ALGEBRA

Exponents and Radicals

$$x^a x^b = x^{a+b} \qquad \frac{x^a}{x^b} = x^{a-b} \qquad x^{-a} = \frac{1}{x^a} \qquad (x^a)^b = x^{ab} \qquad \left(\frac{x}{y}\right)^a = \frac{x^a}{y^a}$$

$$x^{1/n} = \sqrt[n]{x} \qquad x^{m/n} = \sqrt[n]{x^m} = (\sqrt[n]{x})^m \qquad \sqrt[n]{xy} = \sqrt[n]{x}\sqrt[n]{y} \qquad \sqrt[n]{x/y} = \sqrt[n]{x}/\sqrt[n]{y}$$

Factoring Formulas

$$a^2 - b^2 = (a - b)(a + b)$$
$$a^3 - b^3 = (a - b)(a^2 + ab + b^2)$$
$$a^n - b^n = (a - b)(a^{n-1} + a^{n-2}b + a^{n-3}b^2 + \cdots + ab^{n-2} + b^{n-1})$$

$a^2 + b^2$ does not factor over real numbers.
$$a^3 + b^3 = (a + b)(a^2 - ab + b^2)$$

Binomials

$$(a \pm b)^2 = a^2 \pm 2ab + b^2$$
$$(a \pm b)^3 = a^3 \pm 3a^2b + 3ab^2 \pm b^3$$

Binomial Theorem

$$(a + b)^n = a^n + \binom{n}{1}a^{n-1}b + \binom{n}{2}a^{n-2}b^2 + \cdots + \binom{n}{n-1}ab^{n-1} + b^n,$$

where $\dbinom{n}{k} = \dfrac{n(n-1)(n-2)\cdots(n-k+1)}{k(k-1)(k-2)\cdots 3\cdot 2\cdot 1} = \dfrac{n!}{k!(n-k)!}$

Quadratic Formula

The solutions of $ax^2 + bx + c = 0$ are

$$x = \frac{-b \pm \sqrt{b^2 - 4ac}}{2a}.$$

GEOMETRY

Parallelogram	Triangle	Trapezoid	Circle	Sector

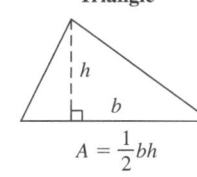 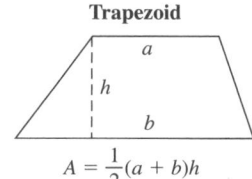 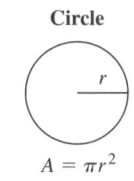 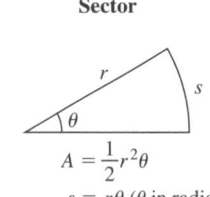

$A = bh$ $\qquad$ $A = \frac{1}{2}bh$ $\qquad$ $A = \frac{1}{2}(a + b)h$ $\qquad$ $A = \pi r^2$ $\qquad$ $A = \frac{1}{2}r^2\theta$

$C = 2\pi r$ $\qquad$ $s = r\theta$ (θ in radians)

Cylinder	Cone	Sphere

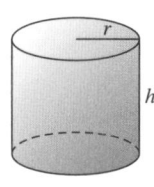 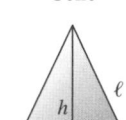

$V = \pi r^2 h$ $\qquad$ $V = \frac{1}{3}\pi r^2 h$ $\qquad$ $V = \frac{4}{3}\pi r^3$

$S = 2\pi rh$ $\qquad$ $S = \pi r \ell$ $\qquad$ $S = 4\pi r^2$

(lateral surface area) $\quad$ (lateral surface area)

Equations of Lines and Circles

$$m = \frac{y_2 - y_1}{x_2 - x_1} \qquad \text{slope of line through } (x_1, y_1) \text{ and } (x_2, y_2)$$

$$y - y_1 = m(x - x_1) \qquad \text{point–slope form of line through } (x_1, y_1) \text{ with slope } m$$

$$y = mx + b \qquad \text{slope–intercept form of line with slope } m \text{ and } y\text{-intercept } (0, b)$$

$$(x - h)^2 + (y - k)^2 = r^2 \qquad \text{circle of radius } r \text{ with center } (h, k)$$

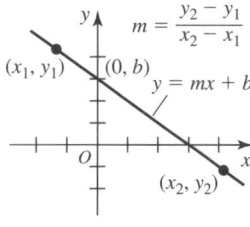

$$m = \frac{y_2 - y_1}{x_2 - x_1}$$
$$y = mx + b$$

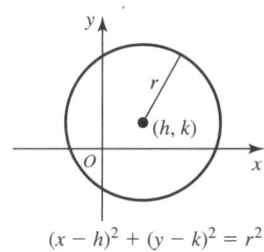

$$(x - h)^2 + (y - k)^2 = r^2$$

TRIGONOMETRY

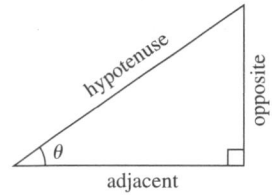

$$\cos \theta = \frac{\text{adj}}{\text{hyp}} \qquad \sin \theta = \frac{\text{opp}}{\text{hyp}} \qquad \tan \theta = \frac{\text{opp}}{\text{adj}}$$

$$\sec \theta = \frac{\text{hyp}}{\text{adj}} \qquad \csc \theta = \frac{\text{hyp}}{\text{opp}} \qquad \cot \theta = \frac{\text{adj}}{\text{opp}}$$

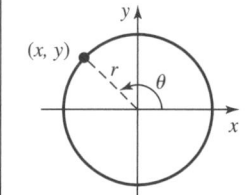

$$\cos \theta = \frac{x}{r} \qquad \sec \theta = \frac{r}{x}$$

$$\sin \theta = \frac{y}{r} \qquad \csc \theta = \frac{r}{y}$$

$$\tan \theta = \frac{y}{x} \qquad \cot \theta = \frac{x}{y}$$

(Continued)

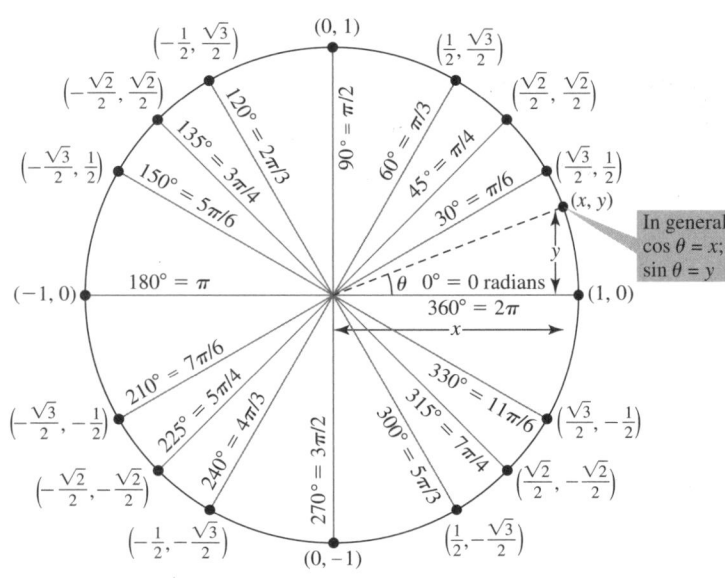

Reciprocal Identities

$$\tan \theta = \frac{\sin \theta}{\cos \theta} \quad \cot \theta = \frac{\cos \theta}{\sin \theta} \quad \sec \theta = \frac{1}{\cos \theta} \quad \csc \theta = \frac{1}{\sin \theta}$$

Pythagorean Identities

$$\sin^2 \theta + \cos^2 \theta = 1 \quad \tan^2 \theta + 1 = \sec^2 \theta \quad 1 + \cot^2 \theta = \csc^2 \theta$$

Sign Identities

$$\sin (-\theta) = -\sin \theta \quad \cos (-\theta) = \cos \theta \quad \tan (-\theta) = -\tan \theta$$
$$\csc (-\theta) = -\csc \theta \quad \sec (-\theta) = \sec \theta \quad \cot (-\theta) = -\cot \theta$$

Double-Angle Identities

$$\sin 2\theta = 2 \sin \theta \cos \theta \qquad \cos 2\theta = \cos^2 \theta - \sin^2 \theta$$
$$= 2 \cos^2 \theta - 1$$
$$\tan 2\theta = \frac{2 \tan \theta}{1 - \tan^2 \theta} \qquad\qquad\quad = 1 - 2 \sin^2 \theta$$

Half-Angle Identities

$$\cos^2 \theta = \frac{1 + \cos 2\theta}{2} \qquad \sin^2 \theta = \frac{1 - \cos 2\theta}{2}$$

Addition Formulas

$$\sin (\alpha + \beta) = \sin \alpha \cos \beta + \cos \alpha \sin \beta \qquad \sin (\alpha - \beta) = \sin \alpha \cos \beta - \cos \alpha \sin \beta$$
$$\cos (\alpha + \beta) = \cos \alpha \cos \beta - \sin \alpha \sin \beta \qquad \cos (\alpha - \beta) = \cos \alpha \cos \beta + \sin \alpha \sin \beta$$
$$\tan (\alpha + \beta) = \frac{\tan \alpha + \tan \beta}{1 - \tan \alpha \tan \beta} \qquad\qquad \tan (\alpha - \beta) = \frac{\tan \alpha - \tan \beta}{1 + \tan \alpha \tan \beta}$$

Law of Sines

$$\frac{\sin \alpha}{a} = \frac{\sin \beta}{b} = \frac{\sin \gamma}{c}$$

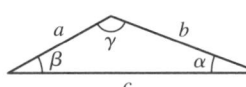

Law of Cosines

$$a^2 = b^2 + c^2 - 2bc \cos \alpha$$

Graphs of Trigonometric Functions and Their Inverses

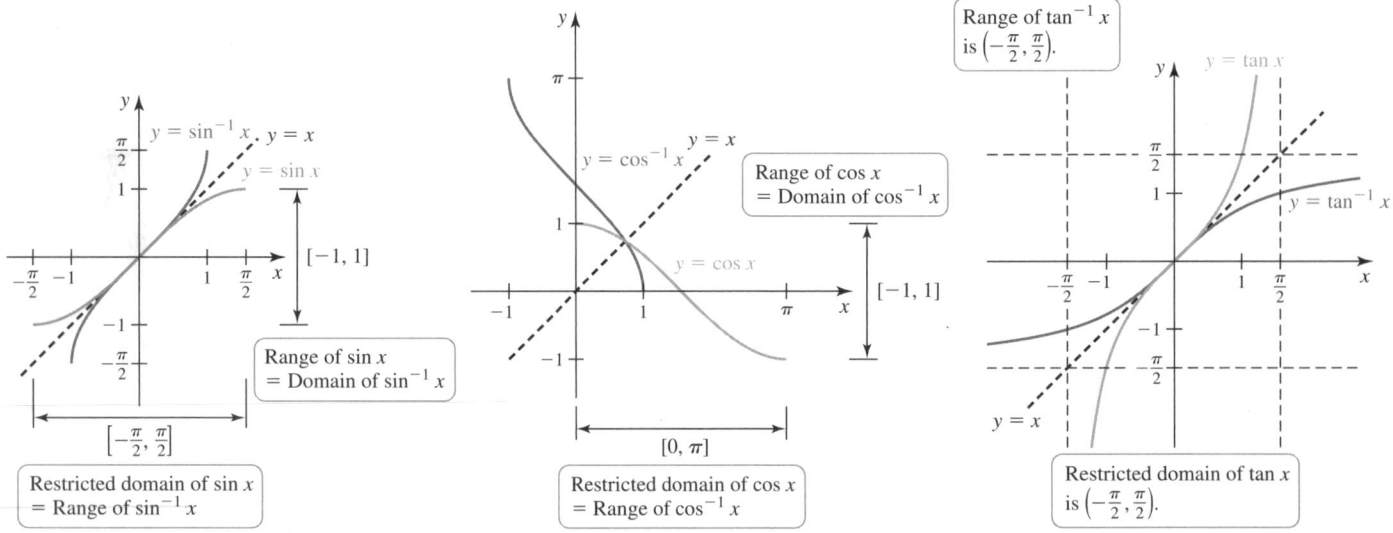

Calculus
EARLY TRANSCENDENTALS
Third Edition

WILLIAM BRIGGS
University of Colorado, Denver

LYLE COCHRAN
Whitworth University

BERNARD GILLETT
University of Colorado, Boulder

ERIC SCHULZ
Walla Walla Community College

 Pearson

Director, Portfolio Management:	Deirdre Lynch
Executive Editor:	Jeff Weidenaar
Content Producer:	Rachel S. Reeve
Managing Producer:	Scott Disanno
Producer:	Stephanie Green
Manager, Courseware QA:	Mary Durnwald
Manager, Content Development:	Kristina Evans
Product Marketing Manager:	Emily Ockay
Field Marketing Manager:	Evan St. Cyr
Marketing Assistants:	Shannon McCormack, Erin Rush
Senior Author Support/Technology Specialist:	Joe Vetere
Manager, Rights and Permissions:	Gina Cheselka
Manufacturing Buyer:	Carol Melville, LSC Communications
Text and Cover Design, Production Coordination, Composition:	Cenveo® Publisher Services
Illustrations:	Network Graphics, Cenveo® Publisher Services
Cover Image:	Nutexzles/Moment/Getty Images

Library of Congress Cataloging-in-Publication Data
Names: Briggs, William L., author. | Cochran, Lyle, author. | Gillett, Bernard, author. | Schulz, Eric P., author.
Title: Calculus. Early transcendentals.
Description: Third edition / William Briggs, University of Colorado, Denver, Lyle Cochran, Whitworth University, Bernard Gillett, University of Colorado, Boulder, Eric Schulz, Walla Walla Community College. | New York, NY : Pearson, [2019] | Includes index.
Identifiers: LCCN 2017046414 | ISBN 9780134763644 (hardcover) | ISBN 0134763645 (hardcover)
Subjects: LCSH: Calculus—Textbooks.
Classification: LCC QA303.2 .B75 2019 | DDC 515—dc23
LC record available at https://lccn.loc.gov/2017046414

2   18

Instructor's Edition
ISBN 13: 978-0-13-476684-3
ISBN 10: 0-13-476684-9

Student Edition
ISBN 13: 978-0-13-476364-4
ISBN 10: 0-13-476364-5

For Julie, Susan, Sally, Sue,
Katie, Jeremy, Elise, Mary, Claire, Katie, Chris, and Annie,
whose support, patience, and encouragement made this book possible.

Contents

Preface

The third edition of *Calculus: Early Transcendentals* supports a three-semester or four-quarter calculus sequence typically taken by students studying mathematics, engineering, the natural sciences, or economics. The third edition has the same goals as the first edition:

• to motivate the essential ideas of calculus with a lively narrative, demonstrating the utility of calculus with applications in diverse fields;

• to introduce new topics through concrete examples, applications, and analogies, appealing to students' intuition and geometric instincts to make calculus natural and believable; and

• once this intuitive foundation is established, to present generalizations and abstractions and to treat theoretical matters in a rigorous way.

The third edition both builds on the success of the previous two editions and addresses the feedback we have received. We have listened to and learned from the instructors who used the text. They have given us wise guidance about how to make the third edition an even more effective learning tool for students and a more powerful resource for instructors. Users of the text continue to tell us that it mirrors the course they teach—and, more important, that students actually read it! Of course, the third edition also benefits from our own experiences using the text, as well as from our experiences teaching mathematics at diverse institutions over the past 30 years.

New to the Third Edition

Exercises

The exercise sets are a major focus of the revision. In response to reviewer and instructor feedback, we've made some significant changes to the exercise sets by rearranging and relabeling exercises, modifying some exercises, and adding many new ones. Of the approximately 10,400 exercises appearing in this edition, 18% are new, and many of the exercises from the second edition were revised for this edition. We analyzed aggregated student usage and performance data from MyLab™ Math for the previous edition of this text. The results of this analysis helped us improve the quality and quantity of exercises that matter the most to instructors and students. We have also simplified the structure of the exercises sets from five parts to the following three:

1. **Getting Started** contains some of the former Review Questions but goes beyond those to include more conceptual exercises, along with new basic skills and short-answer exercises. Our goal in this section is to provide an excellent overall assessment of understanding of the key ideas of a section.

2. **Practice Exercises** consist primarily of exercises from the former Basic Skills, but they also include intermediate-level exercises from the former Further Explorations and Application sections. Unlike previous editions, these exercises are not necessarily organized into groups corresponding to specific examples. For instance, instead of separating out Product Rule exercises from Quotient Rule exercises in Section 3.4, we

have merged these problems into one larger group of exercises. Consequently, specific instructions such as "Use the Product Rule to find the derivative of the following functions" and "Use the Quotient Rule to find the derivative of the given functions" have been replaced with the general instruction "Find the derivative of the following functions." With Product Rule and Quotient Rule exercises mixed together, students must first choose the correct method for evaluating derivatives before solving the problems.

3. **Explorations and Challenges** consist of more challenging problems and those that extend the content of the section.

We no longer have a section of the exercises called "Applications," but (somewhat ironically) in eliminating this section, we feel we are providing better coverage of applications because these exercises have been placed strategically *throughout the exercise sets*. Some are in Getting Started, most are in Practice Exercises, and some are in Explorations and Challenges. The applications nearly always have a boldface heading so that the topic of the application is readily apparent.

Regarding the boldface heads that precede exercises: These heads provide instructors with a quick way to discern the topic of a problem when creating assignments. We heard from users of earlier editions, however, that some of these heads provided too much guidance in how to solve a given problem. In this edition, therefore, we eliminated or reworded run-in heads that provided too much information about the solution method for a problem.

Finally, the **Chapter Review exercises** received a major revamp to provide more exercises (particularly intermediate-level problems) and more opportunities for students to choose a strategy of solution. More than 26% of the Chapter Review exercises are new.

Content Changes

Below are noteworthy changes from the previous edition of the text. Many other detailed changes, not noted here, were made to improve the quality of the narrative and exercises. Bullet points with a 🔑 icon represent major content changes from the previous edition.

Chapter 1 Functions

- Example 2 in Section 1.1 was modified with more emphasis on using algebraic techniques to determine the domain and range of a function. To better illustrate a common feature of limits, we replaced part (c) with a rational function that has a common factor in the numerator and denominator.

- Examples 7 and 8 in Section 1.1 from the second edition (2e) were moved forward in the narrative so that students get an intuitive feel for the composition of two functions using graphs and tables; compositions of functions using algebraic techniques follow.

- Example 10 in Section 1.1, illustrating the importance of secant lines, was made more relevant to students by using real data from a GPS watch during a hike. Corresponding exercises were also added.

- Exercises were added to Section 1.3 to give students practice at finding inverses of functions using the properties of exponential and logarithmic functions.

- New application exercises (investment problems and a biology problem) were added to Section 1.3 to further illustrate the usefulness of logarithmic and exponential functions.

Chapter 2 Limits

- Example 4 in Section 2.2 was revised, emphasizing an algebraic approach to a function with a jump discontinuity, rather than a graphical approach.

- Theorems 2.3 and 2.13 were modified, simplifying the notation to better connect with upcoming material.

- Example 7 in Section 2.3 was added to solidify the notions of left-, right-, and two-sided limits.

- The material explaining the end behavior of exponential and logarithmic functions was reworked, and Example 6 in Section 2.5 was added to show how substitution is used in evaluating limits.

- Exercises were added to Section 2.5 to illustrate the similarities and differences between limits at infinity and infinite limits. We also included some easier exercises in Section 2.5 involving limits at infinity of functions containing square roots.

- Example 5 in Section 2.7 was added to demonstrate an epsilon-delta proof of a limit of a quadratic function.

- We added 17 epsilon-delta exercises to Section 2.7 to provide a greater variety of problems involving limits of quadratic, cubic, trigonometric, and absolute value functions.

Chapter 3 Derivatives

- Chapter 3 now begins with a look back at average and instantaneous velocity, first encountered in Section 2.1, with a corresponding revised example in Section 3.1.

- 🔑 The derivative at a point and the derivative as a function are now treated separately in Sections 3.1 and 3.2.

- After defining the derivative at a point in Section 3.1 with a supporting example, we added a new subsection: Interpreting the Derivative (with two supporting examples).

- Several exercises were added to Section 3.3 that require students to use the Sum and Constant Rules, together with geometry, to evaluate derivatives.

- 🔑 The Power Rule for derivatives in Section 3.4 is stated for all real powers (later proved in Section 3.9). Example 4 in Section 3.4 includes two additional parts to highlight this change, and subsequent examples in upcoming sections rely on the more robust version of the Power Rule. The Power Rule for Rational Exponents in Section 3.8 was deleted because of this change.

- We combined the intermediate-level exercises in Section 3.4 involving the Product Rule and Quotient Rule together under one unified set of directions.

- ⊶ The derivative of e^x still appears early in the chapter, but the derivative of e^{kx} is delayed; it appears only after the Chain Rule is introduced in Section 3.7.

- In Section 3.7, we deleted references to Version 1 and Version 2 of the Chain Rule. Additionally, Chain Rule exercises involving repeated use of the rule were merged with the standard exercises.

- In Section 3.8, we added emphasis on simplifying derivative formulas for implicitly defined functions; see Examples 4 and 5.

- Example 3 in Section 3.11 was replaced; the new version shows how similar triangles are used in solving a related-rates problem.

Chapter 4 Applications of the Derivative

- ⊶ The Mean Value Theorem (MVT) was moved from Section 4.6 to 4.2 so that the proof of Theorem 4.7 is not delayed. We added exercises to Section 4.2 that help students better understand the MVT geometrically, and we included exercises where the MVT is used to prove some well-known identities and inequalities.

- Example 5 in Section 4.5 was added to give guidance on a certain class of optimization problems.

- Example 3b in Section 4.7 was replaced to better drive home the need to simplify after applying l'Hôpital's Rule.

- Most of the intermediate exercises in Section 4.7 are no longer separated out by the type of indeterminate form, and we added some problems in which l'Hôpital's Rule does not apply.

- ⊶ Indefinite integrals of trigonometric functions with argument ax (Table 4.9) were relocated to Section 5.5, where they are derived with the Substitution Rule. A similar change was made to Table 4.10.

- Example 7b in Section 4.9 was added to foreshadow a more complete treatment of the domain of an initial value problem found in Chapter 9.

- We added to Section 4.9 a significant number of intermediate antiderivative exercises that require some preliminary work (e.g., factoring, cancellation, expansion) before the antiderivatives can be determined.

Chapter 5 Integration

- Examples 2 and 3 in Section 5.1 on approximating areas were replaced with a friendlier function where the grid points are more transparent; we return to these approximations in Section 5.3, where an exact result is given (Example 3b).

- Three properties of integrals (bounds on definite integrals) were added in Section 5.2 (Table 5.5); the last of these properties is used in the proof of the Fundamental Theorem (Section 5.3).

- Exercises were added to Sections 5.1 and 5.2 where students are required to evaluate Riemann sums using graphs or tables instead of formulas. These exercises will help students better understand the geometric meaning of Riemann sums.

- We added to Section 5.3 more exercises in which the integrand must be simplified before the integrals can be evaluated.

- A proof of Theorem 5.7 is now offered in Section 5.5.

- Table 5.6 lists the general integration formulas that were relocated from Section 4.9 to Section 5.5; Example 4 in Section 5.5 derives these formulas.

Chapter 6 Applications of Integration
Chapter 7 Logarithmic, Exponential, and Hyperbolic Functions

- ⊶ Chapter 6 from the 2e was split into two chapters in order to match the number of chapters in *Calculus* (Late Transcendentals). The result is a compact Chapter 7.

- Exercises requiring students to evaluate net change using graphs were added to Section 6.1.

- Exercises in Section 6.2 involving area calculations with respect to x and y are now combined under one unified set of directions (so that students must first determine the appropriate variable of integration).

- We increased the number of exercises in Sections 6.3 and 6.4 in which curves are revolved about lines other than the x- and y-axes. We also added introductory exercises that guide students, step by step, through the processes used to find volumes.

- A more gentle introduction to lifting problems (specifically, lifting a chain) was added in Section 6.7 and illustrated in Example 3, accompanied by additional exercises.

- The introduction to exponential growth (Section 7.2) was rewritten to make a clear distinction between the relative growth rate (or percent change) of a quantity and the rate constant k. We revised the narrative so that the equation $y = y_0 e^{kt}$ applies to both growth and decay models. This revision resulted in a small change to the half-life formula.

- The variety of applied exercises in Section 7.2 was increased to further illustrate the utility of calculus in the study of exponential growth and decay.

Chapter 8 Integration Techniques

- Table 8.1 now includes four standard trigonometric integrals that previously appeared in the section Trigonometric Integrals (8.3); these integrals are derived in Examples 1 and 2 in Section 8.1.

- ⊶ A new section (8.6) was added so that students can master integration techniques (that is, choose a strategy) apart from the context given in the previous five sections.

- In Section 8.5 we increased the number and variety of exercises where students must set up the appropriate form of the partial fraction decomposition of a rational function, including more with irreducible quadratic factors.

- A full derivation of Simpson's Rule was added to Section 8.8, accompanied by Example 7, additional figures, and an expanded exercise set.

- 🔑 The Comparison Test for improper integrals was added to Section 8.9, accompanied by Example 7, a two-part example. New exercises in Section 8.9 include some covering doubly infinite improper integrals over infinite intervals.

Chapter 9 Differential Equations

- 🔑 The chapter on differential equations that was available only online in the 2e was converted to a chapter of the text, replacing the single-section coverage found in the 2e.

- More attention was given to the domain of an initial value problem, resulting in the addition and revision of several examples and exercises throughout the chapter.

Chapter 10 Sequences and Infinite Series

- 🔑 The second half of Chapter 10 was reordered: Comparison Tests (Section 10.5), Alternating Series (Section 10.6, which includes the topic of absolute convergence), The Ratio and Root Tests (Section 10.7), and Choosing a Convergence Test (Section 10.8; new section). We split the 2e section that covered the comparison, ratio, and root tests to avoid overwhelming students with too many tests at one time. Section 10.5 focuses entirely on the comparison tests; 39% of the exercises are new. The topic of alternating series now appears before the Ratio and Root Tests so that the latter tests may be stated in their more general form (they now apply to any series rather than only to series with positive terms). The final section (10.8) gives students an opportunity to master convergence tests after encountering each of them separately.

- The terminology associated with sequences (10.2) now includes *bounded above, bounded below,* and *bounded* (rather than only *bounded,* as found in earlier editions).

- Theorem 10.3 (Geometric Sequences) is now developed in the narrative rather than within an example, and an additional example (10.2.3) was added to reinforce the theorem and limit laws from Theorem 10.2.

- Example 5c in Section 10.2 uses mathematical induction to find the limit of a sequence defined recursively; this technique is reinforced in the exercise set.

- Example 3 in Section 10.3 was replaced with telescoping series that are not geometric and that require re-indexing.

- We increased the number and variety of exercises where the student must determine the appropriate series test necessary to determine convergence of a given series.

- We added some easier intermediate-level exercises to Section 10.6, where series are estimated using *n*th partial sums for a given value of *n*.

- Properties of Convergent Series (Theorem 10.8) was expanded (two more properties) and moved to Section 10.3 to better balance the material presented in Sections 10.3 and 10.4. Example 4 in Section 10.3 now has two parts to give students more exposure to the theorem.

Chapter 11 Power Series

- Chapter 11 was revised to mesh with the changes made in Chapter 10.

- We included in Section 11.2 more exercises where the student must find the radius and interval of convergence.

- Example 2 in Section 11.3 was added to illustrate how to choose a different center for a series representation of a function when the original series for the function converges to the function on only part of its domain.

- We addressed an issue with the exercises in Section 11.2 of the previous edition by adding more exercises where the intervals of convergence either are closed or contain one, but not both, endpoints.

- We addressed an issue with exercises in the previous edition by adding many exercises that involve power series centered at locations other than 0.

Chapter 12 Parametric and Polar Curves

- 🔑 The arc length of a two-dimensional curve described by parametric equations was added to Section 12.1, supported by two examples and additional exercises. Area and surfaces of revolution associated with parametric curves were also added to the exercises.

- In Example 3 in Section 12.2, we derive more general polar coordinate equations for circles.

- The arc length of a curve described in polar coordinates is given in Section 12.3.

Chapter 13 Vectors and the Geometry of Space

- 🔑 The material from the 2e chapter Vectors and Vector-Valued Functions is now covered in this chapter and the following chapter.

- Example 5c in Section 13.1 was added to illustrate how to express a vector as a product of its magnitude and its direction.

- We increased the number of applied vector exercises in Section 13.1, starting with some easier exercises, resulting in a wider gradation of exercises.

- 🔑 We adopted a more traditional approach to lines and planes; these topics are now covered together in Section 13.5, followed by cylinders and quadric surfaces in Section 13.6. This arrangement gives students early exposure to all the basic three-dimensional objects that they will encounter throughout the remainder of the text.

- 🔑 A discussion of the distance from a point to a line was moved from the exercises into the narrative, supported with Example 3 in Section 13.5. Example 4 finds the point of intersection of two lines. Several related exercises were added to this section.

- In Section 13.6 there is a larger selection of exercises where the student must identify the quadric surface associated with a given equation. Exercises are also included where students design shapes using quadric surfaces.

Chapter 14 Vector-Valued Functions

- More emphasis was placed on the surface(s) on which a space curve lies in Sections 14.1 and 14.3.

- We added exercises in Section 14.1 where students are asked to find the curve of intersection of two surfaces and where students must verify that a curve lies on a given surface.

- Example 3c in Section 14.3 was added to illustrate how a space curve can be mapped onto a sphere.

- Because the arc length of plane curves (described parametrically in Section 12.1 and with polar coordinates in Section 12.3) was moved to an earlier location in the text, Section 14.4 is now a shorter section.

Chapter 15 Functions of Several Variables

- ⚷ Equations of planes and quadric surfaces were removed from this chapter and now appear in Chapter 13.

- The notation in Theorem 15.2 was simplified to match changes made to Theorem 2.3.

- Example 7 in Section 15.4 was added to illustrate how the Chain Rule is used to compute second partial derivatives.

- We added more challenging partial derivative exercises to Section 15.3 and more challenging Chain Rule exercises to Section 15.4.

- Example 7 in Section 15.5 was expanded to give students more practice finding equations of curves that lie on surfaces.

- Theorem 15.13 was added in Section 15.5; it's a three-dimensional version of Theorem 15.11.

- Example 7 in Section 15.7 was replaced with a more interesting example; the accompanying figure helps tell the story of maximum/minimum problems and can be used to preview Lagrange multipliers.

- We added to Section 15.7 some basic exercises that help students better understand the second derivative test for functions of two variables.

- ⚷ Example 1 in Section 15.8 was modified so that using Lagrange multipliers is the clear path to a solution, rather than eliminating one of the variables and using standard techniques. We also make it clear that care must be taken when using the method of Lagrange multipliers on sets that are not closed and bounded (absolute maximum and minimum values may not exist).

Chapter 16 Multiple Integration

- Example 2 in Section 16.3 was modified because it was too similar to Example 1.

- More care was given to the notation used with polar, cylindrical, and spherical coordinates (see, for example, Theorem 16.3 and the development of integration in different coordinate systems).

- Example 3 in Section 16.4 was modified to make the integration a little more transparent and to show that changing variables to polar coordinates is permissible in more than just the xy-plane.

- More multiple integral exercises were added to Sections 16.1, 16.2, and 16.4, where integration by substitution or integration by parts is needed to evaluate the integrals.

- In Section 16.4 we added more exercises in which the integrals must first be evaluated with respect to x or y instead of z. We also included more exercises that require triple integrals to be expressed in several orderings.

Chapter 17 Vector Calculus

- ⚷ Our approach to scalar line integrals was streamlined; Example 1 in Section 17.2 was modified to reflect this fact.

- We added basic exercises in Section 17.2 emphasizing the geometric meaning of line integrals in a vector field. A subset of exercises was added where line integrals are grouped so that the student must determine the type of line integral before evaluating the integral.

- Theorem 17.5 was added to Section 17.3; it addresses the converse of Theorem 17.4. We also promoted the area of a plane region by a line integral to theorem status (Theorem 17.8 in Section 17.4).

- Example 3 in Section 17.7 was replaced to give an example of a surface whose bounding curve is not a plane curve and to provide an example that buttresses the claims made at the end of the section (that is, Two Final Notes on Stokes' Theorem).

- More line integral exercises were added to Section 17.3 where the student must first find the potential function before evaluating the line integral over a conservative vector field using the Fundamental Theorem of Line Integrals.

- We added to Section 17.7 more challenging surface integrals that are evaluated using Stokes' Theorem.

New to MyLab Math

- **Assignable Exercises** To better support students and instructors, we made the following changes to the assignable exercises:
  - Updated the solution processes in Help Me Solve This and View an Example to better match the techniques used in the text.
  - Added more Setup & Solve exercises to better mirror the types of responses that students are expected to provide on tests. We also added a parallel "standard" version of each Setup & Solve exercise, to allow the instructor to determine which version to assign.
  - Added exercises corresponding to new exercises in the text.

○ Added exercises where MyLab Math users had identified gaps in coverage in the 2e.

○ Added extra practice exercises to each section (clearly labeled EXTRA). These "beyond the text" exercises are perfect for chapter reviews, quizzes, and tests.

○ Analyzed aggregated student usage and performance data from MyLab Math for the previous edition of this text. The results of this analysis helped improve the quality and quantity of exercises that matter the most to instructors and students.

• **Instructional Videos** For each section of the text, there is now a new full-lecture video. Many of these videos make use of Interactive Figures to enhance student understanding of concepts. To make it easier for students to navigate to the specific content they need, each lecture video is segmented into shorter clips (labeled Introduction, Example, or Summary). Both the full lectures and the video segments are assignable within MyLab Math. The videos were created by the following team: Matt Hudelson (Washington State University), Deb Carney and Rebecca Swanson (Colorado School of Mines), Greg Wisloski and Dan Radelet (Indiana University of Pennsylvania), and Nick Ormes (University of Denver).

• **Enhanced Interactive Figures** Incorporating functionality from several standard Interactive Figures makes Enhanced Interactive Figures mathematically richer and ideal for in-class demonstrations. Using a single figure, instructors can illustrate concepts that are difficult for students to visualize and can make important connections to key themes of calculus.

• **Enhanced Sample Assignments** These section-level assignments address gaps in precalculus skills with a personalized review of prerequisites, help keep skills fresh with spaced practice using key calculus concepts, and provide opportunities to work exercises without learning aids so students can check their understanding. They are assignable and editable.

• **Quick Quizzes** have been added to Learning Catalytics™ (an in-class assessment system) for every section of the text.

• **Maple™, Mathematica®, and Texas Instruments® Manuals and Projects** have all been updated to align with the latest software and hardware.

Noteworthy Features

Figures

Given the power of graphics software and the ease with which many students assimilate visual images, we devoted considerable time and deliberation to the figures in this text. Whenever possible, we let the figures communicate essential ideas using annotations reminiscent of an instructor's voice at the board. Readers will quickly find that the figures facilitate learning in new ways.

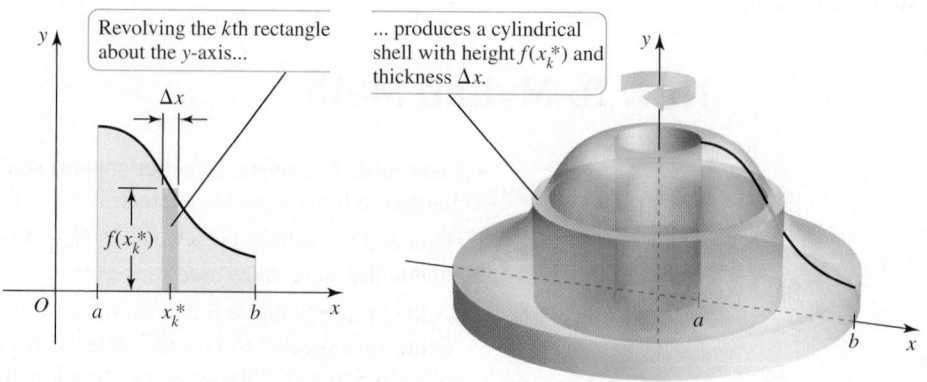

Figure 6.40

Annotated Examples

Worked-out examples feature annotations in blue to guide students through the process of solving the example and to emphasize that each step in a mathematical argument must be rigorously justified. These annotations are designed to echo how instructors "talk through" examples in lecture. They also provide help for students who may struggle with the algebra and trigonometry steps within the solution process.

Quick Checks

The narrative is interspersed with Quick Check questions that encourage students to do the calculus as they are reading about it. These questions resemble the kinds of questions instructors pose in class. Answers to the Quick Check questions are found at the end of the section in which they occur.

Guided Projects

MyLab Math contains 78 Guided Projects that allow students to work in a directed, step-by-step fashion, with various objectives: to carry out extended calculations, to derive physical models, to explore related theoretical topics, or to investigate new applications of calculus. The Guided Projects vividly demonstrate the breadth of calculus and provide a wealth of mathematical excursions that go beyond the typical classroom experience. A list of related Guided Projects is included at the end of each chapter.

Incorporating Technology

We believe that a calculus text should help students strengthen their analytical skills and demonstrate how technology can extend (not replace) those skills. Calculators and graphing utilities are additional tools in the kit, and students must learn when and when not to use them. Our goal is to accommodate the different policies regarding technology adopted by various instructors.

Throughout the text, exercises marked with **T** indicate that the use of technology—ranging from plotting a function with a graphing calculator to carrying out a calculation using a computer algebra system—may be needed. See page xx for information regarding our technology resource manuals covering Maple, Mathematica, and Texas Instruments graphing calculators.

Text Versions

- **eBook with Interactive Figures** The text is supported by a groundbreaking and award-winning electronic book created by Eric Schulz of Walla Walla Community College. This "live book" runs in Wolfram CDF Player (the free version of Mathematica) and contains the complete text of the print book plus interactive versions of approximately 700 figures. Instructors can use these interactive figures in the classroom to illustrate the important ideas of calculus, and students can explore them while they are reading the text. Our experience confirms that the interactive figures help build students' geometric intuition of calculus. The authors have written Interactive Figure Exercises that can be assigned via MyLab Math so that students can engage with the figures outside of class in a directed way. Available only within MyLab Math, the eBook provides instructors with powerful new teaching tools that expand and enrich the learning experience for students.

- **Other eBook Formats** The text is also available in various stand-alone eBook formats. These are listed in the Pearson online catalog: **www.pearson.com**. MyLab Math also contains an HTML eBook that is screen-reader accessible.

- **Other Print Formats** The text is also available in split editions (Single Variable [Chapters 1–12] and Multivariable [Chapters 10–17]) and in unbound (3-hole punched) formats. Again, see the Pearson online catalog for details: **www.pearson.com.**

Acknowledgments

We would like to express our thanks to the following people, who made valuable contributions to this edition as it evolved through its many stages.

Accuracy Checkers

Jennifer Blue

Lisa Collette

Nathan Moyer

Tom Wegleitner

Reviewers

Siham Alfred, *Germanna Community College*

Darry Andrews, *Ohio State University*

Pichmony Anhaouy, *Langara College*

Raul Aparicio, *Blinn College*

Anthony Barcellos, *American River College*

Rajesh K. Barnwal, *Middle Tennessee State University*

Susan Barton, *Brigham Young University, Hawaii*

Aditya Baskota, *Kauai Community College*

Al Batten, *University of Colorado, Colorado Springs*

Jeffrey V. Berg, *Arapahoe Community College*

Martina Bode, *University of Illinois, Chicago*

Kaddour Boukaabar, *California University of Pennsylvania*

Paul W. Britt, *Our Lady of the Lake College*

Steve Brosnan, *Belmont Abbey College*

Brendan Burns Healy, *Tufts University*

MK Choy, *CUNY Hunter College*

Nathan P. Clements, *University of Wyoming*

Gregory M. Coldren, *Frederick Community College*

Brian Crimmel, *U.S. Coast Guard Academy*

Robert D. Crise, Jr, *Crafton Hills College*

Dibyajyoti Deb, *Oregon Institute of Technology*

Elena Dilai, *Monroe Community College*

Johnny I. Duke, *Georgia Highlands College*

Fedor Duzhin, *Nanyang Technological University*

Houssein El Turkey, *University of New Haven*

J. Robson Eby, *Blinn College*

Amy H. Erickson, *Georgia Gwinnett College*

Robert A. Farinelli, *College of Southern Maryland*

Rosemary Farley, *Manhattan College*

Lester A. French, Jr, *University of Maine, Augusta*

Pamela Frost, *Middlesex Community College*

Scott Gentile, *CUNY Hunter College*

Stephen V. Gilmore, *Johnson C. Smith University*

Joseph Hamelman, *Virginia Commonwealth University*

Jayne Ann Harderq, *Oral Roberts University*

Miriam Harris-Botzum, *Lehigh Carbon Community College*

Mako Haruta, *University of Hartford*

Ryan A. Hass, *Oregon State University*

Christy Hediger, *Lehigh Carbon Community College*

Joshua Hertz, *Northeastern University*

Peter Hocking, *Brunswick Community College*

Farhad Jafari, *University of Wyoming*

Yvette Janecek, *Blinn College*

Tom Jerardi, *Howard Community College*

Karen Jones, *Ivy Tech Community College*

Bir Kafle, *Purdue University Northwest*

Mike Kawai, *University of Colorado, Denver*

Mahmoud Khalili, *Oakton Community College*

Lynne E. Kowski, *Raritan Valley Community College*

Tatyana Kravchuk, *Northern Virginia Community College*

Lorraine Krusinski, *Brunswick High School*

Frederic L. Lang, *Howard Community College*

Robert Latta, *University of North Carolina, Charlotte*

Angelica Lyubenko, *University of Colorado, Denver*

Darren E. Mason, *Albion College*

John C. Medcalf, *American River College*

Elaine Merrill, *Brigham Young University, Hawaii*

Robert Miller, *Langara College*

Mishko Mitkovski, *Clemson University*

Carla A. Monticelli, *Camden County College*

Charles Moore, *Washington State University*

Humberto Munoz Barona, *Southern University*

Clark Musselman, *University of Washington, Bothell*

Glenn Newman, *Newbury College*

Daniela Nikolova-Popova, *Florida Atlantic University*

Janis M. Oldham, *North Carolina A&T State University*

Byungdo Park, *CUNY Hunter College*

Denise Race, *Eastfield College*

Hilary Risser, *Montana Tech*

Sylvester Roebuck, *Jr, Olive Harvey College*

John P. Roop, *North Carolina A&T State University*

Paul N. Runnion, *Missouri University of Science and Technology*

Kegan Samuel, *Middlesex Community College*

Steve Scarborough, *Oregon State University*

Vickey Sealy, *West Virginia University*

Ruth Seidel, *Blinn College*

Derek Smith, *Nashville State Community College*

James Talamo, *Ohio State University*

Yan Tang, *Embry-Riddle Aeronautical University*

Kye Taylor, *Tufts University*

Daniel Triolo, *Lake Sumter State College*

Cal Van Niewaal, *Coe College*

Robin Wilson, *California Polytechnic State University, Pomona*

Kurt Withey, *Northampton Community College*

Amy Yielding, *Eastern Oregon University*

Prudence York-Hammons, *Temple College*

Bradley R. Young, *Oakton Community College*

Pablo Zafra, *Kean University*

The following faculty members provided direction on the development of the MyLab Math course for this edition:

Heather Albrecht, *Blinn College*

Terry Barron, *Georgia Gwinnett College*

Jeffrey V. Berg, *Arapahoe Community College*

Joseph Bilello, *University of Colorado, Denver*

Mark Bollman, *Albion College*

Mike Bostick, *Central Wyoming College*

Laurie L. Boudreaux, *Nicholls State University*

Steve Brosnan, *Belmont Abbey College*

Debra S. Carney, *Colorado School of Mines*

Robert D. Crise, Jr, *Crafton Hills College*

Shannon Dingman, *University of Arkansas*

Deland Doscol, *North Georgia Technical College*

J. Robson Eby, *Blinn College*

Stephanie L. Fitch, *Missouri University of Science and Technology*

Dennis Garity, *Oregon State University*

Monica Geist, *Front Range Community College*

Brandie Gilchrist, *Central High School*

Roger M. Goldwyn, *Florida Atlantic University*

Maggie E. Habeeb, *California University of Pennsylvania*

Ryan A. Hass, *Oregon State University*

Danrun Huang, *St. Cloud State University*

Zhongming Huang, *Midland University*

Christa Johnson, *Guilford Technical Community College*

Bir Kafle, *Purdue University Northwest*

Semra Kilic-Bahi, *Colby-Sawyer College*

Kimberly Kinder, *Missouri University of Science and Technology*

Joseph Kotowski, *Oakton Community College*

Lynne E. Kowski, *Raritan Valley Community College*

Paula H. Krone, *Georgia Gwinnett College*

Daniel Lukac, *Miami University*

Jeffrey W. Lyons, *Nova Southeastern University*

James Magee, *Diablo Valley College*

Erum Marfani, *Frederick Community College*

Humberto Munoz Barona, *Southern University*

James A. Mendoza Alvarez, *University of Texas, Arlington*

Glenn Newman, *Newbury College*

Peter Nguyen, *Coker College*

Mike Nicholas, *Colorado School of Mines*

Seungly Oh, *Western New England University*

Denise Race, *Eastfield College*

Paul N. Runnion, *Missouri University of Science and Technology*

Polly Schulle, *Richland College*

Rebecca Swanson, *Colorado School of Mines*

William Sweet, *Blinn College*

M. Velentina Vega-Veglio, *William Paterson University*

Nick Wahle, *Cincinnati State Technical and Community College*

Patrick Wenkanaab, *Morgan State University*

Amanda Wheeler, *Amarillo College*

Diana White, *University of Colorado, Denver*

Gregory A. White, *Linn Benton Community College*

Joseph White, *Olympic College*

Robin T. Wilson, *California Polytechnic State University, Pomona*

Deborah A. Zankofski, *Prince George's Community College*

The following faculty were members of the Engineering Review Panel. This panel made recommendations to improve the text for engineering students.

Al Batten, *University of Colorado, Colorado Springs*

Josh Hertz, *Northeastern University*

Daniel Latta, *University of North Carolina, Charlotte*

Yan Tang, *Embry-Riddle Aeronautical University*

MyLab Math Online Course
for *Calculus: Early Transcendentals, 3e*

(access code required)

MyLab™ Math is available to accompany Pearson's market-leading text offerings. To give students a consistent tone, voice, and teaching method, each text's flavor and approach are tightly integrated throughout the accompanying MyLab Math course, making learning the material as seamless as possible.

PREPAREDNESS

One of the biggest challenges in calculus courses is making sure students are adequately prepared with the prerequisite skills needed to successfully complete their course work. MyLab Math supports students with just-in-time remediation and review of key concepts.

Integrated Review Course

These special MyLab courses contain pre-made, assignable quizzes to assess the prerequisite skills needed for each chapter, plus personalized remediation for any gaps in skills that are identified. Each student, therefore, receives the appropriate level of help—no more, no less.

DEVELOPING DEEPER UNDERSTANDING

MyLab Math provides content and tools that help students build a deeper understanding of course content than would otherwise be possible.

eBook with Interactive Figures

The eBook includes approximately 700 figures that can be manipulated by students to provide a deeper geometric understanding of key concepts and examples as they read and learn new material. Students get unlimited access to the eBook within any MyLab Math course using that edition of the text. The authors have written Interactive Figure Exercises that can be assigned for homework so that students can engage with the figures outside of the classroom.

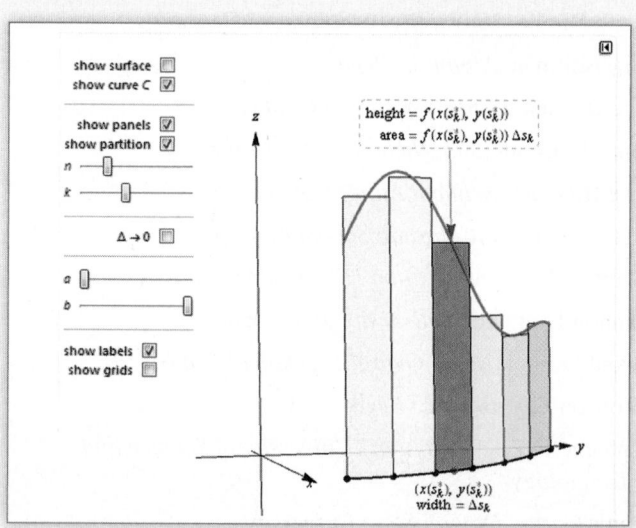

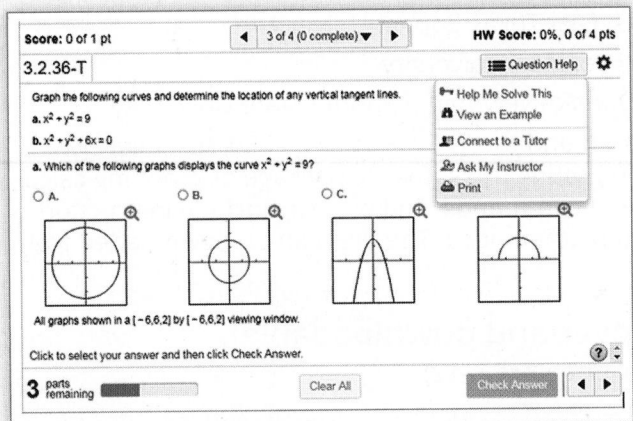

Exercises with Immediate Feedback

The over 8000 homework and practice exercises for this text regenerate algorithmically to give students unlimited opportunity for practice and mastery. MyLab Math provides helpful feedback when students enter incorrect answers and includes the optional learning aids Help Me Solve This, View an Example, videos, and/or the eBook.

NEW! Enhanced Sample Assignments

These section-level assignments include just-in-time review of prerequisites, help keep skills fresh with spaced practice of key concepts, and provide opportunities to work exercises without learning aids so students can check their understanding. They are assignable and editable within MyLab Math.

Additional Conceptual Questions

Additional Conceptual Questions focus on deeper, theoretical understanding of the key concepts in calculus. These questions were written by faculty at Cornell University under an NSF grant and are also assignable through Learning Catalytics™.

Setup & Solve Exercises

These exercises require students to show how they set up a problem, as well as the solution, thus better mirroring what is required on tests. This new type of exercise was widely praised by users of the second edition, so more were added to the third edition.

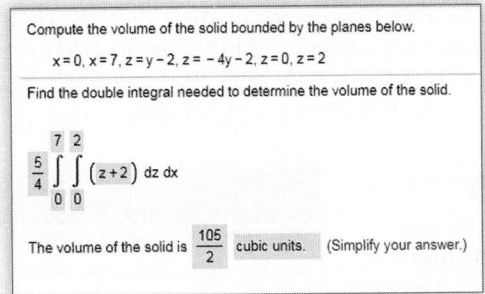

ALL NEW! Instructional Videos

For each section of the text, there is now a new full-lecture video. Many of these videos make use of Interactive Figures to enhance student understanding of concepts. To make it easier for students to navigate to the content they need, each lecture video is segmented into shorter clips (labeled Introduction, Example, or Summary). Both the video lectures and the video segments are assignable within MyLab Math. The Guide to Video-Based Assignments makes it easy to assign videos for homework by showing which MyLab Math exercises correspond to each video.

UPDATED! Technology Manuals (downloadable)

- Maple™ Manual and Projects by Kevin Reeves, East Texas Baptist University
- Mathematica® Manual and Projects by Todd Lee, Elon University
- TI-Graphing Calculator Manual by Elaine McDonald-Newman, Sonoma State University

These manuals cover *Maple* 2017, *Mathematica 11,* and the TI-84 Plus and TI-89, respectively. Each manual provides detailed guidance for integrating the software package or graphing calculator throughout the course, including syntax and commands. The projects include instructions and ready-made application files for Maple and Mathematica. The files can be downloaded from within MyLab Math.

Student's Solutions Manuals (softcover and downloadable)

Single Variable Calculus: Early Transcendentals (Chapters 1–12)
  ISBN: 0-13-477048-X | 978-0-13-477048-2
Multivariable Calculus (Chapters 10–17)
  ISBN: 0-13-476682-2 | 978-0-13-476682-9
Written by Mark Woodard (Furman University), the Student's Solutions Manual contains worked-out solutions to all the odd-numbered exercises. This manual is available in print and can be downloaded from within MyLab Math.

SUPPORTING INSTRUCTION

MyLab Math comes from an experienced partner with educational expertise and an eye on the future. It provides resources to help you assess and improve student results at every turn and unparalleled flexibility to create a course tailored to you and your students.

NEW! Enhanced Interactive Figures

Incorporating functionality from several standard Interactive Figures makes Enhanced Interactive Figures mathematically richer and ideal for in-class demonstrations. Using a single enhanced figure, instructors can illustrate concepts that are difficult for students to visualize and can make important connections to key themes of calculus.

Learning Catalytics

Now included in all MyLab Math courses, this student response tool uses students' smartphones, tablets, or laptops to engage them in more interactive tasks and thinking during lecture. Learning Catalytics™ fosters student engagement and peer-to-peer learning with real-time analytics. Access pre-built exercises created specifically for calculus, including Quick Quiz exercises for each section of the text.

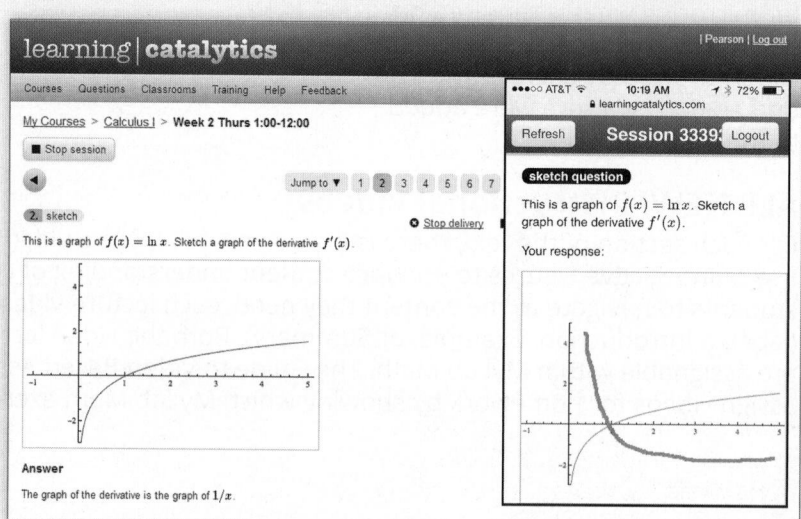

pearson.com/mylab/math

PowerPoint Lecture Resources (downloadable)

Slides contain presentation resources such as key concepts, examples, definitions, figures, and tables from this text. They can be downloaded from within MyLab Math or from Pearson's online catalog at **www.pearson.com**.

Comprehensive Gradebook

The gradebook includes enhanced reporting functionality, such as item analysis and a reporting dashboard to enable you to efficiently manage your course. Student performance data are presented at the class, section, and program levels in an accessible, visual manner so you'll have the information you need to keep your students on track.

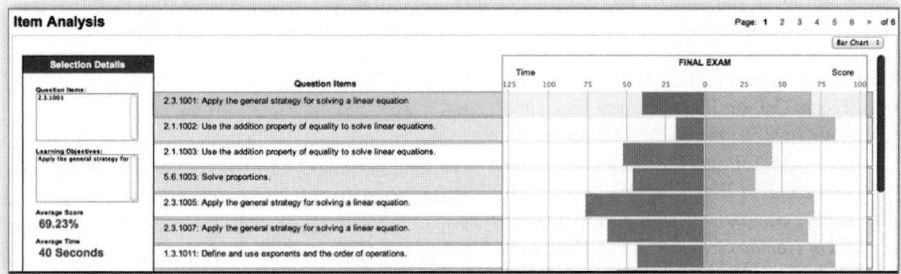

TestGen

TestGen® (**www.pearson.com/testgen**) enables instructors to build, edit, print, and administer tests using a computerized bank of questions developed to cover all the objectives of the text. TestGen is algorithmically based, allowing instructors to create multiple, but equivalent, versions of the same question or test with the click of a button. Instructors can also modify test bank questions and/or add new questions. The software and test bank are available for download from Pearson's online catalog, **www.pearson.com**. The questions are also assignable in MyLab Math.

Instructor's Solutions Manual (downloadable)

Written by Mark Woodard (Furman University), the Instructor's Solutions Manual contains complete solutions to all the exercises in the text. It can be downloaded from within MyLab Math or from Pearson's online catalog, **www.pearson.com**.

Instructor's Resource Guide (downloadable)

This resource includes Guided Projects that require students to make connections between concepts and applications from different parts of the calculus course. They are correlated to specific chapters of the text and can be assigned as individual or group work. The files can be downloaded from within MyLab Math or from Pearson's online catalog, **www.pearson.com**.

Accessibility

Pearson works continuously to ensure that our products are as accessible as possible to all students. We are working toward achieving WCAG 2.0 Level AA and Section 508 standards, as expressed in the Pearson Guidelines for Accessible Educational Web Media, **www.pearson.com/mylab/math/accessibility**.

pearson.com/mylab/math

Credits

Chapter opener art: Andrey Pavlov/123RF

Chapter 1

Page 23, Rivas, E.M. et al (2014). A simple mathematical model that describes the growth of the area and the number of total and viable cells in yeast colonies. *Letters in Applied Microbiology*, 594(59). **Page 25,** Tucker, V. A. (2000, December). The Deep Fovea, Sideways Vision and Spiral Flight Paths in Raptors. *The Journal of Experimental Biology*, 203, 3745–3754. **Page 26,** Collings, B. J. (2007, January). Tennis (and Volleyball) Without Geometric Series. *The College Mathematics Journal*, 38(1). **Page 37,** Murray, I. W. & Wolf, B. O. (2012). Tissue Carbon Incorporation Rates and Diet-to-Tissue Discrimination in Ectotherms: Tortoises Are Really Slow. *Physiological and Biochemical Zoology*, 85(1). **Page 50,** Isaksen, D. C. (1996, September). How to Kick a Field Goal. *The College Mathematics Journal*, 27(4).

Chapter 3

Page 186, Perloff, J. (2012). *Microeconomics*. Prentice Hall. **Page 198,** Murray, I. W. & Wolf, B. O. (2012). Tissue Carbon Incorporation Rates and Diet-to-Tissue Discrimination in Ectotherms: Tortoises Are Really Slow. *Physiological and Biochemical Zoology*, 85(1). **Page 211,** Hook, E. G. & Lindsjo, A. (1978, January). Down syndrome in live births by single year maternal age interval in a Swedish study: comparison with results from a New York State study. *American Journal of Human Genetics*, 30(1). **Page 218,** David, D. (1997, March). Problems and Solutions. *The College Mathematics Journal*, 28(2).

Chapter 4

Page 249, Yu, C. Chau, K.T. & Jiang J. Z. (2009). A flux-mnemonic permanent magnet brushless machine for wind power generation. *Journal of Applied Physics*, 105. **Page 255,** Bragg, L. (2001, September). Arctangent sums. *The College Mathematics Journal*, 32(4). **Page 289,** Adam, J. A. (2011, June). Blood Vessel Branching: Beyond the Standard Calculus Problem. *Mathematics Magazine*, 84(3), 196–207; Apostol T. M. (1967). *Calculus*, Vol. 1. John Wiley & Sons. **Page 291,** Halmos, P. R. Problems For Mathematicians Young And Old. Copyright © Mathematical Association of America, 1991. All rights reserved; Dodge, W. & Viktora, S. (2002, November). Thinking out of the Box … Problem. *Mathematics Teacher*, 95(8). **Page 319,** Ledder, G. (2013). Undergraduate Mathematics for the Life Sciences. *Mathematical Association of America*, Notes No. 81. **Pages 287–288,** Pennings, T. (2003, May). Do Dogs Know Calculus? *The College Mathematics Journal*, 34(6). **Pages 289–290,** Dial, R. (2003). Energetic Savings and The Body Size Distributions of Gliding Mammals. *Evolutionary Ecology Research*, 5, 1151–1162.

Chapter 5

Page 367, Barry, P. (2001, September). Integration from First Principles. *The College Mathematics Journal*, 32(4), 287–289. **Page 387,** Chen, H. (2005, December). Means Generated by an Integral. *Mathematics Magazine*, 78(5), 397–399; Tong, J. (2002, November). A Generalization of the Mean Value Theorem for Integrals. *The College Mathematics Journal*, 33(5). **Page 402,** Plaza, Á. (2008, December). Proof Without Words: Exponential Inequalities. *Mathematics Magazine*, 81(5).

Chapter 6

Page 414, Zobitz, J. M. (2013, November). Forest Carbon Uptake and the Fundamental Theorem of Calculus. *The College Mathematics Journal*, 44(5), 421–424. **Page 424,** Cusick, W. L. (2008, April). Archimedean Quadrature Redux. *Mathematics Magazine*, 81(2), 83–95.

Chapter 7

Page 499, Murray, I. W. & Wolf, B. O. (2012). Tissue Carbon Incorporation Rates and Diet-to-Tissue Discrimination in Ectotherms: Tortoises Are Really Slow. *Physiological and Biochemical Zoology*, 85(1). **Page 501,** Keller, J. (1973, September). A Theory of Competitive Running. *Physics Today*, 26(9); Schreiber, J. S. (2013). Motivating Calculus with Biology. *Mathematical Association of America*, Notes No. 81.

Chapter 8

Page 592, Weidman, P. & Pinelis, I. (2004). Model equations for the Eiffel Tower profile: historical perspective and new results. *C. R. Mécanique*, 332, 571–584; Feuerman, M. etal. (1986, February). Problems. *Mathematics Magazine*, 59(1). **Page 596,** Galperin, G. & Ronsse, G. (2008, April). Lazy Student Integrals. *Mathematics Magazine*, 81(2), 152–154. **Pages 545–546,** Osler, J. T. (2003, May). Visual Proof of Two Integrals. *The College Mathematics Journal*, 34(3), 231–232.

Chapter 10

Page 682, Fleron, J. F. (1999, January). Gabriel's Wedding Cake. *The College Mathematics Journal*, 30(1), 35–38; Selden, A. & Selden, J. (1993, November). Collegiate Mathematics Education Research: What Would That Be Like? *The College Mathematics Journal*, 24(5), 431–445. **Pages 682–683,** Chen, H. & Kennedy, C. (2012, May). Harmonic Series Meets Fibonacci Sequence. *The College Mathematics Journal*, 43(3), 237–243.

Chapter 12

Page 766, Wagon, S. (2010). *Mathematica in Action*. Springer; created by Norton Starr, Amherst College. **Page 767,** Brannen, N. S. (2001, September). The Sun, the Moon, and Convexity. *The College Mathematics Journal*, 32(4), 268–272. **Page 774,** Fray, T. H. (1989). The butterfly curve. *American Mathematical Monthly*, 95(5), 442–443; revived in Wagon, S. & Packel, E. (1994). *Animating Calculus*. Freeman. **Page 778,** Wagon, S. & Packel, E. (1994). *Animating Calculus*. Freeman.

Chapter 13

Page 816, Strang, G. (1991). Calculus. Wellesley-Cambridge Press. **Page 864,** Model Based 3D Tracking of an Articulated Hand, B. Stenger, P. R. S. Mendonça, R. Cipolla, CVPR, Vol. II, p. 310–315, December 2001. CVPR 2001: PROCEEDINGS OF THE 2001 IEEE COMPUTER SOCIETY CONFERENCE ON COMPUTER VISION AND PATTERN RECOGNITION by IEEE Computer Society Reproduced with permission of IEEE COMPUTER SOCIETY PRESS in the format Republish in a book via Copyright Clearance Center.

Chapter 15

Pages 929–930, www.nfl.com. **Page 994,** Karim, R. (2014, December). Optimization of pectin isolation method from pineapple (ananas comosus l.) waste. *Carpathian Journal of Food Science and Technology*, 6(2), 116–122. **Page 996,** Rosenholtz, I. (1985, May). "The Only Critical Point in Town" Test. *Mathematics Magazine*, 58(3), 149–150 and Gillett, P. (1984). *Calculus and Analytical Geometry*, 2nd edition; Rosenholtz, I. (1987, February). Two mountains without a valley. *Mathematics Magazine*, 60(1); Math Horizons, Vol. 11, No. 4, April 2004.

Chapter 16

Page 1072, Glaister, P. (1996). Golden Earrings. *Mathematical Gazette*, 80, 224–225.

1

Functions

Chapter Preview Mathematics is a language with an alphabet, a vocabulary, and many rules. Before beginning your calculus journey, you should be familiar with the elements of this language. Among these elements are algebra skills; the notation and terminology for various sets of real numbers; and the descriptions of lines, circles, and other basic sets in the coordinate plane. A review of this material is found in Appendix B, online at goo.gl/6DCbbM. This chapter begins with the fundamental concept of a function and then presents the entire cast of functions needed for calculus: polynomials, rational functions, algebraic functions, exponential and logarithmic functions, and the trigonometric functions, along with their inverses. Before you begin studying calculus, it is important that you master the ideas in this chapter.

1.1 Review of Functions

Everywhere around us we see relationships among quantities, or **variables**. For example, the consumer price index changes in time and the temperature of the ocean varies with latitude. These relationships can often be expressed by mathematical objects called *functions*. Calculus is the study of functions, and because we use functions to describe the world around us, calculus is a universal language for human inquiry.

> **DEFINITION Function**
>
> A **function** f is a rule that assigns to each value x in a set D a *unique* value denoted $f(x)$. The set D is the **domain** of the function. The **range** is the set of all values of $f(x)$ produced as x varies over the entire domain (**Figure 1.1**).

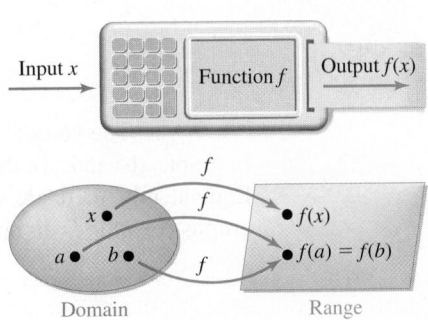

Figure 1.1

1

The **independent variable** is the variable associated with the domain; the **dependent variable** belongs to the range. The **graph** of a function f is the set of all points (x, y) in the xy-plane that satisfy the equation $y = f(x)$. The **argument** of a function is the expression on which the function works. For example, x is the argument when we write $f(x)$. Similarly, 2 is the argument in $f(2)$ and $x^2 + 4$ is the argument in $f(x^2 + 4)$.

QUICK CHECK 1 If $f(x) = x^2 - 2x$, find $f(-1)$, $f(x^2)$, $f(t)$, and $f(p - 1)$. ◄

The requirement that a function assigns a *unique* value of the dependent variable to each value in the domain is expressed in the vertical line test (**Figure 1.2a**). For example, the outside temperature as it varies over the course of a day is a function of time (**Figure 1.2b**).

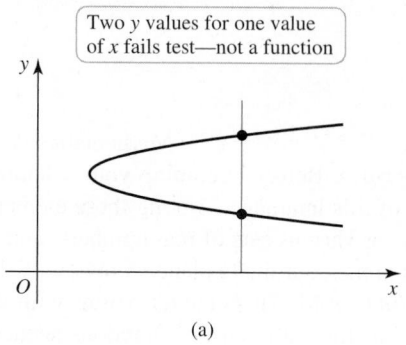

(a)   (b)

Figure 1.2

▶ If the domain is not specified, we take it to be the set of all values of x for which f is defined. We will see shortly that the domain and range of a function may be restricted by the context of the problem.

▶ A set of points or a graph that does *not* correspond to a function represents a **relation** between the variables. All functions are relations, but not all relations are functions.

Vertical Line Test

A graph represents a function if and only if it passes the **vertical line test**: Every vertical line intersects the graph at most once. A graph that fails this test does not represent a function.

EXAMPLE 1   Identifying functions State whether each graph in **Figure 1.3** represents a function.

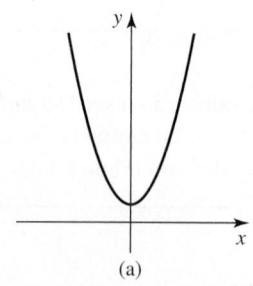

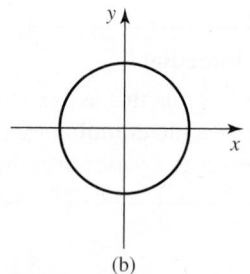
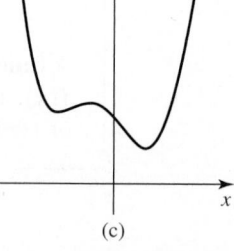
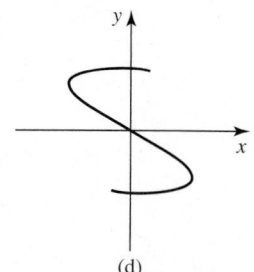

(a)   (b)   (c)   (d)

Figure 1.3

SOLUTION The vertical line test indicates that only graphs (a) and (c) represent functions. In graphs (b) and (d), there are vertical lines that intersect the graph more than once. Equivalently, there are values of x that correspond to more than one value of y. Therefore, graphs (b) and (d) do not pass the vertical line test and do not represent functions.

Related Exercise 3 ◄

EXAMPLE 2   Domain and range Determine the domain and range of each function.

a. $f(x) = x^2 + 1$ **b.** $g(x) = \sqrt{4 - x^2}$ **c.** $h(x) = \dfrac{x^2 - 3x + 2}{x - 1}$

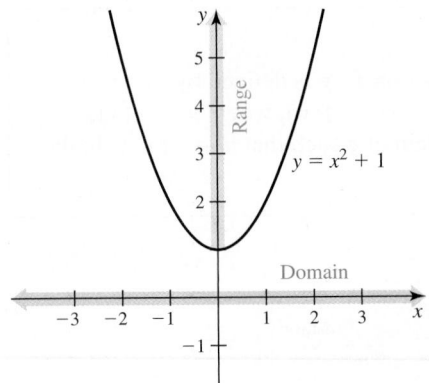

Figure 1.4

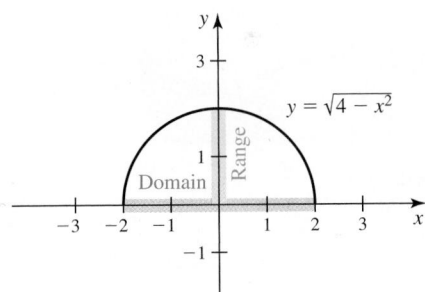

Figure 1.5

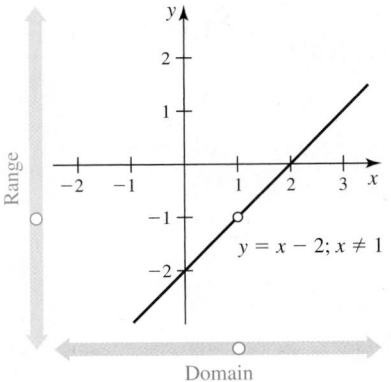

Figure 1.6

SOLUTION

a. Note that f is defined for all values of x; therefore, its domain is the set of all real numbers, written $(-\infty, \infty)$ or $\mathbb{R}$. Because $x^2 \geq 0$ for all x, it follows that $x^2 + 1 \geq 1$, which implies that the range of f is $[1, \infty)$. **Figure 1.4** shows the graph of f along with its domain and range.

b. Functions involving square roots are defined provided the quantity under the root is nonnegative (additional restrictions may also apply). In this case, the function g is defined provided $4 - x^2 \geq 0$, which means $x^2 \leq 4$, or $-2 \leq x \leq 2$. Therefore, the domain of g is $[-2, 2]$. The graph of $g(x) = \sqrt{4 - x^2}$ is the upper half of a circle centered at the origin with radius 2 (**Figure 1.5**; see Appendix B, online at goo.gl/6DCbbM). From the graph we see that the range of g is $[0, 2]$.

c. The function h is defined for all values of $x \neq 1$, so its domain is $\{x : x \neq 1\}$. Factoring the numerator, we find that

$$h(x) = \frac{x^2 - 3x + 2}{x - 1} = \frac{(x - 1)(x - 2)}{x - 1} = x - 2, \text{ provided } x \neq 1.$$

The graph of $y = h(x)$, shown in **Figure 1.6**, is identical to the graph of the line $y = x - 2$ except that it has a hole at $(1, -1)$ because h is undefined at $x = 1$. Therefore, the range of h is $\{y : y \neq -1\}$.   *Related Exercises 23, 25* ◄

EXAMPLE 3 **Domain and range in context** At time $t = 0$, a stone is thrown vertically upward from the ground at a speed of 30 m/s. Its height h above the ground in meters (neglecting air resistance) is approximated by the function $f(t) = 30t - 5t^2$, where t is measured in seconds. Find the domain and range of f in the context of this particular problem.

SOLUTION Although f is defined for all values of t, the only relevant times are between the time the stone is thrown ($t = 0$) and the time it strikes the ground, when $h = 0$. Solving the equation $h = 30t - 5t^2 = 0$, we find that

$$30t - 5t^2 = 0$$
$$5t(6 - t) = 0 \qquad \text{Factor.}$$
$$5t = 0 \quad \text{or} \quad 6 - t = 0 \quad \text{Set each factor equal to 0.}$$
$$t = 0 \quad \text{or} \quad t = 6. \qquad \text{Solve.}$$

Therefore, the stone leaves the ground at $t = 0$ and returns to the ground at $t = 6$. An appropriate domain that fits the context of this problem is $\{t : 0 \leq t \leq 6\}$. The range consists of all values of $h = 30t - 5t^2$ as t varies over $[0, 6]$. The largest value of h occurs when the stone reaches its highest point at $t = 3$ (halfway through its flight), which is $h = f(3) = 45$. Therefore, the range is $[0, 45]$. These observations are confirmed by the graph of the height function (**Figure 1.7**). Note that this graph is *not* the trajectory of the stone; the stone moves vertically.

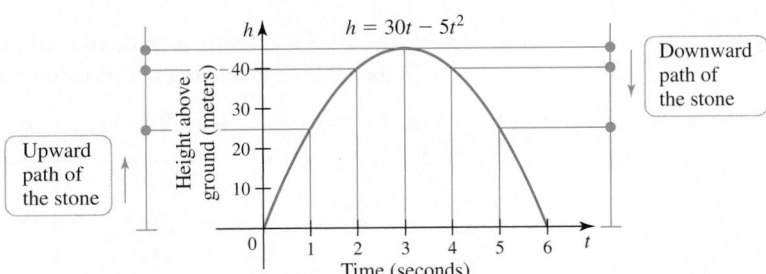

Figure 1.7   *Related Exercises 8–9* ◄

QUICK CHECK 2 State the domain and range of $f(x) = (x^2 + 1)^{-1}$. ◄

Composite Functions

Functions may be combined using sums $(f + g)$, differences $(f - g)$, products (fg), or quotients (f/g). The process called *composition* also produces new functions.

➤ In the composition $y = f(g(x))$, f is the *outer function* and g is the *inner function*.

> **DEFINITION Composite Functions**
>
> Given two functions f and g, the composite function $f \circ g$ is defined by $(f \circ g)(x) = f(g(x))$. It is evaluated in two steps: $y = f(u)$, where $u = g(x)$. The domain of $f \circ g$ consists of all x in the domain of g such that $u = g(x)$ is in the domain of f (**Figure 1.8**).

➤ Three different notations for intervals on the real number line will be used throughout the text:

- $[-2, 3)$ is an example of interval notation,
- $-2 \leq x < 3$ is inequality notation, and
- $\{x: -2 \leq x < 3\}$ is set notation.

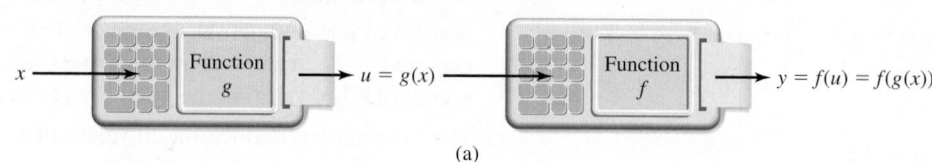

(a)

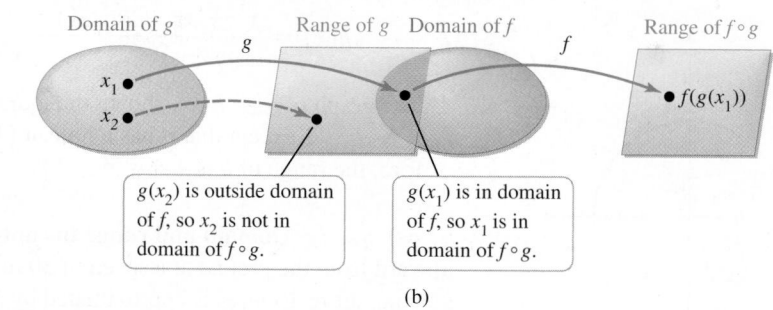

$g(x_2)$ is outside domain of f, so x_2 is not in domain of $f \circ g$.

$g(x_1)$ is in domain of f, so x_1 is in domain of $f \circ g$.

(b)

Figure 1.8

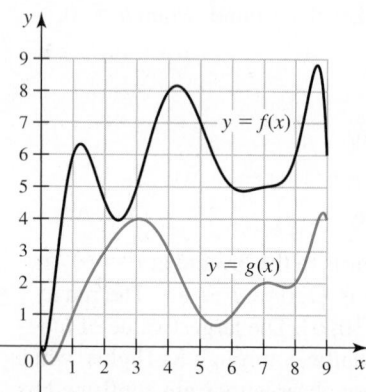

Figure 1.9

EXAMPLE 4 Using graphs to evaluate composite functions Use the graphs of f and g in **Figure 1.9** to find the following values.

a. $f(g(3))$ **b.** $g(f(3))$ **c.** $f(f(4))$ **d.** $f(g(f(8)))$

SOLUTION

a. The graphs indicate that $g(3) = 4$ and $f(4) = 8$, so $f(g(3)) = f(4) = 8$.

b. We see that $g(f(3)) = g(5) = 1$. Observe that $f(g(3)) \neq g(f(3))$.

c. In this case, $f(\underbrace{f(4)}_{8}) = f(8) = 6$.

d. Starting on the inside,

$$f(g(\underbrace{f(8)}_{6})) = f(\underbrace{g(6)}_{1}) = f(1) = 6.$$

Related Exercise 15 ◄

EXAMPLE 5 Using a table to evaluate composite functions Use the function values in the table to evaluate the following composite functions.

a. $(f \circ g)(0)$ **b.** $g(f(-1))$ **c.** $f(g(g(-1)))$

x	-2	-1	0	1	2
$f(x)$	0	1	3	4	2
$g(x)$	-1	0	-2	-3	-4

SOLUTION

a. Using the table, we see that $g(0) = -2$ and $f(-2) = 0$. Therefore, $(f \circ g)(0) = 0$.

b. Because $f(-1) = 1$ and $g(1) = -3$, it follows that $g(f(-1)) = -3$.

c. Starting with the inner function,

$$f(g(\underbrace{g(-1)}_{0})) = f(\underbrace{g(0)}_{-2}) = f(-2) = 0.$$

Related Exercise 16 ◄

EXAMPLE 6 **Composite functions and notation** Let $f(x) = 3x^2 - x$ and $g(x) = 1/x$. Simplify the following expressions.

a. $f(5p + 1)$ **b.** $g(1/x)$ **c.** $f(g(x))$ **d.** $g(f(x))$

SOLUTION In each case, the functions work on their arguments.

a. The argument of f is $5p + 1$, so

$$f(5p + 1) = 3(5p + 1)^2 - (5p + 1) = 75p^2 + 25p + 2.$$

b. Because g requires taking the reciprocal of the argument, we take the reciprocal of $1/x$ and find that $g(1/x) = 1/(1/x) = x$.

c. The argument of f is $g(x)$, so

$$f(g(x)) = f\left(\frac{1}{x}\right) = 3\left(\frac{1}{x}\right)^2 - \left(\frac{1}{x}\right) = \frac{3}{x^2} - \frac{1}{x} = \frac{3 - x}{x^2}.$$

➤ Examples 6c and 6d demonstrate that, in general,

$$f(g(x)) \neq g(f(x)).$$

d. The argument of g is $f(x)$, so

$$g(f(x)) = g(3x^2 - x) = \frac{1}{3x^2 - x}.$$

Related Exercises 33–37 ◄

EXAMPLE 7 **Working with composite functions** Identify possible choices for the inner and outer functions in the following composite functions. Give the domain of the composite function.

a. $h(x) = \sqrt{9x - x^2}$ **b.** $h(x) = \dfrac{2}{(x^2 - 1)^3}$

SOLUTION

➤ Techniques for solving inequalities are discussed in Appendix B, online at goo.gl/6DCbbM.

a. An obvious outer function is $f(x) = \sqrt{x}$, which works on the inner function $g(x) = 9x - x^2$. Therefore, h can be expressed as $h = f \circ g$ or $h(x) = f(g(x))$. The domain of $f \circ g$ consists of all values of x such that $9x - x^2 \geq 0$. Solving this inequality gives $\{x: 0 \leq x \leq 9\}$ as the domain of $f \circ g$.

b. A good choice for an outer function is $f(x) = 2/x^3 = 2x^{-3}$, which works on the inner function $g(x) = x^2 - 1$. Therefore, h can be expressed as $h = f \circ g$ or $h(x) = f(g(x))$. The domain of $f \circ g$ consists of all values of $g(x)$ such that $g(x) \neq 0$, which is $\{x: x \neq \pm 1\}$. *Related Exercises 44–45* ◄

EXAMPLE 8 **More composite functions** Given $f(x) = \sqrt[3]{x}$ and $g(x) = x^2 - x - 6$, find the following composite functions and their domains.

a. $g \circ f$ **b.** $g \circ g$

SOLUTION

a. We have

$$(g \circ f)(x) = g(f(x)) = g(\underbrace{\sqrt[3]{x}}_{f(x)}) = (\underbrace{\sqrt[3]{x}}_{f(x)})^2 - \sqrt[3]{x} - 6 = x^{2/3} - x^{1/3} - 6.$$

Because the domains of f and g are $(-\infty, \infty)$, the domain of $f \circ g$ is also $(-\infty, \infty)$.

b. In this case, we have the composition of two polynomials:

$$(g \circ g)(x) = g(g(x))$$
$$= g(x^2 - x - 6)$$
$$= (\underbrace{x^2 - x - 6}_{g(x)})^2 - (\underbrace{x^2 - x - 6}_{g(x)}) - 6$$
$$= x^4 - 2x^3 - 12x^2 + 13x + 36.$$

QUICK CHECK 3 If $f(x) = x^2 + 1$ and $g(x) = x^2$, find $f \circ g$ and $g \circ f$. ◄

The domain of the composition of two polynomials is $(-\infty, \infty)$.

Related Exercises 47–48 ◄

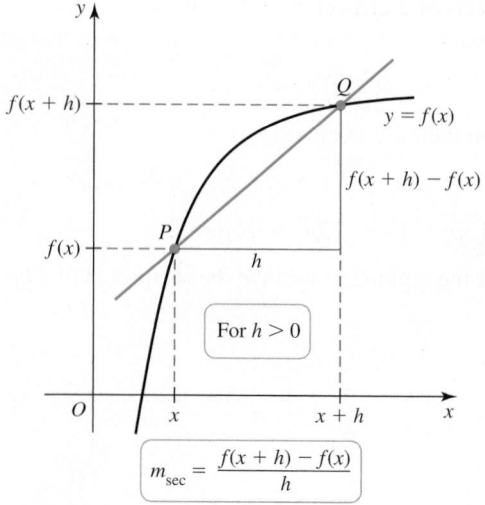

$$m_{sec} = \frac{f(x+h) - f(x)}{h}$$

Figure 1.10

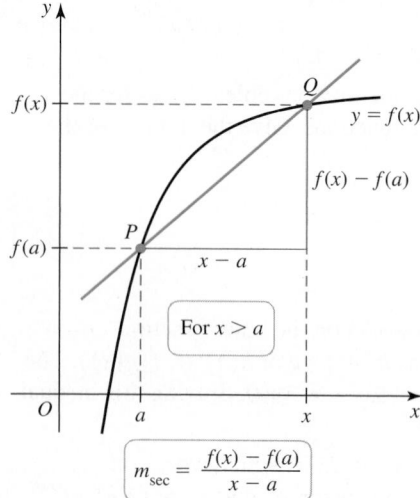

$$m_{sec} = \frac{f(x) - f(a)}{x - a}$$

Figure 1.11

▶ Treat $f(x + h)$ like the composition $f(g(x))$, where $x + h$ plays the role of $g(x)$. It may help to establish a pattern in your mind before evaluating $f(x + h)$. For instance, using the function in Example 9a, we have

$$f(x) = 3x^2 - x;$$
$$f(12) = 3 \cdot 12^2 - 12;$$
$$f(b) = 3b^2 - b;$$
$$f(\text{math}) = 3 \cdot \text{math}^2 - \text{math};$$

therefore,

$$f(x + h) = 3(x + h)^2 - (x + h).$$

▶ See the front papers of this text for a review of factoring formulas.

Secant Lines and the Difference Quotient

As you will see shortly, slopes of lines and curves play a fundamental role in calculus. **Figure 1.10** shows two points $P(x, f(x))$ and $Q(x + h, f(x + h))$ on the graph of $y = f(x)$ in the case that $h > 0$. A line through any two points on a curve is called a **secant line**; its importance in the study of calculus is explained in Chapters 2 and 3. For now, we focus on the slope of the secant line through P and Q, which is denoted m_{sec} and is given by

$$m_{sec} = \frac{\text{change in } y}{\text{change in } x} = \frac{f(x+h) - f(x)}{(x+h) - x} = \frac{f(x+h) - f(x)}{h}.$$

The slope formula $\dfrac{f(x+h) - f(x)}{h}$ is also known as a **difference quotient**, and it can be expressed in several ways depending on how the coordinates of P and Q are labeled. For example, given the coordinates $P(a, f(a))$ and $Q(x, f(x))$ (**Figure 1.11**), the difference quotient is

$$m_{sec} = \frac{f(x) - f(a)}{x - a}.$$

We interpret the slope of the secant line in this form as the **average rate of change** of f over the interval $[a, x]$.

EXAMPLE 9 **Working with difference quotients**

a. Simplify the difference quotient $\dfrac{f(x + h) - f(x)}{h}$, for $f(x) = 3x^2 - x$.

b. Simplify the difference quotient $\dfrac{f(x) - f(a)}{x - a}$, for $f(x) = x^3$.

SOLUTION

a. First note that $f(x + h) = 3(x + h)^2 - (x + h)$. We substitute this expression into the difference quotient and simplify:

$$\frac{f(x+h) - f(x)}{h} = \frac{\overbrace{3(x+h)^2 - (x+h)}^{f(x+h)} - \overbrace{(3x^2 - x)}^{f(x)}}{h}$$

$$= \frac{3(x^2 + 2xh + h^2) - (x + h) - (3x^2 - x)}{h} \quad \text{Expand } (x + h)^2.$$

$$= \frac{3x^2 + 6xh + 3h^2 - x - h - 3x^2 + x}{h} \quad \text{Distribute.}$$

$$= \frac{6xh + 3h^2 - h}{h} \quad \text{Simplify.}$$

$$= \frac{h(6x + 3h - 1)}{h} = 6x + 3h - 1. \quad \text{Factor and simplify.}$$

b. The factoring formula for the difference of perfect cubes is needed:

$$\frac{f(x) - f(a)}{x - a} = \frac{x^3 - a^3}{x - a}$$

$$= \frac{(x - a)(x^2 + ax + a^2)}{x - a} \quad \text{Factoring formula}$$

$$= x^2 + ax + a^2. \quad \text{Simplify.}$$

Related Exercises 66, 72 ◀

EXAMPLE 10   Interpreting the slope of the secant line The position of a hiker on a trail at various times t is recorded by a GPS watch worn by the hiker. These data are then uploaded to a computer to produce the graph of the distance function $d = f(t)$ shown in **Figure 1.12**, where d measures the distance traveled on the trail in miles and t is the elapsed time in hours from the beginning of the hike.

a. Find the slope of the secant line that passes through the points on the graph corresponding to the trail segment between milepost 3 and milepost 5, and interpret the result.

b. Estimate the slope of the secant line that passes through points A and B in Figure 1.12, and compare it to the slope of the secant line found in part (a).

> Figure 1.12 contains actual GPS data collected in Rocky Mountain National Park. See Exercises 75–76 for another look at the data set.

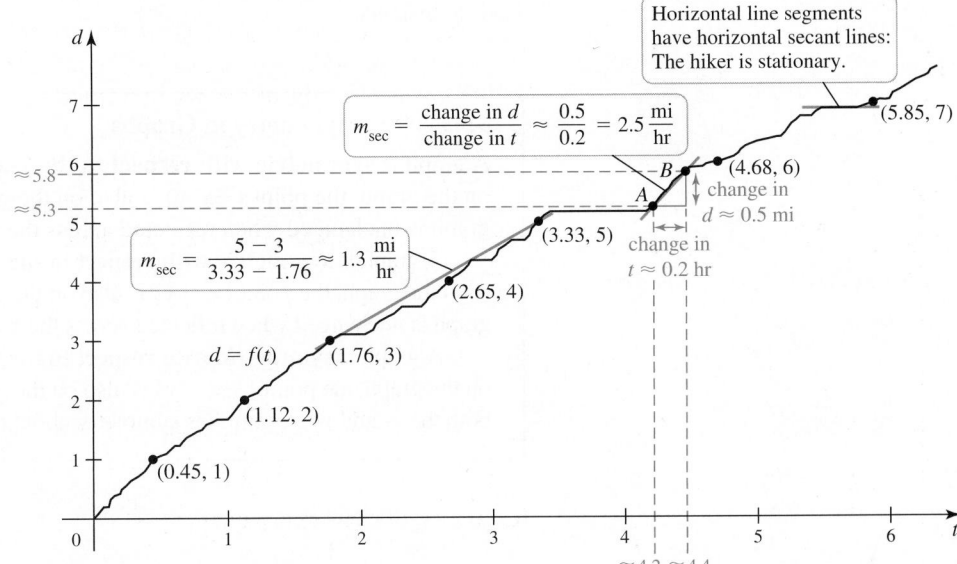

Figure 1.12

SOLUTION

a. We see from the graph of $d = f(t)$ that 1.76 hours (about 1 hour and 46 minutes) has elapsed when the hiker arrives at milepost 3, while milepost 5 is reached 3.33 hours into the hike. This information is also expressed as $f(1.76) = 3$ and $f(3.33) = 5$. To find the slope of the secant line through these points, we compute the change in distance divided by the change in time:

$$m_{\text{sec}} = \frac{f(3.33) - f(1.76)}{3.33 - 1.76} = \frac{5 - 3}{3.33 - 1.76} \approx 1.3 \, \frac{\text{mi}}{\text{hr}}.$$

The units provide a clue about the physical meaning of the slope: It measures the average rate at which the distance changes per hour, which is the average speed of the hiker. In this case, the hiker walks with an average speed of approximately 1.3 mi/hr between mileposts 3 and 5.

b. From the graph we see that the coordinates of points A and B are approximately $(4.2, 5.3)$ and $(4.4, 5.8)$, respectively, which implies the hiker walks $5.8 - 5.3 = 0.5$ mi in $4.4 - 4.2 = 0.2$ hr. The slope of the secant line through A and B is

$$m_{\text{sec}} = \frac{\text{change in } d}{\text{change in } t} \approx \frac{0.5}{0.2} = 2.5 \, \frac{\text{mi}}{\text{hr}}.$$

For this segment of the trail, the hiker walks at an average speed of about 2.5 mi/hr, nearly twice as fast as the average speed computed in part (a). Expressed another way, steep sections of the distance curve yield steep secant lines, which correspond to faster average hiking speeds. Conversely, any secant line with slope equal to 0 corresponds to an average speed of 0. Looking one last time at Figure 1.12, we can identify the time intervals during which the hiker was resting alongside the trail—whenever the distance curve is horizontal, the hiker is not moving.

Related Exercise 75 ◄

QUICK CHECK 4 Refer to Figure 1.12. Find the hiker's average speed during the first mile of the trail and then determine the hiker's average speed in the time interval from 3.9 to 4.1 hours. ◄

Symmetry

The word *symmetry* has many meanings in mathematics. Here we consider symmetries of graphs and the relations they represent. Taking advantage of symmetry often saves time and leads to insights.

DEFINITION Symmetry in Graphs

A graph is **symmetric with respect to the y-axis** if whenever the point (x, y) is on the graph, the point $(-x, y)$ is also on the graph. This property means that the graph is unchanged when reflected across the y-axis (**Figure 1.13a**).

A graph is **symmetric with respect to the x-axis** if whenever the point (x, y) is on the graph, the point $(x, -y)$ is also on the graph. This property means that the graph is unchanged when reflected across the x-axis (**Figure 1.13b**).

A graph is **symmetric with respect to the origin** if whenever the point (x, y) is on the graph, the point $(-x, -y)$ is also on the graph (**Figure 1.13c**). Symmetry about both the x- and y-axes implies symmetry about the origin, but not vice versa.

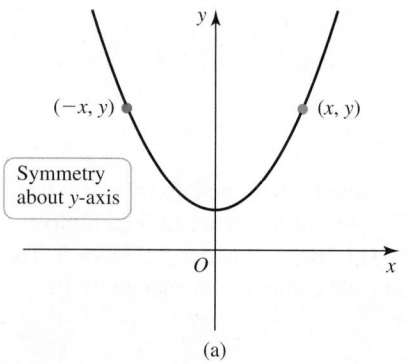

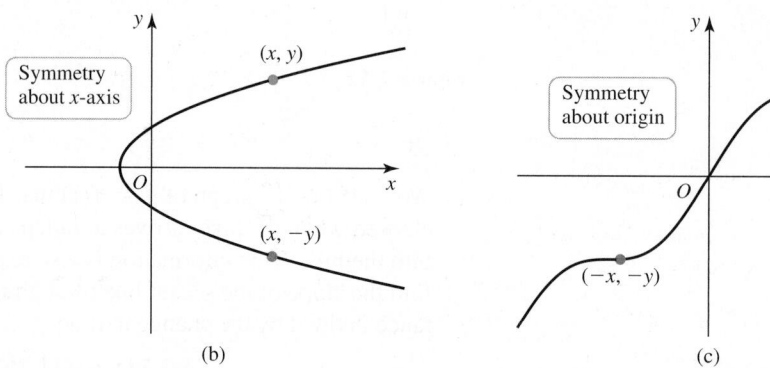

Figure 1.13

DEFINITION Symmetry in Functions

An **even function** f has the property that $f(-x) = f(x)$, for all x in the domain. The graph of an even function is symmetric about the y-axis.

An **odd function** f has the property that $f(-x) = -f(x)$, for all x in the domain. The graph of an odd function is symmetric about the origin.

Polynomials consisting of only even powers of the variable (of the form x^{2n}, where n is a nonnegative integer) are even functions. Polynomials consisting of only odd powers of the variable (of the form x^{2n+1}, where n is a nonnegative integer) are odd functions.

QUICK CHECK 5 Explain why the graph of a nonzero function is never symmetric with respect to the x-axis. ◄

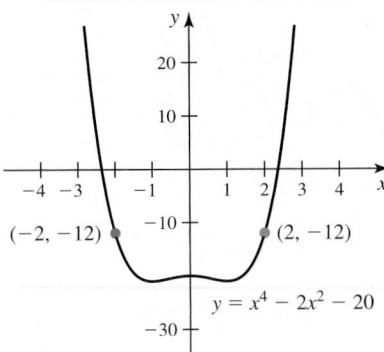

Even function: If (x, y) is on the graph, then $(-x, y)$ is on the graph.

$$y = x^4 - 2x^2 - 20$$

$(-2, -12)$ $(2, -12)$

Figure 1.14

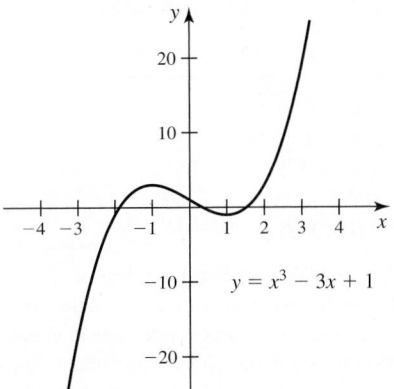

No symmetry: neither even nor odd function.

$$y = x^3 - 3x + 1$$

Figure 1.15

▶ The symmetry of compositions of even and odd functions is considered in Exercises 101–104.

EXAMPLE 11   Identifying symmetry in functions Identify the symmetry, if any, in the following functions.

a. $f(x) = x^4 - 2x^2 - 20$ **b.** $g(x) = x^3 - 3x + 1$ **c.** $h(x) = \dfrac{1}{x^3 - x}$

SOLUTION

a. The function f consists of only even powers of x (where $20 = 20 \cdot 1 = 20x^0$ and x^0 is considered an even power). Therefore, f is an even function (**Figure 1.14**). This fact is verified by showing that $f(-x) = f(x)$:

$$f(-x) = (-x)^4 - 2(-x)^2 - 20 = x^4 - 2x^2 - 20 = f(x).$$

b. The function g consists of two odd powers and one even power (again, $1 = x^0$ is an even power). Therefore, we expect that g has no symmetry about the y-axis or the origin (**Figure 1.15**). Note that

$$g(-x) = (-x)^3 - 3(-x) + 1 = -x^3 + 3x + 1,$$

so $g(-x)$ equals neither $g(x)$ nor $-g(x)$; therefore, g has no symmetry.

c. In this case, h is a composition of an odd function $f(x) = 1/x$ with an odd function $g(x) = x^3 - x$. Note that

$$h(-x) = \frac{1}{(-x)^3 - (-x)} = -\frac{1}{x^3 - x} = -h(x).$$

Because $h(-x) = -h(x)$, h is an odd function (**Figure 1.16**).

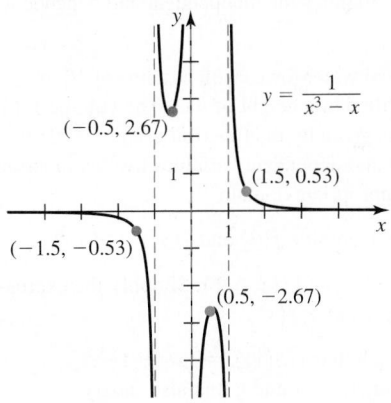

Odd function: If (x, y) is on the graph, then $(-x, -y)$ is on the graph.

$$y = \frac{1}{x^3 - x}$$

$(-0.5, 2.67)$

$(1.5, 0.53)$

$(-1.5, -0.53)$

$(0.5, -2.67)$

Figure 1.16

Related Exercises 79–81 ◀

SECTION 1.1 EXERCISES

Getting Started

1. Use the terms *domain, range, independent variable,* and *dependent variable* to explain how a function relates one variable to another variable.

2. Is the independent variable of a function associated with the domain or range? Is the dependent variable associated with the domain or range?

3. Decide whether graph A, graph B, or both represent functions.

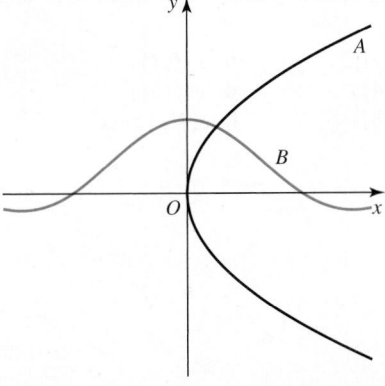

4.   The entire graph of f is given. State the domain and range of f.

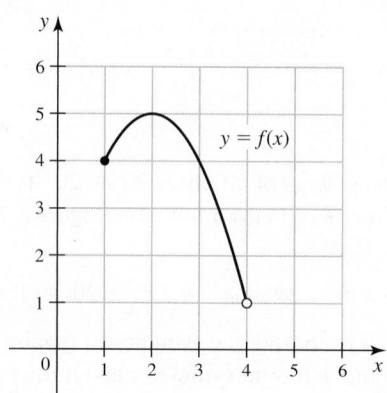

5.   Which statement about a function is true? (i) For each value of x in the domain, there corresponds one unique value of y in the range; (ii) for each value of y in the range, there corresponds one unique value of x in the domain. Explain.

6.   Determine the domain and range of $g(x) = \dfrac{x^2 - 1}{x - 1}$. Sketch a graph of g.

7.   Determine the domain and range of $f(x) = 3x^2 - 10$.

8.   **Throwing a stone** A stone is thrown vertically upward from the ground at a speed of 40 m/s at time $t = 0$. Its distance d (in meters) above the ground (neglecting air resistance) is approximated by the function $f(t) = 40t - 5t^2$. Determine an appropriate domain for this function. Identify the independent and dependent variables.

9.   **Water tower** A cylindrical water tower with a radius of 10 m and a height of 50 m is filled to a height of h m. The volume V of water (in cubic meters) is given by the function $g(h) = 100\pi h$. Identify the independent and dependent variables for this function, and then determine an appropriate domain.

10.   Let $f(x) = 1/(x^3 + 1)$. Compute $f(2)$ and $f(y^2)$.

11.   Let $f(x) = 2x + 1$ and $g(x) = 1/(x - 1)$. Simplify the expressions $f(g(1/2))$, $g(f(4))$, and $g(f(x))$.

12.   Find functions f and g such that $f(g(x)) = (x^2 + 1)^5$. Find a different pair of functions f and g that also satisfy $f(g(x)) = (x^2 + 1)^5$.

13.   Explain how to find the domain of $f \circ g$ if you know the domain and range of f and g.

14.   If $f(x) = \sqrt{x}$ and $g(x) = x^3 - 2$, simplify the expressions $(f \circ g)(3)$, $(f \circ f)(64)$, $(g \circ f)(x)$, and $(f \circ g)(x)$.

15.   Use the graphs of f and g in the figure to determine the following function values.

    a. $(f \circ g)(2)$ **b.** $g(f(2))$
    c. $f(g(4))$ **d.** $g(f(5))$
    e. $f(f(8))$ **f.** $g(f(g(5)))$

16.   Use the table to evaluate the given compositions.

x	-1	0	1	2	3	4
$f(x)$	3	1	0	-1	-3	-1
$g(x)$	-1	0	2	3	4	5
$h(x)$	0	-1	0	3	0	4

    a. $h(g(0))$ **b.** $g(f(4))$
    c. $h(h(0))$ **d.** $g(h(f(4)))$
    e. $f(f(f(1)))$ **f.** $h(h(h(0)))$
    g. $f(h(g(2)))$ **h.** $g(f(h(4)))$
    i. $g(g(g(1)))$ **j.** $f(f(h(3)))$

17.   **Rising radiosonde** The National Weather Service releases approximately 70,000 radiosondes every year to collect data from the atmosphere. Attached to a balloon, a radiosonde rises at about 1000 ft/min until the balloon bursts in the upper atmosphere. Suppose a radiosonde is released from a point 6 ft above the ground and that 5 seconds later, it is 83 ft above the ground. Let $f(t)$ represent the height (in feet) that the radiosonde is above the ground t seconds after it is released. Evaluate $\dfrac{f(5) - f(0)}{5 - 0}$ and interpret the meaning of this quotient.

18.   **World record free fall** On October 14, 2012, Felix Baumgartner stepped off a balloon capsule at an altitude of 127,852.4 feet and began his free fall. It is claimed that Felix reached the speed of sound 34 seconds into his fall at an altitude of 109,731 feet and that he continued to fall at supersonic speed for 30 seconds until he was at an altitude of 75,330.4 feet. Let $f(t)$ equal the distance that Felix had fallen t seconds after leaving his capsule. Calculate $f(0)$, $f(34)$, $f(64)$, and his average supersonic speed $\dfrac{f(64) - f(34)}{64 - 34}$ (in ft/s) over the time interval $[34, 64]$.
(*Source:* http://www.redbullstratos.com)

19.   Suppose f is an even function with $f(2) = 2$ and g is an odd function with $g(2) = -2$. Evaluate $f(-2)$, $g(-2)$, $f(g(2))$, and $g(f(-2))$.

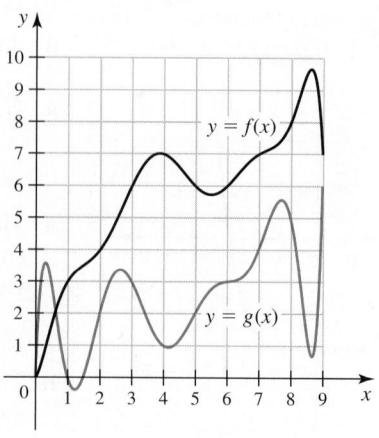

20. Complete the left half of the graph of g if g is an odd function.

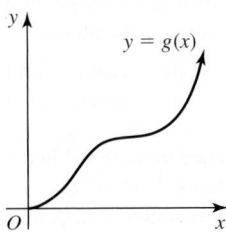

21. State whether the functions represented by graphs A, B, and C in the figure are even, odd, or neither.

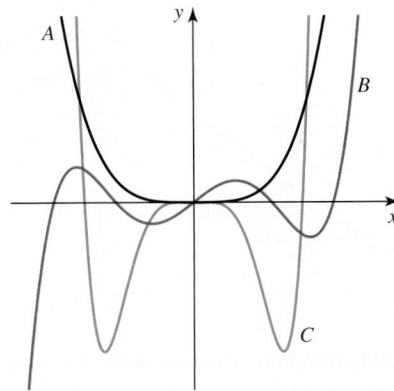

22. State whether the functions represented by graphs A, B, and C in the figure are even, odd, or neither.

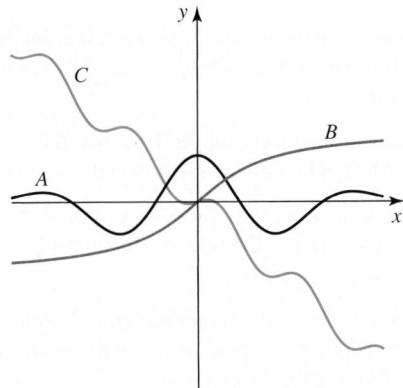

Practice Exercises

23–26. Domain and range *State the domain and range of the function.*

23. $f(x) = \dfrac{x^2 - 5x + 6}{x - 2}$

24. $f(x) = \dfrac{x - 2}{2 - x}$

25. $f(x) = \sqrt{7 - x^2}$

26. $f(x) = -\sqrt{25 - x^2}$

27–30. Domain *State the domain of the function.*

27. $h(u) = \sqrt[3]{u - 1}$

28. $F(w) = \sqrt[4]{2 - w}$

29. $f(x) = (9 - x^2)^{3/2}$

30. $g(t) = \dfrac{1}{1 + t^2}$

T 31. Launching a rocket A small rocket is launched vertically upward from the edge of a cliff 80 ft above the ground at a speed of 96 ft/s. Its height (in feet) above the ground is given by $h(t) = -16t^2 + 96t + 80$, where t represents time measured in seconds.

   a. Assuming the rocket is launched at $t = 0$, what is an appropriate domain for h?

   b. Graph h and determine the time at which the rocket reaches its highest point. What is the height at that time?

32. Draining a tank (Torricelli's law) A cylindrical tank with a cross-sectional area of 10 m^2 is filled to a depth of 25 m with water. At $t = 0$ s, a drain in the bottom of the tank with an area of 1 m^2 is opened, allowing water to flow out of the tank. The depth of water in the tank (in meters) at time $t \geq 0$ is $d(t) = (5 - 0.22t)^2$.

   a. Check that $d(0) = 25$, as specified.

   b. At what time is the tank empty?

   c. What is an appropriate domain for d?

33–42. Composite functions and notation *Let $f(x) = x^2 - 4$, $g(x) = x^3$, and $F(x) = 1/(x - 3)$. Simplify or evaluate the following expressions.*

33. $g(1/z)$

34. $F(y^4)$

35. $F(g(y))$

36. $f(g(w))$

37. $g(f(u))$

38. $\dfrac{f(2 + h) - f(2)}{h}$

39. $F(F(x))$

40. $g(F(f(x)))$

41. $f(\sqrt{x + 4})$

42. $F\left(\dfrac{3x + 1}{x}\right)$

43–46. Working with composite functions *Find possible choices for outer and inner functions f and g such that the given function h equals $f \circ g$.*

43. $h(x) = (x^3 - 5)^{10}$

44. $h(x) = \dfrac{2}{(x^6 + x^2 + 1)^2}$

45. $h(x) = \sqrt{x^4 + 2}$

46. $h(x) = \dfrac{1}{\sqrt{x^3 - 1}}$

47–54. More composite functions *Let $f(x) = |x|$, $g(x) = x^2 - 4$, $F(x) = \sqrt{x}$, and $G(x) = 1/(x - 2)$. Determine the following composite functions and give their domains.*

47. $f \circ g$

48. $g \circ f$

49. $f \circ G$

50. $f \circ g \circ G$

51. $G \circ g \circ f$

52. $g \circ F \circ F$

53. $g \circ g$

54. $G \circ G$

55–60. Missing piece *Let $g(x) = x^2 + 3$. Find a function f that produces the given composition.*

55. $(f \circ g)(x) = x^2$

56. $(f \circ g)(x) = \dfrac{1}{x^2 + 3}$

57. $(f \circ g)(x) = x^4 + 6x^2 + 9$

58. $(f \circ g)(x) = x^4 + 6x^2 + 20$

59. $(g \circ f)(x) = x^4 + 3$

60. $(g \circ f)(x) = x^{2/3} + 3$

61. Explain why or why not Determine whether the following statements are true and give an explanation or counterexample.

    a. The range of $f(x) = 2x - 38$ is all real numbers.

    b. The relation $y = x^6 + 1$ is *not* a function because $y = 2$ for both $x = -1$ and $x = 1$.

    c. If $f(x) = x^{-1}$, then $f(1/x) = 1/f(x)$.

    d. In general, $f(f(x)) = (f(x))^2$.

    e. In general, $f(g(x)) = g(f(x))$.

    f. By definition, $f(g(x)) = (f \circ g)(x)$.

    g. If $f(x)$ is an even function, then $cf(ax)$ is an even function, where a and c are nonzero real numbers.

    h. If $f(x)$ is an odd function, then $f(x) + d$ is an odd function, where d is a nonzero real number.

    i. If f is both even *and* odd, then $f(x) = 0$ for all x.

62–68. Working with difference quotients *Simplify the difference quotient* $\dfrac{f(x + h) - f(x)}{h}$ *for the following functions.*

62. $f(x) = 10$ **63.** $f(x) = 3x$

64. $f(x) = 4x - 3$ **65.** $f(x) = x^2$

66. $f(x) = 2x^2 - 3x + 1$ **67.** $f(x) = \dfrac{2}{x}$

68. $f(x) = \dfrac{x}{x + 1}$

69–74. Working with difference quotients *Simplify the difference quotient* $\dfrac{f(x) - f(a)}{x - a}$ *for the following functions.*

69. $f(x) = x^2 + x$ **70.** $f(x) = 4 - 4x - x^2$

71. $f(x) = x^3 - 2x$ **72.** $f(x) = x^4$

73. $f(x) = -\dfrac{4}{x^2}$ **74.** $f(x) = \dfrac{1}{x} - x^2$

75. GPS data A GPS device tracks the elevation E (in feet) of a hiker walking in the mountains. The elevation t hours after beginning the hike is given in the figure.

    a. Find the slope of the secant line that passes through points A and B. Interpret your answer as an average rate of change over the interval $1 \le t \le 3$.

    b. Repeat the procedure outlined in part (a) for the secant line that passes through points P and Q.

    c. Notice that the curve in the figure is horizontal for an interval of time near $t = 5.5$ hr. Give a plausible explanation for the horizontal line segment.

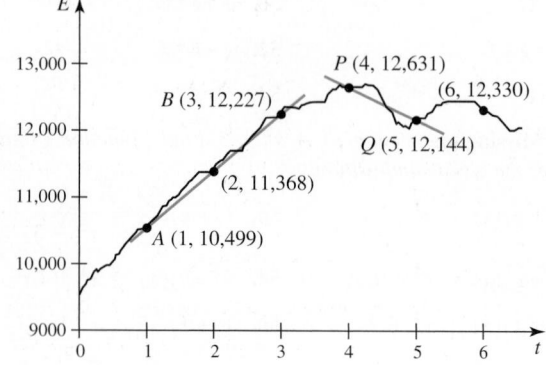

76. Elevation vs. Distance The following graph, obtained from GPS data, shows the elevation of a hiker as a function of the distance d from the starting point of the trail.

    a. Find the slope of the secant line that passes through points A and B. Interpret your answer as an average rate of change over the interval $1 \le d \le 3$.

    b. Repeat the procedure outlined in part (a) for the secant line that passes through points P and Q.

    c. Notice that the elevation function is nearly constant over the segment of the trail from mile $d = 4.5$ to mile $d = 5$. Give a plausible explanation for the horizontal line segment.

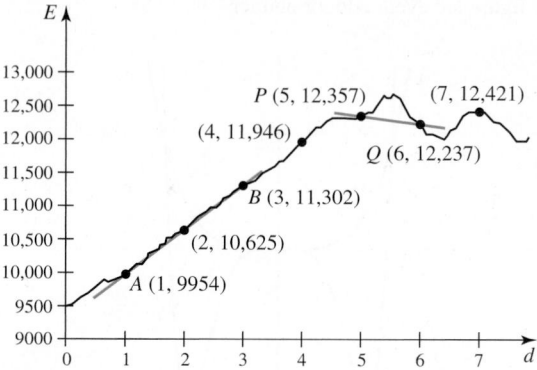

77–78. Interpreting the slope of secant lines *In each exercise, a function and an interval of its independent variable are given. The endpoints of the interval are associated with points P and Q on the graph of the function.*

    a. Sketch a graph of the function and the secant line through P and Q.

    b. Find the slope of the secant line in part (a), and interpret your answer in terms of an average rate of change over the interval. Include units in your answer.

77. After t seconds, an object dropped from rest falls a distance $d = 16t^2$, where d is measured in feet and $2 \le t \le 5$.

78. The volume V of an ideal gas in cubic centimeters is given by $V = 2/p$, where p is the pressure in atmospheres and $0.5 \le p \le 2$.

79–86. Symmetry *Determine whether the graphs of the following equations and functions are symmetric about the x-axis, the y-axis, or the origin. Check your work by graphing.*

79. $f(x) = x^4 + 5x^2 - 12$ **80.** $f(x) = 3x^5 + 2x^3 - x$

81. $f(x) = x^5 - x^3 - 2$ **82.** $f(x) = 2|x|$

83. $x^{2/3} + y^{2/3} = 1$ **84.** $x^3 - y^5 = 0$

85. $f(x) = x|x|$ **86.** $|x| + |y| = 1$

Explorations and Challenges

87. Composition of even and odd functions from graphs Assume f is an even function and g is an odd function. Use the (incomplete) graphs of f and g in the figure to determine the following function values.

    a. $f(g(-2))$ **b.** $g(f(-2))$

    c. $f(g(-4))$ **d.** $g(f(5) - 8)$

    e. $g(g(-7))$ **f.** $f(1 - f(8))$

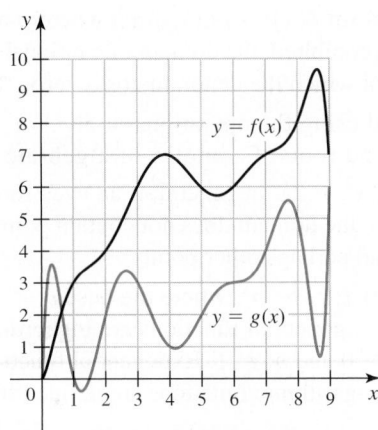

88. Composition of even and odd functions from tables Assume f is an even function, g is an odd function, and both are defined at 0. Use the (incomplete) table to evaluate the given compositions.

x	1	2	3	4
$f(x)$	2	-1	3	-4
$g(x)$	-3	-1	-4	-2

a. $f(g(-1))$ b. $g(f(-4))$
c. $f(g(-3))$ d. $f(g(-2))$
e. $g(g(-1))$ f. $f(g(0) - 1)$
g. $f(g(g(-2)))$ h. $g(f(f(-4)))$
i. $g(g(g(-1)))$

89. Absolute value graphs Use the definition of absolute value (see Appendix B, online at goo.gl/6DCbbM) to graph the equation $|x| - |y| = 1$. Use a graphing utility to check your work.

90. Graphing semicircles Show that the graph of $f(x) = 10 + \sqrt{-x^2 + 10x - 9}$ is the upper half of a circle. Then determine the domain and range of the function.

91. Graphing semicircles Show that the graph of $g(x) = 2 - \sqrt{-x^2 + 6x + 16}$ is the lower half of a circle. Then determine the domain and range of the function.

92. Even and odd at the origin
     a. If $f(0)$ is defined and f is an even function, is it necessarily true that $f(0) = 0$? Explain.
     b. If $f(0)$ is defined and f is an odd function, is it necessarily true that $f(0) = 0$? Explain.

▣ 93–96. Polynomial calculations *Find a polynomial f that satisfies the following properties. (Hint: Determine the degree of f; then substitute a polynomial of that degree and solve for its coefficients.)*

93. $f(f(x)) = 9x - 8$ **94.** $(f(x))^2 = 9x^2 - 12x + 4$

95. $f(f(x)) = x^4 - 12x^2 + 30$ **96.** $(f(x))^2 = x^4 - 12x^2 + 36$

97–100. Difference quotients *Simplify the difference quotients* $\dfrac{f(x+h) - f(x)}{h}$ *and* $\dfrac{f(x) - f(a)}{x - a}$ *by rationalizing the numerator.*

97. $f(x) = \sqrt{x}$ **98.** $f(x) = \sqrt{1 - 2x}$

99. $f(x) = -\dfrac{3}{\sqrt{x}}$ **100.** $f(x) = \sqrt{x^2 + 1}$

101–104. Combining even and odd functions *Let E be an even function and O be an odd function. Determine the symmetry, if any, of the following functions.*

101. $E + O$ **102.** $E \cdot O$

103. $O \circ E$ **104.** $E \circ O$

QUICK CHECK ANSWERS

1. $3, x^4 - 2x^2, t^2 - 2t, p^2 - 4p + 3$ **2.** Domain is all real numbers; range is $\{y: 0 < y \le 1\}$.   **3.** $(f \circ g)(x) = x^4 + 1$ and $(g \circ f)(x) = (x^2 + 1)^2$ **4.** Average speed ≈ 2.2 mi/hr for first mile; average speed $= 0$ on $3.9 \le t \le 4.1$.   **5.** If the graph were symmetric with respect to the x-axis, it would not pass the vertical line test. ◄

1.2 Representing Functions

We consider four approaches to defining and representing functions: formulas, graphs, tables, and words.

Using Formulas

The following list is a brief catalog of the families of functions that are introduced in this chapter and studied systematically throughout this text; they are all defined by *formulas*.

> ➤ One version of the Fundamental Theorem of Algebra states that a nonzero polynomial of degree n has exactly n (possibly complex) roots, counting each root up to its multiplicity.

1. Polynomials are functions of the form

$$p(x) = a_n x^n + a_{n-1} x^{n-1} + \cdots + a_1 x + a_0,$$

where the **coefficients** $a_0, a_1, \ldots, a_n$ are real numbers with $a_n \ne 0$ and the nonnegative integer n is the **degree** of the polynomial. The domain of any polynomial is the set of all real numbers. An nth-degree polynomial can have as many as n real **zeros** or **roots**—values of x at which $p(x) = 0$; the zeros are points at which the graph of p intersects the x-axis.

2. **Rational functions** are ratios of the form $f(x) = p(x)/q(x)$, where p and q are polynomials. Because division by zero is prohibited, the domain of a rational function is the set of all real numbers except those for which the denominator is zero.

3. **Algebraic functions** are constructed using the operations of algebra: addition, subtraction, multiplication, division, and roots. Examples of algebraic functions are $f(x) = \sqrt{2x^3 + 4}$ and $g(x) = x^{1/4}(x^3 + 2)$. In general, if an even root (square root, fourth root, and so forth) appears, then the domain does not contain points at which the quantity under the root is negative (and perhaps other points).

> Exponential and logarithmic functions are introduced in Section 1.3.

4. **Exponential functions** have the form $f(x) = b^x$, where the base $b \neq 1$ is a positive real number. Closely associated with exponential functions are **logarithmic functions** of the form $f(x) = \log_b x$, where $b > 0$ and $b \neq 1$. Exponential functions have a domain consisting of all real numbers. Logarithmic functions are defined for positive real numbers.

   The **natural exponential function** is $f(x) = e^x$, with base $b = e$, where $e \approx 2.71828\ldots$ is one of the fundamental constants of mathematics. Associated with the natural exponential function is the **natural logarithm function** $f(x) = \ln x$, which also has the base $b = e$.

> Trigonometric functions and their inverses are introduced in Section 1.4.

5. The **trigonometric functions** are $\sin x$, $\cos x$, $\tan x$, $\cot x$, $\sec x$, and $\csc x$; they are fundamental to mathematics and many areas of application. Also important are their relatives, the **inverse trigonometric functions**.

6. Trigonometric, exponential, and logarithmic functions are a few examples of a large family called **transcendental functions**. **Figure 1.17** shows the organization of these functions, which are explored in detail in upcoming chapters.

QUICK CHECK 1 Are all polynomials rational functions? Are all algebraic functions polynomials? ◄

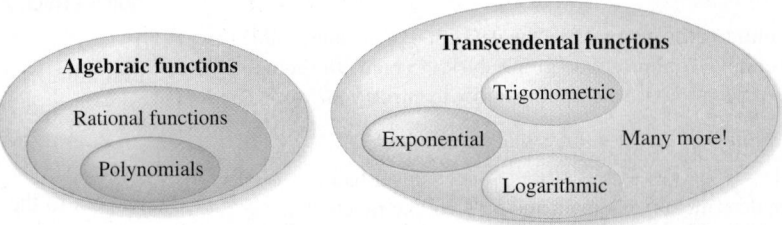

Figure 1.17

Using Graphs

Although formulas are the most compact way to represent many functions, graphs often provide the most illuminating representations. Two of countless examples of functions and their graphs are shown in **Figure 1.18**. Much of this text is devoted to creating and analyzing graphs of functions.

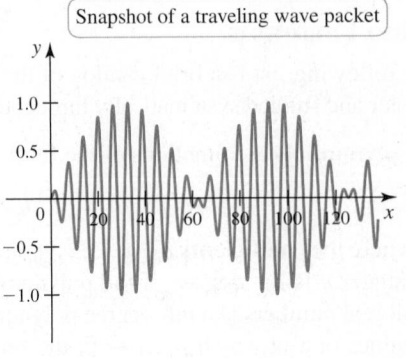

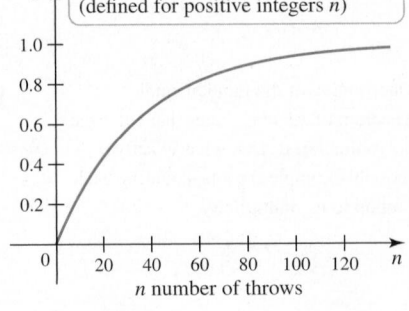

Figure 1.18

There are two approaches to graphing functions.

- Graphing calculators and graphing software are easy to use and powerful. Such **technology** easily produces graphs of most functions encountered in this text. We assume you know how to use a graphing utility.

- Graphing utilities, however, are not infallible. Therefore, you should also strive to master **analytical methods** (pencil-and-paper methods) in order to analyze functions and make accurate graphs by hand. Analytical methods rely heavily on calculus and are presented throughout this text.

The important message is this: Both technology and analytical methods are essential and must be used together in an integrated way to produce accurate graphs.

Linear Functions   One form of the equation of a line (see Appendix B, online at goo.gl/6DCbbM) is $y = mx + b$, where m and b are constants. Therefore, the function $f(x) = mx + b$ has a straight-line graph and is called a **linear function**.

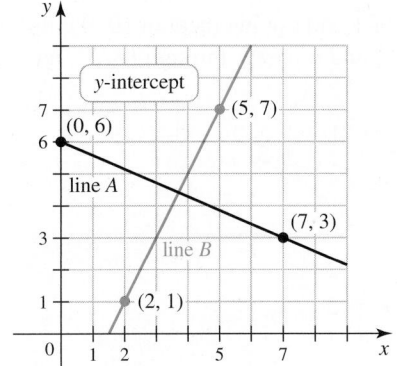

Figure 1.19

EXAMPLE 1   Linear functions and their graphs Determine the function represented by **(a)** line A in **Figure 1.19** and **(b)** line B in Figure 1.19.

SOLUTION

a. From the graph, we see that the y-intercept of line A is $(0, 6)$. Using the points $(0, 6)$ and $(7, 3)$, we find that the slope of the line is

$$m = \frac{3 - 6}{7 - 0} = -\frac{3}{7}.$$

Therefore, the line is described by the function $f(x) = -\dfrac{3}{7}x + 6$.

b. Line B passes through the points $(2, 1)$ and $(5, 7)$, so the slope of this line is

$$m = \frac{7 - 1}{5 - 2} = 2,$$

which implies that the line is described by a function of the form $g(x) = 2x + b$. To determine the value of b, note that $g(2) = 1$:

$$g(2) = 2 \cdot 2 + b = 1.$$

Solving for b, we find that $b = 1 - 4 = -3$. Therefore, the line is described by the function $g(x) = 2x - 3$.

Related Exercises 3, 15–16 ◀

> In the solution to Example 1b, we used the point $(2, 1)$ to determine the value of b. Using the point $(5, 7)$—or, equivalently, the fact that $g(5) = 7$—leads to the same result.

EXAMPLE 2   Demand function for pizzas After studying sales for several months, the owner of a pizza chain knows that the number of two-topping pizzas sold in a week (called the *demand*) decreases as the price increases. Specifically, her data indicate that at a price of $14 per pizza, an average of 400 pizzas are sold per week, while at a price of $17 per pizza, an average of 250 pizzas are sold per week. Assume the demand d is a *linear* function of the price p.

a. Find the constants m and b in the demand function $d = f(p) = mp + b$. Then graph f.

b. According to this model, how many pizzas (on average) are sold per week at a price of $20?

SOLUTION

a. Two points on the graph of the demand function are given: $(p, d) = (14, 400)$ and $(17, 250)$. Therefore, the slope of the demand line is

$$m = \frac{400 - 250}{14 - 17} = -50 \text{ pizzas per dollar.}$$

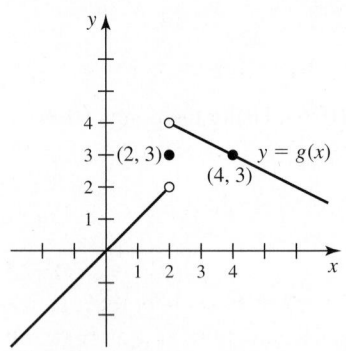

The demand function $d = -50p + 1100$ is defined on the interval $0 \le p \le 22$.

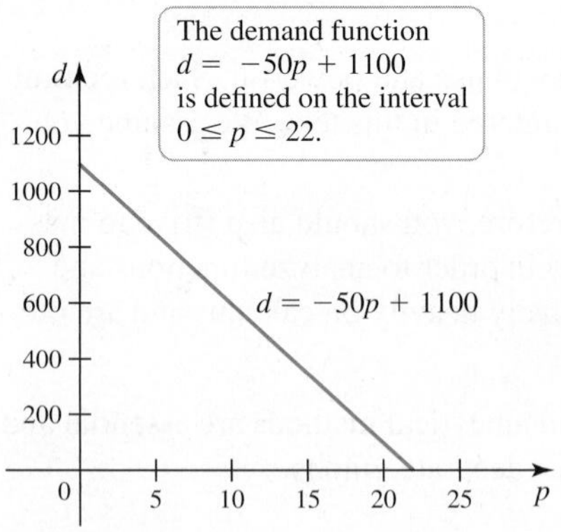

Figure 1.20

➤ The units of the slope have meaning: For every dollar the price is reduced, an average of 50 more pizzas can be sold.

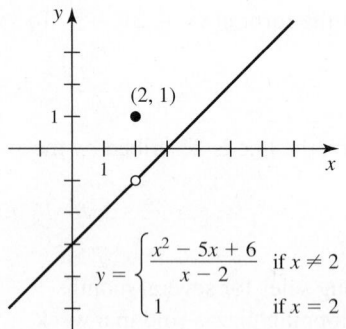

Figure 1.21

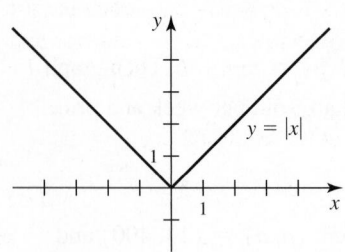

Figure 1.22

It follows that the equation of the linear demand function is

$$d - 250 = -50(p - 17).$$

Expressing d as a function of p, we have $d = f(p) = -50p + 1100$ (**Figure 1.20**).

b. Using the demand function with a price of \$20, the average number of pizzas that could be sold per week is $f(20) = 100$.

Related Exercise 21 ◄

Piecewise Functions   A function may have different definitions on different parts of its domain. For example, income tax is levied in tax brackets that have different tax rates. Functions that have different definitions on different parts of their domain are called **piece-wise functions**. If all the pieces are linear, the function is **piecewise linear**. Here are some examples.

EXAMPLE 3   Defining a piecewise function   The graph of a piecewise linear function g is shown in **Figure 1.21**. Find a formula for the function.

SOLUTION   For $x < 2$, the graph is linear with a slope of 1 and a y-intercept of $(0, 0)$; its equation is $y = x$. For $x > 2$, the slope of the line is $-\frac{1}{2}$ and it passes through $(4, 3)$; an equation of this piece of the function is

$$y - 3 = -\frac{1}{2}(x - 4) \quad \text{or} \quad y = -\frac{1}{2}x + 5.$$

For $x = 2$, we have $g(2) = 3$. Therefore,

$$g(x) = \begin{cases} x & \text{if } x < 2 \\ 3 & \text{if } x = 2 \\ -\dfrac{1}{2}x + 5 & \text{if } x > 2. \end{cases}$$

Related Exercises 25–26 ◄

EXAMPLE 4   Graphing piecewise functions   Graph the following functions.

a. $f(x) = \begin{cases} \dfrac{x^2 - 5x + 6}{x - 2} & \text{if } x \neq 2 \\ 1 & \text{if } x = 2 \end{cases}$

b. $f(x) = |x|$, the **absolute value** function

SOLUTION

a. The function f is simplified by factoring and then canceling $x - 2$, assuming $x \neq 2$:

$$\frac{x^2 - 5x + 6}{x - 2} = \frac{(x - 2)(x - 3)}{x - 2} = x - 3.$$

Therefore, the graph of f is identical to the graph of the line $y = x - 3$ when $x \neq 2$. We are given that $f(2) = 1$ (**Figure 1.22**).

b. The absolute value of a real number is defined as

$$f(x) = |x| = \begin{cases} x & \text{if } x \geq 0 \\ -x & \text{if } x < 0. \end{cases}$$

Graphing $y = -x$, for $x < 0$, and $y = x$, for $x \geq 0$, produces the graph in **Figure 1.23**.

Related Exercises 29–30 ◄

Figure 1.23

Power Functions   Power functions are a special case of polynomials; they have the form $f(x) = x^n$, where n is a positive integer. When n is an even integer, the function values are nonnegative and the graph passes through the origin, opening upward (**Figure 1.24**). For odd integers, the power function $f(x) = x^n$ has values that are positive when x is positive and negative when x is negative (**Figure 1.25**).

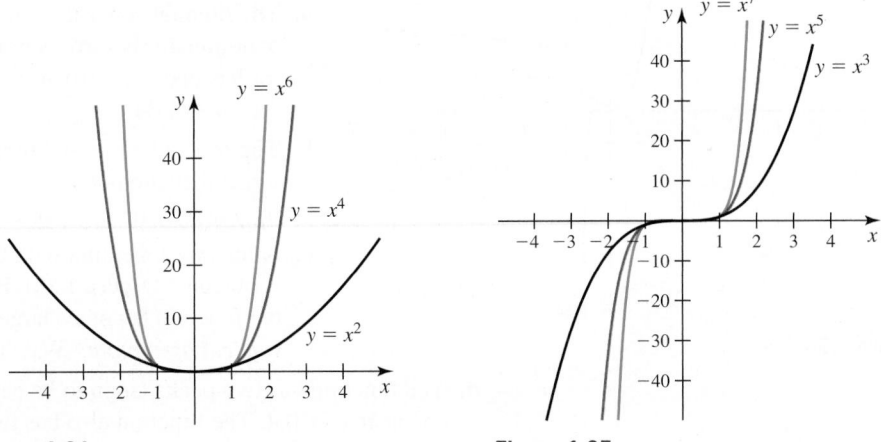

Figure 1.24   Figure 1.25

QUICK CHECK 2 What is the range of $f(x) = x^7$? What is the range of $f(x) = x^8$? ◄

Root Functions   Root functions are a special case of algebraic functions; they have the form $f(x) = x^{1/n}$, where $n > 1$ is a positive integer. Notice that when n is even (square roots, fourth roots, and so forth), the domain and range consist of nonnegative numbers. Their graphs begin steeply at the origin and flatten out as x increases (**Figure 1.26**).

By contrast, the odd root functions (cube roots, fifth roots, and so forth) are defined for all real values of x, and their range is all real numbers. Their graphs pass through the origin, open upward for $x < 0$ and downward for $x > 0$, and flatten out as x increases in magnitude (**Figure 1.27**).

➤ Recall that if n is a positive integer, then $x^{1/n}$ is the nth root of x; that is, $f(x) = x^{1/n} = \sqrt[n]{x}$.

QUICK CHECK 3 What are the domain and range of $f(x) = x^{1/7}$? What are the domain and range of $f(x) = x^{1/10}$? ◄

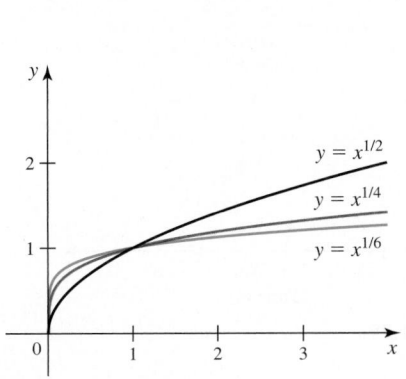

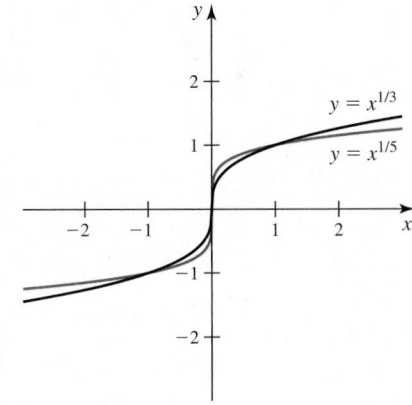

Figure 1.26   Figure 1.27

Rational Functions   Rational functions appear frequently in this text, and much is said later about graphing rational functions. The following example illustrates how analysis and technology work together.

EXAMPLE 5   Technology and analysis Consider the rational function

$$f(x) = \frac{3x^3 - x - 1}{x^3 + 2x^2 - 6}.$$

a. What is the domain of f?

b. Find the roots (zeros) of f.

c. Graph the function using a graphing utility.

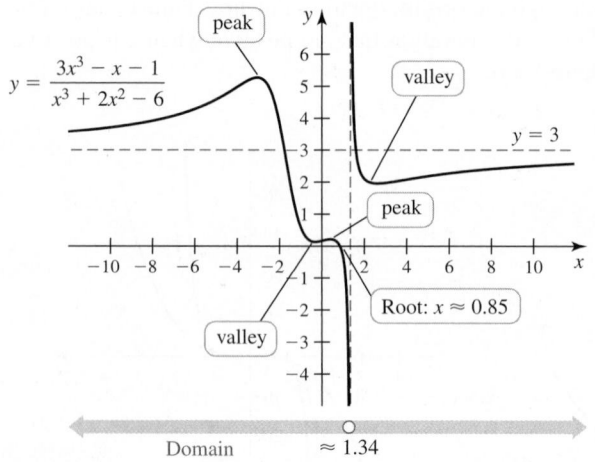

$$y = \frac{3x^3 - x - 1}{x^3 + 2x^2 - 6}$$

Figure 1.28

➤ In Chapter 4, we show how calculus is used to locate the local maximum and local minimum values of a function.

d. At what points does the function have peaks and valleys?

e. How does f behave as x grows large in magnitude?

SOLUTION

a. The domain consists of all real numbers except those at which the denominator is zero. A graphing utility shows that the denominator has one real zero at $x \approx 1.34$ and therefore, the domain of f is $\{x: x \neq 1.34\}$.

b. The roots of a rational function are the roots of the numerator, provided they are not also roots of the denominator. Using a graphing utility, the only real root of the numerator is $x \approx 0.85$.

c. After experimenting with the graphing window, a reasonable graph of f is obtained (**Figure 1.28**). Because the denominator is zero at $x \approx 1.34$, the function becomes large in magnitude at nearby points, and f has a *vertical asymptote*." Watch page break. Maintain current page break.

d. The function has two peaks (soon to be called *local maxima*), one near $x = -3.0$ and one near $x = 0.4$. The function also has two valleys (soon to be called *local minima*), one near $x = -0.3$ and one near $x = 2.6$.

e. By zooming out, it appears that as x increases in the positive direction, the graph approaches the *horizontal asymptote* $y = 3$ from below, and as x becomes large and negative, the graph approaches $y = 3$ from above.

Related Exercises 35–36 ◄

Using Tables

Sometimes functions do not originate as formulas or graphs; they may start as numbers or data. For example, suppose you do an experiment in which a marble is dropped into a cylinder filled with heavy oil and is allowed to fall freely. You measure the total distance d, in centimeters, that the marble falls at times $t = 0, 1, 2, 3, 4, 5, 6$, and 7 seconds after it is dropped (Table 1.1). The first step might be to plot the data points (**Figure 1.29**).

Table 1.1

t (s)	d (cm)
0	0
1	2
2	6
3	14
4	24
5	34
6	44
7	54

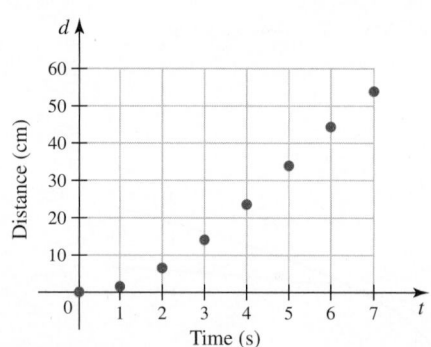

Figure 1.29

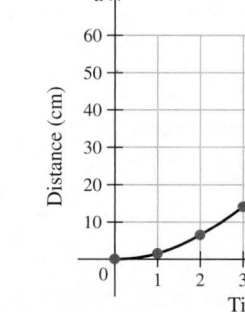

Figure 1.30

The data points suggest that there is a function $d = f(t)$ that gives the distance that the marble falls at *all* times of interest. Because the marble falls through the oil without abrupt changes, a smooth graph passing through the data points (**Figure 1.30**) is reasonable. Finding the function that best fits the data is a more difficult problem, which we discuss later in the text.

Using Words

Using words may be the least mathematical way to define functions, but it is often the way in which functions originate. Once a function is defined in words, it can often be tabulated, graphed, or expressed as a formula.

EXAMPLE 6   A slope function Let S be the **slope function** for a given function f. In words, this means that $S(x)$ is the slope of the curve $y = f(x)$ at the point $(x, f(x))$. Find and graph the slope function for the function f in **Figure 1.31**.

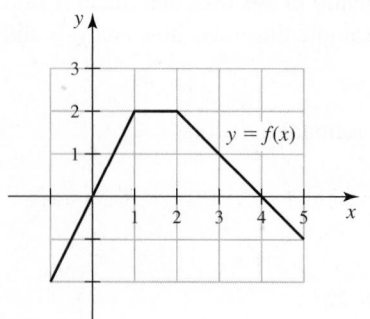

Figure 1.31

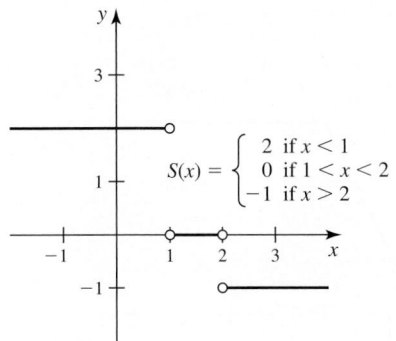

$$S(x) = \begin{cases} 2 & \text{if } x < 1 \\ 0 & \text{if } 1 < x < 2 \\ -1 & \text{if } x > 2 \end{cases}$$

Figure 1.32

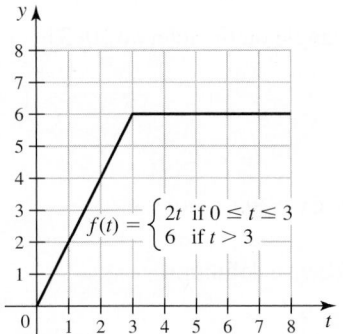

$$f(t) = \begin{cases} 2t & \text{if } 0 \le t \le 3 \\ 6 & \text{if } t > 3 \end{cases}$$

Figure 1.33

> ▶ Slope functions and area functions reappear in upcoming chapters and play an essential part in calculus.

SOLUTION For $x < 1$, the slope of $y = f(x)$ is 2. The slope is 0 for $1 < x < 2$, and the slope is -1 for $x > 2$. At $x = 1$ and $x = 2$, the graph of f has a corner, so the slope is undefined at these points. Therefore, the domain of S is the set of all real numbers except $x = 1$ and $x = 2$, and the slope function (**Figure 1.32**) is defined by the piecewise function

$$S(x) = \begin{cases} 2 & \text{if } x < 1 \\ 0 & \text{if } 1 < x < 2 \\ -1 & \text{if } x > 2. \end{cases}$$

Related Exercises 47–48 ◀

EXAMPLE 7   An area function Let A be an **area function** for a positive function f. In words, this means that $A(x)$ is the area of the region bounded by the graph of f and the t-axis from $t = 0$ to $t = x$. Consider the function (**Figure 1.33**)

$$f(t) = \begin{cases} 2t & \text{if } 0 \le t \le 3 \\ 6 & \text{if } t > 3. \end{cases}$$

a. Find $A(2)$ and $A(5)$.

b. Find a piecewise formula for the area function for f.

SOLUTION

a. The value of $A(2)$ is the area of the shaded region between the graph of f and the t-axis from $t = 0$ to $t = 2$ (**Figure 1.34a**). Using the formula for the area of a triangle,

$$A(2) = \frac{1}{2}(2)(4) = 4.$$

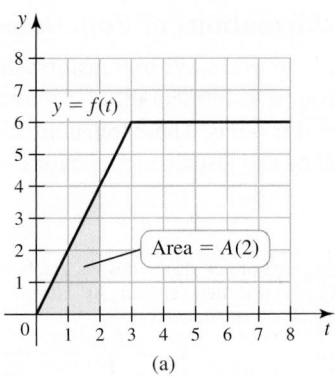

(a)

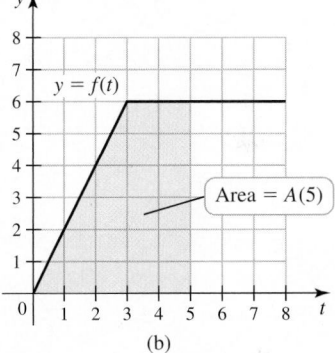

(b)

Figure 1.34

The value of $A(5)$ is the area of the shaded region between the graph of f and the t-axis on the interval $[0, 5]$ (**Figure 1.34b**). This area equals the area of the triangle whose base is the interval $[0, 3]$ plus the area of the rectangle whose base is the interval $[3, 5]$:

$$A(5) = \underbrace{\frac{1}{2}(3)(6)}_{\substack{\text{area of the} \\ \text{triangle}}} + \underbrace{(2)(6)}_{\substack{\text{area of the} \\ \text{rectangle}}} = 21.$$

b. For $0 \le x \le 3$ (**Figure 1.35a**), $A(x)$ is the area of the triangle whose base is the interval $[0, x]$. Because the height of the triangle at $t = x$ is $f(x)$,

$$A(x) = \frac{1}{2}x\,f(x) = \frac{1}{2}x\underbrace{(2x)}_{f(x)} = x^2.$$

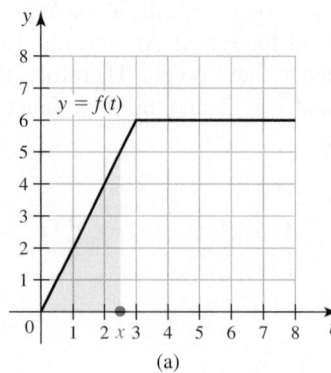

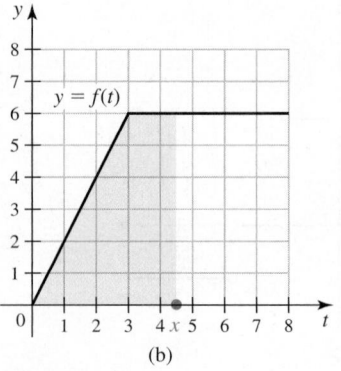

Figure 1.35

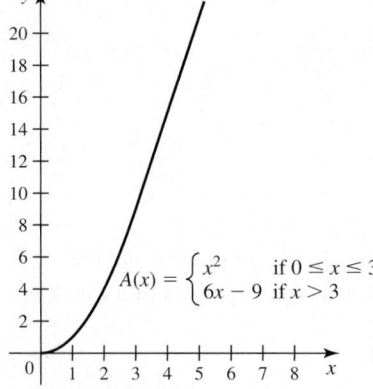

$$A(x) = \begin{cases} x^2 & \text{if } 0 \le x \le 3 \\ 6x - 9 & \text{if } x > 3 \end{cases}$$

Figure 1.36

For $x > 3$ (**Figure 1.35b**), $A(x)$ is the area of the triangle on the interval $[0, 3]$ plus the area of the rectangle on the interval $[3, x]$:

$$A(x) = \underbrace{\frac{1}{2}(3)(6)}_{\substack{\text{area of the} \\ \text{triangle}}} + \underbrace{(x - 3)(6)}_{\substack{\text{area of the} \\ \text{rectangle}}} = 6x - 9.$$

Therefore, the area function A (**Figure 1.36**) has the piecewise definition

$$A(x) = \begin{cases} x^2 & \text{if } 0 \le x \le 3 \\ 6x - 9 & \text{if } x > 3. \end{cases}$$

Related Exercises 51–52 ◄

Transformations of Functions and Graphs

There are several ways to transform the graph of a function to produce graphs of new functions. Four transformations are common: *shifts* in the x- and y-directions and *scalings* in the x- and y-directions. These transformations, summarized in **Figures 1.37–1.42**, can save time in graphing and visualizing functions.

The graph of $y = f(x) + d$ is the graph of $y = f(x)$ shifted vertically by d units: up if $d > 0$ and down if $d < 0$.

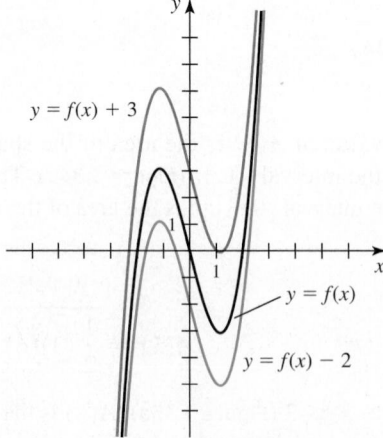

Figure 1.37

The graph of $y = f(x - b)$ is the graph of $y = f(x)$ shifted horizontally by b units: right if $b > 0$ and left if $b < 0$.

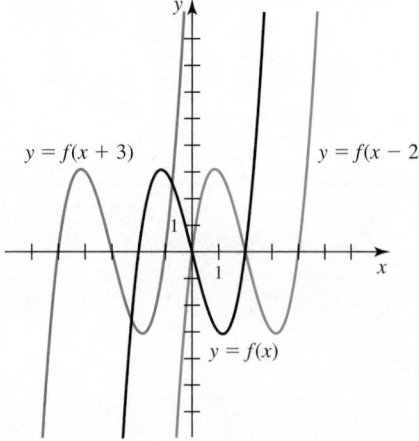

Figure 1.38

For $c > 0$ and $c \neq 1$, the graph of $y = cf(x)$ is the graph of $y = f(x)$ scaled vertically by a factor of c: compressed if $0 < c < 1$ and stretched if $c > 1$.

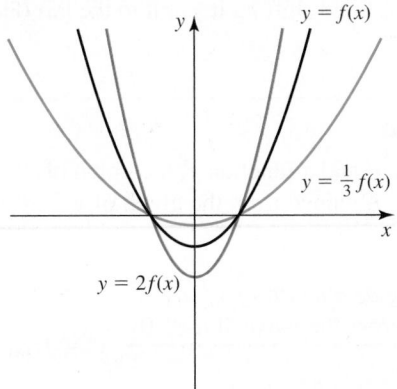

Figure 1.39

For $c < 0$, the graph of $y = cf(x)$ is the graph of $y = f(x)$ reflected across the x-axis and scaled vertically (if $c \neq -1$) by a factor of $|c|$: compressed if $0 < |c| < 1$ and stretched if $|c| > 1$.

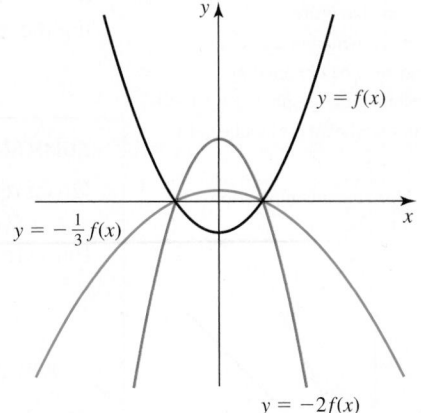

Figure 1.40

For $a > 0$ and $a \neq 1$, the graph of $y = f(ax)$ is the graph $y = f(x)$ scaled horizontally by a factor of $1/a$: compressed if $0 < 1/a < 1$ and stretched if $1/a > 1$.

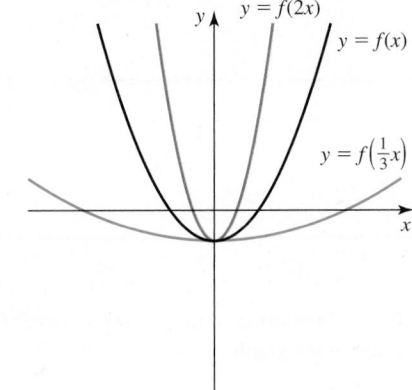

Figure 1.41

For $a < 0$, the graph of $y = f(ax)$ is the graph of $y = f(x)$ reflected across the y-axis and scaled horizontally (if $a \neq -1$) by a factor of $1/|a|$: compressed if $0 < 1/|a| < 1$ and stretched if $1/|a| > 1$.

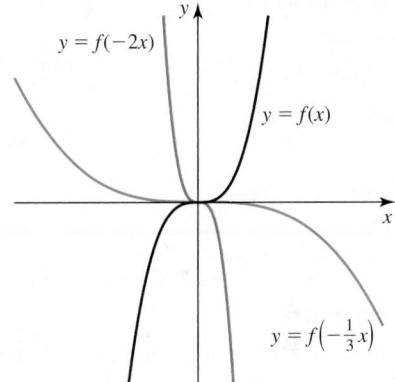

Figure 1.42

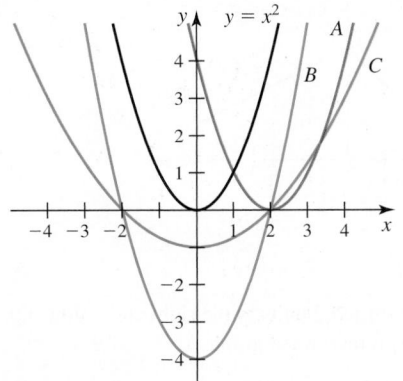

Figure 1.43

➤ You should verify that graph C also corresponds to a horizontal stretch and a vertical shift. It has the equation $y = f(ax) - 1$, where $a = \frac{1}{2}$.

EXAMPLE 8   Transforming parabolas The graphs A, B, and C in **Figure 1.43** are obtained from the graph of $f(x) = x^2$ using shifts and scalings. Find the function that describes each graph.

SOLUTION

a. Graph A is the graph of f shifted to the right by 2 units. It represents the function

$$f(x - 2) = (x - 2)^2 = x^2 - 4x + 4.$$

b. Graph B is the graph of f shifted down by 4 units. It represents the function

$$f(x) - 4 = x^2 - 4.$$

c. Graph C is a vertical scaling of the graph of f shifted down by 1 unit. Therefore, it represents $cf(x) - 1 = cx^2 - 1$, for some value of c, with $0 < c < 1$ (because the graph is vertically compressed). Using the fact that graph C passes through the points $(\pm 2, 0)$, we find that $c = \frac{1}{4}$. Therefore, the graph represents

$$y = \frac{1}{4} f(x) - 1 = \frac{1}{4} x^2 - 1.$$

Related Exercises 55, 58 ◀

QUICK CHECK 4 How do you modify the graph of $f(x) = 1/x$ to produce the graph of $g(x) = 1/(x + 4)$? ◄

➤ Note that we can also write $g(x) = 2\left|x + \frac{1}{2}\right|$, which means the graph of g may also be obtained by vertically stretching the graph of f by a factor of 2, followed by a horizontal shift.

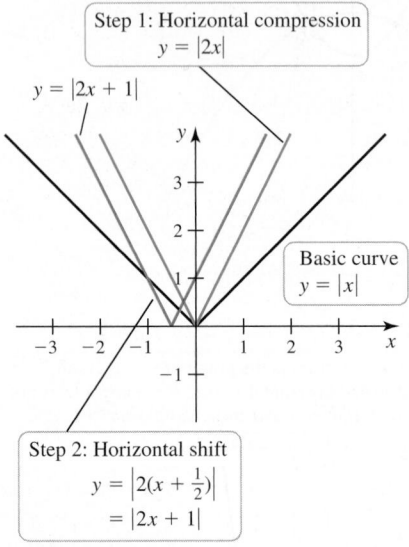

Step 1: Horizontal compression
$y = |2x|$

$y = |2x + 1|$

Basic curve
$y = |x|$

Step 2: Horizontal shift
$y = \left|2\left(x + \frac{1}{2}\right)\right|$
$= |2x + 1|$

Figure 1.44

EXAMPLE 9   Scaling and shifting Graph $g(x) = |2x + 1|$.

SOLUTION We write the function as $g(x) = \left|2\left(x + \frac{1}{2}\right)\right|$. Letting $f(x) = |x|$, we have $g(x) = f\left(2\left(x + \frac{1}{2}\right)\right)$. Therefore, the graph of g is obtained by horizontally compressing the graph of f by a factor of $\frac{1}{2}$ and shifting it $\frac{1}{2}$ unit to the left (**Figure 1.44**).

Related Exercise 64 ◄

SUMMARY   Transformations

Given real numbers a, b, c, and d and a function f, the graph of $y = cf(a(x - b)) + d$ can be obtained from the graph of $y = f(x)$ in the following steps.

$$y = f(x) \xrightarrow[\text{(and reflection across the y-axis if $a < 0$)}]{\text{horizontal scaling by a factor of } 1/|a|} y = f(ax)$$

$$\xrightarrow[\text{by b units}]{\text{horizontal shift}} y = f(a(x - b))$$

$$\xrightarrow[\text{reflection across the x-axis if $c < 0$)}]{\text{vertical scaling by a factor of } |c| \text{ (and}} y = cf(a(x - b))$$

$$\xrightarrow[\text{by d units}]{\text{vertical shift}} y = cf(a(x - b)) + d$$

SECTION 1.2 EXERCISES

Getting Started

1. Give four ways in which functions may be defined and represented.

2. What is the domain of a polynomial?

3. Determine the function f represented by the graph of the line $y = f(x)$ in the figure.

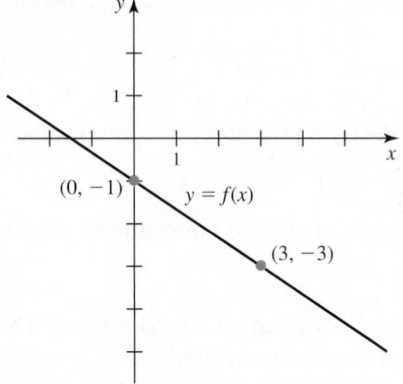

$(0, -1)$ $y = f(x)$

$(3, -3)$

4. Determine the linear function g whose graph is parallel to the line $y = 2x + 1$ and passes through the point $(5, 0)$.

5. What is the domain of a rational function?

6. What is a piecewise linear function?

7. Write a definition of the piecewise linear function $y = f(x)$ that is given in the graph.

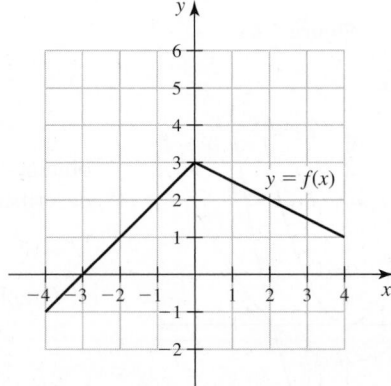

$y = f(x)$

8. The graph of $y = \sqrt{x}$ is shifted 2 units to the right and 3 units up. Write an equation for this transformed graph.

9. How do you obtain the graph of $y = f(x + 2)$ from the graph of $y = f(x)$?

10. How do you obtain the graph of $y = -3f(x)$ from the graph of $y = f(x)$?

11. How do you obtain the graph of $y = f(3x)$ from the graph of $y = f(x)$?

12. How do you obtain the graph of $y = 4(x + 3)^2 + 6$ from the graph of $y = x^2$?

13. The graphs of the functions f and g in the figure are obtained by vertical and horizontal shifts and scalings of $y = |x|$. Find formulas for f and g. Verify your answers with a graphing utility.

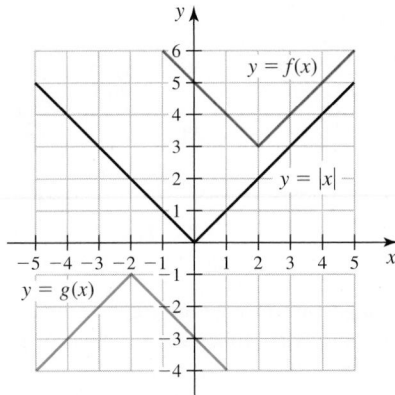

14. **Transformations** Use the graph of f in the figure to plot the following functions.

   a. $y = -f(x)$ **b.** $y = f(x + 2)$
   c. $y = f(x - 2)$ **d.** $y = f(2x)$
   e. $y = f(x - 1) + 2$ **f.** $y = 2f(x)$

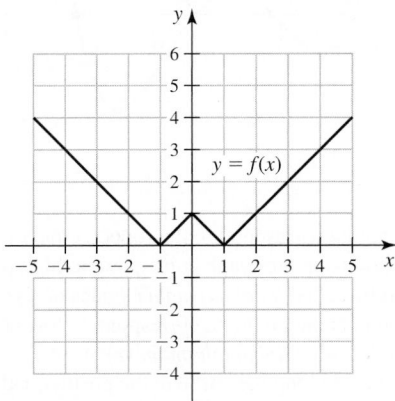

Practice Exercises

15. **Graph of a linear function** Find and graph the linear function that passes through the points $(1, 3)$ and $(2, 5)$.

16. **Graph of a linear function** Find and graph the linear function that passes through the points $(2, -3)$ and $(5, 0)$.

17. **Linear function** Find the linear function whose graph passes though the point $(3, 2)$ and is parallel to the line $y = 3x + 8$.

18. **Linear function** Find the linear function whose graph passes though the point $(-1, 4)$ and is perpendicular to the line $y = \frac{1}{4}x - 7$.

19–20. Yeast growth *Consider a colony of yeast cells that has the shape of a cylinder. As the number of yeast cells increases, the cross-sectional area A (in mm^2) of the colony increases but the height of the colony remains constant. If the colony starts from a single cell, the number of yeast cells (in millions) is approximated by the linear function $N(A) = C_s A$, where the constant C_s is known as the cell-surface*

coefficient. Use the given information to determine the cell-surface coefficient for each of the following colonies of yeast cells, and find the number of yeast cells in the colony when the cross-sectional area A reaches 150 mm^2. (Source: Letters in Applied Microbiology, 594, 59, 2014)

19. The scientific name of baker's or brewer's yeast (used in making bread, wine, and beer) is *Saccharomyces cerevisiae*. When the cross-sectional area of a colony of this yeast reaches 100 mm^2, there are 571 million yeast cells.

20. The yeast *Rhodotorula glutinis* is a laboratory contaminant. When the cross-sectional area of a colony reaches 100 mm^2, there are 226 million yeast cells.

21. **Demand function** Sales records indicate that if Blu-ray players are priced at \$250, then a large store sells an average of 12 units per day. If they are priced at \$200, then the store sells an average of 15 units per day. Find and graph the linear demand function for Blu-ray sales. For what prices is the demand function defined?

22. **Fundraiser** The Biology Club plans to have a fundraiser for which \$8 tickets will be sold. The cost of room rental and refreshments is \$175. Find and graph the function $p = f(n)$ that gives the profit from the fundraiser when n tickets are sold. Notice that $f(0) = -\$175$; that is, the cost of room rental and refreshments must be paid regardless of how many tickets are sold. How many tickets must be sold for the fundraiser to break even (zero profit)?

23. **Bald eagle population** After DDT was banned and the Endangered Species Act was passed in 1973, the number of bald eagles in the United States increased dramatically. In the lower 48 states, the number of breeding pairs of bald eagles increased at a nearly linear rate from 1875 pairs in 1986 to 6471 pairs in 2000.

   a. Use the data points for 1986 and 2000 to find a linear function p that models the number of breeding pairs from 1986 to 2000 ($0 \le t \le 14$).

   b. Using the function in part (a), approximately how many breeding pairs were in the lower 48 states in 1995?

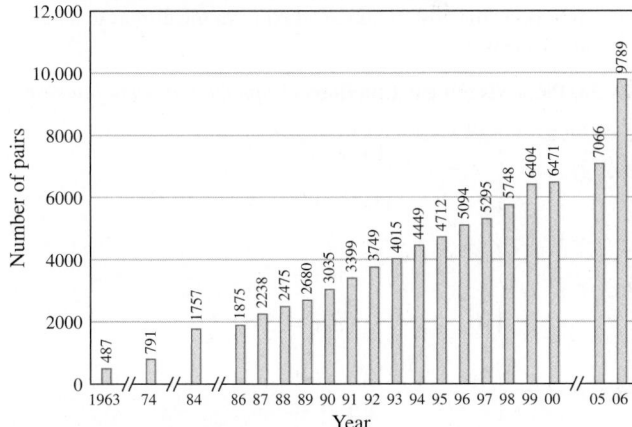

(*Source:* U.S. Fish and Wildlife Service)

24. **Taxicab fees** A taxicab ride costs \$3.50 plus \$2.50 per mile. Let m be the distance (in miles) from the airport to a hotel. Find and graph the function $c(m)$ that represents the cost of taking a taxi from the airport to the hotel. Also determine how much it will cost if the hotel is 9 miles from the airport.

25–26. Defining piecewise functions *Write a definition of the function whose graph is given.*

25.

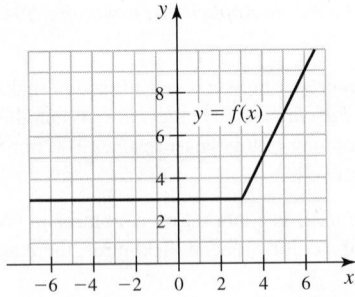

26.

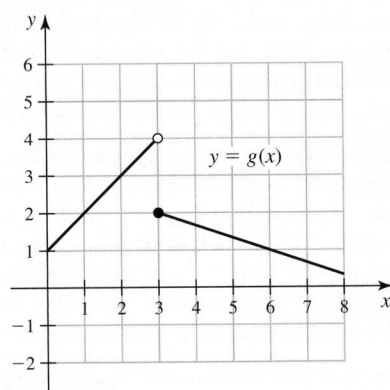

27. Parking fees Suppose that it costs 5¢ per minute to park at the airport, with the rate dropping to 3¢ per minute after 9 P.M. Find and graph the cost function $c(t)$ for values of t satisfying $0 \le t \le 120$. Assume that t is the number of minutes after 8 P.M.

28. Taxicab fees A taxicab ride costs $3.50 plus $2.50 per mile for the first 5 miles, with the rate dropping to $1.50 per mile after the fifth mile. Let m be the distance (in miles) from the airport to a hotel. Find and graph the piecewise linear function $c(m)$ that represents the cost of taking a taxi from the airport to a hotel m miles away.

29–34. Piecewise linear functions *Graph the following functions.*

29. $f(x) = \begin{cases} \dfrac{x^2 - x}{x - 1} & \text{if } x \ne 1 \\ 2 & \text{if } x = 1 \end{cases}$

30. $f(x) = \begin{cases} \dfrac{x^2 - x - 2}{x - 2} & \text{if } x \ne 2 \\ 4 & \text{if } x = 2 \end{cases}$

31. $f(x) = \begin{cases} 3x - 1 & \text{if } x \le 0 \\ -2x + 1 & \text{if } x > 0 \end{cases}$

32. $f(x) = \begin{cases} 3x - 1 & \text{if } x < 1 \\ x + 1 & \text{if } x \ge 1 \end{cases}$

33. $f(x) = \begin{cases} -2x - 1 & \text{if } x < -1 \\ 1 & \text{if } -1 \le x \le 1 \\ 2x - 1 & \text{if } x > 1 \end{cases}$

34. $f(x) = \begin{cases} 2x + 2 & \text{if } x < 0 \\ x + 2 & \text{if } 0 \le x \le 2 \\ 3 - \dfrac{x}{2} & \text{if } x > 2 \end{cases}$

T 35–40. Graphs of functions

a. *Use a graphing utility to produce a graph of the given function. Experiment with different windows to see how the graph changes on different scales. Sketch an accurate graph by hand after using the graphing utility.*
b. *Give the domain of the function.*
c. *Discuss interesting features of the function, such as peaks, valleys, and intercepts (as in Example 5).*

35. $f(x) = x^3 - 2x^2 + 6$

36. $f(x) = \sqrt[3]{2x^2 - 8}$

37. $g(x) = \left| \dfrac{x^2 - 4}{x + 3} \right|$

38. $f(x) = \dfrac{\sqrt{3x^2 - 12}}{x + 1}$

39. $f(x) = 3 - |2x - 1|$

40. $f(x) = \begin{cases} \dfrac{|x - 1|}{x - 1} & \text{if } x \ne 1 \\ 0 & \text{if } x = 1 \end{cases}$

41. Features of a graph Consider the graph of the function f shown in the figure. Answer the following questions by referring to the points A–I.

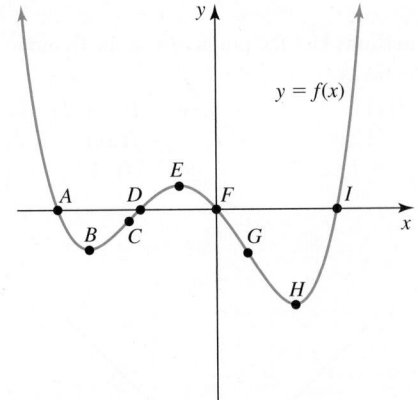

a. Which points correspond to the roots (zeros) of f?
b. Which points on the graph correspond to high points or peaks (soon to be called *local maximum* values of f)?
c. Which points on the graph correspond to low points or valleys (soon to be called *local minimum* values of f)?
d. As you move along the curve in the positive x-direction, at which point is the graph rising most rapidly?
e. As you move along the curve in the positive x-direction, at which point is the graph falling most rapidly?

42. Features of a graph Consider the graph of the function g shown in the figure.

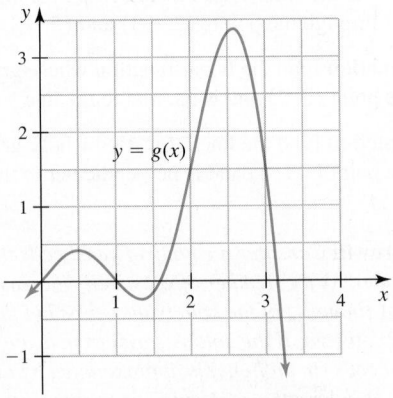

a. Give the approximate roots (zeros) of g.
b. Give the approximate coordinates of the high points or peaks (soon to be called *local maximum* values of g).
c. Give the approximate coordinates of the low points or valleys (soon to be called *local minimum* values of g).

d. Imagine moving along the curve in the positive x-direction on the interval $[0, 3]$. Give the approximate coordinates of the point at which the graph is rising most rapidly.

e. Imagine moving along the curve in the positive x-direction on the interval $[0, 3]$. Give the approximate coordinates of the point at which the graph is falling most rapidly.

T 43. Relative acuity of the human eye The **fovea centralis** (or **fovea**) is responsible for the sharp central vision that humans use for reading and other detail-oriented eyesight. The relative acuity of a human eye, which measures the sharpness of vision, is modeled by the function

$$R(\theta) = \frac{0.568}{0.331|\theta| + 0.568},$$

where θ (in degrees) is the angular deviation of the line of sight from the center of the fovea (see figure).

a. Graph R, for $-15 \le \theta \le 15$.

b. For what value of θ is R maximized? What does this fact indicate about our eyesight?

c. For what values of θ do we maintain at least 90% of our maximum relative acuity? (*Source: The Journal of Experimental Biology*, 203, 24, Dec 2000)

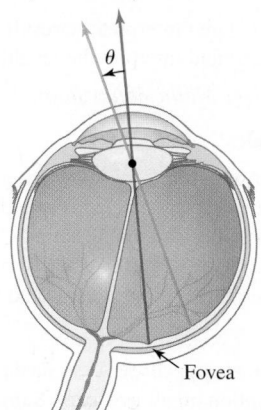
Fovea

44–48. Slope functions *Determine the slope function $S(x)$ for the following functions.*

44. $f(x) = 3$ **45.** $f(x) = 2x + 1$ **46.** $f(x) = |x|$

47. Use the figure for Exercise 7.

48. Use the figure for Exercise 26.

49–52. Area functions *Let $A(x)$ be the area of the region bounded by the t-axis and the graph of $y = f(t)$ from $t = 0$ to $t = x$. Consider the following functions and graphs.*

a. Find $A(2)$.   *b. Find $A(6)$.*   *c. Find a formula for $A(x)$.*

49. $f(t) = 6$

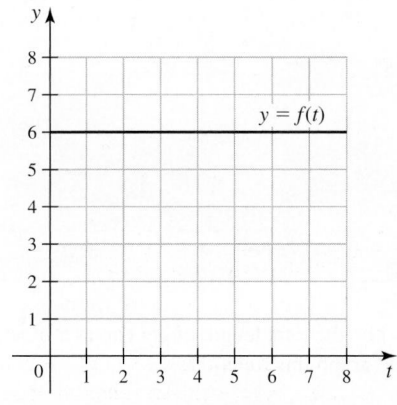
$y = f(t)$

50. $f(t) = \dfrac{t}{2}$

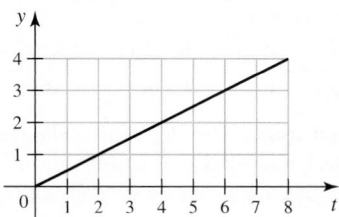

51. $f(t) = \begin{cases} -2t + 8 & \text{if } t \le 3 \\ 2 & \text{if } t > 3 \end{cases}$

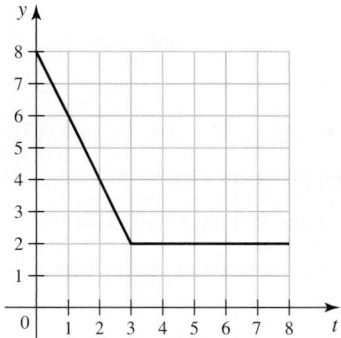

52. $f(t) = |t - 2| + 1$

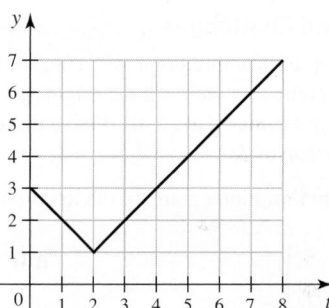

53. Explain why or why not Determine whether the following statements are true and give an explanation or counterexample.

a. All polynomials are rational functions, but not all rational functions are polynomials.

b. If f is a linear polynomial, then $f \circ f$ is a quadratic polynomial.

c. If f and g are polynomials, then the degrees of $f \circ g$ and $g \circ f$ are equal.

d. The graph of $g(x) = f(x + 2)$ is the graph of f shifted 2 units to the right.

54. Shifting a graph Use a shift to explain how the graph of $f(x) = \sqrt{-x^2 + 8x + 9}$ is obtained from the graph of $g(x) = \sqrt{25 - x^2}$. Sketch a graph of f.

55. Transformations of $f(x) = x^2$ Use shifts and scalings to transform the graph of $f(x) = x^2$ into the graph of g. Use a graphing utility to check your work.

a. $g(x) = f(x - 3)$

b. $g(x) = f(2x - 4)$

c. $g(x) = -3f(x - 2) + 4$

d. $g(x) = 6f\left(\dfrac{x - 2}{3}\right) + 1$

56. Transformations of $f(x) = \sqrt{x}$ Use shifts and scalings to transform the graph of $f(x) = \sqrt{x}$ into the graph of g. Use a graphing utility to check your work.

a. $g(x) = f(x + 4)$ **b.** $g(x) = 2f(2x - 1)$
c. $g(x) = \sqrt{x - 1}$ **d.** $g(x) = 3\sqrt{x - 1} - 5$

T 57–64. Shifting and scaling *Use shifts and scalings to graph the given functions. Then check your work with a graphing utility. Be sure to identify an original function on which the shifts and scalings are performed.*

57. $f(x) = (x - 2)^2 + 1$

58. $f(x) = x^2 - 2x + 3$
(*Hint:* Complete the square first.)

59. $g(x) = -3x^2$

60. $g(x) = 2x^3 - 1$

61. $g(x) = 2(x + 3)^2$

62. $p(x) = x^2 + 3x - 5$

63. $h(x) = -4x^2 - 4x + 12$

64. $h(x) = |3x - 6| + 1$

65–67. Intersection problems *Find the following points of intersection.*

65. The point(s) of intersection of the curves $y = 4\sqrt{2x}$ and $y = 2x^2$

66. The point(s) of intersection of the parabola $y = x^2 + 2$ and the line $y = x + 4$

67. The point(s) of intersection of the parabolas $y = x^2$ and $y = -x^2 + 8x$

Explorations and Challenges

68. Two semicircles The entire graph of f consists of the upper half of a circle of radius 2 centered at the origin and the lower half of a circle of radius 3 centered at $(5, 0)$. Find a piecewise function for f and plot a graph of f.

69. Piecewise function Plot a graph of the function

$$f(x) = \begin{cases} \frac{3}{2}x & \text{if } 0 \le x \le 2 \\ 3 + \sqrt{x - 2} & \text{if } 2 < x \le 6 \\ \sqrt{25 - (x - 6)^2} & \text{if } 6 < x \le 11. \end{cases}$$

70. Floor function The floor function, or greatest integer function, $f(x) = \lfloor x \rfloor$, gives the greatest integer less than or equal to x. Graph the floor function for $-3 \le x \le 3$.

71. Ceiling function The ceiling function, or smallest integer function, $f(x) = \lceil x \rceil$, gives the smallest integer greater than or equal to x. Graph the ceiling function for $-3 \le x \le 3$.

72. Sawtooth wave Graph the sawtooth wave defined by

$$f(x) = \begin{cases} \vdots \\ x + 1 & \text{if } -1 \le x < 0 \\ x & \text{if } 0 \le x < 1 \\ x - 1 & \text{if } 1 \le x < 2 \\ x - 2 & \text{if } 2 \le x < 3 \\ \vdots \end{cases}$$

73. Square wave Graph the square wave defined by

$$f(x) = \begin{cases} 0 & \text{if } x < 0 \\ 1 & \text{if } 0 \le x < 1 \\ 0 & \text{if } 1 \le x < 2 \\ 1 & \text{if } 2 \le x < 3 \\ \vdots \end{cases}$$

74–76. Roots and powers *Sketch a graph of the given pairs of functions. Be sure to draw the graphs accurately relative to each other.*

74. $y = x^4$ and $y = x^6$

75. $y = x^3$ and $y = x^7$

76. $y = x^{1/3}$ and $y = x^{1/5}$

77. Tennis probabilities Suppose the probability of a server winning any given point in a tennis match is a constant p, with $0 \le p \le 1$. Then the probability of the server winning a game when serving from deuce is

$$f(p) = \frac{p^2}{1 - 2p(1 - p)}.$$

a. Evaluate $f(0.75)$ and interpret the result.
b. Evaluate $f(0.25)$ and interpret the result.

(*Source: The College Mathematics Journal*, 38, 1, Jan 2007)

78. Temperature scales

a. Find the linear function $C = f(F)$ that gives the reading on the Celsius temperature scale corresponding to a reading on the Fahrenheit scale. Use the facts that $C = 0$ when $F = 32$ (freezing point) and $C = 100$ when $F = 212$ (boiling point).
b. At what temperature are the Celsius and Fahrenheit readings equal?

79. Automobile lease vs. purchase A car dealer offers a purchase option and a lease option on all new cars. Suppose you are interested in a car that can be bought outright for $25,000 or leased for a start-up fee of $1200 plus monthly payments of $350.

a. Find the linear function $y = f(m)$ that gives the total amount you have paid on the lease option after m months.
b. With the lease option, after a 48-month (4-year) term, the car has a residual value of $10,000, which is the amount that you could pay to purchase the car. Assuming no other costs, should you lease or buy?

T 80. Walking and rowing Kelly has finished a picnic on an island that is 200 m off shore (see figure). She wants to return to a beach house that is 600 m from the point P on the shore closest to the island. She plans to row a boat to a point on shore x meters from P and then jog along the (straight) shore to the house.

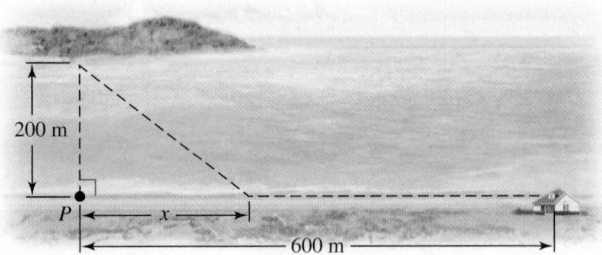

a. Let $d(x)$ be the total length of her trip as a function of x. Find and graph this function.

b. Suppose that Kelly can row at 2 m/s and jog at 4 m/s. Let $T(x)$ be the total time for her trip as a function of x. Find and graph $y = T(x)$.

c. Based on your graph in part (b), estimate the point on the shore at which Kelly should land to minimize the total time of her trip. What is that minimum time?

T 81. Optimal boxes Imagine a lidless box with height h and a square base whose sides have length x. The box must have a volume of 125 ft³.

a. Find and graph the function $S(x)$ that gives the surface area of the box, for all values of $x > 0$.

b. Based on your graph in part (a), estimate the value of x that produces the box with a minimum surface area.

82. Composition of polynomials Let f be an nth-degree polynomial and let g be an mth-degree polynomial. What is the degree of the following polynomials?

a. $f \cdot f$ **b.** $f \circ f$ **c.** $f \cdot g$ **d.** $f \circ g$

83. Parabola vertex property Prove that if a parabola crosses the x-axis twice, the x-coordinate of the vertex of the parabola is halfway between the x-intercepts.

84. Parabola properties Consider the general quadratic function $f(x) = ax^2 + bx + c$, with $a \neq 0$.

a. Find the coordinates of the vertex of the graph of the parabola $y = f(x)$ in terms of a, b, and c.

b. Find the conditions on a, b, and c that guarantee that the graph of f crosses the x-axis twice.

T 85. Factorial function The factorial function is defined for positive integers as $n! = n(n - 1)(n - 2) \cdots 3 \cdot 2 \cdot 1$.

a. Make a table of the factorial function, for $n = 1, 2, 3, 4, 5$.

b. Graph these data points and then connect them with a smooth curve.

c. What is the least value of n for which $n! > 10^6$?

QUICK CHECK ANSWERS

1. Yes; no   **2.** $(-\infty, \infty)$; $[0, \infty)$ **3.** Domain and range are $(-\infty, \infty)$. Domain and range are $[0, \infty)$.   **4.** Shift the graph of f horizontally 4 units to the left. ◄

1.3 Inverse, Exponential, and Logarithmic Functions

Exponential functions are fundamental to all of mathematics. Many processes in the world around us are modeled by *exponential functions*—they appear in finance, medicine, ecology, biology, economics, anthropology, and physics (among other disciplines). Every exponential function has an inverse function, which is a member of the family of *logarithmic functions*, also discussed in this section.

Exponential Functions

Exponential functions have the form $f(x) = b^x$, where the base $b \neq 1$ is a positive real number. An important question arises immediately: For what values of x can b^x be evaluated? We certainly know how to compute b^x when x is an integer. For example, $2^3 = 8$ and $2^{-4} = 1/2^4 = 1/16$. When x is rational, the numerator and denominator are interpreted as a power and root, respectively:

$$16^{3/4} = 16^{3/4} = \left(\sqrt[4]{16}\right)^3 = 8.$$

But what happens when x is irrational? For example, how should 2^π be understood? Your calculator provides an approximation to 2^π, but where does the approximation come from? These questions will be answered eventually. For now, we assume that b^x can be defined for all real numbers x and that it can be approximated as closely as desired by using rational numbers as close to x as needed. In Section 7.1, we prove that the domain of an exponential function is all real numbers.

Properties of Exponential Functions $f(x) = b^x$

1. Because b^x is defined for all real numbers, the domain of f is $\{x: -\infty < x < \infty\}$. Because $b^x > 0$ for all values of x, the range of f is $\{y: 0 < y < \infty\}$.

2. For all $b > 0$, $b^0 = 1$, and therefore $f(0) = 1$.

3. If $b > 1$, then f is an increasing function of x (**Figure 1.45**). For example, if $b = 2$, then $2^x > 2^y$ whenever $x > y$.

> **Exponent Rules**
> For any base $b > 0$ and real numbers x and y, the following relations hold:
>
> **E1.** $b^x b^y = b^{x+y}$
>
> **E2.** $\dfrac{b^x}{b^y} = b^{x-y}$
>
> $\left(\text{which includes } \dfrac{1}{b^y} = b^{-y}\right)$
>
> **E3.** $(b^x)^y = b^{xy}$
>
> **E4.** $b^x > 0$, for all x

> $16^{3/4}$ can also be computed as $\sqrt[4]{16^3} = \sqrt[4]{4096} = 8$.

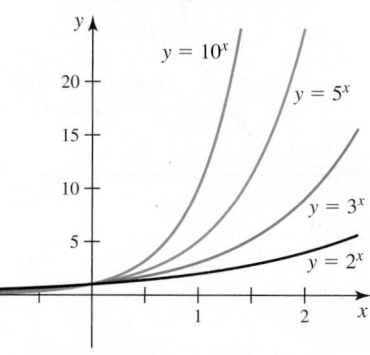

Larger values of b produce greater rates of increase in b^x if $b > 1$.

$y = 10^x$
$y = 5^x$
$y = 3^x$
$y = 2^x$

Figure 1.45

QUICK CHECK 1 Is it possible to raise a positive number b to a power and obtain a negative number? Is it possible to obtain zero? ◄

QUICK CHECK 2 Explain why
$f(x) = \left(\dfrac{1}{3}\right)^x$ is a decreasing
function. ◄

> The notation e was proposed by the Swiss mathematician Leonhard Euler (pronounced *oiler*) (1707–1783).

4. If $0 < b < 1$, then f is a decreasing function of x. For example, if $b = \frac{1}{2}$, then

$$f(x) = \left(\frac{1}{2}\right)^x = \frac{1}{2^x} = 2^{-x},$$

and because 2^x increases with x, 2^{-x} decreases with x (Figure 1.46).

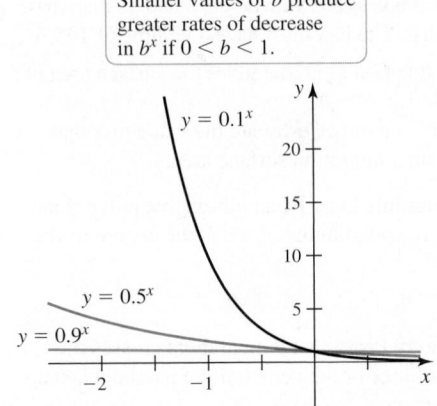

Figure 1.46

The Natural Exponential Function One of the bases used for exponential functions is special. For reasons that will become evident in upcoming chapters, the special base is e, one of the fundamental constants of mathematics. It is an irrational number with a value of $e = 2.718281828459\dots$.

> DEFINITION **The Natural Exponential Function**
>
> The **natural exponential function** is $f(x) = e^x$, which has the base $e = 2.718281828459\dots$.

The base e gives an exponential function that has a valuable property. As shown in Figure 1.47a, the graph of $y = e^x$ lies between the graphs of $y = 2^x$ and $y = 3^x$ (because $2 < e < 3$). At every point on the graph of $y = e^x$, it is possible to draw a *tangent line* (discussed in Chapters 2 and 3) that touches the graph only at that point. The natural exponential function is the only exponential function with the property that the slope of the tangent line at $x = 0$ is 1 (Figure 1.47b); therefore, e^x has both value and slope equal to 1 at $x = 0$. This property—minor as it may seem—leads to many simplifications when we do calculus with exponential functions.

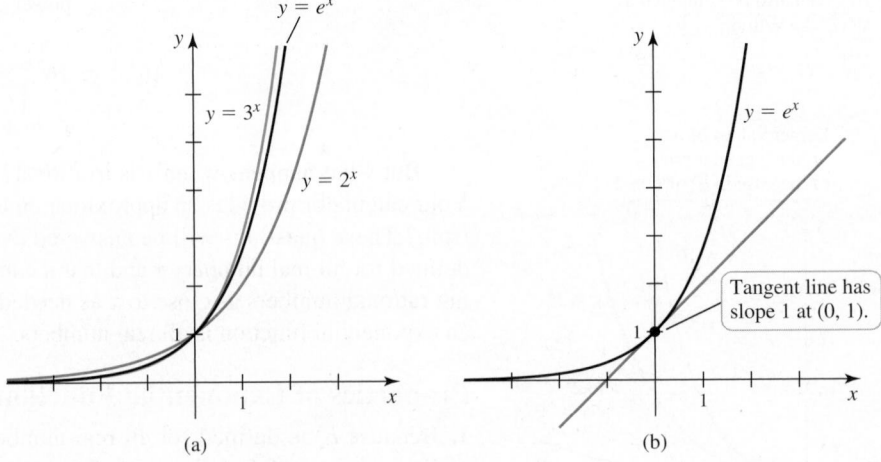

Figure 1.47

Inverse Functions

Consider the linear function $f(x) = 2x$, which takes any value of x and doubles it. The function that reverses this process by taking any value of $f(x) = 2x$ and mapping it back to x is called the *inverse function* of f, denoted f^{-1}. In this case, the inverse function is $f^{-1}(x) = x/2$. The effect of applying these two functions in succession looks like this:

$$x \xrightarrow{\quad f \quad} 2x \xrightarrow{\quad f^{-1} \quad} x.$$

We now generalize this idea.

QUICK CHECK 3 What is the inverse of $f(x) = \frac{1}{3}x$? What is the inverse of $f(x) = x - 7$? ◄

> **DEFINITION Inverse Function**
>
> Given a function f, its inverse (if it exists) is a function f^{-1} such that whenever $y = f(x)$, then $f^{-1}(y) = x$ (**Figure 1.48**).

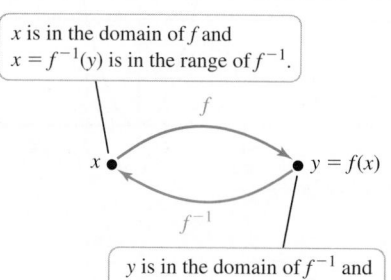

x is in the domain of f and $x = f^{-1}(y)$ is in the range of f^{-1}.

y is in the domain of f^{-1} and $y = f(x)$ is in the range of f.

Figure 1.48

▶ The notation f^{-1} for the inverse can be confusing. The inverse is not the reciprocal; that is, $f^{-1}(x)$ is not $1/f(x) = (f(x))^{-1}$. We adopt the common convention of using simply *inverse* to mean *inverse function*.

Because the inverse "undoes" the original function, if we start with a value of x, apply f to it, and then apply f^{-1} to the result, we recover the original value of x; that is,

$$f^{-1}(f(x)) = x.$$

Similarly, if we apply f^{-1} to a value of y and then apply f to the result, we recover the original value of y; that is,

$$f(f^{-1}(y)) = y.$$

One-to-One Functions   We have defined the inverse of a function, but said nothing about when it exists. To ensure that f has an inverse on a domain, f must be *one-to-one* on that domain. This property means that every output of the function f corresponds to exactly one input. The one-to-one property is checked graphically by using the *horizontal line test*.

> **DEFINITION One-to-One Functions and the Horizontal Line Test**
>
> A function f is **one-to-one** on a domain D if each value of $f(x)$ corresponds to exactly one value of x in D. More precisely, f is one-to-one on D if $f(x_1) \neq f(x_2)$ whenever $x_1 \neq x_2$, for x_1 and x_2 in D. The **horizontal line test** says that every horizontal line intersects the graph of a one-to-one function at most once (**Figure 1.49**).

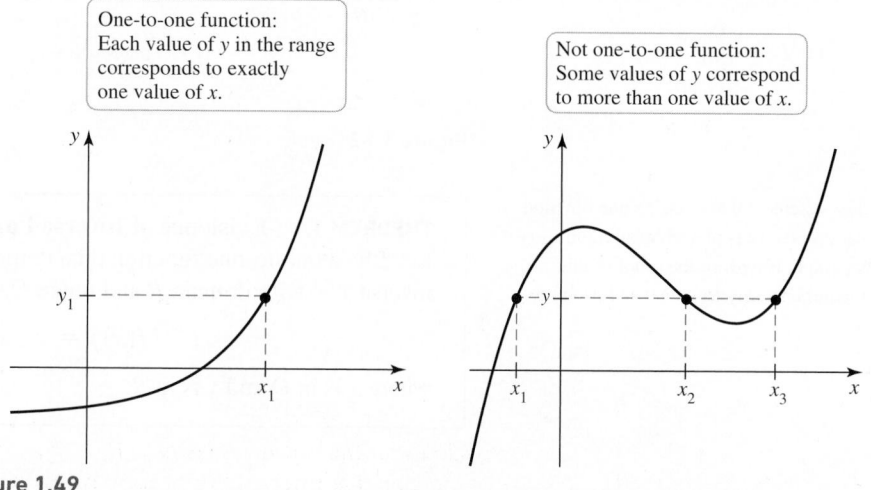

One-to-one function: Each value of y in the range corresponds to exactly one value of x.

Not one-to-one function: Some values of y correspond to more than one value of x.

Figure 1.49

For example, in **Figure 1.50**, some horizontal lines intersect the graph of $f(x) = x^2$ twice. Therefore, f does not have an inverse function on the interval $(-\infty, \infty)$. However, if the domain of f is restricted to one of the intervals $(-\infty, 0]$ or $[0, \infty)$, then the graph of f passes the horizontal line test, and f is one-to-one on these intervals.

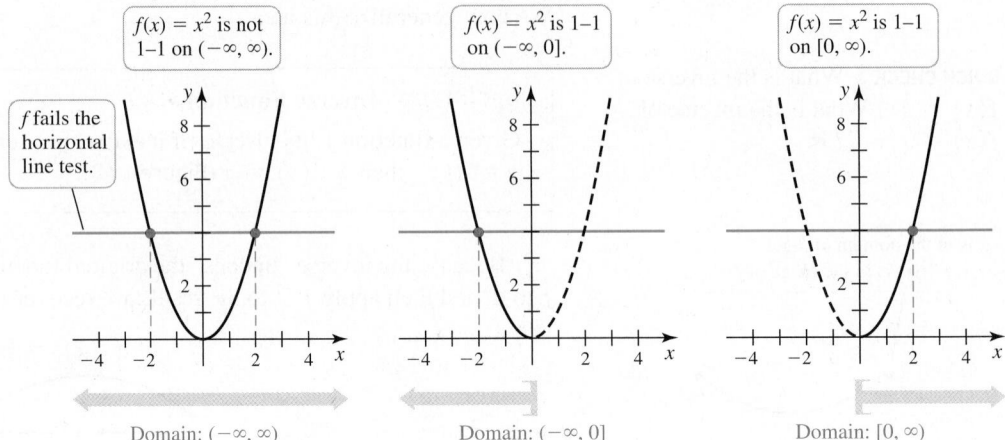

Figure 1.50

EXAMPLE 1   One-to-one functions   Determine the (largest possible) intervals on which the function $f(x) = 2x^2 - x^4$ (**Figure 1.51**) is one-to-one.

SOLUTION   The function is not one-to-one on the entire real line because it fails the horizontal line test. However, on the intervals $(-\infty, -1], [-1, 0], [0, 1],$ and $[1, \infty), f$ is one-to-one. The function is also one-to-one on any subinterval of these four intervals.

Related Exercises 5–6 ◄

Existence of Inverse Functions   Figure 1.52a illustrates the actions of a one-to-one function f and its inverse f^{-1}. We see that f maps a value of x to a unique value of y. In turn, f^{-1} maps that value of y back to the original value of x. This procedure cannot be carried out if f is *not* one-to-one (**Figure 1.52b**).

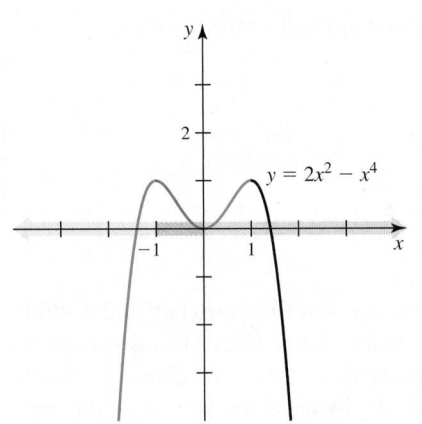

Figure 1.51

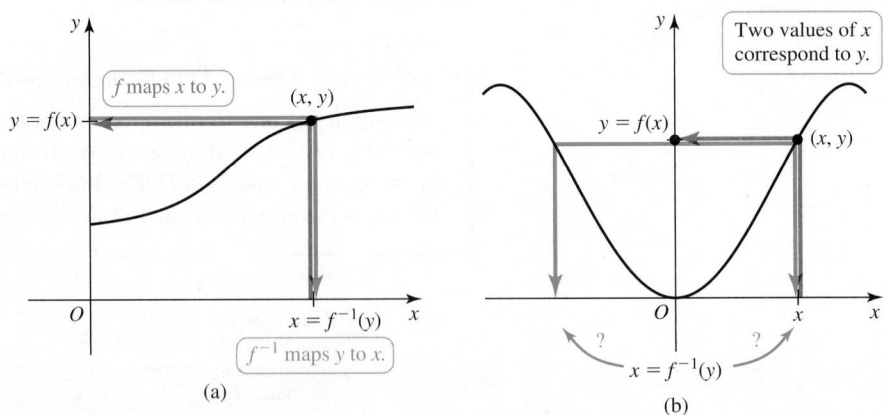

Figure 1.52

▶ The statement that a one-to-one function has an inverse is plausible based on its graph. However, the proof of this theorem is fairly technical and is omitted.

THEOREM 1.1   Existence of Inverse Functions

Let f be a one-to-one function on a domain D with a range R. Then f has a unique inverse f^{-1} with domain R and range D such that

$$f^{-1}(f(x)) = x \quad \text{and} \quad f(f^{-1}(y)) = y,$$

where x is in D and y is in R.

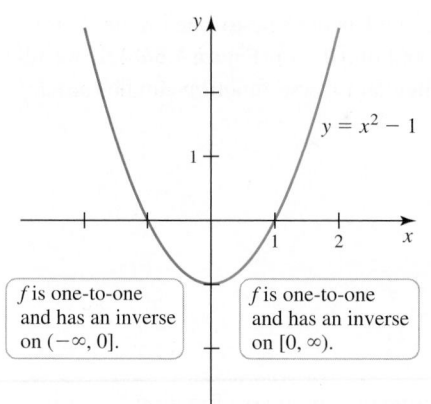

$y = x^2 - 1$

f is one-to-one and has an inverse on $(-\infty, 0]$.

f is one-to-one and has an inverse on $[0, \infty)$.

Figure 1.53

➤ Once you find a formula for f^{-1}, you can check your work by verifying that $f^{-1}(f(x)) = x$ and $f(f^{-1}(x)) = x$.

➤ A constant function (whose graph is a horizontal line) fails the horizontal line test and does not have an inverse.

QUICK CHECK 4 The function that gives degrees Fahrenheit in terms of degrees Celsius is $F = 9C/5 + 32$. Why does this function have an inverse? ◄

EXAMPLE 2   Does an inverse exist? Determine the largest intervals on which $f(x) = x^2 - 1$ has an inverse function.

SOLUTION On the interval $(-\infty, \infty)$ the function does not pass the horizontal line test and is not one-to-one (**Figure 1.53**). However, if the domain of f is restricted to the interval $(-\infty, 0]$ or $[0, \infty)$, then f is one-to-one and an inverse exists.

Related Exercise 25 ◄

Finding Inverse Functions   The crux of finding an inverse for a function f is solving the equation $y = f(x)$ for x in terms of y. If it is possible to do so, then we have found a relationship of the form $x = f^{-1}(y)$. Interchanging x and y in $x = f^{-1}(y)$ so that x is the independent variable (which is the customary role for x), the inverse has the form $y = f^{-1}(x)$. Notice that if f is not one-to-one, this process leads to more than one inverse function.

PROCEDURE   Finding an Inverse Function

Suppose f is one-to-one on an interval I. To find f^{-1}, use the following steps.

1. Solve $y = f(x)$ for x. If necessary, choose the function that corresponds to I.
2. Interchange x and y and write $y = f^{-1}(x)$.

EXAMPLE 3   Finding inverse functions Find the inverse(s) of the following functions. Restrict the domain of f if necessary.

a. $f(x) = 2x + 6$ **b.** $f(x) = x^2 - 1$

SOLUTION

a. Linear functions (except constant linear functions) are one-to-one on the entire real line. Therefore, an inverse function for f exists for all values of x.

Step 1: Solve $y = f(x)$ for x: We see that $y = 2x + 6$ implies that $2x = y - 6$, or

$$x = \frac{1}{2}y - 3.$$

Step 2: Interchange x and y, and write $y = f^{-1}(x)$:

$$y = f^{-1}(x) = \frac{1}{2}x - 3.$$

It is instructive to verify that the inverse relations $f(f^{-1}(x)) = x$ and $f^{-1}(f(x)) = x$ are satisfied:

$$f(f^{-1}(x)) = f\left(\frac{1}{2}x - 3\right) = \underbrace{2\left(\frac{1}{2}x - 3\right) + 6}_{f(x) = 2x + 6} = x - 6 + 6 = x,$$

$$f^{-1}(f(x)) = f^{-1}(2x + 6) = \underbrace{\frac{1}{2}(2x + 6) - 3}_{f^{-1}(x) = \frac{1}{2}x - 3} = x + 3 - 3 = x.$$

b. As shown in Example 2, the function $f(x) = x^2 - 1$ is not one-to-one on the entire real line; however, it is one-to-one on $(-\infty, 0]$ and on $[0, \infty)$ (**Figure 1.54a**). If we restrict our attention to either of these intervals, then an inverse function can be found.

Step 1: Solve $y = f(x)$ for x:

$$y = x^2 - 1$$

$$x^2 = y + 1$$

$$x = \begin{cases} \sqrt{y + 1} \\ -\sqrt{y + 1}. \end{cases}$$

Each branch of the square root corresponds to an inverse function.

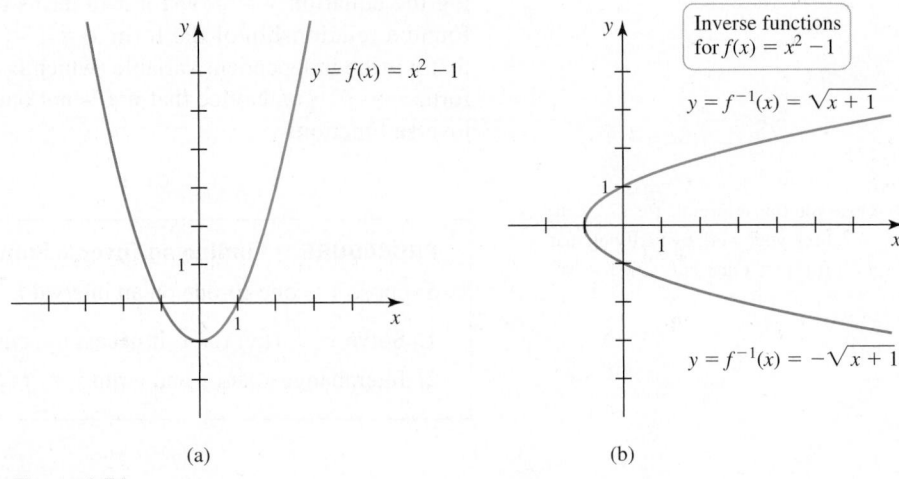

(a)   (b)

Figure 1.54

Step 2: Interchange x and y and write $y = f^{-1}(x)$:

$$y = f^{-1}(x) = \sqrt{x + 1} \quad \text{or} \quad y = f^{-1}(x) = -\sqrt{x + 1}.$$

The interpretation of this result is important. Taking the positive branch of the square root, the inverse function $y = f^{-1}(x) = \sqrt{x + 1}$ gives positive values of y; it corresponds to the branch of $f(x) = x^2 - 1$ on the interval $[0, \infty)$ (**Figure 1.54b**). The negative branch of the square root, $y = f^{-1}(x) = -\sqrt{x + 1}$, is another inverse function that gives negative values of y; it corresponds to the branch of $f(x) = x^2 - 1$ on the interval $(-\infty, 0]$.

Related Exercises 9–10, 33–34 ◄

QUICK CHECK 5 On what interval(s) does the function $f(x) = x^3$ have an inverse? ◄

Graphing Inverse Functions

The graphs of a function and its inverse have a special relationship, which is illustrated in the following example.

EXAMPLE 4   Graphing inverse functions Find f^{-1}, and then plot f and f^{-1} on the same coordinate axes.

a. $f(x) = 2x + 6$

b. $f(x) = \sqrt{x - 1}$

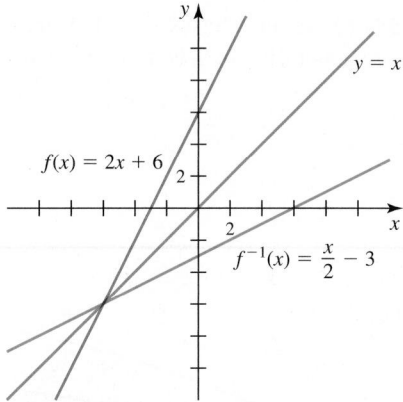

The function $f(x) = 2x + 6$ and its inverse $f^{-1}(x) = \frac{x}{2} - 3$ are symmetric about the line $y = x$.

Figure 1.55

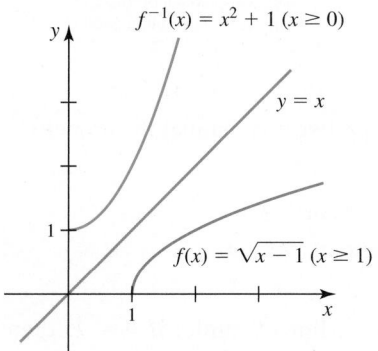

The function $f(x) = \sqrt{x - 1}$ $(x \geq 1)$ and its inverse $f^{-1}(x) = x^2 + 1$ $(x \geq 0)$ are symmetric about $y = x$.

Figure 1.56

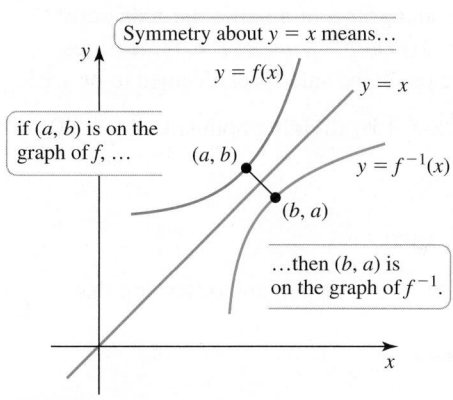

Symmetry about $y = x$ means…

$y = f(x)$

if (a, b) is on the graph of f, …

(a, b)

$y = f^{-1}(x)$

(b, a)

…then (b, a) is on the graph of f^{-1}.

Figure 1.57

➤ Logarithms were invented around 1600 for calculating purposes by the Scotsman John Napier and the Englishman Henry Briggs. Unfortunately, the word *logarithm*, derived from the Greek for reasoning (*logos*) with numbers (*arithmos*), doesn't help with the meaning of the word. **When you see logarithm, you should think exponent.**

SOLUTION

a. The inverse of $f(x) = 2x + 6$, found in Example 3, is

$$y = f^{-1}(x) = \frac{1}{2}x - 3.$$

The graphs of f and f^{-1} are shown in **Figure 1.55**. Notice that both f and f^{-1} are increasing linear functions that intersect at $(-6, -6)$.

b. The domain of $f(x) = \sqrt{x - 1}$ is $[1, \infty)$ and its range is $[0, \infty)$. On this domain, f is one-to-one and has an inverse. It can be found in two steps:

Step 1: Solve $y = \sqrt{x - 1}$ for x:

$$y^2 = x - 1 \quad \text{or} \quad x = y^2 + 1.$$

Step 2: Interchange x and y and write $y = f^{-1}(x)$:

$$y = f^{-1}(x) = x^2 + 1.$$

The graphs of f and f^{-1} are shown in **Figure 1.56**. Notice that the domain of f^{-1} (which is $x \geq 0$) corresponds to the range of f (which is $y \geq 0$).

Related Exercises 28–29 ◄

Looking closely at the graphs in Figure 1.55 and Figure 1.56, you see a symmetry that always occurs when a function and its inverse are plotted on the same set of axes. In each figure, one curve is the reflection of the other curve across the line $y = x$. These curves have *symmetry about the line* $y = x$, which means that the point (a, b) is on one curve whenever the point (b, a) is on the other curve (**Figure 1.57**).

The explanation for the symmetry comes directly from the definition of the inverse. Suppose that the point (a, b) is on the graph of $y = f(x)$, which means that $b = f(a)$. By the definition of the inverse function, we know that $a = f^{-1}(b)$, which means that the point (b, a) is on the graph of $y = f^{-1}(x)$. This argument applies to all relevant points (a, b), so whenever (a, b) is on the graph of f, (b, a) is on the graph of f^{-1}. As a consequence, the graphs are symmetric about the line $y = x$.

Logarithmic Functions

Everything we learned about inverse functions is now applied to the exponential function $f(x) = b^x$. For any $b > 0$, with $b \neq 1$, this function is one-to-one on the interval $(-\infty, \infty)$. Therefore, it has an inverse.

DEFINITION Logarithmic Function Base b

For any base $b > 0$, with $b \neq 1$, the **logarithmic function base b**, denoted $y = \log_b x$, is the inverse of the exponential function $y = b^x$. The inverse of the natural exponential function with base $b = e$ is the **natural logarithm function**, denoted $y = \ln x$.

The inverse relationship between logarithmic and exponential functions may be stated concisely in several ways. First, we have

$$y = \log_b x \quad \text{if and only if} \quad b^y = x.$$

Combining these two conditions results in two important relations.

Inverse Relations for Exponential and Logarithmic Functions

For any base $b > 0$, with $b \neq 1$, the following inverse relations hold.

I1. $b^{\log_b x} = x$, for $x > 0$

I2. $\log_b b^x = x$, for real values of x

➤ **Logarithm Rules**

For any base $b > 0$ $(b \neq 1)$, positive real numbers x and y, and real numbers z, the following relations hold:

L1. $\log_b xy = \log_b x + \log_b y$

L2. $\log_b \dfrac{x}{y} = \log_b x - \log_b y$

$\left(\text{includes } \log_b \dfrac{1}{y} = -\log_b y \right)$

L3. $\log_b x^z = z \log_b x$

L4. $\log_b b = 1$

Specifically, these properties hold with $b = e$, in which case $\log_b x = \log_e x = \ln x$. See Exercises 92–94.

Properties of Logarithmic Functions   The graph of the logarithmic function is generated using the symmetry of the graphs of a function and its inverse. **Figure 1.58** shows how the graph of $y = b^x$, for $b > 1$, is reflected across the line $y = x$ to obtain the graph of $y = \log_b x$.

The graphs of $y = \log_b x$ are shown (**Figure 1.59**) for several bases $b > 1$. Logarithms with bases $0 < b < 1$, although well defined, are generally not used (and they can be expressed in terms of bases with $b > 1$).

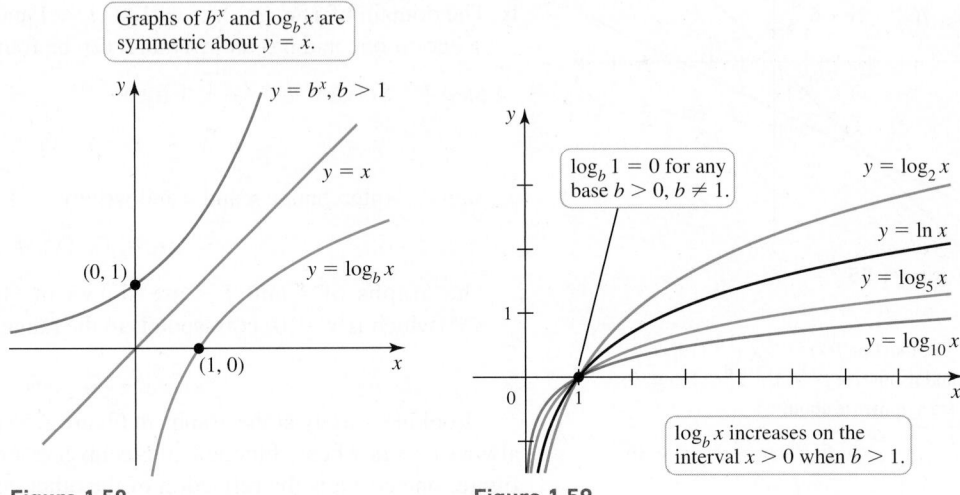

Graphs of b^x and $\log_b x$ are symmetric about $y = x$.

$y = b^x, b > 1$

$y = x$

$(0, 1)$

$y = \log_b x$

$(1, 0)$

Figure 1.58

$\log_b 1 = 0$ for any base $b > 0, b \neq 1$.

$y = \log_2 x$

$y = \ln x$

$y = \log_5 x$

$y = \log_{10} x$

$\log_b x$ increases on the interval $x > 0$ when $b > 1$.

Figure 1.59

Logarithmic functions with base $b > 0$ satisfy properties that parallel the properties of the exponential functions on pp. 27–28.

1. Because the range of b^x is $(0, \infty)$, the domain of $\log_b x$ is $(0, \infty)$.

2. The domain of b^x is $(-\infty, \infty)$, which implies that the range of $\log_b x$ is $(-\infty, \infty)$.

3. Because $b^0 = 1$, it follows that $\log_b 1 = 0$.

4. If $b > 1$, then $\log_b x$ is an increasing function of x. For example, if $b = e$, then $\ln x > \ln y$ whenever $x > y$ (Figure 1.59).

QUICK CHECK 6 What is the domain of $f(x) = \log_b x^2$? What is the range of $f(x) = \log_b x^2$? ◄

EXAMPLE 5   Using inverse relations   One thousand grams of a particular radioactive substance decays according to the function $m(t) = 1000e^{-t/850}$, where $t \geq 0$ measures time in years. When does the mass of the substance reach the safe level, deemed to be 1 g?

SOLUTION   Setting $m(t) = 1$, we solve $1000e^{-t/850} = 1$ by dividing both sides by 1000 and taking the natural logarithm of both sides:

$$\ln \left(e^{-t/850} \right) = \ln \left(\frac{1}{1000} \right).$$

This equation is simplified by calculating $\ln(1/1000) \approx -6.908$ and observing that

$$\ln \left(e^{-t/850} \right) = -\frac{t}{850} \text{ (inverse property I2). Therefore,}$$

➤ Provided the arguments are positive, we can take the $\log_b$ of both sides of an equation and produce an equivalent equation.

$$-\frac{t}{850} \approx -6.908.$$

Solving for t, we find that $t \approx (-850)(-6.908) \approx 5872$ years.

Related Exercise 61 ◄

Change of Base

When working with logarithms and exponentials, it doesn't matter *in principle* which base is used. However, there are practical reasons for switching between bases. For example, most calculators have built-in logarithmic functions in just one or two bases. If you need to use a different base, then the change-of-base rules are essential.

Consider changing bases with exponential functions. Specifically, suppose you want to express b^x (base b) in the form e^y (base e), where y must be determined. Taking the natural logarithm of both sides of $e^y = b^x$, we have

$$\underbrace{\ln e^y}_{y} = \underbrace{\ln b^x}_{x \ln b} \quad \text{which implies that } y = x \ln b.$$

It follows that $b^x = e^y = e^{x \ln b}$. For example, $4^x = e^{x \ln 4}$. This result is derived rigorously in Section 7.1.

> ➤ A similar argument is used to derive more general formulas for changing from base b to any other positive base c.

The formula for changing from $\log_b x$ to $\ln x$ is derived in a similar way. We let $y = \log_b x$, which implies that $x = b^y$. Taking the natural logarithm of both sides of $x = b^y$ gives $\ln x = \ln b^y = y \ln b$. Solving for y gives

$$y = \log_b x = \frac{\ln x}{\ln b}.$$

Change-of-Base Rules

Let b be a positive real number with $b \neq 1$. Then

$$b^x = e^{x \ln b}, \text{ for all } x \quad \text{and} \quad \log_b x = \frac{\ln x}{\ln b}, \text{ for } x > 0.$$

More generally, if c is a positive real number with $c \neq 1$, then

$$b^x = c^{x \log_c b}, \text{ for all } x \quad \text{and} \quad \log_b x = \frac{\log_c x}{\log_c b}, \text{ for } x > 0.$$

EXAMPLE 6   Changing bases

a. Express 2^{x+4} as an exponential function with base e.

b. Express $\log_2 x$ using base e and base 32.

SOLUTION

a. Using the change-of-base rule for exponential functions, we have

$$2^{x+4} = e^{(x+4)\ln 2}.$$

b. Using the change-of-base rule for logarithmic functions, we have

$$\log_2 x = \frac{\ln x}{\ln 2}.$$

To change from base 2 to base 32, we use the general change-of-base formula:

$$\log_2 x = \frac{\log_{32} x}{\log_{32} 2} = \frac{\log_{32} x}{1/5} = 5 \log_{32} x.$$

The second step follows from the fact that $2 = 32^{1/5}$, so $\log_{32} 2 = \frac{1}{5}$.

Related Exercises 67, 71 ◄

SECTION 1.3 EXERCISES

Getting Started

1. For $b > 0$, what are the domain and range of $f(x) = b^x$?

2. Give an example of a function that is one-to-one on the entire real number line.

3. Sketch a graph of a function that is one-to-one on the interval $(-\infty, 0]$ but is not one-to-one on $(-\infty, \infty)$.

4. Sketch a graph of a function that is one-to-one on the intervals $(-\infty, -2]$ and $[-2, \infty)$ but is not one-to-one on $(-\infty, \infty)$.

5. Find three intervals on which f is one-to-one, making each interval as large as possible.

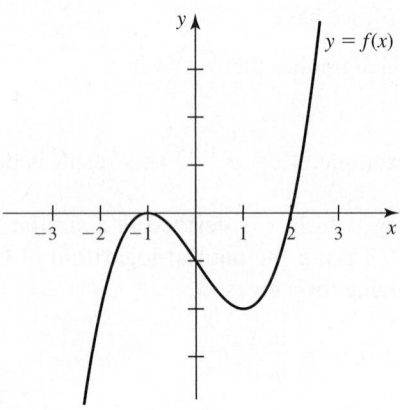

6. Find four intervals on which f is one-to-one, making each interval as large as possible.

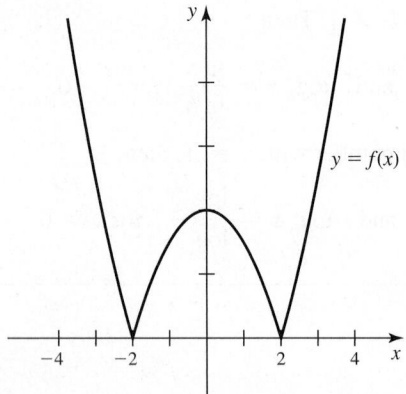

7. Explain why a function that is not one-to-one on an interval I cannot have an inverse function on I.

8. Use the graph of f to find $f^{-1}(2)$, $f^{-1}(9)$, and $f^{-1}(12)$.

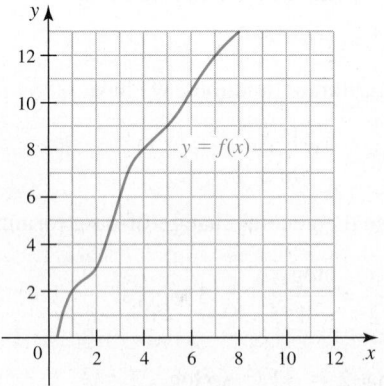

9. Find the inverse of the function $f(x) = 2x$. Verify that $f(f^{-1}(x)) = x$ and $f^{-1}(f(x)) = x$.

10. Find the inverse of the function $f(x) = \sqrt{x}$, for $x \geq 0$. Verify that $f(f^{-1}(x)) = x$ and $f^{-1}(f(x)) = x$.

11. Sketch the graph of the inverse of f.

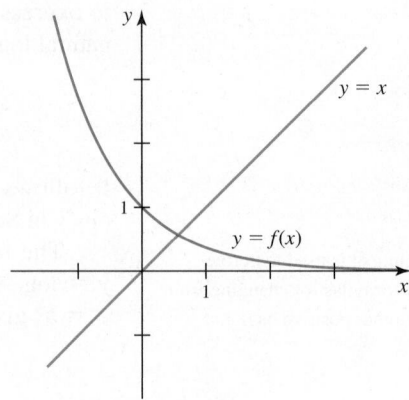

12. Sketch the graph of the inverse of f.

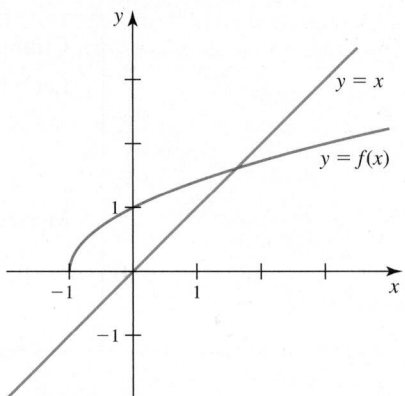

13–14. *The parabola* $y = x^2 + 1$ *consists of two one-to-one functions,* $g_1(x)$ *and* $g_2(x)$. *Complete each exercise and confirm that your answers are consistent with the graphs displayed in the figure.*

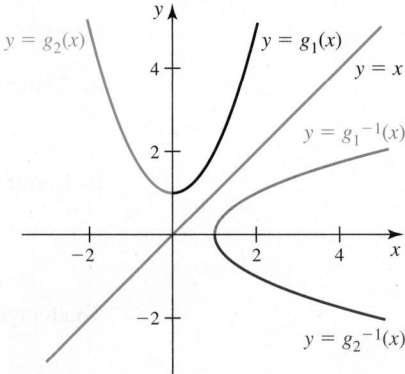

13. Find formulas for $g_1(x)$ and $g_1^{-1}(x)$. State the domain and range of each function.

14. Find formulas for $g_2(x)$ and $g_2^{-1}(x)$. State the domain and range of each function.

15. Explain the meaning of $\log_b x$.

16. How is the property $b^{x+y} = b^x b^y$ related to the property $\log_b(xy) = \log_b x + \log_b y$?

17. For $b > 0$ with $b \neq 1$, what are the domain and range of $f(x) = \log_b x$ and why?

18. Express 2^5 using base e.

19. Evaluate each expression without a calculator.

    a. $\log_{10} 1000$ **b.** $\log_2 16$ **c.** $\log_{10} 0.01$ **d.** $\ln e^3$ **e.** $\ln \sqrt{e}$

20. For a certain constant $a > 1$, $\ln a \approx 3.8067$. Find approximate values of $\log_2 a$ and $\log_a 2$ using the fact that $\ln 2 \approx 0.6931$.

Practice Exercises

21–26. Where do inverses exist? *Use analytical and/or graphical methods to determine the largest possible sets of points on which the following functions have an inverse.*

21. $f(x) = 3x + 4$ **22.** $f(x) = |2x + 1|$

T 23. $f(x) = 1/(x - 5)$ **24.** $f(x) = -(6 - x)^2$

T 25. $f(x) = 1/x^2$ **26.** $f(x) = x^2 - 2x + 8$
                                   (*Hint:* Complete the square.)

27–32. Graphing inverse functions *Find the inverse function (on the given interval, if specified) and graph both f and f^{-1} on the same set of axes. Check your work by looking for the required symmetry in the graphs.*

27. $f(x) = 8 - 4x$ **28.** $f(x) = 3x + 5$

29. $f(x) = \sqrt{x + 2}$, for $x \geq -2$

30. $f(x) = \sqrt{3 - x}$, for $x \leq 3$

31. $f(x) = (x - 2)^2 - 1$, for $x \geq 2$

32. $f(x) = x^2 + 4$, for $x \geq 0$

33–42. Finding inverse functions *Find the inverse $f^{-1}(x)$ of each function (on the given interval, if specified).*

33. $f(x) = 2/(x^2 + 1)$, for $x \geq 0$

34. $f(x) = 6/(x^2 - 9)$, for $x > 3$

35. $f(x) = e^{2x+6}$ **36.** $f(x) = 4e^{5x}$

37. $f(x) = \ln(3x + 1)$ **38.** $f(x) = \log_{10} 4x$

39. $f(x) = 10^{-2x}$ **40.** $f(x) = 1/(e^x + 1)$

41. $f(x) = e^x/(e^x + 2)$ **42.** $f(x) = x/(x - 2)$, for $x > 2$

43. Splitting up curves The unit circle $x^2 + y^2 = 1$ consists of four one-to-one functions, $f_1(x)$, $f_2(x)$, $f_3(x)$, and $f_4(x)$ (see figure).

    a. Find the domain and a formula for each function.
    b. Find the inverse of each function and write it as $y = f^{-1}(x)$.

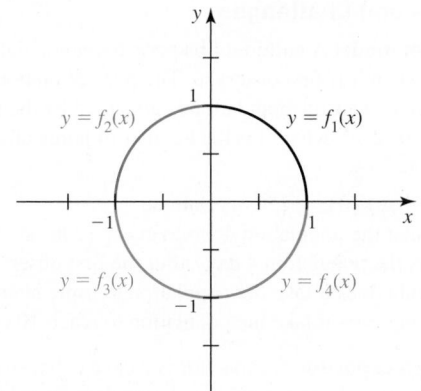

44. Splitting up curves The equation $y^4 = 4x^2$ is associated with four one-to-one functions, $f_1(x)$, $f_2(x)$, $f_3(x)$, and $f_4(x)$ (see figure).

    a. Find the domain and a formula for each function.
    b. Find the inverse of each function and write it as $y = f^{-1}(x)$.

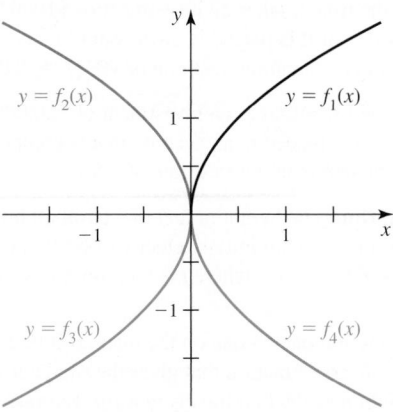

45–50. Properties of logarithms *Assume $\log_b x = 0.36$, $\log_b y = 0.56$, and $\log_b z = 0.83$. Evaluate the following expressions.*

45. $\log_b \dfrac{x}{y}$ **46.** $\log_b x^2$

47. $\log_b xz$ **48.** $\log_b \dfrac{\sqrt{xy}}{z}$

49. $\log_b \dfrac{\sqrt{x}}{\sqrt[3]{z}}$ **50.** $\log_b \dfrac{b^2 x^{5/2}}{\sqrt{y}}$

51–60. Solving equations *Solve the following equations.*

51. $\log_{10} x = 3$ **52.** $\log_5 x = -1$

53. $\log_8 x = \frac{1}{3}$ **54.** $\log_b 125 = 3$

55. $\ln x = -1$ **56.** $\ln y = 3$

57. $7^x = 21$ **58.** $2^x = 55$

59. $3^{3x-4} = 15$ **60.** $5^{3x} = 29$

T 61. Using inverse relations One hundred grams of a particular radioactive substance decays according to the function $m(t) = 100e^{-t/650}$, where $t > 0$ measures time in years. When does the mass reach 50 grams?

T 62. Mass of juvenile desert tortoises In a study conducted at the University of New Mexico, it was found that the mass $m(t)$ (in grams) of juvenile desert tortoises t days after a switch to a particular diet is accurately described by the function $m(t) = m_0 e^{0.004t}$, where m_0 is the mass of the tortoise at the time of the diet switch. According to this function, how long does it take a juvenile desert tortoise to reach a mass of 150 g if it had a mass of 64 g when its diet was switched? (*Source: Physiological and Biochemical Zoology*, 85, 1, 2012)

T 63–64. Investment Problems *An investment of P dollars is deposited in a savings account that is compounded monthly with an annual interest rate of r, where r is expressed as a decimal. The amount of money A in the account after t years is given by $A = P(1 + r/12)^{12t}$. Use this equation to complete the following exercises.*

63. Determine the time it takes an investment of $1000 to increase to $1100 dollars if it is placed in an account that is compounded monthly with an annual interest rate of 1% ($r = 0.01$).

64. Determine the time it takes an investment of $20,000 to increase to $22,000 if it is placed in an account that is compounded monthly with an annual interest rate of 2.5%.

T 65. Height and time The height in feet of a baseball hit straight up from the ground with an initial velocity of 64 ft/s is given by $h = f(t) = 64t - 16t^2$, where t is measured in seconds after the hit.

   a. Is this function one-to-one on the interval $0 \le t \le 4$?

   b. Find the inverse function that gives the time t at which the ball is at height h as the ball travels *upward.* Express your answer in the form $t = f^{-1}(h)$.

   c. Find the inverse function that gives the time t at which the ball is at height h as the ball travels *downward.* Express your answer in the form $t = f^{-1}(h)$.

   d. At what time is the ball at a height of 30 ft on the way up?

   e. At what time is the ball at a height of 10 ft on the way down?

T 66. Velocity of a skydiver The velocity of a skydiver (in m/s) t seconds after jumping from the plane is $v(t) = 600(1 - e^{-kt/60})/k$, where $k > 0$ is a constant. The *terminal velocity* of the skydiver is the value that $v(t)$ approaches as t becomes large. Graph v with $k = 11$ and estimate the terminal velocity.

T 67–70. Calculator base change *Write the following logarithms in terms of the natural logarithm. Then use a calculator to find the value of the logarithm, rounding your result to four decimal places.*

67. $\log_2 15$ **68.** $\log_3 30$

69. $\log_4 40$ **70.** $\log_6 60$

71–76. Changing bases *Convert the following expressions to the indicated base.*

71. 2^x using base e

72. $3^{\sin x}$ using base e

73. $\ln |x|$ using base 5

74. $\log_2(x^2 + 1)$ using base e

75. $a^{1/\ln a}$ using base e, for $a > 0$ and $a \ne 1$

76. $a^{1/\log_{10} a}$ using base 10, for $a > 0$ and $a \ne 1$

77. Explain why or why not Determine whether the following statements are true and give an explanation or counterexample.

   a. If $y = 3^x$, then $x = \sqrt[3]{y}$.

   b. $\dfrac{\log_b x}{\log_b y} = \log_b x - \log_b y$

   c. $\log_5 4^6 = 4 \log_5 6$

   d. $2 = 10^{\log_{10} 2}$

   e. $2 = \ln 2^e$

   f. If $f(x) = x^2 + 1$, then $f^{-1}(x) = \dfrac{1}{x^2 + 1}$.

   g. If $f(x) = 1/x$, then $f^{-1}(x) = 1/x$.

78. Graphs of exponential functions The following figure shows the graphs of $y = 2^x$, $y = 3^x$, $y = 2^{-x}$, and $y = 3^{-x}$. Match each curve with the correct function.

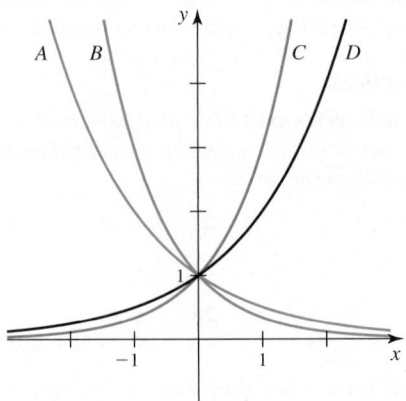

79. Graphs of logarithmic functions The following figure shows the graphs of $y = \log_2 x$, $y = \log_4 x$, and $y = \log_{10} x$. Match each curve with the correct function.

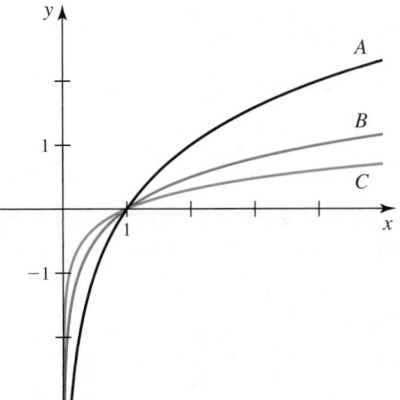

80. Graphs of modified exponential functions Without using a graphing utility, sketch the graph of $y = 2^x$. Then on the same set of axes, sketch the graphs of $y = 2^{-x}$, $y = 2^{x-1}$, $y = 2^x + 1$, and $y = 2^{2x}$.

81. Graphs of modified logarithmic functions Without using a graphing utility, sketch the graph of $y = \log_2 x$. Then on the same set of axes, sketch the graphs of $y = \log_2(x - 1)$, $y = \log_2 x^2$, $y = (\log_2 x)^2$, and $y = \log_2 x + 1$.

Explorations and Challenges

82. Population model A culture of bacteria has a population of 150 cells when it is first observed. The population doubles every 12 hr, which means its population is governed by the function $p(t) = 150 \cdot 2^{t/12}$, where t is the number of hours after the first observation.

   a. Verify that $p(0) = 150$, as claimed.

   b. Show that the population doubles every 12 hr, as claimed.

   c. What is the population 4 days after the first observation?

   d. How long does it take the population to triple in size?

   e. How long does it take the population to reach 10,000?

T 83. Charging a capacitor A capacitor is a device that stores electrical charge. The charge on a capacitor accumulates according to the function $Q(t) = a(1 - e^{-t/c})$, where t is measured in seconds, and a and $c > 0$ are physical constants. The *steady-state charge* is the value that $Q(t)$ approaches as t becomes large.

a. Graph the charge function for $t \geq 0$ using $a = 1$ and $c = 10$. Find a graphing window that shows the full range of the function.

b. Vary the value of a while holding c fixed. Describe the effect on the curve. How does the steady-state charge vary with a?

c. Vary the value of c while holding a fixed. Describe the effect on the curve. How does the steady-state charge vary with c?

d. Find a formula that gives the steady-state charge in terms of a and c.

T 84. Large intersection point Use any means to approximate the intersection point(s) of the graphs of $f(x) = e^x$ and $g(x) = x^{123}$. (*Hint:* Consider using logarithms.)

85–90. Finding all inverses *Find all the inverses associated with the following functions, and state their domains.*

85. $f(x) = x^2 - 2x + 6$ (*Hint:* Complete the square first.)

86. $f(x) = -x^2 - 4x - 3$ (*Hint:* Complete the square first.)

87. $f(x) = (x + 1)^3$ **88.** $f(x) = (x - 4)^2$

89. $f(x) = 2/(x^2 + 2)$ **90.** $f(x) = 2x/(x + 2)$

91. Reciprocal bases Assume that $b > 0$ and $b \neq 1$. Show that $\log_{1/b} x = -\log_b x$.

92. Proof of rule L1 Use the following steps to prove that $\log_b xy = \log_b x + \log_b y$.

a. Let $x = b^p$ and $y = b^q$. Solve these expressions for p and q, respectively.

b. Use property E1 for exponents (p. 27) to express xy in terms of b, p, and q.

c. Compute $\log_b xy$ and simplify.

93. Proof of rule L2 Modify the proof outlined in Exercise 92 and use property E2 for exponents to prove that $\log_b \dfrac{x}{y} = \log_b x - \log_b y$.

94. Proof of rule L3 Use the following steps to prove that $\log_b x^z = z \log_b x$.

a. Let $x = b^p$. Solve this expression for p.

b. Use property E3 for exponents to express x^z in terms of b and p.

c. Compute $\log_b x^z$ and simplify.

T 95. Inverses of a quartic Consider the quartic polynomial $y = f(x) = x^4 - x^2$.

a. Graph f and find the largest intervals on which it is one-to-one. The goal is to find the inverse function on each of these intervals.

b. Make the substitution $u = x^2$ to solve the equation $y = f(x)$ for x in terms of y. Be sure you have included all possible solutions.

c. Write each inverse function in the form $y = f^{-1}(x)$ for each of the intervals found in part (a).

96. Inverse of composite functions

a. Let $g(x) = 2x + 3$ and $h(x) = x^3$. Consider the composite function $f(x) = g(h(x))$. Find f^{-1} directly and then express it in terms of g^{-1} and h^{-1}.

b. Let $g(x) = x^2 + 1$ and $h(x) = \sqrt{x}$. Consider the composite function $f(x) = g(h(x))$. Find f^{-1} directly and then express it in terms of g^{-1} and h^{-1}.

c. Explain why if g and h are one-to-one, the inverse of $f(x) = g(h(x))$ exists.

97. Nice property Prove that $(\log_b c)(\log_c b) = 1$, for $b > 0, c > 0$, $b \neq 1$, and $c \neq 1$.

QUICK CHECK ANSWERS

1. b^x is always positive (and never zero) for all x and for positive bases b.   **2.** Because $(1/3)^x = 1/3^x$ and 3^x increases as x increases, it follows that $(1/3)^x$ decreases as x increases.   **3.** $f^{-1}(x) = 3x$; $f^{-1}(x) = x + 7$.
4. For every Fahrenheit temperature, there is exactly one Celsius temperature, and vice versa. The given relation is also a linear function. It is one-to-one, so it has an inverse function.   **5.** The function $f(x) = x^3$ is one-to-one on $(-\infty, \infty)$, so it has an inverse for all values of x.   **6.** The domain of $\log_b x^2$ is all real numbers except zero (because x^2 is positive for $x \neq 0$). The range of $\log_b x^2$ is all real numbers. ◄

1.4 Trigonometric Functions and Their Inverses

This section is a review of what you need to know in order to study the calculus of trigonometric functions. Once the trigonometric functions are on stage, it makes sense to present the inverse trigonometric functions and their basic properties.

Radian Measure

Calculus typically requires that angles be measured in **radians** (rad). Working with a circle of radius r, the radian measure of an angle θ is the length of the arc associated with θ, denoted s, divided by the radius of the circle r (**Figure 1.60a**). Working on a unit circle ($r = 1$), the radian measure of an angle is simply the length of the arc associated with θ (**Figure 1.60b**). For example, the length of a full unit circle is 2π; therefore, an angle with a radian measure

Degrees	Radians
0	0
30	$\pi/6$
45	$\pi/4$
60	$\pi/3$
90	$\pi/2$
120	$2\pi/3$
135	$3\pi/4$
150	$5\pi/6$
180	π

of π corresponds to a half circle ($\theta = 180°$) and an angle with a radian measure of $\pi/2$ corresponds to a quarter circle ($\theta = 90°$).

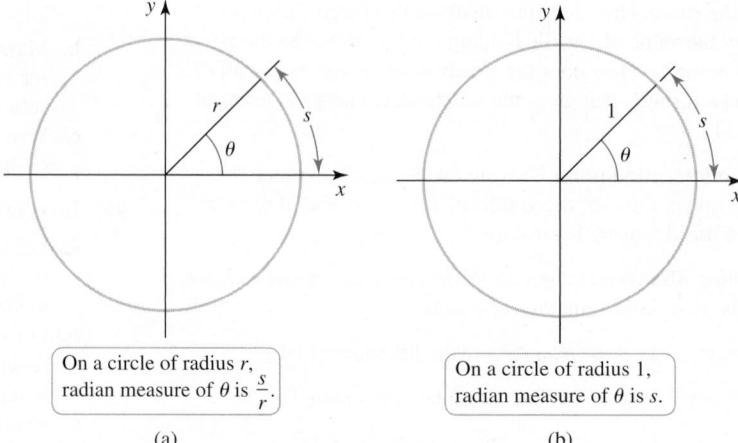

On a circle of radius r, radian measure of θ is $\dfrac{s}{r}$.

(a)

On a circle of radius 1, radian measure of θ is s.

(b)

Figure 1.60

QUICK CHECK 1 What is the radian measure of a 270° angle? What is the degree measure of a $5\pi/4$-rad angle? ◄

Trigonometric Functions

For acute angles, the trigonometric functions are defined as ratios of the lengths of the sides of a right triangle (**Figure 1.61**). To extend these definitions to include all angles, we work in an xy-coordinate system with a circle of radius r centered at the origin. Suppose that $P(x, y)$ is a point on the circle. An angle θ is in **standard position** if its initial side is on the positive x-axis and its terminal side is the line segment OP between the origin and P. An angle is positive if it is obtained by a counterclockwise rotation from the positive x-axis (**Figure 1.62**). When the right-triangle definitions of Figure 1.61 are used with the right triangle in Figure 1.62, the trigonometric functions may be expressed in terms of x, y, and the radius of the circle, $r = \sqrt{x^2 + y^2}$.

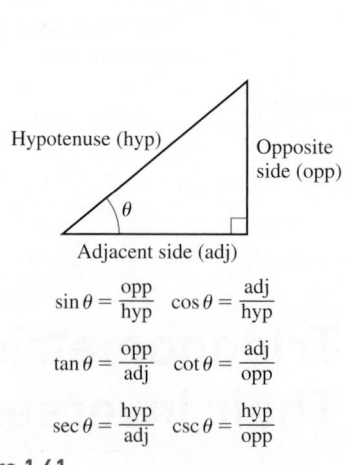

$$\sin\theta = \frac{\text{opp}}{\text{hyp}} \quad \cos\theta = \frac{\text{adj}}{\text{hyp}}$$

$$\tan\theta = \frac{\text{opp}}{\text{adj}} \quad \cot\theta = \frac{\text{adj}}{\text{opp}}$$

$$\sec\theta = \frac{\text{hyp}}{\text{adj}} \quad \csc\theta = \frac{\text{hyp}}{\text{opp}}$$

Figure 1.61

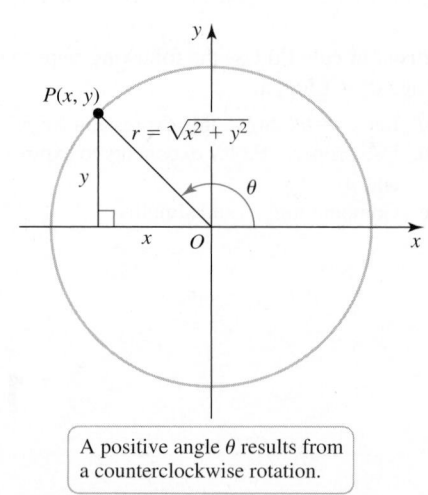

A positive angle θ results from a counterclockwise rotation.

Figure 1.62

▶ When working on a unit circle ($r = 1$), these definitions become

$$\sin\theta = y \quad \cos\theta = x$$
$$\tan\theta = \frac{y}{x} \quad \cot\theta = \frac{x}{y}$$
$$\sec\theta = \frac{1}{x} \quad \csc\theta = \frac{1}{y}$$

DEFINITION   Trigonometric Functions

Let $P(x, y)$ be a point on a circle of radius r associated with the angle θ. Then

$$\sin\theta = \frac{y}{r}, \quad \cos\theta = \frac{x}{r}, \quad \tan\theta = \frac{y}{x},$$

$$\cot\theta = \frac{x}{y}, \quad \sec\theta = \frac{r}{x}, \quad \csc\theta = \frac{r}{y}.$$

> **Standard triangles**

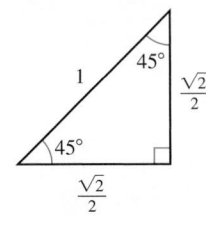

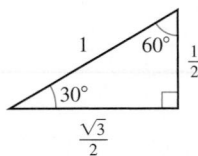

To evaluate the trigonometric functions at the standard angles (multiples of 30° and 45°), it is helpful to know the radian measure of those angles and the coordinates of the associated points on the unit circle (**Figure 1.63**).

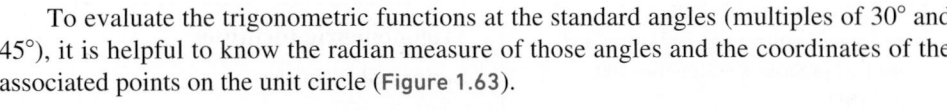

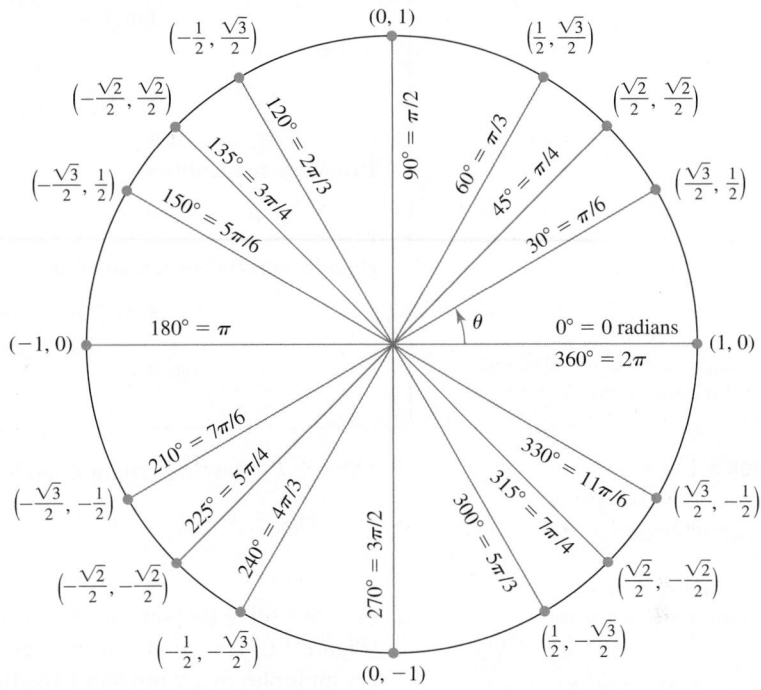

Figure 1.63

Combining the definitions of the trigonometric functions with the coordinates shown in Figure 1.63, we may evaluate these functions at any standard angle. For example,

$$\sin\frac{2\pi}{3} = \frac{\sqrt{3}}{2}, \qquad \cos\frac{5\pi}{6} = -\frac{\sqrt{3}}{2}, \qquad \tan\frac{7\pi}{6} = \frac{1}{\sqrt{3}}, \qquad \tan\frac{3\pi}{2} \text{ is undefined,}$$

$$\cot\frac{5\pi}{3} = -\frac{1}{\sqrt{3}}, \qquad \sec\frac{7\pi}{4} = \sqrt{2}, \qquad \csc\frac{3\pi}{2} = -1, \qquad \sec\frac{\pi}{2} \text{ is undefined.}$$

EXAMPLE 1   Evaluating trigonometric functions Evaluate the following expressions.

a. $\sin(8\pi/3)$ **b.** $\csc(-11\pi/3)$

SOLUTION

a. The angle $\theta = 8\pi/3 = 2\pi + 2\pi/3$ corresponds to a *counterclockwise* revolution of one full circle (2π radians) plus an additional $2\pi/3$ radians (**Figure 1.64**). Therefore, this angle has the same terminal side as the angle $2\pi/3$, and the corresponding point on the unit circle is $\left(-1/2, \sqrt{3}/2\right)$. It follows that $\sin(8\pi/3) = y = \sqrt{3}/2$.

b. The angle $\theta = -11\pi/3 = -2\pi - 5\pi/3$ corresponds to a *clockwise* revolution of one full circle (2π radians) plus an additional $5\pi/3$ radians (**Figure 1.65**). Therefore, this angle has the same terminal side as the angle $\pi/3$. The co-ordinates of the corresponding point on the unit circle are $\left(1/2, \sqrt{3}/2\right)$, so $\csc(-11\pi/3) = 1/y = 2/\sqrt{3}$.

Related Exercises 20, 24 ◀

QUICK CHECK 2 Evaluate $\cos(11\pi/6)$ and $\sin(5\pi/4)$. ◀

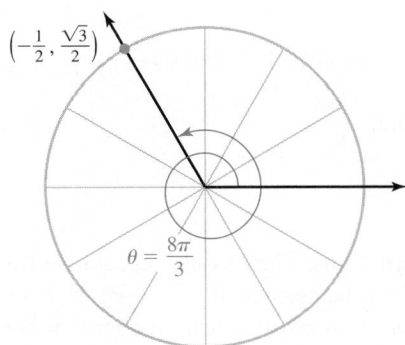

Figure 1.64

Figure 1.65

Trigonometric Identities

Trigonometric functions have a variety of properties, called identities, that are true for all angles in their domain. Here is a list of some commonly used identities.

➤ In addition, to these identities, you should be familiar with the Law of Cosines and the Law of Sines. See Exercises 100 and 101.

Trigonometric Identities

Reciprocal Identities

$$\tan \theta = \frac{\sin \theta}{\cos \theta} \qquad \cot \theta = \frac{1}{\tan \theta} = \frac{\cos \theta}{\sin \theta}$$

$$\csc \theta = \frac{1}{\sin \theta} \qquad \sec \theta = \frac{1}{\cos \theta}$$

Pythagorean Identities

$$\sin^2 \theta + \cos^2 \theta = 1 \qquad 1 + \cot^2 \theta = \csc^2 \theta \qquad \tan^2 \theta + 1 = \sec^2 \theta$$

Double- and Half-Angle Identities

$$\sin 2\theta = 2 \sin \theta \cos \theta \qquad \cos 2\theta = \cos^2 \theta - \sin^2 \theta$$

$$\cos^2 \theta = \frac{1 + \cos 2\theta}{2} \qquad \sin^2 \theta = \frac{1 - \cos 2\theta}{2}$$

➤ The half-angle identities for cosine and sine are also known as *Power-Reducing Identities*.

QUICK CHECK 3 Use $\sin^2 \theta + \cos^2 \theta = 1$ to prove that $1 + \cot^2 \theta = \csc^2 \theta$. ◄

➤ By rationalizing the denominator, observe that $\dfrac{1}{\sqrt{2}} = \dfrac{1}{\sqrt{2}} \cdot \dfrac{\sqrt{2}}{\sqrt{2}} = \dfrac{\sqrt{2}}{2}$.

EXAMPLE 2   Solving trigonometric equations Solve the following equations.

a. $\sqrt{2} \sin x + 1 = 0$ **b.** $\cos 2x = \sin 2x$, where $0 \le x < 2\pi$

SOLUTION

a. First, we solve for $\sin x$ to obtain $\sin x = -1/\sqrt{2} = -\sqrt{2}/2$. From the unit circle (Figure 1.63), we find that $\sin x = -\sqrt{2}/2$ if $x = 5\pi/4$ or $x = 7\pi/4$. Adding integer multiples of 2π produces additional solutions. Therefore, the set of all solutions is

$$x = \frac{5\pi}{4} + 2n\pi \quad \text{and} \quad x = \frac{7\pi}{4} + 2n\pi, \qquad \text{for } n = 0, \pm 1, \pm 2, \pm 3, \ldots.$$

b. Dividing both sides of the equation by $\cos 2x$ (assuming $\cos 2x \ne 0$), we obtain $\tan 2x = 1$. Letting $\theta = 2x$ gives us the equivalent equation $\tan \theta = 1$. This equation is satisfied by

$$\theta = \frac{\pi}{4}, \frac{5\pi}{4}, \frac{9\pi}{4}, \frac{13\pi}{4}, \frac{17\pi}{4}, \ldots.$$

➤ Notice that the assumption $\cos 2x \ne 0$ is valid for these values of x.

Dividing by two and using the restriction $0 \le x < 2\pi$ gives the solutions

$$x = \frac{\theta}{2} = \frac{\pi}{8}, \frac{5\pi}{8}, \frac{9\pi}{8}, \text{ and } \frac{13\pi}{8}.$$

Related Exercises 39, 44 ◄

Graphs of the Trigonometric Functions

Trigonometric functions are examples of **periodic functions**: Their values repeat over every interval of some fixed length. A function f is said to be periodic if $f(x + P) = f(x)$, for all x in the domain, where the **period** P is the smallest positive real number that has this property.

Period of Trigonometric Functions

The functions $\sin \theta$, $\cos \theta$, $\sec \theta$, and $\csc \theta$ have a period of 2π:

$$\sin (\theta + 2\pi) = \sin \theta \qquad \cos (\theta + 2\pi) = \cos \theta$$

$$\sec (\theta + 2\pi) = \sec \theta \qquad \csc (\theta + 2\pi) = \csc \theta,$$

for all θ in the domain.

The functions $\tan \theta$ and $\cot \theta$ have a period of π:

$$\tan (\theta + \pi) = \tan \theta \qquad \cot (\theta + \pi) = \cot \theta,$$

for all θ in the domain.

The graph of $y = \sin\theta$ is shown in **Figure 1.66a**. Because $\csc\theta = 1/\sin\theta$, these two functions have the same sign, but $y = \csc\theta$ is undefined with vertical asymptotes at $\theta = 0, \pm\pi, \pm 2\pi, \ldots$. The functions $\cos\theta$ and $\sec\theta$ have a similar relationship (**Figure 1.66b**).

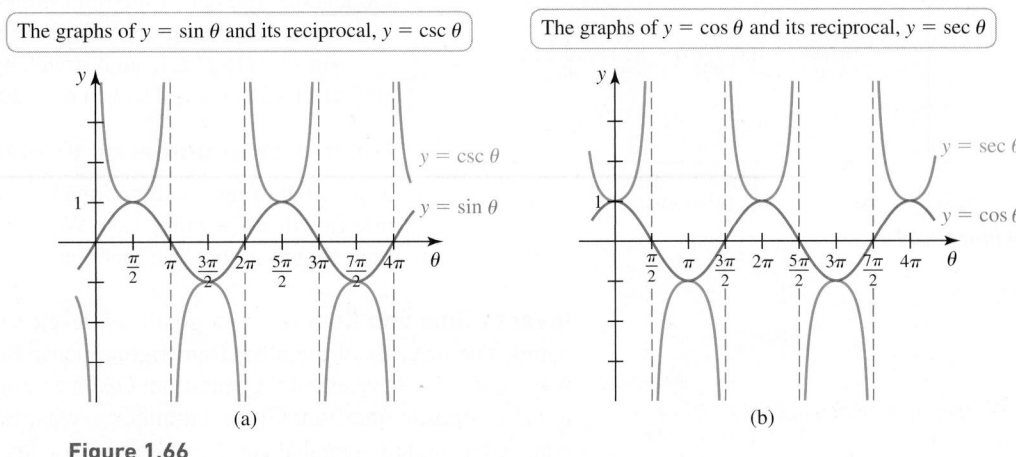

The graphs of $y = \sin\theta$ and its reciprocal, $y = \csc\theta$

The graphs of $y = \cos\theta$ and its reciprocal, $y = \sec\theta$

(a)   (b)

Figure 1.66

The graphs of $\tan\theta$ and $\cot\theta$ are shown in **Figure 1.67**. Each function has points, separated by π units, at which it is undefined.

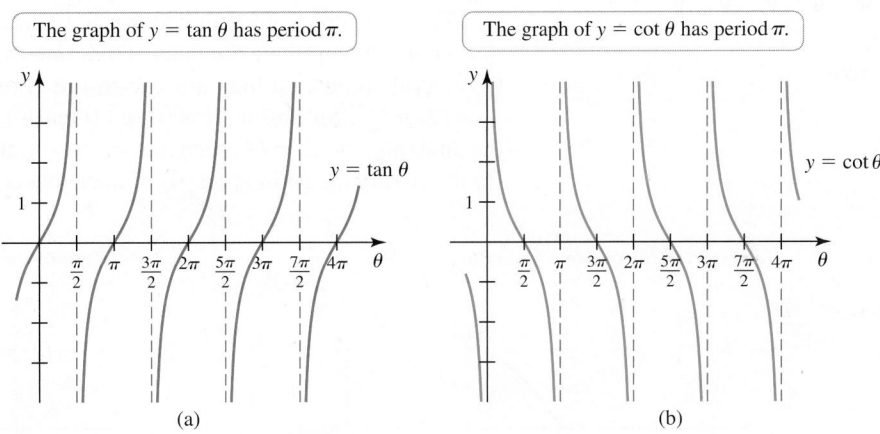

The graph of $y = \tan\theta$ has period π.

The graph of $y = \cot\theta$ has period π.

(a)   (b)

Figure 1.67

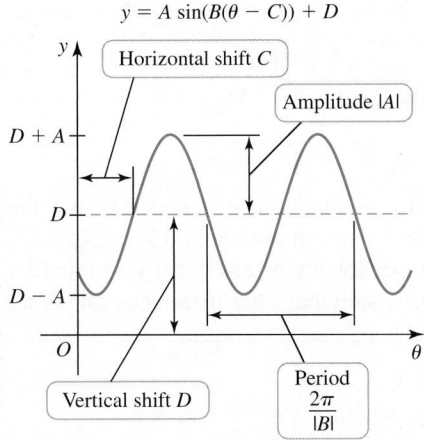

$y = A\sin(B(\theta - C)) + D$

Horizontal shift C

Amplitude $|A|$

Vertical shift D

Period $\dfrac{2\pi}{|B|}$

Figure 1.68

Transforming Graphs

Many physical phenomena, such as the motion of waves and the rising and setting of the sun, can be modeled using trigonometric functions; the sine and cosine functions are especially useful. Using the transformation methods introduced in Section 1.2, we can show that the functions

$$y = A\sin\left(B(\theta - C)\right) + D \quad \text{and} \quad y = A\cos\left(B(\theta - C)\right) + D,$$

when compared to the graphs of $y = \sin\theta$ and $y = \cos\theta$, have a vertical scaling (or **amplitude**) of $|A|$, a period of $2\pi/|B|$, a horizontal shift (or **phase shift**) of C, and a **vertical shift** of D (**Figure 1.68**).

For example, at latitude 40° north (Beijing, Madrid, Philadelphia), there are 12 hours of daylight on the equinoxes (approximately March 21 and September 21), with a maximum of 14.8 hours of daylight on the summer solstice (approximately June 21) and a

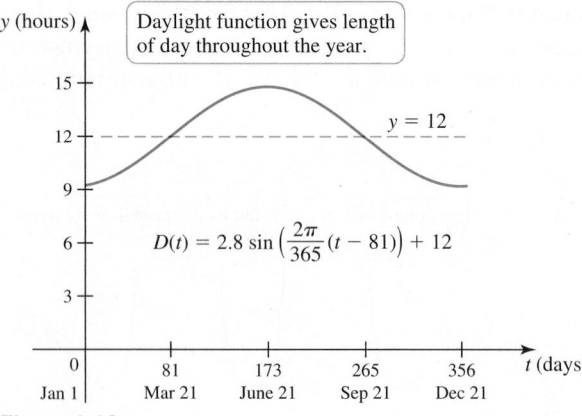

Figure 1.69

minimum of 9.2 hours of daylight on the winter solstice (approximately December 21). Using this information, it can be shown that the function

$$D(t) = 2.8 \sin\left(\frac{2\pi}{365}(t - 81)\right) + 12$$

models the number of daylight hours t days after January 1 (**Figure 1.69**; Exercise 112). The graph of this function is obtained from the graph of $y = \sin t$ by (1) a horizontal stretch by a factor of $365/2\pi$, (2) a horizontal shift of 81, (3) a vertical stretch by a factor of 2.8, and (4) a vertical shift of 12.

Inverse Trigonometric Functions

The notion of inverse functions led from exponential functions to logarithmic functions (Section 1.3). We now carry out a similar procedure—this time with trigonometric functions.

Inverse Sine and Cosine   Our goal is to develop the inverses of the sine and cosine in detail. The inverses of the other four trigonometric functions then follow in an analogous way. So far, we have asked this question: Given an angle x, what is $\sin x$ or $\cos x$? Now we ask the opposite question: Given a number y, what is the angle x such that $\sin x = y$? Or what is the angle x such that $\cos x = y$? These are inverse questions.

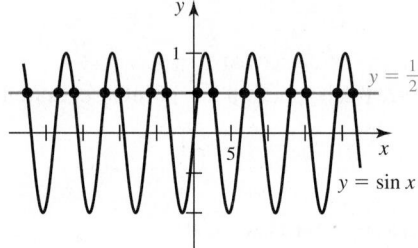

Figure 1.70

There are a few things to notice right away. First, these questions don't make sense if $|y| > 1$, because $-1 \le \sin x \le 1$ and $-1 \le \cos x \le 1$. Next, let's select an acceptable value of y, say, $y = \frac{1}{2}$, and find the angle x that satisfies $\sin x = y = \frac{1}{2}$. It is apparent that infinitely many angles satisfy $\sin x = \frac{1}{2}$; all angles of the form $\pi/6 \pm 2n\pi$ and $5\pi/6 \pm 2n\pi$, where n is an integer, answer the inverse question (**Figure 1.70**). A similar situation occurs with the cosine function.

These inverse questions do not have unique answers because $\sin x$ and $\cos x$ are not one-to-one on their domains. To define their inverses, these functions are restricted to intervals on which they are one-to-one. For the sine function, the standard choice is $[-\pi/2, \pi/2]$; for cosine, it is $[0, \pi]$ (**Figure 1.71**). Now when we ask for the angle x on the interval $[-\pi/2, \pi/2]$ such that $\sin x = \frac{1}{2}$, there is one answer: $x = \pi/6$. When we ask for the angle x on the interval $[0, \pi]$ such that $\cos x = -\frac{1}{2}$, there is one answer: $x = 2\pi/3$.

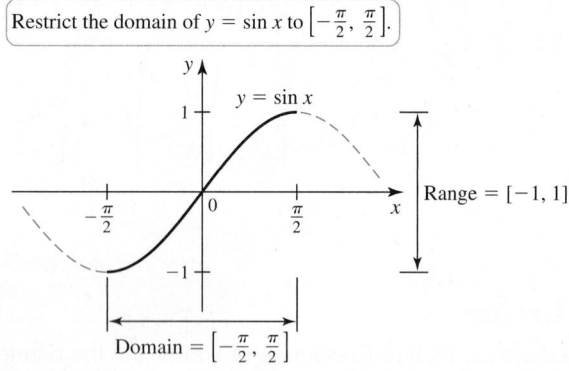

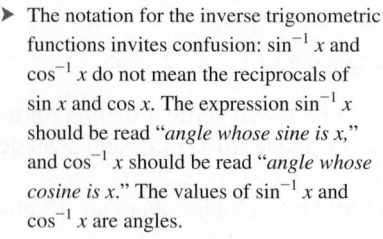

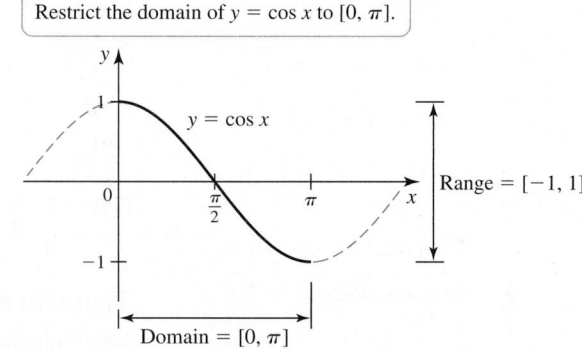

Figure 1.71

▶ The notation for the inverse trigonometric functions invites confusion: $\sin^{-1} x$ and $\cos^{-1} x$ do not mean the reciprocals of $\sin x$ and $\cos x$. The expression $\sin^{-1} x$ should be read "*angle whose sine is x*," and $\cos^{-1} x$ should be read "*angle whose cosine is x*." The values of $\sin^{-1} x$ and $\cos^{-1} x$ are angles.

We define the **inverse sine**, or **arcsine**, denoted $y = \sin^{-1} x$ or $y = \arcsin x$, such that y is the angle whose sine is x, with the provision that y lies in the interval $[-\pi/2, \pi/2]$. Similarly, we define the **inverse cosine**, or **arccosine**, denoted $y = \cos^{-1} x$ or $y = \arccos x$, such that y is the angle whose cosine is x, with the provision that y lies in the interval $[0, \pi]$.

DEFINITION   Inverse Sine and Cosine

$y = \sin^{-1} x$ is the value of y such that $x = \sin y$, where $-\pi/2 \le y \le \pi/2$.
$y = \cos^{-1} x$ is the value of y such that $x = \cos y$, where $0 \le y \le \pi$.
The domain of both $\sin^{-1} x$ and $\cos^{-1} x$ is $\{x: -1 \le x \le 1\}$.

Recall that any invertible function and its inverse satisfy the properties

$$f(f^{-1}(y)) = y \quad \text{and} \quad f^{-1}(f(x)) = x.$$

These properties apply to the inverse sine and cosine, as long as we observe the restrictions on the domains. Here is what we can say:

- $\sin(\sin^{-1} x) = x$ and $\cos(\cos^{-1} x) = x$, for $-1 \le x \le 1$.
- $\sin^{-1}(\sin y) = y$, for $-\pi/2 \le y \le \pi/2$.
- $\cos^{-1}(\cos y) = y$, for $0 \le y \le \pi$.

QUICK CHECK 4 Explain why $\sin^{-1}(\sin 0) = 0$, but $\sin^{-1}(\sin 2\pi) \ne 2\pi$. ◄

EXAMPLE 3 **Working with inverse sine and cosine** Evaluate the following expressions.

a. $\sin^{-1}\left(\sqrt{3}/2\right)$ **b.** $\cos^{-1}\left(-\sqrt{3}/2\right)$ **c.** $\cos^{-1}(\cos 3\pi)$ **d.** $\sin^{-1}\left(\sin \frac{3\pi}{4}\right)$

SOLUTION

a. $\sin^{-1}\left(\sqrt{3}/2\right) = \pi/3$ because $\sin(\pi/3) = \sqrt{3}/2$ and $\pi/3$ is in the interval $[-\pi/2, \pi/2]$.

b. $\cos^{-1}\left(-\sqrt{3}/2\right) = 5\pi/6$ because $\cos(5\pi/6) = -\sqrt{3}/2$ and $5\pi/6$ is in the interval $[0, \pi]$.

c. It's tempting to conclude that $\cos^{-1}(\cos 3\pi) = 3\pi$, but the result of an inverse cosine operation must lie in the interval $[0, \pi]$. Because $\cos(3\pi) = -1$ and $\cos^{-1}(-1) = \pi$, we have

$$\cos^{-1}\underbrace{(\cos 3\pi)}_{-1} = \cos^{-1}(-1) = \pi.$$

d. $\sin^{-1}\left(\underbrace{\sin \frac{3\pi}{4}}_{\sqrt{2}/2}\right) = \sin^{-1}\frac{\sqrt{2}}{2} = \frac{\pi}{4}$

Related Exercises 49–56 ◄

Graphs and Properties   Recall from Section 1.3 that the graph of f^{-1} is obtained by reflecting the graph of f about the line $y = x$. This operation produces the graphs of the inverse sine (**Figure 1.72**) and inverse cosine (**Figure 1.73**). The graphs make it easy to compare the domain and range of each function and its inverse.

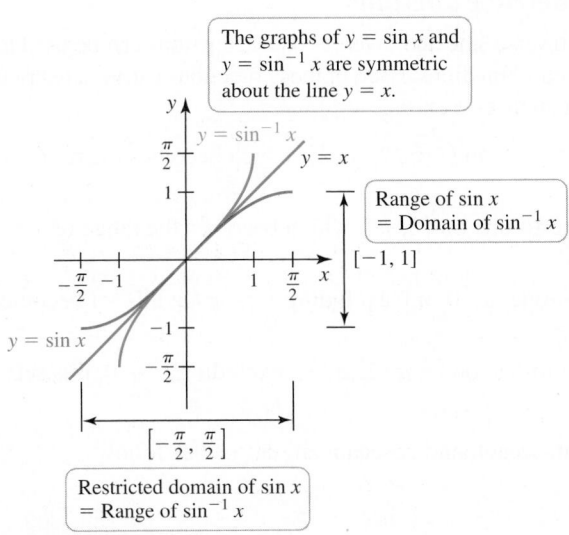

Figure 1.72

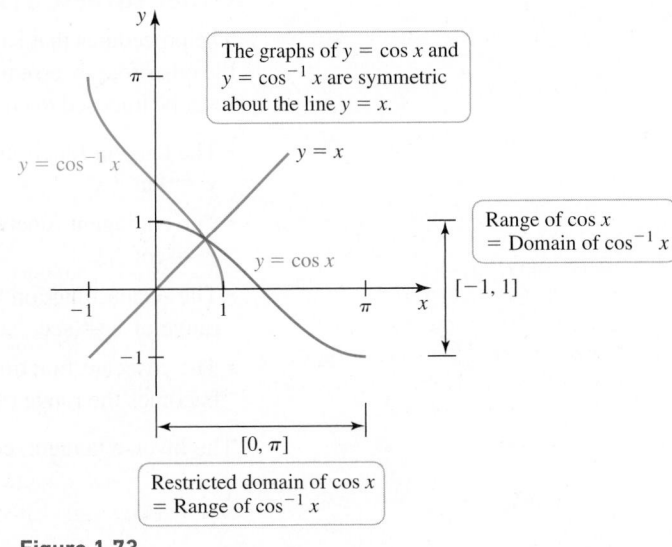

Figure 1.73

EXAMPLE 4 Right-triangle relationships

a. Suppose $\theta = \sin^{-1}(2/5)$. Find $\cos\theta$ and $\tan\theta$.

b. Find an alternative form for $\cot(\cos^{-1}(x/4))$ in terms of x.

SOLUTION

a. Relationships between the trigonometric functions and their inverses can often be simplified using a right-triangle sketch. Notice that $0 < \theta < \pi/2$. The right triangle in **Figure 1.74** satisfies the relationship $\sin\theta = \frac{2}{5}$, or, equivalently, $\theta = \sin^{-1}\frac{2}{5}$. We label the angle θ and the lengths of two sides; then the length of the third side is $\sqrt{21}$ (by the Pythagorean theorem). Now it is easy to read directly from the triangle:

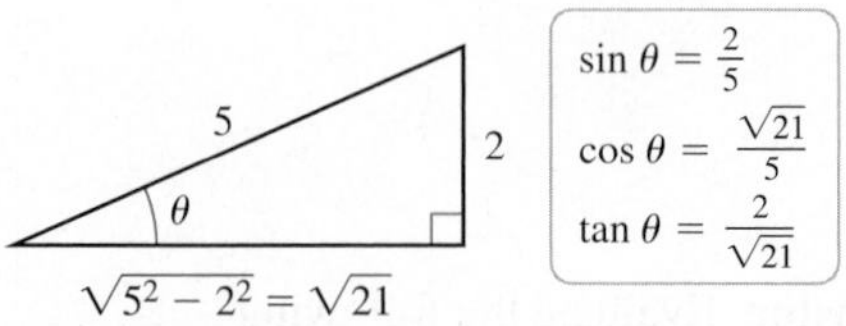

$$\cos\theta = \frac{\sqrt{21}}{5} \quad \text{and} \quad \tan\theta = \frac{2}{\sqrt{21}}.$$

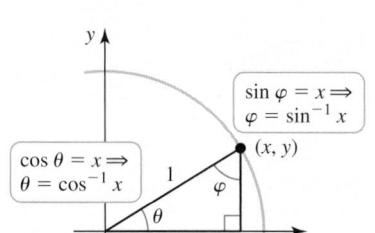

$\sin\theta = \frac{2}{5}$

$\cos\theta = \frac{\sqrt{21}}{5}$

$\tan\theta = \frac{2}{\sqrt{21}}$

$\sqrt{5^2 - 2^2} = \sqrt{21}$

Figure 1.74

b. We draw a right triangle with an angle θ satisfying $\cos\theta = x/4$, or, equivalently, $\theta = \cos^{-1}(x/4)$, where $0 < x < 4$ (**Figure 1.75**). The length of the third side of the triangle is $\sqrt{16 - x^2}$. It now follows that

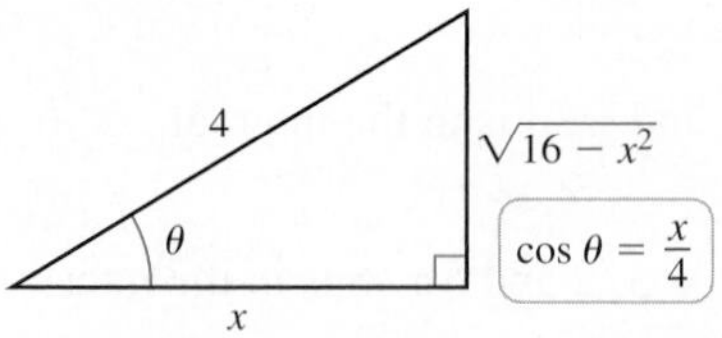

$$\cot\left(\underbrace{\cos^{-1}\frac{x}{4}}_{\theta}\right) = \frac{x}{\sqrt{16 - x^2}}, \text{ for } 0 < x < 4 \text{ and } \theta \text{ in the first quadrant.}$$

$\sqrt{16 - x^2}$

$\cos\theta = \frac{x}{4}$

Figure 1.75

A similar argument with θ in the second quadrant shows this relationship is valid for $|x| < 4$.

Related Exercises 59, 62 ◀

EXAMPLE 5 A useful identity Use right triangles to explain why $\cos^{-1}x + \sin^{-1}x = \pi/2$.

SOLUTION We draw a right triangle in a unit circle and label the acute angles θ and φ (**Figure 1.76**). These angles satisfy $\cos\theta = x$, or $\theta = \cos^{-1}x$, and $\sin\varphi = x$, or $\varphi = \sin^{-1}x$. Because θ and φ are complementary angles, we have

$$\frac{\pi}{2} = \theta + \varphi = \cos^{-1}x + \sin^{-1}x.$$

This result holds for $0 \le x \le 1$. An analogous argument extends the property to $-1 \le x \le 1$.

Related Exercises 71–72 ◀

$\sin\varphi = x \Rightarrow$
$\varphi = \sin^{-1}x$

$\cos\theta = x \Rightarrow$
$\theta = \cos^{-1}x$

Figure 1.76

Other Inverse Trigonometric Functions

The procedures that led to the inverse sine and inverse cosine functions can be used to obtain the other four inverse trigonometric functions. Each of these functions carries a restriction that must be imposed to ensure that an inverse exists.

• The tangent function is one-to-one on $(-\pi/2, \pi/2)$, which becomes the range of $y = \tan^{-1}x$.

• The cotangent function is one-to-one on $(0, \pi)$, which becomes the range of $y = \cot^{-1}x$.

• The secant function is one-to-one on $[0, \pi]$, excluding $x = \pi/2$; this set becomes the range of $y = \sec^{-1}x$.

• The cosecant function is one-to-one on $[-\pi/2, \pi/2]$, excluding $x = 0$; this set becomes the range of $y = \csc^{-1}x$.

The inverse tangent, cotangent, secant, and cosecant are defined as follows.

▶ Tables, books, and computer algebra systems differ on the definition of the inverses of the secant, cosecant, and cotangent. For example, in some books, $\sec^{-1} x$ is defined to lie in the interval $[-\pi, -\pi/2)$ when $x < 0$.

DEFINITION Other Inverse Trigonometric Functions

$y = \tan^{-1} x$ is the value of y such that $x = \tan y$, where $-\pi/2 < y < \pi/2$.
$y = \cot^{-1} x$ is the value of y such that $x = \cot y$, where $0 < y < \pi$.
The domain of both $\tan^{-1} x$ and $\cot^{-1} x$ is $\{x: -\infty < x < \infty\}$.
$y = \sec^{-1} x$ is the value of y such that $x = \sec y$, where $0 \le y \le \pi$, with $y \ne \pi/2$.
$y = \csc^{-1} x$ is the value of y such that $x = \csc y$, where $-\pi/2 \le y \le \pi/2$, with $y \ne 0$.
The domain of both $\sec^{-1} x$ and $\csc^{-1} x$ is $\{x: |x| \ge 1\}$.

The graphs of these inverse functions are obtained by reflecting the graphs of the original trigonometric functions about the line $y = x$ (**Figures 1.77–1.80**). The inverse secant and cosecant are somewhat irregular. The domain of the secant function (**Figure 1.79**) is restricted to the set $[0, \pi]$, excluding $x = \pi/2$, where the secant has a vertical asymptote. This asymptote splits the range of the secant into two disjoint intervals $(-\infty, -1]$ and $[1, \infty)$, which, in turn, splits the domain of the inverse secant into the same two intervals. A similar situation occurs with the cosecant function.

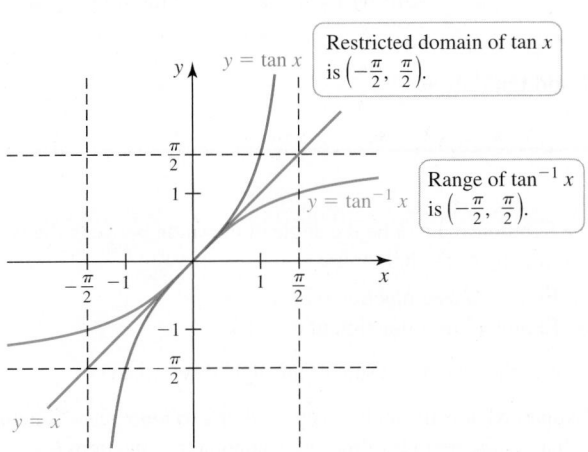

Figure 1.77

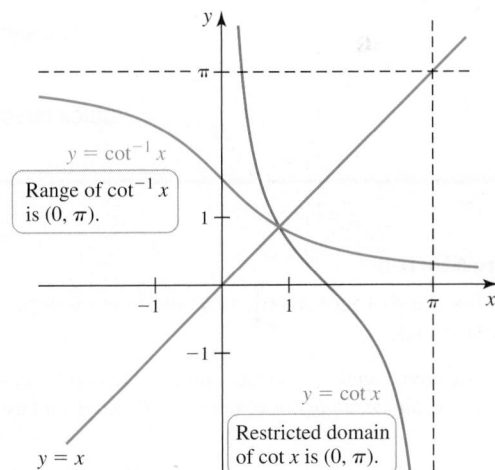

Figure 1.78

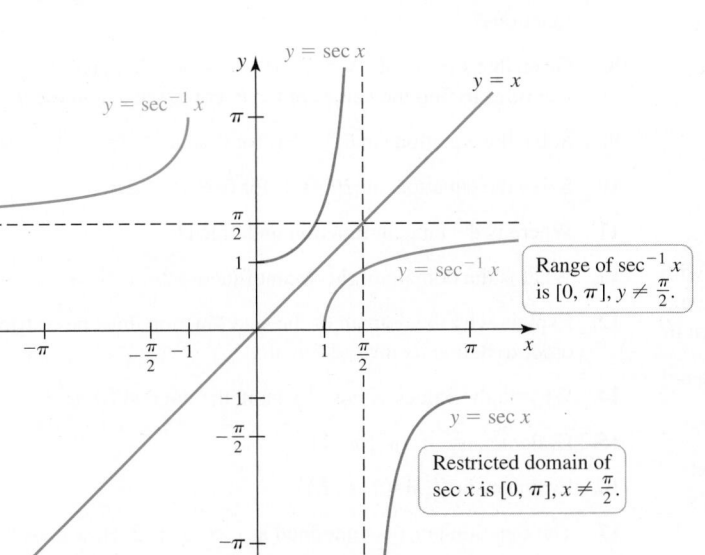

Figure 1.79

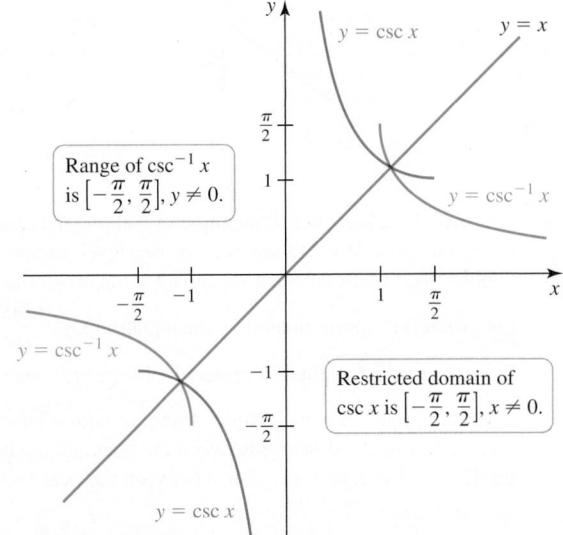

Figure 1.80

EXAMPLE 6 Working with inverse trigonometric functions Evaluate or simplify the following expressions.

a. $\tan^{-1}\left(-1/\sqrt{3}\right)$ **b.** $\sec^{-1}(-2)$ **c.** $\sin\left(\tan^{-1}x\right)$

SOLUTION

a. The result of an inverse tangent operation must lie in the interval $(-\pi/2, \pi/2)$. Therefore,

$$\tan^{-1}\left(-\frac{1}{\sqrt{3}}\right) = -\frac{\pi}{6} \quad \text{because} \quad \tan\left(-\frac{\pi}{6}\right) = -\frac{1}{\sqrt{3}}.$$

b. The result of an inverse secant operation when $x \le -1$ must lie in the interval $(\pi/2, \pi]$. Therefore,

$$\sec^{-1}(-2) = \frac{2\pi}{3} \quad \text{because} \quad \sec\frac{2\pi}{3} = -2.$$

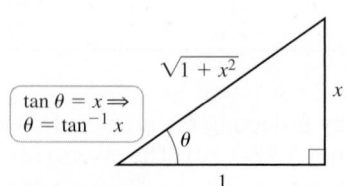

$$\tan\theta = x \Rightarrow$$
$$\theta = \tan^{-1}x$$

Figure 1.81

c. Figure 1.81 shows a right triangle with the relationship $x = \tan\theta$ or $\theta = \tan^{-1}x$, in the case that $0 \le \theta < \pi/2$. We see that

$$\sin\left(\underbrace{\tan^{-1}x}_{\theta}\right) = \frac{x}{\sqrt{1 + x^2}}.$$

The same result follows if $-\pi/2 < \theta < 0$, in which case $x < 0$ and $\sin\theta < 0$.

Related Exercises 75, 77, 83 ◄

QUICK CHECK 5 Evaluate $\sec^{-1}1$ and $\tan^{-1}1$. ◄

SECTION 1.4 EXERCISES

Getting Started

1. Define the six trigonometric functions in terms of the sides of a right triangle.

2. For the given angle θ corresponding to the point $P(-4, -3)$ in the figure, evaluate $\sin\theta$, $\cos\theta$, $\tan\theta$, $\cot\theta$, $\sec\theta$, and $\csc\theta$.

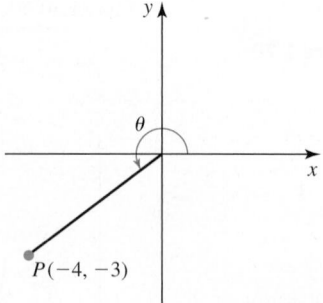

3. A projectile is launched at an angle of θ above the horizontal with an initial speed of v ft/s and travels over level ground. The time of flight t (the time it takes, in seconds, for the projectile to return to the ground) is approximated by the equation $t = \dfrac{v\sin\theta}{16}$. Determine the time of flight of a projectile if $\theta = \pi/6$ and $v = 96$.

4. A boat approaches a 50-ft-high lighthouse whose base is at sea level. Let d be the distance between the boat and the base of the lighthouse. Let L be the distance between the boat and the top of the lighthouse. Let θ be the angle of elevation between the boat and the top of the lighthouse.
  a. Express d as a function of θ.
  b. Express L as a function of θ.

5. How is the radian measure of an angle determined?

6. Explain what is meant by the period of a trigonometric function. What are the periods of the six trigonometric functions?

7. What are the three Pythagorean identities for the trigonometric functions?

8. Given that $\sin\theta = 1/\sqrt{5}$ and $\cos\theta = -2/\sqrt{5}$, use trigonometric identities to find the values of $\tan\theta$, $\cot\theta$, $\sec\theta$, and $\csc\theta$.

9. Solve the equation $\sin\theta = -1$, for $0 \le \theta < 2\pi$.

10. Solve the equation $\sin 2\theta = 1$, for $0 \le \theta < 2\pi$.

11. Where is the tangent function undefined?

12. What is the domain of the secant function?

13. Explain why the domain of the sine function must be restricted in order to define its inverse function.

14. Why do the values of $\cos^{-1}x$ lie in the interval $[0, \pi]$?

15. Evaluate $\cos^{-1}(\cos(5\pi/4))$.

16. Evaluate $\sin^{-1}(\sin(11\pi/6))$.

17. The function $\tan x$ is undefined at $x = \pm\pi/2$. How does this fact appear in the graph of $y = \tan^{-1}x$?

18. State the domain and range of $\sec^{-1}x$.

Practice Exercises

19–34. Evaluating trigonometric functions *Without using a calculator, evaluate the following expressions or state that the quantity is undefined.*

19. $\cos\left(2\pi/3\right)$

20. $\sin\left(2\pi/3\right)$

21. $\tan\left(-3\pi/4\right)$

22. $\tan\left(15\pi/4\right)$

23. $\cot\left(-13\pi/3\right)$

24. $\sec\left(7\pi/6\right)$

25. $\cot\left(-17\pi/3\right)$

26. $\sin\left(16\pi/3\right)$

27. $\cos 0$

28. $\sin\left(-\pi/2\right)$

29. $\cos\left(-\pi\right)$

30. $\tan 3\pi$

31. $\sec\left(5\pi/2\right)$

32. $\cot \pi$

33. $\cos\left(\pi/12\right)$ *(Hint: Use a half-angle formula.)*

34. $\sin\left(3\pi/8\right)$

35–46. Solving trigonometric equations *Solve the following equations.*

35. $\tan x = 1$

36. $2\theta\cos\theta + \theta = 0$

37. $\sin^2\theta = \dfrac{1}{4}, 0 \le \theta < 2\pi$

38. $\cos^2\theta = \dfrac{1}{2}, 0 \le \theta < 2\pi$

39. $\sqrt{2}\sin x - 1 = 0$

40. $\sin^2\theta - 1 = 0$

41. $\sin\theta\cos\theta = 0, 0 \le \theta < 2\pi$

42. $\sin 3x = \dfrac{\sqrt{2}}{2}, 0 \le x < 2\pi$

43. $\cos 3x = \sin 3x, 0 \le x < 2\pi$

44. $\tan^2 2\theta = 1, 0 \le \theta < \pi$

▪ 45. $\sin 2\theta = \dfrac{1}{5}, 0 \le \theta \le \dfrac{\pi}{2}$

▪ 46. $\cos 3\theta = \dfrac{3}{7}, 0 \le \theta \le \pi$

▪ 47–48. Projectile range *A projectile is launched from the ground at an angle θ above the horizontal with an initial speed v in ft/s. The range (the horizontal distance traveled by the projectile over level ground) is approximated by the equation $x = \dfrac{v^2}{32}\sin 2\theta$. Find all launch angles that satisfy the following conditions; express your answers in degrees.*

47. Initial speed of 150 ft/s; range of 400 ft

48. Initial speed of 160 ft/s; range of 350 ft

49–58. Inverse sines and cosines *Without using a calculator, evaluate the following expressions or state that the quantity is undefined.*

49. $\sin^{-1} 1$

50. $\cos^{-1}(-1)$

51. $\sin^{-1}\left(-\dfrac{1}{2}\right)$

52. $\cos^{-1}\left(-\dfrac{\sqrt{2}}{2}\right)$

53. $\sin^{-1}\dfrac{\sqrt{3}}{2}$

54. $\cos^{-1} 2$

55. $\cos^{-1}\left(-\dfrac{1}{2}\right)$

56. $\sin^{-1}(-1)$

57. $\cos\left(\cos^{-1}(-1)\right)$

58. $\cos^{-1}\left(\cos(7\pi/6)\right)$

59–60. Using right triangles *Use a right-triangle sketch to complete the following exercises.*

59. Suppose $\theta = \cos^{-1}(5/13)$. Find $\sin\theta$ and $\tan\theta$.

60. Suppose $\theta = \tan^{-1}(4/3)$. Find $\sec\theta$ and $\csc\theta$.

61–66. Right-triangle relationships *Draw a right triangle to simplify the given expressions. Assume $x > 0$.*

61. $\cos\left(\sin^{-1} x\right)$

62. $\cos\left(\sin^{-1}(x/3)\right)$

63. $\sin\left(\cos^{-1}(x/2)\right)$

64. $\sin^{-1}(\cos\theta)$, for $0 \le \theta \le \dfrac{\pi}{2}$

65. $\sin\left(2\cos^{-1} x\right)$ *(Hint: Use $\sin 2\theta = 2\sin\theta\cos\theta$.)*

66. $\cos\left(2\sin^{-1} x\right)$ *(Hint: Use $\cos 2\theta = \cos^2\theta - \sin^2\theta$.)*

67–74. Identities *Prove the following identities.*

67. $\sec\theta = \dfrac{1}{\cos\theta}$

68. $\tan\theta = \dfrac{\sin\theta}{\cos\theta}$

69. $\tan^2\theta + 1 = \sec^2\theta$

70. $\dfrac{\sin\theta}{\csc\theta} + \dfrac{\cos\theta}{\sec\theta} = 1$

71. $\sec\left(\pi/2 - \theta\right) = \csc\theta$

72. $\sec\left(x + \pi\right) = -\sec x$

73. $\cos^{-1} x + \cos^{-1}(-x) = \pi$

74. $\sin^{-1} y + \sin^{-1}(-y) = 0$

75–82. Evaluating inverse trigonometric functions *Without using a calculator, evaluate the following expressions.*

75. $\tan^{-1}\sqrt{3}$

76. $\cot^{-1}\left(-1/\sqrt{3}\right)$

77. $\sec^{-1} 2$

78. $\csc^{-1}(-1)$

79. $\tan^{-1}\left(\tan\left(\pi/4\right)\right)$

80. $\tan^{-1}\left(\tan\left(3\pi/4\right)\right)$

81. $\csc^{-1}\left(\sec 2\right)$

82. $\tan\left(\tan^{-1} 1\right)$

83–88. Right-triangle relationships *Use a right triangle to simplify the given expressions. Assume $x > 0$.*

83. $\cos\left(\tan^{-1} x\right)$

84. $\tan\left(\cos^{-1} x\right)$

85. $\cos\left(\sec^{-1} x\right)$

86. $\cot\left(\tan^{-1} 2x\right)$

87. $\sin\left(\sec^{-1}\left(\dfrac{\sqrt{x^2 + 16}}{4}\right)\right)$

88. $\cos\left(\tan^{-1}\left(\dfrac{x}{\sqrt{9 - x^2}}\right)\right)$

89–90. Right-triangle pictures *Express θ in terms of x using the inverse sine, inverse tangent, and inverse secant functions.*

89.

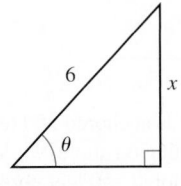

90.

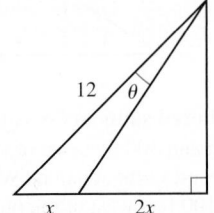

91. Explain why or why not Determine whether the following statements are true and give an explanation or counterexample.

a. $\sin\left(a + b\right) = \sin a + \sin b$

b. The equation $\cos\theta = 2$ has multiple solutions.

c. The equation $\sin\theta = \dfrac{1}{2}$ has exactly one solution.

d. The function $\sin\left(\pi x/12\right)$ has a period of 12.

e. Of the six basic trigonometric functions, only tangent and cotangent have a range of $(-\infty, \infty)$.

f. $\dfrac{\sin^{-1} x}{\cos^{-1} x} = \tan^{-1} x$

g. $\cos^{-1}\left(\cos\left(15\pi/16\right)\right) = 15\pi/16$

h. $\sin^{-1} x = 1/\sin x$

92–95. One function gives all six *Given the following information about one trigonometric function, evaluate the other five functions.*

92. $\sin\theta = -\dfrac{4}{5}$ and $\pi < \theta < \dfrac{3\pi}{2}$

93. $\cos\theta = \dfrac{5}{13}$ and $0 < \theta < \dfrac{\pi}{2}$

94. $\sec\theta = \dfrac{5}{3}$ and $\dfrac{3\pi}{2} < \theta < 2\pi$

95. $\csc\theta = \dfrac{13}{12}$ and $0 < \theta < \dfrac{\pi}{2}$

96–99. Amplitude and period *Identify the amplitude and period of the following functions.*

96. $f(\theta) = 2\sin 2\theta$

97. $g(\theta) = 3\cos\left(\theta/3\right)$

98. $p(t) = 2.5\sin\left(\dfrac{1}{2}\left(t - 3\right)\right)$

99. $q(x) = 3.6\cos\left(\pi x/24\right)$

Explorations and Challenges

100. Law of Cosines Use the figure to prove the Law of Cosines (which is a generalization of the Pythagorean theorem): $c^2 = a^2 + b^2 - 2ab\cos\theta$.

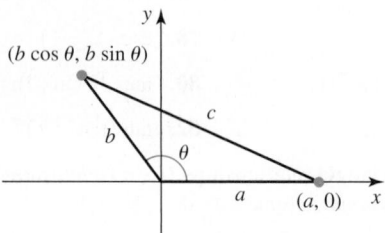

101. Law of Sines Use the figure to prove the Law of Sines:
$\dfrac{\sin A}{a} = \dfrac{\sin B}{b} = \dfrac{\sin C}{c}$.

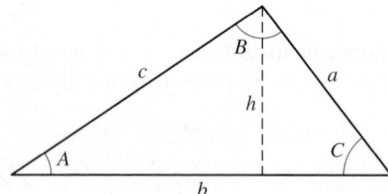

102. Anchored sailboats A sailboat named Ditl is anchored 200 feet north and 300 feet east of an observer standing on shore, while a second sailboat named Windborne is anchored 250 feet north and 100 feet west of the observer. Find the angle between the two sailboats as determined by the observer on shore.

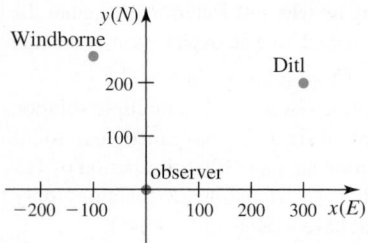

103. Area of a circular sector Prove that the area of a sector of a circle of radius r associated with a central angle θ (measured in radians) is $A = \frac{1}{2}r^2\theta$.

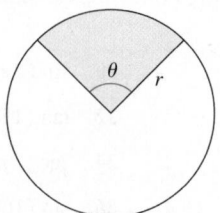

104–107. Graphing sine and cosine functions *Beginning with the graphs of $y = \sin x$ or $y = \cos x$, use shifting and scaling transformations to sketch the graph of the following functions. Use a graphing utility to check your work.*

104. $f(x) = 3\sin 2x$

105. $g(x) = -2\cos\left(x/3\right)$

106. $p(x) = 3\sin\left(2x - \pi/3\right) + 1$

107. $q(x) = 3.6\cos\left(\pi x/24\right) + 2$

108–109. Designer functions *Design a sine function with the given properties.*

108. It has a period of 12 with a minimum value of -4 at $t = 0$ and a maximum value of 4 at $t = 6$.

109. It has a period of 24 with a minimum value of 10 at $t = 3$ and a maximum value of 16 at $t = 15$.

T 110. Field goal attempt During the 1950 Rose Bowl football game, The Ohio State University prepared to kick a field goal at a distance of 23 yd from the end line at point A on the edge of the kicking region (see figure). But before the kick, Ohio State committed a penalty and the ball was backed up 5 yd to point B. After the game, the Ohio State coach claimed that his team deliberately committed a penalty to improve the kicking angle. Given that a successful kick must go between the goal posts G_1 and G_2, is $\angle G_1 B G_2$ greater than $\angle G_1 A G_2$? (In 1950, the goal posts were 23 ft 4 in apart, equidistant from the origin on the end line.) The boundaries of the kicking region are 53 ft 4 in apart and are equidistant from the y-axis. (*Source: The College Mathematics Journal*, 27, 4, Sep 1996)

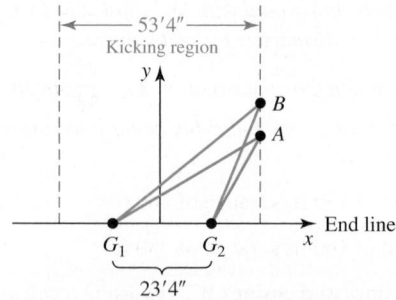

T 111. A surprising result Earth is approximately circular in cross section, with a circumference at the equator of 24,882 miles. Suppose we use two ropes to create two concentric circles: one by wrapping a rope around the equator and another using a rope 38 ft longer (see figure). How much space is between the ropes?

112. Daylight function for 40° N Verify that the function

$$D(t) = 2.8 \sin\left(\frac{2\pi}{365}(t - 81)\right) + 12$$

has the following properties, where t is measured in days and D is the number of hours between sunrise and sunset.

a. It has a period of 365 days.
b. Its maximum and minimum values are 14.8 and 9.2, respectively, which occur approximately at $t = 172$ and $t = 355$, respectively (corresponding to the solstices).
c. $D(81) = 12$ and $D(264) \approx 12$ (corresponding to the equinoxes).

113. Block on a spring A light block hangs at rest from the end of a spring when it is pulled down 10 cm and released (see figure). Assume the block oscillates with an amplitude of 10 cm on either side of its rest position with a period of 1.5 s. Find a trigonometric function $d(t)$ that gives the displacement of the block t seconds after it is released, where $d(t) > 0$ represents downward displacement.

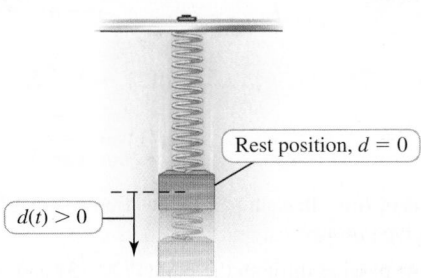

114. Viewing angles An auditorium with a flat floor has a large flat-panel television on one wall. The lower edge of the television is 3 ft above the floor and the upper edge is 10 ft above the floor (see figure). Express θ in terms of x.

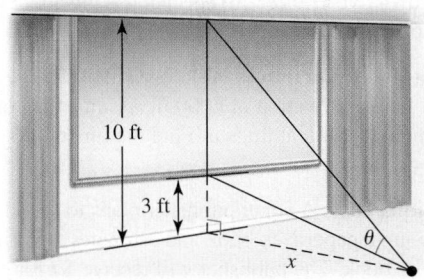

115. Ladders Two ladders of length a lean against opposite walls of an alley with their feet touching (see figure). One ladder extends h feet up the wall and makes a 75° angle with the ground. The other ladder extends k feet up the opposite wall and makes a 45° angle with the ground. Find the width of the alley in terms of h. Assume the ground is horizontal and perpendicular to both walls.

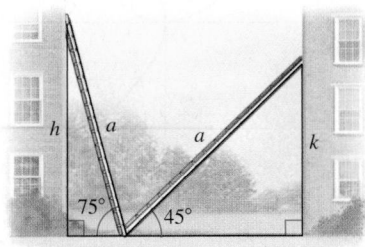

116. Pole in a corner A pole of length L is carried horizontally around a corner where a 3-ft-wide hallway meets a 4-ft-wide hallway. For $0 < \theta < \pi/2$, find the relationship between L and θ at the moment when the pole simultaneously touches both walls and the corner P. Estimate θ when $L = 10$ ft.

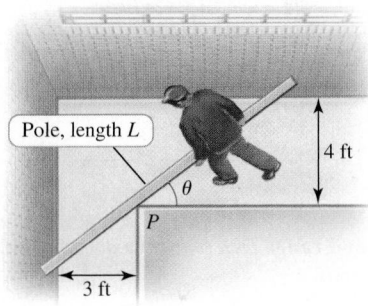

QUICK CHECK ANSWERS

1. $3\pi/2$; 225°   **2.** $\sqrt{3}/2$; $-\sqrt{2}/2$ **3.** Divide both sides of $\sin^2\theta + \cos^2\theta = 1$ by $\sin^2\theta$.   **4.** $\sin^{-1}(\sin 0) = \sin^{-1} 0 = 0$ and $\sin^{-1}(\sin(2\pi)) = \sin^{-1} 0 = 0$
5. $0, \pi/4$ ◄

CHAPTER 1   REVIEW EXERCISES

1. Explain why or why not Determine whether the following statements are true and give an explanation or counterexample.

a. A function could have the property that $f(-x) = f(x)$, for all x.
b. $\cos(a + b) = \cos a + \cos b$, for all a and b in $[0, 2\pi]$.
c. If f is a linear function of the form $f(x) = mx + b$, then $f(u + v) = f(u) + f(v)$, for all u and v.
d. The function $f(x) = 1 - x$ has the property that $f(f(x)) = x$.
e. The set $\{x: |x + 3| > 4\}$ can be drawn on the number line without lifting your pencil.
f. $\log_{10}(xy) = (\log_{10} x)(\log_{10} y)$.
g. $\sin^{-1}(\sin 2\pi) = 0$.

2.   Functions Decide whether graph A, graph B, or both represent functions.

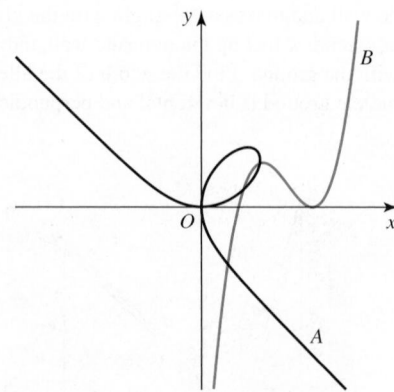

3.   One-to-one functions Decide whether f, g, or both represent one-to-one functions.

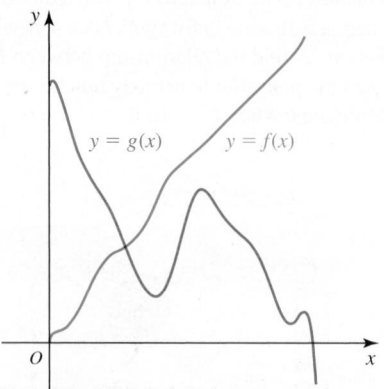

4–7. Domain and range *Determine the domain and range of the following functions.*

4. $f(x) = x^5 + \sqrt{x}$

5. $f(w) = \dfrac{2w^2 - 3w - 2}{w - 2}$

6. $g(x) = \ln(x + 6)$

7. $h(z) = \sqrt{z^2 - 2z - 3}$

8.   Suppose f and g are even functions with $f(2) = 2$ and $g(2) = -2$. Evaluate $f(g(2))$ and $g(f(-2))$.

9.   Is it true that $\tan(\tan^{-1} x) = x$ for all x? Is it true that $\tan^{-1}(\tan x) = x$ for all x?

10–18. Evaluating functions from graphs *Assume f is an odd function and that both f and g are one-to-one. Use the (incomplete) graph of f and the graph of g to find the following function values.*

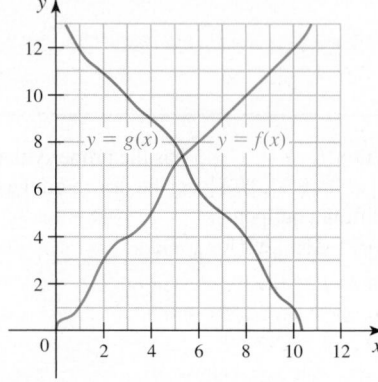

10. $f(g(4))$

11. $g(f(4))$

12. $f^{-1}(10)$

13. $g^{-1}(5)$

14. $f^{-1}(g^{-1}(4))$

15. $g^{-1}(f(3))$

16. $f^{-1}(-8)$

17. $f^{-1}(1 + f(-3))$

18. $g(1 - f(f^{-1}(-7)))$

19.   Composite functions Let $f(x) = x^3$, $g(x) = \sin x$, and $h(x) = \sqrt{x}$.

    a.   Evaluate $h(g(\pi/2))$.
    b.   Find $h(f(x))$.
    c.   Find $f(g(h(x)))$.
    d.   Find the domain of $g \circ f$.
    e.   Find the range of $f \circ g$.

20.   Composite functions Find functions f and g such that $h = f \circ g$.

    a. $h(x) = \sin(x^2 + 1)$ **b.** $h(x) = (x^2 - 4)^{-3}$
    c. $h(x) = e^{\cos 2x}$

21–24. Simplifying difference quotients *Evaluate and simplify the difference quotients $\dfrac{f(x + h) - f(x)}{h}$ and $\dfrac{f(x) - f(a)}{x - a}$ for each function.*

21. $f(x) = x^2 - 2x$

22. $f(x) = 4 - 5x$

23. $f(x) = x^3 + 2$

24. $f(x) = \dfrac{7}{x + 3}$

25.   Equations of lines In each part below, find an equation of the line with the given properties.

    a.   The line passing through the points $(2, -3)$ and $(4, 2)$
    b.   The line with slope $\frac{3}{4}$ and x-intercept $(-4, 0)$
    c.   The line with intercepts $(4, 0)$ and $(0, -2)$

26.   Population function The population of a small town was 500 in 2018 and is growing at a rate of 24 people per year. Find and graph the linear population function $p(t)$ that gives the population of the town t years after 2018. Then use this model to predict the population in 2033.

27.   Boiling-point function Water boils at $212°$ F at sea level and at $200°$ F at an elevation of 6000 ft. Assume the boiling point B varies linearly with altitude a. Find the function $B = f(a)$ that describes the dependence.

28.   Publishing costs A small publisher plans to spend $1000 for advertising a paperback book and estimates the printing cost is $2.50 per book. The publisher will receive $7 for each book sold.

    a.   Find the function $C = f(x)$ that gives the cost of producing x books.
    b.   Find the function $R = g(x)$ that gives the revenue from selling x books.
    c.   Graph the cost and revenue functions; then find the number of books that must be sold for the publisher to break even.

29.   Graphing equations Graph the following equations.

    a. $2x - 3y + 10 = 0$
    b. $y = x^2 + 2x - 3$
    c. $x^2 + 2x + y^2 + 4y + 1 = 0$
    d. $x^2 - 2x + y^2 - 8y + 5 = 0$

30–32. Graphing functions *Sketch a graph of each function.*

30. $f(x) = \begin{cases} 2x & \text{if } x \le 1 \\ 3 - x & \text{if } x > 1 \end{cases}$

31. $g(x) = \begin{cases} 4 - 2x & \text{if } x \leq 1 \\ (x - 1)^2 + 2 & \text{if } x > 1 \end{cases}$

32. $h(x) = \begin{cases} \dfrac{3x^2 - 7x + 2}{x - 2} & \text{if } x \neq 2 \\ 6 & \text{if } x = 2 \end{cases}$

33. Let $f(t) = \begin{cases} t & \text{if } 0 \leq t \leq 4 \\ 8 - t & \text{if } 4 < t \leq 8 \end{cases}$ and let $A(x)$ equal the area of the region bounded by the graph of f and the t-axis from $t = 0$ to $t = x$, where $0 \leq x \leq 8$.

    a. Plot a graph of f.
    b. Find $A(2)$ and $A(6)$.
    c. Find a piecewise formula for $A(x)$.

34. Piecewise linear functions The parking costs in a city garage are \$2 for the first half hour and \$1 for each additional half hour. Graph the function $C = f(t)$ that gives the cost of parking for t hours, where $0 \leq t \leq 3$.

35. Graphing absolute value Consider the function $f(x) = 2(x - |x|)$. Express the function in two pieces without using the absolute value. Then graph the function.

36. Root functions Graph the functions $f(x) = x^{1/3}$ and $g(x) = x^{1/4}$. Find all points where the two graphs intersect. For $x > 1$, is $f(x) > g(x)$ or is $g(x) > f(x)$?

37. Root functions Find the domain and range of the functions $f(x) = x^{1/7}$ and $g(x) = x^{1/4}$.

38. Intersection points Graph the equations $y = x^2$ and $x^2 + y^2 - 7y + 8 = 0$. At what point(s) do the curves intersect?

39. Transformation of graphs How is the graph of $y = x^2 + 6x - 3$ obtained from the graph of $y = x^2$?

40. Shifting and scaling The graph of f is shown in the figure. Graph the following functions.

    a. $f(x + 1)$ **b.** $2f(x - 1)$ **c.** $-f(x/2)$ **d.** $f(2(x - 1))$

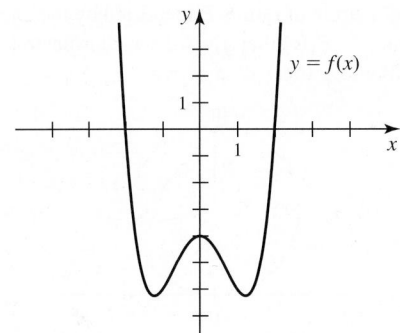

41. Symmetry Identify the symmetry (if any) in the graphs of the following equations.

    a. $y = \cos 3x$ **b.** $y = 3x^4 - 3x^2 + 1$ **c.** $y^2 - 4x^2 = 4$

42–50. Solving equations *Solve each equation.*

42. $48 = 6e^{4k}$

43. $\log_{10} x^2 + 3 \log_{10} x = \log_{10} 32$

44. $\ln 3x + \ln(x + 2) = 0$

45. $3 \ln(5t + 4) = 12$

46. $7^{y-3} = 50$

47. $1 - 2 \sin^2 \theta = 0, 0 \leq \theta < 2\pi$

48. $\sin^2 2\theta = 1/2, -\pi/2 \leq \theta \leq \pi/2$

49. $4 \cos^2 2\theta = 3, -\pi/2 \leq \theta \leq \pi/2$

50. $\sqrt{2} \sin 3\theta + 1 = 2, 0 \leq \theta \leq \pi$

T 51. Using inverse relations The population P of a small town grows according to the function $P(t) = 100\, e^{t/50}$, where t is the number of years after 2010. How long does it take the population to double?

52. Graphs of logarithmic and exponential functions The figure shows the graphs of $y = 2^x$, $y = 3^{-x}$, and $y = -\ln x$. Match each curve with the correct function.

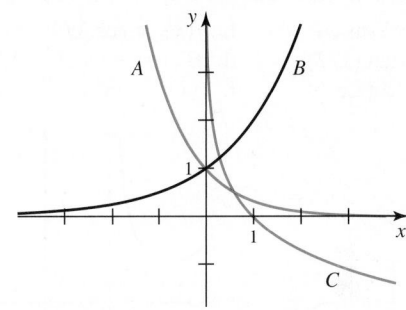

T 53–54. Existence of inverses *Determine the largest intervals on which the following functions have an inverse.*

53. $f(x) = x^3 - 3x^2$ **54.** $g(t) = 2 \sin(t/3)$

55–62. Finding inverses *Find the inverse function.*

55. $f(x) = 6 - 4x$ **56.** $f(x) = 3x - 4$

57. $f(x) = x^2 - 4x + 5$, for $x \geq 2$

58. $f(x) = \dfrac{4x^2}{x^2 + 10}$, for $x \geq 0$

59. $f(x) = 3x^2 + 1$, for $x \leq 0$

60. $f(x) = 1/x^2$, for $x > 0$

61. $f(x) = e^{x^2 + 1}$, for $x \geq 0$

62. $f(x) = \ln(x^2 + 1)$, for $x \geq 0$

T 63. Domain and range of an inverse Find the inverse of $f(x) = \dfrac{6\sqrt{x}}{\sqrt{x} + 2}$. Graph both f and f^{-1} on the same set of axes. (*Hint:* The range of $f(x)$ is $[0, 6)$.)

64. Graphing sine and cosine functions Use shifts and scalings to graph the following functions, and identify the amplitude and period.

    a. $f(x) = 4 \cos(x/2)$ **b.** $g(\theta) = 2 \sin(2\pi\, \theta/3)$
    c. $h(\theta) = -\cos(2(\theta - \pi/4))$

65. Designing functions Find a trigonometric function f that satisfies each set of properties. Answers are not unique.

    a. It has a period of 6 with a minimum value of -2 at $t = 0$ and a maximum value of 2 at $t = 3$.
    b. It has a period of 24 with a maximum value of 20 at $t = 6$ and a minimum value of 10 at $t = 18$.

66. Graph to function Find a trigonometric function f represented by the graph in the figure.

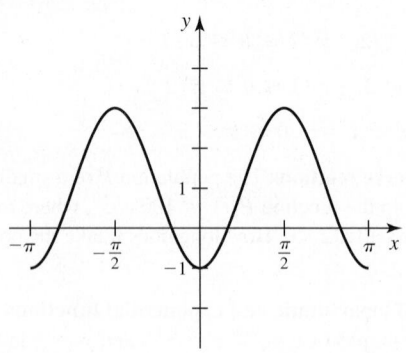

67. Matching Match each function a–f with one of the graphs A–F.

a. $f(x) = -\sin x$ **b.** $f(x) = \cos 2x$
c. $f(x) = \tan(x/2)$ **d.** $f(x) = -\sec x$
e. $f(x) = \cot 2x$ **f.** $f(x) = \sin^2 x$

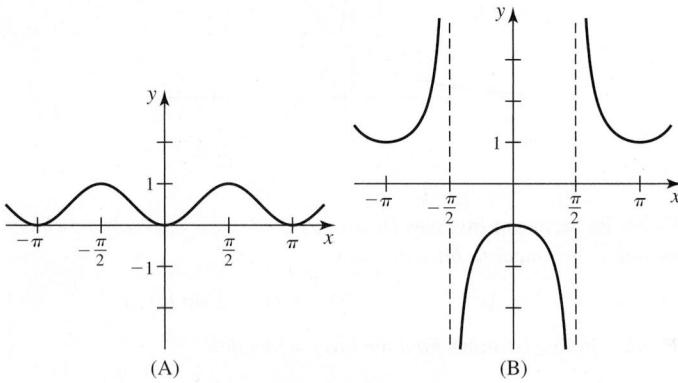

(A)   (B)

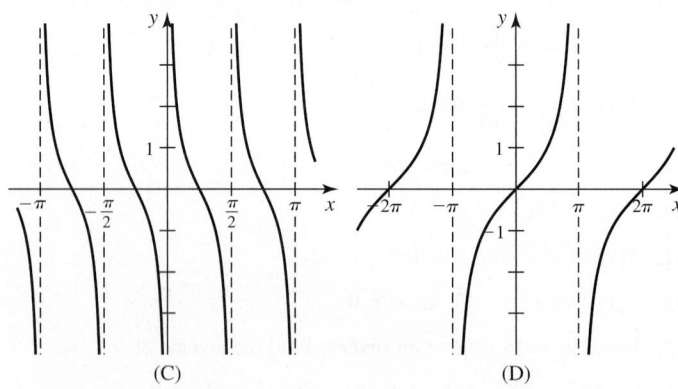

(C)   (D)

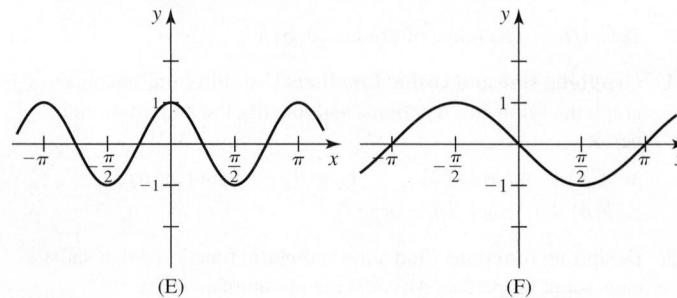

(E)   (F)

68–69. Intersection points *Find the points at which the curves intersect on the given interval.*

68. $y = \sec x$ and $y = 2$ on $(-\pi/2, \pi/2)$

69. $y = \sin x$ and $y = -\dfrac{1}{2}$ on $(0, 2\pi)$

70. Evaluating sine Find the exact value of $\sin \dfrac{5\pi}{8}$.

71. Evaluating cosine Find the exact value of $\cos \dfrac{7\pi}{8}$.

72–78. Inverse sines and cosines *Evaluate or simplify the following expressions without using a calculator.*

72. $\sin^{-1} \dfrac{\sqrt{3}}{2}$ **73.** $\cos^{-1} \dfrac{\sqrt{3}}{2}$

74. $\cos^{-1}\left(-\dfrac{1}{2}\right)$ **75.** $\sin^{-1}(-1)$

76. $\cos(\cos^{-1}(-1))$ **77.** $\sin(\sin^{-1} x)$

78. $\cos^{-1}(\sin 3\pi)$

79. Right triangles Given that $\theta = \sin^{-1} \dfrac{12}{13}$, evaluate $\cos\theta$, $\tan\theta$, $\cot\theta$, $\sec\theta$, and $\csc\theta$.

80–85. Right-triangle relationships *Draw a right triangle to simplify the given expression. Assume $x > 0$ and $0 \le \theta \le \pi/2$.*

80. $\csc(\cot^{-1} x)$ **81.** $\sin(\cos^{-1}(x/4))$

82. $\tan(\sec^{-1}(x/2))$ **83.** $\cot^{-1}(\tan\theta)$

84. $\csc^{-1}(\sec\theta)$ **85.** $\sin^{-1}x + \sin^{-1}(-x)$

86–87. Identities *Prove the following identities.*

86. $\dfrac{\sin\theta}{1 + \cos\theta} = \dfrac{1 - \cos\theta}{\sin\theta}$ **87.** $\tan 2\theta = \dfrac{2\tan\theta}{1 - \tan^2\theta}$

88. Stereographic projections A common way of displaying a sphere (such as Earth) on a plane (such as a map) is to use a *stereographic projection*. Here is the two-dimensional version of the method, which maps a circle to a line. Let P be a point on the right half of a circle of radius R identified by the angle φ. Find the function $x = F(\varphi)$ that gives the x-coordinate ($x \ge 0$) corresponding to φ for $0 \le \varphi \le \pi$.

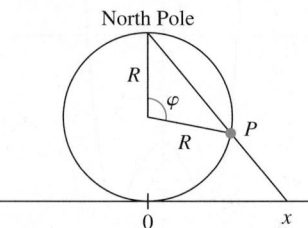

T 89. Sum of squared integers Let $T(n) = 1^2 + 2^2 + \cdots + n^2$, where n is a positive integer. It can be shown that
$$T(n) = \frac{n(n + 1)(2n + 1)}{6}.$$

a. Make a table of $T(n)$, for $n = 1, 2, \ldots, 10$.
b. How would you describe the domain of this function?
c. What is the least value of n for which $T(n) > 1000$?

T 90. Sum of integers Let $S(n) = 1 + 2 + \cdots + n$, where n is a positive integer. It can be shown that $S(n) = n(n + 1)/2$.

a. Make a table of $S(n)$, for $n = 1, 2, \ldots, 10$.
b. How would you describe the domain of this function?
c. What is the least value of n for which $S(n) > 1000$?

91. **Little-known fact** The shortest day of the year occurs on the winter solstice (near December 21) and the longest day of the year occurs on the summer solstice (near June 21). However, the latest sunrise and the earliest sunset do not occur on the winter solstice, and the earliest sunrise and the latest sunset do not occur on the summer solstice. At latitude 40° north, the latest sunrise occurs on January 4 at 7:25 A.M. (14 days after the solstice), and the earliest sunset occurs on December 7 at 4:37 P.M. (14 days before the solstice). Similarly, the earliest sunrise occurs on July 2 at 4:30 A.M. (14 days after the solstice) and the latest sunset occurs on June 7 at 7:32 P.M. (14 days before the solstice). Using sine functions, devise a function $s(t)$ that gives the time of sunrise t days after January 1 and a function $S(t)$ that gives the time of sunset t days after January 1. Assume s and S are measured in minutes and that $s = 0$ and $S = 0$ correspond to 4:00 A.M. Graph the functions. Then graph the length of the day function $D(t) = S(t) - s(t)$ and show that the longest and shortest days occur on the solstices.

Chapter 1 Guided Projects

Applications of the material in this chapter and related topics can be found in the following Guided Projects. For additional information, see the Preface.

- Problem-solving skills
- Constant rate problems
- Functions in action I
- Functions in action II

- Supply and demand
- Phase and amplitude
- Atmospheric CO_2
- Acid, noise, and earthquakes

2

Limits

Chapter Preview   All of calculus is based on the idea of a *limit*. Not only are limits important in their own right but they also underlie the two fundamental operations of calculus: differentiation (calculating derivatives) and integration (evaluating integrals). Derivatives enable us to talk about the instantaneous rate of change of a function, which, in turn, leads to concepts such as velocity and acceleration, population growth rates, marginal cost, and flow rates. Integrals enable us to compute areas under curves, surface areas, and volumes. Because of the incredible reach of this single idea, it is essential to develop a solid understanding of limits. We first present limits intuitively by showing how they arise in computing instantaneous velocities and finding slopes of tangent lines. As the chapter progresses, we build more rigor into the definition of the limit and examine the different ways in which limits arise. The chapter concludes by introducing the important property of *continuity* and by giving the formal definition of a limit.

2.1 The Idea of Limits

This brief opening section illustrates how limits arise in two seemingly unrelated problems: finding the instantaneous velocity of a moving object and finding the slope of a line tangent to a curve. These two problems provide important insights into limits on an intuitive level. In the remainder of the chapter, we develop limits carefully and fill in the mathematical details.

Average Velocity

Suppose you want to calculate your average velocity as you travel along a straight highway. If you pass milepost 100 at noon and milepost 130 at 12:30 P.M., you travel 30 miles in a half hour, so your **average velocity** over this time interval is $(30 \text{ mi})/(0.5 \text{ hr}) = 60 \text{ mi/hr}$. By contrast, even though your average velocity may be 60 mi/hr, it's almost certain that your **instantaneous velocity**, the speed indicated by the speedometer, varies from one moment to the next.

EXAMPLE 1   Average velocity A rock is launched vertically upward from the ground with a speed of 96 ft/s. Neglecting air resistance, a well-known formula from physics states that the position of the rock after t seconds is given by the function

$$s(t) = -16t^2 + 96t.$$

The position s is measured in feet with $s = 0$ corresponding to the ground. Find the average velocity of the rock between each pair of times.

a. $t = 1 \text{ s and } t = 3 \text{ s}$ **b.** $t = 1 \text{ s and } t = 2 \text{ s}$

SOLUTION Figure 2.1 shows the position function of the rock on the time interval $0 \leq t \leq 3$. The graph is *not* the path of the rock. The rock travels up and down on a vertical line.

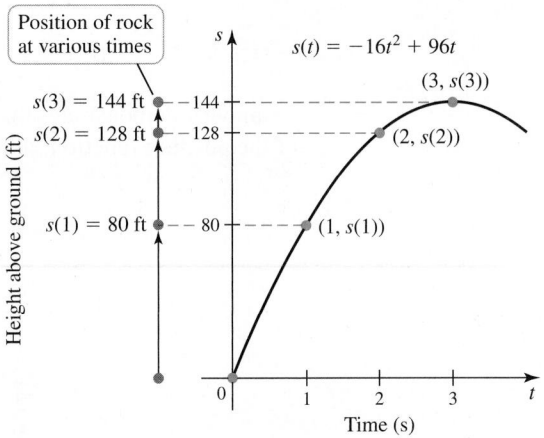

Figure 2.1

a. The average velocity of the rock over any time interval $[t_0, t_1]$ is the change in position divided by the elapsed time:

$$v_{av} = \frac{s(t_1) - s(t_0)}{t_1 - t_0}.$$

Therefore, the average velocity over the interval $[1, 3]$ is

$$v_{av} = \frac{s(3) - s(1)}{3 - 1} = \frac{144 \text{ ft} - 80 \text{ ft}}{3 \text{ s} - 1 \text{ s}} = \frac{64 \text{ ft}}{2 \text{ s}} = 32 \text{ ft/s}.$$

Here is an important observation: As shown in **Figure 2.2a**, the average velocity is simply the slope of the line joining the points $(1, s(1))$ and $(3, s(3))$ on the graph of the position function.

b. The average velocity of the rock over the interval $[1, 2]$ is

$$v_{av} = \frac{s(2) - s(1)}{2 - 1} = \frac{128 \text{ ft} - 80 \text{ ft}}{2 \text{ s} - 1 \text{ s}} = \frac{48 \text{ ft}}{1 \text{ s}} = 48 \text{ ft/s}.$$

Again, the average velocity is the slope of the line joining the points $(1, s(1))$ and $(2, s(2))$ on the graph of the position function (**Figure 2.2b**).

QUICK CHECK 1 In Example 1, what is the average velocity between $t = 2$ and $t = 3$? ◄

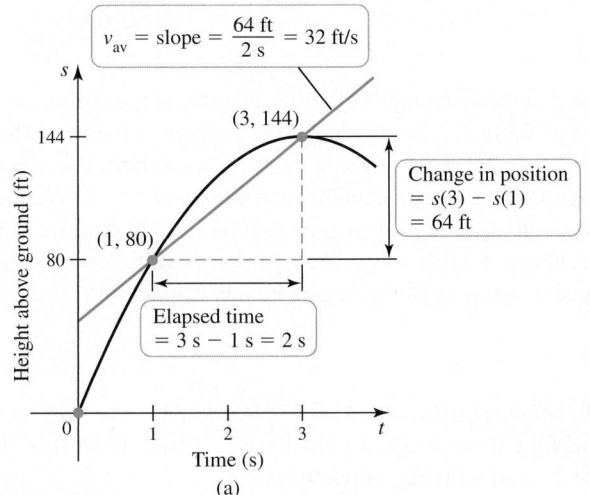

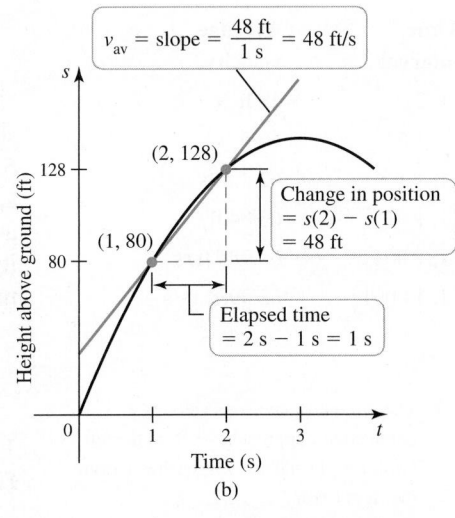

Figure 2.2

Related Exercises 13, 15 ◄

▶ See Section 1.1 for a discussion of secant lines.

In Example 1, we computed slopes of lines passing through two points on a curve. Any such line joining two points on a curve is called a **secant line**. The slope of the secant line, denoted m_{sec}, for the position function in Example 1 on the interval $[t_0, t_1]$ is

$$m_{\text{sec}} = \frac{s(t_1) - s(t_0)}{t_1 - t_0}.$$

Example 1 demonstrates that the average velocity is the slope of a secant line on the graph of the position function; that is, $v_{\text{av}} = m_{\text{sec}}$ (**Figure 2.3**).

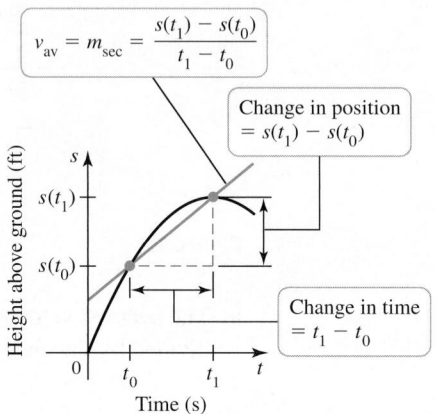

Figure 2.3

Instantaneous Velocity

To compute the average velocity, we use the position of the object at *two* distinct points in time. How do we compute the instantaneous velocity at a *single* point in time? As illustrated in Example 2, the instantaneous velocity at a point $t = t_0$ is determined by computing average velocities over intervals $[t_0, t_1]$ that decrease in length. As t_1 approaches t_0, the average velocities typically approach a unique number, which is the instantaneous velocity. This single number is called a **limit**.

QUICK CHECK 2 Explain the difference between average velocity and instantaneous velocity. ◀

EXAMPLE 2   Instantaneous velocity Estimate the instantaneous velocity of the rock in Example 1 at the *single* point $t = 1$.

SOLUTION We are interested in the instantaneous velocity at $t = 1$, so we compute the average velocity over smaller and smaller time intervals $[1, t]$ using the formula

$$v_{\text{av}} = \frac{s(t) - s(1)}{t - 1}.$$

Notice that these average velocities are also slopes of secant lines, several of which are shown in Table 2.1. For example, the average velocity on the interval $[1, 1.0001]$ is 63.9984 ft/s. Because this time interval is so short, the average velocity gives a good approximation to the instantaneous velocity at $t = 1$. We see that as t approaches 1, the average velocities appear to approach 64 ft/s. In fact, we could make the average velocity as close to 64 ft/s as we like by taking t sufficiently close to 1. Therefore, 64 ft/s is a reasonable estimate of the instantaneous velocity at $t = 1$.

Related Exercises 17, 19 ◀

Table 2.1

Time interval	Average velocity
$[1, 2]$	48 ft/s
$[1, 1.5]$	56 ft/s
$[1, 1.1]$	62.4 ft/s
$[1, 1.01]$	63.84 ft/s
$[1, 1.001]$	63.984 ft/s
$[1, 1.0001]$	63.9984 ft/s

▶ The same instantaneous velocity is obtained as t approaches 1 from the left (with $t < 1$) and as t approaches 1 from the right (with $t > 1$).

In language to be introduced in Section 2.2, we say that the limit of v_{av} as t approaches 1 equals the instantaneous velocity v_{inst}, which is 64 ft/s. This statement is illustrated in **Figure 2.4** and written compactly as

$$v_{\text{inst}} = \lim_{t \to 1} v_{\text{av}} = \lim_{t \to 1} \frac{s(t) - s(1)}{t - 1} = 64 \text{ ft/s}.$$

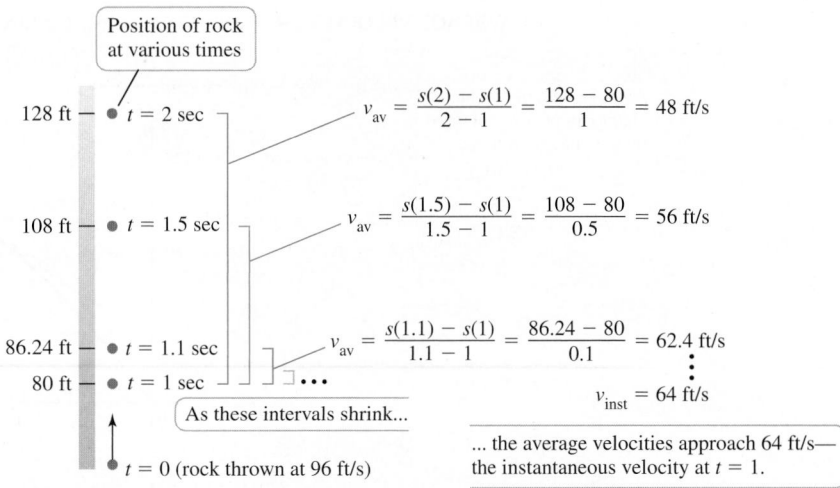

Figure 2.4

> ▶ We define tangent lines carefully in Section 3.1. For the moment, imagine zooming in on a point P on a smooth curve. As you zoom in, the curve appears more and more like a line passing through P. This line is the *tangent line* at P. Because a smooth curve approaches a line as we zoom in on a point, a smooth curve is said to be *locally linear* at any given point.

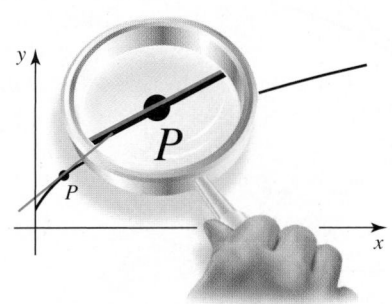

Slope of the Tangent Line

Several important conclusions follow from Examples 1 and 2. Each average velocity in Table 2.1 corresponds to the slope of a secant line on the graph of the position function (**Figure 2.5**). Just as the average velocities approach a limit as t approaches 1, the slopes of the secant lines approach the same limit as t approaches 1. Specifically, as t approaches 1, two things happen:

1. The secant lines approach a unique line called the **tangent line**.
2. The slopes of the secant lines m_{sec} approach the slope of the tangent line m_{tan} at the point $(1, s(1))$. Therefore, the slope of the tangent line is also expressed as a limit:

$$m_{\text{tan}} = \lim_{t \to 1} m_{\text{sec}} = \lim_{t \to 1} \frac{s(t) - s(1)}{t - 1} = 64.$$

This limit is the same limit that defines the instantaneous velocity. Therefore, the instantaneous velocity at $t = 1$ is the slope of the line tangent to the position curve at $t = 1$.

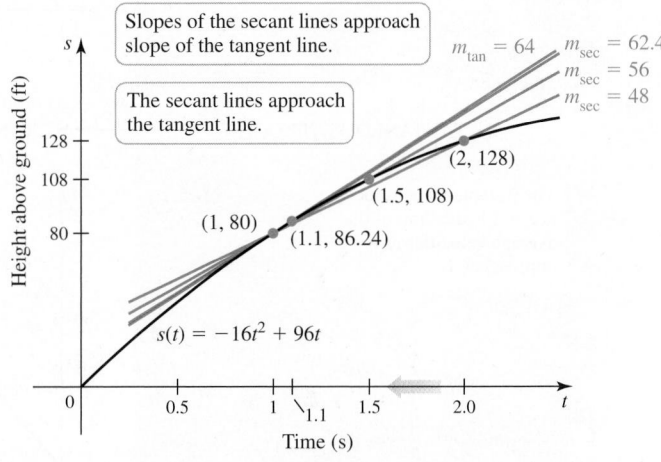

QUICK CHECK 3 In Figure 2.5, is m_{tan} at $t = 2$ greater than or less than m_{tan} at $t = 1$? ◀

Figure 2.5

The parallels between average and instantaneous velocities, on one hand, and between slopes of secant lines and tangent lines, on the other, illuminate the power behind the idea of a limit. As $t \to 1$, slopes of secant lines approach the slope of a tangent line. And as $t \to 1$, average velocities approach an instantaneous velocity. **Figure 2.6** summarizes these two parallel limit processes. These ideas lie at the foundation of what follows in the coming chapters.

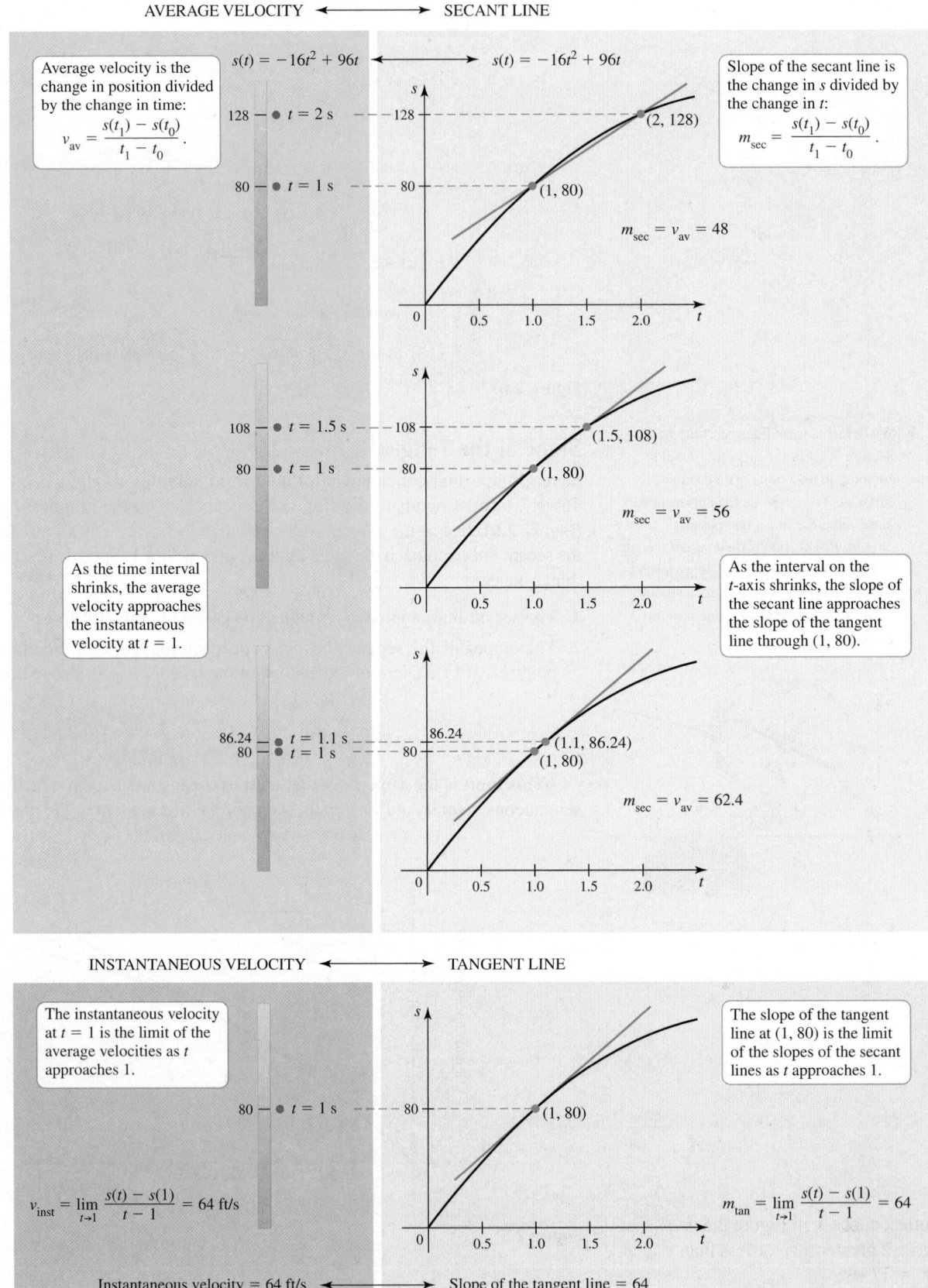

AVERAGE VELOCITY ⟷ **SECANT LINE**

Average velocity is the change in position divided by the change in time:
$$v_{av} = \frac{s(t_1) - s(t_0)}{t_1 - t_0}.$$

$s(t) = -16t^2 + 96t$ ⟷ $s(t) = -16t^2 + 96t$

Slope of the secant line is the change in s divided by the change in t:
$$m_{sec} = \frac{s(t_1) - s(t_0)}{t_1 - t_0}.$$

128 — ● $t = 2$ s — — 128 — — — (2, 128)

80 — ● $t = 1$ s — — 80 — — — (1, 80)

$m_{sec} = v_{av} = 48$

108 — ● $t = 1.5$ s — — 108 — — — (1.5, 108)

80 — ● $t = 1$ s — — 80 — — — (1, 80)

$m_{sec} = v_{av} = 56$

As the time interval shrinks, the average velocity approaches the instantaneous velocity at $t = 1$.

As the interval on the t-axis shrinks, the slope of the secant line approaches the slope of the tangent line through (1, 80).

86.24 — ● $t = 1.1$ s — — 86.24 — (1.1, 86.24)
80 — ● $t = 1$ s — — 80 — (1, 80)

$m_{sec} = v_{av} = 62.4$

INSTANTANEOUS VELOCITY ⟷ **TANGENT LINE**

The instantaneous velocity at $t = 1$ is the limit of the average velocities as t approaches 1.

The slope of the tangent line at (1, 80) is the limit of the slopes of the secant lines as t approaches 1.

80 — ● $t = 1$ s — — 80 — (1, 80)

$$v_{inst} = \lim_{t \to 1} \frac{s(t) - s(1)}{t - 1} = 64 \text{ ft/s}$$

$$m_{tan} = \lim_{t \to 1} \frac{s(t) - s(1)}{t - 1} = 64$$

Instantaneous velocity = 64 ft/s ⟷ Slope of the tangent line = 64

Figure 2.6

SECTION 2.1 EXERCISES

Getting Started

1. Suppose $s(t)$ is the position of an object moving along a line at time $t \geq 0$. What is the average velocity between the times $t = a$ and $t = b$?

2. Suppose $s(t)$ is the position of an object moving along a line at time $t \geq 0$. Describe a process for finding the instantaneous velocity at $t = a$.

3. The function $s(t)$ represents the position of an object at time t moving along a line. Suppose $s(2) = 136$ and $s(3) = 156$. Find the average velocity of the object over the interval of time $[2, 3]$.

4. The function $s(t)$ represents the position of an object at time t moving along a line. Suppose $s(1) = 84$ and $s(4) = 144$. Find the average velocity of the object over the interval of time $[1, 4]$.

5. The table gives the position $s(t)$ of an object moving along a line at time t, over a two-second interval. Find the average velocity of the object over the following intervals.

   a. $[0, 2]$ **b.** $[0, 1.5]$ **c.** $[0, 1]$ **d.** $[0, 0.5]$

t	0	0.5	1	1.5	2
$s(t)$	0	30	52	66	72

6. The graph gives the position $s(t)$ of an object moving along a line at time t, over a 2.5-second interval. Find the average velocity of the object over the following intervals.

   a. $[0.5, 2.5]$ **b.** $[0.5, 2]$ **c.** $[0.5, 1.5]$ **d.** $[0.5, 1]$

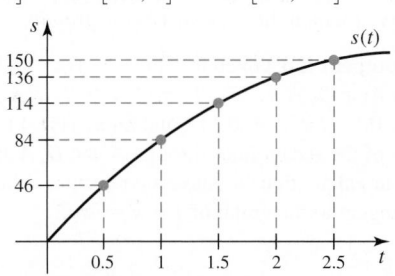

T 7. The following table gives the position $s(t)$ of an object moving along a line at time t. Determine the average velocities over the time intervals $[1, 1.01]$, $[1, 1.001]$, and $[1, 1.0001]$. Then make a conjecture about the value of the instantaneous velocity at $t = 1$.

t	1	1.0001	1.001	1.01
$s(t)$	64	64.00479984	64.047984	64.4784

T 8. The following table gives the position $s(t)$ of an object moving along a line at time t. Determine the average velocities over the time intervals $[2, 2.01]$, $[2, 2.001]$, and $[2, 2.0001]$. Then make a conjecture about the value of the instantaneous velocity at $t = 2$.

t	2	2.0001	2.001	2.01
$s(t)$	56	55.99959984	55.995984	55.9584

9. What is the slope of the secant line that passes through the points $(a, f(a))$ and $(b, f(b))$ on the graph of f?

10. Describe a process for finding the slope of the line tangent to the graph of f at $(a, f(a))$.

11. Describe the parallels between finding the instantaneous velocity of an object at a point in time and finding the slope of the line tangent to the graph of a function at a point on the graph.

T 12. Given the function $f(x) = -16x^2 + 64x$, complete the following.

   a. Find the slopes of the secant lines that pass though the points $(x, f(x))$ and $(2, f(2))$, for $x = 1.5, 1.9, 1.99, 1.999$, and 1.9999 (see figure).

   b. Make a conjecture about the value of the limit of the slopes of the secant lines that pass through $(x, f(x))$ and $(2, f(2))$ as x approaches 2.

   c. What is the relationship between your answer to part (b) and the slope of the line tangent to the curve at $x = 2$ (see figure)?

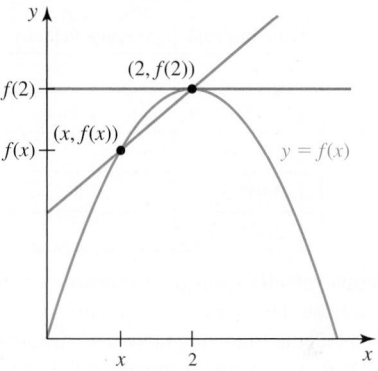

Practice Exercises

T 13. **Average velocity** The position of an object moving vertically along a line is given by the function $s(t) = -16t^2 + 128t$. Find the average velocity of the object over the following intervals.

   a. $[1, 4]$ **b.** $[1, 3]$ **c.** $[1, 2]$
   d. $[1, 1 + h]$, where $h > 0$ is a real number

T 14. **Average velocity** The position of an object moving vertically along a line is given by the function $s(t) = -4.9t^2 + 30t + 20$. Find the average velocity of the object over the following intervals.

   a. $[0, 3]$ **b.** $[0, 2]$ **c.** $[0, 1]$
   d. $[0, h]$, where $h > 0$ is a real number

T 15. **Average velocity** Consider the position function $s(t) = -16t^2 + 100t$ representing the position of an object moving vertically along a line. Sketch a graph of s with the secant line passing through $(0.5, s(0.5))$ and $(2, s(2))$. Determine the slope of the secant line and explain its relationship to the moving object.

16. **Average velocity** Consider the position function $s(t) = \sin \pi t$ representing the position of an object moving along a line on the end of a spring. Sketch a graph of s together with the secant line passing through $(0, s(0))$ and $(0.5, s(0.5))$. Determine the slope of the secant line and explain its relationship to the moving object.

T 17. **Instantaneous velocity** Consider the position function $s(t) = -16t^2 + 128t$ (Exercise 13). Complete the following table with the appropriate average velocities. Then make a conjecture about the value of the instantaneous velocity at $t = 1$.

Time interval	$[1, 2]$	$[1, 1.5]$	$[1, 1.1]$	$[1, 1.01]$	$[1, 1.001]$
Average velocity					

T 18. Instantaneous velocity Consider the position function $s(t) = -4.9t^2 + 30t + 20$ (Exercise 14). Complete the following table with the appropriate average velocities. Then make a conjecture about the value of the instantaneous velocity at $t = 2$.

Time interval	[2, 3]	[2, 2.5]	[2, 2.1]	[2, 2.01]	[2, 2.001]
Average velocity					

T 19. Instantaneous velocity Consider the position function $s(t) = -16t^2 + 100t$. Complete the following table with the appropriate average velocities. Then make a conjecture about the value of the instantaneous velocity at $t = 3$.

Time interval	Average velocity
[2, 3]	
[2.9, 3]	
[2.99, 3]	
[2.999, 3]	
[2.9999, 3]	

T 20. Instantaneous velocity Consider the position function $s(t) = 3 \sin t$ that describes a block bouncing vertically on a spring. Complete the following table with the appropriate average velocities. Then make a conjecture about the value of the instantaneous velocity at $t = \pi/2$.

Time interval	Average velocity
$[\pi/2, \pi]$	
$[\pi/2, \pi/2 + 0.1]$	
$[\pi/2, \pi/2 + 0.01]$	
$[\pi/2, \pi/2 + 0.001]$	
$[\pi/2, \pi/2 + 0.0001]$	

T 21–24. Instantaneous velocity *For the following position functions, make a table of average velocities similar to those in Exercises 19–20 and make a conjecture about the instantaneous velocity at the indicated time.*

21. $s(t) = -16t^2 + 80t + 60$ at $t = 3$

22. $s(t) = 20 \cos t$ at $t = \pi/2$

23. $s(t) = 40 \sin 2t$ at $t = 0$

24. $s(t) = 20/(t + 1)$ at $t = 0$

T 25–28. Slopes of tangent lines *For the following functions, make a table of slopes of secant lines and make a conjecture about the slope of the tangent line at the indicated point.*

25. $f(x) = 2x^2$ at $x = 2$

26. $f(x) = 3 \cos x$ at $x = \pi/2$

27. $f(x) = e^x$ at $x = 0$

28. $f(x) = x^3 - x$ at $x = 1$

Explorations and Challenges

T 29. Tangent lines with zero slope
  a. Graph the function $f(x) = x^2 - 4x + 3$.
  b. Identify the point $(a, f(a))$ at which the function has a tangent line with zero slope.

  c. Confirm your answer to part (b) by making a table of slopes of secant lines to approximate the slope of the tangent line at this point.

T 30. Tangent lines with zero slope
  a. Graph the function $f(x) = 4 - x^2$.
  b. Identify the point $(a, f(a))$ at which the function has a tangent line with zero slope.
  c. Consider the point $(a, f(a))$ found in part (b). Is it true that the secant line between $(a - h, f(a - h))$ and $(a + h, f(a + h))$ has slope zero for any value of $h \neq 0$?

T 31. Zero velocity A projectile is fired vertically upward and has a position given by $s(t) = -16t^2 + 128t + 192$, for $0 \leq t \leq 9$.
  a. Graph the position function, for $0 \leq t \leq 9$.
  b. From the graph of the position function, identify the time at which the projectile has an instantaneous velocity of zero; call this time $t = a$.
  c. Confirm your answer to part (b) by making a table of average velocities to approximate the instantaneous velocity at $t = a$.
  d. For what values of t on the interval $[0, 9]$ is the instantaneous velocity positive (the projectile moves upward)?
  e. For what values of t on the interval $[0, 9]$ is the instantaneous velocity negative (the projectile moves downward)?

T 32. Impact speed A rock is dropped off the edge of a cliff, and its distance s (in feet) from the top of the cliff after t seconds is $s(t) = 16t^2$. Assume the distance from the top of the cliff to the ground is 96 ft.
  a. When will the rock strike the ground?
  b. Make a table of average velocities and approximate the velocity at which the rock strikes the ground.

T 33. Slope of tangent line Given the function $f(x) = 1 - \cos x$ and the points $A(\pi/2, f(\pi/2))$, $B(\pi/2 + 0.05, f(\pi/2 + 0.05))$, $C(\pi/2 + 0.5, f(\pi/2 + 0.5))$, and $D(\pi, f(\pi))$ (see figure), find the slopes of the secant lines through A and D, A and C, and A and B. Use your calculations to make a conjecture about the slope of the line tangent to the graph of f at $x = \pi/2$.

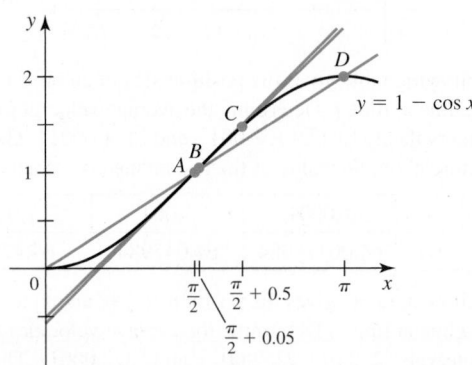

QUICK CHECK ANSWERS

1. 16 ft/s   **2.** Average velocity is the velocity over an interval of time. Instantaneous velocity is the velocity at one point of time.   **3.** Less than ◄

2.2 Definitions of Limits

Computing slopes of tangent lines and instantaneous velocities (Section 2.1) are just two of many important calculus problems that rely on limits. We now put these two problems aside until Chapter 3 and begin with a preliminary definition of the limit of a function.

> **DEFINITION Limit of a Function (Preliminary)**
>
> Suppose the function f is defined for all x near a except possibly at a. If $f(x)$ is arbitrarily close to L (as close to L as we like) for all x sufficiently close (but not equal) to a, we write
>
> $$\lim_{x \to a} f(x) = L$$
>
> and say the limit of $f(x)$ as x approaches a equals L.

➤ The terms *arbitrarily close* and *sufficiently close* will be made precise when rigorous definitions of limits are given in Section 2.7.

Informally, we say that $\lim_{x \to a} f(x) = L$ if $f(x)$ gets closer and closer to L as x gets closer and closer to a from both sides of a. The value of $\lim_{x \to a} f(x)$ (if it exists) depends on the values of f near a, but it does not depend on the value of $f(a)$. In some cases, the limit $\lim_{x \to a} f(x)$ equals $f(a)$. In other instances, $\lim_{x \to a} f(x)$ and $f(a)$ differ, or $f(a)$ may not even be defined.

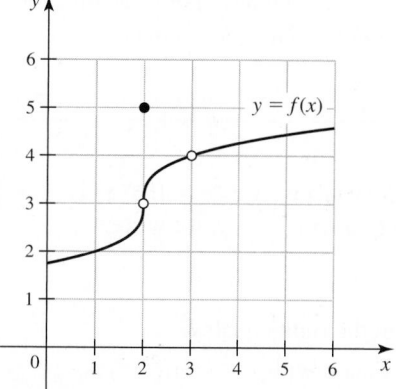

Figure 2.7

EXAMPLE 1 Finding limits from a graph Use the graph of f (Figure 2.7) to determine the following values, if possible.

a. $f(1)$ and $\lim_{x \to 1} f(x)$ **b.** $f(2)$ and $\lim_{x \to 2} f(x)$ **c.** $f(3)$ and $\lim_{x \to 3} f(x)$

SOLUTION

a. We see that $f(1) = 2$. As x approaches 1 from either side, the values of $f(x)$ approach 2 (Figure 2.8). Therefore, $\lim_{x \to 1} f(x) = 2$.

b. We see that $f(2) = 5$. However, as x approaches 2 from either side, $f(x)$ approaches 3 because the points on the graph of f approach the open circle at $(2, 3)$ (Figure 2.9). Therefore, $\lim_{x \to 2} f(x) = 3$ even though $f(2) = 5$.

c. In this case, $f(3)$ is undefined. We see that $f(x)$ approaches 4 as x approaches 3 from either side (Figure 2.10). Therefore, $\lim_{x \to 3} f(x) = 4$ even though $f(3)$ does not exist.

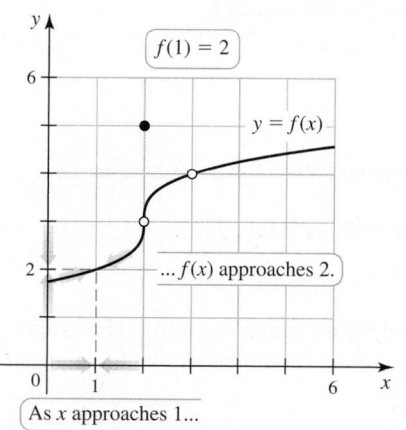

Figure 2.8

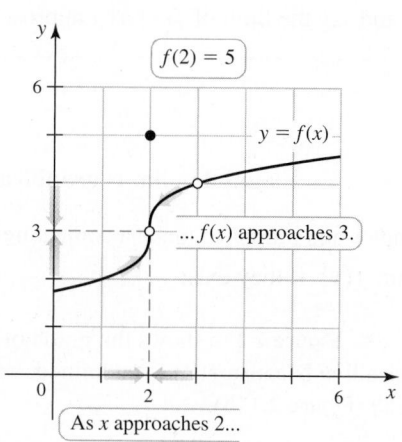

Figure 2.9

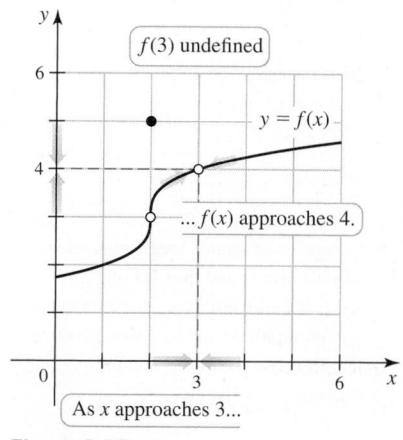

Figure 2.10

Related Exercises 3–4 ◄

QUICK CHECK 1 In Example 1, suppose we redefine the function at one point so that $f(1) = 1$. Does this change the value of $\lim_{x \to 1} f(x)$? ◄

In Example 1, we worked with the graph of a function to estimate limits. Let's now estimate limits using tabulated values of a function.

EXAMPLE 2   Finding limits from a table Create a table of values of $f(x) = \dfrac{\sqrt{x} - 1}{x - 1}$

➤ In Example 2, we have not stated with certainty that $\lim_{x \to 1} f(x) = 0.5$. But this is a reasonable conjecture based on the numerical evidence. Methods for calculating limits precisely are introduced in Section 2.3.

corresponding to values of x near 1. Then make a conjecture about the value of $\lim_{x \to 1} f(x)$.

SOLUTION Table 2.2 lists values of f corresponding to values of x approaching 1 from both sides. The numerical evidence suggests that $f(x)$ approaches 0.5 as x approaches 1. Therefore, we make the conjecture that $\lim_{x \to 1} f(x) = 0.5$.

Table 2.2

x	0.9	0.99	0.999	0.9999	1.0001	1.001	1.01	1.1
$f(x) = \dfrac{\sqrt{x} - 1}{x - 1}$	0.5131670	0.5012563	0.5001251	0.5000125	0.4999875	0.4998751	0.4987562	0.4880885

Related Exercises 7–8 ◄

One-Sided Limits

The limit $\lim_{x \to a} f(x) = L$ is referred to as a *two-sided* limit because $f(x)$ approaches L as x approaches a for values of x less than a *and* for values of x greater than a. For some functions, it makes sense to examine *one-sided* limits called *right-sided* and *left-sided* limits.

➤ As with two-sided limits, the value of a one-sided limit (if it exists) depends on the values of $f(x)$ near a but not on the value of $f(a)$.

DEFINITION One-Sided Limits

1. **Right-sided limit** Suppose f is defined for all x near a with $x > a$. If $f(x)$ is arbitrarily close to L for all x sufficiently close to a with $x > a$, we write

$$\lim_{x \to a^+} f(x) = L$$

and say the limit of $f(x)$ as x approaches a from the right equals L.

2. **Left-sided limit** Suppose f is defined for all x near a with $x < a$. If $f(x)$ is arbitrarily close to L for all x sufficiently close to a with $x < a$, we write

$$\lim_{x \to a^-} f(x) = L$$

and say the limit of $f(x)$ as x approaches a from the left equals L.

EXAMPLE 3   Examining limits graphically and numerically Let $f(x) = \dfrac{x^3 - 8}{4(x - 2)}$.

Use tables and graphs to make a conjecture about the values of $\lim_{x \to 2^+} f(x)$, $\lim_{x \to 2^-} f(x)$,

➤ Computer-generated graphs and tables help us understand the idea of a limit. Keep in mind, however, that computers are not infallible and they may produce incorrect results, even for simple functions (see Example 5).

and $\lim_{x \to 2} f(x)$, if they exist.

SOLUTION Figure 2.11a shows the graph of f obtained with a graphing utility. The graph is misleading because $f(2)$ is undefined, which means there should be a hole in the graph at $(2, 3)$ (**Figure 2.11b**).

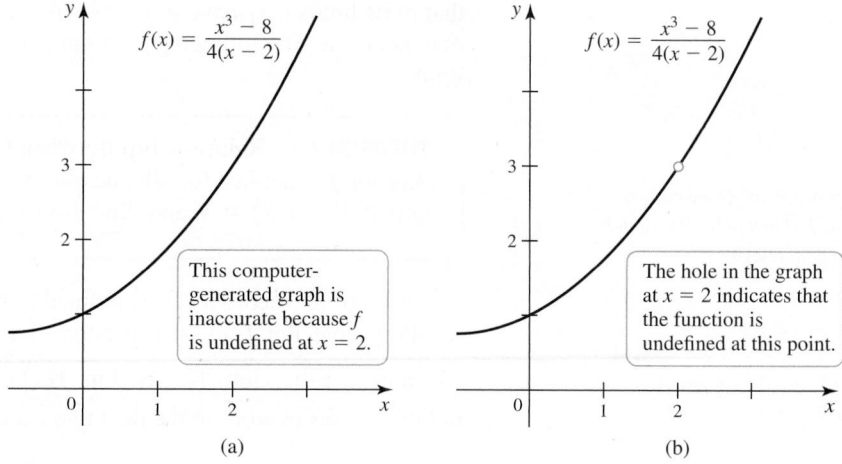

Figure 2.11

The graph in **Figure 2.12a** and the function values in Table 2.3 suggest that $f(x)$ approaches 3 as x approaches 2 from the right. Therefore, we write the right-sided limit

$$\lim_{x \to 2^+} f(x) = 3,$$

which says the limit of $f(x)$ as x approaches 2 from the right equals 3.

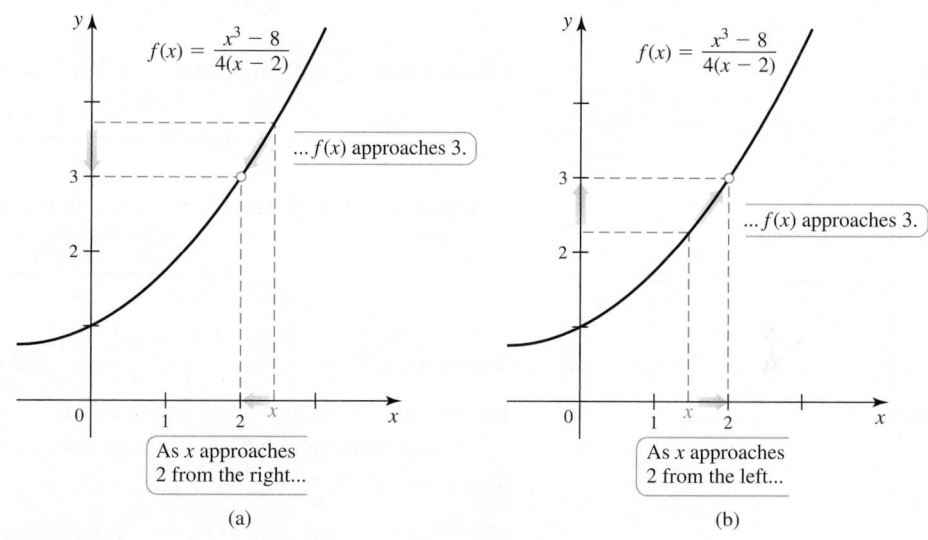

Figure 2.12

Table 2.3

x	1.9	1.99	1.999	1.9999	2.0001	2.001	2.01	2.1
$f(x) = \dfrac{x^3 - 8}{4(x - 2)}$	2.8525	2.985025	2.99850025	2.99985000	3.00015000	3.00150025	3.015025	3.1525

▶ Remember that the value of the limit does not depend on the value of $f(2)$. In Example 3, $\lim_{x \to 2} f(x) = 3$ despite the fact that $f(2)$ is undefined.

Similarly, **Figure 2.12b** and Table 2.3 suggest that as x approaches 2 from the left, $f(x)$ approaches 3. So we write the left-sided limit

$$\lim_{x \to 2^-} f(x) = 3,$$

which says the limit of $f(x)$ as x approaches 2 from the left equals 3. Because $f(x)$ approaches 3 as x approaches 2 from either side, we write the two-sided limit $\lim_{x \to 2} f(x) = 3$.

Related Exercises 27–28 ◀

Based on the previous example, you might wonder whether the limits $\lim_{x \to a^-} f(x)$, $\lim_{x \to a^+} f(x)$, and $\lim_{x \to a} f(x)$ always exist and are equal. The remaining examples demonstrate that these limits may have different values, and in some cases, one or more of these limits may not exist. The following theorem is useful when comparing one-sided and two-sided limits.

> ➤ Suppose P and Q are statements. We write P *if and only if* Q when P implies Q and Q implies P.

THEOREM 2.1   Relationship Between One-Sided and Two-Sided Limits
Assume f is defined for all x near a except possibly at a. Then $\lim\limits_{x \to a} f(x) = L$ if and only if $\lim\limits_{x \to a^+} f(x) = L$ and $\lim\limits_{x \to a^-} f(x) = L$.

A proof of Theorem 2.1 is outlined in Exercise 56 of Section 2.7. Using this theorem, it follows that $\lim\limits_{x \to a} f(x) \neq L$ if either $\lim\limits_{x \to a^+} f(x) \neq L$ or $\lim\limits_{x \to a^-} f(x) \neq L$ (or both). Furthermore, if either $\lim\limits_{x \to a^+} f(x)$ or $\lim\limits_{x \to a^-} f(x)$ does not exist, then $\lim\limits_{x \to a} f(x)$ does not exist. We put these ideas to work in the next two examples.

EXAMPLE 4   A function with a jump Sketch the graph of $g(x) = \dfrac{2x^2 - 6x + 4}{|x - 1|}$ and use the graph to find the values of $\lim\limits_{x \to 1^-} g(x)$, $\lim\limits_{x \to 1^+} g(x)$, and $\lim\limits_{x \to 1} g(x)$, if they exist.

SOLUTION Sketching the graph of g is straightforward if we first rewrite g as a piecewise function. Factoring the numerator of g, we obtain

$$g(x) = \frac{2x^2 - 6x + 4}{|x - 1|} = \frac{2(x - 1)(x - 2)}{|x - 1|}.$$

Observe that g is undefined at $x = 1$. For $x > 1$, $|x - 1| = x - 1$ and

$$g(x) = \frac{2(x - 1)(x - 2)}{x - 1} = 2x - 4.$$

Note that $x - 1 < 0$ when $x < 1$, which implies that $|x - 1| = -(x - 1)$ and

$$g(x) = \frac{2(x - 1)(x - 2)}{-(x - 1)} = -2x + 4.$$

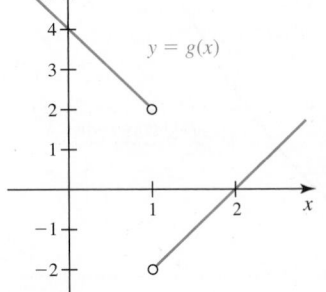

Figure 2.13

Therefore, $g(x) = \begin{cases} 2x - 4 & \text{if } x > 1 \\ -2x + 4 & \text{if } x < 1 \end{cases}$ and the graph of g consists of two linear pieces with a jump at $x = 1$ (Figure 2.13).

Examining the graph of g, we see that as x approaches 1 from the left, $g(x)$ approaches 2. Therefore, $\lim\limits_{x \to 1^-} g(x) = 2$. As x approaches 1 from the right, $g(x)$ approaches -2 and $\lim\limits_{x \to 1^+} g(x) = -2$. By Theorem 2.1, $\lim\limits_{x \to 1} g(x)$ does not exist because $\lim\limits_{x \to 1^-} g(x) \neq \lim\limits_{x \to 1^+} g(x)$.

Related Exercises 19–20 ◄

EXAMPLE 5   Some strange behavior Examine $\lim\limits_{x \to 0} \cos(1/x)$.

SOLUTION From the first three values of $\cos(1/x)$ in Table 2.4, it is tempting to conclude that $\lim\limits_{x \to 0^+} \cos(1/x) = -1$. But this conclusion is not confirmed when we evaluate $\cos(1/x)$ for values of x closer to 0.

Table 2.4

x	$\cos(1/x)$
0.001	0.56238
0.0001	−0.95216
0.00001	−0.99936
0.000001	0.93675
0.0000001	−0.90727
0.00000001	−0.36338

> We might *incorrectly* conclude that $\cos(1/x)$ approaches -1 as x approaches 0 from the right.

The behavior of $\cos{(1/x)}$ near 0 is better understood by letting $x = 1/(n\pi)$, where n is a positive integer. By making this substitution, we can sample the function at discrete points that approach zero. In this case,

$$\cos{\frac{1}{x}} = \cos{n\pi} = \begin{cases} 1 & \text{if } n \text{ is even} \\ -1 & \text{if } n \text{ is odd.} \end{cases}$$

As n increases, the values of $x = 1/(n\pi)$ approach zero, while the values of $\cos{(1/x)}$ oscillate between -1 and 1 (**Figure 2.14**). Therefore, $\cos{(1/x)}$ does not approach a single number as x approaches 0 from the right. We conclude that $\lim\limits_{x \to 0^+} \cos{(1/x)}$ does *not* exist, which implies that $\lim\limits_{x \to 0} \cos{(1/x)}$ does not exist.

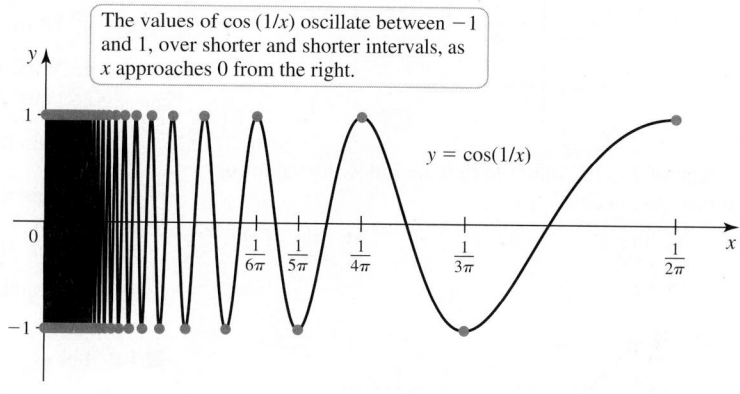

The values of $\cos{(1/x)}$ oscillate between -1 and 1, over shorter and shorter intervals, as x approaches 0 from the right.

$y = \cos(1/x)$

QUICK CHECK 2 Why is the graph of $y = \cos{(1/x)}$ difficult to plot near $x = 0$, as suggested by Figure 2.14? ◀

Figure 2.14

Related Exercise 43 ◀

Using tables and graphs to make conjectures for the values of limits worked well until Example 5. The limitation of technology in this example is not an isolated incident. For this reason, analytical techniques (paper-and-pencil methods) for finding limits are developed in the next section.

SECTION 2.2 EXERCISES

Getting Started

1. Explain the meaning of $\lim\limits_{x \to a} f(x) = L$.

2. True or false: When $\lim\limits_{x \to a} f(x)$ exists, it always equals $f(a)$. Explain.

3. Use the graph of h in the figure to find the following values or state that they do not exist.

   a. $h(2)$ **b.** $\lim\limits_{x \to 2} h(x)$ **c.** $h(4)$ **d.** $\lim\limits_{x \to 4} h(x)$ **e.** $\lim\limits_{x \to 5} h(x)$

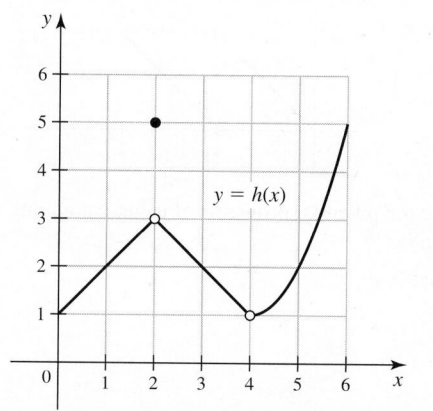

$y = h(x)$

4. Use the graph of g in the figure to find the following values or state that they do not exist.

   a. $g(0)$ **b.** $\lim\limits_{x \to 0} g(x)$ **c.** $g(1)$ **d.** $\lim\limits_{x \to 1} g(x)$

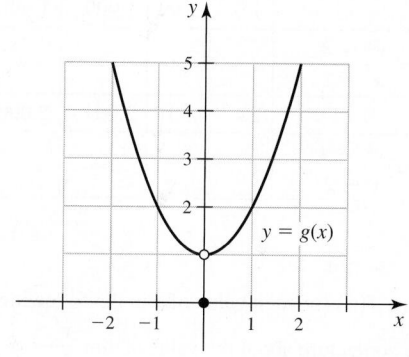

$y = g(x)$

5. Use the graph of f in the figure to find the following values or state that they do not exist.

   a. $f(1)$ **b.** $\lim\limits_{x\to 1} f(x)$ **c.** $f(0)$ **d.** $\lim\limits_{x\to 0} f(x)$

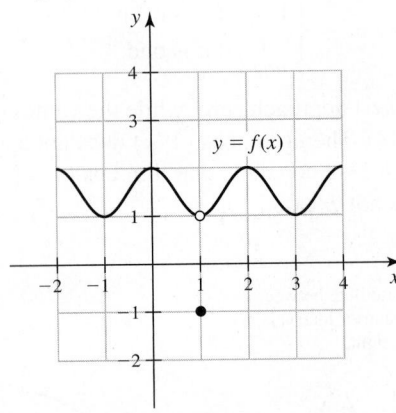

6. Use the graph of f in the figure to find the following values or state that they do not exist.

   a. $f(2)$ **b.** $\lim\limits_{x\to 2} f(x)$ **c.** $\lim\limits_{x\to 4} f(x)$ **d.** $\lim\limits_{x\to 5} f(x)$

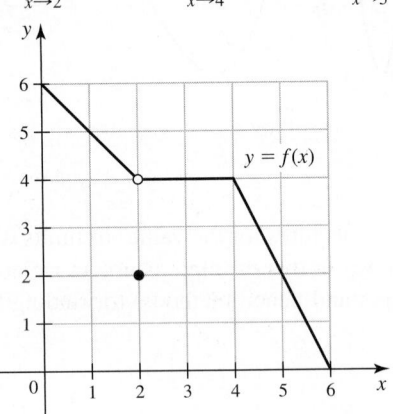

T 7. Let $f(x) = \dfrac{x^2 - 4}{x - 2}$.

   a. Calculate $f(x)$ for each value of x in the following table.

   b. Make a conjecture about the value of $\lim\limits_{x\to 2} \dfrac{x^2 - 4}{x - 2}$.

x	1.9	1.99	1.999	1.9999
$f(x) = \dfrac{x^2 - 4}{x - 2}$				
x	2.1	2.01	2.001	2.0001
$f(x) = \dfrac{x^2 - 4}{x - 2}$				

T 8. Let $f(x) = \dfrac{x^3 - 1}{x - 1}$.

   a. Calculate $f(x)$ for each value of x in the following table.

   b. Make a conjecture about the value of $\lim\limits_{x\to 1} \dfrac{x^3 - 1}{x - 1}$.

x	0.9	0.99	0.999	0.9999
$f(x) = \dfrac{x^3 - 1}{x - 1}$				
x	1.1	1.01	1.001	1.0001
$f(x) = \dfrac{x^3 - 1}{x - 1}$				

T 9. Let $g(t) = \dfrac{t - 9}{\sqrt{t} - 3}$.

   a. Make two tables, one showing values of g for $t = 8.9, 8.99,$ and 8.999 and one showing values of g for $t = 9.1, \ 9.01,$ and 9.001.

   b. Make a conjecture about the value of $\lim\limits_{t\to 9} \dfrac{t - 9}{\sqrt{t} - 3}$.

T 10. Let $f(x) = (1 + x)^{1/x}$.

   a. Make two tables, one showing values of f for $x = 0.01,$ $0.001, 0.0001,$ and 0.00001 and one showing values of f for $x = -0.01, \ -0.001, \ -0.0001,$ and -0.00001. Round your answers to five digits.

   b. Estimate the value of $\lim\limits_{x\to 0} (1 + x)^{1/x}$.

   c. What mathematical constant does $\lim\limits_{x\to 0} (1 + x)^{1/x}$ appear to equal?

11. Explain the meaning of $\lim\limits_{x\to a^+} f(x) = L$.

12. Explain the meaning of $\lim\limits_{x\to a^-} f(x) = L$.

13. If $\lim\limits_{x\to a^-} f(x) = L$ and $\lim\limits_{x\to a^+} f(x) = M$, where L and M are finite real numbers, then how are L and M related if $\lim\limits_{x\to a} f(x)$ exists?

T 14. Let $g(x) = \dfrac{x^3 - 4x}{8|x - 2|}$.

   a. Calculate $g(x)$ for each value of x in the following table.

   b. Make a conjecture about the values of $\lim\limits_{x\to 2^-} g(x)$, $\lim\limits_{x\to 2^+} g(x)$, and $\lim\limits_{x\to 2} g(x)$ or state that they do not exist.

x	1.9	1.99	1.999	1.9999
$g(x) = \dfrac{x^3 - 4x}{8\|x - 2\|}$				
x	2.1	2.01	2.001	2.0001
$g(x) = \dfrac{x^3 - 4x}{8\|x - 2\|}$				

15. Use the graph of f in the figure to find the following values or state that they do not exist. If a limit does not exist, explain why.

   a. $f(1)$ **b.** $\lim\limits_{x\to 1^-} f(x)$ **c.** $\lim\limits_{x\to 1^+} f(x)$ **d.** $\lim\limits_{x\to 1} f(x)$

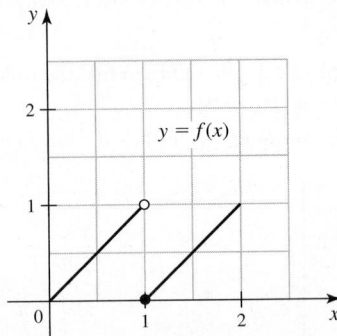

16. What are the potential problems of using a graphing utility to estimate $\lim\limits_{x\to a} f(x)$?

17. Finding limits from a graph Use the graph of f in the figure to find the following values or state that they do not exist. If a limit does not exist, explain why.

a. $f(1)$
b. $\lim_{x \to 1^-} f(x)$
c. $\lim_{x \to 1^+} f(x)$
d. $\lim_{x \to 1} f(x)$
e. $f(3)$
f. $\lim_{x \to 3^-} f(x)$
g. $\lim_{x \to 3^+} f(x)$
h. $\lim_{x \to 3} f(x)$
i. $f(2)$
j. $\lim_{x \to 2^-} f(x)$
k. $\lim_{x \to 2^+} f(x)$
l. $\lim_{x \to 2} f(x)$

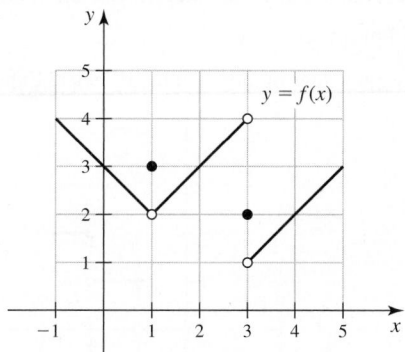

18. One-sided and two-sided limits Use the graph of g in the figure to find the following values or state that they do not exist. If a limit does not exist, explain why.

a. $g(2)$
b. $\lim_{x \to 2^-} g(x)$
c. $\lim_{x \to 2^+} g(x)$
d. $\lim_{x \to 2} g(x)$
e. $g(3)$
f. $\lim_{x \to 3^-} g(x)$
g. $\lim_{x \to 3^+} g(x)$
h. $g(4)$
i. $\lim_{x \to 4} g(x)$

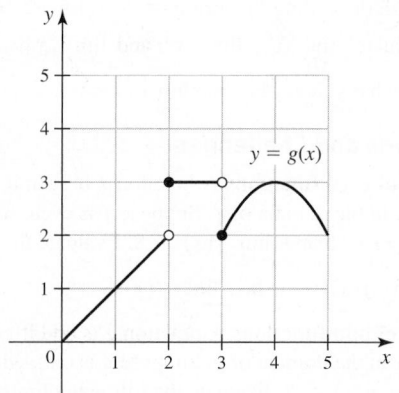

Practice Exercises

19–26. Evaluating limits graphically *Sketch a graph of f and use it to make a conjecture about the values of $f(a)$, $\lim_{x \to a^-} f(x)$, $\lim_{x \to a^+} f(x)$, and $\lim_{x \to a} f(x)$ or state that they do not exist.*

19. $f(x) = \begin{cases} x^2 + 1 & \text{if } x \le -1 \\ 3 & \text{if } x > -1 \end{cases}; a = -1$

20. $f(x) = \begin{cases} 3 - x & \text{if } x < 2 \\ x - 1 & \text{if } x > 2 \end{cases}; a = 2$

21. $f(x) = \begin{cases} \sqrt{x} & \text{if } x < 4 \\ 3 & \text{if } x = 4; \; a = 4 \\ x + 1 & \text{if } x > 4 \end{cases}$

22. $f(x) = |x + 2| + 2; a = -2$

23. $f(x) = \dfrac{x^2 - 25}{x - 5}; a = 5$

T 24. $f(x) = \dfrac{x - 100}{\sqrt{x} - 10}; a = 100$

25. $f(x) = \dfrac{x^2 + x - 2}{x - 1}; a = 1$

26. $f(x) = \dfrac{1 - x^4}{x^2 - 1}; a = 1$

T 27–32. Estimating limits graphically and numerically *Use a graph of f to estimate $\lim_{x \to a} f(x)$ or to show that the limit does not exist. Evaluate $f(x)$ near $x = a$ to support your conjecture.*

27. $f(x) = \dfrac{x - 2}{\ln|x - 2|}; a = 2$

28. $f(x) = \dfrac{e^{2x} - 2x - 1}{x^2}; a = 0$

29. $f(x) = \dfrac{1 - \cos(2x - 2)}{(x - 1)^2}; a = 1$

30. $f(x) = \dfrac{3 \sin x - 2 \cos x + 2}{x}; a = 0$

31. $f(x) = \dfrac{\sin(x + 1)}{|x + 1|}; a = -1$

32. $f(x) = \dfrac{x^3 - 4x^2 + 3x}{|x - 3|}; a = 3$

33. Explain why or why not Determine whether the following statements are true and give an explanation or counterexample.

a. The value of $\lim_{x \to 3} \dfrac{x^2 - 9}{x - 3}$ does not exist.

b. The value of $\lim_{x \to a} f(x)$ is always found by computing $f(a)$.

c. The value of $\lim_{x \to a} f(x)$ does not exist if $f(a)$ is undefined.

d. $\lim_{x \to 0} \sqrt{x} = 0$. (*Hint:* Graph $y = \sqrt{x}$.)

e. $\lim_{x \to \pi/2} \cot x = 0$. (*Hint:* Graph $y = \cot x$.)

34. The Heaviside function The Heaviside function is used in engineering applications to model flipping a switch. It is defined as

$$H(x) = \begin{cases} 0 & \text{if } x < 0 \\ 1 & \text{if } x \ge 0. \end{cases}$$

a. Sketch a graph of H on the interval $[-1, 1]$.
b. Does $\lim_{x \to 0} H(x)$ exist?

35. Postage rates Assume postage for sending a first-class letter in the United States is \$0.47 for the first ounce (up to and including 1 oz) plus \$0.21 for each additional ounce (up to and including each additional ounce).

a. Graph the function $p = f(w)$ that gives the postage p for sending a letter that weighs w ounces, for $0 < w \le 3.5$.
b. Evaluate $\lim_{w \to 2.3} f(w)$.
c. Does $\lim_{w \to 3} f(w)$ exist? Explain.

T 36–42. Calculator limits *Estimate the following limits using graphs or tables.*

36. $\lim\limits_{h \to 0} \dfrac{(1 + 2h)^{1/h}}{2e^{2+h}}$

37. $\lim\limits_{x \to \pi/2} \dfrac{\cot 3x}{\cos x}$

38. $\lim\limits_{x \to 1} \dfrac{18\left(\sqrt[3]{x} - 1\right)}{x^3 - 1}$

39. $\lim\limits_{x \to 1} \dfrac{9\left(\sqrt{2x - x^4} - \sqrt[3]{x}\right)}{1 - x^{3/4}}$

40. $\lim\limits_{x \to 0} \dfrac{6^x - 3^x}{x \ln 16}$

41. $\lim\limits_{h \to 0} \dfrac{\ln(1 + h)}{h}$

42. $\lim\limits_{h \to 0} \dfrac{4^h - 1}{h \ln(h + 2)}$

T 43.   Strange behavior near $x = 0$

    a. Create a table of values of $\sin(1/x)$, for $x = 2/\pi, 2/(3\pi)$, $2/(5\pi), 2/(7\pi), 2/(9\pi)$, and $2/(11\pi)$. Describe the pattern of values you observe.

    b. Why does a graphing utility have difficulty plotting the graph of $y = \sin(1/x)$ near $x = 0$ (see figure)?

    c. What do you conclude about $\lim\limits_{x \to 0} \sin(1/x)$?

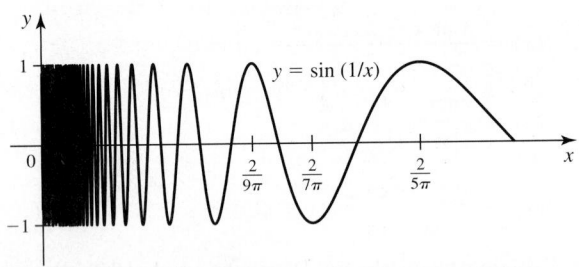

T 44.   Strange behavior near $x = 0$

    a. Create a table of values of $\tan(3/x)$, for $x = 12/\pi, 12/(3\pi)$, $12/(5\pi), \ldots, 12/(11\pi)$. Describe the general pattern in the values you observe.

    b. Use a graphing utility to graph $y = \tan(3/x)$. Why do graphing utilities have difficulty plotting the graph near $x = 0$?

    c. What do you conclude about $\lim\limits_{x \to 0} \tan(3/x)$?

45–49. Sketching graphs of functions *Sketch the graph of a function with the given properties. You do not need to find a formula for the function.*

45. $f(2) = 1,\ \lim\limits_{x \to 2} f(x) = 3$

46. $f(1) = 0,\ f(2) = 4,\ f(3) = 6,\ \lim\limits_{x \to 2^-} f(x) = -3,\ \lim\limits_{x \to 2^+} f(x) = 5$

47. $g(1) = 0,\ g(2) = 1,\ g(3) = -2,\ \lim\limits_{x \to 2} g(x) = 0,$
$\lim\limits_{x \to 3^-} g(x) = -1,\ \lim\limits_{x \to 3^+} g(x) = -2$

48. $h(-1) = 2,\ \lim\limits_{x \to -1^-} h(x) = 0,\ \lim\limits_{x \to -1^+} h(x) = 3,$
$h(1) = \lim\limits_{x \to 1^-} h(x) = 1,\ \lim\limits_{x \to 1^+} h(x) = 4$

49. $p(0) = 2,\ \lim\limits_{x \to 0} p(x) = 0,\ \lim\limits_{x \to 2} p(x)$ does not exist,
$p(2) = \lim\limits_{x \to 2^+} p(x) = 1$

50.   A step function Let $f(x) = \dfrac{|x|}{x}$, for $x \neq 0$.

    a. Sketch a graph of f on the interval $[-2, 2]$.

    b. Does $\lim\limits_{x \to 0} f(x)$ exist? Explain your reasoning after first examining $\lim\limits_{x \to 0^-} f(x)$ and $\lim\limits_{x \to 0^+} f(x)$.

51.   The floor function For any real number x, the *floor function* (or *greatest integer function*) $\lfloor x \rfloor$ is the greatest integer less than or equal to x (see figure).

    a. Compute $\lim\limits_{x \to -1^-} \lfloor x \rfloor$, $\lim\limits_{x \to -1^+} \lfloor x \rfloor$, $\lim\limits_{x \to 2^-} \lfloor x \rfloor$, and $\lim\limits_{x \to 2^+} \lfloor x \rfloor$.

    b. Compute $\lim\limits_{x \to 2.3^-} \lfloor x \rfloor$, $\lim\limits_{x \to 2.3^+} \lfloor x \rfloor$, and $\lim\limits_{x \to 2.3} \lfloor x \rfloor$.

    c. For a given integer a, state the values of $\lim\limits_{x \to a^-} \lfloor x \rfloor$ and $\lim\limits_{x \to a^+} \lfloor x \rfloor$.

    d. In general, if a is not an integer, state the values of $\lim\limits_{x \to a^-} \lfloor x \rfloor$ and $\lim\limits_{x \to a^+} \lfloor x \rfloor$.

    e. For what values of a does $\lim\limits_{x \to a} \lfloor x \rfloor$ exist? Explain?

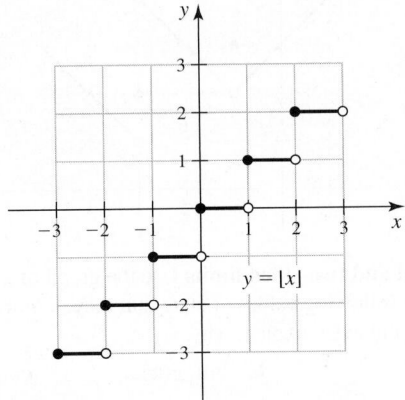

52.   The ceiling function For any real number x, the *ceiling function* $\lceil x \rceil$ is the smallest integer greater than or equal to x.

    a. Graph the ceiling function $y = \lceil x \rceil$, for $-2 \le x \le 3$.

    b. Evaluate $\lim\limits_{x \to 2^-} \lceil x \rceil$, $\lim\limits_{x \to 1^+} \lceil x \rceil$, and $\lim\limits_{x \to 1.5} \lceil x \rceil$.

    c. For what values of a does $\lim\limits_{x \to a} \lceil x \rceil$ exist? Explain.

Explorations and Challenges

53.   Limits of even functions A function f is even if $f(-x) = f(x)$, for all x in the domain of f. Suppose f is even, with $\lim\limits_{x \to 2^+} f(x) = 5$ and $\lim\limits_{x \to 2^-} f(x) = 8$. Evaluate the following limits.

    a. $\lim\limits_{x \to -2^+} f(x)$ **b.** $\lim\limits_{x \to -2^-} f(x)$

54.   Limits of odd functions A function g is odd if $g(-x) = -g(x)$, for all x in the domain of g. Suppose g is odd, with $\lim\limits_{x \to 2^+} g(x) = 5$ and $\lim\limits_{x \to 2^-} g(x) = 8$. Evaluate the following limits.

    a. $\lim\limits_{x \to -2^+} g(x)$ **b.** $\lim\limits_{x \to -2^-} g(x)$

T 55.   Limits by graphs

    a. Use a graphing utility to estimate $\lim\limits_{x \to 0} \dfrac{\tan 2x}{\sin x}$, $\lim\limits_{x \to 0} \dfrac{\tan 3x}{\sin x}$, and $\lim\limits_{x \to 0} \dfrac{\tan 4x}{\sin x}$.

    b. Make a conjecture about the value of $\lim\limits_{x \to 0} \dfrac{\tan px}{\sin x}$, for any real constant p.

T 56.   Limits by graphs Graph $f(x) = \dfrac{\sin nx}{x}$, for $n = 1, 2, 3$, and 4 (four graphs). Use the window $[-1, 1] \times [0, 5]$.

    a. Estimate $\lim\limits_{x \to 0} \dfrac{\sin x}{x}$, $\lim\limits_{x \to 0} \dfrac{\sin 2x}{x}$, $\lim\limits_{x \to 0} \dfrac{\sin 3x}{x}$, and $\lim\limits_{x \to 0} \dfrac{\sin 4x}{x}$.

    b. Make a conjecture about the value of $\lim\limits_{x \to 0} \dfrac{\sin px}{x}$, for any real constant p.

T 57. Limits by graphs Use a graphing utility to plot $y = \dfrac{\sin px}{\sin qx}$ for at least three different pairs of nonzero constants p and q of your choice. Estimate $\lim\limits_{x\to 0} \dfrac{\sin px}{\sin qx}$ in each case. Then use your work to make a conjecture about the value of $\lim\limits_{x\to 0} \dfrac{\sin px}{\sin qx}$ for any nonzero values of p and q.

2.3 Techniques for Computing Limits

Graphical and numerical techniques for estimating limits, like those presented in the previous section, provide intuition about limits. These techniques, however, occasionally lead to incorrect results. Therefore, we turn our attention to analytical methods for evaluating limits precisely.

Limits of Linear Functions

The graph of $f(x) = mx + b$ is a line with slope m and y-intercept b. From **Figure 2.15**, we see that $f(x)$ approaches $f(a)$ as x approaches a. Therefore, if f is a linear function, we have $\lim\limits_{x\to a} f(x) = f(a)$. It follows that for linear functions, $\lim\limits_{x\to a} f(x)$ is found by direct substitution of $x = a$ into $f(x)$. This observation leads to the following theorem, which is proved in Exercise 39 of Section 2.7.

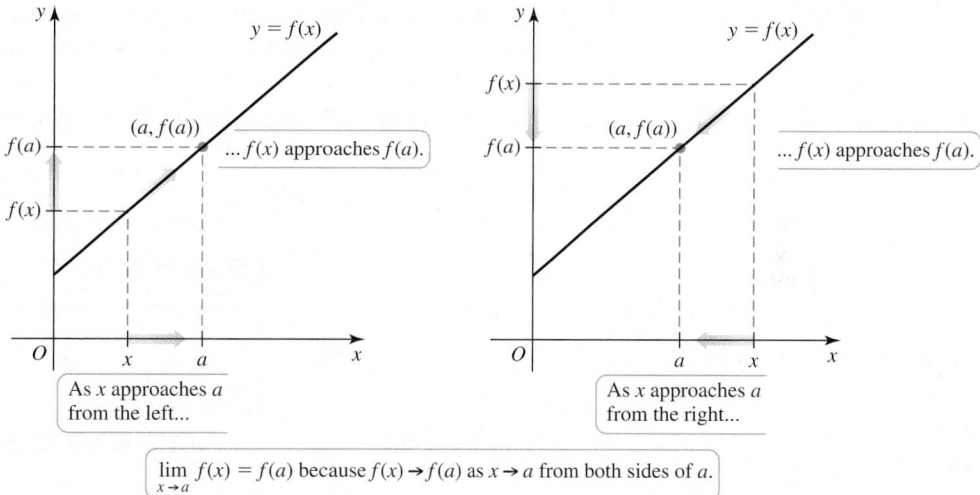

$$\lim_{x\to a} f(x) = f(a) \text{ because } f(x) \to f(a) \text{ as } x \to a \text{ from both sides of } a.$$

Figure 2.15

THEOREM 2.2   Limits of Linear Functions

Let a, b, and m be real numbers. For linear functions $f(x) = mx + b$,

$$\lim_{x\to a} f(x) = f(a) = ma + b.$$

EXAMPLE 1   Limits of linear functions Evaluate the following limits.

a. $\lim\limits_{x\to 3} f(x)$, where $f(x) = \frac{1}{2}x - 7$ **b.** $\lim\limits_{x\to 2} g(x)$, where $g(x) = 6$

SOLUTION

a. $\lim\limits_{x\to 3} f(x) = \lim\limits_{x\to 3} \left(\frac{1}{2}x - 7\right) = f(3) = -\frac{11}{2}$ **b.** $\lim\limits_{x\to 2} g(x) = \lim\limits_{x\to 2} 6 = g(2) = 6$

Related Exercises 19, 22 ◄

Limit Laws

The following limit laws greatly simplify the evaluation of many limits.

THEOREM 2.3   Limit Laws

Assume $\lim\limits_{x \to a} f(x)$ and $\lim\limits_{x \to a} g(x)$ exist. The following properties hold, where c is a real number, and $n > 0$ is an integer.

1. Sum $\lim\limits_{x \to a} (f(x) + g(x)) = \lim\limits_{x \to a} f(x) + \lim\limits_{x \to a} g(x)$

2. Difference $\lim\limits_{x \to a} (f(x) - g(x)) = \lim\limits_{x \to a} f(x) - \lim\limits_{x \to a} g(x)$

3. Constant multiple $\lim\limits_{x \to a} (cf(x)) = c \lim\limits_{x \to a} f(x)$

4. Product $\lim\limits_{x \to a} (f(x)g(x)) = \left(\lim\limits_{x \to a} f(x) \right) \left(\lim\limits_{x \to a} g(x) \right)$

5. Quotient $\lim\limits_{x \to a} \left(\dfrac{f(x)}{g(x)} \right) = \dfrac{\lim\limits_{x \to a} f(x)}{\lim\limits_{x \to a} g(x)}$, provided $\lim\limits_{x \to a} g(x) \neq 0$

6. Power $\lim\limits_{x \to a} (f(x))^n = \left(\lim\limits_{x \to a} f(x) \right)^n$

7. Root $\lim\limits_{x \to a} (f(x))^{1/n} = \left(\lim\limits_{x \to a} f(x) \right)^{1/n}$, provided $f(x) > 0$, for x near a, if n is even

---

A proof of Law 1 is given in Example 6 of Section 2.7; the proofs of Laws 2 and 3 are asked for in Exercises 43 and 44 of the same section. Laws 4 and 5 are proved in Appendix A. Law 6 is proved from Law 4 as follows.

For a positive integer n, if $\lim\limits_{x \to a} f(x)$ exists, we have

$$\lim\limits_{x \to a} (f(x))^n = \lim\limits_{x \to a} \underbrace{(f(x) \, f(x) \cdots f(x))}_{n \text{ factors of } f(x)}$$

$$= \underbrace{\left(\lim\limits_{x \to a} f(x) \right) \left(\lim\limits_{x \to a} f(x) \right) \cdots \left(\lim\limits_{x \to a} f(x) \right)}_{n \text{ factors of } \lim\limits_{x \to a} f(x)} \qquad \text{Repeated use of Law 4}$$

$$= \left(\lim\limits_{x \to a} f(x) \right)^n.$$

Law 7 is a direct consequence of Theorem 2.12 (Section 2.6).

EXAMPLE 2   Evaluating limits Suppose $\lim\limits_{x \to 2} f(x) = 4$, $\lim\limits_{x \to 2} g(x) = 5$, and $\lim\limits_{x \to 2} h(x) = 8$. Use the limit laws in Theorem 2.3 to compute each limit.

a. $\lim\limits_{x \to 2} \dfrac{f(x) - g(x)}{h(x)}$ **b.** $\lim\limits_{x \to 2} (6f(x)g(x) + h(x))$ **c.** $\lim\limits_{x \to 2} (g(x))^3$

SOLUTION

a. $\lim\limits_{x \to 2} \dfrac{f(x) - g(x)}{h(x)} = \dfrac{\lim\limits_{x \to 2} (f(x) - g(x))}{\lim\limits_{x \to 2} h(x)} \qquad$ Law 5

$$= \dfrac{\lim\limits_{x \to 2} f(x) - \lim\limits_{x \to 2} g(x)}{\lim\limits_{x \to 2} h(x)} \qquad \text{Law 2}$$

$$= \dfrac{4 - 5}{8} = -\dfrac{1}{8}$$

b. $\lim\limits_{x \to 2} (6f(x)g(x) + h(x)) = \lim\limits_{x \to 2} (6f(x)g(x)) + \lim\limits_{x \to 2} h(x)$ Law 1

$= 6 \cdot \lim\limits_{x \to 2} (f(x)g(x)) + \lim\limits_{x \to 2} h(x)$ Law 3

$= 6 \left(\lim\limits_{x \to 2} f(x) \right) \left(\lim\limits_{x \to 2} g(x) \right) + \lim\limits_{x \to 2} h(x)$ Law 4

$= 6 \cdot 4 \cdot 5 + 8 = 128$

c. $\lim\limits_{x \to 2} (g(x))^3 = \left(\lim\limits_{x \to 2} g(x) \right)^3 = 5^3 = 125$ Law 6

Related Exercises 11–12 ◄

Limits of Polynomial and Rational Functions

The limit laws are now used to find the limits of polynomial and rational functions. For example, to evaluate the limit of the polynomial $p(x) = 7x^3 + 3x^2 + 4x + 2$ at an arbitrary point a, we proceed as follows:

$\lim\limits_{x \to a} p(x) = \lim\limits_{x \to a} (7x^3 + 3x^2 + 4x + 2)$

$= \lim\limits_{x \to a} (7x^3) + \lim\limits_{x \to a} (3x^2) + \lim\limits_{x \to a} (4x + 2)$ Law 1

$= 7 \lim\limits_{x \to a} (x^3) + 3 \lim\limits_{x \to a} (x^2) + \lim\limits_{x \to a} (4x + 2)$ Law 3

$= 7 \underbrace{\left(\lim\limits_{x \to a} x \right)^3}_{a} + 3 \underbrace{\left(\lim\limits_{x \to a} x \right)^2}_{a} + \underbrace{\lim\limits_{x \to a} (4x + 2)}_{4a + 2}$ Law 6

$= 7a^3 + 3a^2 + 4a + 2 = p(a).$ Theorem 2.2

As in the case of linear functions, the limit of a polynomial is found by direct substitution; that is, $\lim\limits_{x \to a} p(x) = p(a)$ (Exercise 107).

It is now a short step to evaluating limits of rational functions of the form $f(x) = p(x)/q(x)$, where p and q are polynomials. Applying Law 5, we have

$$\lim\limits_{x \to a} \frac{p(x)}{q(x)} = \frac{\lim\limits_{x \to a} p(x)}{\lim\limits_{x \to a} q(x)} = \frac{p(a)}{q(a)}, \text{ provided } q(a) \neq 0,$$

which shows that limits of rational functions are also evaluated by direct substitution.

> ► The conditions under which direct substitution $\left(\lim\limits_{x \to a} f(x) = f(a) \right)$ can be used to evaluate a limit become clear in Section 2.6, when we discuss the important property of *continuity*.

THEOREM 2.4   Limits of Polynomial and Rational Functions

Assume p and q are polynomials and a is a constant.

a. Polynomial functions: $\lim\limits_{x \to a} p(x) = p(a)$

b. Rational functions: $\lim\limits_{x \to a} \dfrac{p(x)}{q(x)} = \dfrac{p(a)}{q(a)}$, provided $q(a) \neq 0$

QUICK CHECK 1 Use Theorem 2.4 to evaluate $\lim\limits_{x \to 2} (2x^4 - 8x - 16)$ and $\lim\limits_{x \to -1} \dfrac{x - 1}{x}$. ◄

EXAMPLE 3   Limit of a rational function Evaluate $\lim\limits_{x \to 2} \dfrac{3x^2 - 4x}{5x^3 - 36}$.

SOLUTION Notice that the denominator of this function is nonzero at $x = 2$. Using Theorem 2.4b, we find that

$$\lim\limits_{x \to 2} \frac{3x^2 - 4x}{5x^3 - 36} = \frac{3(2^2) - 4(2)}{5(2^3) - 36} = 1.$$

Related Exercise 25 ◄

QUICK CHECK 2 Use Theorem 2.4 to compute $\lim\limits_{x \to 1} \dfrac{5x^4 - 3x^2 + 8x - 6}{x + 1}$. ◄

EXAMPLE 4   An algebraic function Evaluate $\lim\limits_{x \to 2} \dfrac{\sqrt{2x^3 + 9} + 3x - 1}{4x + 1}$.

SOLUTION Using Theorems 2.3 and 2.4, we have

$$\lim_{x \to 2} \frac{\sqrt{2x^3 + 9} + 3x - 1}{4x + 1} = \frac{\lim\limits_{x \to 2} \left(\sqrt{2x^3 + 9} + 3x - 1\right)}{\lim\limits_{x \to 2} (4x + 1)} \qquad \text{Law 5}$$

$$= \frac{\sqrt{\lim\limits_{x \to 2} (2x^3 + 9)} + \lim\limits_{x \to 2} (3x - 1)}{\lim\limits_{x \to 2} (4x + 1)} \qquad \text{Laws 1 and 7}$$

$$= \frac{\sqrt{(2(2)^3 + 9)} + (3(2) - 1)}{(4(2) + 1)} \qquad \text{Theorem 2.4}$$

$$= \frac{\sqrt{25} + 5}{9} = \frac{10}{9}.$$

Notice that the limit at $x = 2$ equals the value of the function at $x = 2$.

Related Exercises 26–27 ◄

One-Sided Limits

Theorem 2.2, Limit Laws 1–6, and Theorem 2.4 also hold for left-sided and right-sided limits. In other words, these laws remain valid if we replace $\lim\limits_{x \to a}$ with $\lim\limits_{x \to a^+}$ or $\lim\limits_{x \to a^-}$. Law 7 must be modified slightly for one-sided limits, as shown in the next theorem.

THEOREM 2.3 (CONTINUED)   Limit Laws for One-Sided Limits

Laws 1–6 hold with $\lim\limits_{x \to a}$ replaced with $\lim\limits_{x \to a^+}$ or $\lim\limits_{x \to a^-}$. Law 7 is modified as follows. Assume $n > 0$ is an integer.

7. Root

a. $\lim\limits_{x \to a^+} (f(x))^{1/n} = \left(\lim\limits_{x \to a^+} f(x)\right)^{1/n}$, provided $f(x) \geq 0$, for x near a with
$x > a$, if n is even

b. $\lim\limits_{x \to a^-} (f(x))^{1/n} = \left(\lim\limits_{x \to a^-} f(x)\right)^{1/n}$, provided $f(x) \geq 0$, for x near a with
$x < a$, if n is even

EXAMPLE 5   Calculating left- and right-sided limits Let

$$f(x) = \begin{cases} -2x + 4 & \text{if } x \leq 1 \\ \sqrt{x - 1} & \text{if } x > 1. \end{cases}$$

Find the values of $\lim\limits_{x \to 1^-} f(x)$, $\lim\limits_{x \to 1^+} f(x)$, and $\lim\limits_{x \to 1} f(x)$, or state that they do not exist.

SOLUTION Notice that $f(x) = -2x + 4$, for $x \leq 1$. Therefore,

$$\lim_{x \to 1^-} f(x) = \lim_{x \to 1^-} (-2x + 4) = 2. \quad \text{Theorem 2.2}$$

For $x > 1$, note that $x - 1 > 0$; it follows that

$$\lim_{x \to 1^+} f(x) = \lim_{x \to 1^+} \sqrt{x - 1} = \sqrt{\lim_{x \to 1^+} (x - 1)} = 0. \quad \text{Law 7 for one-sided limits}$$

Because $\lim\limits_{x \to 1^-} f(x) \neq \lim\limits_{x \to 1^+} f(x)$, $\lim\limits_{x \to 1} f(x)$ does not exist by Theorem 2.1. The graph of f (Figure 2.16) is consistent with these findings.

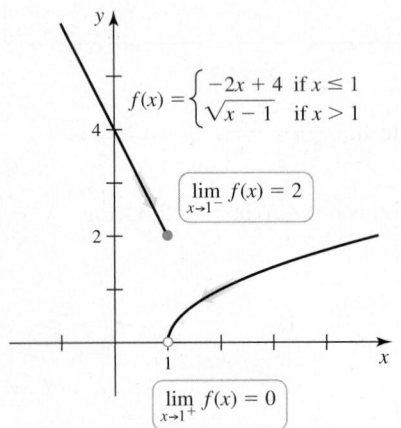

Figure 2.16

Related Exercises 72–73 ◄

Other Techniques

So far, we have evaluated limits by direct substitution. A more challenging problem is finding $\lim\limits_{x \to a} f(x)$ when the limit exists, but $\lim\limits_{x \to a} f(x) \neq f(a)$. Two typical cases are shown in **Figure 2.17**. In the first case, $f(a)$ is defined, but it is not equal to $\lim\limits_{x \to a} f(x)$; in the second case, $f(a)$ is not defined at all. In both cases, direct substitution does not work—we need a new strategy. One way to evaluate a challenging limit is to replace it with an equivalent limit that *can* be evaluated by direct substitution. Example 6 illustrates two common scenarios.

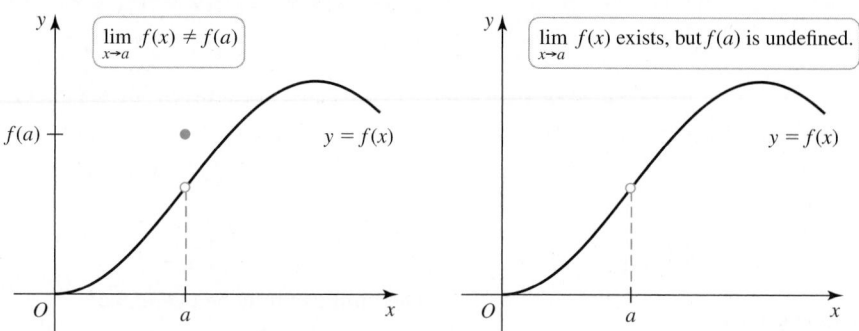

Figure 2.17

EXAMPLE 6 **Other techniques** Evaluate the following limits.

a. $\displaystyle\lim_{x \to 2} \frac{x^2 - 6x + 8}{x^2 - 4}$ **b.** $\displaystyle\lim_{x \to 1} \frac{\sqrt{x} - 1}{x - 1}$

SOLUTION

a. Factor and cancel This limit cannot be found by direct substitution because the denominator is zero when $x = 2$. Instead, the numerator and denominator are factored; then, assuming $x \neq 2$, we cancel like factors:

$$\frac{x^2 - 6x + 8}{x^2 - 4} = \frac{(x - 2)(x - 4)}{(x - 2)(x + 2)} = \frac{x - 4}{x + 2}.$$

> ▶ The argument used in Example 6 relies on the fact that in the limit process, x approaches 2, but $x \neq 2$. Therefore, we may cancel like factors.

Because $\dfrac{x^2 - 6x + 8}{x^2 - 4} = \dfrac{x - 4}{x + 2}$ whenever $x \neq 2$, the two functions have the same limit as x approaches 2 (**Figure 2.18**). Therefore,

$$\lim_{x \to 2} \frac{x^2 - 6x + 8}{x^2 - 4} = \lim_{x \to 2} \frac{x - 4}{x + 2} = \frac{2 - 4}{2 + 2} = -\frac{1}{2}.$$

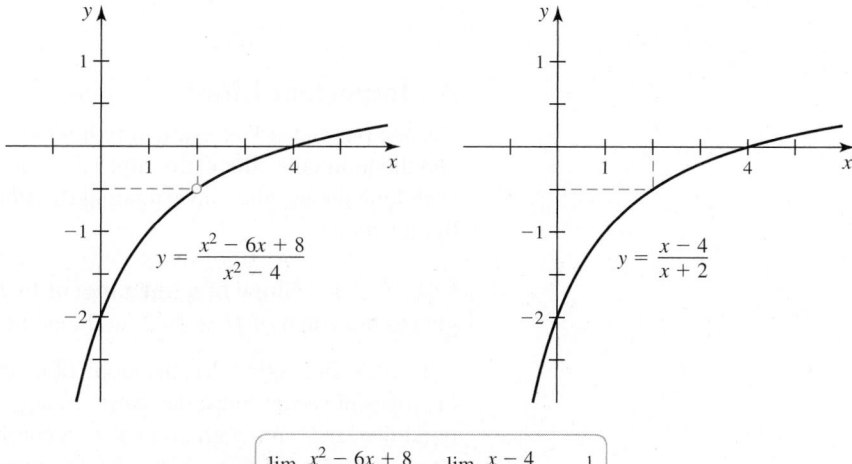

$$\lim_{x \to 2} \frac{x^2 - 6x + 8}{x^2 - 4} = \lim_{x \to 2} \frac{x - 4}{x + 2} = -\frac{1}{2}$$

Figure 2.18

b. Use conjugates This limit was approximated numerically in Example 2 of Section 2.2; we made a conjecture that the value of the limit is $1/2$. Using direct substitution to verify this conjecture fails in this case, because the denominator is zero at $x = 1$. Instead, we first simplify the function by multiplying the numerator and denominator by the *algebraic conjugate* of the numerator. The conjugate of $\sqrt{x} - 1$ is $\sqrt{x} + 1$; therefore,

> We multiply the given function by
> $$1 = \frac{\sqrt{x} + 1}{\sqrt{x} + 1}.$$

$$\frac{\sqrt{x} - 1}{x - 1} = \frac{(\sqrt{x} - 1)(\sqrt{x} + 1)}{(x - 1)(\sqrt{x} + 1)} \qquad \text{Rationalize the numerator; multiply by 1.}$$

$$= \frac{x + \sqrt{x} - \sqrt{x} - 1}{(x - 1)(\sqrt{x} + 1)} \qquad \text{Expand the numerator.}$$

$$= \frac{x - 1}{(x - 1)(\sqrt{x} + 1)} \qquad \text{Simplify.}$$

$$= \frac{1}{\sqrt{x} + 1}. \qquad \text{Cancel like factors assuming } x \neq 1.$$

The limit can now be evaluated:

$$\lim_{x \to 1} \frac{\sqrt{x} - 1}{x - 1} = \lim_{x \to 1} \frac{1}{\sqrt{x} + 1} = \frac{1}{1 + 1} = \frac{1}{2}.$$

Related Exercises 34, 41 ◀

QUICK CHECK 3 Evaluate

$$\lim_{x \to 5} \frac{x^2 - 7x + 10}{x - 5}.$$ ◀

EXAMPLE 7   Finding limits Let $f(x) = \dfrac{x^3 - 6x^2 + 8x}{\sqrt{x - 2}}$. Find the values of $\lim\limits_{x \to 2^-} f(x)$, $\lim\limits_{x \to 2^+} f(x)$, and $\lim\limits_{x \to 2} f(x)$, or state that they do not exist.

SOLUTION Because the denominator of f is $\sqrt{x - 2}$, $f(x)$ is defined only when $x - 2 > 0$. Therefore, the domain of f is $x > 2$ and it follows that $\lim\limits_{x \to 2^-} f(x)$ does not exist, which in turn implies that $\lim\limits_{x \to 2} f(x)$ does not exist (Theorem 2.1). To evaluate $\lim\limits_{x \to 2^+} f(x)$, factor the numerator of f and simplify:

$$\lim_{x \to 2^+} f(x) = \lim_{x \to 2^+} \frac{x(x - 2)(x - 4)}{(x - 2)^{1/2}} \qquad \begin{array}{l} \text{Factor numerator;} \\ \sqrt{x - 2} = (x - 2)^{1/2}. \end{array}$$

$$= \lim_{x \to 2^+} x(x - 4)(x - 2)^{1/2} \qquad \begin{array}{l} \text{Simplify;} \\ (x - 2)/(x - 2)^{1/2} = (x - 2)^{1/2}. \end{array}$$

$$= \lim_{x \to 2^+} x \cdot \lim_{x \to 2^+} (x - 4) \cdot \lim_{x \to 2^+} (x - 2)^{1/2} \qquad \text{Theorem 2.3}$$

$$= 2(-2)(0) = 0. \qquad \begin{array}{l} \text{Theorem 2.2, Law 7 for} \\ \text{one-sided limits} \end{array}$$

Related Exercises 69–70 ◀

An Important Limit

Despite our success in evaluating limits using direct substitution, algebraic manipulation, and the limit laws, there are important limits for which these techniques do not work. One such limit arises when investigating the slope of a line tangent to the graph of an exponential function.

EXAMPLE 8   Slope of a line tangent to $f(x) = 2^x$ Estimate the slope of the line tangent to the graph of $f(x) = 2^x$ at the point $P(0, 1)$.

SOLUTION In Section 2.1, the slope of a tangent line was obtained by finding the limit of slopes of secant lines; the same strategy is employed here. We begin by selecting a point Q near P on the graph of f with coordinates $(x, 2^x)$. The secant line joining the points $P(0, 1)$ and $Q(x, 2^x)$ is an approximation to the tangent line. To estimate the slope of the tangent line (denoted m_{tan}) at $x = 0$, we compute the slope of the secant line $m_{\text{sec}} = (2^x - 1)/x$ and then let x approach 0.

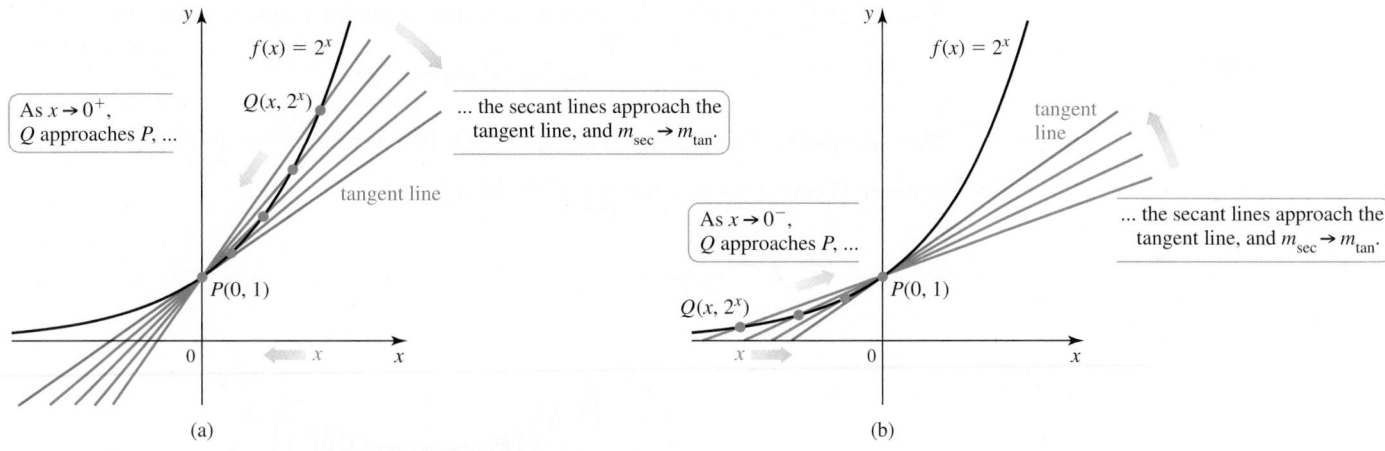

(a)

(b)

Figure 2.19

The limit $\lim\limits_{x \to 0} \dfrac{2^x - 1}{x}$ exists only if it has the same value as $x \to 0^+$ (**Figure 2.19a**) and as $x \to 0^-$ (**Figure 2.19b**). Because it is not an elementary limit, it cannot be evaluated using the limit laws introduced in this section. Instead, we investigate the limit using numerical evidence. Choosing positive values of x near 0 results in Table 2.5.

▶ Example 8 shows that
$\lim\limits_{x \to 0} \dfrac{2^x - 1}{x} \approx 0.693$. The exact value of the limit is ln 2. The connection between the natural logarithm and slopes of lines tangent to exponential curves is made clear in Chapters 3 and 7.

Table 2.5

x	1.0	0.1	0.01	0.001	0.0001	0.00001
$m_{sec} = \dfrac{2^x - 1}{x}$	1.000000	0.7177	0.6956	0.6934	0.6932	0.6931

We see that as x approaches 0 from the right, the slopes of the secant lines approach the slope of the tangent line, which is approximately 0.693. A similar calculation (Exercise 95) gives the same approximation for the limit as x approaches 0 from the left.

Because the left-sided and right-sided limits are the same, we conclude that $\lim\limits_{x \to 0} (2^x - 1)/x \approx 0.693$ (Theorem 2.1). Therefore, the slope of the line tangent to $f(x) = 2^x$ at $x = 0$ is approximately 0.693.

Related Exercises 95–96 ◀

The Squeeze Theorem

▶ The Squeeze Theorem is also called the Pinching Theorem or the Sandwich Theorem.

The *Squeeze Theorem* provides another useful method for calculating limits. Suppose the functions f and h have the same limit L at a and assume the function g is trapped between f and h (**Figure 2.20**). The Squeeze Theorem says that g must also have the limit L at a. A proof of this theorem is assigned in Exercise 68 of Section 2.7.

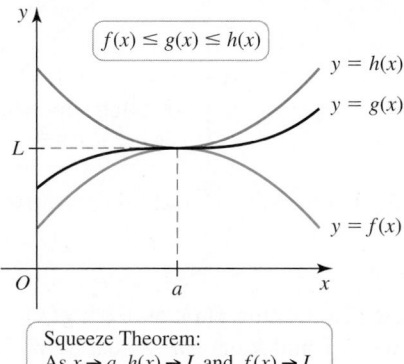

Figure 2.20

THEOREM 2.5   The Squeeze Theorem

Assume the functions f, g, and h satisfy $f(x) \le g(x) \le h(x)$ for all values of x near a, except possibly at a. If $\lim\limits_{x \to a} f(x) = \lim\limits_{x \to a} h(x) = L$, then $\lim\limits_{x \to a} g(x) = L$.

EXAMPLE 9   Applying the Squeeze Theorem Use the Squeeze Theorem to verify that $\lim\limits_{x \to 0} x^2 \sin (1/x) = 0$.

SOLUTION For any real number θ, $-1 \le \sin \theta \le 1$. Letting $\theta = 1/x$ for $x \ne 0$, it follows that

$$-1 \le \sin \frac{1}{x} \le 1.$$

Noting that $x^2 > 0$ for $x \neq 0$, each term in this inequality is multiplied by x^2:

$$-x^2 \leq x^2 \sin \frac{1}{x} \leq x^2.$$

These inequalities are illustrated in **Figure 2.21**. Because $\lim\limits_{x \to 0} x^2 = \lim\limits_{x \to 0} (-x^2) = 0$, the Squeeze Theorem implies that $\lim\limits_{x \to 0} x^2 \sin (1/x) = 0$.

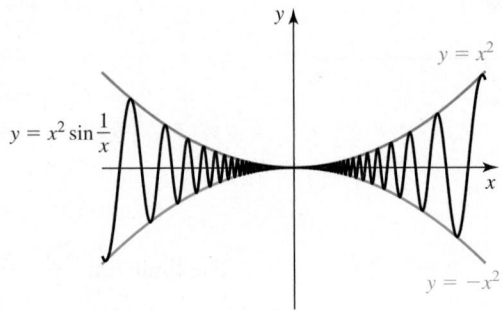

QUICK CHECK 4 Suppose f satisfies $1 \leq f(x) \leq 1 + \dfrac{x^2}{6}$ for all values of x near zero. Find $\lim\limits_{x \to 0} f(x)$, if possible. ◄

Figure 2.21

Related Exercises 81–82 ◄

Trigonometric Limits

The Squeeze Theorem is used to evaluate two important limits that play a crucial role in establishing fundamental properties of the trigonometric functions in Section 2.6. These limits are

$$\lim_{x \to 0} \sin x = 0 \quad \text{and} \quad \lim_{x \to 0} \cos x = 1.$$

To verify these limits, the following inequalities, both valid on $-\pi/2 < x < \pi/2$, are used:

$$-|x| \leq \sin x \leq |x| \text{ (Figure 2.22a)} \quad \text{and} \quad 1 - |x| \leq \cos x \leq 1 \text{ (Figure 2.22b)}.$$

➤ See Exercise 106 for a geometric derivation of the inequalities illustrated in Figure 2.22.

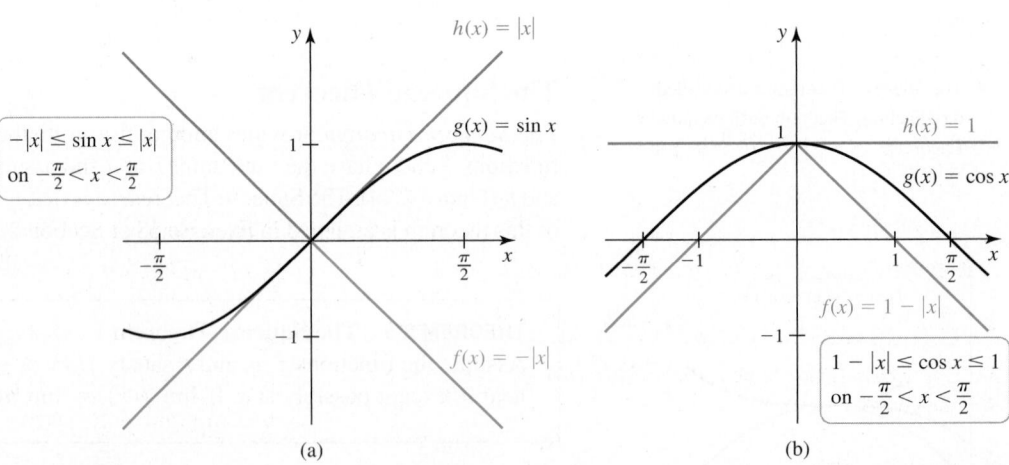

Figure 2.22

Let's begin with the inequality illustrated in Figure 2.22a. Letting $f(x) = -|x|$, $g(x) = \sin x$, and $h(x) = |x|$, we see that g is trapped between f and h on $-\pi/2 < x < \pi/2$. Because $\lim\limits_{x \to 0} f(x) = \lim\limits_{x \to 0} h(x) = 0$ (Exercise 85), the Squeeze Theorem implies that $\lim\limits_{x \to 0} g(x) = \lim\limits_{x \to 0} \sin x = 0$.

To evaluate $\lim\limits_{x\to 0} \cos x$, let $f(x) = 1 - |x|$, $g(x) = \cos x$, and $h(x) = 1$, and notice that g is again trapped between f and h on $-\pi/2 < x < \pi/2$ (Figure 2.22b). Because $\lim\limits_{x\to 0} f(x) = \lim\limits_{x\to 0} h(x) = 1$, the Squeeze Theorem implies that $\lim\limits_{x\to 0} g(x) = \lim\limits_{x\to 0} \cos x = 1$.

Having established that $\lim\limits_{x\to 0} \sin x = 0$ and $\lim\limits_{x\to 0} \cos x = 1$, we can evaluate more complicated limits involving trigonometric functions.

▶ Notice that direct substitution can be used to evaluate the important trigonometric limits just derived. That is, $\lim\limits_{x\to 0} \sin x = \sin 0 = 0$ and $\lim\limits_{x\to 0} \cos x = \cos 0 = 1$. In Section 2.6, these limits are used to show that

$$\lim_{x\to a} \sin x = \sin a \text{ and}$$
$$\lim_{x\to a} \cos x = \cos a.$$

In other words, direct substitution may be used to evaluate limits of the sine and cosine functions for any value of a.

EXAMPLE 10 **Trigonometric limits** Evaluate the following trigonometric limits.

a. $\lim\limits_{x\to 0} \dfrac{\sin^2 x}{1 - \cos x}$ **b.** $\lim\limits_{x\to 0} \dfrac{1 - \cos 2x}{\sin x}$

SOLUTION Direct substitution does not work for either limit because the denominator is zero at $x = 0$ in both parts (a) and (b). Instead, trigonometric identities and limit laws are used to simplify the trigonometric function.

a. The Pythagorean identity $\sin^2 x + \cos^2 x = 1$ is used to simplify the function.

$$
\begin{aligned}
\lim_{x\to 0} \frac{\sin^2 x}{1 - \cos x} &= \lim_{x\to 0} \frac{1 - \cos^2 x}{1 - \cos x} && \sin^2 x = 1 - \cos^2 x \\
&= \lim_{x\to 0} \frac{(1 - \cos x)(1 + \cos x)}{1 - \cos x} && \text{Factor the numerator.} \\
&= \lim_{x\to 0} (1 + \cos x) && \text{Simplify.} \\
&= \underbrace{\lim_{x\to 0} 1}_{1} + \underbrace{\lim_{x\to 0} \cos x}_{1} = 2 && \text{Theorem 2.3 (Law 1)}
\end{aligned}
$$

b. We use the identity $\cos 2x = \cos^2 x - \sin^2 x$ to simplify the function.

$$
\begin{aligned}
\lim_{x\to 0} \frac{1 - \cos 2x}{\sin x} &= \lim_{x\to 0} \frac{1 - (\cos^2 x - \sin^2 x)}{\sin x} && \cos 2x = \cos^2 x - \sin^2 x \\
&= \lim_{x\to 0} \frac{1 - \cos^2 x + \sin^2 x}{\sin x} && \text{Distribute.} \\
&= \lim_{x\to 0} \frac{2 \sin^2 x}{\sin x} && 1 - \cos^2 x = \sin^2 x \\
&= 2 \underbrace{\lim_{x\to 0} \sin x}_{0} = 0 && \text{Simplify; Theorem 2.3 (Law 3).}
\end{aligned}
$$

Related Exercises 60–61 ◀

SECTION 2.3 EXERCISES

Getting Started

1. How is $\lim\limits_{x\to a} p(x)$ calculated if p is a polynomial function?

2. Evaluate $\lim\limits_{x\to 1} (x^3 + 3x^2 - 3x + 1)$.

3. For what values of a does $\lim\limits_{x\to a} r(x) = r(a)$ if r is a rational function?

4. Evaluate $\lim\limits_{x\to 4}\left(\dfrac{x^2 - 4x - 1}{3x - 1}\right)$.

5. Explain why $\lim\limits_{x\to 3} \dfrac{x^2 - 7x + 12}{x - 3} = \lim\limits_{x\to 3} (x - 4)$ and then evaluate $\lim\limits_{x\to 3} \dfrac{x^2 - 7x + 12}{x - 3}$.

6. Evaluate $\lim\limits_{x\to 5}\left(\dfrac{4x^2 - 100}{x - 5}\right)$.

7–13. Assume $\lim\limits_{x\to 1} f(x) = 8$, $\lim\limits_{x\to 1} g(x) = 3$, and $\lim\limits_{x\to 1} h(x) = 2$. Compute the following limits and state the limit laws used to justify your computations.

7. $\lim\limits_{x\to 1} (4f(x))$

8. $\lim\limits_{x\to 1} \dfrac{f(x)}{h(x)}$

9. $\lim\limits_{x\to 1} (f(x) - g(x))$

10. $\lim\limits_{x\to 1} (f(x)h(x))$

11. $\lim\limits_{x\to 1} \dfrac{f(x)}{g(x) - h(x)}$

12. $\lim\limits_{x\to 1} \sqrt[3]{f(x)g(x) + 3}$

13. $\lim\limits_{x\to 1} (f(x))^{2/3}$.

14. How are $\lim\limits_{x\to a^-} p(x)$ and $\lim\limits_{x\to a^+} p(x)$ calculated if p is a polynomial function?

15. Suppose

$$g(x) = \begin{cases} 2x + 1 & \text{if } x \neq 0 \\ 5 & \text{if } x = 0. \end{cases}$$

Compute $g(0)$ and $\lim_{x \to 0} g(x)$.

16. Suppose

$$f(x) = \begin{cases} 4 & \text{if } x \leq 3 \\ x + 2 & \text{if } x > 3. \end{cases}$$

Compute $\lim_{x \to 3^-} f(x)$ and $\lim_{x \to 3^+} f(x)$. Then explain why $\lim_{x \to 3} f(x)$ does not exist.

17. Suppose p and q are polynomials. If $\lim_{x \to 0} \dfrac{p(x)}{q(x)} = 10$ and $q(0) = 2$, find $p(0)$.

18. Suppose $\lim_{x \to 2} f(x) = \lim_{x \to 2} h(x) = 5$. Find $\lim_{x \to 2} g(x)$, where $f(x) \leq g(x) \leq h(x)$, for all x.

Practice Exercises

19–70. Evaluating limits *Find the following limits or state that they do not exist. Assume a, b, c, and k are fixed real numbers.*

19. $\lim_{x \to 4} (3x - 7)$

20. $\lim_{x \to 1} (-2x + 5)$

21. $\lim_{x \to -9} 5x$

22. $\lim_{x \to 6} 4$

23. $\lim_{x \to 1} (2x^3 - 3x^2 + 4x + 5)$

24. $\lim_{t \to -2} (t^2 + 5t + 7)$

25. $\lim_{x \to 1} \dfrac{5x^2 + 6x + 1}{8x - 4}$

26. $\lim_{t \to 3} \sqrt[3]{t^2 - 10}$

27. $\lim_{p \to 2} \dfrac{3p}{\sqrt{4p + 1} - 1}$

28. $\lim_{x \to 2} (x^2 - x)^5$

29. $\lim_{x \to 3} \dfrac{-5x}{\sqrt{4x - 3}}$

30. $\lim_{h \to 0} \dfrac{3}{\sqrt{16 + 3h} + 4}$

31. $\lim_{x \to 2} (5x - 6)^{3/2}$

32. $\lim_{h \to 0} \dfrac{100}{(10h - 1)^{11} + 2}$

33. $\lim_{x \to 1} \dfrac{x^2 - 1}{x - 1}$

34. $\lim_{x \to 3} \dfrac{x^2 - 2x - 3}{x - 3}$

35. $\lim_{x \to 4} \dfrac{x^2 - 16}{4 - x}$

36. $\lim_{t \to 2} \dfrac{3t^2 - 7t + 2}{2 - t}$

37. $\lim_{x \to b} \dfrac{(x - b)^{50} - x + b}{x - b}$

38. $\lim_{x \to -b} \dfrac{(x + b)^7 + (x + b)^{10}}{4(x + b)}$

39. $\lim_{x \to -1} \dfrac{(2x - 1)^2 - 9}{x + 1}$

40. $\lim_{h \to 0} \dfrac{\dfrac{1}{5 + h} - \dfrac{1}{5}}{h}$

41. $\lim_{x \to 9} \dfrac{\sqrt{x} - 3}{x - 9}$

42. $\lim_{w \to 1} \left(\dfrac{1}{w^2 - w} - \dfrac{1}{w - 1} \right)$

43. $\lim_{t \to 5} \left(\dfrac{1}{t^2 - 4t - 5} - \dfrac{1}{6(t - 5)} \right)$

44. $\lim_{t \to 3} \left(\left(4t - \dfrac{2}{t - 3} \right) (6 + t - t^2) \right)$

45. $\lim_{x \to a} \dfrac{x - a}{\sqrt{x} - \sqrt{a}}, a > 0$

46. $\lim_{x \to a} \dfrac{x^2 - a^2}{\sqrt{x} - \sqrt{a}}, a > 0$

47. $\lim_{h \to 0} \dfrac{\sqrt{16 + h} - 4}{h}$

48. $\lim_{x \to c} \dfrac{x^2 - 2cx + c^2}{x - c}$

49. $\lim_{x \to 4} \dfrac{\dfrac{1}{x} - \dfrac{1}{4}}{x - 4}$

50. $\lim_{x \to 3} \dfrac{\dfrac{1}{x^2 + 2x} - \dfrac{1}{15}}{x - 3}$

51. $\lim_{x \to 1} \dfrac{\sqrt{10x - 9} - 1}{x - 1}$

52. $\lim_{x \to 2} \left(\dfrac{1}{x - 2} - \dfrac{2}{x^2 - 2x} \right)$

53. $\lim_{h \to 0} \dfrac{(5 + h)^2 - 25}{h}$

54. $\lim_{w \to -k} \dfrac{w^2 + 5kw + 4k^2}{w^2 + kw}, k \neq 0$

55. $\lim_{x \to 1} \dfrac{x - 1}{\sqrt{x} - 1}$

56. $\lim_{x \to 1} \dfrac{x - 1}{\sqrt{4x + 5} - 3}$

57. $\lim_{x \to 4} \dfrac{3(x - 4)\sqrt{x + 5}}{3 - \sqrt{x + 5}}$

58. $\lim_{x \to 0} \dfrac{x}{\sqrt{cx + 1} - 1}, c \neq 0$

59. $\lim_{x \to 0} x \cos x$

60. $\lim_{x \to 0} \dfrac{\sin 2x}{\sin x}$

61. $\lim_{x \to 0} \dfrac{1 - \cos x}{\cos^2 x - 3 \cos x + 2}$

62. $\lim_{x \to 0} \dfrac{\cos x - 1}{\cos^2 x - 1}$

63. $\lim_{x \to 0^-} \dfrac{x^2 - x}{|x|}$

64. $\lim_{w \to 3^-} \dfrac{|w - 3|}{w^2 - 7w + 12}$

65. $\lim_{t \to 2^+} \dfrac{|2t - 4|}{t^2 - 4}$

66. $\lim_{x \to -1} g(x)$, where $g(x) = \begin{cases} \dfrac{x^2 - 1}{x + 1} & \text{if } x < -1 \\ -2 & \text{if } x \geq -1 \end{cases}$

67. $\lim_{x \to 3} \dfrac{x - 3}{|x - 3|}$

68. $\lim_{x \to 5} \dfrac{|x - 5|}{x^2 - 25}$

69. $\lim_{x \to 1^-} \dfrac{x^3 + 1}{\sqrt{x - 1}}$

70. $\lim_{x \to 1^+} \dfrac{x - 1}{\sqrt{x^2 - 1}}$

71. Explain why or why not Determine whether the following statements are true and give an explanation or counterexample. Assume a and L are finite numbers.

a. If $\lim_{x \to a} f(x) = L$, then $f(a) = L$.

b. If $\lim_{x \to a^-} f(x) = L$, then $\lim_{x \to a^+} f(x) = L$.

c. If $\lim_{x \to a} f(x) = L$ and $\lim_{x \to a} g(x) = L$, then $f(a) = g(a)$.

d. The limit $\lim_{x \to a} \dfrac{f(x)}{g(x)}$ does not exist if $g(a) = 0$.

e. If $\lim_{x \to 1^+} \sqrt{f(x)} = \sqrt{\lim_{x \to 1^+} f(x)}$, it follows that

$$\lim_{x \to 1} \sqrt{f(x)} = \sqrt{\lim_{x \to 1} f(x)}.$$

72. One-sided limits Let

$$g(x) = \begin{cases} 5x - 15 & \text{if } x < 4 \\ \sqrt{6x + 1} & \text{if } x \geq 4. \end{cases}$$

Compute the following limits or state that they do not exist.

a. $\lim\limits_{x \to 4^-} g(x)$ **b.** $\lim\limits_{x \to 4^+} g(x)$ **c.** $\lim\limits_{x \to 4} g(x)$

73. One-sided limits Let

$$f(x) = \begin{cases} x^2 + 1 & \text{if } x < -1 \\ \sqrt{x + 1} & \text{if } x \geq -1. \end{cases}$$

Compute the following limits or state that they do not exist.

a. $\lim\limits_{x \to -1^-} f(x)$ **b.** $\lim\limits_{x \to -1^+} f(x)$ **c.** $\lim\limits_{x \to -1} f(x)$

74. One-sided limits Let

$$f(x) = \begin{cases} 0 & \text{if } x \leq -5 \\ \sqrt{25 - x^2} & \text{if } -5 < x < 5 \\ 3x & \text{if } x \geq 5. \end{cases}$$

Compute the following limits or state that they do not exist.

a. $\lim\limits_{x \to -5^-} f(x)$ **b.** $\lim\limits_{x \to -5^+} f(x)$ **c.** $\lim\limits_{x \to -5} f(x)$
d. $\lim\limits_{x \to 5^-} f(x)$ **e.** $\lim\limits_{x \to 5^+} f(x)$ **f.** $\lim\limits_{x \to 5} f(x)$

75. One-sided limits

a. Evaluate $\lim\limits_{x \to 2^+} \sqrt{x - 2}$.
b. Explain why $\lim\limits_{x \to 2^-} \sqrt{x - 2}$ does not exist.

76. One-sided limits

a. Evaluate $\lim\limits_{x \to 3^-} \sqrt{\dfrac{x - 3}{2 - x}}$.
b. Explain why $\lim\limits_{x \to 3^+} \sqrt{\dfrac{x - 3}{2 - x}}$ does not exist.

T 77. Electric field The magnitude of the electric field at a point x meters from the midpoint of a 0.1-m line of charge is given by

$$E(x) = \frac{4.35}{x\sqrt{x^2 + 0.01}} \text{ (in units of newtons per coulomb, N/C).}$$

Evaluate $\lim\limits_{x \to 10} E(x)$.

78. Torricelli's law A cylindrical tank is filled with water to a depth of 9 meters. At $t = 0$, a drain in the bottom of the tank is opened, and water flows out of the tank. The depth of water in the tank (measured from the bottom of the tank) t seconds after the drain is opened is approximated by $d(t) = (3 - 0.015t)^2$, for $0 \leq t \leq 200$. Evaluate and interpret $\lim\limits_{t \to 200^-} d(t)$.

79. Limit of the radius of a cylinder A right circular cylinder with a height of 10 cm and a surface area of S cm^2 has a radius given by

$$r(S) = \frac{1}{2}\left(\sqrt{100 + \frac{2S}{\pi}} - 10\right).$$

Find $\lim\limits_{S \to 0^+} r(S)$ and interpret your result.

80. A problem from relativity theory Suppose a spaceship of length L_0 travels at a high speed v relative to an observer. To the observer, the ship appears to have a smaller length given by the *Lorentz contraction formula*

$$L = L_0\sqrt{1 - \frac{v^2}{c^2}},$$

where c is the speed of light.

a. What is the observed length L of the ship if it is traveling at 50% of the speed of light?
b. What is the observed length L of the ship if it is traveling at 75% of the speed of light?
c. In parts (a) and (b), what happens to L as the speed of the ship increases?
d. Find $\lim\limits_{v \to c^-} L_0\sqrt{1 - \dfrac{v^2}{c^2}}$ and explain the significance of this limit.

T 81. a. Show that $-|x| \leq x \sin\dfrac{1}{x} \leq |x|$, for $x \neq 0$.

b. Illustrate the inequalities in part (a) with a graph.

c. Show that $\lim\limits_{x \to 0} x \sin\dfrac{1}{x} = 0$.

T 82. A cosine limit It can be shown that $1 - \dfrac{x^2}{2} \leq \cos x \leq 1$, for x near 0.

a. Illustrate these inequalities with a graph.
b. Use these inequalities to evaluate $\lim\limits_{x \to 0} \cos x$.

T 83. A sine limit It can be shown that $1 - \dfrac{x^2}{6} \leq \dfrac{\sin x}{x} \leq 1$, for x near 0.

a. Illustrate these inequalities with a graph.
b. Use these inequalities to evaluate $\lim\limits_{x \to 0} \dfrac{\sin x}{x}$.

T 84. A logarithm limit

a. Draw a graph to verify that $-|x| \leq x^2 \ln x^2 \leq |x|$, for $-1 \leq x \leq 1$, where $x \neq 0$.
b. Use the inequality in part (a) to evaluate $\lim\limits_{x \to 0} x^2 \ln x^2$.

85. Absolute value Show that $\lim\limits_{x \to 0} |x| = 0$ by first evaluating $\lim\limits_{x \to 0^-} |x|$ and $\lim\limits_{x \to 0^+} |x|$. Recall that

$$|x| = \begin{cases} x & \text{if } x \geq 0 \\ -x & \text{if } x < 0. \end{cases}$$

86. Absolute value limit Show that $\lim\limits_{x \to a} |x| = |a|$, for any real number a. (*Hint:* Consider the cases $a < 0$, $a = 0$, and $a > 0$.)

87. Finding a constant Suppose

$$f(x) = \begin{cases} \dfrac{x^2 - 5x + 6}{x - 3} & \text{if } x \neq 3 \\ a & \text{if } x = 3. \end{cases}$$

Determine a value of the constant a for which $\lim\limits_{x \to 3} f(x) = f(3)$.

88. Finding a constant Suppose

$$f(x) = \begin{cases} 3x + b & \text{if } x \leq 2 \\ x - 2 & \text{if } x > 2. \end{cases}$$

Determine a value of the constant b for which $\lim\limits_{x \to 2} f(x)$ exists and state the value of the limit, if possible.

89. Finding a constant Suppose

$$g(x) = \begin{cases} x^2 - 5x & \text{if } x \leq -1 \\ ax^3 - 7 & \text{if } x > -1. \end{cases}$$

Determine a value of the constant a for which $\lim\limits_{x \to -1} g(x)$ exists and state the value of the limit, if possible.

90–94. Useful factorization formula *Calculate the following limits using the factorization formula*

$$x^n - a^n = (x - a)(x^{n-1} + ax^{n-2} + a^2 x^{n-3} + \cdots + a^{n-2}x + a^{n-1}),$$

where n is a positive integer and a is a real number.

90. $\lim\limits_{x \to 2} \dfrac{x^5 - 32}{x - 2}$

91. $\lim\limits_{x \to 1} \dfrac{x^6 - 1}{x - 1}$

92. $\lim\limits_{x \to -1} \dfrac{x^7 + 1}{x + 1}$ (*Hint:* Use the formula for $x^7 - a^7$ with $a = -1$.)

93. $\lim\limits_{x \to a} \dfrac{x^5 - a^5}{x - a}$

94. $\lim\limits_{x \to a} \dfrac{x^n - a^n}{x - a}$, for any positive integer n

T 95. Slope of a tangent line

  a. Sketch a graph of $y = 2^x$ and carefully draw three secant lines connecting the points $P(0, 1)$ and $Q(x, 2^x)$, for $x = -3, -2$, and -1.
  b. Find the slope of the line that passes through $P(0, 1)$ and $Q(x, 2^x)$, for $x \neq 0$.
  c. Complete the table and make a conjecture about the value of
  $$\lim\limits_{x \to 0^-} \frac{2^x - 1}{x}.$$

x	-1	-0.1	-0.01	-0.001	-0.0001	-0.00001
$\dfrac{2^x - 1}{x}$						

T 96. Slope of a tangent line

  a. Sketch a graph of $y = 3^x$ and carefully draw four secant lines connecting the points $P(0, 1)$ and $Q(x, 3^x)$, for $x = -2, -1, 1$, and 2.
  b. Find the slope of the line that passes through $P(0, 1)$ and $Q(x, 3^x)$, for $x \neq 0$.
  c. Complete the table and make a conjecture about the value of
  $$\lim\limits_{x \to 0} \frac{3^x - 1}{x}.$$

x	-0.1	-0.01	-0.001	-0.0001	0.0001	0.001	0.01	0.1
$\dfrac{3^x - 1}{x}$								

Explorations and Challenges

97. Even function limits Suppose f is an even function where $\lim\limits_{x \to 1^-} f(x) = 5$ and $\lim\limits_{x \to 1^+} f(x) = 6$. Find $\lim\limits_{x \to -1^-} f(x)$ and $\lim\limits_{x \to -1^+} f(x)$.

98. Odd function limits Suppose g is an odd function where $\lim\limits_{x \to 1^-} g(x) = 5$ and $\lim\limits_{x \to 1^+} g(x) = 6$. Find $\lim\limits_{x \to -1^-} g(x)$ and $\lim\limits_{x \to -1^+} g(x)$.

99. Evaluate $\lim\limits_{x \to 1} \dfrac{\sqrt[3]{x} - 1}{x - 1}$. (*Hint:* $x - 1 = \left(\sqrt[3]{x}\right)^3 - 1^3$.)

100. Evaluate $\lim\limits_{x \to 16} \dfrac{\sqrt[4]{x} - 2}{x - 16}$.

101. Creating functions satisfying given limit conditions Find functions f and g such that $\lim\limits_{x \to 1} f(x) = 0$ and $\lim\limits_{x \to 1} (f(x)g(x)) = 5$.

102. Creating functions satisfying given limit conditions Find a function f satisfying $\lim\limits_{x \to 1} \dfrac{f(x)}{x - 1} = 2$.

103. Finding constants Find constants b and c in the polynomial $p(x) = x^2 + bx + c$ such that $\lim\limits_{x \to 2} \dfrac{p(x)}{x - 2} = 6$. Are the constants unique?

104. If $\lim\limits_{x \to 1} f(x) = 4$, find $\lim\limits_{x \to -1} f(x^2)$.

105. Suppose $g(x) = f(1 - x)$ for all x, $\lim\limits_{x \to 1^+} f(x) = 4$, and $\lim\limits_{x \to 1^-} f(x) = 6$. Find $\lim\limits_{x \to 0^+} g(x)$ and $\lim\limits_{x \to 0^-} g(x)$.

106. Two trigonometric inequalities Consider the angle θ in standard position in a unit circle where $0 \leq \theta < \pi/2$ or $-\pi/2 < \theta < 0$ (use both figures).

  a. Show that $|AC| = |\sin \theta|$, for $-\pi/2 < \theta < \pi/2$. (*Hint:* Consider the cases $0 < \theta < \pi/2$ and $-\pi/2 < \theta < 0$ separately.)
  b. Show that $|\sin \theta| < |\theta|$, for $-\pi/2 < \theta < \pi/2$. (*Hint:* The length of arc AB is θ if $0 \leq \theta < \pi/2$, and is $-\theta$ if $-\pi/2 < \theta < 0$.)
  c. Conclude that $-|\theta| \leq \sin \theta \leq |\theta|$, for $-\pi/2 < \theta < \pi/2$.
  d. Show that $0 \leq 1 - \cos \theta \leq |\theta|$, for $-\pi/2 < \theta < \pi/2$.
  e. Show that $1 - |\theta| \leq \cos \theta \leq 1$, for $-\pi/2 < \theta < \pi/2$.

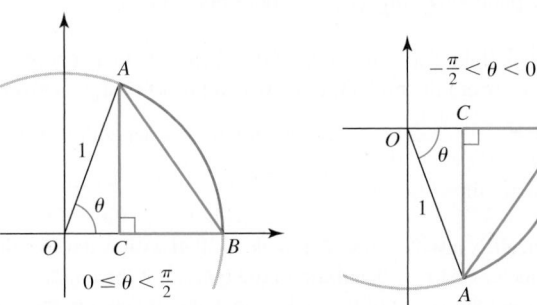

107. Theorem 2.4a Given the polynomial
$$p(x) = b_n x^n + b_{n-1} x^{n-1} + \cdots + b_1 x + b_0,$$
prove that $\lim\limits_{x \to a} p(x) = p(a)$ for any value of a.

QUICK CHECK ANSWERS

1. $0, 2$ **2.** 2 **3.** 3 **4.** 1 ◄

2.4 Infinite Limits

Two more limit scenarios are frequently encountered in calculus and are discussed in this section and the following section. An *infinite limit* occurs when function values increase or decrease without bound near a point. The other type of limit, known as a *limit at infinity,* occurs when the independent variable x increases or decreases without bound. The ideas behind infinite limits and limits at infinity are quite different. Therefore, it is important to distinguish between these limits and the methods used to calculate them.

An Overview

To illustrate the differences between infinite limits and limits at infinity, consider the values of $f(x) = 1/x^2$ in Table 2.6. As x approaches 0 from either side, $f(x)$ grows larger and larger. Because $f(x)$ does not approach a finite number as x approaches 0, $\lim\limits_{x\to 0} f(x)$ does not exist. Nevertheless, we use limit notation and write $\lim\limits_{x\to 0} f(x) = \infty$. The infinity symbol indicates that $f(x)$ grows arbitrarily large as x approaches 0. This is an example of an *infinite limit*; in general, the *dependent variable* becomes arbitrarily large in magnitude as the *independent variable* approaches a finite number.

Table 2.6

x	$f(x) = 1/x^2$
± 0.1	100
± 0.01	10,000
± 0.001	1,000,000
$\downarrow$	$\downarrow$
0	∞

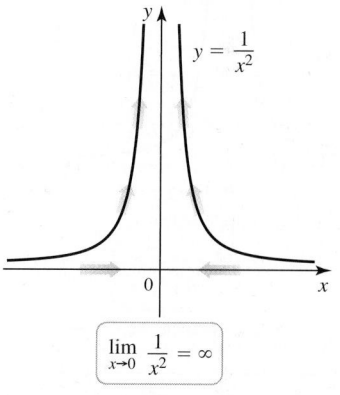

$$\lim_{x\to 0} \frac{1}{x^2} = \infty$$

With *limits at infinity*, the opposite occurs: The *dependent variable* approaches a finite number as the *independent variable* becomes arbitrarily large in magnitude. In Table 2.7 we see that $f(x) = 1/x^2$ approaches 0 as x increases. In this case, we write $\lim\limits_{x\to\infty} f(x) = 0$.

Table 2.7

x	$f(x) = 1/x^2$
10	0.01
100	0.0001
1000	0.000001
$\downarrow$	$\downarrow$
∞	0

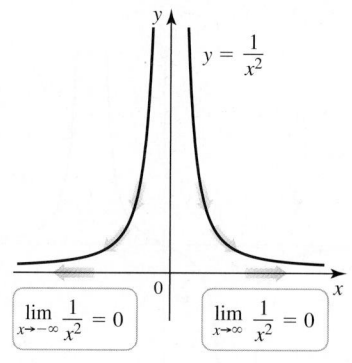

$$\lim_{x\to -\infty} \frac{1}{x^2} = 0 \qquad \lim_{x\to\infty} \frac{1}{x^2} = 0$$

A general picture of these two limit scenarios—occurring with the same function—is shown in **Figure 2.23**.

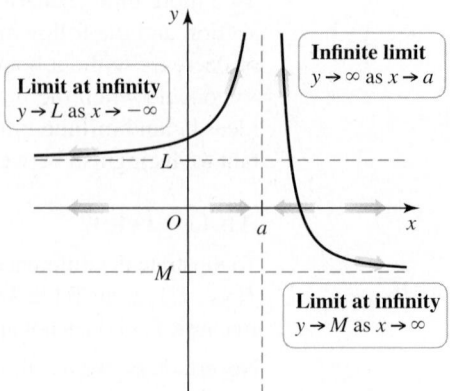

Figure 2.23

Infinite Limits

The following definition of infinite limits is informal, but it is adequate for most functions encountered in this book. A precise definition is given in Section 2.7.

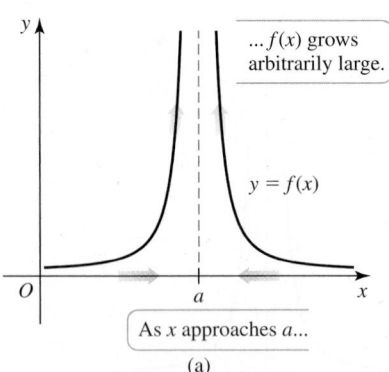

(a)

(b)

Figure 2.24

DEFINITION Infinite Limits

Suppose f is defined for all x near a. If $f(x)$ grows arbitrarily large for all x sufficiently close (but not equal) to a (**Figure 2.24a**), we write

$$\lim_{x \to a} f(x) = \infty$$

and say the limit of $f(x)$ as x approaches a is infinity.

   If $f(x)$ is negative and grows arbitrarily large in magnitude for all x sufficiently close (but not equal) to a (**Figure 2.24b**), we write

$$\lim_{x \to a} f(x) = -\infty$$

and say the limit of $f(x)$ as x approaches a is negative infinity. *In both cases, the limit does not exist.*

EXAMPLE 1   Infinite limits Analyze $\displaystyle\lim_{x \to 1} \frac{x}{\left(x^2 - 1\right)^2}$ and $\displaystyle\lim_{x \to -1} \frac{x}{\left(x^2 - 1\right)^2}$ using the graph of the function.

SOLUTION The graph of $f(x) = \dfrac{x}{\left(x^2 - 1\right)^2}$ (**Figure 2.25**) shows that as x approaches 1 (from either side), the values of f grow arbitrarily large. Therefore, the limit does not exist and we write

$$\lim_{x \to 1} \frac{x}{\left(x^2 - 1\right)^2} = \infty.$$

As x approaches -1, the values of f are negative and grow arbitrarily large in magnitude; therefore,

$$\lim_{x \to -1} \frac{x}{\left(x^2 - 1\right)^2} = -\infty.$$

Related Exercise 6 ◄

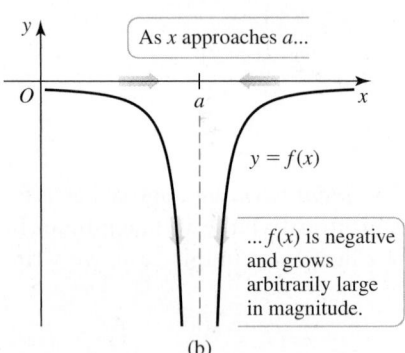

Figure 2.25

   Example 1 illustrates *two-sided* infinite limits. As with finite limits, we also need to work with right-sided and left-sided infinite limits.

As $x \to -4^+$, we find that

$$\lim_{x \to -4^+} \frac{-x^3 + 5x^2 - 6x}{-x^3 - 4x^2} = \lim_{x \to -4^+} \frac{\overbrace{(x-2)(x-3)}^{\text{approaches 42}}}{\underbrace{x(x+4)}_{\substack{\text{negative and} \\ \text{approaches 0}}}} = -\infty.$$

This limit implies that the given function has a vertical asymptote at $x = -4$.

Related Exercises 28, 31 ◀

QUICK CHECK 3 Verify that $x(x+4) \to 0$ through negative values as $x \to -4^+$. ◀

EXAMPLE 5 Location of vertical asymptotes Let $f(x) = \dfrac{x^2 - 4x + 3}{x^2 - 1}$. Determine the following limits and find the vertical asymptotes of f. Verify your work with a graphing utility.

a. $\displaystyle\lim_{x \to 1} f(x)$ **b.** $\displaystyle\lim_{x \to -1^-} f(x)$ **c.** $\displaystyle\lim_{x \to -1^+} f(x)$

SOLUTION

➤ Example 5 illustrates that $f(x)/g(x)$ might not grow arbitrarily large in magnitude if *both* $f(x)$ and $g(x)$ approach 0. Such limits are called *indeterminate forms*; they are examined in detail in Section 4.7.

a. Notice that as $x \to 1$, both the numerator and denominator of f approach 0, and the function is undefined at $x = 1$. To compute $\displaystyle\lim_{x \to 1} f(x)$, we first factor:

$$\lim_{x \to 1} f(x) = \lim_{x \to 1} \frac{x^2 - 4x + 3}{x^2 - 1}$$

➤ It is permissible to cancel the factor $x - 1$ in $\displaystyle\lim_{x \to 1} \frac{(x-1)(x-3)}{(x-1)(x+1)}$ because x approaches 1 but is not equal to 1. Therefore, $x - 1 \neq 0$.

$$= \lim_{x \to 1} \frac{(x-1)(x-3)}{(x-1)(x+1)} \qquad \text{Factor.}$$

$$= \lim_{x \to 1} \frac{(x-3)}{(x+1)} \qquad \text{Cancel like factors, } x \neq 1.$$

$$= \frac{1-3}{1+1} = -1. \qquad \text{Substitute } x = 1.$$

Therefore, $\displaystyle\lim_{x \to 1} f(x) = -1$ (even though $f(1)$ is undefined). The line $x = 1$ is *not* a vertical asymptote of f.

b. In part (a) we showed that

$$f(x) = \frac{x^2 - 4x + 3}{x^2 - 1} = \frac{x - 3}{x + 1}, \qquad \text{provided } x \neq 1.$$

We use this fact again. As x approaches -1 from the left, the one-sided limit is

$$\lim_{x \to -1^-} f(x) = \lim_{x \to -1^-} \frac{\overbrace{x - 3}^{\text{approaches } -4}}{\underbrace{x + 1}_{\substack{\text{negative and} \\ \text{approaches 0}}}} = \infty.$$

c. As x approaches -1 from the right, the one-sided limit is

$$\lim_{x \to -1^+} f(x) = \lim_{x \to -1^+} \frac{\overbrace{x - 3}^{\text{approaches } -4}}{\underbrace{x + 1}_{\substack{\text{positive and} \\ \text{approaches 0}}}} = -\infty.$$

The infinite limits $\displaystyle\lim_{x \to -1^+} f(x) = -\infty$ and $\displaystyle\lim_{x \to -1^-} f(x) = \infty$ each imply that the line $x = -1$ is a vertical asymptote of f. The graph of f generated by a graphing utility *may* appear as shown in **Figure 2.28a**. If so, two corrections must be made. A hole should appear in the graph at $(1, -1)$ because $\lim\limits_{x \to 1} f(x) = -1$, but $f(1)$ is undefined.

It is also a good idea to replace the solid vertical line with a dashed line to emphasize that the vertical asymptote is not a part of the graph of f (**Figure 2.28b**).

> ➤ Graphing utilities vary in how they display vertical asymptotes. The errors shown in Figure 2.28a do not occur on all graphing utilities.

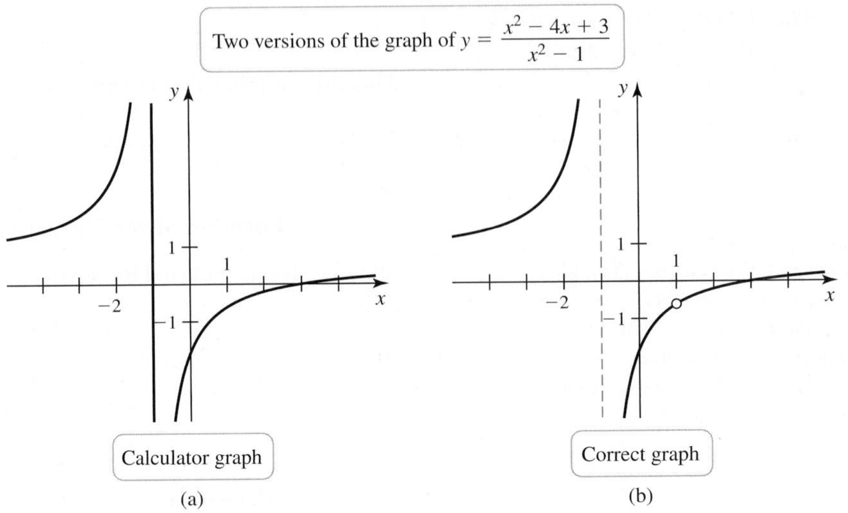

Two versions of the graph of $y = \dfrac{x^2 - 4x + 3}{x^2 - 1}$

Calculator graph

(a)

Correct graph

(b)

Figure 2.28

QUICK CHECK 4 The line $x = 2$ is not a vertical asymptote of $y = \dfrac{(x - 1)(x - 2)}{x - 2}$. Why not? ◄

Related Exercises 45–46 ◄

EXAMPLE 6 Limits of trigonometric functions Analyze the following limits.

a. $\lim\limits_{\theta \to 0^+} \cot \theta$ **b.** $\lim\limits_{\theta \to 0^-} \cot \theta$

SOLUTION

a. Recall that $\cot \theta = \cos \theta / \sin \theta$. Furthermore, it is shown in Section 2.3 that $\lim\limits_{\theta \to 0^+} \cos \theta = 1$ and that $\sin \theta$ is positive and approaches 0 as $\theta \to 0^+$. Therefore, as $\theta \to 0^+$, $\cot \theta$ becomes arbitrarily large and positive, which means $\lim\limits_{\theta \to 0^+} \cot \theta = \infty$. This limit is consistent with the graph of $\cot \theta$ (**Figure 2.29**), which has a vertical asymptote at $\theta = 0$.

b. In this case, $\lim\limits_{\theta \to 0^-} \cos \theta = 1$ and as $\theta \to 0^-$, $\sin \theta \to 0$ with $\sin \theta < 0$. Therefore, as $\theta \to 0^-$, $\cot \theta$ is negative and becomes arbitrarily large in magnitude. It follows that $\lim\limits_{\theta \to 0^-} \cot \theta = -\infty$, as confirmed by the graph of $\cot \theta$.

Related Exercises 39–40 ◄

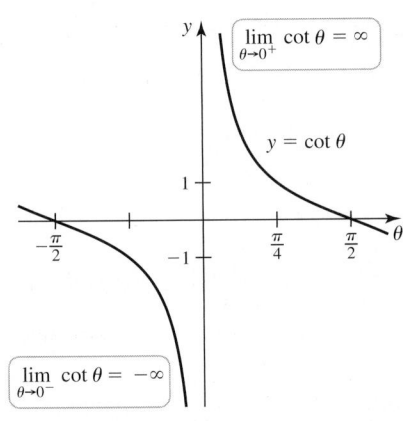

$\lim\limits_{\theta \to 0^+} \cot \theta = \infty$

$y = \cot \theta$

$\lim\limits_{\theta \to 0^-} \cot \theta = -\infty$

Figure 2.29

SECTION 2.4 EXERCISES

Getting Started

1. Explain the meaning of $\lim\limits_{x \to a^+} f(x) = -\infty$.

2. Explain the meaning of $\lim\limits_{x \to a} f(x) = \infty$.

3. What is a vertical asymptote?

4. Consider the function $F(x) = f(x)/g(x)$ with $g(a) = 0$. Does F necessarily have a vertical asymptote at $x = a$? Explain your reasoning.

T 5. Compute the values of $f(x) = \dfrac{x + 1}{(x - 1)^2}$ in the following table and use them to determine $\lim\limits_{x \to 1} f(x)$.

x	$\dfrac{x + 1}{(x - 1)^2}$	x	$\dfrac{x + 1}{(x - 1)^2}$
1.1		0.9	
1.01		0.99	
1.001		0.999	
1.0001		0.9999	

6. Use the graph of $f(x) = \dfrac{x}{(x^2 - 2x - 3)^2}$ to determine $\lim\limits_{x \to -1} f(x)$ and $\lim\limits_{x \to 3} f(x)$.

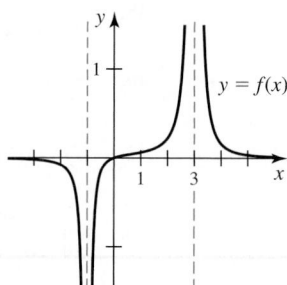

7. The graph of f in the figure has vertical asymptotes at $x = 1$ and $x = 2$. Analyze the following limits.

a. $\lim\limits_{x \to 1^-} f(x)$ **b.** $\lim\limits_{x \to 1^+} f(x)$ **c.** $\lim\limits_{x \to 1} f(x)$

d. $\lim\limits_{x \to 2^-} f(x)$ **e.** $\lim\limits_{x \to 2^+} f(x)$ **f.** $\lim\limits_{x \to 2} f(x)$

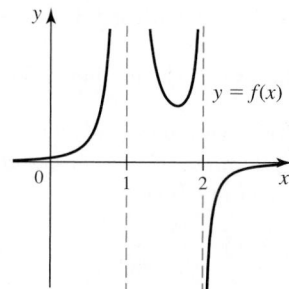

8. The graph of g in the figure has vertical asymptotes at $x = 2$ and $x = 4$. Analyze the following limits.

a. $\lim\limits_{x \to 2^-} g(x)$ **b.** $\lim\limits_{x \to 2^+} g(x)$ **c.** $\lim\limits_{x \to 2} g(x)$

d. $\lim\limits_{x \to 4^-} g(x)$ **e.** $\lim\limits_{x \to 4^+} g(x)$ **f.** $\lim\limits_{x \to 4} g(x)$

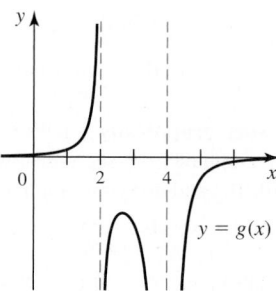

9. The graph of h in the figure has vertical asymptotes at $x = -2$ and $x = 3$. Analyze the following limits.

a. $\lim\limits_{x \to -2^-} h(x)$ **b.** $\lim\limits_{x \to -2^+} h(x)$ **c.** $\lim\limits_{x \to -2} h(x)$

d. $\lim\limits_{x \to 3^-} h(x)$ **e.** $\lim\limits_{x \to 3^+} h(x)$ **f.** $\lim\limits_{x \to 3} h(x)$

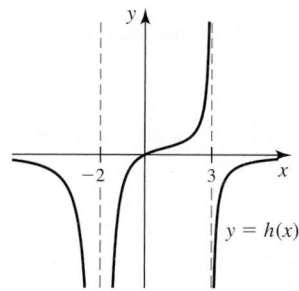

10. The graph of p in the figure has vertical asymptotes at $x = -2$ and $x = 3$. Analyze the following limits.

a. $\lim\limits_{x \to -2^-} p(x)$ **b.** $\lim\limits_{x \to -2^+} p(x)$ **c.** $\lim\limits_{x \to -2} p(x)$

d. $\lim\limits_{x \to 3^-} p(x)$ **e.** $\lim\limits_{x \to 3^+} p(x)$ **f.** $\lim\limits_{x \to 3} p(x)$

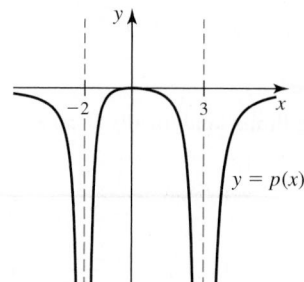

T 11. Graph the function $f(x) = \dfrac{1}{x^2 - x}$ using a graphing utility with the window $[-1, 2] \times [-10, 10]$. Use your graph to determine the following limits.

a. $\lim\limits_{x \to 0^-} f(x)$ **b.** $\lim\limits_{x \to 0^+} f(x)$ **c.** $\lim\limits_{x \to 1^-} f(x)$ **d.** $\lim\limits_{x \to 1^+} f(x)$

T 12. Graph the function $f(x) = \dfrac{e^{-x}}{x(x + 2)^2}$ using a graphing utility. (Experiment with your choice of a graphing window.) Use your graph to determine the following limits.

a. $\lim\limits_{x \to -2^+} f(x)$ **b.** $\lim\limits_{x \to -2} f(x)$ **c.** $\lim\limits_{x \to 0^-} f(x)$ **d.** $\lim\limits_{x \to 0^+} f(x)$

13. Suppose $f(x) \to 100$ and $g(x) \to 0$, with $g(x) < 0$ as $x \to 2$. Determine $\lim\limits_{x \to 2} \dfrac{f(x)}{g(x)}$.

14. Evaluate $\lim\limits_{x \to 3^-} \dfrac{1}{x - 3}$ and $\lim\limits_{x \to 3^+} \dfrac{1}{x - 3}$.

15. Verify that the function $f(x) = \dfrac{x^2 - 4x + 3}{x^2 - 3x + 2}$ is undefined at $x = 1$ and at $x = 2$. Does the graph of f have vertical asymptotes at both these values of x? Explain.

16. Evaluate $\lim\limits_{x \to 0} \dfrac{x + 1}{1 - \cos x}$.

17. Sketch a possible graph of a function f, together with vertical asymptotes, satisfying all the following conditions on $[0, 4]$.

$f(1) = 0, \quad f(3)$ is undefined, $\quad \lim\limits_{x \to 3} f(x) = 1,$

$\lim\limits_{x \to 0^+} f(x) = -\infty, \quad \lim\limits_{x \to 2} f(x) = \infty, \quad \lim\limits_{x \to 4^-} f(x) = \infty$

18. Sketch a possible graph of a function g, together with vertical asymptotes, satisfying all the following conditions.

$g(2) = 1, \quad g(5) = -1, \quad \lim\limits_{x \to 4} g(x) = -\infty,$

$\lim\limits_{x \to 7^-} g(x) = \infty, \quad \lim\limits_{x \to 7^+} g(x) = -\infty$

19. Which of the following statements are correct? Choose all that apply.

a. $\lim\limits_{x \to 1} \dfrac{1}{(x - 1)^2}$ does not exist   **b.** $\lim\limits_{x \to 1} \dfrac{1}{(x - 1)^2} = \infty$

c. $\lim\limits_{x \to 1} \dfrac{1}{(x - 1)^2} = -\infty$

20. Which of the following statements are correct? Choose all that apply.

   a. $\lim_{x \to 1^+} \dfrac{1}{1 - x}$ does not exist   **b.** $\lim_{x \to 1^+} \dfrac{1}{1 - x} = \infty$

   c. $\lim_{x \to 1^+} \dfrac{1}{1 - x} = -\infty$

Practice Exercises

21–44. Determining limits analytically *Determine the following limits.*

21. a. $\lim_{x \to 2^+} \dfrac{1}{x - 2}$ **b.** $\lim_{x \to 2^-} \dfrac{1}{x - 2}$ **c.** $\lim_{x \to 2} \dfrac{1}{x - 2}$

22. a. $\lim_{x \to 3^+} \dfrac{2}{(x - 3)^3}$ **b.** $\lim_{x \to 3^-} \dfrac{2}{(x - 3)^3}$ **c.** $\lim_{x \to 3} \dfrac{2}{(x - 3)^3}$

23. a. $\lim_{x \to 4^+} \dfrac{x - 5}{(x - 4)^2}$ **b.** $\lim_{x \to 4^-} \dfrac{x - 5}{(x - 4)^2}$ **c.** $\lim_{x \to 4} \dfrac{x - 5}{(x - 4)^2}$

24. a. $\lim_{x \to 1^+} \dfrac{x}{|x - 1|}$ **b.** $\lim_{x \to 1^-} \dfrac{x}{|x - 1|}$ **c.** $\lim_{x \to 1} \dfrac{x}{|x - 1|}$

25. a. $\lim_{z \to 3^+} \dfrac{(z - 1)(z - 2)}{(z - 3)}$ **b.** $\lim_{z \to 3^-} \dfrac{(z - 1)(z - 2)}{(z - 3)}$

   c. $\lim_{z \to 3} \dfrac{(z - 1)(z - 2)}{(z - 3)}$

26. a. $\lim_{x \to -2^+} \dfrac{(x - 4)}{x(x + 2)}$ **b.** $\lim_{x \to -2^-} \dfrac{(x - 4)}{x(x + 2)}$ **c.** $\lim_{x \to -2} \dfrac{(x - 4)}{x(x + 2)}$

27. a. $\lim_{x \to 2^+} \dfrac{x^2 - 4x + 3}{(x - 2)^2}$ **b.** $\lim_{x \to 2^-} \dfrac{x^2 - 4x + 3}{(x - 2)^2}$

   c. $\lim_{x \to 2} \dfrac{x^2 - 4x + 3}{(x - 2)^2}$

28. a. $\lim_{t \to -2^+} \dfrac{t^3 - 5t^2 + 6t}{t^4 - 4t^2}$ **b.** $\lim_{t \to -2^-} \dfrac{t^3 - 5t^2 + 6t}{t^4 - 4t^2}$

   c. $\lim_{t \to -2} \dfrac{t^3 - 5t^2 + 6t}{t^4 - 4t^2}$ **d.** $\lim_{t \to 2} \dfrac{t^3 - 5t^2 + 6t}{t^4 - 4t^2}$

29. a. $\lim_{x \to 2^+} \dfrac{1}{\sqrt{x(x - 2)}}$ **b.** $\lim_{x \to 2^-} \dfrac{1}{\sqrt{x(x - 2)}}$

   c. $\lim_{x \to 2} \dfrac{1}{\sqrt{x(x - 2)}}$

30. a. $\lim_{x \to 1^+} \dfrac{x - 3}{\sqrt{x^2 - 5x + 4}}$ **b.** $\lim_{x \to 1^-} \dfrac{x - 3}{\sqrt{x^2 - 5x + 4}}$

   c. $\lim_{x \to 1} \dfrac{x - 3}{\sqrt{x^2 - 5x + 4}}$

31. a. $\lim_{x \to 0} \dfrac{x - 3}{x^4 - 9x^2}$ **b.** $\lim_{x \to 3} \dfrac{x - 3}{x^4 - 9x^2}$ **c.** $\lim_{x \to -3} \dfrac{x - 3}{x^4 - 9x^2}$

32. a. $\lim_{x \to 0} \dfrac{x - 2}{x^5 - 4x^3}$ **b.** $\lim_{x \to 2} \dfrac{x - 2}{x^5 - 4x^3}$ **c.** $\lim_{x \to -2} \dfrac{x - 2}{x^5 - 4x^3}$

33. $\lim_{x \to 0} \dfrac{x^3 - 5x^2}{x^2}$ **34.** $\lim_{t \to 5} \dfrac{4t^2 - 100}{t - 5}$

35. $\lim_{x \to 1^+} \dfrac{x^2 - 5x + 6}{x - 1}$ **36.** $\lim_{z \to 4} \dfrac{z - 5}{(z^2 - 10z + 24)^2}$

37. $\lim_{x \to 6^+} \dfrac{x - 7}{\sqrt{x - 6}}$ **38.** $\lim_{x \to 2^-} \dfrac{x - 1}{\sqrt{(x - 3)(x - 2)}}$

39. $\lim_{\theta \to 0^+} \csc \theta$ **40.** $\lim_{x \to 0^-} \csc x$

41. $\lim_{x \to 0^+} (-10 \cot x)$ **42.** $\lim_{\theta \to \pi/2^+} \dfrac{1}{3} \tan \theta$

43. $\lim_{\theta \to 0} \dfrac{2 + \sin \theta}{1 - \cos^2 \theta}$ **44.** $\lim_{\theta \to 0^-} \dfrac{\sin \theta}{\cos^2 \theta - 1}$

45. Location of vertical asymptotes Analyze the following limits and find the vertical asymptotes of $f(x) = \dfrac{x - 5}{x^2 - 25}$.

   a. $\lim_{x \to 5} f(x)$ **b.** $\lim_{x \to -5^-} f(x)$ **c.** $\lim_{x \to -5^+} f(x)$

46. Location of vertical asymptotes Analyze the following limits and find the vertical asymptotes of $f(x) = \dfrac{x + 7}{x^4 - 49x^2}$.

   a. $\lim_{x \to 7^-} f(x)$ **b.** $\lim_{x \to 7^+} f(x)$ **c.** $\lim_{x \to -7} f(x)$ **d.** $\lim_{x \to 0} f(x)$

47–50. Finding vertical asymptotes *Find all vertical asymptotes $x = a$ of the following functions. For each value of a, determine* $\lim_{x \to a^+} f(x)$, $\lim_{x \to a^-} f(x)$, *and* $\lim_{x \to a} f(x)$.

47. $f(x) = \dfrac{x^2 - 9x + 14}{x^2 - 5x + 6}$ **48.** $f(x) = \dfrac{\cos x}{x^2 + 2x}$

49. $f(x) = \dfrac{x + 1}{x^3 - 4x^2 + 4x}$ **50.** $f(x) = \dfrac{x^3 - 10x^2 + 16x}{x^2 - 8x}$

51. Checking your work graphically Analyze the following limits. Then sketch a graph of $y = \tan x$ with the window $[-\pi, \pi] \times [-10, 10]$ and use your graph to check your work.

   a. $\lim_{x \to \pi/2^+} \tan x$ **b.** $\lim_{x \to \pi/2^-} \tan x$
   c. $\lim_{x \to -\pi/2^+} \tan x$ **d.** $\lim_{x \to -\pi/2^-} \tan x$

T 52. Checking your work graphically Analyze the following limits. Then sketch a graph of $y = \sec x \tan x$ with the window $[-\pi, \pi] \times [-10, 10]$ and use your graph to check your work.

   a. $\lim_{x \to \pi/2^+} \sec x \tan x$ **b.** $\lim_{x \to \pi/2^-} \sec x \tan x$
   c. $\lim_{x \to -\pi/2^+} \sec x \tan x$ **d.** $\lim_{x \to -\pi/2^-} \sec x \tan x$

53. Explain why or why not Determine whether the following statements are true and give an explanation or counterexample.

   a. The line $x = 1$ is a vertical asymptote of the function $f(x) = \dfrac{x^2 - 7x + 6}{x^2 - 1}$.

   b. The line $x = -1$ is a vertical asymptote of the function $f(x) = \dfrac{x^2 - 7x + 6}{x^2 - 1}$.

   c. If g has a vertical asymptote at $x = 1$ and $\lim_{x \to 1^+} g(x) = \infty$, then $\lim_{x \to 1^-} g(x) = \infty$.

54. Matching Match functions a–f with graphs A–F in the figure without using a graphing utility.

a. $f(x) = \dfrac{x}{x^2 + 1}$ b. $f(x) = \dfrac{x}{x^2 - 1}$

c. $f(x) = \dfrac{1}{x^2 - 1}$ d. $f(x) = \dfrac{x}{(x - 1)^2}$

e. $f(x) = \dfrac{1}{(x - 1)^2}$ f. $f(x) = \dfrac{x}{x + 1}$

A.   B.

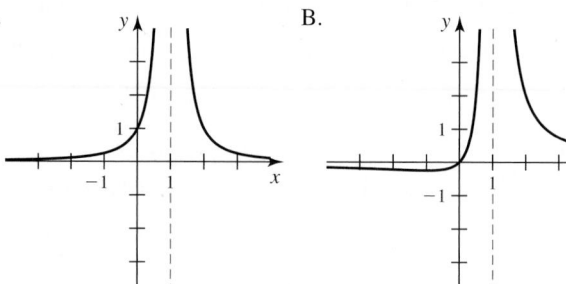

C.   D.

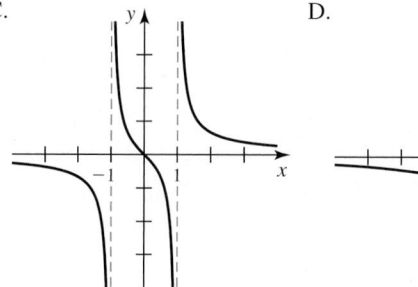
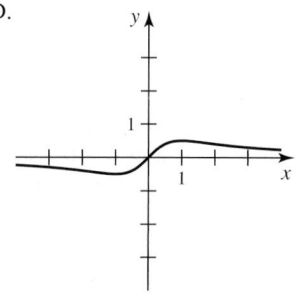

E.   F.

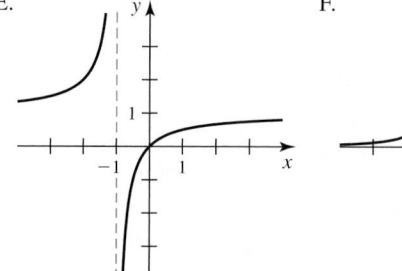
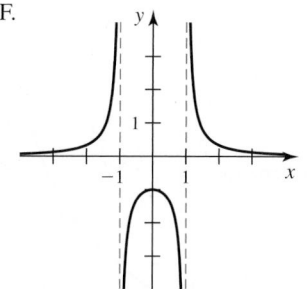

Explorations and Challenges

55. Finding a rational function Find a rational function $r(x)$ such that $r(1)$ is undefined, $\lim\limits_{x \to 1} r(x) = 0$, and $\lim\limits_{x \to 2} r(x) = \infty$.

56. Finding a function with vertical asymptotes Find polynomials p and q such that $f = p/q$ is undefined at 1 and 2, but f has a vertical asymptote only at 2. Sketch a graph of your function.

57. Finding a function with infinite limits Give a formula for a function f that satisfies $\lim\limits_{x \to 6^+} f(x) = \infty$ and $\lim\limits_{x \to 6^-} f(x) = -\infty$.

T 58–66. Asymptotes *Use analytical methods and/or a graphing utility to identify the vertical asymptotes (if any) of the following functions.*

58. $f(x) = \dfrac{x^2 - 1}{x^4 - 1}$ **59.** $f(x) = \dfrac{x^2 - 3x + 2}{x^{10} - x^9}$

60. $g(x) = 2 - \ln x^2$ **61.** $h(x) = \dfrac{e^x}{(x + 1)^3}$

62. $p(x) = \sec \dfrac{\pi x}{2}$, for $|x| < 2$

63. $g(\theta) = \tan \dfrac{\pi \theta}{10}$ **64.** $q(s) = \dfrac{\pi}{s - \sin s}$

65. $f(x) = \dfrac{1}{\sqrt{x} \sec x}$ **66.** $g(x) = e^{1/x}$

67. Limits with a parameter Let $f(x) = \dfrac{x^2 - 7x + 12}{x - a}$.

  a. For what values of a, if any, does $\lim\limits_{x \to a^+} f(x)$ equal a finite number?

  b. For what values of a, if any, does $\lim\limits_{x \to a^+} f(x) = \infty$?

  c. For what values of a, if any, does $\lim\limits_{x \to a^+} f(x) = -\infty$?

68–69. Steep secant lines

  a. *Given the graph of f in the following figures, find the slope of the secant line that passes through $(0, 0)$ and $(h, f(h))$ in terms of h, for $h > 0$ and $h < 0$.*

  b. *Analyze the limit of the slope of the secant line found in part (a) as $h \to 0^+$ and $h \to 0^-$. What does this tell you about the line tangent to the curve at $(0, 0)$?*

68. $f(x) = x^{1/3}$ **69.** $f(x) = x^{2/3}$

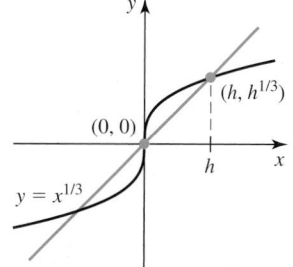
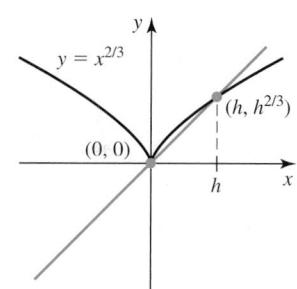

2.5 Limits at Infinity

Limits at infinity—as opposed to infinite limits—occur when the independent variable becomes large in magnitude. For this reason, limits at infinity determine what is called the *end behavior* of a function. An application of these limits is to determine whether a system (such as an ecosystem or a large oscillating structure) reaches a steady state as time increases.

Limits at Infinity and Horizontal Asymptotes

Figure 2.30 shows the graph of $y = \tan x$ (black curve) with vertical asymptotes at $x = \pm\dfrac{\pi}{2}$. Recall from Section 1.3 that on the interval $\left(-\dfrac{\pi}{2}, \dfrac{\pi}{2}\right)$, the graph of $\tan^{-1} x$ is obtained by reflecting the graph of $\tan x$ across the line $y = x$. Notice that when we do this reflection across the line $y = x$, the vertical asymptotes of $\tan x$ become the horizontal lines $y = \pm\dfrac{\pi}{2}$, which are associated with the graph of $y = \tan^{-1} x$ (blue curve). The figure shows that as x becomes arbitrarily large (denoted $x \to \infty$), the graph of $\tan^{-1} x$ approaches the horizontal line $y = \dfrac{\pi}{2}$, and as x becomes arbitrarily large in magnitude and negative (denoted $x \to -\infty$), the graph of $\tan^{-1} x$ approaches the horizontal line $y = -\dfrac{\pi}{2}$. Observe that the limits for $\tan^{-1} x$ as $x \to \pm\infty$ correspond perfectly with the one-sided limits for $\tan x$ as $x \to \pm\dfrac{\pi}{2}$. We use limit notation to summarize the behavior of these two functions concisely:

$$\lim_{x \to \infty} \tan^{-1} x = \frac{\pi}{2} \qquad \text{corresponds to} \qquad \lim_{x \to \pi/2^-} \tan x = \infty \text{ and}$$

$$\lim_{x \to -\infty} \tan^{-1} x = -\frac{\pi}{2} \qquad \text{corresponds to} \qquad \lim_{x \to -\pi/2^+} \tan x = -\infty.$$

The one-sided limits for $\tan x$ are infinite limits from Section 2.4; they indicate vertical asymptotes. The limits we have written for $\tan^{-1} x$ are called *limits at infinity*, and the horizontal lines $y = \pm\dfrac{\pi}{2}$ approached by the graph of $\tan^{-1} x$ are *horizontal asymptotes*.

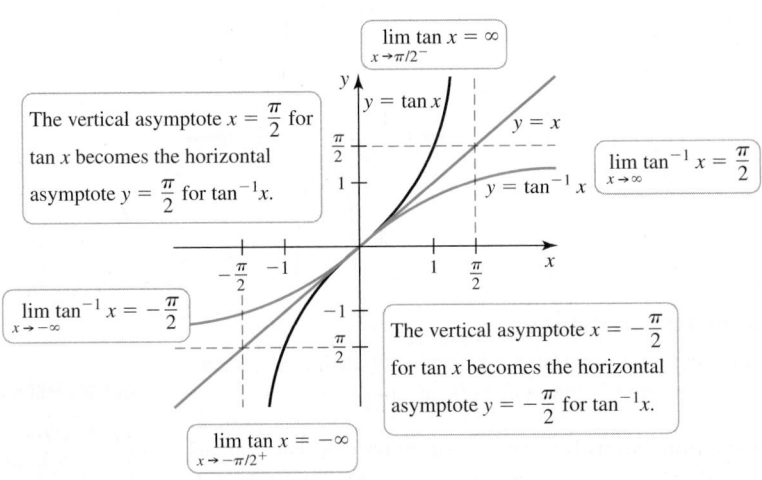

Figure 2.30

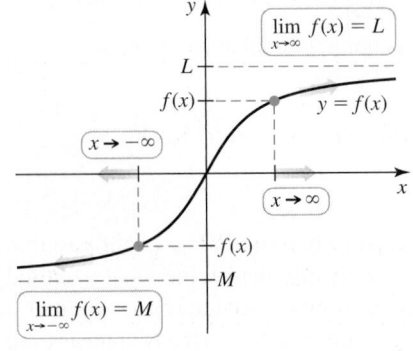

Figure 2.31

> **DEFINITION Limits at Infinity and Horizontal Asymptotes**
>
> If $f(x)$ becomes arbitrarily close to a finite number L for all sufficiently large and positive x, then we write
>
> $$\lim_{x \to \infty} f(x) = L.$$
>
> We say the limit of $f(x)$ as x approaches infinity is L. In this case, the line $y = L$ is a **horizontal asymptote** of f (Figure 2.31). The limit at negative infinity, $\lim_{x \to -\infty} f(x) = M$, is defined analogously. When this limit exists, $y = M$ is a horizontal asymptote.

QUICK CHECK 1 Evaluate $x/(x + 1)$ for $x = 10, 100$, and 1000. What is $\lim\limits_{x \to \infty} \dfrac{x}{x + 1}$? ◄

▶ The limit laws of Theorem 2.3 and the Squeeze Theorem apply if $x \to a$ is replaced with $x \to \infty$ or $x \to -\infty$.

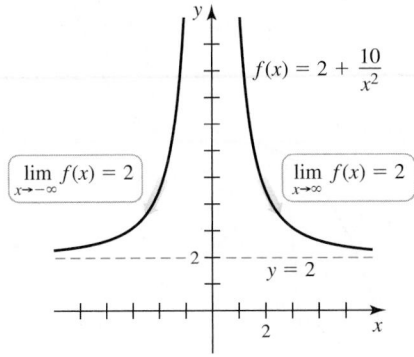

$$\lim_{x \to -\infty} f(x) = 2 \qquad \lim_{x \to \infty} f(x) = 2$$

$$f(x) = 2 + \frac{10}{x^2}$$

$$y = 2$$

Figure 2.32

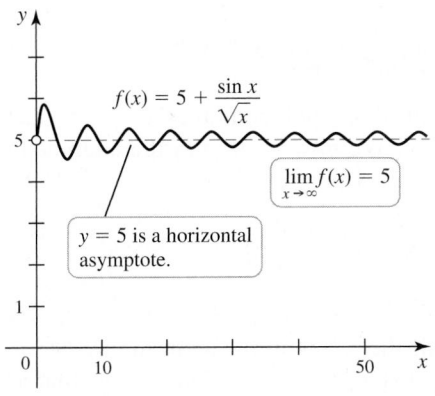

$$f(x) = 5 + \frac{\sin x}{\sqrt{x}}$$

$$\lim_{x \to \infty} f(x) = 5$$

$y = 5$ is a horizontal asymptote.

Figure 2.33

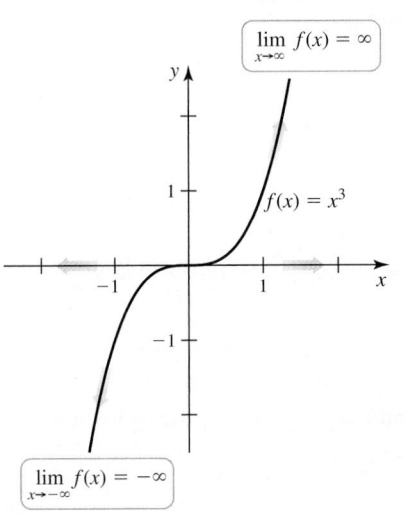

$$\lim_{x \to \infty} f(x) = \infty$$

$$f(x) = x^3$$

$$\lim_{x \to -\infty} f(x) = -\infty$$

Figure 2.34

EXAMPLE 1   Limits at infinity Evaluate the following limits.

a. $\lim\limits_{x \to -\infty} \left(2 + \dfrac{10}{x^2}\right)$ **b.** $\lim\limits_{x \to \infty} \left(5 + \dfrac{\sin x}{\sqrt{x}}\right)$

SOLUTION

a. As x becomes large and negative, x^2 becomes large and positive; in turn, $10/x^2$ approaches 0. By the limit laws of Theorem 2.3,

$$\lim_{x \to -\infty} \left(2 + \frac{10}{x^2}\right) = \underbrace{\lim_{x \to -\infty} 2}_{\text{equals 2}} + \underbrace{\lim_{x \to -\infty} \left(\frac{10}{x^2}\right)}_{\text{equals 0}} = 2 + 0 = 2.$$

Therefore, the graph of $y = 2 + 10/x^2$ approaches the horizontal asymptote $y = 2$ as $x \to -\infty$ (**Figure 2.32**). Notice that $\lim\limits_{x \to \infty} \left(2 + \dfrac{10}{x^2}\right)$ is also equal to 2, which implies that the graph has a single horizontal asymptote.

b. The numerator of $\sin x / \sqrt{x}$ is bounded between -1 and 1; therefore, for $x > 0$,

$$-\frac{1}{\sqrt{x}} \le \frac{\sin x}{\sqrt{x}} \le \frac{1}{\sqrt{x}}.$$

As $x \to \infty$, $\sqrt{x}$ becomes arbitrarily large, which means that

$$\lim_{x \to \infty} \frac{-1}{\sqrt{x}} = \lim_{x \to \infty} \frac{1}{\sqrt{x}} = 0.$$

It follows by the Squeeze Theorem (Theorem 2.5) that $\lim\limits_{x \to \infty} \dfrac{\sin x}{\sqrt{x}} = 0$.

Using the limit laws of Theorem 2.3,

$$\lim_{x \to \infty} \left(5 + \frac{\sin x}{\sqrt{x}}\right) = \underbrace{\lim_{x \to \infty} 5}_{\text{equals 5}} + \underbrace{\lim_{x \to \infty} \frac{\sin x}{\sqrt{x}}}_{\text{equals 0}} = 5.$$

The graph of $y = 5 + \dfrac{\sin x}{\sqrt{x}}$ approaches the horizontal asymptote $y = 5$ as x becomes large (**Figure 2.33**). Note that the curve intersects its asymptote infinitely many times.

Related Exercises 10, 19 ◄

Infinite Limits at Infinity

It is possible for a limit to be *both* an infinite limit and a limit at infinity. This type of limit occurs if $f(x)$ becomes arbitrarily large in magnitude as x becomes arbitrarily large in magnitude. Such a limit is called an *infinite limit at infinity* and is illustrated by the function $f(x) = x^3$ (**Figure 2.34**).

DEFINITION   Infinite Limits at Infinity

If $f(x)$ becomes arbitrarily large as x becomes arbitrarily large, then we write

$$\lim_{x \to \infty} f(x) = \infty.$$

The limits $\lim\limits_{x \to \infty} f(x) = -\infty$, $\lim\limits_{x \to -\infty} f(x) = \infty$, and $\lim\limits_{x \to -\infty} f(x) = -\infty$ are defined similarly.

Infinite limits at infinity tell us about the behavior of polynomials for large-magnitude values of x. First, consider power functions $f(x) = x^n$, where n is a positive integer. Figure 2.35 shows that when n is even, $\displaystyle\lim_{x \to \pm\infty} x^n = \infty$, and when n is odd, $\displaystyle\lim_{x \to \infty} x^n = \infty$ and $\displaystyle\lim_{x \to -\infty} x^n = -\infty$.

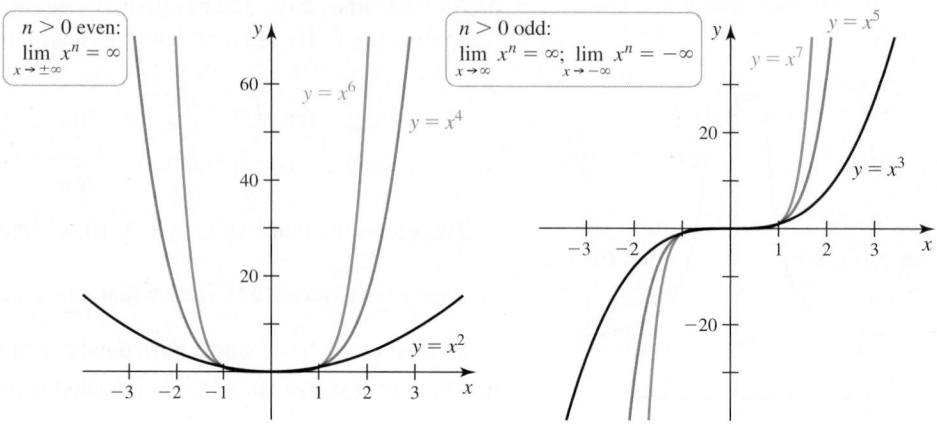

Figure 2.35

It follows that reciprocals of power functions $f(x) = 1/x^n = x^{-n}$, where n is a positive integer, behave as follows:

$$\lim_{x \to \infty} \frac{1}{x^n} = \lim_{x \to \infty} x^{-n} = 0 \quad \text{and} \quad \lim_{x \to -\infty} \frac{1}{x^n} = \lim_{x \to -\infty} x^{-n} = 0.$$

QUICK CHECK 2 Describe the behavior of $p(x) = -3x^3$ as $x \to \infty$ and as $x \to -\infty$. ◄

From here, it is a short step to finding the behavior of any polynomial as $x \to \pm\infty$. Let $p(x) = a_n x^n + a_{n-1} x^{n-1} + \cdots + a_2 x^2 + a_1 x + a_0$. We now write p in the equivalent form

$$p(x) = x^n \left(a_n + \underbrace{\frac{a_{n-1}}{x}}_{\to 0} + \underbrace{\frac{a_{n-2}}{x^2}}_{\to 0} + \cdots + \underbrace{\frac{a_0}{x^n}}_{\to 0} \right).$$

Notice that as x becomes large in magnitude, all the terms in p except the first term approach zero. Therefore, as $x \to \pm\infty$, we see that $p(x) \approx a_n x^n$. This means that as $x \to \pm\infty$, the behavior of p is determined by the term $a_n x^n$ with the highest power of x.

THEOREM 2.6   Limits at Infinity of Powers and Polynomials
Let n be a positive integer and let p be the polynomial
$p(x) = a_n x^n + a_{n-1} x^{n-1} + \cdots + a_2 x^2 + a_1 x + a_0$, where $a_n \ne 0$.

1. $\displaystyle\lim_{x \to \pm\infty} x^n = \infty$ when n is even.

2. $\displaystyle\lim_{x \to \infty} x^n = \infty$ and $\displaystyle\lim_{x \to -\infty} x^n = -\infty$ when n is odd.

3. $\displaystyle\lim_{x \to \pm\infty} \frac{1}{x^n} = \lim_{x \to \pm\infty} x^{-n} = 0.$

4. $\displaystyle\lim_{x \to \pm\infty} p(x) = \lim_{x \to \pm\infty} a_n x^n = \pm\infty$, depending on the degree of the polynomial and the sign of the leading coefficient a_n.

EXAMPLE 2   Limits at infinity Determine the limits as $x \to \pm\infty$ of the following functions.

a. $p(x) = 3x^4 - 6x^2 + x - 10$

b. $q(x) = -2x^3 + 3x^2 - 12$

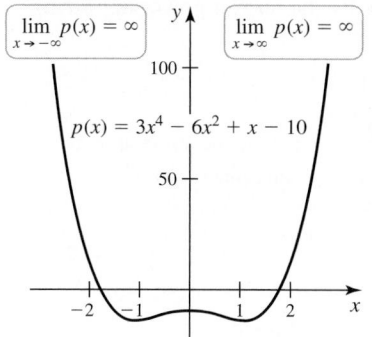

Figure 2.36

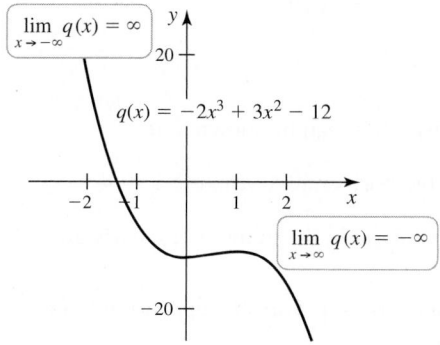

Figure 2.37

SOLUTION

a. We use the fact that the limit is determined by the behavior of the leading term:

$$\lim_{x\to\infty}(3x^4 - 6x^2 + x - 10) = \lim_{x\to\infty} 3\underbrace{x^4}_{\to\,\infty} = \infty.$$

Similarly,

$$\lim_{x\to-\infty}(3x^4 - 6x^2 + x - 10) = \lim_{x\to-\infty} 3\underbrace{x^4}_{\to\,\infty} = \infty.$$

Figure 2.36 illustrates these limits.

b. Noting that the leading coefficient is negative, we have

$$\lim_{x\to\infty}(-2x^3 + 3x^2 - 12) = \lim_{x\to\infty}(-2\underbrace{x^3}_{\to\,\infty}) = -\infty$$

$$\lim_{x\to-\infty}(-2x^3 + 3x^2 - 12) = \lim_{x\to-\infty}(-2\underbrace{x^3}_{\to\,-\infty}) = \infty.$$

The graph of q (Figure 2.37) confirms these results.

Related Exercises 21, 23 ◄

End Behavior

The behavior of polynomials as $x \to \pm\infty$ is an example of what is often called *end behavior*. Having treated polynomials, we now turn to the end behavior of rational, algebraic, and transcendental functions.

EXAMPLE 3 End behavior of rational functions Use limits at infinity to determine the end behavior of the following rational functions.

a. $f(x) = \dfrac{3x + 2}{x^2 - 1}$ 　　**b.** $g(x) = \dfrac{40x^4 + 4x^2 - 1}{10x^4 + 8x^2 + 1}$ 　　**c.** $h(x) = \dfrac{x^3 - 2x + 1}{2x + 4}$

SOLUTION

➤ Recall that the *degree* of a polynomial is the highest power of x that appears.

a. An effective approach for determining limits of rational functions at infinity is to divide both the numerator and denominator by x^n, where n is the degree of the polynomial in the denominator. This strategy forces the terms corresponding to lower powers of x to approach 0 in the limit. In this case, we divide by x^2:

$$\lim_{x\to\infty}\frac{3x+2}{x^2-1} = \lim_{x\to\infty}\frac{\dfrac{3x+2}{x^2}}{\dfrac{x^2-1}{x^2}} = \lim_{x\to\infty}\frac{\overbrace{\dfrac{3}{x}+\dfrac{2}{x^2}}^{\text{approaches }0}}{1-\underbrace{\dfrac{1}{x^2}}_{\text{approaches }0}} = \frac{0}{1} = 0.$$

A similar calculation gives $\displaystyle\lim_{x\to-\infty}\frac{3x+2}{x^2-1} = 0$; therefore, the graph of f has the horizontal asymptote $y = 0$. You should confirm that the zeros of the denominator are -1 and 1, which correspond to vertical asymptotes (Figure 2.38). In this example, the degree of the polynomial in the numerator is *less than* the degree of the polynomial in the denominator.

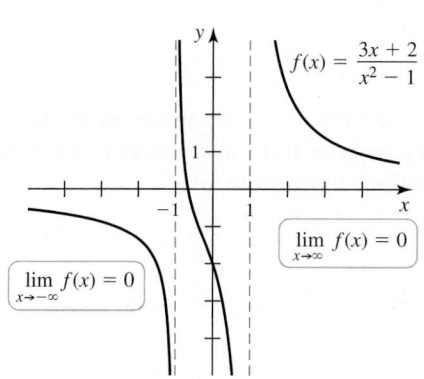

Figure 2.38

b. Again we divide both the numerator and denominator by the largest power appearing in the denominator, which is x^4:

$$\lim_{x\to\infty} \frac{40x^4 + 4x^2 - 1}{10x^4 + 8x^2 + 1} = \lim_{x\to\infty} \frac{\dfrac{40x^4}{x^4} + \dfrac{4x^2}{x^4} - \dfrac{1}{x^4}}{\dfrac{10x^4}{x^4} + \dfrac{8x^2}{x^4} + \dfrac{1}{x^4}}$$
Divide the numerator and denominator by x^4.

approaches 0 approaches 0

$$= \lim_{x\to\infty} \frac{40 + \dfrac{4}{x^2} - \dfrac{1}{x^4}}{10 + \dfrac{8}{x^2} + \dfrac{1}{x^4}}$$
Simplify.

approaches 0 approaches 0

$$= \frac{40 + 0 + 0}{10 + 0 + 0} = 4.$$
Evaluate limits.

Using the same steps (dividing each term by x^4), it can be shown that $\lim_{x\to-\infty} \dfrac{40x^4 + 4x^2 - 1}{10x^4 + 8x^2 + 1} = 4$. This function has the horizontal asymptote $y = 4$

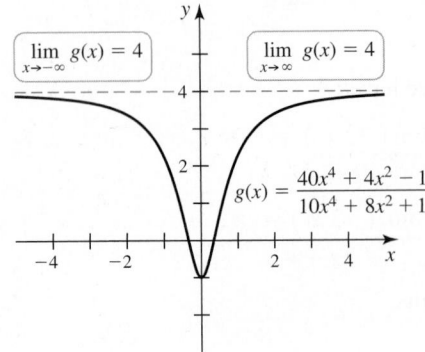

$$\lim_{x\to-\infty} g(x) = 4 \qquad \lim_{x\to\infty} g(x) = 4$$

$$g(x) = \frac{40x^4 + 4x^2 - 1}{10x^4 + 8x^2 + 1}$$

Figure 2.39

(Figure 2.39). Notice that the degree of the polynomial in the numerator *equals* the degree of the polynomial in the denominator.

c. We divide the numerator and denominator by the largest power of x appearing in the denominator, which is x, and then take the limit:

$$\lim_{x\to\infty} \frac{x^3 - 2x + 1}{2x + 4} = \lim_{x\to\infty} \frac{\dfrac{x^3}{x} - \dfrac{2x}{x} + \dfrac{1}{x}}{\dfrac{2x}{x} + \dfrac{4}{x}}$$
Divide the numerator and denominator by x.

arbitrarily large constant   approaches 0

$$= \lim_{x\to\infty} \frac{x^2 - 2 + \dfrac{1}{x}}{2 + \dfrac{4}{x}}$$
Simplify.

constant approaches 0

$$= \infty.$$
Take limits.

As $x \to \infty$, all the terms in this function either approach zero or are constant—except the x^2-term in the numerator, which becomes arbitrarily large. Therefore, the limit of the function does not exist. Using a similar analysis, we find that $\lim_{x\to-\infty} \dfrac{x^3 - 2x + 1}{2x + 4} = \infty$.

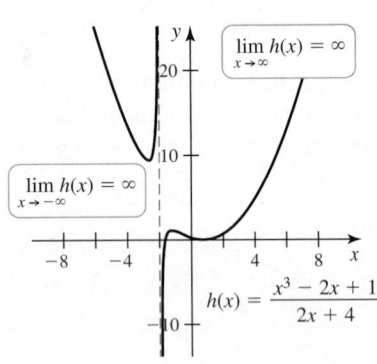

$$\lim_{x\to\infty} h(x) = \infty$$

$$\lim_{x\to-\infty} h(x) = \infty$$

$$h(x) = \frac{x^3 - 2x + 1}{2x + 4}$$

Figure 2.40

These limits are not finite, so the graph of the function has no horizontal asymptote (Figure 2.40). There is, however, a vertical asymptote due to the fact that $x = -2$ is a zero of the denominator. In this case, the degree of the polynomial in the numerator is *greater than* the degree of the polynomial in the denominator.

Related Exercises 38, 41, 43 ◄

A special case of end behavior arises with rational functions. As shown in the next example, if the graph of a function f approaches a line (with finite and nonzero slope) as $x \to \pm\infty$, then that line is a **slant asymptote**, or **oblique asymptote**, of f.

EXAMPLE 4   Slant asymptotes Determine the end behavior of the function
$$f(x) = \frac{2x^2 + 6x - 2}{x + 1}.$$

SOLUTION We first divide the numerator and denominator by the largest power of x appearing in the denominator, which is x:

$$\lim_{x \to \infty} \frac{2x^2 + 6x - 2}{x + 1} = \lim_{x \to \infty} \frac{\dfrac{2x^2}{x} + \dfrac{6x}{x} - \dfrac{2}{x}}{\dfrac{x}{x} + \dfrac{1}{x}} \qquad \text{Divide the numerator and denominator by } x.$$

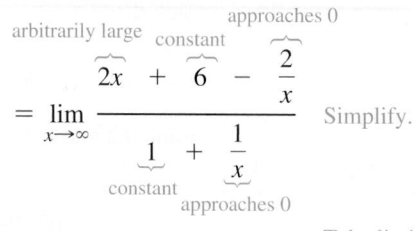

$$= \lim_{x \to \infty} \frac{\overbrace{2x}^{\text{arbitrarily large}} + \overbrace{6}^{\text{constant}} - \overbrace{\dfrac{2}{x}}^{\text{approaches 0}}}{\underbrace{1}_{\text{constant}} + \underbrace{\dfrac{1}{x}}_{\text{approaches 0}}} \qquad \text{Simplify.}$$

$$= \infty. \qquad \text{Take limits.}$$

A similar analysis shows that $\displaystyle\lim_{x \to -\infty} \frac{2x^2 + 6x - 2}{x + 1} = -\infty$. Because these limits are not finite, f has no horizontal asymptote.

However, there is more to be learned about the end behavior of this function. Using long division, the function f is written

$$f(x) = \frac{2x^2 + 6x - 2}{x + 1} = \underbrace{2x + 4}_{\ell(x)} - \underbrace{\frac{6}{x + 1}}_{\substack{\text{approaches 0 as} \\ x \to \infty}}.$$

As $x \to \infty$, the term $6/(x + 1)$ approaches 0, and we see that the function f behaves like the linear function $\ell(x) = 2x + 4$. For this reason, the graph of f approaches the graph of ℓ as $x \to \infty$ (**Figure 2.41**). A similar argument shows that the graph of f also approaches the graph of ℓ as $x \to -\infty$. The line described by ℓ is a slant asymptote; it occurs with rational functions only when the degree of the polynomial in the numerator exceeds the degree of the polynomial in the denominator by exactly 1.

Related Exercises 51–52 ◄

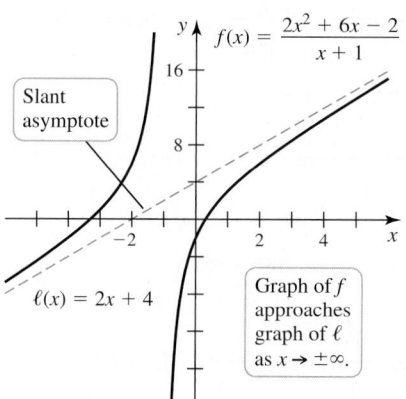

Figure 2.41

▶ More generally, a line $y = \ell(x)$ (with finite and nonzero slope) is a slant asymptote of a function f if $\displaystyle\lim_{x \to \infty}(f(x) - \ell(x)) = 0$ or $\displaystyle\lim_{x \to -\infty}(f(x) - \ell(x)) = 0.$

QUICK CHECK 3 Use Theorem 2.7 to find the vertical and horizontal asymptotes of $y = \dfrac{10x}{3x - 1}$. ◄

The conclusions reached in Examples 3 and 4 can be generalized for all rational functions. These results are summarized in Theorem 2.7; its proof is assigned in Exercise 92.

THEOREM 2.7   End Behavior and Asymptotes of Rational Functions

Suppose $f(x) = \dfrac{p(x)}{q(x)}$ is a rational function, where

$$p(x) = a_m x^m + a_{m-1} x^{m-1} + \cdots + a_2 x^2 + a_1 x + a_0 \quad \text{and}$$
$$q(x) = b_n x^n + b_{n-1} x^{n-1} + \cdots + b_2 x^2 + b_1 x + b_0,$$

with $a_m \neq 0$ and $b_n \neq 0$.

a. Degree of numerator less than degree of denominator If $m < n$, then $\displaystyle\lim_{x \to \pm\infty} f(x) = 0$, and $y = 0$ is a horizontal asymptote of f.

b. Degree of numerator equals degree of denominator If $m = n$, then $\displaystyle\lim_{x \to \pm\infty} f(x) = a_m/b_n$, and $y = a_m/b_n$ is a horizontal asymptote of f.

c. Degree of numerator greater than degree of denominator If $m > n$, then $\displaystyle\lim_{x \to \pm\infty} f(x) = \infty$ or $-\infty$, and f has no horizontal asymptote.

d. Slant asymptote If $m = n + 1$, then $\lim\limits_{x \to \pm\infty} f(x) = \infty$ or $-\infty$, and f has no horizontal asymptote, but f has a slant asymptote.

e. Vertical asymptotes Assuming f is in reduced form (p and q share no common factors), vertical asymptotes occur at the zeros of q.

Although it isn't stated explicitly, Theorem 2.7 implies that a rational function can have at most one horizontal asymptote, and whenever there is a horizontal asymptote, $\lim\limits_{x \to \infty} \dfrac{p(x)}{q(x)} = \lim\limits_{x \to -\infty} \dfrac{p(x)}{q(x)}$. The same cannot be said of other functions, as the next examples show.

EXAMPLE 5 End behavior of an algebraic function Use limits at infinity to determine the end behavior of $f(x) = \dfrac{10x^3 - 3x^2 + 8}{\sqrt{25x^6 + x^4 + 2}}$.

SOLUTION The square root in the denominator forces us to revise the strategy used with rational functions. First, consider the limit as $x \to \infty$. The highest power of the polynomial in the denominator is 6. However, the polynomial is under a square root, so effectively, the term with the highest power in the denominator is $\sqrt{x^6} = x^3$. When we divide the numerator and denominator by x^3, for $x > 0$, the limit becomes

$$\lim_{x \to \infty} \frac{10x^3 - 3x^2 + 8}{\sqrt{25x^6 + x^4 + 2}} = \lim_{x \to \infty} \frac{\dfrac{10x^3}{x^3} - \dfrac{3x^2}{x^3} + \dfrac{8}{x^3}}{\sqrt{\dfrac{25x^6}{x^6} + \dfrac{x^4}{x^6} + \dfrac{2}{x^6}}} \qquad \text{Divide by } \sqrt{x^6} = x^3.$$

$$= \lim_{x \to \infty} \frac{10 - \overbrace{\dfrac{3}{x}}^{\text{approaches } 0} + \overbrace{\dfrac{8}{x^3}}^{\text{approaches } 0}}{\sqrt{25 + \underbrace{\dfrac{1}{x^2}}_{\text{approaches } 0} + \underbrace{\dfrac{2}{x^6}}_{\text{approaches } 0}}} \qquad \text{Simplify.}$$

$$= \frac{10}{\sqrt{25}} = 2. \qquad \text{Evaluate limits.}$$

> Recall that
> $$\sqrt{x^2} = |x| = \begin{cases} x & \text{if } x \geq 0 \\ -x & \text{if } x < 0. \end{cases}$$
> Therefore,
> $$\sqrt{x^6} = |x^3| = \begin{cases} x^3 & \text{if } x \geq 0 \\ -x^3 & \text{if } x < 0. \end{cases}$$
> Because x is negative as $x \to -\infty$, we have $\sqrt{x^6} = -x^3$.

As $x \to -\infty$, x^3 is negative, so we divide the numerator and denominator by $\sqrt{x^6} = -x^3$ (which is positive):

$$\lim_{x \to -\infty} \frac{10x^3 - 3x^2 + 8}{\sqrt{25x^6 + x^4 + 2}} = \lim_{x \to -\infty} \frac{\dfrac{10x^3}{-x^3} - \dfrac{3x^2}{-x^3} + \dfrac{8}{-x^3}}{\sqrt{\dfrac{25x^6}{x^6} + \dfrac{x^4}{x^6} + \dfrac{2}{x^6}}} \qquad \text{Divide by } \sqrt{x^6} = -x^3 > 0.$$

$$= \lim_{x \to -\infty} \frac{-10 + \overbrace{\dfrac{3}{x}}^{\text{approaches } 0} - \overbrace{\dfrac{8}{x^3}}^{\text{approaches } 0}}{\sqrt{25 + \underbrace{\dfrac{1}{x^2}}_{\text{approaches } 0} + \underbrace{\dfrac{2}{x^6}}_{\text{approaches } 0}}} \qquad \text{Simplify.}$$

$$= -\frac{10}{\sqrt{25}} = -2. \qquad \text{Evaluate limits.}$$

The limits reveal two asymptotes, $y = 2$ and $y = -2$. In this particular case, the graph crosses both horizontal asymptotes (**Figure 2.42**).   *Related Exercises 46–47* ◄

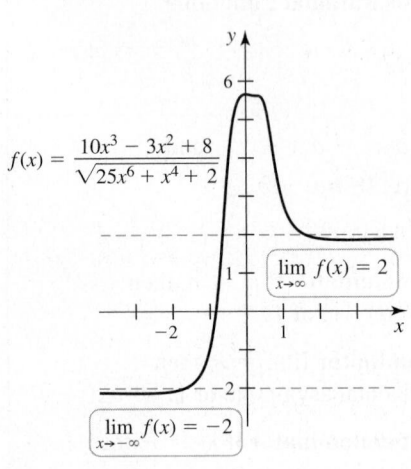

$$f(x) = \frac{10x^3 - 3x^2 + 8}{\sqrt{25x^6 + x^4 + 2}}$$

$\lim\limits_{x \to \infty} f(x) = 2$

$\lim\limits_{x \to -\infty} f(x) = -2$

Figure 2.42

End Behavior for e^x, e^{-x}, and $\ln x$

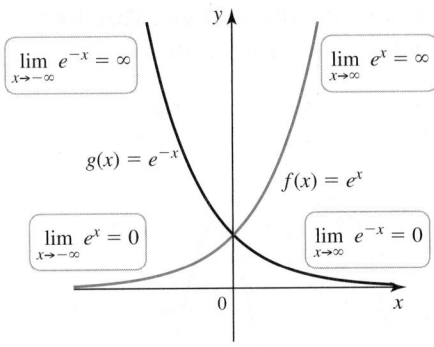

Figure 2.43

To determine the end behavior of e^x and e^{-x}, we begin by examining their graphs (**Figure 2.43**). The graph of e^x makes it clear that as $x \to \infty$, e^x increases without bound. All exponential functions b^x with $b > 1$ behave this way, because raising a number greater than 1 to ever-larger powers produces numbers that increase without bound. We conclude that

$$\lim_{x \to \infty} e^x = \infty \text{ and it immediately follows that } \lim_{x \to \infty} e^{-x} = \lim_{x \to \infty} \frac{1}{e^x} = 0.$$

denominator approaches ∞

The graph of e^{-x} is the reflection of the graph of e^x across the y-axis. Appealing to this symmetry, we have

$$\lim_{x \to -\infty} e^x = \lim_{x \to \infty} e^{-x} = 0 \quad \text{and} \quad \lim_{x \to -\infty} e^{-x} = \lim_{x \to \infty} e^x = \infty.$$

These limits imply that $y = 0$ is a horizontal asymptote of both e^x and e^{-x}.

The domain of $\ln x$ is $\{x : x > 0\}$, so we evaluate $\lim\limits_{x \to 0^+} \ln x$ and $\lim\limits_{x \to \infty} \ln x$ to determine end behavior. For the first limit, recall that $\ln x$ is the inverse of e^x, and the graph of $\ln x$ is a reflection of the graph of e^x across the line $y = x$ (**Figure 2.44**). The horizontal asymptote ($y = 0$) of e^x is also reflected across $y = x$, becoming a vertical asymptote ($x = 0$) for $\ln x$. These observations imply that $\lim\limits_{x \to 0^+} \ln x = -\infty$.

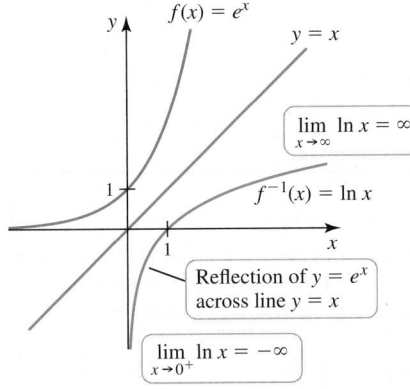

Figure 2.44

To determine $\lim\limits_{x \to \infty} \ln x$, recall that $y = e^x$ implies that $x = \ln y$. Examining once more the graph of e^x (Figure 2.44), we see that $x \to \infty$ if and only if $y \to \infty$; therefore, $x = \ln y \to \infty$ as $y \to \infty$. Written in limit notation, we have $\lim\limits_{y \to \infty} \ln y = \infty$, which, when y is replaced with x, yields $\lim\limits_{x \to \infty} \ln x = \infty$.

The end behavior of exponential and logarithmic functions is important. We summarize these results in the following theorem.

THEOREM 2.8   End Behavior of e^x, e^{-x}, and $\ln x$

The end behavior for e^x and e^{-x} on $(-\infty, \infty)$ and $\ln x$ on $(0, \infty)$ is given by the following limits:

$$\lim_{x \to \infty} e^x = \infty \qquad \text{and} \qquad \lim_{x \to -\infty} e^x = 0,$$

$$\lim_{x \to \infty} e^{-x} = 0 \qquad \text{and} \qquad \lim_{x \to -\infty} e^{-x} = \infty,$$

$$\lim_{x \to 0^+} \ln x = -\infty \qquad \text{and} \qquad \lim_{x \to \infty} \ln x = \infty.$$

EXAMPLE 6   End behavior of transcendental functions Determine the end behavior of the following functions.

a. $f(x) = e^{5x}$ **b.** $f(x) = 4 - \ln 3x$ **c.** $f(x) = \cos x$

SOLUTION

> ▶ The technique of changing variables introduced in the solution to Example 6a could also be used for Example 6b by letting $t = 3x$.

a. Let $t = 5x$. Observe that as $x \to \infty$, $t \to \infty$ and that as $x \to -\infty$, $t \to -\infty$. It follows that $\lim\limits_{x \to \infty} e^{5x} = \lim\limits_{t \to \infty} e^t = \infty$ and $\lim\limits_{x \to -\infty} e^{5x} = \lim\limits_{t \to -\infty} e^t = 0$. The second of these limits implies that the graph of e^{5x} has the horizontal asymptote $y = 0$.

QUICK CHECK 4 How do the functions e^{10x} and e^{-10x} behave as $x \to \infty$ and as $x \to -\infty$? ◀

b. Using a property of logarithms, we have $f(x) = 4 - (\ln 3 + \ln x) = 4 - \ln 3 - \ln x$. Because $\ln x \to \infty$ as $x \to \infty$, it follows that $-\ln x \to -\infty$ as $x \to \infty$. Consequently, $4 - \ln 3 - \ln x$ is negative and grows arbitrarily large in magnitude as $x \to \infty$. Therefore, $\lim\limits_{x \to \infty} (4 - \ln 3x) = -\infty$. A similar argument shows that $\lim\limits_{x \to 0^+} (4 - \ln 3x) = \infty$.

c. The cosine function oscillates between -1 and 1 as $x \to \infty$ (Figure 2.45). Therefore, $\lim\limits_{x \to \infty} \cos x$ does not exist. For the same reason, $\lim\limits_{x \to -\infty} \cos x$ does not exist.

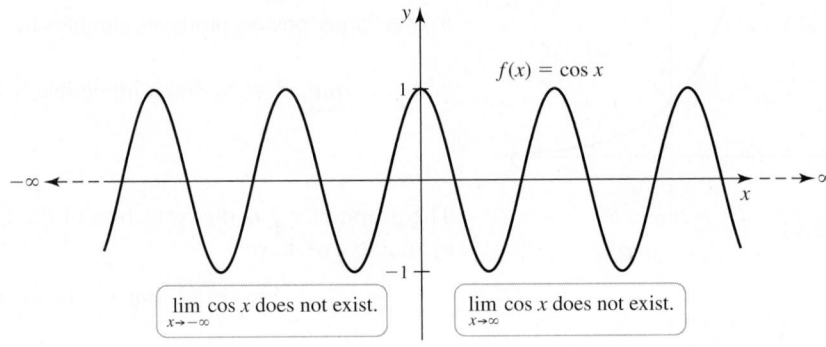

Figure 2.45

Related Exercises 57, 59, 62 ◀

SECTION 2.5 EXERCISES

Getting Started

1. Explain the meaning of $\lim\limits_{x \to -\infty} f(x) = 10$.

2. Evaluate $\lim\limits_{x \to \infty} f(x)$ and $\lim\limits_{x \to -\infty} f(x)$ using the figure.

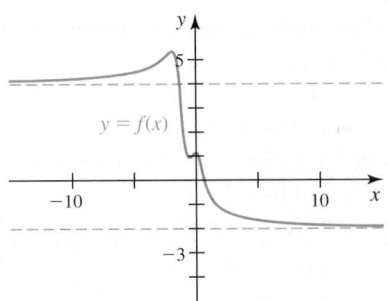

3–13. *Determine the following limits at infinity.*

3. $\lim\limits_{x \to \infty} x^{12}$

4. $\lim\limits_{x \to -\infty} 3x^{11}$

5. $\lim\limits_{x \to \infty} x^{-6}$

6. $\lim\limits_{x \to -\infty} x^{-11}$

7. $\lim\limits_{t \to \infty} (-12t^{-5})$

8. $\lim\limits_{x \to -\infty} 2x^{-8}$

9. $\lim\limits_{x \to \infty} \left(3 + \dfrac{10}{x^2}\right)$

10. $\lim\limits_{x \to \infty} \left(5 + \dfrac{1}{x} + \dfrac{10}{x^2}\right)$

11. $\lim\limits_{x \to \infty} \dfrac{f(x)}{g(x)}$ if $f(x) \to 100{,}000$ and $g(x) \to \infty$ as $x \to \infty$

12. $\lim\limits_{x \to \infty} \dfrac{4x^2 + 2x + 3}{x^2}$

13. $\lim\limits_{t \to \infty} e^t$, $\lim\limits_{t \to -\infty} e^t$, and $\lim\limits_{t \to \infty} e^{-t}$

14. Describe the end behavior of $g(x) = e^{-2x}$.

15. Suppose the function g satisfies the inequality

$$3 - \frac{1}{x^2} \le g(x) \le 3 + \frac{1}{x^2}, \text{ for all nonzero values of } x. \text{ Evaluate}$$

$\lim\limits_{x \to \infty} g(x)$ and $\lim\limits_{x \to -\infty} g(x)$.

16. The graph of g has a vertical asymptote at $x = 2$ and horizontal asymptotes at $y = -1$ and $y = 3$ (see figure). Determine the following limits: $\lim\limits_{x \to -\infty} g(x)$, $\lim\limits_{x \to \infty} g(x)$, $\lim\limits_{x \to 2^-} g(x)$, and $\lim\limits_{x \to 2^+} g(x)$.

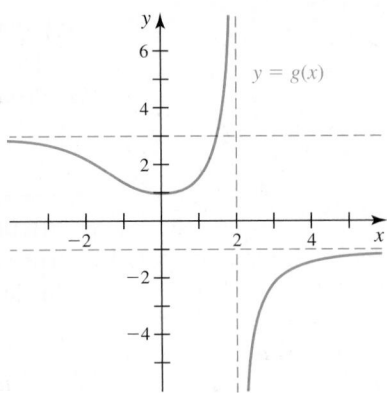

Practice Exercises

17–36. Limits at infinity *Determine the following limits.*

17. $\lim\limits_{\theta \to \infty} \dfrac{\cos \theta}{\theta^2}$

18. $\lim\limits_{t \to \infty} \dfrac{5t^2 + t \sin t}{t^2}$

19. $\lim\limits_{x \to \infty} \dfrac{\cos x^5}{\sqrt{x}}$

20. $\lim\limits_{x \to -\infty} \left(5 + \dfrac{100}{x} + \dfrac{\sin^4 x^3}{x^2}\right)$

21. $\lim\limits_{x \to \infty} (3x^{12} - 9x^7)$

22. $\lim\limits_{x \to -\infty} (3x^7 + x^2)$

23. $\lim\limits_{x \to -\infty} (-3x^{16} + 2)$

24. $\lim\limits_{x \to -\infty} (2x^{-8} + 4x^3)$

25. $\lim\limits_{x \to \infty} \dfrac{14x^3 + 3x^2 - 2x}{21x^3 + x^2 + 2x + 1}$

26. $\lim\limits_{x \to \infty} \dfrac{9x^3 + x^2 - 5}{3x^4 + 4x^2}$

27. $\lim\limits_{x \to -\infty} \dfrac{3x^2 + 3x}{x + 1}$

28. $\lim\limits_{x \to \infty} \dfrac{x^4 + 7}{x^5 + x^2 - x}$

29. $\lim\limits_{w \to \infty} \dfrac{15w^2 + 3w + 1}{\sqrt{9w^4 + w^3}}$

30. $\lim\limits_{x \to -\infty} \dfrac{40x^4 + x^2 + 5x}{\sqrt{64x^8 + x^6}}$

31. $\lim\limits_{x \to -\infty} \dfrac{\sqrt{16x^2 + x}}{x}$

32. $\lim\limits_{x \to \infty} \dfrac{6x^2}{4x^2 + \sqrt{16x^4 + x^2}}$

33. $\lim\limits_{x \to \infty} (x^2 - \sqrt{x^4 + 3x^2})$ (*Hint:* Multiply by $\dfrac{x^2 + \sqrt{x^4 + 3x^2}}{x^2 + \sqrt{x^4 + 3x^2}}$ first.)

34. $\lim\limits_{x \to -\infty} \left(x + \sqrt{x^2 - 5x}\right)$ **35.** $\lim\limits_{x \to \infty} \dfrac{\sin x}{e^x}$

36. $\lim\limits_{x \to -\infty} (e^x \cos x + 3)$

37–50. Horizontal asymptotes *Determine* $\lim\limits_{x \to \infty} f(x)$ *and* $\lim\limits_{x \to -\infty} f(x)$ *for the following functions. Then give the horizontal asymptotes of f (if any).*

37. $f(x) = \dfrac{4x}{20x + 1}$

38. $f(x) = \dfrac{3x^2 - 7}{x^2 + 5x}$

39. $f(x) = \dfrac{6x^2 - 9x + 8}{3x^2 + 2}$

40. $f(x) = \dfrac{12x^8 - 3}{3x^8 - 2x^7}$

41. $f(x) = \dfrac{3x^3 - 7}{x^4 + 5x^2}$

42. $f(x) = \dfrac{2x + 1}{3x^4 - 2}$

43. $f(x) = \dfrac{40x^5 + x^2}{16x^4 - 2x}$

44. $f(x) = \dfrac{6x^2 + 1}{\sqrt{4x^4 + 3x + 1}}$

45. $f(x) = \dfrac{1}{2x^4 - \sqrt{4x^8 - 9x^4}}$

46. $f(x) = \dfrac{\sqrt{x^2 + 1}}{2x + 1}$

47. $f(x) = \dfrac{4x^3 + 1}{2x^3 + \sqrt{16x^6 + 1}}$

48. $f(x) = x - \sqrt{x^2 - 9x}$

49. $f(x) = \dfrac{\sqrt[3]{x^6 + 8}}{4x^2 + \sqrt{3x^4 + 1}}$

50. $f(x) = 4x(3x - \sqrt{9x^2 + 1})$

▣ 51–56. Slant (oblique) asymptotes *Complete the following steps for the given functions.*

  a. *Find the slant asymptote of f.*
  b. *Find the vertical asymptotes of f (if any).*
  c. *Graph f and all of its asymptotes with a graphing utility. Then sketch a graph of the function by hand, correcting any errors appearing in the computer-generated graph.*

51. $f(x) = \dfrac{x^2 - 3}{x + 6}$

52. $f(x) = \dfrac{x^2 - 1}{x + 2}$

53. $f(x) = \dfrac{x^2 - 2x + 5}{3x - 2}$

54. $f(x) = \dfrac{5x^2 - 4}{5x - 5}$

55. $f(x) = \dfrac{4x^3 + 4x^2 + 7x + 4}{x^2 + 1}$

56. $f(x) = \dfrac{3x^2 - 2x + 5}{3x + 4}$

57–62. Transcendental functions *Determine the end behavior of the following transcendental functions by analyzing appropriate limits. Then provide a simple sketch of the associated graph, showing asymptotes if they exist.*

57. $f(x) = -3e^{-x}$ **58.** $f(x) = 2^x$ **59.** $f(x) = 1 - \ln x$

60. $f(x) = |\ln x|$ **61.** $f(x) = \sin x$ **62.** $f(x) = \dfrac{50}{e^{2x}}$

63. Explain why or why not Determine whether the following statements are true and give an explanation or counterexample.

  a. The graph of a function can never cross one of its horizontal asymptotes.
  b. A rational function f has both $\lim\limits_{x \to \infty} f(x) = L$ (where L is finite) and $\lim\limits_{x \to -\infty} f(x) = \infty$.
  c. The graph of a function can have any number of vertical asymptotes but at most two horizontal asymptotes.
  d. $\lim\limits_{x \to \infty} (x^3 - x) = \lim\limits_{x \to \infty} x^3 - \lim\limits_{x \to \infty} x = \infty - \infty = 0$

64–69. Steady states *If a function f represents a system that varies in time, the existence of* $\lim\limits_{t \to \infty} f(t)$ *means that the system reaches a steady state (or equilibrium). For the following systems, determine whether a steady state exists and give the steady-state value.*

64. The population of a bacteria culture is given by $p(t) = \dfrac{2500}{t + 1}$.

65. The population of a culture of tumor cells is given by
$$p(t) = \dfrac{3500t}{t + 1}.$$

66. The amount of drug (in milligrams) in the blood after an IV tube is inserted is given by $m(t) = 200(1 - 2^{-t})$.

67. The value of an investment is given by $v(t) = 1000e^{0.065t}$.

68. The population of a colony of squirrels is given by
$$p(t) = \dfrac{1500}{3 + 2e^{-0.1t}}.$$

69. The amplitude of an oscillator is given by $a(t) = 2\left(\dfrac{t + \sin t}{t}\right)$.

70–81. Horizontal and vertical asymptotes

  a. *Analyze* $\lim\limits_{x \to \infty} f(x)$ *and* $\lim\limits_{x \to -\infty} f(x)$, *and then identify any horizontal asymptotes.*
  b. *Find the vertical asymptotes. For each vertical asymptote $x = a$, analyze* $\lim\limits_{x \to a^-} f(x)$ *and* $\lim\limits_{x \to a^+} f(x)$.

70. $f(x) = \dfrac{x^2 - 4x + 3}{x - 1}$

71. $f(x) = \dfrac{2x^3 + 10x^2 + 12x}{x^3 + 2x^2}$

72. $f(x) = \dfrac{\sqrt{16x^4 + 64x^2} + x^2}{2x^2 - 4}$

73. $f(x) = \dfrac{3x^4 + 3x^3 - 36x^2}{x^4 - 25x^2 + 144}$

74. $f(x) = x^2\left(4x^2 - \sqrt{16x^4 + 1}\right)$

75. $f(x) = \dfrac{x^2 - 9}{x(x - 3)}$ **76.** $f(x) = \dfrac{x^4 - 1}{x^2 - 1}$

77. $f(x) = \dfrac{\sqrt{x^2 + 2x + 6} - 3}{x - 1}$

78. $f(x) = \dfrac{|1 - x^2|}{x(x + 1)}$

79. $f(x) = \sqrt{|x|} - \sqrt{|x - 1|}$

80. $f(x) = \dfrac{3e^x + 10}{e^x}$ **81.** $f(x) = \dfrac{\cos x + 2\sqrt{x}}{\sqrt{x}}$.

82–85. End behavior for transcendental functions

82. Consider the graph of $y = \cot^{-1} x$ (see Section 1.4) and determine the following limits using the graph.

   a. $\lim\limits_{x \to \infty} \cot^{-1} x$ **b.** $\lim\limits_{x \to -\infty} \cot^{-1} x$

83. Consider the graph of $y = \sec^{-1} x$ (see Section 1.4) and evaluate the following limits using the graph. Assume the domain is $\{x : |x| \geq 1\}$.

   a. $\lim\limits_{x \to \infty} \sec^{-1} x$ **b.** $\lim\limits_{x \to -\infty} \sec^{-1} x$

84. The **hyperbolic cosine function**, denoted $\cosh x$, is used to model the shape of a hanging cable (a telephone wire, for example). It is defined as $\cosh x = \dfrac{e^x + e^{-x}}{2}$.

   a. Determine its end behavior by analyzing $\lim\limits_{x \to \infty} \cosh x$ and $\lim\limits_{x \to -\infty} \cosh x$.

   b. Evaluate $\cosh 0$. Use symmetry and part (a) to sketch a plausible graph for $y = \cosh x$.

85. The **hyperbolic sine function** is defined as $\sinh x = \dfrac{e^x - e^{-x}}{2}$.

   a. Determine its end behavior.

   b. Evaluate $\sinh 0$. Use symmetry and part (a) to sketch a plausible graph for $y = \sinh x$.

86–87. Sketching graphs *Sketch a possible graph of a function f that satisfies all of the given conditions. Be sure to identify all vertical and horizontal asymptotes.*

86. $f(-1) = -2$, $f(1) = 2$, $f(0) = 0$, $\lim\limits_{x \to \infty} f(x) = 1$, $\lim\limits_{x \to -\infty} f(x) = -1$

87. $\lim\limits_{x \to 0^+} f(x) = \infty$, $\lim\limits_{x \to 0^-} f(x) = -\infty$, $\lim\limits_{x \to \infty} f(x) = 1$, $\lim\limits_{x \to -\infty} f(x) = -2$

Explorations and Challenges

88–91. Looking ahead to sequences *A sequence is an infinite, ordered list of numbers that is often defined by a function. For example, the sequence $\{2, 4, 6, 8, \ldots\}$ is specified by the function $f(n) = 2n$, where $n = 1, 2, 3, \ldots$. The limit of such a sequence is $\lim\limits_{n \to \infty} f(n)$, provided the limit exists. All the limit laws for limits at infinity may be applied to limits of sequences. Find the limit of the following sequences or state that the limit does not exist.*

88. $\left\{4, 2, \dfrac{4}{3}, 1, \dfrac{4}{5}, \dfrac{2}{3}, \ldots\right\}$, which is defined by $f(n) = \dfrac{4}{n}$, for $n = 1, 2, 3, \ldots$

89. $\left\{0, \dfrac{1}{2}, \dfrac{2}{3}, \dfrac{3}{4}, \ldots\right\}$, which is defined by $f(n) = \dfrac{n - 1}{n}$, for $n = 1, 2, 3, \ldots$

90. $\left\{\dfrac{1}{2}, \dfrac{4}{3}, \dfrac{9}{4}, \dfrac{16}{5}, \ldots\right\}$, which is defined by $f(n) = \dfrac{n^2}{n + 1}$, for $n = 1, 2, 3, \ldots$

91. $\left\{2, \dfrac{3}{4}, \dfrac{4}{9}, \dfrac{5}{16}, \ldots\right\}$, which is defined by $f(n) = \dfrac{n + 1}{n^2}$, for $n = 1, 2, 3, \ldots$

92. End behavior of rational functions Suppose $f(x) = \dfrac{p(x)}{q(x)}$ is a rational function, where

$$p(x) = a_m x^m + a_{m-1} x^{m-1} + \cdots + a_2 x^2 + a_1 x + a_0,$$
$$q(x) = b_n x^n + b_{n-1} x^{n-1} + \cdots + b_2 x^2 + b_1 x + b_0, \; a_m \neq 0,$$

and $b_n \neq 0$.

   a. Prove that if $m = n$, then $\lim\limits_{x \to \pm\infty} f(x) = \dfrac{a_m}{b_n}$.

   b. Prove that if $m < n$, then $\lim\limits_{x \to \pm\infty} f(x) = 0$.

93. Horizontal and slant asymptotes

   a. Is it possible for a rational function to have both slant and horizontal asymptotes? Explain.

   b. Is it possible for an algebraic function to have two distinct slant asymptotes? Explain or give an example.

T 94. End behavior of exponentials Use the following instructions to determine the end behavior of $f(x) = \dfrac{4e^x + 2e^{2x}}{8e^x + e^{2x}}$.

   a. Evaluate $\lim\limits_{x \to \infty} f(x)$ by first dividing the numerator and denominator by e^{2x}.

   b. Evaluate $\lim\limits_{x \to -\infty} f(x)$ by first dividing the numerator and denominator by e^x.

   c. Give the horizontal asymptote(s).

   d. Graph f to confirm your work in parts (a)–(c).

95–96. *Find the horizontal asymptotes of each function using limits at infinity.*

95. $f(x) = \dfrac{2e^x + 3}{e^x + 1}$ **96.** $f(x) = \dfrac{3e^{5x} + 7e^{6x}}{9e^{5x} + 14e^{6x}}$

T 97. Use analytical methods to identify all the asymptotes of $f(x) = \dfrac{\ln x^6}{\ln x^3 - 1}$. Plot a graph of the function with a graphing utility and then sketch a graph by hand, correcting any errors in the computer-generated graph.

QUICK CHECK ANSWERS

1. $10/11$, $100/101$, $1000/1001$; 1 **2.** $p(x) \to -\infty$ as $x \to \infty$ and $p(x) \to \infty$ as $x \to -\infty$ **3.** Horizontal asymptote is $y = \dfrac{10}{3}$; vertical asymptote is $x = \dfrac{1}{3}$.

4. $\lim\limits_{x \to \infty} e^{10x} = \infty$, $\lim\limits_{x \to -\infty} e^{10x} = 0$, $\lim\limits_{x \to \infty} e^{-10x} = 0$, $\lim\limits_{x \to -\infty} e^{-10x} = \infty$ ◄

2.6 Continuity

The graphs of many functions encountered in this text contain no holes, jumps, or breaks. For example, if $L = f(t)$ represents the length (in inches) of a fish t years after it is hatched, then the length of the fish changes gradually as t increases. Consequently, the graph of $L = f(t)$ contains no breaks (**Figure 2.46a**). Some functions, however, do contain abrupt changes in their values. Consider a parking meter that accepts only quarters, where each quarter buys 15 minutes of parking. Letting $c(t)$ be the cost (in dollars) of parking for t minutes, the graph of c has breaks at integer multiples of 15 minutes (**Figure 2.46b**).

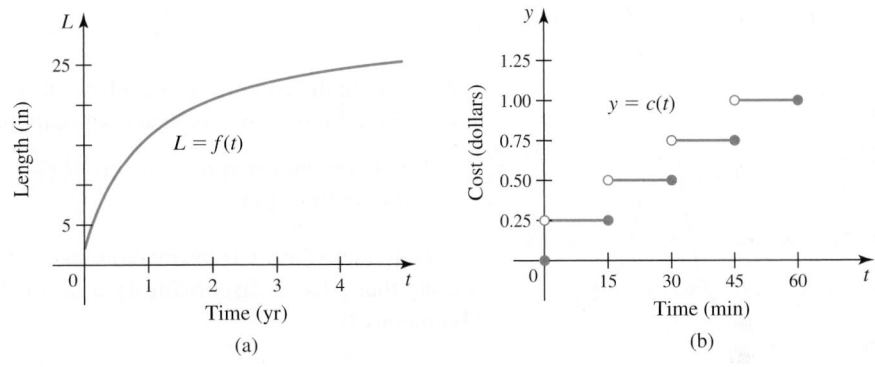

Figure 2.46

Informally, we say that a function f is *continuous* at a if the graph of f does not have a hole or break at a (that is, if the graph near a can be drawn without lifting the pencil).

Continuity at a Point

This informal description of continuity is sufficient for determining the continuity of simple functions, but it is not precise enough to deal with more complicated functions such as

$$h(x) = \begin{cases} x \sin \dfrac{1}{x} & \text{if } x \neq 0 \\ 0 & \text{if } x = 0. \end{cases}$$

It is difficult to determine whether the graph of h has a break at 0 because it oscillates rapidly as x approaches 0 (**Figure 2.47**). We need a better definition.

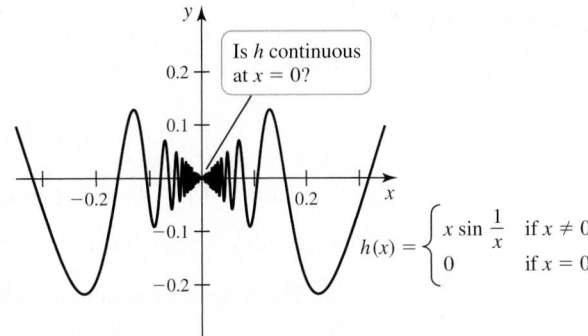

Figure 2.47

DEFINITION Continuity at a Point

A function f is **continuous** at a if $\lim\limits_{x \to a} f(x) = f(a)$.

There is more to this definition than first appears. If $\lim\limits_{x \to a} f(x) = f(a)$, then $f(a)$ and $\lim\limits_{x \to a} f(x)$ must both exist, and they must be equal. The following checklist is helpful in determining whether a function is continuous at a.

Continuity Checklist

In order for f to be continuous at a, the following three conditions must hold.

1. $f(a)$ is defined (a is in the domain of f).
2. $\lim\limits_{x \to a} f(x)$ exists.
3. $\lim\limits_{x \to a} f(x) = f(a)$ (the value of f equals the limit of f at a).

If *any* item in the continuity checklist fails to hold, the function fails to be continuous at a. From this definition, we see that continuity has an important practical consequence:

> *If f is continuous at a, then $\lim\limits_{x \to a} f(x) = f(a)$, and direct substitution may be used to evaluate $\lim\limits_{x \to a} f(x)$.*

Note that when f is defined on an open interval containing a (except possibly at a), we say that f has a **discontinuity** at a (or that a is a **point of discontinuity**) if f is not continuous at a.

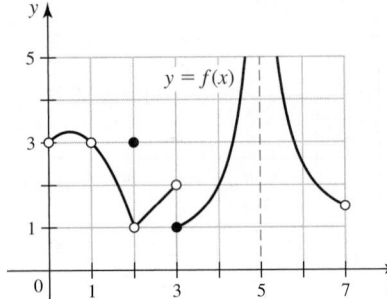

Figure 2.48

➤ In Example 1, the discontinuities at $x = 1$ and $x = 2$ are called **removable discontinuities** because they can be removed by redefining the function at these points (in this case, $f(1) = 3$ and $f(2) = 1$). The discontinuity at $x = 3$ is called a **jump discontinuity**. The discontinuity at $x = 5$ is called an **infinite discontinuity**. These terms are discussed in Exercises 95–101.

EXAMPLE 1 **Points of discontinuity** Use the graph of f in **Figure 2.48** to identify values of x on the interval $(0, 7)$ at which f has a discontinuity.

SOLUTION The function f has discontinuities at $x = 1, 2, 3,$ and 5 because the graph contains holes or breaks at these locations. The continuity checklist tells us why f is not continuous at these points.

- $f(1)$ is not defined.
- $f(2) = 3$ and $\lim\limits_{x \to 2} f(x) = 1$. Therefore, $f(2)$ and $\lim\limits_{x \to 2} f(x)$ exist but are not equal.
- $\lim\limits_{x \to 3} f(x)$ does not exist because the left-sided limit $\lim\limits_{x \to 3^-} f(x) = 2$ differs from the right-sided limit $\lim\limits_{x \to 3^+} f(x) = 1$.
- Neither $\lim\limits_{x \to 5} f(x)$ nor $f(5)$ exists.

Related Exercises 5–6 ◄

QUICK CHECK 1 For what values of t in $(0, 60)$ does the graph of $y = c(t)$ in Figure 2.46b have a discontinuity? ◄

EXAMPLE 2 **Continuity at a point** Determine whether the following functions are continuous at a. Justify each answer using the continuity checklist.

a. $f(x) = \dfrac{3x^2 + 2x + 1}{x - 1}; \quad a = 1$

b. $g(x) = \dfrac{3x^2 + 2x + 1}{x - 1}; \quad a = 2$

c. $h(x) = \begin{cases} x \sin \dfrac{1}{x} & \text{if } x \neq 0 \\ 0 & \text{if } x = 0 \end{cases}; \quad a = 0$

SOLUTION

a. The function f is not continuous at 1 because $f(1)$ is undefined.

b. Because g is a rational function and the denominator is nonzero at 2, it follows by Theorem 2.4 that $\lim\limits_{x \to 2} g(x) = g(2) = 17$. Therefore, g is continuous at 2.

c. By definition, $h(0) = 0$. In Exercise 81 of Section 2.3, we used the Squeeze Theorem to show that $\lim\limits_{x \to 0} x \sin \dfrac{1}{x} = 0$. Therefore, $\lim\limits_{x \to 0} h(x) = h(0)$, which implies that h is continuous at 0.

Related Exercises 17, 22 ◄

The following theorems make it easier to test various combinations of functions for continuity at a point.

THEOREM 2.9   Continuity Rules
If f and g are continuous at a, then the following functions are also continuous at a. Assume c is a constant and $n > 0$ is an integer.

a. $f + g$ **b.** $f - g$

c. cf **d.** fg

e. f/g, provided $g(a) \neq 0$ **f.** $(f(x))^n$

To prove the first result, note that if f and g are continuous at a, then $\lim\limits_{x \to a} f(x) = f(a)$ and $\lim\limits_{x \to a} g(x) = g(a)$. From the limit laws of Theorem 2.3, it follows that

$$\lim_{x \to a} (f(x) + g(x)) = f(a) + g(a).$$

Therefore, $f + g$ is continuous at a. Similar arguments lead to the continuity of differences, products, quotients, and powers of continuous functions. The next theorem is a direct consequence of Theorem 2.9.

THEOREM 2.10   Polynomial and Rational Functions

a. A polynomial function is continuous for all x.

b. A rational function (a function of the form $\dfrac{p}{q}$, where p and q are polynomials) is continuous for all x for which $q(x) \neq 0$.

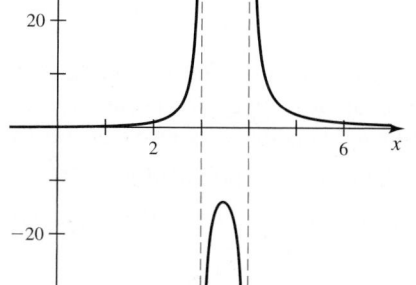

$$f(x) = \frac{x}{x^2 - 7x + 12}$$

Continuous everywhere except $x = 3$ and $x = 4$

Figure 2.49

EXAMPLE 3   Applying the continuity theorems For what values of x is the function

$$f(x) = \frac{x}{x^2 - 7x + 12} \text{ continuous?}$$

SOLUTION Because f is rational, Theorem 2.10b implies it is continuous for all x at which the denominator is nonzero. The denominator factors as $(x - 3)(x - 4)$, so it is zero at $x = 3$ and $x = 4$. Therefore, f is continuous for all x except $x = 3$ and $x = 4$ (Figure 2.49).

Related Exercises 26–27 ◄

The following theorem enables us to determine when a composition of two functions is continuous at a point. Its proof is informative and is outlined in Exercise 102.

THEOREM 2.11   Continuity of Composite Functions at a Point
If g is continuous at a and f is continuous at $g(a)$, then the composite function $f \circ g$ is continuous at a.

Theorem 2.11 is useful because it allows us to conclude that the composition of two continuous functions is continuous at a point. For example, the composite function $\left(\dfrac{x}{x - 1}\right)^3$ is continuous for all $x \neq 1$. Furthermore, under the stated conditions on f and g, the limit of $f \circ g$ is evaluated by direct substitution; that is,

$$\lim_{x \to a} f(g(x)) = f(g(a)).$$

EXAMPLE 4   **Limit of a composition**   Evaluate $\displaystyle\lim_{x \to 0} \left(\dfrac{x^4 - 2x + 2}{x^6 + 2x^4 + 1} \right)^{10}$.

SOLUTION   The rational function $\dfrac{x^4 - 2x + 2}{x^6 + 2x^4 + 1}$ is continuous for all x because its denominator is always positive (Theorem 2.10b). Therefore, $\left(\dfrac{x^4 - 2x + 2}{x^6 + 2x^4 + 1} \right)^{10}$, which is the composition of the continuous function $f(x) = x^{10}$ and a continuous rational function, is continuous for all x by Theorem 2.11. By direct substitution,

$$\lim_{x \to 0} \left(\frac{x^4 - 2x + 2}{x^6 + 2x^4 + 1} \right)^{10} = \left(\frac{0^4 - 2 \cdot 0 + 2}{0^6 + 2 \cdot 0^4 + 1} \right)^{10} = 2^{10} = 1024.$$

Related Exercises 31–32 ◄

Closely related to Theorem 2.11 are two useful results dealing with limits of composite functions. We present these results—one a more general version of the other—in a single theorem.

THEOREM 2.12   Limits of Composite Functions

1. If g is continuous at a and f is continuous at $g(a)$, then

$$\lim_{x \to a} f(g(x)) = f\left(\lim_{x \to a} g(x) \right).$$

2. If $\displaystyle\lim_{x \to a} g(x) = L$ and f is continuous at L, then

$$\lim_{x \to a} f(g(x)) = f\left(\lim_{x \to a} g(x) \right).$$

Proof: The first statement follows directly from Theorem 2.11, which states that $\displaystyle\lim_{x \to a} f(g(x)) = f(g(a))$. If g is continuous at a, then $\displaystyle\lim_{x \to a} g(x) = g(a)$, and it follows that

$$\lim_{x \to a} f(g(x)) = f(\underbrace{g(a)}_{\lim_{x \to a} g(x)}) = f\left(\lim_{x \to a} g(x) \right).$$

The proof of the second statement (see Appendix A) relies on the formal definition of a limit, which is discussed in Section 2.7.   ◄

Both statements of Theorem 2.12 justify interchanging the order of a limit and a function evaluation. By the second statement, the inner function of the composition needn't be continuous at the point of interest, but it must have a limit at that point. Note also that $\displaystyle\lim_{x \to a}$ can be replaced with $\displaystyle\lim_{x \to a^+}$ or $\displaystyle\lim_{x \to a^-}$ in statement (1) of Theorem 2.12, provided g is *right-* or *left-continuous* (see p. 107) at a, respectively. In statement (2), $\displaystyle\lim_{x \to a}$ can be replaced with $\displaystyle\lim_{x \to \infty}$ or $\displaystyle\lim_{x \to -\infty}$.

EXAMPLE 5   **Limits of a composite functions**   Evaluate the following limits.

a. $\displaystyle\lim_{x \to -1} \sqrt{2x^2 - 1}$ **b.** $\displaystyle\lim_{x \to 2} \cos\left(\dfrac{x^2 - 4}{x - 2} \right)$

SOLUTION

a. The inner function of the composite function $\sqrt{2x^2 - 1}$ is $g(x) = 2x^2 - 1$; it is continuous and positive at -1, and $g(-1) = 1$. Because $f(x) = \sqrt{x}$ is continuous at $g(-1) = 1$ (a consequence of Law 7, Theorem 2.3), we have, by the first statement of Theorem 2.12,

$$\lim_{x \to -1} \sqrt{2x^2 - 1} = \sqrt{\underbrace{\lim_{x \to -1} (2x^2 - 1)}_{1}} = \sqrt{1} = 1.$$

b. We show later in this section that $\cos x$ is continuous at all points of its domain. The inner function of the composite function $\cos\left(\dfrac{x^2 - 4}{x - 2}\right)$ is $\dfrac{x^2 - 4}{x - 2}$, which is not continuous at 2. However,

$$\lim_{x\to 2} \frac{x^2 - 4}{x - 2} = \lim_{x\to 2} \frac{(x - 2)(x + 2)}{x - 2} = \lim_{x\to 2} (x + 2) = 4.$$

Therefore, by the second statement of Theorem 2.12,

$$\lim_{x\to 2} \cos\left(\frac{x^2 - 4}{x - 2}\right) = \cos\left(\underbrace{\lim_{x\to 2} \frac{x^2 - 4}{x - 2}}_{4}\right) = \cos 4 \approx -0.654.$$

Related Exercises 33–34 ◄

Continuity on an Interval

A function is *continuous on an interval* if it is continuous at every point in that interval. Consider the functions f and g whose graphs are shown in **Figure 2.50**. Both these functions are continuous for all x in (a, b), but what about the endpoints? To answer this question, we introduce the ideas of *left-continuity* and *right-continuity*.

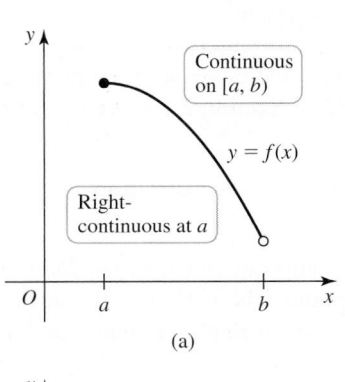

(a)

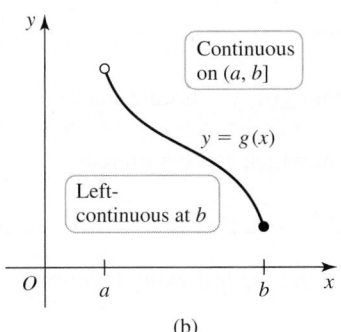

(b)

Figure 2.50

> **DEFINITION** **Continuity at Endpoints**
>
> A function f is **continuous from the right** (or **right-continuous**) at a if $\lim\limits_{x\to a^+} f(x) = f(a)$, and f is **continuous from the left** (or **left-continuous**) at b if $\lim\limits_{x\to b^-} f(x) = f(b)$.

Combining the definitions of left-continuous and right-continuous with the definition of continuity at a point, we define what it means for a function to be continuous on an interval.

> **DEFINITION** **Continuity on an Interval**
>
> A function f is **continuous on an interval I** if it is continuous at all points of I. If I contains its endpoints, continuity on I means continuous from the right or left at the endpoints.

To illustrate these definitions, consider again the functions in Figure 2.50. In **Figure 2.50a**, f is continuous from the right at a because $\lim\limits_{x\to a^+} f(x) = f(a)$, but it is not continuous from the left at b because $f(b)$ is not defined. Therefore, f is continuous on the interval $[a, b)$. The behavior of the function g in **Figure 2.50b** is the opposite: It is continuous from the left at b, but it is not continuous from the right at a. Therefore, g is continuous on $(a, b]$.

QUICK CHECK 2 Modify the graphs of the functions f and g in Figure 2.50 to obtain functions that are continuous on $[a, b]$. ◄

EXAMPLE 6 **Intervals of continuity** Determine the intervals of continuity for

$$f(x) = \begin{cases} x^2 + 1 & \text{if } x \le 0 \\ 3x + 5 & \text{if } x > 0. \end{cases}$$

SOLUTION This piecewise function consists of two polynomials that describe a parabola and a line (**Figure 2.51**). By Theorem 2.10, f is continuous for all $x \ne 0$. From its graph, it appears that f is left-continuous at 0. This observation is verified by noting that

$$\lim_{x\to 0^-} f(x) = \lim_{x\to 0^-} (x^2 + 1) = 1,$$

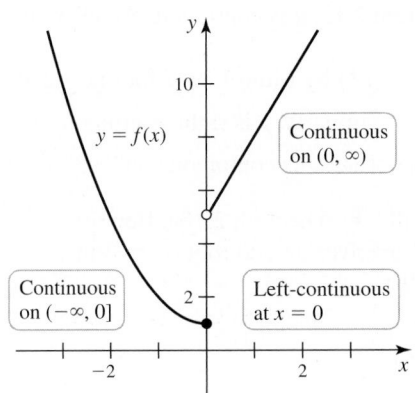

Figure 2.51

which means that $\lim\limits_{x \to 0^-} f(x) = f(0)$. However, because

$$\lim_{x \to 0^+} f(x) = \lim_{x \to 0^+} (3x + 5) = 5 \neq f(0),$$

we see that f is not right-continuous at 0. Therefore, f is continuous on $(-\infty, 0]$ and on $(0, \infty)$.

Related Exercises 39–40 ◄

Functions Involving Roots

Recall that Limit Law 7 of Theorem 2.3 states

$$\lim_{x \to a} (f(x))^{1/n} = \left(\lim_{x \to a} f(x) \right)^{1/n},$$

provided $f(x) > 0$, for x near a, if n is even. Therefore, if n is odd and f is continuous at a, then $(f(x))^{1/n}$ is continuous at a, because

$$\lim_{x \to a} (f(x))^{1/n} = \left(\lim_{x \to a} f(x) \right)^{1/n} = (f(a))^{1/n}.$$

When n is even, the continuity of $(f(x))^{1/n}$ must be handled more carefully because this function is defined only when $f(x) \geq 0$. Exercise 73 of Section 2.7 establishes an important fact:

> If f is continuous at a, and $f(a) > 0$, then $f(x) > 0$ for all values of x in some interval containing a.

Combining this fact with Theorem 2.11 (the continuity of composite functions), it follows that $(f(x))^{1/n}$ is continuous at a provided $f(a) > 0$. At points where $f(a) = 0$, the behavior of $(f(x))^{1/n}$ varies. Often we find that $(f(x))^{1/n}$ is left- or right-continuous at that point, or it may be continuous from both sides.

THEOREM 2.13   Continuity of Functions with Roots

Assume n is a positive integer. If n is an odd integer, then $(f(x))^{1/n}$ is continuous at all points at which f is continuous.

If n is even, then $(f(x))^{1/n}$ is continuous at all points a at which f is continuous and $f(a) > 0$.

EXAMPLE 7   Continuity with roots For what values of x are the following functions continuous?

a. $g(x) = \sqrt{9 - x^2}$

b. $f(x) = (x^2 - 2x + 4)^{2/3}$

SOLUTION

a. The graph of g is the upper half of the circle $x^2 + y^2 = 9$ (which can be verified by solving $x^2 + y^2 = 9$ for y). From **Figure 2.52**, it appears that g is continuous on $[-3, 3]$. To verify this fact, note that g involves an even root ($n = 2$ in Theorem 2.13). If $-3 < x < 3$, then $9 - x^2 > 0$ and by Theorem 2.13, g is continuous for all x on $(-3, 3)$.

At the right endpoint, $\lim\limits_{x \to 3^-} \sqrt{9 - x^2} = 0 = g(3)$ by Limit Law 7 for one-sided limits, which implies that g is left-continuous at 3. Similarly, g is right-continuous at -3 because $\lim\limits_{x \to -3^+} \sqrt{9 - x^2} = 0 = g(-3)$. Therefore, g is continuous on $[-3, 3]$.

b. The polynomial $x^2 - 2x + 4$ is continuous for all x by Theorem 2.10a. Rewriting f as $f(x) = ((x^2 - 2x + 4)^{1/3})^2$, we see that f involves an odd root ($n = 3$ in Theorem 2.13). Therefore, f is continuous for all x.

Related Exercises 44–45 ◄

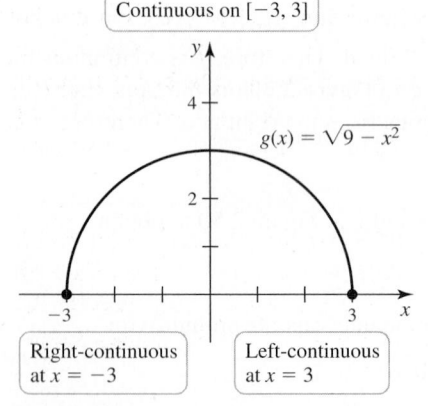

Continuous on $[-3, 3]$

$g(x) = \sqrt{9 - x^2}$

Right-continuous at $x = -3$

Left-continuous at $x = 3$

Figure 2.52

QUICK CHECK 3 On what interval is $f(x) = x^{1/4}$ continuous? On what interval is $f(x) = x^{2/5}$ continuous? ◄

Continuity of Transcendental Functions

The understanding of continuity that we have developed with algebraic functions may now be applied to transcendental functions.

Trigonometric Functions   In Section 2.3, we used the Squeeze Theorem to show that $\lim_{x \to 0} \sin x = 0$ and $\lim_{x \to 0} \cos x = 1$. Because $\sin 0 = 0$ and $\cos 0 = 1$, these limits imply that $\sin x$ and $\cos x$ are continuous at 0. The graph of $y = \sin x$ (**Figure 2.53**) suggests that $\lim_{x \to a} \sin x = \sin a$ for any value of a, which means that $\sin x$ is continuous on $(-\infty, \infty)$. The graph of $y = \cos x$ also indicates that $\cos x$ is continuous for all x. Exercise 105 outlines a proof of these results.

With these facts in hand, we appeal to Theorem 2.9e to discover that the remaining trigonometric functions are continuous on their domains. For example, because $\sec x = 1/\cos x$, the secant function is continuous for all x for which $\cos x \neq 0$ (for all x except odd multiples of $\pi/2$) (**Figure 2.54**). Likewise, the tangent, cotangent, and cosecant functions are continuous at all points of their domains.

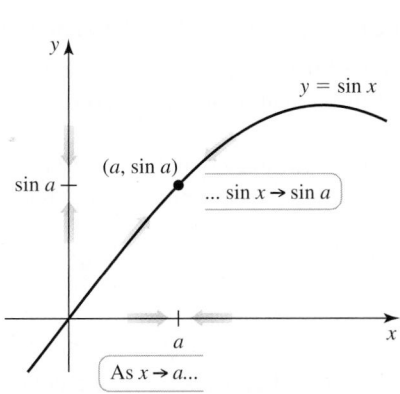

Figure 2.53

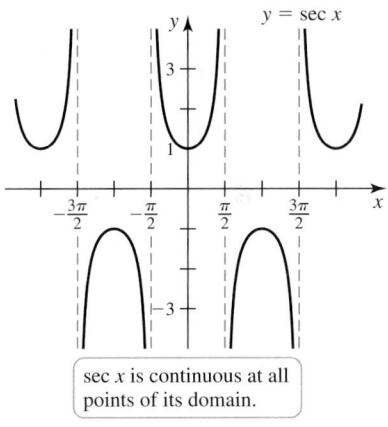

sec x is continuous at all points of its domain.

Figure 2.54

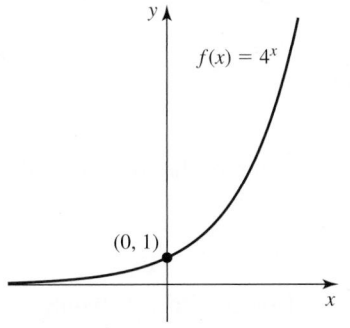

Exponential functions are defined for all real numbers and are continuous on $(-\infty, \infty)$, as shown in Chapter 7.

Figure 2.55

Exponential Functions   The continuity of exponential functions of the form $f(x) = b^x$, with $0 < b < 1$ or $b > 1$, raises an important question. Consider the function $f(x) = 4^x$ (**Figure 2.55**). Evaluating f is routine if x is rational:

$$4^3 = 4 \cdot 4 \cdot 4 = 64; \quad 4^{-2} = \frac{1}{4^2} = \frac{1}{16}; \quad 4^{3/2} = \sqrt{4^3} = 8; \quad \text{and} \quad 4^{-1/3} = \frac{1}{\sqrt[3]{4}}.$$

But what is meant by 4^x when x is an irrational number, such as $\sqrt{2}$? In order for $f(x) = 4^x$ to be continuous for all real numbers, it must also be defined when x is an irrational number. Providing a working definition for an expression such as $4^{\sqrt{2}}$ requires mathematical results that don't appear until Chapter 7. Until then, we assume without proof that the domain of $f(x) = b^x$ is the set of all real numbers and that f is continuous at all points of its domain.

Inverse Functions   Suppose a function f is continuous and one-to-one on an interval I. Reflecting the graph of f through the line $y = x$ generates the graph of f^{-1}. The reflection process introduces no discontinuities in the graph of f^{-1}, so it is plausible (and indeed true) that f^{-1} is continuous on the interval corresponding to I. We state this fact without a formal proof.

THEOREM 2.14   Continuity of Inverse Functions

If a function f is continuous on an interval I and has an inverse on I, then its inverse f^{-1} is also continuous (on the interval consisting of the points $f(x)$, where x is in I).

Because all the trigonometric functions are continuous on their domains, they are also continuous when their domains are restricted for the purpose of defining inverse functions. Therefore, by Theorem 2.14, the inverse trigonometric functions are continuous at all points of their domains.

Logarithmic functions of the form $f(x) = \log_b x$ are continuous at all points of their domains for the same reason: They are inverses of exponential functions, which are one-to-one and continuous. Collecting all these facts, we have the following theorem.

THEOREM 2.15 Continuity of Transcendental Functions
The following functions are continuous at all points of their domains.

Trigonometric		Inverse Trigonometric		Exponential	
$\sin x$	$\cos x$	$\sin^{-1} x$	$\cos^{-1} x$	b^x	e^x
$\tan x$	$\cot x$	$\tan^{-1} x$	$\cot^{-1} x$	**Logarithmic**	
$\sec x$	$\csc x$	$\sec^{-1} x$	$\csc^{-1} x$	$\log_b x$	$\ln x$

For each function listed in Theorem 2.15, we have $\lim\limits_{x \to a} f(x) = f(a)$, provided a is in the domain of the function. This means that limits of these functions may be evaluated by direct substitution at points in the domain.

EXAMPLE 8 Limits involving transcendental functions Analyze the following limits after determining the continuity of the functions involved.

a. $\lim\limits_{x \to 0} \dfrac{\cos^2 x - 1}{\cos x - 1}$

b. $\lim\limits_{x \to 1} \left(\sqrt[4]{\ln x} + \tan^{-1} x \right)$

SOLUTION

> ► Limits like the one in Example 8a are denoted 0/0 and are known as *indeterminate forms*, to be studied further in Section 4.7.

a. Both $\cos^2 x - 1$ and $\cos x - 1$ are continuous for all x by Theorems 2.9 and 2.15. However, the ratio of these functions is continuous only when $\cos x - 1 \ne 0$, which occurs when x is not an integer multiple of 2π. Note that both the numerator and denominator of $\dfrac{\cos^2 x - 1}{\cos x - 1}$ approach 0 as $x \to 0$. To evaluate the limit, we factor and simplify:

$$\lim_{x \to 0} \frac{\cos^2 x - 1}{\cos x - 1} = \lim_{x \to 0} \frac{(\cos x - 1)(\cos x + 1)}{\cos x - 1} = \lim_{x \to 0} (\cos x + 1)$$

(where $\cos x - 1$ may be canceled because it is nonzero as x approaches 0). The limit on the right is now evaluated using direct substitution:

$$\lim_{x \to 0} (\cos x + 1) = \cos 0 + 1 = 2.$$

> **QUICK CHECK 4** Show that $f(x) = \sqrt[4]{\ln x}$ is right-continuous at $x = 1$. ◄

b. By Theorem 2.15, $\ln x$ is continuous on its domain $(0, \infty)$. However, $\ln x > 0$ only when $x > 1$, so Theorem 2.13 implies that $\sqrt[4]{\ln x}$ is continuous on $(1, \infty)$. At $x = 1$, $\sqrt[4]{\ln x}$ is right-continuous (Quick Check 4). The domain of $\tan^{-1} x$ is all real numbers, and it is continuous on $(-\infty, \infty)$. Therefore, $f(x) = \sqrt[4]{\ln x} + \tan^{-1} x$ is continuous on $[1, \infty)$. Because the domain of f does not include points with $x < 1$, $\lim\limits_{x \to 1^-} \left(\sqrt[4]{\ln x} + \tan^{-1} x \right)$ does not exist, which implies that $\lim\limits_{x \to 1} \left(\sqrt[4]{\ln x} + \tan^{-1} x \right)$ does not exist.

Related Exercises 62–63 ◄

We close this section with an important theorem that has both practical and theoretical uses.

Intermediate Value Theorem

A common problem in mathematics is finding solutions to equations of the form $f(x) = L$. Before attempting to find values of x satisfying this equation, it is worthwhile to determine whether a solution exists.

The existence of solutions is often established using a result known as the *Intermediate Value Theorem*. Given a function f and a constant L, we assume L lies strictly between $f(a)$ and $f(b)$. The Intermediate Value Theorem says that if f is continuous on $[a, b]$, then the graph of f must cross the horizontal line $y = L$ at least once (**Figure 2.56**). Although this theorem is easily illustrated, its proof is beyond the scope of this text.

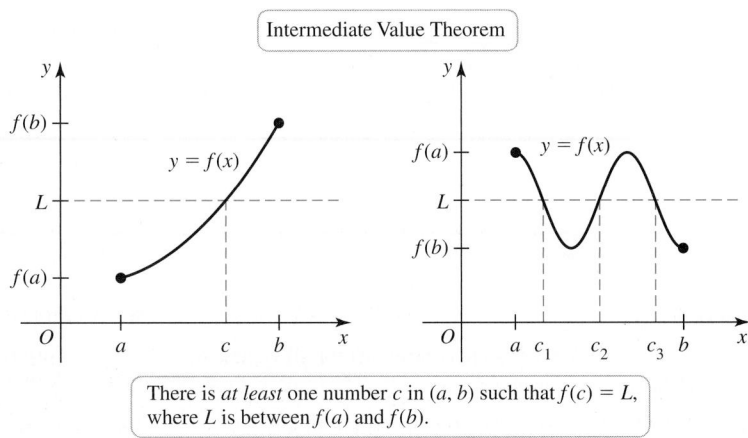

Intermediate Value Theorem

There is *at least* one number c in (a, b) such that $f(c) = L$, where L is between $f(a)$ and $f(b)$.

Figure 2.56

> **THEOREM 2.16   Intermediate Value Theorem**
> Suppose f is continuous on the interval $[a, b]$ and L is a number strictly between $f(a)$ and $f(b)$. Then there exists at least one number c in (a, b) satisfying $f(c) = L$.

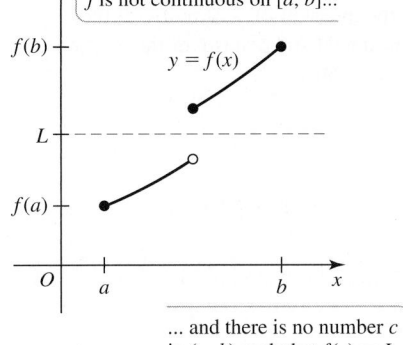

f is not continuous on $[a, b]$...

... and there is no number c in (a, b) such that $f(c) = L$.

Figure 2.57

The importance of continuity in Theorem 2.16 is illustrated in **Figure 2.57**, where we see a function f that is not continuous on $[a, b]$. For the value of L shown in the figure, there is no value of c in (a, b) satisfying $f(c) = L$. The next example illustrates a practical application of the Intermediate Value Theorem.

EXAMPLE 9   Finding an interest rate Suppose you invest $1000 in a special 5-year savings account with a fixed annual interest rate r, with monthly compounding. The amount of money A in the account after 5 years (60 months) is

$$A(r) = 1000\left(1 + \frac{r}{12}\right)^{60}.$$ Your goal is to have $1400 in the account after 5 years.

a. Use the Intermediate Value Theorem to show there is a value of r in $(0, 0.08)$—that is, an interest rate between 0% and 8%—for which $A(r) = 1400$.

b. Use a graphing utility to illustrate your explanation in part (a) and then estimate the interest rate required to reach your goal.

SOLUTION

a. As a polynomial in r (of degree 60), $A(r) = 1000\left(1 + \dfrac{r}{12}\right)^{60}$ is continuous for all r.

Evaluating $A(r)$ at the endpoints of the interval $[0, 0.08]$, we have $A(0) = 1000$ and $A(0.08) \approx 1489.85$. Therefore,

$$A(0) < 1400 < A(0.08),$$

and it follows, by the Intermediate Value Theorem, that there is a value of r in $(0, 0.08)$ for which $A(r) = 1400$.

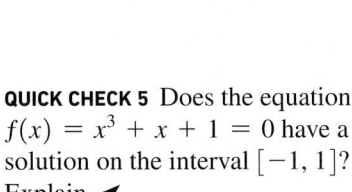

QUICK CHECK 5 Does the equation $f(x) = x^3 + x + 1 = 0$ have a solution on the interval $[-1, 1]$? Explain. ◀

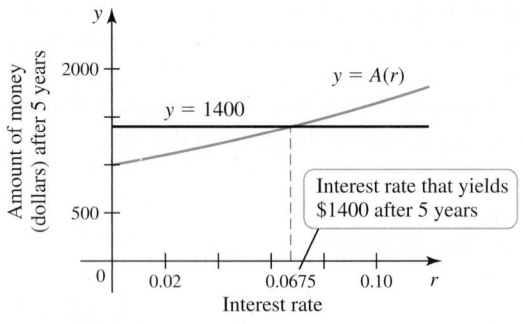

Figure 2.58

b. The graphs of $y = A(r)$ and the horizontal line $y = 1400$ are shown in **Figure 2.58**; it is evident that they intersect between $r = 0$ and $r = 0.08$. Solving $A(r) = 1400$ algebraically or using a root finder reveals that the curve and line intersect at $r \approx 0.0675$. Therefore, an interest rate of approximately 6.75% is required for the investment to be worth $1400 after 5 years.

Related Exercises 67, 75 ◀

SECTION 2.6 EXERCISES

Getting Started

1. Which of the following functions are continuous for all values in their domain? Justify your answers.

    a. $a(t) =$ altitude of a skydiver t seconds after jumping from a plane

    b. $n(t) =$ number of quarters needed to park legally in a metered parking space for t minutes

    c. $T(t) =$ temperature t minutes after midnight in Chicago on January 1

    d. $p(t) =$ number of points scored by a basketball player after t minutes of a basketball game

2. Give the three conditions that must be satisfied by a function to be continuous at a point.

3. What does it mean for a function to be continuous on an interval?

4. We informally describe a function f to be continuous at a if its graph contains no holes or breaks at a. Explain why this is not an adequate definition of continuity.

5–8. *Determine the points on the interval $(0, 5)$ at which the following functions f have discontinuities. At each point of discontinuity, state the conditions in the continuity checklist that are violated.*

5.
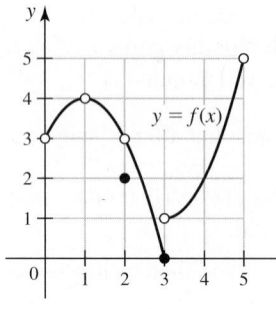

6.
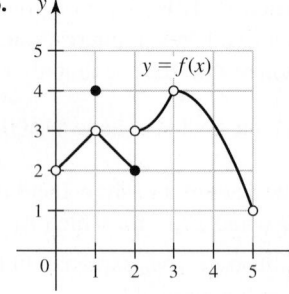

7.
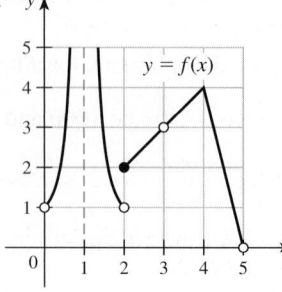

8.
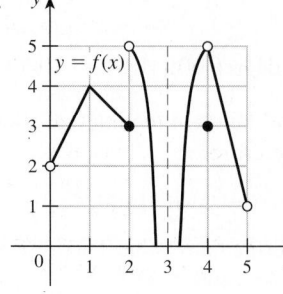

9. Complete the following sentences in terms of a limit.

    a. A function is continuous from the left at a if _____.

    b. A function is continuous from the right at a if _____.

10. Evaluate $f(3)$ if $\lim\limits_{x \to 3^-} f(x) = 5$, $\lim\limits_{x \to 3^+} f(x) = 6$, and f is right-continuous at $x = 3$.

11–14. *Determine the intervals of continuity for the following functions. At which endpoints of these intervals of continuity is f continuous from the left or continuous from the right?*

11. The graph of Exercise 5   **12.** The graph of Exercise 6

13. The graph of Exercise 7   **14.** The graph of Exercise 8

15. What is the domain of $f(x) = e^x/x$ and where is f continuous?

16. **Parking costs** Determine the intervals of continuity for the parking cost function c introduced at the outset of this section (see figure). Consider $0 \le t \le 60$.

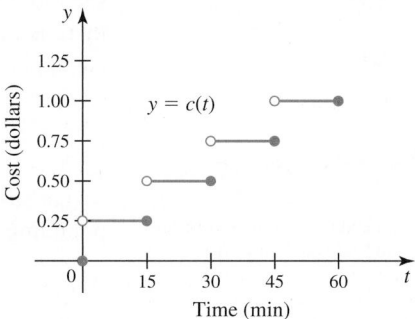

Practice Exercises

17–24. Continuity at a point *Determine whether the following functions are continuous at a. Use the continuity checklist to justify your answer.*

17. $f(x) = \dfrac{2x^2 + 3x + 1}{x^2 + 5x}; a = -5$

18. $f(x) = \dfrac{2x^2 + 3x + 1}{x^2 + 5x}; a = 5$

19. $f(x) = \sqrt{x - 2}; a = 1$

20. $g(x) = \dfrac{1}{x - 3}; a = 3$

21. $f(x) = \begin{cases} \dfrac{x^2 - 1}{x - 1} & \text{if } x \neq 1 \\ 3 & \text{if } x = 1 \end{cases}; a = 1$

22. $f(x) = \begin{cases} \dfrac{x^2 - 4x + 3}{x - 3} & \text{if } x \neq 3 \\ 2 & \text{if } x = 3 \end{cases}; a = 3$

23. $f(x) = \dfrac{5x - 2}{x^2 - 9x + 20}; a = 4$

24. $f(x) = \begin{cases} \dfrac{x^2 + x}{x + 1} & \text{if } x \neq -1 \\ 2 & \text{if } x = -1 \end{cases}; a = -1$

25–30. Continuity *Determine the interval(s) on which the following functions are continuous.*

25. $p(x) = 4x^5 - 3x^2 + 1$

26. $g(x) = \dfrac{3x^2 - 6x + 7}{x^2 + x + 1}$

27. $f(x) = \dfrac{x^5 + 6x + 17}{x^2 - 9}$

28. $s(x) = \dfrac{x^2 - 4x + 3}{x^2 - 1}$

29. $f(x) = \dfrac{1}{x^2 - 4}$

30. $f(t) = \dfrac{t + 2}{t^2 - 4}$

31–38. Limits *Evaluate each limit and justify your answer.*

31. $\lim_{x \to 0} (x^8 - 3x^6 - 1)^{40}$

32. $\lim_{x \to 2} \left(\dfrac{3}{2x^5 - 4x^2 - 50} \right)^4$

33. $\lim_{x \to 4} \sqrt{\dfrac{x^3 - 2x^2 - 8x}{x - 4}}$

34. $\lim_{t \to 4} \dfrac{t - 4}{\sqrt{t} - 2}$

35. $\lim_{x \to 1} \left(\dfrac{x + 5}{x + 2} \right)^4$

36. $\lim_{x \to \infty} \left(\dfrac{2x + 1}{x} \right)^3$

37. $\lim_{x \to 5} \ln \dfrac{6(\sqrt{x^2 - 16} - 3)}{5x - 25}$

38. $\lim_{x \to 0} \left(\dfrac{x}{\sqrt{16x + 1} - 1} \right)^{1/3}$

39–40. Intervals of continuity *Complete the following steps for each function.*

a. *Use the continuity checklist to show that f is not continuous at the given value of a.*

b. *Determine whether f is continuous from the left or the right at a.*

c. *State the interval(s) of continuity.*

39. $f(x) = \begin{cases} 2x & \text{if } x < 1 \\ x^2 + 3x & \text{if } x \geq 1 \end{cases}; a = 1$

40. $f(x) = \begin{cases} x^3 + 4x + 1 & \text{if } x \leq 0 \\ 2x^3 & \text{if } x > 0 \end{cases}; a = 0$

41–48. Functions with roots *Determine the interval(s) on which the following functions are continuous. At which finite endpoints of the intervals of continuity is f continuous from the left or continuous from the right?*

41. $f(x) = \sqrt{5 - x}$

42. $f(x) = \sqrt{25 - x^2}$

43. $f(x) = \sqrt{2x^2 - 16}$

44. $f(x) = \sqrt{x^2 - 3x + 2}$

45. $f(x) = \sqrt[3]{x^2 - 2x - 3}$

46. $f(t) = (t^2 - 1)^{3/2}$

47. $f(x) = (2x - 3)^{2/3}$

48. $f(z) = (z - 1)^{3/4}$

49–60. *Evaluate each limit.*

49. $\lim_{x \to 2} \sqrt{\dfrac{4x + 10}{2x - 2}}$

50. $\lim_{x \to -1} (x^2 - 4 + \sqrt[3]{x^2 - 9})$

51. $\lim_{x \to \pi} \dfrac{\cos^2 x + 3 \cos x + 2}{\cos x + 1}$

52. $\lim_{x \to 3\pi/2} \dfrac{\sin^2 x + 6 \sin x + 5}{\sin^2 x - 1}$

53. $\lim_{x \to 3} \sqrt{x^2 + 7}$

54. $\lim_{t \to 2} \dfrac{t^2 + 5}{1 + \sqrt{t^2 + 5}}$

55. $\lim_{x \to \pi/2} \dfrac{\sin x - 1}{\sqrt{\sin x} - 1}$

56. $\lim_{\theta \to 0} \dfrac{\dfrac{1}{2 + \sin \theta} - \dfrac{1}{2}}{\sin \theta}$

57. $\lim_{x \to 0} \dfrac{\cos x - 1}{\sin^2 x}$

58. $\lim_{x \to 0^+} \dfrac{1 - \cos^2 x}{\sin x}$

59. $\lim_{x \to 0} \dfrac{e^{4x} - 1}{e^x - 1}$

60. $\lim_{x \to e^2} \dfrac{\ln^2 x - 5 \ln x + 6}{\ln x - 2}$

61–66. Continuity and limits with transcendental functions *Determine the interval(s) on which the following functions are continuous; then analyze the given limits.*

61. $f(x) = \csc x;\ \lim_{x \to \pi/4} f(x);\ \lim_{x \to 2\pi^-} f(x)$

62. $f(x) = e^{\sqrt{x}};\ \lim_{x \to 4} f(x);\ \lim_{x \to 0^+} f(x)$

63. $f(x) = \dfrac{1 + \sin x}{\cos x};\ \lim_{x \to \pi/2^-} f(x);\ \lim_{x \to 4\pi/3} f(x)$

64. $f(x) = \dfrac{\ln x}{\sin^{-1} x};\ \lim_{x \to 1^-} f(x)$

65. $f(x) = \dfrac{e^x}{1 - e^x};\ \lim_{x \to 0^-} f(x);\ \lim_{x \to 0^+} f(x)$

66. $f(x) = \dfrac{e^{2x} - 1}{e^x - 1};\ \lim_{x \to 0} f(x)$

T 67–72. Applying the Intermediate Value Theorem

a. *Use the Intermediate Value Theorem to show that the following equations have a solution on the given interval.*

b. *Use a graphing utility to find all the solutions to the equation on the given interval.*

c. *Illustrate your answers with an appropriate graph.*

67. $2x^3 + x - 2 = 0;\ (-1, 1)$

68. $\sqrt{x^4 + 25x^3 + 10} = 5;\ (0, 1)$

69. $x^3 - 5x^2 + 2x = -1;\ (-1, 5)$

70. $-x^5 - 4x^2 + 2\sqrt{x} + 5 = 0;\ (0, 3)$

71. $x + e^x = 0;\ (-1, 0)$ **72.** $x \ln x - 1 = 0;\ (1, e)$

73. Explain why or why not *Determine whether the following statements are true and give an explanation or counterexample.*

a. *If a function is left-continuous and right-continuous at a, then it is continuous at a.*

b. *If a function is continuous at a, then it is left-continuous and right-continuous at a.*

c. *If $a < b$ and $f(a) \leq L \leq f(b)$, then there is some value of c in (a, b) for which $f(c) = L$.*

d. *Suppose f is continuous on $[a, b]$. Then there is a point c in (a, b) such that $f(c) = (f(a) + f(b))/2$.*

T 74. Mortgage payments You are shopping for a $250,000, 30-year (360-month) loan to buy a house. The monthly payment is given by

$$m(r) = \frac{250{,}000(r/12)}{1 - (1 + r/12)^{-360}},$$

where r is the annual interest rate. Suppose banks are currently offering interest rates between 4% and 5%.

a. Show there is a value of r in $(0.04, 0.05)$—an interest rate between 4% and 5%—that allows you to make monthly payments of $1300 per month.

b. Use a graph to illustrate your explanation to part (a). Then determine the interest rate you need for monthly payments of $1300.

T 75. Interest rates Suppose $5000 is invested in a savings account for 10 years (120 months), with an annual interest rate of r, compounded monthly. The amount of money in the account after 10 years is given by $A(r) = 5000(1 + r/12)^{120}$.

a. Show there is a value of r in $(0, 0.08)$—an interest rate between 0% and 8%—that allows you to reach your savings goal of $7000 in 10 years.

b. Use a graph to illustrate your explanation in part (a). Then approximate the interest rate required to reach your goal.

T 76. Investment problem Assume you invest $250 at the end of each year for 10 years at an annual interest rate of r. The amount of money in your account after 10 years is given by $A(r) = \dfrac{250((1 + r)^{10} - 1)}{r}$. Assume your goal is to have $3500 in your account after 10 years.

a. Show that there is an interest rate r in the interval $(0.01, 0.10)$—between 1% and 10%—that allows you to reach your financial goal.

b. Use a calculator to estimate the interest rate required to reach your financial goal.

T 77. Find an interval containing a solution to the equation $2x = \cos x$. Use a graphing utility to approximate the solution.

Explorations and Challenges

78. Continuity of the absolute value function Prove that the absolute value function $|x|$ is continuous for all values of x. (*Hint:* Using the definition of the absolute value function, compute $\lim\limits_{x \to 0^-} |x|$ and $\lim\limits_{x \to 0^+} |x|$.)

79–82. Continuity of functions with absolute values *Use the continuity of the absolute value function (Exercise 78) to determine the interval(s) on which the following functions are continuous.*

79. $f(x) = |x^2 + 3x - 18|$ **80.** $g(x) = \left| \dfrac{x + 4}{x^2 - 4} \right|$

81. $h(x) = \left| \dfrac{1}{\sqrt{x} - 4} \right|$

82. $h(x) = |x^2 + 2x + 5| + \sqrt{x}$

83. Pitfalls using technology The graph of the *sawtooth function* $y = x - \lfloor x \rfloor$, where $\lfloor x \rfloor$ is the greatest integer function or floor function (Exercise 51, Section 2.2), was obtained using a graphing utility (see figure). Identify any inaccuracies appearing in the graph and then plot an accurate graph by hand.

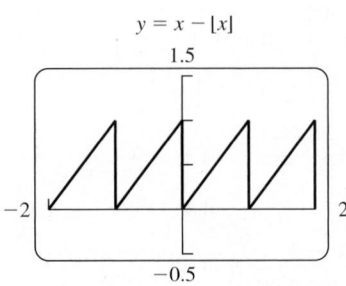

$y = x - \lfloor x \rfloor$

T 84. Pitfalls using technology Graph the function $f(x) = \dfrac{\sin x}{x}$ using a graphing window of $[-\pi, \pi] \times [0, 2]$.

a. Sketch a copy of the graph obtained with your graphing device and describe any inaccuracies appearing in the graph.

b. Sketch an accurate graph of the function. Is f continuous at 0?

c. What is the value of $\lim\limits_{x \to 0} \dfrac{\sin x}{x}$?

85. Sketching functions

a. Sketch the graph of a function that is not continuous at 1, but is defined at 1.

b. Sketch the graph of a function that is not continuous at 1, but has a limit at 1.

86. An unknown constant Determine the value of the constant a for which the function

$$f(x) = \begin{cases} \dfrac{x^2 + 3x + 2}{x + 1} & \text{if } x \neq -1 \\ a & \text{if } x = -1 \end{cases}$$

is continuous at -1.

87. An unknown constant Let

$$g(x) = \begin{cases} x^2 + x & \text{if } x < 1 \\ a & \text{if } x = 1 \\ 3x + 5 & \text{if } x > 1. \end{cases}$$

a. Determine the value of a for which g is continuous from the left at 1.

b. Determine the value of a for which g is continuous from the right at 1.

c. Is there a value of a for which g is continuous at 1? Explain.

T 88. Asymptotes of a function containing exponentials Let $f(x) = \dfrac{2e^x + 5e^{3x}}{e^{2x} - e^{3x}}$. Analyze $\lim\limits_{x \to 0^-} f(x)$, $\lim\limits_{x \to 0^+} f(x)$, $\lim\limits_{x \to -\infty} f(x)$, and $\lim\limits_{x \to \infty} f(x)$. Then give the horizontal and vertical asymptotes of f. Plot f to verify your results.

T 89. Asymptotes of a function containing exponentials Let $f(x) = \dfrac{2e^x + 10e^{-x}}{e^x + e^{-x}}$. Analyze $\lim\limits_{x \to 0} f(x)$, $\lim\limits_{x \to -\infty} f(x)$, and $\lim\limits_{x \to \infty} f(x)$. Then give the horizontal and vertical asymptotes of f. Plot f to verify your results.

T 90–91. Applying the Intermediate Value Theorem *Use the Intermediate Value Theorem to verify that the following equations have three solutions on the given interval. Use a graphing utility to find the approximate roots.*

90. $x^3 + 10x^2 - 100x + 50 = 0;\ (-20, 10)$

91. $70x^3 - 87x^2 + 32x - 3 = 0;\ (0, 1)$

92. Walk in the park Suppose you park your car at a trailhead in a national park and begin a 2-hr hike to a lake at 7 A.M. on a Friday morning. On Sunday morning, you leave the lake at 7 A.M. and start the 2-hr hike back to your car. Assume the lake is 3 mi from your car. Let $f(t)$ be your distance from the car t hours after 7 A.M. on Friday morning, and let $g(t)$ be your distance from the car t hours after 7 A.M. on Sunday morning.

   a. Evaluate $f(0)$, $f(2)$, $g(0)$, and $g(2)$.
   b. Let $h(t) = f(t) - g(t)$. Find $h(0)$ and $h(2)$.
   c. Show that there is some point along the trail that you will pass at exactly the same time on both days.

93. The monk and the mountain A monk set out from a monastery in the valley at dawn. He walked all day up a winding path, stopping for lunch and taking a nap along the way. At dusk, he arrived at a temple on the mountaintop. The next day the monk made the return walk to the valley, leaving the temple at dawn, walking the same path for the entire day, and arriving at the monastery in the evening. Must there be one point along the path that the monk occupied at the same time of day on both the ascent and the descent? Explain. (*Hint:* The question can be answered without the Intermediate Value Theorem.) (*Source:* Arthur Koestler, *The Act of Creation*)

94. Does continuity of $|f|$ imply continuity of f? Let

$$g(x) = \begin{cases} 1 & \text{if } x \geq 0 \\ -1 & \text{if } x < 0. \end{cases}$$

   a. Write a formula for $|g(x)|$.
   b. Is g continuous at $x = 0$? Explain.
   c. Is $|g|$ continuous at $x = 0$? Explain.
   d. For any function f, if $|f|$ is continuous at a, does it necessarily follow that f is continuous at a? Explain.

95–101. Classifying discontinuities *The discontinuities in graphs (a) and (b) are* **removable discontinuities** *because they disappear if we define or redefine f at a such that $f(a) = \lim\limits_{x \to a} f(x)$. The function in graph (c) has a* **jump discontinuity** *because left and right limits exist at a but are unequal. The discontinuity in graph (d) is an* **infinite discontinuity** *because the function has a vertical asymptote at a.*

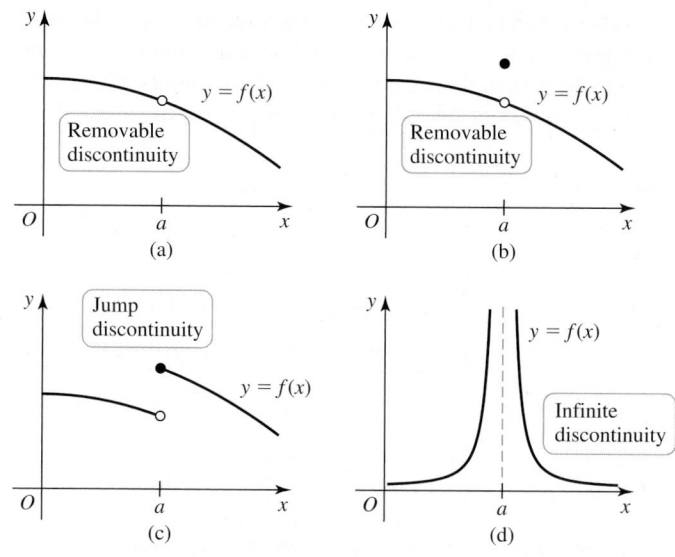

(a)   (b)

(c)   (d)

95. Is the discontinuity at a in graph (c) removable? Explain.

96. Is the discontinuity at a in graph (d) removable? Explain.

T 97–100. *Classify the discontinuities in the following functions at the given points.*

97. $f(x) = \dfrac{x^2 - 7x + 10}{x - 2}; x = 2$

98. $g(x) = \begin{cases} \dfrac{x^2 - 1}{1 - x} & \text{if } x \neq 1 \\ 3 & \text{if } x = 1 \end{cases}; x = 1$

99. $h(x) = \dfrac{x^3 - 4x^2 + 4x}{x(x - 1)}; x = 0 \text{ and } x = 1$

100. $f(x) = \dfrac{|x - 2|}{x - 2}; x = 2$

101. Do removable discontinuities exist?

   a. Does the function $f(x) = x \sin(1/x)$ have a removable discontinuity at $x = 0$? Explain.
   b. Does the function $g(x) = \sin(1/x)$ have a removable discontinuity at $x = 0$? Explain.

102. Continuity of composite functions Prove Theorem 2.11: If g is continuous at a and f is continuous at $g(a)$, then the composition $f \circ g$ is continuous at a. (*Hint:* Write the definition of continuity for f and g separately; then combine them to form the definition of continuity for $f \circ g$.)

103. Continuity of compositions

   a. Find functions f and g such that each function is continuous at 0 but $f \circ g$ is not continuous at 0.
   b. Explain why examples satisfying part (a) do not contradict Theorem 2.11.

104. Violation of the Intermediate Value Theorem? Let $f(x) = \dfrac{|x|}{x}$. Then $f(-2) = -1$ and $f(2) = 1$. Therefore, $f(-2) < 0 < f(2)$, but there is no value of c between -2 and 2 for which $f(c) = 0$. Does this fact violate the Intermediate Value Theorem? Explain.

105. Continuity of $\sin x$ and $\cos x$

   a. Use the identity $\sin(a + h) = \sin a \cos h + \cos a \sin h$ with the fact that $\lim\limits_{x \to 0} \sin x = 0$ to prove that $\lim\limits_{x \to a} \sin x = \sin a$, thereby establishing that $\sin x$ is continuous for all x. (*Hint:* Let $h = x - a$ so that $x = a + h$ and note that $h \to 0$ as $x \to a$.)
   b. Use the identity $\cos(a + h) = \cos a \cos h - \sin a \sin h$ with the fact that $\lim\limits_{x \to 0} \cos x = 1$ to prove that $\lim\limits_{x \to a} \cos x = \cos a$.

QUICK CHECK ANSWERS

1. $t = 15, 30, 45$ **2.** Fill in the endpoints.
3. $[0, \infty); (-\infty, \infty)$ **4.** Note that $\lim\limits_{x \to 1^+} \sqrt[4]{\ln x} = \sqrt[4]{\lim\limits_{x \to 1^+} \ln x} = 0$ and $f(1) = \sqrt[4]{\ln 1} = 0$. Because $\lim\limits_{x \to 1^+} \sqrt[4]{\ln x} = \sqrt[4]{\ln 1}$, the function is right-continuous at $x = 1$.   **5.** The equation has a solution on the interval $[-1, 1]$ because f is continuous on $[-1, 1]$ and $f(-1) < 0 < f(1)$. ◄

2.7 Precise Definitions of Limits

The limit definitions already encountered in this chapter are adequate for most elementary limits. However, some of the terminology used, such as *sufficiently close* and *arbitrarily large*, needs clarification. The goal of this section is to give limits a solid mathematical foundation by transforming the previous limit definitions into precise mathematical statements.

Moving Toward a Precise Definition

> The phrase *for all x near a* means for all x in an open interval containing a.

> The Greek letters δ (delta) and ε (epsilon) represent small positive numbers in the discussion of limits.

> The two conditions $|x - a| < \delta$ and $x \neq a$ are written concisely as $0 < |x - a| < \delta$.

Assume the function f is defined for all x near a, except possibly at a. Recall that $\lim_{x \to a} f(x) = L$ means that $f(x)$ is arbitrarily close to L for all x sufficiently close (but not equal) to a. This limit definition is made precise by observing that the distance between $f(x)$ and L is $|f(x) - L|$ and that the distance between x and a is $|x - a|$. Therefore, we write $\lim_{x \to a} f(x) = L$ if we can make $|f(x) - L|$ arbitrarily small for any x, distinct from a, with $|x - a|$ sufficiently small. For instance, if we want $|f(x) - L|$ to be less than 0.1, then we must find a number $\delta > 0$ such that

$$|f(x) - L| < 0.1 \quad \text{whenever} \quad |x - a| < \delta \quad \text{and} \quad x \neq a.$$

If, instead, we want $|f(x) - L|$ to be less than 0.001, then we must find *another* number $\delta > 0$ such that

$$|f(x) - L| < 0.001 \quad \text{whenever} \quad 0 < |x - a| < \delta.$$

For the limit to exist, it must be true that for *any* $\varepsilon > 0$, we can always find a $\delta > 0$ such that

$$|f(x) - L| < \varepsilon \quad \text{whenever} \quad 0 < |x - a| < \delta.$$

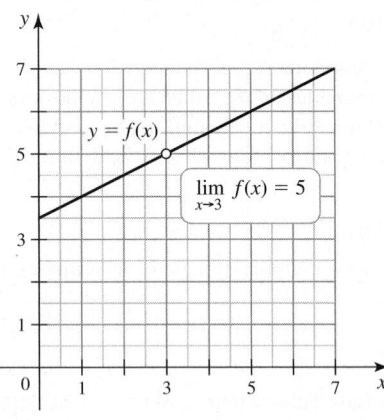

Figure 2.59

$y = f(x)$

$\lim_{x \to 3} f(x) = 5$

> The founders of calculus, Isaac Newton (1642–1727) and Gottfried Leibniz (1646–1716), developed the core ideas of calculus without using a precise definition of a limit. It was not until the 19th century that a rigorous definition was introduced by Augustin-Louis Cauchy (1789–1857) and later refined by Karl Weierstrass (1815–1897).

EXAMPLE 1 **Determining values of δ from a graph** Figure 2.59 shows the graph of a linear function f with $\lim_{x \to 3} f(x) = 5$. For each value of $\varepsilon > 0$, determine a value of $\delta > 0$ satisfying the statement

$$|f(x) - 5| < \varepsilon \quad \text{whenever} \quad 0 < |x - 3| < \delta.$$

a. $\varepsilon = 1$ **b.** $\varepsilon = \frac{1}{2}$

SOLUTION

a. With $\varepsilon = 1$, we want $f(x)$ to be less than 1 unit from 5, which means $f(x)$ is between 4 and 6. To determine a corresponding value of δ, draw the horizontal lines $y = 4$ and $y = 6$ (**Figure 2.60a**). Then sketch vertical lines passing through the points where the horizontal lines and the graph of f intersect (**Figure 2.60b**). We see that the vertical

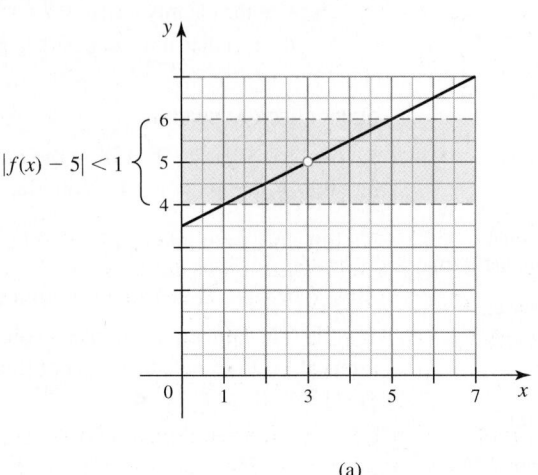

$|f(x) - 5| < 1$

(a)

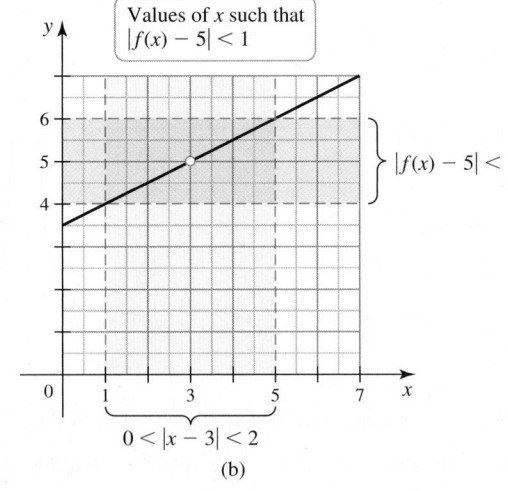

Values of x such that $|f(x) - 5| < 1$

$|f(x) - 5| < 1$

$0 < |x - 3| < 2$

(b)

Figure 2.60

lines intersect the *x*-axis at $x = 1$ and $x = 5$. Note that $f(x)$ is less than 1 unit from 5 on the *y*-axis if *x* is within 2 units of 3 on the *x*-axis. So for $\varepsilon = 1$, we let $\delta = 2$ or any smaller positive value.

b. With $\varepsilon = \frac{1}{2}$, we want $f(x)$ to lie within a half-unit of 5, or equivalently, $f(x)$ must lie between 4.5 and 5.5. Proceeding as in part (a), we see that $f(x)$ is within a half-unit of 5 on the *y*-axis (**Figure 2.61a**) if *x* is less than 1 unit from 3 (**Figure 2.61b**). So for $\varepsilon = \frac{1}{2}$, we let $\delta = 1$ or any smaller positive number.

> Once an acceptable value of δ is found satisfying the statement
> $$|f(x) - L| < \varepsilon \quad \text{whenever}$$
> $$0 < |x - a| < \delta,$$
> any smaller positive value of δ also works.

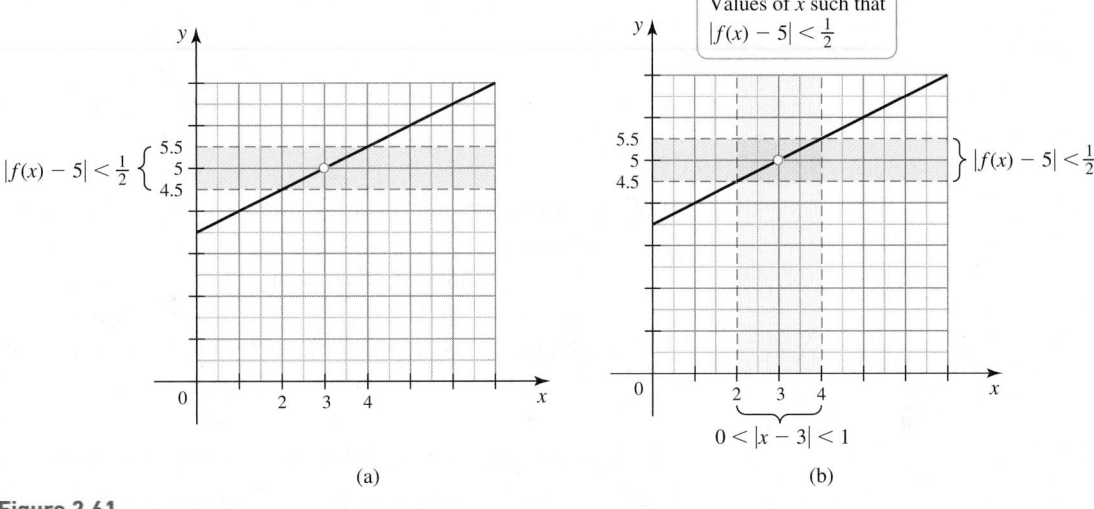

Figure 2.61

Related Exercises 9–10 ◄

The idea of a limit, as illustrated in Example 1, may be described in terms of a contest between two people named Epp and Del. First, Epp picks a particular number $\varepsilon > 0$; then he challenges Del to find a corresponding value of $\delta > 0$ such that

$$|f(x) - 5| < \varepsilon \quad \text{whenever} \quad 0 < |x - 3| < \delta. \tag{1}$$

To illustrate, suppose Epp chooses $\varepsilon = 1$. From Example 1, we know that Del will satisfy (1) by choosing $0 < \delta \le 2$. If Epp chooses $\varepsilon = \frac{1}{2}$, then (by Example 1) Del responds by letting $0 < \delta \le 1$. If Epp lets $\varepsilon = \frac{1}{8}$, then Del chooses $0 < \delta \le \frac{1}{4}$ (**Figure 2.62**). In fact, there is a pattern: For *any* $\varepsilon > 0$ that Epp chooses, no matter how small, Del will satisfy (1) by choosing a positive value of δ satisfying $0 < \delta \le 2\varepsilon$. Del has discovered a mathematical relationship: If $0 < \delta \le 2\varepsilon$ and $0 < |x - 3| < \delta$, then $|f(x) - 5| < \varepsilon$, for *any* $\varepsilon > 0$. This conversation illustrates the general procedure for proving that $\lim_{x \to a} f(x) = L$.

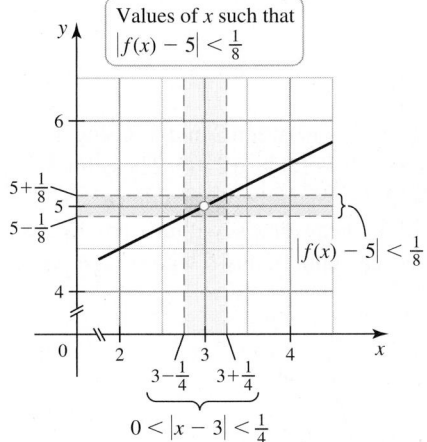

Figure 2.62

QUICK CHECK 1 In Example 1, find a positive number δ satisfying the statement

$$|f(x) - 5| < \frac{1}{100} \quad \text{whenever} \quad 0 < |x - 3| < \delta. \blacktriangleleft$$

A Precise Definition

Example 1 dealt with a linear function, but it points the way to a precise definition of a limit for any function. As shown in **Figure 2.63**, $\lim_{x \to a} f(x) = L$ means that for *any* positive number ε, there is another positive number δ such that

$$|f(x) - L| < \varepsilon \quad \text{whenever} \quad 0 < |x - a| < \delta.$$

In all limit proofs, the goal is to find a relationship between ε and δ that gives an admissible value of δ, in terms of ε only. This relationship must work for any positive value of ε.

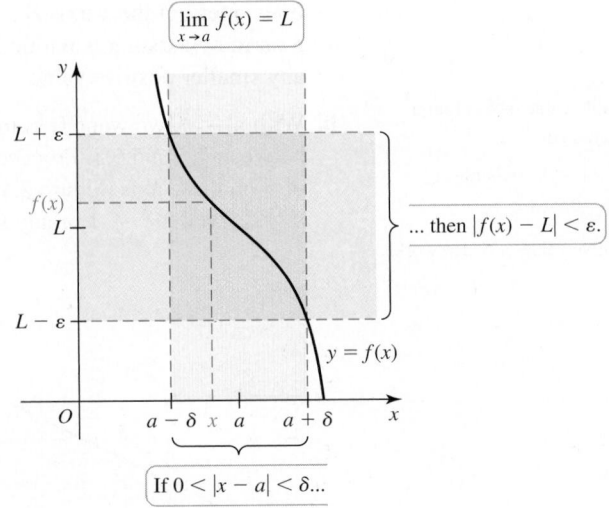

Figure 2.63

DEFINITION Limit of a Function

Assume $f(x)$ is defined for all x in some open interval containing a, except possibly at a. We say **the limit of $f(x)$ as x approaches a is L**, written

$$\lim_{x \to a} f(x) = L,$$

if for *any* number $\varepsilon > 0$ there is a corresponding number $\delta > 0$ such that

$$|f(x) - L| < \varepsilon \quad \text{whenever} \quad 0 < |x - a| < \delta.$$

EXAMPLE 2 Finding δ for a given ε using a graphing utility Let $f(x) = x^3 - 6x^2 + 12x - 5$ and demonstrate that $\lim_{x \to 2} f(x) = 3$ as follows. For the given values of ε, use a graphing utility to find a value of $\delta > 0$ such that

$$|f(x) - 3| < \varepsilon \quad \text{whenever} \quad 0 < |x - 2| < \delta.$$

> ➤ The value of δ in the precise definition of a limit depends only on ε.

a. $\varepsilon = 1$ **b.** $\varepsilon = \frac{1}{2}$

SOLUTION

> ➤ Definitions of the one-sided limits $\lim_{x \to a^+} f(x) = L$ and $\lim_{x \to a^-} f(x) = L$ are discussed in Exercises 51–55.

a. The condition $|f(x) - 3| < \varepsilon = 1$ implies that $f(x)$ lies between 2 and 4. Using a graphing utility, we graph f and the lines $y = 2$ and $y = 4$ (**Figure 2.64**). These lines intersect the graph of f at $x = 1$ and at $x = 3$. We now sketch the vertical lines $x = 1$ and $x = 3$ and observe that $f(x)$ is within 1 unit of 3 whenever x is within 1 unit of 2 on the x-axis (Figure 2.64). Therefore, with $\varepsilon = 1$, we can choose any δ with $0 < \delta \le 1$.

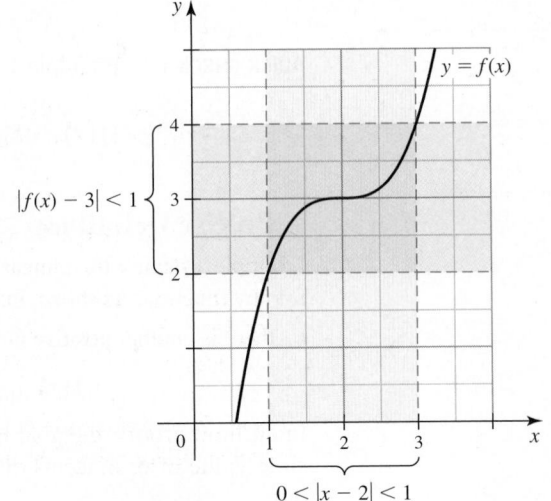

Figure 2.64

b. The condition $|f(x) - 3| < \varepsilon = \frac{1}{2}$ implies that $f(x)$ lies between 2.5 and 3.5 on the y-axis. We now find that the lines $y = 2.5$ and $y = 3.5$ intersect the graph of f at $x \approx 1.21$ and $x \approx 2.79$ (**Figure 2.65**). Observe that if x is less than 0.79 unit from 2 on the x-axis, then $f(x)$ is less than a half-unit from 3 on the y-axis. Therefore, with $\varepsilon = \frac{1}{2}$ we can choose any δ with $0 < \delta \leq 0.79$.

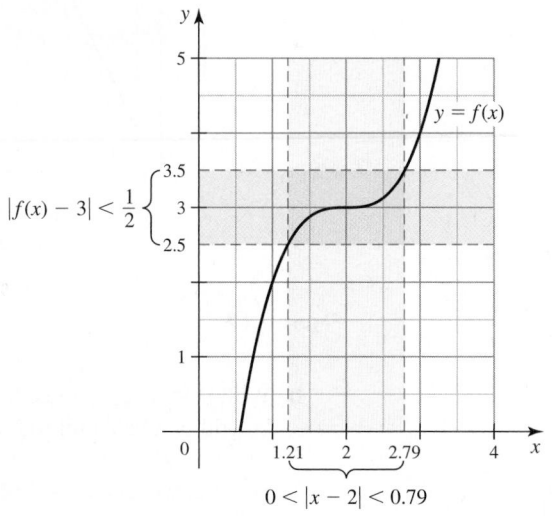

Figure 2.65

This procedure could be repeated for smaller and smaller values of $\varepsilon > 0$. For each value of ε, there exists a corresponding value of δ, proving that the limit exists.

Related Exercise 13 ◄

QUICK CHECK 2 For the function f given in Example 2, estimate a value of $\delta > 0$ satisfying $|f(x) - 3| < 0.25$ whenever $0 < |x - 2| < \delta$. ◄

The inequality $0 < |x - a| < \delta$ means that x lies between $a - \delta$ and $a + \delta$ with $x \neq a$. We say that the interval $(a - \delta, a + \delta)$ is **symmetric about** a because a is the midpoint of the interval. Symmetric intervals are convenient, but Example 3 demonstrates that we don't always get symmetric intervals without a bit of extra work.

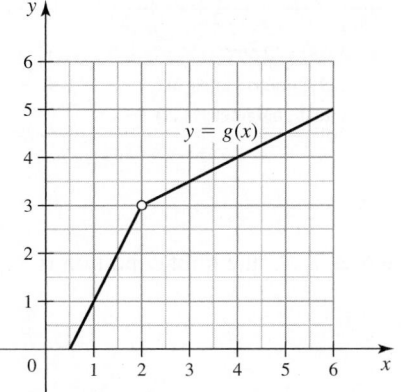

Figure 2.66

EXAMPLE 3   Finding a symmetric interval Figure 2.66 shows the graph of g with $\lim\limits_{x \to 2} g(x) = 3$. For each value of ε, find the corresponding values of $\delta > 0$ that satisfy the condition

$$|g(x) - 3| < \varepsilon \quad \text{whenever} \quad 0 < |x - 2| < \delta.$$

a. $\varepsilon = 2$

b. $\varepsilon = 1$

c. For any given value of ε, make a conjecture about the corresponding values of δ that satisfy the limit condition.

SOLUTION

a. With $\varepsilon = 2$, we need a value of $\delta > 0$ such that $g(x)$ is within 2 units of 3, which means $g(x)$ is between 1 and 5, whenever x is less than δ units from 2. The horizontal lines $y = 1$ and $y = 5$ intersect the graph of g at $x = 1$ and $x = 6$. Therefore, $|g(x) - 3| < 2$ if x lies in the interval $(1, 6)$ with $x \neq 2$ (**Figure 2.67a**). However, we want x to lie in an interval that is *symmetric* about 2. We can guarantee that $|g(x) - 3| < 2$ in an interval symmetric about 2 only if x is less than 1 unit away from 2, on either side of 2 (**Figure 2.67b**). Therefore, with $\varepsilon = 2$, we take $\delta = 1$ or any smaller positive number.

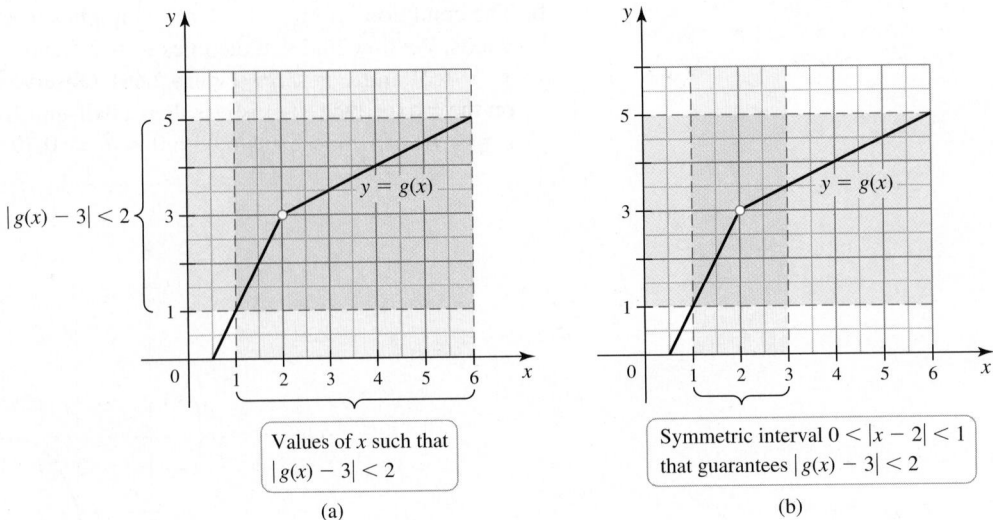

Figure 2.67

b. With $\varepsilon = 1$, $g(x)$ must lie between 2 and 4 (**Figure 2.68a**). This implies that x must be within a half-unit to the left of 2 and within 2 units to the right of 2. Therefore, $|g(x) - 3| < 1$ provided x lies in the interval $(1.5, 4)$. To obtain a symmetric interval about 2, we take $\delta = \frac{1}{2}$ or any smaller positive number. Then we are still guaranteed that $|g(x) - 3| < 1$ when $0 < |x - 2| < \frac{1}{2}$ (**Figure 2.68b**).

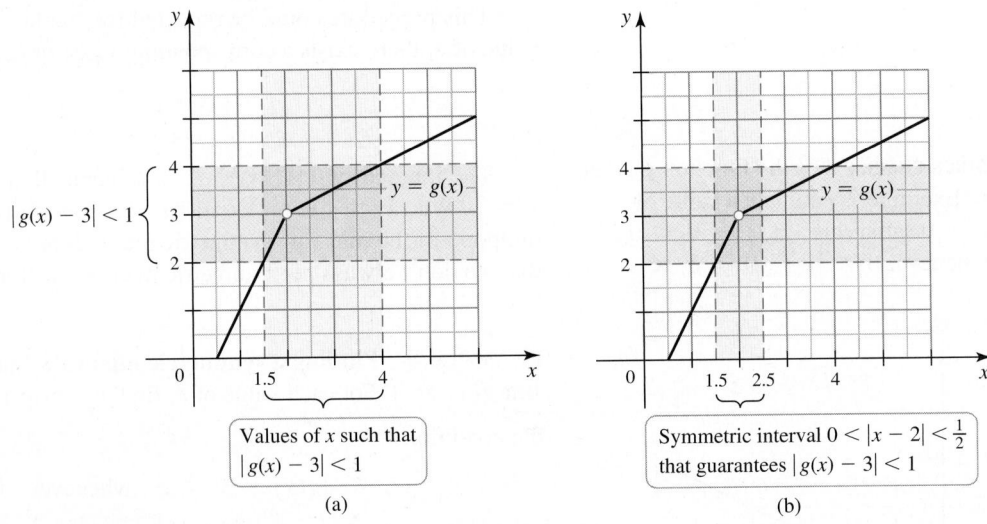

Figure 2.68

c. From parts (a) and (b), it appears that if we choose $\delta \le \varepsilon/2$, the limit condition is satisfied for any $\varepsilon > 0$.

Related Exercises 15–16 ◄

In Examples 2 and 3, we showed that a limit exists by discovering a relationship between ε and δ that satisfies the limit condition. We now generalize this procedure.

Limit Proofs

We use the following two-step process to prove that $\lim_{x \to a} f(x) = L$.

➤ The first step of the limit-proving process is the preliminary work of finding a candidate for δ. The second step verifies that the δ found in the first step actually works.

Steps for proving that $\lim\limits_{x \to a} f(x) = L$

1. **Find δ.** Let ε be an arbitrary positive number. Use the inequality $|f(x) - L| < \varepsilon$ to find a condition of the form $|x - a| < \delta$, where δ depends only on the value of ε.

2. **Write a proof.** For any $\varepsilon > 0$, assume $0 < |x - a| < \delta$ and use the relationship between ε and δ found in Step 1 to prove that $|f(x) - L| < \varepsilon$.

EXAMPLE 4 **Limit of a linear function** Prove that $\lim\limits_{x \to 4} (4x - 15) = 1$ using the precise definition of a limit.

SOLUTION

Step 1: *Find δ.* In this case, $a = 4$ and $L = 1$. Assuming $\varepsilon > 0$ is given, we use $|(4x - 15) - 1| < \varepsilon$ to find an inequality of the form $|x - 4| < \delta$. If $|(4x - 15) - 1| < \varepsilon$, then

$$|4x - 16| < \varepsilon$$

$$4|x - 4| < \varepsilon \qquad \text{Factor } 4x - 16.$$

$$|x - 4| < \frac{\varepsilon}{4}. \qquad \text{Divide by 4 and identify } \delta = \varepsilon/4.$$

We have shown that $|(4x - 15) - 1| < \varepsilon$ implies that $|x - 4| < \varepsilon/4$. Therefore, a plausible relationship between δ and ε is $\delta = \varepsilon/4$. We now write the actual proof.

Step 2: *Write a proof.* Let $\varepsilon > 0$ be given and assume $0 < |x - 4| < \delta$, where $\delta = \varepsilon/4$. The aim is to show that $|(4x - 15) - 1| < \varepsilon$ for all x such that $0 < |x - 4| < \delta$. We simplify $|(4x - 15) - 1|$ and isolate the $|x - 4|$ term:

$$|(4x - 15) - 1| = |4x - 16|$$

$$= 4 \underbrace{|x - 4|}_{\text{less than } \delta \,=\, \varepsilon/4}$$

$$< 4\left(\frac{\varepsilon}{4}\right) = \varepsilon.$$

We have shown that for any $\varepsilon > 0$,

$$|f(x) - L| = |(4x - 15) - 1| < \varepsilon \quad \text{whenever} \quad 0 < |x - 4| < \delta,$$

provided $0 < \delta \le \varepsilon/4$. Therefore, $\lim\limits_{x \to 4} (4x - 15) = 1$.

Related Exercises 19–20 ◄

EXAMPLE 5 **Limit of a quadratic function** Prove that $\lim\limits_{x \to 5} x^2 = 25$ using the precise definition of a limit.

SOLUTION

Step 1: *Find δ.* Given $\varepsilon > 0$, our task is to find an expression for $\delta > 0$ that depends only on ε, such that $|x^2 - 25| < \varepsilon$ whenever $0 < |x - 5| < \delta$. We begin by factoring $|x^2 - 25|$:

$$|x^2 - 25| = |(x + 5)(x - 5)| \qquad \text{Factor.}$$

$$= |x + 5||x - 5|. \qquad |ab| = |a||b|$$

Because the value of $\delta > 0$ in the inequality $0 < |x - 5| < \delta$ typically represents a small positive number, let's assume $\delta \le 1$ so that $|x - 5| < 1$, which implies that $-1 < x - 5 < 1$ or $4 < x < 6$. It follows that x is positive, $|x + 5| < 11$, and

$$|x^2 - 25| = |x + 5||x - 5| < 11|x - 5|.$$

> The minimum value of a and b is denoted min $\{a, b\}$. If $x = $ min $\{a, b\}$, then x is the smaller of a and b. If $a = b$, then x equals the common value of a and b. In either case, $x \leq a$ and $x \leq b$.

Using this inequality, we have $|x^2 - 25| < \varepsilon$, provided $11|x - 5| < \varepsilon$ or $|x - 5| < \varepsilon/11$. Note that two restrictions have been placed on $|x - 5|$:

$$|x - 5| < 1 \quad \text{and} \quad |x - 5| < \frac{\varepsilon}{11}.$$

To ensure that both these inequalities are satisfied, let $\delta = $ min $\{1, \varepsilon/11\}$ so that δ equals the smaller of 1 and $\varepsilon/11$.

Step 2: *Write a proof.* Let $\varepsilon > 0$ be given and assume $0 < |x - 5| < \delta$, where $\delta = $ min$\{1, \varepsilon/11\}$. By factoring $x^2 - 25$, we have

$$|x^2 - 25| = |x + 5||x - 5|.$$

Because $0 < |x - 5| < \delta$ and $\delta \leq \varepsilon/11$, we have $|x - 5| < \varepsilon/11$. It is also the case that $|x - 5| < 1$ because $\delta \leq 1$, which implies that $-1 < x - 5 < 1$ or $4 < x < 6$. Therefore, $|x + 5| < 11$ and

$$|x^2 - 25| = |x + 5||x - 5| < 11\left(\frac{\varepsilon}{11}\right) = \varepsilon.$$

We have shown that for any $\varepsilon > 0$, $|x^2 - 25| < \varepsilon$ whenever $0 < |x - 5| < \delta$, provided $0 < \delta = $ min$\{1, \varepsilon/11\}$. Therefore, $\lim_{x \to 5} x^2 = 25$.

Related Exercises 27, 29 ◄

Justifying Limit Laws

The precise definition of a limit is used to prove the limit laws in Theorem 2.3. Essential in several of these proofs is the **triangle inequality**, which states that

$$|x + y| \leq |x| + |y|, \quad \text{for all real numbers } x \text{ and } y.$$

EXAMPLE 6   Proof of Limit Law 1 Prove that if $\lim_{x \to a} f(x)$ and $\lim_{x \to a} g(x)$ exist, then

$$\lim_{x \to a} (f(x) + g(x)) = \lim_{x \to a} f(x) + \lim_{x \to a} g(x).$$

> Because $\lim_{x \to a} f(x)$ exists, if there exists a $\delta > 0$ for any given $\varepsilon > 0$, then there also exists a $\delta > 0$ for any given $\frac{\varepsilon}{2}$.

SOLUTION Assume $\varepsilon > 0$ is given. Let $\lim_{x \to a} f(x) = L$, which implies there exists a $\delta_1 > 0$ such that

$$|f(x) - L| < \frac{\varepsilon}{2} \quad \text{whenever} \quad 0 < |x - a| < \delta_1.$$

Similarly, let $\lim_{x \to a} g(x) = M$, which implies there exists a $\delta_2 > 0$ such that

$$|g(x) - M| < \frac{\varepsilon}{2} \quad \text{whenever} \quad 0 < |x - a| < \delta_2.$$

Let $\delta = $ min $\{\delta_1, \delta_2\}$ and suppose $0 < |x - a| < \delta$. Because $\delta \leq \delta_1$, it follows that $0 < |x - a| < \delta_1$ and $|f(x) - L| < \varepsilon/2$. Similarly, because $\delta \leq \delta_2$, it follows that $0 < |x - a| < \delta_2$ and $|g(x) - M| < \varepsilon/2$. Therefore,

$$
\begin{aligned}
|(f(x) + g(x)) - (L + M)| &= |(f(x) - L) + (g(x) - M)| &&\text{Rearrange terms.}\\
&\leq |f(x) - L| + |g(x) - M| &&\text{Triangle inequality}\\
&< \frac{\varepsilon}{2} + \frac{\varepsilon}{2} = \varepsilon.
\end{aligned}
$$

> Proofs of other limit laws are outlined in Exercises 43–44.

We have shown that given any $\varepsilon > 0$, if $0 < |x - a| < \delta$, then $|(f(x) + g(x)) - (L + M)| < \varepsilon$, which implies that $\lim_{x \to a} (f(x) + g(x)) = L + M = \lim_{x \to a} f(x) + \lim_{x \to a} g(x)$.

Related Exercises 43–44 ◄

Infinite Limits

> Notice that for infinite limits, N plays the role that ε plays for regular limits. It sets a tolerance or bound for the function values $f(x)$.

In Section 2.4, we stated that $\lim_{x \to a} f(x) = \infty$ if $f(x)$ grows *arbitrarily large* as x approaches a. More precisely, this means that for any positive number N (no matter how large), $f(x)$ is larger than N if x is sufficiently close to a but not equal to a.

> DEFINITION **Two-Sided Infinite Limit**
>
> The **infinite limit** $\lim_{x \to a} f(x) = \infty$ means that for any positive number N, there exists a corresponding $\delta > 0$ such that
>
> $$f(x) > N \quad \text{whenever} \quad 0 < |x - a| < \delta.$$

▸ Precise definitions for $\lim_{x \to a} f(x) = -\infty$, $\lim_{x \to a^-} f(x) = -\infty$, $\lim_{x \to a^+} f(x) = \infty$, $\lim_{x \to a^-} f(x) = -\infty$, and $\lim_{x \to a^-} f(x) = \infty$ are given in Exercises 57–63.

As shown in **Figure 2.69**, to prove that $\lim_{x \to a} f(x) = \infty$, we let N represent *any* positive number. Then we find a value of $\delta > 0$, depending only on N, such that

$$f(x) > N \quad \text{whenever} \quad 0 < |x - a| < \delta.$$

This process is similar to the two-step process for finite limits.

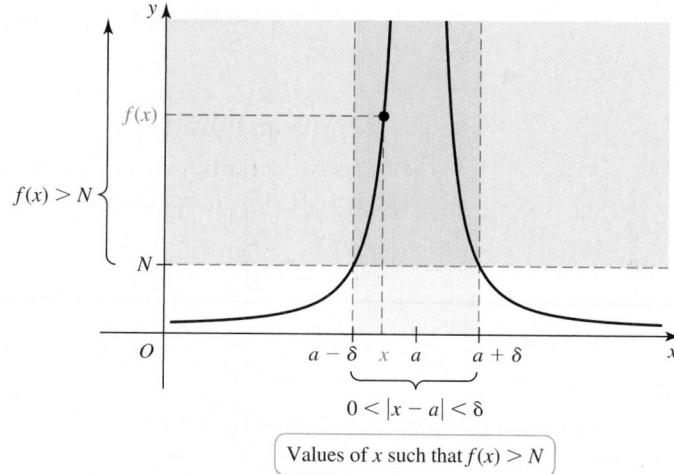

Figure 2.69

> **Steps for proving that $\lim_{x \to a} f(x) = \infty$**
>
> 1. **Find δ.** Let N be an arbitrary positive number. Use the statement $f(x) > N$ to find an inequality of the form $|x - a| < \delta$, where δ depends only on N.
> 2. **Write a proof.** For any $N > 0$, assume $0 < |x - a| < \delta$ and use the relationship between N and δ found in Step 1 to prove that $f(x) > N$.

EXAMPLE 7 **An Infinite Limit Proof** Let $f(x) = \dfrac{1}{(x - 2)^2}$. Prove that $\lim_{x \to 2} f(x) = \infty$.

SOLUTION

Step 1: *Find $\delta > 0$.* Assuming $N > 0$, we use the inequality $\dfrac{1}{(x - 2)^2} > N$ to find δ, where δ depends only on N. Taking reciprocals of this inequality, it follows that

$$(x - 2)^2 < \frac{1}{N}$$

$$|x - 2| < \frac{1}{\sqrt{N}}. \quad \text{Take the square root of both sides.}$$

▸ Recall that $\sqrt{x^2} = |x|$.

The inequality $|x - 2| < \dfrac{1}{\sqrt{N}}$ has the form $|x - 2| < \delta$ if we let $\delta = \dfrac{1}{\sqrt{N}}$. We now write a proof based on this relationship between δ and N.

Step 2:   *Write a proof.* Suppose $N > 0$ is given. Let $\delta = \dfrac{1}{\sqrt{N}}$ and assume

$$0 < |x - 2| < \delta = \frac{1}{\sqrt{N}}.$$ Squaring both sides of the inequality

$|x - 2| < \dfrac{1}{\sqrt{N}}$ and taking reciprocals, we have

$$(x - 2)^2 < \frac{1}{N} \qquad \text{Square both sides.}$$

$$\frac{1}{(x - 2)^2} > N. \qquad \text{Take reciprocals of both sides.}$$

QUICK CHECK 3   In Example 7, if N is increased by a factor of 100, how must δ change? ◄

We see that for any positive N, if $0 < |x - 2| < \delta = \dfrac{1}{\sqrt{N}}$, then

$f(x) = \dfrac{1}{(x - 2)^2} > N$. It follows that $\displaystyle\lim_{x \to 2} \dfrac{1}{(x - 2)^2} = \infty$. Note that because

$\delta = \dfrac{1}{\sqrt{N}}$, δ decreases as N increases.

Related Exercises 45–46 ◄

Limits at Infinity

Precise definitions can also be written for the limits at infinity $\displaystyle\lim_{x \to \infty} f(x) = L$ and $\displaystyle\lim_{x \to -\infty} f(x) = L$. For discussion and examples, see Exercises 64–65.

SECTION 2.7 EXERCISES

Getting Started

1. Suppose x lies in the interval $(1, 3)$ with $x \neq 2$. Find the smallest positive value of δ such that the inequality $0 < |x - 2| < \delta$ is true.

2. Suppose $f(x)$ lies in the interval $(2, 6)$. What is the smallest value of ε such that $|f(x) - 4| < \varepsilon$?

3. Which one of the following intervals is not symmetric about $x = 5$?

   a. $(1, 9)$ **b.** $(4, 6)$ **c.** $(3, 8)$ **d.** $(4.5, 5.5)$

4. Suppose a is a constant and δ is a positive constant. Give a geometric description of the sets $\{x \colon |x - a| < \delta\}$ and $\{x \colon 0 < |x - a| < \delta\}$.

5. State the precise definition of $\displaystyle\lim_{x \to a} f(x) = L$.

6. Interpret $|f(x) - L| < \varepsilon$ in words.

7. Suppose $|f(x) - 5| < 0.1$ whenever $0 < x < 5$. Find all values of $\delta > 0$ such that $|f(x) - 5| < 0.1$ whenever $0 < |x - 2| < \delta$.

8. Give the definition of $\displaystyle\lim_{x \to a} f(x) = \infty$ and interpret it using pictures.

9. **Determining values of δ from a graph** The function f in the figure satisfies $\displaystyle\lim_{x \to 2} f(x) = 5$. Determine the largest value of $\delta > 0$ satisfying each statement.

   a. If $0 < |x - 2| < \delta$, then $|f(x) - 5| < 2$.

   b. If $0 < |x - 2| < \delta$, then $|f(x) - 5| < 1$.

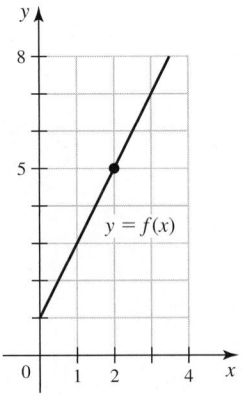

10. **Determining values of δ from a graph** The function f in the figure satisfies $\displaystyle\lim_{x \to 2} f(x) = 4$. Determine the largest value of $\delta > 0$ satisfying each statement.

    a. If $0 < |x - 2| < \delta$, then $|f(x) - 4| < 1$.

    b. If $0 < |x - 2| < \delta$, then $|f(x) - 4| < 1/2$.

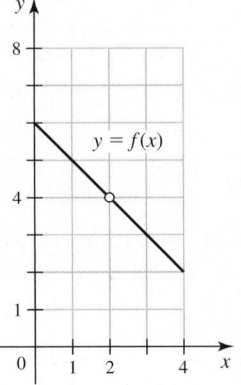

Practice Exercises

11. Determining values of δ from a graph The function f in the figure satisfies $\lim_{x \to 3} f(x) = 6$. Determine the largest value of $\delta > 0$ satisfying each statement.

  a. If $0 < |x - 3| < \delta$, then $|f(x) - 6| < 3$.

  b. If $0 < |x - 3| < \delta$, then $|f(x) - 6| < 1$.

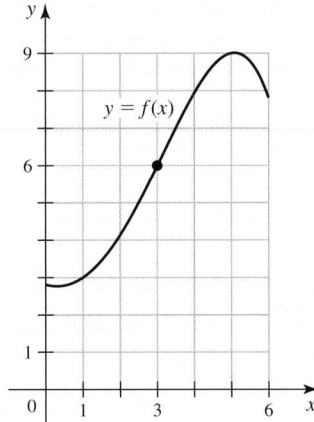

12. Determining values of δ from a graph The function f in the figure satisfies $\lim_{x \to 4} f(x) = 5$. Determine the largest value of $\delta > 0$ satisfying each statement.

  a. If $0 < |x - 4| < \delta$, then $|f(x) - 5| < 1$.

  b. If $0 < |x - 4| < \delta$, then $|f(x) - 5| < 0.5$.

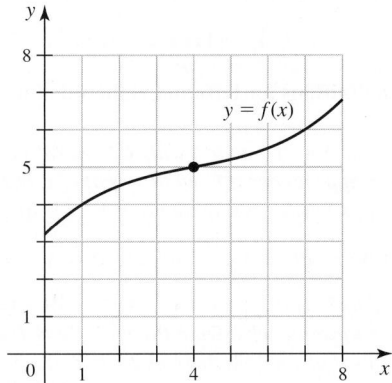

T 13. Finding δ for a given ε using a graph Let $f(x) = x^3 + 3$ and note that $\lim_{x \to 0} f(x) = 3$. For each value of ε, use a graphing utility to find all values of $\delta > 0$ such that $|f(x) - 3| < \varepsilon$ whenever $0 < |x - 0| < \delta$. Sketch graphs illustrating your work.

  a. $\varepsilon = 1$ **b.** $\varepsilon = 0.5$

T 14. Finding δ for a given ε using a graph Let $g(x) = 2x^3 - 12x^2 + 26x + 4$ and note that $\lim_{x \to 2} g(x) = 24$. For each value of ε, use a graphing utility to find all values of $\delta > 0$ such that $|g(x) - 24| < \varepsilon$ whenever $0 < |x - 2| < \delta$. Sketch graphs illustrating your work.

  a. $\varepsilon = 1$ **b.** $\varepsilon = 0.5$

15. Finding a symmetric interval The function f in the figure satisfies $\lim_{x \to 2} f(x) = 3$. For each value of ε, find all values of $\delta > 0$ such that

$$|f(x) - 3| < \varepsilon \quad \text{whenever} \quad 0 < |x - 2| < \delta. \quad (2)$$

  a. $\varepsilon = 1$

  b. $\varepsilon = \frac{1}{2}$

  c. For any $\varepsilon > 0$, make a conjecture about the corresponding values of δ satisfying (2).

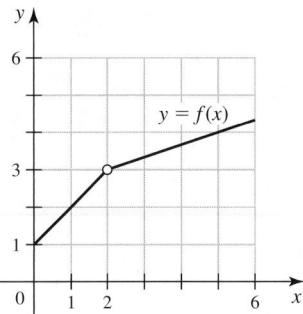

16. Finding a symmetric interval The function f in the figure satisfies $\lim_{x \to 4} f(x) = 5$. For each value of ε, find all values of $\delta > 0$ such that

$$|f(x) - 5| < \varepsilon \quad \text{whenever} \quad 0 < |x - 4| < \delta. \quad (3)$$

  a. $\varepsilon = 2$ **b.** $\varepsilon = 1$

  c. For any $\varepsilon > 0$, make a conjecture about the corresponding values of δ satisfying (3).

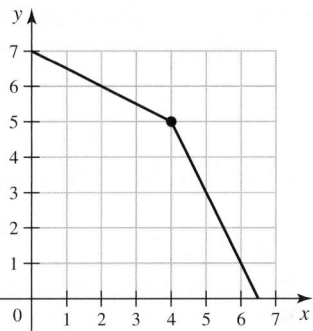

T 17. Finding a symmetric interval Let $f(x) = \dfrac{2x^2 - 2}{x - 1}$ and note that $\lim_{x \to 1} f(x) = 4$. For each value of ε, use a graphing utility to find all values of $\delta > 0$ such that $|f(x) - 4| < \varepsilon$ whenever $0 < |x - 1| < \delta$.

  a. $\varepsilon = 2$ **b.** $\varepsilon = 1$

  c. For any $\varepsilon > 0$, make a conjecture about the value of δ that satisfies the preceding inequality.

T 18. Finding a symmetric interval Let $f(x) = \begin{cases} \frac{1}{3}x + 1 & \text{if } x \le 3 \\ \frac{1}{2}x + \frac{1}{2} & \text{if } x > 3 \end{cases}$

and note that $\lim_{x \to 3} f(x) = 2$. For each value of ε, use a graphing utility to find all values of $\delta > 0$ such that $|f(x) - 2| < \varepsilon$ whenever $0 < |x - 3| < \delta$.

  a. $\varepsilon = \frac{1}{2}$ **b.** $\varepsilon = \frac{1}{4}$

  c. For any $\varepsilon > 0$, make a conjecture about the value of δ that satisfies the preceding inequality.

19–42. Limit proofs *Use the precise definition of a limit to prove the following limits. Specify a relationship between ε and δ that guarantees the limit exists.*

19. $\lim_{x \to 1} (8x + 5) = 13$ **20.** $\lim_{x \to 3} (-2x + 8) = 2$

21. $\lim\limits_{x \to 4} \dfrac{x^2 - 16}{x - 4} = 8$ (*Hint:* Factor and simplify.)

22. $\lim\limits_{x \to 3} \dfrac{x^2 - 7x + 12}{x - 3} = -1$

23. $\lim\limits_{x \to 0} |x| = 0$ 24. $\lim\limits_{x \to 0} |5x| = 0$

25. $\lim\limits_{x \to 7} f(x) = 9$, where $f(x) = \begin{cases} 3x - 12 & \text{if } x \le 7 \\ x + 2 & \text{if } x > 7 \end{cases}$

26. $\lim\limits_{x \to 5} f(x) = 4$, where $f(x) = \begin{cases} 2x - 6 & \text{if } x \le 5 \\ -4x + 24 & \text{if } x > 5 \end{cases}$

27. $\lim\limits_{x \to 0} x^2 = 0$ (*Hint:* Use the identity $\sqrt{x^2} = |x|$.)

28. $\lim\limits_{x \to 3} (x - 3)^2 = 0$ (*Hint:* Use the identity $\sqrt{x^2} = |x|$.)

29. $\lim\limits_{x \to 2} (x^2 + 3x) = 10$ 30. $\lim\limits_{x \to 4} (2x^2 - 4x + 1) = 17$

31. $\lim\limits_{x \to -3} |2x| = 6$ (*Hint:* Use the inequality $\|a\| - |b\| \le |a - b|$, which holds for all constants a and b (see Exercise 74).)

32. $\lim\limits_{x \to 25} \sqrt{x} = 5$ (*Hint:* The factorization $x - 25 = (\sqrt{x} - 5)(\sqrt{x} + 5)$ implies that $\sqrt{x} - 5 = \dfrac{x - 25}{\sqrt{x} + 5}$.)

33. $\lim\limits_{x \to 3} \dfrac{1}{x} = \dfrac{1}{3}$ (*Hint:* As $x \to 3$, eventually the distance between x and 3 is less than 1. Start by assuming $|x - 3| < 1$ and show $\dfrac{1}{|x|} < \dfrac{1}{2}$.)

34. $\lim\limits_{x \to 4} \dfrac{x - 4}{\sqrt{x} - 2} = 4$ (*Hint:* Multiply the numerator and denominator by $\sqrt{x} + 2$.)

35. $\lim\limits_{x \to 1/10} \dfrac{1}{x} = 10$ (*Hint:* To find δ, you need to bound x away from 0. So let $\left| x - \dfrac{1}{10} \right| < \dfrac{1}{20}$.)

36. $\lim\limits_{x \to 0} x \sin \dfrac{1}{x} = 0$

37. $\lim\limits_{x \to 0} (x^2 + x^4) = 0$ (*Hint:* You may use the fact that if $|x| < c$, then $x^2 < c^2$.)

38. $\lim\limits_{x \to a} b = b$, for any constants a and b

39. $\lim\limits_{x \to a} (mx + b) = ma + b$, for any constants a, b, and m

40. $\lim\limits_{x \to 3} x^3 = 27$ 41. $\lim\limits_{x \to 1} x^4 = 1$

42. $\lim\limits_{x \to 5} \dfrac{1}{x^2} = \dfrac{1}{25}$

43. **Proof of Limit Law 2** Suppose $\lim\limits_{x \to a} f(x) = L$ and $\lim\limits_{x \to a} g(x) = M$. Prove that $\lim\limits_{x \to a} (f(x) - g(x)) = L - M$.

44. **Proof of Limit Law 3** Suppose $\lim\limits_{x \to a} f(x) = L$. Prove that $\lim\limits_{x \to a} (cf(x)) = cL$, where c is a constant.

45–48. Limit proofs for infinite limits *Use the precise definition of infinite limits to prove the following limits.*

45. $\lim\limits_{x \to 4} \dfrac{1}{(x - 4)^2} = \infty$ 46. $\lim\limits_{x \to -1} \dfrac{1}{(x + 1)^4} = \infty$

47. $\lim\limits_{x \to 0} \left(\dfrac{1}{x^2} + 1 \right) = \infty$ 48. $\lim\limits_{x \to 0} \left(\dfrac{1}{x^4} - \sin x \right) = \infty$

49. **Explain why or why not** Determine whether the following statements are true and give an explanation or counterexample. Assume a and L are finite numbers and assume $\lim\limits_{x \to a} f(x) = L$.

    a. For a given $\varepsilon > 0$, there is one value of $\delta > 0$ for which $|f(x) - L| < \varepsilon$ whenever $0 < |x - a| < \delta$.

    b. The limit $\lim\limits_{x \to a} f(x) = L$ means that given an arbitrary $\delta > 0$, we can always find an $\varepsilon > 0$ such that $|f(x) - L| < \varepsilon$ whenever $0 < |x - a| < \delta$.

    c. The limit $\lim\limits_{x \to a} f(x) = L$ means that for any arbitrary $\varepsilon > 0$, we can always find a $\delta > 0$ such that $|f(x) - L| < \varepsilon$ whenever $0 < |x - a| < \delta$.

    d. If $|x - a| < \delta$, then $a - \delta < x < a + \delta$.

50. **Finding δ algebraically** Let $f(x) = x^2 - 2x + 3$.

    a. For $\varepsilon = 0.25$, find the largest value of $\delta > 0$ satisfying the statement
    $$|f(x) - 2| < \varepsilon \quad \text{whenever} \quad 0 < |x - 1| < \delta.$$

    b. Verify that $\lim\limits_{x \to 1} f(x) = 2$ as follows. For any $\varepsilon > 0$, find the largest value of $\delta > 0$ satisfying the statement
    $$|f(x) - 2| < \varepsilon \quad \text{whenever} \quad 0 < |x - 1| < \delta.$$

51–55. Precise definitions for left- and right-sided limits *Use the following definitions.*

*Assume f exists for all x near a with $x > a$. We say that **the limit of $f(x)$ as x approaches a from the right of a is L** and write $\lim\limits_{x \to a^+} f(x) = L$, if for any $\varepsilon > 0$ there exists $\delta > 0$ such that*
$$|f(x) - L| < \varepsilon \quad \text{whenever} \quad 0 < x - a < \delta.$$

*Assume f exists for all x near a with $x < a$. We say that **the limit of $f(x)$ as x approaches a from the left of a is L** and write $\lim\limits_{x \to a^-} f(x) = L$, if for any $\varepsilon > 0$ there exists $\delta > 0$ such that*
$$|f(x) - L| < \varepsilon \quad \text{whenever} \quad 0 < a - x < \delta.$$

51. **Comparing definitions** Why is the last inequality in the definition of $\lim\limits_{x \to a} f(x) = L$, namely, $0 < |x - a| < \delta$, replaced with $0 < x - a < \delta$ in the definition of $\lim\limits_{x \to a^+} f(x) = L$?

52. **Comparing definitions** Why is the last inequality in the definition of $\lim\limits_{x \to a} f(x) = L$, namely, $0 < |x - a| < \delta$, replaced with $0 < a - x < \delta$ in the definition of $\lim\limits_{x \to a^-} f(x) = L$?

53. **One-sided limit proofs** Prove the following limits for
$$f(x) = \begin{cases} 3x - 4 & \text{if } x < 0 \\ 2x - 4 & \text{if } x \ge 0. \end{cases}$$

    a. $\lim\limits_{x \to 0^+} f(x) = -4$ b. $\lim\limits_{x \to 0^-} f(x) = -4$ c. $\lim\limits_{x \to 0} f(x) = -4$

54. Determining values of δ from a graph The function f in the figure satisfies $\lim_{x \to 2^+} f(x) = 0$ and $\lim_{x \to 2^-} f(x) = 1$. Determine all values of $\delta > 0$ that satisfy each statement.

a. $|f(x) - 0| < 2$ whenever $0 < x - 2 < \delta$
b. $|f(x) - 0| < 1$ whenever $0 < x - 2 < \delta$
c. $|f(x) - 1| < 2$ whenever $0 < 2 - x < \delta$
d. $|f(x) - 1| < 1$ whenever $0 < 2 - x < \delta$

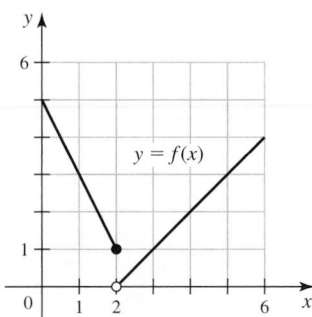

55. One-sided limit proof Prove that $\lim_{x \to 0^+} \sqrt{x} = 0$.

Explorations and Challenges

56. The relationship between one-sided and two-sided limits Prove the following statements to establish the fact that $\lim_{x \to a} f(x) = L$ if and only if $\lim_{x \to a^-} f(x) = L$ and $\lim_{x \to a^+} f(x) = L$.

a. If $\lim_{x \to a^-} f(x) = L$ and $\lim_{x \to a^+} f(x) = L$, then $\lim_{x \to a} f(x) = L$.
b. If $\lim_{x \to a} f(x) = L$, then $\lim_{x \to a^-} f(x) = L$ and $\lim_{x \to a^+} f(x) = L$.

57. Definition of one-sided infinite limits We write $\lim_{x \to a^+} f(x) = -\infty$ if for any negative number N, there exists $\delta > 0$ such that

$$f(x) < N \quad \text{whenever} \quad 0 < x - a < \delta.$$

a. Write an analogous formal definition for $\lim_{x \to a^+} f(x) = \infty$.
b. Write an analogous formal definition for $\lim_{x \to a^-} f(x) = -\infty$.
c. Write an analogous formal definition for $\lim_{x \to a^-} f(x) = \infty$.

58–59. One-sided infinite limits *Use the definitions given in Exercise 57 to prove the following infinite limits.*

58. $\lim_{x \to 1^+} \dfrac{1}{1 - x} = -\infty$ **59.** $\lim_{x \to 1^-} \dfrac{1}{1 - x} = \infty$

60–61. Definition of an infinite limit *We write* $\lim_{x \to a} f(x) = -\infty$ *if for any negative number M, there exists $\delta > 0$ such that*

$$f(x) < M \quad \text{whenever} \quad 0 < |x - a| < \delta.$$

Use this definition to prove the following statements.

60. $\lim_{x \to 1} \dfrac{-2}{(x - 1)^2} = -\infty$ **61.** $\lim_{x \to -2} \dfrac{-10}{(x + 2)^4} = -\infty$

62. Suppose $\lim_{x \to a} f(x) = \infty$. Prove that $\lim_{x \to a} (f(x) + c) = \infty$ for any constant c.

63. Suppose $\lim_{x \to a} f(x) = \infty$ and $\lim_{x \to a} g(x) = \infty$. Prove that $\lim_{x \to a} (f(x) + g(x)) = \infty$.

64–65. Definition of a limit at infinity *The limit at infinity* $\lim_{x \to \infty} f(x) = L$ *means that for any $\varepsilon > 0$, there exists $N > 0$ such that*

$$|f(x) - L| < \varepsilon \quad \text{whenever} \quad x > N.$$

Use this definition to prove the following statements.

64. $\lim_{x \to \infty} \dfrac{10}{x} = 0$ **65.** $\lim_{x \to \infty} \dfrac{2x + 1}{x} = 2$

66–67. Definition of infinite limits at infinity *We write* $\lim_{x \to \infty} f(x) = \infty$ *if for any positive number M, there is a corresponding $N > 0$ such that*

$$f(x) > M \quad \text{whenever} \quad x > N.$$

Use this definition to prove the following statements.

66. $\lim_{x \to \infty} \dfrac{x}{100} = \infty$ **67.** $\lim_{x \to \infty} \dfrac{x^2 + x}{x} = \infty$

68. Proof of the Squeeze Theorem Assume the functions f, g, and h satisfy the inequality $f(x) \le g(x) \le h(x)$, for all x near a, except possibly at a. Prove that if $\lim_{x \to a} f(x) = \lim_{x \to a} h(x) = L$, then $\lim_{x \to a} g(x) = L$.

69. Limit proof Suppose f is defined for all x near a, except possibly at a. Assume for any integer $N > 0$, there is another integer $M > 0$ such that $|f(x) - L| < 1/N$ whenever $|x - a| < 1/M$. Prove that $\lim_{x \to a} f(x) = L$ using the precise definition of a limit.

70–72. Proving that $\lim_{x \to a} f(x) \ne L$ *Use the following definition for the nonexistence of a limit. Assume f is defined for all x near a, except possibly at a. We write* $\lim_{x \to a} f(x) \ne L$ *if for some $\varepsilon > 0$, there is no value of $\delta > 0$ satisfying the condition*

$$|f(x) - L| < \varepsilon \quad \text{whenever} \quad 0 < |x - a| < \delta.$$

70. For the following function, note that $\lim_{x \to 2} f(x) \ne 3$. Find all values of $\varepsilon > 0$ for which the preceding condition for nonexistence is satisfied.

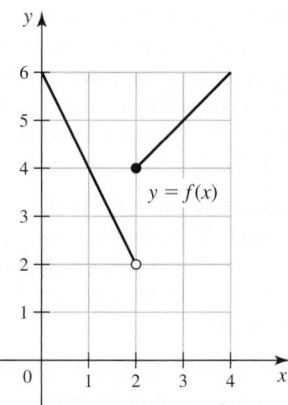

71. Prove that $\lim_{x \to 0} \dfrac{|x|}{x}$ does not exist.

72. Let

$$f(x) = \begin{cases} 0 & \text{if } x \text{ is rational} \\ 1 & \text{if } x \text{ is irrational.} \end{cases}$$

Prove that $\lim_{x \to a} f(x)$ does not exist for any value of a. (*Hint:* Assume $\lim_{x \to a} f(x) = L$ for some values of a and L, and let $\varepsilon = \dfrac{1}{2}$.)

73. A continuity proof Suppose f is continuous at a and defined for all x near a. If $f(a) > 0$, show that there is a positive number $\delta > 0$ for which $f(x) > 0$ for all x in $(a - \delta, a + \delta)$. (In other words, f is positive for all x in some interval containing a.)

74. Show that $\|a| - |b\| \leq |a - b|$ for all constants a and b. (*Hint:* Write $|a| = |(a - b) + b|$ and apply the triangle inequality to $|(a - b) + b|$.)

QUICK CHECK ANSWERS

1. $\delta \leq \frac{1}{50}$ **2.** $\delta \leq 0.62$ **3.** δ must decrease by a factor of $\sqrt{100} = 10$ (at least). ◄

CHAPTER 2   REVIEW EXERCISES

1. Explain why or why not Determine whether the following statements are true and give an explanation or counterexample.

a. The rational function $\dfrac{x - 1}{x^2 - 1}$ has vertical asymptotes at $x = -1$ and $x = 1$.

b. Numerical or graphical methods always produce good estimates of $\lim_{x \to a} f(x)$.

c. The value of $\lim_{x \to a} f(x)$, if it exists, is found by calculating $f(a)$.

d. If $\lim_{x \to a} f(x) = \infty$ or $\lim_{x \to a} f(x) = -\infty$, then $\lim_{x \to a} f(x)$ does not exist.

e. If $\lim_{x \to a} f(x)$ does not exist, then either $\lim_{x \to a} f(x) = \infty$ or $\lim_{x \to a} f(x) = -\infty$.

f. The line $y = 2x + 1$ is a slant asymptote of the function $f(x) = \dfrac{2x^2 + x}{x - 3}$.

g. If a function is continuous on the intervals (a, b) and $[b, c)$, where $a < b < c$, then the function is also continuous on (a, c).

h. If $\lim_{x \to a} f(x)$ can be calculated by direct substitution, then f is continuous at $x = a$.

2. The height above the ground of a stone thrown upwards is given by $s(t)$, where t is measured in seconds. After 1 second, the height of the stone is 48 feet above the ground, and after 1.5 seconds, the height of the stone is 60 feet above the ground. Evaluate $s(1)$ and $s(1.5)$, and then find the average velocity of the stone over the time interval $[1, 1.5]$.

⊤ 3. A baseball is thrown upwards into the air; its distance above the ground after t seconds is given by $s(t) = -16t^2 + 60t + 6$. Make a table of average velocities to make a conjecture about the instantaneous velocity of the baseball at $t = 1.5$ seconds after it was thrown into the air.

4. Estimating limits graphically Use the graph of f in the figure to evaluate the function or analyze the limit.

a. $f(-1)$ **b.** $\lim_{x \to -1^-} f(x)$ **c.** $\lim_{x \to -1^+} f(x)$ **d.** $\lim_{x \to -1} f(x)$

e. $f(1)$ **f.** $\lim_{x \to 1} f(x)$ **g.** $\lim_{x \to 2} f(x)$ **h.** $\lim_{x \to 3^-} f(x)$

i. $\lim_{x \to 3^+} f(x)$ **j.** $\lim_{x \to 3} f(x)$

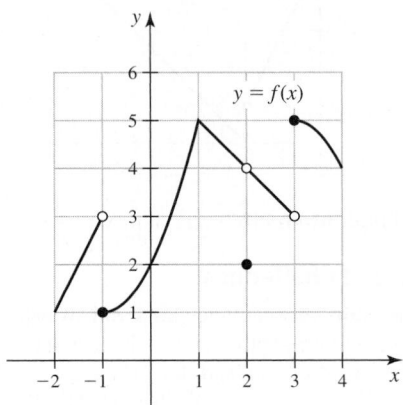

5. Points of discontinuity Use the graph of f in the figure to determine the values of x in the interval $(-3, 5)$ at which f fails to be continuous. Justify your answers using the continuity checklist.

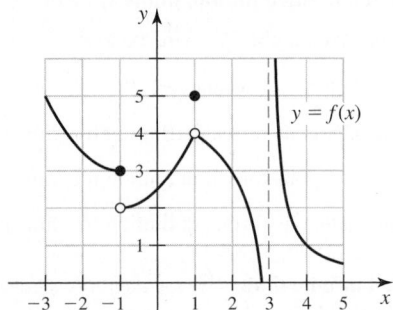

⊤ 6. Computing a limit graphically and analytically

a. Graph $y = \dfrac{\sin 2\theta}{\sin \theta}$ with a graphing utility. Comment on any inaccuracies in the graph and then sketch an accurate graph of the function.

b. Estimate $\lim_{\theta \to 0} \dfrac{\sin 2\theta}{\sin \theta}$ using the graph in part (a).

c. Verify your answer to part (b) by finding the value of $\lim_{\theta \to 0} \dfrac{\sin 2\theta}{\sin \theta}$ analytically using the trigonometric identity $\sin 2\theta = 2 \sin \theta \cos \theta$.

⊤ 7. Computing a limit numerically and analytically

a. Estimate $\lim_{x \to \pi/4} \dfrac{\cos 2x}{\cos x - \sin x}$ by making a table of values of $\dfrac{\cos 2x}{\cos x - \sin x}$ for values of x approaching $\pi/4$. Round your estimate to four digits.

b. Use analytic methods to find the value of $\lim_{x \to \pi/4} \dfrac{\cos 2x}{\cos x - \sin x}$.

8. Snowboard rental Suppose the rental cost for a snowboard is $25 for the first day (or any part of the first day) plus $15 for each additional day (or any part of a day).

  a. Graph the function $c = f(t)$ that gives the cost of renting a snowboard for t days, for $0 \le t \le 5$.

  b. Evaluate $\lim_{t \to 2.9} f(t)$.

  c. Evaluate $\lim_{x \to 3^-} f(t)$ and $\lim_{t \to 3^+} f(t)$.

  d. Interpret the meaning of the limits in part (c).

  e. For what values of t is f continuous? Explain.

9. Sketching a graph Sketch the graph of a function f with all the following properties.

$$\lim_{x \to -2^-} f(x) = \infty \quad \lim_{x \to -2^+} f(x) = -\infty \quad \lim_{x \to 0} f(x) = \infty$$

$$\lim_{x \to 3^-} f(x) = 2 \quad \lim_{x \to 3^+} f(x) = 4 \quad f(3) = 1$$

10–51. Calculating limits *Determine the following limits.*

10. $\lim_{x \to 1000} 18\pi^2$

11. $\lim_{x \to 1} \sqrt{5x + 6}$

12. $\lim_{h \to 0} \dfrac{\sqrt{5x + 5h} - \sqrt{5x}}{h}$, where $x > 0$ is constant

13. $\lim_{h \to 0} \dfrac{(h + 6)^2 + (h + 6) - 42}{h}$

14. $\lim_{x \to a} \dfrac{(3x + 1)^2 - (3a + 1)^2}{x - a}$, where a is constant

15. $\lim_{x \to 1} \dfrac{x^3 - 7x^2 + 12x}{4 - x}$

16. $\lim_{x \to 4} \dfrac{x^3 - 7x^2 + 12x}{4 - x}$

17. $\lim_{x \to 1} \dfrac{1 - x^2}{x^2 - 8x + 7}$

18. $\lim_{x \to 3} \dfrac{\sqrt{3x + 16} - 5}{x - 3}$

19. $\lim_{x \to 3} \dfrac{1}{x - 3}\left(\dfrac{1}{\sqrt{x + 1}} - \dfrac{1}{2}\right)$

20. $\lim_{t \to 1/3} \dfrac{t - 1/3}{(3t - 1)^2}$

21. $\lim_{x \to 3} \dfrac{x^4 - 81}{x - 3}$

22. $\lim_{p \to 1} \dfrac{p^5 - 1}{p - 1}$

23. $\lim_{x \to 81} \dfrac{\sqrt[4]{x} - 3}{x - 81}$

24. $\lim_{\theta \to \pi/2} \dfrac{\sin^2 \theta - 5 \sin \theta + 4}{\sin^2 \theta - 1}$

25. $\lim_{x \to \pi/2} \dfrac{\dfrac{1}{\sqrt{\sin x}} - 1}{x + \pi/2}$

26. $\lim_{x \to 1^+} \sqrt{\dfrac{x - 1}{x - 3}}$ and $\lim_{x \to 1^-} \sqrt{\dfrac{x - 1}{x - 3}}$

27. $\lim_{x \to 5} \dfrac{x - 7}{x(x - 5)^2}$

28. $\lim_{x \to -5^+} \dfrac{x - 5}{x + 5}$

29. $\lim_{x \to 3^-} \dfrac{x - 4}{x^2 - 3x}$

30. $\lim_{u \to 0^+} \dfrac{u - 1}{\sin u}$

31. $\lim_{x \to 1^+} \dfrac{4x^3 - 4x^2}{|x - 1|}$

32. $\lim_{x \to 2^-} \dfrac{|2x - 4|}{x^2 - 5x + 6}$

33. $\lim_{x \to 0^-} \dfrac{2}{\tan x}$

34. $\lim_{x \to \infty} \dfrac{4x^2 + 3x + 1}{\sqrt{8x^4 + 2}}$

35. $\lim_{x \to \infty} \dfrac{2x - 3}{4x + 10}$

36. $\lim_{x \to \infty} \dfrac{x^4 - 1}{x^5 + 2}$

37. $\lim_{x \to -\infty} \dfrac{3x + 1}{\sqrt{ax^2 + 2}}$, where a is a positive constant

38. $\lim_{x \to \infty} \left(\sqrt{x^2 + ax} - \sqrt{x^2 - b}\right)$, where a and b are constants

39. $\lim_{x \to \infty} \left(\dfrac{1}{\sqrt{x^2 - ax} - \sqrt{x^2 - x}}\right)$, where a is constant and $a \ne 1$

40. $\lim_{z \to \infty} \left(e^{-2z} + \dfrac{2}{z}\right)$

41. $\lim_{x \to \infty} (3 \tan^{-1} x + 2)$

42. $\lim_{x \to -\infty} (-3x^3 + 5)$

43. $\lim_{x \to -\infty} (|x - 1| + x)$ and $\lim_{x \to \infty} (|x - 1| + x)$

44. $\lim_{x \to -\infty} \dfrac{(|x - 2| + x)}{x}$ and $\lim_{x \to \infty} \dfrac{(|x - 2| + x)}{x}$

45. $\lim_{w \to \infty} \dfrac{\ln w^2}{\ln w^3 + 1}$

46. $\lim_{r \to -\infty} \dfrac{1}{2 + e^r}$ and $\lim_{r \to \infty} \dfrac{1}{2 + e^r}$

47. $\lim_{r \to \infty} \dfrac{2e^{4r} + 3e^{5r}}{7e^{4r} - 9e^{5r}}$ and $\lim_{r \to -\infty} \dfrac{2e^{4r} + 3e^{5r}}{7e^{4r} - 9e^{5r}}$

48. $\lim_{x \to -\infty} e^x \sin x$

49. $\lim_{x \to \infty} \left(5 + \dfrac{\cos^4 x}{x^2 + x + 1}\right)$

50. $\lim_{t \to \infty} \dfrac{\cos t}{e^{3t}}$

51. $\lim_{x \to 1^-} \dfrac{x}{\ln x}$

52. Assume the function g satisfies the inequality $1 \le g(x) \le \sin^2 x + 1$, for all values of x near 0. Find $\lim_{x \to 0} g(x)$.

T 53. An important limit

  a. Use a graphing utility to illustrate the inequalities

$$\cos x \le \dfrac{\sin x}{x} \le \dfrac{1}{\cos x}$$

  on $[-1, 1]$.

  b. Use part (a) to explain why $\lim_{x \to 0} \dfrac{\sin x}{x} = 1$.

T 54. Finding vertical asymptotes Let $f(x) = \dfrac{x^2 - 5x + 6}{x^2 - 2x}$.

  a. Analyze $\lim_{x \to 0^-} f(x)$, $\lim_{x \to 0^+} f(x)$, $\lim_{x \to 2^-} f(x)$, and $\lim_{x \to 2^+} f(x)$.

  b. Does the graph of f have any vertical asymptotes? Explain.

  c. Graph f using a graphing utility and then sketch the graph with paper and pencil, correcting any errors obtained with the graphing utility.

55–60. End behavior *Evaluate $\lim_{x \to \infty} f(x)$ and $\lim_{x \to -\infty} f(x)$.*

55. $f(x) = \dfrac{4x^3 + 1}{1 - x^3}$

56. $f(x) = \dfrac{x^6 + 1}{\sqrt{16x^{14} + 1}}$

57. $f(x) = 1 - e^{-2x}$

58. $f(x) = \dfrac{1}{\ln x^2}$

59. $f(x) = \dfrac{6e^x + 20}{3e^x + 4}$

60. $f(x) = \dfrac{x + 1}{\sqrt{9x^2 + x}}$

61–65. Slant asymptotes

a. *Analyze $\lim_{x \to \infty} f(x)$ and $\lim_{x \to -\infty} f(x)$ for each function.*

b. *Determine whether f has a slant asymptote. If so, write the equation of the slant asymptote.*

61. $f(x) = \dfrac{3x^2 + 5x + 7}{x + 1}$

62. $f(x) = \dfrac{9x^2 + 4}{(2x - 1)^2}$

63. $f(x) = \dfrac{1 + x - 2x^2 - x^3}{x^2 + 1}$ **64.** $f(x) = \dfrac{x(x + 2)^3}{3x^2 - 4x}$

65. $f(x) = \dfrac{4x^3 + x^2 + 7}{x^2 - x + 1}$

66–68. Finding asymptotes *Find all the asymptotes of the following functions.*

66. $f(x) = \dfrac{2x^2 + 6}{2x^2 + 3x - 2}$ **67.** $f(x) = \dfrac{1}{\tan^{-1} x}$

68. $f(x) = \dfrac{2x^2 - 7}{x - 2}$

T 69. **Two slant asymptotes** Explain why the function

$f(x) = \dfrac{x + xe^x + 10e^x}{2(e^x + 1)}$ has two slant asymptotes, $y = \dfrac{1}{2}x$ and

$y = \dfrac{1}{2}x + 5$. Plot a graph of f together with its two slant asymptotes.

T 70. **Finding all asymptotes** Find all the asymptotes of

$f(x) = \dfrac{x^2 + x + 3}{|x|}$. Plot a graph of f together with its asymp-

totes. (*Hint:* Consider the cases $x = 0$, $x > 0$, and $x < 0$.)

71–74. Continuity at a point *Use the continuity checklist to determine whether the following functions are continuous at the given value of a.*

71. $f(x) = \dfrac{1}{x - 5}; a = 5$

72. $g(x) = \begin{cases} \dfrac{x^2 - 16}{x - 4} & \text{if } x \neq 4 \\ 9 & \text{if } x = 4 \end{cases}; a = 4$

73. $h(x) = \begin{cases} -2x + 14 & \text{if } x \leq 5 \\ \sqrt{x^2 - 9} & \text{if } x > 5 \end{cases}; a = 5$

74. $g(x) = \begin{cases} \dfrac{x^3 - 5x^2 + 6x}{x - 2} & \text{if } x \neq 2 \\ -2 & \text{if } x = 2 \end{cases}; a = 2$

75–78. Continuity on intervals *Find the intervals on which the following functions are continuous. Specify right- or left-continuity at the finite endpoints.*

75. $f(x) = \sqrt{x^2 - 5}$ **76.** $g(x) = e^{\sqrt{x - 2}}$

77. $h(x) = \dfrac{2x}{x^3 - 25x}$ **78.** $g(x) = \cos e^x$

79. **Determining unknown constants** Let

$$g(x) = \begin{cases} 5x - 2 & \text{if } x < 1 \\ a & \text{if } x = 1 \\ ax^2 + bx & \text{if } x > 1. \end{cases}$$

Determine values of the constants a and b, if possible, for which g is continuous at $x = 1$.

80. **Left- and right-continuity**
   a. Is $h(x) = \sqrt{x^2 - 9}$ left-continuous at $x = 3$? Explain.
   b. Is $h(x) = \sqrt{x^2 - 9}$ right-continuous at $x = 3$? Explain.

81. Sketch the graph of a function that is continuous on $(0, 1]$ and on $(1, 2)$ but is not continuous on $(0, 2)$.

T 82–83. Intermediate Value Theorem
   a. *Use the Intermediate Value Theorem to show that the equation has a solution in the given interval.*
   b. *Estimate a solution to the equation in the given interval using a root finder.*

T 82. $x^5 + 7x + 5 = 0; (-1, 0)$

T 83. $x = \cos x; \left(0, \dfrac{\pi}{2}\right)$

84. Suppose on a certain day, the low temperature was $32°$ at midnight, the high temperature was $65°$ at noon, and then the temperature dropped to $32°$ the following midnight. Prove there were two times during that day, which were 12 hours apart, when the temperatures were equal. (*Hint:* Let $T(t)$ equal the temperature t hours after midnight and examine the function $f(t) = T(t) - T(t + 12)$, for $0 \leq t \leq 12$.)

T 85. **Antibiotic dosing** The amount of an antibiotic (in milligrams) in the blood t hours after an intravenous line is opened is given by

$$m(t) = 100(e^{-0.1t} - e^{-0.3t}).$$

   a. Use the Intermediate Value Theorem to show that the amount of drug is 30 mg at some time in the interval $[0, 5]$ and again at some time in the interval $[5, 15]$.
   b. Estimate the times at which $m = 30$ mg.
   c. Is the amount of drug in the blood ever 50 mg?

86–91. Limit proofs *Use an appropriate limit definition to prove the following limits.*

86. $\lim\limits_{x \to 1} (5x - 2) = 3$ **87.** $\lim\limits_{x \to 5} \dfrac{x^2 - 25}{x - 5} = 10$

88. $\lim\limits_{x \to 3} f(x) = 5$, where $f(x) = \begin{cases} 3x - 4 & \text{if } x \leq 3 \\ -4x + 17 & \text{if } x > 3 \end{cases}$

89. $\lim\limits_{x \to 2} (3x^2 - 4) = 8$ **90.** $\lim\limits_{x \to 2^+} \sqrt{4x - 8} = 0$

91. $\lim\limits_{x \to 2} \dfrac{1}{(x - 2)^4} = \infty$

92. **Limit proofs**
   a. Assume $|f(x)| \leq L$ for all x near a and $\lim\limits_{x \to a} g(x) = 0$. Give a formal proof that $\lim\limits_{x \to a} (f(x)g(x)) = 0$.
   b. Find a function f for which $\lim\limits_{x \to 2} (f(x)(x - 2)) \neq 0$. Why doesn't this violate the result stated in part (a)?
   c. The Heaviside function is defined as

$$H(x) = \begin{cases} 0 & \text{if } x < 0 \\ 1 & \text{if } x \geq 0. \end{cases}$$

   Explain why $\lim\limits_{x \to 0} (xH(x)) = 0$.

Chapter 2 Guided Projects

Applications of the material in this chapter and related topics can be found in the following Guided Projects. For additional information, see the Preface.

- Fixed-point iteration
- Local linearity

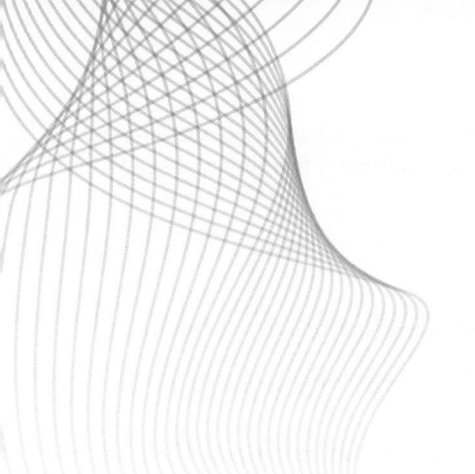

3

Derivatives

Chapter Preview   Now that you are familiar with limits, the door to calculus stands open. The first task is to introduce the fundamental concept of the *derivative*. Suppose a function f represents a quantity of interest—for example, the variable cost of manufacturing an item, the population of a country, or the position of an orbiting satellite. The derivative of f is another function, denoted f', that gives the slope of the curve $y = f(x)$ as it changes with respect to x. Equivalently, the derivative of f gives the *instantaneous rate of change* of f with respect to the independent variable. We use limits not only to define the derivative but also to develop efficient rules for finding derivatives. The applications of the derivative—which we introduce along the way—are endless because almost everything around us is in a state of change, and derivatives describe change.

3.1 Introducing the Derivative

In this section, we return to the problem of finding the slope of a line tangent to a curve, introduced at the beginning of Chapter 2. This problem is important for several reasons.

- We identify the slope of the tangent line with the *instantaneous rate of change* of a function (Figure 3.1).
- The slopes of the tangent lines as they change along a curve are the values of a new function called the *derivative*.
- Looking farther ahead, if a curve represents the trajectory of a moving object, the tangent line at a point on the curve indicates the direction of motion at that point (Figure 3.2).

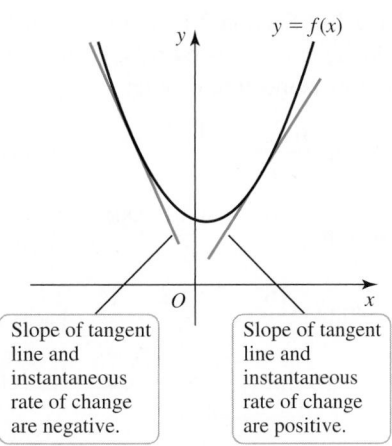

Slope of tangent line and instantaneous rate of change are negative.

Slope of tangent line and instantaneous rate of change are positive.

Figure 3.1

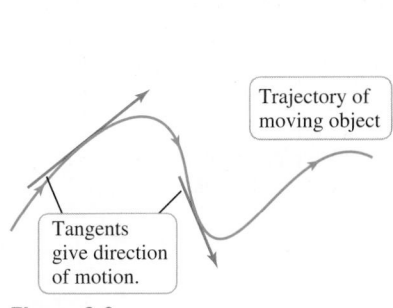

Trajectory of moving object

Tangents give direction of motion.

Figure 3.2

In Section 2.1, we used a limit to define the instantaneous velocity of an object that moves along a line. Recall that if $s(t)$ is the position of the object at time t, then the average velocity of the object over the time interval $[a, t]$ is

$$v_{av} = \frac{s(t) - s(a)}{t - a}.$$

The instantaneous velocity at time $t = a$ is the limit of the average velocity as $t \to a$:

$$v_{inst} = \lim_{t \to a} \frac{s(t) - s(a)}{t - a}.$$

We also learned that these quantities have important geometric interpretations. The average velocity is the slope of the secant line through the points $(a, s(a))$ and $(t, s(t))$ on the graph of s, and the instantaneous velocity is the slope of the tangent line through the point $(a, s(a))$ (**Figure 3.3**).

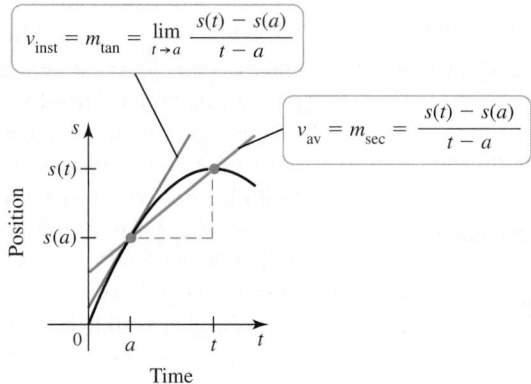

Figure 3.3

EXAMPLE 1   Instantaneous velocity and tangent lines A rock is launched vertically upward from the ground with an initial speed of 96 ft/s. The position of the rock in feet above the ground after t seconds is given by the function $s(t) = -16t^2 + 96t$. Consider the point $P(1, 80)$ on the curve $y = s(t)$.

a. Find the instantaneous velocity of the rock 1 second after launch and find the slope of the line tangent to the graph of s at P.

b. Find an equation of the tangent line in part (a).

SOLUTION

a. In Example 2 of Section 2.1, we used numerical evidence to estimate that the instantaneous velocity at $t = 1$ is 64 ft/s. Using limit techniques developed in Chapter 2, we can verify this conjectured value:

$$
\begin{aligned}
v_{inst} &= \lim_{t \to 1} \frac{s(t) - s(1)}{t - 1} && \text{Definition of instantaneous velocity} \\[1em]
&= \lim_{t \to 1} \frac{-16t^2 + 96t - 80}{t - 1} && s(t) = -16t^2 + 96t;\ s(1) = 80 \\[1em]
&= \lim_{t \to 1} \frac{-16(t - 5)(t - 1)}{t - 1} && \text{Factor the numerator.} \\[1em]
&= -16 \underbrace{\lim_{t \to 1} (t - 5)}_{-4} = 64. && \text{Cancel factors } (t \neq 1) \text{ and evaluate the limit.}
\end{aligned}
$$

We see that the instantaneous velocity at $t = 1$ is 64 ft/s, which also equals the slope of the line tangent to the graph of s at the point $P(1, 80)$.

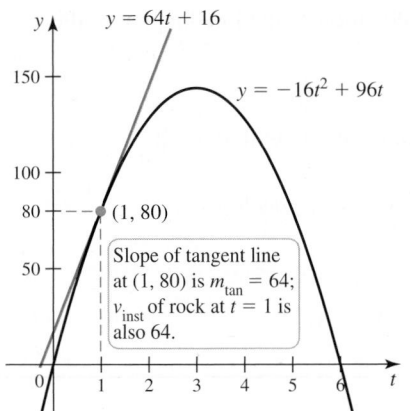

Figure 3.4

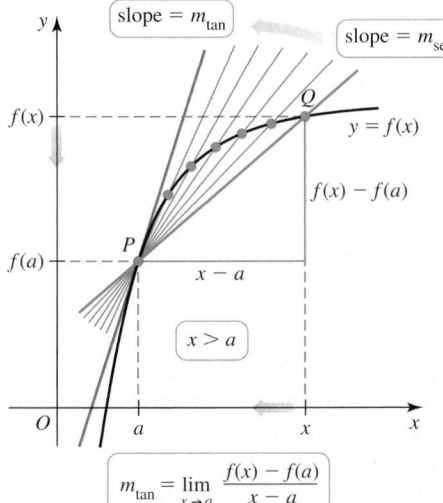

Figure 3.5

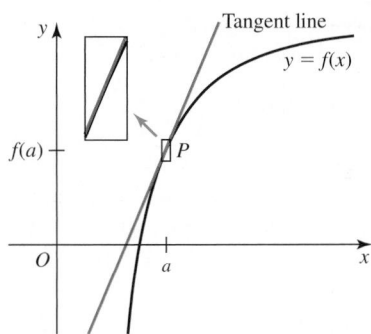

Figure 3.6

▶ The definition of m_{sec} involves a *difference quotient*, introduced in Section 1.1.

▶ If x and y have physical units, then the average and instantaneous rates of change have units of (units of y)/(units of x). For example, if y has units of meters and x has units of seconds, the units of the rates of change are meters/second (m/s).

b. An equation of the line passing through $(1, 80)$ with slope 64 is $y - 80 = 64(t - 1)$ or $y = 64t + 16$ (**Figure 3.4**).

Related Exercises 13–14 ◀

QUICK CHECK 1 In Example 1, is the slope of the tangent line at $(2, 128)$ greater than or less than the slope at $(1, 80)$? ◀

The connection between the instantaneous rate of change of an object's position and the slope of a tangent line on the graph of the position function can be extended far beyond a discussion of velocity. In fact, the slope of a tangent line is one of the central concepts in calculus because it measures the instantaneous rate of change of a function. Whether a given function describes the position of an object, the population of a city, the concentration of a reactant in a chemical reaction, or the weight of a growing child, the slopes of tangent lines associated with these functions measure rates at which the quantities change.

Tangent Lines and Rates of Change

Consider the curve $y = f(x)$ and a secant line intersecting the curve at the points $P(a, f(a))$ and $Q(x, f(x))$ (**Figure 3.5**). The difference $f(x) - f(a)$ is the change in the value of f on the interval $[a, x]$, while $x - a$ is the change in x. As discussed in Chapters 1 and 2, the slope of the secant line $\overleftrightarrow{PQ}$ is

$$m_{sec} = \frac{f(x) - f(a)}{x - a},$$

and it gives the *average rate of change* in f on the interval $[a, x]$.

Figure 3.5 also shows what happens as the variable point x approaches the fixed point a. If the curve is smooth at $P(a, f(a))$—it has no kinks or corners—the secant lines approach a *unique* line that intersects the curve at P; this line is the *tangent line* at P. As x approaches a, the slopes m_{sec} of the secant lines approach a unique number m_{tan} that we call the *slope of the tangent line*; that is,

$$m_{tan} = \lim_{x \to a} \frac{f(x) - f(a)}{x - a}.$$

The slope of the tangent line at P is also called the *instantaneous rate of change* in f at a because it measures how quickly f changes at a.

The tangent line has another geometric interpretation. As discussed in Section 2.1, if the curve $y = f(x)$ is smooth at a point $P(a, f(a))$, then the curve looks more like a line as we zoom in on P. The line that is approached as we zoom in on P is the tangent line (**Figure 3.6**). A smooth curve has the property of *local linearity*, which means that if we look at a point on the curve locally (by zooming in), then the curve appears linear.

DEFINITION Rate of Change and the Slope of the Tangent Line

The **average rate of change** in f on the interval $[a, x]$ is the slope of the corresponding secant line:

$$m_{sec} = \frac{f(x) - f(a)}{x - a}.$$

The **instantaneous rate of change** in f at a is

$$m_{tan} = \lim_{x \to a} \frac{f(x) - f(a)}{x - a}, \tag{1}$$

which is also the **slope of the tangent line** at $(a, f(a))$, provided this limit exists. The **tangent line** is the unique line through $(a, f(a))$ with slope m_{tan}. Its equation is

$$y - f(a) = m_{tan}(x - a).$$

QUICK CHECK 2 Sketch the graph of a function f near a point a. As in Figure 3.5, draw a secant line that passes through $(a, f(a))$ and a neighboring point $(x, f(x))$ with $x < a$. Use arrows to show how the secant lines approach the tangent line as x approaches a. ◀

EXAMPLE 2 Equation of a tangent line Find an equation of the line tangent to the graph of $f(x) = \dfrac{3}{x}$ at $\left(2, \dfrac{3}{2}\right)$.

SOLUTION We use the definition of the slope of the tangent line with $a = 2$:

$$
\begin{aligned}
m_{\text{tan}} &= \lim_{x \to 2} \frac{f(x) - f(2)}{x - 2} && \text{Definition of slope of tangent line} \\[2mm]
&= \lim_{x \to 2} \frac{\dfrac{3}{x} - \dfrac{3}{2}}{x - 2} && f(x) = 3/x;\ f(2) = 3/2 \\[2mm]
&= \lim_{x \to 2} \frac{\dfrac{6 - 3x}{2x}}{x - 2} && \text{Combine fractions with common denominator.} \\[2mm]
&= \lim_{x \to 2} \frac{-3(x - 2)}{2x(x - 2)} && \text{Simplify.} \\[2mm]
&= \lim_{x \to 2} \left(-\frac{3}{2x}\right) = -\frac{3}{4}. && \text{Cancel factors } (x \neq 2) \text{ and evaluate the limit.}
\end{aligned}
$$

The tangent line has slope $m_{\text{tan}} = -\dfrac{3}{4}$ and passes through the point $\left(2, \dfrac{3}{2}\right)$ (**Figure 3.7**). Its equation is $y - \dfrac{3}{2} = -\dfrac{3}{4}(x - 2)$ or $y = -\dfrac{3}{4}x + 3$. We could also say that the instantaneous rate of change in f at $x = 2$ is $-3/4$.

Related Exercises 17–18 ◄

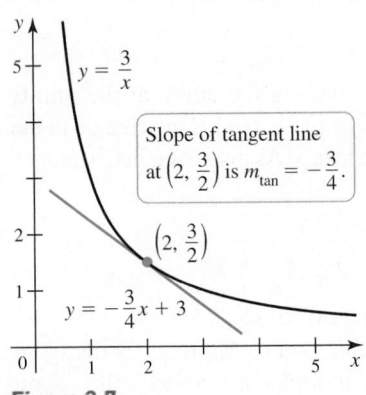

$y = \dfrac{3}{x}$

Slope of tangent line at $\left(2, \dfrac{3}{2}\right)$ is $m_{\text{tan}} = -\dfrac{3}{4}$.

$\left(2, \dfrac{3}{2}\right)$

$y = -\dfrac{3}{4}x + 3$

Figure 3.7

An alternative formula for the slope of the tangent line is helpful for future work. Consider again the curve $y = f(x)$ and the secant line intersecting the curve at the points P and Q. We now let $(a, f(a))$ and $(a + h, f(a + h))$ be the coordinates of P and Q, respectively (**Figure 3.8**). The difference in the x-coordinates of P and Q is $(a + h) - a = h$. Note that Q is located to the right of P if $h > 0$ and to the left of P if $h < 0$.

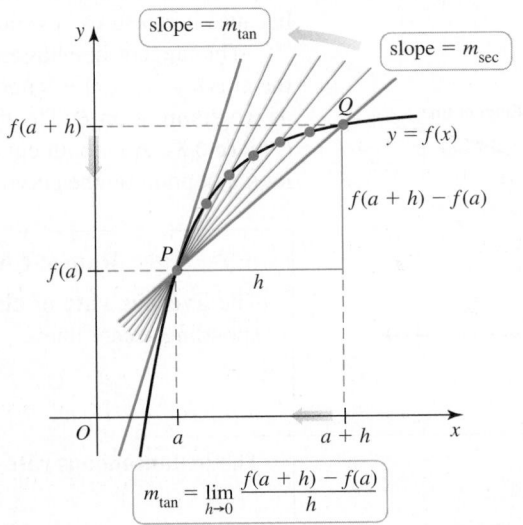

slope $= m_{\text{tan}}$

slope $= m_{\text{sec}}$

$y = f(x)$

$f(a + h) - f(a)$

$f(a + h)$

$f(a)$

h

$$m_{\text{tan}} = \lim_{h \to 0} \frac{f(a + h) - f(a)}{h}$$

Figure 3.8

The slope of the secant line $\overleftrightarrow{PQ}$ using the new notation is $m_{\text{sec}} = \dfrac{f(a + h) - f(a)}{h}$. As h approaches 0, the variable point Q approaches P, and the slopes of the secant lines approach the slope of the tangent line. Therefore, the slope of the tangent line at $(a, f(a))$, which is also the instantaneous rate of change in f at a, is

$$m_{\text{tan}} = \lim_{h \to 0} \frac{f(a + h) - f(a)}{h}.$$

ALTERNATIVE DEFINITION **Rate of Change and the Slope of the Tangent Line**

The **average rate of change** in f on the interval $[a, a + h]$ is the slope of the corresponding secant line:

$$m_{\text{sec}} = \frac{f(a + h) - f(a)}{h}.$$

The **instantaneous rate of change** in f at a is

$$m_{\text{tan}} = \lim_{h \to 0} \frac{f(a + h) - f(a)}{h}, \tag{2}$$

which is also the **slope of the tangent line** at $(a, f(a))$, provided this limit exists.

> ▶ By the definition of the limit as $h \to 0$, notice that h approaches 0 but $h \neq 0$. Therefore, it is permissible to cancel h from the numerator and denominator of $\dfrac{h(h^2 - 3h + 7)}{h}$.

EXAMPLE 3 **Equation of a tangent line** Find an equation of the line tangent to the graph of $f(x) = x^3 + 4x$ at $(1, 5)$.

SOLUTION We let $a = 1$ in definition (2) and first find $f(1 + h)$. After expanding and collecting terms, we have

$$f(1 + h) = (1 + h)^3 + 4(1 + h) = h^3 + 3h^2 + 7h + 5.$$

Substituting $f(1 + h)$ and $f(1) = 5$, the slope of the tangent line is

$$
\begin{aligned}
m_{\text{tan}} &= \lim_{h \to 0} \frac{f(1 + h) - f(1)}{h} && \text{Definition of } m_{\text{tan}} \\
&= \lim_{h \to 0} \frac{(h^3 + 3h^2 + 7h + 5) - 5}{h} && \text{Substitute } f(1 + h) \text{ and } f(1) = 5. \\
&= \lim_{h \to 0} \frac{h(h^2 + 3h + 7)}{h} && \text{Simplify.} \\
&= \lim_{h \to 0} (h^2 + 3h + 7) && \text{Cancel } h, \text{ noting } h \neq 0. \\
&= 7. && \text{Evaluate the limit.}
\end{aligned}
$$

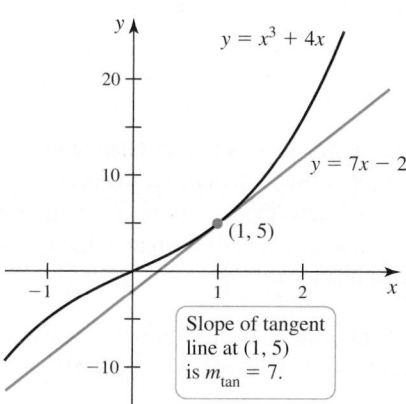

$y = x^3 + 4x$

$y = 7x - 2$

$(1, 5)$

Slope of tangent line at $(1, 5)$ is $m_{\text{tan}} = 7$.

Figure 3.9

The tangent line has slope $m_{\text{tan}} = 7$ and passes through the point $(1, 5)$ (**Figure 3.9**); its equation is $y - 5 = 7(x - 1)$ or $y = 7x - 2$. We could also say that the instantaneous rate of change in f at $x = 1$ is 7.

Related Exercises 23, 27 ◀

QUICK CHECK 3 Set up the calculation in Example 3 using definition (1) for the slope of the tangent line rather than definition (2). Does the calculation appear more difficult using definition (1)? ◀

The Derivative

Computing the slope of the line tangent to the graph of a function f at a given point a gives us the instantaneous rate of change in f at a. This information about the behavior of a function is so important that it has its own name and notation.

> ▶ The derivative notation $f'(a)$ is read f *prime of a*. A minor modification in the notation for the derivative is necessary when the name of the function changes. For example, given the function $y = g(x)$, its derivative at the point a is $g'(a)$.

DEFINITION **The Derivative of a Function at a Point**

The **derivative of f at a**, denoted $f'(a)$, is given by either of the two following limits, provided the limits exist and a is in the domain of f:

$$f'(a) = \lim_{x \to a} \frac{f(x) - f(a)}{x - a} \quad (1) \quad \text{or} \quad f'(a) = \lim_{h \to 0} \frac{f(a + h) - f(a)}{h}. \tag{2}$$

If $f'(a)$ exists, we say that f is **differentiable** at a.

The limits that define the derivative of a function at a point are exactly the same limits used to compute the slope of a tangent line and the instantaneous rate of change of a function at a point. When you compute a derivative, remember that you are also finding a rate of change and the slope of a tangent line.

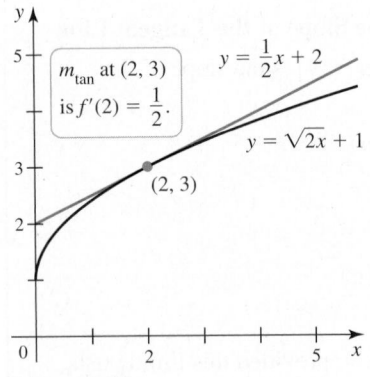

Figure 3.10

▶ Definition (2) could also be used to find $f'(2)$ in Example 4.

QUICK CHECK 4 Verify that the derivative of the function f in Example 4 at the point $(8, 5)$ is $f'(8) = 1/4$. Then find an equation of the line tangent to the graph of f at the point $(8, 5)$. ◀

EXAMPLE 4 **Derivatives and tangent lines** Let $f(x) = \sqrt{2x} + 1$. Compute $f'(2)$, the derivative of f at $x = 2$, and use the result to find an equation of the line tangent to the graph of f at $(2, 3)$.

SOLUTION Definition (1) for $f'(a)$ is used here, with $a = 2$.

$$f'(2) = \lim_{x \to 2} \frac{f(x) - f(2)}{x - 2} \qquad \text{Definition of the derivative at a point}$$

$$= \lim_{x \to 2} \frac{\sqrt{2x} + 1 - 3}{x - 2} \qquad \begin{aligned} &\text{Substitute: } f(x) = \sqrt{2x} + 1 \text{ and} \\ &f(2) = 3. \end{aligned}$$

$$= \lim_{x \to 2} \frac{\sqrt{2x} - 2}{x - 2} \cdot \frac{\sqrt{2x} + 2}{\sqrt{2x} + 2} \qquad \begin{aligned} &\text{Multiply numerator and denominator} \\ &\text{by } \sqrt{2x} + 2. \end{aligned}$$

$$= \lim_{x \to 2} \frac{2x - 4}{(x - 2)(\sqrt{2x} + 2)} \qquad \text{Simplify.}$$

$$= \lim_{x \to 2} \frac{2}{\sqrt{2x} + 2} = \frac{1}{2} \qquad \begin{aligned} &\text{Cancel factors } (x \neq 2) \text{ and evaluate} \\ &\text{the limit.} \end{aligned}$$

Because $f'(2) = 1/2$ is the slope of the line tangent to the graph of f at $(2, 3)$, an equation of the tangent line (**Figure 3.10**) is

$$y - 3 = \frac{1}{2}(x - 2) \quad \text{or} \quad y = \frac{1}{2}x + 2.$$

Related Exercises 39–40 ◀

Interpreting the Derivative

The derivative of a function f at a point a measures the instantaneous rate of change in f at a. When the variables associated with a function represent physical quantities, the derivative takes on special meaning. For instance, suppose $T = g(d)$ describes the water temperature T at depth d in the ocean. Then $g'(d)$ measures the instantaneous rate of change in the water temperature at depth d. The following examples illustrate this idea.

EXAMPLE 5 **Instantaneous rate of change** Sound intensity I, measured in watts per square meter (W/m^2) at a point x meters from a sound source with acoustic power P, is given by $I(x) = \dfrac{P}{4\pi x^2}$. A typical sound system at a rock concert produces an acoustic power of about $P = 3$ W. Compute $I'(3)$ and interpret the result.

SOLUTION With $P = 3$, the sound intensity function is $I(x) = \dfrac{3}{4\pi x^2}$. A useful trick is to write I as $I(x) = \dfrac{3}{4\pi} \cdot \dfrac{1}{x^2}$. Using definition (2) of the derivative at a point (p. 135), with $a = 3$, we have

$$I'(3) = \lim_{h \to 0} \frac{I(3 + h) - I(3)}{h} \qquad \begin{aligned} &\text{Definition of } I'(3) \\ &\text{Numerator units: } \text{W/m}^2; \\ &\text{denominator units: m} \end{aligned}$$

$$= \lim_{h \to 0} \frac{\dfrac{3}{4\pi} \cdot \dfrac{1}{(3 + h)^2} - \dfrac{3}{4\pi} \cdot \dfrac{1}{9}}{h} \qquad \text{Evaluate } I(3 + h) \text{ and } I(3).$$

$$= \frac{3}{4\pi} \lim_{h \to 0} \frac{\dfrac{9 - (3 + h)^2}{9(3 + h)^2}}{h} \qquad \begin{aligned} &\text{Factor out } \dfrac{3}{4\pi} \text{ and simplify} \\ &\text{remaining fraction.} \end{aligned}$$

$$= \frac{3}{4\pi} \lim_{h \to 0} \frac{-h(6 + h)}{9(3 + h)^2} \cdot \frac{1}{h} \qquad \text{Simplify.}$$

$$= -\frac{3}{4\pi} \lim_{h \to 0} \frac{6 + h}{9(3 + h)^2} \qquad \text{Cancel } h; h \neq 0.$$

$$= -\frac{3}{4\pi} \cdot \frac{6}{81} = -\frac{1}{18\pi}. \qquad \text{Evaluate limit.}$$

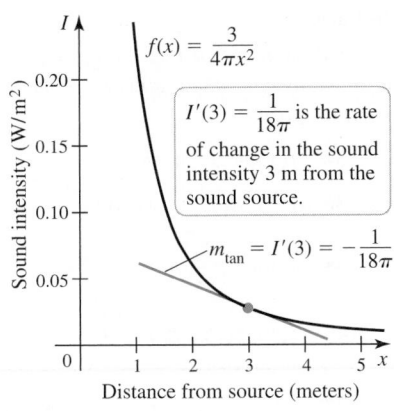

$f(x) = \dfrac{3}{4\pi x^2}$

$I'(3) = \dfrac{1}{18\pi}$ is the rate of change in the sound intensity 3 m from the sound source.

$m_{tan} = I'(3) = -\dfrac{1}{18\pi}$

Figure 3.11

The units associated with the numerator and denominator help to interpret this result. The change in sound intensity I is measured in W/m^2 while the distance from the sound source is measured in meters (**Figure 3.11**). Therefore, the result $I'(3) = -1/(18\pi)$ means that the sound intensity decreases with an instantaneous rate of change of $1/(18\pi)$ W/m^2 *per meter* at a point 3 m from the sound source. Another useful interpretation is that when the distance from the source increases from 3 m to 4 m, the sound intensity decreases by about $1/(18\pi)$ W/m^2.

Related Exercises 49–50 ◄

EXAMPLE 6 **Growth rates of Indian spotted owlets** The Indian spotted owlet is a small owl that is indigenous to Southeast Asia. The body mass $m(t)$ (in grams) of an owl at an age of t weeks is modeled by the graph in **Figure 3.12**. Estimate $m'(2)$ and state the physical meaning of this quantity. (*Source: ZooKeys, 132, 2011*)

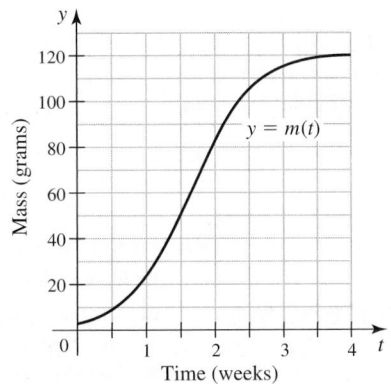

Figure 3.12

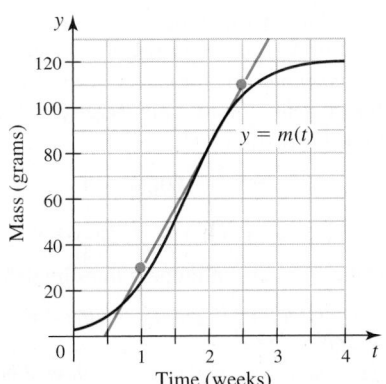

Figure 3.13

➤ The graphical method used in Example 6 to estimate $m'(2)$ provides only a rough approximation to the value of the derivative: The answer depends on the accuracy with which the tangent line is sketched.

SOLUTION Recall that the derivative $m'(2)$ equals the slope of the line tangent to the graph of $y = m(t)$ at $t = 2$. One method for estimating $m'(2)$ is to sketch the line tangent to the curve at $t = 2$ and then estimate its slope (**Figure 3.13**). Searching for two convenient points on this line, we see that the tangent line passes through—or close to—the points $(1, 30)$ and $(2.5, 110)$. Therefore,

$$m'(2) \approx \frac{110 - 30}{2.5 - 1} = \frac{80 \text{ g}}{1.5 \text{ weeks}} \approx 53.3 \text{ g/week,}$$

which means that the owl is growing at 53.3 g/week 2 weeks after birth.

Related Exercises 53–54 ◄

SECTION 3.1 EXERCISES

Getting Started

1. Use definition (1) (p. 133) for the slope of a tangent line to explain how slopes of secant lines approach the slope of the tangent line at a point.

2. Explain why the slope of a secant line can be interpreted as an average rate of change.

3. Explain why the slope of the tangent line can be interpreted as an instantaneous rate of change.

4. Explain the relationships among the slope of a tangent line, the instantaneous rate of change, and the value of the derivative at a point.

5. Given a function f and a point a in its domain, what does $f'(a)$ represent?

6. The following figure shows the graph of f and a line tangent to the graph of f at $x = 6$. Find $f(6)$ and $f'(6)$.

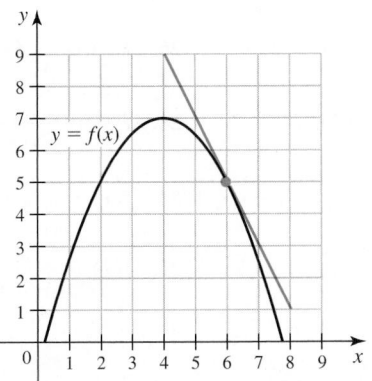

7. An equation of the line tangent to the graph of f at the point $(2, 7)$ is $y = 4x - 1$. Find $f(2)$ and $f'(2)$.

8. An equation of the line tangent to the graph of g at $x = 3$ is $y = 5x + 4$. Find $g(3)$ and $g'(3)$.

9. If $h(1) = 2$ and $h'(1) = 3$, find an equation of the line tangent to the graph of h at $x = 1$.

10. If $f'(-2) = 7$, find an equation of the line tangent to the graph of f at the point $(-2, 4)$.

11. Use definition (1) (p. 133) to find the slope of the line tangent to the graph of $f(x) = -5x + 1$ at the point $(1, -4)$.

12. Use definition (2) (p. 135) to find the slope of the line tangent to the graph of $f(x) = 5$ at $P(1, 5)$.

Practice Exercises

13–14. Velocity functions *A projectile is fired vertically upward into the air; its position (in feet) above the ground after t seconds is given by the function s(t). For the following functions, use limits to determine the instantaneous velocity of the projectile at t = a seconds for the given value of a.*

13. $s(t) = -16t^2 + 100t; a = 1$

14. $s(t) = -16t^2 + 128t + 192; a = 2$

15–20. Equations of tangent lines by definition (1)

a. Use definition (1) (p. 133) to find the slope of the line tangent to the graph of f at P.
b. Determine an equation of the tangent line at P.
c. Plot the graph of f and the tangent line at P.

15. $f(x) = x^2 - 5; P(3, 4)$

T 16. $f(x) = -3x^2 - 5x + 1; P(1, -7)$

17. $f(x) = \dfrac{1}{x}; P(-1, -1)$

T 18. $f(x) = \dfrac{4}{x^2}; P(-1, 4)$

T 19. $f(x) = \sqrt{3x + 3}; P(2, 3)$

T 20. $f(x) = \dfrac{2}{\sqrt{x}}; P(4, 1)$

21–32. Equations of tangent lines by definition (2)

a. Use definition (2) (p. 135) to find the slope of the line tangent to the graph of f at P.
b. Determine an equation of the tangent line at P.

21. $f(x) = 2x + 1; P(0, 1)$ **22.** $f(x) = -7x; P(-1, 7)$

23. $f(x) = 3x^2 - 4x; P(1, -1)$ **24.** $f(x) = 8 - 2x^2; P(0, 8)$

25. $f(x) = x^2 - 4; P(2, 0)$ **26.** $f(x) = \dfrac{1}{x}; P(1, 1)$

27. $f(x) = x^3; P(1, 1)$ **28.** $f(x) = \dfrac{1}{2x + 1}; P(0, 1)$

29. $f(x) = \dfrac{1}{3 - 2x}; P\left(-1, \dfrac{1}{5}\right)$ **30.** $f(x) = \sqrt{x - 1}; P(2, 1)$

31. $f(x) = \sqrt{x + 3}; P(1, 2)$ **32.** $f(x) = \dfrac{x}{x + 1}; P(-2, 2)$

33–42. Derivatives and tangent lines

a. For the following functions and values of a, find f'(a).
b. Determine an equation of the line tangent to the graph of f at the point (a, f(a)) for the given value of a.

33. $f(x) = 8x; a = -3$ **34.** $f(x) = x^2; a = 3$

35. $f(x) = 4x^2 + 2x; a = -2$ **36.** $f(x) = 2x^3; a = 10$

37. $f(x) = \dfrac{1}{\sqrt{x}}; a = \dfrac{1}{4}$ **38.** $f(x) = \dfrac{1}{x^2}; a = 1$

39. $f(x) = \sqrt{2x + 1}; a = 4$ **40.** $f(x) = \sqrt{3x}; a = 12$

41. $f(x) = \dfrac{1}{x + 5}; a = 5$ **42.** $f(x) = \dfrac{1}{3x - 1}; a = 2$

43–46. Derivative calculations *Evaluate the derivative of the following functions at the given point.*

43. $f(t) = \dfrac{1}{t + 1}; a = 1$

44. $f(t) = t - t^2; a = 2$

45. $f(s) = 2\sqrt{s} - 1; a = 25$

46. $f(r) = \pi r^2; a = 3$

47. **Explain why or why not** Determine whether the following statements are true and give an explanation or counterexample.

  a. For linear functions, the slope of any secant line always equals the slope of any tangent line.

  b. The slope of the secant line passing through the points P and Q is less than the slope of the tangent line at P.

  c. Consider the graph of the parabola $f(x) = x^2$. For $a > 0$ and $h > 0$, the secant line through $(a, f(a))$ and $(a + h, f(a + h))$ always has a greater slope than the tangent line at $(a, f(a))$.

48–51. Interpreting the derivative *Find the derivative of each function at the given point and interpret the physical meaning of this quantity. Include units in your answer.*

48. When a faucet is turned on to fill a bathtub, the volume of water in gallons in the tub after t minutes is $V(t) = 3t$. Find $V'(12)$.

49. An object dropped from rest falls $d(t) = 16t^2$ feet in t seconds. Find $d'(4)$.

50. The gravitational force of attraction between two masses separated by a distance of x meters is inversely proportional to the square of the distance between them, which implies that the force is described by the function $F(x) = k/x^2$, for some constant k, where $F(x)$ is measured in newtons. Find $F'(10)$, expressing your answer in terms of k.

51. Suppose the speed of a car approaching a stop sign is given by $v(t) = (t - 5)^2$, for $0 \le t \le 5$, where t is measured in seconds and $v(t)$ is measured in meters per second. Find $v'(3)$.

T 52. **Population of Las Vegas** Let $p(t)$ represent the population of the Las Vegas metropolitan area t years after 1970, as shown in the table and figure.

  a. Compute the average rate of growth of Las Vegas from 1970 to 1980.

  b. Explain why the average rate of growth calculated in part (a) is a good estimate of the instantaneous rate of growth of Las Vegas in 1975.

c. Compute the average rate of growth of Las Vegas from 1970 to 2000. Is the average rate of growth an overestimate or an underestimate of the instantaneous rate of growth of Las Vegas in 2000? Approximate the growth rate in 2000.

Year	1970	1980	1990	2000	2010	2020
t	0	10	20	30	40	50
$p(t)$	305,000	528,000	853,000	1,563,000	1,951,000	2,223,000

Source: U.S. Bureau of Census

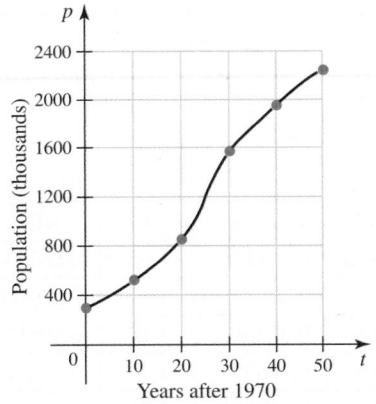

53. Owlet talons Let $L(t)$ equal the average length (in mm) of the middle talon on an Indian spotted owlet that is t weeks old, as shown in the figure.

a. Estimate $L'(1.5)$ and state the physical meaning of this quantity.

b. Estimate the value of $L'(a)$, for $a \geq 4$. What does this tell you about the talon lengths on these birds? (*Source: ZooKeys,* 132, 2011)

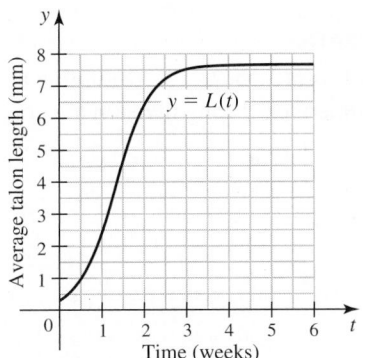

54. Caffeine levels Let $A(t)$ be the amount of caffeine (in mg) in the bloodstream t hours after a cup of coffee has been consumed (see figure). Estimate the values of $A'(7)$ and $A'(15)$, rounding answers to the nearest whole number. Include units in your answers and interpret the physical meaning of these values.

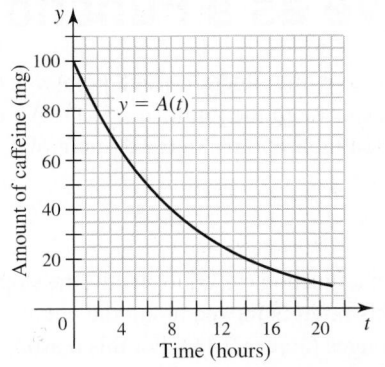

55. Let $D(t)$ equal the number of daylight hours at a latitude of 40° N, t days after January 1. Assuming $D(t)$ is approximated by a continuous function (see figure), estimate the values of $D'(60)$ and $D'(170)$. Include units in your answers and interpret your results.

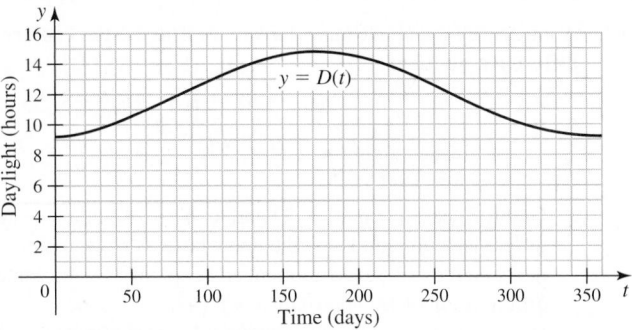

Explorations and Challenges

56–61. Find the function *The following limits represent the slope of a curve $y = f(x)$ at the point $(a, f(a))$. Determine a possible function f and number a; then calculate the limit.*

56. $\lim\limits_{x \to 1} \dfrac{3x^2 + 4x - 7}{x - 1}$

57. $\lim\limits_{x \to 2} \dfrac{5x^2 - 20}{x - 2}$

58. $\lim\limits_{x \to 2} \dfrac{\dfrac{1}{x + 1} - \dfrac{1}{3}}{x - 2}$

59. $\lim\limits_{h \to 0} \dfrac{(2 + h)^4 - 16}{h}$

60. $\lim\limits_{h \to 0} \dfrac{\sqrt{2 + h} - \sqrt{2}}{h}$

61. $\lim\limits_{h \to 0} \dfrac{|-1 + h| - 1}{h}$

⊤ 62–65. Approximating derivatives *Assuming the limit exists, the definition of the derivative $f'(a) = \lim\limits_{h \to 0} \dfrac{f(a + h) - f(a)}{h}$ implies that if h is small, then an approximation to $f'(a)$ is given by*

$$f'(a) \approx \dfrac{f(a + h) - f(a)}{h}.$$

*If $h > 0$, then this approximation is called a **forward difference quotient**; if $h < 0$, it is a **backward difference quotient**. As shown in the following exercises, these formulas are used to approximate f' at a point when f is a complicated function or when f is represented by a set of data points.*

62. Let $f(x) = \sqrt{x}$.

a. Find the exact value of $f'(4)$.

b. Show that $f'(4) \approx \dfrac{f(4 + h) - f(4)}{h} = \dfrac{\sqrt{4 + h} - 2}{h}$.

c. Complete columns 2 and 5 of the following table and describe how $\dfrac{\sqrt{4 + h} - 2}{h}$ behaves as h approaches 0.

h	$(\sqrt{4 + h} - 2)/h$	Error	h	$(\sqrt{4 + h} - 2)/h$	Error
0.1			−0.1		
0.01			−0.01		
0.001			−0.001		
0.0001			−0.0001		

d. The accuracy of an approximation is measured by

$$\text{error} = |\text{exact value} - \text{approximate value}|.$$

Use the exact value of $f'(4)$ in part (a) to complete columns 3 and 6 in the table. Describe the behavior of the errors as h approaches 0.

63. Another way to approximate derivatives is to use the **centered difference quotient:**

$$f'(a) \approx \frac{f(a + h) - f(a - h)}{2h}.$$

Again, consider $f(x) = \sqrt{x}$.

a. Graph f near the point $(4, 2)$ and let $h = 1/2$ in the centered difference quotient. Draw the line whose slope is computed by the centered difference quotient and explain why the centered difference quotient approximates $f'(4)$.

b. Use the centered difference quotient to approximate $f'(4)$ by completing the table.

h	Approximation	Error
0.1		
0.01		
0.001		

c. Explain why it is not necessary to use negative values of h in the table of part (b).

d. Compare the accuracy of the derivative estimates in part (b) with those found in Exercise 62.

64. The following table gives the distance $f(t)$ fallen by a smoke jumper t seconds after she opens her chute.

a. Use the forward difference quotient with $h = 0.5$ to estimate the velocity of the smoke jumper at $t = 2$ seconds.

b. Repeat part (a) using the centered difference quotient.

t (seconds)	$f(t)$ (feet)
0	0
0.5	4
1.0	15
1.5	33
2.0	55
2.5	81
3.0	109
3.5	138
4.0	169

65. The *error function* (denoted erf(x)) is an important function in statistics because it is related to the normal distribution. Its graph is shown in the figure, and values of erf(x) at several points are shown in the table.

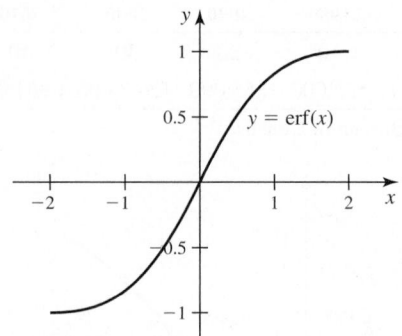

x	erf(x)	x	erf(x)
0.75	0.711156	1.05	0.862436
0.8	0.742101	1.1	0.880205
0.85	0.770668	1.15	0.896124
0.9	0.796908	1.2	0.910314
0.95	0.820891	1.25	0.922900
1.0	0.842701	1.3	0.934008

a. Use forward and centered difference quotients to find approximations to the derivative erf$'(1)$.

b. Given that erf$'(1) = \dfrac{2}{e\sqrt{\pi}}$, compute the error in the approximations in part (a).

QUICK CHECK ANSWERS

1. The slope is less at $x = 2$.   **3.** Definition (1) requires factoring the numerator or long division to cancel $(x - 1)$.

4. $y - 5 = \dfrac{1}{4}(x - 8)$ ◄

3.2 The Derivative as a Function

In Section 3.1, we learned that the derivative of a function f at a point a is the slope of the line tangent to the graph of f that passes through the point $(a, f(a))$. We now extend the concept of a derivative at a point to all (suitable) points in the domain of f to create a new function called the derivative of f.

The Derivative Function

So far we have computed the derivative of a function (or, equivalently, the slope of the tangent line) at one fixed point on a curve. If this point is moved along the curve, the tangent line also moves, and, in general, its slope changes (**Figure 3.14**). For this reason, the slope of the tangent line for the function f is itself a function, called the *derivative* of f.

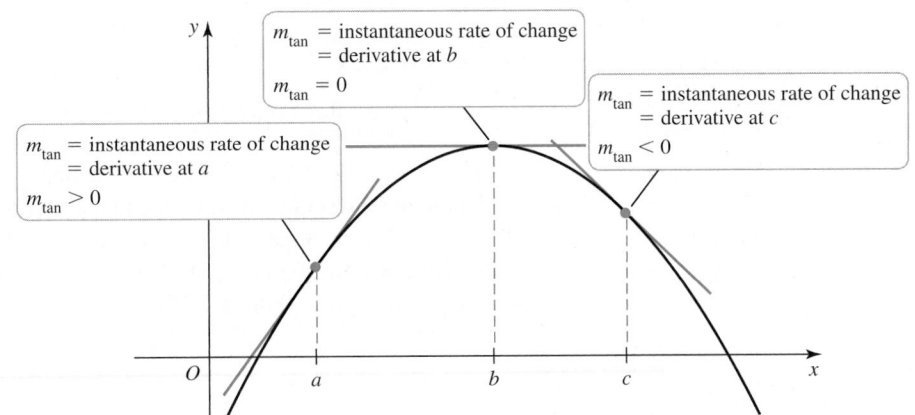

Figure 3.14

We let f' (read f *prime*) denote the derivative function for f, which means that $f'(a)$, when it exists, is the slope of the line tangent to the graph of f at $(a, f(a))$. Using definition (2) for the derivative of f at the point a from Section 3.1, we have

$$f'(a) = \lim_{h \to 0} \frac{f(a + h) - f(a)}{h}.$$

We now take an important step. The derivative is a special function, but it works just like any other function. For example, if the graph of f is smooth and 2 is in the domain of f, then $f'(2)$ is the slope of the line tangent to the graph of f at the point $(2, f(2))$. Similarly, if -2 is in the domain of f, then $f'(-2)$ is the slope of the tangent line at the point $(-2, f(-2))$. In fact, if x is *any* point in the domain of f, then $f'(x)$ is the slope of the tangent line at the point $(x, f(x))$. When we introduce a variable point x, the expression $f'(x)$ becomes the *derivative function*.

▶ To emphasize an important point, $f'(2)$ or $f'(-2)$ or $f'(a)$, for a real number a, are real numbers, whereas f' and $f'(x)$ refer to the derivative *function*.

▶ The process of finding f' is called *differentiation*, and to *differentiate* f means to find f'.

DEFINITION The Derivative Function

The **derivative** of f is the function

$$f'(x) = \lim_{h \to 0} \frac{f(x + h) - f(x)}{h},$$

provided the limit exists and x is in the domain of f. If $f'(x)$ exists, we say that f is **differentiable** at x. If f is differentiable at every point of an open interval I, we say that f is differentiable on I.

Notice that the definition of f' applies only at points in the domain of f. Therefore, the domain of f' is no larger than the domain of f. If the limit in the definition of f' fails to exist at some points, then the domain of f' is a subset of the domain of f. Let's use this definition to compute a derivative function.

EXAMPLE 1 Computing a derivative Find the derivative of $f(x) = -x^2 + 6x$.

SOLUTION

▶ Notice that this argument applies for $h > 0$ and for $h < 0$; that is, the limit as $h \to 0^+$ and the limit as $h \to 0^-$ are equal.

$$f'(x) = \lim_{h \to 0} \frac{f(x + h) - f(x)}{h} \qquad \text{Definition of } f'(x)$$

$$= \lim_{h \to 0} \frac{\overbrace{-(x + h)^2 + 6(x + h)}^{f(x+h)} - \overbrace{(-x^2 + 6x)}^{f(x)}}{h} \qquad \text{Substitute.}$$

$$= \lim_{h \to 0} \frac{-(x^2 + 2xh + h^2) + 6x + 6h + x^2 - 6x}{h} \qquad \text{Expand the numerator.}$$

$$= \lim_{h \to 0} \frac{h(-2x - h + 6)}{h}$$

Simplify and factor
out h.

Cancel h and evaluate
the limit.

$$= \lim_{h \to 0} (-2x - h + 6) = -2x + 6$$

The derivative is $f'(x) = -2x + 6$, which gives the slope of the tangent line (equivalently, the instantaneous rate of change) at *any* point on the curve. For example, at the point $(1, 5)$, the slope of the tangent line is $f'(1) = -2(1) + 6 = 4$. The slope of the tangent line at $(3, 9)$ is $f'(3) = -2(3) + 6 = 0$, which means the tangent line is horizontal at that point (**Figure 3.15**).

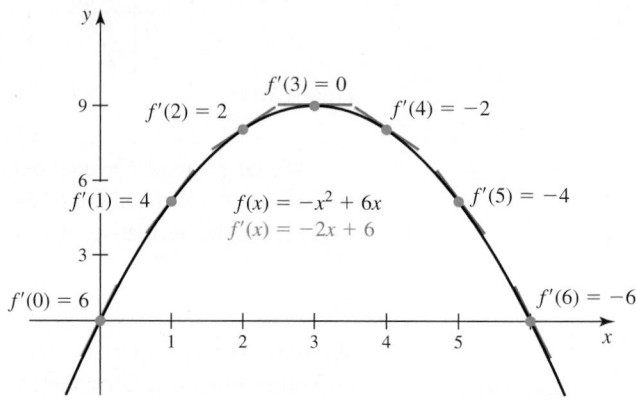

QUICK CHECK 1 In Example 1, determine the slope of the tangent line at $x = 2$. ◀

Figure 3.15

Related Exercises 23–24 ◀

Derivative Notation

For historical and practical reasons, several notations for the derivative are used. To see the origin of one notation, recall that the slope of the secant line through two points $P(x, f(x))$ and $Q(x + h, f(x + h))$ on the curve $y = f(x)$ is $\dfrac{f(x + h) - f(x)}{h}$. The quantity h is the change in the x-coordinate in moving from P to Q. A standard notation for change is the symbol Δ (uppercase Greek letter delta). So we replace h with Δx to represent the change in x. Similarly, $f(x + h) - f(x) = f(x + \Delta x) - f(x)$ is the change in y, denoted Δy (**Figure 3.16**). Therefore, the slope of the secant line is

$$m_{\text{sec}} = \frac{f(x + \Delta x) - f(x)}{\Delta x} = \frac{\Delta y}{\Delta x}.$$

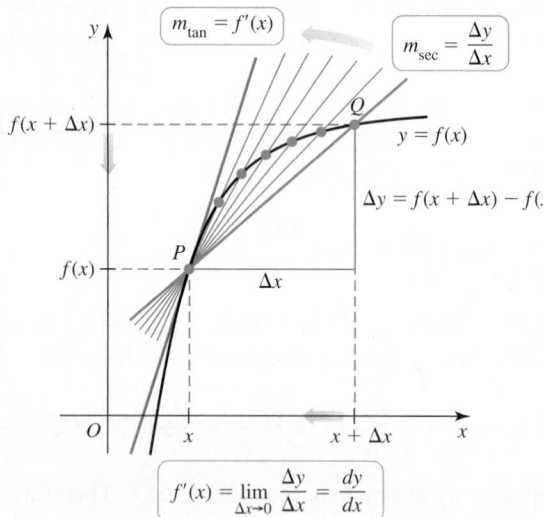

Figure 3.16

By letting $\Delta x \to 0$, the slope of the tangent line at $(x, f(x))$ is

$$f'(x) = \lim_{\Delta x \to 0} \frac{f(x + \Delta x) - f(x)}{\Delta x} = \lim_{\Delta x \to 0} \frac{\Delta y}{\Delta x} = \frac{dy}{dx}.$$

The new notation for the derivative is $\dfrac{dy}{dx}$; it reminds us that $f'(x)$ is the limit of $\dfrac{\Delta y}{\Delta x}$ as $\Delta x \to 0$.

➤ The notation $\dfrac{dy}{dx}$ is read *the derivative of y with respect to x or dy dx*. It does not mean dy divided by dx, but it is a reminder of the limit of the quotient $\dfrac{\Delta y}{\Delta x}$.

In addition to the notation $f'(x)$ and $\dfrac{dy}{dx}$, other common ways of writing the derivative include

$$\frac{df}{dx}, \qquad \frac{d}{dx}(f(x)), \qquad D_x(f(x)), \quad \text{and} \quad y'(x).$$

➤ The derivative notation dy/dx was introduced by Gottfried Wilhelm von Leibniz (1646–1716), one of the coinventors of calculus. His notation is used today in its original form. The notation used by Sir Isaac Newton (1642–1727), the other coinventor of calculus, is rarely used.

The following notations represent the derivative of f evaluated at a.

$$f'(a), \qquad y'(a), \qquad \left.\frac{df}{dx}\right|_{x=a}, \quad \text{and} \quad \left.\frac{dy}{dx}\right|_{x=a}$$

QUICK CHECK 2 What are some other ways to write $f'(3)$, where $y = f(x)$? ◀

EXAMPLE 2　A derivative calculation Let $y = f(x) = \sqrt{x}$.

a. Compute $\dfrac{dy}{dx}$.

b. Find an equation of the line tangent to the graph of f at $(4, 2)$.

SOLUTION

> ▶ Example 2 gives the first of many derivative formulas to be presented in the text:
>
> $$\frac{d}{dx}(\sqrt{x}) = \frac{1}{2\sqrt{x}}.$$
>
> Remember this result. It will be used often.

a.

$$\frac{dy}{dx} = \lim_{h \to 0} \frac{f(x + h) - f(x)}{h} \qquad \text{Definition of } \frac{dy}{dx} = f'(x)$$

$$= \lim_{h \to 0} \frac{\sqrt{x + h} - \sqrt{x}}{h} \qquad \text{Substitute } f(x) = \sqrt{x}.$$

$$= \lim_{h \to 0} \frac{\left(\sqrt{x + h} - \sqrt{x}\right)}{h} \frac{\left(\sqrt{x + h} + \sqrt{x}\right)}{\left(\sqrt{x + h} + \sqrt{x}\right)} \qquad \begin{array}{l}\text{Multiply the numerator and} \\ \text{denominator by } \sqrt{x + h} + \sqrt{x}.\end{array}$$

$$= \lim_{h \to 0} \frac{1}{\sqrt{x + h} + \sqrt{x}} = \frac{1}{2\sqrt{x}} \qquad \text{Simplify and evaluate the limit.}$$

b. The slope of the tangent line at $(4, 2)$ is

$$\left.\frac{dy}{dx}\right|_{x=4} = \frac{1}{2\sqrt{4}} = \frac{1}{4}.$$

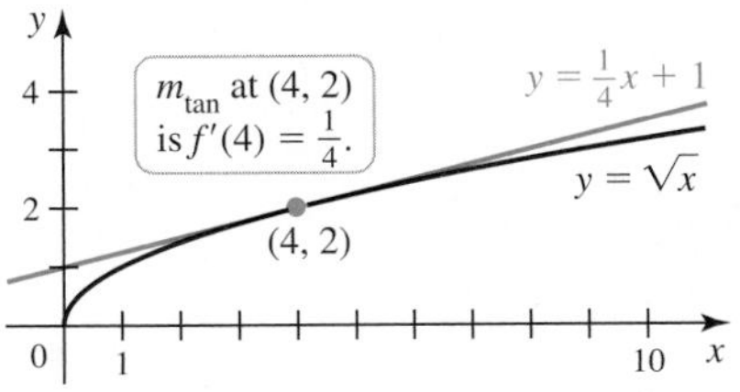

Figure 3.17

Therefore, an equation of the tangent line (**Figure 3.17**) is

$$y - 2 = \frac{1}{4}(x - 4) \quad \text{or} \quad y = \frac{1}{4}x + 1.$$

Related Exercises 37–38 ◀

QUICK CHECK 3 In Example 2, do the slopes of the tangent lines increase or decrease as x increases? Explain. ◀

If a function is given in terms of variables other than x and y, we make an adjustment to the derivative definition. For example, if $y = g(t)$, we replace f with g and x with t to obtain the *derivative of g with respect to t*:

$$g'(t) = \lim_{h \to 0} \frac{g(t + h) - g(t)}{h}.$$

QUICK CHECK 4 Express the derivative of $p = q(r)$ in three ways. ◀

Other notations for $g'(t)$ include $\dfrac{dg}{dt}, \dfrac{d}{dt}(g(t)), D_t(g(t)),$ and $y'(t)$.

EXAMPLE 3　Another derivative calculation Let $g(t) = 1/t^2$ and compute $g'(t)$.

SOLUTION

$$g'(t) = \lim_{h \to 0} \frac{g(t + h) - g(t)}{h} \qquad \text{Definition of } g'$$

$$= \lim_{h \to 0} \frac{1}{h}\left(\frac{1}{(t + h)^2} - \frac{1}{t^2}\right) \qquad \text{Substitute } g(t) = 1/t^2.$$

$$= \lim_{h \to 0} \frac{1}{h}\left(\frac{t^2 - (t + h)^2}{t^2(t + h)^2}\right) \qquad \text{Common denominator}$$

$$= \lim_{h \to 0} \frac{1}{h}\left(\frac{-2ht - h^2}{t^2(t + h)^2}\right) \qquad \text{Expand the numerator and simplify.}$$

$$= \lim_{h \to 0} \left(\frac{-2t - h}{t^2(t + h)^2}\right) \qquad h \neq 0; \text{ cancel } h.$$

$$= -\frac{2}{t^3} \qquad \text{Evaluate the limit.}$$

Related Exercises 25–27 ◀

Graphs of Derivatives

The function f' is called the derivative of f because it is *derived* from f. The following examples illustrate how to *derive* the graph of f' from the graph of f.

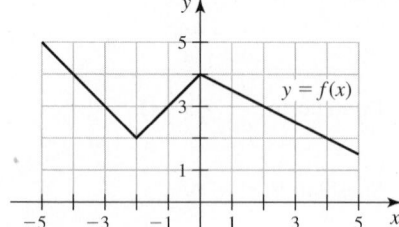

Figure 3.18

EXAMPLE 4 **Graph of the derivative** Sketch the graph of f' from the graph of f (Figure 3.18).

SOLUTION The graph of f consists of line segments, which are their own tangent lines. Therefore, the slope of the curve $y = f(x)$, for $x < -2$, is -1; that is, $f'(x) = -1$, for $x < -2$. Similarly, $f'(x) = 1$, for $-2 < x < 0$, and $f'(x) = -\frac{1}{2}$, for $x > 0$. Figure 3.19 shows the graph of f in black and the graph of f' in red.

➤ In terms of limits at $x = -2$, we can write $\lim\limits_{h \to 0^-} \dfrac{f(-2 + h) - f(-2)}{h} = -1$ and $\lim\limits_{h \to 0^+} \dfrac{f(-2 + h) - f(-2)}{h} = 1$. Because the one-sided limits are not equal, $f'(-2)$ does not exist. The analogous one-sided limits at $x = 0$ are also unequal.

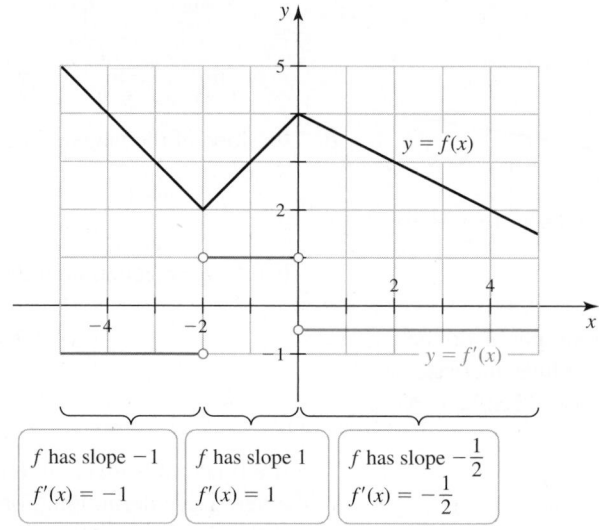

| f has slope -1 | f has slope 1 | f has slope $-\frac{1}{2}$ |
| $f'(x) = -1$ | $f'(x) = 1$ | $f'(x) = -\frac{1}{2}$ |

Figure 3.19

Notice that the slopes of the tangent lines change abruptly at $x = -2$ and $x = 0$. As a result, $f'(-2)$ and $f'(0)$ are undefined and the graph of the derivative has a discontinuity at these points.

Related Exercises 17–18 ◄

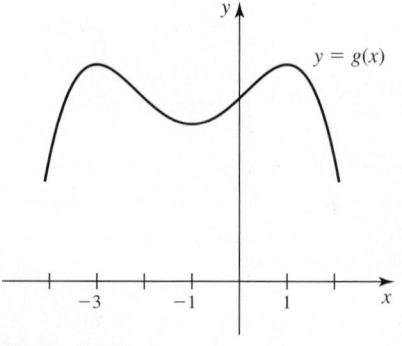

Figure 3.20

QUICK CHECK 5 Is it true that if $f(x) > 0$ at a point, then $f'(x) > 0$ at that point? Is it true that if $f'(x) > 0$ at a point, then $f(x) > 0$ at that point? Explain. ◄

EXAMPLE 5 **Graph of the derivative** Sketch the graph of g' using the graph of g (Figure 3.20).

SOLUTION Without an equation for g, the best we can do is to find the general shape of the graph of g'. Here are the key observations.

1. Note that the lines tangent to the graph of g at $x = -3, -1$, and 1 have a slope of 0. Therefore,

$$g'(-3) = g'(-1) = g'(1) = 0,$$

which means the graph of g' has x-intercepts at these points (Figure 3.21a).

2. For $x < -3$, the slopes of the tangent lines are positive and decrease to 0 as x approaches -3 from the left. Therefore, $g'(x)$ is positive for $x < -3$ and decreases to 0 as x approaches -3.

3. For $-3 < x < -1$, $g'(x)$ is negative; it initially decreases as x increases and then increases to 0 at $x = -1$. For $-1 < x < 1$, $g'(x)$ is positive; it initially increases as x increases and then returns to 0 at $x = 1$.

4. Finally, $g'(x)$ is negative and decreasing for $x > 1$. Because the slope of g changes gradually, the graph of g' is continuous with no jumps or breaks (Figure 3.21b).

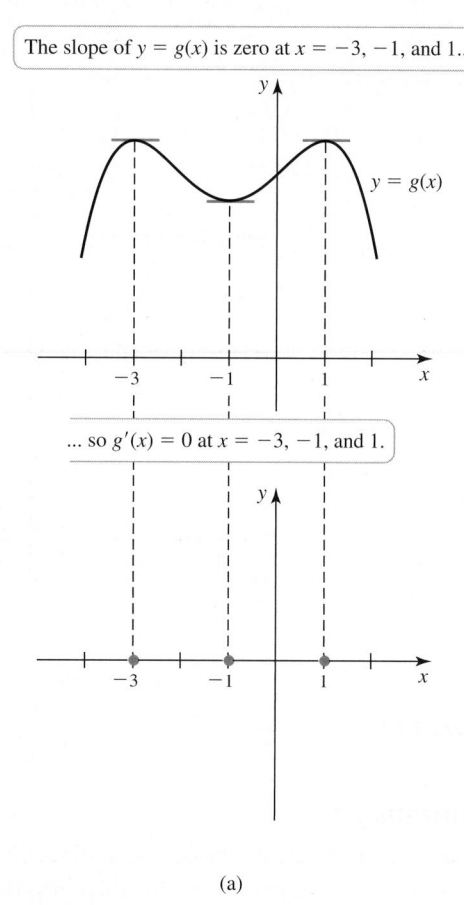

The slope of $y = g(x)$ is zero at $x = -3$, -1, and 1...

$y = g(x)$

... so $g'(x) = 0$ at $x = -3$, -1, and 1.

(a)

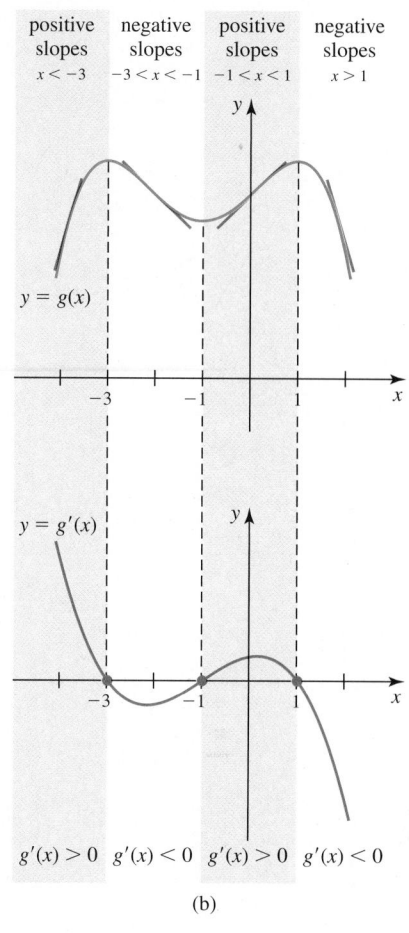

positive slopes	negative slopes	positive slopes	negative slopes
$x < -3$	$-3 < x < -1$	$-1 < x < 1$	$x > 1$

$y = g(x)$

$y = g'(x)$

$g'(x) > 0$ $g'(x) < 0$ $g'(x) > 0$ $g'(x) < 0$

(b)

Figure 3.21

Related Exercises 48–50 ◄

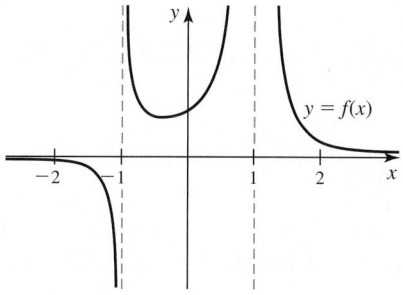

$y = f(x)$

Figure 3.22

➤ Although it is the case in Example 6, a function and its derivative do not always share the same vertical asymptotes.

EXAMPLE 6   Graphing the derivative with asymptotes The graph of the function f is shown in **Figure 3.22**. Sketch a graph of its derivative.

SOLUTION Identifying intervals on which the slopes of tangent lines are zero, positive, and negative, we make the following observations:

• A horizontal tangent line occurs at approximately $\left(-\frac{1}{3}, f\left(-\frac{1}{3}\right)\right)$. Therefore, $f'\left(-\frac{1}{3}\right) = 0$.

• On the interval $(-\infty, -1)$, slopes of tangent lines are negative and increase in magnitude without bound as we approach -1 from the left.

• On the interval $\left(-1, -\frac{1}{3}\right)$, slopes of tangent lines are negative and increase to zero at $-\frac{1}{3}$.

• On the interval $\left(-\frac{1}{3}, 1\right)$, slopes of tangent lines are positive and increase without bound as we approach 1 from the left.

• On the interval $(1, \infty)$, slopes of tangent lines are negative and increase to zero.

Assembling all this information, we obtain a graph of f' shown in **Figure 3.23**. Notice that f and f' have the same vertical asymptotes. However, as we pass through -1, the sign of f changes, but the sign of f' does not change. As we pass through 1, the sign of f does not change, but the sign of f' does change.

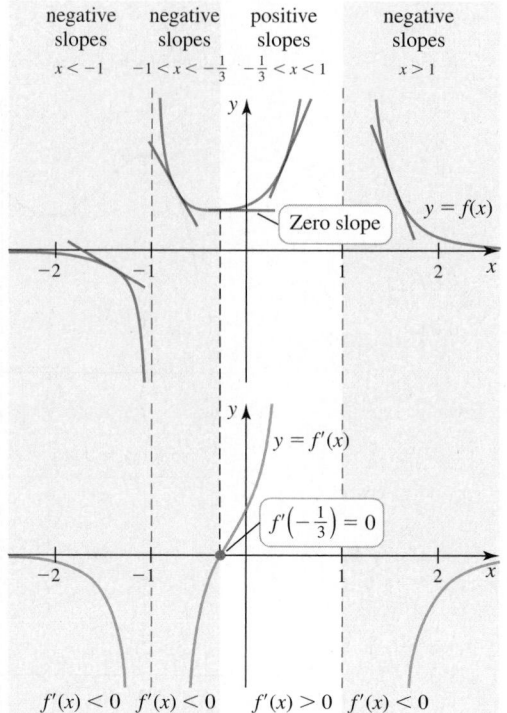

Figure 3.23

Related Exercises 51–52 ◄

Continuity

We now return to the discussion of continuity (Section 2.6) and investigate the relationship between continuity and differentiability. Specifically, we show that if a function is differentiable at a point, then it is also continuous at that point.

> **THEOREM 3.1   Differentiable Implies Continuous**
> If f is differentiable at a, then f is continuous at a.

Proof: Because f is differentiable at a point a, we know that

$$f'(a) = \lim_{x \to a} \frac{f(x) - f(a)}{x - a}$$

exists. To show that f is continuous at a, we must show that $\lim\limits_{x \to a} f(x) = f(a)$. The key is the identity

$$f(x) = \frac{f(x) - f(a)}{x - a}(x - a) + f(a), \quad \text{for } x \neq a. \tag{1}$$

▶ Expression (1) is an identity because it holds for all $x \neq a$, which can be seen by canceling $x - a$ and simplifying.

Taking the limit as x approaches a on both sides of (1) and simplifying, we have

$$\lim_{x \to a} f(x) = \lim_{x \to a} \left(\frac{f(x) - f(a)}{x - a}(x - a) + f(a) \right) \qquad \text{Use identity.}$$

$$= \underbrace{\lim_{x \to a} \left(\frac{f(x) - f(a)}{x - a} \right)}_{f'(a)} \underbrace{\lim_{x \to a} (x - a)}_{0} + \underbrace{\lim_{x \to a} f(a)}_{f(a)} \qquad \text{Theorem 2.3}$$

$$= f'(a) \cdot 0 + f(a) \qquad\qquad\qquad \text{Evaluate limits.}$$

$$= f(a). \qquad\qquad\qquad\qquad\qquad \text{Simplify.}$$

QUICK CHECK 6 Verify that the right-hand side of (1) equals $f(x)$ if $x \neq a$. ◄

Therefore, $\lim\limits_{x \to a} f(x) = f(a)$, which means that f is continuous at a.   ◄

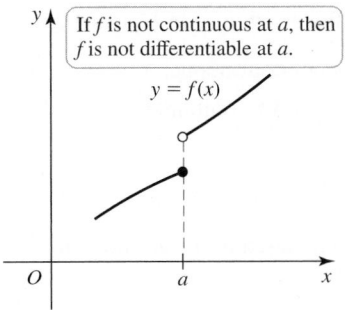

Figure 3.24

▶ The alternative version of Theorem 3.1 is called the *contrapositive* of the first statement of Theorem 3.1. A statement and its contrapositive are two equivalent ways of expressing the same statement. For example, the statement *If I live in Denver, then I live in Colorado* is logically equivalent to its contrapositive: *If I do not live in Colorado, then I do not live in Denver.*

▶ To avoid confusion about continuity and differentiability, it helps to think about the function $f(x) = |x|$: It is continuous everywhere but not differentiable at 0.

▶ Continuity requires that $\lim_{x \to a} (f(x) - f(a)) = 0$. Differentiability requires more: $\lim_{x \to a} \dfrac{f(x) - f(a)}{x - a}$ must exist.

▶ See Exercises 73–76 for a formal definition of a vertical tangent line.

Theorem 3.1 tells us that if f is differentiable at a point, then it is necessarily continuous at that point. Therefore, if f is *not* continuous at a point, then f is *not* differentiable there (**Figure 3.24**). So Theorem 3.1 can be stated in another way.

THEOREM 3.1 (ALTERNATIVE VERSION)   Not Continuous Implies Not Differentiable
If f is not continuous at a, then f is not differentiable at a.

It is tempting to read more into Theorem 3.1 than what it actually states. If f is continuous at a point, f is *not* necessarily differentiable at that point. For example, consider the continuous function in **Figure 3.25** and note the **corner point** at a. Ignoring the portion of the graph for $x > a$, we might be tempted to conclude that ℓ_1 is the line tangent to the curve at a. Ignoring the part of the graph for $x < a$, we might incorrectly conclude that ℓ_2 is the line tangent to the curve at a. The slopes of ℓ_1 and ℓ_2 are not equal. Because of the abrupt change in the slope of the curve at a, f is not differentiable at a: The limit that defines f' does not exist at a.

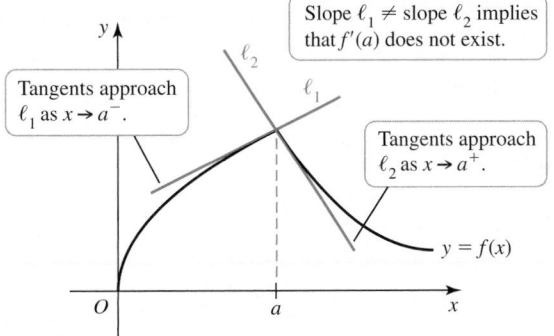

Figure 3.25

Another common situation occurs when the graph of a function f has a vertical tangent line at a. In this case, $f'(a)$ is undefined because the slope of a vertical line is undefined. A vertical tangent line may occur at a sharp point on the curve called a **cusp** (for example, the function $f(x) = \sqrt{|x|}$ in **Figure 3.26a**). In other cases, a vertical tangent line may occur without a cusp (for example, the function $f(x) = \sqrt[3]{x}$ in **Figure 3.26b**).

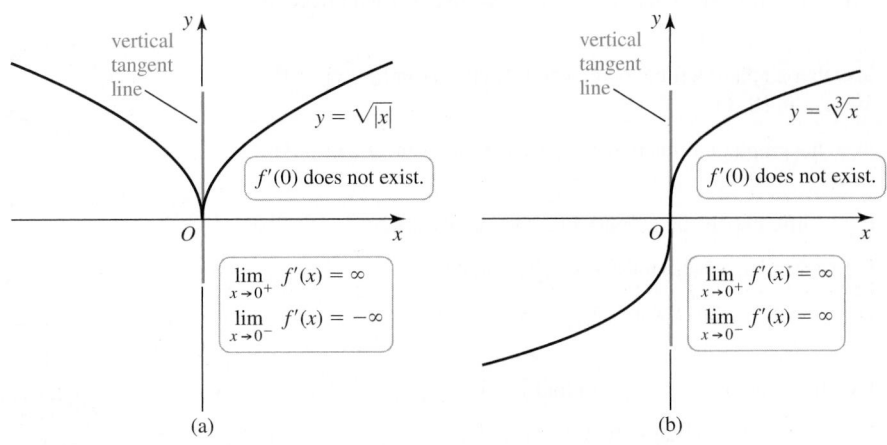

Figure 3.26

When Is a Function Not Differentiable at a Point?

A function f is *not* differentiable at a if at least one of the following conditions holds:

a. f is not continuous at a (Figure 3.24).

b. f has a corner at a (Figure 3.25).

c. f has a vertical tangent at a (Figure 3.26).

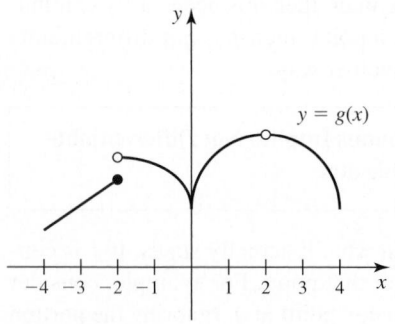

Figure 3.27

EXAMPLE 7 Continuous and differentiable Consider the graph of g in **Figure 3.27**.

a. Find the values of x in the interval $(-4, 4)$ at which g is not continuous.

b. Find the values of x in the interval $(-4, 4)$ at which g is not differentiable.

c. Sketch a graph of the derivative of g.

SOLUTION

a. The function g fails to be continuous at -2 (where the one-sided limits are not equal) and at 2 (where g is not defined).

b. Because it is not continuous at ± 2, g is not differentiable at those points. Furthermore, g is not differentiable at 0, because the graph has a cusp at that point.

c. A sketch of the derivative (**Figure 3.28**) has the following features:

- $g'(x) > 0$, for $-4 < x < -2$ and $0 < x < 2$
- $g'(x) < 0$, for $-2 < x < 0$ and $2 < x < 4$
- $g'(x)$ approaches $-\infty$ as $x \to 0^-$ and as $x \to 4^-$, and $g'(x)$ approaches ∞ as $x \to 0^+$
- $g'(x)$ approaches 0 as $x \to 2$ from either side, although $g'(2)$ does not exist.

Related Exercises 53–54 ◄

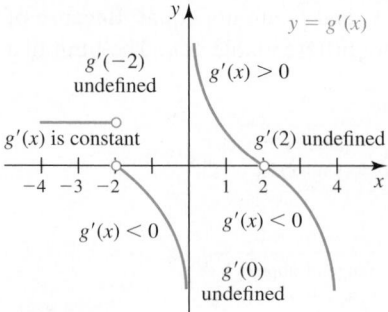

Figure 3.28

SECTION 3.2 EXERCISES

Getting Started

1. For a given function f, what does f' represent?

2. If $f'(x) = 3x + 2$, find the slope of the line tangent to the curve $y = f(x)$ at $x = 1, 2$, and 3.

3. Why is the notation $\dfrac{dy}{dx}$ used to represent the derivative?

4. Give three different notations for the derivative of f with respect to x.

5. Sketch a graph of a function f, where $f(x) < 0$ and $f'(x) > 0$ for all x in $(0, 1)$.

6. Sketch a graph of a function f, where $f(x) > 0$ and $f'(x) < 0$ for all x in $(0, 2)$.

7. If f is differentiable at a, must f be continuous at a?

8. If f is continuous at a, must f be differentiable at a?

9. Describe the graph of f if $f(0) = 1$ and $f'(x) = 3$, for $-\infty < x < \infty$.

10. Use the graph of $f(x) = |x|$ to find $f'(x)$.

11. Use limits to find $f'(x)$ if $f(x) = 7x$.

12. Use limits to find $f'(x)$ if $f(x) = -3x$.

13. Use limits to find $\dfrac{dy}{dx}$ if $y = x^2$. Then evaluate $\dfrac{dy}{dx}\Big|_{x=3}$ and $\dfrac{dy}{dx}\Big|_{x=-2}$.

14. The weight $w(x)$ (in pounds) of an Atlantic salmon can be estimated from its length x (in inches). If $33 \le x \le 48$, the estimated weight is $w(x) = 1.5x - 35.8$. Use limits to find $w'(x)$ and interpret its meaning. (*Source:* www.atlanticsalmonfederation.org)

15. **Matching functions with derivatives** Match graphs a–d of functions with graphs A–C of their derivatives.

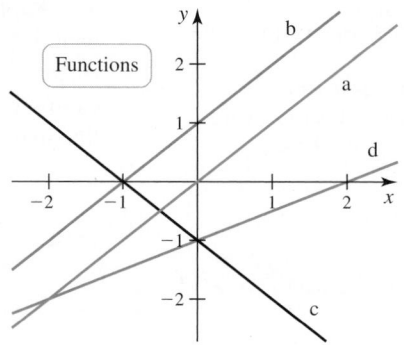

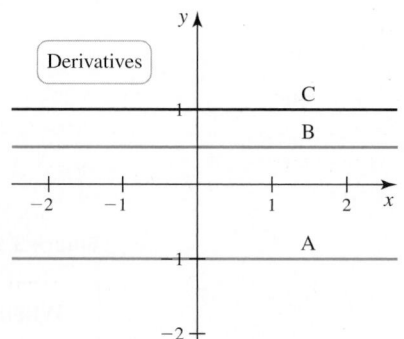

16. Matching derivatives with functions Match graphs a–d of derivative functions with possible graphs A–D of the corresponding functions.

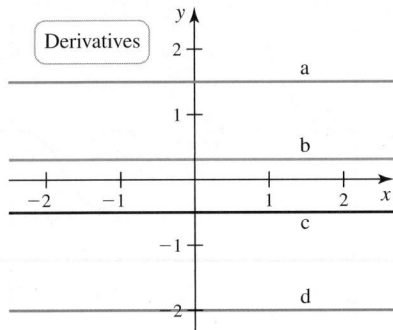

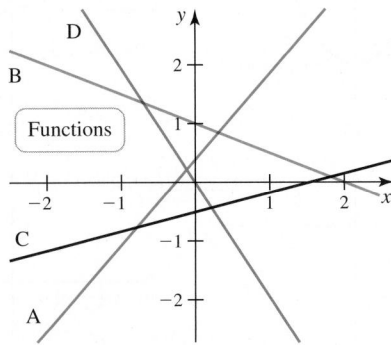

17–18. Sketching derivatives *Reproduce the graph of f and then sketch a graph of f' on the same axes.*

17.

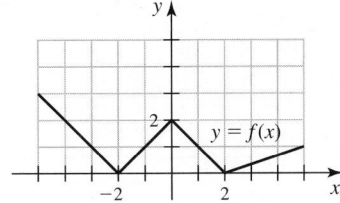

18.

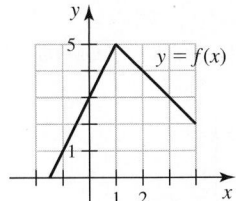

19. Use the graph of *f* in the figure to do the following.

a. Find the values of *x* in (−2, 2) at which *f* is not continuous.

b. Find the values of *x* in (−2, 2) at which *f* is not differentiable.

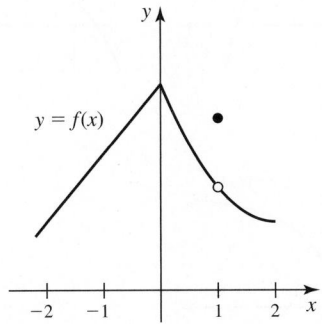

20. Use the graph of *g* in the figure to do the following.

a. Find the values of *x* in (−2, 2) at which *g* is not continuous.

b. Find the values of *x* in (−2, 2) at which *g* is not differentiable.

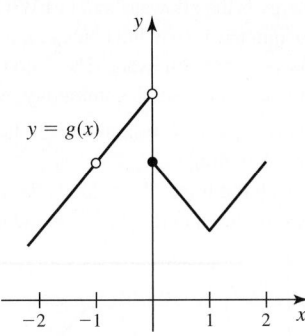

Practice Exercises

21–30. Derivatives

a. *Use limits to find the derivative function f' for the following functions f.*

b. *Evaluate f'(a) for the given values of a.*

21. $f(x) = 5x + 2; a = 1, 2$ **22.** $f(x) = 7; a = -1, 2$

23. $f(x) = 4x^2 + 1; a = 2, 4$ **24.** $f(x) = x^2 + 3x; a = -1, 4$

25. $f(x) = \dfrac{1}{x + 1}; a = -\dfrac{1}{2}, 5$ **26.** $f(x) = \dfrac{x}{x + 2}; a = -1, 0$

27. $f(t) = \dfrac{1}{\sqrt{t}}; a = 9, \dfrac{1}{4}$ **28.** $f(w) = \sqrt{4w - 3}; a = 1, 3$

29. $f(s) = 4s^3 + 3s; a = -3, -1$

30. $f(t) = 3t^4; a = -2, 2$

31–32. Velocity functions *A projectile is fired vertically upward into the air, and its position (in feet) above the ground after t seconds is given by the function s(t).*

a. *For the following functions s(t), find the instantaneous velocity function v(t). (Recall that the velocity function v is the derivative of the position function s.)*

b. *Determine the instantaneous velocity of the projectile at t = 1 and t = 2 seconds.*

31. $s(t) = -16t^2 + 100t$ **32.** $s(t) = -16t^2 + 128t + 192$

33. Evaluate $\dfrac{dy}{dx}$ and $\dfrac{dy}{dx}\bigg|_{x=2}$ if $y = \dfrac{x + 1}{x + 2}$.

34. Evaluate $\dfrac{ds}{dt}$ and $\dfrac{ds}{dt}\bigg|_{t=-1}$ if $s = 11t^3 + t + 1$.

35–40. Tangent lines

a. *Find the derivative function f' for the following functions f.*

b. *Find an equation of the line tangent to the graph of f at (a, f(a)) for the given value of a.*

35. $f(x) = 3x^2 + 2x - 10; a = 1$

36. $f(x) = 5x^2 - 6x + 1; a = 2$

37. $f(x) = \sqrt{3x + 1}; a = 8$ **38.** $f(x) = \sqrt{x + 2}; a = 7$

39. $f(x) = \dfrac{2}{3x + 1}; a = -1$ **40.** $f(x) = \dfrac{1}{x}; a = -5$

41. Power and energy Energy is the capacity to do work, and power is the rate at which energy is used or consumed. Therefore, if $E(t)$ is the energy function for a system, then $P(t) = E'(t)$ is the power function. A unit of energy is the kilowatt-hour (1 kWh is the amount of energy needed to light ten 100-W light bulbs for an hour); the corresponding units for power are kilowatts. The following figure shows the energy demand function for a small community over a 25-hour period.

a. Estimate the power at $t = 10$ and $t = 20$ hr. Be sure to include units in your calculation.

b. At what times on the interval $[0, 25]$ is the power zero?

c. At what times on the interval $[0, 25]$ is the power a maximum?

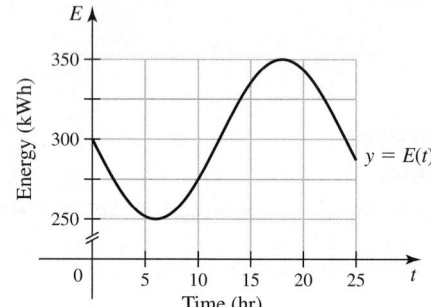

42. Slope of a line Consider the line $f(x) = mx + b$, where m and b are constants. Show that $f'(x) = m$ for all x. Interpret this result.

43. A derivative formula

a. Use the definition of the derivative to determine $\dfrac{d}{dx}(ax^2 + bx + c)$, where a, b, and c are constants.

b. Let $f(x) = 4x^2 - 3x + 10$ and use part (a) to find $f'(x)$.

c. Use part (b) to find $f'(1)$.

44. A derivative formula

a. Use the definition of the derivative to determine $\dfrac{d}{dx}\left(\sqrt{ax + b}\right)$, where a and b are constants.

b. Let $f(x) = \sqrt{5x + 9}$ and use part (a) to find $f'(x)$.

c. Use part (b) to find $f'(-1)$.

45–46. Analyzing slopes *Use the points A, B, C, D, and E in the following graphs to answer these questions.*

a. *At which points is the slope of the curve negative?*

b. *At which points is the slope of the curve positive?*

c. *Using A–E, list the slopes in decreasing order.*

45.

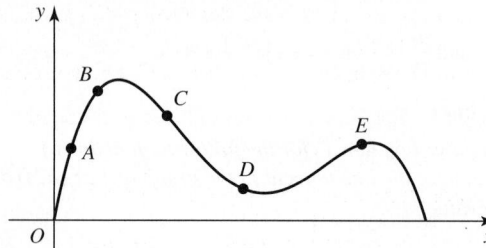

46.

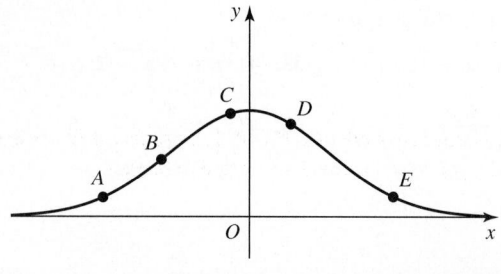

47. Matching functions with derivatives Match the functions a–d in the first set of figures with the derivative functions A–D in the next set of figures.

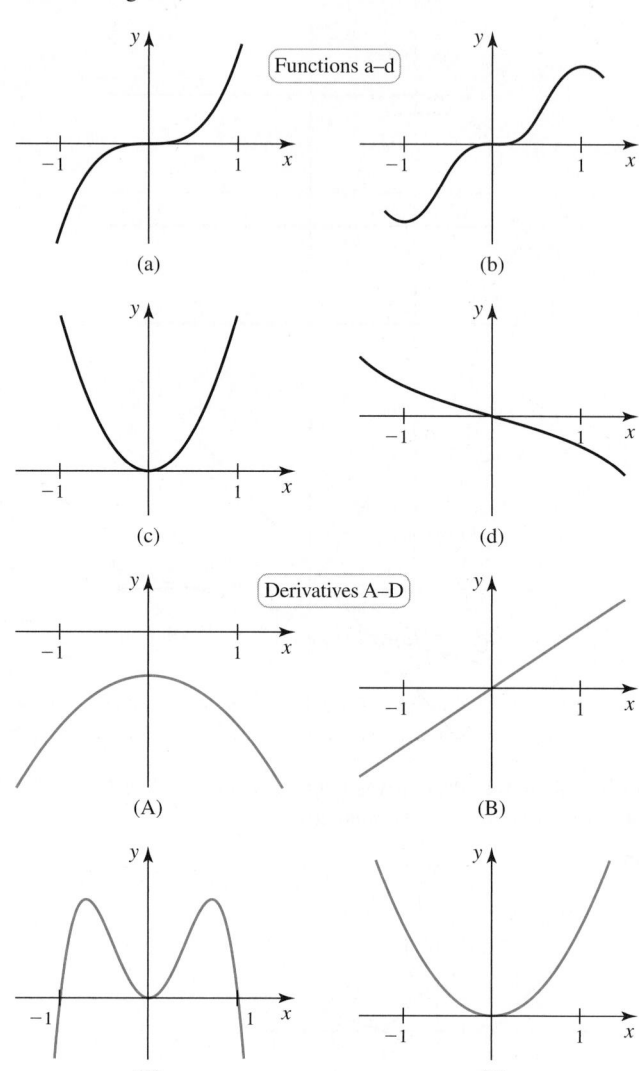

48–52. Sketching derivatives *Reproduce the graph of f and then plot a graph of f′ on the same axes.*

48.

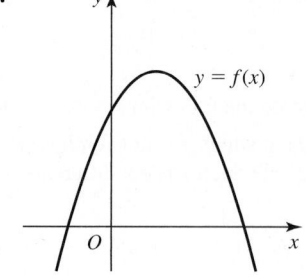

49.

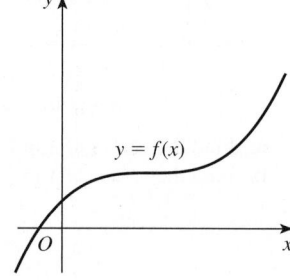

50.

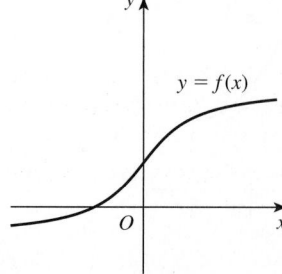

51.

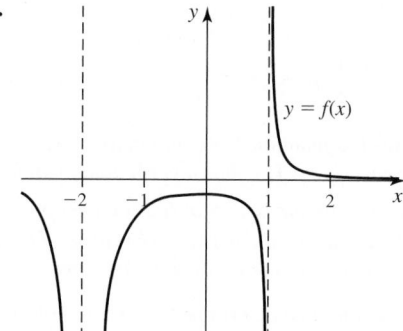

52.

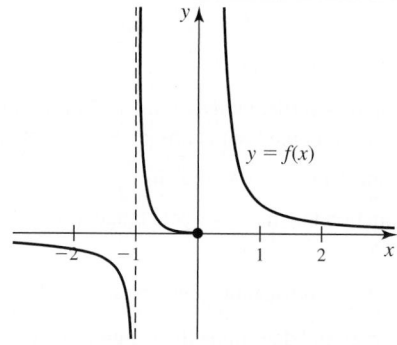

53. Where is the function continuous? Differentiable? Use the graph of f in the figure to do the following.

  a. Find the values of x in $(0, 3)$ at which f is not continuous.
  b. Find the values of x in $(0, 3)$ at which f is not differentiable.
  c. Sketch a graph of f'.

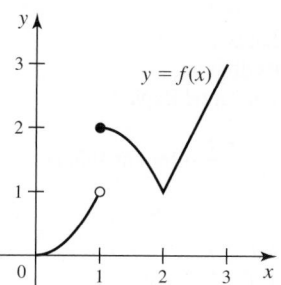

54. Where is the function continuous? Differentiable? Use the graph of g in the figure to do the following.

  a. Find the values of x in $(0, 4)$ at which g is not continuous.
  b. Find the values of x in $(0, 4)$ at which g is not differentiable.
  c. Sketch a graph of g'.

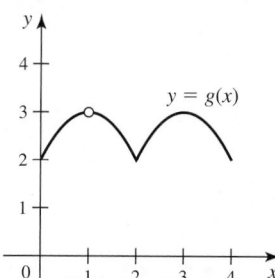

55. Voltage on a capacitor A capacitor is a device in an electrical circuit that stores charge. In one particular circuit, the charge on the capacitor Q varies in time as shown in the figure.

  a. At what time is the rate of change of the charge Q' the greatest?
  b. Is Q' positive or negative for $t \geq 0$?

  c. Is Q' an increasing or decreasing function of time (or neither)?
  d. Sketch the graph of Q'. You do not need a scale on the vertical axis.

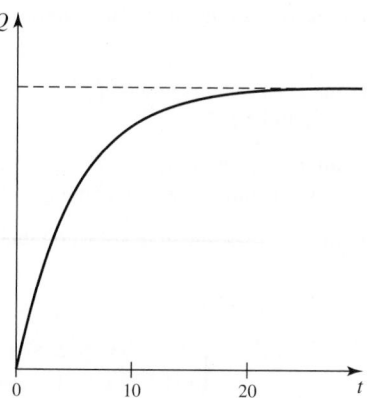

56. Logistic growth A common model for population growth uses the logistic (or sigmoid) curve. Consider the logistic curve in the figure, where $P(t)$ is the population at time $t \geq 0$.

  a. At approximately what time is the rate of growth P' the greatest?
  b. Is P' positive or negative for $t \geq 0$?
  c. Is P' an increasing or decreasing function of time (or neither)?
  d. Sketch the graph of P'. You do not need a scale on the vertical axis.

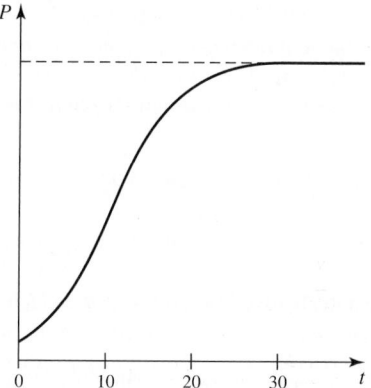

57. Explain why or why not Determine whether the following statements are true and give an explanation or counterexample.

  a. If the function f is differentiable for all values of x, then f is continuous for all values of x.
  b. The function $f(x) = |x + 1|$ is continuous for all x, but not differentiable for all x.
  c. It is possible for the domain of f to be (a, b) and the domain of f' to be $[a, b]$.

Explorations and Challenges

58. Looking ahead: Derivative of x^n Use the definition
$$f'(x) = \lim_{h \to 0} \frac{f(x + h) - f(x)}{h}$$ to find $f'(x)$ for the following functions.

  a. $f(x) = x^2$
  b. $f(x) = x^3$
  c. $f(x) = x^4$
  d. Based on your answers to parts (a)–(c), propose a formula for $f'(x)$ if $f(x) = x^n$, where n is a positive integer.

59. Determining the unknown constant Let

$$f(x) = \begin{cases} 2x^2 & \text{if } x \le 1 \\ ax - 2 & \text{if } x > 1. \end{cases}$$

Determine a value of a (if possible) for which f' is continuous at $x = 1$.

60. Finding f from f' Sketch the graph of $f'(x) = 2$. Then sketch three possible graphs of f.

61. Finding f from f' Sketch the graph of $f'(x) = x$. Then sketch a possible graph of f. Is more than one graph possible?

62. Finding f from f' Create the graph of a continuous function f such that

$$f'(x) = \begin{cases} 1 & \text{if } x < 0 \\ 0 & \text{if } 0 < x < 1 \\ -1 & \text{if } x > 1. \end{cases}$$

Is more than one graph possible?

63–66. Normal lines *A line perpendicular to another line or to a tangent line is often called a **normal line**. Find an equation of the line perpendicular to the line that is tangent to the following curves at the given point P.*

63. $y = 3x - 4; P(1, -1)$ **64.** $y = \sqrt{x}; P(4, 2)$

65. $y = \dfrac{2}{x}; P(1, 2)$ **66.** $y = x^2 - 3x; P(3, 0)$

67–70. Aiming a tangent line *Given the function f and the point Q, find all points P on the graph of f such that the line tangent to f at P passes through Q. Check your work by graphing f and the tangent lines.*

67. $f(x) = x^2 + 1; Q(3, 6)$ **68.** $f(x) = -x^2 + 4x - 3; Q(0, 6)$

69. $f(x) = \dfrac{1}{x}; Q(-2, 4)$ **70.** $f(x) = 3\sqrt{4x + 1}; Q(0, 5)$

71–72. One-sided derivatives *The **right-sided** and **left-sided** derivatives of a function at a point a are given by $f'_+(a) = \lim_{h \to 0^+} \dfrac{f(a + h) - f(a)}{h}$ and $f'_-(a) = \lim_{h \to 0^-} \dfrac{f(a + h) - f(a)}{h}$, respectively, provided these limits exist. The derivative $f'(a)$ exists if and only if $f'_+(a) = f'_-(a)$.*

a. *Sketch the following functions.*
b. *Compute $f'_+(a)$ and $f'_-(a)$ at the given point a.*
c. *Is f continuous at a? Is f differentiable at a?*

71. $f(x) = |x - 2|; a = 2$

72. $f(x) = \begin{cases} 4 - x^2 & \text{if } x \le 1 \\ 2x + 1 & \text{if } x > 1 \end{cases}; a = 1$

73–76. Vertical tangent lines *If a function f is continuous at a and $\lim_{x \to a} |f'(x)| = \infty$, then the curve $y = f(x)$ has a vertical tangent line at a, and the equation of the tangent line is $x = a$. If a is an endpoint of a domain, then the appropriate one-sided derivative (Exercises 71–72) is used. Use this information to answer the following questions.*

▣ 73. Graph the following functions and determine the location of the vertical tangent lines.

a. $f(x) = (x - 2)^{1/3}$ **b.** $f(x) = (x + 1)^{2/3}$
c. $f(x) = \sqrt{|x - 4|}$ **d.** $f(x) = x^{5/3} - 2x^{1/3}$

74. The preceding definition of a vertical tangent line includes four cases: $\lim_{x \to a^+} f'(x) = \pm\infty$ combined with $\lim_{x \to a^-} f'(x) = \pm\infty$ (for example, one case is $\lim_{x \to a^+} f'(x) = -\infty$ and $\lim_{x \to a^-} f'(x) = \infty$). Sketch a continuous function that has a vertical tangent line at a in each of the four cases.

75. Verify that $f(x) = x^{1/3}$ has a vertical tangent line at $x = 0$.

76. Graph the following curves and determine the location of any vertical tangent lines.

a. $x^2 + y^2 = 9$ **b.** $x^2 + y^2 + 2x = 0$

77. a. Graph the function $f(x) = \begin{cases} x & \text{if } x \le 0 \\ x + 1 & \text{if } x > 0. \end{cases}$

b. For $x < 0$, what is $f'(x)$?
c. For $x > 0$, what is $f'(x)$?
d. Graph f' on its domain.
e. Is f differentiable at 0? Explain.

78. Is $f(x) = \dfrac{x^2 - 5x + 6}{x - 2}$ differentiable at $x = 2$? Justify your answer.

QUICK CHECK ANSWERS

1. 2   **2.** $\dfrac{df}{dx}\Big|_{x=3}, \dfrac{dy}{dx}\Big|_{x=3}, y'(3)$ **3.** The slopes of the tangent lines decrease as x increases because the values of $f'(x) = \dfrac{1}{2\sqrt{x}}$ decrease as x increases.   **4.** $\dfrac{dq}{dr}, \dfrac{dp}{dr}$, $D_r(q(r)), q'(r), p'(r)$ **5.** No; no ◄

3.3 Rules of Differentiation

If you always had to use limits to evaluate derivatives, as we did in Section 3.2, calculus would be a tedious affair. The goal of this chapter is to establish rules and formulas for quickly evaluating derivatives—not just for individual functions but for entire families of functions. By the end of the chapter, you will have learned many time-saving rules and formulas, all of which are listed in the endpapers of the text.

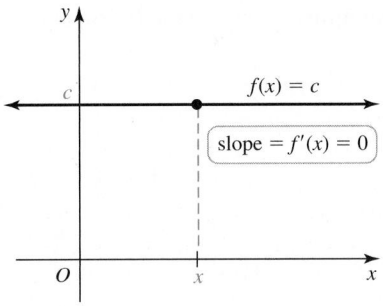

Figure 3.29

➤ The derivative of a constant function is 0 at every point because the values of a constant function do not change. This means the instantaneous rate of change is 0 at every point.

QUICK CHECK 1 Find the values of $\dfrac{d}{dx}(11)$ and $\dfrac{d}{dx}(\pi)$. ◄

➤ Note that this factoring formula agrees with familiar factoring formulas for differences of squares and cubes:

$$x^2 - a^2 = (x - a)(x + a)$$
$$x^3 - a^3 = (x - a)(x^2 + ax + a^2)$$

The Constant and Power Rules for Derivatives

The graph of the **constant function** $f(x) = c$ is a horizontal line with a slope of 0 at every point (**Figure 3.29**). It follows that $f'(x) = 0$ or, equivalently, $\dfrac{d}{dx}(c) = 0$ (Exercise 92). This observation leads to the *Constant Rule* for derivatives.

THEOREM 3.2 Constant Rule

If c is a real number, then $\dfrac{d}{dx}(c) = 0$.

Next, consider power functions of the form $f(x) = x^n$, where n is a nonnegative integer. If you completed Exercise 58 in Section 3.2, you used the limit definition of the derivative to discover that

$$\frac{d}{dx}(x^2) = 2x, \quad \frac{d}{dx}(x^3) = 3x^2, \quad \text{and} \quad \frac{d}{dx}(x^4) = 4x^3.$$

In each case, the derivative of x^n could be evaluated by placing the exponent n in front of x as a coefficient and decreasing the exponent by 1. Based on these observations, we state and prove the following theorem.

THEOREM 3.3 Power Rule

If n is a nonnegative integer, then $\dfrac{d}{dx}(x^n) = nx^{n-1}$.

Proof: We let $f(x) = x^n$ and use the definition of the derivative in the form

$$f'(a) = \lim_{x \to a} \frac{f(x) - f(a)}{x - a}.$$

With $n = 1$ and $f(x) = x$, we have

$$f'(a) = \lim_{x \to a} \frac{f(x) - f(a)}{x - a} = \lim_{x \to a} \frac{x - a}{x - a} = 1,$$

as given by the Power Rule.

With $n \geq 2$ and $f(x) = x^n$, note that $f(x) - f(a) = x^n - a^n$. A factoring formula gives

$$x^n - a^n = (x - a)(x^{n-1} + x^{n-2}a + \cdots + xa^{n-2} + a^{n-1}).$$

Therefore,

$$
\begin{aligned}
f'(a) &= \lim_{x \to a} \frac{x^n - a^n}{x - a} && \text{Definition of } f'(a) \\
&= \lim_{x \to a} \frac{(x - a)(x^{n-1} + x^{n-2}a + \cdots + xa^{n-2} + a^{n-1})}{x - a} && \text{Factor } x^n - a^n. \\
&= \lim_{x \to a} (x^{n-1} + x^{n-2}a + \cdots + xa^{n-2} + a^{n-1}) && \text{Cancel common factors.} \\
&= \underbrace{a^{n-1} + a^{n-2} \cdot a + \cdots + a \cdot a^{n-2} + a^{n-1}}_{n \text{ terms}} = na^{n-1}. && \text{Evaluate the limit.}
\end{aligned}
$$

Replacing a with the variable x in $f'(a) = na^{n-1}$, we obtain the result given in the Power Rule for $n \geq 2$. Finally, note that the Constant Rule is consistent with the Power Rule with $n = 0$. ◄

EXAMPLE 1   Derivatives of power and constant functions Evaluate the following derivatives.

a. $\dfrac{d}{dx}(x^9)$ **b.** $\dfrac{d}{dx}(x)$ **c.** $\dfrac{d}{dx}(2^8)$

SOLUTION

a. $\dfrac{d}{dx}(x^9) = 9x^{9-1} = 9x^8$ Power Rule

QUICK CHECK 2 Use the graph of $y = x$ to give a geometric explanation of why $\dfrac{d}{dx}(x) = 1$. ◄

b. $\dfrac{d}{dx}(x) = \dfrac{d}{dx}(x^1) = 1x^0 = 1$ Power Rule

c. You might be tempted to use the Power Rule here, but $2^8 = 256$ is a constant. So by the Constant Rule, $\dfrac{d}{dx}(2^8) = 0$.

Related Exercises 19–22 ◄

Constant Multiple Rule

Consider the problem of finding the derivative of a constant c multiplied by a function f (assuming f' exists). We apply the definition of the derivative in the form

$$f'(x) = \lim_{h \to 0} \frac{f(x+h) - f(x)}{h}$$

to the function cf:

$$\frac{d}{dx}(cf(x)) = \lim_{h \to 0} \frac{cf(x+h) - cf(x)}{h}$$ Definition of the derivative of cf

$$= \lim_{h \to 0} \frac{c(f(x+h) - f(x))}{h}$$ Factor out c.

$$= c \lim_{h \to 0} \frac{f(x+h) - f(x)}{h}$$ Theorem 2.3

$$= cf'(x).$$ Definition of $f'(x)$

This calculation leads to the *Constant Multiple Rule* for derivatives.

> Theorem 3.4 says that the derivative of a constant multiplied by a function is the constant multiplied by the derivative of the function.

THEOREM 3.4   Constant Multiple Rule
If f is differentiable at x and c is a constant, then

$$\frac{d}{dx}(cf(x)) = cf'(x).$$

EXAMPLE 2   Derivatives of constant multiples of functions Evaluate the following derivatives.

a. $\dfrac{d}{dx}\left(-\dfrac{7x^{11}}{8}\right)$ **b.** $\dfrac{d}{dt}\left(\dfrac{3}{8}\sqrt{t}\right)$

SOLUTION

a. $\dfrac{d}{dx}\left(-\dfrac{7x^{11}}{8}\right) = -\dfrac{7}{8} \cdot \dfrac{d}{dx}(x^{11})$ Constant Multiple Rule

$$= -\frac{7}{8} \cdot 11x^{10}$$ Power Rule

$$= -\frac{77}{8}x^{10}$$ Simplify.

> In Example 2 of Section 3.2, we proved that $\dfrac{d}{dt}(\sqrt{t}) = \dfrac{1}{2\sqrt{t}}$.

b. $\dfrac{d}{dt}\left(\dfrac{3}{8}\sqrt{t}\right) = \dfrac{3}{8} \cdot \dfrac{d}{dt}\left(\sqrt{t}\right)$ Constant Multiple Rule

$\qquad\qquad = \dfrac{3}{8} \cdot \dfrac{1}{2\sqrt{t}} \qquad \dfrac{d}{dt}\left(\sqrt{t}\right) = \dfrac{1}{2\sqrt{t}}$

$\qquad\qquad = \dfrac{3}{16\sqrt{t}}$

Related Exercises 23, 24, 28 ◄

Sum Rule

Many functions are sums of simpler functions. Therefore, it is useful to establish a rule for calculating the derivative of the sum of two or more functions.

> In words, Theorem 3.5 states that the derivative of a sum is the sum of the derivatives.

THEOREM 3.5 Sum Rule

If f and g are differentiable at x, then

$$\frac{d}{dx}(f(x) + g(x)) = f'(x) + g'(x).$$

Proof: Let $F = f + g$, where f and g are differentiable at x, and use the definition of the derivative:

$\dfrac{d}{dx}(f(x) + g(x)) = F'(x)$

$\qquad\qquad = \lim\limits_{h \to 0} \dfrac{F(x + h) - F(x)}{h}$ Definition of derivative

$\qquad\qquad = \lim\limits_{h \to 0} \dfrac{(f(x + h) + g(x + h)) - (f(x) + g(x))}{h}$ $F = f + g$

$\qquad\qquad = \lim\limits_{h \to 0} \left(\dfrac{f(x + h) - f(x)}{h} + \dfrac{g(x + h) - g(x)}{h}\right)$ Regroup.

$\qquad\qquad = \lim\limits_{h \to 0} \dfrac{f(x + h) - f(x)}{h} + \lim\limits_{h \to 0} \dfrac{g(x + h) - g(x)}{h}$ Theorem 2.3

$\qquad\qquad = f'(x) + g'(x).$ Definition of f' and g' ◄

QUICK CHECK 3 If $f(x) = x^2$ and $g(x) = 2x$, what is the derivative of $f(x) + g(x)$? ◄

The Sum Rule can be extended to three or more differentiable functions, $f_1, f_2, \ldots, f_n$, to obtain the **Generalized Sum Rule**:

$$\frac{d}{dx}(f_1(x) + f_2(x) + \cdots + f_n(x)) = f_1{}'(x) + f_2{}'(x) + \cdots + f_n{}'(x).$$

The difference of two functions $f - g$ can be rewritten as the sum $f + (-g)$. By combining the Sum Rule with the Constant Multiple Rule, the **Difference Rule** is established:

$$\frac{d}{dx}(f(x) - g(x)) = f'(x) - g'(x).$$

Let's put the Sum and Difference Rules to work on one of the more common problems: differentiating polynomials.

EXAMPLE 3 Derivative of a polynomial Determine $\dfrac{d}{dw}(2w^3 + 9w^2 - 6w + 4)$.

SOLUTION

$$\dfrac{d}{dw}(2w^3 + 9w^2 - 6w + 4)$$

$$= \dfrac{d}{dw}(2w^3) + \dfrac{d}{dw}(9w^2) - \dfrac{d}{dw}(6w) + \dfrac{d}{dw}(4) \qquad \text{Generalized Sum Rule and Difference Rule}$$

$$= 2\dfrac{d}{dw}(w^3) + 9\dfrac{d}{dw}(w^2) - 6\dfrac{d}{dw}(w) + \dfrac{d}{dw}(4) \qquad \text{Constant Multiple Rule}$$

$$= 2 \cdot 3w^2 + 9 \cdot 2w - 6 \cdot 1 + 0 \qquad \text{Power Rule and Constant Rule}$$

$$= 6w^2 + 18w - 6 \qquad \text{Simplify.}$$

Related Exercises 31–33 ◄

The technique used to differentiate the polynomial in Example 3 may be used for *any* polynomial. Much of the remainder of this chapter is devoted to discovering differentiation rules for the families of functions introduced in Chapter 1: rational, exponential, logarithmic, algebraic, and trigonometric functions.

The Derivative of the Natural Exponential Function

The exponential function $f(x) = b^x$ was introduced in Chapter 1. Let's begin by looking at the graphs of two members of this family, $y = 2^x$ and $y = 3^x$ (**Figure 3.30**). The slope of the line tangent to the graph of $f(x) = b^x$ at $x = 0$ is given by

$$f'(0) = \lim_{h \to 0} \frac{f(0 + h) - f(0)}{h} = \lim_{h \to 0} \frac{b^h - b^0}{h} = \lim_{h \to 0} \frac{b^h - 1}{h}.$$

We investigate this limit numerically for $b = 2$ and $b = 3$. Table 3.1 shows values of $\dfrac{2^h - 1}{h}$ and $\dfrac{3^h - 1}{h}$ (which are slopes of secant lines) for values of h approaching 0 from the right.

► The limit $\lim\limits_{h \to 0} \dfrac{2^h - 1}{h}$ was explored in Example 8 of Section 2.3.

Table 3.1

h	$\dfrac{2^h - 1}{h}$	$\dfrac{3^h - 1}{h}$
1.0	1.000000	2.000000
0.1	0.717735	1.161232
0.01	0.695555	1.104669
0.001	0.693387	1.099216
0.0001	0.693171	1.098673
0.00001	0.693150	1.098618

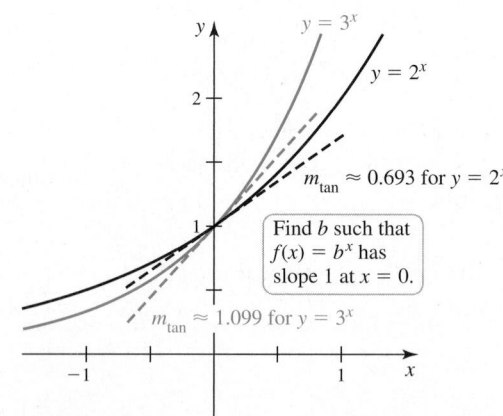

Figure 3.30

Exercise 88 gives similar approximations for the limit as h approaches 0 from the left. These numerical values suggest that

$$\lim_{h \to 0} \frac{2^h - 1}{h} \approx 0.693 \quad \text{Less than 1}$$

$$\lim_{h \to 0} \frac{3^h - 1}{h} \approx 1.099. \quad \text{Greater than 1}$$

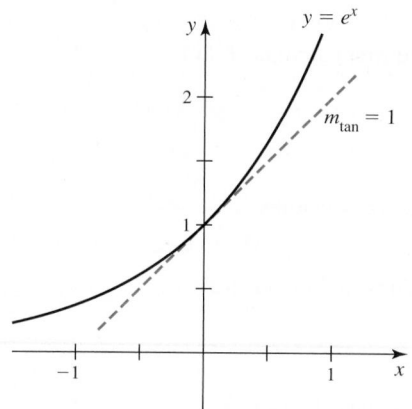

Figure 3.31

▶ The constant e was identified and named by the Swiss mathematician Leonhard Euler (1707–1783) (pronounced "oiler").

These two facts, together with the graphs in Figure 3.30, suggest that there is a number b with $2 < b < 3$ such that the graph of $y = b^x$ has a tangent line with slope 1 at $x = 0$. This number b has the property that

$$\lim_{h \to 0} \frac{b^h - 1}{h} = 1.$$

It can be shown that, indeed, such a number b exists. In fact, it is the number $e = 2.718281828459\ldots$ that was introduced in Chapter 1. Therefore, the exponential function whose tangent line has slope 1 at $x = 0$ is the *natural exponential function* $f(x) = e^x$ (Figure 3.31).

DEFINITION The Number e

The number $e = 2.718281828459\ldots$ satisfies

$$\lim_{h \to 0} \frac{e^h - 1}{h} = 1.$$

It is the base of the natural exponential function $f(x) = e^x$.

With the preceding facts in mind, the derivative of $f(x) = e^x$ is computed as follows:

$$\frac{d}{dx}(e^x) = \lim_{h \to 0} \frac{e^{x+h} - e^x}{h} \qquad \text{Definition of the derivative}$$

$$= \lim_{h \to 0} \frac{e^x \cdot e^h - e^x}{h} \qquad \text{Property of exponents}$$

$$= \lim_{h \to 0} \frac{e^x(e^h - 1)}{h} \qquad \text{Factor out } e^x.$$

$$= e^x \cdot \underbrace{\lim_{h \to 0} \frac{e^h - 1}{h}}_{1} \qquad e^x \text{ is constant as } h \to 0; \text{ definition of } e.$$

$$= e^x \cdot 1 = e^x.$$

We have proved a remarkable fact: The derivative of the exponential function is itself; it is the only function (other than constant multiples of e^x and $f(x) = 0$) with this property.

▶ The Power Rule *cannot* be applied to exponential functions; that is, $\frac{d}{dx}(e^x) \neq xe^{x-1}$. Also note that $\frac{d}{dx}(e^{10}) \neq e^{10}$. Instead, $\frac{d}{dx}(e^c) = 0$, for any real number c, because e^c is a constant.

QUICK CHECK 4 Find the derivative of $f(x) = 4e^x - 3x^2$. ◀

THEOREM 3.6 The Derivative of e^x

The function $f(x) = e^x$ is differentiable for all real numbers x, and

$$\frac{d}{dx}(e^x) = e^x.$$

Slopes of Tangent Lines

The derivative rules presented in this section allow us to determine slopes of tangent lines and rates of change for many functions.

EXAMPLE 4 Finding tangent lines

a. Write an equation of the line tangent to the graph of $f(x) = 2x - \dfrac{e^x}{2}$ at the point $\left(0, -\dfrac{1}{2}\right)$.

b. Find the point(s) on the graph of f at which the tangent line is horizontal.

SOLUTION

a. To find the slope of the tangent line at $\left(0, -\frac{1}{2}\right)$, we first calculate $f'(x)$:

$$f'(x) = \frac{d}{dx}\left(2x - \frac{e^x}{2}\right)$$

$$= \frac{d}{dx}(2x) - \frac{d}{dx}\left(\frac{1}{2}e^x\right) \quad \text{Difference Rule}$$

$$= 2\underbrace{\frac{d}{dx}(x)}_{1} - \frac{1}{2} \cdot \underbrace{\frac{d}{dx}(e^x)}_{e^x} \quad \text{Constant Multiple Rule}$$

$$= 2 - \frac{1}{2}e^x. \quad \text{Evaluate derivatives.}$$

It follows that the slope of the tangent line at $\left(0, -\frac{1}{2}\right)$ is

$$f'(0) = 2 - \frac{1}{2}e^0 = \frac{3}{2}.$$

Figure 3.32 shows the tangent line passing through $\left(0, -\frac{1}{2}\right)$; it has the equation

$$y - \left(-\frac{1}{2}\right) = \frac{3}{2}(x - 0) \quad \text{or} \quad y = \frac{3}{2}x - \frac{1}{2}.$$

b. Because the slope of a horizontal tangent line is 0, our goal is to solve $f'(x) = 2 - \frac{1}{2}e^x = 0$. We multiply both sides of this equation by 2 and rearrange to arrive at the equation $e^x = 4$. Taking the natural logarithm of both sides, we find that $x = \ln 4$. Therefore, $f'(x) = 0$ at $x = \ln 4 \approx 1.39$, and f has a horizontal tangent at $(\ln 4, f(\ln 4)) \approx (1.39, 0.77)$ (Figure 3.32).

Related Exercises 60–61 ◄

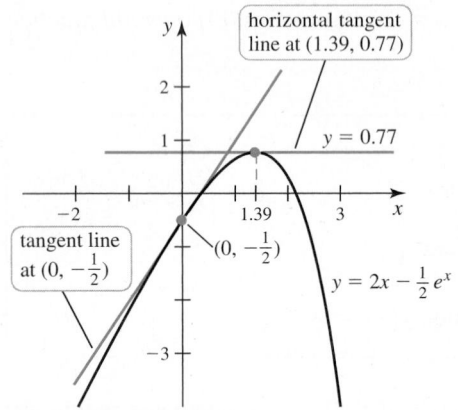

Figure 3.32

➤ Observe that the function has a maximum value of approximately 0.77 at the point where the tangent line has a slope of 0. We explore the importance of horizontal tangent lines in Chapter 4.

EXAMPLE 5   Slope of a tangent line Let $f(x) = 2x^3 - 15x^2 + 24x$. For what values of x does the line tangent to the graph of f have a slope of 6?

SOLUTION The tangent line has a slope of 6 when

$$f'(x) = 6x^2 - 30x + 24 = 6.$$

Subtracting 6 from both sides of the equation and factoring, we have

$$6(x^2 - 5x + 3) = 0.$$

Using the quadratic formula, the roots are

$$x = \frac{5 - \sqrt{13}}{2} \approx 0.697 \quad \text{and} \quad x = \frac{5 + \sqrt{13}}{2} \approx 4.303.$$

Therefore, the slope of the curve at these points is 6 (**Figure 3.33**).

Related Exercises 63–64 ◄

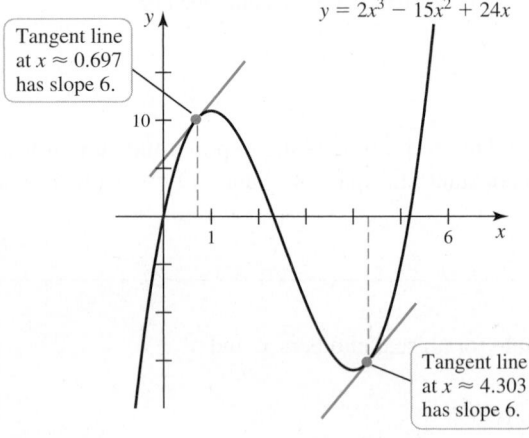

Figure 3.33

QUICK CHECK 5 Determine the point(s) at which $f(x) = x^3 - 12x$ has a horizontal tangent line. ◄

Higher-Order Derivatives

➤ The prime notation, f', f'', and f''', is typically used only for the first, second, and third derivatives.

Because the derivative of a function f is a function in its own right, we can take the derivative of f'. The result is the *second derivative of f*, denoted f'' (read f *double prime*). The derivative of the second derivative is the *third derivative of f*, denoted f''' or (read f *triple prime*). In general, derivatives of order $n \geq 2$ are called *higher-order derivatives*.

> Parentheses are placed around n to distinguish a derivative from a power. Therefore, $f^{(n)}$ is the nth derivative of f, and f^n is the function f raised to the nth power. By convention, $f^{(0)}$ is the function f itself.

> The notation $\dfrac{d^2 f}{dx^2}$ comes from $\dfrac{d}{dx}\left(\dfrac{df}{dx}\right)$ and is read $d\, 2\, f\, dx$ squared.

DEFINITION Higher-Order Derivatives

Assuming $y = f(x)$ can be differentiated as often as necessary, the **second derivative** of f is

$$f''(x) = \frac{d}{dx}(f'(x)).$$

For integers $n \geq 1$, the **nth derivative** of f is

$$f^{(n)}(x) = \frac{d}{dx}(f^{(n-1)}(x)).$$

Other common notations for the second derivative of $y = f(x)$ include $\dfrac{d^2 y}{dx^2}$ and $\dfrac{d^2 f}{dx^2}$; the notations $\dfrac{d^n y}{dx^n}$, $\dfrac{d^n f}{dx^n}$, and $y^{(n)}$ are used for the nth derivative of f.

EXAMPLE 6   Finding higher-order derivatives Find the third derivative of the following functions.

a. $f(x) = 3x^3 - 5x + 12$

b. $y = 3t + 2e^t$

SOLUTION

a.
$$f'(x) = 9x^2 - 5$$

$$f''(x) = \frac{d}{dx}(9x^2 - 5) = 18x$$

$$f'''(x) = 18$$

> In Example 6a, note that $f^{(4)}(x) = 0$, which means that all successive derivatives are also 0. In general, the nth derivative of an nth-degree polynomial is a constant, which implies that derivatives of order $k > n$ are 0.

b. Here we use an alternative notation for higher-order derivatives:

$$\frac{dy}{dt} = \frac{d}{dt}(3t + 2e^t) = 3 + 2e^t$$

$$\frac{d^2 y}{dt^2} = \frac{d}{dt}(3 + 2e^t) = 2e^t$$

$$\frac{d^3 y}{dt^3} = \frac{d}{dt}(2e^t) = 2e^t.$$

QUICK CHECK 6 With $f(x) = x^5$, find $f^{(5)}(x)$ and $f^{(6)}(x)$. With $g(x) = e^x$, find $g^{(100)}(x)$. ◄

In this case, $\dfrac{d^n y}{dt^n} = 2e^t$, for $n \geq 2$.

Related Exercises 69–70 ◄

SECTION 3.3 EXERCISES

Getting Started

1. If the limit definition of a derivative can be used to find f', then what is the purpose of using other rules to find f'?

2. In this section, we showed that the rule $\dfrac{d}{dx}(x^n) = nx^{n-1}$ is valid for what values of n?

3. Give a nonzero function that is its own derivative.

4. How do you find the derivative of the sum of two functions $f + g$?

5. How do you find the derivative of a constant multiplied by a function?

6. How do you find the fifth derivative of a function?

7. Given that $f'(3) = 6$ and $g'(3) = -2$, find $(f + g)'(3)$.

8. If $f'(0) = 6$ and $g(x) = f(x) + e^x + 1$, find $g'(0)$.

9–11. Let $F(x) = f(x) + g(x)$, $G(x) = f(x) - g(x)$, and $H(x) = 3f(x) + 2g(x)$, where the graphs of f and g are shown in the figure. Find each of the following.

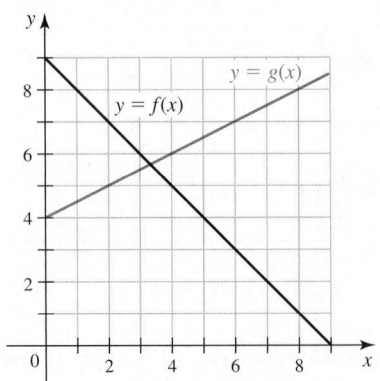

9. $F'(2)$ **10.** $G'(6)$ **11.** $H'(2)$

12–14. Use the table to find the following derivatives.

x	1	2	3	4	5
$f'(x)$	3	5	2	1	4
$g'(x)$	2	4	3	1	5

12. $\dfrac{d}{dx}(f(x) + g(x))\Big|_{x=1}$ **13.** $\dfrac{d}{dx}(1.5f(x))\Big|_{x=2}$

14. $\dfrac{d}{dx}(2x - 3g(x))\Big|_{x=4}$

15. If $f(t) = t^{10}$, find $f'(t)$, $f''(t)$, and $f'''(t)$.

16. Find an equation of the line tangent to the graph of $y = e^x$ at $x = 0$.

17. The line tangent to the graph of f at $x = 5$ is $y = \dfrac{1}{10}x - 2$. Find $\dfrac{d}{dx}(4f(x))\Big|_{x=5}$.

18. The line tangent to the graph of f at $x = 3$ is $y = 4x - 2$ and the line tangent to the graph of g at $x = 3$ is $y = -5x + 1$. Find the values of $(f + g)(3)$ and $(f + g)'(3)$.

Practice Exercises

19–40. Derivatives Find the derivative of the following functions. See Example 2 of Section 3.2 for the derivative of $\sqrt{x}$.

19. $y = x^5$

20. $f(t) = t$

21. $f(x) = 5$

22. $g(x) = e^3$

23. $f(x) = 5x^3$

24. $g(w) = \dfrac{5}{6}w^{12}$

25. $h(t) = \dfrac{t^2}{2} + 1$

26. $f(v) = v^{100} + e^v + 10$

27. $p(x) = 8x$

28. $g(t) = 6\sqrt{t}$

29. $g(t) = 100t^2$

30. $f(s) = \dfrac{\sqrt{s}}{4}$

31. $f(x) = 3x^4 + 7x$

32. $g(x) = 6x^5 - \dfrac{5}{2}x^2 + x + 5$

33. $f(x) = 10x^4 - 32x + e^2$

34. $f(t) = 6\sqrt{t} - 4t^3 + 9$

35. $g(w) = 2w^3 + 3w^2 + 10w$ **36.** $s(t) = 4\sqrt{t} - \dfrac{1}{4}t^4 + t + 1$

37. $f(x) = 3e^x + 5x + 5$ **38.** $g(w) = e^w - e^2 + 8$

39. $f(x) = \begin{cases} x^2 + 1 & \text{if } x \le 0 \\ 2x^2 + x + 1 & \text{if } x > 0 \end{cases}$

40. $g(w) = \begin{cases} w + 5e^w & \text{if } w \le 1 \\ 2w^3 + 4w + 5 & \text{if } w > 1 \end{cases}$

41. Height estimate The distance an object falls (when released from rest, under the influence of Earth's gravity, and with no air resistance) is given by $d(t) = 16t^2$, where d is measured in feet and t is measured in seconds. A rock climber sits on a ledge on a vertical wall and carefully observes the time it takes for a small stone to fall from the ledge to the ground.

  a. Compute $d'(t)$. What units are associated with the derivative and what does it measure?

  b. If it takes 6 s for a stone to fall to the ground, how high is the ledge? How fast is the stone moving when it strikes the ground (in miles per hour)?

T 42. Projectile trajectory The position of a small rocket that is launched vertically upward is given by $s(t) = -5t^2 + 40t + 100$, for $0 \le t \le 10$, where t is measured in seconds and s is measured in meters above the ground.

  a. Find the rate of change in the position (instantaneous velocity) of the rocket, for $0 \le t \le 10$.

  b. At what time is the instantaneous velocity zero?

  c. At what time does the instantaneous velocity have the greatest magnitude, for $0 \le t \le 10$?

  d. Graph the position and instantaneous velocity, for $0 \le t \le 10$.

43. City urbanization City planners model the size of their city using the function $A(t) = -\dfrac{1}{50}t^2 + 2t + 20$, for $0 \le t \le 50$, where A is measured in square miles and t is the number of years after 2010.

  a. Compute $A'(t)$. What units are associated with this derivative and what does the derivative measure?

  b. How fast will the city be growing when it reaches a size of 38 mi^2?

  c. Suppose the population density of the city remains constant from year to year at 1000 people/mi^2. Determine the growth rate of the population in 2030.

44. Cell growth When observations begin at $t = 0$, a cell culture has 1200 cells and continues to grow according to the function $p(t) = 1200e^t$, where p is the number of cells and t is measured in days.

  a. Compute $p'(t)$. What units are associated with the derivative and what does it measure?

  b. On the interval $[0, 4]$, when is the growth rate $p'(t)$ the least? When is it the greatest?

45. Weight of Atlantic salmon The weight $w(x)$ (in pounds) of an Atlantic salmon that is x inches long can be estimated by the function

$$w(x) = \begin{cases} 0.4x - 5 & \text{if } 19 \le x \le 21 \\ 0.8x - 13.4 & \text{if } 21 < x \le 32 \\ 1.5x - 35.8 & \text{if } x > 32 \end{cases}$$

Calculate $w'(x)$ and explain the physical meaning of this derivative. (*Source:* www.atlanticsalmonfederation.org)

46–58. Derivatives of products and quotients *Find the derivative of the following functions by first expanding or simplifying the expression. Simplify your answers.*

46. $f(x) = (\sqrt{x} + 1)(\sqrt{x} - 1)$

47. $f(x) = (2x + 1)(3x^2 + 2)$

48. $g(r) = (5r^3 + 3r + 1)(r^2 + 3)$

49. $f(w) = \dfrac{w^3 - w}{w}$ **50.** $y = \dfrac{12s^3 - 8s^2 + 12s}{4s}$

51. $h(x) = (x^2 + 1)^2$ **52.** $h(x) = \sqrt{x}\left(\sqrt{x} - x^{3/2}\right)$

53. $g(x) = \dfrac{x^2 - 1}{x - 1}$ **54.** $h(x) = \dfrac{x^3 - 6x^2 + 8x}{x^2 - 2x}$

55. $y = \dfrac{x - a}{\sqrt{x} - \sqrt{a}}$; a is a positive constant.

56. $y = \dfrac{x^2 - 2ax + a^2}{x - a}$; a is a constant.

57. $g(w) = \dfrac{e^{2w} + e^{w}}{e^{w}}$ **58.** $r(t) = \dfrac{e^{2t} + 3e^{t} + 2}{e^{t} + 2}$

⊤ 59–62. Equations of tangent lines

a. *Find an equation of the line tangent to the given curve at a.*
b. *Use a graphing utility to graph the curve and the tangent line on the same set of axes.*

59. $y = -3x^2 + 2$; $a = 1$

60. $y = x^3 - 4x^2 + 2x - 1$; $a = 2$

61. $y = e^x$; $a = \ln 3$ **62.** $y = \dfrac{e^x}{4} - x$; $a = 0$

63. Finding slope locations Let $f(x) = x^2 - 6x + 5$.

     a. Find the values of x for which the slope of the curve $y = f(x)$ is 0.

     b. Find the values of x for which the slope of the curve $y = f(x)$ is 2.

64. Finding slope locations Let $f(t) = t^3 - 27t + 5$.

     a. Find the values of t for which the slope of the curve $y = f(t)$ is 0.

     b. Find the values of t for which the slope of the curve $y = f(t)$ is 21.

65. Finding slope locations Let $f(x) = 2x^3 - 3x^2 - 12x + 4$.

     a. Find all points on the graph of f at which the tangent line is horizontal.

     b. Find all points on the graph of f at which the tangent line has slope 60.

66. Finding slope locations Let $f(x) = 2e^x - 6x$.

     a. Find all points on the graph of f at which the tangent line is horizontal.

     b. Find all points on the graph of f at which the tangent line has slope 12.

67. Finding slope locations Let $f(x) = 4\sqrt{x} - x$.

     a. Find all points on the graph of f at which the tangent line is horizontal.

     b. Find all points on the graph of f at which the tangent line has slope $-\dfrac{1}{2}$.

68–72. Higher-order derivatives *Find $f'(x)$, $f''(x)$, and $f'''(x)$ for the following functions.*

68. $f(x) = 3x^3 + 5x^2 + 6x$ **69.** $f(x) = 5x^4 + 10x^3 + 3x + 6$

70. $f(x) = 3x^2 + 5e^x$ **71.** $f(x) = \dfrac{x^2 - 7x - 8}{x + 1}$

72. $f(x) = 10e^x$

73. Explain why or why not Determine whether the following statements are true and give an explanation or counterexample.

     a. $\dfrac{d}{dx}(10^5) = 5 \cdot 10^4$.

     b. The slope of a line tangent to $f(x) = e^x$ is never 0.

     c. $\dfrac{d}{dx}(e^3) = e^3$.   **d.** $\dfrac{d}{dx}(e^x) = xe^{x-1}$.

     e. $\dfrac{d^n}{dx^n}(5x^3 + 2x + 5) = 0$, for any integer $n \geq 3$.

74. Tangent lines Suppose $f(3) = 1$ and $f'(3) = 4$. Let $g(x) = x^2 + f(x)$ and $h(x) = 3f(x)$.

     a. Find an equation of the line tangent to $y = g(x)$ at $x = 3$.

     b. Find an equation of the line tangent to $y = h(x)$ at $x = 3$.

75. Derivatives from tangent lines Suppose the line tangent to the graph of f at $x = 2$ is $y = 4x + 1$ and suppose the line tangent to the graph of g at $x = 2$ has slope 3 and passes through $(0, -2)$. Find an equation of the line tangent to the following curves at $x = 2$.

     a. $y = f(x) + g(x)$ **b.** $y = f(x) - 2g(x)$

     c. $y = 4f(x)$

76. Tangent line Find the equation of the line tangent to the curve $y = x + \sqrt{x}$ that has slope 2.

77. Tangent line given Determine the constants b and c such that the line tangent to $f(x) = x^2 + bx + c$ at $x = 1$ is $y = 4x + 2$.

78–81. Derivatives from a graph *Let $F = f + g$ and $G = 3f - g$, where the graphs of f and g are shown in the figure. Find the following derivatives.*

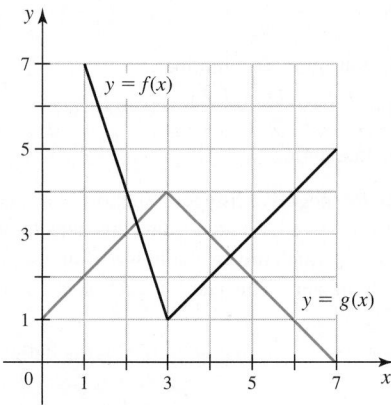

78. $F'(2)$ **79.** $G'(2)$ **80.** $F'(5)$ **81.** $G'(5)$

82–87. Derivatives from limits *The following limits represent $f'(a)$ for some function f and some real number a.*

a. *Find a possible function f and number a.*
b. *Evaluate the limit by computing $f'(a)$.*

82. $\displaystyle\lim_{x \to 0} \dfrac{e^x - 1}{x}$ **83.** $\displaystyle\lim_{x \to 0} \dfrac{x + e^x - 1}{x}$

84. $\lim\limits_{x\to 1}\dfrac{x^{100}-1}{x-1}$

85. $\lim\limits_{h\to 0}\dfrac{\sqrt{9+h}-\sqrt{9}}{h}$

86. $\lim\limits_{h\to 0}\dfrac{(1+h)^8+(1+h)^3-2}{h}$

87. $\lim\limits_{h\to 0}\dfrac{e^{3+h}-e^3}{h}$

Explorations and Challenges

T 88. Important limits Complete the following table and give approximations for $\lim\limits_{h\to 0^-}\dfrac{2^h-1}{h}$ and $\lim\limits_{h\to 0^-}\dfrac{3^h-1}{h}$.

h	$\dfrac{2^h-1}{h}$	$\dfrac{3^h-1}{h}$
-1.0		
-0.1		
-0.01		
-0.001		
-0.0001		
-0.00001		

T 89–91. Calculator limits *Use a calculator to approximate the following limits.*

89. $\lim\limits_{x\to 0}\dfrac{e^{3x}-1}{x}$

90. $\lim\limits_{n\to\infty}\left(1+\dfrac{1}{n}\right)^n$

91. $\lim\limits_{x\to 0^+}x^x$

92. Constant Rule proof For the constant function $f(x)=c$, use the definition of the derivative to show that $f'(x)=0$.

93. Alternative proof of the Power Rule The Binomial Theorem states that for any positive integer n,

$$(a+b)^n = a^n + na^{n-1}b + \dfrac{n(n-1)}{2\cdot 1}a^{n-2}b^2$$
$$+ \dfrac{n(n-1)(n-2)}{3\cdot 2\cdot 1}a^{n-3}b^3 + \cdots + nab^{n-1} + b^n.$$

Use this formula and the definition
$$f'(x)=\lim\limits_{h\to 0}\dfrac{f(x+h)-f(x)}{h}$$ to show that $\dfrac{d}{dx}(x^n)=nx^{n-1}$, for any positive integer n.

94. Power Rule for negative integers Suppose n is a negative integer and $f(x)=x^n$. Use the following steps to prove that $f'(a)=na^{n-1}$, which means the Power Rule for positive integers extends to all integers. This result is proved in Section 3.4 by a different method.

a. Assume $m=-n$, so that $m>0$. Use the definition
$$f'(a)=\lim\limits_{x\to a}\dfrac{x^n-a^n}{x-a}=\lim\limits_{x\to a}\dfrac{x^{-m}-a^{-m}}{x-a}.$$
Simplify using the factoring rule (which is valid for $n>0$)
$$x^n-a^n=(x-a)(x^{n-1}+x^{n-2}a+\cdots+xa^{n-2}+a^{n-1})$$
until it is possible to take the limit.

b. Use this result to find $\dfrac{d}{dx}(x^{-7})$ and $\dfrac{d}{dx}\left(\dfrac{1}{x^{10}}\right)$.

95. Extending the Power Rule to $n=\dfrac{1}{2},\dfrac{3}{2}$, and $\dfrac{5}{2}$ With Theorem 3.3 and Exercise 94, we have shown that the Power Rule, $\dfrac{d}{dx}(x^n)=nx^{n-1}$, applies to any integer n. Later in the chapter, we extend this rule so that it applies to any real number n.

a. Explain why the Power Rule is consistent with the formula
$$\dfrac{d}{dx}(\sqrt{x})=\dfrac{1}{2\sqrt{x}}.$$

b. Prove that the Power Rule holds for $n=\dfrac{3}{2}$. (*Hint:* Use the definition of the derivative: $\dfrac{d}{dx}(x^{3/2})=\lim\limits_{h\to 0}\dfrac{(x+h)^{3/2}-x^{3/2}}{h}$.)

c. Prove that the Power Rule holds for $n=\dfrac{5}{2}$.

d. Propose a formula for $\dfrac{d}{dx}(x^{n/2})$ for any positive integer n.

96. Computing the derivative of $f(x)=e^{-x}$

a. Use the definition of the derivative to show that
$$\dfrac{d}{dx}(e^{-x})=e^{-x}\cdot\lim\limits_{h\to 0}\dfrac{e^{-h}-1}{h}.$$

b. Show that the limit in part (a) is equal to -1. (*Hint:* Use the facts that $\lim\limits_{h\to 0}\dfrac{e^h-1}{h}=1$ and e^x is continuous for all x.)

c. Use parts (a) and (b) to find the derivative of $f(x)=e^{-x}$.

97. Computing the derivative of $f(x)=e^{2x}$

a. Use the definition of the derivative to show that
$$\dfrac{d}{dx}(e^{2x})=e^{2x}\cdot\lim\limits_{h\to 0}\dfrac{e^{2h}-1}{h}.$$

b. Show that the limit in part (a) is equal to 2. (*Hint:* Factor $e^{2h}-1$.)

c. Use parts (a) and (b) to find the derivative of $f(x)=e^{2x}$.

98. Computing the derivative of $f(x)=x^2e^x$

a. Use the definition of the derivative to show that
$$\dfrac{d}{dx}(x^2e^x)=e^x\cdot\lim\limits_{h\to 0}\dfrac{(x^2+2xh+h^2)e^h-x^2}{h}.$$

b. Manipulate the limit in part (a) to arrive at
$$f'(x)=e^x(x^2+2x).\ (Hint:\ \lim\limits_{h\to 0}\dfrac{e^h-1}{h}=1.)$$

QUICK CHECK ANSWERS

1. $\dfrac{d}{dx}(11)=0$ and $\dfrac{d}{dx}(\pi)=0$ because 11 and π are constants.
2. The slope of the curve $y=x$ is 1 at any point; therefore, $\dfrac{d}{dx}(x)=1$.   **3.** $2x+2$ **4.** $f'(x)=4e^x-6x$
5. $x=2$ and $x=-2$ **6.** $f^{(5)}(x)=120, f^{(6)}(x)=0$, $g^{(100)}(x)=e^x$ ◄

3.4 The Product and Quotient Rules

The derivative of a sum of functions is the sum of the derivatives. So you might assume the derivative of a product of functions is the product of the derivatives. Consider, however, the functions $f(x) = x^3$ and $g(x) = x^4$. In this case, $\dfrac{d}{dx}(f(x)g(x)) = \dfrac{d}{dx}(x^7) = 7x^6$, but $f'(x)g'(x) = 3x^2 \cdot 4x^3 = 12x^5$. Therefore, $\dfrac{d}{dx}(f \cdot g) \neq f' \cdot g'$. Similarly, the derivative of a quotient is *not* the quotient of the derivatives. The purpose of this section is to develop rules for differentiating products and quotients of functions.

Product Rule

Here is an anecdote that suggests the formula for the Product Rule. Imagine running along a road at a constant speed. Your speed is determined by two factors: the length of your stride and the number of strides you take each second. Therefore,

$$\text{running speed} = \text{stride length} \cdot \text{stride rate}.$$

For example, if your stride length is 3 ft per stride and you take 2 strides/s, then your speed is 6 ft/s.

Now suppose your stride length increases by 0.5 ft, from 3 to 3.5 ft. Then the change in speed is calculated as follows:

$$\begin{aligned} \textit{change} \text{ in speed} &= \text{change in stride length} \cdot \text{stride rate} \\ &= 0.5 \cdot 2 = 1 \text{ ft/s}. \end{aligned}$$

Alternatively, suppose your stride length remains constant but your stride rate increases by 0.25 stride/s, from 2 to 2.25 strides/s. Then

$$\begin{aligned} \textit{change} \text{ in speed} &= \text{stride length} \cdot \text{change in stride rate} \\ &= 3 \cdot 0.25 = 0.75 \text{ ft/s}. \end{aligned}$$

If both your stride rate and stride length change simultaneously, we expect two contributions to the change in your running speed:

$$\begin{aligned} \textit{change} \text{ in speed} &= (\text{change in stride length} \cdot \text{stride rate}) \\ &\quad + (\text{stride length} \cdot \text{change in stride rate}) \\ &= 1 \text{ ft/s} + 0.75 \text{ ft/s} = 1.75 \text{ ft/s}. \end{aligned}$$

This argument correctly suggests that the derivative (or rate of change) of a product of two functions has *two components*, as shown by the following rule.

> ▶ In words, Theorem 3.7 states that the derivative of the product of two functions equals the derivative of the first function multiplied by the second function, plus the first function multiplied by the derivative of the second function.

THEOREM 3.7   Product Rule

If f and g are differentiable at x, then

$$\frac{d}{dx}(f(x)g(x)) = f'(x)g(x) + f(x)g'(x).$$

Proof: We apply the definition of the derivative to the function fg:

$$\frac{d}{dx}(f(x)g(x)) = \lim_{h \to 0} \frac{f(x+h)g(x+h) - f(x)g(x)}{h}.$$

A useful tactic is to add $-f(x)g(x+h) + f(x)g(x+h)$ (which equals 0) to the numerator, so that

$$\frac{d}{dx}(f(x)g(x)) = \lim_{h \to 0} \frac{f(x+h)g(x+h) - f(x)g(x+h) + f(x)g(x+h) - f(x)g(x)}{h}.$$

The fraction is now split and the numerators are factored:

$$\frac{d}{dx}(f(x)g(x))$$

$$= \lim_{h \to 0} \frac{f(x+h)g(x+h) - f(x)g(x+h)}{h} + \lim_{h \to 0} \frac{f(x)g(x+h) - f(x)g(x)}{h}$$

$$= \lim_{h \to 0} \left(\underbrace{\frac{f(x+h) - f(x)}{h}}_{\substack{\text{approaches } f'(x) \\ \text{as } h \to 0}} \cdot \underbrace{g(x+h)}_{\substack{\text{approaches} \\ g(x) \\ \text{as } h \to 0}} \right) + \lim_{h \to 0} \left(\underbrace{f(x)}_{\substack{\text{equals} \\ f(x) \text{ as} \\ h \to 0}} \cdot \underbrace{\frac{g(x+h) - g(x)}{h}}_{\substack{\text{approaches } g'(x) \\ \text{as } h \to 0}} \right)$$

$$= f'(x) \cdot g(x) + f(x) \cdot g'(x).$$

The continuity of g is used to conclude that $\lim_{h \to 0} g(x+h) = g(x)$.   ◄

> As $h \to 0$, $f(x)$ does not change in value because it is independent of h.

EXAMPLE 1   Using the Product Rule Find and simplify the following derivatives.

a. $\dfrac{d}{dv}\left(v^2(2\sqrt{v} + 1)\right)$ **b.** $\dfrac{d}{dx}(x^2 e^x)$

> In Example 2 of Section 3.2, we proved that $\dfrac{d}{dv}(\sqrt{v}) = \dfrac{1}{2\sqrt{v}}$.

SOLUTION

a. $\dfrac{d}{dv}\left(v^2(2\sqrt{v} + 1)\right) = \left(\dfrac{d}{dv}(v^2)\right)(2\sqrt{v} + 1) + v^2\left(\dfrac{d}{dv}(2\sqrt{v} + 1)\right)$ Product Rule

$$= 2v(2\sqrt{v} + 1) + v^2\left(2 \cdot \frac{1}{2\sqrt{v}}\right) \qquad \text{Evaluate the derivatives.}$$

$$= 4v^{3/2} + 2v + v^{3/2} = 5v^{3/2} + 2v \qquad \text{Simplify.}$$

QUICK CHECK 1 Find the derivative of $f(x) = x^5$. Then find the same derivative using the Product Rule with $f(x) = x^2 x^3$. ◄

b. $\dfrac{d}{dx}(x^2 e^x) = \underbrace{2x}_{\frac{d}{dx}(x^2)} \cdot e^x + x^2 \cdot \underbrace{e^x}_{\frac{d}{dx}(e^x)} = xe^x(2 + x)$

Related Exercises 19–20 ◄

Quotient Rule

Consider the quotient $q(x) = \dfrac{f(x)}{g(x)}$ and note that $f(x) = g(x)q(x)$. By the Product Rule, we have

$$f'(x) = g'(x)q(x) + g(x)q'(x).$$

Solving for $q'(x)$, we find that

$$q'(x) = \frac{f'(x) - g'(x)q(x)}{g(x)}.$$

Substituting $q(x) = \dfrac{f(x)}{g(x)}$ produces a rule for finding $q'(x)$:

$$q'(x) = \frac{f'(x) - g'(x)\dfrac{f(x)}{g(x)}}{g(x)} \qquad \text{Replace } q(x) \text{ with } \frac{f(x)}{g(x)}.$$

$$= \frac{g(x)\left(f'(x) - g'(x)\dfrac{f(x)}{g(x)}\right)}{g(x) \cdot g(x)} \qquad \text{Multiply numerator and denominator by } g(x).$$

$$= \frac{g(x)f'(x) - f(x)g'(x)}{(g(x))^2}. \qquad \text{Simplify.}$$

➤ In words, Theorem 3.8 states that the derivative of the quotient of two functions equals the denominator multiplied by the derivative of the numerator minus the numerator multiplied by the derivative of the denominator, all divided by the denominator squared. An easy way to remember the Quotient Rule is with

$$\frac{LoD(Hi) - HiD(Lo)}{(Lo)^2}.$$

This calculation produces the correct result for the derivative of a quotient. However, there is one subtle point: How do we know that the derivative of f/g exists in the first place? A complete proof of the Quotient Rule is outlined in Exercise 96.

THEOREM 3.8   Quotient Rule
If f and g are differentiable at x and $g(x) \neq 0$, then the derivative of f/g at x exists and

$$\frac{d}{dx}\left(\frac{f(x)}{g(x)}\right) = \frac{g(x)f'(x) - f(x)g'(x)}{(g(x))^2}.$$

EXAMPLE 2   Using the Quotient Rule Find and simplify the following derivatives.

a. $\dfrac{d}{dx}\left(\dfrac{x^2 + 3x + 4}{x^2 - 1}\right)$ **b.** $\dfrac{d}{dx}(e^{-x})$

SOLUTION

➤ The Product and Quotient Rules are used on a regular basis throughout this text. It is a good idea to memorize these rules (along with the other derivative rules and formulas presented in this chapter) so that you can evaluate derivatives quickly.

a.
$$\frac{d}{dx}\left(\frac{x^2 + 3x + 4}{x^2 - 1}\right) = \frac{\overbrace{(x^2 - 1)(2x + 3)}^{\substack{(x^2-1)\,\cdot\,\text{the derivative}\\ \text{of }(x^2+3x+4)}} - \overbrace{(x^2 + 3x + 4)2x}^{\substack{(x^2+3x+4)\,\cdot\,\text{the}\\ \text{derivative of }(x^2-1)}}}{\underbrace{(x^2 - 1)^2}_{\substack{\text{the denominator}\\ (x^2-1)\text{ squared}}}} \qquad \text{Quotient Rule}$$

$$= \frac{2x^3 - 2x + 3x^2 - 3 - 2x^3 - 6x^2 - 8x}{(x^2 - 1)^2} \qquad \text{Expand.}$$

$$= -\frac{3x^2 + 10x + 3}{(x^2 - 1)^2} \qquad \text{Simplify.}$$

b. We rewrite e^{-x} as $\dfrac{1}{e^x}$ and use the Quotient Rule:

$$\frac{d}{dx}\left(\frac{1}{e^x}\right) = \frac{e^x \cdot 0 - 1 \cdot e^x}{(e^x)^2} = -\frac{1}{e^x} = -e^{-x}.$$

Related Exercises 22, 27 ◄

QUICK CHECK 2 Find the derivative of $f(x) = x^5$. Then find the same derivative using the Quotient Rule with $f(x) = x^8/x^3$. ◄

EXAMPLE 3   Finding tangent lines Find an equation of the line tangent to the graph of $f(x) = \dfrac{x^2 + 1}{x^2 - 4}$ at the point $(3, 2)$. Plot the curve and tangent line.

SOLUTION To find the slope of the tangent line, we compute f' using the Quotient Rule:

$$f'(x) = \frac{(x^2 - 4)\,2x - (x^2 + 1)\,2x}{(x^2 - 4)^2} \qquad \text{Quotient Rule}$$

$$= \frac{2x^3 - 8x - 2x^3 - 2x}{(x^2 - 4)^2} = -\frac{10x}{(x^2 - 4)^2}. \qquad \text{Simplify.}$$

The slope of the tangent line at $(3, 2)$ is

$$m_{\text{tan}} = f'(3) = -\frac{10(3)}{(3^2 - 4)^2} = -\frac{6}{5}.$$

Therefore, an equation of the tangent line is

$$y - 2 = -\frac{6}{5}(x - 3), \quad \text{or} \quad y = -\frac{6}{5}x + \frac{28}{5}.$$

The graphs of f and the tangent line are shown in **Figure 3.34**.

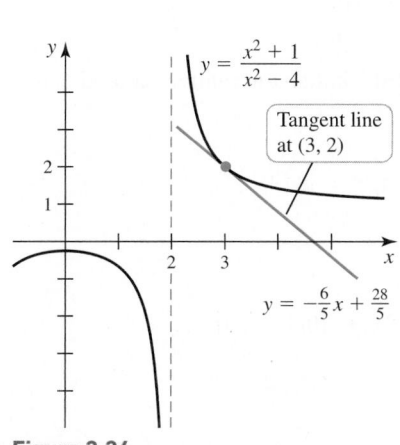

Figure 3.34

Related Exercises 61–62 ◄

Extending the Power Rule to Negative Integers

The Power Rule in Section 3.3 says that $\dfrac{d}{dx}(x^n) = nx^{n-1}$, for nonnegative integers n. Using the Quotient Rule, we show that the Power Rule also holds if n is a negative integer. Assume n is a negative integer and let $m = -n$, so that $m > 0$. Then

$$\frac{d}{dx}(x^n) = \frac{d}{dx}\left(\frac{1}{x^m}\right) \qquad x^n = \frac{1}{x^{-n}} = \frac{1}{x^m}$$

$$= \frac{x^m\overbrace{\left(\dfrac{d}{dx}(1)\right)}^{\substack{\text{derivative of a}\\\text{constant is }0}} - 1\overbrace{\left(\dfrac{d}{dx}x^m\right)}^{\substack{\text{equals}\\ mx^{m-1}}}}{(x^m)^2} \qquad \text{Quotient Rule}$$

$$= -\frac{mx^{m-1}}{x^{2m}} \qquad \text{Simplify.}$$

$$= -mx^{-m-1} \qquad \frac{x^{m-1}}{x^{2m}} = x^{m-1-2m}$$

$$= nx^{n-1}. \qquad \text{Replace } -m \text{ with } n.$$

This calculation leads to the first extension of the Power Rule; the rule now applies to all integers. In Theorem 3.9, we assert that, in fact, the Power Rule is valid for all real powers. A proof of this theorem appears in Section 3.9.

> ▶ In Theorem 3.9, it is necessary to restrict the domain for some values of n. If n is irrational, then the Power Rule holds, provided that $x > 0$. If n is a rational number of the form p/q, where p and q are integers with no common factors and q is even, then the rule holds, provided that $x > 0$.

THEOREM 3.9   Power Rule (general form)
If n is any real number, then

$$\frac{d}{dx}(x^n) = nx^{n-1}.$$

QUICK CHECK 3 Find the derivative of $f(x) = 1/x^5$ in two different ways: using the Power Rule and using the Quotient Rule. ◄

EXAMPLE 4   Using the Power Rule Find the following derivatives.

a. $\dfrac{d}{dx}\left(\dfrac{9}{x^5}\right)$ **b.** $\dfrac{d}{dt}\left(\dfrac{3t^{16} - 4}{t^6}\right)$

c. $\dfrac{d}{dz}\left(\sqrt[3]{z}\,e^z\right)$ **d.** $\dfrac{d}{dx}\left(\dfrac{3x^{5/2}}{2x^2 + 4}\right)$

SOLUTION

a. $\dfrac{d}{dx}\left(\dfrac{9}{x^5}\right) = \dfrac{d}{dx}(9x^{-5}) = 9(-5x^{-6}) = -45x^{-6} = -\dfrac{45}{x^6}$

b. This derivative can be evaluated by the Quotient Rule, but an alternative method is to rewrite the expression using negative powers:

$$\frac{3t^{16} - 4}{t^6} = \frac{3t^{16}}{t^6} - \frac{4}{t^6} = 3t^{10} - 4t^{-6}.$$

We now differentiate using the Power Rule:

$$\frac{d}{dt}\left(\frac{3t^{16} - 4}{t^6}\right) = \frac{d}{dt}(3t^{10} - 4t^{-6}) = 30t^9 + 24t^{-7}.$$

c. Express the cube root as a power and then apply the Product and Power Rules:

$$\frac{d}{dz}\left(\sqrt[3]{z}\, e^z\right) = \frac{d}{dz}(z^{1/3}e^z) \qquad\qquad \sqrt[3]{z} = z^{1/3}$$

$$= \frac{1}{3}z^{-2/3}e^z + z^{1/3}e^z \qquad \text{Product and Power Rules}$$

$$= e^z\left(\frac{1}{3z^{2/3}} + z^{1/3}\right) \qquad \text{Factor.}$$

$$= e^z\left(\frac{1}{3z^{2/3}} + \frac{3z}{3z^{2/3}}\right) \qquad \text{Common denominator}$$

$$= \frac{e^z(1 + 3z)}{3z^{2/3}}. \qquad \text{Combine fractions.}$$

d. The Quotient and Power Rules are required:

$$\frac{d}{dx}\left(\frac{3x^{5/2}}{2x^2 + 4}\right) = \frac{(2x^2 + 4)\cdot\frac{15}{2}x^{3/2} - 3x^{5/2}\cdot 4x}{(2x^2 + 4)^2} \qquad \text{Quotient and Power Rules}$$

$$= \frac{15x^{7/2} + 30x^{3/2} - 12x^{7/2}}{(2(x^2 + 2))^2} \qquad \text{Simplify.}$$

$$= \frac{3x^{7/2} + 30x^{3/2}}{4(x^2 + 2)^2} \qquad \text{Combine like terms.}$$

$$= \frac{3x^{3/2}(x^2 + 10)}{4(x^2 + 2)^2}. \qquad \text{Factor.}$$

Related Exercises 28, 39, 43, 51 ◄

Combining Derivative Rules

Some situations call for the use of multiple differentiation rules. This section concludes with one such example.

EXAMPLE 5   Combining derivative rules Find the derivative of

$$y = \frac{4xe^x}{x^2 + 1}.$$

SOLUTION In this case, we have the quotient of two functions, with a product $(4xe^x)$ in the numerator.

$$\frac{dy}{dx} = \frac{(x^2 + 1)\cdot\dfrac{d}{dx}(4xe^x) - (4xe^x)\cdot\dfrac{d}{dx}(x^2 + 1)}{(x^2 + 1)^2} \qquad \text{Quotient Rule}$$

$$= \frac{(x^2 + 1)(4e^x + 4xe^x) - (4xe^x)(2x)}{(x^2 + 1)^2} \qquad \begin{array}{l}\frac{d}{dx}(4xe^x) = 4e^x + 4xe^x \\ \text{by the Product Rule}\end{array}$$

$$= \frac{4e^x(x^3 - x^2 + x + 1)}{(x^2 + 1)^2} \qquad \text{Simplify.}$$

Related Exercises 45–46 ◄

SECTION 3.4 EXERCISES

Getting Started

1. How do you find the derivative of the product of two functions that are differentiable at a point?

2. How do you find the derivative of the quotient of two functions that are differentiable at a point?

3. Use the Product Rule to evaluate and simplify
$$\frac{d}{dx}((x + 1)(3x + 2)).$$

4. Use the Product Rule to find $f'(1)$ given that $f(x) = x^4 e^x$.

5. Use the Quotient Rule to evaluate and simplify $\dfrac{d}{dx}\left(\dfrac{x - 1}{3x + 2}\right)$.

6. Use the Quotient Rule to find $g'(1)$ given that $g(x) = \dfrac{x^2}{x + 1}$.

7–14. Find the derivative the following ways:

a. Using the Product Rule (Exercises 7–10) or the Quotient Rule (Exercises 11–14). Simplify your result.

b. By expanding the product first (Exercises 7–10) or by simplifying the quotient first (Exercises 11–14). Verify that your answer agrees with part (a).

7. $f(x) = x(x - 1)$

8. $g(t) = (t + 1)(t^2 - t + 1)$

9. $f(x) = (x - 1)(3x + 4)$

10. $h(z) = (z^3 + 4z^2 + z)(z - 1)$

11. $f(w) = \dfrac{w^3 - w}{w}$

12. $g(s) = \dfrac{4s^3 - 8s^2 + 4s}{4s}$

13. $y = \dfrac{x^2 - a^2}{x - a}$, where a is a constant

14. $y = \dfrac{x^2 - 2ax + a^2}{x - a}$, where a is a constant

15. Given that $f(1) = 5$, $f'(1) = 4$, $g(1) = 2$, and $g'(1) = 3$, find
$$\frac{d}{dx}(f(x)g(x))\Big|_{x=1} \text{ and } \frac{d}{dx}\left(\frac{f(x)}{g(x)}\right)\Big|_{x=1}.$$

16. Show two ways to differentiate $f(x) = \dfrac{1}{x^{10}}$.

17. Find the slope of the line tangent to the graph of $f(x) = \dfrac{x}{x + 6}$ at the point $(3, 1/3)$ and at $(-2, -1/2)$.

18. Find the slope of the graph of $f(x) = 2 + xe^x$ at the point $(0, 2)$.

Practice Exercises

19–60. Derivatives *Find and simplify the derivative of the following functions.*

19. $f(x) = 3x^4(2x^2 - 1)$

20. $g(x) = 6x - 2xe^x$

21. $f(x) = \dfrac{x}{x + 1}$

22. $f(x) = \dfrac{x^3 - 4x^2 + x}{x - 2}$

23. $f(t) = t^{5/3}e^t$

24. $g(w) = e^w(5w^2 + 3w + 1)$

25. $f(x) = \dfrac{e^x}{e^x + 1}$

26. $f(x) = \dfrac{2e^x - 1}{2e^x + 1}$

27. $f(x) = xe^{-x}$

28. $f(x) = e^x \sqrt[3]{x}$

29. $y = (3t - 1)(2t - 2)^{-1}$

30. $h(w) = \dfrac{w^2 - 1}{w^2 + 1}$

31. $h(x) = (x - 1)(x^3 + x^2 + x + 1)$

32. $f(x) = \left(1 + \dfrac{1}{x^2}\right)(x^2 + 1)$

33. $g(w) = e^w(w^3 - 1)$

34. $s(t) = \dfrac{t^{4/3}}{e^t}$

35. $f(t) = e^t(t^2 - 2t + 2)$

36. $f(x) = e^x(x^3 - 3x^2 + 6x - 6)$

37. $g(x) = \dfrac{e^x}{x^2 - 1}$

38. $y = (2\sqrt{x} - 1)(4x + 1)^{-1}$

39. $f(x) = 3x^{-9}$

40. $y = \dfrac{4}{p^3}$

41. $g(t) = 3t^2 + \dfrac{6}{t^7}$

42. $y = \dfrac{w^4 + 5w^2 + w}{w^2}$

43. $g(t) = \dfrac{t^3 + 3t^2 + t}{t^3}$

44. $p(x) = \dfrac{4x^3 + 3x + 1}{2x^5}$

45. $g(x) = \dfrac{(x + 1)e^x}{x - 2}$

46. $h(x) = \dfrac{(x - 1)(2x^2 - 1)}{(x^3 - 1)}$

47. $h(x) = \dfrac{xe^x}{x + 1}$

48. $h(x) = \dfrac{x + 1}{x^2 e^x}$

49. $g(w) = \dfrac{\sqrt{w} + w}{\sqrt{w} - w}$

50. $f(x) = \dfrac{4 - x^2}{x - 2}$

51. $h(w) = \dfrac{w^{5/3}}{w^{5/3} + 1}$

52. $g(x) = \dfrac{x^{4/3} - 1}{x^{4/3} + 1}$

53. $f(x) = 4x^2 - \dfrac{2x}{5x + 1}$

54. $f(z) = \left(\dfrac{z^2 + 1}{z}\right)e^z$

55. $h(r) = \dfrac{2 - r - \sqrt{r}}{r + 1}$

56. $y = \dfrac{x - a}{\sqrt{x} - \sqrt{a}}$, where a is a positive constant

57. $h(x) = (5x^7 + 5x)(6x^3 + 3x^2 + 3)$

58. $s(t) = (t + 1)(t + 2)(t + 3)$

59. $f(x) = \sqrt{e^{2x} + 8x^2 e^x + 16x^4}$ (*Hint:* Factor the function under the square root first.)

60. $g(x) = \dfrac{e^{2x} - 1}{e^x - 1}$

T 61–64. Equations of tangent lines

a. Find an equation of the line tangent to the given curve at a.

b. Use a graphing utility to graph the curve and the tangent line on the same set of axes.

61. $y = \dfrac{x + 5}{x - 1}$; $a = 3$

62. $y = \dfrac{2x^2}{3x - 1}$; $a = 1$

63. $y = 1 + 2x + xe^x$; $a = 0$

64. $y = \dfrac{e^x}{x}$; $a = 1$

T 65–66. Population growth *Consider the following population functions.*

a. *Find the instantaneous growth rate of the population, for t ≥ 0.*
b. *What is the instantaneous growth rate at t = 5?*
c. *Estimate the time when the instantaneous growth rate is greatest.*
d. *Evaluate and interpret* $\lim_{t \to \infty} p(t)$.
e. *Use a graphing utility to graph the population and its growth rate.*

65. $p(t) = \dfrac{200t}{t + 2}$ **66.** $p(t) = 600\left(\dfrac{t^2 + 3}{t^2 + 9}\right)$

67. Electrostatic force The magnitude of the electrostatic force between two point charges Q and q of the same sign is given by $F(x) = \dfrac{kQq}{x^2}$, where x is the distance (measured in meters) between the charges and $k = 9 \times 10^9$ N-m²/C² is a physical constant (C stands for coulomb, the unit of charge; N stands for newton, the unit of force).

   a. Find the instantaneous rate of change of the force with respect to the distance between the charges.
   b. For two identical charges with $Q = q = 1$ C, what is the instantaneous rate of change of the force at a separation of $x = 0.001$ m?
   c. Does the magnitude of the instantaneous rate of change of the force increase or decrease with the separation? Explain.

68. Gravitational force The magnitude of the gravitational force between two objects of mass M and m is given by $F(x) = -\dfrac{GMm}{x^2}$, where x is the distance between the centers of mass of the objects and $G = 6.7 \times 10^{-11}$ N-m²/kg² is the gravitational constant (N stands for newton, the unit of force; the negative sign indicates an attractive force).

   a. Find the instantaneous rate of change of the force with respect to the distance between the objects.
   b. For two identical objects of mass $M = m = 0.1$ kg, what is the instantaneous rate of change of the force at a separation of $x = 0.01$ m?
   c. Does the magnitude of the instantaneous rate of change of the force increase or decrease with the separation? Explain.

69. Explain why or why not Determine whether the following statements are true and give an explanation or counterexample.

   a. $\dfrac{d}{dx}(e^5) = 5e^4$

   b. The Quotient Rule must be used to evaluate $\dfrac{d}{dx}\left(\dfrac{x^2 + 3x + 2}{x}\right)$.

   c. $\dfrac{d}{dx}\left(\dfrac{1}{x^5}\right) = \dfrac{1}{5x^4}$.

   d. $\dfrac{d}{dx}(x^3 e^x) = 3x^2 e^x$.

70–71. Higher-order derivatives *Find $f'(x)$, $f''(x)$, and $f'''(x)$.*

70. $f(x) = \dfrac{1}{x}$ **71.** $f(x) = x^2(2 + x^{-3})$

72–73. First and second derivatives *Find $f'(x)$ and $f''(x)$.*

72. $f(x) = \dfrac{x}{x + 2}$ **73.** $f(x) = \dfrac{x^2 - 7x}{x + 1}$

74. Tangent lines Suppose $f(2) = 2$ and $f'(2) = 3$. Let $g(x) = x^2 f(x)$ and $h(x) = \dfrac{f(x)}{x - 3}$.

   a. Find an equation of the line tangent to $y = g(x)$ at $x = 2$.
   b. Find an equation of the line tangent to $y = h(x)$ at $x = 2$.

T 75. The Witch of Agnesi The graph of $y = \dfrac{a^3}{x^2 + a^2}$, where a is a constant, is called the *witch of Agnesi* (named after the 18th-century Italian mathematician Maria Agnesi).

   a. Let $a = 3$ and find an equation of the line tangent to $y = \dfrac{27}{x^2 + 9}$ at $x = 2$.
   b. Plot the function and the tangent line found in part (a).

76–81. Derivatives from a table *Use the following table to find the given derivatives.*

x	1	2	3	4
$f(x)$	5	4	3	2
$f'(x)$	3	5	2	1
$g(x)$	4	2	5	3
$g'(x)$	2	4	3	1

76. $\dfrac{d}{dx}(f(x)g(x))\Big|_{x=1}$ **77.** $\dfrac{d}{dx}\left(\dfrac{f(x)}{g(x)}\right)\Big|_{x=2}$

78. $\dfrac{d}{dx}(xf(x))\Big|_{x=3}$ **79.** $\dfrac{d}{dx}\left(\dfrac{f(x)}{x + 2}\right)\Big|_{x=4}$

80. $\dfrac{d}{dx}\left(\dfrac{xf(x)}{g(x)}\right)\Big|_{x=4}$ **81.** $\dfrac{d}{dx}\left(\dfrac{f(x)g(x)}{x}\right)\Big|_{x=4}$

82–83. Flight formula for Indian spotted owlets *The following table shows the average body mass $m(t)$ (in g) and average wing chord length $w(t)$ (in mm), along with the derivatives $m'(t)$ and $w'(t)$, of t-week-old Indian spotted owlets. The **flight formula** function $f(t) = w(t)/m(t)$, which is the ratio of wing chord length to mass, is used to predict when these fledglings are old enough to fly. The values of f are less than 1, but approach 1 as t increases. When f is close to 1, the fledglings are capable of flying, which is important for determining when rescued fledglings can be released back into the wild. (Source: ZooKeys, 132, 2011)*

t	$m(t)$	$m'(t)$	$w(t)$	$w'(t)$
1	23.32	41.45	10.14	14.5
1.5	50.59	64.94	20.13	26.17
2	82.83	57.95	36.7	39.86
2.5	105.13	31.08	58.92	47.11
3	115.48	12.48	81.55	41.38
3.5	119.4	4.44	98.99	27.94
4	120.76	1.51	109.75	15.74
4.5	121.22	0.51	115.5	7.99
5	121.37	0.17	118.34	3.85
5.5	121.42	0.06	119.69	1.8
6	121.44	0.02	120.32	0.84
6.5	121.45	0.01	120.61	0.38

82. State the units associated with the following derivatives and state the physical meaning of each derivative.

   a. $m'(t)$ **b.** $w'(t)$ **c.** $f'(t)$

T 83. Complete the following steps to examine the behavior of the flight formula.

    a. Sketch an approximate graph of $y = f(t)$ by plotting and connecting the points $(1, f(1)), (1.5, f(1.5)), \ldots, (6.5, f(6.5))$ with a smooth curve.

    b. For what value of t does f appear to be changing most rapidly? Round t to the nearest whole number.

    c. For the value of t found in part (b), use the table and the Quotient Rule to find $f'(t)$. Describe what is happening to the bird at this stage in its life.

    d. Use your graph of f to predict what happens to $f'(t)$ as t grows larger and confirm your prediction by evaluating $f'(6.5)$ using the Quotient Rule. Describe what is happening in the physical development of the fledglings as t grows larger.

84–85. *Assume both the graphs of f and g pass through the point $(3, 2), f'(3) = 5, $ and $g'(3) = -10$. If $p(x) = f(x)g(x)$ and $q(x) = f(x)/g(x)$, find the following derivatives.*

84. $p'(3)$ **85.** $q'(3)$

86. Given that $f(1) = 2$ and $f'(1) = 2$, find the slope of the curve $y = xf(x)$ at the point $(1, 2)$.

87–90. Derivatives from graphs *Use the figure to find the following derivatives.*

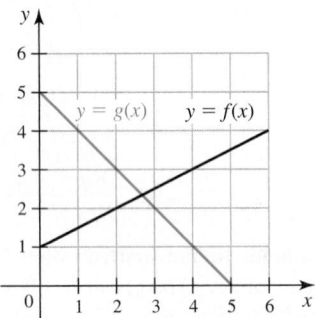

87. $\dfrac{d}{dx}(f(x)g(x))\Big|_{x=4}$ **88.** $\dfrac{d}{dx}\left(\dfrac{f(x)}{g(x)}\right)\Big|_{x=4}$

89. $\dfrac{d}{dx}(xg(x))\Big|_{x=2}$ **90.** $\dfrac{d}{dx}\left(\dfrac{x^2}{f(x)}\right)\Big|_{x=2}$

91. Tangent lines The line tangent to the curve $y = h(x)$ at $x = 4$ is $y = -3x + 14$. Find an equation of the line tangent to the following curves at $x = 4$.

    a. $y = (x^2 - 3x)h(x)$ **b.** $y = h(x)/(x + 2)$

92. Derivatives from tangent lines Suppose the line tangent to the graph of f at $x = 2$ is $y = 4x + 1$ and suppose $y = 3x - 2$ is the line tangent to the graph of g at $x = 2$. Find an equation of the line tangent to the following curves at $x = 2$.

    a. $y = f(x)g(x)$ **b.** $y = f(x)/g(x)$

Explorations and Challenges

93. Avoiding tedious work Given that
$$q(x) = \frac{5x^8 + 6x^5 + 5x^4 + 3x^2 + 20x + 100}{10x^{10} + 8x^9 + 6x^5 + 6x^2 + 4x + 2}, \text{ find } q'(0)$$
without computing $q'(x)$. (*Hint:* Evaluate $f(0), f'(0), g(0),$ and $g'(0)$ where f is the numerator of q and g is the denominator of q.)

94. Given that $p(x) = (5e^x + 10x^5 + 20x^3 + 100x^2 + 5x + 20) \cdot (10x^5 + 40x^3 + 20x^2 + 4x + 10)$, find $p'(0)$ without computing $p'(x)$.

95. Means and tangents Suppose f is differentiable on an interval containing a and b, and let $P(a, f(a))$ and $Q(b, f(b))$ be distinct points on the graph of f. Let c be the x-coordinate of the point at which the lines tangent to the curve at P and Q intersect, assuming the tangent lines are not parallel (see figure).

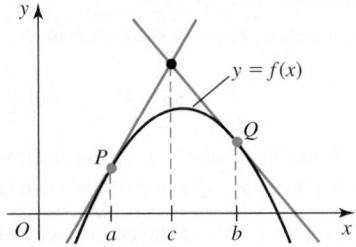

    a. If $f(x) = x^2$, show that $c = (a + b)/2$, the arithmetic mean of a and b, for real numbers a and b.

    b. If $f(x) = \sqrt{x}$, show that $c = \sqrt{ab}$, the geometric mean of a and b, for $a > 0$ and $b > 0$.

    c. If $f(x) = 1/x$, show that $c = 2ab/(a + b)$, the harmonic mean of a and b, for $a > 0$ and $b > 0$.

    d. Find an expression for c in terms of a and b for any (differentiable) function f whenever c exists.

96. Proof of the Quotient Rule Let $F = f/g$ be the quotient of two functions that are differentiable at x.

    a. Use the definition of F' to show that
$$\frac{d}{dx}\left(\frac{f(x)}{g(x)}\right) = \lim_{h\to 0}\frac{f(x + h)g(x) - f(x)g(x + h)}{hg(x + h)g(x)}.$$

    b. Now add $-f(x)g(x) + f(x)g(x)$ (which equals 0) to the numerator in the preceding limit to obtain
$$\lim_{h\to 0}\frac{f(x + h)g(x) - f(x)g(x) + f(x)g(x) - f(x)g(x + h)}{hg(x + h)g(x)}.$$
Use this limit to obtain the Quotient Rule.

    c. Explain why $F' = (f/g)'$ exists, whenever $g(x) \neq 0$.

97. Product Rule for the second derivative Assuming the first and second derivatives of f and g exist at x, find a formula for $d^2/dx^2(f(x)g(x))$.

98. One of the Leibniz Rules One of several Leibniz Rules in calculus deals with higher-order derivatives of products. Let $(fg)^{(n)}$ denote the nth derivative of the product fg, for $n \geq 1$.

    a. Prove that $(fg)^{(2)} = f''g + 2f'g' + fg''$.

    b. Prove that, in general,
$$(fg)^{(n)} = \sum_{k=0}^{n}\binom{n}{k}f^{(k)}g^{(n-k)},$$
where $\binom{n}{k} = \dfrac{n!}{k!(n - k)!}$ are the binomial coefficients.

    c. Compare the result of (b) to the expansion of $(a + b)^n$.

99. Product Rule for three functions Assume $f, g,$ and h are differentiable at x.

    a. Use the Product Rule (twice) to find a formula for $d/dx(f(x)g(x)h(x))$.

    b. Use the formula in (a) to find $d/dx(e^x(x - 1)(x + 3))$.

QUICK CHECK ANSWERS

1. $f'(x) = 5x^4$ by either method   **2.** $f'(x) = 5x^4$ by either method   **3.** $f'(x) = -5x^{-6}$ by either method ◄

3.5 Derivatives of Trigonometric Functions

> ▸ Results stated in this section assume angles are measured in *radians*.

From variations in market trends and ocean temperatures to daily fluctuations in tides and hormone levels, change is often cyclical or periodic. Trigonometric functions are well suited for describing such cyclical behavior. In this section, we investigate the derivatives of trigonometric functions and their many uses.

Two Special Limits

Our principal goal is to determine derivative formulas for $\sin x$ and $\cos x$. To do this, we use two special limits.

Table 3.2

x	$\dfrac{\sin x}{x}$
± 0.1	0.9983341665
± 0.01	0.9999833334
± 0.001	0.9999998333

THEOREM 3.10   Trigonometric Limits

$$\lim_{x \to 0} \frac{\sin x}{x} = 1 \qquad \lim_{x \to 0} \frac{\cos x - 1}{x} = 0$$

Note that these limits cannot be evaluated by direct substitution because in both cases, the numerator and denominator approach zero as $x \to 0$. We first examine numerical and graphical evidence supporting Theorem 3.10, and then we offer an analytic proof.

The values of $\dfrac{\sin x}{x}$, rounded to 10 digits, appear in Table 3.2. As x approaches zero from both sides, it appears that $\dfrac{\sin x}{x}$ approaches 1. **Figure 3.35** shows a graph of $y = \dfrac{\sin x}{x}$, with a hole at $x = 0$, where the function is undefined. The graphical evidence also strongly suggests (but does not prove) that $\lim\limits_{x \to 0} \dfrac{\sin x}{x} = 1$. Similar evidence indicates that $\dfrac{\cos x - 1}{x}$ approaches 0 as x approaches 0.

Using a geometric argument and the methods of Chapter 2, we now prove that $\lim\limits_{x \to 0} \dfrac{\sin x}{x} = 1$. The proof that $\lim\limits_{x \to 0} \dfrac{\cos x - 1}{x} = 0$ is found in Exercise 81.

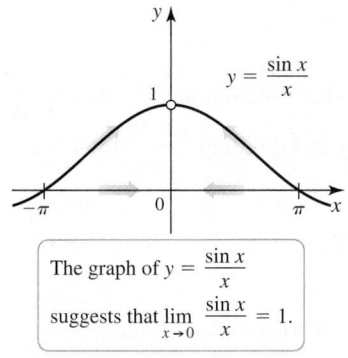

The graph of $y = \dfrac{\sin x}{x}$ suggests that $\lim\limits_{x \to 0} \dfrac{\sin x}{x} = 1$.

Figure 3.35

Proof: Consider **Figure 3.36**, in which $\triangle OAD$, $\triangle OBC$, and the sector OAC of the unit circle (with central angle x) are shown. Observe that with $0 < x < \pi/2$,

$$\text{area of } \triangle OAD < \text{area of sector } OAC < \text{area of } \triangle OBC. \qquad (1)$$

Because the circle in Figure 3.36 is a unit circle, $OA = OC = 1$. It follows that $\sin x = \dfrac{AD}{OA} = AD$, $\cos x = \dfrac{OD}{OA} = OD$, and $\tan x = \dfrac{BC}{OC} = BC$. From these observations, we conclude that

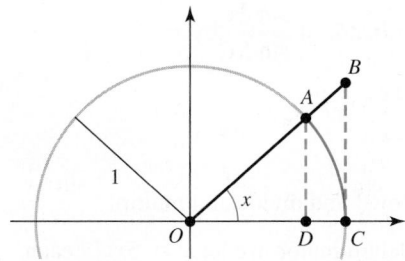

Figure 3.36

• the area of $\triangle OAD = \dfrac{1}{2}(OD)(AD) = \dfrac{1}{2}\cos x \sin x$,

• the area of sector $OAC = \dfrac{1}{2} \cdot 1^2 \cdot x = \dfrac{x}{2}$, and

• the area of $\triangle OBC = \dfrac{1}{2}(OC)(BC) = \dfrac{1}{2}\tan x$.

> ▸ Area of a sector of a circle of radius r formed by a central angle θ:

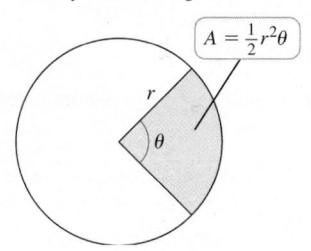

$$A = \tfrac{1}{2}r^2\theta$$

Substituting these results into (1), we have

$$\frac{1}{2}\cos x \sin x < \frac{x}{2} < \frac{1}{2}\tan x.$$

Replacing $\tan x$ with $\dfrac{\sin x}{\cos x}$ and multiplying the inequalities by $\dfrac{2}{\sin x}$ (which is positive) leads to the inequalities

$$\cos x < \frac{x}{\sin x} < \frac{1}{\cos x}.$$

When we take reciprocals and reverse the inequalities, we have

$$\cos x < \frac{\sin x}{x} < \frac{1}{\cos x}, \tag{2}$$

for $0 < x < \pi/2$.

A similar argument shows that the inequalities in (2) also hold for $-\pi/2 < x < 0$. Taking the limit as $x \to 0$ in (2), we find that

> $\displaystyle \lim_{x \to 0} \frac{\sin x}{x} = 1$ implies that if $|x|$ is small, then $\sin x \approx x$.

$$\underbrace{\lim_{x \to 0} \cos x}_{1} \le \lim_{x \to 0} \frac{\sin x}{x} \le \underbrace{\lim_{x \to 0} \frac{1}{\cos x}}_{1}.$$

The Squeeze Theorem (Theorem 2.5) now implies that $\displaystyle \lim_{x \to 0} \frac{\sin x}{x} = 1$.   ◄

EXAMPLE 1   Calculating trigonometric limits Evaluate the following limits.

a. $\displaystyle \lim_{x \to 0} \frac{\sin 4x}{x}$ **b.** $\displaystyle \lim_{x \to 0} \frac{\sin 3x}{\sin 5x}$

SOLUTION

a. To use the fact that $\displaystyle \lim_{x \to 0} \frac{\sin x}{x} = 1$, the argument of the sine function in the numerator must be the same as in the denominator. Multiplying and dividing $\dfrac{\sin 4x}{x}$ by 4, we evaluate the limit as follows:

$$\lim_{x \to 0} \frac{\sin 4x}{x} = \lim_{x \to 0} \frac{4 \sin 4x}{4x} \qquad \text{Multiply and divide by 4.}$$

$$= 4 \lim_{t \to 0} \underbrace{\frac{\sin t}{t}}_{1} \qquad \text{Factor out 4 and let } t = 4x; \, t \to 0 \text{ as } x \to 0.$$

$$= 4(1) = 4. \qquad \text{Theorem 3.10}$$

b. The first step is to divide the numerator and denominator of $\dfrac{\sin 3x}{\sin 5x}$ by x:

$$\frac{\sin 3x}{\sin 5x} = \frac{(\sin 3x)/x}{(\sin 5x)/x}.$$

As in part (a), we now divide and multiply $\dfrac{\sin 3x}{x}$ by 3 and divide and multiply $\dfrac{\sin 5x}{x}$ by 5. In the numerator, we let $t = 3x$, and in the denominator, we let $u = 5x$. In each case, $t \to 0$ and $u \to 0$ as $x \to 0$. Therefore,

$$\lim_{x \to 0} \frac{\sin 3x}{\sin 5x} = \lim_{x \to 0} \frac{\dfrac{3 \sin 3x}{3x}}{\dfrac{5 \sin 5x}{5x}} \qquad \text{Multiply and divide by 3 and 5.}$$

$$= \frac{3}{5} \frac{\displaystyle \lim_{t \to 0} (\sin t)/t}{\displaystyle \lim_{u \to 0} (\sin u)/u} \qquad \text{Let } t = 3x \text{ in numerator and } u = 5x \text{ in denominator.}$$

$$= \frac{3}{5} \cdot \frac{1}{1} = \frac{3}{5}. \qquad \text{Both limits equal 1.}$$

Related Exercises 12–13 ◄

QUICK CHECK 1 Evaluate $\displaystyle \lim_{x \to 0} \frac{\tan 2x}{x}$. ◄

We now use the important limits of Theorem 3.10 to establish the derivatives of $\sin x$ and $\cos x$.

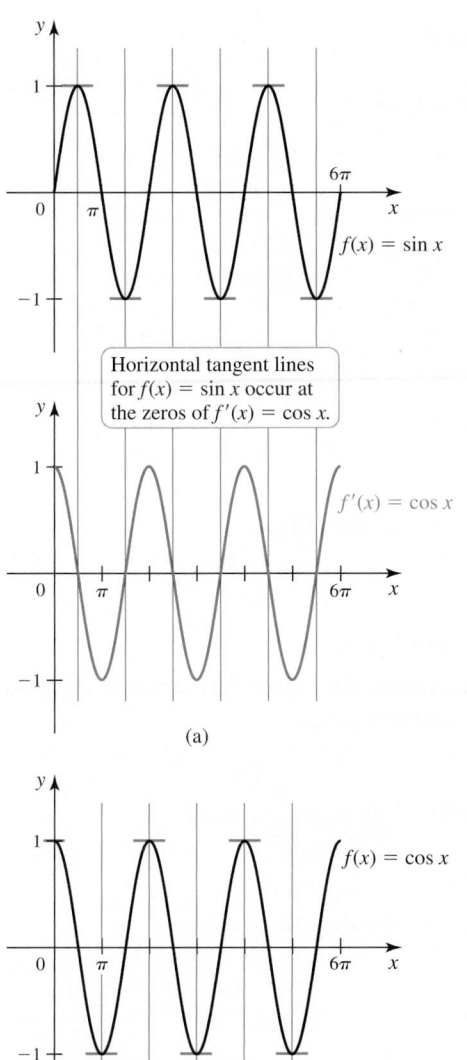

Horizontal tangent lines for $f(x) = \sin x$ occur at the zeros of $f'(x) = \cos x$.

(a)

Horizontal tangent lines for $f(x) = \cos x$ occur at the zeros of $f'(x) = -\sin x$.

(b)

Figure 3.37

QUICK CHECK 2 At what points on the interval $[0, 2\pi]$ does the graph of $f(x) = \sin x$ have tangent lines with positive slopes? At what points on the interval $[0, 2\pi]$ is $\cos x > 0$? Explain the connection. ◄

Derivatives of Sine and Cosine Functions

We start with the definition of the derivative,

$$f'(x) = \lim_{h \to 0} \frac{f(x + h) - f(x)}{h},$$

with $f(x) = \sin x$, and then appeal to the sine addition identity

$$\sin(x + h) = \sin x \cos h + \cos x \sin h.$$

The derivative is

$$
\begin{aligned}
f'(x) &= \lim_{h \to 0} \frac{\sin(x + h) - \sin x}{h} && \text{Definition of derivative} \\
&= \lim_{h \to 0} \frac{\sin x \cos h + \cos x \sin h - \sin x}{h} && \text{Sine addition identity} \\
&= \lim_{h \to 0} \frac{\sin x \,(\cos h - 1) + \cos x \sin h}{h} && \text{Factor } \sin x. \\
&= \lim_{h \to 0} \frac{\sin x \,(\cos h - 1)}{h} + \lim_{h \to 0} \frac{\cos x \sin h}{h} && \text{Theorem 2.3} \\
&= \sin x \left(\underbrace{\lim_{h \to 0} \frac{\cos h - 1}{h}}_{0} \right) + \cos x \left(\underbrace{\lim_{h \to 0} \frac{\sin h}{h}}_{1} \right) && \begin{array}{l}\text{Both } \sin x \text{ and } \cos x \text{ are} \\ \text{independent of } h.\end{array} \\
&= (\sin x)(0) + \cos x \,(1) && \text{Theorem 3.10} \\
&= \cos x. && \text{Simplify.}
\end{aligned}
$$

We have proved the important result that $\dfrac{d}{dx}(\sin x) = \cos x$.

The fact that $\dfrac{d}{dx}(\cos x) = -\sin x$ is proved in a similar way using a cosine addition identity (Exercise 83).

THEOREM 3.11   Derivatives of Sine and Cosine

$$\frac{d}{dx}(\sin x) = \cos x \qquad \frac{d}{dx}(\cos x) = -\sin x$$

From a geometric point of view, these derivative formulas make sense. Because $f(x) = \sin x$ is a periodic function, we expect its derivative to be periodic. Observe that the horizontal tangent lines on the graph of $f(x) = \sin x$ (Figure 3.37a) occur at the zeros of $f'(x) = \cos x$. Similarly, the horizontal tangent lines on the graph of $f(x) = \cos x$ occur at the zeros of $f'(x) = -\sin x$ (Figure 3.37b).

EXAMPLE 2   Derivatives involving trigonometric functions Calculate dy/dx for the following functions.

a. $y = e^x \cos x$ **b.** $y = \sin x - x \cos x$ **c.** $y = \dfrac{1 + \sin x}{1 - \sin x}$

SOLUTION

a. $\dfrac{dy}{dx} = \dfrac{d}{dx}(e^x \cos x) = \overbrace{e^x \cos x}^{\substack{\text{(derivative of } e^x) \cdot \cos x \\ e^x) \cdot \cos x}} + \overbrace{e^x(-\sin x)}^{\substack{e^x \cdot \text{(derivative} \\ \text{of } \cos x)}}$ Product Rule

$= e^x(\cos x - \sin x)$ Simplify.

b. $\dfrac{dy}{dx} = \dfrac{d}{dx}(\sin x) - \dfrac{d}{dx}(x \cos x)$ Difference Rule

$= \cos x - \underbrace{((1) \cos x}_{\substack{(\text{derivative of } x) \\ \cdot \, \cos x}} + \underbrace{x \, (-\sin x))}_{\substack{x \cdot (\text{derivative} \\ \text{of } \cos x)}}$ Product Rule

$= x \sin x$ Simplify.

c. $\dfrac{dy}{dx} = \dfrac{(1 - \sin x)\overbrace{(\cos x)}^{\substack{\text{derivative of} \\ 1 \, + \, \sin x}} - (1 + \sin x)\overbrace{(-\cos x)}^{\substack{\text{derivative of} \\ 1 \, - \, \sin x}}}{(1 - \sin x)^2}$ Quotient Rule

$= \dfrac{\cos x - \cos x \sin x + \cos x + \sin x \cos x}{(1 - \sin x)^2}$ Expand.

$= \dfrac{2 \cos x}{(1 - \sin x)^2}$ Simplify.

Related Exercises 25, 27, 29 ◄

Derivatives of Other Trigonometric Functions

The derivatives of $\tan x$, $\cot x$, $\sec x$, and $\csc x$ are obtained using the derivatives of $\sin x$ and $\cos x$ together with the Quotient Rule and trigonometric identities.

➤ Recall that $\tan x = \dfrac{\sin x}{\cos x}$, $\cot x = \dfrac{\cos x}{\sin x}$, $\sec x = \dfrac{1}{\cos x}$, and $\csc x = \dfrac{1}{\sin x}$.

EXAMPLE 3 **Derivative of the tangent function** Calculate $\dfrac{d}{dx}(\tan x)$.

SOLUTION Using the identity $\tan x = \dfrac{\sin x}{\cos x}$ and the Quotient Rule, we have

$$\dfrac{d}{dx}(\tan x) = \dfrac{d}{dx}\left(\dfrac{\sin x}{\cos x}\right)$$

➤ One way to remember Theorem 3.12 is to learn the derivatives of the sine, tangent, and secant functions. Then replace each function with its corresponding **cofunction** and put a negative sign on the right-hand side of the new derivative formula.

$\dfrac{d}{dx}(\sin x) = \cos x \quad \leftrightarrow$

$\dfrac{d}{dx}(\cos x) = -\sin x$

$\dfrac{d}{dx}(\tan x) = \sec^2 x \quad \leftrightarrow$

$\dfrac{d}{dx}(\cot x) = -\csc^2 x$

$\dfrac{d}{dx}(\sec x) = \sec x \tan x \quad \leftrightarrow$

$\dfrac{d}{dx}(\csc x) = -\csc x \cot x$

QUICK CHECK 3 The formulas for $\dfrac{d}{dx}(\cot x)$, $\dfrac{d}{dx}(\sec x)$, and $\dfrac{d}{dx}(\csc x)$ can be determined using the Quotient Rule. Why? ◄

$= \dfrac{\cos x \overbrace{\cos x}^{\substack{\text{derivative of} \\ \sin x}} - \sin x \overbrace{(-\sin x)}^{\substack{\text{derivative of} \\ \cos x}}}{\cos^2 x}$ Quotient Rule

$= \dfrac{\cos^2 x + \sin^2 x}{\cos^2 x}$ Simplify numerator.

$= \dfrac{1}{\cos^2 x} = \sec^2 x.$ $\cos^2 x + \sin^2 x = 1$

Therefore, $\dfrac{d}{dx}(\tan x) = \sec^2 x.$ *Related Exercises 52–54* ◄

The derivatives of $\cot x$, $\sec x$, and $\csc x$ are given in Theorem 3.12 (Exercises 52–54).

THEOREM 3.12 Derivatives of the Trigonometric Functions

$$\dfrac{d}{dx}(\sin x) = \cos x \qquad \dfrac{d}{dx}(\cos x) = -\sin x$$

$$\dfrac{d}{dx}(\tan x) = \sec^2 x \qquad \dfrac{d}{dx}(\cot x) = -\csc^2 x$$

$$\dfrac{d}{dx}(\sec x) = \sec x \tan x \qquad \dfrac{d}{dx}(\csc x) = -\csc x \cot x$$

EXAMPLE 4   **Derivatives involving sec x and csc x** Find the derivative of $y = \sec x \csc x$.

SOLUTION

$$\frac{dy}{dx} = \frac{d}{dx}(\sec x \cdot \csc x)$$

$$= \underbrace{\sec x \tan x \csc x}_{\text{derivative of } \sec x} + \sec x \underbrace{(-\csc x \cot x)}_{\text{derivative of } \csc x} \qquad \text{Product Rule}$$

$$= \underbrace{\frac{1}{\cos x}}_{\sec x} \cdot \underbrace{\frac{\sin x}{\cos x}}_{\tan x} \cdot \underbrace{\frac{1}{\sin x}}_{\csc x} - \underbrace{\frac{1}{\cos x}}_{\sec x} \cdot \underbrace{\frac{1}{\sin x}}_{\csc x} \cdot \underbrace{\frac{\cos x}{\sin x}}_{\cot x} \qquad \begin{array}{l}\text{Write functions in terms} \\ \text{of } \sin x \text{ and } \cos x.\end{array}$$

$$= \frac{1}{\cos^2 x} - \frac{1}{\sin^2 x} \qquad \text{Cancel and simplify.}$$

$$= \sec^2 x - \csc^2 x \qquad \text{Definition of } \sec x \text{ and } \csc x$$

Related Exercises 43–44 ◄

Higher-Order Trigonometric Derivatives

Higher-order derivatives of the sine and cosine functions are important in many applications, particularly in problems that involve oscillations, vibrations, or waves. A few higher-order derivatives of $y = \sin x$ reveal a pattern.

$$\frac{dy}{dx} = \cos x \qquad\qquad \frac{d^2 y}{dx^2} = \frac{d}{dx}(\cos x) = -\sin x$$

$$\frac{d^3 y}{dx^3} = \frac{d}{dx}(-\sin x) = -\cos x \qquad \frac{d^4 y}{dx^4} = \frac{d}{dx}(-\cos x) = \sin x$$

We see that the higher-order derivatives of $\sin x$ cycle back periodically to $\pm \sin x$. In general, it can be shown that $\dfrac{d^{2n} y}{dx^{2n}} = (-1)^n \sin x$, with a similar result for $\cos x$ (Exercise 88). This cyclic behavior in the derivatives of $\sin x$ and $\cos x$ does not occur with the other trigonometric functions.

QUICK CHECK 4 Find $\dfrac{d^2 y}{dx^2}$ and $\dfrac{d^4 y}{dx^4}$ when $y = \cos x$. Find $\dfrac{d^{40} y}{dx^{40}}$ and $\dfrac{d^{42} y}{dx^{42}}$ when $y = \sin x$. ◄

EXAMPLE 5   **Second-order derivatives** Find the second derivative of $y = \csc x$.

SOLUTION By Theorem 3.12, $\dfrac{dy}{dx} = -\csc x \cot x$.

Applying the Product Rule gives the second derivative:

$$\frac{d^2 y}{dx^2} = \frac{d}{dx}(-\csc x \cot x)$$

$$= \left(\frac{d}{dx}(-\csc x)\right)\cot x - \csc x \frac{d}{dx}(\cot x) \qquad \text{Product Rule}$$

$$= (\csc x \cot x)\cot x - \csc x(-\csc^2 x) \qquad \text{Calculate derivatives.}$$

$$= \csc x (\cot^2 x + \csc^2 x). \qquad \text{Factor.}$$

Related Exercises 61–62 ◄

SECTION 3.5 EXERCISES

Getting Started

1. Why is it not possible to evaluate $\lim\limits_{x \to 0} \dfrac{\sin x}{x}$ by direct substitution?

2. How is $\lim\limits_{x \to 0} \dfrac{\sin x}{x}$ used in this section?

3. Explain why the Quotient Rule is used to determine the derivative of $\tan x$ and $\cot x$.

4. How can you use the derivatives $\dfrac{d}{dx}(\sin x) = \cos x$, $\dfrac{d}{dx}(\tan x) = \sec^2 x$, and $\dfrac{d}{dx}(\sec x) = \sec x \tan x$ to remember the derivatives of $\cos x$, $\cot x$, and $\csc x$?

5. Let $f(x) = \sin x$. What is the value of $f'(\pi)$?

6. Find the value of $\dfrac{d}{dx}(\tan x)\Big|_{x=\frac{\pi}{3}}$.

7. Find an equation of the line tangent to the curve $y = \sin x$ at $x = 0$.

8. Where does $\sin x$ have a horizontal tangent line? Where does $\cos x$ have a value of zero? Explain the connection between these two observations.

9. Find $\dfrac{d^2}{dx^2}(\sin x + \cos x)$.   **10.** Find $\dfrac{d^2}{dx^2}(\sec x)$.

Practice Exercises

11–22. Trigonometric limits *Use Theorem 3.10 to evaluate the following limits.*

11. $\lim\limits_{x \to 0} \dfrac{\sin 3x}{x}$

12. $\lim\limits_{x \to 0} \dfrac{\sin 5x}{3x}$

13. $\lim\limits_{x \to 0} \dfrac{\sin 7x}{\sin 3x}$

14. $\lim\limits_{x \to 0} \dfrac{\sin 3x}{\tan 4x}$

15. $\lim\limits_{x \to 0} \dfrac{\tan 5x}{x}$

16. $\lim\limits_{\theta \to 0} \dfrac{\cos^2 \theta - 1}{\theta}$

17. $\lim\limits_{x \to 0} \dfrac{\tan 7x}{\sin x}$

18. $\lim\limits_{\theta \to 0} \dfrac{\sec \theta - 1}{\theta}$

19. $\lim\limits_{x \to 2} \dfrac{\sin(x - 2)}{x^2 - 4}$

20. $\lim\limits_{x \to -3} \dfrac{\sin(x + 3)}{x^2 + 8x + 15}$

21. $\lim\limits_{x \to 0} \dfrac{\sin ax}{\sin bx}$, where a and b are constants with $b \neq 0$

22. $\lim\limits_{x \to 0} \dfrac{\sin ax}{bx}$, where a and b are constants with $b \neq 0$

23–51. Calculating derivatives *Find the derivative of the following functions.*

23. $y = \sin x + \cos x$

24. $y = 5x^2 + \cos x$

25. $y = e^{-x} \sin x$

26. $y = \sin x + 4e^x$

27. $y = x \sin x$

28. $y = e^x(\cos x + \sin x)$

29. $y = \dfrac{\cos x}{\sin x + 1}$

30. $y = \dfrac{1 - \sin x}{1 + \sin x}$

31. $y = \sin x \cos x$

32. $y = \dfrac{a \sin x + b \cos x}{a \sin x - b \cos x}$; a and b are nonzero constants

33. $y = \cos^2 x$

34. $y = \dfrac{x \sin x}{1 + \cos x}$

35. $y = w^2 \sin w + 2w \cos w - 2 \sin w$

36. $y = -x^3 \cos x + 3x^2 \sin x + 6x \cos x - 6 \sin x$

37. $y = x \cos x \sin x$

38. $y = \dfrac{1}{2 + \sin x}$

39. $y = \dfrac{\sin x}{1 + \cos x}$

40. $y = \dfrac{\sin x + \cos x}{e^x}$

41. $y = \dfrac{1 - \cos x}{1 + \cos x}$

42. $y = \tan x + \cot x$

43. $y = \sec x + \csc x$

44. $y = \sec x \tan x$

45. $y = e^x \csc x$

46. $y = \dfrac{\tan w}{1 + \tan w}$

47. $y = \dfrac{\cot x}{1 + \csc x}$

48. $y = \dfrac{\tan t}{1 + \sec t}$

49. $y = \dfrac{1}{\sec z \csc z}$

50. $y = \csc^2 \theta - 1$

51. $y = x - \cos x \sin x$

52–54. Verifying derivative formulas *Verify the following derivative formulas using the Quotient Rule.*

52. $\dfrac{d}{dx}(\cot x) = -\csc^2 x$

53. $\dfrac{d}{dx}(\sec x) = \sec x \tan x$

54. $\dfrac{d}{dx}(\csc x) = -\csc x \cot x$

T 55. Velocity of an oscillator An object oscillates along a vertical line, and its position in centimeters is given by $y(t) = 30(\sin t - 1)$, where $t \geq 0$ is measured in seconds and y is positive in the upward direction.

  a. Graph the position function, for $0 \leq t \leq 10$.
  b. Find the velocity of the oscillator, $v(t) = y'(t)$.
  c. Graph the velocity function, for $0 \leq t \leq 10$.
  d. At what times and positions is the velocity zero?
  e. At what times and positions is the velocity a maximum?
  f. The acceleration of the oscillator is $a(t) = v'(t)$. Find and graph the acceleration function.

T 56. Damped sine wave The graph of $f(t) = e^{-t} \sin t$ is an example of a *damped* sine wave; it is used in a variety of applications, such as modeling the vibrations of a shock absorber.

  a. Use a graphing utility to graph f and explain why this curve is called a damped sine wave.
  b. Compute $f'(t)$ and use it to determine where the graph of f has a horizontal tangent.
  c. Evaluate $\lim\limits_{t \to \infty} e^{-t} \sin t$ by using the Squeeze Theorem. What does the result say about the oscillations of this damped sine wave?

57–64. Second derivatives *Find y'' for the following functions.*

57. $y = x \sin x$

58. $y = x^2 \cos x$

59. $y = e^x \sin x$

60. $y = \dfrac{1}{2} e^x \cos x$

61. $y = \cot x$

62. $y = \tan x$

63. $y = \sec x \csc x$

64. $y = \cos \theta \sin \theta$

65. Explain why or why not Determine whether the following statements are true and give an explanation or counterexample.

  a. $\dfrac{d}{dx}(\sin^2 x) = \cos^2 x$.   **b.** $\dfrac{d^2}{dx^2}(\sin x) = \sin x$.

  c. $\dfrac{d^4}{dx^4}(\cos x) = \cos x$.

  d. The function $\sec x$ is not differentiable at $x = \pi/2$.

66–71. Trigonometric limits *Evaluate the following limits or state that they do not exist. (Hint: Identify each limit as the derivative of a function at a point.)*

66. $\lim\limits_{x \to \pi/2} \dfrac{\cos x}{x - \pi/2}$

67. $\lim\limits_{x \to \pi/4} \dfrac{\tan x - 1}{x - \pi/4}$

68. $\displaystyle\lim_{h\to 0}\frac{\sin\left(\dfrac{\pi}{6}+h\right)-\dfrac{1}{2}}{h}$

69. $\displaystyle\lim_{h\to 0}\frac{\cos\left(\dfrac{\pi}{6}+h\right)-\dfrac{\sqrt{3}}{2}}{h}$

70. $\displaystyle\lim_{x\to\pi/4}\frac{\cot x-1}{x-\dfrac{\pi}{4}}$

71. $\displaystyle\lim_{h\to 0}\frac{\tan\left(\dfrac{5\pi}{6}+h\right)+\dfrac{1}{\sqrt{3}}}{h}$

T 72–75. Equations of tangent lines

a. *Find an equation of the line tangent to the following curves at the given value of x.*

b. *Use a graphing utility to plot the curve and the tangent line.*

72. $y=4\sin x\cos x;\ x=\dfrac{\pi}{3}$ **73.** $y=1+2\sin x;\ x=\dfrac{\pi}{6}$

74. $y=\csc x;\ x=\dfrac{\pi}{4}$ **75.** $y=\dfrac{\cos x}{1-\cos x};\ x=\dfrac{\pi}{3}$

76. Locations of tangent lines

  a. For what values of x does $g(x)=x-\sin x$ have a horizontal tangent line?

  b. For what values of x does $g(x)=x-\sin x$ have a slope of 1?

77. Locations of horizontal tangent lines For what values of x does $f(x)=x-2\cos x$ have a horizontal tangent line?

78. Matching Match the graphs of the functions in a–d with the graphs of their derivatives in A–D.

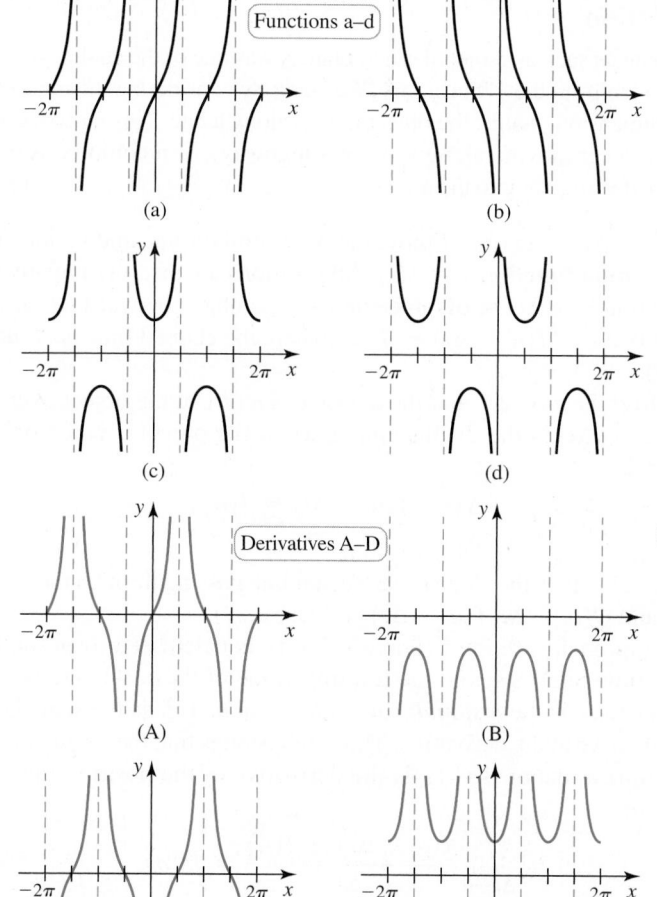

Functions a–d

(a) (b) (c) (d)

Derivatives A–D

(A) (B) (C) (D)

79. A differential equation A differential equation is an equation involving an unknown function and its derivatives. Consider the differential equation $y''(t)+y(t)=0$.

  a. Show that $y=A\sin t$ satisfies the equation for any constant A.

  b. Show that $y=B\cos t$ satisfies the equation for any constant B.

  c. Show that $y=A\sin t+B\cos t$ satisfies the equation for any constants A and B.

Explorations and Challenges

80. Using identities Use the identity $\sin 2x=2\sin x\cos x$ to find $\dfrac{d}{dx}(\sin 2x)$. Then use the identity $\cos 2x=\cos^2 x-\sin^2 x$ to express the derivative of $\sin 2x$ in terms of $\cos 2x$.

81. Proof of $\displaystyle\lim_{x\to 0}\frac{\cos x-1}{x}=0$ Use the trigonometric identity $\cos^2 x+\sin^2 x=1$ to prove that $\displaystyle\lim_{x\to 0}\frac{\cos x-1}{x}=0$. (*Hint:* Begin by multiplying the numerator and denominator by $\cos x+1$.)

82. Another method for proving $\displaystyle\lim_{x\to 0}\frac{\cos x-1}{x}=0$ Use the half-angle formula $\sin^2 x=\dfrac{1-\cos 2x}{2}$ to prove that $\displaystyle\lim_{x\to 0}\frac{\cos x-1}{x}=0$.

83. Proof of $\dfrac{d}{dx}(\cos x)=-\sin x$ Use the definition of the derivative and the trigonometric identity

$$\cos(x+h)=\cos x\cos h-\sin x\sin h$$

to prove that $\dfrac{d}{dx}(\cos x)=-\sin x$.

84. Continuity of a piecewise function Let

$$f(x)=\begin{cases}\dfrac{3\sin x}{x} & \text{if } x\neq 0\\ a & \text{if } x=0.\end{cases}$$

For what values of a is f continuous?

85. Continuity of a piecewise function Let

$$g(x)=\begin{cases}\dfrac{1-\cos x}{2x} & \text{if } x\neq 0\\ a & \text{if } x=0.\end{cases}$$

For what values of a is g continuous?

T 86. Computing limits with angles in degrees Suppose your graphing calculator has two functions, one called $\sin x$, which calculates the sine of x when x is in radians, and the other called $s(x)$, which calculates the sine of x when x is in degrees.

  a. Explain why $s(x)=\sin\left(\dfrac{\pi}{180}x\right)$.

  b. Evaluate $\displaystyle\lim_{x\to 0}\frac{s(x)}{x}$. Verify your answer by estimating the limit on your calculator.

87. Derivatives of $\sin^n x$ Calculate the following derivatives using the Product Rule.

  a. $\dfrac{d}{dx}(\sin^2 x)$ **b.** $\dfrac{d}{dx}(\sin^3 x)$ **c.** $\dfrac{d}{dx}(\sin^4 x)$

d. Based on your answers to parts (a)–(c), make a conjecture about $\dfrac{d}{dx}(\sin^n x)$, where n is a positive integer. Then prove the result by induction.

88. Prove that

$$\frac{d^{2n}}{dx^{2n}}(\sin x) = (-1)^n \sin x \quad \text{and} \quad \frac{d^{2n}}{dx^{2n}}(\cos x) = (-1)^n \cos x.$$

▣ 89–90. Difference quotients *Suppose f is differentiable for all x and consider the function*

$$D(x) = \frac{f(x + 0.01) - f(x)}{0.01}.$$

For the following functions, graph D on the given interval, and explain why the graph appears as it does. What is the relationship between the functions f and D?

89. $f(x) = \sin x$ on $[-\pi, \pi]$ **90.** $f(x) = \dfrac{x^3}{3} + 1$ on $[-2, 2]$

3.6 Derivatives as Rates of Change

The theme of this section is the *derivative as a rate of change*. Observing the world around us, we see that almost everything is in a state of change: The size of the Internet is increasing; your blood pressure fluctuates; as supply increases, prices decrease; and the universe is expanding. This section explores a few of the many applications of this idea and demonstrates why calculus is called the mathematics of change.

One-Dimensional Motion

Describing the motion of objects such as projectiles and planets was one of the challenges that led to the development of calculus in the 17th century. We begin by considering the motion of an object confined to one dimension; that is, the object moves along a line. This motion could be horizontal (for example, a car moving along a straight highway), or it could be vertical (such as a projectile launched vertically into the air).

> When describing the motion of objects, it is customary to use t as the independent variable to represent time. Generally, motion is assumed to begin at $t = 0$.

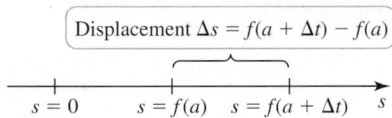

Figure 3.38

Position and Velocity Suppose an object moves along a straight line and its location at time t is given by the **position function** $s = f(t)$. All positions are measured relative to a reference point $s = 0$. The **displacement** of the object between $t = a$ and $t = a + \Delta t$ is $\Delta s = f(a + \Delta t) - f(a)$, where the elapsed time is Δt units (**Figure 3.38**).

Recall from Section 2.1 that the *average velocity* of the object over the interval $[a, a + \Delta t]$ is the displacement Δs of the object divided by the elapsed time Δt:

$$v_{av} = \frac{\Delta s}{\Delta t} = \frac{f(a + \Delta t) - f(a)}{\Delta t}.$$

The average velocity is the slope of the secant line passing through the points $P(a, f(a))$ and $Q(a + \Delta t, f(a + \Delta t))$ (**Figure 3.39**).

As Δt approaches 0, the average velocity is calculated over smaller and smaller time intervals, and the limiting value of these average velocities, when it exists, is the *instantaneous velocity* at a. This is the same argument used to arrive at the derivative. The conclusion is that the instantaneous velocity at time a, denoted $v(a)$, is the derivative of the position function evaluated at a:

$$v(a) = \lim_{\Delta t \to 0} \frac{f(a + \Delta t) - f(a)}{\Delta t} = f'(a).$$

Equivalently, the instantaneous velocity at a is the rate of change in the position function at a; it also equals the slope of the line tangent to the curve $s = f(t)$ at $P(a, f(a))$.

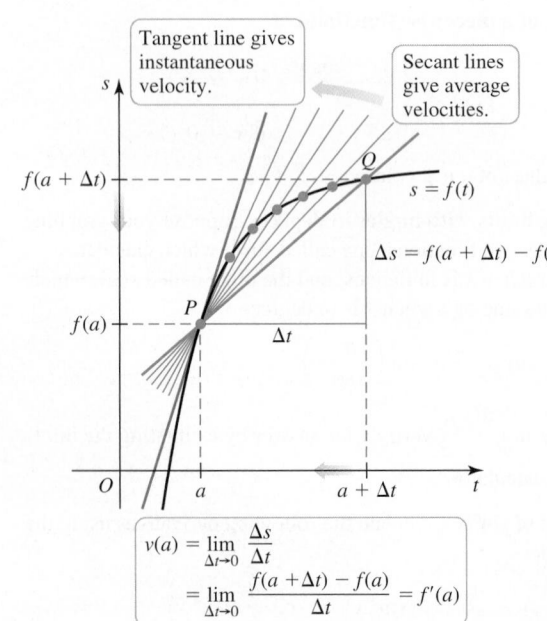

Figure 3.39

▶ Using the various derivative notations, the velocity is also written $v(t) = s'(t) = ds/dt$. If *average* or *instantaneous* is not specified, *velocity* is understood to mean instantaneous velocity.

QUICK CHECK 1 Does the speedometer in your car measure average or instantaneous velocity? ◄

DEFINITION Average and Instantaneous Velocity

Let $s = f(t)$ be the position function of an object moving along a line. The **average velocity** of the object over the time interval $[a, a + \Delta t]$ is the slope of the secant line between $(a, f(a))$ and $(a + \Delta t, f(a + \Delta t))$:

$$v_{av} = \frac{f(a + \Delta t) - f(a)}{\Delta t}.$$

The **instantaneous velocity** at a is the slope of the line tangent to the position curve, which is the derivative of the position function:

$$v(a) = \lim_{\Delta t \to 0} \frac{f(a + \Delta t) - f(a)}{\Delta t} = f'(a).$$

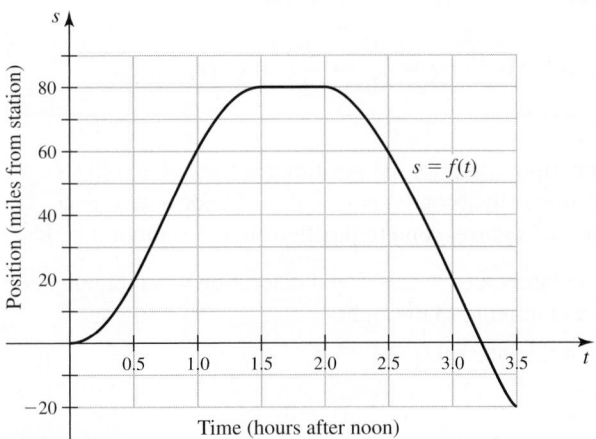

Figure 3.40

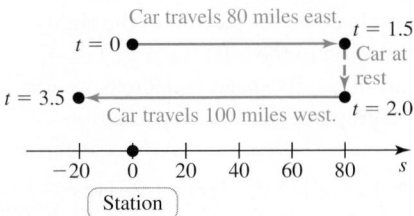

Figure 3.41

EXAMPLE 1   Position and velocity of a patrol car Assume a police station is located along a straight east-west freeway. At noon ($t = 0$), a patrol car leaves the station heading east. The position function of the car $s = f(t)$ gives the location of the car in miles east ($s > 0$) or west ($s < 0$) of the station t hours after noon (**Figure 3.40**).

a. Describe the location of the patrol car during the first 3.5 hr of the trip.

b. Calculate the displacement and average velocity of the car between 2:00 P.M. and 3:30 P.M. ($2 \le t \le 3.5$).

c. At what time(s) is the instantaneous velocity greatest *as the car travels east*?

SOLUTION

a. The graph of the position function indicates the car travels 80 miles east between $t = 0$ (noon) and $t = 1.5$ (1:30 P.M.). The car is at rest and its position does not change from $t = 1.5$ to $t = 2$ (that is, from 1:30 P.M. to 2:00 P.M.). Starting at $t = 2$, the car's distance from the station decreases, which means the car travels west, eventually ending up 20 miles west of the station at $t = 3.5$ (3:30 P.M.) (**Figure 3.41**).

b. The position of the car at 3:30 P.M. is $f(3.5) = -20$ (Figure 3.40; the negative sign indicates the car is 20 miles *west* of the station), and the position of the car at 2:00 P.M. is $f(2) = 80$. Therefore, the displacement is

$$\Delta s = f(3.5) - f(2) = -20 \text{ mi} - 80 \text{ mi} = -100 \text{ mi}$$

during an elapsed time of $\Delta t = 3.5 - 2 = 1.5$ hr (the *negative* displacement indicates that the car moved 100 miles *west*). The average velocity is

$$v_{av} = \frac{\Delta s}{\Delta t} = \frac{-100 \text{ mi}}{1.5 \text{ hr}} \approx -66.7 \text{ mi/hr}.$$

c. The greatest eastward instantaneous velocity corresponds to points at which the graph of the position function has the greatest positive slope. The greatest slope occurs between $t = 0.5$ and $t = 1$. During this time interval, the car also has a nearly constant velocity because the curve is approximately linear. We conclude that the eastward velocity is largest from 12:30 P.M. to 1:00 P.M.

Related Exercises 11–12 ◄

Speed and Acceleration   When only the magnitude of the velocity is of interest, we use *speed*, which is the absolute value of the velocity:

$$\text{speed} = |v|.$$

For example, a car with an instantaneous velocity of -30 mi/hr has a speed of 30 mi/hr.

▶ Newton's First Law of Motion says that in the absence of external forces, a moving object has no acceleration, which means the magnitude and direction of the velocity are constant.

A more complete description of an object moving along a line includes its *acceleration*, which is the rate of change of the velocity; that is, acceleration is the derivative of the velocity function with respect to time t. If the acceleration is positive, the object's velocity increases; if it is negative, the object's velocity decreases. Because velocity is the derivative of the position function, acceleration is the second derivative of the position. Therefore,

$$a = \frac{dv}{dt} = \frac{d^2s}{dt^2}.$$

DEFINITION Velocity, Speed, and Acceleration

Suppose an object moves along a line with position $s = f(t)$. Then

the **velocity** at time t is $v = \dfrac{ds}{dt} = f'(t)$,

the **speed** at time t is $|v| = |f'(t)|$, and

the **acceleration** at time t is $a = \dfrac{dv}{dt} = \dfrac{d^2s}{dt^2} = f''(t)$.

QUICK CHECK 2 For an object moving along a line, is it possible for its velocity to increase while its speed decreases? Is it possible for its velocity to decrease while its speed increases? Give an example to support your answers. ◀

▶ The units of derivatives are consistent with the notation. If s is measured in meters and t is measured in seconds, the units of the velocity $\dfrac{ds}{dt}$ are m/s. The units of the acceleration $\dfrac{d^2s}{dt^2}$ are m/s².

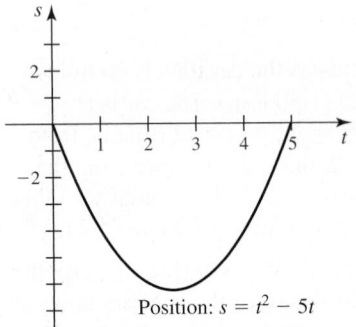

Position: $s = t^2 - 5t$

Figure 3.42

▶ Figure 3.42 gives the graph of the position function, not the path of the object. The motion is along a horizontal line.

EXAMPLE 2 Velocity and acceleration Suppose the position (in feet) of an object moving horizontally along a line at time t (in seconds) is $s = t^2 - 5t$, for $0 \le t \le 5$ (**Figure 3.42**). Assume positive values of s correspond to positions to the right of $s = 0$.

a. Graph the velocity function on the interval $0 \le t \le 5$ and determine when the object is stationary, moving to the left, and moving to the right.

b. Graph the acceleration function on the interval $0 \le t \le 5$.

c. Describe the motion of the object.

SOLUTION

a. The velocity is $v = s'(t) = 2t - 5$. The object is stationary when $v = 2t - 5 = 0$, or at $t = 2.5$. Solving $v = 2t - 5 > 0$, the velocity is positive (motion to the right) for $\frac{5}{2} < t \le 5$. Similarly, the velocity is negative (motion to the left) for $0 \le t < \frac{5}{2}$. Though the velocity of the object is increasing at all times, its speed is decreasing for $0 \le t < \frac{5}{2}$, and then increasing for $\frac{5}{2} < t \le 5$. The graph of the velocity function (**Figure 3.43**) confirms these observations.

b. The acceleration is the derivative of the velocity or $a = v'(t) = s''(t) = 2$. This means that the acceleration is 2 ft/s², for $0 \le t \le 5$ (**Figure 3.44**).

c. Starting at an initial position of $s(0) = 0$, the object moves in the negative direction (to the left) with decreasing speed until it comes to rest momentarily at $s\left(\frac{5}{2}\right) = -\frac{25}{4}$. The object then moves in the positive direction (to the right) with increasing speed, reaching its initial position at $t = 5$. During this time interval, the acceleration is constant.

QUICK CHECK 3 Describe the velocity of an object that has a positive constant acceleration. Could an object have a positive acceleration and a decreasing speed? ◀

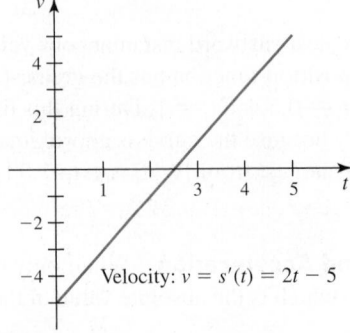

Velocity: $v = s'(t) = 2t - 5$

Figure 3.43

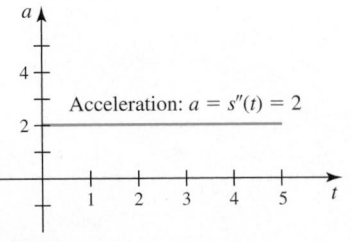

Acceleration: $a = s''(t) = 2$

Figure 3.44

Related Exercises 15–16 ◀

> ➤ The acceleration due to Earth's gravitational field is denoted g. In metric units, $g \approx 9.8 \text{ m/s}^2$ on the surface of Earth; in the U.S. Customary System (USCS), $g \approx 32 \text{ ft/s}^2$.

Free Fall   We now consider a problem in which an object moves vertically in Earth's gravitational field, assuming no other forces (such as air resistance) are at work.

EXAMPLE 3   Motion in a gravitational field   Suppose a stone is thrown vertically upward with an initial velocity of 64 ft/s from a bridge 96 ft above a river. By Newton's laws of motion, the position of the stone (measured as the height above the river) after t seconds is

$$s(t) = -16t^2 + 64t + 96,$$

where $s = 0$ is the level of the river (**Figure 3.45a**).

> ➤ The position function in Example 3 is derived in Section 6.1. Once again we mention that the graph of the position function is not the path of the stone.

a. Find the velocity and acceleration functions.

b. What is the highest point above the river reached by the stone?

c. With what velocity will the stone strike the river?

SOLUTION

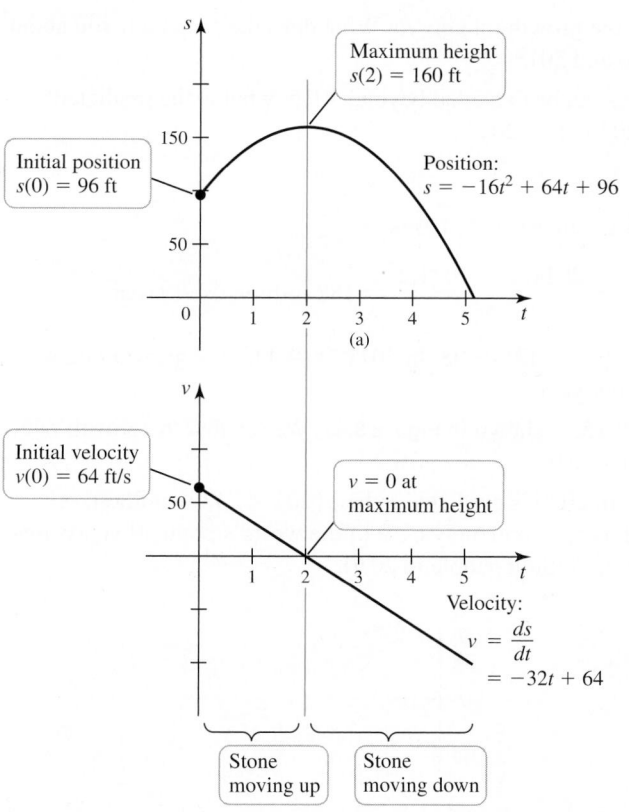

Figure 3.45

a. The velocity of the stone is the derivative of the position function, and its acceleration is the derivative of the velocity function. Therefore,

$$v = \frac{ds}{dt} = -32t + 64 \quad \text{and} \quad a = \frac{dv}{dt} = -32.$$

b. When the stone reaches its high point, its velocity is zero (**Figure 3.45b**). Solving $v(t) = -32t + 64 = 0$ yields $t = 2$; therefore, the stone reaches its maximum height 2 seconds after it is thrown. Its height (in feet) at that instant is

$$s(2) = -16(2)^2 + 64(2) + 96 = 160.$$

c. To determine the velocity at which the stone strikes the river, we first determine *when* it reaches the river. The stone strikes the river when $s(t) = -16t^2 + 64t + 96 = 0$. Dividing both sides of the equation by -16, we obtain $t^2 - 4t - 6 = 0$. Using the quadratic formula, the solutions are $t \approx 5.16$ and $t \approx -1.16$. Because the stone is thrown at $t = 0$, only positive values of t are of interest; therefore, the relevant root is $t \approx 5.16$. The velocity of the stone (in ft/s) when it strikes the river is approximately

$$v(5.16) = -32(5.16) + 64 = -101.12.$$

Related Exercises 24–25 ◄

QUICK CHECK 4 In Example 3, does the rock have a greater speed at $t = 1$ or $t = 3$? ◄

Growth Models

Much of the change in the world around us can be classified as *growth*: Populations, prices, and computer networks all tend to increase in size. Modeling growth is important because it often leads to an understanding of underlying processes and allows for predictions.

We let $p = f(t)$ be the measure of a quantity of interest (for example, the population of a species or the consumer price index), where $t \geq 0$ represents time. The average growth rate of p between time $t = a$ and a later time $t = a + \Delta t$ is the change Δp divided by elapsed time Δt. Therefore, the **average growth rate** of p on the interval $[a, a + \Delta t]$ is

$$\frac{\Delta p}{\Delta t} = \frac{f(a + \Delta t) - f(a)}{\Delta t}.$$

If we now let $\Delta t \to 0$, then $\dfrac{\Delta p}{\Delta t}$ approaches the derivative $\dfrac{dp}{dt}$, which is the **instantaneous growth rate** (or simply **growth rate**) of p with respect to time:

$$\frac{dp}{dt} = \lim_{\Delta t \to 0} \frac{\Delta p}{\Delta t}.$$

Once again, we see the derivative appearing as an instantaneous rate of change. In the next example, a growth function and its derivative are approximated using real data.

EXAMPLE 4   Internet growth The number of worldwide Internet users between 2000 and 2015 is shown in Figure 3.46. A reasonable fit to the data is given by the function $p(t) = 6t^2 + 98t + 431.2$, where t is measured in years after 2000.

a. Use the function p to approximate the average growth rate of Internet users from 2005 $(t = 5)$ to 2010 $(t = 10)$.

b. What was the instantaneous growth rate of Internet users in 2011?

c. Use a graphing utility to plot the growth rate dp/dt. What does the graph tell you about the growth rate between 2000 and 2015?

d. Assuming the growth function can be extended beyond 2015, what is the predicted number of Internet users in 2020 $(t = 20)$?

SOLUTION

a. The average growth rate over the interval $[5, 10]$ is

$$\frac{\Delta P}{\Delta t} = \frac{p(10) - p(5)}{10 - 5} \approx \frac{2011.2 - 1071.2}{5} = 188 \text{ million users/year.}$$

b. The growth rate at time t is $p'(t) = 12t + 98$. In 2011 $(t = 11)$, the growth rate was $p'(11) = 230$ million users per year.

QUICK CHECK 5 Using the growth function in Example 4, compare the growth rates in 2001 and 2012. ◄

c. The graph of p', for $0 \le t \le 15$, is shown in Figure 3.47. We see that the growth rate is positive and increasing, for $t \ge 0$.

d. The number of Internet users in 2020 is predicted to be $p(20) \approx 4791$ million, or about 4.8 billion users, which is approximately 62% of the world's population, assuming a projected population of 7.7 billion people in 2020.

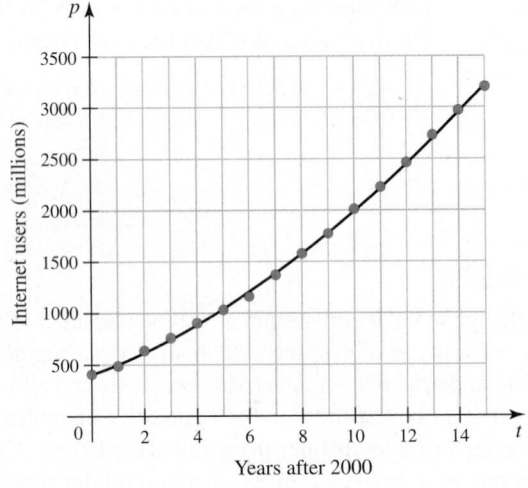

Figure 3.46

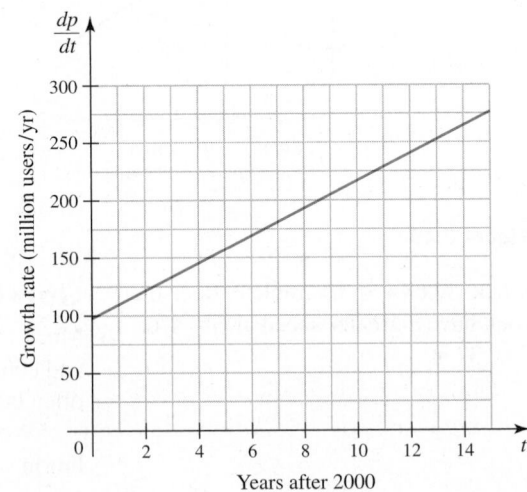

Figure 3.47

Related Exercise 28 ◄

Economics and Business

Our final examples illustrate how derivatives arise in economics and business. As you will see, the mathematics of derivatives is the same as it is in other applications. However, the vocabulary and interpretation are quite different.

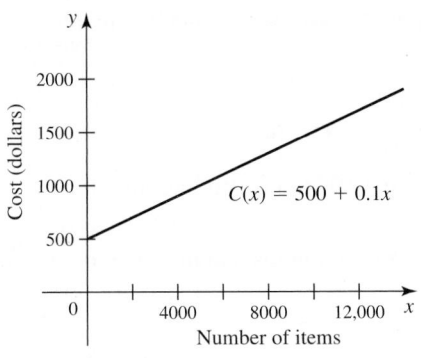

Figure 3.48

➤ Although x is a whole number of units, we treat it as a continuous variable, which is reasonable if x is large.

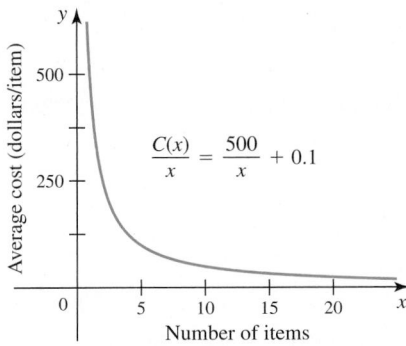

Figure 3.49

➤ The average describes the past; the marginal describes the future.
—Old saying

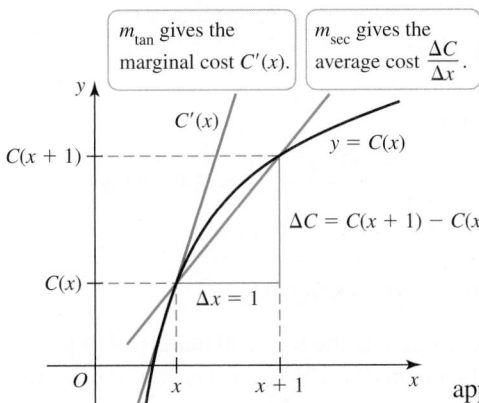

Figure 3.50

➤ The approximation $\Delta C/\Delta x \approx C'(x)$ says that the slope of the secant line between $(x, C(x))$ and $(x + 1, C(x + 1))$ is approximately equal to the slope of the tangent line at $(x, C(x))$. This approximation is good if the cost curve is nearly linear over a 1-unit interval.

Average and Marginal Cost Imagine a company that manufactures large quantities of a product such as mousetraps, DVD players, or snowboards. Associated with the manufacturing process is a *cost function* $C(x)$ that gives the cost of manufacturing the first x items of the product. A simple cost function might have the form $y = C(x) = 500 + 0.1x$, as shown in **Figure 3.48**. It includes a **fixed cost** of $500 (setup costs and overhead) that is independent of the number of items produced. It also includes a **unit cost**, or **variable cost**, of $0.10 per item produced. For example, the cost of producing the first 1000 items is $C(1000) = \$600$.

If the company produces x items at a cost of $C(x)$, then the *average cost* is $\dfrac{C(x)}{x}$ per item. For the cost function $C(x) = 500 + 0.1x$, the average cost is

$$\frac{C(x)}{x} = \frac{500 + 0.1x}{x} = \frac{500}{x} + 0.1.$$

For example, the average cost of manufacturing the first 1000 items is

$$\frac{C(1000)}{1000} = \frac{\$600}{1000} = \$0.60/\text{unit}.$$

Plotting $C(x)/x$, we see that the average cost decreases as the number of items produced increases (**Figure 3.49**).

The average cost gives the cost of items already produced. But what about the cost of producing additional items? Once x items have been produced, the cost of producing another Δx items is $C(x + \Delta x) - C(x)$. Therefore, the average cost per item of producing those Δx additional items is

$$\frac{C(x + \Delta x) - C(x)}{\Delta x} = \frac{\Delta C}{\Delta x}.$$

If we let $\Delta x \to 0$, we see that

$$\lim_{\Delta x \to 0} \frac{\Delta C}{\Delta x} = C'(x),$$

which is called the *marginal cost*. In reality, we cannot let $\Delta x \to 0$ because Δx represents whole numbers of items. However, there is a useful interpretation of the marginal cost. Suppose $\Delta x = 1$. Then $\Delta C = C(x + 1) - C(x)$ is the cost to produce *one* additional item. In this case, we write

$$\frac{\Delta C}{\Delta x} = \frac{C(x + 1) - C(x)}{1}.$$

If the *slope* of the cost curve does not vary significantly near the point x, then—as shown in **Figure 3.50**—we have

$$\frac{\Delta C}{\Delta x} \approx \lim_{\Delta x \to 0} \frac{\Delta C}{\Delta x} = C'(x).$$

Therefore, the cost of producing one additional item, having already produced x items, is approximated by the marginal cost $C'(x)$. In the preceding example, we have $C'(x) = 0.1$; so if $x = 1000$ items have been produced, then the cost of producing the 1001st item is approximately $C'(1000) = \$0.10$. With this simple linear cost function, the marginal cost tells us what we already know: The cost of producing one additional item is the variable cost of $0.10. With more realistic cost functions, the marginal cost may be variable.

DEFINITION Average and Marginal Cost

The **cost function** $C(x)$ gives the cost to produce the first x items in a manufacturing process. The **average cost** to produce x items is $\overline{C}(x) = C(x)/x$. The **marginal cost** $C'(x)$ is the approximate cost to produce one additional item after producing x items.

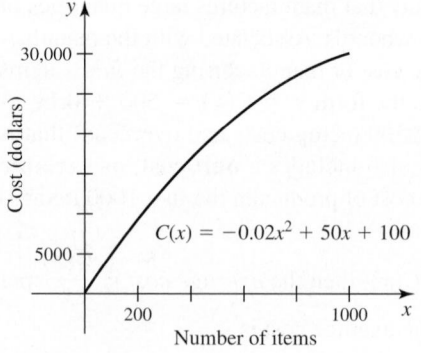

Cost (dollars)

$C(x) = -0.02x^2 + 50x + 100$

Number of items

Figure 3.51

EXAMPLE 5   Average and marginal costs Suppose the cost of producing x items is given by the function (Figure 3.51)

$$C(x) = -0.02x^2 + 50x + 100, \quad \text{for} \quad 0 \le x \le 1000.$$

a. Determine the average and marginal cost functions.

b. Determine the average and marginal cost for the first 100 items and interpret these values.

c. Determine the average and marginal cost for the first 900 items and interpret these values.

SOLUTION

a. The average cost is

$$\overline{C}(x) = \frac{C(x)}{x} = \frac{-0.02x^2 + 50x + 100}{x} = -0.02x + 50 + \frac{100}{x}$$

and the marginal cost is

$$\frac{dC}{dx} = -0.04x + 50.$$

The average cost decreases as the number of items produced increases (Figure 3.52a). The marginal cost decreases linearly with a slope of -0.04 (Figure 3.52b).

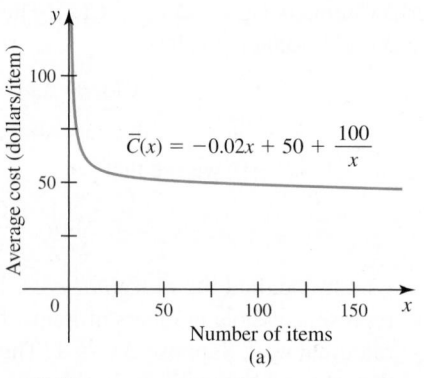

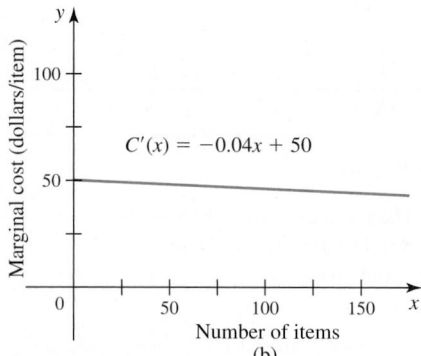

Figure 3.52

b. To produce $x = 100$ items, the average cost is

$$\overline{C}(100) = \frac{C(100)}{100} = \frac{-0.02(100)^2 + 50(100) + 100}{100} = \$49/\text{item}$$

and the marginal cost is

$$C'(100) = -0.04(100) + 50 = \$46/\text{item}.$$

These results mean that the average cost of producing the first 100 items is \$49 per item, but the cost of producing one additional item (the 101st item) is approximately \$46. Therefore, producing one more item is less expensive than the average cost of producing the first 100 items.

c. To produce $x = 900$ items, the average cost is

$$\overline{C}(900) = \frac{C(900)}{900} = \frac{-0.02(900)^2 + 50(900) + 100}{900} \approx \$32/\text{item}$$

and the marginal cost is

$$C'(900) = -0.04(900) + 50 = \$14/\text{item}.$$

QUICK CHECK 6 In Example 5, what happens to the average cost as the number of items produced increases from $x = 1$ to $x = 100$? ◄

The comparison with part (b) is revealing. The average cost of producing 900 items has dropped to \$32 per item. More striking is that the marginal cost (the cost of producing the 901st item) has dropped to \$14.

Related Exercises 29–30 ◄

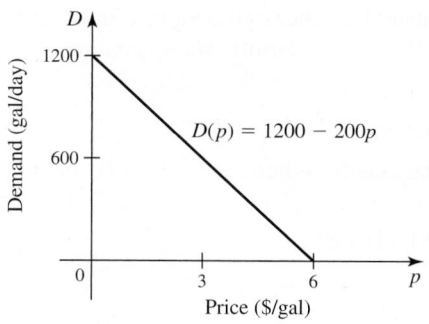

Figure 3.53

Elasticity in Economics Economists apply the term *elasticity* to prices, income, capital, labor, and other variables in systems with input and output. Elasticity describes how changes in the input to a system are related to changes in the output. Because elasticity involves change, it also involves derivatives.

A general rule is that as the price p of an item increases, the number of sales of that item decreases. This relationship is expressed in a demand function. For example, suppose sales at a gas station have the linear demand function $D(p) = 1200 - 200p$ (Figure 3.53), where $D(p)$ is the number of gallons sold per day at a price p (measured in dollars per gallon). According to this function, if gas sells at \$3.60/gal, then the owner can expect to sell $D(3.6) = 480$ gallons. If the price is increased, sales decrease.

Suppose the price of a gallon of gasoline increases from \$3.60 to \$3.96 per gallon; call this change $\Delta p = \$0.36$. The resulting change in the number of gallons sold is $\Delta D = D(3.96) - D(3.60) = -72$. (The change is a decrease, so it is negative.) Comparisons of the variables are more meaningful if we work with percentages. Increasing the price from \$3.60 to \$3.96 per gallon is a percentage change of $\dfrac{\Delta p}{p} = \dfrac{\$0.36}{\$3.60} = 10\%$. Similarly, the corresponding percentage change in the gallons sold is $\dfrac{\Delta D}{D} = \dfrac{-72}{480} = -15\%$.

The *price elasticity of the demand* (or simply, *elasticity*) is the ratio of the percentage change in demand to the percentage change in price; that is, $E = \dfrac{\Delta D/D}{\Delta p/p}$. In the case of the gas demand function, the elasticity of this particular price change is $\dfrac{-15\%}{10\%} = -1.5$.

The elasticity is more useful when it is expressed as a function of the price. To do this, we consider small changes in p and assume the corresponding changes in D are also small. Using the definition of the derivative, the elasticity *function* is

$$E(p) = \lim_{\Delta p \to 0} \frac{\Delta D/D}{\Delta p/p} = \lim_{\Delta p \to 0} \frac{\Delta D}{\Delta p}\left(\frac{p}{D}\right) = \frac{dD}{dp}\frac{p}{D}.$$

Applying this definition to the gas demand function, we find that

$$E(p) = \frac{dD}{dp}\frac{p}{D}$$

$$= \frac{d}{dp}\underbrace{(1200 - 200p)}_{D}\frac{p}{\underbrace{1200 - 200p}_{D}} \qquad \text{Substitute } D = 1200 - 200p.$$

$$= -200\left(\frac{p}{1200 - 200p}\right) \qquad \text{Differentiate.}$$

$$= \frac{p}{p - 6}. \qquad \text{Simplify.}$$

Given a particular price, the elasticity is interpreted as the percentage change in the demand that results for every 1% change in the price. For example, in the gas demand case, with $p = \$3.60$, the elasticity is $E(3.6) = -1.5$; therefore, a 2% increase in the price results in a change of $-1.5 \cdot 2\% = -3\%$ (a decrease) in the number of gallons sold.

▶ Some texts define the elasticity as

$$E(p) = -\frac{dD}{dp}\frac{p}{D} \text{ to make } E(p) \text{ positive.}$$

DEFINITION Elasticity

If the demand for a product varies with the price according to the function

$D = f(p)$, then the **price elasticity of the demand** is $E(p) = \dfrac{dD}{dp}\dfrac{p}{D}$.

EXAMPLE 6   Elasticity in pork prices The demand for processed pork in Canada is described by the function $D(p) = 286 - 20p$. (*Source:* J. Perloff, *Microeconomics,* Prentice Hall, 2012)

a. Compute and graph the price elasticity of the demand.

b. When $-\infty < E < -1$, the demand is said to be **elastic**. When $-1 < E < 0$, the demand is said to be **inelastic**. Interpret these terms.

c. For what prices is the demand for pork elastic? Inelastic?

SOLUTION

a. Substituting the demand function into the definition of elasticity, we find that

$$E(p) = \frac{dD}{dp}\frac{p}{D}$$

$$= \frac{d}{dp}\underbrace{(286 - 20p)}_{D}\underbrace{\frac{p}{286 - 20p}}_{D} \quad \text{Substitute } D = 286 - 20p.$$

$$= -20\left(\frac{p}{286 - 20p}\right) \quad \text{Differentiate.}$$

$$= -\frac{10p}{143 - 10p}. \quad \text{Simplify.}$$

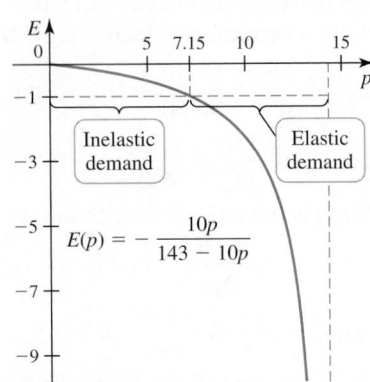

Figure 3.54

➤ When $E(p) = 0$, the demand for a good is said to be *perfectly inelastic*. In Example 6, this case occurs when $p = 0$.

Notice that the elasticity is undefined at $p = 14.3$, which is the price at which the demand reaches zero. (According to the model, no pork can be sold at prices above $14.30.) Therefore, the domain of the elasticity function is $[0, 14.3)$, and on the interval $(0, 14.3)$, the elasticity is negative (**Figure 3.54**).

b. For prices with an elasticity in the interval $-\infty < E(p) < -1$, a $P\%$ increase in the price results in *more* than a $P\%$ decrease in the demand; this is the case of elastic (sensitive) demand. If a price has an elasticity in the interval $-1 < E(p) < 0$, a $P\%$ increase in the price results in *less* than a $P\%$ decrease in the demand; this is the case of inelastic (insensitive) demand.

c. Solving $E(p) = -\dfrac{10p}{143 - 10p} = -1$, we find that $E(p) < -1$, for $p > 7.15$.

For prices in this interval, the demand is elastic (Figure 3.54). For prices with $0 < p < 7.15$, the demand is inelastic.

Related Exercises 33–34 ◄

SECTION 3.6 EXERCISES

Getting Started

1. Explain the difference between the average rate of change and the instantaneous rate of change of a function f.

2. Complete the following statement. If $\dfrac{dy}{dx}$ is large, then small changes in x will result in relatively _____ changes in the value of y.

3. Complete the following statement. If $\dfrac{dy}{dx}$ is small, then small changes in x will result in relatively _____ changes in the value of y.

4. Suppose the function $s(t)$ represents the position (in feet) of a stone t seconds after it is thrown directly upward from 6 feet above Earth's surface.

    a. Is the stone's acceleration positive or negative when the stone is moving upward? Explain.

    b. What is the value of the instantaneous velocity $v(t)$ when the stone reaches its highest point? Explain.

5. Suppose $w(t)$ is the weight (in pounds) of a golden retriever puppy t weeks after it is born. Interpret the meaning of $w'(15) = 1.75$.

6. What is the difference between the *velocity* and the *speed* of an object moving in a straight line?

7. Define the acceleration of an object moving in a straight line.

8. An object moving along a line has a constant negative acceleration. Describe the velocity of the object.

9. The speed of sound (in m/s) in dry air is approximated by the function $v(T) = 331 + 0.6T$, where T is the air temperature (in degrees Celsius). Evaluate $v'(T)$ and interpret its meaning.

10. At noon, a city park manager starts filling a swimming pool. If $V(t)$ is the volume of water (in ft³) in the swimming pool t hours after noon, then what does dV/dt represent?

11. Highway travel A state patrol station is located on a straight north-south freeway. A patrol car leaves the station at 9:00 A.M. heading north with position function $s = f(t)$ that gives its location in miles t hours after 9:00 A.M. (see figure). Assume s is positive when the car is north of the patrol station.

a. Determine the average velocity of the car during the first 45 minutes of the trip.

b. Find the average velocity of the car over the interval $[0.25, 0.75]$. Is the average velocity a good estimate of the velocity at 9:30 A.M.?

c. Find the average velocity of the car over the interval $[1.75, 2.25]$. Estimate the velocity of the car at 11:00 A.M. and determine the direction in which the patrol car is moving.

d. Describe the motion of the patrol car relative to the patrol station between 9:00 A.M. and noon.

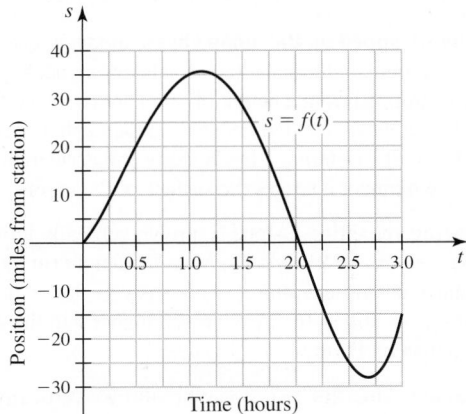

Time (hours)

12. Airline travel The following figure shows the position function of an airliner on an out-and-back trip from Seattle to Minneapolis, where $s = f(t)$ is the number of ground miles from Seattle t hours after take-off at 6:00 A.M. The plane returns to Seattle 8.5 hours later at 2:30 P.M.

a. Calculate the average velocity of the airliner during the first 1.5 hours of the trip ($0 \le t \le 1.5$).

b. Calculate the average velocity of the airliner between 1:30 P.M. and 2:30 P.M. ($7.5 \le t \le 8.5$).

c. At what time(s) is the velocity 0? Give a plausible explanation.

d. Determine the velocity of the airliner at noon ($t = 6$) and explain why the velocity is negative.

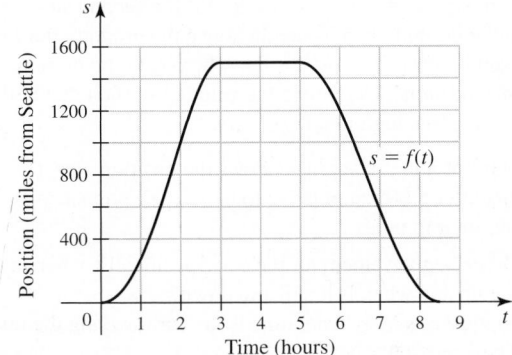

Time (hours)

13. Suppose the average cost of producing 200 gas stoves is $70 per stove and the marginal cost at $x = 200$ is $65 per stove. Interpret these costs.

14. Explain why a decreasing demand function has a negative elasticity function.

Practice Exercises

15–20. Position, velocity, and acceleration *Suppose the position of an object moving horizontally along a line after t seconds is given by the following functions $s = f(t)$, where s is measured in feet, with $s > 0$ corresponding to positions right of the origin.*

a. *Graph the position function.*

b. *Find and graph the velocity function. When is the object stationary, moving to the right, and moving to the left?*

c. *Determine the velocity and acceleration of the object at $t = 1$.*

d. *Determine the acceleration of the object when its velocity is zero.*

e. *On what intervals is the speed increasing?*

15. $f(t) = t^2 - 4t; 0 \le t \le 5$

16. $f(t) = -t^2 + 4t - 3; 0 \le t \le 5$

17. $f(t) = 2t^2 - 9t + 12; 0 \le t \le 3$

18. $f(t) = 18t - 3t^2; 0 \le t \le 8$

19. $f(t) = 2t^3 - 21t^2 + 60t; 0 \le t \le 6$

20. $f(t) = -6t^3 + 36t^2 - 54t; 0 \le t \le 4$

21. A dropped stone on Earth The height (in feet) of a stone dropped from a bridge 64 feet above a river at $t = 0$ seconds is given by $s(t) = -16t^2 + 64$. Find the velocity of the stone and its speed when it hits the water.

22. A dropped stone on Mars A stone is dropped off the edge of a 54-ft cliff on Mars, where the acceleration due to gravity is about 12 ft/s². The height (in feet) of the stone above the ground t seconds after it is dropped is $s(t) = -6t^2 + 54$. Find the velocity of the stone and its speed when it hits the ground.

23. Throwing a stone Suppose a stone is thrown vertically upward from the edge of a cliff on Earth with an initial velocity of 32 ft/s from a height of 48 ft above the ground. The height (in feet) of the stone above the ground t seconds after it is thrown is $s(t) = -16t^2 + 32t + 48$.

a. Determine the velocity v of the stone after t seconds.

b. When does the stone reach its highest point?

c. What is the height of the stone at the highest point?

d. When does the stone strike the ground?

e. With what velocity does the stone strike the ground?

f. On what intervals is the speed increasing?

24. Suppose a stone is thrown vertically upward from the edge of a cliff on Earth with an initial velocity of 19.6 m/s from a height of 24.5 m above the ground. The height (in meters) of the stone above the ground t seconds after it is thrown is $s(t) = -4.9t^2 + 19.6t + 24.5$.

a. Determine the velocity v of the stone after t seconds.

b. When does the stone reach its highest point?

c. What is the height of the stone at the highest point?

d. When does the stone strike the ground?

e. With what velocity does the stone strike the ground?

f. On what intervals is the speed increasing?

25. Suppose a stone is thrown vertically upward from the edge of a cliff on Earth with an initial velocity of 64 ft/s from a height of 32 ft above the ground. The height (in feet) of the stone above the ground t seconds after it is thrown is $s(t) = -16t^2 + 64t + 32$.

a. Determine the velocity v of the stone after t seconds.

b. When does the stone reach its highest point?

c. What is the height of the stone at the highest point?

d. When does the stone strike the ground?

e. With what velocity does the stone strike the ground?

f. On what intervals is the speed increasing?

26. Maximum height Suppose a baseball is thrown vertically upward from the ground with an initial velocity of v_0 ft/s. The approximate height of the ball (in feet) above the ground after t seconds is given by $s(t) = -16t^2 + v_0 t$.

a. What is the height of the ball at its highest point?

b. With what velocity does the ball strike the ground?

27. Initial velocity Suppose a baseball is thrown vertically upward from the ground with an initial velocity of v_0 ft/s. Its height above the ground after t seconds is given by $s(t) = -16t^2 + v_0 t$. Determine the initial velocity of the ball if it reaches a high point of 128 ft.

T 28. Population growth in Washington The population of the state of Washington (in millions) from 2010 ($t = 0$) to 2016 ($t = 6$) is modeled by the polynomial $p(t) = 0.0078t^2 + 0.028t + 6.73$.

a. Determine the average growth rate from 2010 to 2016.

b. What was the growth rate for Washington in 2011 ($t = 1$) and 2015 ($t = 5$)?

c. Use a graphing utility to graph p' for $0 \le t \le 6$. What does this graph tell you about population growth in Washington during the period of time from 2010 to 2016?

29–32. Average and marginal cost *Consider the following cost functions.*

a. Find the average cost and marginal cost functions.

b. Determine the average cost and the marginal cost when $x = a$.

c. Interpret the values obtained in part (b).

29. $C(x) = 1000 + 0.1x, 0 \le x \le 5000, a = 2000$

30. $C(x) = 500 + 0.02x, 0 \le x \le 2000, a = 1000$

31. $C(x) = -0.01x^2 + 40x + 100, 0 \le x \le 1500, a = 1000$

32. $C(x) = -0.04x^2 + 100x + 800, 0 \le x \le 1000, a = 500$

33. Demand and elasticity Based on sales data over the past year, the owner of a DVD store devises the demand function $D(p) = 40 - 2p$, where $D(p)$ is the number of DVDs that can be sold in one day at a price of p dollars.

a. According to the model, how many DVDs can be sold in a day at a price of $10?

b. According to the model, what is the maximum price that can be charged (above which no DVDs can be sold)?

c. Find the elasticity function for this demand function.

d. For what prices is the demand elastic? Inelastic?

e. If the price of DVDs is raised from $10.00 to $10.25, what is the exact percentage decrease in demand (using the demand function)?

f. If the price of DVDs is raised from $10.00 to $10.25, what is the approximate percentage decrease in demand (using the elasticity function)?

34. Demand and elasticity The economic advisor of a large tire store proposes the demand function $D(p) = \dfrac{1800}{p - 40}$, where $D(p)$ is the number of tires of one brand and size that can be sold in one day at a price p.

a. Recalling that the demand must be positive, what is the domain of this function?

b. According to the model, how many tires can be sold in a day at a price of $60 per tire?

c. Find the elasticity function on the domain of the demand function.

d. For what prices is the demand elastic? Inelastic?

e. If the price of tires is raised from $60 to $62, what is the approximate percentage decrease in demand (using the elasticity function)?

35. Explain why or why not Determine whether the following statements are true and give an explanation or counterexample.

a. If the acceleration of an object remains constant, then its velocity is constant.

b. If the acceleration of an object moving along a line is always 0, then its velocity is constant.

c. It is impossible for the instantaneous velocity at all times $a \le t \le b$ to equal the average velocity over the interval $a \le t \le b$.

d. A moving object can have negative acceleration and increasing speed.

36. A feather dropped on the moon On the moon, a feather will fall to the ground at the same rate as a heavy stone. Suppose a feather is dropped from a height of 40 m above the surface of the moon. Its height (in meters) above the ground after t seconds is $s = 40 - 0.8t^2$. Determine the velocity and acceleration of the feather the moment it strikes the surface of the moon.

37. Comparing velocities A stone is thrown vertically into the air at an initial velocity of 96 ft/s. On Mars the height (in feet) of the stone above the ground after t seconds is $s = 96t - 6t^2$, and on Earth it is $s = 96t - 16t^2$. How much higher will the stone travel on Mars than on Earth?

38. Comparing velocities Two stones are thrown vertically upward, each with an initial velocity of 48 ft/s at time $t = 0$. One stone is thrown from the edge of a bridge that is 32 feet above the ground, and the other stone is thrown from ground level. The height above the ground of the stone thrown from the bridge after t seconds is $f(t) = -16t^2 + 48t + 32$, and the height of the stone thrown from the ground after t seconds is $g(t) = -16t^2 + 48t$.

a. Show that the stones reach their high points at the same time.

b. How much higher does the stone thrown from the bridge go than the stone thrown from the ground?

c. When do the stones strike the ground and with what velocities?

39. Matching heights A stone is thrown with an initial velocity of 32 ft/s from the edge of a bridge that is 48 ft above the ground. The height of this stone above the ground t seconds after it is thrown is $f(t) = -16t^2 + 32t + 48$. If a second stone is thrown from the ground, then its height above the ground after t seconds is given by $g(t) = -16t^2 + v_0 t$, where v_0 is the initial velocity of the second stone. Determine the value of v_0 such that both stones reach the same high point.

40. Velocity of a car The graph shows the position $s = f(t)$ of a car t hours after 5:00 P.M. relative to its starting point $s = 0$, where s is measured in miles.

a. Describe the velocity of the car. Specifically, when is it speeding up and when is it slowing down?

b. At approximately what time is the car traveling the fastest? The slowest?

c. What is the approximate maximum velocity of the car? The approximate minimum velocity?

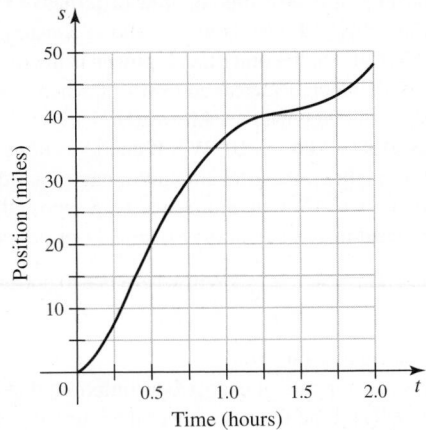

Time (hours)

41. Velocity from position The graph of $s = f(t)$ represents the position of an object moving along a line at time $t \geq 0$.

    a. Assume the velocity of the object is 0 when $t = 0$. For what other values of t is the velocity of the object zero?

    b. When is the object moving in the positive direction and when is it moving in the negative direction?

    c. Sketch a graph of the velocity function.

    d. On what intervals is the speed increasing?

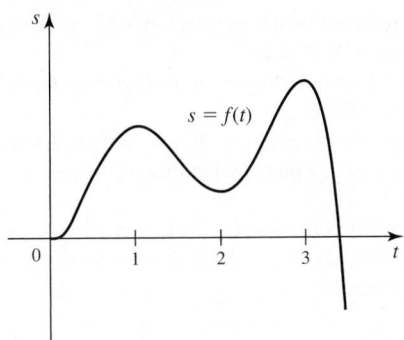

42. Fish length Assume the length L (in centimeters) of a particular species of fish after t years is modeled by the following graph.

    a. What does dL/dt represent and what happens to this derivative as t increases?

    b. What does the derivative tell you about how this species of fish grows?

    c. Sketch a graph of L' and L''.

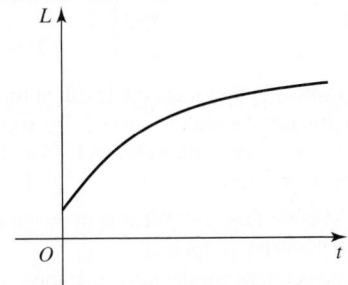

43–46. Average and marginal profit *Let $C(x)$ represent the cost of producing x items and $p(x)$ be the sale price per item if x items are sold. The profit $P(x)$ of selling x items is $P(x) = xp(x) - C(x)$ (revenue minus costs). The **average profit per item** when x items are sold is $P(x)/x$ and the **marginal profit** is dP/dx. The marginal profit approximates the profit obtained by selling one more item, given that x*

items have already been sold. Consider the following cost functions C and price functions p.

    a. *Find the profit function P.*

    b. *Find the average profit function and the marginal profit function.*

    c. *Find the average profit and the marginal profit if $x = a$ units are sold.*

    d. *Interpret the meaning of the values obtained in part (c).*

43. $C(x) = -0.02x^2 + 50x + 100$, $p(x) = 100$, $a = 500$

44. $C(x) = -0.02x^2 + 50x + 100$, $p(x) = 100 - 0.1x$, $a = 500$

45. $C(x) = -0.04x^2 + 100x + 800$, $p(x) = 200$, $a = 1000$

46. $C(x) = -0.04x^2 + 100x + 800$, $p(x) = 200 - 0.1x$, $a = 1000$

47. U.S. population growth The population $p(t)$ (in millions) of the United States t years after the year 1900 is shown in the figure. Approximately when (in what year) was the U.S. population growing most slowly between 1925 and 2020? Estimate the growth rate in that year.

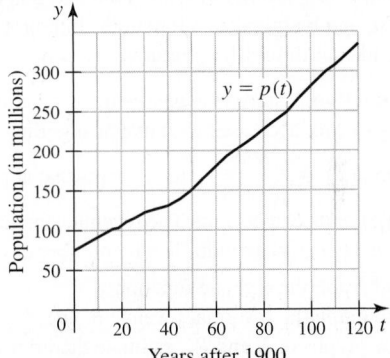

Years after 1900

T 48. Average of marginal production Economists use *production functions* to describe how the output of a system varies with respect to another variable such as labor or capital. For example, the production function $P(L) = 200L + 10L^2 - L^3$ gives the output of a system as a function of the number of laborers L. The *average product $A(L)$* is the average output per laborer when L laborers are working; that is, $A(L) = P(L)/L$. The *marginal product $M(L)$* is the approximate change in output when one additional laborer is added to L laborers; that is, $M(L) = dP/dL$.

    a. For the given production function, compute and graph P, A, and L.

    b. Suppose the peak of the average product curve occurs at $L = L_0$, so that $A'(L_0) = 0$. Show that for a general production function, $M(L_0) = A(L_0)$.

T 49. Velocity of a marble The position (in meters) of a marble, given an initial velocity and rolling up a long incline, is given by $s = \dfrac{100t}{t + 1}$, where t is measured in seconds and $s = 0$ is the starting point.

    a. Graph the position function.

    b. Find the velocity function for the marble.

    c. Graph the velocity function and give a description of the motion of the marble.

    d. At what time is the marble 80 m from its starting point?

    e. At what time is the velocity 50 m/s?

T 50. Tree growth Let b represent the base diameter of a conifer tree and let h represent the height of the tree, where b is measured in centimeters and h is measured in meters. Assume the height is related to the base diameter by the function $h = 5.67 + 0.70b + 0.0067b^2$.

    a. Graph the height function.

    b. Plot and interpret the meaning of dh/db.

Explorations and Challenges

T 51. **A different interpretation of marginal cost** Suppose a large company makes 25,000 gadgets per year in batches of x items at a time. After analyzing setup costs to produce each batch and taking into account storage costs, planners have determined that the total cost $C(x)$ of producing 25,000 gadgets in batches of x items at a time is given by

$$C(x) = 1{,}250{,}000 + \frac{125{,}000{,}000}{x} + 1.5x.$$

   a. Determine the marginal cost and average cost functions. Graph and interpret these functions.

   b. Determine the average cost and marginal cost when $x = 5000$.

   c. The meaning of average cost and marginal cost here is different from earlier examples and exercises. Interpret the meaning of your answer in part (b).

52. **Diminishing returns** A cost function of the form $C(x) = \frac{1}{2}x^2$ reflects *diminishing returns to scale*. Find and graph the cost, average cost, and marginal cost functions. Interpret the graphs and explain the idea of diminishing returns.

T 53. **Revenue function** A store manager estimates that the demand for an energy drink decreases with increasing price according to the function $d(p) = \dfrac{100}{p^2 + 1}$, which means that at price p (in dollars), $d(p)$ units can be sold. The revenue generated at price p is $R(p) = p \cdot d(p)$ (price multiplied by number of units).

   a. Find and graph the revenue function.

   b. Find and graph the marginal revenue $R'(p)$.

   c. From the graphs of R and R', estimate the price that should be charged to maximize the revenue.

T 54. **Fuel economy** Suppose you own a fuel-efficient hybrid automobile with a monitor on the dashboard that displays the mileage and gas consumption. The number of miles you can drive with g gallons of gas remaining in the tank on a particular stretch of highway is given by $m(g) = 50g - 25.8g^2 + 12.5g^3 - 1.6g^4$, for $0 \le g \le 4$.

   a. Graph and interpret the mileage function.

   b. Graph and interpret the gas mileage $m(g)/g$.

   c. Graph and interpret dm/dg.

T 55. **Spring oscillations** A spring hangs from the ceiling at equilibrium with a mass attached to its end. Suppose you pull downward on the mass and release it 10 inches below its equilibrium position with an upward push. The distance x (in inches) of the mass from its equilibrium position after t seconds is given by the function $x(t) = 10 \sin t - 10 \cos t$, where x is positive when the mass is above the equilibrium position.

   a. Graph and interpret this function.

   b. Find dx/dt and interpret the meaning of this derivative.

   c. At what times is the velocity of the mass zero?

   d. The function given here is a model for the motion of an object on a spring. In what ways is this model unrealistic?

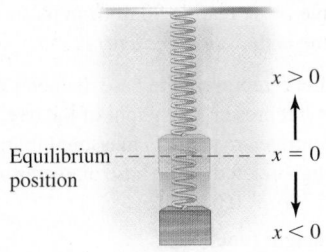

T 56. **Power and energy** Power and energy are often used interchangeably, but they are quite different. **Energy** is what makes matter move or heat up. It is measured in units of **joules** or **Calories**, where 1 Cal = 4184 J. One hour of walking consumes roughly 10^6 J, or 240 Cal. On the other hand, **power** is the rate at which energy is used, which is measured in **watts**, where 1 W = 1 J/s. Other useful units of power are **kilowatts** (1 kW = 10^3 W) and **megawatts** (1 MW = 10^6 W). If energy is used at a rate of 1 kW for one hour, the total amount of energy used is 1 **kilowatt-hour** (1 kWh = 3.6×10^6 J). Suppose the cumulative energy used in a large building over a 24-hr period is given by $E(t) = 100t + 4t^2 - \dfrac{t^3}{9}$ kWh, where $t = 0$ corresponds to midnight.

   a. Graph the energy function.

   b. The power is the rate of energy consumption; that is, $P(t) = E'(t)$. Find the power over the interval $0 \le t \le 24$.

   c. Graph the power function and interpret the graph. What are the units of power in this case?

57. **A race** Jean and Juan run a one-lap race on a circular track. Their angular positions on the track during the race are given by the functions $\theta(t)$ and $\varphi(t)$, respectively, where $0 \le t \le 4$ and t is measured in minutes (see figure). These angles are measured in radians, where $\theta = \varphi = 0$ represent the starting position and $\theta = \varphi = 2\pi$ represent the finish position. The angular velocities of the runners are $\theta'(t)$ and $\varphi'(t)$.

   a. Compare in words the angular velocity of the two runners and the progress of the race.

   b. Which runner has the greater average angular velocity?

   c. Who wins the race?

   d. Jean's position is given by $\theta(t) = \pi t^2/8$. What is her angular velocity at $t = 2$ and at what time is her angular velocity the greatest?

   e. Juan's position is given by $\varphi(t) = \pi t(8 - t)/8$. What is his angular velocity at $t = 2$ and at what time is his angular velocity the greatest?

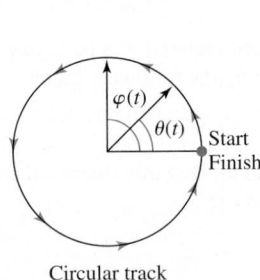

Circular track

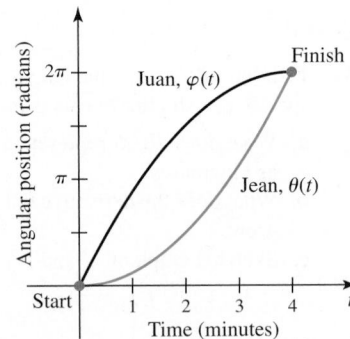

T 58. **Flow from a tank** A cylindrical tank is full at time $t = 0$ when a valve in the bottom of the tank is opened. By Torricelli's law, the volume of water in the tank after t hours is $V = 100(200 - t)^2$, measured in cubic meters.

   a. Graph the volume function. What is the volume of water in the tank before the valve is opened?

   b. How long does it take for the tank to empty?

   c. Find the rate at which water flows from the tank and plot the flow rate function.

   d. At what time is the magnitude of the flow rate a minimum? A maximum?

T 59. Bungee jumper A woman attached to a bungee cord jumps from a bridge that is 30 m above a river. Her height in meters above the river t seconds after the jump is $y(t) = 15(1 + e^{-t} \cos t)$, for $t \geq 0$.

    a. Determine her velocity at $t = 1$ and $t = 3$.

    b. Use a graphing utility to determine when she is moving downward and when she is moving upward during the first 10 s.

    c. Use a graphing utility to estimate the maximum upward velocity.

T 60. Spring runoff The flow of a small stream is monitored for 90 days between May 1 and August 1. The total water that flows past a gauging station is given by

$$V(t) = \begin{cases} \dfrac{4}{5}t^2 & \text{if } 0 \leq t < 45 \\ -\dfrac{4}{5}(t^2 - 180t + 4050) & \text{if } 45 \leq t < 90, \end{cases}$$

where V is measured in cubic feet and t is measured in days, with $t = 0$ corresponding to May 1.

    a. Graph the volume function.

    b. Find the flow rate function $V'(t)$ and graph it. What are the units of the flow rate?

    c. Describe the flow of the stream over the 3-month period. Specifically, when is the flow rate a maximum?

61. Temperature distribution A thin copper rod, 4 m in length, is heated at its midpoint, and the ends are held at a constant temperature of $0°$. When the temperature reaches equilibrium, the temperature profile is given by $T(x) = 40x(4 - x)$, where $0 \leq x \leq 4$ is the position along the rod. The **heat flux** at a point on the rod equals $-kT'(x)$, where $k > 0$ is a constant. If the heat flux is positive at a point, heat moves in the positive x-direction at that point, and if the heat flux is negative, heat moves in the negative x-direction.

    a. With $k = 1$, what is the heat flux at $x = 1$? At $x = 3$?

    b. For what values of x is the heat flux negative? Positive?

    c. Explain the statement that heat flows out of the rod at its ends.

QUICK CHECK ANSWERS

1. Instantaneous velocity **2.** Yes; yes **3.** If an object has positive acceleration, then its velocity is increasing. If the velocity is negative but increasing, then the acceleration is positive and the speed is decreasing. For example, the velocity may increase from -2 m/s to -1 m/s to 0 m/s. **4.** $v(1) = 32$ ft/s and $v(3) = -32$ ft/s, so the speed is 32 ft/s at both times. **5.** The growth rate in 2001 ($t = 1$) is 110 million users/year. It is less than half of the growth rate in 2012 ($t = 12$), which is 242 million users/year. **6.** As x increases from 1 to 100, the average cost decreases from \$149.98/item to \$49/item. ◄

3.7 The Chain Rule

QUICK CHECK 1 Explain why it is not practical to calculate $\dfrac{d}{dx}(5x + 4)^{100}$ by first expanding $(5x + 4)^{100}$. ◄

The differentiation rules presented so far allow us to find derivatives of many functions. However, these rules are inadequate for finding the derivatives of most *composite functions*. Here is a typical situation. If $f(x) = x^3$ and $g(x) = 5x + 4$, then their composition is $f(g(x)) = (5x + 4)^3$. One way to find the derivative is by expanding $(5x + 4)^3$ and differentiating the resulting polynomial. Unfortunately, this strategy becomes prohibitive for functions such as $(5x + 4)^{100}$. We need a better approach.

Chain Rule Formulas

An efficient method for differentiating composite functions, called the *Chain Rule*, is motivated by the following example. Suppose Yancey, Uri, and Xan pick apples. Let y, u, and x represent the number of apples picked in some period of time by Yancey, Uri, and Xan, respectively. Yancey picks apples three times faster than Uri, which means the rate at which Yancey picks apples with respect to Uri is $dy/du = 3$. Uri picks apples twice as fast as Xan, so $du/dx = 2$. Therefore, Yancey picks apples at a rate that is $3 \cdot 2 = 6$ times greater than Xan's rate, which means that $dy/dx = 6$ (**Figure 3.55**). Observe that

$$\frac{dy}{dx} = \frac{dy}{du} \cdot \frac{du}{dx} = 3 \cdot 2 = 6.$$

➤ Expressions such as dy/dx should not be treated as fractions. Nevertheless, you can check symbolically that you have written the Chain Rule correctly by noting that du appears in the "numerator" and "denominator." If it were "canceled," the Chain Rule would have dy/dx on both sides.

The equation $\dfrac{dy}{dx} = \dfrac{dy}{du} \cdot \dfrac{du}{dx}$ is one form of the Chain Rule.

Alternatively, the Chain Rule may be expressed in terms of composite functions. Let $y = f(u)$ and $u = g(x)$, which means y is related to x through the composite function $y = f(u) = f(g(x))$. The derivative $\dfrac{dy}{dx}$ is now expressed as the product

$$\underbrace{\frac{d}{dx}\big(f(g(x))\big)}_{\frac{dy}{dx}} = \underbrace{f'(u)}_{\frac{dy}{du}} \cdot \underbrace{g'(x)}_{\frac{du}{dx}}.$$

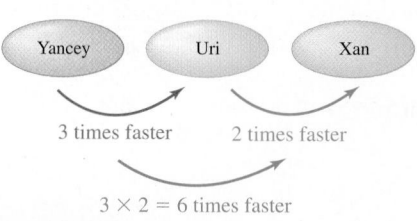

3 times faster   2 times faster

$3 \times 2 = 6$ times faster

Figure 3.55

Replacing u with $g(x)$ results in

$$\frac{d}{dx}\left(f(g(x))\right) = f'(g(x)) \cdot g'(x).$$

▶ The two equations in Theorem 3.13 differ only in notation (Leibniz notation for the derivative versus function notation). Mathematically, they are identical. The second equation states that the derivative of $y = f(g(x))$ is the derivative of f evaluated at $g(x)$ multiplied by the derivative of g evaluated at x.

▶ There may be several ways to choose an inner function $u = g(x)$ and an outer function $y = f(u)$. Nevertheless, we often refer to *the* inner and *the* outer function for the most obvious choices.

THEOREM 3.13   The Chain Rule

Suppose $y = f(u)$ is differentiable at $u = g(x)$ and $u = g(x)$ is differentiable at x. The composite function $y = f(g(x))$ is differentiable at x, and its derivative can be expressed in two equivalent ways.

$$\frac{dy}{dx} = \frac{dy}{du} \cdot \frac{du}{dx} \tag{1}$$

$$\frac{d}{dx}\left(f(g(x))\right) = f'(g(x)) \cdot g'(x) \tag{2}$$

A proof of the Chain Rule is given at the end of this section. For now, it's important to learn how to use it. With the composite function $f(g(x))$, we refer to g as the *inner function* and to f as the *outer function* of the composition. The key to using the Chain Rule is identifying the inner and outer functions. The following four steps outline the differentiation process, although you will soon find that the procedure can be streamlined.

PROCEDURE   Using the Chain Rule

Assume the differentiable function $y = f(g(x))$ is given.

1. Identify an outer function f and an inner function g, and let $u = g(x)$.
2. Replace $g(x)$ with u to express y in terms of u:
$$y = f(\underbrace{g(x)}_{u}) \Rightarrow y = f(u).$$
3. Calculate the product $\dfrac{dy}{du} \cdot \dfrac{du}{dx}$.
4. Replace u with $g(x)$ in $\dfrac{dy}{du}$ to obtain $\dfrac{dy}{dx}$.

QUICK CHECK 2 Identify an inner function (call it g) of $y = (5x + 4)^3$. Let $u = g(x)$ and express the outer function f in terms of u. ◀

EXAMPLE 1   The Chain Rule For each of the following composite functions, find the inner function $u = g(x)$ and the outer function $y = f(u)$. Use equation (1) of the Chain Rule to find $\dfrac{dy}{dx}$.

a. $y = (5x + 4)^{3/2}$ **b.** $y = \sin^3 x$ **c.** $y = \sin x^3$

SOLUTION

a. The inner function of $y = (5x + 4)^{3/2}$ is $u = 5x + 4$, and the outer function is $y = u^{3/2}$. Using equation (1) of the Chain Rule, we have

$$\begin{aligned}
\frac{dy}{dx} &= \frac{dy}{du} \cdot \frac{du}{dx} \\
&= \frac{3}{2}u^{1/2} \cdot 5 \qquad\qquad y = u^{3/2} \Rightarrow \frac{dy}{du} = \frac{3}{2}u^{1/2};\ u = 5x + 4 \Rightarrow \frac{du}{dx} = 5 \\
&= \frac{3}{2}(5x + 4)^{1/2} \cdot 5 \quad \text{Replace } u \text{ with } 5x + 4. \\
&= \frac{15}{2}\sqrt{5x + 4}.
\end{aligned}$$

▶ With trigonometric functions, expressions such as $\sin^n x$ always mean $(\sin x)^n$, except when $n = -1$. In Example 1, $\sin^3 x = (\sin x)^3$.

b. Replacing the shorthand form $y = \sin^3 x$ with $y = (\sin x)^3$, we identify the inner function as $u = \sin x$. Letting $y = u^3$, we have

$$\frac{dy}{dx} = \frac{dy}{du} \cdot \frac{du}{dx} = 3u^2 \cdot \cos x = \underbrace{3 \sin^2 x}_{3u^2} \cos x.$$

c. Although $y = \sin x^3$ appears to be similar to the function $y = \sin^3 x$ in part (b), the inner function in this case is $u = x^3$ and the outer function is $y = \sin u$. Therefore,

$$\frac{dy}{dx} = \frac{dy}{du} \cdot \frac{du}{dx} = (\cos u) \cdot 3x^2 = 3x^2 \cos x^3.$$

Related Exercises 15–17 ◀

When using equation (2) of the Chain Rule, $\frac{d}{dx}\big(f(g(x))\big) = f'(g(x)) \cdot g'(x)$, we identify an outer function $y = f(u)$ and an inner function $u = g(x)$. Then $\frac{d}{dx}\big(f(g(x))\big)$ is the product of $f'(u)$ evaluated at $u = g(x)$ and $g'(x)$.

EXAMPLE 2   The Chain Rule Use equation (2) of the Chain Rule to calculate the derivatives of the following functions.

a. $(6x^3 + 3x + 1)^{10}$ **b.** $\sqrt{5x^2 + 1}$ **c.** $\left(\dfrac{5t^2}{3t^2 + 2}\right)^3$ **d.** e^{-3x}

SOLUTION

a. The inner function of $(6x^3 + 3x + 1)^{10}$ is $g(x) = 6x^3 + 3x + 1$, and the outer function is $f(u) = u^{10}$. The derivative of the outer function is $f'(u) = 10u^9$, which, when evaluated at $g(x)$, is $10(6x^3 + 3x + 1)^9$. The derivative of the inner function is $g'(x) = 18x^2 + 3$. Multiplying the derivatives of the outer and inner functions, we have

$$\frac{d}{dx}\big((6x^3 + 3x + 1)^{10}\big) = \underbrace{10(6x^3 + 3x + 1)^9}_{f'(u) \text{ evaluated at } g(x)} \cdot \underbrace{(18x^2 + 3)}_{g'(x)}$$

$$= 30(6x^2 + 1)(6x^3 + 3x + 1)^9. \qquad \text{Factor and simplify.}$$

b. The inner function of $\sqrt{5x^2 + 1}$ is $g(x) = 5x^2 + 1$, and the outer function is $f(u) = \sqrt{u}$. The derivatives of these functions are $f'(u) = \dfrac{1}{2\sqrt{u}}$ and $g'(x) = 10x$. Therefore,

$$\frac{d}{dx}\sqrt{5x^2 + 1} = \underbrace{\frac{1}{2\sqrt{5x^2 + 1}}}_{\substack{f'(u) \text{ evaluated} \\ \text{at } g(x)}} \cdot \underbrace{10x}_{g'(x)} = \frac{5x}{\sqrt{5x^2 + 1}}.$$

c. The inner function of $\left(\dfrac{5t^2}{3t^2 + 2}\right)^3$ is $g(t) = \dfrac{5t^2}{3t^2 + 2}$. The outer function is $f(u) = u^3$, whose derivative is $f'(u) = 3u^2$. The derivative of the inner function requires the Quotient Rule. Applying the Chain Rule, we have

$$\frac{d}{dt}\left(\frac{5t^2}{3t^2 + 2}\right)^3 = \underbrace{3\left(\frac{5t^2}{3t^2 + 2}\right)^2}_{\substack{f'(u) \text{ evaluated} \\ \text{at } g(t)}} \cdot \underbrace{\frac{(3t^2 + 2)10t - 5t^2(6t)}{(3t^2 + 2)^2}}_{g'(t) \text{ by the Quotient Rule}} = \frac{1500t^5}{(3t^2 + 2)^4}.$$

d. Identifying the inner function of e^{-3x} as $g(x) = -3x$ and the outer function as $f(u) = e^u$, we have

$$\frac{d}{dx}(e^{-3x}) = \underbrace{e^{-3x}}_{\substack{f'(u) \\ \text{evaluated} \\ \text{at } g(x)}} \cdot \underbrace{(-3)}_{g'(x)} = -3e^{-3x}.$$

Related Exercises 28, 29, 41 ◄

The Chain Rule is also used to calculate the derivative of a composite function for a specific value of the variable. If $h(x) = f(g(x))$ and a is a real number, then $h'(a) = f'(g(a))g'(a)$, provided the necessary derivatives exist. Therefore, $h'(a)$ is the derivative of f evaluated at $g(a)$ multiplied by the derivative of g evaluated at a.

EXAMPLE 3   Calculating derivatives at a point Let $h(x) = f(g(x))$. Use the values in Table 3.3 to calculate $h'(1)$ and $h'(2)$.

SOLUTION We use $h'(a) = f'(g(a))g'(a)$ with $a = 1$:

$$h'(1) = f'(g(1))g'(1) = f'(2)g'(1) = 7 \cdot 3 = 21.$$

With $a = 2$, we have

$$h'(2) = f'(g(2))g'(2) = f'(1)g'(2) = 5 \cdot 4 = 20.$$

Related Exercises 25–26 ◄

Table 3.3

x	$f'(x)$	$g(x)$	$g'(x)$
1	5	2	3
2	7	1	4

EXAMPLE 4   Applying the Chain Rule A trail runner programs her GPS unit to record her altitude a (in feet) every 10 minutes during a training run in the mountains; the resulting data are shown in Table 3.4. Meanwhile, at a nearby weather station, a weather probe records the atmospheric pressure p (in hectopascals, or hPa) at various altitudes, shown in Table 3.5.

Table 3.4

t (minutes)	0	10	20	30	40	50	60	70	80
$a(t)$ (altitude)	10,000	10,220	10,510	10,980	11,660	12,330	12,710	13,330	13,440

Table 3.5

a (altitude)	5485	7795	10,260	11,330	12,330	13,330	14,330	15,830	16,230
$p(a)$ (pressure)	1000	925	840	821	793	765	738	700	690

Use the Chain Rule to estimate the rate of change in pressure per unit time experienced by the trail runner when she is 50 minutes into her run.

> The difference quotient $\dfrac{p(a + h) - p(a)}{h}$ is a *forward difference quotient* when $h > 0$ (see Exercises 62–65 in Section 3.1)

SOLUTION We seek the rate of change in the pressure $\dfrac{dp}{dt}$, which is given by the Chain Rule:

$$\frac{dp}{dt} = \frac{dp}{da}\frac{da}{dt}.$$

The runner is at an altitude of 12,330 feet 50 minutes into her run, so we must compute dp/da when $a = 12{,}330$ and da/dt when $t = 50$. These derivatives can be approximated using the following forward difference quotients.

$$\left.\frac{dp}{da}\right|_{a=12,330} \approx \frac{p(12{,}330 + 1000) - p(12{,}330)}{1000}$$

$$= \frac{765 - 793}{1000}$$

$$= -0.028\ \frac{\text{hPa}}{\text{ft}}$$

$$\left.\frac{da}{dt}\right|_{t=50} \approx \frac{a(50 + 10) - a(50)}{10}$$

$$= \frac{12{,}710 - 12{,}330}{10}$$

$$= 38.0\ \frac{\text{ft}}{\text{min}}$$

We now compute the rate of change of the pressure with respect to time:

$$\frac{dp}{dt} = \frac{dp}{da}\frac{da}{dt}$$

$$\approx -0.028\,\frac{\text{hPa}}{\text{ft}} \cdot 38.0\,\frac{\text{ft}}{\text{min}} = -1.06\,\frac{\text{hPa}}{\text{min}}.$$

As expected, dp/dt is negative because the pressure decreases with increasing altitude (Table 3.5) as the runner ascends the trail. Note also that the units are consistent.

Related Exercises 79–80 ◄

Chain Rule for Powers

The Chain Rule leads to a general derivative rule for powers of differentiable functions. In fact, we have already used it in several examples. Consider the function $f(x) = (g(x))^p$, where p is a real number. Letting $f(u) = u^p$ be the outer function and $u = g(x)$ be the inner function, we obtain the Chain Rule for powers of functions.

THEOREM 3.14   Chain Rule for Powers
If g is differentiable for all x in its domain and p is a real number, then

$$\frac{d}{dx}\left((g(x))^p\right) = p(g(x))^{p-1}g'(x).$$

EXAMPLE 5   Chain Rule for powers Find $\dfrac{d}{dx}(\tan x + 10)^{21}$.

SOLUTION With $g(x) = \tan x + 10$, the Chain Rule gives

$$\frac{d}{dx}(\tan x + 10)^{21} = 21(\tan x + 10)^{20}\frac{d}{dx}(\tan x + 10)$$

$$= 21(\tan x + 10)^{20}\sec^2 x.$$

Related Exercises 45–46 ◄

The Composition of Three or More Functions

We can differentiate the composition of three or more functions by applying the Chain Rule repeatedly, as shown in the following example.

EXAMPLE 6   Composition of three functions Calculate the derivative of $\sin(e^{\cos x})$.

SOLUTION The inner function of $\sin(e^{\cos x})$ is $e^{\cos x}$. Because $e^{\cos x}$ is also a composition of two functions, the Chain Rule is used again to calculate $\dfrac{d}{dx}(e^{\cos x})$, where $\cos x$ is the inner function:

$$\frac{d}{dx}\left(\sin\left(e^{\cos x}\right)\right) = \cos\left(e^{\cos x}\right)\underbrace{\frac{d}{dx}\left(e^{\cos x}\right)}_{} \qquad \text{Chain Rule}$$
$$\underbrace{\phantom{\sin(e^{\cos x})}}_{\text{outer}}\ \underbrace{\phantom{e^{\cos x}}}_{\text{inner}}$$

$$= \cos\left(e^{\cos x}\right)\underbrace{e^{\cos x}\cdot\frac{d}{dx}(\cos x)}_{\frac{d}{dx}(e^{\cos x})} \qquad \text{Chain Rule}$$

$$= \cos\left(e^{\cos x}\right)\cdot e^{\cos x}\,(-\sin x) \qquad \text{Differentiate } \cos x.$$
$$= -\sin x \cdot e^{\cos x} \cdot \cos\left(e^{\cos x}\right). \qquad \text{Simplify.}$$

Related Exercises 53–54 ◄

> ► Before dismissing the function in Example 6 as merely a tool to teach the Chain Rule, consider the graph of a related function, $y = \sin\left(e^{1.3\cos x}\right) + 1$ (Figure 3.56). This periodic function has two peaks per cycle and could be used as a simple model of traffic flow (two rush hours followed by light traffic in the middle of the night), tides (high tide, medium tide, high tide, low tide, . . .), or the presence of certain allergens in the air (peaks in the spring and fall).

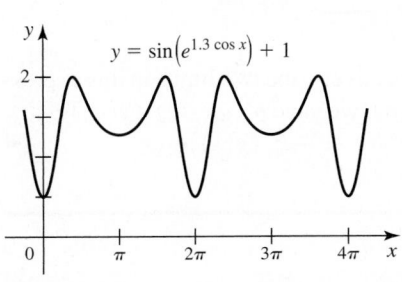

Figure 3.56

QUICK CHECK 3 Let $y = \tan^{10}(x^5)$. Find f, g, and h such that $y = f(u)$, where $u = g(v)$ and $v = h(x)$. ◄

The Chain Rule is often used together with other derivative rules. Example 7 illustrates how several differentiation rules are combined.

EXAMPLE 7   Combining rules Find $\dfrac{d}{dx}\left(x^2\sqrt{x^2+1}\right)$.

SOLUTION The given function is the product of x^2 and $\sqrt{x^2+1}$, and $\sqrt{x^2+1}$ is a composite function. We apply the Product Rule and then the Chain Rule:

$$\frac{d}{dx}\left(x^2\sqrt{x^2+1}\right) = \underbrace{\frac{d}{dx}(x^2)}_{2x}\cdot\sqrt{x^2+1} + x^2\cdot\underbrace{\frac{d}{dx}\left(\sqrt{x^2+1}\right)}_{\text{Use Chain Rule}} \qquad \text{Product Rule}$$

$$= 2x\sqrt{x^2+1} + x^2\cdot\frac{1}{2\sqrt{x^2+1}}\cdot 2x \qquad \text{Chain Rule}$$

$$= 2x\sqrt{x^2+1} + \frac{x^3}{\sqrt{x^2+1}} \qquad \text{Simplify.}$$

$$= \frac{3x^3+2x}{\sqrt{x^2+1}}. \qquad \text{Simplify.}$$

Related Exercises 68–69 ◄

Proof of the Chain Rule

Suppose f is differentiable at $u = g(a)$, g is differentiable at a, and $h(x) = f(g(x))$. By the definition of the derivative of h,

$$h'(a) = \lim_{x\to a}\frac{h(x)-h(a)}{x-a} = \lim_{x\to a}\frac{f(g(x))-f(g(a))}{x-a}. \qquad (3)$$

We assume $g(a) \neq g(x)$ for values of x near a but not equal to a. This assumption holds for most functions encountered in this text. For a proof of the Chain Rule without this assumption, see Exercise 115.

We multiply the right side of equation (3) by $\dfrac{g(x)-g(a)}{g(x)-g(a)}$, which equals 1, and let $v = g(x)$ and $u = g(a)$. The result is

$$h'(a) = \lim_{x\to a}\frac{f(g(x))-f(g(a))}{g(x)-g(a)}\cdot\frac{g(x)-g(a)}{x-a}$$

$$= \lim_{x\to a}\frac{f(v)-f(u)}{v-u}\cdot\frac{g(x)-g(a)}{x-a}.$$

By assumption, g is differentiable at a; therefore, it is continuous at a. This means that $\lim_{x\to a} g(x) = g(a)$, so $v \to u$ as $x \to a$. Consequently,

$$h'(a) = \underbrace{\lim_{v\to u}\frac{f(v)-f(u)}{v-u}}_{f'(u)}\cdot\underbrace{\lim_{x\to a}\frac{g(x)-g(a)}{x-a}}_{g'(a)} = f'(u)g'(a).$$

Because f and g are differentiable at u and a, respectively, the two limits in this expression exist; therefore, $h'(a)$ exists. Noting that $u = g(a)$, we have $h'(a) = f'(g(a))g'(a)$. Replacing a with the variable x gives the Chain Rule: $h'(x) = f'(g(x))g'(x)$.   ◄

SECTION 3.7 EXERCISES

Getting Started

1. Two equivalent forms of the Chain Rule for calculating the derivative of $y = f(g(x))$ are presented in this section. State both forms.

2. Identify the inner and outer functions in the composition $(x^2 + 10)^{-5}$.

3. Identify an inner function $u = g(x)$ and an outer function $y = f(u)$ of $y = (x^3 + x + 1)^4$. Then calculate $\dfrac{dy}{dx}$ using $\dfrac{dy}{dx} = \dfrac{dy}{du}\cdot\dfrac{du}{dx}$.

4. Identify an inner function $u = g(x)$ and an outer function $y = f(u)$ of $y = e^{x^3 + 2x}$. Then calculate $\dfrac{dy}{dx}$ using $\dfrac{dy}{dx} = \dfrac{dy}{du} \cdot \dfrac{du}{dx}$.

5. The two composite functions $y = \cos^3 x$ and $y = \cos x^3$ look similar, but in fact are quite different. For each function, identify the inner function $u = g(x)$ and the outer function $y = f(u)$; then evaluate $\dfrac{dy}{dx}$ using the Chain Rule.

6. Let $h(x) = f(g(x))$, where f and g are differentiable on their domains. If $g(1) = 3$ and $g'(1) = 5$, what else do you need to know to calculate $h'(1)$?

7. Fill in the blanks. The derivative of $f(g(x))$ equals f' evaluated at _____ multiplied by g' evaluated at _____.

8. Evaluate the derivative of $y = (x^2 + 2x + 1)^2$ using $d/dx\, (f(g(x))) = f'(g(x)) \cdot g'(x)$.

9. Evaluate the derivative of $y = \sqrt{4x + 1}$ using $d/dx\, (f(g(x))) = f'(g(x)) \cdot g'(x)$.

10. Express $Q(x) = \cos^4(x^2 + 1)$ as the composition of three functions; that is, identify f, g, and h such that $Q(x) = f(g(h(x)))$.

11. Given that $h(x) = f(g(x))$, find $h'(3)$ if $g(3) = 4$, $g'(3) = 5$, $f(4) = 9$, and $f'(4) = 10$.

12. Given that $h(x) = f(g(x))$, use the graphs of f and g to find $h'(4)$.

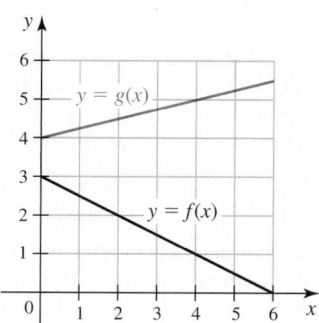

13. What is the derivative of $y = e^{kx}$?

14. Find $f'(x)$ if $f(x) = 15e^{3x}$.

Practice Exercises

15–24. For each of the following composite functions, find an inner function $u = g(x)$ and an outer function $y = f(u)$ such that $y = f(g(x))$. Then calculate dy/dx.

15. $y = (3x + 7)^{10}$

16. $y = (5x^2 + 11x)^{4/3}$

17. $y = \sin^5 x$

18. $y = \sin x^5$

19. $y = \sqrt{x^2 + 1}$

20. $y = \sqrt{7x - 1}$

21. $y = e^{4x^2 + 1}$

22. $y = e^{\sqrt{x}}$

23. $y = \tan 5x^2$

24. $y = \sin \dfrac{x}{4}$

25. **Derivatives using tables** Let $h(x) = f(g(x))$ and $p(x) = g(f(x))$. Use the table to compute the following derivatives.

   a. $h'(3)$ **b.** $h'(2)$ **c.** $p'(4)$ **d.** $p'(2)$ **e.** $h'(5)$

x	1	2	3	4	5
$f(x)$	0	3	5	1	0
$f'(x)$	5	2	-5	-8	-10
$g(x)$	4	5	1	3	2
$g'(x)$	2	10	20	15	20

26. **Derivatives using tables** Let $h(x) = f(g(x))$ and $k(x) = g(g(x))$. Use the table to compute the following derivatives.

   a. $h'(1)$ **b.** $h'(2)$ **c.** $h'(3)$ **d.** $k'(3)$ **e.** $k'(1)$ **f.** $k'(5)$

x	1	2	3	4	5
$f'(x)$	-6	-3	8	7	2
$g(x)$	4	1	5	2	3
$g'(x)$	9	7	3	-1	-5

27–76. Calculate the derivative of the following functions.

27. $y = (3x^2 + 7x)^{10}$

28. $y = (x^2 + 2x + 7)^8$

29. $y = \sqrt{10x + 1}$

30. $y = \sqrt[3]{x^2 + 9}$

31. $y = 5(7x^3 + 1)^{-3}$

32. $y = \cos 5t$

33. $y = \sec(3x + 1)$

34. $y = \csc e^x$

35. $y = \tan e^x$

36. $y = e^{\tan t}$

37. $y = \sin(4x^3 + 3x + 1)$

38. $y = \csc(t^2 + t)$

39. $y = (5x + 1)^{2/3}$

40. $y = x(x + 1)^{1/3}$

41. $y = \sqrt[4]{\dfrac{2x}{4x - 3}}$

42. $y = \cos^4 \theta + \sin^4 \theta$

43. $y = (\sec x + \tan x)^5$

44. $y = \sin(4 \cos z)$

45. $y = (2x^6 - 3x^3 + 3)^{25}$

46. $y = (\cos x + 2 \sin x)^8$

47. $y = (1 + 2 \tan u)^{4.5}$

48. $y = (1 - e^x)^4$

49. $y = \sqrt{1 + \cot^2 x}$

50. $g(x) = \dfrac{x}{e^{3x}}$

51. $y = \dfrac{2e^x + 3e^{-x}}{3}$

52. $f(x) = xe^{7x}$

53. $y = \sin(\sin(e^x))$

54. $y = \sin^2(e^{3x + 1})$

55. $y = \sin^5(\cos 3x)$

56. $y = \cos^{7/4}(4x^3)$

57. $y = \dfrac{e^{2t}}{1 + e^{2t}}$

58. $y = (1 - e^{-0.05x})^{-1}$

59. $y = \sqrt{x + \sqrt{x}}$

60. $y = \sqrt{x + \sqrt{x + \sqrt{x}}}$

61. $y = f(g(x^2))$, where f and g are differentiable for all real numbers

62. $y = (f(g(x^m)))^n$, where f and g are differentiable for all real numbers and m and n are constants

63. $y = \left(\dfrac{x}{x + 1}\right)^5$

64. $y = \left(\dfrac{e^x}{x + 1}\right)^8$

65. $y = e^{x^2 + 1} \sin x^3$

66. $y = \tan(xe^x)$

67. $y = \theta^2 \sec 5\theta$

68. $y = \left(\dfrac{3x}{4x + 2}\right)^5$

69. $y = ((x + 2)(x^2 + 1))^4$

70. $y = e^{2x}(2x - 7)^5$

71. $y = \sqrt[5]{x^4 + \cos 2x}$ **72.** $y = \dfrac{te^t}{t + 1}$

73. $y = (p + 3)^2 \sin p^2$ **74.** $y = (2z + 5)^{1.75} \tan z$

75. $y = \sqrt{f(x)}$, where f is differentiable and nonnegative at x

76. $y = \sqrt[5]{f(x)g(x)}$, where f and g are differentiable at x

77. Explain why or why not Determine whether the following statements are true and give an explanation or counterexample.

   a. The function $f(x) = x \sin x$ can be differentiated without using the Chain Rule.

   b. The function $f(x) = e^{\sqrt{x+1}}$ should be differentiated using the Chain Rule.

   c. The derivative of a product is *not* the product of the derivatives, but the derivative of a composition is a product of derivatives.

   d. $\dfrac{d}{dx} P(Q(x)) = P'(x)Q'(x)$

T 78. Smartphones From 2007 to 2014, there was a dramatic increase in smartphone sales. The number of smartphones (in millions) sold to end users from 2007 to 2014 (see figure) is modeled by the function $c(t) = 114.9e^{0.345t}$, where t represents the number of years after 2007.

   a. Determine the average growth rate in smartphone sales between the years 2007 and 2009 and between 2012 and 2014. During which of these two time intervals was the growth rate greater?

   b. Find the instantaneous growth rate in smartphone sales at $t = 1$ (2008) and $t = 6$ (2013)? At which of these times was the instantaneous growth rate greater?

   c. Use a graphing utility to graph the growth rate, for $0 \le t \le 7$. What does the graph tell you about growth of smartphone sales to end users from 2007 to 2014?

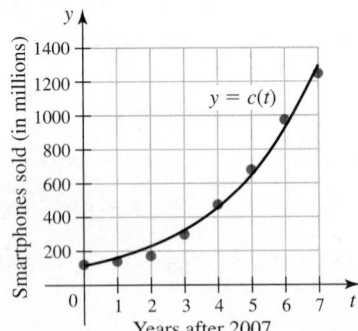

79. Applying the Chain Rule Use the data in Tables 3.4 and 3.5 of Example 4 to estimate the rate of change in pressure with respect to time experienced by the runner when she is at an altitude of 13,330 ft. Make use of a forward difference quotient when estimating the required derivatives.

80. Changing temperature The *lapse rate* is the rate at which the temperature in Earth's atmosphere decreases with altitude. For example, a lapse rate of 6.5° Celsius/km means the temperature *decreases* at a rate of 6.5°C per kilometer of altitude. The lapse rate varies with location and with other variables such as humidity. However, at a given time and location, the lapse rate is often nearly constant in the first 10 kilometers of the atmosphere. A radiosonde (weather balloon) is released from Earth's surface, and

its altitude (measured in kilometers above sea level) at various times (measured in hours) is given in the table below.

Time (hr)	0	0.5	1	1.5	2	2.5
Altitude (km)	0.5	1.2	1.7	2.1	2.5	2.9

   a. Assuming a lapse rate of 6.5°C/km, what is the approximate rate of change of the temperature with respect to time as the balloon rises 1.5 hours into the flight? Specify the units of your result and use a forward difference quotient when estimating the required derivative.

   b. How does an increase in lapse rate change your answer in part (a)?

   c. Is it necessary to know the actual temperature to carry out the calculation in part (a)? Explain.

T 81. Mass of juvenile desert tortoises A study conducted at the University of New Mexico found that the mass $m(t)$ (in grams) of a juvenile desert tortoise t days after a switch to a particular diet is described by the function $m(t) = m_0 e^{0.004t}$, where m_0 is the mass of the tortoise at the time of the diet switch. If $m_0 = 64$, evaluate $m'(65)$ and interpret the meaning of this result. (*Source: Physiological and Biochemical Zoology*, 85, 1, 2012)

T 82. Cell population The population of a culture of cells after t days is approximated by the function $P(t) = \dfrac{1600}{1 + 7e^{-0.02t}}$, for $t \ge 0$.

   a. Graph the population function.

   b. What is the average growth rate during the first 10 days?

   c. Looking at the graph, when does the growth rate appear to be a maximum?

   d. Differentiate the population function to determine the growth rate function $P'(t)$.

   e. Graph the growth rate. When is it a maximum and what is the population at the time that the growth rate is a maximum?

T 83. Bank account A \$200 investment in a savings account grows according to $A(t) = 200e^{0.0398t}$, for $t \ge 0$, where t is measured in years.

   a. Find the balance of the account after 10 years.

   b. How fast is the account growing (in dollars/year) at $t = 10$?

   c. Use your answers to parts (a) and (b) to write the equation of the line tangent to the curve $A = 200e^{0.0398t}$ at the point $(10, A(10))$.

T 84. Pressure and altitude Earth's atmospheric pressure decreases with altitude from a sea level pressure of 1000 millibars (a unit of pressure used by meteorologists). Letting z be the height above Earth's surface (sea level) in kilometers, the atmospheric pressure is modeled by $p(z) = 1000e^{-z/10}$.

   a. Compute the pressure at the summit of Mt. Everest, which has an elevation of roughly 10 km. Compare the pressure on Mt. Everest to the pressure at sea level.

   b. Compute the average change in pressure in the first 5 km above Earth's surface.

   c. Compute the rate of change of the pressure at an elevation of 5 km.

   d. Does $p'(z)$ increase or decrease with z? Explain.

   e. What is the meaning of $\lim_{z \to \infty} p(z) = 0$?

85. Finding slope locations Let $f(x) = xe^{2x}$.

   a. Find the values of x for which the slope of the curve $y = f(x)$ is 0.

   b. Explain the meaning of your answer to part (a) in terms of the graph of f.

86–89. Second derivatives *Find $\dfrac{d^2y}{dx^2}$ for the following functions.*

86. $y = x \cos x^2$

87. $y = \sin x^2$

88. $y = \sqrt{x^2 + 2}$

89. $y = e^{-2x^2}$

90. Derivatives by different methods

a. Calculate $\dfrac{d}{dx}(x^2 + x)^2$ using the Chain Rule. Simplify your answer.

b. Expand $(x^2 + x)^2$ first and then calculate the derivative. Verify that your answer agrees with part (a).

91. Tangent lines Determine an equation of the line tangent to the graph of $y = \dfrac{(x^2 - 1)^2}{x^3 - 6x - 1}$ at the point $(0, -1)$.

■ 92. Tangent lines Determine equations of the lines tangent to the graph of $y = x\sqrt{5 - x^2}$ at the points $(1, 2)$ and $(-2, -2)$. Graph the function and the tangent lines.

93. Tangent lines Assume f and g are differentiable on their domains with $h(x) = f(g(x))$. Suppose the equation of the line tangent to the graph of g at the point $(4, 7)$ is $y = 3x - 5$ and the equation of the line tangent to the graph of f at $(7, 9)$ is $y = -2x + 23$.

a. Calculate $h(4)$ and $h'(4)$.

b. Determine an equation of the line tangent to the graph of h at $(4, h(4))$.

94. Tangent lines Assume f is a differentiable function whose graph passes through the point $(1, 4)$. Suppose $g(x) = f(x^2)$ and the line tangent to the graph of f at $(1, 4)$ is $y = 3x + 1$. Find each of the following.

a. $g(1)$ b. $g'(x)$ c. $g'(1)$

d. An equation of the line tangent to the graph of g when $x = 1$

95. Tangent lines Find the equation of the line tangent to $y = e^{2x}$ at $x = \dfrac{1}{2} \ln 3$.

96. Composition containing $\sin x$ Suppose f is differentiable on $[-2, 2]$ with $f'(0) = 3$ and $f'(1) = 5$. Let $g(x) = f(\sin x)$. Evaluate the following expressions.

a. $g'(0)$ b. $g'\left(\dfrac{\pi}{2}\right)$ c. $g'(\pi)$

97. Composition containing $\sin x$ Suppose f is differentiable for all real numbers with $f(0) = -3$, $f(1) = 3$, $f'(0) = 3$, and $f'(1) = 5$. Let $g(x) = \sin(\pi f(x))$. Evaluate the following expressions.

a. $g'(0)$ b. $g'(1)$

98–100. Vibrations of a spring *Suppose an object of mass m is attached to the end of a spring hanging from the ceiling. The mass is at its equilibrium position $y = 0$ when the mass hangs at rest. Suppose you push the mass to a position y_0 units above its equilibrium position and release it. As the mass oscillates up and down (neglecting any friction in the system), the position y of the mass after t seconds is*

$$y = y_0 \cos\left(t\sqrt{\frac{k}{m}}\right), \qquad (4)$$

where $k > 0$ is a constant measuring the stiffness of the spring (the larger the value of k, the stiffer the spring) and y is positive in the upward direction.

98. Use equation (4) to answer the following questions.

a. Find dy/dt, the velocity of the mass. Assume k and m are constant.

b. How would the velocity be affected if the experiment were repeated with four times the mass on the end of the spring?

c. How would the velocity be affected if the experiment were repeated with a spring having four times the stiffness (k is increased by a factor of 4)?

d. Assume y has units of meters, t has units of seconds, m has units of kg, and k has units of kg/s^2. Show that the units of the velocity in part (a) are consistent.

99. Use equation (4) to answer the following questions.

a. Find the second derivative $\dfrac{d^2y}{dt^2}$.

b. Verify that $\dfrac{d^2y}{dt^2} = -\dfrac{k}{m}y$.

100. Use equation (4) to answer the following questions.

a. The *period T* is the time required by the mass to complete one oscillation. Show that $T = 2\pi\sqrt{\dfrac{m}{k}}$.

b. Assume k is constant and calculate $\dfrac{dT}{dm}$.

c. Give a physical explanation of why $\dfrac{dT}{dm}$ is positive.

■ 101. A damped oscillator The displacement of a mass on a spring suspended from the ceiling is given by $y = 10e^{-t/2}\cos\dfrac{\pi t}{8}$.

a. Graph the displacement function.

b. Compute and graph the velocity of the mass, $v(t) = y'(t)$.

c. Verify that the velocity is zero when the mass reaches the high and low points of its oscillation.

102. Oscillator equation A mechanical oscillator (such as a mass on a spring or a pendulum) subject to frictional forces satisfies the equation (called a differential equation)

$$y''(t) + 2y'(t) + 5y(t) = 0,$$

where y is the displacement of the oscillator from its equilibrium position. Verify by substitution that the function $y(t) = e^{-t}(\sin 2t - 2\cos 2t)$ satisfies this equation.

■ 103. Hours of daylight The number of hours of daylight at any point on Earth fluctuates throughout the year. In the Northern Hemisphere, the shortest day is on the winter solstice and the longest day is on the summer solstice. At 40° north latitude, the length of a day is approximated by

$$D(t) = 12 - 3\cos\left(\frac{2\pi(t + 10)}{365}\right),$$

where D is measured in hours and $0 \le t \le 365$ is measured in days, with $t = 0$ corresponding to January 1.

a. Approximately how much daylight is there on March 1 ($t = 59$)?

b. Find the rate at which the daylight function changes.

c. Find the rate at which the daylight function changes on March 1. Convert your answer to units of min/day and explain what this result means.

d. Graph the function $y = D'(t)$ using a graphing utility.

e. At what times of the year is the length of day changing most rapidly? Least rapidly?

Explorations and Challenges

T 104. A mixing tank A 500-liter (L) tank is filled with pure water. At time $t = 0$, a salt solution begins flowing into the tank at a rate of 5 L/min. At the same time, the (fully mixed) solution flows out of the tank at a rate of 5.5 L/min. The mass of salt in grams in the tank at any time $t \geq 0$ is given by

$$M(t) = 250(1000 - t)(1 - 10^{-30}(1000 - t)^{10})$$

and the volume of solution in the tank is given by

$$V(t) = 500 - 0.5t.$$

a. Graph the mass function and verify that $M(0) = 0$.
b. Graph the volume function and verify that the tank is empty when $t = 1000$ min.
c. The concentration of the salt solution in the tank (in g/L) is given by $C(t) = M(t)/V(t)$. Graph the concentration function and comment on its properties. Specifically, what are $C(0)$ and $\lim\limits_{t \to 1000^-} C(t)$?
d. Find the rate of change of the mass $M'(t)$, for $0 \leq t \leq 1000$.
e. Find the rate of change of the concentration $C'(t)$, for $0 \leq t \leq 1000$.
f. For what times is the concentration of the solution increasing? Decreasing?

T 105. Power and energy The total energy in megawatt-hr (MWh) used by a town is given by

$$E(t) = 400t + \frac{2400}{\pi} \sin \frac{\pi t}{12},$$

where $t \geq 0$ is measured in hours, with $t = 0$ corresponding to noon.

a. Find the power, or rate of energy consumption, $P(t) = E'(t)$ in units of megawatts (MW).
b. At what time of day is the rate of energy consumption a maximum? What is the power at that time of day?
c. At what time of day is the rate of energy consumption a minimum? What is the power at that time of day?
d. Sketch a graph of the power function reflecting the times when energy use is a minimum or a maximum.

106. Deriving trigonometric identities

a. Differentiate both sides of the identity $\cos 2t = \cos^2 t - \sin^2 t$ to prove that $\sin 2t = 2 \sin t \cos t$.
b. Verify that you obtain the same identity for $\sin 2t$ as in part (a) if you differentiate the identity $\cos 2t = 2 \cos^2 t - 1$.
c. Differentiate both sides of the identity $\sin 2t = 2 \sin t \cos t$ to prove that $\cos 2t = \cos^2 t - \sin^2 t$.

107. Quotient Rule derivation Suppose you forgot the Quotient Rule for calculating $\dfrac{d}{dx}\left(\dfrac{f(x)}{g(x)}\right)$. Use the Chain Rule and Product Rule with the identity $\dfrac{f(x)}{g(x)} = f(x)(g(x))^{-1}$ to derive the Quotient Rule.

108. The Chain Rule for second derivatives

a. Derive a formula for the second derivative, $\dfrac{d^2}{dx^2}(f(g(x)))$.
b. Use the formula in part (a) to calculate
$$\frac{d^2}{dx^2}(\sin(3x^4 + 5x^2 + 2)).$$

T 109–112. Calculating limits *The following limits are the derivatives of a composite function g at a point a.*

a. *Find a possible function g and number a.*
b. *Use the Chain Rule to find each limit. Verify your answer by using a calculator.*

109. $\lim\limits_{x \to 2} \dfrac{(x^2 - 3)^5 - 1}{x - 2}$

110. $\lim\limits_{x \to 0} \dfrac{\sqrt{4 + \sin x} - 2}{x}$

111. $\lim\limits_{h \to 0} \dfrac{\sin(\pi/2 + h)^2 - \sin(\pi^2/4)}{h}$

112. $\lim\limits_{h \to 0} \dfrac{\dfrac{1}{3((1 + h)^5 + 7)^{10}} - \dfrac{1}{3(8)^{10}}}{h}$

113. Limit of a difference quotient Assuming f is differentiable for all x, simplify $\lim\limits_{x \to 5} \dfrac{f(x^2) - f(25)}{x - 5}$.

114. Derivatives of even and odd functions Recall that f is even if $f(-x) = f(x)$, for all x in the domain of f, and f is odd if $f(-x) = -f(x)$, for all x in the domain of f.

a. If f is a differentiable, even function on its domain, determine whether f' is even, odd, or neither.
b. If f is a differentiable, odd function on its domain, determine whether f' is even, odd, or neither.

115. A general proof of the Chain Rule Let f and g be differentiable functions with $h(x) = f(g(x))$. For a given constant a, let $u = g(a)$ and $v = g(x)$, and define

$$H(v) = \begin{cases} \dfrac{f(v) - f(u)}{v - u} - f'(u) & \text{if } v \neq u \\ 0 & \text{if } v = u. \end{cases}$$

a. Show that $\lim\limits_{v \to u} H(v) = 0$.
b. For any value of u, show that
$$f(v) - f(u) = (H(v) + f'(u))(v - u).$$
c. Show that
$$h'(a) = \lim\limits_{x \to a}\left((H(g(x)) + f'(g(a))) \cdot \frac{g(x) - g(a)}{x - a}\right).$$
d. Show that $h'(a) = f'(g(a))g'(a)$.

QUICK CHECK ANSWERS

1. The expansion of $(5x + 4)^{100}$ contains 101 terms. It would take too much time to calculate the expansion and the derivative.   **2.** The inner function is $u = 5x + 4$, and the outer function is $y = u^3$.   **3.** $f(u) = u^{10}$; $u = g(v) = \tan v$; $v = h(x) = x^5$ ◄

3.8 Implicit Differentiation

This chapter has been devoted to calculating derivatives of functions of the form $y = f(x)$, where y is defined *explicitly* as a function of x. However, relations between variables are often expressed *implicitly*. For example, the equation of the unit circle $x^2 + y^2 = 1$ does not specify y directly, but rather defines y implicitly. This equation does not represent a single function because its graph fails the vertical line test (**Figure 3.57a**). If, however, the equation $x^2 + y^2 = 1$ is solved for y, then *two* functions emerge: $y = -\sqrt{1 - x^2}$ and $y = \sqrt{1 - x^2}$ (**Figure 3.57b**). Having identified two explicit functions that describe the circle, we can find their derivatives using the Chain Rule.

$$\text{If } y = \sqrt{1 - x^2}, \text{ then } \frac{dy}{dx} = -\frac{x}{\sqrt{1 - x^2}}. \tag{1}$$

$$\text{If } y = -\sqrt{1 - x^2}, \text{ then } \frac{dy}{dx} = \frac{x}{\sqrt{1 - x^2}}. \tag{2}$$

We use equation (1) to find the slope of the curve at any point on the upper half of the unit circle and equation (2) to find the slope of the curve at any point on the lower half of the circle.

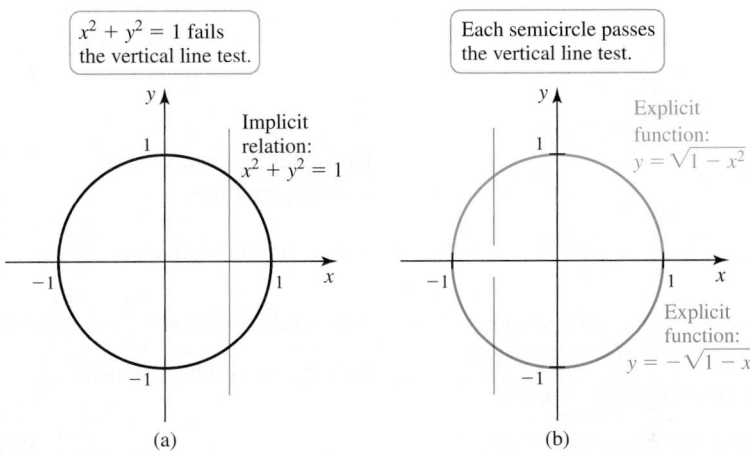

(a)   (b)

Figure 3.57

QUICK CHECK 1 The equation $x - y^2 = 0$ implicitly defines what two functions? ◄

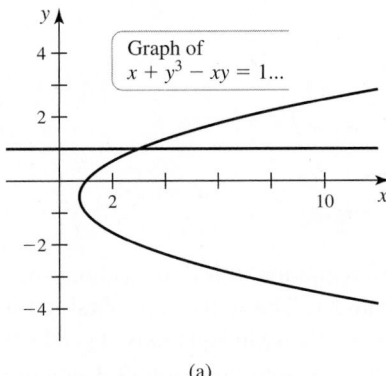

(a)

Figure 3.58

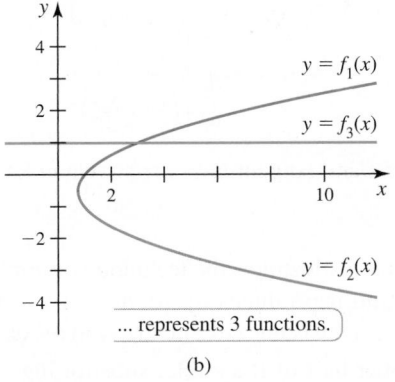

(b)

Although it is straightforward to solve some implicit equations for y (such as $x^2 + y^2 = 1$ or $x - y^2 = 0$), it is difficult or impossible to solve other equations for y. For example, the graph of $x + y^3 - xy = 1$ (**Figure 3.58a**) represents three functions: the upper half of a parabola $y = f_1(x)$, the lower half of a parabola $y = f_2(x)$, and the horizontal line $y = f_3(x)$ (**Figure 3.58b**). Solving for y to obtain these three functions is challenging (Exercise 69), and even after solving for y, we would have to calculate derivatives for each of the three functions separately. The goal of this section is to find a *single* expression for the derivative *directly* from an equation without first solving for y. This technique, called **implicit differentiation**, is demonstrated through examples.

EXAMPLE 1   Implicit differentiation

a. Calculate $\dfrac{dy}{dx}$ directly from the equation for the unit circle $x^2 + y^2 = 1$.

b. Find the slope of the unit circle at $\left(\dfrac{1}{2}, \dfrac{\sqrt{3}}{2}\right)$ and $\left(\dfrac{1}{2}, -\dfrac{\sqrt{3}}{2}\right)$.

SOLUTION

a. To indicate the choice of x as the independent variable, it is helpful to replace the variable y with $y(x)$:

$$x^2 + (y(x))^2 = 1. \quad \text{Replace } y \text{ with } y(x).$$

We now take the derivative of each term in the equation *with respect to x*:

$$\underbrace{\frac{d}{dx}(x^2)}_{2x} + \underbrace{\frac{d}{dx}(y(x))^2}_{\text{Use the Chain Rule}} = \underbrace{\frac{d}{dx}(1)}_{0}.$$

By the Chain Rule, $\dfrac{d}{dx}(y(x))^2 = 2y(x)y'(x)$, or $\dfrac{d}{dx}(y^2) = 2y\dfrac{dy}{dx}$. Substituting this result, we have

$$2x + 2y\frac{dy}{dx} = 0.$$

The last step is to solve for $\dfrac{dy}{dx}$:

$$2y\frac{dy}{dx} = -2x \quad \text{Subtract } 2x \text{ from both sides.}$$

$$\frac{dy}{dx} = -\frac{x}{y}. \quad \text{Divide by } 2y \text{ and simplify.}$$

This result holds provided $y \neq 0$. At the points $(1, 0)$ and $(-1, 0)$, the circle has vertical tangent lines.

b. Notice that the derivative $\dfrac{dy}{dx} = -\dfrac{x}{y}$ depends on *both x* and *y*. Therefore, to find the slope of the circle at $\left(\dfrac{1}{2}, \dfrac{\sqrt{3}}{2}\right)$, we substitute both $x = 1/2$ and $y = \sqrt{3}/2$ into the derivative formula. The result is

$$\frac{dy}{dx}\bigg|_{\left(\frac{1}{2}, \frac{\sqrt{3}}{2}\right)} = -\frac{1/2}{\sqrt{3}/2} = -\frac{1}{\sqrt{3}}.$$

The slope of the curve at $\left(\dfrac{1}{2}, -\dfrac{\sqrt{3}}{2}\right)$ is

$$\frac{dy}{dx}\bigg|_{\left(\frac{1}{2}, -\frac{\sqrt{3}}{2}\right)} = -\frac{1/2}{-\sqrt{3}/2} = \frac{1}{\sqrt{3}}.$$

The curve and tangent lines are shown in **Figure 3.59**.

Related Exercises 13, 15 ◄

> ➤ Implicit differentiation usually produces an expression for dy/dx in terms of both x and y. The notation $\dfrac{dy}{dx}\bigg|_{(a, b)}$ tells us to replace x with a and y with b in the expression for dy/dx.

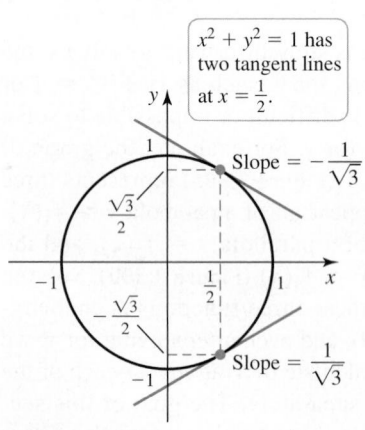

$x^2 + y^2 = 1$ has two tangent lines at $x = \frac{1}{2}$.

Slope $= -\dfrac{1}{\sqrt{3}}$

Slope $= \dfrac{1}{\sqrt{3}}$

Figure 3.59

Example 1 illustrates the technique of implicit differentiation. It is done without solving for y, and it produces dy/dx in terms of both x and y. The derivative obtained in Example 1 is consistent with the derivatives calculated explicitly in equations (1) and (2). For the upper half of the circle, substituting $y = \sqrt{1 - x^2}$ into the implicit derivative $dy/dx = -x/y$ gives

$$\frac{dy}{dx} = -\frac{x}{y} = -\frac{x}{\sqrt{1 - x^2}},$$

which agrees with equation (1). For the lower half of the circle, substituting $y = -\sqrt{1 - x^2}$ into $dy/dx = -x/y$ gives

$$\frac{dy}{dx} = -\frac{x}{y} = \frac{x}{\sqrt{1 - x^2}},$$

which is consistent with equation (2). Therefore, implicit differentiation gives a single unified derivative $dy/dx = -x/y$.

EXAMPLE 2 **Implicit differentiation** Find $y'(x)$ when $\sin xy = x^2 + y$.

SOLUTION It is impossible to solve this equation for y in terms of x, so we differentiate implicitly. Differentiating both sides of the equation with respect to x, using the Chain Rule and the Product Rule on the left side, gives

$$(\cos xy)(y + xy') = 2x + y'.$$

We now solve for y':

$$xy'\cos xy - y' = 2x - y \cos xy \quad \text{Rearrange terms.}$$
$$y'(x \cos xy - 1) = 2x - y \cos xy \quad \text{Factor on left side.}$$
$$y' = \frac{2x - y \cos xy}{x \cos xy - 1}. \quad \text{Solve for } y'.$$

Notice that the final result gives y' in terms of both x and y. *Related Exercises 31, 33* ◄

Slopes of Tangent Lines

Derivatives obtained by implicit differentiation typically depend on x and y. Therefore, the slope of a curve at a particular point (a, b) requires both the x- and y-coordinates of the point. These coordinates are also needed to find an equation of the tangent line at that point.

EXAMPLE 3 **Finding tangent lines with implicit functions** Find an equation of the line tangent to the curve $x^2 + xy - y^3 = 7$ at $(3, 2)$.

SOLUTION We calculate the derivative of each term of the equation $x^2 + xy - y^3 = 7$ with respect to x:

$$\frac{d}{dx}(x^2) + \frac{d}{dx}(xy) - \frac{d}{dx}(y^3) = \frac{d}{dx}(7) \quad \text{Differentiate each term.}$$
$$2x + \underbrace{y + xy'}_{\text{Product Rule}} - \underbrace{3y^2 y'}_{\text{Chain Rule}} = 0 \quad \text{Calculate the derivatives.}$$
$$3y^2 y' - xy' = 2x + y \quad \text{Group the terms containing } y'.$$
$$y' = \frac{2x + y}{3y^2 - x}. \quad \text{Factor and solve for } y'.$$

To find the slope of the tangent line at $(3, 2)$, we substitute $x = 3$ and $y = 2$ into the derivative formula:

$$\left.\frac{dy}{dx}\right|_{(3, 2)} = \left.\frac{2x + y}{3y^2 - x}\right|_{(3, 2)} = \frac{8}{9}.$$

An equation of the line passing through $(3, 2)$ with slope $\frac{8}{9}$ is

$$y - 2 = \frac{8}{9}(x - 3) \quad \text{or} \quad y = \frac{8}{9}x - \frac{2}{3}.$$

Figure 3.60 shows the graphs of the curve $x^2 + xy - y^3 = 7$ and the tangent line.
Related Exercises 47–48 ◄

EXAMPLE 4 **Slope of a curve** Find the slope of the curve $2(x + y)^{1/3} = y$ at the point $(4, 4)$.

SOLUTION We begin by differentiating both sides of the given equation with respect to x:

$$\frac{2}{3}(x + y)^{-2/3}\left(1 + \frac{dy}{dx}\right) = \frac{dy}{dx} \quad \begin{array}{l}\text{Implicit differentiation,}\\ \text{Chain Rule, Power Rule}\end{array}$$
$$\frac{2}{3}(x + y)^{-2/3} = \frac{dy}{dx} - \frac{2}{3}(x + y)^{-2/3}\frac{dy}{dx} \quad \text{Expand and collect like terms.}$$
$$\frac{2}{3}(x + y)^{-2/3} = \frac{dy}{dx}\left(1 - \frac{2}{3}(x + y)^{-2/3}\right). \quad \text{Factor out } \frac{dy}{dx}.$$

QUICK CHECK 2 Use implicit differentiation to find $\frac{dy}{dx}$ for $x - y^2 = 3$. ◄

QUICK CHECK 3 If a function is defined explicitly in the form $y = f(x)$, which coordinates are needed to find the slope of a tangent line—the x-coordinate, the y-coordinate, or both? ◄

➤ Because y is a function of x, we have
$$\frac{d}{dx}(x) = 1 \quad \text{and}$$
$$\frac{d}{dx}(y) = y'.$$
To differentiate y^3 with respect to x, we need the Chain Rule.

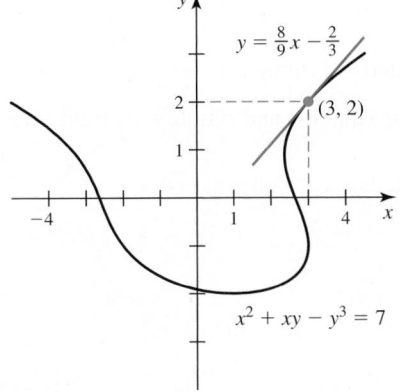

Figure 3.60

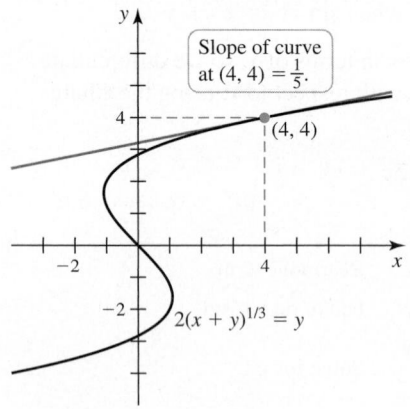

Figure 3.61

▶ The procedure outlined in Example 4 for simplifying dy/dx is a useful trick. When computing derivatives using implicit differentiation, look back to the original equation to determine whether a substitution can be made to simplify the derivative. This technique is illustrated again in Example 5.

Solving for dy/dx, we find that

$$\frac{dy}{dx} = \frac{\frac{2}{3}(x+y)^{-2/3}}{1 - \frac{2}{3}(x+y)^{-2/3}} \qquad \text{Divide by } 1 - \frac{2}{3}(x+y)^{-2/3}.$$

$$\frac{dy}{dx} = \frac{2}{3(x+y)^{2/3} - 2}. \qquad \text{Multiply by } 3(x+y)^{2/3} \text{ and simplify.}$$

At this stage, we could evaluate dy/dx at the point $(4, 4)$ to determine the slope of the curve. However, the derivative can be simplified by employing a subtle maneuver. The original equation $2(x+y)^{1/3} = y$ implies that $(x+y)^{2/3} = y^2/4$. Therefore,

$$\frac{dy}{dx} = \frac{2}{\underbrace{3(x+y)^{2/3}}_{y^2/4} - 2} = \frac{8}{3y^2 - 8}.$$

With this simplified formula, finding the slope of the curve at the point $(4, 4)$ **(Figure 3.61)** requires only the y-coordinate:

$$\left.\frac{dy}{dx}\right|_{(4,4)} = \left.\frac{8}{3y^2 - 8}\right|_{y=4} = \frac{8}{3(4)^2 - 8} = \frac{1}{5}.$$

Related Exercises 25–26 ◀

Higher-Order Derivatives of Implicit Functions

In previous sections of this chapter, we found higher-order derivatives $\dfrac{d^n y}{dx^n}$ by first calculating $\dfrac{dy}{dx}, \dfrac{d^2 y}{dx^2}, \ldots,$ and $\dfrac{d^{n-1} y}{dx^{n-1}}$. The same approach is used with implicit differentiation.

EXAMPLE 5   **A second derivative**   Find $\dfrac{d^2 y}{dx^2}$ if $x^2 + y^2 = 1$.

SOLUTION The first derivative $\dfrac{dy}{dx} = -\dfrac{x}{y}$ was computed in Example 1.

We now calculate the derivative of each side of this equation and simplify the right side:

$$\frac{d}{dx}\left(\frac{dy}{dx}\right) = \frac{d}{dx}\left(-\frac{x}{y}\right) \qquad \text{Take derivatives with respect to } x.$$

$$\frac{d^2 y}{dx^2} = -\frac{y \cdot 1 - x\dfrac{dy}{dx}}{y^2} \qquad \text{Quotient Rule}$$

$$= -\frac{y - x\left(-\dfrac{x}{y}\right)}{y^2} \qquad \text{Substitute for } \frac{dy}{dx}.$$

$$= -\frac{x^2 + y^2}{y^3} \qquad \text{Simplify.}$$

$$= -\frac{1}{y^3}. \qquad x^2 + y^2 = 1$$

In the last step of this calculation we used the original equation $x^2 + y^2 = 1$ to simplify the formula for d^2y/dx^2.

Related Exercises 51–52 ◀

SECTION 3.8 EXERCISES

Getting Started

1. For some equations, such as $x^2 + y^2 = 1$ or $x - y^2 = 0$, it is possible to solve for y and then calculate $\dfrac{dy}{dx}$. Even in these cases, explain why implicit differentiation is usually a more efficient method for calculating the derivative.

2. Explain the differences between computing the derivatives of functions that are defined implicitly and explicitly.

3. Why are both the x-coordinate and the y-coordinate generally needed to find the slope of the tangent line at a point for an implicitly defined function?

4. Identify and correct the error in the following argument. Suppose $y^2 + 2y = 2x^3 - 7$. Differentiating both sides with respect to x to find $\dfrac{dy}{dx}$, we have $2y + 2\dfrac{dy}{dx} = 6x^2$, which implies that $\dfrac{dy}{dx} = 3x^2 - y$.

5–8. Calculate $\dfrac{dy}{dx}$ using implicit differentiation.

5. $x = y^2$

6. $3x + 4y^3 = 7$

7. $\sin y + 2 = x$

8. $e^y - e^x = C$, where C is constant

9. Consider the curve defined by $2x - y + y^3 = 0$ (see figure).

   a. Find the coordinates of the y-intercepts of the curve.

   b. Verify that $\dfrac{dy}{dx} = \dfrac{2}{1 - 3y^2}$.

   c. Find the slope of the curve at each point where $x = 0$.

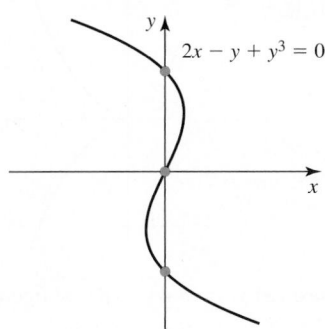

10. Find the slope of the curve $x^2 + y^3 = 2$ at each point where $y = 1$ (see figure).

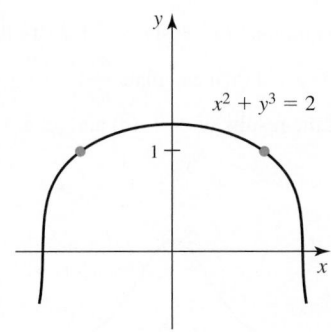

11. Consider the curve $x = y^3$. Use implicit differentiation to verify that $\dfrac{dy}{dx} = \dfrac{1}{3y^2}$ and then find $\dfrac{d^2y}{dx^2}$.

12. Consider the curve $x = e^y$. Use implicit differentiation to verify that $\dfrac{dy}{dx} = e^{-y}$ and then find $\dfrac{d^2y}{dx^2}$.

Practice Exercises

13–26. Implicit differentiation *Carry out the following steps.*

a. Use implicit differentiation to find $\dfrac{dy}{dx}$.

b. Find the slope of the curve at the given point.

13. $x^4 + y^4 = 2;\ (1, -1)$

14. $x = e^y;\ (2, \ln 2)$

15. $y^2 = 4x;\ (1, 2)$

16. $y^2 + 3x = 8;\ \left(1, \sqrt{5}\right)$

17. $\sin y = 5x^4 - 5;\ (1, \pi)$

18. $\sqrt{x} - 2\sqrt{y} = 0;\ (4, 1)$

19. $\cos y = x;\ \left(0, \dfrac{\pi}{2}\right)$

20. $\tan xy = x + y;\ (0, 0)$

21. $xy = 7;\ (1, 7)$

22. $\dfrac{x}{y^2 + 1} = 1;\ (10, 3)$

23. $\sqrt[3]{x} + \sqrt[3]{y^4} = 2;\ (1, 1)$

24. $x^{2/3} + y^{2/3} = 2;\ (1, 1)$

25. $x\sqrt[3]{y} + y = 10;\ (1, 8)$

26. $(x + y)^{2/3} = y;\ (4, 4)$

27–40. Implicit differentiation *Use implicit differentiation to find* $\dfrac{dy}{dx}$.

27. $\sin x + \sin y = y$

28. $y = xe^y$

29. $x + y = \cos y$

30. $x + 2y = \sqrt{y}$

31. $\sin xy = x + y$

32. $e^{xy} = 2y$

33. $\cos y^2 + x = e^y$

34. $y = \dfrac{x + 1}{y - 1}$

35. $x^3 = \dfrac{x + y}{x - y}$

36. $(xy + 1)^3 = x - y^2 + 8$

37. $6x^3 + 7y^3 = 13xy$

38. $\sin x \cos y = \sin x + \cos y$

39. $\sqrt{x^4 + y^2} = 5x + 2y^3$

40. $\sqrt{x + y^2} = \sin y$

41. **Cobb-Douglas production function** The output of an economic system Q, subject to two inputs, such as labor L and capital K, is often modeled by the Cobb-Douglas production function $Q = cL^aK^b$. When $a + b = 1$, the case is called *constant returns to scale*. Suppose $Q = 1280$, $a = \dfrac{1}{3}$, $b = \dfrac{2}{3}$, and $c = 40$.

    a. Find the rate of change of capital with respect to labor, dK/dL.

    b. Evaluate the derivative in part (a) with $L = 8$ and $K = 64$.

42. **Surface area of a cone** The lateral surface area of a cone of radius r and height h (the surface area excluding the base) is $A = \pi r \sqrt{r^2 + h^2}$.

    a. Find dr/dh for a cone with a lateral surface area of $A = 1500\pi$.

    b. Evaluate this derivative when $r = 30$ and $h = 40$.

43. Volume of a spherical cap Imagine slicing through a sphere with a plane (sheet of paper). The smaller piece produced is called a spherical cap. Its volume is $V = \pi h^2(3r - h)/3$, where r is the radius of the sphere and h is the thickness of the cap.

    a. Find dr/dh for a spherical cap with a volume of $5\pi/3$.

    b. Evaluate this derivative when $r = 2$ and $h = 1$.

44. Volume of a torus The volume of a torus (doughnut or bagel) with an inner radius of a and an outer radius of b is $V = \pi^2(b + a)(b - a)^2/4$.

    a. Find db/da for a torus with a volume of $64\pi^2$.

    b. Evaluate this derivative when $a = 6$ and $b = 10$.

45–50. Tangent lines *Carry out the following steps.*

a. *Verify that the given point lies on the curve.*

b. *Determine an equation of the line tangent to the curve at the given point.*

45. $\sin y + 5x = y^2$; $(0, 0)$ **46.** $x^3 + y^3 = 2xy$; $(1, 1)$

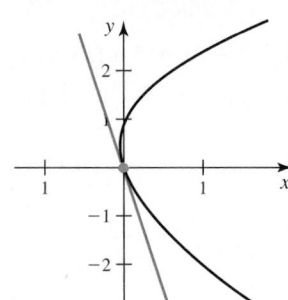 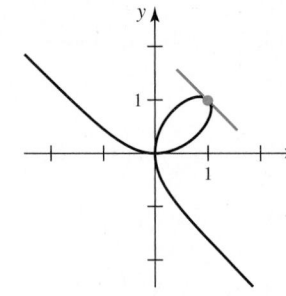

47. $x^2 + xy + y^2 = 7$; $(2, 1)$ **48.** $x^4 - x^2y + y^4 = 1$; $(-1, 1)$

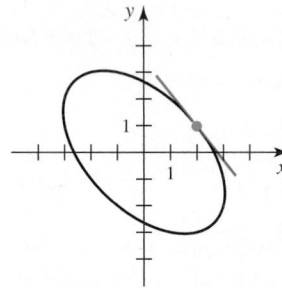 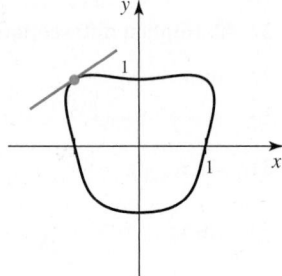

49. $\cos(x - y) + \sin y = \sqrt{2}$;   **50.** $(x^2 + y^2)^2 = \dfrac{25}{4}xy^2$; $(1, 2)$
$\left(\dfrac{\pi}{2}, \dfrac{\pi}{4}\right)$

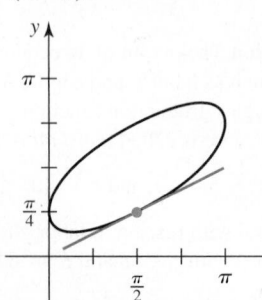 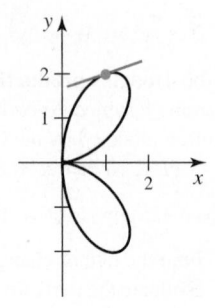

51–56. Second derivatives *Find d^2y/dx^2.*

51. $x + y^2 = 1$ **52.** $2x^2 + y^2 = 4$

53. $x + y = \sin y$ **54.** $x^4 + y^4 = 64$

55. $e^{2y} + x = y$ **56.** $\sin x + x^2y = 10$

57. Explain why or why not Determine whether the following statements are true and give an explanation or counterexample.

    a. For any equation containing the variables x and y, the derivative dy/dx can be found by first using algebra to rewrite the equation in the form $y = f(x)$.

    b. For the equation of a circle of radius r, $x^2 + y^2 = r^2$, we have $\dfrac{dy}{dx} = -\dfrac{x}{y}$, for $y \neq 0$ and any real number $r > 0$.

    c. If $x = 1$, then by implicit differentiation, $1 = 0$.

    d. If $xy = 1$, then $y' = 1/x$.

58–59. *Carry out the following steps.*

a. *Use implicit differentiation to find $\dfrac{dy}{dx}$.*

b. *Find the slope of the curve at the given point.*

58. $xy^{5/2} + x^{3/2}y = 12$; $(4, 1)$ **59.** $xy + x^{3/2}y^{-1/2} = 2$; $(1, 1)$

[T] 60–62. Multiple tangent lines *Complete the following steps.*

a. *Find equations of all lines tangent to the curve at the given value of x.*

b. *Graph the tangent lines on the given graph.*

60. $x + y^3 - y = 1$; $x = 1$ **61.** $x + y^2 - y = 1$; $x = 1$

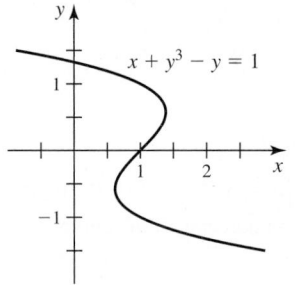 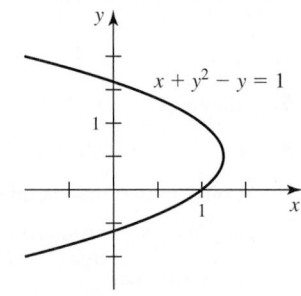

62. $4x^3 = y^2(4 - x)$; $x = 2$
(cissoid of Diocles)

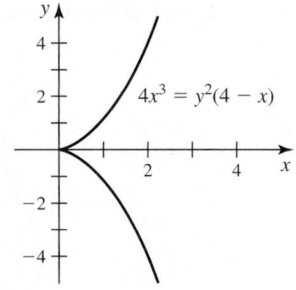

63. Witch of Agnesi Let $y(x^2 + 4) = 8$ (see figure).

    a. Use implicit differentiation to find $\dfrac{dy}{dx}$.

    b. Find equations of all lines tangent to the curve $y(x^2 + 4) = 8$ when $y = 1$.

    c. Solve the equation $y(x^2 + 4) = 8$ for y to find an explicit expression for y and then calculate $\dfrac{dy}{dx}$.

    d. Verify that the results of parts (a) and (c) are consistent.

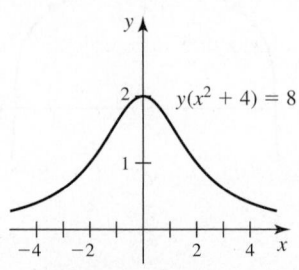

64. Vertical tangent lines

a. Determine the points where the curve $x + y^3 - y = 1$ has a vertical tangent line (see Exercise 60).

b. Does the curve have any horizontal tangent lines? Explain.

65. Vertical tangent lines

a. Determine the points where the curve $x + y^2 - y = 1$ has a vertical tangent line (Exercise 61).

b. Does the curve have any horizontal tangent lines? Explain.

T 66–67. Tangent lines for ellipses *Find the equations of the vertical and horizontal tangent lines of the following ellipses.*

66. $x^2 + 4y^2 + 2xy = 12$

67. $9x^2 + y^2 - 36x + 6y + 36 = 0$

Explorations and Challenges

T 68–72. Identifying functions from an equation *The following equations implicitly define one or more functions.*

a. *Find $\dfrac{dy}{dx}$ using implicit differentiation.*

b. *Solve the given equation for y to identify the implicitly defined functions $y = f_1(x), y = f_2(x), \ldots$.*

c. *Use the functions found in part (b) to graph the given equation.*

68. $y^3 = ax^2$ (Neile's semicubical parabola)

69. $x + y^3 - xy = 1$ (*Hint:* Rewrite as $y^3 - 1 = xy - x$ and then factor both sides.)

70. $y^2 = \dfrac{x^2(4 - x)}{4 + x}$ (right strophoid)

71. $x^4 = 2(x^2 - y^2)$ (eight curve)

72. $y^2(x + 2) = x^2(6 - x)$ (trisectrix)

T 73–78. Normal lines *A normal line at a point P on a curve passes through P and is perpendicular to the line tangent to the curve at P (see figure). Use the following equations and graphs to determine an equation of the normal line at the given point. Illustrate your work by graphing the curve with the normal line.*

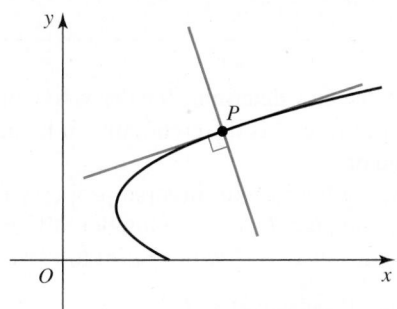

73. Exercise 45   **74.** Exercise 46

75. Exercise 47   **76.** Exercise 48

77. Exercise 49   **78.** Exercise 50

T 79–82. Visualizing tangent and normal lines

a. *Determine an equation of the tangent line and the normal line at the given point (x_0, y_0) on the following curves. (See instructions for Exercises 73–78.)*

b. *Graph the tangent and normal lines on the given graph.*

79. $3x^3 + 7y^3 = 10y$; $(x_0, y_0) = (1, 1)$

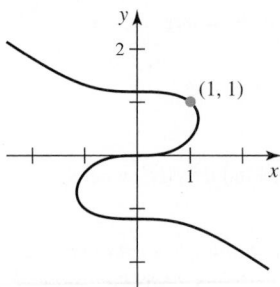

80. $x^4 = 2x^2 + 2y^2$; $(x_0, y_0) = (2, 2)$ (kampyle of Eudoxus)

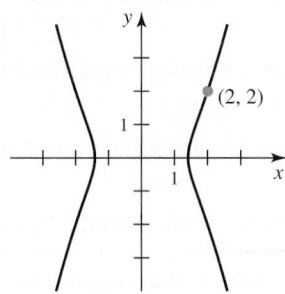

81. $(x^2 + y^2 - 2x)^2 = 2(x^2 + y^2)$; $(x_0, y_0) = (2, 2)$ (limaçon of Pascal)

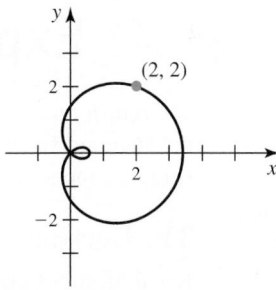

82. $(x^2 + y^2)^2 = \dfrac{25}{3}(x^2 - y^2)$; $(x_0, y_0) = (2, -1)$ (lemniscate of Bernoulli)

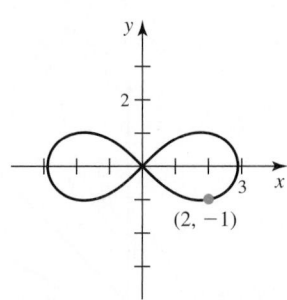

83–85. Orthogonal trajectories *Two curves are **orthogonal** to each other if their tangent lines are perpendicular at each point of intersection (recall that two lines are perpendicular to each other if their slopes are negative reciprocals). A family of curves forms **orthogonal trajectories** with another family of curves if each curve in one family is orthogonal to each curve in the other family. For example, the parabolas $y = cx^2$ form orthogonal trajectories with the family of ellipses $x^2 + 2y^2 = k$, where c and k are constants (see figure).*

Find dy/dx for each equation of the following pairs. Use the derivatives to explain why the families of curves form orthogonal trajectories.

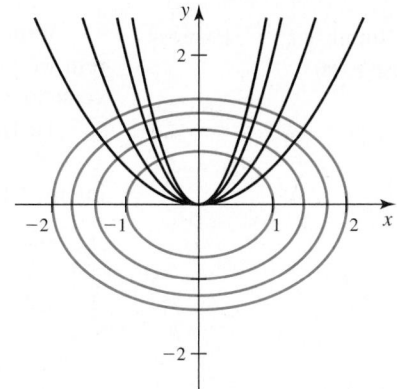

83. $y = mx$; $x^2 + y^2 = a^2$, where m and a are constants

84. $y = cx^2$; $x^2 + 2y^2 = k$, where c and k are constants

85. $xy = a$; $x^2 - y^2 = b$, where a and b are constants

86. Finding slope Find the slope of the curve $5\sqrt{x} - 10\sqrt{y} = \sin x$ at the point $(4\pi, \pi)$.

87. A challenging derivative Find dy/dx, where $(x^2 + y^2)(x^2 + y^2 + x) = 8xy^2$.

88. A challenging derivative Find dy/dx, where $\sqrt{3x^7 + y^2} = \sin^2 y + 100xy$.

89. A challenging second derivative Find d^2y/dx^2, where $\sqrt{y} + xy = 1$.

▣ 90–93. Work carefully *Proceed with caution when using implicit differentiation to find points at which a curve has a specified slope. For the following curves, find the points on the curve (if they exist) at which the tangent line is horizontal or vertical. Once you have found possible points, make sure that they actually lie on the curve. Confirm your results with a graph.*

90. $y^2 - 3xy = 2$ **91.** $x^2(3y^2 - 2y^3) = 4$

92. $x^2(y - 2) - e^y = 0$ **93.** $x(1 - y^2) + y^3 = 0$

QUICK CHECK ANSWERS

1. $y = \sqrt{x}$ and $y = -\sqrt{x}$ **2.** $\dfrac{dy}{dx} = \dfrac{1}{2y}$

3. Only the x-coordinate is needed. ◄

3.9 Derivatives of Logarithmic and Exponential Functions

We return now to the major theme of this chapter: developing rules of differentiation for the standard families of functions. First, we discover how to differentiate the natural logarithmic function. From there, we treat general exponential and logarithmic functions.

The Derivative of $y = \ln x$

Recall from Section 1.3 that the natural exponential function $f(x) = e^x$ is a one-to-one function on the interval $(-\infty, \infty)$. Therefore, it has an inverse, which is the natural logarithmic function $f^{-1}(x) = \ln x$. The domain of f^{-1} is the range of f, which is $(0, \infty)$. The graphs of f and f^{-1} are symmetric about the line $y = x$ (**Figure 3.62**). This inverse relationship has several important consequences, summarized as follows.

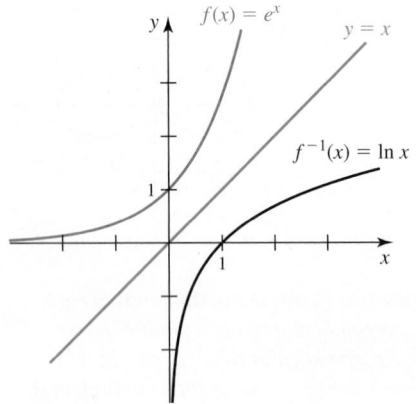

Figure 3.62

➤ A careful development of property (3) is given in Section 7.1.

➤ Figure 3.62 also provides evidence that $\ln x$ is differentiable for $x > 0$: Its graph is smooth with no jumps or cusps.

QUICK CHECK 1 Simplify $e^{2 \ln x}$. Express 5^x using the base e. ◄

Inverse Properties for e^x and $\ln x$

1. $e^{\ln x} = x$, for $x > 0$, and $\ln(e^x) = x$, for all x.

2. $y = \ln x$ if and only if $x = e^y$.

3. For real numbers x and $b > 0$, $b^x = e^{\ln b^x} = e^{x \ln b}$.

With these preliminary observations, we now determine the derivative of $\ln x$. A theorem we prove in Section 3.10 says that because e^x is differentiable on its domain, its inverse $\ln x$ is also differentiable on its domain.

To find the derivative of $y = \ln x$, we begin with inverse property (2) and write $x = e^y$, where $x > 0$. The key step is to compute dy/dx with implicit differentiation. Using the Chain Rule to differentiate both sides of $x = e^y$ with respect to x, we have

$$x = e^y \qquad \text{\small $y = \ln x$ if and only if $x = e^y$}$$

$$1 = e^y \cdot \frac{dy}{dx} \qquad \text{\small Differentiate both sides with respect to x.}$$

$$\frac{dy}{dx} = \frac{1}{e^y} = \frac{1}{x}. \qquad \text{\small Solve for dy/dx and use $x = e^y$.}$$

Therefore,

$$\frac{d}{dx}(\ln x) = \frac{1}{x}.$$

Because the domain of the natural logarithm is $(0, \infty)$, this rule is limited to positive values of x (**Figure 3.63a**).

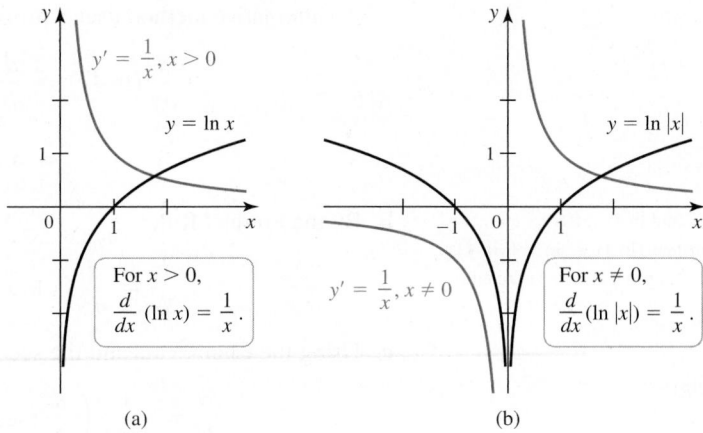

Figure 3.63

▶ Recall that

$$|x| = \begin{cases} x & \text{if } x \geq 0 \\ -x & \text{if } x < 0. \end{cases}$$

An important extension is obtained by considering the function $\ln |x|$, which is defined for all $x \neq 0$. By the definition of absolute value,

$$\ln |x| = \begin{cases} \ln x & \text{if } x > 0 \\ \ln (-x) & \text{if } x < 0. \end{cases}$$

For $x > 0$, it follows immediately that

$$\frac{d}{dx} (\ln |x|) = \frac{d}{dx} (\ln x) = \frac{1}{x}.$$

When $x < 0$, a similar calculation using the Chain Rule reveals that

$$\frac{d}{dx} (\ln |x|) = \frac{d}{dx} (\ln (-x)) = \frac{1}{(-x)} (-1) = \frac{1}{x}.$$

Therefore, we have the result that the derivative of $\ln |x|$ is $\frac{1}{x}$, for $x \neq 0$ (**Figure 3.63b**).

Taking these results one step further and using the Chain Rule to differentiate $\ln |u(x)|$, we obtain the following theorem.

THEOREM 3.15 Derivative of ln x

$$\frac{d}{dx} (\ln x) = \frac{1}{x}, \text{ for } x > 0 \qquad \frac{d}{dx} (\ln |x|) = \frac{1}{x}, \text{ for } x \neq 0$$

If u is differentiable at x and $u(x) \neq 0$, then

$$\frac{d}{dx} (\ln |u(x)|) = \frac{u'(x)}{u(x)}.$$

EXAMPLE 1 Derivatives involving ln x Find $\dfrac{dy}{dx}$ for the following functions.

a. $y = \ln 4x$ **b.** $y = x \ln x$ **c.** $y = \ln |\sec x|$ **d.** $y = \dfrac{\ln x^2}{x^2}$

SOLUTION

a. Using the Chain Rule,

$$\frac{dy}{dx} = \frac{d}{dx} (\ln 4x) = \frac{1}{4x} \cdot 4 = \frac{1}{x}.$$

An alternative method uses a property of logarithms before differentiating:

$$\frac{d}{dx}(\ln 4x) = \frac{d}{dx}(\ln 4 + \ln x) \quad \ln xy = \ln x + \ln y$$

$$= 0 + \frac{1}{x} = \frac{1}{x}. \quad \ln 4 \text{ is a constant.}$$

> ➤ Because ln x and ln $4x$ differ by an additive constant (ln $4x$ = ln x + ln 4), the derivatives of ln x and ln $4x$ are equal.

b. By the Product Rule,

$$\frac{dy}{dx} = \frac{d}{dx}(x \ln x) = 1 \cdot \ln x + x \cdot \frac{1}{x} = \ln x + 1.$$

c. Using the Chain Rule and the second part of Theorem 3.15,

$$\frac{dy}{dx} = \frac{1}{\sec x}\left(\frac{d}{dx}(\sec x)\right) = \frac{1}{\sec x}(\sec x \tan x) = \tan x.$$

QUICK CHECK 2 Find $\dfrac{d}{dx}(\ln x^p)$, where $x > 0$ and p is a real number, in two ways: (1) using the Chain Rule and (2) by first using a property of logarithms. ◄

d. The Quotient Rule and Chain Rule give

$$\frac{dy}{dx} = \frac{x^2\left(\frac{1}{x^2} \cdot 2x\right) - (\ln x^2)\,2x}{(x^2)^2} = \frac{2x - 2x \ln x^2}{x^4} = \frac{2(1 - \ln x^2)}{x^3}.$$

Related Exercises 15, 16, 19 ◄

The Derivative of b^x

A rule similar to $\dfrac{d}{dx}(e^x) = e^x$ exists for computing the derivative of b^x, where $b > 0$. Because $b^x = e^{x \ln b}$ by inverse property (3), its derivative is

$$\frac{d}{dx}(b^x) = \frac{d}{dx}(e^{x \ln b}) = \underbrace{e^{x \ln b}}_{b^x} \cdot \ln b. \quad \text{Chain Rule with } \frac{d}{dx}(x \ln b) = \ln b$$

Noting that $e^{x \ln b} = b^x$ results in the following theorem.

> ➤ Check that when $b = e$, Theorem 3.16 becomes
> $$\frac{d}{dx}(e^x) = e^x.$$

THEOREM 3.16   Derivative of b^x
If $b > 0$ and $b \neq 1$, then for all x,

$$\frac{d}{dx}(b^x) = b^x \ln b.$$

Notice that when $b > 1$, ln $b > 0$ and the graph of $y = b^x$ has tangent lines with positive slopes for all x. When $0 < b < 1$, ln $b < 0$ and the graph of $y = b^x$ has tangent lines with negative slopes for all x. In either case, the tangent line at $(0, 1)$ has slope ln b (Figure 3.64).

EXAMPLE 2   Derivatives with b^x Find the derivative of the following functions.

a. $f(x) = 3^x$ **b.** $g(t) = 108 \cdot 2^{t/12}$

SOLUTION

a. Using Theorem 3.16, $f'(x) = 3^x \ln 3$.

b.
$$g'(t) = 108\,\frac{d}{dt}(2^{t/12}) \quad\quad \text{Constant Multiple Rule}$$

$$= 108 \cdot \ln 2 \cdot 2^{t/12} \underbrace{\frac{d}{dt}\left(\frac{t}{12}\right)}_{1/12} \quad \text{Chain Rule}$$

$$= 9 \ln 2 \cdot 2^{t/12} \quad\quad \text{Simplify.}$$

Related Exercises 37, 39 ◄

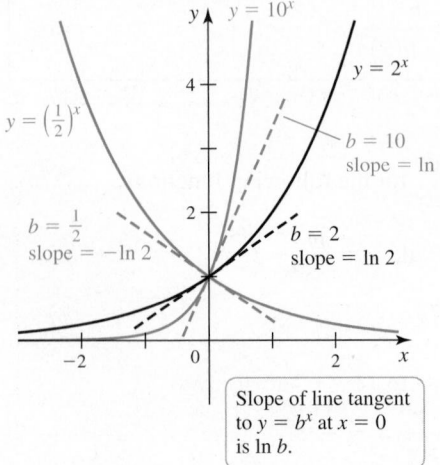

Figure 3.64

Table 3.6

Mother's Age	Incidence of Down Syndrome	Decimal Equivalent
30	1 in 900	0.00111
35	1 in 400	0.00250
36	1 in 300	0.00333
37	1 in 230	0.00435
38	1 in 180	0.00556
39	1 in 135	0.00741
40	1 in 105	0.00952
42	1 in 60	0.01667
44	1 in 35	0.02857
46	1 in 20	0.05000
48	1 in 16	0.06250
49	1 in 12	0.08333

(*Source:* E.G. Hook and A. Lindsjo, *The American Journal of Human Genetics,* 30, Jan 1978)

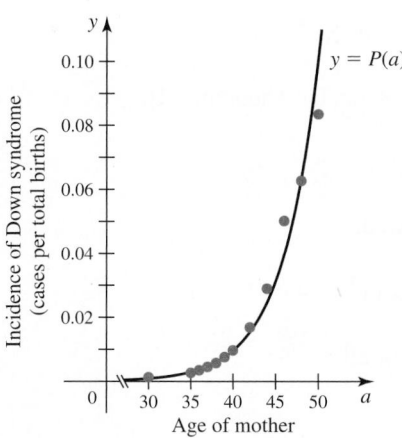

Figure 3.65

➤ The model in Example 3 was created using a method called *exponential regression*. The parameters A and B are chosen so that the function $P(a) = A \cdot B^a$ fits the data as closely as possible.

QUICK CHECK 3 Suppose $A = 500(1.045)^t$. Compute $\dfrac{dA}{dt}$. ◄

EXAMPLE 3   An exponential model Table 3.6 and Figure 3.65 show how the incidence of Down syndrome in newborn infants increases with the age of the mother. The data can be modeled with the exponential function $P(a) = \dfrac{1}{1,613,000} 1.2733^a$, where a is the age of the mother (in years) and $P(a)$ is the incidence (number of Down syndrome children per total births).

a. According to the model, at what age is the incidence of Down syndrome equal to 0.01 (that is, 1 in 100)?

b. Compute $P'(a)$.

c. Find $P'(35)$ and $P'(46)$, and interpret each.

SOLUTION

a. We let $P(a) = 0.01$ and solve for a:

$$0.01 = \frac{1}{1,613,000} 1.2733^a$$

$\ln 16,130 = \ln(1.2733^a)$ Multiply both sides by 1,613,000 and take logarithms of both sides.

$\ln 16,130 = a \ln 1.2733$ Property of logarithms

$a = \dfrac{\ln 16,130}{\ln 1.2733} \approx 40$ (years old).   Solve for a.

b. $P'(a) = \dfrac{1}{1,613,000} \dfrac{d}{da}(1.2733^a)$

$= \dfrac{1}{1,613,000} 1.2733^a \ln 1.2733$

$\approx \dfrac{1}{6,676,000} 1.2733^a$

c. The derivative measures the rate of change of the incidence with respect to age. For a 35-year-old woman,

$$P'(35) = \frac{1}{6,676,000} 1.2733^{35} \approx 0.0007,$$

which means the incidence increases at a rate of about 0.0007/year. By age 46, the rate of change is

$$P'(46) = \frac{1}{6,676,000} 1.2733^{46} \approx 0.01,$$

which is a significant increase over the rate of change of the incidence at age 35.

Related Exercises 56–57 ◄

The General Power Rule

In Theorem 3.9 of Section 3.4, we claimed that the Power Rule for derivatives, $\dfrac{d}{dx}(x^p) = px^{p-1}$, is valid for real powers p. We now have the tools to prove this important result.

THEOREM 3.17   General Power Rule

For real numbers p and for $x > 0$,

$$\frac{d}{dx}(x^p) = px^{p-1}.$$

Furthermore, if u is a positive differentiable function on its domain, then

$$\frac{d}{dx}(u(x)^p) = p(u(x))^{p-1} \cdot u'(x).$$

Proof: For $x > 0$ and real numbers p, we have $x^p = e^{p \ln x}$ by inverse property (3). Therefore, the derivative of x^p is computed as follows:

$$\frac{d}{dx}(x^p) = \frac{d}{dx}(e^{p \ln x}) \quad \text{Inverse property (3)}$$

$$= e^{p \ln x} \cdot \frac{p}{x} \quad \text{Chain Rule, } \frac{d}{dx}(p \ln x) = \frac{p}{x}$$

$$= x^p \cdot \frac{p}{x} \quad e^{p \ln x} = x^p$$

$$= px^{p-1}. \quad \text{Simplify.}$$

We see that $\dfrac{d}{dx}(x^p) = px^{p-1}$ for all real powers p. The second part of the General Power Rule follows from the Chain Rule. ◄

EXAMPLE 4 **Computing derivatives** Find the derivative of the following functions.

a. $y = x^\pi$ **b.** $y = \pi^x$ **c.** $y = (x^2 + 4)^e$

SOLUTION

➤ Recall that power functions have the variable in the base, while exponential functions have the variable in the exponent.

a. With $y = x^\pi$, we have a power function with an irrational exponent; by the General Power Rule,

$$\frac{dy}{dx} = \pi x^{\pi - 1}, \text{ for } x > 0.$$

b. Here we have an exponential function with base $b = \pi$. By Theorem 3.16,

$$\frac{dy}{dx} = \pi^x \ln \pi.$$

c. The Chain Rule and General Power Rule are required:

$$\frac{dy}{dx} = e(x^2 + 4)^{e-1} \cdot 2x = 2ex\,(x^2 + 4)^{e-1}.$$

Because $x^2 + 4 > 0$, for all x, the result is valid for all x.

Related Exercises 10, 33, 35 ◄

Functions of the form $f(x) = (g(x))^{h(x)}$, where both g and h are nonconstant functions, are neither exponential functions nor power functions (they are sometimes called *tower functions*). To compute their derivatives, we use the identity $b^x = e^{x \ln b}$ to rewrite f with base e:

$$f(x) = (g(x))^{h(x)} = e^{h(x) \ln g(x)}.$$

This function carries the restriction $g(x) > 0$. The derivative of f is then computed using the methods developed in this section. A specific case is illustrated in the following example.

EXAMPLE 5 **Tower function** Let $f(x) = x^{\sin x}$, for $x \geq 0$.

a. Find $f'(x)$. **b.** Evaluate $f'\left(\dfrac{\pi}{2}\right)$.

SOLUTION

a. The key step is to use $b^x = e^{x \ln b}$ to write f in the form

$$f(x) = x^{\sin x} = e^{(\sin x) \ln x}.$$

We now differentiate:

$$f'(x) = e^{(\sin x)\ln x}\frac{d}{dx}\left((\sin x)\ln x\right) \qquad \text{Chain Rule}$$

$$= \underbrace{e^{(\sin x)\ln x}}_{x^{\sin x}}\left((\cos x)\ln x + \frac{\sin x}{x}\right) \qquad \text{Product Rule}$$

$$= x^{\sin x}\left((\cos x)\ln x + \frac{\sin x}{x}\right).$$

b. Letting $x = \dfrac{\pi}{2}$, we find that

$$f'\left(\frac{\pi}{2}\right) = \left(\frac{\pi}{2}\right)^{\sin(\pi/2)}\left(\underbrace{\left(\cos\frac{\pi}{2}\right)}_{0}\ln\frac{\pi}{2} + \underbrace{\frac{\sin(\pi/2)}{\pi/2}}_{2/\pi}\right) \qquad \text{Substitute } x = \frac{\pi}{2}.$$

$$= \frac{\pi}{2}\left(0 + \frac{2}{\pi}\right) = 1.$$

Related Exercises 49, 59 ◄

EXAMPLE 6 Finding a horizontal tangent line Determine whether the graph of $f(x) = x^x$, for $x > 0$, has any horizontal tangent lines.

SOLUTION A horizontal tangent occurs when $f'(x) = 0$. To find the derivative, we first write $f(x) = x^x = e^{x\ln x}$:

$$\frac{d}{dx}(x^x) = \frac{d}{dx}(e^{x\ln x})$$

$$= \underbrace{e^{x\ln x}}_{x^x}\left(1\cdot\ln x + x\cdot\frac{1}{x}\right) \qquad \text{Chain Rule; Product Rule}$$

$$= x^x(\ln x + 1). \qquad \text{Simplify; } e^{x\ln x} = x^x.$$

The equation $f'(x) = 0$ implies that $x^x = 0$ or $\ln x + 1 = 0$. The first equation has no solution because $x^x = e^{x\ln x} > 0$, for all $x > 0$. We solve the second equation, $\ln x + 1 = 0$, as follows:

$$\ln x = -1$$
$$e^{\ln x} = e^{-1} \qquad \text{Exponentiate both sides.}$$
$$x = \frac{1}{e}. \qquad e^{\ln x} = x$$

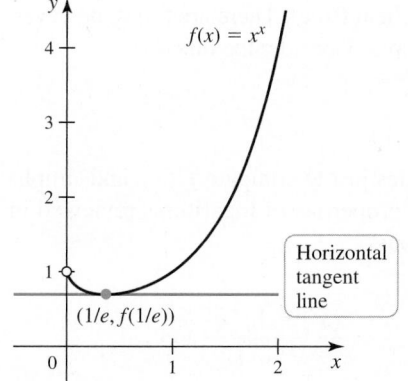
Figure 3.66

Therefore, the graph of $f(x) = x^x$ (**Figure 3.66**) has a single horizontal tangent at $(e^{-1}, f(e^{-1})) \approx (0.368, 0.692)$.

Related Exercises 60–61 ◄

Derivatives of General Logarithmic Functions

The general exponential function $f(x) = b^x$ is one-to-one when $b > 0$ with $b \neq 1$. The inverse function $f^{-1}(x) = \log_b x$ is the logarithmic function with base b. The technique used to differentiate the natural logarithm applies to the general logarithmic function. We begin with the inverse relationship

$$y = \log_b x \Leftrightarrow x = b^y, \text{ where } x > 0.$$

Differentiating both sides of $x = b^y$ with respect to x, we obtain

$$1 = b^y \ln b \cdot \frac{dy}{dx} \qquad \text{Implicit differentiation}$$
$$\frac{dy}{dx} = \frac{1}{b^y \ln b} \qquad \text{Solve for } \frac{dy}{dx}.$$
$$\frac{dy}{dx} = \frac{1}{x \ln b}. \qquad b^y = x.$$

A similar argument applied to $y = \log_b |x|$ with $x < 0$ leads to the following theorem.

▶ An alternative proof of Theorem 3.18 uses the change-of-base formula

$$\log_b x = \frac{\ln x}{\ln b} \text{ (Section 1.3).}$$

Differentiating both sides of this equation gives the same result.

> **THEOREM 3.18 Derivative of $\log_b x$**
> If $b > 0$ and $b \neq 1$, then
> $$\frac{d}{dx}(\log_b x) = \frac{1}{x \ln b}, \text{ for } x > 0 \quad \text{and} \quad \frac{d}{dx}(\log_b |x|) = \frac{1}{x \ln b}, \text{ for } x \neq 0.$$

EXAMPLE 7 Derivatives with general logarithms Compute the derivative of the following functions.

a. $f(x) = \log_5(2x + 1)$ **b.** $T(n) = n \log_2 n$

QUICK CHECK 4 Compute dy/dx for $y = \log_3 x$. ◀

SOLUTION

a. We use Theorem 3.18 with the Chain Rule assuming $2x + 1 > 0$:

$$f'(x) = \frac{1}{(2x + 1)\ln 5} \cdot 2 = \frac{2}{\ln 5} \cdot \frac{1}{2x + 1}. \quad \text{Chain Rule}$$

b.
$$T'(n) = \log_2 n + n \cdot \frac{1}{n \ln 2} = \log_2 n + \frac{1}{\ln 2} \quad \text{Product Rule}$$

We can change bases and write the result in base e:

$$T'(n) = \frac{\ln n}{\ln 2} + \frac{1}{\ln 2} = \frac{\ln n + 1}{\ln 2}.$$

Related Exercises 63–64 ◀

▶ The function in Example 7b is used in computer science as an estimate of the computing time needed to carry out a *sorting algorithm* on a list of n items.

QUICK CHECK 5 Show that the derivative computed in Example 7b can be expressed in base 2 as $T'(n) = \log_2(en)$. ◀

Logarithmic Differentiation

Products, quotients, and powers of functions are usually differentiated using the derivative rules of the same name (perhaps combined with the Chain Rule). There are times, however, when the direct computation of a derivative is very tedious. Consider the function

$$f(x) = \frac{(x^3 - 1)^4 \sqrt{3x - 1}}{x^2 + 4}.$$

We would need the Quotient, Product, and Chain Rules just to compute $f'(x)$, and simplifying the result would require additional work. The properties of logarithms reviewed in Section 1.3 are useful for differentiating such functions.

▶ The properties of logarithms needed for logarithmic differentiation (where $x > 0$, $y > 0$, and z is any real number):
1. $\ln xy = \ln x + \ln y$
2. $\ln(x/y) = \ln x - \ln y$
3. $\ln x^z = z \ln x$
All three properties are used in Example 8.

EXAMPLE 8 Logarithmic differentiation Let $f(x) = \dfrac{(x^2 + 1)^4 e^x}{x^2 + 4}$ and compute $f'(x)$.

SOLUTION We begin by taking the natural logarithm of both sides and simplifying the result:

$$\ln f(x) = \ln \left(\frac{(x^2 + 1)^4 e^x}{x^2 + 4} \right)$$

$$= \ln(x^2 + 1)^4 + \ln e^x - \ln(x^2 + 4) \quad \ln xy = \ln x + \ln y$$

$$= 4 \ln(x^2 + 1) + x - \ln(x^2 + 4). \quad \ln x^y = y \ln x; \ln e^x = x$$

▶ In the event that $f \leq 0$ for some values of x, $\ln f(x)$ is not defined. In that case, we generally find the derivative of $|y| = |f(x)|$.

We now differentiate both sides using the Chain Rule; specifically, the derivative of the left side is $\dfrac{d}{dx}(\ln f(x)) = \dfrac{f'(x)}{f(x)}$. Therefore,

$$\frac{f'(x)}{f(x)} = 4 \cdot \frac{1}{x^2 + 1} \cdot 2x + 1 - \frac{1}{x^2 + 4} \cdot 2x.$$

Solving for $f'(x)$, we have

$$f'(x) = f(x) \left(\frac{8x}{x^2 + 1} + 1 - \frac{2x}{x^2 + 4} \right).$$

Finally, we replace $f(x)$ with the original function:

$$f'(x) = \frac{(x^2 + 1)^4 e^x}{x^2 + 4}\left(\frac{8x}{x^2 + 1} - \frac{2x}{x^2 + 4} + 1\right).$$

Related Exercises 77, 80 ◄

Logarithmic differentiation also provides an alternative method for finding derivatives of tower functions, which are functions of the form $g(x)^{h(x)}$. The derivative of $f(x) = x^x$ (Example 6) is computed as follows, assuming $x > 0$:

$$f(x) = x^x$$

$$\ln f(x) = \ln x^x = x \ln x \qquad \text{Take logarithms of both sides; use properties.}$$

$$\frac{1}{f(x)} f'(x) = 1 \cdot \ln x + x \cdot \frac{1}{x} \qquad \text{Differentiate both sides.}$$

$$f'(x) = f(x)(\ln x + 1) \qquad \text{Solve for } f'(x) \text{ and simplify.}$$

$$f'(x) = x^x (\ln x + 1). \qquad \text{Replace } f(x) \text{ with } x^x.$$

This result agrees with Example 6. Which method to use is largely a matter of personal preference.

SECTION 3.9 EXERCISES

Getting Started

1. Use $x = e^y$ to explain why $\dfrac{d}{dx}(\ln x) = \dfrac{1}{x}$, for $x > 0$.

2. Sketch the graph of $f(x) = \ln |x|$ and explain how the graph shows that $f'(x) = \dfrac{1}{x}$.

3. Show that $\dfrac{d}{dx}(\ln kx) = \dfrac{d}{dx}(\ln x)$, where $x > 0$ and k is a positive real number.

4. State the derivative rule for the exponential function $f(x) = b^x$. How does it differ from the derivative formula for e^x?

5. State the derivative rule for the logarithmic function $f(x) = \log_b x$. How does it differ from the derivative formula for $\ln x$?

6. Explain why $b^x = e^{x \ln b}$.

7. Simplify the expression $e^{x \ln (x^2 + 1)}$.

8. Find $\dfrac{d}{dx}\left(\ln\left(\dfrac{x}{x^2 + 1}\right)\right)$ without using the Quotient Rule.

9. Find $\dfrac{d}{dx}\left(\ln \sqrt{x^2 + 1}\right)$.

10. Evaluate $\dfrac{d}{dx}(x^e + e^x)$

11. Express the function $f(x) = g(x)^{h(x)}$ in terms of the natural logarithm and natural exponential functions (base e).

12. Explain why $\dfrac{d}{dx}\left(\ln \sqrt{f(x)}\right) = \dfrac{1}{2} f'(x)/f(x)$, where $f(x) \geq 0$ for all x.

13. Find $\dfrac{d}{dx}(\ln (xe^x))$ without using the Chain Rule and the Product Rule.

14. Find $\dfrac{d}{dx}(\ln x^{101})$ without using the Chain Rule.

Practice Exercises

15–48. Derivatives *Find the derivative of the following functions.*

15. $y = \ln 7x$

16. $y = x^2 \ln x$

17. $y = \ln x^2$

18. $y = \ln 2x^8$

19. $y = \ln |\sin x|$

20. $y = \dfrac{1 + \ln x^3}{4x^3}$

21. $y = \ln (x^4 + 1)$

22. $y = \ln \sqrt{x^4 + x^2}$

23. $y = \ln\left(\dfrac{x + 1}{x - 1}\right)$

24. $y = x \ln x - x$

25. $y = (x^2 + 1)\ln x$

26. $y = \ln |x^2 - 1|$

27. $y = x^2(1 - \ln x^2)$

28. $y = 3x^3 \ln x - x^3$

29. $y = \ln (\ln x)$

30. $y = \ln (\cos^2 x)$

31. $y = \dfrac{\ln x}{\ln x + 1}$

32. $y = \ln (e^x + e^{-x})$

33. $y = x^e$

34. $y = e^x x^e$

35. $y = (2^x + 1)^\pi$

36. $y = \ln (x^3 + 1)^\pi$

37. $y = 8^x$

38. $y = 5^{3t}$

39. $y = 5 \cdot 4^x$

40. $y = 4^{-x} \sin x$

41. $y = 2^{3 + \sin x}$

42. $y = 10^{\ln 2x}$

43. $y = x^3 3^x$

44. $P = \dfrac{40}{1 + 2^{-t}}$

45. $A = 250(1.045)^{4t}$

46. $y = 10^x(\ln 10^x - 1)$

47. $f(x) = \dfrac{2^x}{2^x + 1}$

48. $s(t) = \cos 2^t$

49–55. Derivatives of tower functions (or g^h) *Find the derivative of each function and evaluate the derivative at the given value of a.*

49. $f(x) = x^{\cos x}; a = \pi/2$

50. $g(x) = x^{\ln x}; a = e$

51. $h(x) = x^{\sqrt{x}};\ a = 4$

52. $f(x) = (x^2 + 1)^x;\ a = 1$

53. $f(x) = (\sin x)^{\ln x};\ a = \pi/2$ **54.** $f(x) = (\tan x)^{x-1};\ a = \pi/4$

55. $f(x) = (4\sin x + 2)^{\cos x};\ a = \pi$

56. Magnitude of an earthquake The energy (in joules) released by an earthquake of magnitude M is given by the equation $E = 25{,}000 \cdot 10^{1.5M}$. (This equation can be solved for M to define the magnitude of a given earthquake; it is a refinement of the original Richter scale created by Charles Richter in 1935.)

   a. Compute the energy released by earthquakes of magnitude 1, 2, 3, 4, and 5. Plot the points on a graph and join them with a smooth curve.

   b. Compute dE/dM and evaluate it for $M = 3$. What does this derivative mean? (M has no units, so the units of the derivative are J per change in magnitude.)

57. Exponential model The following table shows the *time of useful consciousness* at various altitudes in the situation where a pressurized airplane suddenly loses pressure. The change in pressure drastically reduces available oxygen, and hypoxia sets in. The upper value of each time interval is roughly modeled by $T = 10 \cdot 2^{-0.274a}$, where T measures time in minutes and a is the altitude over 22,000 in thousands of feet ($a = 0$ corresponds to 22,000 ft).

Altitude (in ft)	Time of useful consciousness
22,000	5 to 10 min
25,000	3 to 5 min
28,000	2.5 to 3 min
30,000	1 to 2 min
35,000	30 to 60 s
40,000	15 to 20 s
45,000	9 to 15 s

   a. A Learjet flying at 38,000 ft ($a = 16$) suddenly loses pressure when the seal on a window fails. According to this model, how long do the pilot and passengers have to deploy oxygen masks before they become incapacitated?

   b. What is the average rate of change of T with respect to a over the interval from 24,000 to 30,000 ft (include units)?

   c. Find the instantaneous rate of change dT/da, compute it at 30,000 ft, and interpret its meaning.

58. Diagnostic scanning Iodine-123 is a radioactive isotope used in medicine to test the function of the thyroid gland. If a 350-microcurie (μCi) dose of iodine-123 is administered to a patient, the quantity Q left in the body after t hours is approximately $Q = 350(1/2)^{t/13.1}$.

   a. How long does it take for the level of iodine-123 to drop to 10 μCi?

   b. Find the rate of change of the quantity of iodine-123 at 12 hours, 1 day, and 2 days. What do your answers say about the rate at which iodine decreases as time increases?

59–62. Tangent lines

59. Find an equation of the line tangent to $y = x^{\sin x}$ at the point $x = 1$.

60. Determine whether the graph of $y = x^{\sqrt{x}}$ has any horizontal tangent lines.

61. The graph of $y = (x^2)^x$ has two horizontal tangent lines. Find equations for both of them.

62. The graph of $y = x^{\ln x}$ has one horizontal tangent line. Find an equation for it.

63–74. Derivatives of logarithmic functions *Calculate the derivative of the following functions. In some cases, it is useful to use the properties of logarithms to simplify the functions before computing $f'(x)$.*

63. $y = 4\log_3 (x^2 - 1)$

64. $y = \log_{10} x$

65. $y = (\cos x)\ln \cos^2 x$

66. $y = \log_8 |\tan x|$

67. $y = \dfrac{1}{\log_4 x}$

68. $y = \log_2 (\log_2 x)$

69. $f(x) = \ln (3x + 1)^4$

70. $f(x) = \ln \dfrac{2x}{(x^2 + 1)^3}$

71. $f(x) = \ln \sqrt{10x}$

72. $f(x) = \log_2 \dfrac{8}{\sqrt{x+1}}$

73. $f(x) = \ln \dfrac{(2x - 1)(x + 2)^3}{(1 - 4x)^2}$

74. $f(x) = \ln (\sec^4 x \tan^2 x)$

75–86. Logarithmic differentiation *Use logarithmic differentiation to evaluate $f'(x)$.*

75. $f(x) = x^{10x}$

76. $f(x) = (2x)^{2x}$

77. $f(x) = \dfrac{(x + 1)^{10}}{(2x - 4)^8}$

78. $f(x) = x^2 \cos x$

79. $f(x) = x^{\ln x}$

80. $f(x) = \dfrac{\tan^{10} x}{(5x + 3)^6}$

81. $f(x) = \dfrac{(x + 1)^{3/2}(x - 4)^{5/2}}{(5x + 3)^{2/3}}$

82. $f(x) = \dfrac{x^8 \cos^3 x}{\sqrt{x - 1}}$

83. $f(x) = (\sin x)^{\tan x}$

84. $f(x) = (1 + x^2)^{\sin x}$

85. $f(x) = \left(1 + \dfrac{1}{x}\right)^x$

86. $f(x) = x^{(x^{10})}$

87. Explain why or why not Determine whether the following statements are true and give an explanation or counterexample.

   a. The derivative of $\log_2 9$ is $1/(9\ln 2)$.

   b. $\ln (x + 1) + \ln (x - 1) = \ln (x^2 - 1)$, for all x.

   c. The exponential function 2^{x+1} can be written in base e as $e^{2\ln (x+1)}$.

   d. $\dfrac{d}{dx}(\sqrt{2}^x) = x\sqrt{2}^{x-1}$.

   e. $\dfrac{d}{dx}(x^{\sqrt{2}}) = \sqrt{2}x^{\sqrt{2}-1}$.

   f. $(4x + 1)^{\ln x} = x^{\ln(4x+1)}$.

88–91. Higher-order derivatives *Find the following higher-order derivatives.*

88. $\dfrac{d^2}{dx^2}(\ln (x^2 + 1))$

89. $\dfrac{d^2}{dx^2}(\log_{10} x)$

90. $\dfrac{d^n}{dx^n}(2^x)$

91. $\dfrac{d^3}{dx^3}(x^2 \ln x)$

92–94. Derivatives by different methods *Calculate the derivative of the following functions (i) using the fact that $b^x = e^{x \ln b}$ and (ii) using logarithmic differentiation. Verify that both answers are the same.*

92. $y = (x^2 + 1)^x$　　**93.** $y = 3^x$　　**94.** $y = (4x + 1)^{\ln x}$

T 95. Tangent line Find the equation of the line tangent to $y = 2^{\sin x}$ at $x = \pi/2$. Graph the function and the tangent line.

Explorations and Challenges

T 96. Horizontal tangents The graph of $y = (\cos x) \ln \cos^2 x$ has seven horizontal tangent lines on the interval $[0, 2\pi]$. Find the approximate x-coordinates of all points at which these tangent lines occur.

97–100. Logistic growth *Scientists often use the logistic growth function $P(t) = \dfrac{P_0 K}{P_0 + (K - P_0)e^{-r_0 t}}$ to model population growth, where P_0 is the initial population at time $t = 0$, K is the **carrying capacity**, and r_0 is the base growth rate. The carrying capacity is a theoretical upper bound on the total population that the surrounding environment can support. The figure shows the sigmoid (S-shaped) curve associated with a typical logistic model.*

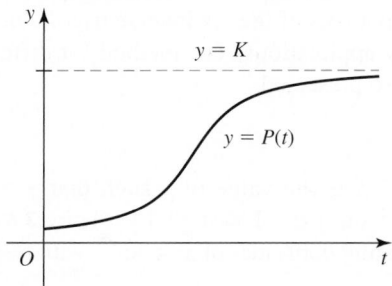

T 97. Gone fishing When a reservoir is created by a new dam, 50 fish are introduced into the reservoir, which has an estimated carrying capacity of 8000 fish. A logistic model of the fish population is
$$P(t) = \frac{400{,}000}{50 + 7950e^{-0.5t}},$$ where t is measured in years.

   a. Graph P using a graphing utility. Experiment with different windows until you produce an S-shaped curve characteristic of the logistic model. What window works well for this function?

   b. How long does it take for the population to reach 5000 fish? How long does it take for the population to reach 90% of the carrying capacity?

   c. How fast (in fish per year) is the population growing at $t = 0$? At $t = 5$?

   d. Graph P' and use the graph to estimate the year in which the population is growing fastest.

T 98. World population (part 1) The population of the world reached 6 billion in 1999 ($t = 0$). Assume Earth's carrying capacity is 15 billion and the base growth rate is $r_0 = 0.025$ per year.

   a. Write a logistic growth function for the world's population (in billions), and graph your equation on the interval $0 \le t \le 200$ using a graphing utility.

   b. What will the population be in the year 2020? When will it reach 12 billion?

99. World population (part 2) The *relative growth rate* r of a function f measures the rate of change of the function compared to its value at a particular point. It is computed as $r(t) = f'(t)/f(t)$.

   a. Confirm that the relative growth rate in 1999 ($t = 0$) for the logistic model in Exercise 98 is $r(0) = P'(0)/P(0) = 0.015$.

This means the world's population was growing at 1.5% per year in 1999.

   b. Compute the relative growth rate of the world's population in 2010 and 2020. What appears to be happening to the relative growth rate as time increases?

   c. Evaluate $\displaystyle\lim_{t \to \infty} r(t) = \lim_{t \to \infty} \frac{P'(t)}{P(t)}$, where $P(t)$ is the logistic growth function from Exercise 98. What does your answer say about populations that follow a logistic growth pattern?

T 100. Population crash The logistic model can be used for situations in which the initial population P_0 is above the carrying capacity K. For example, consider a deer population of 1500 on an island where a large fire has reduced the carrying capacity to 1000 deer.

   a. Assuming a base growth rate of $r_0 = 0.1$ and an initial population of $P(0) = 1500$, write a logistic growth function for the deer population and graph it. Based on the graph, what happens to the deer population in the long run?

   b. How fast (in deer per year) is the population declining immediately after the fire at $t = 0$?

   c. How long does it take for the deer population to decline to 1200 deer?

101. Savings plan Beginning at age 30, a self-employed plumber saves $250 per month in a retirement account until he reaches age 65. The account offers 6% interest, compounded monthly. The balance in the account after t years is given by $A(t) = 50{,}000(1.005^{12t} - 1)$.

   a. Compute the balance in the account after 5, 15, 25, and 35 years. What is the average rate of change in the value of the account over the intervals $[5, 15]$, $[15, 25]$, and $[25, 35]$?

   b. Suppose the plumber started saving at age 25 instead of age 30. Find the balance at age 65 (after 40 years of investing).

   c. Use the derivative dA/dt to explain the surprising result in part (b) and the advice: Start saving for retirement as early as possible.

T 102. Tangency question It is easily verified that the graphs of $y = x^2$ and $y = e^x$ have no point of intersection (for $x > 0$), while the graphs of $y = x^3$ and $y = e^x$ have two points of intersection. It follows that for some real number $2 < p < 3$, the graphs of $y = x^p$ and $y = e^x$ have exactly one point of intersection (for $x > 0$). Using analytical and/or graphical methods, determine p and the coordinates of the single point of intersection.

T 103. Tangency question It is easily verified that the graphs of $y = 1.1^x$ and $y = x$ have two points of intersection, and the graphs of $y = 2^x$ and $y = x$ have no point of intersection. It follows that for some real number $1.1 < p < 2$, the graphs of $y = p^x$ and $y = x$ have exactly one point of intersection. Using analytical and/or graphical methods, determine p and the coordinates of the single point of intersection.

T 104. Triple intersection Graph the functions $f(x) = x^3$, $g(x) = 3^x$, and $h(x) = x^x$ and find their common intersection point (exactly).

105–108. Calculating limits exactly *Use the definition of the derivative to evaluate the following limits.*

105. $\displaystyle\lim_{x \to e} \frac{\ln x - 1}{x - e}$

106. $\displaystyle\lim_{h \to 0} \frac{\ln(e^8 + h) - 8}{h}$

107. $\displaystyle\lim_{h \to 0} \frac{(3 + h)^{3+h} - 27}{h}$

108. $\displaystyle\lim_{x \to 2} \frac{5^x - 25}{x - 2}$

109. Derivative of $u(x)^{v(x)}$ Use logarithmic differentiation to prove that

$$\frac{d}{dx}\left(u(x)^{v(x)}\right) = u(x)^{v(x)}\left(\frac{dv}{dx}\ln u(x) + \frac{v(x)}{u(x)}\frac{du}{dx}\right).$$

110. Tangent lines and exponentials Assume b is given with $b > 0$ and $b \neq 1$. Find the y-coordinate of the point on the curve $y = b^x$ at which the tangent line passes through the origin. (*Source: The College Mathematics Journal*, 28, Mar 1997)

QUICK CHECK ANSWERS

1. x^2 (for $x > 0$); $e^{x\ln 5}$ **2.** Either way, $\dfrac{d}{dx}(\ln x^p) = \dfrac{p}{x}$.

3. $\dfrac{dA}{dt} = 500(1.045)^t \cdot \ln 1.045 \approx 22(1.045)^t$ **4.** $\dfrac{dy}{dx} = \dfrac{1}{x\ln 3}$

5. $T'(n) = \log_2 n + \dfrac{1}{\ln 2} = \log_2 n + \dfrac{1}{\dfrac{\log_2 2}{\log_2 e}} = $

$$\log_2 n + \log_2 e = \log_2(en) \blacktriangleleft$$

3.10 Derivatives of Inverse Trigonometric Functions

The inverse trigonometric functions, introduced in Section 1.4, are major players in calculus. In this section, we develop the derivatives of the six inverse trigonometric functions and begin an exploration of their many applications. The method for differentiating the inverses of more general functions is also presented.

Inverse Sine and Its Derivative

Recall from Section 1.4 that $y = \sin^{-1} x$ is the value of y such that $x = \sin y$, where $-\pi/2 \le y \le \pi/2$. The domain of $\sin^{-1} x$ is $\{x: -1 \le x \le 1\}$ (**Figure 3.67**). The derivative of $y = \sin^{-1} x$ follows by differentiating both sides of $x = \sin y$ with respect to x, simplifying, and solving for dy/dx:

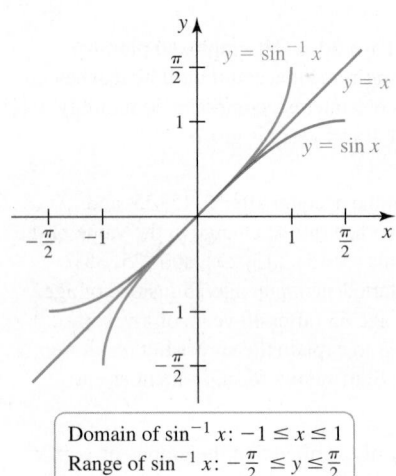

Domain of $\sin^{-1} x$: $-1 \le x \le 1$
Range of $\sin^{-1} x$: $-\frac{\pi}{2} \le y \le \frac{\pi}{2}$

Figure 3.67

$$x = \sin y \qquad y = \sin^{-1} x \iff x = \sin y$$

$$\frac{d}{dx}(x) = \frac{d}{dx}(\sin y) \quad \text{Differentiate with respect to } x.$$

$$1 = (\cos y)\frac{dy}{dx} \quad \text{Chain Rule on the right side}$$

$$\frac{dy}{dx} = \frac{1}{\cos y}. \quad \text{Solve for } \frac{dy}{dx}.$$

The identity $\sin^2 y + \cos^2 y = 1$ is used to express this derivative in terms of x. Solving for $\cos y$ yields

$$\cos y = \pm\sqrt{1 - \underbrace{\sin^2 y}_{x^2}} \quad x = \sin y \implies x^2 = \sin^2 y$$

$$= \pm\sqrt{1 - x^2}.$$

Because y is restricted to the interval $-\pi/2 \le y \le \pi/2$, we have $\cos y \ge 0$. Therefore, we choose the positive branch of the square root, and it follows that

$$\frac{dy}{dx} = \frac{d}{dx}(\sin^{-1} x) = \frac{1}{\sqrt{1 - x^2}}.$$

This result is consistent with the graph of $f(x) = \sin^{-1} x$ (**Figure 3.68**).

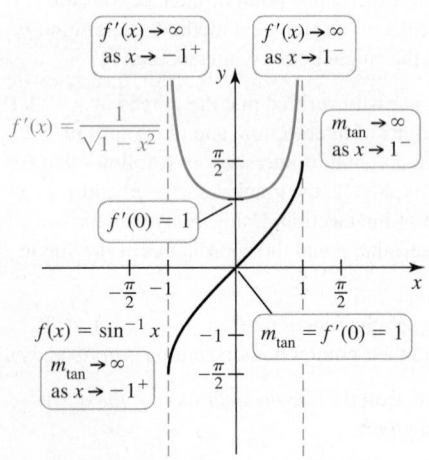

$f'(x) = \dfrac{1}{\sqrt{1 - x^2}}$

$f'(x) \to \infty$ as $x \to -1^+$

$f'(x) \to \infty$ as $x \to 1^-$

$f'(0) = 1$

$m_{\tan} \to \infty$ as $x \to 1^-$

$f(x) = \sin^{-1} x$

$m_{\tan} = f'(0) = 1$

$m_{\tan} \to \infty$ as $x \to -1^+$

Figure 3.68

QUICK CHECK 1 Is $f(x) = \sin^{-1} x$ an even or odd function? Is $f'(x)$ an even or odd function? ◄

THEOREM 3.19   Derivative of Inverse Sine

$$\frac{d}{dx}(\sin^{-1} x) = \frac{1}{\sqrt{1 - x^2}}, \text{ for } -1 < x < 1$$

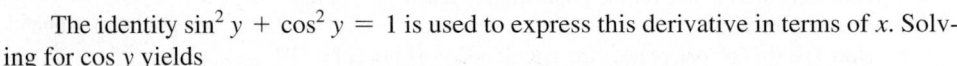

EXAMPLE 1 **Derivatives involving the inverse sine** Compute the following derivatives.

a. $\dfrac{d}{dx}\left(\sin^{-1}(x^2 - 1)\right)$ **b.** $\dfrac{d}{dx}\left(\cos(\sin^{-1} x)\right)$

SOLUTION We apply the Chain Rule for both derivatives.

a. $\dfrac{d}{dx}\left(\sin^{-1}\underbrace{(x^2 - 1)}_{u}\right) = \underbrace{\dfrac{1}{\sqrt{1 - (x^2 - 1)^2}}}_{\text{derivative of } \sin^{-1} u \text{ evaluated at } u = x^2 - 1} \cdot \underbrace{2x}_{u'(x)} = \dfrac{2x}{\sqrt{2x^2 - x^4}}$

▶ The result in Example 1b can also be obtained by noting that $\cos(\sin^{-1} x) = \sqrt{1 - x^2}$ and differentiating this expression (Exercise 87).

b. $\dfrac{d}{dx}\left(\cos(\underbrace{\sin^{-1} x}_{u})\right) = \underbrace{-\sin(\sin^{-1} x)}_{\substack{\text{derivative of the} \\ \text{outer function } \cos u \\ \text{evaluated at } u = \sin^{-1} x}} \cdot \underbrace{\dfrac{1}{\sqrt{1 - x^2}}}_{\substack{\text{derivative of the} \\ \text{inner function } \sin^{-1} x}} = -\dfrac{x}{\sqrt{1 - x^2}}$

This result is valid for $-1 < x < 1$, where $\sin(\sin^{-1} x) = x$.

Related Exercises 13, 15 ◀

Derivatives of Inverse Tangent and Secant

The derivatives of the inverse tangent and inverse secant are derived using a method similar to that used for the inverse sine. Once these three derivative results are known, the derivatives of the inverse cosine, cotangent, and cosecant follow immediately.

Inverse Tangent Recall from Section 1.4 that $y = \tan^{-1} x$ is the value of y such that $x = \tan y$, where $-\pi/2 < y < \pi/2$. The domain of $y = \tan^{-1} x$ is $\{x : -\infty < x < \infty\}$ (Figure 3.69). To find $\dfrac{dy}{dx}$, we differentiate both sides of $x = \tan y$ with respect to x and simplify:

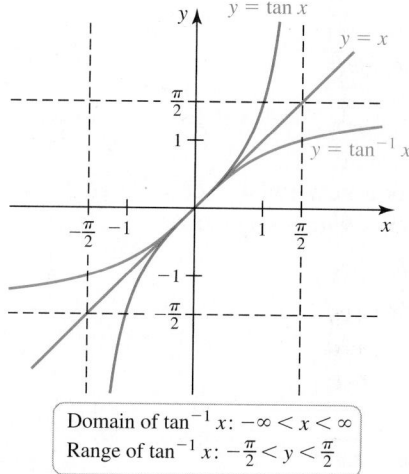

Domain of $\tan^{-1} x$: $-\infty < x < \infty$
Range of $\tan^{-1} x$: $-\frac{\pi}{2} < y < \frac{\pi}{2}$

Figure 3.69

$$x = \tan y \qquad\qquad y = \tan^{-1} x \iff x = \tan y$$

$$\dfrac{d}{dx}(x) = \dfrac{d}{dx}(\tan y) \quad \text{Differentiate with respect to } x.$$

$$1 = \sec^2 y \cdot \dfrac{dy}{dx} \quad \text{Chain Rule}$$

$$\dfrac{dy}{dx} = \dfrac{1}{\sec^2 y}. \quad \text{Solve for } \dfrac{dy}{dx}.$$

To express this derivative in terms of x, we combine the trigonometric identity $\sec^2 y = 1 + \tan^2 y$ with $x = \tan y$ to obtain $\sec^2 y = 1 + x^2$. Substituting this result into the expression for dy/dx, it follows that

$$\dfrac{dy}{dx} = \dfrac{d}{dx}(\tan^{-1} x) = \dfrac{1}{1 + x^2}.$$

The graphs of the inverse tangent and its derivative (Figure 3.70) are informative. Letting $f(x) = \tan^{-1} x$ and $f'(x) = \dfrac{1}{1 + x^2}$, we see that $f'(0) = 1$, which is the maximum value of the derivative; that is, $\tan^{-1} x$ has its maximum slope at $x = 0$. As $x \to \infty$, $f'(x)$ approaches zero; likewise, as $x \to -\infty$, $f'(x)$ approaches zero.

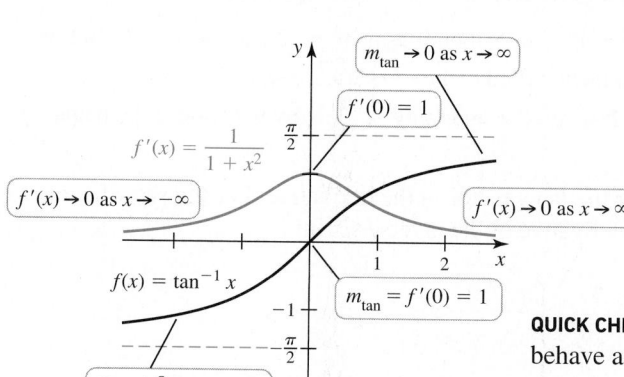

Figure 3.70

QUICK CHECK 2 How do the slopes of the lines tangent to the graph of $y = \tan^{-1} x$ behave as $x \to \infty$? ◀

Inverse Secant   Recall from Section 1.4 that $y = \sec^{-1} x$ is the value of y such that $x = \sec y$, where $0 \le y \le \pi$, with $y \ne \pi/2$. The domain of $y = \sec^{-1} x$ is $\{x: |x| \ge 1\}$ (Figure 3.71).

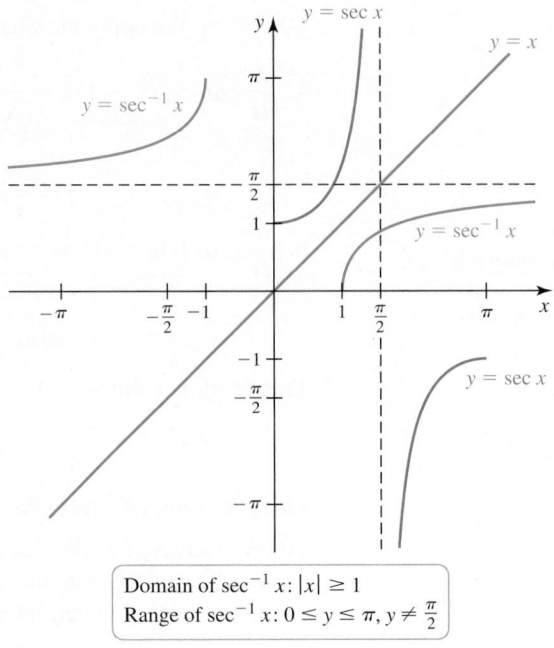

Domain of $\sec^{-1} x: |x| \ge 1$
Range of $\sec^{-1} x: 0 \le y \le \pi, y \ne \dfrac{\pi}{2}$

Figure 3.71

The derivative of the inverse secant presents a new twist. Let $y = \sec^{-1} x$, or $x = \sec y$, and then differentiate both sides of $x = \sec y$ with respect to x:

$$1 = \sec y \tan y \frac{dy}{dx}.$$

Solving for $\dfrac{dy}{dx}$ produces

$$\frac{dy}{dx} = \frac{d}{dx}\left(\sec^{-1} x\right) = \frac{1}{\sec y \tan y}.$$

The final step is to express $\sec y \tan y$ in terms of x by using the identity $\sec^2 y = 1 + \tan^2 y$. Solving this equation for $\tan y$, we have

$$\tan y = \pm \sqrt{\underbrace{\sec^2 y}_{x^2} - 1} = \pm \sqrt{x^2 - 1}.$$

Two cases must be examined to resolve the sign on the square root:

• By the definition of $y = \sec^{-1} x$, if $x \ge 1$, then $0 \le y < \pi/2$ and $\tan y > 0$. In this case, we choose the positive branch and take $\tan y = \sqrt{x^2 - 1}$.

• However, if $x \le -1$, then $\pi/2 < y \le \pi$ and $\tan y < 0$. Now we choose the negative branch.

This argument accounts for the $\tan y$ factor in the derivative. For the $\sec y$ factor, we have $\sec y = x$. Therefore, the derivative of the inverse secant is

$$\frac{d}{dx}\left(\sec^{-1} x\right) = \begin{cases} \dfrac{1}{x\sqrt{x^2 - 1}} & \text{if } x > 1 \\[2ex] -\dfrac{1}{x\sqrt{x^2 - 1}} & \text{if } x < -1, \end{cases}$$

which is an awkward result. The absolute value helps here: Recall that $|x| = x$, if $x > 0$, and $|x| = -x$, if $x < 0$. It follows that

$$\frac{d}{dx}(\sec^{-1} x) = \frac{1}{|x|\sqrt{x^2 - 1}}, \text{ for } |x| > 1.$$

We see that the slope of the inverse secant function is always positive, which is consistent with this derivative result (Figure 3.71).

Derivatives of Other Inverse Trigonometric Functions   The hard work is complete. The derivative of the inverse cosine results from the identity

$$\cos^{-1} x + \sin^{-1} x = \frac{\pi}{2}.$$

▶ This identity was proved in Example 5 of Section 1.4.

Differentiating both sides of this equation with respect to x, we find that

$$\frac{d}{dx}(\cos^{-1} x) + \underbrace{\frac{d}{dx}(\sin^{-1} x)}_{1/\sqrt{1-x^2}} = \underbrace{\frac{d}{dx}\left(\frac{\pi}{2}\right)}_{0}.$$

Solving for $\dfrac{d}{dx}(\cos^{-1} x)$ reveals that the required derivative is

$$\frac{d}{dx}(\cos^{-1} x) = -\frac{1}{\sqrt{1 - x^2}}.$$

In a similar manner, the analogous identities

$$\cot^{-1} x + \tan^{-1} x = \frac{\pi}{2} \quad \text{and} \quad \csc^{-1} x + \sec^{-1} x = \frac{\pi}{2}$$

are used to show that the derivatives of $\cot^{-1} x$ and $\csc^{-1} x$ are the negative of the derivatives of $\tan^{-1} x$ and $\sec^{-1} x$, respectively (Exercise 85).

QUICK CHECK 3 Summarize how the derivatives of inverse trigonometric functions are related to the derivatives of the corresponding inverse cofunctions (for example, inverse tangent and inverse cotangent). ◀

> **THEOREM 3.20   Derivatives of Inverse Trigonometric Functions**
>
> $$\frac{d}{dx}(\sin^{-1} x) = \frac{1}{\sqrt{1-x^2}} \qquad \frac{d}{dx}(\cos^{-1} x) = -\frac{1}{\sqrt{1-x^2}}, \text{ for } -1 < x < 1$$
>
> $$\frac{d}{dx}(\tan^{-1} x) = \frac{1}{1+x^2} \qquad \frac{d}{dx}(\cot^{-1} x) = -\frac{1}{1+x^2}, \text{ for } -\infty < x < \infty$$
>
> $$\frac{d}{dx}(\sec^{-1} x) = \frac{1}{|x|\sqrt{x^2-1}} \qquad \frac{d}{dx}(\csc^{-1} x) = -\frac{1}{|x|\sqrt{x^2-1}}, \text{ for } |x| > 1$$

EXAMPLE 2   Derivatives of inverse trigonometric functions

a. Evaluate $f'(2\sqrt{3})$, where $f(x) = x \tan^{-1}(x/2)$.

b. Find an equation of the line tangent to the graph of $g(x) = \sec^{-1} 2x$ at the point $(1, \pi/3)$.

SOLUTION

a. $f'(x) = 1 \cdot \tan^{-1}\dfrac{x}{2} + x \underbrace{\dfrac{1}{1 + (x/2)^2} \cdot \dfrac{1}{2}}_{\frac{d}{dx}(\tan^{-1}(x/2))}$ Product Rule and Chain Rule

$$= \tan^{-1}\frac{x}{2} + \frac{2x}{4 + x^2} \qquad \text{Simplify.}$$

We evaluate f' at $x = 2\sqrt{3}$ and note that $\tan^{-1}\sqrt{3} = \pi/3$:

$$f'(2\sqrt{3}) = \tan^{-1}\sqrt{3} + \frac{2(2\sqrt{3})}{4 + (2\sqrt{3})^2} = \frac{\pi}{3} + \frac{\sqrt{3}}{4}.$$

b. The slope of the tangent line at $(1, \pi/3)$ is $g'(1)$. Using the Chain Rule, we have

$$g'(x) = \frac{d}{dx}(\sec^{-1} 2x) = \frac{2}{|2x|\sqrt{4x^2 - 1}} = \frac{1}{|x|\sqrt{4x^2 - 1}}.$$

It follows that $g'(1) = 1/\sqrt{3}$. An equation of the tangent line is

$$y - \frac{\pi}{3} = \frac{1}{\sqrt{3}}(x - 1) \quad \text{or} \quad y = \frac{1}{\sqrt{3}}x + \frac{\pi}{3} - \frac{1}{\sqrt{3}}.$$

Related Exercises 27, 41 ◄

Figure 3.72

EXAMPLE 3 Shadows in a ballpark As the sun descends behind the 150-ft grandstand of a baseball stadium, the shadow of the stadium moves across the field (**Figure 3.72**). Let ℓ be the line segment between the edge of the shadow and the sun, and let θ be the angle of elevation of the sun—the angle between ℓ and the horizontal. The shadow length s is the distance between the edge of the shadow and the base of the grandstand.

a. Express θ as a function of the shadow length s.

b. Compute $d\theta/ds$ when $s = 200$ ft and explain what this rate of change measures.

SOLUTION

a. The tangent of θ is

$$\tan \theta = \frac{150}{s},$$

where $s > 0$. Taking the inverse tangent of both sides of this equation, we find that

$$\theta = \tan^{-1}\frac{150}{s}.$$

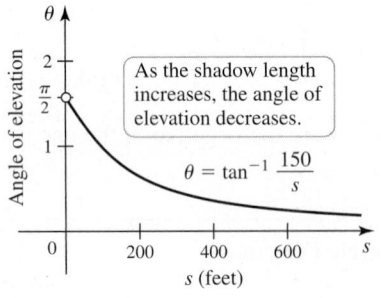

Figure 3.73

Figure 3.73 illustrates how the sun's angle of elevation θ approaches $\pi/2$ as the shadow length approaches zero ($\theta = \pi/2$ means the sun is overhead). As the shadow length increases, θ decreases and approaches zero.

b. Using the Chain Rule, we have

$$\frac{d\theta}{ds} = \frac{1}{1 + (150/s)^2}\frac{d}{ds}\left(\frac{150}{s}\right) \qquad \text{Chain Rule; } \frac{d}{du}(\tan^{-1} u) = \frac{1}{1 + u^2}$$

$$= \frac{1}{1 + (150/s)^2}\left(-\frac{150}{s^2}\right) \qquad \text{Evaluate the derivative.}$$

$$= -\frac{150}{s^2 + 22{,}500}. \qquad \text{Simplify.}$$

Notice that $d\theta/ds$ is negative for all values of s, which means longer shadows are associated with smaller angles of elevation (Figure 3.73). At $s = 200$ ft, we have

$$\left.\frac{d\theta}{ds}\right|_{s=200} = -\frac{150}{200^2 + 150^2} = -0.0024\,\frac{\text{rad}}{\text{ft}}.$$

QUICK CHECK 4 Example 3 makes the claim that $d\theta/ds = -0.0024$ rad/ft is equivalent to $-0.138°$/ft. Verify this claim. ◄

When the shadow length is $s = 200$ ft, the angle of elevation is changing at a rate of -0.0024 rad/ft, or $-0.138°$/ft.

Related Exercises 45–46 ◄

Derivatives of Inverse Functions in General

We found the derivatives of the inverse trigonometric functions using implicit differentiation. However, this approach does not always work. For example, suppose we know only f and its derivative f' and want to evaluate the derivative of f^{-1}. The key to finding the derivative of the inverse function lies in the symmetry of the graphs of f and f^{-1}.

EXAMPLE 4 Linear functions, inverses, and derivatives Consider the general linear function $y = f(x) = mx + b$, where $m \neq 0$ and b are constants.

a. Write the inverse of f in the form $y = f^{-1}(x)$.

b. Find the derivative of the inverse $\dfrac{d}{dx}(f^{-1}(x))$.

c. Consider the specific case $f(x) = 2x - 6$. Graph f and f^{-1}, and find the slope of each line.

SOLUTION

a. Solving $y = mx + b$ for x, we find that $mx = y - b$, or

$$x = \frac{y}{m} - \frac{b}{m}.$$

Writing this function in the form $y = f^{-1}(x)$ (by reversing the roles of x and y), we have

$$y = f^{-1}(x) = \frac{x}{m} - \frac{b}{m},$$

which describes a line with slope $1/m$.

b. The derivative of f^{-1} is

$$(f^{-1})'(x) = \frac{1}{m}.$$

Notice that $f'(x) = m$, so the derivative of f^{-1} is the reciprocal of f'.

c. In the case that $f(x) = 2x - 6$, we have $f^{-1}(x) = x/2 + 3$. The graphs of these two lines are symmetric about the line $y = x$ (Figure 3.74). Furthermore, the slope of the line $y = f(x)$ is 2 and the slope of $y = f^{-1}(x)$ is $\frac{1}{2}$; that is, the slopes (and therefore the derivatives) are reciprocals of each other.

Related Exercises 47–48 ◄

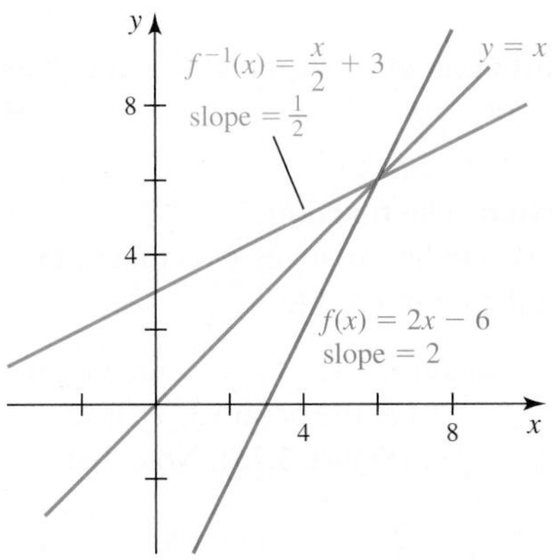

Figure 3.74

The reciprocal property obeyed by f' and $(f^{-1})'$ in Example 4 holds for all differentiable functions with inverses. **Figure 3.75** shows the graphs of a typical one-to-one function and its inverse. It also shows a pair of symmetric points—(x_0, y_0) on the graph of f and (y_0, x_0) on the graph of f^{-1}—along with the tangent lines at these points. Notice that as the lines tangent to the graph of f get steeper (as x increases), the corresponding lines tangent to the graph of f^{-1} get less steep. The next theorem makes this relationship precise.

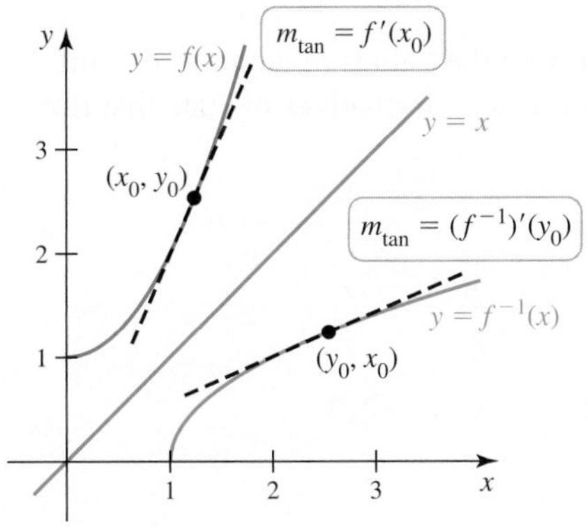

Figure 3.75

► The result of Theorem 3.21 is also written in the form

$$(f^{-1})'(f(x_0)) = \frac{1}{f'(x_0)}$$

or

$$(f^{-1})'(y_0) = \frac{1}{f'(f^{-1}(y_0))}.$$

THEOREM 3.21 Derivative of the Inverse Function

Let f be differentiable and have an inverse on an interval I. If x_0 is a point of I at which $f'(x_0) \neq 0$, then f^{-1} is differentiable at $y_0 = f(x_0)$ and

$$(f^{-1})'(y_0) = \frac{1}{f'(x_0)}, \quad \text{where} \quad y_0 = f(x_0).$$

To understand this theorem, suppose (x_0, y_0) is a point on the graph of f, which means that (y_0, x_0) is the corresponding point on the graph of f^{-1}. Then the slope of the line tangent to the graph of f^{-1} at the point (y_0, x_0) is the reciprocal of the slope of the line tangent to the graph of f at the point (x_0, y_0). Importantly, the theorem says that we can evaluate the derivative of the inverse function without finding the inverse function itself.

Proof: Before doing a short calculation, we note two facts:

• At a point x_0 where f is differentiable, $y_0 = f(x_0)$ and $x_0 = f^{-1}(y_0)$.

• As a differentiable function, f is continuous at x_0 (Theorem 3.1), which implies that f^{-1} is continuous at y_0 (Theorem 2.14). Therefore, as $y \to y_0$, $x \to x_0$.

Using the definition of the derivative, we have

$$(f^{-1})'(y_0) = \lim_{y \to y_0} \frac{f^{-1}(y) - f^{-1}(y_0)}{y - y_0} \qquad \text{Definition of derivative of } f^{-1}$$

$$= \lim_{x \to x_0} \frac{x - x_0}{f(x) - f(x_0)} \qquad y = f(x) \text{ and } x = f^{-1}(y); \ x \to x_0 \text{ as } y \to y_0$$

$$= \lim_{x \to x_0} \frac{1}{\dfrac{f(x) - f(x_0)}{x - x_0}} \qquad \frac{a}{b} = \frac{1}{b/a}$$

$$= \frac{1}{f'(x_0)}. \qquad \text{Definition of derivative of } f$$

We have shown that $(f^{-1})'(y_0)$ exists (f^{-1} is differentiable at y_0) and that it equals the reciprocal of $f'(x_0)$. ◄

QUICK CHECK 5 Sketch the graphs of $y = \sin x$ and $y = \sin^{-1} x$. Then verify that Theorem 3.21 holds at the point $(0, 0)$. ◄

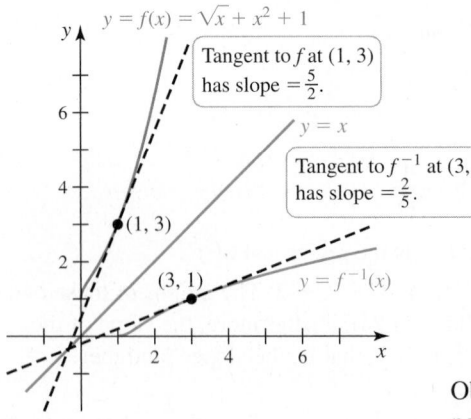

Figure 3.76

EXAMPLE 5 **Derivative of an inverse function** The function $f(x) = \sqrt{x} + x^2 + 1$ is one-to-one, for $x \geq 0$, and has an inverse on that interval. Find the slope of the curve $y = f^{-1}(x)$ at the point $(3, 1)$.

SOLUTION The point $(1, 3)$ is on the graph of f; therefore, $(3, 1)$ is on the graph of f^{-1}. In this case, the slope of the curve $y = f^{-1}(x)$ at the point $(3, 1)$ is the reciprocal of the slope of the curve $y = f(x)$ at $(1, 3)$ (**Figure 3.76**). Note that

$$f'(x) = \frac{1}{2\sqrt{x}} + 2x, \text{ which means that } f'(1) = \frac{1}{2} + 2 = \frac{5}{2}. \text{ Therefore,}$$

$$(f^{-1})'(3) = \frac{1}{f'(1)} = \frac{1}{5/2} = \frac{2}{5}.$$

Observe that it is not necessary to find a formula for f^{-1} to evaluate its derivative at a point.

Related Exercises 50–51 ◄

EXAMPLE 6 **Derivatives of an inverse function** Use the values of a one-to-one differentiable function f in Table 3.7 to compute the indicated derivatives or state that the derivative cannot be determined.

Table 3.7

x	-1	0	1	2	3
$f(x)$	2	3	5	6	7
$f'(x)$	$1/2$	2	$3/2$	1	$2/3$

a. $(f^{-1})'(5)$ **b.** $(f^{-1})'(2)$ **c.** $(f^{-1})'(1)$

SOLUTION We use the relationship $(f^{-1})'(y_0) = \dfrac{1}{f'(x_0)}$, where $y_0 = f(x_0)$.

a. In this case, $y_0 = f(x_0) = 5$. Using Table 3.7, we see that $x_0 = 1$ and $f'(1) = \dfrac{3}{2}$. Therefore, $(f^{-1})'(5) = \dfrac{1}{f'(1)} = \dfrac{2}{3}$.

b. In this case, $y_0 = f(x_0) = 2$, which implies that $x_0 = -1$ and $f'(-1) = \dfrac{1}{2}$. Therefore, $(f^{-1})'(2) = \dfrac{1}{f'(-1)} = 2$.

c. With $y_0 = f(x_0) = 1$, Table 3.7 does not supply a value of x_0. Therefore, neither $f'(x_0)$ nor $(f^{-1})'(1)$ can be determined.

Related Exercises 7–8 ◄

SECTION 3.10 EXERCISES

Getting Started

1. State the derivatives of $\sin^{-1} x$, $\tan^{-1} x$, and $\sec^{-1} x$.

2. Find the slope of the line tangent to the graph of $y = \sin^{-1} x$ at $x = 0$.

3. Find the slope of the line tangent to the graph of $y = \tan^{-1} x$ at $x = -2$.

4. How are the derivatives of $\sin^{-1} x$ and $\cos^{-1} x$ related?

5. Suppose f is a one-to-one function with $f(2) = 8$ and $f'(2) = 4$. What is the value of $(f^{-1})'(8)$?

6. Explain how to find $(f^{-1})'(y_0)$, given that $y_0 = f(x_0)$.

7–8. Derivatives of inverse functions from a table *Use the following tables to determine the indicated derivatives or state that the derivative cannot be determined.*

7.

x	-2	-1	0	1	2
$f(x)$	2	3	4	6	7
$f'(x)$	1	$1/2$	2	$3/2$	1

a. $(f^{-1})'(4)$ b. $(f^{-1})'(6)$ c. $(f^{-1})'(1)$ d. $f'(1)$

8.

x	-4	-2	0	2	4
$f(x)$	0	1	2	3	4
$f'(x)$	5	4	3	2	1

a. $f'(f(0))$ b. $(f^{-1})'(0)$ c. $(f^{-1})'(1)$ d. $(f^{-1})'(f(4))$

9. If f is a one-to-one function with $f(3) = 8$ and $f'(3) = 7$, find the equation of the line tangent to $y = f^{-1}(x)$ at $x = 8$.

10. The line tangent to the graph of the one-to-one function $y = f(x)$ at $x = 3$ is $y = 5x + 1$. Find $f^{-1}(16)$ and $(f^{-1})'(16)$.

11. Find the slope of the curve $y = \sin^{-1} x$ at $\left(\dfrac{1}{2}, \dfrac{\pi}{6}\right)$ without calculating the derivative of $\sin^{-1} x$.

12. Find the slope of the curve $y = \tan^{-1} x$ at $\left(1, \dfrac{\pi}{4}\right)$ without calculating the derivative of $\tan^{-1} x$.

Practice Exercises

13–40. *Evaluate the derivative of the following functions.*

13. $f(x) = \sin^{-1} 2x$

14. $f(x) = x \sin^{-1} x$

15. $f(w) = \cos(\sin^{-1} 2w)$

16. $f(x) = \sin^{-1}(\ln x)$

17. $f(x) = \sin^{-1}(e^{-2x})$

18. $f(x) = \sin^{-1}(e^{\sin x})$

19. $f(x) = \tan^{-1} 10x$

20. $f(x) = 2x \tan^{-1} x - \ln(1 + x^2)$

21. $f(y) = \tan^{-1}(2y^2 - 4)$

22. $g(z) = \tan^{-1}(1/z)$

23. $f(z) = \cot^{-1}\sqrt{z}$

24. $f(x) = \sec^{-1}\sqrt{x}$

25. $f(x) = x^2 + 2x^3 \cot^{-1} x - \ln(1 + x^2)$

26. $f(x) = x \cos^{-1} x - \sqrt{1 - x^2}$

27. $f(w) = w^2 - \tan^{-1} w^2$

28. $f(t) = \ln(\sin^{-1} t^2)$

29. $f(x) = \cos^{-1}(1/x)$

30. $f(t) = (\cos^{-1} t)^2$

31. $f(u) = \csc^{-1}(2u + 1)$

32. $f(t) = \ln(\tan^{-1} t)$

33. $f(y) = \cot^{-1}(1/(y^2 + 1))$

34. $f(w) = \sin(\sec^{-1} 2w)$

35. $f(x) = \sec^{-1}(\ln x)$

36. $f(x) = \tan^{-1}(e^{4x})$

37. $f(x) = \csc^{-1}(\tan e^x)$

38. $f(x) = \sin(\tan^{-1}(\ln x))$

39. $f(s) = \cot^{-1}(e^s)$

40. $f(x) = 1/\tan^{-1}(x^2 + 4)$

41–44. Tangent lines *Find an equation of the line tangent to the graph of f at the given point.*

41. $f(x) = \tan^{-1} 2x$; $(1/2, \pi/4)$

42. $f(x) = \sin^{-1}\dfrac{x}{4}$; $(2, \pi/6)$

43. $f(x) = \cos^{-1} x^2$; $(1/\sqrt{2}, \pi/3)$

44. $f(x) = \sec^{-1}(e^x)$; $(\ln 2, \pi/3)$

T 45. Angular size A boat sails directly toward a 150-meter skyscraper that stands on the edge of a harbor. The angular size θ of the building is the angle formed by lines from the top and bottom of the building to the observer (see figure).

a. What is the rate of change of the angular size $d\theta/dx$ when the boat is $x = 500$ m from the building?

b. Graph $d\theta/dx$ as a function of x and determine the point at which the angular size changes most rapidly.

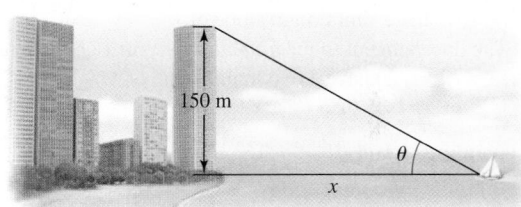

T 46. Angle of elevation A small plane, moving at 70 m/s, flies horizontally on a line 400 meters directly above an observer. Let θ be the angle of elevation of the plane (see figure).

a. What is the rate of change of the angle of elevation $d\theta/dx$ when the plane is $x = 500$ m past the observer?

b. Graph $d\theta/dx$ as a function of x and determine the point at which θ changes most rapidly.

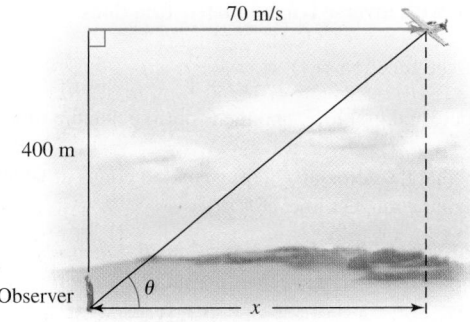

47–56. Derivatives of inverse functions at a point *Consider the following functions. In each case, without finding the inverse, evaluate the derivative of the inverse at the given point.*

47. $f(x) = 3x + 4$; $(16, 4)$ **48.** $f(x) = \frac{1}{2}x + 8$; $(10, 4)$

49. $f(x) = \ln(5x + e)$; $(1, 0)$

50. $f(x) = x^2 + 1$, for $x \geq 0$; $(5, 2)$

51. $f(x) = \tan x$; $(1, \pi/4)$

52. $f(x) = x^2 - 2x - 3$, for $x \leq 1$; $(12, -3)$

53. $f(x) = \sqrt{x}$; $(2, 4)$ **54.** $f(x) = 4e^{10x}$; $(4, 0)$

55. $f(x) = (x + 2)^2$; $(36, 4)$ **56.** $f(x) = \log_{10} 3x$; $\left(0, \frac{1}{3}\right)$

57. Find $(f^{-1})'(3)$, where $f(x) = x^3 + x + 1$.

58. Suppose the slope of the curve $y = f(x)$ at $(7, 4)$ is $2/3$. Find the slope of the curve $y = f^{-1}(x)$ at $(4, 7)$.

59. Suppose the slope of the curve $y = f^{-1}(x)$ at $(4, 7)$ is $4/5$. Find $f'(7)$.

60. Suppose the slope of the curve $y = f(x)$ at $(4, 7)$ is $1/5$. Find $(f^{-1})'(7)$.

61. **Explain why or why not** Determine whether the following statements are true and give an explanation or counterexample.

    a. $\frac{d}{dx}(\sin^{-1}x + \cos^{-1}x) = 0$

    b. $\frac{d}{dx}(\tan^{-1}x) = \sec^2 x$

    c. The lines tangent to the graph of $y = \sin^{-1}x$ on the interval $[-1, 1]$ have a minimum slope of 1.

    d. The lines tangent to the graph of $y = \sin x$ on the interval $[-\pi/2, \pi/2]$ have a maximum slope of 1.

    e. If $f(x) = 1/x$, then $(f^{-1})'(x) = -1/x^2$.

■ 62–65. Graphing f and f'

    a. *Graph f with a graphing utility.*

    b. *Compute and graph f'.*

    c. *Verify that the zeros of f' correspond to points at which f has a horizontal tangent line.*

62. $f(x) = (x - 1)\sin^{-1} x$ on $[-1, 1]$

63. $f(x) = (x^2 - 1)\sin^{-1} x$ on $[-1, 1]$

64. $f(x) = (\sec^{-1}x)/x$ on $[1, \infty)$

65. $f(x) = e^{-x}\tan^{-1} x$ on $[0, \infty)$

■ 66. **Graphing with inverse trigonometric functions**

    a. Graph the function $f(x) = \dfrac{\tan^{-1}x}{x^2 + 1}$.

    b. Compute and graph f', and then approximate the roots of $f'(x)$.

    c. Verify that the zeros of f' correspond to points at which f has a horizontal tangent line.

67–78. Derivatives of inverse functions *Consider the following functions (on the given interval, if specified). Find the derivative of the inverse function.*

67. $f(x) = 3x - 4$ **68.** $f(x) = x^3 + 3$

69. $f(x) = x^2 - 4$, for $x > 0$ **70.** $f(x) = \dfrac{x}{x + 5}$

71. $f(x) = e^{3x+1}$ **72.** $f(x) = \ln(5x + 4)$

73. $f(x) = 10^{12x-6}$ **74.** $f(x) = \log_{10}(2x + 6)$

75. $f(x) = \sqrt{x + 2}$

76. $f(x) = x^{2/3}$, for $x > 0$

77. $f(x) = x^{-1/2}$, for $x > 0$

78. $f(x) = |x + 2|$

Explorations and Challenges

■ 79. **Towing a boat** A boat is towed toward a dock by a cable attached to a winch that stands 10 feet above the water level (see figure). Let θ be the angle of elevation of the winch and let ℓ be the length of the cable as the boat is towed toward the dock.

    a. Show that the rate of change of θ with respect to ℓ is
$$\frac{d\theta}{d\ell} = -\frac{10}{\ell\sqrt{\ell^2 - 100}}.$$

    b. Compute $\dfrac{d\theta}{d\ell}$ when $\ell = 50, 20,$ and 11 ft.

    c. Find $\lim\limits_{\ell \to 10^+} \dfrac{d\theta}{d\ell}$, and explain what happens as the last foot of cable is reeled in (note that the boat is at the dock when $\ell = 10$).

    d. It is evident from the figure that θ increases as the boat is towed to the dock. Why, then, is $d\theta/d\ell$ negative?

80. **Tracking a dive** A biologist standing at the bottom of an 80-foot vertical cliff watches a peregrine falcon dive from the top of the cliff at a 45° angle from the horizontal (see figure).

    a. Express the angle of elevation θ from the biologist to the falcon as a function of the height h of the bird above the ground. (*Hint:* The vertical distance between the top of the cliff and the falcon is $80 - h$.)

    b. What is the rate of change of θ with respect to the bird's height when it is 60 ft above the ground?

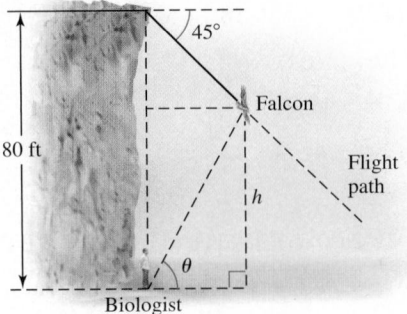

81. Angle to a particle (part 1) A particle travels clockwise on a circular path of diameter D, monitored by a sensor on the circle at point P; the other endpoint of the diameter on which the sensor lies is Q (see figure). Let θ be the angle between the diameter PQ and the line from the sensor to the particle. Let c be the length of the chord from the particle's position to Q.

a. Calculate $\dfrac{d\theta}{dc}$. **b.** Evaluate $\dfrac{d\theta}{dc}\Big|_{c=0}$.

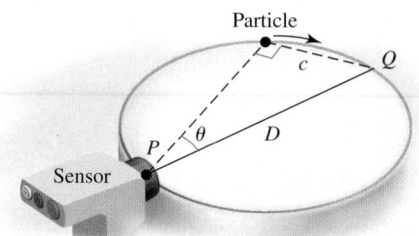

82. Angle to a particle (part 2) The figure in Exercise 81 shows the particle traveling away from the sensor, which may have influenced your solution (we expect you used the inverse sine function). Suppose instead that the particle approaches the sensor (see figure). How would this change the solution? Explain the differences in the two answers.

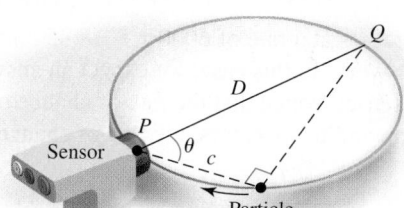

83. Derivative of the inverse sine Find the derivative of the inverse sine function using Theorem 3.21.

84. Derivative of the inverse cosine Find the derivative of the inverse cosine function in the following two ways.
  a. Using Theorem 3.21
  b. Using the identity $\sin^{-1}x + \cos^{-1}x = \pi/2$

85. Derivative of $\cot^{-1}x$ and $\csc^{-1}x$ Use a trigonometric identity to show that the derivatives of the inverse cotangent and inverse cosecant differ from the derivatives of the inverse tangent and inverse secant, respectively, by a multiplicative factor of -1.

86. Tangents and inverses Suppose $L(x) = ax + b$ (with $a \neq 0$) is the equation of the line tangent to the graph of a one-to-one function f at (x_0, y_0). Also, suppose $M(x) = cx + d$ is the equation of the line tangent to the graph of f^{-1} at (y_0, x_0).
  a. Express a and b in terms of x_0 and y_0.
  b. Express c in terms of a, and d in terms of a, x_0, and y_0.
  c. Prove that $L^{-1}(x) = M(x)$.

87–90. Identity proofs *Prove the following identities and give the values of x for which they are true.*

87. $\cos(\sin^{-1}x) = \sqrt{1-x^2}$ **88.** $\cos(2\sin^{-1}x) = 1 - 2x^2$

89. $\tan(2\tan^{-1}x) = \dfrac{2x}{1-x^2}$ **90.** $\sin(2\sin^{-1}x) = 2x\sqrt{1-x^2}$

QUICK CHECK ANSWERS

1. $f(x) = \sin^{-1}x$ is odd and $f'(x) = 1/\sqrt{1-x^2}$ is even. **2.** The slopes of the tangent lines approach 0. **3.** One is the negative of the other. **4.** Recall that $1° = \pi/180$ rad. So 0.0024 rad/ft is equivalent to 0.138°/ft. **5.** Both curves have a slope of 1 at $(0, 0)$. ◄

3.11 Related Rates

We now return to the theme of derivatives as rates of change in problems in which the variables change with respect to *time*. The essential feature of these problems is that two or more variables, which are related in a known way, are themselves changing in time. Here are two examples illustrating this type of problem.

- An oil rig springs a leak and the oil spreads in a (roughly) circular patch around the rig. If the radius of the oil patch increases at a known rate, how fast is the area of the patch changing (Example 1)?

- Two airliners approach an airport with known speeds, one flying west and one flying north. How fast is the distance between the airliners changing (Example 2)?

In the first problem, the two related variables are the radius and the area of the oil patch. Both are changing in time. The second problem has three related variables: the positions of the two airliners and the distance between them. Again, the three variables change in time. The goal in both problems is to determine the rate of change of one of the variables at a specific moment of time—hence the name *related rates*.

We present a progression of examples in this section. After the first example, a general procedure is given for solving related-rate problems.

EXAMPLE 1 **Spreading oil** An oil rig springs a leak in calm seas, and the oil spreads in a circular patch around the rig. If the radius of the oil patch increases at a rate of 30 m/hr, how fast is the area of the patch increasing when the patch has a radius of 100 meters (**Figure 3.77**)?

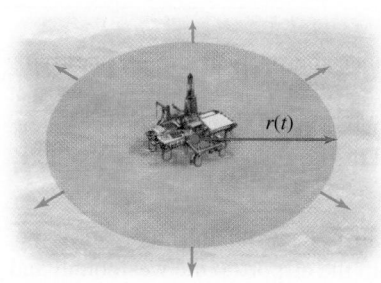

Figure 3.77

SOLUTION Two variables change simultaneously: the radius of the circle and its area. The key relationship between the radius and area is $A = \pi r^2$. It helps to rewrite the basic relationship showing explicitly which quantities vary in time. In this case, we rewrite A and r as $A(t)$ and $r(t)$ to emphasize that they change with respect to t (time). The general expression relating the radius and area at any time t is $A(t) = \pi r(t)^2$.

The goal is to find the rate of change of the area of the circle, which is $A'(t)$, given that $r'(t) = 30$ m/hr. To introduce derivatives into the problem, we differentiate the area relation $A(t) = \pi r(t)^2$ with respect to t:

$$A'(t) = \frac{d}{dt}\left(\pi r(t)^2\right)$$

$$= \pi \frac{d}{dt}\left(r(t)^2\right)$$

$$= \pi \left(2r(t)\right) r'(t) \qquad \text{Chain Rule}$$

$$= 2\pi r(t)\, r'(t). \qquad \text{Simplify.}$$

Substituting the given values $r(t) = 100$ m and $r'(t) = 30$ m/hr, we have (including units)

$$A'(t) = 2\pi r(t)\, r'(t)$$

$$= 2\pi(100 \text{ m})\left(30\,\frac{\text{m}}{\text{hr}}\right)$$

$$= 6000\pi\,\frac{\text{m}^2}{\text{hr}}.$$

> It is important to remember that substitution of specific values of the variables occurs *after* differentiating.

We see that the area of the oil spill increases at a rate of $6000\pi \approx 18{,}850$ m²/hr. Including units is a simple way to check your work. In this case, we expect an answer with units of area per unit time, so m²/hr makes sense. Notice that the rate of change of the area depends on the radius of the spill. As the radius increases, the rate of change of the area also increases.

QUICK CHECK 1 In Example 1, what is the rate of change of the area when the radius is 200 m? 300 m? ◄

Related Exercises 5, 15 ◄

Using Example 1 as a template, we offer a general procedure for solving related-rate problems with the understanding that there are always variations that arise for individual problems.

PROCEDURE   Steps for Related-Rate Problems

1. Read the problem carefully, making a sketch to organize the given information. Identify the rates that are given and the rate that is to be determined.

2. Write one or more equations that express the basic relationships among the variables.

3. Introduce rates of change by differentiating the appropriate equation(s) with respect to time t.

4. Substitute known values and solve for the desired quantity.

5. Check that units are consistent and the answer is reasonable. (For example, does it have the correct sign?)

EXAMPLE 2   Converging airplanes Two small planes approach an airport, one flying due west at 120 mi/hr and the other flying due north at 150 mi/hr. Assuming they fly at the same constant elevation, how fast is the distance between the planes changing when the westbound plane is 180 miles from the airport and the northbound plane is 225 miles from the airport?

SOLUTION A sketch such as **Figure 3.78** helps us visualize the problem and organize the information. Let $x(t)$ and $y(t)$ denote the distance from the airport to the westbound and northbound planes, respectively. The paths of the two planes form the legs of a right

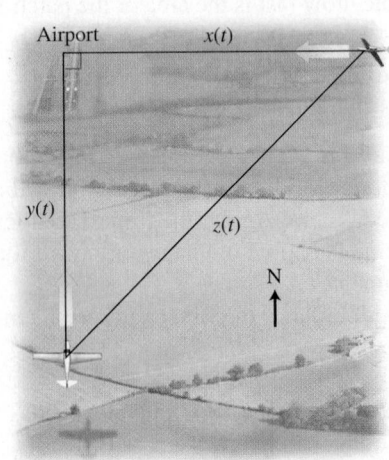

Figure 3.78

triangle, and the distance between them, denoted $z(t)$, is the hypotenuse. By the Pythagorean theorem, $z^2 = x^2 + y^2$.

Our aim is to find dz/dt, the rate of change of the distance between the planes. We first differentiate both sides of $z^2 = x^2 + y^2$ with respect to t:

$$\frac{d}{dt}(z^2) = \frac{d}{dt}(x^2 + y^2) \implies 2z\frac{dz}{dt} = 2x\frac{dx}{dt} + 2y\frac{dy}{dt}.$$

Notice that the Chain Rule is needed because x, y, and z are functions of t. Solving for dz/dt results in

$$\frac{dz}{dt} = \frac{2x\frac{dx}{dt} + 2y\frac{dy}{dt}}{2z} = \frac{x\frac{dx}{dt} + y\frac{dy}{dt}}{z}.$$

This equation relates the unknown rate dz/dt to the known quantities x, y, z, dx/dt, and dy/dt. For the westbound plane, $dx/dt = -120$ mi/hr (negative because the distance is decreasing), and for the northbound plane, $dy/dt = -150$ mi/hr. At the moment of interest, when $x = 180$ mi and $y = 225$ mi, the distance between the planes is

$$z = \sqrt{x^2 + y^2} = \sqrt{180^2 + 225^2} \approx 288 \text{ mi}.$$

Substituting these values gives

$$\frac{dz}{dt} = \frac{x\frac{dx}{dt} + y\frac{dy}{dt}}{z} \approx \frac{(180 \text{ mi})(-120 \text{ mi/hr}) + (225 \text{ mi})(-150 \text{ mi/hr})}{288 \text{ mi}}$$

$$\approx -192 \text{ mi/hr}.$$

Notice that $dz/dt < 0$, which means the distance between the planes is *decreasing* at a rate of about 192 mi/hr.

Related Exercises 22–23 ◄

> In Example 1, we replaced A and r with $A(t)$ and $r(t)$, respectively, to remind us of the independent variable. After some practice, this replacement is not necessary.

> One could solve the equation $z^2 = x^2 + y^2$ for z, with the result
> $$z = \sqrt{x^2 + y^2},$$
> and then differentiate. However, it is easier to differentiate implicitly as shown in the example.

QUICK CHECK 2 Assuming the same plane speeds as in Example 2, how fast is the distance between the planes changing if $x = 60$ mi and $y = 75$ mi? ◄

EXAMPLE 3 Morning coffee Coffee is draining out of a conical filter at a rate of 2.25 in^3/min. If the cone is 5 in tall and has a radius of 2 in, how fast is the coffee level dropping when the coffee is 3 in deep?

SOLUTION A sketch of the problem (**Figure 3.79a**) shows the three relevant variables: the volume V, the radius r, and the height h of the coffee in the filter. The aim is to find the rate of change of the height dh/dt at the instant that $h = 3$ in, given that $dV/dt = -2.25$ in^3/min.

> Because coffee is draining out of the filter, the volume of coffee in the filter is decreasing, which implies that dV/dt is negative.

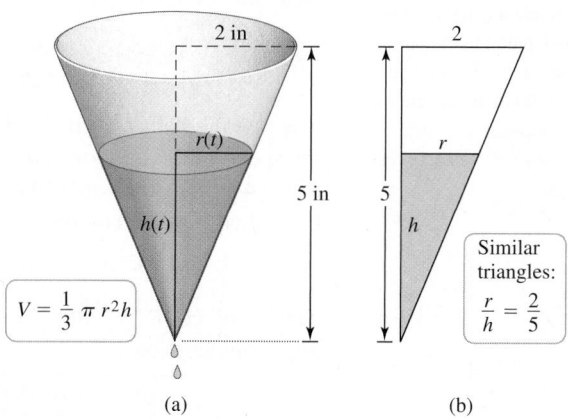

Figure 3.79

The volume formula for a cone, $V = \dfrac{1}{3}\pi r^2 h$, expresses the basic relationship among the relevant variables. Using similar triangles (**Figure 3.79b**), we see that the ratio of the radius of the coffee in the cone to the height of the coffee is 2/5 at all times; that is,

$$\frac{r}{h} = \frac{2}{5} \quad \text{or} \quad r = \frac{2}{5}h.$$

Substituting $r = \dfrac{2}{5} h$ into the volume formula gives V in terms of h:

$$V = \frac{1}{3} \pi r^2 h = \frac{1}{3} \pi \left(\frac{2}{5} h \right)^2 h = \frac{4\pi}{75} h^3.$$

Rates of change are introduced by differentiating both sides of $V = \dfrac{4\pi}{75} h^3$ with respect to t:

$$\frac{dV}{dt} = \frac{4\pi}{25} h^2 \frac{dh}{dt}.$$

Now we find dh/dt at the instant when $h = 3$, given that $dV/dt = -2.25 \text{ in}^3/\text{min}$. Solving for dh/dt and substituting these values, we have

$$\frac{dh}{dt} = \frac{25 \, dV/dt}{4\pi h^2} \qquad \text{Solve for } \frac{dh}{dt}.$$

$$= \frac{25(-2.25 \text{ in}^3/\text{min})}{4\pi(3 \text{ in})^2} \approx -0.497 \frac{\text{in}}{\text{min}}. \qquad \text{Substitute for } \frac{dV}{dt} \text{ and } h.$$

> **QUICK CHECK 3** In Example 3, what is the rate of change of the height when $h = 2$ in? ◄

At the instant that the coffee is 3 inches deep, the height decreases at a rate of 0.497 in/min (almost half an inch per minute). Notice that the units work out consistently.

Related Exercises 36–37 ◄

EXAMPLE 4   Observing a launch An observer stands 200 meters from the launch site of a hot-air balloon at an elevation equal to the elevation of the launch site. The balloon rises vertically at a constant rate of 4 m/s. How fast is the angle of elevation of the balloon increasing 30 seconds after the launch? (The angle of elevation is the angle between the ground and the observer's line of sight to the balloon.)

SOLUTION Figure 3.80 shows the geometry of the launch. As the balloon rises, its distance from the ground y and its angle of elevation θ change simultaneously. An equation expressing the relationship between these variables is $\tan\theta = y/200$. To find $d\theta/dt$, we differentiate both sides of this relationship using the Chain Rule:

$$\sec^2\theta \frac{d\theta}{dt} = \frac{1}{200} \frac{dy}{dt}.$$

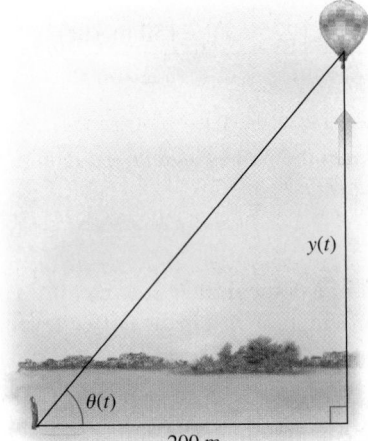

Figure 3.80

> ➤ The solution to Example 4 is reported in units of rad/s. Where did radians come from? Because a radian has no physical dimension (it is the ratio of an arc length to a radius), no unit appears. We write rad/s for clarity because $d\theta/dt$ is the rate of change of an angle.

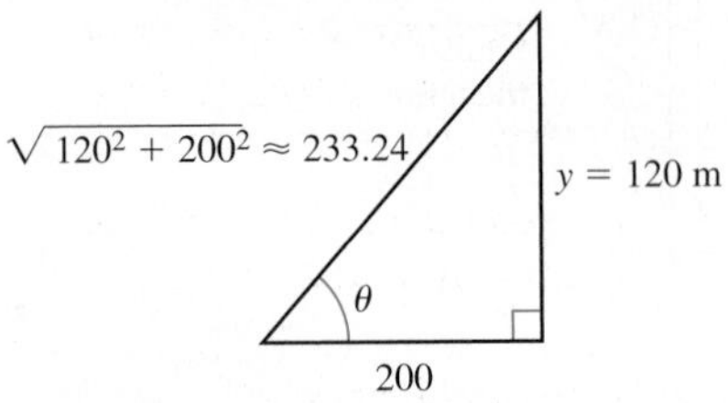

$$\sqrt{120^2 + 200^2} \approx 233.24$$

$y = 120$ m

θ

200

$$\cos\theta \approx \frac{200}{233.24} \approx 0.86$$

Figure 3.81

> ➤ Recall that to convert radians to degrees, we use
> $$\text{degrees} = \frac{180}{\pi} \cdot \text{radians}.$$

Next we solve for $\dfrac{d\theta}{dt}$:

$$\frac{d\theta}{dt} = \frac{dy/dt}{200 \sec^2\theta} = \frac{(dy/dt) \cdot \cos^2\theta}{200}.$$

The rate of change of the angle of elevation depends on the angle of elevation and the speed of the balloon. Thirty seconds after the launch, the balloon has risen $y = (4 \text{ m/s})(30 \text{ s}) = 120$ m. To complete the problem, we need the value of $\cos\theta$. Note that when $y = 120$ m, the distance between the observer and the balloon is

$$d = \sqrt{120^2 + 200^2} \approx 233.24 \text{ m}.$$

Therefore, $\cos\theta \approx 200/233.24 \approx 0.86$ (Figure 3.81), and the rate of change of the angle of elevation is

$$\frac{d\theta}{dt} = \frac{(dy/dt) \cdot \cos^2\theta}{200} \approx \frac{(4 \text{ m/s})(0.86^2)}{200 \text{ m}} = 0.015 \text{ rad/s}.$$

At this instant, the balloon is rising at an angular rate of 0.015 rad/s, or slightly less than 1°/s, as seen by the observer.

Related Exercise 46 ◄

> **QUICK CHECK 4** In Example 4, notice that as the balloon rises (as θ increases), the rate of change of the angle of elevation decreases to zero. When does the maximum value of $\theta'(t)$ occur, and what is it? ◄

SECTION 3.11 EERCISES

Getting Started

1. Give an example in which one dimension of a geometric figure changes and produces a corresponding change in the area or volume of the figure.

2. Charles' law states that for a fixed mass of gas under constant pressure, the volume V and temperature T of the gas (in kelvins) satisfy the equation $V = kT$, where k is constant. Find an equation relating dV/dt to dT/dt.

3. If two opposite sides of a rectangle increase in length, how must the other two opposite sides change if the area of the rectangle is to remain constant?

4. The temperature F in degrees Fahrenheit is related to the temperature C in degrees Celsius by the equation $F = \dfrac{9}{5}C + 32$.

   a. Find an equation relating dF/dt to dC/dt.
   b. How fast is the temperature in an oven changing in degrees Fahrenheit per minute if it is rising at $10°$ Celsius per min?

5. A rectangular swimming pool 10 ft wide by 20 ft long and of uniform depth is being filled with water.

   a. If t is elapsed time, h is the height of the water, and V is the volume of the water, find equations relating V to h and dV/dt to dh/dt.
   b. At what rate is the volume of the water increasing if the water level is rising at $\frac{1}{4}$ ft/min?
   c. At what rate is the water level rising if the pool is filled at a rate of 10 ft^3/min?

6. At all times, the length of a rectangle is twice the width w of the rectangle as the area of the rectangle changes with respect to time t.

   a. Find an equation relating A to w.
   b. Find an equation relating dA/dt to dw/dt.

7. The volume V of a sphere of radius r changes over time t.

   a. Find an equation relating dV/dt to dr/dt.
   b. At what rate is the volume changing if the radius increases at 2 in/min when the radius is 4 inches?
   c. At what rate is the radius changing if the volume increases at 10 in^3/min when the radius is 5 inches?

8. At all times, the length of the long leg of a right triangle is 3 times the length x of the short leg of the triangle. If the area of the triangle changes with respect to time t, find equations relating the area A to x and dA/dt to dx/dt.

9. Assume x, y, and z are functions of t with $z = x + y^3$. Find dz/dt when $dx/dt = -1$, $dy/dt = 5$, and $y = 2$.

10. Assume $w = x^2 y^4$, where x and y are functions of t. Find dw/dt when $x = 3$, $dx/dt = 2$, $dy/dt = 4$, and $y = 1$.

Practice Exercises

11. **Expanding square** The sides of a square increase in length at a rate of 2 m/s.

    a. At what rate is the area of the square changing when the sides are 10 m long?
    b. At what rate is the area of the square changing when the sides are 20 m long?

12. **Shrinking square** The sides of a square decrease in length at a rate of 1 m/s.

    a. At what rate is the area of the square changing when the sides are 5 m long?
    b. At what rate are the lengths of the diagonals of the square changing?

13. **Expanding isosceles triangle** The legs of an isosceles right triangle increase in length at a rate of 2 m/s.

    a. At what rate is the area of the triangle changing when the legs are 2 m long?
    b. At what rate is the area of the triangle changing when the hypotenuse is 1 m long?
    c. At what rate is the length of the hypotenuse changing?

14. **Shrinking isosceles triangle** The hypotenuse of an isosceles right triangle decreases in length at a rate of 4 m/s.

    a. At what rate is the area of the triangle changing when the legs are 5 m long?
    b. At what rate are the lengths of the legs of the triangle changing?
    c. At what rate is the area of the triangle changing when the area is 4 m^2?

15. **Expanding circle** The area of a circle increases at a rate of 1 cm^2/s.

    a. How fast is the radius changing when the radius is 2 cm?
    b. How fast is the radius changing when the circumference is 2 cm?

16. **Expanding cube** The edges of a cube increase at a rate of 2 cm/s. How fast is the volume changing when the length of each edge is 50 cm?

17. **Shrinking circle** A circle has an initial radius of 50 ft when the radius begins decreasing at a rate of 2 ft/min. What is the rate of change of the area at the instant the radius is 10 ft?

18. **Shrinking cube** The volume of a cube decreases at a rate of 0.5 ft^3/min. What is the rate of change of the side length when the side lengths are 12 ft?

19. **Balloons** A spherical balloon is inflated and its volume increases at a rate of 15 in^3/min. What is the rate of change of its radius when the radius is 10 in?

20. **Expanding rectangle** A rectangle initially has dimensions 2 cm by 4 cm. All sides begin increasing in length at a rate of 1 cm/s. At what rate is the area of the rectangle increasing after 20 s?

21. **Melting snowball** A spherical snowball melts at a rate proportional to its surface area. Show that the rate of change of the radius is constant. (*Hint:* Surface area $= 4\pi r^2$.)

22. **Divergent paths** Two boats leave a port at the same time; one travels west at 20 mi/hr and the other travels south at 15 mi/hr.

    a. After 30 minutes, how far is each boat from port?
    b. At what rate is the distance between the boats changing 30 minutes after they leave the port?

23. **Time-lagged flights** An airliner passes over an airport at noon traveling 500 mi/hr due west. At 1:00 P.M., another airliner passes over the same airport at the same elevation traveling due north at 550 mi/hr. Assuming both airliners maintain their (equal) elevations, how fast is the distance between them changing at 2:30 P.M.?

24. **Flying a kite** Once Kate's kite reaches a height of 50 ft (above her hands), it rises no higher but drifts due east in a wind blowing 5 ft/s. How fast is the string running through Kate's hands at the moment when she has released 120 ft of string?

25. **Rope on a boat** A rope passing through a capstan on a dock is attached to a boat offshore. The rope is pulled in at a constant rate of 3 ft/s and the capstan is 5 ft vertically above the water. How fast is the boat traveling when it is 10 ft from the dock?

26. **Bug on a parabola** A bug is moving along the right side of the parabola $y = x^2$ at a rate such that its distance from the origin is increasing at 1 cm/min.

   a. At what rate is the x-coordinate of the bug increasing at the point $(2, 4)$?

   b. Use the equation $y = x^2$ to find an equation relating $\dfrac{dy}{dt}$ to $\dfrac{dx}{dt}$.

   c. At what rate is the y-coordinate of the bug increasing at the point $(2, 4)$?

27. **Balloons and motorcycles** A hot-air balloon is 150 ft above the ground when a motorcycle (traveling in a straight line on a horizontal road) passes directly beneath it going 40 mi/hr (58.67 ft/s). If the balloon rises vertically at a rate of 10 ft/s, what is the rate of change of the distance between the motorcycle and the balloon 10 seconds later?

28. **Baseball runners** Runners stand at first and second base in a baseball game. At the moment a ball is hit, the runner at first base runs to second base at 18 ft/s; simultaneously, the runner on second runs to third base at 20 ft/s. How fast is the distance between the runners changing 1 s after the ball is hit (see figure)? (*Hint:* The distance between consecutive bases is 90 ft and the bases lie at the corners of a square.)

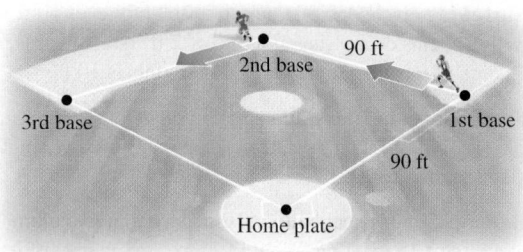

29. **Fishing story** An angler hooks a trout and reels in his line at 4 in/s. Assume the tip of the fishing rod is 12 ft above the water and directly above the angler, and the fish is pulled horizontally directly toward the angler (see figure). Find the horizontal speed of the fish when it is 20 ft from the angler.

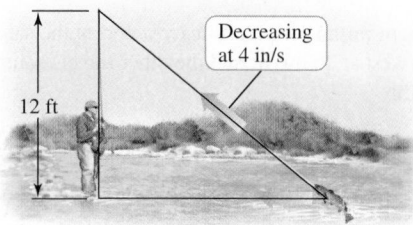

30. **Parabolic motion** An arrow is shot into the air and moves along the parabolic path $y = x(50 - x)$ (see figure). The horizontal

component of velocity is always 30 ft/s. What is the vertical component of velocity when (a) $x = 10$ and (b) $x = 40$?

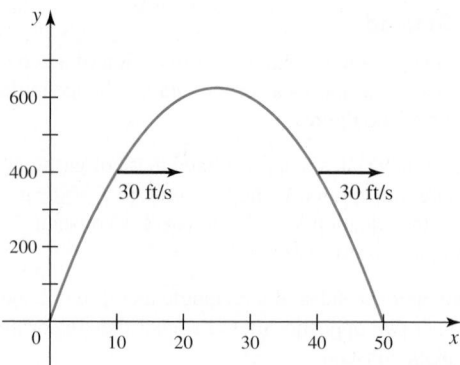

31. **Draining a water heater** A water heater that has the shape of a right cylindrical tank with a radius of 1 ft and a height of 4 ft is being drained. How fast is water draining out of the tank (in ft³/min) if the water level is dropping at 6 in/min?

32. **Drinking a soda** At what rate (in in³/s) is soda being sucked out of a cylindrical glass that is 6 in tall and has a radius of 2 in? The depth of the soda decreases at a constant rate of 0.25 in/s.

33. **Piston compression** A piston is seated at the top of a cylindrical chamber with radius 5 cm when it starts moving into the chamber at a constant speed of 3 cm/s (see figure). What is the rate of change of the volume of the cylinder when the piston is 2 cm from the base of the chamber?

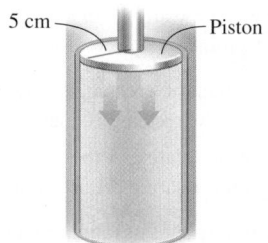

34. **Filling two pools** Two cylindrical swimming pools are being filled simultaneously at the same rate (in m³/min; see figure). The smaller pool has a radius of 5 m, and the water level rises at a rate of 0.5 m/min. The larger pool has a radius of 8 m. How fast is the water level rising in the larger pool?

35. **Growing sandpile** Sand falls from an overhead bin and accumulates in a conical pile with a radius that is always three times its height. Suppose the height of the pile increases at a rate of 2 cm/s when the pile is 12 cm high. At what rate is the sand leaving the bin at that instant?

36. Draining a tank An inverted conical water tank with a height of 12 ft and a radius of 6 ft is drained through a hole in the vertex at a rate of 2 ft³/s (see figure). What is the rate of change of the water depth when the water depth is 3 ft? (*Hint:* Use similar triangles.)

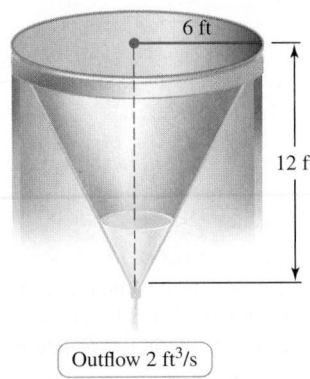

6 ft

12 ft

Outflow 2 ft³/s

37. Draining a cone Water is drained out of an inverted cone that has the same dimensions as the cone depicted in Exercise 36. If the water level drops at 1 ft/min, at what rate is water (in ft³/min) draining from the tank when the water depth is 6 ft?

38. Two tanks A conical tank with an upper radius of 4 m and a height of 5 m drains into a cylindrical tank with a radius of 4 m and a height of 5 m (see figure). If the water level in the conical tank drops at a rate of 0.5 m/min, at what rate does the water level in the cylindrical tank rise when the water level in the conical tank is 3 m? 1 m?

4 m

5 m

4 m

5 m

39. Filling a hemispherical tank A hemispherical tank with a radius of 10 m is filled from an inflow pipe at a rate of 3 m³/min (see figure). How fast is the water level rising when the water level is 5 m from the bottom of the tank? (*Hint:* The volume of a cap of thickness h sliced from a sphere of radius r is $\pi h^2 (3r - h)/3$.)

Inflow 3 m³/min

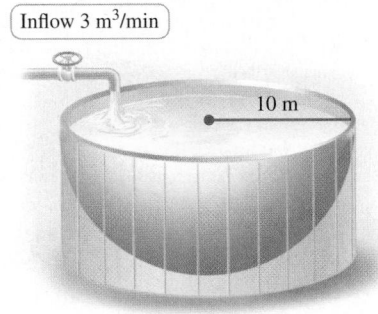

10 m

40. Surface area of hemispherical tank For the situation described in Exercise 39, what is the rate of change of the area of the exposed surface of the water when the water is 5 m deep?

41. Ladder against the wall A 13-ft ladder is leaning against a vertical wall (see figure) when Jack begins pulling the foot of the ladder away from the wall at a rate of 0.5 ft/s. How fast is the top of the ladder sliding down the wall when the foot of the ladder is 5 ft from the wall?

13 ft

42. Ladder against the wall again A 12-ft ladder is leaning against a vertical wall when Jack begins pulling the foot of the ladder away from the wall at a rate of 0.2 ft/s. What is the configuration of the ladder at the instant when the vertical speed of the top of the ladder equals the horizontal speed of the foot of the ladder?

43. Moving shadow A 5-ft-tall woman walks at 8 ft/s toward a streetlight that is 20 ft above the ground. What is the rate of change of the length of her shadow when she is 15 ft from the streetlight? At what rate is the tip of her shadow moving?

44. Another moving shadow A landscape light at ground level lights up the side of a tall building that is 15 feet from the light. A 6-ft-tall man starts walking (on flat terrain) from the light directly toward the building. How fast is he walking when he is 9 feet from the light if his shadow on the building is shrinking at 2 ft/s at that instant?

45. Watching an elevator An observer is 20 m above the ground floor of a large hotel atrium looking at a glass-enclosed elevator shaft that is 20 m horizontally from the observer (see figure). The angle of elevation of the elevator is the angle that the observer's line of sight makes with the horizontal (it may be positive or negative). Assuming the elevator rises at a rate of 5 m/s, what is the rate of change of the angle of elevation when the elevator is 10 m above the ground? When the elevator is 40 m above the ground?

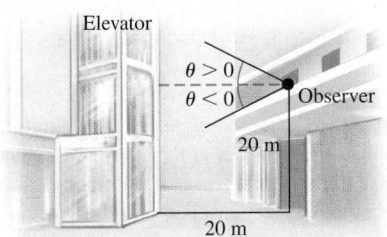

Elevator

$\theta > 0$
$\theta < 0$ Observer

20 m

20 m

46. Observing a launch An observer stands 300 ft from the launch site of a hot-air balloon at an elevation equal to the elevation of the launch site. The balloon is launched vertically and maintains a constant upward velocity of 20 ft/s. What is the rate of change of the angle of elevation of the balloon when it is 400 ft from the ground? (*Hint:* The angle of elevation is the angle θ between the observer's line of sight to the balloon and the ground.)

47. Viewing angle The bottom of a large theater screen is 3 ft above your eye level and the top of the screen is 10 ft above your eye level. Assume you walk away from the screen (perpendicular to the screen) at a rate of 3 ft/s while looking at the screen. What is the rate of change of the viewing angle θ when you are 30 ft from the wall on which the screen hangs, assuming the floor is horizontal (see figure)?

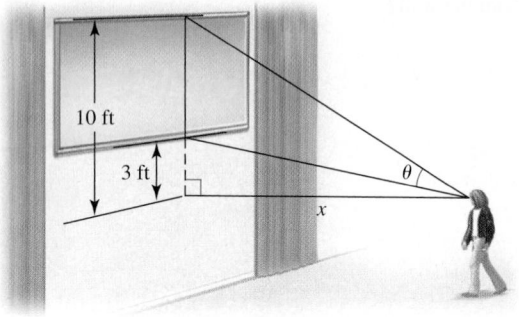

48. Altitude of a jet A jet ascends at a 10° angle from the horizontal with an airspeed of 550 mi/hr (its speed along its line of flight is 550 mi/hr). How fast is the altitude of the jet increasing? If the sun is directly overhead, how fast is the shadow of the jet moving on the ground?

49. Rate of dive of a submarine A surface ship is moving (horizontally) in a straight line at 10 km/hr. At the same time, an enemy submarine maintains a position directly below the ship while diving at an angle that is 20° below the horizontal. How fast is the submarine's altitude decreasing?

50. Revolving light beam A lighthouse stands 500 m off a straight shore, and the focused beam of its light revolves (at a constant rate) four times each minute. As shown in the figure, P is the point on shore closest to the lighthouse and Q is a point on the shore 200 m from P. What is the speed of the beam along the shore when it strikes the point Q? Describe how the speed of the beam along the shore varies with the distance between P and Q. Neglect the height of the lighthouse.

51. Filming a race A camera is set up at the starting line of a drag race 50 ft from a dragster at the starting line (camera 1 in the figure). Two seconds after the start of the race, the dragster has traveled 100 ft and the camera is turning at 0.75 rad/s while filming the dragster.

a. What is the speed of the dragster at this point?
b. A second camera (camera 2 in the figure) filming the dragster is located on the starting line 100 ft away from the dragster at the start of the race. How fast is this camera turning 2 s after the start of the race?

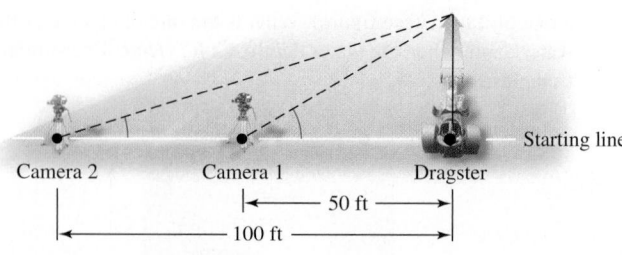

52. Fishing reel An angler hooks a trout and begins turning her circular reel at 1.5 rev/s. Assume the radius of the reel (and the fishing line on it) is 2 inches.

a. Let R equal the number of revolutions the angler has turned her reel and suppose L is the amount of line that she has reeled in. Find an equation for L as a function of R.
b. How fast is she reeling in her fishing line?

53. Wind energy The kinetic energy E (in joules) of a mass in motion satisfies the equation $E = \dfrac{1}{2}mv^2$, where mass m is measured in kg and velocity v is measured in m/s.

a. Power P is defined to be dE/dt, the rate of change in energy with respect to time. Power is measured in units of watts (W), where 1 W = 1 joule/s. If the velocity v is constant, use implicit differentiation to find an equation for power P in terms of the derivative dm/dt.
b. Wind turbines use kinetic energy in the wind to create electrical power. In this case, the derivative dm/dt is called the **mass flow rate** and it satisfies the equation $\dfrac{dm}{dt} = \rho Av$, where ρ is the density of the air in kg/m³, A is the sweep area in m² of the wind turbine (see figure), and v is the velocity of the wind in m/s. Show that $P = \dfrac{1}{2}\rho Av^3$.
c. Suppose a blade on a small wind turbine has a length of 3 m. Find the available power P if the wind is blowing at 10 m/s. (*Hint:* Use $\rho = 1.23$ kg/m³ for the density of air. The density of air varies, but this is a reasonable average value.)
d. Wind turbines convert only a small percentage of the available wind power into electricity. Assume the wind turbine described in this exercise converts only 25% of the available wind power into electricity. How much electrical power is produced?

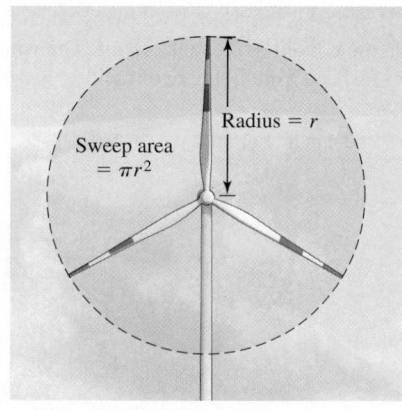

54. Boyle's law Robert Boyle (1627–1691) found that for a given quantity of gas at a constant temperature, the pressure P (in kPa) and volume V of the gas (in m^3) are accurately approximated by the equation $V = k/P$, where $k > 0$ is constant. Suppose the volume of an expanding gas is increasing at a rate of $0.15 \ m^3/min$ when the volume $V = 0.5 \ m^3$ and the pressure is $P = 50$ kPa. At what rate is pressure changing at this moment?

Explorations and Challenges

55. Clock hands The hands of the clock in the tower of the Houses of Parliament in London are approximately 3 m and 2.5 m in length. How fast is the distance between the tips of the hands changing at 9:00? (*Hint:* Use the Law of Cosines.)

56. Divergent paths Two boats leave a port at the same time, one traveling west at 20 mi/hr and the other traveling southwest (45° south of west) at 15 mi/hr. After 30 minutes, how far apart are the boats and at what rate is the distance between them changing? (*Hint:* Use the Law of Cosines.)

57. Filling a pool A swimming pool is 50 m long and 20 m wide. Its depth decreases linearly along the length from 3 m to 1 m (see figure). It was initially empty and is being filled at a rate of $1 \ m^3/min$. How fast is the water level rising 250 min after the filling begins? How long will it take to fill the pool?

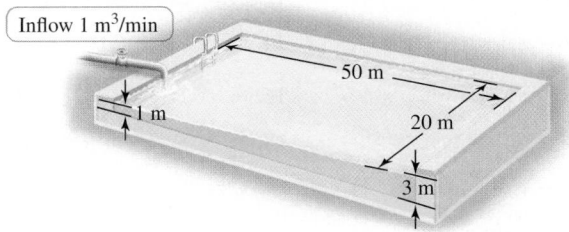

58. Disappearing triangle An equilateral triangle initially has sides of length 20 ft when each vertex moves toward the midpoint of the opposite side at a rate of 1.5 ft/min. Assuming the triangle remains equilateral, what is the rate of change of the area of the triangle at the instant the triangle disappears?

59. Oblique tracking A port and a radar station are 2 mi apart on a straight shore running east and west (see figure). A ship leaves the port at noon traveling northeast at a rate of 15 mi/hr. If the ship maintains its speed and course, what is the rate of change of the tracking angle θ between the shore and the line between the radar station and the ship at 12:30 P.M.? (*Hint:* Use the Law of Sines.)

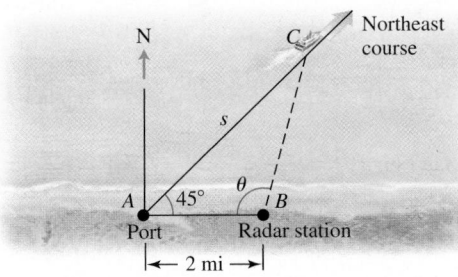

60. Oblique tracking A ship leaves port traveling southwest at a rate of 12 mi/hr. At noon, the ship reaches its closest approach to a radar station, which is on the shore 1.5 mi from the port. If the

ship maintains its speed and course, what is the rate of change of the tracking angle θ between the radar station and the ship at 1:30 P.M. (see figure)? (*Hint:* Use the Law of Sines.)

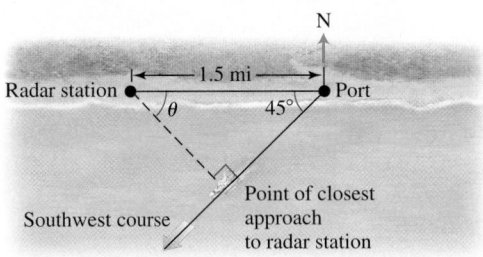

61. Navigation A boat leaves a port traveling due east at 12 mi/hr. At the same time, another boat leaves the same port traveling northeast at 15 mi/hr. The angle θ of the line between the boats is measured relative to due north (see figure). What is the rate of change of this angle 30 min after the boats leave the port? 2 hr after the boats leave the port?

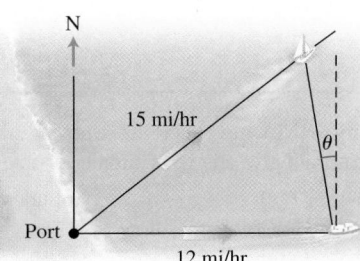

62. Watching a Ferris wheel An observer stands 20 m from the bottom of a 10-m-tall Ferris wheel on a line that is perpendicular to the face of the Ferris wheel. The wheel revolves at a rate of π rad/min and the observer's line of sight with a specific seat on the wheel makes an angle θ with the ground (see figure). Forty seconds after that seat leaves the lowest point on the wheel, what is the rate of change of θ? Assume the observer's eyes are level with the bottom of the wheel.

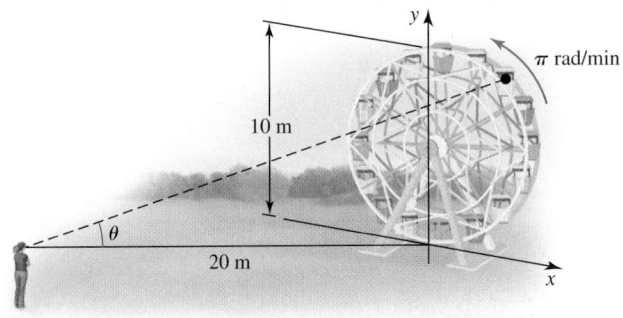

63. Draining a trough A trough in the shape of a half cylinder has length 5 m and radius 1 m. The trough is full of water when a valve is opened, and water flows out of the bottom of the trough at a rate of $1.5 \ m^3/hr$ (see figure). (*Hint:* The area of a sector of a circle of radius r subtended by an angle θ is $r^2\theta/2$.)

   a. How fast is the water level changing when the water level is 0.5 m from the bottom of the trough?

b. What is the rate of change of the surface area of the water when the water is 0.5 m deep?

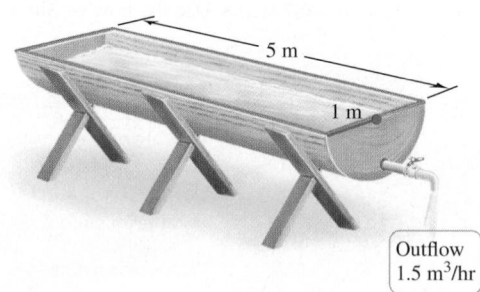

Outflow
1.5 m³/hr

64. Searchlight—wide beam A revolving searchlight, which is 100 m from the nearest point on a straight highway, casts a horizontal beam along a highway (see figure). The beam leaves the spotlight at an angle of $\pi/16$ rad and revolves at a rate $\pi/6$ rad/s. Let w be the width of the beam as it sweeps along the highway

and let θ be the angle that the center of the beam makes with the perpendicular to the highway. What is the rate of change of w when $\theta = \pi/3$? Neglect the height of the searchlight.

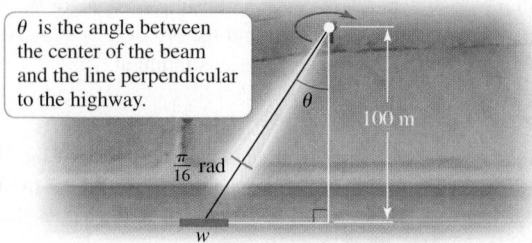

θ is the angle between the center of the beam and the line perpendicular to the highway.

100 m

$\frac{\pi}{16}$ rad

w

QUICK CHECK ANSWERS

1. $12{,}000\pi$ m²/hr, $18{,}000\pi$ m²/hr   **2.** -192 mi/hr
3. -1.12 in/min   **4.** $t = 0, \theta = 0, \theta'(0) = 0.02$ rad/s ◄

CHAPTER 3   REVIEW EXERCISES

1. Explain why or why not Determine whether the following statements are true and give an explanation or counterexample.

  a. The function $f(x) = |2x + 1|$ is continuous for all x; therefore, it is differentiable for all x.

  b. If $\dfrac{d}{dx}(f(x)) = \dfrac{d}{dx}(g(x))$, then $f = g$.

  c. For any function f, $\dfrac{d}{dx}|f(x)| = |f'(x)|$.

  d. The value of $f'(a)$ fails to exist only if the curve $y = f(x)$ has a vertical tangent line at $x = a$.

  e. An object can have negative acceleration and increasing speed.

2–4. *Evaluate the derivative of each of the following functions using a limit definition of the derivative. Check your work by evaluating the derivatives using the derivative rules given in this chapter.*

2. $f(x) = x^2 + 2x + 9$ **3.** $g(x) = \dfrac{1}{x^2 + 5}$

4. $h(t) = \sqrt{3t + 5}$

5–8. *Use differentiation to verify each equation.*

5. $\dfrac{d}{dx}(\tan^3 x - 3\tan x + 3x) = 3\tan^4 x.$

6. $\dfrac{d}{dx}\left(\dfrac{x}{\sqrt{1 - x^2}}\right) = \dfrac{1}{(1 - x^2)^{3/2}}.$

7. $\dfrac{d}{dx}(x^4 - \ln(x^4 + 1)) = \dfrac{4x^7}{1 + x^4}.$

8. $\dfrac{d}{dx}\left(\ln\left(\dfrac{1 - \sqrt{x}}{1 + \sqrt{x}}\right)\right) = \dfrac{1}{\sqrt{x}(x - 1)}.$

9–61. *Evaluate and simplify y'.*

9. $y = \dfrac{2}{3}x^3 + \pi x^2 + 7x + 1$ **10.** $y = 4x^4\ln x - x^4$

11. $y = 2^x$

12. $y = 2x^{\sqrt{2}}$

13. $y = e^{2\theta}$

14. $y = (2x - 3)x^{3/2}$

15. $y = (1 + x^4)^{3/2}$

16. $y = 2x\sqrt{x^2 - 2x + 2}$

17. $y = 5t^2 \sin t$

18. $y = 5x + \sin^3 x + \sin x^3$

19. $y = e^{-x}(x^2 + 2x + 2)$

20. $y = \dfrac{\ln x}{\ln x + a}$, where a is constant

21. $y = \dfrac{\sec 2w}{\sec 2w + 1}$

22. $y = \left(\dfrac{\sin x}{\cos x + 1}\right)^{1/3}$

23. $y = \ln|\sec 3x|$

24. $y = \ln|\csc 7x + \cot 7x|$

25. $y = (5t^2 + 10)^{100}$

26. $y = e^{\sin x + 2x + 1}$

27. $y = \ln(\sin x^3)$

28. $y = e^{\tan x}(\tan x - 1)$

29. $y = \tan^{-1}\sqrt{t^2 - 1}$

30. $y = x^{\sqrt{x + 1}}$

31. $y = 4\tan(\theta^2 + 3\theta + 2)$

32. $y = \csc^5 3x$

33. $y = \dfrac{\ln w}{w^5}$

34. $y = \dfrac{s}{e^{as}}$, where a is a constant

35. $y = \dfrac{4u^2 + u}{8u + 1}$

36. $y = \left(\dfrac{3t^2 - 1}{3t^2 + 1}\right)^{-3}$

37. $y = \tan(\sin \theta)$

38. $y = \left(\dfrac{v}{v + 1}\right)^{4/3}$

39. $y = \sin\sqrt{\cos^2 x + 1}$

40. $y = e^{\sin(\cos x)}$

41. $y = \ln\sqrt{e^t + 1}$

42. $y = xe^{-10x}$

43. $y = x^2 + 2x\tan^{-1}(\cot x)$

44. $y = \sqrt{1 - x^4} + x^2\sin^{-1}x^2$

45. $y = x\ln^2 x$

46. $y = e^{6x}\sin x$

47. $y = 2^{x^2 - x}$

48. $y = 10^{\sin x} + \sin^{10}x$

49. $y = (x^2 + 1)^{\ln x}$

50. $y = x^{\cos 2x}$

51. $y = \sin^{-1}\dfrac{1}{x}$

52. $y = \log_3(x + 8)$

53. $y = 6x \cot^{-1} 3x + \ln(9x^2 + 1)$

54. $y = 2x^2 \cos^{-1} x + \sin^{-1} x$ **55.** $x = \cos(x - y)$

56. $xy^4 + x^4 y = 1$

57. $y = \dfrac{e^y}{1 + \sin x}$

58. $\sin x \cos(y - 1) = \dfrac{1}{2}$

59. $y\sqrt{x^2 + y^2} = 15$

60. $y = \dfrac{(x^2 + 1)^3}{(x^4 + 7)^8(2x + 1)^7}$ **61.** $y = \dfrac{(3x + 5)^{10}\sqrt{x^2 + 5}}{(x^3 + 1)^{50}}$

62. Find $f'(1)$ when $f(x) = \tan^{-1}(4x^2)$.

63. Evaluate $\dfrac{d}{dx}(x \sec^{-1} x)\Big|_{x=\frac{2}{\sqrt{3}}}$.

64. Evaluate $\dfrac{d}{dx}(\tan^{-1} e^{-x})\Big|_{x=0}$.

65. Find $f'(1)$ when $f(x) = x^{1/x}$.

66–71. Higher-order derivatives *Find and simplify y''.*

66. $y = e^{x^2+1}$

67. $y = 2^x x$

68. $y = \dfrac{3x - 1}{x + 1}$

69. $y = \dfrac{\ln x}{x^2}$

70. $x + \sin y = y$

71. $xy + y^2 = 1$

72–76. Tangent lines *Find an equation of the line tangent to each of the following curves at the given point.*

72. $y = 2^{3x-6}; \ (2, 1)$

73. $y = 3x^3 + \sin x; \ (0, 0)$

74. $y = \dfrac{4x}{x^2 + 3}; \ (1, 1)$

75. $y + \sqrt{xy} = 6; \ (1, 4)$

76. $x^2 y + y^3 = 5; \ (2, 1)$

77–80. Derivative formulas *Evaluate the following derivatives. Express your answers in terms of f, g, f', and g'.*

77. $\dfrac{d}{dx}(x^2 f(x))$

78. $\dfrac{d}{dx}\sqrt{\dfrac{f(x)}{g(x)}}, \dfrac{f(x)}{g(x)} \geq 0$

79. $\dfrac{d}{dx}\left(\dfrac{xf(x)}{g(x)}\right)$

80. $\dfrac{d}{dx} f\left(\sqrt{g(x)}\right), g(x) \geq 0$

81. Matching functions and derivatives Match the functions a–d with the derivatives A–D.

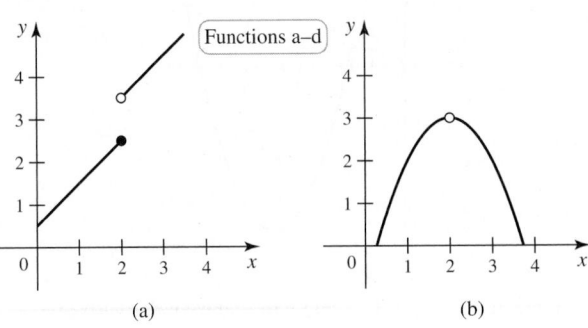

(a)   (b)

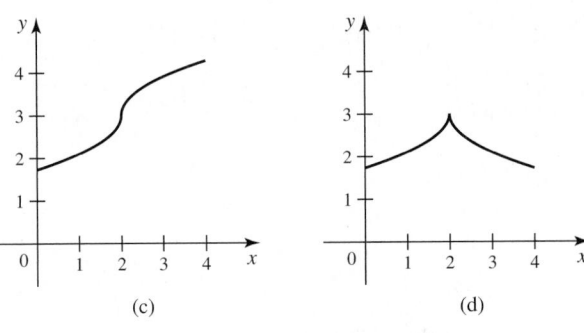

(c)   (d)

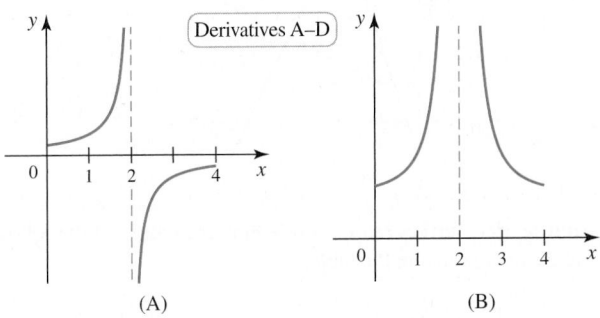

(A)   (B)

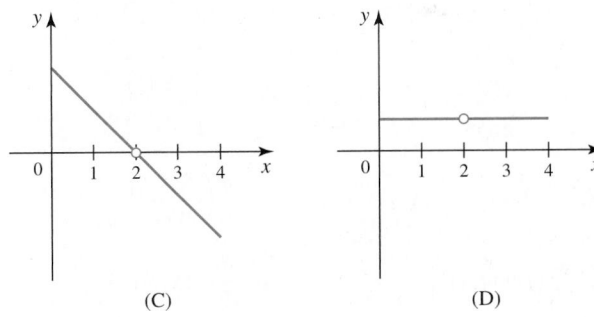

(C)   (D)

82. Sketch a graph of f' for the function f shown in the figure.

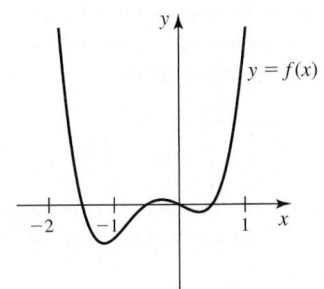

83. Sketch a graph of g' for the function g shown in the figure.

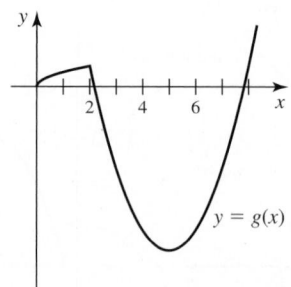

84. Use the given graphs of f and g to find each derivative.

a. $\dfrac{d}{dx}(5f(x) + 3g(x))\Big|_{x=1}$ **b.** $\dfrac{d}{dx}(f(x)g(x))\Big|_{x=1}$

c. $\dfrac{d}{dx}\left(\dfrac{f(x)}{g(x)}\right)\Big|_{x=3}$ **d.** $\dfrac{d}{dx}(f(f(x)))\Big|_{x=4}$

e. $\dfrac{d}{dx}(g(f(x)))\Big|_{x=1}$

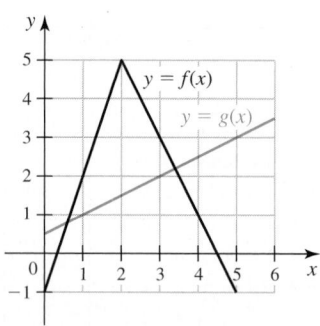

85. Finding derivatives from a table Find the values of the follow-ing derivatives using the table.

x	1	3	5	7	9
$f(x)$	3	1	9	7	5
$f'(x)$	7	9	5	1	3
$g(x)$	9	7	5	3	1
$g'(x)$	5	9	3	1	7

a. $\dfrac{d}{dx}(f(x) + 2g(x))\Big|_{x=3}$ **b.** $\dfrac{d}{dx}\left(\dfrac{f(x)}{g(x)}\right)\Big|_{x=1}$

c. $\dfrac{d}{dx}(f(x)g(x))\Big|_{x=3}$ **d.** $\dfrac{d}{dx}(f(x)^3)\Big|_{x=5}$

e. $(g^{-1})'(7)$

86–87. Derivatives of the inverse at a point *Consider the following functions. In each case, without finding the inverse, evaluate the de-rivative of the inverse at the given point.*

86. $f(x) = 1/(x + 1)$ at $f(0)$

87. $y = \sqrt{x^3 + x - 1}$ at $y = 3$

88–89. Derivative of the inverse *Find the derivative of the inverse of the following functions. Express the result with x as the independent variable.*

88. $f(x) = 12x - 16$ **89.** $f(x) = x^{-1/3}$

90–91. Derivatives from a graph *If possible, evaluate the following derivatives using the graphs of f and f'.*

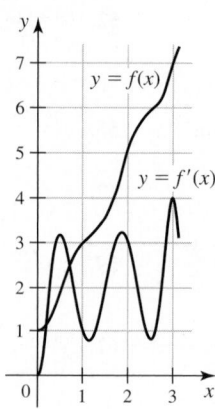

90. a. $\dfrac{d}{dx}(xf(x))\Big|_{x=2}$ **b.** $\dfrac{d}{dx}(f(x^2))\Big|_{x=1}$

c. $\dfrac{d}{dx}(f(f(x)))\Big|_{x=1}$

91. a. $(f^{-1})'(7)$ **b.** $(f^{-1})'(3)$

c. $(f^{-1})'(f(2))$

92–93. *The line tangent to $y = f(x)$ at $x = 3$ is $y = 4x - 10$ and the line tangent to $y = g(x)$ at $x = 5$ is $y = 6x - 27$.*

92. Compute $f(3)$, $f'(3)$, $g(5)$, and $g'(5)$.

93. Find an equation of the line tangent to the graph of $y = f(g(x))$ at $x = 5$.

94. Horizontal motion The position of an object moving horizontally after t seconds is given by the function $s = 27t - t^3$, for $t > 0$, where s is measured in feet, with $s > 0$ corresponding to posi-tions right of the origin.

a. When is the object stationary, moving to the right, and moving to the left?

b. Determine the velocity and acceleration of the object at $t = 2$.

c. Determine the acceleration of the object when its velocity is zero.

d. On what time intervals is the speed decreasing?

95. Projectile on Mars Suppose an object is fired vertically upward from the ground on Mars with an initial velocity of 96 ft/s. The height s (in feet) of the object above the ground after t seconds is given by $s = 96t - 6t^2$.

a. Determine the instantaneous velocity of the object at $t = 1$.

b. When will the object have an instantaneous velocity of 12 ft/s?

c. What is the height of the object at the highest point of its tra-jectory?

d. With what speed does the object strike the ground?

T 96. Beak length The length of the culmen (the upper ridge of a bird's bill) of a t-week-old Indian spotted owlet is modeled by the func-tion $L(t) = \dfrac{11.94}{1 + 4e^{-1.65t}}$, where L is measured in millimeters.

a. Find $L'(1)$ and interpret the meaning of this value.

b. Use a graph of $L'(t)$ to describe how the culmen grows over the first 5 weeks of life.

T 97. Retirement account A self-employed software engineer starts saving $300 a month in an index fund that returns 9% annual interest, compounded monthly. The balance in the account after t years is given by $A(t) = 40{,}000(1.0075^{12t} - 1)$.

a. How much money will be in her account after 20 years and how fast is her account growing at this point?

b. How long will it take until the balance in her account reaches $100,000, and how fast is her account growing at this point?

98. Antibiotic decay The half-life of an antibiotic in the bloodstream is 10 hours. If an initial dose of 20 milligrams is administered, the quantity left after t hours is modeled by $Q(t) = 20e^{-0.0693t}$, for $t \geq 0$.

a. Find the instantaneous rate of change of the amount of antibiotic in the bloodstream, for $t \geq 0$.

b. How fast is the amount of antibiotic changing at $t = 0$? At $t = 2$?

c. Evaluate and interpret $\lim\limits_{t \to \infty} Q(t)$ and $\lim\limits_{t \to \infty} Q'(t)$.

99. Population of the United States The population of the United States (in millions) by decade is given in the table, where t is the number of years after 1910. These data are plotted and fitted with a smooth curve $y = p(t)$ in the figure.

a. Compute the average rate of population growth from 1950 to 1960.

b. Explain why the average rate of growth from 1950 to 1960 is a good approximation to the (instantaneous) rate of growth in 1955.

c. Estimate the instantaneous rate of growth in 1985.

Year	1910	1920	1930	1940	1950	1960
t	0	10	20	30	40	50
$p(t)$	92	106	123	132	152	179

Year	1970	1980	1990	2000	2010	2020
t	60	70	80	90	100	110
$p(t)$	203	227	249	281	309	335

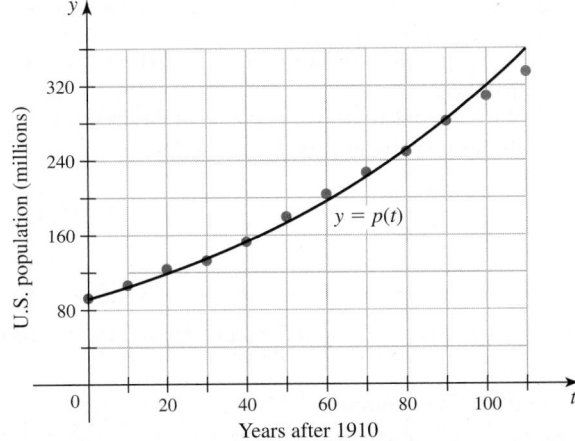

100. Growth rate of bacteria Suppose the following graph represents the number of bacteria in a culture t hours after the start of an experiment.

a. At approximately what time is the instantaneous growth rate the greatest, for $0 \leq t \leq 36$? Estimate the growth rate at this time.

b. At approximately what time in the interval $0 \leq t \leq 36$ is the instantaneous growth rate the least? Estimate the instantaneous growth rate at this time.

c. What is the average growth rate over the interval $0 \leq t \leq 36$?

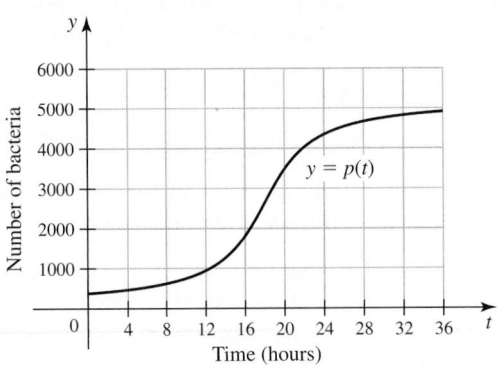

101. Velocity of a skydiver Assume the graph represents the distance (in meters) that a skydiver has fallen t seconds after jumping out of a plane.

a. Estimate the velocity of the skydiver at $t = 15$.

b. Estimate the velocity of the skydiver at $t = 70$.

c. Estimate the average velocity of the skydiver between $t = 20$ and $t = 90$.

d. Sketch a graph of the velocity function, for $0 \leq t \leq 120$.

e. What significant event occurred at $t = 30$?

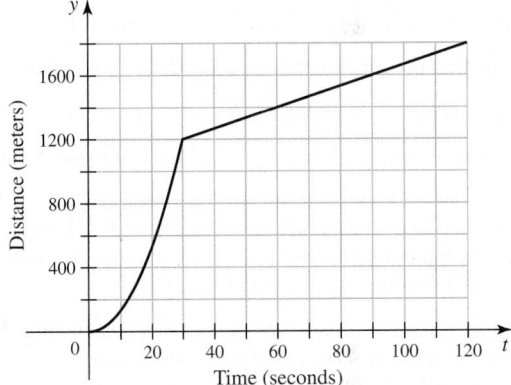

T 102. A function and its inverse function The function $f(x) = \dfrac{x}{x + 1}$ is one-to-one, for $x > -1$, and has an inverse on that interval.

a. Graph f, for $x > -1$.

b. Find the inverse function f^{-1} corresponding to the function graphed in part (a). Graph f^{-1} on the same set of axes as in part (a).

c. Evaluate the derivative of f^{-1} at the point $\left(\dfrac{1}{2}, 1\right)$.

d. Sketch the tangent lines on the graphs of f and f^{-1} at $\left(1, \dfrac{1}{2}\right)$ and $\left(\dfrac{1}{2}, 1\right)$, respectively.

103. For what value(s) of x is the line tangent to the curve $y = x\sqrt{6 - x}$ horizontal? Vertical?

104–105. Limits *The following limits represent the derivative of a function f at a point a. Find a possible f and a, and then evaluate the limit.*

104. $\lim\limits_{h \to 0} \dfrac{\sin^2\left(\dfrac{\pi}{4} + h\right) - \dfrac{1}{2}}{h}$

105. $\lim\limits_{x \to 5} \dfrac{\tan\left(\pi\sqrt{3x - 11}\right)}{x - 5}$

T 106. Velocity of a probe A small probe is launched vertically from the ground. After it reaches its high point, a parachute deploys and the probe descends to Earth. The height of the probe above ground is

$$s(t) = \frac{300t - 50t^2}{t^3 + 2}, \text{ for } 0 \leq t \leq 6.$$

a. Graph the height function and describe the motion of the probe.
b. Find the velocity of the probe.
c. Graph the velocity function and determine the approximate time at which the velocity is a maximum.

107. Marginal and average cost Suppose the cost of producing x lawn mowers is $C(x) = -0.02x^2 + 400x + 5000$.

a. Determine the average and marginal costs for $x = 3000$ lawn mowers.
b. Interpret the meaning of your results in part (a).

108. Marginal and average cost Assume $C(x) = -0.0001x^3 + 0.05x^2 + 60x + 800$ is the cost of making x fly rods.

a. Determine the average and marginal costs for $x = 400$ fly rods.
b. Interpret the meaning of your results in part (a).

T 109. Population growth Suppose $p(t) = -1.7t^3 + 72t^2 + 7200t + 80{,}000$ is the population of a city t years after 1950.

a. Determine the average rate of growth of the city from 1950 to 2000.
b. What was the rate of growth of the city in 1990?

T 110. Position of a piston The distance between the head of a piston and the end of a cylindrical chamber is given by $x(t) = \dfrac{8t}{t + 1}$ cm, for $t \geq 0$ (measured in seconds). The radius of the cylinder is 4 cm.

a. Find the volume of the chamber, for $t \geq 0$.
b. Find the rate of change of the volume $V'(t)$, for $t \geq 0$.
c. Graph the derivative of the volume function. On what intervals is the volume increasing? Decreasing?

111. Boat rates Two boats leave a dock at the same time. One boat travels south at 30 mi/hr and the other travels east at 40 mi/hr. After half an hour, how fast is the distance between the boats increasing?

112. Rate of inflation of a balloon A spherical balloon is inflated at a rate of 10 cm³/min. At what rate is the diameter of the balloon increasing when the balloon has a diameter of 5 cm?

113. Rate of descent of a hot-air balloon A rope is attached to the bottom of a hot-air balloon that is floating above a flat field. If the angle of the rope to the ground remains 65° and the rope is pulled in at 5 ft/s, how quickly is the elevation of the balloon changing?

114. Filling a tank Water flows into a conical tank at a rate of 2 ft³/min. If the radius of the top of the tank is 4 ft and the height is 6 ft, determine how quickly the water level is rising when the water is 2 ft deep in the tank.

115. Angle of elevation A jet flying at 450 mi/hr and traveling in a straight line at a constant elevation of 500 ft passes directly over a spectator at an air show. How quickly is the angle of elevation (between the ground and the line from the spectator to the jet) changing 2 seconds later?

116. Viewing angle A man whose eye level is 6 ft above the ground walks toward a billboard at a rate of 2 ft/s. The bottom of the billboard is 10 ft above the ground, and it is 15 ft high. The man's viewing angle is the angle formed by the lines between the man's eyes and the top and bottom of the billboard. At what rate is the viewing angle changing when the man is 30 ft from the billboard?

117. Shadow length A street light is fastened to the top of a 15-ft-high pole. If a 5-ft-tall women walks away from the pole in a straight line over level ground at a rate of 3 ft/s, how fast is the length of her shadow changing when she is 10 ft away from the pole?

118. Quadratic functions

a. Show that if $(a, f(a))$ is any point on the graph of $f(x) = x^2$, then the slope of the tangent line at that point is $m = 2a$.
b. Show that if $(a, f(a))$ is any point on the graph of $f(x) = bx^2 + cx + d$, then the slope of the tangent line at that point is $m = 2ab + c$.

119. Derivative of the inverse in two ways Let $f(x) = \sin x$, $f^{-1}(x) = \sin^{-1}(x)$, and $(x_0, y_0) = \left(\pi/4, 1/\sqrt{2}\right)$.

a. Evaluate $(f^{-1})'\left(1/\sqrt{2}\right)$ using Theorem 3.21.
b. Evaluate $(f^{-1})'\left(1/\sqrt{2}\right)$ directly by differentiating f^{-1}. Check for agreement with part (a).

120. A parabola property Let $f(x) = x^2$.

a. Show that $\dfrac{f(x) - f(y)}{x - y} = f'\left(\dfrac{x + y}{2}\right)$, for all $x \neq y$.
b. Is this property true for $f(x) = ax^2$, where a is a nonzero real number?
c. Give a geometrical interpretation of this property.
d. Is this property true for $f(x) = ax^3$?

Chapter 3 Guided Projects

Applications of the material in this chapter and related topics can be found in the following Guided Projects. For additional information, see the Preface.

• Numerical differentiation

• Enzyme kinetics

• Elasticity in economics

• Pharmacokinetics—drug metabolism

4

Applications of the Derivative

Chapter Preview   Much of the previous chapter was devoted to the basic mechanics of derivatives: evaluating them and interpreting them as rates of change. We now apply derivatives to a variety of mathematical questions, many of which concern properties of functions and their graphs. One outcome of this work is a set of analytical curve-sketching methods that produce accurate graphs of functions. Equally important, derivatives allow us to formulate and solve a wealth of practical problems. For example, an asteroid passes perilously close to Earth: At what point along their trajectories is the distance separating them smallest, and what is the minimum distance? An economist has a mathematical model that relates the demand for a product to its price: What price maximizes the revenue? In this chapter, we develop the tools needed to answer such questions. In addition, we begin an ongoing discussion about approximating functions, we present an important result called the Mean Value Theorem, and we work with a powerful method that enables us to evaluate a new kind of limit. The chapter concludes with two important topics: a numerical approach to approximating roots of functions, called Newton's method, and a preview of integral calculus, which is the subject of Chapter 5.

4.1 Maxima and Minima

With a working understanding of derivatives, we now undertake one of the fundamental tasks of calculus: analyzing the behavior of functions and producing accurate graphs of them. An important question associated with any function concerns its maximum and minimum values: On a given interval (perhaps the entire domain), where does the function assume its largest and smallest values? Questions about maximum and minimum values take on added significance when a function represents a practical quantity, such as the profits of a company, the surface area of a container, or the speed of a space vehicle.

Absolute Maxima and Minima

Imagine taking a long hike through varying terrain from west to east. Your elevation changes as you walk over hills, through valleys, and across plains, and you reach several high and low points along the journey. Analogously, when we examine a function f over an interval on the x-axis, its values increase and decrease, reaching high points and low points (**Figure 4.1**). You can view our study of functions in this chapter as an exploratory hike along the x-axis.

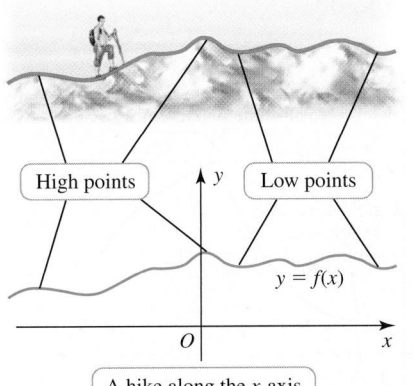

Figure 4.1

> **DEFINITION**   **Absolute Maximum and Minimum**
> Let f be defined on a set D containing c. If $f(c) \geq f(x)$ for every x in D, then $f(c)$ is an **absolute maximum** value of f on D. If $f(c) \leq f(x)$ for every x in D, then $f(c)$ is an **absolute minimum** value of f on D. An **absolute extreme value** is either an absolute maximum value or an absolute minimum value.

The existence and location of absolute extreme values depend on both the function and the interval of interest. **Figure 4.2** shows various cases for the function $f(x) = 4 - x^2$. Notice that if the interval of interest is not closed, a function might not attain absolute extreme values (Figure 4.2a, c, and d).

➤ Absolute maximum and minimum values are also called *global* maximum and minimum values. The plural of *maximum* is *maxima*; the plural of *minimum* is *minima*.

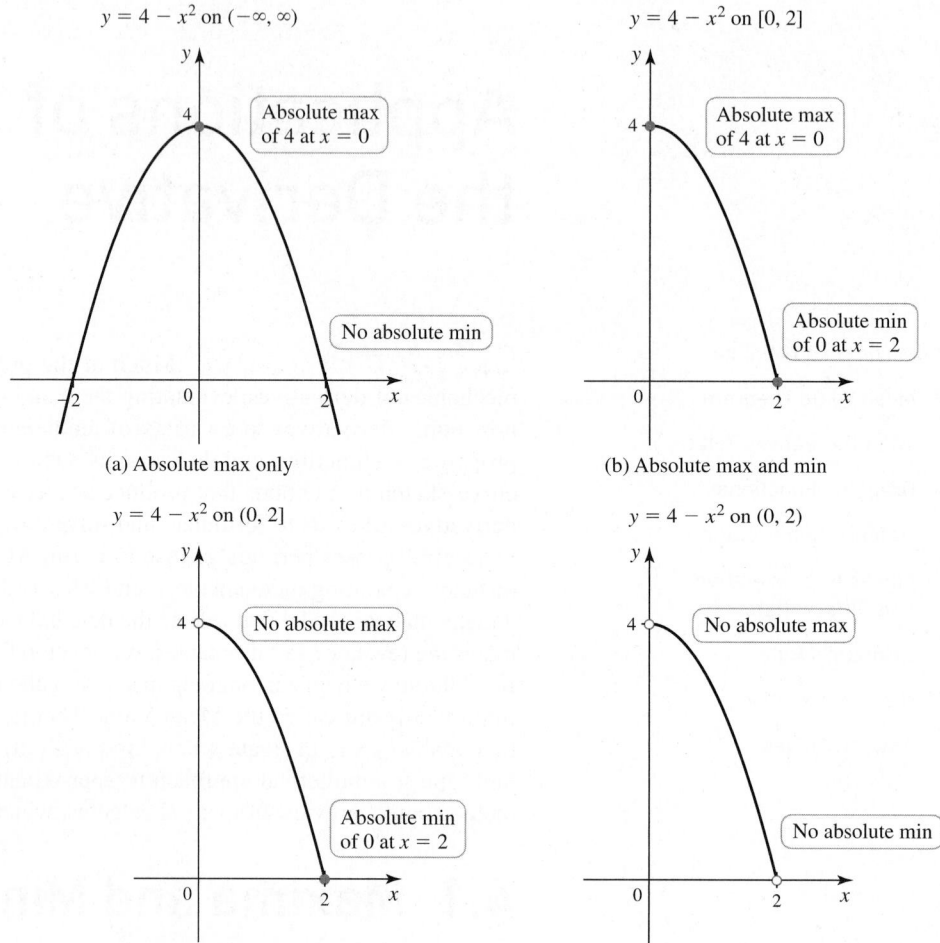

$y = 4 - x^2$ on $(-\infty, \infty)$

Absolute max of 4 at $x = 0$

No absolute min

(a) Absolute max only

$y = 4 - x^2$ on $[0, 2]$

Absolute max of 4 at $x = 0$

Absolute min of 0 at $x = 2$

(b) Absolute max and min

$y = 4 - x^2$ on $(0, 2]$

No absolute max

Absolute min of 0 at $x = 2$

(c) Absolute min only

$y = 4 - x^2$ on $(0, 2)$

No absolute max

No absolute min

(d) No absolute max or min

Figure 4.2

However, defining a function on a closed interval is not enough to guarantee the existence of absolute extreme values. Both functions in **Figure 4.3** are defined at every point of a closed interval, but neither function attains an absolute maximum—the discontinuity in each function prevents it from happening.

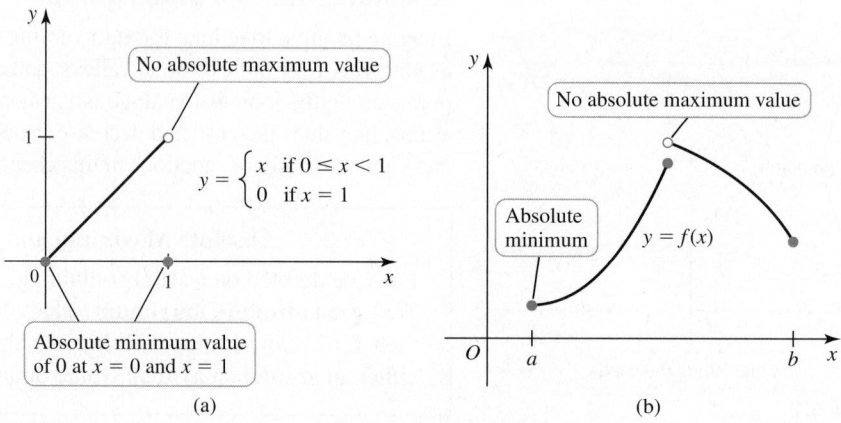

No absolute maximum value

$y = \begin{cases} x & \text{if } 0 \le x < 1 \\ 0 & \text{if } x = 1 \end{cases}$

Absolute minimum value of 0 at $x = 0$ and $x = 1$

(a)

No absolute maximum value

Absolute minimum

$y = f(x)$

(b)

Figure 4.3

It turns out that *two* conditions ensure the existence of absolute maximum and minimum values on an interval: The function must be continuous on the interval, and the interval must be closed and bounded.

> ▶ The proof of the Extreme Value Theorem relies on properties of the real numbers, found in advanced books.

> **THEOREM 4.1   Extreme Value Theorem**
> A function that is continuous on a closed interval $[a, b]$ has an absolute maximum value and an absolute minimum value on that interval.

QUICK CHECK 1 Sketch the graph of a function that is continuous on an interval but does not have an absolute minimum value. Sketch the graph of a function that is defined on a closed interval but does not have an absolute minimum value. ◄

EXAMPLE 1   Locating absolute maximum and minimum values For the functions in Figure 4.4, identify the location of the absolute maximum value and the absolute minimum value on the interval $[a, b]$. Do the functions meet the conditions of the Extreme Value Theorem?

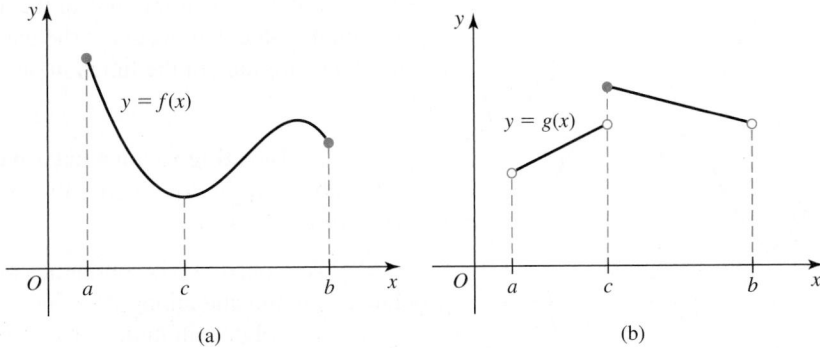

Figure 4.4

SOLUTION

a. The function f is continuous on the closed interval $[a, b]$, so the Extreme Value Theorem guarantees an absolute maximum (which occurs at a) and an absolute minimum (which occurs at c).

b. The function g does not satisfy the conditions of the Extreme Value Theorem because it is not continuous, and it is defined only on the open interval (a, b). It does not have an absolute minimum value. It does, however, have an absolute maximum at c. Therefore, a function may violate the conditions of the Extreme Value Theorem and still have an absolute maximum or minimum (or both).

Related Exercises 11–14 ◄

Local Maxima and Minima

Figure 4.5 shows a function f defined on the interval $[a, b]$. It has an absolute minimum at the endpoint b and an absolute maximum at the interior point s. In addition, the function has

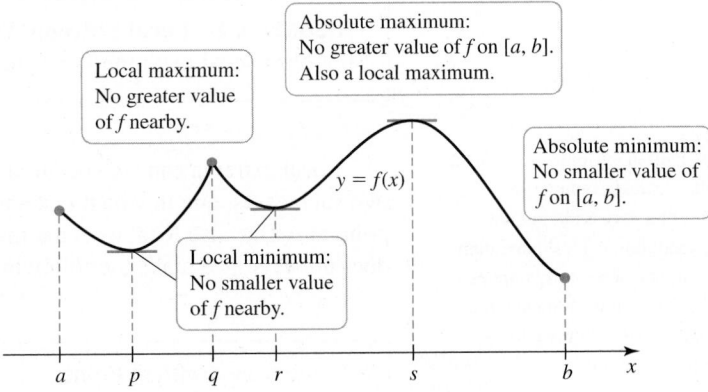

Figure 4.5

special behavior at q, where its value is greatest *among values at nearby points*, and at p and r, where its value is least *among values at nearby points*. A point at which a function takes on the maximum or minimum value among values at nearby points is important.

> Local maximum and minimum values are also called *relative maximum* and *minimum values*. Local extrema (plural) and *local extremum* (singular) refer to either local maxima or local minima.

DEFINITION Local Maximum and Minimum Values

Suppose c is an interior point of some interval I on which f is defined. If $f(c) \geq f(x)$ for all x in I, then $f(c)$ is a **local maximum** value of f. If $f(c) \leq f(x)$ for all x in I, then $f(c)$ is a **local minimum** value of f.

In this text, we adopt the convention that local maximum values and local minimum values occur only at interior points of the interval(s) of interest. For example, in Figure 4.5, the minimum value that occurs at the endpoint b is not a local minimum. However, it is the absolute minimum of the function on $[a, b]$.

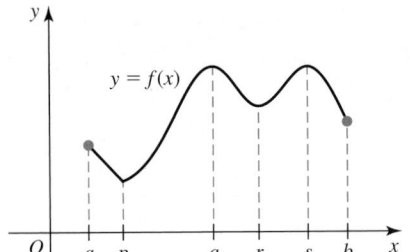

Figure 4.6

EXAMPLE 2 Locating various maxima and minima Figure 4.6 shows the graph of a function defined on $[a, b]$. Identify the location of the various maxima and minima using the terms *absolute* and *local*.

SOLUTION The function f is continuous on a closed interval; by Theorem 4.1, it has absolute maximum and minimum values on $[a, b]$. The function has a local minimum value and its absolute minimum value at p. It has another local minimum value at r. The absolute maximum value of f occurs at both q and s (which also correspond to local maximum values).

Related Exercises 15–18 ◀

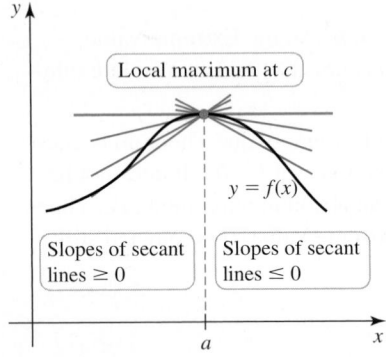

Figure 4.7

Critical Points Another look at Figure 4.6 shows that local maxima and minima occur at points in the open interval (a, b) where the derivative is zero ($x = q, r,$ and s) and at points where the derivative fails to exist ($x = p$). We now make this observation precise.

Figure 4.7 illustrates a function that is differentiable at c with a local maximum at c. For x near c with $x < c$, the secant lines through $(x, f(x))$ and $(c, f(c))$ have nonnegative slopes. For x near c with $x > c$, the secant lines through $(x, f(x))$ and $(c, f(c))$ have nonpositive slopes. As $x \to c$, the slopes of these secant lines approach the slope of the tangent line at $(c, f(c))$. These observations imply that the slope of the tangent line must be both nonnegative and nonpositive, which happens only if $f'(c) = 0$. Similar reasoning leads to the same conclusion for a function with a local minimum at c: if $f'(c)$ exists, then $f'(c)$ must be zero. This argument is an outline of the proof (Exercise 91) of the following theorem.

THEOREM 4.2 Local Extreme Value Theorem

If f has a local maximum or minimum value at c and $f'(c)$ exists, then $f'(c) = 0$.

> Theorem 4.2, often attributed to Fermat, is one of the clearest examples in mathematics of a necessary, but not sufficient, condition. A local maximum (or minimum) at c necessarily implies a critical point at c, but a critical point at c is not sufficient to imply a local maximum (or minimum)

Local extrema can also occur at points c where $f'(c)$ does not exist. Figure 4.8 shows two such cases, one in which c is a point of discontinuity and one in which f has a corner point at c. Because local extrema may occur at points c where $f'(c) = 0$ or where $f'(c)$ does not exist, we make the following definition.

DEFINITION Critical Point

An interior point c of the domain of f at which $f'(c) = 0$ or $f'(c)$ fails to exist is called a **critical point** of f.

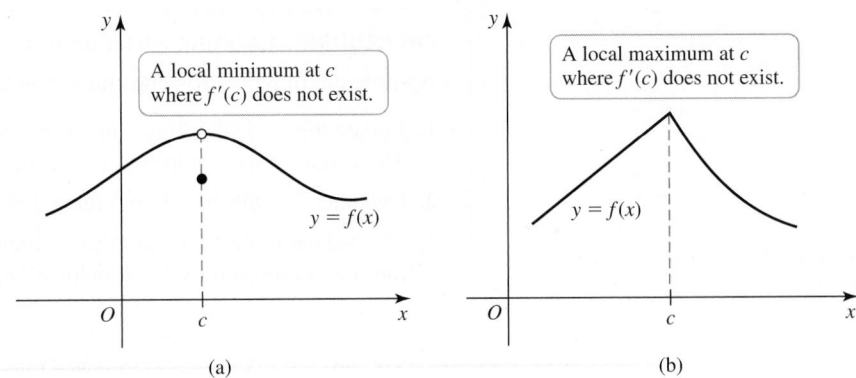

Figure 4.8

Note that the converse of Theorem 4.2 is not necessarily true. It is possible that $f'(c) = 0$ at a point without a local maximum or local minimum value occurring there (Figure 4.9a). It is also possible that $f'(c)$ fails to exist, with no local extreme value occurring at c (Figure 4.9b). Therefore, critical points are *candidates* for the location of local extreme values, but you must determine whether they actually correspond to local maxima or minima. This procedure is discussed in Section 4.3.

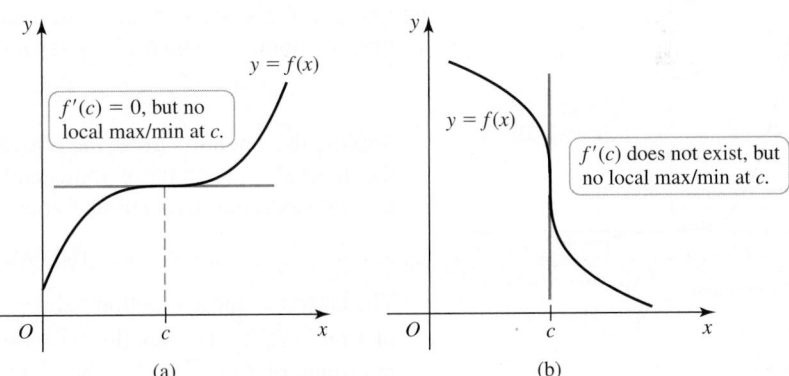

Figure 4.9

EXAMPLE 3   Locating critical points Find the critical points of $f(x) = x^2 \ln x$.

SOLUTION Note that f is differentiable on its domain, which is $(0, \infty)$. By the Product Rule,

$$f'(x) = 2x \cdot \ln x + x^2 \cdot \frac{1}{x} = x(2 \ln x + 1).$$

Setting $f'(x) = 0$ gives $x(2 \ln x + 1) = 0$, which has the solution $x = e^{-1/2} = 1/\sqrt{e}$. Because $x = 0$ is not in the domain of f, it is not a critical point. Therefore, the only critical point is $x = 1/\sqrt{e} \approx 0.61$. A graph of f (Figure 4.10) reveals that a local (and, indeed, absolute) minimum value occurs at $1/\sqrt{e}$, where the value of the function is $-1/(2e)$.

Related Exercises 35–36 ◄

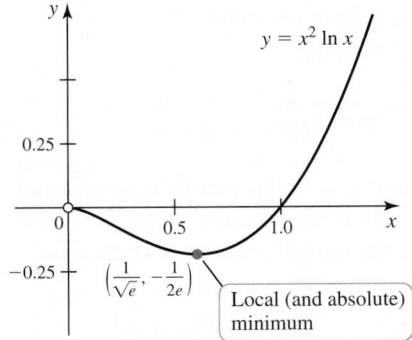

Figure 4.10

QUICK CHECK 2 Consider the function $f(x) = x^3$. Where is the critical point of f? Does f have a local maximum or minimum at the critical point? ◄

Locating Absolute Maxima and Minima

Theorem 4.1 guarantees the existence of absolute extreme values of a continuous function on a closed interval $[a, b]$, but it doesn't say where these values are located. Two observations lead to a procedure for locating absolute extreme values.

- An absolute extreme value in the interior of an interval is also a local extreme value, and we know that local extreme values occur at the critical points of f.

- Absolute extreme values may also occur at the endpoints of the interval of interest.

These two facts suggest the following procedure for locating the absolute extreme values of a continuous function on a closed interval.

> **PROCEDURE**   **Locating Absolute Extreme Values on a Closed Interval**
>
> Assume the function f is continuous on the closed interval $[a, b]$.
>
> 1. Locate the critical points c in (a, b), where $f'(c) = 0$ or $f'(c)$ does not exist. These points are candidates for absolute maxima and minima.
> 2. Evaluate f at the critical points and at the endpoints of $[a, b]$.
> 3. Choose the largest and smallest values of f from Step 2 for the absolute maximum and minimum values, respectively.

Note that the preceding procedure box does not address the case in which f is continuous on an open interval. If the interval of interest is an open interval, then absolute extreme values—if they exist—occur at interior points.

EXAMPLE 4   **Absolute extreme values** Find the absolute maximum and minimum values of the following functions.

a. $f(x) = x^4 - 2x^3$ on the interval $[-2, 2]$

b. $g(x) = x^{2/3}(2 - x)$ on the interval $[-1, 2]$

SOLUTION

a. Because f is a polynomial, its derivative exists everywhere. So if f has critical points, they are points at which $f'(x) = 0$. Computing f' and setting it equal to zero, we have

$$f'(x) = 4x^3 - 6x^2 = 2x^2(2x - 3) = 0.$$

Solving this equation gives the critical points $x = 0$ and $x = \frac{3}{2}$, both of which lie in the interval $[-2, 2]$; these points and the endpoints are *candidates* for the location of absolute extrema. Evaluating f at each of these points, we have

$$f(-2) = 32, \quad f(0) = 0, \quad f\left(\tfrac{3}{2}\right) = -\tfrac{27}{16}, \quad \text{and} \quad f(2) = 0.$$

The largest of these function values is $f(-2) = 32$, which is the absolute maximum of f on $[-2, 2]$. The smallest of these values is $f\left(\frac{3}{2}\right) = -\frac{27}{16}$, which is the absolute minimum of f on $[-2, 2]$. The graph of f **(Figure 4.11)** shows that the critical point $x = 0$ corresponds to neither a local maximum nor a local minimum.

b. Differentiating $g(x) = x^{2/3}(2 - x) = 2x^{2/3} - x^{5/3}$, we have

$$g'(x) = \frac{4}{3}x^{-1/3} - \frac{5}{3}x^{2/3} = \frac{4 - 5x}{3x^{1/3}}.$$

Because $g'(0)$ is undefined and 0 is in the domain of g, $x = 0$ is a critical point. In addition, $g'(x) = 0$ when $4 - 5x = 0$, so $x = \frac{4}{5}$ is also a critical point. These two critical points and the endpoints are *candidates* for the location of absolute extrema. The next step is to evaluate g at the critical points and endpoints:

$$g(-1) = 3, \quad g(0) = 0, \quad g\left(\tfrac{4}{5}\right) \approx 1.03, \quad \text{and} \quad g(2) = 0.$$

The largest of these function values is $g(-1) = 3$, which is the absolute maximum value of g on $[-1, 2]$. The least of these values is 0, which occurs twice. Therefore, g has its absolute minimum value on $[-1, 2]$ at the critical point $x = 0$ and the endpoint $x = 2$ **(Figure 4.12)**.

Related Exercises 46, 52 ◀

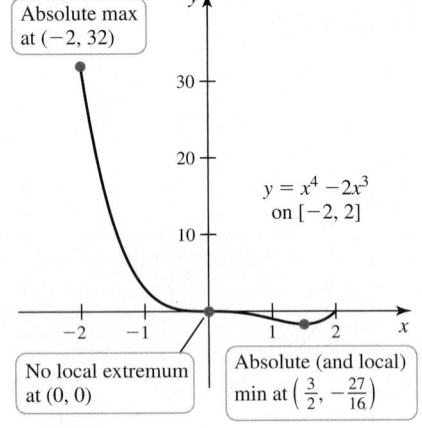

Absolute max at $(-2, 32)$

$y = x^4 - 2x^3$ on $[-2, 2]$

No local extremum at $(0, 0)$

Absolute (and local) min at $\left(\frac{3}{2}, -\frac{27}{16}\right)$

Figure 4.11

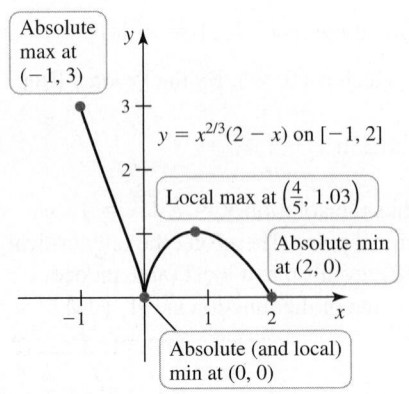

Absolute max at $(-1, 3)$

$y = x^{2/3}(2 - x)$ on $[-1, 2]$

Local max at $\left(\frac{4}{5}, 1.03\right)$

Absolute min at $(2, 0)$

Absolute (and local) min at $(0, 0)$

Figure 4.12

We now apply these ideas to a practical situation.

EXAMPLE 5   **Trajectory high point** A stone is launched vertically upward from a bridge 80 ft above the ground at a speed of 64 ft/s. Its height above the ground t seconds after the launch is given by

$$f(t) = -16t^2 + 64t + 80, \quad \text{for } 0 \le t \le 5.$$

When does the stone reach its maximum height?

➤ The derivation of the position function for an object moving in a gravitational field is given in Section 6.1.

SOLUTION We must evaluate the height function at the critical points and at the endpoints. The critical points satisfy the equation

$$f'(t) = -32t + 64 = -32(t - 2) = 0,$$

so the only critical point is $t = 2$. We now evaluate f at the endpoints and at the critical point:

$$f(0) = 80, \quad f(2) = 144, \quad \text{and} \quad f(5) = 0.$$

On the interval $[0, 5]$, the absolute maximum occurs at $t = 2$, at which time the stone reaches a height of 144 ft. Because $f'(t)$ is the velocity of the stone, the maximum height occurs at the instant the velocity is zero.

Related Exercise 73 ◄

SECTION 4.1 EXERCISES

Getting Started

1. What does it mean for a function to have an absolute extreme value at a point c of an interval $[a, b]$?

2. What are local maximum and minimum values of a function?

3. What conditions must be met to ensure that a function has an absolute maximum value and an absolute minimum value on an interval?

4. Sketch the graph of a function that is continuous on an open interval (a, b) but has neither an absolute maximum nor an absolute minimum value on (a, b).

5. Sketch the graph of a function that has an absolute maximum, a local minimum, but no absolute minimum on $[0, 3]$.

6. What is a critical point of a function?

7. Sketch the graph of a function f that has a local maximum value at a point c where $f'(c) = 0$.

8. Sketch the graph of a function f that has a local minimum value at a point c where $f'(c)$ is undefined.

9. How do you determine the absolute maximum and minimum values of a continuous function on a closed interval?

10. Explain how a function can have an absolute minimum value at an endpoint of an interval.

11–14. *Use the following graphs to identify the points (if any) on the interval* $[a, b]$ *at which the function has an absolute maximum or an absolute minimum value.*

11.

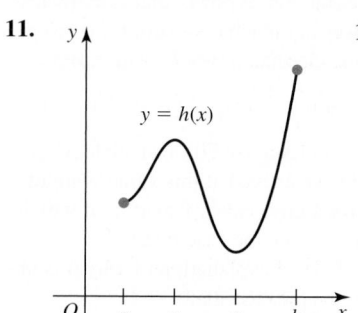

12.

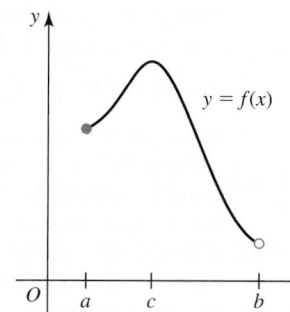

13.   **14.**

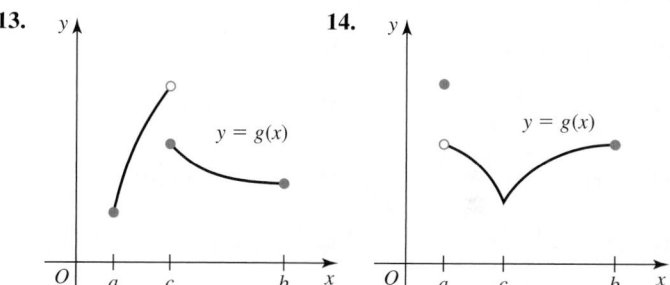

15–18. *Use the following graphs to identify the points on the interval* $[a, b]$ *at which local and absolute extreme values occur.*

15.   **16.**

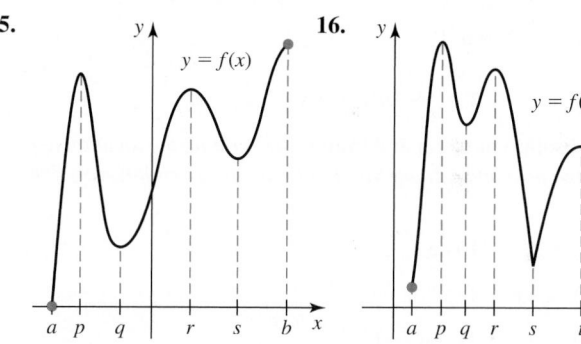

17.   **18.**

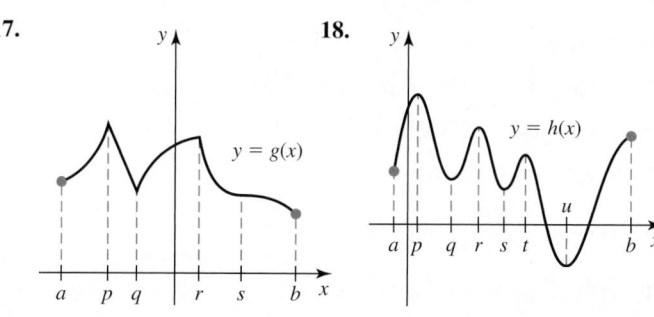

19–22. *Sketch the graph of a continuous function f on* $[0, 4]$ *satisfying the given properties.*

19. $f'(x) = 0$ for $x = 1$ and 2; f has an absolute maximum at $x = 4$; f has an absolute minimum at $x = 0$; and f has a local minimum at $x = 2$.

20. $f'(x) = 0$ for $x = 1$, 2, and 3; f has an absolute minimum at $x = 1$; f has no local extremum at $x = 2$; and f has an absolute maximum at $x = 3$.

21. $f'(1)$ and $f'(3)$ are undefined; $f'(2) = 0$; f has a local maximum at $x = 1$; f has a local minimum at $x = 2$; f has an absolute maximum at $x = 3$; and f has an absolute minimum at $x = 4$.

22. $f'(x) = 0$ at $x = 1$ and 3; $f'(2)$ is undefined; f has an absolute maximum at $x = 2$; f has neither a local maximum nor a local minimum at $x = 1$; and f has an absolute minimum at $x = 3$.

Practice Exercises

23–42. Locating critical points *Find the critical points of the following functions. Assume a is a nonzero constant.*

23. $f(x) = 3x^2 - 4x + 2$

24. $f(x) = \frac{1}{8}x^3 - \frac{1}{2}x$

25. $f(x) = \frac{x^3}{3} - 9x$

26. $f(x) = \frac{x^4}{4} - \frac{x^3}{3} - 3x^2 + 10$

27. $f(x) = 3x^3 + \frac{3x^2}{2} - 2x$

28. $f(x) = \frac{4x^5}{5} - 3x^3 + 5$

29. $f(x) = x^3 - 4a^2 x$

30. $f(x) = x - 5\tan^{-1} x$

31. $f(t) = \frac{t}{t^2 + 1}$

32. $f(x) = 12x^5 - 20x^3$

33. $f(x) = \frac{e^x + e^{-x}}{2}$

34. $f(x) = \sin x \cos x$

35. $f(x) = \frac{1}{x} + \ln x$

36. $f(t) = t^2 - 2\ln(t^2 + 1)$

37. $f(x) = x^2 \sqrt{x + 5}$

38. $f(x) = (\sin^{-1} x)(\cos^{-1} x)$

39. $f(x) = x\sqrt{x - a}$

40. $f(x) = \frac{x}{\sqrt{x - a}}$

41. $f(t) = \frac{1}{5}t^5 - a^4 t$

42. $f(x) = x^3 - 3ax^2 + 3a^2 x - a^3$

43–68. Absolute maxima and minima *Determine the location and value of the absolute extreme values of f on the given interval, if they exist.*

43. $f(x) = x^2 - 10$ on $[-2, 3]$

44. $f(x) = (x + 1)^{4/3}$ on $[-9, 7]$

45. $f(x) = x^3 - 3x^2$ on $[-1, 3]$

46. $f(x) = x^4 - 4x^3 + 4x^2$ on $[-1, 3]$

47. $f(x) = 3x^5 - 25x^3 + 60x$ on $[-2, 3]$

48. $f(x) = 2ex - x^2$ on $[0, 2e]$

49. $f(x) = \cos^2 x$ on $[0, \pi]$

50. $f(x) = \frac{x}{(x^2 + 3)^2}$ on $[-2, 2]$

51. $f(x) = \sin 3x$ on $[-\pi/4, \pi/3]$

52. $f(x) = 3x^{2/3} - x$ on $[0, 27]$

53. $f(x) = (2x)^x$ on $[0.1, 1]$

54. $f(x) = xe^{1-x/2}$ on $[0, 5]$

55. $f(x) = x^2 + \cos^{-1} x$ on $[-1, 1]$

56. $f(x) = x\sqrt{2 - x^2}$ on $[-\sqrt{2}, \sqrt{2}]$

57. $f(x) = 2x^3 - 15x^2 + 24x$ on $[0, 5]$

58. $f(x) = e^x - 2x$ on $[0, 2]$

59. $f(x) = \frac{4x^3}{3} + 5x^2 - 6x$ on $[-4, 1]$

60. $f(x) = 2x^6 - 15x^4 + 24x^2$ on $[-2, 2]$

61. $f(x) = \frac{x}{(x^2 + 9)^5}$ on $[-2, 2]$

62. $f(x) = x^{1/2}\left(\frac{x^2}{5} - 4\right)$ on $[0, 4]$

63. $f(x) = \sec x$ on $\left[-\frac{\pi}{4}, \frac{\pi}{4}\right]$

64. $f(x) = x^{1/3}(x + 4)$ on $[-27, 27]$

65. $f(x) = x^3 e^{-x}$ on $[-1, 5]$ **66.** $f(x) = x \ln \frac{x}{5}$ on $[0.1, 5]$

67. $f(x) = x^{2/3}(4 - x^2)$ on $[-2, 2]$

68. $f(t) = \frac{3t}{t^2 + 1}$ on $[-2, 2]$

69. Efficiency of wind turbines A wind turbine converts wind energy into electrical power. Let v_1 equal the upstream velocity of the wind before it encounters the wind turbine, and let v_2 equal the downstream velocity of the wind after it passes through the area swept out by the turbine blades.

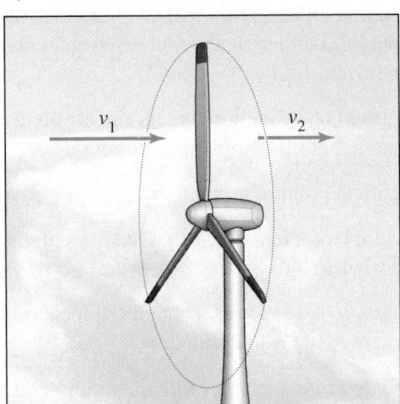

a. Assuming that $v_1 > 0$, give a physical explanation to show that $0 \le \frac{v_2}{v_1} \le 1$.

b. The amount of power extracted from the wind depends on the ratio $r = \frac{v_2}{v_1}$, the ratio of the downstream velocity to upstream velocity. Let $R(r)$ equal the fraction of power that is extracted from the total available power in the wind stream, for a given value of r. In about 1920, the German physicist Albert Betz showed that $R(r) = \frac{1}{2}(1 + r)(1 - r^2)$, where $0 \le r \le 1$ (a derivation of R is outlined in Exercise 70). Calculate $R(1)$ and explain how you could have arrived at this value without using the formula for R. Give a physical explanation of why it is unlikely or impossible for it to be the case that $r = 1$.

c. Calculate $R(0)$ and give a physical explanation of why it is unlikely or impossible for it to be the case that $r = 0$.

d. The maximum value of R is called the *Betz limit*. It represents the theoretical maximum amount of power that can be extracted from the wind. Find this value and explain its physical meaning.

70. Derivation of wind turbine formula A derivation of the function R in Exercise 69, based on three equations from physics, is outlined here. Consider again the figure given in Exercise 69, where v_1 equals the upstream velocity of the wind just before the wind stream encounters the wind turbine, and v_2 equals the downstream velocity of the wind just after the wind stream passes through the area swept out by the turbine blades. An equation for the power extracted by the rotor blades, based on conservation of momentum, is $P = v^2 \rho A(v_1 - v_2)$, where v is the velocity of the wind (in m/s) as it passes through the turbine blades, ρ is the density of air (in kg/m^3), and A is the area (in m^2) of the circular region swept out by the rotor blades.

a. Another expression for the power extracted by the rotor blades, based on conservation of energy, is $P = \dfrac{1}{2} \rho v A(v_1{}^2 - v_2{}^2)$. Equate the two power equations and solve for v.

b. Show that $P = \dfrac{\rho A}{4}(v_1 + v_2)(v_1{}^2 - v_2{}^2)$.

c. If the wind were to pass through the same area A without being disturbed by rotor blades, the amount of available power would be $P_0 = \dfrac{\rho A v_1{}^3}{2}$. Let $r = \dfrac{v_2}{v_1}$ and simplify the ratio $\dfrac{P}{P_0}$ to obtain the function $R(r)$ given in Exercise 69. (*Source: Journal of Applied Physics*, 105, 2009)

71. Suppose the position of an object moving horizontally after t seconds is given by the function $s(t) = 32t - t^4$, where $0 \le t \le 3$ and s is measured in feet, with $s > 0$ corresponding to positions to the right of the origin. When is the object farthest to the right?

72. Minimum-surface-area box All boxes with a square base and a volume of 50 ft^3 have a surface area given by $S(x) = 2x^2 + \dfrac{200}{x}$, where x is the length of the sides of the base. Find the absolute minimum of the surface area function on the interval $(0, \infty)$. What are the dimensions of the box with minimum surface area?

73. Trajectory high point A stone is launched vertically upward from a cliff 192 ft above the ground at a speed of 64 ft/s. Its height above the ground t seconds after the launch is given by $s = -16t^2 + 64t + 192$, for $0 \le t \le 6$. When does the stone reach its maximum height?

74. Maximizing revenue A sales analyst determines that the revenue from sales of fruit smoothies is given by $R(x) = -60x^2 + 300x$, where x is the price in dollars charged per item, for $0 \le x \le 5$.

a. Find the critical points of the revenue function.

b. Determine the absolute maximum value of the revenue function and the price that maximizes the revenue.

75. Maximizing profit Suppose a tour guide has a bus that holds a maximum of 100 people. Assume his profit (in dollars) for taking n people on a city tour is $P(n) = n(50 - 0.5n) - 100$. (Although P is defined only for positive integers, treat it as a continuous function.)

a. How many people should the guide take on a tour to maximize the profit?

b. Suppose the bus holds a maximum of 45 people. How many people should be taken on a tour to maximize the profit?

76. Minimizing rectangle perimeters All rectangles with an area of 64 have a perimeter given by $P(x) = 2x + \dfrac{128}{x}$, where x is the length of one side of the rectangle. Find the absolute minimum value of the perimeter function on the interval $(0, \infty)$. What are the dimensions of the rectangle with minimum perimeter?

77. Explain why or why not Determine whether the following statements are true and give an explanation or counterexample.

a. The function $f(x) = \sqrt{x}$ has a local maximum on the interval $[0, \infty)$.

b. If a function has an absolute maximum on a closed interval, then the function must be continuous on that interval.

c. A function f has the property that $f'(2) = 0$. Therefore, f has a local extreme value at $x = 2$.

d. Absolute extreme values of a function on a closed interval always occur at a critical point or an endpoint of the interval.

Explorations and Challenges

T 78–79. Absolute maxima and minima

a. *Find the critical points of f on the given interval.*

b. *Determine the absolute extreme values of f on the given interval.*

c. *Use a graphing utility to confirm your conclusions.*

78. $f(x) = \dfrac{x}{\sqrt{x-4}}$ on $[6, 12]$ **79.** $f(x) = 2^x \sin x$ on $[-2, 6]$

T 80–83. Critical points and extreme values

a. *Find the critical points of the following functions on the given interval. Use a root finder, if necessary.*

b. *Use a graphing utility to determine whether the critical points correspond to local maxima, local minima, or neither.*

c. *Find the absolute maximum and minimum values on the given interval, if they exist.*

80. $f(x) = 6x^4 - 16x^3 - 45x^2 + 54x + 23$ on $[-5, 5]$

81. $f(\theta) = 2 \sin \theta + \cos \theta$ on $[-2\pi, 2\pi]$

82. $g(x) = (x - 3)^{5/3}(x + 2)$ on $[-4, 4]$

83. $h(x) = \dfrac{5 - x}{x^2 + 2x - 3}$ on $[-10, 10]$

T 84–85. Absolute value functions *Graph the following functions and determine the local and absolute extreme values on the given interval.*

84. $f(x) = |x - 3| + |x + 2|$ on $[-4, 4]$

85. $g(x) = |x - 3| - 2|x + 1|$ on $[-2, 3]$

T 86. Dancing on a parabola Two people, A and B, walk along the parabola $y = x^2$ in such a way that the line segment L between them is always perpendicular to the line tangent to the parabola at A's position. The goal of this exercise is to determine the positions of A and B when L has minimum length. Assume the coordinates of A are (a, a^2).

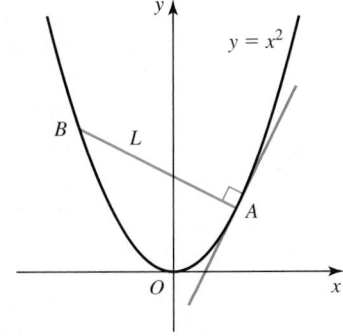

a. Find the slope of the line tangent to the parabola at A, and find the slope of the line that is perpendicular to the tangent line at A.

b. Find the equation of the line joining A and B.

c. Find the position of B on the parabola.

d. Write the function $F(a)$ that gives the *square* of the distance between A and B as it varies with a. (The square of the distance is minimized at the same point that the distance is minimized; it is easier to work with the square of the distance.)

e. Find the critical point of F on the interval $a > 0$.

f. Evaluate F at the critical point and verify that it corresponds to an absolute minimum. What are the positions of A and B that minimize the length of L? What is the minimum length?

g. Graph the function F to check your work.

T 87. Every second counts You must get from a point P on the straight shore of a lake to a stranded swimmer who is 50 m from a point Q on the shore that is 50 m from you (see figure). Assuming that you can swim at a speed of 2 m/s and run at a speed of 4 m/s, the goal of this exercise is to determine the point along the shore, x meters from Q, where you should stop running and start swimming to reach the swimmer in the minimum time.

a. Find the function T that gives the travel time as a function of x, where $0 \le x \le 50$.

b. Find the critical point of T on $(0, 50)$.

c. Evaluate T at the critical point and the endpoints ($x = 0$ and $x = 50$) to verify that the critical point corresponds to an absolute minimum. What is the minimum travel time?

d. Graph the function T to check your work.

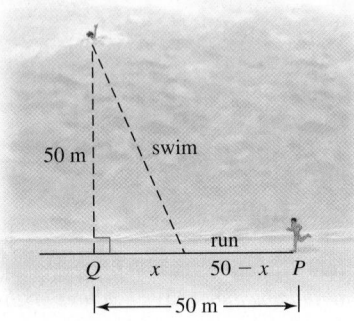

88. Extreme values of parabolas Consider the function $f(x) = ax^2 + bx + c$, with $a \ne 0$. Explain geometrically why f has exactly one absolute extreme value on $(-\infty, \infty)$. Find the critical point to determine the value of x at which f has an extreme value.

89. Values of related functions Suppose f is differentiable on $(-\infty, \infty)$ and assume it has a local extreme value at the point $x = 2$, where $f(2) = 0$. Let $g(x) = xf(x) + 1$ and let $h(x) = xf(x) + x + 1$, for all values of x.

a. Evaluate $g(2)$, $h(2)$, $g'(2)$, and $h'(2)$.

b. Does either g or h have a local extreme value at $x = 2$? Explain.

T 90. A family of double-humped functions Consider the functions $f(x) = \dfrac{x}{(x^2 + 1)^n}$, where n is a positive integer.

a. Show that these functions are odd for all positive integers n.

b. Show that the critical points of these functions are $x = \pm \dfrac{1}{\sqrt{2n - 1}}$, for all positive integers n. (Start with the special cases $n = 1$ and $n = 2$.)

c. Show that as n increases, the absolute maximum values of these functions decrease.

d. Use a graphing utility to verify your conclusions.

91. Proof of the Local Extreme Value Theorem Prove Theorem 4.2 for a local maximum: If f has a local maximum value at the point c and $f'(c)$ exists, then $f'(c) = 0$. Use the following steps.

a. Suppose f has a local maximum at c. What is the sign of $f(x) - f(c)$ if x is near c and $x > c$? What is the sign of $f(x) - f(c)$ if x is near c and $x < c$?

b. If $f'(c)$ exists, then it is defined by $\displaystyle\lim_{x \to c} \dfrac{f(x) - f(c)}{x - c}$. Examine this limit as $x \to c^+$ and conclude that $f'(c) \le 0$.

c. Examine the limit in part (b) as $x \to c^-$ and conclude that $f'(c) \ge 0$.

d. Combine parts (b) and (c) to conclude that $f'(c) = 0$.

92. Even and odd functions

a. Suppose a nonconstant even function f has a local minimum at c. Does f have a local maximum or minimum at $-c$? Explain. (An even function satisfies $f(-x) = f(x)$.)

b. Suppose a nonconstant odd function f has a local minimum at c. Does f have a local maximum or minimum at $-c$? Explain. (An odd function satisfies $f(-x) = -f(x)$.)

QUICK CHECK ANSWERS

1. The continuous function $f(x) = x$ does not have an absolute minimum on the open interval $(0, 1)$. The function $f(x) = -x$ on $\left[0, \frac{1}{2}\right)$ and $f(x) = 0$ on $\left[\frac{1}{2}, 1\right]$ does not have an absolute minimum on $[0, 1]$. **2.** The critical point is $x = 0$. Although $f'(0) = 0$, the function has neither a local maximum nor minimum at $x = 0$. ◄

4.2 Mean Value Theorem

In Section 4.1, we learned how to find the absolute extrema of a function. This information is needed to produce an accurate graph of a function (Sections 4.3 and 4.4) and to solve optimization problems (Section 4.5). The procedures used to solve these types of problems depend on several results developed over the next two sections.

We begin with the *Mean Value Theorem*, a cornerstone in the theoretical framework of calculus. Several critical theorems rely on the Mean Value Theorem; this theorem also appears in practical applications. The proof of the Mean Value Theorem relies on a preliminary result known as Rolle's Theorem.

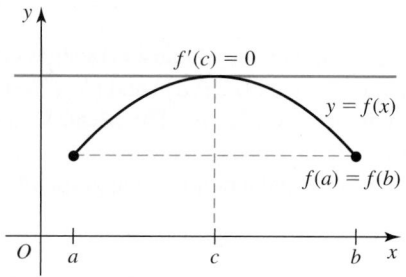

Figure 4.13

Rolle's Theorem

Consider a function f that is continuous on a closed interval $[a, b]$ and differentiable on the open interval (a, b). Furthermore, assume f has the special property that $f(a) = f(b)$ (**Figure 4.13**). The statement of Rolle's Theorem is not surprising: It says that somewhere between a and b, there is at least one point at which f has a horizontal tangent line.

> **THEOREM 4.3   Rolle's Theorem**
> Let f be continuous on a closed interval $[a, b]$ and differentiable on (a, b) with $f(a) = f(b)$. There is at least one point c in (a, b) such that $f'(c) = 0$.

Proof: The function f satisfies the conditions of Theorem 4.1 (Extreme Value Theorem); therefore, it attains its absolute maximum and minimum values on $[a, b]$. Those values are attained either at an endpoint or at an interior point c.

> ► The Extreme Value Theorem, discussed in Section 4.1, states that a function that is continuous on a closed bounded interval attains its absolute maximum and minimum values on that interval.

Case 1: First suppose f attains both its absolute maximum and minimum values at the endpoints. Because $f(a) = f(b)$, the maximum and minimum values are equal, and it follows that f is a constant function on $[a, b]$. Therefore, $f'(x) = 0$ for all x in (a, b), and the conclusion of the theorem holds.

Case 2: Assume at least one of the absolute extreme values of f does not occur at an endpoint. Then f must attain an absolute extreme value at an interior point of $[a, b]$; therefore, f must have either a local maximum or a local minimum at a point c in (a, b). We know from Theorem 4.2 that at a local extremum, the derivative is zero. Therefore, $f'(c) = 0$ for at least one point c of (a, b), and again the conclusion of the theorem holds. ◄

Why does Rolle's Theorem require continuity? A function that is not continuous on $[a, b]$ may have identical values at both endpoints and still not have a horizontal tangent line at any point on the interval (**Figure 4.14a**). Similarly, a function that is continuous on $[a, b]$ but not differentiable at a point of (a, b) may also fail to have a horizontal tangent line (**Figure 4.14b**).

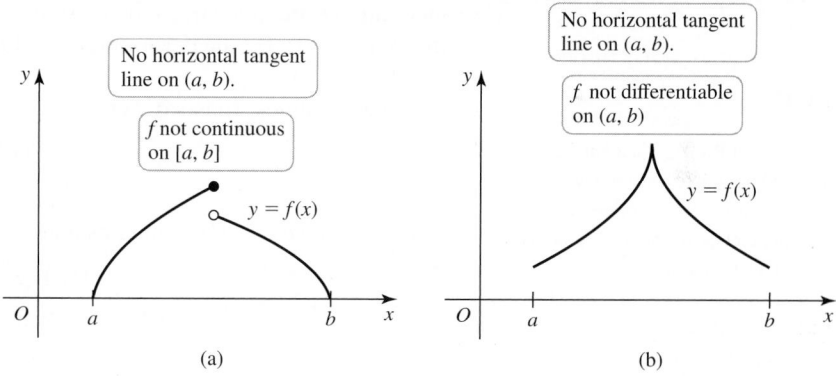

Figure 4.14

QUICK CHECK 1 Where on the interval $[0, 4]$ does $f(x) = 4x - x^2$ have a horizontal tangent line? ◄

EXAMPLE 1   Verifying Rolle's Theorem Find an interval I on which Rolle's Theorem applies to $f(x) = x^3 - 7x^2 + 10x$. Then find all points c in I at which $f'(c) = 0$.

SOLUTION Because f is a polynomial, it is everywhere continuous and differentiable. We need an interval $[a, b]$ with the property that $f(a) = f(b)$. Noting that $f(x) = x(x - 2)(x - 5)$, we choose the interval $[0, 5]$, because $f(0) = f(5) = 0$ (other intervals are possible). The goal is to find points c in the interval $(0, 5)$ at which $f'(c) = 0$, which amounts to the familiar task of finding the critical points of f. The critical points satisfy

$$f'(x) = 3x^2 - 14x + 10 = 0.$$

Using the quadratic formula, the roots are

$$x = \frac{7 \pm \sqrt{19}}{3}, \quad \text{or} \quad x \approx 0.88 \quad \text{and} \quad x \approx 3.79.$$

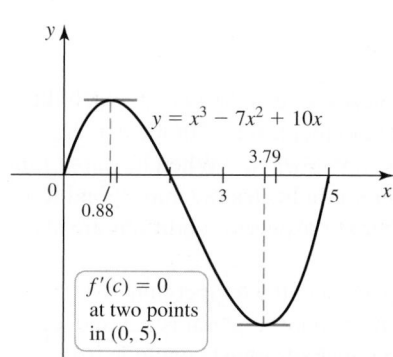

Figure 4.15

As shown in **Figure 4.15**, the graph of f has two points at which the tangent line is horizontal.

Related Exercises 11, 16 ◄

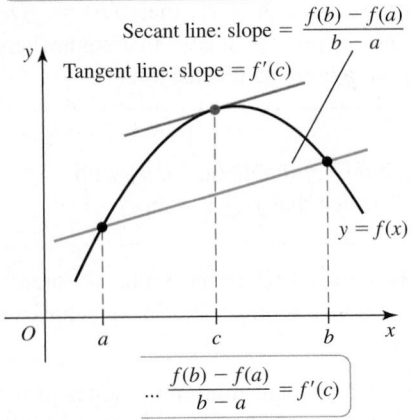

These lines are parallel and their slopes are equal, that is...

Secant line: slope $= \dfrac{f(b) - f(a)}{b - a}$

Tangent line: slope $= f'(c)$

$y = f(x)$

$\dfrac{f(b) - f(a)}{b - a} = f'(c)$

Figure 4.16

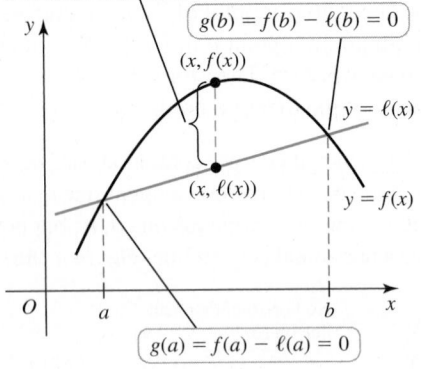

$g(x) = f(x) - \ell(x)$: The difference between the y-coordinates of the points $(x, \ell(x))$ and $(x, f(x))$

$g(b) = f(b) - \ell(b) = 0$

$(x, f(x))$

$y = \ell(x)$

$(x, \ell(x))$

$y = f(x)$

$g(a) = f(a) - \ell(a) = 0$

Figure 4.17

➤ The proofs of Rolle's Theorem and the Mean Value Theorem are nonconstructive: The theorems claim that a certain point exists, but their proofs do not say how to find it.

QUICK CHECK 2 Sketch the graph of a function that illustrates why the continuity condition of the Mean Value Theorem is needed. Sketch the graph of a function that illustrates why the differentiability condition of the Mean Value Theorem is needed. ◄

Mean Value Theorem

The Mean Value Theorem is easily understood with the aid of a picture. **Figure 4.16** shows a function f differentiable on (a, b) with a secant line passing through $(a, f(a))$ and $(b, f(b))$; the slope of the secant line is the average rate of change of f over $[a, b]$. The Mean Value Theorem claims that there exists a point c in (a, b) at which the slope of the tangent line at c is equal to the slope of the secant line. In other words, we can find a point on the graph of f where the tangent line is parallel to the secant line.

> **THEOREM 4.4 Mean Value Theorem**
> If f is continuous on the closed interval $[a, b]$ and differentiable on (a, b), then there is at least one point c in (a, b) such that
> $$\frac{f(b) - f(a)}{b - a} = f'(c).$$

Proof: The strategy of the proof is to use the function f of the Mean Value Theorem to form a new function g that satisfies Rolle's Theorem. Notice that the continuity and differentiability conditions of Rolle's Theorem and the Mean Value Theorem are the same. We devise g so that it satisfies the conditions $g(a) = g(b) = 0$.

As shown in **Figure 4.17**, the secant line passing through $(a, f(a))$ and $(b, f(b))$ is described by a function ℓ. We now define a new function g that measures the difference between the function values of f and ℓ. This function is simply $g(x) = f(x) - \ell(x)$. Because f and ℓ are continuous on $[a, b]$ and differentiable on (a, b), it follows that g is also continuous on $[a, b]$ and differentiable on (a, b). Furthermore, because the graphs of f and ℓ intersect at $x = a$ and $x = b$, we have $g(a) = f(a) - \ell(a) = 0$ and $g(b) = f(b) - \ell(b) = 0$.

We now have a function g that satisfies the conditions of Rolle's Theorem. By that theorem, we are guaranteed the existence of at least one point c in the interval (a, b) such that $g'(c) = 0$. By the definition of g, this condition implies that $f'(c) - \ell'(c) = 0$, or $f'(c) = \ell'(c)$.

We are almost finished. What is $\ell'(c)$? It is just the slope of the secant line, which is

$$\frac{f(b) - f(a)}{b - a}.$$

Therefore, $f'(c) = \ell'(c)$ implies that

$$\frac{f(b) - f(a)}{b - a} = f'(c).$$ ◄

The following situation offers an interpretation of the Mean Value Theorem. Imagine driving for 2 hours to a town 100 miles away. While your average speed is $100 \text{ mi}/2 \text{ hr} = 50 \text{ mi/hr}$, your instantaneous speed (measured by the speedometer) almost certainly varies. The Mean Value Theorem says that at some point during the trip, your instantaneous speed equals your average speed, which is 50 mi/hr. In Example 2, we apply these ideas to the science of weather forecasting.

EXAMPLE 2 Mean Value Theorem in action The *lapse rate* is the rate at which the temperature T decreases in the atmosphere with respect to increasing altitude z. It is typically reported in units of °C/km and is defined by $\gamma = -dT/dz$. When the lapse rate rises above 7°C/km in a certain layer of the atmosphere, it indicates favorable conditions for thunderstorm and tornado formation, provided other atmospheric conditions are also present.

Suppose the temperature at $z = 2.9$ km is $T = 7.6$°C and the temperature at $z = 5.6$ km is $T = -14.3$°C. Assume also that the temperature function is continuous and differentiable at all altitudes of interest. What can a meteorologist conclude from these data?

SOLUTION Figure 4.18 shows the two data points plotted on a graph of altitude and temperature. The slope of the line joining these points is

$$\frac{-14.3°C - 7.6°C}{5.6 \text{ km} - 2.9 \text{ km}} = -8.1°C/\text{km},$$

which means, on average, the temperature is decreasing at 8.1°C/km in the layer of air between 2.9 km and 5.6 km. With only two data points, we cannot know the entire temperature profile. The Mean Value Theorem, however, guarantees that there is at least one altitude at which $dT/dz = -8.1°C/\text{km}$. At each such altitude, the lapse rate is $\gamma = -dT/dz = 8.1°C/\text{km}$. Because this lapse rate is above the 7°C/km threshold associated with unstable weather, the meteorologist might expect an increased likelihood of severe storms.

> Meteorologists look for "steep" lapse rates in the layer of the atmosphere where the pressure is between 700 and 500 hPa (hectopascals). This range of pressure typically corresponds to altitudes between 3 km and 5.5 km. The data in Example 2 were recorded in Denver at nearly the same time a tornado struck 50 mi to the north.

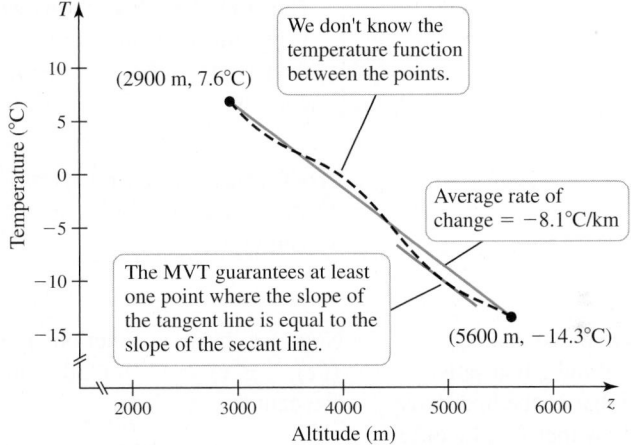

We don't know the temperature function between the points.

(2900 m, 7.6°C)

Average rate of change = −8.1°C/km

The MVT guarantees at least one point where the slope of the tangent line is equal to the slope of the secant line.

(5600 m, −14.3°C)

Figure 4.18

Related Exercises 19, 42 ◄

EXAMPLE 3   Verifying the Mean Value Theorem Determine whether the function $f(x) = 2x^3 - 3x + 1$ satisfies the conditions of the Mean Value Theorem on the interval $[-2, 2]$. If so, find the point(s) guaranteed to exist by the theorem.

SOLUTION The polynomial f is everywhere continuous and differentiable, so it satisfies the conditions of the Mean Value Theorem. The average rate of change of the function on the interval $[-2, 2]$ is

$$\frac{f(2) - f(-2)}{2 - (-2)} = \frac{11 - (-9)}{4} = 5.$$

The goal is to find points in $(-2, 2)$ at which the line tangent to the curve has a slope of 5—that is, to find points at which $f'(x) = 5$. Differentiating f, this condition becomes

$$f'(x) = 6x^2 - 3 = 5 \quad \text{or} \quad x^2 = \frac{4}{3}.$$

Therefore, the points guaranteed to exist by the Mean Value Theorem are $x = \pm 2/\sqrt{3} \approx \pm 1.15$. The tangent lines have slope 5 at the corresponding points on the curve (**Figure 4.19**).

Related Exercises 21–22 ◄

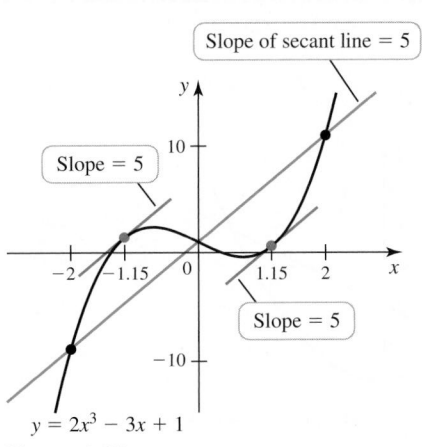

Slope of secant line = 5

Slope = 5

Slope = 5

$y = 2x^3 - 3x + 1$

Figure 4.19

Consequences of the Mean Value Theorem

We close with two results that follow from the Mean Value Theorem.

We already know that the derivative of a constant function is zero; that is, if $f(x) = C$, then $f'(x) = 0$ (Theorem 3.2). Theorem 4.5 states the converse of this result.

THEOREM 4.5   Zero Derivative Implies Constant Function
If f is differentiable and $f'(x) = 0$ at all points of an open interval I, then f is a constant function on I.

Proof: Suppose $f'(x) = 0$ on $[a, b]$, where a and b are distinct points of I. By the Mean Value Theorem, there exists a point c in (a, b) such that

$$\frac{f(b) - f(a)}{b - a} = \underbrace{f'(c) = 0.}_{\substack{f'(x) = 0 \text{ for} \\ \text{all } x \text{ in } I}}$$

Multiplying both sides of this equation by $b - a \neq 0$, it follows that $f(b) = f(a)$, and this is true for every pair of points a and b in I. If $f(b) = f(a)$ for every pair of points in an interval, then f is a constant function on that interval. ◀

Theorem 4.6 builds on the conclusion of Theorem 4.5.

> **THEOREM 4.6 Functions with Equal Derivatives Differ by a Constant**
> If two functions have the property that $f'(x) = g'(x)$, for all x of an open interval I, then $f(x) - g(x) = C$ on I, where C is a constant; that is, f and g differ by a constant.

Proof: The fact that $f'(x) = g'(x)$ on I implies that $f'(x) - g'(x) = 0$ on I. Recall that the derivative of a difference of two functions equals the difference of the derivatives, so we can write

$$f'(x) - g'(x) = (f - g)'(x) = 0.$$

QUICK CHECK 3 Give two distinct linear functions f and g that satisfy $f'(x) = g'(x)$; that is, the lines have equal slopes. Show that f and g differ by a constant. ◀

Now we have a function $f - g$ whose derivative is zero on I. By Theorem 4.5, $f(x) - g(x) = C$, for all x in I, where C is a constant; that is, f and g differ by a constant. ◀

The utility of Theorems 4.5 and 4.6 will become apparent in Section 4.9, where we establish a pivotal result that has far-reaching consequences. These theorems are also useful in verifying identities (see Exercises 34, 35, and 51).

SECTION 4.2 EXERCISES

Getting Started

1. Explain Rolle's Theorem with a sketch.

2. Draw the graph of a function for which the conclusion of Rolle's Theorem does not hold.

3. Explain why Rolle's Theorem cannot be applied to the function $f(x) = |x|$ on the interval $[-a, a]$ for any $a > 0$.

4. Explain the Mean Value Theorem with a sketch.

5–7. *For each function f and interval $[a, b]$, a graph of f is given along with the secant line that passes though the graph of f at $x = a$ and $x = b$.*

a. *Use the graph to make a conjecture about the value(s) of c satisfying the equation $\dfrac{f(b) - f(a)}{b - a} = f'(c)$.*

b. *Verify your answer to part (a) by solving the equation $\dfrac{f(b) - f(a)}{b - a} = f'(c)$ for c.*

5. $f(x) = \dfrac{x^2}{4} + 1; [-2, 4]$

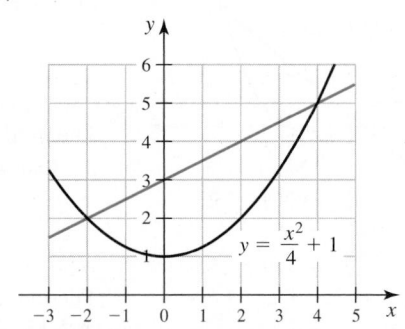

6. $f(x) = 2\sqrt{x}; [0, 4]$

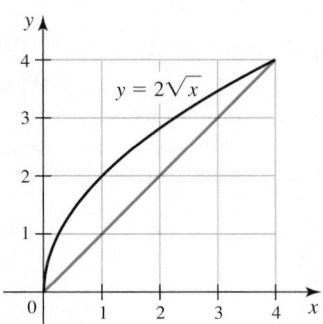

7. $f(x) = \dfrac{x^5}{16}; [-2, 2]$

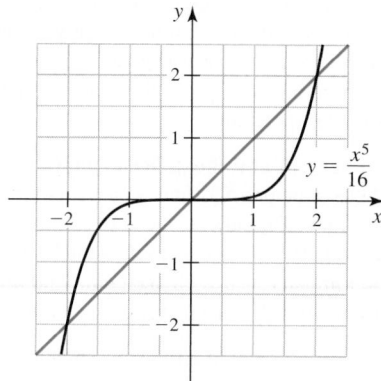

$y = \dfrac{x^5}{16}$

8. At what points c does the conclusion of the Mean Value Theorem hold for $f(x) = x^3$ on the interval $[-10, 10]$?

9. Draw the graph of a function for which the conclusion of the Mean Value Theorem does not hold.

10. Let $f(x) = x^{2/3}$. Show that there is no value of c in the interval $(-1, 8)$ for which $f'(c) = \dfrac{f(8) - f(-1)}{8 - (-1)}$ and explain why this does not violate the Mean Value Theorem.

Practice Exercises

11–18. Rolle's Theorem *Determine whether Rolle's Theorem applies to the following functions on the given interval. If so, find the point(s) guaranteed to exist by Rolle's Theorem.*

11. $f(x) = x(x - 1)^2; [0, 1]$ **12.** $f(x) = \sin 2x; [0, \pi/2]$

13. $f(x) = \cos 4x; [\pi/8, 3\pi/8]$ **14.** $f(x) = 1 - |x|; [-1, 1]$

15. $f(x) = 1 - x^{2/3}; [-1, 1]$

16. $f(x) = x^3 - 2x^2 - 8x; [-2, 4]$

17. $g(x) = x^3 - x^2 - 5x - 3; [-1, 3]$

18. $h(x) = e^{-x^2}; [-a, a]$, where $a > 0$

19. Lapse rates in the atmosphere Refer to Example 2. Concurrent measurements indicate that at an elevation of 6.1 km, the temperature is $-10.3°C$, and at an elevation of 3.2 km, the temperature is $8.0°C$. Based on the Mean Value Theorem, can you conclude that the lapse rate exceeds the threshold value of $7°C/km$ at some intermediate elevation? Explain.

20. Drag racer acceleration The fastest drag racers can reach a speed of 330 mi/hr over a quarter-mile strip in 4.45 seconds (from a standing start). Complete the following sentence about such a drag racer: At some point during the race, the maximum acceleration of the drag racer is at least _____ mi/hr/s.

21–32. Mean Value Theorem *Consider the following functions on the given interval $[a, b]$.*

a. *Determine whether the Mean Value Theorem applies to the following functions on the given interval $[a, b]$.*
b. *If so, find the point(s) that are guaranteed to exist by the Mean Value Theorem.*

21. $f(x) = 7 - x^2; [-1, 2]$ **22.** $f(x) = x^3 - 2x^2; [0, 1]$

23. $f(x) = \begin{cases} -2x & \text{if } x < 0 \\ x & \text{if } x \geq 0 \end{cases}; [-1, 1]$

24. $f(x) = \dfrac{1}{(x - 1)^2}; [0, 2]$ **25.** $f(x) = e^x; [0, 1]$

26. $f(x) = \ln 2x; [1, e]$ **T 27.** $f(x) = \sin x; \left[-\dfrac{\pi}{2}, \dfrac{\pi}{2}\right]$

T 28. $f(x) = \tan x; \left[0, \dfrac{\pi}{4}\right]$ **29.** $f(x) = \sin^{-1} x; \left[0, \dfrac{1}{2}\right]$

30. $f(x) = x + \dfrac{1}{x}; [1, 3]$ **31.** $f(x) = 2x^{1/3}; [-8, 8]$

32. $f(x) = \dfrac{x}{x + 2}; [-1, 2]$

33. Explain why or why not Determine whether the following statements are true and give an explanation or counterexample.

a. The continuous function $f(x) = 1 - |x|$ satisfies the conditions of the Mean Value Theorem on the interval $[-1, 1]$.
b. Two differentiable functions that differ by a constant always have the same derivative.
c. If $f'(x) = 0$, then $f(x) = 10$.

34. An inverse tangent identity

a. Use derivatives to show that $\tan^{-1}x + \tan^{-1}(1/x)$ is constant, for $x > 0$ and for $x < 0$.
b. Prove that $\tan^{-1}x + \tan^{-1}(1/x) = \pi/2$, for $x > 0$.
c. Prove that $\tan^{-1}x + \tan^{-1}(1/x) = -\pi/2$, for $x < 0$, and conclude that $\tan^{-1}x + \tan^{-1}(1/x) = \begin{cases} \dfrac{\pi}{2} & \text{if } x > 0 \\ -\dfrac{\pi}{2} & \text{if } x < 0. \end{cases}$

35. Another inverse tangent identity

a. Use derivatives to show that $\tan^{-1}\dfrac{2}{x^2}$ and $\tan^{-1}(x + 1) - \tan^{-1}(x - 1)$ differ by a constant.
b. Prove that $\tan^{-1}\dfrac{2}{x^2} = \tan^{-1}(x + 1) - \tan^{-1}(x - 1)$, for $x \neq 0$.
(Source: The College Mathematics Journal, 32, 4, Sep 2001)

36. Without evaluating derivatives, determine which of the functions $f(x) = \ln x$, $g(x) = \ln 2x$, $h(x) = \ln x^2$, and $p(x) = \ln 10x^2$ have the same derivative.

37. Without evaluating derivatives, determine which of the functions $g(x) = 2x^{10}$, $h(x) = x^{10} + 2$, and $p(x) = x^{10} - \ln 2$ have the same derivative as $f(x) = x^{10}$.

38. Find all functions f whose derivative is $f'(x) = x + 1$.

39. Mean Value Theorem and graphs By visual inspection, locate all points on the interval $(-4, 4)$ at which the slope of the tangent line equals the average rate of change of the function on the interval $[-4, 4]$.

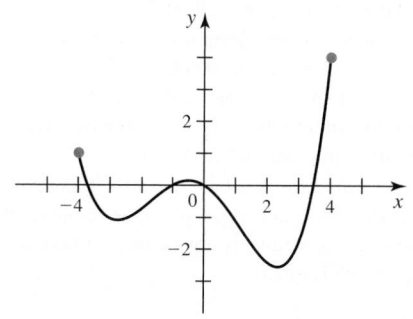

40–41. Mean Value Theorem and graphs *Find all points on the interval* $(1, 3)$ *at which the slope of the tangent line equals the average rate of change of f on* $[1, 3]$. *Reconcile your results with the Mean Value Theorem.*

40.

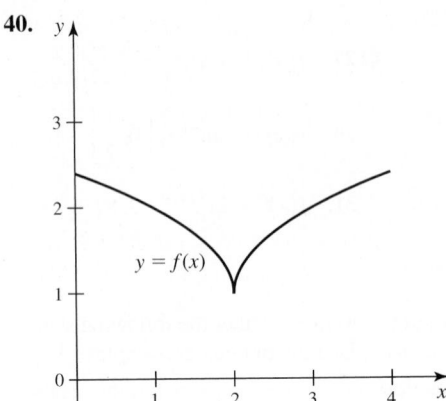

41.

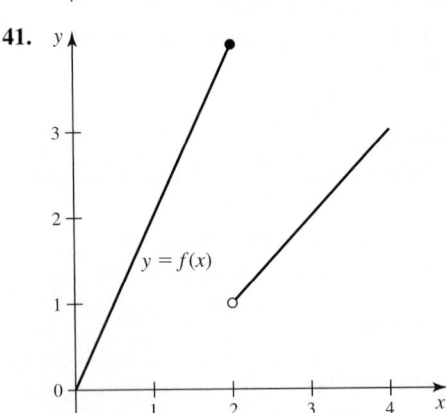

42. Avalanche forecasting Avalanche forecasters measure the *temperature gradient* $\dfrac{dT}{dh}$, which is the rate at which the temperature in a snowpack T changes with respect to its depth h. A large temperature gradient may lead to a weak layer in the snowpack. When these weak layers collapse, avalanches occur. Avalanche forecasters use the following rule of thumb: If $\dfrac{dT}{dh}$ exceeds 10°C/m anywhere in the snowpack, conditions are favorable for weak-layer formation, and the risk of avalanche increases. Assume the temperature function is continuous and differentiable.

    a. An avalanche forecaster digs a snow pit and takes two temperature measurements. At the surface $(h = 0)$, the temperature is $-16°C$. At a depth of 1.1 m, the temperature is $-2°C$. Using the Mean Value Theorem, what can he conclude about the temperature gradient? Is the formation of a weak layer likely?

    b. One mile away, a skier finds that the temperature at a depth of 1.4 m is $-1°C$, and at the surface it is $-12°C$. What can be concluded about the temperature gradient? Is the formation of a weak layer in her location likely?

    c. Because snow is an excellent insulator, the temperature of snow-covered ground is near 0°C. Furthermore, the surface temperature of snow in a particular area does not vary much from one location to the next. Explain why a weak layer is more likely to form in places where the snowpack is not too deep.

    d. The term *isothermal* is used to describe the situation where all layers of the snowpack are at the same temperature (typically near the freezing point). Is a weak layer likely to form in isothermal snow? Explain.

43. Mean Value Theorem and the police A state patrol officer saw a car start from rest at a highway on-ramp. She radioed ahead to a patrol officer 30 mi along the highway. When the car reached the location of the second officer 28 min later, it was clocked going 60 mi/hr. The driver of the car was given a ticket for exceeding the 60-mi/hr speed limit. Why can the officer conclude that the driver exceeded the speed limit?

44. Mean Value Theorem and the police again Compare carefully to Exercise 43. A state patrol officer saw a car start from rest at a highway on-ramp. She radioed ahead to another officer 30 mi along the highway. When the car reached the location of the second officer 30 min later, it was clocked going 60 mi/hr. Can the patrol officer conclude that the driver exceeded the speed limit?

45. Running pace Explain why if a runner completes a 6.2-mi (10-km) race in 32 min, then he must have been running at exactly 11 mi/hr at least twice in the race. Assume the runner's speed at the finish line is zero.

46. Mean Value Theorem for linear functions Interpret the Mean Value Theorem when it is applied to any linear function.

47. Mean Value Theorem for quadratic functions Consider the quadratic function $f(x) = Ax^2 + Bx + C$, where A, B, and C are real numbers with $A \neq 0$. Show that when the Mean Value Theorem is applied to f on the interval $[a, b]$, the number c guaranteed by the theorem is the midpoint of the interval.

48. Means

    a. Show that the point c guaranteed to exist by the Mean Value Theorem for $f(x) = x^2$ on $[a, b]$ is the arithmetic mean of a and b; that is, $c = \dfrac{a + b}{2}$.

    b. Show that the point c guaranteed to exist by the Mean Value Theorem for $f(x) = \dfrac{1}{x}$ on $[a, b]$, where $0 < a < b$, is the geometric mean of a and b; that is, $c = \sqrt{ab}$.

49. Equal derivatives Verify that the functions $f(x) = \tan^2 x$ and $g(x) = \sec^2 x$ have the same derivative. What can you say about the difference $f - g$? Explain.

50. 100-m speed The Jamaican sprinter Usain Bolt set a world record of 9.58 s in the 100-meter dash in the summer of 2009. Did his speed ever exceed 37 km/hr during the race? Explain.

Explorations and Challenges

51. Verify the identity $\sec^{-1} x = \cos^{-1}(1/x)$, for $x \neq 0$.

T 52. Let $f(x) = \ln\left(\dfrac{x + 1}{x - 1}\right)$ and $g(x) = \ln\left(\dfrac{x + 1}{1 - x}\right)$.

    a. Show that f and g have the same derivative.

    b. Sketch graphs of f and g to show that these functions do not differ by a constant.

    c. Explain why parts (a) and (b) do not contradict Theorem 4.6.

53. Suppose $f'(x) < 2$, for all $x \geq 2$, and $f(2) = 7$. Show that $f(4) < 11$.

54. Suppose $f'(x) > 1$, for all $x > 0$, and $f(0) = 0$. Show that $f(x) > x$, for all $x > 0$.

55. Use the Mean Value Theorem to prove that $1 + \dfrac{a}{2} > \sqrt{1 + a}$, for $a > 0$. (*Hint:* For a given value of $a > 0$, let $f(x) = \sqrt{1 + x}$ on $[0, a]$ and use the fact that $\sqrt{1 + c} > 1$, for $c > 0$.)

56. Prove the following statements.

    a. $|\sin a - \sin b| \le |a - b|$, for any real numbers a and b

    b. $|\sin a| \le |a|$, for any real number a

57. **Generalized Mean Value Theorem** Suppose the functions f and g are continuous on $[a, b]$ and differentiable on (a, b), where $g(a) \ne g(b)$. Then there is a point c in (a, b) at which

$$\frac{f(b) - f(a)}{g(b) - g(a)} = \frac{f'(c)}{g'(c)}.$$

This result is known as the **Generalized (or Cauchy's) Mean Value Theorem**.

    a. If $g(x) = x$, then show that the Generalized Mean Value Theorem reduces to the Mean Value Theorem.

    b. Suppose $f(x) = x^2 - 1$, $g(x) = 4x + 2$, and $[a, b] = [0, 1]$. Find a value of c satisfying the Generalized Mean Value Theorem.

58. **Condition for nondifferentiability** Suppose $f'(x) < 0 < f''(x)$, for $x < a$, and $f'(x) > 0 > f''(x)$, for $x > a$. Prove that f is not differentiable at a. (*Hint:* Assume f is differentiable at a, and apply the Mean Value Theorem to f'.) More generally, show that if f' and f'' change sign at the same point, then f is not differentiable at that point.

QUICK CHECK ANSWERS

1. $x = 2$ **2.** The functions shown in Figure 4.14 provide examples.   **3.** The graphs of $f(x) = 3x$ and $g(x) = 3x + 2$ have the same slope. Note that $f(x) - g(x) = -2$, a constant. ◄

4.3 What Derivatives Tell Us

In Section 4.1, we saw that the derivative is a tool for finding critical points, which are related to local maxima and minima. As we show in this section, derivatives (first *and* second derivatives) tell us much more about the behavior of functions.

Increasing and Decreasing Functions

We have used the terms *increasing* and *decreasing* informally in earlier sections to describe a function or its graph. For example, the graph in **Figure 4.20a** rises as x increases, so the corresponding function is increasing. In **Figure 4.20b**, the graph falls as x increases, so the corresponding function is decreasing. The following definition makes these ideas precise.

> ► A function is called **monotonic** if it is either increasing or decreasing. We can make a further distinction by defining **nondecreasing** ($f(x_2) \ge f(x_1)$ whenever $x_2 > x_1$) and **nonincreasing** ($f(x_2) \le f(x_1)$ whenever $x_2 > x_1$).

> **DEFINITION** **Increasing and Decreasing Functions**
>
> Suppose a function f is defined on an interval I. We say that f is **increasing** on I if $f(x_2) > f(x_1)$ whenever x_1 and x_2 are in I and $x_2 > x_1$. We say that f is **decreasing** on I if $f(x_2) < f(x_1)$ whenever x_1 and x_2 are in I and $x_2 > x_1$.

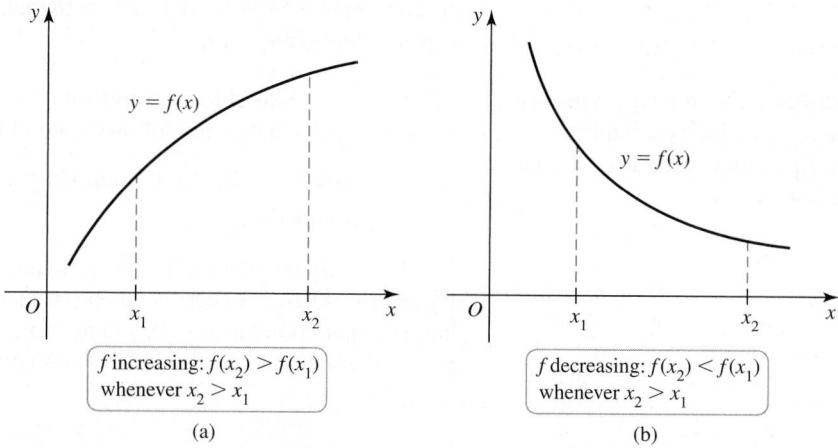

Figure 4.20

Intervals of Increase and Decrease The graph of a function f gives us an idea of the intervals on which f is increasing and decreasing. But how do we determine those intervals precisely? This question is answered by making a connection to the derivative.

    Recall that the derivative of a function gives the slopes of tangent lines. If the derivative is positive on an interval, the tangent lines on that interval have positive slopes, and the function is increasing on the interval (**Figure 4.21a**). Said differently, positive derivatives on an interval imply positive rates of change on the interval, which, in turn, indicate an increase in function values.

Similarly, if the derivative is negative on an interval, the tangent lines on that interval have negative slopes, and the function is decreasing on that interval (**Figure 4.21b**). These observations lead to Theorem 4.7, whose proof relies on the Mean Value Theorem.

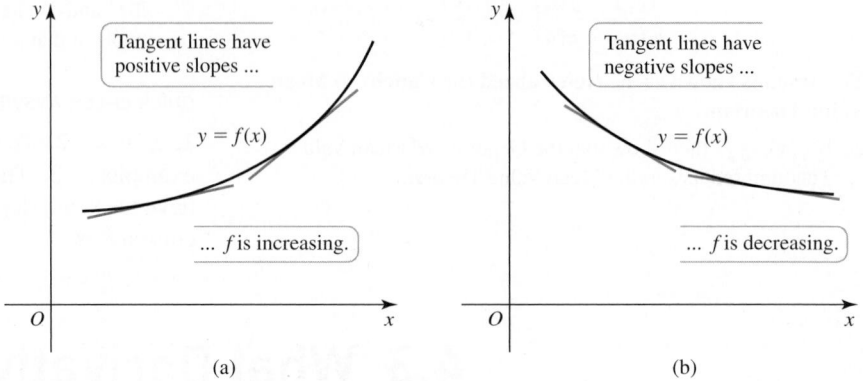

Figure 4.21

> The converse of Theorem 4.7 may not be true. According to the definition, $f(x) = x^3$ is increasing on $(-\infty, \infty)$ but it is not true that $f'(x) > 0$ on $(-\infty, \infty)$ (because $f'(0) = 0$).

THEOREM 4.7   Test for Intervals of Increase and Decrease

Suppose f is continuous on an interval I and differentiable at all interior points of I. If $f'(x) > 0$ at all interior points of I, then f is increasing on I. If $f'(x) < 0$ at all interior points of I, then f is decreasing on I.

Proof: Let a and b be any two distinct points in the interval I with $b > a$. By the Mean Value Theorem,

$$\frac{f(b) - f(a)}{b - a} = f'(c),$$

for some c between a and b. Equivalently,

$$f(b) - f(a) = f'(c)(b - a).$$

Notice that $b - a > 0$ by assumption. So if $f'(c) > 0$, then $f(b) - f(a) > 0$. Therefore, for all a and b in I with $b > a$, we have $f(b) > f(a)$, which implies that f is increasing on I. Similarly, if $f'(c) < 0$, then $f(b) - f(a) < 0$ or $f(b) < f(a)$. It follows that f is decreasing on I.   ◄

QUICK CHECK 1 Explain why a positive derivative on an interval implies that the function is increasing on the interval. ◄

EXAMPLE 1   Sketching a function Sketch a graph of a function f that is continuous on $(-\infty, \infty)$ and satisfies the following conditions.

1. $f' > 0$ on $(-\infty, 0)$, $(4, 6)$, and $(6, \infty)$.   **2.** $f' < 0$ on $(0, 4)$.

3. $f'(0)$ is undefined.   **4.** $f'(4) = f'(6) = 0$.

SOLUTION By condition (1), f is increasing on the intervals $(-\infty, 0)$, $(4, 6)$, and $(6, \infty)$. By condition (2), f is decreasing on $(0, 4)$. Continuity of f and condition (3) imply that f has a cusp or corner at $x = 0$, and by condition (4), the graph has a horizontal tangent line at $x = 4$ and $x = 6$. It is useful to summarize these results using a *sign graph* (**Figure 4.22**).

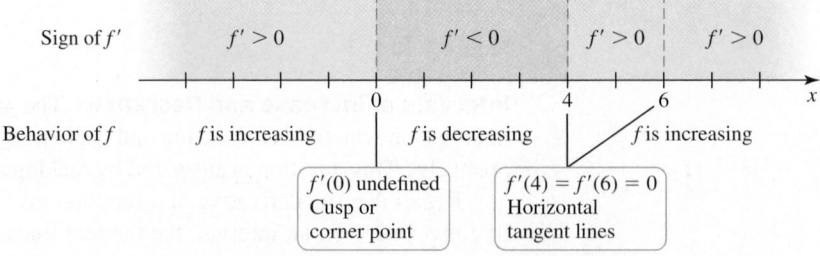

Figure 4.22

One possible graph satisfying these conditions is shown in **Figure 4.23**. Notice that the graph has a cusp at $x = 0$. Furthermore, although $f'(4) = f'(6) = 0$, f has a local minimum at $x = 4$ but no local extremum at $x = 6$.

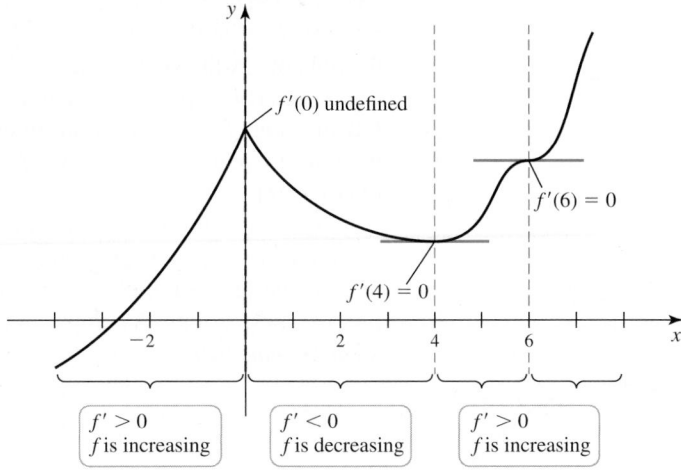

$f'(0)$ undefined

$f'(6) = 0$

$f'(4) = 0$

$f' > 0$
f is increasing

$f' < 0$
f is decreasing

$f' > 0$
f is increasing

Figure 4.23

Related Exercises 9–10 ◄

EXAMPLE 2 Intervals of increase and decrease Find the intervals on which the following functions are increasing and the intervals on which they are decreasing.

a. $f(x) = xe^{-x}$ **b.** $f(x) = 2x^3 + 3x^2 + 1$

SOLUTION

a. By the Product Rule, $f'(x) = e^{-x} + x(-e^{-x}) = (1 - x)e^{-x}$. Solving $f'(x) = 0$ and noting that $e^{-x} \neq 0$ for all x, the sole critical point is $x = 1$. Therefore, if f' changes sign, then it does so at $x = 1$ and nowhere else, which implies f' has the same sign throughout each of the intervals $(-\infty, 1)$ and $(1, \infty)$. We determine the sign of f' on each interval by evaluating f' at selected points in each interval:

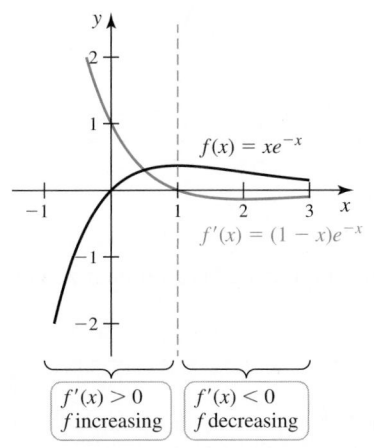

$f(x) = xe^{-x}$

$f'(x) = (1-x)e^{-x}$

$f'(x) > 0$
f increasing

$f'(x) < 0$
f decreasing

Figure 4.24

➤ Appendix B, online at goo.gl/6DCbbM, shows how to solve inequalities using test values.

- At $x = 0$, $f'(0) = 1 > 0$. So $f' > 0$ on $(-\infty, 1)$, which means that f is increasing on $(-\infty, 1)$.
- At $x = 2$, $f'(2) = -e^{-2} < 0$. So $f' < 0$ on $(1, \infty)$, which means that f is decreasing on $(1, \infty)$.

Note also that the graph of f has a horizontal tangent line at $x = 1$. We verify these conclusions by plotting f and f' (**Figure 4.24**).

b. In this case, $f'(x) = 6x^2 + 6x = 6x(x + 1)$. To find the intervals of increase, we first solve $6x(x + 1) = 0$ and determine that the critical points are $x = 0$ and $x = -1$. If f' changes sign, then it does so at these points and nowhere else; that is, f' has the same sign throughout each of the intervals $(-\infty, -1)$, $(-1, 0)$, and $(0, \infty)$. Evaluating f' at selected points of each interval determines the sign of f' on that interval.

- At $x = -2$, $f'(-2) = 12 > 0$, so $f' > 0$ and f is increasing on $(-\infty, -1)$.
- At $x = -\frac{1}{2}$, $f'(-\frac{1}{2}) = -\frac{3}{2} < 0$, so $f' < 0$ and f is decreasing on $(-1, 0)$.
- At $x = 1$, $f'(1) = 12 > 0$, so $f' > 0$ and f is increasing on $(0, \infty)$.

The graph of f has a horizontal tangent line at $x = -1$ and $x = 0$. **Figure 4.25** shows the graph of f superimposed on the graph of f', confirming our conclusions.

Related Exercises 22, 27, 29 ◄

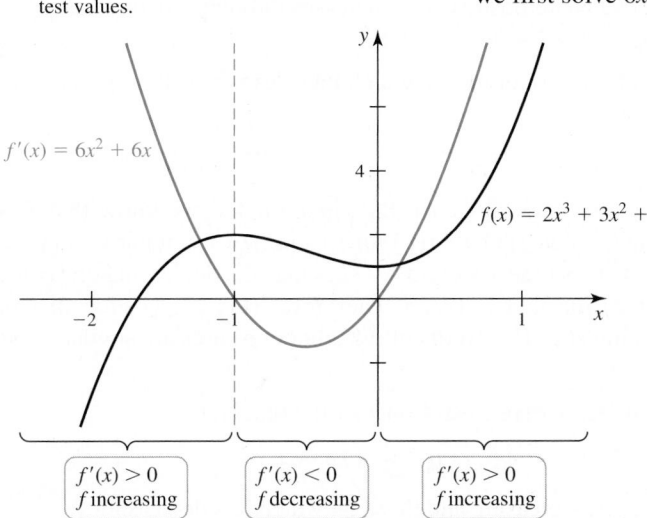

$f'(x) = 6x^2 + 6x$

$f(x) = 2x^3 + 3x^2 + 1$

$f'(x) > 0$
f increasing

$f'(x) < 0$
f decreasing

$f'(x) > 0$
f increasing

Figure 4.25

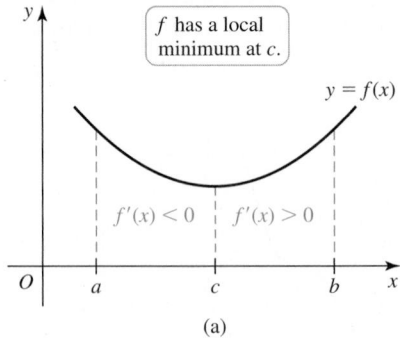

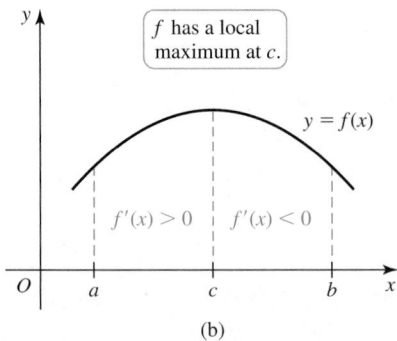

Figure 4.26

QUICK CHECK 2 Sketch a function f that is differentiable on $(-\infty, \infty)$ with the following properties: (i) $x = 0$ and $x = 2$ are critical points; (ii) f is increasing on $(-\infty, 2)$; (iii) f is decreasing on $(2, \infty)$. ◀

Identifying Local Maxima and Minima

Using what we know about increasing and decreasing functions, we can now identify local extrema. Suppose $x = c$ is a critical point of f, where $f'(c) = 0$. Suppose also that f' changes sign at c with $f'(x) < 0$ on an interval (a, c) to the left of c and $f'(x) > 0$ on an interval (c, b) to the right of c. In this case f' is decreasing to the left of c and increasing to the right of c, which means that f has a local minimum at c, as shown in Figure 4.26a.

Similarly, suppose f' changes sign at c with $f'(x) > 0$ on an interval (a, c) to the left of c and $f'(x) < 0$ on an interval (c, b) to the right of c. Then f is increasing to the left of c and decreasing to the right of c, so f has a local maximum at c, as shown in Figure 4.26b.

Figure 4.27 shows typical features of a function on an interval $[a, b]$. At local maxima or minima (c_2, c_3, and c_4), f' changes sign. Although c_1 and c_5 are critical points, the sign of f' is the same on both sides near these points, so there is no local maximum or minimum at these points. As emphasized earlier, *critical points do not always correspond to local extreme values.*

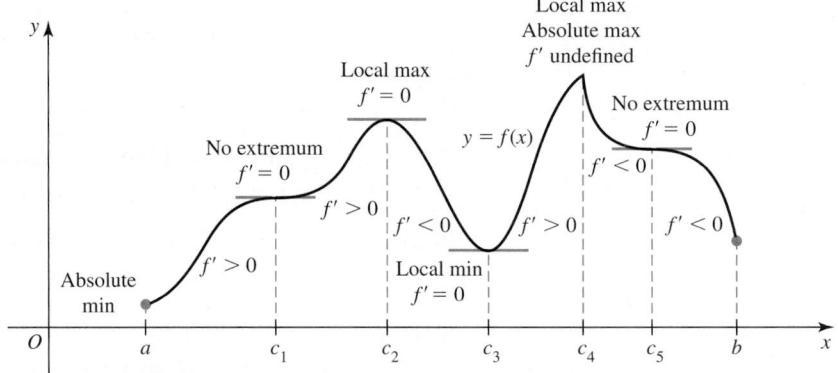

Figure 4.27

First Derivative Test The observations used to interpret Figure 4.27 are summarized in a powerful test for identifying local maxima and minima.

THEOREM 4.8 First Derivative Test

Assume f is continuous on an interval that contains a critical point c, and assume f is differentiable on an interval containing c, except perhaps at c itself.

- If f' changes sign from positive to negative as x increases through c, then f has a **local maximum** at c.

- If f' changes sign from negative to positive as x increases through c, then f has a **local minimum** at c.

- If f' is positive on both sides near c or negative on both sides near c, then f has no local extreme value at c.

Proof: Suppose $f'(x) > 0$ on an interval (a, c). By Theorem 4.7, we know that f is increasing on (a, c), which implies that $f(x) < f(c)$ for all x in (a, c). Similarly, suppose $f'(x) < 0$ on an interval (c, b). This time Theorem 4.7 says that f is decreasing on (c, b), which implies that $f(x) < f(c)$ for all x in (c, b). Therefore, $f(x) \leq f(c)$ for all x in (a, b) and f has a local maximum at c. The proofs of the other two cases are similar. ◀

EXAMPLE 3 Using the First Derivative Test Consider the function

$$f(x) = 3x^4 - 4x^3 - 6x^2 + 12x + 1.$$

a. Find the intervals on which f is increasing and those on which it is decreasing.

b. Identify the local extrema of f.

SOLUTION

a. Differentiating f, we find that

$$f'(x) = 12x^3 - 12x^2 - 12x + 12$$
$$= 12(x^3 - x^2 - x + 1)$$
$$= 12(x + 1)(x - 1)^2.$$

Solving $f'(x) = 0$ gives the critical points $x = -1$ and $x = 1$. The critical points determine the intervals $(-\infty, -1)$, $(-1, 1)$, and $(1, \infty)$ on which f' does not change sign. Choosing a test point in each interval, a sign graph of f' is constructed (**Figure 4.28**) that summarizes the behavior of f.

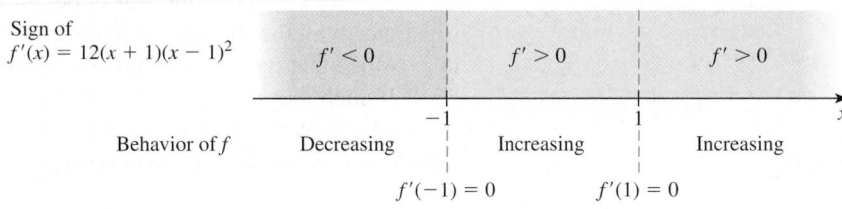

Figure 4.28

b. Note that f is a polynomial, so it is continuous on $(-\infty, \infty)$. Because f' changes sign from negative to positive as x passes through the critical point $x = -1$, it follows by the First Derivative Test that f has a local minimum value of $f(-1) = -10$ at $x = -1$. Observe that f' is positive on both sides near $x = 1$, so f does not have a local extreme value at $x = 1$ (**Figure 4.29**).

Related Exercises 49–50 ◄

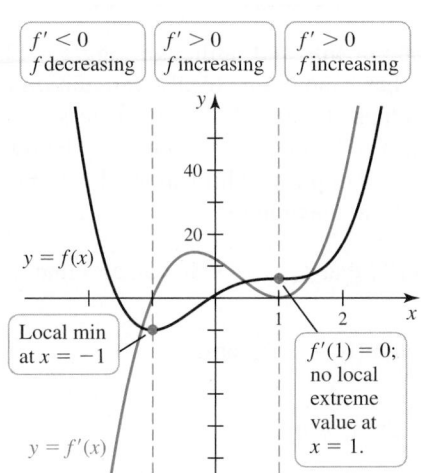

Figure 4.29

EXAMPLE 4 Extreme values Find the local extrema of the function $g(x) = x^{2/3}(2 - x)$.

SOLUTION In Example 4b of Section 4.1, we found that

$$g'(x) = \frac{4}{3}x^{-1/3} - \frac{5}{3}x^{2/3} = \frac{4 - 5x}{3x^{1/3}}$$

and that the critical points of g are $x = 0$ and $x = \frac{4}{5}$. These two critical points are *candidates* for local extrema, and Theorem 4.8 is used to classify each as a local maximum, local minimum, or neither.

On the interval $(-\infty, 0)$, the numerator of g' is positive and the denominator is negative (**Figure 4.30**). Therefore, $g'(x) < 0$ on $(-\infty, 0)$. On the interval $\left(0, \frac{4}{5}\right)$, the numerator of g' is positive, as is the denominator. Therefore, $g'(x) > 0$ on $\left(0, \frac{4}{5}\right)$. We see that as x passes through 0, g' changes sign from negative to positive, which means g has a local minimum at 0. A similar argument shows that g' changes sign from positive to negative as x passes through $\frac{4}{5}$, so g has a local maximum at $\frac{4}{5}$. These observations are confirmed by the graphs of g and g' (**Figure 4.31**).

QUICK CHECK 3 Explain how the First Derivative Test determines whether $f(x) = x^2$ has a local maximum or local minimum at the critical point $x = 0$. ◄

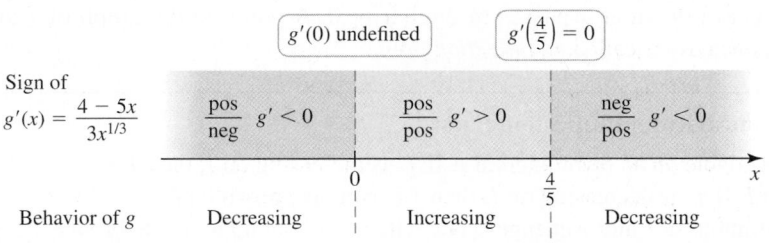

Figure 4.30

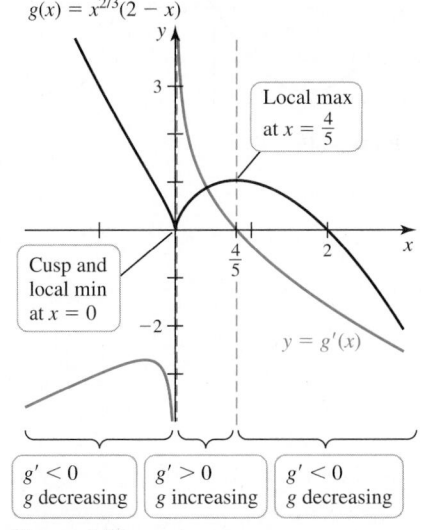

Figure 4.31

Related Exercises 51, 53 ◄

Absolute Extreme Values on Any Interval   Theorem 4.1 guarantees the existence of absolute extreme values only on closed intervals. What can be said about absolute extrema on intervals that are not closed? The following theorem provides a valuable test.

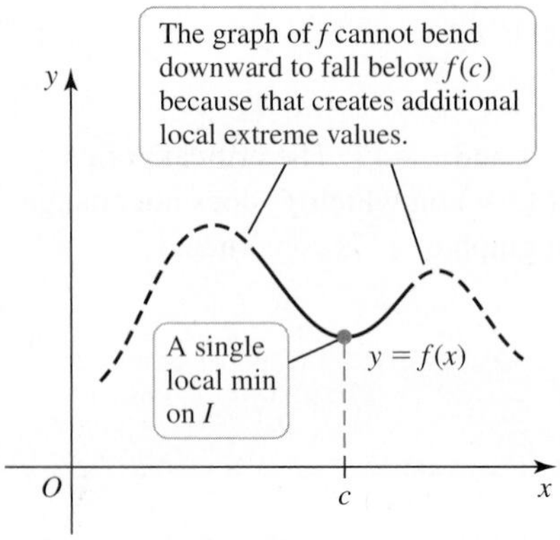

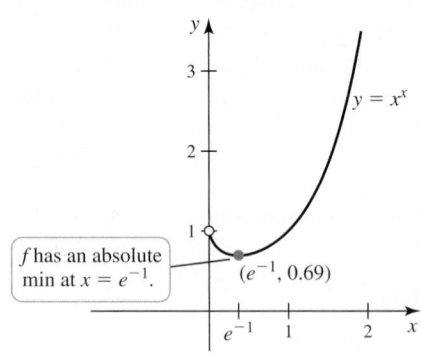

The graph of f cannot bend downward to fall below $f(c)$ because that creates additional local extreme values.

A single local min on I

$y = f(x)$

Figure 4.32

> **THEOREM 4.9   One Local Extremum Implies Absolute Extremum**
> Suppose f is continuous on an interval I that contains exactly one local extremum at c.
> • If a local maximum occurs at c, then $f(c)$ is the absolute maximum of f on I.
> • If a local minimum occurs at c, then $f(c)$ is the absolute minimum of f on I.

The proof of Theorem 4.9 is beyond the scope of this text, although **Figure 4.32** illustrates why the theorem is plausible. Assume f has exactly one local minimum on I at c. Notice that there can be no other point on the graph at which f has a value less than $f(c)$. If such a point did exist, the graph would have to bend downward to drop below $f(c)$, which, by the continuity of f, cannot happen as it implies additional local extreme values on I. A similar argument applies to a solitary local maximum.

EXAMPLE 5   Finding an absolute extremum   Verify that $f(x) = x^x$ has an absolute extreme value on its domain.

SOLUTION   First note that f is continuous on its domain $(0, \infty)$. Because $f(x) = x^x = e^{x \ln x}$, it follows that

$$f'(x) = e^{x \ln x}(1 + \ln x) = x^x(1 + \ln x).$$

Solving $f'(x) = 0$ gives a single critical point $x = e^{-1}$; there is no point in the domain at which $f'(x)$ does not exist. The critical point splits the domain of f into the intervals $(0, e^{-1})$ and (e^{-1}, ∞). Evaluating the sign of f' on each interval gives $f'(x) < 0$ on $(0, e^{-1})$ and $f'(x) > 0$ on (e^{-1}, ∞); therefore, by the First Derivative Test, a local minimum occurs at $x = e^{-1}$. Because it is the only local extremum on $(0, \infty)$, it follows from Theorem 4.9 that the absolute minimum of f occurs at $x = e^{-1}$ (**Figure 4.33**). Its value is $f(e^{-1}) \approx 0.69$.

Related Exercises 55–56 ◄

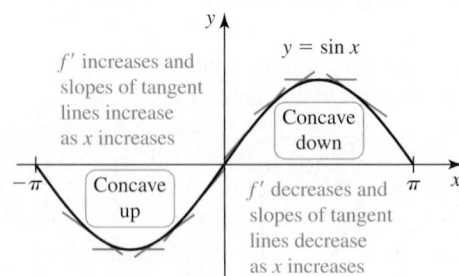

$y = x^x$

f has an absolute min at $x = e^{-1}$.

$(e^{-1}, 0.69)$

Figure 4.33

Concavity and Inflection Points

Just as the first derivative is related to the slope of tangent lines, the second derivative also has geometric meaning. Consider $f(x) = \sin x$, shown in **Figure 4.34**. Its graph bends upward for $-\pi < x < 0$, reflecting the fact that the slopes of the tangent lines increase as x increases. It follows that the first derivative is increasing for $-\pi < x < 0$. A function with the property that f' is increasing on an interval is *concave up* on that interval.

Similarly, $f(x) = \sin x$ bends downward for $0 < x < \pi$ because it has a decreasing first derivative on that interval. A function with the property that f' is decreasing as x increases on an interval is *concave down* on that interval.

Here is another useful characterization of concavity. If a function is concave up at a point (any point in $(-\pi, 0)$, Figure 4.34), then its graph near that point lies *above* the tangent line at that point. Similarly, if a function is concave down at a point (any point in $(0, \pi)$, Figure 4.34), then its graph near that point lies *below* the tangent line at that point (Exercise 112).

Finally, imagine a function f that changes concavity (from up to down, or vice versa) at a point c in the domain of f. For example, $f(x) = \sin x$ in Figure 4.34 changes from concave up to concave down as x passes through $x = 0$. A point on the graph of f at which f changes concavity is called an *inflection point*.

f' increases and slopes of tangent lines increase as x increases

$y = \sin x$

Concave down

Concave up

f' decreases and slopes of tangent lines decrease as x increases

Figure 4.34

> **DEFINITION   Concavity and Inflection Point**
> Let f be differentiable on an open interval I. If f' is increasing on I, then f is **concave up** on I. If f' is decreasing on I, then f is **concave down** on I.
>   If f is continuous at c and f changes concavity at c (from up to down, or vice versa), then f has an **inflection point** at c.

Applying Theorem 4.7 to f' leads to a test for concavity in terms of the second derivative. Specifically, if $f'' > 0$ on an interval I, then f' is increasing on I and f is concave up on I. Similarly, if $f'' < 0$ on I, then f is concave down on I. In addition, if the values of f'' change sign at a point c (from positive to negative, or vice versa), then the concavity of f changes at c and f has an inflection point at c (**Figure 4.35a**). We now have a useful interpretation of the second derivative: It measures *concavity*.

THEOREM 4.10   Test for Concavity
Suppose f'' exists on an open interval I.

- If $f'' > 0$ on I, then f is concave up on I.

- If $f'' < 0$ on I, then f is concave down on I.

- If c is a point of I at which f'' changes sign at c (from positive to negative, or vice versa), then f has an inflection point at c.

There are a few important but subtle points here. The fact that $f''(c) = 0$ does not necessarily imply that f has an inflection point at c. A good example is $f(x) = x^4$. Although $f''(0) = 0$, the concavity does not change at $x = 0$ (a similar function is shown in **Figure 4.35b**).

Typically, if f has an inflection point at c, then $f''(c) = 0$, reflecting the smooth change in concavity. However, an inflection point may also occur at a point where f'' does not exist. For example, the function $f(x) = x^{1/3}$ has a vertical tangent line and an inflection point at $x = 0$ (a similar function is shown in **Figure 4.35c**). Finally, note that the function shown in **Figure 4.35d**, with behavior similar to that of $f(x) = x^{2/3}$, does not have an inflection point at c despite the fact that $f''(c)$ does not exist. In summary, if $f''(c) = 0$ or $f''(c)$ does not exist, then $(c, f(c))$ is a candidate for an inflection point. To be certain an inflection point occurs at c, we must show that the concavity of f changes at c.

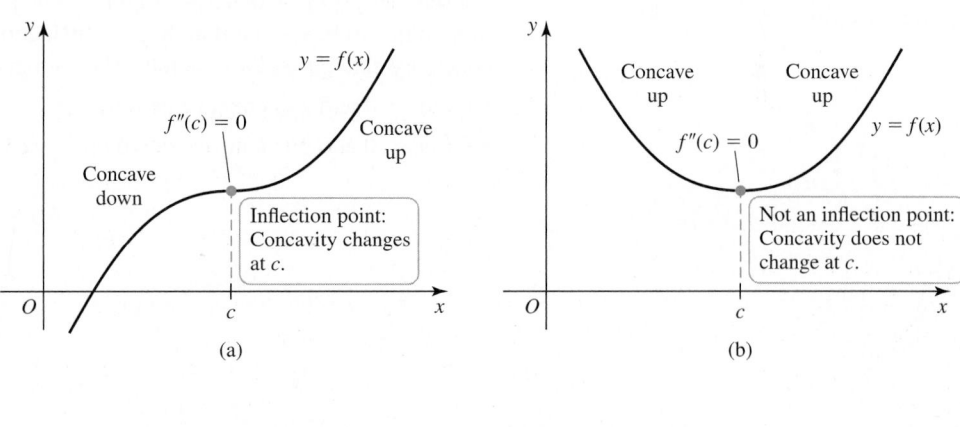

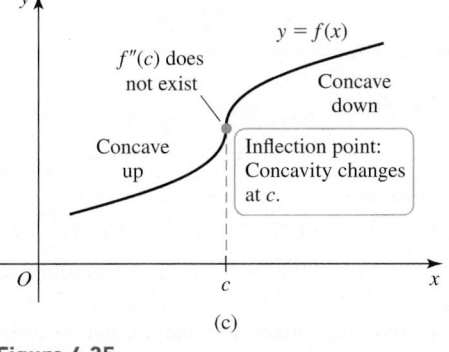

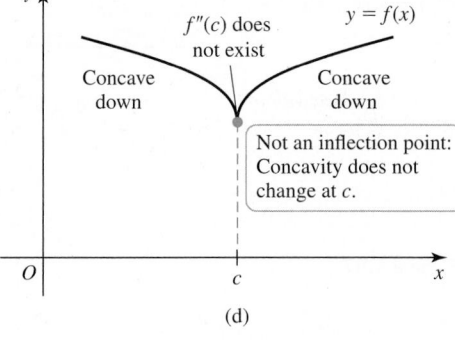

QUICK CHECK 4 Verify that the function $f(x) = x^4$ is concave up for $x > 0$ and for $x < 0$. Is $x = 0$ an inflection point? Explain. ◄

Figure 4.35

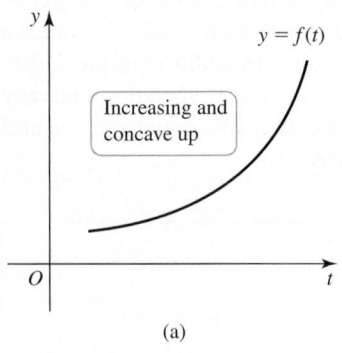

(a)

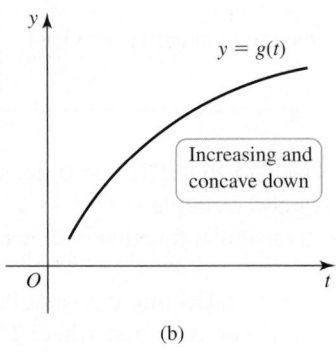

(b)

Figure 4.36

EXAMPLE 6   Interpreting concavity   Sketch a function satisfying each set of conditions on some interval.

a. $f'(t) > 0$ and $f''(t) > 0$ **b.** $g'(t) > 0$ and $g''(t) < 0$

c. Which of the functions, f or g, could describe a population that increases and approaches a steady state as $t \to \infty$?

SOLUTION

a. Figure 4.36a shows the graph of a function that is increasing $(f'(t) > 0)$ and concave up $(f''(t) > 0)$.

b. Figure 4.36b shows the graph of a function that is increasing $(g'(t) > 0)$ and concave down $(g''(t) < 0)$.

c. Because f increases at an *increasing* rate, the graph of f could not approach a horizontal asymptote, so f could not describe a population that approaches a steady state. On the other hand, g increases at a *decreasing* rate, so its graph could approach a horizontal asymptote, depending on the rate at which g increases.

Related Exercises 59–62 ◄

EXAMPLE 7   Detecting concavity   Identify the intervals on which the following functions are concave up or concave down. Then locate the inflection points.

a. $f(x) = 3x^4 - 4x^3 - 6x^2 + 12x + 1$

b. $f(x) = \sin^{-1} x$ on $(-1, 1)$

SOLUTION

a. This function was considered in Example 3, where we found that

$$f'(x) = 12(x + 1)(x - 1)^2.$$

It follows that

$$f''(x) = 12(x - 1)(3x + 1).$$

We see that $f''(x) = 0$ at $x = 1$ and $x = -\frac{1}{3}$. These points are *candidates* for inflection points; to be certain that they are inflection points, we must determine whether the concavity changes at these points. The sign graph in Figure 4.37 shows the following:

• $f''(x) > 0$ and f is concave up on $\left(-\infty, -\frac{1}{3}\right)$ and $(1, \infty)$.

• $f''(x) < 0$ and f is concave down on $\left(-\frac{1}{3}, 1\right)$.

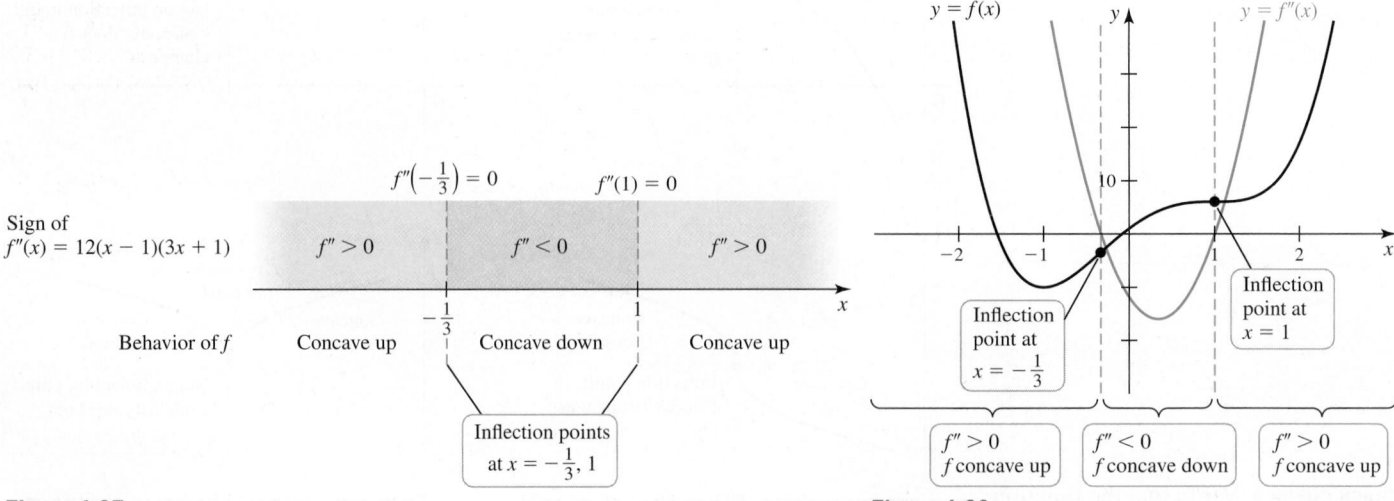

Figure 4.37

Figure 4.38

We see that the sign of f'' changes at $x = -\frac{1}{3}$ and at $x = 1$, so the concavity of f also changes at these points. Therefore, inflection points occur at $x = -\frac{1}{3}$ and $x = 1$. The graphs of f and f'' (Figure 4.38) show that the concavity of f changes at the zeros of f''.

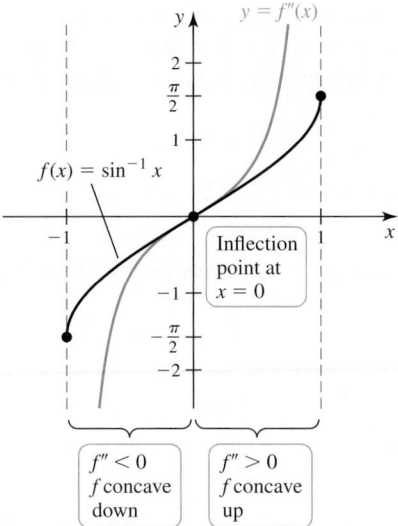

$f(x) = \sin^{-1} x$

Inflection point at $x = 0$

$f'' < 0$ f concave down

$f'' > 0$ f concave up

Figure 4.39

▶ In the inconclusive case of Theorem 4.11 in which $f''(c) = 0$, it is usually best to use the First Derivative Test.

QUICK CHECK 5 Sketch a graph of a function with $f'(x) > 0$ and $f''(x) > 0$ on an interval. Sketch a graph of a function with $f'(x) < 0$ and $f''(x) < 0$ on an interval. ◀

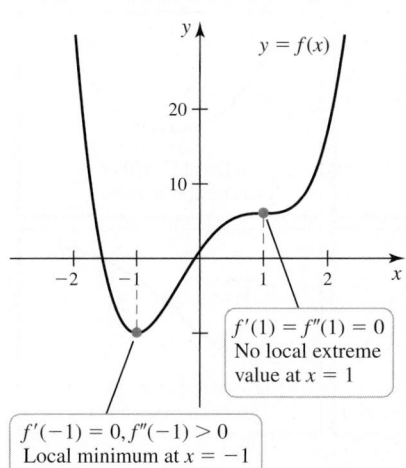

$y = f(x)$

$f'(1) = f''(1) = 0$ No local extreme value at $x = 1$

$f'(-1) = 0, f''(-1) > 0$ Local minimum at $x = -1$

Figure 4.41

b. The first derivative of $f(x) = \sin^{-1} x$ is $f'(x) = 1/\sqrt{1 - x^2}$. We use the Chain Rule to compute the second derivative:

$$f''(x) = -\frac{1}{2}(1 - x^2)^{-3/2} \cdot (-2x) = \frac{x}{(1 - x^2)^{3/2}}.$$

The only zero of f'' is $x = 0$, and because its denominator is positive on $(-1, 1)$, f'' changes sign at $x = 0$ from negative to positive. Therefore, f is concave down on $(-1, 0)$ and concave up on $(0, 1)$, with an inflection point at $x = 0$ (Figure 4.39).

Related Exercises 64, 67 ◀

Second Derivative Test   It is now a short step to a test that uses the second derivative to identify local maxima and minima.

THEOREM 4.11   Second Derivative Test for Local Extrema
Suppose f'' is continuous on an open interval containing c with $f'(c) = 0$.

- If $f''(c) > 0$, then f has a local minimum at c (Figure 4.40a).
- If $f''(c) < 0$, then f has a local maximum at c (Figure 4.40b).
- If $f''(c) = 0$, then the test is inconclusive; f may have a local maximum, a local minimum, or neither at c.

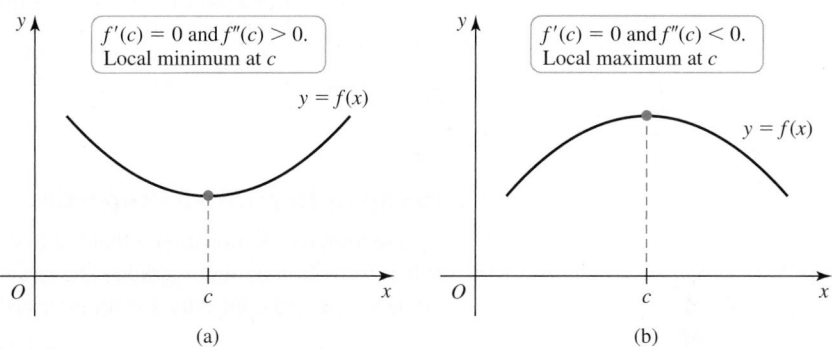

$f'(c) = 0$ and $f''(c) > 0$. Local minimum at c

$y = f(x)$

$f'(c) = 0$ and $f''(c) < 0$. Local maximum at c

$y = f(x)$

(a)   (b)

Figure 4.40

Proof: Assume $f''(c) > 0$. Because f'' is continuous on an interval containing c, it follows that $f'' > 0$ on some open interval I containing c and that f' is increasing on I. Because $f'(c) = 0$, it follows that f' changes sign at c from negative to positive, which, by the First Derivative Test, implies that f has a local minimum at c. The proofs of the second and third statements are similar. ◀

EXAMPLE 8   The Second Derivative Test Use the Second Derivative Test to locate the local extrema of the following functions.

a. $f(x) = 3x^4 - 4x^3 - 6x^2 + 12x + 1$ on $[-2, 2]$
b. $f(x) = \sin^2 x$

SOLUTION

a. This function was considered in Examples 3 and 7, where we found that

$$f'(x) = 12(x + 1)(x - 1)^2 \quad \text{and} \quad f''(x) = 12(x - 1)(3x + 1).$$

Therefore, the critical points of f are $x = -1$ and $x = 1$. Evaluating f'' at the critical points, we find that $f''(-1) = 48 > 0$. By the Second Derivative Test, f has a local minimum at $x = -1$. At the other critical point, $f''(1) = 0$, so the test is inconclusive. You can check that the first derivative does not change sign at $x = 1$, which means f does not have a local maximum or minimum at $x = 1$ (Figure 4.41).

b. Using the Chain Rule and a trigonometric identity, we have $f'(x) = 2 \sin x \cos x = \sin 2x$ and $f''(x) = 2 \cos 2x$. The critical points occur when $f'(x) = \sin 2x = 0$, or when $x = 0, \pm \pi/2, \pm \pi, \ldots$. To apply the Second Derivative Test, we evaluate f'' at the critical points:

• $f''(0) = 2 > 0$, so f has a local minimum at $x = 0$.

• $f''(\pm \pi/2) = -2 < 0$, so f has a local maximum at $x = \pm \pi/2$.

• $f''(\pm \pi) = 2 > 0$, so f has a local minimum at $x = \pm \pi$.

This pattern continues, and we see that f has alternating local maxima and minima, evenly spaced every $\pi/2$ units (Figure 4.42).

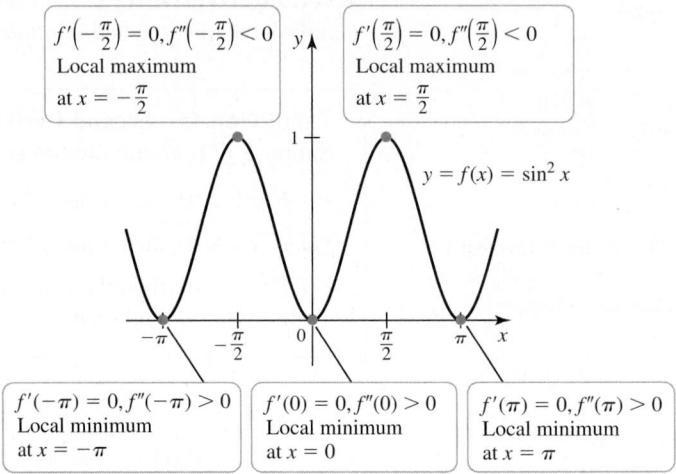

Figure 4.42

Related Exercises 78, 80 ◄

Recap of Derivative Properties

This section has demonstrated that the first and second derivatives of a function provide valuable information about its graph. The relationships among a function's derivatives and its extreme values and concavity are summarized in Figure 4.43.

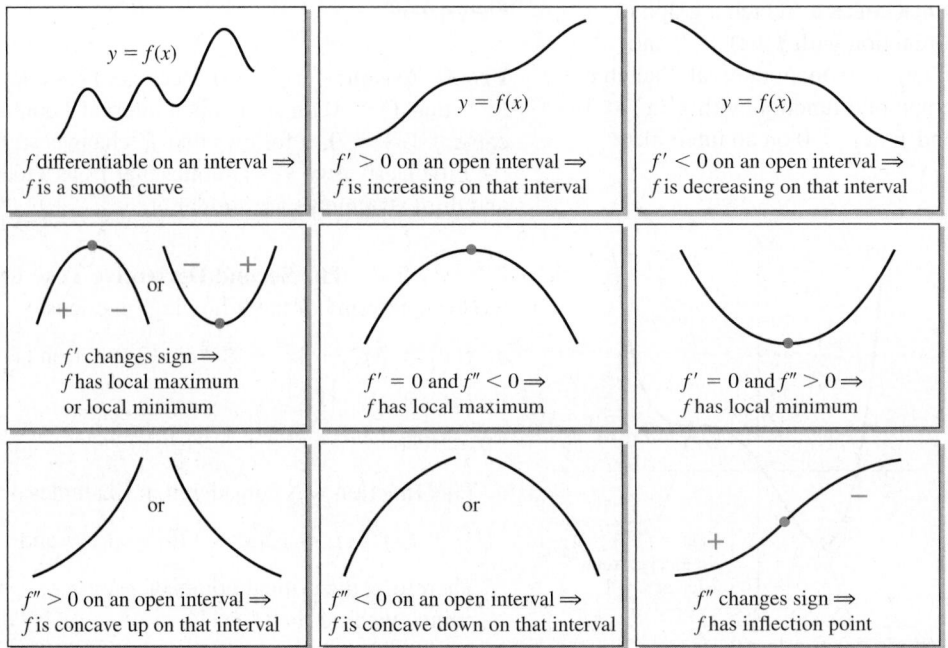

Figure 4.43

SECTION 4.3 EXERCISES

Getting Started

1. Explain how the first derivative of a function determines where the function is increasing or decreasing.

2. Explain how to apply the First Derivative Test.

3. Suppose the derivative of f is $f'(x) = x - 3$.
   a. Find the critical points of f.
   b. On what intervals is f increasing and on what intervals is f decreasing?

4. Suppose the derivative of f is $f'(x) = (x - 1)(x - 2)$.
   a. Find the critical points of f.
   b. On what intervals is f increasing and on what intervals is f decreasing?

5. Sketch the graph of a function that has neither a local maximum nor a local minimum at a point where $f'(x) = 0$.

6. The following graph of the derivative g' has exactly two roots.
   a. Find the critical points of g.
   b. For what values of x in $(0, 3)$ is g increasing? Decreasing?
   c. For what values of x in $(0, 3)$ does g have a local maximum? A local minimum?

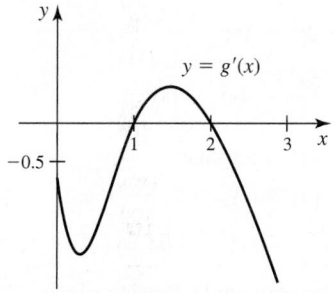

7–8. *The following figures give the graph of the derivative of a continuous function f that passes through the origin. Sketch a possible graph of f on the same set of axes.*

7.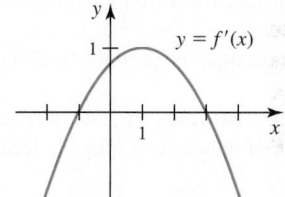

8.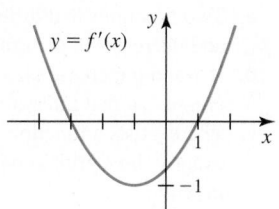

9–12. *Sketch a graph of a function f that is continuous on $(-\infty, \infty)$ and has the following properties. Use a sign graph to summarize information about the function.*

9. $f'(x) < 0$ on $(-\infty, 2)$; $f'(x) > 0$ on $(2, 5)$; $f'(x) < 0$ on $(5, \infty)$

10. $f'(-1)$ is undefined; $f'(x) > 0$ on $(-\infty, -1)$; $f'(x) < 0$ on $(-1, \infty)$

11. $f(0) = f(4) = f'(0) = f'(2) = f'(4) = 0$; $f(x) \geq 0$ on $(-\infty, \infty)$

12. $f'(-2) = f'(2) = f'(6) = 0$; $f'(x) \geq 0$ on $(-\infty, \infty)$

13. Suppose $g''(x) = 2 - x$.
    a. On what intervals is g concave up and on what intervals is g concave down?
    b. State the inflection points of g.

14. The following graph of g'' has exactly three x-intercepts.
    a. For what values of x in $(-4, 3)$ is the graph of g concave up? Concave down?
    b. State the inflection points of g that lie in $(-4, 3)$.

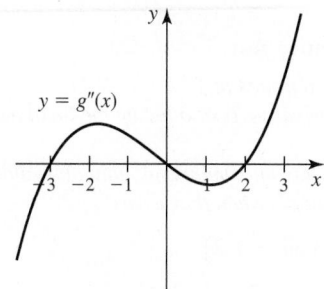

15. Is it possible for a function to satisfy $f(x) > 0$, $f'(x) > 0$, and $f''(x) < 0$ on an interval? Explain.

16. Sketch a graph of a function that changes from concave up to concave down as x increases. Describe how the second derivative of this function changes.

17. Give a function that does not have an inflection point at a point where $f''(x) = 0$.

18. Suppose f is continuous on an interval containing a critical point c and $f''(c) = 0$. How do you determine whether f has a local extreme value at $x = c$?

Practice Exercises

19–44. Increasing and decreasing functions *Find the intervals on which f is increasing and the intervals on which it is decreasing.*

19. $f(x) = 4 - x^2$

20. $f(x) = x^2 - 16$

21. $f(x) = (x - 1)^2$

22. $f(x) = x^3 + 4x$

23. $f(x) = \dfrac{x^3}{3} - \dfrac{5x^2}{2} + 4x$

24. $f(x) = -\dfrac{x^3}{3} + \dfrac{x^2}{2} + 2x$

25. $f(x) = 12 + x - x^2$

26. $f(x) = x^4 - 4x^3 + 4x^2$

27. $f(x) = -\dfrac{x^4}{4} + x^3 - x^2$

28. $f(x) = 2x^5 - \dfrac{15x^4}{4} + \dfrac{5x^3}{3}$

29. $f(x) = x^2 \ln x^2 + 1$

30. $f(x) = \dfrac{e^x}{e^{2x} + 1}$

31. $f(x) = -2 \cos x - x$ on $[0, 2\pi]$

32. $f(x) = \sqrt{2} \sin x - x$ on $[0, 2\pi]$

33. $f(x) = 3 \cos 3x$ on $[-\pi, \pi]$

34. $f(x) = \cos^2 x$ on $[-\pi, \pi]$

35. $f(x) = x^{2/3}(x^2 - 4)$

36. $f(x) = x^2 \sqrt{9 - x^2}$ on $(-3, 3)$

37. $f(x) = \sqrt{9 - x^2} + \sin^{-1}(x/3)$

38. $f(x) = x \ln x - 2x + 3$ on $(0, \infty)$

39. $f(x) = -12x^5 + 75x^4 - 80x^3$

40. $f(x) = x^2 - 2 \ln x$

41. $f(x) = -2x^4 + x^2 + 10$

42. $f(x) = \dfrac{x^4}{4} - \dfrac{8x^3}{3} + \dfrac{15x^2}{2} + 8$

43. $f(x) = xe^{-x^2/2}$

44. $f(x) = \tan^{-1}\left(\dfrac{x}{x^2 + 2}\right)$

45–54. First Derivative Test

a. Locate the critical points of f.
b. Use the First Derivative Test to locate the local maximum and minimum values.
c. Identify the absolute maximum and minimum values of the function on the given interval (when they exist).

45. $f(x) = x^2 + 3$ on $[-3, 2]$

46. $f(x) = -x^2 - x + 2$ on $[-4, 4]$

47. $f(x) = x\sqrt{4 - x^2}$ on $[-2, 2]$

48. $f(x) = 2x^3 + 3x^2 - 12x + 1$ on $[-2, 4]$

49. $f(x) = -x^3 + 9x$ on $[-4, 3]$

50. $f(x) = 2x^5 - 5x^4 - 10x^3 + 4$ on $[-2, 4]$

51. $f(x) = x^{2/3}(x - 5)$ on $[-5, 5]$

52. $f(x) = \dfrac{x^2}{x^2 - 1}$ on $[-4, 4]$

53. $f(x) = \sqrt{x} \ln x$ on $(0, \infty)$

54. $f(x) = x - 2 \tan^{-1} x$ on $[-\sqrt{3}, \sqrt{3}]$

55–58. Absolute extreme values
Verify that the following functions satisfy the conditions of Theorem 4.9 on their domains. Then find the location and value of the absolute extrema guaranteed by the theorem.

55. $f(x) = xe^{-x}$ **56.** $f(x) = 4x + 1/\sqrt{x}$

57. $A(r) = 24/r + 2\pi r^2, r > 0$

58. $f(x) = x\sqrt{3 - x}$

59–62. Sketching curves
Sketch a graph of a function f that is continuous on $(-\infty, \infty)$ and has the following properties.

59. $f'(x) > 0, f''(x) > 0$

60. $f'(x) < 0$ and $f''(x) > 0$ on $(-\infty, 0)$; $f'(x) > 0$ and $f''(x) > 0$ on $(0, \infty)$

61. $f'(x) < 0$ and $f''(x) < 0$ on $(-\infty, 0)$; $f'(x) < 0$ and $f''(x) > 0$ on $(0, \infty)$

62. $f'(x) < 0$ and $f''(x) > 0$ on $(-\infty, 0)$; $f'(x) < 0$ and $f''(x) < 0$ on $(0, \infty)$

63–76. Concavity
Determine the intervals on which the following functions are concave up or concave down. Identify any inflection points.

63. $f(x) = x^4 - 2x^3 + 1$ **64.** $f(x) = -x^4 - 2x^3 + 12x^2$

65. $f(x) = 5x^4 - 20x^3 + 10$ **66.** $f(x) = \dfrac{1}{1 + x^2}$

67. $f(x) = e^x(x - 3)$ **68.** $f(x) = 2x^2 \ln x - 5x^2$

69. $g(t) = \ln(3t^2 + 1)$ **70.** $g(x) = \sqrt[3]{x - 4}$

71. $f(x) = e^{-x^2/2}$ **72.** $p(x) = x^4 e^x + x$

73. $f(x) = \sqrt{x} \ln x$ **74.** $h(t) = 2 + \cos 2t$ on $[0, \pi]$

75. $g(t) = 3t^5 - 30t^4 + 80t^3 + 100$

76. $f(x) = 2x^4 + 8x^3 + 12x^2 - x - 2$

77–94. Second Derivative Test
Locate the critical points of the following functions. Then use the Second Derivative Test to determine (if possible) whether they correspond to local maxima or local minima.

77. $f(x) = x^3 - 3x^2$ **78.** $f(x) = 6x^2 - x^3$

79. $f(x) = 4 - x^2$ **80.** $f(x) = x^3 - \dfrac{3}{2}x^2 - 36x$

81. $f(x) = e^x(x - 7)$ **82.** $f(x) = e^x(x - 2)^2$

83. $f(x) = 2x^3 - 3x^2 + 12$ **84.** $f(x) = \dfrac{e^x}{x + 1}$

85. $f(x) = x^2 e^{-x}$ **86.** $g(x) = \dfrac{x^4}{2 - 12x^2}$

87. $f(x) = 2x^2 \ln x - 11x^2$ **88.** $f(x) = \sqrt{x}\left(\dfrac{12}{7}x^3 - 4x^2\right)$

89. $p(t) = 2t^3 + 3t^2 - 36t$

90. $f(x) = \dfrac{x^4}{4} - \dfrac{5x^3}{3} - 4x^2 + 48x$

91. $h(x) = (x + a)^4$; a constant

92. $f(x) = x^3 - 13x^2 - 9x$

93. $f(x) = 6x^4 \ln x^2 - 7x^4$ **94.** $f(x) = 2x^{-3} - x^{-2}$

95. Explain why or why not Determine whether the following statements are true and give an explanation or counterexample.

a. If $f'(x) > 0$ and $f''(x) < 0$ on an interval, then f is increasing at a decreasing rate on the interval.
b. If $f'(c) > 0$ and $f''(c) = 0$, then f has a local maximum at c.
c. Two functions that differ by an additive constant both increase and decrease on the same intervals.
d. If f and g increase on an interval, then the product fg also increases on that interval.
e. There exists a function f that is continuous on $(-\infty, \infty)$ with exactly three critical points, all of which correspond to local maxima.

96–97. Functions from derivatives Consider the following graphs of f' and f''. On the same set of axes, sketch the graph of a possible function f. The graphs of f are not unique.

96.
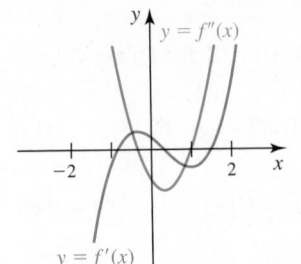

97.
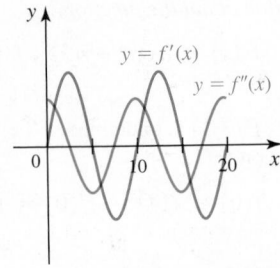

98. Is it possible? Determine whether the following properties can be satisfied by a function that is continuous on $(-\infty, \infty)$. If such a function is possible, provide an example or a sketch of the function. If such a function is not possible, explain why.

a. A function f is concave down and positive everywhere.

b. A function f is increasing and concave down everywhere.

c. A function f has exactly two local extrema and three inflection points.

d. A function f has exactly four zeros and two local extrema.

99. Matching derivatives and functions The following figures show the graphs of three functions (graphs a–c). Match each function with its first derivative (graphs A–C) *and* its second derivative (graphs i–iii).

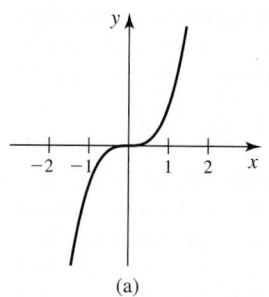

(a)

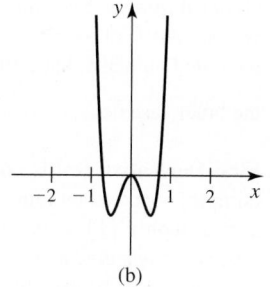

(b)

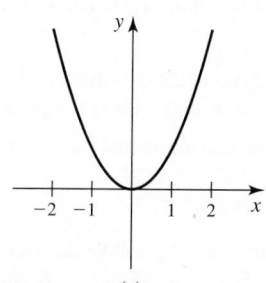

(c)

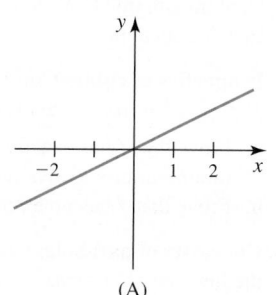

(A)

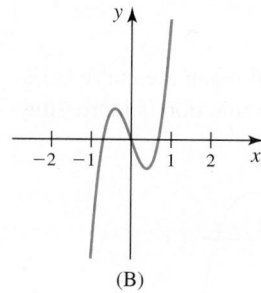

(B)

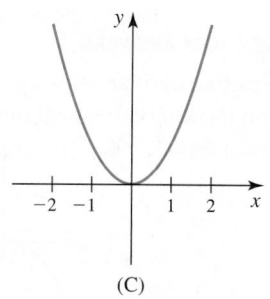

(C)

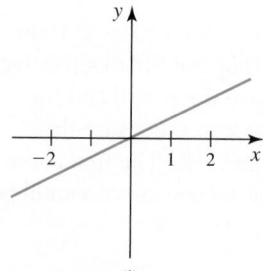

(i)

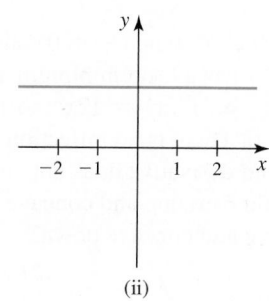

(ii)

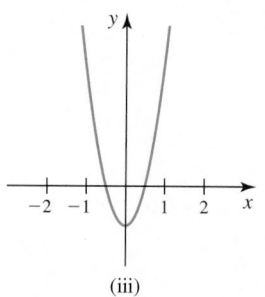

(iii)

100. Graphical analysis The figure shows the graphs of f, f', and f''. Which curve is which?

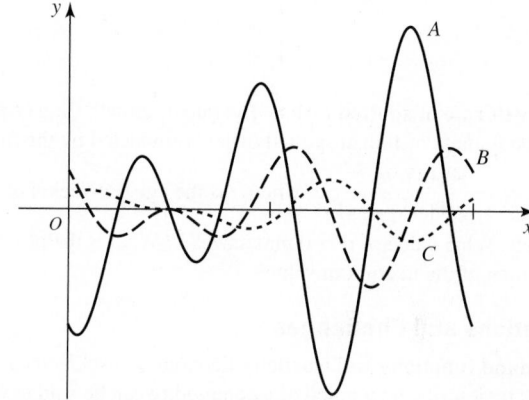

101. Sketching graphs Sketch the graph of a function f that is continuous on $[a, b]$ such that f, f', and f'' have the signs indicated in the following table on $[a, b]$. There are eight different cases lettered A–H that correspond to eight different graphs.

Case	A	B	C	D	E	F	G	H
f	+	+	+	+	−	−	−	−
f'	+	+	−	−	+	+	−	−
f''	+	−	+	−	+	−	+	−

102–105. Designer functions *Sketch the graph of a function f that is continuous on $(-\infty, \infty)$ and satisfies the following sets of conditions.*

102. $f''(x) > 0$ on $(-\infty, -2)$; $f''(-2) = 0$; $f'(-1) = f'(1) = 0$; $f''(2) = 0$; $f'(3) = 0$; $f''(x) > 0$ on $(4, \infty)$

103. $f(-2) = f''(-1) = 0$; $f'\left(-\frac{3}{2}\right) = 0$; $f(0) = f'(0) = 0$; $f(1) = f'(1) = 0$

104. $f'(x) > 0$, for all x in the domain of f'; $f'(-2)$ and $f'(1)$ do not exist; $f''(0) = 0$

105. $f''(x) > 0$ on $(-\infty, -2)$; $f''(x) < 0$ on $(-2, 1)$; $f''(x) > 0$ on $(1, 3)$; $f''(x) < 0$ on $(3, \infty)$

■ 106. Graph carefully Graph the function $f(x) = 60x^5 - 901x^3 + 27x$ in the window $[-4, 4] \times [-10,000, 10,000]$. How many extreme values do you see? Locate *all* the extreme values by analyzing f'.

107. Interpreting the derivative The graph of f' on the interval $[-3, 2]$ is shown in the figure.

a. On what interval(s) is f increasing? Decreasing?

b. Find the critical points of f. Which critical points correspond to local maxima? Local minima? Neither?

c. At what point(s) does f have an inflection point?

d. On what interval(s) is f concave up? Concave down?

e. Sketch the graph of f''.

f. Sketch one possible graph of f.

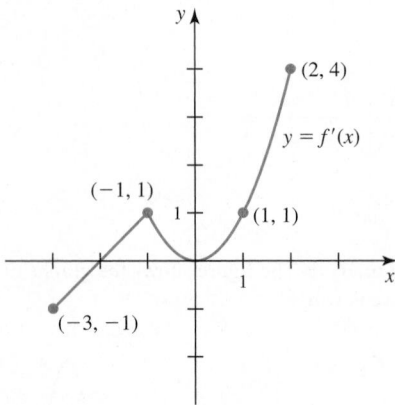

108. Growth rate of spotted owlets The rate of growth (in g/week) of the body mass of Indian spotted owlets is modeled by the function

$$r(t) = \frac{10{,}147.9e^{-2.2t}}{(37.98e^{-2.2t} + 1)^2},$$ where t is the age (in weeks) of the owlets. What value of $t > 0$ maximizes r? What is the physical meaning of the maximum value?

Explorations and Challenges

109. Demand functions and elasticity Economists use *demand functions* to describe how much of a commodity can be sold at varying prices. For example, the demand function $D(p) = 500 - 10p$ says that at a price of $p = 10$, a quantity of $D(10) = 400$ units of the commodity can be sold. The elasticity $E = \dfrac{dD}{dp}\dfrac{p}{D}$ of the demand gives the approximate percent change in the demand for every 1% change in the price. (See Section 3.6 or the Guided Project *Elasticity in Economics* for more on demand functions and elasticity.)

a. Compute the elasticity of the demand function $D(p) = 500 - 10p$.

b. If the price is \$12 and increases by 4.5%, what is the approximate percent change in the demand?

c. Show that for the linear demand function $D(p) = a - bp$, where a and b are positive real numbers, the elasticity is a decreasing function, for $p \geq 0$ and $p \neq a/b$.

d. Show that the demand function $D(p) = a/p^b$, where a and b are positive real numbers, has a constant elasticity for all positive prices.

110. Population models A typical population curve is shown in the figure. The population is small at $t = 0$ and increases toward a steady-state level called the *carrying capacity*. Explain why the maximum growth rate occurs at an inflection point of the population curve.

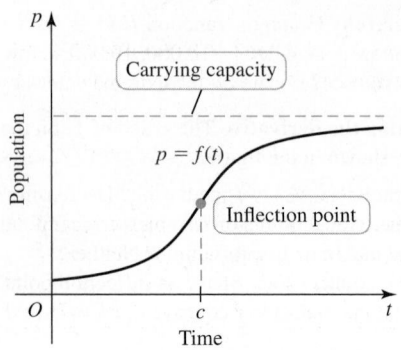

111. Population models The population of a species is given by the function $P(t) = \dfrac{Kt^2}{t^2 + b}$, where $t \geq 0$ is measured in years and K and b are positive real numbers.

a. With $K = 300$ and $b = 30$, what is $\lim_{t \to \infty} P(t)$, the carrying capacity of the population?

b. With $K = 300$ and $b = 30$, when does the maximum growth rate occur?

c. For arbitrary positive values of K and b, when does the maximum growth rate occur (in terms of K and b)?

112. Tangent lines and concavity Give an argument to support the claim that if a function is concave up at a point, then the tangent line at that point lies below the curve near that point.

113. General quartic Show that the general quartic (fourth-degree) polynomial $f(x) = x^4 + ax^3 + bx^2 + cx + d$, where $a, b, c,$ and d are real numbers, has either zero or two inflection points, and the latter case occurs provided $b < \dfrac{3a^2}{8}$.

114. First Derivative Test is not exhaustive Sketch the graph of a (simple) nonconstant function f that has a local maximum at $x = 1$, with $f'(1) = 0$, where f' does not change sign from positive to negative as x increases through 1. Why can't the First Derivative Test be used to classify the critical point at $x = 1$ as a local maximum? How could the test be rephrased to account for such a critical point?

115. Properties of cubics Consider the general cubic polynomial $f(x) = x^3 + ax^2 + bx + c$, where $a, b,$ and c are real numbers.

a. Prove that f has exactly one local maximum and one local minimum provided $a^2 > 3b$.

b. Prove that f has no extreme values if $a^2 < 3b$.

116. Concavity of parabolas Consider the general parabola described by the function $f(x) = ax^2 + bx + c$. For what values of $a, b,$ and c is f concave up? For what values of $a, b,$ and c is f concave down?

QUICK CHECK ANSWERS

1. Positive derivatives on an interval mean the curve is rising on the interval, which means the function is increasing on the interval.   **2.**

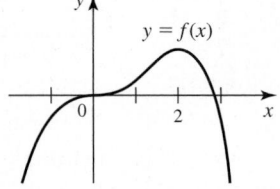

3. $f'(x) < 0$ on $(-\infty, 0)$ and $f'(x) > 0$ on $(0, \infty)$. Therefore, f has a local minimum at $x = 0$ by the First Derivative Test.   **4.** $f''(x) = 12x^2$, so $f''(x) > 0$ for $x < 0$ and for $x > 0$. There is no inflection point at $x = 0$ because the second derivative does not change sign.   **5.** The first curve should be rising and concave up. The second curve should be falling and concave down.

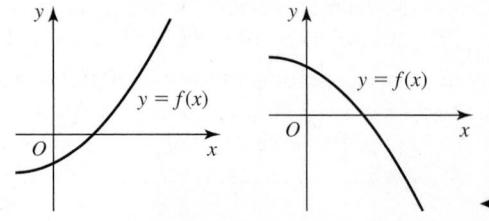

4.4 Graphing Functions

We have now collected the tools required for a comprehensive approach to graphing functions. These *analytical methods* are indispensable, even with the availability of powerful graphing utilities, as illustrated by the following example.

Calculators and Analysis

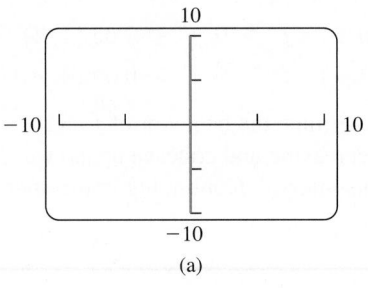

(a)

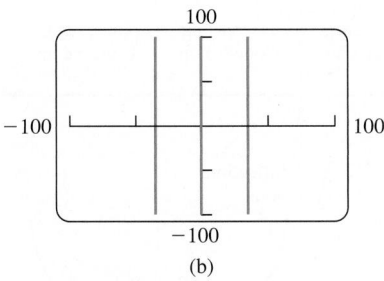

(b)

Figure 4.44

Suppose you want to graph the harmless-looking function $f(x) = x^3/3 - 400x$. The result of plotting f using a graphing calculator with a default window of $[-10, 10] \times [-10, 10]$ is shown in **Figure 4.44a**; one vertical line appears on the screen. Zooming out to the window $[-100, 100] \times [-100, 100]$ produces three vertical lines (**Figure 4.44b**); it is still difficult to understand the behavior of the function using only this graph. Expanding the window even more to $[-1000, 1000] \times [-1000, 1000]$ is no better. So what do we do?

The function $f(x) = x^3/3 - 400x$ has a reasonable graph (see Example 2), but it cannot be found automatically by letting technology do all the work. Here is the message of this section: Graphing utilities are valuable for exploring functions, producing preliminary graphs, and checking your work. But they should not be relied on exclusively because they cannot explain *why* a graph has its shape. Rather, graphing utilities should be used in an interactive way with the analytical methods presented in this chapter.

QUICK CHECK 1 Graph $f(x) = x^3/3 - 400x$ using various windows on a graphing calculator. Find a window that gives a better graph of f than those in Figure 4.44. ◄

Graphing Guidelines

The following set of guidelines need not be followed exactly for every function, and you will find that several steps can often be done at once. Depending on the specific problem, some of the steps are best done analytically, while other steps can be done with a graphing utility. Experiment with both approaches and try to find a good balance. We also present a schematic record-keeping procedure to keep track of discoveries as they are made.

Graphing Guidelines for $y = f(x)$

1. **Identify the domain or interval of interest.** On what interval(s) should the function be graphed? It may be the domain of the function or some subset of the domain.

2. **Exploit symmetry.** Take advantage of symmetry. For example, is the function *even* $(f(-x) = f(x))$, *odd* $(f(-x) = -f(x))$, or neither?

3. **Find the first and second derivatives.** They are needed to determine extreme values, concavity, inflection points, and the intervals on which f is increasing or decreasing. Computing derivatives—particularly second derivatives—may not be practical, so some functions may need to be graphed without complete derivative information.

4. **Find critical points and possible inflection points.** Determine points at which $f'(x) = 0$ or f' is undefined. Determine points at which $f''(x) = 0$ or f'' is undefined.

5. **Find intervals on which the function is increasing/decreasing and concave up/down.** The first derivative determines the intervals on which f is increasing or decreasing. The second derivative determines the intervals on which the function is concave up or concave down.

6. **Identify extreme values and inflection points.** Use either the First or Second Derivative Test to classify the critical points. Both x- and y-coordinates of maxima, minima, and inflection points are needed for graphing.

7. **Locate all asymptotes and determine end behavior.** Vertical asymptotes often occur at zeros of denominators. Horizontal asymptotes require examining limits as $x \to \pm\infty$; these limits determine end behavior. Slant asymptotes occur with rational functions in which the degree of the numerator is one more than the degree of the denominator.

8. **Find the intercepts.** The y-intercept of the graph is found by setting $x = 0$. The x-intercepts are found by solving $f(x) = 0$; they are the real zeros (or roots) of f.

9. **Choose an appropriate graphing window and plot a graph.** Use the results of the previous steps to graph the function. If you use graphing software, check for consistency with your analytical work. Is your graph *complete*—that is, does it show all the essential details of the function?

➤ Limits at infinity determine the end behavior of many functions, though a modification is required for some functions. For example, the end behavior of $f(x) = \dfrac{1 + x}{x + \sqrt{x}}$ is found by computing $\lim\limits_{x \to 0^+} f(x)$ and $\lim\limits_{x \to \infty} f(x)$ because the domain of f is $(0, \infty)$.

EXAMPLE 1 A warm-up Given the following information about the first and second derivatives of a function f that is continuous on $(-\infty, \infty)$, summarize the information using a sign graph, and then sketch a possible graph of f.

$$f' < 0, f'' > 0 \text{ on } (-\infty, 0) \quad f' > 0, f'' > 0 \text{ on } (0, 1) \quad f' > 0, f'' < 0 \text{ on } (1, 2)$$
$$f' < 0, f'' < 0 \text{ on } (2, 3) \quad f' < 0, f'' > 0 \text{ on } (3, 4) \quad f' > 0, f'' > 0 \text{ on } (4, \infty)$$

SOLUTION Figure 4.45 uses the given information to determine the behavior of f and its graph. For example, on the interval $(-\infty, 0)$, f is decreasing and concave up, so we sketch a segment of a curve with these properties on this interval. Continuing in this manner, we obtain a useful summary of the properties of f.

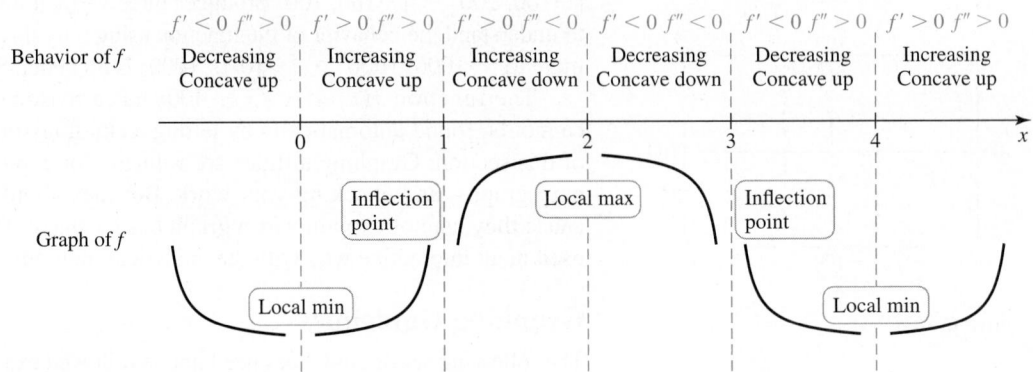

Figure 4.45

Assembling the information shown in Figure 4.45, a possible graph of f is produced (Figure 4.46). Notice that derivative information is not sufficient to determine the y-coordinates of points on the curve.

QUICK CHECK 2 Explain why the functions f and $f + C$, where C is a constant, have the same derivative properties. ◄

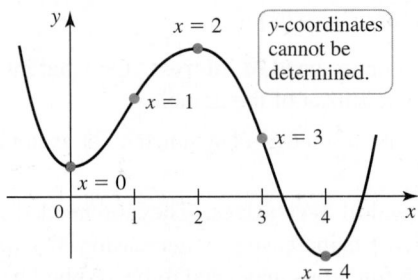

Figure 4.46

Related Exercises 7–8 ◄

EXAMPLE 2 A deceptive polynomial Use the graphing guidelines to graph $f(x) = \dfrac{x^3}{3} - 400x$ on its domain.

SOLUTION

1. **Domain** The domain of any polynomial is $(-\infty, \infty)$.

2. **Symmetry** Because f consists of odd powers of the variable, it is an odd function. Its graph is symmetric about the origin.

➤ Notice that the first derivative of an odd polynomial is an even polynomial and the second derivative is an odd polynomial (assuming the original polynomial is of degree 3 or greater).

3. **Derivatives** The derivatives of f are

$$f'(x) = x^2 - 400 \quad \text{and} \quad f''(x) = 2x.$$

4. **Critical points and possible inflection points** Solving $f'(x) = 0$, we find that the critical points are $x = \pm 20$. Solving $f''(x) = 0$, we see that a possible inflection point occurs at $x = 0$.

➤ See Appendix B, online at goo.gl/6DCbbM, for solving inequalities using test values.

5. **Increasing/decreasing and concavity** Note that

$$f'(x) = x^2 - 400 = (x - 20)(x + 20).$$

Solving the inequality $f'(x) < 0$, we find that f is decreasing on the interval $(-20, 20)$. Solving the inequality $f'(x) > 0$ reveals that f is increasing on the intervals $(-\infty, -20)$ and $(20, \infty)$ (**Figure 4.47**). By the First Derivative Test, we have enough information to conclude that f has a local maximum at $x = -20$ and a local minimum at $x = 20$.

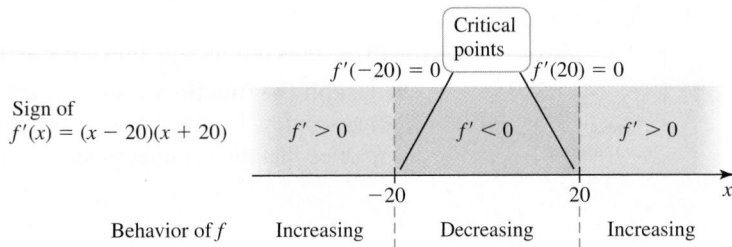

Figure 4.47

Furthermore, $f''(x) = 2x < 0$ on $(-\infty, 0)$, so f is concave down on this interval. Also, $f''(x) > 0$ on $(0, \infty)$, so f is concave up on $(0, \infty)$ (**Figure 4.48**).

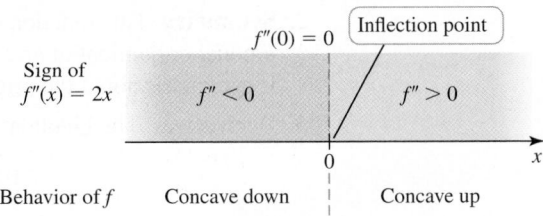

Figure 4.48

The evidence obtained so far is summarized in **Figure 4.49**.

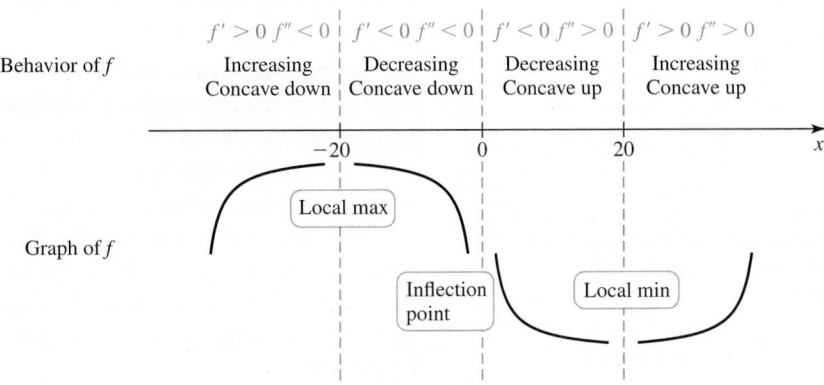

Figure 4.49

6. **Extreme values and inflection points** In this case, the Second Derivative Test is easily applied and it confirms what we have already learned. Because $f''(-20) < 0$ and $f''(20) > 0$, f has a local maximum at $x = -20$ and a local minimum at $x = 20$. The corresponding function values are $f(-20) = 16{,}000/3 = 5333\frac{1}{3}$ and $f(20) = -f(-20) = -5333\frac{1}{3}$. Finally, we see that f'' changes sign at $x = 0$, making $(0, 0)$ an inflection point.

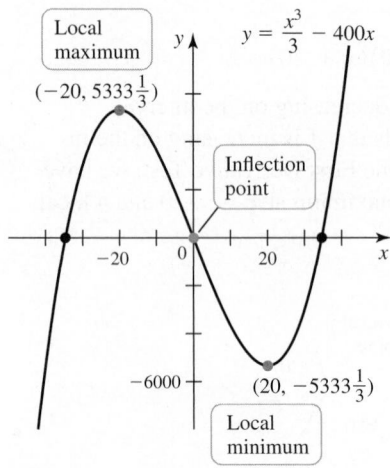

Local maximum

$\left(-20, 5333\tfrac{1}{3}\right)$

$y = \dfrac{x^3}{3} - 400x$

Inflection point

$\left(20, -5333\tfrac{1}{3}\right)$

Local minimum

Figure 4.50

7. **Asymptotes and end behavior** Polynomials have neither vertical nor horizontal asymptotes. Because the highest-power term in the polynomial is x^3 (an odd power) and the leading coefficient is positive, we have the end behavior

$$\lim_{x \to \infty} f(x) = \infty \quad \text{and} \quad \lim_{x \to -\infty} f(x) = -\infty.$$

8. **Intercepts** The y-intercept is $(0, 0)$. We solve the equation $f(x) = 0$ to find the x-intercepts:

$$\frac{x^3}{3} - 400x = x\left(\frac{x^2}{3} - 400\right) = 0.$$

The roots of this equation are $x = 0$ and $x = \pm\sqrt{1200} \approx \pm 34.6$.

9. **Graph the function** Using the information found in Steps 1–8, we choose the graphing window $[-40, 40] \times [-6000, 6000]$ and produce the graph shown in Figure 4.50. Notice that the symmetry detected in Step 2 is evident in this graph.

Related Exercises 17–18 ◄

EXAMPLE 3 **The surprises of a rational function** Use the graphing guidelines to graph $f(x) = \dfrac{10x^3}{x^2 - 1}$ on its domain.

SOLUTION

1. **Domain** The zeros of the denominator are $x = \pm 1$, so the domain is $\{x : x \neq \pm 1\}$.

2. **Symmetry** This function consists of an odd function divided by an even function. The product or quotient of an even function and an odd function is odd. Therefore, the graph is symmetric about the origin.

3. **Derivatives** The Quotient Rule is used to find the first and second derivatives:

$$f'(x) = \frac{10x^2(x^2 - 3)}{(x^2 - 1)^2} \quad \text{and} \quad f''(x) = \frac{20x(x^2 + 3)}{(x^2 - 1)^3}.$$

4. **Critical points and possible inflection points** The solutions of $f'(x) = 0$ occur where the numerator equals 0, provided the denominator is nonzero at those points. Solving $10x^2(x^2 - 3) = 0$ gives the critical points $x = 0$ and $x = \pm\sqrt{3}$. The solutions of $f''(x) = 0$ are found by solving $20x(x^2 + 3) = 0$; we see that the only possible inflection point occurs at $x = 0$.

5. **Increasing/decreasing and concavity** To find the sign of f', first note that the denominator of f' is nonnegative, as is the factor $10x^2$ in the numerator. So the sign of f' is determined by the sign of the factor $x^2 - 3$, which is negative on $(-\sqrt{3}, \sqrt{3})$ (excluding $x = \pm 1$) and positive on $(-\infty, -\sqrt{3})$ and $(\sqrt{3}, \infty)$. Therefore, f is decreasing on $(-\sqrt{3}, \sqrt{3})$ (excluding $x = \pm 1$) and increasing on $(-\infty, -\sqrt{3})$ and $(\sqrt{3}, \infty)$.

   The sign of f'' is a bit trickier. Because $x^2 + 3$ is positive, the sign of f'' is determined by the sign of $20x$ in the numerator and $(x^2 - 1)^3$ in the denominator. When $20x$ and $(x^2 - 1)^3$ have the same sign, $f''(x) > 0$; when $20x$ and $(x^2 - 1)^3$ have opposite signs, $f''(x) < 0$ (Table 4.1). The results of this analysis are shown in Figure 4.51.

> ➤ Sign charts and sign graphs (Table 4.1 and Figure 4.51) must be constructed carefully when vertical asymptotes are involved: The sign of f' and f'' may or may not change at an asymptote.

Table 4.1

	$20x$	$x^2 + 3$	$(x^2 - 1)^3$	**Sign of f''**
$(-\infty, -1)$	$-$	$+$	$+$	$-$
$(-1, 0)$	$-$	$+$	$-$	$+$
$(0, 1)$	$+$	$+$	$-$	$-$
$(1, \infty)$	$+$	$+$	$+$	$+$

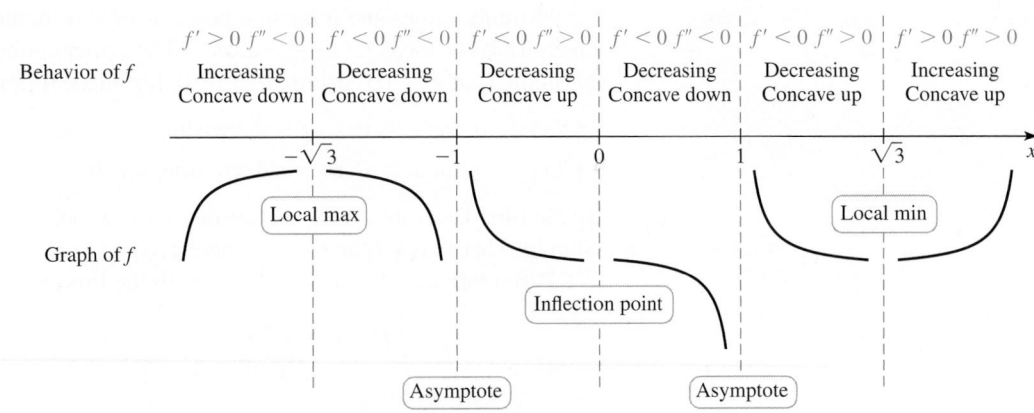

| $f' > 0\ f'' < 0$ | $f' < 0\ f'' < 0$ | $f' < 0\ f'' > 0$ | $f' < 0\ f'' < 0$ | $f' < 0\ f'' > 0$ | $f' > 0\ f'' > 0$ |

Behavior of f — Increasing Concave down | Decreasing Concave down | Decreasing Concave up | Decreasing Concave down | Decreasing Concave up | Increasing Concave up

Local max ... Local min

Inflection point

Asymptote ... Asymptote

Graph of f

Figure 4.51

6. **Extreme values and inflection points** The First Derivative Test is easily applied by looking at Figure 4.51. The function is increasing on $(-\infty, -\sqrt{3})$ and decreasing on $(-\sqrt{3}, -1)$; therefore, f has a local maximum at $x = -\sqrt{3}$, where $f(-\sqrt{3}) = -15\sqrt{3}$. Similarly, f has a local minimum at $x = \sqrt{3}$, where $f(\sqrt{3}) = 15\sqrt{3}$. (These results could also be obtained with the Second Derivative Test.) There is no local extreme value at the critical point $x = 0$, only a horizontal tangent line.

Using Table 4.1 from Step 5, we see that f'' changes sign at $x = \pm 1$ and at $x = 0$. The points $x = \pm 1$ are not in the domain of f, so they cannot correspond to inflection points. However, there is an inflection point at $(0, 0)$.

7. **Asymptotes and end behavior** Recall from Section 2.4 that zeros of the denominator, which in this case are $x = \pm 1$, are candidates for vertical asymptotes. Checking the behavior of f on either side of $x = \pm 1$, we find

$$\lim_{x \to -1^-} f(x) = -\infty, \qquad \lim_{x \to -1^+} f(x) = \infty.$$
$$\lim_{x \to 1^-} f(x) = -\infty, \qquad \lim_{x \to 1^+} f(x) = \infty.$$

It follows that f has vertical asymptotes at $x = \pm 1$. The degree of the numerator is greater than the degree of the denominator, so there is no horizontal asymptote. Using long division, it can be shown that

$$f(x) = 10x + \frac{10x}{x^2 - 1}.$$

Therefore, as $x \to \pm \infty$, the graph of f approaches the line $y = 10x$. This line is a slant asymptote (Section 2.5).

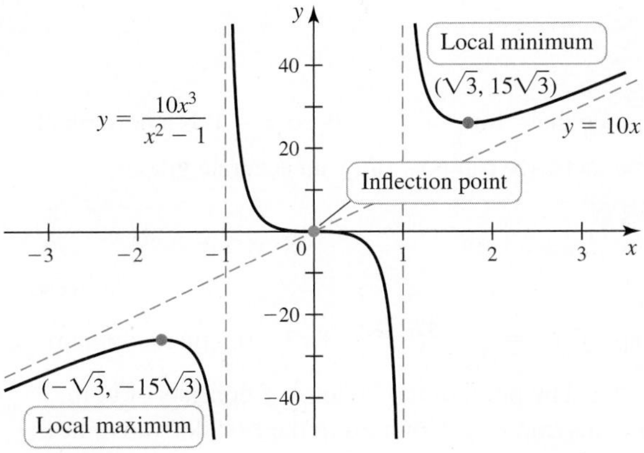

8. **Intercepts** The zeros of a rational function coincide with the zeros of the numerator, provided those points are not also zeros of the denominator. In this case, the zeros of f satisfy $10x^3 = 0$, or $x = 0$ (which is not a zero of the denominator). Therefore, $(0, 0)$ is both the x-intercept and the y-intercept.

9. **Graphing** We now assemble an accurate graph of f, as shown in Figure 4.52. A window of $[-3, 3] \times [-40, 40]$ gives a complete graph of the function. Notice that the symmetry about the origin deduced in Step 2 is apparent in the graph.

Related Exercises 30–31 ◄

$y = \dfrac{10x^3}{x^2 - 1}$

Local minimum $(\sqrt{3}, 15\sqrt{3})$

$y = 10x$

Inflection point

$(-\sqrt{3}, -15\sqrt{3})$

Local maximum

Figure 4.52

QUICK CHECK 3 Verify that the function f in Example 3 is symmetric about the origin by showing that $f(-x) = -f(x)$. ◄

In the next two examples, we show how the guidelines may be streamlined to some extent.

► The function $f(x) = e^{-x^2}$ and the family of functions $f(x) = ce^{-ax^2}$ are central to the study of statistics. They have bell-shaped graphs and describe Gaussian or normal distributions.

EXAMPLE 4 The normal distribution Analyze the function $f(x) = e^{-x^2}$ and draw its graph.

SOLUTION The domain of f is all real numbers, and $f(x) > 0$ for all x. Because $f(-x) = f(x)$, f is an even function and its graph is symmetric about the y-axis.

Extreme values and inflection points follow from the derivatives of f. Using the Chain Rule, we have $f'(x) = -2xe^{-x^2}$. The critical points satisfy $f'(x) = 0$, which has the single root $x = 0$ (because $e^{-x^2} > 0$, for all x). It now follows that

• $f'(x) > 0$, for $x < 0$, so f is increasing on $(-\infty, 0)$.
• $f'(x) < 0$, for $x > 0$, so f is decreasing on $(0, \infty)$.

By the First Derivative Test, we see that f has a local maximum (and an absolute maximum by Theorem 4.9) at $x = 0$, where $f(0) = 1$.

Differentiating $f'(x) = -2xe^{-x^2}$ with the Product Rule yields

$$f''(x) = e^{-x^2}(-2) + (-2x)(-2xe^{-x^2}) \quad \text{Product Rule}$$
$$= 2e^{-x^2}(2x^2 - 1). \quad \text{Simplify.}$$

Again using the fact that $e^{-x^2} > 0$, for all x, we see that $f''(x) = 0$ when $2x^2 - 1 = 0$ or when $x = \pm 1/\sqrt{2}$; these values are candidates for inflection points. Observe that $f''(x) > 0$ and f is concave up on $(-\infty, -1/\sqrt{2})$ and $(1/\sqrt{2}, \infty)$, while $f''(x) < 0$ and f is concave down on $(-1/\sqrt{2}, 1/\sqrt{2})$. Because f'' changes sign at $x = \pm 1/\sqrt{2}$, we have inflection points at $(\pm 1/\sqrt{2}, 1/\sqrt{e})$ (Figure 4.53).

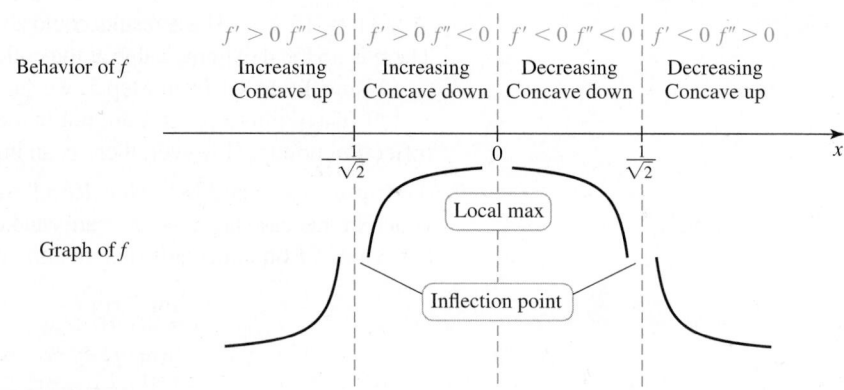

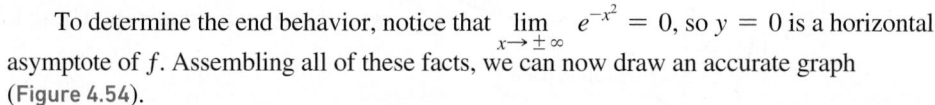

Figure 4.53

To determine the end behavior, notice that $\lim_{x \to \pm \infty} e^{-x^2} = 0$, so $y = 0$ is a horizontal asymptote of f. Assembling all of these facts, we can now draw an accurate graph (Figure 4.54).

Related Exercises 43, 46 ◄

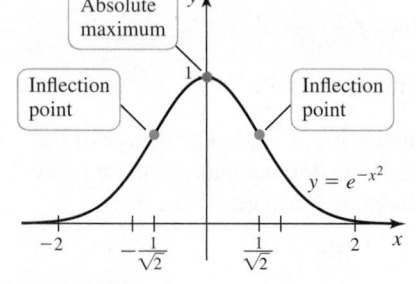

Figure 4.54

EXAMPLE 5   Roots and cusps Graph $f(x) = \frac{1}{8} x^{2/3}(9x^2 - 8x - 16)$ on its domain.

SOLUTION The domain of f is $(-\infty, \infty)$. The polynomial factor in f consists of both even and odd powers, so f has no special symmetry. Computing the first derivative is straightforward if you first expand f as a sum of three terms:

$$f'(x) = \frac{d}{dx}\left(\frac{9x^{8/3}}{8} - x^{5/3} - 2x^{2/3} \right) \quad \text{Expand } f.$$
$$= 3x^{5/3} - \frac{5}{3}x^{2/3} - \frac{4}{3}x^{-1/3} \quad \text{Differentiate.}$$
$$= \frac{(x - 1)(9x + 4)}{3x^{1/3}}. \quad \text{Simplify.}$$

The critical points are now identified: f' is undefined at $x = 0$ (because $x^{-1/3}$ is undefined there) and $f'(x) = 0$ at $x = 1$ and $x = -\frac{4}{9}$. So we have three critical points to analyze. Table 4.2 tracks the signs of the factors in the numerator and denominator of f' and shows the sign of f' on the relevant intervals; this information is recorded in Figure 4.55.

Table 4.2

	$3x^{1/3}$	$9x + 4$	$x - 1$	**Sign of f'**
$\left(-\infty, -\frac{4}{9}\right)$	$-$	$-$	$-$	$-$
$\left(-\frac{4}{9}, 0\right)$	$-$	$+$	$-$	$+$
$(0, 1)$	$+$	$+$	$-$	$-$
$(1, \infty)$	$+$	$+$	$+$	$+$

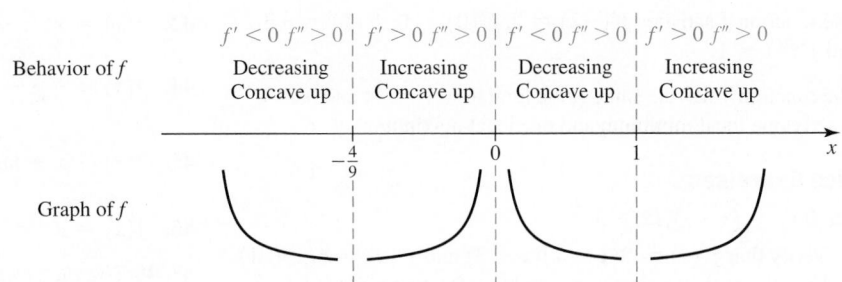

Figure 4.55

> To show that $f''(x) > 0$ (for $x \neq 0$), analyze its numerator and denominator.
> - The graph of $45x^2 - 10x + 4$ is a parabola that opens upward (its leading coefficient is positive) with no x-intercepts (verified by the quadratic formula). Therefore, the numerator is always positive.
> - The denominator is nonnegative because $9x^{4/3} = 9(x^{1/3})^4$ (even powers are nonnegative).

We use the second line in the calculation of f' to compute the second derivative:

$$f''(x) = \frac{d}{dx}\left(3x^{5/3} - \frac{5}{3}x^{2/3} - \frac{4}{3}x^{-1/3}\right)$$

$$= 5x^{2/3} - \frac{10}{9}x^{-1/3} + \frac{4}{9}x^{-4/3} \qquad \text{Differentiate.}$$

$$= \frac{45x^2 - 10x + 4}{9x^{4/3}}. \qquad \text{Simplify.}$$

Solving $f''(x) = 0$, we discover that $f''(x) > 0$, for all x except $x = 0$, where it is undefined. Therefore, f is concave up on $(-\infty, 0)$ and $(0, \infty)$ (Figure 4.55).

By the Second Derivative Test, because $f''(x) > 0$, for $x \neq 0$, the critical points $x = -\frac{4}{9}$ and $x = 1$ correspond to local minima; their y-coordinates are $f(-\frac{4}{9}) \approx -0.78$ and $f(1) = -\frac{15}{8} = -1.875$.

What about the third critical point, $x = 0$? Note that $f(0) = 0$, and f is increasing just to the left of 0 and decreasing just to the right. By the First Derivative Test, f has a local maximum at $x = 0$. Furthermore, $f'(x) \to \infty$ as $x \to 0^-$ and $f'(x) \to -\infty$ as $x \to 0^+$, so the graph of f has a cusp at $x = 0$.

As $x \to \pm\infty$, f is dominated by its highest-power term, which is $9x^{8/3}/8$. This term becomes large and positive as $x \to \pm\infty$; therefore, f has no absolute maximum. Its absolute minimum occurs at $x = 1$ because, comparing the two local minima, $f(1) < f(-\frac{4}{9})$.

The roots of f satisfy $\frac{1}{8}x^{2/3}(9x^2 - 8x - 16) = 0$, which gives $x = 0$ and

$$x = \frac{4}{9}\left(1 \pm \sqrt{10}\right) \approx -0.96 \quad \text{and} \quad 1.85. \quad \text{Use the quadratic formula.}$$

With the information gathered in this analysis, we obtain the graph shown in Figure 4.56.

Related Exercises 38–40 ◄

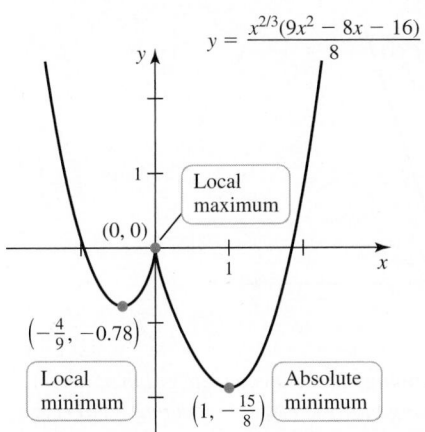

$$y = \frac{x^{2/3}(9x^2 - 8x - 16)}{8}$$

Local maximum
$(0, 0)$
$\left(-\frac{4}{9}, -0.78\right)$
Local minimum
$\left(1, -\frac{15}{8}\right)$ Absolute minimum

Figure 4.56

SECTION 4.4 EXERCISES

Getting Started

1. Why is it important to determine the domain of f before graphing f?

2. Explain why it is useful to know about symmetry in a function.

3. Can the graph of a polynomial have vertical or horizontal asymptotes? Explain.

4. Where are the vertical asymptotes of a rational function located?

5. How do you find the absolute maximum and minimum values of a function that is continuous on a closed interval?

6. Describe the possible end behavior of a polynomial.

7–8. *Sketch a graph of a function f with the following properties.*

7. $f' < 0$ and $f'' < 0$, for $x < 3$
   $f' < 0$ and $f'' > 0$, for $x > 3$

8. $f' < 0$ and $f'' < 0$, for $x < -1$
   $f' < 0$ and $f'' > 0$, for $-1 < x < 2$
   $f' > 0$ and $f'' > 0$, for $2 < x < 8$
   $f' > 0$ and $f'' < 0$, for $8 < x < 10$
   $f' > 0$ and $f'' > 0$, for $x > 10$

9–12. *Sketch a continuous function f on some interval that has the properties described. Answers will vary.*

9. The function f has one inflection point but no local extrema.

10. The function f has three real zeros and exactly two local minima.

11. The function f satisfies $f'(-2) = 2$, $f'(0) = 0$, $f'(1) = -3$, and $f'(4) = 1$.

12. The function f has the same finite limit as $x \to \pm \infty$ and has exactly one local minimum and one local maximum.

Practice Exercises

13. Let $f(x) = (x - 3)(x + 3)^2$.

    a. Verify that $f'(x) = 3(x - 1)(x + 3)$ and $f''(x) = 6(x + 1)$.

    b. Find the critical points and possible inflection points of f.

    c. Find the intervals on which f is increasing or decreasing.

    d. Determine the intervals on which f is concave up or concave down.

    e. Identify the local extreme values and inflection points of f.

    f. State the x- and y-intercepts of the graph of f.

    g. Use your work in parts (a) through (f) to sketch a graph of f.

14. If $f(x) = \dfrac{1}{3x^4 + 5}$, it can be shown that $f'(x) = -\dfrac{12x^3}{(3x^4 + 5)^2}$

and $f''(x) = \dfrac{180x^2(x^2 + 1)(x + 1)(x - 1)}{(3x^4 + 5)^3}$. Use these functions

to complete the following steps.

    a. Find the critical points and possible inflection points of f.

    b. Find the intervals on which f is increasing or decreasing.

    c. Determine the intervals on which f is concave up or concave down.

    d. Identify the local extreme values and inflection points of f.

    e. State the x- and y-intercepts of the graph of f.

    f. Find the asymptotes of f.

    g. Use your work in parts (a) through (f) to sketch a graph of f.

15–46. Graphing functions *Use the guidelines of this section to make a complete graph of f.*

15. $f(x) = x^2 - 6x$

16. $f(x) = x - x^2$

17. $f(x) = x^3 - 6x^2 + 9x$

18. $f(x) = 3x - x^3$

19. $f(x) = x^4 - 6x^2$

20. $f(x) = x^4 + 4x^3$

21. $f(x) = (x - 6)(x + 6)^2$

22. $f(x) = 27(x - 2)^2(x + 2)$

23. $f(x) = x^3 - 6x^2 - 135x$

24. $f(x) = x^4 + 8x^3 - 270x^2 + 1$

25. $f(x) = x^3 - 3x^2 - 144x - 140$

26. $f(x) = x^3 - 147x + 286$

27. $f(x) = x - 2\sqrt{x}$

28. $f(x) = 3\sqrt{x} - x^{3/2}$

29. $f(x) = \dfrac{3x}{x^2 - 1}$

30. $f(x) = \dfrac{2x - 3}{2x - 8}$

31. $f(x) = \dfrac{x^2}{x - 2}$

32. $f(x) = \dfrac{x^2}{x^2 - 4}$

33. $f(x) = \dfrac{x^2 + 12}{2x + 1}$

34. $f(x) = \dfrac{4x}{x^2 + 3}$

35. $f(x) = \tan^{-1}\left(\dfrac{x^2}{\sqrt{3}}\right)$

36. $f(x) = \ln(x^2 + 1)$

37. $f(x) = x + 2\cos x$ on $[-2\pi, 2\pi]$

38. $f(x) = x - 3x^{2/3}$

39. $f(x) = x - 3x^{1/3}$

40. $f(x) = 2 - 2x^{2/3} + x^{4/3}$

41. $f(x) = \sin x - x$ on $[0, 2\pi]$

42. $f(x) = x\sqrt{x + 3}$

43. $f(x) = e^{-x}\sin x$ on $[-\pi, \pi]$

44. $f(x) = \dfrac{1}{e^{-x} - 1}$

45. $f(x) = x + \tan x$ on $\left(-\dfrac{\pi}{2}, \dfrac{\pi}{2}\right)$

46. $f(x) = e^{-x^2/2}$

47–48. *Use the graphs of f' and f'' to find the critical points and inflection points of f, the intervals on which f is increasing or decreasing, and the intervals of concavity. Then graph f assuming $f(0) = 0$.*

47.

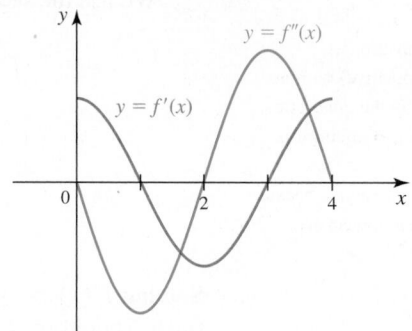

48.

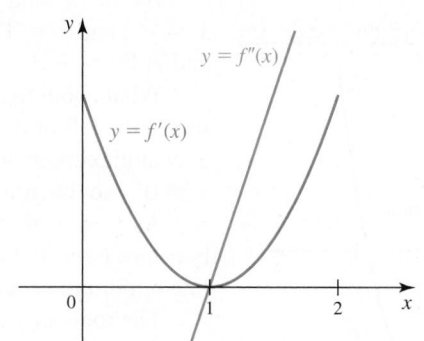

T **49–54. Graphing with technology** *Make a complete graph of the following functions. A graphing utility is useful in locating intercepts, local extreme values, and inflection points.*

49. $f(x) = \dfrac{1}{3}x^3 - 2x^2 - 5x + 2$

50. $f(x) = \dfrac{1}{15}x^3 - x + 1$

51. $f(x) = 3x^4 + 4x^3 - 12x^2$

52. $f(x) = x^3 - 33x^2 + 216x - 2$

53. $f(x) = \dfrac{3x - 5}{x^2 - 1}$

54. $f(x) = x^{1/3}(x - 2)^2$

55. **Explain why or why not** Determine whether the following statements are true and give an explanation or counterexample.

    a. If the zeros of f' are -3, 1, and 4, then the local extrema of f are located at these points.

    b. If the zeros of f'' are -2 and 4, then the inflection points of f are also located at these points.

    c. If the zeros of the denominator of f are -3 and 4, then f has vertical asymptotes at these points.

    d. If a rational function has a finite limit as $x \to \infty$, then it must have a finite limit as $x \to -\infty$.

56–59. Functions from derivatives *Use the derivative f' to determine the x-coordinates of the local maxima and minima of f, and the intervals on which f is increasing or decreasing. Sketch a possible graph of f (f is not unique).*

56. $f'(x) = (x - 1)(x + 2)(x + 4)$

57. $f'(x) = 10 \sin 2x$ on $[-2\pi, 2\pi]$

58. $f'(x) = \frac{1}{6}(x + 1)(x - 2)^2(x - 3)$

59. $f'(x) = x^2(x + 2)(x - 1)$

Explorations and Challenges

60. $e^\pi > \pi^e$ Prove that $e^\pi > \pi^e$ by first finding the maximum value of $f(x) = \frac{\ln x}{x}$.

61. Height vs. volume The figure shows six containers, each of which is filled from the top. Assume water is poured into the containers at a constant rate and each container is filled in 10 s. Assume also that the horizontal cross sections of the containers are always circles. Let $h(t)$ be the depth of water in the container at time t, for $0 \le t \le 10$.

a. For each container, sketch a graph of the function $y = h(t)$, for $0 \le t \le 10$.
b. Explain why h is an increasing function.
c. Describe the concavity of the function. Identify inflection points when they occur.
d. For each container, where does h' (the derivative of h) have an absolute maximum on $[0, 10]$?

(A)   (B)   (C)

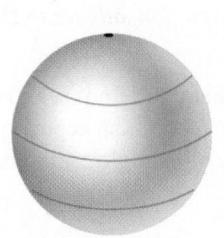  

(D)   (E)   (F)

T 62. Oscillations Consider the function $f(x) = \cos(\ln x)$, for $x > 0$. Use analytical techniques and a graphing utility to complete the following steps.

a. Locate all local extrema on the interval $(0, 4]$.
b. Identify the inflection points on the interval $(0, 4]$.
c. Locate the three smallest zeros of f on the interval $(0.1, \infty)$.
d. Sketch a graph of f.

63. Local max/min of $x^{1/x}$ Use analytical methods to find all local extrema of the function $f(x) = x^{1/x}$, for $x > 0$. Verify your work using a graphing utility.

64. Local max/min of x^x Use analytical methods to find all local extrema of the function $f(x) = x^x$, for $x > 0$. Verify your work using a graphing utility.

65. Derivative information Suppose a continuous function f is concave up on $(-\infty, 0)$ and $(0, \infty)$. Assume f has a local maximum at $x = 0$. What, if anything, do you know about $f'(0)$? Explain with an illustration.

T 66. A pursuit curve A man stands 1 mi east of a crossroads. At noon, a dog starts walking north from the crossroads at 1 mi/hr (see figure). At the same instant, the man starts walking and at all times walks directly toward the dog at $s > 1$ mi/hr. The path in the xy-plane followed by the man as he pursues the dog is given by the function

$$y = f(x) = \frac{s}{2}\left(\frac{x^{(s+1)/s}}{s + 1} - \frac{x^{(s-1)/s}}{s - 1}\right) + \frac{s}{s^2 - 1}.$$

Select various values of $s > 1$ and graph this pursuit curve. Comment on the changes in the curve as s increases.

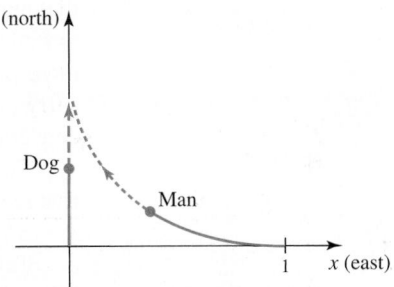

T 67–72. Combining technology with analytical methods *Use a graphing utility together with analytical methods to create a complete graph of the following functions. Be sure to find and label the intercepts, local extrema, inflection points, and asymptotes, and find the intervals on which the function is increasing or decreasing, and the intervals on which the function is concave up or concave down.*

67. $f(x) = \frac{x \sin x}{x^2 + 1}$ on $[-2\pi, 2\pi]$

68. $f(x) = 3\sqrt[4]{x} - \sqrt{x} - 2$

69. $f(x) = 3x^4 - 44x^3 + 60x^2$ (*Hint:* Two different graphing windows may be needed.)

70. $f(x) = \frac{\sin \pi x}{1 + \sin \pi x}$ on $[0, 2]$

71. $f(x) = \frac{\tan^{-1} x}{x^2 + 1}$ **72.** $f(x) = \frac{\sqrt{4x^2 + 1}}{x^2 + 1}$

T 73–77. Special curves *The following classical curves have been studied by generations of mathematicians. Use analytical methods (including implicit differentiation) and a graphing utility to graph the curves. Include as much detail as possible.*

73. $x^{2/3} + y^{2/3} = 1$ (Astroid or hypocycloid with four cusps)

74. $y = \frac{8}{x^2 + 4}$ (Witch of Agnesi)

75. $x^3 + y^3 = 3xy$ (Folium of Descartes)

76. $y^2 = x^3(1 - x)$ (Pear curve)

77. $x^4 - x^2 + y^2 = 0$ (Figure-8 curve)

T 78. Elliptic curves The equation $y^2 = x^3 - ax + 3$, where a is a parameter, defines a well-known family of *elliptic curves*.

a. Plot a graph of the curve when $a = 3$.
b. Plot a graph of the curve when $a = 4$.
c. By experimentation, determine the approximate value of a ($3 < a < 4$) at which the graph separates into two curves.

T 79. Lamé curves The equation $|y/a|^n + |x/a|^n = 1$, where n and a are positive real numbers, defines the family of Lamé curves. Make a complete graph of this function with $a = 1$, for $n = \frac{2}{3}$, 1, 2, and 3. Describe the progression that you observe as n increases.

QUICK CHECK ANSWERS

1. Make the window larger in the y-direction.
2. Notice that f and $f + C$ have the same derivatives.
3. $f(-x) = \dfrac{10(-x)^3}{(-x)^2 - 1} = -\dfrac{10x^3}{x^2 - 1} = -f(x)$ ◄

4.5 Optimization Problems

The theme of this section is *optimization*, a topic arising in many disciplines that rely on mathematics. A structural engineer may seek the dimensions of a beam that maximize strength for a specified cost. A packaging designer may seek the dimensions of a container that maximize the volume of the container for a given surface area. Airline strategists need to find the best allocation of airliners among several hubs to minimize fuel costs and maximize passenger miles. In all these examples, the challenge is to find an *efficient* way to carry out a task, where "efficient" could mean least expensive, most profitable, least time consuming, or, as you will see, many other measures.

To introduce the ideas behind optimization problems, think about pairs of nonnegative real numbers x and y between 0 and 20 with the property that their sum is 20, that is, $x + y = 20$. Of all possible pairs, which has the greatest product?

Table 4.3 displays a few cases showing how the product of two nonnegative numbers varies while their sum remains constant. The condition that $x + y = 20$ is called a **constraint**: It tells us to consider only (nonnegative) values of x and y satisfying this equation.

The quantity that we wish to maximize (or minimize in other cases) is called the **objective function**; in this case, the objective function is the product $P = xy$. From Table 4.3, it appears that the product is greatest if both x and y are near the middle of the interval $[0, 20]$.

This simple problem has all the essential features of optimization problems. At their heart, optimization problems take the following form:

What is the maximum (minimum) value of an objective function subject to the given constraint(s)?

For the problem at hand, this question would be stated as, "What pair of nonnegative numbers maximizes $P = xy$ subject to the constraint $x + y = 20$?" The first step is to use the constraint to express the objective function $P = xy$ in terms of a single variable. In this case, the constraint is

$$x + y = 20, \quad \text{or} \quad y = 20 - x.$$

Substituting for y, the objective function becomes

$$P = xy = x(20 - x) = 20x - x^2,$$

which is a function of the single variable x. Notice that the values of x lie in the interval $0 \le x \le 20$ with $P(0) = P(20) = 0$.

To maximize P, we first find the critical points by solving

$$P'(x) = 20 - 2x = 0$$

to obtain the solution $x = 10$. To find the absolute maximum value of P on the interval $[0, 20]$, we check the endpoints and the critical points. Because $P(0) = P(20) = 0$ and $P(10) = 100$, we conclude that P has its absolute maximum value at $x = 10$. By the constraint $x + y = 20$, the numbers with the greatest product are $x = y = 10$, and their product is $P = 100$.

Table 4.3

x	y	$x + y$	$P = xy$
1	19	20	19
5.5	14.5	20	79.75
9	11	20	99
13	7	20	91
18	2	20	36

➤ In this problem, it is just as easy to eliminate x as y. In other problems, eliminating one variable may result in less work than eliminating other variables.

QUICK CHECK 1 Verify that in the example to the right, the same result is obtained if the constraint $x + y = 20$ is used to eliminate x rather than y. ◄

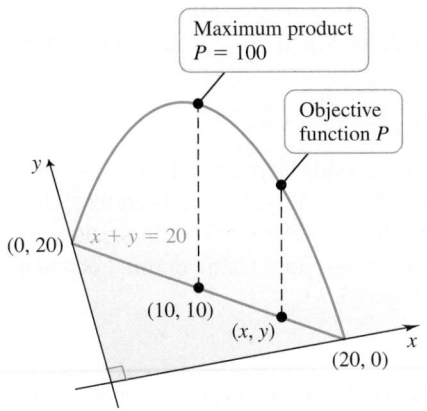

Figure 4.57

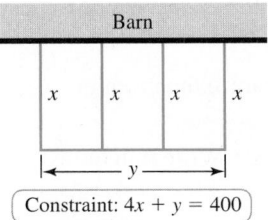

Constraint: $4x + y = 400$

Figure 4.58

> ► Recall from Section 4.1 that, when they exist, the absolute extreme values occur at critical points or endpoints.

QUICK CHECK 2 Find the objective function in Example 1 (in terms of x) (i) if there is no interior fence and (ii) if there is one interior fence that forms a right angle with the barn, as in Figure 4.58. ◄

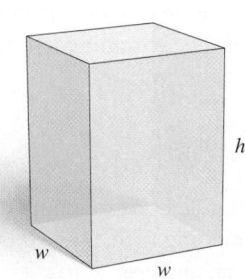

Objective function: $V = w^2h$
Constraint: $2w + h = 64$

Figure 4.60

Figure 4.57 summarizes this problem. We see the constraint line $x + y = 20$ in the xy-plane. Above the line is the objective function $P = xy$. As x and y vary along the constraint line, the objective function changes, reaching a maximum value of 100 when $x = y = 10$.

Most optimization problems have the same basic structure as the preceding example: There is an objective function, which may involve several variables, and one or more constraints. The methods of calculus (Sections 4.1 and 4.3) are used to find the minimum or maximum values of the objective function.

EXAMPLE 1 Rancher's dilemma A rancher has 400 ft of fence for constructing a rectangular corral. One side of the corral will be formed by a barn and requires no fence. Three exterior fences and two interior fences partition the corral into three rectangular regions as shown in **Figure 4.58**. What are the dimensions of the corral that maximize the enclosed area? What is the area of that corral?

SOLUTION We first sketch the corral (Figure 4.58), where $x \geq 0$ is the width and $y \geq 0$ is the length of the corral. The amount of fence required is $4x + y$, so the constraint is $4x + y = 400$, or $y = 400 - 4x$.

The objective function to be maximized is the area of the corral, $A = xy$. Using $y = 400 - 4x$, we eliminate y and express A as a function of x:

$$A = xy = x(400 - 4x) = 400x - 4x^2.$$

Notice that the width of the corral must be at least $x = 0$, and it cannot exceed $x = 100$ (because 400 ft of fence are available). Therefore, we maximize $A(x) = 400x - 4x^2$, for $0 \leq x \leq 100$. The critical points of the objective function satisfy

$$A'(x) = 400 - 8x = 0,$$

which has the solution $x = 50$. To find the absolute maximum value of A, we check the endpoints of $[0, 100]$ and the critical point $x = 50$. Because $A(0) = A(100) = 0$ and $A(50) = 10,000$, the absolute maximum value of A occurs when $x = 50$. Using the constraint, the optimal length of the corral is $y = 400 - 4(50) = 200$. Therefore, the maximum area of 10,000 ft^2 is achieved with dimensions $x = 50$ ft and $y = 200$ ft. The objective function A is shown in **Figure 4.59**.

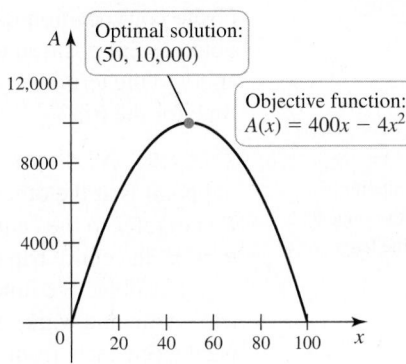

Figure 4.59

Related Exercises 12, 16 ◄

EXAMPLE 2 Airline regulations Suppose an airline policy states that all baggage must be box-shaped with a sum of length, width, and height not exceeding 64 in. What are the dimensions and volume of a square-based box with the greatest volume under these conditions?

SOLUTION We sketch a square-based box whose length and width are w and whose height is h (**Figure 4.60**). By the airline policy, the constraint is $2w + h = 64$. The objective function is the volume, $V = w^2h$. Either w or h may be eliminated from the objective function; the constraint $h = 64 - 2w$ implies that the volume is

$$V = w^2h = w^2(64 - 2w) = 64w^2 - 2w^3.$$

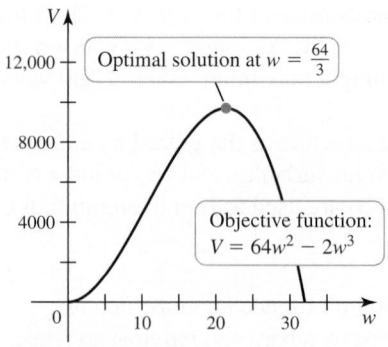

Figure 4.61

QUICK CHECK 3 Find the objective function in Example 2 (in terms of w) if the constraint is that the sum of length and width and height cannot exceed 108 in. ◄

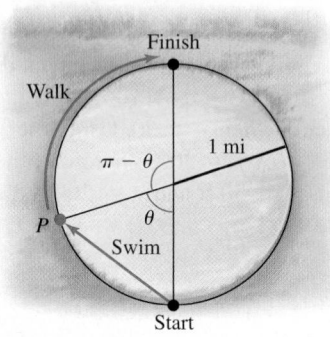

Figure 4.62

➤ To show that the chord length of a circle is $2r \sin(\theta/2)$, draw a line from the center of the circle to the midpoint of the chord. This line bisects the angle θ. Using a right triangle, half the length of the chord is $r \sin(\theta/2)$.

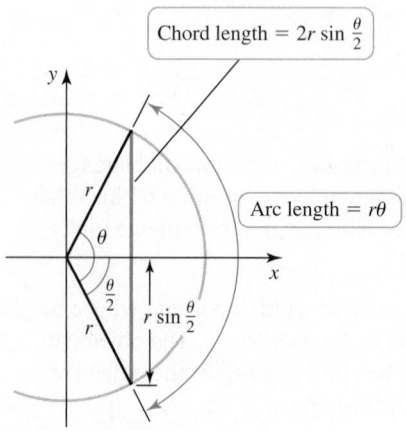

Figure 4.63

The objective function has now been expressed in terms of a single variable. Notice that w is nonnegative and cannot exceed 32, so the domain of V is $0 \le w \le 32$. The critical points satisfy

$$V'(w) = 128w - 6w^2 = 2w(64 - 3w) = 0,$$

which has roots $w = 0$ and $w = \frac{64}{3}$. By the First (or Second) Derivative Test, $w = \frac{64}{3}$ corresponds to a local maximum. At the endpoints, $V(0) = V(32) = 0$. Therefore, the volume function has an absolute maximum of $V(64/3) \approx 9709 \text{ in}^3$. The dimensions of the optimal box are $w = 64/3$ in and $h = 64 - 2w = 64/3$ in, so the optimal box is a cube. A graph of the volume function is shown in **Figure 4.61**.

Related Exercises 19–20 ◄

Optimization Guidelines   With two examples providing some insight, we present a procedure for solving optimization problems. These guidelines provide a general framework, but the details may vary depending on the problem.

Guidelines for Optimization Problems

1. Read the problem carefully, identify the variables, and organize the given information with a picture.

2. Identify the objective function (the function to be optimized). Write it in terms of the variables of the problem.

3. Identify the constraint(s). Write them in terms of the variables of the problem.

4. Use the constraint(s) to eliminate all but one independent variable of the objective function.

5. With the objective function expressed in terms of a single variable, find the interval of interest for that variable.

6. Use methods of calculus to find the absolute maximum or minimum value of the objective function on the interval of interest. If necessary, check the endpoints.

EXAMPLE 3   Walking and swimming   Suppose you are standing on the shore of a circular pond with a radius of 1 mile and you want to get to a point on the shore directly opposite your position (on the other end of a diameter). You plan to swim at 2 mi/hr from your current position to another point P on the shore and then walk at 3 mi/hr along the shore to the terminal point (**Figure 4.62**). How should you choose P to minimize the total time for the trip?

SOLUTION   As shown in Figure 4.62, the initial point is chosen arbitrarily, and the terminal point is at the other end of a diameter. The easiest way to describe the transition point P is to refer to the central angle θ. If $\theta = 0$, then the entire trip is done by walking; if $\theta = \pi$, the entire trip is done by swimming. So the interval of interest is $0 \le \theta \le \pi$.

The objective function is the total travel time as it varies with θ. For each leg of the trip (swim and walk), the travel time is the distance traveled divided by the speed. We need a few facts from circular geometry. The length of the swimming leg is the length of the chord of the circle corresponding to the angle θ. For a circle of radius r, this chord length is given by $2r \sin(\theta/2)$ (**Figure 4.63**). So the time for the swimming leg (with $r = 1$ and a speed of 2 mi/hr) is

$$\text{time} = \frac{\text{distance}}{\text{rate}} = \frac{2 \sin(\theta/2)}{2} = \sin\frac{\theta}{2}.$$

The length of the walking leg is the length of the arc of the circle corresponding to the angle $\pi - \theta$. For a circle of radius r, the arc length corresponding to an angle θ is $r\theta$ (Figure 4.63). Therefore, the time for the walking leg (with an angle $\pi - \theta$, $r = 1$, and a speed of 3 mi/hr) is

$$\text{time} = \frac{\text{distance}}{\text{rate}} = \frac{\pi - \theta}{3}.$$

The total travel time for the trip (in hours) is the objective function

$$T(\theta) = \sin\frac{\theta}{2} + \frac{\pi - \theta}{3}, \quad \text{for} \quad 0 \le \theta \le \pi.$$

We now analyze the objective function. The critical points of T satisfy

$$\frac{dT}{d\theta} = \frac{1}{2}\cos\frac{\theta}{2} - \frac{1}{3} = 0 \quad \text{or} \quad \cos\frac{\theta}{2} = \frac{2}{3}.$$

Using a calculator, the only solution in the interval $[0, \pi]$ is $\theta = 2\cos^{-1}\frac{2}{3} \approx 1.68$ rad $\approx 96°$, which is the critical point.

Evaluating the objective function at the critical point and at the endpoints, we find that $T(1.68) \approx 1.23$ hr, $T(0) = \pi/3 \approx 1.05$ hr, and $T(\pi) = 1$ hr. We conclude that the minimum travel time is $T(\pi) = 1$ hr when the entire trip is done swimming. The *maximum* travel time, corresponding to $\theta \approx 96°$, is $T \approx 1.23$ hr.

The objective function is shown in **Figure 4.64**. In general, the maximum and minimum travel times depend on the walking and swimming speeds (Exercise 26).

Related Exercises 26–27 ◀

> ▶ You can check two special cases: If the entire trip is done walking, the travel time is $(\pi \text{ mi})/(3 \text{ mi/hr}) \approx 1.05$ hr. If the entire trip is done swimming, the travel time is $(2 \text{ mi})/(2 \text{ mi/hr}) = 1$ hr.

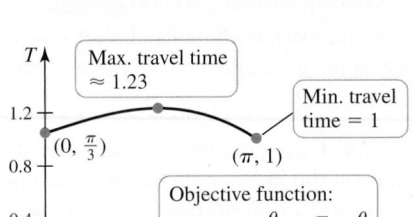

Figure 4.64

EXAMPLE 4 Ladder over the fence An 8-foot-tall fence runs parallel to the side of a house 3 feet away (**Figure 4.65a**). What is the length of the shortest ladder that clears the fence and reaches the house? Assume the vertical wall of the house and the horizontal ground have infinite extent (see Exercise 31 for more realistic assumptions).

SOLUTION Let's first ask why we expect a minimum ladder length. You could put the foot of the ladder far from the fence so that it clears the fence at a shallow angle, but the ladder would be long. Or you could put the foot of the ladder close to the fence so that it clears the fence at a steep angle, but again, the ladder would be long. Somewhere between these extremes is a ladder position that minimizes the ladder length.

The objective function in this problem is the ladder length L. The position of the ladder is specified by x, the distance between the foot of the ladder and the fence (**Figure 4.65b**). The goal is to express L as a function of x, where $x > 0$.

The Pythagorean theorem gives the relationship

$$L^2 = (x + 3)^2 + b^2,$$

where b is the height of the top of the ladder above the ground. Similar triangles give the constraint $8/x = b/(3 + x)$. We now solve the constraint equation for b and substitute to express L^2 in terms of x:

$$L^2 = (x + 3)^2 + \underbrace{\left(\frac{8(x + 3)}{x}\right)^2}_{b} = (x + 3)^2\left(1 + \frac{64}{x^2}\right).$$

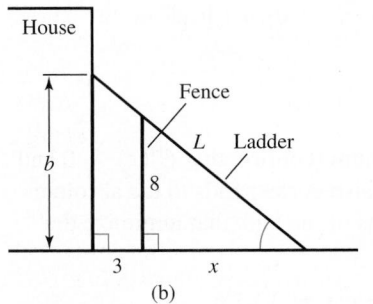

Figure 4.65

At this juncture, we could find the critical points of L by first solving the preceding equation for L and then solving $L' = 0$. However, the solution is simplified considerably if we note that L is a nonnegative function. Therefore, L and L^2 have local extrema at the same points, so we choose to minimize L^2. The derivative of L^2 is

$$\frac{d}{dx}\left((x + 3)^2\left(1 + \frac{64}{x^2}\right)\right) = 2(x + 3)\left(1 + \frac{64}{x^2}\right) + (x + 3)^2\left(-\frac{128}{x^3}\right) \quad \text{Chain Rule and Product Rule}$$

$$= 2(x + 3)\left(1 + \frac{64}{x^2} - (x + 3)\frac{64}{x^3}\right) \quad \text{Factor.}$$

$$= \frac{2(x + 3)(x^3 - 192)}{x^3}. \quad \text{Simplify.}$$

Because $x > 0$, we have $x + 3 \ne 0$; therefore, the condition $\frac{d}{dx}(L^2) = 0$ becomes $x^3 - 192 = 0$, or $x = 4\sqrt[3]{3} \approx 5.77$. By the First Derivative Test, this critical point corresponds to a local minimum. By Theorem 4.9, this solitary local minimum is also the absolute minimum on the interval $(0, \infty)$. Therefore, the minimum ladder length occurs when the foot of the ladder is approximately 5.77 ft from the fence. We find that $L^2(5.77) \approx 224.77$ and the minimum ladder length is $\sqrt{224.77} \approx 15$ ft.

Related Exercises 30–31 ◄

EXAMPLE 5   Water tower   A water storage tank in a small community is built in the shape of a right circular cylinder with a capacity of 32,000 ft³ (about 240,000 gallons). The interior wall and floor of the tank must be cleaned and treated annually. Labor costs for cleaning the wall are twice as high per square foot as the cost to clean the floor. Find the radius and height of the tank that minimize the cleaning cost.

SOLUTION   Figure 4.66 shows a water tank with radius r and height h. The objective function describes the cost to clean the tank, which depends on the interior surface area of the tank. The area of the tank's circular floor is πr^2. Because the circumference of the tank is $2\pi r$, the lateral surface area of the wall is $2\pi rh$. Therefore, the cost to clean the tank is proportional to

$$C = 2 \cdot \underbrace{2\pi rh}_{\text{area of wall}} + \underbrace{\pi r^2}_{\text{area of floor}}.$$

It costs twice as much per square foot to clean the wall, so the area of the wall is multiplied by 2.

To find a constraint equation, we start with the volume of a cylinder $V = \pi r^2 h$ and replace V with 32,000 to obtain $\pi r^2 h = 32,000$, or $h = 32,000/(\pi r^2)$. Substituting this expression for h into the cost equation, we obtain a function of the radius:

$$C(r) = 4\pi r \frac{32,000}{\pi r^2} + \pi r^2 = \frac{128,000}{r} + \pi r^2.$$

To minimize C, we find its critical points on the interval $0 < r < \infty$:

$$C'(r) = -\frac{128,000}{r^2} + 2\pi r = 0 \quad \text{Set the derivative equal to 0.}$$

$$2\pi r = \frac{128,000}{r^2} \quad \text{Rearrange equation.}$$

$$r^3 = \frac{64,000}{\pi} \quad \text{Multiply by } r^2 (r > 0) \text{ and divide by } 2\pi.$$

$$r = \frac{40}{\sqrt[3]{\pi}} \approx 27.3 \text{ ft.} \quad \text{Solve for } r.$$

This solitary critical point corresponds to a local minimum (confirm that $C''(r) > 0$ and use the Second Derivative Test), so by Theorem 4.9, it also corresponds to the absolute minimum. Because $h = 32,000/(\pi r^2)$, the dimensions of the tank that minimize the cleaning cost are

$$r \approx 27.3 \text{ ft} \quad \text{and} \quad h = 32,000/(\pi r^2) \approx 13.7 \text{ ft.}$$

A bit of algebra shows that r is exactly twice as large as h, so the optimal tank is 4 times as wide as it is high.

Related Exercises 21, 35, 42 ◄

Figure 4.66

➤ We don't need to know the actual cost to clean the tank. Whatever it costs (per square foot) to clean the floor, it costs twice as much to clean the wall, which implies that the objective function in Example 5 is a constant multiple of the actual cleaning cost. Because $\frac{d}{dx}(kf(x)) = k\frac{d}{dx}(f(x))$, both $f(x)$ and $kf(x)$ share the same critical points, so either function yields the solution.

SECTION 4.5 EXERCISES

Getting Started

1. Fill in the blanks: The goal of an optimization problem is to find the maximum or minimum value of the _____ function subject to the _____.

2. If the objective function involves more than one independent variable, how are the extra variables eliminated?

3. Suppose the objective function is $Q = x^2 y$ and you know that $x + y = 10$. Write the objective function first in terms of x and then in terms of y.

4. Suppose you wish to minimize a continuous objective function on a closed interval, but you find that it has only a single local maximum. Where should you look for the solution to the problem?

5. Suppose the objective function $P = xy$ is subject to the constraint $10x + y = 100$, where x and y are real numbers.

   a. Eliminate the variable y from the objective function so that P is expressed as a function of one variable x.

   b. Find the absolute maximum value of P subject to the given constraint.

6. Suppose $S = x + 2y$ is an objective function subject to the constraint $xy = 50$, for $x > 0$ and $y > 0$.

   a. Eliminate the variable y from the objective function so that S is expressed as a function of one variable x.

   b. Find the absolute minimum value of S subject to the given constraint.

7. What two nonnegative real numbers with a sum of 23 have the largest possible product?

8. What two nonnegative real numbers a and b whose sum is 23 maximize $a^2 + b^2$? Minimize $a^2 + b^2$?

9. What two positive real numbers whose product is 50 have the smallest possible sum?

10. Find numbers x and y satisfying the equation $3x + y = 12$ such that the product of x and y is as large as possible.

Practice Exercises

11. **Maximum-area rectangles** Of all rectangles with a perimeter of 10, which one has the maximum area? (Give the dimensions.)

12. **Maximum-area rectangles** Of all rectangles with a fixed perimeter of P, which one has the maximum area? (Give the dimensions in terms of P.)

13. **Minimum-perimeter rectangles** Of all rectangles of area 100, which one has the minimum perimeter?

14. **Minimum-perimeter rectangles** Of all rectangles with a fixed area A, which one has the minimum perimeter? (Give the dimensions in terms of A.)

15. **Minimum sum** Find positive numbers x and y satisfying the equation $xy = 12$ such that the sum $2x + y$ is as small as possible.

16. **Pen problems**

   a. A rectangular pen is built with one side against a barn. Two hundred meters of fencing are used for the other three sides of the pen. What dimensions maximize the area of the pen?

   b. A rancher plans to make four identical and adjacent rectangular pens against a barn, each with an area of 100 m² (see figure). What are the dimensions of each pen that minimize the amount of fence that must be used?

Barn			
100	100	100	100

17. **Rectangles beneath a semicircle** A rectangle is constructed with its base on the diameter of a semicircle with radius 5 and its two other vertices on the semicircle. What are the dimensions of the rectangle with maximum area?

18. **Rectangles beneath a parabola** A rectangle is constructed with its base on the x-axis and two of its vertices on the parabola $y = 48 - x^2$. What are the dimensions of the rectangle with the maximum area? What is the area?

19. **Minimum-surface-area box** Of all boxes with a square base and a volume of 8 m³, which one has the minimum surface area? (Give its dimensions.)

20. **Maximum-volume box** Suppose an airline policy states that all baggage must be box-shaped with a sum of length, width, and height not exceeding 108 in. What are the dimensions and volume of a square-based box with the greatest volume under these conditions?

21. **Shipping crates** A square-based, box-shaped shipping crate is designed to have a volume of 16 ft³. The material used to make the base costs twice as much (per square foot) as the material in the sides, and the material used to make the top costs half as much (per square foot) as the material in the sides. What are the dimensions of the crate that minimize the cost of materials?

22. **Closest point on a line** What point on the line $y = 3x + 4$ is closest to the origin?

T 23. **Closest point on a curve** What point on the parabola $y = 1 - x^2$ is closest to the point $(1, 1)$?

24. **Minimum distance** Find the point P on the curve $y = x^2$ that is closest to the point $(18, 0)$. What is the least distance between P and $(18, 0)$? (*Hint:* Use synthetic division.)

25. **Minimum distance** Find the point P on the line $y = 3x$ that is closest to the point $(50, 0)$. What is the least distance between P and $(50, 0)$?

26. **Walking and swimming** A man wishes to get from an initial point on the shore of a circular pond with radius 1 mi to a point on the shore directly opposite (on the other end of the diameter). He plans to swim from the initial point to another point on the shore and then walk along the shore to the terminal point.

   a. If he swims at 2 mi/hr and walks at 4 mi/hr, what are the maximum and minimum times for the trip?

   b. If he swims at 2 mi/hr and walks at 1.5 mi/hr, what are the maximum and minimum times for the trip?

   c. If he swims at 2 mi/hr, what is the minimum walking speed for which it is quickest to walk the entire distance?

27. **Walking and rowing** A boat on the ocean is 4 mi from the nearest point on a straight shoreline; that point is 6 mi from a restaurant on the shore (see figure). A woman plans to row the boat straight to a point on the shore and then walk along the shore to the restaurant.

   a. If she walks at 3 mi/hr and rows at 2 mi/hr, at which point on the shore should she land to minimize the total travel time?

   b. If she walks at 3 mi/hr, what is the minimum speed at which she must row so that the quickest way to the restaurant is to row directly (with no walking)?

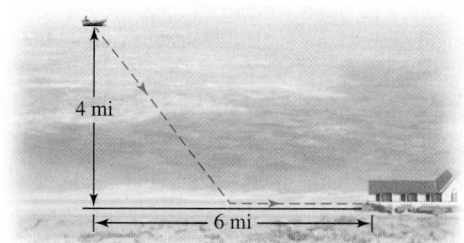

28. Laying cable An island is 3.5 mi from the nearest point on a straight shoreline; that point is 8 mi from a power station (see figure). A utility company plans to lay electrical cable underwater from the island to the shore and then underground along the shore to the power station. Assume it costs $2400/mi to lay underwater cable and $1200/mi to lay underground cable. At what point should the underwater cable meet the shore in order to minimize the cost of the project?

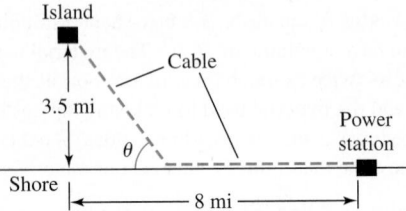

29. Laying cable again Solve the problem in Exercise 28, but this time minimize the cost with respect to the smaller angle θ between the underwater cable and the shore. (You should get the same answer.)

30. Shortest ladder A 10-ft-tall fence runs parallel to the wall of a house at a distance of 4 ft. Find the length of the shortest ladder that extends from the ground to the house without touching the fence. Assume the vertical wall of the house and the horizontal ground have infinite extent.

31. Shortest ladder—more realistic An 8-ft-tall fence runs parallel to the wall of a house at a distance of 5 ft. Find the length of the shortest ladder that extends from the ground to the house without touching the fence. Assume the vertical wall of the house is 20 ft high and the horizontal ground extends 20 ft from the fence.

32. Circle and square A piece of wire of length 60 is cut, and the resulting two pieces are formed to make a circle and a square. Where should the wire be cut to (a) maximize and (b) minimize the combined area of the circle and the square?

33. Maximum-volume cone A cone is constructed by cutting a sector from a circular sheet of metal with radius 20. The cut sheet is then folded up and welded (see figure). Find the radius and height of the cone with maximum volume that can be formed in this way.

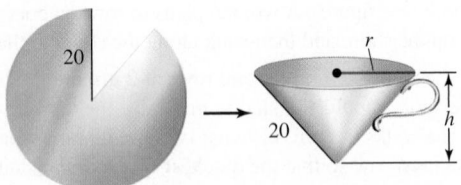

34. Slant height and cones Among all right circular cones with a slant height of 3, what are the dimensions (radius and height) that maximize the volume of the cone? The slant height of a cone is the distance from the outer edge of the base to the vertex.

35. Optimal soda can
   a. Classical problem Find the radius and height of a cylindrical soda can with a volume of 354 cm³ that minimize the surface area.
   b. Real problem Compare your answer in part (a) to a real soda can, which has a volume of 354 cm³, a radius of 3.1 cm, and a height of 12.0 cm, to conclude that real soda cans do not seem

to have an optimal design. Then use the fact that real soda cans have a double thickness in their top and bottom surfaces to find the radius and height that minimize the surface area of a real can (the surface areas of the top and bottom are now twice their values in part (a)). Are these dimensions closer to the dimensions of a real soda can?

36. Covering a marble Imagine a flat-bottomed cylindrical pot with a circular cross section of radius 4. A marble with radius $0 < r < 4$ is placed in the bottom of the pot. What is the radius of the marble that requires the most water to cover it completely?

37. Optimal garden A rectangular flower garden with an area of 30 m² is surrounded by a grass border 1 m wide on two sides and 2 m wide on the other two sides (see figure). What dimensions of the garden minimize the combined area of the garden and borders?

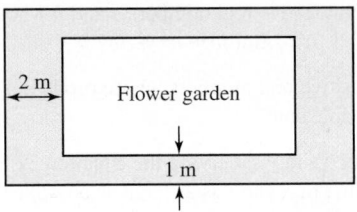

38. Rectangles beneath a line
   a. A rectangle is constructed with one side on the positive x-axis, one side on the positive y-axis, and the vertex opposite the origin on the line $y = 10 - 2x$. What dimensions maximize the area of the rectangle? What is the maximum area?
   b. Is it possible to construct a rectangle with a greater area than that found in part (a) by placing one side of the rectangle on the line $y = 10 - 2x$ and the two vertices not on that line on the positive x- and y-axes? Find the dimensions of the rectangle of maximum area that can be constructed in this way.

39. Designing a box Two squares of length x are cut out of adjacent corners of an 18″ × 18″ piece of cardboard and two rectangles of length 9″ and width x are cut out of the other two corners of the cardboard (see figure). The resulting piece of cardboard is then folded along the dashed lines to form an enclosed box. Find the dimensions and volume of the largest box that can be formed in this way.

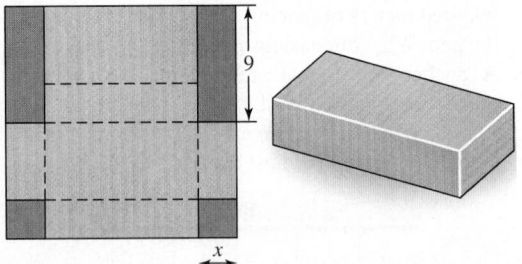

40. Folded boxes
   a. Squares with sides of length x are cut out of each corner of a rectangular piece of cardboard measuring 5 ft by 8 ft. The resulting piece of cardboard is then folded into a box without a lid. Find the volume of the largest box that can be formed in this way.
   b. Squares with sides of length x are cut out of each corner of a square piece of cardboard with sides of length ℓ. Find the volume of the largest open box that can be formed in this way.

41. A window consists of rectangular pane of glass surmounted by a semicircular pane of glass (see figure). If the perimeter of the window is 20 feet, determine the dimensions of the window that maximize the area of the window.

42. **Light transmission** A window consists of a rectangular pane of clear glass surmounted by a semicircular pane of tinted glass. The clear glass transmits twice as much light per unit of surface area as the tinted glass. Of all such windows with a fixed perimeter P, what are the dimensions of the window that transmits the most light?

43. **Kepler's wine barrel** Several mathematical stories originated with the second wedding of the mathematician and astronomer Johannes Kepler. Here is one: While shopping for wine for his wedding, Kepler noticed that the price of a barrel of wine (here assumed to be a cylinder) was determined solely by the length d of a dipstick that was inserted diagonally through a centered hole in the top of the barrel to the edge of the base of the barrel (see figure). Kepler realized that this measurement does not determine the volume of the barrel and that for a fixed value of d, the volume varies with the radius r and height h of the barrel. For a fixed value of d, what is the ratio r/h that maximizes the volume of the barrel?

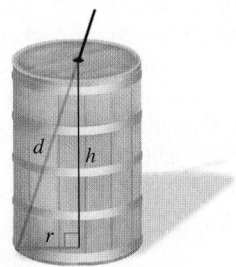

44. **Blood testing** Suppose a blood test for a disease is given to a population of N people, where N is large. At most, N individual blood tests must be done. The following strategy reduces the number of tests. Suppose 100 people are selected from the population and their blood samples are pooled. One test determines whether any of the 100 people test positive. If that test is positive, those 100 people are tested individually, making 101 tests necessary. However, if the pooled sample tests negative, then 100 people have been tested with one test. This procedure is then repeated. Probability theory shows that if the group size is x (for example, $x = 100$, as described here), then the average number of blood tests required to test N people is $N(1 - q^x + 1/x)$, where q is the probability that any one person tests negative. What group size x minimizes the average number of tests in the case that $N = 10,000$ and $q = 0.95$? Assume x is a real number between 1 and 10,000.

45. **Maximum-volume cylinder in a sphere** Find the dimensions of the right circular cylinder of maximum volume that can be placed inside of a sphere of radius R.

46. **Maximizing profit** Suppose you own a tour bus and you book groups of 20 to 70 people for a day tour. The cost per person is $30 minus $0.25 for every ticket sold. If gas and other miscellaneous costs are $200, how many tickets should you sell to maximize your profit? Treat the number of tickets as a nonnegative real number.

47. **Cone in a cone** A right circular cone is inscribed inside a larger right circular cone with a volume of 150 cm³. The axes of the cones coincide, and the vertex of the inner cone touches the center of the base of the outer cone. Find the ratio of the heights of the cones that maximizes the volume of the inner cone.

48. **Cylinder in a sphere** Find the height h, radius r, and volume of a right circular cylinder with maximum volume that is inscribed in a sphere of radius R.

49. **Travel costs** A simple model for travel costs involves the cost of gasoline and the cost of a driver. Specifically, assume gasoline costs $\$p/$gallon and the vehicle gets g miles per gallon. Also assume the driver earns $\$w/$hour.

    a. A plausible function to describe how gas mileage (in mi/gal) varies with speed is $g(v) = v(85 - v)/60$. Evaluate $g(0)$, $g(40)$, and $g(60)$ and explain why these values are reasonable.
    b. At what speed does the gas mileage function have its maximum?
    c. Explain why $C(v) = Lp/g(v) + Lw/v$ gives the cost of the trip in dollars, where L is the length of the trip and v is the constant speed. Show that the dimensions are consistent.
    d. Let $L = 400$ mi, $p = \$4/$gal, and $w = \$20/$hr. At what (constant) speed should the vehicle be driven to minimize the cost of the trip?
    e. Should the optimal speed be increased or decreased (compared with part (d)) if L is increased from 400 mi to 500 mi? Explain.
    f. Should the optimal speed be increased or decreased (compared with part (d)) if p is increased from $\$4/$gal to $\$4.20/$gal? Explain.
    g. Should the optimal speed be increased or decreased (compared with part (d)) if w is decreased from $\$20/$hr to $\$15/$hr? Explain.

50. **Do dogs know calculus?** A mathematician stands on a beach with his dog at point A. He throws a tennis ball so that it hits the water at point B. The dog, wanting to get to the tennis ball as quickly as possible, runs along the straight beach line to point D and then swims from point D to point B to retrieve his ball. Assume C is the point on the edge of the beach closest to the tennis ball (see figure).

    a. Assume the dog runs at speed r and swims at speed s, where $r > s$ and both are measured in meters per second. Also assume the lengths of BC, CD, and AC are x, y, and z, respectively. Find a function $T(y)$ representing the total time it takes for the dog to get to the ball.

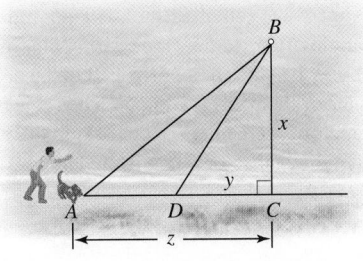

b. Verify that the value of y that minimizes the time it takes to retrieve the ball is $y = \dfrac{x}{\sqrt{r/s + 1}\sqrt{r/s - 1}}$.

c. If the dog runs at 8 m/s and swims at 1 m/s, what ratio y/x produces the fastest retrieving time?

d. A dog named Elvis who runs at 6.4 m/s and swims at 0.910 m/s was found to use an average ratio of y/x of 0.144 to retrieve his ball. Does Elvis appear to know calculus?

(*Source:* T. Pennings, *Do Dogs Know Calculus? The College Mathematics Journal*, 34, 3, May 2003)

51. Viewing angles An auditorium with a flat floor has a large screen on one wall. The lower edge of the screen is 3 ft above eye level and the upper edge of the screen is 10 ft above eye level (see figure). How far from the screen should you stand to maximize your viewing angle?

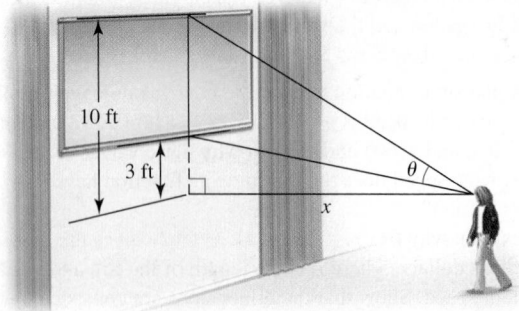

52. Suspension system A load must be suspended 6 m below a high ceiling using cables attached to two supports that are 2 m apart (see figure). How far below the ceiling (x in the figure) should the cables be joined to minimize the total length of cable used?

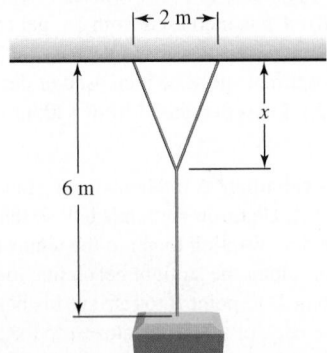

53. Light sources The intensity of a light source at a distance is directly proportional to the strength of the source and inversely proportional to the square of the distance from the source. Two light sources, one twice as strong as the other, are 12 m apart. At what point on the line segment joining the sources is the intensity the weakest?

T 54. Basketball shot A basketball is shot with an initial velocity of v ft/s at an angle of 45° to the floor. The center of the basketball is 8 ft above the floor at a horizontal distance of 18 feet from the center of the basketball hoop when it is released. The height h (in feet) of the center of the basketball after it has traveled a horizontal distance of x feet is modeled by the function

$$h(x) = -\frac{32x^2}{v^2} + x + 8 \text{ (see figure).}$$

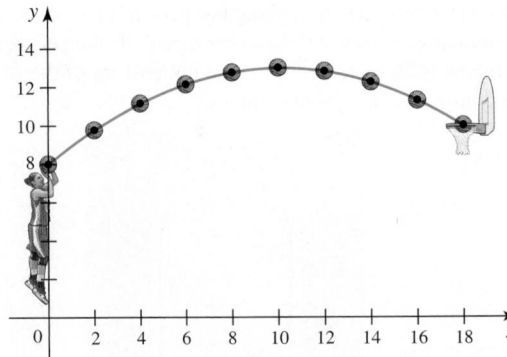

a. Find the initial velocity v if the center of the basketball passes the center of the hoop that is located 10 ft above the floor. Assume the ball does not hit the front of the hoop (otherwise it might not pass though the center of the hoop). The validity of this assumption is explored in the remainder of this exercise.

b. During the flight of the basketball, show that the distance s from the center of the basketball to the front of the hoop is

$$s = \sqrt{(x - 17.25)^2 + \left(-\frac{4x^2}{81} + x - 2\right)^2}.$$

(*Hint:* The diameter of the basketball hoop is 18 inches.)

c. Determine whether the assumption that the basketball does not hit the front of the hoop in part (a) is valid. Use the fact that the diameter of a women's basketball is about 9.23 inches. (*Hint:* The ball will hit the front of the hoop if, during its flight, the distance from the center of the ball to the front of the hoop is less than the radius of the basketball.)

d. A men's basketball has a diameter of about 9.5 inches. Would this larger ball lead to a different conclusion than in part (c)?

55. Fermat's Principle

a. Two poles of heights m and n are separated by a horizontal distance d. A rope is stretched from the top of one pole to the ground and then to the top of the other pole. Show that the configuration that requires the least amount of rope occurs when $\theta_1 = \theta_2$ (see figure).

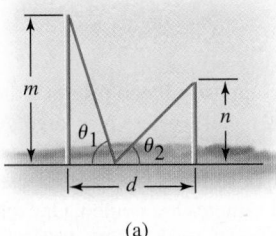

(a)

b. Fermat's Principle states that when light travels between two points in the same medium (at a constant speed), it travels on the path that minimizes the travel time. Show that when light from a source A reflects off a surface and is received at point B, the angle of incidence equals the angle of reflection, or $\theta_1 = \theta_2$ (see figure).

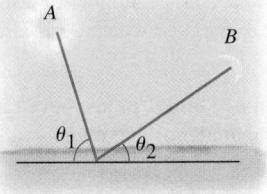

(b)

56. Snell's Law Suppose a light source at A is in a medium in which light travels at a speed v_1 and that point B is in a medium in which light travels at a speed v_2 (see figure). Using Fermat's Principle, which states that light travels along the path that requires the minimum travel time (Exercise 55), show that the path taken between points A and B satisfies $(\sin \theta_1)/v_1 = (\sin \theta_2)/v_2$.

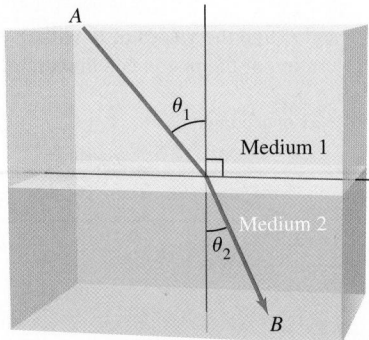

57. Making silos A grain silo consists of a cylindrical concrete tower surmounted by a metal hemispherical dome. The metal in the dome costs 1.5 times as much as the concrete (per unit of surface area). If the volume of the silo is 750 m³, what are the dimensions of the silo (radius and height of the cylindrical tower) that minimize the cost of the materials? Assume the silo has no floor and no flat ceiling under the dome.

58. Blood flow The resistance to blood flow in the circulatory system is one measure of how hard the heart works to pump blood through blood vessels. Lower resistance may correspond to a healthier, higher-efficiency circulatory system. Consider a smaller straight blood vessel of radius r_2 that branches off a larger straight blood vessel of radius r_1 at an angle θ (see figure with given lengths $\ell_1, \ell_2, \ell_3,$ and ℓ_4). Using Poiseuille's Law, it can be shown that the total resistance T to blood flowing along the path from points A to B to D is $T = k\left(\dfrac{\ell_1}{r_1^4} + \dfrac{\ell_2}{r_2^4}\right)$, where $k > 0$ is a constant.

a. Show that $T = k\left(\dfrac{\ell_4 - \ell_3 \cot \theta}{r_1^4} + \dfrac{\ell_3 \csc \theta}{r_2^4}\right)$, assuming that the line segment from A to C is perpendicular to the line segment from C to D and $0 < \theta < \dfrac{\pi}{2}$.

b. Show that the total resistance T is minimized when $\cos \theta = \left(\dfrac{r_2}{r_1}\right)^4$.

c. If the radius of the smaller vessel is 85% of the radius of the larger vessel, then find the value of θ that minimizes T. State your answer in degrees. (*Source: Blood Vessel Branching: Beyond the Standard Calculus Problem, Mathematics Magazine,* 84, 2011)

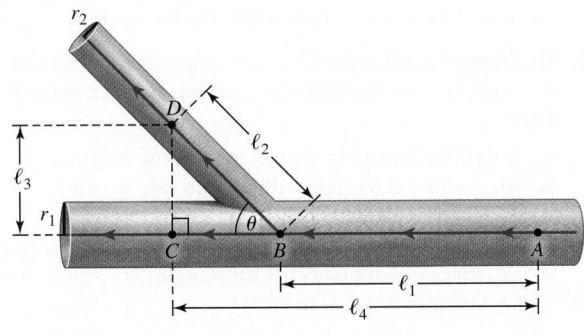

59. Minimizing related functions Complete each of the following parts.

a. What value of x minimizes $f(x) = (x - a_1)^2 + (x - a_2)^2$, for constants a_1 and a_2?

b. What value of x minimizes $f(x) = (x - a_1)^2 + (x - a_2)^2 + (x - a_3)^2$, for constants $a_1, a_2,$ and a_3?

c. Based on the answers to parts (a) and (b), make a conjecture about the values of x that minimize $f(x) = (x - a_1)^2 + (x - a_2)^2 + \cdots + (x - a_n)^2$, for a positive integer n and constants $a_1, a_2, \ldots, a_n$. Use calculus to verify your conjecture.

(*Source:* T. Apostol, *Calculus,* Vol. 1, John Wiley and Sons, 1967)

60. Searchlight problem—narrow beam A searchlight is 100 m from the nearest point on a straight highway (see figure). As it rotates, the searchlight casts a horizontal beam that intersects the highway in a point. If the light revolves at a rate of $\pi/6$ rad/s, find the rate at which the beam sweeps along the highway as a function of θ. For what value of θ is this rate maximized?

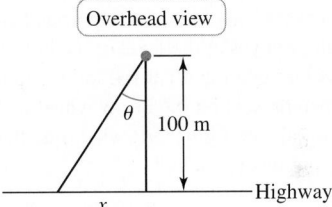

Explorations and Challenges

61. Metal rain gutters A rain gutter is made from sheets of metal 9 in wide. The gutters have a 3-in base and two 3-in sides, folded up at an angle θ (see figure). What angle θ maximizes the cross-sectional area of the gutter?

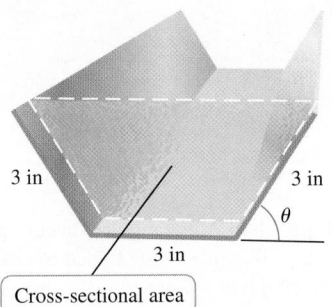

T 62. Gliding mammals Many species of small mammals (such as flying squirrels and marsupial gliders) have the ability to walk and glide. Recent research suggests that these animals choose the most energy-efficient means of travel. According to one empirical model, the energy required for a glider with body mass m to walk a horizontal distance D is $8.46Dm^{2/3}$ (where m is measured in grams, D is measured in meters, and energy is measured in microliters of oxygen consumed in respiration). The energy cost of climbing to a height $D \tan \theta$ and gliding a horizontal distance D at an angle of θ is modeled by $1.36mD \tan \theta$ (where $\theta = 0$ represents horizontal flight and $\theta > 45°$ represents controlled falling). Therefore, the function

$$S(m, \theta) = 8.46m^{2/3} - 1.36m \tan \theta$$

gives the energy difference per horizontal meter traveled between walking and gliding: If $S > 0$ for given values of m and θ, then it is more costly to walk than to glide.

a. For what glide angles is it more efficient for a 200-gram animal to glide rather than walk?

b. Find the threshold function $\theta = g(m)$ that gives the curve along which walking and gliding are equally efficient. Is it an increasing or decreasing function of body mass?

c. To make gliding more efficient than walking, do larger gliders have a larger or smaller selection of glide angles than smaller gliders?

d. Let $\theta = 25°$ (a typical glide angle). Graph S as a function of m, for $0 \le m \le 3000$. For what values of m is gliding more efficient?

e. For $\theta = 25°$, what value of m (call it m^*) maximizes S?

f. Does m^*, as defined in part (e), increase or decrease with increasing θ? That is, as a glider reduces its glide angle, does its optimal size become larger or smaller?

g. Assuming Dumbo is a gliding elephant whose weight is 1 metric ton (10^6 g), what glide angle would Dumbo use to be more efficient at gliding than walking?

(*Source: R. Dial, Energetic savings and the body size distribution of gliding mammals, Evolutionary Ecology Research*, 5, 2003)

T 63. Watching a Ferris wheel An observer stands 20 m from the bottom of a Ferris wheel on a line that is perpendicular to the face of the wheel, with her eyes at the level of the bottom of the wheel. The wheel revolves at a rate of π rad/min, and the observer's line of sight to a specific seat on the Ferris wheel makes an angle θ with the horizontal (see figure). At what time during a full revolution is θ changing most rapidly?

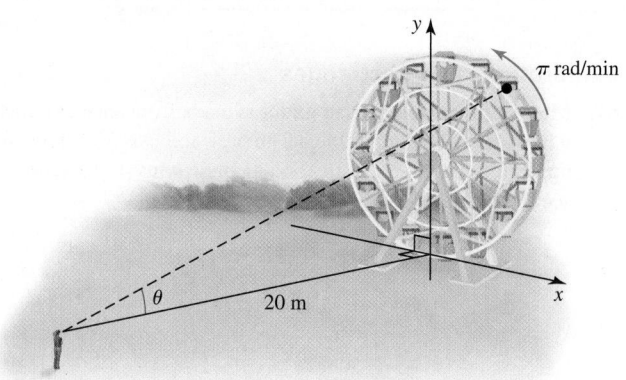

64. Crease-length problem A rectangular sheet of paper of width a and length b, where $0 < a < b$, is folded by taking one corner of the sheet and placing it at a point P on the opposite long side of the sheet (see figure). The fold is then flattened to form a crease across the sheet. Assuming that the fold is made so that there is no flap extending beyond the original sheet, find the point P that produces the crease of minimum length. What is the length of that crease?

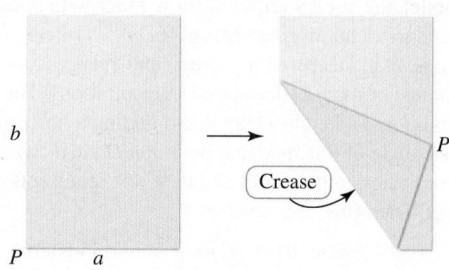

65. Crankshaft A crank of radius r rotates with an angular frequency ω. It is connected to a piston by a connecting rod of length L (see figure). The acceleration of the piston varies with the position of the crank according to the function

$$a(\theta) = \omega^2 r\left(\cos\theta + \frac{r\cos 2\theta}{L}\right).$$

For fixed ω, L, and r, find the values of θ, with $0 \le \theta \le 2\pi$, for which the acceleration of the piston is a maximum and minimum.

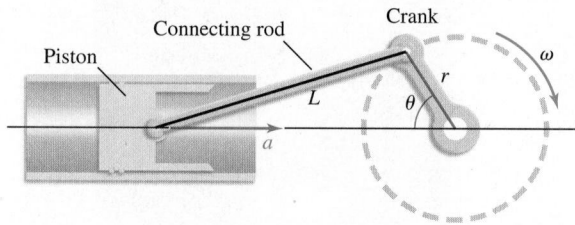

66. Maximum angle Find the value of x that maximizes θ in the figure.

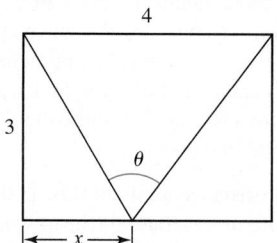

67. Sum of isosceles distances

a. An isosceles triangle has a base of length 4 and two sides of length $2\sqrt{2}$. Let P be a point on the perpendicular bisector of the base. Find the location P that minimizes the sum of the distances between P and the three vertices.

b. Assume in part (a) that the height of the isosceles triangle is $h > 0$ and its base has length 4. Show that the location of P that gives a minimum solution is independent of h for $h \ge \dfrac{2}{\sqrt{3}}$.

68. Cylinder and cones (Putnam Exam 1938) Right circular cones of height h and radius r are attached to each end of a right circular cylinder of height h and radius r, forming a double-pointed object. For a given surface area A, what are the dimensions r and h that maximize the volume of the object?

69. Slowest shortcut Suppose you are standing in a field near a straight section of railroad tracks just as the locomotive of a train passes the point nearest to you, which is $1/4$ mi away. The train, with length $1/3$ mi, is traveling at 20 mi/hr. If you start running in a straight line across the field, how slowly can you run and still catch the train? In which direction should you run?

70. Rectangles in triangles Find the dimensions and area of the rectangle of maximum area that can be inscribed in the following figures.

a. A right triangle with a given hypotenuse length L.

b. An equilateral triangle with a given side length L.

c. A right triangle with a given area A.

d. An arbitrary triangle with a given area A (The result applies to any triangle, but first consider triangles for which all the angles are less than or equal to 90°.)

71. Cylinder in a cone A right circular cylinder is placed inside a cone of radius R and height H so that the base of the cylinder lies on the base of the cone.

    a. Find the dimensions of the cylinder with maximum volume. Specifically, show that the volume of the maximum-volume cylinder is $4/9$ the volume of the cone.

    b. Find the dimensions of the cylinder with maximum lateral surface area (area of the curved surface).

72. Another pen problem A rancher is building a horse pen on the corner of her property using 1000 ft of fencing. Because of the unusual shape of her property, the pen must be built in the shape of a trapezoid (see figure).

    a. Determine the lengths of the sides that maximize the area of the pen.

    b. Suppose there is already a fence along the side of the property opposite the side of length y. Find the lengths of the sides that maximize the area of the pen, using 1000 ft of fencing.

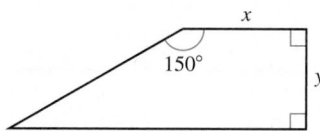

73. Minimum-length roads A house is located at each corner of a square with side lengths of 1 mi. What is the length of the shortest road system with straight roads that connects all of the houses by roads (that is, a road system that allows one to drive from any house to any other house)? (*Hint:* Place two points inside the square at which roads meet.) (*Source:* Paul Halmos, *Problems for Mathematicians Young and Old*, MAA, 1991.)

74. The arbelos An arbelos is the region enclosed by three mutually tangent semicircles; it is the region inside the larger semicircle and outside the two smaller semicircles (see figure).

    a. Given an arbelos in which the diameter of the largest circle is 1, what positions of point B maximize the area of the arbelos?

    b. Show that the area of the arbelos is the area of a circle whose diameter is the distance BD in the figure.

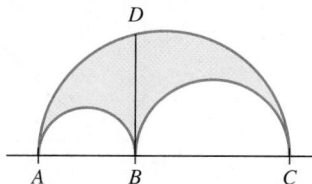

75. Proximity questions Find the point on the graph of $y = \sqrt{x}$ that is nearest the point $(p, 0)$ if (i) $p > \dfrac{1}{2}$; and (ii) $0 < p < \dfrac{1}{2}$. Express the answer in terms of p.

76. Turning a corner with a pole

    a. What is the length of the longest pole that can be carried horizontally around a corner at which a 3-ft corridor and a 4-ft corridor meet at right angles?

    b. What is the length of the longest pole that can be carried horizontally around a corner at which a corridor that is a ft wide and a corridor that is b ft wide meet at right angles?

    c. What is the length of the longest pole that can be carried horizontally around a corner at which a corridor that is $a = 5$ ft wide and a corridor that is $b = 5$ ft wide meet at an angle of $120°$?

    d. What is the length of the longest pole that can be carried around a corner at which a corridor that is a ft wide and a corridor that is b feet wide meet at right angles, assuming there is an 8-foot ceiling and that you may tilt the pole at any angle?

77. Tree notch (Putnam Exam 1938, rephrased) A notch is cut in a cylindrical vertical tree trunk (see figure). The notch penetrates to the axis of the cylinder and is bounded by two half-planes that intersect on a diameter D of the tree. The angle between the two half-planes is θ. Prove that for a given tree and fixed angle θ, the volume of the notch is minimized by taking the bounding planes at equal angles to the horizontal plane that also passes through D.

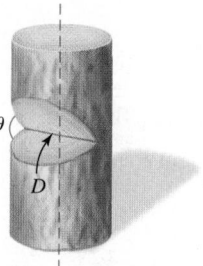

78. Circle in a triangle What are the radius and area of the circle of maximum area that can be inscribed in an isosceles triangle whose two equal sides have length 1?

T 79. A challenging pen problem A farmer uses 200 meters of fencing to build two triangular pens against a barn (see figure); the pens are constructed with three sides and a diagonal dividing fence. What dimensions maximize the area of the pen?

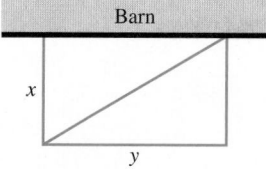

80. Folded boxes Squares with sides of length x are cut out of each corner of a rectangular piece of cardboard with sides of length ℓ and L. Holding ℓ fixed, find the size of the corner squares x that maximizes the volume of the box as $L \to \infty$. (*Source: Mathematics Teacher*, Nov 2002)

QUICK CHECK ANSWERS

2. (i) $A = 400x - 2x^2$ (ii) $A = 400x - 3x^2$
3. $V = 108w^2 - 2w^3$ ◄

4.6 Linear Approximation and Differentials

Imagine plotting a smooth curve with a graphing utility. Now pick a point P on the curve, draw the line tangent to the curve at P, and zoom in on it several times. As you successively enlarge the curve near P, it looks more and more like the tangent line (**Figure 4.67a**). This fundamental observation—that smooth curves appear straighter on smaller scales—is called *local linearity*; it is the basis of many important mathematical ideas, one of which is *linear approximation*.

Now consider a curve with a corner or cusp at a point Q (**Figure 4.67b**). No amount of magnification "straightens out" the curve or removes the corner at Q. The different behavior at P and Q is related to the idea of differentiability: The function in Figure 4.67a is differentiable at P, whereas the function in Figure 4.67b is not differentiable at Q. One of the requirements for the techniques presented in this section is that the function be differentiable at the point in question.

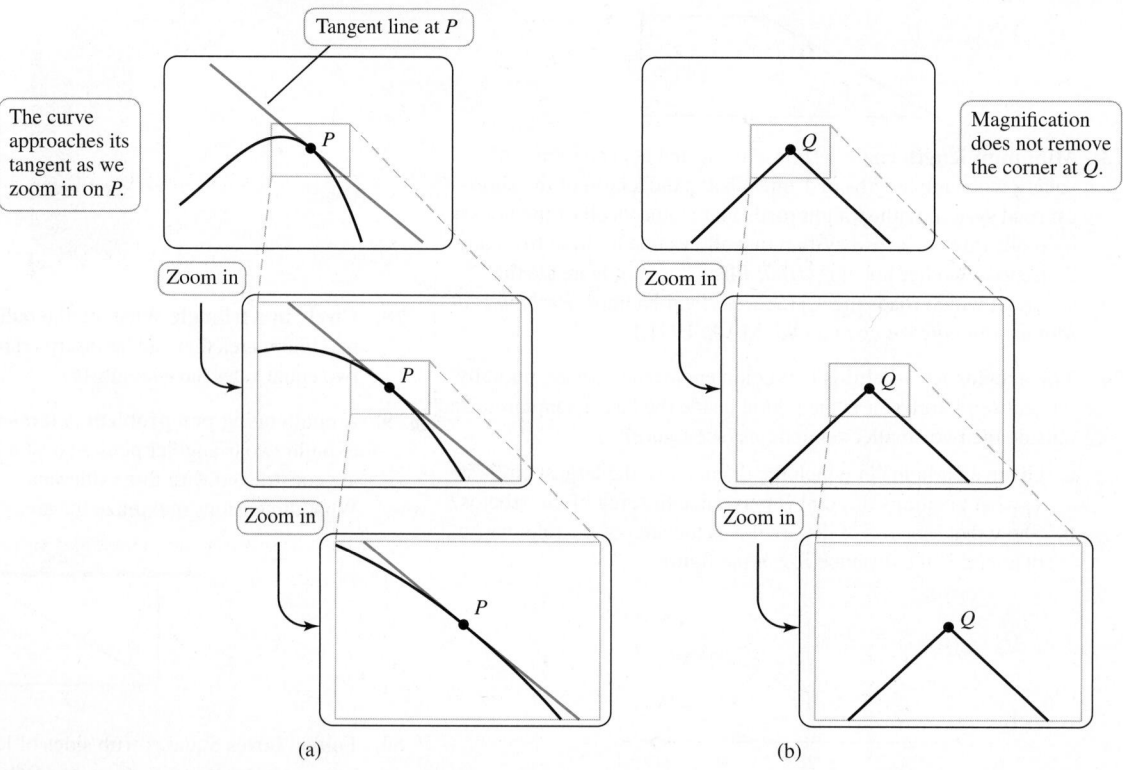

(a)   (b)

Figure 4.67

Linear Approximation

Figure 4.67a suggests that when we zoom in on the graph of a smooth function at a point P, the curve approaches its tangent line at P. This fact is the key to understanding linear approximation. The idea is to use the line tangent to the curve at P to approximate the value of the function at points near P. Here's how it works.

Assume f is differentiable on an interval containing the point a. The slope of the line tangent to the curve at the point $(a, f(a))$ is $f'(a)$. Therefore, an equation of the tangent line is

$$y - f(a) = f'(a)(x - a) \quad \text{or} \quad y = \underbrace{f(a) + f'(a)(x - a)}_{L(x)}.$$

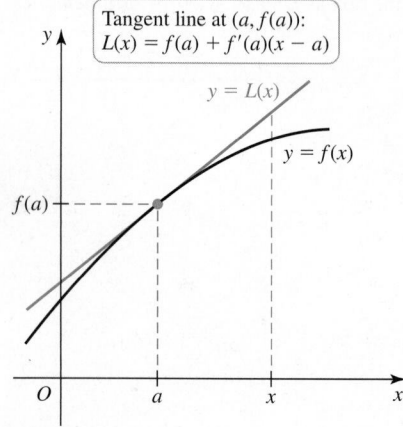

Figure 4.68

This tangent line represents a new function L that we call the *linear approximation* to f at the point a (**Figure 4.68**). If f and f' are easy to evaluate at a, then the value of f at points near a is easily approximated using the linear approximation L. That is,

$$f(x) \approx L(x) = f(a) + f'(a)(x - a).$$

This approximation improves as x approaches a.

QUICK CHECK 1 Sketch the graph of a function f that is concave up on an interval containing the point a. Sketch the linear approximation to f at a. Is the graph of the linear approximation above or below the graph of f? ◀

> DEFINITION **Linear Approximation to f at a**
>
> Suppose f is differentiable on an interval I containing the point a. The **linear approximation** to f at a is the linear function
> $$L(x) = f(a) + f'(a)(x - a), \quad \text{for } x \text{ in } I.$$

EXAMPLE 1   Useful driving math Suppose you are driving along a highway at a constant speed and you record the number of seconds it takes to travel between two consecutive mile markers. If it takes 60 seconds to travel one mile, then your average speed is 1 mi/60 s or 60 mi/hr. Now suppose you travel one mile in $60 + x$ seconds; for example, if it takes 62 seconds, then $x = 2$, and if it takes 57 seconds, then $x = -3$. In this case, your average speed over one mile is 1 mi/$(60 + x)$ s. Because there are 3600 s in 1 hr, the function

▶ In Example 1, notice that when x is positive, you are driving slower than 60 mi/hr; when x is negative, you are driving faster than 60 mi/hr.

$$s(x) = \frac{3600}{60 + x} = 3600(60 + x)^{-1}$$

gives your average speed in mi/hr if you travel one mile in x seconds more or less than 60 seconds. For example, if you travel one mile in 62 seconds, then $x = 2$ and your average speed is $s(2) \approx 58.06$ mi/hr. If you travel one mile in 57 seconds, then $x = -3$ and your average speed is $s(-3) \approx 63.16$ mi/hr. Because you don't want to use a calculator while driving, you need an easy approximation to this function. Use linear approximation to derive such a formula.

SOLUTION The idea is to find the linear approximation to s at the point 0. We first use the Chain Rule to compute

$$s'(x) = -3600(60 + x)^{-2},$$

and then note that $s(0) = 60$ and $s'(0) = -3600 \cdot 60^{-2} = -1$. Using the linear approximation formula, we find that

$$s(x) \approx L(x) = s(0) + s'(0)(x - 0) = 60 - x.$$

QUICK CHECK 2 In Example 1, suppose you travel one mile in 75 seconds. What is the average speed given by the linear approximation formula? What is the exact average speed? Explain the discrepancy between the two values. ◀

For example, if you travel one mile in 62 seconds, then $x = 2$ and your average speed is approximately $L(2) = 58$ mi/hr, which is very close to the exact value given previously. If you travel one mile in 57 seconds, then $x = -3$ and your average speed is approximately $L(-3) = 63$ mi/hr, which again is close to the exact value.

Related Exercises 13–14 ◀

EXAMPLE 2   Linear approximations and errors

a. Find the linear approximation to $f(x) = \sqrt{x}$ at $x = 1$ and use it to approximate $\sqrt{1.1}$.

b. Use linear approximation to estimate the value of $\sqrt{0.1}$.

SOLUTION

a. We construct the linear approximation

$$L(x) = f(a) + f'(a)(x - a),$$

where $f(x) = \sqrt{x}$, $f'(x) = 1/(2\sqrt{x})$, and $a = 1$. Noting that $f(a) = f(1) = 1$ and $f'(a) = f'(1) = \frac{1}{2}$, we have

$$L(x) = 1 + \frac{1}{2}(x - 1) = \frac{1}{2}(x + 1),$$

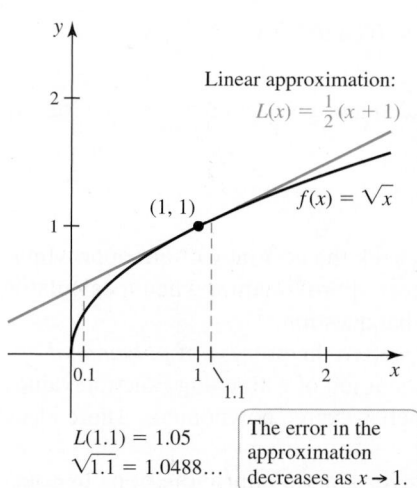

Linear approximation:
$L(x) = \frac{1}{2}(x + 1)$

$(1, 1)$

$f(x) = \sqrt{x}$

$L(1.1) = 1.05$
$\sqrt{1.1} = 1.0488\ldots$

The error in the approximation decreases as $x \to 1$.

Figure 4.69

which describes the line tangent to the curve at the point $(1, 1)$ (Figure 4.69). Because $x = 1.1$ is near $x = 1$, we approximate $\sqrt{1.1}$ by $L(1.1)$:

$$\sqrt{1.1} \approx L(1.1) = \frac{1}{2}(1.1 + 1) = 1.05.$$

Table 4.4

x	$L(x)$	Exact $\sqrt{x}$	Error
1.2	1.1	1.0954...	4.6×10^{-3}
1.1	1.05	1.0488...	1.2×10^{-3}
1.01	1.005	1.0049...	1.2×10^{-5}
1.001	1.0005	1.0005...	1.2×10^{-7}

The exact value is $f(1.1) = \sqrt{1.1} = 1.0488\ldots$; therefore, the linear approximation has an error of about 0.0012. Furthermore, our approximation is an *overestimate* because the tangent line lies above the graph of f. In Table 4.4, we see several approximations to $\sqrt{x}$ for x near 1 and the associated errors $|L(x) - \sqrt{x}|$. Clearly, the errors decrease as x approaches 1.

b. If the linear approximation $L(x) = \frac{1}{2}(x + 1)$ obtained in part (a) is used to approximate $\sqrt{0.1}$, we have

$$\sqrt{0.1} \approx L(0.1) = \frac{1}{2}(0.1 + 1) = 0.55.$$

A calculator gives $\sqrt{0.1} = 0.3162\ldots$, which shows that the approximation is well off the mark. The error arises because the tangent line through $(1, 1)$ is not close to the curve at $x = 0.1$ (Figure 4.69). For this reason, we seek a different value of a, with the requirement that it is near $x = 0.1$, and both $f(a)$ and $f'(a)$ are easily computed. It is tempting to try $a = 0$, but $f'(0)$ is undefined. One choice that works well is $a = \frac{9}{100} = 0.09$. Using the linear approximation $L(x) = f(a) + f'(a)(x - a)$, we have

> ► We choose $a = \frac{9}{100}$ because it is close to 0.1 and its square root is easy to evaluate.

$$\sqrt{0.1} \approx L(0.1) = \underbrace{\sqrt{\frac{9}{100}}}_{f(a)} + \underbrace{\frac{1}{2\sqrt{9/100}}}_{f'(a)} \underbrace{\left(\frac{1}{10} - \frac{9}{100}\right)}_{(x-a)}$$

$$= \frac{3}{10} + \frac{10}{6}\left(\frac{1}{100}\right)$$

$$= \frac{19}{60} \approx 0.3167.$$

This approximation agrees with the exact value to three decimal places.

Related Exercises 31–32 ◄

QUICK CHECK 3 Suppose you want to use linear approximation to estimate $\sqrt{0.18}$. What is a good choice for a? ◄

EXAMPLE 3 **Linear approximation for the sine function** Find the linear approximation to $f(x) = \sin x$ at $x = 0$ and use it to approximate $\sin 2.5°$.

SOLUTION We first construct a linear approximation $L(x) = f(a) + f'(a)(x - a)$, where $f(x) = \sin x$ and $a = 0$. Noting that $f(0) = 0$ and $f'(0) = \cos(0) = 1$, we have

$$L(x) = 0 + 1(x - 0) = x.$$

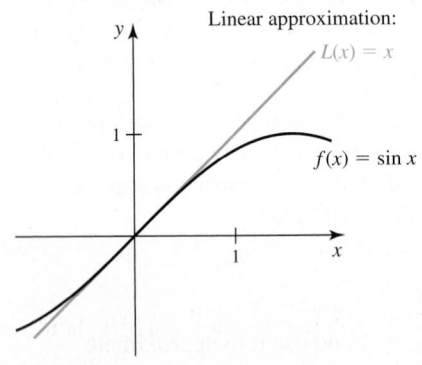

Linear approximation:
$L(x) = x$

$f(x) = \sin x$

Figure 4.70

Again, the linear approximation is the line tangent to the curve at the point $(0, 0)$ (Figure 4.70). Before using $L(x)$ to approximate $\sin 2.5°$, we convert to radian measure (the derivative formulas for trigonometric functions require angles in radians):

$$2.5° = 2.5°\left(\frac{\pi}{180°}\right) = \frac{\pi}{72} \approx 0.04363 \text{ rad}.$$

Therefore, $\sin 2.5° \approx L(0.04363) = 0.04363$. A calculator gives $\sin 2.5° \approx 0.04362$, so the approximation is accurate to four decimal places.

Related Exercises 38, 46 ◄

In Examples 2 and 3, we used a calculator to check the accuracy of our approximations. This raises a question: Why bother with linear approximation when a calculator does a better job? There are some good answers to that question.

Linear approximation is actually just the first step in the process of *polynomial approximation*. While linear approximation does a decent job of estimating function values when x is near a, we can generally do better with higher-degree polynomials. These ideas are explored further in Chapter 11.

Linear approximation also allows us to discover simple approximations to complicated functions. In Example 3, we found the *small-angle approximation to the sine function*: $\sin x \approx x$ for x near 0.

QUICK CHECK 4 Explain why the linear approximation to $f(x) = \cos x$ at $x = 0$ is $L(x) = 1$. ◄

Linear Approximation and Concavity

Additional insight into linear approximation is gained by bringing concavity into the picture. **Figure 4.71a** shows the graph of a function f and its linear approximation (tangent line) at the point $(a, f(a))$. In this particular case, f is concave up on an interval containing a, and the graph of L lies below the graph of f near a. As a result, the linear approximation evaluated at a point near a is less than the exact value of f at that point. In other words, the linear approximation *underestimates* values of f near a.

The contrasting case is shown in **Figure 4.71b**, where we see graphs of f and L when f is concave down on an interval containing a. Now the graph of L lies above the graph of f, which means the linear approximation *overestimates* values of f near a.

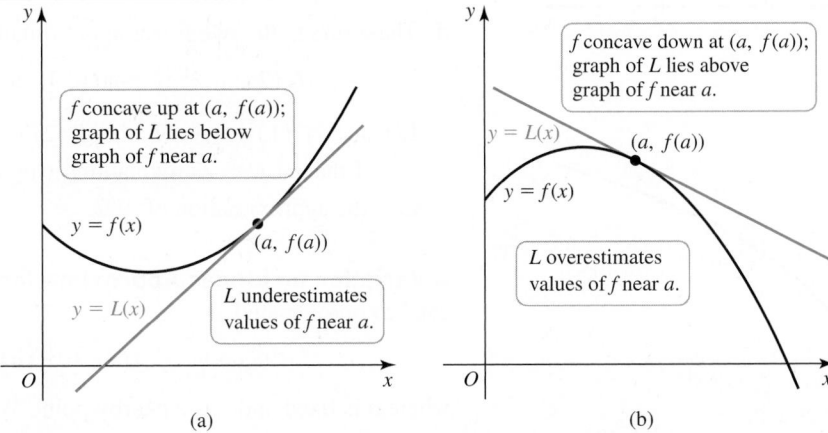

(a)   (b)

Figure 4.71

We can make another observation related to the degree of concavity (also called *curvature*). A large value of $|f''(a)|$ (large curvature) means that near $(a, f(a))$, the slope of the curve changes rapidly and the graph of f separates quickly from the tangent line. A small value of $|f''(a)|$ (small curvature) means the slope of the curve changes slowly and the curve is relatively flat near $(a, f(a))$; therefore, the curve remains close to the tangent line. As a result, absolute errors in linear approximation are larger when $|f''(a)|$ is large.

EXAMPLE 4   Linear approximation and concavity

a. Find the linear approximation to $f(x) = x^{1/3}$ at $x = 1$ and $x = 27$.

b. Use the linear approximations of part (a) to approximate $\sqrt[3]{2}$ and $\sqrt[3]{26}$.

c. Are the approximations in part (b) overestimates or underestimates?

d. Compute the error in each approximation of part (b). Which error is greater? Explain.

SOLUTION

a. Note that

$$f(1) = 1, \quad f(27) = 3, \quad f'(x) = \frac{1}{3x^{2/3}}, \quad f'(1) = \frac{1}{3}, \quad \text{and} \quad f'(27) = \frac{1}{27}.$$

Therefore, the linear approximation at $x = 1$ is

$$L_1(x) = 1 + \frac{1}{3}(x - 1) = \frac{1}{3}x + \frac{2}{3},$$

and the linear approximation at $x = 27$ is

$$L_2(x) = 3 + \frac{1}{27}(x - 27) = \frac{1}{27}x + 2.$$

b. Using the results of part (a), we find that

$$\sqrt[3]{2} \approx L_1(2) = \frac{1}{3} \cdot 2 + \frac{2}{3} = \frac{4}{3} \approx 1.333$$

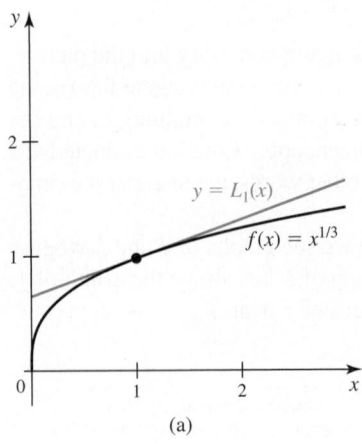

(a)

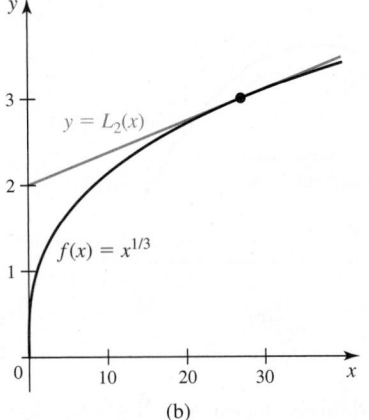

(b)

Figure 4.72

and

$$\sqrt[3]{26} \approx L_2(26) = \frac{1}{27} \cdot 26 + 2 \approx 2.963.$$

c. Figure 4.72 shows the graphs of f and the linear approximations L_1 and L_2 at $x = 1$ and $x = 27$, respectively (note the different scales on the x-axes). We see that f is concave down at both points, which is confirmed by the fact that

$$f''(x) = -\frac{2}{9}x^{-5/3} < 0, \quad \text{for} \quad x > 0.$$

As a result, the linear approximations lie above the graph of f and both approximations are overestimates.

d. The errors in the two linear approximations are

$$|L_1(2) - 2^{1/3}| \approx 0.073 \quad \text{and} \quad |L_2(26) - 26^{1/3}| \approx 0.00047.$$

Because $|f''(1)| \approx 0.22$ and $|f''(27)| \approx 0.00091$, the curvature of f is greater at $x = 1$ than at $x = 27$, explaining why the approximation of $\sqrt[3]{26}$ is more accurate than the approximation of $\sqrt[3]{2}$.

Related Exercises 47–50 ◄

A Variation on Linear Approximation   Linear approximation says that a function f can be approximated as

$$f(x) \approx f(a) + f'(a)(x - a),$$

where a is fixed and x is a nearby point. We first rewrite this expression as

$$\underbrace{f(x) - f(a)}_{\Delta y} \approx f'(a)\underbrace{(x - a)}_{\Delta x}.$$

It is customary to use the notation Δ (capital Greek delta) to denote a change. The factor $x - a$ is the change in the x-coordinate between a and a nearby point x. Similarly, $f(x) - f(a)$ is the corresponding change in the y-coordinate (Figure 4.73). So we write this approximation as

$$\Delta y \approx f'(a)\,\Delta x.$$

In other words, a change in y (the function value) can be approximated by the corresponding change in x magnified or diminished by a factor of $f'(a)$. This interpretation states the familiar fact that $f'(a)$ is the rate of change of y with respect to x.

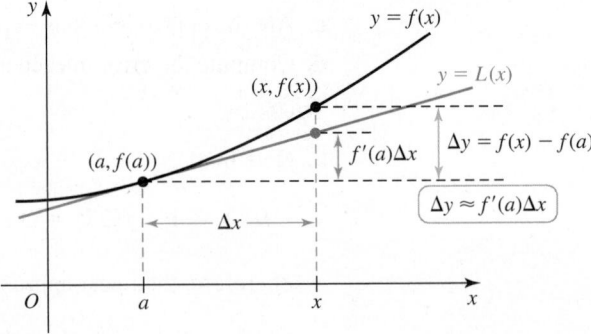

Figure 4.73

Relationship Between Δx and Δy

Suppose f is differentiable on an interval I containing the point a. The change in the value of f between two points a and $a + \Delta x$ is approximately

$$\Delta y \approx f'(a)\,\Delta x,$$

where $a + \Delta x$ is in I.

EXAMPLE 5   Estimating changes with linear approximations

a. Approximate the change in $y = f(x) = x^9 - 2x + 1$ when x changes from 1.00 to 1.05.

b. Approximate the change in the surface area of a spherical hot-air balloon when the radius decreases from 4 m to 3.9 m.

SOLUTION

a. The change in y is $\Delta y \approx f'(a)\,\Delta x$, where $a = 1$, $\Delta x = 0.05$, and $f'(x) = 9x^8 - 2$. Substituting these values, we find that

$$\Delta y \approx f'(a)\,\Delta x = f'(1) \cdot 0.05 = 7 \cdot 0.05 = 0.35.$$

If x increases from 1.00 to 1.05, then y increases by approximately 0.35.

> Notice that the units in these calculations are consistent. If r has units of meters (m), S' has units of $m^2/m = m$, so ΔS has units of m^2, as it should.

b. The surface area of a sphere is $S = 4\pi r^2$, so the change in the surface area when the radius changes by Δr is $\Delta S \approx S'(a)\,\Delta r$. Substituting $S'(r) = 8\pi r$, $a = 4$, and $\Delta r = -0.1$, the approximate change in the surface area is

$$\Delta S \approx S'(a)\,\Delta r = S'(4) \cdot (-0.1) = 32\pi \cdot (-0.1) \approx -10.05.$$

The change in surface area is approximately $-10.05\ m^2$; it is negative, reflecting a decrease.

Related Exercises 55, 59 ◄

QUICK CHECK 5 Given that the volume of a sphere is $V = 4\pi r^3/3$, find an expression for the approximate change in the volume when the radius changes from a to $a + \Delta r$. ◄

SUMMARY Uses of Linear Approximation

• To approximate f near $x = a$, use

$$f(x) \approx L(x) = f(a) + f'(a)(x - a).$$

• To approximate the change Δy in the dependent variable when the independent variable x changes from a to $a + \Delta x$, use

$$\Delta y \approx f'(a)\,\Delta x.$$

Differentials

We now introduce an important concept that allows us to distinguish two related quantities:

• the change in the function $y = f(x)$ as x changes from a to $a + \Delta x$ (which we call Δy, as before), and

• the change in the linear approximation $y = L(x)$ as x changes from a to $a + \Delta x$ (which we call the *differential dy*).

Consider a function $y = f(x)$ differentiable on an interval containing a. If the x-coordinate changes from a to $a + \Delta x$, then the corresponding change in the function is *exactly*

$$\Delta y = f(a + \Delta x) - f(a).$$

Using the linear approximation $L(x) = f(a) + f'(a)(x - a)$, the change in L as x changes from a to $a + \Delta x$ is

$$\Delta L = L(a + \Delta x) - L(a)$$
$$= \underbrace{(f(a) + f'(a)(a + \Delta x - a))}_{L(a + \Delta x)} - \underbrace{(f(a) + f'(a)(a - a))}_{L(a)}$$
$$= f'(a)\,\Delta x.$$

To distinguish Δy and ΔL, we define two new variables called *differentials*. The differential dx is simply Δx; the differential dy is the change in the linear approximation, which is $\Delta L = f'(a)\,\Delta x$. Using this notation,

$$\Delta L = \underbrace{dy}_{\text{same as }\Delta L} = f'(a)\,\Delta x = f'(a)\,\underbrace{dx}_{\text{same as }\Delta x}.$$

Therefore, at the point a, we have $dy = f'(a)\, dx$. More generally, we replace the fixed point a with a variable point x and write

$$dy = f'(x)\, dx.$$

DEFINITION Differentials

Let f be differentiable on an interval containing x. A small change in x is denoted by the **differential** dx. The corresponding change in f is approximated by the **differential** $dy = f'(x)\, dx$; that is,

$$\Delta y = f(x + dx) - f(x) \approx dy = f'(x)\, dx.$$

▶ As one of the two co-inventors of calculus, Gottfried Leibniz relied on the idea of differentials in his development of calculus. Leibniz's notation for differentials is essentially the same as the notation we use today. An Irish philosopher of the day, Bishop Berkeley, called differentials "the ghost of departed quantities."

Figure 4.74 shows that if $\Delta x = dx$ is small, then the change in f, which is Δy, is well approximated by the change in the linear approximation, which is dy. Furthermore, the approximation $\Delta y \approx dy$ improves as dx approaches 0. The notation for differentials is consistent with the notation for the derivative: If we divide both sides of $dy = f'(x)\, dx$ by dx, we have (symbolically)

$$\frac{dy}{dx} = \frac{f'(x)\, dx}{dx} = f'(x).$$

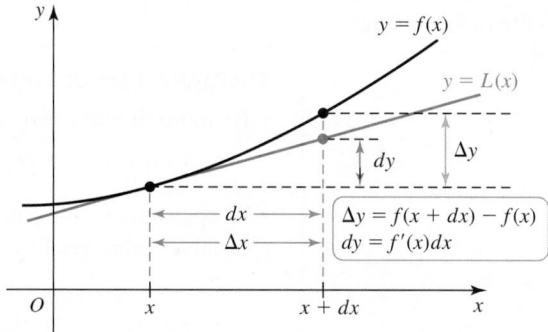

Figure 4.74

EXAMPLE 6   Differentials as change Use the notation of differentials to write the approximate change in $f(x) = 3 \cos^2 x$ given a small change dx.

▶ Recall that $\sin 2x = 2 \sin x \cos x$.

SOLUTION With $f(x) = 3 \cos^2 x$, we have $f'(x) = -6 \cos x \sin x = -3 \sin 2x$. Therefore,

$$dy = f'(x)\, dx = -3 \sin 2x\, dx.$$

The interpretation is that a small change dx in the independent variable x produces an approximate change in the dependent variable of $dy = -3 \sin 2x\, dx$. For example, if x increases from $x = \pi/4$ to $x = \pi/4 + 0.1$, then $dx = 0.1$ and

$$dy = -3 \sin (\pi/2)(0.1) = -0.3.$$

The approximate change in the function is -0.3, which means a decrease of approximately 0.3.

Related Exercises 62, 69 ◀

SECTION 4.6 EXERCISES

Getting Started

1. Sketch the graph of a smooth function f and label a point $P(a, f(a))$ on the curve. Draw the line that represents the linear approximation to f at P.

2. Suppose you find the linear approximation to a differentiable function at a local maximum of that function. Describe the graph of the linear approximation.

3. How is linear approximation used to approximate the value of a function f near a point at which f and f' are easily evaluated?

4. How can linear approximation be used to approximate the change in $y = f(x)$ given a change in x?

5. Suppose f is differentiable on $(-\infty, \infty)$, $f(1) = 2$, and $f'(1) = 3$. Find the linear approximation to f at $x = 1$ and use it to approximate $f(1.1)$.

6. Suppose f is differentiable on $(-\infty, \infty)$ and the equation of the line tangent to the graph of f at $x = 2$ is $y = 5x - 3$. Use the linear approximation to f at $x = 2$ to approximate $f(2.01)$.

7. Use linear approximation to estimate $f(3.85)$ given that $f(4) = 3$ and $f'(4) = 2$.

8. Use linear approximation to estimate $f(5.1)$ given that $f(5) = 10$ and $f'(5) = -2$.

9. Given a function f that is differentiable on its domain, write and explain the relationship between the differentials dx and dy.

10. Does the differential dy represent the change in f or the change in the linear approximation to f? Explain.

11. Suppose f is differentiable on $(-\infty, \infty)$ and $f(5.01) - f(5) = 0.25$. Use linear approximation to estimate the value of $f'(5)$.

12. Suppose f is differentiable on $(-\infty, \infty)$, $f(5.99) = 7$, and $f(6) = 7.002$. Use linear approximation to estimate the value of $f'(6)$.

Practice Exercises

13–14. Estimating speed *Use the linear approximation given in Example 1 to answer the following questions.*

13. If you travel one mile in 59 seconds, what is your approximate average speed? What is your exact speed?

14. If you travel one mile in 63 seconds, what is your approximate average speed? What is your exact speed?

15–18. Estimating time *Suppose you want to travel D miles at a constant speed of $(60 + x)$ mi/hr, where x could be positive or negative. The time in minutes required to travel D miles is $T(x) = 60 D(60 + x)^{-1}$.*

15. Show that the linear approximation to T at the point $x = 0$ is
$$T(x) \approx L(x) = D\left(1 - \frac{x}{60}\right).$$

16. Use the result of Exercise 15 to approximate the amount of time it takes to drive 45 miles at 62 mi/hr. What is the exact time required?

17. Use the result of Exercise 15 to approximate the amount of time it takes to drive 80 miles at 57 mi/hr. What is the exact time required?

18. Use the result of Exercise 15 to approximate the amount of time it takes to drive 93 miles at 63 mi/hr. What is the exact time required?

19–24. Linear approximation *Find the linear approximation to the following functions at the given point a.*

19. $f(x) = 4x^2 + x$; $a = 1$ **20.** $f(x) = x^3 - 5x + 3$; $a = 2$

21. $g(t) = \sqrt{2t + 9}$; $a = -4$ **22.** $h(w) = \sqrt{5w - 1}$; $a = 1$

23. $f(x) = e^{3x-6}$; $a = 2$ **24.** $f(x) = 9(4x + 11)^{2/3}$; $a = 4$

▣ 25–36. Linear approximation

a. *Write the equation of the line that represents the linear approximation to the following functions at the given point a.*

b. *Use the linear approximation to estimate the given quantity.*

c. *Compute the percent error in your approximation, $100 |\text{approximation} - \text{exact}| / |\text{exact}|$, where the exact value is given by a calculator.*

25. $f(x) = 12 - x^2$; $a = 2$; $f(2.1)$

26. $f(x) = \sin x$; $a = \pi/4$; $f(0.75)$

27. $f(x) = \ln(1 + x)$; $a = 0$; $f(0.9)$

28. $f(x) = x/(x + 1)$; $a = 1$; $f(1.1)$

29. $f(x) = \cos x$; $a = 0$; $f(-0.01)$

30. $f(x) = e^x$; $a = 0$; $f(0.05)$

31. $f(x) = (8 + x)^{-1/3}$; $a = 0$; $f(-0.1)$

32. $f(x) = \sqrt[4]{x}$; $a = 81$; $f(85)$

33. $f(x) = 1/(x + 1)$; $a = 0$; $1/1.1$

34. $f(x) = \cos x$; $a = \pi/4$; $\cos 0.8$

35. $f(x) = e^{-x}$; $a = 0$; $e^{-0.03}$

36. $f(x) = \tan x$; $a = 0$; $\tan 3°$

37–46. Estimations with linear approximation *Use linear approximations to estimate the following quantities. Choose a value of a to produce a small error.*

37. $1/203$ **38.** $\tan(-2°)$ **39.** $\sqrt{146}$ **40.** $\sqrt[3]{65}$

41. $\ln 1.05$ **42.** $\sqrt{5/29}$ **43.** $e^{0.06}$ **44.** $1/\sqrt{119}$

45. $1/\sqrt[3]{510}$ **46.** $\cos 31°$

47–50. Linear approximation and concavity *Carry out the following steps for the given functions f and points a.*

a. *Find the linear approximation L to the function f at the point a.*

b. *Graph f and L on the same set of axes.*

c. *Based on the graphs in part (b), state whether linear approximations to f near $x = a$ are underestimates or overestimates.*

d. *Compute $f''(a)$ to confirm your conclusions in part (c).*

47. $f(x) = \dfrac{2}{x}$; $a = 1$ **48.** $f(x) = 5 - x^2$; $a = 2$

49. $f(x) = e^{-x}$; $a = \ln 2$ **50.** $f(x) = \sqrt{2}\cos x$; $a = \dfrac{\pi}{4}$

▣ 51. Error in driving speed Consider again the average speed $s(x)$ and its linear approximation $L(x)$ discussed in Example 1. The error in using $L(x)$ to approximate $s(x)$ is given by $E(x) = |L(x) - s(x)|$. Use a graphing utility to determine the (approximate) values of x for which $E(x) \leq 1$. What does your answer say about the accuracy of the average speeds estimated by $L(x)$ over this interval?

52. Ideal Gas Law The pressure P, temperature T, and volume V of an ideal gas are related by $PV = nRT$, where n is the number of moles of the gas and R is the universal gas constant. For the purposes of this exercise, let $nR = 1$; therefore, $P = T/V$.

a. Suppose the volume is held constant and the temperature increases by $\Delta T = 0.05$. What is the approximate change in the pressure? Does the pressure increase or decrease?

b. Suppose the temperature is held constant and the volume increases by $\Delta V = 0.1$. What is the approximate change in the pressure? Does the pressure increase or decrease?

c. Suppose the pressure is held constant and the volume increases by $\Delta V = 0.1$. What is the approximate change in the temperature? Does the temperature increase or decrease?

53. Explain why or why not Determine whether the following statements are true and give an explanation or counterexample.

a. The linear approximation to $f(x) = x^2$ at $x = 0$ is $L(x) = 0$.

b. Linear approximation at $x = 0$ provides a good approximation to $f(x) = |x|$.

c. If $f(x) = mx + b$, then the linear approximation to f at any point is $L(x) = f(x)$.

d. When linear approximation is used to estimate values of $\ln x$ near $x = e$, the approximations are overestimates of the true values.

54. Time function Show that the function $T(x) = 60D(60 + x)^{-1}$ gives the time in minutes required to drive D miles at $60 + x$ miles per hour.

55–60. Approximating changes

55. Approximate the change in the volume of a sphere when its radius changes from $r = 5$ ft to $r = 5.1$ ft $\left(V(r) = \frac{4}{3}\pi r^3\right)$.

56. Approximate the change in the atmospheric pressure when the altitude increases from $z = 2$ km to $z = 2.01$ km $\left(P(z) = 1000e^{-z/10}\right)$.

57. Approximate the change in the volume of a right circular cylinder of fixed radius $r = 20$ cm when its height decreases from $h = 12$ cm to $h = 11.9$ cm $(V(h) = \pi r^2 h)$.

58. Approximate the change in the volume of a right circular cone of fixed height $h = 4$ m when its radius increases from $r = 3$ m to $r = 3.05$ m $\left(V(r) = \frac{1}{3}\pi r^2 h\right)$.

59. Approximate the change in the lateral surface area (excluding the area of the base) of a right circular cone of fixed height $h = 6$ m when its radius decreases from $r = 10$ m to $r = 9.9$ m $\left(S = \pi r\sqrt{r^2 + h^2}\right)$.

60. Approximate the change in the magnitude of the electrostatic force between two charges when the distance between them increases from $r = 20$ m to $r = 21$ m $(F(r) = 0.01/r^2)$.

61–70. Differentials *Consider the following functions and express the relationship between a small change in x and the corresponding change in y in the form $dy = f'(x)dx$.*

61. $f(x) = 2x + 1$

62. $f(x) = \sin^2 x$

63. $f(x) = \dfrac{1}{x^3}$

64. $f(x) = e^{2x}$

65. $f(x) = 2 - a\cos x,\ a$ constant

66. $f(x) = \dfrac{x + 4}{4 - x}$

67. $f(x) = 3x^3 - 4x$

68. $f(x) = \sin^{-1} x$

69. $f(x) = \tan x$

70. $f(x) = \ln(1 - x)$

Explorations and Challenges

▣ 71. Errors in approximations Suppose $f(x) = \sqrt[3]{x}$ is to be approximated near $x = 8$. Find the linear approximation to f at 8. Then complete the following table, showing the errors in various approximations. Use a calculator to obtain the exact values. The percent error is $100\,|\text{approximation} - \text{exact}|/|\text{exact}|$. Comment on the behavior of the errors as x approaches 8.

x	Linear approx.	Exact value	Percent error
8.1			
8.01			
8.001			
8.0001			
7.9999			
7.999			
7.99			
7.9			

▣ 72. Errors in approximations Suppose $f(x) = 1/(x + 1)$ is to be approximated near $x = 0$. Find the linear approximation to f at 0. Then complete the following table, showing the errors in various approximations. Use a calculator to obtain the exact values. The percent error is $100\,|\text{approximation} - \text{exact}|/|\text{exact}|$. Comment on the behavior of the errors as x approaches 0.

x	Linear approx.	Exact value	Percent error
0.1			
0.01			
0.001			
0.0001			
−0.0001			
−0.001			
−0.01			
−0.1			

73. Linear approximation and the second derivative Draw the graph of a function f such that $f(1) = f'(1) = f''(1) = 1$. Draw the linear approximation to the function at the point $(1, 1)$. Now draw the graph of another function g such that $g(1) = g'(1) = 1$ and $g''(1) = 10$. (It is not possible to represent the second derivative exactly, but your graphs should reflect the fact that $f''(1)$ is relatively small compared to $g''(1)$.) Now suppose linear approximations are used to approximate $f(1.1)$ and $g(1.1)$.

a. Which function has the more accurate linear approximation near $x = 1$ and why?

b. Explain why the error in the linear approximation to f near a point a is proportional to the magnitude of $f''(a)$.

QUICK CHECK ANSWERS

1. The linear approximation lies below the graph of f for x near a.   **2.** $L(15) = 45,\ s(15) = 48;\ x = 15$ is not close to 0.   **3.** $a = 0.16$ **4.** Note that $f(0) = 1$ and $f'(0) = 0$, so $L(x) = 1$ (this is the line tangent to $y = \cos x$ at $(0, 1)$).   **5.** $\Delta V \approx 4\pi a^2\,\Delta r$ ◄

4.7 L'Hôpital's Rule

The study of limits in Chapter 2 was thorough but not exhaustive. Some limits, called *indeterminate forms*, cannot generally be evaluated using the techniques presented in Chapter 2. These limits tend to be the more interesting limits that arise in practice. A powerful result called *l'Hôpital's Rule* enables us to evaluate such limits with relative ease.

Here is how indeterminate forms arise. If f is a *continuous* function at a point a, then we know that $\lim_{x \to a} f(x) = f(a)$, allowing the limit to be evaluated by computing $f(a)$. But there are many limits that cannot be evaluated by substitution. In fact, we encountered such a limit in Section 3.5:

$$\lim_{x \to 0} \frac{\sin x}{x} = 1.$$

If we attempt to substitute $x = 0$ into $(\sin x)/x$, we get $0/0$, which has no meaning. Yet we proved that $(\sin x)/x$ has the limit 1 at $x = 0$ (Theorem 3.10). This limit is an example of an *indeterminate form*; specifically, $\lim_{x \to 0} \dfrac{\sin x}{x}$ has the form $0/0$ because the numerator and denominator both approach 0 as $x \to 0$.

The meaning of an *indeterminate form* is further illustrated by $\lim_{x \to \infty} \dfrac{ax}{x + 1}$, where $a \neq 0$. This limit has the indeterminate form ∞/∞ (meaning that the numerator and denominator become arbitrarily large in magnitude as $x \to \infty$), but the actual value of the limit is $\lim_{x \to \infty} \dfrac{ax}{x + 1} = a \lim_{x \to \infty} \dfrac{x}{x + 1} = a$. In general, a limit with the form ∞/∞ or $0/0$ can have *any* value—which is why these limits must be handled carefully.

L' Hôpital's Rule for the Form 0/0

> ➤ The notations $0/0$ and ∞/∞ are merely symbols used to describe various types of indeterminate forms. The notation $0/0$ does not imply division by 0.

Consider a function of the form $f(x)/g(x)$ and assume $\lim_{x \to a} f(x) = \lim_{x \to a} g(x) = 0$. Then the limit $\lim_{x \to a} \dfrac{f(x)}{g(x)}$ has the indeterminate form $0/0$. We first state l'Hôpital's Rule and then prove a special case.

> ➤ Guillaume François l'Hôpital (lo-pee-tal) (1661–1704) is credited with writing the first calculus textbook. Much of the material in his book, including l'Hôpital's Rule, was provided by the Swiss mathematician Johann Bernoulli (1667–1748).

THEOREM 4.12 L'Hôpital's Rule

Suppose f and g are differentiable on an open interval I containing a with $g'(x) \neq 0$ on I when $x \neq a$. If $\lim_{x \to a} f(x) = \lim_{x \to a} g(x) = 0$, then

$$\lim_{x \to a} \frac{f(x)}{g(x)} = \lim_{x \to a} \frac{f'(x)}{g'(x)},$$

provided the limit on the right exists (or is $\pm \infty$). The rule also applies if $x \to a$ is replaced with $x \to \pm \infty$, $x \to a^+$, or $x \to a^-$.

Proof (special case): The proof of this theorem relies on the Generalized Mean Value Theorem (Exercise 57 of Section 4.2). We prove a special case of the theorem in which we assume f' and g' are continuous at a, $f(a) = g(a) = 0$, and $g'(a) \neq 0$. We have

$$\lim_{x \to a} \frac{f'(x)}{g'(x)} = \frac{f'(a)}{g'(a)} \qquad \text{Continuity of } f' \text{ and } g'$$

$$= \frac{\displaystyle \lim_{x \to a} \frac{f(x) - f(a)}{x - a}}{\displaystyle \lim_{x \to a} \frac{g(x) - g(a)}{x - a}} \qquad \text{Definition of } f'(a) \text{ and } g'(a)$$

▶ The definition of the derivative provides an example of an indeterminate form. Assuming f is differentiable at x,

$$f'(x) = \lim_{h \to 0} \frac{f(x + h) - f(x)}{h}$$

has the form $0/0$.

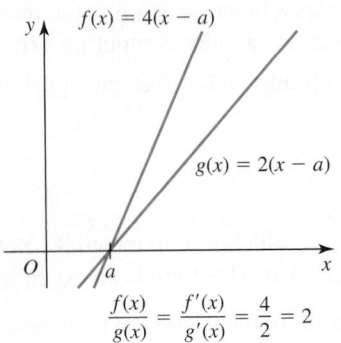

$$\frac{f(x)}{g(x)} = \frac{f'(x)}{g'(x)} = \frac{4}{2} = 2$$

Figure 4.75

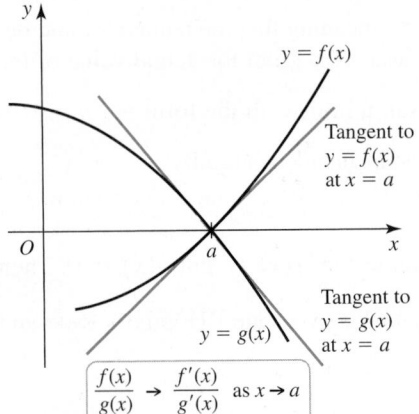

$$\frac{f(x)}{g(x)} \to \frac{f'(x)}{g'(x)} \text{ as } x \to a$$

Figure 4.76

QUICK CHECK 1 Which of the following functions lead to an indeterminate form as $x \to 0$: $f(x) = x^2/(x + 2)$, $g(x) = (\tan 3x)/x$, or $h(x) = (1 - \cos x)/x^2$? ◀

▶ The limit in part (a) can also be evaluated by factoring the numerator and canceling $(x - 1)$:

$$\lim_{x \to 1} \frac{x^3 + x^2 - 2x}{x - 1}$$
$$= \lim_{x \to 1} \frac{x(x - 1)(x + 2)}{x - 1}$$
$$= \lim_{x \to 1} x(x + 2) = 3.$$

$$= \lim_{x \to a} \frac{\dfrac{f(x) - f(a)}{x - a}}{\dfrac{g(x) - g(a)}{x - a}} \qquad \text{Limit of a quotient, } g'(a) \neq 0$$

$$= \lim_{x \to a} \frac{f(x) - f(a)}{g(x) - g(a)} \qquad \text{Cancel } x - a.$$

$$= \lim_{x \to a} \frac{f(x)}{g(x)}. \qquad f(a) = g(a) = 0 \qquad\qquad ◀$$

The geometry of l'Hôpital's Rule offers some insight. First consider two *linear* functions, f and g, whose graphs both pass through the point $(a, 0)$ with slopes 4 and 2, respectively; this means that

$$f(x) = 4(x - a) \quad \text{and} \quad g(x) = 2(x - a).$$

Furthermore, $f(a) = g(a) = 0$, $f'(x) = 4$, and $g'(x) = 2$ (**Figure 4.75**).

Looking at the quotient f/g, we see that

$$\frac{f(x)}{g(x)} = \frac{4(x - a)}{2(x - a)} = \frac{4}{2} = \frac{f'(x)}{g'(x)}.$$

This argument may be generalized, and we find that for any linear functions f and g with $f(a) = g(a) = 0$,

$$\lim_{x \to a} \frac{f(x)}{g(x)} = \lim_{x \to a} \frac{f'(x)}{g'(x)},$$

provided $g'(a) \neq 0$.

If f and g are not linear functions, we replace them with their linear approximations at a (**Figure 4.76**). Zooming in on the point a, the curves are close to their respective tangent lines $y = f'(a)(x - a)$ and $y = g'(a)(x - a)$, which have slopes $f'(a)$ and $g'(a) \neq 0$, respectively. Near $x = a$, we have

$$\frac{f(x)}{g(x)} \approx \frac{f'(a)(x - a)}{g'(a)(x - a)} = \frac{f'(a)}{g'(a)}.$$

Therefore, the ratio of the functions is well approximated by the ratio of the derivatives. In the limit as $x \to a$, we again have

$$\lim_{x \to a} \frac{f(x)}{g(x)} = \lim_{x \to a} \frac{f'(x)}{g'(x)}.$$

EXAMPLE 1 Using l'Hôpital's Rule Evaluate the following limits.

a. $\displaystyle\lim_{x \to 1} \frac{x^3 + x^2 - 2x}{x - 1}$ **b.** $\displaystyle\lim_{x \to 0} \frac{\sqrt{9 + 3x} - 3}{x}$

SOLUTION

a. Direct substitution of $x = 1$ into $\dfrac{x^3 + x^2 - 2x}{x - 1}$ produces the indeterminate form $0/0$.

Applying l'Hôpital's Rule with $f(x) = x^3 + x^2 - 2x$ and $g(x) = x - 1$ gives

$$\lim_{x \to 1} \frac{x^3 + x^2 - 2x}{x - 1} = \lim_{x \to 1} \frac{f'(x)}{g'(x)} = \lim_{x \to 1} \frac{3x^2 + 2x - 2}{1} = 3.$$

b. Substituting $x = 0$ into this function produces the indeterminate form $0/0$. Let $f(x) = \sqrt{9 + 3x} - 3$ and $g(x) = x$, and note that $f'(x) = \dfrac{3}{2\sqrt{9 + 3x}}$ and $g'(x) = 1$. Applying l'Hôpital's Rule, we have

$$\lim_{x \to 0} \underbrace{\frac{\sqrt{9 + 3x} - 3}{x}}_{f/g} = \lim_{x \to 0} \underbrace{\frac{\frac{3}{2\sqrt{9 + 3x}}}{1}}_{f'/g'} = \frac{1}{2}.$$

Related Exercises 17–18 ◄

L'Hôpital's Rule requires evaluating $\lim\limits_{x \to a} f'(x)/g'(x)$. It may happen that this second limit is another indeterminate form to which l'Hôpital's Rule may again be applied.

EXAMPLE 2   L'Hôpital's Rule repeated Evaluate the following limits.

a. $\lim\limits_{x \to 0} \dfrac{e^x - x - 1}{x^2}$ **b.** $\lim\limits_{x \to 2} \dfrac{x^3 - 3x^2 + 4}{x^4 - 4x^3 + 7x^2 - 12x + 12}$

SOLUTION

a. This limit has the indeterminate form $0/0$. Applying l'Hôpital's Rule, we have

$$\lim_{x \to 0} \frac{e^x - x - 1}{x^2} = \lim_{x \to 0} \frac{e^x - 1}{2x},$$

which is another limit of the form $0/0$. Therefore, we apply l'Hôpital's Rule again:

$$\lim_{x \to 0} \frac{e^x - x - 1}{x^2} = \lim_{x \to 0} \frac{e^x - 1}{2x} \qquad \text{L'Hôpital's Rule}$$

$$= \lim_{x \to 0} \frac{e^x}{2} \qquad \text{L'Hôpital's Rule again}$$

$$= \frac{1}{2}. \qquad \text{Evaluate limit.}$$

b. Evaluating the numerator and denominator at $x = 2$, we see that this limit has the form $0/0$. Applying l'Hôpital's Rule twice, we have

$$\lim_{x \to 2} \frac{x^3 - 3x^2 + 4}{x^4 - 4x^3 + 7x^2 - 12x + 12} = \lim_{x \to 2} \underbrace{\frac{3x^2 - 6x}{4x^3 - 12x^2 + 14x - 12}}_{\text{limit of the form } 0/0} \qquad \text{L'Hôpital's Rule}$$

$$= \lim_{x \to 2} \frac{6x - 6}{12x^2 - 24x + 14} \qquad \begin{array}{l}\text{L'Hôpital's Rule}\\\text{again}\end{array}$$

$$= \frac{3}{7}. \qquad \text{Evaluate limit.}$$

It is easy to overlook a crucial step in this computation: After applying l'Hôpital's Rule the first time, you *must* establish that the new limit is an indeterminate form before applying l'Hôpital's Rule a second time.

Related Exercises 36, 39 ◄

Indeterminate Form ∞ / ∞

L'Hôpital's Rule also applies directly to limits of the form $\lim\limits_{x \to a} f(x)/g(x)$, where $\lim\limits_{x \to a} f(x) = \pm\infty$ and $\lim\limits_{x \to a} g(x) = \pm\infty$; this indeterminate form is denoted ∞ / ∞. The proof of this result is found in advanced books.

QUICK CHECK 2 Which of the following functions lead to an indeterminate form as $x \to \infty$: $f(x) = (\sin x)/x$, $g(x) = (3x^2 + 4)/x^2$, or $h(x) = 2^x/x^2$? ◄

> **THEOREM 4.13**   **L'Hôpital's Rule** (∞/∞)
> Suppose f and g are differentiable on an open interval I containing a, with $g'(x) \ne 0$ on I when $x \ne a$. If $\lim\limits_{x \to a} f(x) = \pm\infty$ and $\lim\limits_{x \to a} g(x) = \pm\infty$, then
> $$\lim_{x \to a} \frac{f(x)}{g(x)} = \lim_{x \to a} \frac{f'(x)}{g'(x)},$$
> provided the limit on the right exists (or is $\pm\infty$). The rule also applies for $x \to \pm\infty$, $x \to a^+$, or $x \to a^-$.

EXAMPLE 3   **L'Hôpital's Rule for ∞/∞**   Evaluate the following limits.

a. $\lim\limits_{x \to \infty} \dfrac{4x^3 - 6x^2 + 1}{2x^3 - 10x + 3}$ **b.** $\lim\limits_{x \to 0^+} \dfrac{\ln x}{\csc x}$

SOLUTION

> ➤ As shown in Section 2.5, the limit in Example 3a could also be evaluated by first dividing the numerator and denominator by x^3 or by using Theorem 2.7.

a. This limit has the indeterminate form ∞/∞ because both the numerator and the denominator approach ∞ as $x \to \infty$. Applying l'Hôpital's Rule three times, we have

$$\underbrace{\lim_{x \to \infty} \frac{4x^3 - 6x^2 + 1}{2x^3 - 10x + 3}}_{\infty/\infty} = \underbrace{\lim_{x \to \infty} \frac{12x^2 - 12x}{6x^2 - 10}}_{\infty/\infty} = \underbrace{\lim_{x \to \infty} \frac{24x - 12}{12x}}_{\infty/\infty} = \lim_{x \to \infty} \frac{24}{12} = 2.$$

b. In this limit, the numerator approaches $-\infty$ and the denominator approaches ∞ as $x \to 0^+$. L'Hôpital's Rule gives us

> ➤ In the solution to Example 3b, notice that we simplify $(1/x)/(-\csc x \cot x)$ before using L'Hôpital's Rule again. This step is important.

$$\lim_{x \to 0^+} \frac{\ln x}{\csc x} = \lim_{x \to 0^+} \frac{1/x}{-\csc x \cot x} \qquad \text{L'Hôpital's Rule}$$

$$= -\lim_{x \to 0^+} \frac{\sin^2 x}{x \cos x}. \qquad \text{Simplify.}$$

The new limit has the form $0/0$, so we apply L'Hôpital's Rule again:

$$-\lim_{x \to 0^+} \frac{\sin^2 x}{x \cos x} = -\lim_{x \to 0^+} \frac{2 \sin x \cos x}{\cos x - x \sin x} \qquad \text{L'Hôpital's Rule}$$

$$= -\frac{0}{1} = 0. \qquad \text{Evaluate limit.}$$

Related Exercises 38, 51 ◄

Related Indeterminate Forms: $0 \cdot \infty$ and $\infty - \infty$

We now consider limits of the form $\lim\limits_{x \to a} f(x)g(x)$, where $\lim\limits_{x \to a} f(x) = 0$ and $\lim\limits_{x \to a} g(x) = \pm\infty$; such limits are denoted $0 \cdot \infty$. *L'Hôpital's Rule cannot be directly applied to limits of this form.* Furthermore, it's risky to jump to conclusions about such limits. Suppose $f(x) = x$ and $g(x) = \dfrac{1}{x^2}$, in which case $\lim\limits_{x \to 0} f(x) = 0$, $\lim\limits_{x \to 0} g(x) = \infty$, and $\lim\limits_{x \to 0} f(x)g(x) = \lim\limits_{x \to 0} \dfrac{1}{x}$ does not exist. On the other hand, if $f(x) = x$ and $g(x) = \dfrac{1}{\sqrt{x}}$, we have $\lim\limits_{x \to 0^+} f(x) = 0$, $\lim\limits_{x \to 0^+} g(x) = \infty$, and $\lim\limits_{x \to 0^+} f(x)g(x) = \lim\limits_{x \to 0^+} \sqrt{x} = 0$. So a limit of the form $0 \cdot \infty$, in which the two functions compete with each other, may have any value or may not exist. The following example illustrates how this indeterminate form can be recast in the form $0/0$ or ∞/∞.

EXAMPLE 4 L'Hôpital's Rule for $0 \cdot \infty$ Evaluate $\lim\limits_{x \to \infty} x^2 \sin\left(\dfrac{1}{4x^2}\right)$.

SOLUTION This limit has the form $0 \cdot \infty$. A common technique that converts this form to either $0/0$ or ∞/∞ is to *divide by the reciprocal*. We rewrite the limit and apply l'Hôpital's Rule:

$$\underbrace{\lim_{x \to \infty} x^2 \sin\left(\frac{1}{4x^2}\right)}_{0 \cdot \infty \text{ form}} = \underbrace{\lim_{x \to \infty} \frac{\sin\left(\dfrac{1}{4x^2}\right)}{\dfrac{1}{x^2}}}_{\text{recast in } 0/0 \text{ form}} \qquad x^2 = \frac{1}{1/x^2}$$

$$= \lim_{x \to \infty} \frac{\cos\left(\dfrac{1}{4x^2}\right)\dfrac{1}{4}\left(-2x^{-3}\right)}{-2x^{-3}} \qquad \text{L'Hôpital's Rule}$$

$$= \frac{1}{4} \lim_{x \to \infty} \cos\left(\frac{1}{4x^2}\right) \qquad \text{Simplify.}$$

$$= \frac{1}{4}. \qquad \frac{1}{4x^2} \to 0, \cos 0 = 1$$

Related Exercise 53 ◄

QUICK CHECK 3 What is the form of the limit $\lim\limits_{x \to \pi/2^-} (x - \pi/2)(\tan x)$? Write it in the form $0/0$. ◄

Limits of the form $\lim\limits_{x \to a} (f(x) - g(x))$, where $\lim\limits_{x \to a} f(x) = \infty$ and $\lim\limits_{x \to a} g(x) = \infty$, are indeterminate forms that we denote $\infty - \infty$. L'Hôpital's Rule cannot be applied directly to an $\infty - \infty$ form. It must first be expressed in the form $0/0$ or ∞/∞. With the $\infty - \infty$ form, it is easy to reach erroneous conclusions. For example, if $f(x) = 3x + 5$ and $g(x) = 3x$, then

$$\lim_{x \to \infty} ((3x + 5) - (3x)) = 5.$$

However, if $f(x) = 3x$ and $g(x) = 2x$, then

$$\lim_{x \to \infty} (3x - 2x) = \lim_{x \to \infty} x = \infty.$$

These examples show again why indeterminate forms are deceptive. Before proceeding, we introduce another useful technique.

Occasionally, it helps to convert a limit as $x \to \infty$ to a limit as $t \to 0^+$ (or vice versa) by a *change of variables*. To evaluate $\lim\limits_{x \to \infty} f(x)$, we define $t = 1/x$ and note that as $x \to \infty$, $t \to 0^+$. Then

$$\lim_{x \to \infty} f(x) = \lim_{t \to 0^+} f\left(\frac{1}{t}\right).$$

This idea is illustrated in the next example.

EXAMPLE 5 L'Hôpital's Rule for $\infty - \infty$ Evaluate $\lim\limits_{x \to \infty} \left(x - \sqrt{x^2 - 3x}\right)$.

SOLUTION As $x \to \infty$, both terms in the difference $x - \sqrt{x^2 - 3x}$ approach ∞ and this limit has the form $\infty - \infty$. We first factor x from the expression and form a quotient:

$$\lim_{x \to \infty} \left(x - \sqrt{x^2 - 3x}\right) = \lim_{x \to \infty} \left(x - \sqrt{x^2(1 - 3/x)}\right) \quad \text{Factor } x^2 \text{ under square root.}$$

$$= \lim_{x \to \infty} x(1 - \sqrt{1 - 3/x}) \quad x > 0, \text{ so } \sqrt{x^2} = x$$

$$= \lim_{x \to \infty} \frac{1 - \sqrt{1 - 3/x}}{1/x}. \quad \begin{array}{l}\text{Write } 0 \cdot \infty \text{ form as } 0/0 \text{ form;}\\ x = \dfrac{1}{1/x}.\end{array}$$

This new limit has the form $0/0$, and l'Hôpital's Rule may be applied. One way to proceed is to use the change of variables $t = 1/x$:

$$\lim_{x\to\infty} \frac{1 - \sqrt{1 - 3/x}}{1/x} = \lim_{t\to 0^+} \frac{1 - \sqrt{1 - 3t}}{t} \qquad \text{Let } t = 1/x; \text{ replace } \lim_{x\to\infty} \text{ with } \lim_{t\to 0^+}.$$

$$= \lim_{t\to 0^+} \frac{\dfrac{3}{2\sqrt{1 - 3t}}}{1} \qquad \text{L'Hôpital's Rule}$$

$$= \frac{3}{2}. \qquad \text{Evaluate limit.}$$

Related Exercises 63–64 ◀

Indeterminate Forms 1^∞, 0^0, and ∞^0

The indeterminate forms 1^∞, 0^0, and ∞^0 all arise in limits of the form $\lim\limits_{x\to a} f(x)^{g(x)}$, where $x \to a$ could be replaced with $x \to a^\pm$ or $x \to \pm\infty$. L'Hôpital's Rule cannot be applied directly to the indeterminate forms 1^∞, 0^0, and ∞^0. They must first be expressed in the form $0/0$ or ∞/∞. Here is how we proceed.

The inverse relationship between $\ln x$ and e^x says that $f^g = e^{g \ln f}$, so we first write

$$\lim_{x\to a} f(x)^{g(x)} = \lim_{x\to a} e^{g(x) \ln f(x)}.$$

By the continuity of the exponential function, we switch the order of the limit and the exponential function (Theorem 2.12); therefore,

$$\lim_{x\to a} f(x)^{g(x)} = \lim_{x\to a} e^{g(x) \ln f(x)} = e^{\lim\limits_{x\to a} g(x) \ln f(x)},$$

provided $\lim\limits_{x\to a} g(x) \ln f(x)$ exists. Therefore, $\lim\limits_{x\to a} f(x)^{g(x)}$ is evaluated using the following two steps.

> ► Notice the following:
> - For 1^∞, L has the form $\infty \cdot \ln 1 = \infty \cdot 0$.
> - For 0^0, L has the form $0 \cdot \ln 0 = 0 \cdot -\infty$.
> - For ∞^0, L has the form $0 \cdot \ln \infty = 0 \cdot \infty$.

PROCEDURE **Indeterminate forms 1^∞, 0^0, and ∞^0**

Assume $\lim\limits_{x\to a} f(x)^{g(x)}$ has the indeterminate form 1^∞, 0^0, or ∞^0.

1. Analyze $L = \lim\limits_{x\to a} g(x) \ln f(x)$. This limit can be put in the form $0/0$ or ∞/∞, both of which are handled by l'Hôpital's Rule.

2. When L is finite, $\lim\limits_{x\to a} f(x)^{g(x)} = e^L$. If $L = \infty$ or $-\infty$, then
$$\lim_{x\to a} f(x)^{g(x)} = \infty \quad \text{or} \quad \lim_{x\to a} f(x)^{g(x)} = 0, \text{ respectively.}$$

QUICK CHECK 4 Explain why a limit of the form 0^∞ is not an indeterminate form. ◀

EXAMPLE 6 **Indeterminate forms 0^0 and 1^∞** Evaluate the following limits.

a. $\lim\limits_{x\to 0^+} x^x$ **b.** $\lim\limits_{x\to\infty} \left(1 + \dfrac{1}{x}\right)^x$

SOLUTION

a. This limit has the form 0^0. Using the given two-step procedure, we note that $x^x = e^{x \ln x}$ and first evaluate

$$L = \lim_{x\to 0^+} x \ln x.$$

This limit has the form $0 \cdot \infty$, which may be put in the form ∞/∞ so that l'Hôpital's Rule can be applied:

$$L = \lim_{x\to 0^+} x \ln x = \lim_{x\to 0^+} \frac{\ln x}{1/x} \qquad x = \frac{1}{1/x}$$

$$= \lim_{x\to 0^+} \frac{1/x}{-1/x^2} \qquad \text{L'Hôpital's Rule for } \infty/\infty \text{ form}$$

$$= \lim_{x\to 0^+} (-x) = 0. \qquad \text{Simplify and evaluate the limit.}$$

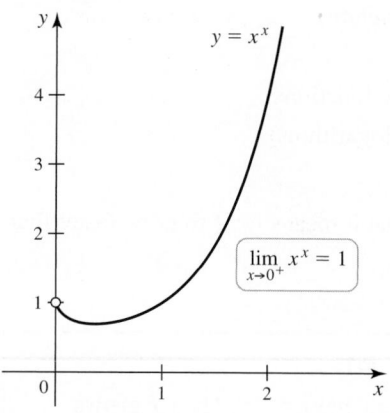

Figure 4.77

> ► The limit in Example 6b is often given as a definition of e. It is a special case of the more general limit
>
> $$\lim_{x \to \infty} \left(1 + \frac{a}{x}\right)^x = e^a.$$
>
> See Exercise 119.

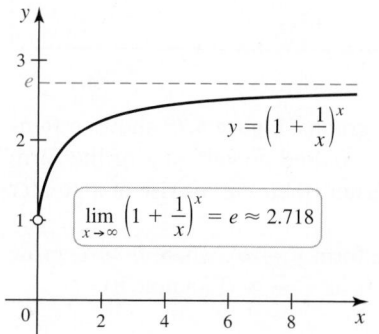

Figure 4.78

> ► Models of epidemics produce more complicated functions than the one given here, but they have the same general features.

> ► The Prime Number Theorem was proved simultaneously (two different proofs) in 1896 by Jacques Hadamard and Charles de la Vallée Poussin, relying on fundamental ideas contributed by Riemann.

The second step is to exponentiate the limit:

$$\lim_{x \to 0^+} x^x = e^L = e^0 = 1.$$

We conclude that $\lim_{x \to 0^+} x^x = 1$ (**Figure 4.77**).

b. This limit has the form 1^∞. Noting that $(1 + 1/x)^x = e^{x \ln (1 + 1/x)}$, the first step is to evaluate

$$L = \lim_{x \to \infty} x \ln \left(1 + \frac{1}{x}\right),$$

which has the form $0 \cdot \infty$. Proceeding as in part (a), we have

$$L = \lim_{x \to \infty} x \ln \left(1 + \frac{1}{x}\right) = \lim_{x \to \infty} \frac{\ln (1 + 1/x)}{1/x} \qquad x = \frac{1}{1/x}$$

$$= \lim_{x \to \infty} \frac{\dfrac{1}{1 + 1/x} \cdot \left(-\dfrac{1}{x^2}\right)}{\left(-\dfrac{1}{x^2}\right)} \qquad \text{L'Hôpital's Rule for } 0/0 \text{ form}$$

$$= \lim_{x \to \infty} \frac{1}{1 + 1/x} = 1. \qquad \text{Simplify and evaluate.}$$

The second step is to exponentiate the limit:

$$\lim_{x \to \infty} \left(1 + \frac{1}{x}\right)^x = e^L = e^1 = e.$$

The function $y = (1 + 1/x)^x$ (**Figure 4.78**) has a horizontal asymptote $y = e \approx 2.71828$.

Related Exercises 75–76 ◄

Growth Rates of Functions

An important use of l'Hôpital's Rule is to compare the growth rates of functions. Here are two questions—one practical and one theoretical—that demonstrate the importance of comparative growth rates of functions.

• A particular theory for modeling the spread of an epidemic predicts that the number of infected people t days after the start of the epidemic is given by the function

$$N(t) = 2.5t^2 e^{-0.01t} = 2.5 \frac{t^2}{e^{0.01t}}.$$

Question: In the long run (as $t \to \infty$), does the epidemic spread or does it die out?

• A prime number is an integer $p \geq 2$ that has only two divisors, 1 and itself. The first few prime numbers are 2, 3, 5, 7, and 11. A celebrated theorem states that the number of prime numbers less than x is approximately

$$P(x) = \frac{x}{\ln x}, \quad \text{for large values of } x.$$

Question: According to this theorem, is the number of prime numbers infinite?

These two questions involve a comparison of two functions. In the first question, if t^2 grows faster than $e^{0.01t}$ as $t \to \infty$, then $\lim_{t \to \infty} N(t) = \infty$ and the epidemic grows. If $e^{0.01t}$ grows faster than t^2 as $t \to \infty$, then $\lim_{t \to \infty} N(t) = 0$ and the epidemic dies out. We will explain what is meant by *grows faster than* in a moment.

In the second question, the comparison is between x and $\ln x$. If x grows faster than $\ln x$ as $x \to \infty$, then $\lim_{x \to \infty} P(x) = \infty$ and the number of prime numbers is infinite.

Our goal is to obtain a ranking of the following families of functions based on their growth rates:

• mx, where $m > 0$ (represents linear functions)

• x^p, where $p > 0$ (represents polynomials and algebraic functions)

▶ Another function with a large growth rate is the factorial function, defined for integers as
$$f(n) = n! = n(n-1) \cdots 2 \cdot 1.$$
See Exercise 116.

QUICK CHECK 5 Before proceeding, use your intuition and rank these classes of functions in order of their growth rates. ◀

- x^x (sometimes called a *superexponential* or *tower function*)
- $\ln x$ (represents logarithmic functions)
- $\ln^q x$, where $q > 0$ (represents powers of logarithmic functions)
- $x^p \ln x$, where $p > 0$ (a combination of powers and logarithms)
- e^x (represents exponential functions).

We need to be precise about growth rates and what it means for f to grow faster than g as $x \to \infty$. We work with the following definitions.

> **DEFINITION Growth Rates of Functions (as $x \to \infty$)**
>
> Suppose f and g are functions with $\lim\limits_{x \to \infty} f(x) = \lim\limits_{x \to \infty} g(x) = \infty$. Then f **grows faster than** g as $x \to \infty$ if
>
> $$\lim_{x \to \infty} \frac{g(x)}{f(x)} = 0 \quad \text{or, equivalently, if} \quad \lim_{x \to \infty} \frac{f(x)}{g(x)} = \infty.$$
>
> The functions f and g have **comparable growth rates** if
>
> $$\lim_{x \to \infty} \frac{f(x)}{g(x)} = M,$$
>
> where $0 < M < \infty$ (M is positive and finite).

The idea of growth rates is illustrated nicely with graphs. **Figure 4.79** shows a family of linear functions of the form $y = mx$, where $m > 0$, and powers of x of the form $y = x^p$, where $p > 1$. We see that powers of x grow faster (their curves rise at a greater rate) than the linear functions as $x \to \infty$.

Figure 4.80 shows that exponential functions of the form $y = b^x$, where $b > 1$, grow faster than powers of x of the form $y = x^p$, where $p > 0$, as $x \to \infty$ (Example 8).

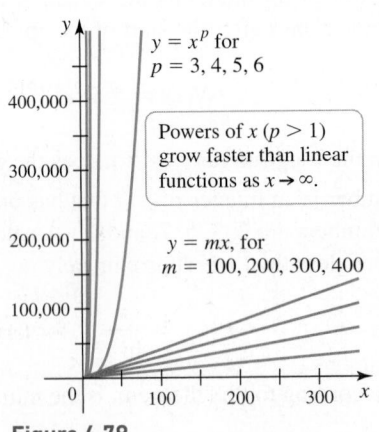

Figure 4.79

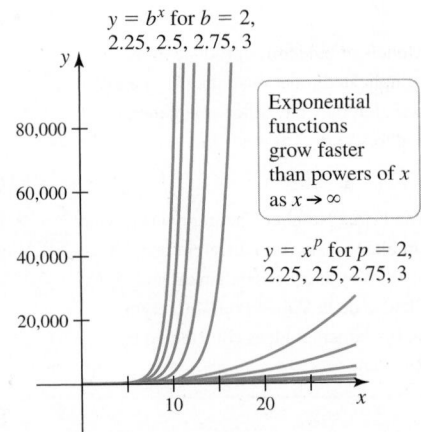

Figure 4.80

QUICK CHECK 6 Compare the growth rates of $f(x) = x^2$ and $g(x) = x^3$ as $x \to \infty$. Compare the growth rates of $f(x) = x^2$ and $g(x) = 10x^2$ as $x \to \infty$. ◀

We now begin a systematic comparison of growth rates. Note that a growth rate limit involves an indeterminate form ∞/∞, so l'Hôpital's Rule is always in the picture.

EXAMPLE 7 Powers of x vs. powers of $\ln x$ Compare the growth rates as $x \to \infty$ of the following pairs of functions.

a. $f(x) = \ln x$ and $g(x) = x^p$, where $p > 0$

b. $f(x) = \ln^q x$ and $g(x) = x^p$, where $p > 0$ and $q > 0$

SOLUTION

a. The limit of the ratio of the two functions is

$$\lim_{x\to\infty} \frac{\ln x}{x^p} = \lim_{x\to\infty} \frac{1/x}{px^{p-1}} \qquad \text{L'Hôpital's Rule}$$

$$= \lim_{x\to\infty} \frac{1}{px^p} \qquad \text{Simplify.}$$

$$= 0. \qquad \text{Evaluate the limit.}$$

We see that any positive power of x grows faster than $\ln x$.

b. We compare $\ln^q x$ and x^p by observing that

$$\lim_{x\to\infty} \frac{\ln^q x}{x^p} = \lim_{x\to\infty} \left(\frac{\ln x}{x^{p/q}}\right)^q = \left(\underbrace{\lim_{x\to\infty} \frac{\ln x}{x^{p/q}}}_{0}\right)^q.$$

By part (a), $\displaystyle\lim_{x\to\infty} \frac{\ln x}{x^{p/q}} = 0$ (because $p/q > 0$). Therefore, $\displaystyle\lim_{x\to\infty} \frac{\ln^q x}{x^p} = 0$ (because $q > 0$). We conclude that any positive power of x grows faster than any positive power of $\ln x$.

Related Exercises 96, 100 ◄

EXAMPLE 8 Powers of x vs. exponentials Compare the rates of growth of $f(x) = x^p$ and $g(x) = e^x$ as $x \to \infty$, where p is a positive real number.

SOLUTION The goal is to evaluate $\displaystyle\lim_{x\to\infty} \frac{x^p}{e^x}$, for $p > 0$. This comparison is most easily done using Example 7 and a change of variables. We let $x = \ln t$ and note that as $x \to \infty$, we also have $t \to \infty$. With this substitution, $x^p = \ln^p t$ and $e^x = e^{\ln t} = t$. Therefore,

$$\lim_{x\to\infty} \frac{x^p}{e^x} = \lim_{t\to\infty} \frac{\ln^p t}{t} = 0. \qquad \text{Example 7}$$

We see that increasing exponential functions grow faster than positive powers of x (Figure 4.80).

Related Exercises 95, 101 ◄

A similar argument applies when comparing the growth rates of $f(x) = x^p$ and $h(x) = b^x$, for $b > 1$ (Exercise 113). These examples, together with the comparison of exponential functions b^x and the superexponential x^x (Exercise 120), establish a ranking of growth rates.

THEOREM 4.14 Ranking Growth Rates as $x \to \infty$

Let $f \ll g$ mean that g grows faster than f as $x \to \infty$. With positive real numbers p, q, r, and s and with $b > 1$,

$$\ln^q x \ll x^p \ll x^p \ln^r x \ll x^{p+s} \ll b^x \ll x^x.$$

You should try to build these relative growth rates into your intuition. They are useful in future chapters and they can be used to evaluate limits at infinity quickly.

Pitfalls in Using l'Hôpital's Rule

We close with a list of common pitfalls to watch out for when using l'Hôpital's Rule.

1. L'Hôpital's Rule says $\displaystyle\lim_{x\to a} \frac{f(x)}{g(x)} = \lim_{x\to a} \frac{f'(x)}{g'(x)}$, not

$$\lim_{x\to a} \frac{f(x)}{g(x)} = \lim_{x\to a} \left(\frac{f(x)}{g(x)}\right)' \quad \text{or} \quad \lim_{x\to a} \frac{f(x)}{g(x)} = \lim_{x\to a} \left(\frac{1}{g(x)}\right)' f'(x).$$

In other words, you should evaluate $f'(x)$ and $g'(x)$, form their quotient, and then take the limit. Don't confuse l'Hôpital's Rule with the Quotient Rule.

2. Be sure that the given limit involves the indeterminate form $0/0$ or ∞/∞ before applying l'Hôpital's Rule. For example, consider the following erroneous use of l'Hôpital's Rule:

$$\lim_{x \to 0} \frac{1 - \sin x}{\cos x} = \lim_{x \to 0} \frac{\cos x}{\sin x},$$

which does not exist. The original limit is not an indeterminate form in the first place. This limit should be evaluated by direct substitution:

$$\lim_{x \to 0} \frac{1 - \sin x}{\cos x} = \frac{1 - \sin 0}{1} = 1.$$

3. When using l'Hôpital's Rule repeatedly, be sure to simplify expressions as much as possible at each step and evaluate the limit as soon as the new limit is no longer an indeterminate form.

4. Repeated use of l'Hôpital's Rule occasionally leads to unending cycles, in which case other methods must be used. For example, limits of the form $\lim\limits_{x \to \infty} \dfrac{\sqrt{ax + 1}}{\sqrt{bx + 1}}$, where a and b are positive real numbers, lead to such behavior (Exercise 111).

5. Be sure that the limit produced by l'Hôpital's Rule exists. Consider $\lim\limits_{x \to \infty} \dfrac{3x + \cos x}{x}$, which has the form ∞/∞. Applying l'Hôpital's Rule, we have

$$\lim_{x \to \infty} \frac{3x + \cos x}{x} = \lim_{x \to \infty} \frac{3 - \sin x}{1}.$$

It is tempting to conclude that because the limit on the right side does not exist, the original limit also does not exist. In fact, the original limit has a value of 3 (divide numerator and denominator by x). To reach a conclusion from l'Hôpital's Rule, the limit produced by l'Hôpital's Rule must exist (or be $\pm \infty$).

SECTION 4.7 EXERCISES

Getting Started

1. Explain with examples what is meant by the indeterminate form $0/0$.

2. Why are special methods such as l'Hôpital's Rule, rather than substitution, needed to evaluate indeterminate forms?

3. Explain the steps used to apply l'Hôpital's Rule to a limit of the form $0/0$.

4. Give examples of each of the following.
   a. A limit with the indeterminate form $0/0$ that equals 3.
   b. A limit with the indeterminate form $0/0$ that equals 4.

5. Give examples of each of the following.
   a. A limit with the indeterminate form $0 \cdot \infty$ that equals 1.
   b. A limit with the indeterminate form $0 \cdot \infty$ that equals 2.

6. Which of the following limits can be evaluated without l'Hôpital's Rule? Evaluate each limit.
   a. $\lim\limits_{x \to 0} \dfrac{\sin x}{x^3 + 2x + 1}$ b. $\lim\limits_{x \to 0} \dfrac{\sin x}{x^3 + 2x}$

7. Explain how to convert a limit of the form $0 \cdot \infty$ to a limit of the form $0/0$ or ∞/∞.

8. Give an example of a limit of the form ∞/∞ as $x \to 0$.

9. The limit $\lim\limits_{x \to 0} (\tan^{-1} x)\left(\dfrac{1}{5x}\right)$ has the indeterminate form $0 \cdot \infty$. Convert the limit to the indeterminate form $0/0$ and then evaluate the limit.

10. Evaluate $\lim\limits_{x \to 2} \dfrac{x^3 - 3x^2 + 2x}{x - 2}$ using l'Hôpital's Rule and then check your work by evaluating the limit using an appropriate Chapter 2 method.

11. Explain why the form 1^{∞} is indeterminate and cannot be evaluated by substitution. Explain how the competing functions behave.

12. Give the two-step method for evaluating a limit of the form $\lim\limits_{x \to a} f(x)^{g(x)}$.

13. In terms of limits, what does it mean for f to grow faster than g as $x \to \infty$?

14. In terms of limits, what does it mean for the rates of growth of f and g to be comparable as $x \to \infty$?

15. Rank the functions x^3, $\ln x$, x^x, and 2^x in order of increasing growth rates as $x \to \infty$.

16. Rank the functions x^{100}, $\ln x^{10}$, x^x, and 10^x in order of increasing growth rates as $x \to \infty$.

Practice Exercises

17–83. Limits *Evaluate the following limits. Use l'Hôpital's Rule when it is convenient and applicable.*

17. $\lim\limits_{x \to 2} \dfrac{x^2 - 2x}{x^2 - 6x + 8}$

18. $\lim\limits_{x \to -1} \dfrac{x^4 + x^3 + 2x + 2}{x + 1}$

19. $\lim\limits_{x \to 1} \dfrac{x^2 + 2x}{x + 3}$

20. $\lim\limits_{x \to 0} \dfrac{e^x - 1}{2x + 5}$

21. $\lim\limits_{x\to 1} \dfrac{\ln x}{4x - x^2 - 3}$

22. $\lim\limits_{x\to 0} \dfrac{e^x - 1}{x^2 + 3x}$

23. $\lim\limits_{x\to\infty} \dfrac{3x^4 - x^2}{6x^4 + 12}$

24. $\lim\limits_{x\to\infty} \dfrac{4x^3 - 2x^2 + 6}{\pi x^3 + 4}$

25. $\lim\limits_{x\to e} \dfrac{\ln x - 1}{x - e}$

26. $\lim\limits_{x\to 1} \dfrac{4\tan^{-1} x - \pi}{x - 1}$

27. $\lim\limits_{x\to 0^+} \dfrac{1 - \ln x}{1 + \ln x}$

28. $\lim\limits_{x\to 0^+} \dfrac{x - 3\sqrt{x}}{x - \sqrt{x}}$

29. $\lim\limits_{x\to 0} \dfrac{3\sin 4x}{5x}$

30. $\lim\limits_{x\to 2\pi} \dfrac{x\sin x + x^2 - 4\pi^2}{x - 2\pi}$

31. $\lim\limits_{u\to\pi/4} \dfrac{\tan u - \cot u}{u - \pi/4}$

32. $\lim\limits_{z\to 0} \dfrac{\tan 4z}{\tan 7z}$

33. $\lim\limits_{x\to 0} \dfrac{1 - \cos 3x}{8x^2}$

34. $\lim\limits_{x\to 0} \dfrac{\sin^2 3x}{x^2}$

35. $\lim\limits_{x\to\pi} \dfrac{\cos x + 1}{(x - \pi)^2}$

36. $\lim\limits_{x\to 0} \dfrac{e^x - x - 1}{5x^2}$

37. $\lim\limits_{x\to\pi/2^-} \dfrac{\tan x}{3/(2x - \pi)}$

38. $\lim\limits_{x\to\infty} \dfrac{e^{3x}}{3e^{3x} + 5}$

39. $\lim\limits_{x\to 0} \dfrac{e^x - \sin x - 1}{x^4 + 8x^3 + 12x^2}$

40. $\lim\limits_{x\to 0} \dfrac{\sin x - x}{7x^3}$

41. $\lim\limits_{x\to\infty} \dfrac{e^{1/x} - 1}{1/x}$

42. $\lim\limits_{x\to\infty} \dfrac{\tan^{-1} x - \pi/2}{1/x}$

43. $\lim\limits_{x\to -1} \dfrac{x^3 - x^2 - 5x - 3}{x^4 + 2x^3 - x^2 - 4x - 2}$

44. $\lim\limits_{x\to 1} \dfrac{x^n - 1}{x - 1}$, n is a positive integer

45. $\lim\limits_{x\to\infty} \dfrac{\ln(3x + 5)}{\ln(7x + 3) + 1}$

46. $\lim\limits_{x\to\infty} \dfrac{\ln(3x + 5e^x)}{\ln(7x + 3e^{2x})}$

47. $\lim\limits_{v\to 3} \dfrac{v - 1 - \sqrt{v^2 - 5}}{v - 3}$

48. $\lim\limits_{y\to 2} \dfrac{y^2 + y - 6}{\sqrt{8 - y^2} - y}$

49. $\lim\limits_{x\to 2} \dfrac{x^2 - 4x + 4}{\sin^2 \pi x}$

50. $\lim\limits_{x\to 2} \dfrac{\sqrt[3]{3x + 2} - 2}{x - 2}$

51. $\lim\limits_{x\to\infty} \dfrac{x^2 - \ln(2/x)}{3x^2 + 2x}$

52. $\lim\limits_{x\to 1^+} \left(\dfrac{1}{x - 1} - \dfrac{1}{\sqrt{x - 1}} \right)$

53. $\lim\limits_{x\to 0} x \csc x$

54. $\lim\limits_{x\to 1^-} (1 - x)\tan\dfrac{\pi x}{2}$

55. $\lim\limits_{x\to 0} \csc 6x \sin 7x$

56. $\lim\limits_{x\to\infty} \csc\left(\dfrac{1}{x}\right)(e^{1/x} - 1)$

57. $\lim\limits_{x\to\pi/2^-} \left(\dfrac{\pi}{2} - x\right)\sec x$

58. $\lim\limits_{x\to 0^+} (\sin x)\sqrt{\dfrac{1 - x}{x}}$

59. $\lim\limits_{x\to 1^+} \left(\dfrac{1}{x - 1} - \dfrac{1}{\ln x} \right)$

60. $\lim\limits_{x\to 1^-} \left(\dfrac{x}{x - 1} - \dfrac{x}{\ln x} \right)$

61. $\lim\limits_{x\to 0^+} \left(\cot x - \dfrac{1}{x} \right)$

62. $\lim\limits_{\theta\to\pi/2^-} (\tan\theta - \sec\theta)$

63. $\lim\limits_{x\to\infty} \left(x^2 - \sqrt{x^4 + 16x^2} \right)$

64. $\lim\limits_{x\to\infty} \left(x - \sqrt{x^2 + 4x} \right)$

65. $\lim\limits_{x\to\infty} \left(\sqrt{x - 2} - \sqrt{x - 4} \right)$

66. $\lim\limits_{x\to\infty} x^2 \ln\left(\cos\dfrac{1}{x} \right)$

67. $\lim\limits_{n\to\infty} \dfrac{1 + 2 + \cdots + n}{n^2}$ $\left(\textit{Hint: } \text{Use } 1 + 2 + \cdots + n = \dfrac{n(n + 1)}{2}.\right)$

68. $\lim\limits_{x\to 0^+} x^{1/\ln x}$

69. $\lim\limits_{x\to\infty} \dfrac{\log_2 x}{\log_3 x}$

70. $\lim\limits_{x\to\infty} (\log_2 x - \log_3 x)$

71. $\lim\limits_{x\to 6} \dfrac{\sqrt[5]{5x + 2} - 2}{1/x - 1/6}$

72. $\lim\limits_{x\to\pi/2} (\pi - 2x)\tan x$

73. $\lim\limits_{x\to\infty} x^3\left(\dfrac{1}{x} - \sin\dfrac{1}{x}\right)$

74. $\lim\limits_{x\to\infty} (x^2 e^{1/x} - x^2 - x)$

75. $\lim\limits_{x\to 0^+} x^{2x}$

76. $\lim\limits_{x\to 0} (1 + 4x)^{3/x}$

77. $\lim\limits_{\theta\to\pi/2^-} (\tan\theta)^{\cos\theta}$

78. $\lim\limits_{\theta\to 0^+} (\sin\theta)^{\tan\theta}$

79. $\lim\limits_{x\to 0^+} (1 + x)^{\cot x}$

80. $\lim\limits_{x\to\infty} \left(1 + \dfrac{a}{x}\right)^x$, for a constant a

81. $\lim\limits_{x\to 0} (e^{ax} + x)^{1/x}$, for a constant a

82. $\lim\limits_{z\to\infty} \left(1 + \dfrac{10}{z^2}\right)^{z^2}$

83. $\lim\limits_{x\to 0} (x + \cos x)^{1/x}$

84. An optics limit The theory of interference of coherent oscillators requires the limit $\lim\limits_{\delta\to 2m\pi} \dfrac{\sin^2(N\delta/2)}{\sin^2(\delta/2)}$, where N is a positive integer and m is any integer. Show that the value of this limit is N^2.

85. Compound interest Suppose you make a deposit of \$$P$ into a savings account that earns interest at a rate of $100r\%$ per year.

  a. Show that if interest is compounded once per year, then the balance after t years is $B(t) = P(1 + r)^t$.

  b. If interest is compounded m times per year, then the balance after t years is $B(t) = P\left(1 + \dfrac{r}{m}\right)^{mt}$. For example, $m = 12$ corresponds to monthly compounding, and the interest rate for each month is $\dfrac{r}{12}$. In the limit $m \to \infty$, the compounding is said to be *continuous*. Show that with continuous compounding, the balance after t years is $B(t) = Pe^{rt}$.

86–87. Two methods *Evaluate the following limits in two different ways: Use the methods of Chapter 2 and use l'Hôpital's Rule.*

86. $\lim\limits_{x\to\infty} \dfrac{2x^3 - x^2 + 1}{5x^3 + 2x}$

87. $\lim\limits_{x\to 0} \dfrac{e^{2x} + 4e^x - 5}{e^{2x} - 1}$

88–94. More limits *Evaluate the following limits.*

88. $\lim\limits_{x\to 0} \left(\dfrac{\sin x}{x}\right)^{1/x^2}$

89. $\lim\limits_{x\to 1} \dfrac{x\ln x - x + 1}{x\ln^2 x}$

90. $\lim\limits_{x\to 1} \dfrac{x\ln x + \ln x - 2x + 2}{x^2 \ln^3 x}$

91. $\lim\limits_{x\to 0^+} x^{1/(1 + \ln x)}$

92. $\lim\limits_{n\to\infty} n^2 \ln\left(n\sin\dfrac{1}{n}\right)$

93. $\lim\limits_{x\to 0} \dfrac{a^x - b^x}{x}$, for positive constants a and b

94. $\lim\limits_{x\to 0} (1 + ax)^{b/x}$, for positive constants a and b

95–104. Comparing growth rates *Use limit methods to determine which of the two given functions grows faster, or state that they have comparable growth rates.*

95. x^{10}; $e^{0.01x}$ **96.** $x^2 \ln x$; $\ln^2 x$ **97.** $\ln x^{20}$; $\ln x$

98. $\ln x$; $\ln(\ln x)$ **99.** 100^x; x^x **100.** $x^2 \ln x$; x^3

101. x^{20}; 1.00001^x **102.** $\ln \sqrt{x}$; $\ln^2 x$ **103.** e^{x^2}; e^{10x}

104. e^{x^2}; $x^{x/10}$

105. Explain why or why not Determine whether the following statements are true and give an explanation or counterexample.

   a. By l'Hôpital's Rule, $\lim\limits_{x \to 2} \dfrac{x-2}{x^2-1} = \lim\limits_{x \to 2} \dfrac{1}{2x} = \dfrac{1}{4}$.

   b. $\lim\limits_{x \to 0} x \sin x = \lim\limits_{x \to 0} f(x)g(x) = \lim\limits_{x \to 0} f'(x) \lim\limits_{x \to 0} g'(x) = \left(\lim\limits_{x \to 0} 1\right)\left(\lim\limits_{x \to 0} \cos x\right) = 1.$

   c. $\lim\limits_{x \to 0^+} x^{1/x}$ is an indeterminate form.

   d. The number 1 raised to any fixed power is 1. Therefore, because $(1 + x) \to 1$ as $x \to 0$, $(1 + x)^{1/x} \to 1$ as $x \to 0$.

   e. The functions $\ln x^{100}$ and $\ln x$ have comparable growth rates as $x \to \infty$.

   f. The function e^x grows faster than 2^x as $x \to \infty$.

106–109. Graphing functions *Make a complete graph of the following functions using the graphing guidelines outlined in Section 4.4.*

106. $g(x) = x^2 \ln x$ **107.** $f(x) = x \ln x$

108. $f(x) = \dfrac{\ln x}{x^2}$ **109.** $p(x) = xe^{-x^2/2}$

Explorations and Challenges

T 110. Algorithm complexity The complexity of a computer algorithm is the number of operations or steps the algorithm needs to complete its tasks assuming there are n pieces of input (for example, the number of steps needed to put n numbers in ascending order). Four algorithms for doing the same task have complexities of A: $n^{3/2}$, B: $n \log_2 n$, C: $n(\log_2 n)^2$, and D: $\sqrt{n} \log_2 n$. Rank the algorithms in order of increasing efficiency for large values of n. Graph the complexities as they vary with n and comment on your observations.

111. L'Hôpital loops Consider the limit $\lim\limits_{x \to \infty} \dfrac{\sqrt{ax+b}}{\sqrt{cx+d}}$, where a, b, c, and d are positive real numbers. Show that l'Hôpital's Rule fails for this limit. Find the limit using another method.

112. Let a and b be positive real numbers. Evaluate $\lim\limits_{x \to \infty} \left(ax - \sqrt{a^2x^2 - bx}\right)$ in terms of a and b.

113. Exponential functions and powers Show that any exponential function b^x, for $b > 1$, grows faster than x^p for $p > 0$.

114. Exponentials with different bases Show that $f(x) = a^x$ grows faster than $g(x) = b^x$ as $x \to \infty$ if $1 < b < a$.

115. Logs with different bases Show that $f(x) = \log_a x$ and $g(x) = \log_b x$, where $a > 1$ and $b > 1$, grow at comparable rates as $x \to \infty$.

T 116. Factorial growth rate The factorial function is defined for positive integers as $n! = n(n-1)(n-2) \cdots 3 \cdot 2 \cdot 1$. For example, $5! = 5 \cdot 4 \cdot 3 \cdot 2 \cdot 1 = 120$. A valuable result that gives good approximations to $n!$ for large values of n is *Stirling's formula*, $n! \approx \sqrt{2\pi n}\, n^n\, e^{-n}$. Use this formula and a calculator to determine where the factorial function appears in the ranking of growth rates given in Theorem 4.14 (see the Guided Project *Stirling's formula and n!*).

117. A geometric limit Let $f(\theta)$ be the area of the triangle *ABP* (see figure), and let $g(\theta)$ be the area of the region between the chord *PB* and the arc *PB*. Evaluate $\lim\limits_{\theta \to 0} \dfrac{g(\theta)}{f(\theta)}$.

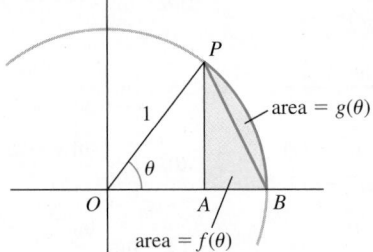

118. Evaluate one of the limits l'Hôpital used in his own textbook in about 1700:

$$\lim_{x \to a} \frac{\sqrt{2a^3 x - x^4} - a\sqrt[3]{a^2 x}}{a - \sqrt[4]{ax^3}}, \text{ where } a \text{ is a real number.}$$

119. Exponential limit Prove that $\lim\limits_{x \to \infty} \left(1 + \dfrac{a}{x}\right)^x = e^a$, for $a \neq 0$.

120. Exponentials vs. superexponentials Show that x^x grows faster than b^x as $x \to \infty$, for $b > 1$.

121. Exponential growth rates

   a. For what values of $b > 0$ does b^x grow faster than e^x as $x \to \infty$?

   b. Compare the growth rates of e^x and e^{ax} as $x \to \infty$, for $a > 0$.

QUICK CHECK ANSWERS

1. g and h **2.** g and h **3.** $0 \cdot \infty$; $(x - \pi/2)/\cot x$
4. The form 0^∞ (for example, $\lim\limits_{x \to 0^+} x^{1/x}$) is not indeterminate, because as the base goes to zero, raising it to larger and larger powers drives the entire function to zero.   **6.** x^3 grows faster than x^2 as $x \to \infty$, whereas x^2 and $10x^2$ have comparable growth rates as $x \to \infty$. ◄

4.8 Newton's Method

➤ Newton's method is attributed to Sir Isaac Newton, who devised the method in 1669. However, similar methods were known prior to Newton's time. A special case of Newton's method for approximating square roots is called the Babylonian method and was probably invented by Greek mathematicians.

A common problem that arises in mathematics is finding the *roots*, or *zeros*, of a function. The roots of a function are the values of x that satisfy the equation $f(x) = 0$. Equivalently, they correspond to the x-intercepts of the graph of f. You have already seen an important example of a root-finding problem. To find the critical points of a function f, we must solve the equation $f'(x) = 0$; that is, we find the roots of f'. Newton's method, which we discuss in this section, is one of the most effective methods for *approximating* the roots of a function.

Why Approximate?

A little background about roots of functions explains why a method is needed to approximate roots. If you are given a linear function, such as $f(x) = 2x - 9$, you know how to use algebraic methods to solve $f(x) = 0$ and find the single root $x = \frac{9}{2}$. Similarly, given the quadratic function $f(x) = x^2 - 6x - 72$, you know how to factor or use the quadratic formula to discover that the roots are $x = 12$ and $x = -6$. It turns out that formulas also exist for finding the roots of cubic (third-degree) and quartic (fourth-degree) polynomials. Methods such as factoring and algebra are called *analytical methods*; when they work, they give the roots of a function *exactly* in terms of arithmetic operations and radicals.

Here is an important fact: Apart from the functions we have listed—polynomials of degree four or less—analytical methods do not give the roots of most functions. To be sure, there are special cases in which analytical methods work. For example, you should verify that the single real root of $f(x) = e^{2x} + 2e^x - 3$ is $x = 0$ and that the two real roots of $f(x) = x^{10} - 1$ are $x = 1$ and $x = -1$. But in general, the roots of even relatively simple functions such as $f(x) = e^{-x} - x$ cannot be found exactly using analytical methods.

When analytical methods do not work, which is the majority of cases, we need another approach. That approach is to approximate roots using *numerical methods*, such as Newton's method.

Deriving Newton's Method

Newton's method is most easily derived geometrically. Assume r is a root of f that we wish to approximate; this means that $f(r) = 0$. We also assume f is differentiable on some interval containing r. Suppose x_0 is an initial approximation to r that is generally obtained by some preliminary analysis. A better approximation to r is often obtained by carrying out the following two steps:

- Draw a line tangent to the curve $y = f(x)$ at the point $(x_0, f(x_0))$.
- Find the point $(x_1, 0)$ at which the tangent line intersects the x-axis, and let x_1 be the new approximation to r.

For the curve shown in **Figure 4.81a**, x_1 is a better approximation to the root r than x_0.

> Sequences are the subject of Chapter 10. An infinite, ordered list of numbers
>
> $$\{x_1, x_2, x_3, \ldots\}$$
>
> is a sequence, and if its values approach a number r, we say that the sequence *converges* to r. If a sequence fails to approach a single number, the sequence *diverges*.

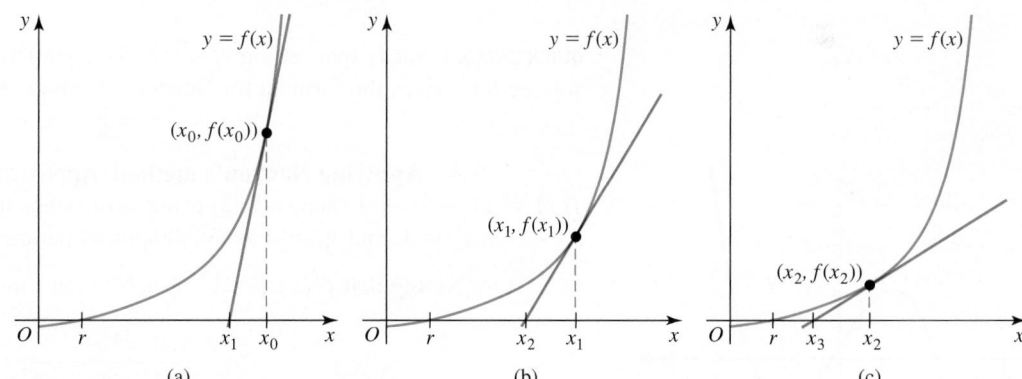

(a) (b) (c)

Figure 4.81

To improve the approximation x_1, we repeat the two-step process, using x_1 to determine the next estimate x_2 (**Figure 4.81b**). Then we use x_2 to obtain x_3 (**Figure 4.81c**), and so forth. Continuing in this fashion, we obtain a *sequence* of approximations $\{x_1, x_2, x_3, \ldots\}$ that ideally get closer and closer, or *converge*, to the root r. Several steps of Newton's method and the convergence of the approximations to the root are shown in **Figure 4.82**.

All that remains is to find a formula that captures the process just described. Assume we have computed the nth approximation x_n to the root r and we want to obtain the next approximation x_{n+1}. We first draw the line tangent to the curve at the point

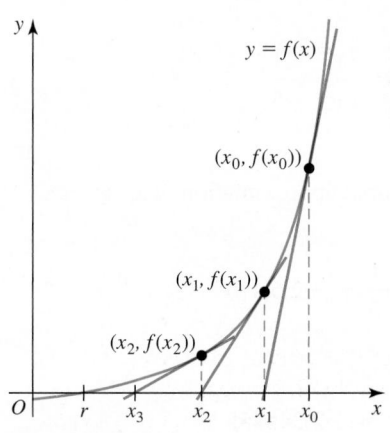

Figure 4.82

▶ Recall that the point-slope form of the equation of a line with slope m passing through (x_n, y_n) is

$$y - y_n = m(x - x_n).$$

$(x_n, f(x_n))$; its slope is $m = f'(x_n)$. Using the point-slope form of the equation of a line, an equation of the tangent line at the point $(x_n, f(x_n))$ is

$$y - f(x_n) = \underbrace{f'(x_n)}_{m}(x - x_n).$$

We find the point at which this line intersects the x-axis by setting $y = 0$ in the equation of the line and solving for x. This value of x becomes the new approximation x_{n+1}:

$$\underbrace{0}_{\text{set } y \text{ to } 0} - f(x_n) = f'(x_n)(x - \underbrace{x_n)}_{\substack{\text{becomes} \\ x_{n+1}}}.$$

Solving for x and calling it x_{n+1}, we find that

$$\underbrace{x_{n+1}}_{\substack{\text{new} \\ \text{approximation}}} = \underbrace{x_n}_{\substack{\text{current} \\ \text{approximation}}} - \frac{f(x_n)}{f'(x_n)}, \quad \text{provided } f'(x_n) \neq 0.$$

We have derived the general step of Newton's method for approximating roots of a function f. This step is repeated for $n = 0, 1, 2, \ldots$, until a termination condition is met (to be discussed).

PROCEDURE   Newton's Method for Approximating Roots of $f(x) = 0$

1. Choose an initial approximation x_0 as close to a root as possible.

2. For $n = 0, 1, 2, \ldots$,

$$x_{n+1} = x_n - \frac{f(x_n)}{f'(x_n)},$$

provided $f'(x_n) \neq 0$.

3. End the calculations when a termination condition is met.

▶ Newton's method is an example of a repetitive loop calculation called an *iteration*. The most efficient way to implement the method is with a calculator or computer. The method is also included in many software packages.

QUICK CHECK 1 Verify that setting $y = 0$ in the equation $y - f(x_n) = f'(x_n)(x - x_n)$ and solving for x gives the formula for Newton's method. ◀

EXAMPLE 1   Applying Newton's method Approximate the roots of $f(x) = x^3 - 5x + 1$ (Figure 4.83) using seven steps of Newton's method. Use $x_0 = -3$, $x_0 = 1$, and $x_0 = 4$ as initial approximations.

SOLUTION Noting that $f'(x) = 3x^2 - 5$, Newton's method takes the form

$$x_{n+1} = x_n - \frac{\overbrace{x_n^3 - 5x_n + 1}^{f(x_n)}}{\underbrace{3x_n^2 - 5}_{f'(x_n)}} = \frac{2x_n^3 - 1}{3x_n^2 - 5},$$

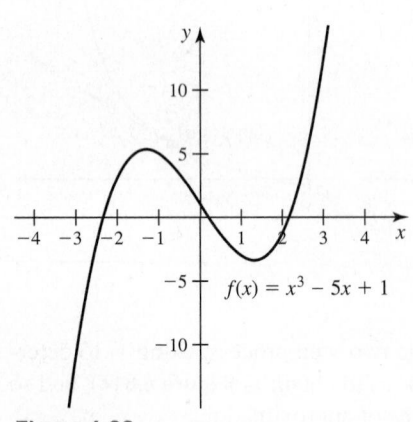

Figure 4.83

where $n = 0, 1, 2, \ldots$, and x_0 is specified. With an initial approximation of $x_0 = -3$, the first approximation is

$$x_1 = \frac{2x_0^3 - 1}{3x_0^2 - 5} = \frac{2(-3)^3 - 1}{3(-3)^2 - 5} = -2.5.$$

The second approximation is

$$x_2 = \frac{2x_1^3 - 1}{3x_1^2 - 5} = \frac{2(-2.5)^3 - 1}{3(-2.5)^2 - 5} \approx -2.345455.$$

Table 4.5

n	x_n	x_n	x_n
0	−3	1	4
1	−2.500000	−0.500000	2.953488
2	−2.345455	0.294118	2.386813
3	−2.330203	0.200215	2.166534
4	−2.330059	0.201639	2.129453
5	−2.330059	0.201640	2.128420
6	−2.330059	0.201640	2.128419
7	−2.330059	0.201640	2.128419

➤ The numbers in Table 4.5 were computed with 16 decimal digits of precision. The results are displayed with 6 digits to the right of the decimal point.

QUICK CHECK 2 What happens if you apply Newton's method to the function $f(x) = x$? ◄

Continuing in this fashion, we generate the first seven approximations shown in Table 4.5. The approximations generated from the initial approximations $x_0 = 1$ and $x_0 = 4$ are also shown in the table.

Notice that with the initial approximation $x_0 = -3$ (second column), the resulting sequence of approximations settles on the value −2.330059 after four iterations, and then there are no further changes in these digits. A similar behavior is seen with the initial approximations $x_0 = 1$ and $x_0 = 4$. Based on this evidence, we conclude that −2.330059, 0.201640, and 2.128419 are approximations to the roots of f with at least six digits (to the right of the decimal point) of accuracy.

A graph of f (**Figure 4.84**) confirms that f has three real roots and that the Newton approximations to the three roots are reasonable. The figure also shows the first three Newton approximations at each root.

Related Exercises 13–14 ◄

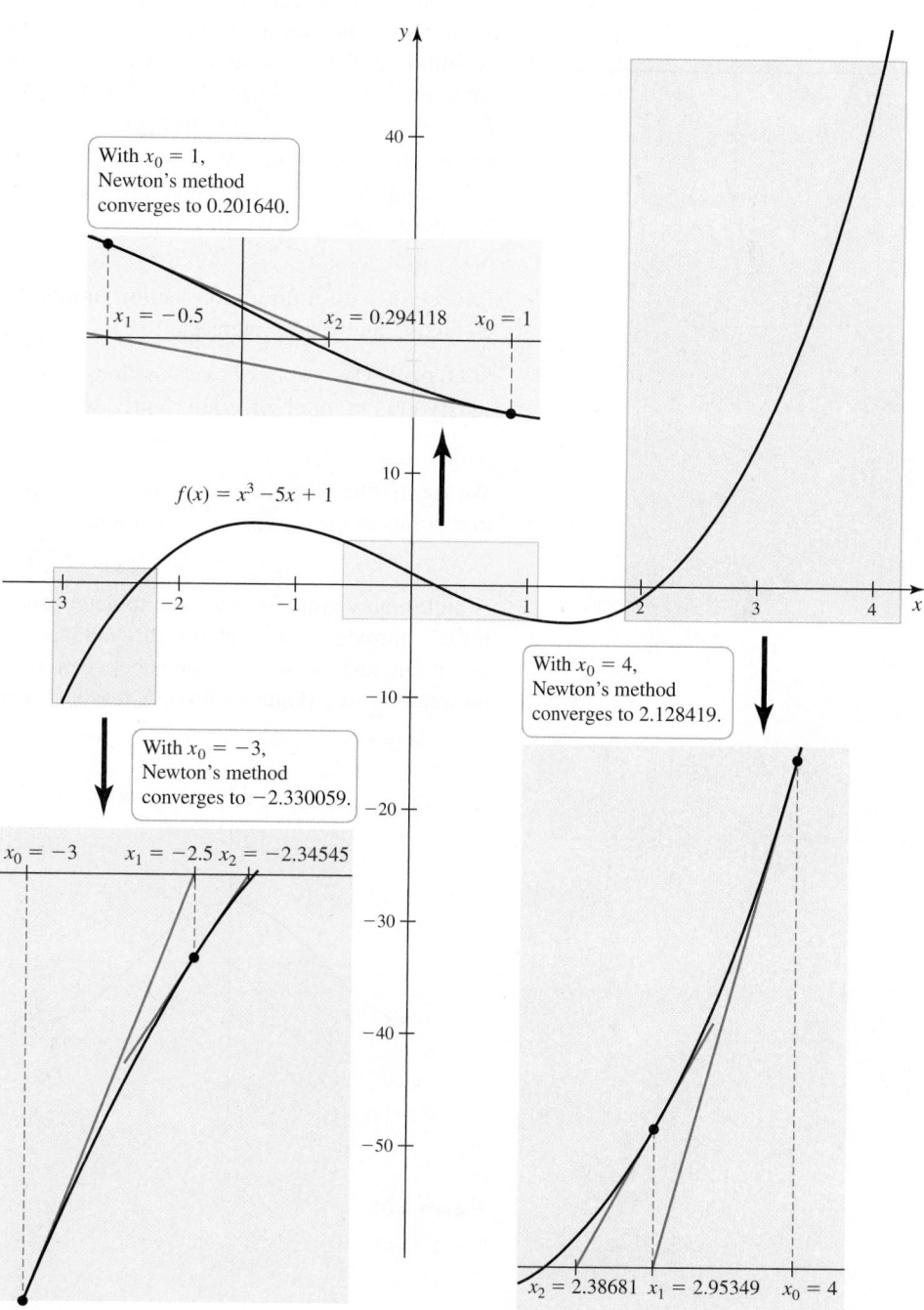

With $x_0 = 1$, Newton's method converges to 0.201640.

$x_1 = -0.5$ $x_2 = 0.294118$ $x_0 = 1$

$f(x) = x^3 - 5x + 1$

With $x_0 = -3$, Newton's method converges to −2.330059.

$x_0 = -3$ $x_1 = -2.5$ $x_2 = -2.34545$

With $x_0 = 4$, Newton's method converges to 2.128419.

$x_2 = 2.38681$ $x_1 = 2.95349$ $x_0 = 4$

Figure 4.84

> ▶ If you write a program for Newton's method, it is a good idea to specify a maximum number of iterations as an escape clause in case the method does not converge.

> ▶ Small residuals do not always imply small errors: The function shown below has a zero at $x = 0$. An approximation such as 0.5 has a small residual but a large error.

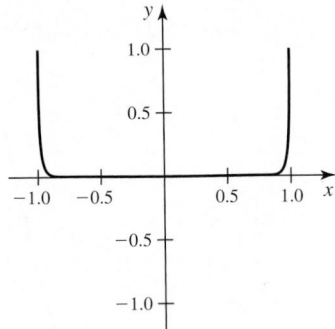

When Do You Stop?

Example 1 raises an important question and gives a practical answer: How many Newton approximations should you compute? Ideally, we would like to compute the **error** in x_n as an approximation to the root r, which is the quantity $|x_n - r|$. Unfortunately, we don't know r in practice; it is the quantity that we are trying to approximate. So we need a practical way to estimate the error.

In the second column of Table 4.5, we see that x_4 and x_5 agree in their seven digits, -2.330059. A general rule of thumb is that if two successive approximations agree to, say, seven digits, then those common digits are accurate (as an approximation to the root). So if you want p digits of accuracy in your approximation, you should compute either until two successive approximations agree to p digits or until some maximum number of iterations is exceeded (in which case Newton's method has failed to find an approximation of the root with the desired accuracy).

There is another practical way to gauge the accuracy of approximations. Because Newton's method generates approximations to a root of f, it follows that as the approximations x_n approach the root, $f(x_n)$ should approach zero. The quantity $f(x_n)$ is called a **residual**, and small residuals usually (but not always) suggest that the approximations have small errors. In Example 1, we find that for the approximations in the second column, $f(x_7) = -1.8 \times 10^{-15}$; for the approximations in the third column, $f(x_7) = 1.1 \times 10^{-16}$; and for the approximations in the fourth column, $f(x_7) = -1.8 \times 10^{-15}$. All these residuals (computed in full precision) are small in magnitude, giving us added confidence that the approximations have small errors.

EXAMPLE 2   Finding intersection points Find the points at which the curves $y = \cos x$ and $y = x$ intersect.

SOLUTION The graphs of two functions g and h intersect at points whose x-coordinates satisfy $g(x) = h(x)$, or, equivalently, where

$$f(x) = g(x) - h(x) = 0.$$

We see that finding intersection points is a root-finding problem. In this case, the intersection points of the curves $y = \cos x$ and $y = x$ satisfy

$$f(x) = \cos x - x = 0.$$

A preliminary graph is advisable to determine the number of intersection points and good initial approximations. From **Figure 4.85a**, we see that the two curves have one intersection point, and its x-coordinate is between 0 and 1. Equivalently, the function f has a zero between 0 and 1 (**Figure 4.85b**). A reasonable initial approximation is $x_0 = 0.5$.

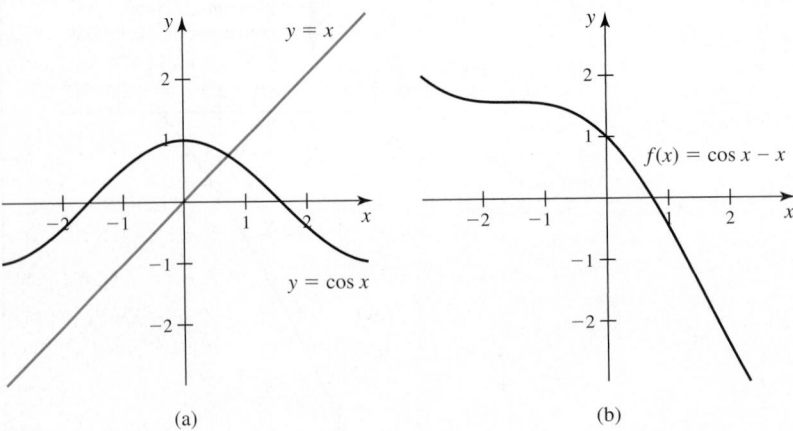

(a)   (b)

Figure 4.85

Table 4.6

n	x_n	Residual
0	0.5	0.377583
1	0.755222	-0.0271033
2	0.739142	-0.0000946154
3	0.739085	-1.18098×10^{-9}
4	0.739085	0
5	0.739085	0

Newton's method takes the form

$$x_{n+1} = x_n - \frac{\overbrace{\cos x_n - x_n}^{f(x_n)}}{\underbrace{-\sin x_n - 1}_{f'(x_n)}} = \frac{x_n \sin x_n + \cos x_n}{\sin x_n + 1}.$$

The results of Newton's method, using an initial approximation of $x_0 = 0.5$, are shown in Table 4.6.

We see that after four iterations, the approximations agree to six digits, so we take 0.739085 as the approximation to the root. Furthermore, the residuals, shown in the last column and computed with full precision, are essentially zero, which confirms the accuracy of the approximation. Therefore, the intersection point is approximately (0.739085, 0.739085) (because the point lies on the line $y = x$).

Related Exercises 27–28 ◄

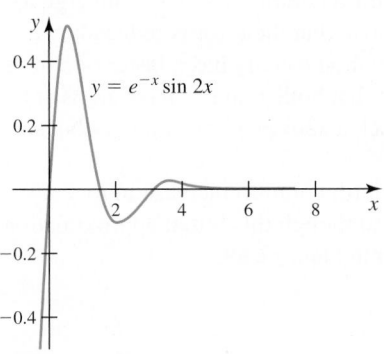

Figure 4.86

Table 4.7

n	x_n	x_n
0	0.200000	2.500000
1	0.499372	1.623915
2	0.550979	2.062202
3	0.553568	2.121018
4	0.553574	2.124360
5	0.553574	2.124371
6	0.553574	2.124371

EXAMPLE 3 **Finding local extrema** Find the x-coordinate of the first local maximum and the first local minimum of the function $f(x) = e^{-x} \sin 2x$ on the interval $(0, \infty)$.

SOLUTION A graph of the function provides some guidance. **Figure 4.86** shows that f has an infinite number of local extrema for $x > 0$. The first local maximum occurs on the interval $[0, 1]$, and the first local minimum occurs on the interval $[2, 3]$.

To locate the local extrema, we must find the critical points by solving

$$f'(x) = e^{-x}(2 \cos 2x - \sin 2x) = 0.$$

To this equation, we apply Newton's method. The results of the calculations, using initial approximations of $x_0 = 0.2$ and $x_0 = 2.5$, are shown in Table 4.7.

Newton's method finds the two critical points quickly, and they are consistent with the graph of f. We conclude that the first local maximum occurs at $x \approx 0.553574$ and the first local minimum occurs at $x \approx 2.124371$.

Related Exercises 43, 46 ◄

▶ A more thorough analysis of the rate at which Newton's method converges and the ways in which it fails to converge is presented in a course in numerical analysis.

Newton's method is widely used because in general, it has a remarkable rate of convergence; the number of digits of accuracy roughly doubles with each iteration.

Pitfalls of Newton's Method

Given a good initial approximation, Newton's method usually converges to a root. In addition, when the method converges, it usually does so quickly. However, when Newton's method fails, it does so in curious and spectacular ways. The formula for Newton's method suggests one way in which the method could encounter difficulties: The term $f'(x_n)$ appears in a denominator, so if at any step $f'(x_n) = 0$, then the method breaks down. Furthermore, if $f'(x_n)$ is close to zero at any step, then the method may converge slowly or may fail to converge. The following example shows three ways in which Newton's method may go awry.

EXAMPLE 4 **Difficulties with Newton's method** Find the root of $f(x) = \dfrac{8x^2}{3x^2 + 1}$ using Newton's method with initial approximations of $x_0 = 1$, $x_0 = 0.15$, and $x_0 = 1.1$.

SOLUTION Notice that f has the single root $x = 0$. So the point of the example is not to find the root, but to investigate the performance of Newton's method. Computing f' and doing a few steps of algebra show that the formula for Newton's method is

$$x_{n+1} = x_n - \frac{f(x_n)}{f'(x_n)} = \frac{x_n}{2}(1 - 3x_n^2).$$

The results of five iterations of Newton's method are displayed in Table 4.8, and they tell three different stories.

Table 4.8

n	x_n	x_n	x_n
0	1	0.15	1.1
1	-1	0.0699375	-1.4465
2	1	0.0344556	3.81665
3	-1	0.0171665	-81.4865
4	1	0.00857564	8.11572×10^5
5	-1	0.00428687	-8.01692×10^{17}

The approximations generated using $x_0 = 1$ (second column) get stuck in a cycle that alternates between $+1$ and -1. The geometry underlying this rare occurrence is illustrated in **Figure 4.87**.

The approximations generated using $x_0 = 0.15$ (third column) actually converge to the root 0, but they converge slowly (**Figure 4.88**). Notice that the error is reduced by a factor of approximately 2 with each step. Newton's method usually has a faster rate of error reduction. The slow convergence is due to the fact that both f and f' have zeros at 0. As mentioned earlier, if the approximations x_n approach a zero of f', the rate of convergence is often compromised.

The approximations generated using $x_0 = 1.1$ (fourth column) increase in magnitude quickly and do not converge to a finite value, even though this initial approximation seems reasonable. The geometry of this case is shown in **Figure 4.89**.

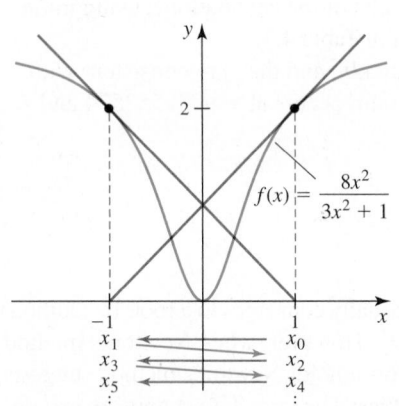

Figure 4.87

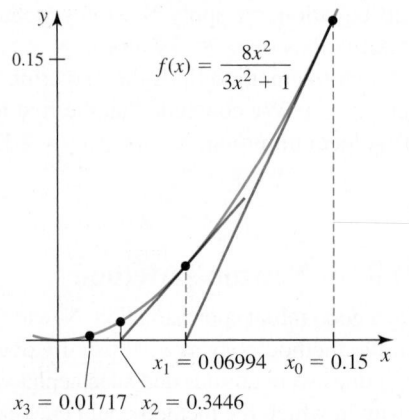

Figure 4.88

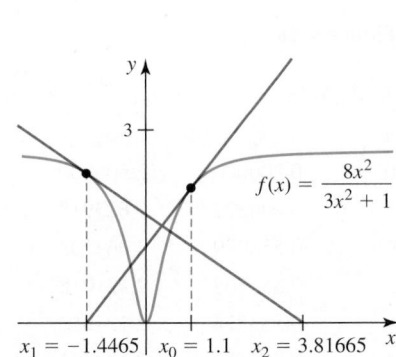

Figure 4.89

The three cases in this example illustrate several ways in which Newton's method may fail to converge at its usual rate: The approximations may cycle or wander, they may converge slowly, or they may diverge (often at a rapid rate).   *Related Exercises 52–53* ◄

SECTION 4.8 EXERCISES

Getting Started

1. Give a geometric explanation of Newton's method.

2. Explain how the iteration formula for Newton's method works.

3. A graph of f and the lines tangent to f at $x = 1, 2,$ and 3 are given. If $x_0 = 3$, find the values of $x_1, x_2,$ and x_3 that are obtained by applying Newton's method.

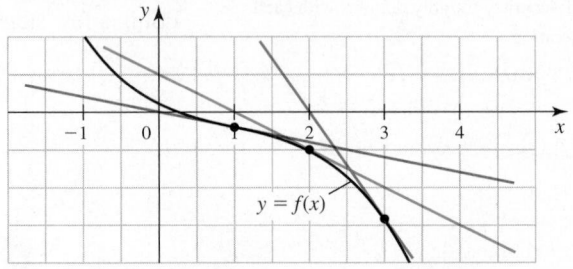

4. A graph of f and the lines tangent to f at $x = -3$, 2, and 3 are given. If $x_0 = 3$, find the values of x_1, x_2, and x_3 that are obtained by applying Newton's method.

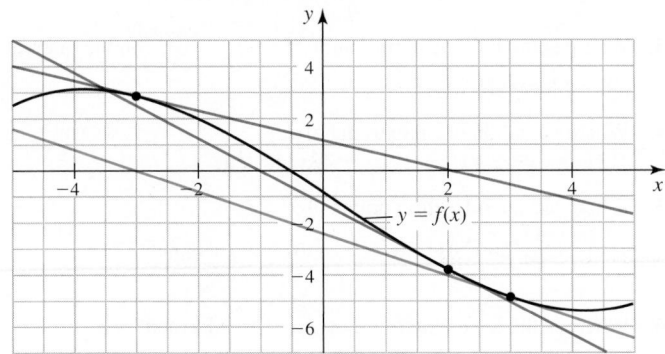

5. Let $f(x) = 2x^3 - 6x^2 + 4x$. Use Newton's method to find x_1 given that $x_0 = 1.4$. Use the graph of f (see figure) and an appropriate tangent line to illustrate how x_1 is obtained from x_0.

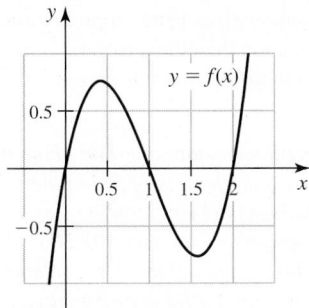

6. The function $f(x) = \dfrac{4x}{x^2 + 4}$ is differentiable and has a local maximum at $x = 2$, where $f'(2) = 0$ (see figure).

   a. Use a graphical explanation to show that Newton's method applied to $f(x) = 0$ fails to produce a value x_1 for the initial approximation $x_0 = 2$.

   b. Use the formula for Newton's method to show that Newton's method fails to produce a value x_1 with the initial approximation $x_0 = 2$.

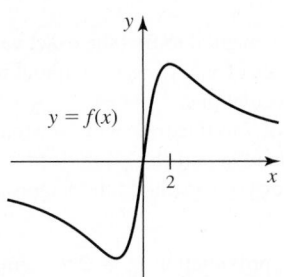

7. How do you decide when to terminate Newton's method?

8. Give the formula for Newton's method for the function $f(x) = x^2 - 5$.

T **9–12.** *Write the formula for Newton's method and use the given initial approximation to compute the approximations x_1 and x_2.*

9. $f(x) = x^2 - 6$; $x_0 = 3$ **10.** $f(x) = x^2 - 2x - 3$; $x_0 = 2$

11. $f(x) = e^{-x} - x$; $x_0 = \ln 2$ **12.** $f(x) = x^3 - 2$; $x_0 = 2$

Practice Exercises

T **13–19. Finding roots with Newton's method** *For the given function f and initial approximation x_0, use Newton's method to approximate a root of f. Stop calculating approximations when two successive approximations agree to five digits to the right of the decimal point after rounding. Show your work by making a table similar to that in Example 1.*

13. $f(x) = x^2 - 10$; $x_0 = 3$

14. $f(x) = x^3 + x^2 + 1$; $x_0 = -1.5$

15. $f(x) = \sin x + x - 1$; $x_0 = 0.5$

16. $f(x) = e^x + x - 5$; $x_0 = 1.6$

17. $f(x) = \tan x - 2x$; $x_0 = 1.2$

18. $f(x) = x \ln(x + 1) - 1$; $x_0 = 1.7$

19. $f(x) = \cos^{-1} x - x$; $x_0 = 0.75$

T **20–26. Finding all roots** *Use Newton's method to find all the roots of the following functions. Use preliminary analysis and graphing to determine good initial approximations.*

20. $f(x) = \cos x - \dfrac{x}{7}$ **21.** $f(x) = \cos 2x - x^2 + 2x$

22. $f(x) = \dfrac{x}{6} - \sec x$ on $[0, 8]$

23. $f(x) = e^{-x} - \dfrac{x + 4}{5}$ **24.** $f(x) = \dfrac{x^5}{5} - \dfrac{x^3}{4} - \dfrac{1}{20}$

25. $f(x) = \ln x - x^2 + 3x - 1$

26. $f(x) = x^2(x - 100) + 1$

T **27–32. Finding intersection points** *Use Newton's method to approximate all the intersection points of the following pairs of curves. Some preliminary graphing or analysis may help in choosing good initial approximations.*

27. $y = \sin x$ and $y = \dfrac{x}{2}$

28. $y = e^x$ and $y = x^3$

29. $y = \dfrac{1}{x}$ and $y = 4 - x^2$

30. $y = x^3$ and $y = x^2 + 1$

31. $y = 4\sqrt{x}$ and $y = x^2 + 1$

32. $y = \ln x$ and $y = x^3 - 2$

T **33–34. Tumor size** *In a study conducted at Dartmouth College, mice with a particular type of cancerous tumor were treated with the chemotherapy drug Cisplatin. If the volume of one of these tumors at the time of treatment is V_0, then the volume of the tumor t days after treatment is modeled by the function $V(t) = V_0(0.99e^{-0.1216t} + 0.01e^{0.239t})$. (Source: Undergraduate Mathematics for the Life Sciences, MAA Notes No. 81, 2013)*

33. Plot a graph of $y = 0.99e^{-0.1216t} + 0.01e^{0.239t}$, for $0 \le t \le 16$, and describe the tumor size over time. Use Newton's method to determine when the tumor decreases to half of its original size.

34. Researchers plan to give a second treatment of Cisplatin just before the tumor starts growing again. Use the derivative of $y = 0.99e^{-0.1216t} + 0.01e^{0.239t}$ and Newton's method to determine when a second treatment should be given.

T 35. Retirement account Assume you invest $10,000 at the end of each year for 30 years at an annual interest rate of r. The amount of money in your account after 30 years is

$$A = \frac{10,000((1 + r)^{30} - 1)}{r}.$$ Assume you want $1,000,000 in your account after 30 years.

a. Show that the minimum value of r required to meet your investment needs satisfies the equation
$1,000,000r - 10,000(1 + r)^{30} + 10,000 = 0.$
b. Apply Newton's method to solve the equation in part (a) to find the interest rate required to meet your investment goal.

T 36. Investment problem A one-time investment of $2500 is deposited in a 5-year savings account paying a fixed annual interest rate r, with monthly compounding. The amount of money in the account after 5 years is $A(r) = 2500\left(1 + \dfrac{r}{12}\right)^{60}$.

a. Use Newton's method to find the value of r if the goal is to have $3200 in the account after 5 years.
b. Verify your answer to part (a) algebraically.

T 37. A damped oscillator The displacement of an object as it bounces vertically up and down on a spring is given by $y(t) = 2.5e^{-t}\cos 2t$, where the initial displacement is $y(0) = 2.5$ and $y = 0$ corresponds to the rest position (see figure).

a. Find the time at which the object first passes the rest position, $y = 0$.
b. Find the time and the displacement when the object reaches its lowest point.
c. Find the time at which the object passes the rest position for the second time.
d. Find the time and the displacement when the object reaches its high point for the second time.

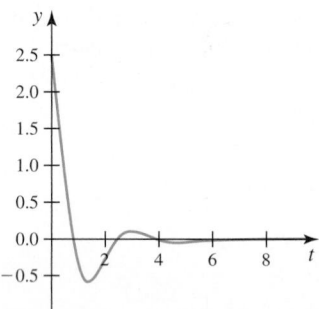

T 38. The sinc function The sinc function, $\text{sinc}(x) = \dfrac{\sin x}{x}$ for $x \neq 0$, $\text{sinc}(0) = 1$, appears frequently in signal-processing applications.

a. Graph the sinc function on $[-2\pi, 2\pi]$.
b. Locate the first local minimum and the first local maximum of $\text{sinc}(x)$, for $x > 0$.

T 39–42. Estimating roots *The values of various roots can be approximated using Newton's method. For example, to approximate the value of $\sqrt[3]{10}$, we let $x = \sqrt[3]{10}$ and cube both sides of the equation to obtain $x^3 = 10$, or $x^3 - 10 = 0$. Therefore, $\sqrt[3]{10}$ is a root of $p(x) = x^3 - 10$, which we can approximate by applying Newton's method. Approximate each value of r by first finding a polynomial with integer coefficients that has a root r. Use an appropriate value of x_0 and stop calculating approximations when two successive approximations agree to five digits to the right of the decimal point after rounding.*

39. $r = 7^{1/4}$ **40.** $r = 2^{1/3}$

41. $r = (-9)^{1/3}$ **42.** $r = (-67)^{1/5}$

T 43–46. Newton's method and curve sketching *Use Newton's method to find approximate answers to the following questions.*

43. Where is the first local minimum of $f(x) = \dfrac{\cos x}{x}$ on the interval $(0, \infty)$ located?

44. Where are all the local extrema of $f(x) = 3x^4 + 8x^3 + 12x^2 + 48x$ located?

45. Where are the inflection points of $f(x) = \dfrac{9}{5}x^5 - \dfrac{15}{2}x^4 + \dfrac{7}{3}x^3 + 30x^2 + 1$ located?

46. Where is the local extremum of $f(x) = \dfrac{e^x}{x}$ located?

47. Explain why or why not Determine whether the following statements are true and give an explanation or counterexample.

a. Newton's method is an example of a numerical method for approximating the roots of a function.
b. Newton's method gives better approximations to the roots of a quadratic equation than the quadratic formula.
c. Newton's method always finds an approximate root of a function.

T 48–51. Fixed points *An important question about many functions concerns the existence and location of fixed points. A **fixed point** of f is a value of x that satisfies the equation $f(x) = x$; it corresponds to a point at which the graph of f intersects the line $y = x$. Find all the fixed points of the following functions. Use preliminary analysis and graphing to determine good initial approximations.*

48. $f(x) = x^3 - 3x^2 + x + 1$ **49.** $f(x) = \cos x$

50. $f(x) = \tan\dfrac{x}{2}$ on $(-\pi, \pi)$ **51.** $f(x) = 2x\cos x$ on $[0, 2]$

Explorations and Challenges

T 52–53. Pitfalls of Newton's method *Let $f(x) = \dfrac{x}{1 + x^2}$, which has just one root, $r = 0$.*

52. Use the initial approximation $x_0 = 1/\sqrt{3}$ to complete the following steps.

a. Use Newton's method to find the exact values of x_1 and x_2.
b. State the values of $x_3, x_4, x_5, \ldots$ without performing any additional calculations.
c. Use a graph of f to illustrate why Newton's method produces the values found in part (b).
d. Why does Newton's method fail to approximate the root $r = 0$ if $x_0 = 1/\sqrt{3}$?

53. Use the initial approximation $x_0 = 2$ to complete the following steps.

a. Use Newton's method to find the values of $x_1, x_2,$ and x_3.
b. Use a graph of f to illustrate why Newton's method produces the values x_1 and x_2 found in part (a).
c. Why does Newton's method fail to approximate the root $r = 0$ if $x_0 = 2$?

T 54. Approximating square roots Let $a > 0$ be given and suppose we want to approximate $\sqrt{a}$ using Newton's method.

a. Explain why the square root problem is equivalent to finding the positive root of $f(x) = x^2 - a$.

b. Show that Newton's method applied to this function takes the form (sometimes called the Babylonian method)

$$x_{n+1} = \frac{1}{2}\left(x_n + \frac{a}{x_n}\right), \text{ for } n = 0, 1, 2, \ldots.$$

c. How would you choose initial approximations to approximate $\sqrt{13}$ and $\sqrt{73}$?

d. Approximate $\sqrt{13}$ and $\sqrt{73}$ with at least ten significant digits.

T 55. Approximating reciprocals To approximate the reciprocal of a number a without using division, we can apply Newton's method to the function $f(x) = \dfrac{1}{x} - a$.

a. Verify that Newton's method gives the formula $x_{n+1} = (2 - ax_n)x_n$.

b. Apply Newton's method with $a = 7$ using a starting value of your choice. Compute an approximation with eight digits of accuracy. What number does Newton's method approximate in this case?

T 56. Modified Newton's method The function f has a root of *multiplicity* 2 at r if $f(r) = f'(r) = 0$ and $f''(r) \ne 0$. In this case, a slight modification of Newton's method, known as the *modified* (or *accelerated*) Newton's method, is given by the formula

$$x_{n+1} = x_n - \frac{2f(x_n)}{f'(x_n)}, \text{ for } n = 0, 1, 2, \ldots.$$

This modified form generally increases the rate of convergence.

a. Verify that 0 is a root of multiplicity 2 of the function $f(x) = e^{2\sin x} - 2x - 1$.

b. Apply Newton's method and the modified Newton's method using $x_0 = 0.1$ to find the value of x_3 in each case. Compare the accuracy of these values of x_3.

c. Consider the function $f(x) = \dfrac{8x^2}{3x^2 + 1}$ given in Example 4. Use the modified Newton's method to find the value of x_3 using $x_0 = 0.15$. Compare this value to the value of x_3 found in Example 4 with $x_0 = 0.15$.

T 57. An eigenvalue problem A certain kind of differential equation (see Chapter 9) leads to the root-finding problem $\tan \pi\lambda = \lambda$, where the roots λ are called **eigenvalues**. Find the first three positive eigenvalues of this problem.

T 58. Fixed points of quadratics and quartics Let $f(x) = ax(1 - x)$, where a is a real number and $0 \le x \le 1$. Recall that the fixed point of a function is a value of x such that $f(x) = x$ (Exercises 48–51).

a. Without using a calculator, find the values of a, with $0 < a \le 4$, such that f has a fixed point. Give the fixed point in terms of a.

b. Consider the polynomial $g(x) = f(f(x))$. Write g in terms of a and powers of x. What is its degree?

c. Graph g for $a = 2, 3$, and 4.

d. Find the number and location of the fixed points of g for $a = 2, 3$, and 4 on the interval $0 \le x \le 1$.

T 59. Basins of attraction Suppose f has a real root r and Newton's method is used to approximate r with an initial approximation x_0. The **basin of attraction** of r is the set of initial approximations that produce a sequence that converges to r. Points near r are often in the basin of attraction of r—but not always. Sometimes an initial approximation x_0 may produce a sequence that doesn't converge, and sometimes an initial approximation x_0 may produce a sequence that converges to a distant root. Let $f(x) = (x + 2)(x + 1)(x - 3)$, which has roots $x = -2, -1$, and 3. Use Newton's method with initial approximations on the interval $[-4, 4]$ and determine (approximately) the basin of each root.

T 60. The functions $f(x) = (x - 1)^2$ and $g(x) = x^2 - 1$ both have a root at $x = 1$. Apply Newton's method to both functions with an initial approximation $x_0 = 2$. Compare the rates at which the method converges in these cases, and give an explanation.

QUICK CHECK ANSWERS

1. $0 - f(x_n) = f'(x_n)(x - x_n) \implies -\dfrac{f(x_n)}{f'(x_n)} = x - x_n \implies$

$x = x_n - \dfrac{f(x_n)}{f'(x_n)}$ **2.** Newton's method will find the root

$x = 0$ exactly in one step. ◄

4.9 Antiderivatives

The goal of differentiation is to find the derivative f' of a given function f. The reverse process, called *antidifferentiation*, is equally important: Given a function f, we look for an *antiderivative* function F whose derivative is f; that is, a function F such that $F' = f$.

> **DEFINITION Antiderivative**
>
> A function F is an **antiderivative** of f on an interval I provided $F'(x) = f(x)$, for all x in I.

In this section, we revisit derivative formulas developed in previous chapters to discover corresponding antiderivative formulas.

Thinking Backward

Consider the derivative formula $\dfrac{d}{dx}(x) = 1$. It implies that an antiderivative of $f(x) = 1$ is $F(x) = x$ because $F'(x) = f(x)$. Using the same logic, we can write

$$\frac{d}{dx}(x^2) = 2x \quad \Rightarrow \quad \text{an antiderivative of } f(x) = 2x \text{ is } F(x) = x^2 \text{ and}$$

$$\frac{d}{dx}(\sin x) = \cos x \quad \Rightarrow \quad \text{an antiderivative of } f(x) = \cos x \text{ is } F(x) = \sin x.$$

QUICK CHECK 1 Verify by differentiation that x^4 is an antiderivative of $4x^3$. ◄

Each of these proposed antiderivative formulas is easily checked by showing that $F' = f$.

An immediate question arises: Does a function have more than one antiderivative? To answer this question, let's focus on $f(x) = 1$ and the antiderivative $F(x) = x$. Because the derivative of a constant C is zero, we see that $F(x) = x + C$ is also an antiderivative of $f(x) = 1$, which is easy to check:

$$F'(x) = \frac{d}{dx}(x + C) = 1 = f(x).$$

Therefore, $f(x) = 1$ actually has an infinite number of antiderivatives. For the same reason, any function of the form $F(x) = x^2 + C$ is an antiderivative of $f(x) = 2x$, and any function of the form $F(x) = \sin x + C$ is an antiderivative of $f(x) = \cos x$, where C is an arbitrary constant.

We might ask whether there are still *more* antiderivatives of a given function. The following theorem provides the answer.

THEOREM 4.15 The Family of Antiderivatives

Let F be any antiderivative of f on an interval I. Then *all* the antiderivatives of f on I have the form $F + C$, where C is an arbitrary constant.

Proof: Suppose F and G are antiderivatives of f on an interval I. Then $F' = f$ and $G' = f$, which implies that $F' = G'$ on I. Theorem 4.6 states that functions with equal derivatives differ by a constant. Therefore, $G = F + C$, and all antiderivatives of f have the form $F + C$, where C is an arbitrary constant. ◄

Theorem 4.15 says that while there are infinitely many antiderivatives of a function, they are all of one family, namely, those functions of the form $F + C$. Because the antiderivatives of a particular function differ by a constant, the antiderivatives are vertical translations of one another (**Figure 4.90**).

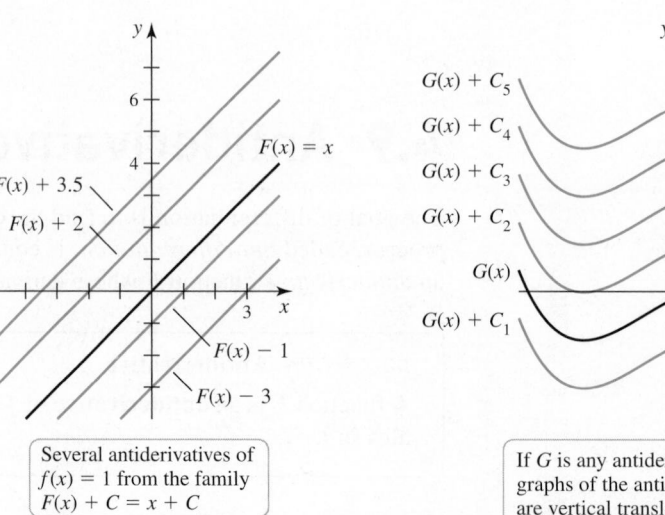

Several antiderivatives of $f(x) = 1$ from the family $F(x) + C = x + C$

If G is any antiderivative of g, the graphs of the antiderivatives $G + C$ are vertical translations of one another.

Figure 4.90

EXAMPLE 1   Finding antiderivatives Use what you know about derivatives to find all antiderivatives of the following functions.

a. $f(x) = 3x^2$ **b.** $f(x) = \dfrac{1}{1 + x^2}$ **c.** $f(t) = \sin t$

SOLUTION

a. Note that $\dfrac{d}{dx}(x^3) = 3x^2$. Therefore, an antiderivative of $f(x) = 3x^2$ is x^3. By Theorem 4.15, the complete family of antiderivatives is $F(x) = x^3 + C$, where C is an arbitrary constant.

b. Because $\dfrac{d}{dx}(\tan^{-1} x) = \dfrac{1}{1 + x^2}$, all antiderivatives of f are of the form $F(x) = \tan^{-1} x + C$, where C is an arbitrary constant.

c. Recall that $\dfrac{d}{dt}(\cos t) = -\sin t$. We seek a function whose derivative is $\sin t$, not $-\sin t$. Observing that $\dfrac{d}{dt}(-\cos t) = \sin t$, it follows that the antiderivatives of $\sin t$ are $F(t) = -\cos t + C$, where C is an arbitrary constant.

Related Exercises 12–13 ◄

> **QUICK CHECK 2** Find the family of antiderivatives for each of $f(x) = e^x$, $g(x) = 4x^3$, and $h(x) = \sec^2 x$. ◄

Indefinite Integrals

The notation $\dfrac{d}{dx}(f(x))$ means *take the derivative of $f(x)$* with respect to x. We need analogous notation for antiderivatives. For historical reasons that become apparent in the next chapter, the notation that means *find the antiderivatives of f* is the **indefinite integral** $\int f(x)\, dx$. Every time an indefinite integral sign $\int$ appears, it is followed by a function called the **integrand**, which in turn is followed by the differential dx. For now, dx simply means that x is the independent variable, or the **variable of integration**. The notation $\int f(x)\, dx$ represents *all* the antiderivatives of f. When the integrand is a function of a variable different from x—say, $g(t)$—we write $\int g(t)\, dt$ to represent the antiderivatives of g.

Using this new notation, the three results of Example 1 are written

$$\int 3x^2\, dx = x^3 + C, \quad \int \frac{1}{1 + x^2}\, dx = \tan^{-1} x + C, \quad \text{and} \quad \int \sin t\, dt = -\cos t + C,$$

where C is an arbitrary constant called a **constant of integration**. The derivative formulas presented earlier in the text may be written in terms of indefinite integrals. We begin with the Power Rule.

> ➤ Notice that if $p = -1$ in Theorem 4.16, then $F(x)$ is undefined. The antiderivative of $f(x) = x^{-1}$ is discussed in Example 5. The case $p = 0$ says that $\int 1\, dx = x + C$.

THEOREM 4.16   Power Rule for Indefinite Integrals

$$\int x^p\, dx = \frac{x^{p+1}}{p + 1} + C,$$

where $p \neq -1$ is a real number and C is an arbitrary constant.

> ➤ Any indefinite integral calculation can be checked by differentiation: The derivative of the alleged indefinite integral must equal the integrand.

Proof: The theorem says that the antiderivatives of $f(x) = x^p$ have the form $F(x) = \dfrac{x^{p+1}}{p + 1} + C$. Differentiating F, we verify that $F'(x) = f(x)$, provided $p \neq -1$:

$$F'(x) = \frac{d}{dx}\left(\frac{x^{p+1}}{p + 1} + C\right)$$

$$= \frac{d}{dx}\left(\frac{x^{p+1}}{p + 1}\right) + \underbrace{\frac{d}{dx}(C)}_{0}$$

$$= \frac{(p + 1)x^{(p+1)-1}}{p + 1} + 0 = x^p.$$

◄

Theorems 3.4 and 3.5 (Section 3.3) state the Constant Multiple and Sum Rules for derivatives. Here are the corresponding antiderivative rules, which are proved by differentiation.

THEOREM 4.17 **Constant Multiple and Sum Rules**

Constant Multiple Rule: $\int cf(x)\,dx = c\int f(x)\,dx$, for real numbers c

Sum Rule: $\int (f(x) + g(x))\,dx = \int f(x)\,dx + \int g(x)\,dx$

The following example shows how Theorems 4.16 and 4.17 are used.

EXAMPLE 2 **Indefinite integrals** Determine the following indefinite integrals.

a. $\int (3x^5 + 2 - 5\sqrt{x})\,dx$ **b.** $\int \left(\dfrac{4x^{19} - 5x^{-8}}{x^2}\right)dx$ **c.** $\int (z^2 + 1)(2z - 5)\,dz$

> $\int dx$ means $\int 1\,dx$, which is the indefinite integral of the constant function $f(x) = 1$, so $\int dx = x + C$.

> Each indefinite integral in Example 2a produces an arbitrary constant, all of which may be combined in one arbitrary constant called C.

SOLUTION

a. $\int (3x^5 + 2 - 5\sqrt{x})\,dx = \int 3x^5\,dx + \int 2\,dx - \int 5x^{1/2}\,dx$ Sum Rule

$\qquad = 3\int x^5\,dx + 2\int dx - 5\int x^{1/2}\,dx$ Constant Multiple Rule

$\qquad = 3\cdot\dfrac{x^6}{6} + 2\cdot x - 5\cdot\dfrac{x^{3/2}}{3/2} + C$ Power Rule

$\qquad = \dfrac{x^6}{2} + 2x - \dfrac{10}{3}x^{3/2} + C$ Simplify.

b. $\int \left(\dfrac{4x^{19} - 5x^{-8}}{x^2}\right)dx = \int (4x^{17} - 5x^{-10})\,dx$ Simplify the integrand.

$\qquad = 4\int x^{17}\,dx - 5\int x^{-10}\,dx$ Sum and Constant Multiple Rules

$\qquad = 4\cdot\dfrac{x^{18}}{18} - 5\cdot\dfrac{x^{-9}}{-9} + C$ Power Rule

$\qquad = \dfrac{2x^{18}}{9} + \dfrac{5x^{-9}}{9} + C$ Simplify.

> Examples 2b and 2c show that in general, the indefinite integral of a product or quotient is not the product or quotient of indefinite integrals.

c. $\int (z^2 + 1)(2z - 5)\,dz = \int (2z^3 - 5z^2 + 2z - 5)\,dz$ Expand integrand.

$\qquad = \dfrac{1}{2}z^4 - \dfrac{5}{3}z^3 + z^2 - 5z + C$ Integrate each term.

All these results should be checked by differentiation.

Related Exercises 24, 25, 31, 35 ◄

Indefinite Integrals of Trigonometric Functions

We used a familiar derivative formula in Example 1c to find the antiderivative of $\sin x$. Our goal in this section is to write the other derivative results for trigonometric functions as indefinite integrals.

EXAMPLE 3 **Indefinite integrals of trigonometric functions** Evaluate the following indefinite integrals.

a. $\int \sec^2 x\,dx$ **b.** $\int (2x + 3\cos x)\,dx$ **c.** $\int \dfrac{\sin x}{\cos^2 x}\,dx$

> Remember the words that go with
> antiderivatives and indefinite integrals.
> The statement $\dfrac{d}{dx}(\tan x) = \sec^2 x$ says
> that $\tan x$ can be differentiated to get
> $\sec^2 x$. Therefore,
> $$\int \sec^2 x \, dx = \tan x + C.$$

SOLUTION

a. The derivative result $\dfrac{d}{dx}(\tan x) = \sec^2 x$ is reversed to produce the indefinite integral

$$\int \sec^2 x \, dx = \tan x + C.$$

b. We first split the integral into two integrals using Theorem 4.17:

$$\int (2x + 3\cos x)\, dx = 2\int x\, dx + 3\int \cos x\, dx. \quad \text{Sum and Constant Multiple Rules}$$

The first of these new integrals is handled by the Power Rule, and the second integral is evaluated by reversing the derivative result $\dfrac{d}{dx}(\sin x) = \cos x$:

$$2\int x\, dx + 3\int \cos x\, dx = 2\cdot\frac{x^2}{2} + 3\sin x + C \quad \begin{array}{l}\text{Power Rule; } \frac{d}{dx}(\sin x) = \cos x \Rightarrow \\ \int \cos x\, dx = \sin x + C\end{array}$$

$$= x^2 + 3\sin x + C. \quad \text{Simplify.}$$

c. When we rewrite the integrand, a familiar derivative formula emerges:

$$\int \frac{\sin x}{\cos^2 x}\, dx = \int \underbrace{\frac{1}{\cos x}}_{\sec x}\cdot \underbrace{\frac{\sin x}{\cos x}}_{\tan x}\, dx$$

$$= \int \sec x \tan x\, dx \quad \text{Rewrite the integrand.}$$

$$= \sec x + C. \quad \begin{array}{l}\frac{d}{dx}(\sec x) = \sec x \tan x \Rightarrow \\ \int \sec x \tan x\, dx = \sec x + C\end{array}$$

Related Exercises 40, 41, 47 ◄

The ideas illustrated in Example 3 are used to obtain the integrals in Table 4.9, where we assume C is an arbitrary constant. Example 4 provides additional integrals involving trigonometric functions.

Table 4.9 Indefinite Integrals of Trigonometric Functions

1. $\dfrac{d}{dx}(\sin x) = \cos x \quad \Rightarrow \quad \int \cos x\, dx = \sin x + C$

2. $\dfrac{d}{dx}(\cos x) = -\sin x \quad \Rightarrow \quad \int \sin x\, dx = -\cos x + C$

3. $\dfrac{d}{dx}(\tan x) = \sec^2 x \quad \Rightarrow \quad \int \sec^2 x\, dx = \tan x + C$

4. $\dfrac{d}{dx}(\cot x) = -\csc^2 x \quad \Rightarrow \quad \int \csc^2 x\, dx = -\cot x + C$

5. $\dfrac{d}{dx}(\sec x) = \sec x \tan x \quad \Rightarrow \quad \int \sec x \tan x\, dx = \sec x + C$

6. $\dfrac{d}{dx}(\csc x) = -\csc x \cot x \quad \Rightarrow \quad \int \csc x \cot x\, dx = -\csc x + C$

QUICK CHECK 3 Use differentiation to verify result 6 in Table 4.9:
$$\int \csc x \cot x\, dx = -\csc x + C. ◄$$

EXAMPLE 4 Indefinite integrals involving trigonometric functions Determine the following indefinite integrals.

a. $\displaystyle\int \left(\frac{2}{\pi}\sin x - 2\csc^2 x\right) dx$

b. $\displaystyle\int \frac{4\cos x + \sin^2 x}{\sin^2 x}\, dx$

SOLUTION

a. Splitting up the integral (Theorem 4.17) and then using Table 4.9, we have

$$\int\left(\frac{2}{\pi}\sin x - 2\csc^2 x\right)dx = \frac{2}{\pi}\int\sin x\,dx - 2\int\csc^2 x\,dx \quad \text{Sum and Constant Multiple Rules}$$

$$= \frac{2}{\pi}(-\cos x) - 2(-\cot x) + C \quad \text{Results (2) and (4), Table 4.9}$$

$$= 2\cot x - \frac{2}{\pi}\cos x + C. \quad \text{Simplify.}$$

b. Again, we split up the integral and then rewrite the first term in the integrand so that it matches result (6) in Table 4.9:

$$\int\frac{4\cos x + \sin^2 x}{\sin^2 x}\,dx = \int\left(\underbrace{\frac{4}{\sin x}\cdot\frac{\cos x}{\sin x}}_{4\csc x\ \cot x} + \underbrace{\frac{\sin^2 x}{\sin^2 x}}_{1}\right)dx$$

$$= 4\int\csc x\cot x\,dx + \int dx \quad \text{Rewrite the integrand; Theorem 4.17.}$$

$$= -4\csc x + x + C. \quad \text{Result (6), Table 4.9; }\int dx = x + C$$

Related Exercises 45, 50 ◄

Other Indefinite Integrals

We now continue the process of rewriting familiar derivative results as indefinite integrals, combining these results with previous techniques.

EXAMPLE 5 **Additional indefinite integrals** Evaluate the following indefinite integrals.

a. $\displaystyle\int\frac{dx}{x}$ **b.** $\displaystyle\int\frac{e^x}{3}\,dx$ **c.** $\displaystyle\int\left(\frac{4}{3\sqrt{1-x^2}} - \frac{3}{x}\right)dx$

SOLUTION

> ► Recognize that evaluating indefinite integrals requires a thorough understanding of all the derivative formulas from Chapter 3. Reviewing the derivative formulas is time well spent.

a. In this case, we know that $\dfrac{d}{dx}(\ln|x|) = \dfrac{1}{x}$, for $x \neq 0$. The corresponding indefinite integral follows immediately:

$$\int\frac{dx}{x} = \ln|x| + C.$$

This result fills the gap in the Power Rule for the case $p = -1$.

b. Because $\dfrac{d}{dx}(e^x) = e^x$, we have

$$\int\frac{e^x}{3}\,dx = \frac{1}{3}\int e^x\,dx = \frac{1}{3}e^x + C. \quad \text{Constant Multiple Rule}$$

c. Recall that $\dfrac{d}{dx}(\sin^{-1}x) = \dfrac{1}{\sqrt{1-x^2}}$. Factoring out constants, we have

$$\int\left(\frac{4}{3\sqrt{1-x^2}} - \frac{3}{x}\right)dx = \frac{4}{3}\int\frac{dx}{\sqrt{1-x^2}} - 3\int\frac{dx}{x} \quad \text{Theorem 4.17}$$

$$= \frac{4}{3}\sin^{-1}x - 3\ln|x| + C.$$

$$\frac{d}{dx}(\sin^{-1}x) = \frac{1}{\sqrt{1-x^2}} \Rightarrow \int\frac{dx}{\sqrt{1-x^2}} = \sin^{-1}x + C;\ \text{part (a)}$$

Related Exercises 51, 53, 59, 61 ◄

The ideas used in Example 5 lead to the results in Table 4.10, where C is an arbitrary constant.

▶ Tables 4.9 and 4.10 are subsets of the table of integrals at the end of the book.

▶ The results in any table of integrals apply regardless of what we call the variable of integration.

Table 4.10 Other Indefinite Integrals

7. $\dfrac{d}{dx}(e^x) = e^x \;\Rightarrow\; \displaystyle\int e^x\,dx = e^x + C$

8. $\dfrac{d}{dx}(\ln|x|) = \dfrac{1}{x} \;\Rightarrow\; \displaystyle\int \dfrac{dx}{x} = \ln|x| + C$

9. $\dfrac{d}{dx}(\sin^{-1} x) = \dfrac{1}{\sqrt{1-x^2}} \;\Rightarrow\; \displaystyle\int \dfrac{dx}{\sqrt{1-x^2}} = \sin^{-1} x + C$

10. $\dfrac{d}{dx}(\tan^{-1} x) = \dfrac{1}{1+x^2} \;\Rightarrow\; \displaystyle\int \dfrac{dx}{1+x^2} = \tan^{-1} x + C$

11. $\dfrac{d}{dx}(\sec^{-1}|x|) = \dfrac{1}{x\sqrt{x^2-1}} \;\Rightarrow\; \displaystyle\int \dfrac{dx}{x\sqrt{x^2-1}} = \sec^{-1}|x| + C$

EXAMPLE 6   Indefinite integrals Determine the following indefinite integrals using Table 4.10.

a. $\displaystyle\int \dfrac{4x^3 + 2x}{3x^2}\,dx$ **b.** $\displaystyle\int \dfrac{5x^2}{4 + 4x^2}\,dx$

SOLUTION

a. We split up the fraction and integrate the resulting terms:

$$\int \frac{4x^3 + 2x}{3x^2}\,dx = \int\left(\frac{4x}{3} + \frac{2}{3x}\right)dx \qquad \text{Split fraction.}$$

$$= \frac{4}{3}\int x\,dx + \frac{2}{3}\int \frac{dx}{x} \qquad \text{Sum and Constant Multiple Rules}$$

$$= \frac{4}{3}\cdot\frac{x^2}{2} + \frac{2}{3}\ln|x| + C \qquad \text{Power Rule; result (8), Table 4.10}$$

$$= \frac{2}{3}\left(x^2 + \ln|x|\right) + C. \qquad \text{Simplify.}$$

b. We factor out a 4 in the denominator and then use the Constant Multiple Rule; the result is

$$\int \frac{5x^2}{4 + 4x^2}\,dx = \frac{5}{4}\int \frac{x^2}{1 + x^2}\,dx.$$

The denominator of the new integrand, $1 + x^2$, looks promising—it matches the denominator in result (10) of Table 4.10—but what about the numerator? Adding and subtracting 1 in the numerator, we have

$$\frac{5}{4}\int \frac{x^2}{1+x^2}\,dx = \frac{5}{4}\int \frac{1 + x^2 - 1}{1 + x^2}\,dx \qquad \text{Add and subtract 1.}$$

$$= \frac{5}{4}\int\left(\frac{1+x^2}{1+x^2} - \frac{1}{1+x^2}\right)dx \qquad \text{Split fraction.}$$

$$= \frac{5}{4}\left(\int dx - \int \frac{1}{1+x^2}\,dx\right) \qquad \text{Sum Rule}$$

$$= \frac{5}{4}\left(x - \tan^{-1} x\right) + C. \qquad \text{Evaluate integrals; result (10), Table 4.10.}$$

Related Exercises 54, 62 ◀

Introduction to Differential Equations

An equation involving an unknown function and its derivatives is called a **differential equation**. Here is an example to get us started.

Suppose you know that the derivative of a function f satisfies the equation

$$f'(x) = 2x + 10.$$

QUICK CHECK 4 Explain why an antiderivative of f' is f. ◄

To find a function f that satisfies this equation, we note that the solutions are antiderivatives of $2x + 10$, which are $x^2 + 10x + C$, where C is an arbitrary constant. So we have found an infinite number of solutions, all of the form $f(x) = x^2 + 10x + C$.

Now consider a more general differential equation of the form $f'(x) = g(x)$, where g is given and f is unknown. The solution f consists of the antiderivatives of g, which involve an arbitrary constant. In most practical cases, the differential equation is accompanied by an **initial condition** that allows us to determine the arbitrary constant. Therefore, we consider problems of the form

$$f'(x) = g(x), \quad \text{where } g \text{ is given, and} \qquad \text{Differential equation}$$
$$f(a) = b, \qquad \text{where } a \text{ and } b \text{ are given.} \quad \text{Initial condition}$$

A differential equation coupled with an initial condition is called an **initial value problem**.

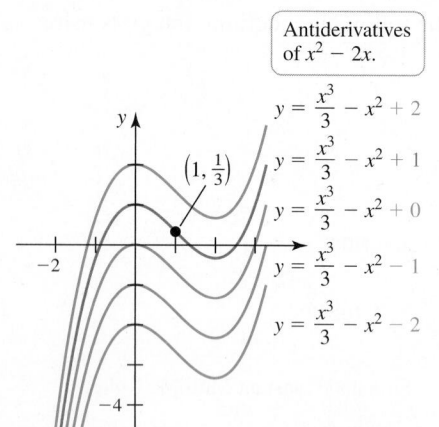

Antiderivatives of $x^2 - 2x$.

$y = \dfrac{x^3}{3} - x^2 + 2$

$y = \dfrac{x^3}{3} - x^2 + 1$

$\left(1, \frac{1}{3}\right)$

$y = \dfrac{x^3}{3} - x^2 + 0$

$y = \dfrac{x^3}{3} - x^2 - 1$

$y = \dfrac{x^3}{3} - x^2 - 2$

Figure 4.91

EXAMPLE 7   Initial value problems Solve the following initial value problems.

a. $f'(x) = x^2 - 2x;\ f(1) = \dfrac{1}{3}$ **b.** $f'(t) = 1 - \dfrac{1}{t^2};\ f(1) = 0$

c. Solve the differential equation in part (b) using the initial condition $f(-1) = -2$.

SOLUTION

a. The solution f is an antiderivative of $x^2 - 2x$. Therefore,

$$f(x) = \frac{x^3}{3} - x^2 + C,$$

where C is an arbitrary constant. We have determined that the solution is a member of a family of functions, all of which differ by a constant. This family of functions, called the **general solution**, is shown in **Figure 4.91**, where we see curves for various choices of C.

Using the initial condition $f(1) = \frac{1}{3}$, we must find the particular function in the general solution whose graph passes through the point $\left(1, \frac{1}{3}\right)$. Imposing the condition $f(1) = \frac{1}{3}$, we reason as follows:

$$f(x) = \frac{x^3}{3} - x^2 + C \quad \text{General solution}$$

$$f(1) = \frac{1}{3} - 1 + C \quad \text{Substitute } x = 1.$$

$$\frac{1}{3} = \frac{1}{3} - 1 + C \quad f(1) = \frac{1}{3}$$

$$C = 1. \qquad \text{Solve for } C.$$

▶ It is advisable to check that the solution satisfies the original problem: $f'(x) = x^2 - 2x$ and $f(1) = \frac{1}{3} - 1 + 1 = \frac{1}{3}$.

Therefore, the solution to the initial value problem is

$$f(x) = \frac{x^3}{3} - x^2 + 1,$$

which is just one of the curves in the family shown in Figure 4.91.

b. By the Power Rule (Theorem 4.16), an antiderivative of $1 - 1/t^2$ is $t + 1/t$. Therefore, the family of functions

$$f(t) = t + \frac{1}{t} + C$$

represents the general solution of the differential equation. **Figure 4.92** shows several members of this family for various values of C; notice that for each value of C, the graph of f has two branches. The initial condition $f(1) = 0$ determines the value of C:

$$f(t) = t + \frac{1}{t} + C \qquad \text{General solution}$$

$$f(1) = 1 + 1 + C \qquad \text{Substitute } t = 1.$$

$$0 = 2 + C \qquad\qquad f(1) = 0$$

$$C = -2 \qquad\qquad \text{Solve for } C.$$

Substituting $C = -2$ into the general solution gives $f(t) = t + \frac{1}{t} - 2$. However, only the right branch of the curve $y = t + \frac{1}{t} - 2$ passes through the point $(1, 0)$ specified by the initial condition (Figure 4.92, quadrant I). Restricting the domain of f to reflect this fact, we find that the solution to the initial value problem is

$$f(t) = t + \frac{1}{t} - 2, \ t > 0.$$

> Because $f(t) = t + \frac{1}{t} - 2$ is undefined at $t = 0$, it cannot be considered a solution to Example 7b over its entire domain. As discussed in Chapter 9, a solution to an initial value problem is a differentiable function defined on a single interval.

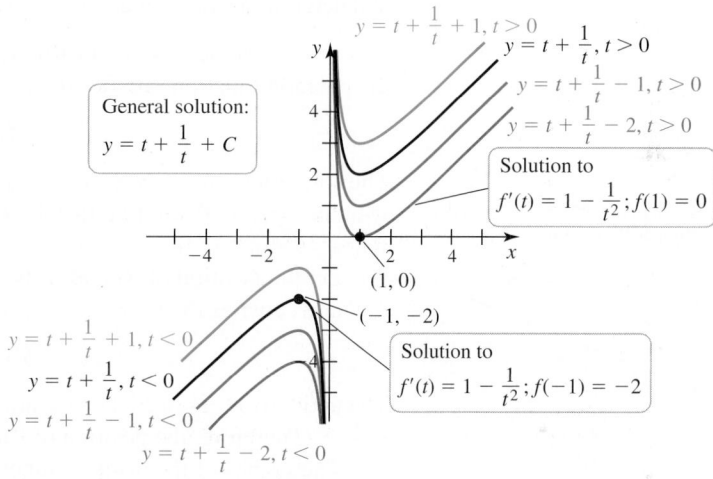

Figure 4.92

c. Using the general solution from part (b) and the initial condition $f(-1) = -2$ leads to $f(-1) = -1 - 1 + C = -2$, which implies that $C = 0$. Substituting this value into the general solution yields $f(t) = t + \frac{1}{t}$, but just as we saw in part (b), an additional restriction is needed because only the left branch of the curve $y = t + \frac{1}{t}$ passes through the point $(-1, -2)$ (Figure 4.92, quadrant III). The solution to this initial value problem is

$$f(t) = t + \frac{1}{t}, \ t < 0.$$

Related Exercises 78, 83 ◄

QUICK CHECK 5 Position is an antiderivative of velocity. But there are infinitely many antiderivatives that differ by a constant. Explain how two objects can have the same velocity function but two different position functions. ◄

> The convention with motion problems is to assume that motion begins at $t = 0$. This means that initial conditions are specified at $t = 0$.

Motion Problems Revisited

Antiderivatives allow us to revisit the topic of one-dimensional motion introduced in Section 3.6. Suppose the position of an object that moves along a line relative to an origin is $s(t)$, where $t \geq 0$ measures elapsed time. The velocity of the object is $v(t) = s'(t)$, which may now be read in terms of antiderivatives: *The position function is an antiderivative of the velocity.* If we are given the velocity function of an object and its position at a particular time, we can determine its position at all future times by solving an initial value problem.

We also know that the acceleration $a(t)$ of an object moving in one dimension is the rate of change of the velocity, which means $a(t) = v'(t)$. In antiderivative terms, this says that the velocity is an antiderivative of the acceleration. Therefore, if we are given the acceleration of an object and its velocity at a particular time, we can determine its velocity at all times. These ideas lie at the heart of modeling the motion of objects.

Initial Value Problems for Velocity and Position

Suppose an object moves along a line with a (known) velocity $v(t)$, for $t \geq 0$. Then its position is found by solving the initial value problem

$$s'(t) = v(t), \quad s(0) = s_0, \text{ where } s_0 \text{ is the (known) initial position.}$$

If the (known) acceleration of the object $a(t)$ is given, then its velocity is found by solving the initial value problem

$$v'(t) = a(t), \quad v(0) = v_0, \text{ where } v_0 \text{ is the (known) initial velocity.}$$

EXAMPLE 8   A race Runner A begins at the point $s(0) = 0$ and runs on a straight and level road with velocity $v(t) = 2t$. Runner B begins with a head start at the point $S(0) = 8$ and runs with velocity $V(t) = 2$. Find the positions of the runners for $t \geq 0$ and determine who is ahead at $t = 6$ time units.

SOLUTION Let the position of Runner A be $s(t)$, with an initial position $s(0) = 0$. Then the position function satisfies the initial value problem

$$s'(t) = 2t, \quad s(0) = 0.$$

The solution is an antiderivative of $s'(t) = 2t$, which has the form $s(t) = t^2 + C$. Substituting $s(0) = 0$, we find that $C = 0$. Therefore, the position of Runner A is given by $s(t) = t^2$, for $t \geq 0$.

Let the position of Runner B be $S(t)$, with an initial position $S(0) = 8$. This position function satisfies the initial value problem

$$S'(t) = 2, \quad S(0) = 8.$$

The antiderivatives of $S'(t) = 2$ are $S(t) = 2t + C$. Substituting $S(0) = 8$ implies that $C = 8$. Therefore, the position of Runner B is given by $S(t) = 2t + 8$, for $t \geq 0$.

The graphs of the position functions are shown in **Figure 4.93**. Runner B begins with a head start but is overtaken when $s(t) = S(t)$, or when $t^2 = 2t + 8$. The solutions of this equation are $t = 4$ and $t = -2$. Only the positive solution is relevant because the race takes place for $t \geq 0$, so Runner A overtakes Runner B at $t = 4$, when $s = S = 16$. When $t = 6$, Runner A has the lead.

Related Exercises 105–106 ◄

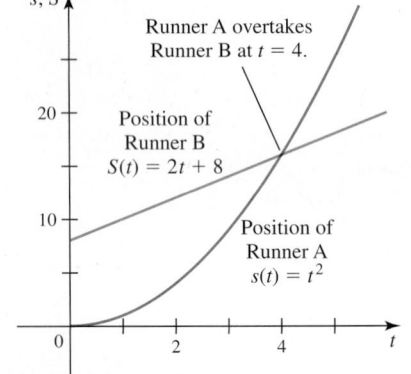

Runner A overtakes Runner B at $t = 4$.

Position of Runner B $S(t) = 2t + 8$

Position of Runner A $s(t) = t^2$

Figure 4.93

EXAMPLE 9   Motion with gravity Neglecting air resistance, the motion of an object moving vertically near Earth's surface is determined by the acceleration due to gravity, which is approximately 9.8 m/s². Suppose a stone is thrown vertically upward at $t = 0$ with a velocity of 40 m/s from the edge of a cliff that is 100 m above a river.

a. Find the velocity $v(t)$ of the object, for $t \geq 0$.

b. Find the position $s(t)$ of the object, for $t \geq 0$.

c. Find the maximum height of the object above the river.

d. With what speed does the object strike the river?

SOLUTION We establish a coordinate system in which the positive s-axis points vertically upward with $s = 0$ corresponding to the river (**Figure 4.94**). Let $s(t)$ be the position of the stone measured relative to the river, for $t \geq 0$. The initial velocity of the stone is $v(0) = 40$ m/s and the initial position of the stone is $s(0) = 100$ m.

a. The acceleration due to gravity points in the *negative* s-direction. Therefore, the initial value problem governing the motion of the object is

$$\text{acceleration} = v'(t) = -9.8, v(0) = 40.$$

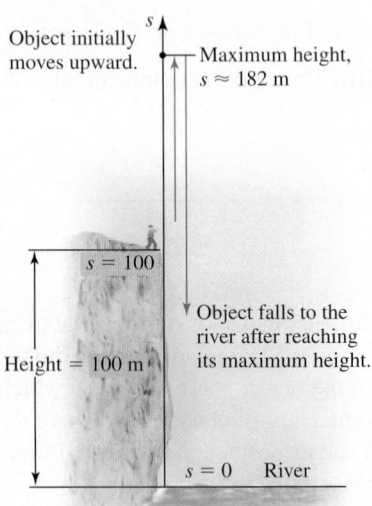

Object initially moves upward.

Maximum height, $s \approx 182$ m

$s = 100$

Object falls to the river after reaching its maximum height.

Height $= 100$ m

$s = 0$ River

Figure 4.94

▶ The acceleration due to gravity at Earth's surface is approximately $g = 9.8 \text{ m/s}^2$, or $g = 32 \text{ ft/s}^2$. It varies (even at sea level) from about 9.8640 at the poles to 9.7982 at the equator. The equation $v'(t) = -g$ is an instance of Newton's Second Law of Motion and assumes no other forces (such as air resistance) are present.

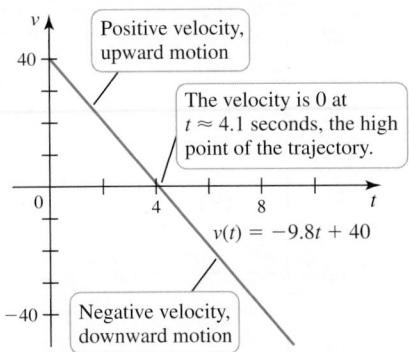

Figure 4.95

The antiderivatives of -9.8 are $v(t) = -9.8t + C$. The initial condition $v(0) = 40$ gives $C = 40$. Therefore, the velocity of the stone is

$$v(t) = -9.8t + 40.$$

As shown in **Figure 4.95**, the velocity decreases from its initial value $v(0) = 40$ until it reaches zero at the high point of the trajectory. This point is reached when

$$v(t) = -9.8t + 40 = 0$$

or when $t \approx 4.1$ s. For $t > 4.1$, the velocity is negative and increases in magnitude as the stone falls to Earth.

b. Knowing the velocity function of the stone, we can determine its position. The position function satisfies the initial value problem

$$v(t) = s'(t) = -9.8t + 40, \quad s(0) = 100.$$

The antiderivatives of $-9.8t + 40$ are

$$s(t) = -4.9t^2 + 40t + C.$$

The initial condition $s(0) = 100$ implies $C = 100$, so the position function of the stone is

$$s(t) = -4.9t^2 + 40t + 100,$$

as shown in **Figure 4.96**. The parabolic graph of the position function is not the actual trajectory of the stone; the stone moves vertically along the s-axis.

c. The position function of the stone increases for $0 < t < 4.1$. At $t \approx 4.1$, the stone reaches a high point of $s(4.1) \approx 182$ m.

d. For $t > 4.1$, the position function decreases, and the stone strikes the river when $s(t) = 0$. The roots of this equation are $t \approx 10.2$ and $t \approx -2.0$. Only the first root is relevant because the motion starts at $t = 0$. Therefore, the stone strikes the ground at $t \approx 10.2$ s. Its speed (in m/s) at this instant is $|v(10.2)| \approx |-60| = 60$.

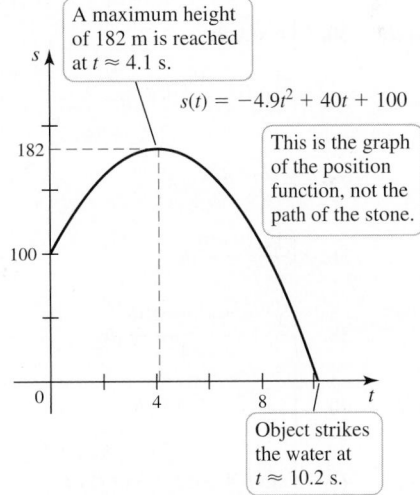

Figure 4.96

Related Exercises 107–108 ◀

SECTION 4.9 EXERCISES

Getting Started

1. Fill in the blanks with either of the words *the derivative* or *an antiderivative*: If $F'(x) = f(x)$, then f is _____ of F, and F is _____ of f.

2. Describe the set of antiderivatives of $f(x) = 0$.

3. Describe the set of antiderivatives of $f(x) = 1$.

4. Why do two different antiderivatives of a function differ by a constant?

5. Give the antiderivatives of x^p. For what values of p does your answer apply?

6. Give the antiderivatives of $a/\sqrt{1-x^2}$, where a is a constant.

7. Give the antiderivatives of $1/x$.

8. Evaluate $\int a \cos x\, dx$ and $\int a \sin x\, dx$, where a is a constant.

9. If $F(x) = x^2 - 3x + C$ and $F(-1) = 4$, what is the value of C?

10. For a given function f, explain the steps used to solve the initial value problem $F'(t) = f(t)$, $F(0) = 10$.

Practice Exercises

11–22. Finding antiderivatives *Find all the antiderivatives of the following functions. Check your work by taking derivatives.*

11. $f(x) = 5x^4$

12. $g(x) = 11x^{10}$

13. $f(x) = 2 \sin x + 1$

14. $g(x) = -4 \cos x - x$

15. $p(x) = 3 \sec^2 x$

16. $q(s) = \csc^2 s$

17. $f(y) = -\dfrac{2}{y^3}$

18. $h(z) = -6z^{-7}$

19. $f(x) = e^x$

20. $h(y) = y^{-1}$

21. $g(s) = \dfrac{1}{s^2 + 1}$

22. $f(t) = \pi$

23–68. Indefinite integrals *Determine the following indefinite integrals. Check your work by differentiation.*

23. $\int (3x^5 - 5x^9)\, dx$

24. $\int (3u^{-2} - 4u^2 + 1)\, du$

25. $\int \left(4\sqrt{x} - \dfrac{4}{\sqrt{x}} \right) dx$

26. $\int \left(\dfrac{5}{t^2} + 4t^2 \right) dt$

27. $\int (5s + 3)^2\, ds$

28. $\int 5m(12m^3 - 10m)\, dm$

29. $\int (3x^{1/3} + 4x^{-1/3} + 6)\, dx$

30. $\int 6\sqrt[3]{x}\, dx$

31. $\int (3x + 1)(4 - x)\, dx$

32. $\int (4z^{1/3} - z^{-1/3})\, dz$

33. $\int \left(\dfrac{3}{x^4} + 2 - \dfrac{3}{x^2} \right) dx$

34. $\int \sqrt[5]{r^2}\, dr$

35. $\int \dfrac{4x^4 - 6x^2}{x}\, dx$

36. $\int \dfrac{12t^8 - t}{t^{3/2}}\, dt$

37. $\int \dfrac{x^2 - 36}{x - 6}\, dx$

38. $\int \dfrac{y^3 - 9y^2 + 20y}{y - 4}\, dy$

39. $\int (\csc^2 \theta + 2\theta^2 - 3\theta)\, d\theta$

40. $\int (\csc^2 \theta + 1)\, d\theta$

41. $\int \dfrac{2 + 3 \cos y}{\sin^2 y}\, dy$

42. $\int \sin t(4 \csc t - \cot t)\, dt$

43. $\int (\sec^2 x - 1)\, dx$

44. $\int \dfrac{\sec^3 v - \sec^2 v}{\sec v - 1}\, dv$

45. $\int (\sec^2 \theta + \sec \theta \tan \theta)\, d\theta$

46. $\int \dfrac{\sin \theta - 1}{\cos^2 \theta}\, d\theta$

47. $\int (3t^2 + 2 \csc^2 t)\, dt$

48. $\int \csc x(\cot x - \csc x)\, dx$

49. $\int \sec \theta(\tan \theta + \sec \theta + \cos \theta)\, d\theta$

50. $\int \dfrac{\csc^3 x + 1}{\csc x}\, dx$

51. $\int \dfrac{1}{2y}\, dy$

52. $\int \dfrac{e^{2t} - 1}{e^t - 1}\, dt$

53. $\int \dfrac{6}{\sqrt{4 - 4x^2}}\, dx$

54. $\int \dfrac{v^3 + v + 1}{1 + v^2}\, dv$

55. $\int \dfrac{4}{x\sqrt{x^2 - 1}}\, dx$

56. $\int \dfrac{2}{25z^2 + 25}\, dz$

57. $\int \dfrac{1}{x\sqrt{36x^2 - 36}}\, dx$

58. $\int (49 - 49x^2)^{-1/2}\, dx$

59. $\int \dfrac{t + 1}{t}\, dt$

60. $\int \dfrac{t^2 - e^{2t}}{t + e^t}\, dt$

61. $\int e^{x+2}\, dx$

62. $\int \dfrac{10t^5 - 3}{t}\, dt$

63. $\int \dfrac{e^{2w} - 5e^w + 4}{e^w - 1}\, dw$

64. $\int (\sqrt[3]{x^2} + \sqrt{x^3})\, dx$

65. $\int \dfrac{1 + \sqrt{x}}{x}\, dx$

66. $\int \dfrac{16 \cos^2 w - 81 \sin^2 w}{4 \cos w - 9 \sin w}\, dw$

67. $\int \sqrt{x}\,(2x^6 - 4\sqrt[3]{x})\, dx$

68. $\int \dfrac{2 + x^2}{1 + x^2}\, dx$

69–76. Particular antiderivatives *For the following functions f, find the antiderivative F that satisfies the given condition.*

69. $f(x) = x^5 - 2x^2 + 1;\ F(0) = 1$

70. $f(x) = 4\sqrt{x} + 6;\ F(1) = 8$

71. $f(x) = 8x^3 + \sin x;\ F(0) = 2$

72. $f(t) = \sec^2 t;\ F(\pi/4) = 1,\ -\pi/2 < t < \pi/2$

73. $f(v) = \sec v \tan v;\ F(0) = 2,\ -\pi/2 < v < \pi/2$

74. $f(u) = 2e^u + 3;\ F(0) = 8$

75. $f(y) = \dfrac{3y^3 + 5}{y};\ F(1) = 3,\ y > 0$

76. $f(\theta) = 2 \sin \theta - 4 \cos \theta;\ F\left(\dfrac{\pi}{4}\right) = 2$

77–86. Solving initial value problems *Find the solution of the following initial value problems.*

77. $f'(x) = 2x - 3;\ f(0) = 4$

78. $g'(x) = 7x^6 - 4x^3 + 12;\ g(1) = 24$

79. $g'(x) = 7x\left(x^6 - \dfrac{1}{7} \right);\ g(1) = 2$

80. $h'(t) = 1 + 6 \sin t;\ h\left(\dfrac{\pi}{3}\right) = -3$

81. $f'(u) = 4(\cos u - \sin u);\ f\left(\dfrac{\pi}{2}\right) = 0$

82. $p'(t) = 10e^t + 70;\ p(0) = 100$

83. $y'(t) = \dfrac{3}{t} + 6; \, y(1) = 8, \, t > 0$

84. $u'(x) = \dfrac{xe^{2x} + 4e^x}{xe^x}; \, u(1) = 0, \, x > 0$

85. $y'(\theta) = \dfrac{\sqrt{2}\cos^3\theta + 1}{\cos^2\theta}; \, y\left(\dfrac{\pi}{4}\right) = 3, \, -\pi/2 < \theta < \pi/2$

86. $v'(x) = 4x^{1/3} + 2x^{-1/3}; \, v(8) = 40, \, x > 0$

T **87–90. Graphing general solutions** *Graph several functions that satisfy each of the following differential equations. Then find and graph the particular function that satisfies the given initial condition.*

87. $f'(x) = 2x - 5; \, f(0) = 4$

88. $f'(x) = 3x^2 - 1; \, f(1) = 2$

89. $f'(x) = 3x + \sin x; \, f(0) = 3$

90. $f'(x) = \cos x - \sin x + 2; \, f(0) = 1$

91–96. Velocity to position *Given the following velocity functions of an object moving along a line, find the position function with the given initial position.*

91. $v(t) = 2t + 4; \, s(0) = 0$ **92.** $v(t) = e^t + 4; \, s(0) = 2$

93. $v(t) = 2\sqrt{t}; \, s(0) = 1$ **94.** $v(t) = 2\cos t; \, s(0) = 0$

95. $v(t) = 6t^2 + 4t - 10; \, s(0) = 0$

96. $v(t) = 4t + \sin t; \, s(0) = 0$

97–102. Acceleration to position *Given the following acceleration functions of an object moving along a line, find the position function with the given initial velocity and position.*

97. $a(t) = -32; \, v(0) = 20, \, s(0) = 0$

98. $a(t) = 4; \, v(0) = -3, \, s(0) = 2$

99. $a(t) = 0.2t; \, v(0) = 0, \, s(0) = 1$

100. $a(t) = 2\cos t; \, v(0) = 1, \, s(0) = 0$

101. $a(t) = 2 + 3\sin t; \, v(0) = 1, \, s(0) = 10$

102. $a(t) = 2e^t - 12; \, v(0) = 1, \, s(0) = 0$

103. A car starting at rest accelerates at 16 ft/s² for 5 seconds on a straight road. How far does it travel during this time?

104. A car is moving at 60 mi/hr (88 ft/s) on a straight road when the driver steps on the brake pedal and begins decelerating at a constant rate of 10 ft/s² for 3 seconds. How far did the car go during this 3-second interval?

T **105–106. Races** *The velocity function and initial position of Runners A and B are given. Analyze the race that results by graphing the position functions of the runners and finding the time and positions (if any) at which they first pass each other.*

105. A: $v(t) = \sin t; \, s(0) = 0$ B: $V(t) = \cos t; \, S(0) = 0$

106. A: $v(t) = e^t; \, s(0) = 0$ B: $V(t) = 2 + \cos t; \, S(0) = 3$

T **107–110. Motion with gravity** *Consider the following descriptions of the vertical motion of an object subject only to the acceleration due to gravity. Begin with the acceleration equation $a(t) = v'(t) = -g$, where $g = 9.8$ m/s².*

a. *Find the velocity of the object for all relevant times.*
b. *Find the position of the object for all relevant times.*
c. *Find the time when the object reaches its highest point. What is the height?*
d. *Find the time when the object strikes the ground.*

107. A softball is popped up vertically (from the ground) with a velocity of 30 m/s.

108. A stone is thrown vertically upward with a velocity of 30 m/s from the edge of a cliff 200 m above a river.

109. A payload is released at an elevation of 400 m from a hot-air balloon that is rising at a rate of 10 m/s.

110. A payload is dropped at an elevation of 400 m from a hot-air balloon that is descending at a rate of 10 m/s.

111. Explain why or why not Determine whether the following statements are true and give an explanation or counterexample.

    a. $F(x) = x^3 - 4x + 100$ and $G(x) = x^3 - 4x - 100$ are antiderivatives of the same function.

    b. If $F'(x) = f(x)$, then f is an antiderivative of F.

    c. If $F'(x) = f(x)$, then $\int f(x)\, dx = F(x) + C$.

    d. $f(x) = x^3 + 3$ and $g(x) = x^3 - 4$ are derivatives of the same function.

    e. If $F'(x) = G'(x)$, then $F(x) = G(x)$.

Explorations and Challenges

112–115. Functions from higher derivatives *Find the function F that satisfies the following differential equations and initial conditions.*

112. $F''(x) = 1; \, F'(0) = 3, \, F(0) = 4$

113. $F''(x) = \cos x; \, F'(0) = 3, \, F(\pi) = 4$

114. $F'''(x) = 4x; \, F''(0) = 0, \, F'(0) = 1, \, F(0) = 3$

115. $F'''(x) = 672x^5 + 24x; \, F''(0) = 0, \, F'(0) = 2, \, F(0) = 1$

116. Mass on a spring A mass oscillates up and down on the end of a spring. Find its position s relative to the equilibrium position if its acceleration is $a(t) = 2\sin t$ and its initial velocity and position are $v(0) = 3$ and $s(0) = 0$, respectively.

117. Flow rate A large tank is filled with water when an outflow valve is opened at $t = 0$. Water flows out at a rate, in gal/min, given by $Q'(t) = 0.1(100 - t^2)$, for $0 \le t \le 10$.

    a. Find the amount of water $Q(t)$ that has flowed out of the tank after t minutes, given the initial condition $Q(0) = 0$.

    b. Graph the flow function Q, for $0 \le t \le 10$.

    c. How much water flows out of the tank in 10 min?

118. General head start problem Suppose object A is located at $s = 0$ at time $t = 0$ and starts moving along the s-axis with a velocity given by $v(t) = 2at$, where $a > 0$. Object B is located at $s = c > 0$ at $t = 0$ and starts moving along the s-axis with a constant velocity given by $V(t) = b > 0$. Show that A always overtakes B at time

$$t = \frac{b + \sqrt{b^2 + 4ac}}{2a}.$$

119–122. Verifying indefinite integrals *Verify the following indefinite integrals by differentiation. These integrals are derived in later chapters.*

119. $\int \dfrac{\cos\sqrt{x}}{\sqrt{x}}\,dx = 2\sin\sqrt{x} + C$

120. $\int \dfrac{x}{\sqrt{x^2+1}}\,dx = \sqrt{x^2+1} + C$

121. $\int x^2 \cos x^3\,dx = \dfrac{1}{3}\sin x^3 + C$

122. $\int \dfrac{x}{(x^2-1)^2}\,dx = -\dfrac{1}{2(x^2-1)} + C$

QUICK CHECK ANSWERS

1. $d/dx(x^4) = 4x^3$ **2.** $e^x + C$, $x^4 + C$, $\tan x + C$
3. $d/dx(-\csc x + C) = \csc x \cot x$ **4.** One function that can be differentiated to get f' is f. Therefore, f is an antiderivative of f'. **5.** The two position functions involve two different initial positions; they differ by a constant. ◄

CHAPTER 4   REVIEW EXERCISES

1. **Explain why or why not** Determine whether the following statements are true and give an explanation or counterexample.

a. If $f'(c) = 0$, then f has a local minimum or maximum at c.
b. If $f''(c) = 0$, then f has an inflection point at c.
c. $F(x) = x^2 + 10$ and $G(x) = x^2 - 100$ are antiderivatives of the same function.
d. Between two local minima of a function continuous on $(-\infty, \infty)$, there must be a local maximum.
e. The linear approximation to $f(x) = \sin x$ at $x = 0$ is $L(x) = x$.
f. If $\lim\limits_{x\to\infty} f(x) = \infty$ and $\lim\limits_{x\to\infty} g(x) = \infty$, then
$$\lim_{x\to\infty}(f(x) - g(x)) = 0.$$

2. **Locating extrema** Consider the graph of a function f on the interval $[-3, 3]$.

a. Give the approximate coordinates of the local maxima and minima of f.
b. Give the approximate coordinates of the absolute maximum and minimum values of f (if they exist).
c. Give the approximate coordinates of the inflection point(s) of f.
d. Give the approximate coordinates of the zero(s) of f.
e. On what intervals (approximately) is f concave up?
f. On what intervals (approximately) is f concave down?

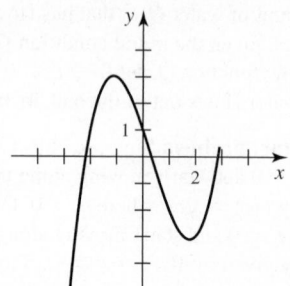

3–4. Designer functions *Sketch the graph of a function continuous on the given interval that satisfies the following conditions.*

3. f is continuous on the interval $[-4, 4]$; $f'(x) = 0$ for $x = -2, 0$, and 3; f has an absolute minimum at $x = 3$; f has a local minimum at $x = -2$; f has a local maximum at $x = 0$; f has an absolute maximum at $x = -4$.

4. f is continuous on $(-\infty, \infty)$; $f'(x) < 0$ and $f''(x) < 0$ on $(-\infty, 0)$; $f'(x) > 0$ and $f''(x) > 0$ on $(0, \infty)$.

5. Use the graphs of f' and f'' to complete the following steps.

a. Find the critical points of f and determine where f is increasing and where it is decreasing.
b. Determine the locations of the inflection points of f and the intervals on which f is concave up or concave down.
c. Determine where f has local maxima and minima.
d. Plot a possible graph of f.

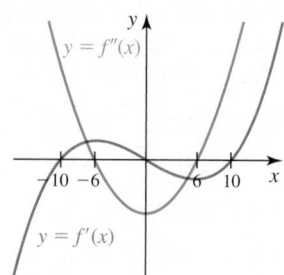

6–16. Critical points *Find the critical points of the following functions on the given intervals. Identify the absolute maximum and absolute minimum values (if they exist).*

6. $f(x) = x^3 - 6x^2$ on $[-1, 5]$

7. $f(x) = 3x^4 - 6x^2 + 9$ on $[-2, 2]$

8. $g(x) = x^4 - 50x^2$ on $[-1, 5]$

9. $f(x) = 2x^3 - 3x^2 - 36x + 12$ on $(-\infty, \infty)$

10. $f(x) = x^3 \ln x$ on $(0, \infty)$

11. $f(x) = \ln(x^2 - 2x + 2)$ on $[0, 2]$

12. $f(x) = \sin 2x + 3$ on $[-\pi, \pi]$

13. $g(x) = -\dfrac{1}{2}\sin x + \dfrac{1}{2}\sin x \cos x$ on $[0, 2\pi]$

14. $f(x) = 4x^{1/2} - x^{5/2}$ on $[0, 4]$

15. $f(x) = 2x \ln x + 10$ on $(0, 4)$

T 16. $g(x) = x \sin^{-1} x$ on $[-1, 1]$

T 17. Absolute values Consider the function $f(x) = |x - 2| + |x + 3|$ on $[-4, 4]$. Graph f, identify the critical points, and give the coordinates of the local and absolute extreme values.

T 18–20. *Use f' and f'' to complete parts (a) and (b).*

a. *Find the intervals on which f is increasing and the intervals on which it is decreasing.*

b. *Find the intervals on which f is concave up and the intervals on which it is concave down.*

18. $f(x) = \dfrac{e^x}{1 + e^x}$ **19.** $f(x) = \dfrac{x^9}{9} + 3x^5 - 16x$

20. $f(x) = x\sqrt{x + 9}$

21. Inflection points Does $f(x) = 2x^5 - 10x^4 + 20x^3 + x + 1$ have any inflection points? If so, identify them.

T 22. Does $f(x) = \dfrac{x^6}{2} + \dfrac{5x^4}{4} - 15x^2$ have any inflection points? If so, identify them.

23. Identify the critical points and the inflection points of $f(x) = (x - a)(x + a)^3$, for $a > 0$.

24–34. Curve sketching *Use the guidelines given in Section 4.4 to make a complete graph of the following functions on their domains or on the given interval. Use a graphing utility to check your work.*

24. $f(x) = x(x + 4)^3$ **25.** $f(x) = \dfrac{x^4}{2} - 3x^2 + 4x + 1$

26. $f(x) = \dfrac{3x}{x^2 + 3}$

27. $f(x) = 4\cos(\pi(x - 1))$ on $[0, 2]$

28. $f(x) = \ln(x^2 + 9)$ **29.** $f(x) = \dfrac{x^2 + 3}{x - 1}$

30. $f(x) = xe^{-x/2}$ **31.** $f(x) = \dfrac{10x^2}{x^2 + 3}$

32. $f(x) = x\sqrt[3]{x + 4}$ (*Hint:* In addition to using the graphing guidelines outlined in Section 4.4, describe the line tangent to the graph of f at $x = -4$ by examining $\lim\limits_{x \to -4} f'(x)$.)

33. $f(x) = \sqrt[3]{x} - \sqrt{x} + 2$ **T 34.** $f(x) = x(x - 1)e^{-x}$

35. Optimal popcorn box A small popcorn box is created from a $12'' \times 12''$ sheet of paperboard by first cutting out four shaded rectangles, each of length x and width $x/2$ (see figure). The remaining paperboard is folded along the solid lines to form a box. What dimensions of the box maximize the volume of the box?

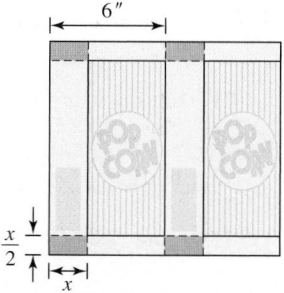 

36. Minimizing time Hannah is standing on the edge of an island, 1 mile from a straight shoreline (see figure). She wants to return to her beach house that is 2 miles from the point P on shore that is closest to where she is standing. Given that she can swim at 2 mi/hr and jog at 6 mi/hr, find the point at which she should come ashore to minimize the total time of her trip. If she starts swimming at noon, can she make it home before 1 p.m.?

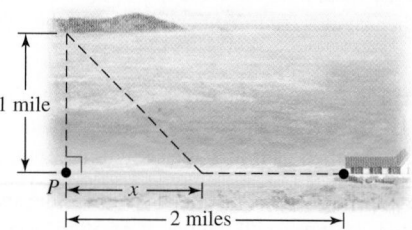

T 37. Minimizing sound intensity Two sound speakers are 100 m apart and one speaker is three times as loud as the other speaker. At what point on a line segment between the speakers is the sound intensity the weakest? (*Hint:* Sound intensity is directly proportional to the sound level and inversely proportional to the square of the distance from the sound source.)

38. Hockey problem A hockey player skates on a line that is perpendicular to the goal line. If the line on which he is skating is 10 feet to the left of the center of the hockey goal and if the hockey goal is 6 ft wide, then where (what value of x) should he shoot the puck to maximize the angle θ on goal (see figure)? (*Hint:* $\theta = \alpha - \beta$.)

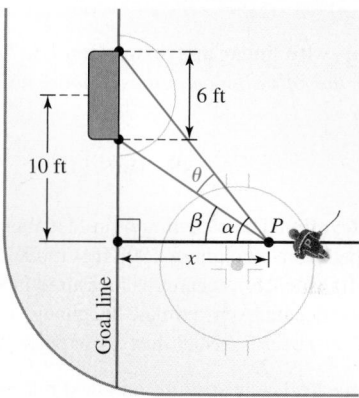

39. Optimization A right triangle has legs of length h and r and a hypotenuse of length 4 (see figure). It is revolved about the leg of length h to sweep out a right circular cone. What values of h and r maximize the volume of the cone? (Volume of a cone $= \dfrac{1}{3}\pi r^2 h$.)

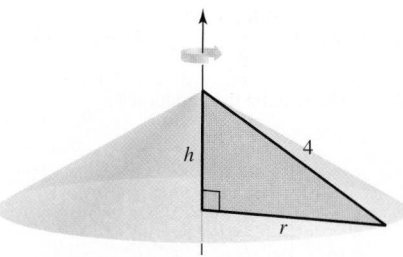

T 40. Rectangles beneath a curve A rectangle is constructed with one side on the positive x-axis, one side on the positive y-axis, and the vertex opposite the origin on the curve $y = \cos x$, for $0 < x < \pi/2$. Approximate the dimensions of the rectangle that maximize the area of the rectangle. What is the area?

41. Maximum printable area A rectangular page in a text (with width x and length y) has an area of 98 in², top and bottom margins set at 1 in, and left and right margins set at $1/2$ in. The printable area of the page is the rectangle that lies within the margins. What are the dimensions of the page that maximize the printable area?

42. Nearest point What point of the graph of $f(x) = 5/2 - x^2$ is closest to the origin? (*Hint:* You can minimize the square of the distance.)

43. Maximum area A line segment of length 10 joins the points $(0, p)$ and $(q, 0)$ to form a triangle in the first quadrant. Find the values of p and q that maximize the area of the triangle.

44. Minimum painting surface A metal cistern in the shape of a right circular cylinder with volume $V = 50$ m³ needs to be painted each year to reduce corrosion. The paint is applied only to surfaces exposed to the elements (the outside cylinder wall and the circular top). Find the dimensions r and h of the cylinder that minimize the area of the painted surfaces.

45–46. Linear approximation

a. *Find the linear approximation to f at the given point a.*
b. *Use your answer from part (a) to estimate the given function value. Does your approximation underestimate or overestimate the exact function value?*

45. $f(x) = x^{2/3}; a = 27; f(29)$

46. $f(x) = \sin^{-1} x; a = \dfrac{1}{2}; f(0.48)$

47–48. Estimations with linear approximation *Use linear approximation to estimate the following quantities. Choose a value of a to produce a small error.*

47. $\dfrac{1}{4.2^2}$ **48.** $\tan^{-1} 1.05$

49. Change in elevation The elevation h (in feet above the ground) of a stone dropped from a height of 1000 ft is modeled by the equation $h(t) = 1000 - 16t^2$, where t is measured in seconds and air resistance is neglected. Approximate the change in elevation over the interval $5 \le t \le 5.7$ (recall that $\Delta h \approx h'(a)\Delta t$).

50. Change in energy The energy E (in joules) released by an earthquake of magnitude M is modeled by the equation $E(M) = 25{,}000 \cdot 10^{1.5M}$. Approximate the change in energy released when the magnitude changes from 7.0 to 7.5 (recall that $\Delta E \approx E'(a)\Delta M$).

51. Mean Value Theorem For the function $f(x) = \sqrt{10x}$ and the interval $[a, b] = [0, 10]$, use the graph to make a conjecture about the value of c for which $\dfrac{f(b) - f(a)}{b - a} = f'(c)$. Then verify your conjectured value by solving the equation $\dfrac{f(b) - f(a)}{b - a} = f'(c)$ for c.

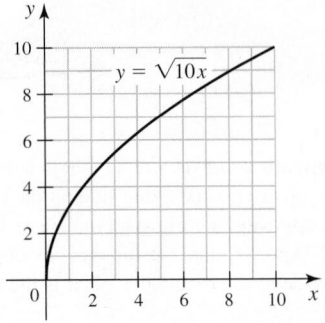

52. Mean Value Theorem Explain why the Mean Value Theorem does not apply to the function $f(x) = |x|$ on $[a, b] = [-1, 2]$.

53. Mean Value Theorem The population of a culture of cells grows according to the function $P(t) = \dfrac{100t}{t + 1}$, where $t \ge 0$ is measured in weeks.

a. What is the average rate of change in the population over the interval $[0, 8]$?
b. At what point of the interval $[0, 8]$ is the instantaneous rate of change equal to the average rate of change?

54. Growth rate of bamboo Bamboo belongs to the grass family and is one of the fastest growing plants in the world.

a. A bamboo shoot was 500 cm tall at 10:00 A.M. and 515 cm tall at 3:00 P.M. Compute the average growth rate of the bamboo shoot in cm/hr over the period of time from 10:00 A.M. to 3:00 P.M.
b. Based on the Mean Value Theorem, what can you conclude about the instantaneous growth rate of bamboo measured in *millimeters per second* between 10:00 A.M. and 3:00 P.M.?

T 55. Newton's method Use Newton's method to approximate the roots of $f(x) = 3x^3 - 4x^2 + 1$ to six digits.

T 56. Newton's method Use Newton's method to approximate the roots of $f(x) = e^{-2x} + 2e^x - 6$ to six digits.

T 57. Newton's method Use Newton's method to approximate the x-coordinates of the inflection points of $f(x) = 2x^5 - 6x^3 - 4x + 2$ to six digits.

58–59. Two methods *Evaluate the following limits in two different ways: with and without l'Hôpital's Rule.*

58. $\displaystyle\lim_{x \to \infty} \dfrac{4x^4 - \sqrt{x}}{2x^4 + x^{-1}}$ **59.** $\displaystyle\lim_{x \to \infty} \dfrac{2x^5 - x + 1}{5x^6 + x}$

60–81. Limits *Evaluate the following limits. Use l'Hôpital's Rule when needed.*

60. $\displaystyle\lim_{t \to 2} \dfrac{t^3 - t^2 - 2t}{t^2 - 4}$ **61.** $\displaystyle\lim_{t \to 0} \dfrac{1 - \cos 6t}{2t}$

62. $\displaystyle\lim_{x \to \infty} \dfrac{5x^2 + 2x - 5}{\sqrt{x^4 - 1}}$ **63.** $\displaystyle\lim_{\theta \to 0} \dfrac{3\sin^2 2\theta}{\theta^2}$

64. $\displaystyle\lim_{x \to \infty} \left(\sqrt{x^2 + x + 1} - \sqrt{x^2 - x} \right)$

65. $\displaystyle\lim_{\theta \to 0} 2\theta \cot 3\theta$ **66.** $\displaystyle\lim_{x \to 0} \dfrac{e^{-2x} - 1 + 2x}{x^2}$

67. $\displaystyle\lim_{y \to 0^+} \dfrac{\ln^{10} y}{\sqrt{y}}$ **68.** $\displaystyle\lim_{\theta \to 0} \dfrac{3\sin 8\theta}{8\sin 3\theta}$

69. $\displaystyle\lim_{x \to 1} \dfrac{x^4 - x^3 - 3x^2 + 5x - 2}{x^3 + x^2 - 5x + 3}$

70. $\displaystyle\lim_{x \to \infty} \dfrac{\ln x^{100}}{\sqrt{x}}$ **71.** $\displaystyle\lim_{x \to 0} \csc x \sin^{-1} x$

72. $\displaystyle\lim_{x \to \infty} \dfrac{\ln^3 x}{\sqrt{x}}$ **73.** $\displaystyle\lim_{x \to \infty} \ln\left(\dfrac{x + 1}{x - 1}\right)$

74. $\displaystyle\lim_{x \to 0^+} (1 + x)^{\ln x}$ **75.** $\displaystyle\lim_{x \to \pi/2^-} (\sin x)^{\tan x}$

76. $\displaystyle\lim_{x \to \infty} (\sqrt{x} + 1)^{1/x}$ **77.** $\displaystyle\lim_{x \to 0^+} |\ln x|^x$

78. $\displaystyle\lim_{x \to \infty} x^{1/x}$ **79.** $\displaystyle\lim_{x \to \infty} \left(1 - \dfrac{3}{x}\right)^x$

80. $\lim\limits_{x\to\infty}\left(\dfrac{2}{\pi}\tan^{-1}x\right)^x$

81. $\lim\limits_{x\to 1}(x-1)^{\sin\pi x}$

82–89. Comparing growth rates *Determine which of the two functions grows faster, or state that they have comparable growth rates.*

82. x^{100} and 1.1^x

83. $x^{1/2}$ and $x^{1/3}$

84. $\ln x$ and $\log_{10}x$

85. $\sqrt{x}$ and $\ln^{10}x$

86. $10x$ and $\ln x^2$

87. e^x and 3^x

88. $\sqrt{x^6+10}$ and x^3

89. 2^x and $4^{x/2}$

90–103. Indefinite integrals *Determine the following indefinite integrals.*

90. $\displaystyle\int(x^8-3x^3+1)\,dx$

91. $\displaystyle\int(2x+1)^2\,dx$

92. $\displaystyle\int\dfrac{x^5-3}{x}\,dx$

93. $\displaystyle\int\left(\dfrac{1}{x^2}-\dfrac{2}{x^{5/2}}\right)dx$

94. $\displaystyle\int\dfrac{x^4-2\sqrt{x}+2}{x^2}\,dx$

95. $\displaystyle\int(1+3\cos\theta)\,d\theta$

96. $\displaystyle\int 2\sec^2\theta\,d\theta$

97. $\displaystyle\int\dfrac{dx}{1-\sin^2 x}$

98. $\displaystyle\int\dfrac{2e^{2x}+e^x}{e^x}\,dx$

99. $\displaystyle\int\dfrac{12}{x}\,dx$

100. $\displaystyle\int\dfrac{dx}{\sqrt{1-x^2}}$

101. $\displaystyle\int\dfrac{x^2}{x^4+x^2}\,dx$

102. $\displaystyle\int\dfrac{1+\tan\theta}{\sec\theta}\,d\theta$

103. $\displaystyle\int\left(\sqrt[4]{x^3}+\sqrt{x^5}\right)dx$

104–107. Functions from derivatives *Find the function f with the following properties.*

104. $f'(x)=3x^2-1;\; f(0)=10$

105. $f'(t)=\sin t+2t;\; f(0)=5$

106. $f'(t)=t^2+t^{-2};\; f(1)=1$, for $t>0$

107. $f'(x)=\dfrac{x^4-2}{1+x^2};\; f(1)=-\dfrac{2}{3}$

T 108. Motion along a line Two objects move along the x-axis with position functions $x_1(t)=2\sin t$ and $x_2(t)=\sin\left(t-\dfrac{\pi}{2}\right)$. At what times on the interval $[0,2\pi]$ are the objects closest to each other and farthest from each other?

109. Vertical motion with gravity A rocket is launched vertically upward with an initial velocity of 120 m/s from a platform that is 125 m above the ground. Assume the only force at work is gravity. Determine the velocity and position functions of the rocket, for $t\ge 0$. Then describe the motion in words.

110. Distance traveled A car starting at rest accelerates at 20 ft/s² for 5 seconds on a straight road. How far does it travel during this time?

111. Projectile motion A ball is thrown vertically upward with a velocity of 64 ft/s from the edge of a cliff 128 ft above a river.

   a. Find the velocity of the ball for all relevant times.

   b. Find the position of the ball above the river for all relevant times.

   c. Find the time when the ball reaches its highest point above the river. What is the height?

   d. Find the velocity at which the ball strikes the ground.

112. Logs of logs Compare the growth rates of $\ln x$, $\ln(\ln x)$, and $\ln(\ln(\ln x))$.

113. Two limits with exponentials Evaluate $\lim\limits_{x\to 0^+}\dfrac{x}{\sqrt{1-e^{-x^2}}}$ and $\lim\limits_{x\to 0^+}\dfrac{x^2}{1-e^{-x^2}}$.

114. Geometric mean Prove that $\lim\limits_{r\to 0}\left(\dfrac{a^r+b^r+c^r}{3}\right)^{1/r}=\sqrt[3]{abc}$, where a, b, and c are positive real numbers.

115. Towers of exponents The functions

$$f(x)=(x^x)^x \quad\text{and}\quad g(x)=x^{(x^x)}$$

are different functions. For example, $f(3)=19{,}683$ and $g(3)\approx 7.6\times 10^{12}$. Determine whether $\lim\limits_{x\to 0^+}f(x)$ and $\lim\limits_{x\to 0^+}g(x)$ are indeterminate forms, and evaluate the limits.

116. Cosine limits Let n be a positive integer. Evaluate the following limits.

   a. $\lim\limits_{x\to 0}\dfrac{1-\cos x^n}{x^{2n}}$

   b. $\lim\limits_{x\to 0}\dfrac{1-\cos^n x}{x^2}$

117. Limits for e Consider the function $g(x)=\left(1+\dfrac{1}{x}\right)^{x+a}$.

   a. Show that if $0\le a<\dfrac{1}{2}$, then $g(x)\to e$ from *below* as $x\to\infty$.

   b. Show that if $\dfrac{1}{2}\le a<1$, then $g(x)\to e$ from *above* as $x\to\infty$.

T 118. A family of superexponential functions Let $f(x)=(a+x)^x$, where $a>0$.

   a. What is the domain of f (in terms of a)?

   b. Describe the end behavior of f (near the left boundary of its domain and as $x\to\infty$).

   c. Compute f'. Then graph f and f' for $a=0.5,\ 1,\ 2,$ and 3.

   d. Show that f has a single local minimum at the point z that satisfies $(z+a)\ln(z+a)+z=0$.

   e. Describe how z (found in part (d)) varies as a increases. Describe how $f(z)$ varies as a increases.

Chapter 4 Guided Projects

Applications of the material in this chapter and related topics can be found in the following Guided Projects. For additional information, see the Preface.

- Oscillators
- Ice cream, geometry, and calculus
- Newton's method

5

Integration

Chapter Preview   We are now at a critical point in the calculus story. Many would argue that this chapter is the cornerstone of calculus because it explains the relationship between the two processes of calculus: differentiation and integration. We begin by explaining why finding the area of regions bounded by the graphs of functions is such an important problem in calculus. Then you will see how antiderivatives lead to definite integrals, which are used to solve the area problem. But there is more to the story. You will also see the remarkable connection between derivatives and integrals, which is expressed in the Fundamental Theorem of Calculus. In this chapter, we develop key properties of definite integrals, investigate a few of their many applications, and present the first of several powerful techniques for evaluating definite integrals.

5.1 Approximating Areas under Curves

The derivative of a function is associated with rates of change and slopes of tangent lines. We also know that antiderivatives (or indefinite integrals) reverse the derivative operation. **Figure 5.1** summarizes our current understanding and raises the question: What is the geometric meaning of the integral? The following example reveals a clue.

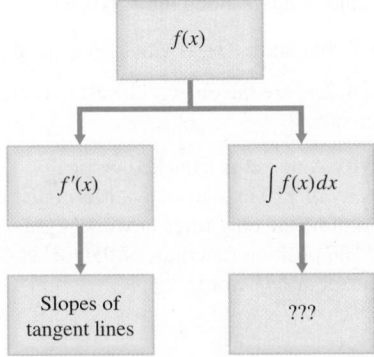

Figure 5.1

Area under a Velocity Curve

Consider an object moving along a line with a known position function. You learned in previous chapters that the slope of the line tangent to the graph of the position function at a certain time gives the velocity v at that time. We now turn the situation around. If we know the velocity function of a moving object, what can we learn about its position function?

▶ Recall from Section 3.6 that the *displacement* of an object moving along a line is given by

    final position − initial position.

If the velocity of an object is positive, its displacement equals the distance traveled.

▶ The side lengths of the rectangle in Figure 5.3 have units mi/hr and hr. Therefore, the units of the area are mi/hr · hr = mi, which is a unit of displacement.

QUICK CHECK 1 What is the displacement of an object that travels at a constant velocity of 10 mi/hr for a half hour, 20 mi/hr for the next half hour, and 30 mi/hr for the next hour? ◀

Imagine a car traveling at a constant velocity of 60 mi/hr along a straight highway over a two-hour period. The graph of the velocity function $v = 60$ on the interval $0 \le t \le 2$ is a horizontal line (**Figure 5.2**). The displacement of the car between $t = 0$ and $t = 2$ is found by a familiar formula:

$$\text{displacement} = \text{rate} \cdot \text{time}$$
$$= 60 \text{ mi/hr} \cdot 2 \text{ hr} = 120 \text{ mi.}$$

This product is the area of the rectangle formed by the velocity curve and the t-axis between $t = 0$ and $t = 2$ (**Figure 5.3**). In this case (constant positive velocity), we see that the area between the velocity curve and the t-axis is the displacement of the moving object.

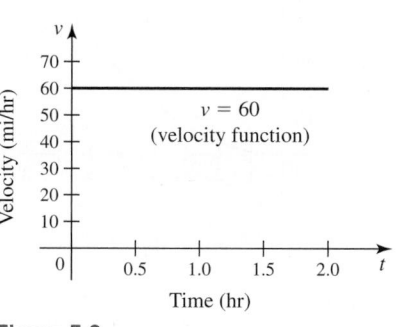

Figure 5.2

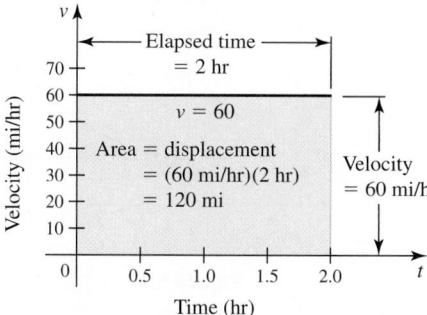

Figure 5.3

Because objects do not necessarily move at a constant velocity, we first extend these ideas to positive velocities that *change* over an interval of time. One strategy is to divide the time interval into many subintervals and approximate the velocity on each subinterval with a constant velocity. Then the displacements on each subinterval are calculated and summed. This strategy produces only an approximation to the displacement; however, this approximation generally improves as the number of subintervals increases.

EXAMPLE 1 **Approximating the displacement** Suppose the velocity in m/s of an object moving along a line is given by the function $v = t^2$, where $0 \le t \le 8$. Approximate the displacement of the object by dividing the time interval $[0, 8]$ into n subintervals of equal length. On each subinterval, approximate the velocity with a constant equal to the value of v evaluated at the midpoint of the subinterval.

a. Begin by dividing $[0, 8]$ into $n = 2$ subintervals: $[0, 4]$ and $[4, 8]$.
b. Divide $[0, 8]$ into $n = 4$ subintervals: $[0, 2]$, $[2, 4]$, $[4, 6]$, and $[6, 8]$.
c. Divide $[0, 8]$ into $n = 8$ subintervals of equal length.

SOLUTION

a. We divide the interval $[0, 8]$ into $n = 2$ subintervals, $[0, 4]$ and $[4, 8]$, each with length 4. The velocity on each subinterval is approximated by evaluating v at the midpoint of that subinterval (**Figure 5.4a**).

- We approximate the velocity on $[0, 4]$ by $v(2) = 2^2 = 4$ m/s. Traveling at 4 m/s for 4 s results in a displacement of 4 m/s · 4 s = 16 m.
- We approximate the velocity on $[4, 8]$ by $v(6) = 6^2 = 36$ m/s. Traveling at 36 m/s for 4 s results in a displacement of 36 m/s · 4 s = 144 m.

Therefore, an approximation to the displacement over the entire interval $[0, 8]$ is

$$(v(2) \cdot 4 \text{ s}) + (v(6) \cdot 4 \text{ s}) = (4 \text{ m/s} \cdot 4 \text{ s}) + (36 \text{ m/s} \cdot 4 \text{ s}) = 160 \text{ m.}$$

b. With $n = 4$ (**Figure 5.4b**), each subinterval has length 2. The approximate displacement over the entire interval is

$$\underbrace{(1 \text{ m/s} \cdot 2 \text{ s})}_{v(1)} + \underbrace{(9 \text{ m/s} \cdot 2 \text{ s})}_{v(3)} + \underbrace{(25 \text{ m/s} \cdot 2 \text{ s})}_{v(5)} + \underbrace{(49 \text{ m/s} \cdot 2 \text{ s})}_{v(7)} = 168 \text{ m.}$$

c. With $n = 8$ subintervals (**Figure 5.4c**), the approximation to the displacement is 170 m. In each case, the approximate displacement is the sum of the areas of the rectangles under the velocity curve.

> The midpoint of each subinterval is used to approximate the velocity over that subinterval.

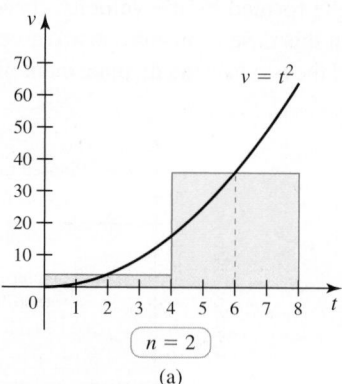

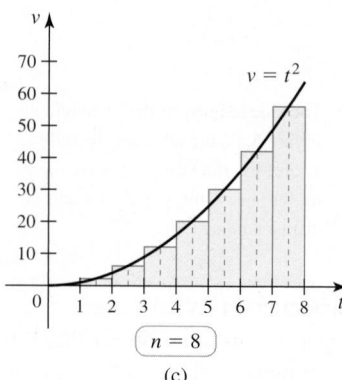

(a)   (b)   (c)

Figure 5.4

Related Exercises 3, 15–16 ◄

QUICK CHECK 2 In Example 1, if we used $n = 32$ subintervals of equal length, what would be the length of each subinterval? Find the midpoint of the first and last subinterval. ◄

The progression in Example 1 may be continued. Larger values of n mean more rectangles; in general, more rectangles give a better fit to the region under the curve (**Figure 5.5**). With the help of a calculator, we can generate the approximations in Table 5.1 using $n = 1, 2, 4, 8, 16, 32,$ and 64 subintervals. Observe that as n increases, the approximations appear to approach a limit of approximately 170.7 m. The limit is the exact displacement, which is represented by the area of the region under the velocity curve. This strategy of taking limits of sums is developed fully in Section 5.2.

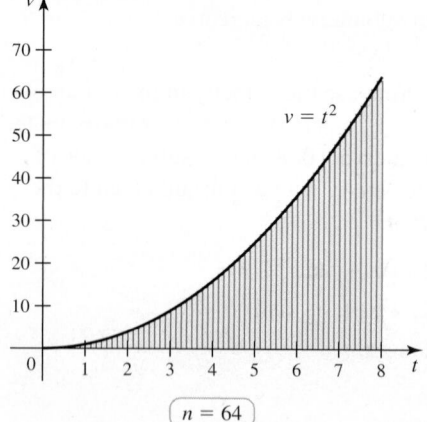

Figure 5.5

> The language "the area of the region bounded by the graph of a function" is often abbreviated as "the area under the curve."

Table 5.1 **Approximations to the area under the velocity curve $v = t^2$ on $[0, 8]$**

Number of subintervals	Length of each subinterval	Approximate displacement (area under curve)
1	8 s	128.0 m
2	4 s	160.0 m
4	2 s	168.0 m
8	1 s	170.0 m
16	0.5 s	170.5 m
32	0.25 s	170.625 m
64	0.125 s	170.65625 m

Approximating Areas by Riemann Sums

We wouldn't spend much time investigating areas under curves if the idea applied only to computing displacements from velocity curves. However, the problem of finding areas under curves arises frequently and turns out to be immensely important—as you will see in the next two chapters. For this reason, we now develop a systematic method for approximating areas under curves. Consider a function f that is continuous and nonnegative on an interval $[a, b]$. The goal is to approximate the area of the region R bounded by the graph of f and the x-axis from $x = a$ to $x = b$ (**Figure 5.6**). We begin by dividing the interval $[a, b]$ into n subintervals of equal length,

$$[x_0, x_1], [x_1, x_2], \ldots, [x_{n-1}, x_n],$$

where $a = x_0$ and $b = x_n$ (**Figure 5.7**). The length of each subinterval, denoted Δx, is found by dividing the length of the entire interval by n:

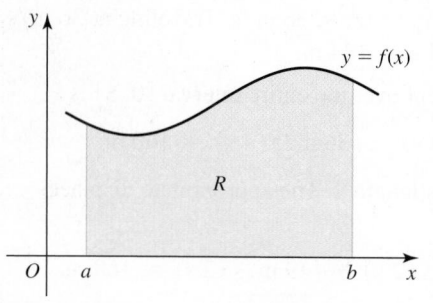

Figure 5.6

$$\Delta x = \frac{b - a}{n}.$$

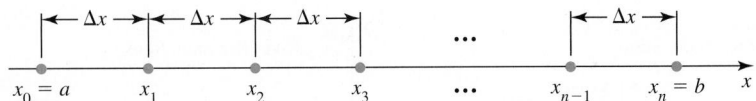

Figure 5.7

DEFINITION **Regular Partition**

Suppose $[a, b]$ is a closed interval containing n subintervals

$$[x_0, x_1], [x_1, x_2], \ldots, [x_{n-1}, x_n]$$

of equal length $\Delta x = \dfrac{b - a}{n}$, with $a = x_0$ and $b = x_n$. The endpoints $x_0, x_1, x_2, \ldots,$ x_{n-1}, x_n of the subintervals are called **grid points**, and they create a **regular partition** of the interval $[a, b]$. In general, the kth grid point is

$$x_k = a + k\Delta x, \text{ for } k = 0, 1, 2, \ldots, n.$$

QUICK CHECK 3 If the interval $[1, 9]$ is partitioned into 4 subintervals of equal length, what is Δx? List the grid points $x_0, x_1, x_2, x_3,$ and x_4. ◀

In the kth subinterval $[x_{k-1}, x_k]$, we choose any point x_k^* and build a rectangle whose height is $f(x_k^*)$, the value of f at x_k^* (Figure 5.8). The area of the rectangle on the kth subinterval is

$$\text{height} \cdot \text{base} = f(x_k^*)\Delta x, \qquad \text{where } k = 1, 2, \ldots, n.$$

Summing the areas of the rectangles in Figure 5.8, we obtain an approximation to the area of R, which is called a **Riemann sum**:

$$f(x_1^*)\Delta x + f(x_2^*)\Delta x + \cdots + f(x_n^*)\Delta x.$$

Three notable Riemann sums are the *left*, *right*, and *midpoint Riemann sums*.

▶ Although the idea of integration was developed in the 17th century, it was almost 200 years later that the German mathematician Bernhard Riemann (1826–1866) worked on the mathematical theory underlying integration.

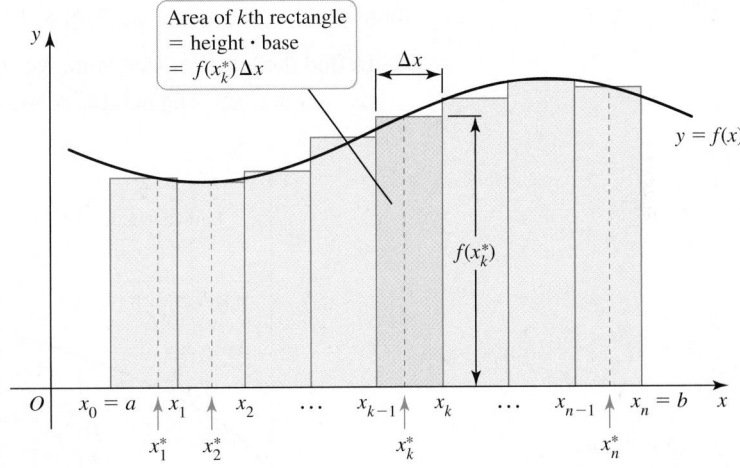

Figure 5.8

DEFINITION **Riemann Sum**

Suppose f is defined on a closed interval $[a, b]$, which is divided into n subintervals of equal length Δx. If x_k^* is any point in the kth subinterval $[x_{k-1}, x_k]$, for $k = 1, 2, \ldots, n$, then

$$f(x_1^*)\Delta x + f(x_2^*)\Delta x + \cdots + f(x_n^*)\Delta x$$

is called a **Riemann sum** for f on $[a, b]$. This sum is called

- a **left Riemann sum** if x_k^* is the left endpoint of $[x_{k-1}, x_k]$ (Figure 5.9);
- a **right Riemann sum** if x_k^* is the right endpoint of $[x_{k-1}, x_k]$ (Figure 5.10); and
- a **midpoint Riemann sum** if x_k^* is the midpoint of $[x_{k-1}, x_k]$ (Figure 5.11), for $k = 1, 2, \ldots, n.$

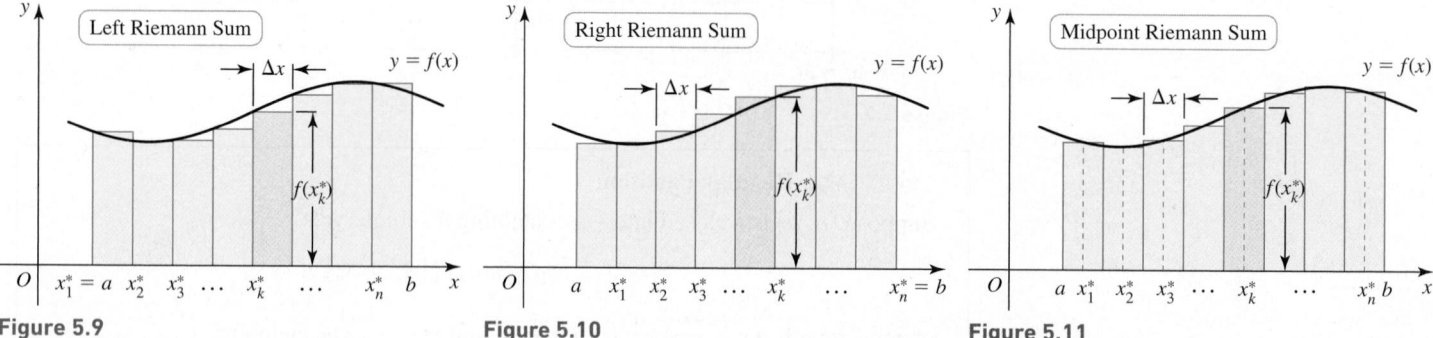

Figure 5.9   **Figure 5.10**   **Figure 5.11**

EXAMPLE 2 Left and right Riemann sums Let R be the region bounded by the graph of $f(x) = 3\sqrt{x}$ and the x-axis between $x = 4$ and $x = 16$.

a. Approximate the area of R using a left Riemann sum with $n = 6$ subintervals. Illustrate the sum with the appropriate rectangles.

b. Approximate the area of R using a right Riemann sum with $n = 6$ subintervals. Illustrate the sum with the appropriate rectangles.

c. Do the area approximations in parts (a) and (b) underestimate or overestimate the actual area under the curve?

SOLUTION Dividing the interval $[a, b] = [4, 16]$ into $n = 6$ subintervals means that the length of each subinterval is

$$\Delta x = \frac{b - a}{n} = \frac{16 - 4}{6} = 2;$$

therefore the grid points are 4, 6, 8, 10, 12, 14, and 16.

a. To find the left Riemann sum, we set $x_1^*, x_2^*, \ldots, x_6^*$ equal to the left endpoints of the six subintervals. The heights of the rectangles are $f(x_k^*)$, for $k = 1, 2, \ldots, 6$.

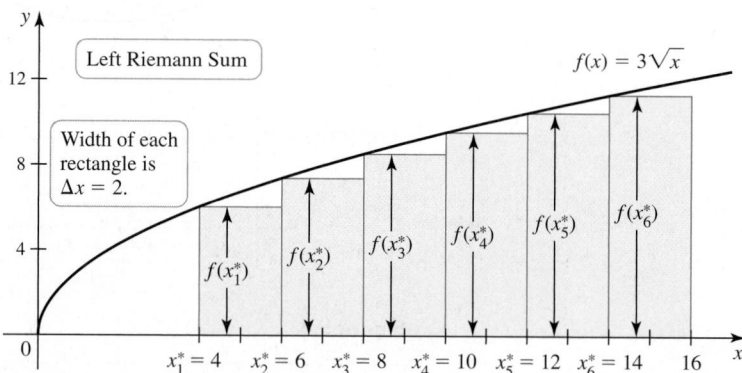

Figure 5.12

The resulting left Riemann sum (**Figure 5.12**) is

$$f(x_1^*)\Delta x + f(x_2^*)\Delta x + \cdots + f(x_6^*)\Delta x$$
$$= f(4) \cdot 2 + f(6) \cdot 2 + f(8) \cdot 2 + f(10) \cdot 2 + f(12) \cdot 2 + f(14) \cdot 2$$
$$= 3\sqrt{4} \cdot 2 + 3\sqrt{6} \cdot 2 + 3\sqrt{8} \cdot 2 + 3\sqrt{10} \cdot 2 + 3\sqrt{12} \cdot 2 + 3\sqrt{14} \cdot 2$$
$$\approx 105.876.$$

b. In a right Riemann sum, the right endpoints are used for $x_1^*, x_2^*, \ldots, x_6^*$, and the heights of the rectangles are $f(x_k^*)$, for $k = 1, 2, \ldots, 6$.

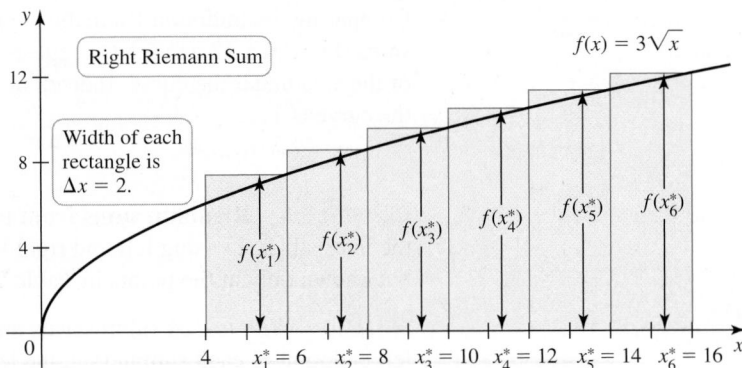

Figure 5.13

The resulting right Riemann sum (**Figure 5.13**) is

$$f(x_1^*)\Delta x + f(x_2^*)\Delta x + \cdots + f(x_6^*)\Delta x$$
$$= f(6)\cdot 2 + f(8)\cdot 2 + f(10)\cdot 2 + f(12)\cdot 2 + f(14)\cdot 2 + f(16)\cdot 2$$
$$= 3\sqrt{6}\cdot 2 + 3\sqrt{8}\cdot 2 + 3\sqrt{10}\cdot 2 + 3\sqrt{12}\cdot 2 + 3\sqrt{14}\cdot 2 + 3\sqrt{16}\cdot 2$$
$$\approx 117.876.$$

QUICK CHECK 4 If the function in Example 2 is replaced with $f(x) = 1/x$, does the left Riemann sum or the right Riemann sum overestimate the area under the curve? ◄

c. Looking at the graphs, we see that the left Riemann sum in part (a) underestimates the actual area of R, whereas the right Riemann sum in part (b) overestimates the area of R. Therefore, the area of R is between 105.876 and 117.876. These approximations improve as the number of rectangles increases.

Related Exercises 23–24, 29 ◄

EXAMPLE 3 A midpoint Riemann sum Let R be the region bounded by the graph of $f(x) = 3\sqrt{x}$ and the x-axis between $x = 4$ and $x = 16$. Approximate the area of R using a midpoint Riemann sum with $n = 6$ subintervals. Illustrate the sum with the appropriate rectangles.

SOLUTION The grid points and the length of the subintervals are the same as in Example 2. To find the midpoint Riemann sum, we set $x_1^*, x_2^*, \ldots, x_6^*$ equal to the midpoints of the subintervals. The midpoint of the first subinterval is the average of x_0 and x_1, which is

$$x_1^* = \frac{x_0 + x_1}{2} = \frac{4 + 6}{2} = 5.$$

The remaining midpoints are also computed by averaging the two nearest grid points.

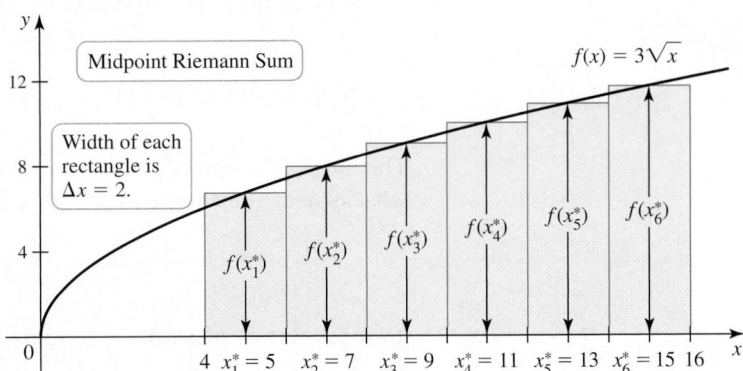

Figure 5.14

The resulting midpoint Riemann sum (**Figure 5.14**) is

$$f(x_1^*)\Delta x + f(x_2^*)\Delta x + \cdots + f(x_6^*)\Delta x$$
$$= f(5)\cdot 2 + f(7)\cdot 2 + f(9)\cdot 2 + f(11)\cdot 2 + f(13)\cdot 2 + f(15)\cdot 2$$
$$= 3\sqrt{5}\cdot 2 + 3\sqrt{7}\cdot 2 + 3\sqrt{9}\cdot 2 + 3\sqrt{11}\cdot 2 + 3\sqrt{13}\cdot 2 + 3\sqrt{15}\cdot 2$$
$$\approx 112.062.$$

Comparing the midpoint Riemann sum (Figure 5.14) with the left and right Riemann sums (Figures 5.12 and 5.13) suggests that the midpoint sum is a more accurate estimate of the area under the curve. Indeed, in Section 5.3, we will learn that the exact area under the curve is 112.

Related Exercises 33–34, 39 ◄

EXAMPLE 4   Riemann sums from tables Estimate the area A under the graph of f on the interval $[0, 2]$ using left and right Riemann sums with $n = 4$, where f is continuous but known only at the points in Table 5.2.

SOLUTION With $n = 4$ subintervals on the interval $[0, 2]$, $\Delta x = 2/4 = 0.5$. Using the left endpoint of each subinterval, the left Riemann sum is

$$A \approx f(0)\Delta x + f(0.5)\Delta x + f(1.0)\Delta x + f(1.5)\Delta x$$
$$= 1 \cdot 0.5 + 3 \cdot 0.5 + 4.5 \cdot 0.5 + 5.5 \cdot 0.5 = 7.0.$$

Using the right endpoint of each subinterval, the right Riemann sum is

$$A \approx f(0.5)\Delta x + f(1.0)\Delta x + f(1.5)\Delta x + f(2.0)\Delta x$$
$$= 3 \cdot 0.5 + 4.5 \cdot 0.5 + 5.5 \cdot 0.5 + 6.0 \cdot 0.5 = 9.5.$$

With only five function values, these estimates of the area are necessarily crude. Better estimates are obtained by using more subintervals and more function values.

Related Exercises 43–44 ◄

Table 5.2

x	$f(x)$
0	1
0.5	3
1.0	4.5
1.5	5.5
2.0	6.0

Sigma (Summation) Notation

Working with Riemann sums is cumbersome with large numbers of subintervals. Therefore, we pause for a moment to introduce some notation that simplifies our work.

Sigma (or **summation**) **notation** is used to express sums in a compact way. For example, the sum $1 + 2 + 3 + \cdots + 10$ is represented in sigma notation as $\sum_{k=1}^{10} k$. Here is how the notation works. The symbol Σ (*sigma*, the Greek capital S) stands for *sum*. The **index** k takes on all integer values from the lower limit ($k = 1$) to the upper limit ($k = 10$). The expression that immediately follows Σ (the **summand**) is evaluated for each value of k, and the resulting values are summed. Here are some examples.

$$\sum_{k=1}^{99} k = 1 + 2 + 3 + \cdots + 99 = 4950 \qquad \sum_{k=1}^{n} k = 1 + 2 + \cdots + n$$

$$\sum_{k=0}^{3} k^2 = 0^2 + 1^2 + 2^2 + 3^2 = 14 \qquad \sum_{k=1}^{4} (2k + 1) = 3 + 5 + 7 + 9 = 24$$

$$\sum_{k=-1}^{2} (k^2 + k) = ((-1)^2 + (-1)) + (0^2 + 0) + (1^2 + 1) + (2^2 + 2) = 8$$

The index in a sum is a *dummy variable*. It is internal to the sum, so it does not matter what symbol you choose as an index. For example,

$$\sum_{k=1}^{99} k = \sum_{n=1}^{99} n = \sum_{p=1}^{99} p.$$

Two properties of sums and sigma notation are useful in upcoming work. Suppose $\{a_1, a_2, \ldots, a_n\}$ and $\{b_1, b_2, \ldots, b_n\}$ are two sets of real numbers, and suppose c is a real number. Then we can factor multiplicative constants out of a sum:

Constant Multiple Rule $$\sum_{k=1}^{n} ca_k = c \sum_{k=1}^{n} a_k.$$

We can also split a sum into two sums:

Addition Rule $$\sum_{k=1}^{n} (a_k + b_k) = \sum_{k=1}^{n} a_k + \sum_{k=1}^{n} b_k.$$

In the coming examples and exercises, the following formulas for sums of powers of integers are essential.

> ▶ Formulas for $\displaystyle\sum_{k=1}^{n} k^p$, where p is a positive integer, have been known for centuries. The formulas for $p = 0, 1, 2,$ and 3 are relatively simple. The formulas become complicated as p increases.

THEOREM 5.1   **Sums of Powers of Integers**

Let n be a positive integer and c a real number.

$$\sum_{k=1}^{n} c = cn \qquad\qquad \sum_{k=1}^{n} k = \frac{n(n+1)}{2}$$

$$\sum_{k=1}^{n} k^2 = \frac{n(n+1)(2n+1)}{6} \qquad \sum_{k=1}^{n} k^3 = \frac{n^2(n+1)^2}{4}$$

Riemann Sums Using Sigma Notation

With sigma notation, a Riemann sum has the convenient compact form

$$f(x_1^*)\Delta x + f(x_2^*)\Delta x + \cdots + f(x_n^*)\Delta x = \sum_{k=1}^{n} f(x_k^*)\Delta x.$$

To express left, right, and midpoint Riemann sums in sigma notation, we must identify the points x_k^*.

- For left Riemann sums, the left endpoints of the subintervals are $x_k^* = a + (k-1)\Delta x$, for $k = 1, \ldots, n$.
- For right Riemann sums, the right endpoints of the subintervals are $x_k^* = a + k\Delta x$, for $k = 1, \ldots, n$.
- For midpoint Riemann sums, the midpoints of the subintervals are $x_k^* = a + \left(k - \frac{1}{2}\right)\Delta x$, for $k = 1, \ldots, n$.

The three Riemann sums are written compactly as follows.

> ▶ For the left Riemann sum,
> $x_1^* = a + 0 \cdot \Delta x, x_2^* = a + 1 \cdot \Delta x,$
> $x_3^* = a + 2 \cdot \Delta x,$
> and in general, $x_k^* = a + (k-1)\Delta x$, for $k = 1, \ldots, n.$
>
> For the right Riemann sum,
> $x_1^* = a + 1 \cdot \Delta x, x_2^* = a + 2 \cdot \Delta x,$
> $x_3^* = a + 3 \cdot \Delta x,$
> and in general, $x_k^* = a + k\Delta x$, for $k = 1, \ldots, n.$
>
> For the midpoint Riemann sum,
> $x_1^* = a + \frac{1}{2}\Delta x, x_2^* = a + \frac{3}{2}\Delta x,$
> and in general,
> $x_k^* = a + \left(k - \frac{1}{2}\right)\Delta x = \dfrac{x_k + x_{k-1}}{2},$
> for $k = 1, \ldots, n.$

DEFINITION   **Left, Right, and Midpoint Riemann Sums in Sigma Notation**

Suppose f is defined on a closed interval $[a, b]$, which is divided into n subintervals of equal length Δx. If x_k^* is a point in the kth subinterval $[x_{k-1}, x_k]$, for $k = 1, 2, \ldots, n,$ then the **Riemann sum** for f on $[a, b]$ is $\displaystyle\sum_{k=1}^{n} f(x_k^*)\Delta x$. Three cases arise in practice.

- $\displaystyle\sum_{k=1}^{n} f(x_k^*)\Delta x$ is a **left Riemann sum** if $x_k^* = a + (k-1)\Delta x$.

- $\displaystyle\sum_{k=1}^{n} f(x_k^*)\Delta x$ is a **right Riemann sum** if $x_k^* = a + k\Delta x$.

- $\displaystyle\sum_{k=1}^{n} f(x_k^*)\Delta x$ is a **midpoint Riemann sum** if $x_k^* = a + \left(k - \frac{1}{2}\right)\Delta x$.

EXAMPLE 5   **Calculating Riemann sums**   Evaluate the left, right, and midpoint Riemann sums for $f(x) = x^3 + 1$ between $a = 0$ and $b = 2$ using $n = 50$ subintervals. Make a conjecture about the exact area of the region under the curve (**Figure 5.15**).

SOLUTION   With $n = 50$, the length of each subinterval is

$$\Delta x = \frac{b - a}{n} = \frac{2 - 0}{50} = \frac{1}{25} = 0.04.$$

The value of x_k^* for the left Riemann sum is

$$x_k^* = a + (k-1)\Delta x = 0 + 0.04(k-1) = 0.04k - 0.04,$$

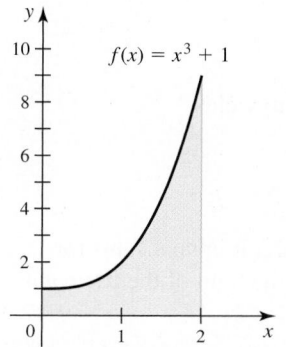

Figure 5.15

$f(x) = x^3 + 1$

for $k = 1, 2, \ldots, 50$. Therefore, the left Riemann sum, evaluated with a calculator, is

$$\sum_{k=1}^{n} f(x_k^*)\Delta x = \sum_{k=1}^{50} f(0.04k - 0.04)0.04 = 5.8416.$$

To evaluate the right Riemann sum, we let $x_k^* = a + k\Delta x = 0.04k$ and find that

$$\sum_{k=1}^{n} f(x_k^*)\Delta x = \sum_{k=1}^{50} f(0.04k)0.04 = 6.1616.$$

For the midpoint Riemann sum, we let

$$x_k^* = a + \left(k - \frac{1}{2}\right)\Delta x = 0 + 0.04\left(k - \frac{1}{2}\right) = 0.04k - 0.02.$$

The value of the sum is

$$\sum_{k=1}^{n} f(x_k^*)\Delta x = \sum_{k=1}^{50} f(0.04k - 0.02)0.04 = 5.9992.$$

Because f is increasing on $[0, 2]$, the left Riemann sum underestimates the area of the shaded region in Figure 5.15, and the right Riemann sum overestimates the area. Therefore, the exact area lies between 5.8416 and 6.1616. The midpoint Riemann sum usually gives the best estimate for increasing or decreasing functions.

Table 5.3 shows the left, right, and midpoint Riemann sum approximations for values of n up to 200. All three sets of approximations approach a value near 6, which is a reasonable estimate of the area under the curve. In Section 5.2, we show rigorously that the limit of all three Riemann sums as $n \to \infty$ is 6.

ALTERNATIVE SOLUTION It is worth examining another approach to Example 5. Consider the right Riemann sum given previously:

$$\sum_{k=1}^{n} f(x_k^*)\Delta x = \sum_{k=1}^{50} f(0.04k)0.04.$$

Rather than evaluating this sum with a calculator, we note that $f(0.04k) = (0.04k)^3 + 1$ and use the properties of sums:

$$\sum_{k=1}^{n} f(x_k^*)\Delta x = \sum_{k=1}^{50} \underbrace{((0.04k)^3 + 1)}_{f(x_k^*)}\underbrace{0.04}_{\Delta x}$$

$$= \sum_{k=1}^{50} (0.04k)^3\, 0.04 + \sum_{k=1}^{50} 1 \cdot 0.04 \qquad \sum(a_k + b_k) = \sum a_k + \sum b_k$$

$$= (0.04)^4 \sum_{k=1}^{50} k^3 + 0.04 \sum_{k=1}^{50} 1. \qquad \sum c a_k = c \sum a_k$$

Using the summation formulas for powers of integers in Theorem 5.1, we find that

$$\sum_{k=1}^{50} 1 = 50 \quad \text{and} \quad \sum_{k=1}^{50} k^3 = \frac{50^2 \cdot 51^2}{4}.$$

Substituting the values of these sums into the right Riemann sum yields

$$\sum_{k=1}^{50} f(x_k^*)\Delta x = \frac{3851}{625} = 6.1616,$$

confirming the result given by a calculator. The idea of evaluating Riemann sums for *arbitrary* values of n is used in Section 5.2, where we evaluate the limit of the Riemann sum as $n \to \infty$.

Related Exercises 51–52 ◄

Table 5.3   Left, right, and midpoint Riemann sum approximations

n	L_n	R_n	M_n
20	5.61	6.41	5.995
40	5.8025	6.2025	5.99875
60	5.86778	6.13444	5.99944
80	5.90063	6.10063	5.99969
100	5.9204	6.0804	5.9998
120	5.93361	6.06694	5.99986
140	5.94306	6.05735	5.9999
160	5.95016	6.05016	5.99992
180	5.95568	6.04457	5.99994
200	5.9601	6.0401	5.99995

SECTION 5.1 EXERCISES

Getting Started

1. Suppose an object moves along a line at 15 m/s, for $0 \leq t < 2$, and at 25 m/s, for $2 \leq t \leq 5$, where t is measured in seconds. Sketch the graph of the velocity function and find the displacement of the object for $0 \leq t \leq 5$.

2. Given the graph of the positive velocity of an object moving along a line, what is the geometrical representation of its displacement over a time interval $[a, b]$?

3. The velocity in ft/s of an object moving along a line is given by $v = f(t)$ on the interval $0 \leq t \leq 8$ (see figure), where t is measured in seconds.

    a. Divide the interval $[0, 8]$ into $n = 2$ subintervals, $[0, 4]$ and $[4, 8]$. On each subinterval, assume the object moves at a constant velocity equal to the value of v evaluated at the midpoint of the subinterval, and use these approximations to estimate the displacement of the object on $[0, 8]$ (see part (a) of the figure).

    b. Repeat part (a) for $n = 4$ subintervals (see part (b) of the figure).

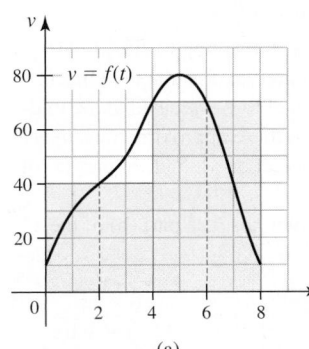

(a)

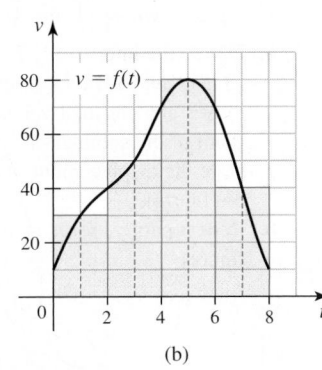

(b)

4. The velocity in ft/s of an object moving along a line is given by $v = f(t)$ on the interval $0 \leq t \leq 6$ (see figure), where t is measured in seconds.

    a. Divide the interval $[0, 6]$ into $n = 3$ subintervals, $[0, 2]$, $[2, 4]$, and $[4, 6]$. On each subinterval, assume the object moves at a constant velocity equal to the value of v evaluated at the left endpoint of the subinterval, and use these approximations to estimate the displacement of the object on $[0, 6]$ (see part (a) of the figure).

    b. Repeat part (a) for $n = 6$ subintervals (see part (b) of the figure).

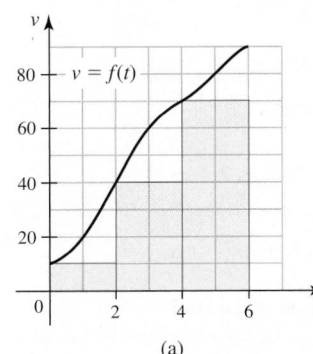

(a)

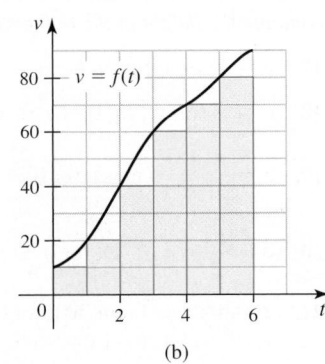

(b)

5. The velocity in ft/s of an object moving along a line is given by $v = f(t)$ on the interval $0 \leq t \leq 6$ (see figure), where t is measured in seconds.

    a. Divide the interval $[0, 6]$ into $n = 3$ subintervals, $[0, 2]$, $[2, 4]$ and $[4, 6]$. On each subinterval, assume the object moves at a constant velocity equal to the value of v evaluated at the right endpoint of the subinterval, and use these approximations to estimate the displacement of the object on $[0, 6]$ (see part (a) of the figure).

    b. Repeat part (a) for $n = 6$ subintervals (see part (b) of the figure).

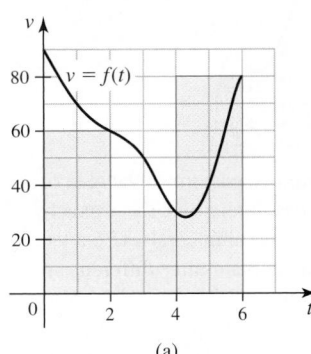

(a)

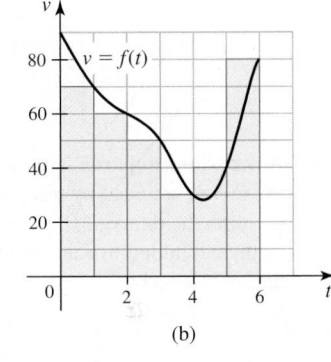

(b)

6. The velocity in ft/s of an object moving along a line is given by $v = f(t)$ on the interval $0 \leq t \leq 10$ (see figure), where t is measured in seconds.

    a. Divide the interval $[0, 10]$ into $n = 5$ subintervals. On each subinterval, assume the object moves at a constant velocity equal to the value of v evaluated at the left endpoint of the subinterval, and use these approximations to estimate the displacement of the object on $[0, 10]$.

    b. Repeat part (a) using the right endpoints to estimate the displacement on $[0, 10]$.

    c. Repeat part (a) using the midpoints of each subinterval to estimate the displacement on $[0, 10]$.

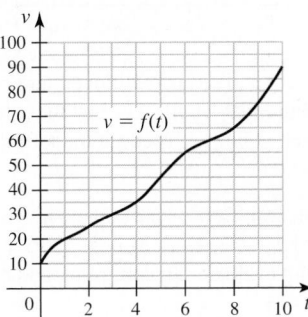

7. Suppose you want to approximate the area of the region bounded by the graph of $f(x) = \cos x$ and the x-axis between $x = 0$ and $x = \frac{\pi}{2}$. Explain a possible strategy.

8. Explain how Riemann sum approximations to the area of a region under a curve change as the number of subintervals increases.

9. **Approximating area from a graph** Approximate the area of the region bounded by the graph (see figure) and the x-axis by dividing the interval $[1, 7]$ into $n = 6$ subintervals. Use

a left and right Riemann sum to obtain two different approximations.

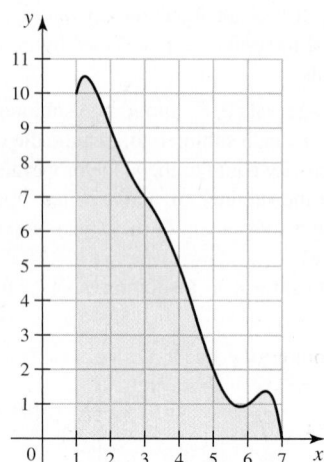

10. Approximating area from a graph Approximate the area of the region bounded by the graph (see figure) and the x-axis by dividing the interval $[0, 6]$ into $n = 3$ subintervals. Use a left, right, and midpoint Riemann sum to obtain three different approximations.

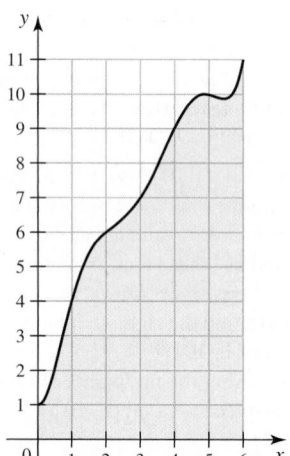

11. Suppose the interval $[1, 3]$ is partitioned into $n = 4$ subintervals. What is the subinterval length Δx? List the grid points x_0, x_1, x_2, x_3, and x_4. Which points are used for the left, right, and midpoint Riemann sums?

12. Suppose the interval $[2, 6]$ is partitioned into $n = 4$ subintervals with grid points $x_0 = 2, x_1 = 3, x_2 = 4, x_3 = 5$, and $x_4 = 6$. Write, but do not evaluate, the left, right, and midpoint Riemann sums for $f(x) = x^2$.

13. Does a right Riemann sum underestimate or overestimate the area of the region under the graph of a function that is positive and decreasing on an interval $[a, b]$? Explain.

14. Does a left Riemann sum underestimate or overestimate the area of the region under the graph of a function that is positive and increasing on an interval $[a, b]$? Explain.

Practice Exercises

15. Approximating displacement The velocity in ft/s of an object moving along a line is given by $v = 3t^2 + 1$ on the interval $0 \le t \le 4$, where t is measured in seconds.

a. Divide the interval $[0, 4]$ into $n = 4$ subintervals, $[0, 1]$, $[1, 2], [2, 3]$, and $[3, 4]$. On each subinterval, assume the object moves at a constant velocity equal to v evaluated at the midpoint of the subinterval, and use these approximations to estimate the displacement of the object on $[0, 4]$ (see part (a) of the figure).

b. Repeat part (a) for $n = 8$ subintervals (see part (b) of the figure).

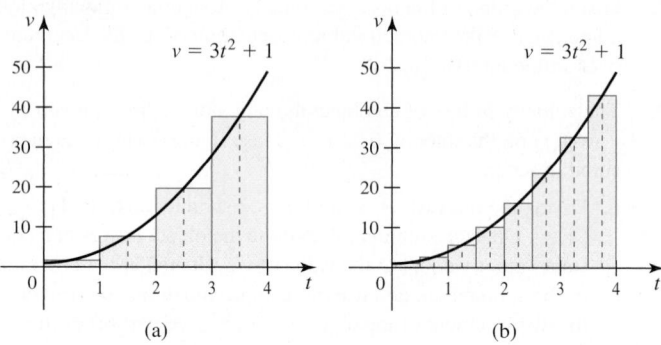

(a)   (b)

T 16. Approximating displacement The velocity in ft/s of an object moving along a line is given by $v = \sqrt{10t}$ on the interval $1 \le t \le 7$, where t is measured in seconds.

a. Divide the interval $[1, 7]$ into $n = 3$ subintervals, $[1, 3]$, $[3, 5]$, and $[5, 7]$. On each subinterval, assume the object moves at a constant velocity equal to v evaluated at the midpoint of the subinterval, and use these approximations to estimate the displacement of the object on $[1, 7]$ (see part (a) of the figure).

b. Repeat part (a) for $n = 6$ subintervals (see part (b) of the figure).

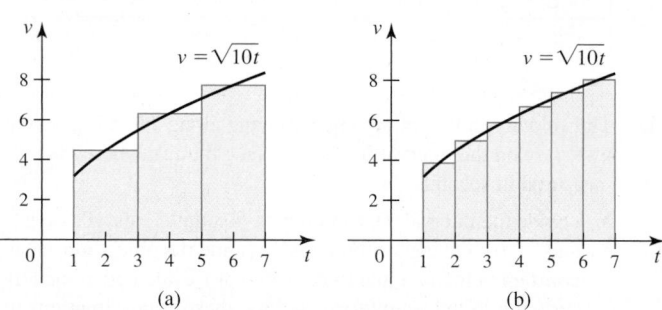

(a)   (b)

17–22. Approximating displacement *The velocity of an object is given by the following functions on a specified interval. Approximate the displacement of the object on this interval by subdividing the interval into n subintervals. Use the left endpoint of each subinterval to compute the height of the rectangles.*

17. $v = 2t + 1 \, (\text{m/s})$, for $0 \le t \le 8; n = 2$

T 18. $v = e^t \, (\text{m/s})$, for $0 \le t \le 3; n = 3$

19. $v = \dfrac{1}{2t + 1} \, (\text{m/s})$, for $0 \le t \le 8; n = 4$

20. $v = \dfrac{t^2}{2} + 4 \, (\text{ft/s})$, for $0 \le t \le 12; n = 6$

T 21. $v = 4\sqrt{t + 1} \, (\text{mi/hr})$, for $0 \le t \le 15; n = 5$

22. $v = \dfrac{t + 3}{6} \, (\text{m/s})$, for $0 \le t \le 4; n = 4$

23–24. Left and right Riemann sums *Use the figures to calculate the left and right Riemann sums for* f *on the given interval and for the given value of* n.

23. $f(x) = x + 1$ on $[1, 6]$; $n = 5$

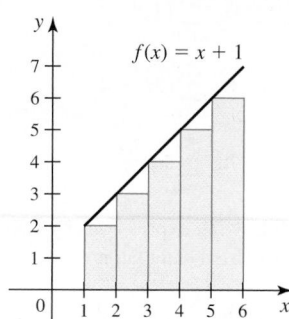
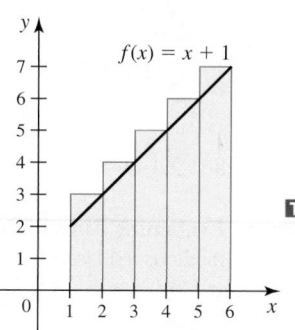

24. $f(x) = \dfrac{1}{x}$ on $[1, 5]$; $n = 4$

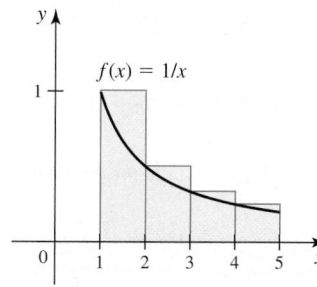
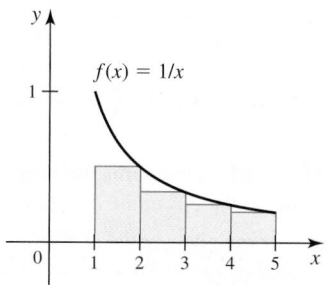

25–32. Left and right Riemann sums *Complete the following steps for the given function, interval, and value of* n.

a. *Sketch the graph of the function on the given interval.*

b. *Calculate* Δx *and the grid points* $x_0, x_1, \ldots, x_n$.

c. *Illustrate the left and right Riemann sums. Then determine which Riemann sum underestimates and which sum overestimates the area under the curve.*

d. *Calculate the left and right Riemann sums.*

25. $f(x) = x + 1$ on $[0, 4]$; $n = 4$

26. $f(x) = 9 - x$ on $[3, 8]$; $n = 5$

T 27. $f(x) = \cos x$ on $\left[0, \dfrac{\pi}{2}\right]$; $n = 4$

T 28. $f(x) = \sin^{-1}\dfrac{x}{3}$ on $[0, 3]$; $n = 6$

29. $f(x) = x^2 - 1$ on $[2, 4]$; $n = 4$

30. $f(x) = 2x^2$ on $[1, 6]$; $n = 5$

T 31. $f(x) = e^{x/2}$ on $[1, 4]$; $n = 6$

T 32. $f(x) = \ln 4x$ on $[1, 3]$; $n = 5$

33. A midpoint Riemann sum Approximate the area of the region bounded by the graph of $f(x) = 100 - x^2$ and the x-axis on $[0, 10]$ with $n = 5$ subintervals. Use the midpoint of each subinterval to determine the height of each rectangle (see figure).

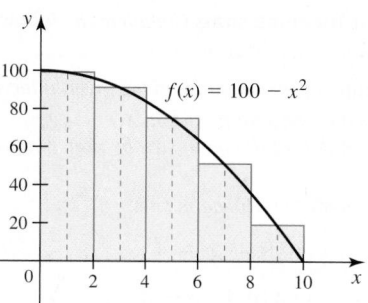

T 34. A midpoint Riemann sum Approximate the area of the region bounded by the graph of $f(t) = \cos \dfrac{t}{2}$ and the t-axis on $[0, \pi]$ with $n = 4$ subintervals. Use the midpoint of each subinterval to determine the height of each rectangle (see figure).

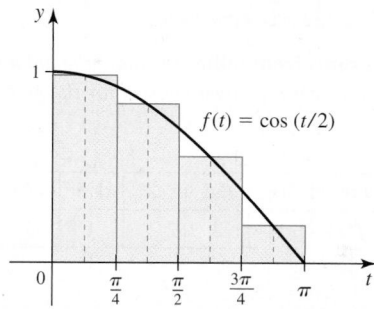

35. Free fall On October 14, 2012, Felix Baumgartner stepped off a balloon capsule at an altitude of almost 39 km above Earth's surface and began his free fall. His velocity in m/s during the fall is given in the figure. It is claimed that Felix reached the speed of sound 34 seconds into his fall and that he continued to fall at supersonic speed for 30 seconds. (*Source:* http://www.redbullstratos.com)

a. Divide the interval $[34, 64]$ into $n = 5$ subintervals with the gridpoints $x_0 = 34$, $x_1 = 40$, $x_2 = 46$, $x_3 = 52$, $x_4 = 58$, and $x_5 = 64$. Use left and right Riemann sums to estimate how far Felix fell while traveling at supersonic speed.

b. It is claimed that the actual distance that Felix fell at supersonic speed was approximately 10,485 m. Which estimate in part (a) produced the more accurate estimate?

c. How could you obtain more accurate estimates of the total distance fallen than those found in part (a)?

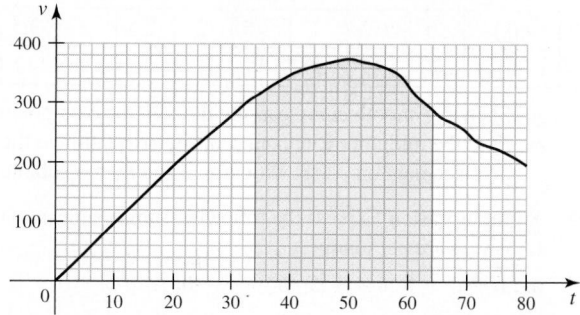

36. Free fall Use geometry and the figure given in Exercise 35 to estimate how far Felix fell in the first 20 seconds of his free fall.

37–42. Midpoint Riemann sums *Complete the following steps for the given function, interval, and value of n.*

a. *Sketch the graph of the function on the given interval.*
b. *Calculate Δx and the grid points $x_0, x_1, \ldots, x_n$.*
c. *Illustrate the midpoint Riemann sum by sketching the appropriate rectangles.*
d. *Calculate the midpoint Riemann sum.*

37. $f(x) = 2x + 1$ on $[0, 4]$; $n = 4$

T 38. $f(x) = 2\cos^{-1} x$ on $[0, 1]$; $n = 5$

T 39. $f(x) = \sqrt{x}$ on $[1, 3]$; $n = 4$

40. $f(x) = x^2$ on $[0, 4]$; $n = 4$

41. $f(x) = \dfrac{1}{x}$ on $[1, 6]$; $n = 5$

42. $f(x) = 4 - x$ on $[-1, 4]$; $n = 5$

43–46. Riemann sums from tables *Evaluate the left and right Riemann sums for f over the given interval for the given value of n.*

43. $n = 4$; $[0, 2]$

x	0	0.5	1	1.5	2
$f(x)$	5	3	2	1	1

44. $n = 8$; $[1, 5]$

x	1	1.5	2	2.5	3	3.5	4	4.5	5
$f(x)$	0	2	3	2	2	1	0	2	3

45. Displacement from a table of velocities The velocities (in mi/hr) of an automobile moving along a straight highway over a two-hour period are given in the following table.

t (hr)	0	0.25	0.5	0.75	1	1.25	1.5	1.75	2
v (mi/hr)	50	50	60	60	55	65	50	60	70

a. Sketch a smooth curve passing through the data points.
b. Find the midpoint Riemann sum approximation to the displacement on $[0, 2]$ with $n = 2$ and $n = 4$.

46. Displacement from a table of velocities The velocities (in m/s) of an automobile moving along a straight freeway over a four-second period are given in the following table.

t (s)	0	0.5	1	1.5	2	2.5	3	3.5	4
v (m/s)	20	25	30	35	30	30	35	40	40

a. Sketch a smooth curve passing through the data points.
b. Find the midpoint Riemann sum approximation to the displacement on $[0, 4]$ with $n = 2$ and $n = 4$ subintervals.

47. Sigma notation Express the following sums using sigma notation. (Answers are not unique.)

a. $1 + 2 + 3 + 4 + 5$
b. $4 + 5 + 6 + 7 + 8 + 9$
c. $1^2 + 2^2 + 3^2 + 4^2$
d. $1 + \dfrac{1}{2} + \dfrac{1}{3} + \dfrac{1}{4}$

48. Sigma notation Express the following sums using sigma notation. (Answers are not unique.)

a. $1 + 3 + 5 + 7 + \cdots + 99$
b. $4 + 9 + 14 + \cdots + 44$
c. $3 + 8 + 13 + \cdots + 63$
d. $\dfrac{1}{1 \cdot 2} + \dfrac{1}{2 \cdot 3} + \dfrac{1}{3 \cdot 4} + \cdots + \dfrac{1}{49 \cdot 50}$

49. Sigma notation Evaluate the following expressions.

a. $\displaystyle\sum_{k=1}^{10} k$
b. $\displaystyle\sum_{k=1}^{6} (2k + 1)$
c. $\displaystyle\sum_{k=1}^{4} k^2$
d. $\displaystyle\sum_{n=1}^{5} (1 + n^2)$
e. $\displaystyle\sum_{m=1}^{3} \frac{2m + 2}{3}$
f. $\displaystyle\sum_{j=1}^{3} (3j - 4)$
g. $\displaystyle\sum_{p=1}^{5} (2p + p^2)$
h. $\displaystyle\sum_{n=0}^{4} \sin \frac{n\pi}{2}$

T 50. Evaluating sums Evaluate the following expressions by two methods. (i) Use Theorem 5.1. (ii) Use a calculator.

a. $\displaystyle\sum_{k=1}^{45} k$
b. $\displaystyle\sum_{k=1}^{45} (5k - 1)$
c. $\displaystyle\sum_{k=1}^{75} 2k^2$
d. $\displaystyle\sum_{n=1}^{50} (1 + n^2)$
e. $\displaystyle\sum_{m=1}^{75} \frac{2m + 2}{3}$
f. $\displaystyle\sum_{j=1}^{20} (3j - 4)$
g. $\displaystyle\sum_{p=1}^{35} (2p + p^2)$
h. $\displaystyle\sum_{n=0}^{40} (n^2 + 3n - 1)$

T 51–54. Riemann sums for larger values of *n* *Complete the following steps for the given function f and interval.*

a. *For the given value of n, use sigma notation to write the left, right, and midpoint Riemann sums. Then evaluate each sum using a calculator.*
b. *Based on the approximations found in part (a), estimate the area of the region bounded by the graph of f and the x-axis on the interval.*

51. $f(x) = 3\sqrt{x}$ on $[0, 4]$; $n = 40$

52. $f(x) = x^2 + 1$ on $[-1, 1]$; $n = 50$

53. $f(x) = x^2 - 1$ on $[2, 5]$; $n = 75$

54. $f(x) = \cos 2x$ on $\left[0, \dfrac{\pi}{4}\right]$; $n = 60$

T 55–58. Approximating areas with a calculator *Use a calculator and right Riemann sums to approximate the area of the given region. Present your calculations in a table showing the approximations for $n = 10, 30, 60,$ and 80 subintervals. Make a conjecture about the limit of Riemann sums as $n \to \infty$.*

55. The region bounded by the graph of $f(x) = 12 - 3x^2$ and the x-axis on the interval $[-1, 1]$.

56. The region bounded by the graph of $f(x) = 3x^2 + 1$ and the x-axis on the interval $[-1, 1]$.

57. The region bounded by the graph of $f(x) = \dfrac{1 - \cos x}{2}$ and the x-axis on the interval $[-\pi, \pi]$.

58. The region bounded by the graph of $f(x) = (2^x + 2^{-x}) \ln 2$ and the x-axis on the interval $[-2, 2]$.

59. Explain why or why not Determine whether the following statements are true and give an explanation or counterexample.

a. Consider the linear function $f(x) = 2x + 5$ and the region bounded by its graph and the x-axis on the interval $[3, 6]$. Suppose the area of this region is approximated using midpoint

Riemann sums. Then the approximations give the exact area of the region for any number of subintervals.

b. A left Riemann sum always overestimates the area of a region bounded by a positive increasing function and the *x*-axis on an interval $[a, b]$.

c. For an increasing or decreasing nonconstant function on an interval $[a, b]$ and a given value of *n*, the value of the midpoint Riemann sum always lies between the values of the left and right Riemann sums.

T 60. Riemann sums for a semicircle Let $f(x) = \sqrt{1 - x^2}$.

a. Show that the graph of *f* is the upper half of a circle of radius 1 centered at the origin.

b. Estimate the area between the graph of *f* and the *x*-axis on the interval $[-1, 1]$ using a midpoint Riemann sum with $n = 25$.

c. Repeat part (b) using $n = 75$ rectangles.

d. What happens to the midpoint Riemann sums on $[-1, 1]$ as $n \to \infty$?

T 61–64. Sigma notation for Riemann sums *Use sigma notation to write the following Riemann sums. Then evaluate each Riemann sum using Theorem 5.1 or a calculator.*

61. The right Riemann sum for $f(x) = x + 1$ on $[0, 4]$ with $n = 50$.

62. The left Riemann sum for $f(x) = e^x$ on $[0, \ln 2]$ with $n = 40$.

63. The midpoint Riemann sum for $f(x) = x^3$ on $[3, 11]$ with $n = 32$.

64. The midpoint Riemann sum for $f(x) = 1 + \cos \pi x$ on $[0, 2]$ with $n = 50$.

65–68. Identifying Riemann sums *Fill in the blanks with an interval and a value of n.*

65. $\sum\limits_{k=1}^{4} f(1 + k) \cdot 1$ is a right Riemann sum for *f* on the interval

$[\underline{\hspace{1cm}}, \underline{\hspace{1cm}}]$ with $n = \underline{\hspace{1cm}}$.

66. $\sum\limits_{k=1}^{4} f(2 + k) \cdot 1$ is a right Riemann sum for *f* on the interval

$[\underline{\hspace{1cm}}, \underline{\hspace{1cm}}]$ with $n = \underline{\hspace{1cm}}$.

67. $\sum\limits_{k=1}^{4} f(1.5 + k) \cdot 1$ is a midpoint Riemann sum for *f* on the interval

$[\underline{\hspace{1cm}}, \underline{\hspace{1cm}}]$ with $n = \underline{\hspace{1cm}}$.

68. $\sum\limits_{k=1}^{8} f\left(1.5 + \dfrac{k}{2}\right) \cdot \dfrac{1}{2}$ is a left Riemann sum for *f* on the interval

$[\underline{\hspace{1cm}}, \underline{\hspace{1cm}}]$ with $n = \underline{\hspace{1cm}}$.

69. Approximating areas Estimate the area of the region bounded by the graph of $f(x) = x^2 + 2$ and the *x*-axis on $[0, 2]$ in the following ways.

a. Divide $[0, 2]$ into $n = 4$ subintervals and approximate the area of the region using a left Riemann sum. Illustrate the solution geometrically.

b. Divide $[0, 2]$ into $n = 4$ subintervals and approximate the area of the region using a midpoint Riemann sum. Illustrate the solution geometrically.

c. Divide $[0, 2]$ into $n = 4$ subintervals and approximate the area of the region using a right Riemann sum. Illustrate the solution geometrically.

70. Displacement from a velocity graph Consider the velocity function for an object moving along a line (see figure).

a. Describe the motion of the object over the interval $[0, 6]$.

b. Use geometry to find the displacement of the object between $t = 0$ and $t = 3$.

c. Use geometry to find the displacement of the object between $t = 3$ and $t = 5$.

d. Assuming the velocity remains 30 m/s, for $t \geq 4$, find the function that gives the displacement between $t = 0$ and any time $t \geq 4$.

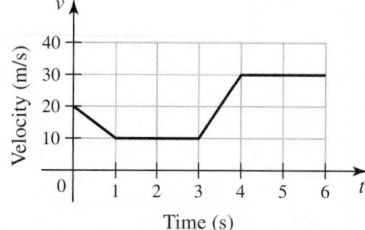

71. Displacement from a velocity graph Consider the velocity function for an object moving along a line (see figure).

a. Describe the motion of the object over the interval $[0, 6]$.

b. Use geometry to find the displacement of the object between $t = 0$ and $t = 2$.

c. Use geometry to find the displacement of the object between $t = 2$ and $t = 5$.

d. Assuming the velocity remains 10 m/s, for $t \geq 5$, find the function that gives the displacement between $t = 0$ and any time $t \geq 5$.

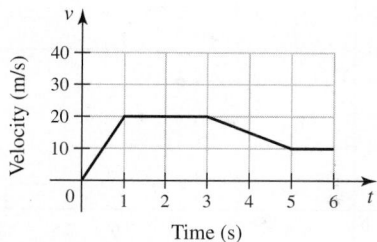

72. Flow rates Suppose a gauge at the outflow of a reservoir measures the flow rate of water in units of ft^3/hr. In Chapter 6, we show that the total amount of water that flows out of the reservoir is the area under the flow rate curve. Consider the flow rate function shown in the figure.

a. Find the amount of water (in units of ft^3) that flows out of the reservoir over the interval $[0, 4]$.

b. Find the amount of water that flows out of the reservoir over the interval $[8, 10]$.

c. Does more water flow out of the reservoir over the interval $[0, 4]$ or $[4, 6]$?

d. Show that the units of your answer are consistent with the units of the variables on the axes.

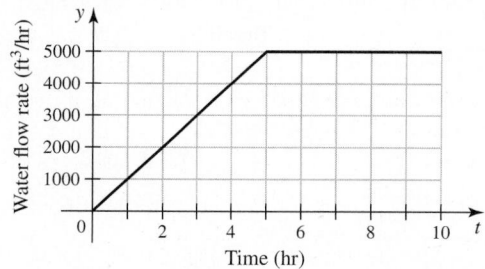

73. Mass from density A thin 10-cm rod is made of an alloy whose density varies along its length according to the function shown in the figure. Assume density is measured in units of g/cm. In Chapter 6, we show that the mass of the rod is the area under the density curve.

  a. Find the mass of the left half of the rod ($0 \le x \le 5$).
  b. Find the mass of the right half of the rod ($5 \le x \le 10$).
  c. Find the mass of the entire rod ($0 \le x \le 10$).
  d. Find the point along the rod at which it will balance (called the center of mass).

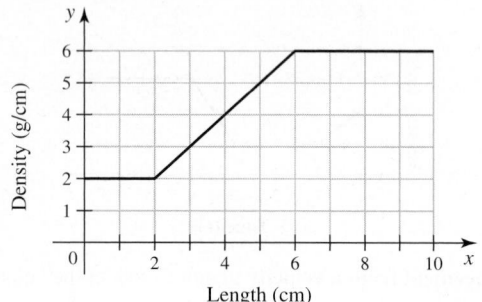

74–75. Displacement from velocity *The following functions describe the velocity of a car (in mi/hr) moving along a straight highway for a 3-hr interval. In each case, find the function that gives the displacement of the car over the interval* $[0, t]$, *where* $0 \le t \le 3$.

74. $v(t) = \begin{cases} 40 & \text{if } 0 \le t \le 1.5 \\ 50 & \text{if } 1.5 < t \le 3 \end{cases}$

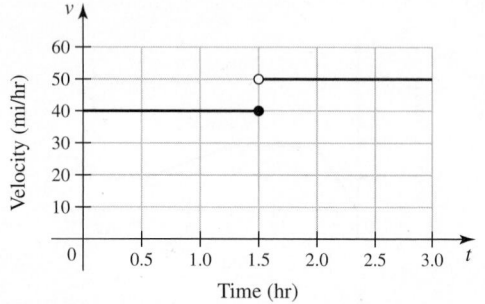

75. $v(t) = \begin{cases} 30 & \text{if } 0 \le t \le 2 \\ 50 & \text{if } 2 < t \le 2.5 \\ 44 & \text{if } 2.5 < t \le 3 \end{cases}$

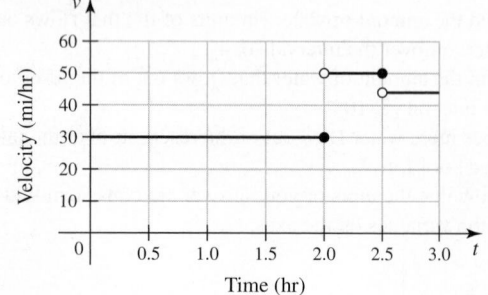

⊤ 76–77. Functions with absolute value *Use a calculator and the method of your choice to approximate the area of the following regions. Present your calculations in a table, showing approximations using* $n = 16, 32,$ *and* 64 *subintervals. Make a conjecture about the limits of the approximations.*

76. The region bounded by the graph of $f(x) = |25 - x^2|$ and the x-axis on the interval $[0, 10]$.

77. The region bounded by the graph of $f(x) = |1 - x^3|$ and the x-axis on the interval $[-1, 2]$.

Explorations and Challenges

78. Riemann sums for constant functions Let $f(x) = c$, where $c > 0$, be a constant function on $[a, b]$. Prove that any Riemann sum for any value of n gives the exact area of the region between the graph of f and the x-axis on $[a, b]$.

79. Riemann sums for linear functions Assume the linear function $f(x) = mx + c$ is positive on the interval $[a, b]$. Prove that the midpoint Riemann sum with any value of n gives the exact area of the region between the graph of f and the x-axis on $[a, b]$.

80. Shape of the graph for left Riemann sums Suppose a left Riemann sum is used to approximate the area of the region bounded by the graph of a positive function and the x-axis on the interval $[a, b]$. Fill in the following table to indicate whether the resulting approximation underestimates or overestimates the exact area in the four cases shown. Use a sketch to explain your reasoning in each case.

	Increasing on $[a, b]$	Decreasing on $[a, b]$
Concave up on $[a, b]$		
Concave down on $[a, b]$		

81. Shape of the graph for right Riemann sums Suppose a right Riemann sum is used to approximate the area of the region bounded by the graph of a positive function and the x-axis on the interval $[a, b]$. Fill in the following table to indicate whether the resulting approximation underestimates or overestimates the exact area in the four cases shown. Use a sketch to explain your reasoning in each case.

	Increasing on $[a, b]$	Decreasing on $[a, b]$
Concave up on $[a, b]$		
Concave down on $[a, b]$		

QUICK CHECK ANSWERS

1. 45 mi   **2.** 0.25, 0.125, 7.875   **3.** $\Delta x = 2$; $\{1, 3, 5, 7, 9\}$
4. The left sum overestimates the area. ◄

5.2 Definite Integrals

We introduced Riemann sums in Section 5.1 as a way to approximate the area of a region bounded by a curve $y = f(x)$ and the x-axis on an interval $[a, b]$. In that discussion, we assumed f to be nonnegative on the interval. Our next task is to discover the geometric meaning of Riemann sums when f is negative on some or all of $[a, b]$. Once this matter is settled, we proceed to the main event of this section, which is to define the *definite integral*. With definite integrals, the approximations given by Riemann sums become exact.

Net Area

How do we interpret Riemann sums when f is negative at some or all points of $[a, b]$? The answer follows directly from the Riemann sum definition.

EXAMPLE 1 **Interpreting Riemann sums** Evaluate and interpret the following Riemann sums for $f(x) = 1 - x^2$ on the interval $[a, b]$ with n equally spaced subintervals.

a. A midpoint Riemann sum with $[a, b] = [1, 3]$ and $n = 4$
b. A left Riemann sum with $[a, b] = [0, 3]$ and $n = 6$

SOLUTION

a. The length of each subinterval is $\Delta x = \dfrac{b - a}{n} = \dfrac{3 - 1}{4} = 0.5$. So the grid points are

$$x_0 = 1, \quad x_1 = 1.5, \quad x_2 = 2, \quad \text{and} \quad x_3 = 2.5, \quad \text{and} \quad x_4 = 3.$$

To compute the midpoint Riemann sum, we evaluate f at the midpoints of the subintervals, which are

$$x_1^* = 1.25, \quad x_2^* = 1.75, \quad x_3^* = 2.25, \quad \text{and} \quad x_4^* = 2.75.$$

The resulting midpoint Riemann sum is

$$\sum_{k=1}^{n} f(x_k^*) \, \Delta x = \sum_{k=1}^{4} f(x_k^*)(0.5)$$
$$= f(1.25)(0.5) + f(1.75)(0.5) + f(2.25)(0.5) + f(2.75)(0.5)$$
$$= (-0.5625 - 2.0625 - 4.0625 - 6.5625)0.5$$
$$= -6.625.$$

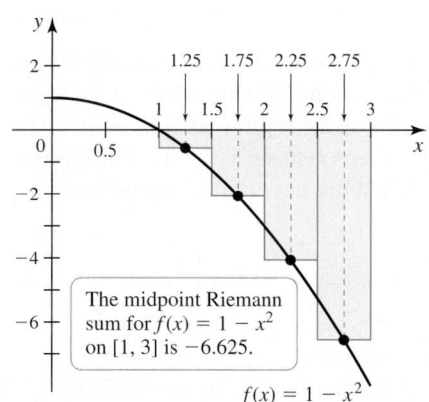

The midpoint Riemann sum for $f(x) = 1 - x^2$ on [1, 3] is −6.625.

Figure 5.16

All values of $f(x_k^*)$ are negative, so the Riemann sum is also negative. Because area is always a nonnegative quantity, this Riemann sum does not approximate the area of the region between the curve and the x-axis on $[1, 3]$. Notice, however, that the values of $f(x_k^*)$ are the *negative* of the heights of the corresponding rectangles (**Figure 5.16**). Therefore, the Riemann sum approximates the *negative* of the area of the region bounded by the curve.

b. The length of each subinterval is $\Delta x = \dfrac{b - a}{n} = \dfrac{3 - 0}{6} = 0.5$, and the grid points are

$$x_0 = 0, \quad x_1 = 0.5, \quad x_2 = 1, \quad x_3 = 1.5, \quad x_4 = 2, \quad x_5 = 2.5, \quad \text{and} \quad x_6 = 3.$$

To calculate the left Riemann sum, we set $x_1^*, x_2^*, \ldots, x_6^*$ equal to the left endpoints of the subintervals:

$$x_1^* = 0, \quad x_2^* = 0.5, \quad x_3^* = 1, \quad x_4^* = 1.5, \quad x_5^* = 2, \quad \text{and} \quad x_6^* = 2.5.$$

The resulting left Riemann sum is

$$\sum_{k=1}^{n} f(x_k^*) \Delta x = \sum_{k=1}^{6} f(x_k^*)(0.5)$$
$$= (\underbrace{f(0) + f(0.5) + f(1)}_{\text{nonnegative contribution}} + \underbrace{f(1.5) + f(2) + f(2.5)}_{\text{negative contribution}})0.5$$
$$= (1 + 0.75 + 0 - 1.25 - 3 - 5.25)0.5$$
$$= -3.875.$$

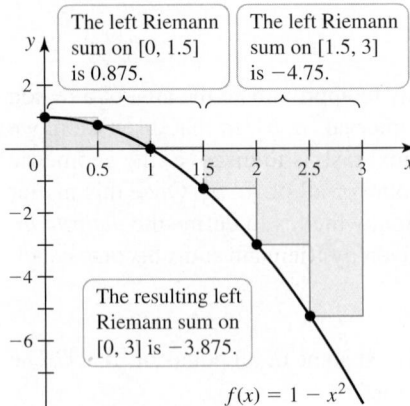

> The left Riemann sum on [0, 1.5] is 0.875.

> The left Riemann sum on [1.5, 3] is −4.75.

> The resulting left Riemann sum on [0, 3] is −3.875.

$f(x) = 1 - x^2$

Figure 5.17

In this case, the values of $f(x_k^*)$ are nonnegative for $k = 1, 2$, and 3, and negative for $k = 4, 5$, and 6 (**Figure 5.17**). Where f is positive, we get positive contributions to the Riemann sum, and where f is negative, we get negative contributions to the sum.

Related Exercises 17, 22 ◄

Let's recap what we learned in Example 1. On intervals where $f(x) < 0$, Riemann sums approximate the *negative* of the area of the region bounded by the curve (**Figure 5.18**).

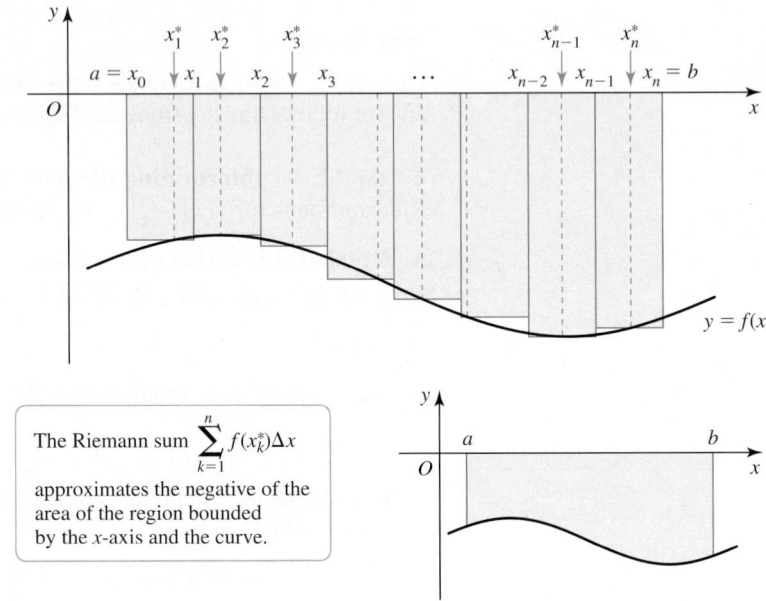

> The Riemann sum $\sum_{k=1}^{n} f(x_k^*)\Delta x$ approximates the negative of the area of the region bounded by the x-axis and the curve.

Figure 5.18

In the more general case that f is positive on part of $[a, b]$, we get positive contributions to the sum where f is positive and negative contributions to the sum where f is negative. In this case, Riemann sums approximate the area of the regions that lie above the x-axis *minus* the area of the regions that lie *below* the x-axis (**Figure 5.19**). This difference between the positive and negative contributions is called the *net area*; it can be positive, negative, or zero.

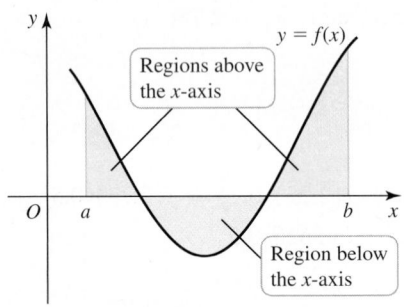

$y = f(x)$

> Regions above the x-axis

> Region below the x-axis

Figure 5.19

➤ Net area suggests the difference between positive and negative contributions much like net change or net profit. Some texts use the term **signed area** for net area.

QUICK CHECK 1 Suppose $f(x) = -5$. What is the net area of the region bounded by the graph of f and the x-axis on the interval $[1, 5]$? Make a sketch of the function and the region. ◄

QUICK CHECK 2 Sketch a continuous function f that is positive over the interval $[0, 1)$ and negative over the interval $(1, 2]$, such that the net area of the region bounded by the graph of f and the x-axis on $[0, 2]$ is zero. ◄

DEFINITION Net Area

Consider the region R bounded by the graph of a continuous function f and the x-axis between $x = a$ and $x = b$. The **net area** of R is the sum of the areas of the parts of R that lie above the x-axis *minus* the sum of the areas of the parts of R that lie below the x-axis on $[a, b]$.

The Definite Integral

Riemann sums for f on $[a, b]$ give *approximations* to the net area of the region bounded by the graph of f and the x-axis between $x = a$ and $x = b$, where $a < b$. How can we make these approximations exact? If f is continuous on $[a, b]$, it is reasonable to expect the Riemann sum approximations to approach the exact value of the net area as the number of subintervals $n \to \infty$ and as the length of the subintervals $\Delta x \to 0$ (**Figure 5.20**). In terms of limits, we write

$$\text{net area} = \lim_{n \to \infty} \sum_{k=1}^{n} f(x_k^*)\Delta x.$$

The Riemann sums we have used so far involve regular partitions in which the subintervals have the same length Δx. We now introduce partitions of $[a, b]$ in which the

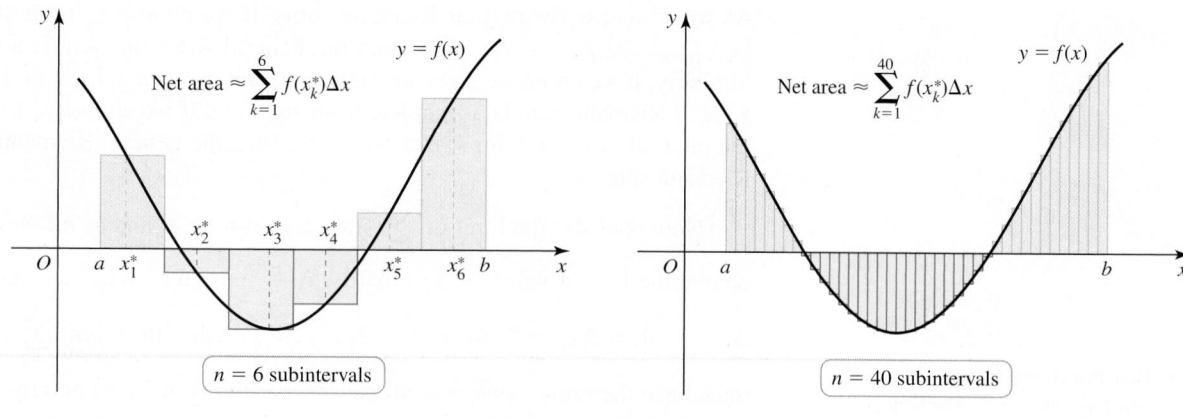

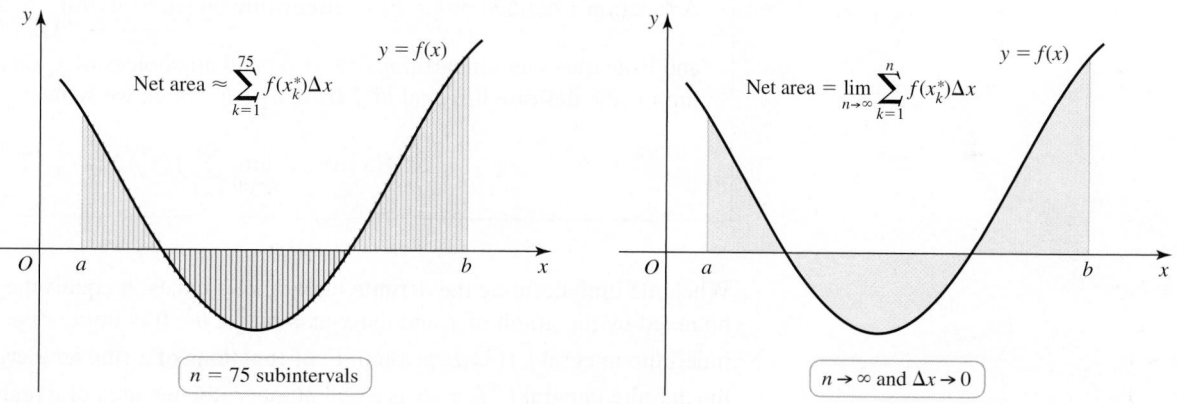

Figure 5.20

lengths of the subintervals are not necessarily equal. A **general partition** of $[a, b]$ consists of the n subintervals

$$[x_0, x_1], [x_1, x_2], \ldots, [x_{n-1}, x_n],$$

where $x_0 = a$ and $x_n = b$. The length of the kth subinterval is $\Delta x_k = x_k - x_{k-1}$, for $k = 1, \ldots, n$. We let x_k^* be any point in the subinterval $[x_{k-1}, x_k]$. This general partition is used to define the *general Riemann sum*.

DEFINITION   **General Riemann Sum**

Suppose $[x_0, x_1], [x_1, x_2], \ldots, [x_{n-1}, x_n]$ are subintervals of $[a, b]$ with

$$a = x_0 < x_1 < x_2 < \cdots < x_{n-1} < x_n = b.$$

Let Δx_k be the length of the subinterval $[x_{k-1}, x_k]$ and let x_k^* be any point in $[x_{k-1}, x_k]$, for $k = 1, 2, \ldots, n$.

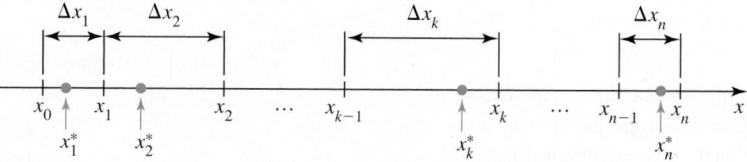

If f is defined on $[a, b]$, the sum

$$\sum_{k=1}^{n} f(x_k^*)\Delta x_k = f(x_1^*)\Delta x_1 + f(x_2^*)\Delta x_2 + \cdots + f(x_n^*)\Delta x_n$$

is called a **general Riemann sum for f on $[a, b]$**.

As was the case for regular Riemann sums, if we choose x_k^* to be the left endpoint of $[x_{k-1}, x_k]$, for $k = 1, 2, \ldots, n$, then the general Riemann sum is a left Riemann sum. Similarly, if we choose x_k^* to be the right endpoint $[x_{k-1}, x_k]$, for $k = 1, 2, \ldots, n$, then the general Riemann sum is a right Riemann sum, and if we choose x_k^* to be the midpoint of the interval $[x_{k-1}, x_k]$, for $k = 1, 2, \ldots, n$, then the general Riemann sum is a midpoint Riemann sum.

Now consider the limit of $\sum_{k=1}^{n} f(x_k^*) \, \Delta x_k$ as $n \to \infty$ and as *all* the $\Delta x_k \to 0$. We let Δ denote the largest value of Δx_k; that is, $\Delta = \max\{\Delta x_1, \Delta x_2, \ldots, \Delta x_n\}$. Observe that if $\Delta \to 0$, then $\Delta x_k \to 0$, for $k = 1, 2, \ldots, n$. For the limit $\lim\limits_{\Delta \to 0} \sum_{k=1}^{n} f(x_k^*) \Delta x_k$ to exist, it must have the same value over all general partitions of $[a, b]$ and for all choices of x_k^* on a partition.

> Note that $\Delta \to 0$ forces all $\Delta x_k \to 0$, which forces $n \to \infty$. Therefore, it suffices to write $\Delta \to 0$ in the limit.

DEFINITION Definite Integral

A function f defined on $[a, b]$ is **integrable** on $[a, b]$ if $\lim\limits_{\Delta \to 0} \sum_{k=1}^{n} f(x_k^*) \Delta x_k$ exists and is unique over all partitions of $[a, b]$ and all choices of x_k^* on a partition. This limit is the **definite integral of f from a to b**, which we write

$$\int_a^b f(x) \, dx = \lim_{\Delta \to 0} \sum_{k=1}^{n} f(x_k^*) \Delta x_k.$$

When the limit defining the definite integral of f exists, it equals the net area of the region bounded by the graph of f and the x-axis on $[a, b]$. It is imperative to remember that the indefinite integral $\int f(x) \, dx$ is a family of functions of x (the antiderivatives of f) and that the definite integral $\int_a^b f(x) \, dx$ is a real number (the net area of a region).

Notation   The notation for the definite integral requires some explanation. There is a direct match between the notation on either side of the equation in the definition (Figure 5.21). In the limit as $\Delta \to 0$, the finite sum, denoted Σ, becomes a sum with an infinite number of terms, denoted $\int$. The integral sign $\int$ is an elongated S for sum. The **limits of integration**, a and b, and the limits of summation also match: The lower limit in the sum, $k = 1$, corresponds to the left endpoint of the interval, $x = a$, and the upper limit in the sum, $k = n$, corresponds to the right endpoint of the interval, $x = b$. The function under the integral sign is called the **integrand**. Finally, the differential dx in the integral (which corresponds to Δx_k in the sum) is an essential part of the notation; it tells us that the **variable of integration** is x.

The variable of integration is a dummy variable that is completely internal to the integral. It does not matter what the variable of integration is called, as long as it does not conflict with other variables that are in use. Therefore, the integrals in Figure 5.22 all have the same meaning.

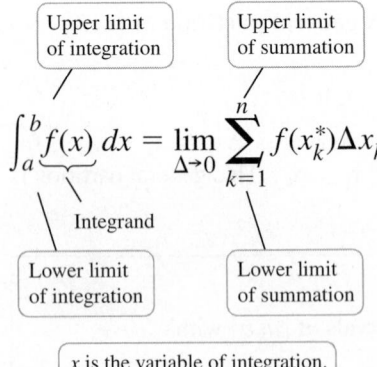

Figure 5.21

> For Leibniz, who introduced this notation in 1675, dx represented the width of an infinitesimally thin rectangle and $f(x) \, dx$ represented the area of such a rectangle. He used $\int_a^b f(x) \, dx$ to denote the sum of these areas from a to b.

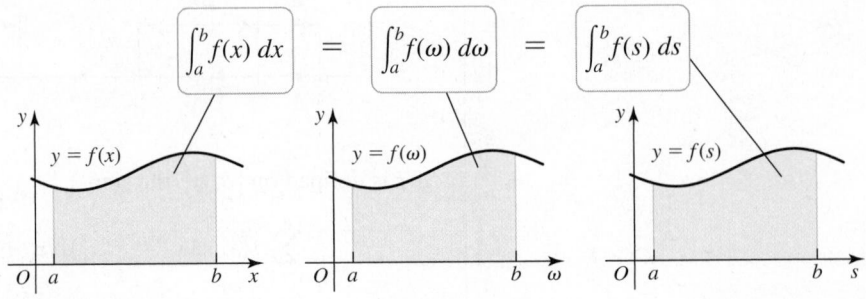

Figure 5.22

The strategy of slicing a region into smaller parts, summing the results from the parts, and taking a limit is used repeatedly in calculus and its applications. We call this strategy the **slice-and-sum method**. It often results in a Riemann sum whose limit is a definite integral.

Evaluating Definite Integrals

> ▶ A function f is bounded on an interval I if there is a number M such that $|f(x)| < M$ for all x in I.

Most of the functions encountered in this text are integrable (see Exercise 95 for an exception). In fact, if f is continuous on $[a, b]$ or if f is bounded on $[a, b]$ with a finite number of discontinuities, then f is integrable on $[a, b]$. The proof of this result lies beyond the scope of this text.

> **THEOREM 5.2   Integrable Functions**
> If f is continuous on $[a, b]$ or bounded on $[a, b]$ with a finite number of discontinuities, then f is integrable on $[a, b]$.

QUICK CHECK 3 Graph $f(x) = x$ and use geometry to evaluate $\int_{-1}^{1} x \, dx$. ◀

When f is continuous on $[a, b]$, we have seen that the definite integral $\int_a^b f(x) \, dx$ is the net area bounded by the graph of f and the x-axis on $[a, b]$. **Figure 5.23** illustrates how the idea of net area carries over to piecewise continuous functions (Exercises 88–92).

> Net area $= \int_a^b f(x) \, dx$
> $=$ area above x-axis (Regions 1 and 3)
> $-$ area below x-axis (Region 2)

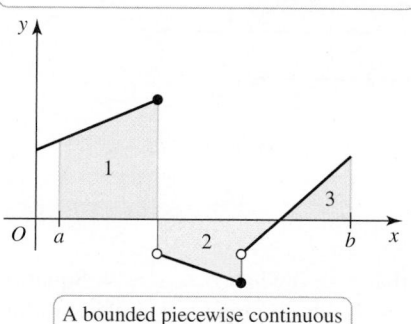

> A bounded piecewise continuous function is integrable.

Figure 5.23

EXAMPLE 2   Identifying the limit of a sum   Assume

$$\lim_{\Delta \to 0} \sum_{k=1}^{n} (3x_k^{*2} + 2x_k^{*} + 1)\Delta x_k$$

is the limit of a Riemann sum for a function f on $[1, 3]$. Identify the function f and express the limit as a definite integral. What does the definite integral represent geometrically?

SOLUTION   By comparing the sum $\sum_{k=1}^{n} (3x_k^{*2} + 2x_k^{*} + 1)\Delta x_k$ to the general Riemann sum $\sum_{k=1}^{n} f(x_k^{*})\Delta x_k$, we see that $f(x) = 3x^2 + 2x + 1$. Because f is a polynomial, it is continuous on $[1, 3]$ and is, therefore, integrable on $[1, 3]$. It follows that

$$\lim_{\Delta \to 0} \sum_{k=1}^{n} (3x_k^{*2} + 2x_k^{*} + 1)\Delta x_k = \int_1^3 (3x^2 + 2x + 1) \, dx.$$

Because f is positive on $[1, 3]$, the definite integral $\int_1^3 (3x^2 + 2x + 1) \, dx$ is the area of the region bounded by the curve $y = 3x^2 + 2x + 1$ and the x-axis on $[1, 3]$ (**Figure 5.24**).

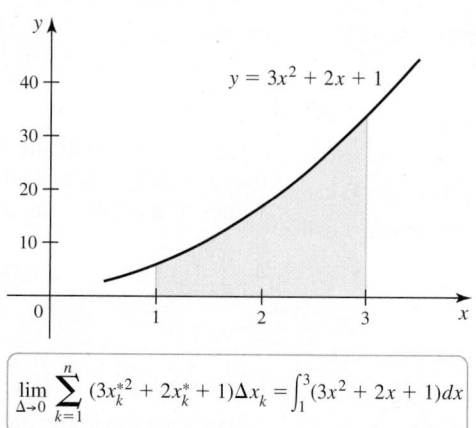

> $$\lim_{\Delta \to 0} \sum_{k=1}^{n} (3x_k^{*2} + 2x_k^{*} + 1)\Delta x_k = \int_1^3 (3x^2 + 2x + 1)dx$$

Figure 5.24

Related Exercises 35–36 ◀

EXAMPLE 3   Evaluating definite integrals using geometry   Use familiar area formulas to evaluate the following definite integrals.

a. $\int_2^4 (2x + 3) \, dx$ **b.** $\int_1^6 (2x - 6) \, dx$ **c.** $\int_3^4 \sqrt{1 - (x - 3)^2} \, dx$

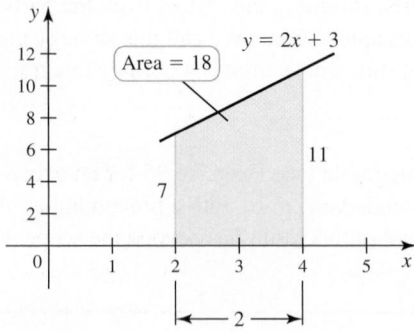

Figure 5.25

▶ **A trapezoid and its area** When $a = 0$, we get the area of a triangle. When $a = b$, we get the area of a rectangle.

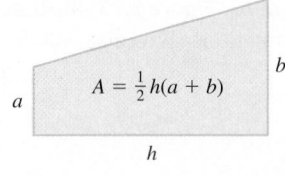

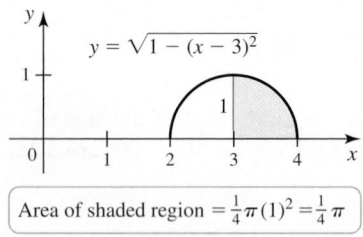

Area of shaded region $= \frac{1}{4}\pi(1)^2 = \frac{1}{4}\pi$

Figure 5.27

QUICK CHECK 4 Let $f(x) = 5$ and use geometry to evaluate $\int_1^3 f(x)\, dx$. What is the value of $\int_a^b c\, dx$, where c is a real number? ◄

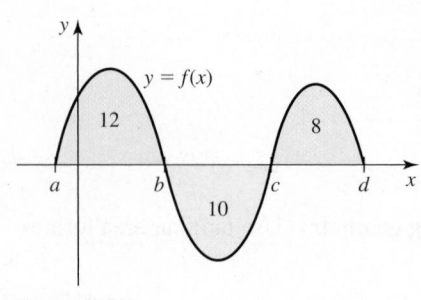

Figure 5.28

SOLUTION To evaluate these definite integrals geometrically, a sketch of the corresponding region is essential.

a. The definite integral $\int_2^4(2x + 3)\, dx$ is the area of the trapezoid bounded by the x-axis and the line $y = 2x + 3$ from $x = 2$ to $x = 4$ (**Figure 5.25**). The width of its base is 2 and the lengths of its two parallel sides are $f(2) = 7$ and $f(4) = 11$. Using the area formula for a trapezoid, we have

$$\int_2^4(2x + 3)\, dx = \frac{1}{2}\cdot 2(11 + 7) = 18.$$

b. A sketch shows that the regions bounded by the line $y = 2x - 6$ and the x-axis are triangles (**Figure 5.26**). The area of the triangle on the interval $[1, 3]$ is $\frac{1}{2}\cdot 2\cdot 4 = 4$. Similarly, the area of the triangle on $[3, 6]$ is $\frac{1}{2}\cdot 3\cdot 6 = 9$. The definite integral is the net area of the entire region, which is the area of the triangle above the x-axis minus the area of the triangle below the x-axis:

$$\int_1^6(2x - 6)\, dx = \text{net area} = 9 - 4 = 5.$$

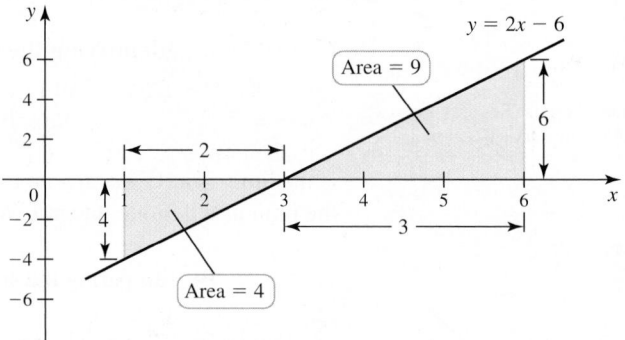

Figure 5.26

c. We first let $y = \sqrt{1 - (x - 3)^2}$ and observe that $y \geq 0$ when $2 \leq x \leq 4$. Squaring both sides leads to the equation $(x - 3)^2 + y^2 = 1$, whose graph is a circle of radius 1 centered at $(3, 0)$. Because $y \geq 0$, the graph of $y = \sqrt{1 - (x - 3)^2}$ is the upper half of the circle. It follows that the integral $\int_3^4\sqrt{1 - (x - 3)^2}\, dx$ is the area of a quarter circle of radius 1 (**Figure 5.27**). Therefore,

$$\int_3^4\sqrt{1 - (x - 3)^2}\, dx = \frac{1}{4}\pi(1)^2 = \frac{\pi}{4}.$$

Related Exercises 39, 40, 43 ◄

EXAMPLE 4 **Definite integrals from graphs** Figure 5.28 shows the graph of a function f with the areas of the regions bounded by its graph and the x-axis given. Find the values of the following definite integrals.

a. $\displaystyle\int_a^b f(x)\, dx$ **b.** $\displaystyle\int_b^c f(x)\, dx$ **c.** $\displaystyle\int_a^c f(x)\, dx$ **d.** $\displaystyle\int_b^d f(x)\, dx$

SOLUTION

a. Because $f(x) \geq 0$ on $[a, b]$, the value of the definite integral is the area of the region between the graph and the x-axis on $[a, b]$; that is, $\int_a^b f(x)\, dx = 12$.

b. Because $f(x) \leq 0$ on $[b, c]$, the value of the definite integral is the negative of the area of the corresponding region; that is, $\int_b^c f(x)\, dx = -10$.

c. The value of the definite integral is the area of the region on $[a, b]$ (where $f(x) \geq 0$) minus the area of the region on $[b, c]$ (where $f(x) \leq 0$). Therefore, $\int_a^c f(x)\, dx = 12 - 10 = 2$.

d. Reasoning as in part (c), we have $\int_b^d f(x)\, dx = -10 + 8 = -2$.

Related Exercises 59–62 ◄

Properties of Definite Integrals

Recall that the definite integral $\int_a^b f(x)\, dx$ was defined assuming $a < b$. There are, however, occasions when it is necessary to reverse the limits of integration. If f is integrable on $[a, b]$, we define

$$\int_b^a f(x)\, dx = -\int_a^b f(x)\, dx.$$

In other words, reversing the limits of integration changes the sign of the integral.

Another fundamental property of integrals is that if we integrate from a point to itself, then the length of the interval of integration is zero, which means the definite integral is also zero.

DEFINITION Reversing Limits and Identical Limits of Integration

Suppose f is integrable on $[a, b]$.

1. $\displaystyle\int_b^a f(x)\, dx = -\int_a^b f(x)\, dx$ **2.** $\displaystyle\int_a^a f(x)\, dx = 0$

QUICK CHECK 5 Evaluate $\int_a^b f(x)\, dx + \int_b^a f(x)\, dx$ assuming f is integrable on $[a, b]$. ◄

Integral of a Sum Definite integrals possess other properties that often simplify their evaluation. Assume f and g are integrable on $[a, b]$. The first property states that their sum $f + g$ is integrable on $[a, b]$ and the integral of their sum is the sum of their integrals:

$$\int_a^b (f(x) + g(x))\, dx = \int_a^b f(x)\, dx + \int_a^b g(x)\, dx.$$

We prove this property assuming f and g are continuous. In this case, $f + g$ is continuous and, therefore, integrable. We then have

$$\int_a^b (f(x) + g(x))\, dx = \lim_{\Delta \to 0} \sum_{k=1}^n (f(x_k^*) + g(x_k^*))\Delta x_k \qquad \text{Definition of definite integral}$$

$$= \lim_{\Delta \to 0} \left(\sum_{k=1}^n f(x_k^*)\Delta x_k + \sum_{k=1}^n g(x_k^*)\Delta x_k \right) \qquad \text{Addition rule for finite sums}$$

$$= \lim_{\Delta \to 0} \sum_{k=1}^n f(x_k^*)\Delta x_k + \lim_{\Delta \to 0} \sum_{k=1}^n g(x_k^*)\Delta x_k \qquad \text{Limit of a sum}$$

$$= \int_a^b f(x)\, dx + \int_a^b g(x)\, dx. \qquad \text{Definition of definite integral}$$

Constants in Integrals Another property of definite integrals is that constants can be factored out of the integral. If f is integrable on $[a, b]$ and c is a constant, then cf is integrable on $[a, b]$ and

$$\int_a^b cf(x)\, dx = c\int_a^b f(x)\, dx.$$

The justification for this property (Exercise 93) is based on the fact that for finite sums,

$$\sum_{k=1}^n cf(x_k^*)\Delta x_k = c\sum_{k=1}^n f(x_k^*)\Delta x_k.$$

Combining the previous two properties, we can also show that

$$\int_a^b (f(x) - g(x))\, dx = \int_a^b f(x)\, dx - \int_a^b g(x)\, dx;$$

that is, the integral of the difference of two functions is the difference of the integrals.

Integrals over Subintervals   If the point p lies between a and b, then the integral on $[a, b]$ may be split into two integrals. As shown in **Figure 5.29**, we have the property

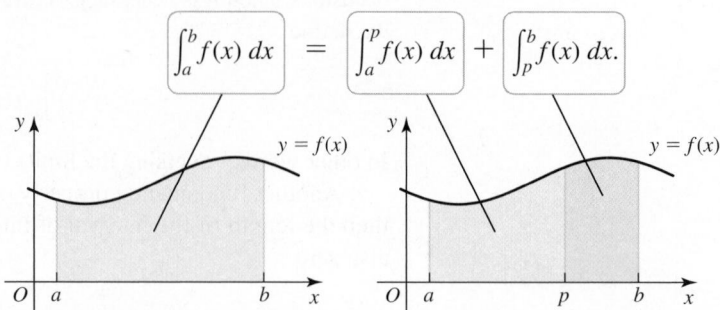

Figure 5.29

It is surprising that this property also holds when p lies outside the interval $[a, b]$. For example, if $a < b < p$ and f is integrable on $[a, p]$, then it follows (**Figure 5.30**) that

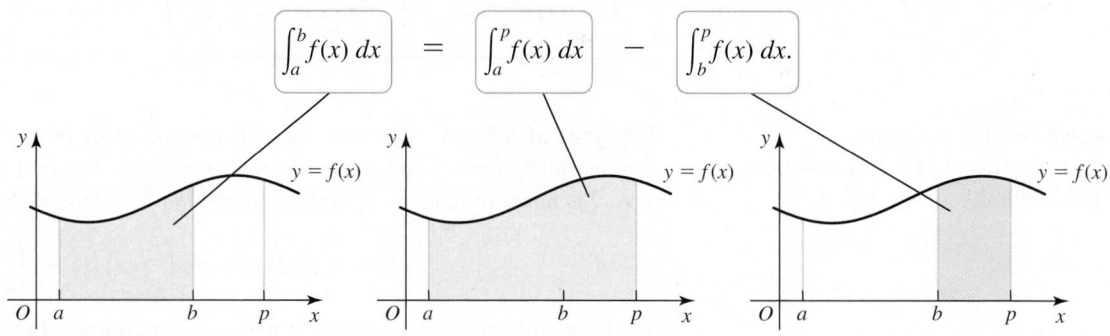

Figure 5.30

Because $\int_p^b f(x)\, dx = -\int_b^p f(x)\, dx$, we have the original property:

$$\int_a^b f(x)\, dx = \int_a^p f(x)\, dx + \int_p^b f(x)\, dx.$$

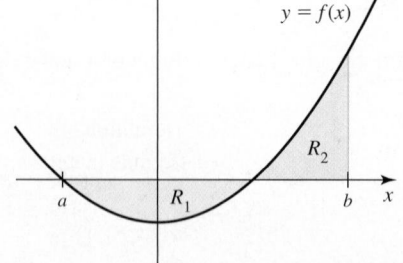

Integrals of Absolute Values   Finally, how do we interpret $\int_a^b |f(x)|\, dx$, the integral of the absolute value of an integrable function? The graphs f and $|f|$ are shown in **Figure 5.31**. The integral $\int_a^b |f(x)|\, dx$ gives the area of regions R_1^* and R_2. But R_1 and R_1^* have the same area; therefore, $\int_a^b |f(x)|\, dx$ also gives the area of R_1 and R_2. The conclusion is that $\int_a^b |f(x)|\, dx$ is the area of the entire region (above and below the x-axis) that lies between the graph of f and the x-axis on $[a, b]$.

All these properties will be used frequently in upcoming work. It's worth collecting them in one table (Table 5.4).

Table 5.4 Properties of definite integrals

Let f and g be integrable functions on an interval that contains a, b, and p.

1. $\int_a^a f(x)\, dx = 0$ Definition

2. $\int_b^a f(x)\, dx = -\int_a^b f(x)\, dx$ Definition

3. $\int_a^b (f(x) \pm g(x))\, dx = \int_a^b f(x)\, dx \pm \int_a^b g(x)\, dx$

4. $\int_a^b c f(x)\, dx = c \int_a^b f(x)\, dx$, for any constant c

5. $\int_a^b f(x)\, dx = \int_a^p f(x)\, dx + \int_p^b f(x)\, dx$

6. The function $|f|$ is integrable on $[a, b]$, and $\int_a^b |f(x)|\, dx$ is the sum of the areas of the regions bounded by the graph of f and the x-axis on $[a, b]$.

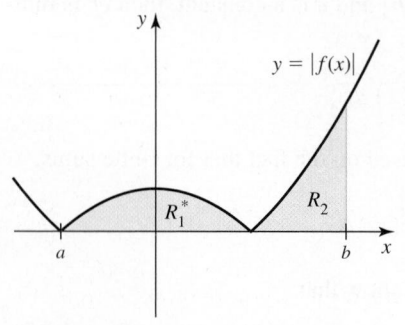

$\int_a^b |f(x)|\, dx = $ area of R_1^* + area of R_2
$= $ area of R_1 + area of R_2

Figure 5.31

EXAMPLE 5 Properties of integrals Assume $\int_0^5 f(x)\,dx = 3$ and $\int_0^7 f(x)\,dx = -10$. Evaluate the following integrals, if possible.

a. $\int_0^7 2f(x)\,dx$ **b.** $\int_5^7 f(x)\,dx$ **c.** $\int_5^0 f(x)\,dx$ **d.** $\int_7^0 6f(x)\,dx$ **e.** $\int_0^7 |f(x)|\,dx$

SOLUTION

a. By Property 4 of Table 5.4,

$$\int_0^7 2f(x)\,dx = 2\int_0^7 f(x)\,dx = 2\cdot(-10) = -20.$$

b. By Property 5 of Table 5.4, $\int_0^7 f(x)\,dx = \int_0^5 f(x)\,dx + \int_5^7 f(x)\,dx$. Therefore,

$$\int_5^7 f(x)\,dx = \int_0^7 f(x)\,dx - \int_0^5 f(x)\,dx = -10 - 3 = -13.$$

c. By Property 2 of Table 5.4,

$$\int_5^0 f(x)\,dx = -\int_0^5 f(x)\,dx = -3.$$

d. Using Properties 2 and 4 of Table 5.4, we have

$$\int_7^0 6f(x)\,dx = -\int_0^7 6f(x)\,dx = -6\int_0^7 f(x)\,dx = (-6)(-10) = 60.$$

e. This integral cannot be evaluated without knowing the intervals on which f is positive and negative. It could have any value greater than or equal to 10.

Related Exercises 51–52 ◄

QUICK CHECK 6 Evaluate $\int_{-1}^2 x\,dx$ and $\int_{-1}^2 |x|\,dx$ using geometry. ◄

Bounds on Definite Integrals

We conclude our discussion of properties of definite integrals with three results that are helpful for upcoming theoretical work. We assume f and g are continuous on $[a, b]$, where $b > a$.

Nonnegative Integrand If $f(x) \geq 0$ on $[a, b]$, it is geometrically apparent that $\int_a^b f(x)\,dx \geq 0$ (**Figure 5.32**). To prove this result, suppose m is the absolute minimum value of f on $[a, b]$, guaranteed to exist by Theorem 4.1 (note that $m \geq 0$). Working with a general partition of $[a, b]$, observe that

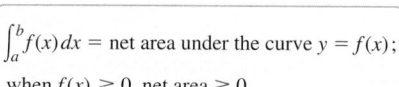

$\int_a^b f(x)\,dx$ = net area under the curve $y = f(x)$; when $f(x) \geq 0$, net area ≥ 0.

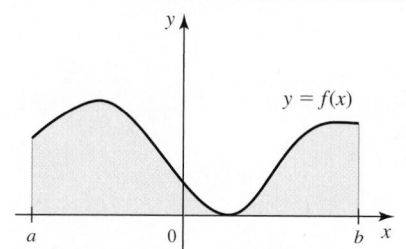

Figure 5.32

$$0 \leq m(b - a) = m\sum_{k=1}^n \Delta x_k \qquad \sum_{k=1}^n \Delta x_k = (b - a)$$

$$= \sum_{k=1}^n m\Delta x_k \qquad \text{Property of finite sums, Section 5.1}$$

$$\leq \underbrace{\sum_{k=1}^n f(x_k^*)\Delta x_k}_{\substack{\text{General Riemann} \\ \text{sum for } f \text{ on } [a, b].}} \qquad m \leq f(x_k^*) \text{ for all } x_k^* \text{ in } [a, b]$$

We have shown that for any general partition of $[a, b]$ and for every choice of x_k^*, the general Riemann sum for f is nonnegative. Taking the limit as $\Delta \to 0$, where $\Delta = \max\{\Delta x_1, \Delta x_2, \ldots, \Delta x_n\}$, we have

$$\int_a^b f(x)\,dx \geq 0.$$

Comparing Definite Integrals A related property of definite integrals says that if $f(x) \geq g(x)$ on $[a, b]$ (**Figure 5.33**), then

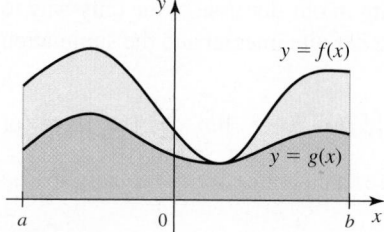

If $f(x) \geq g(x)$ on $[a, b]$, then $\int_a^b f(x)\,dx \geq \int_a^b g(x)\,dx$.

Figure 5.33

$$\int_a^b f(x)\,dx \geq \int_a^b g(x)\,dx.$$

This property follows from the previous result and Property 3 of Table 5.4 (Exercise 98).

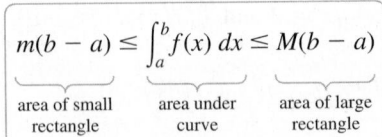

$$m(b - a) \leq \int_a^b f(x)\, dx \leq M(b - a)$$

area of small   area under   area of large
rectangle   curve   rectangle

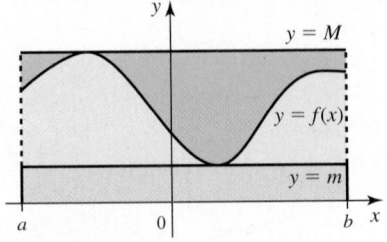

Figure 5.34

Lower and Upper Bounds   Because f is continuous on $[a, b]$, it attains an absolute minimum value m and an absolute maximum value M on $[a, b]$ (Theorem 4.1). Our final property (**Figure 5.34**) claims that

$$m(b - a) \leq \int_a^b f(x)\, dx \leq M(b - a).$$

To prove this result, once again working with a general partition of $[a, b]$, notice that

$$m(b - a) = m \sum_{k=1}^{n} \Delta x_k \qquad \sum_{k=1}^{n} \Delta x_k = b - a$$

$$= \sum_{k=1}^{n} m \Delta x_k \qquad \text{Property of finite sums}$$

$$\leq \sum_{k=1}^{n} f(x_k^*) \Delta x_k \qquad m \leq f(x_k^*) \text{ for all } x_k^* \text{ in } [a, b]$$

$$\leq \sum_{k=1}^{n} M \Delta x_k \qquad f(x_k^*) \leq M \text{ for all } x_k^* \text{ in } [a, b]$$

$$= M \sum_{k=1}^{n} \Delta x_k \qquad \text{Property of finite sums}$$

$$= M(b - a). \qquad \sum_{k=1}^{n} \Delta x_k = b - a$$

We have shown that $m(b - a) \leq \sum_{k=1}^{n} f(x_k^*) \Delta x_k \leq M(b - a)$ for any partition of $[a, b]$ and for every choice of x_k. Letting $\Delta \to 0$, we obtain

$$m(b - a) \leq \int_a^b f(x)\, dx \leq M(b - a).$$

Table 5.5 summarizes our findings.

▶ Although we proved Properties 3, 4, 7, 8, and 9 of Tables 5.4 and 5.5 for continuous functions f and g, these properties hold when f and g are integrable.

Table 5.5 Additional properties of definite integrals

Let f and g be integrable functions on $[a, b]$, where $b > a$.

7. If $f(x) \geq 0$ on $[a, b]$, then $\int_a^b f(x)\, dx \geq 0$.

8. If $f(x) \geq g(x)$ on $[a, b]$, then $\int_a^b f(x)\, dx \geq \int_a^b g(x)\, dx$.

9. If $m \leq f(x) \leq M$, then $m(b - a) \leq \int_a^b f(x)\, dx \leq M(b - a)$.

In the next section, Property 9 is used to prove the central result of this chapter, the Fundamental Theorem of Calculus.

Evaluating Definite Integrals Using Limits

In Example 3, we used area formulas for trapezoids, triangles, and circles to evaluate definite integrals. Regions bounded by more general functions have curved boundaries for which conventional geometrical methods do not work. At this point in our discussion, the only way to handle such integrals is to appeal to the definition of the definite integral and the summation formulas given in Theorem 5.1.

We know that if f is integrable on $[a, b]$, then $\int_a^b f(x)\, dx = \lim\limits_{\Delta \to 0} \sum_{k=1}^{n} f(x_k^*) \Delta x_k$, for any partition of $[a, b]$ and any points x_k^*. To simplify these calculations, we use equally spaced grid points and right Riemann sums. That is, for each value of n, we let $\Delta x_k = \Delta x = \dfrac{b-a}{n}$ and $x_k^* = a + k\,\Delta x$, for $k = 1, 2, \ldots, n$. Then as $n \to \infty$ and $\Delta \to 0$,

$$\int_a^b f(x)\,dx = \lim_{\Delta \to 0} \sum_{k=1}^n f(x_k^*)\Delta x_k = \lim_{n \to \infty} \sum_{k=1}^n f(a + k\Delta x)\Delta x.$$

$$x_1^* = \frac{2}{n}$$
$$x_2^* = \frac{4}{n}$$
$$x_k^* = \frac{2k}{n}$$
$$a = x_0 = 0 \qquad b = x_n = 2$$
$$\Delta x = \frac{2}{n}$$

$$x_k^* = a + k\Delta x = \frac{2k}{n}$$
$$k = 1, \ldots, n$$

EXAMPLE 6 **Evaluating definite integrals** Find the value of $\int_0^2 (x^3 + 1)\,dx$ by evaluating a right Riemann sum and letting $n \to \infty$.

SOLUTION Based on approximations found in Example 5, Section 5.1, we conjectured that the value of this integral is 6. To verify this conjecture, we now evaluate the integral exactly. The interval $[a, b] = [0, 2]$ is divided into n subintervals of length $\Delta x = \dfrac{b-a}{n} = \dfrac{2}{n}$, which produces the grid points

$$x_k^* = a + k\Delta x = 0 + k \cdot \frac{2}{n} = \frac{2k}{n}, \qquad \text{for } k = 1, 2, \ldots, n.$$

Letting $f(x) = x^3 + 1$, the right Riemann sum is

> An analogous calculation could be done using left Riemann sums or midpoint Riemann sums.

$$\sum_{k=1}^n f(x_k^*)\Delta x = \sum_{k=1}^n \left(\left(\frac{2k}{n} \right)^3 + 1 \right) \frac{2}{n}$$

$$= \frac{2}{n} \sum_{k=1}^n \left(\frac{8k^3}{n^3} + 1 \right) \qquad \sum_{k=1}^n ca_k = c\sum_{k=1}^n a_k$$

$$= \frac{2}{n} \left(\frac{8}{n^3} \sum_{k=1}^n k^3 + \sum_{k=1}^n 1 \right) \qquad \sum_{k=1}^n (ca_k + b_k) = c\sum_{k=1}^n a_k + \sum_{k=1}^n b_k$$

$$= \frac{2}{n} \left(\frac{8}{n^3} \left(\frac{n^2(n+1)^2}{4} \right) + n \right) \qquad \sum_{k=1}^n k^3 = \frac{n^2(n+1)^2}{4} \text{ and } \sum_{k=1}^n 1 = n;$$
$$\qquad\qquad\qquad\qquad\qquad\qquad\qquad \text{Theorem 5.1}$$

$$= \frac{4(n^2 + 2n + 1)}{n^2} + 2. \qquad \text{Simplify.}$$

Now we evaluate $\int_0^2 (x^3 + 1)\,dx$ by letting $n \to \infty$ in the Riemann sum:

$$\int_0^2 (x^3 + 1)\,dx = \lim_{n \to \infty} \sum_{k=1}^n f(x_k^*)\Delta x$$

$$= \lim_{n \to \infty} \left(\frac{4(n^2 + 2n + 1)}{n^2} + 2 \right)$$

$$= 4 \underbrace{\lim_{n \to \infty} \left(\frac{n^2 + 2n + 1}{n^2} \right)}_{1} + \lim_{n \to \infty} 2$$

$$= 4(1) + 2 = 6.$$

Therefore, $\int_0^2 (x^3 + 1)\,dx = 6$, confirming our conjecture in Example 5, Section 5.1.

Related Exercises 82, 84 ◄

The Riemann sum calculations in Example 6 are tedious even if f is a simple function. For polynomials of degree 4 and higher, the calculations are more challenging, and for rational and transcendental functions, advanced mathematical results are needed. The next section introduces more efficient methods for evaluating definite integrals.

SECTION 5.2 EXERCISES

Getting Started

1. What does net area measure?

2. Under what conditions does the net area of a region (bounded by a continuous function) equal the area of a region? When does the net area of a region differ from the area of a region?

3. Use the graph of $y = f(x)$ to estimate $\int_0^6 f(x)\, dx$ using a left and right Riemann sum with $n = 6$.

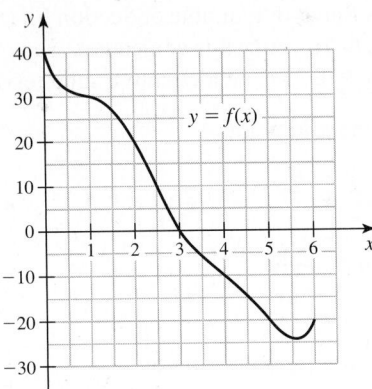

4. Use the graph of $y = g(x)$ to estimate $\int_2^{10} g(x)\, dx$ using a left, right, and midpoint Riemann sum with $n = 4$.

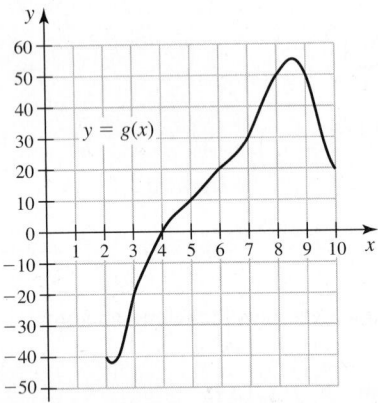

5. Suppose f is continuous on $[2, 8]$. Use the table of values of f to estimate $\int_2^8 f(x)\, dx$ using a left, right, and midpoint Riemann sum with $n = 3$.

x	2	3	4	5	6	7	8
$f(x)$	2	−2	−3	−4	−5	−2	−1

6. Suppose g is continuous on $[1, 9]$. Use the table of values of g to estimate $\int_1^9 g(x)\, dx$ using a left, right, and midpoint Riemann sum with $n = 4$.

x	1	2	3	4	5	6	7	8	9
$g(x)$	−5	−4	−3	0	2	4	5	0	−1

7. Sketch a graph of $y = 2$ on $[-1, 4]$ and use geometry to find the exact value of $\int_{-1}^4 2\, dx$.

8. Sketch a graph of $y = -3$ on $[1, 5]$ and use geometry to find the exact value of $\int_1^5 (-3)\, dx$.

9. Sketch a graph of $y = 2x$ on $[-1, 2]$ and use geometry to find the exact value of $\int_{-1}^2 2x\, dx$.

10. Suppose $\int_1^3 f(x)\, dx = 10$ and $\int_1^3 g(x)\, dx = -20$. Evaluate $\int_1^3 (2f(x) - 4g(x))\, dx$ and $\int_3^1 (2f(x) - 4g(x))\, dx$.

11. Use graphs to evaluate $\int_0^{2\pi} \sin x\, dx$ and $\int_0^{2\pi} \cos x\, dx$.

12. Explain how the notation for Riemann sums, $\sum_{k=1}^{n} f(x_k^*)\Delta x_k$, corresponds to the notation for the definite integral, $\int_a^b f(x)\, dx$.

13. Give a geometrical explanation of why $\int_a^a f(x)\, dx = 0$.

14. Use Table 5.4 to rewrite $\int_1^6 (2x^3 - 4x)\, dx$ as the difference of two integrals.

15. Use geometry to find a formula for $\int_0^a x\, dx$, in terms of a constant $a > 0$.

16. If f is continuous on $[a, b]$ and $\int_a^b |f(x)|\, dx = 0$, what can you conclude about f?

Practice Exercises

17–20. Approximating net area *The following functions are negative on the given interval.*

a. *Sketch the function on the interval.*
b. *Approximate the net area bounded by the graph of f and the x-axis on the interval using a left, right, and midpoint Riemann sum with $n = 4$.*

17. $f(x) = -2x - 1$ on $[0, 4]$ ■ **18.** $f(x) = -4 - x^3$ on $[3, 7]$

■ **19.** $f(x) = \sin 2x$ on $\left[\dfrac{\pi}{2}, \pi\right]$ ■ **20.** $f(x) = x^3 - 1$ on $[-2, 0]$

■ **21–26. Approximating net area** *The following functions are positive and negative on the given interval.*

a. *Sketch the function on the interval.*
b. *Approximate the net area bounded by the graph of f and the x-axis on the interval using a left, right, and midpoint Riemann sum with $n = 4$.*
c. *Use the sketch in part (a) to show which intervals of $[a, b]$ make positive and negative contributions to the net area.*

21. $f(x) = 4 - 2x$ on $[0, 4]$ **22.** $f(x) = 8 - 2x^2$ on $[0, 4]$

23. $f(x) = \sin 2x$ on $\left[0, \dfrac{3\pi}{4}\right]$ **24.** $f(x) = x^3$ on $[-1, 2]$

25. $f(x) = \tan^{-1}(3x - 1)$ on $[0, 1]$

26. $f(x) = xe^{-x}$ on $[-1, 1]$

27–30. Area versus net area *Graph the following functions. Then use geometry (not Riemann sums) to find the area and the net area of the region described.*

27. The region between the graph of $y = -3x$ and the x-axis, for $-2 \le x \le 2$

28. The region between the graph of $y = 4x - 8$ and the x-axis, for $-4 \le x \le 8$

29. The region between the graph of $y = 1 - |x|$ and the x-axis, for $-2 \le x \le 2$

30. The region between the graph of $y = 3x - 6$ and the x-axis, for $0 \le x \le 6$

T 31–34. Approximating definite integrals *Complete the following steps for the given integral and the given value of n.*

a. *Sketch the graph of the integrand on the interval of integration.*
b. *Calculate Δx and the grid points $x_0, x_1, \ldots, x_n$, assuming a regular partition.*
c. *Calculate the left and right Riemann sums for the given value of n.*
d. *Determine which Riemann sum (left or right) underestimates the value of the definite integral and which overestimates the value of the definite integral.*

31. $\int_3^6 (1 - 2x)\, dx$; $n = 6$ **32.** $\int_0^2 (x^2 - 2)\, dx$; $n = 4$

33. $\int_1^7 \frac{1}{x}\, dx$; $n = 6$ **34.** $\int_0^{\pi/2} \cos x\, dx$; $n = 4$

35–38. Identifying definite integrals as limits of sums *Consider the following limits of Riemann sums for a function f on $[a, b]$. Identify f and express the limit as a definite integral.*

35. $\displaystyle\lim_{\Delta \to 0} \sum_{k=1}^n (x_k^{*2} + 1)\Delta x_k$ on $[0, 2]$

36. $\displaystyle\lim_{\Delta \to 0} \sum_{k=1}^n (4 - x_k^{*2})\Delta x_k$ on $[-2, 2]$

37. $\displaystyle\lim_{\Delta \to 0} \sum_{k=1}^n x_k^*(\ln x_k^*)\Delta x_k$ on $[1, 2]$

38. $\displaystyle\lim_{\Delta \to 0} \sum_{k=1}^n |x_k^{*2} - 1|\Delta x_k$ on $[-2, 2]$

39–46. Definite integrals *Use geometry (not Riemann sums) to evaluate the following definite integrals. Sketch a graph of the integrand, show the region in question, and interpret your result.*

39. $\int_0^4 (8 - 2x)\, dx$ **40.** $\int_{-4}^2 (2x + 4)\, dx$

41. $\int_{-1}^2 (-|x|)\, dx$ **42.** $\int_0^2 (1 - x)\, dx$

43. $\int_0^4 \sqrt{16 - x^2}\, dx$ **44.** $\int_{-1}^3 \sqrt{4 - (x - 1)^2}\, dx$

45. $\int_0^4 f(x)\, dx$, where $f(x) = \begin{cases} 5 & \text{if } x \le 2 \\ 3x - 1 & \text{if } x > 2 \end{cases}$

46. $\int_1^{10} g(x)\, dx$, where $g(x) = \begin{cases} 4x & \text{if } 0 \le x \le 2 \\ -8x + 16 & \text{if } 2 < x \le 3 \\ -8 & \text{if } x > 3 \end{cases}$

47–50. *The accompanying figure shows four regions bounded by the graph of $y = x \sin x$: $R_1, R_2, R_3,$ and $R_4,$ whose areas are 1, $\pi - 1$, $\pi + 1$, and $2\pi - 1$, respectively. (We verify these results later in the text.) Use this information to evaluate the following integrals.*

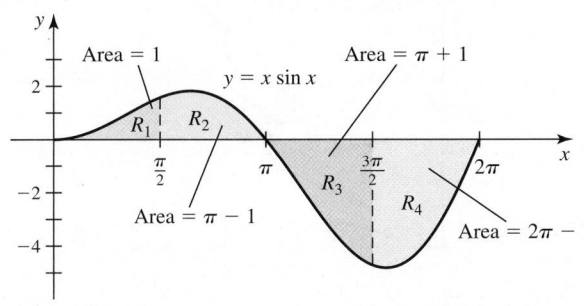

47. $\int_0^{\pi} x \sin x\, dx$ **48.** $\int_0^{3\pi/2} x \sin x\, dx$

49. $\int_0^{2\pi} x \sin x\, dx$ **50.** $\int_{\pi/2}^{2\pi} x \sin x\, dx$

51. Properties of integrals Use only the fact that $\int_0^4 3x(4 - x)\, dx = 32$, and the definitions and properties of integrals, to evaluate the following integrals, if possible.

a. $\int_4^0 3x(4 - x)\, dx$ b. $\int_0^4 x(x - 4)\, dx$

c. $\int_4^0 6x(4 - x)\, dx$ d. $\int_0^8 3x(4 - x)\, dx$

52. Properties of integrals Suppose $\int_1^4 f(x)\, dx = 8$ and $\int_1^6 f(x)\, dx = 5$. Evaluate the following integrals.

a. $\int_1^4 (-3f(x))\, dx$ b. $\int_1^4 3f(x)\, dx$

c. $\int_6^4 12f(x)\, dx$ d. $\int_4^6 3f(x)\, dx$

53. Properties of integrals Suppose $\int_0^3 f(x)\, dx = 2$, $\int_3^6 f(x)\, dx = -5$, and $\int_3^6 g(x)\, dx = 1$. Evaluate the following integrals.

a. $\int_0^3 5f(x)\, dx$ b. $\int_3^6 (-3g(x))\, dx$

c. $\int_3^6 (3f(x) - g(x))\, dx$ d. $\int_6^3 (f(x) + 2g(x))\, dx$

54. Properties of integrals Suppose $f(x) \ge 0$ on $[0, 2]$, $f(x) \le 0$ on $[2, 5]$, $\int_0^2 f(x)\, dx = 6$, and $\int_2^5 f(x)\, dx = -8$. Evaluate the following integrals.

a. $\int_0^5 f(x)\, dx$ b. $\int_0^5 |f(x)|\, dx$

c. $\int_2^5 4|f(x)|\, dx$ d. $\int_0^5 (f(x) + |f(x)|)\, dx$

55. Properties of integrals Consider two functions f and g on $[1, 6]$ such that $\int_1^6 f(x)\, dx = 10$, $\int_1^6 g(x)\, dx = 5$, $\int_4^6 f(x)\, dx = 5$, and $\int_1^4 g(x)\, dx = 2$. Evaluate the following integrals.

a. $\int_1^4 3f(x)\, dx$ b. $\int_1^6 (f(x) - g(x))\, dx$

c. $\int_1^4 (f(x) - g(x))\, dx$ d. $\int_4^6 (g(x) - f(x))\, dx$

e. $\int_4^6 8g(x)\, dx$ f. $\int_4^1 2f(x)\, dx$

56. Suppose f is continuous on $[1, 5]$ and $2 \le f(x) \le 3$, for all x in $[1, 5]$. Find lower and upper bounds for $\int_1^5 f(x)\, dx$.

57–58. Using properties of integrals *Use the value of the first integral I to evaluate the two given integrals.*

57. $I = \int_0^1 (x^3 - 2x)\, dx = -\frac{3}{4}$

a. $\int_0^1 (4x - 2x^3)\, dx$ b. $\int_1^0 (2x - x^3)\, dx$

58. $I = \int_0^{\pi/2} (\cos \theta - 2 \sin \theta)\, d\theta = -1$

a. $\int_0^{\pi/2} (2 \sin \theta - \cos \theta)\, d\theta$ b. $\int_{\pi/2}^0 (4 \cos \theta - 8 \sin \theta)\, d\theta$

59–66. Definite integrals from graphs *The figure shows the areas of regions bounded by the graph of f and the x-axis. Evaluate the following integrals.*

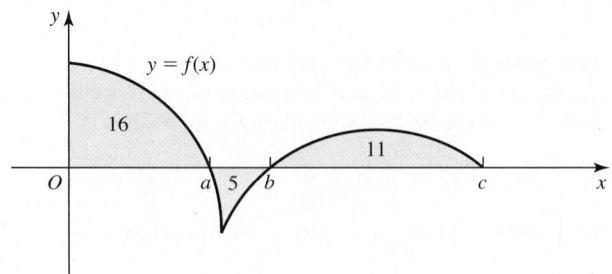

59. $\int_0^a f(x)\, dx$

60. $\int_0^b f(x)\, dx$

61. $\int_a^c f(x)\, dx$

62. $\int_0^c f(x)\, dx$

63. $\int_0^c |f(x)|\, dx$

64. $\int_0^c (2|f(x)| + 3f(x))\, dx$

65. $\int_a^0 f(x)\, dx$

66. $\int_c^0 |f(x)|\, dx$

67. Use geometry and properties of integrals to evaluate
$\int_0^1 (2x + \sqrt{1 - x^2} + 1)\, dx$.

68. Use geometry and properties of integrals to evaluate
$\int_1^5 (|x - 2| + \sqrt{-x^2 + 6x - 5})\, dx$.

69. Explain why or why not Determine whether the following statements are true and give an explanation or counterexample.

   a. If f is a constant function on the interval $[a, b]$, then the right and left Riemann sums give the exact value of $\int_a^b f(x)\, dx$, for any positive integer n.

   b. If f is a linear function on the interval $[a, b]$, then a midpoint Riemann sum gives the exact value of $\int_a^b f(x)\, dx$, for any positive integer n.

   c. $\int_0^{2\pi/a} \sin ax\, dx = \int_0^{2\pi/a} \cos ax\, dx = 0$ (*Hint:* Graph the functions and use properties of trigonometric functions.)

   d. If $\int_a^b f(x)\, dx = \int_b^a f(x)\, dx$, then f is a constant function.

   e. Property 4 of Table 5.4 implies that $\int_a^b xf(x)\, dx = x\int_a^b f(x)\, dx$.

T 70–74. Approximating definite integrals with a calculator *Consider the following definite integrals.*

   a. *Write the left and right Riemann sums in sigma notation for an arbitrary value of n.*

   b. *Evaluate each sum using a calculator with n = 20, 50, and 100. Use these values to estimate the value of the integral.*

70. $\int_4^9 3\sqrt{x}\, dx$

71. $\int_0^1 (x^2 + 1)\, dx$

72. $\int_1^e \ln x\, dx$

73. $\int_0^1 \cos^{-1} x\, dx$

74. $\int_{-1}^1 \pi \cos\left(\frac{\pi x}{2}\right) dx$

T 75–78. Midpoint Riemann sums with a calculator *Consider the following definite integrals.*

   a. *Write the midpoint Riemann sum in sigma notation for an arbitrary value of n.*

   b. *Evaluate each sum using a calculator with n = 20, 50, and 100. Use these values to estimate the value of the integral.*

75. $\int_1^4 2\sqrt{x}\, dx$

76. $\int_{-1}^2 \sin\left(\frac{\pi x}{4}\right) dx$

77. $\int_0^4 (4x - x^2)\, dx$

78. $\int_0^1 \frac{1}{\pi}(\sin^{-1} x + 1)\, dx$

79–85. Limits of sums *Use the definition of the definite integral to evaluate the following definite integrals. Use right Riemann sums and Theorem 5.1.*

79. $\int_0^2 (2x + 1)\, dx$

80. $\int_1^5 (1 - x)\, dx$

81. $\int_3^7 (4x + 6)\, dx$

82. $\int_0^2 (x^2 - 1)\, dx$

83. $\int_1^4 (x^2 - 1)\, dx$

84. $\int_0^2 (x^3 + x + 1)\, dx$

85. $\int_0^1 (4x^3 + 3x^2)\, dx$

Explorations and Challenges

86–87. Area by geometry *Use geometry to evaluate the following integrals.*

86. $\int_1^6 |2x - 4|\, dx$

87. $\int_{-6}^4 \sqrt{24 - 2x - x^2}\, dx$

88. Integrating piecewise continuous functions Suppose f is continuous on the intervals $[a, p]$ and $(p, b]$, where $a < p < b$, with a finite jump at p. Form a uniform partition on the interval $[a, p]$ with n grid points and another uniform partition on the interval $[p, b]$ with m grid points, where p is a grid point of both partitions. Write a Riemann sum for $\int_a^b f(x)\, dx$ and separate it into two pieces for $[a, p]$ and $[p, b]$. Explain why $\int_a^b f(x)\, dx = \int_a^p f(x)\, dx + \int_p^b f(x)\, dx$.

89–90. Integrating piecewise continuous functions *Use geometry and the result of Exercise 88 to evaluate the following integrals.*

89. $\int_0^{10} f(x)\, dx$, where $f(x) = \begin{cases} 2 & \text{if } 0 \le x \le 5 \\ 3 & \text{if } 5 < x \le 10 \end{cases}$

90. $\int_1^6 f(x)\, dx$, where $f(x) = \begin{cases} 2x & \text{if } 1 \le x < 4 \\ 10 - 2x & \text{if } 4 \le x \le 6 \end{cases}$

91–92. Integrating piecewise continuous functions *Recall that the floor function $\lfloor x \rfloor$ is the greatest integer less than or equal to x and that the ceiling function $\lceil x \rceil$ is the least integer greater than or equal to x. Use the result of Exercise 88 and the graphs to evaluate the following integrals.*

91. $\int_1^5 x\lfloor x \rfloor\, dx$

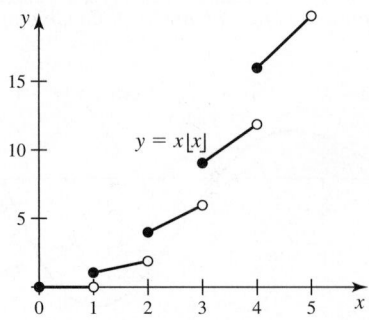

92. $\int_0^4 \dfrac{x}{\lceil x \rceil}\, dx$

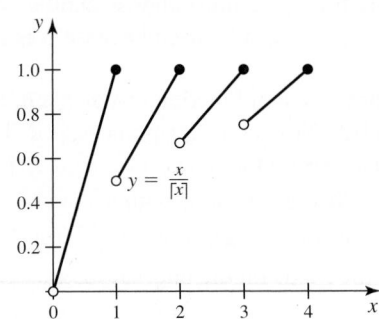

93. Constants in integrals Use the definition of the definite integral to justify the property $\int_a^b c f(x)\, dx = c \int_a^b f(x)\, dx$, where f is continuous and c is a real number.

94. Zero net area Assuming $0 < c < d$, find the value of b (in terms of c and d) for which $\int_c^d (x + b)\, dx = 0$.

95. A nonintegrable function Consider the function defined on $[0, 1]$ such that $f(x) = 1$ if x is a rational number and $f(x) = 0$ if x is irrational. This function has an infinite number of discontinuities, and the integral $\int_0^1 f(x)\, dx$ does not exist. Show that the right, left, and midpoint Riemann sums on *regular* partitions with n subintervals equal 1 for all n. (*Hint:* Between any two real numbers lie a rational and an irrational number.)

96. Powers of x by Riemann sums Consider the integral $I(p) = \int_0^1 x^p\, dx$, where p is a positive integer.

a. Write the left Riemann sum for the integral with n subintervals.

b. It is a fact (proved by the 17th-century mathematicians Fermat and Pascal) that $\displaystyle \lim_{n \to \infty} \frac{1}{n} \sum_{k=0}^{n-1} \left(\frac{k}{n}\right)^p = \frac{1}{p+1}$. Use this fact to evaluate $I(p)$.

97. An exact integration formula Evaluate $\int_a^b \dfrac{dx}{x^2}$, where $0 < a < b$, using the definition of the definite integral and the following steps.

a. Assume $\{x_0, x_1, \ldots, x_n\}$ is a partition of $[a, b]$ with $\Delta x_k = x_k - x_{k-1}$, for $k = 1, 2, \ldots, n$. Show that $x_{k-1} \le \sqrt{x_{k-1} x_k} \le x_k$, for $k = 1, 2, \ldots, n$.

b. Show that $\dfrac{1}{x_{k-1}} - \dfrac{1}{x_k} = \dfrac{\Delta x_k}{x_{k-1} x_k}$, for $k = 1, 2, \ldots, n$.

c. Simplify the general Riemann sum for $\int_a^b \dfrac{dx}{x^2}$ using $x_k^* = \sqrt{x_{k-1} x_k}$.

d. Conclude that $\int_a^b \dfrac{dx}{x^2} = \dfrac{1}{a} - \dfrac{1}{b}$.

(*Source: The College Mathematics Journal, 32, 4, Sep 2001*)

98. Use Property 3 of Table 5.4 and Property 7 of Table 5.5 to prove Property 8 of Table 5.5.

QUICK CHECK ANSWERS

1. -20 **2.** $f(x) = 1 - x$ is one possibility.   **3.** 0
4. $10;\ c(b - a)$ **5.** 0   **6.** $\frac{3}{2};\frac{5}{2}$ ◄

5.3 Fundamental Theorem of Calculus

Evaluating definite integrals using limits of Riemann sums, as described in Section 5.2, is usually not possible or practical. Fortunately, there is a powerful and practical method for evaluating definite integrals, which is developed in this section. Along the way, we discover the inverse relationship between differentiation and integration, expressed in the most important result of calculus, the Fundamental Theorem of Calculus. The first step in this process is to introduce *area functions* (first seen in Section 1.2).

Area Functions

The concept of an area function is crucial to the discussion about the connection between derivatives and integrals. We start with a continuous function $y = f(t)$ defined for $t \ge a$, where a is a fixed number. The *area function* for f with left endpoint a is denoted $A(x)$; it gives the net area of the region bounded by the graph of f and the t-axis between $t = a$ and $t = x$ (**Figure 5.35**). The net area of this region is also given by the definite integral.

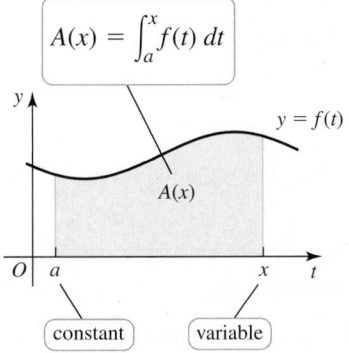

Figure 5.35

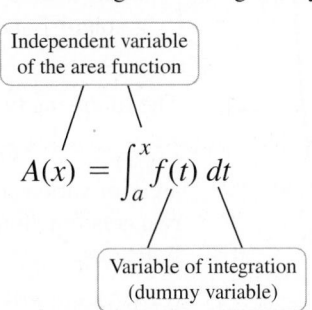

▶ Suppose you want to devise a function A whose value is the sum of the first n positive integers, $1 + 2 + \cdots + n$. The most compact way to write this function is $A(n) = \sum_{k=1}^{n} k$, for $n \geq 1$. In this function, n is the independent variable of the function A, and k, which is internal to the sum, is a dummy variable. This function involving a sum is analogous to an area function involving an integral.

Notice that x is the upper limit of the integral *and* the independent variable of the area function: As x changes, so does the net area under the curve. Because the symbol x is already in use as the independent variable for A, we must choose another symbol for the variable of integration. Any symbol—except x—can be used because it is a *dummy variable*; we have chosen t as the integration variable.

Figure 5.36 gives a general view of how an area function is generated. Suppose f is a continuous function and a is a fixed number. Now choose a point $b > a$. The net area of the region between the graph of f and the t-axis on the interval $[a, b]$ is $A(b)$. Moving the right endpoint to $(c, 0)$ or $(d, 0)$ produces different regions with net areas $A(c)$ and $A(d)$, respectively. In general, if $x > a$ is a variable point, then $A(x) = \int_a^x f(t)\, dt$ is the net area of the region between the graph of f and the t-axis on the interval $[a, x]$.

▶ Notice that t is the independent variable when we plot f, and x is the independent variable when we plot A.

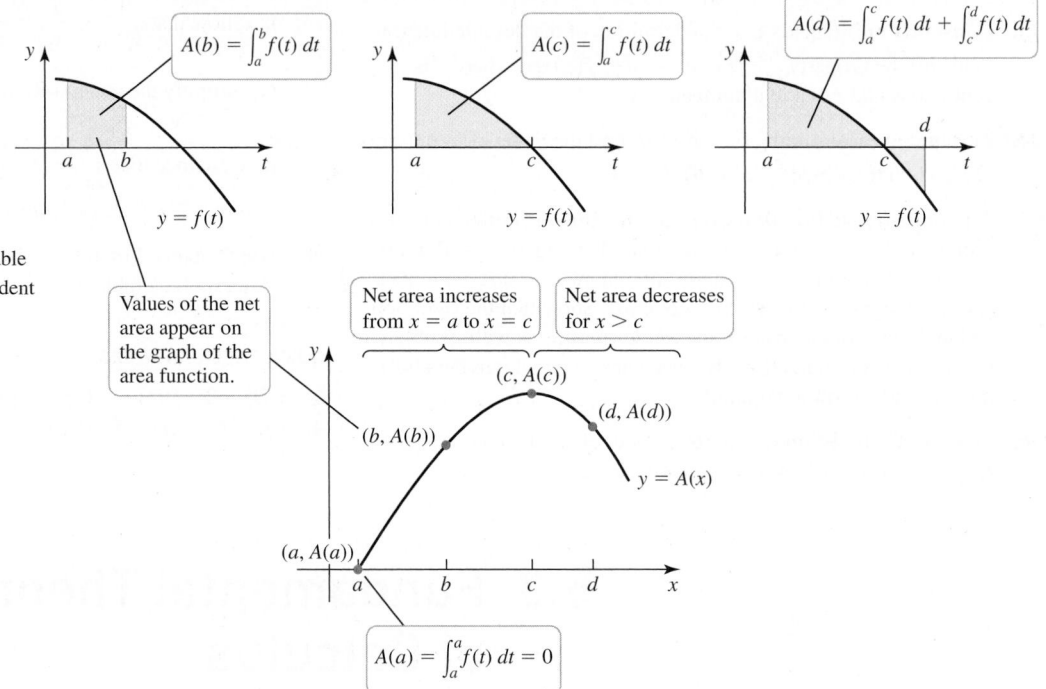

Figure 5.36

Figure 5.36 shows how $A(x)$ varies with respect to x. Notice that $A(a) = \int_a^a f(t)\, dt = 0$. Then for $x > a$, the net area increases for $x < c$, at which point $f(c) = 0$. For $x > c$, the function f is negative, which produces a negative contribution to the area function. As a result, the area function decreases for $x > c$.

DEFINITION Area Function

Let f be a continuous function, for $t \geq a$. The **area function for f with left endpoint a** is

$$A(x) = \int_a^x f(t)\, dt,$$

where $x \geq a$. The area function gives the net area of the region bounded by the graph of f and the t-axis on the interval $[a, x]$.

The following two examples illustrate the idea of area functions.

EXAMPLE 1   Comparing area functions The graph of f is shown in Figure 5.37 with areas of various regions marked. Let $A(x) = \int_{-1}^{x} f(t)\, dt$ and $F(x) = \int_{3}^{x} f(t)\, dt$ be two area functions for f (note the different left endpoints). Evaluate the following area functions.

a. $A(3)$ and $F(3)$ **b.** $A(5)$ and $F(5)$ **c.** $A(9)$ and $F(9)$

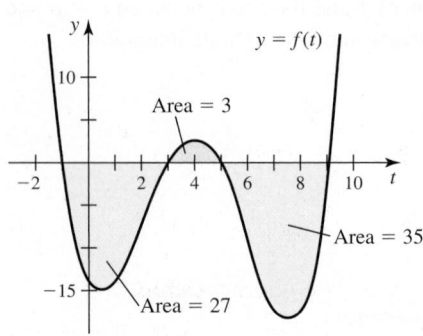

Figure 5.37

SOLUTION

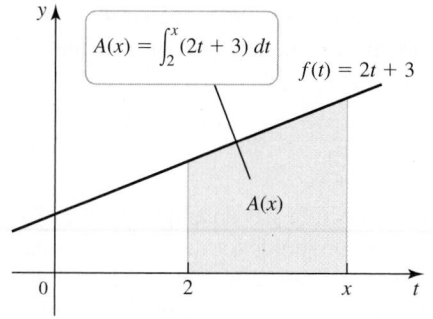

Figure 5.38

a. The value of $A(3) = \int_{-1}^{3} f(t)\, dt$ is the net area of the region bounded by the graph of f and the t-axis on the interval $[-1, 3]$. Using the graph of f, we see that $A(3) = -27$ (because this region has an area of 27 and lies below the t-axis). On the other hand, $F(3) = \int_{3}^{3} f(t)\, dt = 0$ by Property 1 of Table 5.4. Notice that $A(3) - F(3) = -27$.

b. The value of $A(5) = \int_{-1}^{5} f(t)\, dt$ is found by subtracting the area of the region that lies below the t-axis on $[-1, 3]$ from the area of the region that lies above the t-axis on $[3, 5]$. Therefore, $A(5) = 3 - 27 = -24$. Similarly, $F(5)$ is the net area of the region bounded by the graph of f and the t-axis on the interval $[3, 5]$; therefore, $F(5) = 3$. Notice that $A(5) - F(5) = -27$.

c. Reasoning as in parts (a) and (b), we see that $A(9) = -27 + 3 - 35 = -59$ and $F(9) = 3 - 35 = -32$. As before, observe that $A(9) - F(9) = -27$.

Related Exercises 13–14 ◄

QUICK CHECK 1 In Example 1, let $B(x)$ be the area function for f with left endpoint 5. Evaluate $B(5)$ and $B(9)$. ◄

Example 1 illustrates the important fact (to be explained shortly) that two area functions of the same function differ by a constant; in Example 1, the constant is -27.

EXAMPLE 2   Area of a trapezoid Consider the trapezoid bounded by the line $f(t) = 2t + 3$ and the t-axis from $t = 2$ to $t = x$ (Figure 5.38). The area function $A(x) = \int_{2}^{x} f(t)\, dt$ gives the area of the trapezoid, for $x \geq 2$.

a. Evaluate $A(2)$.   **b.** Evaluate $A(5)$.

c. Find and graph the area function $y = A(x)$, for $x \geq 2$.

d. Compare the derivative of A to f.

SOLUTION

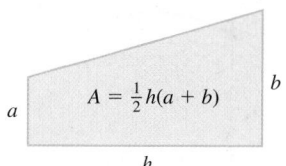

Figure 5.39

a. By Property 1 of Table 5.4, $A(2) = \int_{2}^{2} (2t + 3)\, dt = 0$.

b. Notice that $A(5)$ is the area of the trapezoid (Figure 5.38) bounded by the line $y = 2t + 3$ and the t-axis on the interval $[2, 5]$. Using the area formula for a trapezoid (Figure 5.39), we find that

$$A(5) = \int_{2}^{5} (2t + 3)\, dt = \frac{1}{2}\, \underbrace{(5 - 2)}_{\substack{\text{distance between}\\ \text{parallel sides}}} \cdot \underbrace{(f(2) + f(5))}_{\substack{\text{sum of parallel}\\ \text{side lengths}}} = \frac{1}{2} \cdot 3(7 + 13) = 30.$$

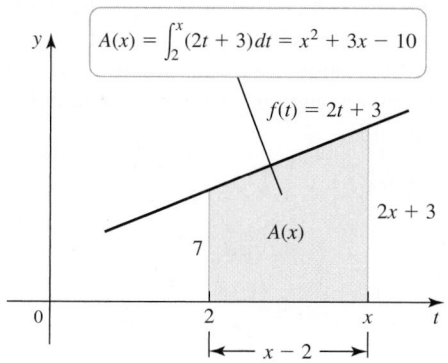

Figure 5.40

c. Now the right endpoint of the base is a variable $x \geq 2$ (Figure 5.40). The distance between the parallel sides of the trapezoid is $x - 2$. By the area formula for a trapezoid, the area of this trapezoid, for any $x \geq 2$, is

$$A(x) = \frac{1}{2}\, \underbrace{(x - 2)}_{\substack{\text{distance between}\\ \text{parallel sides}}} \cdot \underbrace{(f(2) + f(x))}_{\substack{\text{sum of parallel}\\ \text{side lengths}}}$$

$$= \frac{1}{2}(x - 2)(7 + 2x + 3)$$

$$= (x - 2)(x + 5)$$

$$= x^2 + 3x - 10.$$

Expressing the area function in terms of an integral with a variable upper limit, we have

$$A(x) = \int_{2}^{x} (2t + 3)\, dt = x^2 + 3x - 10, \text{ for } x \geq 2.$$

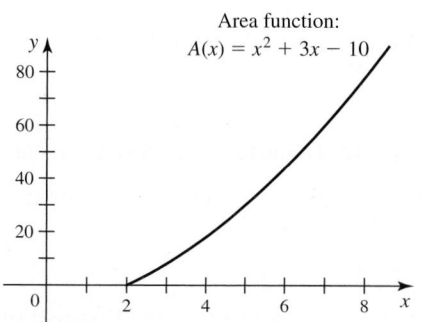

Figure 5.41

Because the line $f(t) = 2t + 3$ is above the t-axis, for $t \geq 2$, the area function $A(x) = x^2 + 3x - 10$ is an increasing function of x with $A(2) = 0$ (Figure 5.41).

➤ Recall that if $A'(x) = f(x)$, then f is the derivative of A; equivalently, A is an antiderivative of f.

QUICK CHECK 2 Verify that the area function in Example 2c gives the correct area when $x = 6$ and $x = 10$. ◄

d. Differentiating the area function, we find that

$$A'(x) = \frac{d}{dx}(x^2 + 3x - 10) = 2x + 3 = f(x).$$

Therefore, $A'(x) = f(x)$, or equivalently, the area function A is an antiderivative of f. We soon show that this relationship is not an accident; it is the first part of the Fundamental Theorem of Calculus.

Related Exercises 18–19 ◄

Fundamental Theorem of Calculus

Example 2 suggests that the area function A for a linear function f is an antiderivative of f; that is, $A'(x) = f(x)$. Our goal is to show that this conjecture holds for more general functions. Let's start with an intuitive argument; a formal proof is given at the end of the section.

Assume f is a continuous function defined on an interval $[a, b]$. As before, $A(x) = \int_a^x f(t)\,dt$ is the area function for f with a left endpoint a: It gives the net area of the region bounded by the graph of f and the t-axis on the interval $[a, x]$, for $x \geq a$. Figure 5.42 is the key to the argument.

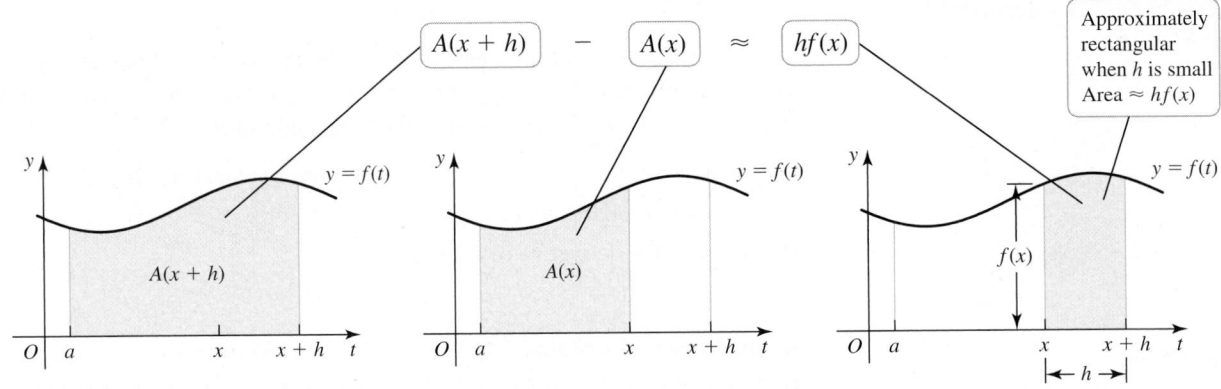

Figure 5.42

Note that with $h > 0$, $A(x + h)$ is the net area of the region whose base is the interval $[a, x + h]$ and $A(x)$ is the net area of the region whose base is the interval $[a, x]$. So the difference $A(x + h) - A(x)$ is the net area of the region whose base is the interval $[x, x + h]$. If h is small, the region in question is nearly rectangular with a base of length h and a height $f(x)$. Therefore, the net area of this region is

$$A(x + h) - A(x) \approx hf(x).$$

Dividing by h, we have

$$\frac{A(x + h) - A(x)}{h} \approx f(x).$$

➤ Recall that

$$f'(x) = \lim_{h \to 0} \frac{f(x + h) - f(x)}{h}.$$

If the function f is replaced with A, then

$$A'(x) = \lim_{h \to 0} \frac{A(x + h) - A(x)}{h}.$$

An analogous argument can be made with $h < 0$. Now observe that as h tends to zero, this approximation improves. In the limit as $h \to 0$, we have

$$\underbrace{\lim_{h \to 0} \frac{A(x + h) - A(x)}{h}}_{A'(x)} = \underbrace{\lim_{h \to 0} f(x)}_{f(x)}.$$

We see that indeed $A'(x) = f(x)$. Because $A(x) = \int_a^x f(t)\,dt$, the result can also be written

$$A'(x) = \frac{d}{dx}\underbrace{\int_a^x f(t)\,dt}_{A(x)} = f(x),$$

which says that the derivative of the integral of f is f. This conclusion is the first part of the Fundamental Theorem of Calculus.

THEOREM 5.3 (PART 1) Fundamental Theorem of Calculus

If f is continuous on $[a, b]$, then the area function

$$A(x) = \int_a^x f(t)\, dt, \quad \text{for} \quad a \le x \le b,$$

is continuous on $[a, b]$ and differentiable on (a, b). The area function satisfies $A'(x) = f(x)$. Equivalently,

$$A'(x) = \frac{d}{dx} \int_a^x f(t)\, dt = f(x),$$

which means that the area function of f is an antiderivative of f on $[a, b]$.

Given that A is an antiderivative of f on $[a, b]$, it is one short step to a powerful method for evaluating definite integrals. Remember (Section 4.9) that any two antiderivatives of f differ by a constant. Assuming F is any other antiderivative of f on $[a, b]$, we have

$$F(x) = A(x) + C, \quad \text{for} \quad a \le x \le b.$$

Noting that $A(a) = 0$, it follows that

$$F(b) - F(a) = (A(b) + C) - (A(a) + C) = A(b).$$

Writing $A(b)$ in terms of a definite integral leads to the remarkable result

$$A(b) = \int_a^b f(x)\, dx = F(b) - F(a).$$

We have shown that to evaluate a definite integral of f, we

- find any antiderivative of f, which we call F; and
- compute $F(b) - F(a)$, the difference in the values of F between the upper and lower limits of integration.

This process comprises the second part of the Fundamental Theorem of Calculus.

THEOREM 5.3 (PART 2) Fundamental Theorem of Calculus

If f is continuous on $[a, b]$ and F is any antiderivative of f on $[a, b]$, then

$$\int_a^b f(x)\, dx = F(b) - F(a).$$

It is customary and convenient to denote the difference $F(b) - F(a)$ by $F(x)\big|_a^b$. Using this shorthand, Part 2 of the Fundamental Theorem is summarized in **Figure 5.43**.

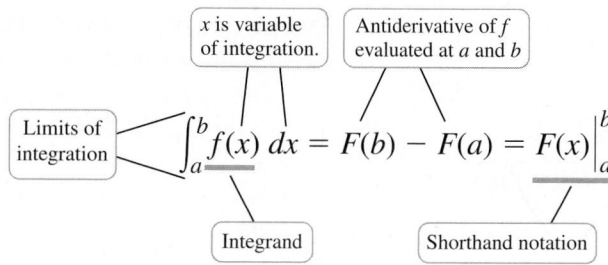

Figure 5.43

QUICK CHECK 3 Evaluate $\left(\dfrac{x}{x+1}\right)\Big|_1^2$. ◄

The Inverse Relationship Between Differentiation and Integration   It is worth pausing to observe that the two parts of the Fundamental Theorem express the inverse relationship between differentiation and integration. Part 1 of the Fundamental Theorem says

$$\frac{d}{dx}\int_a^x f(t)\, dt = f(x),$$

or the derivative of the integral of f is f itself.

Noting that f is an antiderivative of f', Part 2 of the Fundamental Theorem says

$$\int_a^b f'(x)\, dx = f(b) - f(a),$$

QUICK CHECK 4 Explain why f is an antiderivative of f'. ◄

or the definite integral of the derivative of f is given in terms of f evaluated at two points. In other words, the integral "undoes" the derivative.

This last relationship is important because it expresses the integral as an *accumulation* operation. Suppose we know the rate of change of f (which is f') on an interval $[a, b]$. The Fundamental Theorem says that we can integrate (that is, sum or accumulate) the rate of change over that interval, and the result is simply the difference in f evaluated at the endpoints. You will see this accumulation property used many times in the next chapter. Now let's use the Fundamental Theorem to evaluate definite integrals.

EXAMPLE 3   Evaluating definite integrals   Evaluate the following definite integrals using the Fundamental Theorem of Calculus, Part 2. Interpret each result geometrically.

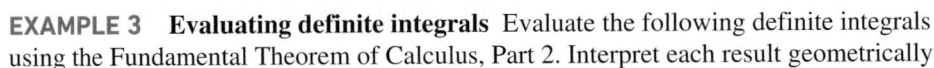

a. $\displaystyle\int_0^{10}(60x - 6x^2)\, dx$ **b.** $\displaystyle\int_4^{16} 3\sqrt{x}\, dx$ **c.** $\displaystyle\int_0^{2\pi} 3\sin x\, dx$ **d.** $\displaystyle\int_{1/16}^{1/4}\frac{\sqrt{t}-1}{t}\, dt$

SOLUTION

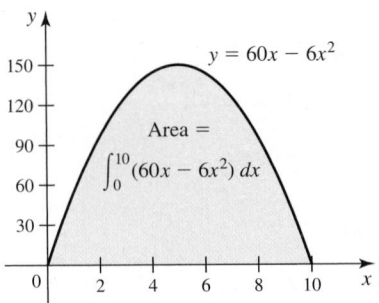

Figure 5.44

a. Using the antiderivative rules of Section 4.9, an antiderivative of $60x - 6x^2$ is $F(x) = 30x^2 - 2x^3$. By the Fundamental Theorem, the value of the definite integral is

$$\int_0^{10}(60x - 6x^2)\, dx = \underbrace{\left(30x^2 - 2x^3\right)\Big|_0^{10}}_{F(x)} \qquad \text{Fundamental Theorem}$$

$$= \underbrace{(30 \cdot 10^2 - 2 \cdot 10^3)}_{F(10)} - \underbrace{(30 \cdot 0^2 - 2 \cdot 0^3)}_{F(0)} \qquad \begin{array}{l}\text{Evaluate at } x = 10 \\ \text{and } x = 0.\end{array}$$

$$= (3000 - 2000) - 0$$

$$= 1000. \qquad \text{Simplify.}$$

Because $f(x) = 60x - 6x^2$ is positive on $[0, 10]$, the definite integral $\int_0^{10}(60x - 6x^2)\, dx$ is the area of the region between the graph of f and the x-axis on the interval $[0, 10]$ **(Figure 5.44)**.

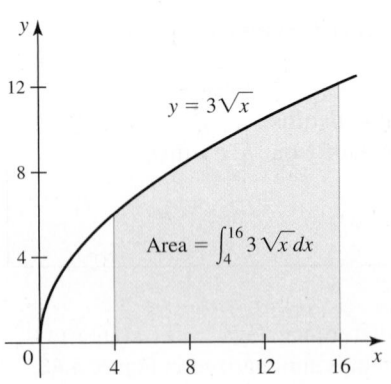

Figure 5.45

b. Because $f(x) = 3\sqrt{x}$ is positive on $[4, 16]$ **(Figure 5.45)**, the definite integral $\int_4^{16} 3\sqrt{x}\, dx$ equals the area of the region under the graph of f on $[4, 16]$. In Example 3 of Section 5.1, we used a midpoint Riemann sum to show that the area is approximately 112.062. Using the Fundamental Theorem, we can compute the exact area. Noting that an antiderivative of $x^{1/2}$ is $\frac{2}{3}x^{3/2}$, we have

$$\int_4^{16} 3\sqrt{x}\, dx = 3\int_4^{16} x^{1/2}\, dx \qquad \text{Property 4, Table 5.4}$$

$$= 3 \cdot \frac{2}{3} x^{3/2}\Big|_4^{16} \qquad \text{Fundamental Theorem}$$

$$= 2(16^{3/2} - 4^{3/2}) \qquad \text{Evaluate at } x = 16 \text{ and } x = 4.$$

$$= 2(64 - 8) = 112. \quad \text{Simplify.}$$

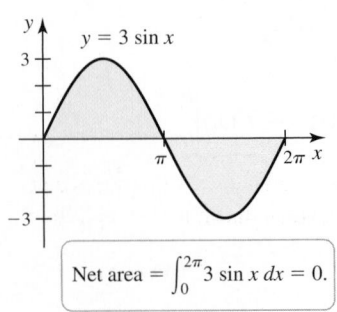

Figure 5.46

c. As shown in **Figure 5.46**, the region bounded by the graph of $f(x) = 3\sin x$ and the x-axis on $[0, 2\pi]$ consists of two parts, one above the x-axis and one below the x-axis. By the symmetry of f, these two regions have the same area, so the definite integral

➤ When evaluating definite integrals, factor out multiplicative constants when possible, as we did in parts (b) and (c) of Example 3. To illustrate: It is better to write

$$-3 \cos x \Big|_0^{2\pi} = -3(\cos 2\pi - \cos 0) \text{ than}$$

$$-3 \cos x \Big|_0^{2\pi} = (-3 \cos 2\pi) - (-3 \cos 0).$$

over $[0, 2\pi]$, which gives the net area of the entire region, is zero. Let's confirm this fact. An antiderivative of $f(x) = 3 \sin x$ is $-3 \cos x$. Therefore, the value of the definite integral is

$$\int_0^{2\pi} 3 \sin x \, dx = -3 \cos x \Big|_0^{2\pi} \qquad \text{Fundamental Theorem}$$

$$= -3(\cos 2\pi - \cos 0) \quad \text{Substitute.}$$

$$= -3(1 - 1) = 0. \qquad \text{Simplify.}$$

d. Although the variable of integration is t, rather than x, we proceed as before, after simplifying the integrand:

$$\frac{\sqrt{t} - 1}{t} = \frac{1}{\sqrt{t}} - \frac{1}{t}.$$

➤ We know that

$$\frac{d}{dt}(t^{1/2}) = \frac{1}{2}t^{-1/2}.$$

Therefore, $\int \frac{1}{2} t^{-1/2} \, dt = t^{1/2} + C$

and $\int \frac{dt}{\sqrt{t}} = \int t^{-1/2} \, dt = 2t^{1/2} + C.$

Finding antiderivatives with respect to t and applying the Fundamental Theorem, we have

$$\int_{1/16}^{1/4} \frac{\sqrt{t} - 1}{t} \, dt = \int_{1/16}^{1/4} \left(t^{-1/2} - \frac{1}{t} \right) dt \qquad \begin{array}{l}\text{Simplify the}\\\text{integrand.}\end{array}$$

$$= (2t^{1/2} - \ln|t|) \Big|_{1/16}^{1/4} \qquad \begin{array}{l}\text{Fundamental}\\\text{Theorem}\end{array}$$

$$= \left(2\left(\frac{1}{4}\right)^{1/2} - \ln\frac{1}{4} \right) - \left(2\left(\frac{1}{16}\right)^{1/2} - \ln\frac{1}{16} \right) \quad \text{Evaluate.}$$

$$= 1 - \ln\frac{1}{4} - \frac{1}{2} + \ln\frac{1}{16} \qquad \text{Simplify.}$$

$$= \frac{1}{2} - \ln 4 \approx -0.8863.$$

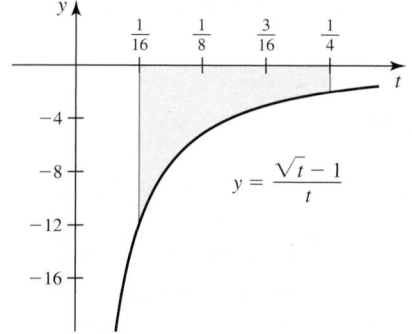

Figure 5.47

The definite integral is negative because the graph of f lies below the t-axis on the interval of integration $\left[\frac{1}{16}, \frac{1}{4}\right]$ (Figure 5.47).

Related Exercises 23, 24, 26, 36 ◀

EXAMPLE 4   Net areas and definite integrals The graph of $f(x) = 6x(x + 1)(x - 2)$ is shown in Figure 5.48. The region R_1 is bounded by the curve and the x-axis on the interval $[-1, 0]$, and R_2 is bounded by the curve and the x-axis on the interval $[0, 2]$.

a. Find the *net area* of the region between the curve and the x-axis on $[-1, 2]$.

b. Find the *area* of the region between the curve and the x-axis on $[-1, 2]$.

SOLUTION

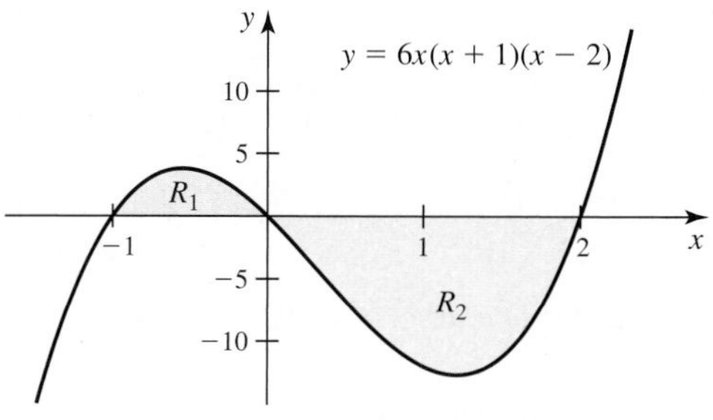

a. The net area of the region is given by a definite integral. The integrand f is first expanded to find an antiderivative:

$$\int_{-1}^{2} f(x) \, dx = \int_{-1}^{2} (6x^3 - 6x^2 - 12x) \, dx. \quad \text{Expand } f.$$

$$= \left(\frac{3}{2}x^4 - 2x^3 - 6x^2 \right) \Big|_{-1}^{2} \quad \text{Fundamental Theorem}$$

$$= -\frac{27}{2}. \qquad \text{Simplify.}$$

The net area of the region between the curve and the x-axis on $[-1, 2]$ is $-\frac{27}{2}$, which is the area of R_1 *minus* the area of R_2 (Figure 5.48). Because R_2 has a larger area than R_1, the net area is negative.

b. The region R_1 lies above the x-axis, so its area is

$$\int_{-1}^{0} (6x^3 - 6x^2 - 12x) \, dx = \left(\frac{3}{2}x^4 - 2x^3 - 6x^2 \right) \Big|_{-1}^{0} = \frac{5}{2}.$$

Figure 5.48

The region R_2 lies below the x-axis, so its net area is negative:

$$\int_0^2 (6x^3 - 6x^2 - 12x)\, dx = \left(\frac{3}{2}x^4 - 2x^3 - 6x^2\right)\Big|_0^2 = -16.$$

Therefore, the *area* of R_2 is $-(-16) = 16$. The combined area of R_1 and R_2 is $\frac{5}{2} + 16 = \frac{37}{2}$. We could also find the area of this region directly by evaluating $\int_{-1}^2 |f(x)|\, dx$.

Related Exercises 68, 70 ◄

Examples 3 and 4 make use of Part 2 of the Fundamental Theorem, which is the most potent tool for evaluating definite integrals. The remaining examples illustrate the use of the equally important Part 1 of the Fundamental Theorem.

EXAMPLE 5 Derivatives of integrals Use Part 1 of the Fundamental Theorem to simplify the following expressions.

a. $\dfrac{d}{dx}\displaystyle\int_1^x \sin^2 t\, dt$
b. $\dfrac{d}{dx}\displaystyle\int_x^5 \sqrt{t^2 + 1}\, dt$
c. $\dfrac{d}{dx}\displaystyle\int_0^{x^2} \cos t^2\, dt$

SOLUTION

a. Using Part 1 of the Fundamental Theorem, we see that

$$\frac{d}{dx}\int_1^x \sin^2 t\, dt = \sin^2 x.$$

b. To apply Part 1 of the Fundamental Theorem, the variable must appear in the upper limit. Therefore, we use the fact that $\int_a^b f(t)\, dt = -\int_b^a f(t)\, dt$ and then apply the Fundamental Theorem:

$$\frac{d}{dx}\int_x^5 \sqrt{t^2 + 1}\, dt = -\frac{d}{dx}\int_5^x \sqrt{t^2 + 1}\, dt = -\sqrt{x^2 + 1}.$$

c. The upper limit of the integral is not x, but a function of x. Therefore, the function to be differentiated is a composite function, which requires the Chain Rule. We let $u = x^2$ to produce

$$y = g(u) = \int_0^u \cos t^2\, dt.$$

By the Chain Rule,

$$\frac{d}{dx}\int_0^{x^2} \cos t^2\, dt = \frac{dy}{dx} = \frac{dy}{du}\frac{du}{dx} \qquad \text{Chain Rule}$$

$$= \left(\frac{d}{du}\int_0^u \cos t^2\, dt\right)(2x) \qquad \text{Substitute for } g\text{; note that } u'(x) = 2x.$$

$$= (\cos u^2)(2x) \qquad \text{Fundamental Theorem}$$

$$= 2x \cos x^4. \qquad \text{Substitute } u = x^2.$$

Related Exercises 73, 75, 78 ◄

➤ Example 5c illustrates one case of Leibniz's Rule:

$$\frac{d}{dx}\int_a^{g(x)} f(t)\, dt = f(g(x))g'(x).$$

EXAMPLE 6 Working with area functions Consider the function f shown in Figure 5.49 and its area function $A(x) = \int_0^x f(t)\, dt$, for $0 \le x \le 17$. Assume the four regions $R_1, R_2, R_3,$ and R_4 have the same area. Based on the graph of f, do the following.

a. Find the zeros of A on $[0, 17]$.

b. Find the points on $[0, 17]$ at which A has a local maximum or a local minimum.

c. Sketch a graph of A, for $0 \le x \le 17$.

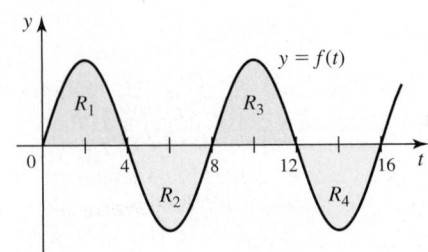

Figure 5.49

SOLUTION

a. The area function $A(x) = \int_0^x f(t)\, dt$ gives the net area bounded by the graph of f and the t-axis on the interval $[0, x]$ (**Figure 5.50a**). Therefore, $A(0) = \int_0^0 f(t)\, dt = 0$. Because R_1 and R_2 have the same area but lie on opposite sides of the t-axis, it follows that $A(8) = \int_0^8 f(t)\, dt = 0$. Similarly, $A(16) = \int_0^{16} f(t)\, dt = 0$. Therefore, the zeros of A are $x = 0$, 8, and 16.

> ▶ Recall that local extrema occur only at interior points of the domain.

b. Observe that the function f is positive, for $0 < t < 4$, which implies that $A(x)$ increases as x increases from 0 to 4 (**Figure 5.50b**). Then as x increases from 4 to 8, $A(x)$ decreases because f is negative, for $4 < t < 8$ (**Figure 5.50c**). Similarly, $A(x)$ increases as x increases from $x = 8$ to $x = 12$ (**Figure 5.50d**) and decreases from $x = 12$ to $x = 16$. By the First Derivative Test, A has local minima at $x = 8$ and $x = 16$ and local maxima at $x = 4$ and $x = 12$ (**Figure 5.50e**).

c. Combining the observations in parts (a) and (b) leads to a qualitative sketch of A (Figure 5.50e). Note that $A(x) \geq 0$, for all $x \geq 0$. It is not possible to determine function values (y-coordinates) on the graph of A.

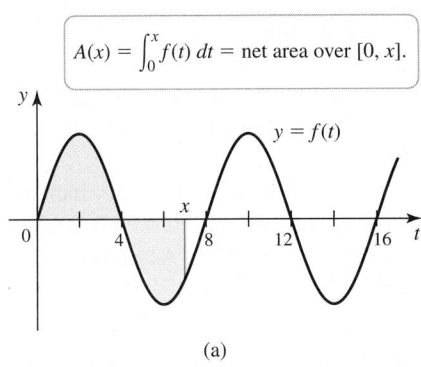

$A(x) = \int_0^x f(t)\, dt = $ net area over $[0, x]$.

(a)

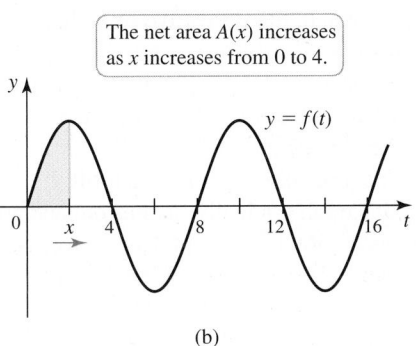

The net area $A(x)$ increases as x increases from 0 to 4.

(b)

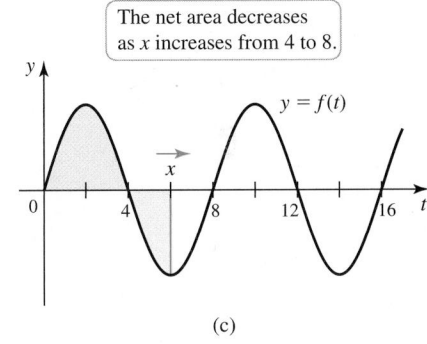

The net area decreases as x increases from 4 to 8.

(c)

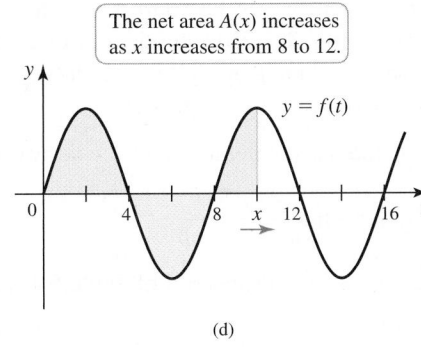

The net area $A(x)$ increases as x increases from 8 to 12.

(d)

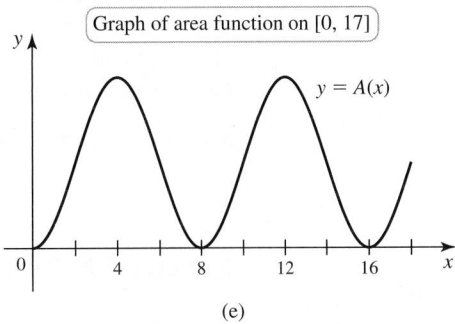

Graph of area function on $[0, 17]$

(e)

Figure 5.50

Related Exercises 71–72 ◀

EXAMPLE 7 The sine integral function Let

$$g(t) = \begin{cases} \dfrac{\sin t}{t} & \text{if } t > 0 \\[2mm] 1 & \text{if } t = 0. \end{cases}$$

Graph the *sine integral function* $S(x) = \int_0^x g(t)\, dt$, for $x \geq 0$.

SOLUTION Notice that S is an area function for g. The independent variable of S is x, and t has been chosen as the (dummy) variable of integration. A good way to start is by graphing the integrand g (**Figure 5.51a**). The function oscillates with a decreasing amplitude with $g(0) = 1$. Beginning with $S(0) = 0$, the area function S increases until $x = \pi$ because g is positive on $(0, \pi)$. However, on $(\pi, 2\pi)$, g is negative and the net area decreases. On $(2\pi, 3\pi)$, g is positive again, so S again increases. Therefore, the graph of S has alternating local maxima and minima. Because the amplitude of g decreases, each maximum of S is less than the previous maximum, and each minimum of S is greater than the previous minimum (**Figure 5.51b**). Determining the exact value of S at these maxima and minima is difficult.

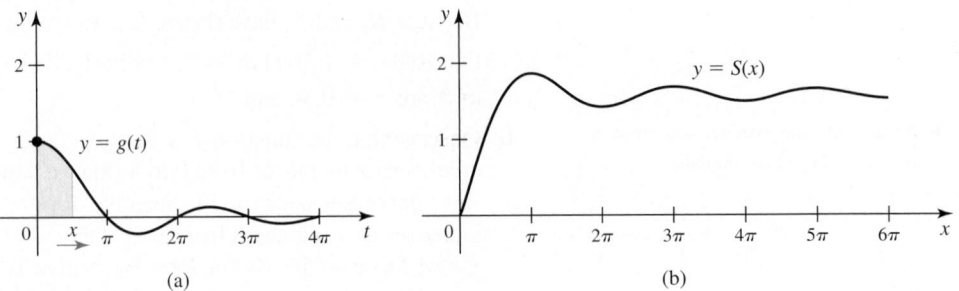

Figure 5.51

Appealing to Part 1 of the Fundamental Theorem, we find that

$$S'(x) = \frac{d}{dx}\int_0^x g(t)\, dt = \frac{\sin x}{x}, \text{ for } x > 0.$$

▶ Note that
$$\lim_{x \to \infty} S'(x) = \lim_{x \to \infty} g(x) = 0.$$

As anticipated, the derivative of S changes sign at integer multiples of π. Specifically, S' is positive and S increases on the intervals $(0, \pi)$, $(2\pi, 3\pi)$, ..., $(2n\pi, (2n + 1)\pi)$, ..., while S' is negative and S decreases on the remaining intervals. Clearly, S has local maxima at $x = \pi, 3\pi, 5\pi, \ldots$, and it has local minima at $x = 2\pi, 4\pi, 6\pi, \ldots$.

One more observation is helpful. It can be shown that although S oscillates for increasing x, its graph gradually flattens out and approaches a horizontal asymptote. (Finding the exact value of this horizontal asymptote is challenging; see Exercise 115.) Assembling all these observations, the graph of the sine integral function emerges (**Figure 5.51b**).

Related Exercises 101–102 ◀

We conclude this section with a formal proof of the Fundamental Theorem of Calculus.

Proof of the Fundamental Theorem: Let f be continuous on $[a, b]$ and let A be the area function for f with left endpoint a. The first step is to prove that A is differentiable on (a, b) and $A'(x) = f(x)$, which is Part 1 of the Fundamental Theorem. The proof of Part 2 then follows.

Step 1. We assume $a < x < b$ and use the definition of the derivative,

$$A'(x) = \lim_{h \to 0} \frac{A(x + h) - A(x)}{h}.$$

First assume $h > 0$. Using **Figure 5.52** and Property 5 of Table 5.4, we have

$$A(x + h) - A(x) = \int_a^{x+h} f(t)\, dt - \int_a^x f(t)\, dt = \int_x^{x+h} f(t)\, dt.$$

That is, $A(x + h) - A(x)$ is the net area of the region bounded by the curve on the interval $[x, x + h]$.

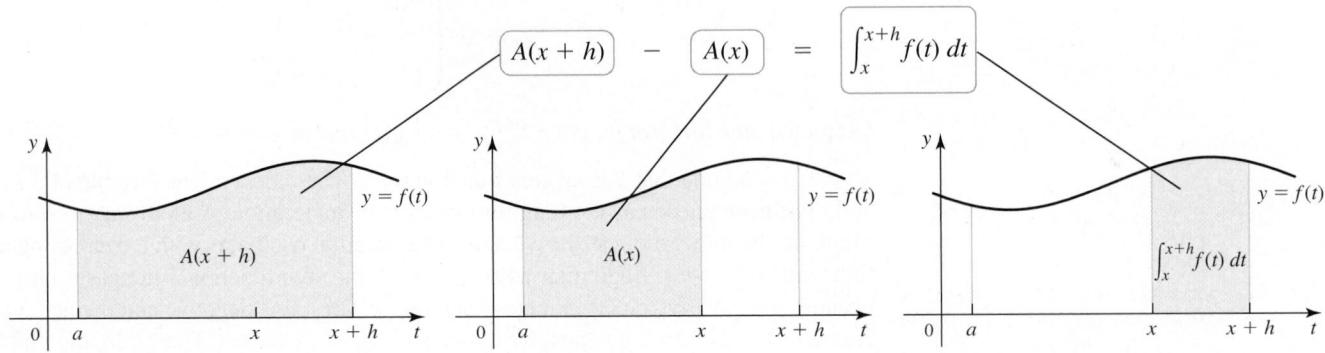

Figure 5.52

> ➤ The quantities m and M exist for any $h > 0$; however, their values depend on h.

Let m and M be the minimum and maximum values of f on $[x, x + h]$, respectively, which exist by the continuity of f. Note that $m \le f(t) \le M$ on $[x, x + h]$ (which is an interval of length h). It follows, by Property 9 of Table 5.5, that

$$mh \le \underbrace{\int_x^{x+h} f(t)\, dt}_{A(x + h) - A(x)} \le Mh.$$

Substituting for the integral, we find that

$$mh \le A(x + h) - A(x) \le Mh.$$

Dividing these inequalities by $h > 0$, we have

$$m \le \frac{A(x + h) - A(x)}{h} \le M.$$

The case $h < 0$ is handled similarly and leads to the same conclusion.

We now take the limit as $h \to 0$ across these inequalities. As $h \to 0$, m and M approach $f(x)$, because f is continuous at x. At the same time, as $h \to 0$, the quotient that is sandwiched between m and M approaches $A'(x)$:

$$\underbrace{\lim_{h \to 0} m}_{f(x)} \le \underbrace{\lim_{h \to 0} \frac{A(x + h) - A(x)}{h}}_{A'(x)} \le \underbrace{\lim_{h \to 0} M}_{f(x)}.$$

By the Squeeze Theorem (Theorem 2.5), we conclude that $A'(x)$ exists and A is differentiable for $a < x < b$. Furthermore, $A'(x) = f(x)$. Finally, because A is differentiable on (a, b), A is continuous on (a, b) by Theorem 3.1. Exercise 118 shows that A is also right- and left-continuous at the endpoints a and b, respectively.

> ➤ Once again we use an important fact: Two antiderivatives of the same function differ by a constant.

Step 2. Having established that the area function A is an antiderivative of f, we know that $F(x) = A(x) + C$, where F is any antiderivative of f and C is a constant. Noting that $A(a) = 0$, it follows that

$$F(b) - F(a) = (A(b) + C) - (A(a) + C) = A(b).$$

Writing $A(b)$ in terms of a definite integral, we have

$$A(b) = \int_a^b f(x)\, dx = F(b) - F(a),$$

which is Part 2 of the Fundamental Theorem.   ◄

SECTION 5.3 EXERCISES

Getting Started

1. Suppose A is an area function of f. What is the relationship between f and A?

2. Suppose F is an antiderivative of f and A is an area function of f. What is the relationship between F and A?

3. Explain mathematically how the Fundamental Theorem of Calculus is used to evaluate definite integrals.

4. Let $f(x) = c$, where c is a positive constant. Explain why an area function of f is an increasing function.

5. The linear function $f(x) = 3 - x$ is decreasing on the interval $[0, 3]$. Is its area function for f (with left endpoint 0) increasing or decreasing on the interval $[0, 3]$? Draw a picture and explain.

6. Evaluate $\int_0^2 3x^2\, dx$ and $\int_{-2}^2 3x^2\, dx$.

7. Explain in words and express mathematically the inverse relationship between differentiation and integration as given by Part 1 of the Fundamental Theorem of Calculus.

8. Why can the constant of integration be omitted from the antiderivative when evaluating a definite integral?

9. Evaluate $\dfrac{d}{dx} \int_a^x f(t)\, dt$ and $\dfrac{d}{dx} \int_a^b f(t)\, dt$, where a and b are constants.

10. Explain why $\int_a^b f'(x)\, dx = f(b) - f(a)$.

11. Evaluate $\int_3^8 f'(t)\, dt$, where f' is continuous on $[3, 8]$, $f(3) = 4$, and $f(8) = 20$.

12. Evaluate $\int_2^7 3\, dx$ using the Fundamental Theorem of Calculus. Check your work by evaluating the integral using geometry.

Practice Exercises

13. Area functions The graph of f is shown in the figure. Let $A(x) = \int_{-2}^{x} f(t)\, dt$ and $F(x) = \int_{4}^{x} f(t)\, dt$ be two area functions for f. Evaluate the following area functions.

  a. $A(-2)$ **b.** $F(8)$ **c.** $A(4)$ **d.** $F(4)$ **e.** $A(8)$

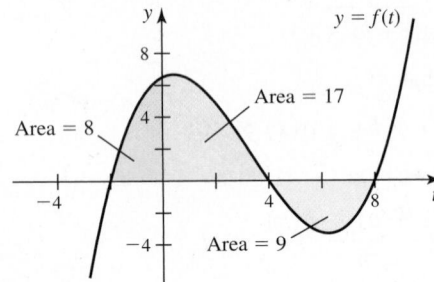

14. Area functions The graph of f is shown in the figure. Let $A(x) = \int_{0}^{x} f(t)\, dt$ and $F(x) = \int_{2}^{x} f(t)\, dt$ be two area functions for f. Evaluate the following area functions.

  a. $A(2)$ **b.** $F(5)$ **c.** $A(0)$ **d.** $F(8)$

  e. $A(8)$ **f.** $A(5)$ **g.** $F(2)$

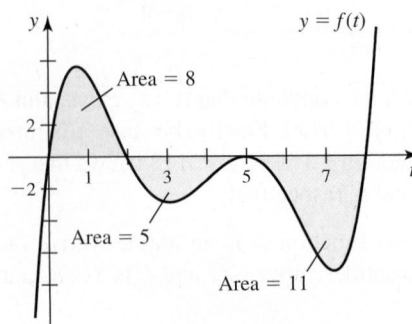

15–16. Area functions for constant functions *Consider the following functions f and real numbers a (see figure).*

a. *Find and graph the area function $A(x) = \int_{a}^{x} f(t)\, dt$ for f.*
b. *Verify that $A'(x) = f(x)$.*

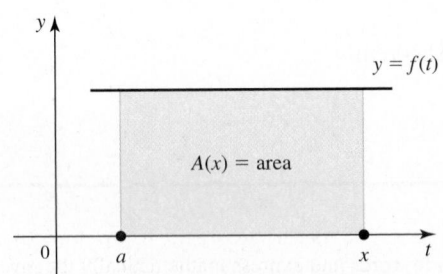

15. $f(t) = 5, a = 0$ **16.** $f(t) = 5, a = -5$

17. Area functions for the same linear function Let $f(t) = t$ and consider the two area functions $A(x) = \int_{0}^{x} f(t)\, dt$ and $F(x) = \int_{2}^{x} f(t)\, dt$.

  a. Evaluate $A(2)$ and $A(4)$. Then use geometry to find an expression for $A(x)$, for $x \geq 0$.
  b. Evaluate $F(4)$ and $F(6)$. Then use geometry to find an expression for $F(x)$, for $x \geq 2$.
  c. Show that $A(x) - F(x)$ is a constant and that $A'(x) = F'(x) = f(x)$.

18. Area functions for the same linear function Let $f(t) = 2t - 2$ and consider the two area functions $A(x) = \int_{1}^{x} f(t)\, dt$ and $F(x) = \int_{4}^{x} f(t)\, dt$.

  a. Evaluate $A(2)$ and $A(3)$. Then use geometry to find an expression for $A(x)$, for $x \geq 1$.
  b. Evaluate $F(5)$ and $F(6)$. Then use geometry to find an expression for $F(x)$, for $x \geq 4$.
  c. Show that $A(x) - F(x)$ is a constant and that $A'(x) = F'(x) = f(x)$.

19–22. Area functions for linear functions *Consider the following functions f and real numbers a (see figure).*

a. *Find and graph the area function $A(x) = \int_{a}^{x} f(t)\, dt$.*
b. *Verify that $A'(x) = f(x)$.*

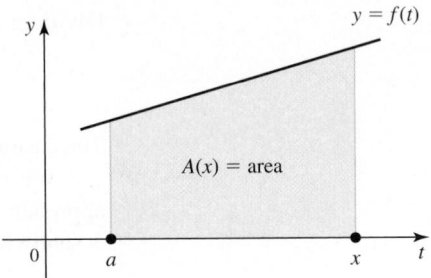

19. $f(t) = t + 5, a = -5$ **20.** $f(t) = 2t + 5, a = 0$

21. $f(t) = 3t + 1, a = 2$ **22.** $f(t) = 4t + 2, a = 0$

23–24. Definite integrals *Evaluate the following integrals using the Fundamental Theorem of Calculus. Explain why your result is consistent with the figure.*

23. $\int_{0}^{1} (x^2 - 2x + 3)\, dx$ **24.** $\int_{-\pi/4}^{7\pi/4} (\sin x + \cos x)\, dx$

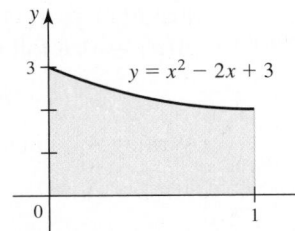   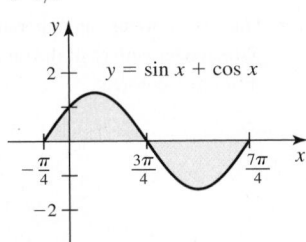

25–28. Definite integrals *Evaluate the following integrals using the Fundamental Theorem of Calculus. Sketch the graph of the integrand and shade the region whose net area you have found.*

25. $\int_{-2}^{3} (x^2 - x - 6)\, dx$ **26.** $\int_{0}^{1} (x - \sqrt{x})\, dx$

27. $\int_{0}^{5} (x^2 - 9)\, dx$ **28.** $\int_{1/2}^{2} \left(1 - \frac{1}{x^2}\right) dx$

29–62. Definite integrals *Evaluate the following integrals using the Fundamental Theorem of Calculus.*

29. $\int_{0}^{2} 4x^3\, dx$ **30.** $\int_{0}^{2} (3x^2 + 2x)\, dx$ **31.** $\int_{1}^{8} 8x^{1/3}\, dx$

32. $\int_{1}^{16} x^{-5/4}\, dx$ **33.** $\int_{0}^{1} (x + \sqrt{x})\, dx$ **34.** $\int_{0}^{\pi/4} 2\cos x\, dx$

35. $\int_{1}^{9} \frac{2}{\sqrt{x}}\, dx$ **36.** $\int_{4}^{9} \frac{2 + \sqrt{t}}{\sqrt{t}}\, dt$ **37.** $\int_{-2}^{2} (x^2 - 4)\, dx$

38. $\int_{0}^{\ln 8} e^x\, dx$ **39.** $\int_{1/2}^{1} (t^{-3} - 8)\, dt$ **40.** $\int_{0}^{4} t(t - 2)(t - 4)\, dt$

41. $\int_1^4 (1 - x)(x - 4)\, dx$

42. $\int_0^{1/2} \frac{dx}{\sqrt{1 - x^2}}$

43. $\int_{-2}^{-1} x^{-3}\, dx$

44. $\int_0^\pi (1 - \sin x)\, dx$

45. $\int_0^{\pi/4} \sec^2 \theta\, d\theta$

46. $\int_{-\pi/2}^{\pi/2} (\cos x - 1)\, dx$

47. $\int_1^2 \frac{3}{t}\, dt$

48. $\int_4^9 \frac{x - \sqrt{x}}{x^2}\, dx$

49. $\int_1^8 \sqrt[3]{y}\, dy$

50. $\frac{1}{2} \int_0^{\ln 2} e^x\, dx$

51. $\int_1^4 \frac{x - 2}{\sqrt{x}}\, dx$

52. $\int_1^2 \frac{2s^2 - 4}{s^3}\, ds$

53. $\int_0^{\pi/3} \sec x \tan x\, dx$

54. $\int_{\pi/4}^{\pi/2} \csc^2 \theta\, d\theta$

55. $\int_{\pi/4}^{3\pi/4} (\cot^2 x + 1)\, dx$

56. $\int_0^1 10 e^{x+3}\, dx$

57. $\int_1^{\sqrt{3}} \frac{1}{1 + x^2}\, dx$

58. $\int_0^{\pi/4} \sec x(\sec x + \cos x)\, dx$

59. $\int_1^2 \frac{z^2 + 4}{z}\, dz$

60. $\int_{\sqrt{2}}^2 \frac{dx}{x\sqrt{x^2 - 1}}$

61. $\int_0^\pi f(x)\, dx$, where $f(x) = \begin{cases} \sin x + 1 & \text{if } x \le \pi/2 \\ 2 \cos x + 2 & \text{if } x > \pi/2 \end{cases}$

62. $\int_1^3 g(x)\, dx$, where $g(x) = \begin{cases} 3x^2 + 4x + 1 & \text{if } x \le 2 \\ 2x + 5 & \text{if } x > 2 \end{cases}$

63–66. Area *Find (i) the net area and (ii) the area of the following regions. Graph the function and indicate the region in question.*

63. The region bounded by $y = x^{1/2}$ and the x-axis between $x = 1$ and $x = 4$

64. The region above the x-axis bounded by $y = 4 - x^2$

65. The region below the x-axis bounded by $y = x^4 - 16$

66. The region bounded by $y = 6 \cos x$ and the x-axis between $x = -\pi/2$ and $x = \pi$

67–72. Areas of regions *Find the area of the region bounded by the graph of f and the x-axis on the given interval.*

67. $f(x) = x^2 - 25$ on $[2, 4]$ **68.** $f(x) = x^3 - 1$ on $[-1, 2]$

69. $f(x) = \frac{1}{x}$ on $[-2, -1]$

70. $f(x) = x(x + 1)(x - 2)$ on $[-1, 2]$

71. $f(x) = \sin x$ on $\left[-\frac{\pi}{4}, \frac{3\pi}{4}\right]$ **72.** $f(x) = \cos x$ on $\left[\frac{\pi}{2}, \pi\right]$

73–86. Derivatives of integrals *Simplify the following expressions.*

73. $\frac{d}{dx} \int_3^x (t^2 + t + 1)\, dt$

74. $\frac{d}{dx} \int_1^x e^{t^2}\, dt$

75. $\frac{d}{dx} \int_x^1 \sqrt{t^4 + 1}\, dt$

76. $\frac{d}{dx} \int_x^0 \frac{dp}{p^2 + 1}$

77. $\frac{d}{dx} \int_2^{x^3} \frac{dp}{p^2}$

78. $\frac{d}{dx} \int_0^{x^2} \frac{dt}{t^2 + 4}$

79. $\frac{d}{dx} \int_0^{\cos x} (t^4 + 6)\, dt$

80. $\frac{d}{dw} \int_0^{\sqrt{w}} \ln (x^2 + 1)\, dx$

81. $\frac{d}{dz} \int_{\sin z}^{10} \frac{dt}{t^4 + 1}$

82. $\frac{d}{dy} \int_{y^3}^{10} \sqrt{x^6 + 1}\, dx$

83. $\frac{d}{dt} \left(\int_1^t \frac{3}{x}\, dx - \int_{t^2}^1 \frac{3}{x}\, dx \right)$

84. $\frac{d}{dt} \left(\int_0^t \frac{dx}{1 + x^2} + \int_1^{1/t} \frac{dx}{1 + x^2} \right)$

85. $\frac{d}{dx} \int_0^x \sqrt{1 + t^2}\, dt$

(*Hint:* $\int_{-x}^x \sqrt{1 + t^2}\, dt = \int_{-x}^0 \sqrt{1 + t^2}\, dt + \int_{-x}^x \sqrt{1 + t^2}\, dt$.)

86. $\frac{d}{dx} \int_{e^x}^{e^{2x}} \ln t^2\, dt$

87. Matching functions with area functions Match the functions f, whose graphs are given in a–d, with the area functions
$$A(x) = \int_0^x f(t)\, dt,$$
whose graphs are given in A–D.

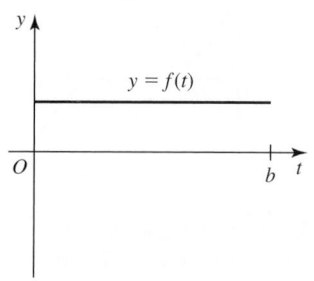

(a)

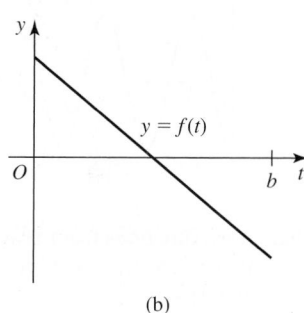

(b)

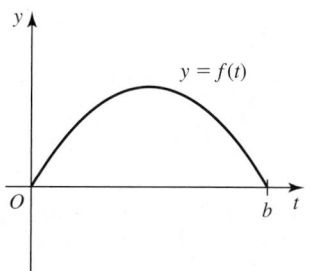

(c)

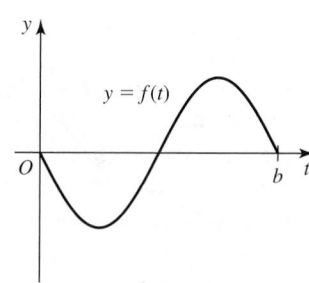

(d)

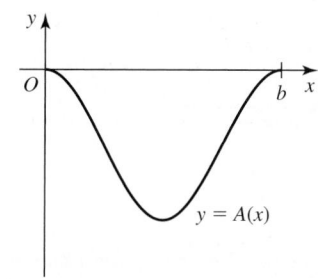

(A)

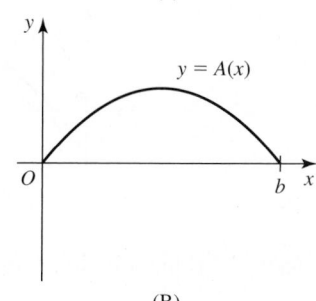

(B)

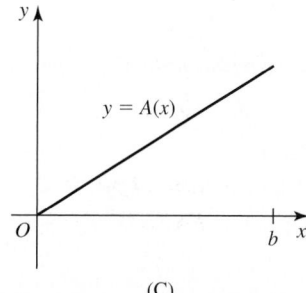

(C)

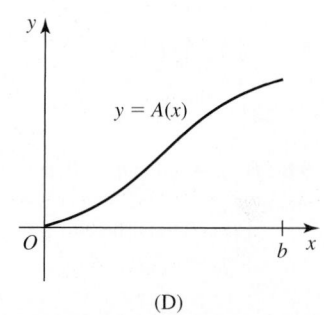

(D)

88–91. Working with area functions *Consider the function f and its graph.*

a. Estimate the zeros of the area function $A(x) = \int_0^x f(t)\,dt$, for $0 \le x \le 10$.

b. Estimate the points (if any) at which A has a local maximum or minimum.

c. Sketch a graph of A, for $0 \le x \le 10$, without a scale on the y-axis.

88.
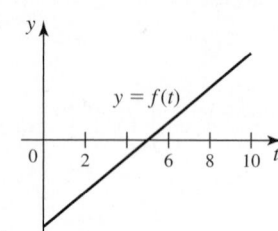

89.
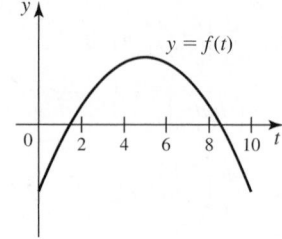

90.
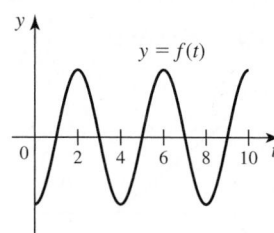

91.
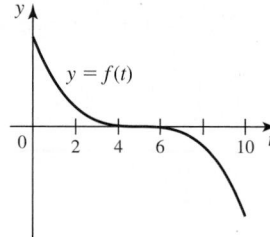

92. Area functions from graphs The graph of f is given in the figure. Let $A(x) = \int_0^x f(t)\,dt$ and evaluate $A(1)$, $A(2)$, $A(4)$, and $A(6)$.

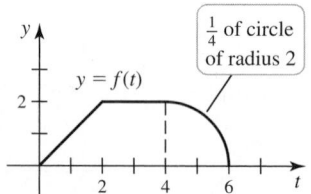

93. Area functions from graphs The graph of f is given in the figure. Let $A(x) = \int_0^x f(t)\,dt$ and evaluate $A(2)$, $A(5)$, $A(8)$, and $A(12)$.

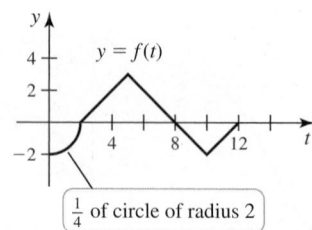

94–98. Working with area functions *Consider the function f and the points a, b, and c.*

a. Find the area function $A(x) = \int_a^x f(t)\,dt$ using the Fundamental Theorem.

b. Graph f and A.

c. Evaluate $A(b)$ and $A(c)$. Interpret the results using the graphs of part (b).

94. $f(x) = \sin x$; $a = 0$, $b = \dfrac{\pi}{2}$, $c = \pi$

95. $f(x) = e^x$; $a = 0$, $b = \ln 2$, $c = \ln 4$

96. $f(x) = -12x(x-1)(x-2)$; $a = 0$, $b = 1$, $c = 2$

97. $f(x) = \cos x$; $a = 0$, $b = \dfrac{\pi}{2}$, $c = \pi$

98. $f(x) = \dfrac{1}{x}$; $a = 1$, $b = 4$, $c = 6$

99. Find the critical points of the function

$$f(x) = \int_{-1}^x t^2(t-3)(t-4)\,dt,$$

and determine the intervals on which f is increasing or decreasing.

100. Determine the intervals on which the function

$$g(x) = \int_x^0 \frac{t}{t^2+1}\,dt \text{ is concave up or concave down.}$$

T 101–102. Functions defined by integrals *Consider the function g, which is given in terms of a definite integral with a variable upper limit.*

a. Graph the integrand.

b. Calculate $g'(x)$.

c. Graph g, showing all your work and reasoning.

101. $g(x) = \int_0^x \sin^2 t\,dt$

102. $g(x) = \int_0^x \sin(\pi t^2)\,dt$ (a Fresnel integral)

T 103–106. Areas of regions *Find the area of the region R bounded by the graph of f and the x-axis on the given interval. Graph f and show the region R.*

103. $f(x) = 2 - |x|$ on $[-2, 4]$

104. $f(x) = (1 - x^2)^{-1/2}$ on $\left[-\dfrac{1}{2}, \dfrac{\sqrt{3}}{2}\right]$

105. $f(x) = x^4 - 4$ on $[1, 4]$

106. $f(x) = x^2(x - 2)$ on $[-1, 3]$

107. Explain why or why not Determine whether the following statements are true and give an explanation or counterexample.

a. Suppose f is a positive decreasing function, for $x > 0$. Then the area function $A(x) = \int_0^x f(t)\,dt$ is an increasing function of x.

b. Suppose f is a negative increasing function, for $x > 0$. Then the area function $A(x) = \int_0^x f(t)\,dt$ is a decreasing function of x.

c. The functions $p(x) = \sin 3x$ and $q(x) = 4 \sin 3x$ are antiderivatives of the same function.

d. If $A(x) = 3x^2 - x - 3$ is an area function for f, then $B(x) = 3x^2 - x$ is also an area function for f.

Explorations and Challenges

108. Evaluate $\displaystyle\lim_{x \to 2} \frac{\displaystyle\int_2^x \sqrt{t^2 + t + 3}\,dt}{x^2 - 4}$.

109. Maximum net area What value of $b > -1$ maximizes the integral $\int_{-1}^b x^2(3 - x)\,dx$?

110. Maximum net area Graph the function $f(x) = 8 + 2x - x^2$ and determine the values of a and b that maximize the value of the integral $\int_a^b f(x)\,dx$.

111. Zero net area Consider the function $f(x) = x^2 - 4x$.

   a. Graph f on the interval $x \geq 0$.

   b. For what value of $b > 0$ is $\int_0^b f(x)\, dx = 0$?

   c. In general, for the function $f(x) = x^2 - ax$, where $a > 0$, for what value of $b > 0$ (as a function of a) is $\int_0^b f(x)\, dx = 0$?

112. Cubic zero net area Consider the graph of the cubic $y = x(x - a)(x - b)$, where $0 < a < b$. Verify that the graph bounds a region above the x-axis, for $0 < x < a$, and bounds a region below the x-axis, for $a < x < b$. What is the relationship between a and b if the areas of these two regions are equal?

113. An integral equation Use the Fundamental Theorem of Calculus, Part 1, to find the function f that satisfies the equation

$$\int_0^x f(t)\, dt = 2\cos x + 3x - 2.$$

Verify the result by substitution into the equation.

114. Max/min of area functions Suppose f is continuous on $[0, \infty)$ and $A(x)$ is the net area of the region bounded by the graph of f and the t-axis on $[0, x]$. Show that the local maxima and minima of A occur at the zeros of f. Verify this fact with the function $f(x) = x^2 - 10x$.

T 115. Asymptote of sine integral Use a calculator to approximate

$$\lim_{x \to \infty} S(x) = \lim_{x \to \infty} \int_0^x \frac{\sin t}{t}\, dt,$$

where S is the sine integral function (see Example 7). Explain your reasoning.

116. Sine integral Show that the sine integral $S(x) = \int_0^x \frac{\sin t}{t}\, dt$ satisfies the (differential) equation $xS'(x) + 2S''(x) + xS'''(x) = 0$.

117. Fresnel integral Show that the Fresnel integral $S(x) = \int_0^x \sin t^2\, dt$ satisfies the (differential) equation $(S'(x))^2 + \left(\dfrac{S''(x)}{2x}\right)^2 = 1$.

118. Continuity at the endpoints Assume f is continuous on $[a, b]$ and let A be the area function for f with left endpoint a. Let m^* and M^* be the absolute minimum and maximum values of f on $[a, b]$, respectively.

   a. Prove that $m^*(x - a) \leq A(x) \leq M^*(x - a)$, for all x in $[a, b]$. Use this result and the Squeeze Theorem to show that A is continuous from the right at $x = a$.

   b. Prove that $m^*(b - x) \leq A(b) - A(x) \leq M^*(b - x)$, for all x in $[a, b]$. Use this result to show that A is continuous from the left at $x = b$.

119. Discrete version of the Fundamental Theorem In this exercise, we work with a discrete problem and show why the relationship $\int_a^b f'(x)\, dx = f(b) - f(a)$ makes sense. Suppose we have a set of equally spaced grid points

$$\{a = x_0 < x_1 < x_2 < \cdots < x_{n-1} < x_n = b\},$$

where the distance between any two grid points is Δx. Suppose also that at each grid point x_k, a function value $f(x_k)$ is defined, for $k = 0, \ldots, n$.

   a. We now replace the integral with a sum and replace the derivative with a difference quotient. Explain why $\int_a^b f'(x)\, dx$ is analogous to $\displaystyle\sum_{k=1}^n \underbrace{\frac{f(x_k) - f(x_{k-1})}{\Delta x}}_{\approx f'(x_k)} \Delta x.$

   b. Simplify the sum in part (a) and show that it is equal to $f(b) - f(a)$.

   c. Explain the correspondence between the integral relationship and the summation relationship.

QUICK CHECK ANSWERS

1. $0, -35$ **2.** $A(6) = 44$; $A(10) = 120$ **3.** $\frac{2}{3} - \frac{1}{2} = \frac{1}{6}$
4. If f is differentiated, we get f'. Therefore, f is an antiderivative of f'. ◀

5.4 Working with Integrals

With the Fundamental Theorem of Calculus in hand, we may begin an investigation of integration and its applications. In this section, we discuss the role of symmetry in integrals, we use the slice-and-sum strategy to define the average value of a function, and we explore a theoretical result called the Mean Value Theorem for Integrals.

Integrating Even and Odd Functions

Symmetry appears throughout mathematics in many different forms, and its use often leads to insights and efficiencies. Here we use the symmetry of a function to simplify integral calculations.

     Section 1.1 introduced the symmetry of even and odd functions. An **even function** satisfies the property $f(-x) = f(x)$, which means that its graph is symmetric about the y-axis (**Figure 5.53a**). Examples of even functions are $f(x) = \cos x$ and $f(x) = x^n$, where n is an even integer. An **odd function** satisfies the property $f(-x) = -f(x)$, which means that its graph is symmetric about the origin (**Figure 5.53b**). Examples of odd functions are $f(x) = \sin x$ and $f(x) = x^n$, where n is an odd integer.

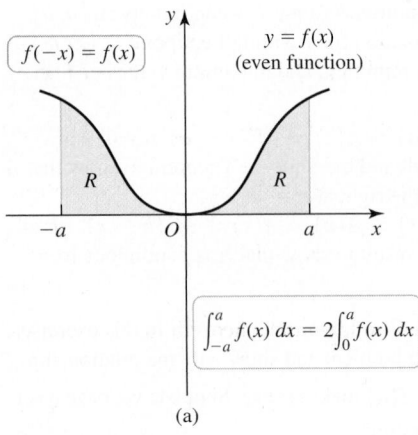

$$f(-x) = f(x)$$

$y = f(x)$ (even function)

$$\int_{-a}^{a} f(x)\, dx = 2\int_{0}^{a} f(x)\, dx$$

(a)

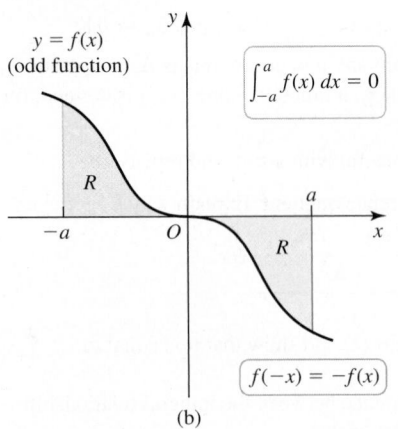

$y = f(x)$ (odd function)

$$\int_{-a}^{a} f(x)\, dx = 0$$

$$f(-x) = -f(x)$$

(b)

Figure 5.53

QUICK CHECK 1 If f and g are both even functions, is the product fg even or odd? Use the facts that $f(-x) = f(x)$ and $g(-x) = g(x)$. ◄

Special things happen when we integrate even and odd functions on intervals centered at the origin. First suppose f is an even function and consider $\int_{-a}^{a} f(x)\, dx$. From Figure 5.53a, we see that the integral of f on $[-a, 0]$ equals the integral of f on $[0, a]$. Therefore, the integral on $[-a, a]$ is twice the integral on $[0, a]$, or

$$\int_{-a}^{a} f(x)\, dx = 2\int_{0}^{a} f(x)\, dx.$$

On the other hand, suppose f is an odd function and consider $\int_{-a}^{a} f(x)\, dx$. As shown in Figure 5.53b, the integral on the interval $[-a, 0]$ is the negative of the integral on $[0, a]$. Therefore, the integral on $[-a, a]$ is zero, or

$$\int_{-a}^{a} f(x)\, dx = 0.$$

We summarize these results in the following theorem.

THEOREM 5.4 Integrals of Even and Odd Functions

Let a be a positive real number and let f be an integrable function on the interval $[-a, a]$.

• If f is even, $\int_{-a}^{a} f(x)\, dx = 2\int_{0}^{a} f(x)\, dx$.

• If f is odd, $\int_{-a}^{a} f(x)\, dx = 0$.

The following example shows how symmetry can simplify integration.

EXAMPLE 1 Integrating symmetric functions Evaluate the following integrals using symmetry arguments.

a. $\displaystyle\int_{-2}^{2} (x^4 - 3x^3)\, dx$ **b.** $\displaystyle\int_{-\pi/2}^{\pi/2} (\cos x - 4\sin^3 x)\, dx$

SOLUTION

a. Note that $x^4 - 3x^3$ is neither odd nor even, so Theorem 5.4 cannot be applied directly. However, we can split the integral and then use symmetry:

$$\int_{-2}^{2} (x^4 - 3x^3)\, dx = \int_{-2}^{2} x^4\, dx - 3\underbrace{\int_{-2}^{2} x^3\, dx}_{0} \quad \text{Properties 3 and 4 of Table 5.4}$$

$$= 2\int_{0}^{2} x^4\, dx - 0 \qquad x^4 \text{ is even; } x^3 \text{ is odd.}$$

$$= 2\left(\frac{x^5}{5}\right)\Big|_{0}^{2} \qquad \text{Fundamental Theorem}$$

$$= 2\left(\frac{32}{5}\right) = \frac{64}{5}. \qquad \text{Simplify.}$$

Notice how the odd-powered term of the integrand is eliminated by symmetry. Integration of the even-powered term is simplified because the lower limit is zero.

> There are a couple of ways to see that $\sin^3 x$ is an odd function. Its graph is symmetric about the origin, indicating that $\sin^3(-x) = -\sin^3 x$. Or by analogy, take an odd power of x and raise it to an odd power. For example, $(x^5)^3 = x^{15}$, which is odd. See Exercises 49–52 for direct proofs of symmetry in composite functions.

b. The $\cos x$ term is an even function, so it can be integrated on the interval $[0, \pi/2]$. What about $\sin^3 x$? It is an odd function raised to an odd power, which results in an odd function; its integral on $[-\pi/2, \pi/2]$ is zero. Therefore,

$$\int_{-\pi/2}^{\pi/2} (\cos x - 4\sin^3 x)\, dx = 2\int_{0}^{\pi/2} \cos x\, dx - 0 \quad \text{Symmetry}$$

$$= 2\sin x \Big|_{0}^{\pi/2} \qquad \text{Fundamental Theorem}$$

$$= 2(1 - 0) = 2. \qquad \text{Simplify.}$$

Related Exercises 15–16 ◄

Average Value of a Function

If five people weigh 155, 143, 180, 105, and 123 lb, their average (mean) weight is

$$\frac{155 + 143 + 180 + 105 + 123}{5} = 141.2 \text{ lb.}$$

This idea generalizes quite naturally to functions. Consider a function f that is continuous on $[a, b]$. Using a regular partition $x_0 = a, x_1, x_2, \ldots, x_n = b$ with $\Delta x = \dfrac{b - a}{n}$, we select a point x_k^* in each subinterval and compute $f(x_k^*)$, for $k = 1, \ldots, n$. The values of $f(x_k^*)$ may be viewed as a sampling of f on $[a, b]$. The average of these function values is

$$\frac{f(x_1^*) + f(x_2^*) + \cdots + f(x_n^*)}{n}.$$

Noting that $n = \dfrac{b - a}{\Delta x}$, we write the average of the n sample values as the Riemann sum

$$\frac{f(x_1^*) + f(x_2^*) + \cdots + f(x_n^*)}{(b - a)/\Delta x} = \frac{1}{b - a} \sum_{k=1}^{n} f(x_k^*) \Delta x.$$

Now suppose we increase n, taking more and more samples of f, while Δx decreases to zero. The limit of this sum is a definite integral that gives the average value $\bar{f}$ on $[a, b]$:

$$\bar{f} = \frac{1}{b - a} \lim_{n \to \infty} \sum_{k=1}^{n} f(x_k^*) \Delta x$$

$$= \frac{1}{b - a} \int_a^b f(x) \, dx.$$

This definition of the average value of a function is analogous to the definition of the average of a finite set of numbers.

> **DEFINITION Average Value of a Function**
>
> The average value of an integrable function f on the interval $[a, b]$ is
>
> $$\bar{f} = \frac{1}{b - a} \int_a^b f(x) \, dx.$$

The average value of a function f on an interval $[a, b]$ has a clear geometrical interpretation. Multiplying both sides of the definition of average value by $(b - a)$, we have

$$\underbrace{(b - a)\bar{f}}_{\substack{\text{net area of} \\ \text{rectangle}}} = \underbrace{\int_a^b f(x) \, dx.}_{\substack{\text{net area of region} \\ \text{bounded by curve}}}$$

We see that $|\bar{f}|$ is the height of a rectangle with base $[a, b]$, and that rectangle has the same net area as the region bounded by the graph of f on the interval $[a, b]$ (**Figure 5.54**). Note that $\bar{f}$ may be zero or negative.

QUICK CHECK 2 What is the average value of a constant function on an interval? What is the average value of an odd function on an interval $[-a, a]$? ◄

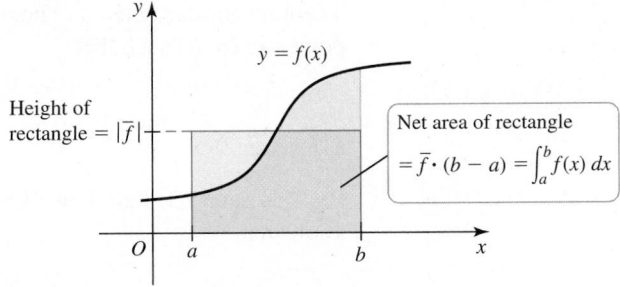

Figure 5.54

EXAMPLE 2   **Average elevation**   A hiking trail has an elevation given by

$$f(x) = 60x^3 - 650x^2 + 1200x + 4500,$$

where f is measured in feet above sea level and x represents horizontal distance along the trail in miles, with $0 \le x \le 5$. What is the average elevation of the trail?

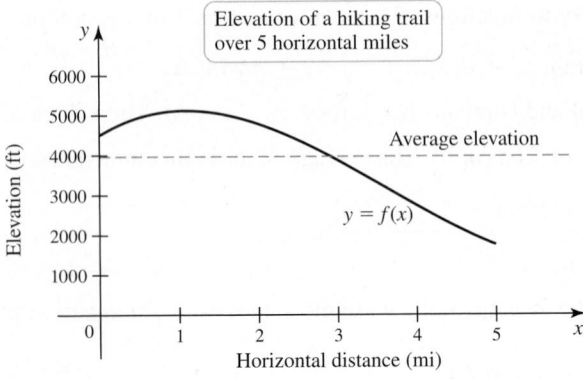

Elevation of a hiking trail over 5 horizontal miles

$y = f(x)$

Figure 5.55

SOLUTION   The trail ranges between elevations of about 2000 and 5000 ft (**Figure 5.55**). If we let the endpoints of the trail correspond to the horizontal distances $a = 0$ and $b = 5$, the average elevation of the trail in feet is

$$\bar{f} = \frac{1}{5}\int_0^5 (60x^3 - 650x^2 + 1200x + 4500)\, dx$$

$$= \frac{1}{5}\left(60\frac{x^4}{4} - 650\frac{x^3}{3} + 1200\frac{x^2}{2} + 4500x\right)\Big|_0^5 \quad \text{Fundamental Theorem}$$

$$= 3958\tfrac{1}{3}. \qquad\qquad\qquad\qquad \text{Simplify.}$$

The average elevation of the trail is slightly less than 3960 ft.

Related Exercises 26, 34 ◄

> Compare this statement to that of the Mean Value Theorem for Derivatives: There is at least one point c in (a, b) such that $f'(c)$ equals the average slope of f.

Mean Value Theorem for Integrals

The average value of a function brings us close to an important theoretical result. The Mean Value Theorem for Integrals says that if f is continuous on $[a, b]$, then there is at least one point c in the interval (a, b) such that $f(c)$ equals the average value of f on (a, b). In other words, the horizontal line $y = \bar{f}$ intersects the graph of f for some point c in (a, b) (**Figure 5.56**). If f were not continuous, such a point might not exist.

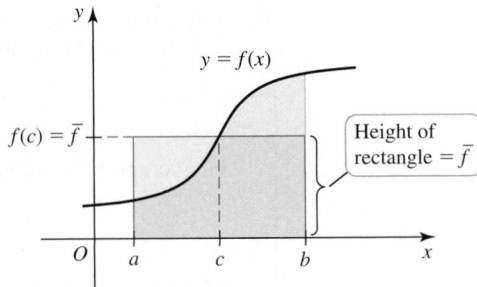

$y = f(x)$

$f(c) = \bar{f}$

Height of rectangle $= \bar{f}$

Figure 5.56

> Theorem 5.5 guarantees a point c in the open interval (a, b) at which f equals its average value. However, f may also equal its average value at an endpoint of that interval.

THEOREM 5.5   Mean Value Theorem for Integrals

Let f be continuous on the interval $[a, b]$. There exists a point c in (a, b) such that

$$f(c) = \bar{f} = \frac{1}{b - a}\int_a^b f(t)\, dt.$$

Proof: We begin by letting $F(x) = \int_a^x f(t)\, dt$ and noting that F is continuous on $[a, b]$ and differentiable on (a, b) (by Theorem 5.3, Part 1). We now apply the Mean Value Theorem for derivatives (Theorem 4.4) to F and conclude that there exists at least one point c in (a, b) such that

$$\underbrace{F'(c)}_{f(c)} = \frac{F(b) - F(a)}{b - a}.$$

By Theorem 5.3, Part 1, we know that $F'(c) = f(c)$, and by Theorem 5.3, Part 2, we know that

$$F(b) - F(a) = \int_a^b f(t)\, dt.$$

▶ A more general form of the Mean Value Theorem states that if f and g are continuous on $[a, b]$ with $g(x) \geq 0$ on $[a, b]$, then there exists a number c in (a, b) such that

$$\int_a^b f(x)g(x)\,dx = f(c)\int_a^b g(x)\,dx.$$

QUICK CHECK 3 Explain why $f(x) = 0$ for at least one point of (a, b) if f is continuous and $\int_a^b f(x)\,dx = 0$. ◀

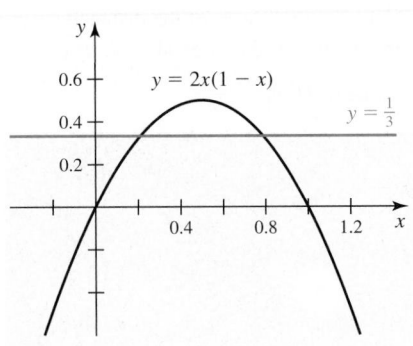

Figure 5.57

Combining these observations, we have

$$f(c) = \frac{1}{b - a}\int_a^b f(t)\,dt,$$

where c is a point in (a, b).   ◀

EXAMPLE 3   Average value equals function value Find the point(s) on the interval $(0, 1)$ at which $f(x) = 2x(1 - x)$ equals its average value on $[0, 1]$.

SOLUTION The average value of f on $[0, 1]$ is

$$\overline{f} = \frac{1}{1 - 0}\int_0^1 2x(1 - x)\,dx = \left(x^2 - \frac{2}{3}x^3\right)\Big|_0^1 = \frac{1}{3}.$$

We must find the points on $(0, 1)$ at which $f(x) = \frac{1}{3}$ (Figure 5.57). Using the quadratic formula, the two solutions of $f(x) = 2x(1 - x) = \frac{1}{3}$ are

$$\frac{1 - \sqrt{1/3}}{2} \approx 0.211 \quad \text{and} \quad \frac{1 + \sqrt{1/3}}{2} \approx 0.789.$$

These two points are located symmetrically on either side of $x = \frac{1}{2}$. The two solutions, 0.211 and 0.789, are the same for $f(x) = ax(1 - x)$ for any nonzero value of a (Exercise 53).

Related Exercises 41–42 ◀

SECTION 5.4 EXERCISES

Getting Started

1. If f is an odd function, why is $\int_{-a}^a f(x)\,dx = 0$?

2. If f is an even function, why is $\int_{-a}^a f(x)\,dx = 2\int_0^a f(x)\,dx$?

3. Suppose f is an even function and $\int_{-8}^8 f(x)\,dx = 18$.
    a. Evaluate $\int_0^8 f(x)\,dx$.   b. Evaluate $\int_{-8}^8 xf(x)\,dx$.

4. Suppose f is an odd function, $\int_0^4 f(x)\,dx = 3$, and $\int_0^8 f(x)\,dx = 9$.
    a. Evaluate $\int_{-4}^8 f(x)\,dx$.   b. Evaluate $\int_{-8}^4 f(x)\,dx$.

5. Use symmetry to explain why
    $$\int_{-4}^4 (5x^4 + 3x^3 + 2x^2 + x + 1)\,dx = 2\int_0^4 (5x^4 + 2x^2 + 1)\,dx.$$

6. Use symmetry to fill in the blanks:
    $$\int_{-\pi}^\pi (\sin x + \cos x)\,dx = \underline{\quad} \int_0^\pi \underline{\quad}\,dx.$$

7. Is x^{12} an even or odd function? Is $\sin x^2$ an even or odd function?

8. Explain how to find the average value of a function on an interval $[a, b]$ and why this definition is analogous to the definition of the average of a set of numbers.

9. Explain the statement that a continuous function on an interval $[a, b]$ equals its average value at some point on (a, b).

10. Sketch the function $y = x$ on the interval $[0, 2]$ and let R be the region bounded by $y = x$ and the x-axis on $[0, 2]$. Now sketch a rectangle in the first quadrant whose base is $[0, 2]$ and whose area equals the area of R.

Practice Exercises

11–24. Symmetry in integrals *Use symmetry to evaluate the following integrals.*

11. $\int_{-2}^2 x^9\,dx$

12. $\int_{-200}^{200} 2x^5\,dx$

13. $\int_{-2}^2 (3x^8 - 2)\,dx$

14. $\int_{-\pi/4}^{\pi/4} \cos x\,dx$

15. $\int_{-2}^2 (x^2 + x^3)\,dx$

16. $\int_{-\pi}^\pi t^2 \sin t\,dt$

17. $\int_{-2}^2 (x^9 - 3x^5 + 2x^2 - 10)\,dx$

18. $\int_{-\pi/2}^{\pi/2} 5 \sin\theta\,d\theta$

19. $\int_{-\pi/4}^{\pi/4} \sin^5 t\,dt$

20. $\int_{-1}^1 (1 - |x|)\,dx$

21. $\int_{-\pi/4}^{\pi/4} \sec^2 x\,dx$

22. $\int_{-\pi/4}^{\pi/4} \tan\theta\,d\theta$

23. $\int_{-2}^2 \frac{x^3 - 4x}{x^2 + 1}\,dx$

24. $\int_{-2}^2 (1 - |x|^3)\,dx$

25–32. Average values *Find the average value of the following functions on the given interval. Draw a graph of the function and indicate the average value.*

25. $f(x) = x^3$ on $[-1, 1]$

26. $f(x) = x^2 + 1$ on $[-2, 2]$

27. $f(x) = \dfrac{1}{x^2 + 1}$ on $[-1, 1]$

28. $f(x) = 1/x$ on $[1, e]$

29. $f(x) = \cos x$ on $\left[-\dfrac{\pi}{2}, \dfrac{\pi}{2}\right]$

30. $f(x) = x(1 - x)$ on $[0, 1]$

31. $f(x) = x^n$ on $[0, 1]$, for any positive integer n

32. $f(x) = x^{1/n}$ on $[0, 1]$, for any positive integer n

33. Average distance on a parabola What is the average distance between the parabola $y = 30x(20 - x)$ and the x-axis on the interval $[0, 20]$?

T 34. Average elevation The elevation of a path is given by $f(x) = x^3 - 5x^2 + 30$, where x measures horizontal distance. Draw a graph of the elevation function and find its average value, for $0 \le x \le 4$.

35. Average velocity The velocity in m/s of an object moving along a line over the time interval $[0, 6]$ is $v(t) = t^2 + 3t$. Find the average velocity of the object over this time interval.

36. Average velocity A rock is launched vertically upward from the ground with a speed of 64 ft/s. The height of the rock (in ft) above the ground after t seconds is given by the function $s(t) = -16t^2 + 64t$. Find its average velocity during its flight.

37. Average height of an arch The height of an arch above the ground is given by the function $y = 10 \sin x$, for $0 \le x \le \pi$. What is the average height of the arch above the ground?

38. Average height of a wave The surface of a water wave is described by $y = 5(1 + \cos x)$, for $-\pi \le x \le \pi$, where $y = 0$ corresponds to a trough of the wave (see figure). Find the average height of the wave above the trough on $[-\pi, \pi]$.

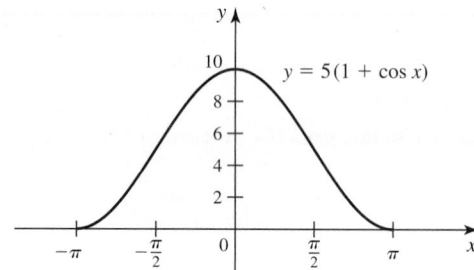

39–44. Mean Value Theorem for Integrals *Find or approximate all points at which the given function equals its average value on the given interval.*

39. $f(x) = 8 - 2x$ on $[0, 4]$ **40.** $f(x) = e^x$ on $[0, 2]$

41. $f(x) = 1 - \dfrac{x^2}{a^2}$ on $[0, a]$, where a is a positive real number

T 42. $f(x) = \dfrac{\pi}{4} \sin x$ on $[0, \pi]$ **43.** $f(x) = 1 - |x|$ on $[-1, 1]$

44. $f(x) = \dfrac{1}{x}$ on $[1, 4]$

45. Explain why or why not Determine whether the following statements are true and give an explanation or counterexample.

   a. If f is symmetric about the line $x = 2$, then $\int_0^4 f(x)\, dx = 2\int_0^2 f(x)\, dx$.

   b. If f has the property $f(a + x) = -f(a - x)$, for all x, where a is constant, then $\int_{a-2}^{a+2} f(x)\, dx = 0$.

   c. The average value of a linear function on an interval $[a, b]$ is the function value at the midpoint of $[a, b]$.

   d. Consider the function $f(x) = x(a - x)$ on the interval $[0, a]$, for $a > 0$. Its average value on $[0, a]$ is $\frac{1}{2}$ of its maximum value.

46. Planetary orbits The planets orbit the Sun in elliptical orbits with the Sun at one focus (see Section 12.4 for more on ellipses). The equation of an ellipse whose dimensions are $2a$ in the x-direction and $2b$ in the y-direction is $\dfrac{x^2}{a^2} + \dfrac{y^2}{b^2} = 1$.

   a. Let d^2 denote the square of the distance from a planet to the center of the ellipse at $(0, 0)$. Integrate over the interval $[-a, a]$ to show that the average value of d^2 is $\dfrac{a^2 + 2b^2}{3}$.

   b. Show that in the case of a circle ($a = b = R$), the average value in part (a) is R^2.

   c. Assuming $0 < b < a$, the coordinates of the Sun are $(\sqrt{a^2 - b^2}, 0)$. Let D^2 denote the square of the distance from the planet to the Sun. Integrate over the interval $[-a, a]$ to show that the average value of D^2 is $\dfrac{4a^2 - b^2}{3}$.

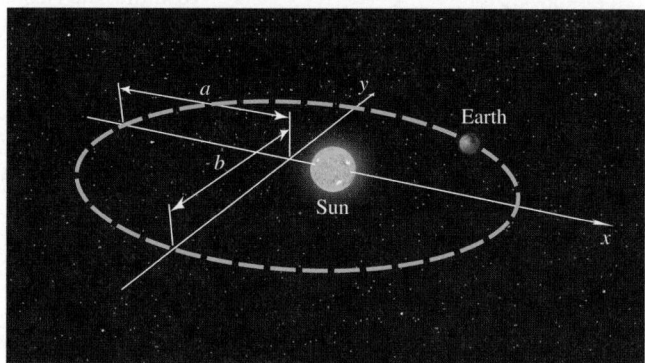

47. Gateway Arch The Gateway Arch in St. Louis is 630 ft high and has a 630-ft base. Its shape can be modeled by the parabola

$$y = 630\left(1 - \left(\frac{x}{315}\right)^2\right).$$

Find the average height of the arch above the ground.

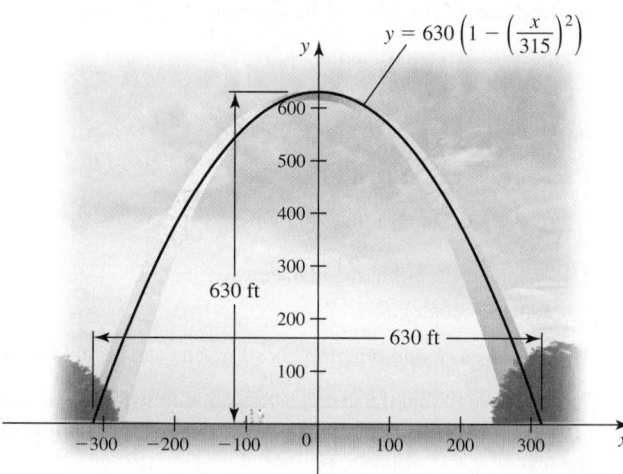

T 48. Comparing a sine and a quadratic function Consider the functions $f(x) = \sin x$ and $g(x) = \dfrac{4}{\pi^2} x(\pi - x)$.

   a. Carefully graph f and g on the same set of axes. Verify that both functions have a single local maximum on the interval $[0, \pi]$ and that they have the same maximum value on $[0, \pi]$.

   b. On the interval $[0, \pi]$, which is true: $f(x) \ge g(x)$, $g(x) \ge f(x)$, or neither?

   c. Compute and compare the average values of f and g on $[0, \pi]$.

49–52. Symmetry of composite functions *Prove that the integrand is either even or odd. Then give the value of the integral or show how it can be simplified. Assume f and g are even functions and p and q are odd functions.*

49. $\int_{-a}^{a} f(g(x))\,dx$

50. $\int_{-a}^{a} f(p(x))\,dx$

51. $\int_{-a}^{a} p(g(x))\,dx$

52. $\int_{-a}^{a} p(q(x))\,dx$

53. Average value with a parameter Consider the function $f(x) = ax(1 - x)$ on the interval $[0, 1]$, where a is a positive real number.

   a. Find the average value of f as a function of a.
   b. Find the points at which the value of f equals its average value, and prove that they are independent of a.

Explorations and Challenges

54. Alternative definitions of means Consider the function

$$f(t) = \frac{\int_{a}^{b} x^{t+1}\,dx}{\int_{a}^{b} x^{t}\,dx}.$$

Show that the following means can be defined in terms of f.

   a. Arithmetic mean: $f(0) = \dfrac{a + b}{2}$

   b. Geometric mean: $f\left(-\dfrac{3}{2}\right) = \sqrt{ab}$

   c. Harmonic mean: $f(-3) = \dfrac{2ab}{a + b}$

   d. Logarithmic mean: $f(-1) = \dfrac{b - a}{\ln b - \ln a}$

(*Source: Mathematics Magazine*, 78, 5, Dec 2005)

55. Problems of antiquity Several calculus problems were solved by Greek mathematicians long before the discovery of calculus. The following problems were solved by Archimedes using methods that predated calculus by 2000 years.

   a. Show that the area of a segment of a parabola is $4/3$ that of its inscribed triangle of greatest area. In other words, the area bounded by the parabola $y = a^2 - x^2$ and the x-axis is $4/3$ the area of the triangle with vertices $(\pm a, 0)$ and $(0, a^2)$. Assume $a > 0$ is unspecified.
   b. Show that the area bounded by the parabola $y = a^2 - x^2$ and the x-axis is $2/3$ the area of the rectangle with vertices $(\pm a, 0)$ and $(\pm a, a^2)$. Assume $a > 0$ is unspecified.

56. Average value of the derivative Suppose f' is a continuous function for all real numbers. Show that the average value of the derivative on an interval $[a, b]$ is $\overline{f}' = \dfrac{f(b) - f(a)}{b - a}$. Interpret this result in terms of secant lines.

57. Symmetry of powers Fill in the following table with either **even** or **odd**, and prove each result. Assume n is a nonnegative integer and f^n means the nth power of f.

	f is even	f is odd
n is even	f^n is _____	f^n is _____
n is odd	f^n is _____	f^n is _____

58. Bounds on an integral Suppose f is continuous on $[a, b]$ with $f''(x) > 0$ on the interval. It can be shown that

$$(b - a)f\left(\frac{a + b}{2}\right) \le \int_{a}^{b} f(x)\,dx \le (b - a)\frac{f(a) + f(b)}{2}.$$

   a. Assuming f is nonnegative on $[a, b]$, draw a figure to illustrate the geometric meaning of these inequalities. Discuss your conclusions.
   b. Divide these inequalities by $(b - a)$ and interpret the resulting inequalities in terms of the average value of f on $[a, b]$.

59. Generalizing the Mean Value Theorem for Integrals Suppose f and g are continuous on $[a, b]$ and let

$$h(x) = (x - b)\int_{a}^{x} f(t)\,dt + (x - a)\int_{x}^{b} g(t)\,dt.$$

   a. Use Rolle's Theorem to show that there is a number c in (a, b) such that

$$\int_{a}^{c} f(t)\,dt + \int_{c}^{b} g(t)\,dt = f(c)(b - c) + g(c)(c - a),$$

   which is a generalization of the Mean Value Theorem for Integrals.
   b. Show that there is a number c in (a, b) such that $\int_{a}^{c} f(t)\,dt = f(c)(b - c)$.
   c. Use a sketch to interpret part (b) geometrically.
   d. Use the result of part (a) to give an alternative proof of the Mean Value Theorem for Integrals.

(*Source: The College Mathematics Journal*, 33, 5, Nov 2002)

60. Evaluating a sine integral by Riemann sums Consider the integral $I = \int_{0}^{\pi/2} \sin x\,dx$.

   a. Write the left Riemann sum for I with n subintervals.
   b. Show that $\displaystyle\lim_{\theta \to 0} \theta\left(\frac{\cos\theta + \sin\theta - 1}{2(1 - \cos\theta)}\right) = 1$.

   c. It is a fact that $\displaystyle\sum_{k=0}^{n-1} \sin\left(\frac{\pi k}{2n}\right) = \frac{\cos\left(\dfrac{\pi}{2n}\right) + \sin\left(\dfrac{\pi}{2n}\right) - 1}{2\left(1 - \cos\left(\dfrac{\pi}{2n}\right)\right)}.$

   Use this fact and part (b) to evaluate I by taking the limit of the Riemann sum as $n \to \infty$.

QUICK CHECK ANSWERS

1. $f(-x)g(-x) = f(x)g(x)$; therefore, fg is even.
2. The average value is the constant; the average value is 0.
3. The average value is zero on the interval; by the Mean Value Theorem for Integrals, $f(x) = 0$ at some point on the interval. ◄

5.5 Substitution Rule

Given just about any differentiable function, with enough know-how and persistence, you can compute its derivative. But the same cannot be said of antiderivatives. Many functions, even relatively simple ones, do not have antiderivatives that can be expressed in terms of familiar functions. Examples are $\sin x^2$, $(\sin x)/x$, and x^x. The immediate goal of this section is to enlarge the family of functions for which we can find antiderivatives. This campaign resumes in Chapter 8, where additional integration methods are developed.

Indefinite Integrals

One way to find new antiderivative rules is to start with familiar derivative rules and work backward. When applied to the Chain Rule, this strategy leads to the Substitution Rule. For example, consider the indefinite integral $\int \cos 2x \, dx$. The closest familiar integral related to this problem is

> ➤ We assume C is an arbitrary constant without stating so each time it appears.

$$\int \cos x \, dx = \sin x + C,$$

which is true because

$$\frac{d}{dx}(\sin x + C) = \cos x.$$

Therefore, we might *incorrectly* conclude that the indefinite integral of $\cos 2x$ is $\sin 2x + C$. However, by the Chain Rule,

$$\frac{d}{dx}(\sin 2x + C) = 2 \cos 2x \neq \cos 2x.$$

Note that $\sin 2x$ fails to be an antiderivative of $\cos 2x$ by a multiplicative factor of 2. A small adjustment corrects this problem. Let's try $\frac{1}{2} \sin 2x$:

$$\frac{d}{dx}\left(\frac{1}{2} \sin 2x\right) = \frac{1}{2} \cdot 2 \cos 2x = \cos 2x.$$

It works! So we have

$$\int \cos 2x \, dx = \frac{1}{2} \sin 2x + C.$$

The trial-and-error approach of the previous example is impractical for complicated integrals. To develop a systematic method, consider a composite function $F(g(x))$, where F is an antiderivative of f; that is, $F' = f$. Using the Chain Rule to differentiate the composite function $F(g(x))$, we find that

$$\frac{d}{dx}(F(g(x))) = \underbrace{F'(g(x))}_{f(g(x))}g'(x) = f(g(x))g'(x).$$

This equation says that $F(g(x))$ is an antiderivative of $f(g(x))g'(x)$, which is written

$$\int f(g(x))g'(x) \, dx = F(g(x)) + C, \tag{1}$$

where F is any antiderivative of f.

> ➤ You can call the new variable anything you want because it is just another variable of integration. Typically, u is the standard choice for the new variable.

Why is this approach called the *Substitution Rule* (or *Change of Variables Rule*)? In the composite function $f(g(x))$ in equation (1), we identify the inner function as $u = g(x)$, which implies that $du = g'(x) \, dx$. Making this identification, the integral in equation (1) is written

$$\int \underbrace{f(g(x))}_{f(u)}\underbrace{g'(x) \, dx}_{du} = \int f(u) \, du = F(u) + C.$$

We see that the integral $\int f(g(x))g'(x) \, dx$ with respect to x is replaced with a new integral $\int f(u) \, du$ with respect to the new variable u. In other words, we have substituted the new variable u for the old variable x. Of course, if the new integral with respect to u is no easier to find than the original integral, then the change of variables has not helped. The Substitution Rule requires plenty of practice until certain patterns become familiar.

THEOREM 5.6   Substitution Rule for Indefinite Integrals
Let $u = g(x)$, where g is differentiable on an interval, and let f be continuous on the corresponding range of g. On that interval,

$$\int f(g(x))g'(x)\, dx = \int f(u)\, du.$$

In practice, Theorem 5.6 is applied using the following procedure.

PROCEDURE   Substitution Rule (Change of Variables)
1. Given an indefinite integral involving a composite function $f(g(x))$, identify an inner function $u = g(x)$ such that a constant multiple of $g'(x)$ appears in the integrand.
2. Substitute $u = g(x)$ and $du = g'(x)\, dx$ in the integral.
3. Evaluate the new indefinite integral with respect to u.
4. Write the result in terms of x using $u = g(x)$.

Disclaimer: Not all integrals yield to the Substitution Rule.

EXAMPLE 1   Perfect substitutions Use the Substitution Rule to find the following indefinite integrals. Check your work by differentiating.

a. $\displaystyle\int 2(2x + 1)^3\, dx$ **b.** $\displaystyle\int 10e^{10x}\, dx$

SOLUTION

a. We identify $u = 2x + 1$ as the inner function of the composite function $(2x + 1)^3$. Therefore, we choose the new variable $u = 2x + 1$, which implies that $\dfrac{du}{dx} = 2$, or $du = 2\, dx$. Notice that $du = 2\, dx$ appears as a factor in the integrand. The change of variables looks like this:

$$\int \underbrace{(2x + 1)^3}_{u^3} \cdot \underbrace{2\, dx}_{du} = \int u^3\, du \qquad \text{Substitute } u = 2x + 1,\, du = 2\, dx.$$

$$= \frac{u^4}{4} + C \qquad \text{Antiderivative}$$

$$= \frac{(2x + 1)^4}{4} + C. \qquad \text{Replace } u \text{ with } 2x + 1.$$

> Use the Chain Rule to check that
> $$\frac{d}{dx}\left(\frac{(2x + 1)^4}{4} + C\right) = 2(2x + 1)^3.$$

Notice that the final step uses $u = 2x + 1$ to return to the original variable.

b. The composite function e^{10x} has the inner function $u = 10x$, which implies that $du = 10\, dx$. The change of variables appears as

$$\int \underbrace{e^{10x}}_{e^u}\, \underbrace{10\, dx}_{du} = \int e^u\, du \qquad \text{Substitute } u = 10x,\, du = 10\, dx.$$

$$= e^u + C \qquad \text{Antiderivative}$$

$$= e^{10x} + C. \qquad \text{Replace } u \text{ with } 10x.$$

To check the result, we compute $\dfrac{d}{dx}(e^{10x} + C) = e^{10x} \cdot 10 = 10e^{10x}$.

QUICK CHECK 1 Find a new variable u so that $\int 4x^3(x^4 + 5)^{10}\, dx = \int u^{10}\, du$. ◄

Related Exercises 17, 20–21 ◄

Most substitutions are not perfect. The remaining examples show more typical situations that require introducing a constant factor.

EXAMPLE 2   Introducing a constant Find the following indefinite integrals.

a. $\int x^4(x^5 + 6)^9 \, dx$ **b.** $\int \cos^3 x \sin x \, dx$

SOLUTION

a. The inner function of the composite function $(x^5 + 6)^9$ is $x^5 + 6$, and its derivative $5x^4$ also appears in the integrand (up to a multiplicative factor). Therefore, we use the substitution $u = x^5 + 6$, which implies that $du = 5x^4 \, dx$, or $x^4 \, dx = \dfrac{1}{5} du$. By the Substitution Rule,

$$\int \underbrace{(x^5 + 6)^9}_{u^9} \underbrace{x^4 \, dx}_{\frac{1}{5} du} = \int u^9 \cdot \frac{1}{5} \, du \qquad \begin{array}{l} \text{Substitute } u = x^5 + 6, \\ du = 5x^4 \, dx \Rightarrow x^4 \, dx = \dfrac{1}{5} du. \end{array}$$

$$= \frac{1}{5} \int u^9 \, du \qquad\qquad \int cf(x) \, dx = c \int f(x) \, dx$$

$$= \frac{1}{5} \cdot \frac{u^{10}}{10} + C \qquad\qquad \text{Antiderivative}$$

$$= \frac{1}{50}(x^5 + 6)^{10} + C. \quad \text{Replace } u \text{ with } x^5 + 6.$$

QUICK CHECK 2 In Example 2a, explain why the same substitution would not work as well for the integral $\int x^3(x^5 + 6)^9 dx$. ◄

b. The integrand can be written as $(\cos x)^3 \sin x$. The inner function in the composition $(\cos x)^3$ is $\cos x$, which suggests the substitution $u = \cos x$. Note that $du = -\sin x \, dx$ or $\sin x \, dx = -du$. The change of variables appears as

$$\int \underbrace{\cos^3 x}_{u^3} \underbrace{\sin x \, dx}_{-du} = -\int u^3 \, du \qquad \text{Substitute } u = \cos x, du = -\sin x \, dx.$$

$$= -\frac{u^4}{4} + C \qquad\qquad \text{Antiderivative}$$

$$= -\frac{\cos^4 x}{4} + C. \qquad \text{Replace } u \text{ with } \cos x.$$

Related Exercises 23–24 ◄

Sometimes the choice for a *u*-substitution is not so obvious *or* more than one *u*-substitution works. The following example illustrates both of these points.

EXAMPLE 3   Variations on the substitution method Find $\displaystyle\int \frac{x}{\sqrt{x + 1}} \, dx$.

SOLUTION

Substitution 1 The composite function $\sqrt{x + 1}$ suggests the new variable $u = x + 1$. You might doubt whether this choice will work because $du = dx$, which leaves the x in the numerator of the integrand unaccounted for. But let's proceed. Letting $u = x + 1$, we have $x = u - 1$, $du = dx$, and

$$\int \frac{x}{\sqrt{x + 1}} \, dx = \int \frac{u - 1}{\sqrt{u}} \, du \qquad \text{Substitute } u = x + 1, du = dx.$$

$$= \int \left(\sqrt{u} - \frac{1}{\sqrt{u}} \right) du \quad \text{Rewrite integrand.}$$

$$= \int (u^{1/2} - u^{-1/2}) \, du. \quad \text{Fractional powers}$$

We integrate each term individually and then return to the original variable x:

$$\int (u^{1/2} - u^{-1/2})\, du = \frac{2}{3}u^{3/2} - 2u^{1/2} + C \qquad \text{Antiderivatives}$$

$$= \frac{2}{3}(x + 1)^{3/2} - 2(x + 1)^{1/2} + C \quad \text{Replace } u \text{ with } x + 1.$$

$$= \frac{2}{3}(x + 1)^{1/2}(x - 2) + C. \qquad \begin{array}{l}\text{Factor out } (x + 1)^{1/2} \\ \text{and simplify.}\end{array}$$

> In Substitution 2, you could also use the fact that
>
> $$u'(x) = \frac{1}{2\sqrt{x + 1}},$$
>
> which implies
>
> $$du = \frac{1}{2\sqrt{x + 1}}\, dx.$$

Substitution 2 Another possible substitution is $u = \sqrt{x + 1}$. Now $u^2 = x + 1$, $x = u^2 - 1$, and $dx = 2u\, du$. Making these substitutions leads to

$$\int \frac{x}{\sqrt{x + 1}}\, dx = \int \frac{u^2 - 1}{u} 2u\, du \qquad \text{Substitute } u = \sqrt{x + 1},\, x = u^2 - 1.$$

$$= 2\int (u^2 - 1)\, du \qquad \text{Simplify the integrand.}$$

$$= 2\left(\frac{u^3}{3} - u\right) + C \qquad \text{Antiderivatives}$$

$$= \frac{2}{3}(x + 1)^{3/2} - 2(x + 1)^{1/2} + C \quad \text{Replace } u \text{ with } \sqrt{x + 1}.$$

$$= \frac{2}{3}(x + 1)^{1/2}(x - 2) + C. \qquad \text{Factor out } (x + 1)^{1/2} \text{ and simplify.}$$

Observe that the same indefinite integral is found using either substitution.

Related Exercises 78–79 ◄

General Formulas for Indefinite Integrals

Integrals of the form $\int f(ax)\, dx$ occur frequently in the remainder of this text, so our aim here is to generalize the integral formulas first introduced in Section 4.9. We encountered an integral of the form $\int f(ax)\, dx$ in the opening of this section, where we used trial and error to discover that $\int \cos 2x\, dx = \frac{1}{2}\sin 2x + C$. Notice that 2 could be replaced by any nonzero constant a to produce the more general result

$$\int \cos ax\, dx = \frac{1}{a}\sin ax + C.$$

Let's verify this result without resorting to trial and error by using substitution. Letting $u = ax$, we have $du = a\, dx$, or $dx = \frac{1}{a}\, du$. By the Substitution Rule,

$$\int \cos ax\, dx = \int \cos u \cdot \frac{1}{a}\, du \quad \text{Substitute } u = ax,\, dx = \frac{1}{a}\, du.$$

$$= \frac{1}{a}\int \cos u\, du \qquad \int cf(x)\, dx = c\int f(x)\, dx$$

$$= \frac{1}{a}\sin u + C \qquad \text{Antiderivative}$$

$$= \frac{1}{a}\sin ax + C. \qquad \text{Replace } u \text{ with } ax.$$

QUICK CHECK 3 Evaluate $\int \cos 6x\, dx$ without using the substitution method. ◄

Now that we have established a general result for $\int \cos ax\, dx$, we can use it to evaluate integrals such as $\int \cos 6x\, dx$ without resorting to the substitution method. Table 5.6 lists additional general formulas for standard integrals; we assume $a \neq 0$ is a real number in results (1)–(7) and (9), and $a > 0$ in results (10) and (11). The derivations of results (2)–(6) in Table 5.6 are similar to the derivation of result (1) just given. In Example 4, we derive several additional results. Note that all these integration formulas can be verified by differentiation.

Table 5.6 General Integration Formulas

1. $\int \cos ax \, dx = \dfrac{1}{a} \sin ax + C$

2. $\int \sin ax \, dx = -\dfrac{1}{a} \cos ax + C$

3. $\int \sec^2 ax \, dx = \dfrac{1}{a} \tan ax + C$

4. $\int \csc^2 ax \, dx = -\dfrac{1}{a} \cot ax + C$

5. $\int \sec ax \tan ax \, dx = \dfrac{1}{a} \sec ax + C$

6. $\int \csc ax \cot ax = -\dfrac{1}{a} \csc ax + C$

7. $\int e^{ax} \, dx = \dfrac{1}{a} e^{ax} + C$

8. $\int b^x \, dx = \dfrac{1}{\ln b} b^x + C, b > 0, b \neq 1$

9. $\int \dfrac{dx}{a^2 + x^2} = \dfrac{1}{a} \tan^{-1} \dfrac{x}{a} + C$

10. $\int \dfrac{dx}{\sqrt{a^2 - x^2}} = \sin^{-1} \dfrac{x}{a} + C, a > 0$

11. $\int \dfrac{dx}{x\sqrt{x^2 - a^2}} = \dfrac{1}{a} \sec^{-1} \left| \dfrac{x}{a} \right| + C, a > 0$

EXAMPLE 4 Deriving integration formulas Find formulas for the following general indefinite integrals to verify results (7), (8), and (10) in Table 5.6.

a. $\int e^{ax} \, dx$ **b.** $\int b^x \, dx, b > 0, b \neq 1$ **c.** $\int \dfrac{dx}{\sqrt{a^2 - x^2}}$, for $a > 0$

SOLUTION

a. Letting $u = ax$, we have $du = a \, dx$, or $dx = \dfrac{1}{a} du$. Substituting these expressions into the integral, we have

$$\int e^{ax} \, dx = \underbrace{\int e^u \frac{1}{a} \, du}_{u \,=\, ax;\ dx \,=\, 1/a \, du} = \frac{1}{a} \int e^u \, du = \frac{1}{a} e^u + C = \frac{1}{a} e^{ax} + C.$$

b. At first glance, $\int b^x \, dx$ does not seem to fit into Table 5.6. However, expressing b^x with base e leads to

$$\int b^x \, dx = \int e^{x \ln b} \, dx,$$

which matches result (7) derived in part (a), with $a = \ln b$. It follows that

$$\int b^x \, dx = \int e^{x \ln b} dx = \frac{1}{\ln b} e^{x \ln b} + C = \frac{1}{\ln b} b^x + C.$$

c. From Section 4.9, recall that $\int \dfrac{dx}{\sqrt{1 - x^2}} = \sin^{-1} x + C$. To handle the more general form, we factor a^2 in the denominator to prepare the integrand for a standard substitution:

$$\int \frac{dx}{\sqrt{a^2 - x^2}} = \int \frac{dx}{\sqrt{a^2(1 - x^2/a^2)}} \quad \text{Factor.}$$

$$= \frac{1}{a} \int \frac{dx}{\sqrt{1 - (x/a)^2}}. \quad \int cf(x) \, dx = c \int f(x) \, dx; \ \sqrt{a^2} = a$$
$$\text{when } a > 0$$

▶ In keeping with the pattern established in results 1–7 of Table 5.6, where we investigate functions of the form $y = f(ax)$, we might look for antiderivatives of $\dfrac{1}{\sqrt{1 - (ax)^2}}$ in Example 4c, rather than antiderivatives of $\dfrac{1}{\sqrt{a^2 - x^2}}$. We work with the latter form because it produces a more common form of the integral.

Now let $u = x/a$, which implies that $du = 1/a \, dx$, or $dx = a \, du$. Making these substitutions, we have

$$\frac{1}{a}\int \frac{dx}{\sqrt{1 - (x/a)^2}} = \frac{1}{a}\int \frac{a}{\sqrt{1 - u^2}}\, du \quad u = x/a,\ dx = a\,du$$

$$= \sin^{-1} u + C \qquad \text{Cancel } a \text{ and integrate.}$$

$$= \sin^{-1}\frac{x}{a} + C. \qquad u = x/a$$

Notice that the constant a cancels in the second step, resulting in an antiderivative without a constant multiple of $1/a$; it is the only formula in Table 5.6 with this form.

Related Exercises 15–16 ◀

QUICK CHECK 4 Evaluate $\displaystyle\int \frac{4}{\sqrt{4 - x^2}}\, dx.$ ◀

Definite Integrals

The Substitution Rule is also used for definite integrals; in fact, there are two ways to proceed.

- You may use the Substitution Rule to find an antiderivative F and then use the Fundamental Theorem to evaluate $F(b) - F(a)$.

- Alternatively, once you have changed variables from x to u, you also may change the limits of integration and complete the integration with respect to u. Specifically, if $u = g(x)$, the lower limit $x = a$ is replaced with $u = g(a)$ and the upper limit $x = b$ is replaced with $u = g(b)$.

The second option tends to be more efficient, and we use it whenever possible. This approach is summarized in the following theorem, which is then applied to several definite integrals.

THEOREM 5.7   Substitution Rule for Definite Integrals
Let $u = g(x)$, where g' is continuous on $[a, b]$, and let f be continuous on the range of g. Then

$$\int_a^b f(g(x))g'(x)\, dx = \int_{g(a)}^{g(b)} f(u)\, du. \tag{2}$$

Proof: We apply Part 2 of the Fundamental Theorem of Calculus to both sides of equation (2). Let F be an antiderivative of f. Then, by equation (1) (p. 388), we know that $F(g(x))$ is an antiderivative of $f(g(x))g'(x)$, which implies that

$$\int_a^b f(g(x))g'(x)\, dx = F(g(x))\Big|_a^b = F(g(b)) - F(g(a)). \quad \text{Fundamental Theorem, Part 2}$$

Applying the Fundamental Theorem to the right side of equation (2) leads to the same result:

$$\int_{g(a)}^{g(b)} f(u)\, du = F(u)\Big|_{g(a)}^{g(b)} = F(g(b)) - F(g(a)). \qquad ◀$$

EXAMPLE 5   Definite integrals   Evaluate the following integrals.

a. $\displaystyle\int_0^2 \frac{dx}{(x + 3)^3}$
b. $\displaystyle\int_2^3 \frac{x^2}{x^3 - 7}\, dx$
c. $\displaystyle\int_0^{\pi/2} \sin^4 x \cos x \, dx$

SOLUTION

▶ When the integrand has the form $f(ax + b)$, the substitution $u = ax + b$ is often effective.

a. Let the new variable be $u = x + 3$ and then $du = dx$. Because we have changed the variable of integration from x to u, the limits of integration must also be expressed in terms of u. In this case,

$$x = 0 \text{ implies } u = 0 + 3 = 3, \quad \text{Lower limit}$$

$$x = 2 \text{ implies } u = 2 + 3 = 5. \quad \text{Upper limit}$$

The entire integration is carried out as follows:

$$\int_0^2 \frac{dx}{(x+3)^3} = \int_3^5 u^{-3}\, du \qquad\qquad \text{Substitute } u = x+3,\, du = dx.$$

$$= -\frac{u^{-2}}{2}\bigg|_3^5 \qquad\qquad \text{Fundamental Theorem}$$

$$= -\frac{1}{2}\left(5^{-2} - 3^{-2}\right) = \frac{8}{225}. \quad \text{Simplify.}$$

b. Notice that a multiple of the derivative of the denominator appears in the numerator; therefore, we let $u = x^3 - 7$, which implies that $du = 3x^2\, dx$, or $x^2\, dx = \frac{1}{3}\, du$. Changing limits of integration,

$$x = 2 \text{ implies } u = 2^3 - 7 = 1, \quad \text{Lower limit}$$
$$x = 3 \text{ implies } u = 3^3 - 7 = 20. \quad \text{Upper limit}$$

Changing variables, we have

$$\int_2^3 \frac{x^2}{x^3 - 7}\, dx = \frac{1}{3}\int_1^{20} \frac{du}{u} \qquad \text{Substitute } u = x^3 - 7,\, du = 3x^2\, dx.$$

$$= \frac{1}{3}\ln|u|\,\bigg|_1^{20} \qquad\qquad \text{Fundamental Theorem}$$

$$= \frac{1}{3}\left(\ln 20 - \ln 1\right) \quad \text{Simplify.}$$

$$= \frac{1}{3}\ln 20 \approx 0.999. \quad \text{ln } 1 = 0$$

c. Let $u = \sin x$, which implies that $du = \cos x\, dx$. The lower limit of integration becomes $u = 0$ and the upper limit becomes $u = 1$. Changing variables, we have

$$\int_0^{\pi/2} \sin^4 x \cos x\, dx = \int_0^1 u^4\, du \qquad u = \sin x,\, du = \cos x\, dx$$

$$= \frac{u^5}{5}\bigg|_0^1 = \frac{1}{5}. \quad \text{Fundamental Theorem}$$

Related Exercises 48, 51, 64 ◄

The Substitution Rule enables us to find two standard integrals that appear frequently in practice, $\int \sin^2 x\, dx$ and $\int \cos^2 x\, dx$. These integrals are handled using the identities

$$\sin^2 x = \frac{1 - \cos 2x}{2} \quad \text{and} \quad \cos^2 x = \frac{1 + \cos 2x}{2}.$$

EXAMPLE 6   Integral of $\cos^2 \theta$ Evaluate $\int_0^{\pi/2} \cos^2 \theta\, d\theta$.

SOLUTION Working with the indefinite integral first, we use the identity for $\cos^2 \theta$:

$$\int \cos^2 \theta\, d\theta = \int \frac{1 + \cos 2\theta}{2}\, d\theta = \frac{1}{2}\int d\theta + \frac{1}{2}\int \cos 2\theta\, d\theta.$$

> ► See Exercise 112 for a generalization of Example 6. Trigonometric integrals involving powers of $\sin x$ and $\cos x$ are explored in greater detail in Section 8.3.

Result (1) of Table 5.6 is used for the second integral, and we have

$$\int \cos^2 \theta\, d\theta = \frac{1}{2}\int d\theta + \frac{1}{2}\underbrace{\int \cos 2\theta\, d\theta}_{\frac{1}{2}\sin 2\theta\, +\, C}$$

$$= \frac{\theta}{2} + \frac{1}{4}\sin 2\theta + C. \qquad \text{Evaluate integrals; Table 5.6}$$

Using the Fundamental Theorem of Calculus, the value of the definite integral is

$$\int_0^{\pi/2} \cos^2 \theta \, d\theta = \left(\frac{\theta}{2} + \frac{1}{4} \sin 2\theta \right) \Big|_0^{\pi/2}$$

$$= \left(\frac{\pi}{4} + \frac{1}{4} \sin \pi \right) - \left(0 + \frac{1}{4} \sin 0 \right) = \frac{\pi}{4}.$$

Related Exercises 87, 91 ◄

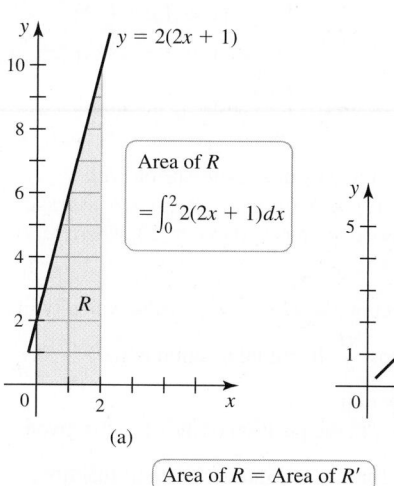

(a)

Area of R = Area of R'

Figure 5.58

Geometry of Substitution

The Substitution Rule has a geometric interpretation. To keep matters simple, consider the integral $\int_0^2 2(2x + 1) \, dx$. The graph of the integrand $y = 2(2x + 1)$ on the interval $[0, 2]$ is shown in **Figure 5.58a**, along with the region R whose area is given by the integral. The change of variables $u = 2x + 1$, $du = 2 \, dx$, $u(0) = 1$, and $u(2) = 5$ leads to the new integral

$$\int_0^2 2(2x + 1) \, dx = \int_1^5 u \, du.$$

Figure 5.58b shows the graph of the new integrand $y = u$ on the interval $[1, 5]$ and the region R' whose area is given by the new integral. You can check that the areas of R and R' are equal. An analogous interpretation may be given to more complicated integrands and substitutions.

QUICK CHECK 5 Changes of variables occur frequently in mathematics. For example, suppose you want to solve the equation $x^4 - 13x^2 + 36 = 0$. If you use the substitution $u = x^2$, what is the new equation that must be solved for u? What are the roots of the original equation? ◄

SECTION 5.5 EXERCISES

Getting Started

1. On which derivative rule is the Substitution Rule based?

2. Why is the Substitution Rule referred to as a change of variables?

3. The composite function $f(g(x))$ consists of an inner function g and an outer function f. If an integrand includes $f(g(x))$, which function is often a likely choice for a new variable u?

4. Find a suitable substitution for evaluating $\int \tan x \sec^2 x \, dx$ and explain your choice.

5. When using a change of variables $u = g(x)$ to evaluate the definite integral $\int_a^b f(g(x))g'(x) \, dx$, how are the limits of integration transformed?

6. If the change of variables $u = x^2 - 4$ is used to evaluate the definite integral $\int_2^4 f(x) \, dx$, what are the new limits of integration?

7–10. *Use the given substitution to evaluate the following indefinite integrals. Check your answer by differentiating.*

7. $\int 2x(x^2 + 1)^4 dx$, $u = x^2 + 1$

8. $\int 8x \cos(4x^2 + 3) \, dx$, $u = 4x^2 + 3$

9. $\int \sin^3 x \cos x \, dx$, $u = \sin x$

10. $\int (6x + 1)\sqrt{3x^2 + x} \, dx$, $u = 3x^2 + x$

11–14. *Use a substitution of the form $u = ax + b$ to evaluate the following indefinite integrals.*

11. $\int (x + 1)^{12} dx$

12. $\int e^{3x+1} dx$

13. $\int \sqrt{2x + 1} \, dx$

14. $\int \cos(2x + 5) \, dx$

15. Use Table 5.6 to evaluate the following indefinite integrals.

  a. $\int e^{10x} dx$

  b. $\int \sec 5x \tan 5x \, dx$

  c. $\int \sin 7x \, dx$

  d. $\int \cos \frac{x}{7} \, dx$

  e. $\int \frac{dx}{81 + 9x^2}$ (*Hint:* Factor a 9 out of the denominator first.)

  f. $\int \frac{dx}{\sqrt{36 - x^2}}$

16. Use Table 5.6 to evaluate the following definite integrals.

  a. $\int_0^1 10^x dx$

  b. $\int_0^{\pi/40} \cos 20x \, dx$

  c. $\int_{3\sqrt{2}}^6 \frac{dx}{x\sqrt{x^2 - 9}}$

  d. $\int_0^{\pi/16} \sec^2 4x \, dx$

Practice Exercises

17–44. Indefinite integrals *Use a change of variables or Table 5.6 to evaluate the following indefinite integrals. Check your work by differentiating.*

17. $\int 2x(x^2 - 1)^{99} dx$

18. $\int xe^{x^2} dx$

19. $\displaystyle\int \frac{2x^2}{\sqrt{1 - 4x^3}}\, dx$

20. $\displaystyle\int \frac{(\sqrt{x} + 1)^4}{2\sqrt{x}}\, dx$

21. $\displaystyle\int (x^2 + x)^{10}(2x + 1)\, dx$

22. $\displaystyle\int \frac{1}{10x - 3}\, dx$

23. $\displaystyle\int x^3(x^4 + 16)^6\, dx$

24. $\displaystyle\int \sin^{10}\theta \cos\theta\, d\theta$

25. $\displaystyle\int \frac{dx}{\sqrt{36 - 4x^2}}$

26. $\displaystyle\int \frac{dx}{\sqrt{1 - 9x^2}}$

27. $\displaystyle\int 6x^2 4^{x^3}\, dx$

28. $\displaystyle\int x^9 \sin x^{10}\, dx$

29. $\displaystyle\int (x^6 - 3x^2)^4(x^5 - x)\, dx$

30. $\displaystyle\int \frac{dx}{1 + 4x^2}$

31. $\displaystyle\int \frac{3}{\sqrt{1 - 25x^2}}\, dx$

32. $\displaystyle\int \frac{2}{x\sqrt{4x^2 - 1}}\, dx,\ x > \frac{1}{2}$

33. $\displaystyle\int \frac{e^w}{36 + e^{2w}}\, dw$

34. $\displaystyle\int \frac{8x + 6}{2x^2 + 3x}\, dx$

35. $\displaystyle\int x \csc x^2 \cot x^2\, dx$

36. $\displaystyle\int \sec 4w \tan 4w\, dw$

37. $\displaystyle\int \sec^2(10x + 7)\, dx$

38. $\displaystyle\int \frac{\tan^{-1} w}{w^2 + 1}\, dw$

39. $\displaystyle\int 10^{4t+1}\, dt$

40. $\displaystyle\int (\sin^5 x + 3\sin^3 x - \sin x)\cos x\, dx$

41. $\displaystyle\int \frac{\csc^2 x}{\cot^3 x}\, dx$

42. $\displaystyle\int (x^{3/2} + 8)^5 \sqrt{x}\, dx$

43. $\displaystyle\int \sin x \sec^8 x\, dx$

44. $\displaystyle\int \frac{e^{2x}}{e^{2x} + 1}\, dx$

45–74. Definite integrals *Use a change of variables or Table 5.6 to evaluate the following definite integrals.*

45. $\displaystyle\int_0^{\pi/8} \cos 2x\, dx$

46. $\displaystyle\int_0^1 2e^{2x}\, dx$

47. $\displaystyle\int_0^1 2x(4 - x^2)\, dx$

48. $\displaystyle\int_0^2 \frac{2x}{(x^2 + 1)^2}\, dx$

49. $\displaystyle\int_1^3 \frac{2^x}{2^x + 4}\, dx$

50. $\displaystyle\int_{-2\pi}^{2\pi} \cos\frac{\theta}{8}\, d\theta$

51. $\displaystyle\int_0^{\pi/2} \sin^2\theta \cos\theta\, d\theta$

52. $\displaystyle\int_0^{\pi/4} \frac{\sin x}{\cos^2 x}\, dx$

53. $\displaystyle\int_{\ln\frac{\pi}{4}}^{\ln\frac{\pi}{2}} e^w \cos e^w\, dw$

54. $\displaystyle\int_{\pi/16}^{\pi/8} 8\csc^2 4x\, dx$

55. $\displaystyle\int_{-1}^2 x^2 e^{x^3+1}\, dx$

56. $\displaystyle\int_0^4 \frac{p}{\sqrt{9 + p^2}}\, dp$

57. $\displaystyle\int_{\pi/4}^{\pi/2} \frac{\cos x}{\sin^2 x}\, dx$

58. $\displaystyle\int_0^{\pi/4} \frac{\sin\theta}{\cos^3\theta}\, d\theta$

59. $\displaystyle\int_{2/(5\sqrt{3})}^{2/5} \frac{dx}{x\sqrt{25x^2 - 1}}$

60. $\displaystyle\int_0^1 \frac{v^3 + 1}{\sqrt{v^4 + 4v + 4}}\, dv$

61. $\displaystyle\int_0^4 \frac{x}{x^2 + 1}\, dx$

62. $\displaystyle\int_0^{1/8} \frac{x}{\sqrt{1 - 16x^2}}\, dx$

63. $\displaystyle\int_{1/3}^{1/\sqrt{3}} \frac{4}{9x^2 + 1}\, dx$

64. $\displaystyle\int_0^{\ln 4} \frac{e^x}{3 + 2e^x}\, dx$

65. $\displaystyle\int_0^1 x\sqrt{1 - x^2}\, dx$

66. $\displaystyle\int_1^{e^2} \frac{\ln p}{p}\, dp$

67. $\displaystyle\int_2^3 \frac{x}{\sqrt[3]{x^2 - 1}}\, dx$

68. $\displaystyle\int_0^{6/5} \frac{dx}{25x^2 + 36}$

69. $\displaystyle\int_0^2 x^3\sqrt{16 - x^4}\, dx$

70. $\displaystyle\int_{-1}^1 (x - 1)(x^2 - 2x)^7\, dx$

71. $\displaystyle\int_{-\pi}^0 \frac{\sin x}{2 + \cos x}\, dx$

72. $\displaystyle\int_0^1 \frac{(v + 1)(v + 2)}{2v^3 + 9v^2 + 12v + 36}\, dv$

73. $\displaystyle\int_1^2 \frac{4}{9x^2 + 6x + 1}\, dx$

74. $\displaystyle\int_0^{\pi/4} e^{\sin^2 x} \sin 2x\, dx$

75. Average velocity An object moves in one dimension with a velocity in m/s given by $v(t) = 8\sin\pi t + 2t$. Find its average velocity over the time interval from $t = 0$ to $t = 10$, where t is measured in seconds.

T 76. Periodic motion An object moves along a line with a velocity in m/s given by $v(t) = 8\cos\dfrac{\pi t}{6}$. Its initial position is $s(0) = 0$.

   a. Graph the velocity function.

   b. As discussed in Chapter 6, the position of the object is given by $s(t) = \displaystyle\int_0^t v(y)\, dy$, for $t \geq 0$. Find the position function, for $t \geq 0$.

   c. What is the period of the motion—that is, starting at any point, how long does it take the object to return to that position?

77. Population models The population of a culture of bacteria has a growth rate given by $p'(t) = \dfrac{200}{(t + 1)^r}$ bacteria per hour, for $t \geq 0$, where $r > 1$ is a real number. In Chapter 6 it is shown that the increase in the population over the time interval $[0, t]$ is given by $\int_0^t p'(s)\, ds$. (Note that the growth rate decreases in time, reflecting competition for space and food.)

   a. Using the population model with $r = 2$, what is the increase in the population over the time interval $0 \leq t \leq 4$?

   b. Using the population model with $r = 3$, what is the increase in the population over the time interval $0 \leq t \leq 6$?

   c. Let ΔP be the increase in the population over a fixed time interval $[0, T]$. For fixed T, does ΔP increase or decrease with the parameter r? Explain.

   d. A lab technician measures an increase in the population of 350 bacteria over the 10-hr period $[0, 10]$. Estimate the value of r that best fits this data point.

   e. Looking ahead: Use the population model in part (b) to find the increase in population over the time interval $[0, T]$, for any $T > 0$. If the culture is allowed to grow indefinitely ($T \to \infty$), does the bacteria population increase without bound? Or does it approach a finite limit?

78–86. Variations on the substitution method *Evaluate the following integrals.*

78. $\displaystyle\int \frac{x}{x - 2}\, dx$

79. $\displaystyle\int \frac{x}{\sqrt{x - 4}}\, dx$

80. $\displaystyle\int \frac{y^2}{(y + 1)^4}\, dy$

81. $\displaystyle\int \frac{x}{\sqrt[3]{x + 4}}\, dx$

82. $\displaystyle\int \frac{e^x - e^{-x}}{e^x + e^{-x}}\, dx$

83. $\displaystyle\int x\sqrt[3]{2x + 1}\, dx$

84. $\displaystyle\int (z + 1)\sqrt{3z + 2}\, dz$ **85.** $\displaystyle\int x(x + 10)^9 dx$

86. $\displaystyle\int_0^{\sqrt{3}} \frac{3\, dx}{9 + x^2}$

87–94. Integrals with $\sin^2 x$ and $\cos^2 x$ *Evaluate the following integrals.*

87. $\displaystyle\int_{-\pi}^{\pi} \cos^2 x\, dx$ **88.** $\displaystyle\int \sin^2 x\, dx$

89. $\displaystyle\int \sin^2\left(\theta + \frac{\pi}{6}\right) d\theta$ **90.** $\displaystyle\int_0^{\pi/4} \cos^2 8\theta\, d\theta$

91. $\displaystyle\int_{-\pi/4}^{\pi/4} \sin^2 2\theta\, d\theta$ **92.** $\displaystyle\int x\cos^2 x^2 dx$

93. $\displaystyle\int_0^{\pi/6} \frac{\sin 2y}{\sin^2 y + 2}\, dy$ *(Hint: $\sin 2y = 2 \sin y \cos y$.)*

94. $\displaystyle\int_0^{\pi/2} \sin^4 \theta\, d\theta$

95. Explain why or why not Determine whether the following statements are true and give an explanation or counterexample. Assume f, f', and f'' are continuous functions for all real numbers.

a. $\displaystyle\int f(x)f'(x)\, dx = \frac{1}{2}(f(x))^2 + C.$

b. $\displaystyle\int (f(x))^n f'(x)\, dx = \frac{1}{n + 1}(f(x))^{n+1} + C, n \neq -1.$

c. $\displaystyle\int \sin 2x\, dx = 2\int \sin x\, dx.$

d. $\displaystyle\int (x^2 + 1)^9 dx = \frac{(x^2 + 1)^{10}}{10} + C.$

e. $\displaystyle\int_a^b f'(x)f''(x)\, dx = f'(b) - f'(a).$

96–98. Areas of regions *Find the area of the following regions.*

96. The region bounded by the graph of $f(x) = \dfrac{x}{\sqrt{x^2 - 9}}$ and the x-axis between $x = 4$ and $x = 5$

97. The region bounded by the graph of $f(x) = x \sin x^2$ and the x-axis between $x = 0$ and $x = \sqrt{\pi}$

98. The region bounded by the graph of $f(x) = (x - 4)^4$ and the x-axis between $x = 2$ and $x = 6$

Explorations and Challenges

99. Morphing parabolas The family of parabolas $y = \dfrac{1}{a} - \dfrac{x^2}{a^3}$, where $a > 0$, has the property that for $x \geq 0$, the x-intercept is $(a, 0)$ and the y-intercept is $(0, 1/a)$. Let $A(a)$ be the area of the region in the first quadrant bounded by the parabola and the x-axis. Find $A(a)$ and determine whether it is an increasing, decreasing, or constant function of a.

100. Substitutions Suppose f is an even function with $\int_0^8 f(x)\, dx = 9$. Evaluate each integral.

a. $\displaystyle\int_{-1}^{1} xf(x^2)\, dx.$ **b.** $\displaystyle\int_{-2}^{2} x^2 f(x^3)\, dx.$

101. Substitutions Suppose p is a nonzero real number and f is an odd function with $\int_0^1 f(x)\, dx = \pi$. Evaluate each integral.

a. $\displaystyle\int_0^{\pi/(2p)} (\cos px)f(\sin px)\, dx$ **b.** $\displaystyle\int_{-\pi/2}^{\pi/2} (\cos x)f(\sin x)\, dx$

102. Average distance on a triangle Consider the right triangle with vertices $(0, 0)$, $(0, b)$, and $(a, 0)$, where $a > 0$ and $b > 0$. Show that the average vertical distance from points on the x-axis to the hypotenuse is $b/2$, for all $a > 0$.

T 103. Average value of sine functions Use a graphing utility to verify that the functions $f(x) = \sin kx$ have a period of $2\pi/k$, where $k = 1, 2, 3, \ldots$. Equivalently, the first "hump" of $f(x) = \sin kx$ occurs on the interval $[0, \pi/k]$. Verify that the average value of the first hump of $f(x) = \sin kx$ is independent of k. What is the average value?

104. Equal areas The area of the shaded region under the curve $y = 2 \sin 2x$ in part (a) of the figure equals the area of the shaded region under the curve $y = \sin x$ in part (b) of the figure. Explain why this is true without computing areas.

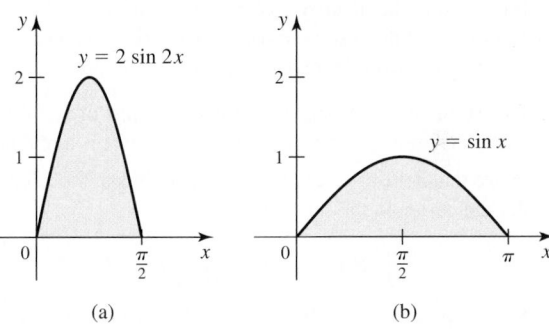

(a)   (b)

105. Equal areas The area of the shaded region under the curve $y = \dfrac{(\sqrt{x} - 1)^2}{2\sqrt{x}}$ on the interval $[4, 9]$ in part (a) of the following figure equals the area of the shaded region under the curve $y = x^2$ on the interval $[1, 2]$ in part (b) of the figure. Without computing areas, explain why.

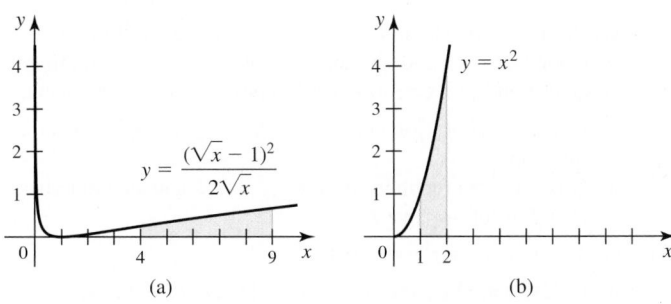

(a)   (b)

106–108. General results *Evaluate the following integrals in which the function f is unspecified. Note that $f^{(p)}$ is the pth derivative of f and f^p is the pth power of f. Assume f and its derivatives are continuous for all real numbers.*

106. $\displaystyle\int (5f^3(x) + 7f^2(x) + f(x))f'(x)\, dx$

107. $\displaystyle\int_1^2 (5f^3(x) + 7f^2(x) + f(x))f'(x)\, dx$, where $f(1) = 4$, $f(2) = 5$

108. $\displaystyle\int (f^{(p)}(x))^n f^{(p+1)}(x)\, dx$, where p is a positive integer, $n \neq -1$

109–111. More than one way *Occasionally, two different substitutions do the job. Use each substitution to evaluate the following integrals.*

109. $\int_0^1 x\sqrt{x+a}\,dx;\ a > 0$ $(u = \sqrt{x+a}$ and $u = x + a)$

110. $\int_0^1 x\sqrt[p]{x+a}\,dx;\ a > 0$ $(u = \sqrt[p]{x+a}$ and $u = x + a)$

111. $\int \sec^3\theta \tan\theta\,d\theta$ $(u = \cos\theta$ and $u = \sec\theta)$

112. $\sin^2 ax$ and $\cos^2 ax$ integrals Use the Substitution Rule to prove that

$$\int \sin^2 ax\,dx = \frac{x}{2} - \frac{\sin(2ax)}{4a} + C \text{ and}$$

$$\int \cos^2 ax\,dx = \frac{x}{2} + \frac{\sin(2ax)}{4a} + C.$$

113. Integral of $\sin^2 x \cos^2 x$ Consider the integral $I = \int \sin^2 x \cos^2 x\,dx$.

a. Find I using the identity $\sin 2x = 2\sin x \cos x$.
b. Find I using the identity $\cos^2 x = 1 - \sin^2 x$.
c. Confirm that the results in parts (a) and (b) are consistent and compare the work involved in the two methods.

114. Substitution: shift Perhaps the simplest change of variables is the shift or translation given by $u = x + c$, where c is a real number.

a. Prove that shifting a function does not change the net area under the curve, in the sense that

$$\int_a^b f(x+c)\,dx = \int_{a+c}^{b+c} f(u)\,du.$$

b. Draw a picture to illustrate this change of variables in the case where $f(x) = \sin x$, $a = 0$, $b = \pi$, and $c = \pi/2$.

115. Substitution: scaling Another change of variables that can be interpreted geometrically is the scaling $u = cx$, where c is a real number. Prove and interpret the fact that

$$\int_a^b f(cx)\,dx = \frac{1}{c}\int_{ac}^{bc} f(u)\,du.$$

Draw a picture to illustrate this change of variables in the case where $f(x) = \sin x$, $a = 0$, $b = \pi$, and $c = 1/2$.

116–119. Multiple substitutions *If necessary, use two or more substitutions to find the following integrals.*

116. $\int x \sin^4 x^2 \cos x^2\,dx$ (*Hint:* Begin with $u = x^2$, and then use $v = \sin u$.)

117. $\int \dfrac{dx}{\sqrt{1 + \sqrt{1 + x}}}$ (*Hint:* Begin with $u = \sqrt{1+x}$.)

118. $\int \tan^{10} 4x \sec^4 4x\,dx$ (*Hint:* Begin with $u = 4x$.)

119. $\int_0^{\pi/2} \dfrac{\cos\theta \sin\theta}{\sqrt{\cos^2\theta + 16}}\,d\theta$ (*Hint:* Begin with $u = \cos\theta$.)

QUICK CHECK ANSWERS

1. $u = x^4 + 5$ 2. With $u = x^5 + 6$, we have $du = 5x^4$, and x^4 does not appear in the integrand. 3. $\dfrac{1}{6}\sin 6x + C$
4. $4\sin^{-1}\dfrac{x}{2} + C$ 5. New equation: $u^2 - 13u + 36 = 0$; roots: $x = \pm 2, \pm 3$ ◄

CHAPTER 5   REVIEW EXERCISES

1. **Explain why or why not** Determine whether the following statements are true and give an explanation or counterexample. Assume f and f' are continuous functions for all real numbers.

   a. If $A(x) = \int_a^x f(t)\,dt$ and $f(t) = 2t - 3$, then A is a quadratic function.
   b. Given an area function $A(x) = \int_a^x f(t)\,dt$ and an antiderivative F of f, it follows that $A'(x) = F(x)$.
   c. $\int_a^b f'(x)\,dx = f(b) - f(a)$.
   d. If f is continuous on $[a, b]$ and $\int_a^b |f(x)|\,dx = 0$, then $f(x) = 0$ on $[a, b]$.
   e. If the average value of f on $[a, b]$ is zero, then $f(x) = 0$ on $[a, b]$.
   f. $\int_a^b (2f(x) - 3g(x))\,dx = 2\int_a^b f(x)\,dx + 3\int_b^a g(x)\,dx$.
   g. $\int f'(g(x))g'(x)\,dx = f(g(x)) + C$.

2. The velocity in ft/s of an object moving along a line is given by the function $v = f(t)$ on the interval $0 \le t \le 12$, where t is in seconds (see figure). Estimate the displacement of the object on $[0, 12]$ using left, right, and midpoint Riemann sums with $n = 3$ subintervals and a regular partition.

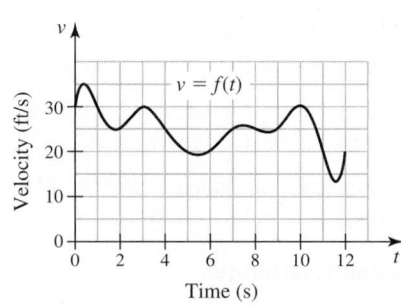

Time (s)

3. **Ascent rate of a scuba diver** Divers who ascend too quickly in the water risk *decompression illness*. A common recommendation for a maximum rate of ascent is 30 feet/minute with a 5-minute safety stop 15 feet below the surface of the water. Suppose a diver ascends to the surface in 8 minutes according to the velocity function

$$v(t) = \begin{cases} 30 & \text{if } 0 \le t \le 2 \\ 0 & \text{if } 2 < t \le 7 \\ 15 & \text{if } 7 < t \le 8. \end{cases}$$

   a. Graph the velocity function v.
   b. Compute the area under the velocity curve.
   c. Interpret the physical meaning of the area under the velocity curve.

4. Use the tabulated values of f to estimate the value of $\int_0^6 f(x)\,dx$ by evaluating the left, right, and midpoint Riemann sums using a regular partition with $n = 3$ subintervals.

x	0	1	2	3	4	5	6
$f(x)$	20	22	26	30	34	38	40

5. Estimate $\int_1^4 \sqrt{4x + 1}\,dx$ by evaluating the left, right, and midpoint Riemann sums using a regular partition with $n = 6$ subintervals.

6. **Evaluating Riemann sums** Consider the function $f(x) = 3x + 4$ on the interval $[3, 7]$. Show that the midpoint Riemann sum with $n = 4$ gives the exact area of the region bounded by the graph. Assume a regular partition.

▼ 7. **Estimating a definite integral** Use a calculator and midpoint Riemann sums to approximate $\int_1^{25} \sqrt{2x - 1}\,dx$. Present your calculations in a table showing the approximations for $n = 10$, 30, and 60 subintervals, assuming a regular partition. Make a conjecture about the exact value of the integral and verify your conjecture using the Fundamental Theorem of Calculus.

8. Suppose the expression $\displaystyle\lim_{\Delta \to 0} \sum_{k=1}^{n} (x_k^{*3} + x_k^{*})\Delta x_k$ is the limit of a Riemann sum of a function f on $[3, 8]$. Identify a possible function f and express the limit as a definite integral.

9. **Integration by Riemann sums** Consider the integral $\int_1^4 (3x - 2)\,dx$.

   a. Evaluate the right Riemann sum for the integral with $n = 3$.
   b. Use summation notation to express the right Riemann sum in terms of a positive integer n.
   c. Evaluate the definite integral by taking the limit as $n \to \infty$ of the Riemann sum in part (b).
   d. Confirm the result from part (c) by graphing $y = 3x - 2$ and using geometry to evaluate the integral. Then evaluate $\int_1^4 (3x - 2)\,dx$ with the Fundamental Theorem of Calculus.

10–13. Limit definition of the definite integral *Use the limit definition of the definite integral with right Riemann sums and a regular partition to evaluate the following definite integrals. Use the Fundamental Theorem of Calculus to check your answer.*

10. $\int_0^1 (4x - 2)\,dx$ **11.** $\int_0^2 (x^2 - 4)\,dx$

12. $\int_1^2 (3x^2 + x)\,dx$ **13.** $\int_0^4 (x^3 - x)\,dx$

14. **Sum to integral** Evaluate the following limit by identifying the integral that it represents:

$$\lim_{n \to \infty} \sum_{k=1}^{n} \left(\left(\frac{4k}{n}\right)^5 + 1\right)\frac{4}{n}.$$

15. **Symmetry properties** Suppose $\int_0^4 f(x)\,dx = 10$ and $\int_0^4 g(x)\,dx = 20$. Furthermore, suppose f is an even function and g is an odd function. Evaluate the following integrals.

   a. $\int_{-4}^4 f(x)\,dx$ **b.** $\int_{-4}^4 3g(x)\,dx$

   c. $\int_{-4}^4 (4f(x) - 3g(x))\,dx$ **d.** $\int_0^1 8xf(4x^2)\,dx$

   e. $\int_{-2}^2 3xf(x)\,dx$

16. **Properties of integrals** The figure shows the areas of regions bounded by the graph of f and the x-axis. Evaluate the following integrals.

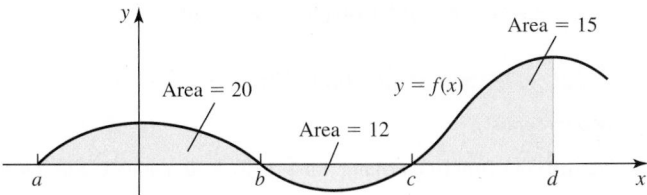

   a. $\int_a^c f(x)\,dx$ **b.** $\int_b^d f(x)\,dx$

   c. $\int_c^b 2f(x)\,dx$ **d.** $\int_a^d 4f(x)\,dx$

   e. $\int_a^b 3f(x)\,dx$ **f.** $\int_b^d 2f(x)\,dx$

   g. $\int_a^c |f(x)|\,dx$ **h.** $\int_d^a |f(x)|\,dx$

17–22. Properties of integrals *Suppose $\int_1^4 f(x)\,dx = 6$, $\int_1^4 g(x)\,dx = 4$, and $\int_3^4 f(x)\,dx = 2$. Evaluate the following integrals or state that there is not enough information.*

17. $\int_1^4 3f(x)\,dx$ **18.** $-\int_4^1 2f(x)\,dx$

19. $\int_1^4 (3f(x) - 2g(x))\,dx$ **20.** $\int_1^4 f(x)g(x)\,dx$

21. $\int_1^3 \frac{f(x)}{g(x)}\,dx$ **22.** $\int_4^1 (f(x) - g(x))\,dx$

23. **Area by geometry** Use geometry to evaluate the following definite integrals, where the graph of f is given in the figure.

   a. $\int_0^4 f(x)\,dx$ **b.** $\int_6^4 f(x)\,dx$

   c. $\int_5^7 f(x)\,dx$ **d.** $\int_0^7 f(x)\,dx$

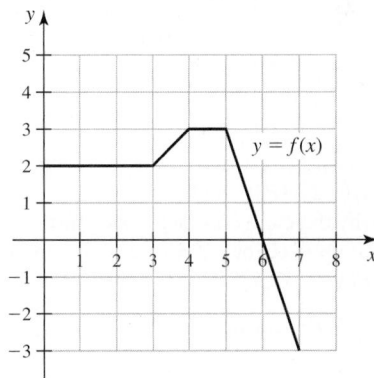

24. Use geometry to find the displacement of an object moving along a line for the time intervals (i) $0 \le t \le 5$, (ii) $3 \le t \le 7$, and (iii) $0 \le t \le 8$, where the graph of its velocity $v = g(t)$ is given in the figure.

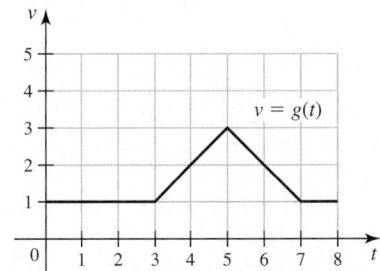

25–26. *Use geometry and properties of integrals to evaluate the following definite integrals.*

25. $\displaystyle\int_0^4 \sqrt{8x - x^2}\, dx$. (*Hint:* Complete the square.)

26. $\displaystyle\int_{-4}^0 \left(2x + \sqrt{16 - x^2}\right) dx$ (*Hint:* Write the integral as a sum of two integrals.)

27. Identifying functions Match the graphs A, B, and C in the figure with the functions $f(x)$, $f'(x)$, and $\int_0^x f(t)\, dt$.

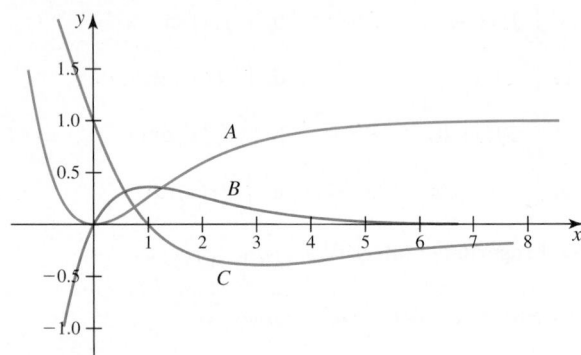

28. Area functions Consider the graph of the function f in the figure and let $F(x) = \int_0^x f(t)\, dt$ and $G(x) = \int_1^x f(t)\, dt$. Assume the graph consists of a line segment from $(0, -2)$ to $(2, 2)$ and two quarter circles of radius 2.

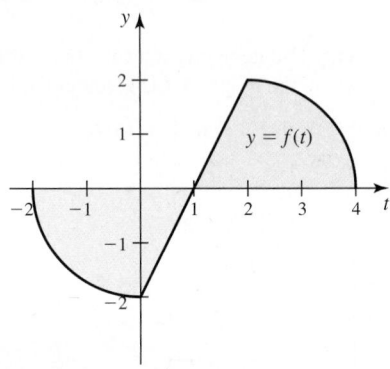

a. Evaluate $F(2)$, $F(-2)$, and $F(4)$.
b. Evaluate $G(-2)$, $G(0)$, and $G(4)$.
c. Explain why there is a constant C such that $F(x) = G(x) + C$, for $-2 \le x \le 4$. Fill in the blank with a number: $F(x) = G(x) + $ _____ , for $-2 \le x \le 4$.

29–34. *Evaluate the following derivatives.*

29. $\displaystyle\frac{d}{dx}\int_7^x \sqrt{1 + t^4 + t^6}\, dt$

30. $\displaystyle\frac{d}{dx}\int_3^{e^x} \cos t^2\, dt$

31. $\displaystyle\frac{d}{dx}\int_x^5 \sin w^6\, dw$

32. $\displaystyle\frac{d}{dx}\int_{x^2}^5 \sin w^6\, dw$

33. $\displaystyle\frac{d}{dx}\int_{-x}^x \frac{dt}{t^{10} + 1}$

34. $\displaystyle\frac{d}{dx}\int_{x^2}^{e^x} \sin^3 t\, dt$

35. Find the intervals on which $f(x) = \int_x^1 (t - 3)(t - 6)^{11} dt$ is increasing and the intervals on which it is decreasing.

36. Area function by geometry Use geometry to find the area $A(x)$ that is bounded by the graph of $f(t) = 2t - 4$ and the t-axis between the point $(2, 0)$ and the variable point $(x, 0)$, where $x \ge 2$. Verify that $A'(x) = f(x)$.

37. Given that $F' = f$, use the substitution method to show that $\displaystyle\int f(ax + b)\, dx = \frac{1}{a} F(ax + b) + C$, for nonzero constants a and b.

38–86. Evaluating integrals *Evaluate the following integrals.*

38. $\displaystyle\int_1^5 dx$

39. $\displaystyle\int_{-2}^2 (3x^4 - 2x + 1)\, dx$

40. $\displaystyle\int_0^1 (4x^{21} - 2x^{16} + 1)\, dx$

41. $\displaystyle\int (9x^8 - 7x^6)\, dx$

42. $\displaystyle\int \frac{\sqrt{x} + 1}{\sqrt{x}}\, dx$

43. $\displaystyle\int_0^1 \sqrt{x}\left(\sqrt{x} + 1\right) dx$

44. $\displaystyle\int (3x + 1)(3x^2 + 2x + 1)^3\, dx$

45. $\displaystyle\int_{\pi/6}^{\pi/3} (\sec^2 t + \csc^2 t)\, dt$

46. $\displaystyle\int_{\pi/12}^{\pi/9} (\csc 3x \cot 3x + \sec 3x \tan 3x)\, dx$

47. $\displaystyle\int_{\sqrt{2}}^2 \frac{dx}{x\sqrt{x^2 - 1}}$

48. $\displaystyle\int_1^4 \left(\frac{\sqrt{v} + v}{v}\right) dv$

49. $\displaystyle\int \frac{\cos x}{\sin^{7/4} x}\, dx$

50. $\displaystyle\int_1^e \frac{dx}{x(1 + \ln x)}$

51. $\displaystyle\int x^2 \cos x^3\, dx$

52. $\displaystyle\int \frac{\cos 3t}{1 + \sin 3t}\, dt$

53. $\displaystyle\int \frac{\cos 7w}{16 + \sin^2 7w}\, dw$

54. $\displaystyle\int \sqrt{1 + \tan 2t}\,\sec^2 2t\, dt$

55. $\displaystyle\int_0^1 x \cdot 2^{x^2 + 1}\, dx$

56. $\displaystyle\int \cos 3x\, dx$

57. $\displaystyle\int_0^2 (2x + 1)^3\, dx$

58. $\displaystyle\int_{-2}^2 e^{4x + 8}\, dx$

59. $\displaystyle\int_0^1 5r e^{3r^2 + 2}\, dr$

60. $\displaystyle\int \sin z \sin(\cos z)\, dz$

61. $\displaystyle\int e^{x + e^x}\, dx$

62. $\displaystyle\int \frac{y^2}{y^3 + 27}\, dy$

63. $\displaystyle\int \frac{dx}{\sqrt{1 - 4x^2}}$

64. $\displaystyle\int y^2 (3y^3 + 1)^4\, dy$

65. $\displaystyle\int_0^{2\pi} \cos^2 \frac{x}{6}\, dx$

66. $\displaystyle\int x \sin x^2 \cos^8 x^2\, dx$

67. $\displaystyle\int_0^\pi \sin^2 5\theta\, d\theta$

68. $\displaystyle\int_0^\pi (1 - \cos^2 3\theta)\, d\theta$

69. $\displaystyle\int_2^3 \frac{x^2 + 2x - 2}{x^3 + 3x^2 - 6x}\, dx$

70. $\displaystyle\int_0^{\ln 2} \frac{e^x}{1 + e^{2x}}\, dx$

71. $\displaystyle\int_{-5}^5 \frac{w^3}{\sqrt{w^{50} + w^{20} + 1}}\, dw$ (*Hint:* Use symmetry.)

72. $\displaystyle\int_{-3}^3 (511x^{17} + 302x^{13} + 117x^9 + 303x^3 + x^2)\, dx$

73. $\displaystyle\int \frac{1}{x^2} \sin \frac{1}{x}\, dx$

74. $\displaystyle\int \frac{(\tan^{-1} x)^5}{1 + x^2}\, dx$

75. $\int \dfrac{dx}{(\tan^{-1}x)(1 + x^2)}$

76. $\int \dfrac{\sin^{-1}x}{\sqrt{1 - x^2}}\, dx$

77. $\int x(x + 3)^{10}\, dx$

78. $\int x^7 \sqrt{x^4 + 1}\, dx$

79. $\int_0^3 \dfrac{x}{\sqrt{25 - x^2}}\, dx$

80. $\int_0^1 \dfrac{dx}{\sqrt{4 - x^2}}$

81. $\int_{\sqrt{2}/5}^{2/5} \dfrac{dx}{x\sqrt{25x^2 - 1}}$

82. $\int \dfrac{\sin 2x}{1 + \cos^2 x}\, dx$ (*Hint:* $\sin 2x = 2 \sin x \cos x$.)

83. $\int_{-10}^{10} \dfrac{x}{\sqrt{200 - x^2}}\, dx$

84. $\int_{-\pi/2}^{\pi/2} (\cos 2x + \cos x \sin x - 3 \sin x^5)\, dx$

85. $\int_0^4 f(x)\, dx$ for $f(x) = \begin{cases} 2x + 1 & \text{if } x \le 3 \\ 3x^2 + 2x - 8 & \text{if } x > 3 \end{cases}$

86. $\int_0^5 |2x - 8|\, dx$

87–90. Area of regions *Compute the area of the region bounded by the graph of f and the x-axis on the given interval. You may find it useful to sketch the region.*

87. $f(x) = 16 - x^2$ on $[-4, 4]$

88. $f(x) = x^3 - x$ on $[-1, 0]$

89. $f(x) = 2 \sin \dfrac{x}{4}$ on $[0, 2\pi]$

90. $f(x) = \dfrac{1}{x^2 + 1}$ on $[-1, \sqrt{3}]$

91–92. Area versus net area *Find (i) the net area and (ii) the area of the region bounded by the graph of f and the x-axis on the given interval. You may find it useful to sketch the region.*

91. $f(x) = x^4 - x^2$ on $[-1, 1]$

92. $f(x) = x^2 - x$ on $[0, 3]$

93. Gateway Arch The Gateway Arch in St. Louis is 630 ft high and has a 630-ft base. Its shape can be modeled by the function

$$y = 1260 - 315(e^{0.00418x} + e^{-0.00418x}),$$

where the base of the arch is $[-315, 315]$ and x and y are measured in feet. Find the average height of the arch above the ground.

94. Root mean square The root mean square (or RMS) is another measure of average value, often used with oscillating functions (for example, sine and cosine functions that describe the current, voltage, or power in an alternating circuit). The RMS of a function f on the interval $[0, T]$ is

$$\overline{f}_{RMS} = \sqrt{\dfrac{1}{T}\int_0^T f(t)^2\, dt}.$$

Compute the RMS of $f(t) = A \sin(\omega t)$, where A and ω are positive constants and T is any integer multiple of the period of f, which is $2\pi/\omega$.

95. Displacement from velocity A particle moves along a line with a velocity given by $v(t) = 5 \sin \pi t$, starting with an initial position $s(0) = 0$. Find the displacement of the particle between $t = 0$ and $t = 2$, which is given by $s(t) = \int_0^2 v(t)\, dt$. Find the distance traveled by the particle during this interval, which is $\int_0^2 |v(t)|\, dt$.

96. Velocity to displacement An object travels on the x-axis with a velocity given by $v(t) = 2t + 5$, for $0 \le t \le 4$.

   a. How far does the object travel, for $0 \le t \le 4$?

   b. What is the average velocity of the object on the interval $[0, 4]$?

   c. True or false: The object would travel as far as in part (a) if it traveled at its average velocity (a constant), for $0 \le t \le 4$.

97. Find the average value of $f(x) = e^{2x}$ on $[0, \ln 2]$.

98. Average height A baseball is launched into the outfield on a parabolic trajectory given by $y = 0.01x(200 - x)$. Find the average height of the baseball over the horizontal extent of its flight.

99. Average values Integration is not needed.

   a. Find the average value of f shown in the figure on the interval $[1, 6]$ and then find the point(s) c in $(1, 6)$ guaranteed to exist by the Mean Value Theorem for Integrals.

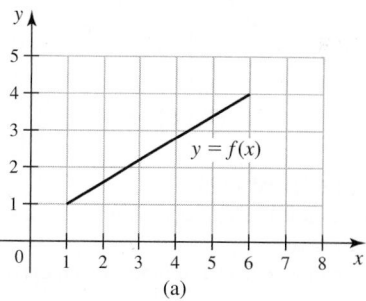

(a)

   b. Find the average value of f shown in the figure on the interval $[2, 6]$ and then find the point(s) c in $(2, 6)$ guaranteed to exist by the Mean Value Theorem for Integrals.

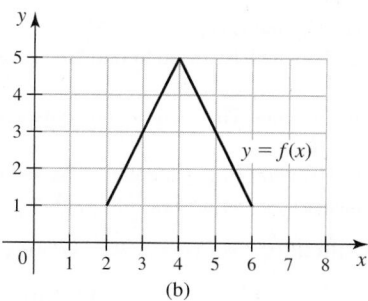

(b)

100. An unknown function The function f satisfies the equation $3x^4 - 48 = \int_2^x f(t)\, dt$. Find f and check your answer by substitution.

101. An unknown function Assume f' is a continuous on $[2, 4]$, $\int_1^2 f'(2x)\, dx = 10$, and $f(2) = 4$. Evaluate $f(4)$.

102. Function defined by an integral Let $H(x) = \int_0^x \sqrt{4 - t^2}\, dt$, for $-2 \le x \le 2$.

   a. Evaluate $H(0)$.

   b. Evaluate $H'(1)$.

   c. Evaluate $H'(2)$.

   d. Use geometry to evaluate $H(2)$.

   e. Find the value of s such that $H(x) = sH(-x)$.

103. Function defined by an integral Make a graph of the function
$$f(x) = \int_1^x \frac{dt}{t}, \text{ for } x \ge 1. \text{ Be sure to include all of the evidence}$$
you used to arrive at the graph.

104. Change of variables Use the change of variables $u^3 = x^2 - 1$ to evaluate the integral $\int_1^3 x \sqrt[3]{x^2 - 1} \, dx$.

105–106. Area functions and the Fundamental Theorem *Consider the function*
$$f(t) = \begin{cases} t & \text{if } -2 \le t < 0 \\ t^2/2 & \text{if } 0 \le t \le 2 \end{cases}$$

and its graph shown below. Let $F(x) = \int_{-1}^x f(t) \, dt$ and $G(x) = \int_{-2}^x f(t) \, dt$.

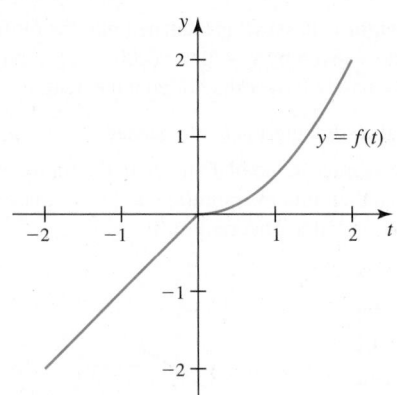

105. a. Evaluate $F(-2)$ and $F(2)$.
  b. Use the Fundamental Theorem to find an expression for $F'(x)$, for $-2 \le x < 0$.
  c. Use the Fundamental Theorem to find an expression for $F'(x)$, for $0 \le x \le 2$.
  d. Evaluate $F'(-1)$ and $F'(1)$. Interpret these values.
  e. Evaluate $F''(-1)$ and $F''(1)$.
  f. Find a constant C such that $F(x) = G(x) + C$.

106. a. Evaluate $G(-1)$ and $G(1)$.
  b. Use the Fundamental Theorem to find an expression for $G'(x)$, for $-2 \le x < 0$.
  c. Use the Fundamental Theorem to find an expression for $G'(x)$, for $0 \le x \le 2$.
  d. Evaluate $G'(0)$ and $G'(1)$. Interpret these values.
  e. Find a constant C such that $F(x) = G(x) + C$.

107–108. Limits with integrals *Evaluate the following limits.*

107. $\lim\limits_{x \to 2} \dfrac{\int_2^x e^{t^2} dt}{x - 2}$
  108. $\lim\limits_{x \to 1} \dfrac{\int_1^{x^2} e^{t^3} dt}{x - 1}$

109. Geometry of integrals Without evaluating the integrals, explain why the following statement is true for positive integers n:
$$\int_0^1 x^n \, dx + \int_0^1 \sqrt[n]{x} \, dx = 1.$$

110. Area with a parameter Let $a > 0$ be a real number and consider the family of functions $f(x) = \sin ax$ on the interval $[0, \pi/a]$.
  a. Graph f, for $a = 1, 2,$ and 3.
  b. Let $g(a)$ be the area of the region bounded by the graph of f and the x-axis on the interval $[0, \pi/a]$. Graph g for $0 < a < \infty$. Is g an increasing function, a decreasing function, or neither?

111. Inverse tangent integral Prove that for nonzero constants a and b, $\int \dfrac{dx}{a^2 x^2 + b^2} = \dfrac{1}{ab} \tan^{-1}\left(\dfrac{ax}{b}\right) + C$.

▣ 112. Area function properties Consider the function
  $f(t) = t^2 - 5t + 4$ and the area function $A(x) = \int_0^x f(t) \, dt$.
  a. Graph f on the interval $[0, 6]$.
  b. Compute and graph A on the interval $[0, 6]$.
  c. Show that the local extrema of A occur at the zeros of f.
  d. Give a geometric and analytical explanation for the observation in part (c).
  e. Find the approximate zeros of A, other than 0, and call them x_1 and x_2, where $x_1 < x_2$.
  f. Find b such that the area bounded by the graph of f and the t-axis on the interval $[0, x_1]$ equals the area bounded by the graph of f and the t-axis on the interval $[x_1, b]$.
  g. If f is an integrable function and $A(x) = \int_a^x f(t) \, dt$, is it always true that the local extrema of A occur at the zeros of f? Explain.

113. Function defined by an integral
  Let $f(x) = \int_0^x (t - 1)^{15}(t - 2)^9 dt$.
  a. Find the intervals on which f is increasing and the intervals on which f is decreasing.
  b. Find the intervals on which f is concave up and the intervals on which f is concave down.
  c. For what values of x does f have local minima? Local maxima?
  d. Where are the inflection points of f?

114. Exponential inequalities Sketch a graph of $f(t) = e^t$ on an arbitrary interval $[a, b]$. Use the graph and compare areas of regions to prove that
$$e^{(a+b)/2} < \frac{e^b - e^a}{b - a} < \frac{e^a + e^b}{2}.$$

(*Source: Mathematics Magazine*, 81, 5, Dec 2008)

115. Equivalent equations Explain why, if a function u satisfies the equation $u(x) + 2\int_0^x u(t) \, dt = 10$, then it also satisfies the equation $u'(x) + 2u(x) = 0$. Is it true that if u satisfies the second equation, then it satisfies the first equation?

116. Unit area sine curve Find the value of c such that the region bounded by $y = c \sin x$ and the x-axis on the interval $[0, \pi]$ has area 1.

117. Unit area cubic Find the value of $c > 0$ such that the region bounded by the cubic $y = x(x - c)^2$ and the x-axis on the interval $[0, c]$ has area 1.

Chapter 5 Guided Projects

Applications of the material in this chapter and related topics can be found in the following Guided Projects. For additional information, see the Preface.

- Limits of sums
- Distribution of wealth
- Symmetry in integrals

6

Applications of Integration

Chapter Preview Now that we have some basic techniques for evaluating integrals, we turn our attention to the uses of integration, which are virtually endless. We first illustrate the general rule that if the rate of change of a quantity is known, then integration can be used to determine the net change or the future value of that quantity over a certain time interval. Next, we explore some rich geometric applications of integration: computing the area of regions bounded by several curves, the volume and surface area of three-dimensional solids, and the length of curves. We end the chapter with an examination of a variety of physical applications of integration, such as finding the work done by a variable force and computing the total force exerted by water behind a dam. All these applications are unified by their use of the *slice-and-sum* strategy.

6.1 Velocity and Net Change

In previous chapters, we established the relationship between the position and velocity of an object moving along a line. With integration, we can now say much more about this relationship. Once we relate velocity and position through integration, we can make analogous observations about a variety of other practical problems, which include fluid flow, population growth, manufacturing costs, and production and consumption of natural resources. The ideas in this section come directly from the Fundamental Theorem of Calculus, and they are among the most powerful applications of calculus.

Velocity, Position, and Displacement

Suppose you are driving along a straight highway and your position relative to a reference point or origin is $s(t)$ for times $t \geq 0$. Your *displacement* over a time interval $[a, b]$ is the change in the position $s(b) - s(a)$ (**Figure 6.1**). If $s(b) > s(a)$, then your displacement is positive; when $s(b) < s(a)$, your displacement is negative.

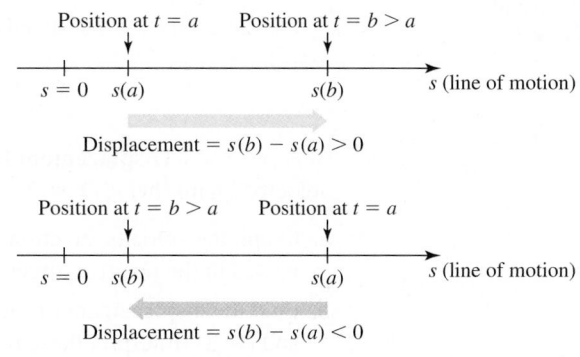

Figure 6.1

Now assume $v(t)$ is the velocity of the object at a particular time t. Recall from Chapter 3 that $v(t) = s'(t)$, which means that s is an antiderivative of v. From the Fundamental Theorem of Calculus, it follows that

$$\int_a^b v(t)\, dt = \int_a^b s'(t)\, dt = s(b) - s(a) = \text{displacement}.$$

We see that the definite integral $\int_a^b v(t)\, dt$ is the displacement (change in position) between times $t = a$ and $t = b$. Equivalently, the displacement over the time interval $[a, b]$ is the net area under the velocity curve over $[a, b]$ (Figure 6.2a).

Not to be confused with the displacement is the *distance traveled* over a time interval, which is the total distance traveled by the object, independent of the direction of motion. If the velocity is positive, the object moves in the positive direction and the displacement equals the distance traveled. However, if the velocity changes sign, then the displacement and the distance traveled are not generally equal.

QUICK CHECK 1 A police officer leaves his station on a north-south freeway at 9 A.M., traveling north (the positive direction) for 40 mi between 9 A.M. and 10 A.M. From 10 A.M. to 11 A.M., he travels south to a point 20 mi south of the station. What are the distance traveled and the displacement between 9 A.M. and 11 A.M.? ◄

To compute the distance traveled, we need the magnitude, but not the sign, of the velocity. The magnitude of the velocity $|v(t)|$ is called the *speed*. The distance traveled over a small time interval dt is $|v(t)|\, dt$ (speed multiplied by elapsed time). Summing these distances, the distance traveled over the time interval $[a, b]$ is the integral of the speed; that is,

$$\text{distance traveled} = \int_a^b |v(t)|\, dt.$$

As shown in Figure 6.2b, integrating the speed produces the area (not net area) bounded by the velocity curve and the t-axis, which corresponds to the distance traveled. The distance traveled is always nonnegative.

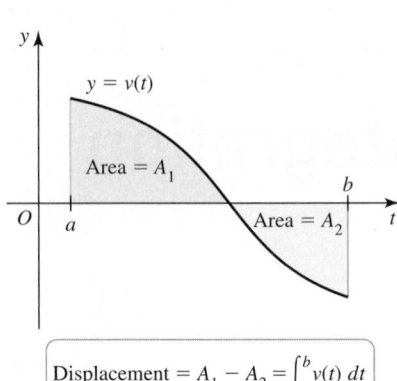

$$\boxed{\text{Displacement} = A_1 - A_2 = \int_a^b v(t)\, dt}$$

(a)

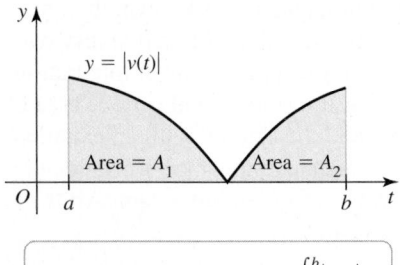

$$\boxed{\text{Distance traveled} = A_1 + A_2 = \int_a^b |v(t)|\, dt}$$

(b)

Figure 6.2

DEFINITION Position, Velocity, Displacement, and Distance

1. The **position** of an object moving along a line at time t, denoted $s(t)$, is the location of the object relative to the origin.

2. The **velocity** of an object at time t is $v(t) = s'(t)$.

3. The **displacement** of the object between $t = a$ and $t = b > a$ is

$$s(b) - s(a) = \int_a^b v(t)\, dt.$$

4. The **distance traveled** by the object between $t = a$ and $t = b > a$ is

$$\int_a^b |v(t)|\, dt,$$

where $|v(t)|$ is the **speed** of the object at time t.

QUICK CHECK 2 Describe a possible motion of an object along a line, for $0 \le t \le 5$, for which the displacement and the distance traveled are different. ◄

EXAMPLE 1 Displacement from velocity A jogger runs along a straight road with velocity (in mi/hr) $v(t) = 2t^2 - 8t + 6$, for $0 \le t \le 3$, where t is measured in hours.

a. Graph the velocity function over the interval $[0, 3]$. Determine when the jogger moves in the positive direction and when she moves in the negative direction.

b. Find the displacement of the jogger (in miles) on the time intervals $[0, 1]$, $[1, 3]$, and $[0, 3]$. Interpret these results.

c. Find the distance traveled over the interval $[0, 3]$.

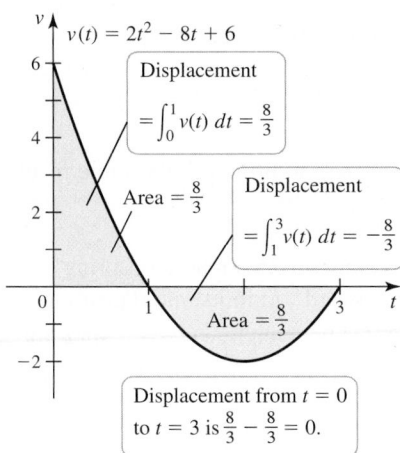

$v(t) = 2t^2 - 8t + 6$

Displacement
$= \int_0^1 v(t)\, dt = \dfrac{8}{3}$

Area $= \dfrac{8}{3}$

Displacement
$= \int_1^3 v(t)\, dt = -\dfrac{8}{3}$

Area $= \dfrac{8}{3}$

Displacement from $t = 0$
to $t = 3$ is $\dfrac{8}{3} - \dfrac{8}{3} = 0$.

(a)

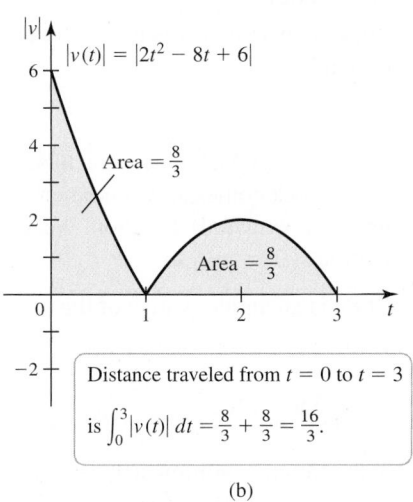

$|v(t)| = |2t^2 - 8t + 6|$

Area $= \dfrac{8}{3}$

Area $= \dfrac{8}{3}$

Distance traveled from $t = 0$ to $t = 3$

is $\int_0^3 |v(t)|\, dt = \dfrac{8}{3} + \dfrac{8}{3} = \dfrac{16}{3}$.

(b)

Figure 6.3

▶ Note that t is the independent variable of the position function. Therefore, another (dummy) variable, in this case x, must be used as the variable of integration.

▶ Theorem 6.1 is a consequence (actually a statement) of the Fundamental Theorem of Calculus.

SOLUTION

a. By solving $v(t) = 2t^2 - 8t + 6 = 2(t - 1)(t - 3) = 0$, we find that the velocity is zero at $t = 1$ and $t = 3$; these values are the t-intercepts of the graph of v, which is an upward-opening parabola with a v-intercept of 6 (**Figure 6.3a**). The velocity is positive on the interval $0 \le t < 1$, which means the jogger moves in the positive s direction. For $1 < t < 3$, the velocity is negative and the jogger moves in the negative s direction.

b. The displacement (in miles) over the interval $[0, 1]$ is

$$s(1) - s(0) = \int_0^1 v(t)\, dt$$

$$= \int_0^1 (2t^2 - 8t + 6)\, dt \qquad \text{Substitute for } v.$$

$$= \left(\frac{2}{3} t^3 - 4t^2 + 6t \right) \Big|_0^1 = \frac{8}{3}. \qquad \text{Evaluate integral.}$$

A similar calculation shows that the displacement over the interval $[1, 3]$ is

$$s(3) - s(1) = \int_1^3 v(t)\, dt = -\frac{8}{3}.$$

Over the interval $[0, 3]$, the displacement is $\frac{8}{3} + \left(-\frac{8}{3}\right) = 0$, which means the jogger returns to the starting point after three hours.

c. From part (b), we can deduce the total distance traveled by the jogger. On the interval $[0, 1]$, the distance traveled is $\frac{8}{3}$ mi; on the interval $[1, 3]$, the distance traveled is also $\frac{8}{3}$ mi. Therefore, the distance traveled on $[0, 3]$ is $\frac{16}{3}$ mi. Alternatively (**Figure 6.3b**), we can integrate the speed and get the same result:

$$\int_0^3 |v(t)|\, dt = \int_0^1 (2t^2 - 8t + 6)\, dt + \int_1^3 (-(2t^2 - 8t + 6))\, dt \quad \text{Definition of } |v(t)|$$

$$= \left(\frac{2}{3} t^3 - 4t^2 + 6t \right) \Big|_0^1 + \left(-\frac{2}{3} t^3 + 4t^2 - 6t \right) \Big|_1^3 \quad \text{Evaluate integrals.}$$

$$= \frac{16}{3}. \qquad\qquad \text{Simplify.}$$

Related Exercises 7–8 ◀

Future Value of the Position Function

To find the displacement of an object, we do not need to know its initial position. For example, whether an object moves from $s = -20$ to $s = -10$ or from $s = 50$ to $s = 60$, its displacement is 10 units. What happens if we are interested in the actual *position* of the object at some future time?

Suppose we know the velocity of an object and its initial position $s(0)$. The goal is to find the position $s(t)$ at some future time $t \ge 0$. The Fundamental Theorem of Calculus gives us the answer directly. Because the position s is an antiderivative of the velocity v, we have

$$\int_0^t v(x)\, dx = \int_0^t s'(x)\, dx = s(x) \Big|_0^t = s(t) - s(0).$$

Rearranging this expression leads to the following result.

THEOREM 6.1 Position from Velocity
Given the velocity $v(t)$ of an object moving along a line and its initial position $s(0)$, the position function of the object for future times $t \ge 0$ is

$$\underbrace{s(t)}_{\substack{\text{position} \\ \text{at } t}} = \underbrace{s(0)}_{\substack{\text{initial} \\ \text{position}}} + \underbrace{\int_0^t v(x)\, dx}_{\substack{\text{displacement} \\ \text{over } [0, t]}}.$$

QUICK CHECK 3 Is the position $s(t)$ a number or a function? For fixed times $t = a$ and $t = b$, is the displacement $s(b) - s(a)$ a number or a function? ◄

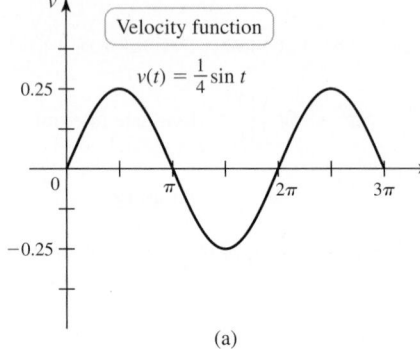

Figure 6.4

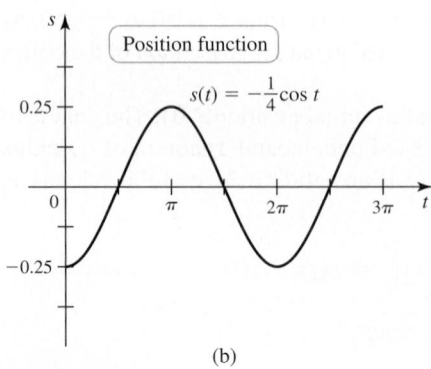

(a)

(b)

Figure 6.5

➤ It is worth repeating that to find the displacement, we need to know only the velocity. To find the position, we must know both the velocity and the initial position $s(0)$.

Theorem 6.1 says that to find the position $s(t)$, we add the displacement over the interval $[0, t]$ to the initial position $s(0)$.

There are two *equivalent* ways to determine the position function:

• Using antiderivatives (Section 4.9)
• Using Theorem 6.1

The latter method is usually more efficient, but either method produces the same result. The following example illustrates both approaches.

EXAMPLE 2   Position from velocity A block hangs at rest from a massless spring at the origin ($s = 0$). At $t = 0$, the block is pulled downward $\frac{1}{4}$ m to its initial position $s(0) = -\frac{1}{4}$ and released (**Figure 6.4**). Its velocity (in m/s) is given by $v(t) = \frac{1}{4}\sin t$, for $t \geq 0$. Assume the upward direction is positive.

a. Find the position of the block, for $t \geq 0$.
b. Graph the position function, for $0 \leq t \leq 3\pi$.
c. When does the block move through the origin for the first time?
d. When does the block reach its highest point for the first time and what is its position at that time? When does the block return to its lowest point?

SOLUTION

a. The velocity function (**Figure 6.5a**) is positive for $0 < t < \pi$, which means the block moves in the positive (upward) direction. At $t = \pi$, the block comes to rest momentarily; for $\pi < t < 2\pi$, the block moves in the negative (downward) direction. We let $s(t)$ be the position at time $t \geq 0$ with the initial position $s(0) = -\frac{1}{4}$ m.

Method 1: Using antiderivatives Because the position is an antiderivative of the velocity, we have

$$s(t) = \int v(t)\, dt = \int \frac{1}{4}\sin t\, dt = -\frac{1}{4}\cos t + C.$$

To determine the arbitrary constant C, we substitute the initial condition $s(0) = -\frac{1}{4}$ into the expression for $s(t)$:

$$-\frac{1}{4} = -\frac{1}{4}\cos 0 + C.$$

Solving for C, we find that $C = 0$. Therefore, the position for any time $t \geq 0$ is

$$s(t) = -\frac{1}{4}\cos t.$$

Method 2: Using Theorem 6.1 Alternatively, we may use the relationship

$$s(t) = s(0) + \int_0^t v(x)\, dx.$$

Substituting $v(x) = \frac{1}{4}\sin x$ and $s(0) = -\frac{1}{4}$, the position function is

$$s(t) = \underbrace{-\frac{1}{4}}_{s(0)} + \int_0^t \underbrace{\frac{1}{4}\sin x}_{v(x)}\, dx$$

$$= -\frac{1}{4} - \left(\frac{1}{4}\cos x\right)\Big|_0^t \quad \text{Evaluate integral.}$$

$$= -\frac{1}{4} - \frac{1}{4}(\cos t - 1) \quad \text{Simplify.}$$

$$= -\frac{1}{4}\cos t. \quad \text{Simplify.}$$

b. The graph of the position function is shown in **Figure 6.5b**. We see that $s(0) = -\frac{1}{4}$ m, as prescribed.

c. The block initially moves in the positive s direction (upward), reaching the origin $(s = 0)$ when $s(t) = -\frac{1}{4}\cos t = 0$. So the block arrives at the origin for the first time when $t = \pi/2$.

QUICK CHECK 4 Without doing further calculations, what are the displacement and distance traveled by the block in Example 2 over the interval $[0, 2\pi]$? ◄

d. The block moves in the positive direction and reaches its high point for the first time when $t = \pi$; the position at that moment is $s(\pi) = \frac{1}{4}$ m. The block then reverses direction and moves in the negative (downward) direction, reaching its low point at $t = 2\pi$. This motion repeats every 2π seconds.

Related Exercises 17, 20 ◄

▶ The terminal velocity of an object depends on its density, shape, and size and on the medium through which it falls. Estimates for humans in free fall in the lower atmosphere vary from 120 mi/hr (54 m/s) to 180 mi/hr (80 m/s).

EXAMPLE 3 Skydiving Suppose a skydiver leaps from a hovering helicopter and falls in a straight line. Assume he reaches a terminal velocity of 80 m/s immediately at $t = 0$ and falls for 19 seconds, at which time he opens his parachute. The velocity decreases linearly to 6 m/s over a two-second period and then remains constant until he reaches the ground at $t = 40$ s. The motion is described by the velocity function

$$v(t) = \begin{cases} 80 & \text{if } 0 \le t < 19 \\ 783 - 37t & \text{if } 19 \le t < 21 \\ 6 & \text{if } 21 \le t \le 40. \end{cases}$$

Determine the height above the ground from which the skydiver jumped.

SOLUTION We let the position of the skydiver increase *downward* with the origin $(s = 0)$ corresponding to the position of the helicopter. The velocity is positive, so the distance traveled by the skydiver equals the displacement, which is

$$\int_0^{40} |v(t)|\, dt = \int_0^{19} 80\, dt + \int_{19}^{21} (783 - 37t)\, dt + \int_{21}^{40} 6\, dt$$

$$= 80t \Big|_0^{19} + \left(783t - \frac{37t^2}{2}\right)\Big|_{19}^{21} + 6t \Big|_{21}^{40} \qquad \text{Fundamental Theorem}$$

$$= 1720. \qquad \text{Evaluate and simplify.}$$

The skydiver jumped from 1720 m above the ground. Notice that the displacement of the skydiver is the area under the velocity curve (**Figure 6.6**).

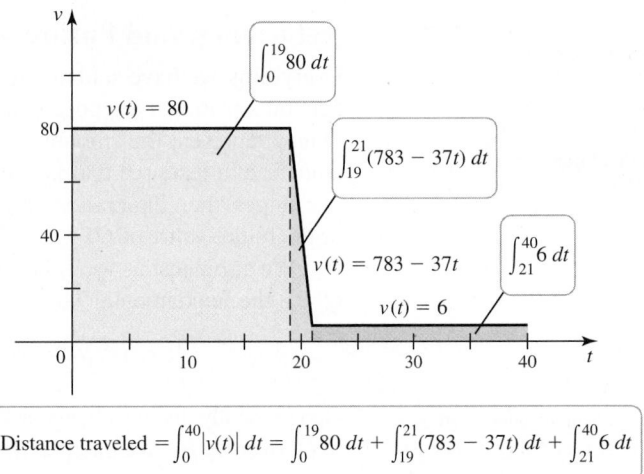

QUICK CHECK 5 Suppose (unrealistically) in Example 3 that the velocity of the skydiver is 80 m/s, for $0 \le t < 20$, and then it changes instantaneously to 6 m/s, for $20 \le t \le 40$. Sketch the velocity function and, without integrating, find the distance the skydiver falls in 40 s. ◄

Distance traveled $= \int_0^{40} |v(t)|\, dt = \int_0^{19} 80\, dt + \int_{19}^{21} (783 - 37t)\, dt + \int_{21}^{40} 6\, dt$

Figure 6.6

Related Exercises 27–28 ◄

Acceleration

Because the acceleration of an object moving along a line is given by $a(t) = v'(t)$, the relationship between velocity and acceleration is the same as the relationship between position and velocity. Given the acceleration of an object, the change in velocity over an interval $[a, b]$ is

$$\text{change in velocity} = v(b) - v(a) = \int_a^b v'(t)\, dt = \int_a^b a(t)\, dt.$$

Furthermore, if we know the acceleration and initial velocity $v(0)$, then we can also find the velocity at future times.

> ➤ Theorem 6.2 is a consequence of the Fundamental Theorem of Calculus.

THEOREM 6.2   Velocity from Acceleration

Given the acceleration $a(t)$ of an object moving along a line and its initial velocity $v(0)$, the velocity of the object for future times $t \geq 0$ is

$$v(t) = v(0) + \int_0^t a(x)\, dx.$$

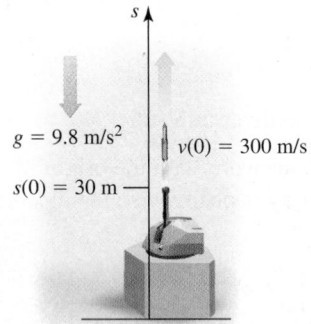

$g = 9.8 \text{ m/s}^2$

$v(0) = 300 \text{ m/s}$

$s(0) = 30 \text{ m}$

Figure 6.7

EXAMPLE 4   Motion in a gravitational field   An artillery shell is fired directly upward with an initial velocity of 300 m/s from a point 30 m above the ground (**Figure 6.7**). Assume only the force of gravity acts on the shell and it produces an acceleration of 9.8 m/s². Find the velocity of the shell while it is in the air.

SOLUTION   We let the positive direction be upward with the origin $(s = 0)$ corresponding to the ground. The initial velocity of the shell is $v(0) = 300$ m/s. The acceleration due to gravity is downward; therefore, $a(t) = -9.8$ m/s². Integrating the acceleration, the velocity is

$$v(t) = \underbrace{v(0)}_{300 \text{ m/s}} + \int_0^t \underbrace{a(x)}_{-9.8 \text{ m/s}^2}\, dx = 300 + \int_0^t (-9.8)\, dx = 300 - 9.8t.$$

The velocity decreases from its initial value of 300 m/s, reaching zero at the high point of the trajectory when $v(t) = 300 - 9.8t = 0$, or at $t \approx 30.6$ s (**Figure 6.8**). At this point, the velocity becomes negative, and the shell begins its descent to Earth.

Knowing the velocity function, you could now find the position function using the methods of Example 3.

Related Exercises 30–31 ◄

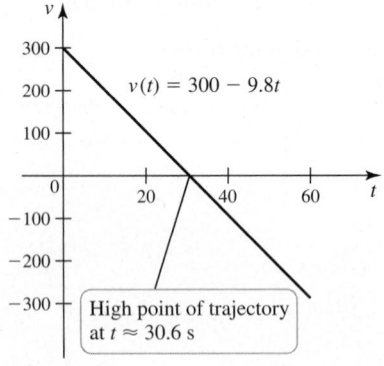

$v(t) = 300 - 9.8t$

High point of trajectory at $t \approx 30.6$ s

Figure 6.8

Net Change and Future Value

Everything we have said about velocity, position, and displacement carries over to more general situations. Suppose you are interested in some quantity Q that changes over *time*; Q may represent the amount of water in a reservoir, the population of a cell culture, or the amount of a resource that is consumed or produced. If you are given the rate Q' at which Q changes, then integration allows you to calculate either the net change in the quantity Q or the future value of Q.

We argue just as we did for velocity and position: Because $Q(t)$ is an antiderivative of $Q'(t)$, the Fundamental Theorem of Calculus tells us that

$$\int_a^b Q'(t)\, dt = Q(b) - Q(a) = \text{net change in } Q \text{ over } [a, b].$$

> ➤ Note that the units in the integral are consistent. For example, if Q' has units of gallons/second, and t and x have units of seconds, then $Q'(x)\, dx$ has units of (gallons/second)(seconds) = gallons, which are the units of Q.

Geometrically, the net change in Q over the time interval $[a, b]$ is the net area under the graph of Q' over $[a, b]$. We interpret the product $Q'(t)\, dt$ as a change in Q over a small increment of time. Integrating $Q'(t)$ accumulates, or adds up, these small changes over the interval $[a, b]$. The result is the net change in Q between $t = a$ and $t = b$. We see that accumulating the rate of change of a quantity over the interval gives the net change in that quantity over the interval.

Alternatively, suppose we are given both the rate of change Q' and the initial value $Q(0)$. Integrating over the interval $[0, t]$, where $t \geq 0$, we have

$$\int_0^t Q'(x)\, dx = Q(t) - Q(0).$$

Rearranging this equation, we write the value of Q at any future time $t \geq 0$ as

$$\underbrace{Q(t)}_{\substack{\text{future} \\ \text{value}}} = \underbrace{Q(0)}_{\substack{\text{initial} \\ \text{value}}} + \underbrace{\int_0^t Q'(x)\, dx}_{\substack{\text{net change} \\ \text{over } [0,\, t]}}.$$

> At the risk of being repetitious, Theorem 6.3 is also a consequence of the Fundamental Theorem of Calculus. We assume Q' is an integrable function.

THEOREM 6.3   Net Change and Future Value

Suppose a quantity Q changes over time at a known rate Q'. Then the **net change** in Q between $t = a$ and $t = b > a$ is

$$\underbrace{Q(b) - Q(a)}_{\text{net change in } Q} = \int_a^b Q'(t)\, dt.$$

Given the initial value $Q(0)$, the **future value** of Q at time $t \geq 0$ is

$$Q(t) = Q(0) + \int_0^t Q'(x)\, dx.$$

The correspondences between velocity–displacement problems and more general problems are shown in Table 6.1.

Table 6.1

Velocity–Displacement Problems	General Problems
Position $s(t)$	Quantity $Q(t)$ (such as volume or population)
Velocity: $s'(t) = v(t)$	Rate of change: $Q'(t)$
Displacement: $s(b) - s(a) = \int_a^b v(t)\, dt$	Net change: $Q(b) - Q(a) = \int_a^b Q'(t)\, dt$
Future position: $s(t) = s(0) + \int_0^t v(x)\, dx$	Future value of Q: $Q(t) = Q(0) + \int_0^t Q'(x)\, dx$

We now consider two general applications that involve integrating rates of change.

EXAMPLE 5   Cell growth A culture of cells in a lab has a population of 100 cells when nutrients are added at time $t = 0$. Suppose the population $N(t)$ (in cells/hr) increases at a rate given by

$$N'(t) = 90e^{-0.1t}.$$

Find $N(t)$, for $t \geq 0$.

SOLUTION As shown in **Figure 6.9**, the growth rate is large when t is small (plenty of food and space) and decreases as t increases. Knowing that the initial population is $N(0) = 100$ cells, we can find the population $N(t)$ at any future time $t \geq 0$ using Theorem 6.3:

$$
\begin{aligned}
N(t) &= N(0) + \int_0^t N'(x)\, dx \\
&= \underbrace{100}_{N(0)} + \int_0^t \underbrace{90e^{-0.1x}}_{N'(x)}\, dx \\
&= 100 + \left[\left(\frac{90}{-0.1} \right) e^{-0.1x} \right]_0^t \quad \text{Fundamental Theorem} \\
&= 1000 - 900e^{-0.1t}. \quad \text{Simplify.}
\end{aligned}
$$

> Although N is a positive integer (the number of cells), we treat it as a continuous variable in this example.

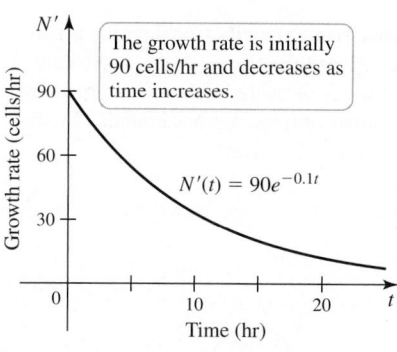

The growth rate is initially 90 cells/hr and decreases as time increases.

$N'(t) = 90e^{-0.1t}$

Figure 6.9

The graph of the population function (**Figure 6.10**) shows that the population increases, but at a decreasing rate. Note that the initial condition $N(0) = 100$ cells is satisfied and that the population size approaches 1000 cells as $t \to \infty$.

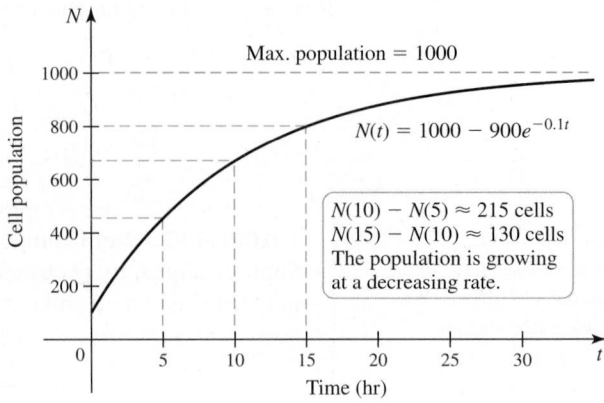

Figure 6.10

Related Exercises 43–44 ◄

EXAMPLE 6   **Production costs**   A book publisher estimates that the marginal cost of producing a particular title (in dollars/book) is given by

$$C'(x) = 12 - 0.0002x,$$

where $0 \le x \le 50{,}000$ is the number of books printed. What is the cost of producing the 12,001st through the 15,000th book?

> Although x is a positive integer (the number of books produced), we treat it as a continuous variable in this example.

SOLUTION   Recall from Section 3.6 that the cost function $C(x)$ is the cost required to produce x units of a product. The marginal cost $C'(x)$ is the approximate cost of producing one additional unit after x units have already been produced. The cost of producing books $x = 12{,}001$ through $x = 15{,}000$ is the cost of producing 15,000 books minus the cost of producing the first 12,000 books. Therefore, the cost in dollars of producing books 12,001 through 15,000 is

$$
\begin{aligned}
C(15{,}000) - C(12{,}000) &= \int_{12{,}000}^{15{,}000} C'(x)\,dx \\
&= \int_{12{,}000}^{15{,}000} (12 - 0.0002x)\,dx \quad \text{Substitute for } C'(x). \\
&= \left. (12x - 0.0001x^2) \right|_{12{,}000}^{15{,}000} \quad \text{Fundamental Theorem} \\
&= 27{,}900. \quad \text{Simplify.}
\end{aligned}
$$

Related Exercises 55–56 ◄

QUICK CHECK 6 Is the cost of increasing production from 9000 books to 12,000 books in Example 6 more or less than the cost of increasing production from 12,000 books to 15,000 books? Explain. ◄

SECTION 6.1 EXERCISES

Getting Started

1. Explain the meaning of position, displacement, and distance traveled as they apply to an object moving along a line.

2. Suppose the velocity of an object moving along a line is positive. Are displacement and distance traveled equal? Explain.

3. Given the velocity function v of an object moving along a line, explain how definite integrals can be used to find the displacement of the object.

4. Explain how to use definite integrals to find the net change in a quantity, given the rate of change of that quantity.

5. Given the rate of change of a quantity Q and its initial value $Q(0)$, explain how to find the value of Q at a future time $t \ge 0$.

6. What is the result of integrating a population growth rate between times $t = a$ and $t = b$, where $b > a$?

7. **Displacement and distance from velocity** Consider the graph shown in the figure, which gives the velocity of an object moving along a line. Assume time is measured in hours and distance is measured in miles. The areas of three regions bounded by the velocity curve and the t-axis are also given.

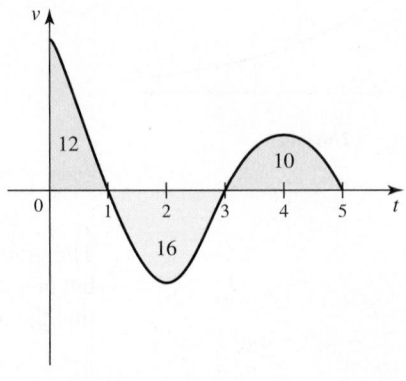

a. On what intervals is the object moving in the positive direction?
b. What is the displacement of the object over the interval $[0, 3]$?
c. What is the total distance traveled by the object over the interval $[1, 5]$?
d. What is the displacement of the object over the interval $[0, 5]$?
e. Describe the position of the object relative to its initial position after 5 hours.

8. **Displacement and distance from velocity** Consider the velocity function shown below of an object moving along a line. Assume time is measured in seconds and distance is measured in meters. The areas of four regions bounded by the velocity curve and the t-axis are also given.

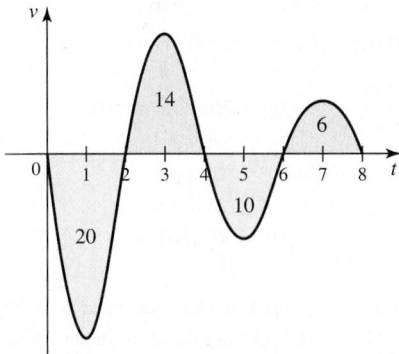

a. On what intervals is the object moving in the negative direction?
b. What is the displacement of the object over the interval $[2, 6]$?
c. What is the total distance traveled by the object over the interval $[0, 6]$?
d. What is the displacement of the object over the interval $[0, 8]$?
e. Describe the position of the object relative to its initial position after 8 seconds.

9–10. Velocity graphs *The figures show velocity functions for motion along a line. Assume the motion begins with an initial position of $s(0) = 0$. Determine the following.*

a. *The displacement between $t = 0$ and $t = 5$*
b. *The distance traveled between $t = 0$ and $t = 5$*
c. *The position at $t = 5$*
d. *A piecewise function for $s(t)$*

9.

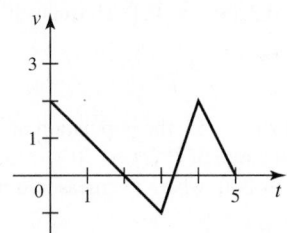

10.

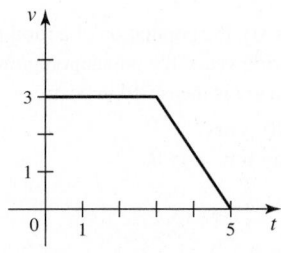

Practice Exercises

11. **Distance traveled and displacement** Suppose an object moves along a line with velocity (in m/s) $v(t) = 3 \sin 2t$, for $0 \le t \le 2\pi$, where t is measured in seconds (see figure).

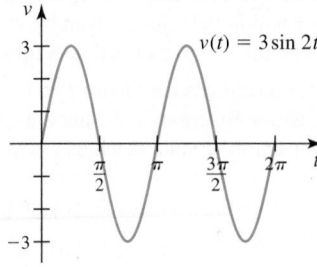

a. Find the distance traveled by the object on the time interval $[0, \pi/2]$.
b. Find the displacement of the object on the time intervals $[0, \pi/2], [0, \pi], [0, 3\pi/2],$ and $[0, 2\pi]$. (*Hint:* Use your answer to part (a) together with the symmetry of the graph to find the displacement values.)
c. Find the distance traveled by the object on the time interval $[0, 2\pi]$.

12. **Distance traveled and displacement** Suppose an object moves along a line with velocity (in ft/s) $v(t) = 6 - 2t$, for $0 \le t \le 6$, where t is measured in seconds.

a. Graph the velocity function on the interval $0 \le t \le 6$. Determine when the motion is in the positive direction and when it is in the negative direction on $0 \le t \le 6$.
b. Find the displacement of the object on the interval $0 \le t \le 6$.
c. Find the distance traveled by the object on the interval $0 \le t \le 6$.

13–16. Displacement from velocity *Consider an object moving along a line with the given velocity v. Assume time t is measured in seconds and velocities have units of m/s.*

a. *Determine when the motion is in the positive direction and when it is in the negative direction.*
b. *Find the displacement over the given interval.*
c. *Find the distance traveled over the given interval.*

13. $v(t) = 3t^2 - 6t$ on $[0, 3]$

14. $v(t) = 4t^3 - 24t^2 + 20t$ on $[0, 5]$

15. $v(t) = 3t^2 - 18t + 24$ on $[0, 5]$

16. $v(t) = 50e^{-2t}$ on $[0, 4]$

17–22. Position from velocity *Consider an object moving along a line with the given velocity v and initial position.*

a. *Determine the position function, for $t \ge 0$, using the antiderivative method*
b. *Determine the position function, for $t \ge 0$, using the Fundamental Theorem of Calculus (Theorem 6.1). Check for agreement with the answer to part (a).*

17. $v(t) = \sin t$ on $[0, 2\pi]$; $s(0) = 1$

18. $v(t) = -t^3 + 3t^2 - 2t$ on $[0, 3]$; $s(0) = 4$

19. $v(t) = 6 - 2t$ on $[0, 5]$; $s(0) = 0$

20. $v(t) = 3 \sin \pi t$ on $[0, 4]$; $s(0) = 1$

21. $v(t) = 9 - t^2$ on $[0, 4]$; $s(0) = -2$

22. $v(t) = \dfrac{1}{t + 1}$ on $[0, 8]$; $s(0) = -4$

T 23. Oscillating motion A mass hanging from a spring is set in motion, and its ensuing velocity is given by $v(t) = 2\pi \cos \pi t$, for $t \geq 0$. Assume the positive direction is upward and $s(0) = 0$.

   a. Determine the position function, for $t \geq 0$.
   b. Graph the position function on the interval $[0, 4]$.
   c. At what times does the mass reach its low point the first three times?
   d. At what times does the mass reach its high point the first three times?

24. Cycling distance A cyclist rides down a long straight road with a velocity (in m/min) given by $v(t) = 400 - 20t$, for $0 \leq t \leq 10$, where t is measured in minutes.

   a. How far does the cyclist travel in the first 5 min?
   b. How far does the cyclist travel in the first 10 min?
   c. How far has the cyclist traveled when her velocity is 250 m/min?

25. Flying into a headwind The velocity (in mi/hr) of an airplane flying into a headwind is given by $v(t) = 30(16 - t^2)$, for $0 \leq t \leq 3$. Assume $s(0) = 0$ and t is measured in hours.

   a. Determine the position function, for $0 \leq t \leq 3$.
   b. How far does the airplane travel in the first 2 hr?
   c. How far has the airplane traveled at the instant its velocity reaches 400 mi/hr?

26. Day hike The velocity (in mi/hr) of a hiker walking along a straight trail is given by $v(t) = 3 \sin^2 \dfrac{\pi t}{2}$, for $0 \leq t \leq 4$.

Assume $s(0) = 0$ and t is measured in hours.

   a. Determine the position function, for $0 \leq t \leq 4$.
   (*Hint:* $\sin^2 t = \dfrac{1 - \cos 2t}{2}$.)
   b. What is the distance traveled by the hiker in the first 15 min of the hike?
   c. What is the hiker's position at $t = 3$?

27. Piecewise velocity The velocity of a (fast) automobile on a straight highway is given by the function

$$v(t) = \begin{cases} 3t & \text{if } 0 \leq t < 20 \\ 60 & \text{if } 20 \leq t < 45 \\ 240 - 4t & \text{if } t \geq 45, \end{cases}$$

where t is measured in seconds and v has units of m/s.

   a. Graph the velocity function, for $0 \leq t \leq 70$. When is the velocity a maximum? When is the velocity zero?
   b. What is the distance traveled by the automobile in the first 30 s?
   c. What is the distance traveled by the automobile in the first 60 s?
   d. What is the position of the automobile when $t = 75$?

28. Probe speed A data collection probe is dropped from a stationary balloon, and it falls with a velocity (in m/s) given by $v(t) = 9.8t$, neglecting air resistance. After 10 s, a chute deploys and the probe immediately slows to a constant speed of 10 m/s, which it maintains until it enters the ocean.

   a. Graph the velocity function.
   b. How far does the probe fall in the first 30 s after it is released?
   c. If the probe was released from an altitude of 3 km, when does it enter the ocean?

29–36. Position and velocity from acceleration *Find the position and velocity of an object moving along a straight line with the given acceleration, initial velocity, and initial position. Use the Fundamental Theorem of Calculus (Theorems 6.1 and 6.2).*

29. $a(t) = -32$; $v(0) = 70$; $s(0) = 10$

30. $a(t) = -32$; $v(0) = 50$; $s(0) = 0$

31. $a(t) = -9.8$; $v(0) = 20$; $s(0) = 0$

32. $a(t) = e^{-t}$; $v(0) = 60$; $s(0) = 40$

33. $a(t) = -0.01t$; $v(0) = 10$; $s(0) = 0$

34. $a(t) = \dfrac{20}{(t + 2)^2}$; $v(0) = 20$; $s(0) = 10$

35. $a(t) = \cos 2t$; $v(0) = 5$; $s(0) = 7$

36. $a(t) = \dfrac{2t}{(t^2 + 1)^2}$; $v(0) = 0$; $s(0) = 0$

37. Acceleration A drag racer accelerates at $a(t) = 88$ ft/s^2. Assume $v(0) = 0$, $s(0) = 0$, and t is measured in seconds.

   a. Determine the position function, for $t \geq 0$.
   b. How far does the racer travel in the first 4 seconds?
   c. At this rate, how long will it take the racer to travel $\frac{1}{4}$ mi?
   d. How long does it take the racer to travel 300 ft?
   e. How far has the racer traveled when it reaches a speed of 178 ft/s?

38. Deceleration A car slows down with an acceleration of $a(t) = -15$ ft/s^2. Assume $v(0) = 60$ ft/s, $s(0) = 0$, and t is measured in seconds.

   a. Determine the position function, for $t \geq 0$.
   b. How far does the car travel in the time it takes to come to rest?

39. Approaching a station At $t = 0$, a train approaching a station begins decelerating from a speed of 80 mi/hr according to the acceleration function $a(t) = -1280(1 + 8t)^{-3}$, where $t \geq 0$ is measured in hours. How far does the train travel between $t = 0$ and $t = 0.2$? Between $t = 0.2$ and $t = 0.4$? The units of acceleration are mi/hr^2.

40–43. Population growth

40. Starting with an initial value of $P(0) = 55$, the population of a prairie dog community grows at a rate of $P'(t) = 20 - t/5$ (prairie dogs/month), for $0 \leq t \leq 200$, where t is measured in months.

   a. What is the population 6 months later?
   b. Find the population $P(t)$, for $0 \leq t \leq 200$.

41. When records were first kept ($t = 0$), the population of a rural town was 250 people. During the following years, the population grew at a rate of $P'(t) = 30(1 + \sqrt{t})$, where t is measured in years.

   a. What is the population after 20 years?
   b. Find the population $P(t)$ at any time $t \geq 0$.

42. The population of a community of foxes is observed to fluctuate on a 10-year cycle due to variations in the availability of prey. When population measurements began ($t = 0$), the population was 35 foxes. The growth rate in units of foxes/year was observed to be

$$P'(t) = 5 + 10 \sin \frac{\pi t}{5}.$$

a. What is the population 15 years later? 35 years later?
b. Find the population $P(t)$ at any time $t \geq 0$.

43. A culture of bacteria in a Petri dish has an initial population of 1500 cells and grows at a rate (in cells/day) of $N'(t) = 100e^{-0.25t}$. Assume t is measured in days.

a. What is the population after 20 days? After 40 days?
b. Find the population $N(t)$ at any time $t \geq 0$.

T 44. Cancer treatment A cancerous tumor in a mouse is treated with a chemotherapy drug. After treatment, the rate of change in the size of the tumor (in cm^3/day) is given by the function $r(t) = 0.0025e^{0.25t} - 0.1485e^{-0.15t}$, where t is measured in days.

a. Find the value of t_0 for which $r(t_0) = 0$.
b. Plot the function $r(t)$, for $0 \leq t \leq 15$, and describe what happens to the tumor over the 15-day period after treatment.
c. Evaluate $\int_0^{t_0} r(t)\, dt$ and interpret the physical meaning of this integral.

45. Oil production An oil refinery produces oil at a variable rate given by

$$Q'(t) = \begin{cases} 800 & \text{if } 0 \leq t < 30 \\ 2600 - 60t & \text{if } 30 \leq t < 40 \\ 200 & \text{if } t \geq 40, \end{cases}$$

where t is measured in days and Q is measured in barrels.

a. How many barrels are produced in the first 35 days?
b. How many barrels are produced in the first 50 days?
c. Without using integration, determine the number of barrels produced over the interval $[60, 80]$.

46. Flow rates in the Spokane River The daily discharge of the Spokane River as it flows through Spokane, Washington, in April and June is modeled by the functions

$$r_1(t) = 0.25t^2 + 37.46t + 722.47 \text{ (April) and}$$
$$r_2(t) = 0.90t^2 - 69.06t + 2053.12 \text{ (June)},$$

where the discharge is measured in millions of cubic feet per day, and $t = 0$ corresponds to the beginning of the first day of the month (see figure).

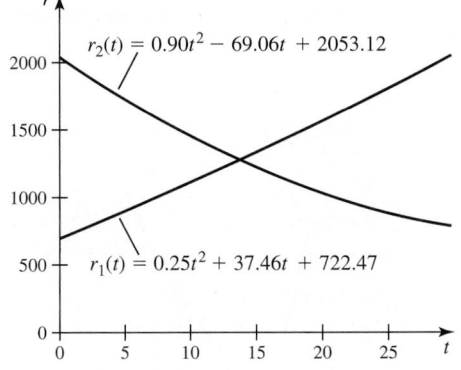

a. Determine the total amount of water that flows through Spokane in April (30 days).
b. Determine the total amount of water that flows through Spokane in June (30 days).
c. The Spokane River flows out of Lake Coeur d'Alene, which contains approximately 0.67 mi^3 of water. Determine the percentage of Lake Coeur d'Alene's volume that flows through Spokane in April and June.

47. Depletion of natural resources Suppose $r(t) = r_0 e^{-kt}$, with $k > 0$, is the rate at which a nation extracts oil, where $r_0 = 10^7$ barrels/yr is the current rate of extraction. Suppose also that the estimate of the total oil reserve is 2×10^9 barrels.

a. Find $Q(t)$, the total amount of oil extracted by the nation after t years.
b. Evaluate $\lim_{t \to \infty} Q(t)$ and explain the meaning of this limit.
c. Find the minimum decay constant k for which the total oil reserves will last forever.
d. Suppose $r_0 = 2 \times 10^7$ barrels/yr and the decay constant k is the minimum value found in part (c). How long will the total oil reserves last?

48. Filling a tank A 2000-liter cistern is empty when water begins flowing into it (at $t = 0$) at a rate (in L/min) given by $Q'(t) = 3\sqrt{t}$, where t is measured in minutes.

a. How much water flows into the cistern in 1 hour?
b. Find the function that gives the amount of water in the tank at any time $t \geq 0$.
c. When will the tank be full?

T 49. Filling a reservoir A reservoir with a capacity of 2500 m^3 is filled with a single inflow pipe. The reservoir is empty when the inflow pipe is opened at $t = 0$. Letting $Q(t)$ be the amount of water in the reservoir at time t, the flow rate of water into the reservoir (in m^3/hr) oscillates on a 24-hr cycle (see figure) and is given by

$$Q'(t) = 20\left(1 + \cos \frac{\pi t}{12}\right).$$

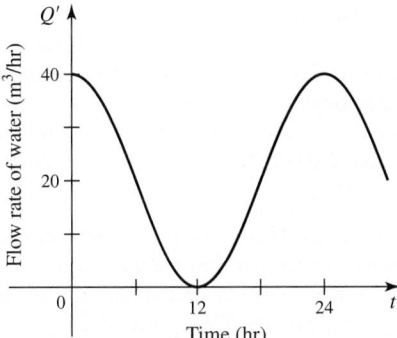

a. How much water flows into the reservoir in the first 2 hr?
b. Find the function that gives the amount of water in the reservoir over the interval $[0, t]$, where $t \geq 0$.
c. When is the reservoir full?

50. Blood flow A typical human heart pumps 70 mL of blood (the stroke volume) with each beat. Assuming a heart rate of 60 beats/min (1 beat/s), a reasonable model for the outflow rate of the heart is $V'(t) = 70(1 + \sin 2\pi t)$, where $V(t)$ is the amount of blood (in milliliters) pumped over the interval $[0, t]$, $V(0) = 0$, and t is measured in seconds.

a. Verify that the amount of blood pumped over a one-second interval is 70 mL.
b. Find the function that gives the total blood pumped between $t = 0$ and a future time $t > 0$.
c. What is the cardiac output over a period of 1 min? (Use calculus; then check your answer with algebra.)

51. Air flow in the lungs A simple model (with different parameters for different people) for the flow of air in and out of the lungs is

$$V'(t) = -\frac{\pi}{2} \sin \frac{\pi t}{2},$$

where $V(t)$ (measured in liters) is the volume of air in the lungs at time $t \geq 0$, t is measured in seconds, and $t = 0$ corresponds to a time at which the lungs are full and exhalation begins. Only a fraction of the air in the lungs is exchanged with each breath. The amount that is exchanged is called the *tidal volume*.

a. Find the volume function V assuming $V(0) = 6$ L.
b. What is the breathing rate in breaths/min?
c. What is the tidal volume and what is the total capacity of the lungs?

52. Oscillating growth rates Some species have growth rates that oscillate with an (approximately) constant period P. Consider the growth rate function

$$N'(t) = r + A \sin \frac{2\pi t}{P},$$

where A and r are constants with units of individuals/yr, and t is measured in years. A species becomes extinct if its population ever reaches 0 after $t = 0$.

a. Suppose $P = 10, A = 20$, and $r = 0$. If the initial population is $N(0) = 10$, does the population ever become extinct? Explain.
b. Suppose $P = 10, A = 20$, and $r = 0$. If the initial population is $N(0) = 100$, does the population ever become extinct? Explain.
c. Suppose $P = 10, A = 50$, and $r = 5$. If the initial population is $N(0) = 10$, does the population ever become extinct? Explain.
d. Suppose $P = 10, A = 50$, and $r = -5$. Find the initial population $N(0)$ needed to ensure that the population never becomes extinct.

53. Power and energy The terms *power* and *energy* are often used interchangeably, but they are quite different. **Energy** is what makes matter move or heat up and is measured in units of **joules** (J) or **Calories** (Cal), where 1 Cal = 4184 J. One hour of walking consumes roughly 10^6 J, or 250 Cal. On the other hand, **power** is the rate at which energy is used and is measured in **watts** (W; 1 W = 1 J/s). Other useful units of power are **kilowatts** (1 kW = 10^3 W) and **megawatts** (1 MW = 10^6 W). If energy is used at a rate of 1 kW for 1 hr, the total amount of energy used is 1 **kilowatt-hour** (kWh), which is 3.6×10^6 J.

Suppose the power function of a large city over a 24-hr period is given by

$$P(t) = E'(t) = 300 - 200 \sin \frac{\pi t}{12},$$

where P is measured in megawatts and $t = 0$ corresponds to 6:00 P.M. (see figure).

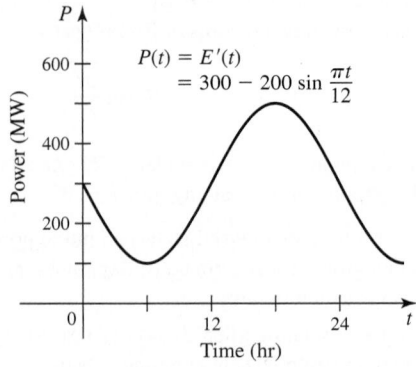

a. How much energy is consumed by this city in a typical 24-hr period? Express the answer in megawatt-hours and in joules.
b. Burning 1 kg of coal produces about 450 kWh of energy. How many kilograms of coal are required to meet the energy needs of the city for 1 day? For 1 year?
c. Fission of 1 gram of uranium-235 (U-235) produces about 16,000 kWh of energy. How many grams of uranium are needed to meet the energy needs of the city for 1 day? For 1 year?
d. A typical wind turbine generates electrical power at a rate of about 200 kW. Approximately how many wind turbines are needed to meet the average energy needs of the city?

54. Carbon uptake An important process in the study of global warming and greenhouse gases is the *net ecosystem exchange*, which is the rate at which carbon leaves an ecosystem and enters the atmosphere in a particular geographic region. Let $N(t)$ equal the net ecosystem exchange on an average July day in a high-altitude coniferous forest, where $N(t)$ is measured in grams of carbon per square meter per hour and t is the number of hours past midnight so that $0 \leq t \leq 24$ (see figure). Negative values of N correspond to times when the amount of carbon in the atmosphere decreases, and positive values of N occur when the amount of carbon in the atmosphere increases.

a. Trees and other plants help reduce carbon emissions in the atmosphere by using photosynthesis to absorb carbon dioxide and release oxygen. Give a possible explanation why N is negative on the interval $5 < t < 17$.
b. The *cumulative net carbon uptake*, $\int_0^{24} N(t)\, dt$, is the net change in the amount of carbon in the atmosphere over a 24-hour period in the coniferous forest. Use a midpoint Riemann sum with $n = 12$ subintervals to estimate the cumulative net carbon uptake in the coniferous forest and interpret the result. (*Source: Forest Carbon Uptake and the Fundamental Theorem of Calculus, The College Mathematics Journal, 44, 5, Nov 2013*)

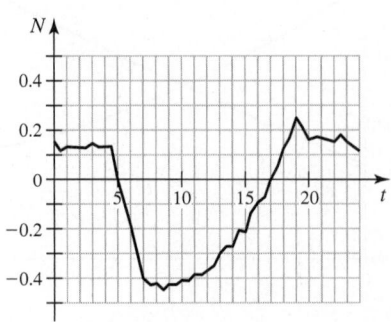

55–58. Marginal cost *Consider the following marginal cost functions.*

a. *Find the additional cost incurred in dollars when production is increased from* 100 *units to* 150 *units.*

b. *Find the additional cost incurred in dollars when production is increased from* 500 *units to* 550 *units.*

55. $C'(x) = 2000 - 0.5x$ **56.** $C'(x) = 200 - 0.05x$

57. $C'(x) = 300 + 10x - 0.01x^2$

58. $C'(x) = 3000 - x - 0.001x^2$

59. Explain why or why not Determine whether the following statements are true and give an explanation or counterexample.

   a. The distance traveled by an object moving along a line is the same as the displacement of the object.
   b. When the velocity is positive on an interval, the displacement and the distance traveled on that interval are equal.
   c. Consider a tank that is filled and drained at a flow rate of $V'(t) = 1 - \dfrac{t^2}{100}$ (gal/min), for $t \geq 0$, where t is measured in minutes. It follows that the volume of water in the tank increases for 10 min and then decreases until the tank is empty.
   d. A particular marginal cost function has the property that it is positive and decreasing. The cost of increasing production from A units to $2A$ units is greater than the cost of increasing production from $2A$ units to $3A$ units.

60–63. Equivalent constant velocity *Consider the following velocity functions. In each case, complete the sentence: The same distance could have been traveled over the given time period at a <u>constant</u> velocity of _____.*

60. $v(t) = 2t + 6$, for $0 \leq t \leq 8$

61. $v(t) = 1 - \dfrac{t^2}{16}$, for $0 \leq t \leq 4$

62. $v(t) = 2 \sin t$, for $0 \leq t \leq \pi$

63. $v(t) = t(25 - t^2)^{1/2}$, for $0 \leq t \leq 5$

Explorations and Challenges

64. Where do they meet? Kelly started at noon ($t = 0$) riding a bike from Niwot to Berthoud, a distance of 20 km, with velocity $v(t) = \dfrac{15}{(t + 1)^2}$ (decreasing because of fatigue). Sandy started at noon ($t = 0$) riding a bike in the opposite direction from Berthoud to Niwot with velocity $u(t) = \dfrac{20}{(t + 1)^2}$ (also decreasing because of fatigue). Assume distance is measured in kilometers and time is measured in hours.

   a. Make a graph of Kelly's distance from Niwot as a function of time.
   b. Make a graph of Sandy's distance from Berthoud as a function of time.
   c. When do they meet? How far has each person traveled when they meet?
   d. More generally, if the riders' speeds are $v(t) = \dfrac{A}{(t + 1)^2}$ and $u(t) = \dfrac{B}{(t + 1)^2}$ and the distance between the towns is D, what conditions on A, B, and D must be met to ensure that the riders will pass each other?
   e. Looking ahead: With the velocity functions given in part (d), make a conjecture about the maximum distance each person can ride (given unlimited time).

65. Bike race Theo and Sasha start at the same place on a straight road, riding bikes with the following velocities (measured in mi/hr). Assume t is measured in hours.

   Theo: $v_T(t) = 10$, for $t \geq 0$

   Sasha: $v_S(t) = 15t$, for $0 \leq t \leq 1$, and $v_S(t) = 15$, for $t > 1$

   a. Graph the velocity function for both riders.
   b. If the riders ride for 1 hr, who rides farther? Interpret your answer geometrically using the graphs of part (a).
   c. If the riders ride for 2 hr, who rides farther? Interpret your answer geometrically using the graphs of part (a).
   d. Which rider arrives first at the 10-, 15-, and 20-mile markers of the race? Interpret your answers geometrically using the graphs of part (a).
   e. Suppose Sasha gives Theo a head start of 0.2 mi and the riders ride for 20 mi. Who wins the race?
   f. Suppose Sasha gives Theo a head start of 0.2 hr and the riders ride for 20 mi. Who wins the race?

66. Two runners At noon ($t = 0$), Alicia starts running along a long straight road at 4 mi/hr. Her velocity decreases according to the function $v(t) = \dfrac{4}{t + 1}$, for $t \geq 0$. At noon, Boris also starts running along the same road with a 2-mi head start on Alicia; his velocity is given by $u(t) = \dfrac{2}{t + 1}$, for $t \geq 0$. Assume t is measured in hours.

   a. Find the position functions for Alicia and Boris, where $s = 0$ corresponds to Alicia's starting point.
   b. When, if ever, does Alicia overtake Boris?

67. Snowplow problem With snow on the ground and falling at a constant rate, a snowplow began plowing down a long straight road at noon. The plow traveled twice as far in the first hour as it did in the second hour. At what time did the snow start falling? Assume the plowing rate is inversely proportional to the depth of the snow.

68. Variable gravity At Earth's surface, the acceleration due to gravity is approximately $g = 9.8$ m/s² (with local variations). However, the acceleration decreases with distance from the surface according to Newton's law of gravitation. At a distance of y meters from Earth's surface, the acceleration is given by

$$a(y) = -\frac{g}{(1 + y/R)^2},$$

where $R = 6.4 \times 10^6$ m is the radius of Earth.

   a. Suppose a projectile is launched upward with an initial velocity of v_0 m/s. Let $v(t)$ be its velocity and $y(t)$ its height (in meters) above the surface t seconds after the launch. Neglecting forces such as air resistance, explain why $\dfrac{dv}{dt} = a(y)$ and $\dfrac{dy}{dt} = v(t)$.
   b. Use the Chain Rule to show that $\dfrac{dv}{dt} = \dfrac{1}{2} \dfrac{d}{dy}(v^2)$.
   c. Show that the equation of motion for the projectile is $\dfrac{1}{2} \dfrac{d}{dy}(v^2) = a(y)$, where $a(y)$ is given previously.
   d. Integrate both sides of the equation in part (c) with respect to y using the fact that when $y = 0$, $v = v_0$. Show that

$$\frac{1}{2}(v^2 - v_0^2) = gR\left(\frac{1}{1 + y/R} - 1\right).$$

e. When the projectile reaches its maximum height, $v = 0$. Use this fact to determine that the maximum height is

$$y_{\max} = \frac{Rv_0^2}{2gR - v_0^2}.$$

f. Graph $y_{\max}$ as a function of v_0. What is the maximum height when $v_0 = 500$ m/s, 1500 m/s, and 5 km/s?

g. Show that the value of v_0 needed to put the projectile into orbit (called the escape velocity) is $\sqrt{2gR}$.

69. Suppose f and g have continuous derivatives on an interval $[a, b]$. Prove that if $f(a) = g(a)$ and $f(b) = g(b)$, then

$$\int_a^b f'(x)\, dx = \int_a^b g'(x)\, dx.$$

70. Use Exercise 69 to prove that if two runners start and finish at the same time and place, then *regardless of the velocities at which they run*, their displacements are equal.

71. Use Exercise 69 to prove that if two trails start at the same place and finish at the same place, then *regardless of the ups and downs of the trails*, they have the same net change in elevation.

72. Without evaluating integrals, prove that

$$\int_0^2 \frac{d}{dx}(12 \sin \pi x^2)\, dx = \int_0^2 \frac{d}{dx}(x^{10}(2 - x)^3)\, dx.$$

QUICK CHECK ANSWERS

1. Displacement $= -20$ mi (20 mi south); distance traveled $= 100$ mi   **2.** Suppose the object moves in the positive direction, for $0 \le t \le 3$, and then moves in the negative direction, for $3 < t \le 5$.   **3.** A function; a number   **4.** Displacement $= 0$; distance traveled $= 1$ **5.** 1720 m   **6.** The production cost would increase more between 9000 and 12,000 books than between 12,000 and 15,000 books. Graph C' and look at the area under the curve. ◄

6.2 Regions Between Curves

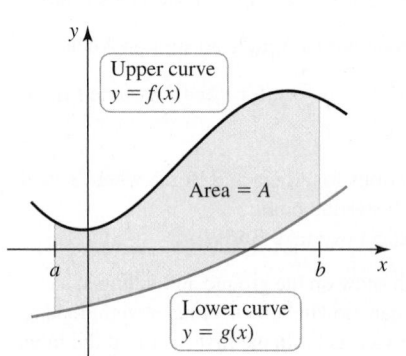

Figure 6.11

In this section, the method for finding the area of a region bounded by a single curve is generalized to regions bounded by two or more curves. Consider two functions f and g continuous on an interval $[a, b]$ on which $f(x) \ge g(x)$ (**Figure 6.11**). The goal is to find the area A of the region bounded by the two curves and the vertical lines $x = a$ and $x = b$.

Once again, we rely on the *slice-and-sum* strategy (Section 5.2) for finding areas by Riemann sums. The interval $[a, b]$ is partitioned into n subintervals using uniformly spaced grid points separated by a distance $\Delta x = (b - a)/n$ (**Figure 6.12**). On each subinterval, we build a rectangle extending from the lower curve to the upper curve. On the kth subinterval, a point x_k^* is chosen, and the height of the corresponding rectangle is taken to be $f(x_k^*) - g(x_k^*)$. Therefore, the area of the kth rectangle is $(f(x_k^*) - g(x_k^*))\Delta x$ (**Figure 6.13**). Summing the areas of the n rectangles gives an approximation to the area of the region between the curves:

$$A \approx \sum_{k=1}^{n} (f(x_k^*) - g(x_k^*))\Delta x.$$

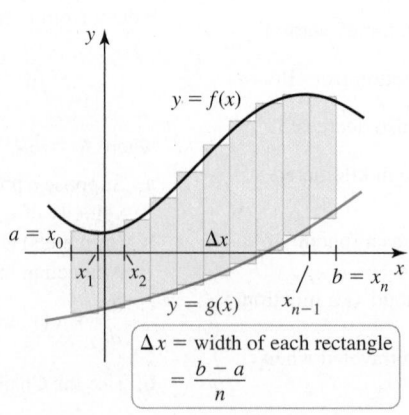

Figure 6.12

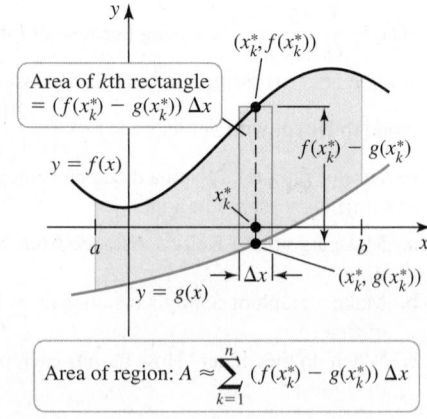

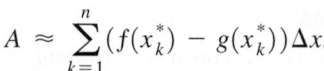

Figure 6.13

As the number of grid points increases, Δx approaches zero and these sums approach the area of the region between the curves; that is,

$$A = \lim_{n \to \infty} \sum_{k=1}^{n} (f(x_k^*) - g(x_k^*))\Delta x.$$

The limit of these Riemann sums is a definite integral of the function $f - g$.

➤ It is helpful to interpret the area formula: $f(x) - g(x)$ is the height of a rectangle and dx represents its width. We sum (integrate) the areas of the rectangles $(f(x) - g(x))\, dx$ to obtain the area of the region.

QUICK CHECK 1 In the area formula for a region between two curves, verify that if the lower curve is $g(x) = 0$, the formula becomes the usual formula for the area of the region bounded by $y = f(x)$ and the x-axis. ◄

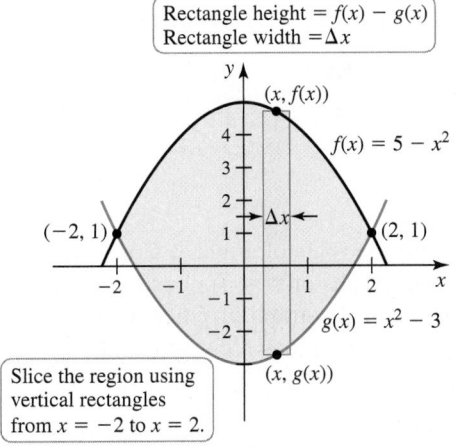

Rectangle height $= f(x) - g(x)$
Rectangle width $= \Delta x$

Slice the region using vertical rectangles from $x = -2$ to $x = 2$.

Figure 6.14

QUICK CHECK 2 Interpret the area formula when it is written in the form $A = \int_a^b f(x)\, dx - \int_a^b g(x)\, dx$, where $f(x) \geq g(x) \geq 0$ on $[a, b]$. ◄

> **DEFINITION Area of a Region Between Two Curves**
>
> Suppose f and g are continuous functions with $f(x) \geq g(x)$ on the interval $[a, b]$. The area of the region bounded by the graphs of f and g on $[a, b]$ is
>
> $$A = \int_a^b (f(x) - g(x))\, dx.$$

EXAMPLE 1   Area between curves Find the area of the region bounded by the graphs of $f(x) = 5 - x^2$ and $g(x) = x^2 - 3$ (Figure 6.14).

SOLUTION A key step in the solution of many area problems is finding the intersection points of the boundary curves, which often determine the limits of integration. The intersection points of these two curves satisfy the equation $5 - x^2 = x^2 - 3$. The solutions of this equation are $x = -2$ and $x = 2$, which become the lower and upper limits of integration, respectively. The graph of f is the upper curve and the graph of g is the lower curve on the interval $[-2, 2]$. Therefore, the area of the region is

$$A = \int_{-2}^{2} (\underbrace{(5 - x^2)}_{f(x)} - \underbrace{(x^2 - 3)}_{g(x)})\, dx \quad \text{Substitute for } f \text{ and } g.$$

$$= 2\int_{0}^{2} (8 - 2x^2)\, dx \qquad\qquad \text{Simplify and use symmetry.}$$

$$= 2\left(8x - \frac{2}{3}x^3\right)\Big|_0^2 \qquad\qquad \text{Fundamental Theorem}$$

$$= \frac{64}{3}. \qquad\qquad\qquad\qquad \text{Simplify.}$$

Notice how the symmetry of the problem simplifies the integration. Additionally, note that the area formula $A = \int_a^b (f(x) - g(x))\, dx$ is valid even if one or both curves lie below the x-axis, as long as $f(x) \geq g(x)$ on $[a, b]$.

Related Exercises 9–10 ◄

EXAMPLE 2   Compound region Find the area of the region bounded by the graphs of $f(x) = -x^2 + 3x + 6$ and $g(x) = |2x|$ (Figure 6.15a).

SOLUTION The lower boundary of the region consists of two different branches of the absolute value function. In situations like this, the region is divided into two (or more) subregions whose areas are found independently and then summed; these subregions are labeled R_1 and R_2 (Figure 6.15b). By the definition of absolute value,

$$g(x) = |2x| = \begin{cases} 2x & \text{if } x \geq 0 \\ -2x & \text{if } x < 0. \end{cases}$$

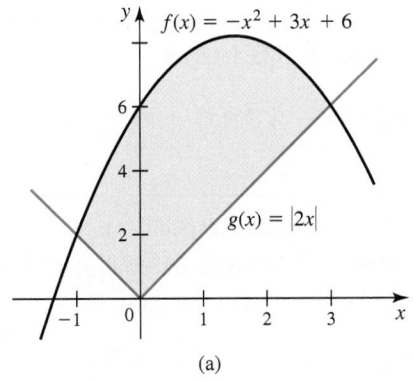

(a)

Figure 6.15

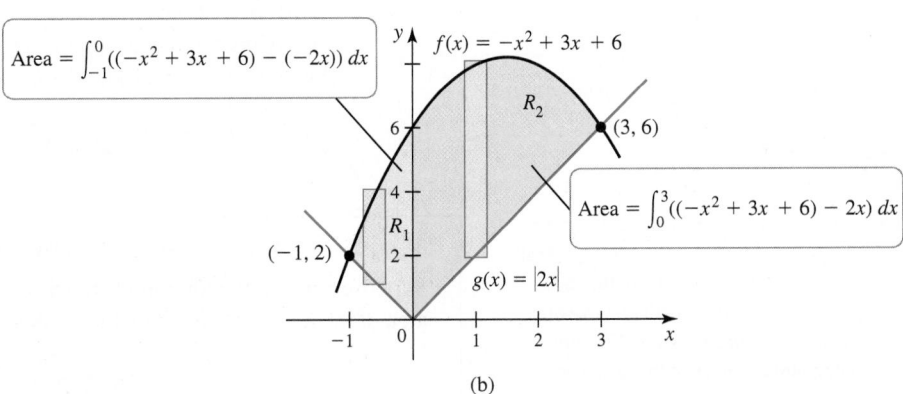

(b)

▶ The solution $x = 6$ corresponds to the intersection of the parabola $y = -x^2 + 3x + 6$ and the line $y = -2x$ in the fourth quadrant, not shown in Figure 6.15 because $g(x) = -2x$ only when $x < 0$.

The left intersection point of f and g satisfies $-2x = -x^2 + 3x + 6$, or $x^2 - 5x - 6 = 0$. Solving for x, we find that $(x + 1)(x - 6) = 0$, which implies $x = -1$ or $x = 6$; only the first solution is relevant. The right intersection point of f and g satisfies $2x = -x^2 + 3x + 6$; you should verify that the relevant solution in this case is $x = 3$.

Given these points of intersection, we see that the region R_1 is bounded by $y = -x^2 + 3x + 6$ and $y = -2x$ on the interval $[-1, 0]$. Similarly, region R_2 is bounded by $y = -x^2 + 3x + 6$ and $y = 2x$ on $[0, 3]$ (Figure 6.15b). Therefore,

$$A = \underbrace{\int_{-1}^{0} ((-x^2 + 3x + 6) - (-2x)) \, dx}_{\text{area of region } R_1} + \underbrace{\int_{0}^{3} ((-x^2 + 3x + 6) - 2x) \, dx}_{\text{area of region } R_2}$$

$$= \int_{-1}^{0} (-x^2 + 5x + 6) \, dx + \int_{0}^{3} (-x^2 + x + 6) \, dx \qquad \text{Simplify.}$$

$$= \left(-\frac{x^3}{3} + \frac{5}{2}x^2 + 6x \right)\Big|_{-1}^{0} + \left(-\frac{x^3}{3} + \frac{1}{2}x^2 + 6x \right)\Big|_{0}^{3} \qquad \text{Fundamental Theorem}$$

$$= 0 - \left(\frac{1}{3} + \frac{5}{2} - 6 \right) + \left(-9 + \frac{9}{2} + 18 \right) - 0 = \frac{50}{3}. \qquad \text{Simplify.}$$

Related Exercises 15–16 ◀

Integrating with Respect to y

There are occasions when it is convenient to reverse the roles of x and y. Consider the regions shown in **Figure 6.16** that are bounded by the graphs of $x = f(y)$ and $x = g(y)$, where $f(y) \geq g(y)$, for $c \leq y \leq d$ (which implies that the graph of f lies to the right of the graph of g). The lower and upper boundaries of the regions are $y = c$ and $y = d$, respectively.

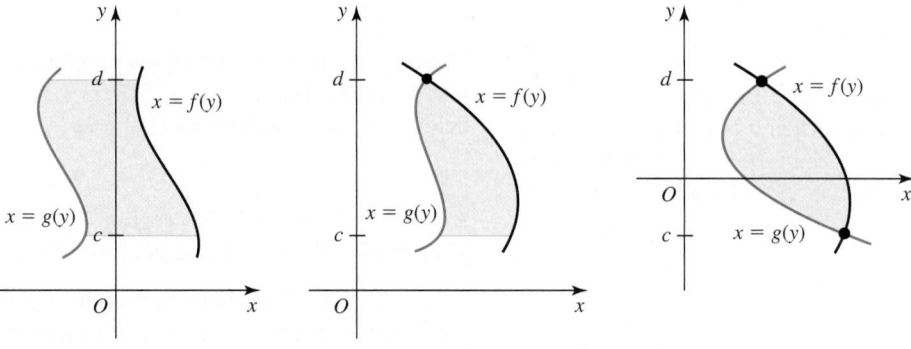

Figure 6.16

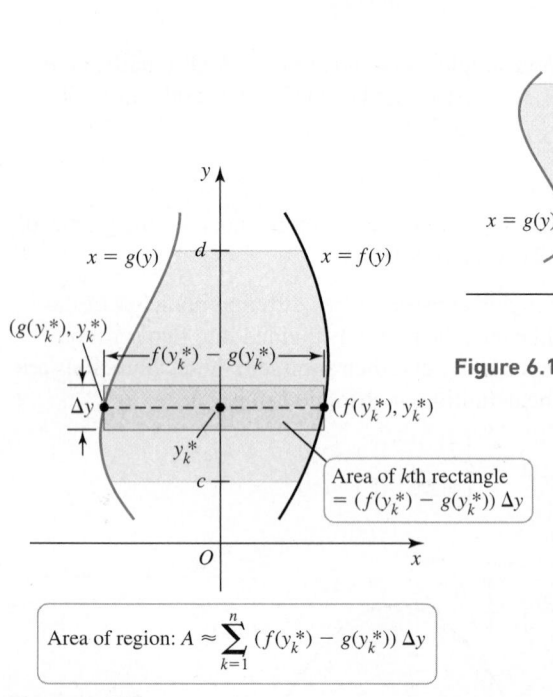

Area of region: $A \approx \sum_{k=1}^{n} (f(y_k^*) - g(y_k^*)) \, \Delta y$

Figure 6.17

In cases such as these, we treat y as the independent variable and divide the interval $[c, d]$ into n subintervals of width $\Delta y = (d - c)/n$ (**Figure 6.17**). On the kth subinterval, a point y_k^* is selected, and we construct a rectangle that extends from the left curve to the right curve. The kth rectangle has length $f(y_k^*) - g(y_k^*)$, so the area of the kth rectangle is $(f(y_k^*) - g(y_k^*))\Delta y$. The area of the region is approximated by the sum of the areas of the rectangles. In the limit as $n \to \infty$ and $\Delta y \to 0$, the area of the region is given as the definite integral

$$A = \lim_{n \to \infty} \sum_{k=1}^{n} (f(y_k^*) - g(y_k^*))\Delta y = \int_{c}^{d} (f(y) - g(y)) \, dy.$$

▶ This area formula is analogous to the one given on p. 417; it is now expressed with respect to the y-axis. In this case, $f(y) - g(y)$ is the length of a rectangle and dy represents its width. We sum (integrate) the areas of the rectangles $(f(y) - g(y)) \, dy$ to obtain the area of the region.

DEFINITION Area of a Region Between Two Curves with Respect to y

Suppose f and g are continuous functions with $f(y) \geq g(y)$ on the interval $[c, d]$. The area of the region bounded by the graphs $x = f(y)$ and $x = g(y)$ on $[c, d]$ is

$$A = \int_{c}^{d} (f(y) - g(y)) \, dy.$$

EXAMPLE 3 Integrating with respect to y Find the area of the region R bounded by the graphs of $y = x^3$, $y = x + 6$, and the x-axis.

SOLUTION The area of this region could be found by integrating with respect to x. But this approach requires splitting the region into two pieces (Figure 6.18). Alternatively, we can view y as the independent variable, express the bounding curves as functions of y, and make horizontal slices parallel to the x-axis (Figure 6.19).

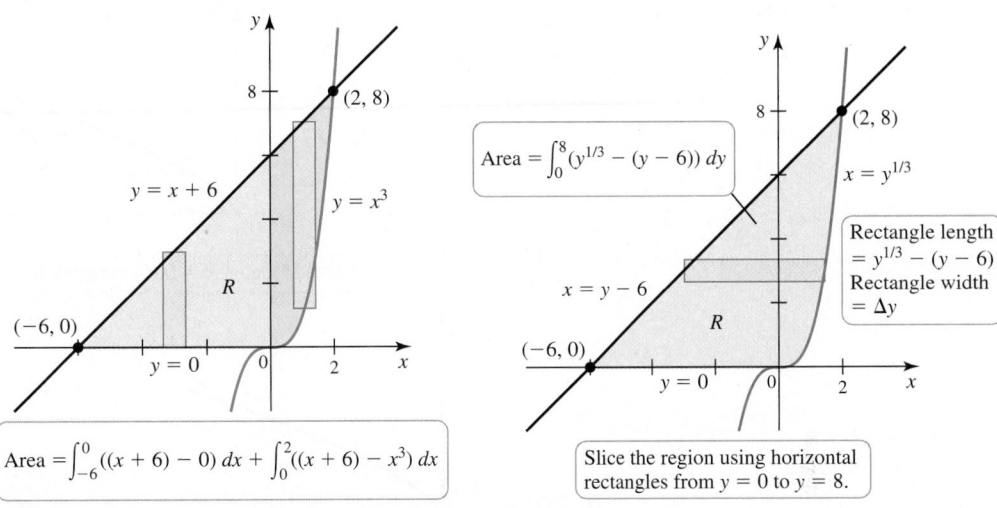

$$\text{Area} = \int_{-6}^{0}((x+6) - 0)\,dx + \int_{0}^{2}((x+6) - x^3)\,dx$$

Figure 6.18

$$\text{Area} = \int_{0}^{8}(y^{1/3} - (y-6))\,dy$$

$x = y^{1/3}$

Rectangle length
$= y^{1/3} - (y - 6)$
Rectangle width
$= \Delta y$

$x = y - 6$

Slice the region using horizontal rectangles from $y = 0$ to $y = 8$.

Figure 6.19

> You may use synthetic division or a root finder to factor the cubic polynomial in Example 3. Then the quadratic formula shows that the equation
>
> $$y^2 - 10y + 27 = 0$$
>
> has no real roots.

Solving for x in terms of y, the right curve $y = x^3$ becomes $x = f(y) = y^{1/3}$. The left curve $y = x + 6$ becomes $x = g(y) = y - 6$. The intersection point of the curves satisfies the equation $y^{1/3} = y - 6$, or $y = (y - 6)^3$. Expanding this equation gives the cubic equation

$$y^3 - 18y^2 + 107y - 216 = (y - 8)(y^2 - 10y + 27) = 0,$$

whose only real root is $y = 8$. As shown in Figure 6.19, the areas of the slices through the region are summed from $y = 0$ to $y = 8$. Therefore, the area of the region is given by

$$\int_{0}^{8}(y^{1/3} - (y - 6))\,dy = \left(\frac{3}{4}y^{4/3} - \frac{y^2}{2} + 6y\right)\Big|_{0}^{8} \qquad \text{Fundamental Theorem}$$

$$= \left(\frac{3}{4}\cdot 16 - 32 + 48\right) - 0 = 28. \qquad \text{Simplify.}$$

Related Exercises 19–20 ◄

QUICK CHECK 3 The region R is bounded by the curve $y = \sqrt{x}$, the line $y = x - 2$, and the x-axis. Express the area of R in terms of (a) integral(s) with respect to x and (b) integral(s) with respect to y. ◄

EXAMPLE 4 Calculus and geometry Find the area of the region R in the first quadrant bounded by the curves $y = x^{2/3}$ and $y = x - 4$ (Figure 6.20).

SOLUTION Slicing the region vertically and integrating with respect to x requires two integrals. Slicing the region horizontally requires a single integral with respect to y. The second approach appears to involve less work.

Slicing horizontally, the right bounding curve is $x = y + 4$ and the left bounding curve is $x = y^{3/2}$. The two curves intersect at $(8, 4)$, so the limits of integration are $y = 0$ and $y = 4$. The area of R is

$$\int_{0}^{4}(\underbrace{(y + 4)}_{\text{right curve}} - \underbrace{y^{3/2}}_{\text{left curve}})\,dy = \left(\frac{y^2}{2} + 4y - \frac{2}{5}y^{5/2}\right)\Big|_{0}^{4} = \frac{56}{5}.$$

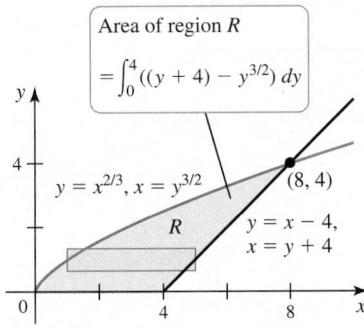

Figure 6.20

Can this area be found using a different approach? Sometimes it helps to use geometry. Notice that the region R can be formed by taking the entire region under the curve

▶ To find the point of intersection in Example 4, solve $y^{3/2} = y + 4$ by first squaring both sides of the equation.

$y = x^{2/3}$ on the interval $[0, 8]$ and then removing a triangle whose base is the interval $[4, 8]$ (Figure 6.21). The area of the region R_1 under the curve $y = x^{2/3}$ is

$$\int_0^8 x^{2/3}\, dx = \frac{3}{5} x^{5/3} \Big|_0^8 = \frac{96}{5}.$$

The triangle R_2 has a base of length 4 and a height of 4, so its area is $\frac{1}{2} \cdot 4 \cdot 4 = 8$. Therefore, the area of R is $\frac{96}{5} - 8 = \frac{56}{5}$, which agrees with the first calculation.

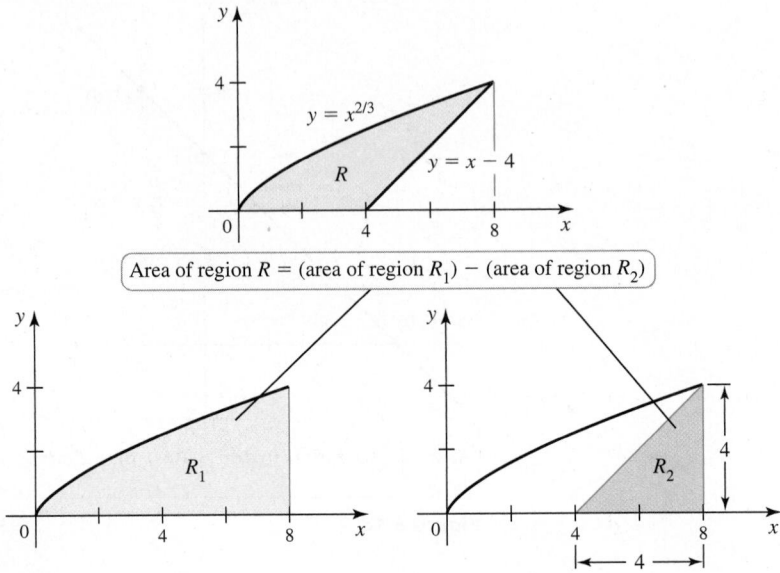

Area of region R = (area of region R_1) − (area of region R_2)

QUICK CHECK 4 An alternative way to determine the area of the region in Example 3 (Figure 6.18) is to compute $18 + \int_0^2 (x + 6 - x^3)\, dx$. Why? ◀

Figure 6.21

Related Exercises 34–36 ◀

SECTION 6.2 EXERCISES

Getting Started

1. Set up a sum of two integrals that equals the area of the shaded region bounded by the graphs of the functions f and g on $[a, c]$ (see figure). Assume the curves intersect at $x = b$.

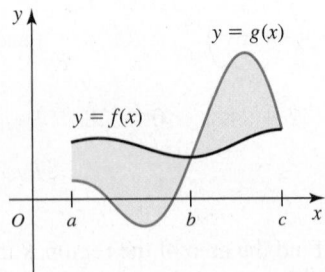

2. Set up an integral that equals the area of the region (see figure) in the following two ways. Do not evaluate the integrals.

   a. Using integration with respect to x.
   b. Using integration with respect to y.

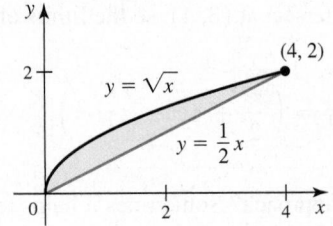

3. Make a sketch to show a case in which the area bounded by two curves is most easily found by integrating with respect to x.

4. Make a sketch to show a case in which the area bounded by two curves is most easily found by integrating with respect to y.

5. Find the area of the region (see figure) in two ways.

   a. Using integration with respect to x.
   b. Using geometry.

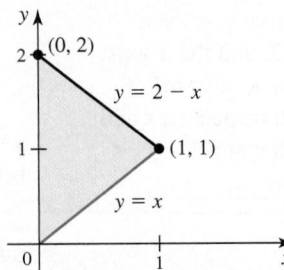

6. Find the area of the region (see figure) in two ways.

   a. By integrating with respect to y.
   b. Using geometry.

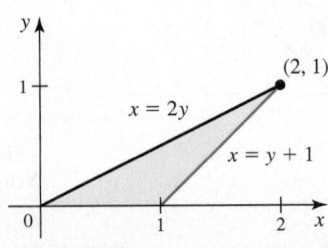

7. Express the area of the shaded region in Exercise 5 as the sum of two integrals with respect to y. Do not evaluate the integrals.

8. Express the area of the shaded region in Exercise 6 as the sum of two integrals with respect to x. Do not evaluate the integrals.

Practice Exercises

9–30. Finding area *Determine the area of the shaded region in the following figures.*

9.
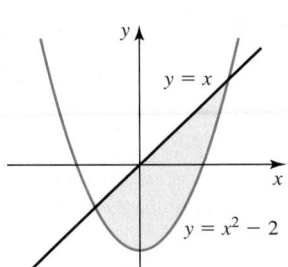

10.
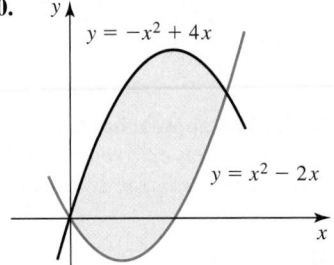

11.
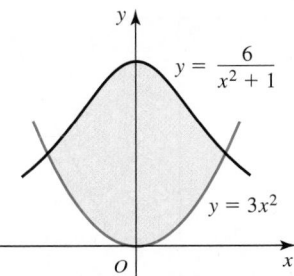

12.
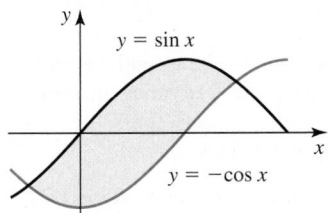

13. (*Hint:* Find the intersection point by inspection.)
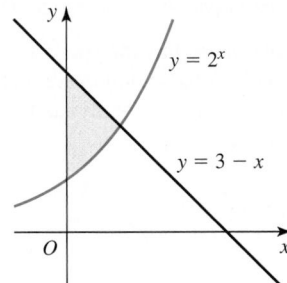

14.
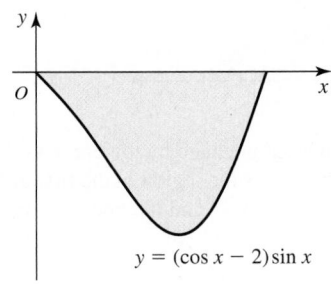

15.
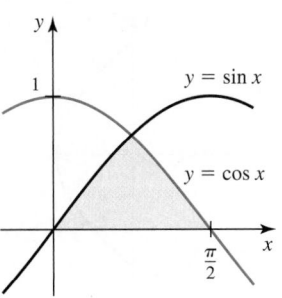

16.
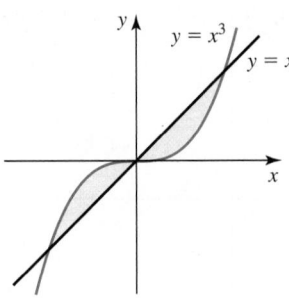

17.
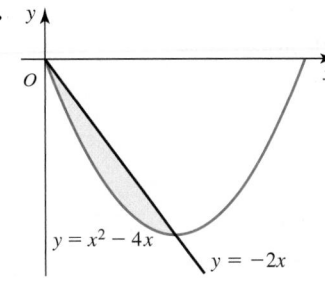

18.
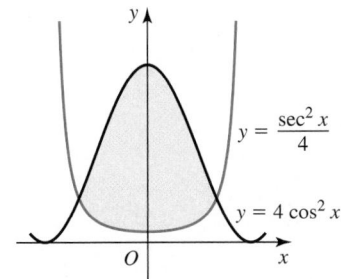

19.
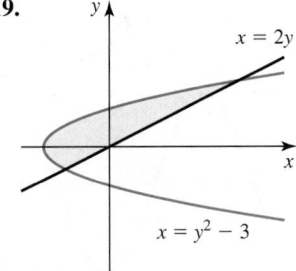

20.
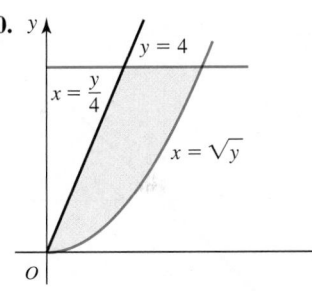

21.
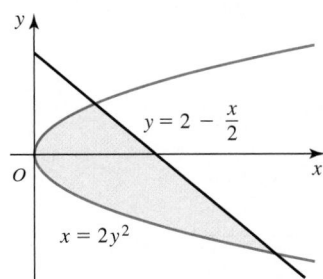

22.
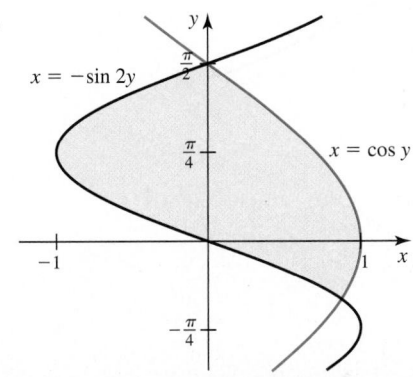

23.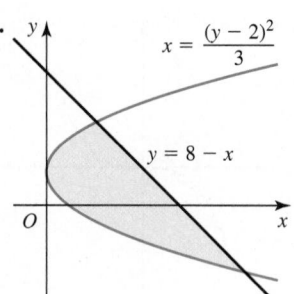

$x = \dfrac{(y-2)^2}{3}$

$y = 8 - x$

24.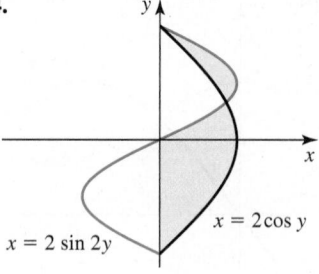

$x = 2\sin 2y$

$x = 2\cos y$

25.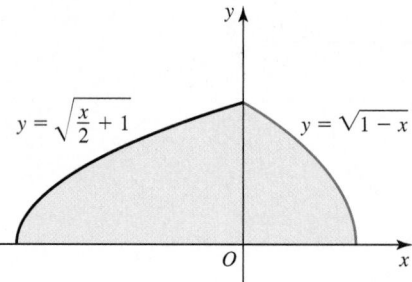

$y = \sqrt{\dfrac{x}{2} + 1}$ $y = \sqrt{1 - x}$

26.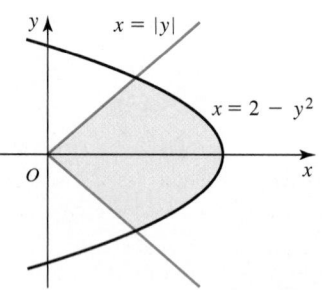

$x = |y|$

$x = 2 - y^2$

27.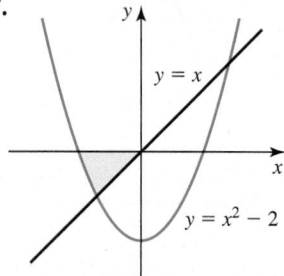

$y = x$

$y = x^2 - 2$

28.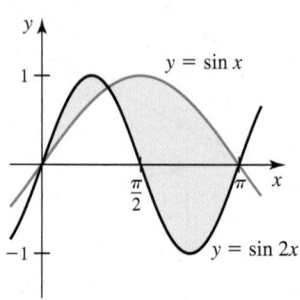

$y = \sin x$

$y = \sin 2x$

29.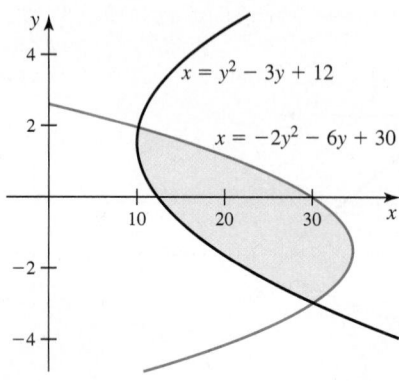

$x = y^2 - 3y + 12$

$x = -2y^2 - 6y + 30$

30.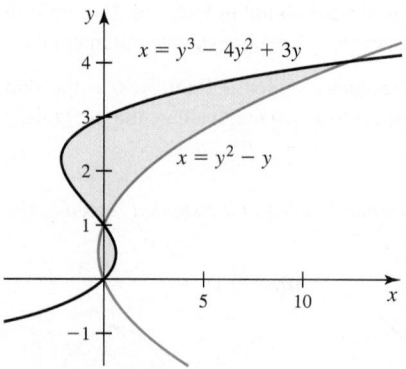

$x = y^3 - 4y^2 + 3y$

$x = y^2 - y$

31–32. Two approaches *Express the area of the following shaded regions in terms of (a) one or more integrals with respect to x and (b) one or more integrals with respect to y. You do not need to evaluate the integrals.*

31.   **32.**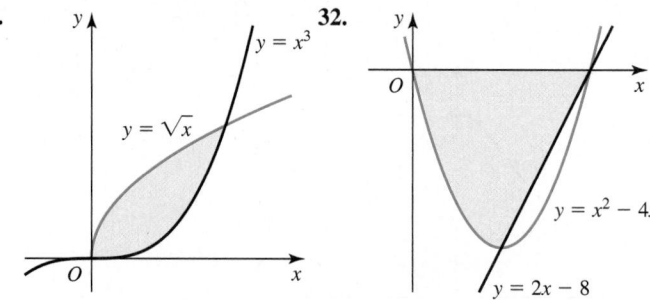

$y = x^3$

$y = \sqrt{x}$

$y = x^2 - 4x$

$y = 2x - 8$

33. Area between velocity curves Two runners, starting at the same location, run along a straight road for 1 hour. The velocity of one runner is $v_1(t) = 7t$ and the velocity of the other runner is $v_2(t) = 10\sqrt{t}$. Assume t is measured in hours and the velocities $v_1(t)$ and $v_2(t)$ are measured in km/hr. Determine the area between the curves $y = v_1(t)$ and $y = v_2(t)$, for $0 \le t \le 1$. Interpret the physical meaning of this area.

34–36. Calculus and geometry *For the given regions R_1 and R_2, complete the following steps.*

a. *Find the area of region R_1.*
b. *Find the area of region R_2 using geometry and the answer to part (a).*

34. R_1 is the region in the first quadrant bounded by the y-axis and the curves $y = 2x^2$ and $y = 3 - x$; R_2 is the region in the first quadrant bounded by the x-axis and the curves $y = 2x^2$ and $y = 3 - x$ (see figure).

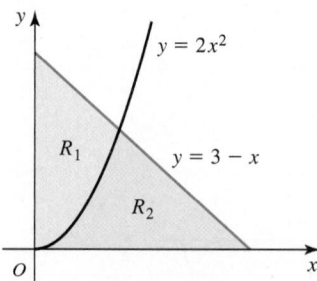

$y = 2x^2$

R_1 $y = 3 - x$

R_2

T 35. R_1 is the region in the first quadrant bounded by the line $x = 1$ and the curve $y = 6x(2 - x^2)^2$; R_2 is the region in the first quadrant bounded the curve $y = 6x(2 - x^2)^2$ and the line $y = 6x$.

36. R_1 is the region in the first quadrant bounded by the coordinate axes and the curve $y = \cos^{-1} x$; R_2 is the region bounded by the lines $y = \dfrac{\pi}{2}$ and $x = 1$, and the curve $y = \cos^{-1} x$.

37–62. Regions between curves *Find the area of the region described in the following exercises.*

37. The region bounded by $y = 4x + 4$, $y = 6x + 6$, and $x = 4$

38. The region bounded by $y = \cos x$ and $y = \sin x$ between $x = \dfrac{\pi}{4}$ and $x = \dfrac{5\pi}{4}$

39. The region bounded by $y = e^x$, $y = e^{-2x}$, and $x = \ln 4$

40. The region bounded by $y = 6x$ and $y = 3x^2 - 6x$

41. The region bounded by $y = \dfrac{2}{1 + x^2}$ and $y = 1$

42. The region bounded by $y = 24\sqrt{x}$ and $y = 3x^2$

43. The region bounded by $y = x$, $y = \dfrac{1}{x}$, $y = 0$, and $x = 2$

44. The region in the first quadrant on the interval $[0, 2]$ bounded by $y = 4x - x^2$ and $y = 4x - 4$

45. The region bounded by $y = 2 - |x|$ and $y = x^2$

46. The region bounded by $y = x^3$ and $y = 9x$

47. The region bounded by $y = |x - 3|$ and $y = \dfrac{x}{2}$

48. The region bounded by $y = 3x - x^3$ and $y = -x$

49. The region in the first quadrant bounded by $y = x^{2/3}$ and $y = 4$

50. The region in the first quadrant bounded by $y = 2$ and $y = 2\sin x$ on the interval $[0, \pi/2]$

51. The region bounded by $y = e^x$, $y = 2e^{-x} + 1$, and $x = 0$

T 52. The region below the line $y = 2$ and above the curve $y = \sec^2 x$ on the interval $[0, \pi/4]$

T 53. The region between the line $y = x$ and the curve $y = 2x\sqrt{1 - x^2}$ in the first quadrant

54. The region bounded by $x = y^2 - 4$ and $y = \dfrac{x}{3}$

55. The region bounded by $y = \sqrt{x}$, $y = 2x - 15$, and $y = 0$

T 56. The region bounded by $y = 2$ and $y = \dfrac{1}{\sqrt{1 - x^2}}$

57. The region bounded by $y = x^2 - 2x + 1$ and $y = 5x - 9$

58. The region bounded by $x = y(y - 1)$ and $x = -y(y - 1)$

59. The region bounded by $x = y(y - 1)$ and $y = \dfrac{x}{3}$

60. The region bounded by $y = \sin x$ and $y = x(x - \pi)$, for $0 \le x \le \pi$

61. The region in the first quadrant bounded by $y = \dfrac{5}{2} - \dfrac{1}{x}$ and $y = x$

62. The region in the first quadrant bounded by $y = x^{-1}$, $y = 4x$, and $y = \dfrac{x}{4}$

63–64. Complicated regions *Find the area of the shaded regions in the following figures.*

63.

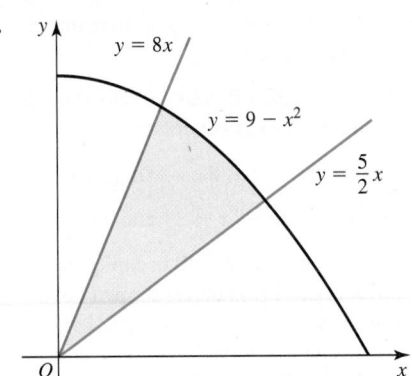

64.

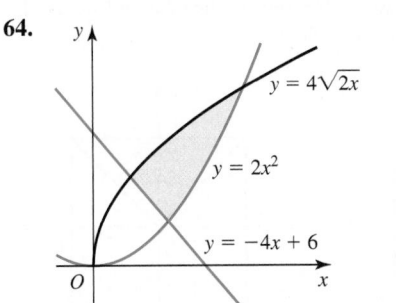

65. Explain why or why not Determine whether the following statements are true and give an explanation or counterexample.

    a. The area of the region bounded by $y = x$ and $x = y^2$ can be found only by integrating with respect to x.

    b. The area of the region between $y = \sin x$ and $y = \cos x$ on the interval $[0, \pi/2]$ is $\int_0^{\pi/2}(\cos x - \sin x)\, dx$.

    c. $\int_0^1 (x - x^2)\, dx = \int_0^1 (\sqrt{y} - y)\, dy$.

Explorations and Challenges

66. Differences of even functions Assume f and g are even, integrable functions on $[-a, a]$, where $a > 1$. Suppose $f(x) > g(x) > 0$ on $[-a, a]$ and that the area bounded by the graphs of f and g on $[-a, a]$ is 10. What is the value of $\int_0^{\sqrt{a}} x(f(x^2) - g(x^2))\, dx$?

67. Area of a curve defined implicitly Determine the area of the shaded region bounded by the curve $x^2 = y^4(1 - y^3)$ (see figure).

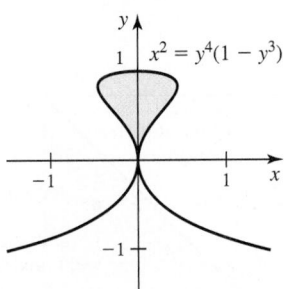

68–71. Roots and powers *Find the area of the following regions, expressing your results in terms of the positive integer $n \ge 2$.*

68. The region bounded by $f(x) = x$ and $g(x) = x^n$, for $x \ge 0$

69. The region bounded by $f(x) = x$ and $g(x) = x^{1/n}$, for $x \ge 0$

70. The region bounded by $f(x) = x^{1/n}$ and $g(x) = x^n$, for $x \geq 0$

71. Let A_n be the area of the region bounded by $f(x) = x^{1/n}$ and $g(x) = x^n$ on the interval $[0, 1]$, where n is a positive integer. Evaluate $\lim_{n \to \infty} A_n$ and interpret the result.

72–73. Bisecting regions *For each region R, find the horizontal line* $y = k$ *that divides R into two subregions of equal area.*

72. R is the region bounded by $y = 1 - x$, the x-axis, and the y-axis.

73. R is the region bounded by $y = 1 - |x - 1|$ and the x-axis.

74. Geometric probability Suppose a dartboard occupies the square $\{(x, y): 0 \leq |x| \leq 1, 0 \leq |y| \leq 1\}$. A dart is thrown randomly at the board many times (meaning it is equally likely to land at any point in the square). What fraction of the dart throws land closer to the edge of the board than to the center? Equivalently, what is the probability that the dart lands closer to the edge of the board than to the center? Proceed as follows.

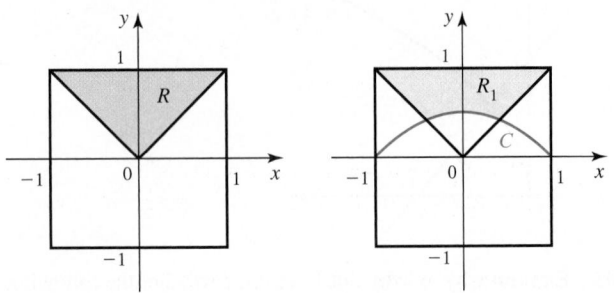

a. Argue that by symmetry it is necessary to consider only one quarter of the board, say the region $R: \{(x,y): |x| \leq y \leq 1\}$.

b. Find the curve C in this region that is equidistant from the center of the board and the top edge of the board (see figure).

c. The probability that the dart lands closer to the edge of the board than to the center is the ratio of the area of the region R_1 above C to the area of the entire region R. Compute this probability.

▣ 75. Lorenz curves and the Gini index A **Lorenz curve** is given by $y = L(x)$, where $0 \leq x \leq 1$ represents the lowest fraction of the population of a society in terms of wealth, and $0 \leq y \leq 1$ represents the fraction of the total wealth that is owned by that fraction of the society. For example, the Lorenz curve in the figure shows that $L(0.5) = 0.2$, which means that the lowest 0.5 (50%) of the society owns 0.2 (20%) of the wealth. (See Guided Project *Distribution of Wealth* for more on Lorenz curves.)

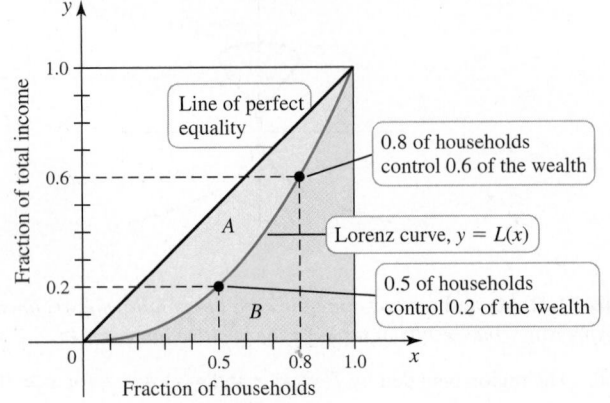

a. A Lorenz curve $y = L(x)$ is accompanied by the line $y = x$, called the **line of perfect equality**. Explain why this line is given this name.

b. Explain why a Lorenz curve satisfies the conditions $L(0) = 0$, $L(1) = 1$, $L(x) \leq x$, and $L'(x) \geq 0$ on $[0, 1]$.

c. Graph the Lorenz curves $L(x) = x^p$ corresponding to $p = 1.1$, 1.5, 2, 3, and 4. Which value of p corresponds to the *most* equitable distribution of wealth (closest to the line of perfect equality)? Which value of p corresponds to the *least* equitable distribution of wealth? Explain.

d. The information in the Lorenz curve is often summarized in a single measure called the **Gini index**, which is defined as follows. Let A be the area of the region between $y = x$ and $y = L(x)$ (see figure) and let B be the area of the region between $y = L(x)$ and the x-axis. Then the Gini index is

$$G = \frac{A}{A + B}.$$ Show that $G = 2A = 1 - 2\int_0^1 L(x)\, dx$.

e. Compute the Gini index for the cases $L(x) = x^p$ and $p = 1.1$, 1.5, 2, 3, and 4.

f. What is the smallest interval $[a, b]$ on which values of the Gini index lie for $L(x) = x^p$ with $p \geq 1$? Which endpoints of $[a, b]$ correspond to the least and most equitable distribution of wealth?

g. Consider the Lorenz curve described by $L(x) = \dfrac{5x^2}{6} + \dfrac{x}{6}$. Show that it satisfies the conditions $L(0) = 0$, $L(1) = 1$, and $L'(x) \geq 0$ on $[0, 1]$. Find the Gini index for this function.

76. Equal area properties for parabolas Consider the parabola $y = x^2$. Let P, Q, and R be points on the parabola with R between P and Q on the curve. Let ℓ_P, ℓ_Q, and ℓ_R be the lines tangent to the parabola at P, Q, and R, respectively (see figure). Let P' be the intersection point of ℓ_Q and ℓ_R, let Q' be the intersection point of ℓ_P and ℓ_R, and let R' be the intersection point of ℓ_P and ℓ_Q. Prove that Area $\triangle PQR = 2 \cdot$ Area $\triangle P'Q'R'$ in the following cases. (In fact, the property holds for any three points on any parabola.) (*Source: Mathematics Magazine*, 81, 2, Apr 2008)

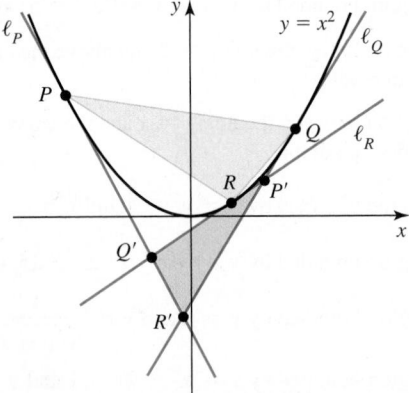

a. $P(-a, a^2)$, $Q(a, a^2)$, and $R(0, 0)$, where a is a positive real number

b. $P(-a, a^2)$, $Q(b, b^2)$, and $R(0, 0)$, where a and b are positive real numbers

c. $P(-a, a^2)$, $Q(b, b^2)$, and R is any point between P and Q on the curve

▣ 77. Roots and powers Consider the functions $f(x) = x^n$ and $g(x) = x^{1/n}$, where $n \geq 2$ is a positive integer.

a. Graph f and g for $n = 2$, 3, and 4, for $x \geq 0$.

b. Give a geometric interpretation of the area function
$A_n(x) = \int_0^x (f(s) - g(s))\, ds$, for $n = 2, 3, 4, \ldots$ and $x > 0$.

c. Find the positive root of $A_n(x) = 0$ in terms of n. Does the root increase or decrease with n?

T 78. **Shifting sines** Consider the functions $f(x) = a \sin 2x$ and $g(x) = \dfrac{\sin x}{a}$, where $a > 0$ is a real number.

a. Graph the two functions on the interval $[0, \frac{\pi}{2}]$, for $a = \frac{1}{2}, 1$, and 2.

b. Show that the curves have an intersection point x^* (other than $x = 0$) on $[0, \frac{\pi}{2}]$ that satisfies $\cos x^* = \frac{1}{2a^2}$, provided $a > 1/\sqrt{2}$.

c. Find the area of the region between the two curves on $[0, x^*]$ when $a = 1$.

d. Show that as $a \to 1/\sqrt{2}^+$, the area of the region between the two curves on $[0, x^*]$ approaches zero.

6.3 Volume by Slicing

We have seen that integration is used to compute the area of two-dimensional regions bounded by curves. Integrals are also used to find the volume of three-dimensional regions (or solids). Once again, the slice-and-sum method is the key to solving these problems.

General Slicing Method

Consider a solid object that extends in the x-direction from $x = a$ to $x = b$. Imagine making a vertical cut through the solid, perpendicular to the x-axis at a particular point x, and suppose the area of the cross section created by the cut is given by a known integrable function A (Figure 6.22).

To find the volume of this solid, we first divide $[a, b]$ into n subintervals of length $\Delta x = (b - a)/n$. The endpoints of the subintervals are the grid points $x_0 = a, x_1, x_2, \ldots, x_n = b$. We now make vertical cuts through the solid perpendicular to the x-axis at each grid point, which produces n slices of thickness Δx. (Imagine cutting a loaf of bread to create n slices of equal width.) On each subinterval, an arbitrary point x_k^* is identified. The kth slice through the solid has a thickness Δx, and we take $A(x_k^*)$ as a representative cross-sectional area of the slice. Therefore, the volume of the kth slice is approximately $A(x_k^*)\Delta x$ (Figure 6.23). Summing the volumes of the slices, the approximate volume of the solid is

$$V \approx \sum_{k=1}^n A(x_k^*)\Delta x.$$

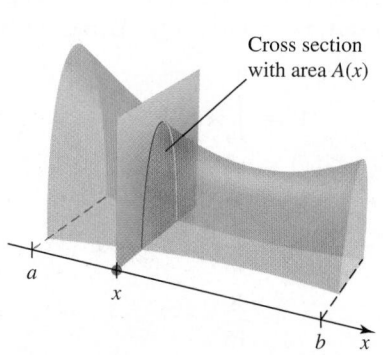

Cross section with area $A(x)$

Figure 6.22

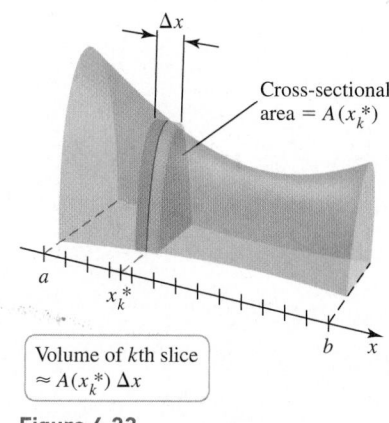

Cross-sectional area $= A(x_k^*)$

Volume of kth slice $\approx A(x_k^*)\, \Delta x$

Figure 6.23

As the number of slices increases $(n \rightarrow \infty)$ and the thickness of each slice goes to zero $(\Delta x \rightarrow 0)$, the exact volume V is obtained in terms of a definite integral (Figure 6.24):

$$V = \lim_{n \to \infty} \sum_{k=1}^{n} A(x_k^*) \Delta x = \int_a^b A(x)\, dx.$$

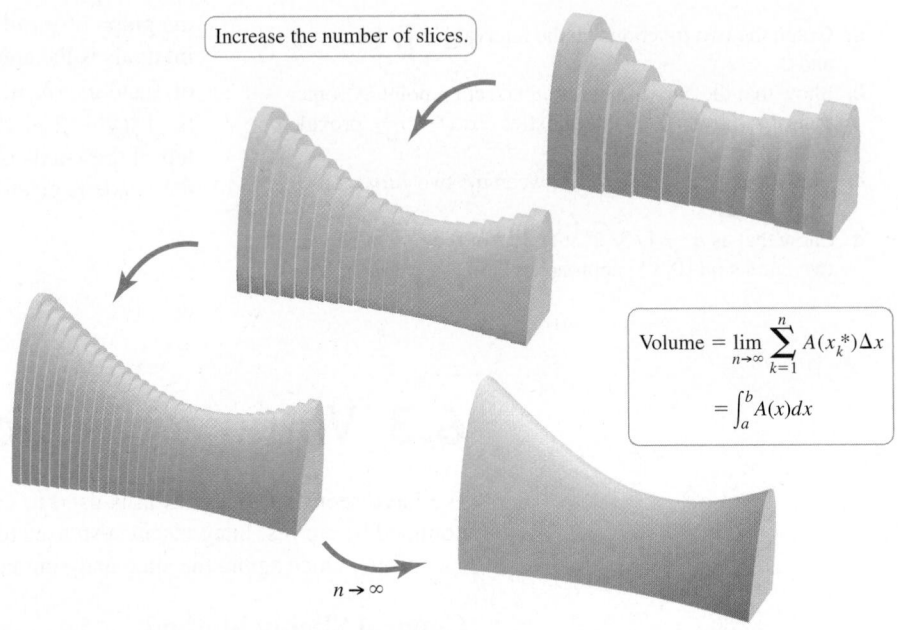

Increase the number of slices.

$$\text{Volume} = \lim_{n \to \infty} \sum_{k=1}^{n} A(x_k^*) \Delta x$$
$$= \int_a^b A(x)\,dx$$

$n \to \infty$

Figure 6.24

We summarize the important general slicing method, which is also the basis of other volume formulas to follow.

> ➤ The factors in this volume integral have meaning: $A(x)$ is the cross-sectional area of a slice and dx represents its thickness. Summing (integrating) the volumes of the slices $A(x)\, dx$ gives the volume of the solid.

QUICK CHECK 1 Why is the volume, as given by the general slicing method, equal to the average value of the area function A on $[a, b]$ multiplied by $b - a$? ◄

General Slicing Method

Suppose a solid object extends from $x = a$ to $x = b$, and the cross section of the solid perpendicular to the x-axis has an area given by a function A that is integrable on $[a, b]$. The volume of the solid is

$$V = \int_a^b A(x)\, dx.$$

EXAMPLE 1   Volume of a "parabolic cube" Let R be the region in the first quadrant bounded by the coordinate axes and the curve $y = 1 - x^2$. A solid has a base R, and cross sections through the solid perpendicular to the base and parallel to the y-axis are squares (Figure 6.25a). Find the volume of the solid.

SOLUTION Focus on a cross section through the solid at a point x, where $0 \le x \le 1$. That cross section is a square with sides of length $1 - x^2$. Therefore, the area of a typical cross section is $A(x) = (1 - x^2)^2$. Using the general slicing method, the volume of the solid is

$$V = \int_0^1 A(x)\, dx \qquad \text{General slicing method}$$

$$= \int_0^1 (1 - x^2)^2\, dx \qquad \text{Substitute for } A(x).$$

$$= \int_0^1 (1 - 2x^2 + x^4)\, dx \qquad \text{Expand integrand.}$$

$$= \frac{8}{15}. \qquad \text{Evaluate.}$$

The actual solid with a square cross section is shown in **Figure 6.25b**.

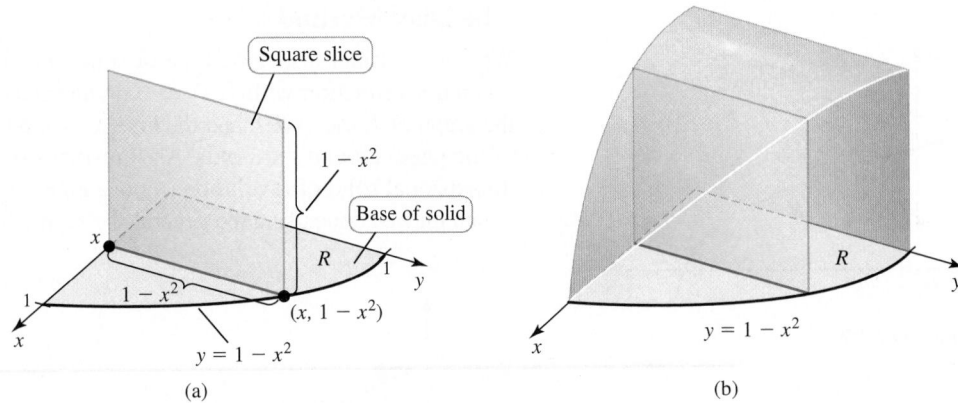

Figure 6.25

Related Exercises 11–12 ◄

EXAMPLE 2   Volume of a "parabolic hemisphere" A solid has a base that is bounded by the curves $y = x^2$ and $y = 2 - x^2$ in the xy-plane. Cross sections through the solid perpendicular to the base and parallel to the y-axis are semicircular disks. Find the volume of the solid.

SOLUTION Because a typical cross section perpendicular to the x-axis is a semicircular disk (Figure 6.26), the area of a cross section is $\frac{1}{2}\pi r^2$, where r is the radius of the cross section.

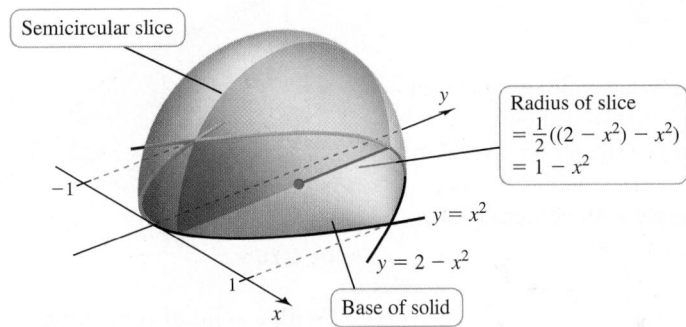

Figure 6.26

The key observation is that this radius is one-half of the distance between the upper bounding curve $y = 2 - x^2$ and the lower bounding curve $y = x^2$. So the radius at the point x is

$$r = \frac{1}{2}\left((2 - x^2) - x^2\right) = 1 - x^2.$$

This means that the area of the semicircular cross section at the point x is

$$A(x) = \frac{1}{2}\pi r^2 = \frac{\pi}{2}(1 - x^2)^2.$$

The intersection points of the two bounding curves satisfy $2 - x^2 = x^2$, which has solutions $x = \pm 1$. Therefore, the cross sections lie between $x = -1$ and $x = 1$. Integrating the cross-sectional areas, the volume of the solid is

$$
\begin{aligned}
V &= \int_{-1}^{1} A(x)\, dx && \text{General slicing method} \\
&= \int_{-1}^{1} \frac{\pi}{2}(1 - x^2)^2\, dx && \text{Substitute for } A(x). \\
&= \frac{\pi}{2}\int_{-1}^{1}(1 - 2x^2 + x^4)\, dx && \text{Expand integrand.} \\
&= \frac{8\pi}{15}. && \text{Evaluate.}
\end{aligned}
$$

QUICK CHECK 2 In Example 2, what is the cross-sectional area function $A(x)$ if cross sections perpendicular to the base are squares rather than semicircles? ◄

Related Exercise 15 ◄

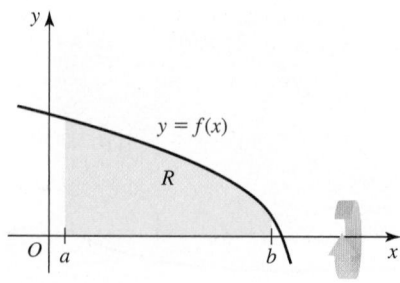

Figure 6.27

The Disk Method

We now consider a specific type of solid known as a **solid of revolution**. Suppose f is a continuous function with $f(x) \geq 0$ on an interval $[a, b]$. Let R be the region bounded by the graph of f, the x-axis, and the lines $x = a$ and $x = b$ (Figure 6.27). Now revolve R (out of the page) around the x-axis. As R revolves once about the x-axis, it sweeps out a three-dimensional solid of revolution (Figure 6.28). The goal is to find the volume of this solid, which may be done using the general slicing method.

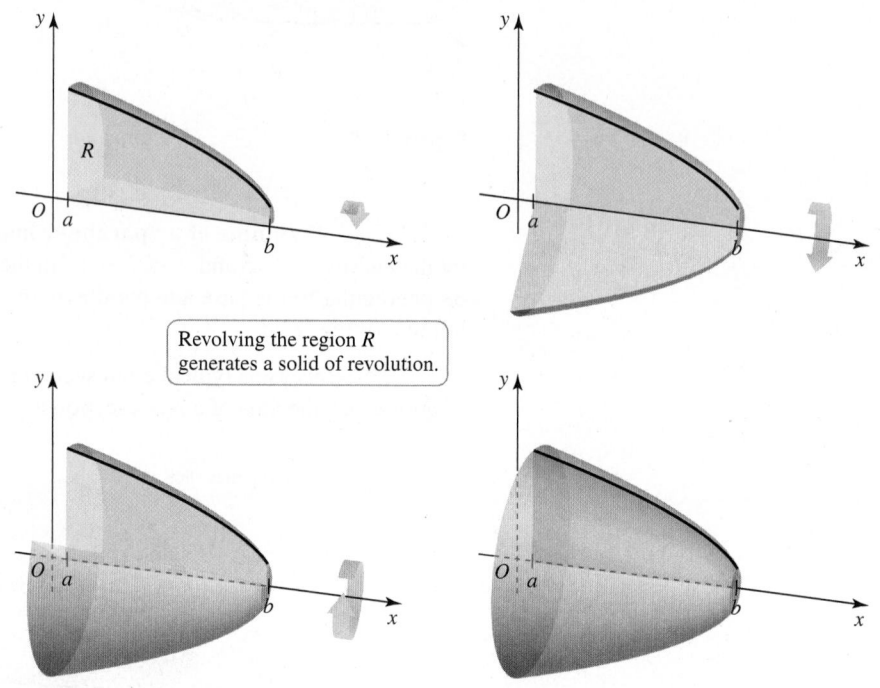

Revolving the region R generates a solid of revolution.

QUICK CHECK 3 What solid results when the region R is revolved about the x-axis if (a) R is a square with vertices $(0, 0)$, $(0, 2)$, $(2, 0)$, and $(2, 2)$, and (b) R is a triangle with vertices $(0, 0)$, $(0, 2)$, and $(2, 0)$? ◄

Figure 6.28

With a solid of revolution, the cross-sectional area function has a special form because all cross sections perpendicular to the x-axis are *circular disks* with radius $f(x)$ (Figure 6.29). Therefore, the cross section at the point x, where $a \leq x \leq b$, has area

$$A(x) = \pi(\text{radius})^2 = \pi f(x)^2.$$

By the general slicing method, the volume of the solid is

$$V = \int_a^b A(x)\, dx = \int_a^b \pi f(x)^2\, dx.$$

Because each slice through the solid is a circular disk, the resulting method is called the *disk method*.

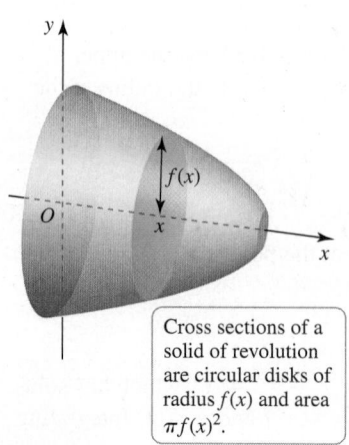

Cross sections of a solid of revolution are circular disks of radius $f(x)$ and area $\pi f(x)^2$.

Figure 6.29

Disk Method about the x-Axis

Let f be continuous with $f(x) \geq 0$ on the interval $[a, b]$. If the region R bounded by the graph of f, the x-axis, and the lines $x = a$ and $x = b$ is revolved about the x-axis, the volume of the resulting solid of revolution is

$$V = \int_a^b \pi \underbrace{f(x)^2}_{\substack{\text{disk} \\ \text{radius}}}\, dx.$$

EXAMPLE 3 **Disk method at work** Let R be the region bounded by the curve $f(x) = (x + 1)^2$, the x-axis, and the lines $x = 0$ and $x = 2$ (Figure 6.30a). Find the volume of the solid of revolution obtained by revolving R about the x-axis.

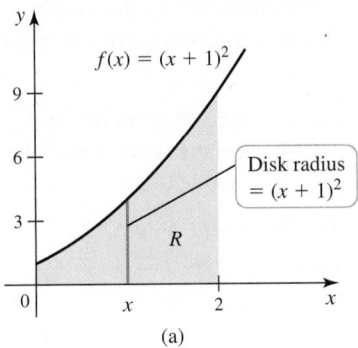

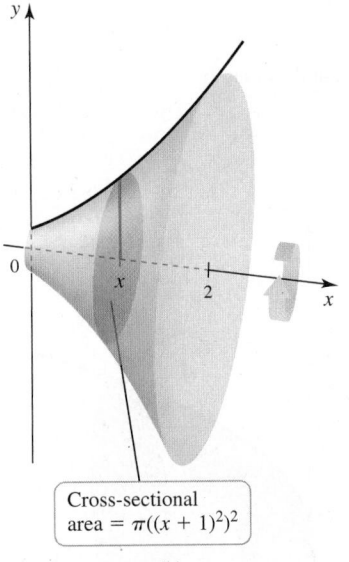

Figure 6.30

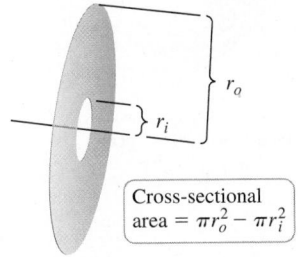

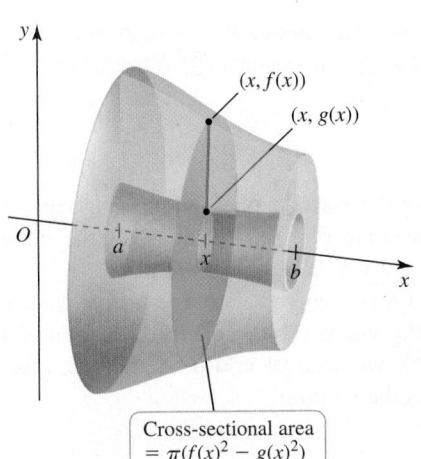

Figure 6.32

SOLUTION When the region R is revolved about the x-axis, it generates a solid of revolution (Figure 6.30b). A cross section perpendicular to the x-axis at the point $0 \le x \le 2$ is a circular disk of radius $f(x)$. Therefore, a typical cross section has area

$$A(x) = \pi f(x)^2 = \pi((x+1)^2)^2.$$

Integrating these cross-sectional areas between $x = 0$ and $x = 2$ gives the volume of the solid:

$$V = \int_0^2 A(x)\, dx = \int_0^2 \pi(\underbrace{(x+1)^2}_{\text{disk radius}})^2\, dx \quad \text{Substitute for } A(x).$$

$$= \int_0^2 \pi(x+1)^4\, dx \qquad\qquad \text{Simplify.}$$

$$= \pi \left.\frac{u^5}{5}\right|_1^3 = \frac{242\,\pi}{5}. \qquad \text{Let } u = x+1 \text{ and evaluate.}$$

Related Exercises 17, 19 ◄

Washer Method

A slight variation on the disk method enables us to compute the volume of more exotic solids of revolution. Suppose R is the region bounded by the graphs of f and g between $x = a$ and $x = b$, where $f(x) \ge g(x) \ge 0$ (Figure 6.31). If R is revolved about the x-axis to generate a solid of revolution, the resulting solid generally has a hole through it.

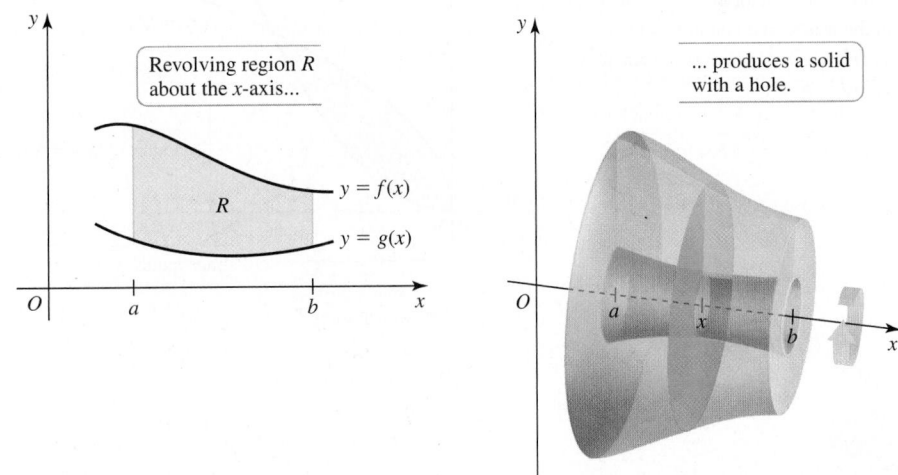

Figure 6.31

Once again we apply the general slicing method. In this case, a cross section through the solid perpendicular to the x-axis is a circular *washer* with an outer radius of $r_o = f(x)$ and an inner radius of $r_i = g(x)$, where $a \le x \le b$. The area of the cross section is the area of the entire disk minus the area of the hole, or

$$A(x) = \pi(r_o^2 - r_i^2) = \pi(f(x)^2 - g(x)^2)$$

(Figure 6.32). The general slicing method gives the volume of the solid.

Washer Method about the x-Axis

Let f and g be continuous functions with $f(x) \ge g(x) \ge 0$ on $[a, b]$. Let R be the region bounded by $y = f(x)$, $y = g(x)$, and the lines $x = a$ and $x = b$. When R is revolved about the x-axis, the volume of the resulting solid of revolution is

$$V = \int_a^b \pi(\underbrace{f(x)^2}_{\substack{\text{outer} \\ \text{radius}}} - \underbrace{g(x)^2}_{\substack{\text{inner} \\ \text{radius}}})\, dx.$$

QUICK CHECK 4 Show that when $g(x) = 0$ in the washer method, the result is the disk method. ◄

▶ The washer method is really two applications of the disk method. We compute the volume of the entire solid without the hole (by the disk method) and then subtract the volume of the hole (also computed by the disk method).

▶ Ignoring the factor of π, the integrand in the washer method integral is $f(x)^2 - g(x)^2$, which is not equal to $(f(x) - g(x))^2$.

EXAMPLE 4   **Volume by the washer method**   The region R is bounded by the graphs of $f(x) = \sqrt{x}$ and $g(x) = x^2$ between $x = 0$ and $x = 1$. What is the volume of the solid that results when R is revolved about the x-axis?

SOLUTION   The region R is bounded by the graphs of f and g with $f(x) \geq g(x)$ on $[0, 1]$, so the washer method is applicable (**Figure 6.33**). The area of a typical cross section at the point x is

$$A(x) = \pi(\,f(x)^2 - g(x)^2) = \pi((\sqrt{x})^2 - (x^2)^2) = \pi(x - x^4).$$

$$\underbrace{\phantom{(\sqrt{x})^2}}_{\substack{\text{outer} \\ \text{radius}}} \quad \underbrace{}_{\substack{\text{inner} \\ \text{radius}}}$$

Therefore, the volume of the solid is

$$V = \int_0^1 \pi(x - x^4)\,dx \qquad \text{Washer method}$$

$$= \pi\left(\frac{x^2}{2} - \frac{x^5}{5}\right)\Big|_0^1 = \frac{3\pi}{10}. \qquad \text{Fundamental Theorem}$$

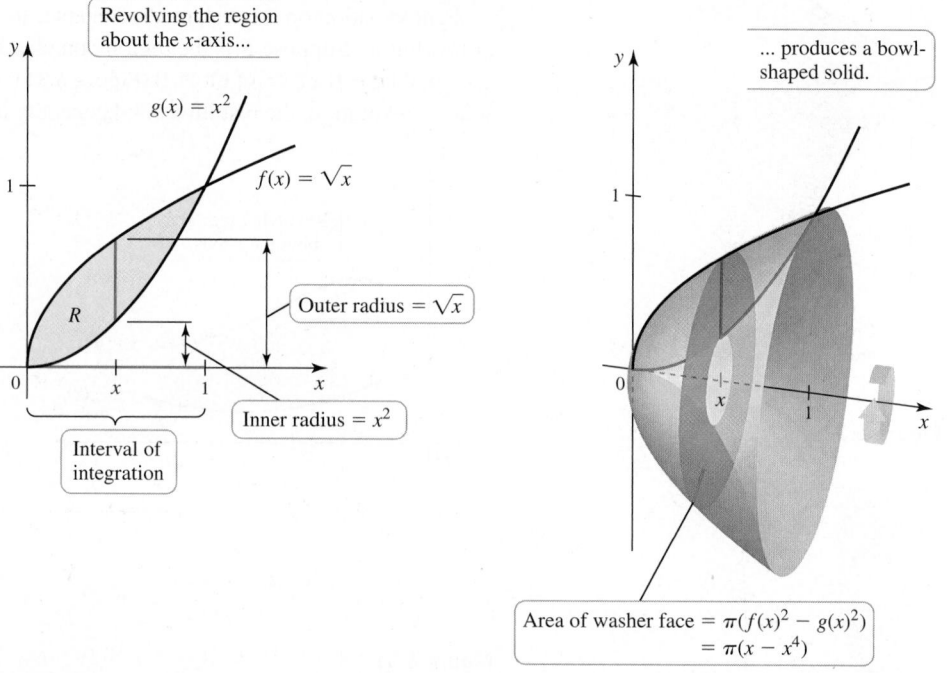

Revolving the region about the x-axis...

$g(x) = x^2$

$f(x) = \sqrt{x}$

R

Outer radius $= \sqrt{x}$

Inner radius $= x^2$

Interval of integration

... produces a bowl-shaped solid.

Area of washer face $= \pi(f(x)^2 - g(x)^2)$
$= \pi(x - x^4)$

Figure 6.33

Related Exercises 23–24 ◄

QUICK CHECK 5 Suppose the region in Example 4 is revolved about the line $y = -1$ instead of the x-axis. (a) What is the inner radius of a typical washer? (b) What is the outer radius of a typical washer? ◄

Revolving about the y-Axis

Everything you learned about revolving regions about the x-axis applies to revolving regions about the y-axis. Consider a region R bounded by the curve $x = p(y)$ on the right, the curve $x = q(y)$ on the left, and the horizontal lines $y = c$ and $y = d$ (**Figure 6.34a**).

To find the volume of the solid generated when R is revolved about the y-axis, we use the general slicing method—now with respect to the y-axis (**Figure 6.34b**). The area of a typical cross section is $A(y) = \pi(p(y)^2 - q(y)^2)$, where $c \leq y \leq d$. As before, integrating these cross-sectional areas of the solid gives the volume.

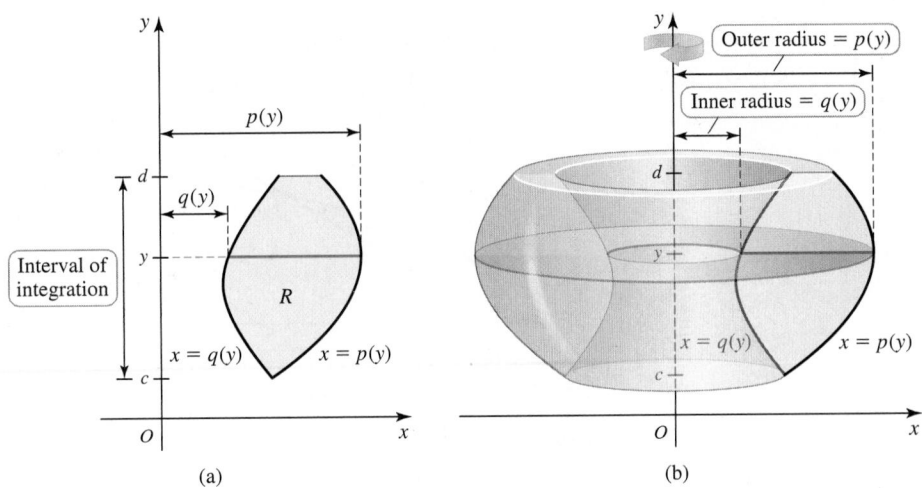

Figure 6.34

▶ The disk/washer method about the y-axis is the disk/washer method about the x-axis with x replaced with y.

Disk and Washer Methods about the y-Axis

Let p and q be continuous functions with $p(y) \geq q(y) \geq 0$ on $[c, d]$. Let R be the region bounded by $x = p(y)$, $x = q(y)$, and the lines $y = c$ and $y = d$. When R is revolved about the y-axis, the volume of the resulting solid of revolution is given by

$$V = \int_c^d \pi \big(\underbrace{p(y)^2}_{\substack{\text{outer} \\ \text{radius}}} - \underbrace{q(y)^2}_{\substack{\text{inner} \\ \text{radius}}} \big) \, dy.$$

If $q(y) = 0$, the disk method results:

$$V = \int_c^d \pi \underbrace{p(y)^2}_{\substack{\text{disk} \\ \text{radius}}} \, dy.$$

EXAMPLE 5 Which solid has greater volume? Let R be the region in the first quadrant bounded by the graphs of $x = y^3$ and $x = 4y$. Which is greater, the volume of the solid generated when R is revolved about the x-axis or the y-axis?

SOLUTION Solving $y^3 = 4y$, or equivalently, $y(y^2 - 4) = 0$, we find that the bounding curves of R intersect at the points $(0, 0)$ and $(8, 2)$. When the region R (Figure 6.35a) is revolved about the y-axis, it generates a funnel with a curved inner surface (Figure 6.35b). Washer-shaped cross sections perpendicular to the y-axis extend from $y = 0$ to $y = 2$. The outer radius of the cross section at the point y is determined by the line $x = p(y) = 4y$. The inner radius of the cross section at the point y is determined by the curve $x = q(y) = y^3$. Applying the washer method, the volume of this solid is

$$V = \int_0^2 \pi(p(y)^2 - q(y)^2) \, dy \qquad \text{Washer method}$$

$$= \int_0^2 \pi(16y^2 - y^6) \, dy \qquad \text{Substitute for } p \text{ and } q.$$

$$= \pi\left(\frac{16}{3} y^3 - \frac{y^7}{7}\right)\Big|_0^2 \qquad \text{Fundamental Theorem}$$

$$= \frac{512\pi}{21}. \qquad \text{Evaluate.}$$

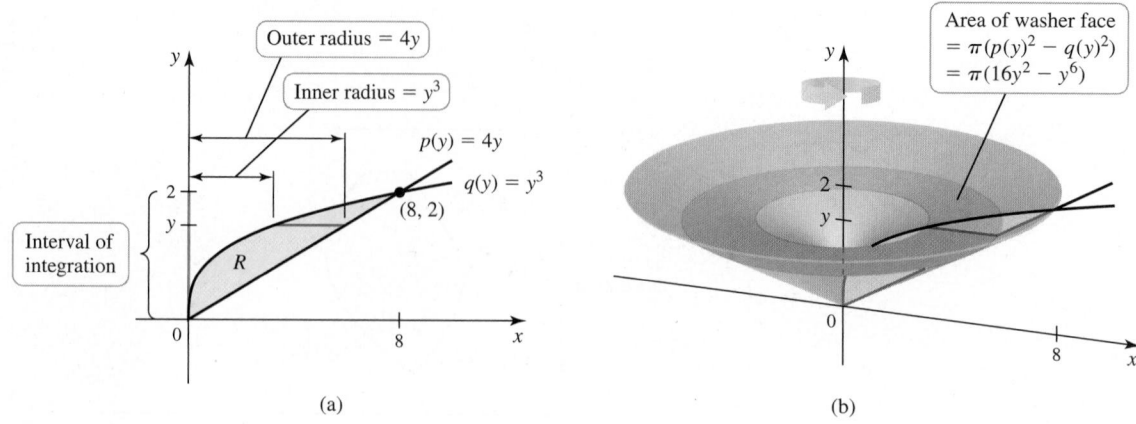

Figure 6.35

When the region R is revolved about the x-axis, it generates a different funnel (Figure 6.36). Vertical slices through the solid between $x = 0$ and $x = 8$ produce washers. The outer radius of the washer at the point x is determined by the curve $x = y^3$, or $y = f(x) = x^{1/3}$. The inner radius is determined by $x = 4y$, or $y = g(x) = x/4$. The volume of the resulting solid is

$$V = \int_0^8 \pi (f(x)^2 - g(x)^2)\, dx \quad \text{Washer method}$$

$$= \int_0^8 \pi \left(x^{2/3} - \frac{x^2}{16} \right) dx \quad \text{Substitute for } f \text{ and } g.$$

$$= \pi \left(\frac{3}{5} x^{5/3} - \frac{x^3}{48} \right) \Bigg|_0^8 \quad \text{Fundamental Theorem}$$

$$= \frac{128\pi}{15}. \quad \text{Evaluate.}$$

We see that revolving the region about the y-axis produces a solid of greater volume.

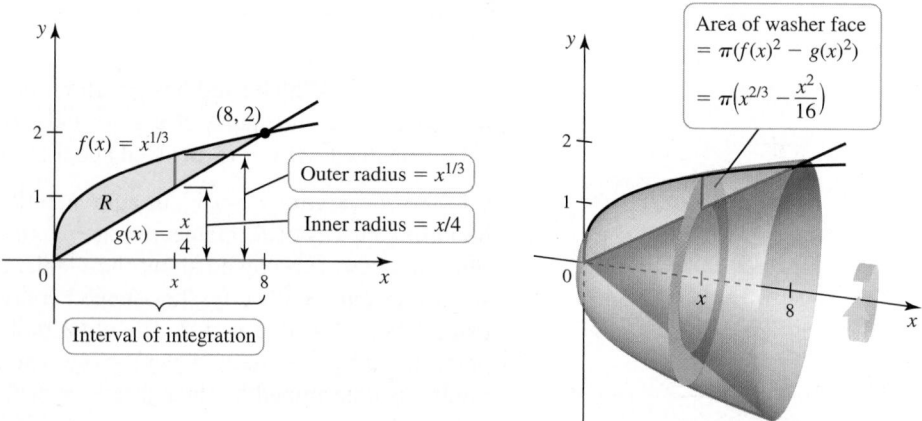

Figure 6.36

Related Exercises 47–48 ◄

QUICK CHECK 6 The region in the first quadrant bounded by $y = x$ and $y = x^3$ is revolved about the y-axis. Give the integral for the volume of the solid that is generated. ◄

The disk and washer methods may be generalized to handle situations in which a region R is revolved about a line parallel to one of the coordinate axes. The next example discusses three such cases.

EXAMPLE 6 **Revolving about other lines** Let $f(x) = \sqrt{x} + 1$ and $g(x) = x^2 + 1$.

a. Find the volume of the solid generated when the region R_1 bounded by the graph of f and the line $y = 2$ on the interval $[0, 1]$ is revolved about the line $y = 2$.

b. Find the volume of the solid generated when the region R_2 bounded by the graphs of f and g on the interval $[0, 1]$ is revolved about the line $y = -1$.

c. Find the volume of the solid generated when the region R_2 bounded by the graphs of f and g on the interval $[0, 1]$ is revolved about the line $x = 2$.

SOLUTION

a. Figure 6.37a shows the region R_1 and the axis of revolution. Applying the disk method, we see that a disk located at a point x has a radius of $2 - f(x) = 2 - (\sqrt{x} + 1) = 1 - \sqrt{x}$. Therefore, the volume of the solid generated when R_1 is revolved about $y = 2$ is

$$\int_0^1 \pi \underbrace{(1 - \sqrt{x})^2}_{\substack{\text{disk} \\ \text{radius}}} dx = \pi \int_0^1 (1 - 2\sqrt{x} + x)\, dx = \frac{\pi}{6}.$$

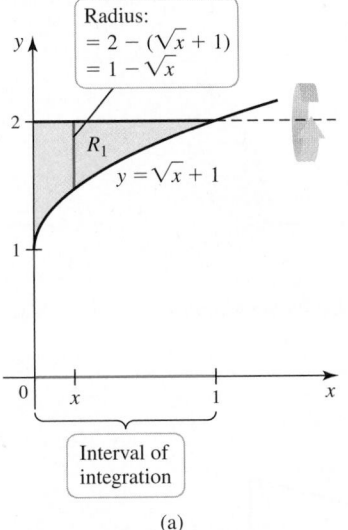

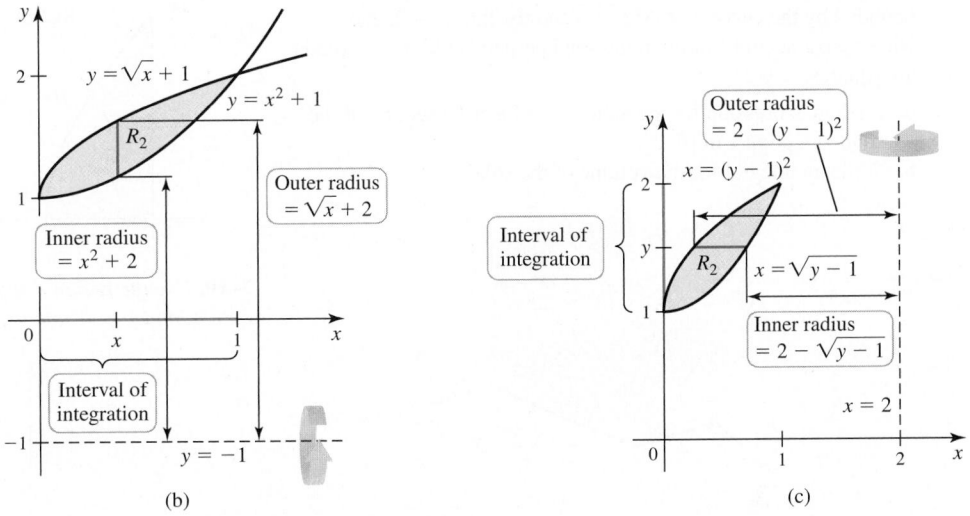

Figure 6.37

b. When the graph of f is revolved about $y = -1$, it sweeps out a solid of revolution whose radius at a point x is $f(x) + 1 = \sqrt{x} + 2$. Similarly, when the graph of g is revolved about $y = -1$, it sweeps out a solid of revolution whose radius at a point x is $g(x) + 1 = x^2 + 2$ (Figure 6.37b). Using the washer method, the volume of the solid generated when R_2 is revolved about $y = -1$ is

$$\int_0^1 \pi (\underbrace{(\sqrt{x} + 2)^2}_{\substack{\text{outer} \\ \text{radius}}} - \underbrace{(x^2 + 2)^2}_{\substack{\text{inner} \\ \text{radius}}})\, dx \qquad \text{Washer method}$$

$$= \pi \int_0^1 (-x^4 - 4x^2 + x + 4\sqrt{x})\, dx \qquad \text{Simplify integrand.}$$

$$= \frac{49\pi}{30}. \qquad \text{Evaluate integral.}$$

c. When the region R_2 is revolved about the line $x = 2$, we use the washer method and integrate in the y-direction. First note that the graph of f is described by $y = \sqrt{x} + 1$, or equivalently, $x = (y - 1)^2$, for $y \geq 1$. Also, the graph of g is described by $y = x^2 + 1$, or equivalently, $x = \sqrt{y - 1}$ for $y \geq 1$ (Figure 6.37c). When the graph of f is revolved about the line $x = 2$, the radius of a typical disk at a point y is $2 - (y - 1)^2$. Similarly, when the graph of g is revolved about $x = 2$, the radius of a typical disk at a point y is $2 - \sqrt{y - 1}$. Finally, observe that the extent of the region R_2 in the y-direction is the interval $1 \leq y \leq 2$.

Applying the washer method, simplifying the integrand, and integrating powers of y, the volume of the solid of revolution is

$$\int_1^2 \pi \Big(\underbrace{(2 - (y-1)^2)^2}_{\text{outer radius}} - \underbrace{(2 - \sqrt{y-1})^2}_{\text{inner radius}} \Big)\, dy = \frac{31\pi}{30}.$$

Related Exercises 49, 52 ◄

SECTION 6.3 EXERCISES

Getting Started

1. Suppose a cut is made through a solid object perpendicular to the x-axis at a particular point x. Explain the meaning of $A(x)$.

2. A solid has a circular base; cross sections perpendicular to the base are squares. What method should be used to find the volume of the solid?

3. Consider a solid whose base is the region in the first quadrant bounded by the curve $y = \sqrt{3 - x}$ and the line $x = 2$, and whose cross sections through the solid perpendicular to the x-axis are squares.

   a. Find an expression for the area $A(x)$ of a cross section of the solid at a point x in $[0, 2]$.
   b. Write an integral for the volume of the solid.

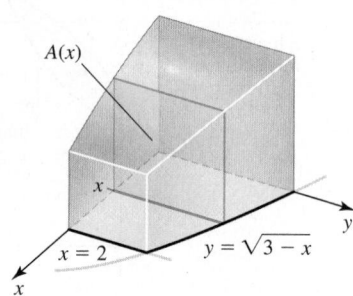

4. Why is the disk method a special case of the general slicing method?

5. Let R be the region bounded by the curve $y = \sqrt{\cos x}$ and the x-axis on $[0, \pi/2]$. A solid of revolution is obtained by revolving R about the x-axis (see figures).

   a. Find an expression for the radius of a cross section of the solid of revolution at a point x in $[0, \pi/2]$.
   b. Find an expression for the area $A(x)$ of a cross section of the solid at a point x in $[0, \pi/2]$.
   c. Write an integral for the volume of the solid.

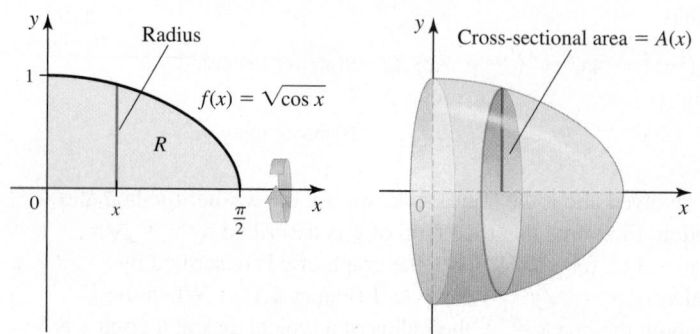

6. Let R be the region bounded by the curve $y = \cos^{-1} x$ and the x-axis on $[0, 1]$. A solid of revolution is obtained by revolving R about the y-axis (see figures).

   a. Find an expression for the radius of a cross section of the solid at a point y in $[0, \pi/2]$.
   b. Find an expression for the area $A(y)$ of a cross section of the solid at a point y in $[0, \pi/2]$.
   c. Write an integral for the volume of the solid.

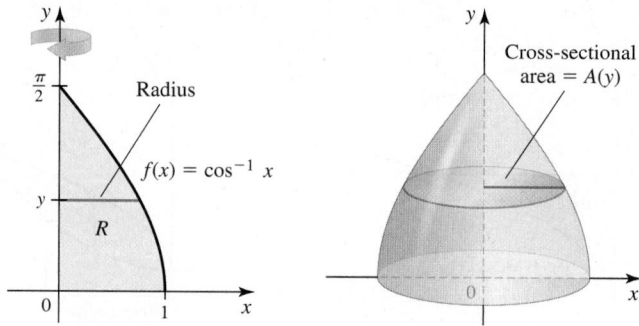

7–10. *Use the region R that is bounded by the graphs of $y = 1 + \sqrt{x}$, $x = 4$, and $y = 1$ to complete the exercises.*

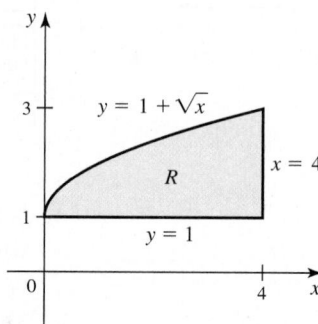

7. Region R is revolved about the x-axis to form a solid of revolution whose cross sections are washers.

   a. What is the outer radius of a cross section of the solid at a point x in $[0, 4]$?
   b. What is the inner radius of a cross section of the solid at a point x in $[0, 4]$?
   c. What is the area $A(x)$ of a cross section of the solid at a point x in $[0, 4]$?
   d. Write an integral for the volume of the solid.

8. Region R is revolved about the y-axis to form a solid of revolution whose cross sections are washers.

   a. What is the outer radius of a cross section of the solid at a point y in $[1, 3]$?
   b. What is the inner radius of a cross section of the solid at a point y in $[1, 3]$?
   c. What is the area $A(y)$ of a cross section of the solid at a point y in $[1, 3]$?
   d. Write an integral for the volume of the solid.

9. Region R is revolved about the line $y = 1$ to form a solid of revolution.

    a. What is the radius of a cross section of the solid at a point x in $[0, 4]$?

    b. What is the area $A(x)$ of a cross section of the solid at a point x in $[0, 4]$?

    c. Write an integral for the volume of the solid.

10. Region R is revolved about the line $x = 4$ to form a solid of revolution.

    a. What is the radius of a cross section of the solid at a point y in $[1, 3]$?

    b. What is the area $A(y)$ of a cross section of the solid at a point y in $[1, 3]$?

    c. Write an integral for the volume of the solid.

Practice Exercises

11–16. General slicing method *Use the general slicing method to find the volume of the following solids.*

11. The solid whose base is the region bounded by the semicircle $y = \sqrt{1 - x^2}$ and the x-axis, and whose cross sections through the solid perpendicular to the x-axis are squares

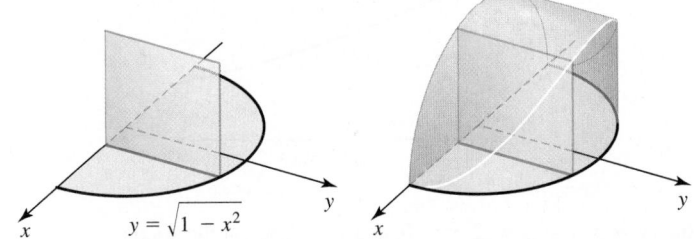

12. The solid whose base is the region bounded by the curves $y = x^2$ and $y = 2 - x^2$, and whose cross sections through the solid perpendicular to the x-axis are squares

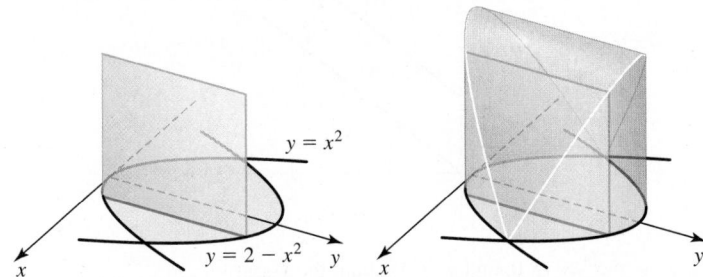

13. The solid whose base is the region bounded by the curve $y = \sqrt{\cos x}$ and the x-axis on $[-\pi/2, \pi/2]$, and whose cross sections through the solid perpendicular to the x-axis are isosceles right triangles with a horizontal leg in the xy-plane and a vertical leg above the x-axis

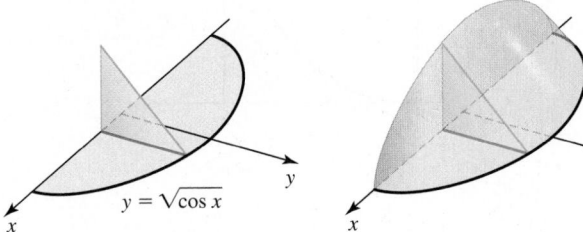

14. The solid with a semicircular base of radius 5 whose cross sections perpendicular to the base and parallel to the diameter are squares

15. The solid whose base is the triangle with vertices $(0, 0)$, $(2, 0)$, and $(0, 2)$, and whose cross sections perpendicular to the base and parallel to the y-axis are semicircles

16. The solid whose base is the region bounded by $y = x^2$ and the line $y = 1$, and whose cross sections perpendicular to the base and parallel to the y-axis are squares

17–44. Solids of revolution *Let R be the region bounded by the following curves. Find the volume of the solid generated when R is revolved about the given axis.*

17. $y = 2x, y = 0$, and $x = 3$; about the x-axis (Verify that your answer agrees with the volume formula for a cone.)

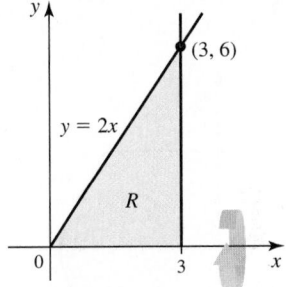

18. $y = 2 - 2x, y = 0$, and $x = 0$; about the x-axis (Verify that your answer agrees with the volume formula for a cone.)

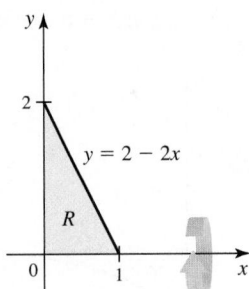

19. $y = e^{-x}, y = 0, x = 0$, and $x = \ln 4$; about the x-axis

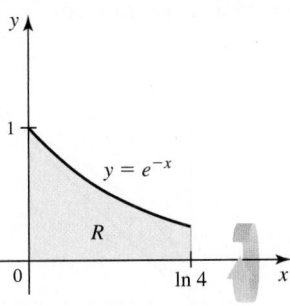

20. $y = \dfrac{1}{\sqrt[4]{1 - x^2}}$, $y = 0$, $x = 0$, and $x = \dfrac{1}{2}$; about the x-axis

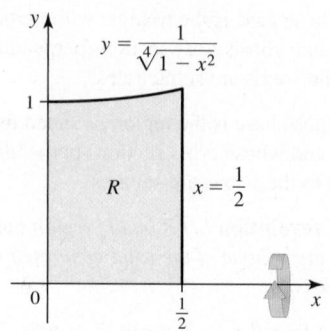

21. $x = \dfrac{1}{\sqrt{1 + y^2}}$, $x = 0$, $y = -1$, and $y = 1$; about the y-axis

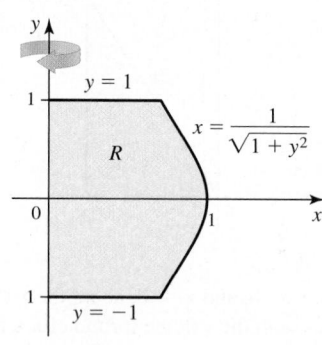

22. $y = 0$, $y = \ln x$, $y = 2$, and $x = 0$; about the y-axis

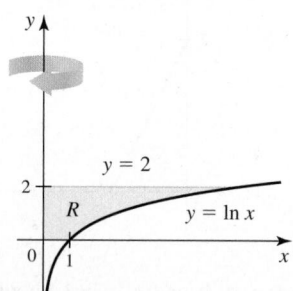

23. $y = x$ and $y = 2\sqrt{x}$; about the x-axis

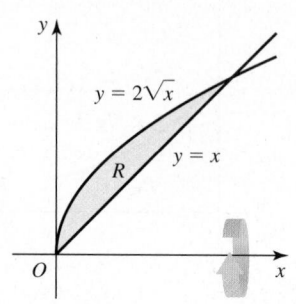

24. $y = x$ and $y = \sqrt[4]{x}$; about the x-axis

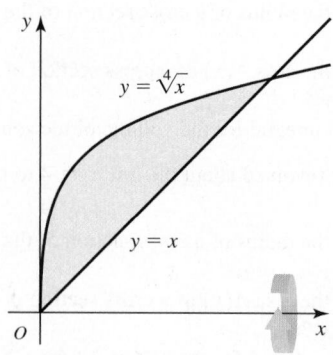

25. $y = e^{x/2}$, $y = e^{-x/2}$, $x = \ln 2$, and $x = \ln 3$; about the x-axis

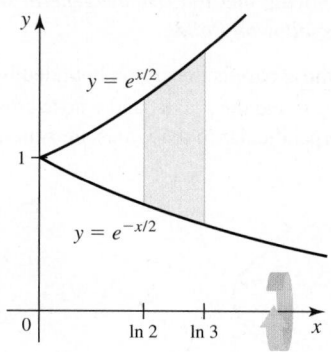

26. $y = x$, $y = x + 2$, $x = 0$, and $x = 4$; about the x-axis

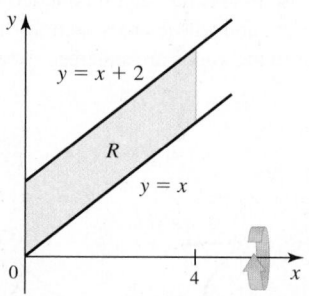

27. $y = x^3$, $y = 0$, and $x = 1$; about the y-axis

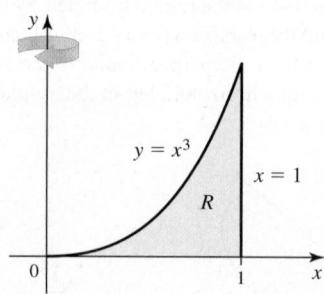

28. $y = x$, $y = 2x$, and $y = 6$; about the y-axis

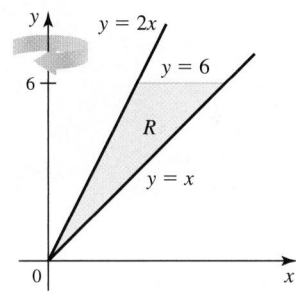

29. $y = \cos x$ on $[0, \pi/2]$, $y = 0$, and $x = 0$; about the x-axis

(*Hint:* Recall that $\cos^2 x = \dfrac{1 + \cos 2x}{2}$.)

30. $y = \sqrt{25 - x^2}$ and $y = 0$; about the x-axis (Verify that your answer agrees with the volume formula for a sphere.)

31. $y = \sin x$ on $[0, \pi]$ and $y = 0$; about the x-axis

(*Hint:* Recall that $\sin^2 x = \dfrac{1 - \cos 2x}{2}$.)

32. $y = \sec^{-1} x$, $x = 0$, $y = 0$, and $y = \frac{\pi}{4}$; about the y-axis

33. $y = \sin^{-1} x$, $x = 0$, $y = \frac{\pi}{4}$; about the y-axis

34. $y = \sqrt{\sin x}$, $y = 1$, and $x = 0$; about the x-axis

35. $y = \sin x$ and $y = \sqrt{\sin x}$, for $0 \le x \le \frac{\pi}{2}$; about the x-axis

36. $y = |x|$ and $y = 2 - x^2$; about the x-axis

37. $y = \sqrt{x}$, $y = 0$, and $x = 4$; about the y-axis

38. $y = 4 - x^2$, $x = 2$, and $y = 4$; about the y-axis

39. $y = 1/\sqrt{x}$, $y = 0$, $x = 2$, and $x = 6$; about the x-axis

40. $y = x^2$, $y = 2 - x$, and $y = 0$; about the y-axis

41. $y = e^x$, $y = 0$, $x = 0$, and $x = 2$; about the x-axis

42. $y = e^{-x}$, $y = e^x$, $x = 0$, and $x = \ln 4$; about the x-axis

43. $y = \ln x$, $y = \ln x^2$, and $y = \ln 8$; about the y-axis

44. $y = e^{-x}$, $y = 0$, $x = 0$, and $x = p > 0$; about the x-axis (Is the volume bounded as $p \to \infty$?)

45–48. Which is greater? *For the following regions R, determine which is greater—the volume of the solid generated when R is revolved about the x-axis or about the y-axis.*

45. R is bounded by $y = 2x$, the x-axis, and $x = 5$.

46. R is bounded by $y = 4 - 2x$, the x-axis, and the y-axis.

47. R is bounded by $y = 1 - x^3$, the x-axis, and the y-axis.

48. R is bounded by $y = x^2$ and $y = \sqrt{8x}$.

49–59. Revolution about other axes *Let R be the region bounded by the following curves. Find the volume of the solid generated when R is revolved about the given line.*

49. $y = 1 - \sqrt{x}$, $y = 1$, and $x = 1$; about $y = 1$

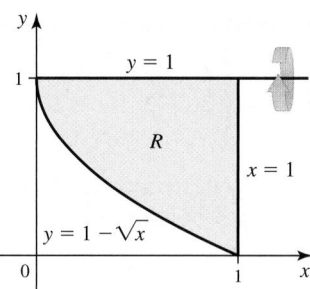

50. $y = e^{-x/2} + 2$, $y = 2$, $x = 0$, and $x = 1$; about $y = 2$

51. $x = 2 - \sec y$, $x = 2$, $y = \frac{\pi}{3}$, and $y = 0$; about $x = 2$

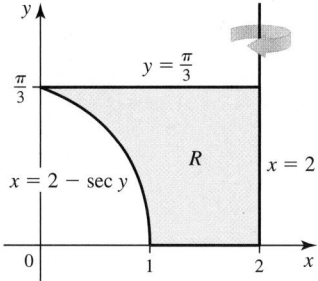

52. $y = 1 - \sqrt{x}$, $x = 1$, and $y = 1$; about $x = 1$

53. $x = 0$, $y = \sqrt{x}$, and $y = 1$; about $y = 1$

54. $x = 0$, $y = \sqrt{x}$, and $y = 2$; about $x = 4$

55. $y = 2 \sin x$ and $y = 0$ on $[0, \pi]$; about $y = -2$

56. $y = \ln x$ and $x = 0$ on the interval $0 \le y \le 1$; about $x = -1$

57. $y = \sin x$ and $y = 1 - \sin x$ on the interval $\frac{\pi}{6} \le x \le \frac{5\pi}{6}$; about $y = -1$

58. $y = x$ and $y = 1 + \frac{x}{2}$; about $y = 3$

59. $y = 2 - x$ and $y = 2 - 2x$; about $x = 3$

60. Comparing volumes The region R is bounded by the graph of $f(x) = 2x(2 - x)$ and the x-axis. Which is greater, the volume of the solid generated when R is revolved about the line $y = 2$ or the volume of the solid generated when R is revolved about the line $y = 0$? Use integration to justify your answer.

61. Explain why or why not Determine whether the following statements are true and give an explanation or counterexample.

a. A pyramid is a solid of revolution.

b. The volume of a hemisphere can be computed using the disk method.

c. Let R_1 be the region bounded by $y = \cos x$ and the x-axis on $[-\pi/2, \pi/2]$. Let R_2 be the region bounded by $y = \sin x$ and the x-axis on $[0, \pi]$. The volumes of the solids generated when R_1 and R_2 are revolved about the x-axis are equal.

Explorations and Challenges

62. Use calculus to find the volume of a tetrahedron (pyramid with four triangular faces), all of whose edges have length 4.

63. Fermat's volume calculation (1636) Let R be the region bounded by the curve $y = \sqrt{x + a}$ (with $a > 0$), the y-axis, and the x-axis. Let S be the solid generated by rotating R about the y-axis. Let T be the inscribed cone that has the same circular base as S and height $\sqrt{a}$. Show that $\dfrac{\text{volume}(S)}{\text{volume}(T)} = \dfrac{8}{5}$.

64. Solid from a piecewise function Let

$$f(x) = \begin{cases} x & \text{if } 0 \le x \le 2 \\ 2x - 2 & \text{if } 2 < x \le 5 \\ -2x + 18 & \text{if } 5 < x \le 6. \end{cases}$$

Find the volume of the solid formed when the region bounded by the graph of f, the x-axis, and the line $x = 6$ is revolved about the x-axis.

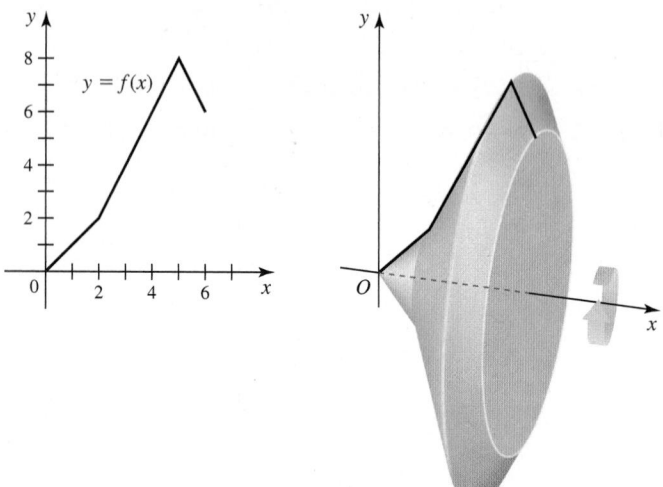

65. Solids from integrals Sketch a solid of revolution whose volume by the disk method is given by the following integrals. Indicate the function that generates the solid. Solutions are not unique.

a. $\displaystyle\int_0^\pi \pi \sin^2 x \, dx$ **b.** $\displaystyle\int_0^2 \pi (x^2 + 2x + 1) \, dx$

66. Volume of a cup A 6-inch-tall plastic cup is shaped like a surface obtained by rotating a line segment in the first quadrant about the x-axis. Given that the radius of the base of the cup is 1 inch, the radius of the top of the cup is 2 inches, and the cup is filled to the brim with water, use integration to approximate the volume of the water in the cup.

67. Estimating volume Suppose the region bounded by the curve $y = f(x)$ from $x = 0$ to $x = 4$ (see figure) is revolved about the x-axis to form a solid of revolution. Use left, right, and midpoint Riemann sums, with $n = 4$ subintervals of equal length, to estimate the volume of the solid of revolution.

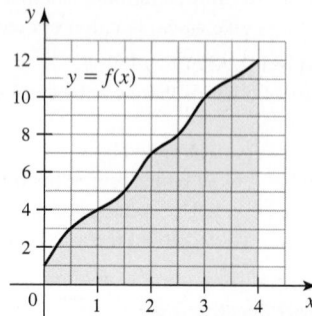

T 68. Volume of a wooden object A solid wooden object turned on a lathe has a length of 50 cm and the diameters (measured in cm) shown in the figure. (A lathe is a tool that spins and cuts a block of wood so that it has circular cross sections.) Use left Riemann sums with uniformly spaced grid points to estimate the volume of the object.

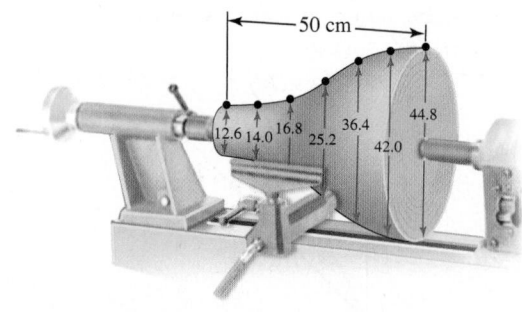

69. Cylinder, cone, hemisphere A right circular cylinder with height R and radius R has a volume of $V_C = \pi R^3$ (height = radius).

  a. Find the volume of the cone that is inscribed in the cylinder with the same base as the cylinder and height R. Express the volume in terms of V_C.
  b. Find the volume of the hemisphere that is inscribed in the cylinder with the same base as the cylinder. Express the volume in terms of V_C.

70. Water in a bowl A hemispherical bowl of radius 8 inches is filled to a depth of h inches, where $0 \le h \le 8$. Find the volume of water in the bowl as a function of h. (Check the special cases $h = 0$ and $h = 8$.)

T 71. A torus (doughnut) Find the volume of the torus formed when the circle of radius 2 centered at $(3, 0)$ is revolved about the y-axis. Use geometry to evaluate the integral.

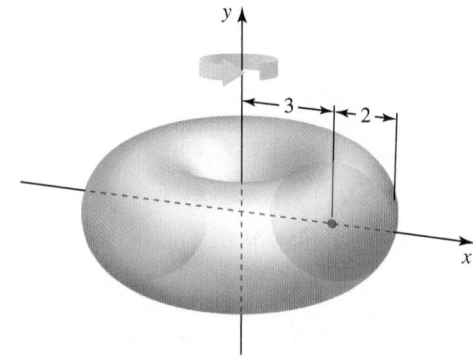

72. Which is greater? Let R be the region bounded by $y = x^2$ and $y = \sqrt{x}$. Use integration to determine which is greater, the volume of the solid generated when R is revolved about the x-axis or about the line $y = 1$.

73. Cavalieri's principle *Cavalieri's principle* states that if two solids with equal altitudes have the same cross-sectional areas at every height, then they have equal volumes (see figure).

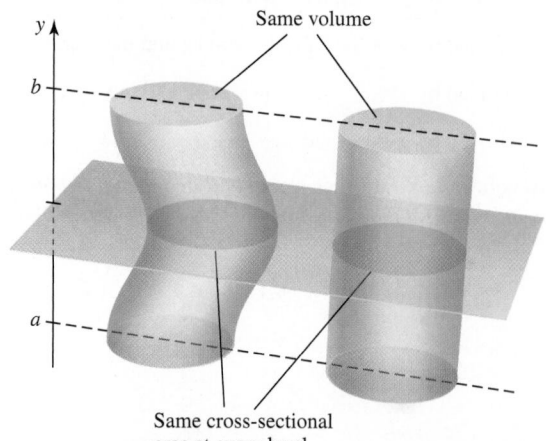

a. Use the general slicing method to justify Cavalieri's principle.

b. Use Cavalieri's principle to find the volume of a circular cylinder of radius r and height h whose axis is at an angle of $\frac{\pi}{4}$ to the base (see figure).

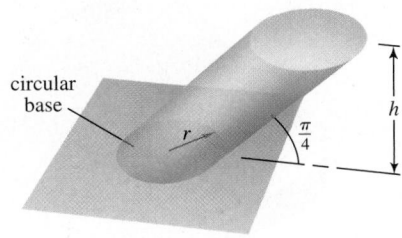

74. Limiting volume Consider the region R in the first quadrant bounded by $y = x^{1/n}$ and $y = x^n$, where $n > 1$ is a positive number.

a. Find the volume $V(n)$ of the solid generated when R is revolved about the x-axis. Express your answer in terms of n.

b. Evaluate $\lim_{n \to \infty} V(n)$. Interpret this limit geometrically.

QUICK CHECK ANSWERS

1. The average value of A on $[a, b]$ is $\overline{A} = \dfrac{1}{b - a}\displaystyle\int_a^b A(x)\, dx$. Therefore, $V = (b - a)\overline{A}$.   **2.** $A(x) = (2 - 2x^2)^2$

3. **(a)** A cylinder with height 2 and radius 2; **(b)** a cone with height 2 and base radius 2   **4.** When $g(x) = 0$, the washer method $V = \displaystyle\int_a^b \pi(f(x)^2 - g(x)^2)\, dx$ reduces to the disk method $V = \displaystyle\int_a^b \pi(f(x)^2)\, dx$.   **5.** **(a)** Inner radius $= x^2 + 1$;

(b) outer radius $= \sqrt{x} + 1$ **6.** $\displaystyle\int_0^1 \pi(y^{2/3} - y^2)\, dy$ ◄

6.4 Volume by Shells

You can solve many challenging volume problems using the disk/washer method. There are, however, some volume problems that are difficult to solve with this method. For this reason, we extend our discussion of volume problems to the *shell method*, which—like the disk/washer method—is used to compute the volume of solids of revolution.

Cylindrical Shells

Let R be a region bounded by the graph of f, the x-axis, and the lines $x = a$ and $x = b$, where $0 \le a < b$ and $f(x) \ge 0$ on $[a, b]$. When R is revolved about the y-axis, a solid is generated (Figure 6.39) whose volume is computed with the slice-and-sum strategy.

▶ Why another method? Suppose R is the region in the first quadrant bounded by the graph of $y = x^2 - x^3$ and the x-axis (**Figure 6.38**). When R is revolved about the y-axis, the resulting solid has a volume that is difficult to compute using the washer method. The volume is much easier to compute using the shell method.

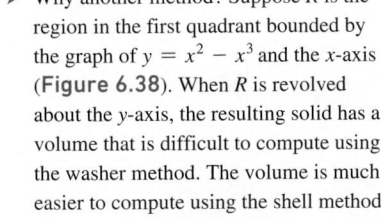

Figure 6.38

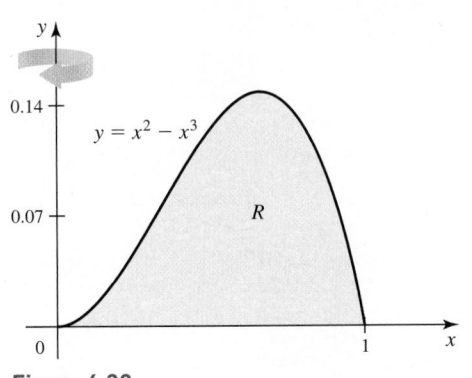

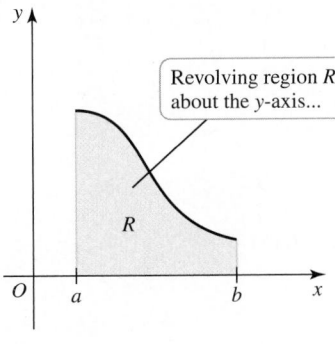

Figure 6.39

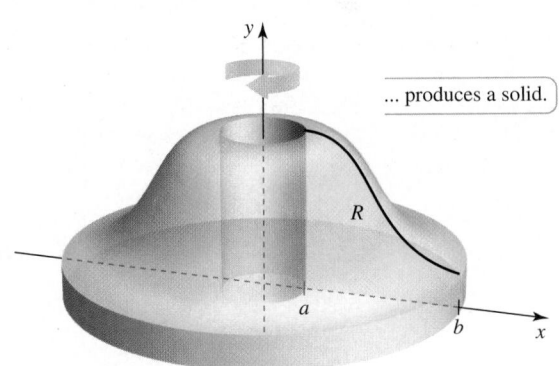

We divide $[a, b]$ into n subintervals of length $\Delta x = (b - a)/n$ and identify an arbitrary point x_k^* on the kth subinterval, for $k = 1, \ldots, n$. Now observe the rectangle built on the kth subinterval with a height of $f(x_k^*)$ and a width Δx (**Figure 6.40**). As it revolves about the y-axis, this rectangle sweeps out a thin *cylindrical shell*.

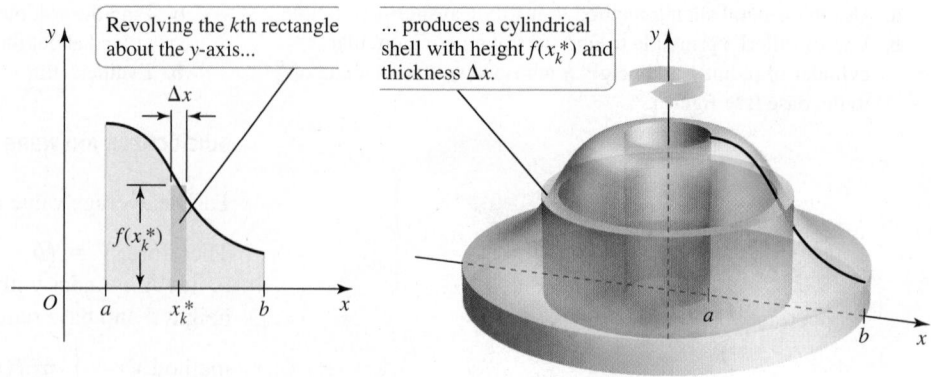

Figure 6.40

When the kth cylindrical shell is unwrapped (Figure 6.41), it approximates a thin rectangular slab. The approximate length of the slab is the circumference of a circle with radius x_k^*, which is $2\pi x_k^*$. The height of the slab is the height of the original rectangle $f(x_k^*)$ and its thickness is Δx; therefore, the volume of the kth shell is approximately

$$\underbrace{2\pi x_k^*}_{\text{length}} \cdot \underbrace{f(x_k^*)}_{\text{height}} \cdot \underbrace{\Delta x}_{\text{thickness}} = 2\pi x_k^* f(x_k^*)\Delta x.$$

Summing the volumes of the n cylindrical shells gives an approximation to the volume of the entire solid:

$$V \approx \sum_{k=1}^{n} 2\pi x_k^* f(x_k^*)\Delta x.$$

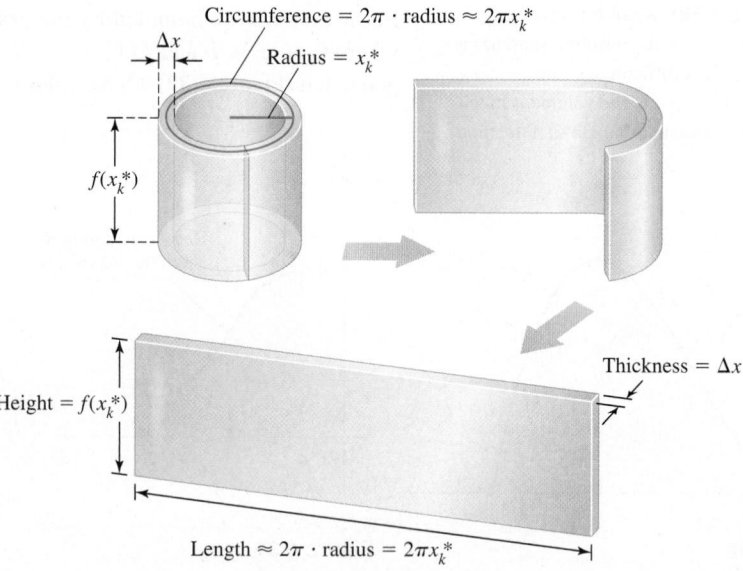

Figure 6.41

As n increases and Δx approaches 0 (Figure 6.42), we obtain the exact volume of the solid as a definite integral:

$$V = \lim_{n\to\infty} \sum_{k=1}^{n} \underbrace{2\pi x_k^*}_{\substack{\text{shell} \\ \text{circumference}}} \overbrace{f(x_k^*)}^{\substack{\text{shell} \\ \text{height}}} \underbrace{\Delta x}_{\substack{\text{shell} \\ \text{thickness}}} = \int_a^b 2\pi x f(x)\, dx.$$

▶ Rather than memorizing, think of the meaning of the factors in this formula: $f(x)$ is the height of a single cylindrical shell, $2\pi x$ is the circumference of the shell, and dx corresponds to the thickness of a shell. Therefore, $2\pi x\, f(x)\, dx$ represents the volume of a single shell, and we sum the volumes from $x = a$ to $x = b$. Notice that the integrand for the shell method is the function $A(x)$ that gives the surface area of the shell of radius x, for $a \le x \le b$.

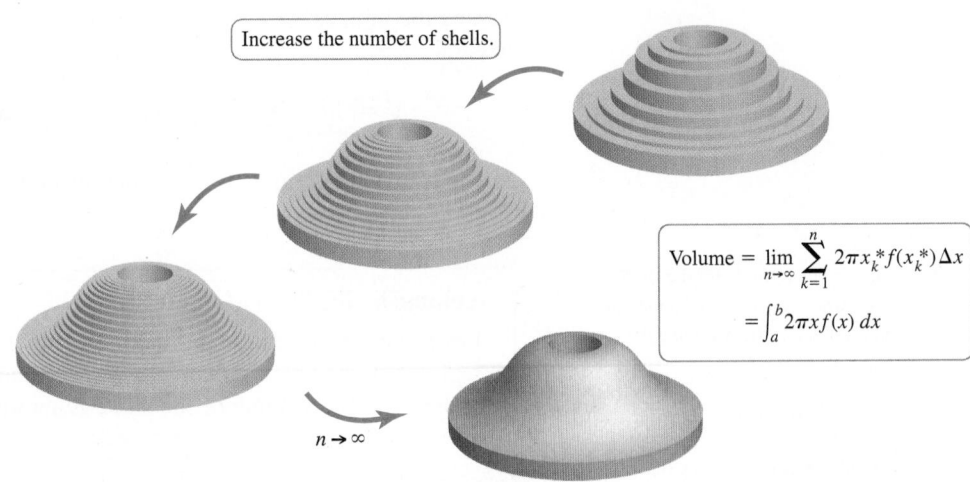

Increase the number of shells.

$$\text{Volume} = \lim_{n \to \infty} \sum_{k=1}^{n} 2\pi x_k^* f(x_k^*)\Delta x$$
$$= \int_a^b 2\pi x f(x)\, dx$$

$n \to \infty$

Figure 6.42

Before doing examples, we generalize this method as we did for the disk method. Suppose the region R is bounded by two curves, $y = f(x)$ and $y = g(x)$, where $f(x) \ge g(x)$ on $[a, b]$ (Figure 6.43). What is the volume of the solid generated when R is revolved about the y-axis?

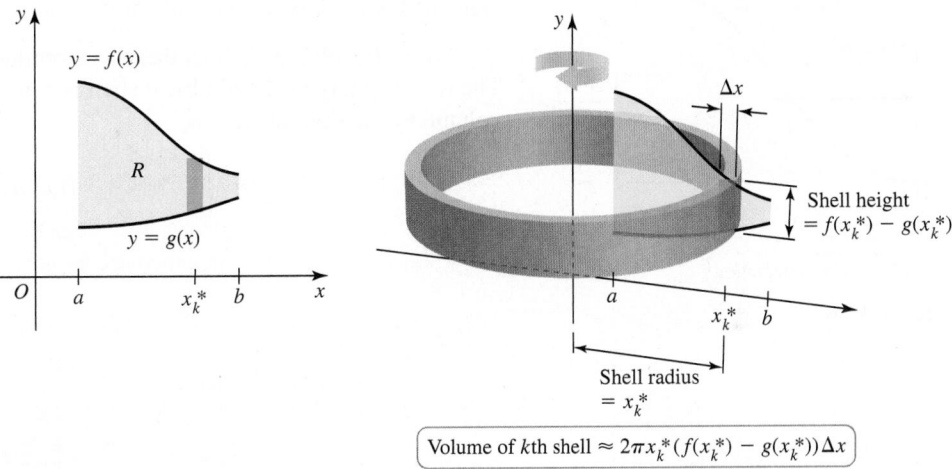

Shell height $= f(x_k^*) - g(x_k^*)$

Shell radius $= x_k^*$

Volume of kth shell $\approx 2\pi x_k^*(f(x_k^*) - g(x_k^*))\Delta x$

Figure 6.43

The situation is similar to the case we just considered. A typical rectangle in R sweeps out a cylindrical shell, but now the height of the kth shell is $f(x_k^*) - g(x_k^*)$, for $k = 1, \ldots, n$. As before, we take the radius of the kth shell to be x_k^*, which means the volume of the kth shell is approximated by $2\pi x_k^*(f(x_k^*) - g(x_k^*))\Delta x$ (Figure 6.43). Summing the volumes of all the shells gives an approximation to the volume of the entire solid:

$$V \approx \sum_{k=1}^{n} \underbrace{2\pi x_k^*}_{\substack{\text{shell} \\ \text{circumference}}} \underbrace{(f(x_k^*) - g(x_k^*))}_{\substack{\text{shell} \\ \text{height}}} \Delta x.$$

Taking the limit as $n \to \infty$ (which implies that $\Delta x \to 0$), the exact volume is the definite integral

$$V = \lim_{n \to \infty} \sum_{k=1}^{n} 2\pi x_k^*(f(x_k^*) - g(x_k^*))\Delta x = \int_a^b 2\pi x(f(x) - g(x))\, dx.$$

We now have the formula for the shell method.

▶ An analogous formula for the shell method when R is revolved about the x-axis is obtained by reversing the roles of x and y:

$$V = \int_c^d 2\pi y\,(p(y) - q(y))\, dy.$$

We assume R is bounded by the curves $x = p(y)$ and $x = q(y)$ where $p(y) \geq q(y)$ on $[c, d]$.

Volume by the Shell Method

Let f and g be continuous functions with $f(x) \geq g(x)$ on $[a, b]$. If R is the region bounded by the curves $y = f(x)$ and $y = g(x)$ between the lines $x = a$ and $x = b$, the volume of the solid generated when R is revolved about the y-axis is

$$V = \int_a^b 2\pi x\underbrace{(f(x) - g(x))}_{\substack{\text{shell} \quad \text{shell} \\ \text{circumference} \; \text{height}}} dx.$$

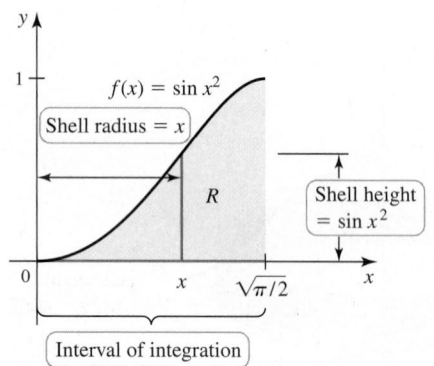

Figure 6.44

EXAMPLE 1   A sine bowl Let R be the region bounded by the graph of $f(x) = \sin x^2$, the x-axis, and the vertical line $x = \sqrt{\pi/2}$ (Figure 6.44). Find the volume of the solid generated when R is revolved about the y-axis.

SOLUTION Revolving R about the y-axis produces a bowl-shaped region (Figure 6.45). The radius of a typical cylindrical shell is x and its height is $f(x) = \sin x^2$. Therefore, the volume by the shell method is

$$V = \int_a^b \underbrace{2\pi x}_{\substack{\text{shell} \\ \text{circumference}}} \underbrace{f(x)}_{\substack{\text{shell} \\ \text{height}}} dx = \int_0^{\sqrt{\pi/2}} 2\pi x \sin x^2\, dx.$$

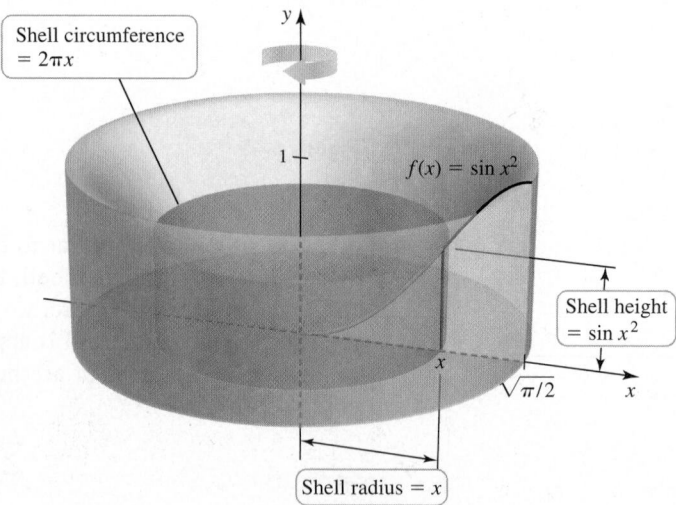

Figure 6.45

▶ When computing volumes using the shell method, it is best to sketch the region R in the xy-plane and draw a slice through the region that generates a typical shell.

Now we make the change of variables $u = x^2$, which means that $du = 2x \, dx$. The lower limit $x = 0$ becomes $u = 0$, and the upper limit $x = \sqrt{\pi/2}$ becomes $u = \pi/2$. The volume of the solid is

$$V = \int_0^{\sqrt{\pi/2}} \underbrace{2\pi x}_{\substack{\text{shell} \\ \text{circumference}}} \underbrace{\sin x^2}_{\substack{\text{shell} \\ \text{height}}} dx = \pi \int_0^{\pi/2} \sin u \, du \qquad u = x^2, \, du = 2x \, dx$$

$$= \pi(-\cos u)\Big|_0^{\pi/2} \qquad \text{Fundamental Theorem}$$

$$= \pi(0 - (-1)) = \pi. \quad \text{Simplify.}$$

Related Exercises 10, 17 ◀

QUICK CHECK 1 The triangle bounded by the x-axis, the line $y = 2x$, and the line $x = 1$ is revolved about the y-axis. Give an integral that equals the volume of the resulting solid using the shell method. ◀

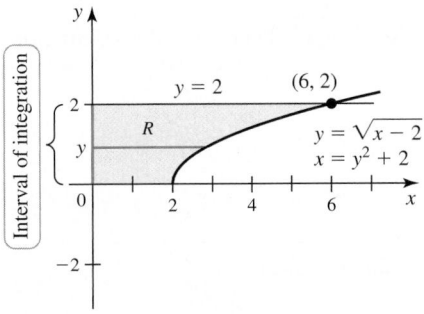

Figure 6.46

▶ In Example 2, we could use the disk/washer method to compute the volume, but notice that this approach requires splitting the region into two subregions. A better approach is to use the shell method and integrate along the y-axis.

EXAMPLE 2 Shells about the x-axis Let R be the region in the first quadrant bounded by the graph of $y = \sqrt{x - 2}$ and the line $y = 2$. Find the volume of the solid generated when R is revolved about the x-axis.

SOLUTION The revolution is about the x-axis, so the integration in the shell method is with respect to y. A typical shell runs parallel to the x-axis and has radius y, where $0 \le y \le 2$; the shells extend from the y-axis to the curve $y = \sqrt{x - 2}$ (**Figure 6.46**). Solving $y = \sqrt{x - 2}$ for x, we have $x = y^2 + 2$, which is the height of the shell at the point y (**Figure 6.47**). Integrating with respect to y, the volume of the solid is

$$V = \int_0^2 \underbrace{2\pi y}_{\substack{\text{shell} \\ \text{circumference}}} \underbrace{(y^2 + 2)}_{\substack{\text{shell} \\ \text{height}}} dy = 2\pi \int_0^2 (y^3 + 2y) \, dy = 16\pi.$$

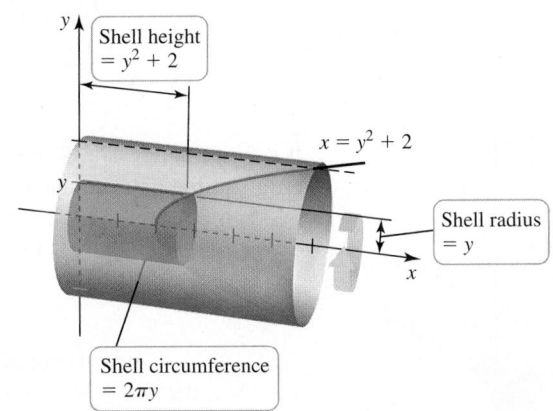

Figure 6.47

Related Exercises 13, 21 ◀

EXAMPLE 3 Volume of a drilled sphere A cylindrical hole with radius r is drilled symmetrically through the center of a sphere with radius a, where $0 \le r \le a$. What is the volume of the remaining material?

SOLUTION The y-axis is chosen to coincide with the axis of the cylindrical hole. We let R be the region in the xy-plane bounded above by $f(x) = \sqrt{a^2 - x^2}$, the upper half of a circle of radius a, and bounded below by $g(x) = -\sqrt{a^2 - x^2}$, the lower half of a circle of radius a, for $r \le x \le a$ (**Figure 6.48a**). Slices are taken perpendicular to the x-axis from $x = r$ to $x = a$. When a slice is revolved about the y-axis, it sweeps out a cylindrical shell that is concentric with the hole through the sphere (**Figure 6.48b**). The radius of a

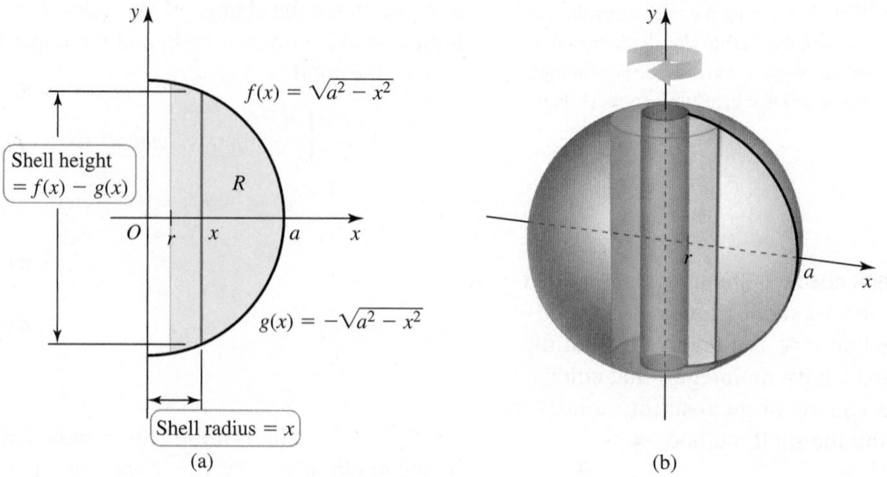

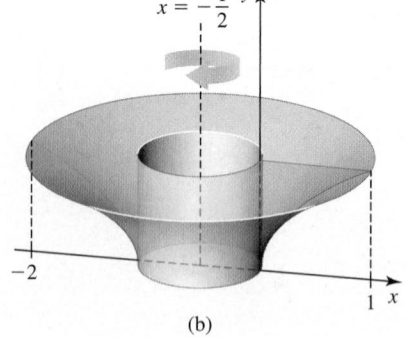

Figure 6.48

typical shell is x and its height is $f(x) - g(x) = 2\sqrt{a^2 - x^2}$. Therefore, the volume of the material that remains in the sphere is

$$V = \int_r^a \underbrace{2\pi x}_{\substack{\text{shell} \\ \text{circumference}}} \underbrace{\left(2\sqrt{a^2 - x^2}\right)}_{\substack{\text{shell} \\ \text{height}}} dx$$

$$= -2\pi \int_{a^2 - r^2}^0 \sqrt{u}\, du \quad u = a^2 - x^2,\, du = -2x\, dx$$

$$= 2\pi \left(\frac{2}{3} u^{3/2}\right)\Big|_0^{a^2 - r^2} \qquad \text{Fundamental Theorem}$$

$$= \frac{4\pi}{3}\left(a^2 - r^2\right)^{3/2}. \qquad \text{Simplify.}$$

It is important to check the result by examining special cases. In the case that $r = a$ (the radius of the hole equals the radius of the sphere), our calculation gives a volume of 0, which is correct. In the case that $r = 0$ (no hole in the sphere), our calculation gives the correct volume of a sphere, $\frac{4}{3}\pi a^3$.

Related Exercises 66–67 ◄

EXAMPLE 4   Revolving about other lines Let R be the region bounded by the curve $y = \sqrt{x}$, the line $y = 1$, and the y-axis (Figure 6.49a).

a. Use the shell method to find the volume of the solid generated when R is revolved about the line $x = -\frac{1}{2}$ (Figure 6.49b).

b. Use the disk/washer method to find the volume of the solid generated when R is revolved about the line $y = 1$ (Figure 6.49c).

SOLUTION

a. Using the shell method, we must imagine taking slices through R parallel to the y-axis. A typical slice through R at a point x, where $0 \le x \le 1$, has height $1 - \sqrt{x}$. When that slice is revolved about the line $x = -\frac{1}{2}$, it sweeps out a cylindrical shell with a radius of $x + \frac{1}{2}$ and a height of $1 - \sqrt{x}$ (Figure 6.50). A slight modification of the standard shell method gives the volume of the solid:

$$\int_0^1 \underbrace{2\pi\left(x + \frac{1}{2}\right)}_{\substack{\text{shell} \\ \text{circumference}}} \underbrace{(1 - \sqrt{x})}_{\substack{\text{shell} \\ \text{height}}} dx = 2\pi \int_0^1 \left(x - x^{3/2} + \frac{1}{2} - \frac{x^{1/2}}{2}\right) dx \qquad \begin{array}{l}\text{Expand} \\ \text{integrand.}\end{array}$$

$$= 2\pi \left(\frac{1}{2}x^2 - \frac{2}{5}x^{5/2} + \frac{1}{2}x - \frac{1}{3}x^{3/2}\right)\Big|_0^1 = \frac{8\pi}{15}. \qquad \begin{array}{l}\text{Evaluate} \\ \text{integral.}\end{array}$$

Figure 6.49

> ► If we instead revolved about the *y*-axis
> $(x = 0)$, the radius of the shell would be x.
> Because we are revolving about the line
> $x = -\frac{1}{2}$, the radius of the shell is $x + \frac{1}{2}$.

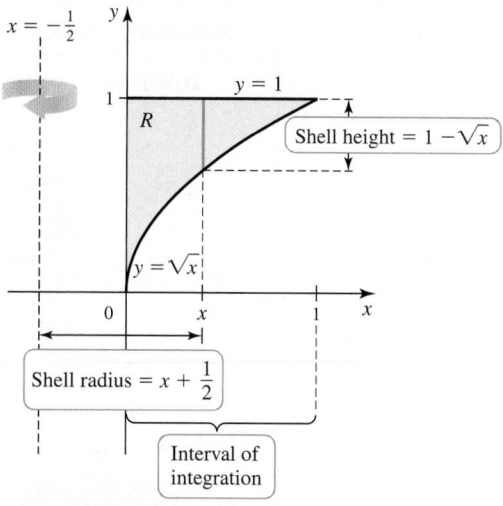

Figure 6.50

> ► The disk/washer method can also be
> used for part (a), and the shell method
> can also be used for part (b).

b. Using the disk/washer method, we take slices through R parallel to the *y*-axis. Consider a typical slice at a point x, where $0 \leq x \leq 1$. Its length, now measured with respect to the line $y = 1$, is $1 - \sqrt{x}$. When that slice is revolved about the line $y = 1$, it sweeps out a disk of radius $1 - \sqrt{x}$ (**Figure 6.51**). By the disk/washer method, the volume of the solid is

$$\int_0^1 \pi \underbrace{(1 - \sqrt{x})^2}_{\text{disk radius}} dx = \pi \int_0^1 (1 - 2\sqrt{x} + x)\, dx \qquad \text{Expand integrand.}$$

$$= \pi \left(x - \frac{4}{3} x^{3/2} + \frac{1}{2} x^2 \right) \Big|_0^1 \qquad \text{Evaluate integral.}$$

$$= \frac{\pi}{6}.$$

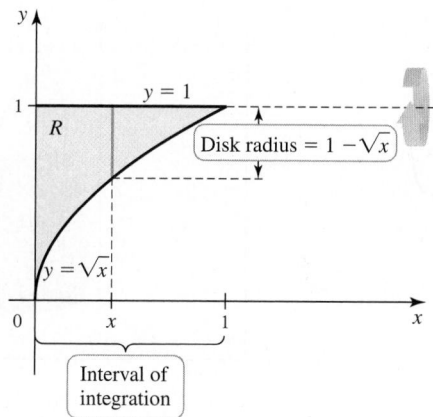

Figure 6.51

QUICK CHECK 2 Write the volume integral in Example 4b in the case that R is revolved about the line $y = -5$. ◄

Related Exercises 36–37 ◄

Restoring Order

After working with slices, disks, washers, and shells, you may feel somewhat overwhelmed. How do you choose a method, and which method is best?

Notice that the disk method is just a special case of the washer method. So for solids of revolution, the choice is between the washer method and the shell method. In *principle*, either method can be used. In *practice*, one method usually produces an integral that is easier to evaluate than the other method. The following table summarizes these methods.

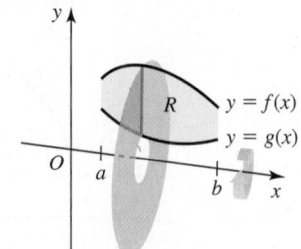

SUMMARY Disk/Washer and Shell Methods	
Integration with respect to x 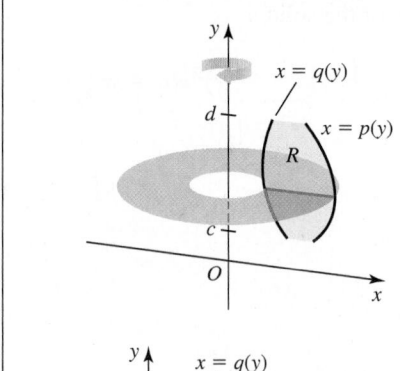	**Disk/washer method about the x-axis** Disks/washers are *perpendicular* to the x-axis. $$\int_a^b \pi\big(\underbrace{f(x)^2}_{\substack{\text{outer} \\ \text{radius}}} - \underbrace{g(x)^2}_{\substack{\text{inner} \\ \text{radius}}}\big)\,dx$$
	Shell method about the y-axis Shells are *parallel* to the y-axis. $$\int_a^b 2\pi x\big(\underbrace{x}_{\substack{\text{shell} \\ \text{circumference}}}\big)\big(\underbrace{f(x) - g(x)}_{\substack{\text{shell} \\ \text{height}}}\big)\,dx$$
Integration with respect to y	**Disk/washer method about the y-axis** Disks/washers are *perpendicular* to the y-axis. $$\int_c^d \pi\big(\underbrace{p(y)^2}_{\substack{\text{outer} \\ \text{radius}}} - \underbrace{q(y)^2}_{\substack{\text{inner} \\ \text{radius}}}\big)\,dy$$
	Shell method about the x-axis Shells are *parallel* to the x-axis. $$\int_c^d 2\pi y\big(\underbrace{y}_{\substack{\text{shell} \\ \text{circumference}}}\big)\big(\underbrace{p(y) - q(y)}_{\substack{\text{shell} \\ \text{height}}}\big)\,dy$$

The following example shows that while two methods may be used on the same problem, one of them may be preferable.

EXAMPLE 5   Volume by which method? The region R is bounded by the graphs of $f(x) = 2x - x^2$ and $g(x) = x$ on the interval $[0, 1]$ (**Figure 6.52**). Use the washer method and the shell method to find the volume of the solid formed when R is revolved about the x-axis.

SOLUTION Solving $f(x) = g(x)$, we find that the curves intersect at the points $(0, 0)$ and $(1, 1)$. Using the washer method, the upper bounding curve is the graph of f, the lower

Figure 6.52

$f(x) = 2x - x^2$ (1, 1)

R

$g(x) = x$

bounding curve is the graph of g, and a typical washer is perpendicular to the x-axis (Figure 6.53). Therefore, the volume is

$$V = \int_0^1 \pi (\underbrace{(2x - x^2)^2}_{\substack{\text{outer} \\ \text{radius}}} - \underbrace{x^2}_{\substack{\text{inner} \\ \text{radius}}}) \, dx \qquad \text{Washer method}$$

$$= \pi \int_0^1 (x^4 - 4x^3 + 3x^2) \, dx \qquad \text{Expand integrand.}$$

$$= \pi \left(\frac{x^5}{5} - x^4 + x^3 \right) \Big|_0^1 = \frac{\pi}{5}. \qquad \text{Evaluate integral.}$$

▶ To solve $y = 2x - x^2$ for x, write the equation as $x^2 - 2x + y = 0$ and complete the square or use the quadratic formula.

The shell method requires expressing the bounding curves in the form $x = p(y)$ for the right curve and $x = q(y)$ for the left curve. The right curve is $x = y$. Solving $y = 2x - x^2$ for x, we find that $x = 1 - \sqrt{1 - y}$ describes the left curve. A typical shell is parallel to the x-axis (Figure 6.54). Therefore, the volume is

$$V = \int_0^1 \underbrace{2\pi y}_{\substack{\text{shell} \\ \text{circumference}}} \underbrace{(y - (1 - \sqrt{1 - y}))}_{\substack{\text{shell} \\ \text{height}}} \, dy.$$

This integral equals $\frac{\pi}{5}$, but it is more difficult to evaluate than the integral required by the washer method. In this case, the washer method is preferable. Of course, the shell method may be preferable for other problems.

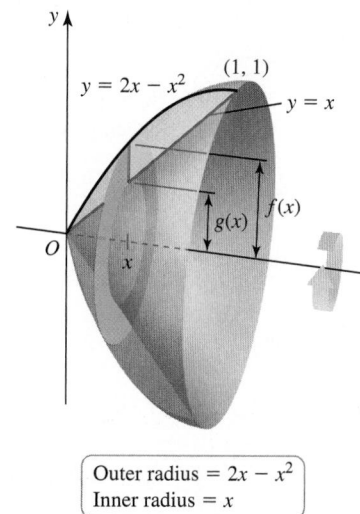

$y = 2x - x^2$
$(1, 1)$
$y = x$
$g(x)$ $f(x)$

Outer radius $= 2x - x^2$
Inner radius $= x$

Figure 6.53

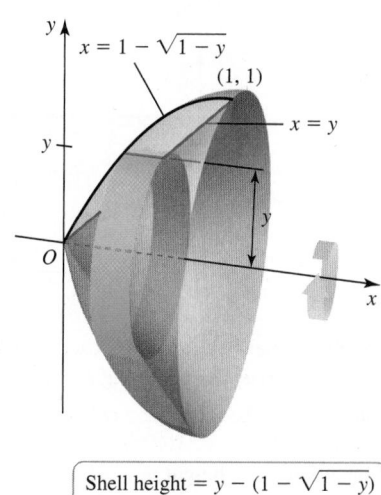

$x = 1 - \sqrt{1 - y}$
$(1, 1)$
$x = y$
y

Shell height $= y - (1 - \sqrt{1 - y})$
Shell radius $= y$

Figure 6.54

QUICK CHECK 3 Suppose the region in Example 5 is revolved about the y-axis. Which method (washer or shell) leads to an easier integral? ◀

Related Exercises 53–54 ◀

SECTION 6.4 EXERCISES

Getting Started

1. Assume f and g are continuous, with $f(x) \geq g(x) \geq 0$ on $[a, b]$. The region bounded by the graphs of f and g and the lines $x = a$ and $x = b$ is revolved about the y-axis. Write the integral given by the shell method that equals the volume of the resulting solid.

2. Fill in the blanks: A region R is revolved about the y-axis. The volume of the resulting solid could (in principle) be found by using the disk/washer method and integrating with respect

to _____ or using the shell method and integrating with respect to _____.

3. Fill in the blanks: A region R is revolved about the x-axis. The volume of the resulting solid could (in principle) be found by using the disk/washer method and integrating with respect to _____ or using the shell method and integrating with respect to _____.

4. Look again at the region R in Figure 6.38 (p. 439). Explain why it would be difficult to use the washer method to find the volume of the solid of revolution that results when R is revolved about the y-axis.

5. Let R be the region in the first quadrant bounded above by the curve $y = 2 - x^2$ and bounded below by the line $y = x$. Suppose the shell method is used to determine the volume of the solid generated by revolving R about the y-axis.

   a. What is the radius of a cylindrical shell at a point x in $[0, 2]$?

   b. What is the height of a cylindrical shell at a point x in $[0, 2]$?

   c. Write an integral for the volume of the solid using the shell method.

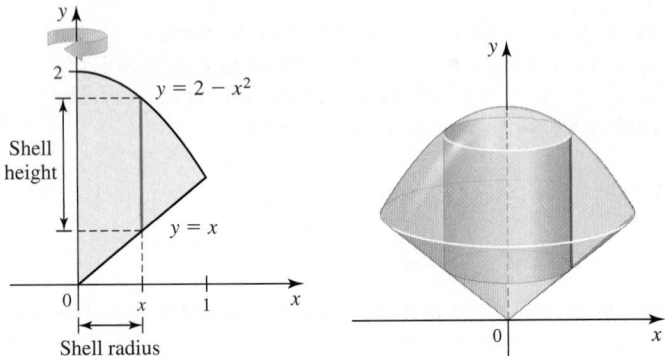

6–8. Let R be the region bounded by the curves $y = 2 - \sqrt{x}, y = 2$, and $x = 4$ in the first quadrant.

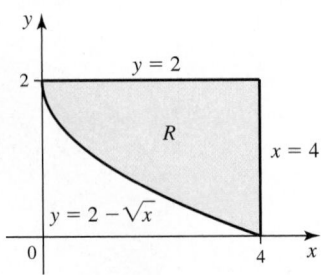

6. Suppose the shell method is used to determine the volume of the solid generated by revolving R about the x-axis.

   a. What is the radius of a cylindrical shell at a point y in $[0, 2]$?

   b. What is the height of a cylindrical shell at a point y in $[0, 2]$?

   c. Write an integral for the volume of the solid using the shell method.

7. Suppose the shell method is used to determine the volume of the solid generated by revolving R about the line $y = 2$.

   a. What is the radius of a cylindrical shell at a point y in $[0, 2]$?

   b. What is the height of a cylindrical shell at a point y in $[0, 2]$?

   c. Write an integral for the volume of the solid using the shell method.

8. Suppose the shell method is used to determine the volume of the solid generated by revolving R about the line $x = 4$.

   a. What is the radius of a cylindrical shell at a point x in $[0, 4]$?

   b. What is the height of a cylindrical shell at a point x in $[0, 4]$?

   c. Write an integral for the volume of the solid using the shell method.

Practice Exercises

9–34. Shell method *Let R be the region bounded by the following curves. Use the shell method to find the volume of the solid generated when R is revolved about indicated axis.*

9. $y = x - x^2, y = 0$; about the y-axis

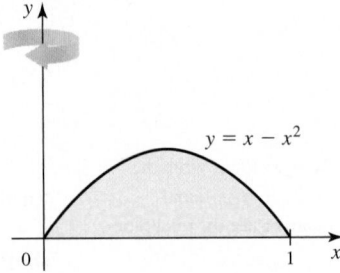

10. $y = (1 + x^2)^{-1}, y = 0, x = 0$, and $x = 2$; about the y-axis

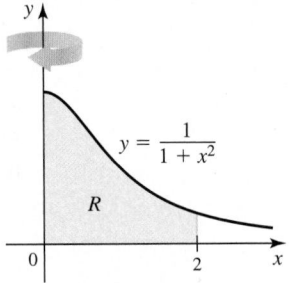

11. $y = 3x, y = 3$, and $x = 0$; about the y-axis (Use integration and check your answer using the volume formula for a cone.)

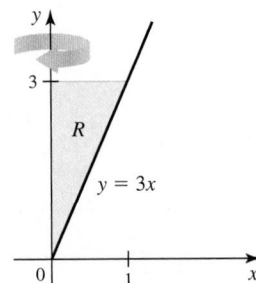

12. $y = -x^2 + 4x + 2, y = x^2 - 6x + 10$; about the y-axis

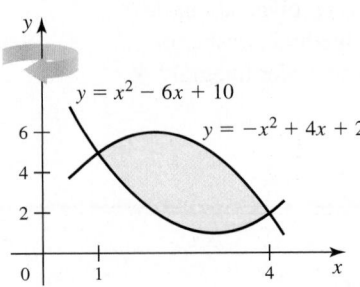

13. $y = \sqrt{x}, y = 0$, and $x = 4$; about the x-axis

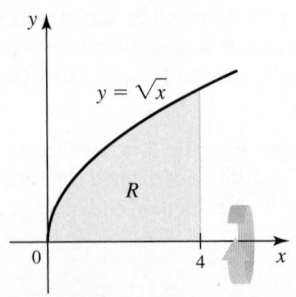

14. $y = \sqrt{x}, y = 2 - x,$ and $y = 0$; about the x-axis

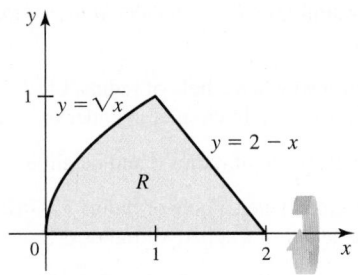

15. $y = 4 - x, y = 2,$ and $x = 0$; about the x-axis

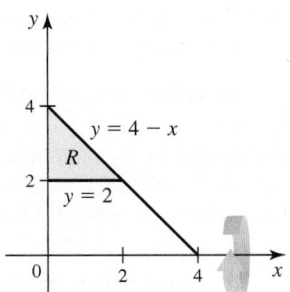

16. $x = \dfrac{4}{y + y^3}, x = \dfrac{1}{\sqrt{3}},$ and $y = 1$; about the x-axis

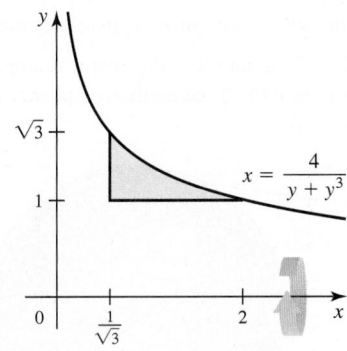

17. $y = \cos x^2, y = 0,$ for $0 \le x \le \sqrt{\pi/2}$; about the y-axis

18. $y = 6 - x, y = 0, x = 2,$ and $x = 4$; about the y-axis

19. $y = 1 - x^2, x = 0,$ and $y = 0,$ in the first quadrant; about the y-axis

20. $y = \sqrt{x}, y = 0,$ and $x = 1$; about the y-axis

21. $x = y^2, x = 0,$ and $y = 3$; about the x-axis

22. $y = x^3, y = 1,$ and $x = 0$; about the x-axis

23. $y = x, y = 2 - x,$ and $y = 0$; about the x-axis

24. $y = \sqrt{4 - 2x^2}, y = 0,$ and $x = 0,$ in the first quadrant; about the y-axis

T 25. $y = \dfrac{1}{(x^2 + 1)^2}, y = 0, x = 1,$ and $x = 2$; about the y-axis

26. $x = y^2, x = 4,$ and $y = 0$; about the x-axis

T 27. $y = 2x^{-3/2}, y = 2, y = 16,$ and $x = 0$; about the x-axis

T 28. $y = \sqrt{\sin^{-1} x}, y = \sqrt{\pi/2},$ and $x = 0$; about the x-axis

T 29. $y = \dfrac{e^x}{x}, y = 0, x = 1,$ and $x = 2$; about the y-axis

T 30. $y = \dfrac{\ln x}{x^2}, y = 0,$ and $x = 3$; about the y-axis

T 31. $y = \sqrt{\cos^{-1} x},$ in the first quadrant; about the x-axis

T 32. $y = \sqrt{50 - 2x^2},$ in the first quadrant; about the x-axis

33. $y = x^3 - x^8 + 1, y = 1$; about the y-axis

T 34. $y^2 = \ln x, y^2 = \ln x^3,$ and $y = 2$; about the x-axis

35–38. Shell and washer methods *Let R be the region bounded by the following curves. Use both the shell method and the washer method to find the volume of the solid generated when R is revolved about the indicated axis.*

35. $y = x, y = x^{1/3},$ in the first quadrant; about the x-axis

36. $y = \dfrac{1}{x + 1}, y = 1 - \dfrac{x}{3}$; about the x-axis

37. $y = (x - 2)^3 - 2, x = 0,$ and $y = 25$; about the y-axis

38. $y = 8, y = 2x + 2, x = 0,$ and $x = 2$; about the y-axis

39–44. Shell method about other lines *Let R be the region bounded by $y = x^2, x = 1,$ and $y = 0$. Use the shell method to find the volume of the solid generated when R is revolved about the following lines.*

39. $x = -2$ **40.** $x = 1$ **41.** $x = 2$

42. $y = 1$ **43.** $y = -2$ **44.** $y = 2$

45–48. Shell and washer methods about other lines *Use both the shell method and the washer method to find the volume of the solid that is generated when the region in the first quadrant bounded by $y = x^2, y = 1,$ and $x = 0$ is revolved about the following lines.*

45. $y = -2$ **46.** $x = -1$ **47.** $y = 6$ **48.** $x = 2$

49. Volume of a sphere Let R be the region bounded by the upper half of the circle $x^2 + y^2 = r^2$ and the x-axis. A sphere of radius r is obtained by revolving R about the x-axis.

  a. Use the shell method to verify that the volume of a sphere of radius r is $\frac{4}{3}\pi r^3$.
  b. Repeat part (a) using the disk method.

50. Comparing American and rugby union footballs An ellipse centered at the origin is described by the equation $\dfrac{x^2}{a^2} + \dfrac{y^2}{b^2} = 1$, where a and b are positive constants. If the upper half of such an ellipse is revolved about the x-axis, the resulting surface is an *ellipsoid*.

  a. Use the washer method to find the volume of an ellipsoid (in terms of a and b). Check your work using the shell method.
  b. Both American and rugby union footballs have the shape of ellipsoids. The maximum regulation size of a rugby union football corresponds to parameters of $a = 6$ in and $b = 3.82$ in, and the maximum regulation size of an American football corresponds to parameters of $a = 5.62$ in and $b = 3.38$ in. Find the volume of each football.
  c. Fill in the blank: At their maximum regulation sizes, the volume of a rugby union football has approximately _____ times the volume of an American football.

51. A torus (doughnut) A torus is formed when a circle of radius 2 centered at $(3, 0)$ is revolved about the y-axis.

   a. Use the shell method to write an integral for the volume of the torus.

   b. Use the washer method to write an integral for the volume of the torus.

   c. Find the volume of the torus by evaluating one of the two integrals obtained in parts (a) and (b). (*Hint:* Both integrals can be evaluated without using the Fundamental Theorem of Calculus.)

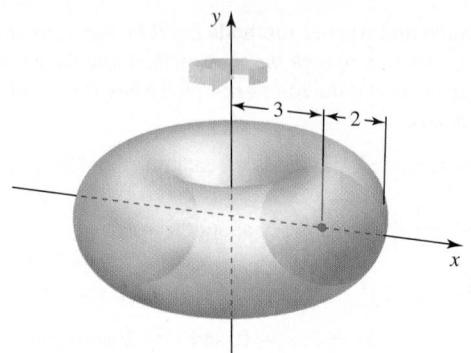

52. A cone by two methods Verify that the volume of a right circular cone with a base radius of r and a height of h is $\pi r^2 h/3$. Use the region bounded by the line $y = rx/h$, the x-axis, and the line $x = h$, where the region is rotated around the x-axis. Then (a) use the disk method and integrate with respect to x, and (b) use the shell method and integrate with respect to y.

53–62. Choose your method *Let R be the region bounded by the following curves. Use the method of your choice to find the volume of the solid generated when R is revolved about the given axis.*

53. $y = x - x^4$, $y = 0$; about the x-axis.

54. $y = x - x^4$, $y = 0$; about the y-axis.

55. $y = x^2$ and $y = 2 - x^2$; about the x-axis

56. $y = \sin x$ and $y = 1 - \sin x$, for $\pi/6 \le x \le 5\pi/6$; about the x-axis

57. $y = x$, $y = 2x + 2$, $x = 2$, and $x = 6$; about the y-axis

58. $y = x^3$, $y = 0$, and $x = 2$; about the x-axis

T 59. $y = \sqrt{\ln x}$, $y = \sqrt{\ln x^2}$, and $y = 1$; about the x-axis

60. $y = 2$, $y = 2x + 2$, and $x = 6$; about the y-axis

61. $y = x^2$, $y = 2 - x$, and $x = 0$, in the first quadrant; about the y-axis

62. $y = \sqrt{x}$, the x-axis, and $x = 4$; about the x-axis

63. Explain why or why not Determine whether the following statements are true and give an explanation or counterexample.

   a. When using the shell method, the axis of the cylindrical shells is parallel to the axis of revolution.

   b. If a region is revolved about the y-axis, then the shell method must be used.

   c. If a region is revolved about the x-axis, then in principle, it is possible to use the disk/washer method and integrate with respect to x or to use the shell method and integrate with respect to y.

Explorations and Challenges

64–68. Shell method *Use the shell method to find the volume of the following solids.*

64. The solid formed when a hole of radius 2 is drilled symmetrically along the axis of a right circular cylinder of height 6 and radius 4

65. A right circular cone of radius 3 and height 8

66. The solid formed when a hole of radius 3 is drilled symmetrically through the center of a sphere of radius 6

67. The solid formed when a hole of radius 3 is drilled symmetrically along the axis of a right circular cone of radius 6 and height 9

68. A hole of radius $r \le R$ is drilled symmetrically along the axis of a bullet. The bullet is formed by revolving the parabola
$y = 6\left(1 - \dfrac{x^2}{R^2}\right)$ about the y-axis, where $0 \le x \le R$.

T 69. Equal volumes Consider the region R bounded by the curves $y = ax^2 + 1$, $y = 0$, $x = 0$, and $x = 1$, for $a \ge -1$. Let S_1 and S_2 be solids generated when R is revolved about the x- and y-axes, respectively.

   a. Find V_1 and V_2, the volumes of S_1 and S_2, as functions of a.

   b. What are the values of $a \ge -1$ for which $V_1(a) = V_2(a)$?

70. A spherical cap Consider the cap of thickness h that has been sliced from a sphere of radius r (see figure). Verify that the volume of the cap is $\dfrac{\pi}{3}h^2(3r - h)$ using (a) the washer method, (b) the shell method, and (c) the general slicing method. Check for consistency among the three methods and check the special cases $h = r$ and $h = 0$.

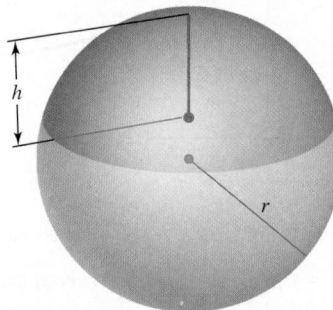

71. Change of variables Suppose $f(x) > 0$, for all x, and $\int_0^4 f(x)\, dx = 10$. Let R be the region in the first quadrant bounded by the coordinate axes, $y = f(x^2)$, and $x = 2$. Find the volume of the solid generated by revolving R around the y-axis.

72. Equal integrals Without evaluating integrals, explain the following equalities. (*Hint:* Draw pictures.)

   a. $\pi \displaystyle\int_0^4 (8 - 2x)^2\, dx = 2\pi \int_0^8 y\left(4 - \dfrac{y}{2}\right) dy$

   b. $\displaystyle\int_0^2 (25 - (x^2 + 1)^2)\, dx = 2\int_1^5 y\sqrt{y - 1}\, dy$

73. Volumes without calculus Solve the following problems with *and* without calculus. A good picture helps.

   a. A cube with side length r is inscribed in a sphere, which is inscribed in a right circular cone, which is inscribed in a right circular cylinder. The side length (slant height) of the cone is equal to its diameter. What is the volume of the cylinder?

b. A cube is inscribed in a right circular cone with a radius of 1 and a height of 3. What is the volume of the cube?

c. A cylindrical hole 10 in long is drilled symmetrically through the center of a sphere. How much material is left in the sphere? (Enough information *is* given).

74. Wedge from a tree Imagine a cylindrical tree of radius a. A wedge is cut from the tree by making two cuts: one in a horizontal plane P perpendicular to the axis of the cylinder, and one that makes an angle θ with P, intersecting P along a diameter of the tree (see figure). What is the volume of the wedge?

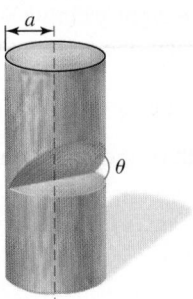

75. Different axes of revolution Suppose R is the region bounded by $y = f(x)$ and $y = g(x)$ on the interval $[a, b]$, where $f(x) \geq g(x) \geq 0$.

a. Show that if R is revolved about the horizontal line $y = y_0$ that lies below R, then by the washer method, the volume of the resulting solid is

$$V = \int_a^b \pi \left((f(x) - y_0)^2 - (g(x) - y_0)^2 \right) dx.$$

b. How is this formula changed if the line $y = y_0$ lies above R?

76. Different axes of revolution Suppose R is the region bounded by $y = f(x)$ and $y = g(x)$ on the interval $[a, b]$, where $f(x) \geq g(x)$.

a. Show that if R is revolved about the vertical line $x = x_0$, where $x_0 < a$, then by the shell method, the volume of the resulting solid is $V = \int_a^b 2\pi (x - x_0)(f(x) - g(x)) \, dx$.

b. How is this formula changed if $x_0 > b$?

QUICK CHECK ANSWERS

1. $\int_0^1 2\pi x(2x) \, dx$ 2. $V = \int_0^1 \pi (36 - (\sqrt{x} + 5)^2) \, dx$
3. The shell method is easier. ◄

6.5 Length of Curves

The space station orbits Earth in an elliptical path. How far does it travel in one orbit? A baseball slugger launches a home run into the upper deck and the sportscaster claims it landed 480 feet from home plate. But how far did the ball actually travel along its flight path? These questions deal with the length of trajectories or, more generally, with *arc length*. As you will see, their answers can be found by integration.

There are two common ways to formulate problems about arc length: The curve may be given explicitly in the form $y = f(x)$ or it may be defined *parametrically*. In this section, we deal with the first case. Parametric curves and the associated arc length problem are discussed in Section 12.1.

Arc Length for $y = f(x)$

Suppose a curve is given by $y = f(x)$, where f is a function with a continuous first derivative on the interval $[a, b]$. The goal is to determine how far you would travel if you walked along the curve from $(a, f(a))$ to $(b, f(b))$. This distance is the arc length, which we denote L.

As shown in **Figure 6.55**, we divide $[a, b]$ into n subintervals of length $\Delta x = (b - a)/n$, where x_k is the right endpoint of the kth subinterval, for $k = 1, \ldots, n$.

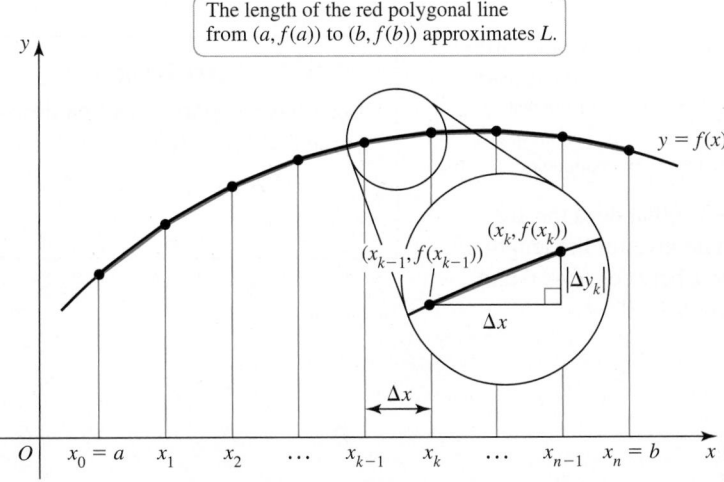

Figure 6.55

▶ More generally, we may choose any point in the kth subinterval and Δx may vary from one subinterval to the next. Using right endpoints, as we do here, simplifies the discussion and leads to the same result.

Joining the corresponding points on the curve by line segments, we obtain a polygonal line with n line segments. If n is large and Δx is small, the length of the polygonal line is a good approximation to the length of the actual curve. The strategy is to find the length of the polygonal line and then let n increase, while Δx goes to zero, to get the exact length of the curve.

Consider the kth subinterval $[x_{k-1}, x_k]$ and the line segment between the points $(x_{k-1}, f(x_{k-1}))$ and $(x_k, f(x_k))$. We let the change in the y-coordinate between these points be

$$\Delta y_k = f(x_k) - f(x_{k-1}).$$

The kth line segment is the hypotenuse of a right triangle with sides of length Δx and $|\Delta y_k| = |f(x_k) - f(x_{k-1})|$. The length of each line segment is

$$\sqrt{(\Delta x)^2 + |\Delta y_k|^2}, \quad \text{for} \quad k = 1, 2, \ldots, n.$$

▶ Notice that Δx is the same for each subinterval, but Δy_k depends on the subinterval.

Summing these lengths, we obtain the length of the polygonal line, which approximates the length L of the curve:

$$L \approx \sum_{k=1}^{n} \sqrt{(\Delta x)^2 + |\Delta y_k|^2}.$$

In previous applications of the integral, we would, at this point, take the limit as $n \to \infty$ and $\Delta x \to 0$ to obtain a definite integral. However, because of the presence of the Δy_k term, we must complete one additional step before taking a limit. Notice that the slope of the line segment on the kth subinterval is $\Delta y_k / \Delta x$ (rise over run). By the Mean Value Theorem (see the margin figure and Section 4.2), this slope equals $f'(x_k^*)$ for some point x_k^* on the kth subinterval. Therefore,

$$L \approx \sum_{k=1}^{n} \sqrt{(\Delta x)^2 + |\Delta y_k|^2}$$

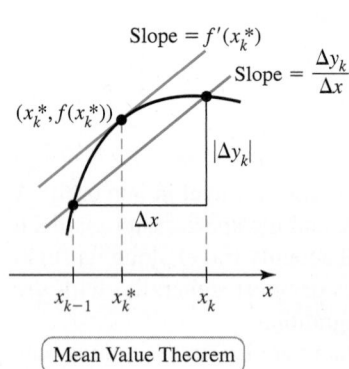

Mean Value Theorem

$$= \sum_{k=1}^{n} \sqrt{(\Delta x)^2 \left(1 + \left(\frac{\Delta y_k}{\Delta x}\right)^2\right)} \qquad \text{Factor out } (\Delta x)^2.$$

$$= \sum_{k=1}^{n} \sqrt{1 + \left(\frac{\Delta y_k}{\Delta x}\right)^2} \, \Delta x \qquad \text{Bring } \Delta x \text{ out of the square root.}$$

$$= \sum_{k=1}^{n} \sqrt{1 + f'(x_k^*)^2} \, \Delta x. \qquad \text{Mean Value Theorem}$$

Now we have a Riemann sum. As n increases and as Δx approaches zero, the sum approaches a definite integral, which is also the exact length of the curve. We have

$$L = \lim_{n \to \infty} \sum_{k=1}^{n} \sqrt{1 + f'(x_k^*)^2} \, \Delta x = \int_a^b \sqrt{1 + f'(x)^2} \, dx.$$

▶ Note that $1 + f'(x)^2$ is positive, so the square root in the integrand is defined whenever f' exists. To ensure that $\sqrt{1 + f'(x)^2}$ is integrable on $[a, b]$, we require that f' be continuous.

QUICK CHECK 1 What does the arc length formula give for the length of the line $y = x$ between $x = 0$ and $x = a$, where $a \geq 0$? ◄

DEFINITION   Arc Length for $y = f(x)$

Let f have a continuous first derivative on the interval $[a, b]$. The length of the curve from $(a, f(a))$ to $(b, f(b))$ is

$$L = \int_a^b \sqrt{1 + f'(x)^2} \, dx.$$

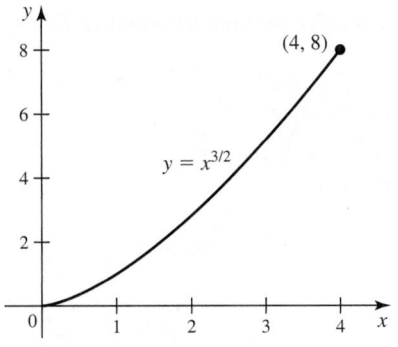

Figure 6.56

EXAMPLE 1 Arc length Find the length of the curve $f(x) = x^{3/2}$ between $x = 0$ and $x = 4$ (Figure 6.56).

SOLUTION Notice that $f'(x) = \frac{3}{2}x^{1/2}$, which is continuous on the interval $[0, 4]$. Using the arc length formula, we have

$$L = \int_a^b \sqrt{1 + f'(x)^2}\, dx = \int_0^4 \sqrt{1 + \left(\frac{3}{2}x^{1/2}\right)^2}\, dx \qquad \text{Substitute for } f'(x).$$

$$= \int_0^4 \sqrt{1 + \frac{9}{4}x}\, dx \qquad \text{Simplify.}$$

$$= \frac{4}{9}\int_1^{10} \sqrt{u}\, du \qquad u = 1 + \frac{9x}{4},\ du = \frac{9}{4}dx$$

$$= \frac{4}{9}\left(\frac{2}{3}u^{3/2}\right)\Big|_1^{10} \qquad \text{Fundamental Theorem}$$

$$= \frac{8}{27}(10^{3/2} - 1). \qquad \text{Simplify.}$$

The length of the curve is $\frac{8}{27}(10^{3/2} - 1) \approx 9.1$ units.

Related Exercise 11 ◄

EXAMPLE 2 Arc length of an exponential curve Find the length of the curve $f(x) = 2e^x + \frac{1}{8}e^{-x}$ on the interval $[0, \ln 2]$.

SOLUTION We first calculate $f'(x) = 2e^x - \frac{1}{8}e^{-x}$ and $f'(x)^2 = 4e^{2x} - \frac{1}{2} + \frac{1}{64}e^{-2x}$. The length of the curve on the interval $[0, \ln 2]$ is

$$L = \int_0^{\ln 2}\sqrt{1 + f'(x)^2}\, dx = \int_0^{\ln 2}\sqrt{1 + \left(4e^{2x} - \frac{1}{2} + \frac{1}{64}e^{-2x}\right)}\, dx$$

$$= \int_0^{\ln 2}\sqrt{4e^{2x} + \frac{1}{2} + \frac{1}{64}e^{-2x}}\, dx \qquad \text{Simplify.}$$

$$= \int_0^{\ln 2}\sqrt{\left(2e^x + \frac{1}{8}e^{-x}\right)^2}\, dx \qquad \text{Factor.}$$

$$= \int_0^{\ln 2}\left(2e^x + \frac{1}{8}e^{-x}\right) dx \qquad \text{Simplify.}$$

$$= \left(2e^x - \frac{1}{8}e^{-x}\right)\Big|_0^{\ln 2} = \frac{33}{16}. \qquad \begin{array}{l}\text{Evaluate the} \\ \text{integral.}\end{array}$$

Related Exercise 10 ◄

EXAMPLE 3 Circumference of a circle Confirm that the circumference of a circle of radius a is $2\pi a$.

SOLUTION The upper half of a circle of radius a centered at $(0, 0)$ is given by the function $f(x) = \sqrt{a^2 - x^2}$ for $|x| \le a$ (Figure 6.57). So we might consider using the arc length formula on the interval $[-a, a]$ to find the length of a semicircle. However, the circle has vertical tangent lines at $x = \pm a$ and $f'(\pm a)$ is undefined, which prevents us from using the arc length formula. An alternative approach is to use symmetry and avoid the points $x = \pm a$. For example, let's compute the length of one-eighth of the circle on the interval $[0, a/\sqrt{2}]$ (Figure 6.57).

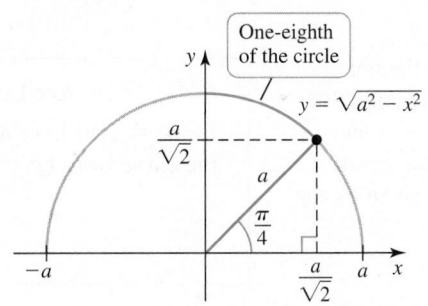

Figure 6.57

We first determine that $f'(x) = -\dfrac{x}{\sqrt{a^2 - x^2}}$, which is continuous on $[0, a/\sqrt{2}]$. The length of one-eighth of the circle is

$$
\begin{aligned}
\int_0^{a/\sqrt{2}} \sqrt{1 + f'(x)^2}\, dx &= \int_0^{a/\sqrt{2}} \sqrt{1 + \left(-\frac{x}{\sqrt{a^2 - x^2}}\right)^2}\, dx \\
&= \int_0^{a/\sqrt{2}} \sqrt{\frac{a^2}{a^2 - x^2}}\, dx && \text{Simplify.} \\
&= a \int_0^{a/\sqrt{2}} \frac{dx}{\sqrt{a^2 - x^2}} && \text{Simplify; } a > 0. \\
&= a \sin^{-1} \frac{x}{a} \Big|_0^{a/\sqrt{2}} && \text{Integrate.} \\
&= a \left(\sin^{-1} \frac{1}{\sqrt{2}} - 0 \right) && \text{Evaluate.} \\
&= \frac{\pi a}{4}. && \text{Simplify.}
\end{aligned}
$$

> The arc length integral for the semicircle on $[-a, a]$ is an example of an *improper integral*, a topic considered in Section 8.9.

It follows that the circumference of the full circle is $8(\pi a/4) = 2\pi a$ units.

Related Exercise 13 ◄

EXAMPLE 4 Looking ahead Consider the segment of the parabola $f(x) = x^2$ on the interval $[0, 2]$.

a. Write the integral for the length of the curve.

b. Use a calculator to evaluate the integral.

SOLUTION

a. Noting that $f'(x) = 2x$, the arc length integral is

$$\int_0^2 \sqrt{1 + f'(x)^2}\, dx = \int_0^2 \sqrt{1 + 4x^2}\, dx.$$

> When relying on technology, it is a good idea to check whether an answer is plausible. In Example 4, we found that the arc length of $y = x^2$ on $[0, 2]$ is approximately 4.647. The straight-line distance between $(0, 0)$ and $(2, 4)$ is $\sqrt{20} \approx 4.472$, so our answer is reasonable.

b. Using integration techniques presented so far, this integral cannot be evaluated (the required method is given in Section 8.4). This is typical of arc length integrals—even simple functions can lead to arc length integrals that are difficult to evaluate analytically. Without an analytical method, we may use numerical integration to *approximate* the value of a definite integral (Section 8.8). Many calculators have built-in functions for this purpose. For this integral, the approximate arc length is

$$\int_0^2 \sqrt{1 + 4x^2}\, dx \approx 4.647.$$

Related Exercises 21, 24 ◄

Arc Length for $x = g(y)$

Sometimes it is advantageous to describe a curve as a function of y—that is, $x = g(y)$. The arc length formula in this case is derived exactly as in the case of $y = f(x)$, switching the roles of x and y. The result is the following arc length formula.

QUICK CHECK 2 What does the arc length formula give for the length of the line $x = y$ between $y = c$ and $y = d$, where $d \geq c$? Is the result consistent with the result given by the Pythagorean theorem? ◄

DEFINITION Arc Length for $x = g(y)$

Let $x = g(y)$ have a continuous first derivative on the interval $[c, d]$. The length of the curve from $(g(c), c)$ to $(g(d), d)$ is

$$L = \int_c^d \sqrt{1 + g'(y)^2}\, dy.$$

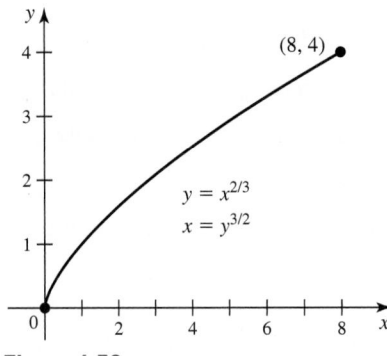

Figure 6.58

EXAMPLE 5 **Arc length** Find the length of the curve $y = f(x) = x^{2/3}$ between $x = 0$ and $x = 8$ (Figure 6.58).

SOLUTION The derivative of $f(x) = x^{2/3}$ is $f'(x) = \frac{2}{3}x^{-1/3}$, which is undefined at $x = 0$. Therefore, the arc length formula with respect to x cannot be used, yet the curve certainly appears to have a well-defined length.

The key is to describe the curve with y as the independent variable. Solving $y = x^{2/3}$ for x, we have $x = g(y) = \pm y^{3/2}$. Notice that when $x = 8$, $y = 8^{2/3} = 4$, which says that we should use the positive branch of $\pm y^{3/2}$. Therefore, finding the length of the curve $y = f(x) = x^{2/3}$ from $x = 0$ to $x = 8$ is equivalent to finding the length of the curve $x = g(y) = y^{3/2}$ from $y = 0$ to $y = 4$. This is precisely the problem solved in Example 1. The arc length is $\frac{8}{27}(10^{3/2} - 1) \approx 9.1$ units.

Related Exercise 19 ◄

QUICK CHECK 3 Write the integral for the length of the curve $x = \sin y$ on the interval $0 \le y \le \pi$. ◄

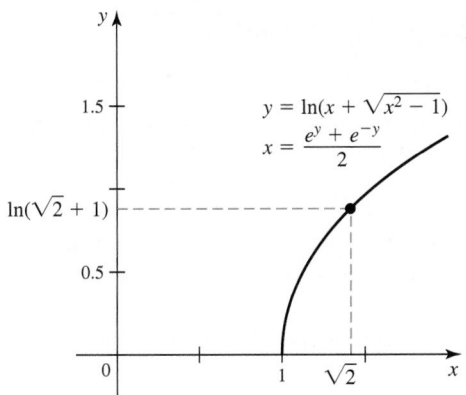

Figure 6.59

➤ The function $\frac{1}{2}(e^y + e^{-y})$ is the **hyperbolic cosine**, denoted cosh y. The function $\frac{1}{2}(e^y - e^{-y})$ is the **hyperbolic sine**, denoted sinh y. See Section 7.3.

EXAMPLE 6 **Ingenuity required** Find the length of the curve $y = f(x) = \ln(x + \sqrt{x^2 - 1})$ on the interval $[1, \sqrt{2}]$ (Figure 6.59).

SOLUTION Calculating f' shows that the graph of f has a vertical tangent line at $x = 1$. Therefore, the integrand in the arc length integral is undefined at $x = 1$. An alternative strategy is to express the function in the form $x = g(y)$ and evaluate the arc length integral with respect to y. Noting that $x \ge 1$ and $y \ge 0$, we solve $y = \ln(x + \sqrt{x^2 - 1})$ for x in the following steps:

$$e^y = x + \sqrt{x^2 - 1} \qquad \text{Exponentiate both sides.}$$
$$e^y - x = \sqrt{x^2 - 1} \qquad \text{Subtract } x \text{ from both sides.}$$
$$e^{2y} - 2e^y x = -1 \qquad \text{Square both sides and cancel } x^2.$$
$$x = \frac{e^{2y} + 1}{2e^y} = \frac{e^y + e^{-y}}{2}. \qquad \text{Solve for } x.$$

We conclude that the given curve is also described by the function $x = g(y) = \dfrac{e^y + e^{-y}}{2}$. The interval $1 \le x \le \sqrt{2}$ corresponds to the interval $0 \le y \le \ln(\sqrt{2} + 1)$ (Figure 6.59). Note that $g'(y) = \dfrac{e^y - e^{-y}}{2}$ is continuous on $[0, \ln(\sqrt{2} + 1)]$. The arc length is

$$\int_0^{\ln(\sqrt{2}+1)} \sqrt{1 + g'(y)^2}\, dy = \int_0^{\ln(\sqrt{2}+1)} \sqrt{1 + \left(\frac{e^y - e^{-y}}{2}\right)^2}\, dy \qquad \text{Substitute for } g'(y).$$
$$= \frac{1}{2}\int_0^{\ln(\sqrt{2}+1)} (e^y + e^{-y})\, dy \qquad \text{Expand and simplify.}$$
$$= \frac{1}{2}(e^y - e^{-y})\Big|_0^{\ln(\sqrt{2}+1)} = 1. \qquad \text{Fundamental Theorem}$$

Related Exercise 20 ◄

SECTION 6.5 EXERCISES

Getting Started

1. Explain the steps required to find the length of a curve $y = f(x)$ between $x = a$ and $x = b$.

2. Explain the steps required to find the length of a curve $x = g(y)$ between $y = c$ and $y = d$.

3–6. Setting up arc length integrals *Write and simplify, but do not evaluate, an integral with respect to x that gives the length of the following curves on the given interval.*

3. $y = x^3 + 2$ on $[-2, 5]$

4. $y = 2\cos 3x$ on $[-\pi, \pi]$

5. $y = e^{-2x}$ on $[0, 2]$

6. $y = \ln x$ on $[1, 10]$

7. Find the arc length of the line $y = 2x + 1$ on $[1, 5]$ using calculus and verify your answer using geometry.

8. Find the arc length of the line $y = 4 - 3x$ on $[-3, 2]$ using calculus and verify your answer using geometry.

Practice Exercises

9–20. Arc length calculations *Find the arc length of the following curves on the given interval.*

9. $y = -8x - 3$ on $[-2, 6]$ (Use calculus.)

10. $y = \dfrac{1}{2}(e^x + e^{-x})$ on $[-\ln 2, \ln 2]$

11. $y = \dfrac{1}{3}x^{3/2}$ on $[0, 60]$

12. $y = 3 \ln x - \dfrac{x^2}{24}$ on $[1, 6]$

13. $y = \dfrac{(x^2 + 2)^{3/2}}{3}$ on $[0, 1]$

14. $y = \dfrac{x^{3/2}}{3} - x^{1/2}$ on $[4, 16]$

15. $y = \dfrac{x^4}{4} + \dfrac{1}{8x^2}$ on $[1, 2]$

16. $y = \dfrac{2}{3}x^{3/2} - \dfrac{1}{2}x^{1/2}$ on $[1, 9]$

17. $x = \dfrac{y^4}{4} + \dfrac{1}{8y^2}$, for $1 \le y \le 2$

18. $x = 2e^{\sqrt{2}y} + \dfrac{1}{16}e^{-\sqrt{2}y}$, for $0 \le y \le \dfrac{\ln 2}{\sqrt{2}}$

19. $x = 2y - 4$, for $-3 \le y \le 4$ (Use calculus, but check your work using geometry.)

20. $y = \ln(x - \sqrt{x^2 - 1})$, for $1 \le x \le \sqrt{2}$ (*Hint:* Integrate with respect to y.)

⊤ 21–30. Arc length by calculator

a. *Write and simplify the integral that gives the arc length of the following curves on the given interval.*

b. *If necessary, use technology to evaluate or approximate the integral.*

21. $y = x^2$, for $-1 \le x \le 1$

22. $y = \sin x$, for $0 \le x \le \pi$

23. $y = \ln x$, for $1 \le x \le 4$

24. $y = \dfrac{x^3}{3}$, for $-1 \le x \le 1$

25. $x = \sqrt{y - 2}$, for $3 \le y \le 4$

26. $y = \dfrac{8}{x^2}$, for $1 \le x \le 4$

27. $y = \cos 2x$, for $0 \le x \le \pi$

28. $y = 4x - x^2$, for $0 \le x \le 4$

29. $y = \dfrac{1}{x}$, for $1 \le x \le 10$

30. $x = \dfrac{1}{y^2 + 1}$, for $-5 \le y \le 5$

⊤ 31. Golden Gate cables The profile of the cables on a suspension bridge may be modeled by a parabola. The central span of the Golden Gate Bridge (see figure) is 1280 m long and 152 m high. The parabola $y = 0.00037x^2$ gives a good fit to the shape of the cables, where $|x| \le 640$, and x and y are measured in meters. Approximate the length of the cables that stretch between the tops of the two towers.

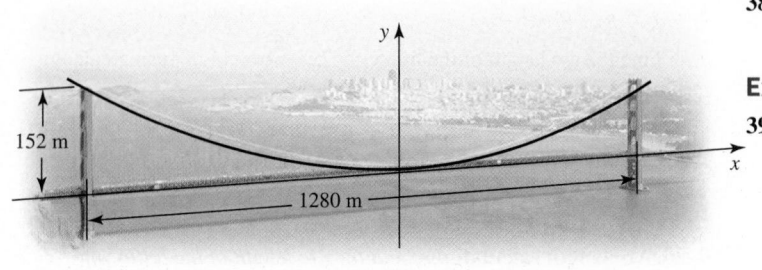

⊤ 32. Gateway Arch The shape of the Gateway Arch in St. Louis (with a height and a base length of 630 ft) is modeled by the function

$$y = -630 \cosh \dfrac{x}{239.2} + 1260, \text{ where } |x| \le 315,$$

and x and y are measured in feet (see figure). The function $\cosh x$ is the **hyperbolic cosine,** defined by $\cosh x = \dfrac{e^x + e^{-x}}{2}$ (see Section 7.3 for more on hyperbolic functions). Estimate the length of the Gateway Arch.

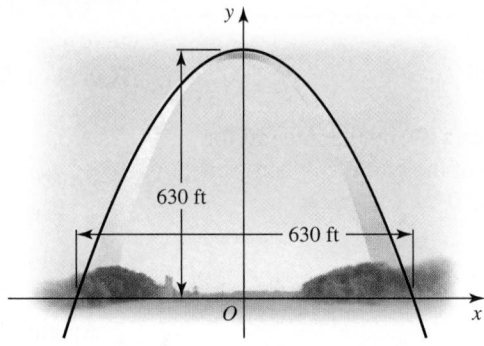

630 ft

630 ft

33. **Explain why or why not** Determine whether the following statements are true and give an explanation or counterexample.

a. $\displaystyle\int_a^b \sqrt{1 + f'(x)^2}\, dx = \int_a^b (1 + f'(x))\, dx$

b. Assuming f' is continuous on the interval $[a, b]$, the length of the curve $y = f(x)$ on $[a, b]$ is the area under the curve $y = \sqrt{1 + f'(x)^2}$ on $[a, b]$.

c. Arc length may be negative if $f(x) < 0$ on part of the interval in question.

34. **Arc length for a line** Consider the segment of the line $y = mx + c$ on the interval $[a, b]$. Use the arc length formula to show that the length of the line segment is $(b - a)\sqrt{1 + m^2}$. Verify this result by computing the length of the line segment using the distance formula.

35. **Functions from arc length** What differentiable functions have an arc length on the interval $[a, b]$ given by the following integrals? Note that the answers are not unique. Give a family of functions that satisfy the conditions.

a. $\displaystyle\int_a^b \sqrt{1 + 16x^4}\, dx$

b. $\displaystyle\int_a^b \sqrt{1 + 36 \cos^2 2x}\, dx$

36. **Function from arc length** Find a curve that passes through the point $(1, 5)$ and has an arc length on the interval $[2, 6]$ given by $\displaystyle\int_2^6 \sqrt{1 + 16x^{-6}}\, dx$.

⊤ 37. Cosine vs. parabola Which curve has the greater length on the interval $[-1, 1]$, $y = 1 - x^2$ or $y = \cos \dfrac{\pi x}{2}$?

38. **Function defined as an integral** Write the integral that gives the length of the curve $y = f(x) = \displaystyle\int_0^x \sin t\, dt$ on the interval $[0, \pi]$.

Explorations and Challenges

39. **Lengths of related curves** Suppose the graph of f on the interval $[a, b]$ has length L, where f' is continuous on $[a, b]$. Evaluate the following integrals in terms of L.

a. $\displaystyle\int_{a/2}^{b/2} \sqrt{1 + f'(2x)^2}\, dx$

b. $\displaystyle\int_{a/c}^{b/c} \sqrt{1 + f'(cx)^2}\, dx$ if $c \ne 0$

40. Lengths of symmetric curves Suppose a curve is described by $y = f(x)$ on the interval $[-b, b]$, where f' is continuous on $[-b, b]$. Show that if f is odd *or* f is even, then the length of the curve $y = f(x)$ from $x = -b$ to $x = b$ is twice the length of the curve from $x = 0$ to $x = b$. Use a geometric argument and prove it using integration.

41. A family of exponential functions

a. Show that the arc length integral for the function

$$f(x) = Ae^{ax} + \frac{1}{4Aa^2}e^{-ax}, \text{ where } a > 0 \text{ and } A > 0, \text{ may be}$$

integrated using methods you already know.

b. Verify that the arc length of the curve $y = f(x)$ on the interval $[0, \ln 2]$ is

$$A(2^a - 1) - \frac{1}{4a^2A}(2^{-a} - 1).$$

T 42. Bernoulli's "parabolas" Johann Bernoulli (1667−1748) evaluated the arc length of curves of the form $y = x^{(2n+1)/2n}$, where n is a positive integer, on the interval $[0, a]$.

a. Write the arc length integral.

b. Make the change of variables $u^2 = 1 + \left(\dfrac{2n+1}{2n}\right)^2 x^{1/n}$ to obtain a new integral with respect to u.

c. Use the Binomial Theorem to expand this integrand and evaluate the integral.

d. The case $n = 1$ ($y = x^{3/2}$) was done in Example 1. With $a = 1$, compute the arc length in the cases $n = 2$ and $n = 3$. Does the arc length increase or decrease with n?

e. Graph the arc length of the curves for $a = 1$ as a function of n.

QUICK CHECK ANSWERS

1. $\sqrt{2}a$ (the length of the line segment joining the points)
2. $\sqrt{2}(d - c)$ (the length of the line segment joining the points)   **3.** $L = \int_0^\pi \sqrt{1 + \cos^2 y}\, dy$ ◀

6.6 Surface Area

In Sections 6.3 and 6.4, we introduced solids of revolution and presented methods for computing the volume of such solids. We now consider a related problem: computing the *area* of the surface of a solid of revolution. Surface area calculations are important in aerodynamics (computing the lift on an airplane wing) and biology (computing transport rates across cell membranes), to name just two applications. Here is an interesting observation: A surface area problem is "between" a volume problem (which is three-dimensional) and an arc length problem (which is one-dimensional). For this reason, you will see ideas that appear in both volume and arc length calculations as we develop the surface area integral.

Some Preliminary Calculations

Consider a curve $y = f(x)$ on an interval $[a, b]$, where f is a nonnegative function with a continuous first derivative on $[a, b]$. Now imagine revolving the curve about the x-axis to generate a *surface of revolution* (Figure 6.60). Our objective is to find the area of this surface.

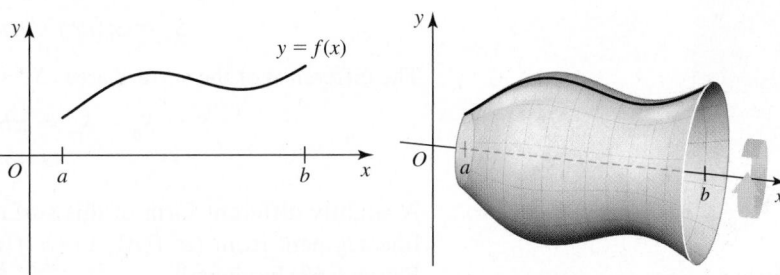

Figure 6.60

Before tackling this problem, we consider a preliminary problem upon which we build a general surface area formula. First consider the graph of $f(x) = rx/h$ on the interval $[0, h]$, where $h > 0$ and $r > 0$. When this line segment is revolved about the x-axis, it generates the surface of a cone of radius r and height h (Figure 6.61). A formula from geometry states that the surface area of a right circular cone of radius r and height h (excluding the base) is $\pi r \sqrt{r^2 + h^2} = \pi r \ell$, where ℓ is the slant height of the cone (the length of the slanted "edge" of the cone).

➤ One way to derive the formula for the surface area of a cone (not including the base) is to cut the cone on a line from its base to its vertex. When the cone is unfolded, it forms a sector of a circular disk of radius ℓ with a curved edge of length $2\pi r$. This sector is a fraction

$$\frac{2\pi r}{2\pi\ell} = \frac{r}{\ell}$$ of a full circular disk of

radius ℓ. So the area of the sector, which is also the surface area of the cone, is

$$\pi\ell^2 \cdot \frac{r}{\ell} = \pi r\ell.$$

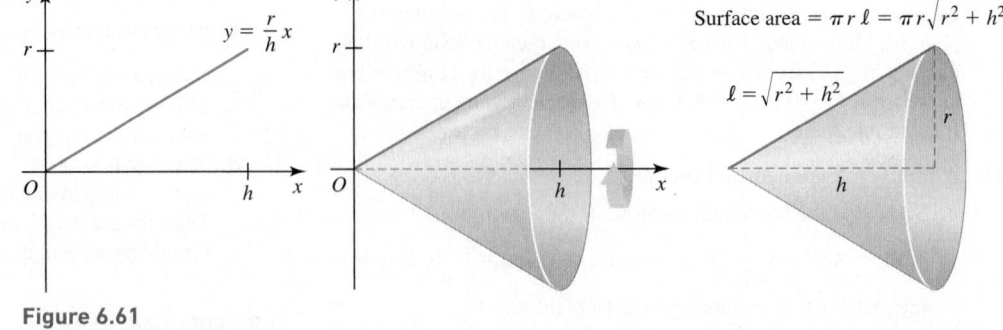

Figure 6.61

Curved edge length $= 2\pi r$

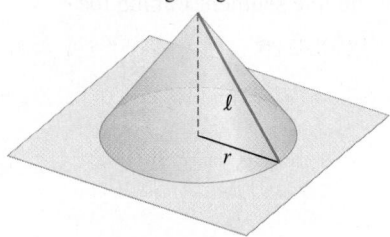

Curved edge length $= 2\pi r$

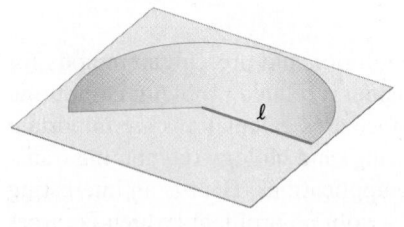

QUICK CHECK 1 Which is greater, the surface area of a cone of height 10 and radius 20 or the surface area of a cone of height 20 and radius 10 (excluding the bases)? ◄

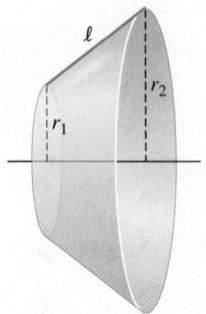

Surface area of frustum:
$$S = \pi(f(b) + f(a))\ell$$
$$= \pi(r_2 + r_1)\ell$$

With this result, we can solve a preliminary problem that will be useful. Consider the linear function $f(x) = cx$ on the interval $[a, b]$, where $0 < a < b$ and $c > 0$. When this line segment is revolved about the x-axis, it generates a *frustum of a cone* (a cone whose top has been sliced off). The goal is to find S, the surface area of the frustum. Figure 6.62 shows that S is the difference between the surface area S_b of the cone that extends over the interval $[0, b]$ and the surface area S_a of the cone that extends over the interval $[0, a]$.

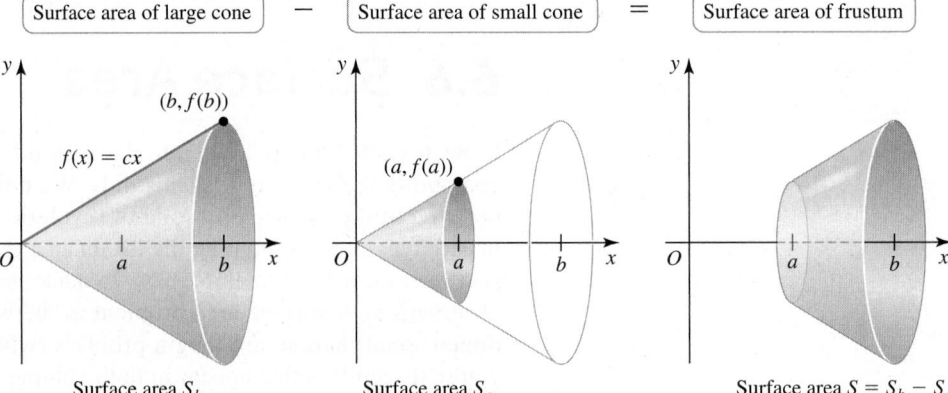

Figure 6.62

Notice that the radius of the cone on $[0, b]$ is $r = f(b) = cb$, and its height is $h = b$. Therefore, this cone has surface area

$$S_b = \pi r\sqrt{r^2 + h^2} = \pi(bc)\sqrt{(bc)^2 + b^2} = \pi b^2 c\sqrt{c^2 + 1}.$$

Similarly, the cone on $[0, a]$ has radius $r = f(a) = ca$ and height $h = a$, so its surface area is

$$S_a = \pi(ac)\sqrt{(ac)^2 + a^2} = \pi a^2 c\sqrt{c^2 + 1}.$$

The difference of the surface areas $S_b - S_a$ is the surface area S of the frustum on $[a, b]$:

$$S = S_b - S_a = \pi b^2 c\sqrt{c^2 + 1} - \pi a^2 c\sqrt{c^2 + 1}$$
$$= \pi c(b^2 - a^2)\sqrt{c^2 + 1}.$$

A slightly different form of this surface area formula will be useful. Observe that the line segment from $(a, f(a))$ to $(b, f(b))$ (which is the slant height of the frustum in Figure 6.62) has length

$$\ell = \sqrt{(b - a)^2 + (bc - ac)^2} = (b - a)\sqrt{c^2 + 1}.$$

Therefore, the surface area of the frustum can also be written

$$S = \pi c(b^2 - a^2)\sqrt{c^2 + 1}$$
$$= \pi c(b + a)(b - a)\sqrt{c^2 + 1} \qquad \text{Factor } b^2 - a^2.$$
$$= \pi\underbrace{(cb + ca)}_{f(b)\ \ f(a)}\underbrace{(b - a)\sqrt{c^2 + 1}}_{\ell} \qquad \text{Expand } c(b + a).$$
$$= \pi(f(b) + f(a))\ell.$$

QUICK CHECK 2 What is the surface area of the frustum of a cone generated when the graph of $f(x) = 3x$ on the interval $[2, 5]$ is revolved about the x-axis? ◄

This result can be generalized to *any* linear function $g(x) = cx + d$ that is positive on the interval $[a, b]$. That is, the surface area of the frustum generated by revolving the line segment between $(a, g(a))$ and $(b, g(b))$ about the x-axis is given by $\pi(g(b) + g(a))\ell$ (Exercise 40).

Surface Area Formula

With the surface area formula for a frustum of a cone, we now derive a general area formula for a surface of revolution. We assume f is a nonnegative function with a continuous first derivative on $[a, b]$. The surface is generated by revolving the graph of f on the interval $[a, b]$ about the x-axis. We begin by subdividing the interval $[a, b]$ into n subintervals of equal length $\Delta x = \dfrac{b - a}{n}$. The grid points in this partition are

$$x_0 = a, x_1, x_2, \ldots, x_{n-1}, x_n = b.$$

Now consider the kth subinterval $[x_{k-1}, x_k]$ and the line segment between the points $(x_{k-1}, f(x_{k-1}))$ and $(x_k, f(x_k))$ (Figure 6.63). We let the change in the y-coordinates between these points be $\Delta y_k = f(x_k) - f(x_{k-1})$.

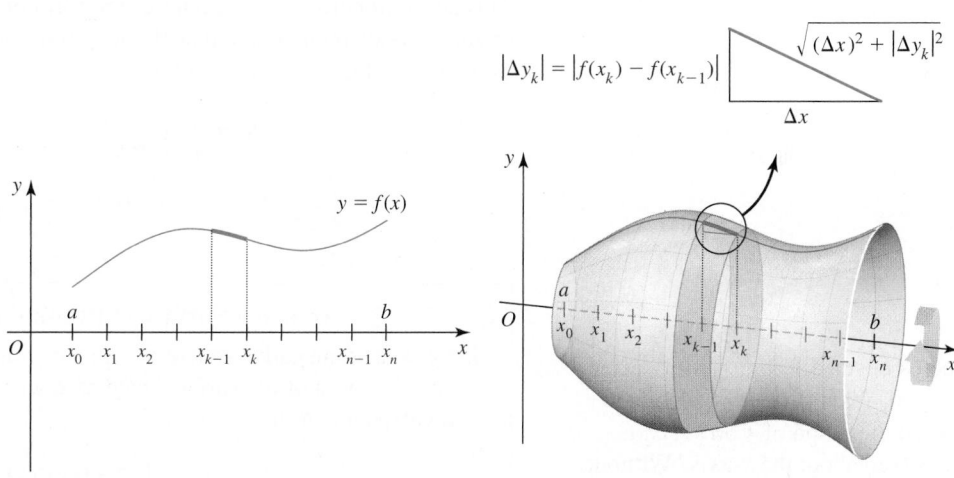

Figure 6.63   **Figure 6.64**

When this line segment is revolved about the x-axis, it generates a frustum of a cone (Figure 6.64). The slant height of this frustum is the length of the hypotenuse of a right triangle whose sides have lengths Δx and $|\Delta y_k|$. Therefore, the slant height of the kth frustum is

$$\sqrt{(\Delta x)^2 + |\Delta y_k|^2} = \sqrt{(\Delta x)^2 + (\Delta y_k)^2}$$

and its surface area is

$$S_k = \pi(f(x_k) + f(x_{k-1}))\sqrt{(\Delta x)^2 + (\Delta y_k)^2}.$$

It follows that the area S of the entire surface of revolution is approximately the sum of the surface areas of the individual frustums S_k, for $k = 1, \ldots, n$; that is,

$$S \approx \sum_{k=1}^{n} S_k = \sum_{k=1}^{n} \pi(f(x_k) + f(x_{k-1}))\sqrt{(\Delta x)^2 + (\Delta y_k)^2}.$$

We would like to identify this sum as a Riemann sum. However, one more step is required to put it in the correct form. We apply the Mean Value Theorem on the kth subinterval $[x_{k-1}, x_k]$ and observe that

$$\frac{f(x_k) - f(x_{k-1})}{\Delta x} = f'(x_k^*),$$

for some number x_k^* in the interval (x_{k-1}, x_k), for $k = 1, \ldots, n$. It follows that $\Delta y_k = f(x_k) - f(x_{k-1}) = f'(x_k^*)\Delta x$.

➤ Notice that f is assumed to be differentiable on $[a, b]$; therefore, it satisfies the conditions of the Mean Value Theorem. Recall that a similar argument was used to derive the arc length formula in Section 6.5.

We now replace Δy_k with $f'(x_k^*)\Delta x$ in the expression for the approximate surface area. The result is

$$
\begin{aligned}
S &\approx \sum_{k=1}^{n} S_k = \sum_{k=1}^{n} \pi(f(x_k) + f(x_{k-1}))\sqrt{(\Delta x)^2 + (\Delta y_k)^2} \\
&= \sum_{k=1}^{n} \pi(f(x_k) + f(x_{k-1}))\sqrt{(\Delta x)^2(1 + f'(x_k^*)^2)} \qquad \text{Mean Value Theorem} \\
&= \sum_{k=1}^{n} \pi(f(x_k) + f(x_{k-1}))\sqrt{1 + f'(x_k^*)^2}\,\Delta x. \qquad \text{Factor out } \Delta x.
\end{aligned}
$$

When Δx is small, we have $x_{k-1} \approx x_k \approx x_k^*$, and by the continuity of f, it follows that $f(x_{k-1}) \approx f(x_k) \approx f(x_k^*)$, for $k = 1, \ldots, n$. These observations allow us to write

$$
\begin{aligned}
S &\approx \sum_{k=1}^{n} \pi(f(x_k^*) + f(x_k^*))\sqrt{1 + f'(x_k^*)^2}\,\Delta x \\
&= \sum_{k=1}^{n} 2\pi f(x_k^*)\sqrt{1 + f'(x_k^*)^2}\,\Delta x.
\end{aligned}
$$

This approximation to S, which has the form of a Riemann sum, improves as the number of subintervals increases and as the length of the subintervals approaches 0. Specifically, as $n \to \infty$ and as $\Delta x \to 0$, we obtain an integral for the surface area:

$$
\begin{aligned}
S &= \lim_{n \to \infty} \sum_{k=1}^{n} 2\pi f(x_k^*)\sqrt{1 + f'(x_k^*)^2}\,\Delta x \\
&= \int_a^b 2\pi f(x)\sqrt{1 + f'(x)^2}\,dx.
\end{aligned}
$$

DEFINITION **Area of a Surface of Revolution**

Let f be a nonnegative function with a continuous first derivative on the interval $[a, b]$. The area of the surface generated when the graph of f on the interval $[a, b]$ is revolved about the x-axis is

$$
S = \int_a^b 2\pi f(x)\sqrt{1 + f'(x)^2}\,dx.
$$

QUICK CHECK 3 Let $f(x) = c$, where $c > 0$. What surface is generated when the graph of f on $[a, b]$ is revolved about the x-axis? Without using calculus, what is the area of the surface? ◄

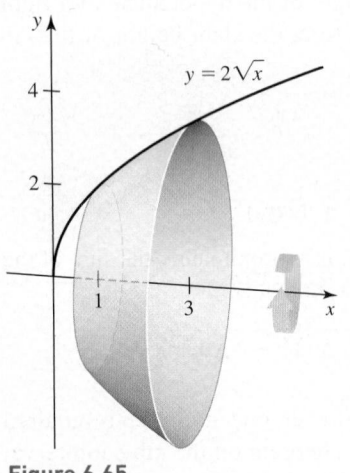

Figure 6.65

EXAMPLE 1 **Using the surface area formula** The graph of $f(x) = 2\sqrt{x}$ on the interval $[1, 3]$ is revolved about the x-axis. What is the area of the surface generated (Figure 6.65)?

SOLUTION Noting that $f'(x) = \dfrac{1}{\sqrt{x}}$, the surface area formula gives

$$
\begin{aligned}
S &= \int_a^b 2\pi f(x)\sqrt{1 + f'(x)^2}\,dx \\
&= 2\pi \int_1^3 2\sqrt{x}\sqrt{1 + \frac{1}{x}}\,dx \qquad \text{Substitute for } f \text{ and } f'. \\
&= 4\pi \int_1^3 \sqrt{x + 1}\,dx \qquad \text{Simplify.} \\
&= \frac{8\pi}{3}(x+1)^{3/2}\Big|_1^3 = \frac{16\pi}{3}(4 - \sqrt{2}). \qquad \text{Integrate and simplify.}
\end{aligned}
$$

Related Exercise 9 ◄

EXAMPLE 2 **Surface area of a spherical zone** A spherical zone is produced when a sphere of radius a is sliced by two parallel planes. In this example, we compute the surface area of the spherical zone that results when the first plane is oriented vertically and cuts the sphere in half while the second plane lies h units to the right, where $0 \le h < a$

(Figure 6.66a). Show that the area of this spherical zone of width h cut from a sphere of radius a is $2\pi ah$, and use the result to show that the surface area of the sphere is $4\pi a^2$.

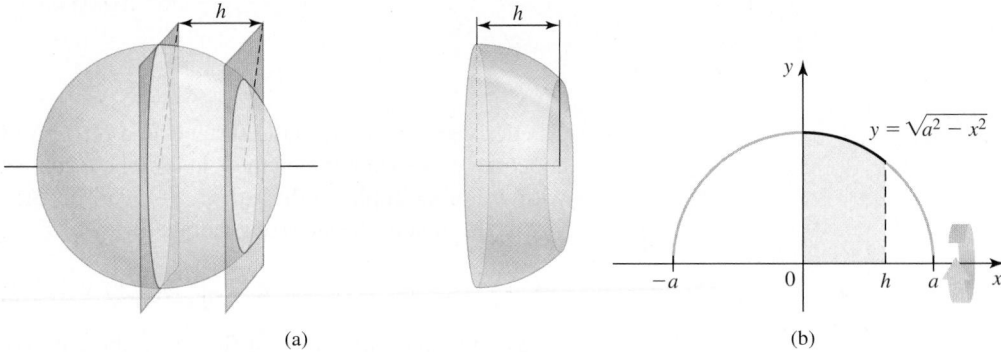

(a)   (b)

Figure 6.66

SOLUTION To generate the spherical zone, we revolve the curve $f(x) = \sqrt{a^2 - x^2}$ on the interval $[0, h]$ about the x-axis (Figure 6.66b). Noting that $f'(x) = -x(a^2 - x^2)^{-1/2}$, the surface area of the spherical zone of width h is

$$S = \int_a^b 2\pi f(x)\sqrt{1 + f'(x)^2}\, dx$$

$$= 2\pi \int_0^h \sqrt{a^2 - x^2}\sqrt{1 + (-x(a^2 - x^2)^{-1/2})^2}\, dx \quad \text{Substitute for } f \text{ and } f'.$$

$$= 2\pi \int_0^h \sqrt{a^2 - x^2}\sqrt{\frac{a^2}{a^2 - x^2}}\, dx \quad \text{Simplify.}$$

$$= 2\pi \int_0^h a\, dx = 2\pi ah. \quad \text{Simplify and integrate.}$$

To find the surface area of the entire sphere, it is tempting to integrate not over the interval $[0, h]$, but rather over the interval $[-a, a]$. Notice, however, that f is not differentiable at $\pm a$, so we cannot use the surface area formula on p. 460. Instead, we use the formula for the area of the spherical zone and let h approach a from the left to find the area of the associated hemisphere:

$$\text{Area of zone} = 2\pi ah, \text{ so}$$

$$\text{Area of hemisphere} = \lim_{h \to a^-} 2\pi ah = 2\pi a^2.$$

Therefore, the surface area of a sphere of radius a is $2(2\pi a^2) = 4\pi a^2$.

Related Exercise 13 ◄

EXAMPLE 3 **Painting a funnel** The curved surface of a funnel is generated by revolving the graph of $y = f(x) = x^3 + \dfrac{1}{12x}$ on the interval $[1, 2]$ about the x-axis (Figure 6.67). Approximately what volume of paint is needed to cover the outside of the funnel with a layer of paint 0.05 cm thick? Assume x and y are measured in centimeters.

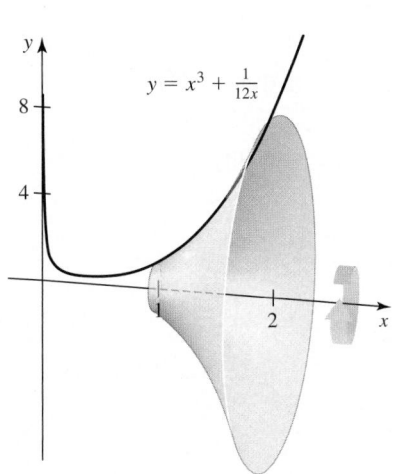

Figure 6.67

SOLUTION Note that $f'(x) = 3x^2 - \dfrac{1}{12x^2}$. Therefore, the surface area of the funnel in cm^2 is

$$S = \int_a^b 2\pi f(x)\sqrt{1 + f'(x)^2}\, dx$$

$$= 2\pi \int_1^2 \left(x^3 + \frac{1}{12x}\right)\sqrt{1 + \left(3x^2 - \frac{1}{12x^2}\right)^2}\, dx \quad \text{Substitute for } f \text{ and } f'.$$

$$= 2\pi \int_1^2 \left(x^3 + \frac{1}{12x}\right)\sqrt{\left(3x^2 + \frac{1}{12x^2}\right)^2}\, dx \quad \text{Expand and factor under square root.}$$

$$= 2\pi \int_1^2 \left(x^3 + \frac{1}{12x}\right)\left(3x^2 + \frac{1}{12x^2}\right) dx \quad \text{Simplify.}$$

$$= \frac{12{,}289}{192}\pi. \quad \text{Evaluate integral.}$$

Because the paint layer is 0.05 cm thick, the volume of paint needed is approximately

$$\left(\frac{12{,}289\pi}{192}\ \text{cm}^2\right)(0.05\ \text{cm}) \approx 10.1\ \text{cm}^3.$$

Related Exercises 21–22 ◄

The derivation that led to the surface area integral may be used when a curve is revolved about the *y*-axis (rather than the *x*-axis). The result is the same integral, where *x* is replaced with *y*. For example, if the curve $x = g(y)$ on the interval $[c, d]$ is revolved about the *y*-axis, the area of the surface generated is

$$S = \int_c^d 2\pi g(y)\sqrt{1 + g'(y)^2}\,dy.$$

To use this integral, we must first describe the given curve as a differentiable function of *y*.

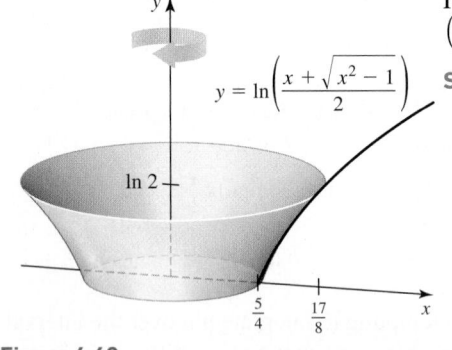

$$y = \ln\left(\frac{x + \sqrt{x^2 - 1}}{2}\right)$$

ln 2

$\frac{5}{4}$ $\frac{17}{8}$

Figure 6.68

EXAMPLE 4   Change of perspective Consider the function $y = \ln\left(\dfrac{x + \sqrt{x^2 - 1}}{2}\right)$.

Find the area of the surface generated when the part of the curve between the points $\left(\frac{5}{4}, 0\right)$ and $\left(\frac{17}{8}, \ln 2\right)$ is revolved about the *y*-axis (**Figure 6.68**).

SOLUTION We solve for *x* in terms of *y* in the following steps:

$$y = \ln\left(\frac{x + \sqrt{x^2 - 1}}{2}\right)$$

$$e^y = \frac{x + \sqrt{x^2 - 1}}{2} \qquad \text{Exponentiate both sides.}$$

$$2e^y - x = \sqrt{x^2 - 1} \qquad \text{Rearrange terms.}$$

$$4e^{2y} - 4xe^y + x^2 = x^2 - 1 \qquad \text{Square both sides.}$$

$$x = g(y) = e^y + \frac{1}{4}e^{-y}. \qquad \text{Solve for } x.$$

Note that $g'(y) = e^y - \frac{1}{4}e^{-y}$ and that the interval of integration on the *y*-axis is $[0, \ln 2]$. The area of the surface is

$$S = \int_c^d 2\pi g(y)\sqrt{1 + g'(y)^2}\,dy$$

$$= 2\pi \int_0^{\ln 2}\left(e^y + \frac{1}{4}e^{-y}\right)\sqrt{1 + \left(e^y - \frac{1}{4}e^{-y}\right)^2}\,dy \quad \text{Substitute for } g \text{ and } g'.$$

$$= 2\pi \int_0^{\ln 2}\left(e^y + \frac{1}{4}e^{-y}\right)\sqrt{\left(e^y + \frac{1}{4}e^{-y}\right)^2}\,dy \qquad \text{Expand and factor.}$$

$$= 2\pi \int_0^{\ln 2}\left(e^y + \frac{1}{4}e^{-y}\right)^2\,dy \qquad \text{Simplify.}$$

$$= 2\pi \int_0^{\ln 2}\left(e^{2y} + \frac{1}{2} + \frac{1}{16}e^{-2y}\right)\,dy \qquad \text{Expand.}$$

$$= \pi\left(\frac{195}{64} + \ln 2\right). \qquad \text{Integrate.}$$

Related Exercise 34 ◄

SECTION 6.6 EXERCISES

Getting Started

1. What is the area of the curved surface of a right circular cone of radius 3 and height 4?

2. A frustum of a cone is generated by revolving the graph of $y = 4x$ on the interval $[2, 6]$ about the x-axis. What is the area of the surface of the frustum?

3. Assume f is a nonnegative function with a continuous first derivative on $[a, b]$. The curve $y = f(x)$ on $[a, b]$ is revolved about the x-axis. Explain how to find the area of the surface that is generated.

4. Assume g is a nonnegative function with a continuous first derivative on $[c, d]$. The curve $x = g(y)$ on $[c, d]$ is revolved about the y-axis. Explain how to find the area of the surface that is generated.

5. A surface is generated by revolving the line $f(x) = 2 - x$, for $0 \le x \le 2$, about the x-axis. Find the area of the resulting surface in the following ways.

   a. Using calculus
   b. Using geometry, after first determining the shape and dimensions of the surface

6. A surface is generated by revolving the line $x = 3$, for $0 \le y \le 8$, about the y-axis. Find the area of the resulting surface in the following ways.

   a. Using calculus with $g(y) = 3$
   b. Using geometry, after first determining the shape and dimensions of the surface

Practice Exercises

7–20. Computing surface areas *Find the area of the surface generated when the given curve is revolved about the given axis.*

7. $y = 3x + 4$, for $0 \le x \le 6$; about the x-axis

8. $y = 12 - 3x$, for $1 \le x \le 3$; about the x-axis

9. $y = 8\sqrt{x}$, for $9 \le x \le 20$; about the x-axis

10. $y = x^3$, for $0 \le x \le 1$; about the x-axis

11. $y = (3x)^{1/3}$, for $0 \le x \le \dfrac{8}{3}$; about the y-axis

12. $y = \dfrac{x^2}{4}$, for $2 \le x \le 4$; about the y-axis

13. $y = \sqrt{1 - x^2}$, for $-\dfrac{1}{2} \le x \le \dfrac{1}{2}$; about the x-axis

14. $x = \sqrt{-y^2 + 6y - 8}$, for $3 \le y \le \dfrac{7}{2}$; about the y-axis

15. $y = 4x - 1$, for $1 \le x \le 4$; about the y-axis (*Hint:* Integrate with respect to y.)

16. $y = \sqrt{4x + 6}$, for $0 \le x \le 5$; about the x-axis

17. $y = \dfrac{1}{4}(e^{2x} + e^{-2x})$, for $-2 \le x \le 2$; about the x-axis

18. $y = \sqrt{5x - x^2}$, for $1 \le x \le 4$; about the x-axis

19. $x = \sqrt{12y - y^2}$, for $2 \le y \le 10$; about the y-axis

20. $y = 1 + \sqrt{1 - x^2}$ between the points $(1, 1)$ and $\left(\dfrac{\sqrt{3}}{2}, \dfrac{3}{2}\right)$; about the y-axis

21–22. Painting surfaces *A 1.5-mm layer of paint is applied to one side of the following surfaces. Find the approximate volume of paint needed. Assume x and y are measured in meters.*

21. The spherical zone generated when the curve $y = \sqrt{8x - x^2}$ on the interval $1 \le x \le 7$ is revolved about the x-axis

22. The spherical zone generated when the upper portion of the circle $x^2 + y^2 = 100$ on the interval $-8 \le x \le 8$ is revolved about the x-axis

23. **Explain why or why not** Determine whether the following statements are true and give an explanation or counterexample.

   a. If the curve $y = f(x)$ on the interval $[a, b]$ is revolved about the y-axis, the area of the surface generated is
   $$\int_{f(a)}^{f(b)} 2\pi f(y)\sqrt{1 + f'(y)^2}\, dy.$$

   b. If f is not one-to-one on the interval $[a, b]$, then the area of the surface generated when the graph of f on $[a, b]$ is revolved about the x-axis is not defined.

   c. Let $f(x) = 12x^2$. The area of the surface generated when the graph of f on $[-4, 4]$ is revolved about the x-axis is twice the area of the surface generated when the graph of f on $[0, 4]$ is revolved about the x-axis.

   d. Let $f(x) = 12x^2$. The area of the surface generated when the graph of f on $[-4, 4]$ is revolved about the y-axis is twice the area of the surface generated when the graph of f on $[0, 4]$ is revolved about the y-axis.

T 24–28. Surface area using technology *Consider the following curves on the given intervals.*

   a. *Write the integral that gives the area of the surface generated when the curve is revolved about the given axis.*
   b. *Use a calculator or software to approximate the surface area.*

24. $y = x^5$, for $0 \le x \le 1$; about the x-axis

25. $y = \cos x$, for $0 \le x \le \dfrac{\pi}{2}$; about the x-axis

26. $y = e^x$, for $0 \le x \le 1$; about the y-axis

27. $y = \tan x$, for $0 \le x \le \dfrac{\pi}{4}$; about the x-axis

28. $y = \ln x^2$, for $1 \le x \le \sqrt{e}$; about the x-axis

29. **Revolving an astroid** Consider the upper half of the astroid described by $x^{2/3} + y^{2/3} = a^{2/3}$, where $a > 0$ and $|x| \le a$. Find the area of the surface generated when this curve is revolved about the x-axis. Note that the function describing the curve is not differentiable at 0. However, the surface area integral can be evaluated using symmetry and methods you know.

30. **Cones and cylinders** The volume of a cone of radius r and height h is one-third the volume of a cylinder with the same radius and height. Does the surface area of a cone of radius r and height h equal one-third the surface area of a cylinder with the same radius and height? If not, find the correct relationship. Exclude the bases of the cone and cylinder.

Explorations and Challenges

31–35. Challenging surface area calculations *Find the area of the surface generated when the given curve is revolved about the given axis.*

31. $y = x^{3/2} - \dfrac{x^{1/2}}{3}$, for $1 \le x \le 2$; about the x-axis

32. $y = \dfrac{x^4}{8} + \dfrac{1}{4x^2}$, for $1 \le x \le 2$; about the x-axis

33. $y = \dfrac{x^3}{3} + \dfrac{1}{4x}$, for $\dfrac{1}{2} \le x \le 2$; about the x-axis

34. $y = \dfrac{1}{2}\ln\left(2x + \sqrt{4x^2 - 1}\right)$ between the points $\left(\dfrac{1}{2}, 0\right)$ and $\left(\dfrac{17}{16}, \ln 2\right)$; about the y-axis

35. $x = 4y^{3/2} - \dfrac{y^{1/2}}{12}$, for $1 \le y \le 4$; about the y-axis

36. Surface area of a torus When the circle $x^2 + (y - a)^2 = r^2$ on the interval $[-r, r]$ is revolved about the x-axis, the result is the surface of a torus, where $0 < r < a$. Show that the surface area of the torus is $S = 4\pi^2 ar$.

37. Zones of a sphere Suppose a sphere of radius r is sliced by two horizontal planes h units apart (see figure). Show that the surface area of the resulting zone on the sphere is $2\pi r h$, independent of the location of the cutting planes.

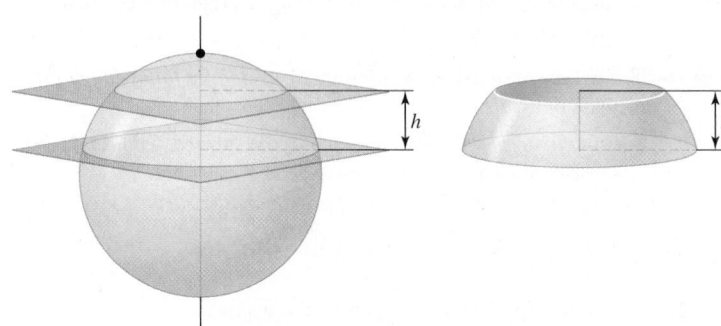

38. Surface area of an ellipsoid If the top half of the ellipse $\dfrac{x^2}{a^2} + \dfrac{y^2}{b^2} = 1$ is revolved about the x-axis, the result is an *ellipsoid* whose axis along the x-axis has length $2a$, whose axis along the y-axis has length $2b$, and whose axis perpendicular to the xy-plane has length $2b$. We assume $0 < b < a$ (see figure). Use the following steps to find the surface area S of this ellipsoid.

a. Use the surface area formula to show that
$$S = \frac{4\pi b}{a}\int_0^a \sqrt{a^2 - c^2 x^2}\, dx, \text{ where } c^2 = 1 - \frac{b^2}{a^2}.$$

b. Use the change of variables $u = cx$ to show that
$$S = \frac{4\pi b}{\sqrt{a^2 - b^2}}\int_0^{\sqrt{a^2 - b^2}} \sqrt{a^2 - u^2}\, du.$$

c. A table of integrals shows that
$$\int \sqrt{a^2 - u^2}\, du = \frac{1}{2}\left(u\sqrt{a^2 - u^2} + a^2 \sin^{-1}\frac{u}{a}\right) + C.$$
Use this fact to show that the surface area of the ellipsoid is
$$S = 2\pi b\left(b + \frac{a^2}{\sqrt{a^2 - b^2}}\sin^{-1}\frac{\sqrt{a^2 - b^2}}{a}\right).$$

d. If a and b have units of length (say, meters), what are the units of S according to this formula?

e. Use part (a) to show that if $a = b$, then $S = 4\pi a^2$, which is the surface area of a sphere of radius a.

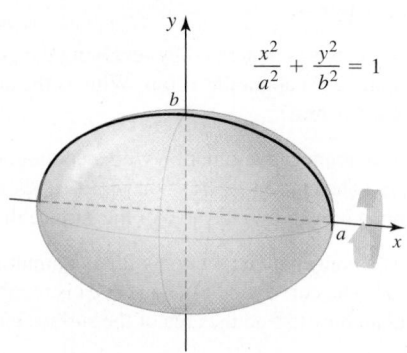

39. Surface-area-to-volume ratio (SAV) In the design of solid objects (both artificial and natural), the ratio of the surface area to the volume of the object is important. Animals typically generate heat at a rate proportional to their volume and lose heat at a rate proportional to their surface area. Therefore, animals with a low SAV ratio tend to retain heat, whereas animals with a high SAV ratio (such as children and hummingbirds) lose heat relatively quickly.

a. What is the SAV ratio of a cube with side lengths R?

b. What is the SAV ratio of a ball with radius R?

c. Use the result of Exercise 38 to find the SAV ratio of an ellipsoid whose long axis has length $2R\sqrt[3]{4}$, for $R \ge 1$, and whose other two axes have half the length of the long axis. (This scaling is used so that, for a given value of R, the volumes of the ellipsoid and the ball of radius R are equal.) The volume of a general ellipsoid is $V = \dfrac{4\pi}{3}ABC$, where the axes have lengths $2A$, $2B$, and $2C$.

d. Graph the SAV ratio of the ball of radius $R \ge 1$ as a function of R (part (b)) and graph the SAV ratio of the ellipsoid described in part (c) on the same set of axes. Which object has the smaller SAV ratio?

e. Among all ellipsoids of a fixed volume, which one would you choose for the shape of an animal if the goal were to minimize heat loss?

40. Surface area of a frustum Show that the surface area of the frustum of a cone generated by revolving the line segment between $(a, g(a))$ and $(b, g(b))$ about the x-axis is $\pi(g(b) + g(a))\ell$, for any linear function $g(x) = cx + d$ that is positive on the interval $[a, b]$, where ℓ is the slant height of the frustum.

41. Scaling surface area Let f be a nonnegative function with a continuous first derivative on $[a, b]$ and suppose $g(x) = c\, f(x)$ and $h(x) = f(cx)$, where $c > 0$. When the curve $y = f(x)$ on $[a, b]$ is revolved about the x-axis, the area of the resulting surface is A. Evaluate the following integrals in terms of A and c.

a. $\displaystyle\int_a^b 2\pi g(x)\sqrt{c^2 + g'(x)^2}\, dx$

b. $\displaystyle\int_{a/c}^{b/c} 2\pi h(x)\sqrt{c^2 + h'(x)^2}\, dx$

42. Surface plus cylinder Suppose f is a nonnegative function with a continuous first derivative on $[a, b]$. Let L equal the length of the graph of f on $[a, b]$ and let S be the area of the surface generated by revolving the graph of f on $[a, b]$ about the x-axis. For a positive constant C, assume the curve $y = f(x) + C$ is revolved about the x-axis. Show that the area of the resulting surface equals the sum of S and the surface area of a right circular cylinder of radius C and height L.

QUICK CHECK ANSWERS

1. The surface area of the first cone $(200\sqrt{5}\pi)$ is twice as great as the surface area of the second cone $(100\sqrt{5}\pi)$.
2. The surface area is $63\sqrt{10}\pi$.   **3.** The surface is a cylinder of radius c and height $b - a$. The area of the curved surface is $2\pi c(b - a)$. ◄

6.7 Physical Applications

We conclude this chapter on applications of integration with several problems from physics and engineering. The physical themes in these problems are mass, work, force, and pressure. The common mathematical theme is the use of the slice-and-sum strategy, which always leads to a definite integral.

Density and Mass

➤ In Chapter 16, we return to mass calculations for two- and three-dimensional objects (plates and solids).

$x = a$ $x = b$

Figure 6.69

Density is the concentration of mass in an object and is usually measured in units of mass per volume (for example, g/cm^3). An object with *uniform* density satisfies the basic relationship

$$\text{mass} = \text{density} \cdot \text{volume}.$$

When the density of an object *varies*, this formula no longer holds, and we must appeal to calculus.

In this section, we introduce mass calculations for thin objects that can be viewed as line segments (such as wires or thin bars). The bar shown in **Figure 6.69** has a density ρ that varies along its length. For one-dimensional objects, we use *linear density* with units of mass per length (for example, g/cm). What is the mass of such an object?

QUICK CHECK 1 In Figure 6.69, suppose $a = 0, b = 3$, and the density of the rod in g/cm is $\rho(x) = (4 - x)$. (a) Where is the rod lightest and heaviest? (b) What is the density at the middle of the bar? ◄

We begin by dividing the bar, represented by the interval $a \le x \le b$, into n subintervals of equal length $\Delta x = (b - a)/n$ (**Figure 6.70**). Let x_k^* be any point in the kth subinterval, for $k = 1, \ldots, n$. The mass of the kth segment of the bar m_k is approximately the density at x_k^* multiplied by the length of the interval, or $m_k \approx \rho(x_k^*)\Delta x$. So the approximate mass of the entire bar is

Mass of kth subinterval:
$m_k \approx \rho(x_k^*)\Delta x$

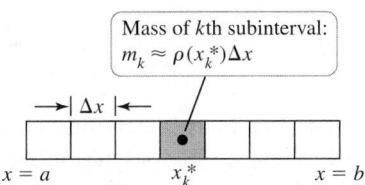

$x = a$ x_k^* $x = b$

Figure 6.70

$$\sum_{k=1}^{n} m_k \approx \sum_{k=1}^{n} \underbrace{\rho(x_k^*)\Delta x}_{m_k}.$$

The exact mass is obtained by taking the limit as $n \to \infty$ and as $\Delta x \to 0$, which produces a definite integral.

➤ Note that the units of the integral work out as they should: ρ has units of mass per length and dx has units of length, so $\rho(x)\,dx$ has units of mass.

DEFINITION Mass of a One-Dimensional Object

Suppose a thin bar or wire is represented by the interval $a \le x \le b$ with a density function ρ (with units of mass per length). The **mass** of the object is

$$m = \int_a^b \rho(x)\,dx.$$

QUICK CHECK 2 A thin bar occupies the interval $0 \le x \le 2$ and has a density in kg/m of $\rho(x) = (1 + x^2)$. Using the minimum value of the density, what is a lower bound for the mass of the object? Using the maximum value of the density, what is an upper bound for the mass of the object? ◄

EXAMPLE 1 Mass from variable density A thin, two-meter bar, represented by the interval $0 \le x \le 2$, is made of an alloy whose density in units of kg/m is given by $\rho(x) = (1 + x^2)$. What is the mass of the bar?

SOLUTION The mass of the bar in kilograms is

$$m = \int_a^b \rho(x)\,dx = \int_0^2 (1 + x^2)\,dx = \left(x + \frac{x^3}{3}\right)\Big|_0^2 = \frac{14}{3}.$$

Related Exercises 14–15 ◄

Work

Work can be described as the change in energy when a force causes a displacement of an object. When you carry a basket of laundry up a flight of stairs or push a stalled car, you apply a force that results in the displacement of an object, and work is done. If a *constant* force F displaces an object a distance d in the direction of the force, the work done is the force multiplied by the distance:

$$\text{work} = \text{force} \cdot \text{distance}.$$

It is easiest to use metric units for force and work. A newton (N) is the force required to give a 1-kg mass an acceleration of 1 m/s^2. A joule (J) is 1 newton-meter (N-m), the work done by a 1-N force acting over a distance of 1 m.

Calculus enters the picture with *variable* forces. Suppose an object is moved along the x-axis by a variable force F that is directed along the x-axis (Figure 6.71). How much work is done in moving the object between $x = a$ and $x = b$? Once again, we use the slice-and-sum strategy.

The interval $[a, b]$ is divided into n subintervals of equal length $\Delta x = (b - a)/n$. We let x_k^* be any point in the kth subinterval, for $k = 1, \ldots, n$. On that subinterval, the force is approximately constant with a value of $F(x_k^*)$. Therefore, the work done in moving the object across the kth subinterval is approximately $F(x_k^*)\Delta x$ (force · distance). Summing the work done over each of the n subintervals, the total work over the interval $[a, b]$ is approximately

$$W \approx \sum_{k=1}^{n} F(x_k^*)\Delta x.$$

This approximation becomes exact when we take the limit as $n \to \infty$ and $\Delta x \to 0$. The total work done is the integral of the force over the interval $[a, b]$ (or, equivalently, the net area under the force curve in Figure 6.71).

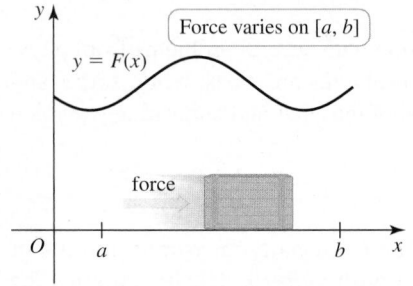

Force varies on $[a, b]$

$y = F(x)$

force

Figure 6.71

QUICK CHECK 3 Explain why the sum of the work over n subintervals is only an approximation of the total work. ◄

DEFINITION Work

The work done by a variable force F moving an object along a line from $x = a$ to $x = b$ in the direction of the force is

$$W = \int_{a}^{b} F(x)\, dx.$$

An application of force and work that is easy to visualize is the stretching and compression of a spring. Suppose an object is attached to a spring on a frictionless horizontal surface; the object slides back and forth under the influence of the spring. We say that the spring is at *equilibrium* when it is neither compressed nor stretched. It is convenient to let x be the position of the object, where $x = 0$ is the equilibrium position (Figure 6.72).

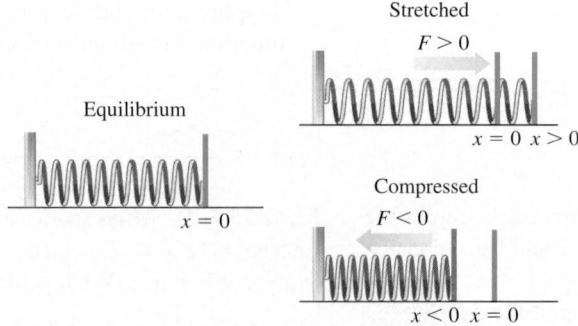

Stretched

$F > 0$

$x = 0 \quad x > 0$

Equilibrium

$x = 0$

Compressed

$F < 0$

$x < 0 \quad x = 0$

Figure 6.72

According to **Hooke's law**, the force required to keep the spring in a compressed or stretched position x units from the equilibrium position is $F(x) = kx$, where the positive spring constant k measures the stiffness of the spring. Note that to stretch the spring

Figure 6.73

> ▶ Hooke's law was proposed by the English scientist Robert Hooke (1635–1703), who also coined the biological term *cell*. Hooke's law works well for springs made of many common materials. However, some springs obey more complicated spring laws (see Exercise 59).

> ▶ Notice again that the units in the integral are consistent. If F has units of N and x has units of m, then W has units of $F\,dx$, or N-m, which are the units of work (1 N-m = 1 J).

QUICK CHECK 4 In Example 2, explain why more work is needed in part (d) than in part (c), even though the displacement is the same. ◀

to a position $x > 0$, a force $F > 0$ (in the positive direction) is required. To compress the spring to a position $x < 0$, a force $F < 0$ (in the negative direction) is required (**Figure 6.73**). In other words, the force required to displace the spring is always in the direction of the displacement.

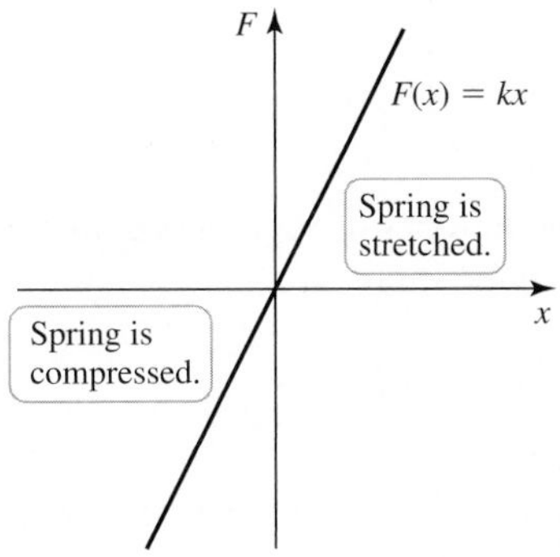

EXAMPLE 2   Compressing a spring Suppose a force of 10 N is required to stretch a spring 0.1 m from its equilibrium position and hold it in that position.

a. Assuming the spring obeys Hooke's law, find the spring constant k.

b. How much work is needed to *compress* the spring 0.5 m from its equilibrium position?

c. How much work is needed to *stretch* the spring 0.25 m from its equilibrium position?

d. How much additional work is required to stretch the spring 0.25 m if it has already been stretched 0.1 m from its equilibrium position?

SOLUTION

a. The fact that a force of 10 N is required to keep the spring stretched at $x = 0.1$ m means (by Hooke's law) that $F(0.1) = k(0.1 \text{ m}) = 10$ N. Solving for the spring constant, we find that $k = 100$ N/m. Therefore, Hooke's law for this spring is $F(x) = 100x$.

b. The work in joules required to compress the spring from $x = 0$ to $x = -0.5$ is

$$W = \int_a^b F(x)\,dx = \int_0^{-0.5} 100x\,dx = 50x^2 \Big|_0^{-0.5} = 12.5.$$

c. The work in joules required to stretch the spring from $x = 0$ to $x = 0.25$ is

$$W = \int_a^b F(x)\,dx = \int_0^{0.25} 100x\,dx = 50x^2 \Big|_0^{0.25} = 3.125.$$

d. The work in joules required to stretch the spring from $x = 0.1$ to $x = 0.35$ is

$$W = \int_a^b F(x)\,dx = \int_{0.1}^{0.35} 100x\,dx = 50x^2 \Big|_{0.1}^{0.35} = 5.625.$$

Comparing parts (c) and (d), we see that more work is required to stretch the spring 0.25 m starting at $x = 0.1$ than starting at $x = 0$.

Related Exercises 23–24 ◀

Lifting Problems Another common work problem arises when the motion is vertical and the force is due to gravity. The gravitational force exerted on an object with mass m (measured in kg) is $F = mg$, where $g = 9.8$ m/s^2 is the acceleration due to gravity near the surface of Earth. The work in joules required to lift an object of mass m a vertical distance of y meters is

$$\text{work} = \text{force} \cdot \text{distance} = mgy.$$

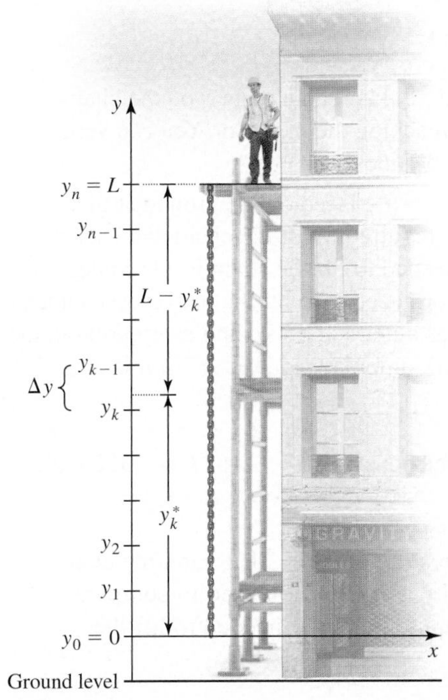

This type of problem becomes interesting when the object being lifted is a rope, a chain, or a cable. In these situations, different parts of the object are lifted different distances, so integration is necessary.

   Imagine a chain of length L meters with a constant density of ρ kg/m hanging vertically from a scaffolding platform at a construction site. To compute the work done in lifting the chain to the platform, we introduce a coordinate system where $y = 0$ corresponds to the bottom of the chain and $y = L$ corresponds to the top of the chain (**Figure 6.74**). We then divide the interval $[0, L]$ into n subintervals of equal length Δy and choose a point y_k^* from each subinterval $[y_{k-1}, y_k]$, for $k = 1, \ldots, n$.

   The mass of each segment of the chain is $m = \rho \Delta y$, and the point y_k^* in the kth segment is lifted a distance of $L - y_k^*$. Therefore, the work required to lift the kth segment is approximately

$$W_k = \underbrace{\rho \Delta y g}_{\text{force}} \cdot \underbrace{(L - y_k^*)}_{\text{distance}},$$

Figure 6.74

▶ The units of ρ, Δy, g, and $L - y_k^*$ are kg/m, m, m/s², and m, respectively. When multiplied, they yield $\dfrac{\text{kg m}^2}{\text{s}^2}$, which are the units of a joule.

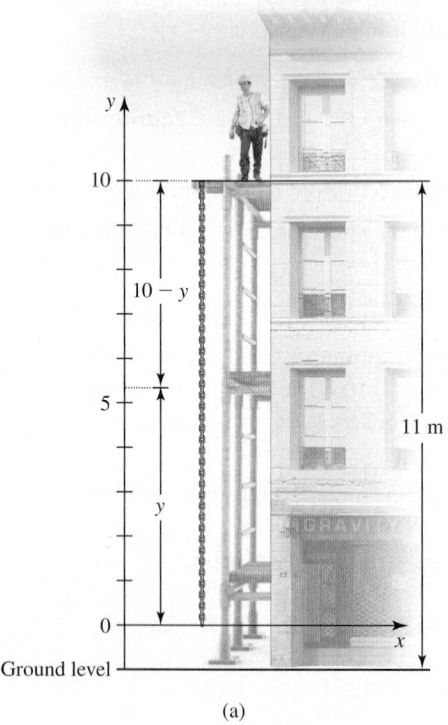

(a)

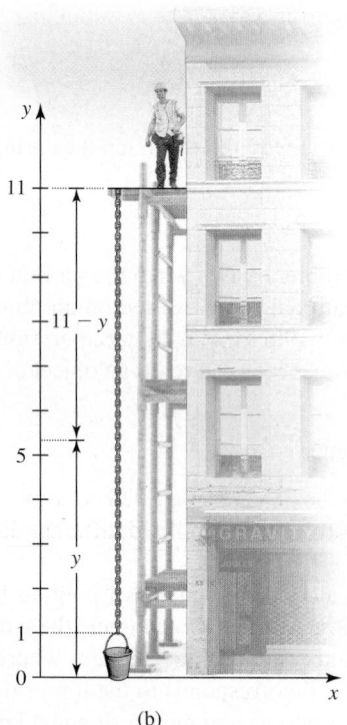

(b)

Figure 6.75

so the total work required to lift the chain is

$$W \approx \sum_{k=1}^{n} \rho g (L - y_k^*) \Delta y.$$

As the length of each segment Δy tends to zero and the number of segments tends to infinity, we obtain a definite integral for the total work in the limit:

$$W = \lim_{n \to \infty} \sum_{k=1}^{n} \rho g (L - y_k^*) \Delta y = \int_0^L \rho g (L - y)\, dy. \qquad (1)$$

The function $L - y$ in equation (1) measures the distance through which a point y on the chain moves. Recognize that this function changes depending on the location of the origin relative to the chain, as illustrated in Example 3 (and Quick Check 6).

EXAMPLE 3 Lifting a chain and bucket A ten-meter chain with density of 1.5 kg/m hangs from a platform at a construction site that is 11 meters above the ground (Figure 6.75a).

a. Compute the work required to lift the chain to the platform.

b. Several packages of nails are placed in a one-meter-tall bucket that rests on the ground; the mass of the bucket and nails together is 15 kg, and the chain is attached to the bucket (Figure 6.75b). How much work is required to lift the bucket to the platform?

SOLUTION

a. We use a coordinate system with $y = 0$ placed at the bottom of the chain. By equation (1), the work in joules required to lift the chain to the platform is

$$W = \int_0^L \rho g (L - y)\, dy = 1.5g \int_0^{10} (10 - y)\, dy \quad L = 10 \text{ m}; \rho = 1.5 \text{ kg/m}$$

$$= 1.5g \left(10y - \frac{y^2}{2} \right) \Big|_0^{10} \quad \text{Integrate.}$$

$$= 1.5g \left(100 - \frac{100}{2} \right) \quad \text{Evaluate.}$$

$$= 735. \quad \text{Simplify; } g = 9.8 \text{ m/s.}$$

Had we instead chosen a coordinate system with $y = 0$ corresponding to the ground, the work integral would be

$$W = \int_1^{11} 1.5g (11 - y)\, dy$$

because the chain would then lie in the interval $[1, 11]$, and a point y on the chain would move a distance of $(11 - y)$ m before reaching the platform. You can verify that the value of this definite integral is also 735 joules.

b. This time we use a coordinate system with $y = 0$ placed on the ground at the bottom of the bucket; we compute the work required to lift the chain and bucket separately. As shown in part (a), the work needed to lift the chain is 735 joules. Integration is not necessary to compute the work needed to lift the bucket (see Quick Check 5). We simply note that the bucket has a mass of 15 kg and every point in the bucket moves a distance of 11 m, so the work in joules required to lift it is

$$W = mgy = 15(9.8)(11) = 1617.$$

The work required to lift both the chain and the bucket is $735 + 1617 = 2352$ joules.
Related Exercises 31–32 ◄

QUICK CHECK 5 In Example 3b, the bucket occupies the interval $[0, 1]$ and the chain occupies the interval $[1, 11]$ (Figure 6.75b). Why is integration used to compute the work needed to lift the chain but not to compute the work needed to lift the bucket? ◄

> The choice of a coordinate system is somewhat arbitrary and may depend on the geometry of the problem. You can let the y-axis point upward or downward, and there are usually several logical choices for the location of $y = 0$. You should experiment with different coordinate systems.

Pumping Problems   The chains and cables encountered in the lifting problems from the previous pages can be modeled as one-dimensional objects. An extra layer of complexity is added when investigating the work required to pump fluid from a tank, although the same principles remain in play.

Suppose a fluid such as water is pumped out of a tank to a height h above the bottom of the tank. How much work is required, assuming the tank is full of water? Three key observations lead to the solution.

- Water from different levels of the tank is lifted different vertical distances, requiring different amounts of work.
- Two equal volumes of water from the same horizontal plane are lifted the same distance and require the same amount of work.
- A volume V of water has mass ρV, where $\rho = 1 \text{ g/cm}^3 = 1000 \text{ kg/m}^3$ is the density of water.

To solve this problem, we let the y-axis point upward with $y = 0$ at the bottom of the tank. The body of water that must be lifted extends from $y = 0$ to $y = b$ (which *may* be the top of the tank). The level to which the water must be raised is $y = h$, where $h \geq b$ (**Figure 6.76**). We now slice the water into n horizontal layers, each having thickness Δy. The kth layer occupying the interval $[y_{k-1}, y_k]$, for $k = 1, \ldots, n$, is approximately y_k^* units above the bottom of the tank, where y_k^* is any point in $[y_{k-1}, y_k]$.

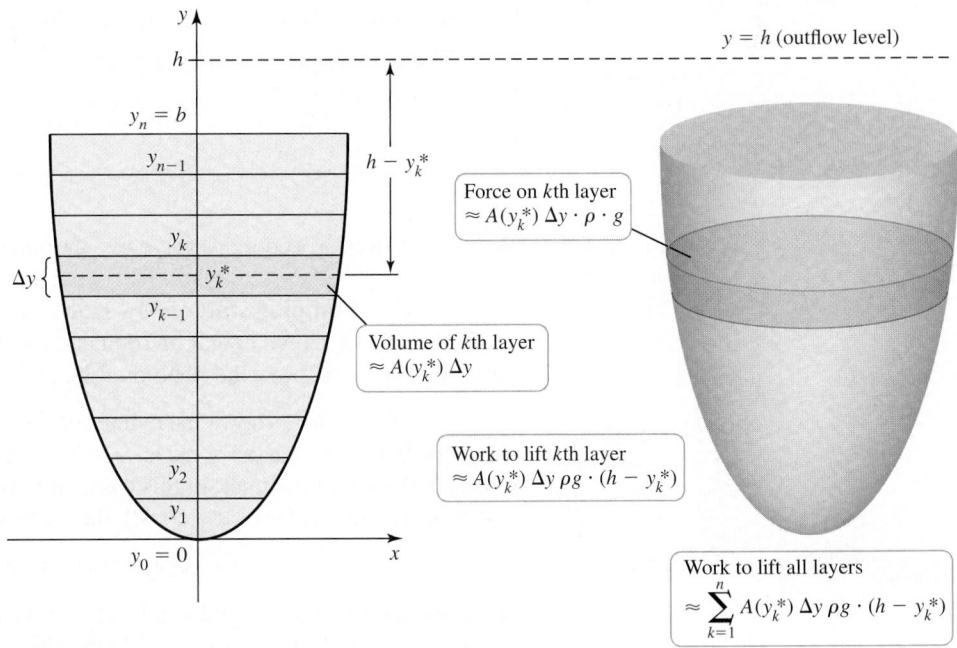

Figure 6.76

The cross-sectional area of the kth layer at y_k^*, denoted $A(y_k^*)$, is determined by the shape of the tank; the solution depends on being able to find A for all values of y. Because the volume of the kth layer is approximately $A(y_k^*)\Delta y$, the force on the kth layer (its weight) is

$$F_k = mg \approx \underbrace{\underbrace{A(y_k^*)\Delta y}_{\text{volume}} \cdot \underbrace{\rho}_{\text{density}} \cdot g}_{\text{mass}}.$$

To reach the level $y = h$, the kth layer is lifted an approximate distance $(h - y_k^*)$ (**Figure 6.76**). So the work in lifting the kth layer to a height h is approximately

$$W_k = \underbrace{A(y_k^*)\Delta y \rho g}_{\text{force}} \cdot \underbrace{(h - y_k^*)}_{\text{distance}}.$$

Summing the work required to lift all the layers to a height h, the total work is

$$W \approx \sum_{k=1}^{n} W_k = \sum_{k=1}^{n} A(y_k^*)\rho g(h - y_k^*)\Delta y.$$

This approximation becomes more accurate as the width of the layers Δy tends to zero and the number of layers tends to infinity. In this limit, we obtain a definite integral from $y = 0$ to $y = b$. The total work required to empty the tank is

$$W = \lim_{n \to \infty} \sum_{k=1}^{n} A(y_k^*)\rho g(h - y_k^*)\Delta y = \int_0^b \rho g A(y)(h - y)\, dy.$$

This derivation assumes the *bottom* of the tank is at $y = 0$, in which case the distance that the slice at level y must be lifted is $D(y) = h - y$. If you choose a different location for the origin, the function D will be different. Here is a general procedure for any choice of origin.

> Notice that the work integral for pumping problems reduces to the work integral for lifting problems given in equation (1) on p. 468 when we take $A(y) = 1$.

PROCEDURE   Solving Pumping Problems

1. Draw a y-axis in the vertical direction (parallel to gravity) and choose a convenient origin. Assume the interval $[a, b]$ corresponds to the vertical extent of the fluid.
2. For $a \le y \le b$, find the cross-sectional area $A(y)$ of the horizontal slices and the distance $D(y)$ the slices must be lifted.
3. The work required to lift the water is

$$W = \int_a^b \rho g A(y) D(y)\, dy.$$

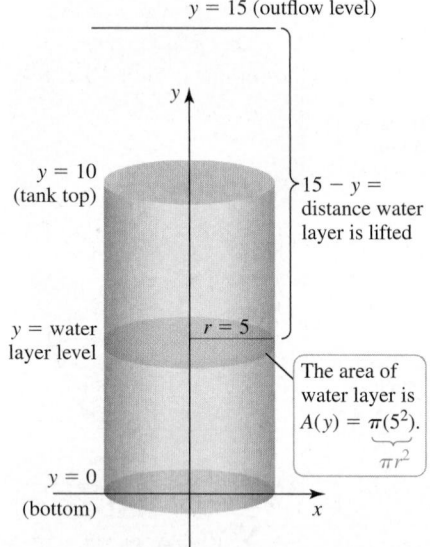

Figure 6.77

> Recall that $g \approx 9.8$ m/s². You should verify that the units are consistent in this calculation: The units of ρ, g, $A(y)$, $D(y)$, and dy are kg/m³, m/s², m², m, and m, respectively. The resulting units of W are kg m²/s², or J. A more convenient unit for large amounts of work and energy is the kilowatt-hour, which is 3.6 million joules.

QUICK CHECK 7 In Example 4, how would the integral change if the outflow pipe were at the top of the tank? ◄

We now use this procedure to solve two pumping problems.

EXAMPLE 4   Pumping water How much work is needed to pump all the water out of a cylindrical tank with a height of 10 m and a radius of 5 m? The water is pumped to an outflow pipe 15 m above the bottom of the tank.

SOLUTION Figure 6.77 shows the cylindrical tank filled to capacity and the outflow 15 m above the bottom of the tank. We let $y = 0$ represent the bottom of the tank and $y = 10$ represent the top of the tank. In this case, all horizontal slices are circular disks of radius $r = 5$ m. Therefore, for $0 \le y \le 10$, the cross-sectional area is

$$A(y) = \pi r^2 = \pi 5^2 = 25\pi.$$

Note that the water is pumped to a level $h = 15$ m above the bottom of the tank, so the lifting distance is $D(y) = 15 - y$. The resulting work integral is

$$W = \int_0^{10} \rho g \underbrace{A(y)}_{25\pi} \underbrace{D(y)}_{15 - y}\, dy = 25\pi\rho g \int_0^{10} (15 - y)\, dy.$$

Substituting $\rho = 1000$ kg/m³ and $g = 9.8$ m/s², the total work in joules is

$$\begin{aligned} W &= 25\pi\rho g \int_0^{10} (15 - y)\, dy \\ &= 25\pi \underbrace{(1000)}_{\rho}\underbrace{(9.8)}_{g}\left(15y - \frac{1}{2}y^2\right)\Big|_0^{10} \\ &\approx 7.7 \times 10^7. \end{aligned}$$

The work required to pump the water out of the tank is approximately 77 million joules.

Related Exercises 36–37 ◄

EXAMPLE 5 **Pumping gasoline** A cylindrical tank with a length of 10 m and a radius of 5 m is on its side and half full of gasoline (Figure 6.78). How much work is required to empty the tank through an outlet pipe at the top of the tank? (The density of gasoline is $\rho = 737 \text{ kg/m}^3$.)

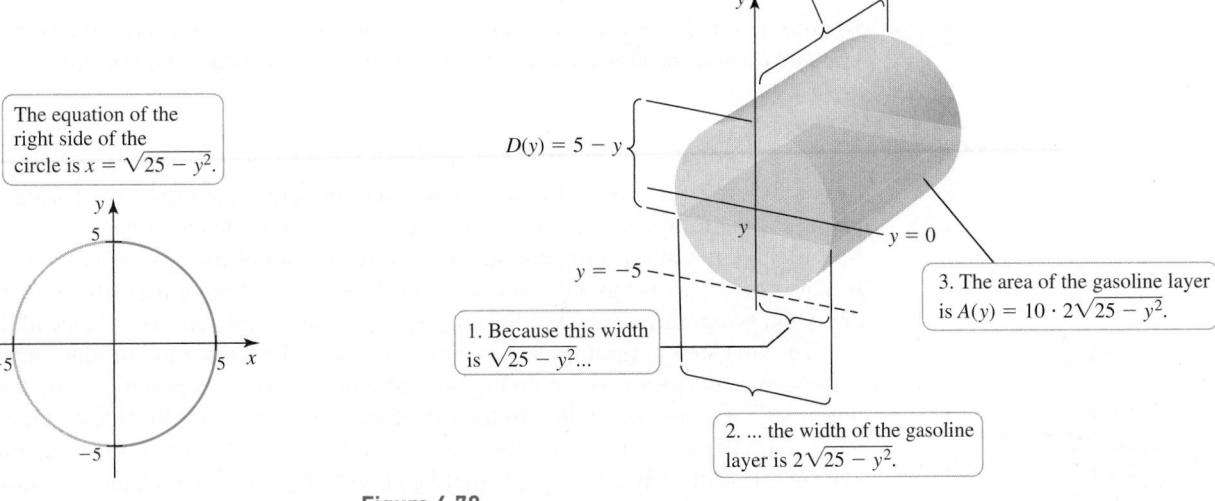

The equation of the right side of the circle is $x = \sqrt{25 - y^2}$.

length = 10

$D(y) = 5 - y$

$y = -5$

1. Because this width is $\sqrt{25 - y^2}$...

2. ... the width of the gasoline layer is $2\sqrt{25 - y^2}$.

3. The area of the gasoline layer is $A(y) = 10 \cdot 2\sqrt{25 - y^2}$.

Figure 6.78

▶ Again, there are several choices for the location of the origin. The location in this example makes $A(y)$ easy to compute.

SOLUTION In this problem, we choose a different origin by letting $y = 0$ and $y = -5$ correspond to the center and the bottom of the tank, respectively. For $-5 \le y \le 0$, a horizontal layer of gasoline located at a depth y is a rectangle with a length of 10 and a width of $2\sqrt{25 - y^2}$ (Figure 6.78). Therefore, the cross-sectional area of the layer at depth y is

$$A(y) = 20\sqrt{25 - y^2}.$$

The distance the layer at level y must be lifted to reach the top of the tank is $D(y) = 5 - y$, where $-5 \le y \le 0$; note that $5 \le D(y) \le 10$. The resulting work integral is

$$W = \underbrace{737}_{\rho}\underbrace{(9.8)}_{g} \int_{-5}^{0} \underbrace{20\sqrt{25 - y^2}}_{A(y)} \underbrace{(5 - y)}_{D(y)} \, dy = 144{,}452 \int_{-5}^{0} \sqrt{25 - y^2} \, (5 - y) \, dy.$$

This integral is evaluated by splitting the integrand into two pieces and recognizing that one piece is the area of a quarter circle of radius 5:

$$\int_{-5}^{0} \sqrt{25 - y^2} \, (5 - y) \, dy = 5 \underbrace{\int_{-5}^{0} \sqrt{25 - y^2} \, dy}_{\text{area of quarter circle}} - \underbrace{\int_{-5}^{0} y\sqrt{25 - y^2} \, dy}_{\text{let } u = 25 - y^2; \; du = -2y \, dy}$$

$$= 5 \cdot \frac{25\pi}{4} + \frac{1}{2}\int_{0}^{25} \sqrt{u} \, du$$

$$= \frac{125\pi}{4} + \frac{1}{3}u^{3/2}\Big|_{0}^{25} = \frac{375\pi + 500}{12}.$$

Multiplying this result by $20\,\rho g = 144{,}452$, we find that the work required is approximately 20.2 million joules.

Related Exercises 42–44 ◀

Force and Pressure

Another application of integration deals with the force exerted on a surface by a body of water. Again, we need a few physical principles.

Pressure is a force per unit area, measured in units such as newtons per square meter (N/m²). For example, the pressure of the atmosphere on the surface of Earth is about 14 lb/in² (approximately 100 kilopascals, or 10^5 N/m²). As another example, if you stood on the bottom of a swimming pool, you would feel pressure due to the weight (force) of the column of water above your head. If your head is flat and has surface area A m^2 and it is h meters below the surface, then the column of water above your head has volume Ah m^3. That column of water exerts a force (its weight)

$$F = \text{mass} \cdot \text{acceleration} = \underbrace{\text{volume} \cdot \text{density}}_{\text{mass}} \cdot g = Ah\rho g,$$

where ρ is the density of water and g is the acceleration due to gravity. Therefore, the pressure on your head is the force divided by the surface area of your head:

$$\text{pressure} = \frac{\text{force}}{A} = \frac{Ah\rho g}{A} = \rho g h.$$

This pressure is called **hydrostatic pressure** (meaning the pressure of *water at rest*), and it has the following important property: *It has the same magnitude in all directions.* Specifically, the hydrostatic pressure on a vertical wall of the swimming pool at a depth h is also $\rho g h$. This is the only fact needed to find the total force on vertical walls such as dams and swimming pools. We assume the water completely covers the face of the dam.

The first step in finding the force on the face of the dam is to introduce a coordinate system. We choose a y-axis pointing upward with $y = 0$ corresponding to the base of the dam and $y = a$ corresponding to the top of the dam (**Figure 6.79**). Because the pressure varies with depth (y-direction), the dam is sliced horizontally into n strips of equal thickness Δy. The kth strip corresponds to the interval $[y_{k-1}, y_k]$, and we let y_k^* be any point in that interval. The depth of that strip is approximately $h = a - y_k^*$, so the hydrostatic pressure on that strip is approximately $\rho g(a - y_k^*)$.

The crux of any dam problem is finding the width of the strips as a function of y, which we denote $w(y)$. Each dam has its own width function; however, once the width function is known, the solution follows directly. The approximate area of the kth strip is its width multiplied by its thickness, or $w(y_k^*)\Delta y$. The force on the kth strip (which is the area of the strip multiplied by the pressure) is approximately

$$F_k = \underbrace{\rho g(a - y_k^*)}_{\text{pressure}}\underbrace{w(y_k^*)\Delta y}_{\text{area of strip}}.$$

Summing the forces over the n strips, the total force is

$$F \approx \sum_{k=1}^{n} F_k = \sum_{k=1}^{n} \rho g(a - y_k^*)w(y_k^*)\Delta y.$$

To find the exact force, we let the thickness of the strips tend to zero and the number of strips tend to infinity, which produces a definite integral. The limits of integration correspond to the base ($y = 0$) and top ($y = a$) of the dam. Therefore, the total force on the dam is

$$F = \lim_{n \to \infty} \sum_{k=1}^{n} \rho g(a - y_k^*)w(y_k^*)\Delta y = \int_0^a \rho g(a - y)w(y) \, dy.$$

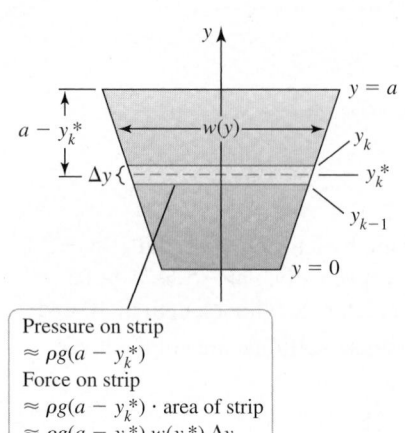

Pressure on strip
$\approx \rho g(a - y_k^*)$
Force on strip
$\approx \rho g(a - y_k^*) \cdot$ area of strip
$\approx \rho g(a - y_k^*) \, w(y_k^*) \, \Delta y$

Figure 6.79

➤ We have chosen $y = 0$ to be the base of the dam. Depending on the geometry of the problem, it may be more convenient (less computation) to let $y = 0$ be at the top of the dam. Note that the depth function $D(y) = a - y$ varies with the chosen coordinate system. Experiment with different choices.

PROCEDURE   **Solving Force-on-Dam Problems**

1. Draw a y-axis on the face of the dam in the vertical direction and choose a convenient origin (often taken to be the base of the dam).

2. Find the width function $w(y)$ for each value of y on the face of the dam.

3. If the base of the dam is at $y = 0$ and the top of the dam is at $y = a$, then the total force on the dam is

$$F = \int_0^a \rho g \underbrace{(a - y)}_{\text{depth}}\underbrace{w(y)}_{\text{width}} \, dy.$$

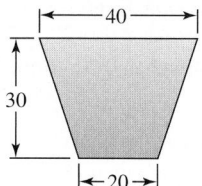

Figure 6.80

EXAMPLE 6 Force on a dam A large vertical dam in the shape of a symmetric trapezoid has a height of 30 m, a width of 20 m at its base, and a width of 40 m at the top (Figure 6.80). What is the total force on the face of the dam when the reservoir is full?

SOLUTION We place the origin at the center of the base of the dam (Figure 6.81). The right slanted edge of the dam is a segment of the line that passes through the points $(10, 0)$ and $(20, 30)$. An equation of that line is

$$y - 0 = \frac{30}{10}(x - 10) \quad \text{or} \quad y = 3x - 30 \quad \text{or} \quad x = \frac{1}{3}(y + 30).$$

➤ You should check the width function: $w(0) = 20$ (the width of the dam at its base) and $w(30) = 40$ (the width of the dam at its top).

Notice that at a depth of y, where $0 \le y \le 30$, the width of the dam is

$$w(y) = 2x = \frac{2}{3}(y + 30).$$

Using $\rho = 1000 \text{ kg/m}^3$ and $g = 9.8 \text{ m/s}^2$, the total force on the dam (in newtons) is

$$F = \int_0^a \rho g(a - y)w(y)\,dy \qquad \text{Force integral}$$

$$= \rho g \int_0^{30} \underbrace{(30 - y)}_{a\,-\,y} \underbrace{\frac{2}{3}(y + 30)}_{w(y)}\,dy \qquad \text{Substitute.}$$

$$= \frac{2}{3}\rho g \int_0^{30} (900 - y^2)\,dy \qquad \text{Simplify.}$$

$$= \frac{2}{3}\rho g\left(900y - \frac{y^3}{3}\right)\Big|_0^{30} \qquad \text{Fundamental Theorem}$$

$$\approx 1.18 \times 10^8.$$

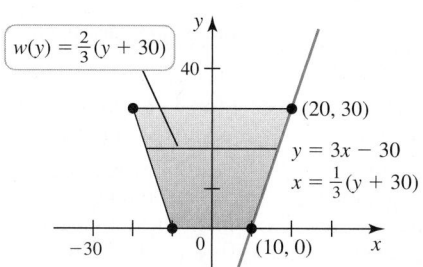

Figure 6.81

The force of 1.18×10^8 N on the dam amounts to about 26 million pounds, or 13,000 tons.

Related Exercises 46–47 ◄

SECTION 6.7 EXERCISES

Getting Started

1. Suppose a 1-m cylindrical bar has a constant density of 1 g/cm for its left half and a constant density of 2 g/cm for its right half. What is its mass?

2. Explain how to find the mass of a one-dimensional object with a variable density ρ.

3. How much work is required to move an object from $x = 0$ to $x = 5$ (measured in meters) in the presence of a constant force of 5 N acting along the x-axis?

4. Why is integration used to find the work done by a variable force?

5. Why is integration used to find the work required to pump water out of a tank?

6. Why is integration used to find the total force on the face of a dam?

7. What is the pressure on a horizontal surface with an area of 2 m² that is 4 m underwater?

8. Explain why you integrate in the vertical direction (parallel to the acceleration due to gravity) rather than the horizontal direction to find the force on the face of a dam.

9–12. *Consider the cylindrical tank in Example 4 that has a height of 10 m and a radius of 5 m. Recall that if the tank is full of water, then $\int_0^{10} 25\,\pi\rho g(15 - y)\,dy$ equals the work required to pump all the water out of the tank, through an outflow pipe that is 15 m above the bottom of the tank. Revise this work integral for the following scenarios. (Do not evaluate the integrals.)*

9. The work required to empty the top half of the tank

10. The work required to empty the tank if it is half full

11. The work required to empty the tank through an outflow pipe at the top of the tank

12. The work required to empty the tank if the water in the tank is only 3 m deep and the outflow pipe is at the top of the tank

Practice Exercises

13–20. Mass of one-dimensional objects *Find the mass of the following thin bars with the given density function.*

13. $\rho(x) = 1 + \sin x$, for $0 \le x \le \pi$

14. $\rho(x) = 1 + x^3$, for $0 \le x \le 1$

15. $\rho(x) = 2 - \dfrac{x}{2}$, for $0 \le x \le 2$

16. $\rho(x) = 5\,e^{-2x}$, for $0 \le x \le 4$

17. $\rho(x) = x\sqrt{2 - x^2}$, for $0 \le x \le 1$

18. $\rho(x) = \begin{cases} 1 & \text{if } 0 \le x \le 2 \\ 2 & \text{if } 2 < x \le 3 \end{cases}$

19. $\rho(x) = \begin{cases} 1 & \text{if } 0 \le x \le 2 \\ 1 + x & \text{if } 2 < x \le 4 \end{cases}$

20. $\rho(x) = \begin{cases} x^2 & \text{if } 0 \le x \le 1 \\ x(2 - x) & \text{if } 1 < x \le 2 \end{cases}$

21. **Work from force** How much work is required to move an object from $x = 0$ to $x = 3$ (measured in meters) in the presence of a force (in N) given by $F(x) = 2x$ acting along the x-axis?

22. **Work from force** How much work is required to move an object from $x = 1$ to $x = 3$ (measured in meters) in the presence of a force (in N) given by $F(x) = \dfrac{2}{x^2}$ acting along the x-axis?

23. **Compressing and stretching a spring** Suppose a force of 30 N is required to stretch and hold a spring 0.2 m from its equilibrium position.

    a. Assuming the spring obeys Hooke's law, find the spring constant k.
    b. How much work is required to compress the spring 0.4 m from its equilibrium position?
    c. How much work is required to stretch the spring 0.3 m from its equilibrium position?
    d. How much additional work is required to stretch the spring 0.2 m if it has already been stretched 0.2 m from its equilibrium position?

24. **Compressing and stretching a spring** Suppose a force of 15 N is required to stretch and hold a spring 0.25 m from its equilibrium position.

    a. Assuming the spring obeys Hooke's law, find the spring constant k.
    b. How much work is required to compress the spring 0.2 m from its equilibrium position?
    c. How much additional work is required to stretch the spring 0.3 m if it has already been stretched 0.25 m from its equilibrium position?

25. **Work done by a spring** A spring on a horizontal surface can be stretched and held 0.5 m from its equilibrium position with a force of 50 N.

    a. How much work is done in stretching the spring 1.5 m from its equilibrium position?
    b. How much work is done in compressing the spring 0.5 m from its equilibrium position?

26. **Shock absorber** A heavy-duty shock absorber is compressed 2 cm from its equilibrium position by a mass of 500 kg. How much work is required to compress the shock absorber 4 cm from its equilibrium position? (A mass of 500 kg exerts a force (in newtons) of $500g$, where $g = 9.8 \text{ m/s}^2$.)

27. **Calculating work for different springs** Calculate the work required to stretch the following springs 0.5 m from their equilibrium positions. Assume Hooke's law is obeyed.

    a. A spring that requires a force of 50 N to be stretched 0.2 m from its equilibrium position
    b. A spring that requires 50 J of work to be stretched 0.2 m from its equilibrium position

28. **Calculating work for different springs** Calculate the work required to stretch the following springs 0.4 m from their equilibrium positions. Assume Hooke's law is obeyed.

    a. A spring that requires a force of 50 N to be stretched 0.1 m from its equilibrium position
    b. A spring that requires 2 J of work to be stretched 0.1 m from its equilibrium position

29. **Calculating work for different springs** Calculate the work required to stretch the following springs 1.25 m from their equilibrium positions. Assume Hooke's law is obeyed.

    a. A spring that requires 100 J of work to be stretched 0.5 m from its equilibrium position
    b. A spring that requires a force of 250 N to be stretched 0.5 m from its equilibrium position

30. **Work function** A spring has a restoring force given by $F(x) = 25x$. Let $W(x)$ be the work required to stretch the spring from its equilibrium position $(x = 0)$ to a variable distance x. Find and graph the work function. Compare the work required to stretch the spring x units from equilibrium to the work required to compress the spring x units from equilibrium.

31. **Winding a chain** A 30-m-long chain hangs vertically from a cylinder attached to a winch. Assume there is no friction in the system and the chain has a density of 5 kg/m.

    a. How much work is required to wind the entire chain onto the cylinder using the winch?
    b. How much work is required to wind the chain onto the cylinder if a 50-kg block is attached to the end of the chain?

32. **Coiling a rope** A 60-m-long, 9.4-mm-diameter rope hangs freely from a ledge. The density of the rope is 55 g/m. How much work is needed to pull the entire rope to the ledge?

33. **Winding part of a chain** A 20-m-long, 50-kg chain hangs vertically from a cylinder attached to a winch. How much work is needed to wind the upper half of the chain onto the winch?

34. **Leaky Bucket** A 1-kg bucket resting on the ground contains 3 kg of water. How much work is required to raise the bucket vertically a distance of 10 m if water leaks out of the bucket at a constant rate of $\frac{1}{5}$ kg/m? Assume the weight of the rope used to raise the bucket is negligible. (*Hint:* Use the definition of work, $W = \int_a^b F(y)\,dy$, where F is the variable force required to lift an object along a vertical line from $y = a$ to $y = b$.)

35. **Emptying a swimming pool** A swimming pool has the shape of a box with a base that measures 25 m by 15 m and a uniform depth of 2.5 m. How much work is required to pump the water out of the pool (to the level of the top of the pool) when it is full?

36. **Emptying a cylindrical tank** A cylindrical water tank has height 8 m and radius 2 m (see figure).

    a. If the tank is full of water, how much work is required to pump the water to the level of the top of the tank and out of the tank?

b. Is it true that it takes half as much work to pump the water out of the tank when it is half full as when it is full? Explain.

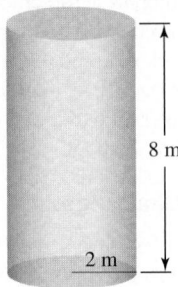

8 m

2 m

37. Emptying a half-full cylindrical tank Suppose the water tank in Exercise 36 is half full of water. Determine the work required to empty the tank by pumping the water to a level 2 m above the top of the tank.

38. Emptying a partially filled swimming pool If the water in the swimming pool in Exercise 35 is 2 m deep, then how much work is required to pump all the water to a level 3 m above the bottom of the pool?

39. Emptying a conical tank A water tank is shaped like an inverted cone with height 6 m and base radius 1.5 m (see figure).

   a. If the tank is full, how much work is required to pump the water to the level of the top of the tank and out of the tank?

   b. Is it true that it takes half as much work to pump the water out of the tank when it is filled to half its depth as when it is full? Explain.

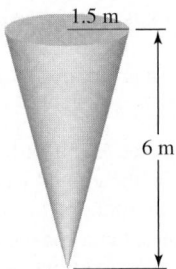

1.5 m

6 m

40. Upper and lower half A cylinder with height 8 m and radius 3 m is filled with water and must be emptied through an outlet pipe 2 m above the top of the cylinder.

   a. Compute the work required to empty the water in the top half of the tank.

   b. Compute the work required to empty the (equal amount of) water in the lower half of the tank.

   c. Interpret the results of parts (a) and (b).

41. Filling a spherical tank A spherical water tank with an inner radius of 8 m has its lowest point 2 m above the ground. It is filled by a pipe that feeds the tank at its lowest point (see figure). Neglecting the volume of the inflow pipe, how much work is required to fill the tank if it is initially empty?

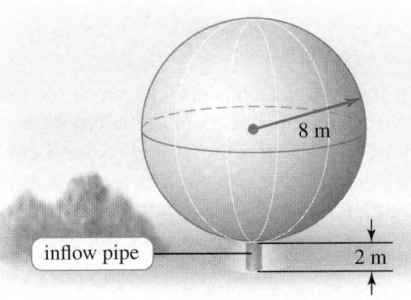

8 m

inflow pipe   2 m

42. Emptying a water trough A water trough has a semicircular cross section with a radius of 0.25 m and a length of 3 m (see figure).

   a. How much work is required to pump the water out of the trough (to the level of the top of the trough) when it is full?

   b. If the length is doubled, is the required work doubled? Explain.

   c. If the radius is doubled, is the required work doubled? Explain.

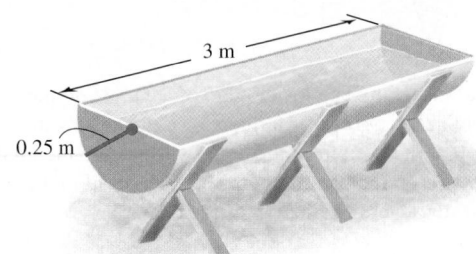

3 m

0.25 m

43. Emptying a water trough A cattle trough has a trapezoidal cross section with a height of 1 m and horizontal sides of length $\frac{1}{2}$ m and 1 m. Assume the length of the trough is 10 m (see figure).

   a. How much work is required to pump the water out of the trough (to the level of the top of the trough) when it is full?

   b. If the length is doubled, is the required work doubled? Explain.

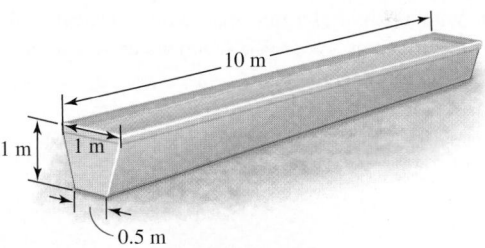

10 m

1 m   1 m

0.5 m

44. Pumping water Suppose the tank in Example 5 is full of water (rather than half full of gasoline). Determine the work required to pump all the water to an outlet pipe 15 m above the bottom of the tank.

45. Emptying a conical tank An inverted cone (base above the vertex) is 2 m high and has a base radius of $\frac{1}{2}$ m. If the tank is full, how much work is required to pump the water to a level 1 m above the top of the tank?

46–50. Force on dams *The following figures show the shapes and dimensions of small dams. Assuming the water level is at the top of the dam, find the total force on the face of the dam.*

46.

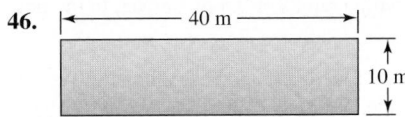

40 m

10 m

47.

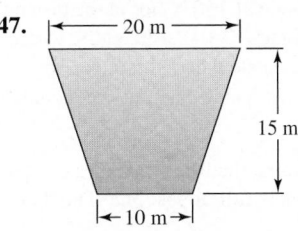

20 m

15 m

10 m

48.

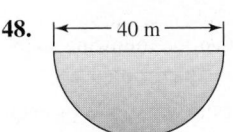

40 m

49.

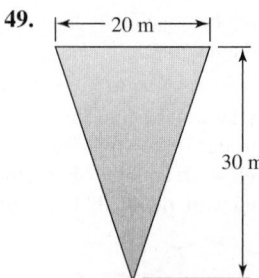

50. Parabolic dam The lower edge of a dam is defined by the parabola $y = x^2/16$ (see figure). Use a coordinate system with $y = 0$ at the bottom of the dam to determine the total force of the water on the dam. Lengths are measured in meters. Assume the water level is at the top of the dam.

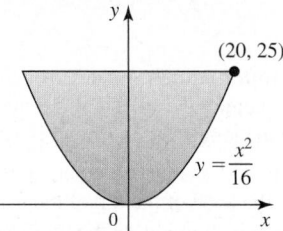

51. Force on a triangular plate A plate shaped like an isosceles triangle with a height of 1 m is placed on a vertical wall 1 m below the surface of a pool filled with water (see figure). Compute the force on the plate.

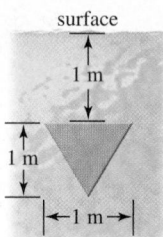

52–54. Force on a window *A diving pool that is 4 m deep and full of water has a viewing window on one of its vertical walls. Find the force on the following windows.*

52. The window is a square, 0.5 m on a side, with the lower edge of the window on the bottom of the pool.

53. The window is a square, 0.5 m on a side, with the lower edge of the window 1 m from the bottom of the pool.

54. The window is circular, with a radius of 0.5 m, tangent to the bottom of the pool.

55. Force on a building A large building shaped like a box is 50 m high with a face that is 80 m wide. A strong wind blows directly at the face of the building, exerting a pressure of 150 N/m² at the ground and increasing with height according to $P(y) = 150 + 2y$, where y is the height above the ground. Calculate the total force on the building, which is a measure of the resistance that must be included in the design of the building.

56. Force on the end of a tank Determine the force on a circular end of the tank in Figure 6.78 if the tank is full of gasoline. The density of gasoline is $\rho = 737 \text{ kg/m}^3$.

57. Explain why or why not Determine whether the following statements are true and give an explanation or counterexample.

a. The mass of a thin wire is the length of the wire times its average density over its length.

b. The work required to stretch a linear spring (that obeys Hooke's law) 100 cm from equilibrium is the same as the work required to compress it 100 cm from equilibrium.

c. The work required to lift a 10-kg object vertically 10 m is the same as the work required to lift a 20-kg object vertically 5 m.

d. The total force on a 10-ft² region on the (horizontal) floor of a pool is the same as the total force on a 10-ft² region on a (vertical) wall of the pool.

Explorations and Challenges

58. Mass of two bars Two bars of length L have densities $\rho_1(x) = 4e^{-x}$ and $\rho_2(x) = 6e^{-2x}$, for $0 \le x \le L$.

a. For what values of L is bar 1 heavier than bar 2?

b. As the lengths of the bars increase, do their masses increase without bound? Explain.

59. A nonlinear spring Hooke's law is applicable to idealized (linear) springs that are not stretched or compressed too far from their equilibrium positions. Consider a nonlinear spring whose restoring force is given by $F(x) = 16x - 0.1x^3$, for $|x| \le 7$.

a. Graph the restoring force and interpret it.

b. How much work is done in stretching the spring from its equilibrium position $(x = 0)$ to $x = 1.5$?

c. How much work is done in compressing the spring from its equilibrium position $(x = 0)$ to $x = -2$?

60. A vertical spring A 10-kg mass is attached to a spring that hangs vertically and is stretched 2 m from the equilibrium position of the spring. Assume a linear spring with $F(x) = kx$.

a. How much work is required to compress the spring and lift the mass 0.5 m?

b. How much work is required to stretch the spring and lower the mass 0.5 m?

61. Leaky cement bucket A 350 kg-bucket containing 4650 kg of cement is resting on the ground when a crane begins lifting it at a constant rate of 5 m/min. As the crane raises the bucket, cement leaks out of the bucket at a constant rate of 100 kg/min. How much work is required to lift the bucket a distance of 30 m if we ignore the weight of the crane cable attached to the bucket?

62. Emptying a real swimming pool A swimming pool is 20 m long and 10 m wide, with a bottom that slopes uniformly from a depth of 1 m at one end to a depth of 2 m at the other end (see figure). Assuming the pool is full, how much work is required to pump the water to a level 0.2 m above the top of the pool?

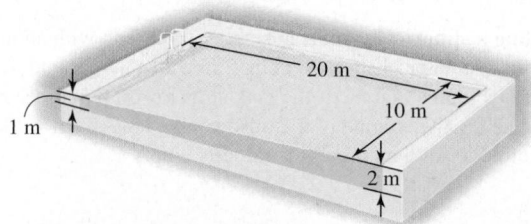

63. Drinking juice A glass has circular cross sections that taper (linearly) from a radius of 5 cm at the top of the glass to a radius of 4 cm at the bottom. The glass is 15 cm high and full of orange juice. How much work is required to drink all the juice through a straw if your mouth is 5 cm above the top of the glass? Assume the density of orange juice equals the density of water.

64. Lifting a pendulum A body of mass m is suspended by a rod of length L that pivots without friction (see figure). The mass is slowly lifted along a circular arc to a height h.

a. Assuming the only force acting on the mass is the gravitational force, show that the component of this force acting along the arc of motion is $F = mg \sin \theta$.

b. Noting that an element of length along the path of the pendulum is $ds = L\,d\theta$, evaluate an integral in θ to show that the work done in lifting the mass to a height h is mgh.

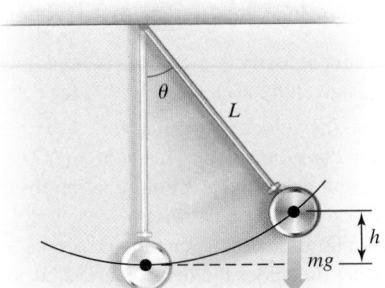

65. Critical depth A large tank has a plastic window on one wall that is designed to withstand a force of 90,000 N. The square window is 2 m on a side, and its lower edge is 1 m from the bottom of the tank.

a. If the tank is filled to a depth of 4 m, will the window withstand the resulting force?

b. What is the maximum depth to which the tank can be filled without the window failing?

66. Orientation and force A plate shaped like an equilateral triangle 1 m on a side is placed on a vertical wall 1 m below the surface of a pool filled with water. On which plate in the figure is the force greater? Try to anticipate the answer and then compute the force on each plate.

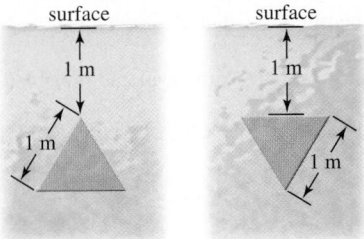

67. Orientation and force A square plate 1 m on a side is placed on a vertical wall 1 m below the surface of a pool filled with water. On which plate in the figure is the force greater? Try to anticipate the answer and then compute the force on each plate.

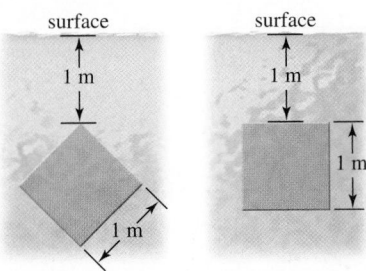

68. Work by two different integrals A rigid body with a mass of 2 kg moves along a line due to a force that produces a position function $x(t) = 4t^2$, where x is measured in meters and t is measured in seconds. Find the work done during the first 5 s in two ways.

a. Note that $x''(t) = 8$; then use Newton's second law $(F = ma = mx''(t))$ to evaluate the work integral $W = \int_{x_0}^{x_f} F(x)\,dx$, where x_0 and x_f are the initial and final positions, respectively.

b. Change variables in the work integral and integrate with respect to t. Be sure your answer agrees with part (a).

69. Work in a gravitational field For large distances from the surface of Earth, the gravitational force is given by $F(x) = \dfrac{GMm}{(x + R)^2}$, where $G = 6.7 \times 10^{-11}\,\text{N m}^2/\text{kg}^2$ is the gravitational constant, $M = 6 \times 10^{24}$ kg is the mass of Earth, m is the mass of the object in the gravitational field, $R = 6.378 \times 10^6$ m is the radius of Earth, and $x \geq 0$ is the distance above the surface of Earth (in meters).

a. How much work is required to launch a rocket with a mass of 500 kg in a vertical flight path to a height of 2500 km (from Earth's surface)?

b. Find the work required to launch the rocket to a height of x kilometers, for $x > 0$.

c. How much work is required to reach outer space $(x \rightarrow \infty)$?

d. Equate the work in part (c) to the initial kinetic energy of the rocket, $\frac{1}{2}mv^2$, to compute the escape velocity of the rocket.

70. Buoyancy Archimedes' principle says that the buoyant force exerted on an object that is (partially or totally) submerged in water is equal to the weight of the water displaced by the object (see figure). Let $\rho_w = 1\,\text{g/cm}^3 = 1000\,\text{kg/m}^3$ be the density of water and let ρ be the density of an object in water. Let $f = \rho/\rho_w$. If $0 < f \leq 1$, then the object floats with a fraction f of its volume submerged; if $f > 1$, then the object sinks. Consider a cubical box with sides 2 m long floating in water with one-half of its volume submerged $(\rho = \rho_w/2)$. Find the force required to fully submerge the box (so its top surface is at the water level). (See the Guided Project *Buoyancy and Archimedes' Principle* for further explorations of buoyancy problems.)

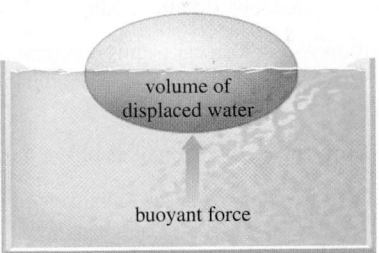

QUICK CHECK ANSWERS

1. a. The bar is heaviest at the left end and lightest at the right end. **b.** $\rho = 2.5\,\text{g/cm}$. **2.** Minimum mass = 2 kg; maximum mass = 10 kg **3.** We assume the force is constant over each subinterval, when, in fact, it varies over each subinterval. **4.** The restoring force of the spring increases as the spring is stretched $(f(x) = 100x)$. Greater restoring forces are encountered on the interval $[0.1, 0.35]$ than on the interval $[0, 0.25]$. **5.** The chain is a flexible object, and different points on the chain are lifted different distances as it is pulled to the platform. The bucket, however, is a rigid object, and every point in the bucket is lifted 11 m when it goes from resting on the ground to resting on the platform (in other words, it can be treated as a point mass.) **6.** $\int_0^{10} 1.5gy\,dy = 735$ joules. **7.** The factor $(15 - y)$ in the integral is replaced with $(10 - y)$. ◄

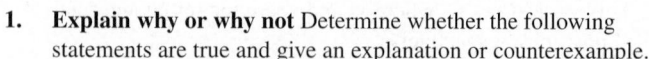

CHAPTER 6   REVIEW EXERCISES

1. **Explain why or why not** Determine whether the following statements are true and give an explanation or counterexample.

   a. A region R is revolved about the y-axis to generate a solid S. To find the volume of S, you could either use the disk/washer method and integrate with respect to y or use the shell method and integrate with respect to x.

   b. Given only the velocity of an object moving on a line, it is possible to find its displacement, but not its position.

   c. If water flows into a tank at a constant rate (for example, 6 gal/min), the volume of water in the tank increases according to a linear function of time.

2–3. Displacement, distance, and position *Consider an object moving along a line with the following velocities and initial positions. Assume time t is measured in seconds and velocities have units of m/s.*

   a. *Over the given interval, determine when the object is moving in the positive direction and when it is moving in the negative direction.*

   b. *Find the displacement over the given interval.*

   c. *Find the distance traveled over the given interval.*

   d. *Determine the position function $s(t)$ using the Fundamental Theorem of Calculus (Theorem 6.1). Check your answer by finding the position function using the antiderivative method.*

2. $v(t) = 1 - 2\cos(\pi t/3)$, for $0 \le t \le 3$; $s(0) = 0$

3. $v(t) = 12t^2 - 30t + 12$, for $0 \le t \le 3$; $s(0) = 1$

4. **Displacement from velocity** The velocity of an object moving along a line is given by $v(t) = 20\cos\pi t$ (in ft/s). What is the displacement of the object after 1.5 s?

5. **Position, displacement, and distance** A projectile is launched vertically from the ground at $t = 0$, and its velocity in flight (in m/s) is given by $v(t) = 20 - 10t$. Find the position, displacement, and distance traveled after t seconds, for $0 \le t \le 4$.

6. **Deceleration** At $t = 0$, a car begins decelerating from a velocity of 80 ft/s at a constant rate of 5 ft/s². Find its position function assuming $s(0) = 0$.

7. **An oscillator** The acceleration of an object moving along a line is given by $a(t) = 2\sin\dfrac{\pi t}{4}$. The initial velocity and position are $v(0) = -\dfrac{8}{\pi}$ and $s(0) = 0$.

   a. Find the velocity and position for $t \ge 0$.

   b. What are the minimum and maximum values of s?

   c. Find the average velocity and average position over the interval $[0, 8]$.

8. **A race** Starting at the same point on a straight road, Anna and Benny begin running with velocities (in mi/hr) given by $v_A(t) = 2t + 1$ and $v_B(t) = 4 - t$, respectively.

   a. Graph the velocity functions, for $0 \le t \le 4$.

   b. If the runners run for 1 hr, who runs farther? Interpret your conclusion geometrically using the graph in part (a).

   c. If the runners run for 6 mi, who wins the race? Interpret your conclusion geometrically using the graph in part (a).

9. **Fuel consumption** A small plane in flight consumes fuel at a rate (in gal/min) given by

   $$R'(t) = \begin{cases} 4t^{1/3} & \text{if } 0 \le t \le 8 \quad \text{(take-off)} \\ 2 & \text{if } t > 8 \quad \text{(cruising)}. \end{cases}$$

   a. Find a function R that gives the total fuel consumed, for $0 \le t \le 8$.

   b. Find a function R that gives the total fuel consumed, for $t \ge 0$.

   c. If the fuel tank capacity is 150 gal, when does the fuel run out?

10. **Variable flow rate** Water flows out of a tank at a rate (in m³/hr) given by $V'(t) = \dfrac{15}{t+1}$. If the tank initially holds 75 m³ of water, when will the tank be empty?

T 11. Decreasing velocity A projectile is fired upward, and its velocity in m/s is given by $v(t) = 200e^{-t/10}$, for $t \ge 0$.

   a. Graph the velocity function, for $t \ge 0$.

   b. When does the velocity reach 50 m/s?

   c. Find and graph the position function for the projectile, for $t \ge 0$, assuming $s(0) = 0$.

   d. Given unlimited time, can the projectile travel 2500 m? If so, at what time does the distance traveled equal 2500 m?

T 12. Decreasing velocity A projectile is fired upward, and its velocity (in m/s) is given by $v(t) = \dfrac{200}{\sqrt{t+1}}$, for $t \ge 0$.

   a. Graph the velocity function, for $t \ge 0$.

   b. Find and graph the position function for the projectile, for $t \ge 0$, assuming $s(0) = 0$.

   c. Given unlimited time, can the projectile travel 2500 m? If so, at what time does the distance traveled equal 2500 m?

13. **An exponential bike ride** Tom and Sue took a bike ride, both starting at the same time and position. Tom started riding at 20 mi/hr, and his velocity decreased according to the function $v(t) = 20e^{-2t}$, for $t \ge 0$. Sue started riding at 15 mi/hr, and her velocity decreased according to the function $u(t) = 15e^{-t}$, for $t \ge 0$.

   a. Find and graph the position functions of Tom and Sue.

   b. Find the times at which the riders had the same position at the same time.

   c. Who ultimately took the lead and remained in the lead?

14–25. Areas of regions *Determine the area of the given region.*

14.
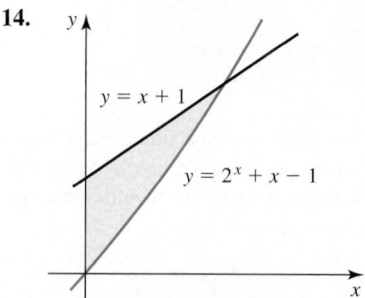

15.
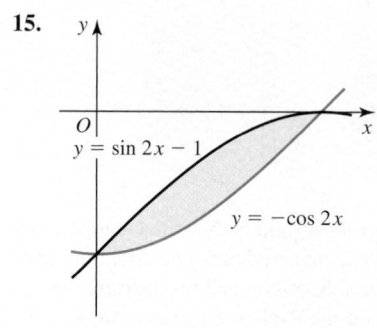

16.

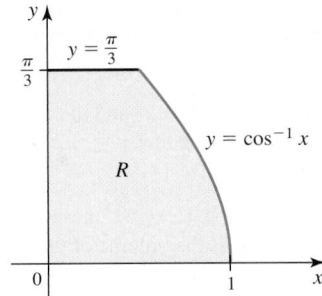

17. The region bounded by $y = \ln x$, $y = 1$, and $x = 1$

18. The region bounded by $y = \sin \dfrac{x}{2}$ and $y = \cos \dfrac{x}{2}$ on $\left[\dfrac{-3\pi}{2}, \dfrac{5\pi}{2}\right]$

19. The region in the first quadrant bounded by $y = 4x$ and $y = x\sqrt{25 - x^2}$

20. The region bounded by $y = 6 - 3x^2$ and $y = 6x - 3$

21. The region bounded by $y = x^2$, $y = 2x^2 - 4x$, and $y = 0$

22. The region in the first quadrant bounded by the curve $\sqrt{x} + \sqrt{y} = 1$

23. The region in the first quadrant bounded by $y = x/6$ and $y = 1 - |x/2 - 1|$

24. The region in the first quadrant bounded by $y = x^p$ and $y = \sqrt[p]{x}$, where $p = 100$ and $p = 1000$ (two area calculations)

25. The region bounded by $y = 2x^2 - 6x + 5$ and $y = 1$

26. Determine the area of the region bounded by the curves $x = y^2$ and $x = (2 - y^2)^2$ (see figure).

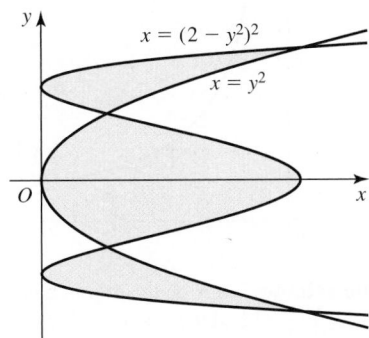

27–33. Multiple regions *The regions R_1, R_2, and R_3 (see figure) are formed by the graphs of $y = 2\sqrt{x}$, $y = 3 - x$, and $x = 3$.*

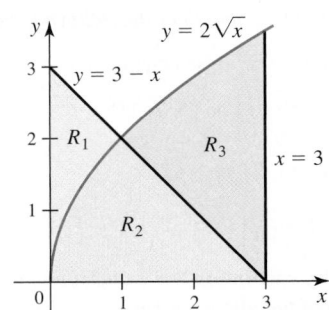

27. Find the area of each of the regions R_1, R_2, and R_3.

28. Find the volume of the solid obtained by revolving region R_1 about the x-axis.

29. Find the volume of the solid obtained by revolving region R_1 about the y-axis.

30. Find the volume of the solid obtained by revolving region R_2 about the y-axis.

31. Find the volume of the solid obtained by revolving region R_2 about the x-axis.

32. Use the disk method to find an integral, or sum of integrals, that equals the volume of the solid obtained by revolving region R_3 about the line $x = 3$. Do not evaluate the integral.

33. Use the shell method to find an integral, or sum of integrals, that equals the volume of the solid obtained by revolving region R_3 about the line $x = 3$. Do not evaluate the integral.

34. Area and volume The region R is bounded by the curves $x = y^2 + 2$, $y = x - 4$, and $y = 0$ (see figure).

   a. Write a single integral that gives the area of R.

   b. Write a single integral that gives the volume of the solid generated when R is revolved about the x-axis.

   c. Write a single integral that gives the volume of the solid generated when R is revolved about the y-axis.

   d. Suppose S is a solid whose base is R and whose cross sections perpendicular to R and parallel to the x-axis are semicircles. Write a single integral that gives the volume of S.

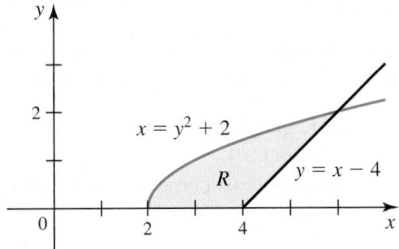

35–38. Area and volume *Let R be the region in the first quadrant bounded by the graph of $f(x) = \begin{cases} 1 & \text{if } 0 \le x \le 1 \\ 2 - \sqrt{x} & \text{if } 1 < x \le 4. \end{cases}$*

35. Find the area of the region R.

36. Find the volume of the solid generated when R is revolved about the x-axis.

37. Find the volume of the solid generated when R is revolved about the y-axis.

38. Find the area of the surface generated when the curve $y = f(x)$, for $0 \le x \le 4$, is revolved about the x-axis.

39. Find the area of the shaded regions R_1 and R_2 shown in the figure.

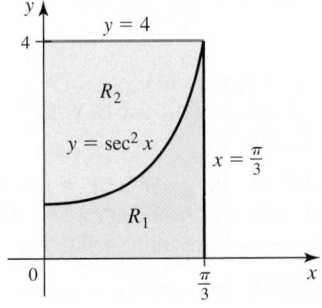

40. An area function Let $R(x)$ be the area of the shaded region between the graphs of $y = f(t)$ and $y = g(t)$ on the interval $[a, x]$ (see figure).

a. Sketch a plausible graph of R, for $a \le x \le c$.
b. Give expressions for $R(x)$ and $R'(x)$, for $a \le x \le c$.

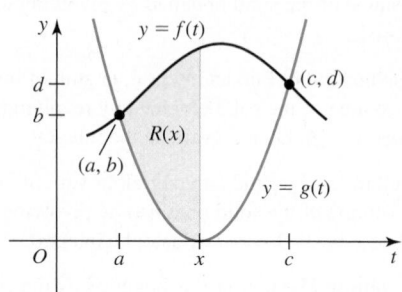

41. An area function Consider the functions $y = \dfrac{x^2}{a}$ and $y = \sqrt{\dfrac{x}{a}}$, where $a > 0$. Find $A(a)$, the area of the region between the curves.

42. Two methods The region R in the first quadrant bounded by the parabola $y = 4 - x^2$ and coordinate axes is revolved about the y-axis to produce a dome-shaped solid. Find the volume of the solid in the following ways:

a. Apply the disk method and integrate with respect to y.
b. Apply the shell method and integrate with respect to x.

43–55. Volumes of solids *Choose the general slicing method, the disk/washer method, or the shell method to answer the following questions.*

43. What is the volume of the solid whose base is the region in the first quadrant bounded by $y = \sqrt{x}, y = 2 - x$, and the x-axis, and whose cross sections perpendicular to the base and parallel to the y-axis are squares?

44. What is the volume of the solid whose base is the region in the first quadrant bounded by $y = \sqrt{x}, y = 2 - x$, and the x-axis, and whose cross sections perpendicular to the base and parallel to the y-axis are semicircles?

45. What is the volume of the solid whose base is the region in the first quadrant bounded by $y = \sqrt{x}, y = 2 - x$, and the y-axis, and whose cross sections perpendicular to the base and parallel to the x-axis are squares?

46. The region bounded by the curves $y = -x^2 + 2x + 2$ and $y = 2x^2 - 4x + 2$ is revolved about the x-axis. What is the volume of the solid that is generated?

47. The region bounded by the curve $y = 1 + \sqrt{x}$, the curve $y = 1 - \sqrt{x}$, and the line $x = 1$ is revolved about the y-axis. Find the volume of the resulting solid by (a) integrating with respect to x and (b) integrating with respect to y. Be sure your answers agree.

48. The region bounded by the curve $y = 2e^{-x}$, the curve $y = e^x$, and the y-axis is revolved about the x-axis. What is the volume of the solid that is generated?

49. The region bounded by the graphs of $x = 0, x = \sqrt{\ln y}$, and $x = \sqrt{2 - \ln y}$ in the first quadrant is revolved about the y-axis. What is the volume of the resulting solid?

50. The region bounded by the curves $y = \sec x$ and $y = 2$, for $0 \le x \le \frac{\pi}{3}$, is revolved about the x-axis. What is the volume of the solid that is generated?

51. The region bounded by $y = (1 - x^2)^{-1/2}$ and the x-axis over the interval $[0, \sqrt{3}/2]$ is revolved about the y-axis. What is the volume of the solid that is generated?

52. The region bounded by the graph of $y = 4 - x^2$ and the x-axis on the interval $[-2, 2]$ is revolved about the line $x = -2$. What is the volume of the solid that is generated?

53. The region bounded by the graphs of $y = (x - 2)^2$ and $y = 4$ is revolved about the line $y = 4$. What is the volume of the resulting solid?

54. The region bounded by the graphs of $y = 6x$ and $y = x^2 + 5$ is revolved about the line $y = -1$ and the line $x = -1$. Find the volumes of the resulting solids. Which one is greater?

55. The region bounded by the graphs of $y = 2x, y = 6 - x$, and $y = 0$ is revolved about the line $y = -2$ and the line $x = -2$. Find the volumes of the resulting solids. Which one is greater?

56. Comparing volumes Let R be the region bounded by $y = 1/x^p$ and the x-axis on the interval $[1, a]$, where $p > 0$ and $a > 1$ (see figure). Let V_x and V_y be the volumes of the solids generated when R is revolved about the x- and y-axes, respectively.

a. With $a = 2$ and $p = 1$, which is greater, V_x or V_y?
b. With $a = 4$ and $p = 3$, which is greater, V_x or V_y?
c. Find a general expression for V_x in terms of a and p. Note that $p = \frac{1}{2}$ is a special case. What is V_x when $p = \frac{1}{2}$?
d. Find a general expression for V_y in terms of a and p. Note that $p = 2$ is a special case. What is V_y when $p = 2$?
e. Explain how parts (c) and (d) demonstrate that $\lim\limits_{h \to 0} \dfrac{a^h - 1}{h} = \ln a$.
f. Find any values of a and p for which $V_x > V_y$.

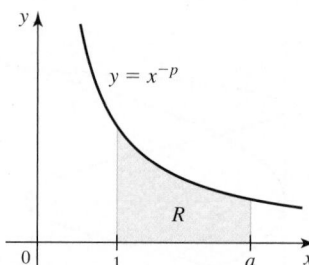

57. Comparing volumes Let R be the region bounded by the graph of $f(x) = cx(1 - x)$ and the x-axis on $[0, 1]$. Find the positive value of c such that the volume of the solid generated by revolving R about the x-axis equals the volume of the solid generated by revolving R about the y-axis.

58–61. Arc length *Find the length of the following curves.*

58. $y = 2x + 4$ on $[-2, 2]$ (Use calculus.)

59. $y = \ln(x + \sqrt{x^2 - 1})$ on $[\sqrt{2}, \sqrt{5}]$

60. $y = \dfrac{x^3}{6} + \dfrac{1}{2x}$ on $[1, 2]$

61. $y = x^{1/2} - x^{3/2}/3$ on $[1, 3]$

T 62–65. Arc length by calculator *Write and simplify the integral that gives the arc length of the following curves on the given interval. Then use technology to approximate the integral.*

62. $y = \sin \pi x$ on $[0, 1]$

63. $y = (x + 1)^2$ on $[2, 4]$

64. $y = \dfrac{x^3}{3} + \dfrac{x^2}{2}$ on $[0, 2]$

65. $y = \ln x$ between $x = 1$ and $x = b > 1$ given that

$$\int \sqrt{\frac{x^2 + a^2}{x}} \, dx = \sqrt{x^2 + a^2} - a \ln\left(\frac{a + \sqrt{x^2 + a^2}}{x}\right) + C.$$

Use any means to approximate the value of b for which the curve has length 2.

66. Surface area and volume Let $f(x) = \frac{1}{3}x^3$ and let R be the region bounded by the graph of f and the x-axis on the interval $[0, 2]$.

    a. Find the area of the surface generated when the graph of f on $[0, 2]$ is revolved about the x-axis.

    b. Find the volume of the solid generated when R is revolved about the y-axis.

    c. Find the volume of the solid generated when R is revolved about the x-axis.

67. Surface area and volume Let $f(x) = \sqrt{3x - x^2}$ and let R be the region bounded by the graph of f and the x-axis on the interval $[0, 3]$.

    a. Find the area of the surface generated when the graph of f on $[0, 3]$ is revolved about the x-axis.

    b. Find the volume of the solid generated when R is revolved about the x-axis.

68. Surface area of a cone Find the surface area of a cone (excluding the base) with radius 4 and height 8 using integration and a surface area integral.

T 69. Surface area and more Let $f(x) = \dfrac{x^4}{2} + \dfrac{1}{16x^2}$ and let R be the region bounded by the graph of f and the x-axis on the interval $[1, 2]$.

    a. Find the area of the surface generated when the graph of f on $[1, 2]$ is revolved about the x-axis. Use technology to evaluate or approximate the integral.

    b. Find the length of the curve $y = f(x)$ on $[1, 2]$.

    c. Find the volume of the solid generated when R is revolved about the y-axis.

    d. Find the volume of the solid generated when R is revolved about the x-axis. Use technology to evaluate or approximate the integral.

70–72. Variable density in one dimension *Find the mass of the following thin bars.*

70. A bar on the interval $0 \le x \le 9$ with a density (in g/m) given by $\rho(x) = 3 + 2\sqrt{x}$

71. A 3-m bar with a density (in g/m) of $\rho(x) = 150\,e^{-x/3}$, for $0 \le x \le 3$

72. A bar on the interval $0 \le x \le 6$ with a density

$$\rho(x) = \begin{cases} 1 & \text{if } 0 \le x < 2 \\ 2 & \text{if } 2 \le x < 4 \\ 4 & \text{if } 4 \le x \le 6 \end{cases}$$

73. Spring work

    a. It takes 50 J of work to stretch a spring 0.2 m from its equilibrium position. How much work is needed to stretch it an additional 0.5 m?

    b. It takes 50 N of force to stretch a spring 0.2 m from its equilibrium position. How much work is needed to stretch it an additional 0.5 m?

74. Leaky bucket A 1-kg bucket resting on the ground initially contains 6 kg of water. The bucket is lifted from the ground at a constant rate, and while it is rising, water leaks out of the bucket at a constant rate of $\frac{1}{4}$ kg/m. How much work is required to lift the bucket a vertical distance of 8 m? Assume the weight of the rope used to raise the bucket is negligible.

75. Lifting problem A 10-m, 20-kg chain hangs vertically from a cylinder attached to a winch.

    a. How much work is required to wind the entire chain onto the cylinder using the winch?

    b. How much work is required to wind the upper 4 m of the chain onto the cylinder using the winch?

76. Lifting problem A 4-kg mass is attached to the bottom of a 5-m, 15-kg chain. If the chain hangs from a platform, how much work is required to pull the chain and the mass onto the platform?

77. Pumping water A water tank has the shape of a box that is 2 m wide, 4 m long, and 6 m high.

    a. If the tank is full, how much work is required to pump the water to the level of the top of the tank?

    b. If the water in the tank is 2 m deep, how much work is required to pump the water to a level of 1 m above the top of the tank?

78. Pumping water A cylindrical water tank has a height of 6 m and a radius of 4 m. How much work is required to empty the full tank by pumping the water to an outflow pipe at the top of the tank?

79. Pumping water A water tank that is full of water has the shape of an inverted cone with a height of 6 m and a radius of 4 m (see figure). Assume the water is pumped out to the level of the top of the tank.

    a. How much work is required to pump out the water?

    b. Which requires more work, pumping out the top 3 m of water or the bottom 3 m of water?

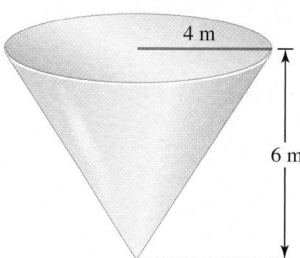

80. Pumping water A water tank that has the shape of a sphere with a radius of 2 m is half full of water. How much work is required to pump the water to the level of the top of the tank?

81. Pumping water A tank has the shape of the surface obtained by revolving the curve $y = x^2$, for $0 \le x \le 3$, about the y-axis. (Assume x and y are in meters.) If the tank is full of water, how much work is required to pump the water to an outflow pipe 1 m above the top of the tank?

82–84. Fluid Forces *Suppose the following plates are placed on a vertical wall so that the top of the plate is* 2 m *below the surface of a pool that is filled with water. Compute the force on each plate.*

82. A square plate with side length of 1 m, oriented so that the top and bottom sides of the square are horizontal

83. A plate shaped like an isosceles triangle with a height of 1 m and a base of length $\frac{1}{2}$ m (see figure)

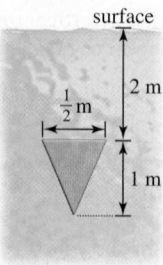

surface

$\frac{1}{2}$ m

2 m

1 m

84. A circular plate with a radius of 2 m

85. Force on a dam Find the total force on the face of a semicircular dam with a radius of 20 m when its reservoir is full of water. The diameter of the semicircle is the top of the dam.

86. Equal area property for parabolas Let $f(x) = ax^2 + bx + c$ be an arbitrary quadratic function and choose two points $x = p$ and $x = q$. Let L_1 be the line tangent to the graph of f at the point $(p, f(p))$ and let L_2 be the line tangent to the graph at the point $(q, f(q))$. Let $x = s$ be the vertical line through the intersection point of L_1 and L_2. Finally, let R_1 be the region bounded by $y = f(x)$, L_1, and the vertical line $x = s$, and let R_2 be the region bounded by $y = f(x)$, L_2, and the vertical line $x = s$. Prove that the area of R_1 equals the area of R_2.

Chapter 6 Guided Projects

Applications of the material in this chapter and related topics can be found in the following Guided Projects. For additional information, see the Preface.

- Means and tangent lines
- Landing an airliner
- Geometric probability
- Mathematics of the CD player

- Designing a water clock
- Buoyancy and Archimedes' principle
- Dipstick problems
- Inverse sine from geometry

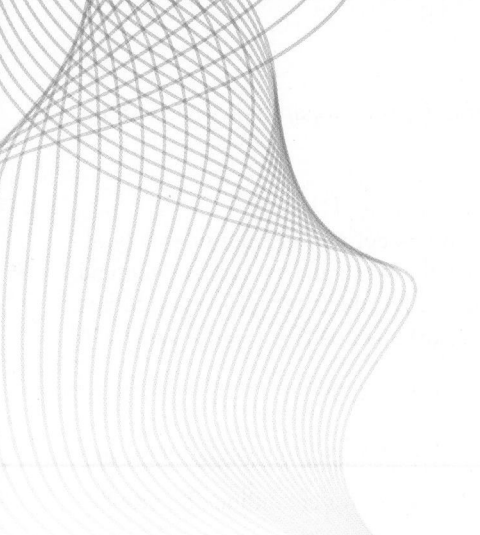

7

Logarithmic, Exponential, and Hyperbolic Functions

Chapter Preview   In this brief chapter we revisit the logarithmic and exponential functions and reexamine these functions and their properties through the lens of calculus. Although we have been working with these functions since Chapter 1, we have not yet established some of their properties in a rigorous manner. Once this task is complete, we explore a variety of applications of exponential functions. The chapter concludes with an introduction to the hyperbolic functions, a family of functions related to both exponential and trigonometric functions.

7.1 Logarithmic and Exponential Functions Revisited

Earlier in the text, we made several claims about exponential and logarithmic functions, but we did not prove them. (For example, these functions are continuous and differentiable on their domains.) Our objective in this section is to place these important functions on a solid foundation by presenting a more rigorous development of their properties.

Before embarking on this program, we offer a roadmap to help guide you through the section. We carry out the following four steps.

1. We define the natural logarithm function in terms of an integral and derive the properties of $\ln x$ directly from this new definition.

2. The natural exponential function e^x is introduced as the inverse of $\ln x$, and the properties of e^x are developed by appealing to this inverse relationship. We also present derivative and integral formulas associated with these functions.

3. Next, we define the general exponential function b^x in terms of e^x, and the general logarithmic function $\log_b x$ in terms of $\ln x$. The derivative and integral results stated in Section 3.9 follow immediately from these definitions.

4. Finally, we revisit the General Power Rule (Section 3.9) and derive a limit that can be used to approximate e.

Step 1: The Natural Logarithm

Our aim is to develop the properties of the natural logarithm using definite integrals. It all begins with the following definition.

DEFINITION **The Natural Logarithm**

The **natural logarithm** of a number $x > 0$ is $\ln x = \displaystyle\int_1^x \frac{1}{t}\, dt$.

All the familiar geometric and algebraic properties of the natural logarithmic function follow directly from this new integral definition.

Properties of the Natural Logarithm

Domain, range, and sign Because the natural logarithm is defined as a definite integral, its value is the net area under the curve $y = 1/t$ between $t = 1$ and $t = x$. The integrand is undefined at $t = 0$, so the domain of $\ln x$ is $(0, \infty)$. On the interval $(1, \infty)$, $\ln x$ is positive because the net area of the region under the curve is positive (**Figure 7.1a**). On $(0, 1)$, we have $\int_1^x \frac{1}{t}\, dt = -\int_x^1 \frac{1}{t}\, dt$, which implies $\ln x$ is negative (**Figure 7.1b**). As expected, when $x = 1$, we have $\ln 1 = \int_1^1 \frac{1}{t}\, dt = 0$. The net area interpretation of $\ln x$ also implies that the range of $\ln x$ is $(-\infty, \infty)$ (see Exercise 74 for an outline of a proof).

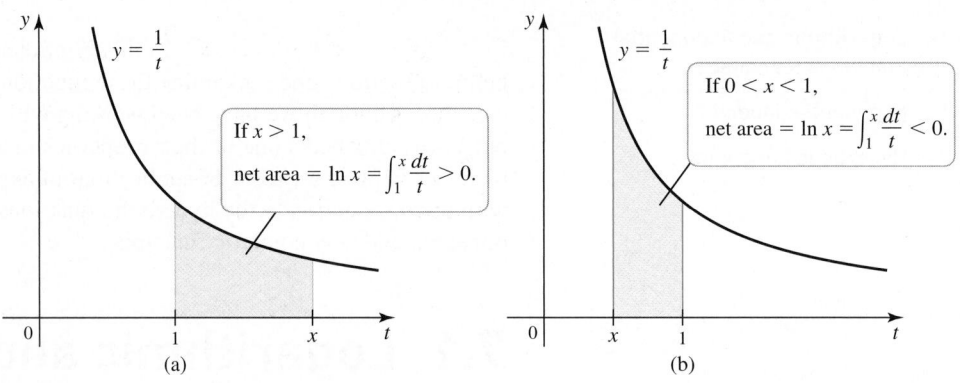

Figure 7.1

> By the Fundamental Theorem of Calculus,
>
> $$\frac{d}{dx} \int_a^x f(t)\, dt = f(x).$$

Derivative The derivative of the natural logarithm follows immediately from its definition and the Fundamental Theorem of Calculus:

$$\frac{d}{dx}(\ln x) = \frac{d}{dx} \int_1^x \frac{dt}{t} = \frac{1}{x}, \text{ for } x > 0.$$

We have two important consequences.

• Because its derivative is defined for $x > 0$, $\ln x$ is a differentiable function for $x > 0$, which means it is continuous on its domain (Theorem 3.1).

• Because $1/x > 0$, for $x > 0$, $\ln x$ is strictly increasing and one-to-one on its domain; therefore, it has a well-defined inverse.

The Chain Rule allows us to extend the derivative property to all nonzero real numbers (Exercise 72). By differentiating $\ln(-x)$, for $x < 0$, we find that

$$\frac{d}{dx}(\ln |x|) = \frac{1}{x}, \quad \text{for } x \neq 0.$$

QUICK CHECK 1 What is the domain of $\ln |x|$? ◄

More generally, by the Chain Rule,

$$\frac{d}{dx}(\ln |u(x)|) = \frac{1}{u(x)} u'(x) = \frac{u'(x)}{u(x)}.$$

Graph of $\ln x$ As noted before, $\ln x$ is continuous and strictly increasing, for $x > 0$. The second derivative, $\frac{d^2}{dx^2}(\ln x) = -\frac{1}{x^2}$, is negative, for $x > 0$, which implies the graph of $\ln x$ is concave down, for $x > 0$. As demonstrated in Exercise 74,

$$\lim_{x \to \infty} \ln x = \infty, \quad \text{and} \quad \lim_{x \to 0^+} \ln x = -\infty.$$

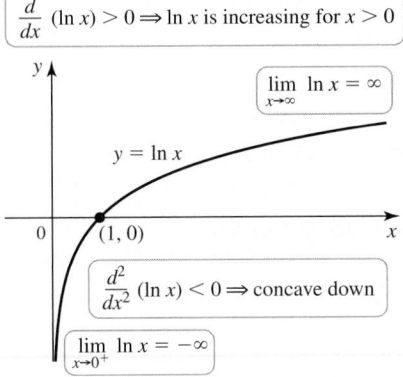

$\dfrac{d}{dx}(\ln x) > 0 \Rightarrow \ln x$ is increasing for $x > 0$

$\displaystyle\lim_{x\to\infty} \ln x = \infty$

$y = \ln x$

$(1, 0)$

$\dfrac{d^2}{dx^2}(\ln x) < 0 \Rightarrow$ concave down

$\displaystyle\lim_{x\to 0^+} \ln x = -\infty$

Figure 7.2

This information, coupled with the fact that $\ln 1 = 0$, gives the graph of $y = \ln x$ (Figure 7.2).

Logarithm of a product The familiar logarithm property

$$\ln xy = \ln x + \ln y, \qquad \text{for } x > 0, \quad y > 0,$$

may be proved using the integral definition:

$$\ln xy = \int_1^{xy} \frac{dt}{t} \qquad\qquad \text{Definition of } \ln xy$$

$$= \int_1^{x} \frac{dt}{t} + \int_x^{xy} \frac{dt}{t} \qquad \text{Additive property of integrals}$$

$$= \int_1^{x} \frac{dt}{t} + \int_1^{y} \frac{du}{u} \qquad \text{Substitute } u = t/x \text{ in second integral.}$$

$$= \ln x + \ln y. \qquad\qquad \text{Definition of } \ln$$

Logarithm of a quotient Assuming $x > 0$ and $y > 0$, the product property and a bit of algebra give

$$\ln x = \ln\left(y \cdot \frac{x}{y}\right) = \ln y + \ln \frac{x}{y}.$$

Solving for $\ln (x/y)$, we have

$$\ln \frac{x}{y} = \ln x - \ln y,$$

which is the quotient property for logarithms.

Logarithm of a power Assuming $x > 0$ and p is rational, we have

$$\ln x^p = \int_1^{x^p} \frac{dt}{t} \qquad \text{Definition of } \ln x^p$$

$$= p \int_1^{x} \frac{du}{u} \qquad \text{Let } t = u^p;\ dt = pu^{p-1}\, du.$$

$$= p \ln x. \qquad \text{By definition, } \ln x = \int_1^{x} \frac{du}{u}.$$

This argument relies on the Power Rule ($dt = pu^{p-1}\, du$), which we proved only for rational exponents. Later in this section, we prove the Power Rule for all real values of p.

Integrals Because $\dfrac{d}{dx}(\ln|x|) = \dfrac{1}{x}$, we have

$$\int \frac{1}{x}\, dx = \ln|x| + C.$$

We have shown that the familiar properties of $\ln x$ follow from its new integral definition.

THEOREM 7.1 Properties of the Natural Logarithm

1. The domain and range of $\ln x$ are $(0, \infty)$ and $(-\infty, \infty)$, respectively.

2. $\ln xy = \ln x + \ln y$, for $x > 0$ and $y > 0$

3. $\ln (x/y) = \ln x - \ln y$, for $x > 0$ and $y > 0$

4. $\ln x^p = p \ln x$, for $x > 0$ and p a rational number

5. $\dfrac{d}{dx}(\ln|x|) = \dfrac{1}{x}$, for $x \neq 0$

6. $\dfrac{d}{dx}(\ln|u(x)|) = \dfrac{u'(x)}{u(x)}$, for $u(x) \neq 0$

7. $\displaystyle\int \frac{1}{x}\, dx = \ln|x| + C$

EXAMPLE 1   Integrals with ln x   Evaluate $\int_0^4 \dfrac{x}{x^2 + 9}\,dx$.

SOLUTION

$$\int_0^4 \frac{x}{x^2 + 9}\,dx = \frac{1}{2}\int_9^{25} \frac{du}{u} \qquad \text{Let } u = x^2 + 9;\ du = 2x\,dx.$$

$$= \frac{1}{2}\ln |u|\Big|_9^{25} \qquad \text{Fundamental Theorem}$$

$$= \frac{1}{2}\left(\ln 25 - \ln 9\right) \quad \text{Evaluate.}$$

$$= \ln \frac{5}{3} \qquad \text{Properties of logarithms}$$

Related Exercises 30, 32 ◄

Step 2: The Exponential Function

We have established that $f(x) = \ln x$ is a continuous, increasing function on the interval $(0, \infty)$. Therefore, it is one-to-one and its inverse function exists on $(0, \infty)$. We denote the inverse function $f^{-1}(x) = \exp(x)$. Its graph is obtained by reflecting the graph of $f(x) = \ln x$ about the line $y = x$ (**Figure 7.3**). The domain of $\exp(x)$ is $(-\infty, \infty)$ because the range of $\ln x$ is $(-\infty, \infty)$, and the range of $\exp(x)$ is $(0, \infty)$ because the domain of $\ln x$ is $(0, \infty)$.

The usual relationships between a function and its inverse also hold:

- $y = \exp(x)$ if and only if $x = \ln y$;
- $\exp(\ln x) = x$, for $x > 0$, and $\ln(\exp(x)) = x$, for all x.

We now appeal to the properties of $\ln x$ and use the inverse relations between $\ln x$ and $\exp(x)$ to show that $\exp(x)$ satisfies the expected properties of any exponential function. For example, if $x_1 = \ln y_1$ and $x_2 = \ln y_2$, then it follows that $y_1 = \exp(x_1)$, $y_2 = \exp(x_2)$, and

$$\exp(x_1 + x_2) = \exp(\underbrace{\ln y_1 + \ln y_2}_{\ln y_1 y_2}) \quad \text{Substitute } x_1 = \ln y_1, x_2 = \ln y_2.$$

$$= \exp(\ln y_1 y_2) \qquad \text{Properties of logarithms}$$

$$= y_1 y_2 \qquad \text{Inverse property of } \exp(x) \text{ and } \ln x$$

$$= \exp(x_1)\exp(x_2). \qquad y_1 = \exp(x_1), y_2 = \exp(x_2)$$

Therefore, $\exp(x)$ satisfies the property of exponential functions $b^{x_1 + x_2} = b^{x_1} b^{x_2}$. Similar arguments show that $\exp(x)$ satisfies other characteristic properties of exponential functions (Exercise 73):

$$\exp(0) = 1,$$

$$\exp(x_1 - x_2) = \frac{\exp(x_1)}{\exp(x_2)}, \quad \text{and}$$

$$(\exp(x))^p = \exp(px), \text{ for rational numbers } p.$$

Suspecting that $\exp(x)$ is an exponential function, we proceed to identify its base. Let's consider the real number $\exp(1)$, and with a bit of forethought, call it e. The inverse relationship between $\ln x$ and $\exp(x)$ implies that

$$\text{if } e = \exp(1), \text{ then } \ln e = \ln(\exp(1)) = 1.$$

Using the fact that $\ln e = 1$ and the integral definition of $\ln x$, we now formally define e.

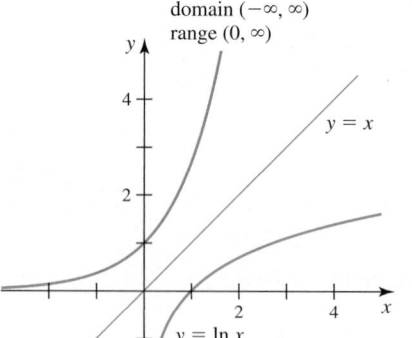

$y = \exp(x)$
domain $(-\infty, \infty)$
range $(0, \infty)$

$y = x$

$y = \ln x$
domain $(0, \infty)$
range $(-\infty, \infty)$

Figure 7.3

DEFINITION   The Number e

The number e is the real number that satisfies $\ln e = \displaystyle\int_1^e \frac{dt}{t} = 1$.

➤ We give a limit definition that provides a good approximation to e at the end of this section.

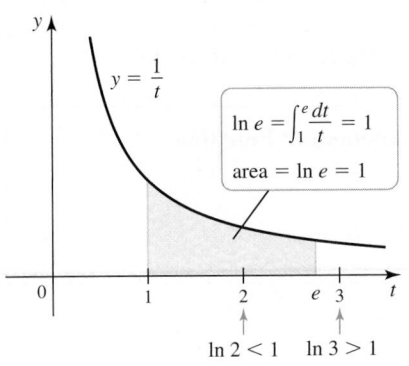

$$\ln e = \int_1^e \frac{dt}{t} = 1$$

area $= \ln e = 1$

Figure 7.4

The number e has the property that the area of the region bounded by the graph of $y = \dfrac{1}{t}$ and the t-axis on the interval $[1, e]$ is 1 (**Figure 7.4**). Note that $\ln 2 < 1$ and $\ln 3 > 1$ (Exercise 75). Because $\ln x$ is continuous on its domain, the Intermediate Value Theorem ensures that there is a number e with $2 < e < 3$ such that $\ln e = 1$.

We can now show that indeed $\exp(x)$ is the exponential function e^x. Assume p is a rational number and note that $e^p > 0$. By property 4 of Theorem 7.1 we have

$$\ln e^p = p \underbrace{\ln e}_{1} = p.$$

Using the inverse relationship between $\ln x$ and $\exp(x)$, we also know that

$$\ln \exp(p) = p.$$

Equating these two expressions for p, we conclude that $\ln e^p = \ln \exp(p)$. Because $\ln x$ is a one-to-one function, it follows that

$$e^p = \exp(p), \text{ for rational numbers } p,$$

and we conclude that $\exp(x)$ is the exponential function with base e.

We already know how to evaluate e^x when x is rational. For example, $e^3 = e \cdot e \cdot e$, $e^{-2} = \dfrac{1}{e \cdot e}$, and $e^{1/2} = \sqrt{e}$. But how do we evaluate e^x when x is irrational? We proceed as follows. The function $x = \ln y$ is defined for $y > 0$ and its range is all real numbers. Therefore, the domain of its inverse $y = \exp(x)$ is all real numbers; that is, $\exp(x)$ is defined for all real numbers. We now define e^x to be $\exp(x)$ when x is irrational.

DEFINITION The Exponential Function

For any real number x, $y = e^x = \exp(x)$, where $x = \ln y$.

We may now dispense with the notation $\exp(x)$ and use e^x as the inverse of $\ln x$. The usual inverse relationships between e^x and $\ln x$ hold, and the properties of $\exp(x)$ can now be written for e^x.

THEOREM 7.2 Properties of e^x

The exponential function e^x satisfies the following properties, all of which result from the integral definition of $\ln x$. Let x and y be real numbers.

1. $e^{x+y} = e^x e^y$
2. $e^{x-y} = e^x / e^y$
3. $(e^x)^p = e^{xp}$, where p is a rational number
4. $\ln(e^x) = x$
5. $e^{\ln x} = x$, for $x > 0$

➤ The restriction on p in property 3 will be lifted shortly.

QUICK CHECK 2 Simplify $e^{\ln 2x}$, $\ln(e^{2x})$, $e^{2 \ln x}$, and $\ln(2e^x)$. ◄

Derivatives and Integrals By Theorem 3.21 (derivatives of inverse functions), the derivative of the exponential function exists for all x. To compute $\dfrac{d}{dx}(e^x)$, we observe that $\ln(e^x) = x$ and then differentiate both sides with respect to x:

$$\frac{d}{dx}(\ln e^x) = \underbrace{\frac{d}{dx}(x)}_{1}$$

$$\frac{1}{e^x} \frac{d}{dx}(e^x) = 1 \qquad \frac{d}{dx}(\ln u(x)) = \frac{u'(x)}{u(x)} \text{ (Chain Rule)}$$

$$\frac{d}{dx}(e^x) = e^x. \qquad \text{Solve for } \frac{d}{dx}(e^x).$$

QUICK CHECK 3 What is the slope of the curve $y = e^x$ at $x = \ln 2$? What is the area of the region bounded by the graph of $y = e^x$ and the x-axis between $x = 0$ and $x = \ln 2$? ◄

Once again, we obtain the remarkable result that the exponential function is its own derivative. It follows that e^x is its own antiderivative up to a constant; that is,

$$\int e^x \, dx = e^x + C.$$

Extending these results using the Chain Rule, we have the following theorem.

THEOREM 7.3 **Derivative and Integral of the Exponential Function**
For real numbers x,

$$\frac{d}{dx}\left(e^{u(x)}\right) = e^{u(x)}u'(x) \quad \text{and} \quad \int e^x \, dx = e^x + C.$$

EXAMPLE 2 **Integrals with e^x** Evaluate $\int \dfrac{e^x}{1 + e^x} \, dx$.

SOLUTION The change of variables $u = 1 + e^x$ implies $du = e^x dx$:

$$\int \underbrace{\frac{1}{1 + e^x}}_{u} \underbrace{e^x \, dx}_{du} = \int \frac{1}{u} \, du \qquad u = 1 + e^x, du = e^x \, dx$$

$$= \ln |u| + C \qquad \text{Antiderivative of } u^{-1}$$

$$= \ln (1 + e^x) + C. \quad \text{Replace } u \text{ with } 1 + e^x.$$

Note that the absolute value may be removed from $\ln |u|$ because $1 + e^x > 0$, for all x.

Related Exercises 39–40 ◄

Step 3: General Logarithmic and Exponential Functions

We now turn to exponential and logarithmic functions with a general positive base b. The first step is to define the exponential function b^x for positive bases with $b \neq 1$ and for all real numbers x. We use property 3 of Theorem 7.2 and the fact that $b = e^{\ln b}$. If x is a rational number, then

$$b^x = \underbrace{(e^{\ln b})^x}_{b} = e^{x \ln b};$$

this important relationship expresses b^x in terms of e^x. Because e^x is defined for all real x, we use this relationship to define b^x for all real x.

DEFINITION **Exponential Functions with General Bases**
Let b be a positive real number with $b \neq 1$. Then for all real x,

$$b^x = e^{x \ln b}.$$

This definition fills the gap in property 4 of Theorem 7.1 ($\ln x^p = p \ln x$). We use the definition of b^x to write

$$x^p = e^{p \ln x}, \text{ for } x > 0 \text{ and } p \text{ real.}$$

Taking the natural logarithm of both sides and using the inverse relationship between e^x and $\ln x$, we find that

$$\ln x^p = \ln e^{p \ln x} = p \ln x, \text{ for } x > 0 \text{ and } p \text{ real.}$$

In this way, we extend property 4 of Theorem 7.1 to real powers.

Just as b^x is defined in terms of e^x, logarithms with base $b > 0$ and $b \neq 1$ may be expressed in terms of $\ln x$. All that is needed is the change of base formula (Section 1.3)

$$\log_b x = \frac{\ln x}{\ln b}.$$

► Knowing that $\ln x^p = p \ln x$ for real p, we can also extend property 3 of Theorem 7.2 to real numbers. For real x and y, we take the natural logarithm of both sides of $z = (e^x)^y$, which gives $\ln z = y \ln e^x = xy$, or $z = e^{xy}$. Therefore, $(e^x)^y = e^{xy}$.

Theorems 3.16 and 3.18 give us the derivative results for exponential and logarithmic functions with a general base $b > 0$. Extending those results with the Chain Rule, we have the following derivatives and integrals.

SUMMARY Derivatives and Integrals with Other Bases

Let $b > 0$ and $b \neq 1$. Then

$$\frac{d}{dx}\left(\log_b |u(x)|\right) = \frac{1}{\ln b}\frac{u'(x)}{u(x)}, \text{ for } u(x) \neq 0 \text{ and } \frac{d}{dx}\left(b^{u(x)}\right) = (\ln b)b^{u(x)}u'(x).$$

For $b > 0$ and $b \neq 1$, $\displaystyle\int b^x dx = \frac{1}{\ln b} b^x + C.$

QUICK CHECK 4 Verify that the derivative and integral results for a general base b reduce to the expected results when $b = e$. ◄

EXAMPLE 3 Integrals involving exponentials with other bases Evaluate the following integrals.

a. $\displaystyle\int x\, 3^{x^2}\, dx$
b. $\displaystyle\int_1^4 \frac{6^{-\sqrt{x}}}{\sqrt{x}}\, dx$

SOLUTION

a. $\displaystyle\int x\, 3^{x^2}\, dx = \frac{1}{2}\int 3^u\, du$ $u = x^2, du = 2x\, dx$

$\qquad\qquad = \frac{1}{2}\frac{1}{\ln 3} 3^u + C$ Integrate.

$\qquad\qquad = \frac{1}{2\ln 3} 3^{x^2} + C$ Substitute $u = x^2$.

b. $\displaystyle\int_1^4 \frac{6^{-\sqrt{x}}}{\sqrt{x}}\, dx = -2\int_{-1}^{-2} 6^u\, du$ $u = -\sqrt{x}, du = -\frac{1}{2\sqrt{x}}\, dx$

$\qquad\qquad = -\frac{2}{\ln 6} 6^u \Big|_{-1}^{-2}$ Fundamental Theorem

$\qquad\qquad = \frac{5}{18\ln 6}$ Simplify.

Related Exercises 44, 47 ◄

Step 4: General Power Rule

With the identity $x^p = e^{p\ln x}$, we can state and prove the final version of the Power Rule. In Chapter 3, we showed that

$$\frac{d}{dx}(x^p) = px^{p-1}$$

when p is a rational number. This result is extended to all real values of p by differentiating $x^p = e^{p\ln x}$:

$$\frac{d}{dx}(x^p) = \frac{d}{dx}\left(e^{p\ln x}\right) \quad x^p = e^{p\ln x}$$

$$= \underbrace{e^{p\ln x}}_{x^p}\frac{p}{x} \qquad \text{Chain Rule}$$

$$= x^p\frac{p}{x} \qquad e^{p\ln x} = x^p$$

$$= px^{p-1}. \qquad \text{Simplify.}$$

> **THEOREM 7.4   General Power Rule**
> For any real number p,
>
> $$\frac{d}{dx}(x^p) = px^{p-1} \quad \text{and} \quad \frac{d}{dx}(u(x)^p) = pu(x)^{p-1}u'(x).$$

EXAMPLE 4   Derivative of a tower function Evaluate the derivative of $f(x) = x^{2x}$.

SOLUTION We use the inverse relationship $e^{\ln x} = x$ to write $x^{2x} = e^{\ln(x^{2x})} = e^{2x \ln x}$. It follows that

$$\frac{d}{dx}(x^{2x}) = \frac{d}{dx}(e^{2x \ln x})$$

$$= \underbrace{e^{2x \ln x}}_{x^{2x}} \frac{d}{dx}(2x \ln x) \qquad \frac{d}{dx}(e^{u(x)}) = e^{u(x)}u'(x)$$

$$= x^{2x}\left(2 \ln x + 2x \cdot \frac{1}{x}\right) \qquad \text{Product Rule}$$

$$= 2x^{2x}(1 + \ln x). \qquad \text{Simplify.}$$

Related Exercises 13, 17 ◄

Approximating e

We have shown that the number e serves as a base for both $\ln x$ and e^x, but how do we approximate its value? Recall that the derivative of $\ln x$ at $x = 1$ is 1. By the definition of the derivative, it follows that

> ➤ Because $\frac{d}{dx}(\ln x) = \frac{1}{x}$,
>
> $\frac{d}{dx}(\ln x)\Big|_{x=1} = \frac{1}{1} = 1.$

$$1 = \frac{d}{dx}(\ln x)\Big|_{x=1} = \lim_{h \to 0} \frac{\ln(1+h) - \ln 1}{h} \qquad \text{Derivative of } \ln x \text{ at } x = 1$$

$$= \lim_{h \to 0} \frac{\ln(1+h)}{h} \qquad \ln 1 = 0$$

$$= \lim_{h \to 0} \ln(1+h)^{1/h}. \qquad p \ln x = \ln x^p$$

> ➤ Here we rely on Theorem 2.12 of Section 2.6. If f is continuous at $g(a)$ and $\lim_{x \to a} g(x)$ exists, then $\lim_{x \to a} f(g(x)) = f(\lim_{x \to a} g(x))$.

The natural logarithm is continuous for $x > 0$, so it is permissible to interchange the order of $\lim_{h \to 0}$ and the evaluation of $\ln(1+h)^{1/h}$. The result is that

$$\ln\left(\underbrace{\lim_{h \to 0}(1+h)^{1/h}}_{e}\right) = 1.$$

Observe that the limit within the brackets is e because $\ln e = 1$ and only one number satisfies this equation. Therefore, we have isolated e as a limit:

$$e = \lim_{h \to 0}(1+h)^{1/h}.$$

Table 7.1

h	$(1+h)^{1/h}$	h	$(1+h)^{1/h}$
10^{-1}	2.593742	-10^{-1}	2.867972
10^{-2}	2.704814	-10^{-2}	2.731999
10^{-3}	2.716924	-10^{-3}	2.719642
10^{-4}	2.718146	-10^{-4}	2.718418
10^{-5}	2.718268	-10^{-5}	2.718295
10^{-6}	2.718280	-10^{-6}	2.718283
10^{-7}	2.718282	-10^{-7}	2.718282

It is evident from Table 7.1 that $(1+h)^{1/h} \to 2.718282\ldots$ as $h \to 0$. The value of this limit is e, and it has been computed to millions of digits. A better approximation,

$$e \approx 2.718281828459045,$$

is obtained by methods introduced in Chapter 11.

SECTION 7.1 EXERCISES

Getting Started

1. What are the domain and range of $\ln x$?

2. Give a geometric interpretation of the function $\ln x = \int_1^x \frac{dt}{t}$.

3. Evaluate $\int 4^x \, dx$.

4. What is the inverse function of $\ln x$, and what are its domain and range?

5. Express 3^x, x^π, and $x^{\sin x}$ using the base e.

6. Evaluate $\frac{d}{dx}(3^x)$.

Practice Exercises

7–28. Derivatives *Evaluate the following derivatives.*

7. $\dfrac{d}{dx}(x \ln x^3)$

8. $\dfrac{d}{dx}(\ln(\ln x))$

9. $\dfrac{d}{dx}(\sin(\ln x))$

10. $\dfrac{d}{dx}(\ln(\cos^2 x))$

11. $\dfrac{d}{dx}((\ln 2x)^{-5})$

12. $\dfrac{d}{dx}(\ln^3(3x^2 + 2))$

13. $\dfrac{d}{dx}((2x)^{4x})$

14. $\dfrac{d}{dx}(x^\pi)$

15. $\dfrac{d}{dx}(2^{(x^2)})$

16. $\dfrac{d}{dt}((\sin t)^{\sqrt{t}})$

17. $\dfrac{d}{dx}((x + 1)^{2x})$

18. $\dfrac{d}{dx}(x^{-\ln x})$

19. $\dfrac{d}{dy}(y^{\sin y})$

20. $\dfrac{d}{dt}(t^{1/t})$

21. $\dfrac{d}{dx}(e^{-10x^2})$

22. $\dfrac{d}{dx}(x^e + e^x)$

23. $\dfrac{d}{dx}(x^{2x})$

24. $\dfrac{d}{dx}(x^{\tan x})$

25. $\dfrac{d}{dx}\left(\left(\dfrac{1}{x}\right)^x\right)$

26. $\dfrac{d}{dx}(x^{(x^{10})})$

27. $\dfrac{d}{dx}\left(1 + \dfrac{4}{x}\right)^x$

28. $\dfrac{d}{dx}(\cos(x^2 \sin x))$

29–62. Integrals *Evaluate the following integrals. Include absolute values only when needed.*

29. $\displaystyle\int_0^3 \dfrac{2x - 1}{x + 1}\,dx$

30. $\displaystyle\int \dfrac{x^2}{4x^3 + 7}\,dx$

31. $\displaystyle\int_e^{e^2} \dfrac{dx}{x \ln^3 x}$

32. $\displaystyle\int_0^{\pi/2} \dfrac{\sin x}{1 + \cos x}\,dx$

33. $\displaystyle\int \dfrac{e^{2x}}{4 + e^{2x}}\,dx$

34. $\displaystyle\int \dfrac{dx}{x \ln x \ln(\ln x)}$

35. $\displaystyle\int_{e^2}^{e^3} \dfrac{dx}{x \ln x \ln^2(\ln x)}$

36. $\displaystyle\int_0^1 \dfrac{y \ln^4(y^2 + 1)}{y^2 + 1}\,dy$

37. $\displaystyle\int_0^2 4xe^{-x^2/2}\,dx$

38. $\displaystyle\int \dfrac{e^{\sin x}}{\sec x}\,dx$

39. $\displaystyle\int \dfrac{e^{\sqrt{x}}}{\sqrt{x}}\,dx$

40. $\displaystyle\int_{-2}^2 \dfrac{e^{z/2}}{e^{z/2} + 1}\,dz$

41. $\displaystyle\int \dfrac{e^x + e^{-x}}{e^x - e^{-x}}\,dx$

42. $\displaystyle\int_{\ln 2}^{\ln 3} \dfrac{e^x + e^{-x}}{e^{2x} - 2 + e^{-2x}}\,dx$

43. $\displaystyle\int_{-1}^1 10^x\,dx$

44. $\displaystyle\int_0^{\pi/2} 4^{\sin x} \cos x\,dx$

45. $\displaystyle\int_1^2 (1 + \ln x)x^x\,dx$

46. $\displaystyle\int_{1/3}^{1/2} \dfrac{10^{1/p}}{p^2}\,dp$

47. $\displaystyle\int x^2 6^{x^3 + 8}\,dx$

48. $\displaystyle\int \dfrac{4^{\cot x}}{\sin^2 x}\,dx$

49. $\displaystyle\int xe^{3x^2 + 1}\,dx$

50. $\displaystyle\int 7^{2x}\,dx$

51. $\displaystyle\int 3^{-2x}\,dx$

52. $\displaystyle\int_0^5 5^{5x}\,dx$

53. $\displaystyle\int x^2 10^{x^3}\,dx$

54. $\displaystyle\int_0^\pi 2^{\sin x} \cos x\,dx$

55. $\displaystyle\int_1^{2e} \dfrac{3^{\ln x}}{x}\,dx$

56. $\displaystyle\int \dfrac{\sin(\ln x)}{4x}\,dx$

57. $\displaystyle\int_1^{e^2} \dfrac{(\ln x)^5}{x}\,dx$

58. $\displaystyle\int \dfrac{\ln^2 x + 2 \ln x - 1}{x}\,dx$

59. $\displaystyle\int_0^{\ln 2} \dfrac{e^{3x} - e^{-3x}}{e^{3x} + e^{-3x}}\,dx$

60. $\displaystyle\int \dfrac{e^{2x}}{(e^{2x} + 1)^2}\,dx$

61. $\displaystyle\int \dfrac{e^{5 + \sqrt{x}}}{\sqrt{x}}\,dx$

62. $\displaystyle\int_0^1 \dfrac{16^x}{4^{2x}}\,dx$

T 63–66. Calculator limits *Use a calculator to make a table similar to Table 7.1 to approximate the following limits. Confirm your result with l'Hôpital's Rule.*

63. $\displaystyle\lim_{h \to 0}(1 + 2h)^{1/h}$

64. $\displaystyle\lim_{h \to 0}(1 + 3h)^{2/h}$

65. $\displaystyle\lim_{x \to 0} \dfrac{2^x - 1}{x}$

66. $\displaystyle\lim_{x \to 0} \dfrac{\ln(1 + x)}{x}$

67. Explain why or why not Determine whether the following statements are true and give an explanation or counterexample. Assume $x > 0$ and $y > 0$.

a. $\ln xy = \ln x + \ln y$.

b. $\ln 0 = 1$.

c. $\ln(x + y) = \ln x + \ln y$.

d. $2^x = e^{2 \ln x}$.

e. The area under the curve $y = 1/x$ and the x-axis on the interval $[1, e]$ is 1.

Explorations and Challenges

68. Logarithm properties Use the integral definition of the natural logarithm to prove that $\ln(x/y) = \ln x - \ln y$.

69. Average value What is the average value of $f(x) = \dfrac{1}{x}$ on the interval $[1, p]$ for $p > 1$? What is the average value of f as $p \to \infty$?

T 70. Behavior at the origin Using calculus and accurate sketches, explain how the graphs of $f(x) = x^p \ln x$ differ as $x \to 0^+$ for $p = 1/2, 1,$ and 2.

71. Zero net area Consider the function $f(x) = \dfrac{1 - x}{x}$.

a. Are there numbers $0 < a < 1$ such that $\displaystyle\int_{1-a}^{1+a} f(x)\,dx = 0$?

b. Are there numbers $a > 1$ such that $\displaystyle\int_{1/a}^a f(x)\,dx = 0$?

72. Derivative of $\ln|x|$ Differentiate $\ln x$, for $x > 0$, and differentiate $\ln(-x)$, for $x < 0$, to conclude that $\dfrac{d}{dx}(\ln|x|) = \dfrac{1}{x}$.

73. Properties of $\exp(x)$ Use the inverse relations between $\ln x$ and $\exp(x)$, and the properties of $\ln x$, to prove the following properties.

a. $\exp(0) = 1$

b. $\exp(x - y) = \dfrac{\exp(x)}{\exp(y)}$

c. $(\exp(x))^p = \exp(px), p$ rational

74. ln x is unbounded Use the following argument to show that $\lim_{x\to\infty} \ln x = \infty$ and $\lim_{x\to 0^+} \ln x = -\infty$.

a. Make a sketch of the function $f(x) = 1/x$ on the interval $[1, 2]$. Explain why the area of the region bounded by $y = f(x)$ and the x-axis on $[1, 2]$ is ln 2.

b. Construct a rectangle over the interval $[1, 2]$ with height $1/2$. Explain why $\ln 2 > 1/2$.

c. Show that $\ln 2^n > n/2$ and $\ln 2^{-n} < -n/2$.

d. Conclude that $\lim_{x\to\infty} \ln x = \infty$ and $\lim_{x\to 0^+} \ln x = -\infty$.

75. Bounds on e Use a left Riemann sum with at least $n = 2$ subintervals of equal length to approximate $\ln 2 = \int_1^2 \frac{dt}{t}$ and show that $\ln 2 < 1$. Use a right Riemann sum with $n = 7$ subintervals of equal length to approximate $\ln 3 = \int_1^3 \frac{dt}{t}$ and show that $\ln 3 > 1$.

76. Alternative proof of product property Assume $y > 0$ is fixed and $x > 0$. Show that $\frac{d}{dx}(\ln xy) = \frac{d}{dx}(\ln x)$. Recall that if two functions have the same derivative, then they differ by an additive constant. Set $x = 1$ to evaluate the constant and prove that $\ln xy = \ln x + \ln y$.

77. Harmonic sum In Chapter 10, we will encounter the harmonic sum $1 + \frac{1}{2} + \frac{1}{3} + \cdots + \frac{1}{n}$. Use a left Riemann sum to approximate $\int_1^{n+1} \frac{dx}{x}$ (with unit spacing between the grid points) to show

that $1 + \frac{1}{2} + \frac{1}{3} + \cdots + \frac{1}{n} > \ln(n + 1)$. Use this fact to conclude that $\lim_{n\to\infty} \left(1 + \frac{1}{2} + \frac{1}{3} + \cdots + \frac{1}{n}\right)$ does not exist.

78. Probability as an integral Two points P and Q are chosen randomly, one on each of two adjacent sides of a unit square (see figure). What is the probability that the area of the triangle formed by the sides of the square and the line segment PQ is less than one-fourth the area of the square? Begin by showing that x and y must satisfy $xy < \frac{1}{2}$ in order for the area condition to be met. Then argue that the required probability is $\frac{1}{2} + \int_{1/2}^1 \frac{dx}{2x}$ and evaluate the integral.

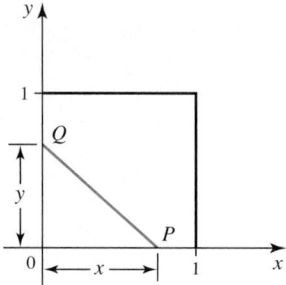

QUICK CHECK ANSWERS

1. $\{x : x \neq 0\}$ **2.** $2x, 2x, x^2, \ln 2 + x$ **3.** Slope = 2; area = 1   **4.** Note that when $b = e$, we have $\ln b = 1$. ◄

7.2 Exponential Models

The uses of exponential functions are wide-ranging. In this section, you will see them applied to problems in finance, medicine, ecology, biology, economics, pharmacokinetics, anthropology, and physics.

Exponential Growth

Imagine a quantity that grows in the following manner: The percent change in the quantity over the span of one year—or over any unit of time—remains constant. For example, suppose the value of a real estate investment increases at 7% per year, or the number of yeast cells in a batch of bread dough grows at 1.3% per minute. In both cases, we can create a simple function that models the growth of the quantity.

Let's focus on the real estate investment growing at 7% per year; we assume its value at time $t = 0$ is y_0. To compute its value one year later, we simply take 7% of y_0 and add it to y_0:

$$\text{value at } t = 1: \quad y_0 + 0.07y_0 = y_0(1 + 0.07) = y_0(1.07).$$

The previous calculation shows that multiplication of a quantity by 1.07 is equivalent to increasing the quantity by 7%. Therefore, the value of the real estate investment after two years is

$$\text{value at } t = 2: \underbrace{y_0(1.07)}_{\text{value at } t = 1}(1.07) = y_0(1.07)^2.$$

Already a pattern is emerging. To find the value of the investment at $t = 3$ years, we multiply the value of the investment at $t = 2$ years by 1.07 to arrive at

$$\text{value at } t = 3: \underbrace{y_0(1.07)^2}_{\text{value at } t = 2}(1.07) = y_0(1.07)^3.$$

Building on this pattern, we find that the value of the investment at any time t is given by the exponential function

$$y(t) = y_0(1.07)^t. \qquad (1)$$

While the exponential function in equation (1) with base 1.07 adequately describes the value of the real estate investment at any time t, it is generally easier to work with functions expressed in base e. Recall from Section 7.1 that $b^x = e^{x \ln b}$. Therefore, equation (1) can be written as

$$y(t) = y_0(1.07)^t = y_0 e^{t \ln 1.07} \approx y_0 e^{0.0677t}. \qquad \ln 1.07 \approx 0.0677$$

It is evident that we could carry out this procedure for any quantity whose percent change from one time unit to the next remains constant; the result would be

$$y(t) = y_0 e^{kt},$$

where the constant k, called the **rate constant**, is determined by the constant percent change in the quantity over one time unit. When $k > 0$, the function $y(t) = y_0 e^{kt}$ is called an *exponential growth function* (**Figure 7.5**), and it has a special property. Notice that the derivative of y is

$$y'(t) = y_0 k e^{kt} = k(\underbrace{y_0 e^{kt}}_{y(t)}); \qquad (2)$$

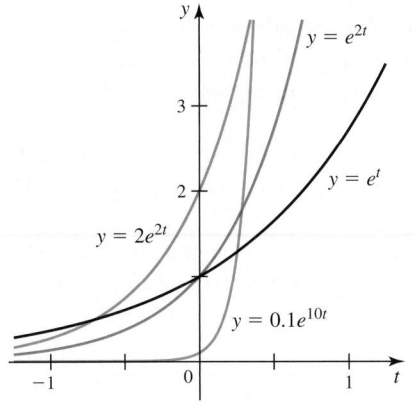

Figure 7.5

> The derivative $\dfrac{dy}{dt}$ is the *absolute* growth rate, but it is usually called the *growth rate*.

that is, $y'(t) = ky(t)$. Equation (2) says that the rate of change of an exponential growth function is proportional to its value. If $y(t)$ represents a population, then $y'(t)$ is the **(absolute) growth rate** of the population, with units such as people/month or cells/hr. Because the growth rate is proportional to the value of the function, the larger the population, the faster its growth.

Another way to talk about growth rates is to use the **relative growth rate**, which is the growth rate divided by the current value of the quantity, or $y'(t)/y(t)$. When equation (2) is written in the form $y'(t)/y(t) = k$, it has another interpretation. It says that *a quantity that grows exponentially has a constant relative growth rate*. Constant relative growth rate, or constant percent change, is the hallmark of exponential growth. Returning to equation (1), we see that a constant percent change of 7% per year is equivalent to a rate constant of $k = \ln 1.07 \approx 0.0677$. Example 1 illustrates the difference between exponential growth and linear growth.

> A consumer price index that increases at a constant rate of 4% per year increases exponentially. A currency that is devalued at a constant rate of 3% per month decreases exponentially. By contrast, linear growth is characterized by constant absolute growth rates, such as 500 people per year or $400 per month.

EXAMPLE 1 **Linear versus exponential growth** Suppose the population of the town of Pine is given by $P(t) = 1500 + 125t$, while the population of the town of Spruce is given by $S(t) = 1500e^{0.1t}$, where $t \geq 0$ is measured in years. Find the growth rate and the relative growth rate of each town.

SOLUTION Note that Pine grows according to a linear function, while Spruce grows exponentially (**Figure 7.6**). The growth rate of Pine is $P'(t) = 125$ people/year, which is constant for all times. The growth rate of Spruce is

$$S'(t) = 0.1(\underbrace{1500e^{0.1t}}_{S(t)}) = 0.1S(t),$$

showing that the growth rate is proportional to the population. The relative growth rate of Pine is $\dfrac{P'(t)}{P(t)} = \dfrac{125}{1500 + 125t}$, which decreases in time. The relative growth rate of Spruce is

$$\frac{S'(t)}{S(t)} = \frac{0.1 \cdot 1500e^{0.1t}}{1500e^{0.1t}} = 0.1,$$

which is constant for all times. In summary, the linear population function has a *constant absolute growth rate* and the exponential population function has a *constant relative growth rate*.

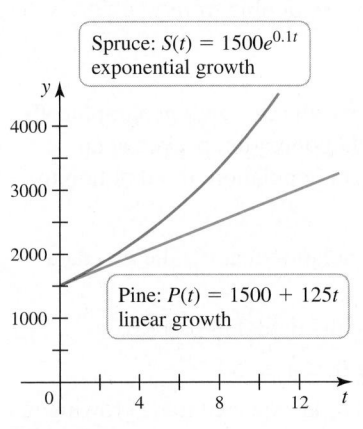

Figure 7.6

Related Exercises 13–14 ◄

QUICK CHECK 1 Population A increases at a constant rate of $4\%/\text{yr}$. Population B increases at a constant rate of 500 people/yr. Which population exhibits exponential growth? What kind of growth is exhibited by the other population? ◄

> The unit time^{-1} is read *per unit time*. For example, month^{-1} is read *per month*.

The rate constant k in $y(t) = y_0 e^{kt}$ determines the growth rate of the exponential function. For problems that involve time, the units of k are time^{-1}; for example, if t is measured in months, the units of k are month^{-1}. In this way, the exponent kt is dimensionless (without units).

Unless there is good reason to do otherwise, it is customary to take $t = 0$ as the reference point for time. Notice that with $y(t) = y_0 e^{kt}$, we have $y(0) = y_0$. Therefore, y_0 has a simple meaning: It is the *initial value* of the quantity of interest. In the examples that follow, two pieces of information are typically given: the initial value and clues for determining the rate constant k. The initial value and the rate constant determine an exponential growth function completely.

Exponential Growth Functions

Exponential growth is described by functions of the form $y(t) = y_0 e^{kt}$. The **initial value** of y at $t = 0$ is $y(0) = y_0$, and the **rate constant** $k > 0$ determines the rate of growth. Exponential growth is characterized by a constant relative growth rate.

Because exponential growth is characterized by a constant relative growth rate, the time required for a quantity to double (a 100% increase) is constant. Therefore, one way to describe an exponentially growing quantity is to give its *doubling time*. To compute the time it takes the function $y(t) = y_0 e^{kt}$ to double in value, say from y_0 to $2y_0$, we find the value of t that satisfies

$$2y_0 = y_0 e^{kt}.$$

> Note that the initial value y_0 appears on both sides of this equation. It may be canceled, meaning that the doubling time is independent of the initial condition: *The doubling time is constant for all t.*

Canceling y_0 from the equation $2y_0 = y_0 e^{kt}$ leaves the equation $2 = e^{kt}$. Taking logarithms of both sides, we have $\ln 2 = \ln e^{kt}$, or $\ln 2 = kt$, which has the solution $t = \dfrac{\ln 2}{k}$. We denote this doubling time T_2 so that $T_2 = \dfrac{\ln 2}{k}$. If y increases exponentially, the time it takes to double from 100 to 200 is the same as the time it takes to double from 1000 to 2000.

DEFINITION Doubling Time

The quantity described by the function $y(t) = y_0 e^{kt}$, for $k > 0$, has a constant **doubling time** of $T_2 = \dfrac{\ln 2}{k}$, with the same units as t.

QUICK CHECK 2 Verify that the time needed for $y(t) = y_0 e^{kt}$ to double from y_0 to $2y_0$ is the same as the time needed to double from $2y_0$ to $4y_0$. ◄

> **World population**
>
> | | |
> |---|---|
> | 1804 | 1 billion |
> | 1927 | 2 billion |
> | 1960 | 3 billion |
> | 1974 | 4 billion |
> | 1987 | 5 billion |
> | 1999 | 6 billion |
> | 2012 | 7 billion |
> | 2050 | 9 billion (proj.) |

EXAMPLE 2 World population Human population growth rates vary geographically and fluctuate over time. The overall growth rate for world population peaked at an annual rate of 2.1% per year in the 1960s. Assume a world population of 6.0 billion in 1999 ($t = 0$) and 7.4 billion in 2017 ($t = 18$).

a. Find an exponential growth function for the world population that fits the two data points.

b. Find the doubling time for the world population using the model in part (a).

c. Find the (absolute) growth rate $y'(t)$ and graph it, for $0 \le t \le 50$.

d. According to the growth model, how fast is the population expected to be growing in 2020 ($t = 21$)?

SOLUTION

a. Let $y(t)$ be world population measured in billions of people t years after 1999. We use the growth function $y(t) = y_0 e^{kt}$, where y_0 and k must be determined. The initial value is $y_0 = 6$ (billion). To determine the rate constant k, we use the fact that $y(18) = 7.4$. Substituting $t = 18$ into the growth function with $y_0 = 6$ implies

$$y(18) = 6e^{18k} = 7.4.$$

Solving for k yields the rate constant $k = \dfrac{\ln(7.4/6)}{18} \approx 0.01165 \text{ year}^{-1}$.

Therefore, the growth function is

$$y(t) = 6e^{0.01165t}.$$

b. The doubling time of the population is

$$T_2 = \frac{\ln 2}{k} \approx \frac{\ln 2}{0.01165} \approx 59.5 \text{ years}.$$

c. Working with the growth function $y(t) = 6e^{0.01165t}$, we find that

$$y'(t) = 6(0.01165)e^{0.01165t} = 0.0699e^{0.01165t},$$

which has units of *billions of people/year*. As shown in **Figure 7.7** the growth rate itself increases exponentially.

d. In 2020 ($t = 21$), the growth rate is expected to be

$$y'(21) = 0.0699e^{(0.01165)(21)} \approx 0.0893 \text{ billion people/year},$$

or roughly 89 million people/year.

Related Exercises 16–17 ◄

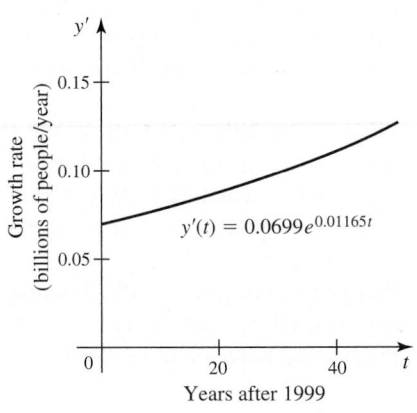

Figure 7.7

> ➤ Converted to a daily rate (dividing by 365), the world population in 2020 is expected to increase at a rate of roughly 244,600 people per day.

QUICK CHECK 3 Assume $y(t) = 100e^{0.05t}$. By (exactly) what percentage does y increase when t increases by 1 unit? ◄

A Financial Model Exponential functions are used in many financial applications, several of which are explored in the exercises. For now, consider a simple savings account in which an initial deposit earns interest that is reinvested in the account. Interest payments are made on a regular basis (for example, annually, monthly, daily), or interest may be compounded continuously. In all cases, the balance in the account increases exponentially at a rate that can be determined from the advertised **annual percentage yield** (or **APY**) of the account. Assuming no additional deposits are made, the balance in the account is given by the exponential growth function $y(t) = y_0 e^{kt}$, where y_0 is the initial deposit, t is measured in years, and k is determined by the annual percentage yield.

> ➤ The concept of continuous compounding was introduced in Exercise 85 of Section 4.7.

EXAMPLE 3 Compounding The APY of a savings account is the percentage increase in the balance over the course of a year. Suppose you deposit $500 in a savings account that has an APY of 6.18% per year. Assume the interest rate remains constant and no additional deposits or withdrawals are made. How long will it take the balance to reach $2500?

SOLUTION Because the balance grows by a fixed percentage every year, it grows exponentially. Letting $y(t)$ be the balance t years after the initial deposit of $y_0 = \$500$, we have $y(t) = y_0 e^{kt}$, where the rate constant k must be determined. Note that if the initial balance is y_0, one year later the balance is 6.18% more, or

$$y(1) = 1.0618y_0 = y_0 e^k.$$

> ➤ If the balance increases by 6.18% in one year, it increases by a factor of 1.0618 in one year.

Solving for k, we find that the rate constant is

$$k = \ln 1.0618 \approx 0.060 \text{ yr}^{-1}.$$

Therefore, the balance at any time $t \geq 0$ is $y(t) = 500e^{0.060t}$. To determine the time required for the balance to reach $2500, we solve the equation

$$y(t) = 500e^{0.060t} = 2500.$$

Dividing by 500 and taking the natural logarithm of both sides yields

$$0.060t = \ln 5.$$

The balance reaches \$2500 in $t = (\ln 5)/0.060 \approx 26.8$ yr.   *Related Exercises 18–19* ◄

Resource Consumption   Among the many resources that people use, energy is certainly one of the most important. The basic unit of energy is the **joule** (J), roughly the energy needed to lift a 0.1-kg object (say an orange) 1 m. The *rate* at which energy is consumed is called **power.** The basic unit of power is the **watt** (W), where 1 W = 1 J/s. If you turn on a 100-W lightbulb for 1 min, the bulb consumes energy at a rate of 100 J/s, and it uses a total of 100 J/s · 60 s = 6000 J of energy.

A more useful measure of energy for large quantities is the **kilowatt-hour** (kWh). A kilowatt is 1000 W, or 1000 J/s. So if you consume energy at the rate of 1 kW for 1 hr (3600 s), you use a total of 1000 J/s · 3600 s = 3.6×10^6 J, which is 1 kWh. A person running for one hour consumes roughly 1 kWh of energy. A typical house uses on the order of 1000 kWh of energy in a month.

Assume the total energy used (by a person, machine, or city) is given by the function $E(t)$. Because the power $P(t)$ is the rate at which energy is used, we have $P(t) = E'(t)$. Using the ideas of Section 6.1, the total amount of energy used between the times $t = a$ and $t = b$ is

$$\text{total energy used} = \int_a^b E'(t) \, dt = \int_a^b P(t) \, dt.$$

We see that energy is the area under the power curve. With this background, we can investigate a situation in which the rate of energy consumption increases exponentially.

EXAMPLE 4   Energy consumption   At the beginning of 2010, the rate of energy consumption for the city of Denver was 7000 megawatts (MW), where 1 MW = 10^6 W. That rate is expected to increase at an annual growth rate of 2% per year.

a. Find the function that gives the power or rate of energy consumption for all times after the beginning of 2010.

b. Find the total amount of energy used during 2018.

c. Find the function that gives the total (cumulative) amount of energy used by the city between 2010 and any time $t \geq 0$.

SOLUTION

a. Let $t \geq 0$ be the number of years after the beginning of 2010 and let $P(t)$ be the power function that gives the rate of energy consumption at time t. Because P increases at a constant rate of 2% per year, it increases exponentially. Therefore, $P(t) = P_0 e^{kt}$, where $P_0 = 7000$ MW. We determine k as before by setting $t = 1$; after one year the power is

$$P(1) = P_0 e^k = 1.02 P_0. \quad \text{Increase by 2\%, or by a factor of 1.02.}$$

Canceling P_0 and solving for k, we find that $k = \ln 1.02 \approx 0.0198$. Therefore, the power function (**Figure 7.8**) is

$$P(t) = 7000 e^{0.0198t}, \quad \text{for } t \geq 0.$$

b. The entire year 2018 corresponds to the interval $8 \leq t \leq 9$. Substituting $P(t) = 7000 e^{0.0198t}$ reveals that the total energy used in 2018 was

$$\int_8^9 P(t) \, dt = \int_8^9 7000 e^{0.0198t} \, dt \quad \text{Substitute for } P(t).$$

$$= \frac{7000}{0.0198} e^{0.0198t} \Big|_8^9 \quad \text{Fundamental Theorem}$$

$$\approx 8283. \quad \text{Evaluate.}$$

Because the units of P are MW and t is measured in years, the units of energy are MW-yr. To convert to MWh, we multiply by 8760 hr/yr to get the total energy of about 7.26×10^7 MWh (or 7.26×10^{10} kWh).

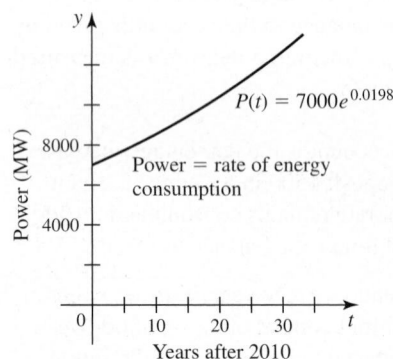

Power (MW)

8000

Power = rate of energy
consumption

4000

$P(t) = 7000 e^{0.0198t}$

0   10   20   30 t

Years after 2010

Figure 7.8

➤ It is a common mistake to assume that if the annual growth rate is 2% per year, then $k = 2\% = 0.02$ year^{-1}. The rate constant k must be calculated, as it is in Example 4a, to give $k \approx 0.0198$. For larger growth rates, the difference between k and the actual growth rate is greater.

c. The total energy used between $t = 0$ and any future time t is given by the future value formula (Section 6.1):

$$E(t) = E(0) + \int_0^t E'(s)\,ds = E(0) + \int_0^t P(s)\,ds.$$

Assuming $t = 0$ corresponds to the beginning of 2010, we take $E(0) = 0$. Substituting again for the power function P, the total energy (in MW-yr) at time t is

$$E(t) = E(0) + \int_0^t P(s)\,ds$$

$$= 0 + \int_0^t 7000 e^{0.0198s}\,ds \quad \text{Substitute for } P(s) \text{ and } E(0).$$

$$= \frac{7000}{0.0198} e^{0.0198s}\bigg|_0^t \quad \text{Fundamental Theorem}$$

$$\approx 353{,}535(e^{0.0198t} - 1). \quad \text{Evaluate.}$$

As shown in **Figure 7.9**, when the rate of energy consumption increases exponentially, the total amount of energy consumed also increases exponentially.

Related Exercises 24, 26 ◄

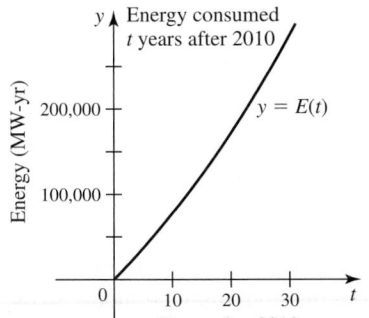

Figure 7.9

Exponential Decay

Everything you have learned about exponential growth carries over directly to exponential decay. A function that decreases exponentially has the form $y(t) = y_0 e^{kt}$, where $y_0 = y(0)$ is the initial value and $k < 0$ is the rate constant.

Exponential decay is characterized by a constant relative decay rate and by a constant *half-life*. For example, radioactive plutonium has a half-life of 24,000 years. An initial sample of 1 mg decays to 0.5 mg after 24,000 years and to 0.25 mg after 48,000 years. To compute the half-life, denoted $T_{1/2}$, we determine the time required for the quantity $y(t) = y_0 e^{kt}$ to reach one-half of its current value; that is, we solve $y_0/2 = y_0 e^{kt}$ for t. Canceling y_0 and taking logarithms of both sides, we find that

$$\frac{1}{2} = e^{kt} \quad \Rightarrow \quad \ln\frac{1}{2} = kt \quad \Rightarrow \quad t = \frac{\ln(1/2)}{k}.$$

QUICK CHECK 4 If a quantity decreases by a factor of 8 every 30 years, what is its half-life? ◄

> ► Because $\ln(1/2) < 0$ and $k < 0$, it follows that $T_{1/2} > 0$, as it should be.

Exponential Decay Functions

Exponential decay is described by functions of the form $y(t) = y_0 e^{kt}$. The initial value of y is $y(0) = y_0$, and the rate constant $k < 0$ determines the rate of decay. Exponential decay is characterized by a constant relative decay rate. The constant **half-life** is $T_{1/2} = \dfrac{\ln(1/2)}{k}$, with the same units as t.

Radiometric Dating   A powerful method for estimating the age of ancient objects (for example, fossils, bones, meteorites, and cave paintings) relies on the radioactive decay of certain elements. A common version of radiometric dating uses the carbon isotope C-14, which is present in all living matter. When a living organism dies, it ceases to replace C-14, and the C-14 that is present decays with a half-life of about $T_{1/2} = 5730$ years. Comparing the C-14 in a living organism to the amount in a dead sample provides an estimate of its age.

EXAMPLE 5   Radiometric dating Researchers determine that a fossilized bone has 30% of the C-14 of a live bone. Estimate the age of the bone. Assume a half-life for C-14 of 5730 years.

SOLUTION The exponential decay function $y(t) = y_0 e^{kt}$ represents the amount of C-14 in the bone t years after the animal died. By the half-life formula, $T_{1/2} = \ln(1/2)/k$. Substituting $T_{1/2} = 5730$ yr, the rate constant is

$$k = \frac{\ln(1/2)}{T_{1/2}} = \frac{\ln(1/2)}{5730 \text{ yr}} \approx -0.000121 \text{ yr}^{-1}$$

Assume the amount of C-14 in a living bone is y_0. Over t years, the amount of C-14 in the fossilized bone decays to 30% of its initial value, or $0.3y_0$. Using the decay function, we have

$$0.3y_0 = y_0 e^{-0.000121t}.$$

Solving for t, the age of the bone (in years) is

$$t = \frac{\ln 0.3}{-0.000121} \approx 9950.$$

Related Exercises 34–35 ◄

Pharmacokinetics   Pharmacokinetics describes the processes by which drugs are assimilated by the body. The elimination of most drugs from the body may be modeled by an exponential decay function with a known half-life (alcohol is a notable exception). The simplest models assume an entire drug dose is immediately absorbed into the blood. This assumption is a bit of an idealization; more refined mathematical models account for the absorption process.

▶ **Approximate half-lives of common drugs**

Penicillin	1 hr
Amoxicillin	1 hr
Nicotine	2 hr
Morphine	3 hr
Tetracycline	9 hr
Digitalis	33 hr
Phenobarbitol	2–6 days

EXAMPLE 6   Pharmacokinetics   An exponential decay function $y(t) = y_0 e^{kt}$ models the amount of drug in the blood t hr after an initial dose of $y_0 = 100$ mg is administered. Assume the half-life of the drug is 16 hours.

a. Find the exponential decay function that governs the amount of drug in the blood.

b. How much time is required for the drug to reach 1% of the initial dose (1 mg)?

c. If a second 100-mg dose is given 12 hr after the first dose, how much time is required for the drug level to reach 1 mg?

SOLUTION

a. Knowing that the half-life is 16 hr, the rate constant is

$$k = \frac{\ln(1/2)}{T_{1/2}} = \frac{\ln(1/2)}{16 \text{ hr}} \approx -0.0433 \text{ hr}^{-1}.$$

Therefore, the decay function is $y(t) = 100e^{-0.0433t}$.

b. The time required for the drug to reach 1 mg is the solution of

$$100e^{-0.0433t} = 1.$$

Solving for t, we have

$$t = \frac{\ln 0.01}{-0.0433 \text{ hr}^{-1}} \approx 106 \text{ hr}.$$

It takes more than 4 days for the drug to be reduced to 1% of the initial dose.

c. Using the exponential decay function of part (a), the amount of drug in the blood after 12 hr is

$$y(12) = 100e^{-0.0433 \cdot 12} \approx 59.5 \text{ mg}.$$

The second 100-mg dose given after 12 hr increases the amount of drug (assuming instantaneous absorption) to 159.5 mg. This amount becomes the new initial value for another exponential decay process (**Figure 7.10**). Measuring t from the time of the second dose, the amount of drug in the blood is

$$y(t) = 159.5e^{-0.0433t}.$$

The amount of drug reaches 1 mg when

$$y(t) = 159.5e^{-0.0433t} = 1,$$

which implies that

$$t = \frac{-\ln 159.5}{-0.0433 \text{ hr}^{-1}} = 117.1 \text{ hr}.$$

$$\ln \frac{1}{159.5} = -\ln 159.5$$

Approximately 117 hr after the second dose (or 129 hr after the first dose), the amount of drug reaches 1 mg.

Related Exercises 29, 38 ◄

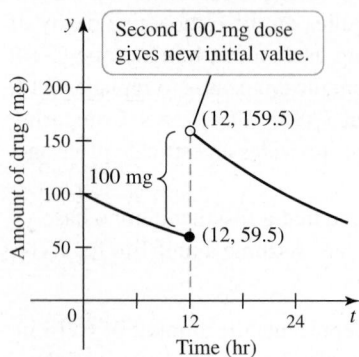

Figure 7.10

SECTION 7.2 EXERCISES

Getting Started

1. In terms of relative growth rate, what is the defining property of exponential growth?

2. Give two pieces of information that may be used to formulate an exponential growth or decay function.

3. Explain the meaning of doubling time.

4. Explain the meaning of half-life.

5. How are the rate constant and the doubling time related?

6. How are the rate constant and the half-life related?

7. Suppose a quantity described by the function $y(t) = y_0e^{kt}$, where t is measured in years, has a doubling time of 20 years. Find the rate constant k.

8. Suppose a quantity is described by the function $y(t) = 30{,}000e^{-0.05t}$, where t is measured in years. Find the half-life of the quantity.

9. Give two examples of processes that are modeled by exponential growth.

10. Give two examples of processes that are modeled by exponential decay.

11. Because of the absence of predators, the number of rabbits on a small island increases at a rate of 11% per month. If $y(t)$ equals the number of rabbits on the island t months from now, find the rate constant k for the growth function $y(t) = y_0e^{kt}$.

12. After the introduction of foxes on an island, the number of rabbits on the island decreases by 4.5% per month. If $y(t)$ equals the number of rabbits on the island t months after foxes were introduced, find the rate constant k for the exponential decay function $y(t) = y_0e^{kt}$.

Practice Exercises

13–14. Absolute and relative growth rates *Two functions f and g are given. Show that the growth rate of the linear function is constant and the relative growth rate of the exponential function is constant.*

13. $f(t) = 100 + 10.5t$, $g(t) = 100e^{t/10}$

14. $f(t) = 2200 + 400t$, $g(t) = 400 \cdot 2^{t/20}$

15–20. Designing exponential growth functions *Complete the following steps for the given situation.*

a. *Find the rate constant k and use it to devise an exponential growth function that fits the given data.*

b. *Answer the accompanying question.*

15. **Population** The population of a town with a 2016 population of 90,000 grows at a rate of 2.4%/yr. In what year will the population reach 120,000?

16. **Population** The population of Clark County, Nevada, was about 2.115 million in 2015. Assuming an annual growth rate of 1.5%/yr, what will the county population be in 2025?

17. **Population** The current population of a town is 50,000 and is growing exponentially. If the population is projected to be 55,000 in 10 years, then what will be the population 20 years from now?

18. **Savings account** An initial deposit of $1500 is placed in a savings account with an APY of 3.1%. How long will it take until the balance of the account is $2500? Assume the interest rate remains constant and no additional deposits or withdrawals are made.

19. **Rising costs** Between 2010 and 2016, the average rate of inflation was about 1.6%/yr. If a cart of groceries cost $100 in 2010, what will it cost in 2025, assuming the rate of inflation remains constant at 1.6%?

20. **Cell growth** The number of cells in a tumor doubles every 6 weeks starting with 8 cells. After how many weeks does the tumor have 1500 cells?

21. **Determining APY** Suppose $1000 is deposited in a savings account that increases exponentially. Determine the APY if the account increases to $1200 in 5 years. Assume the interest rate remains constant and no additional deposits or withdrawals are made.

22. **Tortoise growth** In a study conducted at University of New Mexico, it was found that the mass (weight) of juvenile desert tortoises exhibited exponential growth after a diet switch. One of these tortoises had a mass of about 64 g at the time of the diet switch, and 33 days later the mass was 73 g. How many days after the diet switch did the tortoise have a mass of 100 g? (*Source: Physiological and Biochemical Zoology*, 85, 1, 2012)

23. **Projection sensitivity** According to the 2014 national population projections published by the U.S. Census Bureau, the U.S. population is projected to be 334.4 million in 2020 with an estimated growth rate of 0.79%/yr.

    a. Based on these figures, find the doubling time and the projected population in 2050. Assume the growth rate remains constant.

    b. Suppose the actual growth rate is instead 0.7%. What are the resulting doubling time and projected 2050 population?

24. **Energy consumption** On the first day of the year ($t = 0$), a city uses electricity at a rate of 2000 MW. That rate is projected to increase at a rate of 1.3% per year.

    a. Based on these figures, find an exponential growth function for the power (rate of electricity use) for the city.

    b. Find the total energy (in MW-yr) used by the city over four full years beginning at $t = 0$.

    c. Find a function that gives the total energy used (in MW-yr) between $t = 0$ and any future time $t > 0$.

25. **Population of Texas** Texas was the third fastest growing state in the United States in 2016. Texas grew from 25.1 million in 2010 to 26.47 million in 2016. Use an exponential growth model to predict the population of Texas in 2025.

26. **Oil consumption** Starting in 2018 ($t = 0$), the rate at which oil is consumed by a small country increases at a rate of 1.5%/yr, starting with an initial rate of 1.2 million barrels/yr.

    a. How much oil is consumed over the course of the year 2018 (between $t = 0$ and $t = 1$)?

    b. Find the function that gives the amount of oil consumed between $t = 0$ and any future time t.

    c. How many years after 2018 will the amount of oil consumed since 2018 reach 10 million barrels?

27–30. Designing exponential decay functions *Devise an exponential decay function that fits the following data; then answer the accompanying questions. Be sure to identify the reference point (t = 0) and units of time.*

27. **Crime rate** The homicide rate decreases at a rate of 3%/yr in a city that had 800 homicides/yr in 2018. At this rate, when will the homicide rate reach 600 homicides/yr?

28. **Drug metabolism** A drug is eliminated from the body at a rate of 15%/hr. After how many hours does the amount of drug reach 10% of the initial dose?

29. **Valium metabolism** The drug Valium is eliminated from the bloodstream with a half-life of 36 hr. Suppose a patient receives an initial dose of 20 mg of Valium at midnight. How much Valium is in the patient's blood at noon the next day? When will the Valium concentration reach 10% of its initial level?

30. **China's population** China's one-child policy was implemented with a goal of reducing China's population to 700 million by 2050 (from 1.2 billion in 2000). Suppose China's population then declined at a rate of 0.5%/yr. Would this rate of decline be sufficient to meet the goal?

31. **Population of West Virginia** The population of West Virginia decreased from about 1.853 million in 2010 to 1.831 million in 2016. Use an exponential model to predict the population in 2025. Explain why an exponential (decay) model might not be an appropriate long-term model of the population of West Virginia.

32. **Depreciation of equipment** A large die-casting machine used to make automobile engine blocks is purchased for $2.5 million. For tax purposes, the value of the machine can be depreciated by 6.8% of its current value each year.

    a. What is the value of the machine after 10 years?
    b. After how many years is the value of the machine 10% of its original value?

33. **Atmospheric pressure** The pressure of Earth's atmosphere at sea level is approximately 1000 millibars and decreases exponentially with elevation. At an elevation of 30,000 ft (approximately the altitude of Mt. Everest), the pressure is one-third the sea-level pressure. At what elevation is the pressure half the sea-level pressure? At what elevation is it 1% of the sea-level pressure?

34. **Carbon dating** The half-life of C-14 is about 5730 yr.

    a. Archaeologists find a piece of cloth painted with organic dyes. Analysis of the dye in the cloth shows that only 77% of the C-14 originally in the dye remains. When was the cloth painted?
    b. A well-preserved piece of wood found at an archaeological site has 6.2% of the C-14 that it had when it was alive. Estimate when the wood was cut.

35. **Uranium dating** Uranium-238 (U-238) has a half-life of 4.5 billion years. Geologists find a rock containing a mixture of U-238 and lead, and they determine that 85% of the original U-238 remains; the other 15% has decayed into lead. How old is the rock?

36. **Radioiodine treatment** Roughly 12,000 Americans are diagnosed with thyroid cancer every year, which accounts for 1% of all cancer cases. It occurs in women three times as frequently as in men. Fortunately, thyroid cancer can be treated successfully in many cases with radioactive iodine, or I-131. This unstable form of iodine has a half-life of 8 days and is given in small doses measured in millicuries.

    a. Suppose a patient is given an initial dose of 100 millicuries. Find the function that gives the amount of I-131 in the body after $t \geq 0$ days.
    b. How long does it take the amount of I-131 to reach 10% of the initial dose?
    c. Finding the initial dose to give a particular patient is a critical calculation. How does the time to reach 10% of the initial dose change if the initial dose is increased by 5%?

37–38. Caffeine *After an individual drinks a beverage containing caffeine, the amount of caffeine in the bloodstream can be modeled by an exponential decay function, with a half-life that depends on several factors, including age and body weight. For the sake of simplicity, assume the caffeine in the following drinks immediately enters the bloodstream upon consumption.*

37. An individual consumes an energy drink that contains caffeine. If 80% of the caffeine from the energy drink is still in the bloodstream 2 hours later, find the half-life of caffeine for this individual.

38. An individual consumes two cups of coffee, each containing 90 mg of caffeine, two hours apart. Assume the half-life of caffeine for this individual is 5.7 hours.

    a. Determine the amount of caffeine in the bloodstream 1 hour after drinking the first cup of coffee.
    b. Determine the amount of caffeine in the bloodstream 1 hour after drinking the second cup of coffee.

39–40. LED lighting *LED (light-emitting diode) bulbs are rapidly decreasing in cost, and they are more energy-efficient than standard incandescent light bulbs and CFL (compact fluorescent light) bulbs. By some estimates, LED bulbs last more than 40 times longer than incandescent bulbs and more than 8 times longer than CFL bulbs. Haitz's law, which is explored in the following two exercises, predicts that over time, LED bulbs will exponentially increase in efficiency and exponentially decrease in cost.*

39. Haitz's law predicts that the cost per lumen of an LED bulb decreases by a factor of 10 every 10 years. This means that 10 years from now, the cost of an LED bulb will be 1/10 of its current cost. Predict the cost of a particular LED bulb in 2021 if it costs 4 dollars in 2018.

40. LED packages are assemblies that house LED chips. Haitz's law predicts that the amount of light produced by an LED package increases by a factor of 20 every 10 years. Compared to its current light output, how much more light will an LED package produce 3 years from now?

41. **Tumor growth** Suppose the cells of a tumor are idealized as spheres, each with a radius of 5 μm (micrometers). The number of cells has a doubling time of 35 days. Approximately how long will it take a single cell to grow into a multi-celled spherical tumor with a volume of 0.5 cm^3 (1 cm = 10,000 μm)? Assume the tumor spheres are tightly packed.

42. **Tripling time** A quantity increases according to the exponential function $y(t) = y_0 e^{kt}$. What is the tripling time for the quantity? What is the time required for the quantity to increase p-fold?

43. **Explain why or why not** Determine whether the following statements are true and give an explanation or counterexample.

    a. A quantity that increases at 6%/yr obeys the growth function $y(t) = y_0 e^{0.06t}$.
    b. If a quantity increases by 10%/yr, it increases by 30% over 3 years.

c. A quantity decreases by one-third every month. Therefore, it decreases exponentially.

d. If the rate constant of an exponential growth function is increased, its doubling time is decreased.

e. If a quantity increases exponentially, the time required to increase by a factor of 10 remains constant for all time.

Explorations and Challenges

T 44. A running model A model for the startup of a runner in a short race results in the velocity function $v(t) = a(1 - e^{-t/c})$, where a and c are positive constants, t is measured in seconds, and v has units of m/s. (*Source:* Joe Keller, *A Theory of Competitive Running, Physics Today,* 26, Sep 1973)

  a. Graph the velocity function for $a = 12$ and $c = 2$. What is the runner's maximum velocity?

  b. Using the velocity in part (a) and assuming $s(0) = 0$, find the position function $s(t)$, for $t \geq 0$.

  c. Graph the position function and estimate the time required to run 100 m.

T 45. Chemotherapy In an experimental study at Dartmouth College, mice with tumors were treated with the chemotherapeutic drug Cisplatin. Before treatment, the tumors consisted entirely of *clonogenic* cells that divide rapidly, causing the tumors to double in size every 2.9 days. Immediately after treatment, 99% of the cells in the tumor became *quiescent* cells which do not divide and lose 50% of their volume every 5.7 days. For a particular mouse, assume the tumor size is 0.5 cm³ at the time of treatment.

  a. Find an exponential decay function $V_1(t)$ that equals the total volume of the quiescent cells in the tumor t days after treatment.

  b. Find an exponential growth function $V_2(t)$ that equals the total volume of the clonogenic cells in the tumor t days after treatment.

  c. Use parts (a) and (b) to find a function $V(t)$ that equals the volume of the tumor t days after treatment.

  d. Plot a graph of $V(t)$ for $0 \leq t \leq 15$. What happens to the size of the tumor, assuming there are no follow-up treatments with Cisplatin?

  e. In cases where more than one chemotherapy treatment is required, it is often best to give a second treatment just before the tumor starts growing again. For the mice in this exercise, when should the second treatment be given? (*Source: Motivating Calculus with Biology, MAA Notes,* 81, 2013)

46. Overtaking City A has a current population of 500,000 people and grows at a rate of 3%/yr. City B has a current population of 300,000 and grows at a rate of 5%/yr.

  a. When will the cities have the same population?

  b. Suppose City C has a current population of $y_0 < 500,000$ and a growth rate of $p > 3\%$/yr. What is the relationship between y_0 and p such that Cities A and C have the same population in 10 years?

T 47. A slowing race Starting at the same time and place, Abe and Bob race, running at velocities $u(t) = \dfrac{4}{t + 1}$ mi/hr and $v(t) = 4e^{-t/2}$ mi/hr, respectively, for $t \geq 0$.

  a. Who is ahead after $t = 5$ hr? After $t = 10$ hr?

  b. Find and graph the position functions of both runners. Which runner can run only a finite distance in an unlimited amount of time?

48. Rule of 70 Bankers use the Rule of 70, which says that if an account increases at a fixed rate of $p\%$/yr, its doubling time is approximately $70/p$. Use linear approximation to explain why and when this is true.

49. Compounded inflation The U.S. government reports the rate of inflation (as measured by the consumer price index) both monthly and annually. Suppose for a particular month, the *monthly* rate of inflation is reported as 0.8%. Assuming this rate remains constant, what is the corresponding *annual* rate of inflation? Is the annual rate 12 times the monthly rate? Explain.

50. Acceleration, velocity, position Suppose the acceleration of an object moving along a line is given by $a(t) = -kv(t)$, where k is a positive constant and v is the object's velocity. Assume the initial velocity and position are given by $v(0) = 10$ and $s(0) = 0$, respectively.

  a. Use $a(t) = v'(t)$ to find the velocity of the object as a function of time.

  b. Use $v(t) = s'(t)$ to find the position of the object as a function of time.

  c. Use the fact that $\dfrac{dv}{dt} = \dfrac{dv}{ds}\dfrac{ds}{dt}$ (by the Chain Rule) to find the velocity as a function of position.

T 51. Air resistance (adapted from Putnam Exam, 1939) An object moves freely in a straight line, acted on by air resistance, which is proportional to its speed; this means its acceleration is $a(t) = -kv(t)$. The velocity of the object decreases from 1000 ft/s to 900 ft/s over a distance of 1200 ft. Approximate the time required for this deceleration to occur. (Exercise 50 may be useful.)

52. General relative growth rates Define the relative growth rate of the function f over the time interval T to be the relative change in f over an interval of length T:

$$R_T = \frac{f(t + T) - f(t)}{f(t)}.$$

Show that for the exponential function $y(t) = y_0 e^{kt}$, the relative growth rate R_T, for fixed T, is constant for all t.

53. Equivalent growth functions The same exponential growth function can be written in the forms $y(t) = y_0 e^{kt}$, $y(t) = y_0(1 + r)^t$, and $y(t) = y_0 2^{t/T_2}$. Write k as a function of r, r as a function of T_2, and T_2 as a function of k.

54. Geometric means A quantity grows exponentially according to $y(t) = y_0 e^{kt}$. What is the relationship among m, n, and p such that $y(p) = \sqrt{y(m)y(n)}$?

55. Constant doubling time Prove that the doubling time for an exponentially increasing quantity is constant for all time.

QUICK CHECK ANSWERS

1. Population A grows exponentially; population B grows linearly. **3.** The function $100e^{0.05t}$ increases by a factor of 1.0513, or by 5.13%, in 1 unit of time. **4.** 10 years ◄

7.3 Hyperbolic Functions

In this section, we introduce a new family of functions called the *hyperbolic* functions, which are closely related to both trigonometric functions and exponential functions. Hyperbolic functions find widespread use in fluid dynamics, projectile motion, architecture, and electrical engineering, to name just a few areas. Hyperbolic functions are also important in the development of many theoretical results in mathematics.

Relationship Between Trigonometric and Hyperbolic Functions

The trigonometric functions defined in Chapter 1 are based on relationships involving a circle—for this reason, trigonometric functions are also known as *circular* functions. Specifically, $\cos t$ and $\sin t$ are equal to the x- and y-coordinates, respectively, of the point $P(x, y)$ on the unit circle that corresponds to an angle of t radians (**Figure 7.11**). We can also regard t as the length of the arc from $(1, 0)$ to the point $P(x, y)$.

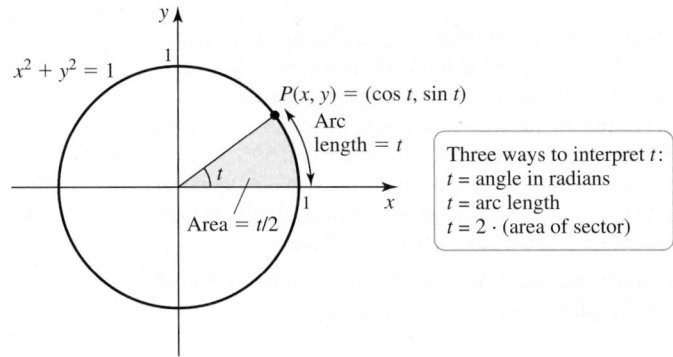

Figure 7.11

> Recall that the area of a circular sector of radius r and angle θ is $A = \frac{1}{2}r^2\theta$. With $r = 1$ and $\theta = t$, we have $A = \frac{1}{2}t$, which implies $t = 2A$.

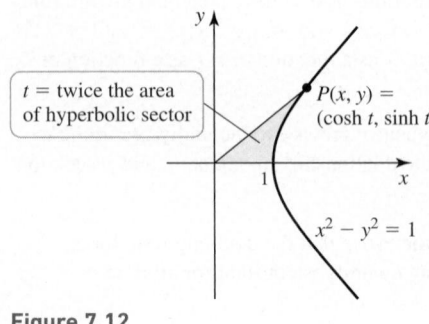

Figure 7.12

There is yet another way to interpret the number t, and it is this third interpretation that links the trigonometric and hyperbolic functions. Observe that t is twice the area of the circular sector in Figure 7.11. The functions $\cos t$ and $\sin t$ are still defined as the x- and y-coordinates of the point P, but now we associate P with a sector whose area is one-half of t.

The *hyperbolic cosine* and *hyperbolic sine* are defined in an analogous fashion using the hyperbola $x^2 - y^2 = 1$ instead of the circle $x^2 + y^2 = 1$. Consider the region bounded by the x-axis, the right branch of the unit hyperbola $x^2 - y^2 = 1$, and a line segment from the origin to a point $P(x, y)$ on the hyperbola (**Figure 7.12**); let t equal twice the area of this region.

The hyperbolic cosine of t, denoted $\cosh t$, is the x-coordinate of P, and the hyperbolic sine of t, denoted $\sinh t$, is the y-coordinate of P. Expressing x and y in terms of t leads to the standard definitions of the hyperbolic functions. We accomplish this task by writing t, which is an area, as an integral that depends on the coordinates of P. In Exercise 112, we ask you to carry out the calculations to show that

$$x = \cosh t = \frac{e^t + e^{-t}}{2} \quad \text{and} \quad y = \sinh t = \frac{e^t - e^{-t}}{2}.$$

Everything that follows in this section is based on these two definitions.

Definitions, Identities, and Graphs of the Hyperbolic Functions

Once the hyperbolic cosine and hyperbolic sine are defined, the four remaining hyperbolic functions follow in a manner analogous to the trigonometric functions.

> There is no universally accepted pronunciation of the names of the hyperbolic functions. In the United States, *cohsh x* (long *oh* sound) and *sinch x* are common choices for cosh *x* and sinh *x*. The pronunciations *tanch x*, *cotanch x*, *seech x* or *sech x*, and *coseech x* or *cosech x* are used for the other functions. International pronunciations vary as well.

DEFINITION Hyperbolic Functions

Hyperbolic cosine

$$\cosh x = \frac{e^x + e^{-x}}{2}$$

Hyperbolic sine

$$\sinh x = \frac{e^x - e^{-x}}{2}$$

Hyperbolic tangent

$$\tanh x = \frac{\sinh x}{\cosh x} = \frac{e^x - e^{-x}}{e^x + e^{-x}}$$

Hyperbolic cotangent

$$\coth x = \frac{\cosh x}{\sinh x} = \frac{e^x + e^{-x}}{e^x - e^{-x}}$$

Hyperbolic secant

$$\operatorname{sech} x = \frac{1}{\cosh x} = \frac{2}{e^x + e^{-x}}$$

Hyperbolic cosecant

$$\operatorname{csch} x = \frac{1}{\sinh x} = \frac{2}{e^x - e^{-x}}$$

The hyperbolic functions satisfy many important identities. Let's begin with the fundamental identity for hyperbolic functions, which is analogous to the familiar trigonometric identity $\cos^2 x + \sin^2 x = 1$:

$$\cosh^2 x - \sinh^2 x = 1.$$

This identity is verified by appealing to the definitions:

> The fundamental identity for hyperbolic functions can also be understood in terms of the geometric definition of the hyperbolic functions. Because the point $P(\cosh t, \sinh t)$ is on the hyperbola $x^2 - y^2 = 1$, the coordinates of P satisfy the equation of the hyperbola, which leads immediately to
>
> $$\cosh^2 t - \sinh^2 t = 1.$$

$$\cosh^2 x - \sinh^2 x = \left(\frac{e^x + e^{-x}}{2}\right)^2 - \left(\frac{e^x - e^{-x}}{2}\right)^2 \quad \text{Definition of } \cosh x \text{ and } \sinh x$$

$$= \frac{e^{2x} + 2 + e^{-2x} - (e^{2x} - 2 + e^{-2x})}{4} \quad \text{Expand and combine fractions.}$$

$$= \frac{4}{4} = 1. \quad \text{Simplify.}$$

EXAMPLE 1 Deriving hyperbolic identities

a. Use the fundamental identity $\cosh^2 x - \sinh^2 x = 1$ to prove that $1 - \tanh^2 x = \operatorname{sech}^2 x$.

b. Derive the identity $\sinh 2x = 2\sinh x \cosh x$.

SOLUTION

a. Dividing both sides of the fundamental identity $\cosh^2 x - \sinh^2 x = 1$ by $\cosh^2 x$ leads to the desired result:

$$\cosh^2 x - \sinh^2 x = 1 \quad \text{Fundamental identity}$$

$$\underbrace{\frac{\cosh^2 x}{\cosh^2 x}}_{} - \underbrace{\frac{\sinh^2 x}{\cosh^2 x}}_{\tanh^2 x} = \underbrace{\frac{1}{\cosh^2 x}}_{\operatorname{sech}^2 x} \quad \text{Divide both sides by } \cosh^2 x.$$

$$1 - \tanh^2 x = \operatorname{sech}^2 x. \quad \text{Identify functions.}$$

b. Using the definition of the hyperbolic sine, we have

$$\sinh 2x = \frac{e^{2x} - e^{-2x}}{2} \quad \text{Definition of sinh}$$

$$= \frac{(e^x - e^{-x})(e^x + e^{-x})}{2} \quad \text{Factor; difference of perfect squares}$$

$$= 2\sinh x \cosh x. \quad \text{Identify functions.}$$

Related Exercises 16–18 ◄

The identities in Example 1 are just two of many useful hyperbolic identities, some of which we list next.

> **Hyperbolic Identities**
>
> $\cosh^2 x - \sinh^2 x = 1$ $\qquad$ $\cosh(-x) = \cosh x$
>
> $1 - \tanh^2 x = \operatorname{sech}^2 x$ $\qquad$ $\sinh(-x) = -\sinh x$
>
> $\coth^2 x - 1 = \operatorname{csch}^2 x$ $\qquad$ $\tanh(-x) = -\tanh x$
>
> $\cosh(x + y) = \cosh x \cosh y + \sinh x \sinh y$
>
> $\sinh(x + y) = \sinh x \cosh y + \cosh x \sinh y$
>
> $\cosh 2x = \cosh^2 x + \sinh^2 x$ $\qquad$ $\sinh 2x = 2 \sinh x \cosh x$
>
> $\cosh^2 x = \dfrac{\cosh 2x + 1}{2}$ $\qquad$ $\sinh^2 x = \dfrac{\cosh 2x - 1}{2}$

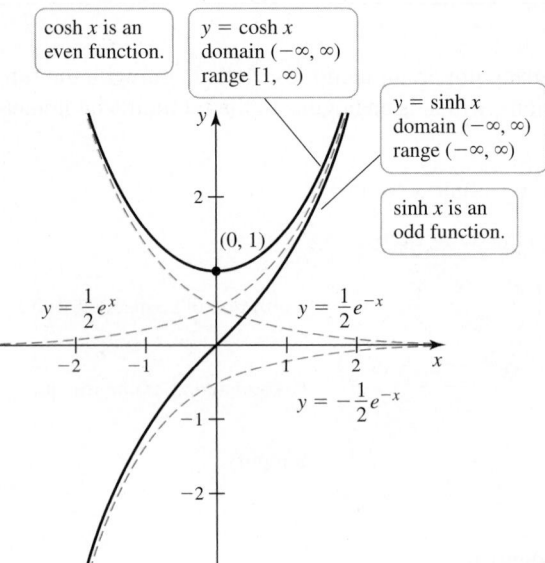

> cosh x is an even function.

> $y = \cosh x$
> domain $(-\infty, \infty)$
> range $[1, \infty)$

> $y = \sinh x$
> domain $(-\infty, \infty)$
> range $(-\infty, \infty)$

> sinh x is an odd function.

$y = \dfrac{1}{2}e^x$

$y = \dfrac{1}{2}e^{-x}$

$y = -\dfrac{1}{2}e^{-x}$

Figure 7.13

Graphs of the hyperbolic functions are relatively easy to produce because they are based on the familiar graphs of e^x and e^{-x}. Recall that $\lim\limits_{x \to \infty} e^{-x} = 0$ and that $\lim\limits_{x \to -\infty} e^x = 0$. With these facts in mind, we see that the graph of $\cosh x$ (**Figure 7.13**) approaches the graph of $y = \dfrac{1}{2}e^x$ as $x \to \infty$ because $\cosh x = \dfrac{e^x + e^{-x}}{2} \approx \dfrac{e^x}{2}$ for large values of x. A similar argument shows that as $x \to -\infty$, $\cosh x$ approaches $y = \dfrac{1}{2}e^{-x}$. Note also that $\cosh x$ is an even function:

$$\cosh(-x) = \frac{e^{-x} + e^{-(-x)}}{2} = \frac{e^x + e^{-x}}{2} = \cosh x.$$

Finally, $\cosh 0 = \dfrac{e^0 + e^0}{2} = 1$, so its y-intercept is $(0, 1)$. The behavior of $\sinh x$, an odd function also shown in Figure 7.13, can be explained in much the same way.

QUICK CHECK 1 Use the definition of the hyperbolic sine to show that $\sinh x$ is an odd function. ◄

The graphs of the other four hyperbolic functions are shown in **Figure 7.14**. As a consequence of their definitions, we see that the domain of $\cosh x$, $\sinh x$, $\tanh x$, and $\operatorname{sech} x$ is $(-\infty, \infty)$, whereas the domain of $\coth x$ and $\operatorname{csch} x$ is the set of all real numbers excluding 0.

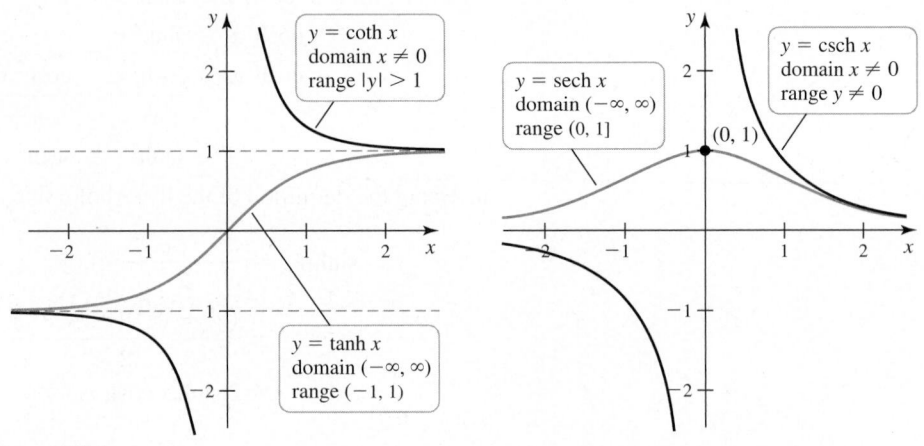

> $y = \coth x$
> domain $x \neq 0$
> range $|y| > 1$

> $y = \operatorname{sech} x$
> domain $(-\infty, \infty)$
> range $(0, 1]$

> $y = \operatorname{csch} x$
> domain $x \neq 0$
> range $y \neq 0$

> $y = \tanh x$
> domain $(-\infty, \infty)$
> range $(-1, 1)$

Figure 7.14

QUICK CHECK 2 Explain why the graph of $\tanh x$ has the horizontal asymptotes $y = 1$ and $y = -1$. ◄

Derivatives and Integrals of Hyperbolic Functions

Because the hyperbolic functions are defined in terms of e^x and e^{-x}, computing their derivatives is straightforward. The derivatives of the hyperbolic functions are given in Theorem 7.5—reversing these formulas produces corresponding integral formulas.

> ▶ The identities, derivative formulas, and integral formulas for the hyperbolic functions are similar to the corresponding formulas for the trigonometric functions, which makes them easy to remember. However, be aware of some subtle differences in the signs associated with these formulas. For instance,
>
> $$d/dx(\cos x) = -\sin x,$$
>
> whereas
>
> $$d/dx(\cosh x) = \sinh x.$$

THEOREM 7.5   Derivative and Integral Formulas

1. $\dfrac{d}{dx}(\cosh x) = \sinh x \quad \Rightarrow \quad \displaystyle\int \sinh x \, dx = \cosh x + C$

2. $\dfrac{d}{dx}(\sinh x) = \cosh x \quad \Rightarrow \quad \displaystyle\int \cosh x \, dx = \sinh x + C$

3. $\dfrac{d}{dx}(\tanh x) = \operatorname{sech}^2 x \quad \Rightarrow \quad \displaystyle\int \operatorname{sech}^2 x \, dx = \tanh x + C$

4. $\dfrac{d}{dx}(\coth x) = -\operatorname{csch}^2 x \quad \Rightarrow \quad \displaystyle\int \operatorname{csch}^2 x \, dx = -\coth x + C$

5. $\dfrac{d}{dx}(\operatorname{sech} x) = -\operatorname{sech} x \tanh x \quad \Rightarrow \quad \displaystyle\int \operatorname{sech} x \tanh x \, dx = -\operatorname{sech} x + C$

6. $\dfrac{d}{dx}(\operatorname{csch} x) = -\operatorname{csch} x \coth x \quad \Rightarrow \quad \displaystyle\int \operatorname{csch} x \coth x \, dx = -\operatorname{csch} x + C$

Proof: Using the definitions of $\cosh x$ and $\sinh x$, we have

$$\frac{d}{dx}(\cosh x) = \frac{d}{dx}\left(\frac{e^x + e^{-x}}{2}\right) = \frac{e^x - e^{-x}}{2} = \sinh x \quad \text{and}$$

$$\frac{d}{dx}(\sinh x) = \frac{d}{dx}\left(\frac{e^x - e^{-x}}{2}\right) = \frac{e^x + e^{-x}}{2} = \cosh x.$$

To prove formula (3), we begin with $\tanh x = \sinh x / \cosh x$ and then apply the Quotient Rule:

$$\frac{d}{dx}(\tanh x) = \frac{d}{dx}\left(\frac{\sinh x}{\cosh x}\right) \qquad \text{Definition of } \tanh x$$

$$= \frac{(\cosh x)\cosh x - (\sinh x)\sinh x}{\cosh^2 x} \qquad \text{Quotient Rule}$$

$$= \frac{1}{\cosh^2 x} \qquad \cosh^2 x - \sinh^2 x = 1$$

$$= \operatorname{sech}^2 x. \qquad \operatorname{sech} x = 1/\cosh x$$

The proofs of the remaining derivative formulas are assigned in Exercises 19–21. The integral formulas are a direct consequence of their corresponding derivative formulas. ◀

EXAMPLE 2   Derivatives and integrals of hyperbolic functions Evaluate the following derivatives and integrals.

a. $\dfrac{d}{dx}(\operatorname{sech} 3x)$

b. $\dfrac{d^2}{dx^2}(\operatorname{sech} 3x)$

c. $\displaystyle\int \frac{\operatorname{csch}^2 \sqrt{x}}{\sqrt{x}} \, dx$

d. $\displaystyle\int_0^{\ln 3} \sinh^3 x \cosh x \, dx$

SOLUTION

a. Combining formula (5) of Theorem 7.5 with the Chain Rule gives

$$\frac{d}{dx}(\text{sech } 3x) = -3 \text{ sech } 3x \tanh 3x.$$

b. Applying the Product Rule and Chain Rule to the result of part (a), we have

$$\frac{d^2}{dx^2}(\text{sech } 3x) = \frac{d}{dx}(-3 \text{ sech } 3x \tanh 3x)$$

$$= \underbrace{\frac{d}{dx}(-3 \text{ sech } 3x)}_{9 \text{ sech } 3x \tanh 3x} \cdot \tanh 3x + (-3 \text{ sech } 3x) \cdot \underbrace{\frac{d}{dx}(\tanh 3x)}_{3 \text{ sech}^2 3x} \qquad \text{Product Rule}$$

$$= 9 \text{ sech } 3x \tanh^2 3x - 9 \text{ sech}^3 3x \qquad\qquad\qquad\qquad \text{Chain Rule}$$

$$= 9 \text{ sech } 3x(\tanh^2 3x - \text{sech}^2 3x). \qquad\qquad\qquad\qquad \text{Simplify.}$$

c. The integrand suggests the substitution $u = \sqrt{x}$:

$$\int \frac{\text{csch}^2 \sqrt{x}}{\sqrt{x}} dx = 2 \int \text{csch}^2 u \, du \qquad \text{Let } u = \sqrt{x}; du = \frac{1}{2\sqrt{x}} dx.$$

$$= -2 \coth u + C \qquad \text{Formula (4), Theorem 7.5}$$

$$= -2 \coth \sqrt{x} + C. \quad u = \sqrt{x}$$

d. The derivative formula $\dfrac{d}{dx}(\sinh x) = \cosh x$ suggests the substitution $u = \sinh x$:

$$\int_0^{\ln 3} \sinh^3 x \cosh x \, dx = \int_0^{4/3} u^3 \, du. \quad \text{Let } u = \sinh x; du = \cosh x \, dx.$$

The new limits of integration are determined by the calculations

$$x = 0 \quad \Rightarrow \quad u = \sinh 0 = 0 \quad \text{and}$$

$$x = \ln 3 \quad \Rightarrow \quad u = \sinh(\ln 3) = \frac{e^{\ln 3} - e^{-\ln 3}}{2} = \frac{3 - 1/3}{2} = \frac{4}{3}.$$

We now evaluate the integral in the variable u:

$$\int_0^{4/3} u^3 \, du = \frac{1}{4} u^4 \Big|_0^{4/3}$$

$$= \frac{1}{4}\left(\left(\frac{4}{3}\right)^4 - 0^4\right) = \frac{64}{81}.$$

Related Exercises 22, 28, 43, 44 ◄

QUICK CHECK 3 Find both the derivative and indefinite integral of $f(x) = 4 \cosh 2x$. ◄

Theorem 7.6 presents integral formulas for the four hyperbolic functions not covered in Theorem 7.5.

THEOREM 7.6   Integrals of Hyperbolic Functions

1. $\displaystyle\int \tanh x \, dx = \ln \cosh x + C$ **2.** $\displaystyle\int \coth x \, dx = \ln|\sinh x| + C$

3. $\displaystyle\int \text{sech } x \, dx = \tan^{-1}(\sinh x) + C$ **4.** $\displaystyle\int \text{csch } x \, dx = \ln|\tanh(x/2)| + C$

Proof: Formula (1) is derived by first writing $\tanh x$ in terms of $\sinh x$ and $\cosh x$:

$$\int \tanh x \, dx = \int \frac{\sinh x}{\cosh x} \, dx \qquad \text{Definition of } \tanh x$$

$$= \int \frac{1}{u} \, du \qquad \text{Let } u = \cosh x; \, du = \sinh x \, dx.$$

$$= \ln|u| + C \qquad \text{Evaluate integral.}$$

$$= \ln \cosh x + C. \qquad u = \cosh x > 0$$

Formula (2) is derived in a similar fashion (Exercise 60). The more challenging proofs of formulas (3) and (4) are considered in Exercises 107 and 108. ◄

EXAMPLE 3 **Integrals involving hyperbolic functions** Determine the indefinite integral $\int x \coth x^2 \, dx$.

SOLUTION The integrand suggests the substitution $u = x^2$:

$$\int x \coth x^2 \, dx = \frac{1}{2} \int \coth u \, du \qquad \text{Let } u = x^2; \, du = 2x \, dx.$$

$$= \frac{1}{2} \ln|\sinh u| + C \qquad \text{Evaluate integral; use Theorem 7.6.}$$

$$= \frac{1}{2} \ln(\sinh x^2) + C. \qquad u = x^2; \sinh x^2 \ge 0$$

Related Exercises 45, 58 ◄

QUICK CHECK 4 Determine the indefinite integral $\int \text{csch } 2x \, dx$. ◄

Inverse Hyperbolic Functions

At present, we don't have the tools for evaluating an integral such as $\int \dfrac{dx}{\sqrt{x^2 + 4}}$. By studying inverse hyperbolic functions, we can discover new integration formulas. Inverse hyperbolic functions are also useful for solving equations involving hyperbolic functions.

Figures 7.13 and 7.14 show that the functions $\sinh x$, $\tanh x$, $\coth x$, and $\text{csch } x$ are all one-to-one on their respective domains. This observation implies that each of these functions has a well-defined inverse. However, the function $y = \cosh x$ is not one-to-one on $(-\infty, \infty)$, so its inverse, denoted $y = \cosh^{-1} x$, exists only if we restrict the domain of $\cosh x$. Specifically, when $y = \cosh x$ is restricted to the interval $[0, \infty)$, it is one-to-one, and its inverse is defined as follows:

$$y = \cosh^{-1} x \quad \text{if and only if} \quad x = \cosh y, \text{ for } x \ge 1 \text{ and } 0 \le y < \infty.$$

Figure 7.15a shows the graph of $y = \cosh^{-1} x$, obtained by reflecting the graph of $y = \cosh x$ on $[0, \infty)$ over the line $y = x$. The definitions and graphs of the other five inverse hyperbolic functions are also shown in Figure 7.15. Notice that the domain of $y = \text{sech } x$ (Figure 7.15d) must be restricted to $[0, \infty)$ to ensure the existence of its inverse.

Because hyperbolic functions are defined in terms of exponential functions, we can find explicit formulas for their inverses in terms of logarithms. For example, let's start with the definition of the inverse hyperbolic sine. For all real x and y, we have

$$y = \sinh^{-1} x \quad \Longleftrightarrow \quad x = \sinh y.$$

Following the procedure outlined in Section 1.3, we solve

$$x = \sinh y = \frac{e^y - e^{-y}}{2}$$

for y to give a formula for $\sinh^{-1} x$:

$$x = \frac{e^y - e^{-y}}{2} \quad \Longrightarrow \quad e^y - 2x - e^{-y} = 0 \qquad \text{Rearrange equation.}$$

$$\Longrightarrow \quad (e^y)^2 - 2xe^y - 1 = 0. \qquad \text{Multiply by } e^y.$$

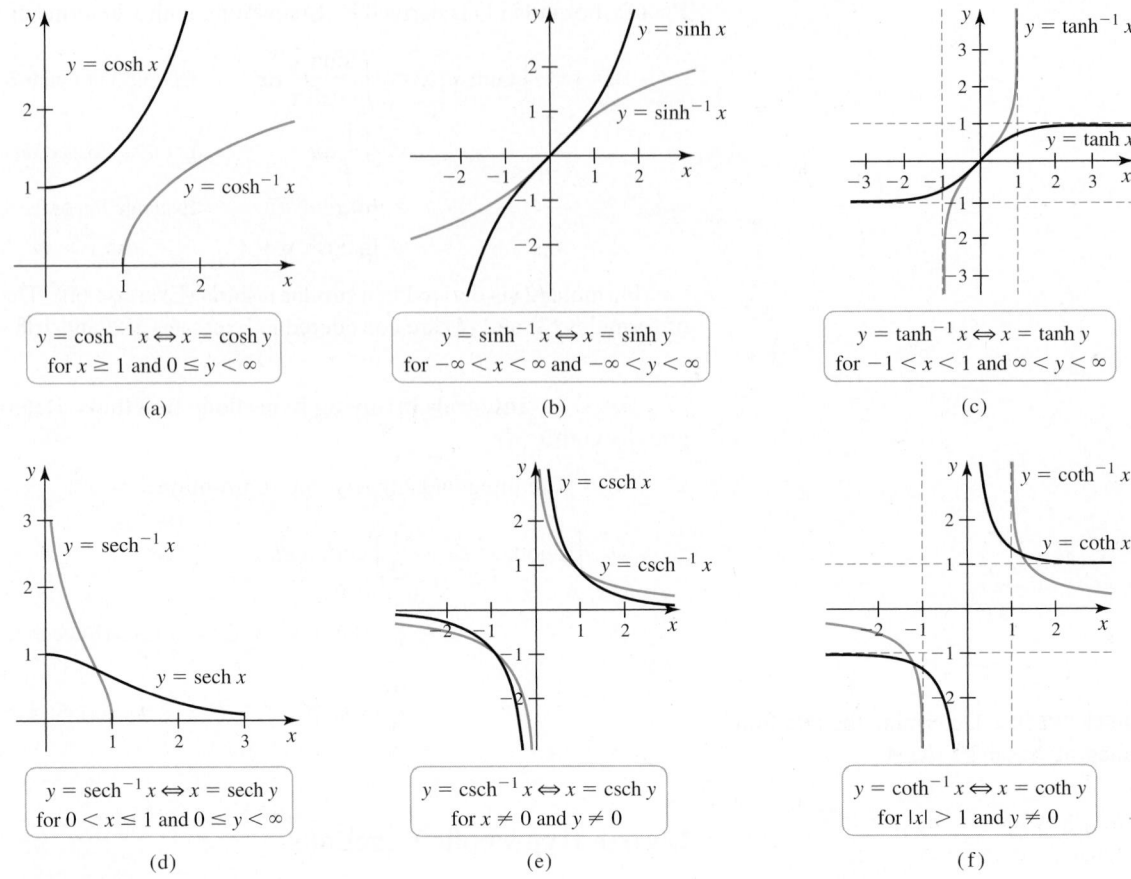

Figure 7.15

At this stage, we recognize a quadratic equation in e^y and solve for e^y using the quadratic formula, with $a = 1$, $b = -2x$, and $c = -1$:

$$e^y = \frac{2x \pm \sqrt{4x^2 + 4}}{2} = x \pm \sqrt{x^2 + 1} = \underbrace{x + \sqrt{x^2 + 1}}_{\text{choose positive root}}.$$

Because $e^y > 0$ and $\sqrt{x^2 + 1} > x$, the positive root must be chosen. We now solve for y by taking the natural logarithm of both sides:

$$e^y = x + \sqrt{x^2 + 1} \quad \Rightarrow \quad y = \ln(x + \sqrt{x^2 + 1}).$$

Therefore, the formula we seek is $\sinh^{-1} x = \ln(x + \sqrt{x^2 + 1})$.

A similar procedure can be carried out for the other inverse hyperbolic functions (Exercise 110). Theorem 7.7 lists the results of these calculations.

▶ Most calculators allow for the direct evaluation of the hyperbolic sine, cosine, and tangent, along with their inverses, but are not programmed to evaluate $\text{sech}^{-1} x$, $\text{csch}^{-1} x$, and $\text{coth}^{-1} x$. The formulas in Theorem 7.7 are useful for evaluating these functions on a calculator.

THEOREM 7.7   Inverses of the Hyperbolic Functions Expressed as Logarithms

$$\cosh^{-1} x = \ln(x + \sqrt{x^2 - 1}) \; (x \geq 1) \qquad \text{sech}^{-1} x = \cosh^{-1} \frac{1}{x} \; (0 < x \leq 1)$$

$$\sinh^{-1} x = \ln(x + \sqrt{x^2 + 1}) \qquad\qquad \text{csch}^{-1} x = \sinh^{-1} \frac{1}{x} \; (x \neq 0)$$

$$\tanh^{-1} x = \frac{1}{2} \ln\left(\frac{1 + x}{1 - x}\right) (|x| < 1) \qquad \text{coth}^{-1} x = \tanh^{-1} \frac{1}{x} \; (|x| > 1)$$

Notice that the formulas in Theorem 7.7 for the inverse hyperbolic secant, cosecant, and cotangent are given in terms of the inverses of their corresponding reciprocal functions. Justification for these formulas follows from the definitions of the inverse functions. For example, from the definition of $\mathrm{csch}^{-1} x$, we have

$$y = \mathrm{csch}^{-1} x \quad \Leftrightarrow \quad x = \mathrm{csch}\, y \quad \Leftrightarrow \quad 1/x = \sinh y.$$

Applying the inverse hyperbolic sine to both sides of $1/x = \sinh y$ yields

$$\sinh^{-1}(1/x) = \underbrace{\sinh^{-1}(\sinh y)}_{y} \quad \text{or} \quad y = \mathrm{csch}^{-1} x = \sinh^{-1}(1/x).$$

Similar derivations yield the other two formulas.

EXAMPLE 4   Points of intersection   Find the points at which the curves $y = \cosh x$ and $y = \frac{5}{3}$ intersect (**Figure 7.16**).

SOLUTION The x-coordinates of the points of intersection satisfy the equation $\cosh x = \frac{5}{3}$, which is solved by applying $\cosh^{-1}$ to both sides of the equation. However, evaluating $\cosh^{-1}(\cosh x)$ requires care—in Exercise 105, you are asked to show that $\cosh^{-1}(\cosh x) = |x|$. With this fact, the points of intersection can be found:

$$\cosh x = \frac{5}{3} \qquad\qquad \text{Set equations equal to one another.}$$

$$\cosh^{-1}(\cosh x) = \cosh^{-1}\frac{5}{3} \qquad\qquad \text{Apply } \cosh^{-1} \text{ to both sides.}$$

$$|x| = \ln\left(\frac{5}{3} + \sqrt{\left(\frac{5}{3}\right)^2 - 1}\right) \qquad \text{Simplify; use Theorem 7.7.}$$

$$x = \pm\ln 3. \qquad\qquad \text{Simplify.}$$

The points of intersection lie on the line $y = \frac{5}{3}$, so the points are $\left(-\ln 3, \frac{5}{3}\right)$ and $\left(\ln 3, \frac{5}{3}\right)$.

Related Exercises 61–62 ◄

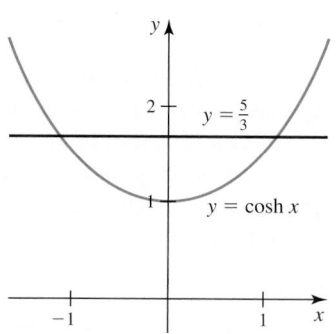

Figure 7.16

QUICK CHECK 5 Use the results of Example 4 to write an integral for the area of the region bounded by $y = \cosh x$ and $y = \frac{5}{3}$ (Figure 7.16) and then evaluate the integral. ◄

Derivatives of the Inverse Hyperbolic Functions and Related Integral Formulas

The derivatives of the inverse hyperbolic functions can be computed directly from the logarithmic formulas given in Theorem 7.7. However, it is more efficient to use the definitions in Figure 7.15.

Recall that the inverse hyperbolic sine is defined by

$$y = \sinh^{-1} x \quad \Leftrightarrow \quad x = \sinh y.$$

We differentiate both sides of $x = \sinh y$ with respect to x and solve for dy/dx:

$$x = \sinh y \qquad\qquad y = \sinh^{-1} x \quad \Leftrightarrow \quad x = \sinh y$$

$$1 = (\cosh y)\frac{dy}{dx} \qquad\qquad \text{Use implicit differentiation.}$$

$$\frac{dy}{dx} = \frac{1}{\cosh y} \qquad\qquad \text{Solve for } dy/dx.$$

$$\frac{dy}{dx} = \frac{1}{\pm\sqrt{\sinh^2 y + 1}} \qquad \cosh^2 y - \sinh^2 y = 1$$

$$\frac{dy}{dx} = \frac{1}{\sqrt{x^2 + 1}}. \qquad\qquad x = \sinh y$$

In the last step, the positive root is chosen because $\cosh y > 0$ for all y.

The derivatives of the other inverse hyperbolic functions, listed in Theorem 7.8, are derived in a similar way (Exercise 106).

THEOREM 7.8   Derivatives of the Inverse Hyperbolic Functions

$$\frac{d}{dx}(\cosh^{-1} x) = \frac{1}{\sqrt{x^2 - 1}} \ (x > 1) \qquad \frac{d}{dx}(\sinh^{-1} x) = \frac{1}{\sqrt{x^2 + 1}}$$

$$\frac{d}{dx}(\tanh^{-1} x) = \frac{1}{1 - x^2} \ (|x| < 1) \qquad \frac{d}{dx}(\coth^{-1} x) = \frac{1}{1 - x^2} \ (|x| > 1)$$

$$\frac{d}{dx}(\operatorname{sech}^{-1} x) = -\frac{1}{x\sqrt{1 - x^2}} \ (0 < x < 1) \quad \frac{d}{dx}(\operatorname{csch}^{-1} x) = -\frac{1}{|x|\sqrt{1 + x^2}} \ (x \neq 0)$$

The restrictions associated with the formulas in Theorem 7.8 are a direct consequence of the domains of the inverse functions (Figure 7.15). Note that the derivative of both $\tanh^{-1} x$ and $\coth^{-1} x$ is $1/(1 - x^2)$, although this result is valid on different domains ($|x| < 1$ for $\tanh^{-1} x$ and $|x| > 1$ for $\coth^{-1} x$). These facts have a bearing on formula (3) in the next theorem, which is a reversal of the derivative formulas in Theorem 7.8. Here we list integral results, where a is a positive constant; each formula can be verified by differentiation.

▶ The integrals in Theorem 7.9 appear again in Chapter 8 in terms of logarithms and with fewer restrictions on the variable of integration.

THEOREM 7.9   Integral Formulas

1. $\displaystyle\int \frac{dx}{\sqrt{x^2 - a^2}} = \cosh^{-1} \frac{x}{a} + C, \text{ for } x > a$

2. $\displaystyle\int \frac{dx}{\sqrt{x^2 + a^2}} = \sinh^{-1} \frac{x}{a} + C, \text{ for all } x$

3. $\displaystyle\int \frac{dx}{a^2 - x^2} = \begin{cases} \dfrac{1}{a} \tanh^{-1} \dfrac{x}{a} + C, \text{ for } |x| < a \\[2mm] \dfrac{1}{a} \coth^{-1} \dfrac{x}{a} + C, \text{ for } |x| > a \end{cases}$

4. $\displaystyle\int \frac{dx}{x\sqrt{a^2 - x^2}} = -\frac{1}{a} \operatorname{sech}^{-1} \frac{x}{a} + C, \text{ for } 0 < x < a$

5. $\displaystyle\int \frac{dx}{x\sqrt{a^2 + x^2}} = -\frac{1}{a} \operatorname{csch}^{-1} \frac{|x|}{a} + C, \text{ for } x \neq 0$

EXAMPLE 5   Derivatives of inverse hyperbolic functions Compute dy/dx for each function.

a. $y = \tanh^{-1} 3x$ **b.** $y = x^2 \sinh^{-1} x$

SOLUTION

a. Using the Chain Rule, we have

$$\frac{dy}{dx} = \frac{d}{dx}(\tanh^{-1} 3x) = \frac{1}{1 - (3x)^2} \cdot 3 = \frac{3}{1 - 9x^2}.$$

▶ The function $3/(1 - 9x^2)$ in the solution to Example 5a is defined for all $x \neq \pm 1/3$. However, the derivative formula $dy/dx = 3/(1 - 9x^2)$ is valid only on $-1/3 < x < 1/3$ because $\tanh^{-1} 3x$ is defined only on $-1/3 < x < 1/3$. The result of computing $\frac{d}{dx}(\coth^{-1} 3x)$ is the same, but valid on $(-\infty, -1/3) \cup (1/3, \infty)$.

b. $\dfrac{dy}{dx} = 2x \sinh^{-1} x + x^2 \cdot \dfrac{1}{\sqrt{x^2 + 1}}$ *Product Rule; Theorem 7.8*

$$= x \left(\frac{2\sqrt{x^2 + 1} \cdot \sinh^{-1} x + x}{\sqrt{x^2 + 1}} \right) \qquad \textit{Simplify.}$$

Related Exercises 31, 35 ◀

EXAMPLE 6   **Integral computations**

a. Compute the area of the region bounded by $y = 1/\sqrt{x^2 + 16}$ over the interval $[0, 3]$.

b. Evaluate $\displaystyle\int_9^{25} \frac{dx}{\sqrt{x}(4 - x)}$.

SOLUTION

a. The region in question is shown in **Figure 7.17**, and its area is given by $\displaystyle\int_0^3 \frac{dx}{\sqrt{x^2 + 16}}$.

Using formula (2) in Theorem 7.9 with $a = 4$, we have

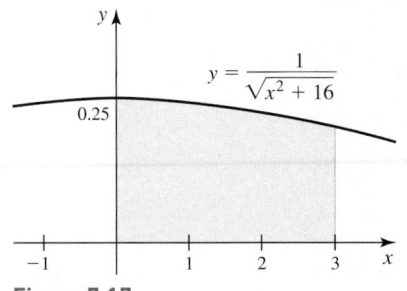

Figure 7.17

$$\int_0^3 \frac{dx}{\sqrt{x^2 + 16}} = \sinh^{-1} \frac{x}{4} \Big|_0^3 \qquad \text{Theorem 7.9}$$

$$= \sinh^{-1} \frac{3}{4} - \sinh^{-1} 0 \quad \text{Evaluate.}$$

$$= \sinh^{-1} \frac{3}{4}. \qquad \sinh^{-1} 0 = 0$$

A calculator gives an approximate result of $\sinh^{-1}(3/4) \approx 0.693$. The exact result can be written in terms of logarithms using Theorem 7.7:

$$\sinh^{-1}(3/4) = \ln(3/4 + \sqrt{(3/4)^2 + 1}) = \ln 2.$$

b. The integral doesn't match any of the formulas in Theorem 7.9, so we use the substitution $u = \sqrt{x}$:

$$\int_9^{25} \frac{dx}{\sqrt{x}(4 - x)} = 2 \int_3^5 \frac{du}{4 - u^2}. \quad \text{Let } u = \sqrt{x}; du = \frac{dx}{2\sqrt{x}}.$$

The new integral now matches formula (3), with $a = 2$. We conclude that

$$2 \int_3^5 \frac{du}{4 - u^2} = 2 \cdot \frac{1}{2} \coth^{-1} \frac{u}{2} \Big|_3^5 \qquad \int \frac{dx}{a^2 - x^2} = \frac{1}{a} \coth^{-1} \frac{x}{a} + C$$

$$= \coth^{-1} \frac{5}{2} - \coth^{-1} \frac{3}{2}. \quad \text{Evaluate.}$$

The antiderivative involving $\coth^{-1} x$ was chosen because the interval of integration ($3 \le u \le 5$) satisfies $|u| > a = 2$. Theorem 7.7 is used to express the result in numerical form in case your calculator cannot evaluate $\coth^{-1} x$:

$$\coth^{-1} \frac{5}{2} - \coth^{-1} \frac{3}{2} = \tanh^{-1} \frac{2}{5} - \tanh^{-1} \frac{2}{3} \approx -0.381.$$

Related Exercises 47, 63 ◄

QUICK CHECK 6 Evaluate $\displaystyle\int_0^1 \frac{du}{4 - u^2}$. ◄

Applications of Hyperbolic Functions

This section concludes with a brief look at two applied problems associated with hyperbolic functions. Additional applications are presented in the exercises.

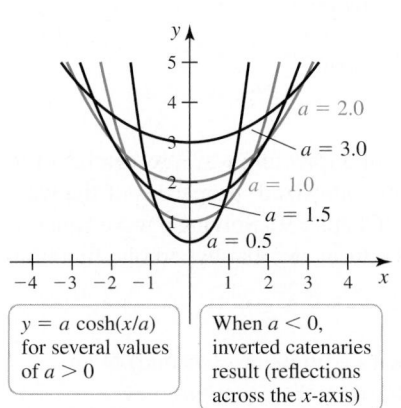

$y = a \cosh(x/a)$ for several values of $a > 0$

When $a < 0$, inverted catenaries result (reflections across the x-axis)

Figure 7.18

The Catenary   When a free-hanging rope or flexible cable supporting only its own weight is attached to two points of equal height, it takes the shape of a curve known as a *catenary*. You can see catenaries in telephone wires, ropes strung across chasms for Tyrolean traverses (Example 7), and spider webs.

The equation for a general catenary is $y = a \cosh(x/a)$, where $a \ne 0$ is a real number. When $a < 0$, the curve is called an inverted catenary, sometimes used in the design of arches. **Figure 7.18** illustrates catenaries for several values of a.

▶ A Tyrolean traverse is used to pass over difficult terrain, such as a chasm between two cliffs or a raging river. A rope is strung between two anchor points, the climber clips onto the rope and then traverses the gap by pulling on the rope.

EXAMPLE 7   Length of a catenary A climber anchors a rope at two points of equal height, separated by a distance of 100 ft, in order to perform a *Tyrolean traverse.* The rope follows the catenary $f(x) = 200 \cosh(x/200)$ over the interval $[-50, 50]$ (**Figure 7.19**). Find the length of the unweighted rope between the two anchor points.

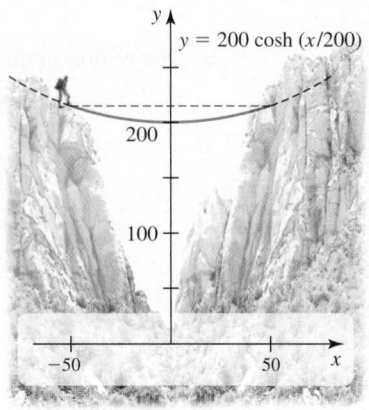

Figure 7.19

SOLUTION Recall from Section 6.5 that the arc length of the curve $y = f(x)$ over the interval $[a, b]$ is $L = \int_a^b \sqrt{1 + f'(x)^2}\, dx$. Also note that

$$f'(x) = 200 \sinh\left(\frac{x}{200}\right) \cdot \frac{1}{200} = \sinh\frac{x}{200}.$$

Therefore, the length of the rope is

$$
\begin{aligned}
L &= \int_{-50}^{50} \sqrt{1 + \sinh^2\left(\frac{x}{200}\right)}\, dx & \text{Arc length formula} \\[2mm]
&= 2\int_{0}^{50} \sqrt{1 + \sinh^2\left(\frac{x}{200}\right)}\, dx & \text{Use symmetry.} \\[2mm]
&= 400 \int_{0}^{1/4} \sqrt{1 + \sinh^2 u}\, du & \text{Let } u = \frac{x}{200}. \\[2mm]
&= 400 \int_{0}^{1/4} \cosh u\, du & 1 + \sinh^2 u = \cosh^2 u \\[2mm]
&= 400 \sinh u\Big|_{0}^{1/4} & \text{Evaluate integral.} \\[2mm]
&= 400\left(\sinh\frac{1}{4} - \sinh 0\right) & \text{Simplify.} \\[2mm]
&\approx 101 \text{ ft.} & \text{Evaluate.}
\end{aligned}
$$

Related Exercises 69–70 ◀

▶ Using the principles of vector analysis introduced in Chapter 13, the tension in the rope and the forces acting upon the anchors in Example 7 can be computed. This is crucial information for anyone setting up a Tyrolean traverse; the *sag angle* (Exercise 72) figures into these calculations. Similar calculations are important for catenary lifelines used in construction and for rigging camera shots in Hollywood movies.

Velocity of a Wave To describe the characteristics of a traveling wave, researchers formulate a *wave equation* that reflects the known (or hypothesized) properties of the wave and that often takes the form of a differential equation (Chapter 9). Solving a wave equation produces additional information about the wave, and it turns out that hyperbolic functions may arise naturally in this context.

EXAMPLE 8   Velocity of an ocean wave The velocity v (in meters/second) of an idealized surface wave traveling on the ocean is modeled by the equation

$$v = \sqrt{\frac{g\lambda}{2\pi} \tanh\left(\frac{2\pi d}{\lambda}\right)},$$

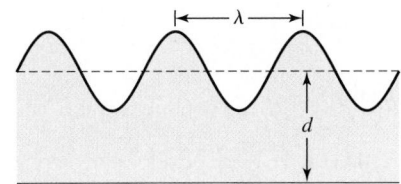

Figure 7.20

▶ In fluid dynamics, *water depth* is often discussed in terms of the depth-to-wavelength ratio d/λ, not the actual depth of the water. Three classifications are generally used:

shallow water: $d/\lambda < 0.05$

intermediate depth: $0.05 < d/\lambda < 0.5$

deep water: $d/\lambda > 0.5$

where $g = 9.8 \text{ m/s}^2$ is the acceleration due to gravity, λ is the wavelength measured in meters from crest to crest, and d is the depth of the undisturbed water, also measured in meters (**Figure 7.20**).

a. A sea kayaker observes several waves that pass beneath her kayak, and she estimates that $\lambda = 12$ m and $v = 4$ m/s. How deep is the water in which she is kayaking?

b. The *deep-water* equation for wave velocity is $v = \sqrt{\dfrac{g\lambda}{2\pi}}$, which is an approximation to the velocity formula given above. Waves are said to be in deep water if the depth-to-wavelength ratio d/λ is greater than $\dfrac{1}{2}$. Explain why $v = \sqrt{\dfrac{g\lambda}{2\pi}}$ is a good approximation when $d/\lambda > \dfrac{1}{2}$.

SOLUTION

a. We substitute $\lambda = 12$ and $v = 4$ into the velocity equation and solve for d.

$$4 = \sqrt{\frac{g \cdot 12}{2\pi} \tanh\left(\frac{2\pi d}{12}\right)} \quad \Rightarrow \quad 16 = \frac{6g}{\pi} \tanh\left(\frac{\pi d}{6}\right) \quad \text{Square both sides.}$$

$$\Rightarrow \quad \frac{8\pi}{3g} = \tanh\left(\frac{\pi d}{6}\right) \quad \text{Multiply by } \frac{\pi}{6g}.$$

In order to extract d from the argument of tanh, we apply $\tanh^{-1}$ to both sides of the equation and then use the property $\tanh^{-1}(\tanh x) = x$, for all x.

$$\tanh^{-1}\left(\frac{8\pi}{3g}\right) = \tanh^{-1}\left(\tanh\left(\frac{\pi d}{6}\right)\right) \quad \text{Apply } \tanh^{-1} \text{ to both sides.}$$

$$\tanh^{-1}\left(\frac{8\pi}{29.4}\right) = \frac{\pi d}{6} \quad \text{Simplify; } 3g = 29.4.$$

$$d = \frac{6}{\pi} \tanh^{-1}\left(\frac{8\pi}{29.4}\right) \approx 2.4 \text{ m} \quad \text{Solve for } d.$$

Therefore, the kayaker is in water that is about 2.4 m deep.

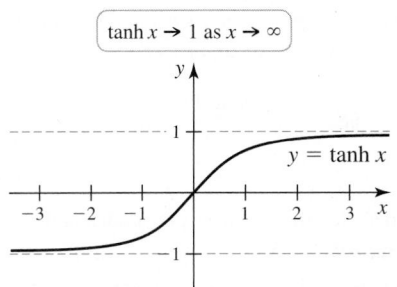

tanh $x \to 1$ as $x \to \infty$

Figure 7.21

QUICK CHECK 7 Explain why longer waves travel faster than shorter waves in deep water. ◀

b. Recall that $y = \tanh x$ is an increasing function $(dy/dx = \text{sech}^2 x > 0)$ whose values approach 1 as $x \to \infty$ (**Figure 7.21**). Also notice that when $\dfrac{d}{\lambda} = \dfrac{1}{2}$,

$$\tanh\left(\frac{2\pi d}{\lambda}\right) = \tanh \pi \approx 0.996,$$ which is nearly equal to 1. These facts imply that

whenever $\dfrac{d}{\lambda} > \dfrac{1}{2}$, we can replace $\tanh\left(\dfrac{2\pi d}{\lambda}\right)$ with 1 in the velocity formula, resulting

in the deep-water velocity function $v = \sqrt{\dfrac{g\lambda}{2\pi}}$.

Related Exercises 73–74 ◀

SECTION 7.3 EXERCISES

Getting Started

1. State the definitions of the hyperbolic cosine and hyperbolic sine functions.

2. Sketch the graphs of $y = \cosh x$, $y = \sinh x$, and $y = \tanh x$ (include asymptotes), and state whether each function is even, odd, or neither.

3. What is the fundamental identity for hyperbolic functions?

4. How are the derivative formulas for the hyperbolic functions and the trigonometric functions alike? How are they different?

5. Express $\sinh^{-1} x$ in terms of logarithms.

6. What is the domain of $\text{sech}^{-1} x$? How is $\text{sech}^{-1} x$ defined in terms of the inverse hyperbolic cosine?

7. A calculator has a built-in $\sinh^{-1} x$ function, but no $\text{csch}^{-1} x$ function. How do you evaluate $\text{csch}^{-1} 5$ on such a calculator?

8. On what interval is the formula $\dfrac{d}{dx}(\tanh^{-1} x) = \dfrac{1}{1 - x^2}$ valid?

9. When evaluating the definite integral $\int_6^8 \dfrac{dx}{16 - x^2}$, why must you choose the antiderivative $\dfrac{1}{4}\coth^{-1}\dfrac{x}{4}$ rather than $\dfrac{1}{4}\tanh^{-1}\dfrac{x}{4}$?

10. How does the graph of the catenary $y = a\cosh\dfrac{x}{a}$ change as $a > 0$ increases?

Practice Exercises

11–15. Identities *Prove each identity using the definitions of the hyperbolic functions.*

11. $\tanh x = \dfrac{e^{2x} - 1}{e^{2x} + 1}$ **12.** $\tanh(-x) = -\tanh x$

13. $\cosh 2x = \cosh^2 x + \sinh^2 x$ (*Hint:* Begin with the right side of the equation.)

14. $2\sinh(\ln(\sec x)) = \sin x \tan x$

15. $\cosh x + \sinh x = e^x$

16–18. Identities *Use the given identity to prove the related identity.*

16. Use the fundamental identity $\cosh^2 x - \sinh^2 x = 1$ to prove the identity $\coth^2 x - 1 = \operatorname{csch}^2 x$.

17. Use the identity $\cosh 2x = \cosh^2 x + \sinh^2 x$ to prove the identities $\cosh^2 x = \dfrac{\cosh 2x + 1}{2}$ and $\sinh^2 x = \dfrac{\cosh 2x - 1}{2}$.

18. Use the identity $\cosh(x + y) = \cosh x \cosh y + \sinh x \sinh y$ to prove the identity $\cosh 2x = \cosh^2 x + \sinh^2 x$.

19–21. Derivative formulas *Derive the following derivative formulas given that* $\dfrac{d}{dx}(\cosh x) = \sinh x$ *and* $\dfrac{d}{dx}(\sinh x) = \cosh x$.

19. $\dfrac{d}{dx}(\coth x) = -\operatorname{csch}^2 x$ **20.** $\dfrac{d}{dx}(\operatorname{sech} x) = -\operatorname{sech} x \tanh x$

21. $\dfrac{d}{dx}(\operatorname{csch} x) = -\operatorname{csch} x \coth x$

22–36. Derivatives *Find the derivatives of the following functions.*

22. $f(x) = \sinh 4x$ **23.** $f(x) = \cosh^2 x$

24. $f(x) = -\sinh^3 4x$ **25.** $f(x) = \tanh^2 x$

26. $f(x) = \sqrt{\coth 3x}$ **27.** $f(x) = \ln \operatorname{sech} 2x$

28. $f(x) = x \tanh x$ **29.** $f(x) = x^2 \cosh^2 3x$

30. $f(x) = \dfrac{x}{\operatorname{csch} x}$ **31.** $f(x) = \cosh^{-1} 4x$

32. $f(t) = 2\tanh^{-1}\sqrt{t}$ **33.** $f(v) = \sinh^{-1} v^2$

34. $f(x) = \operatorname{csch}^{-1}\left(\dfrac{2}{x}\right)$ **35.** $f(x) = x\sinh^{-1} x - \sqrt{x^2 + 1}$

36. $f(u) = \sinh^{-1}(\tan u)$

37–56. Integrals *Evaluate each integral.*

37. $\displaystyle\int \cosh 2x\, dx$ **38.** $\displaystyle\int \operatorname{sech}^2 w \tanh w\, dw$

39. $\displaystyle\int \dfrac{\sinh x}{1 + \cosh x}\, dx$ **40.** $\displaystyle\int \coth^2 x \operatorname{csch}^2 x\, dx$

41. $\displaystyle\int \tanh^2 x\, dx$ (*Hint:* Use an identity.)

42. $\displaystyle\int \sinh^2 z\, dz$ (*Hint:* Use an identity.)

43. $\displaystyle\int_0^1 \cosh^3 3x \sinh 3x\, dx$ **44.** $\displaystyle\int_0^4 \dfrac{\operatorname{sech}^2 \sqrt{x}}{\sqrt{x}}\, dx$

45. $\displaystyle\int_0^{\ln 2} \tanh x\, dx$ **46.** $\displaystyle\int_{\ln 2}^{\ln 3} \operatorname{csch} y\, dy$

47. $\displaystyle\int \dfrac{dx}{8 - x^2},\ x > 2\sqrt{2}$ **48.** $\displaystyle\int \dfrac{dx}{\sqrt{x^2 - 16}},\ x > 4$

49. $\displaystyle\int \dfrac{e^x}{36 - e^{2x}}\, dx,\ x < \ln 6$ **50.** $\displaystyle\int \dfrac{dx}{x\sqrt{16 + x^2}}$

51. $\displaystyle\int \dfrac{dx}{x\sqrt{4 - x^8}}$ **52.** $\displaystyle\int \dfrac{dx}{x\sqrt{1 + x^4}}$

53. $\displaystyle\int \dfrac{\cosh z}{\sinh^2 z}\, dz$ **54.** $\displaystyle\int \dfrac{\cos \theta}{9 - \sin^2 \theta}\, d\theta$

55. $\displaystyle\int_{5/12}^{3/4} \dfrac{\sinh^{-1} x}{\sqrt{x^2 + 1}}\, dx$

56. $\displaystyle\int_{25}^{225} \dfrac{dx}{\sqrt{x^2 + 25x}}$ (*Hint:* $\sqrt{x^2 + 25x} = \sqrt{x}\sqrt{x + 25}$.)

57–58. Two ways *Evaluate the following integrals two ways.*

a. *Simplify the integrand first and then integrate.*

b. *Change variables (let $u = \ln x$), integrate, and then simplify your answer. Verify that both methods give the same answer.*

57. $\displaystyle\int \dfrac{\sinh(\ln x)}{x}\, dx$ **58.** $\displaystyle\int_1^{\sqrt{3}} \dfrac{\operatorname{sech}(\ln x)}{x}\, dx$

T 59. Visual approximation

a. Use a graphing utility to sketch the graph of $y = \coth x$ and then explain why $\int_5^{10} \coth x\, dx \approx 5$.

b. Evaluate $\int_5^{10} \coth x\, dx$ analytically and use a calculator to arrive at a decimal approximation to the answer. How large is the error in the approximation in part (a)?

60. Integral proof Prove the formula $\int \coth x\, dx = \ln|\sinh x| + C$ of Theorem 7.6.

61–62. Points of intersection and area

a. *Sketch the graphs of the functions f and g and find the x-coordinate of the points at which they intersect.*

b. *Compute the area of the region described.*

61. $f(x) = \operatorname{sech} x$, $g(x) = \tanh x$; the region bounded by the graphs of f, g, and the y-axis

62. $f(x) = \sinh x$, $g(x) = \tanh x$; the region bounded by the graphs of f, g, and $x = \ln 3$

63–68. Definite integrals *Evaluate the following definite integrals. Use Theorem 7.7 to express your answer in terms of logarithms.*

63. $\displaystyle\int_1^{e^2} \dfrac{dx}{x\sqrt{\ln^2 x + 1}}$ **64.** $\displaystyle\int_5^{3\sqrt{5}} \dfrac{dx}{\sqrt{x^2 - 9}}$

65. $\displaystyle\int_{-2}^2 \dfrac{dt}{t^2 - 9}$ **66.** $\displaystyle\int_{1/6}^{1/4} \dfrac{dt}{t\sqrt{1 - 4t^2}}$

67. $\displaystyle\int_{1/8}^{1} \frac{dx}{x\sqrt{1 + x^{2/3}}}$

68. $\displaystyle\int_{\ln 5}^{\ln 9} \frac{\cosh x}{4 - \sinh^2 x}\, dx$

69. Catenary arch The portion of the curve $y = \dfrac{17}{15} - \cosh x$ that lies above the x-axis forms a catenary arch. Find the average height of the arch above the x-axis.

70. Length of a catenary Show that the arc length of the catenary $y = \cosh x$ over the interval $[0, a]$ is $L = \sinh a$.

⊤ 71. Power lines A power line is attached at the same height to two utility poles that are separated by a distance of 100 ft; the power line follows the curve $f(x) = a \cosh \dfrac{x}{a}$. Use the following steps to find the value of a that produces a sag of 10 ft midway between the poles. Use a coordinate system that places the poles at $x = \pm 50$.

   a. Show that a satisfies the equation $\cosh \dfrac{50}{a} - 1 = \dfrac{10}{a}$.

   b. Let $t = \dfrac{10}{a}$, confirm that the equation in part (a) reduces to $\cosh 5t - 1 = t$, and solve for t using a graphing utility. Report your answer accurate to two decimal places.

   c. Use your answer in part (b) to find a, and then compute the length of the power line.

72. Sag angle Imagine a climber clipping onto the rope described in Example 7 and pulling himself to the rope's midpoint. Because the rope is supporting the weight of the climber, it no longer takes the shape of the catenary $y = 200 \cosh \dfrac{x}{200}$. Instead, the rope (nearly) forms two sides of an isosceles triangle. Compute the *sag angle* θ illustrated in the figure, assuming the rope does not stretch when weighted. Recall from Example 7 that the length of the rope is 101 ft.

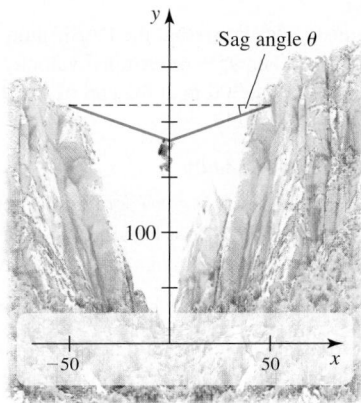

⊤ 73. Wavelength The velocity of a surface wave on the ocean is given by $v = \sqrt{\dfrac{g\lambda}{2\pi}\tanh\left(\dfrac{2\pi d}{\lambda}\right)}$ (Example 8). Use a graphing utility or root finder to approximate the wavelength λ of an ocean wave traveling at $v = 7$ m/s in water that is $d = 10$ m deep.

74. Wave velocity Use Exercise 73 to do the following calculations.

   a. Find the velocity of a wave where $\lambda = 50$ m and $d = 20$ m.

   b. Determine the depth of the water if a wave with $\lambda = 15$ m is traveling at $v = 4.5$ m/s.

75. Shallow-water velocity equation

   a. Confirm that the linear approximation to $f(x) = \tanh x$ at $a = 0$ is $L(x) = x$.

   b. Recall that the velocity of a surface wave on the ocean is $v = \sqrt{\dfrac{g\lambda}{2\pi}\tanh\left(\dfrac{2\pi d}{\lambda}\right)}$. In fluid dynamics, *shallow water* refers to water where the depth-to-wavelength ratio $\dfrac{d}{\lambda} < 0.05$. Use your answer to part (a) to explain why the approximate shallow-water velocity equation is $v = \sqrt{gd}$.

   c. Use the shallow-water velocity equation to explain why waves tend to slow down as they approach the shore.

76. Tsunamis A tsunami is an ocean wave often caused by earthquakes on the ocean floor; these waves typically have long wavelengths, ranging from 150 to 1000 km. Imagine a tsunami traveling across the Pacific Ocean, which is the deepest ocean in the world, with an average depth of about 4000 m. Explain why the *shallow-water velocity equation* (Exercise 75) applies to tsunamis even though the actual depth of the water is large. What does the shallow-water equation say about the speed of a tsunami in the Pacific Ocean (use $d = 4000$ m)?

77. Explain why or why not Determine whether the following statements are true and give an explanation or counterexample.

   a. $\dfrac{d}{dx}(\sinh(\ln 3)) = \dfrac{\cosh(\ln 3)}{3}$.

   b. $\dfrac{d}{dx}(\sinh x) = \cosh x$ and $\dfrac{d}{dx}(\cosh x) = -\sinh x$.

   c. $\ln(1 + \sqrt{2}) = -\ln(-1 + \sqrt{2})$.

   d. $\displaystyle\int_{0}^{1} \dfrac{dx}{4 - x^2} = \dfrac{1}{2}\left(\coth^{-1}\dfrac{1}{2} - \coth^{-1}0\right)$.

⊤ 78. Evaluating hyperbolic functions Use a calculator to evaluate each expression or state that the value does not exist. Report answers accurate to four decimal places to the right of the decimal point.

   a. $\coth 4$ **b.** $\tanh^{-1} 2$

   c. $\operatorname{csch}^{-1} 5$ **d.** $\operatorname{csch} x\Big|_{1/2}^{2}$

   e. $\ln\left|\tanh\dfrac{x}{2}\right|\,\Big|_{1}^{10}$ **f.** $\tan^{-1}(\sinh x)\Big|_{-3}^{3}$

   g. $\dfrac{1}{4}\coth^{-1}\dfrac{x}{4}\,\Big|_{20}^{36}$

79. Evaluating hyperbolic functions Evaluate each expression without using a calculator or state that the value does not exist. Simplify answers as much as possible.

   a. $\cosh 0$ **b.** $\tanh 0$

   c. $\operatorname{csch} 0$ **d.** $\operatorname{sech}(\sinh 0)$

   e. $\coth(\ln 5)$ **f.** $\sinh(2\ln 3)$

   g. $\cosh^2 1$ **h.** $\operatorname{sech}^{-1}(\ln 3)$

   i. $\cosh^{-1}\dfrac{17}{8}$ **j.** $\sinh^{-1}\dfrac{e^2 - 1}{2e}$

80. Confirming a graph The graph of $f(x) = \sinh x$ is shown in Figure 7.13. Use calculus to find the intervals on which f is increasing or decreasing, and to find the intervals on which f is concave up or concave down to confirm that the graph is correct.

81. Critical points Find the critical points of the function $f(x) = \sinh^2 x \cosh x$.

T 82. Critical points

a. Show that the critical points of $f(x) = \dfrac{\cosh x}{x}$ satisfy $x = \coth x$.

b. Use a root finder to approximate the critical points of f.

83. Points of inflection Find the x-coordinate of the point(s) of inflection of $f(x) = \tanh^2 x$.

84. Points of inflection Find the x-coordinate of the point(s) of inflection of $f(x) = \operatorname{sech} x$. Report exact answers in terms of logarithms (use Theorem 7.7).

85. Area of region Find the area of the region bounded by $y = \operatorname{sech} x$, $x = 1$, and the unit circle (see figure).

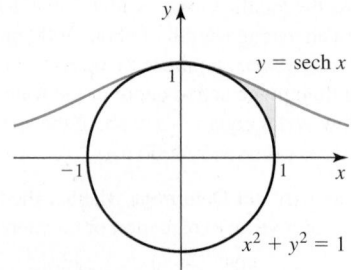

86. Solid of revolution Compute the volume of the solid of revolution that results when the region in Exercise 85 is revolved about the x-axis.

87. L'Hôpital loophole Explain why l'Hôpital's Rule fails when applied to the limit $\lim\limits_{x \to \infty} \dfrac{\sinh x}{\cosh x}$ and then find the limit another way.

88–91. Limits *Use l'Hôpital's Rule to evaluate the following limits.*

88. $\lim\limits_{x \to \infty} \dfrac{1 - \coth x}{1 - \tanh x}$

89. $\lim\limits_{x \to 0} \dfrac{\tanh^{-1} x}{\tan (\pi x / 2)}$

90. $\lim\limits_{x \to 1^-} \dfrac{\tanh^{-1} x}{\tan (\pi x / 2)}$

91. $\lim\limits_{x \to 0^+} (\tanh x)^x$

T 92. Slant asymptote The linear function $\ell(x) = mx + b$, for finite $m \neq 0$, is a slant asymptote of $f(x)$ if $\lim\limits_{x \to \infty} (f(x) - \ell(x)) = 0$.

a. Use a graphing utility to make a sketch that shows $\ell(x) = x$ is a slant asymptote of $f(x) = x \tanh x$. Does f have any other slant asymptotes?

b. Provide an intuitive argument showing that $f(x) = x \tanh x$ behaves like $\ell(x) = x$ as x gets large.

c. Prove that $\ell(x) = x$ is a slant asymptote of f by confirming $\lim\limits_{x \to \infty} (x \tanh x - x) = 0$.

Explorations and Challenges

93. Kiln design Find the volume interior to the inverted catenary kiln (an oven used to fire pottery) shown in the figure.

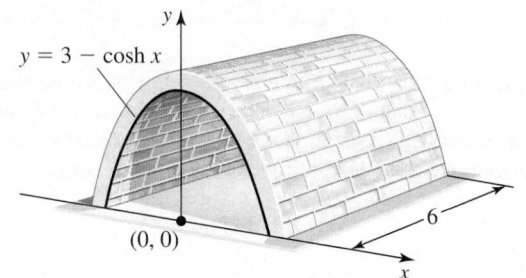

T 94. Newton's method Use Newton's method to find all local extreme values of $f(x) = x \operatorname{sech} x$.

95. Falling body When an object falling from rest encounters air resistance proportional to the square of its velocity, the distance it falls (in meters) after t seconds is given by

$$d(t) = \frac{m}{k} \ln \left(\cosh \left(\sqrt{\frac{kg}{m}}\, t \right) \right),$$ where m is the mass of the object

in kilograms, $g = 9.8 \text{ m/s}^2$ is the acceleration due to gravity, and k is a physical constant.

a. A BASE jumper ($m = 75$ kg) leaps from a tall cliff and performs a ten-second delay (she free-falls for 10 s and then opens her chute). How far does she fall in 10 s? Assume $k = 0.2$.

b. How long does it take for her to fall the first 100 m? The second 100 m? What is her average velocity over each of these intervals?

96. Velocity of falling body Refer to Exercise 95, which gives the position function for a falling body. Use $m = 75$ kg and $k = 0.2$.

a. Confirm that the BASE jumper's velocity t seconds after

jumping is $v(t) = d'(t) = \sqrt{\dfrac{mg}{k}} \tanh \left(\sqrt{\dfrac{kg}{m}}\, t \right)$.

b. How fast is the BASE jumper falling ten seconds after jumping?

c. How long does it take for the BASE jumper to reach a speed of 45 m/s (roughly 100 mi/hr)?

97. Terminal velocity Refer to Exercises 95 and 96.

a. Compute a jumper's *terminal velocity*, which is defined as

$$\lim_{t \to \infty} v(t) = \lim_{t \to \infty} \sqrt{\frac{mg}{k}} \tanh \left(\sqrt{\frac{kg}{m}}\, t \right).$$

b. Find the terminal velocity for the jumper in Exercise 96 ($m = 75$ kg and $k = 0.2$).

c. How long does it take any falling object to reach a speed equal to 95% of its terminal velocity? Leave your answer in terms of k, g, and m.

d. How tall must a cliff be so that the BASE jumper ($m = 75$ kg and $k = 0.2$) reaches 95% of terminal velocity? Assume the jumper needs at least 300 m at the end of free fall to deploy the chute and land safely.

98. Acceleration of a falling body

a. Find the acceleration $a(t) = v'(t)$ of a falling body whose velocity is given in part (a) of Exercise 96.

b. Compute $\lim\limits_{t \to \infty} a(t)$. Explain your answer as it relates to terminal velocity (Exercise 97).

99. Differential equations Hyperbolic functions are useful in solving differential equations (Chapter 9). Show that the functions $y = A \sinh kx$ and $y = B \cosh kx$, where A, B, and k are constants, satisfy the equation $y''(x) - k^2 y(x) = 0$.

100. Surface area of a catenoid When the catenary $y = a \cosh \dfrac{x}{a}$ is revolved about the x-axis, it sweeps out a surface of revolution called a *catenoid*. Find the area of the surface generated when $y = \cosh x$ on $[-\ln 2, \ln 2]$ is rotated about the x-axis.

101–104. Proving identities *Prove the following identities.*

101. $\sinh (\cosh^{-1} x) = \sqrt{x^2 - 1}$, for $x \geq 1$

102. $\cosh (\sinh^{-1} x) = \sqrt{x^2 + 1}$, for all x

103. $\cosh (x + y) = \cosh x \cosh y + \sinh x \sinh y$

104. $\sinh (x + y) = \sinh x \cosh y + \cosh x \sinh y$

105. Inverse identity Show that $\cosh^{-1}(\cosh x) = |x|$ by using the formula $\cosh^{-1} t = \ln\left(t + \sqrt{t^2 - 1}\right)$ and considering the cases $x \geq 0$ and $x < 0$.

106. Theorem 7.8

a. The definition of the inverse hyperbolic cosine is $y = \cosh^{-1} x \Leftrightarrow x = \cosh y$, for $x \geq 1, 0 \leq y < \infty$. Use implicit differentiation to show that $\dfrac{d}{dx}(\cosh^{-1} x) = \dfrac{1}{\sqrt{x^2 - 1}}$.

b. Differentiate $\sinh^{-1} x = \ln\left(x + \sqrt{x^2 + 1}\right)$ to show that $\dfrac{d}{dx}(\sinh^{-1} x) = \dfrac{1}{\sqrt{x^2 + 1}}$.

107. Many formulas There are several ways to express the indefinite integral of sech x.

a. Show that $\int \operatorname{sech} x \, dx = \tan^{-1}(\sinh x) + C$ (Theorem 7.6). (*Hint:* Write $\operatorname{sech} x = \dfrac{1}{\cosh x} = \dfrac{\cosh x}{\cosh^2 x} = \dfrac{\cosh x}{1 + \sinh^2 x}$ and then make a change of variables.)

b. Show that $\int \operatorname{sech} x \, dx = \sin^{-1}(\tanh x) + C$. (*Hint:* Show that $\operatorname{sech} x = \dfrac{\operatorname{sech}^2 x}{\sqrt{1 - \tanh^2 x}}$ and then make a change of variables.)

c. Verify that $\int \operatorname{sech} x \, dx = 2 \tan^{-1} e^x + C$ by proving $\dfrac{d}{dx}(2 \tan^{-1} e^x) = \operatorname{sech} x$.

108. Integral formula Carry out the following steps to derive the formula $\int \operatorname{csch} x \, dx = \ln\left|\tanh \dfrac{x}{2}\right| + C$ (Theorem 7.6).

a. Change variables with the substitution $u = \dfrac{x}{2}$ to show that $\int \operatorname{csch} x \, dx = \int \dfrac{2 \, du}{\sinh 2u}$.

b. Use the identity for $\sinh 2u$ to show that $\dfrac{2}{\sinh 2u} = \dfrac{\operatorname{sech}^2 u}{\tanh u}$.

c. Change variables again to determine $\int \dfrac{\operatorname{sech}^2 u}{\tanh u} \, du$, and then express your answer in terms of x.

109. Arc length Use the result of Exercise 108 to find the arc length of the curve $f(x) = \ln\left|\tanh \dfrac{x}{2}\right|$ on $[\ln 2, \ln 8]$.

110. Inverse hyperbolic tangent Recall that the inverse hyperbolic tangent is defined as $y = \tanh^{-1} x \Leftrightarrow x = \tanh y$, for $-1 < x < 1$ and all real y. Solve $x = \tanh y$ for y to express the formula for $\tanh^{-1} x$ in terms of logarithms.

111. Integral family Use the substitution $u = x^r$ to show that $\int \dfrac{dx}{x\sqrt{1 - x^{2r}}} = -\dfrac{1}{r} \operatorname{sech}^{-1} x^r + C$, for $r > 0$ and $0 < x < 1$.

112. Definitions of hyperbolic sine and cosine Complete the following steps to prove that when the x- and y-coordinates of a point on the hyperbola $x^2 - y^2 = 1$ are defined as $\cosh t$ and $\sinh t$, respectively, where t is twice the area of the shaded region in the figure, x and y can be expressed as

$$x = \cosh t = \frac{e^t + e^{-t}}{2} \quad \text{and} \quad y = \sinh t = \frac{e^t - e^{-t}}{2}.$$

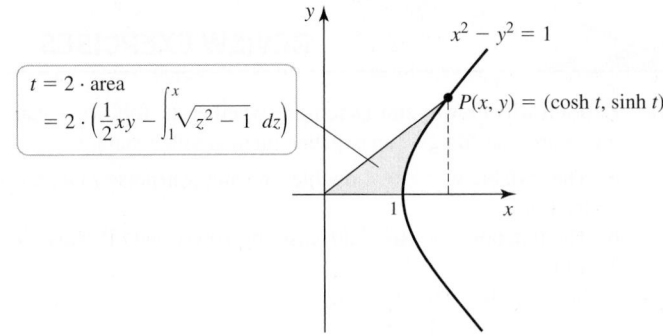

$$t = 2 \cdot \text{area} = 2 \cdot \left(\frac{1}{2}xy - \int_1^x \sqrt{z^2 - 1} \, dz\right)$$

$$x^2 - y^2 = 1$$

$$P(x, y) = (\cosh t, \sinh t)$$

a. Explain why twice the area of the shaded region is given by

$$t = 2 \cdot \left(\frac{1}{2}xy - \int_1^x \sqrt{z^2 - 1} \, dz\right) = x\sqrt{x^2 - 1} - 2\int_1^x \sqrt{z^2 - 1} \, dz.$$

b. In Chapter 8, the formula for the integral in part (a) is derived:

$$\int \sqrt{z^2 - 1} \, dz = \frac{z}{2}\sqrt{z^2 - 1} - \frac{1}{2}\ln\left|z + \sqrt{z^2 - 1}\right| + C.$$

Evaluate this integral on the interval $[1, x]$, explain why the absolute value can be dropped, and combine the result with part (a) to show that

$$t = \ln\left(x + \sqrt{x^2 - 1}\right).$$

c. Solve the final equation from part (b) for x to show that

$$x = \frac{e^t + e^{-t}}{2}.$$

d. Use the fact that $y = \sqrt{x^2 - 1}$, in combination with part (c), to show that $y = \dfrac{e^t - e^{-t}}{2}$.

QUICK CHECK ANSWERS

1. $\sinh(-x) = \dfrac{e^{-x} - e^{-(-x)}}{2} = -\dfrac{e^x - e^{-x}}{2} = -\sinh x$

2. Because $\tanh x = \dfrac{e^x - e^{-x}}{e^x + e^{-x}}$ and $\lim\limits_{x \to \infty} e^{-x} = 0$, $\tanh x \approx \dfrac{e^x}{e^x} = 1$ for large x, which implies that $y = 1$ is a horizontal asymptote. A similar argument shows that $\tanh x \to -1$ as $x \to -\infty$, which means that $y = -1$ is also a horizontal asymptote. 3. $\dfrac{d}{dx}(4 \cosh 2x) = 8 \sinh 2x$; $\int 4 \cosh 2x \, dx = 2 \sinh 2x + C$ 4. $\dfrac{1}{2}\ln|\tanh x| + C$

5. $\text{Area} = 2\int_0^{\ln 3}\left(\dfrac{5}{3} - \cosh x\right) dx = \dfrac{2}{3}(5\ln 3 - 4) \approx 0.995$

6. $\int_0^1 \dfrac{du}{4 - u^2} = \dfrac{1}{2}\tanh^{-1}\dfrac{1}{2} \approx 0.275$ 7. The deep-water velocity formula is $v = \sqrt{\dfrac{g\lambda}{2\pi}}$, which is an increasing function of the wavelength λ. Therefore, larger values of λ correspond to faster waves. ◄

CHAPTER 7 REVIEW EXERCISES

1. Explain why or why not Determine whether the following statements are true and give an explanation or counterexample.

a. The variable $y = t + 1$ doubles in value whenever t increases by 1 unit.

b. The function $y = Ae^{0.1t}$ increases by 10% when t increases by 1 unit.

c. $\ln xy = (\ln x)(\ln y)$.

d. $\sinh(\ln x) = \dfrac{x^2 - 1}{2x}$.

2–9. Integrals *Evaluate the following integrals.*

2. $\displaystyle\int \frac{e^x}{4e^x + 6}\, dx$

3. $\displaystyle\int_{e^2}^{e^8} \frac{dx}{x \ln x}$

4. $\displaystyle\int_1^4 \frac{10^{\sqrt{x}}}{\sqrt{x}}\, dx$

5. $\displaystyle\int \frac{x + 4}{x^2 + 8x + 25}\, dx$

6. $\displaystyle\int_{\ln 2}^{\ln 3} \coth s\, ds$

7. $\displaystyle\int \frac{dx}{\sqrt{x^2 - 9}}, x > 3$

8. $\displaystyle\int \frac{e^x}{\sqrt{e^{2x} + 4}}\, dx$

9. $\displaystyle\int_0^1 \frac{x^2}{9 - x^6}\, dx$

10–19. Derivatives *Find the derivatives of the following functions.*

10. $f(x) = \ln(3 \sin^2 4x)$

11. $g(x) = x^{3x^2 + 1}$

12. $f(x) = \dfrac{\sinh x}{1 + \sinh x}$

13. $f(t) = \cosh t \sinh t$

14. $g(w) = \tanh(4w^2 + w)$

15. $f(x) = \cosh^2(3x - 1)$

16. $h(z) = \ln \sinh z - z \coth z$

17. $f(x) = \tanh^{-1}(\cos x)$

18. $g(t) = \sinh^{-1} \sqrt{t}$

19. $f(x) = \operatorname{sech}^{-1} \dfrac{1}{x^2}$, for $x > 1$

20. Population growth The population of a large city grows exponentially with a current population of 1.3 million and a predicted population of 1.45 million 10 years from now.

a. Use an exponential model to estimate the population in 20 years. Assume the annual growth rate is constant.

b. Find the doubling time of the population.

21. Caffeine An adult consumes an espresso containing 75 mg of caffeine. If the caffeine has a half-life of 5.5 hours, when will the amount of caffeine in her bloodstream equal 30 mg?

22. Two cups of coffee A college student consumed two cups of coffee, each containing 80 mg of caffeine, 90 minutes apart. Suppose that just before he consumed his second cup of coffee, 65 mg of caffeine was still in his system from the first cup of coffee. How much caffeine remains in his system 7 hours after he drank his first cup of coffee?

23. Moore's Law In 1965, Gordon Moore observed that the number of transistors that could be placed on an integrated circuit was approximately doubling each year, and he predicted that this trend would continue for another decade. In 1975, Moore revised the doubling time to every two years, and this prediction became known as *Moore's Law*.

a. In 1979, Intel introduced the Intel 8088 processor; each of its integrated circuits contained 29,000 transistors. Use Moore's revised doubling time to find a function $y(t)$ that approximates the number of transistors on an integrated circuit t years after 1979.

b. In 2000, the Pentium 4 integrated circuit was introduced; it contained 42 million transistors. Compare this value to the value obtained using the function found in part (a).

24. Radioactive decay The mass of radioactive material in a sample has decreased by 30% since the decay began. Assuming a half-life of 1500 years, how long ago did the decay begin?

25. Population growth Growing from an initial population of 150,000 at a constant annual growth rate of 4%/yr, how long will it take a city to reach a population of 1 million?

26. Savings account A savings account advertises an annual percentage yield (APY) of 5.4%, which means that the balance in the account increases at an annual growth rate of 5.4%/yr.

a. Find the balance in the account for $t \geq 0$ with an initial deposit of \$1500, assuming the APY remains fixed and no additional deposits or withdrawals are made.

b. What is the doubling time of the balance?

c. After how many years does the balance reach \$5000?

27–28. Curve sketching *Use the graphing techniques of Section 4.4 to graph the following functions on their domains. Identify local extreme points, inflection points, concavity, and end behavior. Use a graphing utility only to check your work.*

27. $f(x) = e^x(x^2 - x)$

28. $f(x) = \ln x - \ln^2 x$

T 29. Log-normal probability distribution A commonly used distribution in probability and statistics is the log-normal distribution. (If the logarithm of a variable has a normal distribution, then the variable itself has a log-normal distribution.) The distribution function is

$$f(x) = \frac{1}{x\sigma\sqrt{2\pi}} e^{-\ln^2 x/(2\sigma^2)}, \quad \text{for } x \geq 0,$$

where $\ln x$ has zero mean and standard deviation $\sigma > 0$.

a. Graph f for $\sigma = \dfrac{1}{2}$, 1, and 2. Based on your graphs, does $\displaystyle\lim_{x \to 0} f(x)$ appear to exist?

b. Evaluate $\displaystyle\lim_{x \to 0} f(x)$. (*Hint:* Let $x = e^y$.)

c. Show that f has a single local maximum at $x^* = e^{-\sigma^2}$.

d. Evaluate $f(x^*)$ and express the result as a function of σ.

e. For what value of $\sigma > 0$ in part (d) does $f(x^*)$ have a minimum?

30. Area of region Find the area of the region bounded by the curves $f(x) = 8\operatorname{sech}^2 x$ and $g(x) = \cosh x$.

31. Linear approximation Find the linear approximation to $f(x) = \cosh x$ at $a = \ln 3$ and then use it to approximate the value of $\cosh 1$.

32. Limit Evaluate $\lim\limits_{x\to\infty} (\tanh x)^x$.

33. Derivatives of hyperbolic functions Compute the following derivatives.

a. $\dfrac{d^6}{dx^6}(\cosh x)$ **b.** $\dfrac{d}{dx}(x \operatorname{sech} x)$

T 34. Arc length Find the arc length of the curve $y = \ln x$ between $x = 1$ and $x = b > 1$ given that

$$\int \sqrt{\frac{x^2 + a^2}{x}}\, dx = \sqrt{x^2 + a^2} - a \ln\!\left(\frac{a + \sqrt{x^2 + a^2}}{x}\right) + C.$$

Use any means to approximate the value of b for which the curve has length 2.

Chapter 7 Guided Projects

Applications of the material in this chapter and related topics can be found in the following Guided Projects. For additional information, see the Preface.

- Hyperbolic functions

- Optimizing fuel use

8

Integration Techniques

Chapter Preview In this chapter, we return to integration methods and present a variety of new strategies that supplement the substitution (or change of variables) method. The new techniques introduced here are integration by parts, trigonometric substitution, and partial fractions. Taken altogether, these *analytical methods* (pencil-and-paper methods) greatly enlarge the collection of integrals that we can evaluate. Nevertheless, it is important to recognize that these methods are limited because many integrals do not yield to them. For this reason, we also introduce table-based methods, which are used to evaluate many indefinite integrals, and computer-based methods for approximating definite integrals. The discussion then turns to integrals that have either infinite integrands or infinite intervals of integration. These integrals, called *improper integrals*, offer surprising results and have many practical applications.

8.1 Basic Approaches

Before plunging into new integration techniques, we devote this section to two practical goals. The first is to review what you learned about the substitution method in Section 5.5. The other is to introduce several basic simplifying procedures that are worth keeping in mind when evaluating integrals.

▶ Table 8.1 is similar to Tables 4.9 and 4.10 in Section 4.9. It is a subset of the table of integrals at the back of the text.

Table 8.1 includes the basic integration formulas presented earlier in the text. To this list we add four standard trigonometric integrals (formulas 14–17), which are derived in Examples 1 and 2 using substitution.

Table 8.1 Basic Integration Formulas

1. $\int k \, dx = kx + C, k$ real

2. $\int x^p \, dx = \dfrac{x^{p+1}}{p+1} + C, p \neq -1$ real

3. $\int \cos ax \, dx = \dfrac{1}{a} \sin ax + C$

4. $\int \sin ax \, dx = -\dfrac{1}{a} \cos ax + C$

5. $\int \sec^2 ax \, dx = \dfrac{1}{a} \tan ax + C$

6. $\int \csc^2 ax \, dx = -\dfrac{1}{a} \cot ax + C$

7. $\int \sec ax \tan ax \, dx = \dfrac{1}{a} \sec ax + C$

8. $\int \csc ax \cot ax \, dx = -\dfrac{1}{a} \csc ax + C$

9. $\int e^{ax} \, dx = \dfrac{1}{a} e^{ax} + C$

10. $\int \dfrac{dx}{x} = \ln |x| + C$

11. $\int \dfrac{dx}{a^2 + x^2} = \dfrac{1}{a} \tan^{-1} \dfrac{x}{a} + C$

12. $\int \dfrac{dx}{\sqrt{a^2 - x^2}} = \sin^{-1} \dfrac{x}{a} + C, a > 0$

13. $\int \dfrac{dx}{x\sqrt{x^2 - a^2}} = \dfrac{1}{a} \sec^{-1} \left| \dfrac{x}{a} \right| + C, a > 0$

14. $\int \tan ax \, dx = \dfrac{1}{a} \ln |\sec ax| + C$

15. $\int \cot ax \, dx = \dfrac{1}{a} \ln |\sin ax| + C$

16. $\int \sec ax \, dx = \dfrac{1}{a} \ln |\sec ax + \tan ax| + C$

17. $\int \csc ax \, dx = -\dfrac{1}{a} \ln |\csc ax + \cot ax| + C$

In Section 4.9, we introduced formulas for $\int \cos x \, dx$ and $\int \sin x \, dx$ and then derived more general results listed in Table 5.6 of Section 5.5. Missing from that table of integrals are formulas for the antiderivatives of $\tan x$, $\cot x$, $\sec x$, and $\csc x$. Examples 1 and 2 (and their related exercises) fill in these holes.

EXAMPLE 1   **Substitution review**   Derive integral formula 14 in Table 8.1:

$$\int \tan ax \, dx = \frac{1}{a} \ln|\sec ax| + C.$$

SOLUTION   We begin with a simpler form of the integral, $\int \tan x \, dx$, and express $\tan x$ in terms of $\sin x$ and $\cos x$ to prepare for a standard substitution:

$$\int \tan x \, dx = \int \frac{\sin x}{\cos x} dx$$

$$= -\int \frac{1}{u} \, du \qquad\qquad u = \cos x; \, du = -\sin x \, dx$$

$$= -\ln|u| + C = -\ln|\cos x| + C.$$

Using the properties of logarithms, the integral can also be written

$$\int \tan x \, dx = -\ln|\cos x| + C = -\ln|1/\sec x| + C$$

$$= -(\ln 1 - \ln|\sec x|) = \ln|\sec x| + C.$$

With this result in hand, the more general formula given in Table 8.1 is easily derived by using the substitution $u = ax$. The derivation of formula 15 in Table 8.1 is similar (Exercise 66).

Related Exercise 66 ◄

QUICK CHECK 1 What change of variable would you use for the integral $\int \dfrac{\sec^2 x}{\tan^3 x} \, dx$? ◄

EXAMPLE 2   **Multiplication by 1**   Derive integral formula 16 in Table 8.1:

$$\int \sec ax \, dx = \frac{1}{a} \ln|\sec ax + \tan ax| + C.$$

SOLUTION   As in Example 1, it is easier to begin with a simpler form of the integral, $\int \sec x \, dx$. The key to evaluating this integral is admittedly not obvious, and the trick works only on special integrals. The idea is to multiply the integrand by 1, but the challenge is finding the appropriate representation of 1. In this case, we use

$$1 = \frac{\sec x + \tan x}{\sec x + \tan x}.$$

The integral is evaluated as follows:

$$\int \sec x \, dx = \int \sec x \cdot \underbrace{\frac{\sec x + \tan x}{\sec x + \tan x}}_{1} dx \qquad \text{Multiply integrand by 1.}$$

$$= \int \frac{\sec^2 x + \sec x \tan x}{\sec x + \tan x} dx \qquad \text{Expand numerator.}$$

$$= \int \frac{du}{u} \qquad\qquad u = \sec x + \tan x; \, du = (\sec^2 x + \sec x \tan x) dx$$

$$= \ln|u| + C \qquad\qquad \text{Integrate.}$$

$$= \ln|\sec x + \tan x| + C. \qquad u = \sec x + \tan x$$

Using the substitution $u = ax$ with $\int \sec ax \, dx$ leads to the more general result in Table 8.1. The derivation of formula 17 in Table 8.1 employs the same technique used here (Exercise 67).

Related Exercise 67 ◄

EXAMPLE 3 Subtle substitution Evaluate $\int \dfrac{dx}{e^x + e^{-x}}$.

SOLUTION In this case, we see nothing in Table 8.1 that resembles the given integral. In a spirit of trial and error, we try the technique used in Example 2 and multiply the numerator and denominator of the integrand by e^x:

$$\int \frac{dx}{e^x + e^{-x}} = \int \frac{e^x}{e^{2x} + 1}\, dx. \quad e^x \cdot e^x = e^{2x}$$

This form of the integrand suggests the substitution $u = e^x$, which implies that $du = e^x\, dx$. Making these substitutions, the integral becomes

$$\int \frac{e^x}{e^{2x} + 1}\, dx = \int \frac{du}{u^2 + 1} \qquad \text{Substitute } u = e^x,\, du = e^x\, dx.$$

$$= \tan^{-1} u + C \qquad \text{Table 8.1}$$

$$= \tan^{-1} e^x + C. \qquad u = e^x$$

Related Exercises 45–46 ◄

EXAMPLE 4 Split up fractions Evaluate $\int \dfrac{\cos x + \sin^3 x}{\sec x}\, dx$.

SOLUTION Don't overlook the opportunity to split a fraction into two or more fractions. In this case, the integrand is simplified in a useful way:

$$\int \frac{\cos x + \sin^3 x}{\sec x}\, dx = \int \frac{\cos x}{\sec x}\, dx + \int \frac{\sin^3 x}{\sec x}\, dx \qquad \text{Split fraction.}$$

$$= \int \cos^2 x\, dx + \int \sin^3 x \cos x\, dx. \quad \sec x = \frac{1}{\cos x}$$

> **Half-angle formulas**
> $$\cos^2 x = \frac{1 + \cos 2x}{2}$$
> $$\sin^2 x = \frac{1 - \cos 2x}{2}$$

The first of the resulting integrals is evaluated using a half-angle formula (Example 6 of Section 5.5). In the second integral, the substitution $u = \sin x$ is used:

$$\int \frac{\cos x + \sin^3 x}{\sec x}\, dx = \int \cos^2 x\, dx + \int \sin^3 x \cos x\, dx$$

$$= \int \frac{1 + \cos 2x}{2}\, dx + \int \sin^3 x \cos x\, dx \quad \text{Half-angle formula}$$

$$= \frac{1}{2}\int dx + \frac{1}{2}\int \cos 2x\, dx + \int u^3\, du \quad u = \sin x,\, du = \cos x\, dx$$

$$= \frac{x}{2} + \frac{1}{4}\sin 2x + \frac{1}{4}\sin^4 x + C. \qquad \text{Evaluate integrals.}$$

QUICK CHECK 2 Explain how to simplify the integrand of $\int \dfrac{x^3 + \sqrt{x}}{x^{3/2}}\, dx$ before integrating. ◄

Related Exercises 22, 25 ◄

EXAMPLE 5 Division with rational functions Evaluate $\int \dfrac{x^2 + 2x - 1}{x + 4}\, dx$.

SOLUTION When integrating rational functions (polynomials in the numerator and denominator), check to see whether the function is *improper* (the degree of the numerator is greater than or equal to the degree of the denominator). In this example, we have an improper rational function, and long division is used to simplify it. The integration is done as follows:

$$
\begin{array}{r}
x - 2 \\
x + 4\,\overline{)\,x^2 + 2x - 1} \\
\underline{x^2 + 4x} \\
-2x - 1 \\
\underline{-2x - 8} \\
7
\end{array}
$$

$$\int \frac{x^2 + 2x - 1}{x + 4}\, dx = \int (x - 2)\, dx + \int \frac{7}{x + 4}\, dx \quad \text{Long division}$$

$$= \frac{x^2}{2} - 2x + 7 \ln |x + 4| + C. \quad \text{Evaluate integrals.}$$

Related Exercises 34–35 ◄

QUICK CHECK 3 Explain how to simplify the integrand of $\int \dfrac{x + 1}{x - 1}\, dx$ before integrating. ◄

EXAMPLE 6 **Complete the square** Evaluate $\displaystyle\int \dfrac{dx}{\sqrt{-x^2 - 8x - 7}}$.

SOLUTION We don't see an integral in Table 8.1 that looks like the given integral, so some preliminary work is needed. In this case, the key is to complete the square on the polynomial in the denominator. We find that

$$-x^2 - 8x - 7 = -(x^2 + 8x + 7)$$
$$= -(x^2 + 8x \underbrace{+\ 16 - 16}_{\text{add and subtract } 16} + 7) \quad \text{Complete the square.}$$
$$= -((x + 4)^2 - 9) \qquad \text{Factor and combine terms.}$$
$$= 9 - (x + 4)^2. \qquad \text{Rearrange terms.}$$

After a change of variables, the integral is recognizable:

$$\int \dfrac{dx}{\sqrt{-7 - 8x - x^2}} = \int \dfrac{dx}{\sqrt{9 - (x + 4)^2}} \qquad \text{Complete the square.}$$
$$= \int \dfrac{du}{\sqrt{9 - u^2}} \qquad u = x + 4,\ du = dx$$
$$= \sin^{-1}\dfrac{u}{3} + C \qquad \text{Table 8.1}$$
$$= \sin^{-1}\!\left(\dfrac{x + 4}{3}\right) + C. \qquad \text{Replace } u \text{ with } x + 4.$$

Related Exercises 31, 37 ◄

QUICK CHECK 4 Express $x^2 + 6x + 16$ in terms of a perfect square. ◄

The techniques illustrated in this section are designed to transform or simplify an integrand before you apply a specific method. In fact, these ideas may help you recognize the best method to use. Keep them in mind as you learn new integration methods and improve your integration skills.

SECTION 8.1 EXERCISES

Getting Started

1. What change of variables would you use for the integral $\int (4 - 7x)^{-6}\, dx$?

2. Evaluate $\int (\sec x + 1)^2\, dx$. (*Hint:* Expand $(\sec x + 1)^2$ first.)

3. What trigonometric identity is useful in evaluating $\int \sin^2 x\, dx$?

4. Let $f(x) = \dfrac{4x^3 + x^2 + 4x + 2}{x^2 + 1}$. Use long division to show that $f(x) = 4x + 1 + \dfrac{1}{x^2 + 1}$ and use this result to evaluate $\int f(x)\, dx$.

5. Describe a first step in integrating $\displaystyle\int \dfrac{10}{x^2 - 4x - 9}\, dx$.

6. Evaluate $\displaystyle\int \dfrac{2x + 1}{x^2 + 1}\, dx$ using the following steps.
    a. Fill in the blanks: By splitting the integrand into two fractions, we have $\displaystyle\int \dfrac{2x + 1}{x^2 + 1}\, dx = \int \underline{\quad}\, dx + \int \underline{\quad}\, dx.$
    b. Evaluate the two integrals on the right side of the equation in part (a).

Practice Exercises

7–64. Integration review *Evaluate the following integrals.*

7. $\displaystyle\int \dfrac{dx}{(3 - 5x)^4}$

8. $\displaystyle\int (9x - 2)^{-3} dx$

9. $\displaystyle\int_0^{3\pi/8} \sin\!\left(2x - \dfrac{\pi}{4}\right) dx$

10. $\displaystyle\int e^{3 - 4x} dx$

11. $\displaystyle\int \dfrac{\ln 2x}{x}\, dx$

12. $\displaystyle\int_{-5}^0 \dfrac{dx}{\sqrt{4 - x}}$

13. $\displaystyle\int \dfrac{e^x}{e^x + 1}\, dx$

14. $\displaystyle\int_0^1 x\, 3^{x^2 + 1}\, dx$

15. $\displaystyle\int_1^{e^2} \dfrac{\ln^2(x^2)}{x}\, dx$

16. $\displaystyle\int_0^1 \dfrac{t^2}{1 + t^6}\, dt$

17. $\displaystyle\int_1^2 s(s - 1)^9\, ds$

18. $\displaystyle\int_3^7 (t - 6)\sqrt{t - 3}\, dt$

19. $\displaystyle\int \dfrac{(\ln w - 1)^7 \ln w}{w}\, dw$

20. $\displaystyle\int e^x (1 + e^x)^9 (1 - e^x)\, dx$

21. $\displaystyle\int \dfrac{x + 2}{x^2 + 4}\, dx$

22. $\displaystyle\int \dfrac{\sin x + 1}{\cos x}\, dx$

23. $\displaystyle\int e^x \csc(3e^x + 4)\, dx$

24. $\displaystyle\int_0^9 \frac{x^{5/2} - x^{1/2}}{x^{3/2}}\, dx$

25. $\displaystyle\int_0^{\pi/4} \frac{\sec\theta + \csc\theta}{\sec\theta\,\csc\theta}\, d\theta$

26. $\displaystyle\int \frac{4 + e^{-2x}}{e^{3x}}\, dx$

27. $\displaystyle\int \frac{2 - 3x}{\sqrt{1 - x^2}}\, dx$

28. $\displaystyle\int \frac{3x + 1}{\sqrt{4 - x^2}}\, dx$

29. $\displaystyle\int_{\pi/4}^{\pi/2} \sqrt{1 + \cot^2 x}\, dx$

30. $\displaystyle\int_{e^{\pi/4}}^{e^{\pi/3}} \frac{\cot(\ln x)}{x}\, dx$

31. $\displaystyle\int \frac{dx}{x^2 - 2x + 10}$

32. $\displaystyle\int_0^2 \frac{x}{x^2 + 4x + 8}\, dx$

33. $\displaystyle\int \frac{x^3 + 2x^2 + 5x + 3}{x^2 + x + 2}\, dx$

34. $\displaystyle\int_2^4 \frac{x^2 + 2}{x - 1}\, dx$

35. $\displaystyle\int_0^1 \frac{t^4 + t^3 + t^2 + t + 1}{t^2 + 1}\, dt$

36. $\displaystyle\int \frac{t^3 - 2}{t + 1}\, dt$

37. $\displaystyle\int \frac{d\theta}{\sqrt{27 - 6\theta - \theta^2}}$

38. $\displaystyle\int \frac{x}{x^4 + 2x^2 + 1}\, dx$

39. $\displaystyle\int \frac{d\theta}{1 + \sin\theta}$

40. $\displaystyle\int \frac{1 - x}{1 - \sqrt{x}}\, dx$

41. $\displaystyle\int \frac{dx}{\sec x - 1}$

42. $\displaystyle\int \frac{d\theta}{1 - \csc\theta}$

43. $\displaystyle\int \frac{\cosh 3x}{1 + \sinh 3x}\, dx$

44. $\displaystyle\int_0^{\sqrt{3}} \frac{6x^3}{\sqrt{x^2 + 1}}\, dx$

45. $\displaystyle\int \frac{e^x}{e^x - 2e^{-x}}\, dx$

46. $\displaystyle\int \frac{e^{2z}}{e^{2z} - 4e^{-z}}\, dz$

47. $\displaystyle\int \frac{dx}{x^{-1} + 1}$

48. $\displaystyle\int \frac{dy}{y^{-1} + y^{-3}}$

49. $\displaystyle\int \sqrt{9 + \sqrt{t + 1}}\, dt$

50. $\displaystyle\int_4^9 \frac{dx}{1 - \sqrt{x}}$

51. $\displaystyle\int_{-1}^0 \frac{x}{x^2 + 2x + 2}\, dx$

52. $\displaystyle\int_{\pi/6}^{\pi/2} \frac{dy}{\sin y}$

53. $\displaystyle\int e^x \sec(e^x + 1)\, dx$

54. $\displaystyle\int_0^1 \sqrt{1 + \sqrt{x}}\, dx$

55. $\displaystyle\int \sin x \sin 2x\, dx$

56. $\displaystyle\int_0^{\pi/2} \sqrt{1 + \cos 2x}\, dx$

57. $\displaystyle\int \frac{dx}{x^{1/2} + x^{3/2}}$

58. $\displaystyle\int_0^1 \frac{dp}{4 - \sqrt{p}}$

59. $\displaystyle\int \frac{x - 2}{x^2 + 6x + 13}\, dx$

60. $\displaystyle\int_0^{\pi/4} 3\sqrt{1 + \sin 2x}\, dx$

61. $\displaystyle\int \frac{e^x}{e^{2x} + 2e^x + 1}\, dx$

62. $\displaystyle\int \frac{-x^5 - x^4 - 2x^3 + 4x + 3}{x^2 + x + 1}\, dx$

63. $\displaystyle\int_1^3 \frac{2}{x^2 + 2x + 1}\, dx$

64. $\displaystyle\int_0^2 \frac{2}{s^3 + 3s^2 + 3s + 1}\, ds$

65. **Explain why or why not** Determine whether the following statements are true and give an explanation or counterexample.

    a. $\displaystyle\int \frac{3}{x^2 + 4}\, dx = \int \frac{3}{x^2}\, dx + \int \frac{3}{4}\, dx.$

    b. Long division simplifies the evaluation of the integral
$$\int \frac{x^3 + 2}{3x^4 + x}\, dx.$$

    c. $\displaystyle\int \frac{1}{\sin x + 1}\, dx = \ln|\sin x + 1| + C.$

    d. $\displaystyle\int \frac{dx}{e^x} = \ln e^x + C.$

66–67. Integrals of cot x and csc x

66. Use a change of variables to prove that $\int \cot x\, dx = \ln|\sin x| + C$. (*Hint:* See Example 1.)

67. Prove that $\int \csc x\, dx = -\ln|\csc x + \cot x| + C$. (*Hint:* See Example 2.)

68. **Different methods**

    a. Evaluate $\int \cot x \csc^2 x\, dx$ using the substitution $u = \cot x$.

    b. Evaluate $\int \cot x \csc^2 x\, dx$ using the substitution $u = \csc x$.

    c. Reconcile the results in parts (a) and (b).

69. **Different substitutions**

    a. Evaluate $\int \tan x \sec^2 x\, dx$ using the substitution $u = \tan x$.

    b. Evaluate $\int \tan x \sec^2 x\, dx$ using the substitution $u = \sec x$.

    c. Reconcile the results in parts (a) and (b).

70. **Different methods** Let $I = \displaystyle\int \frac{x + 2}{x + 4}\, dx$.

    a. Evaluate I after first performing long division on the integrand.

    b. Evaluate I without performing long division on the integrand.

    c. Reconcile the results in parts (a) and (b).

71. **Different methods** Let $I = \displaystyle\int \frac{x^2}{x + 1}\, dx$.

    a. Evaluate I using the substitution $u = x + 1$.

    b. Evaluate I after first performing long division on the integrand.

    c. Reconcile the results in parts (a) and (b).

72. **Area of a region between curves** Find the area of the entire region bounded by the curves $y = \dfrac{x^3}{x^2 + 1}$ and $y = \dfrac{8x}{x^2 + 1}$.

73. **Area of a region between curves** Find the area of the region bounded by the curves $y = \dfrac{x^2}{x^3 - 3x}$ and $y = \dfrac{1}{x^3 - 3x}$ on the interval $[2, 4]$.

74. **Volume of a solid** Consider the region R bounded by the graph of $f(x) = \dfrac{1}{x + 2}$ and the x-axis on the interval $[0, 3]$. Find the volume of the solid formed when R is revolved about the y-axis.

75. **Volume of a solid** Consider the region R bounded by the graph of $f(x) = \sqrt{x^2 + 1}$ and the x-axis on the interval $[0, 2]$. Find the volume of the solid formed when R is revolved about the y-axis.

Explorations and Challenges

76. **Different substitutions**

    a. Show that $\displaystyle\int \frac{dx}{\sqrt{x - x^2}} = \sin^{-1}(2x - 1) + C$ using either
$$u = 2x - 1 \text{ or } u = x - \frac{1}{2}.$$

b. Show that $\int \dfrac{dx}{\sqrt{x - x^2}} = 2 \sin^{-1}\sqrt{x} + C$ using $u = \sqrt{x}$.

c. Prove the identity $2 \sin^{-1}\sqrt{x} - \sin^{-1}(2x - 1) = \dfrac{\pi}{2}$.

(*Source: The College Mathematics Journal, 32, 5, Nov 2001*)

77. Surface area Let $f(x) = \sqrt{x + 1}$. Find the area of the surface generated when the region bounded by the graph of f and the x-axis on the interval $[0, 1]$ is revolved about the x-axis.

78. Surface area Find the area of the surface generated when the region bounded by the graph of $y = e^x + \dfrac{1}{4}e^{-x}$ and the x-axis on the interval $[0, \ln 2]$ is revolved about the x-axis.

79. Arc length Find the length of the curve $y = x^{5/4}$ on the interval $[0, 1]$. (*Hint:* Write the arc length integral and let $u^2 = 1 + \left(\dfrac{5}{4}\right)^2 \sqrt{x}$.)

80. Skydiving A skydiver in free fall subject to gravitational acceleration and air resistance has a velocity given by

$$v(t) = v_T\left(\dfrac{e^{at} - 1}{e^{at} + 1}\right),$$ where v_T is the terminal velocity and $a > 0$

is a physical constant. Find the distance that the skydiver falls after t seconds, which is $d(t) = \int_0^t v(y)\, dy$.

QUICK CHECK ANSWERS

1. Let $u = \tan x$.   **2.** Write the integrand as $x^{3/2} + x^{-1}$.

3. Use long division to write the integrand as $1 + \dfrac{2}{x - 1}$.

4. $(x + 3)^2 + 7$ ◄

8.2 Integration by Parts

The Substitution Rule (Section 5.5) emerges when we reverse the Chain Rule for derivatives. In this section, we use a similar strategy and reverse the Product Rule for derivatives. The result is an integration technique called *integration by parts*. To illustrate the importance of integration by parts, consider the indefinite integrals

$$\int e^x\, dx = e^x + C \quad \text{and} \quad \int xe^x\, dx.$$

The first integral is an elementary integral that we have already encountered. The second integral is only slightly different—and yet, the appearance of the product xe^x in the integrand makes this integral (at the moment) impossible to evaluate. Integration by parts is ideally suited for evaluating integrals of *products* of functions.

Integration by Parts for Indefinite Integrals

Given two differentiable functions u and v, the Product Rule states that

$$\dfrac{d}{dx}(u(x)v(x)) = u'(x)v(x) + u(x)v'(x).$$

By integrating both sides, we can write this rule in terms of an indefinite integral:

$$u(x)v(x) = \int (u'(x)v(x) + u(x)v'(x))\, dx.$$

Rearranging this expression in the form

$$\int u(x)\underbrace{v'(x)\, dx}_{dv} = u(x)v(x) - \int v(x)\underbrace{u'(x)\, dx}_{du}$$

leads to the basic relationship for *integration by parts*. It is expressed compactly by noting that $du = u'(x)\, dx$ and $dv = v'(x)\, dx$. Suppressing the independent variable x, we have

$$\int u\, dv = uv - \int v\, du.$$

The integral $\int u\, dv$ is viewed as the given integral, and we use integration by parts to express it in terms of a new integral $\int v\, du$. The technique is successful if the new integral can be evaluated.

> **Integration by Parts**
>
> Suppose u and v are differentiable functions. Then
>
> $$\int u\, dv = uv - \int v\, du.$$

(Handwritten margin notes:)

$\int 3x\sqrt{2x+1}\, dx$

$3x(2x^2+1)^{1/2}$

$\int 3x(\sqrt{2}\, x + 1)\, dx$

$3\sqrt{2}\, x^2 + 3x\, dx$

$\sqrt{2}\, x^3 + \dfrac{3}{2}x^2 + C$

▶ The integration by parts calculation may be done without including the constant of integration—as long as it is included in the final result.

EXAMPLE 1   Integration by parts   Evaluate $\int xe^x \, dx$.

SOLUTION   The presence of *products* in the integrand often suggests integration by parts. We split the product xe^x into two factors, one of which must be identified as u and the other as dv (the latter always includes the differential dx). Powers of x are *often* good choices for u. The choice for dv should be easy to integrate. In this case, the choices $u = x$ and $dv = e^x \, dx$ are advisable. It follows that $du = dx$. The relationship $dv = e^x \, dx$ means that v is an antiderivative of e^x, which implies $v = e^x$. A table is helpful for organizing these calculations.

▶ The arrows in the table show how to combine factors in the integration by parts formula. The first arrow indicates the product uv; the second arrow indicates the integrand $v \, du$.

Functions in original integral	$u = x$	$dv = e^x \, dx$
Functions in new integral	$du = dx$	$v = e^x$

The integration by parts rule is now applied:

$$\int \underbrace{x}_{u} \underbrace{e^x \, dx}_{dv} = \underbrace{x}_{u} \underbrace{e^x}_{v} - \int \underbrace{e^x}_{v} \underbrace{dx}_{du}.$$

The original integral $\int xe^x \, dx$ has been replaced with the integral of e^x, which is easier to evaluate: $\int e^x \, dx = e^x + C$. The entire procedure looks like this:

$$\int xe^x \, dx = xe^x - \int e^x \, dx \qquad \text{Integration by parts}$$

$$= xe^x - e^x + C. \qquad \text{Evaluate the new integral.}$$

Related Exercises 11–12 ◀

▶ To make the table, first write the functions in the original integral:

$$u = \underline{\quad}, \, dv = \underline{\quad}.$$

Then find the functions in the new integral by differentiating u and integrating dv:

$$du = \underline{\quad}, \, v = \underline{\quad}.$$

EXAMPLE 2   Integration by parts   Evaluate $\int x \sin x \, dx$.

SOLUTION   Remembering that powers of x are often a good choice for u, we form the following table.

$u = x$	$dv = \sin x \, dx$
$du = dx$	$v = -\cos x$

Applying integration by parts, we have

$$\int \underbrace{x}_{u} \underbrace{\sin x \, dx}_{dv} = \underbrace{x}_{u} \underbrace{(-\cos x)}_{v} - \int \underbrace{(-\cos x)}_{v} \underbrace{dx}_{du} \qquad \text{Integration by parts}$$

$$= -x \cos x + \sin x + C. \qquad \text{Evaluate } \int \cos x \, dx = \sin x.$$

Related Exercises 10, 16 ◀

QUICK CHECK 1 What are the best choices for u and dv in evaluating $\int x \cos x \, dx$? ◀

In general, integration by parts works when we can easily integrate the choice for dv and when the new integral is easier to evaluate than the original. Integration by parts is often used for integrals of the form $\int x^n f(x) \, dx$, where n is a positive integer. Such integrals generally require the repeated use of integration by parts, as shown in the following example.

EXAMPLE 3   Repeated use of integration by parts

a. Evaluate $\int x^2 e^x \, dx$.

b. How would you evaluate $\int x^n e^x \, dx$, where n is a positive integer?

SOLUTION

a. The factor x^2 is a good choice for u, leaving $dv = e^x \, dx$. We then have

$u = x^2$	$dv = e^x \, dx$
$du = 2x \, dx$	$v = e^x$

$$\int \underbrace{x^2}_{u} \underbrace{e^x \, dx}_{dv} = \underbrace{x^2}_{u} \underbrace{e^x}_{v} - \int \underbrace{e^x}_{v} \underbrace{2x \, dx}_{du}.$$

Notice that the new integral on the right side is simpler than the original integral because the power of x has been reduced by one. In fact, the new integral was evaluated in Example 1. Therefore, after using integration by parts twice, we have

$$\int x^2 e^x \, dx = x^2 e^x - 2 \int x e^x \, dx \qquad \text{Integration by parts}$$

$$= x^2 e^x - 2(x e^x - e^x) + C \quad \text{Result of Example 1}$$

$$= e^x (x^2 - 2x + 2) + C. \quad \text{Simplify.}$$

b. We now let $u = x^n$ and $dv = e^x \, dx$. The integration takes the form

$$\int x^n e^x \, dx = x^n e^x - n \int x^{n-1} e^x \, dx.$$

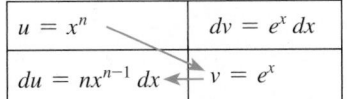

> An integral identity in which the power of a variable is reduced is called a **reduction formula.** Other examples of reduction formulas are explored in Exercises 50–53.

We see that integration by parts reduces the power of the variable in the integrand. The integral in part (a) with $n = 2$ requires two uses of integration by parts. You can probably anticipate that evaluating the integral $\int x^n e^x \, dx$ requires n applications of integration by parts to reach the integral $\int e^x \, dx$, which is easily evaluated.

Related Exercises 24–25 ◀

EXAMPLE 4   Repeated use of integration by parts Evaluate $\int e^{2x} \sin x \, dx$.

SOLUTION The integrand consists of a product, which suggests integration by parts. In this case, there is no obvious choice for u and dv, so let's try the following choices.

$u = e^{2x}$	$dv = \sin x \, dx$
$du = 2e^{2x} \, dx$	$v = -\cos x$

> In Example 4, we could also use $u = \sin x$ and $dv = e^{2x} \, dx$. In general, some trial and error may be required when using integration by parts. Effective choices come with practice.

The integral then becomes

$$\int e^{2x} \sin x \, dx = -e^{2x} \cos x + 2 \int e^{2x} \cos x \, dx. \qquad (1)$$

The original integral has been expressed in terms of a new integral, $\int e^{2x} \cos x \, dx$, which is no easier to evaluate than the original integral. It is tempting to start over with a new choice of u and dv, but a little persistence pays off. Suppose we evaluate $\int e^{2x} \cos x \, dx$ using integration by parts with the following choices.

$u = e^{2x}$	$dv = \cos x \, dx$
$du = 2e^{2x} \, dx$	$v = \sin x$

> When using integration by parts, the acronym LIPET may help. If the integrand is the product of two or more functions, choose u to be the first function type that appears in the list
>
> **L**ogarithmic, **I**nverse trigonometric, **P**olynomial, **E**xponential, **T**rigonometric.

Integrating by parts, we have

$$\int e^{2x} \cos x \, dx = e^{2x} \sin x - 2 \int e^{2x} \sin x \, dx. \qquad (2)$$

Now observe that equation (2) contains the original integral, $\int e^{2x} \sin x \, dx$. Substituting the result of equation (2) into equation (1), we find that

$$\int e^{2x} \sin x \, dx = -e^{2x} \cos x + 2 \int e^{2x} \cos x \, dx$$

> To solve for $\int e^{2x} \sin x \, dx$ in the equation $\int e^{2x} \sin x \, dx = -e^{2x} \cos x + 2e^{2x} \sin x - 4 \int e^{2x} \sin x \, dx$, add $4 \int e^{2x} \sin x \, dx$ to both sides of the equation and then divide both sides by 5.

$$= -e^{2x} \cos x + 2(e^{2x} \sin x - 2 \int e^{2x} \sin x \, dx) \quad \text{Substitute for } \int e^{2x} \cos x \, dx.$$

$$= -e^{2x} \cos x + 2e^{2x} \sin x - 4 \int e^{2x} \sin x \, dx. \quad \text{Simplify.}$$

Now it is a matter of solving for $\int e^{2x} \sin x \, dx$ and including the constant of integration:

$$\int e^{2x} \sin x \, dx = \frac{1}{5} e^{2x} (2 \sin x - \cos x) + C.$$

Related Exercises 27–28 ◀

Integration by Parts for Definite Integrals

Integration by parts with definite integrals presents two options. You can use the method outlined in Examples 1–4 to find an antiderivative and then evaluate it at the upper and lower limits of integration. Alternatively, the limits of integration can be incorporated directly into the integration by parts process. With the second approach, integration by parts for definite integrals has the following form.

> ➤ Integration by parts for definite integrals still has the form
> $$\int u\, dv = uv - \int v\, du.$$
> However, both definite integrals must be written with respect to x.

Integration by Parts for Definite Integrals

Let u and v be differentiable. Then

$$\int_a^b u(x)v'(x)\, dx = u(x)v(x)\Big|_a^b - \int_a^b v(x)u'(x)\, dx.$$

EXAMPLE 5   **A definite integral**   Evaluate $\int_1^2 \ln x\, dx$.

SOLUTION This example is instructive because the integrand does not appear to be a product. The key is to view the integrand as the product $(\ln x)(1\, dx)$. Then the following choices are plausible.

$u = \ln x$	$dv = dx$
$du = \dfrac{1}{x}\, dx$	$v = x$

Using integration by parts, we have

$$\int_1^2 \underbrace{\ln x}_{u}\, \underbrace{dx}_{dv} = \Big(\underbrace{(\ln x)}_{u}\, \underbrace{x}_{v}\Big)\Big|_1^2 - \int_1^2 \underbrace{x}_{v}\, \underbrace{\frac{1}{x}\, dx}_{du} \qquad \text{Integration by parts}$$

$$= (x \ln x - x)\Big|_1^2 \qquad\qquad \text{Integrate and simplify.}$$

$$= (2 \ln 2 - 2) - (0 - 1) \qquad \text{Evaluate.}$$

$$= 2 \ln 2 - 1 \approx 0.386. \qquad \text{Simplify.}$$

Related Exercises 33–34 ◄

In Example 5, we evaluated a definite integral of $\ln x$. The corresponding indefinite integral can be added to our list of integration formulas.

> **QUICK CHECK 2** Verify by differentiation that $\int \ln x\, dx = x \ln x - x + C$. ◄

Integral of $\ln x$

$$\int \ln x\, dx = x \ln x - x + C$$

We now apply integration by parts to a familiar geometry problem.

EXAMPLE 6   **Solids of revolution**   Let R be the region bounded by $y = \ln x$, the x-axis, and the line $x = e$ (**Figure 8.1**). Find the volume of the solid that is generated when the region R is revolved about the x-axis.

SOLUTION Revolving R about the x-axis generates a solid whose volume is computed with the disk method (Section 6.3). Its volume is

$$V = \int_1^e \pi (\ln x)^2\, dx.$$

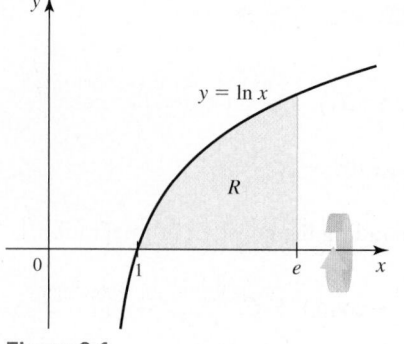

Figure 8.1

➤ Recall that if $f(x) \geq 0$ on $[a, b]$ and the region bounded by the graph of f and the x-axis on $[a, b]$ is revolved about the x-axis, then the volume of the solid generated is $V = \int_a^b \pi f(x)^2 \, dx$.

We integrate by parts with the following assignments.

$u = (\ln x)^2$	$dv = dx$
$du = \dfrac{2 \ln x}{x} dx$	$v = x$

The integration is carried out as follows, using the indefinite integral of $\ln x$ just given:

$$V = \int_1^e \pi (\ln x)^2 \, dx \qquad \text{Disk method}$$

$$= \pi \left(\underbrace{(\ln x)^2}_{u} \underbrace{x}_{v} \Big|_1^e - \int_1^e \underbrace{x}_{v} \underbrace{\frac{2 \ln x}{x}}_{du} dx \right) \qquad \text{Integration by parts}$$

$$= \pi \left(x(\ln x)^2 \Big|_1^e - 2 \int_1^e \ln x \, dx \right) \qquad \text{Simplify.}$$

QUICK CHECK 3 How many times do you need to integrate by parts to reduce $\int_1^e (\ln x)^6 \, dx$ to an integral of $\ln x$? ◄

$$= \pi \left(x(\ln x)^2 \Big|_1^e - 2(x \ln x - x) \Big|_1^e \right) \qquad \int \ln x \, dx = x \ln x - x + C$$

$$= \pi (e(\ln e)^2 - 2e \ln e + 2e - 2) \qquad \text{Evaluate and simplify.}$$

$$= \pi (e - 2) \approx 2.257. \qquad \text{Simplify.}$$

Related Exercises 45, 47 ◄

SECTION 8.2 EXERCISES

Getting Started

1. On which derivative rule is integration by parts based?

2. Use integration by parts to evaluate $\int x \cos x \, dx$ with $u = x$ and $dv = \cos x \, dx$.

3. Use integration by parts to evaluate $\int x \ln x \, dx$ with $u = \ln x$ and $dv = x \, dx$.

4. How is integration by parts used to evaluate a definite integral?

5. What type of integrand is a good candidate for integration by parts?

6. How would you choose dv when evaluating $\int x^n e^{ax} \, dx$ using integration by parts?

7–8. *Use a substitution to reduce the following integrals to $\int \ln u \, du$. Then evaluate the resulting integral using the formula for $\int \ln x \, dx$.*

7. $\int (\sec^2 x) \ln (\tan x + 2) \, dx$ 8. $\int (\cos x) \ln (\sin x) \, dx$

Practice Exercises

9–40. Integration by parts *Evaluate the following integrals using integration by parts.*

9. $\int x \cos 5x \, dx$

10. $\int x \sin 2x \, dx$

11. $\int t e^{6t} \, dt$

12. $\int 2x e^{3x} \, dx$

13. $\int x \ln 10x \, dx$

14. $\int s e^{-2s} \, ds$

15. $\int (2w + 4) \cos 2w \, dw$

16. $\int \theta \sec^2 \theta \, d\theta$

17. $\int x \, 3^x \, dx$

18. $\int x^9 \ln x \, dx$

19. $\int \dfrac{\ln x}{x^{10}} \, dx$

20. $\int \sin^{-1} x \, dx$

21. $\int x \sin x \cos x \, dx$

22. $\int e^{2x} \sin e^x \, dx$

23. $\int x^2 \sin 2x \, dx$

24. $\int x^2 e^{4x} \, dx$

25. $\int t^2 e^{-t} \, dt$

26. $\int t^3 \sin t \, dt$

27. $\int e^x \cos x \, dx$

28. $\int e^{3x} \cos 2x \, dx$

29. $\int e^{-x} \sin 4x \, dx$

30. $\int e^{-2\theta} \sin 6\theta \, d\theta$

31. $\int e^{3x} \sin e^x \, dx$

32. $\int_0^1 x^2 \, 2^x \, dx$

33. $\int_0^\pi x \sin x \, dx$

34. $\int_1^e \ln 2x \, dx$

35. $\int_0^{\pi/2} x \cos 2x \, dx$

36. $\int_0^{\ln 2} x e^x \, dx$

37. $\int_1^{e^2} x^2 \ln x \, dx$

38. $\int x^2 \ln^2 x \, dx$

39. $\int_0^1 \sin^{-1} y \, dy$

40. $\int e^{\sqrt{x}} \, dx$

41. Evaluate the integral in part (a) and then use this result to evaluate the integral in part (b).

   a. $\int \tan^{-1} x \, dx$ **b.** $\int x \tan^{-1} x^2 \, dx$

42–47. Volumes of solids *Find the volume of the solid that is generated when the given region is revolved as described.*

42. The region bounded by $f(x) = \ln x$, $y = 1$, and the coordinate axes is revolved about the x-axis.

43. The region bounded by $f(x) = e^{-x}$, $x = \ln 2$, and the coordinate axes is revolved about the y-axis.

44. The region bounded by $f(x) = \sin x$ and the x-axis on $[0, \pi]$ is revolved about the y-axis.

45. The region bounded by $g(x) = \sqrt{\ln x}$ and the x-axis on $[1, e]$ is revolved about the x-axis.

46. The region bounded by $f(x) = e^{-x}$ and the x-axis on $[0, \ln 2]$ is revolved about the line $x = \ln 2$.

47. The region bounded by $f(x) = x \ln x$ and the x-axis on $[1, e^2]$ is revolved about the x-axis.

48. Integral of $\sec^3 x$ Use integration by parts to show that

$$\int \sec^3 x \, dx = \frac{1}{2} \sec x \tan x + \frac{1}{2} \int \sec x \, dx.$$

49. Explain why or why not Determine whether the following statements are true and give an explanation or counterexample.

   a. $\displaystyle\int uv' dx = \left(\int u \, dx \right)\left(\int v' dx \right).$

   b. $\displaystyle\int uv' dx = uv - \int vu' dx.$

   c. $\displaystyle\int v \, du = uv - \int u \, dv.$

50–53. Reduction formulas *Use integration by parts to derive the following reduction formulas.*

50. $\displaystyle\int x^n e^{ax} \, dx = \frac{x^n e^{ax}}{a} - \frac{n}{a} \int x^{n-1} e^{ax} \, dx$, for $a \neq 0$

51. $\displaystyle\int x^n \cos ax \, dx = \frac{x^n \sin ax}{a} - \frac{n}{a} \int x^{n-1} \sin ax \, dx$, for $a \neq 0$

52. $\displaystyle\int x^n \sin ax \, dx = -\frac{x^n \cos ax}{a} + \frac{n}{a} \int x^{n-1} \cos ax \, dx$, for $a \neq 0$

53. $\displaystyle\int \ln^n x \, dx = x \ln^n x - n \int \ln^{n-1} x \, dx$

54–57. Applying reduction formulas *Use the reduction formulas in Exercises 50–53 to evaluate the following integrals.*

54. $\displaystyle\int x^2 e^{3x} \, dx$ **55.** $\displaystyle\int x^2 \cos 5x \, dx$

56. $\displaystyle\int x^3 \sin x \, dx$ **57.** $\displaystyle\int_1^e \ln^3 x \, dx$

58. Two methods Evaluate $\int_0^{\pi/3} \sin x \ln(\cos x) \, dx$ in the following two ways.

   a. Use integration by parts.   **b.** Use substitution.

59. Two methods

   a. Evaluate $\displaystyle\int \frac{x}{\sqrt{x+1}} \, dx$ using integration by parts.

   b. Evaluate $\displaystyle\int \frac{x}{\sqrt{x+1}} \, dx$ using substitution.

   c. Verify that your answers to parts (a) and (b) are consistent.

60. Two methods

   a. Evaluate $\int x \ln x^2 \, dx$ using the substitution $u = x^2$ and evaluating $\int \ln u \, du$.

   b. Evaluate $\int x \ln x^2 \, dx$ using integration by parts.

   c. Verify that your answers to parts (a) and (b) are consistent.

61. Logarithm base b Prove that $\displaystyle\int \log_b x \, dx = \frac{1}{\ln b}(x \ln x - x) + C.$

62. Two integration methods Evaluate $\int \sin x \cos x \, dx$ using integration by parts. Then evaluate the integral using a substitution. Reconcile your answers.

63. Combining two integration methods Evaluate $\int \cos \sqrt{x} \, dx$ using a substitution followed by integration by parts.

64. Combining two integration methods Evaluate $\int_0^{\pi^2/4} \sin \sqrt{x} \, dx$ using a substitution followed by integration by parts.

65. An identity Show that if f has continuous derivatives on $[a, b]$ and $f'(a) = f'(b) = 0$, then

$$\int_a^b x f''(x) \, dx = f(a) - f(b).$$

66. Integrating derivatives Use integration by parts to show that if f' is continuous on $[a, b]$, then

$$\int_a^b f(x) f'(x) \, dx = \frac{1}{2}(f(b)^2 - f(a)^2).$$

67. Function defined as an integral Find the arc length of the function $f(x) = \int_e^x \sqrt{\ln^2 t - 1} \, dt$ on $[e, e^3]$.

68. Log integrals Use integration by parts to show that for $m \neq -1$,

$$\int x^m \ln x \, dx = \frac{x^{m+1}}{m+1}\left(\ln x - \frac{1}{m+1} \right) + C$$

and for $m = -1$,

$$\int \frac{\ln x}{x} \, dx = \frac{1}{2} \ln^2 x + C.$$

69. Comparing volumes Let R be the region bounded by $y = \sin x$ and the x-axis on the interval $[0, \pi]$. Which is greater, the volume of the solid generated when R is revolved about the x-axis or the volume of the solid generated when R is revolved about the y-axis?

70. A useful integral

   a. Use integration by parts to show that if f' is continuous, then

$$\int x f'(x) \, dx = x f(x) - \int f(x) \, dx.$$

   b. Use part (a) to evaluate $\int x e^{3x} \, dx$.

71. Solid of revolution Find the volume of the solid generated when the region bounded by $y = \cos x$ and the x-axis on the interval $[0, \pi/2]$ is revolved about the y-axis.

Explorations and Challenges

72. Between the sine and inverse sine Find the area of the region bounded by the curves $y = \sin x$ and $y = \sin^{-1} x$ on the interval $[0, 1/2]$.

73. Two useful exponential integrals Use integration by parts to derive the following formulas for real numbers a and b.

$$\int e^{ax} \sin bx \, dx = \frac{e^{ax}(a \sin bx - b \cos bx)}{a^2 + b^2} + C$$

$$\int e^{ax} \cos bx \, dx = \frac{e^{ax}(a \cos bx + b \sin bx)}{a^2 + b^2} + C$$

74. Integrating inverse functions Assume f has an inverse on its domain.

   a. Let $y = f^{-1}(x)$ and show that
   $$\int f^{-1}(x)\, dx = \int y f'(y)\, dy.$$

   b. Use part (a) to show that
   $$\int f^{-1}(x)\, dx = y f(y) - \int f(y)\, dy.$$

   c. Use the result of part (b) to evaluate $\int \ln x\, dx$ (express the result in terms of x).

   d. Use the result of part (b) to evaluate $\int \sin^{-1} x\, dx$.

   e. Use the result of part (b) to evaluate $\int \tan^{-1} x\, dx$.

T 75. Oscillator displacements Suppose a mass on a spring that is slowed by friction has the position function $s(t) = e^{-t} \sin t$.

   a. Graph the position function. At what times does the oscillator pass through the position $s = 0$?

   b. Find the average value of the position on the interval $[0, \pi]$.

   c. Generalize part (b) and find the average value of the position on the interval $[n\pi, (n + 1)\pi]$, for $n = 0, 1, 2, \ldots$.

   d. Let a_n be the absolute value of the average position on the interval $[n\pi, (n + 1)\pi]$, for $n = 0, 1, 2, \ldots$. Describe the pattern in the numbers $a_0, a_1, a_2, \ldots$.

76. Find the error Suppose you evaluate $\int \dfrac{dx}{x}$ using integration by parts. With $u = 1/x$ and $dv = dx$, you find that $du = -1/x^2\, dx$, $v = x$, and
$$\int \frac{dx}{x} = \left(\frac{1}{x}\right) x - \int x\left(-\frac{1}{x^2}\right) dx = 1 + \int \frac{dx}{x}.$$

You conclude that $0 = 1$. Explain the problem with the calculation.

77. Tabular integration Consider the integral $\int f(x) g(x)\, dx$, where f can be differentiated repeatedly and g can be integrated repeatedly. Let G_k represent the result of calculating k indefinite integrals of g, where the constants of integration are omitted.

   a. Show that integration by parts, when applied to $\int f(x) g(x)\, dx$ with the choices $u = f(x)$ and $dv = g(x)\, dx$, leads to $\int f(x) g(x)\, dx = f(x) G_1(x) - \int f'(x) G_1(x)\, dx$. This formula can be remembered by utilizing the following table, where a right arrow represents a product of functions on the right side of the integration by parts formula, and a left arrow represents the *integral* of a product of functions (also appearing on the right side of the formula). Explain the significance of the signs associated with the arrows.

f and its derivatives	g and its integrals
$f(x)$	$g(x)$
$f'(x)$	$G_1(x)$

   b. Perform integration by parts again on $\int f'(x) G_1(x)\, dx$ (from part (a)) with the choices $u = f'(x)$ and $dv = G_1(x)\, dx$ to show that $\int f(x) g(x)\, dx = f(x) G_1(x) - f'(x) G_2(x) + \int f''(x) G_2(x)\, dx$. Explain the connection between this integral formula and the following table, paying close attention to the signs attached to the arrows.

f and its derivatives	g and its integrals
$f(x)$	$g(x)$
$f'(x)$	$G_1(x)$
$f''(x)$	$G_2(x)$

   c. Continue the pattern established in parts (a) and (b) and integrate by parts a third time. Write the integral formula that results from three applications of integration by parts, and construct the associated *tabular integration* table (include signs of the arrows).

   d. The tabular integration table from part (c) is easily extended to allow for as many steps as necessary in the process of integration by parts. Evaluate $\int x^2 e^{x/2}\, dx$ by constructing an appropriate table, and explain why the process terminates after four rows of the table have been filled in.

   e. Use tabular integration to evaluate $\int x^3 \cos x\, dx$. How many rows of the table are necessary? Why?

   f. Explain why tabular integration is particularly suited to integrals of the form $\int p_n(x) g(x)\, dx$, where p_n is a polynomial of degree $n > 0$ (and where, as before, we assume g is easily integrated as many times as necessary).

78. Practice with tabular integration Evaluate the following integrals using tabular integration (refer to Exercise 77).

   a. $\displaystyle\int x^4 e^x\, dx$

   b. $\displaystyle\int 7x e^{3x}\, dx$

   c. $\displaystyle\int_{-1}^{0} 2x^2 \sqrt{x + 1}\, dx$

   d. $\displaystyle\int (x^3 - 2x) \sin 2x\, dx$

   e. $\displaystyle\int \frac{2x^2 - 3x}{(x - 1)^3}\, dx$

   f. $\displaystyle\int \frac{x^2 + 3x + 4}{\sqrt[3]{2x + 1}}\, dx$

   g. Why doesn't tabular integration work well when applied to $\displaystyle\int \frac{x}{\sqrt{1 - x^2}}\, dx$? Evaluate this integral using a different method.

79. Tabular integration extended Refer to Exercise 77.

   a. The following table shows the method of tabular integration applied to $\int e^x \cos x\, dx$. Use the table to express $\int e^x \cos x\, dx$ in terms of the sum of functions and an indefinite integral.

f and its derivatives	g and its integrals
e^x	$\cos x$
e^x	$\sin x$
e^x	$-\cos x$

   b. Solve the equation in part (a) for $\int e^x \cos x\, dx$.

   c. Evaluate $\int e^{-2x} \sin 3x\, dx$ by applying the idea from parts (a) and (b).

80. An identity Show that if f and g have continuous second derivatives and $f(0) = f(1) = g(0) = g(1) = 0$, then
$$\int_0^1 f''(x) g(x)\, dx = \int_0^1 f(x) g''(x)\, dx.$$

81. Possible and impossible integrals Let $I_n = \int x^n e^{-x^2}\, dx$, where n is a nonnegative integer.

   a. $I_0 = \int e^{-x^2}\, dx$ cannot be expressed in terms of elementary functions. Evaluate I_1.

   b. Use integration by parts to evaluate I_3.

c. Use integration by parts and the result of part (b) to evaluate I_5.

d. Show that, in general, if n is odd, then $I_n = -\dfrac{1}{2}e^{-x^2}p_{n-1}(x)$, where p_{n-1} is a polynomial of degree $n - 1$.

e. Argue that if n is even, then I_n cannot be expressed in terms of elementary functions.

82. A family of exponentials The curves $y = xe^{-ax}$ are shown in the figure for $a = 1, 2,$ and 3.

a. Find the area of the region bounded by $y = xe^{-x}$ and the x-axis on the interval $[0, 4]$.

b. Find the area of the region bounded by $y = xe^{-ax}$ and the x-axis on the interval $[0, 4]$, where $a > 0$.

c. Find the area of the region bounded by $y = xe^{-ax}$ and the x-axis on the interval $[0, b]$. Because this area depends on a and b, we call it $A(a, b)$.

d. Use part (c) to show that $A(1, \ln b) = 4A\left(2, \dfrac{\ln b}{2}\right)$.

e. Does this pattern continue? Is it true that $A(1, \ln b) = a^2 A(a, (\ln b)/a)$?

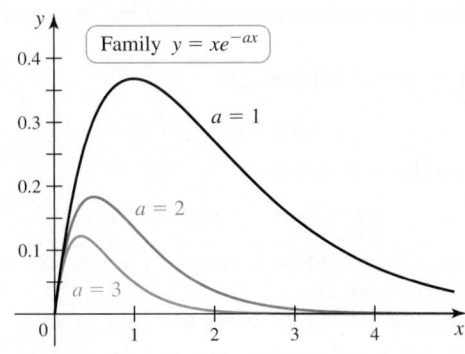

Family $y = xe^{-ax}$

QUICK CHECK ANSWERS

1. Let $u = x$ and $dv = \cos x\, dx$.

2. $\dfrac{d}{dx}(x \ln x - x + C) = \ln x$

3. Integration by parts must be applied five times. ◄

8.3 Trigonometric Integrals

At the moment, our inventory of integrals involving trigonometric functions is rather limited. For example, we can integrate the basic trigonometric functions found in Table 8.1, but have not yet encountered integrals such as $\int \cos^5 x\, dx$ and $\int \cos^2 x \sin^4 x\, dx$. The goal of this section is to develop techniques for evaluating integrals involving powers of trigonometric functions. These techniques are indispensable when we encounter *trigonometric substitutions* in the next section.

Integrating Powers of $\sin x$ or $\cos x$

Two strategies are used when evaluating integrals of the form $\int \sin^m x\, dx$ or $\int \cos^n x\, dx$, where m and n are positive integers. Both strategies use trigonometric identities to recast the integrand, as shown in the first example.

> Some of the techniques described in this section also work for negative powers of trigonometric functions.

EXAMPLE 1 **Powers of sine or cosine** Evaluate the following integrals.

a. $\int \cos^5 x\, dx$ **b.** $\int \sin^4 x\, dx$

SOLUTION

> Pythagorean identities:
> $$\cos^2 x + \sin^2 x = 1$$
> $$1 + \tan^2 x = \sec^2 x$$
> $$\cot^2 x + 1 = \csc^2 x$$

a. Integrals involving odd powers of $\cos x$ (or $\sin x$) are most easily evaluated by splitting off a single factor of $\cos x$ (or $\sin x$). In this case, we rewrite $\cos^5 x$ as $\cos^4 x \cdot \cos x$. Notice that $\cos^4 x$ can be written in terms of $\sin x$ using the identity $\cos^2 x = 1 - \sin^2 x$. The result is an integrand that readily yields to the substitution $u = \sin x$:

$$\int \cos^5 x\, dx = \int \cos^4 x \cos x\, dx \qquad \text{Split off } \cos x.$$

$$= \int (1 - \sin^2 x)^2 \cos x\, dx \qquad \text{Pythagorean identity}$$

$$= \int (1 - u^2)^2\, du \qquad \text{Let } u = \sin x;\ du = \cos x\, dx.$$

$$= \int (1 - 2u^2 + u^4)\, du \qquad \text{Expand.}$$

$$= u - \frac{2}{3}u^3 + \frac{1}{5}u^5 + C \qquad \text{Integrate.}$$

$$= \sin x - \frac{2}{3}\sin^3 x + \frac{1}{5}\sin^5 x + C. \qquad \text{Replace } u \text{ with } \sin x.$$

▶ The half-angle formulas for $\sin^2 x$ and $\cos^2 x$ are easily confused. Use the phrase "sine is minus" to remember that a minus sign is associated with the half-angle formula for $\sin^2 x$, whereas a positive sign is used for $\cos^2 x$.

b. With even positive powers of $\sin x$ or $\cos x$, we use the half-angle formulas

$$\sin^2 x = \frac{1 - \cos 2x}{2} \quad \text{and} \quad \cos^2 x = \frac{1 + \cos 2x}{2}$$

to reduce the powers in the integrand:

$$\int \sin^4 x \, dx = \int \underbrace{\left(\frac{1 - \cos 2x}{2} \right)^2}_{\sin^2 x} dx \qquad \text{Half-angle formula}$$

$$= \frac{1}{4} \int (1 - 2 \cos 2x + \cos^2 2x) \, dx. \quad \text{Expand the integrand.}$$

Using the half-angle formula for $\cos^2 2x$, the evaluation may be completed:

$$\int \sin^4 x \, dx = \frac{1}{4} \int \left(1 - 2 \cos 2x + \underbrace{\frac{1 + \cos 4x}{2}}_{\cos^2 2x} \right) dx \quad \text{Half-angle formula}$$

$$= \frac{1}{4} \int \left(\frac{3}{2} - 2 \cos 2x + \frac{1}{2} \cos 4x \right) dx \qquad \text{Simplify.}$$

$$= \frac{3x}{8} - \frac{1}{4} \sin 2x + \frac{1}{32} \sin 4x + C. \qquad \text{Evaluate the integrals.}$$

Related Exercises 9–11 ◀

QUICK CHECK 1 Evaluate $\int \sin^3 x \, dx$ by splitting off a factor of $\sin x$, rewriting $\sin^2 x$ in terms of $\cos x$, and using an appropriate u-substitution. ◀

Integrating Products of Powers of $\sin x$ and $\cos x$

We now consider integrals of the form $\int \sin^m x \cos^n x \, dx$. If m is an odd, positive integer, we split off a factor of $\sin x$ and write the remaining even power of $\sin x$ in terms of cosine functions. This step prepares the integrand for the substitution $u = \cos x$, and the resulting integral is readily evaluated. A similar strategy is used when n is an odd, positive integer.

If both m and n are even positive integers, the half-angle formulas are used to transform the integrand into a polynomial in $\cos 2x$, each of whose terms can be integrated, as shown in Example 2.

EXAMPLE 2   Products of sine and cosine Evaluate the following integrals.

a. $\int \sin^4 x \cos^2 x \, dx$ **b.** $\int \sin^3 x \cos^{-2} x \, dx$

SOLUTION

a. When both powers are even positive integers, the half-angle formulas are used:

$$\int \sin^4 x \cos^2 x \, dx = \int \underbrace{\left(\frac{1 - \cos 2x}{2} \right)^2}_{\sin^2 x} \underbrace{\left(\frac{1 + \cos 2x}{2} \right)}_{\cos^2 x} dx \qquad \text{Half-angle formulas}$$

$$= \frac{1}{8} \int (1 - \cos 2x - \cos^2 2x + \cos^3 2x) \, dx. \quad \text{Expand.}$$

The third term in the integrand is rewritten with a half-angle formula. For the last term, a factor of $\cos 2x$ is split off, and the resulting even power of $\cos 2x$ is written in terms of $\sin 2x$ to prepare for a u-substitution:

$$\int \sin^4 x \cos^2 x \, dx =$$

$$\frac{1}{8} \int \left(1 - \cos 2x - \left(\overbrace{\frac{1 + \cos 4x}{2}}^{\cos^2 2x} \right) \right) dx + \frac{1}{8} \int \overbrace{(1 - \sin^2 2x)}^{\cos^2 2x} \cdot \cos 2x \, dx.$$

Finally, the integrals are evaluated, using the substitution $u = \sin 2x$ for the second integral. After simplification, we find that

$$\int \sin^4 x \cos^2 x \, dx = \frac{1}{16} x - \frac{1}{64} \sin 4x - \frac{1}{48} \sin^3 2x + C.$$

b. When at least one power is odd and positive, the following approach works:

$$\int \sin^3 x \cos^{-2} x \, dx = \int \sin^2 x \cos^{-2} x \cdot \sin x \, dx \qquad \text{Split off } \sin x.$$

$$= \int (1 - \cos^2 x) \cos^{-2} x \cdot \sin x \, dx \qquad \text{Pythagorean identity}$$

$$= -\int (1 - u^2) u^{-2} \, du \qquad u = \cos x; \, du = -\sin x \, dx$$

$$= \int (1 - u^{-2}) \, du = u + \frac{1}{u} + C \qquad \text{Evaluate the integral.}$$

$$= \cos x + \sec x + C. \qquad \text{Replace } u \text{ with } \cos x.$$

Related Exercises 15–17 ◄

tan²x sec²x (tan x sec x)
(sec²x − 1)(sec²x)(tan x sec x)
(sec⁴x − sec²x)(tan x sec x)

tan⁴x (tan²x) sec²x
(tan⁶x + tan⁴x)sec²x
5(u⁶ + u⁴) du du
u⁷/7 + u⁵/5 ⟹ tan⁷x/7 + tan⁵x/5 +C

sec⁶/... sec⁴/... + c
u² − u
u³/3 − u²/2 + c

QUICK CHECK 2 What strategy would you use to evaluate $\int \sin^3 x \cos^3 x \, dx$? ◄

Table 8.2 summarizes the techniques used to evaluate integrals of the form $\int \sin^m x \cos^n x \, dx$.

Table 8.2

$\int \sin^m x \cos^n x \, dx$	Strategy
m odd and positive, n real	Split off $\sin x$, rewrite the resulting even power of $\sin x$ in terms of $\cos x$, and then use $u = \cos x$.
n odd and positive, m real	Split off $\cos x$, rewrite the resulting even power of $\cos x$ in terms of $\sin x$, and then use $u = \sin x$.
m and n both even, nonnegative integers	Use half-angle formulas to transform the integrand into a polynomial in $\cos 2x$, and apply the preceding strategies once again to powers of $\cos 2x$ greater than 1.

► If both m and n are odd, you may split off $\sin x$ or $\cos x$, though in practice, one choice may be more efficient than the other.

Reduction Formulas

Evaluating an integral such as $\int \sin^8 x \, dx$ using the method of Example 1b is tedious, at best. For this reason, *reduction formulas* have been developed to ease the workload. A reduction formula equates an integral involving a power of a function with another integral in which the power is reduced; several reduction formulas were encountered in Exercises 50–53 of Section 8.2. Here are some frequently used reduction formulas for trigonometric integrals.

∫tan⁴x sec²x dx
tan⁴x(tan²x+1)dx
tan⁶x + tan⁴x dx
tan⁹x tan⁷x + tan⁵x
tan⁴x(tan²x+1)
u⁴ du
u⁵/5 +c du
tan⁵/5 + c

Reduction Formulas

Assume n is a positive integer.

1. $\displaystyle \int \sin^n x \, dx = -\frac{\sin^{n-1} x \cos x}{n} + \frac{n-1}{n} \int \sin^{n-2} x \, dx$

2. $\displaystyle \int \cos^n x \, dx = \frac{\cos^{n-1} x \sin x}{n} + \frac{n-1}{n} \int \cos^{n-2} x \, dx$

3. $\displaystyle \int \tan^n x \, dx = \frac{\tan^{n-1} x}{n-1} - \int \tan^{n-2} x \, dx, \quad n \neq 1$

4. $\displaystyle \int \sec^n x \, dx = \frac{\sec^{n-2} x \tan x}{n-1} + \frac{n-2}{n-1} \int \sec^{n-2} x \, dx, \quad n \neq 1$

Formulas 1, 3, and 4 are derived in Exercises 72–74. The derivation of formula 2 is similar to that of formula 1.

EXAMPLE 3 **Powers of tan x** Evaluate $\int \tan^4 x \, dx$.

SOLUTION Reduction formula 3 gives

$$\int \tan^4 x \, dx = \frac{1}{3} \tan^3 x - \underbrace{\int \tan^2 x \, dx}_{\text{use (3) again}}$$

$$= \frac{1}{3} \tan^3 x - \left(\tan x - \underbrace{\int \tan^0 x \, dx}_{=1} \right)$$

$$= \frac{1}{3} \tan^3 x - \tan x + x + C.$$

An alternative solution uses the identity $\tan^2 x = \sec^2 x - 1$:

$$\int \tan^4 x \, dx = \int \tan^2 x \underbrace{\left(\sec^2 x - 1 \right)}_{\tan^2 x} dx$$

$$= \int \tan^2 x \sec^2 x \, dx - \int \tan^2 x \, dx.$$

The substitution $u = \tan x$, $du = \sec^2 x \, dx$ is used in the first integral, and the identity $\tan^2 x = \sec^2 x - 1$ is used again in the second integral:

$$\int \tan^4 x \, dx = \underbrace{\int \tan^2 x \sec^2 x \, dx}_{u^2 \quad du} - \int \tan^2 x \, dx$$

$$= \int u^2 \, du - \int \left(\sec^2 x - 1 \right) dx \qquad \text{Substitution and identity}$$

$$= \frac{u^3}{3} - \tan x + x + C \qquad \text{Evaluate integrals.}$$

$$= \frac{1}{3} \tan^3 x - \tan x + x + C. \qquad u = \tan x$$

Related Exercises 30–31 ◄

Note that for odd powers of $\tan x$ and $\sec x$, the use of reduction formula 3 or 4 will eventually lead to $\int \tan x \, dx$ or $\int \sec x \, dx$. Table 8.1 in Section 8.1 gives these integrals, along with the integrals of $\cot x$ and $\csc x$.

Integrating Products of Powers of $\tan x$ and $\sec x$

Integrals of the form $\int \tan^m x \sec^n x \, dx$ are evaluated using methods analogous to those used for $\int \sin^m x \cos^n x \, dx$. For example, if n is an even positive integer, we split off a factor of $\sec^2 x$ and write the remaining even power of $\sec x$ in terms of $\tan x$. This step prepares the integral for the substitution $u = \tan x$. If m is odd and positive, we split off a factor of $\sec x \tan x$ (the derivative of $\sec x$), which prepares the integral for the substitution $u = \sec x$. If m is even and n is odd, the integrand is expressed as a polynomial in $\sec x$, each of whose terms is handled by a reduction formula. Example 4 illustrates these techniques.

EXAMPLE 4 **Products of tan x and sec x** Evaluate the following integrals.

a. $\int \tan^3 x \sec^4 x \, dx$ **b.** $\int \tan^2 x \sec x \, dx$

SOLUTION

a. With an even power of $\sec x$, we split off a factor of $\sec^2 x$ and prepare the integral for the substitution $u = \tan x$:

$$\int \tan^3 x \sec^4 x \, dx = \int \tan^3 x \sec^2 x \cdot \sec^2 x \, dx$$

$$= \int \tan^3 x \left(\tan^2 x + 1 \right) \cdot \sec^2 x \, dx \qquad \sec^2 x = \tan^2 x + 1$$

$$= \int u^3 \left(u^2 + 1 \right) du \qquad u = \tan x; \, du = \sec^2 x \, dx$$

$$= \frac{1}{6} \tan^6 x + \frac{1}{4} \tan^4 x + C. \qquad \text{Evaluate; } u = \tan x.$$

▶ In Example 4a, the two methods produce results that look different but are equivalent. This is common when evaluating trigonometric integrals. For instance, evaluate $\int \sin^4 x \, dx$ using reduction formula 1, and compare your answer to

$$\frac{3x}{8} - \frac{1}{4} \sin 2x + \frac{1}{32} \sin 4x + C,$$

the solution found in Example 1b.

Because the integrand also has an odd power of tan x, an alternative solution is to split off a factor of sec x tan x and prepare the integral for the substitution $u = \sec x$:

$$\int \tan^3 x \sec^4 x \, dx = \int \underbrace{\tan^2 x}_{\sec^2 x - 1} \sec^3 x \cdot \sec x \tan x \, dx$$

$$= \int (\sec^2 x - 1) \sec^3 x \cdot \sec x \tan x \, dx$$

$$= \int (u^2 - 1) u^3 \, du \qquad \begin{array}{l} u = \sec x; \\ du = \sec x \tan x \, dx \end{array}$$

$$= \frac{1}{6} \sec^6 x - \frac{1}{4} \sec^4 x + C. \qquad \text{Evaluate; } u = \sec x.$$

The apparent difference in the two solutions given here is reconciled by using the identity $1 + \tan^2 x = \sec^2 x$ to transform the second result into the first, the only difference being an additive constant, which is part of C.

b. In this case, we write the even power of tan x in terms of sec x:

$$\int \tan^2 x \sec x \, dx = \int (\sec^2 x - 1) \sec x \, dx \qquad \tan^2 x = \sec^2 x - 1$$

$$= \int \sec^3 x \, dx - \int \sec x \, dx \qquad \text{Expand.}$$

$$= \overbrace{\frac{1}{2} \sec x \tan x + \frac{1}{2} \int \sec x \, dx}^{\text{reduction formula 4}} - \int \sec x \, dx \qquad \text{Use reduction formula.}$$

$$= \frac{1}{2} \sec x \tan x - \frac{1}{2} \ln |\sec x + \tan x| + C. \qquad \begin{array}{l} \text{Add secant integrals; use} \\ \text{Table 8.1 in Section 8.1.} \end{array}$$

Related Exercises 33–35 ◀

Table 8.3 summarizes the methods used to integrate $\int \tan^m x \sec^n x \, dx$. Analogous techniques are used for $\int \cot^m x \csc^n x \, dx$.

Table 8.3

$\int \tan^m x \sec^n x \, dx$	Strategy
n even and positive, m real	Split off $\sec^2 x$, rewrite the remaining even power of sec x in terms of tan x, and use $u = \tan x$.
m odd and positive, n real	Split off sec x tan x, rewrite the remaining even power of tan x in terms of sec x, and use $u = \sec x$.
m even and positive, n odd and positive	Rewrite the even power of tan x in terms of sec x to produce a polynomial in sec x; apply reduction formula 4 to each term.

SECTION 8.3 EXERCISES

Getting Started

1. State the half-angle identities used to integrate $\sin^2 x$ and $\cos^2 x$.

2. State the three Pythagorean identities.

3. Describe the method used to integrate $\sin^3 x$.

4. Describe the method used to integrate $\sin^m x \cos^n x$, for m even and n odd.

5. What is a reduction formula?

6. How would you evaluate $\int \cos^2 x \sin^3 x \, dx$?

7. How would you evaluate $\int \tan^{10} x \sec^2 x \, dx$?

8. How would you evaluate $\int \sec^{12} x \tan x \, dx$?

Practice Exercises

9–61. Trigonometric integrals *Evaluate the following integrals.*

9. $\displaystyle\int \cos^3 x \, dx$

10. $\displaystyle\int \sin^3 x \, dx$

11. $\displaystyle\int \sin^2 3x \, dx$

12. $\displaystyle\int \cos^4 2\theta \, d\theta$

13. $\displaystyle\int \sin^5 x \, dx$

14. $\displaystyle\int \cos^3 20x \, dx$

15. $\displaystyle\int \sin^3 x \cos^2 x \, dx$

16. $\displaystyle\int \sin^2 \theta \cos^5 \theta \, d\theta$

17. $\displaystyle\int \cos^3 x \sqrt{\sin x} \, dx$

18. $\displaystyle\int \sin^3 \theta \cos^{-2} \theta \, d\theta$

19. $\displaystyle\int_0^{\pi/3} \sin^5 x \cos^{-2} x \, dx$

20. $\displaystyle\int \sin^{-3/2} x \cos^3 x \, dx$

21. $\int_0^{\pi/2} \cos^3 x \sqrt{\sin^3 x}\, dx$ **22.** $\int_{\pi/4}^{\pi/2} \sin^2 2x \cos^3 2x\, dx$

23. $\int \sin^2 x \cos^2 x\, dx$ **24.** $\int \sin^3 x \cos^5 x\, dx$

25. $\int \sin^2 x \cos^4 x\, dx$ **26.** $\int \sin^3 x \cos^{3/2} x\, dx$

27. $\int \tan^2 x\, dx$ **28.** $\int 6 \sec^4 x\, dx$

29. $\int \cot^4 x\, dx$ **30.** $\int \tan^3 \theta\, d\theta$

31. $\int 20 \tan^6 x\, dx$ **32.** $\int \cot^5 3x\, dx$

33. $\int 10 \tan^9 x \sec^2 x\, dx$ **34.** $\int \tan^9 x \sec^4 x\, dx$

35. $\int \tan x \sec^3 x\, dx$ **36.** $\int \tan 4x \sec^{3/2} 4x\, dx$

37. $\int \dfrac{\sec^4(\ln \theta)}{\theta}\, d\theta$ **38.** $\int \tan^5 \theta \sec^4 \theta\, d\theta$

39. $\int_{-\pi/3}^{\pi/3} \sqrt{\sec^2 \theta - 1}\, d\theta$ **40.** $\int_0^{\pi/6} \tan^5 2x \sec 2x\, dx$

41. $\int_0^{\pi/4} \sec^7 x \sin x\, dx$ **42.** $\int \sqrt{\tan x}\, \sec^4 x\, dx$

43. $\int \tan^3 4x\, dx$ **44.** $\int \dfrac{\sec^2 x}{\tan^5 x}\, dx$

45. $\int \sec^2 x \tan^{1/2} x\, dx$ **46.** $\int \sec^{-2} x \tan^3 x\, dx$

47. $\int \dfrac{\csc^4 x}{\cot^2 x}\, dx$ **48.** $\int \csc^{10} x \cot x\, dx$

49. $\int_{\pi/20}^{\pi/10} \csc^2 5w \cot^4 5w\, dw$ **50.** $\int \csc^{10} x \cot^3 x\, dx$

51. $\int (\csc^2 x + \csc^4 x)\, dx$ **52.** $\int_0^{\pi/8} (\tan 2x + \tan^3 2x)\, dx$

53. $\int_0^{\pi/4} \sec^4 \theta\, d\theta$ **54.** $\int_0^{\sqrt{\pi/2}} x \sin^3(x^2)\, dx$

55. $\int_{\pi/6}^{\pi/3} \cot^3 \theta\, d\theta$ **56.** $\int_0^{\pi/4} \tan^3 \theta \sec^2 \theta\, d\theta$

57. $\int_0^{\pi} (1 - \cos 2x)^{3/2}\, dx$ **58.** $\int_{-\pi/4}^{\pi/4} \sqrt{1 + \cos 4x}\, dx$

59. $\int_0^{\pi/2} \sqrt{1 - \cos 2x}\, dx$ **60.** $\int_0^{\pi/8} \sqrt{1 - \cos 8x}\, dx$

61. $\int_0^{\pi/4} (1 + \cos 4x)^{3/2}\, dx$

62. Arc length Find the length of the curve $y = \ln(\sec x)$, for $0 \le x \le \pi/4$.

63. Explain why or why not Determine whether the following statements are true and give an explanation or counterexample.
 a. If m is a positive integer, then $\int_0^{\pi} \cos^{2m+1} x\, dx = 0$.
 b. If m is a positive integer, then $\int_0^{\pi} \sin^m x\, dx = 0$.

64. Sine football Find the volume of the solid generated when the region bounded by $y = \sin x$ and the x-axis on the interval $[0, \pi]$ is revolved about the x-axis.

65. Volume Find the volume of the solid generated when the region bounded by $y = \sin^2 x \cos^{3/2} x$ and the x-axis on the interval $[0, \pi/2]$ is revolved about the x-axis.

66. Particle position A particle moves along a line with a velocity (in m/s) given by $v(t) = \sec^4 \dfrac{\pi t}{12}$, for $0 \le t \le 5$, where t is measured in seconds. Determine the position function $s(t)$, for $0 \le t \le 5$. Assume $s(0) = 0$.

67–70. Integrals of the form $\int \sin mx \cos nx\, dx$ *Use the following three identities to evaluate the given integrals.*

$$\sin mx \sin nx = \frac{1}{2}(\cos((m-n)x) - \cos((m+n)x))$$
$$\sin mx \cos nx = \frac{1}{2}(\sin((m-n)x) + \sin((m+n)x))$$
$$\cos mx \cos nx = \frac{1}{2}(\cos((m-n)x) + \cos((m+n)x))$$

67. $\int \sin 3x \cos 7x\, dx$ **68.** $\int \sin 5x \sin 7x\, dx$

69. $\int \sin 3x \sin 2x\, dx$ **70.** $\int \cos x \cos 2x\, dx$

Explorations and Challenges

71. Prove the following **orthogonality relations** (which are used to generate *Fourier series*). Assume m and n are integers with $m \ne n$.
 a. $\int_0^{\pi} \sin mx \sin nx\, dx = 0$ **b.** $\int_0^{\pi} \cos mx \cos nx\, dx = 0$
 c. $\int_0^{\pi} \sin mx \cos nx\, dx = 0$, for $|m + n|$ even

72. A sine reduction formula Use integration by parts to obtain the reduction formula for positive integers n:
$$\int \sin^n x\, dx = -\sin^{n-1} x \cos x + (n-1) \int \sin^{n-2} x \cos^2 x\, dx.$$
Then use an identity to obtain the reduction formula
$$\int \sin^n x\, dx = -\frac{\sin^{n-1} x \cos x}{n} + \frac{n-1}{n} \int \sin^{n-2} x\, dx.$$
Use this reduction formula to evaluate $\int \sin^6 x\, dx$.

73. A tangent reduction formula Prove that for positive integers $n \ne 1$,
$$\int \tan^n x\, dx = \frac{\tan^{n-1} x}{n-1} - \int \tan^{n-2} x\, dx.$$
Use the formula to evaluate $\int_0^{\pi/4} \tan^3 x\, dx$.

74. A secant reduction formula Prove that for positive integers $n \ne 1$,
$$\int \sec^n x\, dx = \frac{\sec^{n-2} x \tan x}{n-1} + \frac{n-2}{n-1} \int \sec^{n-2} x\, dx.$$
(*Hint:* Integrate by parts with $u = \sec^{n-2} x$ and $dv = \sec^2 x\, dx$.)

75. Exploring powers of sine and cosine

a. Graph the functions $f_1(x) = \sin^2 x$ and $f_2(x) = \sin^2 2x$ on the interval $[0, \pi]$. Find the area under these curves on $[0, \pi]$.

b. Graph a few more of the functions $f_n(x) = \sin^2 nx$ on the interval $[0, \pi]$, where n is a positive integer. Find the area under these curves on $[0, \pi]$. Comment on your observations.

c. Prove that $\int_0^\pi \sin^2 nx \, dx$ has the same value for all positive integers n.

d. Does the conclusion of part (c) hold if sine is replaced by cosine?

e. Repeat parts (a), (b), and (c) with $\sin^2 x$ replaced by $\sin^4 x$. Comment on your observations.

QUICK CHECK ANSWERS

1. $\frac{1}{3} \cos^3 x - \cos x + C$ **2.** Write $\int \sin^3 x \cos^3 x \, dx = \int \sin^2 x \cos^3 x \sin x \, dx = \int (1 - \cos^2 x) \cos^3 x \sin x \, dx$. Then use the substitution $u = \cos x$. Or begin by writing $\int \sin^3 x \cos^3 x \, dx = \int \sin^3 x \cos^2 x \cos x \, dx$. ◄

8.4 Trigonometric Substitutions

In Example 4 of Section 6.5, we wrote the arc length integral for the segment of the parabola $y = x^2$ on the interval $[0, 2]$ as

$$\int_0^2 \sqrt{1 + 4x^2} \, dx = \int_0^2 2\sqrt{\tfrac{1}{4} + x^2} \, dx.$$

At the time, we did not have the analytical methods needed to evaluate this integral. The difficulty with $\int_0^2 \sqrt{1 + 4x^2} \, dx$ is that the square root of a sum (or difference) of two squares is not easily simplified. On the other hand, the square root of a product of two squares is easily simplified: $\sqrt{A^2 B^2} = |AB|$. If we could somehow replace $1 + 4x^2$ with a product of squares, this integral might be easier to evaluate. The goal of this section is to introduce techniques that transform sums of squares $a^2 + x^2$ (and the difference of squares $a^2 - x^2$ and $x^2 - a^2$) into products of squares.

Integrals similar to the arc length integral for the parabola arise in many different situations. For example, electrostatic, magnetic, and gravitational forces obey an inverse square law (their strength is proportional to $1/r^2$, where r is a distance). Computing these force fields in two dimensions leads to integrals such as $\int \dfrac{dx}{\sqrt{x^2 + a^2}}$ or $\int \dfrac{dx}{(x^2 + a^2)^{3/2}}$.

It turns out that integrals containing the terms $a^2 \pm x^2$ or $x^2 - a^2$, where a is a constant, can be simplified using somewhat unexpected substitutions involving trigonometric functions. The new integrals produced by these substitutions are often trigonometric integrals of the variety studied in the preceding section.

Integrals Involving $a^2 - x^2$

Suppose you are faced with an integral whose integrand contains the term $a^2 - x^2$, where a is a positive constant. Observe what happens when x is replaced with $a \sin \theta$:

$$
\begin{aligned}
a^2 - x^2 &= a^2 - (a \sin \theta)^2 && \text{Replace } x \text{ with } a \sin \theta. \\
&= a^2 - a^2 \sin^2 \theta && \text{Simplify.} \\
&= a^2 (1 - \sin^2 \theta) && \text{Factor.} \\
&= a^2 \cos^2 \theta. && 1 - \sin^2 \theta = \cos^2 \theta
\end{aligned}
$$

This calculation shows that the change of variables $x = a \sin \theta$ turns the difference $a^2 - x^2$ into the product $a^2 \cos^2 \theta$. The resulting integral—now with respect to θ—is often easier to evaluate than the original integral. The details of this procedure are spelled out in the following examples.

> ▶ To understand how a difference of squares is rewritten as a product of squares, think of the Pythagorean Theorem: $a^2 + b^2 = c^2$. A rearrangement of this theorem leads to the standard substitution for integrals involving the difference of squares $a^2 - x^2$. The term $\sqrt{a^2 - x^2}$ is the length of one side of a right triangle whose hypotenuse has length a and whose other side has length x. Labeling one acute angle θ, we see that the substitution $x = a \sin \theta$ transforms $a^2 - x^2$ into $a^2 \cos^2 \theta$.

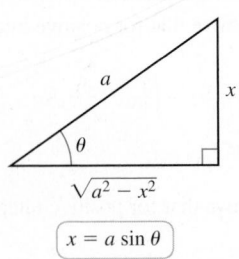

$$x = a \sin \theta$$

QUICK CHECK 1 Use a substitution of the form $x = a \sin \theta$ to transform $9 - x^2$ into a product. ◄

EXAMPLE 1   Area of a circle   Verify that the area of a circle of radius a is πa^2.

SOLUTION   The function $f(x) = \sqrt{a^2 - x^2}$ describes the upper half of a circle centered at the origin with radius a (**Figure 8.2**). The region under this curve on the interval $[0, a]$ is a quarter-circle. Therefore, the area of the full circle is $4 \int_0^a \sqrt{a^2 - x^2} \, dx$.

Because the integrand contains the expression $a^2 - x^2$, we use the trigonometric substitution $x = a \sin \theta$, which can also be written $\theta = \sin^{-1}(x/a)$. This substitution is well defined only when the angle θ is restricted to the interval $-\dfrac{\pi}{2} \le \theta \le \dfrac{\pi}{2}$, which is the range of $\sin^{-1}(x/a)$ (**Figure 8.3**). As with all substitutions, the differential associated with the substitution must be computed:

$$x = a \sin \theta \quad \text{implies that} \quad dx = a \cos \theta \, d\theta.$$

Notice that the new variable θ plays the role of an angle. Replacing x with $a \sin \theta$ in the integrand, we have

$$\sqrt{a^2 - x^2} = \sqrt{a^2 - (a \sin \theta)^2} \quad \text{Replace } x \text{ with } a \sin \theta.$$
$$= \sqrt{a^2 (1 - \sin^2 \theta)} \quad \text{Factor.}$$
$$= \sqrt{a^2 \cos^2 \theta} \quad 1 - \sin^2 \theta = \cos^2 \theta$$
$$= |a \cos \theta| \quad \sqrt{x^2} = |x|$$
$$= a \cos \theta. \quad a > 0, \cos \theta \ge 0, \text{ for } -\frac{\pi}{2} \le \theta \le \frac{\pi}{2}$$

We also change the limits of integration: When $x = 0$, $\theta = \sin^{-1} 0 = 0$; when $x = a$, $\theta = \sin^{-1}(a/a) = \sin^{-1} 1 = \pi/2$. Making these substitutions, we evaluate the integral as follows:

$$4 \int_0^a \sqrt{a^2 - x^2} \, dx = 4 \int_0^{\pi/2} \underbrace{a \cos \theta}_{\substack{\text{integrand} \\ \text{simplified}}} \cdot \underbrace{a \cos \theta \, d\theta}_{dx} \quad x = a \sin \theta, dx = a \cos \theta \, d\theta$$

$$= 4a^2 \int_0^{\pi/2} \cos^2 \theta \, d\theta \quad \text{Simplify.}$$

$$= 4a^2 \left(\frac{\theta}{2} + \frac{\sin 2\theta}{4} \right) \Big|_0^{\pi/2} \quad \cos^2 \theta = \frac{1 + \cos 2\theta}{2}$$

$$= 4a^2 \left(\left(\frac{\pi}{4} + 0 \right) - (0 + 0) \right) = \pi a^2. \quad \text{Simplify.}$$

A similar calculation (Exercise 58) gives the area of an ellipse.

Related Exercises 7, 9 ◀

EXAMPLE 2 **Sine substitution** Evaluate $\displaystyle \int \frac{dx}{(16 - x^2)^{3/2}}$.

SOLUTION The factor $16 - x^2$ has the form $a^2 - x^2$ with $a = 4$, so we use the substitution $x = 4 \sin \theta$. It follows that $dx = 4 \cos \theta \, d\theta$. We now simplify $(16 - x^2)^{3/2}$:

$$(16 - x^2)^{3/2} = (16 - (4 \sin \theta)^2)^{3/2} \quad \text{Substitute } x = 4 \sin \theta.$$
$$= (16 (1 - \sin^2 \theta))^{3/2} \quad \text{Factor.}$$
$$= (16 \cos^2 \theta)^{3/2} \quad 1 - \sin^2 \theta = \cos^2 \theta$$
$$= 64 \cos^3 \theta. \quad \text{Simplify.}$$

Replacing the factors $(16 - x^2)^{3/2}$ and dx of the original integral with appropriate expressions in θ, we have

$$\int \frac{\overbrace{dx}^{4 \cos \theta \, d\theta}}{\underbrace{(16 - x^2)^{3/2}}_{64 \cos^3 \theta}} = \int \frac{4 \cos \theta}{64 \cos^3 \theta} \, d\theta$$

$$= \frac{1}{16} \int \frac{d\theta}{\cos^2 \theta}$$

$$= \frac{1}{16} \int \sec^2 \theta \, d\theta \quad \text{Simplify.}$$

$$= \frac{1}{16} \tan \theta + C. \quad \text{Evaluate the integral.}$$

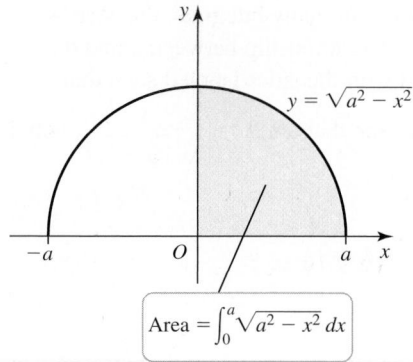

Figure 8.2

$$\text{Area} = \int_0^a \sqrt{a^2 - x^2} \, dx$$

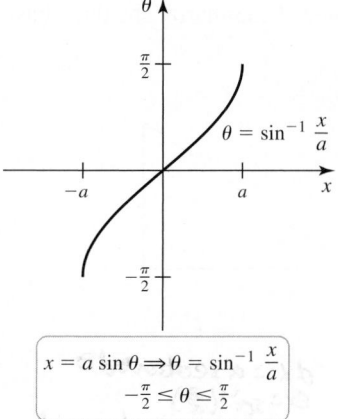

$$x = a \sin \theta \Rightarrow \theta = \sin^{-1} \frac{x}{a}$$
$$-\frac{\pi}{2} \le \theta \le \frac{\pi}{2}$$

Figure 8.3

▶ The key identities for integrating $\sin^2 \theta$ and $\cos^2 \theta$ are

$$\sin^2 \theta = \frac{1 - \cos 2\theta}{2} \quad \text{and}$$
$$\cos^2 \theta = \frac{1 + \cos 2\theta}{2}.$$

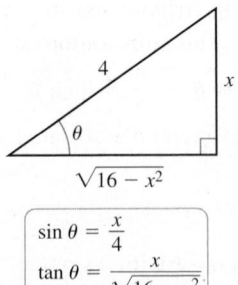

Figure 8.4

The final step is to express this result in terms of x. In many integrals, this step is most easily done with a reference triangle showing the relationship between x and θ. Figure 8.4 shows a right triangle with an angle θ and with the sides labeled such that $x = 4 \sin \theta$ (or $\sin \theta = x/4$). Using this triangle, we see that $\tan \theta = \dfrac{x}{\sqrt{16 - x^2}}$, which implies that

$$\int \frac{dx}{(16 - x^2)^{3/2}} = \frac{1}{16} \tan \theta + C = \frac{x}{16\sqrt{16 - x^2}} + C.$$

Related Exercises 8, 11 ◄

Integrals Involving $a^2 + x^2$ or $x^2 - a^2$

The additional trigonometric substitutions involving tangent and secant use a procedure similar to that used for the sine substitution. Figure 8.5 and Table 8.4 summarize the three basic trigonometric substitutions for real numbers $a > 0$.

$SIN\theta = \frac{x}{a}$

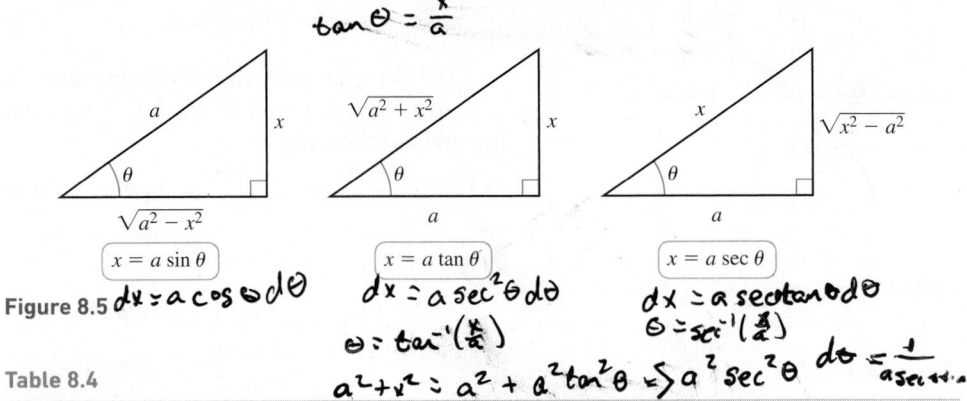

Figure 8.5

$dx = a\cos\theta\, d\theta$ $dx = a\sec^2\theta\, d\theta$ $dx = a\sec\theta\tan\theta\, d\theta$

$\theta = \tan^{-1}\left(\frac{x}{a}\right)$ $\theta = \sec^{-1}\left(\frac{x}{a}\right)$

$a^2 + x^2 = a^2 + a^2\tan^2\theta \Rightarrow a^2\sec^2\theta$ $d\theta = \frac{1}{a\sec\theta\tan\theta}$

Table 8.4

The Integral Contains . . .	Corresponding Substitution	Useful Identity
$a^2 - x^2$	$x = a \sin \theta, -\dfrac{\pi}{2} \le \theta \le \dfrac{\pi}{2}$, for $\lvert x \rvert \le a$	$a^2 - a^2 \sin^2 \theta = a^2 \cos^2 \theta$
$a^2 + x^2$	$x = a \tan \theta, -\dfrac{\pi}{2} < \theta < \dfrac{\pi}{2}$	$a^2 + a^2 \tan^2 \theta = a^2 \sec^2 \theta$
$x^2 - a^2$	$x = a \sec \theta, \begin{cases} 0 \le \theta < \dfrac{\pi}{2}, \text{ for } x \ge a \\ \dfrac{\pi}{2} < \theta \le \pi, \text{ for } x \le -a \end{cases}$	$a^2 \sec^2 \theta - a^2 = a^2 \tan^2 \theta$

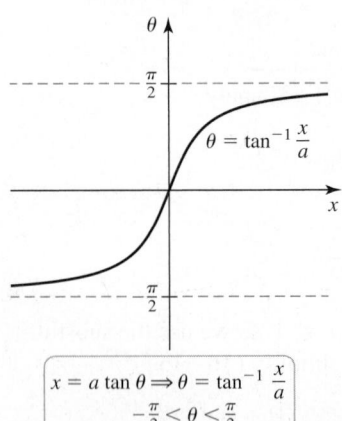

$x = a \tan \theta \Rightarrow \theta = \tan^{-1} \dfrac{x}{a}$
$-\dfrac{\pi}{2} < \theta < \dfrac{\pi}{2}$

Figure 8.6

In order for the tangent substitution $x = a \tan \theta$ to be well defined, the angle θ must be restricted to the interval $-\pi/2 < \theta < \pi/2$, which is the range of $\tan^{-1}(x/a)$ (Figure 8.6). On this interval, $\sec \theta > 0$ and with $a > 0$, it is valid to write

$$\sqrt{a^2 + x^2} = \sqrt{a^2 + (a \tan \theta)^2} = \sqrt{a^2 \underbrace{(1 + \tan^2 \theta)}_{\sec^2 \theta}} = a \sec \theta.$$

With the secant substitution, there is a technicality. As discussed in Section 1.4, $\theta = \sec^{-1}(x/a)$ is defined for $x \ge a$, in which case $0 \le \theta < \pi/2$, *and for* $x \le -a$, in which case $\pi/2 < \theta \le \pi$ (Figure 8.7). These restrictions on θ must be treated carefully when simplifying integrands with a factor of $\sqrt{x^2 - a^2}$. Because $\tan \theta$ is positive in the first quadrant but negative in the second, we have

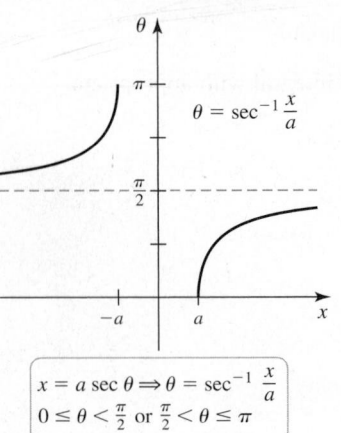

$x = a \sec \theta \Rightarrow \theta = \sec^{-1} \dfrac{x}{a}$
$0 \le \theta < \dfrac{\pi}{2}$ or $\dfrac{\pi}{2} < \theta \le \pi$

Figure 8.7

$$\sqrt{x^2 - a^2} = \sqrt{a^2 \underbrace{(\sec^2 \theta - 1)}_{\tan^2 \theta}} = \lvert a \tan \theta \rvert = \begin{cases} a \tan \theta & \text{if } 0 \le \theta < \dfrac{\pi}{2} \\ -a \tan \theta & \text{if } \dfrac{\pi}{2} < \theta \le \pi. \end{cases}$$

QUICK CHECK 2 Suggest a trigonometric substitution for each integral, and then write the new integral that results after carrying out the substitution.

a. $\displaystyle\int \frac{x^2}{\sqrt{x^2 + 9}}\, dx$

b. $\displaystyle\int \frac{3}{\sqrt{x^2 - 16}}\, dx, x > 4$ ◀

> ▶ Because we are evaluating a definite integral in Example 3, we could change the limits of integration to $\theta = 0$ and $\theta = \tan^{-1} 4$. However, $\tan^{-1} 4$ is not a standard angle, so it is easier to express the antiderivative in terms of x and use the original limits of integration.

$\tan \theta = 2x$

$\sec \theta = \sqrt{1 + 4x^2}$

Figure 8.8

When evaluating a definite integral, you should check the limits of integration to see which of these two cases applies. For indefinite integrals, a piecewise formula is often needed, unless a restriction on the variable is given in the problem (see Exercises 79–80).

EXAMPLE 3 **Arc length of a parabola** Evaluate $\int_0^2 \sqrt{1 + 4x^2}\, dx$, the arc length of the segment of the parabola $y = x^2$ on $[0, 2]$.

SOLUTION Removing a factor of 4 from the square root, we have

$$\int_0^2 \sqrt{1 + 4x^2}\, dx = 2\int_0^2 \sqrt{\tfrac{1}{4} + x^2}\, dx = 2\int_0^2 \sqrt{\left(\tfrac{1}{2}\right)^2 + x^2}\, dx.$$

The integrand contains the expression $a^2 + x^2$, with $a = \tfrac{1}{2}$, which suggests the substitution $x = \tfrac{1}{2}\tan \theta$. It follows that $dx = \tfrac{1}{2}\sec^2 \theta\, d\theta$, and

$$\sqrt{\left(\tfrac{1}{2}\right)^2 + x^2} = \sqrt{\left(\tfrac{1}{2}\right)^2 + \left(\tfrac{1}{2}\tan \theta\right)^2} = \frac{1}{2}\underbrace{\sqrt{1 + \tan^2 \theta}}_{\sec^2 \theta} = \frac{1}{2}\sec \theta.$$

Setting aside the limits of integration for the moment, we compute the antiderivative:

$$2\int \sqrt{\left(\tfrac{1}{2}\right)^2 + x^2}\, dx = 2\int \tfrac{1}{2}\sec \theta \underbrace{\tfrac{1}{2}\sec^2 \theta\, d\theta}_{dx}$$

$$= \frac{1}{2}\int \sec^3 \theta\, d\theta \qquad \text{Simplify.}$$

$$= \frac{1}{4}\left(\sec \theta \tan \theta + \ln\left|\sec \theta + \tan \theta\right|\right) + C. \qquad \begin{array}{l}\text{Reduction formula 4,}\\\text{Section 8.3}\end{array}$$

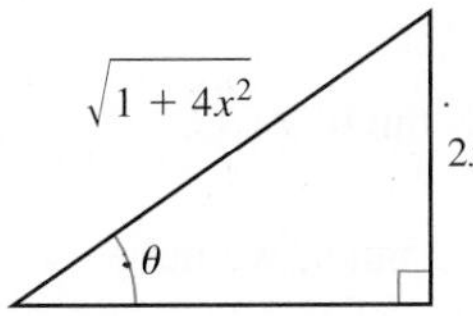

> $x = \tfrac{1}{2}\tan \theta,$
> $dx = \tfrac{1}{2}\sec^2 \theta\, d\theta$

Using a reference triangle (**Figure 8.8**), we express the antiderivative in terms of the original variable x and evaluate the definite integral:

$$2\int_0^2 \sqrt{\left(\tfrac{1}{2}\right)^2 + x^2}\, dx = \frac{1}{4}\left(\underbrace{\sqrt{1 + 4x^2}}_{\sec \theta}\,\underbrace{2x}_{\tan \theta} + \ln\left|\underbrace{\sqrt{1 + 4x^2}}_{\sec \theta} + \underbrace{2x}_{\tan \theta}\right|\right)\Bigg|_0^2$$

$$= \frac{1}{4}\left(4\sqrt{17} + \ln\left(\sqrt{17} + 4\right)\right) \approx 4.65.$$

Related Exercises 41, 77 ◀

QUICK CHECK 3 The integral $\displaystyle\int \frac{dx}{a^2 + x^2} = \frac{1}{a}\tan^{-1}\frac{x}{a} + C$ is given in Section 5.5. Verify this result with the appropriate trigonometric substitution. ◀

EXAMPLE 4 **Another tangent substitution** Evaluate $\displaystyle\int \frac{dx}{(1 + x^2)^2}$.

SOLUTION The factor $1 + x^2$ suggests the substitution $x = \tan \theta$. It follows that $\theta = \tan^{-1} x$, $dx = \sec^2 \theta\, d\theta$, and

$$(1 + x^2)^2 = \underbrace{(1 + \tan^2 \theta)^2}_{\sec^2 \theta} = \sec^4 \theta.$$

Substituting these factors leads to

$$\int \frac{dx}{(1 + x^2)^2} = \int \frac{\sec^2 \theta}{\sec^4 \theta}\, d\theta \qquad x = \tan \theta, dx = \sec^2 \theta\, d\theta$$

$$= \int \cos^2 \theta\, d\theta \qquad \text{Simplify.}$$

$$= \frac{\theta}{2} + \frac{\sin 2\theta}{4} + C. \qquad \text{Integrate } \cos^2 \theta = \frac{1 + \cos 2\theta}{2}$$

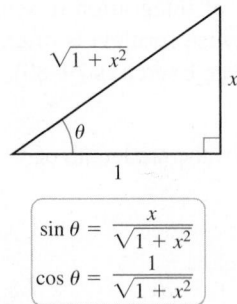

Figure 8.9

$$\sin \theta = \frac{x}{\sqrt{1 + x^2}}$$

$$\cos \theta = \frac{1}{\sqrt{1 + x^2}}$$

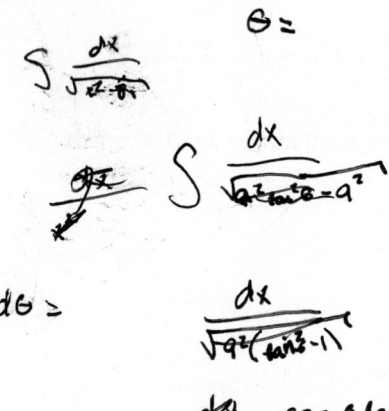

The final step is to return to the original variable x. The first term $\theta/2$ is replaced with $\frac{1}{2}\tan^{-1} x$. The second term involving $\sin 2\theta$ requires the identity $\sin 2\theta = 2 \sin \theta \cos \theta$. The reference triangle (**Figure 8.9**) tells us that

$$\frac{1}{4}\sin 2\theta = \frac{1}{2}\sin \theta \cos \theta = \frac{1}{2} \cdot \frac{x}{\sqrt{1 + x^2}} \cdot \frac{1}{\sqrt{1 + x^2}} = \frac{x}{2(1 + x^2)}.$$

The integration can now be completed:

$$\int \frac{dx}{(1 + x^2)^2} = \frac{\theta}{2} + \frac{\sin 2\theta}{4} + C$$

$$= \frac{1}{2}\tan^{-1} x + \frac{x}{2(1 + x^2)} + C.$$

Related Exercises 28–29 ◄

EXAMPLE 5   Multiple approaches Evaluate the integral $\int \frac{dx}{\sqrt{x^2 + 4}}$.

SOLUTION Our goal is to show that several different methods lead to the same end.

<u>Solution 1:</u> The term $x^2 + 4$ suggests the substitution $x = 2 \tan \theta$, which implies that $dx = 2 \sec^2 \theta \, d\theta$ and

$$\sqrt{x^2 + 4} = \sqrt{4 \tan^2 \theta + 4} = \sqrt{4(\tan^2 \theta + 1)} = 2\sqrt{\sec^2 \theta} = 2 \sec \theta.$$

Making these substitutions, the integral becomes

$$\int \frac{dx}{\sqrt{x^2 + 4}} = \int \frac{2 \sec^2 \theta}{2 \sec \theta} \, d\theta = \int \sec \theta \, d\theta = \ln |\sec \theta + \tan \theta| + C.$$

To express the indefinite integral in terms of x, notice that with $x = 2 \tan \theta$, we have

$$\tan \theta = \frac{x}{2} \quad \text{and} \quad \sec \theta = \sqrt{\tan^2 \theta + 1} = \frac{1}{2}\sqrt{x^2 + 4}.$$

Therefore,

$$\int \frac{dx}{\sqrt{x^2 + 4}} = \ln |\sec \theta + \tan \theta| + C$$

$$= \ln \left| \frac{1}{2}\sqrt{x^2 + 4} + \frac{x}{2} \right| + C \qquad \text{Substitute for } \sec \theta \text{ and } \tan \theta.$$

$$= \ln \left(\frac{1}{2}\left(\sqrt{x^2 + 4} + x\right) \right) + C \qquad \text{Factor; } \sqrt{x^2 + 4} + x > 0.$$

$$= \ln \frac{1}{2} + \ln(\sqrt{x^2 + 4} + x) + C \qquad \ln ab = \ln a + \ln b$$

$$= \ln(\sqrt{x^2 + 4} + x) + C. \qquad \text{Absorb constant in } C.$$

<u>Solution 2:</u> Using Theorem 7.9 of Section 7.3, we see that

$$\int \frac{dx}{\sqrt{x^2 + 4}} = \sinh^{-1} \frac{x}{2} + C.$$

By Theorem 7.7 of Section 7.3, we also know that

$$\sinh^{-1} \frac{x}{2} = \ln \left(\frac{x}{2} + \sqrt{\left(\frac{x}{2}\right)^2 + 1} \right) = \ln \left(\frac{1}{2}\left(\sqrt{x^2 + 4} + x\right) \right),$$

which leads to the same result as in Solution 1.

<u>Solution 3:</u> Yet another approach is to use the substitution $x = 2 \sinh t$, which implies that $dx = 2 \cosh t \, dt$ and

$$\sqrt{x^2 + 4} = \sqrt{4 \sinh^2 t + 4} = \sqrt{4(\sinh^2 t + 1)} = 2\sqrt{\cosh^2 t} = 2 \cosh t.$$

The original integral now becomes

$$\int \frac{dx}{\sqrt{x^2 + 4}} = \int \frac{2 \cosh t}{2 \cosh t} \, dt = \int dt = t + C.$$

Because $x = 2 \sinh t$, we have $t = \sinh^{-1} \frac{x}{2}$, which, by Theorem 7.7, leads to the result found in Solution 2.

This example shows that some integrals may be evaluated by more than one method. With practice, you will learn to identify the best method for a given integral.

Related Exercises 44, 49 ◄

▶ Recall that to complete the square with $x^2 + bx + c$, you add and subtract $(b/2)^2$ to and from the expression, and then factor to form a perfect square. You could also make the single substitution $x + 2 = 3 \sec \theta$ in Example 6.

EXAMPLE 6 A secant substitution Evaluate $\int_1^4 \frac{\sqrt{x^2 + 4x - 5}}{x + 2} \, dx$.

SOLUTION This example illustrates a useful preliminary step first encountered in Section 8.1. The integrand does not contain any of the patterns in Table 8.4 that suggest a trigonometric substitution. Completing the square does, however, lead to one of those patterns. Noting that $x^2 + 4x - 5 = (x + 2)^2 - 9$, we change variables with $u = x + 2$ and write the integral as

$$\int_1^4 \frac{\sqrt{x^2 + 4x - 5}}{x + 2} \, dx = \int_1^4 \frac{\sqrt{(x + 2)^2 - 9}}{x + 2} \, dx \qquad \text{Complete the square.}$$

$$= \int_3^6 \frac{\sqrt{u^2 - 9}}{u} \, du. \qquad \begin{array}{l} u = x + 2, du = dx \\ \text{Change limits of integration.} \end{array}$$

▶ The substitution $u = 3 \sec \theta$ can be rewritten as $\theta = \sec^{-1}(u/3)$. Because $u \geq 3$ in the integral $\int_3^6 \frac{\sqrt{u^2 - 9}}{u} \, du$, we have $0 \leq \theta < \frac{\pi}{2}$.

This new integral calls for the secant substitution $u = 3 \sec \theta$ (where $0 \leq \theta < \pi/2$), which implies that $du = 3 \sec \theta \tan \theta \, d\theta$ and $\sqrt{u^2 - 9} = 3 \tan \theta$. We also change the limits of integration: When $u = 3, \theta = 0$, and when $u = 6, \theta = \pi/3$. The complete integration can now be done:

$$\int_1^4 \frac{\sqrt{x^2 + 4x - 5}}{x + 2} \, dx = \int_3^6 \frac{\sqrt{u^2 - 9}}{u} \, du \qquad u = x + 2, du = dx$$

$$= \int_0^{\pi/3} \frac{3 \tan \theta}{3 \sec \theta} 3 \sec \theta \tan \theta \, d\theta \qquad u = 3 \sec \theta, du = 3 \sec \theta \tan \theta \, d\theta$$

$$= 3 \int_0^{\pi/3} \tan^2 \theta \, d\theta \qquad \text{Simplify.}$$

$$= 3 \int_0^{\pi/3} (\sec^2 \theta - 1) \, d\theta \qquad \tan^2 \theta = \sec^2 \theta - 1$$

$$= 3 (\tan \theta - \theta) \Big|_0^{\pi/3} \qquad \text{Evaluate the integral.}$$

$$= 3\sqrt{3} - \pi. \qquad \text{Simplify.}$$

Related Exercises 40, 60 ◄

SECTION 8.4 EXERCISES

Getting Started

1. What change of variables is suggested by an integral containing $\sqrt{x^2 - 9}$?

2. What change of variables is suggested by an integral containing $\sqrt{x^2 + 36}$?

3. What change of variables is suggested by an integral containing $\sqrt{100 - x^2}$?

4. If $x = 4 \tan \theta$, express $\sin \theta$ in terms of x.

5. Using the trigonometric substitution $x = 2 \sin \theta$, for $-\frac{\pi}{2} \leq \theta \leq \frac{\pi}{2}$, express $\cot \theta$ in terms of x.

6. Using the trigonometric substitution $x = 8 \sec \theta$, $x \geq 8$ and $0 < \theta \leq \frac{\pi}{2}$, express $\tan \theta$ in terms of x.

Practice Exercises

7–56. Trigonometric substitutions *Evaluate the following integrals using trigonometric substitution.*

7. $\int_0^{5/2} \frac{dx}{\sqrt{25 - x^2}}$ (*Hint:* Check your answer without using trigonometric substitution.)

8. $\displaystyle\int_0^{3/2} \frac{dx}{(9-x^2)^{3/2}}$

9. $\displaystyle\int_5^{5\sqrt{3}} \sqrt{100-x^2}\,dx$

10. $\displaystyle\int_0^{\sqrt{2}} \frac{x^2}{\sqrt{4-x^2}}\,dx$

11. $\displaystyle\int_{1/2}^{\sqrt{3}/2} \frac{x^2}{\sqrt{1-x^2}}\,dx$

12. $\displaystyle\int_{1/2}^{1} \frac{\sqrt{1-x^2}}{x^2}\,dx$

13. $\displaystyle\int \frac{dx}{\sqrt{16-x^2}}$

14. $\displaystyle\int \sqrt{36-t^2}\,dt$

15. $\displaystyle\int \frac{dx}{x^2\sqrt{x^2+9}}$

16. $\displaystyle\int \frac{x^2}{(25+x^2)^2}\,dx$

17. $\displaystyle\int_0^2 \frac{x^2}{x^2+4}\,dx$

18. $\displaystyle\int \frac{dx}{(1+x^2)^{3/2}}$

19. $\displaystyle\int \frac{dx}{\sqrt{x^2-81}},\, x>9$

20. $\displaystyle\int \frac{dx}{\sqrt{x^2-49}},\, x>7$

21. $\displaystyle\int \sqrt{64-x^2}\,dx$

22. $\displaystyle\int \frac{dt}{t^2\sqrt{9-t^2}}$

23. $\displaystyle\int \frac{dx}{(25-x^2)^{3/2}}$

24. $\displaystyle\int \frac{\sqrt{9-x^2}}{x^2}\,dx$

25. $\displaystyle\int \frac{\sqrt{9-x^2}}{x}\,dx$

26. $\displaystyle\int_{\sqrt{2}}^2 \frac{\sqrt{x^2-1}}{x}\,dx$

27. $\displaystyle\int_0^{1/3} \frac{dx}{(9x^2+1)^{3/2}}$

28. $\displaystyle\int_0^6 \frac{z^2}{(z^2+36)^2}\,dz$

29. $\displaystyle\int \frac{dx}{(4+x^2)^2}$

30. $\displaystyle\int x^3\sqrt{1-x^2}\,dx$

31. $\displaystyle\int \frac{x^2}{\sqrt{16-x^2}}\,dx$

32. $\displaystyle\int \frac{dx}{(x^2-36)^{3/2}},\, x>6$

33. $\displaystyle\int \frac{\sqrt{x^2-9}}{x}\,dx,\, x>3$

34. $\displaystyle\int \frac{dx}{x^3\sqrt{x^2-1}},\, x>1$

35. $\displaystyle\int \frac{dx}{x(x^2-1)^{3/2}},\, x>1$

36. $\displaystyle\int_{8\sqrt{2}}^{16} \frac{dx}{\sqrt{x^2-64}}$

37. $\displaystyle\int_{1/\sqrt{3}}^{1} \frac{dx}{x^2\sqrt{1+x^2}}$

38. $\displaystyle\int_1^{\sqrt{2}} \frac{dx}{x^2\sqrt{4-x^2}}$

39. $\displaystyle\int \frac{x^2}{(100-x^2)^{3/2}}\,dx$

40. $\displaystyle\int_{10/\sqrt{3}}^{10} \frac{dy}{\sqrt{y^2-25}}$

41. $\displaystyle\int \frac{dx}{(1+4x^2)^{3/2}}$

42. $\displaystyle\int \frac{dx}{x^2\sqrt{9x^2-1}},\, x>\frac{1}{3}$

43. $\displaystyle\int_0^{4/\sqrt{3}} \frac{dx}{\sqrt{x^2+16}}$

44. $\displaystyle\int \frac{dx}{\sqrt{16+4x^2}}$

45. $\displaystyle\int \frac{x^3}{(81-x^2)^2}\,dx$

46. $\displaystyle\int \frac{dx}{\sqrt{1-2x^2}}$

47. $\displaystyle\int_{4/\sqrt{3}}^{4} \frac{dx}{x^2(x^2-4)}$

48. $\displaystyle\int \sqrt{9-4x^2}\,dx$

49. $\displaystyle\int_0^{1/\sqrt{3}} \sqrt{x^2+1}\,dx$

50. $\displaystyle\int (36-9x^2)^{-3/2}\,dx$

51. $\displaystyle\int \frac{x^2}{\sqrt{4+x^2}}\,dx$

52. $\displaystyle\int \frac{\sqrt{4x^2-1}}{x^2}\,dx,\, x>\frac{1}{2}$

53. $\displaystyle\int \frac{\sqrt{9x^2-25}}{x^3}\,dx,\, x>\frac{5}{3}$

54. $\displaystyle\int \frac{y^4}{1+y^2}\,dy$

55. $\displaystyle\int \frac{dx}{x^3\sqrt{x^2-100}},\, x>10$

56. $\displaystyle\int \frac{x^3}{(x^2-16)^{3/2}}\,dx,\, x<-4$

57. Explain why or why not Determine whether the following statements are true and give an explanation or counterexample.

a. If $x=4\tan\theta$, then $\csc\theta=4/x$.

b. The integral $\displaystyle\int_1^2 \sqrt{1-x^2}\,dx$ does not have a finite real value.

c. The integral $\displaystyle\int_1^2 \sqrt{x^2-1}\,dx$ does not have a finite real value.

d. The integral $\displaystyle\int \frac{dx}{x^2+4x+9}$ cannot be evaluated using a trigonometric substitution.

58. Area of an ellipse The upper half of the ellipse centered at the origin with axes of length $2a$ and $2b$ is described by $y=\dfrac{b}{a}\sqrt{a^2-x^2}$ (see figure). Find the area of the ellipse in terms of a and b.

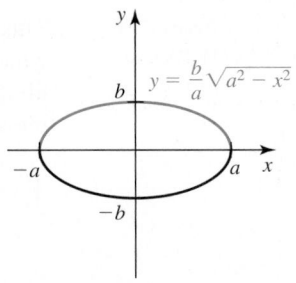

59. Area of a segment of a circle Use two approaches to show that the area of a cap (or segment) of a circle of radius r subtended by an angle θ (see figure) is given by

$$A_{\text{seg}}=\frac{1}{2}r^2(\theta-\sin\theta).$$

a. Find the area using geometry (no calculus).
b. Find the area using calculus.

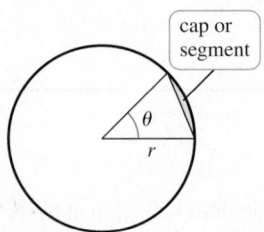

60–69. Completing the square *Evaluate the following integrals.*

60. $\displaystyle\int \frac{dx}{x^2-6x+34}$

61. $\displaystyle\int \frac{dx}{\sqrt{3-2x-x^2}}$

62. $\displaystyle\int \frac{du}{2u^2-12u+36}$

63. $\displaystyle\int \frac{dx}{x^2+6x+18}$

64. $\displaystyle\int \frac{x^2-2x+1}{\sqrt{x^2-2x+10}}\,dx$

65. $\displaystyle\int_{1/2}^{(\sqrt{2}+3)/(2\sqrt{2})} \frac{dx}{8x^2-8x+11}$

66. $\displaystyle\int_{1}^{4} \frac{dt}{t^2 - 2t + 10}$

67. $\displaystyle\int \frac{x^2 - 8x + 16}{(9 + 8x - x^2)^{3/2}}\, dx$

68. $\displaystyle\int \frac{dx}{\sqrt{(x-1)(3-x)}}$

69. $\displaystyle\int_{2+\sqrt{2}}^{4} \frac{dx}{\sqrt{(x-1)(x-3)}}$

⊤ 70–74. Using the integral of $\sec^3 u$ *By reduction formula 4 in Section 8.3,*

$$\int \sec^3 u \, du = \frac{1}{2}(\sec u \tan u + \ln|\sec u + \tan u|) + C.$$

Graph the following functions and find the area under the curve on the given interval.

70. $f(x) = (4 + x^2)^{1/2}, [0, 2]$ **71.** $f(x) = (9 - x^2)^{-2}, \left[0, \dfrac{3}{2}\right]$

72. $f(x) = (x^2 - 25)^{1/2}, [5, 10]$

73. $f(x) = \sqrt{x^2 - 9}/x, [3, 6]$

74. $f(x) = \dfrac{1}{x\sqrt{x^2 - 36}}, \left[\dfrac{12}{\sqrt{3}}, 12\right]$

75. Find the area of the region bounded by the curve $f(x) = (16 + x^2)^{-3/2}$ and the x-axis on the interval $[0, 3]$.

Explorations and Challenges

76. Area and volume Consider the function $f(x) = (9 + x^2)^{-1/2}$ and the region R on the interval $[0, 4]$ (see figure).

a. Find the area of R.
b. Find the volume of the solid generated when R is revolved about the x-axis.
c. Find the volume of the solid generated when R is revolved about the y-axis.

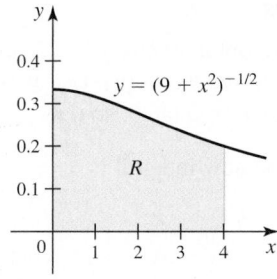

77. Arc length of a parabola Find the length of the curve $y = ax^2$ from $x = 0$ to $x = 10$, where $a > 0$ is a real number.

78. Computing areas On the interval $[0, 2]$, the graphs of $f(x) = x^2/3$ and $g(x) = x^2(9 - x^2)^{-1/2}$ have similar shapes.

a. Find the area of the region bounded by the graph of f and the x-axis on the interval $[0, 2]$.
b. Find the area of the region bounded by the graph of g and the x-axis on the interval $[0, 2]$.
c. Which region has greater area?

⊤ 79–80. Care with the secant substitution *Recall that the substitution $x = a \sec\theta$ implies either $x \geq a$ (in which case $0 \leq \theta < \pi/2$ and $\tan\theta \geq 0$) or $x \leq -a$ (in which case $\pi/2 < \theta \leq \pi$ and $\tan\theta \leq 0$).*

79. Show that $\displaystyle\int \frac{dx}{x\sqrt{x^2 - 1}} = \begin{cases} \sec^{-1} x + C & \text{if } x > 1 \\ -\sec^{-1} x + C & \text{if } x < -1. \end{cases}$

80. Evaluate for $\displaystyle\int \frac{\sqrt{x^2 - 1}}{x^3}\, dx$, for $x > 1$ and for $x < -1$.

81. Electric field due to a line of charge A total charge of Q is distributed uniformly on a line segment of length $2L$ along the y-axis (see figure). The x-component of the electric field at a point $(a, 0)$ is given by

$$E_x(a) = \frac{kQa}{2L}\int_{-L}^{L} \frac{dy}{(a^2 + y^2)^{3/2}},$$

where k is a physical constant and $a > 0$.

a. Confirm that $E_x(a) = \dfrac{kQ}{a\sqrt{a^2 + L^2}}$.
b. Letting $\rho = Q/2L$ be the charge density on the line segment, show that if $L \to \infty$, then $E_x(a) = 2k\rho/a$.

(See the Guided Project *Electric field integrals* for a derivation of this and other similar integrals.)

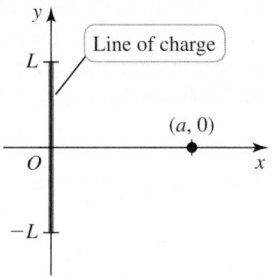

82. Magnetic field due to current in a straight wire A long straight wire of length $2L$ on the y-axis carries a current I. According to the Biot-Savart Law, the magnitude of the magnetic field due to the current at a point $(a, 0)$ is given by

$$B(a) = \frac{\mu_0 I}{4\pi}\int_{-L}^{L} \frac{\sin\theta}{r^2}\, dy,$$

where μ_0 is a physical constant, $a > 0$, and θ, r, and y are related as shown in the figure.

a. Show that the magnitude of the magnetic field at $(a, 0)$ is
$$B(a) = \frac{\mu_0 I L}{2\pi a\sqrt{a^2 + L^2}}.$$
b. What is the magnitude of the magnetic field at $(a, 0)$ due to an infinitely long wire $(L \to \infty)$?

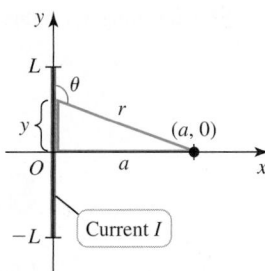

83. Visual proof Let $F(x) = \int_0^x \sqrt{a^2 - t^2}\, dt$. The figure shows that $F(x) = $ area of sector $OAB + $ area of triangle OBC.

a. Use the figure to prove that
$$F(x) = \frac{a^2 \sin^{-1}(x/a)}{2} + \frac{x\sqrt{a^2 - x^2}}{2}.$$

b. Conclude that

$$\int \sqrt{a^2 - x^2}\, dx = \frac{a^2 \sin^{-1}(x/a)}{2} + \frac{x\sqrt{a^2 - x^2}}{2} + C.$$

(*Source: The College Mathematics Journal*, 34, 3, May 2003)

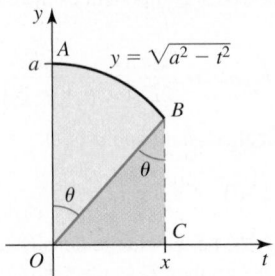

T 84. Maximum path length of a projectile (Adapted from Putnam Exam 1940) A projectile is launched from the ground with an initial speed V at an angle θ from the horizontal. Assume the x-axis is the horizontal ground and y is the height above the ground. Neglecting air resistance and letting g be the acceleration due to gravity, it can be shown that the trajectory of the projectile is given by

$$y = -\frac{1}{2} kx^2 + y_{max}, \quad \text{where } k = \frac{g}{(V \cos \theta)^2}$$

and $y_{max} = \dfrac{(V \sin \theta)^2}{2g}$.

a. Note that the high point of the trajectory occurs at $(0, y_{max})$. If the projectile is on the ground at $(-a, 0)$ and $(a, 0)$, what is a?

b. Show that the length of the trajectory (arc length) is

$$2 \int_0^a \sqrt{1 + k^2 x^2}\, dx.$$

c. Evaluate the arc length integral and express your result in terms of V, g, and θ.

d. For a fixed value of V and g, show that the launch angle θ that maximizes the length of the trajectory satisfies $(\sin \theta)\ln(\sec \theta + \tan \theta) = 1$.

e. Use a graphing utility to approximate the optimal launch angle.

85. Fastest descent time The cycloid is the curve traced by a point on the rim of a rolling wheel. Imagine a wire shaped like an inverted cycloid (see figure). A bead sliding down this wire without

friction has some remarkable properties. Among all wire shapes, the cycloid is the shape that produces the fastest descent time (see the Guided Project *The amazing cycloid* for more about the *brachistochrone property*). It can be shown that the descent time between any two points $0 \le a < b \le \pi$ on the curve is

$$\text{descent time} = \int_a^b \sqrt{\frac{1 - \cos t}{g(\cos a - \cos t)}}\, dt,$$

where g is the acceleration due to gravity, $t = 0$ corresponds to the top of the wire, and $t = \pi$ corresponds to the lowest point on the wire.

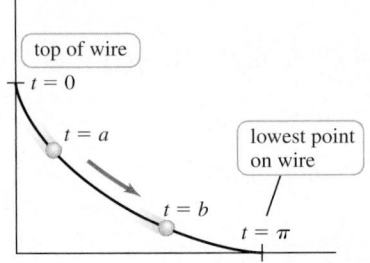

a. Find the descent time on the interval $[a, b]$ by making the substitution $u = \cos t$.

b. Show that when $b = \pi$, the descent time is the same for all values of a; that is, the descent time to the bottom of the wire is the same for all starting points.

86. Clever substitution Evaluate $\displaystyle\int \frac{dx}{1 + \sin x + \cos x}$ using the substitution $x = 2 \tan^{-1} \theta$. The identities $\sin x = 2 \sin \dfrac{x}{2} \cos \dfrac{x}{2}$ and $\cos x = \cos^2 \dfrac{x}{2} - \sin^2 \dfrac{x}{2}$ are helpful.

QUICK CHECK ANSWERS

1. Use $x = 3 \sin \theta$ to obtain $9 \cos^2 \theta$.　**2.** (a) Use $x = 3 \tan \theta$; $\int 9 \tan^2 \theta \sec \theta\, d\theta$. (b) Use $x = 4 \sec \theta$; $\int 3 \sec \theta\, d\theta$.　**3.** Let $x = a \tan \theta$, so that $dx = a \sec^2 \theta\, d\theta$. The new integral is $\displaystyle\int \frac{a \sec^2 \theta\, d\theta}{a^2(1 + \tan^2 \theta)} = \frac{1}{a}\int d\theta = \frac{1}{a}\theta + C = \frac{1}{a}\tan^{-1}\frac{x}{a} + C.$ ◀

8.5 Partial Fractions

> Recall that a rational function has the form p/q, where p and q are polynomials.

Later in this chapter, we will see that finding the velocity of a skydiver requires evaluating an integral of the form $\displaystyle\int \frac{dv}{a - bv^2}$ and finding the population of a species that is limited in size involves an integral of the form $\displaystyle\int \frac{dP}{aP(1 - bP)}$, where a and b are constants in both cases. These integrals have the common feature that their integrands are rational functions. Similar integrals result from modeling mechanical and electrical networks. The goal of this section is to introduce the *method of partial fractions* for integrating rational functions. When combined with standard and trigonometric substitutions (Section 8.4), this method allows us (in principle) to integrate any rational function.

Method of Partial Fractions

Given a function such as

$$f(x) = \frac{1}{x - 2} + \frac{2}{x + 4},$$

it is a straightforward task to find a common denominator and write the equivalent expression

$$f(x) = \frac{(x + 4) + 2(x - 2)}{(x - 2)(x + 4)} = \frac{3x}{(x - 2)(x + 4)} = \frac{3x}{x^2 + 2x - 8}.$$

The purpose of partial fractions is to reverse this process. Given a rational function that is difficult to integrate, the method of partial fractions produces an equivalent function that is much easier to integrate.

<table>
<tr><td>Rational function</td><td></td><td>Partial fraction decomposition</td></tr>
<tr><td>$\dfrac{3x}{x^2 + 2x - 8}$</td><td>method of partial fractions →</td><td>$\dfrac{1}{x - 2} + \dfrac{2}{x + 4}$</td></tr>
<tr><td>Difficult to integrate</td><td></td><td>Easy to integrate</td></tr>
<tr><td>$\displaystyle\int \dfrac{3x}{x^2 + 2x - 8}\, dx$</td><td></td><td>$\displaystyle\int \left(\dfrac{1}{x - 2} + \dfrac{2}{x + 4}\right) dx$</td></tr>
</table>

QUICK CHECK 1 Find an antiderivative of $f(x) = \dfrac{1}{x - 2} + \dfrac{2}{x + 4}$. ◄

The Key Idea Working with the same function, $f(x) = \dfrac{3x}{(x - 2)(x + 4)}$, our objective is to write it in the form

$$\frac{A}{x - 2} + \frac{B}{x + 4},$$

where A and B are constants to be determined. This expression is called the **partial fraction decomposition** of the original function; in this case, it has two terms, one for each factor in the denominator of the original function.

▶ Notice that the numerator of the original rational function does not affect the form of the partial fraction decomposition. The constants A and B are called *undetermined coefficients*.

The constants A and B are determined using the condition that the original function f and its partial fraction decomposition must be equal for all values of x in the domain of f; that is,

$$\frac{3x}{(x - 2)(x + 4)} = \frac{A}{x - 2} + \frac{B}{x + 4}. \tag{1}$$

▶ This step requires that $x \neq 2$ and $x \neq -4$; both values are outside the domain of f.

Multiplying both sides of equation (1) by $(x - 2)(x + 4)$ gives

$$3x = A(x + 4) + B(x - 2).$$

Collecting like powers of x results in

$$3x = (A + B)x + (4A - 2B). \tag{2}$$

If equation (2) is to hold for all values of x, then

• the coefficients of x^1 on both sides of the equation must be equal, and

• the coefficients of x^0 (that is, the constants) on both sides of the equation must be equal.

$$3x + 0 = \underbrace{(A + B)}x + \overbrace{(4A - 2B)}$$

These observations lead to two equations for A and B.

$$\text{Equate coefficients of } x^1: \quad 3 = A + B$$
$$\text{Equate coefficients of } x^0: \quad 0 = 4A - 2B$$

The first equation says that $A = 3 - B$. Substituting $A = 3 - B$ into the second equation gives the equation $0 = 4(3 - B) - 2B$. Solving for B, we find that $6B = 12$, or $B = 2$. The value of A now follows; we have $A = 3 - B = 1$.

Substituting these values of A and B into equation (1), the partial fraction decomposition is

$$\frac{3x}{(x - 2)(x + 4)} = \frac{1}{x - 2} + \frac{2}{x + 4}.$$

> ➤ Like an ordinary fraction, a rational function is said to be in **reduced form** if the numerator and denominator have no common factors, and it is said to be **proper** if the degree of the numerator is less than the degree of the denominator.

Simple Linear Factors

The previous calculation illustrates the method of partial fractions with **simple linear factors**, meaning the denominator of the original function consists only of linear factors of the form $(x - r)$, which appear to the first power and no higher power. Here is the general procedure for this case.

> ➤ For the sake of simplicity, the factors of the denominator q in Step 1 are written with a leading coefficient of 1. The same procedure may be applied to factors whose leading coefficients are not equal to 1 as long as no factor is a constant multiple of any other factor. For example, $(x + 2)(2x + 4) = 2(x + 2)^2$ should be recognized as a repeated linear factor (p. 550), not two simple linear factors. See Example 2.

QUICK CHECK 2 If the denominator of a reduced proper rational function is $(x - 1)(x + 5)(x - 10)$, what is the general form of its partial fraction decomposition? ◄

PROCEDURE   Partial Fractions with Simple Linear Factors

Suppose $f(x) = p(x)/q(x)$, where p and q are polynomials with no common factors and with the degree of p less than the degree of q. Assume q is the product of simple linear factors. The partial fraction decomposition is obtained as follows.

Step 1.   **Factor the denominator q** in the form $(x - r_1)(x - r_2) \cdots (x - r_n)$, where $r_1, \ldots, r_n$ are distinct real numbers.

Step 2.   **Partial fraction decomposition** Form the partial fraction decomposition by writing

$$\frac{p(x)}{q(x)} = \frac{A_1}{(x - r_1)} + \frac{A_2}{(x - r_2)} + \cdots + \frac{A_n}{(x - r_n)}.$$

Step 3.   **Clear denominators** Multiply both sides of the equation in Step 2 by $q(x) = (x - r_1)(x - r_2) \cdots (x - r_n)$, which produces conditions for $A_1, \ldots, A_n$.

Step 4.   **Solve for coefficients** Equate like powers of x in Step 3 to solve for the undetermined coefficients $A_1, \ldots, A_n$.

EXAMPLE 1   Integrating with partial fractions

a. Find the partial fraction decomposition for $f(x) = \dfrac{3x^2 + 7x - 2}{x^3 - x^2 - 2x}$.

b. Evaluate $\displaystyle\int f(x)\, dx$.

SOLUTION

a. The partial fraction decomposition is done in four steps.

Step 1:   Factoring the denominator, we find that

$$x^3 - x^2 - 2x = x(x + 1)(x - 2),$$

in which only simple linear factors appear.

> ➤ You can call the undetermined coefficients $A_1, A_2, A_3, \ldots$ or $A, B, C, \ldots$. The latter choice avoids subscripts.

Step 2:   The partial fraction decomposition has one term for each factor in the denominator:

$$\frac{3x^2 + 7x - 2}{x(x + 1)(x - 2)} = \frac{A}{x} + \frac{B}{x + 1} + \frac{C}{x - 2}. \tag{3}$$

The goal is to find the undetermined coefficients A, B, and C.

Step 3:   We multiply both sides of equation (3) by $x(x + 1)(x - 2)$:

$$3x^2 + 7x - 2 = A(x + 1)(x - 2) + Bx(x - 2) + Cx(x + 1)$$
$$= (A + B + C)x^2 + (-A - 2B + C)x - 2A.$$

Step 4: We now equate coefficients of x^2, x^1, and x^0 on both sides of the equation in Step 3.

Equate coefficients of x^2: $\qquad A + B + C = 3$

Equate coefficients of x^1: $\qquad -A - 2B + C = 7$

Equate coefficients of x^0: $\qquad\qquad -2A = -2$

The third equation implies that $A = 1$, which is substituted into the first two equations to give

$$B + C = 2 \quad \text{and} \quad -2B + C = 8.$$

Solving for B and C, we conclude that $A = 1$, $B = -2$, and $C = 4$. Substituting the values of A, B, and C into equation (3), the partial fraction decomposition is

$$f(x) = \frac{1}{x} - \frac{2}{x + 1} + \frac{4}{x - 2}.$$

b. Integration is now straightforward:

$$\int \frac{3x^2 + 7x - 2}{x^3 - x^2 - 2x}\, dx = \int \left(\frac{1}{x} - \frac{2}{x + 1} + \frac{4}{x - 2} \right) dx \qquad \text{Partial fractions}$$

$$= \ln |x| - 2 \ln |x + 1| + 4 \ln |x - 2| + K \qquad \begin{array}{l}\text{Integrate; arbitrary}\\ \text{constant } K.\end{array}$$

$$= \ln \frac{|x|(x - 2)^4}{(x + 1)^2} + K. \qquad \begin{array}{l}\text{Properties of}\\ \text{logarithms}\end{array}$$

Related Exercises 30, 35 ◄

A Shortcut (Convenient Values) Solving for more than three unknown coefficients in a partial fraction decomposition may be difficult. In the case of simple linear factors, a shortcut saves work. In Example 1, Step 3 led to the equation

$$3x^2 + 7x - 2 = A(x + 1)(x - 2) + Bx(x - 2) + Cx(x + 1).$$

▶ In cases other than simple linear factors, the shortcut can be used to determine some, but not all the coefficients, which reduces the work required to find the remaining coefficients. A modified shortcut can be utilized to find all the coefficients; see the margin note next to Example 3.

Because this equation holds for *all* values of x, it must hold for any particular value of x. By choosing values of x judiciously, it is easy to solve for A, B, and C. For example, setting $x = 0$ in this equation results in $-2 = -2A$, or $A = 1$. Setting $x = -1$ results in $-6 = 3B$, or $B = -2$, and setting $x = 2$ results in $24 = 6C$, or $C = 4$. In each case, we choose a value of x that eliminates all but one term on the right side of the equation.

EXAMPLE 2 **Using the shortcut**

a. Find the partial fraction decomposition for $f(x) = \dfrac{9x^2 + 2x - 1}{(x - 1)(2x^2 + 7x - 4)}$.

b. Evaluate $\displaystyle\int_2^{14} f(x)\, dx$.

SOLUTION

a. We use four steps to obtain the partial fraction decomposition.

Step 1: Factoring the denominator of f results in $(x - 1)(2x - 1)(x + 4)$, so the integrand has three simple linear factors.

Step 2: We form the partial fraction decomposition with one term for each factor in the denominator:

$$\frac{9x^2 + 2x - 1}{(x - 1)(2x - 1)(x + 4)} = \frac{A}{x - 1} + \frac{B}{2x - 1} + \frac{C}{x + 4}. \tag{4}$$

The goal is to find the undetermined coefficients A, B, and C.

Step 3: We now multiply both sides of equation (4) by $(x - 1)(2x - 1)(x + 4)$:

$$9x^2 + 2x - 1 = A(2x - 1)(x + 4) + B(x - 1)(x + 4) + C(x - 1)(2x - 1). \tag{5}$$

Step 4: The shortcut is now used to determine A, B, and C. Substituting $x = 1$, $1/2$, and -4 in equation (5) allows us to solve directly for the coefficients:

$$\text{Letting } x = 1 \Rightarrow 10 = 5A + 0 \cdot B + 0 \cdot C \Rightarrow A = 2;$$
$$\text{Letting } x = 1/2 \Rightarrow 9/4 = 0 \cdot A - 9B/4 + 0 \cdot C \Rightarrow B = -1;$$
$$\text{Letting } x = -4 \Rightarrow 135 = 0 \cdot A + 0 \cdot B + 45C \Rightarrow C = 3.$$

Substituting the values of A, B, and C into equation (4) gives the partial fraction decomposition

$$f(x) = \frac{2}{x - 1} - \frac{1}{2x - 1} + \frac{3}{x + 4}.$$

b. We now carry out the integration.

$$\int_2^{14} f(x)\, dx = \int_2^{14} \left(\frac{2}{x - 1} - \frac{1}{2x - 1} + \frac{3}{x + 4} \right) dx \qquad \text{Partial fractions}$$

$$= \left(2 \ln |x - 1| - \frac{1}{2} \ln |2x - 1| + 3 \ln |x + 4| \right) \Big|_2^{14} \qquad \begin{array}{l}\text{Integrate;}\\ u = 2x - 1,\\ du = 2dx.\end{array}$$

$$= 2 \ln 13 - \frac{1}{2} \ln 27 + 3 \ln 18 - \left(\underbrace{2 \ln 1}_{0} - \frac{1}{2} \ln 3 + 3 \ln 6 \right) \qquad \text{Evaluate.}$$

$$= \ln 13^2 + \underbrace{\ln 18^3 - \ln 6^3}_{\ln 27} - \frac{1}{2} \underbrace{(\ln 27 - \ln 3)}_{\ln 3} \qquad \text{Log properties}$$

$$= \ln 1521 \approx 7.327 \qquad \text{Simplify.}$$

Related Exercises 27–28 ◄

Repeated Linear Factors

> *Simple* means the factor is raised to the first power; *repeated* means the factor is raised to an integer power greater than 1.

The preceding discussion relies on the assumption that the denominator of the rational function can be factored into simple linear factors of the form $(x - r)$. But what about denominators such as $x^2(x - 3)$ or $(x + 2)^2(x - 4)^3$, in which linear factors are raised to integer powers greater than 1? In these cases we have *repeated linear factors*, and a modification to the previous procedure must be made.

Here is the modification: Suppose the factor $(x - r)^m$ appears in the denominator, where $m > 1$ is an integer. Then there must be a partial fraction for each power of $(x - r)$ up to and including the mth power. For example, if $x^2(x - 3)^4$ appears in the denominator, then the partial fraction decomposition includes the terms

> Think of x^2 as the repeated linear factor $(x - 0)^2$.

$$\frac{A}{x} + \frac{B}{x^2} + \frac{C}{(x - 3)} + \frac{D}{(x - 3)^2} + \frac{E}{(x - 3)^3} + \frac{F}{(x - 3)^4}.$$

The rest of the partial fraction procedure remains the same, although the amount of work increases as the number of coefficients increases.

PROCEDURE   Partial Fractions for Repeated Linear Factors

Suppose the repeated linear factor $(x - r)^m$ appears in the denominator of a proper rational function in reduced form. The partial fraction decomposition has a partial fraction for each power of $(x - r)$ up to and including the mth power; that is, the partial fraction decomposition contains the sum

$$\frac{A_1}{(x - r)} + \frac{A_2}{(x - r)^2} + \frac{A_3}{(x - r)^3} + \cdots + \frac{A_m}{(x - r)^m},$$

where $A_1, \ldots, A_m$ are constants to be determined.

EXAMPLE 3   Integrating with repeated linear factors Evaluate $\int f(x)\, dx$, where

$$f(x) = \frac{5x^2 - 3x + 2}{x^3 - 2x^2}.$$

SOLUTION The denominator factors as $x^3 - 2x^2 = x^2(x - 2)$, so it has one simple linear factor $(x - 2)$ and one repeated linear factor x^2. The partial fraction decomposition has the form

$$\frac{5x^2 - 3x + 2}{x^2(x - 2)} = \frac{A}{x} + \frac{B}{x^2} + \frac{C}{(x - 2)}.$$

Multiplying both sides of the partial fraction decomposition by $x^2(x - 2)$, we find

$$5x^2 - 3x + 2 = Ax(x - 2) + B(x - 2) + Cx^2$$
$$= (A + C)x^2 + (-2A + B)x - 2B.$$

The coefficients A, B, and C are determined by equating the coefficients of x^2, x^1, and x^0.

> ➤ The shortcut can be used to obtain two of the three coefficients easily. Choosing $x = 0$ allows B to be determined. Choosing $x = 2$ determines C. To find A, any other value of x may be substituted.

Equate coefficients of x^2: $A + C = 5$

Equate coefficients of x^1: $-2A + B = -3$

Equate coefficients of x^0: $-2B = 2$

Solving these three equations in three unknowns results in the solution $A = 1$, $B = -1$, and $C = 4$. When A, B, and C are substituted, the partial fraction decomposition is

$$f(x) = \frac{1}{x} - \frac{1}{x^2} + \frac{4}{x - 2}.$$

Integration is now straightforward:

$$\int \frac{5x^2 - 3x + 2}{x^3 - 2x^2}\, dx = \int \left(\frac{1}{x} - \frac{1}{x^2} + \frac{4}{x - 2} \right) dx \qquad \text{Partial fractions}$$

QUICK CHECK 3 State the form of the partial fraction decomposition of the reduced proper rational function $p(x)/q(x)$ if $q(x) = x^2(x - 3)^2(x - 1)$. ◄

$$= \ln |x| + \frac{1}{x} + 4 \ln |x - 2| + K \quad \text{Integrate; arbitrary constant } K.$$

$$= \frac{1}{x} + \ln \left(|x|(x - 2)^4 \right) + K. \qquad \text{Properties of logarithms}$$

Related Exercises 40–41 ◄

Irreducible Quadratic Factors

By the Fundamental Theorem of Algebra, we know that a polynomial with real-valued coefficients can be written as the product of linear factors of the form $x - r$ and *irreducible quadratic factors* of the form $ax^2 + bx + c$, where r, a, b, and c are real numbers. By irreducible, we mean that $ax^2 + bx + c$ cannot be factored over the real numbers. For example, the polynomial

> ➤ The quadratic $ax^2 + bx + c$ has no real roots and cannot be factored over the real numbers if $b^2 - 4ac < 0$.

$$x^9 + 4x^8 + 6x^7 + 34x^6 + 64x^5 - 84x^4 - 287x^3 - 500x^2 - 354x - 180$$

factors as

$$\underbrace{(x - 2)}_{\substack{\text{linear}\\\text{factor}}}\underbrace{(x + 3)^2}_{\substack{\text{repeated}\\\text{linear}\\\text{factor}}}\underbrace{(x^2 - 2x + 10)}_{\substack{\text{irreducible}\\\text{quadratic}\\\text{factor}}}\underbrace{(x^2 + x + 1)^2}_{\substack{\text{repeated}\\\text{irreducible}\\\text{quadratic factor}}}.$$

In this factored form, we see linear factors (simple and repeated) and irreducible quadratic factors (simple and repeated).

With irreducible quadratic factors, two cases must be considered: simple and repeated factors. Simple quadratic factors are examined in the following examples, and repeated quadratic factors (which generally involve long computations) are explored in the exercises.

PROCEDURE   Partial Fractions with Simple Irreducible Quadratic Factors

Suppose a simple irreducible factor $ax^2 + bx + c$ appears in the denominator of a proper rational function in reduced form. The partial fraction decomposition contains a term of the form

$$\frac{Ax + B}{ax^2 + bx + c},$$

where A and B are unknown coefficients to be determined.

EXAMPLE 4   Setting up partial fractions Give the appropriate form of the partial fraction decomposition for the following functions.

a. $\dfrac{x^2 + 1}{x^4 - 4x^3 - 32x^2}$ **b.** $\dfrac{10}{(x - 2)^2(x^2 + 2x + 2)}$

SOLUTION

a. The denominator factors as $x^2(x^2 - 4x - 32) = x^2(x - 8)(x + 4)$. Therefore, x is a repeated linear factor, and $(x - 8)$ and $(x + 4)$ are simple linear factors. The required form of the decomposition is

$$\frac{A}{x} + \frac{B}{x^2} + \frac{C}{x - 8} + \frac{D}{x + 4}.$$

We see that the factor $x^2 - 4x - 32$ is quadratic, but it can be factored as $(x - 8)(x + 4)$, so it is not irreducible.

> In Example 4b, the factor $(x - 2)^2$ cannot be treated as an irreducible quadratic factor; it is a repeated linear factor.

b. The denominator is already fully factored. The quadratic factor $x^2 + 2x + 2$ cannot be factored using real numbers; therefore, it is irreducible. The form of the decomposition is

$$\frac{A}{x - 2} + \frac{B}{(x - 2)^2} + \frac{Cx + D}{x^2 + 2x + 2}.$$

Related Exercises 8, 10 ◄

EXAMPLE 5   Integrating with quadratic factors Evaluate

$$\int \frac{7x^2 - 13x + 13}{(x - 2)(x^2 - 2x + 3)}\, dx.$$

SOLUTION Note that the quadratic factor is irreducible. The appropriate form of the partial fraction decomposition is

$$\frac{7x^2 - 13x + 13}{(x - 2)(x^2 - 2x + 3)} = \frac{A}{x - 2} + \frac{Bx + C}{x^2 - 2x + 3}.$$

Multiplying both sides of this equation by $(x - 2)(x^2 - 2x + 3)$ leads to

$$7x^2 - 13x + 13 = A(x^2 - 2x + 3) + (Bx + C)(x - 2)$$
$$= (A + B)x^2 + (-2A - 2B + C)x + (3A - 2C).$$

Equating coefficients of equal powers of x results in the equations

$$A + B = 7, \quad -2A - 2B + C = -13, \quad \text{and} \quad 3A - 2C = 13.$$

Solving this system of equations gives $A = 5$, $B = 2$, and $C = 1$; therefore, the original integral can be written as

$$\int \frac{7x^2 - 13x + 13}{(x - 2)(x^2 - 2x + 3)}\, dx = \int \frac{5}{x - 2}\, dx + \int \frac{2x + 1}{x^2 - 2x + 3}\, dx.$$

Let's work on the second (more difficult) integral. The substitution $u = x^2 - 2x + 3$ would work if $du = (2x - 2)\,dx$ appeared in the numerator. For this reason, we write the numerator as $2x + 1 = (2x - 2) + 3$ and split the integral:

$$\int \frac{2x+1}{x^2-2x+3}\,dx = \int \frac{2x-2}{x^2-2x+3}\,dx + \int \frac{3}{x^2-2x+3}\,dx.$$

Assembling all the pieces, we have

$$\int \frac{7x^2 - 13x + 13}{(x-2)(x^2-2x+3)}\,dx$$

$$= \int \frac{5}{x-2}\,dx + \underbrace{\int \frac{2x-2}{x^2-2x+3}\,dx}_{\text{let } u = x^2 - 2x + 3} + \underbrace{\int \frac{3}{x^2-2x+3}\,dx}_{(x-1)^2 + 2} \qquad \text{Complete the square in the third integral.}$$

$$= 5\ln|x-2| + \ln|x^2-2x+3| + \frac{3}{\sqrt{2}}\tan^{-1}\left(\frac{x-1}{\sqrt{2}}\right) + K \qquad \text{Integrate.}$$

$$= \ln\left|(x-2)^5(x^2-2x+3)\right| + \frac{3}{\sqrt{2}}\tan^{-1}\left(\frac{x-1}{\sqrt{2}}\right) + K. \qquad \text{Property of logarithms}$$

To evaluate the last integral $\int \frac{3}{x^2-2x+3}\,dx$, we completed the square in the denominator and used the substitution $u = x - 1$ to produce $3\int \frac{du}{u^2+2}$, which is a standard form.

Related Exercises 53, 55 ◄

Long Division The preceding discussion of partial fraction decomposition assumes $f(x) = p(x)/q(x)$ is a proper rational function. If this is not the case and we are faced with an improper rational function f, we divide the denominator into the numerator and express f in two parts. One part will be a polynomial, and the other will be a proper rational function. For example, given the function

$$f(x) = \frac{2x^3 + 11x^2 + 28x + 33}{x^2 - x - 6},$$

we perform long division.

$$
\begin{array}{r}
2x + 13 \\
x^2 - x - 6 \overline{)\,2x^3 + 11x^2 + 28x + 33} \\
\underline{2x^3 - 2x^2 - 12x} \\
13x^2 + 40x + 33 \\
\underline{13x^2 - 13x - 78} \\
53x + 111
\end{array}
$$

It follows that

$$f(x) = \underbrace{2x + 13}_{\substack{\text{polynomial;}\\\text{easy to}\\\text{integrate}}} + \underbrace{\frac{53x + 111}{x^2 - x - 6}}_{\substack{\text{apply partial fraction}\\\text{decomposition}}}.$$

The first piece is easily integrated, and the second piece now qualifies for the methods described in this section.

QUICK CHECK 4 What is the result of simplifying $\dfrac{x}{x+1}$ by long division? ◄

SUMMARY Partial Fraction Decompositions

Let $f(x) = p(x)/q(x)$ be a proper rational function in reduced form. Assume the denominator q has been factored completely over the real numbers and m is a positive integer.

1. **Simple linear factor** A factor $x - r$ in the denominator requires the partial fraction $\dfrac{A}{x - r}$.

2. **Repeated linear factor** A factor $(x - r)^m$ with $m > 1$ in the denominator requires the partial fractions
$$\frac{A_1}{(x - r)} + \frac{A_2}{(x - r)^2} + \frac{A_3}{(x - r)^3} + \cdots + \frac{A_m}{(x - r)^m}.$$

3. **Simple irreducible quadratic factor** An irreducible factor $ax^2 + bx + c$ in the denominator requires the partial fraction
$$\frac{Ax + B}{ax^2 + bx + c}.$$

4. **Repeated irreducible quadratic factor** (See Exercises 61–64.) An irreducible factor $(ax^2 + bx + c)^m$ with $m > 1$ in the denominator requires the partial fractions
$$\frac{A_1x + B_1}{ax^2 + bx + c} + \frac{A_2x + B_2}{(ax^2 + bx + c)^2} + \cdots + \frac{A_mx + B_m}{(ax^2 + bx + c)^m}.$$

SECTION 8.5 EXERCISES

Getting Started

1. What kinds of functions can be integrated using partial fraction decomposition?

2. Give an example of each of the following.

   a. A simple linear factor
   b. A repeated linear factor
   c. A simple irreducible quadratic factor
   d. A repeated irreducible quadratic factor

3. What term(s) should appear in the partial fraction decomposition of a proper rational function with each of the following?

   a. A factor of $x - 3$ in the denominator
   b. A factor of $(x - 4)^3$ in the denominator
   c. A factor of $x^2 + 2x + 6$ in the denominator

4. What is the first step in integrating $\dfrac{x^2 + 2x - 3}{x + 1}$?

5–16. *Set up the appropriate form of the partial fraction decomposition for the following expressions. Do not find the values of the unknown constants.*

5. $\dfrac{4x}{x^2 - 9x + 20}$

6. $\dfrac{4x + 1}{4x^2 - 1}$

7. $\dfrac{x + 3}{(x - 5)^2}$

8. $\dfrac{2}{x^3 - 2x^2 + x}$

9. $\dfrac{4}{x^5 - 5x^3 + 4x}$

10. $\dfrac{20x}{(x - 1)^2(x^2 + 1)}$

11. $\dfrac{1}{x(x^2 + 1)}$

12. $\dfrac{2x^2 + 3}{(x^2 - 8x + 16)(x^2 + 3x + 4)}$

13. $\dfrac{x^4 + 12x^2}{(x - 2)^2(x^2 + x + 2)^2}$

14. $\dfrac{6x^4 - 4x^3 + 15x^2 - 5x + 7}{(x - 2)(2x^2 + 3)^2}$

15. $\dfrac{x}{(x^4 - 16)^2}$

16. $\dfrac{x^2 + 2x + 6}{x^3(x^2 + x + 1)^2}$

Practice Exercises

17–22. *Give the partial fraction decomposition for the following expressions.*

17. $\dfrac{5x - 7}{x^2 - 3x + 2}$

18. $\dfrac{11x - 10}{x^2 - x}$

19. $\dfrac{6}{x^2 - 2x - 8}$

20. $\dfrac{x^2 - 4x + 11}{(x - 3)(x - 1)(x + 1)}$

21. $\dfrac{2x^2 + 5x + 6}{x^2 + 3x + 2}$ (*Hint:* Use long division first.)

22. $\dfrac{x^4 + 2x^3 + x}{x^2 - 1}$

23–64. Integration *Evaluate the following integrals.*

23. $\displaystyle\int \frac{3}{(x - 1)(x + 2)}\, dx$

24. $\displaystyle\int \frac{8}{(x - 2)(x + 6)}\, dx$

25. $\displaystyle\int \frac{6}{x^2 - 1}\, dx$

26. $\displaystyle\int_0^1 \frac{dt}{t^2 - 9}$

27. $\displaystyle\int \frac{8x-5}{3x^2-5x+2}\,dx$

28. $\displaystyle\int_1^2 \frac{7x-2}{3x^2-2x}\,dx$

29. $\displaystyle\int_{-1}^2 \frac{5x}{x^2-x-6}\,dx$

30. $\displaystyle\int \frac{21x^2}{x^3-x^2-12x}\,dx$

31. $\displaystyle\int \frac{6x^2}{x^4-5x^2+4}\,dx$

32. $\displaystyle\int \frac{4x-2}{x^3-x}\,dx$

33. $\displaystyle\int \frac{3x^2+4x-6}{x^2-3x+2}\,dx$

34. $\displaystyle\int \frac{2z^3+z^2-6z+7}{z^2+z-6}\,dz$

35. $\displaystyle\int \frac{x^2+12x-4}{x^3-4x}\,dx$

36. $\displaystyle\int \frac{z^2+20z-15}{z^3+4z^2-5z}\,dz$

37. $\displaystyle\int \frac{dx}{x^4-10x^2+9}$

38. $\displaystyle\int_0^5 \frac{2}{x^2-4x-32}\,dx$

39. $\displaystyle\int \frac{81}{x^3-9x^2}\,dx$

40. $\displaystyle\int \frac{16x^2}{(x-6)(x+2)^2}\,dx$

41. $\displaystyle\int_{-1}^1 \frac{x}{(x+3)^2}\,dx$

42. $\displaystyle\int \frac{dx}{x^3-2x^2-4x+8}$

43. $\displaystyle\int \frac{2}{x^3+x^2}\,dx$

44. $\displaystyle\int_1^2 \frac{2}{t^3(t+1)}\,dt$

45. $\displaystyle\int \frac{x-5}{x^2(x+1)}\,dx$

46. $\displaystyle\int \frac{x^2}{(x-2)^3}\,dx$

47. $\displaystyle\int \frac{x^3-10x^2+27x}{x^2-10x+25}\,dx$

48. $\displaystyle\int \frac{x^3+2}{x^3-2x^2+x}\,dx$

49. $\displaystyle\int \frac{x^2-4}{x^3-2x^2+x}\,dx$

50. $\displaystyle\int \frac{8(x^2+4)}{x(x^2+8)}\,dx$

51. $\displaystyle\int \frac{x^2+x+2}{(x+1)(x^2+1)}\,dx$

52. $\displaystyle\int \frac{x^2+3x+2}{x(x^2+2x+2)}\,dx$

53. $\displaystyle\int \frac{2x^2+5x+5}{(x+1)(x^2+2x+2)}\,dx$

54. $\displaystyle\int \frac{z+1}{z(z^2+4)}\,dz$

55. $\displaystyle\int \frac{20x}{(x-1)(x^2+4x+5)}\,dx$

56. $\displaystyle\int \frac{2x+1}{x^2+4}\,dx$

57. $\displaystyle\int \frac{x^3+5x}{(x^2+3)^2}\,dx$

58. $\displaystyle\int \frac{x^4+1}{x^3+9x}\,dx$

59. $\displaystyle\int \frac{x^3+6x^2+12x+6}{(x^2+6x+10)^2}\,dx$

60. $\displaystyle\int \frac{dy}{(y^2+1)(y^2+2)}$

61. $\displaystyle\int \frac{2}{x(x^2+1)^2}\,dx$

62. $\displaystyle\int \frac{dx}{(x+1)(x^2+2x+2)^2}$

63. $\displaystyle\int \frac{9x^2+x+21}{(3x^2+7)^2}\,dx$

64. $\displaystyle\int \frac{9x^5+6x^3}{(3x^2+1)^3}\,dx$

65. Explain why or why not Determine whether the following statements are true and give an explanation or counterexample.

   a. To evaluate $\displaystyle\int \frac{4x^6}{x^4+3x^2}\,dx$, the first step is to find the partial fraction decomposition of the integrand.

   b. The easiest way to evaluate $\displaystyle\int \frac{6x+1}{3x^2+x}\,dx$ is with a partial fraction decomposition of the integrand.

   c. The rational function $f(x) = \dfrac{1}{x^2-13x+42}$ has an irreducible quadratic denominator.

   d. The rational function $f(x) = \dfrac{1}{x^2-13x+43}$ has an irreducible quadratic denominator.

T 66–68. Areas of regions *Find the area of the following regions.*

66. The region bounded by the curve $y = \dfrac{x-x^2}{(x+1)(x^2+1)}$ and the x-axis from $x=0$ to $x=1$

67. The region bounded by the curve $y = \dfrac{10}{x^2-2x-24}$, the x-axis, and the lines $x=-2$ and $x=2$

68. The region in the first quadrant bounded by the curves $y = \dfrac{3x^2+2x+1}{x(x^2+x+1)}$, $y = \dfrac{2}{x}$, and $x=2$

69–72. Volumes of solids *Find the volume of the following solids.*

69. The region bounded by $y = \dfrac{x}{x+1}$, the x-axis, and $x=4$ is revolved about the x-axis.

70. The region bounded by $y = \dfrac{1}{x^2(x^2+2)^2}$, $y=0$, $x=1$, and $x=2$ is revolved about the y-axis.

71. The region bounded by $y = \dfrac{1}{\sqrt{x(3-x)}}$, $y=0$, $x=1$, and $x=2$ is revolved about the x-axis.

72. The region bounded by $y = \dfrac{3x^2+25}{x^2(x^2+25)}$, $y=0$, $x=5$, and $x=10$ is revolved about the y-axis.

73. Two methods Evaluate $\displaystyle\int \frac{dx}{x^2-1}$, for $x>1$, in two ways: using partial fractions and a trigonometric substitution. Reconcile your two answers.

T 74–75. Finding constants with a computer algebra system *Give the appropriate form of the partial fraction decomposition of the expression, and then use a computer algebra system to find the unknown constants.*

74. $\dfrac{3x^2+2x+1}{(x+1)^3(x^2+x+1)^2}$

75. $\dfrac{x^4+3x^2+1}{x(x^2+1)^2(x^2+x+4)^2}$

76–83. Preliminary steps *The following integrals require a preliminary step such as a change of variables before using the method of partial fractions. Evaluate these integrals.*

76. $\displaystyle\int \frac{\cos\theta}{(\sin^3\theta - 4\sin\theta)}\,d\theta$

77. $\displaystyle\int \frac{e^x}{(e^x-1)(e^x+2)}\,dx$

78. $\displaystyle\int \frac{dy}{y(\sqrt{a}-\sqrt{y})}$, $a>0$ (*Hint: Let $u=\sqrt{y}$.*)

79. $\displaystyle\int \frac{\sec t}{1+\sin t}\,dt$

80. $\displaystyle\int \sqrt{e^x+1}\,dx$ (*Hint: Let $u=\sqrt{e^x+1}$.*)

81. $\displaystyle\int \frac{(e^{3x}+e^{2x}+e^x)}{(e^{2x}+1)^2}\,dx$

82. $\displaystyle\int \frac{dx}{x\sqrt{1+2x}}$

83. $\displaystyle\int \frac{dx}{1 + e^x}$

84. What's wrong? Why are there no constants A and B satisfying

$$\frac{x^2}{(x - 4)(x + 5)} = \frac{A}{x - 4} + \frac{B}{x + 5}?$$

85. Another form of $\displaystyle\int \sec x \, dx$

a. Verify the identity $\sec x = \dfrac{\cos x}{1 - \sin^2 x}$.

b. Use the identity in part (a) to verify that

$$\int \sec x \, dx = \frac{1}{2} \ln \left| \frac{1 + \sin x}{1 - \sin x} \right| + C.$$

(*Source: The College Mathematics Journal, 32, 5, Nov 2001*)

Explorations and Challenges

86. Partial fractions from an integral A computer algebra system was used to obtain the following result.

$$\int \frac{2x^4 + 14x^3 + 14x^2 + 51x + 19}{x^5 + x^4 + 3x^3 + 3x^2 - 4x - 4} \, dx$$

$$= 5 \ln (x - 1) - 3 \ln (x + 1) - \frac{3}{x + 1} + \frac{1}{2} \tan^{-1} \frac{x}{2} + C$$

Use this fact to find the partial fraction decomposition of the integrand on the left side of the equation.

87–92. *An integrand with trigonometric functions in the numerator and denominator can often be converted to a rational function using the substitution $u = \tan (x/2)$ or, equivalently, $x = 2 \tan^{-1} u$. The following relations are used in making this change of variables.*

$$A: \ dx = \frac{2}{1 + u^2} \, du \quad B: \ \sin x = \frac{2u}{1 + u^2} \quad C: \ \cos x = \frac{1 - u^2}{1 + u^2}$$

87. Verify relation A by differentiating $x = 2 \tan^{-1} u$. Verify relations B and C using a right-triangle diagram and the double-angle formulas

$$\sin x = 2 \sin \frac{x}{2} \cos \frac{x}{2} \quad \text{and} \quad \cos x = 2 \cos^2 \frac{x}{2} - 1.$$

88. Evaluate $\displaystyle\int \frac{dx}{2 + \cos x}$.

89. Evaluate $\displaystyle\int \frac{dx}{1 - \cos x}$.

90. Evaluate $\displaystyle\int \frac{dx}{1 + \sin x + \cos x}$.

91. Evaluate $\displaystyle\int_0^{\pi/2} \frac{d\theta}{\cos \theta + \sin \theta}$.

92. Evaluate $\displaystyle\int_0^{\pi/3} \frac{\sin \theta}{1 - \sin \theta} \, d\theta$.

93. Three start-ups Three cars, A, B, and C, start from rest and accelerate along a line according to the following velocity functions:

$$v_A(t) = \frac{88t}{t + 1}, \quad v_B(t) = \frac{88t^2}{(t + 1)^2}, \quad \text{and} \quad v_C(t) = \frac{88t^2}{t^2 + 1}.$$

a. Which car travels farthest on the interval $0 \leq t \leq 1$?

b. Which car travels farthest on the interval $0 \leq t \leq 5$?

c. Find the position functions for each car assuming each car starts at the origin.

d. Which car ultimately gains the lead and remains in front?

T 94. Skydiving A skydiver has a downward velocity given by

$$v(t) = V_T \left(\frac{1 - e^{-2gt/V_T}}{1 + e^{-2gt/V_T}} \right),$$

where $t = 0$ is the instant the skydiver starts falling, $g = 9.8 \ \text{m/s}^2$ is the acceleration due to gravity, and V_T is the terminal velocity of the skydiver.

a. Evaluate $v(0)$ and $\lim_{t \to \infty} v(t)$ and interpret these results.

b. Graph the velocity function.

c. Verify by integration that the position function is given by

$$s(t) = V_T t + \frac{V_T^2}{g} \ln \left(\frac{1 + e^{-2gt/V_T}}{2} \right),$$

where $s'(t) = v(t)$ and $s(0) = 0$.

d. Graph the position function.
(See the Guided Project *Terminal velocity* for more details on free fall and terminal velocity.)

95. $\pi < 22/7$ One of the earliest approximations to π is $22/7$. Verify that $0 < \displaystyle\int_0^1 \frac{x^4(1 - x)^4}{1 + x^2} \, dx = \frac{22}{7} - \pi$. Why can you conclude that $\pi < 22/7$?

96. Challenge Show that with the change of variables $u = \sqrt{\tan x}$, the integral $\int \sqrt{\tan x} \, dx$ can be converted to an integral amenable to partial fractions. Evaluate $\int_0^{\pi/4} \sqrt{\tan x} \, dx$.

QUICK CHECK ANSWERS

1. $\ln |x - 2| + 2 \ln |x + 4| = \ln |(x - 2)(x + 4)^2|$
2. $A/(x - 1) + B/(x + 5) + C/(x - 10)$
3. $A/x + B/x^2 + C/(x - 3) + D/(x - 3)^2 + E/(x - 1)$
4. $1 - \dfrac{1}{x + 1}$ ◄

8.6 Integration Strategies

In the previous five sections, we introduced various methods for evaluating integrals, though always in the context of learning a specific technique of integration. For instance, in Section 8.3, we focused on integrals involving trigonometric functions. In this section, the chief goal is to illustrate how to attack a generic integration problem when the appropriate method is not obvious.

Integration Strategies

As a starting point, we assume you know all the integrals listed in Table 8.1 at the beginning of this chapter, and that you are familiar with the reduction formulas presented in

Sections 8.2 and 8.3. Your instructor may also require that you know the formulas presented in Section 7.3 that involve hyperbolic functions.

We noted in Section 5.5 that computing the derivative of a (differentiable) function usually requires only the proper application of the derivative rules from Chapter 3. Finding the antiderivative of a function is, generally speaking, more difficult. In what follows, we offer a collection of strategies to use when confronted with an unfamiliar integral, illustrated with brief examples.

Rewrite the integrand Several techniques can be used to transform an integrand. We can split up fractions, use long division or trigonometric identities, complete the square in an integrand that includes a quadratic function, perform algebraic manipulations (such as multiplying numerator and denominator by the same expression), and cancel common factors.

EXAMPLE 1 **Rewrite the integrand** Evaluate $\int \dfrac{\sin x + 1}{\cos^2 x} \, dx$.

QUICK CHECK 1 Use Table 8.1 (p. 520) to complete the process of evaluating $\int \dfrac{\sin x + 1}{\cos^2 x} \, dx$ given in Example 1. ◄

SOLUTION Split the fraction and use trigonometric identities to rewrite the integral as the sum of two integrals:

$$\int \frac{\sin x + 1}{\cos^2 x} \, dx = \int \left(\frac{1}{\cos x} \cdot \frac{\sin x}{\cos x} + \frac{1}{\cos^2 x} \right) dx = \int \sec x \tan x \, dx + \int \sec^2 x \, dx.$$

Both integrals on the right are now easily integrated using Table 8.1.

Related Exercises 4, 11 ◄

Try substitution As a general rule, if an integrand contains the function $f(x)$ and its derivative $f'(x)$, the substitution $u = f(x)$ often results in an easier integral. For instance, $\int \dfrac{\sin (1/x)}{x^2} \, dx$ suggests the substitution $u = 1/x$ because $\dfrac{du}{dx} = -\dfrac{1}{x^2}$ appears in the integrand (up to a constant multiplicative factor). When attempting to evaluate an integral using the substitution method, it is important to understand that there is no list of "rules" one must follow—any substitution is fair game. On occasion, a creative substitution does the trick.

QUICK CHECK 2 Use the substitution $u = \dfrac{1}{x}$ to evaluate $\int \dfrac{\sin (1/x)}{x^2} \, dx$. ◄

EXAMPLE 2 **Substitution after rewriting the integrand** Evaluate $\int \dfrac{x^2}{\sqrt{4 - x^6}} \, dx$.

SOLUTION Expressing the denominator of the integrand as $\sqrt{4 - (x^3)^2}$ looks promising, particularly because a constant multiple of the derivative of x^3 appears in the numerator. Therefore, we try the substitution $u = x^3$:

$$\int \frac{x^2}{\sqrt{4 - x^6}} \, dx = \int \frac{x^2}{\sqrt{4 - (x^3)^2}} \, dx = \frac{1}{3} \int \frac{du}{\sqrt{4 - u^2}}. \quad \text{Let } u = x^3; \, du = 3x^2 \, dx.$$

QUICK CHECK 3 Complete the evaluation of $\int \dfrac{x^2}{\sqrt{4 - x^6}} \, dx$ in Example 2; make sure to express your answer in terms of x. ◄

Observe that $\int \dfrac{du}{\sqrt{4 - u^2}}$ is a standard integral, found in Table 8.1.

Related Exercises 14, 43 ◄

Recognize a common pattern present in the integrand Here we list certain patterns to look for in the integrand and the associated integration techniques.

1. Integrating the product of two functions often calls for integration by parts (Section 8.2), particularly if the factors of the product are a polynomial and a transcendental function (a logarithmic, exponential, trigonometric, or inverse trigonometric function). In some cases, integration by parts is successful when the integrand is a single function (for example, $\int \ln x \, dx$ or $\int \sin^{-1} x \, dx$).

2. Powers and products of $\sin x$ and $\cos x$, $\tan x$ and $\sec x$, or $\cot x$ and $\csc x$ are integrated using identities, reduction formulas, or the substitutions given in Tables 8.2 and 8.3 of Section 8.3.

3. Integrals containing the expression $a^2 - x^2$, $a^2 + x^2$, or $x^2 - a^2$ may yield to trigonometric substitution. Use the substitutions given in Table 8.4 of Section 8.4 to tackle such integrals, but don't overlook a simpler solution. For example, $\int \dfrac{x}{1 - x^2} \, dx$ is most easily evaluated using the substitution $u = 1 - x^2$ rather than $x = \sin \theta$.

4. Use partial fraction decomposition (Section 8.5) to integrate rational functions. Remember to first cancel common factors and to apply long division if the rational function is not proper.

EXAMPLE 3   **Identify a strategy**   Suggest a technique of integration for each of the following integrals based on patterns in the integrand.

a. $\displaystyle\int \dfrac{4 - 3x^2}{x(x^2 - 4)} \, dx$ b. $\displaystyle\int xe^{\sqrt{1+x^2}} \, dx$ c. $\displaystyle\int \ln(1 + x^2) \, dx$

SOLUTION

a. The integrand is a rational function whose denominator factors easily, so partial fraction decomposition is an obvious choice. However, it is worth considering other options that may save time. Expanding the denominator, we recognize the integral admits an almost perfect u-substitution:

$$\int \frac{4 - 3x^2}{x(x^2 - 4)} \, dx = \int \frac{4 - 3x^2}{x^3 - 4x} \, dx. \qquad \begin{array}{l} \text{Let } u = x^3 - 4x; \, du = (3x^2 - 4)\,dx, \\ \text{or } -du = (4 - 3x^2)\,dx. \end{array}$$

This substitution leads to $-\displaystyle\int \dfrac{du}{u}$, which is a standard integral.

b. The integrand suggests several techniques. The substitution $u = 1 + x^2$ looks promising given that $du/dx = 2x$ appears in the integrand (up to a constant multiple). This choice leads to $\dfrac{1}{2} \displaystyle\int e^{\sqrt{u}} \, du$, which requires a crafty substitution followed by integration by parts.

    Given that the integrand is a product of two functions, we might try integration by parts as a first step. We leave it to the reader to conclude that this path is not fruitful.

    Finally, the presence of $1 + x^2$ in the integrand suggests the substitution $x = \tan \theta$. This choice leads to

$$\int xe^{\sqrt{1+x^2}} \, dx = \int \tan\theta \cdot e^{\sqrt{1+\tan^2\theta}} \sec^2\theta \, d\theta \qquad x = \tan\theta, \, dx = \sec^2\theta \, d\theta$$

$$= \int \sec\theta \cdot e^{\sec\theta} \sec\theta \tan\theta \, d\theta \qquad \begin{array}{l}\text{Rewrite the integrand to prepare} \\ \text{for a substitution.}\end{array}$$

$$= \int u e^u \, du. \qquad \text{Let } u = \sec\theta; \, du = \sec\theta \tan\theta \, d\theta.$$

> ➤ The substitution $u = \sqrt{1 + x^2}$ is yet another option for the integral in Example 3b; this choice leads immediately (though not transparently) to $\int ue^u \, du$. It takes a lot of practice to find creative solutions such as this one.

We encountered the last integral in Example 1 of Section 8.2 and evaluated it using integration by parts.

c. Because $1 + x^2$ is present in the integrand, we could try the substitution $x = \tan \theta$, which yields $\int \ln(\sec^2\theta) \sec^2\theta \, d\theta$. Using integration by parts on this new integral eventually leads to the answer, but it is more efficient to use integration by parts from the beginning. We try the following choices for u and dv.

$u = \ln(1 + x^2)$	$dv = dx$
$du = \dfrac{2x}{1 + x^2} \, dx$	$v = x$

The integral then becomes

$$\int \ln(1 + x^2) \, dx = x \ln(1 + x^2) - 2 \int \frac{x^2}{1 + x^2} \, dx,$$

and the new integral is handled with long division followed by routine integration.

Related Exercises 55, 70, 96 ◄

The integrals in Example 3 bring to light two important concepts. First, recognize that more than one technique may work for a given integral. In practice, one method is usually more efficient than the others, so be ready to try another method if your first choice does not work. Second, to determine which choice is best, you may need to carry out the first few steps of several techniques in order to recognize the best approach.

Be persistent   Be prepared to experience dead ends when attempting to evaluate an integral. Look again at the integral given in Example 1. Instead of splitting the fraction, suppose we used the identity $\cos^2 x = 1 - \sin^2 x$:

$$\int \frac{\sin x + 1}{\cos^2 x}\, dx = \int \frac{\sin x + 1}{1 - \sin^2 x}\, dx = \int \frac{\sin x + 1}{(1 + \sin x)(1 - \sin x)}\, dx = \int \frac{1}{1 - \sin x}\, dx.$$

No error has been committed in the previous calculation, but the final integral is not easily integrated in its present form. In situations such as this one, start again using a different approach.

EXAMPLE 4   Integration by parts? Evaluate $\int x e^{\sqrt{x}}\, dx$.

SOLUTION The integrand is the product of a power of x and a transcendental function; integrals of this form often yield to integration by parts. We try the following choices for u and dv.

$u = e^{\sqrt{x}}$	$dv = x\, dx$
$du = \dfrac{e^{\sqrt{x}}}{2\sqrt{x}}\, dx$	$v = \dfrac{x^2}{2}$

The integral then becomes

$$\int x e^{\sqrt{x}}\, dx = \frac{x^2 e^{\sqrt{x}}}{2} - \frac{1}{4}\int x^{3/2} e^{\sqrt{x}}\, dx.$$

> ▶ The integral that results in Example 4 when applying integration by parts, $\int x^{3/2} e^{\sqrt{x}}\, dx$, can be evaluated. For instance, the substitution $u = e^{\sqrt{x}}$ leads to $2 \int \ln^4 u\, du$, and the substitution $u = \sqrt{x}$ leads to $2 \int u^4 e^u\, du$, both of which can be handled by reduction formulas given in the exercise set of Section 8.2. However, the second solution, using the substitution $u = \sqrt{x}$, is more efficient.

The new integral appears to be more difficult than the original. Because integration by parts was not successful, we return to the original integral. Let's try the substitution $u = \sqrt{x}$, which implies that $du = \dfrac{1}{2\sqrt{x}}\, dx$. At this point, there are two paths we might follow.

Option 1: Our substitution would succeed if $2\sqrt{x}$ appeared in the denominator of the integrand. Using the technique of multiplication by 1, we multiply the numerator and denominator of the integrand by $2\sqrt{x}$ to prepare for the substitution:

$$\int x e^{\sqrt{x}}\, dx = \int \frac{2\sqrt{x} \cdot x e^{\sqrt{x}}}{2\sqrt{x}}\, dx \quad \text{Multiply by 1.}$$

$$= 2 \int u^3 e^u\, du. \qquad u = \sqrt{x};\ du = \frac{dx}{2\sqrt{x}};\ \sqrt{x} \cdot x = (\sqrt{x})^3 = u^3$$

Option 2: Because $u = \sqrt{x}$, we have $x = u^2$, which implies that $dx = 2u\, du$. Making these substitutions in the integral leads to

$$\int x e^{\sqrt{x}}\, dx = \int u^2 e^u \cdot 2u\, du \quad u = \sqrt{x};\ x = u^2;\ dx = 2u\, du$$

$$= 2 \int u^3 e^u\, du. \qquad \text{Simplify.}$$

In both cases we arrive at the same integral, $2 \int u^3 e^u\, du$, which is handled by applying repeated integration by parts (see tabular integration, Exercises 77–79, Section 8.2) or by using a reduction formula (Exercise 50, Section 8.2).

Related Exercise 98 ◀

Multiple techniques   More than one technique may be necessary for complicated integrals. We conclude this section with a full evaluation of an integral that requires several techniques of integration.

EXAMPLE 5   Multiple techniques needed Find the area of the surface of revolution that results when the curve $f(x) = e^x$ on $[0, \ln 2]$ is revolved about the x-axis.

SOLUTION Recall from Section 6.6 that the integral $\int_a^b 2\pi f(x)\sqrt{1 + f'(x)^2}\, dx$ gives the area of the surface generated when $y = f(x)$ on $[a, b]$ is revolved about the x-axis. Because $f'(x) = e^x$, the area of the surface is

$$\int_0^{\ln 2} 2\pi e^x \sqrt{1 + (e^x)^2}\, dx.$$

The presence of e^x and its derivative in the integrand suggests the substitution $u = e^x$:

$$\int_0^{\ln 2} 2\pi e^x \sqrt{1 + (e^x)^2}\, dx = 2\pi \int_1^2 \sqrt{1 + u^2}\, du \qquad \begin{array}{l} u = e^x;\ du = e^x\, dx \\ x = 0 \Rightarrow u = 1;\ x = \ln 2 \Rightarrow u = 2 \end{array}$$

The new integrand contains $1 + u^2$, so we try the trigonometric substitution $u = \tan\theta$. Setting the definite integral aside for the moment, we first focus on evaluating the indefinite integral:

$$2\pi \int \sqrt{1 + u^2}\, du = 2\pi \int \sqrt{1 + \tan^2\theta}\, \sec^2\theta\, d\theta \qquad \text{Let } u = \tan\theta;\ du = \sec^2\theta\, d\theta.$$

$$= 2\pi \int \sec^3\theta\, d\theta. \qquad 1 + \tan^2\theta = \sec^2\theta$$

To complete the solution, we rely on the secant reduction formula from Section 8.3 and return to the variable u:

$$2\pi \int \sec^3\theta\, d\theta = 2\pi \left(\frac{1}{2} \sec\theta \tan\theta + \frac{1}{2}\int \sec\theta\, d\theta \right) \qquad \text{Secant reduction formula}$$

$$= \pi(\sec\theta \tan\theta + \ln|\sec\theta + \tan\theta|) + C \qquad \text{Evaluate } \int \sec\theta\, d\theta.$$

$$= \pi\left(u\sqrt{1 + u^2} + \ln|u + \sqrt{1 + u^2}|\right) + C. \qquad \tan\theta = u;\ \sec\theta = \sqrt{1 + u^2}$$

Evaluating the antiderivative using limits of integration from the integral in u gives the area of the surface:

$$\text{Area} = \pi\left(u\sqrt{1 + u^2} + \ln\left(u + \sqrt{1 + u^2}\right)\right)\Big|_1^2$$

$$= \pi\left(2\sqrt{5} - \sqrt{2} + \ln\frac{2 + \sqrt{5}}{1 + \sqrt{2}}\right) \approx 11.37.$$

Related Exercise 87 ◄

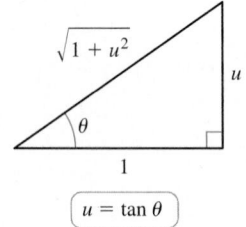

$u = \tan\theta$

SECTION 8.6 EXERCISES

Getting Started

1–6. Choosing an integration strategy *Identify a technique of integration for evaluating the following integrals. If necessary, explain how to first simplify the integrand before applying the suggested technique of integration. You do not need to evaluate the integrals.*

1. $\displaystyle\int 4x \sin 5x\, dx$

2. $\displaystyle\int (1 + \tan x)\sec^2 x\, dx$

3. $\displaystyle\int \frac{x^3}{\sqrt{64 - x^2}}\, dx$

4. $\displaystyle\int \frac{\tan^2 x + 1}{\tan x}\, dx$

5. $\displaystyle\int \frac{5x^2 + 18x + 20}{(2x + 3)(x^2 + 4x + 8)}\, dx$

6. $\displaystyle\int \frac{\cos^5 x \sin^4 x}{1 - \sin^2 x}\, dx$

Practice Exercises

7–84. *Evaluate the following integrals.*

7. $\displaystyle\int_0^{\pi/2} \frac{\sin\theta}{1 + \cos^2\theta}\, d\theta$

8. $\displaystyle\int \cos^2 10x\, dx$

9. $\displaystyle\int_4^6 \frac{dx}{\sqrt{8x - x^2}}$

10. $\displaystyle\int \sin^9 x \cos^3 x\, dx$

11. $\displaystyle\int_0^{\pi/4} (\sec x - \cos x)^2\, dx$

12. $\displaystyle\int \frac{e^x}{\sqrt{1 - e^{2x}}}\, dx$

13. $\displaystyle\int \frac{dx}{e^x\sqrt{1 - e^{2x}}}$

14. $\displaystyle\int \frac{x^{-2} + x^{-3}}{x^{-1} + 16x^{-3}}\, dx$

15. $\displaystyle\int_1^4 \frac{2^{\sqrt{x}}}{\sqrt{x}}\, dx$

16. $\displaystyle\int \frac{dx}{x^4 - 1}$

17. $\displaystyle\int_1^2 w^3 e^{w^2}\, dw$

18. $\displaystyle\int_5^6 x(x-5)^{10}\, dx$

19. $\displaystyle\int_0^{\pi/2} \sin^7 x\, dx$

20. $\displaystyle\int_1^3 \frac{dt}{\sqrt{t}(t+1)}$

21. $\displaystyle\int x^9 \ln 3x\, dx$

22. $\displaystyle\int \frac{dx}{(x-a)(x-b)},\ a \neq b$

23. $\displaystyle\int \frac{\sin x}{\cos^2 x + \cos x}\, dx$

24. $\displaystyle\int \frac{3w^5 + 2w^4 - 12w^3 - 12w - 32}{w^3 - 4w}\, dw$

25. $\displaystyle\int \frac{dx}{x\sqrt{1-x^2}}$

26. $\displaystyle\int_{1/e}^1 \frac{dx}{x(\ln^2 x + 2\ln x + 2)}$

27. $\displaystyle\int \sin^4 \frac{x}{2}\, dx$

28. $\displaystyle\int \frac{3x^2 + 2x + 3}{x^4 + 2x^2 + 1}\, dx$

29. $\displaystyle\int \frac{2\cos x + \cot x}{1 + \sin x}\, dx$

30. $\displaystyle\int_{5/2}^{5\sqrt{3}/2} \frac{dv}{v^2\sqrt{25 - v^2}}$

31. $\displaystyle\int \sqrt{36 - 9x^2}\, dx$

32. $\displaystyle\int \frac{dx}{\sqrt{36x^2 - 25}},\ x > 5/6$

33. $\displaystyle\int \frac{e^x}{a^2 + e^{2x}}\, dx,\ a \neq 0$

34. $\displaystyle\int_0^{\pi/9} \frac{\sin 3x}{\cos 3x + 1}\, dx$

35. $\displaystyle\int_0^{\pi/4} (\tan^2\theta + \tan\theta + 1)\sec^2\theta\, d\theta$

36. $\displaystyle\int x\, 10^x\, dx$

37. $\displaystyle\int_0^{\pi/6} \frac{dx}{1 - \sin 2x}$

38. $\displaystyle\int_{\pi/6}^{\pi/2} \cos x \ln(\sin x)\, dx$

39. $\displaystyle\int \sin x \ln(\sin x)\, dx$

40. $\displaystyle\int \sin 2x \ln(\sin x)\, dx$

41. $\displaystyle\int \cot^{3/2} x\, \csc^4 x\, dx$

42. $\displaystyle\int_0^{1/2} \frac{\sin^{-1} x}{\sqrt{1-x^2}}\, dx$

43. $\displaystyle\int \frac{x^9}{\sqrt{1 - x^{20}}}\, dx$

44. $\displaystyle\int \frac{dx}{x^3 - x^2}$

45. $\displaystyle\int_0^{\ln 2} \frac{1}{(1 + e^x)^2}\, dx$

46. $\displaystyle\int \frac{dx}{e^{2x} + 1}$

47. $\displaystyle\int \frac{2x^3 + x^2 - 2x - 4}{x^2 - x - 2}\, dx$

48. $\displaystyle\int \frac{\sqrt{16 - x^2}}{x^2}\, dx$

49. $\displaystyle\int \tan^3 x \sec^9 x\, dx$

50. $\displaystyle\int \tan^7 x \sec^4 x\, dx$

51. $\displaystyle\int_0^{\pi/3} \tan x \sec^{7/4} x\, dx$

52. $\displaystyle\int t^2 e^{3t}\, dt$

53. $\displaystyle\int e^x \cot^3 e^x\, dx$

54. $\displaystyle\int \frac{2x^2 + 3x + 26}{(x-2)(x^2 + 16)}\, dx$

55. $\displaystyle\int \frac{3x^2 + 3x + 1}{x^3 + x}\, dx$

56. $\displaystyle\int_\pi^{3\pi/2} \sin 2x\, e^{\sin^2 x}\, dx$

57. $\displaystyle\int \sin\sqrt{x}\, dx$

58. $\displaystyle\int w^2 \tan^{-1} w\, dw$

59. $\displaystyle\int \frac{dx}{x^4 + x^2}$

60. $\displaystyle\int_0^{\pi/2} e^{-3x} \cos x\, dx$

61. $\displaystyle\int_0^{\sqrt{2}/2} e^{\sin^{-1} x}\, dx$

62. $\displaystyle\int_0^{\pi/2} \sqrt{1 + \cos\theta}\, d\theta$

63. $\displaystyle\int x^a \ln x\, dx,\ a \neq -1$

64. $\displaystyle\int \frac{\ln ax}{x}\, dx,\ a \neq 0$

65. $\displaystyle\int_0^{1/6} \frac{dx}{\sqrt{1 - 9x^2}}$

66. $\displaystyle\int \frac{x}{\sqrt{1 - 9x^2}}\, dx$

67. $\displaystyle\int \frac{x^2}{\sqrt{1 - 9x^2}}\, dx$

68. $\displaystyle\int \frac{e^x}{e^{2x} + 2e^x + 17}\, dx$

69. $\displaystyle\int \frac{dx}{1 - x^2 + \sqrt{1 - x^2}}$

70. $\displaystyle\int \ln(x^2 + a^2)\, dx,\ a \neq 0$

71. $\displaystyle\int \frac{1 - \cos x}{1 + \cos x}\, dx$

72. $\displaystyle\int x^2 \sinh x\, dx$

73. $\displaystyle\int_9^{16} \sqrt{1 + \sqrt{x}}\, dx$

74. $\displaystyle\int \frac{e^{3x}}{e^x - 1}\, dx$

75. $\displaystyle\int_1^3 \frac{\tan^{-1}\sqrt{x}}{x^{1/2} + x^{3/2}}\, dx$

76. $\displaystyle\int \frac{x}{x^2 + 6x + 18}\, dx$

77. $\displaystyle\int \cos^{-1} x\, dx$

78. $\displaystyle\int (\cos^{-1} x)^2\, dx$

79. $\displaystyle\int \frac{\sin^{-1} x}{x^2}\, dx$

80. $\displaystyle\int_{-2}^{-1} \sqrt{-4x - x^2}\, dx$

81. $\displaystyle\int \frac{x^4 + 2x^3 + 5x^2 + 2x + 1}{x^5 + 2x^3 + x}\, dx$

82. $\displaystyle\int \frac{dx}{1 + \tan x}$

83. $\displaystyle\int e^x \sin^{998}(e^x)\cos^3(e^x)\, dx$

84. $\displaystyle\int \frac{\tan\theta + \tan^3\theta}{(1 + \tan\theta)^{50}}\, d\theta$

85. **Explain why or why not** Determine whether the following statements are true and give an explanation or counterexample.

   a. More than one integration method can be used to evaluate
   $$\int \frac{dx}{1 - x^2}.$$

   b. Using the substitution $u = \sqrt[3]{x}$ in $\displaystyle\int \sin\sqrt[3]{x}\, dx$ leads to
   $$\int 3u^2 \sin u\, du.$$

   c. The most efficient way to evaluate $\displaystyle\int \tan 3x \sec^2 3x\, dx$ is to first rewrite the integrand in terms of $\sin 3x$ and $\cos 3x$.

   d. Using the substitution $u = \tan x$ in $\displaystyle\int \frac{\tan^2 x}{\tan x - 1}\, dx$ leads to
   $$\int \frac{u^2}{u - 1}\, du.$$

86. **Area** Find the area of the region bounded by the curves
$$y = \frac{x}{x^2 - 2x + 2},\ y = \frac{2}{x^2 - 2x + 2},\ \text{and } x = 0.$$

87. **Surface area** Find the area of the surface generated when the curve $f(x) = \sin x$ on $[0, \pi/2]$ is revolved about the x-axis.

88. Volume Find the volume of the solid obtained by revolving the region bounded by the curve $y = \dfrac{x}{\sqrt{4 - x^2}}$ on $[0, 1]$ about the y-axis.

89. Volume Find the volume of the solid obtained by revolving the region bounded by the curve $y = \dfrac{1}{1 - \sin x}$ on $[0, \pi/4]$ about the x-axis.

90. Work Let R be the region in the first quadrant bounded by the curve $y = \sqrt{x^4 - 4}$, and the lines $y = 0$ and $y = 2$. Suppose a tank that is full of water has the shape of a solid of revolution obtained by revolving region R about the y-axis. How much work is required to pump all the water to the top of the tank? Assume x and y are in meters.

91. Work Let R be the region in the first quadrant bounded by the curve $y = \sec^{-1} x$ and the line $y = \pi/3$. Suppose a tank that is full of water has the shape of a solid of revolution obtained by revolving region R about the y-axis. How much work is required to pump all the water to the top of the tank? Assume x and y are in meters.

Explorations and Challenges

92–98. *Evaluate the following integrals.*

92. $\displaystyle\int_{1}^{\sqrt[3]{2}} y^8 e^{y^3}\, dy$

93. $\displaystyle\int \dfrac{68}{e^{2x} + 2e^x + 17}\, dx$

94. $\displaystyle\int \dfrac{dt}{t^3 + 1}$

95. $\displaystyle\int \dfrac{dx}{1 - \tan^2 x}$

96. $\displaystyle\int e^{\sqrt[3]{x}}\, dx$

97. $\displaystyle\int \tan^{-1} \sqrt[3]{x}\, dx$

98. $\displaystyle\int e^{\sqrt{\sin x}} \cos x\, dx$

99. Surface area Find the area of the surface generated when the curve $f(x) = \tan x$ on $[0, \pi/4]$ is revolved about the x-axis.

QUICK CHECK ANSWERS

1. $\sec x + \tan x + C$ **2.** $\cos(1/x) + C$

3. $\dfrac{1}{3}\sin^{-1}\dfrac{x^3}{2} + C$ ◄

8.7 Other Methods of Integration

The integration methods studied so far—various substitutions, integration by parts, and partial fractions—are examples of *analytical methods*; they are done with pencil and paper, and they give exact results. While many important integrals can be evaluated with analytical methods, many more integrals lie beyond their reach. For example, the following integrals cannot be evaluated in terms of familiar functions:

$$\int e^{x^2}\, dx, \qquad \int \sin x^2\, dx, \qquad \int \dfrac{\sin x}{x}\, dx, \qquad \int \dfrac{e^{-x}}{x}\, dx, \quad \text{and} \quad \int \ln(\ln x)\, dx.$$

The next two sections survey alternative strategies for evaluating integrals when standard analytical methods do not work. These strategies fall into three categories.

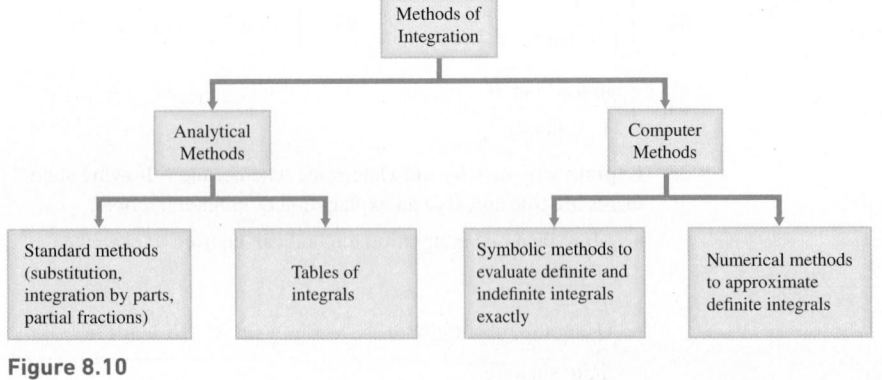

Figure 8.10

1. **Tables of integrals** The endpapers of this text contain a table of many standard integrals. Because these integrals were evaluated analytically, using tables is considered an analytical method. Tables of integrals also contain reduction formulas like those discussed in Sections 8.2 and 8.3.

2. **Symbolic methods** Computer algebra systems have sophisticated algorithms to evaluate difficult integrals. Many definite and indefinite integrals can be evaluated exactly using these symbolic methods.

3. **Numerical methods** The value of a definite integral can be approximated accurately using numerical methods introduced in the next section. *Numerical* means that these methods compute numbers rather than manipulate symbols. Computers and calculators often have built-in functions to carry out numerical calculations.

Figure 8.10 is a chart of the various integration strategies and how they are related.

Using Tables of Integrals

Given a specific integral, you *may* be able to find the identical integral in a table of integrals. More often, some preliminary work is needed to convert the given integral into one that appears in a table. Most tables give only indefinite integrals, although some tables include special definite integrals. The following examples illustrate various ways in which tables of integrals are used.

> A short table of integrals is found at the end of the text. Longer tables of integrals are found online and in venerable collections such as the *CRC Mathematical Tables* and *Handbook of Mathematical Functions*, by Abramowitz and Stegun.

> Letting $u^2 = 2x - 9$, we have $u \, du = dx$ and $x = \frac{1}{2}(u^2 + 9)$. Therefore,
> $$\int \frac{dx}{x\sqrt{2x - 9}} = 2\int \frac{du}{u^2 + 9}.$$

EXAMPLE 1   Using tables of integrals Evaluate the integral $\displaystyle \int \frac{dx}{x\sqrt{2x - 9}}$.

SOLUTION It is worth noting that this integral may be evaluated with the change of variables $u^2 = 2x - 9$. Alternatively, a table of integrals includes the integral

$$\int \frac{dx}{x\sqrt{ax - b}} = \frac{2}{\sqrt{b}}\tan^{-1}\sqrt{\frac{ax - b}{b}} + C, \quad \text{where} \quad b > 0,$$

which matches the given integral. Letting $a = 2$ and $b = 9$, we find that

$$\int \frac{dx}{x\sqrt{2x - 9}} = \frac{2}{\sqrt{9}}\tan^{-1}\sqrt{\frac{2x - 9}{9}} + C = \frac{2}{3}\tan^{-1}\frac{\sqrt{2x - 9}}{3} + C.$$

Related Exercises 9–10 ◄

EXAMPLE 2   Preliminary work Evaluate $\int \sqrt{x^2 + 6x}\,dx$.

SOLUTION Most tables of integrals do not include this integral. The nearest integral you are likely to find is $\int \sqrt{x^2 \pm a^2}\,dx$. The given integral can be put into this form by completing the square and using a substitution:

$$x^2 + 6x = x^2 + 6x + 9 - 9 = (x + 3)^2 - 9.$$

With the change of variables $u = x + 3$, the evaluation appears as follows:

$$\int \sqrt{x^2 + 6x}\,dx = \int \sqrt{(x + 3)^2 - 9}\,dx \qquad \text{Complete the square.}$$

$$= \int \sqrt{u^2 - 9}\,du \qquad u = x + 3,\, du = dx$$

$$= \frac{u}{2}\sqrt{u^2 - 9} - \frac{9}{2}\ln|u + \sqrt{u^2 - 9}| + C \qquad \text{Table of integrals}$$

$$= \frac{x + 3}{2}\sqrt{(x + 3)^2 - 9} - \frac{9}{2}\ln|x + 3 + \sqrt{(x + 3)^2 - 9}| + C$$

$$\text{Replace } u \text{ with } x + 3.$$

$$= \frac{x + 3}{2}\sqrt{x^2 + 6x} - \frac{9}{2}\ln|x + 3 + \sqrt{x^2 + 6x}| + C. \quad \text{Simplify.}$$

Related Exercises 31–33 ◄

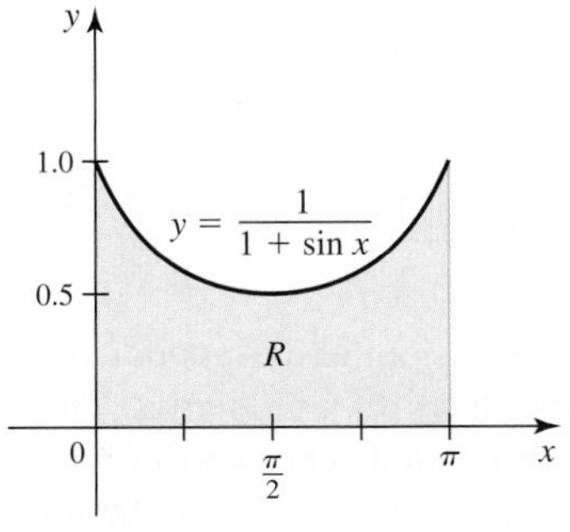

Figure 8.11

QUICK CHECK 1 Use the result of Example 3 to evaluate $\displaystyle \int_0^{\pi/2} \frac{dx}{1 + \sin x}$. ◄

EXAMPLE 3   Using tables of integrals for area Find the area of the region bounded by the curve $y = \dfrac{1}{1 + \sin x}$ and the x-axis between $x = 0$ and $x = \pi$.

SOLUTION The region in question (Figure 8.11) lies entirely above the x-axis, so its area is $\displaystyle \int_0^\pi \frac{dx}{1 + \sin x}$. A matching integral in a table of integrals is

$$\int \frac{dx}{1 + \sin ax} = -\frac{1}{a}\tan\left(\frac{\pi}{4} - \frac{ax}{2}\right) + C.$$

Evaluating the definite integral with $a = 1$, we have

$$\int_0^\pi \frac{dx}{1 + \sin x} = -\tan\left(\frac{\pi}{4} - \frac{x}{2}\right)\Big|_0^\pi = -\tan\left(-\frac{\pi}{4}\right) - \left(-\tan\frac{\pi}{4}\right) = 2.$$

Related Exercises 45–46 ◄

Symbolic Methods

Computer algebra systems evaluate many integrals exactly using symbolic methods, and they approximate many definite integrals using numerical methods. Different software packages may produce different results for the same indefinite integral, but ultimately, they must agree. The discussion that follows does not rely on one particular computer algebra system. Rather, it illustrates results from different systems and shows some of the idiosyncrasies of using a computer algebra system.

QUICK CHECK 2 Using one computer algebra system, it was found that $\int \sin x \cos x \, dx = \frac{1}{2} \sin^2 x + C$; using another computer algebra system, it was found that $\int \sin x \cos x \, dx = -\frac{1}{2} \cos^2 x + C$. Reconcile the two answers. ◄

> Some computer algebra systems do not include the constant of integration after evaluating an indefinite integral. But it should always be included when reporting a result.

EXAMPLE 4   Apparent discrepancies Evaluate $\int \dfrac{dx}{\sqrt{e^x + 1}}$ using tables and a computer algebra system.

SOLUTION Using one particular computer algebra system, we find that

$$\int \frac{dx}{\sqrt{e^x + 1}} = -2 \tanh^{-1} \sqrt{e^x + 1} + C,$$

> Recall that the *hyperbolic tangent* is defined as
> $$\tanh x = \frac{e^x - e^{-x}}{e^x + e^{-x}}.$$
> Its inverse is the *inverse hyperbolic tangent*, written $\tanh^{-1} x$.

where $\tanh^{-1}$ is the *inverse hyperbolic tangent* function (Section 7.3). However, we can obtain a result in terms of more familiar functions by first using the substitution $u = e^x$, which implies that $du = e^x \, dx$ or $dx = du/e^x = du/u$. The integral becomes

$$\int \frac{dx}{\sqrt{e^x + 1}} = \int \frac{du}{u\sqrt{u + 1}}.$$

Using a computer algebra system again, we obtain

$$\int \frac{dx}{\sqrt{e^x + 1}} = \int \frac{du}{u\sqrt{u + 1}} = \ln\left(\sqrt{1 + u} - 1\right) - \ln\left(\sqrt{1 + u} + 1\right)$$

$$= \ln\left(\sqrt{1 + e^x} - 1\right) - \ln\left(\sqrt{1 + e^x} + 1\right).$$

> Some computer algebra systems use $\log x$ for $\ln x$.

A table of integrals leads to a third equivalent form of the integral:

$$\int \frac{dx}{\sqrt{e^x + 1}} = \int \frac{du}{u\sqrt{u + 1}} = \ln\left(\frac{\sqrt{u + 1} - 1}{\sqrt{u + 1} + 1}\right) + C$$

$$= \ln\left(\frac{\sqrt{e^x + 1} - 1}{\sqrt{e^x + 1} + 1}\right) + C.$$

Often the difference between two results is a few steps of algebra or a trigonometric identity. In this case, the final two results are reconciled using logarithm properties. This example illustrates that some computer algebra systems do not include constants of integration and may omit absolute values when logarithms appear. It is important for the user to determine whether integration constants and absolute values are needed.

Related Exercises 49–50 ◄

QUICK CHECK 3 Using partial fractions, we know that $\int \dfrac{dx}{x(x + 1)} = \ln\left|\dfrac{x}{x + 1}\right| + C$.

Using a computer algebra system, we find that $\int \dfrac{dx}{x(x + 1)} = \ln x - \ln (x + 1)$. What is wrong with the result from the computer algebra system? ◄

EXAMPLE 5   **Symbolic versus numerical integration**   Use a computer algebra system to evaluate $\int_0^1 \sin x^2 \, dx$.

SOLUTION   Sometimes a computer algebra system gives the exact value of an integral in terms of an unfamiliar function, or it may not be able to evaluate the integral exactly. For example, one particular computer algebra system returns the result

$$\int_0^1 \sin x^2 \, dx = \sqrt{\frac{\pi}{2}} \, S\left(\sqrt{\frac{2}{\pi}}\right),$$

where S is known as the *Fresnel integral function* $\left(S(x) = \int_0^x \sin\left(\frac{\pi t^2}{2}\right) dt \right)$. However, if the computer algebra system is instructed to approximate the integral, the result is

$$\int_0^1 \sin x^2 \, dx \approx 0.3102683017,$$

which is an excellent approximation.

Related Exercises 61, 63 ◄

SECTION 8.7 EXERCISES

Getting Started

1. Give some examples of analytical methods for evaluating integrals.

2. Does a computer algebra system give an exact result for an indefinite integral? Explain.

3. Why might an integral found in a table differ from the same integral evaluated by a computer algebra system?

4. Is a reduction formula an analytical method or a numerical method? Explain.

5. Evaluate $\int e^x \cos^3(e^x) \, dx$ using tables after performing the substitution $u = e^x$.

6. Evaluate $\int \cos x \sqrt{100 - \sin^2 x} \, dx$ using tables after performing the substitution $u = \sin x$.

Practice Exercises

7–40. Table look-up integrals *Use a table of integrals to evaluate the following indefinite integrals. Some of the integrals require preliminary work, such as completing the square or changing variables, before they can be found in a table.*

7. $\int \cos^{-1} x \, dx$

8. $\int \sin 3x \cos 2x \, dx$

9. $\int \dfrac{dx}{\sqrt{x^2 + 16}}$

10. $\int \dfrac{dx}{\sqrt{x^2 - 25}}$

11. $\int \dfrac{3u}{2u + 7} \, du$

12. $\int \dfrac{dy}{y(2y + 9)}$

13. $\int \dfrac{dx}{1 - \cos 4x}$

14. $\int \dfrac{dx}{x\sqrt{81 - x^2}}$

15. $\int \dfrac{x}{\sqrt{4x + 1}} \, dx$

16. $\int t\sqrt{4t + 12} \, dt$

17. $\int \dfrac{dx}{\sqrt{9x^2 - 100}}, x > \dfrac{10}{3}$

18. $\int \dfrac{dx}{225 - 16x^2}$

19. $\int \dfrac{e^x}{\sqrt{e^{2x} + 4}} \, dx$

20. $\int \dfrac{\sqrt{\ln^2 x + 4}}{x} \, dx$

21. $\int \dfrac{\cos x}{\sin^2 x + 2 \sin x} \, dx$

22. $\int \dfrac{\cos^{-1} \sqrt{x}}{\sqrt{x}} \, dx$

23. $\int \dfrac{(\ln x)\sin^{-1}(\ln x)}{x} \, dx$

24. $\int \dfrac{dt}{\sqrt{1 + 4e^t}}$

25. $\int \dfrac{dx}{(16 + 9x^2)^{3/2}}$

26. $\int \sqrt{4x^2 - 9} \, dx, x > \dfrac{3}{2}$

27. $\int \dfrac{dx}{x\sqrt{144 - x^2}}$

28. $\int \dfrac{dv}{v(v^2 + 8)}$

29. $\int \ln^2 x \, dx$

30. $\int x^2 e^{5x} \, dx$

31. $\int \sqrt{x^2 + 10x} \, dx, x > 0$

32. $\int \sqrt{x^2 - 8x} \, dx, x > 8$

33. $\int \dfrac{dx}{x^2 + 2x + 10}$

34. $\int \sqrt{x^2 - 4x + 8} \, dx$

35. $\int \dfrac{dx}{x(x^{10} + 1)}$

36. $\int \dfrac{dt}{t(t^8 - 256)}$

37. $\int \dfrac{dx}{\sqrt{x^2 - 6x}}, x > 6$

38. $\int \dfrac{dx}{\sqrt{x^2 + 10x}}, x > 0$

39. $\int \dfrac{\tan^{-1} x^3}{x^4} \, dx$

40. $\int \dfrac{e^{3t}}{\sqrt{4 + e^{2t}}} \, dt$

41–48. Geometry problems *Use a table of integrals to solve the following problems.*

41. Find the length of the curve $y = \dfrac{x^2}{4}$ on the interval $[0, 8]$.

42. Find the length of the curve $y = x^{3/2} + 8$ on the interval $[0, 2]$.

43. Find the length of the curve $y = e^x$ on the interval $[0, \ln 2]$.

44. The region bounded by the graph of $y = x^2\sqrt{\ln x}$ and the x-axis on the interval $[1, e]$ is revolved about the x-axis. What is the volume of the solid that is formed?

45. The region bounded by the graph of $y = 1/\sqrt{x + 4}$ and the x-axis on the interval $[0, 12]$ is revolved about the y-axis. What is the volume of the solid that is formed?

46. Find the area of the region bounded by the graph of
$$y = \frac{1}{\sqrt{x^2 - 2x + 2}}$$ and the x-axis between $x = 0$ and $x = 3$.

47. The region bounded by the graphs of $y = \pi/2$, $y = \sin^{-1} x$, and the y-axis is revolved about the y-axis. What is the volume of the solid that is formed?

48. The graphs of $f(x) = \dfrac{2}{x^2 + 1}$ and $g(x) = \dfrac{7}{4\sqrt{x^2 + 1}}$ are shown in the figure. Which is greater, the average value of f or the average value of g on the interval $[-1, 1]$?

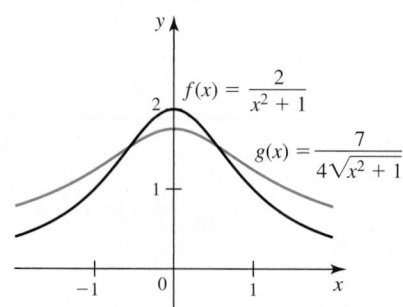

T 49–63. Integrating with a CAS *Use a computer algebra system to evaluate the following integrals. Find both an exact result and an approximate result for each definite integral. Assume a is a positive real number.*

49. $\displaystyle\int \frac{x}{\sqrt{2x + 3}}\, dx$

50. $\displaystyle\int \sqrt{4x^2 + 36}\, dx$

51. $\displaystyle\int \tan^2 3x\, dx$

52. $\displaystyle\int_0^{\pi/2} \cos^6 x\, dx$

53. $\displaystyle\int_0^4 (9 + x^2)^{3/2}\, dx$

54. $\displaystyle\int (a^2 - t^2)^{-2}\, dt$

55. $\displaystyle\int \frac{(x^2 - a^2)^{3/2}}{x}\, dx$

56. $\displaystyle\int \frac{dx}{x(a^2 - x^2)^2}$

57. $\displaystyle\int_0^{\pi/2} \frac{dt}{1 + \tan^2 t}$

58. $\displaystyle\int_0^{2\pi} \frac{dt}{(4 + 2\sin t)^2}$

59. $\displaystyle\int (a^2 - x^2)^{3/2}\, dx$

60. $\displaystyle\int (y^2 + a^2)^{-5/2}\, dy$

61. $\displaystyle\int_0^1 (\ln x)\ln(1 + x)\, dx$

62. $\displaystyle\int_0^1 x^{31} e^{x^2}\, dx$

63. $\displaystyle\int_1^2 \frac{x^{19}}{\sqrt{x^4 - 1}}\, dx$

64. Using a computer algebra system, it was determined that
$$\int x(x + 1)^8\, dx = \frac{x^{10}}{10} + \frac{8x^9}{9} + \frac{7x^8}{2} + 8x^7 + \frac{35x^6}{3} + \frac{56x^5}{5} +$$
$$7x^4 + \frac{8x^3}{3} + \frac{x^2}{2} + C.$$ Use integration by substitution to evaluate
$$\int x(x + 1)^8\, dx.$$

65–68. Reduction formulas *Use the reduction formulas in a table of integrals to evaluate the following integrals.*

65. $\displaystyle\int x^3 e^{2x}\, dx$

66. $\displaystyle\int p^2 e^{-3p}\, dp$

67. $\displaystyle\int \tan^4 3y\, dy$

68. $\displaystyle\int \sec^4 4t\, dt$

T 69–70. Using a computer algebra system *Use a computer algebra system to solve the following problems.*

69. Find the exact area of the region bounded by the curves
$$y = \sqrt{x + \sqrt{x}} \text{ and } y = \frac{2}{\sqrt{1 + \sqrt{x}}}$$ in the first quadrant.

70. Find the approximate area of the surface generated when the curve $y = 1 + \sin x + \cos x$, for $0 \le x \le \pi$, is revolved about the x-axis.

71–74. Deriving formulas *Evaluate the following integrals. Assume a and b are real numbers and n is a positive integer.*

71. $\displaystyle\int \frac{x}{ax + b}\, dx$ *(Hint: $u = ax + b$.)*

72. $\displaystyle\int \frac{x}{\sqrt{ax + b}}\, dx$ *(Hint: $u^2 = ax + b$.)*

73. $\displaystyle\int x(ax + b)^n\, dx$ *(Hint: $u = ax + b$.)*

74. $\displaystyle\int x^n \sin^{-1} x\, dx$ *(Hint: integration by parts.)*

75. Explain why or why not Determine whether the following statements are true and give an explanation or counterexample.

a. It is possible for a computer algebra system to give the result
$$\int \frac{dx}{x(x - 1)} = \ln(x - 1) - \ln x$$ and a table of integrals to give the result $\displaystyle\int \frac{dx}{x(x - 1)} = \ln\left|\frac{x - 1}{x}\right| + C.$

b. A computer algebra system working in symbolic mode could give the result $\displaystyle\int_0^1 x^8\, dx = \frac{1}{9}$, and a computer algebra system working in approximate (numerical) mode could give the result $\int_0^1 x^8\, dx = 0.11111111$.

Explorations and Challenges

76. Apparent discrepancy Three different computer algebra systems give the following results:
$$\int \frac{dx}{x\sqrt{x^4 - 1}} = \frac{1}{2}\cos^{-1}\sqrt{x^{-4}} = \frac{1}{2}\cos^{-1} x^{-2} = \frac{1}{2}\tan^{-1}\sqrt{x^4 - 1}.$$
Explain how all three can be correct.

77. Reconciling results Using one computer algebra system, it was found that $\displaystyle\int \frac{dx}{1 + \sin x} = \frac{\sin x - 1}{\cos x}$, and using another computer algebra system, it was found that
$$\int \frac{dx}{1 + \sin x} = \frac{2\sin(x/2)}{\cos(x/2) + \sin(x/2)}.$$ Reconcile the two answers.

78. Apparent discrepancy Resolve the apparent discrepancy between $\displaystyle\int \frac{dx}{x(x - 1)(x + 2)} = \frac{1}{6}\ln\frac{(x - 1)^2 |x + 2|}{|x|^3} + C$ and
$$\int \frac{dx}{x(x - 1)(x + 2)} = \frac{\ln|x - 1|}{3} + \frac{\ln|x + 2|}{6} - \frac{\ln|x|}{2} + C.$$

T 79–82. Double table look-up *The following integrals may require more than one table look-up. Evaluate the integrals using a table of integrals, and then check your answer with a computer algebra system.*

79. $\displaystyle\int x \sin^{-1} 2x \, dx$

80. $\displaystyle\int 4x \cos^{-1} 10x \, dx$

81. $\displaystyle\int \frac{\tan^{-1} x}{x^2} \, dx$

82. $\displaystyle\int \frac{\sin^{-1} ax}{x^2} \, dx, \, a > 0$

83. Evaluating an integral without the Fundamental Theorem of Calculus Evaluate $\int_0^{\pi/4} \ln(1 + \tan x)\, dx$ using the following steps.

a. If f is integrable on $[0, b]$, use substitution to show that
$$\int_0^b f(x)\, dx = \int_0^{b/2} (f(x) + f(b - x))\, dx.$$

b. Use part (a) and the identity $\tan(\alpha + \beta) = \dfrac{\tan \alpha + \tan \beta}{1 - \tan \alpha \tan \beta}$
to evaluate $\int_0^{\pi/4} \ln(1 + \tan x)\, dx$.

(*Source: The College Mathematics Journal, 33, 4, Sep 2004*)

84. Two integration approaches Evaluate $\int \cos(\ln x)\, dx$ two different ways:

a. Use tables after first using the substitution $u = \ln x$.
b. Use integration by parts twice to verify your answer to part (a).

T 85. Period of a pendulum Consider a pendulum of length L meters that swings under the influence of gravity alone. Suppose the pendulum starts swinging with an initial displacement of θ_0 radians (see figure). The period (time to complete one full cycle) is given by
$$T = \frac{4}{\omega} \int_0^{\pi/2} \frac{d\varphi}{\sqrt{1 - k^2 \sin^2 \varphi}},$$
where $\omega^2 = \dfrac{g}{L}$, $g = 9.8 \text{ m/s}^2$ is the acceleration due to gravity, and $k^2 = \sin^2 \dfrac{\theta_0}{2}$. Assume $L = 9.8$ m, which means $\omega = 1\text{s}^{-1}$.

a. Use a computer algebra system to find the period of the pendulum for $\theta_0 = 0.1, 0.2, \ldots, 0.9, 1.0$ rad.
b. For small values of θ_0, the period should be approximately 2π s. For what values of θ_0 are your computed values within 10% of 2π (relative error less than 0.1)?

86. Arc length of a parabola Let $L(c)$ be the length of the parabola $f(x) = x^2$ from $x = 0$ to $x = c$, where $c \geq 0$ is a constant.

a. Find an expression for L.
b. Is L concave up or concave down on $[0, \infty)$?
c. Show that as c becomes large and positive, the arc length function increases as c^2; that is, $L(c) \approx kc^2$, where k is a constant.

T 87. Powers of sine and cosine It can be shown that
$$\int_0^{\pi/2} \sin^n x \, dx = \int_0^{\pi/2} \cos^n x \, dx =$$
$$\begin{cases} \dfrac{1 \cdot 3 \cdot 5 \cdots (n-1)}{2 \cdot 4 \cdot 6 \cdots n} \cdot \dfrac{\pi}{2} & \text{if } n \geq 2 \text{ is an even integer} \\[2mm] \dfrac{2 \cdot 4 \cdot 6 \cdots (n-1)}{3 \cdot 5 \cdot 7 \cdots n} & \text{if } n \geq 3 \text{ is an odd integer.} \end{cases}$$

a. Use a computer algebra system to confirm this result for $n = 2, 3, 4,$ and 5.
b. Evaluate the integrals with $n = 10$ and confirm the result.
c. Using graphing and/or symbolic computation, determine whether the values of the integrals increase or decrease as n increases.

T 88. A remarkable integral It is a fact that $\int_0^{\pi/2} \dfrac{dx}{1 + \tan^m x} = \dfrac{\pi}{4}$, for *all* real numbers m.

a. Graph the integrand for $m = -2, -3/2, -1, -1/2, 0, 1/2, 1, 3/2,$ and 2, and explain geometrically how the area under the curve on the interval $[0, \pi/2]$ remains constant as m varies.
b. Use a computer algebra system to confirm that the integral is constant for all m.

QUICK CHECK ANSWERS

1. 1 **2.** Because $\sin^2 x = 1 - \cos^2 x$, the two results differ by a constant, which can be absorbed in the arbitrary constant C.
3. The second result agrees with the first for $x > 0$ after using $\ln a - \ln b = \ln(a/b)$. The second result should have absolute values and an arbitrary constant. ◄

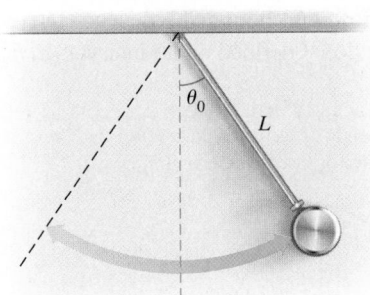

8.8 Numerical Integration

Situations arise in which the analytical methods we have developed so far cannot be used to evaluate a definite integral. For example, an integrand may not have an obvious antiderivative (such as $\cos x^2$ and $1/\ln x$), or perhaps the value of the integrand is known only at a finite set of points, which makes finding an antiderivative impossible.

When analytical methods fail, we often turn to *numerical methods*, which are typically done on a calculator or computer. These methods do not produce exact values of definite integrals, but they provide approximations that are generally quite accurate. Many calculators, software packages, and computer algebra systems have built-in numerical integration methods. In this section, we explore some of these methods.

Absolute and Relative Error

Because numerical methods do not typically produce exact results, we should be concerned about the accuracy of approximations, which leads to the ideas of *absolute* and *relative error.*

> **DEFINITION Absolute and Relative Error**
>
> Suppose c is a computed numerical solution to a problem having an exact solution x. There are two common measures of the error in c as an approximation to x:
>
> $$\textbf{absolute error} = |c - x|$$
>
> and
>
> $$\textbf{relative error} = \frac{|c - x|}{|x|} \quad (\text{if } x \neq 0).$$

▶ Because the exact solution is usually not known, the goal in practice is to estimate the maximum size of the error.

EXAMPLE 1 Absolute and relative error The ancient Greeks used $\frac{22}{7}$ to approximate the value of π. Determine the absolute and relative error in this approximation to π.

SOLUTION Letting $c = \frac{22}{7}$ be the approximate value of $x = \pi$, we find that

$$\text{absolute error} = \left| \frac{22}{7} - \pi \right| \approx 0.00126$$

and

$$\text{relative error} = \frac{|22/7 - \pi|}{|\pi|} \approx 0.000402 \approx 0.04\%.$$

Related Exercises 11–14 ◀

Midpoint Rule

Many numerical integration methods are based on the ideas that underlie Riemann sums; these methods approximate the net area of regions bounded by curves. A typical problem is shown in **Figure 8.12**, where we see a function f defined on an interval $[a, b]$. The goal is to

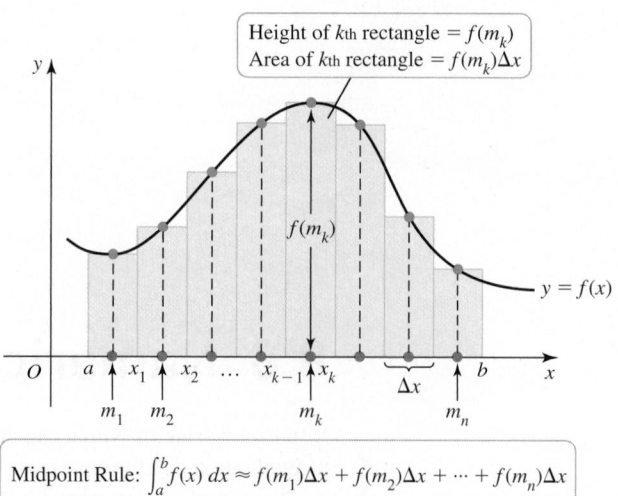

Midpoint Rule: $\int_a^b f(x)\,dx \approx f(m_1)\Delta x + f(m_2)\Delta x + \cdots + f(m_n)\Delta x$

Figure 8.12

approximate the value of $\int_a^b f(x)\,dx$. As with Riemann sums, we first partition the interval $[a, b]$ into n subintervals of equal length $\Delta x = (b - a)/n$. This partition establishes $n + 1$ grid points

$$x_0 = a, \quad x_1 = a + \Delta x, \quad x_2 = a + 2\Delta x, \dots, \quad x_k = a + k\Delta x, \dots, \quad x_n = b.$$

The kth subinterval is $[x_{k-1}, x_k]$, for $k = 1, 2, \dots, n$.

The Midpoint Rule approximates the region under the curve using rectangles. The bases of the rectangles have width Δx. The height of the kth rectangle is $f(m_k)$, where $m_k = (x_{k-1} + x_k)/2$ is the midpoint of the kth subinterval (Figure 8.12). Therefore, the net area of the kth rectangle is $f(m_k)\Delta x$.

Let $M(n)$ be the Midpoint Rule approximation to the integral using n rectangles. Summing the net areas of the rectangles, we have

$$\begin{aligned}
\int_a^b f(x)\,dx &\approx M(n) \\
&= f(m_1)\Delta x + f(m_2)\Delta x + \cdots + f(m_n)\Delta x \\
&= \sum_{k=1}^{n} f\left(\frac{x_{k-1} + x_k}{2}\right)\Delta x.
\end{aligned}$$

> If $f(m_k) < 0$ for some k, then the net area of the corresponding rectangle is negative, which makes a negative contribution to the approximation (Section 5.2).

Just as with Riemann sums, the Midpoint Rule approximations to $\int_a^b f(x)\,dx$ generally improve as n increases.

> The Midpoint Rule is a midpoint Riemann sum, introduced in Section 5.1.

DEFINITION Midpoint Rule

Suppose f is defined and integrable on $[a, b]$. The **Midpoint Rule approximation** to $\int_a^b f(x)\,dx$ using n equally spaced subintervals on $[a, b]$ is

$$\begin{aligned}
M(n) &= f(m_1)\Delta x + f(m_2)\Delta x + \cdots + f(m_n)\Delta x \\
&= \sum_{k=1}^{n} f\left(\frac{x_{k-1} + x_k}{2}\right)\Delta x,
\end{aligned}$$

where $\Delta x = (b - a)/n$, $x_0 = a$, $x_k = a + k\Delta x$, and $m_k = (x_{k-1} + x_k)/2$ is the midpoint of $[x_{k-1}, x_k]$, for $k = 1, \dots, n$.

QUICK CHECK 1 To apply the Midpoint Rule on the interval $[3, 11]$ with $n = 4$, at what points must the integrand be evaluated? ◄

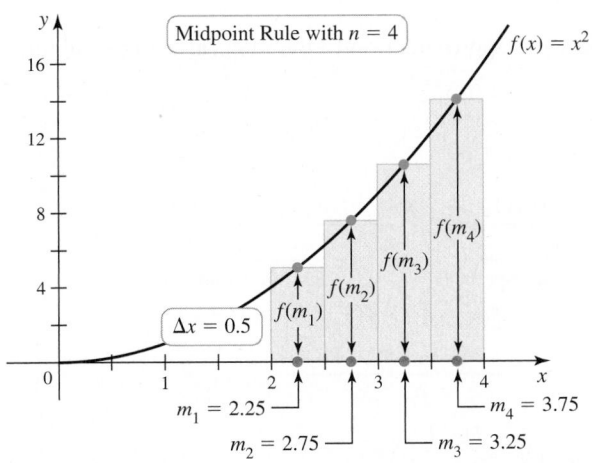

Figure 8.13

EXAMPLE 2 Applying the Midpoint Rule Approximate $\int_2^4 x^2\,dx$ using the Midpoint Rule with $n = 4$ and $n = 8$ subintervals.

SOLUTION With $a = 2$, $b = 4$, and $n = 4$ subintervals, the length of each subinterval is $\Delta x = (b - a)/n = 2/4 = 0.5$. The grid points are

$$x_0 = 2, \quad x_1 = 2.5, \quad x_2 = 3, \quad x_3 = 3.5, \quad \text{and} \quad x_4 = 4.$$

The integrand must be evaluated at the midpoints (Figure 8.13)

$$m_1 = 2.25, \quad m_2 = 2.75, \quad m_3 = 3.25, \quad \text{and} \quad m_4 = 3.75.$$

With $f(x) = x^2$ and $n = 4$, the Midpoint Rule approximation is

$$\begin{aligned}
M(4) &= f(m_1)\Delta x + f(m_2)\Delta x + f(m_3)\Delta x + f(m_4)\Delta x \\
&= (m_1^2 + m_2^2 + m_3^2 + m_4^2)\Delta x \\
&= (2.25^2 + 2.75^2 + 3.25^2 + 3.75^2) \cdot 0.5 \\
&= 18.625.
\end{aligned}$$

The exact area of the region is $\frac{56}{3}$, so this Midpoint Rule approximation has an absolute error of

$$|18.625 - 56/3| \approx 0.0417$$

and a relative error of

$$\left|\frac{18.625 - 56/3}{56/3}\right| \approx 0.00223 = 0.223\%.$$

Using $n = 8$ subintervals, the midpoint approximation is

$$M(8) = \sum_{k=1}^{8} f(m_k)\Delta x = 18.65625,$$

which has an absolute error of about 0.0104 and a relative error of about 0.0558%. We see that increasing n and using more rectangles decreases the error in the approximations.

Related Exercises 15–16 ◄

The Trapezoid Rule

Another numerical method for estimating $\int_a^b f(x)\,dx$ is the Trapezoid Rule, which uses the same partition of the interval $[a, b]$ described for the Midpoint Rule. Instead of approximating the region under the curve by rectangles, the Trapezoid Rule uses (what else?) trapezoids. The bases of the trapezoids have length Δx. The sides of the kth trapezoid have lengths $f(x_{k-1})$ and $f(x_k)$, for $k = 1, 2, \ldots, n$ (Figure 8.14). Therefore, the net area of the kth trapezoid is $\dfrac{f(x_{k-1}) + f(x_k)}{2}\Delta x$.

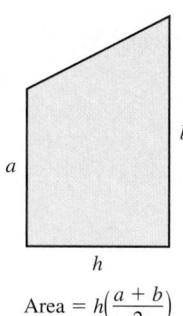

Area of a trapezoid

$$\text{Area} = h\left(\frac{a + b}{2}\right)$$

▶ This derivation of the Trapezoid Rule assumes f is nonnegative on $[a, b]$. However, the same argument can be used if f is negative on all or part of $[a, b]$. In fact, the argument illustrates how negative contributions to the net area arise when f is negative.

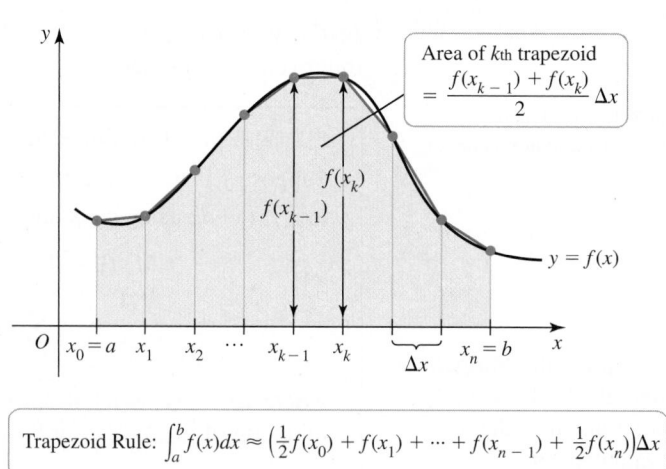

Trapezoid Rule: $\int_a^b f(x)dx \approx \left(\frac{1}{2}f(x_0) + f(x_1) + \cdots + f(x_{n-1}) + \frac{1}{2}f(x_n)\right)\Delta x$

Figure 8.14

Letting $T(n)$ be the Trapezoid Rule approximation to the integral using n subintervals, we have

$$\int_a^b f(x)\,dx \approx T(n)$$

$$= \underbrace{\frac{f(x_0) + f(x_1)}{2}\Delta x}_{\text{area of first trapezoid}} + \underbrace{\frac{f(x_1) + f(x_2)}{2}\Delta x}_{\text{area of second trapezoid}} + \cdots + \underbrace{\frac{f(x_{n-1}) + f(x_n)}{2}\Delta x}_{\text{area of }n\text{th trapezoid}}$$

$$= \left(\frac{f(x_0)}{2} + \underbrace{\frac{f(x_1)}{2} + \frac{f(x_1)}{2}}_{f(x_1)} + \cdots + \underbrace{\frac{f(x_{n-1})}{2} + \frac{f(x_{n-1})}{2}}_{f(x_{n-1})} + \frac{f(x_n)}{2}\right)\Delta x$$

$$= \left(\frac{f(x_0)}{2} + \underbrace{f(x_1) + \cdots + f(x_{n-1})}_{\sum_{k=1}^{n-1} f(x_k)} + \frac{f(x_n)}{2}\right)\Delta x.$$

As with the Midpoint Rule, the Trapezoid Rule approximations generally improve as n increases.

DEFINITION Trapezoid Rule

Suppose f is defined and integrable on $[a, b]$. The **Trapezoid Rule approximation** to $\int_a^b f(x)\, dx$ using n equally spaced subintervals on $[a, b]$ is

$$T(n) = \left(\frac{1}{2} f(x_0) + \sum_{k=1}^{n-1} f(x_k) + \frac{1}{2} f(x_n) \right) \Delta x,$$

where $\Delta x = (b - a)/n$ and $x_k = a + k\Delta x$, for $k = 0, 1, \ldots, n$.

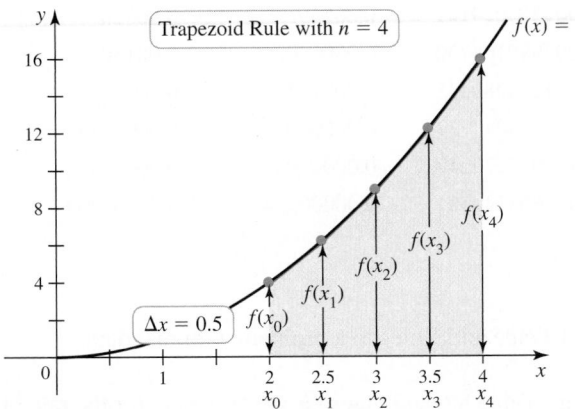

Figure 8.15

EXAMPLE 3 Applying the Trapezoid Rule Approximate $\int_2^4 x^2\, dx$ using the Trapezoid Rule with $n = 4$ subintervals.

SOLUTION As in Example 2, the grid points are

$$x_0 = 2, \quad x_1 = 2.5, \quad x_2 = 3, \quad x_3 = 3.5, \quad \text{and} \quad x_4 = 4.$$

With $f(x) = x^2$ and $n = 4$, the Trapezoid Rule approximation is

$$
\begin{aligned}
T(4) &= \tfrac{1}{2} f(x_0)\Delta x + f(x_1)\Delta x + f(x_2)\Delta x + f(x_3)\Delta x + \tfrac{1}{2} f(x_4)\Delta x \\
&= \left(\tfrac{1}{2} x_0^2 + x_1^2 + x_2^2 + x_3^2 + \tfrac{1}{2} x_4^2 \right) \Delta x \\
&= \left(\tfrac{1}{2} \cdot 2^2 + 2.5^2 + 3^2 + 3.5^2 + \tfrac{1}{2} \cdot 4^2 \right) \cdot 0.5 \\
&= 18.75.
\end{aligned}
$$

Figure 8.15 shows the approximation with $n = 4$ trapezoids. The exact area of the region is $56/3$, so the Trapezoid Rule approximation has an absolute error of about 0.0833 and a relative error of approximately 0.00446, or 0.446%. Increasing n decreases this error.

Related Exercises 19–20 ◄

EXAMPLE 4 Errors in the Midpoint and Trapezoid Rules Given that

$$\int_0^1 xe^{-x}\, dx = 1 - 2e^{-1},$$

find the absolute errors in the Midpoint Rule and Trapezoid Rule approximations to the integral with $n = 4, 8, 16, 32, 64,$ and 128 subintervals.

SOLUTION Because the exact value of the integral is known (which does *not* happen in practice), we can compute the error in various approximations. For example, if $n = 16$, then

$$\Delta x = \frac{1}{16} \quad \text{and} \quad x_k = \frac{k}{16}, \quad \text{for } k = 0, 1, \ldots, n.$$

Using sigma notation and a calculator, we have

$$M(16) = \sum_{k=1}^{16} f\!\left(\overbrace{\frac{(k-1)/16}{2}}^{x_{k-1}} + \overbrace{k/16}^{x_k} \right) \overbrace{\frac{1}{16}}^{\Delta x} = \sum_{k=1}^{16} f\!\left(\frac{2k-1}{32} \right) \frac{1}{16} \approx 0.26440383609318$$

and

$$T(16) = \left(\frac{1}{2} \underbrace{f(0)}_{x_0 = a} + \sum_{k=1}^{15} \underbrace{f(k/16)}_{x_k} + \frac{1}{2} \underbrace{f(1)}_{x_{16} = b} \right) \frac{1}{16} \approx 0.26391564480235.$$

The absolute error in the Midpoint Rule approximation with $n = 16$ is $\left| M(16) - (1 - 2e^{-1}) \right| \approx 0.000163$. The absolute error in the Trapezoid Rule approximation with $n = 16$ is $\left| T(16) - (1 - 2e^{-1}) \right| \approx 0.000325$.

The Midpoint Rule and Trapezoid Rule approximations to the integral, together with the associated absolute errors, are shown in Table 8.5 for various values of n. Notice that as n increases, the errors in both methods decrease, as expected. With $n = 128$ subintervals, the approximations $M(128)$ and $T(128)$ agree to four decimal places. Based on these approximations, a good approximation to the integral is 0.2642. The way in which the errors decrease is also worth noting. If you look carefully at both error columns in Table 8.5, you will see that each time n is doubled (or Δx is halved), the error decreases by a factor of approximately 4.

QUICK CHECK 3 Compute the approximate factor by which the error decreases in Table 8.5 between $T(16)$ and $T(32)$, and between $T(32)$ and $T(64)$. ◄

Table 8.5

n	$M(n)$	$T(n)$	Error in $M(n)$	Error in $T(n)$
4	0.26683456310319	0.25904504019141	0.00259	0.00520
8	0.26489148795740	0.26293980164730	0.000650	0.00130
16	0.26440383609318	0.26391564480235	0.000163	0.000325
32	0.26428180513718	0.26415974044777	0.0000407	0.0000814
64	0.26425129001915	0.26422077279247	0.0000102	0.0000203
128	0.26424366077837	0.26423603140581	0.00000254	0.00000509

Related Exercises 31–32 ◄

We now apply the Midpoint and Trapezoid Rules to a problem with real data.

Table 8.6

	World Crude Oil Production
Year	**(billions of barrels/yr)**
1995	21.9
1996	22.3
1997	23.0
1998	23.7
1999	24.5
2000	23.7
2001	25.2
2002	24.8
2003	24.5
2004	25.2
2005	25.9
2006	26.3
2007	27.0
2008	26.9
2009	26.4
2010	27.0
2011	27.0

(*Source:* U.S. Energy Information Administration)

EXAMPLE 5   World oil production Table 8.6 and **Figure 8.16** show data for the rate of world crude oil production (in billions of barrels/yr) over a 16-year period. If the rate of oil production is given by the (assumed to be integrable) function R, then the total amount of oil produced in billions of barrels over the time period $a \le t \le b$ is $Q = \int_a^b R(t)\,dt$ (Section 6.1). Use the Midpoint and Trapezoid Rules to approximate the total oil produced between 1995 and 2011.

SOLUTION For convenience, let $t = 0$ represent 1995 and $t = 16$ represent 2011. We let $R(t)$ be the rate of oil production in the year corresponding to t (for example, $R(6) = 25.2$ is the rate in 2001). The goal is to approximate $Q = \int_0^{16} R(t)\,dt$. If we use $n = 4$ subintervals, then $\Delta t = 4$ yr. The resulting Midpoint and Trapezoid Rule approximations (in billions of barrels) are

$$
\begin{aligned}
Q \approx M(4) &= (R(2) + R(6) + R(10) + R(14))\Delta t \\
&= (23.0 + 25.2 + 25.9 + 26.4)4 \\
&= 402.0
\end{aligned}
$$

and

$$
\begin{aligned}
Q \approx T(4) &= \left(\frac{1}{2}R(0) + R(4) + R(8) + R(12) + \frac{1}{2}R(16)\right)\Delta t \\
&= \left(\frac{1}{2}\cdot 21.9 + 24.5 + 24.5 + 27.0 + \frac{1}{2}\cdot 27.0\right)4 \\
&= 401.8.
\end{aligned}
$$

The two methods give reasonable agreement. Using $n = 8$ subintervals, with $\Delta t = 2$ yr, similar calculations give the approximations

$$
Q \approx M(8) = 399.8 \quad \text{and} \quad Q \approx T(8) = 401.9.
$$

The given data do not allow us to compute the next Midpoint Rule approximation $M(16)$. However, we can compute the next Trapezoid Rule approximation $T(16)$, and here is a good way to do it. If $T(n)$ and $M(n)$ are known, then the next Trapezoid Rule approximation is (Exercise 76)

$$
T(2n) = \frac{T(n) + M(n)}{2}.
$$

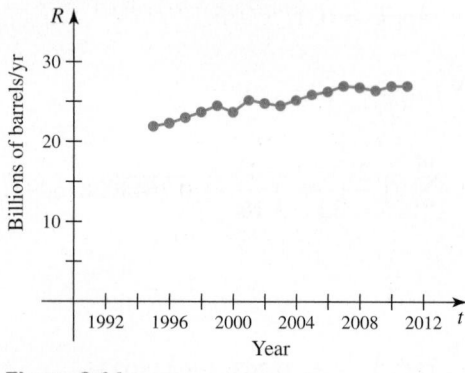

Figure 8.16
(*Source:* U.S. Energy Information Administration)

Using this identity, we find that

$$T(16) = \frac{T(8) + M(8)}{2} = \frac{401.9 + 399.8}{2} \approx 400.9.$$

Based on these calculations, the best approximation to the total oil produced between 1995 and 2011 is 400.9 billion barrels.   *Related Exercises 37–40* ◄

The Midpoint and Trapezoid Rules, as well as left and right Riemann sums, can be applied to problems in which data are given on a nonuniform grid (that is, the lengths of the subintervals vary). In the case of the Trapezoid Rule, the net areas of the approximating trapezoids must be computed individually and then summed, as shown in the next example.

EXAMPLE 6 **Net change in sea level** Table 8.7 lists rates of change $s'(t)$ in global sea level $s(t)$ in various years from 1995 ($t = 0$) to 2011 ($t = 16$), with rates of change reported in mm/yr.

Table 8.7

t (years from 1995)	0 (1995)	3 (1998)	5 (2000)	7 (2002)	8 (2003)	12 (2007)	14 (2009)	16 (2011)
$s'(t)$ (mm/yr)	0.51	5.19	4.39	2.21	5.24	0.63	4.19	2.38

(*Source:* Collecte Localisation Satellites/Centre national d'études spatiales/Legos)

> ► The rate of change in sea level varies from one location on Earth to the next; sea level also varies seasonally and is influenced by ocean currents. The data in Table 8.7 reflect approximate rates of change at the beginning of each year listed, averaged over the entire globe.

a. Assuming s' is continuous on $[0, 16]$, explain how a definite integral can be used to find the net change in sea level from 1995 to 2011; then write the definite integral.

b. Use the data in Table 8.7 and generalize the Trapezoid Rule to estimate the value of the integral from part (a).

SOLUTION

a. The net change in any quantity Q over the interval $[a, b]$ is $Q(b) - Q(a)$ (Section 6.1). When the rate of change Q' is known, the net change in Q is found by integrating Q' over the same interval; that is,

$$\text{net change in } Q = Q(b) - Q(a) = \int_a^b Q'(t)\, dt. \quad \text{Fundamental Theorem}$$

Therefore, the net change in sea level from 1995 to 2011 is $\int_0^{16} s'(t)\, dt$.

b. The values from Table 8.7 are plotted in **Figure 8.17**, accompanied by seven trapezoids whose area approximates $\int_0^{16} s'(t)\, dt$. Notice that the grid points (the t-values in Table 8.7) do not form a regular partition of the interval $[0, 16]$. Therefore, we must generalize the standard Trapezoid Rule and compute the area of each trapezoid separately.

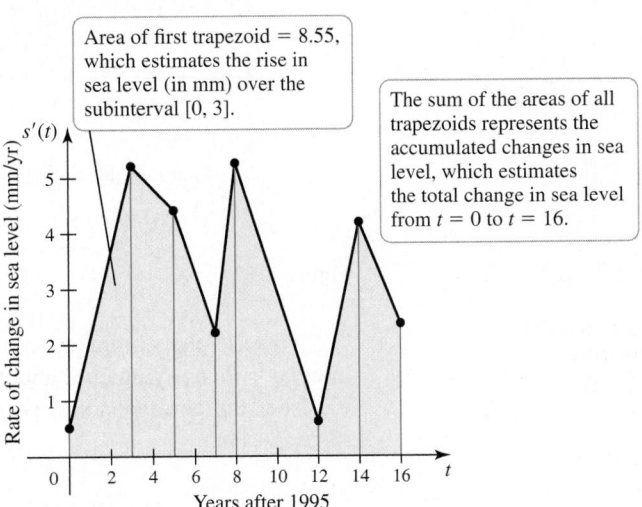

Area of first trapezoid = 8.55, which estimates the rise in sea level (in mm) over the subinterval [0, 3].

The sum of the areas of all trapezoids represents the accumulated changes in sea level, which estimates the total change in sea level from $t = 0$ to $t = 16$.

Figure 8.17

Focusing on the first trapezoid over the subinterval $[0, 3]$, we find that its area is

$$\underbrace{\text{area of first trapezoid}}_{A = \frac{1}{2}(b_1 + b_2)h} = \frac{1}{2} \cdot \underbrace{(s'(0) + s'(3))}_{\text{measured in mm/yr}} \cdot \underbrace{3}_{\text{yr}} = \frac{1}{2} \cdot (0.51 + 5.19) \cdot 3 = 8.55.$$

Because s' is measured in mm/yr and t is measured in yr, the area of this trapezoid (8.55) is interpreted as the approximate net change in sea level from 1995 to 1998, measured in mm. As we add new trapezoid areas to the ongoing sum that approximates $\int_0^{16} s'(t)\, dt$, the changes in sea level accumulate, resulting in the total change in sea level on $[0, 16]$. The sum of the areas of all seven trapezoids is

$$\underbrace{\frac{1}{2}(s'(0) + s'(3)) \cdot 3}_{\text{first trapezoid}} + \underbrace{\frac{1}{2}(s'(3) + s'(5)) \cdot 2}_{\text{second trapezoid...}} + \frac{1}{2}(s'(5) + s'(7)) \cdot 2 + \frac{1}{2}(s'(7) + s'(8)) \cdot 1$$

$$+ \frac{1}{2}(s'(8) + s'(12)) \cdot 4 + \frac{1}{2}(s'(12) + s'(14)) \cdot 2 + \underbrace{\frac{1}{2}(s'(14) + s'(16)) \cdot 2}_{\text{... last trapezoid}} = 51.585.$$

We conclude that an estimate of the rise in sea level from 1995 to 2011 is 51.585 mm.

Related Exercises 41, 43 ◄

Simpson's Rule

An improvement over the Midpoint Rule and the Trapezoid Rule results when the graph of f is approximated with curves rather than line segments. Returning to the partition of the interval $[a, b]$ used by the Midpoint and Trapezoid Rules, suppose we work with three neighboring points on the curve $y = f(x)$, say $(x_0, f(x_0))$, $(x_1, f(x_1))$, and $(x_2, f(x_2))$. These three points determine a parabola, and it is easy to find the net area bounded by a parabola on the interval $[x_0, x_2]$. When this idea is applied to every group of three consecutive points along the interval of integration, the result is *Simpson's Rule* (Figure 8.18). Our immediate goal is to determine a formula for this rule.

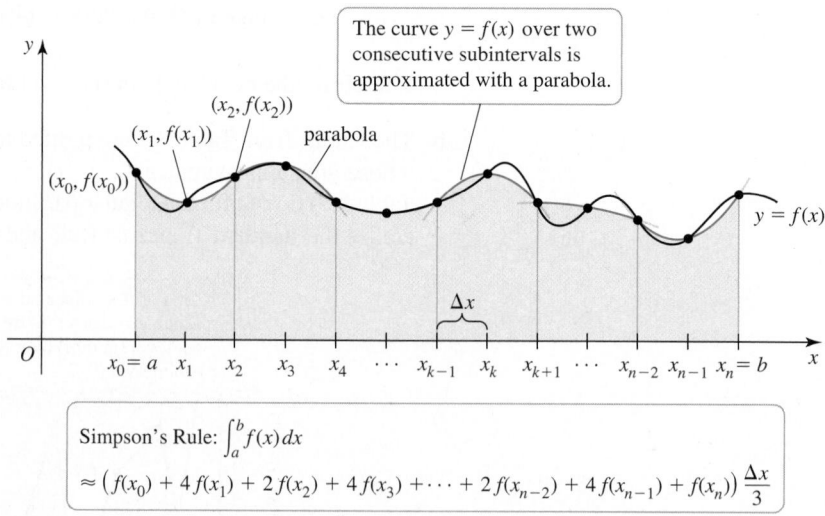

Simpson's Rule: $\int_a^b f(x)\, dx$

$$\approx \left(f(x_0) + 4f(x_1) + 2f(x_2) + 4f(x_3) + \cdots + 2f(x_{n-2}) + 4f(x_{n-1}) + f(x_n)\right) \frac{\Delta x}{3}$$

Figure 8.18

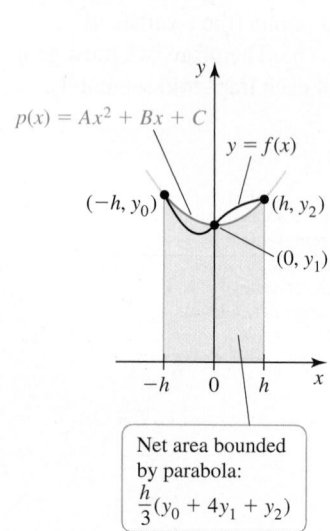

Net area bounded by parabola:
$$\frac{h}{3}(y_0 + 4y_1 + y_2)$$

Figure 8.19

To ease the computation involved in deriving Simpson's Rule, we consider an interval $[-h, h]$ symmetric about the origin with grid points $-h$, 0, and h, where $h = \Delta x$; we label the corresponding points on the graph of f as $(-h, y_0)$, $(0, y_1)$, and (h, y_2) (Figure 8.19).

If these points are to lie on the parabola $p(x) = Ax^2 + Bx + C$, they must satisfy the following conditions:

$$p(-h) = A(-h)^2 + B(-h) + C = y_0, \quad \text{or} \quad Ah^2 - Bh + C = y_0;$$
$$p(0) = A(0)^2 + B(0) + C = y_1, \quad \text{or} \quad C = y_1; \text{ and}$$
$$p(h) = A(h)^2 + B(h) + C = y_2, \quad \text{or} \quad Ah^2 + Bh + C = y_2.$$

Solving this system of equations for A, B, and C yields

$$A = \frac{y_0 - 2y_1 + y_2}{2h^2}, \quad B = \frac{y_2 - y_0}{2h}, \quad \text{and } C = y_1.$$

Therefore, the net area bounded by the parabola is given by

> ► Recall that if $f(x)$ is an even function and $g(x)$ is odd, then
>
> $$\int_{-h}^{h} f(x)\,dx = 2\int_{0}^{h} f(x)\,dx \quad \text{and}$$
>
> $$\int_{-h}^{h} g(x)\,dx = 0.$$

$$\int_{-h}^{h}\left(\underbrace{\frac{y_0 - 2y_1 + y_2}{2h^2}x^2}_{A} + \overbrace{\underbrace{\frac{y_2 - y_0}{2h}x}_{B}}^{\text{odd function}} + \underbrace{y_1}_{C}\right)dx = 2\int_{0}^{h}\left(\frac{y_0 - 2y_1 + y_2}{2h^2}x^2 + y_1\right)dx$$

Use symmetry.

$$= 2\left(\frac{y_0 - 2y_1 + y_2}{6h^2}x^3 + y_1 x\right)\Big|_{0}^{h}$$

Integrate.

> ► In the event that the three points $(-h, y_0)$, $(0, y_1)$, and (h, y_2) are collinear, there is no parabola that passes through all the points. However, it can be shown that the net area bounded by the line segment through these points is also given by $\frac{h}{3}(y_0 + 4y_1 + y_2)$.

$$= 2\left(\frac{(y_0 - 2y_1 + y_2)h}{6} + y_1 h\right)$$

Evaluate.

$$= \frac{h}{3}(y_0 + 4y_1 + y_2).$$

Simplify.

Notice that a horizontal shift of the parabola shown in Figure 8.19 does not change its associated net area, which implies that the net area bounded by the parabola on $[x_0, x_2]$ shown in Figure 8.18 is also given by

$$\frac{h}{3}(y_0 + 4y_1 + y_2), \quad \text{or} \quad (f(x_0) + 4f(x_1) + f(x_2))\frac{\Delta x}{3}.$$

Similarly, the net area bounded by the parabola on the interval $[x_2, x_4]$ is

$$(f(x_2) + 4f(x_3) + f(x_4))\frac{\Delta x}{3}.$$

When we sum the net area of all such parabolas, we arrive at an approximation to the net area bounded by the graph of f on $[a, b]$. With n subintervals, Simpson's Rule is denoted $S(n)$ and is given by

$$S(n) = \underbrace{(f(x_0) + 4f(x_1) + f(x_2))\frac{\Delta x}{3}}_{\text{net area bounded by first parabola}} + \underbrace{(f(x_2) + 4f(x_3) + f(x_4))\frac{\Delta x}{3}}_{\text{net area bounded by second parabola}} + \cdots$$

$$+ \underbrace{(f(x_{n-2}) + 4f(x_{n-1}) + f(x_n))\frac{\Delta x}{3}}_{\text{net area bounded by } n\text{th parabola}}$$

$$= (f(x_0) + 4f(x_1) + 2f(x_2) + 4f(x_3) + \cdots + 2f(x_{n-2}) + 4f(x_{n-1}) + f(x_n))\frac{\Delta x}{3}.$$

Notice that apart from the first and last terms, the coefficients alternate between 4 and 2; to apply this rule, *n must be an even integer.*

DEFINITION **Simpson's Rule**

Suppose f is defined and integrable on $[a, b]$ and $n \geq 2$ is an even integer. The **Simpson's Rule approximation** to $\int_a^b f(x)\, dx$ using n equally spaced subintervals on $[a, b]$ is

$$S(n) = (f(x_0) + 4f(x_1) + 2f(x_2) + 4f(x_3) + \cdots + 4f(x_{n-1}) + f(x_n))\frac{\Delta x}{3},$$

where n is an even integer, $\Delta x = (b - a)/n$, and $x_k = a + k\Delta x$, for $k = 0, 1, \ldots, n$.

EXAMPLE 7 **Applying Simpson's Rule** Approximate $\int_{\pi/2}^{5\pi/2} \dfrac{\sin x}{x}\, dx$ using Simpson's Rule with $n = 4$ subintervals.

SOLUTION We have $\Delta x = \dfrac{5\pi/2 - \pi/2}{4} = \pi/2$, and therefore the grid points are

$$x_0 = \pi/2, \ x_1 = \pi, \ x_2 = 3\pi/2, \ x_3 = 2\pi, \ \text{ and } \ x_4 = 5\pi/2.$$

With $f(x) = \dfrac{\sin x}{x}$ and $n = 4$, the approximation given by Simpson's Rule is

$$S(4) = (f(x_0) + 4f(x_1) + 2f(x_2) + 4f(x_3) + f(x_4))\frac{\pi/2}{3}$$

$$= \left(\frac{\sin x_0}{x_0} + \frac{4 \sin x_1}{x_1} + \frac{2 \sin x_2}{x_2} + \frac{4 \sin x_3}{x_3} + \frac{\sin x_4}{x_4}\right)\frac{\pi}{6}$$

$$= \left(\frac{1}{\pi/2} + \frac{4(0)}{\pi} + \frac{2(-1)}{3\pi/2} + \frac{4(0)}{2\pi} + \frac{1}{5\pi/2}\right)\frac{\pi}{6}$$

$$= \left(\frac{2}{\pi} - \frac{4}{3\pi} + \frac{2}{5\pi}\right)\frac{\pi}{6}$$

$$= \frac{8}{45} \approx 0.178.$$

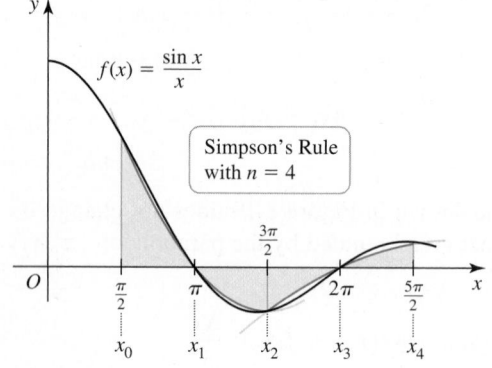

Figure 8.20

QUICK CHECK 4 Write the Simpson's Rule formula for $\int_0^3 g(x)\, dx$ with $n = 6$ subintervals. ◄

Figure 8.20 shows the approximation with two parabolas. To perform error analysis on our approximation, we need to look ahead to Theorem 8.1 (p. 577); this theorem involves finding an upper bound of $|f^{(4)}(x)|$ on the interval $\left[\frac{\pi}{2}, \frac{5\pi}{2}\right]$.

Related Exercises 23, 26 ◄

You can use the formula for Simpson's Rule, but here is an easier way. If you already have the Trapezoid Rule approximations $T(n)$ and $T(2n)$, the next Simpson's Rule approximation follows immediately with a simple calculation (Exercise 78):

$$S(2n) = \frac{4T(2n) - T(n)}{3}.$$

EXAMPLE 8 **Errors in the Trapezoid Rule and Simpson's Rule** Given that $\int_0^1 xe^{-x}\, dx = 1 - 2e^{-1}$, find the absolute errors in the Trapezoid Rule and Simpson's Rule approximations to the integral with $n = 8, 16, 32, 64,$ and 128 subintervals.

SOLUTION Because the shortcut formula for Simpson's Rule is based on values generated by the Trapezoid Rule, it is best to calculate the Trapezoid Rule approximations first. The second column of Table 8.8 shows the Trapezoid Rule approximations computed in Example 4. Having a column of Trapezoid Rule approximations, we can readily find the corresponding Simpson's Rule approximations. For example, if $n = 4$, we have

$$S(8) = \frac{4T(8) - T(4)}{3} \approx 0.26423805546593.$$

The table also shows the absolute errors in the approximations. The Simpson's Rule errors decrease more rapidly than the Trapezoid Rule errors. By careful inspection, you will see that the Simpson's Rule errors decrease with a clear pattern: Each time n is doubled (or Δx is halved), the errors decrease by a factor of approximately 16, which makes Simpson's Rule a more accurate method.

QUICK CHECK 5 Compute the approximate factor by which the error decreases in Table 8.8 between $S(16)$ and $S(32)$ and between $S(32)$ and $S(64)$. ◄

Table 8.8

n	$T(n)$	$S(n)$	Error in $T(n)$	Error in $S(n)$
4	0.25904504019141		0.00520	
8	0.26293980164730	0.26423805546593	0.00130	0.00000306
16	0.26391564480235	0.26424092585404	0.000325	0.000000192
32	0.26415974044777	0.26424110566291	0.0000814	0.0000000120
64	0.26422077279247	0.26424111690738	0.0000203	0.000000000750
128	0.26423603140581	0.26424111761026	0.00000509	0.0000000000469

Related Exercises 45–48 ◄

Errors in Numerical Integration

A detailed analysis of the errors in the three methods we have discussed goes beyond the scope of the text. We state without proof the standard error theorems for the methods and note that Examples 2, 3, 4, and 8 are consistent with these results.

> Because $\Delta x = \dfrac{b-a}{n}$, the error bounds in Theorem 8.1 can also be written as
>
> $$E_M \le \frac{k(b-a)^3}{24n^2},$$
>
> $$E_T \le \frac{k(b-a)^3}{12n^2}, \text{ and}$$
>
> $$E_S \le \frac{K(b-a)^5}{180n^4}.$$

THEOREM 8.1 Errors in Numerical Integration

Assume f'' is continuous on the interval $[a, b]$ and k is a real number such that $|f''(x)| \le k$, for all x in $[a, b]$. The absolute errors in approximating the integral $\int_a^b f(x)\, dx$ by the Midpoint Rule and Trapezoid Rule with n subintervals satisfy the inequalities

$$E_M \le \frac{k(b-a)}{24}(\Delta x)^2 \quad \text{and} \quad E_T \le \frac{k(b-a)}{12}(\Delta x)^2,$$

respectively, where $\Delta x = (b-a)/n$.

Assume $f^{(4)}$ is continuous on the interval $[a, b]$ and K is a real number such that $|f^{(4)}(x)| \le K$ on $[a, b]$. The absolute error in approximating the integral $\int_a^b f(x)\, dx$ by Simpson's Rule with n subintervals satisfies the inequality

$$E_S \le \frac{K(b-a)}{180}(\Delta x)^4.$$

QUICK CHECK 6 In Example 7, we used Simpson's Rule with $n = 4$ to find that $\int_{\pi/2}^{5\pi/2} \dfrac{\sin x}{x}\, dx \approx 0.178$. Use Theorem 8.1 to find an upper bound on E_S, the absolute error in this approximation, given that $|f^{(4)}(x)| \le 0.2$ on $\left[\frac{\pi}{2}, \frac{5\pi}{2}\right]$. ◄

The absolute errors associated with the Midpoint Rule and Trapezoid Rule are proportional to $(\Delta x)^2$. So if Δx is reduced by a factor of 2, the errors decrease roughly by a factor of 4, as shown in Example 4. Simpson's Rule is a more accurate method; its error is proportional to $(\Delta x)^4$, which means that if Δx is reduced by a factor of 2, the errors decrease roughly by a factor of 16, as shown in Example 8. Computing both the Trapezoid Rule and Simpson's Rule together, as shown in Example 8, is a powerful method that produces accurate approximations with relatively little work.

SECTION 8.8 EXERCISES

Getting Started

1. If the interval $[4, 18]$ is partitioned into $n = 28$ subintervals of equal length, what is Δx?

2. Explain geometrically how the Midpoint Rule is used to approximate a definite integral.

3. Explain geometrically how the Trapezoid Rule is used to approximate a definite integral.

4. If the Midpoint Rule is used on the interval $[-1, 11]$ with $n = 3$ subintervals, at what x-coordinates is the integrand evaluated?

5–8. *Compute the following estimates of $\int_0^8 f(x)\, dx$ using the graph in the figure.*

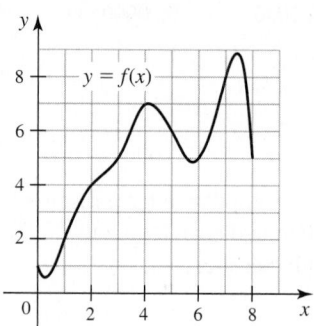

5. $M(4)$ 6. $T(4)$ 7. $S(4)$ 8. $S(8)$

9. If the Trapezoid Rule is used on the interval $[-1, 9]$ with $n = 5$ subintervals, at what x-coordinates is the integrand evaluated?

10. Suppose two Trapezoidal Rule approximations of $\int_a^b f(x)\, dx$ are $T(2) = 6$ and $T(4) = 5.1$. Find the Simpson's Rule approximation $S(4)$.

11–14. *Compute the absolute and relative errors in using c to approximate x.*

11. $x = \pi$; $c = 3.14$ 12. $x = \sqrt{2}$; $c = 1.414$

13. $x = e$; $c = 2.72$ 14. $x = e$; $c = 2.718$

Practice Exercises

15–18. Midpoint Rule approximations *Find the indicated Midpoint Rule approximations to the following integrals.*

15. $\int_2^{10} 2x^2\, dx$ using $n = 1, 2,$ and 4 subintervals

16. $\int_1^9 x^3\, dx$ using $n = 1, 2,$ and 4 subintervals

17. $\int_0^1 \sin \pi x\, dx$ using $n = 6$ subintervals

18. $\int_0^1 e^{-x}\, dx$ using $n = 8$ subintervals

19–22. Trapezoid Rule approximations *Find the indicated Trapezoid Rule approximations to the following integrals.*

19. $\int_2^{10} 2x^2\, dx$ using $n = 2, 4,$ and 8 subintervals

20. $\int_1^9 x^3\, dx$ using $n = 2, 4,$ and 8 subintervals

21. $\int_0^1 \sin \pi x\, dx$ using $n = 6$ subintervals

22. $\int_0^1 e^{-x}\, dx$ using $n = 8$ subintervals

23–26. Simpson's Rule approximations *Find the indicated Simpson's Rule approximations to the following integrals.*

23. $\int_0^\pi \sqrt{\sin x}\, dx$ using $n = 4$ and $n = 6$ subintervals

24. $\int_4^8 \sqrt{x}\, dx$ using $n = 4$ and $n = 8$ subintervals

25. $\int_{-2}^3 e^{-x^2}\, dx$ using $n = 10$ subintervals

26. $\int_2^4 \cos \sqrt{x}\, dx$ using $n = 8$ subintervals

27. **Midpoint Rule, Trapezoid Rule, and relative error** Find the Midpoint and Trapezoid Rule approximations to $\int_0^1 \sin \pi x\, dx$ using $n = 25$ subintervals. Compute the relative error of each approximation.

28. **Midpoint Rule, Trapezoid Rule, and relative error** Find the Midpoint and Trapezoid Rule approximations to $\int_0^1 e^{-x}\, dx$ using $n = 50$ subintervals. Compute the relative error of each approximation.

29–34. Comparing the Midpoint and Trapezoid Rules *Apply the Midpoint and Trapezoid Rules to the following integrals. Make a table similar to Table 8.5 showing the approximations and errors for $n = 4, 8, 16,$ and 32. The exact values of the integrals are given for computing the error.*

29. $\int_1^5 (3x^2 - 2x)\, dx = 100$ 30. $\int_{-2}^6 \left(\dfrac{x^3}{16} - x \right) dx = 4$

31. $\int_0^{\pi/4} 3 \sin 2x\, dx = \dfrac{3}{2}$ 32. $\int_1^e \ln x\, dx = 1$

33. $\int_0^\pi \sin x \cos 3x\, dx = 0$ 34. $\int_0^8 e^{-2x}\, dx = \dfrac{1 - e^{-16}}{2}$

35–36. River flow rates *The following figure shows the discharge rate $r(t)$ of the Snoqualmie River near Carnation, Washington, starting on a February day when the air temperature was rising. The variable t is the number of hours after midnight, $r(t)$ is given in millions of cubic feet per hour, and $\int_0^{24} r(t)\, dt$ equals the total amount of water that flows by the town of Carnation over a 24-hour period. Estimate $\int_0^{24} r(t)\, dt$ using the Trapezoidal Rule and Simpson's Rule with the following values of n.*

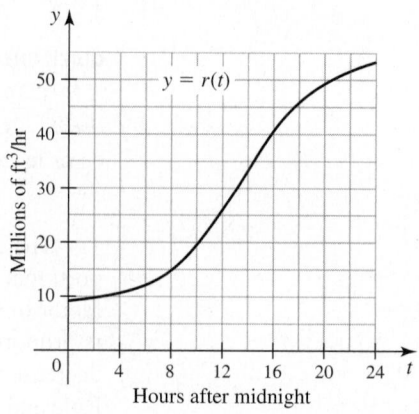

35. $n = 4$ 36. $n = 6$

37–40. Temperature data *Hourly temperature data for Boulder, Colorado; San Francisco, California; Nantucket, Massachusetts; and Duluth, Minnesota, over a 12-hr period on the same day of January are shown in the figure. Assume these data are taken from a continuous temperature function $T(t)$. The average temperature (in °F) over the 12-hr period is $\overline{T} = \dfrac{1}{12}\displaystyle\int_0^{12} T(t)\,dt.$*

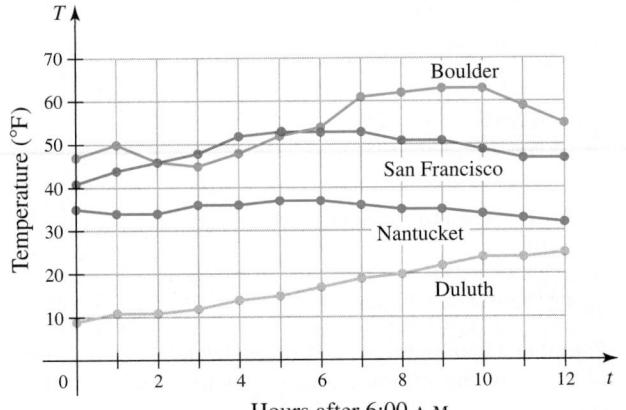

Hours after 6:00 A.M.

t	0	1	2	3	4	5	6	7	8	9	10	11	12
B	47	50	46	45	48	52	54	61	62	63	63	59	55
SF	41	44	46	48	52	53	53	53	51	51	49	47	47
N	35	34	34	36	36	37	37	36	35	35	34	33	32
D	9	11	11	12	14	15	17	19	20	22	24	24	25

37. Find an accurate approximation to the average temperature over the 12-hr period for Boulder. State your method.

38. Find an accurate approximation to the average temperature over the 12-hr period for San Francisco. State your method.

39. Find an accurate approximation to the average temperature over the 12-hr period for Nantucket. State your method.

40. Find an accurate approximation to the average temperature over the 12-hr period for Duluth. State your method.

41–44. Nonuniform grids *Use the indicated methods to solve the following problems with nonuniform grids.*

41. A curling iron is plugged into an outlet at time $t = 0$. Its temperature T in degrees Fahrenheit, assumed to be a continuous function that is strictly increasing and concave down on $0 \le t \le 120$, is given at various times (in seconds) in the table.

t (seconds)	0	20	45	60	90	110	120
$T(t)$ (°F)	70	130	200	239	311	355	375

a. Approximate $\dfrac{1}{120}\displaystyle\int_0^{120} T(t)\,dt$ in three ways: using a left Riemann sum, a right Riemann sum, and the Trapezoid Rule. Interpret the value of $\dfrac{1}{120}\displaystyle\int_0^{120} T(t)\,dt$ in the context of this problem.

b. Which of the estimates in part (a) overestimate the value of $\dfrac{1}{120}\displaystyle\int_0^{120} T(t)\,dt$? Which underestimate it? Justify your answers with a simple sketch of the sums you computed.

c. Evaluate and interpret $\dfrac{1}{120}\displaystyle\int_0^{120} T'(t)\,dt$ in the context of this problem.

42. Approximating integrals The function f is twice differentiable on $(-\infty, \infty)$. Values of f at various points on $[0, 20]$ are given in the table.

x	0	4	7	12	14	18	20
$f(x)$	3	0	-2	-1	2	4	7

a. Approximate $\int_0^{20} f(x)\,dx$ in three ways: using a left Riemann sum, a right Riemann sum, and the Trapezoid Rule.

b. A scatterplot of the data in the table is provided in the figure. Use the scatterplot to illustrate each of the approximations in part (a) by sketching appropriate rectangles for the Riemann sums and by sketching trapezoids for the Trapezoid Rule approximation.

c. Evaluate $\int_4^{12}(3f'(x) + 2)\,dx$.

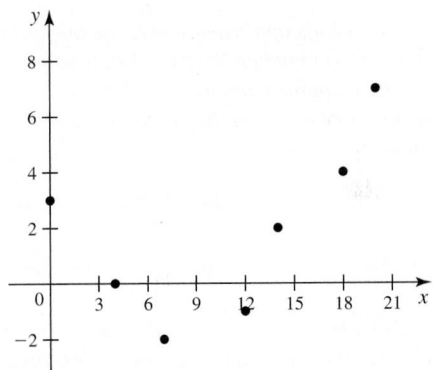

43. A hot-air balloon is launched from an elevation of 5400 ft above sea level. As it rises, the vertical velocity is computed using a device (called a *variometer*) that measures the change in atmospheric pressure. The vertical velocities at selected times are shown in the table (with units of ft/min).

t (min)	0	1	1.5	3	3.5	4	5
Velocity (ft/min)	0	100	120	150	110	90	80

a. Use the Trapezoid Rule to estimate the elevation of the balloon after five minutes. Remember that the balloon starts at an elevation of 5400 ft.

b. Use a right Riemann sum to estimate the elevation of the balloon after five minutes.

c. A polynomial that fits the data reasonably well is
$$g(t) = 3.49t^3 - 43.21t^2 + 142.43t - 1.75.$$
Estimate the elevation of the balloon after five minutes using this polynomial.

44. A piece of wood paneling must be cut in the shape shown in the figure. The coordinates of several points on its curved surface are also shown (with units of inches).

a. Estimate the surface area of the paneling using the Trapezoid Rule.

b. Estimate the surface area of the paneling using a left Riemann sum.

c. Could two identical pieces be cut from a 9-in by 9-in piece of wood? Answer carefully.

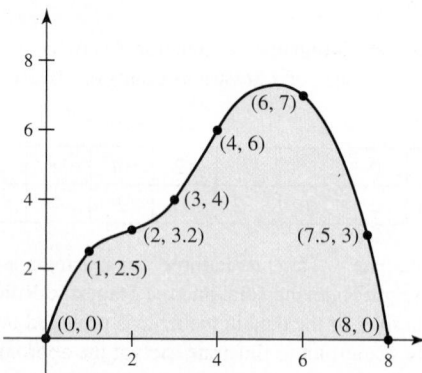

T 45–48. Trapezoid Rule and Simpson's Rule *Consider the following integrals and the given values of n.*

a. *Find the Trapezoid Rule approximations to the integral using n and 2n subintervals.*

b. *Find the Simpson's Rule approximation to the integral using 2n subintervals. It is easiest to obtain Simpson's Rule approximations from the Trapezoid Rule approximations, as in Example 8.*

c. *Compute the absolute errors in the Trapezoid Rule and Simpson's Rule with 2n subintervals.*

45. $\int_0^1 e^{2x}\,dx;\ n=25$ **46.** $\int_0^2 x^4\,dx;\ n=30$

47. $\int_1^e \frac{dx}{x};\ n=50$ **48.** $\int_0^{\pi/4}\frac{dx}{1+x^2};\ n=64$

T 49–52. Simpson's Rule *Apply Simpson's Rule to the following integrals. It is easiest to obtain the Simpson's Rule approximations from the Trapezoid Rule approximations, as in Example 8. Make a table similar to Table 8.8 showing the approximations and errors for n = 4, 8, 16, and 32. The exact values of the integrals are given for computing the error.*

49. $\int_0^4 (3x^5 - 8x^3)\,dx = 1536$ **50.** $\int_1^e \ln x\,dx = 1$

51. $\int_0^\pi e^{-t}\sin t\,dt = \frac{1}{2}(e^{-\pi}+1)$

52. $\int_0^6 3e^{-3x}\,dx = 1 - e^{-18} \approx 1.000000$

53. Explain why or why not Determine whether the following statements are true and give an explanation or counterexample.

a. Suppose $\int_a^b f(x)\,dx$ is approximated with Simpson's Rule using $n=18$ subintervals, where $|f^{(4)}(x)| \le 1$ on $[a,b]$. The absolute error E_S in approximating the integral satisfies $E_S \le \frac{(\Delta x)^5}{10}$.

b. If the number of subintervals used in the Midpoint Rule is increased by a factor of 3, the error is expected to decrease by a factor of 8.

c. If the number of subintervals used in the Trapezoid Rule is increased by a factor of 4, the error is expected to decrease by a factor of 16.

T 54–57. Comparing the Midpoint and Trapezoid Rules *Compare the errors in the Midpoint and Trapezoid Rules with n = 4, 8, 16, and 32 subintervals when they are applied to the following integrals (with their exact values given).*

54. $\int_0^{\pi/2}\sin^6 x\,dx = \frac{5\pi}{32}$ **55.** $\int_0^{\pi/2}\cos^9 x\,dx = \frac{128}{315}$

56. $\int_0^1 (8x^7 - 7x^8)\,dx = \frac{2}{9}$

57. $\int_0^\pi \ln(5 + 3\cos x)\,dx = \pi\ln\frac{9}{2}$

T 58–61. Using Simpson's Rule *Approximate the following integrals using Simpson's Rule. Experiment with values of n to ensure that the error is less than 10^{-3}.*

58. $\int_0^{2\pi}\frac{dx}{(5 + 3\sin x)^2} = \frac{5\pi}{32}$

59. $\int_0^\pi \frac{4\cos x}{5 - 4\cos x}\,dx = \frac{2\pi}{3}$

60. $\int_0^\pi \ln(2 + \cos x)\,dx = \pi\ln\left(\frac{2+\sqrt{3}}{2}\right)$

61. $\int_0^\pi \sin 6x \cos 3x\,dx = \frac{4}{9}$

T 62. Period of a pendulum A standard pendulum of length L that swings under the influence of gravity alone (no resistance) has a period of

$$T = \frac{4}{\omega}\int_0^{\pi/2}\frac{d\varphi}{\sqrt{1 - k^2\sin^2\varphi}},$$

where $\omega^2 = g/L$, $k^2 = \sin^2(\theta_0/2)$, $g = 9.8\ \text{m/s}^2$ is the acceleration due to gravity, and θ_0 is the initial angle from which the pendulum is released (in radians). Use numerical integration to approximate the period of a pendulum with $L = 1$ m that is released from an angle of $\theta_0 = \pi/4$ rad.

T 63. Normal distribution of heights The heights of U.S. men are normally distributed with a mean of 69 in and a standard deviation of 3 in. This means that the fraction of men with a height between a and b (with $a < b$) inches is given by the integral

$$\frac{1}{3\sqrt{2\pi}}\int_a^b e^{-((x-69)/3)^2/2}\,dx.$$

What percentage of American men are between 66 and 72 inches tall? Use the method of your choice, and experiment with the number of subintervals until you obtain successive approximations that differ by less than 10^{-3}.

T 64. Normal distribution of movie lengths A study revealed that the lengths of U.S. movies are normally distributed with a mean of 110 minutes and a standard deviation of 22 minutes. This means that the fraction of movies with lengths between a and b minutes (with $a < b$) is given by the integral

$$\frac{1}{22\sqrt{2\pi}}\int_a^b e^{-((x-110)/22)^2/2}\,dx.$$

What percentage of U.S. movies are between 1 hr and 1.5 hr long (60–90 min)?

T 65. U.S. oil produced and imported The figure shows the rate at which U.S. oil was produced and imported between 1920 and 2005 in units of millions of barrels per day. The total amount of oil produced or imported is given by the area of the region under the corresponding curve. Be careful with units because both days and years are used in this data set.

a. Use numerical integration to estimate the amount of U.S. oil produced between 1940 and 2000. Use the method of your choice, and experiment with values of n.

b. Use numerical integration to estimate the amount of oil imported between 1940 and 2000. Use the method of your choice, and experiment with values of n.

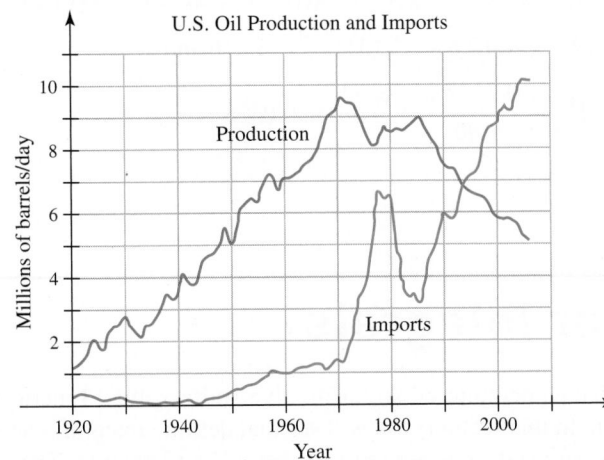

U.S. Oil Production and Imports

(*Source:* U.S. Energy Information Administration)

T 66–71. Estimating error *Refer to Theorem 8.1 in the following exercises.*

66. Let $f(x) = \cos x^2$.

   a. Find a Midpoint Rule approximation to $\int_{-1}^{1} \cos x^2 \, dx$ using $n = 30$ subintervals.
   b. Calculate $f''(x)$.
   c. Explain why $|f''(x)| \le 6$ on $[-1, 1]$.
   d. Use Theorem 8.1 to find an upper bound on the absolute error in the estimate found in part (a).

67. Let $f(x) = \sqrt{x^3 + 1}$.

   a. Find a Midpoint Rule approximation to $\int_{1}^{6} \sqrt{x^3 + 1} \, dx$ using $n = 50$ subintervals.
   b. Calculate $f''(x)$.
   c. Use the fact that f'' is decreasing and positive on $[1, 6]$ to show that $|f''(x)| \le 15/(8\sqrt{2})$ on $[1, 6]$.
   d. Use Theorem 8.1 to find an upper bound on the absolute error in the estimate found in part (a).

68. Let $f(x) = e^{x^2}$.

   a. Find a Trapezoid Rule approximation to $\int_{0}^{1} e^{x^2} \, dx$ using $n = 50$ subintervals.
   b. Calculate $f''(x)$.
   c. Explain why $|f''(x)| < 18$ on $[0, 1]$, given that $e < 3$.
   d. Use Theorem 8.1 to find an upper bound on the absolute error in the estimate found in part (a).

69. Let $f(x) = \sin e^x$.

   a. Find a Trapezoid Rule approximation to $\int_{0}^{1} \sin e^x \, dx$ using $n = 40$ subintervals.
   b. Calculate $f''(x)$.
   c. Explain why $|f''(x)| < 6$ on $[0, 1]$, given that $e < 3$. (*Hint:* Graph f''.)
   d. Find an upper bound on the absolute error in the estimate found in part (a) using Theorem 8.1.

70. Let $f(x) = e^{-x^2}$.

   a. Find a Simpson's Rule approximation to $\int_{0}^{3} e^{-x^2} \, dx$ using $n = 30$ subintervals.

   b. Calculate $f^{(4)}(x)$.
   c. Find an upper bound on the absolute error in the estimate found in part (a) using Theorem 8.1. (*Hint:* Use a graph to find an upper bound for $|f^{(4)}(x)|$ on $[0, 3]$.)

71. Let $f(x) = \sqrt{\sin x}$.

   a. Find a Simpson's Rule approximation to $\int_{1}^{2} \sqrt{\sin x} \, dx$ using $n = 20$ subintervals.
   b. Find an upper bound on the absolute error in the estimate found in part (a) using Theorem 8.1. (*Hint:* Use the fact that $|f^{(4)}(x)| \le 1$ on $[1, 2]$.)

Explorations and Challenges

72. **Exact Trapezoid Rule** Prove that the Trapezoid Rule is exact (no error) when approximating the definite integral of a linear function.

T 73. **Arc length of an ellipse** The length of an ellipse with axes of length $2a$ and $2b$ is

$$\int_{0}^{2\pi} \sqrt{a^2 \cos^2 t + b^2 \sin^2 t} \, dt.$$

Use numerical integration, and experiment with different values of n to approximate the length of an ellipse with $a = 4$ and $b = 8$.

T 74. **Sine integral** The theory of diffraction produces the sine integral function $\text{Si}(x) = \int_{0}^{x} \frac{\sin t}{t} \, dt$. Use the Midpoint Rule to approximate $\text{Si}(1)$ and $\text{Si}(10)$. (Recall that $\lim_{x \to 0} \frac{\sin x}{x} = 1$.) Experiment with the number of subintervals until you obtain approximations that have an error less than 10^{-3}. A rule of thumb is that if two successive approximations differ by less than 10^{-3}, then the error is usually less than 10^{-3}.

75. **Exact Simpson's Rule**

   a. Use Simpson's Rule to approximate $\int_{0}^{4} x^3 \, dx$ using two subintervals ($n = 2$); compare the approximation to the value of the integral.
   b. Use Simpson's Rule to approximate $\int_{0}^{4} x^3 \, dx$ using four subintervals ($n = 4$); compare the approximation to the value of the integral.
   c. Use the error bound associated with Simpson's Rule given in Theorem 8.1 to explain why the approximations in parts (a) and (b) give the exact value of the integral.
   d. Use Theorem 8.1 to explain why a Simpson's Rule approximation using any (even) number of subintervals gives the exact value of $\int_{a}^{b} f(x) \, dx$, where $f(x)$ is a polynomial of degree 3 or less.

76. **Shortcut for the Trapezoid Rule** Given a Midpoint Rule approximation $M(n)$ and a Trapezoid Rule approximation $T(n)$ for a continuous function on $[a, b]$ with n subintervals, show that

$$T(2n) = \frac{T(n) + M(n)}{2}.$$

77. **Trapezoid Rule and concavity** Suppose f is positive and its first two derivatives are continuous on $[a, b]$. If f'' is positive on $[a, b]$, then is a Trapezoid Rule estimate of $\int_{a}^{b} f(x) \, dx$ an underestimate or overestimate of the integral? Justify your answer using Theorem 8.1 and an illustration.

78. Shortcut for Simpson's Rule Using the notation of the text, prove that $S(2n) = \dfrac{4T(2n) - T(n)}{3}$, for $n \geq 1$.

T 79. Another Simpson's Rule formula Another Simpson's Rule formula is $S(2n) = \dfrac{2M(n) + T(n)}{3}$, for $n \geq 1$. Use this rule to estimate $\displaystyle\int_1^e \frac{1}{x}\, dx$ using $n = 10$ subintervals.

QUICK CHECK ANSWERS

1. 4, 6, 8, 10   **2.** Overestimates   **3.** 4 and 4
4. $\displaystyle\int_0^3 g(x)\, dx \approx (g(0) + 4g(0.5) + 2g(1) + 4g(1.5) +$
$2g(2) + 4g(2.5) + g(3))\dfrac{0.5}{3}$ **5.** 16 and 16
6. $E_S \leq \dfrac{0.2\,(2\pi)}{180} \cdot \left(\dfrac{\pi}{2}\right)^4 \approx 0.0425$ ◄

8.9 Improper Integrals

The definite integrals we have encountered so far involve finite-valued functions and finite intervals of integration. In this section, you will see that definite integrals can sometimes be evaluated when these conditions are not met. Here is an example. The energy required to launch a rocket from the surface of Earth ($R = 6370$ km from the center of Earth) to an altitude H is given by an integral of the form $\int_R^{R+H} k/x^2\, dx$, where k is a constant that includes the mass of the rocket, the mass of Earth, and the gravitational constant. This integral may be evaluated for any finite altitude $H > 0$. Now suppose the aim is to launch the rocket to an arbitrarily large altitude H so that it escapes Earth's gravitational field. The energy required is given by the preceding integral as $H \to \infty$, which we write $\int_R^{\infty} k/x^2\, dx$. This integral is an example of an *improper integral*, and it has a finite value (which explains why it is possible to launch rockets to outer space). For historical reasons, the term *improper integral* is used for cases in which

- the interval of integration is infinite, or
- the integrand is unbounded on the interval of integration.

In this section, we explore improper integrals and their many uses.

Infinite Intervals

A simple example illustrates what can happen when we integrate a function over an infinite interval. Consider the integral $\displaystyle\int_1^b \frac{1}{x^2}\, dx$, for any real number $b > 1$. As shown in Figure 8.21, this integral gives the area of the region bounded by the curve $y = x^{-2}$ and the x-axis between $x = 1$ and $x = b$. In fact, the value of the integral is

$$\int_1^b \frac{1}{x^2}\, dx = -\frac{1}{x}\Big|_1^b = 1 - \frac{1}{b}.$$

For example, if $b = 2$, the area under the curve is $\frac{1}{2}$; if $b = 3$, the area under the curve is $\frac{2}{3}$. In general, as b increases, the area under the curve increases.

Now let's ask what happens to the area as b becomes arbitrarily large. Letting $b \to \infty$, the area of the region under the curve is

$$\lim_{b \to \infty}\left(1 - \frac{1}{b}\right) = 1.$$

We have discovered, surprising as it may seem, a curve of *infinite* length that bounds a region with *finite* area (1 square unit).

We express this result as

$$\int_1^{\infty} \frac{1}{x^2}\, dx = 1,$$

which is an improper integral because ∞ appears in the upper limit. In general, to evaluate $\int_a^{\infty} f(x)\, dx$, we first integrate over a finite interval $[a, b]$ and then let $b \to \infty$. Similar procedures are used to evaluate $\int_{-\infty}^b f(x)\, dx$ and $\int_{-\infty}^{\infty} f(x)\, dx$.

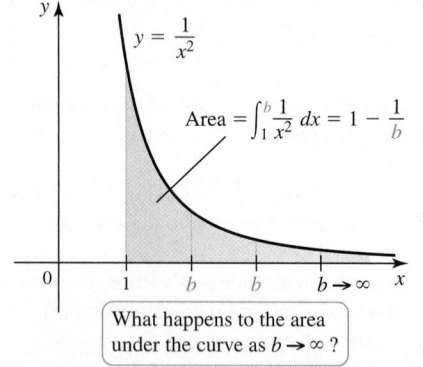

Area $= \displaystyle\int_1^b \frac{1}{x^2}\, dx = 1 - \frac{1}{b}$

$y = \dfrac{1}{x^2}$

What happens to the area under the curve as $b \to \infty$?

Figure 8.21

DEFINITION Improper Integrals over Infinite Intervals

1. If f is continuous on $[a, \infty)$, then
$$\int_a^\infty f(x)\, dx = \lim_{b \to \infty} \int_a^b f(x)\, dx.$$

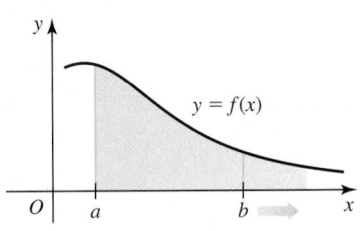

2. If f is continuous on $(-\infty, b]$, then
$$\int_{-\infty}^b f(x)\, dx = \lim_{a \to -\infty} \int_a^b f(x)\, dx.$$

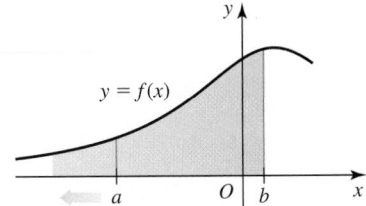

3. If f is continuous on $(-\infty, \infty)$, then
$$\int_{-\infty}^\infty f(x)\, dx = \lim_{a \to -\infty} \int_a^c f(x)\, dx$$
$$+ \lim_{b \to \infty} \int_c^b f(x)\, dx,$$

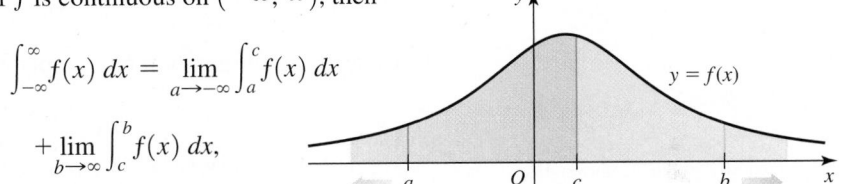

where c is any real number. It can be shown that the choice of c does not affect the value or convergence of the original integral.

If the limits in cases 1–3 exist, then the improper integrals **converge**; otherwise, they **diverge**.

> ▶ Doubly infinite integrals (case 3 in the definition) must be evaluated as two independent limits and not as
> $$\int_{-\infty}^\infty f(x)\, dx = \lim_{b \to \infty} \int_{-b}^b f(x)\, dx.$$

EXAMPLE 1 Infinite intervals Evaluate each integral.

a. $\displaystyle\int_0^\infty e^{-x}\, dx$ **b.** $\displaystyle\int_{-\infty}^\infty \frac{dx}{1 + x^2}$

SOLUTION

a. Using the definition of the improper integral, we have

$$\int_0^\infty e^{-x}\, dx = \lim_{b \to \infty} \int_0^b e^{-x}\, dx \qquad \text{Definition of improper integral}$$

$$= \lim_{b \to \infty} \left. (-e^{-x}) \right|_0^b \qquad \text{Evaluate the integral.}$$

$$= \lim_{b \to \infty} (1 - e^{-b}) \qquad \text{Simplify.}$$

$$= \left(1 - \underbrace{\lim_{b \to \infty} \frac{1}{e^b}}_{\text{equals } 0}\right) = 1. \qquad \text{Evaluate the limit; } e^{-b} = \frac{1}{e^b}.$$

In this case, the limit exists, so the integral converges and the region under the curve has a finite area of 1 (Figure 8.22).

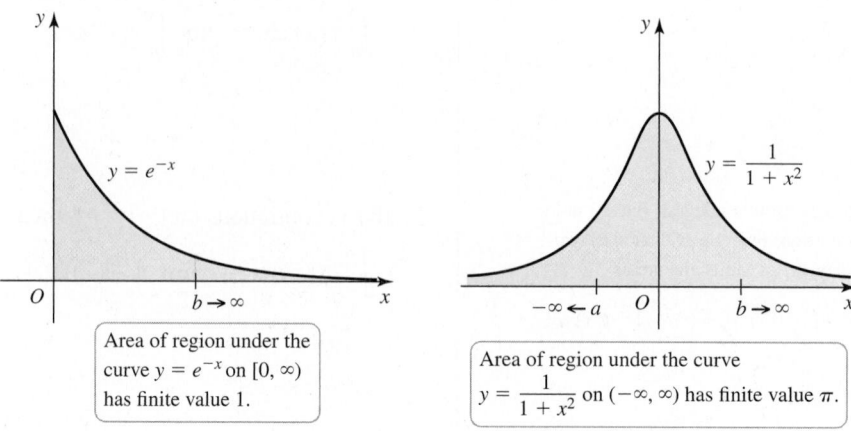

Area of region under the curve $y = e^{-x}$ on $[0, \infty)$ has finite value 1.

Figure 8.22

Area of region under the curve $y = \dfrac{1}{1 + x^2}$ on $(-\infty, \infty)$ has finite value π.

Figure 8.23

▶ Recall that
$$\int \frac{dx}{a^2 + x^2} = \frac{1}{a}\tan^{-1}\frac{x}{a} + C.$$

The graph of $y = \tan^{-1} x$ shows that
$$\lim_{x \to \infty} \tan^{-1} x = \frac{\pi}{2} \text{ and } \lim_{x \to -\infty} \tan^{-1} x = -\frac{\pi}{2}.$$

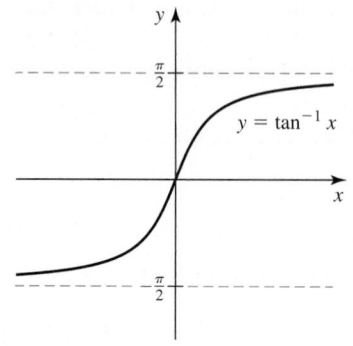

QUICK CHECK 1 The function $f(x) = 1 + x^{-1}$ decreases to 1 as $x \to \infty$. Does $\int_1^\infty f(x)\, dx$ exist? ◀

▶ Recall that for $p \ne 1$,
$$\int \frac{1}{x^p}\, dx = \int x^{-p}\, dx$$
$$= \frac{x^{-p+1}}{-p + 1} + C$$
$$= \frac{x^{1-p}}{1 - p} + C.$$

▶ Example 2 is important in the study of infinite series in Chapter 10. It shows that a continuous function f must do more than simply decrease to zero for its integral on $[a, \infty)$ to converge; it must decrease to zero *sufficiently fast*.

b. Using the definition of the improper integral, we choose $c = 0$ and write

$$\int_{-\infty}^{\infty} \frac{dx}{1 + x^2} = \lim_{a \to -\infty} \int_a^c \frac{dx}{1 + x^2} + \lim_{b \to \infty} \int_c^b \frac{dx}{1 + x^2} \qquad \text{Definition of improper integral}$$

$$= \lim_{a \to -\infty} \tan^{-1} x \Big|_a^0 + \lim_{b \to \infty} \tan^{-1} x \Big|_0^b \qquad \text{Evaluate integral; } c = 0.$$

$$= \lim_{a \to -\infty} (0 - \tan^{-1} a) + \lim_{b \to \infty} (\tan^{-1} b - 0) \quad \text{Simplify.}$$

$$= \frac{\pi}{2} + \frac{\pi}{2} = \pi. \qquad \text{Evaluate limits.}$$

The same result is obtained with any value of the intermediate point c; therefore, the value of the integral is π (Figure 8.23).

Related Exercises 10, 15 ◀

EXAMPLE 2   The family $f(x) = 1/x^p$ Consider the family of functions $f(x) = 1/x^p$, where p is a real number. For what values of p does $\int_1^\infty f(x)\, dx$ converge?

SOLUTION For $p > 0$, the functions $f(x) = 1/x^p$ approach zero as $x \to \infty$, with larger values of p giving greater rates of decrease (Figure 8.24). Assuming $p \ne 1$, the integral is evaluated as follows:

$$\int_1^\infty \frac{1}{x^p}\, dx = \lim_{b \to \infty} \int_1^b x^{-p}\, dx \qquad \text{Definition of improper integral}$$

$$= \frac{1}{1 - p} \lim_{b \to \infty} \left(x^{1-p} \Big|_1^b \right) \qquad \text{Evaluate the integral on a finite interval.}$$

$$= \frac{1}{1 - p} \lim_{b \to \infty} (b^{1-p} - 1). \quad \text{Simplify.}$$

It is easiest to consider three cases.

Case 1: If $p > 1$, then $p - 1 > 0$, and $b^{1-p} = 1/b^{p-1}$ approaches 0 as $b \to \infty$. Therefore, the integral converges and its value is

$$\int_1^\infty \frac{1}{x^p}\, dx = \frac{1}{1 - p} \lim_{b \to \infty} \underbrace{(b^{1-p}}_{\substack{\text{approaches} \\ 0}} - 1) = \frac{1}{p - 1}.$$

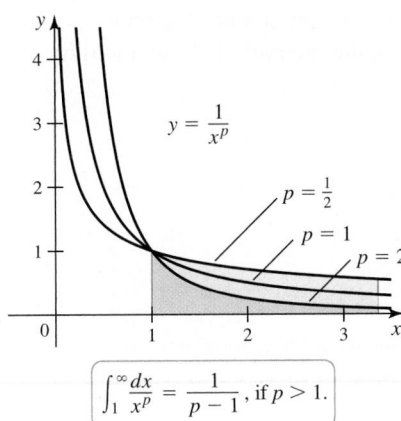

$$\int_1^\infty \frac{dx}{x^p} = \frac{1}{p-1}, \text{ if } p > 1.$$

Figure 8.24

QUICK CHECK 2 Use the result of Example 2 to evaluate $\int_1^\infty \frac{1}{x^4}\,dx.$ ◀

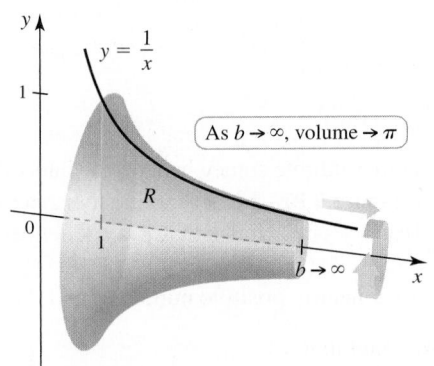

Figure 8.25

▶ The integral in Example 3b can be evaluated directly by using the substitution $u = x^2$ and then consulting a table of integrals.

▶ The solid in Examples 3a and 3b is called *Gabriel's horn* or *Torricelli's trumpet*. We have shown that—quite remarkably—it has finite volume and infinite surface area.

Case 2: If $p < 1$, then $1 - p > 0$, and the integral diverges:

$$\int_1^\infty \frac{1}{x^p}\,dx = \frac{1}{1-p}\lim_{b\to\infty}\underbrace{(b^{1-p}-1)}_{\substack{\text{arbitrarily}\\\text{large}}} = \infty.$$

Case 3: If $p = 1$, then $\int_1^\infty \frac{1}{x}\,dx = \lim_{b\to\infty}\ln b = \infty$, so the integral diverges.

In summary, $\int_1^\infty \frac{1}{x^p}\,dx = \frac{1}{p-1}$ if $p > 1$, and the integral diverges if $p \le 1$.

Related Exercises 8–9 ◀

EXAMPLE 3   Solids of revolution Let R be the region bounded by the graph of $y = x^{-1}$ and the x-axis, for $x \ge 1$.

a. What is the volume of the solid generated when R is revolved about the x-axis?

b. What is the surface area of the solid generated when R is revolved about the x-axis?

c. What is the volume of the solid generated when R is revolved about the y-axis?

SOLUTION

a. The region R and the corresponding solid of revolution are shown in Figure 8.25. We use the disk method (Section 6.3) over the interval $[1, b]$ and then let $b \to \infty$:

$$\text{Volume} = \int_1^\infty \pi(f(x))^2\,dx \qquad \text{Disk method}$$
$$= \pi\lim_{b\to\infty}\int_1^b \frac{1}{x^2}\,dx \qquad \text{Definition of improper integral}$$
$$= \pi\lim_{b\to\infty}\left(1 - \frac{1}{b}\right) = \pi. \quad \text{Evaluate the integral.}$$

The improper integral exists, and the solid has a volume of π cubic units.

b. Using the results of Section 6.6, the area of the surface generated on the interval $[1, b]$, where $b > 1$, is

$$\int_1^b 2\pi f(x)\sqrt{1 + f'(x)^2}\,dx.$$

The area of the surface generated on the interval $[1, \infty)$ is found by letting $b \to \infty$:

$$\text{Surface area} = 2\pi\lim_{b\to\infty}\int_1^b f(x)\sqrt{1 + f'(x)^2}\,dx \quad \text{Surface area formula; let } b \to \infty.$$
$$= 2\pi\lim_{b\to\infty}\int_1^b \frac{1}{x}\sqrt{1 + \left(-\frac{1}{x^2}\right)^2}\,dx \quad \text{Substitute } f \text{ and } f'.$$
$$= 2\pi\lim_{b\to\infty}\int_1^b \frac{1}{x^3}\sqrt{1 + x^4}\,dx. \qquad \text{Simplify.}$$

Notice that on the interval of integration $x \ge 1$, we have $\sqrt{1 + x^4} > \sqrt{x^4} = x^2$, which means that

$$\frac{1}{x^3}\sqrt{1 + x^4} > \frac{x^2}{x^3} = \frac{1}{x}.$$

Therefore, for all b with $1 < b < \infty$,

$$\text{Surface area} = 2\pi\int_1^b \frac{1}{x^3}\sqrt{1 + x^4}\,dx > 2\pi\int_1^b \frac{1}{x}\,dx.$$

Because $2\pi\lim_{b\to\infty}\int_1^b \frac{1}{x}\,dx = \infty$ (by Example 2), the preceding inequality implies that $2\pi\lim_{b\to\infty}\int_1^b \frac{1}{x^3}\sqrt{1 + x^4}\,dx = \infty$. Therefore, the integral diverges and the surface area of the solid is infinite.

➤ Recall that if $f(x) > 0$ on $[a, b]$ and the region bounded by the graph of f and the x-axis on $[a, b]$ is revolved about the y-axis, the volume of the solid generated is

$$V = \int_a^b 2\pi x f(x) \, dx.$$

c. The region in question and the corresponding solid of revolution are shown in Figure 8.26. Using the shell method (Section 6.4) on the interval $[1, b)$ and letting $b \to \infty$, the volume is given by

$$
\begin{aligned}
\text{Volume} &= \int_1^\infty 2\pi x f(x) \, dx && \text{Shell method} \\
&= 2\pi \int_1^\infty 1 \, dx && f(x) = x^{-1} \\
&= 2\pi \lim_{b \to \infty} \int_1^b 1 \, dx && \text{Definition of improper integral} \\
&= 2\pi \lim_{b \to \infty} (b - 1) && \text{Evaluate the integral over a finite interval.} \\
&= \infty. && \text{The improper integral diverges.}
\end{aligned}
$$

Revolving the region R about the y-axis, the volume of the resulting solid is infinite.

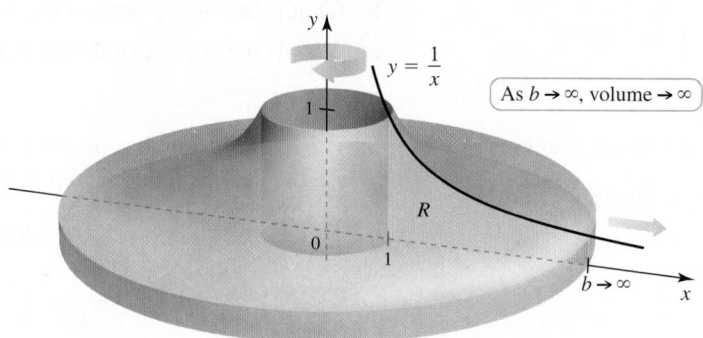

$y = \dfrac{1}{x}$

As $b \to \infty$, volume $\to \infty$

R

$b \to \infty$

Figure 8.26

Related Exercises 65–66 ◄

Unbounded Integrands

Improper integrals also occur when the integrand becomes infinite somewhere on the interval of integration. Consider the function $f(x) = 1/\sqrt{x}$ (Figure 8.27). Let's examine the area of the region bounded by the graph of f between $x = 0$ and $x = 1$. Notice that f is not defined at $x = 0$, and it increases without bound as $x \to 0^+$.

The idea here is to replace the lower limit 0 with a nearby positive number c and then consider the integral $\int_c^1 \dfrac{dx}{\sqrt{x}}$, where $0 < c < 1$. We find that

$$\int_c^1 \frac{dx}{\sqrt{x}} = 2\sqrt{x}\,\Big|_c^1 = 2(1 - \sqrt{c}).$$

To find the area of the region under the curve over the interval $(0, 1]$, we let $c \to 0^+$. The resulting area, which we denote $\int_0^1 \dfrac{dx}{\sqrt{x}}$, is

$$\lim_{c \to 0^+} \int_c^1 \frac{dx}{\sqrt{x}} = \lim_{c \to 0^+} 2(1 - \sqrt{c}) = 2.$$

Once again, we have a surprising result: Although the region in question has a boundary curve with infinite length, the area of the region is finite.

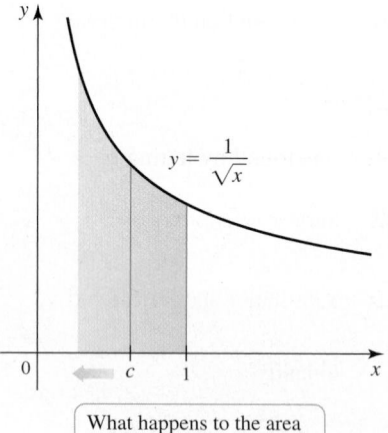

$y = \dfrac{1}{\sqrt{x}}$

What happens to the area under the curve as $c \to 0^+$?

Figure 8.27

➤ The functions $f(x) = 1/x^p$ are unbounded at $x = 0$, for $p > 0$. It can be shown (Exercise 94) that

$$\int_0^1 \frac{dx}{x^p} = \frac{1}{1 - p},$$

provided $p < 1$. Otherwise, the integral diverges.

QUICK CHECK 3 Explain why the one-sided limit $c \to 0^+$ (instead of a two-sided limit) must be used in the previous calculation. ◄

The preceding example shows that if a function is unbounded at a point p, it may be possible to integrate that function over an interval that contains p. The point p may occur at either endpoint or at an interior point of the interval of integration.

DEFINITION **Improper Integrals with an Unbounded Integrand**

1. Suppose f is continuous on $(a, b]$ with $\lim\limits_{x \to a^+} f(x) = \pm\infty$. Then

$$\int_a^b f(x)\, dx = \lim_{c \to a^+} \int_c^b f(x)\, dx.$$

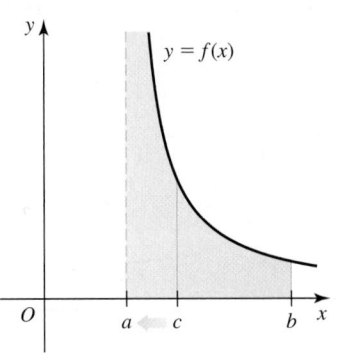

2. Suppose f is continuous on $[a, b)$ with $\lim\limits_{x \to b^-} f(x) = \pm\infty$. Then

$$\int_a^b f(x)\, dx = \lim_{c \to b^-} \int_a^c f(x)\, dx.$$

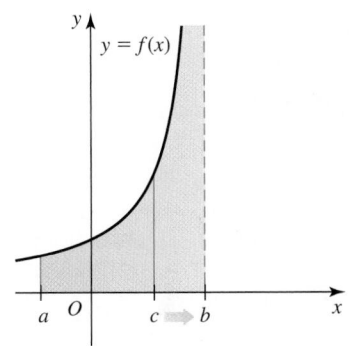

3. Suppose f is continuous on $[a, b]$ except at the interior point p where f is unbounded. Then

$$\int_a^b f(x)\, dx = \lim_{c \to p^-} \int_a^c f(x)\, dx + \lim_{d \to p^+} \int_d^b f(x)\, dx.$$

If the limits in cases 1–3 exist, then the improper integrals **converge**; otherwise, they **diverge**.

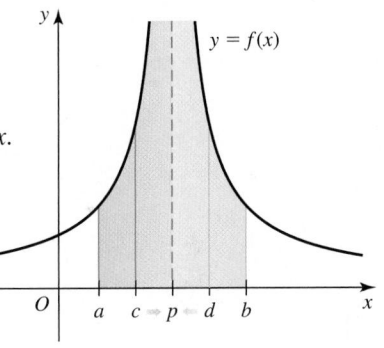

> **Recall that**
> $$\int \frac{dx}{\sqrt{a^2 - x^2}} = \sin^{-1}\frac{x}{a} + C.$$

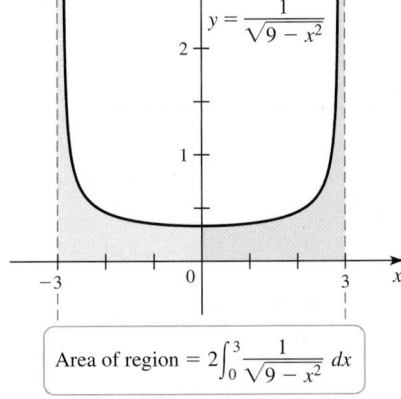

Area of region $= 2\int_0^3 \dfrac{1}{\sqrt{9 - x^2}}\, dx$

Figure 8.28

EXAMPLE 4 **Infinite integrand** Find the area of the region R between the graph of $f(x) = \dfrac{1}{\sqrt{9 - x^2}}$ and the x-axis on the interval $(-3, 3)$ (if it exists).

SOLUTION The integrand is even and has vertical asymptotes at $x = \pm 3$ (Figure 8.28). By symmetry, the area of R is given by

$$\int_{-3}^{3} \frac{1}{\sqrt{9 - x^2}}\, dx = 2\int_0^3 \frac{1}{\sqrt{9 - x^2}}\, dx,$$

assuming these improper integrals exist. Because the integrand is unbounded at $x = 3$, we replace the upper limit with c, evaluate the resulting integral, and then let $c \to 3^-$:

$$2\int_0^3 \frac{dx}{\sqrt{9 - x^2}} = 2\lim_{c \to 3^-} \int_0^c \frac{dx}{\sqrt{9 - x^2}} \qquad \text{Definition of improper integral}$$

$$= 2\lim_{c \to 3^-} \left.\sin^{-1}\frac{x}{3}\right|_0^c \qquad\qquad \text{Evaluate the integral.}$$

$$= 2\lim_{c \to 3^-} \left(\underbrace{\sin^{-1}\frac{c}{3}}_{\text{approaches } \pi/2} - \underbrace{\sin^{-1} 0}_{\text{equals } 0} \right). \quad \text{Simplify.}$$

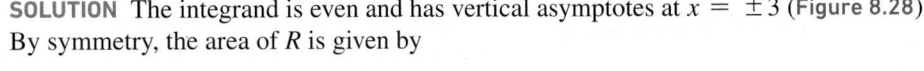

Note that as $c \to 3^-$, $\sin^{-1}(c/3) \to \sin^{-1} 1 = \pi/2$. Therefore, the area of R is

$$2 \int_0^3 \frac{1}{\sqrt{9 - x^2}} \, dx = 2\left(\frac{\pi}{2} - 0\right) = \pi.$$

Related Exercises 37–38 ◀

EXAMPLE 5   Infinite integrand at an interior point Evaluate $\displaystyle\int_1^{10} \frac{dx}{(x - 2)^{1/3}}$.

SOLUTION The integrand is unbounded at $x = 2$, which is an interior point of the interval of integration (**Figure 8.29**). We split the interval into two subintervals and evaluate an improper integral on each subinterval:

$$\int_1^{10} \frac{dx}{(x - 2)^{1/3}} = \lim_{c \to 2^-} \int_1^c \frac{dx}{(x - 2)^{1/3}} + \lim_{d \to 2^+} \int_d^{10} \frac{dx}{(x - 2)^{1/3}} \quad \text{Definition of improper integral}$$

$$= \lim_{c \to 2^-} \frac{3}{2}(x - 2)^{2/3}\Big|_1^c + \lim_{d \to 2^+} \frac{3}{2}(x - 2)^{2/3}\Big|_d^{10} \quad \text{Evaluate integrals.}$$

$$= \frac{3}{2}\left(\lim_{c \to 2^-}(c - 2)^{2/3} - (1 - 2)^{2/3}\right)$$

$$+ \frac{3}{2}\left((10 - 2)^{2/3} - \lim_{d \to 2^+}(d - 2)^{2/3}\right) \quad \text{Simplify.}$$

$$= \frac{3}{2}\left(0 - (-1)^{2/3} + 8^{2/3} - 0\right) = \frac{9}{2}. \quad \text{Evaluate limits.}$$

Related Exercises 48–49 ◀

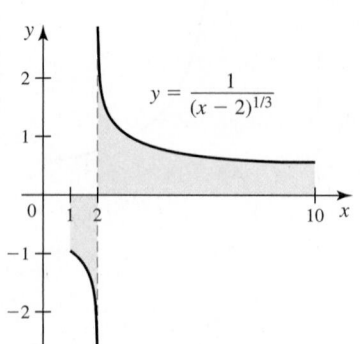

Figure 8.29

➤ We interpret the result of $9/2$ in Example 5 as the net area bounded by the curve $y = 1/(x - 2)^{1/3}$ over the interval $[1, 10]$.

The following example illustrates one of many practical uses of improper integrals.

EXAMPLE 6   Bioavailability The most efficient way to deliver a drug to its intended target site is to administer it intravenously (directly into the blood). If a drug is administered any other way (for example, by injection, orally, by nasal inhalant, or by skin patch), then some of the drug is typically lost to absorption before it gets to the blood. By definition, the *bioavailability* of a drug measures the effectiveness of a nonintravenous method compared to the intravenous method. The bioavailability of intravenous dosing is 100%.

Let the functions $C_i(t)$ and $C_o(t)$ give the concentration of a drug in the blood, for times $t \geq 0$, using intravenous and oral dosing, respectively. (These functions can be determined through clinical experiments.) Assuming the same amount of drug is initially administered by both methods, the bioavailability for an oral dose is defined to be

$$F = \frac{\text{AUC}_o}{\text{AUC}_i} = \frac{\int_0^\infty C_o(t) \, dt}{\int_0^\infty C_i(t) \, dt},$$

where AUC is used in the pharmacology literature to mean *area under the curve*.

Suppose the concentration of a certain drug in the blood in mg/L when given intravenously is $C_i(t) = 100e^{-0.3t}$, where $t \geq 0$ is measured in hours. Suppose also that the concentration of the same drug when delivered orally is $C_o(t) = 90(e^{-0.3t} - e^{-2.5t})$ (**Figure 8.30**). Find the bioavailability of the drug.

SOLUTION Evaluating the integrals of the concentration functions, we find that

$$\text{AUC}_i = \int_0^\infty C_i(t) \, dt = \int_0^\infty 100e^{-0.3t} \, dt$$

$$= \lim_{b \to \infty} \int_0^b 100e^{-0.3t} \, dt \quad \text{Improper integral}$$

$$= \lim_{b \to \infty} \frac{1000}{3}\left(1 - \underbrace{e^{-0.3b}}_{\substack{\text{approaches} \\ \text{zero}}}\right) \quad \text{Evaluate the integral.}$$

$$= \frac{1000}{3}. \quad \text{Evaluate the limit.}$$

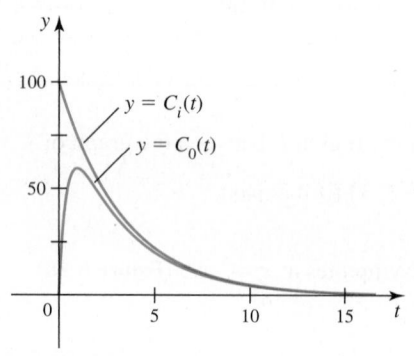

Figure 8.30

Similarly,

$$AUC_o = \int_0^\infty C_o(t)\, dt = \int_0^\infty 90(e^{-0.3t} - e^{-2.5t})\, dt$$

$$= \lim_{b \to \infty} \int_0^b 90(e^{-0.3t} - e^{-2.5t})\, dt \qquad \text{Improper integral}$$

$$= \lim_{b \to \infty} \left(300(1 - \underbrace{e^{-0.3b}}_{\substack{\text{approaches} \\ \text{zero}}}) - 36(1 - \underbrace{e^{-2.5b}}_{\substack{\text{approaches} \\ \text{zero}}})\right) \qquad \text{Evaluate the integral.}$$

$$= 264. \qquad \text{Evaluate the limit.}$$

Therefore, the bioavailability is $F = 264/(1000/3) = 0.792$, which means oral administration of the drug is roughly 80% as effective as intravenous dosing. Notice that F is the ratio of the areas under the two curves on the interval $[0, \infty)$.

Related Exercises 61–62 ◀

The Comparison Test

There are some cases in which we cannot determine whether a given improper integral converges, simply because it is impossible to compute an antiderivative using the catalog of integrals in this text. For example, the antiderivative of $g(x) = e^{-x^2}$ is not expressible in terms of elementary functions, so the procedures used earlier in this section fail to determine whether $\int_a^\infty e^{-x^2}\, dx$ converges, for any real number a. Theorem 8.2 is helpful here.

> ▶ The integral $\int_a^\infty e^{-x^2}\, dx$ plays an essential role in probability theory, and it is important to determine whether it converges—see Example 7a.

THEOREM 8.2   Comparison Test for Improper Integrals
Suppose the functions f and g are continuous on the interval $[a, \infty)$, with $f(x) \geq g(x) \geq 0$, for $x \geq a$.

1. If $\int_a^\infty f(x)\, dx$ converges, then $\int_a^\infty g(x)\, dx$ converges.

2. If $\int_a^\infty g(x)\, dx$ diverges, then $\int_a^\infty f(x)\, dx$ diverges.

We omit a formal proof, although **Figure 8.31** helps to explain statement (1) of Theorem 8.2. If the area beneath the graph of f is finite on $[a, \infty)$ and if $f(x) \geq g(x) \geq 0$, then the area under the graph of g on $[a, \infty)$ must also be finite. Statement (2) follows from (1) because it is the contrapositive of (1). Analogous theorems apply to improper integrals of the form $\int_{-\infty}^b f(x)\, dx$ and to integrals over finite intervals with an unbounded integrand.

If the area under the graph of f on $[a, \infty)$ is finite . . .

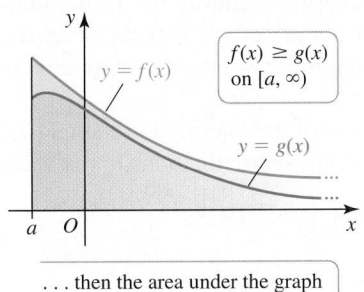

$f(x) \geq g(x)$ on $[a, \infty)$

. . . then the area under the graph of g on $[a, \infty)$ is also finite.

Figure 8.31

QUICK CHECK 4 Show that
$$\int_1^\infty e^{-x}\, dx = \frac{1}{e}. \quad ◀$$

EXAMPLE 7   Comparison Test Use Theorem 8.2 to determine whether the following integrals converge.

a. $\displaystyle\int_1^\infty e^{-x^2}\, dx$ **b.** $\displaystyle\int_1^\infty \frac{1}{\sqrt[3]{x^2 - 0.5}}\, dx$

SOLUTION

a. Before invoking Theorem 8.2, it helps to predict whether a given improper integral converges or diverges so we can decide whether to apply statement (1) or (2) of the Comparison Test. In Example 1a, we learned that $\int_0^\infty e^{-x}\, dx$ converges, which implies that $\int_1^\infty e^{-x}\, dx$ also converges (Quick Check 4). Because e^{-x^2} approaches 0 more rapidly than e^{-x} as $x \to \infty$ ($x^2 > x$ for large values of x), we expect that $\int_1^\infty e^{-x^2}\, dx$ also converges. So we let $f(x) = e^{-x}$ and $g(x) = e^{-x^2}$ in statement (1); our goal is to prove that $f(x) \geq g(x)$ on $[1, \infty)$ (note that both functions are positive for all x). Because $x \leq x^2$ and e^x is increasing on $[1, \infty)$, we have $e^x \leq e^{x^2}$ on $[1, \infty)$, which implies that

$$e^{-x} \geq e^{-x^2} \text{ on } [1, \infty). \quad \text{Figure 8.32}$$

Therefore, by statement (1) of Theorem 8.2, $\int_1^\infty e^{-x^2}\, dx$ converges.

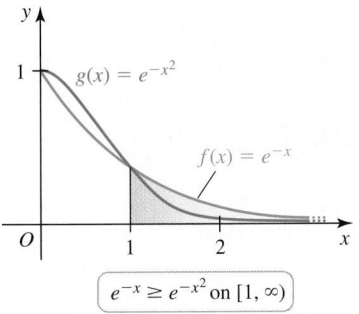

$e^{-x} \geq e^{-x^2}$ on $[1, \infty)$

Figure 8.32

b. One challenge in using Theorem 8.2 is choosing an appropriate improper integral for comparison to the given improper integral. In this case, we see that for large values of x,

$$\frac{1}{\sqrt[3]{x^2 - 0.5}} \approx \frac{1}{\sqrt[3]{x^2}} = \frac{1}{x^{2/3}},$$ which suggests that we compare $\int_1^\infty \frac{1}{\sqrt[3]{x^2 - 0.5}} \, dx$ to

$\int_1^\infty \frac{1}{x^{2/3}} \, dx$. In Example 2 we showed that $\int_1^\infty \frac{1}{x^p} \, dx$ diverges if $p \leq 1$, so we

expect that $\int_1^\infty \frac{1}{\sqrt[3]{x^2 - 0.5}} \, dx$ also diverges. Therefore, we let $f(x) = \sqrt[3]{x^2 - 0.5}$ and

$g(x) = x^{2/3}$ in statement (2) of Theorem 8.2; our goal is to show that $f(x) \geq g(x)$ on $[1, \infty)$. Note that $\sqrt[3]{x^2 - 0.5} < \sqrt[3]{x^2} = x^{2/3}$ on $[1, \infty)$, which implies that

$$\frac{1}{\sqrt[3]{x^2 - 0.5}} > \frac{1}{x^{2/3}} \text{ on } [1, \infty).$$

By statement (2) of Theorem 8.2, $\int_1^\infty \frac{1}{\sqrt[3]{x^2 - 0.5}} \, dx$ diverges.

Related Exercises 77, 79 ◀

We close this section with two important points. First, for improper integrals of the form $\int_a^\infty g(x) \, dx$, where g satisfies the conditions of Theorem 8.2, one need not find a function f that satisfies the theorem's conditions on the entire interval $[a, \infty)$. Rather, it suffices to find a function f such that

$$f(x) \geq g(x) \text{ on } [c, \infty), \text{ where } c > a, \quad \text{and} \quad \int_c^\infty f(x) \, dx \text{ converges.}$$

These two conditions are enough to conclude that $\int_a^\infty g(x) \, dx$ converges because $g(x)$ is bounded on $[a, c]$ and therefore $\int_a^c g(x) \, dx$ is finite. For example, even though e^{-x} is not greater than or equal to e^{-x^2} on $[0, \infty)$ (Figure 8.32), we can use the result of Example 7a to conclude that $\int_0^\infty e^{-x^2} \, dx$ (with lower limit of 0) converges—adding the finite number $\int_0^1 e^{-x^2} \, dx$ to the convergent integral $\int_1^\infty e^{-x^2} \, dx$ does not affect whether it converges. Similar principles can be applied to the other forms of improper integrals encountered in this section.

The second point is that we never discovered the value of the integral in Example 7a. In future chapters, we develop methods that provide accurate approximations to the value of a convergent integral, and in Section 16.3, we show that $\int_0^\infty e^{-x^2} \, dx = \frac{\sqrt{\pi}}{2}$.

SECTION 8.9 EXERCISES

Getting Started

1. What are the two general ways in which an improper integral may occur?

2. Evaluate $\int_2^\infty \frac{dx}{x^3}$ after writing the expression as a limit.

3. Rewrite $\int_2^\infty \frac{dx}{x^{1/5}}$ as a limit and then show that the integral diverges.

4. Evaluate $\int_0^1 \frac{dx}{x^{1/5}}$ after writing the integral as a limit.

5. Write $\lim_{a \to -\infty} \int_a^0 f(x) \, dx + \lim_{b \to \infty} \int_0^b f(x) \, dx$ as an improper integral.

6. For what values of p does $\int_1^\infty x^{-p} \, dx$ converge?

Practice Exercises

7–58. Improper integrals *Evaluate the following integrals or state that they diverge.*

7. $\int_3^\infty \frac{dx}{x^2}$

8. $\int_2^\infty \frac{dx}{x}$

9. $\int_2^\infty \frac{dx}{\sqrt{x}}$

10. $\int_0^\infty e^{-2x} \, dx$

11. $\int_0^\infty e^{-ax} \, dx, a > 0$

12. $\int_{-\infty}^{-1} \frac{dx}{\sqrt[3]{x}}$

13. $\int_0^\infty \cos x \, dx$

14. $\int_{-\infty}^{-1} \frac{dx}{x^3}$

15. $\int_{-\infty}^\infty \frac{dx}{x^2 + 100}$

16. $\int_{-\infty}^\infty \frac{dx}{x^2 + a^2}, a > 0$

17. $\displaystyle\int_7^\infty \frac{dx}{(x+1)^{1/3}}$

18. $\displaystyle\int_2^\infty \frac{dx}{(x+2)^2}$

19. $\displaystyle\int_1^\infty \frac{3x^2+1}{x^3+x}\,dx$

20. $\displaystyle\int_1^\infty 2^{-x}\,dx$

21. $\displaystyle\int_2^\infty \frac{\cos(\pi/x)}{x^2}\,dx$

22. $\displaystyle\int_{-\infty}^{-2} \frac{1}{z^2}\sin\frac{\pi}{z}\,dz$

23. $\displaystyle\int_0^\infty \frac{e^u}{e^{2u}+1}\,du$

24. $\displaystyle\int_{-\infty}^a \sqrt{e^x}\,dx,\, a>0$

25. $\displaystyle\int_{-\infty}^\infty \frac{e^{3x}}{1+e^{6x}}\,dx$

26. $\displaystyle\int_{-\infty}^\infty \frac{x}{(x^2+1)^2}\,dx$

27. $\displaystyle\int_{-\infty}^\infty xe^{-x^2}\,dx$

28. $\displaystyle\int_1^\infty \frac{\tan^{-1}s}{s^2+1}\,ds$

29. $\displaystyle\int_{-\infty}^\infty \frac{(\tan^{-1}t)^2}{t^2+1}\,dt$

30. $\displaystyle\int_{-\infty}^0 e^x\,dx$

31. $\displaystyle\int_1^\infty \frac{dv}{v(v+1)}$

32. $\displaystyle\int_1^\infty \frac{dx}{x^2(x-1)}$

33. $\displaystyle\int_2^\infty \frac{dy}{y\ln y}$

34. $\displaystyle\int_{-\infty}^{-4/\pi} \frac{1}{x^2}\sec^2\!\left(\frac{1}{x}\right)dx$

35. $\displaystyle\int_{-\infty}^0 \frac{dx}{\sqrt[3]{2-x}}$

36. $\displaystyle\int_{e^2}^\infty \frac{dx}{x\ln^p x},\, p>1$

37. $\displaystyle\int_0^8 \frac{dx}{\sqrt[3]{x}}$

38. $\displaystyle\int_1^2 \frac{dx}{\sqrt{x-1}}$

39. $\displaystyle\int_0^{\pi/2} \tan\theta\,d\theta$

40. $\displaystyle\int_{-3}^1 \frac{dx}{(2x+6)^{2/3}}$

41. $\displaystyle\int_0^{\pi/2} \sec x\tan x\,dx$

42. $\displaystyle\int_3^4 \frac{dz}{(z-3)^{3/2}}$

43. $\displaystyle\int_0^1 \frac{e^{\sqrt{x}}}{\sqrt{x}}\,dx$

44. $\displaystyle\int_0^{\ln 3} \frac{e^y}{(e^y-1)^{2/3}}\,dy$

45. $\displaystyle\int_0^1 \frac{x^3}{x^4-1}\,dx$

46. $\displaystyle\int_1^\infty \frac{dx}{\sqrt[3]{x-1}}$

47. $\displaystyle\int_0^{10} \frac{dx}{\sqrt[4]{10-x}}$

48. $\displaystyle\int_1^{11} \frac{dx}{(x-3)^{2/3}}$

49. $\displaystyle\int_0^2 \frac{dx}{(x-1)^2}$

50. $\displaystyle\int_0^9 \frac{dx}{(x-1)^{1/3}}$

51. $\displaystyle\int_{-2}^2 \frac{dp}{\sqrt{4-p^2}}$

52. $\displaystyle\int_0^\infty xe^{-x}\,dx$

53. $\displaystyle\int_0^1 \ln x\,dx$

54. $\displaystyle\int_1^\infty \frac{\ln x}{x^2}\,dx$

55. $\displaystyle\int_0^{\ln 2} \frac{e^x}{\sqrt{e^{2x}-1}}\,dx$

56. $\displaystyle\int_0^1 \frac{dx}{x+\sqrt{x}}$

57. $\displaystyle\int_{-\infty}^\infty e^{-|x|}\,dx$

58. $\displaystyle\int_{-\infty}^\infty \frac{dx}{x^2+2x+5}$

59. Perpetual annuity Imagine that today you deposit $\$B$ in a savings account that earns interest at a rate of $p\%$ per year compounded continuously (see Section 7.2). The goal is to draw an income of $\$I$ per year from the account forever. The amount of money that must be deposited is $B = I\int_0^\infty e^{-rt}\,dt$, where $r = p/100$. Suppose you find an account that earns 12% interest annually and you wish to have an income from the account of \$5000 per year. How much must you deposit today?

60. Draining a pool Water is drained from a swimming pool at a rate given by $R(t) = 100e^{-0.05t}$ gal/hr. If the drain is left open indefinitely, how much water drains from the pool?

61. Bioavailability When a drug is given intravenously, the concentration of the drug in the blood is $C_i(t) = 250e^{-0.08t}$, for $t \geq 0$. When the same drug is given orally, the concentration of the drug in the blood is $C_o(t) = 200(e^{-0.08t} - e^{-1.8t})$, for $t \geq 0$. Compute the bioavailability of the drug.

62. Electronic chips Suppose the probability that a particular computer chip fails after a hours of operation is $0.00005\int_a^\infty e^{-0.00005t}\,dt$.

    a. Find the probability that the computer chip fails after 15,000 hr of operation.

    b. Of the chips that are still operating after 15,000 hr, what fraction of these will operate for at least another 15,000 hr?

    c. Evaluate $0.00005\int_0^\infty e^{-0.00005t}\,dt$ and interpret its meaning.

63. Average lifetime The average time until a computer chip fails (see Exercise 62) is $0.00005\int_0^\infty te^{-0.00005t}\,dt$. Find this value.

64. Maximum distance An object moves on a line with velocity
$$v(t) = \frac{10}{(t+1)^2}\ \text{mi/hr, for } t \geq 0, \text{ where } t \text{ is measured in hours.}$$
What is the maximum distance the object can travel?

65–76. Volumes *Find the volume of the described solid of revolution or state that it does not exist.*

65. The region bounded by $f(x) = x^{-2}$ and the x-axis on the interval $[1, \infty)$ is revolved about the x-axis.

66. The region bounded by $f(x) = (x^2+1)^{-1/2}$ and the x-axis on the interval $[2, \infty)$ is revolved about the x-axis.

67. The region bounded by $f(x) = \sqrt{\dfrac{x+1}{x^3}}$ and the x-axis on the interval $[1, \infty)$ is revolved about the x-axis.

68. The region bounded by $f(x) = (x+1)^{-3}$ and the x-axis on the interval $[0, \infty)$ is revolved about the y-axis.

69. The region bounded by $f(x) = \dfrac{1}{\sqrt{x}\ln x}$ and the x-axis on the interval $[2, \infty)$ is revolved about the x-axis.

70. The region bounded by $f(x) = \dfrac{\sqrt{x}}{\sqrt[3]{x^2+1}}$ and the x-axis on the interval $[0, \infty)$ is revolved about the x-axis.

71. The region bounded by $f(x) = (x-1)^{-1/4}$ and the x-axis on the interval $(1, 2]$ is revolved about the x-axis.

72. The region bounded by $f(x) = (x+1)^{-3/2}$ and the x-axis on the interval $(-1, 1]$ is revolved about the line $y = -1$.

73. The region bounded by $f(x) = \tan x$ and the x-axis on the interval $[0, \pi/2)$ is revolved about the x-axis.

74. The region bounded by $f(x) = -\ln x$ and the x-axis on the interval $(0, 1]$ is revolved about the x-axis.

75. The region bounded by $f(x) = (4-x)^{-1/3}$ and the x-axis on the interval $[0, 4)$ is revolved about the y-axis.

76. The region bounded by $f(x) = (x^2-1)^{-1/4}$ and the x-axis on the interval $(1, 2]$ is revolved about the y-axis.

77–86. Comparison Test *Determine whether the following integrals converge or diverge.*

77. $\int_1^\infty \dfrac{dx}{x^3 + 1}$

78. $\int_0^\infty \dfrac{dx}{e^x + x + 1}$

79. $\int_3^\infty \dfrac{dx}{\ln x}$ *(Hint:* $\ln x \le x$.)

80. $\int_2^\infty \dfrac{x^3}{x^4 - x - 1}\,dx$

81. $\int_1^\infty \dfrac{\sin^2 x}{x^2}\,dx$

82. $\int_1^\infty \dfrac{1}{e^x(1 + x^2)}\,dx$

83. $\int_1^\infty \dfrac{2 + \cos x}{\sqrt{x}}\,dx$

84. $\int_1^\infty \dfrac{2 + \cos x}{x^2}\,dx$

85. $\int_0^1 \dfrac{dx}{\sqrt{x^{1/3} + x}}$

86. $\int_0^1 \dfrac{\sin x + 1}{x^5}\,dx$

87. Explain why or why not Determine whether the following statements are true and give an explanation or counterexample.

  a. If f is continuous and $0 < f(x) < g(x)$ on the interval $[0, \infty)$, and $\int_0^\infty g(x)\,dx = M < \infty$, then $\int_0^\infty f(x)\,dx$ exists.

  b. If $\lim\limits_{x \to \infty} f(x) = 1$, then $\int_0^\infty f(x)\,dx$ exists.

  c. If $\int_0^1 x^{-p}\,dx$ exists, then $\int_0^1 x^{-q}\,dx$ exists, where $q > p$.

  d. If $\int_1^\infty x^{-p}\,dx$ exists, then $\int_1^\infty x^{-q}\,dx$ exists, where $q > p$.

  e. $\int_1^\infty \dfrac{dx}{x^{3p+2}}$ exists, for $p > -\dfrac{1}{3}$.

88. Incorrect calculation

  a. What is wrong with this calculation?
  $$\int_{-1}^1 \dfrac{dx}{x} = \ln|x|\,\Big|_{-1}^1 = \ln 1 - \ln 1 = 0$$

  b. Evaluate $\int_{-1}^1 \dfrac{dx}{x}$ or show that the integral does not exist.

89. Area between curves Let R be the region bounded by the graphs of $y = e^{-ax}$ and $y = e^{-bx}$, for $x \ge 0$, where $a > b > 0$. Find the area of R in terms of a and b.

90. Area between curves Let R be the region bounded by the graphs of $y = x^{-p}$ and $y = x^{-q}$, for $x \ge 1$, where $q > p > 1$. Find the area of R in terms of p and q.

T 91. Regions bounded by exponentials Let $a > 0$ and let R be the region bounded by the graph of $y = e^{-ax}$ and the x-axis on the interval $[b, \infty)$.

  a. Find $A(a, b)$, the area of R as a function of a and b.
  b. Find the relationship $b = g(a)$ such that $A(a, b) = 2$.
  c. What is the minimum value of b (call it b^*) such that when $b > b^*$, $A(a, b) = 2$ for some value of $a > 0$?

Explorations and Challenges

92–93. Improper integrals with infinite intervals and unbounded integrands *For a real number a, suppose* $\lim\limits_{x \to a^+} f(x) = -\infty$ *or* $\lim\limits_{x \to a^+} f(x) = \infty$. *In these cases, the integral* $\int_a^\infty f(x)\,dx$ *is improper for two reasons:* ∞ *appears in the upper limit and* f *is unbounded at* $x = a$. *It can be shown that* $\int_a^\infty f(x)\,dx = \int_a^c f(x)\,dx + \int_c^\infty f(x)\,dx$, *for any* $c > a$. *Use this result to evaluate the following improper integrals.*

92. $\int_0^\infty x^{-x}(\ln x + 1)\,dx$

93. $\int_1^\infty \dfrac{dx}{x\sqrt{x - 1}}$

94. The family $f(x) = \dfrac{1}{x^p}$ **revisited** Consider the family of functions $f(x) = \dfrac{1}{x^p}$, where p is a real number. For what values of p does the integral $\int_0^1 f(x)\,dx$ exist? What is its value?

T 95–98. Numerical methods *Use numerical methods or a calculator to approximate the following integrals as closely as possible. The exact value of each integral is given.*

95. $\int_0^{\pi/2} \ln(\sin x)\,dx = \int_0^{\pi/2} \ln(\cos x)\,dx = -\dfrac{\pi \ln 2}{2}$

96. $\int_0^\infty \dfrac{\sin^2 x}{x^2}\,dx = \dfrac{\pi}{2}$

97. $\int_0^\infty \ln\left(\dfrac{e^x + 1}{e^x - 1}\right)dx = \dfrac{\pi^2}{4}$

98. $\int_0^1 \dfrac{\ln x}{1 + x}\,dx = -\dfrac{\pi^2}{12}$

99. Decaying oscillations Let $a > 0$ and b be real numbers. Use integration to confirm the following identities. (See Exercise 73 of Section 8.2)

  a. $\int_0^\infty e^{-ax} \cos bx\,dx = \dfrac{a}{a^2 + b^2}$

  b. $\int_0^\infty e^{-ax} \sin bx\,dx = \dfrac{b}{a^2 + b^2}$

100. The Eiffel Tower property Let R be the region between the curves $y = e^{-cx}$ and $y = -e^{-cx}$ on the interval $[a, \infty)$, where $a \ge 0$ and $c > 0$. The center of mass of R is located at $(\bar{x}, 0)$, where $\bar{x} = \dfrac{\int_a^\infty x e^{-cx}\,dx}{\int_a^\infty e^{-cx}\,dx}$. (The profile of the Eiffel Tower is modeled by the two exponential curves; see the Guided Project *The exponential Eiffel Tower*.)

  a. For $a = 0$ and $c = 2$, sketch the curves that define R and find the center of mass of R. Indicate the location of the center of mass.
  b. With $a = 0$ and $c = 2$, find equations of the lines tangent to the curves at the points corresponding to $x = 0$.
  c. Show that the tangent lines intersect at the center of mass.
  d. Show that this same property holds for any $a \ge 0$ and any $c > 0$; that is, the tangent lines to the curves $y = \pm e^{-cx}$ at $x = a$ intersect at the center of mass of R.

  (*Source:* P. Weidman and I. Pinelis, *Comptes Rendu, Mechanique*, 332, 571–584, 2004)

101. Many methods needed Show that $\int_0^\infty \dfrac{\sqrt{x}\,\ln x}{(1 + x)^2}\,dx = \pi$ in the following steps.

  a. Integrate by parts with $u = \sqrt{x}\,\ln x$.
  b. Change variables by letting $y = 1/x$.
  c. Show that $\int_0^1 \dfrac{\ln x}{\sqrt{x}(1 + x)}\,dx = -\int_1^\infty \dfrac{\ln x}{\sqrt{x}(1 + x)}\,dx$ (and that both integrals converge). Conclude that $\int_0^\infty \dfrac{\ln x}{\sqrt{x}(1 + x)}\,dx = 0$.
  d. Evaluate the remaining integral using the change of variables $z = \sqrt{x}$.

  (*Source: Mathematics Magazine* 59, 1, Feb 1986)

102–106. Laplace transforms *A powerful tool in solving problems in engineering and physics is the Laplace transform. Given a function* $f(t)$*, the Laplace transform is a new function* $F(s)$ *defined by*

$$F(s) = \int_0^\infty e^{-st} f(t)\, dt,$$

where we assume s is a positive real number. For example, to find the Laplace transform of $f(t) = e^{-t}$*, the following improper integral is evaluated using integration by parts:*

$$F(s) = \int_0^\infty e^{-st} e^{-t}\, dt = \int_0^\infty e^{-(s+1)t}\, dt = \frac{1}{s+1}.$$

Verify the following Laplace transforms, where a is a real number.

102. $f(t) = 1 \rightarrow F(s) = \dfrac{1}{s}$ **103.** $f(t) = e^{at} \rightarrow F(s) = \dfrac{1}{s-a}$

104. $f(t) = t \rightarrow F(s) = \dfrac{1}{s^2}$

105. $f(t) = \sin at \rightarrow F(s) = \dfrac{a}{s^2 + a^2}$

106. $f(t) = \cos at \rightarrow F(s) = \dfrac{s}{s^2 + a^2}$

107. Improper integrals Evaluate the following improper integrals (Putnam Exam, 1939).

a. $\displaystyle\int_1^3 \frac{dx}{\sqrt{(x-1)(3-x)}}$ **b.** $\displaystyle\int_1^\infty \frac{dx}{e^{x+1} + e^{3-x}}$

108. Draining a tank Water is drained from a 3000-gal tank at a rate that starts at 100 gal/hr and decreases continuously by 5%/hr. If the drain is left open indefinitely, how much water drains from the tank? Can a full tank be emptied at this rate?

109. Escape velocity and black holes The work required to launch an object from the surface of Earth to outer space is given by $W = \int_R^\infty F(x)\, dx$, where $R = 6370$ km is the approximate radius of Earth, $F(x) = \dfrac{GMm}{x^2}$ is the gravitational force between Earth and the object, G is the gravitational constant, M is the mass of Earth, m is the mass of the object, and $GM = 4 \times 10^{14}\, \mathrm{m^3/s^2}$.

a. Find the work required to launch an object in terms of m.
b. What escape velocity v_e is required to give the object a kinetic energy $\dfrac{1}{2} m v_e^{\,2}$ equal to W?
c. The French scientist Laplace anticipated the existence of black holes in the 18th century with the following argument: If a body has an escape velocity that equals or exceeds the speed

of light, $c = 300{,}000$ km/s, then light cannot escape the body and it cannot be seen. Show that such a body has a radius $R \le 2GM/c^2$. For Earth to be a black hole, what would its radius need to be?

110. Adding a proton to a nucleus The nucleus of an atom is positively charged because it consists of positively charged protons and uncharged neutrons. To bring a free proton toward a nucleus, a repulsive force $F(r) = kqQ/r^2$ must be overcome, where $q = 1.6 \times 10^{-19}$ C (coulombs) is the charge on the proton, $k = 9 \times 10^9\, \mathrm{N \cdot m^2/C^2}$, Q is the charge on the nucleus, and r is the distance between the center of the nucleus and the proton. Find the work required to bring a free proton (assumed to be a point mass) from a large distance $(r \to \infty)$ to the edge of a nucleus that has a charge $Q = 50q$ and a radius of 6×10^{-11} m.

111. Gamma function The gamma function is defined by $\Gamma(p) = \int_0^\infty x^{p-1} e^{-x}\, dx$, for p not equal to zero or a negative integer.

a. Use the reduction formula

$$\int_0^\infty x^p e^{-x}\, dx = p \int_0^\infty x^{p-1} e^{-x}\, dx \quad \text{for } p = 1, 2, 3, \ldots$$

to show that $\Gamma(p+1) = p!$ (p factorial).

b. Use the substitution $x = u^2$ and the fact that $\displaystyle\int_0^\infty e^{-u^2}\, du = \frac{\sqrt{\pi}}{2}$ to show that $\Gamma\left(\dfrac{1}{2}\right) = \sqrt{\pi}$.

T 112. Gaussians An important function in statistics is the Gaussian (or normal distribution, or bell-shaped curve), $f(x) = e^{-ax^2}$.

a. Graph the Gaussian for $a = 0.5, 1$, and 2.

b. Given that $\displaystyle\int_{-\infty}^\infty e^{-ax^2}\, dx = \sqrt{\dfrac{\pi}{a}}$, compute the area under the curves in part (a).

c. Complete the square to evaluate $\displaystyle\int_{-\infty}^\infty e^{-(ax^2 + bx + c)}\, dx$, where $a > 0$, b, and c are real numbers.

QUICK CHECK ANSWERS

1. The integral diverges. $\displaystyle\lim_{b \to \infty} \int_1^b (1 + x^{-1})\, dx =$

$\displaystyle\lim_{b \to \infty} (x + \ln x)\Big|_1^b$ does not exist.   **2.** $\frac{1}{3}$ **3.** c must approach 0 through values in the interval of integration $(0, 1)$. Therefore, $c \to 0^+$.   **4.** $\displaystyle\int_1^\infty e^{-x}\, dx = \lim_{b \to \infty}\left(-e^{-x}\Big|_1^b\right) = e^{-1}$ ◄

CHAPTER 8   REVIEW EXERCISES

1. Explain why or why not Determine whether the following statements are true and give an explanation or counterexample.

a. The integral $\int x^2 e^{2x}\, dx$ can be evaluated using integration by parts.

b. To evaluate the integral $\displaystyle\int \frac{dx}{\sqrt{x^2 - 100}}$ analytically, it is best to use partial fractions.

c. One computer algebra system produces $\int 2 \sin x \cos x\, dx = \sin^2 x$. Another computer algebra system

produces $\int 2 \sin x \cos x\, dx = -\cos^2 x$. One computer algebra system is wrong (apart from a missing constant of integration).

d. $\displaystyle\int 2 \sin x \cos x\, dx = -\frac{1}{2}\cos 2x + C$.

e. The best approach to evaluating $\displaystyle\int \frac{x^3 + 1}{3x^2}\, dx$ is to use the change of variables $u = x^3 + 1$.

2–74. Integration techniques *Use the methods introduced in Sections 8.1 through 8.5 to evaluate the following integrals.*

2. $\displaystyle\int \cos\left(\frac{x}{2} + \frac{\pi}{3}\right) dx$

3. $\displaystyle\int \frac{3x}{\sqrt{x+4}}\, dx$

4. $\displaystyle\int_{-1}^{\ln 2} \frac{3t}{e^t}\, dt$

5. $\displaystyle\int \frac{x}{2\sqrt{x+2}}\, dx$

6. $\displaystyle\int \frac{2 - \sin 2\theta}{\cos^2 2\theta}\, d\theta$

7. $\displaystyle\int_{-2}^{1} \frac{3}{x^2 + 4x + 13}\, dx$

8. $\displaystyle\int_{\pi}^{2\pi} \cot \frac{x}{3}\, dx$

9. $\displaystyle\int_{0}^{\pi/4} \cos^5 2x \sin^2 2x\, dx$

10. $\displaystyle\int \frac{x^3 + 3x^2 + 1}{x^3 + 1}\, dx$

11. $\displaystyle\int \frac{\sqrt{t} - 1}{2t}\, dt$

12. $\displaystyle\int \frac{8x + 5}{2x^2 + 3x + 1}\, dx$

13. $\displaystyle\int_{0}^{\pi} e^{3x} \sin 6x\, dx$

14. $\displaystyle\int_{2}^{3} (6w - 10)e^{3w}\, dw$

15. $\displaystyle\int_{1}^{2} \frac{3x^5 + 48x^3 + 3x^2 + 16}{x^3 + 16x}\, dx$

16. $\displaystyle\int \frac{x^6 + 2x^4 + x^2 + 1}{(x^2 + 1)^2}\, dx$

17. $\displaystyle\int \frac{2x^2 + 7x + 4}{x^3 + 2x^2 + 2x}\, dx$

18. $\displaystyle\int_{0}^{\sqrt{2}} \frac{x + 1}{3x^2 + 6}\, dx$

19. $\displaystyle\int \frac{x^2 + 4x + 7}{(x + 3)^3}\, dx$

20. $\displaystyle\int \frac{\sqrt{1 - x^2}}{x}\, dx$

21. $\displaystyle\int_{\sqrt{2}}^{2} \frac{\sqrt{x^2 - 1}}{x}\, dx$

22. $\displaystyle\int \tan^3 5\theta\, d\theta$

23. $\displaystyle\int \frac{\sin^4 t}{\cos^6 t}\, dt$

24. $\displaystyle\int \frac{dx}{\sqrt{18x - x^2}}$

25. $\displaystyle\int_{-3/2}^{-1} \frac{dx}{4x^2 + 12x + 10}$

26. $\displaystyle\int \tan 10x\, dx$

27. $\displaystyle\int \frac{dw}{(w + 1)^2 \sqrt{w^2 + 2w - 8}}$

28. $\displaystyle\int \frac{\sin^3 x}{\cos^5 x}\, dx$

29. $\displaystyle\int \frac{\cos^4 x}{\sin^6 x}\, dx$

30. $\displaystyle\int x \tan^{-1} 7x\, dx$

31. $\displaystyle\int x \sinh 2x\, dx$

32. $\displaystyle\int \csc^2 6x \cot 6x\, dx$

33. $\displaystyle\int \tan^3 3\theta \sec^3 3\theta\, d\theta$

34. $\displaystyle\int \frac{w^3}{\sqrt{36 - 9w^2}}\, dw$

35. $\displaystyle\int \frac{x^3}{\sqrt{4x^2 + 16}}\, dx$

36. $\displaystyle\int_{-1/2}^{1/2} \frac{u^2 + 1}{u^2 - 1}\, du$

37. $\displaystyle\int \frac{3x^3 + 4x^2 + 6x}{(x + 1)^2(x^2 + 4)}\, dx$

38. $\displaystyle\int_{\pi/4}^{\pi/2} x \csc^2 x\, dx$

39. $\displaystyle\int \frac{dt}{2 + e^t}$

40. $\displaystyle\int \frac{x^2 - 4}{x + 4}\, dx$

41. $\displaystyle\int \frac{d\theta}{1 + \cos 4\theta}$

42. $\displaystyle\int x^2 \cos x\, dx$

43. $\displaystyle\int e^x \sin x\, dx$

44. $\displaystyle\int_{1}^{e} x^2 \ln x\, dx$

45. $\displaystyle\int \frac{2x^3 + 5x^2 + 13x + 9}{x^2(x^2 + 4x + 9)}\, dx$

46. $\displaystyle\int \frac{x^3 + 4x^2 + 12x + 4}{(x^2 + 4x + 10)^2}\, dx$

47. $\displaystyle\int \cos^2 4\theta\, d\theta$

48. $\displaystyle\int \sin 3x \cos^6 3x\, dx$

49. $\displaystyle\int \sec^{49} 2z \tan 2z\, dz$

50. $\displaystyle\int_{0}^{\pi/6} \cos^4 3x\, dx$

51. $\displaystyle\int_{0}^{\pi/4} \sin^5 4\theta\, d\theta$

52. $\displaystyle\int \tan^4 2u\, du$

53. $\displaystyle\int \frac{dx}{\sqrt{x} + \sqrt[3]{x}}$ (*Hint:* Let $u = x^6$.)

54. $\displaystyle\int \frac{dx}{\sqrt{9x^2 - 25}}, x > \frac{5}{3}$

55. $\displaystyle\int \frac{dy}{y^2 \sqrt{18 - 2y^2}}$

56. $\displaystyle\int_{0}^{\sqrt{3}/2} \frac{x^2}{(1 - x^2)^{3/2}}\, dx$

57. $\displaystyle\int_{0}^{\sqrt{3}/2} \frac{4}{9 + 4x^2}\, dx$

58. $\displaystyle\int \frac{(1 - u^2)^{5/2}}{u^8}\, du$

59. $\displaystyle\int \operatorname{sech}^2 x \sinh x\, dx$

60. $\displaystyle\int x^2 \cosh x\, dx$

61. $\displaystyle\int_{0}^{\ln(\sqrt{3}+2)} \frac{\cosh x}{\sqrt{4 - \sinh^2 x}}\, dx$

62. $\displaystyle\int \sinh^{-1} x\, dx$

63. $\displaystyle\int \frac{dx}{x^2 - 2x - 15}$

64. $\displaystyle\int \frac{dx}{x^3 - 2x^2}$

65. $\displaystyle\int_{0}^{1} \frac{dy}{(y + 1)(y^2 + 1)}$

66. $\displaystyle\int_{0}^{\infty} \frac{6x}{1 + x^6}\, dx$

67. $\displaystyle\int_{0}^{2} \frac{dx}{\sqrt[3]{|x - 1|}}$

68. $\displaystyle\int_{-1}^{1} \frac{dx}{x^2 + 2x + 5}$

69. $\displaystyle\int \frac{dx}{x^2 - x - 2}$

70. $\displaystyle\int \frac{3x^2 + x - 3}{x^2 - 1}\, dx$

71. $\displaystyle\int \frac{2x^2 - 4x}{x^2 - 4}\, dx$

72. $\displaystyle\int_{1/12}^{1/4} \frac{dx}{\sqrt{x}(1 + 4x)}$

73. $\displaystyle\int \frac{e^{2t}}{(1 + e^{4t})^{3/2}}\, dt$

74. $\displaystyle\int \frac{dx}{\sqrt{\sqrt{1 + \sqrt{x}}}}$

75. Evaluate the integral in part (a) and then use this result to evaluate the integral in part (b).

   a. $\displaystyle\int e^x \sec e^x \tan e^x\, dx$ **b.** $\displaystyle\int e^{2x} \sec e^x \tan e^x\, dx$

76–81. Table of integrals *Use a table of integrals to evaluate the following integrals.*

76. $\displaystyle\int x(2x + 3)^5\, dx$

77. $\displaystyle\int \frac{dx}{x\sqrt{4x - 6}}$

78. $\displaystyle\int_0^{\pi/2} \frac{d\theta}{1 + \sin 2\theta}$

79. $\displaystyle\int \sec^5 x \, dx$

80. $\displaystyle\int_0^1 \frac{e^x}{e^{2x}\sqrt{16 - e^{2x}}} \, dx$

81. $\displaystyle\int_0^{\pi/2} \cos x \ln^3(1 + \sin x) \, dx$

82–88. Improper integrals *Evaluate the following integrals or show that the integral diverges.*

82. $\displaystyle\int_{-\infty}^{-1} \frac{dx}{(x - 1)^4}$

83. $\displaystyle\int_0^\infty xe^{-x} \, dx$

84. $\displaystyle\int_0^\pi \sec^2 x \, dx$

85. $\displaystyle\int_0^3 \frac{dx}{\sqrt{9 - x^2}}$

86. $\displaystyle\int_{-\infty}^\infty \frac{x^3}{1 + x^8} \, dx$

87. $\displaystyle\int_1^\infty \frac{dx}{x(x - 1)^{1/3}}$ (*Hint:* First write the integral as the sum of two improper integrals.)

88. $\displaystyle\int_{-\infty}^\infty \frac{2}{x^2 + 2x + 2} \, dx$

89–91. Comparison Test *Determine whether the following integrals converge or diverge.*

89. $\displaystyle\int_1^\infty \frac{dx}{x^5 + x^4 + x^3 + 1}$

90. $\displaystyle\int_0^1 \frac{dx}{\sqrt{x + \sin x}}$

91. $\displaystyle\int_3^\infty \frac{x^3}{\sqrt{x^7 - 1}} \, dx$

92. Integral with a parameter For what values of p does the integral $\displaystyle\int_2^\infty \frac{dx}{x \ln^p x}$ converge and what is its value (in terms of p)?

T 93–94. Approximations *Use a computer algebra system to approximate the value of the following integrals.*

93. $\displaystyle\int_{-1}^1 e^{-2x^2} \, dx$

94. $\displaystyle\int_1^{\sqrt{e}} x^3 \ln^3 x \, dx$

T 95–98. Numerical integration *Estimate the following integrals using the Midpoint Rule $M(n)$, the Trapezoidal Rule $T(n)$, and Simpson's Rule $S(n)$ for the given values of n.*

95. $\displaystyle\int_0^8 f(x) \, dx; n = 4$ (see figure)

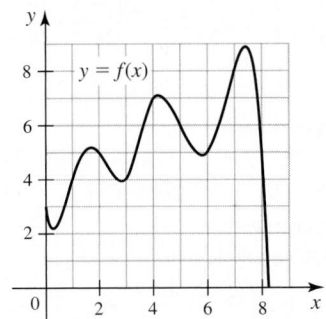

96. $\displaystyle\int_1^3 \frac{dx}{x^3 + x + 1}; n = 4$

97. $\displaystyle\int_0^1 \tan x^2 \, dx; n = 40$

98. $\displaystyle\int_1^8 e^{\sin x} \, dx; n = 60$

99. Improper integrals by numerical methods Use the Trapezoid Rule (Section 8.8) to approximate $\int_0^R e^{-x^2} \, dx$ with $R = 2$, 4, and 8. For each value of R, take $n = 4$, 8, 16, and 32, and compare approximations with successive values of n. Use these approximations to approximate $I = \int_0^\infty e^{-x^2} \, dx$.

100. Comparing areas Show that the area of the region bounded by the graph of $y = ae^{-ax}$ and the x-axis on the interval $[0, \infty)$ is the same for all values of $a > 0$.

101. Comparing volumes Let R be the region bounded by the graph of $y = \sin x$ and the x-axis on the interval $[0, \pi]$. Which is greater, the volume of the solid generated when R is revolved about the x-axis or about the y-axis?

102–105. Volumes *The region R is bounded by the curve $y = \ln x$ and the x-axis on the interval $[1, e]$. Find the volume of the solid that is generated when R is revolved in the following ways.*

102. About the y-axis

103. About the x-axis

104. About the line $y = 1$

105. About the line $x = 1$

106. Arc length Find the length of the curve
$$y = \frac{x}{2}\sqrt{3 - x^2} + \frac{3}{2}\sin^{-1}\frac{x}{\sqrt{3}} \text{ from } x = 0 \text{ to } x = 1.$$

T 107. Zero log integral It is evident from the graph of $y = \ln x$ that for every real number a with $0 < a < 1$, there is a unique real number $b = g(a)$ with $b > 1$, such that $\int_a^b \ln x \, dx = 0$ (the net area bounded by the graph of $y = \ln x$ on $[a, b]$ is 0).

   a. Approximate $b = g\left(\dfrac{1}{2}\right)$.

   b. Approximate $b = g\left(\dfrac{1}{3}\right)$.

   c. Find the equation satisfied by all pairs of numbers (a, b) such that $b = g(a)$.

   d. Is g an increasing or decreasing function of a? Explain.

108. Arc length Find the length of the curve $y = \ln x$ on the interval $[1, e^2]$.

109. Average velocity Find the average velocity of a projectile whose velocity over the interval $0 \le t \le \pi$ is given by $v(t) = 10 \sin 3t$.

110. Comparing distances Suppose two cars started at the same time and place ($t = 0$ and $s = 0$). The velocity of car A (in mi/hr) is given by $u(t) = 40/(t + 1)$ and the velocity of car B (in mi/hr) is given by $v(t) = 40e^{-t/2}$.

   a. After $t = 2$ hr, which car has traveled farther?

   b. After $t = 3$ hr, which car has traveled farther?

   c. If allowed to travel indefinitely ($t \to \infty$), which car will travel a finite distance?

111. Traffic flow When data from a traffic study are fitted to a curve, the flow rate of cars past a point on a highway is approximated by $R(t) = 800te^{-t/2}$ cars/hr. How many cars pass the measuring site during the time interval $0 \le t \le 4$?

T 112. Comparing integrals Graph the functions $f(x) = \pm\dfrac{1}{x^2}$, $g(x) = \dfrac{\cos x}{x^2}$, and $h(x) = \dfrac{\cos^2 x}{x^2}$. Without evaluating integrals and knowing that $\int_1^\infty f(x) \, dx$ has a finite value, determine whether $\int_1^\infty g(x) \, dx$ and $\int_1^\infty h(x) \, dx$ have finite values.

113. A family of logarithm integrals Let $I(p) = \int_1^e \dfrac{\ln x}{x^p} dx$, where p is a real number.

a. Find an expression for $I(p)$, for all real values of p.

b. Evaluate $\lim\limits_{p \to \infty} I(p)$ and $\lim\limits_{p \to -\infty} I(p)$.

c. For what value of p is $I(p) = 1$?

T 114. Arc length of the natural logarithm Consider the curve $y = \ln x$.

a. Find the length of the curve from $x = 1$ to $x = a$ and call it $L(a)$. (*Hint:* The change of variables $u = \sqrt{x^2 + 1}$ allows evaluation by partial fractions.)

b. Graph $L(a)$.

c. As a increases, $L(a)$ increases as what power of a?

T 115. Best approximation Let $I = \int_0^1 \dfrac{x^2 - x}{\ln x} dx$. Use any method you choose to find a good approximation to I. You may use the facts that $\lim\limits_{x \to 0^+} \dfrac{x^2 - x}{\ln x} = 0$ and $\lim\limits_{x \to 1^-} \dfrac{x^2 - x}{\ln x} = 1$.

T 116. Numerical integration Use a calculator to determine the integer n that satisfies $\int_0^{1/2} \dfrac{\ln (1 + 2x)}{x} dx = \dfrac{\pi^2}{n}$.

T 117. Numerical integration Use a calculator to determine the integer n that satisfies $\int_0^1 \dfrac{\sin^{-1} x}{x} dx = \dfrac{\pi \ln 2}{n}$.

118. Two worthy integrals

a. Let $I(a) = \int_0^\infty \dfrac{dx}{(1 + x^a)(1 + x^2)}$, where a is a real number. Evaluate $I(a)$ and show that its value is independent of a. (*Hint:* Split the integral into two integrals over $[0, 1]$ and $[1, \infty)$; then use a change of variables to convert the second integral into an integral over $[0, 1]$.)

b. Let f be any positive continuous function on $[0, \pi/2]$. Evaluate $\int_0^{\pi/2} \dfrac{f(\cos x)}{f(\cos x) + f(\sin x)} dx$. (*Hint:* Use the identity $\cos\left(\dfrac{\pi}{2} - x\right) = \sin x$.)

(*Source: Mathematics Magazine 81, 2, Apr 2008*)

T 119. Comparing volumes Let R be the region bounded by $y = \ln x$, the x-axis, and the line $x = a$, where $a > 1$.

a. Find the volume $V_1(a)$ of the solid generated when R is revolved about the x-axis (as a function of a).

b. Find the volume $V_2(a)$ of the solid generated when R is revolved about the y-axis (as a function of a).

c. Graph V_1 and V_2. For what values of $a > 1$ is $V_1(a) > V_2(a)$?

120. Equal volumes

a. Let R be the region bounded by the graph of $f(x) = x^{-p}$ and the x-axis, for $x \geq 1$. Let V_1 and V_2 be the volumes of the solids generated when R is revolved about the x-axis and the

y-axis, respectively, if they exist. For what values of p (if any) is $V_1 = V_2$?

b. Repeat part (a) on the interval $(0, 1]$.

121. Equal volumes Let R_1 be the region bounded by the graph of $y = e^{-ax}$ and the x-axis on the interval $[0, b]$, where $a > 0$ and $b > 0$. Let R_2 be the region bounded by the graph of $y = e^{-ax}$ and the x-axis on the interval $[b, \infty)$. Let V_1 and V_2 be the volumes of the solids generated when R_1 and R_2 are revolved about the x-axis. Find and graph the relationship between a and b for which $V_1 = V_2$.

122. Comparing areas The region R_1 is bounded by the graph of $y = \tan x$ and the x-axis on the interval $[0, \pi/3]$. The region R_2 is bounded by the graph of $y = \sec x$ and the x-axis on the interval $[0, \pi/6]$. Which region has the greater area?

123. Region between curves Find the area of the region bounded by the graphs of $y = \tan x$ and $y = \sec x$ on the interval $[0, \pi/4]$.

124. Mercator map projection The Mercator map projection was proposed by the Flemish geographer Gerardus Mercator (1512–1594). The stretching factor of the Mercator map as a function of the latitude θ is given by the function

$$G(\theta) = \int_0^\theta \sec x \, dx.$$

Graph G, for $0 \leq \theta < \pi/2$. (See the Guided Project *Mercator projections* for a derivation of this integral.)

125. Wallis products Complete the following steps to prove a well-known formula discovered by the 17th-century English mathematician John Wallis.

a. Use a reduction formula to show that
$$\int_0^\pi \sin^m x \, dx = \frac{m - 1}{m} \int_0^\pi \sin^{m-2} x \, dx,$$
for any integer $m \geq 2$.

b. Show that $\int_0^\pi \sin^{2n} x \, dx = \dfrac{2n - 1}{2n} \cdot \dfrac{2n - 3}{2n - 2} \cdots \dfrac{1}{2} \cdot \pi$, for any integer $n \geq 1$. The product on the right is called a *Wallis product*.

c. Show that $\int_0^\pi \sin^{2n+1} x \, dx = \dfrac{2n}{2n + 1} \cdot \dfrac{2n - 2}{2n - 1} \cdots \dfrac{2}{3} \cdot 2$, for any integer $n \geq 1$. The product on the right is another example of a Wallis product.

d. Use the inequality $0 \leq \sin x \leq 1$ on $[0, \pi]$ to show that $0 \leq \int_0^\pi \sin^{m+1} x \, dx \leq \int_0^\pi \sin^m x \, dx$, for any integer $m \geq 0$.

e. Use part (d) to show that
$$\frac{2n}{2n + 1} \cdot \frac{2n - 2}{2n - 1} \cdots \frac{2}{3} \cdot 2 \leq \frac{2n - 1}{2n} \cdot \frac{2n - 3}{2n - 2} \cdots \frac{1}{2} \cdot \pi$$
$$\leq \frac{2n - 2}{2n - 1} \cdot \frac{2n - 4}{2n - 3} \cdots \frac{2}{3} \cdot 2,$$
for any integer $n \geq 2$.

f. Show that $\dfrac{2n}{2n + 1} \cdot \dfrac{\pi}{2} \leq \dfrac{(2n)^2}{2n + 1} \cdot \dfrac{(2n - 2)^2}{(2n - 1)^2} \cdots \dfrac{2^2}{3^2} \leq \dfrac{\pi}{2}$, for any integer $n \geq 2$.

g. Use part (f) to conclude that
$$\frac{\pi}{2} = \lim_{n \to \infty} \left(\frac{2}{1} \cdot \frac{2}{3} \cdot \frac{4}{3} \cdot \frac{4}{5} \cdots \frac{2n}{2n - 1} \cdot \frac{2n}{2n + 1} \right).$$

This is known as the *Wallis formula*.

Chapter 8 Guided Projects

Applications of the material in this chapter and related topics can be found in the following Guided Projects. For additional information, see the Preface.

• Simpson's Rule

• How long will your iPod last?

• Mercator projections

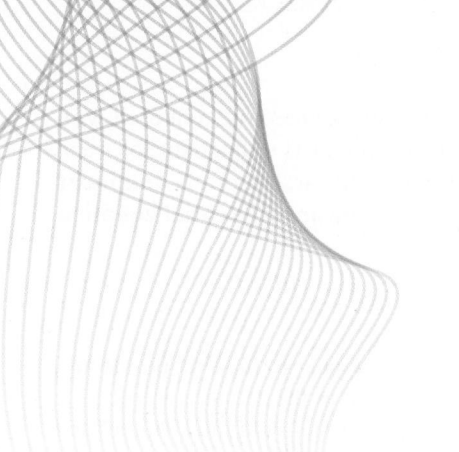

9

Differential Equations

Chapter Preview If you wanted to demonstrate the utility of mathematics to a skeptic, perhaps the most convincing way would be to talk about *differential equations*. This vast subject lies at the heart of mathematical modeling and is used in engineering, physics, chemistry, biology, geophysics, economics and finance, and health sciences. Its many applications in these areas include analyzing the stability of buildings and bridges, simulating planet and satellite orbits, describing chemical reactions, modeling populations and epidemics, predicting weather, locating oil reserves, forecasting financial markets, producing medical images, and simulating drug kinetics. Differential equations rely heavily on calculus and are usually studied in advanced courses that follow calculus. Nevertheless, you have now seen enough calculus to take a brief tour of this rich and powerful subject.

9.1 Basic Ideas

If you studied Section 4.9 or 6.1, then you saw a preview of differential equations. There you learned that if you are given the derivative of a function (for example, a velocity or some other rate of change), you can find the function itself by integration. This process amounts to solving a differential equation.

A differential equation involves an unknown function y and its derivatives. The unknown in a differential equation is not a number (as in an algebraic equation), but rather a *function*. Examples of differential equations are

> ➤ Common choices for the independent variable in a differential equation are x and t, with t being used for time-dependent problems.

$$\text{(A) } \frac{dy}{dx} + 4y = \cos x, \quad \text{(B) } \frac{d^2y}{dx^2} + 16y = 0, \quad \text{and} \quad \text{(C) } y'(t) = 0.1y(100 - y).$$

In each case, the goal is to find functions y that satisfy the equation. Just to be clear about what we mean by a solution, consider equation (B). If we substitute $y = \cos 4x$ and $y'' = -16 \cos 4x$ into this equation, we find that

$$\underbrace{-16 \cos 4x}_{y''} + \underbrace{16 \cos 4x}_{16y} = 0,$$

which implies that $y = \cos 4x$ is a solution of the equation. You should verify that $y = C \cos 4x$ is also a solution, for any real number C (as is $y = C \sin 4x$). Already we see that one differential equation can have many solutions.

Let's begin with a brief discussion of the terminology associated with differential equations. The **order** of a differential equation is the highest order that appears on a derivative in the equation. Of the three differential equations just given, (A) and (C) are first order, and (B) is second order. A differential equation is **linear** if the unknown

▶ A *linear* differential equation cannot have terms such as y^2, yy', or $\sin y$, where y is the unknown function.

▶ To keep matters simple, we will use *general solution* to refer to the most general family of solutions of a differential equation. However, some nonlinear equations may have isolated solutions that are not included in this family of solutions. For example, you should check that for real numbers C, the functions $y = 1/(C - t)$ satisfy the equation $y'(t) = y^2$. Therefore, we call $y = 1/(C - t)$ the general solution of the equation, even though it doesn't include the solution $y = 0$.

function y and its derivatives appear only to the first power and are not composed with other functions. Furthermore, a linear equation cannot have products or quotients of y and its derivatives. Of the equations just given, (A) and (B) are linear, but (C) is **nonlinear** (because the right side contains y^2). The general **first-order linear differential equation** has the form

$$\frac{dy}{dt} + p(t)y = q(t),$$

where p and q are given functions of t. Notice that y and y' appear to the first power and not in products or compositions that involve y or y', which makes the equation linear.

Solving a first-order differential equation requires integration—you must "undo" the derivative y' in order to find y. Integration introduces an arbitrary constant, so the most general solution of a first-order differential equation typically involves one arbitrary constant. Similarly, the most general solution of a second-order differential equation involves two arbitrary constants, and for an nth-order differential equation, the most general solution involves n arbitrary constants. The most general family of functions that solves a differential equation, including the appropriate number of arbitrary constants, is called (not surprisingly) the **general solution**.

A differential equation is often accompanied by **initial conditions** that specify the values of y, and possibly its derivatives, at a particular point. In general, an nth-order equation requires n initial conditions, which can be used to determine the n arbitrary constants in the general solution. A differential equation, together with the appropriate number of initial conditions, is called an **initial value problem**. A typical first-order initial value problem has the form

$$y'(t) = F(t, y) \quad \text{Differential equation}$$
$$y(a) = b, \quad\quad \text{Initial condition}$$

where a and b are given, and F is a given expression that involves t and/or y. A **solution** to the first-order initial value problem $y'(t) = F(t, y)$, $y(a) = b$ is a function $y(t)$ that satisfies the differential equation on an interval I and whose graph includes the point (a, b). We refer to I as the **domain** of the solution.

As we see in upcoming examples, the domain of a solution to an initial value problem is the natural domain of the function y, possibly restricted by the properties of the differential equation, the initial condition(s), or the physical context of the differential equation.

EXAMPLE 1 Verifying solutions

a. Show by substitution that the function $y(t) = Ce^{2.5t}$ is a solution of the differential equation $y'(t) = 2.5y(t)$, where C is an arbitrary constant.

b. Show by substitution that the function $y(t) = 3.2e^{2.5t}$ satisfies the initial value problem

$$y'(t) = 2.5y(t) \quad \text{Differential equation}$$
$$y(0) = 3.2. \quad\quad \text{Initial condition}$$

SOLUTION

a. We differentiate $y(t) = Ce^{2.5t}$ to get $y'(t) = 2.5Ce^{2.5t}$. Substituting into the differential equation, we find that

$$y'(t) = \underbrace{2.5Ce^{2.5t}}_{y'(t)} = 2.5\underbrace{Ce^{2.5t}}_{y(t)} = 2.5y(t).$$

In other words, the function $y(t) = Ce^{2.5t}$ satisfies the equation $y'(t) = 2.5y(t)$, for any value of C. Therefore, $y(t) = Ce^{2.5t}$ is a family of solutions of the differential equation.

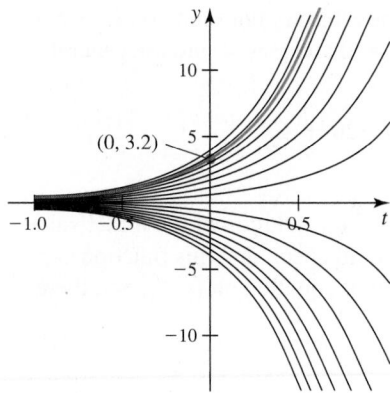

Figure 9.1

b. By part (a) with $C = 3.2$, the function $y(t) = 3.2e^{2.5t}$ satisfies the differential equation $y'(t) = 2.5y(t)$. We can also check that this function satisfies the initial condition $y(0) = 3.2$:

$$y(0) = 3.2e^{2.5 \cdot 0} = 3.2 \cdot e^0 = 3.2.$$

Therefore, $y(t) = 3.2e^{2.5t}$ is a solution of the initial value problem, and its domain is $(-\infty, \infty)$ because $y(t)$ satisfies the differential equation for every real value of t. Figure 9.1 shows the family of curves $y = Ce^{2.5t}$ with several different values of the constant C. It also shows the function $y(t) = 3.2e^{2.5t}$ highlighted in red, which is the solution of the initial value problem.

Related Exercises 7–8 ◄

EXAMPLE 2   General solutions Find the general solution of the following differential equations.

a. $y'(t) = 5 \cos t + 6 \sin 3t$
b. $y''(t) = 10t^3 - 144t^7 + 12t$

SOLUTION

a. The solution of the equation consists of the antiderivatives of $5 \cos t + 6 \sin 3t$. Taking the indefinite integral of both sides of the equation, we have

$$\int y'(t)\, dt = \int (5 \cos t + 6 \sin 3t)\, dt \quad \text{Integrate both sides with respect to } t.$$

$$y(t) = 5 \sin t - 2 \cos 3t + C, \quad \text{Evaluate integrals.}$$

where C is an arbitrary constant. The function $y(t) = 5 \sin t - 2 \cos 3t + C$ is the general solution of the differential equation.

b. In this second-order equation, we are given $y''(t)$ in terms of the independent variable t. Taking the indefinite integral of both sides of the equation yields

$$\int y''(t)\, dt = \int (10t^3 - 144t^7 + 12t)\, dt \quad \text{Integrate both sides with respect to } t.$$

$$y'(t) = \frac{5}{2}t^4 - 18t^8 + 6t^2 + C_1. \quad \text{Evaluate integrals.}$$

Integrating once gives $y'(t)$ and introduces an arbitrary constant that we call C_1. We now integrate again:

$$\int y'(t)\, dt = \int \left(\frac{5}{2}t^4 - 18t^8 + 6t^2 + C_1\right) dt \quad \text{Integrate both sides with respect to } t.$$

$$y(t) = \frac{1}{2}t^5 - 2t^9 + 2t^3 + C_1 t + C_2. \quad \text{Evaluate integrals.}$$

This function, which involves two arbitrary constants, is the general solution of the differential equation.

Related Exercises 22–23 ◄

QUICK CHECK 1 What are the orders of the equations in Example 2? Are they linear or nonlinear? ◄

EXAMPLE 3   An initial value problem Solve the initial value problem

$$y'(t) = 10e^{-t/2}, \quad y(0) = 4.$$

SOLUTION The general solution is found by taking the indefinite integral of both sides of the differential equation with respect to t:

$$\int y'(t)\, dt = \int 10e^{-t/2}\, dt \quad \text{Integrate both sides with respect to } t.$$

$$y(t) = -20e^{-t/2} + C. \quad \text{Evaluate integrals.}$$

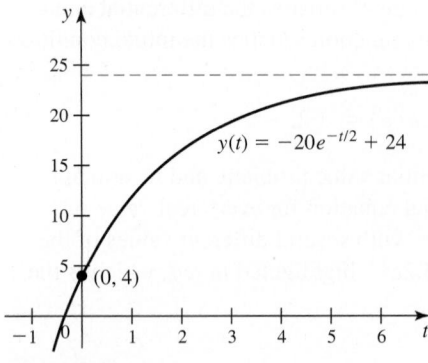

Figure 9.2

QUICK CHECK 2 What is the solution of the initial value problem in Example 3 with the initial condition $y(0) = 16$? ◄

We have found the general solution, which involves one arbitrary constant. To determine its value, we use the initial condition by substituting $t = 0$ and $y = 4$ into the general solution:

$$\underbrace{y(0)}_{4} = -20e^{-0/2} + C = -20 + C,$$

which implies that $4 = -20 + C$ or $C = 24$. Therefore, the solution of the initial value problem is $y(t) = -20e^{-t/2} + 24$ (**Figure 9.2**). You should check that this function satisfies the initial condition and that it satisfies the differential equation on $(-\infty, \infty)$; therefore, the domain of the solution is all real numbers.

Related Exercises 33–34 ◄

In Example 3, the interval on which the solution $y(t) = -20e^{t/2} + 24$ satisfies the differential equation matches the natural domain of the function, which is $(-\infty, \infty)$. In such cases it is not necessary to state the domain of the solution. However, in Example 4, we see that the differential equation and the initial condition may restrict the natural domain of the solution function.

EXAMPLE 4   Determining the domain Solve each initial value problem and determine the domain of the solution.

a. $y'(t) = 1 + \dfrac{2}{\sqrt[3]{t}}$, $y(1) = 4$ **b.** $y'(x) = \csc 2x \cot 2x$, $y(\pi/4) = 0$

SOLUTION

a. Integrating both sides of the differential equation with respect to t gives the general solution:

$$\int y'(t)\, dt = \int (1 + 2t^{-1/3})\, dt \qquad \text{Integrate both sides with respect to } t.$$

$$y(t) = t + 3t^{2/3} + C. \qquad \text{Evaluate integrals.}$$

The initial condition $y(1) = 4$ determines the value of C:

$$y(1) = \underbrace{1 + 3 \cdot 1^{2/3}}_{4} + C = 4 + C = 4,$$

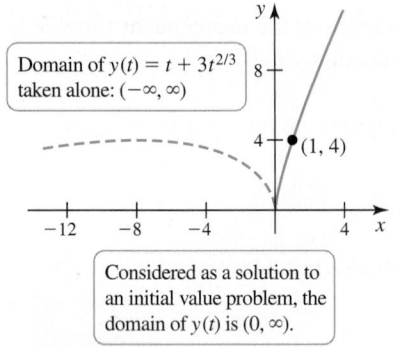

Figure 9.3

QUICK CHECK 3 Solve the initial value problem in Example 4a with an initial condition of $y(-1) = 4$. What is the domain of the solution? ◄

which implies that $C = 0$. Therefore, $y(t) = t + 3t^{2/3}$ is the solution to the initial value problem. When treated as a stand-alone function, the domain of $y(t) = t + 3t^{2/3}$ is all real numbers, as shown in **Figure 9.3**. However, when it is regarded as a solution to the differential equation $y'(t) = 1 + \dfrac{2}{\sqrt[3]{t}}$, the domain of $y(t)$ cannot include $t = 0$ because $y'(t)$ is undefined at 0. Only the right branch of $y(t) = t + 3t^{2/3}$ satisfies both the differential equation and the initial condition $y(1) = 4$, which implies that the domain of the solution to the initial value problem is $(0, \infty)$.

b. To find the general solution, we integrate both sides of the equation with respect to x using the substitution $u = 2x$:

$$\int y'(x)\, dx = \int \csc 2x \cot 2x\, dx \qquad \text{Integrate with respect to } x.$$

$$y(x) = \frac{1}{2}\int \csc u \cot u\, du \qquad u = 2x;\ du = 2\, dx$$

$$= -\frac{1}{2}\csc u + C = -\frac{1}{2}\csc 2x + C. \qquad \text{Evaluate integral.}$$

The initial condition $y(\pi/4) = 0$ implies that

$$y\left(\frac{\pi}{4}\right) = -\frac{1}{2}\csc \underbrace{\frac{\pi}{2}}_{1} + C = 0 \quad \text{or} \quad C = \frac{1}{2},$$

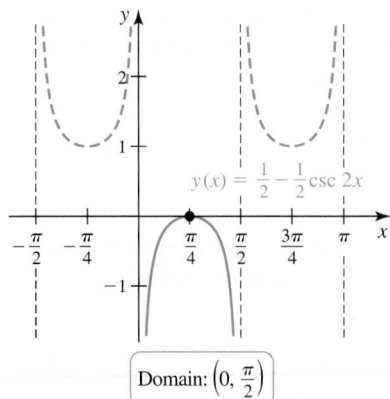

$y(x) = \frac{1}{2} - \frac{1}{2}\csc 2x$

Domain: $\left(0, \frac{\pi}{2}\right)$

Figure 9.4

> The term *initial condition* originates with equations in which the independent variable is *time*. The initial condition $y(0) = a$ gives the initial state of the system. In such cases, the solution is usually graphed only for $t \geq 0$, and the domain of the solution is determined by the physical constraints of the problem.

so the solution to the initial value problem is $y(x) = \dfrac{1}{2} - \dfrac{1}{2}\csc 2x$. The graph of y has infinitely many branches (**Figure 9.4**); we choose the branch that satisfies the initial condition, which implies that the domain of the solution is $(0, \pi/2)$.

Related Exercises 35–36 ◄

Differential Equations in Action

We close this section with three examples of differential equations that are used to model particular physical systems. The first example of one-dimensional motion in a gravitational field was introduced in Example 9 of Section 4.9; it is useful to revisit this problem using the language of differential equations. The equations in Examples 6 and 7 reappear later in the chapter when we show how to solve them. These examples illustrate how the domain of a solution is determined by the physical context of the problem.

EXAMPLE 5   Motion in a gravitational field   A stone is launched vertically upward with a velocity of v_0 m/s from a point s_0 meters above the ground, where $v_0 > 0$ and $s_0 \geq 0$. Assume the stone is launched at time $t = 0$ and that $s(t)$ is the position of the stone at time $t \geq 0$; the positive s-axis points upward with the origin at the ground. By Newton's Second Law of Motion, assuming no air resistance, the position of the stone is governed by the differential equation $s''(t) = -g$, where $g = 9.8$ m/s^2 is the acceleration due to gravity (in the downward direction).

a. Find the position $s(t)$ of the stone for all times at which the stone is above the ground.

b. At what time does the stone reach its highest point and what is its height above the ground?

c. Does the stone go higher if it is launched at $v(0) = v_0 = 39.2$ m/s from the ground $(s_0 = 0)$ or at $v_0 = 19.6$ m/s from a height of $s_0 = 50$ m?

SOLUTION

a. Integrating both sides of the differential equation $s''(t) = -9.8$ gives the velocity $v(t)$:

$$\int s''(t)\, dt = -\int 9.8\, dt \qquad \text{Integrate both sides.}$$

$$s'(t) = v(t) = -9.8t + C_1. \quad \text{Evaluate integrals.}$$

To evaluate the constant C_1, we use the initial condition $v(0) = v_0$, finding that $v(0) = -9.8 \cdot 0 + C_1 = C_1 = v_0$. Therefore, $C_1 = v_0$ and the velocity is $v(t) = s'(t) = -9.8t + v_0$.

Integrating both sides of this velocity equation gives the position function:

$$\int s'(t)\, dt = \int (-9.8t + v_0)\, dt \qquad \text{Integrate both sides.}$$

$$s(t) = -4.9t^2 + v_0 t + C_2. \quad \text{Evaluate integrals.}$$

We now use the initial condition $s(0) = s_0$ to evaluate C_2, finding that

$$s(0) = -4.9 \cdot 0^2 + v_0 \cdot 0 + C_2 = C_2 = s_0.$$

Therefore, $C_2 = s_0$, and the position function is $s(t) = -4.9t^2 + v_0 t + s_0$, where v_0 and s_0 are given. This function is valid while the stone is in flight. To find the domain of the solution, which depends on v_0 and s_0, we solve $s(t) \geq 0$.

> To find the time at which the stone reaches its highest point, we could instead locate the local maximum of the position function, which also requires solving $s'(t) = v(t) = 0$.

b. The stone reaches its highest point when $v(t) = 0$. Solving $v(t) = -9.8t + v_0 = 0$, we find that the stone reaches its highest point when $t = v_0/9.8$, measured in seconds. So the position at the highest point is

$$s_{\max} = s\left(\frac{v_0}{9.8}\right) = -4.9\left(\frac{v_0}{9.8}\right)^2 + v_0\left(\frac{v_0}{9.8}\right) + s_0 = \frac{v_0{}^2}{19.6} + s_0.$$

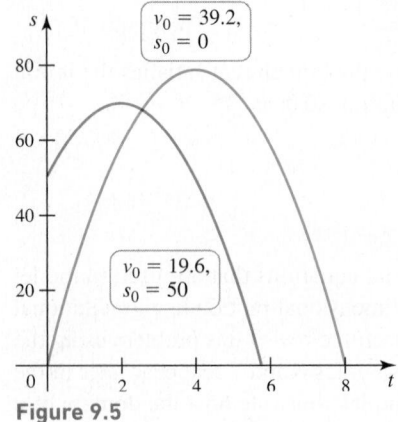

Figure 9.5

▶ The curves in Figure 9.5 are not the trajectories of the stones. The motion is one-dimensional because the stones travel along a vertical line.

c. Now it is a matter of substituting the given values of s_0 and v_0. In the first case, with $v_0 = 39.2$ and $s_0 = 0$, we have $s_{max} = 78.4$ m. In the second case, with $v_0 = 19.6$ and $s_0 = 50$, we have $s_{max} = 69.6$ m. The position functions in the two cases are shown in Figure 9.5. We see that the stone goes higher with $v_0 = 39.2$ and $s_0 = 0$.

Related Exercises 43–44 ◀

QUICK CHECK 4 Suppose the initial conditions in Example 5a are $v_0 = 14.7$ m/s and $s_0 = 49$ m. Write the position function $s(t)$, and state its domain. At what time will the stone reach its maximum height? What is the maximum height at that time? ◀

EXAMPLE 6 **A harvesting model** A simple model of a harvested resource (for example, timber or fish) assumes a competition between the harvesting and the natural growth of the resource. This process may be described by the differential equation

$$\underbrace{p'(t)}_{\substack{\text{rate of}\\\text{change}}} = \underbrace{rp(t)}_{\substack{\text{natural}\\\text{growth rate}}} - \underbrace{H}_{\text{harvesting}}, \quad \text{for} \quad t \ge 0,$$

where $p(t)$ is the amount (or population) of the resource at time $t \ge 0$, $r > 0$ is the natural growth rate of the resource, and $H > 0$ is the constant harvesting rate. An initial condition $p(0) = p_0$ is also specified to create an initial value problem. Notice that the rate of change $p'(t)$ has a positive contribution from the natural growth rate and a negative contribution from the harvesting term.

a. For given constants p_0, r, and H, verify that the function

$$p(t) = \left(p_0 - \frac{H}{r}\right)e^{rt} + \frac{H}{r}$$

is a solution of the initial value problem.

b. Let $p_0 = 1000$ and $r = 0.1$. Graph the solutions for the harvesting rates $H = 50, 90, 130,$ and 170. Describe and interpret the four curves.

c. What value of H gives a constant value of p, for all $t \ge 0$?

SOLUTION

a. Differentiating the given solution, we find that

$$p'(t) = \left(p_0 - \frac{H}{r}\right)re^{rt} = (rp_0 - H)e^{rt}.$$

Simplifying the right side of the differential equation, we find that

$$rp(t) - H = r\underbrace{\left(\left(p_0 - \frac{H}{r}\right)e^{rt} + \frac{H}{r}\right)}_{p(t)} - H = (rp_0 - H)e^{rt}.$$

Therefore, the left and right sides of the equation $p'(t) = rp(t) - H$ are equal, so the equation is satisfied by the given function. You can verify that the given solution also satisfies $p(0) = p_0$, which means p satisfies the initial value problem.

b. Letting $p_0 = 1000$ and $r = 0.1$, the function

$$p(t) = (1000 - 10H)e^{0.1t} + 10H$$

is graphed in Figure 9.6, for $H = 50, 90, 130,$ and 170. We see that for small values of H ($H = 50$ and $H = 90$), the amount of the resource increases with time. On the other hand, for large values of H ($H = 130$ and $H = 170$), the amount of the resource decreases with time, eventually reaching zero. The model predicts that if the harvesting rate is too large, the resource will eventually disappear. The domain of each solution depends on p_0, r, and H. For example, with $p_0 = 1000$, $r = 0.1$, and $H = 170$, we find that the domain is approximately $[0, 8.9]$ (bottom curve in Figure 9.6).

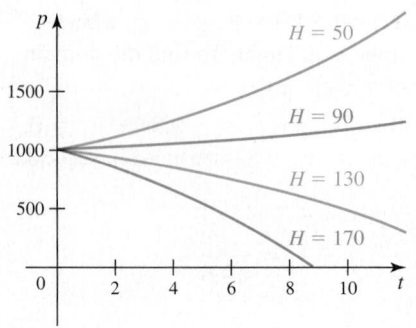

Figure 9.6

c. The solution

$$p(t) = (1000 - 10H)e^{0.1t} + 10H$$

is constant (independent of t) when $1000 - 10H = 0$, which implies $H = 100$. In this case, the solution is

$$p(t) = \underbrace{(1000 - 10H)}_{0} e^{0.1t} + 10H = 1000.$$

Therefore, if the harvesting rate is $H = 100$, then the harvesting exactly balances the natural growth of the resource, and p is constant. This solution is called an *equilibrium solution*. For $H > 100$, the amount of resource decreases in time, and for $H < 100$, it increases in time.

Related Exercises 45–46 ◄

▶ Evangelista Torricelli was an Italian mathematician and physicist who lived from 1608 to 1647. He is credited with inventing the barometer.

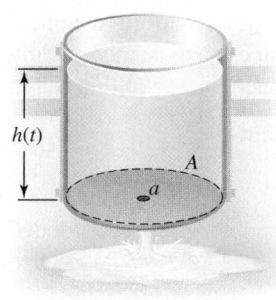

Figure 9.7

EXAMPLE 7   Flow from a tank Imagine a large cylindrical tank with cross-sectional area A. The bottom of the tank has a circular drain with cross-sectional area a. Assume the tank is initially filled with water to a height (in meters) of $h(0) = H$ (Figure 9.7). According to Torricelli's law, the height of the water t seconds after the drain is opened is described by the differential equation

$$h'(t) = -k\sqrt{h}, \quad \text{where } t \geq 0, k = \frac{a}{A}\sqrt{2g},$$

and $g = 9.8 \text{ m/s}^2$ is the acceleration due to gravity.

a. According to the differential equation, is h an increasing or decreasing function of t, for $t \geq 0$?

b. Verify by substitution that the solution of the initial value problem is

$$h(t) = \left(\sqrt{H} - \frac{kt}{2}\right)^2.$$

c. Graph the solution for $H = 1.44$ m, $A = 1 \text{ m}^2$, and $a = 0.05 \text{ m}^2$.

d. After how many seconds is the tank in part (c) empty?

SOLUTION

a. Because $k > 0$, the differential equation implies that $h'(t) < 0$, for $t \geq 0$. Therefore, the height of the water decreases in time, consistent with the fact that the tank is being drained.

b. We first check the initial condition. Substituting $t = 0$ into the proposed solution, we see that

$$h(0) = \left(\sqrt{H} - \frac{k \cdot 0}{2}\right)^2 = (\sqrt{H})^2 = H.$$

Differentiating the proposed solution, we have

$$h'(t) = 2\underbrace{\left(\sqrt{H} - \frac{kt}{2}\right)}_{\sqrt{h(t)}}\left(-\frac{k}{2}\right) = -k\sqrt{h}.$$

Therefore, h satisfies the initial condition and the differential equation.

c. With the given values of the parameters,

$$k = \frac{a}{A}\sqrt{2g} = \frac{0.05 \text{ m}^2}{1 \text{ m}^2}\sqrt{2 \cdot 9.8 \text{ m/s}^2} \approx 0.22 \text{ m}^{1/2}/\text{s},$$

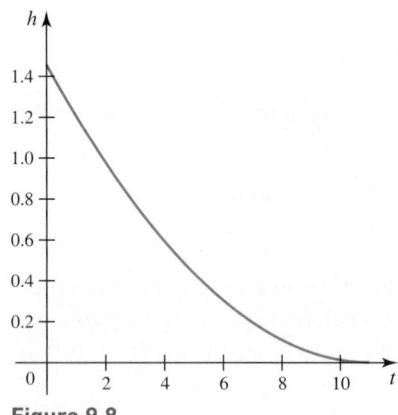

Figure 9.8

and the solution becomes

$$h(t) = \left(\sqrt{H} - \frac{kt}{2}\right)^2 \approx (\sqrt{1.44} - 0.11t)^2 = (1.2 - 0.11t)^2.$$

The graph of the solution (**Figure 9.8**) shows the height of the water decreasing from $h(0) = 1.44$ to zero at approximately $t = 11$ s.

d. Solving the equation

$$h(t) = (1.2 - 0.11t)^2 = 0,$$

we find that the tank is empty at $t \approx 10.9$ s. This calculation determines the domain of the solution, which is approximately $[0, 10.9]$.

Related Exercises 47–48 ◄

QUICK CHECK 5 In Example 7, if the height function were given by $h(t) = (4.2 - 0.14t)^2$, at what time would the tank be empty? What does your answer say about the domain of this solution? ◄

Final Note   Throughout this section, we found solutions to initial value problems without worrying about whether there might be other solutions. Once we find a solution to an initial value problem, how can we be sure there aren't other solutions? More generally, given a particular initial value problem, how do we know whether a solution exists and whether it is unique?

These theoretical questions have answers, and they are provided by powerful *existence and uniqueness theorems*. These theorems and their proofs are quite technical and are handled in advanced courses. Here is an informal statement of an existence and uniqueness theorem for a particular class of initial value problems encountered in this chapter:

The solution of the general first-order initial value problem

$$y'(t) = F(t, y), y(a) = b$$

exists and is unique in some region that contains the point (a, b) provided F is a "well-behaved" function in that region.

The technical challenges arise in defining *well-behaved* in the most general way possible. The initial value problems we consider in this chapter satisfy the conditions of this theorem, and can be assumed to have unique solutions.

SECTION 9.1 EXERCISES

Getting Started

1. Consider the differential equation $y'(t) + 9y(t) = 10$.

   a. How many arbitrary constants appear in the general solution of the differential equation?

   b. Is the differential equation linear or nonlinear?

2. If the general solution of a differential equation is $y(t) = Ce^{-3t} + 10$, what is the solution that satisfies the initial condition $y(0) = 5$?

3. Does the function $y(t) = 2t$ satisfy the differential equation $y'''(t) + y'(t) = 2$?

4. Does the function $y(t) = 6e^{-3t}$ satisfy the initial value problem $y'(t) - 3y(t) = 0, y(0) = 6$?

5. The solution to the initial value problem $y'(t) = 2 \sec t \tan t$, $y(\pi) = -2$ is $y(t) = 2 \sec t$. What is the domain of this solution? (*Hint:* Sketch a graph of y).

6. Explain why the graph of the solution to the initial value problem $y'(t) = \dfrac{t^2}{1 - t}, y(-1) = \ln 2$ cannot cross the line $t = 1$.

Practice Exercises

7–16. Verifying general solutions *Verify that the given function is a solution of the differential equation that follows it. Assume C, C_1, C_2, and C_3 are arbitrary constants.*

7. $y(t) = Ce^{-5t}$; $y'(t) + 5y(t) = 0$

8. $y(t) = Ct^3$; $ty'(t) - 3y(t) = 0$

9. $y(t) = C_1 \sin 4t + C_2 \cos 4t$; $y''(t) + 16y(t) = 0$

10. $y(x) = C_1 e^{-x} + C_2 e^x$; $y''(x) - y(x) = 0$

11. $u(t) = Ce^{1/(4t^4)}$; $u'(t) + \dfrac{1}{t^5} u(t) = 0$

12. $u(t) = C_1 e^t + C_2 t e^t$; $u''(t) - 2u'(t) + u(t) = 0$

13. $g(x) = C_1 e^{-2x} + C_2 x e^{-2x} + 2$; $g''(x) + 4g'(x) + 4g(x) = 8$

14. $u(t) = C_1 t^2 + C_2 t^3$; $t^2 u''(t) - 4tu'(t) + 6u(t) = 0$

15. $u(t) = C_1 t^5 + C_2 t^{-4} - t^3$; $t^2 u''(t) - 20u(t) = 14t^3$

16. $z(t) = C_1 e^{-t} + C_2 e^{2t} + C_3 e^{-3t} - e^t$; $z'''(t) + 2z''(t) - 5z'(t) - 6z(t) = 8e^t$

17–20. Verifying solutions of initial value problems *Verify that the given function y is a solution of the initial value problem that follows it.*

17. $y(t) = 16e^{2t} - 10; y'(t) - 2y(t) = 20, y(0) = 6$

18. $y(t) = 8t^6 - 3; ty'(t) - 6y(t) = 18, y(1) = 5$

19. $y(t) = -3 \cos 3t; y''(t) + 9y(t) = 0, y(0) = -3, y'(0) = 0$

20. $y(x) = \dfrac{1}{4}(e^{2x} - e^{-2x}); y''(x) - 4y(x) = 0, y(0) = 0, y'(0) = 1$

21–32. Finding general solutions *Find the general solution of each differential equation. Use* $C, C_1, C_2, \ldots$ *to denote arbitrary constants.*

21. $y'(t) = 3 + e^{-2t}$

22. $y'(t) = 12t^5 - 20t^4 + 2 - 6t^{-2}$

23. $y'(x) = 4 \tan 2x - 3 \cos x$

24. $p'(x) = \dfrac{16}{x^9} - 5 + 14x^6$

25. $y''(t) = 60t^4 - 4 + 12t^{-3}$

26. $y''(t) = 15e^{3t} + \sin 4t$

27. $u''(x) = 55x^9 + 36x^7 - 21x^5 + 10x^{-3}$

28. $v''(x) = \dfrac{5}{x^2}$

29. $u'(x) = \dfrac{2(x-1)}{x^2 + 4}$

30. $y'(t) = t \ln t + 1$

31. $y''(x) = \dfrac{x}{(1 - x^2)^{3/2}}$

32. $v'(t) = \dfrac{4}{t^2 - 4}$

33–42. Solving initial value problems *Solve the following initial value problems.*

33. $y'(t) = 1 + e^t, y(0) = 4$

34. $y'(t) = \sin t + \cos 2t, y(0) = 4$

35. $y'(x) = 3x^2 - 3x^{-4}, y(1) = 0$

36. $y'(x) = 4 \sec^2 2x, y(0) = 8$

37. $y''(t) = 12t - 20t^3, y(0) = 1, y'(0) = 0$

38. $u''(x) = 4e^{2x} - 8e^{-2x}, u(0) = 1, u'(0) = 3$

39. $y''(t) = te^t, y(0) = 0, y'(0) = 1$

40. $y'(t) = te^t, y(0) = -1$

41. $u'(x) = \dfrac{1}{x^2 + 16} - 4, u(0) = 2$

42. $p'(x) = \dfrac{2}{x^2 + x}, p(1) = 0$

43–44. Motion in a gravitational field *An object is fired vertically upward with an initial velocity* $v(0) = v_0$ *from an initial position* $s(0) = s_0$.

a. *For the following values of* v_0 *and* s_0, *find the position and velocity functions for all times at which the object is above the ground* $(s = 0)$.

b. *Find the time at which the highest point of the trajectory is reached and the height of the object at that time.*

43. $v_0 = 29.4 \text{ m/s}, s_0 = 30 \text{ m}$

44. $v_0 = 49 \text{ m/s}, s_0 = 60 \text{ m}$

45–46. Harvesting problems *Consider the harvesting problem in Example 6.*

45. If $r = 0.05$ and $p_0 = 1500$, for what values of H is the amount of the resource increasing? For what value of H is the amount of the resource constant? If $H = 100$, when does the resource vanish?

46. If $r = 0.05$ and $H = 500$, for what values of p_0 is the amount of the resource decreasing? For what value of p_0 is the amount of the resource constant? If $p_0 = 9000$, when does the resource vanish?

T 47–48. Draining tanks *Consider the tank problem in Example 7. For the following parameter values, find the water height function. Then determine the approximate time at which the tank is first empty and graph the solution.*

47. $H = 1.96 \text{ m}, A = 1.5 \text{ m}^2, a = 0.3 \text{ m}^2$

48. $H = 2.25 \text{ m}, A = 2 \text{ m}^2, a = 0.5 \text{ m}^2$

49. Explain why or why not Determine whether the following statements are true and give an explanation or counterexample.

a. The general solution of the differential equation $y'(t) = 1$ is $y(t) = t$.

b. The differential equation $y''(t) - y(t)y'(t) = 0$ is second order and linear.

c. To find the solution of an initial value problem, we usually begin by finding a general solution of the differential equation.

Explorations and Challenges

50. A second-order equation Consider the differential equation $y''(t) - k^2y(t) = 0$, where $k > 0$ is a real number.

a. Verify by substitution that when $k = 1$, a solution of the equation is $y(t) = C_1e^t + C_2e^{-t}$. You may assume this function is the general solution.

b. Verify by substitution that when $k = 2$, the general solution of the equation is $y(t) = C_1e^{2t} + C_2e^{-2t}$.

c. Give the general solution of the equation for arbitrary $k > 0$ and verify your conjecture.

d. For a positive real number k, verify that the general solution of the equation may also be expressed in the form $y(t) = C_1 \cosh kt + C_2 \sinh kt$, where cosh and sinh are the hyperbolic cosine and hyperbolic sine, respectively (Section 7.3).

51. Another second-order equation Consider the differential equation $y''(t) + k^2y(t) = 0$, where k is a positive real number.

a. Verify by substitution that when $k = 1$, a solution of the equation is $y(t) = C_1 \sin t + C_2 \cos t$. You may assume this function is the general solution.

b. Verify by substitution that when $k = 2$, the general solution of the equation is $y(t) = C_1 \sin 2t + C_2 \cos 2t$.

c. Give the general solution of the equation for arbitrary $k > 0$ and verify your conjecture.

52–56. *In this section, several models are presented and the solution of the associated differential equation is given. Later in the chapter, we present methods for solving these differential equations.*

T 52. Drug infusion The delivery of a drug (such as an antibiotic) through an intravenous line may be modeled by the differential equation $m'(t) + km(t) = I$, where $m(t)$ is the mass of the drug in the blood at time $t \geq 0$, k is a constant that describes the rate at which the drug is absorbed, and I is the infusion rate.

a. Show by substitution that if the initial mass of drug in the blood is zero ($m(0) = 0$), then the solution of the initial value problem is $m(t) = \dfrac{I}{k}(1 - e^{-kt})$.

b. Graph the solution for $I = 10$ mg/hr and $k = 0.05$ hr^{-1}.

c. Evaluate $\lim\limits_{t \to \infty} m(t)$, the steady-state drug level, and verify the result using the graph in part (b).

T 53. Logistic population growth Widely used models for population growth involve the *logistic equation* $P'(t) = rP\left(1 - \dfrac{P}{K}\right)$, where $P(t)$ is the population, for $t \geq 0$, and $r > 0$ and $K > 0$ are given constants.

a. Verify by substitution that the general solution of the equation is $P(t) = \dfrac{K}{1 + Ce^{-rt}}$, where C is an arbitrary constant.

b. Find the value of C that corresponds to the initial condition $P(0) = 50$.

c. Graph the solution for $P(0) = 50$, $r = 0.1$, and $K = 300$.

d. Find $\lim\limits_{t \to \infty} P(t)$ and check that the result is consistent with the graph in part (c).

T 54. Free fall One possible model that describes the free fall of an object in a gravitational field subject to air resistance uses the equation $v'(t) = g - bv$, where $v(t)$ is the velocity of the object for $t \geq 0$, $g = 9.8$ m/s^2 is the acceleration due to gravity, and $b > 0$ is a constant that involves the mass of the object and the air resistance.

a. Verify by substitution that a solution of the equation, subject to the initial condition $v(0) = 0$, is $v(t) = \dfrac{g}{b}(1 - e^{-bt})$.

b. Graph the solution with $b = 0.1$ s^{-1}.

c. Using the graph in part (b), estimate the terminal velocity $\lim\limits_{t \to \infty} v(t)$.

T 55. Chemical rate equations The reaction of certain chemical compounds can be modeled using a differential equation of the form

$y'(t) = -ky^n(t)$, where $y(t)$ is the concentration of the compound, for $t \geq 0$, $k > 0$ is a constant that determines the speed of the reaction, and n is a positive integer called the *order* of the reaction. Assume the initial concentration of the compound is $y(0) = y_0 > 0$.

a. Consider a first-order reaction ($n = 1$) and show that the solution of the initial value problem is $y(t) = y_0 e^{-kt}$.

b. Consider a second-order reaction ($n = 2$) and show that the solution of the initial value problem is $y(t) = \dfrac{y_0}{y_0 kt + 1}$.

c. Let $y_0 = 1$ and $k = 0.1$. Graph the first-order and second-order solutions found in parts (a) and (b). Compare the two reactions.

T 56. Tumor growth The growth of cancer tumors may be modeled by the Gompertz growth equation. Let $M(t)$ be the mass of a tumor, for $t \geq 0$. The relevant initial value problem is

$$\frac{dM}{dt} = -rM(t)\ln\left(\frac{M(t)}{K}\right), \quad M(0) = M_0,$$

where r and K are positive constants and $0 < M_0 < K$.

a. Show by substitution that the solution of the initial value problem is

$$M(t) = K\left(\frac{M_0}{K}\right)^{\exp(-rt)}.$$

b. Graph the solution for $M_0 = 100$ and $r = 0.05$.

c. Using the graph in part (b), estimate $\lim\limits_{t \to \infty} M(t)$, the limiting size of the tumor.

QUICK CHECK ANSWERS

1. The first equation is first order and linear. The second equation is second order and linear.   2. $y(t) = -20e^{-t/2} + 36$
3. $y(t) = t + 3t^{2/3} + 2$; domain is $(-\infty, 0)$
4. $s(t) = -4.9t^2 + 14.7t + 49$; domain is $[0, 5]$; stone reaches maximum height of 60.025 m at $t = 1.5$ s
5. The tank is empty at $t = 30$ s; the domain is $[0, 30]$. ◄

9.2 Direction Fields and Euler's Method

The goal of this chapter is to present methods for finding solutions of various kinds of differential equations. However, before taking up that task, we spend a few pages investigating a remarkable fact: It is possible to visualize and draw approximate graphs of the solutions of a differential equation without ever solving the equation. You might wonder how one can graph a function without knowing a formula for it. It turns out that the differential equation itself contains enough information to draw accurate graphs of its solutions. The tool that makes this visualization possible and allows us to explore the geometry of a differential equation is called the *direction field* (or *slope field*).

Direction Fields

We work with first-order differential equations of the form

$$\frac{dy}{dt} = f(t, y),$$

where the notation $f(t, y)$ means an expression involving the independent variable t and/or the unknown solution y. If a solution of this equation is displayed in the ty-plane, then the

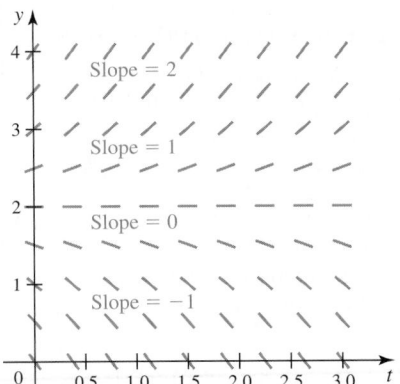

Figure 9.9

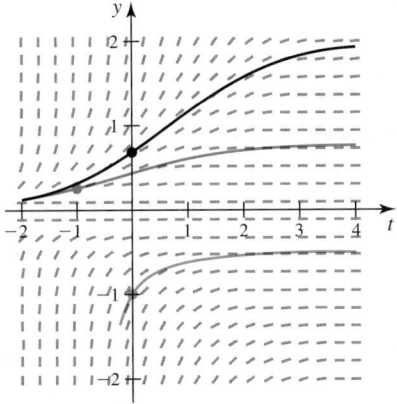

Figure 9.10

▶ If the function f in the differential equation is even slightly complicated, drawing the direction field by hand is tedious. It's best to use a calculator or software. Examples 1 and 2 show some basic steps in plotting fairly simple direction fields by hand.

QUICK CHECK 1 Assuming solutions are unique (at most one solution curve passes through each point), explain why a solution curve cannot cross the line $y = 2$ in Example 1. ◀

▶ A differential equation in which the function f is independent of t is said to be **autonomous**.

differential equation simply says that at each point (t, y) of the solution curve, the slope of the curve is $y'(t) = f(t, y)$ (Figure 9.9). A **direction field** is a picture that shows the slope of the solution at selected points in the ty-plane.

For example, consider the equation $y'(t) = f(t, y) = y^2 e^{-t}$. We choose a regular grid of points in the ty-plane and at each point $P(t, y)$ we make a small line segment with slope $y^2 e^{-t}$. The line segment at a point P gives the slope of the solution curve that passes through P (Figure 9.10). For instance, we see that along the t-axis $(y = 0)$, the slopes of the line segments are $f(t, 0) = 0$, which means the line segments are horizontal. Along the y-axis $(t = 0)$, the slopes of the line segments are $f(0, y) = y^2$, which means the slopes of the line segments increase as we move up or down the y-axis.

Now suppose an initial condition $y(0) = \frac{2}{3}$ is given. We start at the point $(0, \frac{2}{3})$ in the ty-plane and sketch a curve that follows the flow of the direction field (black curve in Figure 9.10). At each point of the solution curve, the slope matches the direction field. Different initial conditions ($y(-1) = \frac{1}{3}$ and $y(0) = -1$ in Figure 9.10) give different solution curves. The collection of solution curves for several different initial conditions is a representation of the general solution of the equation.

EXAMPLE 1   Direction field for a linear differential equation Figure 9.11 shows the direction field for the equation $y'(t) = y - 2$, for $t \geq 0$ and $y \geq 0$. For what initial conditions at $t = 0$ are the solutions constant? Increasing? Decreasing?

SOLUTION The direction field has horizontal line segments (slope zero) for $y = 2$. Therefore, $y'(t) = 0$ when $y = 2$, for all $t \geq 0$. These horizontal line segments correspond to a solution that is constant in time; that is, if the initial condition is $y(0) = 2$, then the solution is $y(t) = 2$, for all $t \geq 0$.

We also see that the direction field has line segments with positive slopes above the line $y = 2$ (with increasing slopes as you move away from $y = 2$). Therefore, $y'(t) > 0$ when $y > 2$, and solutions are increasing in this region.

Similarly, the direction field has line segments with negative slopes below the line $y = 2$ (with increasingly negative slopes as you move away from $y = 2$). Therefore, $y'(t) < 0$ when $y < 2$, and solutions are decreasing in this region.

Combining these observations, we see that if the initial condition satisfies $y(0) > 2$, the resulting solution is increasing, for $t \geq 0$. If the initial condition satisfies $y(0) < 2$, the resulting solution is decreasing, for $t \geq 0$. Figure 9.12 shows the solution curves with initial conditions $y(0) = 2.25$, $y(0) = 2$, and $y(0) = 1.75$.

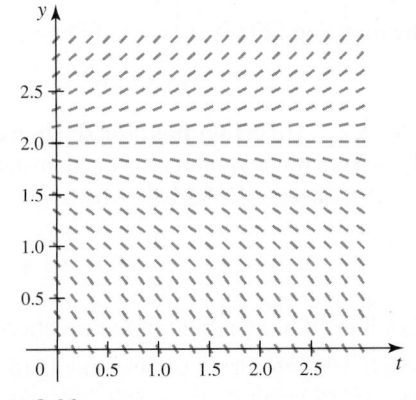

Figure 9.11

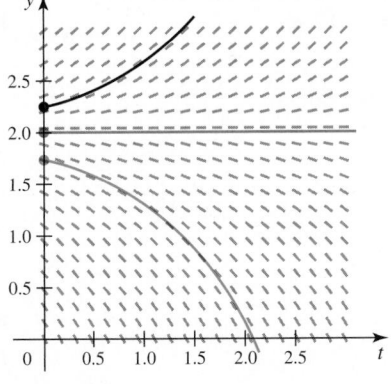

Figure 9.12

Related Exercises 12, 17 ◀

For a differential equation of the form $y'(t) = f(y)$ (that is, the function f depends only on y), the following steps are useful in sketching the direction field. Notice that because the direction field depends only on y, it has the same slope on any given horizontal line. A detailed direction field is usually not required.

PROCEDURE   Sketching a Direction Field by Hand for $y'(t) = f(y)$

1. Find the values of y for which $f(y) = 0$. For example, suppose $f(a) = 0$. Then we have $y'(t) = 0$ whenever $y = a$, and the direction field at all points (t, a) consists of horizontal line segments. If the initial condition is $y(0) = a$, then the solution is $y(t) = a$, for all $t \geq 0$. Such a constant solution is called an **equilibrium solution**.

2. Find the values of y for which $f(y) > 0$. For example, suppose $f(b) > 0$. Then $y'(t) > 0$ whenever $y = b$. It follows that the direction field at all points (t, b) has line segments with positive slopes, and the solution is increasing at those points.

3. Find the values of y for which $f(y) < 0$. For example, suppose $f(c) < 0$. Then $y'(t) < 0$ whenever $y = c$. It follows that the direction field at all points (t, c) has line segments with negative slopes and the solution is decreasing at those points.

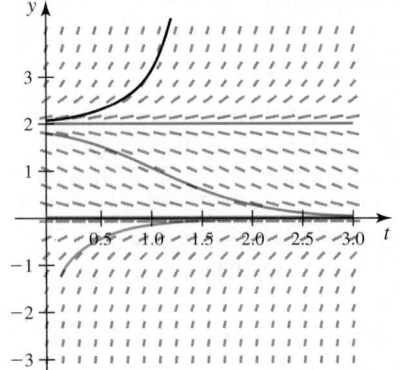

Figure 9.13

QUICK CHECK 2 In Example 2, is the solution to the equation increasing or decreasing if the initial condition is $y(0) = 2.01$? Is it increasing or decreasing if the initial condition is $y(1) = -1$? ◄

EXAMPLE 2   Direction field for a simple nonlinear equation Consider the differential equation $y'(t) = y(y - 2)$, for $t \geq 0$.

a. For what initial conditions $y(0) = a$ is the resulting solution constant? Increasing? Decreasing?

b. Sketch the direction field for the equation.

SOLUTION

a. We follow the steps given in the procedure.

1. Letting $f(y) = y(y - 2)$, we see that $f(y) = 0$ when $y = 0$ or $y = 2$. Therefore, the direction field has horizontal line segments when $y = 0$ and $y = 2$. As a result, the constant functions $y(t) = 0$ and $y(t) = 2$, for $t \geq 0$, are equilibrium solutions (Figure 9.13).

2. The solutions of the inequality $f(y) = y(y - 2) > 0$ are $y < 0$ or $y > 2$. Therefore, below the line $y = 0$ and above the line $y = 2$, the direction field has positive slopes and the solutions are increasing in these regions.

3. The solution of the inequality $f(y) = y(y - 2) < 0$ is $0 < y < 2$. Therefore, between the lines $y = 0$ and $y = 2$, the direction field has negative slopes and the solutions are decreasing in this region.

b. The direction field is shown in Figure 9.13 with several representative solutions.

Related Exercises 18, 20 ◄

EXAMPLE 3   Direction field for the logistic equation The logistic equation is commonly used to model populations with a stable equilibrium solution (called the *carrying capacity*). Consider the logistic equation

$$\frac{dP}{dt} = 0.1P\left(1 - \frac{P}{300}\right), \text{ for } t \geq 0.$$

a. Sketch the direction field of the equation.

b. Sketch solution curves corresponding to the initial conditions $P(0) = 50$, $P(0) = 150$, and $P(0) = 350$.

c. Find and interpret $\lim_{t \to \infty} P(t)$.

SOLUTION

➤ The constant solutions $P = 0$ and $P = 300$ are equilibrium solutions. The solution $P = 0$ is an *unstable* equilibrium because nearby solution curves move away from $P = 0$. By contrast, the solution $P = 300$ is a *stable equilibrium* because nearby solution curves are attracted to $P = 300$.

a. We follow the steps in the summary box for sketching the direction field. Because P represents a population, we assume $P \geq 0$.

1. Notice that $P'(t) = 0$ when $P = 0$ or $P = 300$. Therefore, if the initial population is either $P = 0$ or $P = 300$, then $P'(t) = 0$, for all $t \geq 0$, and the solution is

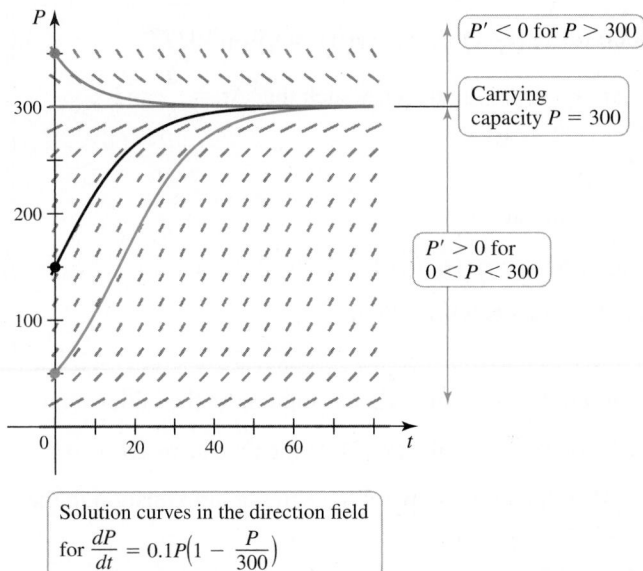

$P' < 0$ for $P > 300$

Carrying capacity $P = 300$

$P' > 0$ for $0 < P < 300$

Solution curves in the direction field

for $\dfrac{dP}{dt} = 0.1P\left(1 - \dfrac{P}{300}\right)$

Figure 9.14

constant. For this reason we expect the direction field to show horizontal lines (with zero slope) at $P = 0$ and $P = 300$.

2. The equation implies that $P'(t) > 0$ provided $0 < P < 300$. Therefore, the direction field has positive slopes and the solutions are increasing, for $t \geq 0$ and $0 < P < 300$.

3. The equation also implies that $P'(t) < 0$ provided $P > 300$ (it was assumed $P \geq 0$). Therefore, the direction field has negative slopes and the solutions are decreasing, for $t \geq 0$ and $P > 300$.

b. **Figure 9.14** shows the direction field with three solution curves corresponding to the three different initial conditions.

c. The horizontal line $P = 300$ corresponds to the carrying capacity of the population. We see that if the initial population is positive and less than 300, the resulting solution increases to the carrying capacity from below. If the initial population is greater than 300, the resulting solution decreases to the carrying capacity from above.

Related Exercises 21–22 ◄

QUICK CHECK 3 According to Figure 9.14, for what approximate value of P is the growth rate of the solution the greatest? ◄

➤ Euler proposed his method for finding approximate solutions to differential equations 200 years before digital computers were invented.

$\Delta t = \dfrac{T}{N}$

$t_0 = 0$ $t_1 = \Delta t$ $t_k = k\Delta t$ $t_N = T$

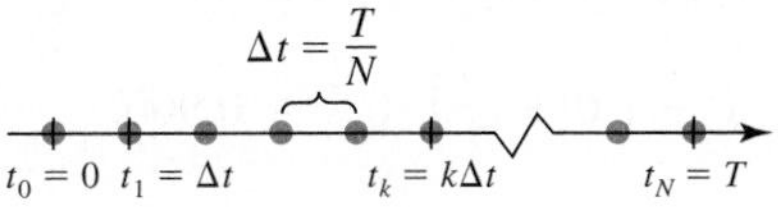

Figure 9.15

➤ See Exercise 45 for setting up Euler's method on a more general interval $[a, b]$.

➤ The argument used to derive the first step of Euler's method is really an application of linear approximation (Section 4.6). We draw a line tangent to the curve at the point (t_0, u_0). The point on that line corresponding to $t = t_1$ is (t_1, u_1), where u_1 is the Euler approximation to $y(t_1)$.

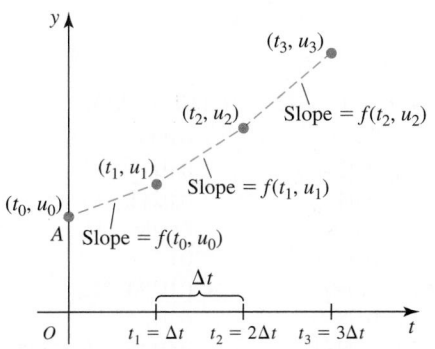

Figure 9.16

Euler's Method

Direction fields are useful for at least two reasons. As shown in previous examples, a direction field provides valuable qualitative information about the solutions of a differential equation *without solving the equation*. In addition, it turns out that direction fields are the basis for many computer-based methods for approximating solutions of a differential equation. The computer begins with the initial condition and advances the solution in small steps, always following the direction field at each time step. The simplest method that uses this idea is called *Euler's method*.

Suppose we wish to approximate the solution to the initial value problem $y'(t) = f(t, y)$, $y(0) = A$ on an interval $[0, T]$. We begin by dividing the interval $[0, T]$ into N **time steps** of equal length $\Delta t = \dfrac{T}{N}$. In so doing, we create a set of grid points (**Figure 9.15**)

$$t_0 = 0, t_1 = \Delta t, t_2 = 2\Delta t, \ldots, t_k = k\Delta t, \ldots, t_N = N\Delta t = T.$$

The exact solution of the initial value problem at the grid points is $y(t_k)$, for $k = 0, 1, 2, \ldots, N$, which is generally unknown unless we are able to solve the original differential equation. The goal is to compute a set of *approximations* to the exact solution at the grid points, which we denote u_k, for $k = 0, 1, 2, \ldots, N$; that is, $u_k \approx y(t_k)$.

The initial condition says that $u_0 = y(0) = A$ (exactly). We now make one step forward in time of length Δt and compute an approximation u_1 to $y(t_1)$. The key observation is that, according to the direction field, the solution at the point (t_0, u_0) has slope $f(t_0, u_0)$. We obtain u_1 from u_0 by drawing a line segment starting at (t_0, u_0) with horizontal extent Δt and slope $f(t_0, u_0)$. The other endpoint of the line segment is (t_1, u_1) (**Figure 9.16**). Applying the slope formula to the two points (t_0, u_0) and (t_1, u_1), we have

$$f(t_0, u_0) = \frac{u_1 - u_0}{t_1 - t_0}.$$

Solving for u_1 and noting that $t_1 - t_0 = \Delta t$, we have

$$u_1 = u_0 + f(t_0, u_0)\Delta t.$$

This basic *Euler step* is now repeated over each time step until we reach $t = T$. That is, having computed u_1, we apply the same argument to obtain u_2. From u_2, we compute u_3. In general, u_{k+1} is computed from u_k, for $k = 0, 1, 2, \ldots, N - 1$. Hand calculations with Euler's method quickly become laborious. The method is usually carried out on a calculator or with a computer program. It is also included in many software packages.

> **PROCEDURE   Euler's Method for** $y'(t) = f(t, y), y(0) = A$ **on** $[0, T]$
>
> 1. Choose either a time step Δt or a positive integer N such that $\Delta t = \dfrac{T}{N}$ and $t_k = k\Delta t$, for $k = 0, 1, 2, \ldots, N - 1$.
> 2. Let $u_0 = y(0) = A$.
> 3. For $k = 0, 1, 2, \ldots, N - 1$, compute
> $$u_{k+1} = u_k + f(t_k, u_k)\Delta t.$$
> Each u_k is an approximation to the exact solution $y(t_k)$.

EXAMPLE 4   Using Euler's method Find an approximate solution to the initial value problem $y'(t) = t - \dfrac{y}{2}, y(0) = 1$, on the interval $[0, 2]$. Use the time steps $\Delta t = 0.2$ ($N = 10$) and $\Delta t = 0.1$ ($N = 20$). Which time step gives a better approximation to the exact solution, which is $y(t) = 5e^{-t/2} + 2t - 4$?

SOLUTION With a time step of $\Delta t = 0.2$, the grid points on the interval $[0, 2]$ are

$$t_0 = 0.0, t_1 = 0.2, t_2 = 0.4, \ldots, t_{10} = 2.0.$$

We identify $f(t, y) = t - \dfrac{y}{2}$ and let u_k be the Euler approximation to $y(t_k)$. Euler's method takes the form

$$u_0 = y(0) = 1, \quad u_{k+1} = u_k + f(t_k, u_k)\Delta t = u_k + \left(t_k - \frac{u_k}{2}\right)\Delta t,$$

where $k = 0, 1, 2, \ldots, 9$. For example, the value of the approximation u_1 is given by

$$u_1 = u_0 + f(t_0, u_0)\Delta t = u_0 + \left(t_0 - \frac{u_0}{2}\right)\Delta t = 1 + \left(0 - \frac{1}{2}\right) \cdot 0.2 = 0.900,$$

and the value of u_2 is given by

$$u_2 = u_1 + f(t_1, u_1)\Delta t = u_1 + \left(t_1 - \frac{u_1}{2}\right)\Delta t = 0.9 + \left(0.2 - \frac{0.9}{2}\right) \cdot 0.2 = 0.850.$$

A similar procedure is used with $\Delta t = 0.1$. In this case, $N = 20$ time steps are needed to cover the interval $[0, 2]$. The results of the two calculations are shown in Figure 9.17, where the exact solution appears as a solid curve and the Euler approximations are shown as points. From these graphs, it appears that the time step $\Delta t = 0.1$ gives better approximations to the solution.

A more detailed account of these calculations is given in Table 9.1, which shows the numerical values of the Euler approximations for $\Delta t = 0.2$ and $\Delta t = 0.1$. Notice that the approximations with $\Delta t = 0.1$ are tabulated at *every other* time step so that they may be compared to the $\Delta t = 0.2$ approximations.

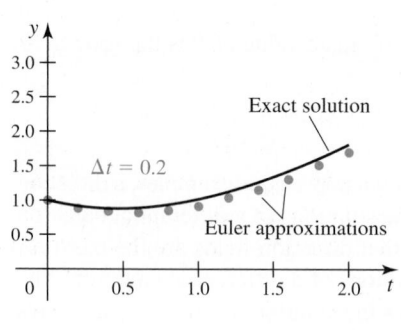

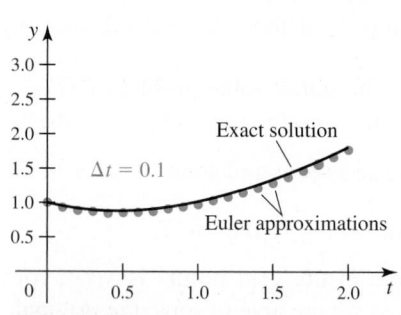

Figure 9.17

> ➤ Because computers produce small errors at each time step, taking a large number of time steps may eventually lead to an unacceptable accumulation of errors. When more accuracy is needed, it may be best to use other methods that require more work per time step, but also give more accurate results.

QUICK CHECK 4 Notice that the errors in Table 9.1 increase in time for both time steps. Give a possible explanation for this increase in the errors. ◄

Table 9.1

t_k	$u_k(\Delta t = 0.2)$	$u_k(\Delta t = 0.1)$	$e_k(\Delta t = 0.2)$	$e_k(\Delta t = 0.1)$
0.0	1.000	1.000	0.000	0.000
0.2	0.900	0.913	0.0242	0.0117
0.4	0.850	0.873	0.0437	0.0211
0.6	0.845	0.875	0.0591	0.0286
0.8	0.881	0.917	0.0711	0.0345
1.0	0.952	0.994	0.0802	0.0390
1.2	1.057	1.102	0.0869	0.0423
1.4	1.191	1.238	0.0914	0.0446
1.6	1.352	1.401	0.0943	0.0460
1.8	1.537	1.586	0.0957	0.0468
2.0	1.743	1.792	0.0960	0.0470

How accurate are these approximations? Although it does not generally happen in practice, we can determine the solution of this particular initial value problem exactly. (You can check that the solution is $y(t) = 5e^{-t/2} + 2t - 4$.) We investigate the accuracy of the Euler approximations by computing the *error*, $e_k = |u_k - y(t_k)|$, at each grid point. The error simply measures the difference between the exact solution and the corresponding approximations. The last two columns of Table 9.1 show the errors associated with the approximations. We see that at every grid point, the approximations with $\Delta t = 0.1$ have errors with roughly half the magnitude of the errors with $\Delta t = 0.2$.

This pattern is typical of Euler's method. If we focus on one point in time, halving the time step roughly halves the errors. However, nothing is free: Halving the time step also requires twice as many time steps and twice the amount of computational work to cover the same time interval.

Related Exercises 29, 32 ◄

Final Note Euler's method is the simplest of a vast collection of *numerical methods* for approximating solutions of differential equations (often studied in courses on *numerical analysis*). As we have seen, Euler's method uses linear approximation; that is, the method follows the direction field using line segments. This idea works well provided the direction field varies smoothly and slowly. In less well behaved cases, Euler's method may encounter difficulties. More robust and accurate methods do a better job of following the direction field (for example, by using parabolas or higher-degree polynomials instead of linear approximation). While these refined methods are generally more accurate than Euler's method, they often require more computational work per time step. As with Euler's method, all methods have the property that their accuracy improves as the time step decreases. The upshot is that there are often trade-offs in choosing a method to approximate the solution of a differential equation. However, Euler's method is a good place to start and may be adequate.

SECTION 9.2 EXERCISES

Getting Started

1. Explain how to sketch the direction field of the equation $y'(t) = f(t, y)$, where f is given.

2. Consider the differential equation $y'(t) = t^2 - 3y^2$ and the solution curve that passes through the point $(3, 1)$. What is the slope of the curve at $(3, 1)$?

3. Consider the initial value problem $y'(t) = t^2 - 3y^2$, $y(3) = 1$. What is the approximation to $y(3.1)$ given by Euler's method with a time step of $\Delta t = 0.1$?

4. Give a geometrical explanation of how Euler's method works.

5. **Matching direction fields** Match equations a–d with direction fields A–D.

   a. $y'(t) = \dfrac{t}{2}$ **b.** $y'(t) = \dfrac{y}{2}$

   c. $y'(t) = \dfrac{t^2 + y^2}{2}$ **d.** $y'(t) = \dfrac{y}{t}$

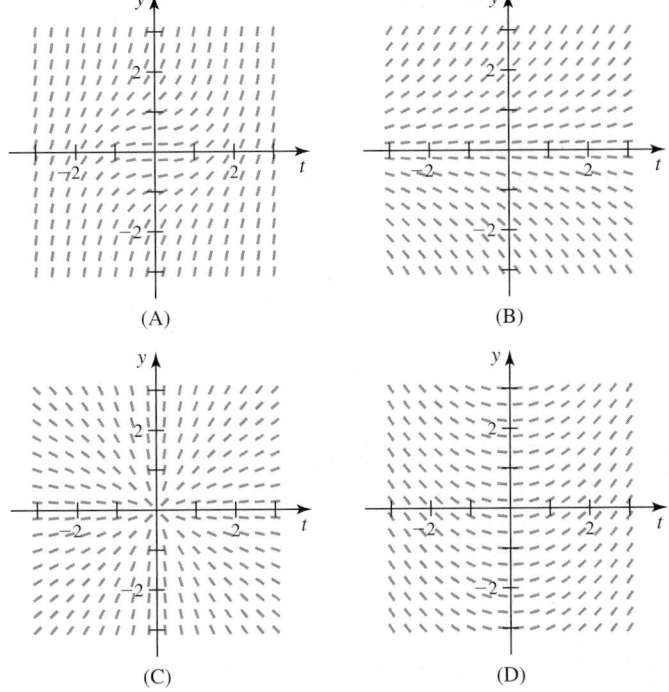

(A)

(B)

(C)

(D)

6. Identifying direction fields Which of the differential equations a–d corresponds to the following direction field? Explain your reasoning.

  a. $y'(t) = 0.5(y + 1)(t - 1)$
  b. $y'(t) = -0.5(y + 1)(t - 1)$
  c. $y'(t) = 0.5(y - 1)(t + 1)$
  d. $y'(t) = -0.5(y - 1)(t + 1)$

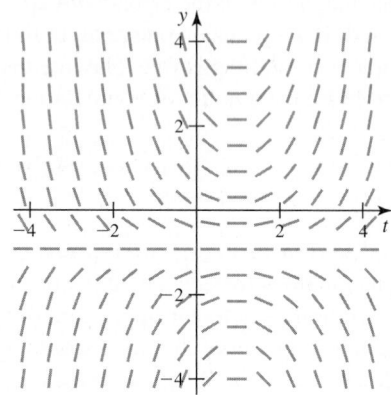

Practice Exercises

7–8. Direction fields *A differential equation and its direction field are shown in the following figures. Sketch a graph of the solution curve that passes through the given initial conditions.*

7. $y'(t) = \dfrac{t^2}{y^2 + 1}, y(0) = -2$ **8.** $y'(t) = \dfrac{\sin t}{y}, y(-2) = -2$
  and $y(-2) = 0$.   and $y(-2) = 2$

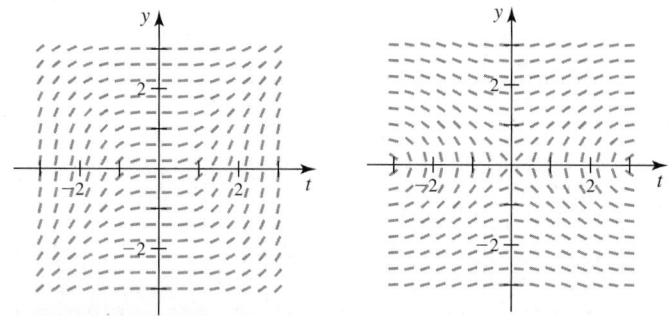

▣ 9–11. Direction fields with technology *Plot a direction field for the following differential equation with a graphing utility. Then find the solutions that are constant and determine which initial conditions $y(0) = A$ lead to solutions that are increasing in time.*

9. $y'(t) = 0.5(y + 1)^2(t - 1)^2, |t| \le 3$ and $|y| \le 3$

10. $y'(t) = (y - 1)\sin \pi t, 0 \le t \le 2, 0 \le y \le 2$

11. $y'(t) = t(y - 1), 0 \le t \le 2, 0 \le y \le 2$

12–16. Sketching direction fields *Use the window $[-2, 2] \times [-2, 2]$ to sketch a direction field for the following equations. Then sketch the solution curve that corresponds to the given initial condition. A detailed direction field is not needed.*

12. $y'(t) = y - 3, y(0) = 1$

13. $y'(t) = 4 - y, y(0) = -1$

14. $y'(t) = y(2 - y), y(0) = 1$

15. $y'(x) = \sin x, y(-2) = 2$

16. $y'(x) = \sin y, y(-2) = \dfrac{1}{2}$

17–20. Increasing and decreasing solutions *Consider the following differential equations. A detailed direction field is not needed.*

  a. *Find the solutions that are constant, for all $t \ge 0$ (the equilibrium solutions).*
  b. *In what regions are solutions increasing? Decreasing?*
  c. *Which initial conditions $y(0) = A$ lead to solutions that are increasing in time? Decreasing?*
  d. *Sketch the direction field and verify that it is consistent with parts (a)–(c).*

17. $y'(t) = (y - 1)(1 + y)$ **18.** $y'(t) = (y - 2)(y + 1)$

19. $y'(t) = \cos y$, for $|y| \le \pi$ **20.** $y'(t) = y(y + 3)(4 - y)$

21–24. Logistic equations *Consider the following logistic equations. In each case, sketch the direction field, draw the solution curve for each initial condition, and find the equilibrium solutions. A detailed direction field is not needed. Assume $t \ge 0$ and $P \ge 0$.*

21. $P'(t) = 0.05P\left(1 - \dfrac{P}{500}\right); P(0) = 100, P(0) = 400,$
    $P(0) = 700$

22. $P'(t) = 0.1P\left(1 - \dfrac{P}{1200}\right); P(0) = 600, P(0) = 800,$
    $P(0) = 1600$

23. $P'(t) = 0.02P\left(4 - \dfrac{P}{800}\right); P(0) = 1600, P(0) = 2400,$
    $P(0) = 4000$

24. $P'(t) = 0.05P - 0.001P^2; P(0) = 10, P(0) = 40, P(0) = 80$

25–28. Two steps of Euler's method *For the following initial value problems, compute the first two approximations u_1 and u_2 given by Euler's method using the given time step.*

25. $y'(t) = 2y, y(0) = 2; \Delta t = 0.5$

26. $y'(t) = -y, y(0) = -1; \Delta t = 0.2$

27. $y'(t) = 2 - y, y(0) = 1; \Delta t = 0.1$

28. $y'(t) = t + y, y(0) = 4; \Delta t = 0.5$

▣ 29–32. Errors in Euler's method *Consider the following initial value problems.*

  a. *Find the approximations to $y(0.2)$ and $y(0.4)$ using Euler's method with time steps of $\Delta t = 0.2, 0.1, 0.05,$ and 0.025.*
  b. *Using the exact solution given, compute the errors in the Euler approximations at $t = 0.2$ and $t = 0.4$.*
  c. *Which time step results in the more accurate approximation? Explain your observations.*
  d. *In general, how does halving the time step affect the error at $t = 0.2$ and $t = 0.4$?*

29. $y'(t) = -y, y(0) = 1; y(t) = e^{-t}$

30. $y'(t) = \dfrac{y}{2}, y(0) = 2; y(t) = 2e^{t/2}$

31. $y'(t) = 4 - y, y(0) = 3; y(t) = 4 - e^{-t}$

32. $y'(t) = 2t + 1, y(0) = 0; y(t) = t^2 + t$

T 33–36. Computing Euler approximations *Use a calculator or computer program to carry out the following steps.*

a. *Approximate the value of $y(T)$ using Euler's method with the given time step on the interval $[0, T]$.*
b. *Using the exact solution (also given), find the error in the approximation to $y(T)$ (only at the right endpoint of the time interval).*
c. *Repeating parts (a) and (b) using half the time step used in those calculations, again find an approximation to $y(T)$.*
d. *Compare the errors in the approximations to $y(T)$.*

33. $y'(t) = -2y, y(0) = 1; \Delta t = 0.2, T = 2; y(t) = e^{-2t}$

34. $y'(t) = 6 - 2y, y(0) = -1; \Delta t = 0.2, T = 3;$
$y(t) = 3 - 4e^{-2t}$

35. $y'(t) = t - y, y(0) = 4; \Delta t = 0.2, T = 4;$
$y(t) = 5e^{-t} + t - 1$

36. $y'(t) = \dfrac{t}{y}, y(0) = 4; \Delta t = 0.1, T = 2; y(t) = \sqrt{t^2 + 16}$

37. Explain why or why not Determine whether the following statements are true and give an explanation or counterexample.

a. A direction field allows you to visualize the solution of a differential equation, but it does not give exact values of the solution at particular points.
b. Euler's method is used to compute exact values of the solution of an initial value problem.

38–43. Equilibrium solutions *A differential equation of the form $y'(t) = f(y)$ is said to be **autonomous** (the function f depends only on y). The constant function $y = y_0$ is an **equilibrium solution** of the equation provided $f(y_0) = 0$ (because then $y'(t) = 0$ and the solution remains constant for all t). Note that equilibrium solutions correspond to horizontal lines in the direction field. Note also that for autonomous equations, the direction field is independent of t. Carry out the following analysis on the given equations.*

a. *Find the equilibrium solutions.*
b. *Sketch the direction field, for $t \geq 0$.*
c. *Sketch the solution curve that corresponds to the initial condition $y(0) = 1$.*

38. $y'(t) = 2y + 4$ **39.** $y'(t) = 6 - 2y$

40. $y'(t) = y(2 - y)$ **41.** $y'(t) = y(y - 3)$

42. $y'(t) = \sin y$ **43.** $y'(t) = y(y - 3)(y + 2)$

Explorations and Challenges

44. Direction field analysis Consider the first-order initial value problem $y'(t) = ay + b, y(0) = A$, for $t \geq 0$, where a, b, and A are real numbers.

a. Explain why $y = -b/a$ is an equilibrium solution and corresponds to a horizontal line in the direction field.
b. Draw a representative direction field in the case that $a > 0$. Show that if $A > -b/a$, then the solution increases for $t \geq 0$, and that if $A < -b/a$, then the solution decreases for $t \geq 0$.
c. Draw a representative direction field in the case that $a < 0$. Show that if $A > -b/a$, then the solution decreases for $t \geq 0$, and that if $A < -b/a$, then the solution increases for $t \geq 0$.

45. Euler's method on more general grids Suppose the solution of the initial value problem $y'(t) = f(t, y), y(a) = A$ is to be approximated on the interval $[a, b]$.

a. If $N + 1$ grid points are used (including the endpoints), what is the time step Δt?
b. Write the first step of Euler's method to compute u_1.
c. Write the general step of Euler's method that applies, for $k = 0, 1, \ldots, N - 1$.

46–48. Analyzing models *The following models were discussed in Section 9.1 and reappear in later sections of this chapter. In each case, carry out the indicated analysis using direction fields.*

46. Drug infusion The delivery of a drug (such as an antibiotic) through an intravenous line may be modeled by the differential equation $m'(t) + km(t) = I$, where $m(t)$ is the mass of the drug in the blood at time $t \geq 0$, k is a constant that describes the rate at which the drug is absorbed, and I is the infusion rate. Let $I = 10$ mg/hr and $k = 0.05$ hr^{-1}.

a. Draw the direction field, for $0 \leq t \leq 100, 0 \leq y \leq 600$.
b. For what initial values $m(0) = A$ are solutions increasing? Decreasing?
c. What is the equilibrium solution?

47. Free fall A model that describes the free fall of an object in a gravitational field subject to air resistance uses the equation $v'(t) = g - bv$, where $v(t)$ is the velocity of the object, for $t \geq 0, g = 9.8$ m/s^2 is the acceleration due to gravity, and $b > 0$ is a constant that involves the mass of the object and the air resistance. Let $b = 0.1$ s^{-1}.

a. Draw the direction field for $0 \leq t \leq 60, 0 \leq y \leq 150$.
b. For what initial values $v(0) = A$ are solutions increasing? Decreasing?
c. What is the equilibrium solution?

48. Chemical rate equations Consider the chemical rate equations $y'(t) = -ky(t)$ and $y'(t) = -ky^2(t)$, where $y(t)$ is the concentration of the compound for $t \geq 0$, and $k > 0$ is a constant that determines the speed of the reaction. Assume the initial concentration of the compound is $y(0) = y_0 > 0$.

a. Let $k = 0.3$ and make a sketch of the direction fields for both equations. What is the equilibrium solution in both cases?
b. According to the direction fields, which reaction approaches its equilibrium solution faster?

49. Convergence of Euler's method Suppose Euler's method is applied to the initial value problem $y'(t) = ay, y(0) = 1$, which has the exact solution $y(t) = e^{at}$. For this exercise, let h denote the time step (rather than Δt). The grid points are then given by $t_k = kh$. We let u_k be the Euler approximation to the exact solution $y(t_k)$, for $k = 0, 1, 2, \ldots$.

a. Show that Euler's method applied to this problem can be written $u_0 = 1, u_{k+1} = (1 + ah)u_k$, for $k = 0, 1, 2, \ldots$.
b. Show by substitution that $u_k = (1 + ah)^k$ is a solution of the equations in part (a), for $k = 0, 1, 2, \ldots$.
c. Recall from Section 4.7 that $\lim\limits_{h \to 0} (1 + ah)^{1/h} = e^a$. Use this fact to show that as the time step goes to zero ($h \to 0$, with $t_k = kh$ fixed), the approximations given by Euler's method approach the exact solution of the initial value problem; that is, $\lim\limits_{h \to 0} u_k = \lim\limits_{h \to 0} (1 + ah)^k = y(t_k) = e^{at_k}$.

50. Stability of Euler's method Consider the initial value problem $y'(t) = -ay$, $y(0) = 1$, where $a > 0$; it has the exact solution $y(t) = e^{-at}$, which is a decreasing function.

a. Show that Euler's method applied to this problem with time step h can be written $u_0 = 1$, $u_{k+1} = (1 - ah)u_k$, for $k = 0, 1, 2, \ldots$.

b. Show by substitution that $u_k = (1 - ah)^k$ is a solution of the equations in part (a), for $k = 0, 1, 2, \ldots$.

c. Explain why, as k increases, the Euler approximations $u_k = (1 - ah)^k$ decrease in magnitude only if $|1 - ah| < 1$.

d. Show that the inequality in part (c) implies that the time step must satisfy the condition $0 < h < \dfrac{2}{a}$. If the time step does not satisfy this condition, then Euler's method is *unstable* and produces approximations that actually increase in time.

QUICK CHECK ANSWERS

1. To cross the line $y = 2$, the solution must have a slope different from zero when $y = 2$. However, according to the direction field, a solution on the line $y = 2$ has zero slope.   **2.** The solutions originating at both initial conditions are increasing.   **3.** The direction field is steepest when $P = 150$.   **4.** Each step of Euler's method introduces an error. With each successive step of the calculation, the errors could accumulate (or propagate). ◄

9.3 Separable Differential Equations

Sketching solutions of a differential equation using its direction field is a powerful technique, and it provides a wealth of information about the solutions. However, valuable as they are, direction fields do not produce the actual solutions of a differential equation. In this section, we examine methods that lead to the solutions of certain differential equations in terms of an algebraic expression (often called an *analytical solution*). The equations we consider are first order and belong to a class called *separable equations*.

Method of Solution

> If the equation has the form $y'(t) = f(t)$ (that is, the right side depends only on t), then solving the equation amounts to finding the antiderivatives of f, a problem discussed in Section 4.9.

The most general first-order differential equation has the form $y'(t) = F(t, y)$, where $F(t, y)$ is an expression that may involve both the independent variable t and the unknown function y. We have a *chance* of solving such an equation if it can be written in the form

$$g(y)y'(t) = h(t).$$

In the equation $g(y)y'(t) = h(t)$, the factor $g(y)$ involves only y, and $h(t)$ involves only t; that is, the variables have been separated. An equation that can be written in this form is said to be **separable**.

In general, we solve a separable differential equation by integrating both sides of the equation with respect to t:

$$\int g(y)\underbrace{y'(t)\,dt}_{dy} = \int h(t)\,dt \qquad \text{Integrate both sides with respect to } t.$$

$$\int g(y)\,dy = \int h(t)\,dt. \qquad \text{Change variables on left; } dy = y'(t)\,dt.$$

The fact that $dy = y'(t)\,dt$ on the left side of the equation leaves us with two integrals to evaluate, one with respect to y and one with respect to t. Finding a solution depends on evaluating these integrals.

QUICK CHECK 1 Which of the following equations are separable? (A) $y'(t) = y + t$, (B) $y'(t) = \dfrac{ty}{t + 1}$, and (C) $y'(x) = e^{x+y}$ ◄

EXAMPLE 1   **A separable equation** Find a function that satisfies the following initial value problem.

$$y'(t) = y^2 e^{-t}, \quad y(0) = \tfrac{1}{2}$$

SOLUTION The equation is written in separable form by dividing both sides of the equation by y^2 to give $\dfrac{y'(t)}{y^2} = e^{-t}$. We now integrate both sides of the equation with respect to t and evaluate the resulting integrals:

$$\int \frac{1}{y^2} y'(t)\, dt = \int e^{-t}\, dt$$

$$\int \frac{dy}{y^2} = \int e^{-t}\, dt \qquad \text{Change variables on left side.}$$

$$-\frac{1}{y} = -e^{-t} + C. \qquad \text{Evaluate integrals.}$$

> In practice, the change of variable on the left side is often omitted, and we go directly to the second step, which is to integrate the left side with respect to y and the right side with respect to t.

> Notice that each integration produces a constant of integration. The two constants of integration may be combined into one.

Solving for y gives the general solution

$$y(t) = \frac{1}{e^{-t} - C}.$$

The initial condition $y(0) = \frac{1}{2}$ implies that

$$y(0) = \frac{1}{e^0 - C} = \frac{1}{1 - C} = \frac{1}{2}.$$

It follows that $C = -1$, so the solution to the initial value problem is

$$y(t) = \frac{1}{e^{-t} + 1}, \text{ for all real } t.$$

The solution (**Figure 9.18**) passes through $\left(0, \frac{1}{2}\right)$ and increases to approach the asymptote $y = 1$ because $\displaystyle\lim_{t\to\infty} \frac{1}{e^{-t} + 1} = 1.$

Related Exercises 17, 20 ◄

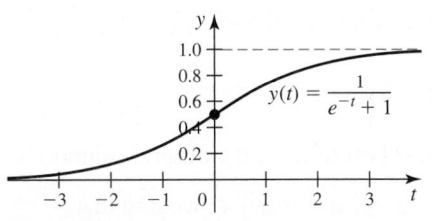

Figure 9.18

QUICK CHECK 2 Write

$$y'(t) = (t^2 + 1)/y^3$$

in separated form. ◄

EXAMPLE 2 **Another separable equation** Find the solutions of the equation $y'(x) = e^{-y} \sin x$ subject to the three different initial conditions

$$y(0) = 1, \quad y\left(\frac{\pi}{2}\right) = \frac{1}{2}, \quad \text{and} \quad y(\pi) = \ln 2.$$

SOLUTION Writing the equation in the form $e^y y'(x) = \sin x$, we see that it is separable. Integrating both sides with respect to x, we have

$$\int e^y y'(x)\, dx = \int \sin x\, dx$$

$$\int e^y\, dy = \int \sin x\, dx \qquad \text{Change variables on left side.}$$

$$e^y = -\cos x + C. \qquad \text{Evaluate integrals.}$$

The general solution y is found by taking logarithms of both sides of this equation:

$$y(x) = \ln(C - \cos x).$$

The three initial conditions are now used to evaluate the constant C for the three solutions:

$$y(0) = 1 \Rightarrow 1 = \ln(C - \cos 0) = \ln(C - 1) \Rightarrow e = C - 1 \Rightarrow C = e + 1,$$

$$y\left(\frac{\pi}{2}\right) = \frac{1}{2} \Rightarrow \frac{1}{2} = \ln\left(C - \cos\frac{\pi}{2}\right) = \ln C \Rightarrow C = e^{1/2}, \text{ and}$$

$$y(\pi) = \ln 2 \Rightarrow \ln 2 = \ln(C - \cos \pi) = \ln(C + 1) \Rightarrow 2 = C + 1 \Rightarrow C = 1.$$

Substituting these values of C into the general solution gives the solutions of the three initial value problems (**Figure 9.19**):

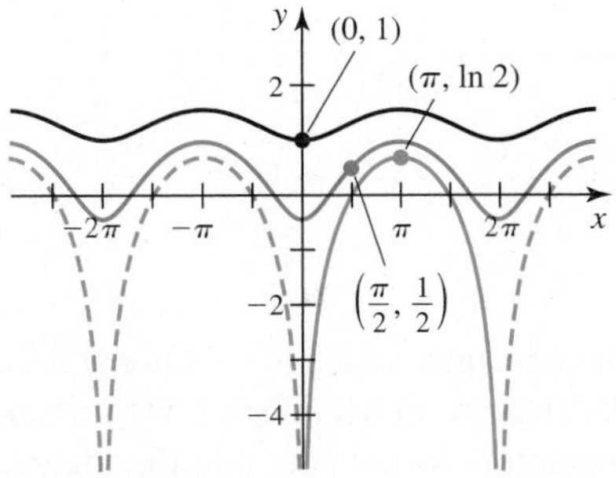

$$y_1(x) = \ln(e + 1 - \cos x), \tag{1}$$

$$y_2(x) = \ln(e^{1/2} - \cos x), \text{ and} \tag{2}$$

$$y_3(x) = \ln(1 - \cos x). \tag{3}$$

Figure 9.19

QUICK CHECK 3 Find the value of the constant C in Example 2 with the initial condition $y(\pi) = 0$, and then explain why the domain of the corresponding solution is $(\pi/2, 3\pi/2)$. ◄

Solutions (1) and (2) have a domain of all real numbers (both $e + 1 - \cos x$ and $e^{1/2} - \cos x$ are greater than 0, for all x). The graph of solution (3) has a vertical asymptote at $x = 0, \pm 2\pi, \pm 4\pi, \ldots$ because $1 - \cos x = 0$ at those points and $\lim_{x \to 0^+} \ln x = -\infty$. Therefore, we choose the branch of the graph that passes through the initial condition $(\pi, \ln 2)$ (Figure 9.19) and conclude that the domain of $y_3(x)$ is $(0, 2\pi)$.

Related Exercises 26–27 ◄

Even if we can evaluate the integrals necessary to solve a separable equation, the solution may not be easily expressed in an explicit form. Here is an example of a solution that is best left in implicit form.

EXAMPLE 3   An implicit solution Find and graph the solution of the initial value problem

$$(\cos y)\, y'(t) = \sin^2 t \cos t, \quad y(0) = \frac{\pi}{6}.$$

SOLUTION The equation is already in separated form. Integrating both sides with respect to t, we have

> ➤ For the integral on the right side, we use the substitution $u = \sin t$. The integral becomes $\int u^2\, du = \frac{1}{3}u^3 + C.$

$$\int \cos y \, dy = \int \sin^2 t \cos t \, dt \quad \text{Integrate both sides.}$$

$$\sin y = \frac{1}{3}\sin^3 t + C. \quad \text{Evaluate integrals.}$$

When imposing the initial condition in this case, it is best to leave the general solution in implicit form. Substituting $t = 0$ and $y = \frac{\pi}{6}$ into the general solution, we find that

> ➤ Care must be used when graphing and interpreting implicit solutions. The graph of
> $$\sin y = \frac{1}{3}\sin^3 t + \frac{1}{2}$$
> is a family of an infinite number of curves.

$$\sin \frac{\pi}{6} = \frac{1}{3}\sin^3 0 + C \quad \text{or} \quad C = \frac{1}{2}.$$

Therefore, the solution of the initial value problem is

$$\sin y = \frac{1}{3}\sin^3 t + \frac{1}{2}.$$

In order to graph the solution in this implicit form, it is easiest to use graphing software. The result is shown in **Figure 9.20**.

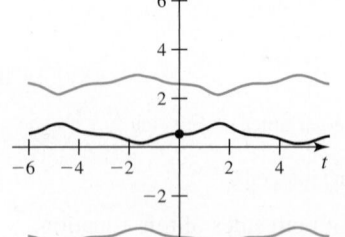

You must choose the curve that satisfies the initial condition, as shown in Figure 9.20. The domain of this solution is $(-\infty, \infty)$.

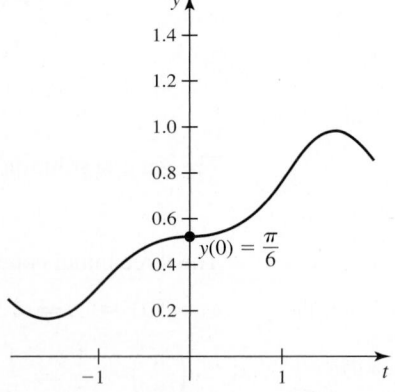

Figure 9.20

Related Exercises 33, 36 ◄

QUICK CHECK 4 Find the value of the constant C in Example 3 with the initial condition $y\left(\dfrac{\pi}{6}\right) = 0$. ◄

Logistic Equation Revisited

> ➤ The derivation of the logistic equation is discussed in Section 9.5.

In Exercise 53 of Section 9.1, we introduced the logistic equation, which is commonly used for modeling populations, epidemics, and the spread of rumors. In Section 9.2, we investigated the direction field associated with the logistic equation. It turns out that the logistic equation is a separable equation, so we now have the tools needed to solve it.

EXAMPLE 4 **Logistic population growth** Assume 50 fruit flies are in a large jar at the beginning of an experiment. Let $P(t)$ be the number of fruit flies in the jar t days later. At first, the population grows exponentially, but due to limited space and food supply, the growth rate decreases and the population is prevented from growing without bound. This experiment is modeled by the *logistic equation*

$$\frac{dP}{dt} = 0.1P\left(1 - \frac{P}{300}\right), \text{ for } t \geq 0,$$

together with the initial condition $P(0) = 50$. Solve this initial value problem.

SOLUTION We see that the equation is separable by writing it in the form

$$\frac{1}{P\left(1 - \dfrac{P}{300}\right)} \cdot \frac{dP}{dt} = 0.1.$$

Integrating both sides with respect to t leads to the equation

$$\int \frac{dP}{P\left(1 - \dfrac{P}{300}\right)} = \int 0.1 \, dt. \tag{4}$$

The integral on the right side of equation (4) is $\int 0.1 \, dt = 0.1t + C$.

Because the integrand on the left side is a rational function in P, we use partial fractions. You should verify that

$$\frac{1}{P\left(1 - \dfrac{P}{300}\right)} = \frac{300}{P(300 - P)} = \frac{1}{P} + \frac{1}{300 - P}$$

and therefore,

$$\int \frac{1}{P\left(1 - \dfrac{P}{300}\right)} \, dP = \int \left(\frac{1}{P} + \frac{1}{300 - P}\right) dP = \ln\left|\frac{P}{300 - P}\right| + C.$$

After integration, equation (4) becomes

$$\ln\left|\frac{P}{300 - P}\right| = 0.1t + C. \tag{5}$$

The next step is to solve for P, which is tangled up inside the logarithm. We first exponentiate both sides of equation (5) to obtain

$$\left|\frac{P}{300 - P}\right| = e^C \cdot e^{0.1t}.$$

We can remove the absolute value on the left side of equation (5) by writing

$$\frac{P}{300 - P} = \pm \, e^C \cdot e^{0.1t}.$$

> ▶ There are not many times in mathematics when we can redefine a constant in the middle of a calculation. When working with arbitrary constants, it may be possible, if it is done carefully.

At this point, a useful trick simplifies matters. Because C is an arbitrary constant, $\pm e^C$ is also an arbitrary constant, so we rename $\pm e^C$ as C. We now have

$$\frac{P}{300 - P} = Ce^{0.1t}. \tag{6}$$

Solving equation (6) for P and replacing $1/C$ by C gives the general solution

$$P(t) = \frac{300}{1 + Ce^{-0.1t}}.$$

➤ We could also use the initial condition in equation (6) to solve for *C*.

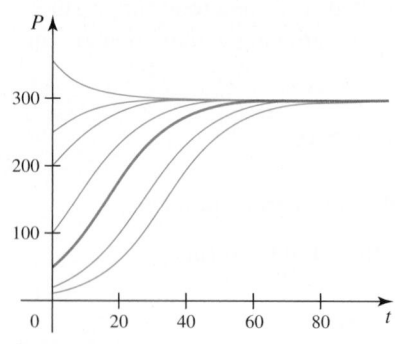

Figure 9.21

Figure 9.21 shows the general solution, with curves corresponding to several different values of *C*. Using the initial condition $P(0) = 50$, we find the value of *C* for our specific problem is $C = 5$. It follows that the solution of the initial value problem is

$$P(t) = \frac{300}{1 + 5e^{-0.1t}}, \text{ for } t \geq 0.$$

Figure 9.21 also shows this particular solution (in red) among the curves in the general solution. A significant feature of this model is that, for $0 < P(0) < 300$, the population increases, but not without bound. Instead, it approaches an **equilibrium**, or **steady-state**, solution with a value of

$$\lim_{t \to \infty} P(t) = \lim_{t \to \infty} \frac{300}{1 + 5e^{-0.1t}} = 300,$$

which is the maximum population that the environment (space and food supply) can sustain. This equilibrium population is called the **carrying capacity**. Notice that all the curves in the general solution approach the carrying capacity as *t* increases.

Related Exercises 39–40 ◄

SECTION 9.3 EXERCISES

Getting Started

1. What is a separable first-order differential equation?

2. Is the equation $t^2 y'(t) = \dfrac{t + 4}{y^2}$ separable?

3. Is the equation $y'(t) = 2y - t$ separable?

4. Explain how to solve a separable differential equation of the form $g(t)y'(t) = h(t)$.

Practice Exercises

5–16. Solving separable equations *Find the general solution of the following equations. Express the solution explicitly as a function of the independent variable.*

5. $t^{-3}y'(t) = 1$

6. $e^{4t}y'(t) = 5$

7. $\dfrac{dy}{dt} = \dfrac{3t^2}{y}$

8. $\dfrac{dy}{dx} = y(x^2 + 1)$

9. $y'(t) = e^{y/2} \sin t$

10. $x^2 \dfrac{dw}{dx} = \sqrt{w}(3x + 1)$

11. $x^2 y'(x) = y^2$

12. $(t^2 + 1)^3 yy'(t) = t(y^2 + 4)$

13. $y'(t)\csc t = -\dfrac{y^3}{2}$

14. $y'(t)e^{t/2} = y^2 + 4$

15. $u'(x) = e^{2x-u}$

16. $xu'(x) = u^2 - 4$

17–32. Solving initial value problems *Determine whether the following equations are separable. If so, solve the initial value problem.*

17. $2yy'(t) = 3t^2, y(0) = 9$

18. $y'(t) = e^{ty}, y(0) = 1$

19. $\dfrac{dy}{dt} = ty + 2, y(1) = 2$

20. $y'(t) = y(4t^3 + 1), y(0) = 4$

21. $y'(t) = ye^t, y(0) = -1$

22. $y'(x) = y \cos x, y(0) = 3$

23. $\dfrac{dy}{dx} = e^{x-y}, y(0) = \ln 3$

24. $y'(t) = \cos^2 y, y(1) = \dfrac{\pi}{4}$

25. $y'(t) = \dfrac{\ln^3 t}{te^y}, y(1) = 0$

26. $ty'(t) = y(y + 1), y(3) = 1$

27. $y'(t) = \dfrac{\sec^2 t}{2y}, y(\pi/4) = 1$

28. $y'(t) = \dfrac{y + 3}{5t + 6}, y(2) = 0$

29. $y'(t) = \dfrac{t}{y}, y(1) = 2$

30. $y'(t) = y^3 \sin t, y(0) = 1$

31. $ty'(t) = 1, y(1) = 2$

32. $(\sec t)y'(t) = 1, y(0) = 1$

⊤ 33–38. Solutions in implicit form *Solve the following initial value problems and leave the solution in implicit form. Use graphing software to plot the solution. If the implicit solution describes more than one function, be sure to indicate which function corresponds to the solution of the initial value problem.*

33. $y'(t) = \dfrac{2t^2}{y^2 - 1}, y(0) = 0$

34. $y'(x) = \dfrac{1 + x}{2 - y}, y(1) = 1$

35. $u'(x) = \csc u \cos \dfrac{x}{2}, u(\pi) = \dfrac{\pi}{2}$

36. $yy'(x) = \dfrac{2x}{(2 + y^2)^2}, y(1) = -1$

37. $y'(x)\sqrt{y + 4} = \sqrt{x + 1}, y(3) = 5$

38. $z'(x) = \dfrac{z^2 + 4}{x^2 + 16}, z(4) = 2$

⊤ 39. Logistic equation for a population A community of hares on an island has a population of 50 when observations begin (at $t = 0$). The population is modeled by the initial value problem

$$\frac{dP}{dt} = 0.08P\left(1 - \frac{P}{200}\right), P(0) = 50.$$

a. Find and graph the solution of the initial value problem, for $t \geq 0$.
b. What is the steady-state population?

⊤ 40. Logistic equation for an epidemic When an infected person is introduced into a closed and otherwise healthy community, the number of people who contract the disease (in the absence of any intervention) may be modeled by the logistic equation

$$\frac{dP}{dt} = kP\left(1 - \frac{P}{A}\right), P(0) = P_0,$$

where k is a positive infection rate, A is the number of people in the community, and P_0 is the number of infected people at $t = 0$. The model also assumes no recovery.

a. Find the solution of the initial value problem, for $t \geq 0$, in terms of k, A, and P_0.

b. Graph the solution in the case that $k = 0.025$, $A = 300$, and $P_0 = 1$.

c. For a fixed value of k and A, describe the long-term behavior of the solutions, for any P_0 with $0 < P_0 < A$.

41. Explain why or why not Determine whether the following statements are true and give an explanation or counterexample.

a. The equation $u'(x) = (x^2 u^7)^{-1}$ is separable.

b. The general solution of the separable equation
$$y'(t) = \frac{t}{y^7 + 10y^4}$$
can be expressed explicitly with y in terms of t.

c. The general solution of the equation $yy'(x) = xe^{-y}$ can be found using integration by parts.

42–43. Implicit solutions for separable equations *For the following separable equations, carry out the indicated analysis.*

a. *Find the general solution of the equation.*

b. *Find the value of the arbitrary constant associated with each initial condition. (Each initial condition requires a different constant.)*

c. *Use the graph of the general solution that is provided to sketch the solution curve for each initial condition.*

42. $y'(t) = \dfrac{t^2}{y^2 + 1}; y(-1) = 1, y(0) = 0, y(-1) = -1$

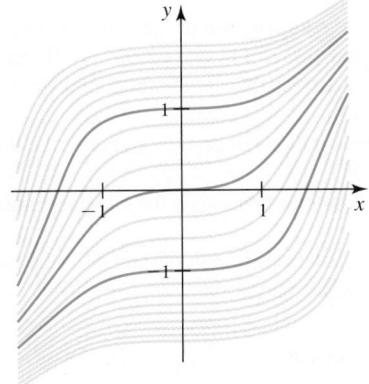

43. $e^{-y/2} y'(x) = 4x \sin x^2 - x; y(0) = 0,$
$$y(0) = \ln\left(\frac{1}{4}\right), y\left(\sqrt{\frac{\pi}{2}}\right) = 0$$

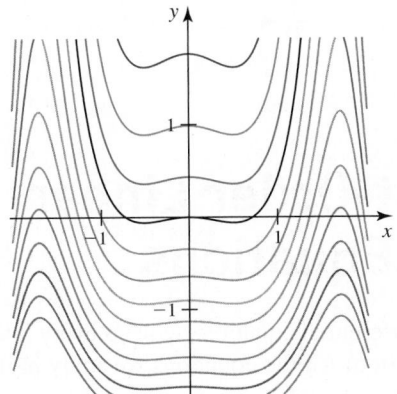

44. Orthogonal trajectories Two curves are orthogonal to each other if their tangent lines are perpendicular at each point of intersection. A family of curves forms **orthogonal trajectories** with another family of curves if each curve in one family is orthogonal to each curve in the other family. Use the following steps to find the orthogonal trajectories of the family of ellipses $2x^2 + y^2 = a^2$.

a. Apply implicit differentiation to $2x^2 + y^2 = a^2$ to show that
$$\frac{dy}{dx} = \frac{-2x}{y}.$$

b. The family of trajectories orthogonal to $2x^2 + y^2 = a^2$ satisfies the differential equation $\dfrac{dy}{dx} = \dfrac{y}{2x}$. Why?

c. Solve the differential equation in part (b) to verify that $y^2 = e^C |x|$ and then explain why it follows that $y^2 = kx$, where k is an arbitrary constant. Therefore, the family of parabolas $y^2 = kx$ forms the orthogonal trajectories of the family of ellipses $2x^2 + y^2 = a^2$.

45. Orthogonal trajectories Use the method in Exercise 44 to find the orthogonal trajectories for the family of circles $x^2 + y^2 = a^2$.

T 46. Logistic equation for spread of rumors Sociologists model the spread of rumors using logistic equations. The key assumption is that at any given time, a fraction y of the population, where $0 \leq y \leq 1$, knows the rumor, while the remaining fraction $1 - y$ does not. Furthermore, the rumor spreads by interactions between those who know the rumor and those who do not. The number of such interactions is proportional to $y(1 - y)$. Therefore, the equation that describes the spread of the rumor is $y'(t) = ky(1 - y)$, for $t \geq 0$, where k is a positive real number and t is measured in weeks. The number of people who initially know the rumor is $y(0) = y_0$, where $0 \leq y_0 \leq 1$.

a. Solve this initial value problem and give the solution in terms of k and y_0.

b. Assume $k = 0.3$ weeks^{-1} and graph the solution for $y_0 = 0.1$ and $y_0 = 0.7$.

c. Describe and interpret the long-term behavior of the rumor function, for any $0 \leq y_0 \leq 1$.

T 47. Free fall An object in free fall may be modeled by assuming the only forces at work are the gravitational force and air resistance. By Newton's Second Law of Motion (mass · acceleration = the sum of external forces), the velocity of the object satisfies the differential equation
$$\underbrace{m}_{\text{mass}} \cdot \underbrace{v'(t)}_{\text{acceleration}} = \underbrace{mg + f(v)}_{\text{external forces}},$$

where f is a function that models the air resistance (assuming the positive direction is downward). One common assumption (often used for motion in air) is that $f(v) = -kv^2$, for $t \geq 0$, where $k > 0$ is a drag coefficient.

a. Show that the equation can be written in the form $v'(t) = g - av^2$, where $a = \dfrac{k}{m}$.

b. For what (positive) value of v is $v'(t) = 0$? (This equilibrium solution is called the *terminal velocity*.)

c. Find the solution of this separable equation assuming $v(0) = 0$ and $0 < v^2 < \dfrac{g}{a}$.

d. Graph the solution found in part (c) with $g = 9.8$ m/s^2, $m = 1$, and $k = 0.1$, and verify that the terminal velocity agrees with the value found in part (b).

T 48. Free fall Using the background given in Exercise 47, assume the resistance is given by $f(v) = -Rv$, for $t \geq 0$, where $R > 0$ is a drag coefficient (an assumption often made for a heavy medium such as water or oil).

a. Show that the equation can be written in the form
$$v'(t) = g - bv, \text{ where } b = \frac{R}{m}.$$

b. For what value of v is $v'(t) = 0$? (This equilibrium solution is called the terminal velocity.)

c. Find the solution of this separable equation assuming $v(0) = 0$ and $0 < v < \dfrac{g}{b}$.

d. Graph the solution found in part (c) with $g = 9.8$ m/s^2, $m = 1$, and $R = 0.1$, and verify that the terminal velocity agrees with the value found in part (b).

T 49. Torricelli's law An open cylindrical tank initially filled with water drains through a hole in the bottom of the tank according to Torricelli's law (see figure). If $h(t)$ is the depth of water in the tank for $t \geq 0$ s, then Torricelli's law implies $h'(t) = -k\sqrt{h}$, where k is a constant that includes $g = 9.8$ m/s^2, the radius of the tank, and the radius of the drain. Assume the initial depth of the water is $h(0) = H$ m.

a. Find the solution of the initial value problem.
b. Find the solution in the case that $k = 0.1$ and $H = 0.5$ m.
c. In part (b), how long does it take for the tank to drain?
d. Graph the solution in part (b) and check that it is consistent with part (c).

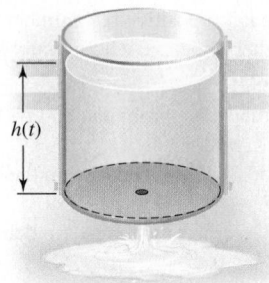

T 50. Chemical rate equations Let $y(t)$ be the concentration of a substance in a chemical reaction (typical units are moles/liter). The change in the concentration, under appropriate conditions, is modeled by the equation $\dfrac{dy}{dt} = -ky^n$, for $t \geq 0$, where $k > 0$ is a rate constant and the positive integer n is the order of the reaction.

a. Show that for a first-order reaction ($n = 1$), the concentration obeys an exponential decay law.
b. Solve the initial value problem for a second-order reaction ($n = 2$) assuming $y(0) = y_0$.
c. Graph the concentration for a first-order and second-order reaction with $k = 0.1$ and $y_0 = 1$.

Explorations and Challenges

T 51. Tumor growth The Gompertz growth equation is often used to model the growth of tumors. Let $M(t)$ be the mass of a tumor at time $t \geq 0$. The relevant initial value problem is
$$\frac{dM}{dt} = -rM \ln\left(\frac{M}{K}\right), M(0) = M_0,$$
where r and K are positive constants and $0 < M_0 < K$.

a. Graph the growth rate function $R(M) = -rM \ln\left(\dfrac{M}{K}\right)$ (which equals $M'(t)$) assuming $r = 1$ and $K = 4$. For what values of M is the growth rate positive? For what value of M is the growth rate a maximum?

b. Solve the initial value problem and graph the solution for $r = 1$, $K = 4$, and $M_0 = 1$. Describe the growth pattern of the tumor. Is the growth unbounded? If not, what is the limiting size of the tumor?

T 52. Technology for an initial value problem Solve $y'(t) = ye^t \cos^3 4t$, $y(0) = 1$, and plot the solution for $0 \leq t \leq \pi$.

53. Blowup in finite time Consider the initial value problem $y'(t) = y^{n+1}$, $y(0) = y_0$, where n is a positive integer.

a. Solve the initial value problem with $n = 1$ and $y_0 = 1$.
b. Solve the initial value problem with $n = 2$ and $y_0 = 1/\sqrt{2}$.
c. Solve the problem for positive integers n and $y_0 = n^{-1/n}$. How do solutions behave as $t \to 1^-$?

T 54. Analysis of a separable equation Consider the differential equation $yy'(t) = \dfrac{1}{2}e^t + t$ and carry out the following analysis.

a. Find the general solution of the equation and express it explicitly as a function of t in two cases: $y > 0$ and $y < 0$.
b. Find the solutions that satisfy the initial conditions $y(-1) = 1$ and $y(-1) = 2$.
c. Graph the solutions in part (b) and describe their behavior as t increases.
d. Find the solutions that satisfy the initial conditions $y(-1) = -1$ and $y(-1) = -2$.
e. Graph the solutions in part (d) and describe their behavior as t increases.

QUICK CHECK ANSWERS

1. B and C are separable.   **2.** $y^3 y'(t) = t^2 + 1$ **3.** The initial condition $y(\pi) = 0$ implies that $C = 0$; the solution $y(x) = \ln(-\cos x)$ is defined only when $-\cos x > 0$, which occurs on the intervals ... $\left(-\frac{3\pi}{2}, -\frac{\pi}{2}\right), \left(\frac{\pi}{2}, \frac{3\pi}{2}\right), \left(\frac{5\pi}{2}, \frac{7\pi}{2}\right), \ldots$. The interval containing the initial condition, $\left(\frac{\pi}{2}, \frac{3\pi}{2}\right)$, is the domain.   **4.** $C = -\frac{1}{24}$ ◄

9.4 Special First-Order Linear Differential Equations

> The exponential growth and decay problems studied in Section 7.2 appear again in this section, but now with a differential equations perspective.

We now focus on a special class of differential equations with so many interesting applications that they warrant special attention. All the equations we study in this section are first order and linear.

Method of Solution

Consider the first-order linear equation $y'(t) = ky + b$, where $k \neq 0$ and b are real numbers. By varying the values of k and b, this versatile equation may be used to model a wide variety of phenomena. Specifically, the terms of the equation have the following general meaning:

$$\underbrace{y'(t)}_{\substack{\text{rate of change} \\ \text{of } y}} = \underbrace{ky(t)}_{\substack{\text{natural growth or} \\ \text{decay rate of } y}} + \underbrace{b.}_{\substack{\text{growth or decay} \\ \text{rate due to} \\ \text{external effects}}}$$

▶ In the most general first-order linear equation, k and b are functions of t. This general first-order linear equation is not separable. See Exercises 45–48 for this more challenging case.

For example, if y represents the number of fish in a hatchery, then $ky(t)$ (with $k > 0$) models exponential growth in the fish population, in the absence of other factors, and $b < 0$ is the harvesting rate at which the population is depleted. As another example, if y represents the amount of a drug in the blood, then $ky(t)$ (with $k < 0$) models exponential decay of the drug through the kidneys, and $b > 0$ is the rate at which the drug is added to the blood intravenously. Because k and b are constants, the equation is separable and we can give an explicit solution.

To solve this equation, we begin by dividing both sides of $y'(t) = ky + b$ by $ky + b$ to express it in separated form:

$$\frac{y'(t)}{ky + b} = 1.$$

We now integrate both sides of this equation with respect to t and observe that $dy = y'(t)\,dt$, which gives

$$\int \frac{dy}{ky + b} = \int dt \qquad \text{Integrate both sides of the equation.}$$

$$\frac{1}{k} \ln |ky + b| = t + C. \qquad \text{Evaluate integrals.}$$

For the moment, we assume $ky + b > 0$, or $y > -b/k$, so the absolute value may be removed. Multiplying through by k and exponentiating both sides of the equation, we have

$$ky + b = e^{kt + kC} = e^{kt} \cdot \underbrace{e^{kC}}_{\text{redefine as } C} = Ce^{kt}.$$

Notice that we use the standard practice of redefining the arbitrary constant C as we solve for y: If C is arbitrary, then e^{kC} and C/k are also arbitrary. We now solve for the general solution:

$$y(t) = Ce^{kt} - \frac{b}{k}.$$

We can also show that if $ky + b < 0$, or $y < -b/k$, then the same solution results (Exercise 36).

SUMMARY Solution of a First-Order Linear Differential Equation

The general solution of the first-order linear equation $y'(t) = ky + b$, where $k \neq 0$ and b are real numbers, is

$$y(t) = Ce^{kt} - \frac{b}{k},$$

where C is an arbitrary constant. Given an initial condition, the value of C may be determined.

QUICK CHECK 1 Verify by substitution that $y(t) = Ce^{kt} - b/k$ is a solution of $y'(t) = ky + b$, for real numbers b and $k \neq 0$. ◀

EXAMPLE 1   An initial value problem for drug dosing A drug is administered to a patient through an intravenous line at a rate of 6 mg/hr. The drug has a half-life that corresponds to a rate constant of $k = 0.03 \text{ hr}^{-1}$. Let $y(t)$ be the amount of drug in the blood, for $t \geq 0$. Solve the initial value problem that governs the process,

$$y'(t) = -0.03y + 6, \; y(0) = 0,$$

and interpret the solution.

SOLUTION The equation has the form $y'(t) = ky + b$, where $k = -0.03$ and $b = 6$. Therefore, the general solution is

$$y(t) = Ce^{-0.03t} + 200.$$

To determine the value of C for this particular problem, we substitute $y(0) = 0$ into the general solution. The result is that $y(0) = C + 200 = 0$, which implies that $C = -200$. Therefore, the solution of the initial value problem is

$$y(t) = -200e^{-0.03t} + 200 = 200(1 - e^{-0.03t}), \text{ for } t \geq 0.$$

The graph of the solution (**Figure 9.22**) reveals an important fact: The amount of drug in the blood increases, but it approaches a steady-state level of

$$\lim_{t \to \infty} y(t) = \lim_{t \to \infty} (200(1 - e^{-0.03t})) = 200 \text{ mg}.$$

A doctor can obtain practical information from this solution. For example, after 100 hours, the drug level reaches 95% of the steady state.

Related Exercise 31 ◄

Figure 9.22

QUICK CHECK 2 If the rate constant in Example 1 were 0.3 instead of 0.03, would the steady-state level of the drug change? If so, to what value? ◄

▶ The idea of stable and unstable equilibrium solutions can be illustrated using a hemispherical bowl and a small ball. When the ball rests at the bottom of the bowl, it is at rest in an equilibrium state. If the ball is moved away from the equilibrium state, it returns to that state.

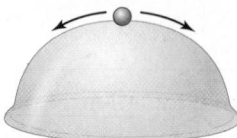

Stable

By contrast, when the ball rests on top of the inverted bowl, it is at rest in an equilibrium state. However, if the ball is moved away from the equilibrium state, it does not return to that state.

Unstable

EXAMPLE 2   Direction field analysis Use direction fields to analyze the behavior of the solutions of the following equations, where $k > 0$ and b is nonzero. Assume $t \geq 0$.

a. $y'(t) = -ky + b$ **b.** $y'(t) = ky + b$

SOLUTION

a. First notice that $y'(t) = 0$ when $y = b/k$. Therefore, the direction field consists of horizontal line segments when $y = b/k$. These horizontal line segments correspond to the constant *equilibrium* solution $y(t) = \dfrac{b}{k}$. Depending on the sign of b, the constant solution could be positive or negative. If $-ky + b > 0$, or equivalently, $y < b/k$, then $y'(t) > 0$, and solutions are increasing in this region. Similarly, if $-ky + b < 0$, or equivalently, $y > b/k$, then $y'(t) < 0$, and solutions are decreasing in this region. **Figure 9.23** shows a typical direction field in the case that $b > 0$. Notice that the solution curves are attracted to the equilibrium solution. For this reason, the equilibrium is said to be *stable*.

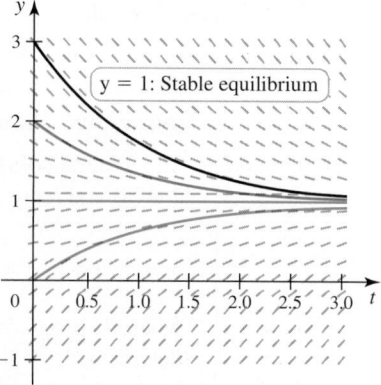

$y = 1$: Stable equilibrium

Figure 9.23

b. The analysis is similar to that in part (a). In this case, we have an equilibrium solution at $y = -b/k$, which may be positive or negative depending on the sign of b. If

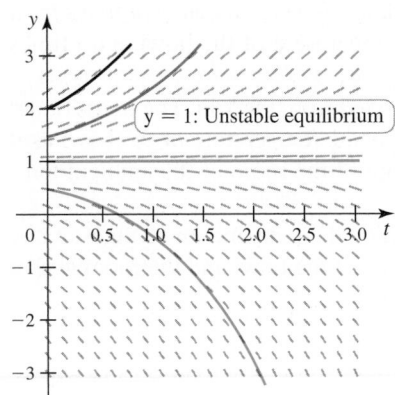

Figure 9.24

$ky + b > 0$, or equivalently, $y > -b/k$, then $y'(t) > 0$, and solutions are increasing in this region. Similarly, if $ky + b < 0$, or equivalently, $y < -b/k$, then $y'(t) < 0$, and solutions are decreasing in this region. Figure 9.24 shows a direction field for $b < 0$. Now the solution curves move away from the equilibrium solution, and the equilibrium is *unstable*.

Related Exercises 17–18 ◄

QUICK CHECK 3 What is the equilibrium solution of the equation $y'(t) = 2y - 4$? Is it stable or unstable? ◄

We give a qualitative summary of the important ideas introduced in Example 2.

SUMMARY   Equilibrium Solutions

The differential equation $y'(t) = f(y)$ has a (constant) **equilibrium** solution $y = a$ when $f(a) = 0$. The equilibrium is **stable** if initial conditions near $y = a$ produce solutions that approach $y = a$ as $t \to \infty$. The equilibrium is **unstable** if initial conditions near $y = a$ produce solutions that do not approach $y = a$ as $t \to \infty$.

EXAMPLE 3   Paying off a loan Suppose you borrow \$60,000 with a monthly interest rate of 0.5% and plan to pay it back with monthly payments of \$600. The balance in the loan is described approximately by the initial value problem

$$B'(t) = \underbrace{0.005B}_{\text{interest}} - \underbrace{600}_{\substack{\text{monthly} \\ \text{payments}}}, B(0) = 60{,}000,$$

where $B(t)$ is the balance in the loan after t months. Notice that the interest increases the loan balance, while the monthly payments decrease the loan balance.

a. Find and graph the loan balance function.

b. After approximately how many months does the loan balance reach zero?

SOLUTION

> ► Loan payments are an example of a discrete process (interest is assessed and payments are made each month). However, they may be modeled as a continuous process using a differential equation because the time interval between payments is small compared to the length of the entire loan process. Discrete processes are often modeled using *difference equations*.

a. The differential equation has the form $y'(t) = ky + b$, where $k = 0.005$ month^{-1} and $b = -\$600/\text{month}$. Using the summary box, the general solution is

$$B(t) = Ce^{kt} - \frac{b}{k} = Ce^{0.005t} + 120{,}000.$$

The initial condition implies that

$$B(0) = C + 120{,}000 = 60{,}000 \Rightarrow C = -60{,}000.$$

Therefore, the solution of the initial value problem is

$$B(t) = Ce^{kt} - \frac{b}{k} = 120{,}000 - 60{,}000e^{0.005t}.$$

b. The graph (Figure 9.25) shows the loan balance decreasing and reaching zero at $t \approx 139$ months (which implies an approximate domain of $[0, 139]$). This fact can be confirmed by solving $B(t) = 0$ algebraically.

Related Exercises 23–26 ◄

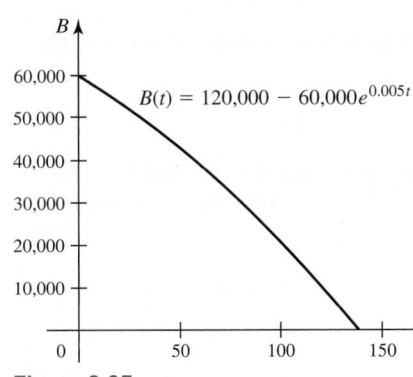

Figure 9.25

$B(t) = 120{,}000 - 60{,}000e^{0.005t}$

Newton's Law of Cooling

Imagine taking a fired bowl out of a hot pottery kiln and putting it on a rack to cool at room temperature. Your intuition tells you that because the temperature of the bowl is greater than the temperature of the room, the pot cools and its temperature approaches the temperature of the room. (We assume the room is sufficiently large that the heating of the room by the bowl is negligible.)

It turns out that this process can be described approximately using a first-order differential equation similar to those studied in this section. That equation is often called

Newton's Law of Cooling, and it is based on the familiar observation that *heat flows from hot to cold*. The solution of the equation gives the temperature of the bowl at all times after it is removed from the kiln.

We let $t = 0$ be the time at which the bowl is removed from the kiln. The temperature of the bowl at any time $t \geq 0$ is $T(t)$, and $T(0) = T_0$ is the temperature of the bowl as it comes out of the kiln. We also let A be the temperature of the room or the ambient temperature. Both T_0 and A are assumed to be known.

Newton's Law of Cooling says that the rate at which the temperature changes at any time is proportional to the temperature *difference* between the bowl and the room at that time; that is,

$$\frac{dT}{dt} = -k(T(t) - A),$$

where $k > 0$ is a constant determined by the thermal properties of the bowl. Notice that the equation makes sense.

• If $T(t) > A$ (the bowl is hotter than the room), then $\dfrac{dT}{dt} < 0$, and the temperature of the bowl decreases (cooling).

• If $T(t) < A$ (the bowl is colder than the room), then $\dfrac{dT}{dt} > 0$, and the temperature of the bowl increases (heating).

We see that Newton's Law of Cooling amounts to a first-order differential equation that we know how to solve. The equation has the form $T'(t) = -kT + b$, where $b = kA$. This equation was studied earlier in the section; its general solution is

$$T(t) = Ce^{-kt} + A.$$

When we use the initial condition $T(0) = T_0$ to determine C, we find that

$$T(0) = C + A = T_0 \quad \Rightarrow \quad C = T_0 - A.$$

Therefore, the solution of the initial value problem is

$$T(t) = (T_0 - A)e^{-kt} + A.$$

Newton's Law of Cooling models the cooling process well when the object is a good conductor of heat and when the temperature is fairly uniform throughout the object.

QUICK CHECK 4 Verify that the solution of the initial value problem satisfies $T(0) = T_0$. What is the solution of the problem if $T_0 = A$? ◄

EXAMPLE 4   Cooling a bowl A bowl is removed from a pottery kiln at a temperature of 200°C and placed on a rack in a room with an ambient temperature of 20°C. Two minutes after the bowl is removed, its temperature is 160°C. Find the temperature of the bowl, for all $t \geq 0$.

SOLUTION Letting $A = 20$, the general solution of the cooling equation is

$$T(t) = Ce^{-kt} + 20.$$

As always, the arbitrary constant is determined using the initial condition $T(0) = 200$. Substituting this condition we find that

$$T(0) = C + 20 = 200 \quad \Rightarrow \quad C = 180.$$

► The value of the thermal constant k is known for common materials. Example 4 illustrates one way to estimate the constant experimentally.

The solution at this point is $T(t) = 180e^{-kt} + 20$, but notice that the constant k is still unknown. It is determined using the additional fact that $T(2) = 160$. We substitute this condition into the solution and solve for k:

$$T(2) = 180e^{-2k} + 20 = 160 \qquad \text{Substitute } t = 2.$$

$$180e^{-2k} = 140 \qquad \text{Rearrange.}$$

$$e^{-2k} = \frac{140}{180} = \frac{7}{9} \qquad \text{Rearrange.}$$

$$k = -\frac{1}{2}\ln\frac{7}{9} \approx 0.126. \qquad \text{Solve for } k.$$

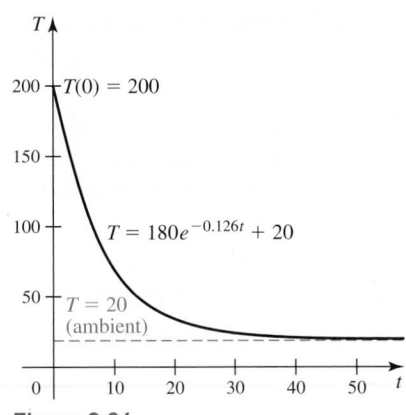

Figure 9.26

Therefore, the solution for $t \geq 0$ is

$$T(t) = 180e^{-kt} + 20 \approx 180e^{-0.126t} + 20.$$

The graph (**Figure 9.26**) confirms that $T(0) = 200$ and that $T(2) = 160$. Notice also that $\lim_{t \to \infty} T(t) = 20$, meaning that the temperature of the bowl approaches the ambient temperature as $t \to \infty$. Equivalently, the solution $T = 20$ is a stable equilibrium of the system.

Related Exercises 27, 30 ◄

QUICK CHECK 5 In general, what is the equilibrium temperature for any Newton cooling problem? Is it a stable or unstable equilibrium? ◄

SECTION 9.4 EXERCISES

Getting Started

1. The general solution of a first-order linear differential equation is $y(t) = Ce^{-10t} - 13$. What solution satisfies the initial condition $y(0) = 4$?

2. What is the general solution of the equation $y'(t) = 3y - 12$?

3. What is the general solution of the equation $y'(t) = -4y + 6$?

4. What is the equilibrium solution of the equation $y'(t) = 3y - 9$? Is it stable or unstable?

Practice Exercises

5–10. First-order linear equations *Find the general solution of the following equations.*

5. $y'(t) = 3y - 4$

6. $y'(x) = -y + 2$

7. $y'(x) + 2y = -4$

8. $y'(x) = 2y + 6$

9. $u'(t) + 12u = 15$

10. $v'(y) - \dfrac{v}{2} = 14$

11–16. Initial value problems *Solve the following initial value problems.*

11. $y'(t) = 3y - 6, y(0) = 9$

12. $y'(x) = -y + 2, y(0) = -2$

13. $y'(t) - 2y = 8, y(0) = 0$

14. $u'(x) = 2u + 6, u(1) = 6$

15. $y'(t) - 3y = 12, y(1) = 4$

16. $z'(t) + \dfrac{z}{2} = 6, z(-1) = 0$

17–22. Stability of equilibrium points *Find the equilibrium solution of the following equations, make a sketch of the direction field, for $t \geq 0$, and determine whether the equilibrium solution is stable. The direction field needs to indicate only whether solutions are increasing or decreasing on either side of the equilibrium solution.*

17. $y'(t) = 12y - 18$

18. $y'(t) = -6y + 12$

19. $y'(t) = -\dfrac{y}{3} - 1$

20. $y'(t) - \dfrac{y}{4} - 1 = 0$

21. $u'(t) + 7u + 21 = 0$

22. $u'(t) - 4u = 3$

23–26. Loan problems *The following initial value problems model the payoff of a loan. In each case, solve the initial value problem, for $t \geq 0$, graph the solution, and determine the first month in which the loan balance is zero.*

23. $B'(t) = 0.005B - 500, B(0) = 50,000$

24. $B'(t) = 0.01B - 750, B(0) = 45,000$

25. $B'(t) = 0.0075B - 1500, B(0) = 100,000$

26. $B'(t) = 0.004B - 800, B(0) = 40,000$

27–30. Newton's Law of Cooling *Solve the differential equation for Newton's Law of Cooling to find the temperature function in the following cases. Then answer any additional questions.*

27. A cup of coffee has a temperature of 90°C when it is poured and allowed to cool in a room with a temperature of 25°C. One minute after the coffee is poured, its temperature is 85°C. How long must you wait until the coffee is cool enough to drink, say 30°C?

28. An iron rod is removed from a blacksmith's forge at a temperature of 900°C. Assume $k = 0.02$ and the rod cools in a room with a temperature of 30°C. When does the temperature of the rod reach 100°C?

29. A glass of milk is moved from a refrigerator with a temperature of 5°C to a room with a temperature of 20°C. One minute later the milk has warmed to a temperature of 7°C. After how many minutes does the milk have a temperature that is 90% of the ambient temperature?

30. A pot of boiling soup (100°C) is put in a cellar with a temperature of 10°C. After 30 minutes, the soup has cooled to 80°C. When will the temperature of the soup reach 30°C?

T 31. Intravenous drug dosing The amount of drug in the blood of a patient (in milligrams) administered via an intravenous line is governed by the initial value problem $y'(t) = -0.02y + 3$, $y(0) = 0$, where t is measured in hours.

a. Find and graph the solution of the initial value problem.

b. What is the steady-state level of the drug?

c. When does the drug level reach 90% of the steady-state value?

T 32. Fish harvesting A fish hatchery has 500 fish at $t = 0$, when harvesting begins at a rate of $b > 0$ fish/year. The fish population is modeled by the initial value problem $y'(t) = 0.01y - b$, $y(0) = 500$, where t is measured in years.

a. Find the fish population, for $t \geq 0$, in terms of the harvesting rate b.

b. Graph the solution in the case that $b = 40$ fish/year. Describe the solution.

c. Graph the solution in the case that $b = 60$ fish/year. Describe the solution.

33. Optimal harvesting rate Let $y(t)$ be the population of a species that is being harvested, for $t \geq 0$. Consider the harvesting model $y'(t) = 0.008y - h$, $y(0) = y_0$, where h is the annual harvesting rate, y_0 is the initial population of the species, and t is measured in years.

a. If $y_0 = 2000$, what harvesting rate should be used to maintain a constant population of $y = 2000$, for $t \geq 0$?

b. If the harvesting rate is $h = 200$/year, what initial population ensures a constant population?

T 34. Endowment model An endowment is an investment account in which the balance ideally remains constant and withdrawals are made on the interest earned by the account. Such an account may be modeled by the initial value problem $B'(t) = rB - m$, for $t \geq 0$, with $B(0) = B_0$. The constant $r > 0$ reflects the annual interest rate, $m > 0$ is the annual rate of withdrawal, B_0 is the initial balance in the account, and t is measured in years.

a. Solve the initial value problem with $r = 0.05$, $m = \$1000$/year, and $B_0 = \$15,000$. Does the balance in the account increase or decrease?

b. If $r = 0.05$ and $B_0 = \$50,000$, what is the annual withdrawal rate m that ensures a constant balance in the account? What is the constant balance?

35. Explain why or why not Determine whether the following statements are true and give an explanation or counterexample.

a. The general solution of $y'(t) = 2y - 18$ is $y(t) = 2e^{2t} + 9$.

b. If $k > 0$ and $b > 0$, then $y(t) = 0$ is never a solution of $y'(t) = ky - b$.

c. The equation $y'(t) = ty(t) + 3$ is separable and can be solved using the methods of this section.

d. According to Newton's Law of Cooling, the temperature of a hot object will reach the ambient temperature after a finite amount of time.

36. Case 2 of the general solution Solve the equation $y'(t) = ky + b$ in the case that $ky + b < 0$ and verify that the general solution is $y(t) = Ce^{kt} - \dfrac{b}{k}$.

37. A bad loan Consider a loan repayment plan described by the initial value problem

$$B'(t) = 0.03B - 600, \quad B(0) = 40,000,$$

where the amount borrowed is $B(0) = \$40,000$, the monthly payments are $\$600$, and $B(t)$ is the unpaid balance in the loan.

a. Find the solution of the initial value problem and explain why B is an increasing function.

b. What is the most that you can borrow under the terms of this loan without going further into debt each month?

c. Now consider the more general loan repayment plan described by the initial value problem

$$B'(t) = rB - m, \quad B(0) = B_0,$$

where $r > 0$ reflects the interest rate, $m > 0$ is the monthly payment, and $B_0 > 0$ is the amount borrowed. In terms of m and r, what is the maximum amount B_0 that can be borrowed without going further into debt each month?

38. Cooling time Suppose an object with an initial temperature of $T_0 > 0$ is put in surroundings with an ambient temperature of A, where $A < \dfrac{T_0}{2}$. Let $t_{1/2}$ be the time required for the object to cool to $\dfrac{T_0}{2}$.

a. Show that $t_{1/2} = -\dfrac{1}{k}\ln\left(\dfrac{T_0 - 2A}{2(T_0 - A)}\right)$.

b. Does $t_{1/2}$ increase or decrease as k increases? Explain.

c. Why is the condition $A < \dfrac{T_0}{2}$ needed?

Explorations and Challenges

39–42. Special equations *A special class of first-order linear equations have the form $a(t)y'(t) + a'(t)y(t) = f(t)$, where a and f are given functions of t. Notice that the left side of this equation can be written as the derivative of a product, so the equation has the form*

$$a(t)y'(t) + a'(t)y(t) = \dfrac{d}{dt}(a(t)y(t)) = f(t).$$

Therefore, the equation can be solved by integrating both sides with respect to t. Use this idea to solve the following initial value problems.

39. $ty'(t) + y = 1 + t$, $y(1) = 4$

40. $t^3y'(t) + 3t^2y = \dfrac{1 + t}{t}$, $y(1) = 6$

41. $e^{-t}y'(t) - e^{-t}y = e^{2t}$, $y(0) = 4$

42. $(t^2 + 1)y'(t) + 2ty = 3t^2$, $y(2) = 8$

43. Change of variables in a Bernoulli equation The equation $y'(t) + ay = by^p$, where a, b, and p are real numbers, is called a *Bernoulli equation*. Unless $p = 1$, the equation is nonlinear and would appear to be difficult to solve—except for a small miracle. Through the change of variables $v(t) = (y(t))^{1-p}$, the equation can be made linear. Carry out the following steps.

a. Letting $v = y^{1-p}$, show that $y'(t) = \dfrac{y(t)^p}{1 - p}v'(t)$.

b. Substitute this expression for $y'(t)$ into the differential equation and simplify to obtain the new (linear) equation $v'(t) = a(1 - p)v = b(1 - p)$, which can be solved using the methods of this section. The solution y of the original equation can then be found from v.

44. Solving Bernoulli equations Use the method outlined in Exercise 43 to solve the following Bernoulli equations.

a. $y'(t) + y = 2y^2$

b. $y'(t) - 2y = 3y^{-1}$

c. $y'(t) + y = \sqrt{y}$

45–48. General first-order linear equations *Consider the general first-order linear equation* $y'(t) + a(t)y(t) = f(t)$. *This equation can be solved, in principle, by defining the* **integrating factor** $p(t) = \exp\left(\int a(t)\,dt\right)$. *Here is how the integrating factor works. Multiply both sides of the equation by p (which is always positive) and show that the left side becomes an exact derivative. Therefore, the equation becomes*

$$p(t)(y'(t) + a(t)y(t)) = \frac{d}{dt}(p(t)y(t)) = p(t)f(t).$$

Now integrate both sides of the equation with respect to t to obtain the solution. Use this method to solve the following initial value problems. Begin by computing the required integrating factor.

45. $y'(t) + \dfrac{1}{t}y(t) = 0,\ y(1) = 6$

46. $y'(t) + \dfrac{3}{t}y(t) = 1 - 2t,\ y(2) = 0$

47. $y'(t) + \dfrac{2t}{t^2 + 1}y(t) = 1 + 3t^2,\ y(1) = 4$

48. $y'(t) + 2ty(t) = 3t,\ y(0) = 1$

QUICK CHECK ANSWERS

1. $y'(t) = Cke^{kt}$, while $ky + b = k(Ce^{kt} - b/k) + b = Cke^{kt}$.   **2.** The steady-state drug level would be $y = 20$. **3.** The equilibrium solution $y = 2$ is unstable. **4.** $T(0) = (T_0 - A) + A = T_0$. If $T_0 = A$, $T(t) = A$ for all $t \geq 0$.   **5.** The ambient temperature is a stable equilibrium. ◄

9.5 Modeling with Differential Equations

Many examples and exercises of this chapter have illustrated the use of differential equations to model various real-world problems. In this concluding section, we focus on three specific applications and explore some of the ideas involved in formulating mathematical models. The first application is the modeling of populations, examples of which we have already encountered. Next we derive the differential equation that governs a mixed-tank reaction. Finally, we introduce and analyze a well-known two-species ecosystem model.

Population Models

So far, we have seen two examples of differential equations that model population growth. Letting $P(t)$ be the population of a species at time $t \geq 0$, both equations have the general form $P'(t) = f(P)$, where $f(P)$ is a function that depends only on the population, and r and K are constants.

$$\text{Exponential growth: } P'(t) = f(P) = rP$$

$$\text{Logistic growth: } P'(t) = f(P) = rP\left(1 - \frac{P}{K}\right)$$

The **growth rate function** f specifies the rate of growth of the population and is chosen to give the best description of the population. Figure 9.27 shows a graph of the growth rate functions for the exponential and logistic models. Note that population P is the variable on the horizontal axis and the growth rate function, which defines P', is on the vertical axis. In both cases, the growth rate function is nonnegative, so both models describe populations that increase. Population values for which $f(P) = 0$ correspond to equilibrium solutions.

For the exponential model, the growth rate function increases linearly with the population size, implying that the larger the population, the larger the growth rate. Therefore, with this model, populations increase (unrealistically) without bound (Figure 9.28).

The growth rate function for the logistic model has zeros at $P = 0$ and $P = K$ (equilibrium solutions) and has a local maximum at $P = K/2$. As a result, the population increases slowly at first, approaches a maximum growth rate, and then grows more slowly as it approaches the **carrying capacity** $P = K$ (Figure 9.28). The important feature of this model is that the population is bounded in size, reflecting the effects of overcrowding or a shortage of resources.

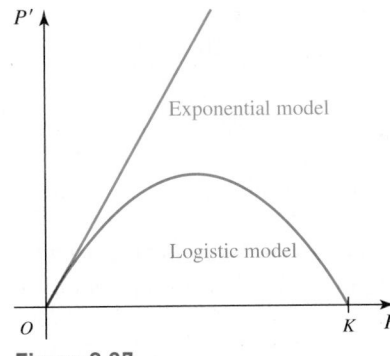

Figure 9.27

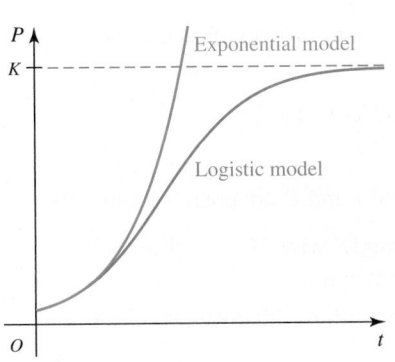

Figure 9.28

An important observation about the logistic model is that when the population is small compared to the carrying capacity (often written $P \ll K$), the population grows exponentially with a rate constant r. We see this fact in the growth rate function:

$$f(P) = rP\left(1 - \frac{P}{K}\right) \approx rP.$$

smal

This fact is also evident in Figure 9.28, where the population curves are nearly identical for small values of t. Therefore, r may be interpreted as the natural growth rate of the species in ideal conditions (unlimited space and resources).

QUICK CHECK 1 Explain why the maximum growth rate for the logistic equation occurs at $P = K/2$. ◄

EXAMPLE 1 Designing a logistic model Wildlife biologists observe a prairie dog community for several years. When observations begin, there are 8 prairie dogs; after one year the population reaches 20 prairie dogs. After 10 years, the population has leveled out at approximately 200 prairie dogs. Assuming a logistic growth model applies to this community, find a function that models the population.

SOLUTION The biologists' measurements suggest that the initial population of the community is $P_0 = 8$ and the carrying capacity is $K = 200$. Using the logistic equation, the resulting initial value problem is

$$P'(t) = rP\left(1 - \frac{P}{200}\right), \quad P(0) = P_0 = 8.$$

Using the methods of Section 9.3, the solution of this problem (Exercise 33) is

$$P(t) = \frac{200}{24e^{-rt} + 1}.$$

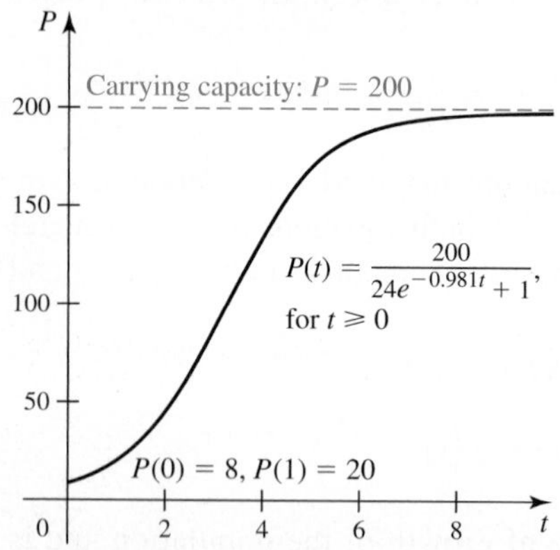

Notice that the natural growth rate r is still undetermined; it is estimated using the fact that the population after one year is 20 prairie dogs. Substituting $P(1) = 20$ into the solution, we have

$$P(1) = \frac{200}{24e^{-r} + 1} = 20 \quad \text{Substitute } t = 1 \text{ and } P = 20.$$

$$e^{-r} = \frac{3}{8} \quad \text{Simplify.}$$

$$r = -\ln\frac{3}{8} \approx 0.981. \quad \text{Take logarithms of both sides.}$$

Substituting this value of r, we obtain the population function shown in **Figure 9.29**. Notice that the initial condition $P(0) = 8$ is satisfied, $P(1) \approx 20$, and the population approaches the carrying capacity of 200 prairie dogs.

Related Exercises 17–18 ◄

Figure 9.29

Carrying capacity: $P = 200$

$P(t) = \dfrac{200}{24e^{-0.981t} + 1}$, for $t \geq 0$

$P(0) = 8, P(1) = 20$

EXAMPLE 2 Gompertz growth model Models of tumor growth often use the Gompertz equation

$$M'(t) = -rM \ln\left(\frac{M}{K}\right),$$

where $M(t)$ is the mass of the tumor at time $t \geq 0$, and r and K are positive constants.

a. Graph the growth rate function for the Gompertz model with $M > 0$, discuss its features, and compare it to the logistic growth rate function.

b. Find the general solution of the Gompertz equation with positive values of r, K, and $M(0) = M_0$, assuming $0 < M_0 < K$.

c. Graph the solution in part (b) when $r = 0.5$, $K = 10$, and $M_0 = 0.01$.

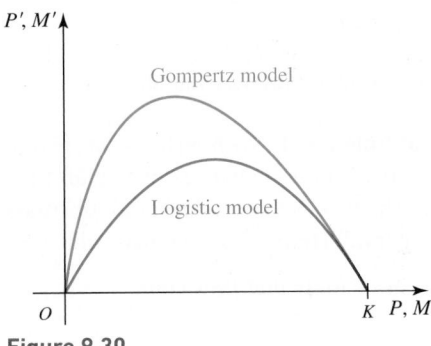

Figure 9.30

> Recall that $\exp(u)$ is another way to write e^u.

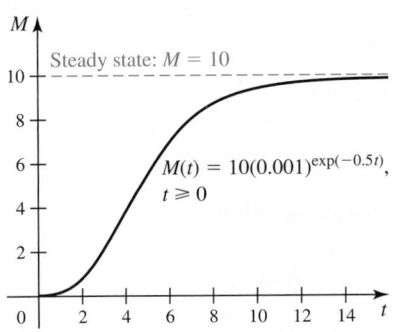

Figure 9.31

SOLUTION

a. The growth rate function $f(M) = -rM \ln\left(\dfrac{M}{K}\right)$ is a skewed version of the logistic growth rate function (**Figure 9.30**). It is left as an exercise (Exercise 32) to show that the Gompertz model has a maximum growth rate of rK/e when $M = K/e$ (compared to the logistic growth rate function, which has a maximum of $rK/4$ when $P = K/2$).

b. The Gompertz equation is separable and is solved using the methods of Section 9.3:

$$\frac{M'(t)}{M \ln\left(\dfrac{M}{K}\right)} = -r \qquad \text{Write equation in separated form.}$$

$$\int \frac{dM}{M \ln\left(\dfrac{M}{K}\right)} = -\int r\, dt \qquad \text{Integrate both sides; } M'(t)\,dt = dM.$$

$$\ln\left|\ln\left(\frac{M}{K}\right)\right| = -rt + C \qquad \text{Integrate with } u = \ln\left(\dfrac{M}{K}\right) \text{ on left side.}$$

$$\ln\left(\frac{M}{K}\right) = Ce^{-rt} \qquad \text{Exponentiate both sides; relabel } e^{\pm C} \text{ as } C.$$

$$M(t) = K \exp\left(Ce^{-rt}\right). \qquad \text{Exponentiate both sides.}$$

We now have the general solution. The initial condition $M(0) = M_0$ implies that $M_0 = Ke^C$. Solving for C, we find that $C = \ln\dfrac{M_0}{K}$.

Substituting this value of C into the general solution gives the solution of the initial value problem (Exercise 19):

$$M(t) = K\left(\frac{M_0}{K}\right)^{\exp(-rt)}.$$

You should check that this unusual solution satisfies the initial condition $M(0) = M_0$. Furthermore, the solution has a steady state given by $\lim\limits_{t\to\infty} M(t) = K$ (Exercise 34).

c. The solution given in part (b) with $r = 0.5$, $K = 10$, and $M_0 = 0.01$ is

$$M(t) = 10(0.001)^{\exp(-0.5t)};$$

its graph is shown in **Figure 9.31**. Notice that the solution approaches the steady-state mass $M = K = 10$.

Related Exercises 20–22 ◄

Stirred Tank Reactions

In their many forms and variations, models of stirred tank reactions are used to simulate industrial and manufacturing processes. They are also adapted so they can be applied to physiological problems, such as the assimilation of drugs by systems of organs. The differential equations governing these reactions can be derived from first principles, and we have the tools to solve them.

A stirred tank reaction takes place in a large tank that is initially filled with a solution of a soluble substance, such as salt or sugar. The solution has a known initial concentration of the substance, measured in grams per liter (g/L). The tank is filled by an inflow pipe at a known rate of R liters per second (L/s) with a solution of the same substance that has a known concentration of C_i g/L. The tank also has an outflow pipe that allows solution to leave the tank at a rate equal to the inflow rate of R (L/s). Therefore, at all times the volume of solution, denoted V and measured in liters, is constant. The configuration of the tank and the names of the various parameters are shown in **Figure 9.32**.

Imagine that at time $t = 0$, the inflow and outflow pipes are opened and solution begins flowing into and out of the tank. We assume at all times the tank is thoroughly stirred, so the solution in the tank has a uniform—but changing—concentration. The goal is to find the mass of the substance (salt or sugar) in the tank at all times.

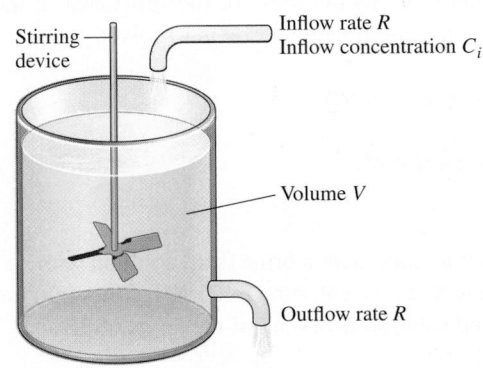

Figure 9.32

QUICK CHECK 2 Suppose the tank is filled with a salt solution that initially has a concentration of 50 g/L and the inflow pipe carries pure water (concentration of 0 g/L). If the stirred tank reaction runs for a long time, what is the eventual concentration of the salt solution in the tank? ◄

A key fact in modeling the stirred tank reaction is that

$$\text{concentration} = \frac{\text{mass}}{\text{volume}} \quad \text{or} \quad \text{mass} = \text{concentration} \cdot \text{volume}.$$

Let $m(t)$ be the mass of the substance in the tank at time $t \geq 0$, with $m(0) = m_0$ given. Assuming the mass $m(t)$ is known at some time t, we ask how it changes in a small time interval $[t, t + \Delta t]$ to give a new mass $m(t + \Delta t)$. The task is to account for all the mass that flows into and out of the tank during this time interval. Here is how the mass changes:

$$\underbrace{m(t + \Delta t)}_{\substack{\text{mass at end} \\ \text{of interval}}} \approx \underbrace{m(t)}_{\substack{\text{current} \\ \text{mass}}} + \underbrace{\text{mass that flows in}}_{C_i R \Delta t} - \underbrace{\text{mass that flows out.}}_{\frac{m(t)}{V} R \Delta t}$$

The crux of the modeling process is to determine the inflow and outflow terms in this equation. Consider the inflow first. Solution flows in at a rate R L/s, so the volume of solution that flows into the tank in Δt seconds is $R\Delta t$ liters (check that the units work out). The solution that flows into the tank has a concentration of C_i g/L; therefore, the mass of substance that flows into the tank in time Δt is

$$\underbrace{C_i}_{\substack{\text{concentration} \\ \text{g/L}}} \cdot \underbrace{R}_{\substack{\text{inflow rate} \\ \text{L/s}}} \cdot \underbrace{\Delta t}_{\substack{\text{time} \\ \text{interval}}} \quad (\text{grams}).$$

(Remember that mass = concentration · volume; check the units.)

Now let's look at the outflow term. At time t, the mass of substance in the tank is $m(t)$, so the concentration of the solution is $m(t)/V$ g/L. As with the inflow, the volume of solution that flows out of the tank in Δt seconds is $R\Delta t$ liters, and the mass of substance that flows out of the tank in time Δt is

$$\underbrace{\frac{m(t)}{V}}_{\substack{\text{concentration} \\ \text{g/L}}} \cdot \underbrace{R}_{\substack{\text{outflow rate} \\ \text{L/s}}} \cdot \underbrace{\Delta t}_{\substack{\text{time} \\ \text{interval}}} \quad (\text{grams}).$$

We now substitute these quantities into the mass change equation:

$$m(t + \Delta t) \approx m(t) + \underbrace{C_i R \Delta t}_{\text{inflow}} - \underbrace{\frac{m(t)}{V} R \Delta t}_{\text{outflow}}.$$

This equation is an approximation because the mass of substance changes during the time interval $[t, t + \Delta t]$. However, the approximation improves as the length of the time interval Δt decreases. We divide through the mass change equation by Δt:

$$\underbrace{\frac{m(t + \Delta t) - m(t)}{\Delta t}}_{\to m'(t) \text{ as } \Delta t \to 0} \approx C_i R - \frac{m(t)}{V} R.$$

Observe that the left side of the equation approaches the derivative $m'(t)$ as Δt approaches zero. The result is a differential equation that governs the mass of the substance in the stirred tank. We have a familiar, linear first-order initial value problem to solve:

$$m'(t) = -\frac{R}{V} m(t) + C_i R, \quad m(0) = m_0.$$

The solution of this equation is analyzed in Exercise 35.

EXAMPLE 3   A stirred tank A 1000-L tank is filled with a brine (salt) solution with an initial concentration of 5 g/L. Brine solution with a concentration of 25 g/L flows into the tank at a rate of 8 L/s, while thoroughly mixed solution flows out of the tank at 8 L/s.

a. Find the mass of salt in the tank, for $t \geq 0$.

b. Find the concentration of the solution in the tank, for $t \geq 0$.

SOLUTION

a. We are given the initial concentration of the solution in the tank. To find the initial mass of salt in the tank, we multiply the concentration by the volume:

$$m_0 = 1000 \text{ L} \cdot 5 \frac{g}{L} = 5000 \text{ g}.$$

The inflow concentration is $C_i = 25$ g/L and inflow rate is $R = 8$ L/s. Therefore, the initial value problem for the reaction is

$$m'(t) = -\frac{8}{1000} m(t) + 25 \cdot 8$$

$$= -0.008m(t) + 200, \quad m(0) = 5000.$$

This is an equation of the form $y'(t) = ky + b$, which was discussed in Section 9.4. Letting $k = -0.008$ and $b = 200$, the general solution is

$$m(t) = Ce^{kt} - \frac{b}{k} = Ce^{-0.008t} - \frac{200}{(-0.008)} = Ce^{-0.008t} + 25,000.$$

The initial condition $m(0) = 5000$, when substituted into the general solution, implies that

$$5000 = C + 25,000 \quad \Rightarrow \quad C = -20,000.$$

Therefore, the solution of the initial value problem is

$$m(t) = 25,000 - 20,000e^{-0.008t}, \text{ for } t \geq 0.$$

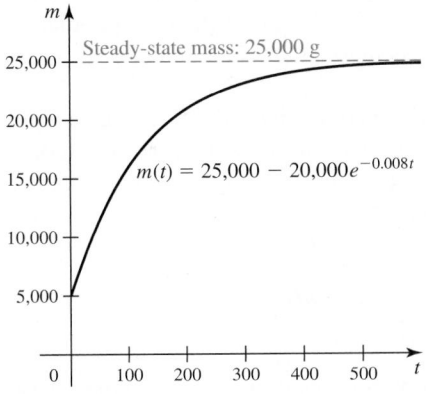

Figure 9.33

The graph of m (**Figure 9.33**) indicates that the mass of salt in the tank approaches 25,000 g as t increases. This mass corresponds to a concentration of 25,000 g/ 1000 L = 25 g/L, which is the concentration of the inflow solution. As time increases, the original solution in the tank is replaced by the inflow solution.

b. The concentration is found by dividing the mass function by the volume of the tank. Therefore, the concentration function is

$$C(t) = 25 - 20e^{-0.008t}, \text{ for } t \geq 0.$$

Related Exercises 24–25 ◄

Predator-Prey Models

Perhaps the best-known graph in wildlife ecology shows 100 years of data, collected by the Hudson Bay Company, of populations of Canadian lynx and snowshoe hare (**Figure 9.34**). The striking features of these graphs are the cyclic fluctuations of the two populations and the fact that the hare population is out of phase with the lynx population. In general, two species may interact in a competitive way, in a cooperative way, or, as in the case of the lynx-hare pair, as predator and prey. In this section, we investigate the fundamental model for describing predator-prey interactions.

Our task is to consider a system consisting of two species—one a predator and one a prey—and to devise a *pair* of differential equations that describes their interactions and whose solutions give their populations. To be specific, let the predator be foxes, whose population at time $t \geq 0$ is $F(t)$, and let the prey be hares, whose population at time $t \geq 0$ is $H(t)$. Here are the assumptions that underlie the model.

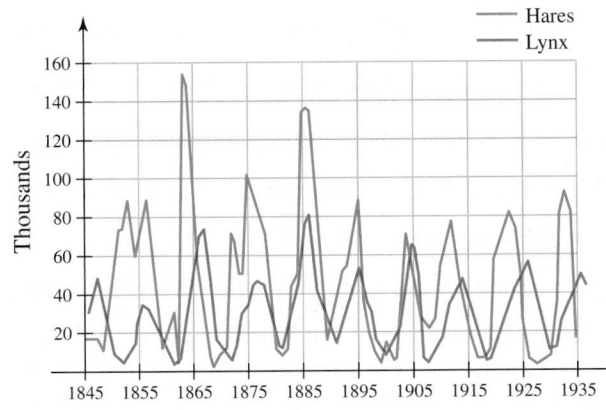

Figure 9.34

➤ The original predator-prey model is attributed to the Belgian mathematician Pierre François Verhulst (1804–1849). The model was further developed independently by the American biophysicist Alfred Lotka and the Italian mathematician Vito Volterra, who used it to study shark populations. These equations are also called the Lotka-Volterra equations.

• When the hare population (prey) is sparse, the fox population decreases exponentially, while encounters between hares and foxes increase the fox population (the hares are the food supply).

• When the fox population (predators) is sparse, the hare population increases exponentially, while encounters between hares and foxes deplete the hare population (the foxes eat the hares).

Here is a set of differential equations that incorporate these assumptions.

$$\underbrace{F'(t)}_{\substack{\text{rate of change} \\ \text{of fox population}}} = \underbrace{-aF(t)}_{\substack{\text{natural decay} \\ \text{of foxes}}} + \underbrace{bF(t)H(t)}_{\substack{\text{increase in foxes} \\ \text{due to fox-hare} \\ \text{encounters}}}$$

$$\underbrace{H'(t)}_{\substack{\text{rate of change} \\ \text{of hare population}}} = \underbrace{cH(t)}_{\substack{\text{natural growth} \\ \text{of hares}}} - \underbrace{dF(t)H(t)}_{\substack{\text{decrease in hares} \\ \text{due to fox-hare} \\ \text{encounters}}}$$

In these equations, a, b, c, and d are positive real numbers.

Notice that in the first equation, the rate of change of the fox population decreases with the size of the fox population and increases with the number of fox-hare interactions. We assume the number of fox-hare interactions is proportional to the product of the fox and hare populations. In the second equation, the rate of change of the hare population increases with the size of the hare population and decreases with the number of fox-hare interactions.

We have no methods for solving such a pair of equations; indeed, finding an analytical solution is challenging (Exercise 39). Fortunately, we can resort to a familiar tool to study the solutions: direction fields. However, in this case, because there are two unknown solutions, the direction field is plotted in the *FH*-plane.

Let's first rewrite the governing equations more compactly.

$$F'(t) = -aF + bFH = F(-a + bH)$$
$$H'(t) = cH - dFH = H(c - dF)$$

As with the direction fields we studied earlier, we look for conditions for which the derivatives are zero, positive, and negative. Because F and H are populations, we assume they have nonnegative values.

The points in the *FH*-plane at which $F'(t) = H'(t) = 0$ are special because they correspond to equilibrium solutions. You should verify that there are two such points: $(F, H) = (0, 0)$ and $(F, H) = \left(\dfrac{c}{d}, \dfrac{a}{b}\right)$. If the two populations have either of these initial values, then they remain constant for all time.

Recalling that $F > 0$, the condition $F'(t) = F(-a + bH) > 0$ is satisfied when $-a + bH > 0$—or, equivalently, when $H > \dfrac{a}{b}$. Using similar reasoning, the condition $F'(t) < 0$ is satisfied when $0 < H < \dfrac{a}{b}$. Repeating this process on the second equation gives $H'(t) = H(c - dF) > 0$ when $0 < F < \dfrac{c}{d}$ and $H'(t) < 0$ when $F > \dfrac{c}{d}$. Figure 9.35 summarizes everything we have learned so far.

We see that the vertical line $F = \dfrac{c}{d}$ and the horizontal line $H = \dfrac{a}{b}$ divide the first quadrant of the *FH*-plane into four regions. In each region, the derivatives of F and H have particular signs. For example, in the region nearest to the origin, we have $F' < 0$ and $H' > 0$, which means that F is decreasing and H is increasing in this region. Therefore, we mark this region with an arrow that points in the direction of decreasing F and increasing H. All solution curves move in the negative F-direction and positive H-direction in this region. Similar arguments explain the arrows in the other three regions of Figure 9.35.

If we stand back and look at Figure 9.35, we can see the general "flow" of the solution curves. They circulate around the equilibrium point in the clockwise direction. While it is not evident from this analysis, it can be shown that the solution curves are actually closed curves; that is, they close on themselves. Therefore, if we choose an initial population of foxes and hares, corresponding to a single point in the *FH*-plane, the resulting solution curve eventually returns to the same point. In the process, both F and H oscillate in a cyclic fashion—as seen in the Hudson Bay data. Figure 9.36 shows two solution curves superimposed on the direction field.

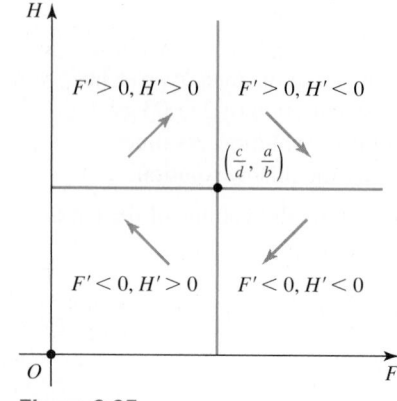

Figure 9.35

▶ Notice that when we plot solutions in the *FH*-plane, the independent variable t does not appear explicitly in the graph. That is, the graph in the *FH*-plane is different from a graph of F as a function of t or of H as a function of t. As a result, it is not possible to determine the period of the oscillations from the graph in the *FH*-plane.

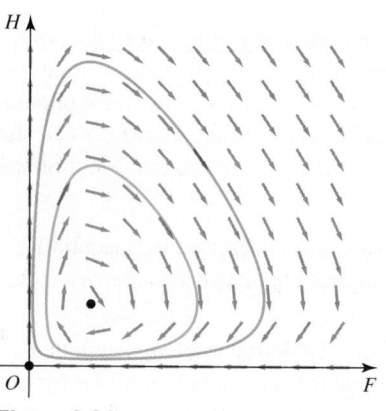

Figure 9.36

QUICK CHECK 3 Explain why a closed solution curve in the *FH*-plane represents fox and hare populations that oscillate in a cyclic way. ◄

▶ The variables in population models are often scaled to some reference quantity. For example, *F* and *H* may be measured in hundreds of individuals, so that $F = 3$ might mean 300 foxes.

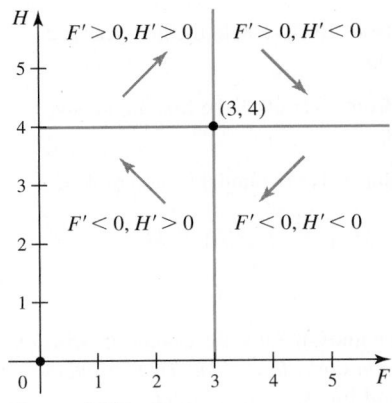

Figure 9.37

EXAMPLE 4 **A predator-prey model** Consider the predator-prey model given by the equations

$$F'(t) = -12F + 3FH,$$
$$H'(t) = 15H - 5FH.$$

a. Find the lines on which $F' = 0$ or $H' = 0$, and the equilibrium points of the system.

b. Make a sketch of four regions in the first quadrant of the *FH*-plane and indicate the directions in which the solution curves move in each region.

c. Sketch a representative solution curve in the *FH*-plane.

SOLUTION

a. Using the first equation and solving $F' = 0$ gives the condition

$$-12F + 3FH = 3F(-4 + H) = 0,$$

which implies that $F' = 0$ when $F = 0$ or when $H = 4$. Using the second equation and solving $H' = 0$ implies that

$$15H - 5FH = 5H(3 - F) = 0.$$

Therefore, $H' = 0$ when $H = 0$ or $F = 3$. The equilibrium points occur when $F' = H' = 0$ (simultaneously). These conditions are satisfied at the points $(0, 0)$ and $(F, H) = (3, 4)$. Therefore, the system has two equilibrium points. These observations are recorded in **Figure 9.37**.

b. The horizontal line $H = 4$ and the vertical line $F = 3$ divide the first quadrant of the *FH*-plane into four regions. The condition

$$F' = -12F + 3FH = 3F(-4 + H) > 0$$

implies that $H > 4$ (recall that $F > 0$). It follows that $F' < 0$ when $0 < H < 4$. Similarly,

$$H' = 15H - 5FH = 5H(3 - F) > 0$$

when $0 < F < 3$, and $H' < 0$ when $F > 3$. These observations are also shown in Figure 9.37, and from them, we can see that the solution curves circulate around the equilibrium point $(3, 4)$ in the clockwise direction.

c. **Figure 9.38** shows the direction field in detail with a typical solution curve superimposed on the direction field.

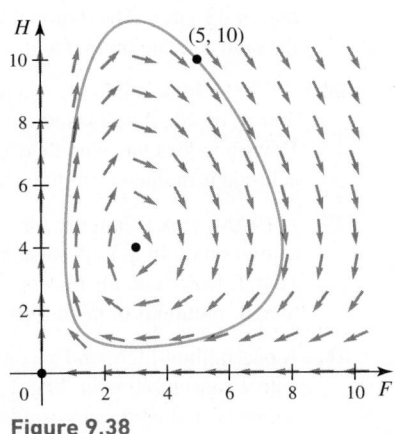

Figure 9.38

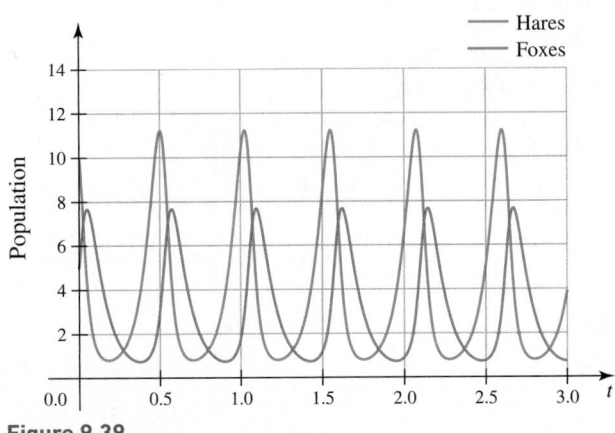

Figure 9.39

A final view of the solutions is obtained by using a numerical method, such as Euler's method, to approximate the solutions of the predator-prey equations. **Figure 9.39** shows the fox and hare populations, now graphed as functions of time. The cyclic behavior is evident, and the period of the oscillations is also seen to be approximately 0.5 time units.

Related Exercises 27–30 ◄

SECTION 9.5 EXERCISES

Getting Started

1. Explain how the growth rate function determines the solution of a population model.

2. What is a carrying capacity? Mathematically, how does it appear on the graph of a population function?

3. Explain how the growth rate function can be decreasing while the population function is increasing.

4. Explain how a stirred tank reaction works.

5. Is the differential equation that describes a stirred tank reaction (as developed in this section) linear or nonlinear? What is its order?

6. What are the assumptions underlying the predator-prey model discussed in this section?

7. Describe the solution curves in a predator-prey model in the FH-plane.

8. Describe the behavior of the two populations in a predator-prey model as functions of time.

9–14. Growth rate functions *Make a sketch of the population function P (as a function of time) that results from the following growth rate functions. Assume the population at time $t = 0$ begins at some positive value.*

9.

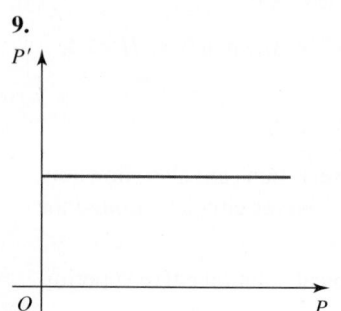

10.

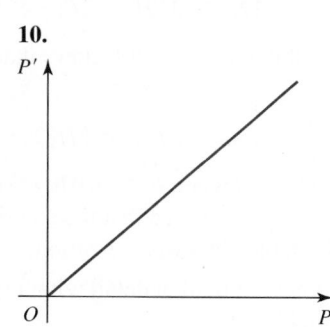

11.

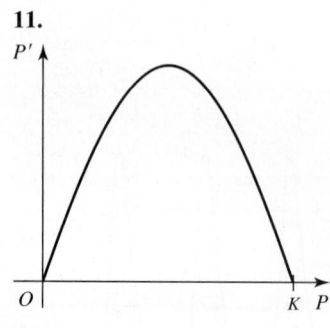

12.

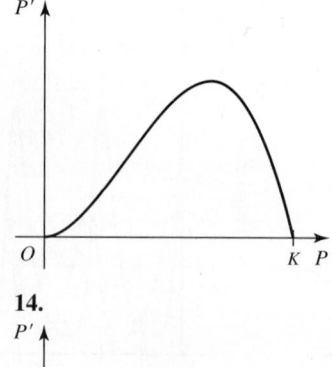

13.

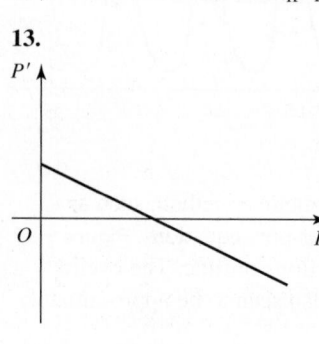

14.

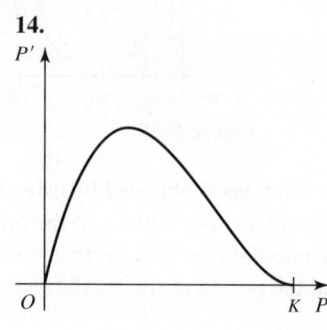

Practice Exercises

T 15–16. Solving logistic equations *Write a logistic equation with the following parameter values. Then solve the initial value problem and graph the solution. Let r be the natural growth rate, K the carrying capacity, and P_0 the initial population.*

15. $r = 0.2, K = 300, P_0 = 50$

16. $r = 0.4, K = 5500, P_0 = 100$

T 17–18. Designing logistic functions *Use the method of Example 1 to find a logistic function that describes the following populations. Graph the population function.*

17. The population increases from 200 to 600 in the first year and eventually levels off at 2000.

18. The population increases from 50 to 60 in the first month and eventually levels off at 150.

19. General Gompertz solution Solve the initial value problem

$$M'(t) = -rM \ln\left(\frac{M}{K}\right), M(0) = M_0$$

with arbitrary positive values of r, K, and M_0.

T 20–22. Solving the Gompertz equation *Solve the Gompertz equation in Exercise 19 with the given values of r, K, and M_0. Then graph the solution to be sure that $M(0)$ and $\lim_{t \to \infty} M(t)$ are correct.*

20. $r = 0.1, K = 500, M_0 = 50$

21. $r = 0.05, K = 1200, M_0 = 90$

22. $r = 0.6, K = 5500, M_0 = 20$

23–26. Stirred tank reactions *For each of the following stirred tank reactions, carry out the following analysis.*

a. Write an initial value problem for the mass of the substance.

b. Solve the initial value problem.

23. A 500-L tank is initially filled with pure water. A copper sulfate solution with a concentration of 20 g/L flows into the tank at a rate of 4 L/min. The thoroughly mixed solution is drained from the tank at a rate of 4 L/min.

24. A 1500-L tank is initially filled with a solution that contains 3000 g of salt. A salt solution with a concentration of 20 g/L flows into the tank at a rate of 3 L/min. The thoroughly mixed solution is drained from the tank at a rate of 3 L/min.

25. A 2000-L tank is initially filled with a sugar solution with a concentration of 40 g/L. A sugar solution with a concentration of 10 g/L flows into the tank at a rate of 10 L/min. The thoroughly mixed solution is drained from the tank at a rate of 10 L/min.

26. A one-million-liter pond is contaminated by a chemical pollutant with a concentration of 20 g/L. The source of the pollutant is removed, and pure water is allowed to flow into the pond at a rate of 1200 L/hr. Assuming the pond is thoroughly mixed and drained at a rate of 1200 L/hr, how long does it take to reduce the concentration of the solution in the pond to 10% of the initial value?

27–30. Predator-prey models *Consider the following pairs of differential equations that model a predator-prey system with populations x and y. In each case, carry out the following steps.*

a. *Identify which equation corresponds to the predator and which corresponds to the prey.*

b. *Find the lines along which $x'(t) = 0$. Find the lines along which $y'(t) = 0$.*

c. *Find the equilibrium points for the system.*

d. *Identify the four regions in the first quadrant of the xy-plane in which x' and y' are positive or negative.*

e. *Sketch a representative solution curve in the xy-plane and indicate the direction in which the solution evolves.*

27. $x'(t) = -3x + 6xy,\ y'(t) = y - 4xy$

28. $x'(t) = 2x - 4xy,\ y'(t) = -y + 2xy$

29. $x'(t) = -3x + xy,\ y'(t) = 2y - xy$

30. $x'(t) = 2x - xy,\ y'(t) = -y + xy$

31. Explain why or why not Determine whether the following statements are true and give an explanation or counterexample.

a. If the growth rate function for a population model is positive, then the population is increasing.

b. The solution of a stirred tank initial value problem always approaches a constant as $t \to \infty$.

c. In the predator-prey models discussed in this section, if the initial predator population is zero and the initial prey population is positive, then the prey population increases without bound.

Explorations and Challenges

32. Growth rate functions

a. Show that the logistic growth rate function $f(P) = rP\left(1 - \dfrac{P}{K}\right)$ has a maximum value of $\dfrac{rK}{4}$ at the point $P = \dfrac{K}{2}$.

b. Show that the Gompertz growth rate function
$f(M) = -rM \ln\left(\dfrac{M}{K}\right)$ has a maximum value of $\dfrac{rK}{e}$ at the point $M = \dfrac{K}{e}$.

33. Solution of the logistic equation Use separation of variables to show that the solution of the initial value problem

$$P'(t) = rP\left(1 - \frac{P}{K}\right), \quad P(0) = P_0$$

is $P(t) = \dfrac{K}{\left(\dfrac{K}{P_0} - 1\right)e^{-rt} + 1}$.

34. Properties of the Gompertz solution Verify that the function

$$M(t) = K\left(\frac{M_0}{K}\right)^{\exp(-rt)}$$

satisfies the properties $M(0) = M_0$ and $\lim\limits_{t \to \infty} M(t) = K$.

35. Properties of stirred tank solutions

a. Show that for general positive values of R, V, C_i, and m_0, the solution of the initial value problem

$$m'(t) = -\frac{R}{V}m(t) + C_i R, \quad m(0) = m_0$$

is $m(t) = (m_0 - C_i V)e^{-Rt/V} + C_i V$.

b. Verify that $m(0) = m_0$.

c. Evaluate $\lim\limits_{t \to \infty} m(t)$ and give a physical interpretation of the result.

d. Suppose m_0 and V are fixed. Describe the effect of increasing R on the graph of the solution.

36. A physiological model A common assumption in modeling drug assimilation is that the blood volume in a person is a single compartment that behaves like a stirred tank. Suppose the blood volume is a four-liter tank that initially has a zero concentration of a particular drug. At time $t = 0$, an intravenous line is inserted into a vein (into the tank) that carries a drug solution with a concentration of 500 mg/L. The inflow rate is 0.06 L/min. Assume the drug is quickly mixed thoroughly in the blood and that the volume of blood remains constant.

a. Write an initial value problem that models the mass of the drug in the blood, for $t \geq 0$.

b. Solve the initial value problem, and graph both the mass of the drug and the concentration of the drug.

c. What is the steady-state mass of the drug in the blood?

d. After how many minutes does the drug mass reach 90% of its steady-state level?

37. RC circuit equation Suppose a battery with voltage V is connected in series to a capacitor (a charge storage device) with capacitance C and a resistor with resistance R. As the charge Q in the capacitor increases, the current I across the capacitor decreases according to the following initial value problems. Solve each initial value problem and interpret the solution. Assume C, R, and V are given.

a. $I'(t) + \dfrac{1}{RC}I(t) = 0,\ I(0) = \dfrac{V}{R}$

b. $Q'(t) + \dfrac{1}{RC}Q(t) = \dfrac{V}{R},\ Q(0) = 0$

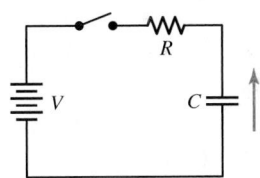

38. U.S. population projections According to the U.S. Census Bureau, the nation's population (to the nearest million) was 296 million in 2005 and 321 million in 2015. The Bureau also projects a 2050 population of 398 million. To construct a logistic model, both the growth rate and the carrying capacity must be estimated. There are several ways to estimate these parameters. Here is one approach:

a. Assume $t = 0$ corresponds to 2005 and that the population growth is exponential for the first ten years; that is, between 2005 and 2015, the population is given by $P(t) = P(0)e^{rt}$. Estimate the growth rate r using this assumption.

b. Write the solution of the logistic equation with the value of r found in part (a). Use the projected value $P(45) = 398$ million to find a value of the carrying capacity K.

c. According to the logistic model determined in parts (a) and (b), when will the U.S. population reach 95% of its carrying capacity?

d. Estimations of this kind must be made and interpreted carefully. Suppose the projected population for 2050 is 410 million rather than 398 million. What is the value of the carrying capacity in this case?

e. Repeat part (d) assuming the projected population for 2050 is 380 million rather than 398 million. What is the value of the carrying capacity in this case?

f. Comment on the sensitivity of the carrying capacity to the 35-year population projection.

39. Analytical solution of the predator-prey equations The solution of the predator-prey equations

$$x'(t) = -ax + bxy, \, y'(t) = cy - dxy$$

can be viewed as parametric equations that describe the solution curves. Assume a, b, c, and d are positive constants and consider solutions in the first quadrant.

a. Recalling that $\dfrac{dy}{dx} = \dfrac{y'(t)}{x'(t)}$, divide the first equation by the second equation to obtain a separable differential equation in terms of x and y.

b. Show that the general solution can be written in the implicit form $e^{dx+by} = Cx^c y^a$, where C is an arbitrary constant.

c. Let $a = 0.8$, $b = 0.4$, $c = 0.9$, and $d = 0.3$. Plot the solution curves for $C = 1.5, 2$, and 2.5, and confirm that they are, in fact, closed curves. Use the graphing window $[0, 9] \times [0, 9]$.

QUICK CHECK ANSWERS

1. The graph of the growth rate function is a parabola with zeros (intercepts on the horizontal axis) at $P = 0$ and $P = K$. The vertex of a parabola occurs at the midpoint of the interval between the zeros, or in this case, at $P = K/2$. **2.** $0 \, \text{g/L}$ **3.** Trace a closed solution curve in the FH-plane, and watch the values of either variable, F or H. As you traverse the curve once, the values of each variable increase, then decrease (or vice versa), and return to their starting values. ◄

CHAPTER 9   REVIEW EXERCISES

1. Explain why or why not Determine whether the following statements are true and give an explanation or counterexample.

a. The differential equation $y' + 2y = t$ is first-order, linear, and separable.

b. The differential equation $y'y = 2t^2$ is first-order, linear, and separable.

c. The function $y = t + 1/t$, for $t > 0$, satisfies the initial value problem $ty' + y = 2t$, $y(1) = 2$.

d. The direction field for the differential equation $y'(t) = t + y(t)$ is plotted in the ty-plane.

e. Euler's method gives the exact solution to the initial value problem $y' = ty^2$, $y(0) = 3$ on the interval $[0, a]$ provided a is not too large.

2–10. General solutions *Use the method of your choice to find the general solution of the following differential equations.*

2. $y'(t) + 3y = 0$

3. $y'(t) + 2y = 6$

4. $p'(x) = 4p + 8$

5. $y'(t) = 2ty$

6. $y'(t) = \sqrt{\dfrac{y}{t}}$

7. $y'(t) = \dfrac{y}{t^2 + 1}$

8. $y'(x) = \dfrac{\sin x}{2y}$

9. $y'(t) = (2t + 1)(y^2 + 1)$

10. $z'(t) = \dfrac{tz}{t^2 + 1}$

11–18. Solving initial value problems *Use the method of your choice to find the solution of the following initial value problems.*

11. $y'(t) = 2t + \cos t$, $y(0) = 1$

12. $y'(t) = -3y + 9$, $y(0) = 4$

13. $Q'(t) = Q - 8$, $Q(1) = 0$

14. $y'(x) = \dfrac{x}{y}$, $y(2) = 4$

15. $u'(t) = \left(\dfrac{u}{t}\right)^{1/3}$, $u(1) = 8$

16. $y'(x) = 4x \csc y$, $y(0) = \dfrac{\pi}{2}$

17. $t(t^2 + 1)s'(t) = s$, $s(1) = 1$

18. $\theta'(x) = 4x \cos^2 \theta$, $\theta(0) = \dfrac{\pi}{4}$

19. Direction fields Consider the direction field for the equation $y' = y(2 - y)$ shown in the figure and initial conditions of the form $y(0) = A$.

a. Sketch a solution on the direction field with the initial condition $y(0) = 1$.

b. Sketch a solution on the direction field with the initial condition $y(0) = 3$.

c. For what values of A are the corresponding solutions increasing, for $t \geq 0$?

d. For what values of A are the corresponding solutions decreasing, for $t \geq 0$?

e. Identify the equilibrium solutions for the differential equation.

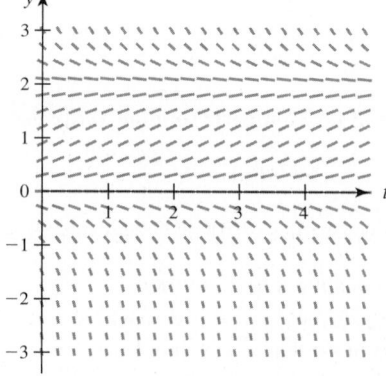

20. Direction fields The direction field for the equation
$y'(t) = t - y$, for $|t| \le 4$ and $|y| \le 4$, is shown in the figure.

  a. Use the direction field to sketch the solution curve that passes through the point $(0, 1/2)$.

  b. Use the direction field to sketch the solution curve that passes through the point $(0, -1/2)$.

  c. In what region of the ty-plane are solutions increasing? Decreasing?

  d. Complete the following sentence. The solution of the differential equation with the initial condition $y(0) = A$, where A is a real number, approaches the line _____ as $t \to \infty$.

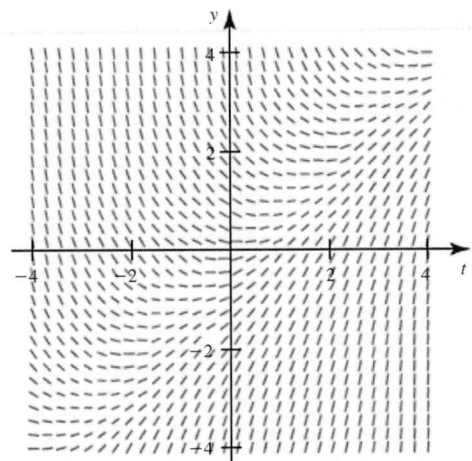

21. Euler's method Consider the initial value problem $y'(t) = \dfrac{1}{2y}$, $y(0) = 1$.

  a. Use Euler's method with $\Delta t = 0.1$ to compute approximations to $y(0.1)$ and $y(0.2)$.

  b. Use Euler's method with $\Delta t = 0.05$ to compute approximations to $y(0.1)$ and $y(0.2)$.

  c. The exact solution of this initial value problem is $y = \sqrt{t + 1}$, for $t > -1$. Compute the errors on the approximations to $y(0.2)$ found in parts (a) and (b). Which approximation gives the smaller error?

22–25. Equilibrium solutions *Find the equilibrium solutions of the following equations and determine whether each solution is stable or unstable.*

22. $y'(t) = y(2 - y)$ **23.** $y'(t) = y(3 + y)(y - 5)$

24. $y'(t) = \sin 2y$, for $|y| < \pi$ **25.** $y'(t) = y^3 - y^2 - 2y$

26. Logistic growth The population of a rabbit community is governed by the initial value problem

$$P'(t) = 0.2P\left(1 - \frac{P}{1200}\right), P(0) = 50.$$

  a. Find the equilibrium solutions.

  b. Find the population, for all times $t \ge 0$.

  c. What is the carrying capacity of the population?

  d. What is the population when the growth rate is a maximum?

27. Logistic growth parameters A cell culture has a population of 20 when a nutrient solution is added at $t = 0$. After 20 hours, the cell population is 80 and the carrying capacity of the culture is estimated to be 1600 cells.

  a. Use the population data at $t = 0$ and $t = 20$ to find the natural growth rate of the population.

  b. Find and solve the logistic equation for the cell population.

  c. After how many hours does the population reach half of the carrying capacity?

28. Logistic growth in India The population of India was 435 million in 1960 ($t = 0$) and 487 million in 1965 ($t = 5$). The projected population for 2050 is 1.57 billion.

  a. Assume the population increased exponentially between 1960 and 1965, and use the populations in these years to determine the natural growth rate in a logistic model.

  b. Use the solution of the logistic equation and the 2050 projected population to determine the carrying capacity.

  c. Based on the values of r and K found in parts (a) and (b), write the logistic growth function for India's population (measured in millions of people).

  d. In approximately what year does the population of India first exceed 2 billion people?

  e. Discuss some possible shortcomings of this model. Why might the carrying capacity be either greater than or less than the value predicted by the model?

29. Stirred tank reaction A 100-L tank is filled with pure water when an inflow pipe is opened and a sugar solution with a concentration of 20 gm/L flows into the tank at a rate of 0.5 L/min. The solution is thoroughly mixed and flows out of the tank at a rate of 0.5 L/min.

  a. Find the mass of sugar in the tank at all times after the inflow pipe is opened.

  b. What is the steady-state mass of sugar in the tank?

  c. At what time does the mass of sugar reach 95% of its steady-state level?

30. Newton's Law of Cooling A cup of coffee is removed from a microwave oven with a temperature of 80°C and allowed to cool in a room with a temperature of 25°C. Five minutes later, the temperature of the coffee is 60°C.

  a. Find the rate constant k for the cooling process.

  b. Find the temperature of the coffee, for $t \ge 0$.

  c. When does the temperature of the coffee reach 50°C?

31. A predator-prey model Consider the predator-prey model
$$x'(t) = -4x + 2xy, \, y'(t) = 5y - xy.$$

  a. Does x represent the population of the predator species or the prey species?

  b. Find the lines along which $x'(t) = 0$. Find the lines along which $y'(t) = 0$.

  c. Find the equilibrium points for the system.

  d. Identify the four regions in the first quadrant of the xy-plane in which x' and y' are positive or negative.

  e. Sketch a typical solution curve in the xy-plane. In which direction does the solution evolve?

32. A first-order equation Consider the equation
$t^2 y'(t) + 2ty(t) = e^{-t}$.

  a. Show that the left side of the equation can be written as the derivative of a single term.

  b. Integrate both sides of the equation to obtain the general solution.

  c. Find the solution that satisfies the condition $y(1) = 0$.

33. A second-order equation Consider the equation
$t^2 y''(t) + 2ty'(t) - 12y(t) = 0.$

    a. Look for solutions of the form $y(t) = t^p$, where p is to be determined. Substitute this trial solution into the equation and find two values of p that give solutions; call them p_1 and p_2.

    b. Assuming the general solution of the equation is $y(t) = C_1 t^{p_1} + C_2 t^{p_2}$, find the solution that satisfies the conditions $y(1) = 0, y'(1) = 7$.

Chapter 9 Guided Projects

Applications of the material in this chapter and related topics can be found in the following Guided Projects. For additional information, see the Preface.

- Cooling coffee
- Euler's method for differential equations
- Predator-prey models
- Period of the pendulum
- Terminal velocity
- Logistic growth
- A pursuit problem

10

Sequences and Infinite Series

Chapter Preview This chapter covers topics that lie at the foundation of calculus—indeed, at the foundation of mathematics. The first task is to make a clear distinction between a *sequence* and an *infinite series*. A sequence is an ordered *list* of numbers, $a_1, a_2, \ldots$, while an infinite series is a *sum* of numbers, $a_1 + a_2 + \cdots$. The idea of convergence to a limit is important for both sequences and series, but convergence is analyzed differently in the two cases. To determine limits of sequences, we use the same tools used for limits of functions at infinity. The analysis of infinite series requires new methods, which are introduced in this chapter. The study of infinite series begins with *geometric series*, which have theoretical importance and are used to answer many practical questions (When is your auto loan paid off? How much antibiotic is in your blood if you take three pills per day?). We then present several tests that are used to determine whether a given series converges. In the final section of the chapter, we offer guidelines to help choose an appropriate convergence test for a given series.

10.1 An Overview

To understand sequences and series, you must understand how they differ and how they are related. The purposes of this opening section are to introduce sequences and series in concrete terms, and to illustrate both their differences and their relationships with each other.

Examples of Sequences

> ▶ In keeping with common practice, the terms *series* and *infinite series* are used interchangeably throughout this chapter.

> ▶ The dots (. . . , an ellipsis) after the last number of a sequence mean that the list continues indefinitely.

Consider the following *list* of numbers:

$$\{1, 4, 7, 10, 13, 16, \ldots\}.$$

Each number in the list is obtained by adding 3 to the previous number in the list. With this rule, we could extend the list indefinitely.

This list is an example of a *sequence*, where each number in the sequence is called a **term** of the sequence. We denote sequences in any of the following forms:

$$\{a_1, a_2, a_3, \ldots, a_n, \ldots\}, \quad \{a_n\}_{n=1}^{\infty}, \quad \text{or} \quad \{a_n\}.$$

The subscript n that appears in a_n is called an **index**, and it indicates the order of terms in the sequence. The choice of a starting index is arbitrary, but sequences usually begin with $n = 0$ or $n = 1$.

The sequence $\{1, 4, 7, 10, \dots\}$ can be defined in two ways. First, we have the rule that each term of the sequence is 3 more than the previous term; that is, $a_2 = a_1 + 3$, $a_3 = a_2 + 3$, $a_4 = a_3 + 3$, and so forth. In general, we see that

$$a_1 = 1 \quad \text{and} \quad a_{n+1} = a_n + 3, \qquad \text{for } n = 1, 2, 3, \dots.$$

This way of defining a sequence is called a *recurrence relation* (or an *implicit formula*). It specifies the initial term of the sequence (in this case, $a_1 = 1$) and gives a general rule for computing the next term of the sequence from previous terms. For example, if you know a_{100}, the recurrence relation can be used to find a_{101}.

Suppose instead you want to find a_{147} directly without computing the first 146 terms of the sequence. The first four terms of the sequence can be written

$$a_1 = 1 + (3 \cdot 0), \quad a_2 = 1 + (3 \cdot 1), \quad a_3 = 1 + (3 \cdot 2), \quad a_4 = 1 + (3 \cdot 3).$$

Observe the pattern: The nth term of the sequence is 1 plus 3 multiplied by $n - 1$, or

$$a_n = 1 + 3(n - 1) = 3n - 2, \qquad \text{for } n = 1, 2, 3, \dots.$$

> **QUICK CHECK 1** Find a_{10} for the sequence $\{1, 4, 7, 10, \dots\}$ using the recurrence relation and then again using the explicit formula for the nth term. ◄

With this *explicit formula*, the nth term of the sequence is determined directly from the value of n. For example, with $n = 147$,

$$a_{147} = 3 \cdot \underbrace{147}_{n} - 2 = 439.$$

> ► When defined by an explicit formula $a_n = f(n)$, it is evident that sequences are functions. The domain is generally a subset of the nonnegative integers, and one real number a_n is assigned to each integer n in the domain.

DEFINITION **Sequence**

A **sequence** $\{a_n\}$ is an ordered list of numbers of the form

$$\{a_1, a_2, a_3, \dots, a_n, \dots\}.$$

A sequence may be generated by a **recurrence relation** of the form $a_{n+1} = f(a_n)$, for $n = 1, 2, 3, \dots$, where a_1 is given. A sequence may also be defined with an **explicit formula** of the form $a_n = f(n)$, for $n = 1, 2, 3, \dots$.

EXAMPLE 1 **Explicit formulas** Use the explicit formula for $\{a_n\}_{n=1}^{\infty}$ to write the first four terms of each sequence. Sketch a graph of the sequence.

a. $a_n = \dfrac{1}{2^n}$ **b.** $a_n = \dfrac{(-1)^n \, n}{n^2 + 1}$

SOLUTION

a. Substituting $n = 1, 2, 3, 4, \dots$ into the explicit formula $a_n = \dfrac{1}{2^n}$, we find that the terms of the sequence are

$$\left\{\frac{1}{2}, \frac{1}{2^2}, \frac{1}{2^3}, \frac{1}{2^4}, \dots\right\} = \left\{\frac{1}{2}, \frac{1}{4}, \frac{1}{8}, \frac{1}{16}, \dots\right\}.$$

The graph of a sequence is the graph of a function that is defined only on a set of integers. In this case, we plot the coordinate pairs (n, a_n), for $n = 1, 2, 3, \dots$, resulting in a graph consisting of individual points. The graph of the sequence $a_n = \dfrac{1}{2^n}$ suggests that the terms of this sequence approach 0 as n increases (Figure 10.1).

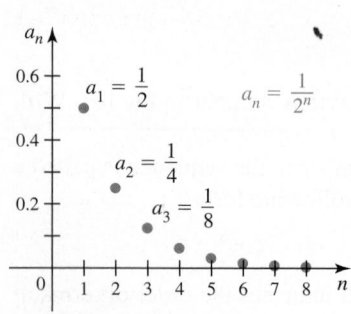

Figure 10.1

b. Substituting $n = 1, 2, 3, 4, \dots$ into the explicit formula, the terms of the sequence are

> ► The "switch" $(-1)^n$ is used frequently to alternate the signs of the terms of sequences and series.

$$\left\{\frac{(-1)^1(1)}{1^2 + 1}, \frac{(-1)^2 2}{2^2 + 1}, \frac{(-1)^3 3}{3^2 + 1}, \frac{(-1)^4 4}{4^2 + 1}, \dots\right\} = \left\{-\frac{1}{2}, \frac{2}{5}, -\frac{3}{10}, \frac{4}{17}, \dots\right\}.$$

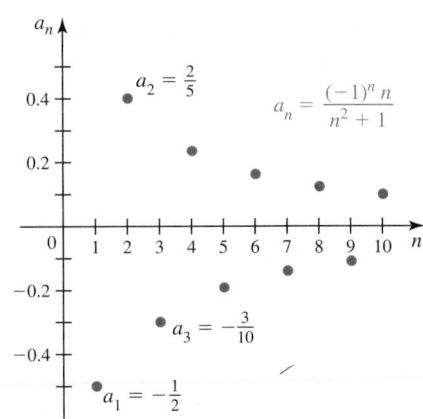

Figure 10.2

From the graph (**Figure 10.2**), we see that the terms of the sequence alternate in sign and appear to approach 0 as n increases.

Related Exercises 13, 16 ◀

EXAMPLE 2   **Recurrence relations**   Use the recurrence relation for $\{a_n\}_{n=1}^{\infty}$ to write the first four terms of the sequences

$$a_{n+1} = 2a_n + 1,\, a_1 = 1 \quad \text{and} \quad a_{n+1} = 2a_n + 1,\, a_1 = -1.$$

SOLUTION   Notice that the recurrence relation is the same for the two sequences; only the first term differs. The first four terms of each sequence are as follows.

n	a_n with $a_1 = 1$	a_n with $a_1 = -1$
1	$a_1 = 1$ (given)	$a_1 = -1$ (given)
2	$a_2 = 2a_1 + 1 = 2 \cdot 1 + 1 = 3$	$a_2 = 2a_1 + 1 = 2(-1) + 1 = -1$
3	$a_3 = 2a_2 + 1 = 2 \cdot 3 + 1 = 7$	$a_3 = 2a_2 + 1 = 2(-1) + 1 = -1$
4	$a_4 = 2a_3 + 1 = 2 \cdot 7 + 1 = 15$	$a_4 = 2a_3 + 1 = 2(-1) + 1 = -1$

QUICK CHECK 2 Find an explicit formula for the sequence $\{1, 3, 7, 15, \ldots\}$ (Example 2). ◀

We see that the terms of the first sequence increase without bound, while all terms of the second sequence are -1. Clearly, the initial term of the sequence may determine the behavior of the entire sequence.

Related Exercises 21–22 ◀

EXAMPLE 3   **Working with sequences**   Consider the following sequences.

a. $\{a_n\} = \{-2, 5, 12, 19, \ldots\}$ **b.** $\{b_n\} = \{3, 6, 12, 24, 48, \ldots\}$

(i)   Find the next two terms of the sequence.

(ii)   Find a recurrence relation that generates the sequence.

(iii)   Find an explicit formula for the nth term of the sequence.

SOLUTION

a. (i)   Each term is obtained by adding 7 to its predecessor. The next two terms are $19 + 7 = 26$ and $26 + 7 = 33$.

➤ In Example 3, we chose the starting index $n = 0$. Other choices are possible.

(ii)   Because each term is seven more than its predecessor, a recurrence relation is

$$a_{n+1} = a_n + 7,\, a_0 = -2, \quad \text{for } n = 0, 1, 2, \ldots.$$

(iii)   Notice that $a_0 = -2$, $a_1 = -2 + (1 \cdot 7)$, and $a_2 = -2 + (2 \cdot 7)$, so an explicit formula is

$$a_n = 7n - 2, \quad \text{for } n = 0, 1, 2, \ldots.$$

b. (i)   Each term is obtained by multiplying its predecessor by 2. The next two terms are $48 \cdot 2 = 96$ and $96 \cdot 2 = 192$.

(ii)   Because each term is two times its predecessor, a recurrence relation is

$$a_{n+1} = 2a_n,\, a_0 = 3, \quad \text{for } n = 0, 1, 2, \ldots.$$

(iii)   To obtain an explicit formula, note that $a_0 = 3$, $a_1 = 3(2^1)$, and $a_2 = 3(2^2)$. In general,

$$a_n = 3(2^n), \quad \text{for } n = 0, 1, 2, \ldots.$$

Related Exercises 27–28 ◀

Limit of a Sequence

Perhaps the most important question about a sequence is this: If you go farther and farther out in the sequence, $a_{100}, \ldots, a_{10,000}, \ldots, a_{100,000}, \ldots$, how do the terms of the sequence behave? Do they approach a specific number, and if so, what is that number? Or do they grow in magnitude without bound? Or do they wander around with or without a pattern?

The long-term behavior of a sequence is described by its **limit**. The limit of a sequence is defined rigorously in the next section. For now, we work with an informal definition.

DEFINITION Limit of a Sequence

If the terms of a sequence $\{a_n\}$ approach a unique number L as n increases—that is, if a_n can be made arbitrarily close to L by taking n sufficiently large—then we say $\lim\limits_{n \to \infty} a_n = L$ exists, and the sequence **converges** to L. If the terms of the sequence do not approach a single number as n increases, the sequence has no limit, and the sequence **diverges**.

EXAMPLE 4   Limits of sequences Write the first four terms of each sequence. If you believe the sequence converges, make a conjecture about its limit. If the sequence appears to diverge, explain why.

a. $\left\{ \dfrac{(-1)^n}{n^2 + 1} \right\}_{n=1}^{\infty}$ Explicit formula

b. $\{ \cos n\pi \}_{n=1}^{\infty}$ Explicit formula

c. $\{ a_n \}_{n=1}^{\infty}$, where $a_{n+1} = -2a_n, a_1 = 1$ Recurrence relation

SOLUTION

a. Beginning with $n = 1$, the first four terms of the sequence are

$$\left\{ \frac{(-1)^1}{1^2 + 1}, \frac{(-1)^2}{2^2 + 1}, \frac{(-1)^3}{3^2 + 1}, \frac{(-1)^4}{4^2 + 1}, \ldots \right\} = \left\{ -\frac{1}{2}, \frac{1}{5}, -\frac{1}{10}, \frac{1}{17}, \ldots \right\}.$$

The terms decrease in magnitude and approach zero with alternating signs. The limit appears to be 0 (**Figure 10.3**).

b. The first four terms of the sequence are

$$\{ \cos \pi, \cos 2\pi, \cos 3\pi, \cos 4\pi, \ldots \} = \{ -1, 1, -1, 1, \ldots \}.$$

In this case, the terms of the sequence alternate between -1 and $+1$, and never approach a single value. Therefore, the sequence diverges (**Figure 10.4**).

c. The first four terms of the sequence are

$$\{ 1, -2a_1, -2a_2, -2a_3, \ldots \} = \{ 1, -2, 4, -8, \ldots \}.$$

Because the magnitudes of the terms increase without bound, the sequence diverges (**Figure 10.5**).

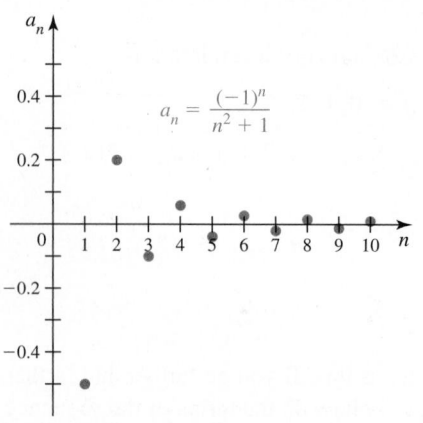

Figure 10.3

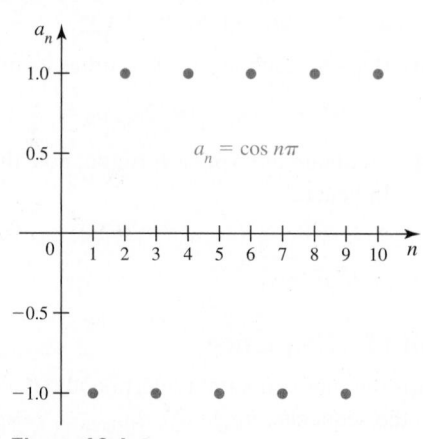

Figure 10.4

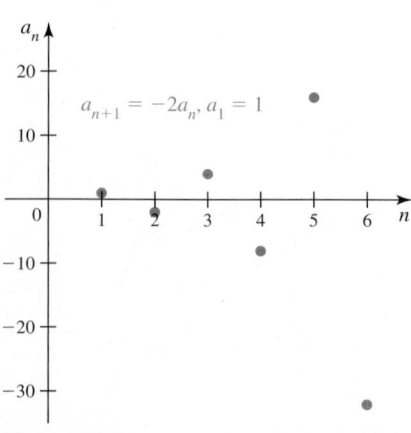

Figure 10.5

Related Exercises 35–36 ◄

EXAMPLE 5   Limit of a sequence Enumerate and graph the terms of the following sequence, and make a conjecture about its limit.

$$a_n = \frac{4n^3}{n^3 + 1}, \qquad \text{for } n = 1, 2, 3, \ldots. \quad \text{Explicit formula}$$

SOLUTION The first 14 terms of the sequence $\{a_n\}$ are tabulated in Table 10.1 and graphed in **Figure 10.6**. The terms appear to approach 4.

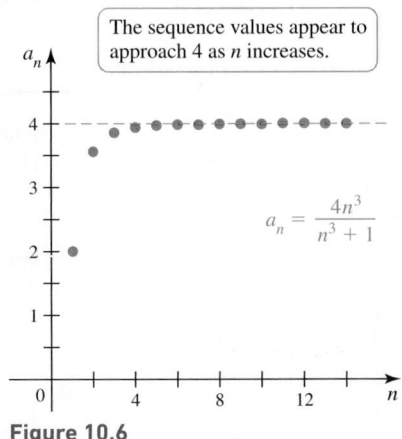

The sequence values appear to approach 4 as n increases.

$$a_n = \frac{4n^3}{n^3 + 1}$$

Figure 10.6

Table 10.1

n	a_n	n	a_n
1	2.000	8	3.992
2	3.556	9	3.995
3	3.857	10	3.996
4	3.938	11	3.997
5	3.968	12	3.998
6	3.982	13	3.998
7	3.988	14	3.999

Related Exercise 45 ◄

EXAMPLE 6   A bouncing ball A basketball tossed straight up in the air reaches a high point and falls to the floor. Each time the ball bounces on the floor it rebounds to 0.8 of its previous height. Let h_n be the high point after the nth bounce, with the initial height being $h_0 = 20$ ft.

a. Find a recurrence relation and an explicit formula for the sequence $\{h_n\}$.

b. What is the high point after the 10th bounce? After the 20th bounce?

c. Speculate on the limit of the sequence $\{h_n\}$.

SOLUTION

a. We first write and graph the heights of the ball for several bounces using the rule that each height is 0.8 of the previous height (**Figure 10.7**). For example, we have

$$h_0 = 20 \text{ ft},$$
$$h_1 = 0.8\, h_0 = 16 \text{ ft},$$
$$h_2 = 0.8\, h_1 = 0.8^2\, h_0 = 12.80 \text{ ft},$$
$$h_3 = 0.8\, h_2 = 0.8^3\, h_0 = 10.24 \text{ ft, and}$$
$$h_4 = 0.8\, h_3 = 0.8^4\, h_0 \approx 8.19 \text{ ft}.$$

Each number in the list is 0.8 of the previous number. Therefore, the recurrence relation for the sequence of heights is

$$h_{n+1} = 0.8\, h_n, \quad h_0 = 20, \qquad \text{for } n = 0, 1, 2, 3, \ldots.$$

To find an explicit formula for the nth term, note that

$$h_1 = h_0 \cdot 0.8, \qquad h_2 = h_0 \cdot 0.8^2, \qquad h_3 = h_0 \cdot 0.8^3, \qquad \text{and} \qquad h_4 = h_0 \cdot 0.8^4.$$

In general, we have

$$h_n = h_0 \cdot 0.8^n = 20 \cdot 0.8^n, \qquad \text{for } n = 0, 1, 2, 3, \ldots,$$

which is an explicit formula for the terms of the sequence.

b. Using the explicit formula for the sequence, we see that after $n = 10$ bounces, the next height is

$$h_{10} = 20 \cdot 0.8^{10} \approx 2.15 \text{ ft}.$$

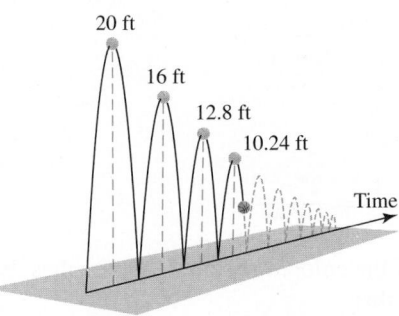

The height of each bounce of the basketball is 0.8 of the height of the previous bounce.

20 ft

16 ft

12.8 ft

10.24 ft

Time

Figure 10.7

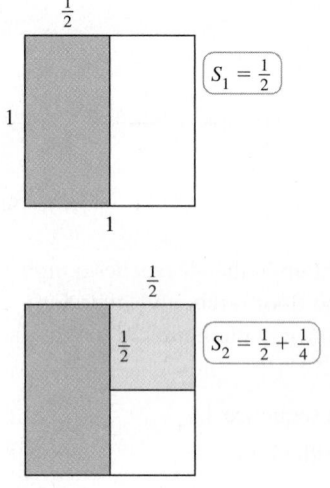

Figure 10.8

After $n = 20$ bounces, the next height is

$$h_{20} = 20 \cdot 0.8^{20} \approx 0.23 \text{ ft.}$$

c. The terms of the sequence (Figure 10.8) appear to decrease and approach 0. A reasonable conjecture is that $\lim_{n \to \infty} h_n = 0$.

Related Exercises 57–60 ◄

Infinite Series and the Sequence of Partial Sums

An infinite series can be viewed as a *sum* of an infinite set of numbers; it has the form

$$a_1 + a_2 + \cdots + a_n + \cdots,$$

where the terms of the series, $a_1, a_2, \ldots$, are real numbers. We first answer the question: How is it possible to sum an infinite set of numbers and produce a finite number? Here is an informative example.

Consider a unit square (sides of length 1) that is subdivided as shown in Figure 10.9. We let S_n be the area of the colored region in the nth figure of the progression. The area of the colored region in the first figure is

$$S_1 = \frac{1}{2} \cdot 1 = \frac{1}{2}. \quad \frac{1}{2} = \frac{2^1 - 1}{2^1}$$

The area of the colored region in the second figure is S_1 plus the area of the smaller blue square, which is $\frac{1}{2} \cdot \frac{1}{2} = \frac{1}{4}$. Therefore,

$$S_2 = \frac{1}{2} + \frac{1}{4} = \frac{3}{4}. \quad \frac{3}{4} = \frac{2^2 - 1}{2^2}$$

The area of the colored region in the third figure is S_2 plus the area of the smaller green rectangle, which is $\frac{1}{2} \cdot \frac{1}{4} = \frac{1}{8}$. Therefore,

$$S_3 = \frac{1}{2} + \frac{1}{4} + \frac{1}{8} = \frac{7}{8}. \quad \frac{7}{8} = \frac{2^3 - 1}{2^3}$$

Continuing in this manner, we find that

$$S_n = \frac{1}{2} + \frac{1}{4} + \frac{1}{8} + \cdots + \frac{1}{2^n} = \frac{2^n - 1}{2^n}.$$

If this process is continued indefinitely, the area of the colored region S_n approaches the area of the unit square, which is 1. So it is plausible that

$$\lim_{n \to \infty} S_n = \underbrace{\frac{1}{2} + \frac{1}{4} + \frac{1}{8} + \cdots}_{\text{sum continues indefinitely}} = 1.$$

The explicit formula $S_n = \dfrac{2^n - 1}{2^n}$ can be analyzed to verify our assertion that $\lim_{n \to \infty} S_n = 1$; we turn to that task in Section 10.2.

This example shows that it is possible to sum an infinite set of numbers and obtain a finite number—in this case, the sum is 1. The sequence $\{S_n\}$ generated in this example is extremely important. It is called a *sequence of partial sums*, and its limit is the value of the infinite series $\frac{1}{2} + \frac{1}{4} + \frac{1}{8} + \cdots$. The idea of a sequence of partial sums is illustrated by the decimal expansion of 1.

EXAMPLE 7   Working with series Consider the infinite series

$$0.9 + 0.09 + 0.009 + 0.0009 + \cdots,$$

where each term of the sum is $\frac{1}{10}$ of the previous term.

a. Find the sum of the first one, two, three, and four terms of the series.

b. What value would you assign to the infinite series $0.9 + 0.09 + 0.009 + \cdots$?

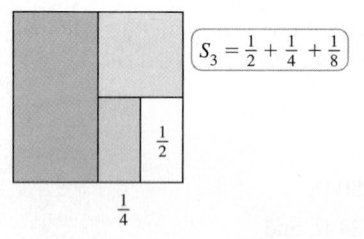

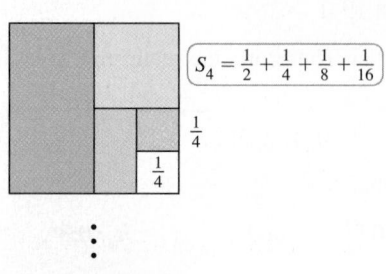

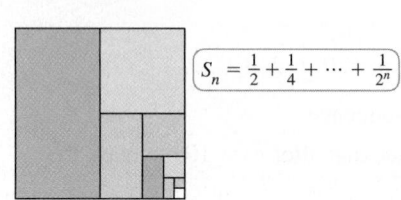

Figure 10.9

SOLUTION

a. Let S_n denote the sum of the first n terms of the given series. Then

$$S_1 = 0.9,$$
$$S_2 = 0.9 + 0.09 = 0.99,$$
$$S_3 = 0.9 + 0.09 + 0.009 = 0.999, \text{ and}$$
$$S_4 = 0.9 + 0.09 + 0.009 + 0.0009 = 0.9999.$$

b. The sums $S_1, S_2, \ldots, S_n$ form a sequence $\{S_n\}$, which is a sequence of partial sums. As more and more terms are included, the values of S_n approach 1. Therefore, a reasonable conjecture for the value of the series is 1:

$$\underbrace{\underbrace{\underbrace{0.9}_{S_1 = 0.9} + 0.09}_{S_2 = 0.99} + 0.009}_{S_3 = 0.999} + 0.0009 + \cdots = 1.$$

Related Exercises 61–62 ◀

> **QUICK CHECK 3** Reasoning as in Example 7, what is the value of $0.3 + 0.03 + 0.003 + \cdots$? ◀

The nth term of the sequence of partial sums is

$$S_n = \underbrace{0.9 + 0.09 + 0.009 + \cdots + 0.0\ldots09}_{n \text{ terms}} = \sum_{k=1}^{n} 9 \cdot 0.1^k.$$

> ► Recall the summation notation introduced in Chapter 5: $\sum_{k=1}^{n} a_k$ means $a_1 + a_2 + \cdots + a_n$.

We observed that $\lim\limits_{n \to \infty} S_n = 1$. For this reason, we write

$$\lim_{n \to \infty} S_n = \lim_{n \to \infty} \underbrace{\sum_{k=1}^{n} 9 \cdot 0.1^k}_{S_n} = \underbrace{\sum_{k=1}^{\infty} 9 \cdot 0.1^k}_{\text{new object}} = 1.$$

By letting $n \to \infty$, a new mathematical object $\sum\limits_{k=1}^{\infty} 9 \cdot 0.1^k$ is created. It is an infinite series, and its value is the *limit* of the sequence of partial sums.

> ► The term *series* is used for historical reasons. When you see *series*, you should think *sum*.

DEFINITION Infinite Series

Given a sequence $\{a_1, a_2, a_3, \ldots\}$, the sum of its terms

$$a_1 + a_2 + a_3 + \cdots = \sum_{k=1}^{\infty} a_k$$

is called an **infinite series**. The **sequence of partial sums** $\{S_n\}$ associated with this series has the terms

$$S_1 = a_1$$
$$S_2 = a_1 + a_2$$
$$S_3 = a_1 + a_2 + a_3$$
$$\vdots$$
$$S_n = a_1 + a_2 + a_3 + \cdots + a_n = \sum_{k=1}^{n} a_k, \quad \text{for } n = 1, 2, 3, \ldots.$$

If the sequence of partial sums $\{S_n\}$ has a limit L, the infinite series **converges** to that limit, and we write

$$\sum_{k=1}^{\infty} a_k = \lim_{n \to \infty} \underbrace{\sum_{k=1}^{n} a_k}_{S_n} = \lim_{n \to \infty} S_n = L.$$

> **QUICK CHECK 4** Do the series $\sum\limits_{k=1}^{\infty} 1$ and $\sum\limits_{k=1}^{\infty} k$ converge or diverge? ◀

If the sequence of partial sums diverges, the infinite series also **diverges**.

EXAMPLE 8   Sequence of partial sums   Consider the infinite series

$$\sum_{k=1}^{\infty} \frac{1}{k(k+1)}.$$

a. Find the first four terms of the sequence of partial sums.

b. Find an expression for S_n and make a conjecture about the value of the series.

SOLUTION

a. The sequence of partial sums can be evaluated explicitly:

$$S_1 = \sum_{k=1}^{1} \frac{1}{k(k+1)} = \frac{1}{2},$$

$$S_2 = \sum_{k=1}^{2} \frac{1}{k(k+1)} = \frac{1}{2} + \frac{1}{6} = \frac{2}{3},$$

$$S_3 = \sum_{k=1}^{3} \frac{1}{k(k+1)} = \frac{1}{2} + \frac{1}{6} + \frac{1}{12} = \frac{3}{4}, \text{ and}$$

$$S_4 = \sum_{k=1}^{4} \frac{1}{k(k+1)} = \frac{1}{2} + \frac{1}{6} + \frac{1}{12} + \frac{1}{20} = \frac{4}{5}.$$

b. Based on the pattern in the sequence of partial sums, a reasonable conjecture is that

$$S_n = \frac{n}{n+1}, \text{ for } n = 1, 2, 3, \ldots, \text{ which produces the sequence } \left\{ \frac{1}{2}, \frac{2}{3}, \frac{3}{4}, \frac{4}{5}, \frac{5}{6}, \ldots \right\}.$$

It appears that the values of S_n approach 1 (**Figure 10.10**), so a reasonable conjecture is that $\lim_{n \to \infty} S_n = \lim_{n \to \infty} \frac{n}{n+1} = 1$. Therefore, we claim that

$$\sum_{k=1}^{\infty} \frac{1}{k(k+1)} = \lim_{n \to \infty} S_n = 1.$$

Related Exercise 67 ◄

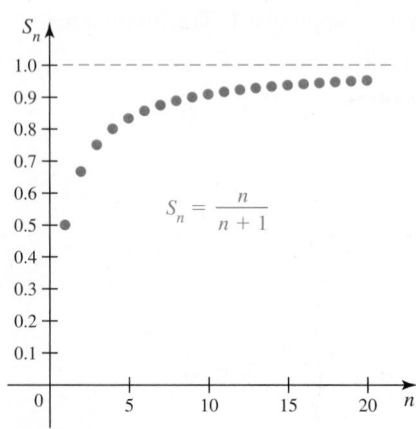

Figure 10.10

QUICK CHECK 5 Find the first four terms of the sequence of partial sums for the series $\sum_{k=1}^{\infty} (-1)^k k$. Does the series converge or diverge? ◄

Summary

This section features three key ideas to keep in mind.

• A *sequence* $\{a_1, a_2, \ldots, a_n, \ldots\}$ is an ordered *list* of numbers.

• An *infinite series* $\sum_{k=1}^{\infty} a_k = a_1 + a_2 + a_3 + \cdots$ is a *sum* of numbers.

• A *sequence of partial sums* $\{S_1, S_2, S_3, \ldots\}$, where $S_n = a_1 + a_2 + \cdots + a_n$, is used to evaluate the series $\sum_{k=1}^{\infty} a_k$.

For sequences, we ask about the behavior of the individual terms as we go out farther and farther in the list; that is, we ask about $\lim_{n \to \infty} a_n$. For infinite series, we examine the sequence of partial sums related to the series. If the sequence of partial sums $\{S_n\}$ has a limit, then the infinite series $\sum_{k=1}^{\infty} a_k$ converges to that limit. If the sequence of partial sums does not have a limit, the infinite series diverges.

Table 10.2 shows the correspondences between sequences/series and functions, and between summation and integration. For a sequence, the index n plays the role of the independent variable and takes on integer values; the terms of the sequence $\{a_n\}$ correspond to the dependent variable.

With sequences $\{a_n\}$, the idea of accumulation corresponds to summation, whereas with functions, accumulation corresponds to integration. A finite sum is analogous to integrating a function over a finite interval. An infinite series is analogous to integrating a function over an infinite interval.

Table 10.2

	Sequences/Series	Functions
Independent variable	n	x
Dependent variable	a_n	$f(x)$
Domain	Integers e.g., $n = 1, 2, 3, \ldots$	Real numbers e.g., $\{x : x \geq 1\}$
Accumulation	Sums	Integrals
Accumulation over a finite interval	$\displaystyle\sum_{k=1}^{n} a_k$	$\displaystyle\int_{1}^{n} f(x)\, dx$
Accumulation over an infinite interval	$\displaystyle\sum_{k=1}^{\infty} a_k$	$\displaystyle\int_{1}^{\infty} f(x)\, dx$

SECTION 10.1 EXERCISES

Getting Started

1. Define *sequence* and give an example.

2. Suppose the sequence $\{a_n\}$ is defined by the explicit formula $a_n = 1/n$, for $n = 1, 2, 3, \ldots$. Write out the first five terms of the sequence.

3. Suppose the sequence $\{a_n\}$ is defined by the recurrence relation $a_{n+1} = na_n$, for $n = 1, 2, 3, \ldots$, where $a_1 = 1$. Write out the first five terms of the sequence.

4. Find two different explicit formulas for the sequence $\{-1, 1, -1, 1, -1, 1, \ldots\}$.

5. Find two different explicit formulas for the sequence $\{1, -2, 3, -4, -5, \ldots\}$.

6. The first ten terms of the sequence $\{2\tan^{-1} 10^n\}_{n=1}^{\infty}$ are rounded to 8 digits right of the decimal point (see table). Make a conjecture about the limit of the sequence.

n	a_n
1	2.94225535
2	3.12159332
3	3.13959265
4	3.14139265
5	3.14157265
6	3.14159065
7	3.14159245
8	3.14159263
9	3.14159265
10	3.14159265

7. The first ten terms of the sequence $\left\{\left(1 + \dfrac{1}{10^n}\right)^{10^n}\right\}_{n=1}^{\infty}$ are rounded to 8 digits right of the decimal point (see table). Make a conjecture about the limit of the sequence.

n	a_n
1	2.59374246
2	2.70481383
3	2.71692393
4	2.71814593
5	2.71826824
6	2.71828047
7	2.71828169
8	2.71828179
9	2.71828204
10	2.71828203

8. Does the sequence $\{1, 2, 1, 2, 1, 2, \ldots\}$ converge? Explain.

9. The terms of a sequence of partial sums are defined by $$S_n = \sum_{k=1}^{n} k^2, \text{ for } n = 1, 2, 3, \ldots.$$ Evaluate the first four terms of the sequence.

10. Given the series $\displaystyle\sum_{k=1}^{\infty} k$, evaluate the first four terms of its sequence of partial sums $S_n = \displaystyle\sum_{k=1}^{n} k$.

11. Use sigma notation to write an infinite series whose first four successive partial sums are 10, 20, 30, and 40.

12. Consider the infinite series $\displaystyle\sum_{k=1}^{\infty} \dfrac{1}{k}$. Evaluate the first four terms of the sequence of partial sums.

Practice Exercises

13–20. Explicit formulas *Write the first four terms of the sequence* $\{a_n\}_{n=1}^{\infty}$.

13. $a_n = \dfrac{1}{10^n}$

14. $a_n = 3n + 1$

15. $a_n = \dfrac{(-1)^n}{2^n}$

16. $a_n = 2 + (-1)^n$

17. $a_n = \dfrac{2^{n+1}}{2^n + 1}$ **18.** $a_n = n + \dfrac{1}{n}$

19. $a_n = 1 + \sin \dfrac{\pi n}{2}$

20. $a_n = n!$ (*Hint:* Recall that $n! = n(n-1)(n-2)\cdots 2 \cdot 1$.)

21–26. Recurrence relations *Write the first four terms of the sequence* $\{a_n\}$ *defined by the following recurrence relations.*

21. $a_{n+1} = 2a_n;\ a_1 = 2$

22. $a_{n+1} = \dfrac{a_n}{2};\ a_1 = 32$

23. $a_{n+1} = 3a_n - 12;\ a_1 = 10$

24. $a_{n+1} = a_n^2 - 1;\ a_1 = 1$

25. $a_{n+1} = \dfrac{1}{1 + a_n};\ a_0 = 1$

26. $a_{n+1} = a_n + a_{n-1};\ a_1 = 1, a_0 = 1$

27–34. Working with sequences *Several terms of a sequence* $\{a_n\}_{n=1}^{\infty}$ *are given.*

a. *Find the next two terms of the sequence.*
b. *Find a recurrence relation that generates the sequence (supply the initial value of the index and the first term of the sequence).*
c. *Find an explicit formula for the nth term of the sequence.*

27. $\left\{1, \dfrac{1}{2}, \dfrac{1}{4}, \dfrac{1}{8}, \dfrac{1}{16}, \ldots\right\}$ **28.** $\{2, 5, 8, 11, \ldots\}$

29. $\{1, 2, 4, 8, 16, \ldots\}$ **30.** $\{64, 32, 16, 8, 4, \ldots\}$

31. $\{1, 3, 9, 27, 81, \ldots\}$ **32.** $\{1, 4, 9, 16, 25, \ldots\}$

33. $\{-5, 5, -5, 5, \ldots\}$ **34.** $\{1, 0, 1, 0, 1, 0, 1, \ldots\}$

35–44. Limits of sequences *Write the terms* $a_1, a_2, a_3,$ *and* a_4 *of the following sequences. If the sequence appears to converge, make a conjecture about its limit. If the sequence diverges, explain why.*

35. $a_n = 10^n - 1;\ n = 1, 2, 3, \ldots$

36. $a_{n+1} = \dfrac{10}{a_n};\ a_1 = 1$

37. $a_n = \dfrac{1}{10^n};\ n = 1, 2, 3, \ldots$

38. $a_{n+1} = \dfrac{a_n}{10};\ a_0 = 1$

39. $a_n = 3 + \cos \pi n;\ n = 1, 2, 3, \ldots$

40. $a_n = 1 - 10^{-n};\ n = 1, 2, 3, \ldots$

41. $a_{n+1} = 1 + \dfrac{a_n}{2};\ a_0 = 2$

42. $a_{n+1} = 1 - \dfrac{a_n}{2};\ a_0 = \dfrac{2}{3}$

T 43. $a_{n+1} = \dfrac{a_n}{11} + 50;\ a_0 = 50$

44. $a_{n+1} = 10a_n - 1;\ a_0 = 0$

T 45–48. Explicit formulas for sequences *Consider the formulas for the following sequences* $\{a_n\}_{n=1}^{\infty}$. *Make a table with at least ten terms and determine a plausible limit of the sequence or state that the sequence diverges.*

45. $a_n = \dfrac{5^n}{5^n + 1}$ **46.** $a_n = 2^n \sin(2^{-n})$

47. $a_n = n^2 + n$ **48.** $a_n = \dfrac{100n - 1}{10n}$

49–50. Limits from graphs *Consider the following sequences.*

a. *Find the first four terms of the sequence.*
b. *Based on part (a) and the figure, determine a plausible limit of the sequence.*

49. $a_n = 2 + 2^{-n};\ n = 1, 2, 3, \ldots$

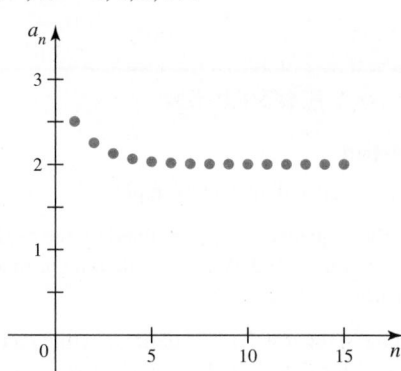

50. $a_n = \dfrac{n^2}{n^2 - 1};\ n = 2, 3, 4, \ldots$

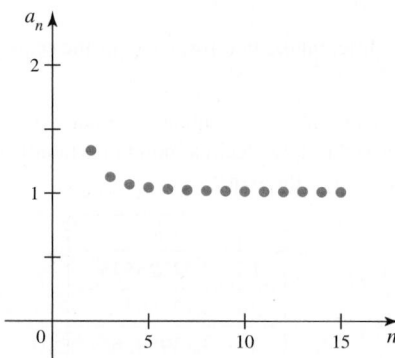

T 51–56. Recurrence relations *Consider the following recurrence relations. Make a table with at least ten terms and determine a plausible limit of the sequence or state that the sequence diverges.*

51. $a_{n+1} = \dfrac{1}{2}a_n + 2;\ a_1 = 3$ **52.** $a_n = \dfrac{1}{4}a_{n-1} - 3;\ a_0 = 1$

53. $a_{n+1} = 4a_n + 1;\ a_0 = 1$ **54.** $a_{n+1} = \dfrac{a_n}{10} + 3;\ a_0 = 10$

55. $a_{n+1} = \dfrac{1}{2}\sqrt{a_n} + 3;\ a_1 = 8$

56. $a_{n+1} = \sqrt{8a_n + 9};\ a_1 = 10$

57–60. Heights of bouncing balls *A ball is thrown upward to a height of h_0 meters. After each bounce, the ball rebounds to a fraction r of its previous height. Let h_n be the height after the nth bounce. Consider the following values of h_0 and r.*

a. *Find the first four terms of the sequence of heights $\{h_n\}$.*
b. *Find an explicit formula for the nth term of the sequence $\{h_n\}$.*

57. $h_0 = 20, \ r = 0.5$ **58.** $h_0 = 10, \ r = 0.9$

59. $h_0 = 30, \ r = 0.25$ **60.** $h_0 = 20, \ r = 0.75$

61–66. Sequences of partial sums *For the following infinite series, find the first four terms of the sequence of partial sums. Then make a conjecture about the value of the infinite series or state that the series diverges.*

61. $0.3 + 0.03 + 0.003 + \cdots$

62. $0.6 + 0.06 + 0.006 + \cdots$

63. $4 + 0.9 + 0.09 + 0.009 + \cdots$

64. $\displaystyle\sum_{k=1}^{\infty} 10^k$ **65.** $\displaystyle\sum_{k=1}^{\infty} \frac{6}{10^k}$

66. $\displaystyle\sum_{k=1}^{\infty} \cos \pi k$

67–70. Formulas for sequences of partial sums *Consider the following infinite series.*

a. *Find the first four partial sums $S_1, S_2, S_3,$ and S_4 of the series.*
b. *Find a formula for the nth partial sum S_n of the infinite series. Use this formula to find the next four partial sums $S_5, S_6, S_7,$ and S_8 of the infinite series.*
c. *Make a conjecture for the value of the series.*

67. $\displaystyle\sum_{k=1}^{\infty} \frac{2}{(2k-1)(2k+1)}$ **68.** $\displaystyle\sum_{k=1}^{\infty} \frac{1}{2^k}$

69. $\displaystyle\sum_{k=1}^{\infty} 90(0.1)^k$ **T 70.** $\displaystyle\sum_{k=1}^{\infty} \frac{2}{3^{k-1}}$

71. Explain why or why not Determine whether the following statements are true and give an explanation or counterexample.

  a. The sequence of partial sums for the series $1 + 2 + 3 + \cdots$ is $\{1, 3, 6, 10, \ldots\}$.
  b. If a sequence of positive numbers converges, then the sequence is decreasing.
  c. If the terms of the sequence $\{a_n\}$ are positive and increasing, then the sequence of partial sums for the series $\displaystyle\sum_{k=1}^{\infty} a_k$ diverges.

T 72–75. Practical sequences *Consider the following situations that generate a sequence.*

a. *Write out the first five terms of the sequence.*
b. *Find an explicit formula for the terms of the sequence.*
c. *Find a recurrence relation that generates the sequence.*
d. *Using a calculator or a graphing utility, estimate the limit of the sequence or state that it does not exist.*

72. Population growth When a biologist begins a study, a colony of prairie dogs has a population of 250. Regular measurements reveal that each month the prairie dog population increases by 3%. Let p_n be the population (rounded to whole numbers) at the end of the nth month, where the initial population is $p_0 = 250$.

73. Radioactive decay A material transmutes 50% of its mass to another element every 10 years due to radioactive decay. Let M_n be the mass of the radioactive material at the end of the nth decade, where the initial mass of the material is $M_0 = 20$ g.

74. Consumer Price Index The Consumer Price Index (the CPI is a measure of the U.S. cost of living) is given a base value of 100 in the year 1984. Assume the CPI has increased by an average of 3% per year since 1984. Let c_n be the CPI n years after 1984, where $c_0 = 100$.

75. Drug elimination Jack took a 200-mg dose of a pain killer at midnight. Every hour, 5% of the drug is washed out of his bloodstream. Let d_n be the amount of drug in Jack's blood n hours after the drug was taken, where $d_0 = 200$ mg.

Explorations and Challenges

T 76–77. Distance traveled by bouncing balls *A ball is thrown upward to a height of h_0 meters. After each bounce, the ball rebounds to a fraction r of its previous height. Let h_n be the height after the nth bounce and let S_n be the total distance the ball has traveled at the moment of the nth bounce.*

a. *Find the first four terms of the sequence $\{S_n\}$.*
b. *Make a table of 20 terms of the sequence $\{S_n\}$ and determine a plausible value for the limit of $\{S_n\}$.*

76. $h_0 = 20, r = 0.5$ **77.** $h_0 = 20, r = 0.75$

T 78. A square root finder A well-known method for approximating $\sqrt{c}$ for positive real numbers c consists of the following recurrence relation (based on Newton's method; see Section 4.8). Let $a_0 = c$ and

$$a_{n+1} = \frac{1}{2}\left(a_n + \frac{c}{a_n}\right), \quad \text{for } n = 0, 1, 2, 3, \ldots.$$

  a. Use this recurrence relation to approximate $\sqrt{10}$. How many terms of the sequence are needed to approximate $\sqrt{10}$ with an error less than 0.01? How many terms of the sequence are needed to approximate $\sqrt{10}$ with an error less than 0.0001? (To compute the error, assume a calculator gives the exact value.)
  b. Use this recurrence relation to approximate $\sqrt{c}$, for $c = 2, 3, \ldots, 10$. Make a table showing the number of terms of the sequence needed to approximate $\sqrt{c}$ with an error less than 0.01.

T 79–80. Fixed-point iteration *A method for estimating a solution to the equation $x = f(x)$, known as **fixed-point iteration**, is based on the following recurrence relation. Let $x_0 = c$ and $x_{n+1} = f(x_n)$, for $n = 1, 2, 3, \ldots$ and a real number c. If the sequence $\{x_n\}_{n=0}^{\infty}$ converges to L, then L is a solution to the equation $x = f(x)$ and L is called a fixed point of f. To estimate L with p digits of accuracy to the right of the decimal point, we can compute the terms of the sequence $\{x_n\}_{n=0}^{\infty}$ until two successive values agree to p digits of accuracy. Use fixed-point iteration to find a solution to the following equations with $p = 3$ digits of accuracy using the given value of x_0.*

79. $x = \cos x; \ x_0 = 0.8$ **80.** $x = \dfrac{\sqrt{x^3 + 1}}{20}; \ x_0 = 5$

QUICK CHECK ANSWERS

1. $a_{10} = 28$ **2.** $a_n = 2^n - 1, \ n = 1, 2, 3, \ldots$
3. $0.33333 \ldots = \frac{1}{3}$ **4.** Both diverge.   **5.** $S_1 = -1, S_2 = 1,$ $S_3 = -2, S_4 = 2$; the series diverges. ◄

10.2 Sequences

The previous section sets the stage for an in-depth investigation of sequences and infinite series. This section is devoted to sequences, and the remainder of the chapter deals with series.

Limit of a Sequence and Limit Laws

A fundamental question about sequences concerns the behavior of the terms as we go out farther and farther in the sequence. For example, in the sequence

$$\{a_n\}_{n=0}^{\infty} = \left\{\frac{1}{n^2 + 1}\right\}_{n=0}^{\infty} = \left\{1, \frac{1}{2}, \frac{1}{5}, \frac{1}{10}, \cdots\right\},$$

the terms remain positive and decrease to 0. We say that this sequence converges and its limit is 0, written $\lim_{n \to \infty} a_n = 0$. Similarly, the terms of the sequence

$$\{b_n\}_{n=1}^{\infty} = \left\{(-1)^n \frac{n(n+1)}{2}\right\}_{n=1}^{\infty} = \{-1, 3, -6, 10, \ldots\}$$

increase in magnitude and do not approach a unique value as n increases. In this case, we say that the sequence diverges.

Limits of sequences are really no different from limits at infinity of functions except that the variable n assumes only integer values as $n \to \infty$. This idea works as follows.

Given a sequence $\{a_n\}$, we define a function f such that $f(n) = a_n$ for all indices n. For example, if $a_n = n/(n+1)$, then we let $f(x) = x/(x+1)$. By the methods of Section 2.5, we know that $\lim_{x \to \infty} f(x) = 1$; because the terms of the sequence lie on the graph of f, it follows that $\lim_{n \to \infty} a_n = 1$ (**Figure 10.11**). This reasoning is the basis of the following theorem.

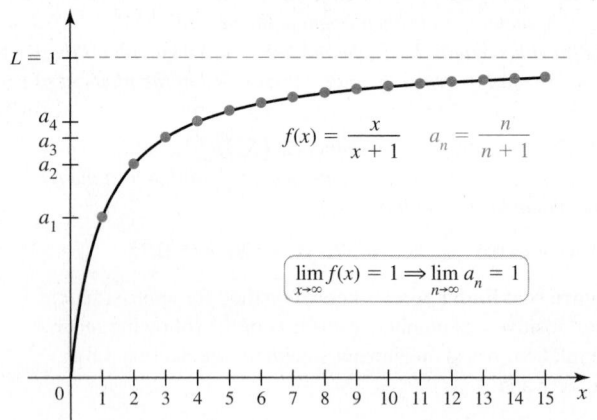

$$f(x) = \frac{x}{x+1} \qquad a_n = \frac{n}{n+1}$$

$$\lim_{x \to \infty} f(x) = 1 \Rightarrow \lim_{n \to \infty} a_n = 1$$

Figure 10.11

▶ The converse of Theorem 10.1 is not true. For example, if $a_n = \cos 2\pi n$, then $\lim_{n \to \infty} a_n = 1$, but $\lim_{x \to \infty} \cos 2\pi x$ does not exist.

> **THEOREM 10.1   Limits of Sequences from Limits of Functions**
> Suppose f is a function such that $f(n) = a_n$, for positive integers n. If $\lim_{x \to \infty} f(x) = L$, then the limit of the sequence $\{a_n\}$ is also L, where L may be $\pm \infty$.

Because of the correspondence between limits of sequences and limits of functions at infinity, we have the following properties that are analogous to those for functions given in Theorem 2.3.

▶ The limit of a sequence $\{a_n\}$ is determined by the terms in the *tail* of the sequence—the terms with large values of n. If the sequences $\{a_n\}$ and $\{b_n\}$ differ in their first 100 terms but have identical terms for $n > 100$, then they have the same limit. For this reason, the initial index of a sequence (for example, $n = 0$ or $n = 1$) is often not specified.

> **THEOREM 10.2   Limit Laws for Sequences**
> Assume the sequences $\{a_n\}$ and $\{b_n\}$ have limits A and B, respectively. Then
>
> **1.** $\lim_{n \to \infty} (a_n \pm b_n) = A \pm B$
>
> **2.** $\lim_{n \to \infty} ca_n = cA$, where c is a real number
>
> **3.** $\lim_{n \to \infty} a_n b_n = AB$
>
> **4.** $\lim_{n \to \infty} \frac{a_n}{b_n} = \frac{A}{B}$, provided $B \neq 0$.

EXAMPLE 1   Limits of sequences   Determine the limits of the following sequences.

a. $a_n = \dfrac{3n^3}{n^3 + 1}$ **b.** $b_n = \left(\dfrac{n+5}{n}\right)^n$ **c.** $c_n = n^{1/n}$

SOLUTION

a. A function with the property that $f(n) = a_n$ is $f(x) = \dfrac{3x^3}{x^3 + 1}$. Dividing numerator and denominator by x^3 (or appealing to Theorem 2.7), we find that $\lim\limits_{x \to \infty} f(x) = 3$. (Alternatively, we can apply l'Hôpital's Rule and obtain the same result.) We conclude that $\lim\limits_{n \to \infty} a_n = 3$.

b. The limit

$$\lim_{n \to \infty} b_n = \lim_{n \to \infty} \left(\frac{n + 5}{n} \right)^n = \lim_{n \to \infty} \left(1 + \frac{5}{n} \right)^n$$

has the indeterminate form 1^∞. Recall that for this limit (Section 4.7), we first evaluate

$$L = \lim_{n \to \infty} \ln \left(1 + \frac{5}{n} \right)^n = \lim_{n \to \infty} n \ln \left(1 + \frac{5}{n} \right),$$

> It is customary to treat n as a continuous variable and differentiate with respect to n, rather than write the sequence as a function of x, as was done in Example 1a.

and then, if L exists, $\lim\limits_{n \to \infty} b_n = e^L$. Using l'Hôpital's Rule for the indeterminate form $0/0$, we have

$$L = \lim_{n \to \infty} n \ln \left(1 + \frac{5}{n} \right) = \lim_{n \to \infty} \frac{\ln \left(1 + (5/n) \right)}{1/n} \qquad \text{Indeterminate form } 0/0$$

> For a review of l'Hôpital's Rule, see Section 4.7, where we showed that
> $$\lim_{x \to \infty} \left(1 + \frac{a}{x} \right)^x = e^a.$$

$$= \lim_{n \to \infty} \frac{\dfrac{1}{1 + (5/n)} \left(-\dfrac{5}{n^2} \right)}{-1/n^2} \qquad \text{L'Hôpital's Rule}$$

$$= \lim_{n \to \infty} \frac{5}{1 + (5/n)} = 5. \qquad \text{Simplify; } 5/n \to 0 \text{ as } n \to \infty.$$

Because $\lim\limits_{n \to \infty} b_n = e^L = e^5$, we have $\lim\limits_{n \to \infty} \left(\dfrac{5 + n}{n} \right)^n = e^5$.

c. The limit has the indeterminate form ∞^0, so we first evaluate $L = \lim\limits_{n \to \infty} \ln n^{1/n} = \lim\limits_{n \to \infty} \dfrac{\ln n}{n}$; if L exists, then $\lim\limits_{n \to \infty} c_n = e^L$. Using either l'Hôpital's Rule or the relative growth rates in Section 4.7, we find that $L = 0$. Therefore, $\lim\limits_{n \to \infty} c_n = e^0 = 1$.

Related Exercises 14, 38, 39 ◄

Terminology for Sequences

We now introduce some terminology for sequences that is similar to that used for functions. The following terms are used to describe sequences $\{a_n\}$.

> Because an increasing sequence is, by definition, nondecreasing, it is also monotonic. Similarly, a decreasing sequence is monotonic.

DEFINITIONS Terminology for Sequences

- $\{a_n\}$ is **increasing** if $a_{n+1} > a_n$; for example, $\{0, 1, 2, 3, \dots\}$.
- $\{a_n\}$ is **nondecreasing** if $a_{n+1} \geq a_n$; for example, $\{1, 1, 2, 2, 3, 3, \dots\}$.
- $\{a_n\}$ is **decreasing** if $a_{n+1} < a_n$; for example, $\{2, 1, 0, -1, \dots\}$.
- $\{a_n\}$ is **nonincreasing** if $a_{n+1} \leq a_n$; for example, $\{0, -1, -1, -2, -2, \dots\}$.
- $\{a_n\}$ is **monotonic** if it is either nonincreasing or nondecreasing (it moves in one direction).
- $\{a_n\}$ is **bounded above** if there is a number M such that $a_n \leq M$, for all relevant values of n, and $\{a_n\}$ is **bounded below** if there is a number N such that $a_n \geq N$, for all relevant values of n.
- If $\{a_n\}$ is bounded above and bounded below, then we say that $\{a_n\}$ is a **bounded** sequence.

For example, the sequence

$$\{a_n\} = \left\{1 - \frac{1}{n}\right\}_{n=1}^{\infty} = \left\{0, \frac{1}{2}, \frac{2}{3}, \frac{3}{4}, \dots\right\}$$

satisfies $0 \leq a_n \leq 1$, for $n \geq 1$, and its terms are increasing in size. Therefore, the sequence is bounded below, bounded above, and increasing; it is also monotonic (Figure 10.12). The sequence

$$\{b_n\} = \left\{1 + \frac{1}{n}\right\}_{n=1}^{\infty} = \left\{2, \frac{3}{2}, \frac{4}{3}, \frac{5}{4}, \dots\right\}$$

satisfies $1 \leq b_n \leq 2$, for $n \geq 1$, and its terms are decreasing in size. Therefore, the sequence is bounded below, bounded above, and decreasing; it is also monotonic (Figure 10.12). Because $\{a_n\}$ and $\{b_n\}$ are bounded below and above, both are bounded sequences.

QUICK CHECK 1 Classify the following sequences as bounded, monotonic, or neither.

a. $\left\{\frac{1}{2}, \frac{3}{4}, \frac{7}{8}, \frac{15}{16}, \dots\right\}$

b. $\left\{1, -\frac{1}{2}, \frac{1}{4}, -\frac{1}{8}, \frac{1}{16}, \dots\right\}$

c. $\{1, -2, 3, -4, 5, \dots\}$

d. $\{1, 1, 1, 1, \dots\}$ ◄

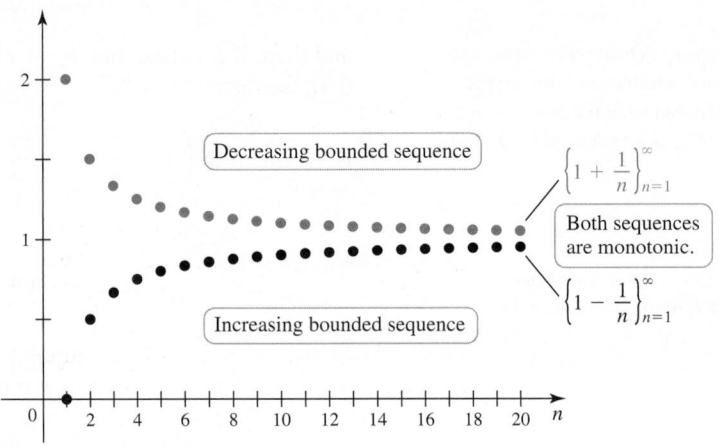

Figure 10.12

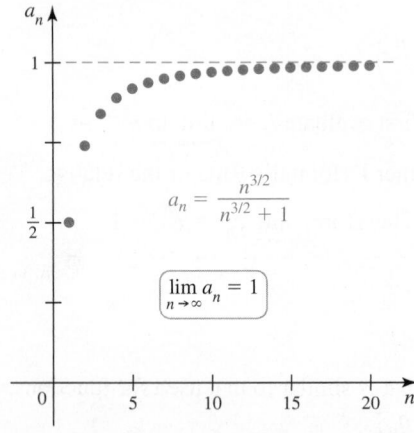

Figure 10.13

EXAMPLE 2   Limits of sequences and graphing   Compare and contrast the behavior of $\{a_n\}$ and $\{b_n\}$ as $n \to \infty$.

a. $a_n = \dfrac{n^{3/2}}{n^{3/2} + 1}$ b. $b_n = \dfrac{(-1)^n \, n^{3/2}}{n^{3/2} + 1}$

SOLUTION

a. The terms of $\{a_n\}$ are positive, increasing, and bounded (Figure 10.13). Dividing the numerator and denominator of a_n by $n^{3/2}$, we see that

$$\lim_{n\to\infty} a_n = \lim_{n\to\infty} \frac{n^{3/2}}{n^{3/2}+1} = \lim_{n\to\infty} \frac{1}{1 + \underbrace{\frac{1}{n^{3/2}}}_{\text{approaches } 0 \text{ as } n \to \infty}} = 1.$$

b. The terms of the bounded sequence $\{b_n\}$ alternate in sign. Using the result of part (a), it follows that the even terms form an increasing sequence that approaches 1, and the odd terms form a decreasing sequence that approaches -1 (Figure 10.14). Therefore, the sequence diverges, illustrating the fact that the presence of $(-1)^n$ may significantly alter the behavior of a sequence.

Related Exercises 55, 60 ◄

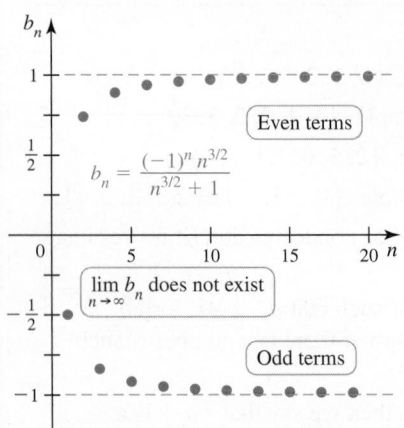

Figure 10.14

Geometric Sequences

Geometric sequences have the property that each term is obtained by multiplying the previous term by a fixed constant, called the **ratio**. They have the form $\{r^n\}$ or $\{ar^n\}$, where the ratio r and $a \neq 0$ are real numbers. The value of r determines the behavior of the sequence.

When a number less than 1 in magnitude is raised to increasing powers, the resulting numbers decrease to zero. For example, the geometric sequence $\{0.75^n\}$, where $r = 0.75$, converges to zero and is monotonic (Figure 10.15). If we choose $r = -0.75$, we have $\{(-0.75)^n\} = \{(-1)^n 0.75^n\}$. Observe that the factor $(-1)^n$ oscillates between 1 and -1, while 0.75^n decreases to zero as n increases. Therefore, the sequence oscillates and converges to zero (Figure 10.16).

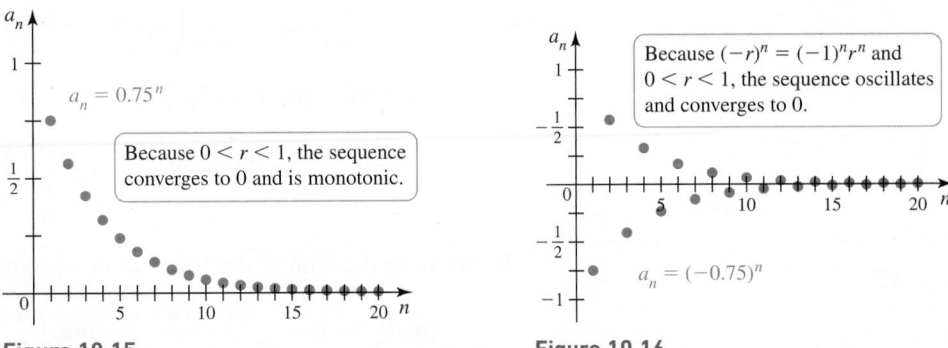

Figure 10.15   **Figure 10.16**

When a number greater than 1 in magnitude is raised to increasing powers, the resulting numbers increase in magnitude. For example, letting $r = 1.15$ results in the sequence $\{1.15^n\}$, whose terms are positive and increase without bound. In this case, the sequence diverges and is monotonic (Figure 10.17). Choosing $r = -1.15$, we have $\{(-1.15)^n\} = \{(-1)^n 1.15^n\}$. As before, the factor $(-1)^n$ oscillates between 1 and -1, while 1.15^n increases without bound as n increases. The terms of the sequence increase in magnitude without bound and alternate in sign. In this case, the sequence oscillates and diverges (Figure 10.18).

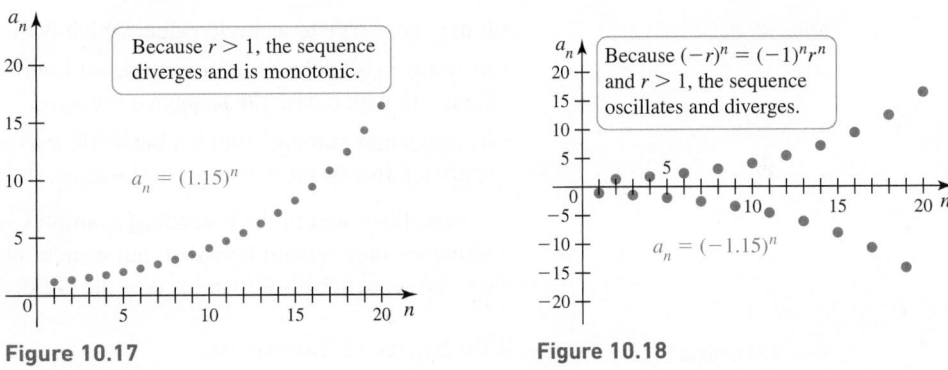

Figure 10.17   **Figure 10.18**

QUICK CHECK 2 Describe the behavior of $\{r^n\}$ in the cases $r = -1$ and $r = 1$. ◄

The preceding discussion and Quick Check 2 are summarized in the following theorem.

THEOREM 10.3 Geometric Sequences

Let r be a real number. Then

$$\lim_{n \to \infty} r^n = \begin{cases} 0 & \text{if } |r| < 1 \\ 1 & \text{if } r = 1 \\ \text{does not exist} & \text{if } r \le -1 \text{ or } r > 1. \end{cases}$$

If $r > 0$, then $\{r^n\}$ is a monotonic sequence. If $r < 0$, then $\{r^n\}$ oscillates.

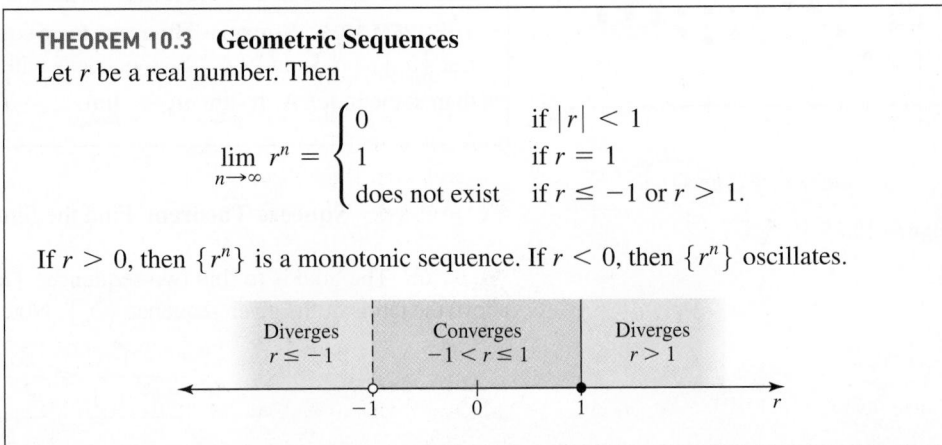

EXAMPLE 3   Using Limit Laws Determine the limits of the following sequences.

a. $a_n = 5(0.6)^n - \dfrac{1}{3^n}$ **b.** $b_n = \dfrac{2n^2 + n}{2^n(3n^2 - 4)}$

SOLUTION

a. Using Laws 1 and 2 from Theorem 10.2, we have

$$\lim_{n\to\infty} a_n = \lim_{n\to\infty}\left(5(0.6)^n - \frac{1}{3^n}\right) = 5\lim_{n\to\infty}(0.6)^n - \lim_{n\to\infty}\left(\frac{1}{3}\right)^n.$$

Both $\{(0.6)^n\}$ and $\{(1/3)^n\}$ are geometric sequences with $|r| < 1$, which implies that

$$\lim_{n\to\infty} a_n = 5\underbrace{\lim_{n\to\infty}(0.6)^n}_{0} - \underbrace{\lim_{n\to\infty}\left(\frac{1}{3}\right)^n}_{0} = 5\cdot 0 - 0 = 0. \quad \text{Theorem 10.3}$$

b. We write the limit of the sequence as a product of limits:

$$\lim_{n\to\infty} b_n = \lim_{n\to\infty}\frac{2n^2 + n}{2^n(3n^2 - 4)} = \lim_{n\to\infty}\left(\frac{1}{2}\right)^n \cdot \lim_{n\to\infty}\frac{2n^2 + n}{3n^2 - 4}. \quad \text{Law 3, Theorem 10.2}$$

The first of the new limits involves a geometric sequence with $r = 1/2$, and the second limit yields to Theorem 2.7:

$$\lim_{n\to\infty} b_n = \underbrace{\lim_{n\to\infty}\left(\frac{1}{2}\right)^n}_{0} \cdot \underbrace{\lim_{n\to\infty}\frac{2n^2 + n}{3n^2 - 4}}_{2/3} = 0\cdot\frac{2}{3} = 0. \quad \text{Evaluate limits.}$$

Related Exercises 26–27 ◄

The previous examples show that a sequence may display any of the following behaviors:

• It may converge to a single value, which is the limit of the sequence.

• Its terms may increase in magnitude without bound (either with one sign or with mixed signs), in which case the sequence diverges.

• Its terms may remain bounded but settle into an oscillating pattern in which the terms approach two or more values; in this case the sequence diverges.

Not illustrated in the preceding examples is one other type of behavior: The terms of a sequence may remain bounded, but wander chaotically forever without a pattern. In this case, the sequence also diverges (see the Guided Project *Chaos!*)

The Squeeze Theorem

We cite two theorems that are used to evaluate limits and to establish that limits exist. The first theorem is a direct analog of the Squeeze Theorem in Section 2.3.

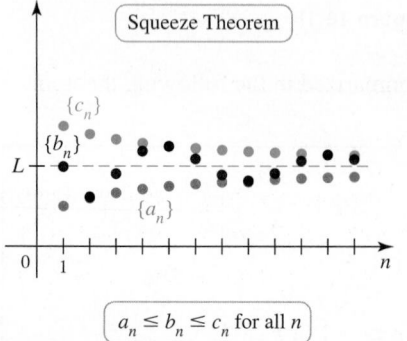

Figure 10.19

> **THEOREM 10.4   Squeeze Theorem for Sequences**
> Let $\{a_n\}$, $\{b_n\}$, and $\{c_n\}$ be sequences with $a_n \le b_n \le c_n$, for all integers n greater than some index N. If $\lim_{n\to\infty} a_n = \lim_{n\to\infty} c_n = L$, then $\lim_{n\to\infty} b_n = L$ (**Figure 10.19**).

EXAMPLE 4   Squeeze Theorem Find the limit of the sequence $b_n = \dfrac{\cos n}{n^2 + 1}$.

SOLUTION The goal is to find two sequences $\{a_n\}$ and $\{c_n\}$ whose terms lie below and above the terms of the given sequence $\{b_n\}$. Note that $-1 \le \cos n \le 1$, for all n. Therefore,

$$\underbrace{-\frac{1}{n^2 + 1}}_{a_n} \le \underbrace{\frac{\cos n}{n^2 + 1}}_{b_n} \le \underbrace{\frac{1}{n^2 + 1}}_{c_n}.$$

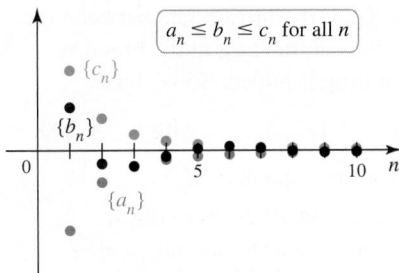

$a_n \le b_n \le c_n$ for all n

$\{c_n\}$

$\{b_n\}$

$\{a_n\}$

Figure 10.20

Letting $a_n = -\dfrac{1}{n^2 + 1}$ and $c_n = \dfrac{1}{n^2 + 1}$, we have $a_n \le b_n \le c_n$, for $n \ge 1$. Furthermore, $\lim\limits_{n \to \infty} a_n = \lim\limits_{n \to \infty} c_n = 0$. By the Squeeze Theorem, $\lim\limits_{n \to \infty} b_n = 0$ (Figure 10.20).

Related Exercises 65, 67 ◄

Bounded Monotonic Sequence Theorem

Suppose you pour a cup of hot coffee and put it on your desk to cool. Assume every minute you measure the temperature of the coffee to create a sequence of temperature readings $\{T_1, T_2, T_3, \dots\}$. This sequence has two notable properties: First, the terms of the sequence are decreasing (because the coffee is cooling); and second, the sequence is bounded below (because the temperature of the coffee cannot be less than the temperature of the surrounding room). In fact, if the measurements continue indefinitely, the sequence of temperatures converges to the temperature of the room. This example illustrates an important theorem that characterizes convergent sequences in terms of boundedness and monotonicity. The theorem is easy to believe, but its proof is beyond the scope of this text.

> **THEOREM 10.5   Bounded Monotonic Sequence**
> A bounded monotonic sequence converges.

> ➤ **Some optional terminology:** M is called an *upper bound* of the sequence in Figure 10.21a, and N is a *lower bound* of the sequence in Figure 10.21b. The number M^* is the *least upper bound* of a sequence (or a set) if it is the smallest of all the upper bounds. It is a fundamental property of the real numbers that if a sequence (or a nonempty set) is bounded above, then it has a least upper bound. It can be shown that an increasing sequence that is bounded above converges to its least upper bound. Similarly, a decreasing sequence that is bounded below converges to its greatest lower bound.

Figure 10.21 shows the two cases of this theorem. In the first case, we see a nondecreasing sequence, all of whose terms are less than M. It must converge to a limit less than or equal to M. Similarly, a nonincreasing sequence, all of whose terms are greater than N, must converge to a limit greater than or equal to N.

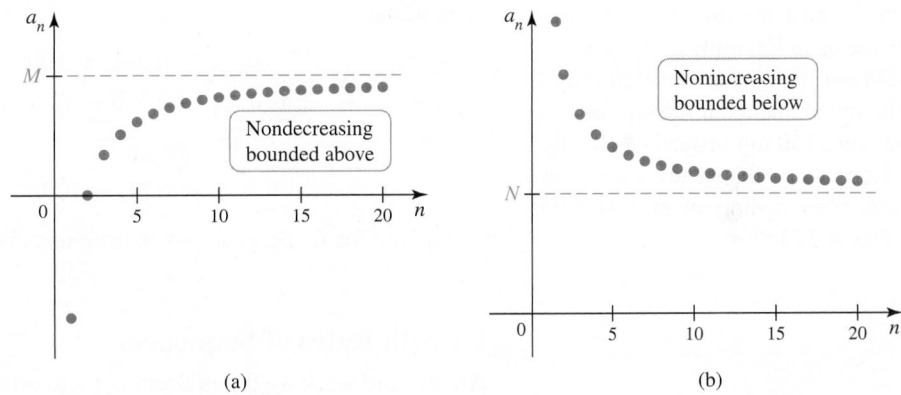

(a)   Nondecreasing bounded above

(b)   Nonincreasing bounded below

Figure 10.21

An Application: Recurrence Relations

> ➤ Most drugs decay exponentially in the bloodstream and have a characteristic half-life assuming the drug absorbs quickly into the blood.

EXAMPLE 5   Sequences for drug doses Suppose your doctor prescribes a 100-mg dose of an antibiotic to be taken every 12 hours. Furthermore, the drug is known to have a half-life of 12 hours; that is, every 12 hours half of the drug in your blood is eliminated.

a. Find the sequence that gives the amount of drug in your blood immediately after each dose.

b. Use a graph to propose the limit of this sequence; that is, in the long run, how much drug do you have in your blood?

c. Find the limit of the sequence directly.

SOLUTION

a. Let d_n be the amount of drug in the blood immediately following the nth dose, where $n = 1, 2, 3, \dots$ and $d_1 = 100$ mg. We want to write a recurrence relation that gives the amount of drug in the blood after the $(n + 1)$st dose (d_{n+1}) in terms of the

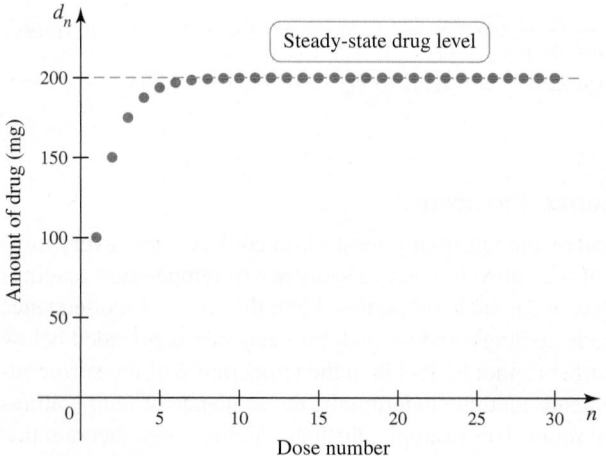

Figure 10.22

▶ *Mathematical induction* is used to prove statements involving natural numbers. In the first step, we show that the statement holds for $n = 1$. In the second step, we show that if the statement is true for $n = k$, then the statement is also true for $n = k + 1$, where k is any natural number greater than 1. If the proof of the second step is successful, it implies that the statement holds for any natural number.

QUICK CHECK 3 If a drug has the same half-life as in Example 5, (i) how would the steady-state level of drug in the blood change if the regular dose were 150 mg instead of 100 mg? (ii) How would the steady-state level change if the dosing interval were 6 hr instead of 12 hr? ◄

amount of drug after the nth dose (d_n). In the 12 hours between the nth dose and the $(n + 1)$st dose, half of the drug in the blood is eliminated *and* another 100 mg of drug is added. So we have

$$d_{n+1} = 0.5\, d_n + 100, \qquad \text{for } n = 1, 2, 3, \ldots, \text{ with } d_1 = 100,$$

which is the recurrence relation for the sequence $\{d_n\}$.

b. We see from **Figure 10.22** that after about 10 doses (5 days) the amount of antibiotic in the blood is close to 200 mg, and— importantly for your body—it never exceeds 200 mg.

c. Figure 10.22 gives evidence that the terms of the sequence are increasing and bounded above by 200. To prove these facts, note that $d_1 = 100 < 200$. Now suppose $d_k < 200$, for any $k > 1$. The recurrence relation in part (a) implies that $d_{k+1} = 0.5 d_k + 100$. Because $d_k < 200$, we have

$$d_{k+1} = 0.5 \underbrace{d_k}_{<200} + 100 < 0.5(200) + 100 = 200.$$

We conclude by mathematical induction that $d_k < 200$, for all k. To prove that $\{d_k\}$ is increasing (and therefore monotonic), we need to show that $d_{k+1} - d_k > 0$. Note that

$$d_{k+1} - d_k = \underbrace{0.5 d_k + 100}_{\text{recurrence relation}} - d_k = 100 - 0.5 d_k > 0;$$

the final inequality follows from the fact that $d_k < 200$. Because $\{d_k\}$ is increasing, it is bounded below by its first term, which is $d_1 = 100$, and therefore $\{d_k\}$ is bounded. By the Bounded Monotonic Sequence Theorem, the sequence has a limit; therefore, $\lim\limits_{n\to\infty} d_n = L$ and $\lim\limits_{n\to\infty} d_{n+1} = L$. We now take the limit of both sides of the recurrence relation:

$$d_{n+1} = 0.5\, d_n + 100 \qquad \text{Recurrence relation}$$

$$\underbrace{\lim_{n\to\infty} d_{n+1}}_{L} = 0.5 \underbrace{\lim_{n\to\infty} d_n}_{L} + \lim_{n\to\infty} 100 \qquad \text{Limits of both sides}$$

$$L = 0.5L + 100. \qquad \text{Substitute } L.$$

Solving for L, the steady-state drug level is $L = 200$.

Related Exercises 71, 89 ◄

Growth Rates of Sequences

All the hard work we did in Section 4.7 to establish the relative growth rates of functions is now applied to sequences. Here is the question: Given two nondecreasing sequences of positive terms $\{a_n\}$ and $\{b_n\}$, which sequence grows faster as $n \to \infty$? As with functions, to compare growth rates, we evaluate $\lim\limits_{n\to\infty} a_n/b_n$. If $\lim\limits_{n\to\infty} a_n/b_n = 0$, then $\{b_n\}$ grows faster than $\{a_n\}$. If $\lim\limits_{n\to\infty} a_n/b_n = \infty$, then $\{a_n\}$ grows faster than $\{b_n\}$.

Using the results of Section 4.7, we immediately arrive at the following ranking of growth rates of sequences as $n \to \infty$, with positive real numbers $p, q, r, s,$ and $b > 1$:

$$\{\ln^q n\} \ll \{n^p\} \ll \{n^p \ln^r n\} \ll \{n^{p+s}\} \ll \{b^n\} \ll \{n^n\}.$$

▶ $0! = 1$ (by definition)
$1! = 1$
$2! = 2 \cdot 1! = 2$
$3! = 3 \cdot 2! = 6$
$4! = 4 \cdot 3! = 24$
$5! = 5 \cdot 4! = 120$
$6! = 6 \cdot 5! = 720$

As before, the notation $\{a_n\} \ll \{b_n\}$ means $\{b_n\}$ *grows faster than* $\{a_n\}$ as $n \to \infty$. Another important sequence that should be added to the list is the **factorial sequence** $\{n!\}$, where $n! = n(n - 1)(n - 2) \cdots 2 \cdot 1$. Where does the factorial sequence $\{n!\}$ appear in the list? The following argument provides some intuition. Notice that

$$n^n = \underbrace{n \cdot n \cdot n \cdots n}_{n \text{ factors}}, \qquad \text{whereas}$$

$$n! = \underbrace{n \cdot (n - 1) \cdot (n - 2) \cdots 2 \cdot 1}_{n \text{ factors}}.$$

The nth term of both sequences involves the product of n factors; however, the factors of $n!$ decrease, while the factors of n^n are the same. Based on this observation, we claim that $\{n^n\}$ grows faster than $\{n!\}$, and we have the ordering $\{n!\} \ll \{n^n\}$. But where does $\{n!\}$ appear in the list relative to $\{b^n\}$? Again, some intuition is gained by noting that

$$b^n = \underbrace{b \cdot b \cdot b \cdots b}_{n \text{ factors}}, \qquad \text{whereas}$$

$$n! = \underbrace{n \cdot (n-1) \cdot (n-2) \cdots 2 \cdot 1}_{n \text{ factors}}.$$

The nth term of both sequences involves a product of n factors; however, the factors of b^n remain constant as n increases, while the factors of $n!$ increase with n. So we claim that $\{n!\}$ grows faster than $\{b^n\}$. This conjecture is supported by computation, although the outcome of the race may not be immediately evident if b is large (Exercise 111).

> **THEOREM 10.6   Growth Rates of Sequences**
> The following sequences are ordered according to increasing growth rates as
> $n \to \infty$; that is, if $\{a_n\}$ appears before $\{b_n\}$ in the list, then $\lim\limits_{n \to \infty} \dfrac{a_n}{b_n} = 0$ and
> $\lim\limits_{n \to \infty} \dfrac{b_n}{a_n} = \infty$:
>
> $$\{\ln^q n\} \ll \{n^p\} \ll \{n^p \ln^r n\} \ll \{n^{p+s}\} \ll \{b^n\} \ll \{n!\} \ll \{n^n\}.$$
>
> The ordering applies for positive real numbers p, q, r, s, and $b > 1$.

QUICK CHECK 4 Which sequence grows faster: $\{\ln n\}$ or $\{n^{1.1}\}$? What is
$$\lim_{n \to \infty} \frac{n^{1,000,000}}{e^n}?$$ ◄

It is worth noting that the rankings in Theorem 10.6 do not change if a sequence is multiplied by a positive constant (Exercise 110).

EXAMPLE 6   Convergence and growth rates Compare growth rates of sequences to determine whether the following sequences converge.

a. $\left\{ \dfrac{\ln n^{10}}{0.00001n} \right\}$ **b.** $\left\{ \dfrac{n^8 \ln n}{n^{8.001}} \right\}$ **c.** $\left\{ \dfrac{n!}{10^n} \right\}$

SOLUTION

a. Because $\ln n^{10} = 10 \ln n$, the sequence in the numerator is a constant multiple of the sequence $\{\ln n\}$. Similarly, the sequence in the denominator is a constant multiple of the sequence $\{n\}$. By Theorem 10.6, $\{n\}$ grows faster than $\{\ln n\}$ as $n \to \infty$; therefore, the sequence $\left\{ \dfrac{\ln n^{10}}{0.00001n} \right\}$ converges to zero.

b. The sequence in the numerator is $\{n^p \ln^r n\}$ of Theorem 10.6 with $p = 8$ and $r = 1$. The sequence in the denominator is $\{n^{p+s}\}$ of Theorem 10.6 with $p = 8$ and $s = 0.001$. Because $\{n^{p+s}\}$ grows faster than $\{n^p \ln^r n\}$ as $n \to \infty$, we conclude that $\left\{ \dfrac{n^8 \ln n}{n^{8.001}} \right\}$ converges to zero.

c. Using Theorem 10.6, we see that $n!$ grows faster than any exponential function as $n \to \infty$. Therefore, $\lim\limits_{n \to \infty} \dfrac{n!}{10^n} = \infty$, and the sequence diverges. Figure 10.23 gives a visual comparison of the growth rates of $\{n!\}$ and $\{10^n\}$. Because these sequences grow so quickly, we plot the logarithm of the terms. The exponential sequence $\{10^n\}$ dominates the factorial sequence $\{n!\}$ until $n = 25$ terms. At that point, the factorial sequence overtakes the exponential sequence.

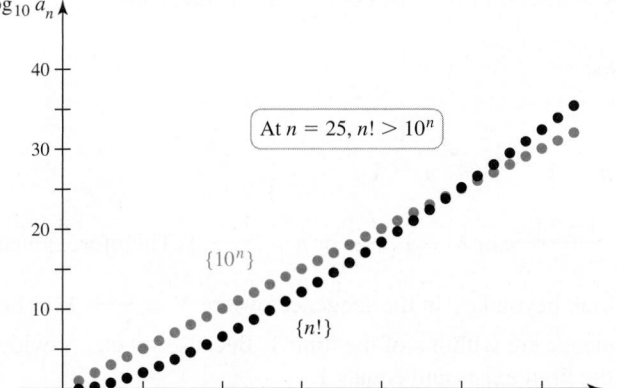

Figure 10.23

Related Exercises 76, 78 ◄

Formal Definition of a Limit of a Sequence

As with limits of functions, there is a formal definition of the limit of a sequence.

DEFINITION Limit of a Sequence

The sequence $\{a_n\}$ converges to L provided the terms of a_n can be made arbitrarily close to L by taking n sufficiently large. More precisely, $\{a_n\}$ has the unique limit L if, given any $\varepsilon > 0$, it is possible to find a positive integer N (depending only on ε) such that

$$|a_n - L| < \varepsilon \qquad \text{whenever } n > N.$$

If the **limit of a sequence** is L, we say the sequence **converges** to L, written

$$\lim_{n \to \infty} a_n = L.$$

A sequence that does not converge is said to **diverge**.

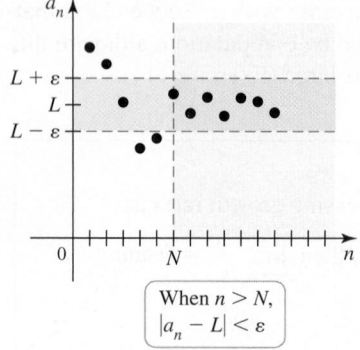

When $n > N$,
$|a_n - L| < \varepsilon$

Figure 10.24

The formal definition of the limit of a convergent sequence is interpreted in much the same way as the limit at infinity of a function. Given a small tolerance $\varepsilon > 0$, how far out in the sequence must you go so that all succeeding terms are within ε of the limit L (**Figure 10.24**)? Given *any* value of $\varepsilon > 0$ (no matter how small), you must find a value of N such that all terms beyond a_N are within ε of L.

EXAMPLE 7 Limits using the formal definition Consider the claim that

$$\lim_{n \to \infty} a_n = \lim_{n \to \infty} \frac{n}{n-1} = 1.$$

a. Given $\varepsilon = 0.01$, find a value of N that satisfies the conditions of the limit definition.

b. Prove that $\lim_{n \to \infty} a_n = 1$.

SOLUTION

a. We must find an integer N such that $|a_n - 1| < \varepsilon = 0.01$ whenever $n > N$. This condition can be written

$$|a_n - 1| = \left| \frac{n}{n-1} - 1 \right| = \left| \frac{1}{n-1} \right| < 0.01.$$

Noting that $n > 1$, the absolute value can be removed. The condition on n becomes $n - 1 > 1/0.01 = 100$, or $n > 101$. Therefore, we take $N = 101$ or any larger number. This means that $|a_n - 1| < 0.01$ whenever $n > 101$.

b. Given *any* $\varepsilon > 0$, we must find a value of N (depending on ε) that guarantees

$|a_n - 1| = \left| \dfrac{n}{n-1} - 1 \right| < \varepsilon$ whenever $n > N$. For $n > 1$, the inequality

$\left| \dfrac{n}{n-1} - 1 \right| < \varepsilon$ implies that

$$\left| \frac{n}{n-1} - 1 \right| = \frac{1}{n-1} < \varepsilon.$$

> In general, $1/\varepsilon + 1$ is not an integer, so N should be the least integer greater than $1/\varepsilon + 1$ or any larger integer.

Solving for n, we find that $\dfrac{1}{n-1} < \varepsilon$ or $n - 1 > \dfrac{1}{\varepsilon}$ or $n > \dfrac{1}{\varepsilon} + 1$. Therefore, given a tolerance $\varepsilon > 0$, we must look beyond a_N in the sequence, where $N \geq \dfrac{1}{\varepsilon} + 1$, to be sure that the terms of the sequence are within ε of the limit 1. Because we can provide a value of N for *any* $\varepsilon > 0$, the limit exists and equals 1.

Related Exercise 105 ◄

SECTION 10.2 EXERCISES

Getting Started

1. Give an example of a nonincreasing sequence with a limit.

2. Give an example of a nondecreasing sequence without a limit.

3. Give an example of a bounded sequence that has a limit.

4. Give an example of a bounded sequence without a limit.

5. For what values of r does the sequence $\{r^n\}$ converge? Diverge?

6–9. *Determine whether the following sequences converge or diverge, and state whether they are monotonic or whether they oscillate. Give the limit when the sequence converges.*

6. $\{0.2^n\}$

7. $\{1.00001^n\}$

8. $\{(-2.5)^n\}$

9. $\{(-0.7)^n\}$

10. Find the limit of the sequence $\{a_n\}$ if $1 - \dfrac{1}{n} < a_n < 1 + \dfrac{1}{n}$, for every integer $n \geq 1$.

11. Compare the growth rates of $\{n^{100}\}$ and $\{e^{n/100}\}$ as $n \to \infty$.

12. Use Theorem 10.6 to evaluate $\displaystyle\lim_{n\to\infty} \dfrac{n^{100}}{n^n}$.

Practice Exercises

13–52. Limits of sequences *Find the limit of the following sequences or determine that the sequence diverges.*

13. $\left\{\dfrac{n^3}{n^4 + 1}\right\}$

14. $\left\{\dfrac{n^{12}}{3n^{12} + 4}\right\}$

15. $\left\{\dfrac{3n^3 - 1}{2n^3 + 1}\right\}$

16. $\left\{\dfrac{n^5 + 3n}{10n^3 + n}\right\}$

17. $\left\{\tan^{-1}\left(\dfrac{10n}{10n + 4}\right)\right\}$

18. $\left\{\cot\left(\dfrac{n\pi}{2n + 2}\right)\right\}$

19. $\left\{1 + \cos\dfrac{1}{n}\right\}$

20. $\{\ln(n^3 + 1) - \ln(3n^3 + 10n)\}$

21. $\left\{\ln\sin\dfrac{1}{n} + \ln n\right\}$

22. $\left\{\dfrac{k}{\sqrt{9k^2 + 1}}\right\}$

23. $\left\{\dfrac{\sqrt{4n^4 + 3n}}{8n^2 + 1}\right\}$

24. $\left\{\dfrac{2e^n + 1}{e^n}\right\}$

25. $\left\{\dfrac{\ln n^2}{\ln 3n}\right\}$

26. $\{5(-1.01)^n\}$

27. $\{2^{n+1}3^{-n}\}$

28. $\{100(-0.003)^n\}$

29. $\{(0.5)^n + 3(0.75)^n\}$

30. $\left\{\dfrac{e^n + \pi^n}{e^n}\right\}$

31. $\left\{\dfrac{3^{n+1} + 3}{3^n}\right\}$

32. $\left\{\dfrac{3^n}{3^n + 4^n}\right\}$

33. $\left\{\dfrac{(n + 1)!}{n!}\right\}$

34. $\left\{\dfrac{(2n)!\, n^2}{(2n + 2)!}\right\}$

35. $\{\tan^{-1} n\}$

36. $\left\{\dfrac{e^{n/10}}{2^n}\right\}$

37. $\left\{\sqrt{n^2 + 1} - n\right\}$

38. $\{n^{2/n}\}$

39. $\left\{\left(1 + \dfrac{2}{n}\right)^n\right\}$

40. $\left\{\dfrac{\ln n}{n^{1.1}}\right\}$

41. $\left\{\sqrt[n]{e^{3n+4}}\right\}$

42. $\left\{\left(\dfrac{n}{n + 5}\right)^n\right\}$

43. $\left\{\sqrt{\left(1 + \dfrac{1}{2n}\right)^n}\right\}$

44. $\left\{\left(1 + \dfrac{4}{n}\right)^{3n}\right\}$

45. $\left\{\dfrac{n}{e^n + 3n}\right\}$

46. $\left\{\dfrac{\ln(1/n)}{n}\right\}$

47. $\left\{\left(\dfrac{1}{n}\right)^{1/n}\right\}$

48. $\left\{\left(1 - \dfrac{4}{n}\right)^n\right\}$

49. $\{b_n\}$, where $b_n = \begin{cases} \dfrac{n}{n + 1} & \text{if } n \leq 5000 \\ ne^{-n} & \text{if } n > 5000 \end{cases}$

50. $\left\{n\left(1 - \cos\dfrac{1}{n}\right)\right\}$

51. $\left\{n\sin\dfrac{6}{n}\right\}$

52. $\left\{\dfrac{n^8 + n^7}{n^7 + n^8 \ln n}\right\}$

53. Plot a graph of the sequence $\{a_n\}$, for $a_n = \sin\dfrac{n\pi}{2}$. Then determine the limit of the sequence or explain why the sequence diverges.

54. Plot a graph of the sequence $\{a_n\}$, for $a_n = \dfrac{(-1)^n n}{n + 1}$. Then determine the limit of the sequence or explain why the sequence diverges.

55–70. More sequences *Find the limit of the following sequences or determine that the sequence diverges.*

55. $\left\{\dfrac{(-1)^n}{2^n}\right\}$

56. $\left\{\dfrac{(-1)^n}{n}\right\}$

57. $a_n = (-1)^n \sqrt[n]{n}$

58. $\left\{\dfrac{\cos(n\pi/2)}{\sqrt{n}}\right\}$

59. $\left\{\dfrac{n\sin^3(n\pi/2)}{n + 1}\right\}$

60. $\left\{\dfrac{(-1)^{n+1} n^2}{2n^3 + n}\right\}$

61. $a_n = e^{-n}\cos n$

62. $a_n = \dfrac{e^{-n}}{2\sin(e^{-n})}$

63. $\left\{\dfrac{\tan^{-1} n}{n}\right\}$

64. $\left\{\dfrac{(2n^3 + n)\tan^{-1} n}{n^3 + 4}\right\}$

65. $\left\{\dfrac{\cos n}{n}\right\}$

66. $\left\{\dfrac{\sin 6n}{5n}\right\}$

67. $\left\{\dfrac{\sin n}{2^n}\right\}$

68. $\left\{\displaystyle\int_1^n x^{-2}\, dx\right\}$

69. $\left\{\dfrac{75^{n-1}}{99^n} + \dfrac{5^n \sin n}{8^n}\right\}$

70. $\left\{\cos(0.99^n) + \dfrac{7^n + 9^n}{63^n}\right\}$

T 71. Periodic dosing Many people take aspirin on a regular basis as a preventive measure for heart disease. Suppose a person takes 80 mg of aspirin every 24 hours. Assume aspirin has a half-life of 24 hours; that is, every 24 hours, half of the drug in the blood is eliminated.

   a. Find a recurrence relation for the sequence $\{d_n\}$ that gives the amount of drug in the blood after the nth dose, where $d_1 = 80$.

   b. Use a calculator to estimate this limit. In the long run, how much drug is in the person's blood?

   c. Assuming the sequence has a limit, confirm the result of part (b) by finding the limit of $\{d_n\}$ directly.

T 72. A car loan Marie takes out a $20,000 loan for a new car. The loan has an annual interest rate of 6% or, equivalently, a monthly interest rate of 0.5%. Each month, the bank adds interest to the loan balance (the interest is always 0.5% of the current balance), and then Marie makes a $200 payment to reduce the loan balance. Let B_n be the loan balance immediately after the nth payment, where $B_0 = \$20,000$.

   a. Write the first five terms of the sequence $\{B_n\}$.

   b. Find a recurrence relation that generates the sequence $\{B_n\}$.

   c. Determine how many months are needed to reduce the loan balance to zero.

T 73. A savings plan James begins a savings plan in which he deposits $100 at the beginning of each month into an account that earns 9% interest annually or, equivalently, 0.75% per month. To be clear, on the first day of each month, the bank adds 0.75% of the current balance as interest, and then James deposits $100. Let B_n be the balance in the account after the nth payment, where $B_0 = \$0$.

   a. Write the first five terms of the sequence $\{B_n\}$.

   b. Find a recurrence relation that generates the sequence $\{B_n\}$.

   c. How many months are needed to reach a balance of $5000?

T 74. Diluting a solution A tank is filled with 100 L of a 40% alcohol solution (by volume). You repeatedly perform the following operation: Remove 2 L of the solution from the tank and replace them with 2 L of 10% alcohol solution.

   a. Let C_n be the concentration of the solution in the tank after the nth replacement, where $C_0 = 40\%$. Write the first five terms of the sequence $\{C_n\}$.

   b. After how many replacements does the alcohol concentration reach 15%?

   c. Determine the limiting (steady-state) concentration of the solution that is approached after many replacements.

75–82. Growth rates of sequences *Use Theorem 10.6 to find the limit of the following sequences or state that they diverge.*

75. $\left\{\dfrac{n!}{n^n}\right\}$ **76.** $\left\{\dfrac{3^n}{n!}\right\}$ **77.** $\left\{\dfrac{n^{10}}{\ln^{20} n}\right\}$

78. $\left\{\dfrac{n^{10}}{\ln^{1000} n}\right\}$ **79.** $\left\{\dfrac{n^{1000}}{2^n}\right\}$ **80.** $a_n = \dfrac{4^n + 5n!}{n! + 2^n}$

81. $a_n = \dfrac{6^n + 3^n}{6^n + n^{100}}$ **82.** $a_n = \dfrac{7^n}{n^7 5^n}$

83. Explain why or why not Determine whether the following statements are true and give an explanation or counterexample.

   a. If $\lim\limits_{n\to\infty} a_n = 1$ and $\lim\limits_{n\to\infty} b_n = 3$, then $\lim\limits_{n\to\infty} \dfrac{b_n}{a_n} = 3$.

   b. If $\lim\limits_{n\to\infty} a_n = 0$ and $\lim\limits_{n\to\infty} b_n = \infty$, then $\lim\limits_{n\to\infty} a_n b_n = 0$.

   c. The convergent sequences $\{a_n\}$ and $\{b_n\}$ differ in their first 100 terms, but $a_n = b_n$, for $n > 100$. It follows that $\lim\limits_{n\to\infty} a_n = \lim\limits_{n\to\infty} b_n$.

   d. If $\{a_n\} = \left\{1, \dfrac{1}{2}, \dfrac{1}{3}, \dfrac{1}{4}, \dfrac{1}{5}, \ldots\right\}$ and $\{b_n\} = \left\{1, 0, \dfrac{1}{2}, 0, \dfrac{1}{3}, 0, \dfrac{1}{4}, 0, \ldots\right\}$, then $\lim\limits_{n\to\infty} a_n = \lim\limits_{n\to\infty} b_n$.

   e. If the sequence $\{a_n\}$ converges, then the sequence $\{(-1)^n a_n\}$ converges.

   f. If the sequence $\{a_n\}$ diverges, then the sequence $\{0.000001 a_n\}$ diverges.

T 84–87. Sequences by recurrence relations *The following sequences, defined by a recurrence relation, are monotonic and bounded, and therefore converge by Theorem 10.5.*

   a. *Examine the first three terms of the sequence to determine whether the sequence is nondecreasing or nonincreasing.*

   b. *Use analytical methods to find the limit of the sequence.*

84. $a_{n+1} = \dfrac{1}{2}a_n + 2;\ a_0 = 1$

85. $a_{n+1} = 2a_n(1 - a_n);\ a_0 = 0.3$

86. $a_{n+1} = \dfrac{1}{2}\left(a_n + \dfrac{2}{a_n}\right);\ a_0 = 2$

T 87. $a_{n+1} = \sqrt{2 + a_n};\ a_0 = 3$

T 88. Fish harvesting A fishery manager knows that her fish population naturally increases at a rate of 1.5% per month, while 80 fish are harvested each month. Let F_n be the fish population after the nth month, where $F_0 = 4000$ fish.

   a. Write out the first five terms of the sequence $\{F_n\}$.

   b. Find a recurrence relation that generates the sequence $\{F_n\}$.

   c. Does the fish population decrease or increase in the long run?

   d. Determine whether the fish population decreases or increases in the long run if the initial population is 5500 fish.

   e. Determine the initial fish population F_0 below which the population decreases.

T 89. Drug Dosing A patient takes 75 mg of a medication every 12 hours; 60% of the medication in the blood is eliminated every 12 hours.

   a. Let d_n equal the amount of medication (in mg) in the bloodstream after n doses, where $d_1 = 75$. Find a recurrence relation for d_n.

   b. Show that $\{d_n\}$ is monotonic and bounded, and therefore converges.

   c. Find the limit of the sequence. What is the physical meaning of this limit?

T 90. Sleep model After many nights of observation, you notice that if you oversleep one night, you tend to undersleep the following night, and vice versa. This pattern of compensation is described by the relationship

$$x_{n+1} = \dfrac{1}{2}(x_n + x_{n-1}), \quad \text{for } n = 1, 2, 3, \ldots,$$

where x_n is the number of hours of sleep you get on the nth night, and $x_0 = 7$ and $x_1 = 6$ are the number of hours of sleep on the first two nights, respectively.

   a. Write out the first six terms of the sequence $\{x_n\}$ and confirm that the terms alternately increase and decrease.

b. Show that the explicit formula

$$x_n = \frac{19}{3} + \frac{2}{3}\left(-\frac{1}{2}\right)^n, \quad \text{for } n \geq 0,$$

generates the terms of the sequence in part (a).

c. Assume the limit of the sequence exists. What is the limit of the sequence?

T 91. Calculator algorithm The CORDIC (COordinate Rotation DIgital Calculation) algorithm is used by most calculators to evaluate trigonometric and logarithmic functions. An important number in the CORDIC algorithm, called the *aggregate constant*, is given by the infinite product $\prod_{n=0}^{\infty} \dfrac{2^n}{\sqrt{1 + 2^{2n}}}$, where $\prod_{n=0}^{N} a_n$ is the *partial product* $a_0 \cdot a_1 \cdots a_N$. This infinite product is the limit of the sequence

$$\left\{ \prod_{n=0}^{0} \frac{2^n}{\sqrt{1 + 2^{2n}}}, \prod_{n=0}^{1} \frac{2^n}{\sqrt{1 + 2^{2n}}}, \prod_{n=0}^{2} \frac{2^n}{\sqrt{1 + 2^{2n}}}, \ldots \right\}.$$

Estimate the value of the aggregate constant. (See the Guided Project *CORDIC algorithms: How your calculator works.*)

92. A sequence of products Find the limit of the sequence

$$\{a_n\}_{n=2}^{\infty} = \left\{ \left(1 - \frac{1}{2}\right)\left(1 - \frac{1}{3}\right) \cdots \left(1 - \frac{1}{n}\right) \right\}_{n=2}^{\infty}.$$

Explorations and Challenges

93. Suppose the sequence $\{a_n\}_{n=0}^{\infty}$ is defined by the recurrence relation $a_{n+1} = \frac{1}{3} a_n + 6$; $a_0 = 3$.

  a. Prove that the sequence is increasing and bounded.
  b. Explain why $\{a_n\}_{n=0}^{\infty}$ converges and find the limit.

94. Suppose the sequence $\{a_n\}_{n=0}^{\infty}$ is defined by the recurrence relation $a_{n+1} = \sqrt{a_n + 20}$; $a_0 = 6$.

  a. Prove that the sequence is decreasing and bounded.
  b. Explain why $\{a_n\}_{n=0}^{\infty}$ converges and find the limit.

T 95. Repeated square roots Consider the sequence defined by $a_{n+1} = \sqrt{2 + a_n}$, $a_0 = \sqrt{2}$, for $n = 0, 1, 2, 3, \ldots$.

  a. Evaluate the first four terms of the sequence $\{a_n\}$. State the exact values first, and then the approximate values.
  b. Show that the sequence is increasing and bounded.
  c. Assuming the limit exists, use the method of Example 5 to determine the limit exactly.

T 96. Towers of powers For a positive real number p, the tower of exponents $p^{p^{p^{\cdot}}}$ continues indefinitely but the expression is ambiguous. The tower could be built from the top as the limit of the sequence $\{p^p, (p^p)^p, ((p^p)^p)^p, \ldots\}$, in which case the sequence is defined recursively as

$$a_{n+1} = a_n^p \text{ (building from the top),} \qquad (1)$$

where $a_1 = p^p$. The tower could also be built from the bottom as the limit of the sequence $\{p^p, p^{(p^p)}, p^{(p^{(p^p)})}, \ldots\}$, in which case the sequence is defined recursively as

$$a_{n+1} = p^{a_n} \text{ (building from the bottom),} \qquad (2)$$

where again $a_1 = p^p$.

  a. Estimate the value of the tower with $p = 0.5$ by building from the top. That is, use tables to estimate the limit of the sequence defined recursively by (1) with $p = 0.5$. Estimate the maximum value of $p > 0$ for which the sequence has a limit.

  b. Estimate the value of the tower with $p = 1.2$ by building from the bottom. That is, use tables to estimate the limit of the sequence defined recursively by (2) with $p = 1.2$. Estimate the maximum value of $p > 1$ for which the sequence has a limit.

T 97. Fibonacci sequence The famous Fibonacci sequence was proposed by Leonardo Pisano, also known as Fibonacci, in about A.D. 1200 as a model for the growth of rabbit populations. It is given by the recurrence relation $f_{n+1} = f_n + f_{n-1}$, for $n = 1, 2, 3, \ldots$, where $f_0 = 1$ and $f_1 = 1$. Each term of the sequence is the sum of its two predecessors.

  a. Write out the first ten terms of the sequence.
  b. Is the sequence bounded?
  c. Estimate or determine $\varphi = \lim\limits_{n \to \infty} \dfrac{f_{n+1}}{f_n}$, the ratio of the successive terms of the sequence. Provide evidence that $\varphi = \dfrac{1 + \sqrt{5}}{2}$, a number known as the *golden mean*.
  d. Use induction to verify the remarkable result that

$$f_n = \frac{1}{\sqrt{5}} (\varphi^n - (-1)^n \varphi^{-n}).$$

98. Arithmetic-geometric mean Pick two positive numbers a_0 and b_0 with $a_0 > b_0$ and write out the first few terms of the two sequences $\{a_n\}$ and $\{b_n\}$:

$$a_{n+1} = \frac{a_n + b_n}{2}, \quad b_{n+1} = \sqrt{a_n b_n}, \quad \text{for } n = 0, 1, 2, \ldots.$$

Recall that the arithmetic mean $A = (p + q)/2$ and the geometric mean $G = \sqrt{pq}$ of two positive numbers p and q satisfy $A \geq G$.

  a. Show that $a_n > b_n$ for all n.
  b. Show that $\{a_n\}$ is a decreasing sequence and $\{b_n\}$ is an increasing sequence.
  c. Conclude that $\{a_n\}$ and $\{b_n\}$ converge.
  d. Show that $a_{n+1} - b_{n+1} < \dfrac{a_n - b_n}{2}$ and conclude that $\lim\limits_{n \to \infty} a_n = \lim\limits_{n \to \infty} b_n$. The common value of these limits is called the arithmetic-geometric mean of a_0 and b_0, denoted $\text{AGM}(a_0, b_0)$.
  e. Estimate $\text{AGM}(12, 20)$. Estimate Gauss' constant $\dfrac{1}{\text{AGM}(1, \sqrt{2})}$.

T 99. Continued fractions The expression

$$1 + \cfrac{1}{1 + \cfrac{1}{1 + \cfrac{1}{1 + \cfrac{1}{1 + \cdots}}}},$$

where the process continues indefinitely, is called a *continued fraction*.

  a. Show that this expression can be built in steps using the recurrence relation $a_0 = 1$, $a_{n+1} = 1 + \dfrac{1}{a_n}$ for $n = 0, 1, 2, 3, \ldots$. Explain why the value of the expression can be interpreted as $\lim\limits_{n \to \infty} a_n$, provided the limit exists.
  b. Evaluate the first five terms of the sequence $\{a_n\}$.
  c. Using computation and/or graphing, estimate the limit of the sequence.

d. Assuming the limit exists, use the method of Example 5 to determine the limit exactly. Compare your estimate with $\dfrac{1 + \sqrt{5}}{2}$, a number known as the *golden mean*.

e. Assuming the limit exists, use the same ideas to determine the value of

$$a + \cfrac{b}{a + \cfrac{b}{a + \cfrac{b}{a + \cfrac{b}{a + \ddots}}}},$$

where a and b are positive real numbers.

100. The hailstone sequence Here is a fascinating (unsolved) problem known as the hailstone problem (or the Ulam Conjecture or the Collatz Conjecture). It involves sequences in two different ways. First, choose a positive integer N and call it a_0. This is the *seed* of a sequence. The rest of the sequence is generated as follows: For $n = 0, 1, 2, \ldots$

$$a_{n+1} = \begin{cases} a_n/2 & \text{if } a_n \text{ is even} \\ 3a_n + 1 & \text{if } a_n \text{ is odd.} \end{cases}$$

However, if $a_n = 1$ for any n, then the sequence terminates.

a. Compute the sequence that results from the seeds $N = 2, 3, 4, \ldots, 10$. You should verify that in all these cases, the sequence eventually terminates. The hailstone conjecture (still unproved) states that for all positive integers N, the sequence terminates after a finite number of terms.

b. Now define the hailstone sequence $\{H_k\}$, which is the number of terms needed for the sequence $\{a_n\}$ to terminate starting with a seed of k. Verify that $H_2 = 1$, $H_3 = 7$, and $H_4 = 2$.

c. Plot as many terms of the hailstone sequence as is feasible. How did the sequence get its name? Does the conjecture appear to be true?

101–106. Formal proofs of limits *Use the formal definition of the limit of a sequence to prove the following limits.*

101. $\displaystyle\lim_{n\to\infty} \frac{1}{n} = 0$

102. $\displaystyle\lim_{n\to\infty} \frac{1}{n^2} = 0$

103. $\displaystyle\lim_{n\to\infty} \frac{3n^2}{4n^2 + 1} = \frac{3}{4}$

104. $\displaystyle\lim_{n\to\infty} b^{-n} = 0$, for $b > 1$

105. $\displaystyle\lim_{n\to\infty} \frac{cn}{bn + 1} = \frac{c}{b}$, for real numbers $b > 0$ and $c > 0$

106. $\displaystyle\lim_{n\to\infty} \frac{n}{n^2 + 1} = 0$

⊞ 107. Comparing sequences with a parameter For what values of a does the sequence $\{n!\}$ grow faster than the sequence $\{n^{an}\}$? (*Hint:* Stirling's formula is useful: $n! \approx \sqrt{2\pi n}\, n^n e^{-n}$, for large values of n.)

108–109. Reindexing *Express each sequence $\{a_n\}_{n=1}^{\infty}$ as an equivalent sequence of the form $\{b_n\}_{n=3}^{\infty}$.*

108. $\{2n + 1\}_{n=1}^{\infty}$

109. $\{n^2 + 6n - 9\}_{n=1}^{\infty}$

110. Prove that if $\{a_n\} \ll \{b_n\}$ (as used in Theorem 10.6), then $\{ca_n\} \ll \{db_n\}$, where c and d are positive real numbers.

⊞ 111. Crossover point The sequence $\{n!\}$ ultimately grows faster than the sequence $\{b^n\}$, for any $b > 1$, as $n \to \infty$. However, b^n is generally greater than $n!$ for small values of n. Use a calculator to determine the smallest value of n such that $n! > b^n$ for each of the cases $b = 2$, $b = e$, and $b = 10$.

QUICK CHECK ANSWERS

1. a. Bounded, monotonic;   **b.** Bounded, not monotonic;   **c.** Not bounded, not monotonic;   **d.** Bounded, monotonic (both nonincreasing and nondecreasing)   **2.** If $r = -1$, the sequence is $\{-1, 1, -1, 1, \ldots\}$, the terms alternate in sign, and the sequence diverges. If $r = 1$, the sequence is $\{1, 1, 1, 1, \ldots\}$, the terms are constant, and the sequence converges to 1.   **3.** Both changes would increase the steady-state level of drug.   **4.** $\{n^{1.1}\}$ grows faster; the limit is 0. ◄

10.3 Infinite Series

We begin our discussion of infinite series with *geometric series*. These series arise more frequently than any other infinite series, they are used in many practical problems, and they illustrate all the essential features of infinite series in general. First let's summarize some important ideas from Section 10.1.

Recall that every infinite series $\displaystyle\sum_{k=1}^{\infty} a_k$ has a sequence of partial sums:

$$S_1 = a_1, \qquad S_2 = a_1 + a_2, \qquad S_3 = a_1 + a_2 + a_3,$$

and in general, $S_n = \displaystyle\sum_{k=1}^{n} a_k$, for $n = 1, 2, 3, \ldots$.

If the sequence of partial sums $\{S_n\}$ converges—that is, if $\displaystyle\lim_{n\to\infty} S_n = L$—then the value of the infinite series is also L. If the sequence of partial sums diverges, then the infinite series also diverges.

➤ The sequence of partial sums may be visualized nicely as follows:

$$a_1 + a_2 + a_3 + a_4 + \cdots$$
$$\underbrace{\quad}_{S_1}$$
$$\underbrace{\qquad}_{S_2}$$
$$\underbrace{\qquad\quad}_{S_3}$$

When the initial value of the index is not 1, the nth partial sum is still the sum of the first n terms of the series. For example, the nth partial sum of $\displaystyle\sum_{k=3}^{\infty} a_k$ is

$$S_n = \underbrace{a_3 + a_4 + \cdots + a_{n+2}}_{n \text{ terms}}.$$

QUICK CHECK 1 Write the nth partial sum of the series $\sum\limits_{k=0}^{\infty} b_n$. ◄

▶ Geometric *sequences* have the form $\{r^k\}$ or $\{ar^k\}$. Geometric *sums* and *series* have the form $\sum r^k$ or $\sum ar^k$.

QUICK CHECK 2 Which of the following sums are not geometric sums?

a. $\sum\limits_{k=0}^{10} \left(\dfrac{1}{2}\right)^k$ **b.** $\sum\limits_{k=0}^{20} \dfrac{1}{k}$

c. $\sum\limits_{k=0}^{30} (2k + 1)$ ◄

▶ The notation $\sum\limits_{k=0}^{\infty} ar^k$ appears to have an undefined first term when $r = 0$. The notation is understood to mean $a + ar + ar^2 + \cdots$ and therefore, the series has a value of a when $r = 0$.

QUICK CHECK 3 Verify that the geometric sum formula gives the correct result for the sums $1 + \frac{1}{2}$ and $\frac{1}{2} + \frac{1}{4} + \frac{1}{8}$. ◄

In summary, to evaluate an infinite series, it is necessary to determine a formula for the sequence of partial sums $\{S_n\}$ and then find its limit. This procedure can be carried out with the series that we discuss in this section: geometric series and telescoping series.

Geometric Sums and Series

As a preliminary step to geometric series, we study geometric sums, which are *finite sums* in which each term in the sum is a constant multiple of the previous term. A **geometric sum** with n terms has the form

$$S_n = a + ar + ar^2 + \cdots + ar^{n-1} = \sum_{k=0}^{n-1} ar^k,$$

where $a \neq 0$ and r are real numbers; r is called the **ratio** of the sum and a is its first term. For example, the geometric sum with $r = 0.1$, $a = 0.9$, and $n = 4$ is

$$0.9 + 0.09 + 0.009 + 0.0009 = 0.9(1 + 0.1 + 0.01 + 0.001)$$

$$= \sum_{k=0}^{3} 0.9(0.1^k).$$

Our goal is to find a formula for the value of the geometric sum

$$S_n = a + ar + ar^2 + \cdots + ar^{n-1}, \tag{1}$$

for any values of $a \neq 0$, r, and the positive integer n. Doing so requires a clever maneuver. The first step is to multiply both sides of equation (1) by the ratio r:

$$rS_n = r(a + ar + ar^2 + ar^3 + \cdots + ar^{n-1})$$
$$= ar + ar^2 + ar^3 + \cdots + ar^{n-1} + ar^n. \tag{2}$$

We now subtract equation (2) from equation (1). Notice how most of the terms on the right sides of these equations cancel, leaving

$$S_n - rS_n = a - ar^n.$$

Assuming $r \neq 1$ and solving for S_n results in a general formula for the value of a geometric sum:

$$S_n = a\frac{1 - r^n}{1 - r}. \tag{3}$$

Having dealt with geometric sums, it is a short step to *geometric series*. We simply note that the geometric sums $S_n = \sum\limits_{k=0}^{n-1} ar^k$ form the sequence of partial sums for the geometric series $\sum\limits_{k=0}^{\infty} ar^k$. The value of the geometric series is the limit of its sequence of partial sums (provided that limit exists). Using equation (3), we have

$$\underbrace{\sum_{k=0}^{\infty} ar^k}_{\text{geometric series}} = \lim_{n\to\infty} \underbrace{\sum_{k=0}^{n-1} ar^k}_{\text{geometric sum } S_n} = \lim_{n\to\infty} a\frac{1 - r^n}{1 - r}.$$

To compute this limit, we must examine the behavior of r^n as $n \to \infty$. Recall from our work with geometric sequences (Section 10.2) that

$$\lim_{n\to\infty} r^n = \begin{cases} 0 & \text{if } |r| < 1 \\ 1 & \text{if } r = 1 \\ \text{does not exist} & \text{if } r \leq -1 \text{ or } r > 1. \end{cases}$$

Case 1: $|r| < 1$ Because $\lim\limits_{n\to\infty} r^n = 0$, we have

$$\lim_{n\to\infty} S_n = \lim_{n\to\infty} a\frac{1 - r^n}{1 - r} = a\frac{1 - \overset{0}{\overbrace{\lim_{n\to\infty} r^n}}}{1 - r} = \frac{a}{1 - r}.$$

In the case that $|r| < 1$, the geometric series *converges* to $\dfrac{a}{1 - r}$.

Case 2: $|r| > 1$ In this case, $\lim\limits_{n \to \infty} r^n$ does not exist, so $\lim\limits_{n \to \infty} S_n$ does not exist and the series *diverges*.

Case 3: $|r| = 1$ If $r = 1$, then the geometric series is $\sum\limits_{k=0}^{\infty} a = a + a + a + \cdots$, which diverges. If $r = -1$, the geometric series is $a \sum\limits_{k=0}^{\infty} (-1)^k = a - a + a - \cdots$, which also diverges (because the sequence of partial sums oscillates between 0 and a). We summarize these results in Theorem 10.7.

QUICK CHECK 4 Evaluate
$\frac{1}{2} + \frac{1}{4} + \frac{1}{8} + \frac{1}{16} + \cdots$. ◄

QUICK CHECK 5 Explain why $\sum\limits_{k=0}^{\infty} 0.2^k$ converges and why $\sum\limits_{k=0}^{\infty} 2^k$ diverges. ◄

> **THEOREM 10.7   Geometric Series**
>
> Let $a \neq 0$ and r be real numbers. If $|r| < 1$, then $\sum\limits_{k=0}^{\infty} ar^k = \dfrac{a}{1 - r}$. If $|r| \geq 1$, then the series diverges.
>
>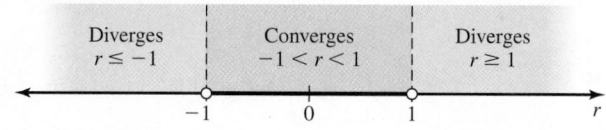

EXAMPLE 1   Geometric series Evaluate the following geometric series or state that the series diverges.

a. $\sum\limits_{k=0}^{\infty} 1.1^k$ **b.** $\sum\limits_{k=0}^{\infty} e^{-k}$ **c.** $\sum\limits_{k=2}^{\infty} 3(-0.75)^k$

SOLUTION

a. The ratio of this geometric series is $r = 1.1$. Because $|r| \geq 1$, the series diverges.

b. Note that $e^{-k} = \dfrac{1}{e^k} = \left(\dfrac{1}{e} \right)^k$. Therefore, the ratio of the series is $r = \dfrac{1}{e}$, and its first term is $a = 1$. Because $|r| < 1$, the series converges and its value is

$$\sum_{k=0}^{\infty} e^{-k} = \sum_{k=0}^{\infty} \left(\frac{1}{e} \right)^k = \frac{1}{1 - (1/e)} = \frac{e}{e - 1} \approx 1.582.$$

> ▶ The series in Example 1c is called an *alternating series* because the terms alternate in sign. Such series are discussed in detail in Section 10.6.

c. Writing out the first few terms of the series is helpful:

$$\sum_{k=2}^{\infty} 3(-0.75)^k = \underbrace{3(-0.75)^2}_{a} + \underbrace{3(-0.75)^3}_{ar} + \underbrace{3(-0.75)^4}_{ar^2} + \cdots.$$

We see that the first term of the series is $a = 3(-0.75)^2$ and that the ratio of the series is $r = -0.75$. Because $|r| < 1$, the series converges, and its value is

$$\sum_{k=2}^{\infty} 3(-0.75)^k = \underbrace{\frac{3(-0.75)^2}{1 - (-0.75)}}_{\frac{a}{1 - r}} = \frac{27}{28}.$$

Related Exercises 24, 27, 29 ◄

EXAMPLE 2   Decimal expansions as geometric series Write $1.0\overline{35} = 1.0353535\ldots$ as a geometric series and express its value as a fraction.

SOLUTION Notice that the decimal part of this number is a convergent geometric series with $a = 0.035$ and $r = 0.01$:

$$1.0353535\ldots = 1 + \underbrace{0.035 + 0.00035 + 0.0000035 + \cdots}_{\text{geometric series with } a = 0.035 \text{ and } r = 0.01}.$$

Evaluating the series, we have

$$1.0353535\ldots = 1 + \frac{a}{1-r} = 1 + \frac{0.035}{1-0.01} = 1 + \frac{35}{990} = \frac{205}{198}.$$

Related Exercises 49–50 ◄

Telescoping Series

With geometric series, we carried out the entire evaluation process by finding a formula for the sequence of partial sums and evaluating the limit of the sequence. Not many infinite series can be subjected to this sort of analysis. With another class of series, called **telescoping series**, it can also be done. Here is an example.

EXAMPLE 3   Telescoping series Evaluate the following series.

a. $\displaystyle\sum_{k=1}^{\infty}\left(\cos\frac{1}{k} - \cos\frac{1}{k+1}\right)$ **b.** $\displaystyle\sum_{k=3}^{\infty}\frac{1}{(k-2)(k-1)}$

SOLUTION

a. The nth term of the sequence of partial sums is

$$S_n = \sum_{k=1}^{n}\left(\cos\frac{1}{k} - \cos\frac{1}{k+1}\right)$$

$$= \left(\cos 1 - \cos\frac{1}{2}\right) + \left(\cos\frac{1}{2} - \cos\frac{1}{3}\right) + \cdots + \left(\cos\frac{1}{n} - \cos\frac{1}{n+1}\right)$$

$$= \cos 1 + \underbrace{\left(-\cos\frac{1}{2} + \cos\frac{1}{2}\right)}_{0} + \cdots + \underbrace{\left(-\cos\frac{1}{n} + \cos\frac{1}{n}\right)}_{0} - \cos\frac{1}{n+1} \qquad \text{Regroup terms.}$$

$$= \cos 1 - \cos\frac{1}{n+1}. \qquad \text{Simplify.}$$

Observe that the interior terms of the sum cancel (or telescope), leaving a simple expression for S_n. Taking the limit, we find that

$$\sum_{k=1}^{\infty}\left(\cos\frac{1}{k} - \cos\frac{1}{k+1}\right) = \lim_{n\to\infty} S_n$$

$$= \lim_{n\to\infty}\left(\cos 1 - \underbrace{\cos\frac{1}{n+1}}_{\to\,\cos 0\,=\,1}\right) = \cos 1 - 1 \approx -0.46.$$

b. Because the initial value of the index is $k = 3$, the nth partial sum S_n involves n terms that run from $k = 3$ to $k = n + 2$, which is somewhat awkward. The nth partial sums are most easily written when the initial value of the index is $k = 1$. We *reindex* the original sum by using the substitution $j = k - 2$, which implies that $j = 1$ when $k = 3$ and that $k - 1 = j + 1$:

$$\sum_{k=3}^{\infty}\frac{1}{(k-2)(k-1)} = \sum_{j=1}^{\infty}\frac{1}{j(j+1)} \qquad \text{Let } j = k - 2.$$

$$= \sum_{k=1}^{\infty}\frac{1}{k(k+1)}.$$

The variable used as an index does not affect the value of a series, so it is permissible to return to the index k.

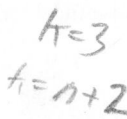

► See Section 8.5 for a review of partial fractions.

Using the method of partial fractions, the sequence of partial sums is

$$S_n = \sum_{k=1}^{n}\frac{1}{k(k+1)} = \sum_{k=1}^{n}\left(\frac{1}{k} - \frac{1}{k+1}\right).$$

Writing out this sum, we see that

$$S_n = \left(1 - \frac{1}{2}\right) + \left(\frac{1}{2} - \frac{1}{3}\right) + \left(\frac{1}{3} - \frac{1}{4}\right) + \cdots + \left(\frac{1}{n} - \frac{1}{n+1}\right)$$

$$= 1 + \underbrace{\left(-\frac{1}{2} + \frac{1}{2}\right)}_{0} + \underbrace{\left(-\frac{1}{3} + \frac{1}{3}\right)}_{0} + \cdots + \underbrace{\left(-\frac{1}{n} + \frac{1}{n}\right)}_{0} - \frac{1}{n+1}$$

$$= 1 - \frac{1}{n+1}.$$

Again, the sum telescopes and all the interior terms cancel. The result is a simple formula for the nth term of the sequence of partial sums. The value of the series is

$$\sum_{k=1}^{\infty} \frac{1}{k(k+1)} = \lim_{n\to\infty} S_n = \lim_{n\to\infty} \left(1 - \frac{1}{n+1}\right) = 1.$$

Related Exercises 59, 69 ◄

Properties of Convergent Series

We close this section with several properties that are useful in upcoming work. The notation $\sum a_k$, without initial and final values of k, is used to refer to a general infinite series whose terms may be positive or negative (or both).

> ▶ The **leading terms** of an infinite series are those at the beginning with a small index. The **tail** of an infinite series consists of the terms at the "end" of the series with a large and increasing index. The convergence or divergence of an infinite series depends on the tail of the series, while the value of a convergent series is determined primarily by the leading terms.

THEOREM 10.8   Properties of Convergent Series

1. Suppose $\sum a_k$ converges to A and c is a real number. The series $\sum ca_k$ converges, and $\sum ca_k = c \sum a_k = cA$.
2. Suppose $\sum a_k$ diverges. Then $\sum ca_k$ also diverges, for any real number $c \neq 0$.
3. Suppose $\sum a_k$ converges to A and $\sum b_k$ converges to B. The series $\sum (a_k \pm b_k)$ converges, and $\sum (a_k \pm b_k) = \sum a_k \pm \sum b_k = A \pm B$.
4. Suppose $\sum a_k$ diverges and $\sum b_k$ converges. Then $\sum (a_k \pm b_k)$ diverges.
5. If M is a positive integer, then $\displaystyle\sum_{k=1}^{\infty} a_k$ and $\displaystyle\sum_{k=M}^{\infty} a_k$ either both converge or both diverge. In general, *whether* a series converges does not depend on a finite number of terms added to or removed from the series. However, the *value* of a convergent series does change if nonzero terms are added or removed.

Proof: These properties are proved using properties of finite sums and limits of sequences. To prove Property 1, assume $\displaystyle\sum_{k=1}^{\infty} a_k$ converges to A and note that

$$\sum_{k=1}^{\infty} ca_k = \lim_{n\to\infty} \sum_{k=1}^{n} ca_k \qquad \text{Definition of infinite series}$$

$$= \lim_{n\to\infty} c \sum_{k=1}^{n} a_k \qquad \text{Property of finite sums}$$

$$= c \lim_{n\to\infty} \sum_{k=1}^{n} a_k \qquad \text{Property of limits}$$

$$= c \sum_{k=1}^{\infty} a_k \qquad \text{Definition of infinite series}$$

$$= cA. \qquad \text{Value of the series}$$

The proofs of Properties 2 and 3 are similar (Exercises 96–97); we offer Property 4 with no proof.

Property 5 follows by noting that for finite sums with $1 < M < n$,

$$\sum_{k=M}^{n} a_k = \sum_{k=1}^{n} a_k - \sum_{k=1}^{M-1} a_k.$$

Letting $n \to \infty$ in this equation and assuming the value of the infinite series is $\sum_{k=1}^{\infty} a_k = A$, it follows that

$$\sum_{k=M}^{\infty} a_k = \underbrace{\sum_{k=1}^{\infty} a_k}_{A} - \underbrace{\sum_{k=1}^{M-1} a_k}_{\text{finite number}}.$$

QUICK CHECK 6 Explain why if $\sum_{k=1}^{\infty} a_k$ converges, then the series $\sum_{k=5}^{\infty} a_k$ (with a different starting index) also converges. Do the two series have the same value? ◄

Because the right side has a finite value, $\sum_{k=M}^{\infty} a_k$ converges. Similarly, if $\sum_{k=M}^{\infty} a_k$ converges, then $\sum_{k=1}^{\infty} a_k$ converges. By an analogous argument, if one of these series diverges, then the other series diverges. ◄

EXAMPLE 4 Using properties of series Evaluate the following infinite series.

a. $\displaystyle\sum_{k=2}^{\infty} \left(\frac{1}{3^k} - \frac{2}{3^{k+2}} \right)$ **b.** $\displaystyle\sum_{k=1}^{\infty} \left(5\left(\frac{2}{3}\right)^k - \frac{7^{k-1}}{6^k} \right)$

SOLUTION

a. As long as both $\displaystyle\sum_{k=2}^{\infty} \frac{1}{3^k}$ and $\displaystyle\sum_{k=2}^{\infty} \frac{2}{3^{k+1}}$ converge, we can use Properties 1 and 3 of Theorem 10.8 to write

$$\sum_{k=2}^{\infty} \left(\frac{1}{3^k} - \frac{2}{3^{k+1}} \right) = \sum_{k=2}^{\infty} \frac{1}{3^k} - 2\sum_{k=2}^{\infty} \frac{1}{3^{k+1}}.$$

Both of the new series are geometric, and the ratio of each series is $r = \dfrac{1}{3} < 1$, so each series converges. The first term of $\displaystyle\sum_{k=2}^{\infty} \frac{1}{3^k}$ is $a = \dfrac{1}{9}$ and the first term of $\displaystyle\sum_{k=2}^{\infty} \frac{1}{3^{k+1}}$ is $a = \dfrac{1}{27}$, so we have

$$\sum_{k=2}^{\infty} \left(\frac{1}{3^k} - \frac{2}{3^{k+1}} \right) = \sum_{k=2}^{\infty} \frac{1}{3^k} - 2\sum_{k=2}^{\infty} \frac{1}{3^{k+1}} = \frac{\frac{1}{9}}{1 - \frac{1}{3}} - 2\frac{\frac{1}{27}}{1 - \frac{1}{3}} = \frac{1}{9} \cdot \frac{3}{2} - \frac{2}{27} \cdot \frac{3}{2} = \frac{1}{18}.$$

b. We examine the two series $\displaystyle\sum_{k=1}^{\infty} 5\left(\frac{2}{3}\right)^k$ and $\displaystyle\sum_{k=1}^{\infty} \frac{7^{k-1}}{6^k}$ individually. The first series is a geometric series and is evaluated using the methods of Section 10.3. Its first few terms are

$$\sum_{k=1}^{\infty} 5\left(\frac{2}{3}\right)^k = 5\left(\frac{2}{3}\right) + 5\left(\frac{2}{3}\right)^2 + 5\left(\frac{2}{3}\right)^3 + \cdots.$$

The first term of the series is $a = 5\left(\frac{2}{3}\right)$ and the ratio is $r = \frac{2}{3} < 1$; therefore,

$$\sum_{k=1}^{\infty} 5\left(\frac{2}{3}\right)^k = \frac{a}{1-r} = \frac{5\left(\frac{2}{3}\right)}{1 - \frac{2}{3}} = 10.$$

Writing out the first few terms of the second series, we see that it, too, is a geometric series:

$$\sum_{k=1}^{\infty} \frac{7^{k-1}}{6^k} = \frac{1}{6} + \frac{7}{6^2} + \frac{7^2}{6^3} + \cdots.$$

QUICK CHECK 7 For a series with positive terms, explain why the sequence of partial sums $\{S_n\}$ is an increasing sequence. ◄

Because the ratio is $r = \frac{7}{6} > 1$, $\displaystyle\sum_{k=1}^{\infty} \frac{7^{k-1}}{6^k}$ diverges. By Property 4 of Theorem 10.8, the original series also diverges.

Related Exercises 81–82 ◄

SECTION 10.3 EXERCISES

Getting Started

1. What is meant by the *ratio* of a geometric series?

2. What is the difference between a geometric sum and a geometric series?

3. Does a geometric series always have a finite value?

4. Does a geometric sum always have a finite value?

5. Find the first term a and the ratio r of each geometric series.

   a. $\sum_{k=0}^{\infty} \frac{2}{3}\left(\frac{1}{5}\right)^k$

   b. $\sum_{k=2}^{\infty} \frac{1}{3}\left(-\frac{1}{3}\right)^k$

6. What is the condition for convergence of the geometric series $\sum_{k=0}^{\infty} ar^k$?

7. Find a formula for the nth partial sum S_n of $\sum_{k=1}^{\infty}\left(\frac{1}{k+3} - \frac{1}{k+4}\right)$. Use your formula to find the sum of the first 36 terms of the series.

8. Reindex the series $\sum_{k=5}^{\infty} \frac{3}{4k^2 - 63}$ so that it starts at $k = 1$.

Practice Exercises

9–15. Geometric sums *Evaluate each geometric sum.*

9. $\sum_{k=0}^{8} 3^k$

T 10. $\sum_{k=0}^{10}\left(\frac{1}{4}\right)^k$

T 11. $\sum_{k=0}^{20}\left(\frac{2}{5}\right)^{2k}$

12. $\sum_{k=4}^{12} 2^k$

T 13. $\sum_{k=0}^{9}\left(-\frac{3}{4}\right)^k$

14. $\sum_{k=0}^{6} \pi^k$

T 15. $\frac{1}{4} + \frac{1}{12} + \frac{1}{36} + \frac{1}{108} + \cdots + \frac{1}{2916}$

T 16–17. Periodic savings *Suppose you deposit m dollars at the beginning of every month in a savings account that earns a monthly interest rate of r, which is the annual interest rate divided by 12 (for example, if the annual interest rate is 2.4%, r = 0.024/12 = 0.002). For an initial investment of m dollars, the amount of money in your account at the beginning of the second month is the sum of your second deposit and your initial deposit plus interest, or m + m(1 + r). Continuing in this fashion, it can be shown that the amount of money in your account after n months is $A_n = m + m(1 + r) + \cdots + m(1 + r)^{n-1}$. Use geometric sums to determine the amount of money in your savings account after 5 years (60 months) using the given monthly deposit and interest rate.*

16. Monthly deposits of $100 at an annual interest rate of 1.8%

17. Monthly deposits of $250 at a monthly interest rate of 0.2%

18–20. Evaluating geometric series two ways *Evaluate each geometric series two ways.*

a. Find the nth partial sum S_n of the series and evaluate $\lim_{n\to\infty} S_n$.
b. Evaluate the series using Theorem 10.7.

18. $\sum_{k=0}^{\infty}\left(\frac{2}{5}\right)^k$

19. $\sum_{k=0}^{\infty}\left(-\frac{2}{7}\right)^k$

20. $2 + \frac{4}{3} + \frac{8}{9} + \frac{16}{27} + \cdots$

21–42. Geometric series *Evaluate each geometric series or state that it diverges.*

21. $\sum_{k=0}^{\infty}\left(\frac{1}{4}\right)^k$

22. $\sum_{k=0}^{\infty}\left(\frac{3}{5}\right)^k$

23. $\sum_{k=0}^{\infty}\left(-\frac{9}{10}\right)^k$

24. $\sum_{k=1}^{\infty}\left(-\frac{2}{3}\right)^k$

25. $\sum_{k=0}^{\infty} 0.9^k$

26. $1 + \frac{2}{7} + \frac{2^2}{7^2} + \frac{2^3}{7^3} + \cdots$

27. $1 + 1.01 + 1.01^2 + 1.01^3 + \cdots$

28. $1 + \frac{1}{\pi} + \frac{1}{\pi^2} + \frac{1}{\pi^3} + \cdots$

29. $\sum_{k=1}^{\infty} e^{-2k}$

30. $\sum_{m=2}^{\infty} \frac{5}{2^m}$

31. $\sum_{k=1}^{\infty} 2^{-3k}$

32. $\sum_{k=3}^{\infty} \frac{3 \cdot 4^k}{7^k}$

33. $\sum_{k=4}^{\infty} \frac{1}{5^k}$

34. $\sum_{k=0}^{\infty}\left(\frac{4}{3}\right)^{-k}$

35. $\sum_{k=0}^{\infty} 3(-\pi)^{-k}$

36. $\sum_{k=1}^{\infty} (-e)^{-k}$

37. $1 + \frac{e}{\pi} + \frac{e^2}{\pi^2} + \frac{e^3}{\pi^3} + \cdots$

38. $\frac{1}{16} + \frac{3}{64} + \frac{9}{256} + \frac{27}{1024} + \cdots$

39. $\sum_{k=2}^{\infty} (-0.15)^k$

40. $\sum_{k=1}^{\infty} 3\left(-\frac{1}{8}\right)^{3k}$

41. $\sum_{k=1}^{\infty} \frac{4}{12^k}$

42. $\sum_{k=2}^{\infty} 3e^{-k}$

43–44. Periodic doses *Suppose you take a dose of m mg of a particular medication once per day. Assume f equals the fraction of the medication that remains in your blood one day later. Just after taking another dose of medication on the second day, the amount of medication in your blood equals the sum of the second dose and the fraction of the first dose remaining in your blood, which is m + mf. Continuing in this fashion, the amount of medication in your blood just after your nth dose is $A_n = m + mf + \cdots + mf^{n-1}$. For the given values of f and m, calculate A_5, A_{10}, A_{30}, and $\lim_{n\to\infty} A_n$. Interpret the meaning of the limit $\lim_{n\to\infty} A_n$.*

43. $f = 0.25, m = 200$ mg

44. $f = 0.4, m = 150$ mg

45. **Periodic doses** Suppose you take 200 mg of an antibiotic every 6 hr. The half-life of the drug (the time it takes for half of the drug to be eliminated from your blood) is 6 hr. Use infinite series to find the long-term (steady-state) amount of antibiotic in your blood. You may assume the steady-state amount is finite.

46–53. Decimal expansions *Write each repeating decimal first as a geometric series and then as a fraction (a ratio of two integers).*

46. $0.\overline{6} = 0.666\ldots$

47. $0.\overline{3} = 0.333\ldots$

48. $0.\overline{09} = 0.090909\ldots$

49. $0.\overline{037} = 0.037037\ldots$

50. $1.\overline{25} = 1.252525\ldots$ **51.** $0.\overline{456} = 0.456456456\ldots$

52. $5.12\overline{83} = 5.12838383\ldots$ **53.** $0.00\overline{952} = 0.00952952\ldots$

54–69. Telescoping series *For the following telescoping series, find a formula for the nth term of the sequence of partial sums $\{S_n\}$. Then evaluate $\lim_{n\to\infty} S_n$ to obtain the value of the series or state that the series diverges.*

54. $\displaystyle\sum_{k=1}^{\infty}\left(\frac{1}{k+2} - \frac{1}{k+3}\right)$ **55.** $\displaystyle\sum_{k=1}^{\infty}\left(\frac{1}{k+1} - \frac{1}{k+2}\right)$

56. $\displaystyle\sum_{k=1}^{\infty}\frac{20}{25k^2 + 15k - 4}$ **57.** $\displaystyle\sum_{k=1}^{\infty}\frac{1}{(k+6)(k+7)}$

58. $\displaystyle\sum_{k=-3}^{\infty}\frac{10}{4k^2 + 32k + 63}$ **59.** $\displaystyle\sum_{k=3}^{\infty}\frac{4}{(4k-3)(4k+1)}$

60. $\displaystyle\sum_{k=0}^{\infty}\frac{1}{(3k+1)(3k+4)}$ **61.** $\displaystyle\sum_{k=1}^{\infty}\ln\frac{k+1}{k}$

62. $\displaystyle\sum_{k=3}^{\infty}\frac{2}{(2k-1)(2k+1)}$

63. $\displaystyle\sum_{k=1}^{\infty}\frac{1}{(k+p)(k+p+1)}$, where p is a positive integer

64. $\displaystyle\sum_{k=1}^{\infty}\frac{1}{(ak+1)(ak+a+1)}$, where a is a positive integer

65. $\displaystyle\sum_{k=1}^{\infty}\left(\frac{1}{\sqrt{k+1}} - \frac{1}{\sqrt{k+3}}\right)$

66. $\displaystyle\sum_{k=1}^{\infty}\frac{6}{k^2 + 2k}$ **67.** $\displaystyle\sum_{k=1}^{\infty}\frac{3}{k^2 + 5k + 4}$

68. $\displaystyle\sum_{k=1}^{\infty}\left(\sqrt{k+1} - \sqrt{k}\right)$ **69.** $\displaystyle\sum_{k=1}^{\infty}\left(\tan^{-1}(k+1) - \tan^{-1}k\right)$

70. Evaluating an infinite series two ways Evaluate the series
$$\sum_{k=1}^{\infty}\left(\frac{1}{2^k} - \frac{1}{2^{k+1}}\right)$$ two ways.

a. Use a telescoping series argument.
b. Use a geometric series argument with Theorem 10.8.

71. Evaluating an infinite series two ways Evaluate the series
$$\sum_{k=1}^{\infty}\left(\frac{4}{3^k} - \frac{4}{3^{k+1}}\right)$$ two ways.

a. Use a telescoping series argument.
b. Use a geometric series argument with Theorem 10.8.

72–86. Evaluating series *Evaluate each series or state that it diverges.*

72. $\displaystyle\sum_{k=1}^{\infty}\frac{1}{9k^2 + 39k + 40}$ **73.** $\displaystyle\sum_{k=0}^{\infty}\frac{1}{16k^2 + 8k - 3}$

74. $\displaystyle\sum_{k=0}^{\infty}\left(\sin\left(\frac{(k+1)\pi}{2k+1}\right) - \sin\left(\frac{k\pi}{2k-1}\right)\right)$

75. $\displaystyle\sum_{k=0}^{\infty}\left(\frac{1}{4}\right)^k 5^{3-k}$ **T 76.** $\displaystyle\sum_{k=2}^{\infty}\left(\frac{3}{8}\right)^{3k}$

77. $\displaystyle\sum_{k=1}^{\infty}\frac{(-2)^k}{3^{k+1}}$

78. $\displaystyle\sum_{k=1}^{\infty}\left(\sin^{-1}\left(\frac{1}{k}\right) - \sin^{-1}\left(\frac{1}{k+1}\right)\right)$

79. $\displaystyle\sum_{k=2}^{\infty}\frac{\ln\left((k+1)k^{-1}\right)}{(\ln k)\ln(k+1)}$ **80.** $\displaystyle\sum_{k=1}^{\infty}\frac{\pi^k}{e^{k+1}}$

81. $\displaystyle\sum_{k=0}^{\infty}\left(3\left(\frac{2}{5}\right)^k - 2\left(\frac{5}{7}\right)^k\right)$ **82.** $\displaystyle\sum_{k=1}^{\infty}\left(2\left(\frac{3}{5}\right)^k + 3\left(\frac{4}{9}\right)^k\right)$

83. $\displaystyle\sum_{k=1}^{\infty}\left(\frac{1}{3}\left(\frac{5}{6}\right)^k + \frac{3}{5}\left(\frac{7}{9}\right)^k\right)$ **84.** $\displaystyle\sum_{k=0}^{\infty}\left(\frac{1}{2}(0.2)^k + \frac{3}{2}(0.8)^k\right)$

85. $\displaystyle\sum_{k=1}^{\infty}\left(\left(\frac{1}{6}\right)^k + \left(\frac{1}{3}\right)^{k-1}\right)$ **86.** $\displaystyle\sum_{k=0}^{\infty}\frac{2 - 3^k}{6^k}$

87. Explain why or why not Determine whether the following statements are true and give an explanation or counterexample.

a. If $\displaystyle\sum_{k=1}^{\infty}a_k$ converges, then $\displaystyle\sum_{k=10}^{\infty}a_k$ converges.

b. If $\displaystyle\sum_{k=1}^{\infty}a_k$ diverges, then $\displaystyle\sum_{k=10}^{\infty}a_k$ diverges.

c. If $\displaystyle\sum a_k$ converges, then $\displaystyle\sum (a_k + 0.0001)$ converges.

d. If $\displaystyle\sum p^k$ diverges, then $\displaystyle\sum (p + 0.001)^k$ diverges, for a fixed real number p.

e. $\displaystyle\sum_{k=1}^{\infty}\left(\frac{\pi}{e}\right)^{-k}$ is a convergent geometric series.

f. If the series $\displaystyle\sum_{k=1}^{\infty}a^k$ converges and $|a| < |b|$, then the series $\displaystyle\sum_{k=1}^{\infty}b^k$ converges.

g. Viewed as a function of r, the series $1 + r + r^2 + r^3 + \cdots$ takes on all values in the interval $\left(\frac{1}{2}, \infty\right)$.

88–89. Binary numbers *Humans use the ten digits 0 through 9 to form base-10 or decimal numbers, whereas computers calculate and store numbers internally as binary numbers—numbers consisting entirely of 0's and 1's. For this exercise, we consider binary numbers that have the form $0.b_1b_2b_3\ldots$, where each of the digits $b_1, b_2, b_3, \ldots$ is either 0 or 1. The base-10 representation of the binary number $0.b_1b_2b_3\ldots$ is the infinite series $\frac{b_1}{2^1} + \frac{b_2}{2^2} + \frac{b_3}{2^3} + \cdots$.*

88. Verify that the base-10 representation of the binary number $0.010101\ldots$ (which can also be written as $0.\overline{01}$) is $\frac{1}{3}$.

89. Computers can store only a finite number of digits and therefore numbers with nonterminating digits must be rounded or truncated before they can be used and stored by a computer.

a. Find the base-10 representation of the binary number $0.\overline{0011}$.
b. Suppose a computer rounds the binary number $0.\overline{0011}$ to $b = 0.0011001100110011$ before storing it. Find the approximate base-10 representation of b, rounding your answer to 8 digits to the right of the decimal place.

90. For what value of r does $1 + r + r^2 + r^3 + \cdots = 10$?

T 91. For what value of r does $\displaystyle\sum_{k=3}^{\infty}r^{2k} = 10$?

92. Value of a series

a. Evaluate the series

$$\sum_{k=1}^{\infty} \frac{3^k}{(3^{k+1} - 1)(3^k - 1)}.$$

(*Hint:* Find constants b and c so that

$$\frac{3^k}{(3^{k+1} - 1)(3^k - 1)} = \frac{b}{3^k - 1} + \frac{c}{3^{k+1} - 1}.)$$

b. For what values of a does the series

$$\sum_{k=1}^{\infty} \frac{a^k}{(a^{k+1} - 1)(a^k - 1)}$$

converge, and in those cases, what is its value?

93–95. Loans *Suppose you borrow P dollars from a bank at a monthly interest rate of r and you make monthly payments of M dollars/month to pay off the loan (for example, if your borrow $30,000 at a rate of 1.5% per month, then P = 30,000 and r = 0.015). Each month, the bank first adds interest to the loan balance and then subtracts your monthly payment to determine the new balance on your loan. So if A_0 is your original loan balance and A_n equals your loan balance after n payments, then your balance after n payments is given by the recursive formula $A_n = A_{n-1}(1 + r) - M, A_0 = P$.*

93. Use the recursive formula and geometric sums to obtain the explicit formula $A_n = \left(A_0 - \dfrac{M}{r}\right)(1 + r)^n + \dfrac{M}{r}$.

T 94. Car loan Suppose you borrow $20,000 for a new car at a monthly interest rate of 0.75%. If you make payments of $600/month, after how many months will the loan balance be zero?

T 95. House loan Suppose you take out a home mortgage for $180,000 at a monthly interest rate of 0.5%. If you make payments of $1000/month, after how many months will the loan balance be zero?

Explorations and Challenges

96. Properties proof Use the ideas in the proof of Property 1 of Theorem 10.8 to prove Property 3 of Theorem 10.8.

97. Property of divergent series Prove Property 2 of Theorem 10.8: If Σa_k diverges, then $\Sigma c a_k$ also diverges, for any real number $c \neq 0$.

98. Zeno's paradox The Greek philosopher Zeno of Elea (who lived about 450 B.C.) invented many paradoxes, the most famous of which tells of a race between the swift warrior Achilles and a tortoise. Zeno argued as follows.

The slower when running will never be overtaken by the quicker; for that which is pursuing must first reach the point from which that which is fleeing started, so that the slower must necessarily always be some distance ahead.

In other words, giving the tortoise a head start ensures that Achilles will never overtake the tortoise because every time Achilles reaches the point where the tortoise was, the tortoise has moved ahead. Resolve this paradox by assuming Achilles gives the tortoise a 1-mi head start and runs 5 mi/hr to the tortoise's 1 mi/hr. How far does Achilles run before he overtakes the tortoise, and how long does it take?

99. Archimedes' quadrature of the parabola The Greeks solved several calculus problems almost 2000 years before the discovery of calculus. One example is Archimedes' calculation of the area of the region R bounded by a segment of a parabola, which he did using the "method of exhaustion." As shown in the figure, the idea was to fill R with an infinite sequence of triangles. Archimedes began with an isosceles triangle inscribed in the parabola, with area A_1, and proceeded in stages, with the number of new triangles doubling at each stage. He was able to show (the key to the solution) that at each stage, the area of a new triangle is $1/8$ of the area of a triangle at the previous stage; for example, $A_2 = 1/8 \, A_1$, and so forth. Show, as Archimedes did, that the area of R is $4/3$ times the area of A_1.

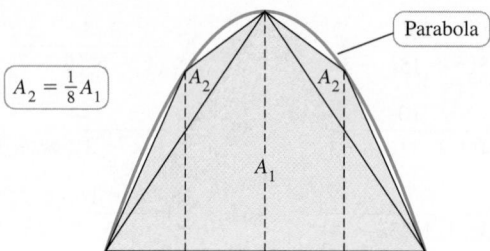

100. Double glass An insulated window consists of two parallel panes of glass with a small spacing between them. Suppose each pane reflects a fraction p of the incoming light and transmits the remaining light. Considering all reflections of light between the panes, what fraction of the incoming light is ultimately transmitted by the window? Assume the amount of incoming light is 1.

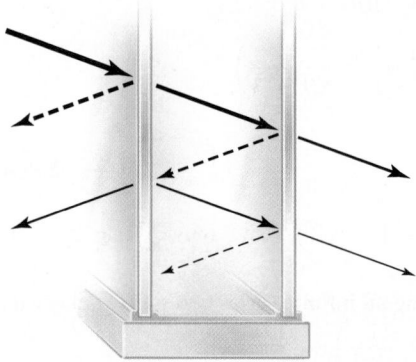

101. Snowflake island fractal The fractal called the *snowflake island* (or *Koch island*) is constructed as follows: Let I_0 be an equilateral triangle with sides of length 1. The figure I_1 is obtained by replacing the middle third of each side of I_0 with a new outward equilateral triangle with sides of length $1/3$ (see figure). The process is repeated, where I_{n+1} is obtained by replacing the middle third of each side of I_n with a new outward equilateral triangle with sides of length $\dfrac{1}{3^{n+1}}$. The limiting figure as $n \to \infty$ is called the snowflake island.

a. Let L_n be the perimeter of I_n. Show that $\lim\limits_{n \to \infty} L_n = \infty$.

b. Let A_n be the area of I_n. Find $\lim\limits_{n \to \infty} A_n$. It exists!

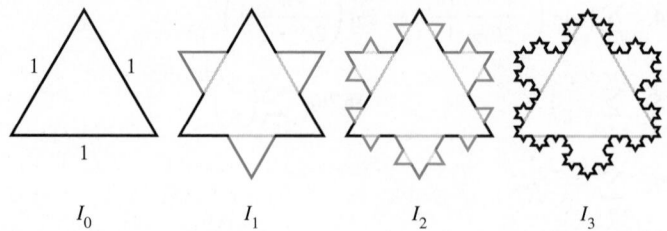

102. Bubbles Imagine a stack of hemispherical soap bubbles with decreasing radii $r_1 = 1, r_2, r_3, \ldots$ (see figure). Let h_n be the distance between the diameters of bubble n and bubble $n + 1$, and let H_n be the total height of the stack with n bubbles.

   a. Use the Pythagorean theorem to show that in a stack with n bubbles, $h_1^2 = r_1^2 - r_2^2, h_2^2 = r_2^2 - r_3^2$, and so forth. Note that for the last bubble $h_n = r_n$.

   b. Use part (a) to show that the height of a stack with n bubbles is

$$H_n = \sqrt{r_1^2 - r_2^2} + \sqrt{r_2^2 - r_3^2} + \cdots + \sqrt{r_{n-1}^2 - r_n^2} + r_n.$$

   c. The height of a stack of bubbles depends on how the radii decrease. Suppose $r_1 = 1, r_2 = a, r_3 = a^2, \ldots, r_n = a^{n-1}$, where $0 < a < 1$ is a fixed real number. In terms of a, find the height H_n of a stack with n bubbles.

   d. Suppose the stack in part (c) is extended indefinitely ($n \to \infty$). In terms of a, how high would the stack be?

   e. Challenge problem: Fix n and determine the sequence of radii $r_1, r_2, r_3, \ldots, r_n$ that maximizes H_n, the height of the stack with n bubbles.

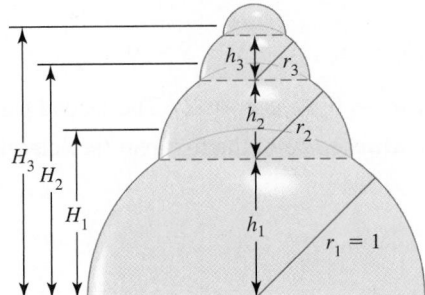

103. Remainder term Consider the geometric series $S = \sum\limits_{k=0}^{\infty} r^k$, which

has the value $\dfrac{1}{1 - r}$ provided $|r| < 1$. Let $S_n = \sum\limits_{k=0}^{n-1} r^k = \dfrac{1 - r^n}{1 - r}$

be the sum of the first n terms. The magnitude of the remainder R_n is the error in approximating S by S_n. Show that

$$R_n = S - S_n = \frac{r^n}{1 - r}.$$

104–107. Comparing remainder terms *Use Exercise 103 to determine how many terms of each series are needed so that the partial sum is within 10^{-6} of the value of the series (that is, to ensure $|R_n| < 10^{-6}$).*

104. a. $\sum\limits_{k=0}^{\infty} 0.6^k$ **b.** $\sum\limits_{k=0}^{\infty} 0.15^k$

105. a. $\sum\limits_{k=0}^{\infty} (-0.8)^k$ **b.** $\sum\limits_{k=0}^{\infty} 0.2^k$

106. a. $\sum\limits_{k=0}^{\infty} 0.72^k$ **b.** $\sum\limits_{k=0}^{\infty} (-0.25)^k$

107. a. $\sum\limits_{k=0}^{\infty} \left(\dfrac{1}{\pi}\right)^k$ **b.** $\sum\limits_{k=0}^{\infty} \left(\dfrac{1}{e}\right)^k$

108. Function defined by a series Suppose a function f is defined by the geometric series $f(x) = \sum\limits_{k=0}^{\infty} x^k$.

   a. Evaluate $f(0), f(0.2), f(0.5), f(1)$, and $f(1.5)$, if possible.

   b. What is the domain of f?

109. Function defined by a series Suppose a function f is defined by the geometric series $f(x) = \sum\limits_{k=0}^{\infty} (-1)^k x^k$.

   a. Evaluate $f(0), f(0.2), f(0.5), f(1)$, and $f(1.5)$, if possible.

   b. What is the domain of f?

110. Function defined by a series Suppose a function f is defined by the geometric series $f(x) = \sum\limits_{k=0}^{\infty} x^{2k}$.

   a. Evaluate $f(0), f(0.2), f(0.5), f(1)$, and $f(1.5)$, if possible.

   b. What is the domain of f?

111. Series in an equation For what values of x does the geometric series

$$f(x) = \sum_{k=0}^{\infty} \left(\frac{1}{1 + x}\right)^k$$

converge? Solve $f(x) = 3$.

QUICK CHECK ANSWERS

1. $S_n = b_0 + b_1 + \cdots + b_{n-1}$ **2.** b and c   **3.** Using the formula, the values are $\frac{3}{2}$ and $\frac{7}{8}$.   **4.** 1   **5.** The first converges because $|r| = 0.2 < 1$; the second diverges because $|r| = 2 > 1$.   **6.** Removing a finite number of terms does not change whether the series converges. It generally changes the value of the series.   **7.** Given the nth term of the sequence of partial sums S_n, the next term is obtained by adding a positive number. So $S_{n+1} > S_n$, which means the sequence is increasing. ◄

10.4 The Divergence and Integral Tests

With geometric series and telescoping series, the sequence of partial sums can be found and its limit can be evaluated (when it exists). Unfortunately, it is difficult or impossible to find an explicit formula for the sequence of partial sums for most infinite series. Therefore, it is difficult to obtain the exact value of most convergent series.

    In light of these observations, we now shift our focus and ask a simple *yes* or *no* question: Given an infinite series, does it converge? If the answer is *no*, the series diverges and there are no more questions to ask. If the answer is *yes*, the series converges and it may be possible to estimate its value.

The Divergence Test

One of the simplest and most useful tests determines whether an infinite series *diverges*. Though our focus in this section and the next is on series with positive terms, the Divergence Test applies to series with arbitrary terms.

THEOREM 10.9 Divergence Test

If $\sum a_k$ converges, then $\lim\limits_{k \to \infty} a_k = 0$. Equivalently, if $\lim\limits_{k \to \infty} a_k \neq 0$, then the series diverges.

Important note: Theorem 10.9 cannot be used to conclude that a series converges.

Proof: Let $\{S_n\}$ be the sequence of partial sums for the series $\sum a_k$. Assuming the series converges, it has a finite value, call it S, where

$$S = \lim_{n \to \infty} S_n = \lim_{n \to \infty} S_{n-1}.$$

Note that $S_n - S_{n-1} = a_n$. Therefore,

$$\lim_{n \to \infty} a_n = \lim_{n \to \infty} (S_n - S_{n-1}) = S - S = 0;$$

> ▶ If the statement *if p, then q* is true, then its contrapositive, *if (not q), then (not p)*, is also true. However its converse, *if q, then p*, is not necessarily true. Try it out on the true statement, *if I live in Paris, then I live in France.*

that is, $\lim\limits_{n \to \infty} a_n = 0$ (which implies $\lim\limits_{k \to \infty} a_k = 0$; **Figure 10.25**). The second part of the test follows immediately because it is the *contrapositive* of the first part (see margin note). ◀

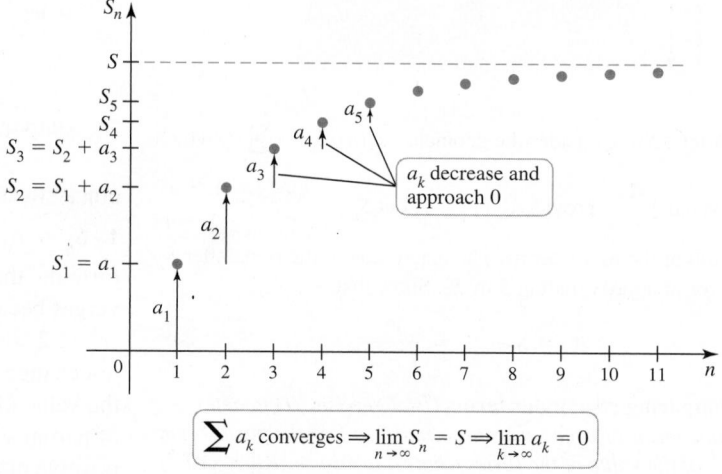

$$\sum a_k \text{ converges} \Rightarrow \lim_{n \to \infty} S_n = S \Rightarrow \lim_{k \to \infty} a_k = 0$$

Figure 10.25

EXAMPLE 1 Using the Divergence Test Determine whether the following series diverge or state that the Divergence Test is inconclusive.

a. $\displaystyle\sum_{k=0}^{\infty} \frac{k}{k+1}$ **b.** $\displaystyle\sum_{k=1}^{\infty} \frac{1 + 3^k}{2^k}$ **c.** $\displaystyle\sum_{k=1}^{\infty} \frac{1}{k}$ **d.** $\displaystyle\sum_{k=1}^{\infty} \frac{1}{k^2}$

SOLUTION By the Divergence Test, if $\lim\limits_{k \to \infty} a_k \neq 0$, then the series $\sum a_k$ diverges.

a. $\lim\limits_{k \to \infty} a_k = \lim\limits_{k \to \infty} \dfrac{k}{k+1} = 1 \neq 0$

The terms of the series do not approach zero, so the series diverges by the Divergence Test.

b. $\lim\limits_{k\to\infty} a_k = \lim\limits_{k\to\infty} \dfrac{1 + 3^k}{2^k}$

$$= \lim\limits_{k\to\infty} \left(\underbrace{2^{-k}}_{\to\, 0} + \underbrace{\left(\dfrac{3}{2}\right)^k}_{\to\, \infty} \right) \quad \text{Simplify.}$$

$$= \infty$$

In this case, $\lim\limits_{k\to\infty} a_k \neq 0$, so the corresponding series $\sum\limits_{k=1}^{\infty} \dfrac{1 + 3^k}{2^k}$ diverges by the Divergence Test.

c. $\lim\limits_{k\to\infty} a_k = \lim\limits_{k\to\infty} \dfrac{1}{k} = 0$

In this case, the terms of the series approach zero, so the Divergence Test is inconclusive. Remember, the Divergence Test cannot be used to prove that a series converges.

QUICK CHECK 1 Apply the Divergence Test to the geometric series $\sum r^k$. For what values of r does the series diverge? ◄

d. $\lim\limits_{k\to\infty} a_k = \lim\limits_{k\to\infty} \dfrac{1}{k^2} = 0$

As in part (c), the terms of the series approach 0, so the Divergence Test is inconclusive.

Related Exercises 9–10 ◄

To summarize: If the terms a_k of a given series do *not* approach zero as $k \to \infty$, then the series diverges. Unfortunately, the test is easy to misuse. It's tempting to conclude that if the terms of the series approach zero, then the series converges. However, look again at the series in Examples 1c and 1d. Although it is true that $\lim\limits_{k\to\infty} a_k = 0$ for both series, we will soon discover that one of them converges while the other diverges. We cannot tell which behavior to expect based only on the observation that $\lim\limits_{k\to\infty} a_k = 0$.

The Harmonic Series

We now look at an example with a surprising result. Consider the infinite series

$$\sum_{k=1}^{\infty} \dfrac{1}{k} = 1 + \dfrac{1}{2} + \dfrac{1}{3} + \dfrac{1}{4} + \dfrac{1}{5} + \cdots,$$

a famous series known as the **harmonic series**. Does it converge? As explained in Example 1c, this question cannot be answered by the Divergence Test, despite the fact that $\lim\limits_{k\to\infty} \dfrac{1}{k} = 0$. Suppose instead you try to answer the convergence question by writing out the terms of the sequence of partial sums:

$$S_1 = 1, \qquad\qquad S_2 = 1 + \dfrac{1}{2} = \dfrac{3}{2},$$

$$S_3 = 1 + \dfrac{1}{2} + \dfrac{1}{3} = \dfrac{11}{6}, \quad S_4 = 1 + \dfrac{1}{2} + \dfrac{1}{3} + \dfrac{1}{4} = \dfrac{25}{12},$$

and in general,

$$S_n = \sum_{k=1}^{n} \dfrac{1}{k} = 1 + \dfrac{1}{2} + \dfrac{1}{3} + \dfrac{1}{4} + \cdots + \dfrac{1}{n}.$$

There is no obvious pattern in this sequence, and in fact, no simple explicit formula for S_n exists, so we analyze the sequence numerically. Have a look at the first 200 terms of the sequence of partial sums shown in **Figure 10.26**. What do you think—does the series converge? The terms of the sequence of partial sums increase, but at a decreasing rate. They could approach a limit or they could increase without bound.

Computing additional terms of the sequence of partial sums does not provide conclusive evidence. Table 10.3 shows that the sum of the first million terms is less than 15; the sum of the first 10^{40} terms—an unimaginably large number of terms—is less than 100. This is a case in which computation alone is not sufficient to determine whether a series converges. We need another way to determine whether the series converges.

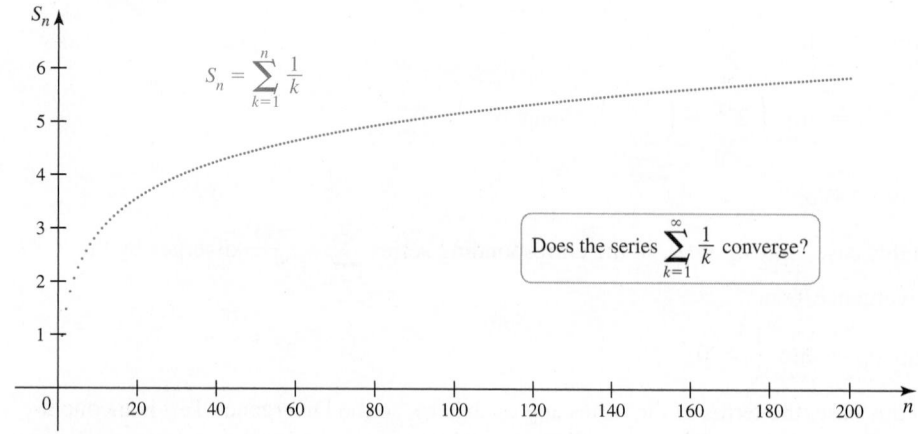

Figure 10.26

Table 10.3

n	S_n	n	S_n
10^3	≈ 7.49	10^{10}	≈ 23.60
10^4	≈ 9.79	10^{20}	≈ 46.63
10^5	≈ 12.09	10^{30}	≈ 69.65
10^6	≈ 14.39	10^{40}	≈ 92.68

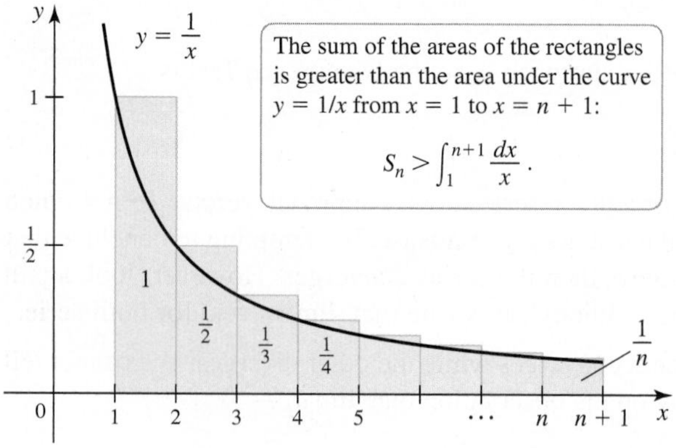

Does the series $\sum_{k=1}^{\infty} \frac{1}{k}$ converge?

$y = \frac{1}{x}$

The sum of the areas of the rectangles is greater than the area under the curve $y = 1/x$ from $x = 1$ to $x = n + 1$:

$$S_n > \int_1^{n+1} \frac{dx}{x}.$$

Figure 10.27

▶ Recall that $\int \frac{dx}{x} = \ln|x| + C$. In Section 8.9, we showed that $\int_1^{\infty} \frac{dx}{x^p}$ diverges for $p \le 1$. Therefore, $\int_1^{\infty} \frac{dx}{x}$ diverges.

Observe that the nth term of the sequence of partial sums,

$$S_n = \sum_{k=1}^{n} \frac{1}{k} = 1 + \frac{1}{2} + \frac{1}{3} + \frac{1}{4} + \cdots + \frac{1}{n},$$

is represented geometrically by a left Riemann sum of the function $y = \frac{1}{x}$ on the interval $[1, n+1]$ (Figure 10.27). This fact follows by noticing that the areas of the rectangles, from left to right, are $1, \frac{1}{2}, \ldots,$ and $\frac{1}{n}$. Comparing the sum of the areas of these n rectangles with the area under the curve, we see that $S_n > \int_1^{n+1} \frac{dx}{x}$. We know that $\int_1^{n+1} \frac{dx}{x} = \ln(n+1)$ increases without bound as n increases. Because S_n exceeds $\int_1^{n+1} \frac{dx}{x}$, S_n also increases without bound; therefore, $\lim_{n \to \infty} S_n = \infty$ and the harmonic series $\sum_{k=1}^{\infty} \frac{1}{k}$ diverges. This argument justifies the following theorem.

THEOREM 10.10 Harmonic Series

The harmonic series $\sum_{k=1}^{\infty} \frac{1}{k} = 1 + \frac{1}{2} + \frac{1}{3} + \frac{1}{4} + \frac{1}{5} + \cdots$ diverges—even though the terms of the series approach zero.

The ideas used to demonstrate that the harmonic series diverges are now used to prove a new and powerful convergence test. This test and those presented in Section 10.5 apply only to series with positive terms.

The Integral Test

The fact that infinite series are sums and that integrals are limits of sums suggests a connection between series and integrals. The Integral Test exploits this connection.

THEOREM 10.11 Integral Test

Suppose f is a continuous, positive, decreasing function, for $x \ge 1$, and let $a_k = f(k)$, for $k = 1, 2, 3, \ldots$. Then

$$\sum_{k=1}^{\infty} a_k \quad \text{and} \quad \int_1^{\infty} f(x)\, dx$$

either both converge or both diverge. In the case of convergence, the value of the integral is *not* equal to the value of the series.

▶ The Integral Test also applies if the terms of the series a_k are decreasing for $k > N$ for some finite number $N > 1$. The proof can be modified to account for this situation.

Proof: By comparing the shaded regions in **Figure 10.28**, it follows that

$$\sum_{k=2}^{n} a_k < \int_1^n f(x)\, dx < \sum_{k=1}^{n-1} a_k. \tag{1}$$

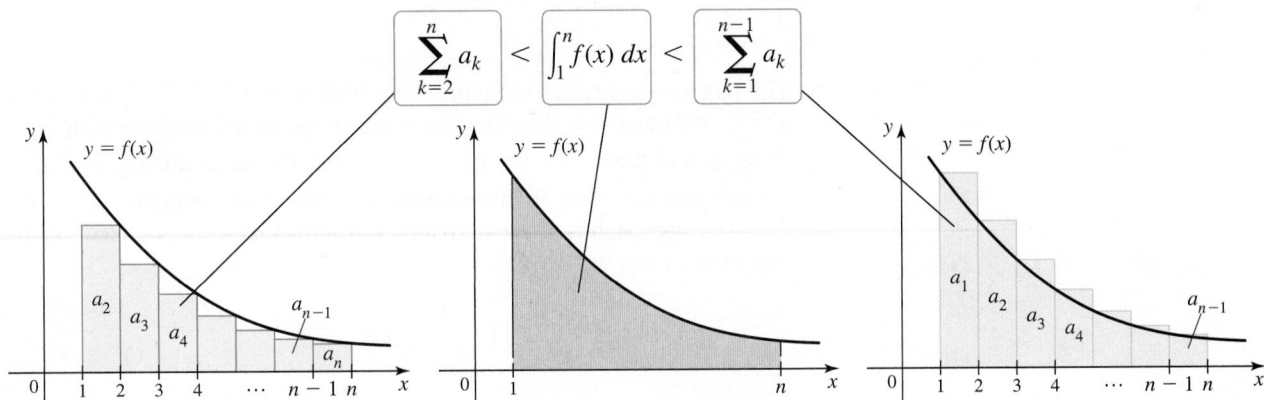

Figure 10.28

The proof must demonstrate two results: If the improper integral $\int_1^\infty f(x)\, dx$ has a finite value, then the infinite series converges, *and* if the infinite series converges, then the improper integral has a finite value. First suppose the improper integral $\int_1^\infty f(x)\, dx$ has a finite value, say I. We have

$$\sum_{k=1}^{n} a_k = a_1 + \sum_{k=2}^{n} a_k \qquad \text{Separate the first term of the series.}$$

$$< a_1 + \int_1^n f(x)\, dx \qquad \text{Left inequality in expression (1)}$$

$$< a_1 + \int_1^\infty f(x)\, dx \qquad f \text{ is positive, so } \int_1^n f(x)\, dx < \int_1^\infty f(x)\, dx.$$

$$= a_1 + I.$$

> In this proof, we rely twice on the Bounded Monotonic Sequence Theorem of Section 10.2: A bounded monotonic sequence converges.

This argument implies that the terms of the sequence of partial sums $S_n = \sum_{k=1}^{n} a_k$ are bounded above by $a_1 + I$. Because $\{S_n\}$ is also increasing (the series consists of positive terms), the sequence of partial sums converges, which means the series $\sum_{k=1}^{\infty} a_k$ converges (to a value less than or equal to $a_1 + I$).

Now suppose the infinite series $\sum_{k=1}^{\infty} a_k$ converges and has a value S. We have

$$\int_1^n f(x)\, dx < \sum_{k=1}^{n-1} a_k \qquad \text{Right inequality in expression (1)}$$

$$< \sum_{k=1}^{\infty} a_k \qquad \text{Terms } a_k \text{ are positive.}$$

$$= S. \qquad \text{Value of infinite series}$$

> An extended version of this proof can be used to show that, in fact,
> $$\int_1^\infty f(x)\, dx < \sum_{k=1}^{\infty} a_k \text{ (strict inequality)}$$
> in all cases.

We see that the sequence $\left\{ \int_1^n f(x)\, dx \right\}$ is increasing (because $f(x) > 0$) and bounded above by a fixed number S. Therefore, the improper integral $\int_1^\infty f(x)\, dx = \lim_{n \to \infty} \int_1^n f(x)\, dx$ has a finite value (less than or equal to S).

We have shown that if $\int_1^\infty f(x)\, dx$ is finite, then $\sum a_k$ converges, and vice versa. The same inequalities imply that $\int_1^\infty f(x)\, dx$ and $\sum a_k$ also diverge together. ◄

The Integral Test is used to determine *whether* a series converges or diverges. For this reason, adding or subtracting a few terms in the series *or* changing the lower limit of integration to another finite point does not change the outcome of the test. Therefore, the test depends on neither the lower index of the series nor the lower limit of the integral.

EXAMPLE 2   Applying the Integral Test Determine whether the following series converge.

a. $\displaystyle\sum_{k=1}^{\infty} \frac{k}{k^2 + 1}$
   b. $\displaystyle\sum_{k=3}^{\infty} \frac{1}{\sqrt{2k - 5}}$
   c. $\displaystyle\sum_{k=0}^{\infty} \frac{1}{k^2 + 4}$

SOLUTION

a. The function associated with this series is $f(x) = x/(x^2 + 1)$, which is positive, for $x \geq 1$. We must also show that the terms of the series are decreasing beyond some fixed term of the series. The first few terms of the series are $\left\{ \frac{1}{2}, \frac{2}{5}, \frac{3}{10}, \frac{4}{17}, \ldots \right\}$, and it appears that the terms are decreasing. When the decreasing property is difficult to confirm, one approach is to use derivatives to show that the associated function is decreasing. In this case, we have

$$f'(x) = \frac{d}{dx}\left(\frac{x}{x^2 + 1} \right) = \underbrace{\frac{x^2 + 1 - 2x^2}{(x^2 + 1)^2}}_{\text{Quotient Rule}} = \frac{1 - x^2}{(x^2 + 1)^2}.$$

For $x > 1$, $f'(x) < 0$, which implies that the function and the terms of the series are decreasing. The integral that determines convergence is

$$\int_1^{\infty} \frac{x}{x^2 + 1}\, dx = \lim_{b \to \infty} \int_1^b \frac{x}{x^2 + 1}\, dx \qquad \text{Definition of improper integral}$$

$$= \lim_{b \to \infty} \frac{1}{2} \ln (x^2 + 1) \Big|_1^b \qquad \text{Evaluate integral.}$$

$$= \frac{1}{2} \lim_{b \to \infty} \left(\ln (b^2 + 1) - \ln 2 \right) \qquad \text{Simplify.}$$

$$= \infty. \qquad \lim_{b \to \infty} \ln (b^2 + 1) = \infty$$

Because the integral diverges, the series diverges.

b. The Integral Test may be modified to accommodate initial indices other than $k = 1$. The terms of this series decrease, for $k \geq 3$. In this case, the relevant integral is

$$\int_3^{\infty} \frac{dx}{\sqrt{2x - 5}} = \lim_{b \to \infty} \int_3^b \frac{dx}{\sqrt{2x - 5}} \qquad \text{Definition of improper integral}$$

$$= \lim_{b \to \infty} \sqrt{2x - 5} \Big|_3^b \qquad \text{Evaluate integral.}$$

$$= \infty. \qquad \lim_{b \to \infty} \sqrt{2b - 5} = \infty$$

Because the integral diverges, the series also diverges.

c. The terms of the series are positive and decreasing, for $k \geq 0$. The relevant integral is

$$\int_0^{\infty} \frac{dx}{x^2 + 4} = \lim_{b \to \infty} \int_0^b \frac{dx}{x^2 + 4} \qquad \text{Definition of improper integral}$$

$$= \lim_{b \to \infty} \frac{1}{2} \tan^{-1} \frac{x}{2} \Big|_0^b \qquad \text{Evaluate integral.}$$

$$= \frac{1}{2} \underbrace{\lim_{b \to \infty} \tan^{-1} \frac{b}{2}}_{\frac{\pi}{2}} - \tan^{-1} 0 \qquad \text{Simplify.}$$

$$= \frac{\pi}{4}. \qquad \tan^{-1} x \to \frac{\pi}{2}, \text{ as } x \to \infty.$$

Because the integral is finite (equivalently, it converges), the infinite series also converges $\left(\text{but not to } \frac{\pi}{4} \right)$.

Related Exercises 17, 20 ◄

The *p*-series

The Integral Test is used to prove Theorem 10.12, which addresses the convergence of an entire family of infinite series known as the *p-series.*

> **THEOREM 10.12   Convergence of the *p*-series**
>
> The ***p*-series** $\displaystyle\sum_{k=1}^{\infty} \frac{1}{k^p}$ converges for $p > 1$ and diverges for $p \le 1$.

QUICK CHECK 2 Which of the following series are *p*-series, and which series converge?

a. $\displaystyle\sum_{k=1}^{\infty} k^{-0.8}$ **b.** $\displaystyle\sum_{k=1}^{\infty} 2^{-k}$ **c.** $\displaystyle\sum_{k=10}^{\infty} k^{-4}$ ◄

Proof: To apply the Integral Test, observe that the terms of the given series are positive and decreasing, for $p > 0$. The function associated with the series is $f(x) = \dfrac{1}{x^p}$. The relevant integral is $\displaystyle\int_1^{\infty} \frac{dx}{x^p}$. Appealing to Example 2 in Section 8.9, recall that this improper integral converges for $p > 1$ and diverges for $p \le 1$. Therefore, by the Integral Test, the *p*-series $\displaystyle\sum_{k=1}^{\infty} \frac{1}{k^p}$ converges for $p > 1$ and diverges for $0 < p \le 1$. For $p \le 0$, the series diverges by the Divergence Test. ◄

EXAMPLE 3   Using the *p*-series test Determine whether the following series converge or diverge.

a. $\displaystyle\sum_{k=1}^{\infty} k^{-3}$ **b.** $\displaystyle\sum_{k=1}^{\infty} \frac{1}{\sqrt[4]{k^3}}$ **c.** $\displaystyle\sum_{k=4}^{\infty} \frac{1}{(k-1)^2}$

SOLUTION

a. Because $\displaystyle\sum_{k=1}^{\infty} k^{-3} = \sum_{k=1}^{\infty} \frac{1}{k^3}$ is a *p*-series with $p = 3$, it converges by Theorem 10.12.

b. This series is a *p*-series with $p = \frac{3}{4}$. By Theorem 10.12, it diverges.

c. The series

$$\sum_{k=4}^{\infty} \frac{1}{(k-1)^2} = \sum_{k=3}^{\infty} \frac{1}{k^2} = \frac{1}{3^2} + \frac{1}{4^2} + \frac{1}{5^2} + \cdots$$

is a convergent *p*-series ($p = 2$) without the first two terms. Recall that adding or removing a finite number of terms does not affect the convergence of a series (Property 5, Theorem 10.8). Therefore, the given series converges.

Related Exercises 23–24 ◄

Estimating the Value of Infinite Series

The Integral Test is powerful in its own right, but it comes with an added bonus. It can be used to estimate the value of a convergent series with positive terms. We define the **remainder** to be the error in approximating a convergent series by the sum of its first n terms; that is,

$$R_n = \underbrace{\sum_{k=1}^{\infty} a_k}_{\substack{\text{value of} \\ \text{series}}} - \underbrace{\sum_{k=1}^{n} a_k}_{\substack{\text{approximation based} \\ \text{on first } n \text{ terms}}} = a_{n+1} + a_{n+2} + a_{n+3} + \cdots.$$

QUICK CHECK 3 If $\sum a_k$ is a convergent series of positive terms, why is $R_n > 0$? ◄

The remainder consists of the *tail* of the series—those terms beyond a_n. For series with positive terms, the remainder is positive.

We now argue much as we did in the proof of the Integral Test. Let f be a continuous, positive, decreasing function such that $f(k) = a_k$, for all relevant k. From Figure 10.29, we see that $\int_{n+1}^{\infty} f(x)\, dx < R_n$.

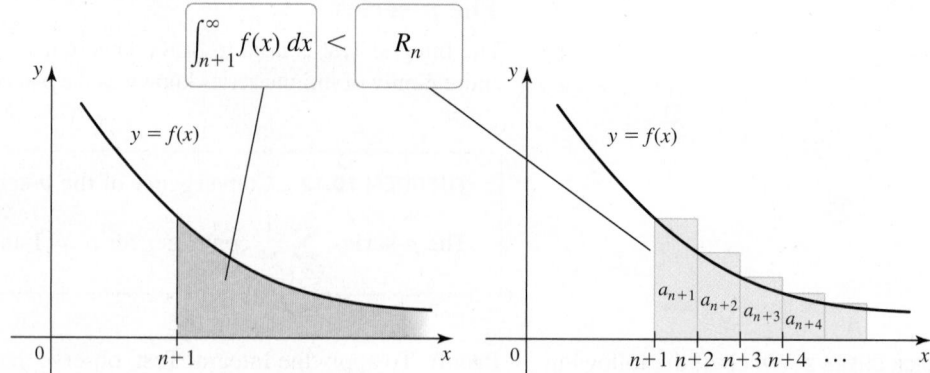

Figure 10.29

Similarly, Figure 10.30 shows that $R_n < \int_n^\infty f(x)\, dx$.

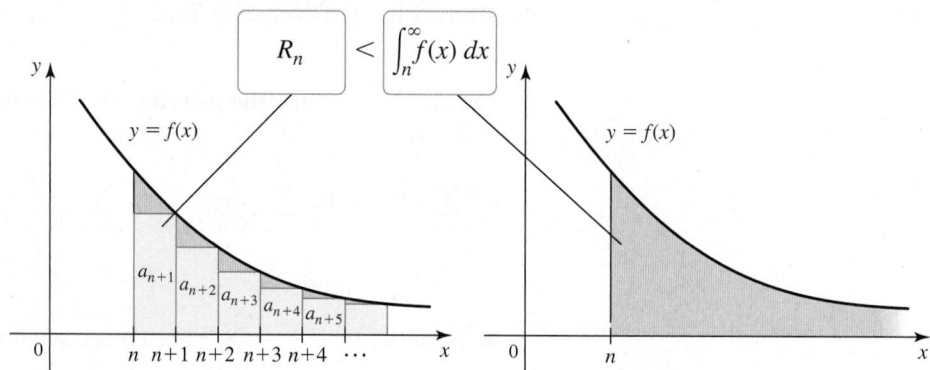

Figure 10.30

Combining these two inequalities, the remainder is squeezed between two integrals:

$$\int_{n+1}^\infty f(x)\, dx < R_n < \int_n^\infty f(x)\, dx. \qquad (2)$$

If the integrals can be evaluated, this result provides an estimate of the remainder.

There is, however, another equally useful way to express this result. Notice that the value of the series is

$$S = \sum_{k=1}^\infty a_k = \underbrace{\sum_{k=1}^n a_k}_{S_n} + R_n,$$

which is the sum of the first n terms S_n and the remainder R_n. Adding S_n to each term of (2), we have

$$\underbrace{S_n + \int_{n+1}^\infty f(x)\, dx}_{L_n} < \underbrace{\sum_{k=1}^\infty a_k}_{S_n + R_n = S} < \underbrace{S_n + \int_n^\infty f(x)\, dx}_{U_n}.$$

These inequalities can be abbreviated as $L_n < S < U_n$, where S is the exact value of the series, and L_n and U_n are lower and upper bounds for S, respectively. If the integrals in these bounds can be evaluated, it is straightforward to compute S_n (by summing the first n terms of the series) and to compute both L_n and U_n.

> **THEOREM 10.13   Estimating Series with Positive Terms**
> Let f be a continuous, positive, decreasing function, for $x \geq 1$, and let $a_k = f(k)$, for $k = 1, 2, 3, \ldots$. Let $S = \sum_{k=1}^{\infty} a_k$ be a convergent series and let $S_n = \sum_{k=1}^{n} a_k$ be the sum of the first n terms of the series. The remainder $R_n = S - S_n$ satisfies
> $$R_n < \int_n^{\infty} f(x) \, dx.$$
> Furthermore, the exact value of the series is bounded as follows:
> $$L_n = S_n + \int_{n+1}^{\infty} f(x) \, dx < \sum_{k=1}^{\infty} a_k < S_n + \int_n^{\infty} f(x) \, dx = U_n.$$

EXAMPLE 4   Approximating a p-series

a. How many terms of the convergent p-series $\sum_{k=1}^{\infty} \dfrac{1}{k^2}$ must be summed to obtain an approximation that is within 10^{-3} of the exact value of the series?

b. Find an approximation to the series using 50 terms of the series.

SOLUTION The function associated with this series is $f(x) = 1/x^2$.

a. Using the bound on the remainder, we have

$$R_n < \int_n^{\infty} f(x) \, dx = \int_n^{\infty} \frac{dx}{x^2} = \frac{1}{n}.$$

To ensure that $R_n < 10^{-3}$, we must choose n so that $1/n < 10^{-3}$, which implies that $n > 1000$. In other words, we must sum at least 1001 terms of the series to be sure that the remainder is less than 10^{-3}.

b. Using the bounds on the series, we have $L_n < S < U_n$, where S is the exact value of the series, and

$$L_n = S_n + \int_{n+1}^{\infty} \frac{dx}{x^2} = S_n + \frac{1}{n+1} \quad \text{and} \quad U_n = S_n + \int_n^{\infty} \frac{dx}{x^2} = S_n + \frac{1}{n}.$$

Therefore, the series is bounded as follows:

$$S_n + \frac{1}{n+1} < S < S_n + \frac{1}{n},$$

> ➤ The values of p-series with even values of p are generally known. For example, with $p = 2$, the series converges to $\pi^2/6$ (a proof is outlined in Exercise 68); with $p = 4$, the series converges to $\pi^4/90$. The values of p-series with odd values of p are not known.

where S_n is the sum of the first n terms. Using a calculator to sum the first 50 terms of the series, we find that $S_{50} \approx 1.625133$. The exact value of the series is in the interval

$$S_{50} + \frac{1}{50+1} < S < S_{50} + \frac{1}{50},$$

or $1.644741 < S < 1.645133$. Taking the average of these two bounds as our approximation of S, we find that $S \approx 1.644937$. This estimate is better than simply using S_{50}. **Figure 10.31a** shows the lower and upper bounds, L_n and U_n, respectively, for $n = 1, 2, \ldots, 50$. **Figure 10.31b** shows these bounds on an enlarged scale for $n = 50, 51, \ldots, 100$. These figures illustrate how the exact value of the series is squeezed into a narrowing interval as n increases.

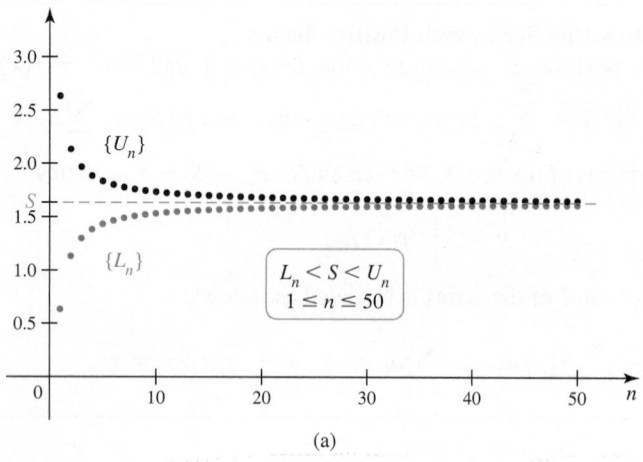

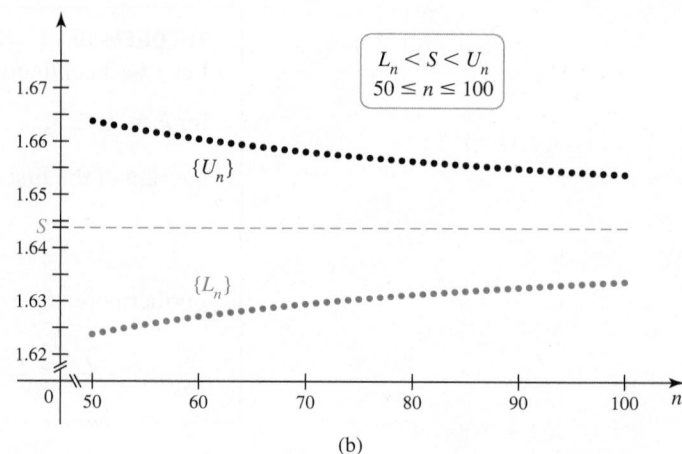

Figure 10.31

Related Exercises 41, 43 ◄

SECTION 10.4 EXERCISES

Getting Started

1. If we know that $\lim\limits_{k \to \infty} a_k = 1$, then what can we say about $\sum\limits_{k=1}^{\infty} a_k$?

2. Is it true that if the terms of a series of positive terms decrease to zero, then the series converges? Explain using an example.

3. If we know that $\sum\limits_{k=1}^{\infty} a_k = 10{,}000$, then what can we say about $\lim\limits_{k \to \infty} a_k$?

4. For what values of p does the series $\sum\limits_{k=1}^{\infty} \dfrac{1}{k^p}$ converge? For what values of p does it diverge?

5. For what values of p does the series $\sum\limits_{k=10}^{\infty} \dfrac{1}{k^p}$ converge (initial index is 10)? For what values of p does it diverge?

6. Explain why the sequence of partial sums for a series of positive terms is an increasing sequence.

7. Define the remainder of an infinite series.

8. If a series of positive terms converges, does it follow that the remainder R_n must decrease to zero as $n \to \infty$? Explain.

Practice Exercises

9–16. Divergence Test *Use the Divergence Test to determine whether the following series diverge or state that the test is inconclusive.*

9. $\sum\limits_{k=0}^{\infty} \dfrac{k}{2k + 1}$

10. $\sum\limits_{k=1}^{\infty} \dfrac{k}{k^2 + 1}$

11. $\sum\limits_{k=0}^{\infty} \dfrac{1}{1000 + k}$

12. $\sum\limits_{k=1}^{\infty} \dfrac{k^3}{k^3 + 1}$

13. $\sum\limits_{k=2}^{\infty} \dfrac{k}{\ln k}$

14. $\sum\limits_{k=1}^{\infty} \dfrac{k^2}{2^k}$

15. $\sum\limits_{k=2}^{\infty} \dfrac{\sqrt{k}}{\ln^{10} k}$

16. $\sum\limits_{k=1}^{\infty} \dfrac{k^3}{k!}$

17–22. Integral Test *Use the Integral Test to determine whether the following series converge after showing that the conditions of the Integral Test are satisfied.*

17. $\sum\limits_{k=0}^{\infty} \dfrac{1}{\sqrt{k + 8}}$

18. $\sum\limits_{k=1}^{\infty} \dfrac{1}{(2k + 4)^2}$

19. $\sum\limits_{k=1}^{\infty} \dfrac{1}{\sqrt[3]{5k + 3}}$

20. $\sum\limits_{k=1}^{\infty} \dfrac{e^k}{1 + e^{2k}}$

21. $\sum\limits_{k=1}^{\infty} k e^{-2k^2}$

22. $\sum\limits_{k=2}^{\infty} \dfrac{1}{k(\ln k)^2}$

23–38. Divergence, Integral, and p-series Tests *Use the Divergence Test, the Integral Test, or the p-series test to determine whether the following series converge.*

23. $\sum\limits_{k=1}^{\infty} k^{-1/5}$

24. $\sum\limits_{k=1}^{\infty} \dfrac{1}{\sqrt{k^3}}$

25. $\sum\limits_{k=1}^{\infty} \dfrac{k^3}{e^{k^4}}$

26. $\sum\limits_{k=1}^{\infty} \dfrac{k^2}{k^3 + 5}$

27. $\sum\limits_{k=1}^{\infty} k^{1/k}$

28. $\sum\limits_{k=1}^{\infty} \dfrac{\sqrt{k^2 + 1}}{k}$

29. $\sum\limits_{k=1}^{\infty} \dfrac{1}{k^{10}}$

30. $\sum\limits_{k=2}^{\infty} \dfrac{k^e}{k^\pi}$

31. $\sum\limits_{k=3}^{\infty} \dfrac{1}{(k - 2)^4}$

32. $\sum\limits_{k=1}^{\infty} 2k^{-3/2}$

33. $\sum\limits_{k=1}^{\infty} \dfrac{k}{e^k}$

34. $\sum\limits_{k=3}^{\infty} \dfrac{1}{k(\ln k) \ln \ln k}$

35. $\sum\limits_{k=1}^{\infty} \left(\dfrac{k}{k + 10}\right)^k$

36. $\sum\limits_{k=1}^{\infty} \dfrac{k}{(k^2 + 1)^3}$

37. $\sum\limits_{k=1}^{\infty} \dfrac{1}{\sqrt[3]{k}}$

38. $\sum\limits_{k=1}^{\infty} \dfrac{1}{\sqrt[3]{27k^2}}$

T **39–40. Lower and upper bounds of a series** *For each convergent series and given value of n, use Theorem 10.13 to complete the following.*

a. *Use S_n to estimate the sum of the series.*
b. *Find an upper bound for the remainder R_n.*
c. *Find lower and upper bounds (L_n and U_n, respectively) for the exact value of the series.*

39. $\displaystyle\sum_{k=1}^{\infty} \frac{1}{k^7}; n = 2$ **40.** $\displaystyle\sum_{k=1}^{\infty} \frac{1}{k^3}; n = 5$

T **41–44. Remainders and estimates** *Consider the following convergent series.*

a. *Find an upper bound for the remainder in terms of n.*
b. *Find how many terms are needed to ensure that the remainder is less than 10^{-3}.*
c. *Find lower and upper bounds (L_n and U_n, respectively) on the exact value of the series.*
d. *Find an interval in which the value of the series must lie if you approximate it using ten terms of the series.*

41. $\displaystyle\sum_{k=1}^{\infty} \frac{1}{k^6}$ **42.** $\displaystyle\sum_{k=2}^{\infty} \frac{1}{k(\ln k)^2}$

43. $\displaystyle\sum_{k=1}^{\infty} \frac{1}{3^k}$ **44.** $\displaystyle\sum_{k=1}^{\infty} ke^{-k^2}$

T **45.** Estimate the series $\displaystyle\sum_{k=1}^{\infty} \frac{1}{k^7}$ to within 10^{-4} of its exact value.

T **46.** Estimate the series $\displaystyle\sum_{k=1}^{\infty} \frac{1}{(3k + 2)^2}$ to within 10^{-3} of its exact value.

47. Explain why or why not Determine whether the following statements are true and give an explanation or counterexample.

a. The sum $\displaystyle\sum_{k=1}^{\infty} \frac{1}{3^k}$ is a *p*-series.
b. The sum $\displaystyle\sum_{k=3}^{\infty} \frac{1}{\sqrt{k - 2}}$ is a *p*-series.
c. Suppose f is a continuous, positive, decreasing function, for $x \geq 1$, and $a_k = f(k)$, for $k = 1, 2, 3, \ldots$. If $\displaystyle\sum_{k=1}^{\infty} a_k$ converges to L, then $\displaystyle\int_1^{\infty} f(x)\, dx$ converges to L.
d. Every partial sum S_n of the series $\displaystyle\sum_{k=1}^{\infty} \frac{1}{k^2}$ underestimates the exact value of $\displaystyle\sum_{k=1}^{\infty} \frac{1}{k^2}$.
e. If $\displaystyle\sum k^{-p}$ converges, then $\displaystyle\sum k^{-p+0.001}$ converges.
f. If $\displaystyle\lim_{k \to \infty} a_k = 0$, then $\displaystyle\sum a_k$ converges.

48–63. Choose your test *Determine whether the following series converge or diverge using the properties and tests introduced in Sections 10.3 and 10.4.*

48. $\displaystyle\sum_{k=1}^{\infty} \ln\left(\frac{2k^6}{1 + k^6}\right)$ **49.** $\displaystyle\sum_{k=2}^{\infty} \frac{1}{e^k}$

50. $\displaystyle\sum_{k=1}^{\infty} \sqrt{\frac{k + 1}{k}}$ **51.** $\displaystyle\sum_{k=1}^{\infty} \frac{1}{(3k + 1)(3k + 4)}$

52. $\displaystyle\sum_{k=0}^{\infty} \frac{10}{k^2 + 9}$ **53.** $\displaystyle\sum_{k=1}^{\infty} \frac{k}{\sqrt{k^2 + 1}}$

54. $\displaystyle\sum_{k=1}^{\infty} \frac{2^k + 3^k}{4^k}$ **55.** $\displaystyle\sum_{k=3}^{\infty} \frac{4}{k\sqrt{\ln k}}$

56. $\displaystyle\sum_{k=1}^{\infty} \frac{1}{k^2 + 7k + 12}$ **57.** $\displaystyle\sum_{k=1}^{\infty} \left(\frac{5}{6}\right)^{-k}$

58. $\displaystyle\sum_{k=1}^{\infty} \left(\frac{k + 6}{k}\right)^k$ **59.** $\displaystyle\sum_{k=4}^{\infty} \frac{3}{(k - 3)^4}$

60. $\displaystyle\sum_{k=1}^{\infty} \frac{3^k}{k^2 + 1}$ **61.** $\dfrac{2}{4^2} + \dfrac{2}{5^2} + \dfrac{2}{6^2} + \cdots$

62. $\dfrac{1}{3} + \dfrac{1}{5} + \dfrac{1}{7} + \cdots$ **63.** $\displaystyle\sum_{k=1}^{\infty} \frac{3^{k+2}}{5^k}$

64. Log *p*-series Consider the series $\displaystyle\sum_{k=2}^{\infty} \frac{1}{k(\ln k)^p}$, where p is a real number.

a. Use the Integral Test to determine the values of p for which this series converges.
b. Does this series converge faster for $p = 2$ or $p = 3$? Explain.

65. Loglog *p*-series Consider the series $\displaystyle\sum_{k=2}^{\infty} \frac{1}{k(\ln k)(\ln \ln k)^p}$, where p is a real number.

a. For what values of p does this series converge?
b. Which of the following series converges faster? Explain.

$$\sum_{k=2}^{\infty} \frac{1}{k(\ln k)^2} \quad \text{or} \quad \sum_{k=3}^{\infty} \frac{1}{k(\ln k)(\ln \ln k)^2}?$$

Explorations and Challenges

66. Prime numbers The prime numbers are those positive integers that are divisible by only 1 and themselves (for example, 2, 3, 5, 7, 11, 13, ...). A celebrated theorem states that the sequence of prime numbers $\{p_k\}$ satisfies $\displaystyle\lim_{k \to \infty} \frac{p_k}{k \ln k} = 1$. Show that $\displaystyle\sum_{k=2}^{\infty} \frac{1}{k \ln k}$ diverges, which implies that the series $\displaystyle\sum_{k=1}^{\infty} \frac{1}{p_k}$ diverges.

T **67. The zeta function** The Riemann zeta function is the subject of extensive research and is associated with several renowned unsolved problems. It is defined by $\displaystyle\zeta(x) = \sum_{k=1}^{\infty} \frac{1}{k^x}$. When x is a real number, the zeta function becomes a *p*-series. For even positive integers p, the value of $\zeta(p)$ is known exactly. For example,

$$\sum_{k=1}^{\infty} \frac{1}{k^2} = \frac{\pi^2}{6}, \quad \sum_{k=1}^{\infty} \frac{1}{k^4} = \frac{\pi^4}{90}, \quad \text{and} \quad \sum_{k=1}^{\infty} \frac{1}{k^6} = \frac{\pi^6}{945}, \ldots.$$

Use the estimation techniques described in the text to approximate $\zeta(3)$ and $\zeta(5)$ (whose values are not known exactly) with a remainder less than 10^{-3}.

68. Showing that $\displaystyle\sum_{k=1}^{\infty} \frac{1}{k^2} = \frac{\pi^2}{6}$ In 1734, Leonhard Euler informally proved that $\displaystyle\sum_{k=1}^{\infty} \frac{1}{k^2} = \frac{\pi^2}{6}$. An elegant proof is outlined here that uses the inequality

$$\cot^2 x < \frac{1}{x^2} < 1 + \cot^2 x \left(\text{provided } 0 < x < \frac{\pi}{2}\right)$$

and the identity

$$\sum_{k=1}^{n} \cot^2 k\theta = \frac{n(2n-1)}{3}, \text{ for } n = 1, 2, 3, \ldots, \text{ where } \theta = \frac{\pi}{2n+1}.$$

a. Show that $\displaystyle\sum_{k=1}^{n} \cot^2 k\theta < \frac{1}{\theta^2} \sum_{k=1}^{n} \frac{1}{k^2} < n + \sum_{k=1}^{n} \cot^2 k\theta.$

b. Use the inequality in part (a) to show that

$$\frac{n(2n-1)\pi^2}{3(2n+1)^2} < \sum_{k=1}^{n} \frac{1}{k^2} < \frac{n(2n+2)\pi^2}{3(2n+1)^2}.$$

c. Use the Squeeze Theorem to conclude that $\displaystyle\sum_{k=1}^{\infty} \frac{1}{k^2} = \frac{\pi^2}{6}.$

(*Source: The College Mathematics Journal*, 24, 5, Nov 1993)

69. **Reciprocals of odd squares** Assume $\displaystyle\sum_{k=1}^{\infty} \frac{1}{k^2} = \frac{\pi^2}{6}$ (Exercises 67 and 68) and the terms of this series may be rearranged without changing the value of the series. Determine the sum of the reciprocals of the squares of the odd positive integers.

70. **Gabriel's wedding cake** Consider a wedding cake of infinite height, each layer of which is a right circular cylinder of height 1. The bottom layer of the cake has a radius of 1, the second layer has a radius of $1/2$, the third layer has a radius of $1/3$, and the nth layer has a radius of $1/n$ (see figure).

a. Find the area of the lateral (vertical) sides of the wedding cake.
b. Determine the volume of the cake. (*Hint:* Use the result of Exercise 68.)
c. Comment on your answers to parts (a) and (b).

(*Source: The College Mathematics Journal*, 30, 1, Jan 1999)

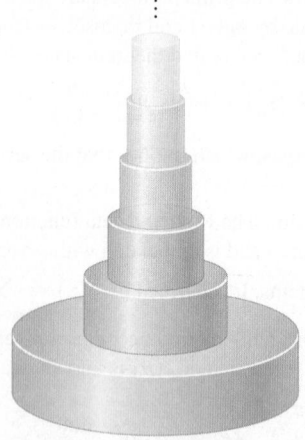

71. **A divergence proof** Give an argument similar to that given in the text for the harmonic series to show that $\displaystyle\sum_{k=1}^{\infty} \frac{1}{\sqrt{k}}$ diverges.

T 72. **Shifted p-series** Consider the sequence $\{F_n\}$ defined by

$$F_n = \sum_{k=1}^{\infty} \frac{1}{k(k+n)},$$

for $n = 0, 1, 2, \ldots$. When $n = 0$, the series is a p-series, and we have $F_0 = \dfrac{\pi^2}{6}$ (Exercises 67 and 68).

a. Explain why $\{F_n\}$ is a decreasing sequence.
b. Plot $\{F_n\}$, for $n = 1, 2, \ldots, 20$.
c. Based on your experiments, make a conjecture about $\displaystyle\lim_{n \to \infty} F_n$.

73. **A sequence of sums** Consider the sequence $\{x_n\}$ defined for $n = 1, 2, 3, \ldots$ by

$$x_n = \sum_{k=n+1}^{2n} \frac{1}{k} = \frac{1}{n+1} + \frac{1}{n+2} + \cdots + \frac{1}{2n}.$$

a. Write out the terms x_1, x_2, x_3.
b. Show that $\dfrac{1}{2} \le x_n < 1$, for $n = 1, 2, 3, \ldots$.
c. Show that x_n is the right Riemann sum for $\displaystyle\int_1^2 \frac{dx}{x}$ using n subintervals.
d. Conclude that $\displaystyle\lim_{n \to \infty} x_n = \ln 2$.

T 74. **The harmonic series and Euler's constant**

a. Sketch the function $f(x) = 1/x$ on the interval $[1, n+1]$, where n is a positive integer. Use this graph to verify that

$$\ln(n+1) < 1 + \frac{1}{2} + \frac{1}{3} + \cdots + \frac{1}{n} < 1 + \ln n.$$

b. Let S_n be the sum of the first n terms of the harmonic series, so part (a) says $\ln(n+1) < S_n < 1 + \ln n$. Define the new sequence $\{E_n\}$ by

$$E_n = S_n - \ln(n+1), \quad \text{for } n = 1, 2, 3, \ldots.$$

Show that $E_n > 0$ for $n = 1, 2, 3, \ldots$.
c. Using a figure similar to that used in part (a), show that

$$\frac{1}{n+1} > \ln(n+2) - \ln(n+1).$$

d. Use parts (a) and (c) to show that $\{E_n\}$ is an increasing sequence ($E_{n+1} > E_n$).
e. Use part (a) to show that $\{E_n\}$ is bounded above by 1.
f. Conclude from parts (d) and (e) that $\{E_n\}$ has a limit less than or equal to 1. This limit is known as **Euler's constant** and is denoted γ (the Greek lowercase gamma).
g. By computing terms of $\{E_n\}$, estimate the value of γ and compare it to the value $\gamma \approx 0.5772$. (It has been conjectured that γ is irrational.)
h. The preceding arguments show that the sum of the first n terms of the harmonic series satisfy $S_n \approx 0.5772 + \ln(n+1)$. How many terms must be summed for the sum to exceed 10?

75. **The harmonic series and the Fibonacci sequence** The Fibonacci sequence $\{1, 1, 2, 3, 5, 8, 13, \ldots\}$ is generated by the recurrence relation

$$f_{n+1} = f_n + f_{n-1}, \text{ for } n = 1, 2, 3, \ldots, \text{ where } f_0 = 1, f_1 = 1.$$

a. It can be shown that the sequence of ratios of successive terms of the sequence $\left\{\dfrac{f_{n+1}}{f_n}\right\}$ has a limit φ. Divide both sides of the recurrence relation by f_n, take the limit as $n \to \infty$, and show that $\varphi = \displaystyle\lim_{n \to \infty} \frac{f_{n+1}}{f_n} = \frac{1 + \sqrt{5}}{2} \approx 1.618.$

b. Show that $\displaystyle\lim_{n \to \infty} \frac{f_{n-1}}{f_{n+1}} = 1 - \frac{1}{\varphi} \approx 0.382.$

c. Now consider the harmonic series and group terms as follows:

$$\sum_{k=1}^{\infty} \frac{1}{k} = 1 + \frac{1}{2} + \frac{1}{3} + \underbrace{\left(\frac{1}{4} + \frac{1}{5}\right)}_{\text{2 terms}} + \underbrace{\left(\frac{1}{6} + \frac{1}{7} + \frac{1}{8}\right)}_{\text{3 terms}}$$

$$+ \underbrace{\left(\frac{1}{9} + \cdots + \frac{1}{13}\right)}_{\text{5 terms}} + \cdots.$$

With the Fibonacci sequence in mind, show that

$$\sum_{k=1}^{\infty} \frac{1}{k} \geq 1 + \frac{1}{2} + \frac{1}{3} + \frac{2}{5} + \frac{3}{8} + \frac{5}{13} + \cdots = 1 + \sum_{k=1}^{\infty} \frac{f_{k-1}}{f_{k+1}}.$$

d. Use part (b) to conclude that the harmonic series diverges.

(*Source: The College Mathematics Journal*, 43, May 2012)

76. **Stacking dominoes** Consider a set of identical dominoes that are 2 in long. The dominoes are stacked on top of each other with their long edges aligned so that each domino overhangs the one beneath it *as far as possible* (see figure).

a. If there are n dominoes in the stack, what is the *greatest* distance that the top domino can be made to overhang the bottom domino? (*Hint:* Put the nth domino beneath the previous $n - 1$ dominoes.)

b. If we allow for infinitely many dominoes in the stack, what is the greatest distance that the top domino can be made to overhang the bottom domino?

10.5 Comparison Tests

The Integral Test from the previous section is easily applied to a series such as $\displaystyle\sum_{k=1}^{\infty} \frac{4}{4 + k^2}$ because $\displaystyle\int_1^{\infty} \frac{4}{4 + x^2}\, dx$ is a standard integral. However, using the Integral Test to determine whether $\displaystyle\sum_{k=1}^{\infty} \frac{5k^4 - 2k^2 + 3}{2k^6 - k + 5}$ converges is decidedly more difficult. For this reason, we develop two additional convergence tests, called *comparison tests*, and we use one of them in Example 2a to investigate the convergence of $\displaystyle\sum_{k=1}^{\infty} \frac{5k^4 - 2k^2 + 3}{2k^6 - k + 5}$. As with the Integral Test, both comparison tests apply only to series with positive terms.

The Comparison Test

Comparison tests use known series to test the convergence of unknown series. The first test is the Direct Comparison Test or simply the Comparison Test.

> ➤ Whether a series converges depends on the behavior of terms in the tail (large values of the index). So the inequalities $a_k \leq b_k$ and $b_k \leq a_k$ need not hold for all terms of the series. They must hold for all $k > N$ for some positive integer N.

THEOREM 10.14   Comparison Test

Let $\sum a_k$ and $\sum b_k$ be series with positive terms.

1. If $a_k \leq b_k$ and $\sum b_k$ converges, then $\sum a_k$ converges.
2. If $b_k \leq a_k$ and $\sum b_k$ diverges, then $\sum a_k$ diverges.

Proof: Assume $\sum b_k$ converges, which means that $\sum b_k$ has a finite value B. The sequence of partial sums for $\sum a_k$ satisfies

$$S_n = \sum_{k=1}^{n} a_k \leq \sum_{k=1}^{n} b_k \qquad a_k \leq b_k$$

$$< \sum_{k=1}^{\infty} b_k \qquad \text{Positive terms are added to a finite sum.}$$

$$= B. \qquad \text{Value of series}$$

Therefore, the sequence of partial sums for $\sum a_k$ is increasing and bounded above by B. By the Bounded Monotonic Sequence Theorem (Theorem 10.5), the sequence of partial sums of $\sum a_k$ has a limit, which implies that $\sum a_k$ converges. The second case of the theorem is proved in a similar way. ◄

The Comparison Test can be illustrated with graphs of sequences of partial sums. Consider the series

$$\sum_{k=1}^{\infty} a_k = \sum_{k=1}^{\infty} \frac{1}{k^2 + 10} \quad \text{and} \quad \sum_{k=1}^{\infty} b_k = \sum_{k=1}^{\infty} \frac{1}{k^2}.$$

Because $\frac{1}{k^2 + 10} < \frac{1}{k^2}$, it follows that $a_k < b_k$, for $k \geq 1$. Furthermore, $\sum b_k$ is a convergent p-series. By the Comparison Test, we conclude that $\sum a_k$ also converges (**Figure 10.32**). The second case of the Comparison Test is illustrated with the series

$$\sum_{k=4}^{\infty} a_k = \sum_{k=4}^{\infty} \frac{1}{\sqrt{k - 3}} \quad \text{and} \quad \sum_{k=4}^{\infty} b_k = \sum_{k=4}^{\infty} \frac{1}{\sqrt{k}}.$$

Now $\frac{1}{\sqrt{k}} < \frac{1}{\sqrt{k - 3}}$, for $k \geq 4$. Therefore, $b_k < a_k$, for $k \geq 4$. Because $\sum b_k$ is a divergent p-series, by the Comparison Test, $\sum a_k$ also diverges. **Figure 10.33** shows that the sequence of partial sums for $\sum a_k$ lies above the sequence of partial sums for $\sum b_k$. Because the sequence of partial sums for $\sum b_k$ diverges, the sequence of partial sums for $\sum a_k$ also diverges.

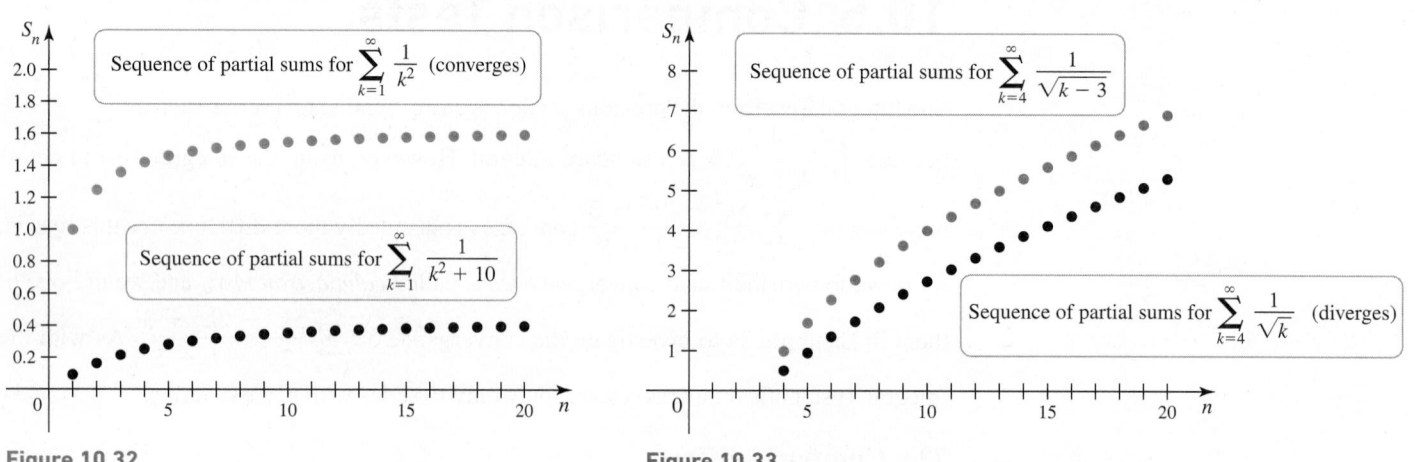

Figure 10.32   **Figure 10.33**

The key in using the Comparison Test is finding an appropriate comparison series. Plenty of practice will enable you to spot patterns and choose good comparison series.

EXAMPLE 1 **Using the Comparison Test** Determine whether the following series converge.

a. $\displaystyle\sum_{k=1}^{\infty} \frac{k^3}{2k^4 - 1}$ **b.** $\displaystyle\sum_{k=2}^{\infty} \frac{\ln k}{k^3}$

SOLUTION In using comparison tests, it's helpful to get a feel for how the terms of the given series behave.

> If $\sum a_k$ diverges, then $\sum c a_k$ also diverges for any constant $c \neq 0$ (Exercise 97 of Section 10.3).

a. As we go farther and farther out in this series ($k \to \infty$), the terms behave like

$$\frac{k^3}{2k^4 - 1} \approx \frac{k^3}{2k^4} = \frac{1}{2k}.$$

So a reasonable choice for a comparison series is the divergent series $\sum \frac{1}{2k}$. We must now show that the terms of the given series are *greater* than the terms of the comparison series. It is done by noting that $2k^4 - 1 < 2k^4$. Inverting both sides, we have

$$\frac{1}{2k^4 - 1} > \frac{1}{2k^4}, \quad \text{which implies that} \quad \frac{k^3}{2k^4 - 1} > \frac{k^3}{2k^4} = \frac{1}{2k}.$$

Because $\sum \frac{1}{2k}$ diverges, case (2) of the Comparison Test implies that the given series also diverges.

b. We note that $\ln k < k$, for $k \geq 2$, and then divide by k^3:

$$\frac{\ln k}{k^3} < \frac{k}{k^3} = \frac{1}{k^2}.$$

QUICK CHECK 1 Explain why it is difficult to use the divergent series $\sum 1/k$ as a comparison series to test $\sum 1/(k+1)$. ◄

Therefore, an appropriate comparison series is the convergent p-series $\sum \dfrac{1}{k^2}$. Because $\sum \dfrac{1}{k^2}$ converges, the given series converges.

Related Exercises 13, 19 ◄

The Limit Comparison Test

The Comparison Test should be tried if there is an obvious comparison series and the necessary inequality is easily established. Notice, however, that if the series in Example 1a were $\displaystyle\sum_{k=1}^{\infty} \frac{k^3}{2k^4 + 10}$ instead of $\displaystyle\sum_{k=1}^{\infty} \frac{k^3}{2k^4 - 1}$, then the comparison to the series $\sum \dfrac{1}{2k}$ would not work. Rather than fiddling with inequalities, it is often easier to use a more refined test called the *Limit Comparison Test*.

QUICK CHECK 2 For case (1) of the Limit Comparison Test, we must have $0 < L < \infty$. Why can either a_k or b_k be chosen as the known comparison series? That is, why can L be the limit of either a_k/b_k or b_k/a_k? ◄

> **THEOREM 10.15 Limit Comparison Test**
> Let $\sum a_k$ and $\sum b_k$ be series with positive terms and let
> $$\lim_{k \to \infty} \frac{a_k}{b_k} = L.$$
> 1. If $0 < L < \infty$ (that is, L is a finite positive number), then $\sum a_k$ and $\sum b_k$ either both converge or both diverge.
> 2. If $L = 0$ and $\sum b_k$ converges, then $\sum a_k$ converges.
> 3. If $L = \infty$ and $\sum b_k$ diverges, then $\sum a_k$ diverges.

➤ Recall that $|x| < a$ is equivalent to $-a < x < a$.

Proof (Case 1): Recall the definition of $\displaystyle\lim_{k \to \infty} \frac{a_k}{b_k} = L$: Given any $\varepsilon > 0$, $\left| \dfrac{a_k}{b_k} - L \right| < \varepsilon$ provided k is sufficiently large. In this case, let's take $\varepsilon = L/2$. It then follows that for sufficiently large k, $\left| \dfrac{a_k}{b_k} - L \right| < \dfrac{L}{2}$, or (removing the absolute value) $-\dfrac{L}{2} < \dfrac{a_k}{b_k} - L < \dfrac{L}{2}$. Adding L to all terms in these inequalities, we have

$$\frac{L}{2} < \frac{a_k}{b_k} < \frac{3L}{2}.$$

These inequalities imply that for sufficiently large k,

$$\frac{Lb_k}{2} < a_k < \frac{3Lb_k}{2}.$$

We see that the terms of $\sum a_k$ are sandwiched between multiples of the terms of $\sum b_k$. By the Comparison Test, it follows that the two series converge or diverge together. Cases (2) and (3) ($L = 0$ and $L = \infty$, respectively) are treated in Exercise 65. ◄

EXAMPLE 2 Using the Limit Comparison Test Determine whether the following series converge.

a. $\displaystyle\sum_{k=1}^{\infty} \frac{5k^4 - 2k^2 + 3}{2k^6 - k + 5}$ **b.** $\displaystyle\sum_{k=1}^{\infty} \frac{\ln k}{k^2}$.

SOLUTION In both cases, we must find a comparison series whose terms behave like the terms of the given series as $k \to \infty$.

a. As $k \to \infty$, a rational function behaves like the ratio of the leading (highest-power) terms. In this case, as $k \to \infty$,

$$\frac{5k^4 - 2k^2 + 3}{2k^6 - k + 5} \approx \frac{5k^4}{2k^6} = \frac{5}{2k^2}.$$

Therefore, a reasonable comparison series is the convergent p-series $\sum\limits_{k=1}^{\infty} \dfrac{1}{k^2}$ (the factor of $5/2$ does not affect whether the given series converges). Having chosen a comparison series, we compute the limit L:

$$
\begin{aligned}
L &= \lim_{k \to \infty} \frac{(5k^4 - 2k^2 + 3)/(2k^6 - k + 5)}{1/k^2} && \text{Ratio of terms of series}\\[2mm]
&= \lim_{k \to \infty} \frac{k^2(5k^4 - 2k^2 + 3)}{2k^6 - k + 5} && \text{Simplify.}\\[2mm]
&= \lim_{k \to \infty} \frac{5k^6 - 2k^4 + 3k^2}{2k^6 - k + 5} = \frac{5}{2}. && \text{Simplify and evaluate the limit.}
\end{aligned}
$$

We see that $0 < L < \infty$; therefore, the given series converges.

b. Why is this series interesting? We know that $\sum\limits_{k=1}^{\infty} \dfrac{1}{k^2}$ converges and that $\sum\limits_{k=1}^{\infty} \dfrac{1}{k}$ diverges. The given series $\sum\limits_{k=1}^{\infty} \dfrac{\ln k}{k^2}$ is "between" these two series. This observation suggests that we use either $\sum\limits_{k=1}^{\infty} \dfrac{1}{k^2}$ or $\sum\limits_{k=1}^{\infty} \dfrac{1}{k}$ as a comparison series. In the first case, letting $a_k = \ln k/k^2$ and $b_k = 1/k^2$, we find that

$$
L = \lim_{k \to \infty} \frac{a_k}{b_k} = \lim_{k \to \infty} \frac{\ln k/k^2}{1/k^2} = \lim_{k \to \infty} \ln k = \infty.
$$

Case (3) of the Limit Comparison Test does not apply here because the comparison series $\sum\limits_{k=1}^{\infty} \dfrac{1}{k^2}$ converges; case (3) is conclusive only when the comparison series *diverges*.

If, instead, we use the comparison series $\sum b_k = \sum \dfrac{1}{k}$, then

$$
L = \lim_{k \to \infty} \frac{a_k}{b_k} = \lim_{k \to \infty} \frac{\ln k/k^2}{1/k} = \lim_{k \to \infty} \frac{\ln k}{k} = 0.
$$

Case (2) of the Limit Comparison Test does not apply here because the comparison series $\sum\limits_{k=1}^{\infty} \dfrac{1}{k}$ diverges; case (2) is conclusive only when the comparison series *converges*.

With a bit more cunning, the Limit Comparison Test becomes conclusive. A series that lies "between" $\sum\limits_{k=1}^{\infty} \dfrac{1}{k^2}$ and $\sum\limits_{k=1}^{\infty} \dfrac{1}{k}$ is the convergent p-series $\sum\limits_{k=1}^{\infty} \dfrac{1}{k^{3/2}}$; we try it as a comparison series. Letting $a_k = \ln k/k^2$ and $b_k = 1/k^{3/2}$, we find that

$$
L = \lim_{k \to \infty} \frac{a_k}{b_k} = \lim_{k \to \infty} \frac{\ln k/k^2}{1/k^{3/2}} = \lim_{k \to \infty} \frac{\ln k}{\sqrt{k}} = 0.
$$

(This limit is evaluated using l'Hôpital's Rule or by recalling that $\ln k$ grows more slowly than any positive power of k.) Now case (2) of the Limit Comparison Test applies; the comparison series $\sum \dfrac{1}{k^{3/2}}$ converges, so the given series converges.

Related Exercises 11, 30 ◄

SECTION 10.5 EXERCISES

Getting Started

1. Explain how the Limit Comparison Test works.

2. Explain why, with a series of positive terms, the sequence of partial sums is an increasing sequence.

3. What comparison series would you use with the Comparison Test to determine whether $\sum_{k=1}^{\infty} \dfrac{1}{k^2 + 1}$ converges?

4. Can the Comparison Test, with the comparison series $\sum_{k=1}^{\infty} \dfrac{1}{k^4}$, be used to show that the series $\sum_{k=1}^{\infty} \dfrac{1}{k^4 - 1}$ converges?

5. What comparison series would you use with the Comparison Test to determine whether $\sum_{k=1}^{\infty} \dfrac{2^k}{3^k + 1}$ converges?

6. Determine whether $\sum_{k=2}^{\infty} \dfrac{1}{\sqrt{k} - 1}$ converges using the Comparison Test with the comparison series $\sum_{k=1}^{\infty} \dfrac{1}{\sqrt{k}}$.

7. What comparison series would you use with the Limit Comparison Test to determine whether $\sum_{k=1}^{\infty} \dfrac{k^2 + k + 5}{k^3 + 3k + 1}$ converges?

8. Determine whether $\sum_{k=1}^{\infty} \dfrac{k + 1}{3k^3 + 2}$ converges using the Limit Comparison Test with the comparison series $\sum_{k=1}^{\infty} \dfrac{1}{k^2}$.

Practice Exercises

9–36. Comparison tests *Use the Comparison Test or the Limit Comparison Test to determine whether the following series converge.*

9. $\sum_{k=1}^{\infty} \dfrac{1}{k^2 + 4}$

10. $\sum_{k=1}^{\infty} \dfrac{k^2 + k - 1}{k^4 + 4k^2 - 3}$

11. $\sum_{k=1}^{\infty} \dfrac{k^2 - 1}{k^3 + 4}$

12. $\sum_{k=1}^{\infty} \dfrac{0.0001}{k + 4}$

13. $\sum_{k=1}^{\infty} \dfrac{\sqrt{k}}{k^2 + 3}$

14. $\sum_{k=1}^{\infty} \dfrac{1}{5^k + 3}$

15. $\sum_{k=1}^{\infty} \dfrac{4^k}{5^k - 3}$

16. $\sum_{k=1}^{\infty} \dfrac{\sin^2 k}{k^2}$

17. $\sum_{k=1}^{\infty} \dfrac{1}{k^{3/2} + 1}$

18. $\sum_{k=1}^{\infty} \dfrac{1}{k \cdot 10^k}$

19. $\sum_{k=4}^{\infty} \dfrac{1 + \cos^2 k}{k - 3}$

20. $\sum_{k=1}^{\infty} \dfrac{k^2 + k + 2}{6^k(k^2 + 1)}$

21. $\sum_{k=1}^{\infty} \dfrac{(3k^3 + 4)(7k^2 + 1)}{(2k^3 + 1)(4k^3 - 1)}$

22. $\sum_{k=1}^{\infty} \sqrt{\dfrac{k}{k^3 + 1}}$

23. $\sum_{k=1}^{\infty} \dfrac{\sin(1/k)}{k^2}$

24. $\sum_{k=1}^{\infty} \dfrac{1}{3^k - 2^k}$

25. $\sum_{k=1}^{\infty} \dfrac{1}{2k - \sqrt{k}}$

26. $\sum_{k=1}^{\infty} \dfrac{1}{k\sqrt{k} + 2}$

27. $\sum_{k=1}^{\infty} \dfrac{2 + (-1)^k}{k^2}$

28. $\sum_{k=1}^{\infty} \dfrac{2 + \sin k}{k}$

29. $\sum_{k=1}^{\infty} \dfrac{\sqrt[3]{k^2 + 1}}{\sqrt{k^3 + 2}}$

30. $\sum_{k=3}^{\infty} \dfrac{1}{(k \ln k)^2}$

31. $\sum_{k=1}^{\infty} \dfrac{20}{\sqrt[3]{k} + \sqrt{k}}$

32. $\sum_{k=3}^{\infty} \dfrac{\sqrt{\ln k}}{k}$

33. $\sum_{k=1}^{\infty} \dfrac{1}{2^{\ln k}}$

34. $\sum_{k=2}^{\infty} \dfrac{\ln^2 k}{k^4}$

T 35. $\sum_{k=1}^{\infty} \dfrac{1}{4^{\ln k}}$

36. $\sum_{k=1}^{\infty} \dfrac{5}{\pi^k + e^k - 2}$

37. **Explain why or why not** Determine whether the following statements are true and give an explanation or counterexample.

a. Suppose $0 < a_k < b_k$. If Σa_k converges, then Σb_k converges.

b. Suppose $0 < a_k < b_k$. If Σa_k diverges, then Σb_k diverges.

c. Suppose $0 < b_k < c_k < a_k$. If Σa_k converges, then Σb_k and Σc_k converge.

d. When applying the Limit Comparison Test, an appropriate comparison series for $\sum_{k=1}^{\infty} \dfrac{k^2 + 2k + 1}{k^5 + 5k + 7}$ is $\sum_{k=1}^{\infty} \dfrac{1}{k^3}$.

38–39. Examining a series two ways *Determine whether the follow series converge using either the Comparison Test or the Limit Comparison Test. Then use another method to check your answer.*

38. $\sum_{k=1}^{\infty} \dfrac{e^{2k}}{e^{4k} + 1}$

39. $\sum_{k=1}^{\infty} \dfrac{1}{k^2 + 2k + 1}$

40–62. Choose your test *Use the test of your choice to determine whether the following series converge.*

40. $\left(\dfrac{1}{2}\right)^2 + \left(\dfrac{2}{3}\right)^3 + \left(\dfrac{3}{4}\right)^4 + \cdots$

41. $\sum_{k=1}^{\infty} \left(1 + \dfrac{2}{k}\right)^k$

42. $\sum_{k=1}^{\infty} \dfrac{e^{2k} + k}{e^{5k} - k^2}$

43. $\sum_{k=1}^{\infty} \dfrac{k^2 + 2k + 1}{3k^2 + 1}$

44. $\sum_{k=1}^{\infty} \dfrac{2^k}{e^k - 1}$

45. $\sum_{k=3}^{\infty} \dfrac{1}{\ln k}$

46. $\sum_{k=1}^{\infty} \dfrac{1}{5^k - 1}$

47. $\sum_{k=1}^{\infty} \dfrac{1}{\sqrt{k^3 - k + 1}}$

48. $\sum_{k=3}^{\infty} \dfrac{1}{5^k - 3^k}$

49. $\sum_{k=1}^{\infty} \left(\dfrac{1}{k} + 2^{-k}\right)$

50. $\sum_{k=2}^{\infty} \dfrac{5 \ln k}{k}$

51. $\sum_{k=1}^{\infty} \dfrac{k^8}{k^{11} + 3}$

52. $\sum_{k=2}^{\infty} \dfrac{1}{k^2 \ln k}$

53. $\sum_{k=1}^{\infty} \dfrac{1}{k^{1+p}}, \; p > 0$

54. $\sum_{k=1}^{\infty} \dfrac{\tan^{-1} k}{k^2}$

55. $\displaystyle\sum_{k=1}^{\infty} \ln\left(\frac{k+2}{k+1}\right)$

56. $\displaystyle\sum_{k=1}^{\infty} k^{-1/k}$

57. $\displaystyle\sum_{k=2}^{\infty} \frac{1}{k^{\ln k}}$

58. $\displaystyle\sum_{k=1}^{\infty} \sin^2 \frac{1}{k}$

59. $\displaystyle\sum_{k=1}^{\infty} \tan \frac{1}{k}$

60. $\displaystyle\sum_{k=1}^{\infty} \sqrt{\frac{1}{2^{2k}} + \frac{1}{k^2}}$

61. $\dfrac{1}{1\cdot 3} + \dfrac{1}{3\cdot 5} + \dfrac{1}{5\cdot 7} + \cdots$

62. $\dfrac{1}{2^2} + \dfrac{2}{3^2} + \dfrac{3}{4^2} + \cdots$

63. Series of squares Prove that if Σa_k is a convergent series of positive terms, then the series Σa_k^2 also converges.

64. Two sine series Determine whether the following series converge.

a. $\displaystyle\sum_{k=1}^{\infty} \sin \frac{1}{k}$

b. $\displaystyle\sum_{k=1}^{\infty} \frac{1}{k} \sin \frac{1}{k}$

Explorations and Challenges

65. Limit Comparison Test proof Use the proof of case (1) of the Limit Comparison Test (Theorem 10.15) to prove cases (2) and (3).

66. Infinite products An infinite product $P = a_1 a_2 a_3 \ldots$, which is denoted $\displaystyle\prod_{k=1}^{\infty} a_k$, is the limit of the *sequence of partial products* $\{a_1, a_1 a_2, a_1 a_2 a_3, \ldots\}$. Assume $a_k > 0$ for all k and L is a finite constant.

a. Evaluate $\displaystyle\prod_{k=1}^{\infty}\left(\frac{k}{k+1}\right) = \frac{1}{2}\cdot\frac{2}{3}\cdot\frac{3}{4}\cdot\frac{4}{5}\cdots.$

b. Show that if $\displaystyle\sum_{k=1}^{\infty} \ln a_k = L$, then $\displaystyle\prod_{k=1}^{\infty} a_k = e^L.$

c. Use the result of part (b) to evaluate
$$\prod_{k=0}^{\infty} e^{1/2^k} = e\cdot e^{1/2}\cdot e^{1/4}\cdot e^{1/8}\cdots.$$

67. An early limit Working in the early 1600s, the mathematicians Wallis, Pascal, and Fermat wanted to calculate the area of the region under the curve $y = x^p$ between $x = 0$ and $x = 1$, where p is a positive integer. Using arguments that predated the Fundamental Theorem of Calculus, they were able to prove that
$$\lim_{n\to\infty} \frac{1}{n} \sum_{k=0}^{n-1} \left(\frac{k}{n}\right)^p = \frac{1}{p+1}.$$
Use what you know about Riemann sums and integrals to verify this limit.

QUICK CHECK ANSWERS

1. To use the Comparison Test, we would need to show that $1/(k+1) > 1/k$, which is not true.   **2.** If $\displaystyle\lim_{k\to\infty} \frac{a_k}{b_k} = L$ for $0 < L < \infty$, then $\displaystyle\lim_{k\to\infty} \frac{b_k}{a_k} = \frac{1}{L}$, where $0 < 1/L < \infty$. ◄

10.6 Alternating Series

Our previous discussion focused on infinite series with positive terms, which is certainly an important part of the entire subject. But there are many interesting series with terms of mixed sign. For example, the series

$$1 + \frac{1}{2} - \frac{1}{3} - \frac{1}{4} + \frac{1}{5} + \frac{1}{6} - \frac{1}{7} - \frac{1}{8} + \cdots$$

has the pattern that two positive terms are followed by two negative terms, and vice versa. Clearly, infinite series could have endless sign patterns, so we need to restrict our attention. Fortunately, the simplest sign pattern is also the most important. We consider **alternating series** in which the signs strictly alternate, as in the series

$$\sum_{k=1}^{\infty} \frac{(-1)^{k+1}}{k} = 1 - \frac{1}{2} + \frac{1}{3} - \frac{1}{4} + \frac{1}{5} - \frac{1}{6} + \frac{1}{7} - \frac{1}{8} + \cdots.$$

The factor $(-1)^{k+1}$ (or $(-1)^k$) has the pattern $\{\ldots, 1, -1, 1, -1, \ldots\}$ and provides the alternating signs.

Alternating Harmonic Series

Let's see what is different about alternating series by working with the series $\displaystyle\sum_{k=1}^{\infty} \frac{(-1)^{k+1}}{k}$, which is called the **alternating harmonic series**. Recall that this series *without* the alternating signs, $\displaystyle\sum_{k=1}^{\infty} \frac{1}{k}$, is the *divergent* harmonic series. So an immediate question is whether the presence of alternating signs affects the convergence of a series.

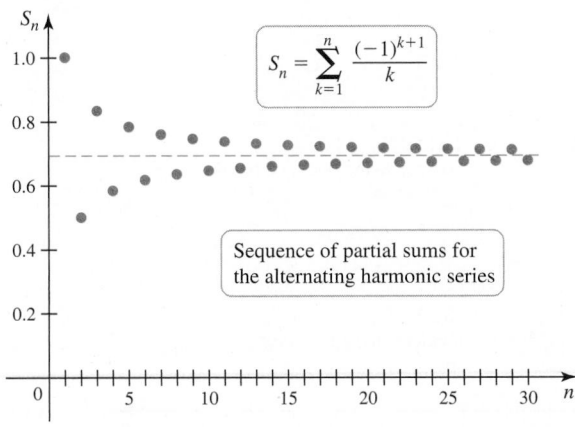

Figure 10.34

$$S_n = \sum_{k=1}^{n} \frac{(-1)^{k+1}}{k}$$

Sequence of partial sums for the alternating harmonic series

We investigate this question by looking at the sequence of partial sums for the series. In this case, the first four terms of the sequence of partial sums are

$$S_1 = 1$$

$$S_2 = 1 - \frac{1}{2} = \frac{1}{2}$$

$$S_3 = 1 - \frac{1}{2} + \frac{1}{3} = \frac{5}{6}$$

$$S_4 = 1 - \frac{1}{2} + \frac{1}{3} - \frac{1}{4} = \frac{7}{12}.$$

Plotting the first 30 terms of the sequence of partial sums results in **Figure 10.34**, which has several noteworthy features.

- The terms of the sequence of partial sums appear to converge to a limit; if they do, it means that, while the harmonic series diverges, the *alternating* harmonic series converges. We will soon learn that taking a divergent series with positive terms and making it an alternating series *may* turn it into a convergent series.

- For series with *positive* terms, the sequence of partial sums is necessarily an increasing sequence. Because the terms of an alternating series alternate in sign, the sequence of partial sums is not increasing (Figure 10.34).

- For the alternating harmonic series, the odd terms of the sequence of partial sums form a decreasing sequence and the even terms form an increasing sequence. As a result, the limit of the sequence of partial sums lies between any two consecutive terms of the sequence.

QUICK CHECK 1 Write out the first few terms of the sequence of partial sums for the alternating series $1 - 2 + 3 - 4 + 5 - 6 + \cdots$. Does this series appear to converge or diverge? ◄

Alternating Series Test

We now consider alternating series in general, which are written $\sum (-1)^{k+1} a_k$, where $a_k > 0$. With the exception of the Divergence Test, none of the convergence tests for series with positive terms applies to alternating series. Fortunately, one test works for most alternating series—and it is easy to use.

► Recall that the Divergence Test of Section 10.4 applies to all series: If the terms of *any* series (including an alternating series) do not tend to zero, then the series diverges.

► Depending on the sign of the first term of the series, an alternating series may be written with $(-1)^k$ or $(-1)^{k+1}$. Theorem 10.16 also applies to series of the form $\sum (-1)^k a_k$.

THEOREM 10.16 Alternating Series Test
The alternating series $\sum (-1)^{k+1} a_k$ converges provided

1. the terms of the series are nonincreasing in magnitude ($0 < a_{k+1} \leq a_k$, for k greater than some index N) and

2. $\lim\limits_{k \to \infty} a_k = 0$.

There is potential for confusion here. *For series of positive terms,* $\lim\limits_{k \to \infty} a_k = 0$ **does not** *imply convergence. For alternating series with nonincreasing terms,* $\lim\limits_{k \to \infty} a_k = 0$ **does** *imply convergence.*

Proof: The proof is short and instructive; it relies on **Figure 10.35**. We consider an alternating series in the form

$$\sum_{k=1}^{\infty} (-1)^{k+1} a_k = a_1 - a_2 + a_3 - a_4 + \cdots.$$

Because the terms of the series are nonincreasing in magnitude, the even terms of the sequence of partial sums $\{S_{2k}\} = \{S_2, S_4, \ldots\}$ form a nondecreasing sequence that is bounded above by S_1. By the Bounded Monotonic Sequence Theorem (Section 10.2), this sequence has a limit; call it L. Similarly, the odd terms of the sequence of partial sums $\{S_{2k-1}\} = \{S_1, S_3, \ldots\}$ form a nonincreasing sequence that is bounded below by S_2. By

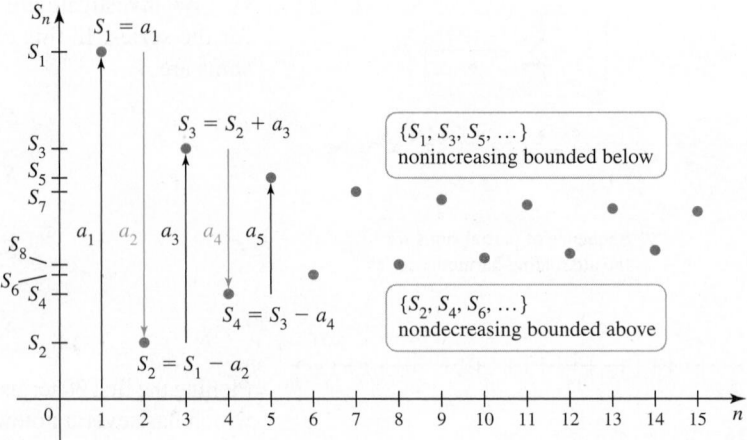

Figure 10.35

the Bounded Monotonic Sequence Theorem, this sequence also has a limit; call it L'. At the moment, we cannot conclude that $L = L'$. However, notice that $S_{2k} = S_{2k-1} - a_{2k}$. By the condition that $\lim_{k \to \infty} a_k = 0$, it follows that

$$\underbrace{\lim_{k \to \infty} S_{2k}}_{L} = \underbrace{\lim_{k \to \infty} S_{2k-1}}_{L'} - \underbrace{\lim_{k \to \infty} a_{2k}}_{0},$$

or $L = L'$. Therefore, the sequence of partial sums converges to a (unique) limit, and the corresponding alternating series converges to that limit.   ◄

Now we can confirm that the alternating harmonic series $\sum_{k=1}^{\infty} \dfrac{(-1)^{k+1}}{k}$ converges.

This fact follows immediately from the Alternating Series Test because the terms $a_k = \dfrac{1}{k}$

decrease and $\lim_{k \to \infty} a_k = 0$.

> $\sum_{k=1}^{\infty} \dfrac{1}{k}$ $\sum_{k=1}^{\infty} \dfrac{(-1)^{k+1}}{k}$
>
> • Diverges   • Converges
> • Partial sums   • Partial sums
>   increase.   bound the
>   series above
>   and below.

> **THEOREM 10.17   Alternating Harmonic Series**
>
> The alternating harmonic series $\sum_{k=1}^{\infty} \dfrac{(-1)^{k+1}}{k}$ converges (even though the harmonic
>
> series $\sum_{k=1}^{\infty} \dfrac{1}{k}$ diverges).

QUICK CHECK 2 Explain why the value of a convergent alternating series, with terms that are nonincreasing in magnitude, is trapped between successive terms of the sequence of partial sums. ◄

EXAMPLE 1   Alternating Series Test Determine whether the following series converge or diverge.

a. $\displaystyle\sum_{k=1}^{\infty} \dfrac{(-1)^{k+1}}{k^2}$ **b.** $2 - \dfrac{3}{2} + \dfrac{4}{3} - \dfrac{5}{4} + \cdots$ **c.** $\displaystyle\sum_{k=2}^{\infty} \dfrac{(-1)^k \ln k}{k}$

SOLUTION

a. The terms of this series decrease in magnitude, for $k \geq 1$. Furthermore,

$$\lim_{k \to \infty} a_k = \lim_{k \to \infty} \frac{1}{k^2} = 0.$$

Therefore, the series converges.

b. Using sigma notation, this series is written as $\displaystyle\sum_{k=1}^{\infty} (-1)^{k+1} \dfrac{k+1}{k}$. Because

$\lim_{k \to \infty} \dfrac{k+1}{k} = 1$, the sequence $\left\{ (-1)^{k+1} \dfrac{k+1}{k} \right\}$ oscillates and diverges. Therefore,

$\displaystyle\sum_{k=1}^{\infty} (-1)^{k+1} \dfrac{k+1}{k}$ diverges by the Divergence Test.

c. The first step is to show that the terms decrease in magnitude after some fixed term of the series. One way to proceed is to look at the function $f(x) = \dfrac{\ln x}{x}$, which generates the terms of the series. By the Quotient Rule, $f'(x) = \dfrac{1 - \ln x}{x^2}$. The fact that $f'(x) < 0$, for $x > e$, implies that the terms $\dfrac{\ln k}{k}$ decrease, for $k \geq 3$. As long as the terms of the series decrease for all k greater than some fixed integer, the first condition of the test is met. Furthermore, using l'Hôpital's Rule or the fact that $\{\ln k\}$ increases more slowly than $\{k\}$ (Section 10.2), we see that

$$\lim_{k \to \infty} a_k = \lim_{k \to \infty} \frac{\ln k}{k} = 0.$$

The conditions of the Alternating Series Test are met and the series converges.

Related Exercises 14, 15, 18 ◄

Remainders in Alternating Series

Recall that if a series converges to a value S, then the remainder is $R_n = S - S_n$, where S_n is the sum of the first n terms of the series. The magnitude of the remainder is the *absolute error* in approximating S by S_n.

An upper bound on the magnitude of the remainder in an alternating series arises from the following observation: When the terms are nonincreasing in magnitude, the value of the series is always trapped between successive terms of the sequence of partial sums. Therefore, as shown in **Figure 10.36**,

$$|R_n| = |S - S_n| \leq |S_{n+1} - S_n| = a_{n+1}.$$

This argument justifies the following theorem.

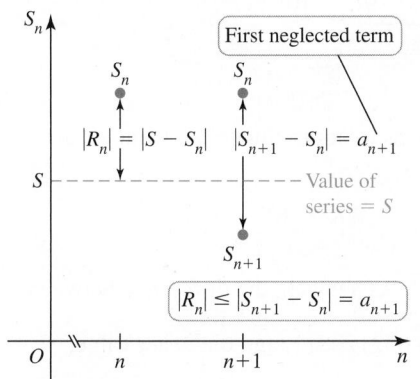

Figure 10.36

> **THEOREM 10.18   Remainder in Alternating Series**
>
> Let $\sum\limits_{k=1}^{\infty} (-1)^{k+1} a_k$ be a convergent alternating series with terms that are nonincreasing in magnitude. Let $R_n = S - S_n$ be the remainder in approximating the value of that series by the sum of its first n terms. Then $|R_n| \leq a_{n+1}$. In other words, the magnitude of the remainder is less than or equal to the magnitude of the first neglected term.

EXAMPLE 2   Remainder in an alternating series

a. In turns out that $\ln 2 = 1 - \dfrac{1}{2} + \dfrac{1}{3} - \dfrac{1}{4} + \cdots = \sum\limits_{k=1}^{\infty} \dfrac{(-1)^{k+1}}{k}$. How many terms of the series are required to approximate $\ln 2$ with an error less than 10^{-6}? The exact value of the series is given but is not needed to answer the question.

b. Consider the series $-1 + \dfrac{1}{2!} - \dfrac{1}{3!} + \dfrac{1}{4!} - \cdots = \sum\limits_{k=1}^{\infty} \dfrac{(-1)^k}{k!}$. Find an upper bound for the magnitude of the error in approximating the value of the series (which is $e^{-1} - 1$) with $n = 9$ terms.

SOLUTION Notice that both series meet the conditions of Theorem 10.18.

a. The series is expressed as the sum of the first n terms plus the remainder:

$$\sum_{k=1}^{\infty} \frac{(-1)^{k+1}}{k} = \underbrace{1 - \frac{1}{2} + \frac{1}{3} - \frac{1}{4} + \cdots + \frac{(-1)^{n+1}}{n}}_{S_n \,=\, \text{the sum of the first } n \text{ terms}} + \overbrace{\frac{(-1)^{n+2}}{n+1} + \cdots}^{R_n}.$$

$|R_n| = |S - S_n|$ is less than the magnitude of this term

In magnitude, the remainder is less than or equal to the magnitude of the $(n + 1)$st term:

$$|R_n| = |S - S_n| \leq a_{n+1} = \frac{1}{n + 1}.$$

To ensure that the error is less than 10^{-6}, we require that

$$a_{n+1} = \frac{1}{n + 1} < 10^{-6}, \quad \text{or} \quad n + 1 > 10^6.$$

Therefore, it takes 1 million terms of the series to approximate $\ln 2$ with an error less than 10^{-6}.

b. The series may be expressed as the sum of the first nine terms plus the remainder:

$$\sum_{k=1}^{\infty} \frac{(-1)^k}{k!} = \underbrace{-1 + \frac{1}{2!} - \frac{1}{3!} + \cdots - \frac{1}{9!}}_{S_9 \,=\, \text{sum of first 9 terms}} + \overbrace{\underbrace{\frac{1}{10!}}_{\substack{|R_9| = |S - S_9| \\ \text{is less than} \\ \text{this term}}} - \cdots}^{R_9}.$$

The error committed when using the first nine terms to approximate the value of the series satisfies

$$|R_9| = |S - S_9| \leq a_{10} = \frac{1}{10!} \approx 2.8 \times 10^{-7}.$$

QUICK CHECK 3 Compare and comment on the speed of convergence of the two series in Example 2. Why does one series converge more rapidly than the other? ◄

Therefore, the error is no greater than 2.8×10^{-7}. As a check, the difference between the sum of the first nine terms, $\displaystyle\sum_{k=1}^{9} \frac{(-1)^k}{k!} \approx -0.632120811$, and the exact value, $S = e^{-1} - 1 \approx -0.632120559$, is approximately 2.5×10^{-7}. Therefore, the actual error satisfies the bound given by Theorem 10.18.

Related Exercises 32–33 ◄

Absolute and Conditional Convergence

In this final segment, we introduce some terminology that is needed in Chapter 11. We now let the notation $\sum a_k$ denote any series—a series of positive terms, an alternating series, or even a more general infinite series.

Look again at the convergent alternating harmonic series $\sum (-1)^{k+1}/k$. The corresponding series of positive terms, $\sum 1/k$, is the divergent harmonic series. In contrast, we saw in Example 1a that the alternating series $\sum (-1)^{k+1}/k^2$ converges, and the corresponding *p*-series of positive terms $\sum 1/k^2$ also converges. These examples illustrate that removing the alternating signs in a convergent series *may* or *may not* result in a convergent series. The terminology that we now introduce distinguishes between these cases.

DEFINITION Absolute and Conditional Convergence

If $\sum |a_k|$ converges, then $\sum a_k$ **converges absolutely**.
If $\sum |a_k|$ diverges and $\sum a_k$ converges, then $\sum a_k$ **converges conditionally**.

The series $\sum (-1)^{k+1}/k^2$ is an example of an absolutely convergent series because the series of absolute values,

$$\sum_{k=1}^{\infty} \left| \frac{(-1)^{k+1}}{k^2} \right| = \sum_{k=1}^{\infty} \frac{1}{k^2},$$

is a convergent *p*-series. In this case, removing the alternating signs in the series does *not* affect its convergence.

On the other hand, the convergent alternating harmonic series $\sum(-1)^{k+1}/k$ has the property that the corresponding series of absolute values,

$$\sum_{k=1}^{\infty}\left|\frac{(-1)^{k+1}}{k}\right| = \sum_{k=1}^{\infty}\frac{1}{k},$$

does *not* converge. In this case, removing the alternating signs in the series *does* affect convergence, so this series does not converge absolutely. Instead, we say it *converges conditionally*. A convergent series (such as $\sum(-1)^{k+1}/k$) may not converge absolutely. However, if a series converges absolutely, then it converges.

> **THEOREM 10.19 Absolute Convergence Implies Convergence**
> If $\sum|a_k|$ converges, then $\sum a_k$ converges (absolute convergence implies convergence). Equivalently, if $\sum a_k$ diverges, then $\sum|a_k|$ diverges.

Proof: Because $|a_k| = a_k$ or $|a_k| = -a_k$, it follows that $0 \le a_k + |a_k| \le 2|a_k|$. By assumption, $\sum|a_k|$ converges, which, in turn, implies that $2\sum|a_k|$ converges. Using the Comparison Test and the inequality $0 \le a_k + |a_k| \le 2|a_k|$, it follows that $\sum(a_k + |a_k|)$ converges. Now note that

$$\sum a_k = \sum(a_k + |a_k| - |a_k|) = \underbrace{\sum(a_k + |a_k|)}_{\text{converges}} - \underbrace{\sum|a_k|}_{\text{converges}}.$$

We see that $\sum a_k$ is the sum of two convergent series, so it also converges. The second statement of the theorem is logically equivalent to the first statement. ◀

Figure 10.37 gives an overview of absolute and conditional convergence. It shows the universe of all infinite series, split first according to whether they converge or diverge. Convergent series are further divided between absolutely and conditionally convergent series. Here are a few more consequences of these definitions.

- The distinction between absolute and conditional convergence is relevant only for series of mixed sign, which includes alternating series. If a series of positive terms converges, it converges absolutely; conditional convergence does not apply.
- To test for absolute convergence, we test the series $\sum|a_k|$, which is a series of positive terms. Therefore, the convergence tests of Sections 10.4 and 10.5 (for positive-term series) are used to determine absolute convergence.

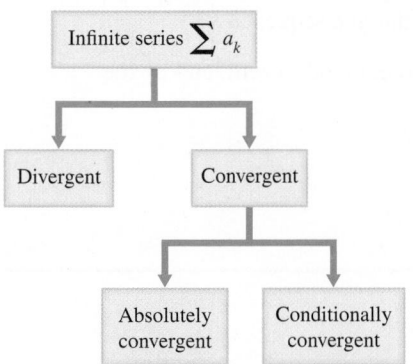

Figure 10.37

QUICK CHECK 4 Explain why a convergent series of positive terms converges absolutely. ◀

EXAMPLE 3 Absolute and conditional convergence Determine whether the following series diverge, converge absolutely, or converge conditionally.

a. $\displaystyle\sum_{k=1}^{\infty}\frac{(-1)^{k+1}}{\sqrt{k}}$ **b.** $\displaystyle\sum_{k=1}^{\infty}\frac{(-1)^{k+1}}{\sqrt{k^3}}$ **c.** $\displaystyle\sum_{k=1}^{\infty}\frac{\sin k}{k^2}$ **d.** $\displaystyle\sum_{k=1}^{\infty}\frac{(-1)^k k}{k+1}$

SOLUTION

a. We examine the series of absolute values,

$$\sum_{k=1}^{\infty}\left|\frac{(-1)^{k+1}}{\sqrt{k}}\right| = \sum_{k=1}^{\infty}\frac{1}{\sqrt{k}},$$

which is a divergent *p*-series (with $p = \frac{1}{2} < 1$). Therefore, the given alternating series does not converge absolutely. To determine whether the series converges conditionally, we look at the original series—with alternating signs. The magnitude of the terms of this series decrease with $\lim_{k\to\infty} 1/\sqrt{k} = 0$, so by the Alternating Series Test, the series converges. Because this series converges, but not absolutely, it converges conditionally.

b. To assess absolute convergence, we look at the series of absolute values,

$$\sum_{k=1}^{\infty} \left| \frac{(-1)^{k+1}}{\sqrt{k^3}} \right| = \sum_{k=1}^{\infty} \frac{1}{k^{3/2}},$$

which is a convergent p-series (with $p = \frac{3}{2} > 1$). Therefore, the original alternating series converges absolutely (and by Theorem 10.19 it converges).

c. The terms of this series do not strictly alternate in sign (the first few signs are $+++---$), so the Alternating Series Test does not apply. Because $|\sin k| \le 1$, the terms of the series of absolute values satisfy

$$\left| \frac{\sin k}{k^2} \right| = \frac{|\sin k|}{k^2} \le \frac{1}{k^2}.$$

The series $\sum \frac{1}{k^2}$ is a convergent p-series. Therefore, by the Comparison Test, the series $\sum \left| \frac{\sin k}{k^2} \right|$ converges, which implies that the series $\sum \frac{\sin k}{k^2}$ converges absolutely (and by Theorem 10.19 it converges).

d. Notice that $\lim_{k \to \infty} k/(k + 1) = 1$, which implies that the sequence $\left\{ \frac{(-1)^k k}{k + 1} \right\}$ oscillates and diverges. The terms of the series do not tend to zero, and by the Divergence Test, the series diverges.

Related Exercises 45, 51, 53 ◄

SECTION 10.6 EXERCISES

Getting Started

1. Explain why the sequence of partial sums for an alternating series is not an increasing sequence.

2. Describe how to apply the Alternating Series Test.

3. Why does the value of a converging alternating series with terms that are nonincreasing in magnitude lie between any two consecutive terms of its sequence of partial sums?

4. Suppose an alternating series with terms that are nonincreasing in magnitude converges to a value L. Explain how to estimate the remainder that occurs when the series is terminated after n terms.

5. Explain why the magnitude of the remainder in an alternating series (with terms that are nonincreasing in magnitude) is less than or equal to the magnitude of the first neglected term.

6. Give an example of a convergent alternating series that fails to converge absolutely.

7. Is it possible for a series of positive terms to converge conditionally? Explain.

8. Does the geometric series $\sum_{k=0}^{\infty} (-0.5)^k$ converge absolutely?

9. Is it possible for an alternating series to converge absolutely but not conditionally?

10. Determine the values of p for which the series $\sum_{k=0}^{\infty} \frac{(-1)^k}{k^p}$ converges conditionally.

Practice Exercises

11–27. Alternating Series Test *Determine whether the following series converge.*

11. $\displaystyle \sum_{k=0}^{\infty} \frac{(-1)^k}{2k + 1}$

12. $\displaystyle \sum_{k=1}^{\infty} \frac{(-1)^k}{\sqrt{k}}$

13. $\displaystyle \sum_{k=1}^{\infty} \frac{(-1)^k k}{3k + 2}$

14. $\displaystyle \sum_{k=1}^{\infty} (-1)^k \left(1 + \frac{1}{k} \right)^k$

15. $\displaystyle \sum_{k=1}^{\infty} \frac{(-1)^{k+1}}{k^3}$

16. $\displaystyle \sum_{k=0}^{\infty} \frac{(-1)^k}{k^2 + 10}$

17. $\displaystyle \sum_{k=1}^{\infty} (-1)^{k+1} \frac{k^2}{k^3 + 1}$

18. $\displaystyle \sum_{k=2}^{\infty} (-1)^k \frac{\ln k}{k^2}$

19. $\displaystyle \sum_{k=2}^{\infty} (-1)^k \frac{k^2 - 1}{k^2 + 3}$

20. $\displaystyle \sum_{k=0}^{\infty} \left(-\frac{1}{5} \right)^k$

21. $\displaystyle \sum_{k=2}^{\infty} (-1)^k \left(1 + \frac{1}{k} \right)$

22. $\displaystyle \sum_{k=1}^{\infty} \frac{\cos \pi k}{k^2}$

23. $\displaystyle \sum_{k=1}^{\infty} (-1)^k \frac{k^{11} + 2k^5 + 1}{4k(k^{10} + 1)}$

24. $\displaystyle \sum_{k=2}^{\infty} \frac{(-1)^k}{k \ln^2 k}$

25. $\displaystyle \sum_{k=1}^{\infty} (-1)^{k+1} k^{1/k}$

26. $\displaystyle \sum_{k=1}^{\infty} (-1)^k k \sin \frac{1}{k}$

27. $\displaystyle \sum_{k=0}^{\infty} \frac{(-1)^k}{\sqrt{k^2 + 4}}$

T 28–32. Estimating errors in partial sums *For each of the following convergent alternating series, evaluate the nth partial sum for the given value of n. Then use Theorem 10.18 to find an upper bound for the error* $|S - S_n|$ *in using the nth partial sum* S_n *to estimate the value of the series S.*

28. $\displaystyle\sum_{k=1}^{\infty} \frac{(-1)^{k+1}}{k^3 + 1}$; $n = 3$

29. $\displaystyle\sum_{k=1}^{\infty} \frac{(-1)^k}{k^4}$; $n = 4$

30. $\displaystyle\sum_{k=0}^{\infty} \frac{(-1)^k}{(3k + 1)^4}$; $n = 2$

31. $\displaystyle\sum_{k=0}^{\infty} \frac{(-1)^k}{k^4 + k^2 + 1}$; $n = 5$

32. $\displaystyle\sum_{k=0}^{\infty} \frac{(-1)^k}{(2k + 1)!}$; $n = 3$

T 33–38. Remainders in alternating series *Determine how many terms of the following convergent series must be summed to be sure that the remainder is less than* 10^{-4} *in magnitude. Although you do not need it, the exact value of the series is given in each case.*

33. $\ln 2 = \displaystyle\sum_{k=1}^{\infty} \frac{(-1)^{k+1}}{k}$

34. $\dfrac{1}{e} = \displaystyle\sum_{k=0}^{\infty} \frac{(-1)^k}{k!}$

35. $\dfrac{\pi}{4} = \displaystyle\sum_{k=0}^{\infty} \frac{(-1)^k}{2k + 1}$

36. $\dfrac{\pi^2}{12} = \displaystyle\sum_{k=1}^{\infty} \frac{(-1)^{k+1}}{k^2}$

37. $\dfrac{7\pi^4}{720} = \displaystyle\sum_{k=1}^{\infty} \frac{(-1)^{k+1}}{k^4}$

38. $\dfrac{\pi^3}{32} = \displaystyle\sum_{k=0}^{\infty} \frac{(-1)^k}{(2k + 1)^3}$

T 39–44. Estimating infinite series *Estimate the value of the following convergent series with an absolute error less than* 10^{-3}.

39. $\displaystyle\sum_{k=1}^{\infty} \frac{(-1)^k}{k^5}$

40. $\displaystyle\sum_{k=1}^{\infty} \frac{(-1)^k}{(2k + 1)^3}$

41. $\displaystyle\sum_{k=1}^{\infty} \frac{(-1)^k k}{k^2 + 1}$

42. $\displaystyle\sum_{k=1}^{\infty} \frac{(-1)^k k}{k^4 + 1}$

43. $\displaystyle\sum_{k=1}^{\infty} \frac{(-1)^k}{k^k}$

44. $\displaystyle\sum_{k=1}^{\infty} \frac{(-1)^{k+1}}{(2k + 1)!}$

45–63. Absolute and conditional convergence *Determine whether the following series converge absolutely, converge conditionally, or diverge.*

45. $\displaystyle\sum_{k=1}^{\infty} \frac{(-1)^k}{k^{2/3}}$

46. $\displaystyle\sum_{k=0}^{\infty} \frac{(-1)^k 2k}{\sqrt{k^2 + 9}}$

47. $\displaystyle\sum_{k=1}^{\infty} \frac{(-1)^{k+1}}{(k + 1)!}$

48. $\displaystyle\sum_{k=1}^{\infty} \left(-\frac{1}{3}\right)^k$

49. $\displaystyle\sum_{k=1}^{\infty} \left(\frac{3}{4}\right)^k$

50. $\displaystyle\sum_{k=1}^{\infty} \frac{(-1)^{k+1}}{k^{0.99}}$

51. $\displaystyle\sum_{k=1}^{\infty} \frac{\cos k}{k^3}$

52. $\displaystyle\sum_{k=1}^{\infty} \frac{(-1)^k k^2}{\sqrt{k^6 + 1}}$

53. $\displaystyle\sum_{k=1}^{\infty} (-1)^k \tan^{-1} k$

54. $\displaystyle\sum_{k=1}^{\infty} (-1)^k e^{-k}$

55. $\displaystyle\sum_{k=1}^{\infty} \frac{(-1)^{k+1}}{2\sqrt{k} - 1}$

56. $\displaystyle\sum_{k=2}^{\infty} \frac{(-1)^k}{3^k - 1}$

57. $\displaystyle\sum_{k=1}^{\infty} \frac{(-1)^k k}{2k + 1}$

58. $\displaystyle\sum_{k=3}^{\infty} \frac{(-1)^k}{\ln k}$

59. $\displaystyle\sum_{k=1}^{\infty} \frac{(-1)^k \tan^{-1} k}{k^3}$

60. $\displaystyle\sum_{k=1}^{\infty} (-1)^k \frac{k^2 + 1}{3k^4 + 3}$

61. $\displaystyle\sum_{k=2}^{\infty} (-1)^k \frac{k^2 + 1}{k^3 - 1}$

62. $\displaystyle\sum_{k=1}^{\infty} \frac{\sin k}{3^k + 4^k}$

63. $\displaystyle\sum_{k=1}^{\infty} (-1)^{k+1} \frac{k!}{k^k}$ (*Hint:* Show that $\dfrac{k!}{k^k} \leq \dfrac{2}{k^2}$, for $k \geq 3$.)

64. Alternating Series Test Show that the series

$$\frac{1}{3} - \frac{2}{5} + \frac{3}{7} - \frac{4}{9} + \cdots = \sum_{k=1}^{\infty} (-1)^{k+1} \frac{k}{2k + 1}$$

diverges. Which condition of the Alternating Series Test is not satisfied?

65. Explain why or why not Determine whether the following statements are true and give an explanation or counterexample.

   a. A series that converges must converge absolutely.
   b. A series that converges absolutely must converge.
   c. A series that converges conditionally must converge.
   d. If $\sum a_k$ diverges, then $\sum |a_k|$ diverges.
   e. If $\sum a_k^2$ converges, then $\sum a_k$ converges.
   f. If $a_k > 0$ and $\sum a_k$ converges, then $\sum a_k^2$ converges.
   g. If $\sum a_k$ converges conditionally, then $\sum |a_k|$ diverges.

Explorations and Challenges

66. Rearranging series It can be proved that if a series converges absolutely, then its terms may be summed in any order without changing the value of the series. However, if a series converges conditionally, then the value of the series depends on the order of summation. For example, the (conditionally convergent) alternating harmonic series has the value

$$1 - \frac{1}{2} + \frac{1}{3} - \frac{1}{4} + \cdots = \ln 2.$$

Show that by rearranging the terms (so the sign pattern is $++-$),

$$1 + \frac{1}{3} - \frac{1}{2} + \frac{1}{5} + \frac{1}{7} - \frac{1}{4} + \cdots = \frac{3}{2} \ln 2.$$

67. Alternating p-series Given that $\displaystyle\sum_{k=1}^{\infty} \frac{1}{k^2} = \frac{\pi^2}{6}$, show that $\displaystyle\sum_{k=1}^{\infty} \frac{(-1)^{k+1}}{k^2} = \frac{\pi^2}{12}$. (Assume the result of Exercise 66.)

68. Alternating p-series Given that $\displaystyle\sum_{k=1}^{\infty} \frac{1}{k^4} = \frac{\pi^4}{90}$, show that $\displaystyle\sum_{k=1}^{\infty} \frac{(-1)^{k+1}}{k^4} = \frac{7\pi^4}{720}$. (Assume the result of Exercise 66.)

69. A fallacy Explain the fallacy in the following argument. Let $x = 1 + \dfrac{1}{3} + \dfrac{1}{5} + \dfrac{1}{7} + \cdots$ and $y = \dfrac{1}{2} + \dfrac{1}{4} + \dfrac{1}{6} + \dfrac{1}{8} + \cdots$. It follows that $2y = x + y$, which implies that $x = y$. On the other hand,

$$x - y = \underbrace{\left(1 - \frac{1}{2}\right)}_{>0} + \underbrace{\left(\frac{1}{3} - \frac{1}{4}\right)}_{>0} + \underbrace{\left(\frac{1}{5} - \frac{1}{6}\right)}_{>0} + \cdots > 0$$

is a sum of positive terms, so $x > y$. Therefore, we have shown that $x = y$ and $x > y$.

70. Conditions of the Alternating Series Test It can be shown that if the sequence $\{a_{2n}\} = \{a_2, a_4, a_6, \ldots\}$ and the sequence $\{a_{2n-1}\} = \{a_1, a_3, a_5, \ldots\}$ both converge to L, then the sequence $\{a_n\} = \{a_1, a_2, a_3, \ldots\}$ converges to L. It is also the case that if $\{a_{2n}\}$ or $\{a_{2n-1}\}$ diverges, then $\{a_n\}$ diverges. Use these results in this exercise. Consider the alternating series

$$\sum_{k=1}^{\infty}(-1)^{k+1}a_k, \text{ where } a_k = \begin{cases} \dfrac{4}{k+1} & \text{if } k \text{ is odd} \\ \dfrac{2}{k} & \text{if } k \text{ is even.} \end{cases}$$

a. Show that $\{a_n\}$ converges to 0.

b. Show that $S_{2n} = \displaystyle\sum_{k=1}^{n}\frac{1}{k}$ and explain why $\displaystyle\lim_{k\to\infty} S_{2n} = \infty$.

c. Explain why the series $\displaystyle\sum_{k=1}^{\infty}(-1)^{k+1}a_k$ diverges even though $\{a_n\}$ converges to 0. Explain why this result does not contradict the Alternating Series Test.

71. A diverging alternating series Consider the alternating series

$$\sum_{k=1}^{\infty}(-1)^{k+1}a_k = \frac{1}{1^2} - \frac{1}{1} + \frac{1}{2^2} - \frac{1}{2} + \frac{1}{3^2} - \frac{1}{3} + \cdots.$$

a. Show that the individual terms of the series converge to 0. (*Hint:* See Exercise 70.)

b. Find a formula for S_{2n}, the sum of the first $2n$ terms of the series.

c. Explain why the alternating series diverges even though individual terms of the series converge to 0. Explain why this result does not contradict the Alternating Series Test.

QUICK CHECK ANSWERS

1. $1, -1, 2, -2, 3, -3, \ldots$; series diverges.   **2.** The even terms of the sequence of partial sums approach the value of the series from one side; the odd terms of the sequence of partial sums approach the value of the series from the other side.   **3.** The second series with $k!$ in the denominator converges much more quickly than the first series because $k!$ increases much faster than k as $k \to \infty$.   **4.** If a series has positive terms, the series of absolute values is the same as the series itself. ◄

10.7 The Ratio and Root Tests

We now consider two additional convergence tests for general series: the Ratio Test and the Root Test. Both tests are used frequently throughout the next chapter. As in previous sections, these tests determine *whether* an infinite series converges, but they do not establish the value of the series.

The Ratio Test

The tests we have developed so far to determine whether a series converges are powerful, but they all have limitations. For example, the Integral Test requires evaluating integrals, and it is difficult to use with a series such as $\displaystyle\sum_{k=1}^{\infty}\frac{1}{k!}$ because integrating $1/k!$ is problematic. Further, the Integral Test and the comparison tests work only with series that have positive terms.

The Ratio Test, which we introduce in the next theorem, significantly enlarges the set of infinite series that we can analyze. It can be applied to any series, though its success depends on evaluating the limit of the ratio of successive terms in a series. The conclusions reached by the Ratio Test rely on the idea of absolute convergence and Theorem 10.19 from Section 10.6. We assume $a_k \neq 0$, for all k, so that the ratio a_{k+1}/a_k is defined.

THEOREM 10.20   Ratio Test

Let $\sum a_k$ be an infinite series, and let $r = \displaystyle\lim_{k\to\infty}\left|\frac{a_{k+1}}{a_k}\right|$.

1. If $r < 1$, the series converges absolutely, and therefore it converges (by Theorem 10.19).

2. If $r > 1$ (including $r = \infty$), the series diverges.

3. If $r = 1$, the test is inconclusive.

> See Appendix A for a formal proof of Theorem 10.20.

Proof (outline): The idea behind the proof provides insight. Let's assume the limit $r \geq 0$ exists. Then as k gets large and the ratio $|a_{k+1}/a_k|$ approaches r, we have $|a_{k+1}| \approx r|a_k|$. Therefore, as one goes farther and farther out in the series, it behaves like

$$|a_k| + |a_{k+1}| + |a_{k+2}| + \cdots \approx |a_k| + r|a_k| + r^2|a_k| + r^3|a_k| + \cdots$$
$$= |a_k|(1 + r + r^2 + r^3 + \cdots).$$

The tail of the series, which determines whether the series converges, behaves like a geometric series with ratio r. We know that if $r < 1$, the geometric series converges, and if $r > 1$, the series diverges, which is the conclusion of the Ratio Test. ◄

EXAMPLE 1 Using the Ratio Test Use the Ratio Test to determine whether the following series converge.

a. $\displaystyle\sum_{k=1}^{\infty} \frac{10^k}{k!}$ **b.** $\displaystyle\sum_{k=1}^{\infty} \frac{(-1)^k k^k}{k!}$ **c.** $\displaystyle\sum_{k=1}^{\infty}(-1)^{k+1} e^{-k}(k^2 + 4)$

SOLUTION In each case, the limit of the ratio of successive terms is determined.

a. In this case, we have a series with positive terms. Therefore,

> Recall that
> $$k! = k \cdot (k-1) \cdots 2 \cdot 1.$$
> Therefore,
> $$(k+1)! = (k+1)\underbrace{k(k-1)\cdots 1}_{k!}$$
> $$= (k+1)k!.$$

$$r = \lim_{k\to\infty}\left|\frac{a_{k+1}}{a_k}\right| = \lim_{k\to\infty}\frac{10^{k+1}/(k+1)!}{10^k/k!} \quad \text{Substitute } a_{k+1} \text{ and } a_k.$$

$$= \lim_{k\to\infty}\frac{10^{k+1}}{10^k}\cdot\frac{k!}{(k+1)k!} \quad \text{Invert and multiply.}$$

$$= \lim_{k\to\infty}\frac{10}{k+1} = 0 \quad \text{Simplify and evaluate the limit.}$$

Because $r = 0 < 1$, the series converges by the Ratio Test.

b. $r = \displaystyle\lim_{k\to\infty}\left|\frac{a_{k+1}}{a_k}\right| = \lim_{k\to\infty}\left|\frac{(-1)^{k+1}(k+1)^{k+1}/(k+1)!}{(-1)^k k^k/k!}\right| \quad \text{Substitute } a_{k+1} \text{ and } a_k.$

$$= \lim_{k\to\infty}\left(\frac{k+1}{k}\right)^k \quad \text{Simplify.}$$

> Recall from Section 4.7 that
> $$\lim_{k\to\infty}\left(1 + \frac{1}{k}\right)^k = e \approx 2.718.$$

$$= \lim_{k\to\infty}\left(1 + \frac{1}{k}\right)^k = e \quad \text{Simplify and evaluate the limit.}$$

Because $r = e > 1$, the series diverges by the Ratio Test. Alternatively, we could have noted that $\displaystyle\lim_{k\to\infty} k^k/k! = \infty$ (Theorem 10.6) and used the Divergence Test to reach the same conclusion.

c. $r = \displaystyle\lim_{k\to\infty}\left|\frac{a_{k+1}}{a_k}\right| = \lim_{k\to\infty}\frac{e^{-(k+1)}((k+1)^2 + 4)}{e^{-k}(k^2 + 4)} \quad \text{Substitute } a_{k+1} \text{ and } a_k.$

$$= \lim_{k\to\infty}\frac{e^{-k}e^{-1}(k^2 + 2k + 5)}{e^{-k}(k^2 + 4)} \quad \text{Simplify.}$$

$$= e^{-1}\underbrace{\lim_{k\to\infty}\frac{k^2 + 2k + 5}{k^2 + 4}}_{1} \quad \text{Simplify.}$$

$$= e^{-1}$$

QUICK CHECK 1 Evaluate $10!/9!$, $(k+2)!/k!$, and $k!/(k+1)!$ ◄

Because $e^{-1} = \dfrac{1}{e} < 1$, the series converges absolutely by the Ratio Test, which implies that it converges.

Related Exercises 10, 14 ◄

▶ In Section 10.8, we offer guidelines to help you to decide which convergence test is best suited for a given series.

The Ratio Test is conclusive for many series. Nevertheless, observe what happens when the Ratio Test is applied to the harmonic series $\sum\limits_{k=1}^{\infty} \dfrac{1}{k}$:

$$r = \lim_{k \to \infty} \left| \frac{a_{k+1}}{a_k} \right| = \lim_{k \to \infty} \frac{1/(k+1)}{1/k} = \lim_{k \to \infty} \frac{k}{k+1} = 1,$$

which means the test is inconclusive. We know the harmonic series diverges, yet the Ratio Test cannot be used to establish this fact. Like all the convergence tests encountered in this chapter, the Ratio Test works only for certain classes of series.

QUICK CHECK 2 Verify that the Ratio Test is inconclusive for $\sum\limits_{k=1}^{\infty} \dfrac{1}{k^2}$. What test could be applied to show that $\sum\limits_{k=1}^{\infty} \dfrac{1}{k^2}$ converges? ◄

The Root Test

The Root Test is similar to the Ratio Test and is well suited to series whose general term involves powers.

THEOREM 10.21   Root Test

Let $\sum a_k$ be an infinite series, and let $\rho = \lim\limits_{k \to \infty} \sqrt[k]{|a_k|}$.

1. If $\rho < 1$, the series converges absolutely, and therefore it converges (by Theorem 10.19).
2. If $\rho > 1$ (including $\rho = \infty$), the series diverges.
3. If $\rho = 1$, the test is inconclusive.

Proof (outline): Assume the limit ρ exists. If k is large, we have $\rho \approx \sqrt[k]{|a_k|}$ or $|a_k| \approx \rho^k$. For large values of k, the tail of the series, which determines whether a series converges, behaves like

$$|a_k| + |a_{k+1}| + |a_{k+2}| + \cdots \approx \rho^k + \rho^{k+1} + \rho^{k+2} + \cdots.$$

▶ See Appendix A for a formal proof of Theorem 10.21.

Therefore, the tail of the series is approximately a geometric series with ratio ρ. If $\rho < 1$, the geometric series converges, and if $\rho > 1$, the series diverges, which is the conclusion of the Root Test. ◄

EXAMPLE 2   Using the Root Test Use the Root Test to determine whether the following series converge.

a. $\sum\limits_{k=1}^{\infty} \left(\dfrac{3 - 4k^2}{7k^2 + 6} \right)^k$ **b.** $\sum\limits_{k=1}^{\infty} \dfrac{(-2)^k}{k^{10}}$

SOLUTION

a. The required limit is

$$\rho = \lim_{k \to \infty} \sqrt[k]{\left| \left(\frac{3 - 4k^2}{7k^2 + 6} \right)^k \right|} = \lim_{k \to \infty} \frac{4k^2 - 3}{7k^2 + 6} = \frac{4}{7}. \qquad |3 - 4k^2| = 4k^2 - 3, \text{ for } k > 1$$

Because $\rho < 1$, the series converges absolutely by the Root Test (and therefore it converges).

b. In this case,

$$\rho = \lim_{k \to \infty} \sqrt[k]{\left| \frac{(-2)^k}{k^{10}} \right|} = \lim_{k \to \infty} \frac{2}{k^{10/k}} = \lim_{k \to \infty} \frac{2}{\left(k^{1/k} \right)^{10}} = 2. \qquad \lim_{k \to \infty} k^{1/k} = 1$$

Because $\rho > 1$, the series diverges by the Root Test.

　　We could have used the Ratio Test for both series in this example, but the Root Test is easier to apply in each case. In part (b), the Divergence Test leads to the same conclusion.

Related Exercises 17, 32 ◄

SECTION 10.7 EXERCISES

Getting Started

1. Explain how the Ratio Test works.

2. Explain how the Root Test works.

3. Evaluate $1000!/998!$ without a calculator.

4. Evaluate $(100!)^2/(99!)^2$ without a calculator.

5. Simplify $\dfrac{k!}{(k+2)!}$, for any integer $k \geq 0$.

6. Simplify $\dfrac{(2k+1)!}{(2k-1)!}$, for any integer $k \geq 1$.

7. What test is advisable if a series involves a factorial term?

8. Can the value of a series be determined using the Root Test or the Ratio Test?

Practice Exercises

9–30. The Ratio and Root Tests *Use the Ratio Test or the Root Test to determine whether the following series converge absolutely or diverge.*

9. $\displaystyle\sum_{k=1}^{\infty} \frac{(-1)^k}{k!}$

10. $\displaystyle\sum_{k=1}^{\infty} \frac{(-2)^k}{k!}$

11. $\displaystyle\sum_{k=1}^{\infty} (-1)^{k+1} \left(\frac{10k^3 + k}{9k^3 + k + 1} \right)^k$

12. $\displaystyle\sum_{k=1}^{\infty} \left(-\frac{2k}{k+1} \right)^k$

13. $\displaystyle\sum_{k=1}^{\infty} \frac{k^2}{4^k}$

14. $\displaystyle\sum_{k=1}^{\infty} \frac{k^k}{2^k}$

15. $\displaystyle\sum_{k=1}^{\infty} (-1)^{k+1} k e^{-k}$

16. $\displaystyle\sum_{k=1}^{\infty} (-1)^k \frac{k!}{k^k}$

17. $\displaystyle\sum_{k=1}^{\infty} \frac{(-7)^k}{k^2}$

18. $\displaystyle\sum_{k=1}^{\infty} \left(1 + \frac{3}{k} \right)^{k^2}$

19. $\displaystyle\sum_{k=1}^{\infty} \frac{2^k}{k^{99}}$

20. $\displaystyle\sum_{k=1}^{\infty} \frac{(-1)^k k!}{k^6}$

21. $\displaystyle\sum_{k=1}^{\infty} \frac{(k!)^2}{(2k)!}$

22. $2 + \dfrac{4}{16} + \dfrac{8}{81} + \dfrac{16}{256} + \cdots$

23. $\displaystyle\sum_{k=1}^{\infty} \left(\frac{k}{k+1} \right)^{2k^2}$

24. $\displaystyle\sum_{k=1}^{\infty} (-1)^{k+1} \frac{3^{k^2}}{k^k}$

25. $\displaystyle\sum_{k=1}^{\infty} (-1)^{k+1} \left(\frac{3k^2 + 4k}{2k^2 + 1} \right)^k$

26. $\displaystyle\sum_{k=1}^{\infty} \left(\frac{1}{\ln (k+1)} \right)^k$

27. $1 + \left(\dfrac{1}{2} \right)^2 + \left(\dfrac{1}{3} \right)^3 + \left(\dfrac{1}{4} \right)^4 + \cdots$

28. $\displaystyle\sum_{k=1}^{\infty} \frac{k}{e^k}$

29. $\displaystyle\sum_{k=1}^{\infty} \frac{(-1)^{k+1} k^{2k}}{k! k!}$

30. $\displaystyle\sum_{k=1}^{\infty} \left(\frac{k}{1 - 5k} \right)^k$

31. **Explain why or why not** Determine whether the following statements are true and give an explanation or counterexample.

  a. $n! n! = (2n)!$, for all positive integers n.

  b. $\dfrac{(2n)!}{(2n-1)!} = 2n$.

  c. If $\displaystyle\lim_{k \to \infty} \sqrt[k]{|a_k|} = \frac{1}{4}$, then $\sum 10 a_k$ converges absolutely.

  d. The Ratio Test is always inconclusive when applied to $\sum a_k$, where a_k is a nonzero rational function of k.

32–49. Choose your test *Use the test of your choice to determine whether the following series converge absolutely, converge conditionally, or diverge.*

32. $\displaystyle\sum_{k=1}^{\infty} \left(\frac{4k^3 + 7}{9k^2 - 1} \right)^k$

33. $\displaystyle\sum_{k=1}^{\infty} (-1)^k \frac{2^{k+1}}{9^{k-1}}$

34. $\displaystyle\sum_{k=1}^{\infty} \frac{k!}{(2k+1)!}$

35. $\displaystyle\sum_{k=1}^{\infty} \frac{(2k+1)!}{(k!)^2}$

36. $\displaystyle\sum_{k=1}^{\infty} \left(\frac{k^2}{2k^2 + 1} \right)^k$

37. $\displaystyle\sum_{k=1}^{\infty} \frac{k^{100}}{(k+1)!}$

38. $\displaystyle\sum_{k=1}^{\infty} k^3 \sin (1/k^3)$

39. $\displaystyle\sum_{k=1}^{\infty} \frac{(-1)^k k^3}{\sqrt{k^8 + 1}}$

40. $\displaystyle\sum_{k=1}^{\infty} \frac{1}{(1 + p)^k}, \ p > 0$

41. $\displaystyle\sum_{k=1}^{\infty} \left(\sqrt[k]{k} - 1 \right)^{2k}$

42. $\displaystyle\sum_{k=1}^{\infty} \frac{(k!)^3}{(3k)!}$

43. $\displaystyle\sum_{k=1}^{\infty} \frac{2^k k!}{k^k}$

44. $\displaystyle\sum_{k=1}^{\infty} \left(1 - \frac{1}{k} \right)^{k^2}$

45. $\displaystyle\sum_{k=1}^{\infty} \frac{(-1)^k}{k^{0.99}}$

46. $\displaystyle\sum_{k=2}^{\infty} 100 k^{-k}$

47. $\dfrac{1}{1!} + \dfrac{4}{2!} + \dfrac{9}{3!} + \dfrac{16}{4!} + \cdots$

48. $\displaystyle\sum_{k=1}^{\infty} \frac{(-1)^k}{\tan^{-1} k}$

49. $\displaystyle\sum_{k=1}^{\infty} \frac{(-1)^k}{\sqrt{k^{3/2} + k}}$

Explorations and Challenges

50–57. Convergence parameter *Find the values of the parameter $p > 0$ for which the following series converge.*

50. $\displaystyle\sum_{k=2}^{\infty} \frac{1}{(\ln k)^p}$

51. $\displaystyle\sum_{k=2}^{\infty} \frac{\ln k}{k^p}$

52. $\displaystyle\sum_{k=2}^{\infty} \frac{1}{k(\ln k)(\ln \ln k)^p}$

53. $\displaystyle\sum_{k=2}^{\infty} \left(\frac{\ln k}{k} \right)^p$

54. $\displaystyle\sum_{k=0}^{\infty} \frac{k! p^k}{(k+1)^k}$

  (*Hint:* Stirling's formula is useful: $k! \approx \sqrt{2\pi k}\, k^k e^{-k}$, for large k.)

55. $\displaystyle\sum_{k=1}^{\infty} \frac{k p^k}{k + 1}$

56. $\displaystyle\sum_{k=1}^{\infty} \ln \left(\frac{k}{k+1} \right)^p$

57. $\displaystyle\sum_{k=1}^{\infty} \left(1 - \frac{p}{k} \right)^k$

58–63. A glimpse ahead to power series *Use the Ratio Test or the Root Test to determine the values of x for which each series converges.*

58. $\displaystyle\sum_{k=1}^{\infty} \frac{x^k}{k!}$ **59.** $\displaystyle\sum_{k=1}^{\infty} x^k$ **60.** $\displaystyle\sum_{k=1}^{\infty} \frac{x^k}{k}$

61. $\displaystyle\sum_{k=1}^{\infty} \frac{x^k}{k^2}$ **62.** $\displaystyle\sum_{k=1}^{\infty} \frac{x^{2k}}{k^2}$ **63.** $\displaystyle\sum_{k=1}^{\infty} \frac{x^k}{2^k}$

64. Stirling's formula Complete the following steps to find the values of $p > 0$ for which the series $\displaystyle\sum_{k=1}^{\infty} \frac{1 \cdot 3 \cdot 5 \cdots (2k-1)}{p^k k!}$ converges.

    a. Use the Ratio Test to show that $\displaystyle\sum_{k=1}^{\infty} \frac{1 \cdot 3 \cdot 5 \cdots (2k-1)}{p^k k!}$ converges for $p > 2$.

    b. Use Stirling's formula, $k! \approx \sqrt{2\pi k}\, k^k e^{-k}$ for large k, to determine whether the series converges with $p = 2$.

    (Hint: $1 \cdot 3 \cdot 5 \cdots (2k-1) = \dfrac{1 \cdot 2 \cdot 3 \cdot 4 \cdot 5 \cdot 6 \cdots (2k-1)2k}{2 \cdot 4 \cdot 6 \cdots 2k}$.*)*

    (See the Guided Project *Stirling's formula and n!* for more on this topic.)

QUICK CHECK ANSWERS

1. $10; (k+2)(k+1); 1/(k+1)$ **2.** The Integral Test or p-series with $p = 2$ ◄

10.8 Choosing a Convergence Test

In previous sections, we presented methods for determining whether an infinite series converges, and in some cases we discovered that it is possible to determine the exact value of a convergent series. In this section, we step back and focus on general strategies for classifying infinite series and deciding which tests are appropriate for a specific series.

Series Strategies

It is helpful if you are already familiar with the convergence tests summarized in Table 10.4. With these tests in mind, we explore various strategies and illustrate them with examples. These strategies can be applied sequentially, more or less in the order in which they are given.

Categorize the Series   The first step in approaching an infinite series is to note some of the distinctive properties that help to classify it. Is it a series with positive terms? An alternating series? Or something more general? Does the general term consist of rational functions, exponential functions, or factorials? Can the series be simplified in any obvious ways before applying convergence tests? Here is a quick suggestive example.

EXAMPLE 1   Categorize the series   Discuss the series $\displaystyle\sum_{k=1}^{\infty} \frac{2^k + \cos(\pi k)\sqrt{k}}{3^{k+1}}$.

SOLUTION   A good first step in analyzing this series would be to split it into two series that can be examined separately. We begin by writing

$$\sum_{k=1}^{\infty} \frac{2^k + \cos(\pi k)\sqrt{k}}{3^{k+1}} = \underbrace{\sum_{k=1}^{\infty} \frac{2^k}{3^{k+1}}}_{\text{geometric series}} + \underbrace{\sum_{k=1}^{\infty} \frac{\cos(\pi k)\sqrt{k}}{3^{k+1}}}_{\text{alternating series}}.$$

QUICK CHECK 1 Evaluate the geometric series in Example 1. Does the alternating series in Example 1 converge? ◄

The general term of the first series is the ratio of exponential functions, so this series is a geometric series that can be evaluated explicitly. Because $\cos(\pi k) = (-1)^k$, we see that the second series is an alternating series, whose convergence can also be tested. Notice that for the original series to converge, both the geometric series and the alternating series must converge.

Related Exercise 8 ◄

Apply the Divergence Test   Almost without exception, the Divergence Test is the easiest test to apply because it requires nothing more than taking a limit. Furthermore, it can lead to a quick conclusion, namely that the given series diverges.

Table 10.4 **Special Series and Convergence Tests**

Series or Test	Form of Series	Condition for Convergence	Condition for Divergence	Comments
Geometric series	$\sum\limits_{k=0}^{\infty} ar^k, a \neq 0$	$\lvert r \rvert < 1$	$\lvert r \rvert \geq 1$	If $\lvert r \rvert < 1$, then $\sum\limits_{k=0}^{\infty} ar^k = \dfrac{a}{1-r}$.
Divergence Test	$\sum\limits_{k=1}^{\infty} a_k$	Does not apply	$\lim\limits_{k \to \infty} a_k \neq 0$	Cannot be used to prove convergence
Integral Test	$\sum\limits_{k=1}^{\infty} a_k$, where $a_k = f(k)$ and f is continuous, positive, and decreasing	$\int_1^{\infty} f(x)\, dx$ converges.	$\int_1^{\infty} f(x)\, dx$ diverges.	The value of the integral is not the value of the series.
p-series	$\sum\limits_{k=1}^{\infty} \dfrac{1}{k^p}$	$p > 1$	$p \leq 1$	Useful for comparison tests
Ratio Test	$\sum\limits_{k=1}^{\infty} a_k$	$\lim\limits_{k \to \infty} \left\lvert \dfrac{a_{k+1}}{a_k} \right\rvert < 1$	$\lim\limits_{k \to \infty} \left\lvert \dfrac{a_{k+1}}{a_k} \right\rvert > 1$	Inconclusive if $\lim\limits_{k \to \infty} \left\lvert \dfrac{a_{k+1}}{a_k} \right\rvert = 1$
Root Test	$\sum\limits_{k=1}^{\infty} a_k$	$\lim\limits_{k \to \infty} \sqrt[k]{\lvert a_k \rvert} < 1$	$\lim\limits_{k \to \infty} \sqrt[k]{\lvert a_k \rvert} > 1$	Inconclusive if $\lim\limits_{k \to \infty} \sqrt[k]{\lvert a_k \rvert} = 1$
Comparison Test	$\sum\limits_{k=1}^{\infty} a_k$, where $a_k > 0$	$a_k \leq b_k$ and $\sum\limits_{k=1}^{\infty} b_k$ converges.	$b_k \leq a_k$ and $\sum\limits_{k=1}^{\infty} b_k$ diverges.	$\sum\limits_{k=1}^{\infty} a_k$ is given; you supply $\sum\limits_{k=1}^{\infty} b_k$.
Limit Comparison Test	$\sum\limits_{k=1}^{\infty} a_k$, where $a_k > 0, b_k > 0$	$0 \leq \lim\limits_{k \to \infty} \dfrac{a_k}{b_k} < \infty$ and $\sum\limits_{k=1}^{\infty} b_k$ converges.	$\lim\limits_{k \to \infty} \dfrac{a_k}{b_k} > 0$ and $\sum\limits_{k=1}^{\infty} b_k$ diverges.	$\sum\limits_{k=1}^{\infty} a_k$ is given; you supply $\sum\limits_{k=1}^{\infty} b_k$.
Alternating Series Test	$\sum\limits_{k=1}^{\infty} (-1)^k a_k$, where $a_k > 0$	$\lim\limits_{k \to \infty} a_k = 0$ and $0 < a_{k+1} \leq a_k$	$\lim\limits_{k \to \infty} a_k \neq 0$	Remainder R_n satisfies $\lvert R_n \rvert \leq a_{n+1}$
Absolute Convergence	$\sum\limits_{k=1}^{\infty} a_k$, a_k arbitrary	$\sum\limits_{k=1}^{\infty} \lvert a_k \rvert$ converges.		Applies to arbitrary series

EXAMPLE 2 **Divergence Test** Does the series $\sum\limits_{k=1}^{\infty} \left(1 - \dfrac{10}{k} \right)^k$ converge or diverge?

SOLUTION The terms of this series alternate in sign for $k < 10$, and then remain positive and close to zero for $k > 10$. Applying the Divergence Test, we evaluate the limit of the general term using l'Hôpital's Rule and find that

$$\lim\limits_{k \to \infty} \left(1 - \dfrac{10}{k} \right)^k = e^{-10} \neq 0.$$

QUICK CHECK 2 Show the steps in evaluating the limit in Example 2. ◄

Because the limit of the general term is non-zero, the series diverges (very slowly).

Related Exercise 18 ◄

Identify Special Series We have studied several special families of series that are readily identified and analyzed for convergence. Specifically, you should look for geometric series, p-series, and (less obvious and less common) telescoping series.

EXAMPLE 3 **Special series** Does the series $\displaystyle\sum_{k=4}^{\infty} \frac{1}{\sqrt[4]{k^2 - 6k + 9}}$ converge or diverge?

SOLUTION The Divergence Test is inconclusive for this series of positive terms. One approach is to rewrite the series in the following steps:

$$\sum_{k=4}^{\infty} \frac{1}{\sqrt[4]{k^2 - 6k + 9}} = \sum_{k=4}^{\infty} \frac{1}{\sqrt[4]{(k-3)^2}} \qquad \text{Factor denominator.}$$

$$= \sum_{j=1}^{\infty} \frac{1}{\sqrt[4]{j^2}} \qquad \text{Shift index; let } j = k - 3.$$

$$= \sum_{k=1}^{\infty} \frac{1}{\sqrt{k}}. \qquad \text{Simplify.}$$

We now identify this series as a p-series with $p = \dfrac{1}{2} < 1$; therefore, it diverges.

Related Exercise 24 ◄

Integral, Ratio, and Root Tests You have probably evaluated enough integrals by now to guess whether the general term of a series can be integrated easily. There are also some immediate clues: If a series involves factorials (something involving $k!$) or terms of the form $f(k)^k$, the Integral Test is not likely to help. On the other hand, series with terms of the form $k!$, k^k, or a^k, where a is a constant, often submit to the Ratio Test or the Root Test.

► When using the Integral Test, make sure the integrand satisfies the conditions given in Theorem 10.11: It must be continuous, positive, and decreasing on $[k, \infty)$ for some finite number k. In Example 4, $f(x) = x^2 e^{-2x}$ satisfies these conditions on $[1, \infty)$.

EXAMPLE 4 **More than one test** Does the series $\displaystyle\sum_{k=1}^{\infty} k^2 e^{-2k}$ converge or diverge?

SOLUTION The general term of this series of positive terms can be integrated using integration by parts, which reveals that $\displaystyle\int_1^{\infty} x^2 e^{-2x}\,dx = \frac{5e^{-2}}{4} < \infty$. Because the integral converges, the infinite series also converges. However, it may be easier to apply the Ratio Test. We find that

$$\lim_{k \to \infty} \frac{(k+1)^2 e^{-2(k+1)}}{k^2 e^{-2k}} = e^{-2} \underbrace{\lim_{k \to \infty} \left(\frac{k+1}{k}\right)^2}_{1} = e^{-2} \cdot 1 \approx 0.135 < 1.$$

QUICK CHECK 3 Verify the limit in the Ratio Test in Example 4. ◄

Because the ratio of successive terms in less than 1, the series converges.

Related Exercise 25 ◄

Rational Terms If a series involves a rational function, or ratios of roots or algebraic functions, then a comparison test is a likely candidate. When the ordinary Comparison Test does not work easily, the more powerful Limit Comparison Test may lead to a conclusion. As shown in the following example, analyzing the end behavior (as $k \to \infty$) of the general term is often the trick needed to find a good comparison series.

EXAMPLE 5 **Comparison tests** Does the series $\displaystyle\sum_{k=2}^{\infty} \sqrt[3]{\frac{k^2 - 1}{k^4 + 4}}$ converge or diverge?

SOLUTION The Divergence Test is inconclusive. Ruling out the Integral, Ratio, and Root Tests as impractical, we are left with a comparison test. The end behavior of the general term is determined by working with the leading terms of the numerator and denominator. We find that for large values of k,

$$\sqrt[3]{\frac{k^2 - 1}{k^4 + 4}} \approx \sqrt[3]{\frac{k^2}{k^4}} = \frac{1}{k^{2/3}}.$$

Therefore, a good comparison series is $\sum k^{-2/3}$, which is a divergent p-series. The ordinary Comparison Test is difficult to apply in this case, so we turn to the Limit

Comparison Test. Taking the limit of the ratio of terms of the given series and the comparison series, we find that

$$L = \lim_{k \to \infty} \frac{\sqrt[3]{\dfrac{k^2 - 1}{k^4 + 4}}}{k^{-2/3}} = \lim_{k \to \infty} \sqrt[3]{\frac{k^2 - 1}{k^4 + 4}} \cdot k^{2/3} = \lim_{k \to \infty} \sqrt[3]{\frac{k^4 - k^2}{k^4 + 4}} = \sqrt[3]{\lim_{k \to \infty} \frac{1 - k^{-2}}{1 + 4k^{-4}}} = 1.$$

Because $0 < L < \infty$ and the comparison series diverges, the given series also diverges.

Related Exercise 49 ◄

The guidelines presented in this section are a good start, but in the end, convergence tests are mastered through practice. It's your turn.

SECTION 10.8 EXERCISES

Getting Started

1–10. Choosing convergence tests *Identify a convergence test for each of the following series. If necessary, explain how to simplify or rewrite the series before applying the convergence test. You do not need to carry out the convergence test.*

1. $\displaystyle\sum_{k=1}^{\infty} (-1)^k \left(2 + \frac{1}{k^2}\right)^k$

2. $\displaystyle\sum_{k=3}^{\infty} \frac{2}{k^2 - k - 2}$

3. $\displaystyle\sum_{k=3}^{\infty} \frac{2k^2}{k^2 - k - 2}$

4. $\displaystyle\sum_{k=3}^{\infty} \frac{1}{k \ln^7 k}$

5. $\displaystyle\sum_{k=10}^{\infty} \frac{1}{(k-9)^5}$

6. $\displaystyle\sum_{k=10}^{\infty} \frac{100^k}{k! \, k^2}$

7. $\displaystyle\sum_{k=1}^{\infty} \frac{k^2}{k^4 + k^3 + 1}$

8. $\displaystyle\sum_{k=1}^{\infty} \frac{(-3)^k}{4^{k+1}}$

9. $\displaystyle\sum_{k=1}^{\infty} \frac{(-1)^{k+1}}{\sqrt{2^k + \ln k}}$

10. $\displaystyle\sum_{k=1}^{\infty} (\tan^{-1} 2k - \tan^{-1}(2k - 2))$

Practice Exercises

11–86. Applying convergence tests *Determine whether the following series converge. Justify your answers.*

11. $\displaystyle\sum_{k=1}^{\infty} \frac{2k^4 + k}{4k^4 - 8k}$

12. $\displaystyle\sum_{k=1}^{\infty} 7^{-2k}$

13. $\displaystyle\sum_{k=3}^{\infty} \frac{5}{2 + \ln k}$

14. $\displaystyle\sum_{k=1}^{\infty} \frac{7k^2 - k - 2}{4k^4 - 3k + 1}$

15. $\displaystyle\sum_{k=1}^{\infty} \frac{(-7)^k}{k!}$

16. $\displaystyle\sum_{k=1}^{\infty} \frac{7^k}{k! + 10}$

17. $\displaystyle\sum_{k=1}^{\infty} \frac{(-k)^3}{3k^3 + 2}$

18. $\displaystyle\sum_{k=1}^{\infty} \left(1 + \frac{a}{k}\right)^k$, *a* is real

19. $\displaystyle\sum_{k=0}^{\infty} \frac{3^{k+4}}{5^{k-2}}$

20. $\displaystyle\sum_{j=1}^{\infty} \frac{4j^{10}}{j^{11} + 1}$

21. $\displaystyle\sum_{k=1}^{\infty} \frac{(-1)^k k}{k^3 + 1}$

22. $\displaystyle\sum_{k=1}^{\infty} \left(\frac{e+1}{\pi}\right)^k$

23. $\displaystyle\sum_{k=1}^{\infty} \frac{k^5}{5^k}$

24. $\displaystyle\sum_{k=1}^{\infty} \frac{4}{(k+3)^3}$

25. $\displaystyle\sum_{k=1}^{\infty} \frac{1}{\sqrt{k} e^{\sqrt{k}}}$

26. $\displaystyle\sum_{k=1}^{\infty} \frac{5 + \sin k}{\sqrt{k}}$

27. $\displaystyle\sum_{k=1}^{\infty} \frac{3 + \cos 5k}{k^3}$

28. $\displaystyle\sum_{k=3}^{\infty} \frac{(-1)^k \ln k}{k^{1/3}}$

29. $\displaystyle\sum_{k=1}^{\infty} \frac{10^k + 1}{k^{10}}$

30. $\displaystyle\sum_{k=3}^{\infty} \frac{1}{k^3 \ln k}$

31. $\displaystyle\sum_{j=1}^{\infty} \frac{5}{j^2 + 4}$

32. $\displaystyle\sum_{k=1}^{\infty} \frac{k^k}{(k+2)^k}$

33. $\displaystyle\sum_{k=3}^{\infty} \frac{1}{k^{1/3} \ln k}$

34. $\displaystyle\sum_{k=1}^{\infty} \frac{(-1)^k 5k^2}{\sqrt{3k^5 + 1}}$

35. $\displaystyle\sum_{k=1}^{\infty} \frac{2^k 3^k}{k^k}$

36. $\displaystyle\sum_{k=1}^{\infty} (-1)^{k+1} \frac{2k + 1}{k!}$

37. $\displaystyle\sum_{k=1}^{\infty} (-1)^k \left(\frac{5k}{3k + 7}\right)^k$

38. $\displaystyle\sum_{k=1}^{\infty} \frac{2^k (k-1)!}{k!}$

39. $\displaystyle\sum_{k=1}^{\infty} \frac{5^k (k!)^2}{(2k)!}$

40. $\displaystyle\sum_{j=1}^{\infty} \frac{\cos((2j + 1)\pi)}{j^2 + 1}$

41. $\displaystyle\sum_{k=1}^{\infty} \frac{2^k}{3^k - 2^k}$

42. $\displaystyle\sum_{k=1}^{\infty} \frac{2 + (-1)^k}{2^k}$

43. $\displaystyle\sum_{k=1}^{\infty} \cos \frac{(3k - 1)\pi}{3}$

44. $\displaystyle\sum_{k=10}^{\infty} \frac{1}{\sqrt{k(k-9)}}$

45. $\displaystyle\sum_{k=1}^{\infty} \frac{k^4}{e^{k^5}}$

46. $\displaystyle\sum_{k=1}^{\infty} \frac{1}{(k+1)! - k!}$

47. $\displaystyle\sum_{k=1}^{\infty} \frac{(4k)!}{(k!)^4}$

48. $\dfrac{2}{3} + \dfrac{3}{8} + \dfrac{4}{15} + \dfrac{5}{24} + \dfrac{6}{35} + \cdots$

49. $\displaystyle\sum_{k=1}^{\infty} \frac{\sqrt[5]{k}}{\sqrt[3]{k^7 + 1}}$

50. $\displaystyle\sum_{k=1}^{\infty} e^{-k^3}$

51. $\displaystyle\sum_{k=1}^{\infty} \frac{7^k + 11^k}{11^k}$

52. $\displaystyle\sum_{k=1}^{\infty} \frac{7^k + 11^k}{13^k}$

53. $\displaystyle\sum_{k=1}^{\infty} \sin \frac{1}{k^9}$

54. $\displaystyle\sum_{j=1}^{\infty} j^9 \sin \frac{1}{j^9}$

55. $\displaystyle\sum_{k=1}^{\infty} \cos \frac{1}{k^9}$

56. $\displaystyle\sum_{k=1}^{\infty} \left(\frac{k}{k+10} \right)^{-k^2}$

57. $\displaystyle\sum_{k=1}^{\infty} 5^{1-2k}$

58. $\displaystyle\sum_{k=1}^{\infty} \frac{(-1)^k}{\sqrt{\ln(k+2)}}$

59. $\displaystyle\sum_{k=1}^{\infty} \frac{k!}{k^k + 3}$

60. $\displaystyle\sum_{k=1}^{\infty} \left(\frac{1}{k^2} + \frac{1}{k^5} \right)$

61. $\displaystyle\sum_{k=1}^{\infty} \frac{1}{\ln(e^k + 1)}$

62. $\displaystyle\sum_{k=0}^{\infty} \left(\frac{\tan^{-1} k}{\pi} \right)^k$

63. $\displaystyle\sum_{k=1}^{\infty} \left(\frac{k+a}{k} \right)^{k^2}, a > 0$

64. $\dfrac{1}{1 \cdot 4} + \dfrac{1}{2 \cdot 7} + \dfrac{1}{3 \cdot 10} + \dfrac{1}{4 \cdot 13} + \cdots$

65. $\displaystyle\sum_{k=1}^{\infty} \left(\cos \frac{1}{k} - \cos \frac{1}{k+1} \right)$

66. $\displaystyle\sum_{k=1}^{\infty} \frac{4^{k^2}}{k!}$

67. $\displaystyle\sum_{j=1}^{\infty} \frac{\cot^{-1} \frac{1}{j}}{2^j}$

68. $\displaystyle\sum_{k=2}^{\infty} \frac{\sqrt{k+1} - \sqrt{k}}{\sqrt{k^2 - k}}$

69. $\displaystyle\sum_{k=1}^{\infty} \left(1 + \frac{1}{2k} \right)^k$

70. $\displaystyle\sum_{k=0}^{\infty} e^{k/100}$

71. $\displaystyle\sum_{k=1}^{\infty} \frac{\ln^2 k}{k^{3/2}}$

72. $\displaystyle\sum_{k=1}^{\infty} (-1)^k \cos \frac{1}{k}$

73. $\displaystyle\sum_{k=0}^{\infty} k^2 \cdot 1.001^{-k}$

74. $\displaystyle\sum_{k=0}^{\infty} k \cdot 0.999^{-k}$

75. $\displaystyle\sum_{k=1}^{\infty} \left(-\frac{1}{k} \right)^k$

76. $\displaystyle\sum_{k=3}^{\infty} \frac{2}{k(2-k)}$

77. $\displaystyle\sum_{k=0}^{\infty} \frac{3k}{\sqrt[4]{k^4 + 3}}$

78. $\displaystyle\sum_{k=0}^{\infty} k \cdot 3^{-k^2}$

79. $\displaystyle\sum_{k=1}^{\infty} \tan^{-1} \frac{1}{\sqrt{k}}$

80. $\displaystyle\sum_{k=1}^{\infty} \left(\frac{3k^8 - 2}{3k^9 + 2} \right)^k$

81. $\displaystyle\sum_{k=1}^{\infty} \left(\frac{1}{\sqrt{k+2}} - \frac{1}{\sqrt{k}} \right)$

82. $\displaystyle\sum_{k=0}^{\infty} \left(\left(\frac{2}{3} \right)^{k+1} - \left(\frac{3}{2} \right)^{-k+1} \right)$

83. $\displaystyle\sum_{j=2}^{\infty} \frac{1}{j \ln^{10} j}$

84. $\displaystyle\sum_{k=1}^{\infty} \tan^{-1} \frac{1}{k^3}$

85. $\dfrac{1}{2 \cdot 3} + \dfrac{1}{4 \cdot 5} + \dfrac{1}{6 \cdot 7} + \dfrac{1}{8 \cdot 9} + \cdots$

86. $\displaystyle\sum_{k=1}^{\infty} \frac{a^k}{k!}, a \neq 0$

87. Explain why or why not Determine whether the following statements are true and give an explanation or counterexample.

   a. If the Limit Comparison Test can be applied successfully to a given series with a certain comparison series, the Comparison Test also works with the same comparison series.

   b. The series $\displaystyle\sum_{k=3}^{\infty} \frac{1}{k \ln^p k}$ converges for the same values of p as the series $\displaystyle\sum_{k=3}^{\infty} \frac{1}{k^p}$.

   c. Both the Ratio Test and the Root Test can be applied conclusively to a geometric series.

   d. The Alternating Series Test can be used to show that some series diverge.

Explorations and Challenges

88–93. A few more series *Determine whether the following series converge. Justify your answers.*

88. $\displaystyle\sum_{k=1}^{\infty} \left(\sqrt{k^4 + 1} - k^2 \right)$

89. $\displaystyle\sum_{k=1}^{\infty} \frac{k}{1 + 2 + \cdots + k}$

90. $\displaystyle\sum_{k=1}^{\infty} \left(\frac{1^3 + 2^3 + 3^3 + \cdots + k^3}{k^5} \right)$

91. $\displaystyle\sum_{k=0}^{\infty} \ln \left(\frac{2k+1}{2k+4} \right)$

92. $\displaystyle\sum_{k=1}^{\infty} \frac{1}{10^{\ln k}}$

93. $\displaystyle\sum_{k=1}^{\infty} \frac{1}{2^{\ln k} + 2}$

QUICK CHECK ANSWERS

1. 2/3; yes ◄

CHAPTER 10 REVIEW EXERCISES

1. **Explain why or why not** Determine whether the following statements are true and give an explanation or counterexample.

   a. The terms of the sequence $\{a_n\}$ increase in magnitude, so the limit of the sequence does not exist.

   b. The terms of the series $\displaystyle\sum \frac{1}{\sqrt{k}}$ approach zero, so the series converges.

   c. The terms of the sequence of partial sums of the series $\displaystyle\sum a_k$ approach $\dfrac{5}{2}$, so the infinite series converges to $\dfrac{5}{2}$.

   d. An alternating series that converges absolutely must converge conditionally.

   e. The sequence $a_n = \dfrac{n^2}{n^2 + 1}$ converges.

   f. The sequence $a_n = \dfrac{(-1)^n n^2}{n^2 + 1}$ converges.

   g. The series $\displaystyle\sum_{k=1}^{\infty} \frac{k^2}{k^2 + 1}$ converges.

   h. The sequence of partial sums associated with the series $\displaystyle\sum_{k=1}^{\infty} \frac{1}{k^2 + 1}$ converges.

2. Related sequences

a. Find the limit of the sequence $\left\{\dfrac{12n^3 + 4n}{6n^3 + 5}\right\}$.

b. Explain why the sequence $\left\{(-1)^n \dfrac{12n^3 + 4n}{6n^3 + 5}\right\}$ diverges.

3. Geometric sums Evaluate the geometric sums $\displaystyle\sum_{k=0}^{9}(0.2)^k$ and $\displaystyle\sum_{k=2}^{9}(0.2)^k$.

4. A savings plan Suppose you open a savings account by depositing $100. The account earns interest at an annual rate of 3% per year (0.25% per month). At the end of each month, you earn interest on the current balance, and then you deposit $100. Let B_n be the balance at the beginning of the nth month, where $B_0 = \$100$.

a. Find a recurrence relation for the sequence $\{B_n\}$.
b. Find an explicit formula that gives B_n, for $n = 0, 1, 2, 3, \ldots$.
c. Find B_{30}, the balance at the beginning of the 30th month.

5. Partial sums Let S_n be the nth partial sum of $\displaystyle\sum_{k=1}^{\infty} a_k = 8$. Find $\displaystyle\lim_{k\to\infty} a_k$ and $\displaystyle\lim_{n\to\infty} S_n$.

6. Find the value of r for which $\displaystyle\sum_{k=0}^{\infty} 3r^k = 6$.

7. Give an example (if possible) of a sequence $\{a_k\}$ that converges, while the series $\displaystyle\sum_{k=1}^{\infty} a_k$ diverges.

8. Give an example (if possible) of a series $\displaystyle\sum_{k=1}^{\infty} a_k$ that converges, while the sequence $\{a_k\}$ diverges.

9. Sequences versus series

a. Find the limit of the sequence $\left\{\left(-\dfrac{4}{5}\right)^k\right\}$.

b. Evaluate $\displaystyle\sum_{k=0}^{\infty}\left(-\dfrac{4}{5}\right)^k$.

10. Sequences versus series

a. Find the limit of the sequence $\left\{\dfrac{1}{k} - \dfrac{1}{k+1}\right\}$.

b. Evaluate $\displaystyle\sum_{k=1}^{\infty}\left(\dfrac{1}{k} - \dfrac{1}{k+1}\right)$.

11. a. Does the sequence $\left\{\dfrac{k}{k+1}\right\}$ converge? Why or why not?

b. Does the series $\displaystyle\sum_{k=1}^{\infty}\dfrac{k}{k+1}$ converge? Why or why not?

12–24. Limits of sequences *Evaluate the limit of the sequence or state that it does not exist.*

12. $a_n = \dfrac{3n^3 + 4n}{6n^3 + 5}$

13. $a_n = (-1)^n \dfrac{3n^3 + 4n}{6n^3 + 5}$

14. $a_n = \dfrac{(2n+2)!}{(2n)!\, n^2}$

15. $a_n = \dfrac{2^n + 5^{n+1}}{5^n}$

16. $a_n = \dfrac{n^2 + 4}{\sqrt{4n^4 + 1}}$

17. $a_n = \dfrac{8^n}{n!}$

18. $a_n = \left(1 + \dfrac{3}{n}\right)^{2n}$

19. $a_n = \dfrac{(3n^2 + 2n + 1)\sin n}{4n^3 + n}$ (*Hint:* Use the Squeeze Theorem.)

20. $a_n = n - \sqrt{n^2 - 1}$

21. $a_n = \left(\dfrac{1}{n}\right)^{1/\ln n}$

22. $a_n = \sin\dfrac{\pi n}{6}$

23. $a_n = \dfrac{(-1)^n}{0.9^n}$

24. $a_n = \dfrac{\pi}{2} - \tan^{-1} n$

25–26. Recursively defined sequences *The following sequences* $\{a_n\}_{n=0}^{\infty}$ *are defined by a recurrence relation. Assume each sequence is monotonic and bounded.*

a. *Find the first five terms* $a_0, a_1, \ldots, a_4$ *of each sequence.*
b. *Determine the limit of each sequence.*

25. $a_{n+1} = \dfrac{1}{2}a_n + 8;\ a_0 = 80$

▉ 26. $a_{n+1} = \sqrt{\dfrac{1}{3}a_n + 34};\ a_0 = 81$

27–37. Evaluating series *Evaluate the following infinite series or state that the series diverges.*

27. $\displaystyle\sum_{k=1}^{\infty} 3(1.001)^k$

28. $\displaystyle\sum_{k=1}^{\infty}\left(\dfrac{9}{10}\right)^k$

29. $\displaystyle\sum_{k=0}^{\infty}\left(\left(\dfrac{1}{3}\right)^k + \left(\dfrac{4}{3}\right)^k\right)$

30. $\displaystyle\sum_{k=1}^{\infty}\dfrac{3^k - 5^k}{10^k}$

31. $\displaystyle\sum_{k=1}^{\infty}\ln\left(\dfrac{2k+1}{2k-1}\right)$

32. $\displaystyle\sum_{k=0}^{\infty}\left(-\dfrac{1}{5}\right)^k$

33. $\displaystyle\sum_{k=0}^{\infty}\left(\tan^{-1}(k+2) - \tan^{-1}k\right)$

34. $\displaystyle\sum_{k=2}^{\infty}\left(\dfrac{1}{\sqrt{k}} - \dfrac{1}{\sqrt{k-1}}\right)$

35. $\displaystyle\sum_{k=1}^{\infty}\dfrac{9}{(3k-2)(3k+1)}$

36. $\displaystyle\sum_{k=1}^{\infty} 4^{-3k}$

37. $\displaystyle\sum_{k=1}^{\infty}\dfrac{2^k}{3^{k+2}}$

38. Express $0.29292929\ldots$ as a ratio of two integers.

39. Express $0.314141414\ldots$ as a ratio of two integers.

40. Finding steady states through sequences Suppose you take 100 mg of aspirin once per day. Assume the aspirin has a half-life of one day; that is, every day, half of the aspirin in your blood is eliminated. Assume d_n is the amount of aspirin in your blood after the nth dose, where $d_1 = 100$.

a. Find a recurrence relation for the sequence $\{d_n\}$.
b. Assuming the sequence $\{d_n\}$ converges, find the long-term (steady-state) amount of aspirin in your blood.

41. Finding steady states using infinite series Solve Exercise 40 by expressing the amount of aspirin in your blood as a geometric series and evaluating the series.

42–76. Convergence or divergence *Use a convergence test of your choice to determine whether the following series converge.*

42. $\displaystyle\sum_{k=1}^{\infty} \frac{k}{k^{5/2} + 1}$

43. $\displaystyle\sum_{k=1}^{\infty} k^{-2/3}$

44. $\displaystyle\sum_{j=1}^{\infty} \frac{2j^4 + j^3 + 1}{3j^6 + 4}$

45. $\displaystyle\sum_{k=1}^{\infty} \frac{2k^4 + k^3 + 1}{3k^4 + 4}$

46. $\displaystyle\sum_{k=1}^{\infty} \frac{k^{k/2} 2^k}{\sqrt{k!}}$

47. $\displaystyle\sum_{k=1}^{\infty} \frac{7 + \sin k}{k^2}$

48. $\displaystyle\sum_{k=1}^{\infty} \frac{2k^4 + 1}{\sqrt{9k^8 + 2}}$

49. $\displaystyle\sum_{k=1}^{\infty} \frac{k^4}{\sqrt{9k^{12} + 2}}$

50. $\displaystyle\sum_{k=1}^{\infty} \sqrt{\frac{k!}{k^k}}$

51. $\displaystyle\sum_{k=1}^{\infty} \frac{2^k}{e^k}$

52. $\displaystyle\sum_{k=1}^{\infty} \left(\frac{k}{k + 3}\right)^{2k}$

53. $\displaystyle\sum_{k=1}^{\infty} \left(\frac{k}{2k + 3}\right)^{k}$

54. $\displaystyle\sum_{k=1}^{\infty} \sin \frac{1}{k^4}$

55. $\displaystyle\sum_{k=1}^{\infty} \frac{k!}{e^k k^k}$

56. $\displaystyle\sum_{k=1}^{\infty} \frac{1}{\sqrt{k}\sqrt{k + 1}}$

57. $\displaystyle\sum_{k=1}^{\infty} \frac{5^k}{2^{2k+1}}$

58. $\displaystyle\sum_{k=1}^{\infty} ke^{-k}$

59. $\displaystyle\sum_{j=0}^{\infty} \frac{2 \cdot 4^j}{(2j + 1)!}$

60. $\displaystyle\sum_{k=0}^{\infty} \frac{9^k}{(2k)!}$

61. $\displaystyle\sum_{k=3}^{\infty} \frac{\ln k}{k^{3/2}}$

62. $\displaystyle\sum_{k=1}^{\infty} \frac{k^2}{(2k - 1)!}$

63. $\displaystyle\sum_{k=1}^{\infty} \frac{3}{2 + e^k}$

64. $\displaystyle\sum_{k=1}^{\infty} k \sin \frac{1}{k}$

65. $\displaystyle\sum_{k=1}^{\infty} \frac{\sqrt[k]{k}}{k^3}$

66. $\displaystyle\sum_{k=1}^{\infty} \frac{1}{1 + \ln k}$

67. $\displaystyle\sum_{k=1}^{\infty} k^5 e^{-k}$

68. $\displaystyle\sum_{k=4}^{\infty} \frac{2}{k^2 - 10}$

69. $\displaystyle\sum_{k=1}^{\infty} \left(1 - \cos \frac{1}{k}\right)$

70. $\displaystyle\sum_{k=1}^{\infty} \left(\frac{\pi}{2} - \tan^{-1} k\right)$

71. $\displaystyle\sum_{k=1}^{\infty} \left(1 - \cos \frac{1}{k}\right)^{2}$

72. $\displaystyle\sum_{k=1}^{\infty} \left(\frac{\pi}{2} - \tan^{-1} k\right)^{2}$

73. $\displaystyle\sum_{k=1}^{\infty} \frac{\coth k}{k}$

74. $\displaystyle\sum_{k=1}^{\infty} \frac{1}{\sinh k}$

75. $\displaystyle\sum_{k=1}^{\infty} \tanh k$

76. $\displaystyle\sum_{k=0}^{\infty} \text{sech } k$

77–87. Absolute or conditional convergence *Determine whether the following series converge absolutely, converge conditionally, or diverge.*

77. $\displaystyle\sum_{k=1}^{\infty} \frac{(-1)^{k+1}}{k^{3/7}}$

78. $\displaystyle\sum_{k=1}^{\infty} \frac{(-1)^{k}}{k^{5/2}}$

79. $\displaystyle\sum_{k=2}^{\infty} \frac{(-1)^{k}}{k^2 - 1}$

80. $\displaystyle\sum_{j=1}^{\infty} \frac{(-1)^{j}}{\ln(j + 3)}$

81. $\displaystyle\sum_{k=1}^{\infty} \frac{(-1)^{k+1}(k^2 + 4)}{2k^2 + 1}$

82. $\displaystyle\sum_{k=1}^{\infty} \frac{(-1)^{k}}{\sqrt{k^2 + 1}}$

83. $\displaystyle\sum_{k=1}^{\infty} (-1)^k k e^{-k}$

84. $\displaystyle\sum_{k=0}^{\infty} \frac{(-1)^{k}}{e^k + e^{-k}}$

85. $\displaystyle\sum_{k=1}^{\infty} \frac{(-1)^{k+1} 10^k}{k!}$

86. $\displaystyle\sum_{k=2}^{\infty} \frac{(-1)^{k}}{k \ln k}$

87. $\displaystyle\sum_{k=1}^{\infty} \frac{(-2)^{k+1}}{k^2}$

88. Logarithmic *p*-series Show that the series $\displaystyle\sum_{k=2}^{\infty} \frac{1}{k(\ln k)^p}$ converges provided $p > 1$.

T 89–90. Lower and upper bounds of a series *For each convergent series and given value of n, complete the following.*

a. *Use the nth partial sum S_n to estimate the value of the series.*
b. *Find an upper bound for the remainder R_n.*
c. *Find lower and upper bounds (L_n and U_n, respectively) for the exact value of the series.*

89. $\displaystyle\sum_{k=1}^{\infty} \frac{1}{k^5}; n = 5$

90. $\displaystyle\sum_{k=1}^{\infty} \frac{1}{k^2 + 1}; n = 20$

91. Estimate the value of the series $\displaystyle\sum_{k=1}^{\infty} \frac{1}{(2k + 5)^3}$ to within 10^{-4} of its exact value.

92. Estimate the value of the series $\displaystyle\sum_{k=1}^{\infty} \frac{(-1)^{k+1}}{(2k + 5)^3}$ using the fifth partial sum S_5 and then find an upper bound for the error $|S - S_5|$.

T 93. Error in a finite alternating sum How many terms of the series $\displaystyle\sum_{k=1}^{\infty} \frac{(-1)^{k+1}}{k^4}$ must be summed to ensure that the approximation error is less than 10^{-8}?

94. Equations involving series Solve the following equations for *x*.

a. $\displaystyle\sum_{k=0}^{\infty} e^{kx} = 2$

b. $\displaystyle\sum_{k=0}^{\infty} (3x)^k = 4$

95. Building a tunnel—first scenario A crew of workers is constructing a tunnel through a mountain. Understandably, the rate of construction decreases because rocks and earth must be removed a greater distance as the tunnel gets longer. Suppose each week the crew digs 0.95 of the distance it dug the previous week. In the first week, the crew constructed 100 m of tunnel.

a. How far does the crew dig in 10 weeks? 20 weeks? *N* weeks?
b. What is the longest tunnel the crew can build at this rate?

96. Building a tunnel—second scenario As in Exercise 95, a crew of workers is constructing a tunnel. The time required to dig 100 m increases by 10% each week, starting with 1 week to dig the first 100 m. Can the crew complete a 1.5-km (1500-m) tunnel in 30 weeks? Explain.

97. Pages of circles On page 1 of a book, there is one circle of radius 1. On page 2, there are two circles of radius $1/2$. On page *n*, there are 2^{n-1} circles of radius 2^{-n+1}.

a. What is the sum of the areas of the circles on page *n* of the book?
b. Assuming the book continues indefinitely ($n \to \infty$), what is the sum of the areas of all the circles in the book?

T 98. Max sine sequence Let $a_n = \max\{\sin 1, \sin 2, \ldots, \sin n\}$, for $n = 1, 2, 3, \ldots$, where $\max\{\ldots\}$ denotes the maximum element of the set. Does $\{a_n\}$ converge? If so, make a conjecture about the limit.

99. Sierpinski triangle The fractal called the *Sierpinski triangle* is the limit of a sequence of figures. Starting with the equilateral triangle with sides of length 1, an inverted equilateral triangle with sides of length $1/2$ is removed. Then, three inverted equilateral triangles with sides of length $1/4$ are removed from this figure (see figure). The process continues in this way. Let T_n be the total area of the removed triangles after stage n of the process. The area of an equilateral triangle with side length L is $A = \dfrac{\sqrt{3}L^2}{4}$.

  a. Find T_1 and T_2, the total area of the removed triangles after stages 1 and 2, respectively.

  b. Find T_n, for $n = 1, 2, 3, \ldots$.

  c. Find $\lim\limits_{n \to \infty} T_n$.

  d. What is the area of the original triangle that remains as $n \to \infty$?

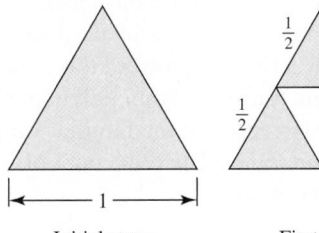 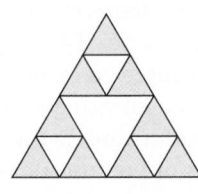

    Initial stage   First stage   Second stage

100. Multiplier effect Imagine that the government of a small community decides to give a total of $\$W$, distributed equally, to all its citizens. Suppose each month each citizen saves a fraction p of his or her new wealth and spends the remaining $1 - p$ in the community. Assume no money leaves or enters the community, and all the spent money is redistributed throughout the community.

  a. If this cycle of saving and spending continues for many months, how much money is ultimately spent? Specifically, by what factor is the initial investment of $\$W$ increased? Economists refer to this increase in the investment as the *multiplier effect*.

  b. Evaluate the limits $p \to 0$ and $p \to 1$, and interpret their meanings. (See the Guided Project *Economic stimulus packages* for more on stimulus packages.)

101. Bouncing ball for time Suppose a rubber ball, when dropped from a given height, returns to a fraction p of that height. In the absence of air resistance, a ball dropped from a height h requires $\sqrt{\dfrac{2h}{g}}$ seconds to fall to the ground, where $g \approx 9.8 \text{ m/s}^2$ is the acceleration due to gravity. The time taken to bounce *up* to a given height equals the time to fall from that height to the ground. How long does it take for a ball dropped from 10 m to come to rest?

Chapter 10 Guided Projects

Applications of the material in this chapter and related topics can be found in the following Guided Projects. For additional information, see the Preface.

- Chaos!
- Financial matters
- Periodic drug dosing
- Economic stimulus packages
- The mathematics of loans
- Archimedes' approximation to π
- Exact values of infinite series
- Conditional convergence in a crystal lattice

11

Power Series

Chapter Preview   Until now, you have worked with infinite series consisting of real numbers. In this chapter, we make a seemingly small, but significant, change by considering infinite series whose terms include powers of a variable. With this change, an infinite series becomes a *power series*. One of the most fundamental ideas in all of calculus is that functions can be represented by power series. As a first step toward this result, we look at approximating functions using polynomials. The transition from polynomials to power series is then straightforward, and we learn how to represent the familiar functions of mathematics in terms of power series called *Taylor series*. The remainder of the chapter is devoted to the properties and many uses of Taylor series.

11.1 Approximating Functions with Polynomials

Power series provide a way to represent familiar functions and to define new functions. For this reason, power series—like sets and functions—are among the most fundamental entities in mathematics.

What Is a Power Series?

A *power series* is an infinite series of the form

$$\sum_{k=0}^{\infty} c_k x^k = \underbrace{c_0 + c_1 x + c_2 x^2 + \cdots + c_n x^n}_{n\text{th-degree polynomial}} + \underbrace{c_{n+1} x^{n+1} + \cdots}_{\text{terms continue}},$$

or, more generally,

$$\sum_{k=0}^{\infty} c_k (x - a)^k = \underbrace{c_0 + c_1(x - a) + \cdots + c_n(x - a)^n}_{n\text{th-degree polynomial}} + \underbrace{c_{n+1}(x - a)^{n+1} + \cdots}_{\text{terms continue}},$$

where the *center* of the series a and the coefficients c_k are constants. This type of series is called a power series because it consists of powers of x or $(x - a)$.

Viewed another way, a power series is built up from polynomials of increasing degree, as shown in the following progression.

$$\left.\begin{array}{l} \text{Degree 0: } c_0 \\[4pt] \text{Degree 1: } c_0 + c_1 x \\[4pt] \text{Degree 2: } c_0 + c_1 x + c_2 x^2 \\[4pt] \quad \vdots \qquad \vdots \qquad \vdots \\[4pt] \text{Degree } n: c_0 + c_1 x + c_2 x^2 + \cdots + c_n x^n = \displaystyle\sum_{k=0}^{n} c_k x^k \\[4pt] \quad \vdots \qquad \vdots \qquad \vdots \end{array}\right\} \text{Polynomials}$$

$$\left. c_0 + c_1 x + c_2 x^2 + \cdots + c_n x^n + \cdots = \displaystyle\sum_{k=0}^{\infty} c_k x^k \right\} \text{Power series}$$

According to this perspective, a power series is a "super-polynomial." Therefore, we begin our exploration of power series by using polynomials to approximate functions.

Polynomial Approximation

An important observation motivates our work. To evaluate a polynomial $\left(\text{such as } f(x) = x^8 - 4x^5 + \frac{1}{2}\right)$, all we need is arithmetic—addition, subtraction, multiplication, and division. However, algebraic functions $\left(\text{say, } f(x) = \sqrt[3]{x^4 - 1}\right)$ and the trigonometric, logarithmic, and exponential functions usually cannot be evaluated exactly using arithmetic. Therefore, it makes practical sense to use the simplest of functions, polynomials, to approximate more complicated functions.

Linear and Quadratic Approximation

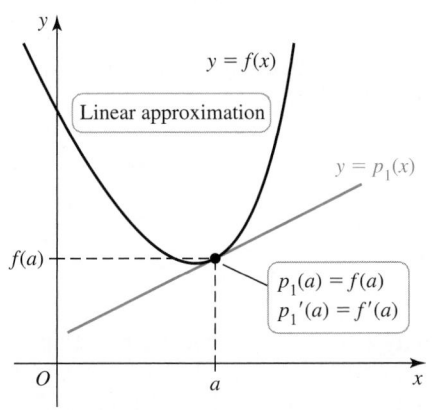

Figure 11.1

In Section 4.6, you learned that if a function f is differentiable at a point a, then it can be approximated near a by its tangent line, which is the linear approximation to f at the point a. The linear approximation at a is given by

$$y - f(a) = f'(a)(x - a) \quad \text{or} \quad y = f(a) + f'(a)(x - a).$$

Because the linear approximation is a first-degree polynomial, we name it p_1:

$$p_1(x) = f(a) + f'(a)(x - a).$$

This polynomial has some important properties: It matches f in *value* and in *slope* at a. In other words (**Figure 11.1**),

$$p_1(a) = f(a) \quad \text{and} \quad p_1'(a) = f'(a).$$

Linear approximation works well if f has a fairly constant slope near a. However, if f has a lot of curvature near a, then the tangent line may not provide an accurate approximation. To remedy this situation, we create a quadratic approximating polynomial by adding one new term to the linear polynomial. Denoting this new polynomial p_2, we let

$$p_2(x) = \underbrace{f(a) + f'(a)(x - a)}_{p_1(x)} + \underbrace{c_2(x - a)^2}_{\text{quadratic term}}.$$

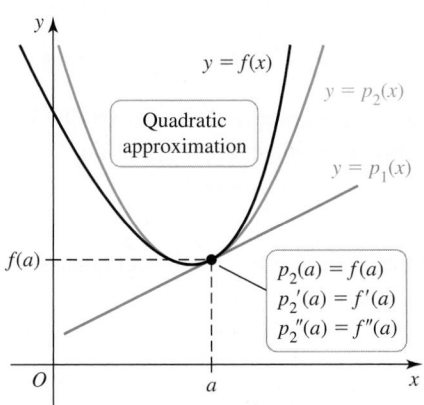

Figure 11.2

▶ Matching concavity (second derivatives) ensures that the graph of p_2 bends in the same direction as the graph of f at a.

The new term consists of a coefficient c_2 that must be determined and a quadratic factor $(x - a)^2$.

To determine c_2 and to ensure that p_2 is a good approximation to f near the point a, we require that p_2 agree with f in value, slope, and concavity at a; that is, p_2 must satisfy the matching conditions

$$p_2(a) = f(a), \quad p_2'(a) = f'(a), \quad \text{and} \quad p_2''(a) = f''(a),$$

where we assume f and its first and second derivatives exist at a (**Figure 11.2**). Substituting $x = a$ into p_2, we see immediately that $p_2(a) = f(a)$, so the first matching condition is met. Differentiating p_2 once, we have

$$p_2'(x) = f'(a) + 2c_2(x - a).$$

So $p_2'(a) = f'(a)$, and the second matching condition is also met. Because $p_2''(a) = 2c_2$, the third matching condition is

$$p_2''(a) = 2c_2 = f''(a).$$

It follows that $c_2 = \frac{1}{2}f''(a)$; therefore, the quadratic approximating polynomial is

$$p_2(x) = \underbrace{f(a) + f'(a)(x - a)}_{p_1(x)} + \frac{f''(a)}{2}(x - a)^2.$$

EXAMPLE 1   **Linear and quadratic approximations for ln x**

a. Find the linear approximation to $f(x) = \ln x$ at $x = 1$.

b. Find the quadratic approximation to $f(x) = \ln x$ at $x = 1$.

c. Use these approximations to estimate ln 1.05.

SOLUTION

a. Note that $f(1) = 0$, $f'(x) = 1/x$, and $f'(1) = 1$. Therefore, the linear approximation to $f(x) = \ln x$ at $x = 1$ is

$$p_1(x) = f(1) + f'(1)(x - 1) = 0 + 1(x - 1) = x - 1.$$

As shown in **Figure 11.3**, p_1 matches f in value ($p_1(1) = f(1)$) and in slope ($p_1'(1) = f'(1)$) at $x = 1$.

b. We first compute $f''(x) = -1/x^2$ and $f''(1) = -1$. Building on the linear approximation found in part (a), the quadratic approximation is

$$p_2(x) = \underbrace{x - 1}_{p_1(x)} + \underbrace{\frac{1}{2}f''(1)(x - 1)^2}_{c_2}$$

$$= x - 1 - \frac{1}{2}(x - 1)^2.$$

Because p_2 matches f in value, slope, and concavity at $x = 1$, it provides a better approximation to f near $x = 1$ (Figure 11.3).

c. To approximate ln 1.05, we substitute $x = 1.05$ into each polynomial approximation:

$$p_1(1.05) = 1.05 - 1 = 0.05 \text{ and} \qquad \text{Linear approximation}$$

$$p_2(1.05) = 1.05 - 1 - \frac{1}{2}(1.05 - 1)^2 = 0.04875. \qquad \text{Quadratic approximation}$$

The value of ln 1.05 given by a calculator, rounded to five decimal places, is 0.04879, showing the improvement in quadratic approximation over linear approximation.

Related Exercises 10, 13 ◄

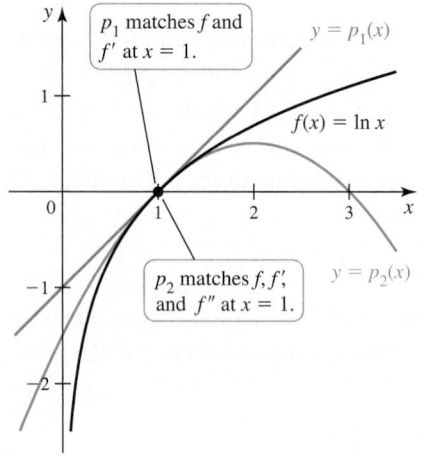

Figure 11.3

We now extend the idea of linear and quadratic approximation to obtain higher-degree polynomials that generally provide better approximations.

Taylor Polynomials

▶ Building on ideas that were already circulating in the early 18th century, Brook Taylor (1685–1731) published Taylor's Theorem in 1715. He is also credited with discovering integration by parts.

Assume f and its first n derivatives exist at a; our goal is to find an nth-degree polynomial that approximates the values of f near a. The first step is to use p_2 to obtain a cubic polynomial p_3 of the form

$$p_3(x) = p_2(x) + c_3(x - a)^3$$

that satisfies the four matching conditions

$$p_3(a) = f(a), \quad p_3'(a) = f'(a), \quad p_3''(a) = f''(a), \quad \text{and} \quad p_3'''(a) = f'''(a).$$

Because p_3 is built "on top of" p_2, the first three matching conditions are met. The last condition, $p_3'''(a) = f'''(a)$, is used to determine c_3. A short calculation shows that $p_3'''(x) = 3 \cdot 2c_3 = 3!c_3$, so the last matching condition is $p_3'''(a) = 3!c_3 = f'''(a)$. Solving for c_3, we have $c_3 = \dfrac{f'''(a)}{3!}$. Therefore, the cubic approximating polynomial is

$$p_3(x) = \underbrace{f(a) + f'(a)(x - a) + \frac{f''(a)}{2!}(x - a)^2}_{p_2(x)} + \frac{f'''(a)}{3!}(x - a)^3.$$

Continuing in this fashion (Exercise 66), building each new polynomial on the previous polynomial, the nth approximating polynomial for f at a is

$$p_n(x) = \underbrace{f(a) + f'(a)(x - a) + \frac{f''(a)}{2!}(x - a)^2 + \cdots}_{p_{n-1}(x)} + \frac{f^{(n)}(a)}{n!}(x - a)^n.$$

It satisfies the $n + 1$ matching conditions

$$p_n(a) = f(a), \quad p_n'(a) = f'(a), \quad p_n''(a) = f''(a), \ldots, \text{ and } p_n^{(n)}(a) = f^{(n)}(a).$$

These conditions ensure that the graph of p_n conforms as closely as possible to the graph of f near a (Figure 11.4).

▶ Recall that $2! = 2 \cdot 1$, $3! = 3 \cdot 2 \cdot 1$, $k! = k \cdot (k - 1)!$, and by definition, $0! = 1$.

QUICK CHECK 1 Verify that p_3 satisfies $p_3^{(k)}(a) = f^{(k)}(a)$, for $k = 0, 1, 2,$ and 3. ◀

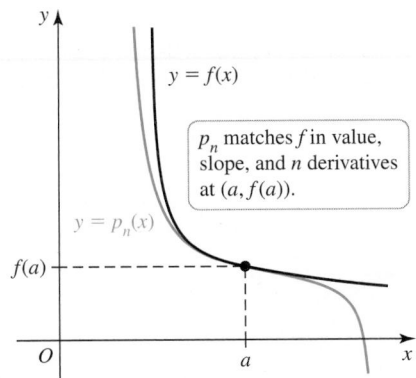

Figure 11.4

▶ Recall that $f^{(n)}$ denotes the nth derivative of f. By convention, the zeroth derivative $f^{(0)}$ is f itself.

> **DEFINITION Taylor Polynomials**
>
> Let f be a function with f', f'', ..., and $f^{(n)}$ defined at a. The **nth-order Taylor polynomial** for f with its **center** at a, denoted p_n, has the property that it matches f in value, slope, and all derivatives up to the nth derivative at a; that is,
>
> $$p_n(a) = f(a), p_n'(a) = f'(a), \ldots, \text{ and } p_n^{(n)}(a) = f^{(n)}(a).$$
>
> The nth-order Taylor polynomial centered at a is
>
> $$p_n(x) = f(a) + f'(a)(x - a) + \frac{f''(a)}{2!}(x - a)^2 + \cdots + \frac{f^{(n)}(a)}{n!}(x - a)^n.$$
>
> More compactly, $p_n(x) = \displaystyle\sum_{k=0}^{n} c_k(x - a)^k$, where the **coefficients** are
>
> $$c_k = \frac{f^{(k)}(a)}{k!}, \quad \text{for } k = 0, 1, 2, \ldots, n.$$

EXAMPLE 2 Taylor polynomials for $\sin x$ Find the Taylor polynomials $p_1, \ldots, p_7$ centered at $x = 0$ for $f(x) = \sin x$.

SOLUTION We begin by differentiating f repeatedly and evaluating the derivatives at 0; these calculations enable us to compute c_k, for $k = 0, 1, \ldots, 7$. Notice that a pattern emerges:

$$\begin{aligned} f(x) &= \sin x \Rightarrow f(0) = 0 \\ f'(x) &= \cos x \Rightarrow f'(0) = 1 \\ f''(x) &= -\sin x \Rightarrow f''(0) = 0 \\ f'''(x) &= -\cos x \Rightarrow f'''(0) = -1 \\ f^{(4)}(x) &= \sin x \Rightarrow f^{(4)}(0) = 0. \end{aligned}$$

The derivatives of $\sin x$ at 0 cycle through the values $\{0, 1, 0, -1\}$. Therefore, $f^{(5)}(0) = 1$, $f^{(6)}(0) = 0$, and $f^{(7)}(0) = -1$.

We now construct the Taylor polynomials that approximate $f(x) = \sin x$ near 0, beginning with the linear polynomial. The polynomial of order $n = 1$ is

$$p_1(x) = f(0) + f'(0)(x - 0) = x,$$

whose graph is the line through the origin with slope 1 (Figure 11.5). Notice that f and p_1 agree in value ($f(0) = p_1(0) = 0$) and in slope ($f'(0) = p_1'(0) = 1$) at 0. We see that p_1 provides a good fit to f near 0, but the graphs diverge visibly for $|x| > \pi/4$.

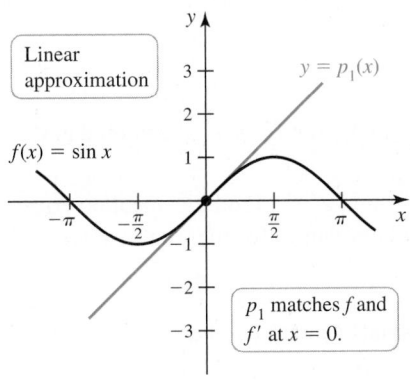

Figure 11.5

▶ It is worth repeating that the next polynomial in the sequence is obtained by adding one new term to the previous polynomial. For example,

$$p_3(x) = p_2(x) + \frac{f'''(a)}{3!}(x-a)^3.$$

The polynomial of order $n = 2$ is

$$p_2(x) = \underbrace{f(0)}_{0} + \underbrace{f'(0)}_{1}x + \underbrace{\frac{f''(0)}{2!}}_{0}x^2 = x,$$

so p_2 is the same as p_1.

The polynomial of order $n = 3$ is

$$p_3(x) = \underbrace{f(0) + f'(0)x + \frac{f''(0)}{2!}x^2}_{p_2(x)\,=\,x} + \underbrace{\frac{f'''(0)}{3!}}_{-1/3!}x^3 = x - \frac{x^3}{6}.$$

QUICK CHECK 2 Verify the following properties for $f(x) = \sin x$ and $p_3(x) = x - x^3/6$:

$$f(0) = p_3(0),$$
$$f'(0) = p_3{}'(0),$$
$$f''(0) = p_3{}''(0), \text{ and}$$
$$f'''(0) = p_3{}'''(0).$$ ◀

We have designed p_3 to agree with f in value, slope, concavity, and third derivative at 0 (Figure 11.6). Consequently, p_3 provides a better approximation to f over a larger interval than p_1.

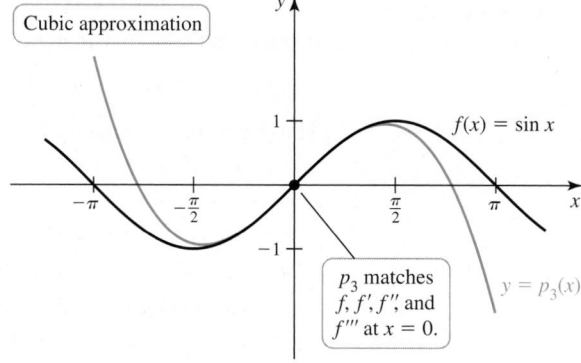

Figure 11.6

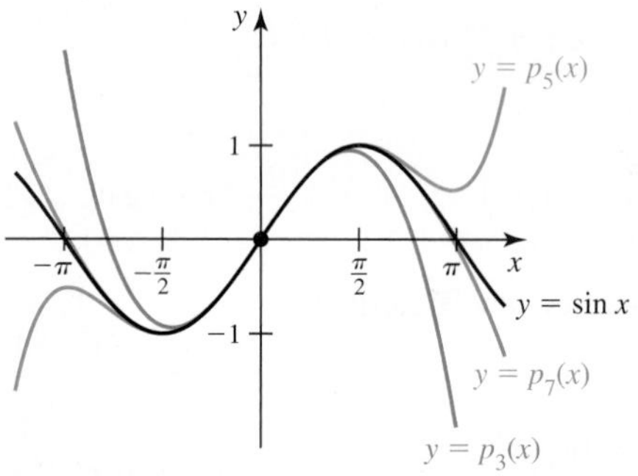

The procedure for finding Taylor polynomials may be extended to polynomials of any order. Because the even derivatives of $f(x) = \sin x$ are zero at $x = 0$, $p_4(x) = p_3(x)$. For the same reason, $p_6(x) = p_5(x)$:

$$p_6(x) = p_5(x) = x - \frac{x^3}{3!} + \frac{x^5}{5!}, \qquad c_5 = \frac{f^{(5)}(0)}{5!} = \frac{1}{5!}$$

Finally, the Taylor polynomial of order $n = 7$ is

$$p_7(x) = x - \frac{x^3}{3!} + \frac{x^5}{5!} - \frac{x^7}{7!}, \qquad c_7 = \frac{f^{(7)}(0)}{7!} = -\frac{1}{7!}$$

Figure 11.7

QUICK CHECK 3 Why do the Taylor polynomials for $\sin x$ centered at 0 consist only of odd powers of x? ◀

From Figure 11.7 we see that as the order of the Taylor polynomials increases, more accurate approximations to $f(x) = \sin x$ are obtained over larger intervals centered at 0. For example, p_7 is a good fit to $f(x) = \sin x$ over the interval $[-\pi, \pi]$. Notice that $\sin x$ and its Taylor polynomials (centered at 0) are all odd functions.   *Related Exercises 17, 20* ◀

Approximations with Taylor Polynomials

Taylor polynomials find widespread use in approximating functions, as illustrated in the following examples.

EXAMPLE 3   Taylor polynomials for e^x

a. Find the Taylor polynomials of order $n = 0, 1, 2,$ and 3 for $f(x) = e^x$ centered at 0. Graph f and the polynomials.

▶ Recall that if c is an approximation to x, the absolute error in c is $|x - c|$ and the relative error in c is $|x - c|/|x|$. We use *error* to refer to *absolute error*.

b. Use the polynomials in part (a) to approximate $e^{0.1}$ and $e^{-0.25}$. Find the absolute errors, $|f(x) - p_n(x)|$, in the approximations. Use calculator values for the exact values of f.

SOLUTION

a. Recall that the coefficients for the Taylor polynomials centered at 0 are

$$c_k = \frac{f^{(k)}(0)}{k!}, \qquad \text{for } k = 0, 1, 2, \ldots, n.$$

With $f(x) = e^x$, we have $f^{(k)}(x) = e^x$, $f^{(k)}(0) = 1$, and $c_k = 1/k!$, for $k = 0, 1, 2, 3, \ldots$. The first four polynomials are

$$p_0(x) = f(0) = 1,$$
$$p_1(x) = \underbrace{f(0)}_{p_0(x)\,=\,1} + \underbrace{f'(0)x}_{1} = 1 + x,$$

$$p_2(x) = \underbrace{f(0) + f'(0)x}_{p_1(x)\,=\,1\,+\,x} + \underbrace{\frac{f''(0)}{2!}}_{1/2} x^2 = 1 + x + \frac{x^2}{2}, \text{ and}$$

$$p_3(x) = \underbrace{f(0) + f'(0)x + \frac{f''(0)}{2!}x^2}_{p_2(x)\,=\,1\,+\,x\,+\,x^2/2} + \underbrace{\frac{f^{(3)}(0)}{3!}}_{1/6} x^3 = 1 + x + \frac{x^2}{2} + \frac{x^3}{6}.$$

Notice that each successive polynomial provides a better fit to $f(x) = e^x$ near 0 (**Figure 11.8**). Continuing the pattern in these polynomials, the nth-order Taylor polynomial for e^x centered at 0 is

$$p_n(x) = 1 + x + \frac{x^2}{2!} + \frac{x^3}{3!} + \cdots + \frac{x^n}{n!} = \sum_{k=0}^{n} \frac{x^k}{k!}.$$

b. We evaluate $p_n(0.1)$ and $p_n(-0.25)$, for $n = 0, 1, 2$, and 3, and compare these values to the calculator values of $e^{0.1} \approx 1.1051709$ and $e^{-0.25} \approx 0.77880078$. The results are shown in Table 11.1. Observe that the errors in the approximations decrease as n increases. In addition, the errors in approximating $e^{0.1}$ are smaller in magnitude than the errors in approximating $e^{-0.25}$ because $x = 0.1$ is closer to the center of the polynomials than $x = -0.25$. Reasonable approximations based on these calculations are $e^{0.1} \approx 1.105$ and $e^{-0.25} \approx 0.78$.

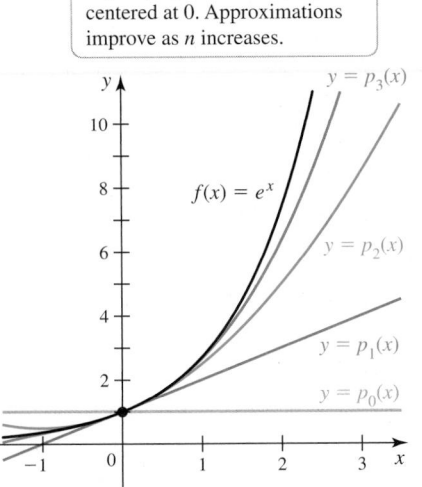

Taylor polynomials for $f(x) = e^x$ centered at 0. Approximations improve as n increases.

$y = p_3(x)$

$f(x) = e^x$

$y = p_2(x)$

$y = p_1(x)$

$y = p_0(x)$

Figure 11.8

▶ A rule of thumb in finding estimates based on several approximations: Keep all the digits that are common to the last two approximations after rounding.

QUICK CHECK 4 Write out the next two Taylor polynomials p_4 and p_5 for $f(x) = e^x$ in Example 3. ◀

Table 11.1

n	Approximation $p_n(0.1)$	Absolute Error $\lvert e^{0.1} - p_n(0.1) \rvert$	Approximation $p_n(-0.25)$	Absolute Error $\lvert e^{-0.25} - p_n(-0.25) \rvert$
0	1	1.1×10^{-1}	1	2.2×10^{-1}
1	1.1	5.2×10^{-3}	0.75	2.9×10^{-2}
2	1.105	1.7×10^{-4}	0.78125	2.4×10^{-3}
3	1.105167	4.3×10^{-6}	0.778646	1.5×10^{-4}

Related Exercise 31 ◀

EXAMPLE 4   Approximating a real number using Taylor polynomials Use polynomials of order $n = 0, 1, 2$, and 3 to approximate $\sqrt{18}$.

SOLUTION Letting $f(x) = \sqrt{x}$, we choose the center $a = 16$ because it is near 18, and f and its derivatives are easy to evaluate at 16. The Taylor polynomials have the form

$$p_n(x) = f(16) + f'(16)(x - 16) + \frac{f''(16)}{2!}(x - 16)^2 + \cdots + \frac{f^{(n)}(16)}{n!}(x - 16)^n.$$

We now evaluate the required derivatives:

$$f(x) = \sqrt{x} \Rightarrow f(16) = 4,$$
$$f'(x) = \frac{1}{2}x^{-1/2} \Rightarrow f'(16) = \frac{1}{8},$$
$$f''(x) = -\frac{1}{4}x^{-3/2} \Rightarrow f''(16) = -\frac{1}{256}, \text{ and}$$
$$f'''(x) = \frac{3}{8}x^{-5/2} \Rightarrow f'''(16) = \frac{3}{8192}.$$

Therefore, the polynomial p_3 (which includes p_0, p_1, and p_2) is

$$p_3(x) = \underbrace{\underbrace{\underbrace{4 + \frac{1}{8}(x - 16)}_{p_0(x)} - \frac{1}{512}(x - 16)^2}_{p_1(x)} + \frac{1}{16{,}384}(x - 16)^3}_{p_2(x)}.$$

The Taylor polynomials (Figure 11.9) give better approximations to f as the order of the approximation increases.

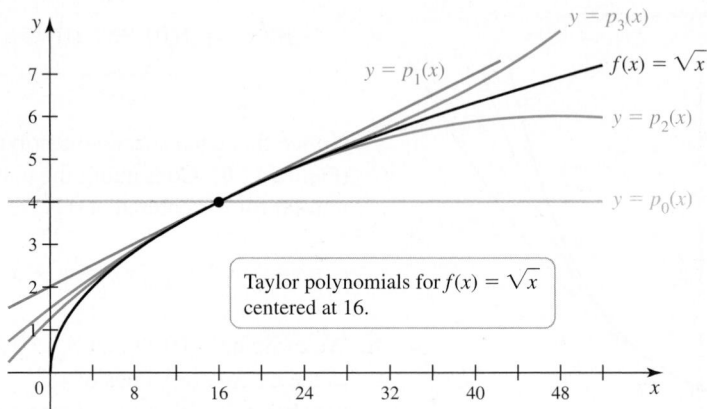

Figure 11.9

Letting $x = 18$, we obtain the approximations to $\sqrt{18}$ and the associated absolute errors shown in Table 11.2. (A calculator is used for the value of $\sqrt{18}$.) As expected, the errors decrease as n increases. Based on these calculations, a reasonable approximation is $\sqrt{18} \approx 4.24$.

Table 11.2

n	Approximation $p_n(18)$	Absolute Error $\lvert\sqrt{18} - p_n(18)\rvert$
0	4	2.4×10^{-1}
1	4.25	7.4×10^{-3}
2	4.242188	4.5×10^{-4}
3	4.242676	3.5×10^{-5}

Related Exercises 37–38 ◄

QUICK CHECK 5 At what point would you center the Taylor polynomials for $\sqrt{x}$ and $\sqrt[4]{x}$ to approximate $\sqrt{51}$ and $\sqrt[4]{15}$, respectively? ◄

Remainder in a Taylor Polynomial

Taylor polynomials provide good approximations to functions near a specific point. But how accurate are the approximations? To answer this question we define the *remainder* in a Taylor polynomial. If p_n is the Taylor polynomial for f of order n, then the remainder at the point x is

$$R_n(x) = f(x) - p_n(x).$$

The absolute value of the remainder is the error made in approximating $f(x)$ by $p_n(x)$. Equivalently, we have $f(x) = p_n(x) + R_n(x)$, which says that f consists of two components: the polynomial approximation and the associated remainder.

DEFINITION Remainder in a Taylor Polynomial

Let p_n be the Taylor polynomial of order n for f. The **remainder** in using p_n to approximate f at the point x is

$$R_n(x) = f(x) - p_n(x).$$

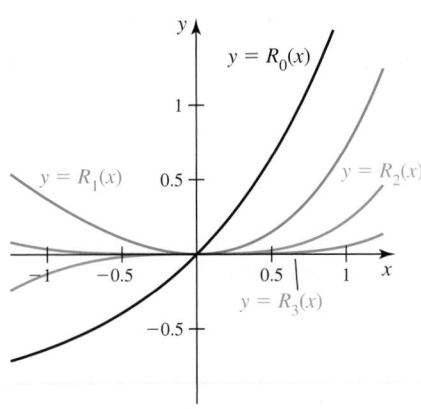

Remainders increase in magnitude as $|x|$ increases. Remainders decrease in magnitude to zero as n increases.

Figure 11.10

▶ The remainder R_n for a Taylor polynomial can be expressed in several different forms. The form stated in Theorem 11.1 is called the *Lagrange form* of the remainder.

The idea of a remainder is illustrated in **Figure 11.10**, where we see the remainders associated with various Taylor polynomials for $f(x) = e^x$ centered at 0 (Example 3). For fixed order n, the remainders tend to increase in magnitude as x moves farther from the center of the polynomials (in this case 0). And for fixed x, remainders decrease in magnitude to zero with increasing n.

The remainder for a Taylor polynomial may be written quite concisely, which enables us to estimate remainders. The following result is known as *Taylor's Theorem* (or the *Remainder Theorem*).

THEOREM 11.1   Taylor's Theorem (Remainder Theorem)

Let f have continuous derivatives up to $f^{(n+1)}$ on an open interval I containing a. For all x in I,

$$f(x) = p_n(x) + R_n(x),$$

where p_n is the nth-order Taylor polynomial for f centered at a and the remainder is

$$R_n(x) = \frac{f^{(n+1)}(c)}{(n+1)!}(x-a)^{n+1},$$

for some point c between x and a.

Discussion: We make two observations about Theorem 11.1 and outline a proof in Exercise 82. First, the case $n = 0$ is the Mean Value Theorem (Section 4.2), which states that

$$\frac{f(x) - f(a)}{x - a} = f'(c),$$

where c is a point between x and a. Rearranging this expression, we have

$$f(x) = \underbrace{f(a)}_{p_0(x)} + \underbrace{f'(c)(x - a)}_{R_0(x)}$$

$$= p_0(x) + R_0(x),$$

which is Taylor's Theorem with $n = 0$. Not surprisingly, the term $f^{(n+1)}(c)$ in Taylor's Theorem comes from a Mean Value Theorem argument.

The second observation makes the remainder easier to remember. If you write the $(n+1)$st Taylor polynomial p_{n+1}, the highest-degree term is $\dfrac{f^{(n+1)}(a)}{(n+1)!}(x-a)^{n+1}$. Replacing $f^{(n+1)}(a)$ with $f^{(n+1)}(c)$ results in the remainder for p_n.

Estimating the Remainder

The remainder has both practical and theoretical importance. We deal with practical matters now and theoretical matters in Section 11.3. The remainder is used to estimate errors in approximations and to determine the number of terms of a Taylor polynomial needed to achieve a prescribed accuracy.

Because c is generally unknown, the difficulty in estimating the remainder is finding a bound for $|f^{(n+1)}(c)|$. Assuming this can be done, the following theorem gives a standard estimate for the remainder term.

THEOREM 11.2   Estimate of the Remainder

Let n be a fixed positive integer. Suppose there exists a number M such that $|f^{(n+1)}(c)| \le M$, for all c between a and x inclusive. The remainder in the nth-order Taylor polynomial for f centered at a satisfies

$$|R_n(x)| = |f(x) - p_n(x)| \le M\frac{|x-a|^{n+1}}{(n+1)!}.$$

Proof: The proof requires taking the absolute value of the remainder in Theorem 11.1, replacing $\left| f^{(n+1)}(c) \right|$ with a larger quantity M, and forming an inequality.   ◄

We now give three examples that demonstrate how an upper bound for the remainder is computed and used in different ways.

EXAMPLE 5   **Estimating the remainder for cos x**   Find a bound for the magnitude of the remainder for the Taylor polynomials of $f(x) = \cos x$ centered at 0.

SOLUTION According to Theorem 11.1 with $a = 0$, we have

$$R_n(x) = \frac{f^{(n+1)}(c)}{(n+1)!} x^{n+1},$$

where c is a point between 0 and x. Notice that $f^{(n+1)}(c) = \pm \sin c$, or $f^{(n+1)}(c) = \pm \cos c$, depending on the value of n. In all cases, $\left| f^{(n+1)}(c) \right| \leq 1$. Therefore, we take $M = 1$ in Theorem 11.2, and the absolute value of the remainder can be bounded as

$$\left| R_n(x) \right| = \left| \frac{f^{(n+1)}(c)}{(n+1)!} x^{n+1} \right| \leq \frac{|x|^{n+1}}{(n+1)!}.$$

For example, if we approximate cos 0.1 using the Taylor polynomial p_{10}, the remainder satisfies

$$\left| R_{10}(0.1) \right| \leq \frac{0.1^{11}}{11!} \approx 2.5 \times 10^{-19}.$$

Related Exercises 41–42 ◄

EXAMPLE 6   **Estimating a remainder**   Consider again Example 4, in which we approximated $\sqrt{18}$ using the Taylor polynomial

$$p_3(x) = 4 + \frac{1}{8}(x-16) - \frac{1}{512}(x-16)^2 + \frac{1}{16,384}(x-16)^3.$$

Find an upper bound on the magnitude of the remainder when using $p_3(x)$ to approximate $\sqrt{18}$.

SOLUTION In Example 4, we computed the error in the approximation knowing the exact value of $\sqrt{18}$ (obtained with a calculator). In the more realistic case in which we do not know the exact value, Theorem 11.2 allows us to estimate remainders (or errors). Applying this theorem with $n = 3$, $a = 16$, and $x = 18$, we find that the remainder in approximating $\sqrt{18}$ by $p_3(18)$ satisfies the bound

$$\left| R_3(18) \right| \leq M \frac{(18-16)^4}{4!} = \frac{2}{3} M,$$

where M is a number that satisfies $\left| f^{(4)}(c) \right| \leq M$, for all c between 16 and 18 inclusive. In this particular problem, we find that $f^{(4)}(c) = -\frac{15}{16} c^{-7/2}$, so M must be chosen (as small as possible) such that $\left| f^{(4)}(c) \right| = \frac{15}{16} c^{-7/2} = \frac{15}{16 c^{7/2}} \leq M$, for $16 \leq c \leq 18$. You can verify that $\frac{15}{16 c^{7/2}}$ is a decreasing function of c on $[16, 18]$ and has a maximum value of approximately 5.7×10^{-5} at $c = 16$ (**Figure 11.11**). Therefore, a bound on the remainder is

$$\left| R_3(18) \right| \leq \frac{2}{3} M \approx \frac{2}{3} \cdot 5.7 \times 10^{-5} \approx 3.8 \times 10^{-5}.$$

Notice that the actual error computed in Example 4 (Table 11.2) is 3.5×10^{-5}, which is less than the bound on the remainder—as it should be.

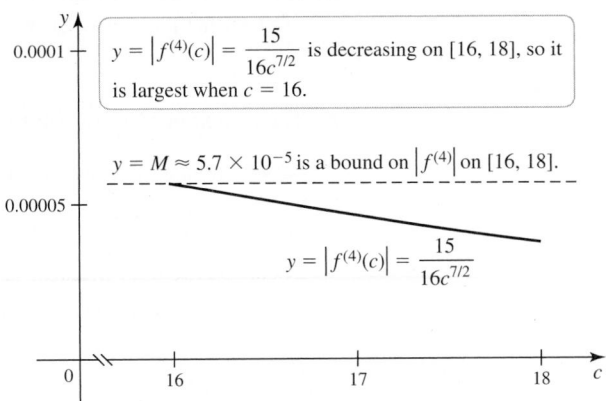

Figure 11.11

Related Exercise 52 ◄

EXAMPLE 7 Estimating the remainder for e^x Find a bound on the remainder in approximating $e^{0.45}$ using the Taylor polynomial of order $n = 6$ for $f(x) = e^x$ centered at 0.

SOLUTION Using Theorem 11.2, a bound on the remainder is given by

$$|R_n(x)| \leq M \frac{|x - a|^{n+1}}{(n + 1)!},$$

where M is chosen such that $|f^{(n+1)}(c)| \leq M$, for all c between a and x inclusive. Notice that $f(x) = e^x$ implies that $f^{(k)}(x) = e^x$, for $k = 0, 1, 2, \ldots$. In this particular problem, we have $n = 6$, $a = 0$, and $x = 0.45$, so the bound on the remainder takes the form

$$|R_6(0.45)| \leq M \frac{|0.45 - 0|^7}{7!} \approx 7.4 \times 10^{-7} M,$$

where M is chosen such that $|f^{(7)}(c)| = e^c \leq M$, for all c in the interval $[0, 0.45]$. Because e^c is an increasing function of c, its maximum value on the interval $[0, 0.45]$ occurs at $c = 0.45$ and is $e^{0.45}$. However, $e^{0.45}$ cannot be evaluated exactly (it is the number we are approximating), so we must find a number M such that $e^{0.45} \leq M$. Here is one of many ways to obtain a bound: We observe that $e^{0.45} < e^{1/2} < 4^{1/2} = 2$ and take $M = 2$ (Figure 11.12). Therefore, a bound on the remainder is

$$|R_6(0.45)| \leq 7.4 \times 10^{-7} M \approx 1.5 \times 10^{-6}.$$

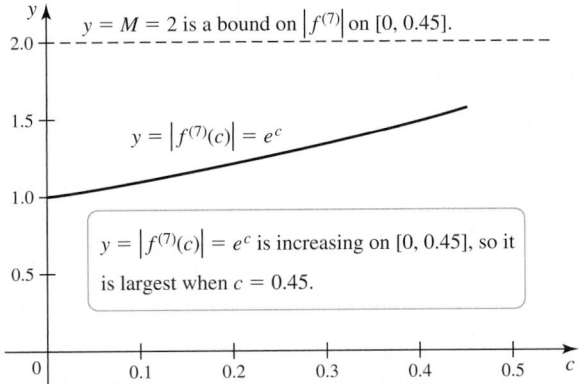

Figure 11.12

► Recall that if $f(x) = e^x$, then

$$p_n(x) = \sum_{k=0}^{n} \frac{x^k}{k!}.$$

Using the Taylor polynomial derived in Example 3 with $n = 6$, the resulting approximation to $e^{0.45}$ is

$$p_6(0.45) = \sum_{k=0}^{6} \frac{0.45^k}{k!} \approx 1.5683114;$$

it has an error that does not exceed 1.5×10^{-6}.

Related Exercise 49 ◄

QUICK CHECK 6 In Example 7, find an approximate upper bound for $R_7(0.45)$. ◄

EXAMPLE 8 Working with the remainder In Example 4b of Section 11.2, we show that the nth-order Taylor polynomial for $f(x) = \ln(1 - x)$ centered at 0 is

$$p_n(x) = -\sum_{k=1}^{n} \frac{x^k}{k} = -x - \frac{x^2}{2} - \frac{x^3}{3} - \cdots - \frac{x^n}{n}.$$

a. Find a bound on the error in approximating $\ln(1 - x)$ by $p_3(x)$ for values of x in the interval $\left[-\frac{1}{2}, \frac{1}{2}\right]$.

b. How many terms of the Taylor polynomial are needed to approximate values of $f(x) = \ln(1 - x)$ with an error less than 10^{-3} on the interval $\left[-\frac{1}{2}, \frac{1}{2}\right]$?

SOLUTION

a. The remainder for the Taylor polynomial p_3 is $R_3(x) = \dfrac{f^{(4)}(c)}{4!} x^4$, where c is between 0 and x. Computing four derivatives of f, we find that $f^{(4)}(x) = -\dfrac{6}{(1 - x)^4}$.

On the interval $\left[-\frac{1}{2}, \frac{1}{2}\right]$, the maximum magnitude of this derivative occurs at $x = \frac{1}{2}$ (because the denominator is smallest at $x = \frac{1}{2}$) and is $6/\left(\frac{1}{2}\right)^4 = 96$. Similarly, the factor x^4 has its maximum magnitude at $x = \pm\frac{1}{2}$, and it is $\left(\frac{1}{2}\right)^4 = \frac{1}{16}$. Therefore,

$$|R_3(x)| \le \frac{96}{4!}\left(\frac{1}{16}\right) = 0.25 \text{ on the interval } \left[-\frac{1}{2}, \frac{1}{2}\right].$$ The error in approximating $f(x)$ by $p_3(x)$, for $-\frac{1}{2} \le x \le \frac{1}{2}$, does not exceed 0.25.

b. For any positive integer n, the remainder is $R_n(x) = \dfrac{f^{(n+1)}(c)}{(n+1)!} x^{n+1}$. Differentiating f several times reveals that

$$f^{(n+1)}(x) = -\frac{n!}{(1 - x)^{n+1}}.$$

On the interval $\left[-\frac{1}{2}, \frac{1}{2}\right]$, the maximum magnitude of this derivative occurs at $x = \frac{1}{2}$ and is $n!/\left(\frac{1}{2}\right)^{n+1}$. Similarly, x^{n+1} has its maximum magnitude at $x = \pm\frac{1}{2}$, and it is $\left(\frac{1}{2}\right)^{n+1}$. Therefore, a bound on the magnitude of the remainder is

$$|R_n(x)| = \frac{1}{(n+1)!} \cdot \underbrace{|f^{(n+1)}(c)|}_{\le\, n!2^{n+1}} \cdot \underbrace{|x|^{n+1}}_{\le\, \left(\frac{1}{2}\right)^{n+1}}$$

$$\le \frac{1}{(n+1)!} \cdot n!2^{n+1} \cdot \frac{1}{2^{n+1}}$$

$$= \frac{1}{n+1}. \qquad\qquad \frac{n!}{(n+1)!} = \frac{1}{n+1}$$

To ensure that the error is less than 10^{-3} on the entire interval $\left[-\frac{1}{2}, \frac{1}{2}\right]$, n must satisfy $|R_n| \le \dfrac{1}{n+1} < 10^{-3}$ or $n > 999$. The error is likely to be significantly less than 10^{-3} if x is near 0.

Related Exercises 57, 62 ◄

SECTION 11.1 EXERCISES

Getting Started

1. Suppose you use a second-order Taylor polynomial centered at 0 to approximate a function f. What matching conditions are satisfied by the polynomial?

2. Does the accuracy of an approximation given by a Taylor polynomial generally increase or decrease with the order of the approximation? Explain.

3. The first three Taylor polynomials for $f(x) = \sqrt{1 + x}$ centered at 0 are $p_0 = 1$, $p_1 = 1 + \dfrac{x}{2}$, and $p_2 = 1 + \dfrac{x}{2} - \dfrac{x^2}{8}$. Find three approximations to $\sqrt{1.1}$.

4. Suppose $f(0) = 1$, $f'(0) = 2$, and $f''(0) = -1$. Find the quadratic approximating polynomial for f centered at 0 and use it to approximate $f(0.1)$.

5. Suppose $f(0) = 1$, $f'(0) = 0$, $f''(0) = 2$, and $f^{(3)}(0) = 6$. Find the third-order Taylor polynomial for f centered at 0 and use it to approximate $f(0.2)$.

6. How is the remainder $R_n(x)$ in a Taylor polynomial defined?

7. Suppose $f(2) = 1$, $f'(2) = 1$, $f''(2) = 0$, and $f^{(3)}(2) = 12$. Find the third-order Taylor polynomial for f centered at 2 and use this polynomial to estimate $f(1.9)$.

8. Suppose you want to estimate $\sqrt{26}$ using a fourth-order Taylor polynomial centered at $x = a$ for $f(x) = \sqrt{x}$. Choose an appropriate value for the center a.

Practice Exercises

T 9–16. Linear and quadratic approximation

a. *Find the linear approximating polynomial for the following functions centered at the given point a.*

b. *Find the quadratic approximating polynomial for the following functions centered at a.*

c. *Use the polynomials obtained in parts (a) and (b) to approximate the given quantity.*

9. $f(x) = 8x^{3/2}, a = 1$; approximate $8 \cdot 1.1^{3/2}$.

10. $f(x) = \dfrac{1}{x}, a = 1$; approximate $\dfrac{1}{1.05}$.

11. $f(x) = e^{-2x}, a = 0$; approximate $e^{-0.2}$.

12. $f(x) = \sqrt{x}, a = 4$; approximate $\sqrt{3.9}$.

13. $f(x) = (1 + x)^{-1}, a = 0$; approximate $\dfrac{1}{1.05}$.

14. $f(x) = \cos x, a = \dfrac{\pi}{4}$; approximate $\cos(0.24\pi)$.

15. $f(x) = x^{1/3}, a = 8$; approximate $7.5^{1/3}$.

16. $f(x) = \tan^{-1} x, a = 0$; approximate $\tan^{-1} 0.1$.

17. Find the Taylor polynomials $p_1, \ldots, p_4$ centered at $a = 0$ for $f(x) = \cos 6x$.

18. Find the Taylor polynomials $p_1, \ldots, p_5$ centered at $a = 0$ for $f(x) = e^{-x}$.

19. Find the Taylor polynomials p_3 and p_4 centered at $a = 0$ for $f(x) = (1 + x)^{-3}$.

20. Find the Taylor polynomials p_4 and p_5 centered at $a = \pi/6$ for $f(x) = \cos x$.

21. Find the Taylor polynomials p_1, p_2, and p_3 centered at $a = 1$ for $f(x) = x^3$.

22. Find the Taylor polynomials p_3 and p_4 centered at $a = 1$ for $f(x) = 8\sqrt{x}$.

23. Find the Taylor polynomial p_3 centered at $a = e$ for $f(x) = \ln x$.

24. Find the Taylor polynomial p_2 centered at $a = 8$ for $f(x) = \sqrt[3]{x}$.

T 25–28. Graphing Taylor polynomials

a. *Find the nth-order Taylor polynomials for the following functions centered at the given point a, for $n = 1$ and $n = 2$.*

b. *Graph the Taylor polynomials and the function.*

25. $f(x) = \ln(1 - x), a = 0$ **26.** $f(x) = (1 + x)^{-1/2}, a = 0$

27. $f(x) = \sin x, a = \dfrac{\pi}{4}$ **28.** $f(x) = \sqrt{x}, a = 9$

T 29–32. Approximations with Taylor polynomials

a. *Use the given Taylor polynomial p_2 to approximate the given quantity.*

b. *Compute the absolute error in the approximation, assuming the exact value is given by a calculator.*

29. Approximate $\sqrt{1.05}$ using $f(x) = \sqrt{1 + x}$ and $p_2(x) = 1 + x/2 - x^2/8$.

30. Approximate $1/\sqrt{1.08}$ using $f(x) = 1/\sqrt{1 + x}$ and $p_2(x) = 1 - x/2 + 3x^2/8$.

31. Approximate $e^{-0.15}$ using $f(x) = e^{-x}$ and $p_2(x) = 1 - x + x^2/2$.

32. Approximate $\ln 1.06$ using $f(x) = \ln(1 + x)$ and $p_2(x) = x - x^2/2$.

T 33–40. Approximations with Taylor polynomials

a. *Approximate the given quantities using Taylor polynomials with $n = 3$.*

b. *Compute the absolute error in the approximation, assuming the exact value is given by a calculator.*

33. $e^{0.12}$ **34.** $\cos(-0.2)$ **35.** $\tan(-0.1)$

36. $\ln 1.05$ **37.** $\sqrt{1.06}$ **38.** $\sqrt[4]{79}$

39. $\sinh 0.5$ **40.** $\tanh 0.5$

41–46. Remainders *Find the remainder R_n for the nth-order Taylor polynomial centered at a for the given functions. Express the result for a general value of n.*

41. $f(x) = \sin x, a = 0$ **42.** $f(x) = \cos 2x, a = 0$

43. $f(x) = e^{-x}, a = 0$ **44.** $f(x) = \cos x, a = \dfrac{\pi}{2}$

45. $f(x) = \sin x, a = \dfrac{\pi}{2}$ **46.** $f(x) = \dfrac{1}{1 - x}, a = 0$

T 47–52. Estimating errors *Use the remainder to find a bound on the error in approximating the following quantities with the nth-order Taylor polynomial centered at 0. Estimates are not unique.*

47. $\sin 0.3, n = 4$ **48.** $\cos 0.45, n = 3$

49. $e^{0.25}, n = 4$ **50.** $\tan 0.3, n = 2$

51. $e^{-0.5}, n = 4$ **52.** $\ln 1.04, n = 3$

T 53–58. Maximum error *Use the remainder term to find a bound on the error in the following approximations on the given interval. Error bounds are not unique.*

53. $\sin x \approx x - \dfrac{x^3}{6}$ on $\left[-\dfrac{\pi}{4}, \dfrac{\pi}{4}\right]$

54. $\cos x \approx 1 - \dfrac{x^2}{2}$ on $\left[-\dfrac{\pi}{4}, \dfrac{\pi}{4}\right]$

55. $e^x \approx 1 + x + \dfrac{x^2}{2}$ on $\left[-\dfrac{1}{2}, \dfrac{1}{2}\right]$

56. $\tan x \approx x$ on $\left[-\dfrac{\pi}{6}, \dfrac{\pi}{6}\right]$

57. $\ln(1 + x) \approx x - \dfrac{x^2}{2}$ on $[-0.2, 0.2]$

58. $\sqrt{1 + x} \approx 1 + \dfrac{x}{2}$ on $[-0.1, 0.1]$

59–64. Number of terms *What is the minimum order of the Taylor polynomial required to approximate the following quantities with an absolute error no greater than 10^{-3}? (The answer depends on your choice of a center.)*

59. $e^{-0.5}$ **60.** $\sin 0.2$ **61.** $\cos(-0.25)$

62. $\ln 0.85$ **63.** $\sqrt{1.06}$ **64.** $\dfrac{1}{\sqrt{0.85}}$

65. Explain why or why not Determine whether the following statements are true and give an explanation or counterexample.

a. Only even powers of x appear in the Taylor polynomials for $f(x) = e^{-2x}$ centered at 0.

b. Let $f(x) = x^5 - 1$. The Taylor polynomial for f of order 10 centered at 0 is f itself.

c. Only even powers of x appear in the nth-order Taylor polynomial for $f(x) = \sqrt{1 + x^2}$ centered at 0.

d. Suppose f'' is continuous on an interval that contains a, where f has an inflection point at a. Then the second-order Taylor polynomial for f at a is linear.

66. Taylor coefficients for $x = a$ Follow the procedure in the text to show that the nth-order Taylor polynomial that matches f and its derivatives up to order n at a has coefficients

$$c_k = \frac{f^{(k)}(a)}{k!}, \quad \text{for } k = 0, 1, 2, \ldots, n.$$

67. Matching functions with polynomials Match functions a–f with Taylor polynomials A–F (all centered at 0). Give reasons for your choices.

a. $\sqrt{1 + 2x}$ A. $p_2(x) = 1 + 2x + 2x^2$

b. $\dfrac{1}{\sqrt{1 + 2x}}$ B. $p_2(x) = 1 - 6x + 24x^2$

c. e^{2x} C. $p_2(x) = 1 + x - \dfrac{x^2}{2}$

d. $\dfrac{1}{1 + 2x}$ D. $p_2(x) = 1 - 2x + 4x^2$

e. $\dfrac{1}{(1 + 2x)^3}$ E. $p_2(x) = 1 - x + \dfrac{3}{2}x^2$

f. e^{-2x} F. $p_2(x) = 1 - 2x + 2x^2$

T 68. Dependence of errors on x Consider $f(x) = \ln(1 - x)$ and its Taylor polynomials given in Example 8.

a. Graph $y = |f(x) - p_2(x)|$ and $y = |f(x) - p_3(x)|$ on the interval $[-1/2, 1/2]$ (two curves).

b. At what points of $[-1/2, 1/2]$ is the error largest? Smallest?

c. Are these results consistent with the theoretical error bounds obtained in Example 8?

T 69–76. Small argument approximations *Consider the following common approximations when x is near zero.*

a. *Estimate $f(0.1)$ and give a bound on the error in the approximation.*

b. *Estimate $f(0.2)$ and give a bound on the error in the approximation.*

69. $f(x) = \sin x \approx x$ **70.** $f(x) = \tan x \approx x$

71. $f(x) = \cos x \approx 1 - \dfrac{x^2}{2}$ **72.** $f(x) = \tan^{-1}x \approx x$

73. $f(x) = \sqrt{1 + x} \approx 1 + \dfrac{x}{2}$ **74.** $f(x) = \ln(1 + x) \approx x - \dfrac{x^2}{2}$

75. $f(x) = e^x \approx 1 + x$ **76.** $f(x) = \sin^{-1}x \approx x$

Explorations and Challenges

T 77. Errors in approximations Suppose you approximate $f(x) = \sec x$ at the points $x = -0.2, -0.1, 0.0, 0.1, 0.2$ using the Taylor polynomials $p_2(x) = 1 + x^2/2$ and $p_4(x) = 1 + x^2/2 + 5x^4/24$. Assume the exact value of $\sec x$ is given by a calculator.

a. Complete the table showing the absolute errors in the approximations at each point. Show three significant digits.

b. In each error column, how do the errors vary with x? For what values of x are the errors largest and smallest in magnitude?

x	$|\sec x - p_2(x)|$	$|\sec x - p_4(x)|$
-0.2		
-0.1		
0.0		
0.1		
0.2		

T 78–79. Errors in approximations *Carry out the procedure described in Exercise 77 with the following functions and Taylor polynomials.*

78. $f(x) = \cos x$, $p_2(x) = 1 - \dfrac{x^2}{2}$, $p_4(x) = 1 - \dfrac{x^2}{2} + \dfrac{x^4}{24}$

79. $f(x) = e^{-x}$, $p_1(x) = 1 - x$, $p_2(x) = 1 - x + \dfrac{x^2}{2}$

T 80. Best center point Suppose you wish to approximate $\cos(\pi/12)$ using Taylor polynomials. Is the approximation more accurate if you use Taylor polynomials centered at 0 or at $\pi/6$? Use a calculator for numerical experiments and check for consistency with Theorem 11.2. Does the answer depend on the order of the polynomial?

T 81. Best center point Suppose you wish to approximate $e^{0.35}$ using Taylor polynomials. Is the approximation more accurate if you use Taylor polynomials centered at 0 or at $\ln 2$? Use a calculator for numerical experiments and check for consistency with Theorem 11.2. Does the answer depend on the order of the polynomial?

82. Proof of Taylor's Theorem There are several proofs of Taylor's Theorem, which lead to various forms of the remainder. The following proof is instructive because it leads to two different forms of the remainder and it relies on the Fundamental Theorem of Calculus, integration by parts, and the Mean Value Theorem for Integrals. Assume f has at least $n + 1$ continuous derivatives on an interval containing a.

a. Show that the Fundamental Theorem of Calculus can be written in the form

$$f(x) = f(a) + \int_a^x f'(t)\, dt.$$

b. Use integration by parts ($u = f'(t)$, $dv = dt$) to show that

$$f(x) = f(a) + (x - a)f'(a) + \int_a^x (x - t)f''(t)\, dt.$$

c. Show that n integrations by parts give

$$f(x) = f(a) + \frac{f'(a)}{1!}(x - a) + \frac{f''(a)}{2!}(x - a)^2 + \cdots$$

$$+ \frac{f^{(n)}(a)}{n!}(x - a)^n + \underbrace{\int_a^x \frac{f^{(n+1)}(t)}{n!}(x - t)^n\, dt}_{R_n(x)}.$$

d. *Challenge:* The result in part (c) has the form $f(x) = p_n(x) + R_n(x)$, where p_n is the nth-order Taylor polynomial and R_n is a new form of the remainder, known as the integral form of the remainder. Use the Mean Value

Theorem for Integrals (Section 5.4) to show that R_n can be expressed in the form

$$R_n(x) = \frac{f^{(n+1)}(c)}{(n+1)!}(x-a)^{n+1},$$

where c is between a and x.

83. Tangent line is p_1 Let f be differentiable at $x = a$.

   a. Find the equation of the line tangent to the curve $y = f(x)$ at $(a, f(a))$.

   b. Verify that the Taylor polynomial p_1 centered at a describes the tangent line found in part (a).

84. Local extreme points and inflection points Suppose f has continuous first and second derivatives at a.

   a. Show that if f has a local maximum at a, then the Taylor polynomial p_2 centered at a also has a local maximum at a.

   b. Show that if f has a local minimum at a, then the Taylor polynomial p_2 centered at a also has a local minimum at a.

   c. Is it true that if f has an inflection point at a, then the Taylor polynomial p_2 centered at a also has an inflection point at a?

   d. Are the converses in parts (a) and (b) true? If p_2 has a local extreme point at a, does f have the same type of point at a?

T 85. Approximating sin x Let $f(x) = \sin x$, and let p_n and q_n be nth-order Taylor polynomials for f centered at 0 and π, respectively.

   a. Find p_5 and q_5.

   b. Graph f, p_5, and q_5 on the interval $[-\pi, 2\pi]$. On what interval is p_5 a better approximation to f than q_5? On what interval is q_5 a better approximation to f than p_5?

   c. Complete the following table showing the errors in the approximations given by p_5 and q_5 at selected points.

x	$\lvert \sin x - p_5(x) \rvert$	$\lvert \sin x - q_5(x) \rvert$
$\pi/4$		
$\pi/2$		
$3\pi/4$		
$5\pi/4$		
$7\pi/4$		

   d. At which points in the table is p_5 a better approximation to f than q_5? At which points do p_5 and q_5 give equal approximations to f? Explain your observations.

T 86. Approximating ln x Let $f(x) = \ln x$, and let p_n and q_n be the nth-order Taylor polynomials for f centered at 1 and e, respectively.

   a. Find p_3 and q_3.

   b. Graph f, p_3, and q_3 on the interval $(0, 4]$.

   c. Complete the following table showing the errors in the approximations given by p_3 and q_3 at selected points.

x	$\lvert \ln x - p_3(x) \rvert$	$\lvert \ln x - q_3(x) \rvert$
0.5		
1.0		
1.5		
2		
2.5		
3		
3.5		

   d. At which points in the table is p_3 a better approximation to f than q_3? Explain your observations.

T 87. Approximating square roots Let p_1 and q_1 be the first-order Taylor polynomials for $f(x) = \sqrt{x}$, centered at 36 and 49, respectively.

   a. Find p_1 and q_1.

   b. Complete the following table showing the errors when using p_1 and q_1 to approximate $f(x)$ at $x = 37, 39, 41, 43, 45,$ and 47. Use a calculator to obtain an exact value of $f(x)$.

x	$\lvert \sqrt{x} - p_1(x) \rvert$	$\lvert \sqrt{x} - q_1(x) \rvert$
37		
39		
41		
43		
45		
47		

   c. At which points in the table is p_1 a better approximation to f than q_1? Explain this result.

T 88. A different kind of approximation When approximating a function f using a Taylor polynomial, we use information about f and its derivative at one point. An alternative approach (called *interpolation*) uses information about f at several different points. Suppose we wish to approximate $f(x) = \sin x$ on the interval $[0, \pi]$.

   a. Write the (quadratic) Taylor polynomial p_2 for f centered at $\pi/2$.

   b. Now consider a quadratic interpolating polynomial $q(x) = ax^2 + bx + c$. The coefficients a, b, and c are chosen such that the following conditions are satisfied:

$$q(0) = f(0), \quad q\left(\frac{\pi}{2}\right) = f\left(\frac{\pi}{2}\right), \quad \text{and } q(\pi) = f(\pi).$$

   Show that $q(x) = -\dfrac{4}{\pi^2}x^2 + \dfrac{4}{\pi}x.$

   c. Graph f, p_2, and q on $[0, \pi]$.

   d. Find the error in approximating $f(x) = \sin x$ at the points $\dfrac{\pi}{4}$, $\dfrac{\pi}{2}$, $\dfrac{3\pi}{4}$, and π using p_2 and q.

   e. Which function, p_2 or q, is a better approximation to f on $[0, \pi]$? Explain.

QUICK CHECK ANSWERS

3. $f(x) = \sin x$ is an odd function, and its even-ordered derivatives are zero at 0, so its Taylor polynomials are also odd functions.   **4.** $p_4(x) = p_3(x) + \dfrac{x^4}{4!}$; $p_5(x) = p_4(x) + \dfrac{x^5}{5!}$

5. $x = 49$ and $x = 16$ are good choices.   **6.** Because $e^{0.45} < 2$, $\lvert R_7(0.45) \rvert < 2\dfrac{0.45^8}{8!} \approx 8.3 \times 10^{-8}$. ◄

11.2 Properties of Power Series

The preceding section demonstrated that Taylor polynomials provide accurate approximations to many functions and that, in general, the approximations improve as the degree of the polynomials increases. In this section, we take the next step and let the degree of the Taylor polynomials increase without bound to produce a *power series*.

Geometric Series as Power Series

A good way to become familiar with power series is to return to *geometric series*, first encountered in Section 10.3. Recall that for a fixed number r,

$$\sum_{k=0}^{\infty} r^k = 1 + r + r^2 + \cdots = \frac{1}{1-r}, \qquad \text{provided } |r| < 1.$$

It's a small change to replace the real number r with the variable x. In doing so, the geometric series becomes a new representation of a familiar function:

$$\sum_{k=0}^{\infty} x^k = 1 + x + x^2 + \cdots = \frac{1}{1-x}, \qquad \text{provided } |x| < 1.$$

This infinite series is a power series and it is a representation of the function $1/(1-x)$ that is valid on the interval $|x| < 1$.

In general, power series are used to represent familiar functions such as trigonometric, exponential, and logarithmic functions. They are also used to define new functions. For example, consider the function defined by

$$g(x) = \sum_{k=1}^{\infty} \frac{(-1)^k k}{4^k} x^{2k}.$$

The term *function* is used advisedly because it's not yet clear whether g really is a function. If so, is it continuous? Does it have a derivative? Judging by its graph (**Figure 11.13**), g appears to be an ordinary continuous and differentiable function on $(-2, 2)$ (which is identified at the end of the chapter). In fact, power series satisfy the defining property of all functions: For each admissible value of x, a power series has at most one value. For this reason, we refer to a power series as a function, although the domain, properties, and identity of the function may need to be discovered.

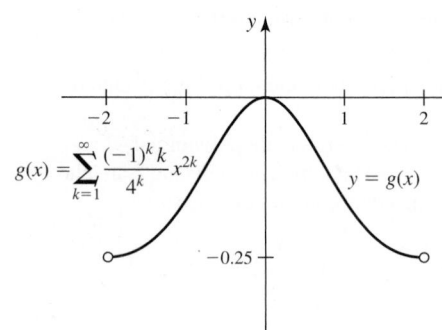

Figure 11.13

➤ Figure 11.13 shows an approximation to the graph of g made by summing the first 500 terms of the power series at selected values of x on the interval $(-2, 2)$.

QUICK CHECK 1 By substituting $x = 0$ in the power series for g, evaluate $g(0)$ for the function in Figure 11.13. ◄

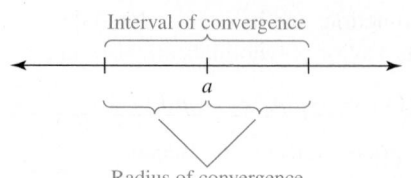

Figure 11.14

Convergence of Power Series

First let's establish some terminology associated with power series.

DEFINITION Power Series

A **power series** has the general form

$$\sum_{k=0}^{\infty} c_k (x - a)^k,$$

where a and c_k are real numbers, and x is a variable. The c_k's are the **coefficients** of the power series, and a is the **center** of the power series. The set of values of x for which the series converges is its **interval of convergence**. The **radius of convergence** of the power series, denoted R, is the distance from the center of the series to the boundary of the interval of convergence (**Figure 11.14**).

How do we determine the interval of convergence for a given power series? The presence of the term x^k or $(x - a)^k$ in a power series suggests using the Ratio Test or the Root Test to determine the values of x for which the series converges absolutely. By Theorem 10.19, if we determine the interval on which the series converges absolutely, we have a set of values for which the series converges.

Theorem 11.3 spells out the ways in which a power series can converge; its proof is given in Appendix A.

➤ Theorem 11.3 implies that the interval of convergence is symmetric about the center of the series; the radius of convergence R is determined by analyzing r from the Ratio Test (or ρ from the Root Test). The theorem says nothing about convergence at the endpoints. For example, the intervals of convergence $(2, 6)$, $(2, 6]$, $[2, 6)$, and $[2, 6]$ all have a radius of convergence of $R = 2$.

THEOREM 11.3   Convergence of Power Series

A power series $\displaystyle\sum_{k=0}^{\infty} c_k(x - a)^k$ centered at a converges in one of three ways:

1. The series converges absolutely for all x. It follows, by Theorem 10.19, that the series converges for all x, in which case the interval of convergence is $(-\infty, \infty)$ and the radius of convergence is $R = \infty$.

2. There is a real number $R > 0$ such that the series converges absolutely (and therefore converges) for $|x - a| < R$ and diverges for $|x - a| > R$, in which case the radius of convergence is R.

3. The series converges only at a, in which case the radius of convergence is $R = 0$.

The following examples illustrate how the Ratio and Root Tests are used to determine the radius and interval of convergence, and how the cases of Theorem 11.3 arise.

EXAMPLE 1   Radius and interval of convergence Find the radius and interval of convergence for each power series.

a. $\displaystyle\sum_{k=0}^{\infty} \frac{x^k}{k!}$ **b.** $\displaystyle\sum_{k=0}^{\infty} \frac{(-1)^k(x - 2)^k}{4^k}$ **c.** $\displaystyle\sum_{k=1}^{\infty} k!\, x^k$

SOLUTION

a. The center of the power series is 0 and the terms of the series are $x^k/k!$. Due to the presence of the factor $k!$, we test the series for absolute convergence using the Ratio Test:

$$r = \lim_{k \to \infty} \frac{|x^{k+1}/(k + 1)!|}{|x^k/k!|} \quad \text{Ratio Test}$$

$$= \lim_{k \to \infty} \frac{|x|^{k+1}}{|x|^k} \cdot \frac{k!}{(k + 1)!} \quad \text{Invert and multiply.}$$

$$= |x| \lim_{k \to \infty} \frac{1}{k + 1} = 0. \quad \text{Simplify and take the limit with } x \text{ fixed.}$$

Notice that in taking the limit as $k \to \infty$, x is held fixed. Because $r = 0$ for all real numbers x, the series converges absolutely for all x, which implies that the series converges for all x. Therefore, the interval of convergence is $(-\infty, \infty)$ **(Figure 11.15)** and the radius of convergence is $R = \infty$.

b. We test for absolute convergence using the Root Test:

$$\rho = \lim_{k \to \infty} \sqrt[k]{\left| \frac{(-1)^k(x - 2)^k}{4^k} \right|} = \frac{|x - 2|}{4}.$$

In this case, ρ depends on the value of x. For absolute convergence, x must satisfy

$$\rho = \frac{|x - 2|}{4} < 1,$$

which implies that $|x - 2| < 4$. Using standard techniques for solving inequalities, the solution set is $-4 < x - 2 < 4$, or $-2 < x < 6$. We conclude that the series converges absolutely (and therefore converges) on $(-2, 6)$, which means the radius of convergence is $R = 4$. When $-\infty < x < -2$ or $6 < x < \infty$, we have $\rho > 1$, so the series diverges on these intervals (the terms of the series do not approach 0 as $k \to \infty$ and the Divergence Test applies).

The Root Test does not give information about convergence at the endpoints $x = -2$ and $x = 6$, because at these points, the Root Test results in $\rho = 1$. To test for

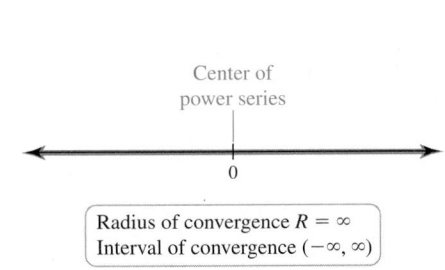

Center of power series

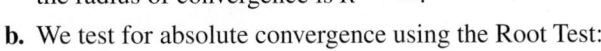

0

Radius of convergence $R = \infty$
Interval of convergence $(-\infty, \infty)$

Figure 11.15

➤ Either the Ratio Test or the Root Test works for the power series in Example 1b.

convergence at the endpoints, we substitute each endpoint into the series and carry out separate tests. At $x = -2$, the power series becomes

$$\sum_{k=0}^{\infty} \frac{(-1)^k(x-2)^k}{4^k} = \sum_{k=0}^{\infty} \frac{4^k}{4^k} \quad \text{Substitute } x = -2 \text{ and simplify.}$$

$$= \sum_{k=0}^{\infty} 1. \quad \text{Diverges by Divergence Test}$$

The series clearly diverges at the left endpoint. At $x = 6$, the power series is

$$\sum_{k=0}^{\infty} \frac{(-1)^k(x-2)^k}{4^k} = \sum_{k=0}^{\infty} (-1)^k \frac{4^k}{4^k} \quad \text{Substitute } x = 6 \text{ and simplify.}$$

$$= \sum_{k=0}^{\infty} (-1)^k. \quad \text{Diverges by Divergence Test}$$

This series also diverges at the right endpoint. Therefore, the interval of convergence is $(-2, 6)$, excluding the endpoints (**Figure 11.16**).

> The Ratio and Root Tests determine the radius of convergence conclusively. However, the interval of convergence is not determined until the endpoints are tested.

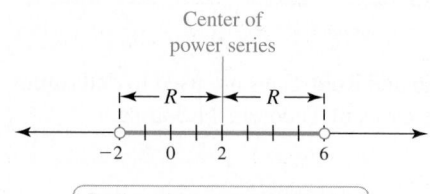

Radius of convergence $R = 4$
Interval of convergence $(-2, 6)$

Figure 11.16

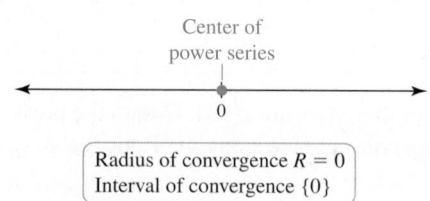

Radius of convergence $R = 0$
Interval of convergence $\{0\}$

Figure 11.17

QUICK CHECK 2 What are the radius and interval of convergence of the geometric series $\sum x^k$? ◄

c. In this case, the Ratio Test is preferable:

$$r = \lim_{k \to \infty} \frac{|(k+1)!\, x^{k+1}|}{|k!\, x^k|} \quad \text{Ratio Test}$$

$$= |x| \lim_{k \to \infty} \frac{(k+1)!}{k!} \quad \text{Simplify.}$$

$$= |x| \lim_{k \to \infty} (k+1) \quad \text{Simplify.}$$

$$= \infty. \quad \text{If } x \neq 0$$

We see that $r > 1$ for all $x \neq 0$, so the series diverges on $(-\infty, 0)$ and $(0, \infty)$.

The only way to satisfy $r < 1$ is to take $x = 0$, in which case the power series has a value of 0. The interval of convergence of the power series consists of the single point $x = 0$ (**Figure 11.17**), and the radius of convergence is $R = 0$.

Related Exercises 10–12 ◄

EXAMPLE 2 **Radius and interval of convergence** Use the Ratio Test to find the radius and interval of convergence of $\displaystyle\sum_{k=1}^{\infty} \frac{(x-2)^k}{\sqrt{k}}$.

SOLUTION

> The power series in Example 2 could also be analyzed using the Root Test.

$$r = \lim_{k \to \infty} \frac{|(x-2)^{k+1}/\sqrt{k+1}|}{|(x-2)^k/\sqrt{k}|} \quad \text{Ratio Test}$$

$$= |x-2| \lim_{k \to \infty} \sqrt{\frac{k}{k+1}} \quad \text{Simplify.}$$

$$= |x-2| \underbrace{\sqrt{\lim_{k \to \infty} \frac{k}{k+1}}}_{1} \quad \text{Limit Law}$$

$$= |x-2| \quad \text{Limit equals 1.}$$

The series converges absolutely (and therefore converges) for all x such that $r < 1$, which implies $|x - 2| < 1$, or $1 < x < 3$; we see that the radius of convergence is $R = 1$. On the intervals $-\infty < x < 1$ and $3 < x < \infty$, we have $r > 1$, so the series diverges.

We now test the endpoints. Substituting $x = 1$ gives the series

$$\sum_{k=1}^{\infty} \frac{(x-2)^k}{\sqrt{k}} = \sum_{k=1}^{\infty} \frac{(-1)^k}{\sqrt{k}}.$$

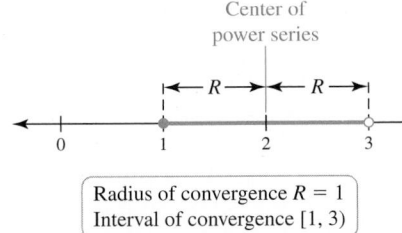

Center of
power series

| Radius of convergence $R = 1$ |
| Interval of convergence $[1, 3)$ |

Figure 11.18

This series converges by the Alternating Series Test (the terms of the series decrease in magnitude and approach 0 as $k \to \infty$). Substituting $x = 3$ gives the series

$$\sum_{k=1}^{\infty} \frac{(x-2)^k}{\sqrt{k}} = \sum_{k=1}^{\infty} \frac{1}{\sqrt{k}},$$

which is a divergent p-series. We conclude that the interval of convergence is $1 \le x < 3$ (**Figure 11.18**).

Related Exercise 18 ◄

The technique used in Examples 1 and 2 to determine the radius and interval of convergence for a given power series is summarized below.

SUMMARY Determining the Radius and Interval of Convergence of $\sum c_k(x - a)^k$

1. Use the Ratio Test or the Root Test to find the interval $(a - R, a + R)$ on which the series converges absolutely; the radius of convergence for the series is R.

2. Use the *radius* of convergence to find the *interval* of convergence:
   • If $R = \infty$, the interval of convergence is $(-\infty, \infty)$.
   • If $R = 0$, the interval of convergence is the single point $x = a$.
   • If $0 < R < \infty$, the interval of convergence consists of the interval $(a - R, a + R)$ and possibly one or both of its endpoints. Determining whether the series $\sum c_k(x - a)^k$ converges at the endpoints $x = a - R$ and $x = a + R$ amounts to analyzing the series $\sum c_k(-R)^k$ and $\sum c_k R^k$.

Although the procedure just outlined can be applied to any power series, it is generally used when nothing is known about the series in question. As we learn shortly in Theorems 11.4 and 11.5, new power series can be built from known power series, and in these cases, other methods may be used to find the interval of convergence.

Combining Power Series

A power series defines a function on its interval of convergence. When power series are combined algebraically, new functions are defined. The following theorem, stated without proof, gives three common ways to combine power series.

➤ New power series can also be defined as the product and quotient of power series. The calculation of the coefficients of such series is more challenging (Exercise 77).

➤ Theorem 11.4 also applies to power series centered at points other than $x = 0$. Property 1 applies directly; Properties 2 and 3 apply with slight modifications.

THEOREM 11.4 Combining Power Series
Suppose the power series $\sum c_k x^k$ and $\sum d_k x^k$ converge to $f(x)$ and $g(x)$, respectively, on an interval I.

1. **Sum and difference:** The power series $\sum (c_k \pm d_k)x^k$ converges to $f(x) \pm g(x)$ on I.
2. **Multiplication by a power:** Suppose m is an integer such that $k + m \ge 0$, for all terms of the power series $x^m \sum c_k x^k = \sum c_k x^{k+m}$. This series converges to $x^m f(x)$, for all $x \ne 0$ in I. When $x = 0$, the series converges to $\lim_{x \to 0} x^m f(x)$.
3. **Composition:** If $h(x) = bx^m$, where m is a positive integer and b is a nonzero real number, the power series $\sum c_k (h(x))^k$ converges to the composite function $f(h(x))$, for all x such that $h(x)$ is in I.

EXAMPLE 3 Combining power series Given the geometric series

$$\frac{1}{1 - x} = \sum_{k=0}^{\infty} x^k = 1 + x + x^2 + x^3 + \cdots, \qquad \text{for } |x| < 1,$$

find the power series and interval of convergence for the following functions.

a. $\dfrac{x^5}{1 - x}$ **b.** $\dfrac{1}{1 - 2x}$ **c.** $\dfrac{1}{1 + x^2}$

SOLUTION

a.
$$\frac{x^5}{1-x} = x^5 \left(1 + x + x^2 + \cdots\right) \quad \text{Theorem 11.4, Property 2}$$
$$= x^5 + x^6 + x^7 + \cdots$$
$$= \sum_{k=0}^{\infty} x^{k+5}$$

This geometric series has a ratio $r = x$ and converges when $|r| = |x| < 1$. The interval of convergence is $|x| < 1$.

b. We substitute $2x$ for x in the power series for $\dfrac{1}{1-x}$:

$$\frac{1}{1-2x} = 1 + (2x) + (2x)^2 + \cdots \quad \text{Theorem 11.4, Property 3}$$
$$= \sum_{k=0}^{\infty} (2x)^k.$$

This geometric series has a ratio $r = 2x$ and converges provided $|r| = |2x| < 1$ or $|x| < \frac{1}{2}$. The interval of convergence is $|x| < \frac{1}{2}$.

c. We substitute $-x^2$ for x in the power series for $\dfrac{1}{1-x}$:

$$\frac{1}{1+x^2} = 1 + (-x^2) + (-x^2)^2 + \cdots \quad \text{Theorem 11.4, Property 3}$$
$$= 1 - x^2 + x^4 - \cdots$$
$$= \sum_{k=0}^{\infty} (-1)^k x^{2k}.$$

This geometric series has a ratio of $r = -x^2$ and converges provided $|r| = |-x^2| = |x^2| < 1$ or $|x| < 1$.

Related Exercises 41–42 ◄

Differentiating and Integrating Power Series

Some properties of polynomials carry over to power series, but others do not. For example, a polynomial is defined for all values of x, whereas a power series is defined only on its interval of convergence. In general, the properties of polynomials carry over to power series when the power series is restricted to its interval of convergence. The following result illustrates this principle.

➤ Theorem 11.5 makes no claim about the convergence of the differentiated or integrated series at the endpoints of the interval of convergence.

THEOREM 11.5 Differentiating and Integrating Power Series

Suppose the power series $\sum c_k(x-a)^k$ converges for $|x-a| < R$ and defines a function f on that interval.

1. Then f is differentiable (which implies continuous) for $|x-a| < R$, and f' is found by differentiating the power series for f term by term; that is,
$$f'(x) = \sum k c_k (x-a)^{k-1},$$
for $|x-a| < R$.

2. The indefinite integral of f is found by integrating the power series for f term by term; that is,
$$\int f(x)\,dx = \sum c_k \frac{(x-a)^{k+1}}{k+1} + C,$$
for $|x-a| < R$, where C is an arbitrary constant.

The proof of this theorem requires advanced ideas and is omitted. However, some discussion is in order before turning to examples. The statements in Theorem 11.5 about term-by-term differentiation and integration say two things. First, the differentiated and integrated power series converge, provided x belongs to the interior of the interval of convergence. But the theorem claims more than convergence. According to the theorem, the differentiated and integrated power series converge to the derivative and indefinite integral of f, respectively, on the interior of the interval of convergence. Let's use this theorem to develop new power series.

EXAMPLE 4   Differentiating and integrating power series   Consider the geometric series

$$f(x) = \frac{1}{1-x} = \sum_{k=0}^{\infty} x^k = 1 + x + x^2 + x^3 + \cdots, \quad \text{for } |x| < 1.$$

a. Differentiate this series term by term to find the power series for f' and identify the function it represents.

b. Integrate this series term by term and identify the function it represents.

SOLUTION

a. We know that $f'(x) = (1-x)^{-2}$. Differentiating the series, we find that

$$f'(x) = \frac{d}{dx}(1 + x + x^2 + x^3 + \cdots) \quad \text{Differentiate the power series for } f.$$

$$= 1 + 2x + 3x^2 + \cdots \quad \text{Differentiate term by term.}$$

$$= \sum_{k=0}^{\infty} (k+1) x^k. \quad \text{Summation notation}$$

Therefore, on the interval $|x| < 1$,

$$f'(x) = (1-x)^{-2} = \sum_{k=0}^{\infty} (k+1) x^k.$$

Theorem 11.5 makes no claim about convergence of the differentiated series to f' at the endpoints. In this case, substituting $x = \pm 1$ into the power series for f' reveals that the series diverges at both endpoints.

b. Integrating f and integrating the power series term by term, we have

$$\int \frac{dx}{1-x} = \int (1 + x + x^2 + x^3 + \cdots) \, dx,$$

which implies that

$$-\ln|1-x| = x + \frac{x^2}{2} + \frac{x^3}{3} + \frac{x^4}{4} + \cdots + C,$$

where C is an arbitrary constant. Notice that the left side is 0 when $x = 0$. The right side is 0 when $x = 0$ provided we choose $C = 0$. Because $|x| < 1$, the absolute value sign on the left side may be removed. Multiplying both sides by -1, we have a series representation for $\ln(1-x)$:

$$\ln(1-x) = -x - \frac{x^2}{2} - \frac{x^3}{3} - \frac{x^4}{4} - \cdots = -\sum_{k=1}^{\infty} \frac{x^k}{k}.$$

It is interesting to test the endpoints of the interval $|x| < 1$. When $x = 1$, the series is (a multiple of) the divergent harmonic series, and when $x = -1$, the series is the convergent alternating harmonic series (Section 10.6). So the interval of convergence is $-1 \le x < 1$. Although we know the series converges at $x = -1$, Theorem 11.5 guarantees convergence to $\ln(1-x)$ only at the interior points. We cannot use Theorem 11.5 to claim that the series converges to $\ln 2$ at $x = -1$. In fact, it does, as shown in Section 11.3.

QUICK CHECK 3 Use the result of Example 4 to write a series representation for $\ln \frac{1}{2} = -\ln 2$. ◄

Related Exercises 51, 55 ◄

EXAMPLE 5 Functions to power series Find power series representations centered at 0 for the following functions and give their intervals of convergence.

a. $\tan^{-1} x$ **b.** $\ln\left(\dfrac{1 + x}{1 - x}\right)$

SOLUTION In both cases, we work with known power series and use differentiation, integration, and other combinations.

a. The key is to recall that

$$\int \frac{dx}{1 + x^2} = \tan^{-1} x + C$$

and that, by Example 3c,

$$\frac{1}{1 + x^2} = 1 - x^2 + x^4 - \cdots, \qquad \text{provided } |x| < 1.$$

We now integrate both sides of this last expression:

$$\int \frac{dx}{1 + x^2} = \int (1 - x^2 + x^4 - \cdots)\, dx,$$

which implies that

$$\tan^{-1} x = x - \frac{x^3}{3} + \frac{x^5}{5} - \cdots + C.$$

Substituting $x = 0$ and noting that $\tan^{-1} 0 = 0$, the two sides of this equation agree provided we choose $C = 0$. Therefore,

$$\tan^{-1} x = x - \frac{x^3}{3} + \frac{x^5}{5} - \cdots = \sum_{k=0}^{\infty} \frac{(-1)^k x^{2k+1}}{2k + 1}.$$

> Again, Theorem 11.5 does not guarantee that the power series in Example 5a converges to $\tan^{-1} x$ at $x = \pm 1$. In fact, it does.

By Theorem 11.5, this power series converges to $\tan^{-1} x$ for $|x| < 1$. Testing the endpoints separately, we find that it also converges at $x = \pm 1$. Therefore, the interval of convergence is $[-1, 1]$.

b. We have already seen (Example 4) that

$$\ln(1 - x) = -x - \frac{x^2}{2} - \frac{x^3}{3} - \cdots.$$

> Nicolaus Mercator (1620–1687) and Sir Isaac Newton (1642–1727) independently derived the power series for $\ln(1 + x)$, which is called the *Mercator series*.

Replacing x with $-x$ (Property 3 of Theorem 11.4), we have

$$\ln(1 - (-x)) = \ln(1 + x) = x - \frac{x^2}{2} + \frac{x^3}{3} - \cdots.$$

Subtracting these two power series gives

$$\ln\left(\frac{1 + x}{1 - x}\right) = \ln(1 + x) - \ln(1 - x) \quad \text{\small Properties of logarithms}$$

$$= \underbrace{\left(x - \frac{x^2}{2} + \frac{x^3}{3} - \cdots\right)}_{\ln(1+x)} - \underbrace{\left(-x - \frac{x^2}{2} - \frac{x^3}{3} - \cdots\right)}_{\ln(1-x)}, \quad \text{for } |x| < 1$$

$$= 2\left(x + \frac{x^3}{3} + \frac{x^5}{5} + \cdots\right) \quad \text{\small Combine; use Property 1 of Theorem 11.4.}$$

$$= 2\sum_{k=0}^{\infty} \frac{x^{2k+1}}{2k + 1}. \quad \text{\small Summation notation}$$

QUICK CHECK 4 Verify that the power series in Example 5b does not converge at the endpoints $x = \pm 1$. ◄

This power series is the difference of two power series, both of which converge on the interval $|x| < 1$. Therefore, by Theorem 11.4, the new series also converges on $|x| < 1$.

Related Exercises 57, 62 ◄

If you look carefully, every example in this section is ultimately based on the geometric series. Using this single series, we were able to develop power series for many other functions. Imagine what we could do with a few more basic power series. The following section accomplishes precisely that end. There, we discover power series for many of the standard functions of calculus.

SECTION 11.2 EXERCISES

Getting Started

1. Write the first four terms of a power series with coefficients c_0, c_1, c_2, and c_3 centered at 0.

2. Is $\displaystyle\sum_{k=0}^{\infty} (5x - 20)^k$ a power series? If so, find the center a of the power series and state a formula for the coefficients c_k of the power series.

3. What tests are used to determine the radius of convergence of a power series?

4. Is $\displaystyle\sum_{k=0}^{\infty} x^{2k}$ a power series? If so, find the center a of the power series and state a formula for the coefficients c_k of the power series.

5. Do the radius and interval of convergence of a power series change when the series is differentiated or integrated? Explain.

6. Suppose a power series converges if $|x - 3| < 4$ and diverges if $|x - 3| \geq 4$. Determine the radius and interval of convergence.

7. Suppose a power series converges if $|4x - 8| \leq 40$ and diverges if $|4x - 8| > 40$. Determine the radius and interval of convergence.

8. Suppose the power series $\displaystyle\sum_{k=0}^{\infty} c_k (x - a)^k$ has an interval of convergence of $(-3, 7]$. Find the center a and the radius of convergence R.

Practice Exercises

9–36. Radius and interval of convergence *Determine the radius and interval of convergence of the following power series.*

9. $\displaystyle\sum_{k=0}^{\infty} (2x)^k$

10. $\displaystyle\sum_{k=0}^{\infty} \frac{(x - 1)^k}{k!}$

11. $\displaystyle\sum_{k=1}^{\infty} (kx)^k$

12. $\displaystyle\sum_{k=0}^{\infty} k!(x - 10)^k$

13. $\displaystyle\sum_{k=1}^{\infty} \sin^k\left(\frac{1}{k}\right) x^k$

14. $\displaystyle\sum_{k=2}^{\infty} \frac{2^k (x - 3)^k}{k}$

15. $\displaystyle\sum_{k=0}^{\infty} \left(\frac{x}{3}\right)^k$

16. $\displaystyle\sum_{k=0}^{\infty} (-1)^k \frac{x^k}{5^k}$

17. $\displaystyle\sum_{k=1}^{\infty} \frac{x^k}{k^k}$

18. $\displaystyle\sum_{k=1}^{\infty} \frac{(-1)^k x^k}{\sqrt{k}}$

19. $\displaystyle\sum_{k=0}^{\infty} \frac{x^k}{2^k + 1}$

20. $\displaystyle\sum_{k=0}^{\infty} \frac{(2x)^k}{k!}$

21. $-\dfrac{x^2}{1!} + \dfrac{x^4}{2!} - \dfrac{x^6}{3!} + \dfrac{x^8}{4!} - \cdots$

22. $x - \dfrac{x^3}{4} + \dfrac{x^5}{9} - \dfrac{x^7}{16} + \cdots$

23. $\displaystyle\sum_{k=1}^{\infty} \frac{(-1)^{k+1}(x - 1)^k}{k}$

24. $\displaystyle\sum_{k=0}^{\infty} (-1)^k \frac{k(x - 4)^k}{2^k}$

25. $\displaystyle\sum_{k=0}^{\infty} \frac{(4x - 1)^k}{k^2 + 4}$

26. $\displaystyle\sum_{k=1}^{\infty} \frac{(3x + 2)^k}{k}$

27. $\displaystyle\sum_{k=0}^{\infty} \frac{k^{10}(2x - 4)^k}{10^k}$

28. $\displaystyle\sum_{k=2}^{\infty} \frac{(x + 3)^k}{k \ln^2 k}$

29. $\displaystyle\sum_{k=0}^{\infty} \frac{k^2 x^{2k}}{k!}$

30. $\displaystyle\sum_{k=0}^{\infty} k(x - 1)^k$

31. $\displaystyle\sum_{k=1}^{\infty} \frac{x^{2k+1}}{3^{k-1}}$

32. $\displaystyle\sum_{k=0}^{\infty} \left(-\frac{x}{10}\right)^{2k}$

33. $\displaystyle\sum_{k=1}^{\infty} \frac{(x - 1)^k k^k}{(k + 1)^k}$

34. $\displaystyle\sum_{k=0}^{\infty} \frac{(-2)^k (x + 3)^k}{3^{k+1}}$

35. $\displaystyle\sum_{k=0}^{\infty} \frac{k^{20} x^k}{(2k + 1)!}$

36. $\displaystyle\sum_{k=0}^{\infty} (-1)^k \frac{x^{3k}}{27^k}$

37–40. Radius of convergence *Find the radius of convergence for the following power series.*

37. $\displaystyle\sum_{k=1}^{\infty} \frac{k! x^k}{k^k}$

38. $\displaystyle\sum_{k=1}^{\infty} \left(1 + \frac{1}{k}\right)^{k^2} x^k$

39. $\displaystyle\sum_{k=1}^{\infty} \left(\frac{k}{k + 4}\right)^{k^2} x^k$

40. $\displaystyle\sum_{k=1}^{\infty} \left(1 - \cos\frac{1}{2^k}\right) x^k$

41–46. Combining power series *Use the geometric series*

$$f(x) = \frac{1}{1 - x} = \sum_{k=0}^{\infty} x^k, \quad \text{for } |x| < 1,$$

to find the power series representation for the following functions (centered at 0). Give the interval of convergence of the new series.

41. $f(3x) = \dfrac{1}{1 - 3x}$

42. $g(x) = \dfrac{x^3}{1 - x}$

43. $h(x) = \dfrac{2x^3}{1 - x}$

44. $f(x^3) = \dfrac{1}{1 - x^3}$

45. $p(x) = \dfrac{4x^{12}}{1 - x}$

46. $f(-4x) = \dfrac{1}{1 + 4x}$

47–50. Combining power series *Use the power series representation*

$$f(x) = \ln(1 - x) = -\sum_{k=1}^{\infty} \frac{x^k}{k}, \quad \text{for } -1 \leq x < 1,$$

to find the power series for the following functions (centered at 0). Give the interval of convergence of the new series.

47. $f(3x) = \ln(1 - 3x)$

48. $g(x) = x^3 \ln(1 - x)$

49. $p(x) = 2x^6 \ln(1 - x)$

50. $f(x^3) = \ln(1 - x^3)$

51–56. Differentiating and integrating power series *Find the power series representation for g centered at 0 by differentiating or integrating the power series for f (perhaps more than once). Give the interval of convergence for the resulting series.*

51. $g(x) = \dfrac{2}{(1 - 2x)^2}$ using $f(x) = \dfrac{1}{1 - 2x}$

52. $g(x) = \dfrac{1}{(1 - x)^3}$ using $f(x) = \dfrac{1}{1 - x}$

53. $g(x) = -\dfrac{1}{(1 + x)^2}$ using $f(x) = \dfrac{1}{1 + x}$

54. $g(x) = \dfrac{x}{(1 + x^2)^2}$ using $f(x) = \dfrac{1}{1 + x^2}$

55. $g(x) = \ln(1 - 3x)$ using $f(x) = \dfrac{1}{1 - 3x}$

56. $g(x) = \ln(1 + x^2)$ using $f(x) = \dfrac{x}{1 + x^2}$

57–62. Functions to power series *Find power series representations centered at 0 for the following functions using known power series. Give the interval of convergence for the resulting series.*

57. $f(x) = \dfrac{2x}{(1 + x^2)^2}$

58. $f(x) = \dfrac{1}{1 - x^4}$

59. $f(x) = \dfrac{3}{3 + x}$

60. $f(x) = \ln\sqrt{1 - x^2}$

61. $f(x) = \ln\sqrt{4 - x^2}$

62. $f(x) = \tan^{-1}(4x^2)$

63. Explain why or why not Determine whether the following statements are true and give an explanation or counterexample.

   a. The interval of convergence of the power series $\sum c_k(x - 3)^k$ could be $(-2, 8)$.

   b. The series $\sum\limits_{k=0}^{\infty}(-2x)^k$ converges on the interval $-\frac{1}{2} < x < \frac{1}{2}$.

   c. If $f(x) = \sum\limits_{k=0}^{\infty} c_k x^k$ on the interval $|x| < 1$, then $f(x^2) = \sum c_k x^{2k}$ on the interval $|x| < 1$.

   d. If $f(x) = \sum\limits_{k=0}^{\infty} c_k x^k = 0$, for all x on an interval $(-a, a)$, then $c_k = 0$, for all k.

64. Scaling power series If the power series $f(x) = \sum c_k x^k$ has an interval of convergence of $|x| < R$, what is the interval of convergence of the power series for $f(ax)$, where $a \neq 0$ is a real number?

65. Shifting power series If the power series $f(x) = \sum c_k x^k$ has an interval of convergence of $|x| < R$, what is the interval of convergence of the power series for $f(x - a)$, where $a \neq 0$ is a real number?

66. A useful substitution Replace x with $x - 1$ in the series $\ln(1 + x) = \sum\limits_{k=1}^{\infty} \dfrac{(-1)^{k+1}x^k}{k}$ to obtain a power series for $\ln x$ centered at $x = 1$. What is the interval of convergence for the new power series?

67–71. Series to functions *Find the function represented by the following series, and find the interval of convergence of the series. (Not all these series are power series.)*

67. $\sum\limits_{k=0}^{\infty}(\sqrt{x} - 2)^k$

68. $\sum\limits_{k=1}^{\infty} \dfrac{x^{2k}}{4^k}$

69. $\sum\limits_{k=0}^{\infty} e^{-kx}$

70. $\sum\limits_{k=1}^{\infty} \dfrac{(x - 2)^k}{3^{2k}}$

71. $\sum\limits_{k=0}^{\infty}\left(\dfrac{x^2 - 1}{3}\right)^k$

72–74. Exponential function *In Section 11.3, we show that the power series for the exponential function centered at 0 is*

$$e^x = \sum_{k=0}^{\infty} \frac{x^k}{k!}, \quad \text{for } -\infty < x < \infty.$$

Use the methods of this section to find the power series centered at 0 for the following functions. Give the interval of convergence for the resulting series.

72. $f(x) = e^{2x}$ **73.** $f(x) = e^{-3x}$ **74.** $f(x) = x^2 e^x$

Explorations and Challenges

75. Powers of x multiplied by a power series Prove that if $f(x) = \sum\limits_{k=0}^{\infty} c_k x^k$ converges with radius of convergence R, then the power series for $x^m f(x)$ also converges with radius of convergence R, for positive integers m.

T 76. Remainders Let

$$f(x) = \sum_{k=0}^{\infty} x^k = \frac{1}{1 - x} \quad \text{and} \quad S_n(x) = \sum_{k=0}^{n-1} x^k.$$

The remainder in truncating the power series after n terms is $R_n = f(x) - S_n(x)$, which depends on x.

   a. Show that $R_n(x) = x^n/(1 - x)$.

   b. Graph the remainder function on the interval $|x| < 1$, for $n = 1, 2,$ and 3. Discuss and interpret the graph. Where on the interval is $|R_n(x)|$ largest? Smallest?

   c. For fixed n, minimize $|R_n(x)|$ with respect to x. Does the result agree with the observations in part (b)?

   d. Let $N(x)$ be the number of terms required to reduce $|R_n(x)|$ to less than 10^{-6}. Graph the function $N(x)$ on the interval $|x| < 1$. Discuss and interpret the graph.

77. Product of power series Let

$$f(x) = \sum_{k=0}^{\infty} c_k x^k \quad \text{and} \quad g(x) = \sum_{k=0}^{\infty} d_k x^k.$$

   a. Multiply the power series together as if they were polynomials, collecting all terms that are multiples of 1, x, and x^2. Write the first three terms of the product $f(x)g(x)$.

   b. Find a general expression for the coefficient of x^n in the product series, for $n = 0, 1, 2, \ldots$.

78. Inverse sine Given the power series

$$\frac{1}{\sqrt{1 - x^2}} = 1 + \frac{1}{2}x^2 + \frac{1 \cdot 3}{2 \cdot 4}x^4 + \frac{1 \cdot 3 \cdot 5}{2 \cdot 4 \cdot 6}x^6 + \cdots,$$

for $-1 < x < 1$, find the power series for $f(x) = \sin^{-1}x$ centered at 0.

QUICK CHECK ANSWERS

1. $g(0) = 0$ **2.** $|x| < 1, R = 1$ **3.** Substituting $x = 1/2$, $\ln(1/2) = -\ln 2 = -\sum\limits_{k=1}^{\infty} \dfrac{1}{2^k k}$ ◄

11.3 Taylor Series

In the preceding section, we saw that a power series represents a function on its interval of convergence. This section explores the opposite question: Given a function, what is its power series representation? We have already made significant progress in answering this question because we know how Taylor polynomials are used to approximate functions. We now extend Taylor polynomials to produce power series—called *Taylor series*—that provide series representations for functions.

Taylor Series for a Function

Suppose a function f has derivatives $f^{(k)}(a)$ of *all* orders at the point a. If we write the nth-order Taylor polynomial for f centered at a and allow n to increase indefinitely, a power series is obtained:

$$\underbrace{c_0 + c_1(x-a) + c_2(x-a)^2 + \cdots + c_n(x-a)^n}_{\text{Taylor polynomial of order } n} + \underset{n\to\infty}{\cdots} = \sum_{k=0}^{\infty} c_k(x-a)^k.$$

The coefficients of the Taylor polynomial are given by

$$c_k = \frac{f^{(k)}(a)}{k!}, \quad \text{for } k = 0, 1, 2, \ldots.$$

These coefficients are also the coefficients of the power series, which is called the *Taylor series for f centered at a*. It is the natural extension of the set of Taylor polynomials for f at a. The special case of a Taylor series centered at 0 is called a *Maclaurin series*.

> ► Maclaurin series are named after the Scottish mathematician Colin Maclaurin (1698–1746), who described them (with credit to Taylor) in a textbook in 1742.

DEFINITION Taylor/Maclaurin Series for a Function

Suppose the function f has derivatives of all orders on an interval centered at the point a. The **Taylor series for f centered at a** is

$$f(a) + f'(a)(x-a) + \frac{f''(a)}{2!}(x-a)^2 + \frac{f^{(3)}(a)}{3!}(x-a)^3 + \cdots$$

$$= \sum_{k=0}^{\infty} \frac{f^{(k)}(a)}{k!}(x-a)^k.$$

A Taylor series centered at 0 is called a **Maclaurin series**.

For the Taylor series to be useful, we need to know two things:

- the values of x for which the Taylor series converges, and
- the values of x for which the Taylor series for f *equals* f.

> ► There are unusual cases in which the Taylor series for a function converges to a different function (Exercise 92).

The second question is subtle and is postponed for a few pages. For now, we find the Taylor series for f centered at a point, but we refrain from saying $f(x)$ equals the power series.

QUICK CHECK 1 Verify that if the Taylor series for f centered at a is evaluated at $x = a$, then the Taylor series equals $f(a)$. ◄

EXAMPLE 1 Maclaurin series and convergence Find the Maclaurin series (which is the Taylor series centered at 0) for the following functions, and then find the interval of convergence.

a. $f(x) = \cos x$ **b.** $f(x) = \dfrac{1}{1-x}$

SOLUTION The procedure for finding the coefficients of a Taylor series is the same as for Taylor polynomials; most of the work is computing the derivatives of f.

a. The Maclaurin series has the form

$$\sum_{k=0}^{\infty} c_k x^k, \quad \text{where } c_k = \frac{f^{(k)}(0)}{k!}, \quad \text{for } k = 0, 1, 2, \ldots.$$

We evaluate derivatives of $f(x) = \cos x$ at $x = 0$.

$$f(x) = \cos x \;\Rightarrow\; f(0) = 1$$
$$f'(x) = -\sin x \;\Rightarrow\; f'(0) = 0$$
$$f''(x) = -\cos x \;\Rightarrow\; f''(0) = -1$$
$$f'''(x) = \sin x \;\Rightarrow\; f'''(0) = 0$$
$$f^{(4)}(x) = \cos x \;\Rightarrow\; f^{(4)}(0) = 1$$
$$\vdots \qquad\qquad \vdots$$

> ➤ In Example 1a, we note that both $\cos x$ and its Maclaurin series are even functions. Be cautious with this observation. A Taylor series for an even function centered at a point different from 0 may be even, odd, or neither. A similar behavior occurs with odd functions.

Because the odd-order derivatives are zero, $c_k = \dfrac{f^{(k)}(0)}{k!} = 0$ when k is odd. Using the even-order derivatives, we have

$$c_0 = f(0) = 1, \qquad c_2 = \frac{f^{(2)}(0)}{2!} = -\frac{1}{2!},$$

$$c_4 = \frac{f^{(4)}(0)}{4!} = \frac{1}{4!}, \quad c_6 = \frac{f^{(6)}(0)}{6!} = -\frac{1}{6!},$$

and in general, $c_{2k} = \dfrac{(-1)^k}{(2k)!}$. Therefore, the Maclaurin series for f is

$$1 - \frac{x^2}{2!} + \frac{x^4}{4!} - \frac{x^6}{6!} + \cdots = \sum_{k=0}^{\infty} \frac{(-1)^k}{(2k)!} x^{2k}.$$

Notice that this series contains all the Taylor polynomials. In this case, it consists only of even powers of x, reflecting the fact that $\cos x$ is an even function.

For what values of x does the series converge? As discussed in Section 11.2, we apply the Ratio Test:

$$r = \lim_{k \to \infty} \left| \frac{(-1)^{k+1} x^{2(k+1)}/(2(k+1))!}{(-1)^k x^{2k}/(2k)!} \right| \qquad r = \lim_{k \to \infty} \left| \frac{a_{k+1}}{a_k} \right|$$

$$= \lim_{k \to \infty} \left| \frac{x^2}{(2k+2)(2k+1)} \right| = 0. \qquad \text{Simplify and take the limit with } x \text{ fixed.}$$

> ➤ Recall that
> $$(2k + 2)! = (2k + 2)(2k + 1)(2k)!.$$
> Therefore, $\dfrac{(2k)!}{(2k+2)!} = \dfrac{1}{(2k+2)(2k+1)}.$

In this case, $r < 1$ for all x, so the Maclaurin series converges absolutely for all x, which implies (by Theorem 10.19) that the series converges for all x. We conclude that the interval of convergence is $-\infty < x < \infty$.

b. We proceed in a similar way with $f(x) = 1/(1 - x)$ by evaluating the derivatives of f at 0:

$$f(x) = \frac{1}{1 - x} \;\Rightarrow\; f(0) = 1,$$

$$f'(x) = \frac{1}{(1 - x)^2} \;\Rightarrow\; f'(0) = 1,$$

$$f''(x) = \frac{2}{(1 - x)^3} \;\Rightarrow\; f''(0) = 2!,$$

$$f'''(x) = \frac{3 \cdot 2}{(1 - x)^4} \;\Rightarrow\; f'''(0) = 3!,$$

$$f^{(4)}(x) = \frac{4 \cdot 3 \cdot 2}{(1 - x)^5} \;\Rightarrow\; f^{(4)}(0) = 4!,$$

and in general, $f^{(k)}(0) = k!$. Therefore, the Maclaurin series coefficients are $c_k = \dfrac{f^{(k)}(0)}{k!} = \dfrac{k!}{k!} = 1$, for $k = 0, 1, 2, \ldots$. The series for f centered at 0 is

$$1 + x + x^2 + x^3 + \cdots = \sum_{k=0}^{\infty} x^k.$$

This power series is familiar! The Maclaurin series for $f(x) = 1/(1 - x)$ is a geometric series. We could apply the Ratio Test, but we have already demonstrated that this series converges for $|x| < 1$.

Related Exercises 12, 18 ◄

QUICK CHECK 2 Based on Example 1b, what is the Taylor series for $f(x) = (1 + x)^{-1}$? ◄

The preceding example has an important lesson. *There is only one power series representation for a given function about a given point; however, there may be several ways to find it.* Another important point is this: The series in Example 1b represents the function $f(x) = 1/(1 - x)$ only on the interval $(-1, 1)$. If we want a power series representation for f on another interval, a new center must be chosen and the coefficients of the power series must be recomputed, as shown in the next example.

EXAMPLE 2   Choosing a different center Find the Taylor series for $f(x) = \dfrac{1}{1 - x}$ centered at 5, and then determine its radius and interval of convergence.

SOLUTION Using the pattern discovered in Example 1b, we find that the kth derivative of f is $f^{(k)}(x) = \dfrac{k!}{(1 - x)^{k+1}}$. Therefore, the Taylor series coefficients are

$$c_k = \frac{f^{(k)}(5)}{k!} = \frac{k!}{k!(1 - 5)^{k+1}} = \frac{(-1)^{k+1}}{4^{k+1}}, \quad \text{for } k = 0, 1, 2, \ldots,$$

and the series for f centered at 5 is

$$\sum_{k=0}^{\infty} \frac{(-1)^{k+1}(x - 5)^k}{4^{k+1}} = -\frac{1}{4} + \frac{x - 5}{4^2} - \frac{(x - 5)^2}{4^3} + \frac{(x - 5)^3}{4^4} \cdots.$$

To determine the interval of convergence for this series, we apply the Ratio Test:

QUICK CHECK 3 Verify that the series $\sum_{k=0}^{\infty} \dfrac{(-1)^{k+1}(x - 5)^k}{4^{k+1}}$ from Example 2 diverges at $x = 1$ and $x = 9$. ◄

$$r = \lim_{k \to \infty} \left| \frac{(-1)^{k+2}(x - 5)^{k+1}/4^{k+2}}{(-1)^{k+1}(x - 5)^k/4^{k+1}} \right| = \lim_{k \to \infty} \frac{|x - 5|}{4} = \frac{|x - 5|}{4}.$$

The series converges absolutely when $r = \dfrac{|x - 5|}{4} < 1$, or when $1 < x < 9$, so the radius of convergence is 4. The series diverges at $x = 1$ and $x = 9$ (Quick Check 3), so we conclude that the interval of convergence is $(1, 9)$.

The graphs of f and the 7th-order Taylor polynomial p_7 centered at 5 (blue curve) are shown in Figure 11.19, along with the graph of p_7 centered at 0 (red curve) found in Example 1b. Notice that in both cases, the interval of convergence extends to the vertical asymptote ($x = 1$) of $f(x) = \dfrac{1}{1 - x}$. In fact, it can be shown that the Taylor series for f, with any choice of center $a \neq 1$, has an interval of convergence symmetric about a that extends to $x = 1$. Remarkably, Taylor series for f "know" the location of the discontinuity of f.

Related Exercises 29–30 ◄

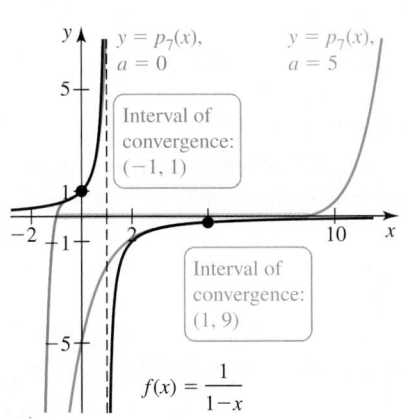

Figure 11.19

EXAMPLE 3   Center other than 0 Find the first four nonzero terms of the Taylor series for $f(x) = \sqrt[3]{x}$ centered at 8.

SOLUTION Notice that f has derivatives of all orders at $x = 8$. The Taylor series centered at 8 has the form

$$\sum_{k=0}^{\infty} c_k(x - 8)^k, \quad \text{where } c_k = \frac{f^{(k)}(8)}{k!}.$$

Next, we evaluate derivatives:

$$f(x) = x^{1/3} \Rightarrow f(8) = 2,$$

$$f'(x) = \frac{1}{3}x^{-2/3} \Rightarrow f'(8) = \frac{1}{12},$$

$$f''(x) = -\frac{2}{9}x^{-5/3} \Rightarrow f''(8) = -\frac{1}{144}, \text{ and}$$

$$f'''(x) = \frac{10}{27}x^{-8/3} \Rightarrow f'''(8) = \frac{5}{3456}.$$

We now assemble the power series:

$$2 + \frac{1}{12}(x - 8) + \frac{1}{2!}\left(-\frac{1}{144}\right)(x - 8)^2 + \frac{1}{3!}\left(\frac{5}{3456}\right)(x - 8)^3 + \cdots$$

$$= 2 + \frac{1}{12}(x - 8) - \frac{1}{288}(x - 8)^2 + \frac{5}{20,736}(x - 8)^3 + \cdots.$$

Related Exercise 32 ◄

EXAMPLE 4   **Manipulating Maclaurin series**   Let $f(x) = e^x$.

a. Find the Maclaurin series for f.

b. Find its interval of convergence.

c. Use the Maclaurin series for e^x to find the Maclaurin series for the functions $x^4 e^x$, e^{-2x}, and e^{-x^2}.

SOLUTION

a. The coefficients of the Taylor polynomials for $f(x) = e^x$ centered at 0 are $c_k = 1/k!$ (Example 3, Section 11.1). They are also the coefficients of the Maclaurin series. Therefore, the Maclaurin series for e^x is

$$1 + \frac{x}{1!} + \frac{x^2}{2!} + \cdots + \frac{x^n}{n!} + \cdots = \sum_{k=0}^{\infty} \frac{x^k}{k!}.$$

b. By the Ratio Test,

$$r = \lim_{k \to \infty} \left| \frac{x^{k+1}/(k+1)!}{x^k/k!} \right| \qquad \text{Substitute } (k+1)\text{st and } k\text{th terms.}$$

$$= \lim_{k \to \infty} \left| \frac{x}{k+1} \right| = 0. \qquad \text{Simplify; take the limit with } x \text{ fixed.}$$

Because $r < 1$ for all x, the interval of convergence is $-\infty < x < \infty$.

c. As stated in Theorem 11.4, power series may be added, multiplied by powers of x, or composed with functions on their intervals of convergence. Therefore, the Maclaurin series for $x^4 e^x$ is

$$x^4 \sum_{k=0}^{\infty} \frac{x^k}{k!} = \sum_{k=0}^{\infty} \frac{x^{k+4}}{k!} = x^4 + \frac{x^5}{1!} + \frac{x^6}{2!} + \cdots + \frac{x^{k+4}}{k!} + \cdots.$$

Similarly, e^{-2x} is the composition $f(-2x)$. Replacing x with $-2x$ in the Maclaurin series for f, the series representation for e^{-2x} is

$$\sum_{k=0}^{\infty} \frac{(-2x)^k}{k!} = \sum_{k=0}^{\infty} \frac{(-1)^k (2x)^k}{k!} = 1 - 2x + 2x^2 - \frac{4}{3}x^3 + \cdots.$$

The Maclaurin series for e^{-x^2} is obtained by replacing x with $-x^2$ in the power series for f. The resulting series is

$$\sum_{k=0}^{\infty} \frac{(-x^2)^k}{k!} = \sum_{k=0}^{\infty} \frac{(-1)^k x^{2k}}{k!} = 1 - x^2 + \frac{x^4}{2!} - \frac{x^6}{3!} + \cdots.$$

QUICK CHECK 4 Find the first three terms of the Maclaurin series for $2xe^x$ and e^{-x}. ◄

► For nonnegative integers p and k with $0 \le k \le p$, the binomial coefficients may also be defined as
$$\binom{p}{k} = \frac{p!}{k!(p-k)!},$$
where $0! = 1$. The coefficients form the rows of Pascal's triangle. The coefficients of $(1 + x)^5$ form the sixth row of the triangle.

```
          1
        1   1
      1   2   1
    1   3   3   1
  1   4   6   4   1
1   5   10  10   5   1
```

QUICK CHECK 5 Evaluate the binomial coefficients $\binom{-3}{2}$ and $\binom{\frac{1}{2}}{3}$. ◄

Because the interval of convergence of $f(x) = e^x$ is $-\infty < x < \infty$, the manipulations used to obtain the series for $x^4 e^x$, e^{-2x}, or e^{-x^2} do not change the interval of convergence. If in doubt about the interval of convergence of a new series, apply the Ratio Test.

Related Exercises 35, 38 ◄

The Binomial Series

We know from algebra that if p is a positive integer, then $(1 + x)^p$ is a polynomial of degree p. In fact,

$$(1 + x)^p = \binom{p}{0} + \binom{p}{1}x + \binom{p}{2}x^2 + \cdots + \binom{p}{p}x^p,$$

where the binomial coefficients $\binom{p}{k}$ are defined as follows.

DEFINITION Binomial Coefficients

For real numbers p and integers $k \ge 1$,

$$\binom{p}{k} = \frac{p(p-1)(p-2)\cdots(p-k+1)}{k!}, \qquad \binom{p}{0} = 1.$$

For example,

$$(1 + x)^5 = \underbrace{\binom{5}{0}}_{1} + \underbrace{\binom{5}{1}}_{5}x + \underbrace{\binom{5}{2}}_{10}x^2 + \underbrace{\binom{5}{3}}_{10}x^3 + \underbrace{\binom{5}{4}}_{5}x^4 + \underbrace{\binom{5}{5}}_{1}x^5$$

$$= 1 + 5x + 10x^2 + 10x^3 + 5x^4 + x^5.$$

Our goal is to extend this idea to the functions $f(x) = (1 + x)^p$, where $p \ne 0$ is a real number. The result is a Taylor series called the *binomial series*.

THEOREM 11.6 Binomial Series

For real numbers $p \ne 0$, the Taylor series for $f(x) = (1 + x)^p$ centered at 0 is the **binomial series**

$$\sum_{k=0}^{\infty} \binom{p}{k} x^k = 1 + \sum_{k=1}^{\infty} \frac{p(p-1)(p-2)\cdots(p-k+1)}{k!} x^k$$

$$= 1 + px + \frac{p(p-1)}{2!}x^2 + \frac{p(p-1)(p-2)}{3!}x^3 + \cdots.$$

The series converges for $|x| < 1$ (and possibly at the endpoints, depending on p). If p is a nonnegative integer, the series terminates and results in a polynomial of degree p.

► To evaluate $\binom{p}{k}$, start with p and successively subtract 1 until k factors are obtained; then take the product of these k factors and divide by $k!$. Recall that
$$\binom{p}{0} = 1.$$

Proof: We seek a power series centered at 0 of the form

$$\sum_{k=0}^{\infty} c_k x^k, \quad \text{where } c_k = \frac{f^{(k)}(0)}{k!}, \quad \text{for } k = 0, 1, 2, \ldots.$$

The job is to evaluate the derivatives of f at 0:

$$f(x) = (1 + x)^p \Rightarrow f(0) = 1,$$
$$f'(x) = p(1 + x)^{p-1} \Rightarrow f'(0) = p,$$
$$f''(x) = p(p-1)(1 + x)^{p-2} \Rightarrow f''(0) = p(p-1), \text{ and}$$
$$f'''(x) = p(p-1)(p-2)(1 + x)^{p-3} \Rightarrow f'''(0) = p(p-1)(p-2).$$

A pattern emerges: The kth derivative $f^{(k)}(0)$ involves the k factors $p(p-1)(p-2)\cdots(p-k+1)$. In general, we have

$$f^{(k)}(0) = p(p-1)(p-2)\cdots(p-k+1).$$

Therefore,

$$c_k = \frac{f^{(k)}(0)}{k!} = \frac{p(p-1)(p-2)\cdots(p-k+1)}{k!} = \binom{p}{k}, \quad \text{for } k = 0, 1, 2, \ldots.$$

The Taylor series for $f(x) = (1+x)^p$ centered at 0 is

$$\binom{p}{0} + \binom{p}{1}x + \binom{p}{2}x^2 + \binom{p}{3}x^3 + \cdots = \sum_{k=0}^{\infty}\binom{p}{k}x^k.$$

This series has the same general form for all values of p. When p is a nonnegative integer, the series terminates and it is a polynomial of degree p.

The radius of convergence for the binomial series is determined by the Ratio Test. Holding p and x fixed, the relevant limit is

$$r = \lim_{k\to\infty}\left|\frac{x^{k+1}p(p-1)\cdots(p-k+1)(p-k)/(k+1)!}{x^k p(p-1)\cdots(p-k+1)/k!}\right| \qquad \text{Ratio of } (k+1)\text{st to } k\text{th term}$$

$$= |x|\lim_{k\to\infty}\underbrace{\left|\frac{p-k}{k+1}\right|}_{\text{approaches } 1} \qquad \text{Cancel factors and simplify.}$$

> In Theorem 11.6, it can be shown that the interval of convergence for the binomial series is
> - $(-1, 1)$ if $p \le -1$,
> - $(-1, 1]$ if $-1 < p < 0$, and
> - $[-1, 1]$ if $p > 0$ and not an integer.

$$= |x|.$$

With p fixed,
$$\lim_{k\to\infty}\left|\frac{p-k}{k+1}\right| = 1.$$

Absolute convergence requires that $r = |x| < 1$. Therefore, the series converges for $|x| < 1$. Depending on the value of p, the interval of convergence may include the endpoints; see margin note. ◄

> A binomial series is a Taylor series. Because the series in Example 5 is centered at 0, it is also a Maclaurin series.

EXAMPLE 5 Binomial series Consider the function $f(x) = \sqrt{1+x}$.

a. Find the first four terms of the binomial series for f centered at 0.
b. Approximate $\sqrt{1.15}$ to three decimal places. Assume the series for f converges to f on its interval of convergence, which is $[-1, 1]$.

SOLUTION

a. We use the formula for the binomial coefficients with $p = \frac{1}{2}$ to compute the first four coefficients:

Table 11.3

n	Approximation $p_n(0.15)$
0	1.0
1	1.075
2	1.0721875
3	1.072398438

$$c_0 = 1, \qquad c_1 = \binom{\frac{1}{2}}{1} = \frac{\frac{1}{2}}{1!} = \frac{1}{2},$$

$$c_2 = \binom{\frac{1}{2}}{2} = \frac{\frac{1}{2}\left(-\frac{1}{2}\right)}{2!} = -\frac{1}{8}, \text{ and } c_3 = \binom{\frac{1}{2}}{3} = \frac{\frac{1}{2}\left(-\frac{1}{2}\right)\left(-\frac{3}{2}\right)}{3!} = \frac{1}{16}.$$

The leading terms of the binomial series are

$$1 + \frac{1}{2}x - \frac{1}{8}x^2 + \frac{1}{16}x^3 - \cdots.$$

> The remainder theorem for alternating series (Section 10.6) could be used in Example 5 to estimate the number of terms of the Maclaurin series needed to achieve a desired accuracy.

b. Truncating the binomial series in part (a) produces Taylor polynomials p_n that may be used to approximate $f(0.15) = \sqrt{1.15}$. With $x = 0.15$, we find the polynomial approximations shown in Table 11.3. Four terms of the power series ($n = 3$) give $\sqrt{1.15} \approx 1.072$. Because the approximations with $n = 2$ and $n = 3$ agree to three decimal places, when rounded, the approximation 1.072 is accurate to three decimal places.

QUICK CHECK 6 Use two and three terms of the binomial series in Example 5 to approximate $\sqrt{1.1}$. ◄

Related Exercises 46–47 ◄

EXAMPLE 6 **Working with binomial series** Consider the functions

$$f(x) = \sqrt[3]{1 + x} \quad \text{and} \quad g(x) = \sqrt[3]{c + x}, \quad \text{where } c > 0 \text{ is a constant.}$$

a. Find the first four terms of the binomial series for f centered at 0.

b. Use part (a) to find the first four terms of the binomial series for g centered at 0.

c. Use part (b) to approximate $\sqrt[3]{23}, \sqrt[3]{24}, \ldots, \sqrt[3]{31}$. Assume the series for g converges to g on its interval of convergence.

SOLUTION

a. Because $f(x) = (1 + x)^{1/3}$, we find the binomial coefficients with $p = \frac{1}{3}$:

$$c_0 = \binom{\frac{1}{3}}{0} = 1, \qquad\qquad c_1 = \binom{\frac{1}{3}}{1} = \frac{\frac{1}{3}}{1!} = \frac{1}{3},$$

$$c_2 = \binom{\frac{1}{3}}{2} = \frac{\left(\frac{1}{3}\right)\left(\frac{1}{3} - 1\right)}{2!} = -\frac{1}{9}, \text{ and } c_3 = \binom{\frac{1}{3}}{3} = \frac{\left(\frac{1}{3}\right)\left(\frac{1}{3} - 1\right)\left(\frac{1}{3} - 2\right)}{3!} = \frac{5}{81}.$$

The first four terms of the binomial series are

$$1 + \frac{1}{3}x - \frac{1}{9}x^2 + \frac{5}{81}x^3.$$

b. To avoid deriving a new series for $g(x) = \sqrt[3]{c + x}$, a few steps of algebra allow us to use part (a). Note that

$$g(x) = \sqrt[3]{c + x} = \sqrt[3]{c\left(1 + \frac{x}{c}\right)} = \sqrt[3]{c} \cdot \sqrt[3]{1 + \frac{x}{c}} = \sqrt[3]{c} \cdot f\left(\frac{x}{c}\right).$$

In other words, g can be expressed in terms of f, for which we already have a binomial series. The binomial series for g is obtained by substituting x/c into the binomial series for f and multiplying by $\sqrt[3]{c}$:

$$g(x) = \sqrt[3]{c}\underbrace{\left(1 + \frac{1}{3}\left(\frac{x}{c}\right) - \frac{1}{9}\left(\frac{x}{c}\right)^2 + \frac{5}{81}\left(\frac{x}{c}\right)^3 - \cdots\right)}_{f(x/c)}.$$

It can be shown that the series for f in part (a) converges to $f(x)$ for $|x| \le 1$. Therefore, the series for $f(x/c)$ converges to $f(x/c)$ provided $|x/c| \le 1$, or, equivalently, for $|x| \le c$.

c. The series of part (b) may be truncated after four terms to approximate cube roots. For example, note that $\sqrt[3]{29} = \sqrt[3]{\underset{c}{\underbrace{27}} + \underset{x}{\underbrace{2}}}$, so we take $c = 27$ and $x = 2$.

The choice $c = 27$ is made because 29 is near 27 and $\sqrt[3]{c} = \sqrt[3]{27} = 3$ is easy to evaluate. Substituting $c = 27$ and $x = 2$, we find that

$$\sqrt[3]{29} \approx \sqrt[3]{27}\left(1 + \frac{1}{3}\left(\frac{2}{27}\right) - \frac{1}{9}\left(\frac{2}{27}\right)^2 + \frac{5}{81}\left(\frac{2}{27}\right)^3\right) \approx 3.0723.$$

The same method is used to approximate the cube roots of $23, 24, \ldots, 30, 31$ (Table 11.4). The absolute error is the difference between the approximation and the value given by a calculator. Notice that the errors increase as we move away from 27.

Related Exercises 52–54 ◄

Table 11.4

	Approximation	Absolute Error
$\sqrt[3]{23}$	2.8439	6.7×10^{-5}
$\sqrt[3]{24}$	2.8845	2.0×10^{-5}
$\sqrt[3]{25}$	2.9240	3.9×10^{-6}
$\sqrt[3]{26}$	2.9625	2.4×10^{-7}
$\sqrt[3]{27}$	3	0
$\sqrt[3]{28}$	3.0366	2.3×10^{-7}
$\sqrt[3]{29}$	3.0723	3.5×10^{-6}
$\sqrt[3]{30}$	3.1072	1.7×10^{-5}
$\sqrt[3]{31}$	3.1414	5.4×10^{-5}

Convergence of Taylor Series

It may seem that the story of Taylor series is over. But there is a technical point that is easily overlooked. Given a function f, we know how to write its Taylor series centered at a point a, and we know how to find its interval of convergence. We still do not know that the series actually converges to f. The remaining task is to determine when the Taylor series for f actually converges to f on its interval of convergence. Fortunately, the necessary tools have already been presented in Taylor's Theorem (Theorem 11.1), which gives the remainder for Taylor polynomials.

Assume f has derivatives of *all* orders on an open interval containing the point a. Taylor's Theorem tells us that

$$f(x) = p_n(x) + R_n(x),$$

where p_n is the nth-order Taylor polynomial for f centered at a, that

$$R_n(x) = \frac{f^{(n+1)}(c)}{(n+1)!}(x-a)^{n+1}$$

is the remainder, and that c is a point between x and a. We see that the remainder, $R_n(x) = f(x) - p_n(x)$, measures the difference between f and the approximating polynomial p_n. When we say the Taylor series converges to f at a point x, we mean the value of the Taylor series at x equals $f(x)$; that is, $\lim_{n \to \infty} p_n(x) = f(x)$. The following theorem makes these ideas precise.

THEOREM 11.7   Convergence of Taylor Series

Let f have derivatives of all orders on an open interval I containing a. The Taylor series for f centered at a converges to f, for all x in I, if and only if $\lim_{n \to \infty} R_n(x) = 0$, for all x in I, where

$$R_n(x) = \frac{f^{(n+1)}(c)}{(n+1)!}(x-a)^{n+1}$$

is the remainder at x, with c between x and a.

Proof: The theorem requires derivatives of *all* orders. Therefore, by Taylor's Theorem (Theorem 11.1), the remainder exists in the given form for all n. Let p_n denote the nth-order Taylor polynomial and note that $\lim_{n \to \infty} p_n(x)$ is the Taylor series for f centered at a, evaluated at a point x in I.

First, assume $\lim_{n \to \infty} R_n(x) = 0$ on the interval I and recall that $p_n(x) = f(x) - R_n(x)$. Taking limits of both sides, we have

$$\underbrace{\lim_{n \to \infty} p_n(x)}_{\text{Taylor series}} = \lim_{n \to \infty}\left(f(x) - R_n(x)\right) = \underbrace{\lim_{n \to \infty} f(x)}_{f(x)} - \underbrace{\lim_{n \to \infty} R_n(x)}_{0} = f(x).$$

We conclude that the Taylor series $\lim_{n \to \infty} p_n(x)$ equals $f(x)$, for all x in I.

Conversely, if the Taylor series converges to f, then $f(x) = \lim_{n \to \infty} p_n(x)$ and

$$0 = f(x) - \lim_{n \to \infty} p_n(x) = \lim_{n \to \infty}\underbrace{\left(f(x) - p_n(x)\right)}_{R_n(x)} = \lim_{n \to \infty} R_n(x).$$

It follows that $\lim_{n \to \infty} R_n(x) = 0$, for all x in I.   ◄

Even with an expression for the remainder, it may be difficult to show that $\lim_{n \to \infty} R_n(x) = 0$. The following examples illustrate cases in which it is possible.

EXAMPLE 7   Remainder in the Maclaurin series for e^x Show that the Maclaurin series for $f(x) = e^x$ converges to $f(x)$, for $-\infty < x < \infty$.

SOLUTION As shown in Example 4, the Maclaurin series for $f(x) = e^x$ is

$$\sum_{k=0}^{\infty} \frac{x^k}{k!} = 1 + x + \frac{x^2}{2!} + \cdots + \frac{x^n}{n!} + \cdots,$$

which converges, for $-\infty < x < \infty$. In Example 7 of Section 11.1 it was shown that the remainder is

$$R_n(x) = \frac{e^c}{(n+1)!} x^{n+1},$$

where c is between 0 and x. Notice that the intermediate point c varies with n, but it is always between 0 and x. Therefore, e^c is between $e^0 = 1$ and e^x; in fact, $e^c \leq e^{|x|}$, for all n. It follows that

$$|R_n(x)| \leq \frac{e^{|x|}}{(n+1)!} |x|^{n+1}.$$

Holding x fixed, we have

$$\lim_{n\to\infty} |R_n(x)| = \lim_{n\to\infty} \frac{e^{|x|}}{(n+1)!}|x|^{n+1} = e^{|x|} \lim_{n\to\infty} \frac{|x|^{n+1}}{(n+1)!} = 0,$$

where we used the fact that $\lim_{n\to\infty} x^n/n! = 0$, for $-\infty < x < \infty$ (Section 10.2). Because $\lim_{n\to\infty} |R_n(x)| = 0$, it follows that for all real numbers x, the Taylor series converges to e^x, or

$$e^x = \sum_{k=0}^{\infty} \frac{x^k}{k!} = 1 + x + \frac{x^2}{2!} + \cdots + \frac{x^n}{n!} + \cdots.$$

The convergence of the Taylor series to e^x is illustrated in Figure 11.20, where Taylor polynomials of increasing degree are graphed together with e^x.

Related Exercise 65 ◄

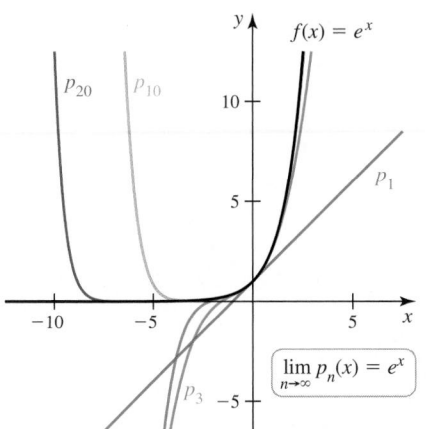

Figure 11.20

EXAMPLE 8   Maclaurin series convergence for cos x Show that the Maclaurin series for cos x,

$$1 - \frac{x^2}{2!} + \frac{x^4}{4!} - \frac{x^6}{6!} + \cdots = \sum_{k=0}^{\infty} (-1)^k \frac{x^{2k}}{(2k)!},$$

converges to $f(x) = \cos x$, for $-\infty < x < \infty$.

SOLUTION To show that the power series converges to f, we must show that $\lim_{n\to\infty} |R_n(x)| = 0$, for $-\infty < x < \infty$. According to Taylor's Theorem with $a = 0$,

$$R_n(x) = \frac{f^{(n+1)}(c)}{(n+1)!} x^{n+1},$$

where c is between 0 and x. Notice that $f^{(n+1)}(c) = \pm \sin c$ or $f^{(n+1)}(c) = \pm \cos c$. In all cases, $|f^{(n+1)}(c)| \leq 1$. Therefore, the absolute value of the remainder term is bounded as

$$|R_n(x)| = \left| \frac{f^{(n+1)}(c)}{(n+1)!} x^{n+1} \right| \leq \frac{|x|^{n+1}}{(n+1)!}.$$

Holding x fixed and using $\lim_{n\to\infty} x^n/n! = 0$, we see that $\lim_{n\to\infty} R_n(x) = 0$ for all x. Therefore, the given power series converges to $f(x) = \cos x$, for all x; that is, $\cos x = \sum_{k=0}^{\infty} \frac{(-1)^k x^{2k}}{(2k)!}$. The convergence of the Taylor series to $\cos x$ is illustrated in Figure 11.21.

Related Exercise 63 ◄

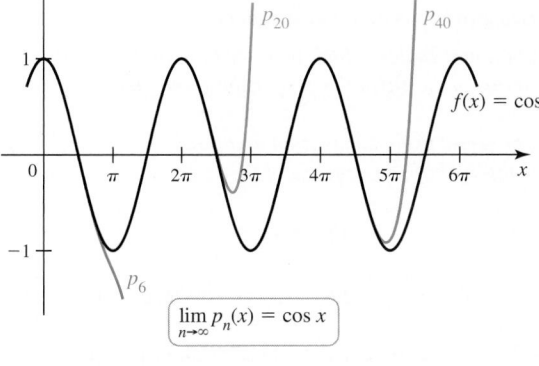

Figure 11.21

The procedure used in Examples 7 and 8 can be carried out for all the Taylor series we have worked with so far (with varying degrees of difficulty). In each case, the Taylor series converges to the function it represents on the interval of convergence. Table 11.5 summarizes commonly used Taylor series centered at 0 and the functions to which they converge.

▶ Table 11.5 asserts, without proof, that in several cases, the Taylor series for f converges to f at the endpoints of the interval of convergence. Proving convergence at the endpoints generally requires advanced techniques. It may also be done using the following theorem:

Suppose the Taylor series for f centered at 0 converges to f on the interval $(-R, R)$. If the series converges at $x = R$, then it converges to $\lim_{x \to R^-} f(x)$. If the series converges at $x = -R$, then it converges to $\lim_{x \to -R^+} f(x)$.

For example, this theorem would allow us to conclude that the series for $\ln(1 + x)$ converges to $\ln 2$ at $x = 1$.

Table 11.5

$$\frac{1}{1 - x} = 1 + x + x^2 + \cdots + x^k + \cdots = \sum_{k=0}^{\infty} x^k, \quad \text{for } |x| < 1$$

$$\frac{1}{1 + x} = 1 - x + x^2 - \cdots + (-1)^k x^k + \cdots = \sum_{k=0}^{\infty} (-1)^k x^k, \quad \text{for } |x| < 1$$

$$e^x = 1 + x + \frac{x^2}{2!} + \cdots + \frac{x^k}{k!} + \cdots = \sum_{k=0}^{\infty} \frac{x^k}{k!}, \quad \text{for } |x| < \infty$$

$$\sin x = x - \frac{x^3}{3!} + \frac{x^5}{5!} - \cdots + \frac{(-1)^k x^{2k+1}}{(2k+1)!} + \cdots = \sum_{k=0}^{\infty} \frac{(-1)^k x^{2k+1}}{(2k+1)!}, \quad \text{for } |x| < \infty$$

$$\cos x = 1 - \frac{x^2}{2!} + \frac{x^4}{4!} - \cdots + \frac{(-1)^k x^{2k}}{(2k)!} + \cdots = \sum_{k=0}^{\infty} \frac{(-1)^k x^{2k}}{(2k)!}, \quad \text{for } |x| < \infty$$

$$\ln(1 + x) = x - \frac{x^2}{2} + \frac{x^3}{3} - \cdots + \frac{(-1)^{k+1} x^k}{k} + \cdots = \sum_{k=1}^{\infty} \frac{(-1)^{k+1} x^k}{k}, \quad \text{for } -1 < x \le 1$$

$$-\ln(1 - x) = x + \frac{x^2}{2} + \frac{x^3}{3} + \cdots + \frac{x^k}{k} + \cdots = \sum_{k=1}^{\infty} \frac{x^k}{k}, \quad \text{for } -1 \le x < 1$$

$$\tan^{-1} x = x - \frac{x^3}{3} + \frac{x^5}{5} - \cdots + \frac{(-1)^k x^{2k+1}}{2k+1} + \cdots = \sum_{k=0}^{\infty} \frac{(-1)^k x^{2k+1}}{2k+1}, \quad \text{for } |x| \le 1$$

$$\sinh x = x + \frac{x^3}{3!} + \frac{x^5}{5!} + \cdots + \frac{x^{2k+1}}{(2k+1)!} + \cdots = \sum_{k=0}^{\infty} \frac{x^{2k+1}}{(2k+1)!}, \quad \text{for } |x| < \infty$$

$$\cosh x = 1 + \frac{x^2}{2!} + \frac{x^4}{4!} + \cdots + \frac{x^{2k}}{(2k)!} + \cdots = \sum_{k=0}^{\infty} \frac{x^{2k}}{(2k)!}, \quad \text{for } |x| < \infty$$

▶ As noted in Theorem 11.6, the binomial series may converge to $(1 + x)^p$ at $x = \pm 1$, depending on the value of p.

$$(1 + x)^p = \sum_{k=0}^{\infty} \binom{p}{k} x^k, \text{ for } |x| < 1 \text{ and } \binom{p}{k} = \frac{p(p-1)(p-2)\cdots(p-k+1)}{k!}, \binom{p}{0} = 1$$

SECTION 11.3 EXERCISES

Getting Started

1. How are the Taylor polynomials for a function f centered at a related to the Taylor series of the function f centered at a?

2. What conditions must be satisfied by a function f to have a Taylor series centered at a?

3. Find a Taylor series for f centered at 2 given that $f^{(k)}(2) = 1$, for all nonnegative integers k.

4. Find a Taylor series for f centered at 0 given that $f^{(k)}(0) = (k + 1)!$, for all nonnegative integers k.

5. Suppose you know the Maclaurin series for f and that it converges to $f(x)$ for $|x| < 1$. How do you find the Maclaurin series for $f(x^2)$ and where does it converge?

6. For what values of p does the Taylor series for $f(x) = (1 + x)^p$ centered at 0 terminate?

7. In terms of the remainder, what does it mean for a Taylor series for a function f to converge to f?

8. Find the Maclaurin series for $\sin(-x)$ using the definition of a Maclaurin series. Check your answer by finding the Maclaurin series for $\sin(-x)$ using Table 11.5.

Practice Exercises

9–26. Taylor series and interval of convergence

a. *Use the definition of a Taylor/Maclaurin series to find the first four nonzero terms of the Taylor series for the given function centered at a.*

b. *Write the power series using summation notation.*

c. *Determine the interval of convergence of the series.*

9. $f(x) = \dfrac{1}{x^2}, a = 1$

10. $f(x) = \dfrac{1}{x^2}, a = -1$

11. $f(x) = e^{-x}, a = 0$

12. $f(x) = \cos 2x, a = 0$

13. $f(x) = \dfrac{2}{(1 - x)^3}, a = 0$

14. $f(x) = x \sin x, a = 0$

15. $f(x) = (1 + x^2)^{-1}, a = 0$

16. $f(x) = \ln(1 + 4x), a = 0$

17. $f(x) = e^{2x}, a = 0$

18. $f(x) = (1 + 2x)^{-1}, a = 0$

19. $f(x) = \tan^{-1}\dfrac{x}{2}, a = 0$

20. $f(x) = \sin 3x, a = 0$

21. $f(x) = 3^x, a = 0$

22. $f(x) = \log_3(x + 1), a = 0$

23. $f(x) = \cosh 3x, a = 0$ **24.** $f(x) = \sinh 2x, a = 0$

25. $f(x) = \ln(x - 2), a = 3$ **26.** $f(x) = e^x, a = 2$

27–34. Taylor series

a. Use the definition of a Taylor series to find the first four nonzero terms of the Taylor series for the given function centered at a.

b. Write the power series using summation notation.

27. $f(x) = \sin x, a = \dfrac{\pi}{2}$ **28.** $f(x) = \cos x, a = \pi$

29. $f(x) = \dfrac{1}{x}, a = 1$ **30.** $f(x) = \dfrac{1}{x}, a = 2$

31. $f(x) = \ln x, a = 3$ **32.** $f(x) = e^x, a = \ln 2$

33. $f(x) = 2^x, a = 1$ **34.** $f(x) = x \ln x - x + 1; a = 1$

35–44. Manipulating Taylor series *Use the Taylor series in Table 11.5 to find the first four nonzero terms of the Taylor series for the following functions centered at 0.*

35. $\ln(1 + x^2)$ **36.** $\sin x^2$

37. $\dfrac{1}{1 - 2x}$ **38.** $\ln(1 + 2x)$

39. $\begin{cases} \dfrac{e^x - 1}{x} & \text{if } x \neq 1 \\ 1 & \text{if } x = 1 \end{cases}$ **40.** $\cos x^3$

41. $(1 + x^4)^{-1}$ **42.** $x \tan^{-1} x^2$

43. $\sinh x^2$ **44.** $\cosh(-2x)$

▣ 45–50. Binomial series

a. Find the first four nonzero terms of the binomial series centered at 0 for the given function.

b. Use the first four terms of the series to approximate the given quantity.

45. $f(x) = (1 + x)^{-2}$; approximate $\dfrac{1}{1.21} = \dfrac{1}{1.1^2}$.

46. $f(x) = \sqrt{1 + x}$; approximate $\sqrt{1.06}$.

47. $f(x) = \sqrt[4]{1 + x}$; approximate $\sqrt[4]{1.12}$.

48. $f(x) = (1 + x)^{-3}$; approximate $\dfrac{1}{1.331} = \dfrac{1}{1.1^3}$.

49. $f(x) = (1 + x)^{-2/3}$; approximate $1.18^{-2/3}$.

50. $f(x) = (1 + x)^{2/3}$; approximate $1.02^{2/3}$.

51–56. Working with binomial series *Use properties of power series, substitution, and factoring to find the first four nonzero terms of the Maclaurin series for the following functions. Give the interval of convergence for the new series (Theorem 11.4 is useful). Use the Maclaurin series*

$$\sqrt{1 + x} = 1 + \frac{x}{2} - \frac{x^2}{8} + \frac{x^3}{16} - \cdots, \quad \text{for } -1 \leq x \leq 1.$$

51. $\sqrt{1 + x^2}$ **52.** $\sqrt{4 + x}$

53. $\sqrt{9 - 9x}$ **54.** $\sqrt{1 - 4x}$

55. $\sqrt{a^2 + x^2}, a > 0$ **56.** $\sqrt{4 - 16x^2}$

57–62. Working with binomial series *Use properties of power series, substitution, and factoring to find the first four nonzero terms of the Maclaurin series for the following functions. Use the Maclaurin series*

$$(1 + x)^{-2} = 1 - 2x + 3x^2 - 4x^3 + \cdots, \quad \text{for } -1 < x < 1.$$

57. $(1 + 4x)^{-2}$ **58.** $\dfrac{1}{(1 - 4x)^2}$

59. $\dfrac{1}{(4 + x^2)^2}$ **60.** $(x^2 - 4x + 5)^{-2}$

61. $\dfrac{1}{(3 + 4x)^2}$ **62.** $\dfrac{1}{(1 + 4x^2)^2}$

63–66. Remainders *Find the remainder in the Taylor series centered at the point a for the following functions. Then show that $\lim_{n \to \infty} R_n(x) = 0$, for all x in the interval of convergence.*

63. $f(x) = \sin x, a = 0$ **64.** $f(x) = \cos 2x, a = 0$

65. $f(x) = e^{-x}, a = 0$ **66.** $f(x) = \cos x, a = \dfrac{\pi}{2}$

67. **Explain why or why not** Determine whether the following statements are true and give an explanation or counterexample.

a. The function $f(x) = \sqrt{x}$ has a Taylor series centered at 0.

b. The function $f(x) = \csc x$ has a Taylor series centered at $\pi/2$.

c. If f has a Taylor series that converges only on $(-2, 2)$, then $f(x^2)$ has a Taylor series that also converges only on $(-2, 2)$.

d. If $p(x)$ is the Taylor series for f centered at 0, then $p(x - 1)$ is the Taylor series for f centered at 1.

e. The Taylor series for an even function centered at 0 has only even powers of x.

68–75. Any method

a. Use any analytical method to find the first four nonzero terms of the Taylor series centered at 0 for the following functions. You do not need to use the definition of the Taylor series coefficients.

b. Determine the radius of convergence of the series.

68. $f(x) = \cos 2x + 2 \sin x$ **69.** $f(x) = \dfrac{e^x + e^{-x}}{2}$

70. $f(x) = \begin{cases} \dfrac{\sin x}{x} & \text{if } x \neq 0 \\ 1 & \text{if } x = 0 \end{cases}$ **71.** $f(x) = (1 + x^2)^{-2/3}$

72. $f(x) = x^2 \cos x^2$ **73.** $f(x) = \sqrt{1 - x^2}$

74. $f(x) = b^x$, for $b > 0, b \neq 1$

75. $f(x) = \dfrac{1}{x^4 + 2x^2 + 1}$

▣ 76–79. Approximating powers *Compute the coefficients for the Taylor series for the following functions about the given point a, and then use the first four terms of the series to approximate the given number.*

76. $f(x) = \sqrt{x}$ with $a = 36$; approximate $\sqrt{39}$.

77. $f(x) = \sqrt[3]{x}$ with $a = 64$; approximate $\sqrt[3]{60}$.

78. $f(x) = \dfrac{1}{\sqrt{x}}$ with $a = 4$; approximate $\dfrac{1}{\sqrt{3}}$.

79. $f(x) = \sqrt[4]{x}$ with $a = 16$; approximate $\sqrt[4]{13}$.

80. Geometric/binomial series Recall that the Taylor series for $f(x) = 1/(1 - x)$ centered at 0 is the geometric series $\sum_{k=0}^{\infty} x^k$. Show that this series can also be found as a binomial series.

81. Integer coefficients Show that the first five nonzero coefficients of the Taylor series (binomial series) for $f(x) = \sqrt{1 + 4x}$ centered at 0 are integers. (In fact, *all* the coefficients are integers.)

Explorations and Challenges

82. Choosing a good center Suppose you want to approximate $\sqrt{72}$ using four terms of a Taylor series. Compare the accuracy of the approximations obtained using the Taylor series for $\sqrt{x}$ centered at 64 and 81.

83. Alternative means By comparing the first four terms, show that the Maclaurin series for $\sin^2 x$ can be found (a) by squaring the Maclaurin series for $\sin x$, (b) by using the identity $\sin^2 x = \dfrac{1 - \cos 2x}{2}$, or (c) by computing the coefficients using the definition.

84. Alternative means By comparing the first four terms, show that the Maclaurin series for $\cos^2 x$ can be found (a) by squaring the Maclaurin series for $\cos x$, (b) by using the identity $\cos^2 x = \dfrac{1 + \cos 2x}{2}$, or (c) by computing the coefficients using the definition.

85. Designer series Find a power series that has $(2, 6)$ as an interval of convergence.

86. Composition of series Use composition of series to find the first three terms of the Maclaurin series for the following functions.

  a. $e^{\sin x}$ **b.** $e^{\tan x}$ **c.** $\sqrt{1 + \sin^2 x}$

▣ 87–88. Approximations *Choose a Taylor series and a center point to approximate the following quantities with an error of 10^{-4} or less.*

87. $\cos 40°$ **88.** $\sin(0.98\pi)$

89. Different approximation strategies Suppose you want to approximate $\sqrt[3]{128}$ to within 10^{-4} of the exact value.

  a. Use a Taylor polynomial for $f(x) = (125 + x)^{1/3}$ centered at 0.
  b. Use a Taylor polynomial for $f(x) = x^{1/3}$ centered at 125.
  c. Compare the two approaches. Are they equivalent?

90. Nonconvergence to f Consider the function

$$f(x) = \begin{cases} e^{-1/x^2} & \text{if } x \neq 0 \\ 0 & \text{if } x = 0. \end{cases}$$

  a. Use the definition of the derivative to show that $f'(0) = 0$.
  b. Assume the fact that $f^{(k)}(0) = 0$, for $k = 1, 2, 3, \ldots$. (You can write a proof using the definition of the derivative.) Write the Taylor series for f centered at 0.
  c. Explain why the Taylor series for f does not converge to f for $x \neq 0$.

91. Version of the Second Derivative Test Assume f has at least two continuous derivatives on an interval containing a with $f'(a) = 0$. Use Taylor's Theorem to prove the following version of the Second Derivative Test:

  a. If $f''(x) > 0$ on some interval containing a, then f has a local minimum at a.
  b. If $f''(x) < 0$ on some interval containing a, then f has a local maximum at a.

QUICK CHECK ANSWERS

1. When evaluated at $x = a$, all terms of the series are zero except the first term, which is $f(a)$. Therefore, the series equals $f(a)$ at this point.   **2.** $1 - x + x^2 - x^3 + x^4 - \cdots$
3. The series diverges at both endpoint by the Divergence Test.   **4.** $2x + 2x^2 + x^3; 1 - x + x^2/2$ **5.** $6, 1/16$
6. $1.05, 1.04875$ ◄

11.4 Working with Taylor Series

We now know the Taylor series for many familiar functions, and we have tools for working with power series. The goal of this final section is to illustrate additional techniques associated with power series. As you will see, power series cover the entire landscape of calculus from limits and derivatives to integrals and approximation. We present five different topics that you can explore selectively.

Limits by Taylor Series

An important use of Taylor series is evaluating limits. Two examples illustrate the essential ideas.

EXAMPLE 1   A limit by Taylor series Evaluate $\lim\limits_{x \to 0} \dfrac{x^2 + 2 \cos x - 2}{3x^4}$.

► L'Hôpital's Rule may be impractical when it must be used more than once on the same limit or when derivatives are difficult to compute.

SOLUTION Because the limit has the indeterminate form $0/0$, l'Hôpital's Rule can be used, which requires four applications of the rule. Alternatively, because the limit involves values of x near 0, we substitute the Maclaurin series for $\cos x$. Recalling that

$$\cos x = 1 - \frac{x^2}{2} + \frac{x^4}{24} - \frac{x^6}{720} + \cdots, \quad \text{Table 11.5}$$

➤ In using series to evaluate limits, it is often not obvious how many terms of the Taylor series to use. When in doubt, include extra (higher-order) terms. The dots in the calculation stand for powers of x greater than the last power that appears.

we have

$$\lim_{x\to 0}\frac{x^2+2\cos x-2}{3x^4}=\lim_{x\to 0}\frac{x^2+2\left(1-\dfrac{x^2}{2}+\dfrac{x^4}{24}-\dfrac{x^6}{720}+\cdots\right)-2}{3x^4}$$ Substitute for $\cos x$.

$$=\lim_{x\to 0}\frac{x^2+\left(2-x^2+\dfrac{x^4}{12}-\dfrac{x^6}{360}+\cdots\right)-2}{3x^4}$$ Simplify.

$$=\lim_{x\to 0}\frac{\dfrac{x^4}{12}-\dfrac{x^6}{360}+\cdots}{3x^4}$$ Simplify.

$$=\lim_{x\to 0}\left(\frac{1}{36}-\frac{x^2}{1080}+\cdots\right)=\frac{1}{36}.$$ Use Theorem 11.4, Property 2; evaluate limit.

Related Exercise 13 ◄

QUICK CHECK 1 Use the Taylor series $\sin x=x-x^3/6+\cdots$ to verify that $\lim\limits_{x\to 0}(\sin x)/x=1$. ◄

EXAMPLE 2   A limit by Taylor series Evaluate

$$\lim_{x\to\infty}\left(6x^5\sin\frac{1}{x}-6x^4+x^2\right).$$

SOLUTION A Taylor series may be centered at any finite point in the domain of the function, but we don't have the tools needed to expand a function about $x=\infty$. Using a technique introduced earlier, we replace x with $1/t$ and note that as $x\to\infty$, $t\to 0^+$. The new limit becomes

$$\lim_{x\to\infty}\left(6x^5\sin\frac{1}{x}-6x^4+x^2\right)=\lim_{t\to 0^+}\left(\frac{6\sin t}{t^5}-\frac{6}{t^4}+\frac{1}{t^2}\right)$$ Replace x with $1/t$.

$$=\lim_{t\to 0^+}\left(\frac{6\sin t-6t+t^3}{t^5}\right).$$ Common denominator

This limit has the indeterminate form $0/0$. We now expand $\sin t$ in a Taylor series centered at $t=0$. Because

$$\sin t=t-\frac{t^3}{6}+\frac{t^5}{120}-\frac{t^7}{5040}+\cdots,$$ Table 11.5

the value of the original limit is

$$\lim_{t\to 0^+}\left(\frac{6\sin t-6t+t^3}{t^5}\right)$$

$$=\lim_{t\to 0^+}\left(\frac{6\left(t-\dfrac{t^3}{6}+\dfrac{t^5}{120}-\dfrac{t^7}{5040}+\cdots\right)-6t+t^3}{t^5}\right)$$ Substitute for $\sin t$.

$$=\lim_{t\to 0^+}\left(\frac{\dfrac{t^5}{20}-\dfrac{t^7}{840}+\cdots}{t^5}\right)$$ Simplify.

$$=\lim_{t\to 0^+}\left(\frac{1}{20}-\frac{t^2}{840}+\cdots\right)=\frac{1}{20}.$$ Use Theorem 11.4, Property 2; evaluate limit.

Related Exercise 19 ◄

Differentiating Power Series

The following examples illustrate ways in which term-by-term differentiation (Theorem 11.5) may be used.

EXAMPLE 3   Power series for derivatives Differentiate the Maclaurin series for $f(x)=\sin x$ to verify that $\dfrac{d}{dx}(\sin x)=\cos x$.

SOLUTION The Maclaurin series for $f(x) = \sin x$ is

$$\sin x = x - \frac{x^3}{3!} + \frac{x^5}{5!} - \frac{x^7}{7!} + \cdots,$$

and it converges for $-\infty < x < \infty$. By Theorem 11.5, the differentiated series also converges for $-\infty < x < \infty$ and it converges to $f'(x)$. Differentiating, we have

$$\frac{d}{dx}\left(x - \frac{x^3}{3!} + \frac{x^5}{5!} - \frac{x^7}{7!} + \cdots\right) = 1 - \frac{x^2}{2!} + \frac{x^4}{4!} - \frac{x^6}{6!} + \cdots = \cos x.$$

The differentiated series is the Maclaurin series for $\cos x$, confirming that $f'(x) = \cos x$.

Related Exercises 25–26 ◄

QUICK CHECK 2 Differentiate the power series for $\cos x$ (given in Example 3) and identify the result. ◄

EXAMPLE 4   **A differential equation**   Find a power series solution of the differential equation $y'(t) = y + 2$, subject to the initial condition $y(0) = 6$. Identify the function represented by the power series.

SOLUTION Because the initial condition is given at $t = 0$, we assume the solution has a Taylor series centered at 0 of the form $y(t) = \sum_{k=0}^{\infty} c_k t^k$, where the coefficients c_k must be determined. Recall that the coefficients of the Taylor series are given by

$$c_k = \frac{y^{(k)}(0)}{k!}, \quad \text{for } k = 0, 1, 2, \ldots.$$

If we can determine $y^{(k)}(0)$, for $k = 0, 1, 2, \ldots$, the coefficients of the series are also determined.

Substituting the initial condition $t = 0$ and $y = 6$ into the power series

$$y(t) = c_0 + c_1 t + c_2 t^2 + \cdots,$$

we find that

$$6 = c_0 + c_1(0) + c_2(0)^2 + \cdots.$$

It follows that $c_0 = 6$. To determine $y'(0)$, we substitute $t = 0$ into the differential equation; the result is $y'(0) = y(0) + 2 = 6 + 2 = 8$. Therefore, $c_1 = y'(0)/1! = 8$.

The remaining derivatives are obtained by successively differentiating the differential equation and substituting $t = 0$. We find that $y''(0) = y'(0) = 8$, $y'''(0) = y''(0) = 8$, and in general, $y^{(k)}(0) = 8$, for $k = 1, 2, 3, 4, \ldots$. Therefore,

$$c_k = \frac{y^{(k)}(0)}{k!} = \frac{8}{k!}, \quad \text{for } k = 1, 2, 3, \ldots,$$

and the Taylor series for the solution is

$$y(t) = c_0 + c_1 t + c_2 t^2 + \cdots$$

$$= 6 + \frac{8}{1!}t + \frac{8}{2!}t^2 + \frac{8}{3!}t^3 + \cdots.$$

To identify the function represented by this series, we write

$$y(t) = \underbrace{-2 + 8}_{6} + \frac{8}{1!}t + \frac{8}{2!}t^2 + \frac{8}{3!}t^3 + \cdots$$

$$= -2 + 8\underbrace{\left(1 + t + \frac{t^2}{2!} + \frac{t^3}{3!} + \cdots\right)}_{e^t}.$$

➤ You should check that $y(t) = -2 + 8e^t$ satisfies $y'(t) = y + 2$ and $y(0) = 6$.

The power series that appears is the Taylor series for e^t. Therefore, the solution is $y = -2 + 8e^t$.

Related Exercise 36 ◄

Integrating Power Series

The following example illustrates the use of power series in approximating integrals that cannot be evaluated by analytical methods.

EXAMPLE 5 **Approximating a definite integral** Approximate the value of the integral $\int_0^1 e^{-x^2}\,dx$ with an error no greater than 5×10^{-4}.

SOLUTION The antiderivative of e^{-x^2} cannot be expressed in terms of familiar functions. The strategy is to write the Maclaurin series for e^{-x^2} and integrate it term by term. Recall that integration of a power series is valid within its interval of convergence (Theorem 11.5). Beginning with the Maclaurin series

$$e^x = 1 + x + \frac{x^2}{2!} + \frac{x^3}{3!} + \cdots + \frac{x^n}{n!} + \cdots,$$

which converges for $-\infty < x < \infty$, we replace x with $-x^2$ to obtain

$$e^{-x^2} = 1 - x^2 + \frac{x^4}{2!} - \frac{x^6}{3!} + \cdots + \frac{(-1)^n x^{2n}}{n!} + \cdots,$$

which also converges for $-\infty < x < \infty$. By the Fundamental Theorem of Calculus,

$$\int_0^1 e^{-x^2}\,dx = \left(x - \frac{x^3}{3} + \frac{x^5}{5 \cdot 2!} - \frac{x^7}{7 \cdot 3!} + \cdots + \frac{(-1)^n x^{2n+1}}{(2n+1)n!} + \cdots \right)\Bigg|_0^1$$

$$= 1 - \frac{1}{3} + \frac{1}{5 \cdot 2!} - \frac{1}{7 \cdot 3!} + \cdots + \frac{(-1)^n}{(2n+1)n!} + \cdots.$$

> The integral in Example 5 is important in statistics and probability theory because of its relationship to the *normal distribution*.

Because the definite integral is expressed as an alternating series, the magnitude of the remainder in truncating the series after n terms is less than the magnitude of the first neglected term, which is $\left| \dfrac{(-1)^{n+1}}{(2n+3)(n+1)!} \right|$. By trial and error, we find that the magnitude of this term is less than 5×10^{-4} if $n \geq 5$ (with $n = 5$, we have $\dfrac{1}{13 \cdot 6!} \approx 1.07 \times 10^{-4}$). The sum of the terms of the series up to $n = 5$ gives the approximation

$$\int_0^1 e^{-x^2}\,dx \approx 1 - \frac{1}{3} + \frac{1}{5 \cdot 2!} - \frac{1}{7 \cdot 3!} + \frac{1}{9 \cdot 4!} - \frac{1}{11 \cdot 5!} \approx 0.747.$$

Related Exercises 37–38 ◄

Representing Real Numbers

When values of x are substituted into a convergent power series, the result may be a series representation of a familiar real number. The following example illustrates some techniques.

EXAMPLE 6 **Evaluating infinite series**

a. Use the Maclaurin series for $f(x) = \tan^{-1}x$ to evaluate

$$1 - \frac{1}{3} + \frac{1}{5} - \cdots = \sum_{k=0}^{\infty} \frac{(-1)^k}{2k+1}.$$

b. Let $f(x) = (e^x - 1)/x$, for $x \neq 0$, and $f(0) = 1$. Use the Maclaurin series for f to evaluate $f'(1)$ and $\displaystyle\sum_{k=1}^{\infty} \frac{k}{(k+1)!}$.

SOLUTION

a. From Table 11.5, we see that for $|x| \leq 1$,

$$\tan^{-1}x = x - \frac{x^3}{3} + \frac{x^5}{5} - \cdots + \frac{(-1)^k x^{2k+1}}{2k+1} + \cdots = \sum_{k=0}^{\infty} \frac{(-1)^k x^{2k+1}}{2k+1}.$$

▶ The series in Example 6a (known as the *Gregory series*) is one of a multitude of series representations of π. Because this series converges slowly, it does not provide an efficient way to approximate π.

Substituting $x = 1$, we have

$$\tan^{-1} 1 = 1 - \frac{1^3}{3} + \frac{1^5}{5} - \cdots = \sum_{k=0}^{\infty} \frac{(-1)^k}{2k + 1}.$$

Because $\tan^{-1} 1 = \pi/4$, the value of the series is $\pi/4$.

b. Using the Maclaurin series for e^x, the series for $f(x) = (e^x - 1)/x$ is

$$f(x) = \frac{e^x - 1}{x} = \frac{1}{x}\left(\left(\underbrace{1 + x + \frac{x^2}{2!} + \frac{x^3}{3!} + \cdots}_{e^x}\right) - 1\right) \quad \text{Substitute series for } e^x.$$

$$= 1 + \frac{x}{2!} + \frac{x^2}{3!} + \frac{x^3}{4!} + \cdots = \sum_{k=1}^{\infty} \frac{x^{k-1}}{k!}, \quad \text{Theorem 11.4, Property 2}$$

which converges for $-\infty < x < \infty$. By the Quotient Rule,

$$f'(x) = \frac{xe^x - (e^x - 1)}{x^2}.$$

Differentiating the series for f term by term (Theorem 11.5), we find that

$$f'(x) = \frac{d}{dx}\left(1 + \frac{x}{2!} + \frac{x^2}{3!} + \frac{x^3}{4!} + \cdots\right)$$

$$= \frac{1}{2!} + \frac{2x}{3!} + \frac{3x^2}{4!} + \cdots = \sum_{k=1}^{\infty} \frac{kx^{k-1}}{(k + 1)!}.$$

QUICK CHECK 3 What value of x would you substitute into the Maclaurin series for $\tan^{-1} x$ to obtain a series representation for $\pi/6$? ◄

We now have two expressions for f'; they are evaluated at $x = 1$ to show that

$$f'(1) = 1 = \sum_{k=1}^{\infty} \frac{k}{(k + 1)!}.$$

Related Exercises 50–51 ◄

Representing Functions as Power Series

Power series have a fundamental role in mathematics in defining functions and providing alternative representations of familiar functions. As an overall review, we close this chapter with two examples that use many techniques for working with power series.

EXAMPLE 7   Identify the series Identify the function represented by the power series $\sum_{k=0}^{\infty} \frac{(1 - 2x)^k}{k!}$ and give its interval of convergence.

SOLUTION The Maclaurin series for the exponential function,

$$e^x = \sum_{k=0}^{\infty} \frac{x^k}{k!},$$

converges for $-\infty < x < \infty$. Replacing x with $1 - 2x$ produces the given series:

$$\sum_{k=0}^{\infty} \frac{(1 - 2x)^k}{k!} = e^{1 - 2x}.$$

This replacement is allowed because $1 - 2x$ is within the interval of convergence of the series for e^x; that is, $-\infty < 1 - 2x < \infty$, for all x. Therefore, the given series represents $e^{1 - 2x}$, for $-\infty < x < \infty$.

Related Exercises 55–56 ◄

EXAMPLE 8   Mystery series The power series $\sum_{k=1}^{\infty} \frac{(-1)^k k}{4^k} x^{2k}$ appeared in the opening of Section 11.2. Determine the interval of convergence of the power series and find the function it represents on this interval.

SOLUTION Applying the Ratio Test to the series, we determine that it converges when $|x^2/4| < 1$, which implies that $|x| < 2$. A quick check of the endpoints of the original series confirms that it diverges at $x = \pm 2$. Therefore, the interval of convergence is $|x| < 2$.

To find the function represented by the series, we apply several maneuvers until we obtain a geometric series. First note that

$$\sum_{k=1}^{\infty} \frac{(-1)^k k}{4^k} x^{2k} = \sum_{k=1}^{\infty} k\left(-\frac{1}{4}\right)^k x^{2k}.$$

The series on the right is not a geometric series because of the presence of the factor k. The key is to realize that k could appear in this way through differentiation; specifically, something like $\frac{d}{dx}(x^{2k}) = 2kx^{2k-1}$. To achieve terms of this form, we write

$$\underbrace{\sum_{k=1}^{\infty} \frac{(-1)^k k}{4^k} x^{2k}}_{\text{original series}} = \sum_{k=1}^{\infty} k\left(-\frac{1}{4}\right)^k x^{2k}$$

$$= \frac{1}{2} \sum_{k=1}^{\infty} 2k\left(-\frac{1}{4}\right)^k x^{2k} \qquad \text{Multiply and divide by 2.}$$

$$= \frac{x}{2} \sum_{k=1}^{\infty} 2k\left(-\frac{1}{4}\right)^k x^{2k-1}. \qquad \text{Remove } x \text{ from the series.}$$

Now we identify the last series as the derivative of another series:

$$\underbrace{\sum_{k=1}^{\infty} \frac{(-1)^k k}{4^k} x^{2k}}_{\text{original series}} = \frac{x}{2} \sum_{k=1}^{\infty} \left(-\frac{1}{4}\right)^k \underbrace{2kx^{2k-1}}_{\frac{d}{dx}(x^{2k})}$$

$$= \frac{x}{2} \sum_{k=1}^{\infty} \left(-\frac{1}{4}\right)^k \frac{d}{dx}(x^{2k}) \qquad \text{Identify a derivative.}$$

$$= \frac{x}{2} \frac{d}{dx}\left(\sum_{k=1}^{\infty} \left(-\frac{x^2}{4}\right)^k \right). \qquad \text{Combine factors; differentiate term by term.}$$

This last series is a geometric series with a ratio $r = -x^2/4$ and first term $-x^2/4$; therefore, its value is $\dfrac{-x^2/4}{1 + (x^2/4)}$, provided $\left|\dfrac{x^2}{4}\right| < 1$, or $|x| < 2$. We now have

$$\underbrace{\sum_{k=1}^{\infty} \frac{(-1)^k k}{4^k} x^{2k}}_{\text{original series}} = \frac{x}{2} \frac{d}{dx}\left(\sum_{k=1}^{\infty} \left(-\frac{x^2}{4}\right)^k \right)$$

$$= \frac{x}{2} \frac{d}{dx}\left(\frac{-x^2/4}{1 + (x^2/4)} \right) \qquad \text{Sum of geometric series}$$

$$= \frac{x}{2} \frac{d}{dx}\left(\frac{-x^2}{4 + x^2} \right) \qquad \text{Simplify.}$$

$$= -\frac{4x^2}{(4 + x^2)^2}. \qquad \text{Differentiate and simplify.}$$

Therefore, the function represented by the power series on $(-2, 2)$ has been uncovered; it is

$$f(x) = -\frac{4x^2}{(4 + x^2)^2}.$$

Notice that f is defined for $-\infty < x < \infty$ (**Figure 11.22**), but its power series centered at 0 converges to f only on $(-2, 2)$.

Related Exercises 61, 63 ◄

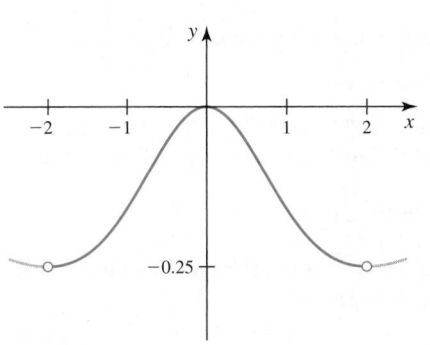

$$\sum_{k=1}^{\infty} \frac{(-1)^k k}{4^k} x^{2k} = -\frac{4x^2}{(4 + x^2)^2} \text{ on } (-2, 2)$$

Figure 11.22

SECTION 11.4 EXERCISES

Getting Started

1. Explain the strategy presented in this section for evaluating a limit of the form $\lim\limits_{x \to a} \dfrac{f(x)}{g(x)}$, where f and g have Taylor series centered at a.

2. Explain the method presented in this section for approximating $\int_a^b f(x)\,dx$, where f has a Taylor series with an interval of convergence centered at a that includes b.

3. How would you approximate $e^{-0.6}$ using the Taylor series for e^x?

4. Use the Taylor series for $\cos x$ centered at 0 to verify that
$$\lim_{x \to 0} \frac{1 - \cos x}{x} = 0.$$

5. Use the Taylor series for $\sinh x$ and $\cosh x$ to verify that
$$\frac{d}{dx} \sinh x = \cosh x.$$

6. What condition must be met by a function f for it to have a Taylor series centered at a?

Practice Exercises

7–24. Limits *Evaluate the following limits using Taylor series.*

7. $\lim\limits_{x \to 0} \dfrac{e^x - 1}{x}$

8. $\lim\limits_{x \to 0} \dfrac{\tan^{-1}x - x}{x^3}$

9. $\lim\limits_{x \to 0} \dfrac{-x - \ln(1 - x)}{x^2}$

10. $\lim\limits_{x \to 0} \dfrac{\sin 2x}{x}$

11. $\lim\limits_{x \to 0} \dfrac{e^x - e^{-x}}{x}$

12. $\lim\limits_{x \to 0} \dfrac{1 + x - e^x}{4x^2}$

13. $\lim\limits_{x \to 0} \dfrac{2 \cos 2x - 2 + 4x^2}{2x^4}$

14. $\lim\limits_{x \to \infty} x \sin \dfrac{1}{x}$

15. $\lim\limits_{x \to 0} \dfrac{\ln(1 + x) - x + x^2/2}{x^3}$

16. $\lim\limits_{x \to 4} \dfrac{x^2 - 16}{\ln(x - 3)}$

17. $\lim\limits_{x \to 0} \dfrac{3 \tan^{-1}x - 3x + x^3}{x^5}$

18. $\lim\limits_{x \to 0} \dfrac{\sqrt{1 + x} - 1 - (x/2)}{4x^2}$

19. $\lim\limits_{x \to 0} \dfrac{12x - 8x^3 - 6 \sin 2x}{x^5}$

20. $\lim\limits_{x \to 1} \dfrac{x - 1}{\ln x}$

21. $\lim\limits_{x \to 2} \dfrac{x - 2}{\ln(x - 1)}$

22. $\lim\limits_{x \to \infty} x(e^{1/x} - 1)$

23. $\lim\limits_{x \to 0} \dfrac{e^{-2x} - 4e^{-x/2} + 3}{2x^2}$

24. $\lim\limits_{x \to 0} \dfrac{(1 - 2x)^{-1/2} - e^x}{8x^2}$

25–32. Power series for derivatives

a. *Differentiate the Taylor series centered at 0 for the following functions.*
b. *Identify the function represented by the differentiated series.*
c. *Give the interval of convergence of the power series for the derivative.*

25. $f(x) = e^x$

26. $f(x) = -\cos x$

27. $f(x) = \ln(1 + x)$

28. $f(x) = \sin x^2$

29. $f(x) = e^{-2x}$

30. $f(x) = (1 - x)^{-1}$

31. $f(x) = \tan^{-1}x$

32. $f(x) = -\ln(1 - x)$

33–36. Differential equations

a. *Find a power series for the solution of the following differential equations, subject to the given initial condition.*
b. *Identify the function represented by the power series.*

33. $y'(t) - y = 0, y(0) = 2$

34. $y'(t) + 4y = 8, y(0) = 0$

35. $y'(t) - 3y = 10, y(0) = 2$

36. $y'(t) = 6y + 9, y(0) = 2$

▣ 37–44. Approximating definite integrals *Use a Taylor series to approximate the following definite integrals. Retain as many terms as needed to ensure the error is less than 10^{-4}.*

37. $\int_0^{0.25} e^{-x^2}\,dx$

38. $\int_0^{0.2} \sin x^2\,dx$

39. $\int_{-0.35}^{0.35} \cos 2x^2\,dx$

40. $\int_0^{0.2} \sqrt{1 + x^4}\,dx$

41. $\int_0^{0.35} \tan^{-1}x\,dx$

42. $\int_0^{0.4} \ln(1 + x^2)\,dx$

43. $\int_0^{0.5} \dfrac{dx}{\sqrt{1 + x^6}}$

44. $\int_0^{0.2} \dfrac{\ln(1 + t)}{t}\,dt$

45–50. Approximating real numbers *Use an appropriate Taylor series to find the first four nonzero terms of an infinite series that is equal to the following numbers.*

45. e^2

46. $\sqrt{e}$

47. $\cos 2$

48. $\sin 1$

49. $\ln \dfrac{3}{2}$

50. $\tan^{-1} \dfrac{1}{2}$

51. **Evaluating an infinite series** Let $f(x) = \dfrac{e^x - 1}{x}$, for $x \neq 0$, and $f(0) = 1$. Use the Taylor series for f centered at 0 to evaluate $f(1)$ and to find the value of $\sum\limits_{k=0}^{\infty} \dfrac{1}{(k + 1)!}$.

52. **Evaluating an infinite series** Let $f(x) = \dfrac{e^x - 1}{x}$, for $x \neq 0$, and $f(0) = 1$. Use the Taylor series for f and f' centered at 0 to evaluate $f'(2)$ and to find the value of $\sum\limits_{k=1}^{\infty} \dfrac{k 2^{k-1}}{(k + 1)!}$.

53. **Evaluating an infinite series** Write the Taylor series for $f(x) = \ln(1 + x)$ centered at 0 and find its interval of convergence. Assume the Taylor series converges to f on the interval of convergence. Evaluate $f(1)$ to find the value of $\sum\limits_{k=1}^{\infty} \dfrac{(-1)^{k+1}}{k}$ (the alternating harmonic series).

54. **Evaluating an infinite series** Write the Maclaurin series for $f(x) = \ln(1 + x)$ and find the interval of convergence. Evaluate $f\left(-\dfrac{1}{2}\right)$ to find the value of $\sum\limits_{k=1}^{\infty} \dfrac{1}{k 2^k}$.

55–64. Representing functions by power series *Identify the functions represented by the following power series.*

55. $\sum\limits_{k=0}^{\infty} \dfrac{x^k}{2^k}$

56. $\sum\limits_{k=0}^{\infty} (-1)^k \dfrac{x^k}{3^k}$

57. $\sum\limits_{k=0}^{\infty} (-1)^k \dfrac{x^{2k}}{4^k}$

58. $\sum\limits_{k=0}^{\infty} 2^k x^{2k+1}$

59. $\sum\limits_{k=1}^{\infty} \dfrac{x^k}{k}$

60. $\sum\limits_{k=0}^{\infty} \dfrac{(-1)^k x^{k+1}}{4^k}$

61. $\displaystyle\sum_{k=1}^{\infty} (-1)^k \frac{kx^{k+1}}{3^k}$

62. $\displaystyle\sum_{k=1}^{\infty} \frac{x^{2k}}{k}$

63. $\displaystyle\sum_{k=2}^{\infty} \frac{k(k-1)x^k}{3^k}$

64. $\displaystyle\sum_{k=2}^{\infty} \frac{x^k}{k(k-1)}$

65. Explain why or why not Determine whether the following statements are true and give an explanation or counterexample.

a. To evaluate $\displaystyle\int_0^2 \frac{dx}{1-x}$, one could expand the integrand in a Taylor series and integrate term by term.

b. To approximate $\pi/3$, one could substitute $x = \sqrt{3}$ into the Taylor series for $\tan^{-1}x$.

c. $\displaystyle\sum_{k=0}^{\infty} \frac{(\ln 2)^k}{k!} = 2.$

66–68. Limits with a parameter *Use Taylor series to evaluate the following limits. Express the result in terms of the nonzero real parameter(s).*

66. $\displaystyle\lim_{x\to 0} \frac{e^{ax} - 1}{x}$

67. $\displaystyle\lim_{x\to 0} \frac{\sin ax}{\sin bx}$

68. $\displaystyle\lim_{x\to 0} \frac{\sin ax - \tan^{-1} ax}{bx^3}$

Explorations and Challenges

69. A limit by Taylor series Use Taylor series to evaluate

$$\lim_{x\to 0} \left(\frac{\sin x}{x}\right)^{1/x^2}.$$

70. Inverse hyperbolic sine The *inverse of hyperbolic sine* is defined in several ways; among them are

$$\sinh^{-1} x = \ln\left(x + \sqrt{x^2 + 1}\right) = \int_0^x \frac{dt}{\sqrt{1 + t^2}}.$$

Find the first four terms of the Taylor series for $\sinh^{-1} x$ using these two definitions (and be sure they agree).

71–74. Derivative trick *Here is an alternative way to evaluate higher derivatives of a function f that may save time. Suppose you can find the Taylor series for f centered at the point a without evaluating derivatives (for example, from a known series). Then $f^{(k)}(a) = k!$ multiplied by the coefficient of $(x - a)^k$. Use this idea to evaluate $f^{(3)}(0)$ and $f^{(4)}(0)$ for the following functions. Use known series and do not evaluate derivatives.*

71. $f(x) = e^{\cos x}$

72. $f(x) = \dfrac{x^2 + 1}{\sqrt[3]{1 + x}}$

73. $f(x) = \displaystyle\int_0^x \sin t^2 \, dt$

74. $f(x) = \displaystyle\int_0^x \frac{1}{1 + t^4} \, dt$

75. Probability: tossing for a head The expected (average) number of tosses of a fair coin required to obtain the first head is $\displaystyle\sum_{k=1}^{\infty} k\left(\frac{1}{2}\right)^k$. Evaluate this series and determine the expected number of tosses. (*Hint:* Differentiate a geometric series.)

76. Probability: sudden-death playoff Teams A and B go into sudden-death overtime after playing to a tie. The teams alternate possession of the ball, and the first team to score wins. Assume each team has a $1/6$ chance of scoring when it has the ball, and Team A has the ball first.

a. The probability that Team A ultimately wins is $\displaystyle\sum_{k=0}^{\infty} \frac{1}{6}\left(\frac{5}{6}\right)^{2k}$. Evaluate this series.

b. The expected number of rounds (possessions by either team) required for the overtime to end is $\displaystyle\frac{1}{6}\sum_{k=1}^{\infty} k\left(\frac{5}{6}\right)^{k-1}$. Evaluate this series.

T 77. Elliptic integrals The period of an undamped pendulum is given by

$$T = 4\sqrt{\frac{\ell}{g}} \int_0^{\pi/2} \frac{d\theta}{\sqrt{1 - k^2 \sin^2 \theta}} = 4\sqrt{\frac{\ell}{g}} F(k),$$

where ℓ is the length of the pendulum, $g = 9.8 \text{ m/s}^2$ is the acceleration due to gravity, $k = \sin\dfrac{\theta_0}{2}$, and θ_0 is the initial angular displacement of the pendulum (in radians). The integral in this formula $F(k)$ is called an *elliptic integral*, and it cannot be evaluated analytically. Approximate $F(0.1)$ by expanding the integrand in a Taylor (binomial) series and integrating term by term.

78. Sine integral function The function $\text{Si}(x) = \displaystyle\int_0^x f(t)\, dt$, where

$$f(t) = \begin{cases} \frac{\sin t}{t} & \text{if } t \neq 0 \\ 1 & \text{if } t = 0 \end{cases}, \text{ is called the } \textit{sine integral function.}$$

a. Expand the integrand in a Taylor series centered at 0.

b. Integrate the series to find a Taylor series for Si.

c. Approximate $\text{Si}(0.5)$ and $\text{Si}(1)$. Use enough terms of the series so the error in the approximation does not exceed 10^{-3}.

T 79. Fresnel integrals The theory of optics gives rise to the two *Fresnel integrals*

$$S(x) = \int_0^x \sin t^2 \, dt \quad \text{and} \quad C(x) = \int_0^x \cos t^2 \, dt.$$

a. Compute $S'(x)$ and $C'(x)$.

b. Expand $\sin t^2$ and $\cos t^2$ in a Maclaurin series, and then integrate to find the first four nonzero terms of the Maclaurin series for S and C.

c. Use the polynomials in part (b) to approximate $S(0.05)$ and $C(-0.25)$.

d. How many terms of the Maclaurin series are required to approximate $S(0.05)$ with an error no greater than 10^{-4}?

e. How many terms of the Maclaurin series are required to approximate $C(-0.25)$ with an error no greater than 10^{-6}?

T 80. Error function An essential function in statistics and the study of the normal distribution is the *error function*

$$\text{erf}(x) = \frac{2}{\sqrt{\pi}} \int_0^x e^{-t^2} \, dt.$$

a. Compute the derivative of $\text{erf}(x)$.

b. Expand e^{-t^2} in a Maclaurin series; then integrate to find the first four nonzero terms of the Maclaurin series for erf.

c. Use the polynomial in part (b) to approximate $\text{erf}(0.15)$ and $\text{erf}(-0.09)$.

d. Estimate the error in the approximations of part (c).

T 81. Bessel functions Bessel functions arise in the study of wave propagation in circular geometries (for example, waves on a circular drum head). They are conveniently defined as power series. One of an infinite family of Bessel functions is

$$J_0(x) = \sum_{k=0}^{\infty} \frac{(-1)^k}{2^{2k}(k!)^2} x^{2k}.$$

a. Write out the first four terms of J_0.

b. Find the radius and interval of convergence of the power series for J_0.

c. Differentiate J_0 twice and show (by keeping terms through x^6) that J_0 satisfies the equation $x^2 y''(x) + xy'(x) + x^2 y(x) = 0$.

T 82. Newton's derivation of the sine and arcsine series Newton discovered the binomial series and then used it ingeniously to obtain many more results. Here is a case in point.

a. Referring to the figure, show that $x = \sin s$ or $s = \sin^{-1} x$.

b. The area of a circular sector of radius r subtended by an angle θ is $1/2 \, r^2\theta$. Show that the area of the circular sector APE is $s/2$, which implies that

$$s = 2\int_0^x \sqrt{1 - t^2} \, dt - x\sqrt{1 - x^2}.$$

c. Use the binomial series for $f(x) = \sqrt{1 - x^2}$ to obtain the first few terms of the Taylor series for $s = \sin^{-1}x$.

d. Newton next inverted the series in part (c) to obtain the Taylor series for $x = \sin s$. He did this by assuming $\sin s = \sum a_k s^k$ and solving $x = \sin(\sin^{-1}x)$ for the coefficients a_k. Find the first few terms of the Taylor series for $\sin s$ using this idea (a computer algebra system might be helpful as well).

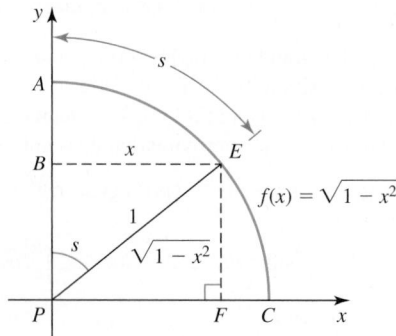

83. L'Hôpital's Rule by Taylor series Suppose f and g have Taylor series about the point a.

a. If $f(a) = g(a) = 0$ and $g'(a) \neq 0$, evaluate $\lim\limits_{x \to a} \dfrac{f(x)}{g(x)}$ by expanding f and g in their Taylor series. Show that the result is consistent with l'Hôpital's Rule.

b. If $f(a) = g(a) = f'(a) = g'(a) = 0$ and $g''(a) \neq 0$, evaluate $\lim\limits_{x \to a} \dfrac{f(x)}{g(x)}$ by expanding f and g in their Taylor series. Show that the result is consistent with two applications of l'Hôpital's Rule.

84. Symmetry

a. Use infinite series to show that $\cos x$ is an even function. That is, show $\cos(-x) = \cos x$.

b. Use infinite series to show that $\sin x$ is an odd function. That is, show $\sin(-x) = -\sin x$.

QUICK CHECK ANSWERS

1. $\dfrac{\sin x}{x} = \dfrac{x - x^3/3! + \cdots}{x} = 1 - \dfrac{x^2}{3!} + \cdots \to 1$ as $x \to 0$

2. The result is the power series for $-\sin x$.   **3.** $x = 1/\sqrt{3}$ (which lies in the interval of convergence) ◄

CHAPTER 11   REVIEW EXERCISES

1. **Explain why or why not** Determine whether the following statements are true and give an explanation or counterexample.

a. Let p_n be the nth-order Taylor polynomial for f centered at 2. The approximation $p_3(2.1) \approx f(2.1)$ is likely to be more accurate than the approximation $p_2(2.2) \approx f(2.2)$.

b. If the Taylor series for f centered at 3 has a radius of convergence of 6, then the interval of convergence is $[-3, 9]$.

c. The interval of convergence of the power series $\sum c_k x^k$ could be $(-7/3, 7/3)$.

d. The Maclaurin series for $f(x) = (1 + x)^{12}$ has a finite number of nonzero terms.

e. If the power series $\sum c_k (x - 3)^k$ has a radius of convergence of $R = 4$ and converges at the endpoints of its interval of convergence, then its interval of convergence is $[-1, 7]$.

2–9. Taylor polynomials *Find the nth-order Taylor polynomial for the following functions centered at the given point a.*

2. $f(x) = \sin 2x, n = 3, a = 0$

3. $f(x) = \cos^3 x, n = 2, a = 0$

4. $f(x) = \cos^{-1} x, n = 2, a = \dfrac{1}{2}$

5. $f(x) = e^{1/x - 1}, n = 2, a = 1$

6. $f(x) = e^{\sin x}, n = 2, a = 0$

7. $f(x) = \cos(\ln x), n = 2, a = 1$

8. $f(x) = \sinh(-3x), n = 3, a = 0$

9. $f(x) = \cosh x, n = 3, a = \ln 2$

T 10–13. Approximations

a. *Find the Taylor polynomials of order $n = 1$ and $n = 2$ for the given functions centered at the given point a.*

b. *Use the Taylor polynomials to approximate the given expression. Make a table showing the approximations and the absolute error in these approximations using a calculator for the exact function value.*

10. $f(x) = \cos x, a = 0; \cos(-0.08)$

11. $f(x) = e^x, a = 0; e^{-0.08}$

12. $f(x) = \sqrt{1 + x}, a = 0; \sqrt{1.08}$

13. $f(x) = \sin x, a = \dfrac{\pi}{4}; \sin\dfrac{\pi}{5}$

14–16. Estimating remainders *Find the remainder term $R_n(x)$ for the Taylor series centered at 0 for the following functions. Find an upper bound for the magnitude of the remainder on the given interval for the given value of n. (The bound is not unique.)*

14. $f(x) = e^x$; bound $R_3(x)$, for $|x| < 1$

15. $f(x) = \sin x$; bound $R_3(x)$, for $|x| < \pi$

16. $f(x) = \ln(1 - x)$; bound $R_3(x)$, for $|x| < \dfrac{1}{2}$

17–26. Radius and interval of convergence *Use the Ratio Test or the Root Test to determine the radius of convergence of the following power series. Test the endpoints to determine the interval of convergence, when appropriate.*

17. $\displaystyle\sum_{k=1}^{\infty} \frac{k^2 x^k}{k!}$

18. $\displaystyle\sum_{k=1}^{\infty} \frac{x^{4k}}{k^2}$

19. $\displaystyle\sum_{k=1}^{\infty} \frac{(-1)^k (x + 1)^{2k}}{k!}$

20. $\displaystyle\sum_{k=1}^{\infty} \frac{(x - 1)^k}{k \, 5^k}$

21. $\displaystyle\sum_{k=0}^{\infty} \left(\frac{x}{9}\right)^{3k}$

22. $\displaystyle\sum_{k=1}^{\infty} \frac{(x + 2)^k}{\sqrt{k}}$

23. $\displaystyle\sum_{k=1}^{\infty} \frac{(x + 2)^k}{2^k \ln k}$

24. $\displaystyle\sum_{k=1}^{\infty} \frac{(2k)!(x - 2)^k}{k!}$

25. $\displaystyle\sum_{k=1}^{\infty} \frac{(-1)^k (2x + 1)^k}{k^2 3^k}$

26. $x + \dfrac{x^3}{3} + \dfrac{x^5}{5} + \dfrac{x^7}{7} + \cdots$

27–28. Radius of convergence *Find the radius of convergence of each series.*

27. $\displaystyle\sum_{k=1}^{\infty} \frac{(3k)! \, x^k}{(k!)^3}.$

28. $\displaystyle\sum_{k=1}^{\infty} \frac{2^k k! (x - 5)^k}{k^k}$

29–34. Power series from the geometric series *Use the geometric series $\displaystyle\sum_{k=0}^{\infty} x^k = \frac{1}{1 - x}$, for $|x| < 1$, to determine the Maclaurin series and the interval of convergence for the following functions.*

29. $f(x) = \dfrac{1}{1 - x^2}$

30. $f(x) = \dfrac{1}{1 + x^3}$

31. $f(x) = \dfrac{1}{1 + 5x}$

32. $f(x) = \dfrac{10x}{1 + x}$

33. $f(x) = \dfrac{1}{(1 - 10x)^2}$

34. $f(x) = \ln(1 - 4x)$

35–42. Taylor series *Write out the first three nonzero terms of the Taylor series for the following functions centered at the given point a. Then write the series using summation notation.*

35. $f(x) = e^{3x}, a = 0$

36. $f(x) = \dfrac{1}{2x + 1}, a = 1$

37. $f(x) = \cos x, a = \dfrac{\pi}{2}$

38. $f(x) = \dfrac{x^2}{1 + x}, a = 0$

39. $f(x) = \tan^{-1} 4x, a = 0$

40. $f(x) = \sin 2x, a = -\dfrac{\pi}{2}$

41. $f(x) = \cosh(2x - 2), a = 1$

42. $f(x) = \dfrac{1}{4 + x^2}, a = 0$

43–46. Binomial series *Write out the first three terms of the Maclaurin series for the following functions.*

43. $f(x) = (1 + x)^{1/3}$

44. $f(x) = (1 + x)^{-1/2}$

45. $f(x) = \left(1 + \dfrac{x}{2}\right)^{-3}$

46. $f(x) = (1 + 2x)^{-5}$

47–48. Convergence *Write the remainder term $R_n(x)$ for the Taylor series for the following functions centered at the given point a. Then show that $\lim_{n \to \infty} |R_n(x)| = 0$, for all x in the given interval.*

47. $f(x) = \sinh x + \cosh x, a = 0, -\infty < x < \infty$

48. $f(x) = \ln(1 + x), a = 0, -\dfrac{1}{2} \leq x \leq \dfrac{1}{2}$

49–54. Limits by power series *Use Taylor series to evaluate the following limits.*

49. $\displaystyle\lim_{x \to 0} \frac{x^2/2 - 1 + \cos x}{x^4}$

50. $\displaystyle\lim_{x \to 0} \frac{2 \sin x - \tan^{-1} x - x}{2x^5}$

51. $\displaystyle\lim_{x \to 4} \frac{\ln(x - 3)}{x^2 - 16}$

52. $\displaystyle\lim_{x \to 0} \frac{\sqrt{1 + 2x} - 1 - x}{x^2}$

53. $\displaystyle\lim_{x \to 0} \frac{\sec x - \cos x - x^2}{x^4}$ *(Hint: The Maclaurin series for $\sec x$ is*
$1 + \dfrac{x^2}{2} + \dfrac{5x^4}{24} + \dfrac{61x^6}{720} + \cdots .)$

54. $\displaystyle\lim_{x \to 0} \frac{(1 + x)^{-2} - \sqrt[3]{1 - 6x}}{2x^2}$

T 55–58. Definite integrals by power series *Use a Taylor series to approximate the following definite integrals. Retain as many terms as necessary to ensure the error is less than 10^{-3}.*

55. $\displaystyle\int_0^{1/2} e^{-x^2} \, dx$

56. $\displaystyle\int_0^{1/2} \tan^{-1} x \, dx$

57. $\displaystyle\int_0^{1} x \cos x \, dx$

58. $\displaystyle\int_0^{1/2} x^2 \tan^{-1} x \, dx$

T 59–62. Approximating real numbers *Use an appropriate Taylor series to find the first four nonzero terms of an infinite series that is equal to the following numbers. There is more than one way to choose the center of the series.*

59. $\sqrt{119}$

60. $\sin 20°$

61. $\tan^{-1}\left(-\dfrac{1}{3}\right)$

62. $\sinh(-1)$

63. A differential equation Find a power series solution of the differential equation $y'(x) - 4y + 12 = 0$, subject to the condition $y(0) = 4$. Identify the solution in terms of known functions.

T 64. Rejected quarters The probability that a random quarter is *not* rejected by a vending machine is given by the integral $11.4 \int_0^{0.14} e^{-102x^2} \, dx$ (assuming the weights of quarters are normally distributed with a mean of 5.670 g and a standard deviation of 0.07 g). Estimate the value of the integral by using the first two terms of the Maclaurin series for e^{-102x^2}.

T 65. Approximating ln 2 Consider the following three ways to approximate ln 2.

a. Use the Taylor series for $\ln(1 + x)$ centered at 0 and evaluate it at $x = 1$ (convergence was asserted in Table 11.5). Write the resulting infinite series.

b. Use the Taylor series for $\ln(1 - x)$ centered at 0 and the identity $\ln 2 = -\ln \dfrac{1}{2}$. Write the resulting infinite series.

c. Use the property $\ln \dfrac{a}{b} = \ln a - \ln b$ and the series of parts (a) and (b) to find the Taylor series for $f(x) = \ln\left(\dfrac{1 + x}{1 - x}\right)$ centered at 0.

d. At what value of x should the series in part (c) be evaluated to approximate ln 2? Write the resulting infinite series for ln 2.

e. Using four terms of the series, which of the three series derived in parts (a)–(d) gives the best approximation to ln 2? Can you explain why?

T 66. Graphing Taylor polynomials Consider the function $f(x) = (1 + x)^{-4}$.

a. Find the Taylor polynomials p_0, p_1, p_2, and p_3 centered at 0.

b. Use a graphing utility to plot the Taylor polynomials and f, for $-1 < x < 1$.

c. For each Taylor polynomial, give the interval on which its graph appears indistinguishable from the graph of f.

Chapter 11 Guided Projects

Applications of the material in this chapter and related topics can be found in the following Guided Projects. For additional information, see the Preface.

- Series approximations to π
- Euler's formula (Taylor series with complex numbers)
- Stirling's formula and $n!$
- Three-sigma quality control
- Fourier series

12

Parametric and Polar Curves

Chapter Preview Up to this point, we have focused our attention on functions of the form $y = f(x)$ and studied their behavior in the setting of the Cartesian coordinate system. There are, however, alternative ways to generate curves and represent functions. We begin by introducing parametric equations, which are featured prominently in upcoming chapters, to represent curves and trajectories in three-dimensional space. When working with objects that have circular, cylindrical, or spherical shapes, other coordinate systems are often advantageous. In this chapter, we introduce the polar coordinate system for circular geometries. Cylindrical and spherical coordinate systems appear in Chapter 16. After working with parametric equations and polar coordinates, the next step is to investigate calculus in these settings. How do we find slopes of tangent lines, rates of change, areas of regions, and arc length when the curves of interest are described with parametric equations or described in polar coordinates? The chapter ends with the related topic of *conic sections*. Ellipses, parabolas, and hyperbolas (all of which are conic sections) can be represented in both Cartesian and polar coordinates. These important families of curves have many fascinating properties and they appear throughout the remainder of the text.

12.1 Parametric Equations

So far, we have used functions of the form $y = F(x)$ to describe curves in the xy-plane. In this section, we look at another way to define curves, known as *parametric equations*. As you will see, parametric curves enable us to describe both common and exotic curves; they are also indispensable for modeling the trajectories of moving objects.

Basic Ideas

A motor boat travels counterclockwise around a circular course with a radius of 4 miles, completing one lap every 2π hours at a constant speed. Suppose we wish to describe the points on the path of the boat $(x(t), y(t))$ at any time $t \geq 0$, where t is measured in hours. We assume the boat starts on the positive x-axis at the point $(4, 0)$ (**Figure 12.1**). Note that the angle θ corresponding to the position of the boat increases by 2π radians every 2π hours beginning with $\theta = 0$ when $t = 0$; therefore, $\theta = t$, for $t \geq 0$. It follows that the x- and y-coordinates of the boat are

$$x = 4\cos\theta = 4\cos t \quad \text{and} \quad y = 4\sin\theta = 4\sin t,$$

where $t \geq 0$. You can confirm that when $t = 0$, the boat is at the starting point $(4, 0)$; when $t = 2\pi$, it returns to the starting point.

The equations $x = 4\cos t$ and $y = 4\sin t$ are examples of **parametric equations**. They specify x and y in terms of a third variable t called a **parameter**, which often represents time.

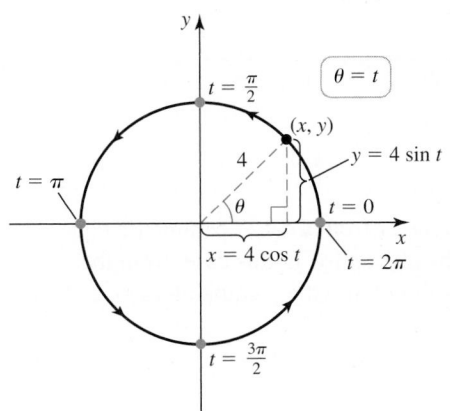

Figure 12.1

753

In general, parametric equations have the form

$$x = f(t), \quad y = g(t),$$

where f and g are given functions and the parameter t typically varies over a specified interval $a \le t \le b$ (**Figure 12.2**). The **parametric curve** described by these equations consists of the points in the plane

$$(x, y) = (f(t), g(t)), \quad \text{for } a \le t \le b.$$

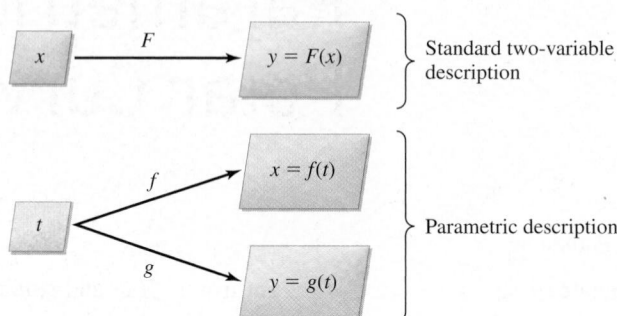

> ➤ With a parametric description, the parameter t is the independent variable. There are two dependent variables, x and y.

Figure 12.2

EXAMPLE 1   **Parametric parabola** Graph and analyze the parametric equations

$$x = f(t) = 2t, \quad y = g(t) = \frac{1}{2}t^2 - 4, \quad \text{for } 0 \le t \le 8.$$

Table 12.1

t	x	y	(x, y)
0	0	-4	$(0, -4)$
1	2	$-\frac{7}{2}$	$\left(2, -\frac{7}{2}\right)$
2	4	-2	$(4, -2)$
3	6	$\frac{1}{2}$	$\left(6, \frac{1}{2}\right)$
4	8	4	$(8, 4)$
5	10	$\frac{17}{2}$	$\left(10, \frac{17}{2}\right)$
6	12	14	$(12, 14)$
7	14	$\frac{41}{2}$	$\left(14, \frac{41}{2}\right)$
8	16	28	$(16, 28)$

SOLUTION Plotting individual points often helps in visualizing a parametric curve. Table 12.1 shows the values of x and y corresponding to several values of t on the interval $[0, 8]$. By plotting the (x, y) pairs in Table 12.1 and connecting them with a smooth curve, we obtain the graph shown in **Figure 12.3**. As t increases from its initial value of $t = 0$ to its final value of $t = 8$, the curve is generated from the initial point $(0, -4)$ to the final point $(16, 28)$. Notice that the values of the parameter do not appear in the graph. The only signature of the parameter is the direction in which the curve is generated: In this case, it unfolds upward and to the right, as indicated by the arrows on the curve.

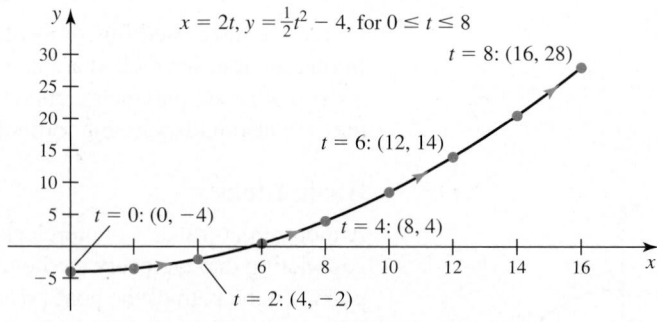

Figure 12.3

Sometimes it is possible to eliminate the parameter from a set of parametric equations and obtain a description of the curve in terms of x and y. In this case, from the x-equation, we have $t = x/2$, which may be substituted into the y-equation to give

$$y = \frac{1}{2}t^2 - 4 = \frac{1}{2}\left(\frac{x}{2}\right)^2 - 4 = \frac{x^2}{8} - 4.$$

QUICK CHECK 1 Identify the graph generated by the parametric equations $x = t^2, y = t$, for $-10 \le t \le 10$. ◄

Expressed in this form, we identify the graph as part of a parabola. Because t lies in the interval $0 \le t \le 8$ and $x = 2t$, it follows that x lies in the interval $0 \le x \le 16$. Therefore, the parametric equations generate the segment of the parabola for $0 \le x \le 16$.

Related Exercise 12 ◄

Given a set of parametric equations, the preceding example shows that as the parameter increases, the corresponding curve unfolds in a particular direction. The following definition captures this fact and is important in upcoming work.

> **DEFINITION Positive Orientation**
>
> The direction in which a parametric curve is generated as the parameter increases is called the **positive orientation** of the curve (and is indicated by arrows on the curve).

EXAMPLE 2 Parametric circle

a. Confirm that the parametric equations

$$x = 4 \cos 2\pi t, \quad y = 4 \sin 2\pi t, \quad \text{for } 0 \le t \le 1$$

describe a circle of radius 4 centered at the origin.

b. Suppose a turtle walks with constant speed in the counterclockwise direction on the circular path from part (a). Starting from the point $(4, 0)$, the turtle completes one lap in 30 minutes. Find a parametric description of the path of the turtle at any time $t \ge 0$, where t is measured in minutes.

SOLUTION

a. For each value of t in Table 12.2, the corresponding ordered pairs (x, y) are recorded. Plotting these points as t increases from $t = 0$ to $t = 1$ results in a graph that appears to be a circle of radius 4; it is generated with positive orientation in the counterclockwise direction, beginning and ending at $(4, 0)$ (**Figure 12.4**). Letting t increase beyond $t = 1$ would simply retrace the same curve.

To identify the curve conclusively, the parameter t is eliminated by observing that

$$
\begin{aligned}
x^2 + y^2 &= (4 \cos 2\pi t)^2 + (4 \sin 2\pi t)^2 \\
&= 16\underbrace{(\cos^2 2\pi t + \sin^2 2\pi t)}_{1} = 16. \quad \cos^2 \theta + \sin^2 \theta = 1
\end{aligned}
$$

We have confirmed that the graph of the parametric equations is the circle $x^2 + y^2 = 16$.

Table 12.2

t	(x, y)
0	$(4, 0)$
$\frac{1}{8}$	$(2\sqrt{2}, 2\sqrt{2})$
$\frac{1}{4}$	$(0, 4)$
$\frac{3}{8}$	$(-2\sqrt{2}, 2\sqrt{2})$
$\frac{1}{2}$	$(-4, 0)$
$\frac{3}{4}$	$(0, -4)$
1	$(4, 0)$

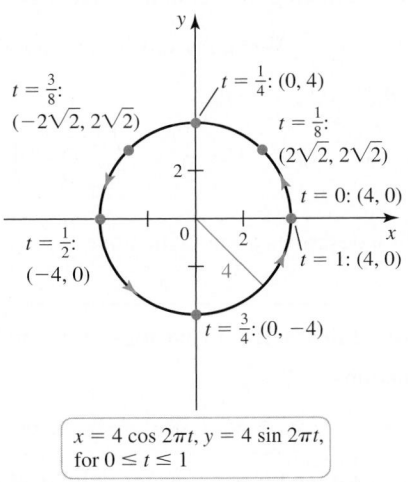

$x = 4 \cos 2\pi t, y = 4 \sin 2\pi t,$ for $0 \le t \le 1$

Figure 12.4

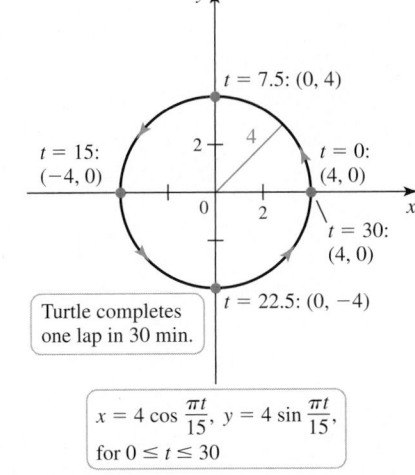

Turtle completes one lap in 30 min.

$x = 4 \cos \dfrac{\pi t}{15}, y = 4 \sin \dfrac{\pi t}{15},$ for $0 \le t \le 30$

Figure 12.5

b. Duplicating the calculations at the end of part (a) with any nonzero real number b, it can be shown that the parametric equations

$$x = 4 \cos bt, \quad y = 4 \sin bt$$

also describe a circle of radius 4. When $b > 0$, the graph is generated in the counterclockwise direction. The *angular frequency b* must be chosen so that as t varies

▶ In Example 2b, the constant $|b|$ is called the *angular frequency* because it is the number of radians the object moves per unit time. The turtle travels 2π rad every 30 min, so the angular frequency is $2\pi/30 = \pi/15$ rad/min. Because radians have no units, the angular frequency in this case has units *per minute*, which is written $\min^{-1}$.

from 0 to 30, the product bt varies from 0 to 2π. Specifically, when $t = 30$, we must have $30b = 2\pi$, or $b = \dfrac{\pi}{15}$ rad/min. Therefore, the parametric equations for the turtle's motion are

$$x = 4 \cos \frac{\pi t}{15}, \quad y = 4 \sin \frac{\pi t}{15}, \quad \text{for } 0 \le t \le 30.$$

You should check that as t varies from 0 to 30, the points (x, y) make one complete circuit of a circle of radius 4 (**Figure 12.5**).

Related Exercises 20, 37, 55 ◀

Generalizing Example 2 for nonzero real numbers a and b in the parametric equations $x = a \cos bt$, $y = a \sin bt$, we find that

$$x^2 + y^2 = (a \cos bt)^2 + (a \sin bt)^2$$
$$= a^2 \underbrace{(\cos^2 bt + \sin^2 bt)}_{1} = a^2.$$

▶ For a nonzero constant b, the functions $\sin bt$ and $\cos bt$ have period $2\pi/|b|$. The equations $x = a \cos bt$, $y = -a \sin bt$ also describe a circle of radius $|a|$, as do the equations $x = \pm a \sin bt$, $y = \pm a \cos bt$, as t varies over an interval of length $2\pi/|b|$.

Therefore, the parametric equations $x = a \cos bt$, $y = a \sin bt$ describe all or part of the circle $x^2 + y^2 = a^2$, centered at the origin with radius $|a|$, for any nonzero value of b. The circle is traversed once as t varies over any interval of length $2\pi/|b|$. If t represents time, the circle is traversed once in $2\pi/|b|$ time units, which means we can vary the speed at which the curve unfolds by varying b. If $b > 0$, the positive orientation is in the counterclockwise direction. If $b < 0$, the curve is generated in the clockwise direction.

More generally, the parametric equations

$$x = x_0 + a \cos bt, \quad y = y_0 + a \sin bt$$

QUICK CHECK 2 Give the center and radius of the circle generated by the equations $x = 3 \sin t$, $y = -3 \cos t$, for $0 \le t \le 2\pi$. Specify the direction of positive orientation. ◀

describe all or part of the circle $(x - x_0)^2 + (y - y_0)^2 = a^2$, centered at (x_0, y_0) with radius $|a|$. If $b > 0$, then the circle is generated in the counterclockwise direction. Example 2 shows that a single curve—for example, a circle of radius 4—may be parameterized in many different ways.

Among the most important of all parametric equations are

$$x = x_0 + at, y = y_0 + bt, \quad \text{for } -\infty < t < \infty,$$

where x_0, y_0, a, and b are constants with $a \ne 0$. The curve described by these equations is found by eliminating the parameter. The first step is to solve the x-equation for t, which gives us $t = \dfrac{x - x_0}{a}$. When t is substituted into the y-equation, the result is an equation for y in terms of x:

$$y = y_0 + bt = y_0 + b\left(\frac{x - x_0}{a}\right) \quad \text{or} \quad y - y_0 = \frac{b}{a}(x - x_0).$$

This equation describes a line with slope $\dfrac{b}{a}$ passing through the point (x_0, y_0).

SUMMARY Parametric Equations of a Line

The equations

$$x = x_0 + at, y = y_0 + bt, \quad \text{for } -\infty < t < \infty,$$

where x_0, y_0, a, and b are constants with $a \ne 0$, describe a line with slope $\dfrac{b}{a}$ passing through the point (x_0, y_0). If $a = 0$ and $b \ne 0$, the line is vertical.

Notice that the parametric description of a given line is not unique: For example, if k is any nonzero constant, the numbers a and b may be replaced with ka and kb, respectively, and the resulting equations describe the same line (although it may be generated in the opposite direction and at a different speed).

EXAMPLE 3   Parametric equations of lines

a. Consider the parametric equations $x = -2 + 3t$, $y = 4 - 6t$, for $-\infty < t < \infty$, which describe a line. Find the slope-intercept form of the line.

b. Find two pairs of parametric equations for the line with slope $\frac{1}{3}$ that passes through the point $(2, 1)$.

c. Find parametric equations for the line segment starting at $P(4, 7)$ and ending at $Q(2, -3)$.

SOLUTION

a. To eliminate the parameter, first solve the x-equation for t to find that $t = \frac{x + 2}{3}$. Replacing t in the y-equation yields

$$y = 4 - 6\left(\frac{x + 2}{3}\right) = 4 - 2x - 4 = -2x.$$

The line $y = -2x$ passes through the origin with slope -2.

b. We use the general parametric equations of a line given in the Summary box. Because the slope of the line is $\frac{1}{3}$, we choose $a = 3$ and $b = 1$. Letting $x_0 = 2$ and $y_0 = 1$, parametric equations for the line are $x = 2 + 3t$, $y = 1 + t$, for $-\infty < t < \infty$. The line passes through $(2, 1)$ when $t = 0$ and rises to the right as t increases (**Figure 12.6**). Notice that other choices for a and b also work. For example, with $a = -6$ and $b = -2$, the equations are $x = 2 - 6t$, $y = 1 - 2t$, for $-\infty < t < \infty$. These equations describe the same line, but now, as t increases, the line is generated in the opposite direction (descending to the left).

c. The slope of this line is $\frac{7 - (-3)}{4 - 2} = 5$. However, notice that when the line segment is traversed from P to Q, both x and y are decreasing (**Figure 12.7**). To account for the direction in which the line segment is generated, we let $a = -1$ and $b = -5$. Because $P(4, 7)$ is the starting point of the line segment, we choose $x_0 = 4$ and $y_0 = 7$. The resulting equations are $x = 4 - t$, $y = 7 - 5t$. Notice that $t = 0$ corresponds to the starting point $(4, 7)$. Because the equations describe a line *segment*, the interval for t must be restricted. What value of t corresponds to the endpoint of the line segment $Q(2, -3)$? Setting $x = 4 - t = 2$, we find that $t = 2$. As a check, we set $y = 7 - 5t = -3$, which also implies that $t = 2$. (If these two calculations do not give the same value of t, it probably means the slope was not computed correctly.) Therefore, the equations for the line segment are $x = 4 - t$, $y = 7 - 5t$, for $0 \le t \le 2$.

Related Exercises 15, 41–42 ◄

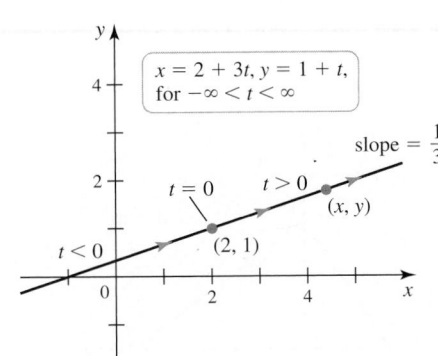

Figure 12.6

➤ The choices $a = -1$ and $b = -5$ result in a slope of $b/a = 5$. These choices also imply that as we move from P to Q, a decrease in x corresponds to a decrease in y.

➤ Lines and line segments may have unexpected parametric representations. For example, the equations $x = \sin t$, $y = 2 \sin t$ represent the line segment $y = 2x$, where $-1 \le x \le 1$ (recall that the range of $x = \sin t$ is $[-1, 1]$).

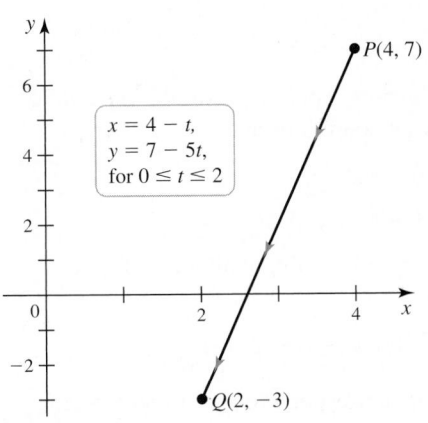

Figure 12.7

QUICK CHECK 3 Describe the curve generated by $x = 3 + 2t$, $y = -12 - 6t$, for $-\infty < t < \infty$. ◄

EXAMPLE 4   Parametric equations of curves A common task (particularly in upcoming chapters) is to parameterize curves given either by Cartesian equations or by graphs. Find a parametric representation of the following curves.

a. The segment of the parabola $y = 9 - x^2$, for $-1 \le x \le 3$

b. The complete curve $x = (y - 5)^2 + \sqrt{y}$

c. The piecewise linear path connecting $P(-2, 0)$ to $Q(0, 3)$ to $R(4, 0)$ (in that order), where the parameter varies over the interval $0 \le t \le 2$

SOLUTION

a. The simplest way to represent a curve $y = f(x)$ parametrically is to let $x = t$ and $y = f(t)$, where t is the parameter. We must then find the appropriate interval for the parameter. Using this approach, the curve $y = 9 - x^2$ has the parametric representation

$$x = t, \quad y = 9 - t^2, \quad \text{for} \quad -1 \le t \le 3.$$

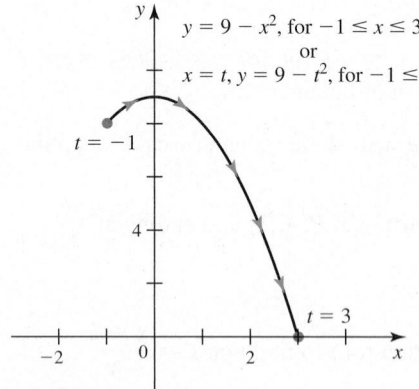

Figure 12.8

This representation is not unique. For example, you can verify that the parametric equations

$$x = 1 - t, \quad y = 9 - (1 - t)^2, \quad \text{for} \quad -2 \le t \le 2$$

also do the job, although these equations trace the parabola from right to left, while the original equations trace the curve from left to right (**Figure 12.8**).

b. In this case, it is easier to let $y = t$. Then a parametric description of the curve is

$$x = (t - 5)^2 + \sqrt{t}, \quad y = t.$$

Notice that t can take values only in the interval $[0, \infty)$. As $t \to \infty$, we see that $x \to \infty$ and $y \to \infty$ (**Figure 12.9**).

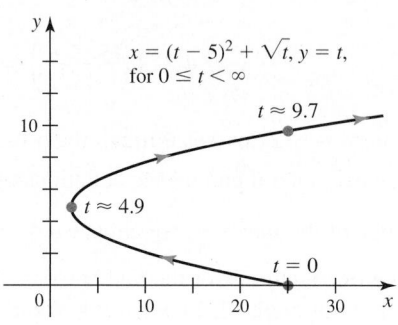

Figure 12.9

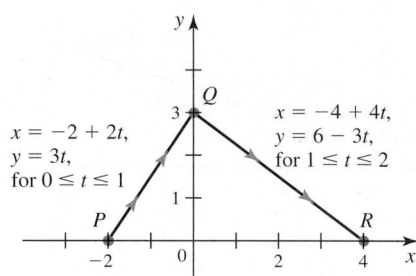

Figure 12.10

c. The path consists of two line segments (**Figure 12.10**) that can be parameterized separately in the form $x = x_0 + at$ and $y = y_0 + bt$. The line segment PQ originates at $P(-2, 0)$ and unfolds in the positive x-direction with slope $\frac{3}{2}$. It can be represented as

$$x = -2 + 2t, \quad y = 3t, \quad \text{for } 0 \le t \le 1.$$

Finding the parametric equations for the line segment QR requires some ingenuity. We want the line segment to originate at $Q(0, 3)$ when $t = 1$ and end at $R(4, 0)$ when $t = 2$. Observe that when $t = 1, x = 0$ and when $t = 2, x = 4$. Substituting these pairs of values into the general x-equation $x = x_0 + at$, we obtain the equations

$$\begin{aligned} x_0 + a &= 0 \quad &x = 0 \text{ when } t = 1 \\ x_0 + 2a &= 4. \quad &x = 4 \text{ when } t = 2 \end{aligned}$$

Solving for x_0 and a, we find that $x_0 = -4$ and $a = 4$. Applying a similar procedure to the general y-equation $y = y_0 + bt$, the relevant conditions are

$$\begin{aligned} y_0 + b &= 3 \quad &y = 3 \text{ when } t = 1 \\ y_0 + 2b &= 0. \quad &y = 0 \text{ when } t = 2 \end{aligned}$$

Solving for y_0 and b, we find that $y_0 = 6$ and $b = -3$. Putting it all together, the equations for the line segment QR are

$$x = -4 + 4t, y = 6 - 3t, \quad \text{for } 1 \le t \le 2.$$

You can verify that the points $Q(0, 3)$ and $R(4, 0)$ correspond to $t = 1$ and $t = 2$, respectively. Furthermore, the slope of the line is $\dfrac{b}{a} = -\dfrac{3}{4}$, which is correct.

Related Exercises 43–44, 47 ◄

QUICK CHECK 4 Find parametric equations for the line segment that goes from $Q(0, 3)$ to $P(-2, 0)$. ◄

Many fascinating curves are generated by points on rolling wheels; Examples 5 and 8 investigate two such curves.

EXAMPLE 5   Rolling wheels The path of a light on the rim of a wheel rolling on a flat surface (**Figure 12.11a**) is a **cycloid**, which has the parametric equations

$$x = a(t - \sin t), \quad y = a(1 - \cos t), \quad \text{for } t \geq 0,$$

where $a > 0$. Use a graphing utility to graph the cycloid with $a = 1$. On what interval does the parameter generate the first arch of the cycloid?

SOLUTION The graph of the cycloid, for $0 \leq t \leq 3\pi$, is shown in **Figure 12.11b**. The wheel completes one full revolution on the interval $0 \leq t \leq 2\pi$, which gives one arch of the cycloid.

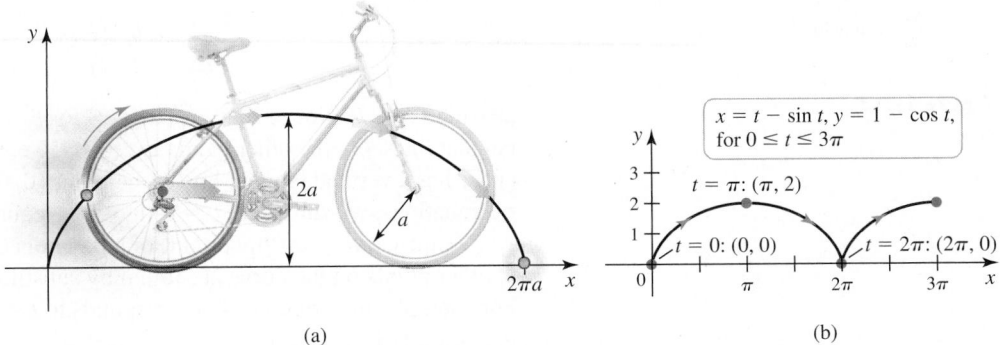

(a)   (b)

Figure 12.11

Related Exercise 58 ◀

Derivatives and Parametric Equations

Parametric equations express a relationship between the variables x and y. Therefore, it makes sense to ask about dy/dx, the rate of change of y with respect to x at a point on a parametric curve. Once we know how to compute dy/dx, it can be used to determine slopes of lines tangent to parametric curves.

Consider the parametric equations $x = f(t)$, $y = g(t)$ on an interval on which both f and g are differentiable. The Chain Rule relates the derivatives $\dfrac{dy}{dt}$, $\dfrac{dx}{dt}$, and $\dfrac{dy}{dx}$:

$$\frac{dy}{dt} = \frac{dy}{dx}\frac{dx}{dt}.$$

Provided $\dfrac{dx}{dt} \neq 0$, we divide both sides of this equation by $\dfrac{dx}{dt}$ and solve for $\dfrac{dy}{dx}$ to obtain the following result.

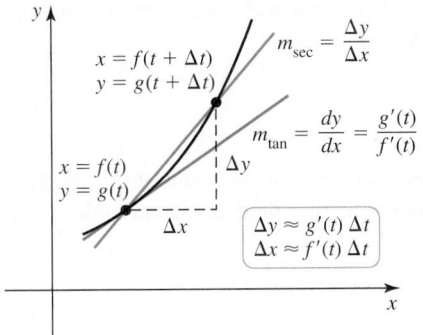

Figure 12.12

QUICK CHECK 5 Use Theorem 12.1 to find the slope of the line $x = 4t$, $y = 2t$, for $-\infty < t < \infty$. ◀

THEOREM 12.1   Derivative for Parametric Curves

Let $x = f(t)$ and $y = g(t)$, where f and g are differentiable on an interval $[a, b]$. Then the slope of the line tangent to the curve at the point corresponding to t is

$$\frac{dy}{dx} = \frac{dy/dt}{dx/dt} = \frac{g'(t)}{f'(t)},$$

provided $f'(t) \neq 0$.

Figure 12.12 gives a geometric explanation of Theorem 12.1. The slope of the line tangent to a curve at a point is $\dfrac{dy}{dx} = \lim\limits_{\Delta x \to 0} \dfrac{\Delta y}{\Delta x}$. Using linear approximation (Section 4.6), we have $\Delta x \approx f'(t)\Delta t$ and $\Delta y \approx g'(t)\Delta t$, with these approximations improving as $\Delta t \to 0$. Notice also that $\Delta t \to 0$ as $\Delta x \to 0$. Therefore, the slope of the tangent line is

$$\frac{dy}{dx} = \lim_{\Delta x \to 0} \frac{\Delta y}{\Delta x} = \lim_{\Delta t \to 0} \frac{g'(t)\Delta t}{f'(t)\Delta t} = \frac{g'(t)}{f'(t)}. \quad \text{Cancel } \Delta t \neq 0.$$

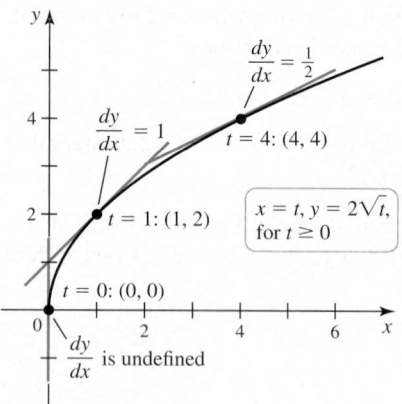

Figure 12.13

> In general, the equations $x = a \cos t$, $y = b \sin t$, for $0 \le t \le 2\pi$, describe an ellipse. The constants a and b can be seen as horizontal and vertical scalings of the unit circle $x = \cos t$, $y = \sin t$. Ellipses are explored in Exercises 91–94 and in Section 12.4.

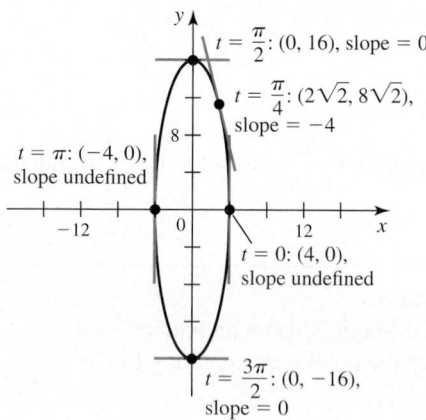

Figure 12.14

EXAMPLE 6 Slopes of tangent lines Find $\dfrac{dy}{dx}$ for the following curves. Interpret the result and determine the points (if any) at which the curve has a horizontal or a vertical tangent line.

a. $x = f(t) = t$, $y = g(t) = 2\sqrt{t}$, for $t \ge 0$

b. $x = f(t) = 4 \cos t$, $y = g(t) = 16 \sin t$, for $0 \le t \le 2\pi$

SOLUTION

a. We find that $f'(t) = 1$ and $g'(t) = 1/\sqrt{t}$. Therefore,

$$\frac{dy}{dx} = \frac{g'(t)}{f'(t)} = \frac{1/\sqrt{t}}{1} = \frac{1}{\sqrt{t}},$$

provided $t \ne 0$. Notice that $dy/dx \ne 0$ for $t > 0$, so the curve has no horizontal tangent lines. On the other hand, as $t \to 0^+$, we see that $dy/dx \to \infty$. Therefore, the curve has a vertical tangent line at the point $(0, 0)$. To eliminate t from the parametric equations, we substitute $t = x$ into the y-equation to find that $y = 2\sqrt{x}$. Because $y \ge 0$, the curve is the upper half of a parabola (Figure 12.13). Slopes of tangent lines at other points on the curve are found by substituting the corresponding values of t. For example, the point $(4, 4)$ corresponds to $t = 4$ and the slope of the tangent line at that point is $1/\sqrt{4} = \frac{1}{2}$.

b. These parametric equations describe an **ellipse** with a major axis of length 32 on the y-axis and a minor axis of length 8 on the x-axis (Figure 12.14). In this case, $f'(t) = -4 \sin t$ and $g'(t) = 16 \cos t$. Therefore,

$$\frac{dy}{dx} = \frac{g'(t)}{f'(t)} = \frac{16 \cos t}{-4 \sin t} = -4 \cot t.$$

At $t = 0$ and $t = \pi$, $\cot t$ is undefined. Notice that

$$\lim_{t \to 0^+} \frac{dy}{dx} = \lim_{t \to 0^+} (-4 \cot t) = -\infty \quad \text{and} \quad \lim_{t \to 0^-} \frac{dy}{dx} = \lim_{t \to 0^-} (-4 \cot t) = \infty.$$

Consequently, a vertical tangent line occurs at the point corresponding to $t = 0$, which is $(4, 0)$ (Figure 12.14). A similar argument shows that a vertical tangent line occurs at the point corresponding to $t = \pi$, which is $(-4, 0)$.

At $t = \pi/2$ and $t = 3\pi/2$, $\cot t = 0$ and the curve has horizontal tangent lines at the corresponding points $(0, \pm 16)$. Slopes of tangent lines at other points on the curve may be found. For example, the point $(2\sqrt{2}, 8\sqrt{2})$ corresponds to $t = \pi/4$; the slope of the tangent line at that point is $-4 \cot \pi/4 = -4$.

Related Exercises 69–70 ◄

In Section 6.5, we derived the formula for the arc length of curves of the form $y = f(x)$. We conclude this section with a similar derivation for the arc length of curves described parametrically.

Arc Length

Suppose a curve C is given by the parametric equations $x = f(t)$, $y = g(t)$, for $a \le t \le b$, where f' and g' are continuous on $[a, b]$. To find the length of the curve between $(f(a), g(a))$ and $(f(b), g(b))$, we first subdivide the interval $[a, b]$ into n subintervals using the grid points

$$a = t_0 < t_1 < t_2 < \cdots < t_n = b.$$

The next step is to connect consecutive points on the curve,

$$(f(t_0), g(t_0)), \ldots, (f(t_k), g(t_k)), \ldots, (f(t_n), g(t_n)),$$

with line segments (Figure 12.15a).

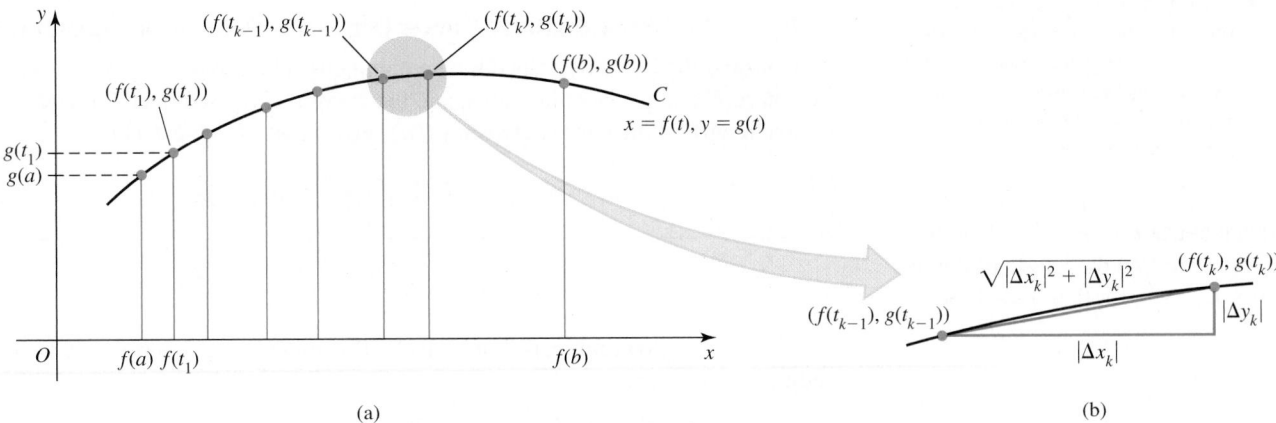

Figure 12.15

The kth line segment is the hypotenuse of a right triangle, whose legs have lengths $|\Delta x_k|$ and $|\Delta y_k|$, where

$$\Delta x_k = f(t_k) - f(t_{k-1}) \quad \text{and} \quad \Delta y_k = g(t_k) - g(t_{k-1}),$$

for $k = 1, 2, \ldots, n$ (**Figure 12.15b**). Therefore, the length of the kth line segment is

$$\sqrt{|\Delta x_k|^2 + |\Delta y_k|^2}.$$

The length of the entire curve L is approximated by the sum of the lengths of the line segments:

$$L \approx \sum_{k=1}^{n} \sqrt{|\Delta x_k|^2 + |\Delta y_k|^2} = \sum_{k=1}^{n} \sqrt{(\Delta x_k)^2 + (\Delta y_k)^2}. \tag{1}$$

The goal is to express this sum as a Riemann sum.

The change in $x = f(t)$ over the kth subinterval is $\Delta x_k = f(t_k) - f(t_{k-1})$. By the Mean Value Theorem, there is a point t_k^* in (t_{k-1}, t_k) such that

$$\underbrace{\frac{\overbrace{f(t_k) - f(t_{k-1})}^{\Delta x_k}}{\underbrace{t_k - t_{k-1}}_{\Delta t_k}}} = f'(t_k^*).$$

So the change in x as t changes by $\Delta t_k = t_k - t_{k-1}$ is

$$\Delta x_k = f(t_k) - f(t_{k-1}) = f'(t_k^*)\Delta t_k.$$

Similarly, the change in y over the kth subinterval is

$$\Delta y_k = g(t_k) - g(t_{k-1}) = g'(\hat{t}_k)\Delta t_k,$$

where $\hat{t}_k$ is also a point in (t_{k-1}, t_k). We substitute these expressions for Δx_k and Δy_k into equation (1):

$$L \approx \sum_{k=1}^{n} \sqrt{(\Delta x_k)^2 + (\Delta y_k)^2}$$

$$= \sum_{k=1}^{n} \sqrt{(f'(t_k^*)\Delta t_k)^2 + (g'(\hat{t}_k)\Delta t_k)^2} \quad \text{Substitute for } \Delta x_k \text{ and } \Delta y_k.$$

$$= \sum_{k=1}^{n} \sqrt{f'(t_k^*)^2 + g'(\hat{t}_k)^2}\,\Delta t_k. \quad \text{Factor } \Delta t_k \text{ out of square root.}$$

The intermediate points t_k^* and $\hat{t}_k$ both approach t_k as n increases and as Δt_k approaches zero. Therefore, given the conditions on f' and g', the limit of this sum as $n \to \infty$ and $\Delta t_k \to 0$, for all k, exists and equals a definite integral:

$$L = \lim_{n \to \infty} \sum_{k=1}^{n} \sqrt{f'(t_k^*)^2 + g'(\hat{t}_k)^2}\,\Delta t_k = \int_a^b \sqrt{f'(t)^2 + g'(t)^2}\, dt.$$

We take this integral as the definition of the length of the curve C.

> ➤ Arc length integrals are usually difficult to evaluate exactly. The few easily evaluated integrals appear in the examples and exercises. Numerical methods are used to approximate challenging integrals.

> **QUICK CHECK 6** Use the arc length formula to find the length of the line $x = t, y = t$, for $0 \le t \le 1$. ◀

> ➤ An important fact is that the arc length of a smooth parameterized curve is independent of the choice of parameter (See Section 14.4).

DEFINITION Arc Length for Curves Defined by Parametric Equations

Consider the curve described by the parametric equations $x = f(t), y = g(t)$, where f' and g' are continuous, and the curve is traversed once for $a \le t \le b$. The **arc length** of the curve between $(f(a), g(a))$ and $(f(b), g(b))$ is

$$L = \int_a^b \sqrt{f'(t)^2 + g'(t)^2}\, dt.$$

EXAMPLE 7   Circumference of a circle Prove that the circumference of a circle of radius $a > 0$ is $2\pi a$.

SOLUTION A circle of radius a is described by

$$x = f(t) = a \cos t, \quad y = g(t) = a \sin t, \quad \text{for } 0 \le t \le 2\pi.$$

Note that $f'(t) = -a \sin t$ and $g'(t) = a \cos t$. The circumference is

$$
\begin{aligned}
L &= \int_0^{2\pi} \sqrt{f'(t)^2 + g'(t)^2}\, dt & &\text{Arc length formula} \\
&= \int_0^{2\pi} \sqrt{(-a \sin t)^2 + (a \cos t)^2}\, dt & &\text{Substitute for } f' \text{ and } g'. \\
&= a \int_0^{2\pi} \sqrt{\sin^2 t + \cos^2 t}\, dt & &\text{Factor } a > 0 \text{ out of square root.} \\
&= a \int_0^{2\pi} 1\, dt & &\sin^2 t + \cos^2 t = 1 \\
&= 2\pi a. & &\text{Integrate a constant.}
\end{aligned}
$$

Related Exercise 82 ◀

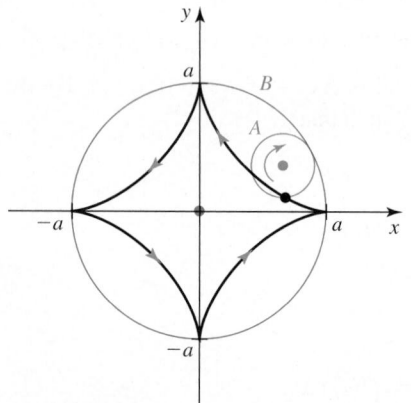

Figure 12.16

EXAMPLE 8   Astroid arc length The path of a point on circle A with radius $a/4$ that rolls on the inside of circle B with radius a (**Figure 12.16**) is an **astroid** or a **hypocycloid**. Its parametric equations are

$$x = a \cos^3 t, \quad y = a \sin^3 t, \quad \text{for } 0 \le t \le 2\pi.$$

a. Graph the astroid with $a = 1$ and find its equation in terms of x and y.
b. Find the arc length of the astroid with $a = 1$.

SOLUTION

a. Because both $\cos^3 t$ and $\sin^3 t$ have a period of 2π, the complete curve is generated on the interval $0 \le t \le 2\pi$ (**Figure 12.17**). To eliminate t from the parametric equations, note that $x^{2/3} = \cos^2 t$ and $y^{2/3} = \sin^2 t$. Therefore,

$$x^{2/3} + y^{2/3} = \cos^2 t + \sin^2 t = 1,$$

where the Pythagorean identity has been used. We see that an alternative description of the astroid is $x^{2/3} + y^{2/3} = 1$.

b. The length of the entire curve is four times the length of the curve in the first quadrant. You should verify that the curve in the first quadrant is generated as the parameter varies from $t = 0$ (corresponding to $(1, 0)$) to $t = \pi/2$ (corresponding to $(0, 1)$; Figure 12.17). Letting $f(t) = \cos^3 t$ and $g(t) = \sin^3 t$, we have

$$f'(t) = -3 \cos^2 t \sin t \quad \text{and} \quad g'(t) = 3 \sin^2 t \cos t.$$

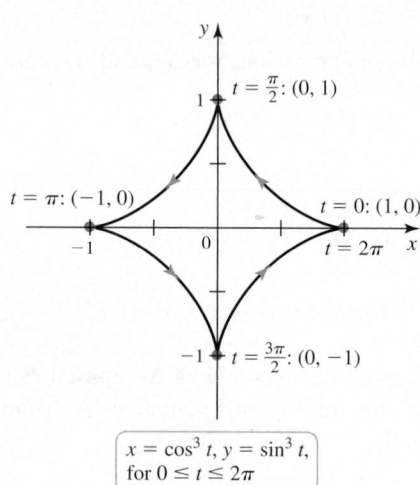

$$x = \cos^3 t,\ y = \sin^3 t,$$
$$\text{for } 0 \le t \le 2\pi$$

Figure 12.17

The arc length of the full curve is

$$L = 4 \int_0^{\pi/2} \sqrt{f'(t)^2 + g'(t)^2} \, dt \qquad \text{Factor of 4 by symmetry}$$

$$= 4 \int_0^{\pi/2} \sqrt{(-3 \cos^2 t \sin t)^2 + (3 \sin^2 t \cos t)^2} \, dt \qquad \text{Substitute for } f' \text{ and } g'.$$

$$= 4 \int_0^{\pi/2} \sqrt{9 \cos^4 t \sin^2 t + 9 \cos^2 t \sin^4 t} \, dt \qquad \text{Simplify terms.}$$

$$= 4 \int_0^{\pi/2} 3 \sqrt{\cos^2 t \sin^2 t \underbrace{(\cos^2 t + \sin^2 t)}_{1}} \, dt \qquad \text{Factor.}$$

$$= 12 \int_0^{\pi/2} \cos t \sin t \, dt. \qquad \cos t \sin t \geq 0, \text{ for } 0 \leq t \leq \frac{\pi}{2}$$

▶ The arc length formula from Section 6.5 could be used to compute the length of the astroid in Example 8 because we found a Cartesian equation for the curve in part (a). However, in cases where it is impossible to eliminate the parameter, the arc length formula derived in this section is required.

Letting $u = \sin t$ with $du = \cos t \, dt$, we have

$$L = 12 \int_0^{\pi/2} \cos t \sin t \, dt = 12 \int_0^1 u \, du = 6.$$

The length of the entire hypocycloid is 6 units.

Related Exercise 87 ◀

SECTION 12.1 EXERCISES

Getting Started

1. Explain how a pair of parametric equations generates a curve in the xy-plane.

2. Give two pairs of parametric equations that generate a circle centered at the origin with radius 6.

3. Give parametric equations that describe a circle of radius R, centered at the origin with clockwise orientation, where the parameter varies over the interval $[0, 10]$.

4. Give parametric equations that generate the line with slope -2 passing through $(1, 3)$.

5. Find parametric equations for the complete parabola $x = y^2$. Answers are not unique.

6. Describe the similarities between the graphs of the parametric equations $x = \sin^2 t, y = \sin t$, for $0 \leq t \leq \pi/2$, and $x = \sin^2 t, y = \sin t$, for $\pi/2 \leq t \leq \pi$. Begin by eliminating the parameter t to obtain an equation in x and y.

7. Find the slope of the parametric curve $x = -2t^3 + 1, y = 3t^2$, for $-\infty < t < \infty$, at the point corresponding to $t = 2$.

8. In which direction is the curve $x = -2 \sin t, y = 2 \cos t$, for $0 \leq t < 2\pi$, generated?

9. Find three different pairs of parametric equations for the line segment that starts at $(0, 0)$ and ends at $(6, 6)$.

10. Use calculus to find the arc length of the line segment $x = 3t + 1$, $y = 4t$, for $0 \leq t \leq 1$. Check your work by finding the distance between the endpoints of the line segment.

Practice Exercises

11–14. Working with parametric equations *Consider the following parametric equations.*

a. *Make a brief table of values of t, x, and y.*
b. *Plot the (x, y) pairs in the table and the complete parametric curve, indicating the positive orientation (the direction of increasing t).*
c. *Eliminate the parameter to obtain an equation in x and y.*
d. *Describe the curve.*

11. $x = 2t, y = 3t - 4; -10 \leq t \leq 10$

12. $x = t^2 + 2, y = 4t; -4 \leq t \leq 4$

13. $x = -t + 6, y = 3t - 3; -5 \leq t \leq 5$

14. $x = t^3 - 1, y = 5t + 1; -3 \leq t \leq 3$

15–30. Working with parametric equations *Consider the following parametric equations.*

a. *Eliminate the parameter to obtain an equation in x and y.*
b. *Describe the curve and indicate the positive orientation.*

15. $x = 3 + t, y = 1 - t; 0 \leq t \leq 1$

16. $x = 4 - 3t, y = -2 + 6t; -1 \leq t \leq 3$

17. $x = \sqrt{t} + 4, y = 3\sqrt{t}; 0 \leq t \leq 16$

18. $x = (t + 1)^2, y = t + 2; -10 \leq t \leq 10$

19. $x = 3 \cos t, y = 3 \sin t; \pi \leq t \leq 2\pi$

20. $x = 3 \cos t, y = 3 \sin t; 0 \leq t \leq \pi/2$

21. $x = \cos t, y = \sin^2 t; 0 \leq t \leq \pi$

22. $x = 1 - \sin^2 s, y = \cos s; \pi \le s \le 2\pi$

23. $x = \cos t, y = 1 + \sin t; 0 \le t \le 2\pi$

24. $x = \sin t, y = 2 \sin t + 1; 0 \le t \le \pi/2$

25. $x = r - 1, y = r^3; -4 \le r \le 4$

26. $x = e^{2t}, y = e^t + 1; 0 \le t \le 25$

27. $x = -7 \cos 2t, y = -7 \sin 2t; 0 \le t \le \pi$

28. $x = 1 - 3 \sin 4\pi t, y = 2 + 3 \cos 4\pi t; 0 \le t \le 1/2$

29. $x = 8 + 2t, y = 1; -\infty < t < \infty$

30. $x = 5, y = 3t; -\infty < t < \infty$

31–36. Eliminating the parameter *Eliminate the parameter to express the following parametric equations as a single equation in x and y.*

31. $x = 2 \sin 8t, y = 2 \cos 8t$ **32.** $x = \sin 8t, y = 2 \cos 8t$

33. $x = t, y = \sqrt{4 - t^2}$ **34.** $x = \sqrt{t + 1}, y = \dfrac{1}{t + 1}$

35. $x = \tan t, y = \sec^2 t - 1$

36. $x = a \sin^n t, y = b \cos^n t$, where a and b are real numbers and n is a positive integer

37–52. Curves to parametric equations *Find parametric equations for the following curves. Include an interval for the parameter values. Answers are not unique.*

37. A circle centered at the origin with radius 4, generated counterclockwise

38. A circle centered at the origin with radius 12, generated clockwise with initial point $(0, 12)$

39. A circle centered at $(2, 3)$ with radius 1, generated counterclockwise

40. A circle centered at $(-2, -3)$ with radius 8, generated clockwise

41. The line segment starting at $P(0, 0)$ and ending at $Q(2, 8)$

42. The line segment starting at $P(-1, -3)$ and ending at $Q(6, -16)$

43. The segment of the parabola $y = 2x^2 - 4$, where $-1 \le x \le 5$

44. The complete curve $x = y^3 - 3y$

45. The vertical line segment starting at $P(2, 3)$ and ending at $Q(2, 9)$

46. The horizontal line segment starting at $P(8, 2)$ and ending at $Q(-2, 2)$

47. The piecewise linear path from $P(-2, 3)$ to $Q(2, -3)$ to $R(3, 5)$, using parameter values $0 \le t \le 2$

48. The path consisting of the line segment from $(-4, 4)$ to $(0, 8)$, followed by the segment of the parabola $y = 8 - 2x^2$ from $(0, 8)$ to $(2, 0)$, using parameter values $0 \le t \le 3$

49. The line that passes through the points $P(1, 1)$ and $Q(3, 5)$, oriented in the direction of increasing x

50. The left half of the parabola $y = x^2 + 1$, originating at $(0, 1)$

51. The upper half of the parabola $x = y^2$, originating at $(0, 0)$

52. The lower half of the circle centered at $(-2, 2)$ with radius 6, oriented in the counterclockwise direction

53–56. Circular motion *Find parametric equations that describe the circular path of the following objects. For Exercises 53–55, assume (x, y) denotes the position of the object relative to the origin at the center of the circle. Use the units of time specified in the problem. There are many ways to describe any circle.*

53. A go-cart moves counterclockwise with constant speed around a circular track of radius 400 m, completing a lap in 1.5 min.

54. The tip of the 15-inch second hand of a clock completes one revolution in 60 seconds.

55. A bicyclist rides counterclockwise with constant speed around a circular velodrome track with a radius of 50 m, completing one lap in 24 seconds.

56. A Ferris wheel has a radius of 20 m and completes a revolution in the clockwise direction at constant speed in 3 min. Assume x and y measure the horizontal and vertical positions of a seat on the Ferris wheel relative to a coordinate system whose origin is at the low point of the wheel. Assume the seat begins moving at the origin.

▦ 57–62. More parametric curves *Use a graphing utility to graph the following curves. Be sure to choose an interval for the parameter that generates all features of interest.*

57. Spiral $x = t \cos t, y = t \sin t; t \ge 0$

58. Witch of Agnesi $x = 2 \cot t, y = 1 - \cos 2t$

59. Folium of Descartes $x = \dfrac{3t}{1 + t^3}, y = \dfrac{3t^2}{1 + t^3}$

60. Involute of a circle $x = \cos t + t \sin t, y = \sin t - t \cos t$

61. Evolute of an ellipse $x = \dfrac{a^2 - b^2}{a} \cos^3 t, y = \dfrac{a^2 - b^2}{b} \sin^3 t$; $a = 4$ and $b = 3$

62. Cissoid of Diocles $x = 2 \sin 2t, y = \dfrac{2 \sin^3 t}{\cos t}$

▦ 63. Implicit function graph Explain and carry out a method for graphing the curve $x = 1 + \cos^2 y - \sin^2 y$ using parametric equations and a graphing utility.

64. Air drop A plane traveling horizontally at 80 m/s over flat ground at an elevation of 3000 m releases an emergency packet. The trajectory of the packet is given by

$$x = 80t, \quad y = -4.9t^2 + 3000, \quad \text{for } t \ge 0,$$

where the origin is the point on the ground directly beneath the plane at the moment of the release (see figure). Graph the trajectory of the packet and find the coordinates of the point where the packet lands.

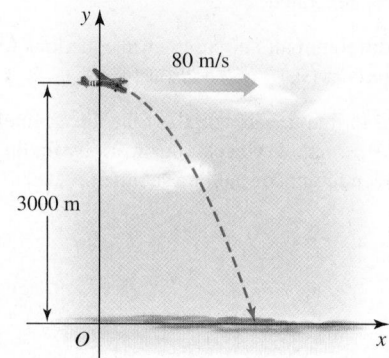

65. Air drop—inverse problem A plane traveling horizontally at 100 m/s over flat ground at an elevation of 4000 m must drop an emergency packet on a target on the ground. The trajectory of the packet is given by

$$x = 100t, \quad y = -4.9t^2 + 4000, \quad \text{for } t \geq 0,$$

where the origin is the point on the ground directly beneath the plane at the moment of the release. How many horizontal meters before the target should the packet be released in order to hit the target?

66. Projectile explorations A projectile launched from the ground with an initial speed of 20 m/s and a launch angle θ follows a trajectory approximated by

$$x = (20 \cos \theta)t, \quad y = -4.9t^2 + (20 \sin \theta)t,$$

where x and y are the horizontal and vertical positions of the projectile relative to the launch point $(0, 0)$.

   a. Graph the trajectory for various values of θ in the range $0 < \theta < \pi/2$.

   b. Based on your observations, what value of θ gives the greatest range (the horizontal distance between the launch and landing points)?

67–72. Derivatives *Consider the following parametric curves.*

a. Determine dy/dx in terms of t and evaluate it at the given value of t.

b. Make a sketch of the curve showing the tangent line at the point corresponding to the given value of t.

67. $x = 2 + 4t, y = 4 - 8t; t = 2$

68. $x = 3 \sin t, y = 3 \cos t; t = \pi/2$

69. $x = \cos t, y = 8 \sin t; t = \pi/2$

70. $x = 2t, y = t^3; t = -1$

T 71. $x = t + 1/t, y = t - 1/t; t = 1$

72. $x = \sqrt{t}, y = 2t; t = 4$

73–76. Tangent lines *Find an equation of the line tangent to the curve at the point corresponding to the given value of t.*

73. $x = t^2 - 1, y = t^3 + t; t = 2$

74. $x = \sin t, y = \cos t; t = \pi/4$

75. $x = \cos t + t \sin t, y = \sin t - t \cos t; t = \pi/4$

76. $x = e^t, y = \ln(t + 1); t = 0$

77–80. Slopes of tangent lines *Find all points at which the following curves have the given slope.*

77. $x = 4 \cos t, y = 4 \sin t;$ slope $= 1/2$

78. $x = 2 \cos t, y = 8 \sin t;$ slope $= -1$

79. $x = t + 1/t, y = t - 1/t;$ slope $= 1$

80. $x = 2 + \sqrt{t}, y = 2 - 4t;$ slope $= -8$

81–88. Arc length *Find the arc length of the following curves on the given interval.*

81. $x = 3t + 1, y = 4t + 2; 0 \leq t \leq 2$

82. $x = 3 \cos t, y = 3 \sin t + 1; 0 \leq t \leq 2\pi$

83. $x = \cos t - \sin t, y = \cos t + \sin t; 0 \leq t \leq \pi$

84. $x = e^t \sin t, y = e^t \cos t; 0 \leq t \leq 2\pi$

85. $x = t^4, y = \dfrac{t^6}{3}; 0 \leq t \leq 1$

86. $x = 2t \sin t - t^2 \cos t, y = 2t \cos t + t^2 \sin t; 0 \leq t \leq \pi$

87. $x = \cos^3 2t, y = \sin^3 2t; 0 \leq t \leq \pi/4$

88. $x = \sin t, y = t - \cos t; 0 \leq t \leq \pi/2$

89. Explain why or why not Determine whether the following statements are true and give an explanation or counterexample.

   a. The equations $x = -\cos t, y = -\sin t$, for $0 \leq t \leq 2\pi$, generate a circle in the clockwise direction.

   b. An object following the parametric curve $x = 2 \cos 2\pi t$, $y = 2 \sin 2\pi t$ circles the origin once every 1 time unit.

   c. The parametric equations $x = t, y = t^2$, for $t \geq 0$, describe the complete parabola $y = x^2$.

   d. The parametric equations $x = \cos t, y = \sin t$, for $-\pi/2 \leq t \leq \pi/2$, describe a semicircle.

   e. There are two points on the curve $x = -4 \cos t, y = \sin t$, for $0 \leq t \leq 2\pi$, at which there is a vertical tangent line.

90. Matching curves and equations Match equations a–d with graphs A–D. Explain your reasoning.

   a. $x = t^2 - 2, y = t^3 - t$

   b. $x = \cos(t + \sin 50t), y = \sin(t + \cos 50t)$

   c. $x = t + \cos 2t, y = t - \sin 4t$

   d. $x = 2 \cos t + \cos 20t, y = 2 \sin t + \sin 20t$

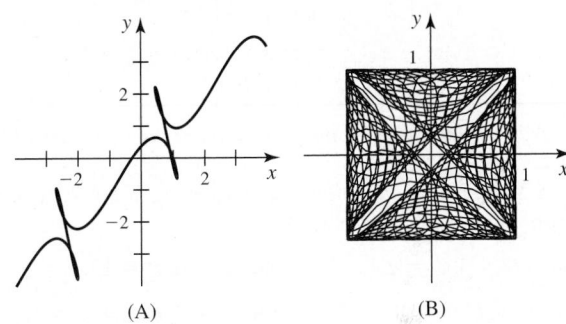

(A)   (B)

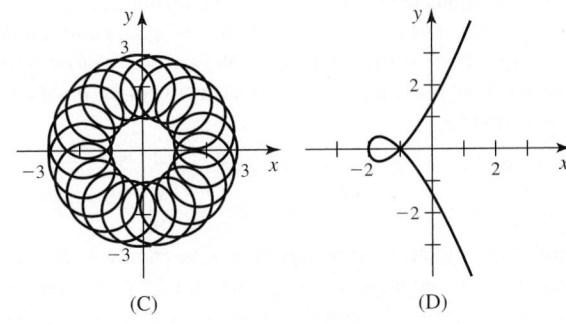

(C)   (D)

91–92. Ellipses *An **ellipse** (discussed in detail in Section 12.4) is generated by the parametric equations $x = a \cos t, y = b \sin t$. If $0 < a < b$, then the long axis (or **major axis**) lies on the y-axis and the short axis (or **minor axis**) lies on the x-axis. If $0 < b < a$, the axes are reversed. The lengths of the axes in the x- and y-directions are 2a and 2b, respectively. Sketch the graph of the following ellipses. Specify an interval in t over which the entire curve is generated.*

91. $x = 4 \cos t, y = 9 \sin t$

92. $x = 12 \sin 2t, y = 3 \cos 2t$

93–94. Parametric equations of ellipses *Find parametric equations (not unique) of the following ellipses (see Exercises 91–92). Graph the ellipse and find a description in terms of x and y.*

93. An ellipse centered at the origin with major axis of length 6 on the *x*-axis and minor axis of length 3 on the *y*-axis, generated counterclockwise

94. An ellipse centered at $(-2, -3)$ with major and minor axes of lengths 30 and 20, parallel to the *x*- and *y*-axes, respectively, generated counterclockwise (*Hint:* Shift the parametric equations.)

95. Intersecting lines Consider the following pairs of lines. Determine whether the lines are parallel or intersecting. If the lines intersect, then determine the point of intersection.

a. $x = 1 + s, y = 2s$ and $x = 1 + 2t, y = 3t$
b. $x = 2 + 5s, y = 1 + s$ and $x = 4 + 10t, y = 3 + 2t$
c. $x = 1 + 3s, y = 4 + 2s$ and $x = 4 - 3t, y = 6 + 4t$

96. Multiple descriptions Which of the following parametric equations describe the same curve?

a. $x = 2t^2, y = 4 + t; -4 \le t \le 4$
b. $x = 2t^4, y = 4 + t^2; -2 \le t \le 2$
c. $x = 2t^{2/3}, y = 4 + t^{1/3}; -64 \le t \le 64$

Explorations and Challenges

T 97–100. Beautiful curves *Consider the family of curves*

$$x = \left(2 + \frac{1}{2}\sin at\right)\cos\left(t + \frac{\sin bt}{c}\right),$$

$$y = \left(2 + \frac{1}{2}\sin at\right)\sin\left(t + \frac{\sin bt}{c}\right).$$

Plot a graph of the curve for the given values of a, b, and c with $0 \le t \le 2\pi$.
(*Source: Mathematica in Action, Stan Wagon, Springer, 2010; created by Norton Starr, Amherst College*)

97. $a = b = 5, c = 2$
98. $a = 6, b = 12, c = 3$
99. $a = 18, b = 18, c = 7$
100. $a = 7, b = 4, c = 1$

T 101–102. Lissajous curves *Consider the following Lissajous curves. Graph the curve, and estimate the coordinates of the points on the curve at which there is (a) a horizontal tangent line or (b) a vertical tangent line. (See the Guided Project* Parametric art *for more on Lissajous curves.)*

101. $x = \sin 2t, y = 2\sin t; 0 \le t \le 2\pi$

102. $x = \sin 4t, y = \sin 3t; 0 \le t \le 2\pi$

103–105. Area under a curve *Suppose the function $y = h(x)$ is nonnegative and continuous on $[\alpha, \beta]$, which implies that the area bounded by the graph of h and the x-axis on $[\alpha, \beta]$ equals $\int_\alpha^\beta h(x)\,dx$ or $\int_\alpha^\beta y\,dx$. If the graph of $y = h(x)$ on $[\alpha, \beta]$ is traced exactly once by the parametric equations $x = f(t), y = g(t)$, for $a \le t \le b$, then it follows by substitution that the area bounded by h is*

$$\int_\alpha^\beta h(x)\,dx = \int_\alpha^\beta y\,dx = \int_a^b g(t)f'(t)\,dt \text{ if } \alpha = f(a) \text{ and } \beta = f(b)$$

$$\left(\text{or } \int_\alpha^\beta h(x)\,dx = \int_b^a g(t)f'(t)\,dt \text{ if } \alpha = f(b) \text{ and } \beta = f(a)\right).$$

103. Find the area under one arch of the cycloid $x = 3(t - \sin t)$, $y = 3(1 - \cos t)$ (see Example 5).

104. Show that the area of the region bounded by the ellipse $x = 3\cos t, y = 4\sin t$, for $0 \le t \le 2\pi$, equals

$$4\int_{\pi/2}^0 4\sin t(-3\sin t)\,dt. \text{ Then evaluate the integral.}$$

105. Find the area of the region bounded by the astroid $x = \cos^3 t, y = \sin^3 t$, for $0 \le t \le 2\pi$ (see Example 8, Figure 12.17).

106–111. Surfaces of revolution *Let C be the curve $x = f(t)$, $y = g(t)$, for $a \le t \le b$, where f' and g' are continuous on $[a, b]$ and C does not intersect itself, except possibly at its endpoints. If g is nonnegative on $[a, b]$, then the area of the surface obtained by revolving C about the x-axis is*

$$S = \int_a^b 2\pi g(t)\sqrt{f'(t)^2 + g'(t)^2}\,dt.$$

Likewise, if f is nonnegative on $[a, b]$, then the area of the surface obtained by revolving C about the y-axis is

$$S = \int_a^b 2\pi f(t)\sqrt{f'(t)^2 + g'(t)^2}\,dt.$$

(These results can be derived in a manner similar to the derivations given in Section 6.6 for surfaces of revolution generated by the curve $y = f(x)$.)

106. Use the parametric equations of a semicircle of radius 1, $x = \cos t$, $y = \sin t$, for $0 \le t \le \pi$, to verify that surface area of a unit sphere is 4π.

107. Consider the curve $x = 3\cos t, y = 3\sin t + 4$, for $0 \le t \le 2\pi$.

a. Describe the curve.
b. If the curve is revolved about the *x*-axis, describe the shape of the surface of revolution and find the area of the surface.

108. Find the area of the surface obtained by revolving the curve $x = \cos^3 t, y = \sin^3 t$, for $0 \le t \le \pi/2$, about the *x*-axis.

109. Find the area of the surface obtained by revolving one arch of the cycloid $x = t - \sin t, y = 1 - \cos t$, for $0 \le t \le 2\pi$, about the *x*-axis.

110. Find the area of the surface obtained by revolving the curve $x = \sqrt{t + 1}, y = t$, for $1 \le t \le 5$, about the *y*-axis.

T 111. A surface is obtained by revolving the curve $x = e^{3t} + 1$, $y = e^{2t}$, for $0 \le t \le 1$, about the *y*-axis. Find an integral that gives the area of the surface and approximate the value of the integral.

T 112. Paths of moons An idealized model of the path of a moon (relative to the Sun) moving with constant speed in a circular orbit around a planet, where the planet in turn revolves around the Sun, is given by the parametric equations

$$x(\theta) = a\cos\theta + \cos n\theta, \quad y(\theta) = a\sin\theta + \sin n\theta.$$

The distance from the moon to the planet is taken to be 1, the distance from the planet to the Sun is a, and n is the number of times the moon orbits the planet for every 1 revolution of the planet around the Sun. Plot the graph of the path of a moon for the given constants; then conjecture which values of n produce loops for a fixed value of a.

a. $a = 4, n = 3$ **b.** $a = 4, n = 4$ **c.** $a = 4, n = 5$

T **113. Paths of the moons of Earth and Jupiter** Use the equations in Exercise 112 to analyze the paths of the following moons in our solar system.

   a. The path of Callisto (one of Jupiter's moons) corresponds to values of $a = 727.5$ and $n = 259.6$. Plot a small portion of the graph to see the detailed behavior of the orbit.

   b. The path of Io (another of Jupiter's moons) corresponds to values of $a = 1846.2$ and $n = 2448.8$. Plot a small portion of the path of Io to see the loops in its orbit.

   (*Source: The Sun, the Moon, and Convexity, The College Mathematics Journal*, 32, Sep 2001)

114. Second derivative Assume a curve is given by the parametric equations $x = f(t)$ and $y = g(t)$, where f and g are twice differentiable. Use the Chain Rule to show that

$$y''(x) = \frac{f'(t)g''(t) - g'(t)f''(t)}{(f'(t))^3}.$$

QUICK CHECK ANSWERS

1. A segment of the parabola $x = y^2$ opening to the right with vertex at the origin   **2.** The circle has center $(0, 0)$ and radius 3; it is generated in the counterclockwise direction (positive orientation) starting at $(0, -3)$.   **3.** The line $y = -3x - 3$ with slope -3 passing through $(3, -12)$ (when $t = 0$)   **4.** One possibility is $x = -2t$, $y = 3 - 3t$, for $0 \le t \le 1$.   **5.** $\dfrac{1}{2}$ **6.** $\sqrt{2}$ ◄

12.2 Polar Coordinates

Suppose you work for a company that designs heat shields for space vehicles. The shields are thin plates that are either rectangular or circular in shape. To solve the heat transfer equations for these two shields, you must choose a coordinate system that best fits the geometry of the problem. A Cartesian (rectangular) coordinate system is a natural choice for the rectangular shields (**Figure 12.18a**). However, it does not provide a good fit for the circular shields (**Figure 12.18b**). On the other hand, a **polar coordinate** system, in which the coordinates are constant on circles and rays, is better suited for the circular shields (**Figure 12.18c**).

> ➤ Recall that the terms *Cartesian* and *rectangular* both describe the usual *xy*-coordinate system.

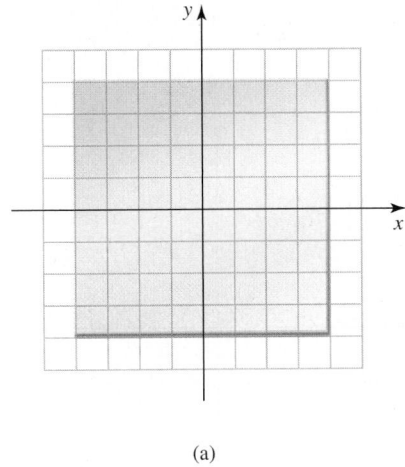

(a)

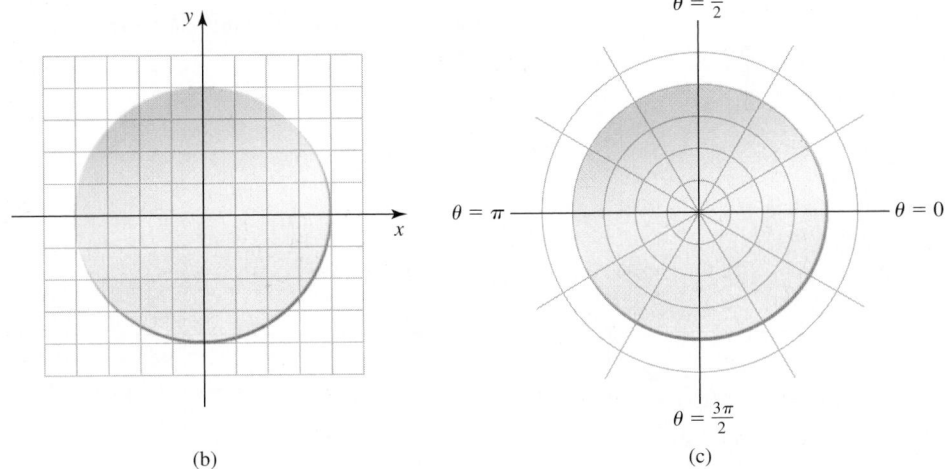

(b)   (c)

Figure 12.18

Defining Polar Coordinates

> ➤ Polar points and curves are plotted on a rectangular coordinate system, with standard "*x*" and "*y*" labels attached to the axes. However, plotting polar points and curves is often easier using polar graph paper, which has concentric circles centered at the origin and rays emanating from the origin (Figure 12.18c).

Like Cartesian coordinates, polar coordinates are used to locate points in the plane. When working in polar coordinates, the origin of the coordinate system is also called the **pole**, and the positive *x*-axis is called the **polar axis**. The polar coordinates for a point P have the form (r, θ). **The radial coordinate** r describes the *signed* (or *directed*) distance from the origin to P. The **angular coordinate** θ describes an angle whose initial side is the positive *x*-axis and whose terminal side lies on the ray passing through the origin and P (**Figure 12.19a**). Positive angles are measured counterclockwise from the positive *x*-axis.

With polar coordinates, points have more than one representation for two reasons. First, angles are determined up to multiples of 2π radians, so the coordinates (r, θ) and

QUICK CHECK 1 Which of the following coordinates represent the same point: $(3, \pi/2)$, $(3, 3\pi/2)$, $(3, 5\pi/2)$, $(-3, -\pi/2)$, and $(-3, 3\pi/2)$? ◄

$(r, \theta \pm 2\pi)$ refer to the same point (**Figure 12.19b**). Second, the radial coordinate may be negative, which is interpreted as follows: The points (r, θ) and $(-r, \theta)$ are reflections of each other through the origin (**Figure 12.19c**). This means that (r, θ), $(-r, \theta + \pi)$, and $(-r, \theta - \pi)$ all refer to the same point. The origin is specified as $(0, \theta)$ in polar coordinates, where θ is any angle.

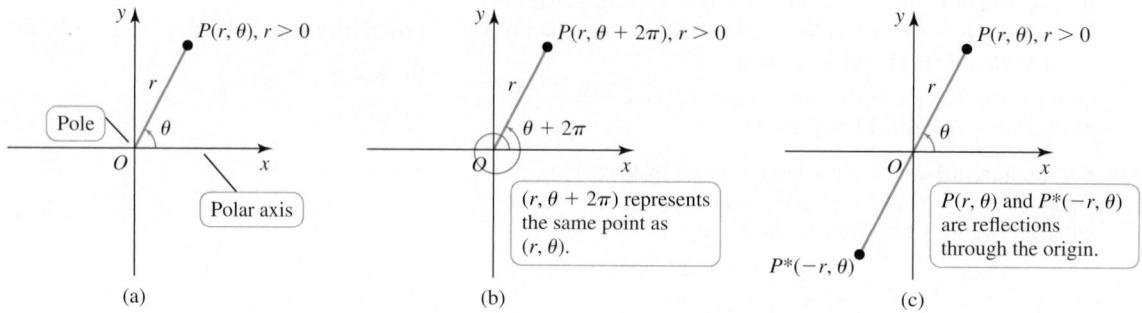

(a)   (b)   (c)

Figure 12.19

EXAMPLE 1   Points in polar coordinates Graph the following points in polar coordinates: $Q\left(1, \frac{5\pi}{4}\right)$, $R\left(-1, \frac{7\pi}{4}\right)$, and $S\left(2, -\frac{3\pi}{2}\right)$. Give two alternative representations for each point.

SOLUTION The point $Q\left(1, \frac{5\pi}{4}\right)$ is one unit from the origin O on a line OQ that makes an angle of $\frac{5\pi}{4}$ with the positive x-axis (**Figure 12.20a**). Subtracting 2π from the angle, the point Q can be represented as $\left(1, -\frac{3\pi}{4}\right)$. Subtracting π from the angle and negating the radial coordinate implies that Q also has the coordinates $\left(-1, \frac{\pi}{4}\right)$.

To locate the point $R\left(-1, \frac{7\pi}{4}\right)$, it is easiest first to find the point $R*\left(1, \frac{7\pi}{4}\right)$ in the fourth quadrant. Then $R\left(-1, \frac{7\pi}{4}\right)$ is the reflection of $R*$ through the origin (**Figure 12.20b**). Other representations of R include $\left(-1, -\frac{\pi}{4}\right)$ and $\left(1, \frac{3\pi}{4}\right)$.

The point $S\left(2, -\frac{3\pi}{2}\right)$ is two units from the origin, found by rotating *clockwise* through an angle of $\frac{3\pi}{2}$ (**Figure 12.20c**). The point S can also be represented as $\left(2, \frac{\pi}{2}\right)$ or $\left(-2, -\frac{\pi}{2}\right)$.

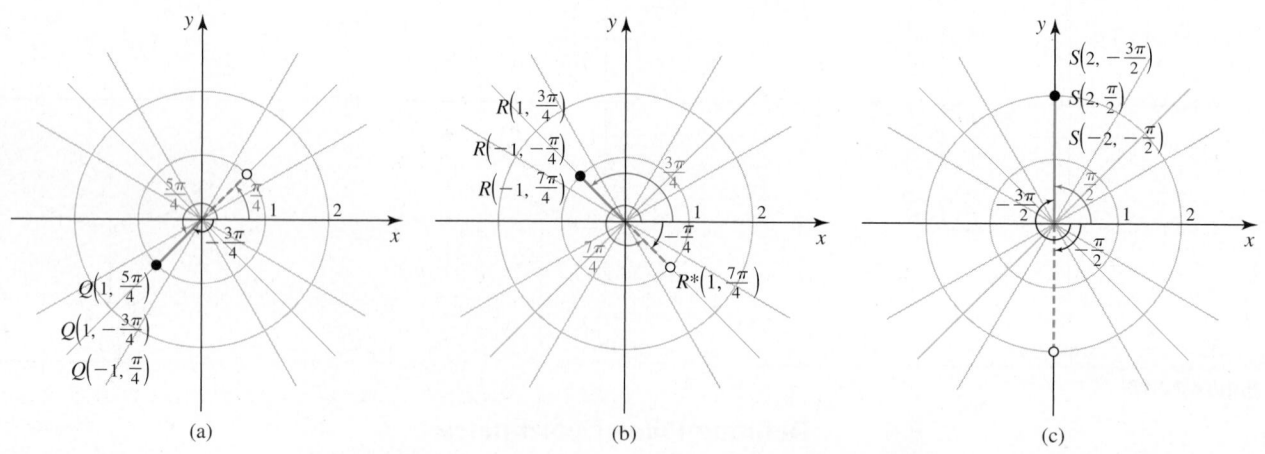

(a)   (b)   (c)

Figure 12.20

Related Exercises 9–13 ◄

Converting Between Cartesian and Polar Coordinates

We often need to convert between Cartesian and polar coordinates. The conversion equations emerge when we look at a right triangle (**Figure 12.21**) in which

$$\cos \theta = \frac{x}{r} \quad \text{and} \quad \sin \theta = \frac{y}{r}.$$

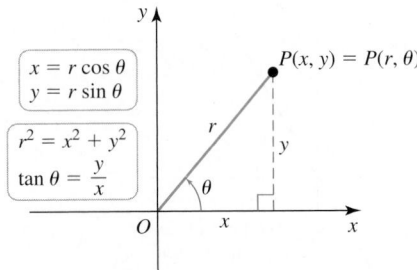

Figure 12.21

QUICK CHECK 2 Draw versions of Figure 12.21 with P in the second, third, and fourth quadrants. Verify that the same conversion formulas hold in all cases. ◄

▶ To determine θ, you may also use the relationships $\cos \theta = x/r$ and $\sin \theta = y/r$. Either method requires checking the signs of x and y to verify that θ is in the correct quadrant.

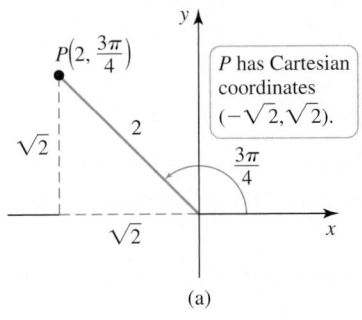

(a)

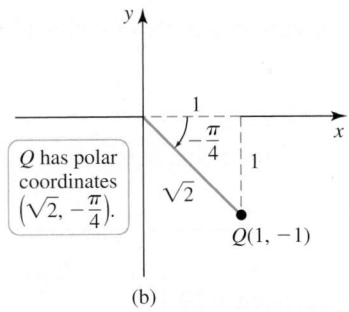

(b)

Figure 12.22

QUICK CHECK 3 Give two polar coordinate descriptions of the point with Cartesian coordinates $(1, 0)$. What are the Cartesian coordinates of the point with polar coordinates $\left(2, \frac{\pi}{2}\right)$? ◄

Given a point with polar coordinates (r, θ), we see that its Cartesian coordinates are $x = r \cos \theta$ and $y = r \sin \theta$. Conversely, given a point with Cartesian coordinates (x, y), its radial polar coordinate satisfies $r^2 = x^2 + y^2$. The coordinate θ is determined using the relation $\tan \theta = y/x$, where the quadrant in which θ lies is determined by the signs of x and y. Figure 12.21 illustrates the conversion formulas for a point P in the first quadrant. The same relationships hold if P is in any of the other three quadrants.

PROCEDURE   Converting Coordinates

A point with polar coordinates (r, θ) has Cartesian coordinates (x, y), where

$$x = r \cos \theta \quad \text{and} \quad y = r \sin \theta.$$

A point with Cartesian coordinates (x, y) has polar coordinates (r, θ), where

$$r^2 = x^2 + y^2 \quad \text{and} \quad \tan \theta = \frac{y}{x}.$$

EXAMPLE 2   Converting coordinates

a. Express the point with polar coordinates $P\left(2, \frac{3\pi}{4}\right)$ in Cartesian coordinates.

b. Express the point with Cartesian coordinates $Q(1, -1)$ in polar coordinates.

SOLUTION

a. The point $P\left(2, \frac{3\pi}{4}\right)$ has Cartesian coordinates

$$x = r \cos \theta = 2 \cos \frac{3\pi}{4} = -\sqrt{2} \quad \text{and}$$

$$y = r \sin \theta = 2 \sin \frac{3\pi}{4} = \sqrt{2}.$$

As shown in **Figure 12.22a**, P is in the second quadrant.

b. It's best to locate this point first to be sure that the angle θ is chosen correctly. As shown in **Figure 12.22b**, the point $Q(1, -1)$ is in the fourth quadrant at a distance $r = \sqrt{1^2 + (-1)^2} = \sqrt{2}$ from the origin. The coordinate θ satisfies

$$\tan \theta = \frac{y}{x} = \frac{-1}{1} = -1.$$

The angle in the fourth quadrant with $\tan \theta = -1$ is $\theta = -\frac{\pi}{4}$ or $\frac{7\pi}{4}$. Therefore, two (of infinitely many) polar representations of Q are $\left(\sqrt{2}, -\frac{\pi}{4}\right)$ and $\left(\sqrt{2}, \frac{7\pi}{4}\right)$.

Related Exercises 25, 28, 35 ◄

Basic Curves in Polar Coordinates

A curve in polar coordinates is the set of points that satisfy an equation in r and θ. Some sets of points are easier to describe in polar coordinates than in Cartesian coordinates. Let's begin by examining polar equations of circles, lines, and spirals.

The polar equation $r = 3$ is satisfied by the set of points whose distance from the origin is 3. The angle θ is arbitrary because it is not specified by the equation, so the graph of $r = 3$ is the circle of radius 3 centered at the origin. In general, the equation $r = a$ describes a circle of radius $|a|$ centered at the origin (**Figure 12.23a**).

The equation $\theta = \pi/3$ is satisfied by the points whose angle with respect to the positive x-axis is $\pi/3$. Because r is unspecified, it is arbitrary (and can be positive or negative). Therefore, $\theta = \pi/3$ describes the line through the origin making an angle of $\pi/3$ with the positive x-axis. More generally, $\theta = \theta_0$ describes the line through the origin making an angle of θ_0 with the positive x-axis (**Figure 12.23b**).

➤ If the equation $\theta = \theta_0$ is accompanied by the condition $r \geq 0$, the resulting set of points is a *ray* emanating from the origin.

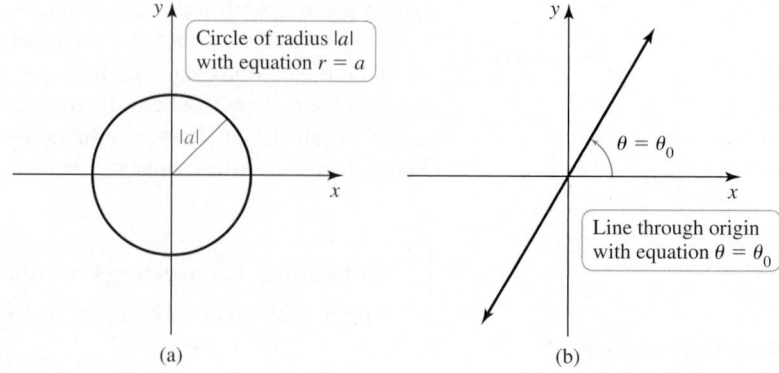

Figure 12.23

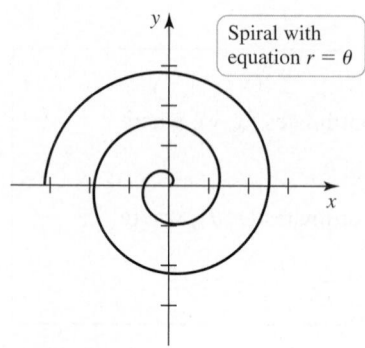

Figure 12.24

The simplest polar equation that involves both r and θ is $r = \theta$. Restricting θ to the interval $\theta \geq 0$, we see that as θ increases, r increases. Therefore, as θ increases, the points on the curve move away from the origin as they circle the origin in a counterclockwise direction, generating a spiral (**Figure 12.24**).

QUICK CHECK 4 Describe the polar curves $r = 12$, $r = 6\theta$, and $r \sin \theta = 10$. ◄

EXAMPLE 3   Polar to Cartesian coordinates Convert the polar equation $r = 10 \cos \theta + 24 \sin \theta$ to Cartesian coordinates and describe the corresponding graph.

SOLUTION Multiplying both sides of the equation by r produces the equation $r^2 = 10r \cos \theta + 24r \sin \theta$. Using the conversion relations $r^2 = x^2 + y^2$, $x = r \cos \theta$, and $y = r \sin \theta$, the equation

$$\underbrace{r^2}_{x^2 + y^2} = \underbrace{10r \cos \theta}_{10x} + \underbrace{24r \sin \theta}_{24y}$$

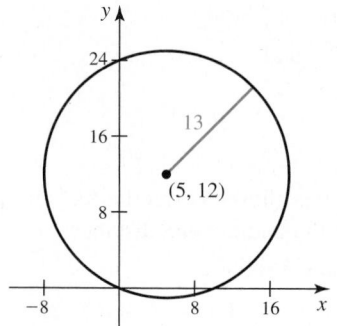

Figure 12.25

becomes $x^2 + y^2 - 10x - 24y = 0$. Completing the square in both variables gives the equation

$$\underbrace{x^2 - 10x + 25}_{(x-5)^2} - 25 + \underbrace{y^2 - 24y + 144}_{(y-12)^2} - 144 = 0, \quad \text{or} \quad (x-5)^2 + (y-12)^2 = 169.$$

We recognize $(x-5)^2 + (y-12)^2 = 169$ as the equation of a circle of radius 13 centered at $(5, 12)$ (**Figure 12.25**).

Related Exercises 41, 43 ◄

Calculations similar to those in Example 3 lead to the following equations of circles in polar coordinates.

SUMMARY   Circles in Polar Coordinates

The equation $r = a$ describes a circle of radius $|a|$ centered at $(0, 0)$.

The equation $r = 2a \cos \theta + 2b \sin \theta$ describes a circle of radius $\sqrt{a^2 + b^2}$ centered at (a, b).

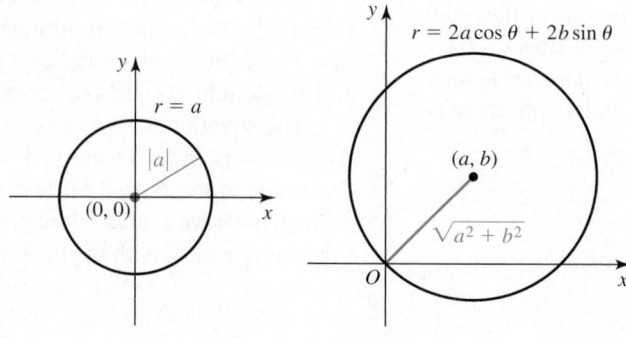

➤ Two important cases deserve mention. The equation $r = 2a \cos \theta$ describes a circle of radius $|a|$ centered at $(a, 0)$. The equation $r = 2b \sin \theta$ describes a circle of radius $|b|$ centered at $(0, b)$.

QUICK CHECK 5 Describe the graph of the equation $r = 2 \cos \theta + 4 \sin \theta$. ◄

Graphing in Polar Coordinates

Equations in polar coordinates often describe curves that are difficult to represent in Cartesian coordinates. Partly for this reason, curve-sketching methods for polar coordinates differ from those used for curves in Cartesian coordinates. Conceptually, the easiest graphing method is to choose several values of θ, calculate the corresponding r-values, and tabulate the coordinates. The points are then plotted and connected with a smooth curve.

▶ When a curve is described as $r = f(\theta)$, it is natural to tabulate points in θ-r format, just as we list points in x-y format for $y = f(x)$. Despite this fact, the standard form for writing an ordered pair in polar coordinates is (r, θ).

EXAMPLE 4   Plotting a polar curve Graph the polar equation $r = f(\theta) = 1 + \sin \theta$.

SOLUTION The domain of f consists of all real values of θ; however, the complete curve is generated by letting θ vary over any interval of length 2π. Table 12.3 shows several (r, θ) pairs, which are plotted in **Figure 12.26**. The resulting curve, called a **cardioid**, is symmetric about the y-axis.

Table 12.3

θ	$r = 1 + \sin \theta$
0	1
$\pi/6$	3/2
$\pi/2$	2
$5\pi/6$	3/2
π	1
$7\pi/6$	1/2
$3\pi/2$	0
$11\pi/6$	1/2
2π	1

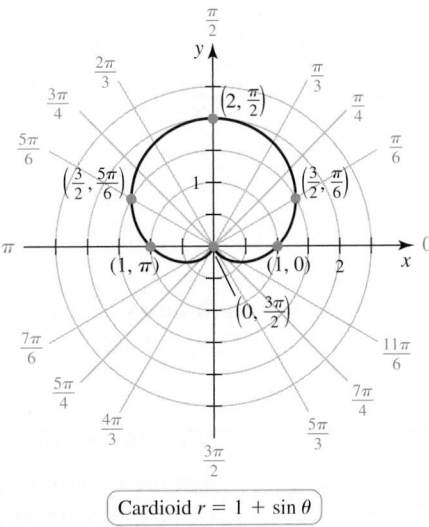

Cardioid $r = 1 + \sin \theta$

Figure 12.26

Related Exercises 53–54 ◀

Cartesian-to-Polar Method Plotting polar curves point by point is time-consuming, and important details may not be revealed. Here is an alternative procedure for graphing polar curves that is usually quicker and more reliable.

PROCEDURE   Cartesian-to-Polar Method for Graphing $r = f(\theta)$

1. Graph $r = f(\theta)$ *as if r and θ were Cartesian coordinates* with θ on the horizontal axis and r on the vertical axis. Be sure to choose an interval for θ on which the entire polar curve is produced.

2. Use the Cartesian graph that you created in Step 1 as a guide to sketch the points (r, θ) on the final *polar* curve.

EXAMPLE 5   Plotting polar graphs Use the Cartesian-to-polar method to graph the polar equation $r = 1 + \sin \theta$ (Example 4).

SOLUTION Viewing r and θ as Cartesian coordinates, the graph of $r = 1 + \sin \theta$ on the interval $[0, 2\pi]$ is a standard sine curve with amplitude 1 shifted up 1 unit (**Figure 12.27**). Notice that the graph begins with $r = 1$ at $\theta = 0$, increases to $r = 2$ at $\theta = \pi/2$, decreases to $r = 0$ at $\theta = 3\pi/2$ (which indicates an intersection with the origin on the polar graph), and increases to $r = 1$ at $\theta = 2\pi$. The second row of Figure 12.27 shows the final polar curve (a cardioid) as it is transferred from the Cartesian curve.

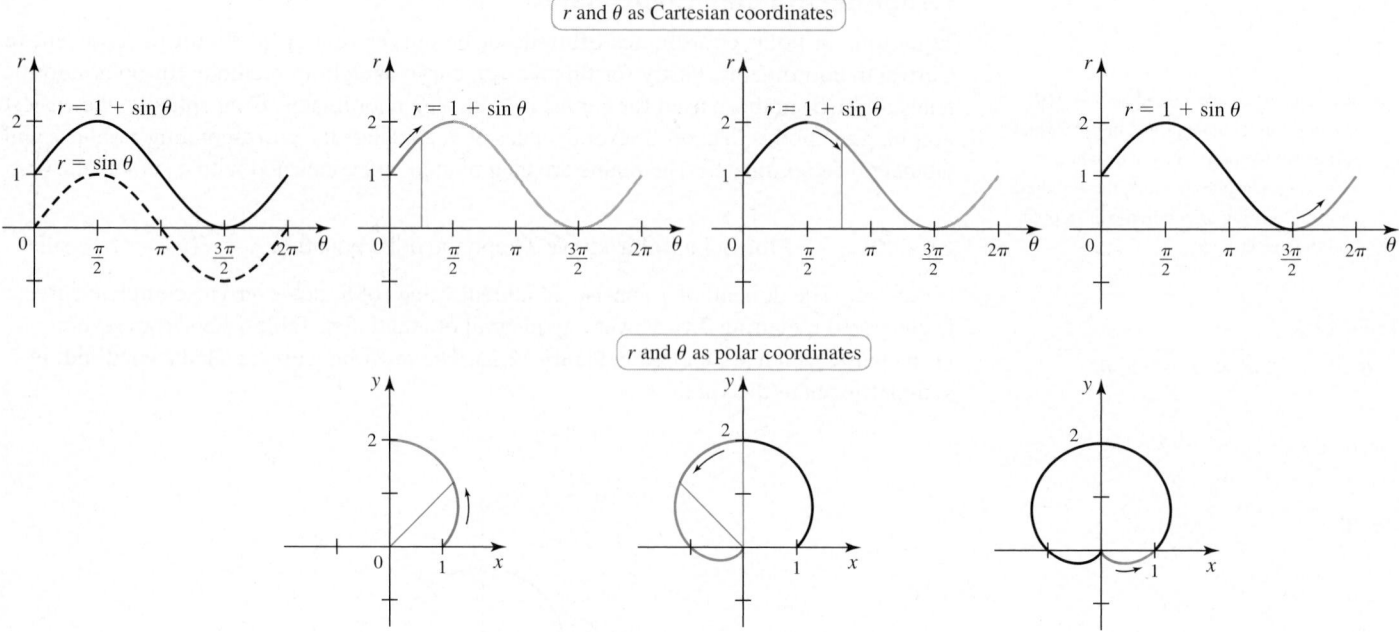

Figure 12.27

Related Exercise 58 ◄

Symmetry   Given a polar equation in r and θ, three types of symmetry are easy to spot (**Figure 12.28**).

➤ Any two of these three symmetries implies the third. For example, if a graph is symmetric about both the x- and y-axes, then it is symmetric about the origin.

> **SUMMARY Symmetry in Polar Equations**
>
> **Symmetry about the x-axis** occurs if the point (r, θ) is on the graph whenever $(r, -\theta)$ is on the graph.
>
> **Symmetry about the y-axis** occurs if the point (r, θ) is on the graph whenever $(r, \pi - \theta) = (-r, -\theta)$ is on the graph.
>
> **Symmetry about the origin** occurs if the point (r, θ) is on the graph whenever $(-r, \theta) = (r, \theta + \pi)$ is on the graph.

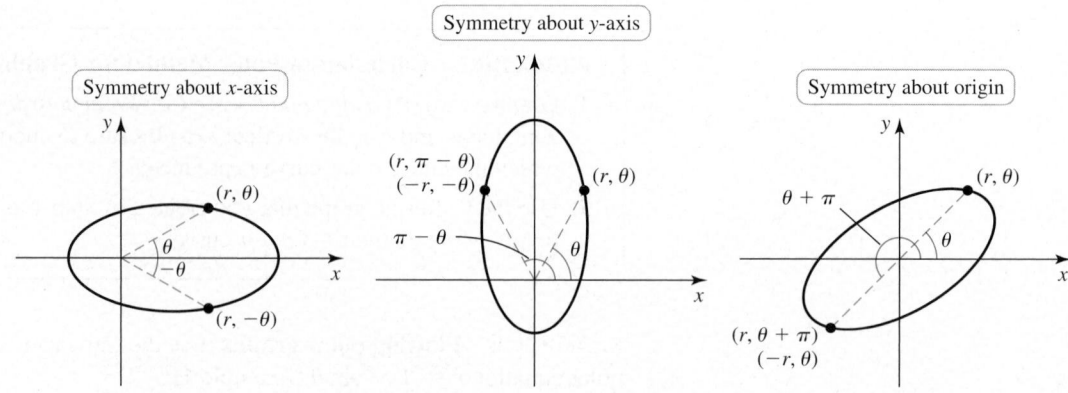

Figure 12.28

QUICK CHECK 6 Identify the symmetry in the graph of (a) $r = 4 + 4 \cos \theta$ and (b) $r = 4 \sin \theta$. ◄

For instance, consider the polar equation $r = 1 + \sin \theta$ in Example 5. If (r, θ) satisfies the equation, then $(r, \pi - \theta)$ also satisfies the equation because $\sin \theta = \sin (\pi - \theta)$. Therefore, the graph is symmetric about the y-axis, as shown in Figure 12.28. Testing for symmetry produces a more accurate graph and often simplifies the task of graphing polar equations.

EXAMPLE 6 **Plotting polar graphs** Graph the polar equation $r = 3 \sin 2\theta$.

SOLUTION The Cartesian graph of $r = 3 \sin 2\theta$ on the interval $[0, 2\pi]$ has amplitude 3 and period π (Figure 12.29a). The θ-intercepts occur at $\theta = 0, \pi/2, \pi, 3\pi/2$, and 2π, which correspond to the intersections with the origin on the polar graph. Furthermore, the arches of the Cartesian curve between θ-intercepts correspond to loops in the polar curve. The resulting polar curve is a **four-leaf rose** (Figure 12.29b).

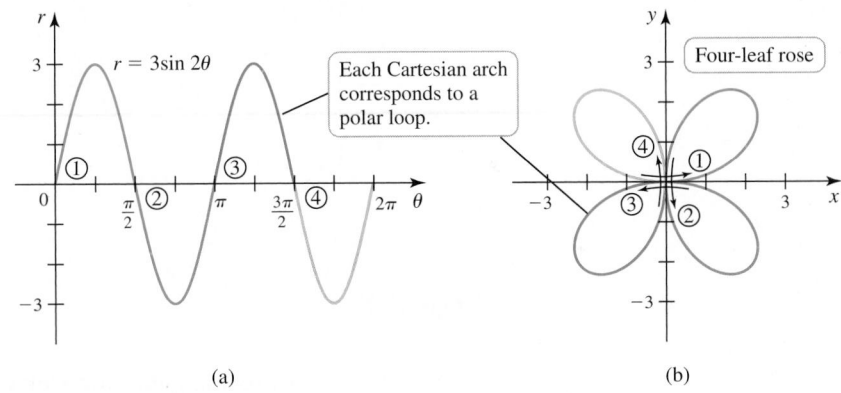

Figure 12.29

The graph is symmetric about the x-axis, the y-axis, and the origin. It is instructive to see how these symmetries are justified. To prove symmetry about the y-axis, notice that

$$(r, \theta) \text{ on the graph} \Rightarrow r = 3 \sin 2\theta$$
$$\Rightarrow r = -3 \sin 2(-\theta) \qquad \sin(-\theta) = -\sin\theta$$
$$\Rightarrow -r = 3 \sin 2(-\theta) \qquad \text{Simplify.}$$
$$\Rightarrow (-r, -\theta) \text{ on the graph.}$$

We see that if (r, θ) is on the graph, then $(-r, -\theta)$ is also on the graph, which implies symmetry about the y-axis. Similarly, to prove symmetry about the origin, notice that

$$(r, \theta) \text{ on the graph} \Rightarrow r = 3 \sin 2\theta$$
$$\Rightarrow r = 3 \sin(2\theta + 2\pi) \qquad \sin(\theta + 2\pi) = \sin\theta$$
$$\Rightarrow r = 3 \sin(2(\theta + \pi)) \qquad \text{Simplify.}$$
$$\Rightarrow (r, \theta + \pi) \text{ on the graph.}$$

We have shown that if (r, θ) is on the graph, then $(r, \theta + \pi)$ is also on the graph, which implies symmetry about the origin. Symmetry about the y-axis and the origin imply symmetry about the x-axis. Had we proved these symmetries in advance, we could have graphed the curve only in the first quadrant—reflections about the x- and y-axes would produce the full curve.

Related Exercises 63–64 ◄

> **Subtle Point**
> The fact that one point has infinitely many representations in polar coordinates presents potential pitfalls. In Example 6, you can show that $(-r, \theta)$ does *not* satisfy the equation $r = 3 \sin 2\theta$ when (r, θ) satisfies the equation. And yet, as shown, the graph is symmetric about the origin because $(r, \theta + \pi)$ satisfies the equation whenever (r, θ) satisfies the equation. Note that $(-r, \theta)$ and $(r, \theta + \pi)$ are the same point.

EXAMPLE 7 **Plotting polar graphs** Graph the polar equation $r^2 = 9 \cos\theta$. Use a graphing utility to check your work.

SOLUTION The graph of this equation has symmetry about the origin (because of the r^2) and about the x-axis (because of $\cos\theta$). These two symmetries imply symmetry about the y-axis.

A preliminary step is required before using the Cartesian-to-polar method for graphing the curve. Solving the given equation for r, we find that $r = \pm 3\sqrt{\cos\theta}$. Notice that $\cos\theta < 0$, for $\pi/2 < \theta < 3\pi/2$, so the curve does not exist on that interval. Therefore, we plot the curve on the intervals $0 \le \theta \le \pi/2$ and $3\pi/2 \le \theta \le 2\pi$ (the interval $\left[-\frac{\pi}{2}, \frac{\pi}{2}\right]$ would also work). Both the positive and negative values of r are included in the Cartesian graph (Figure 12.30a).

Now we are ready to transfer points from the Cartesian graph to the final polar graph (**Figure 12.30b**). The resulting curve is called a **lemniscate**.

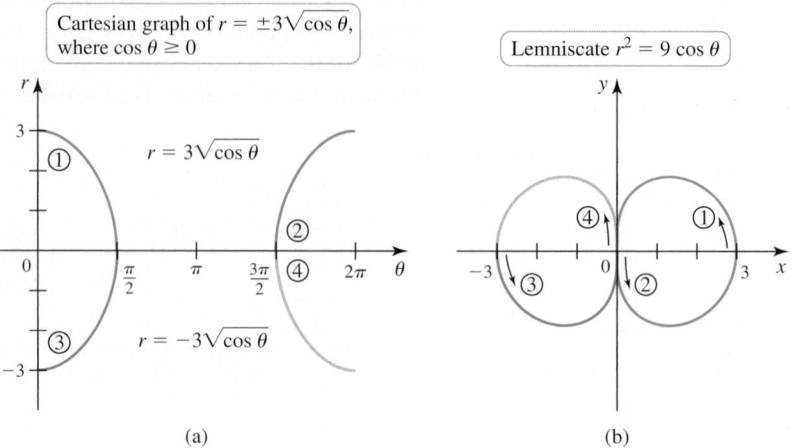

(a)   (b)

Figure 12.30

Related Exercise 61 ◄

EXAMPLE 8   Matching polar and Cartesian graphs The butterfly curve is described by the equation

$$r = e^{\sin\theta} - 2\cos 4\theta, \quad \text{for } 0 \le \theta \le 2\pi,$$

which is plotted in Cartesian and polar coordinates in **Figure 12.31**. Follow the Cartesian graph through the points $A, B, C, \ldots, N, O$ and mark the corresponding points on the polar curve

SOLUTION Point A in **Figure 12.31a** has the Cartesian coordinates $(\theta = 0, r = -1)$. The corresponding point in the polar plot (**Figure 12.31b**) with polar coordinates $(-1, 0)$ is marked A. Point B in the Cartesian plot is on the θ-axis; therefore, $r = 0$. The corresponding point in the polar plot is the origin. The same argument used to locate B applies to $F, H, J, L,$ and N, all of which appear at the origin in the polar plot. In general, the local and endpoint maxima and minima in the Cartesian graph $(A, C, D, E, G, I, K, M,$ and $O)$ correspond to the extreme points of the loops of the polar plot and are marked accordingly in Figure 12.31b.

> ➤ See Exercise 105 for a spectacular enhancement of the butterfly curve.

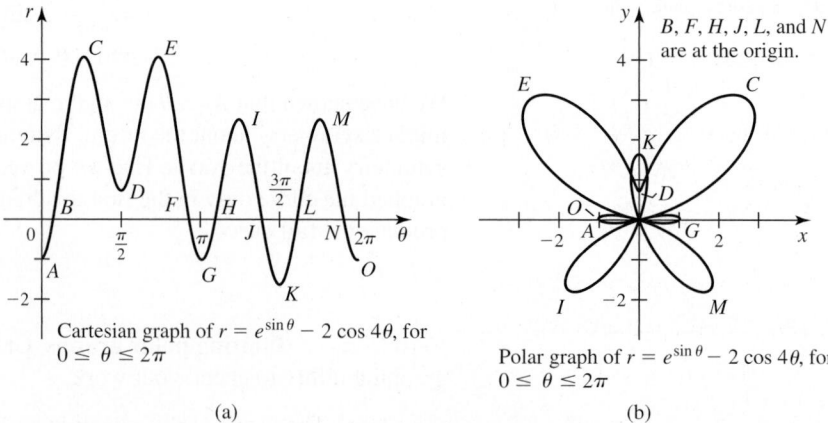

(a)   (b)

Figure 12.31

(*Source:* T. H. Fay, *Amer. Math. Monthly* 96, 1989, revived in Wagon and Packel, *Animating Calculus*, Freeman, 1994)

Related Exercises 65–68 ◄

Using Graphing Utilities

When graphing polar curves that eventually close on themselves, it is useful to specify an interval in θ that generates the entire curve. In some cases, this problem is a challenge in itself.

▶ **Using a parametric equation plotter to graph polar curves**

To graph $r = f(\theta)$, treat θ as a parameter and define the parametric equations

$$x = r \cos \theta = \underbrace{f(\theta)}_{r} \cos \theta$$

$$y = r \sin \theta = \underbrace{f(\theta)}_{r} \sin \theta$$

Then graph $(x(\theta), y(\theta))$ as a parametric curve with θ as the parameter.

▶ The prescription given in Example 9 for finding P when working with functions of the form $f(\theta) = \sin \dfrac{p\theta}{q}$ or $f(\theta) = \cos \dfrac{p\theta}{q}$ ensures that the complete curve is generated. Smaller values of P work in some cases.

EXAMPLE 9 Plotting complete curves Consider the closed curve described by $r = \cos(2\theta/5)$. Give an interval in θ that generates the entire curve and then graph the curve.

SOLUTION Recall that $\cos \theta$ has a period of 2π. Therefore, $\cos(2\theta/5)$ completes one cycle when $2\theta/5$ varies from 0 to 2π, or when θ varies from 0 to 5π. Therefore, it is tempting to conclude that the complete curve $r = \cos(2\theta/5)$ is generated as θ varies from 0 to 5π. But you can check that the point corresponding to $\theta = 0$ is *not* the point corresponding to $\theta = 5\pi$, which means the curve does not close on itself over the interval $[0, 5\pi]$ (Figure 12.32a).

To graph the *complete* curve $r = \cos(2\theta/5)$, we must find an interval $[0, P]$, where P is an integer multiple of 5π (so that $f(0) = f(P)$) and an integer multiple of 2π (so that the points $(0, f(0))$ and $(P, f(P))$ are the same). The smallest number satisfying these conditions is 10π. Graphing $r = \cos(2\theta/5)$ over the interval $[0, 10\pi]$ produces the complete curve (Figure 12.32b).

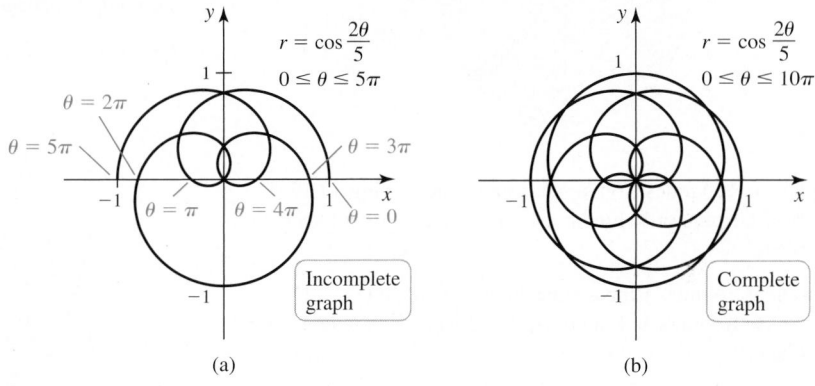

(a)   (b)

Figure 12.32

Related Exercises 72–73 ◀

SECTION 12.2 EXERCISES

Getting Started

1. Plot the points with polar coordinates $(2, \pi/6)$ and $(-3, -\pi/2)$. Give two alternative sets of coordinate pairs for both points.

2. Write the equations that are used to express a point with polar coordinates (r, θ) in Cartesian coordinates.

3. Write the equations that are used to express a point with Cartesian coordinates (x, y) in polar coordinates.

4. What is the polar equation of a circle of radius $\sqrt{a^2 + b^2}$ centered at (a, b)?

5. What is the polar equation of the vertical line $x = 5$?

6. What is the polar equation of the horizontal line $y = 5$?

7. Explain three symmetries in polar graphs and how they are detected in equations.

8. Given three polar coordinate representations for the origin.

Practice Exercises

9–13. *Graph the points with the following polar coordinates. Give two alternative representations of the points in polar coordinates.*

9. $\left(2, \dfrac{\pi}{4}\right)$

10. $\left(3, \dfrac{2\pi}{3}\right)$

11. $\left(-1, -\dfrac{\pi}{3}\right)$

12. $\left(2, \dfrac{7\pi}{4}\right)$

13. $\left(-4, \dfrac{3\pi}{2}\right)$

14. **Points in polar coordinates** Give two sets of polar coordinates for each of the points A–F in the figure.

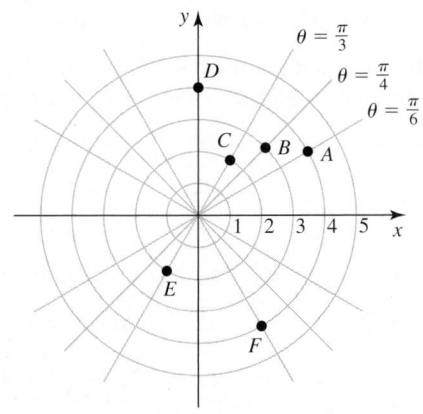

15–22. Sets in polar coordinates *Sketch the following sets of points.*

15. $r \geq 2$

16. $r = 3$

17. $\theta = \dfrac{2\pi}{3}$

18. $2 \leq r \leq 8$

19. $\dfrac{\pi}{2} \le \theta \le \dfrac{3\pi}{4}$

20. $1 < r < 2$ and $\dfrac{\pi}{6} \le \theta \le \dfrac{\pi}{3}$

21. $|\theta| \le \dfrac{\pi}{3}$

22. $0 < r < 3$ and $0 \le \theta \le \pi$

23–24. Radar *Airplanes are equipped with transponders that allow air traffic controllers to see their locations on radar screens. Radar gives the distance of the plane from the radar station (located at the origin) and the angular position of the plane, typically measured in degrees clockwise from north.*

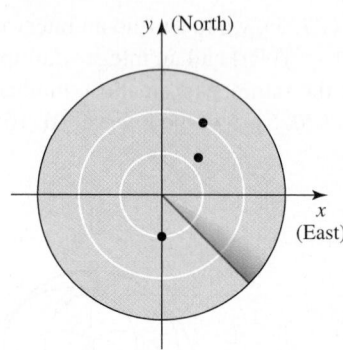

23. A plane is 100 miles from a radar station at an angle of 135° clockwise from north. Find polar coordinates for the location of the plane.

24. A plane is 50 miles from a radar station at an angle of 210° clockwise from north. Find polar coordinates for the location of the plane.

25–30. Converting coordinates *Express the following polar coordinates in Cartesian coordinates.*

25. $\left(3, \dfrac{\pi}{4}\right)$ **26.** $\left(1, \dfrac{2\pi}{3}\right)$ **27.** $\left(1, -\dfrac{\pi}{3}\right)$

28. $\left(2, \dfrac{7\pi}{4}\right)$ **29.** $\left(-4, \dfrac{3\pi}{4}\right)$ **30.** $(4, 5\pi)$

31–36. Converting coordinates *Express the following Cartesian coordinates in polar coordinates in at least two different ways.*

31. $(2, 2)$ **32.** $(-1, 0)$ **33.** $(1, \sqrt{3})$

34. $(0, -9)$ **35.** $(-4, 4\sqrt{3})$ **36.** $(-8\sqrt{3}, -8)$

37–48. Polar-to-Cartesian coordinates *Convert the following equations to Cartesian coordinates. Describe the resulting curve.*

37. $r \cos \theta = -4$ **38.** $r = \cot \theta \csc \theta$

39. $r = 2$ **40.** $r = 3 \csc \theta$

41. $r = 2 \sin \theta + 2 \cos \theta$ **42.** $r = -2 \cos \theta$

43. $r = 6 \cos \theta + 8 \sin \theta$ **44.** $\sin \theta = |\cos \theta|$

45. $r \cos \theta = \sin 2\theta$ **46.** $r = \sin \theta \sec^2 \theta$

47. $r = 8 \sin \theta$ **48.** $r = \dfrac{1}{2 \cos \theta + 3 \sin \theta}$

49–52. Cartesian-to-polar coordinates *Convert the following equations to polar coordinates.*

49. $y = x^2$ **50.** $y = 3$

51. $y = \dfrac{1}{x}$ **52.** $(x - 1)^2 + y^2 = 1$

53–56. Simple curves *Tabulate and plot enough points to sketch a graph of the following equations.*

53. $r = 8 \cos \theta$ **54.** $r = 4 + 4 \cos \theta$

55. $r(\sin \theta - 2 \cos \theta) = 0$ **56.** $r = 1 - \cos \theta$

57–64. Graphing polar curves *Graph the following equations. Use a graphing utility to check your work and produce a final graph.*

57. $r = 2 \sin \theta - 1$ **58.** $r = 2 - 2 \sin \theta$

59. $r = \sin^2 (\theta/2)$ **60.** $r^2 = 4 \sin \theta$

61. $r^2 = 16 \cos \theta$ **62.** $r^2 = 16 \sin 2\theta$

63. $r = \sin 3\theta$ **64.** $r = 2 \sin 5\theta$

65–68. Matching polar and Cartesian curves *Cartesian and polar graphs of* $r = f(\theta)$ *are given in the figures. Mark the points on the polar graph that correspond to the lettered points on the Cartesian graph.*

65. $r = 1 - 2 \sin 3\theta$

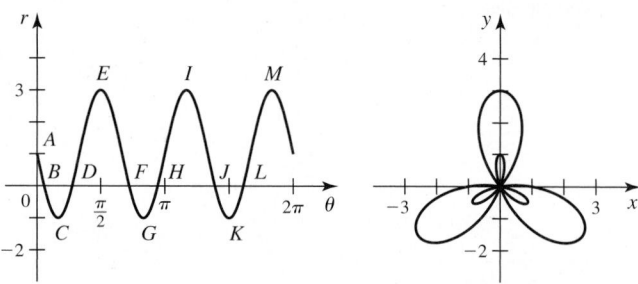

66. $r = \sin (1 + 3 \cos \theta)$

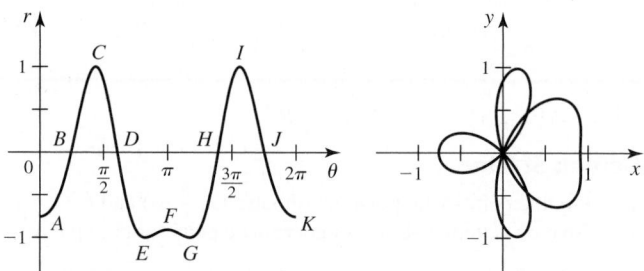

67. $r = \dfrac{1}{4} - \cos 4\theta$

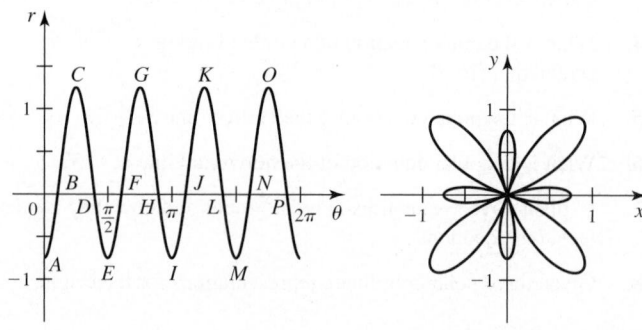

68. $r = \cos \theta + \sin 2\theta$

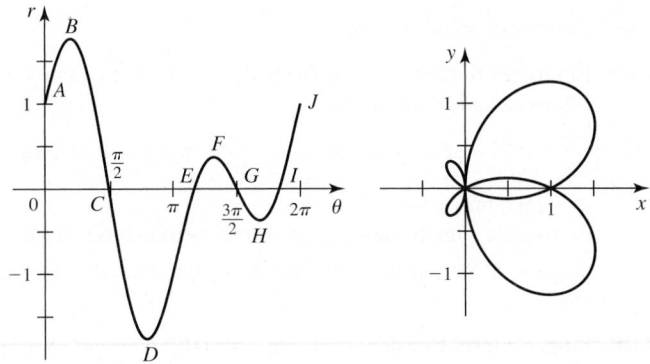

or less than the area of the region inside the square but outside the circles.

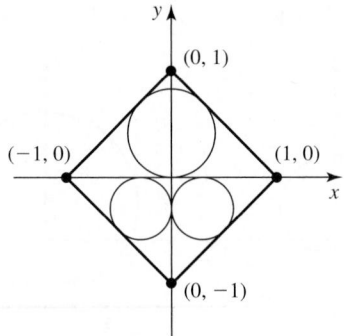

⊤ 69–76. Using a graphing utility *Use a graphing utility to graph the following equations. In each case, give the smallest interval* $[0, P]$ *that generates the entire curve.*

69. $r = \sin \dfrac{\theta}{4}$

70. $r = 2 - 4 \cos 5\theta$

71. $r = \cos 3\theta + \cos^2 2\theta$

72. $r = 2 \sin \dfrac{2\theta}{3}$

73. $r = \cos \dfrac{3\theta}{5}$

74. $r = \sin \dfrac{3\theta}{7}$

75. $r = 1 - 3 \cos 2\theta$

76. $r = 1 - 2 \sin 5\theta$

77. Explain why or why not Determine whether the following statements are true and give an explanation or counterexample.

   a. The point with Cartesian coordinates $(-2, 2)$ has polar coordinates $(2\sqrt{2}, 3\pi/4)$, $(2\sqrt{2}, 11\pi/4)$, $(2\sqrt{2}, -5\pi/4)$, and $(-2\sqrt{2}, -\pi/4)$.

   b. The graphs of $r \cos \theta = 4$ and $r \sin \theta = -2$ intersect exactly once.

   c. The graphs of $r = 2$ and $\theta = \pi/4$ intersect exactly once.

   d. The point $(3, \pi/2)$ lies on the graph of $r = 3 \cos 2\theta$.

   e. The graphs of $r = 2 \sec \theta$ and $r = 3 \csc \theta$ are lines.

78. Circles in general Show that the polar equation

$$r^2 - 2r(a \cos \theta + b \sin \theta) = R^2 - a^2 - b^2$$

describes a circle of radius R centered at (a, b).

79. Circles in general Show that the polar equation

$$r^2 - 2rr_0 \cos (\theta - \theta_0) = R^2 - r_0^2$$

describes a circle of radius R whose center has polar coordinates (r_0, θ_0).

80–83. Equations of circles *Use the results of Exercises 78–79 to describe and graph the following circles.*

80. $r^2 - 6r \cos \theta = 16$

81. $r^2 - 4r \cos (\theta - \pi/3) = 12$

82. $r^2 - 8r \cos (\theta - \pi/2) = 9$

83. $r^2 - 2r(2 \cos \theta + 3 \sin \theta) = 3$

Explorations and Challenges

84. Equations of circles Find equations of the circles in the figure. Determine whether the combined area of the circles is greater than

85. Navigating A plane is 150 miles north of a radar station, and 30 minutes later it is 60° east of north at a distance of 100 miles from the radar station. Assume the plane flies on a straight line and maintains constant altitude during this 30-minute period.

   a. Find the distance traveled during this 30-minute period.

   b. Determine the average velocity of the plane (relative to the ground) during this 30-minute period.

86. Lines in polar coordinates

   a. Show that an equation of the line $y = mx + b$ in polar coordinates is $r = \dfrac{b}{\sin \theta - m \cos \theta}$.

   b. Use the figure to find an alternative polar equation of a line, $r \cos (\theta_0 - \theta) = r_0$. Note that $Q(r_0, \theta_0)$ is a fixed point on the line such that OQ is perpendicular to the line and $r_0 \geq 0$; $P(r, \theta)$ is an arbitrary point on the line.

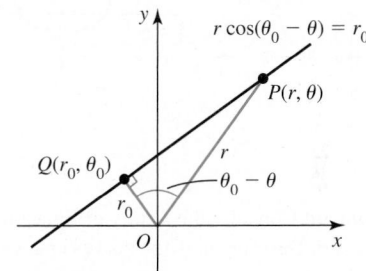

87–90. Equations of lines *Use the result of Exercise 86 to describe and graph the following lines.*

87. $r \cos \left(\dfrac{\pi}{3} - \theta \right) = 3$

88. $r \cos \left(\theta + \dfrac{\pi}{6} \right) = 4$

89. $r(\sin \theta - 4 \cos \theta) - 3 = 0$

90. $r(4 \sin \theta - 3 \cos \theta) = 6$

91. The limaçon family The equations $r = a + b \cos \theta$ and $r = a + b \sin \theta$ describe curves known as *limaçons* (from Latin for *snail*). We have already encountered cardioids, which occur when $|a| = |b|$. The limaçon has an inner loop if $|a| < |b|$. The limaçon has a dent or dimple if $|b| < |a| < 2|b|$. And the limaçon is oval-shaped if $|a| > 2|b|$. Match equations a–f with the limaçons in figures A–F.

a. $r = -1 + \sin \theta$

b. $r = -1 + 2 \cos \theta$

c. $r = 2 + \sin \theta$

d. $r = 1 - 2 \cos \theta$

e. $r = 1 + 2 \sin \theta$

f. $r = 1 + 2/3 \sin \theta$

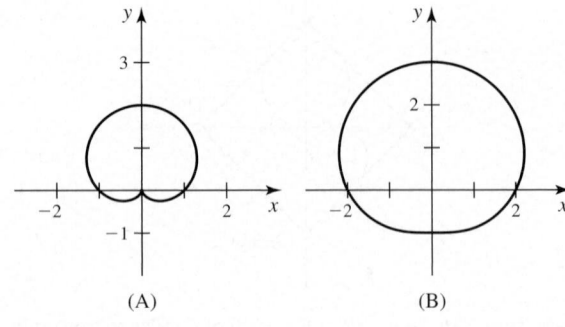

(A)   (B)

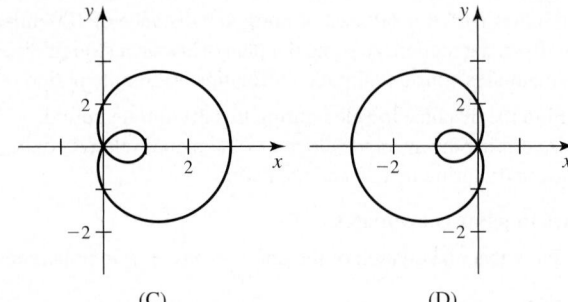

(C)   (D)

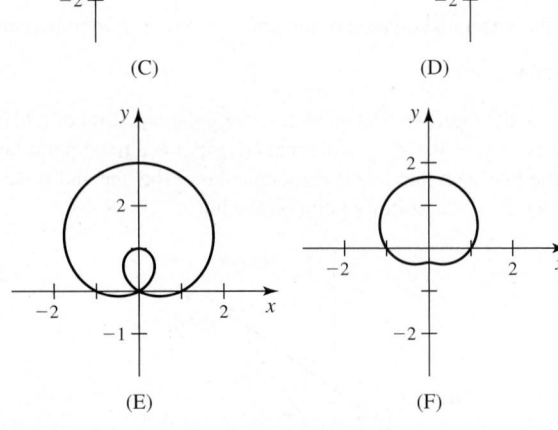

(E)   (F)

92. Limiting limaçon Consider the family of limaçons $r = 1 + b \cos \theta$. Describe how the curves change as $b \to \infty$.

93–96. The lemniscate family *Equations of the form* $r^2 = a \sin 2\theta$ *and* $r^2 = a \cos 2\theta$ *describe lemniscates (see Example 7). Graph the following lemniscates.*

93. $r^2 = \cos 2\theta$

94. $r^2 = 4 \sin 2\theta$

95. $r^2 = -2 \sin 2\theta$

96. $r^2 = -8 \cos 2\theta$

97–100. The rose family *Equations of the form* $r = a \sin m\theta$ *or* $r = a \cos m\theta$, *where a is a real number and m is a positive integer, have graphs known as roses (see Example 6). Graph the following roses.*

97. $r = \sin 2\theta$

98. $r = 4 \cos 3\theta$

99. $r = 2 \sin 4\theta$

100. $r = 6 \sin 5\theta$

101. Number of rose petals Show that the graph of $r = a \sin m\theta$ or $r = a \cos m\theta$ is a rose with m leaves if m is an odd integer and a rose with $2m$ leaves if m is an even integer.

102–104. Spirals *Graph the following spirals. Indicate the direction in which the spiral is generated as θ increases, where $\theta > 0$. Let $a = 1$ and $a = -1$.*

102. Spiral of Archimedes: $r = a\theta$

103. Logarithmic spiral: $r = e^{a\theta}$

104. Hyperbolic spiral: $r = a/\theta$

T 105. Enhanced butterfly curve The butterfly curve of Example 8 is enhanced by adding a term:

$$r = e^{\sin \theta} - 2 \cos 4\theta + \sin^5 \frac{\theta}{12}, \quad \text{for } 0 \le \theta \le 24\pi.$$

a. Graph the curve.

b. Explain why the new term produces the observed effect.

(*Source:* S. Wagon and E. Packel, *Animating Calculus*, Freeman, 1994)

T 106. Finger curves Consider the curve $r = f(\theta) = \cos a^\theta - 1.5$, where $a = (1 + 12\pi)^{1/(2\pi)} \approx 1.78933$ (see figure).

a. Show that $f(0) = f(2\pi)$ and find the point on the curve that corresponds to $\theta = 0$ and $\theta = 2\pi$.

b. Is the same curve produced over the intervals $[-\pi, \pi]$ and $[0, 2\pi]$?

c. Let $f(\theta) = \cos(a^\theta) - b$, where $a = (1 + 2k\pi)^{1/(2\pi)}$, k is an integer, and b is a real number. Show that $f(0) = f(2\pi)$ and the curve closes on itself.

d. Plot the curve with various values of k. How many fingers can you produce?

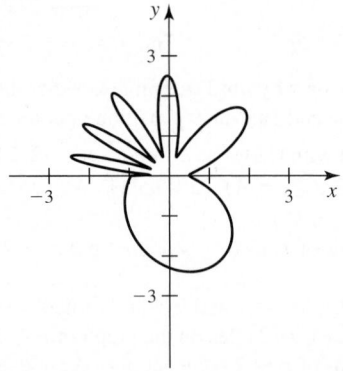

T 107. Earth-Mars system A simplified model assumes the orbits of Earth and Mars around the Sun are circular with radii of 2 and 3, respectively, and that Earth completes one orbit in one year while Mars takes two years. When $t = 0$, Earth is at $(2, 0)$ and Mars is at $(3, 0)$; both orbit the Sun (at $(0, 0)$) in the counterclockwise direction. The position of Mars relative to Earth is given by the parametric equations

$$x = (3 - 4 \cos \pi t) \cos \pi t + 2, \quad y = (3 - 4 \cos \pi t) \sin \pi t.$$

a. Graph the parametric equations, for $0 \le t \le 2$.

b. Letting $r = (3 - 4 \cos \pi t)$, explain why the path of Mars relative to Earth is a limaçon (Exercise 91).

108. Channel flow Water flows in a shallow semicircular channel with inner and outer radii of 1 m and 2 m (see figure). At a point $P(r, \theta)$ in the channel, the flow is in the tangential direction (counterclockwise along circles), and it depends only on r, the distance from the center of the semicircles.

a. Express the region formed by the channel as a set in polar coordinates.

b. Express the inflow and outflow regions of the channel as sets in polar coordinates.

c. Suppose the tangential velocity of the water in meters per second is given by $v(r) = 10r$, for $1 \le r \le 2$. Is the velocity greater at $\left(1.5, \dfrac{\pi}{4}\right)$ or $\left(1.2, \dfrac{3\pi}{4}\right)$? Explain.

d. Suppose the tangential velocity of the water is given by $v(r) = 20/r$, for $1 \le r \le 2$. Is the velocity greater at $(1.8, \pi/6)$ or $(1.3, 2\pi/3)$? Explain.

e. The total amount of water that flows through the channel (across a cross section of the channel $\theta = \theta_0$) is proportional to $\int_1^2 v(r)\, dr$. Is the total flow through the channel greater for the flow in part (c) or part (d)?

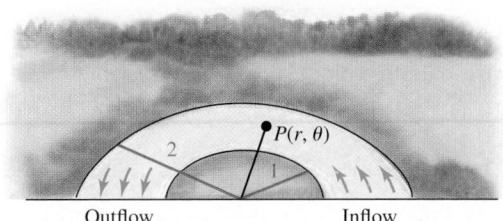

Outflow   Inflow

109. Subtle symmetry Without using a graphing utility, determine the symmetries (if any) of the curve $r = 4 - \sin(\theta/2)$.

110. Complete curves Consider the polar curve $r = \cos(n\theta/m)$, where n and m are integers.

a. Graph the complete curve when $n = 2$ and $m = 3$.

b. Graph the complete curve when $n = 3$ and $m = 7$.

c. Find a general rule in terms of m and n (where m and n have no common factors) for determining the least positive number P such that the complete curve is generated over the interval $[0, P]$.

111. Cartesian lemniscate Find the equation in Cartesian coordinates of the lemniscate $r^2 = a^2 \cos 2\theta$, where a is a real number.

QUICK CHECK ANSWERS

1. All the points are the same except $(3, 3\pi/2)$.   **3.** Polar coordinates: $(1, 0)$, $(1, 2\pi)$; Cartesian coordinates: $(0, 2)$
4. A circle centered at the origin with radius 12; two spirals; the horizontal line $y = 10$ **5.** A circle of radius $\sqrt{5}$ with center $(1, 2)$ **6.** (a) Symmetric about the x-axis; (b) symmetric about the y-axis ◄

12.3 Calculus in Polar Coordinates

Having learned about the *geometry* of polar coordinates, we now have the tools needed to explore *calculus* in polar coordinates. Familiar topics, such as slopes of tangent lines, areas bounded by curves, and arc length, are now revisited in a different setting.

Slopes of Tangent Lines

Given a function $y = f(x)$, the slope of the line tangent to the graph at a given point is $\dfrac{dy}{dx}$ or $f'(x)$. So it is tempting to conclude that the slope of a curve described by the polar equation $r = f(\theta)$ is $\dfrac{dr}{d\theta} = f'(\theta)$. Unfortunately, it's not that simple.

The key observation is that the slope of a tangent line—in any coordinate system—is the rate of change of the vertical coordinate y with respect to the horizontal coordinate x, which is $\dfrac{dy}{dx}$. We begin by writing the polar equation $r = f(\theta)$ in parametric form with θ as a parameter:

> The slope is the change in the vertical coordinate divided by the change in the horizontal coordinate, independent of the coordinate system. In polar coordinates, neither r nor θ corresponds to a vertical or horizontal coordinate.

$$x = r\cos\theta = f(\theta)\cos\theta \quad \text{and} \quad y = r\sin\theta = f(\theta)\sin\theta. \tag{1}$$

From Section 12.1, when x and y are defined parametrically as differentiable functions of θ, the derivative is $\dfrac{dy}{dx} = \dfrac{dy/d\theta}{dx/d\theta}$. Using the Product Rule to compute $\dfrac{dy}{d\theta}$ and $\dfrac{dx}{d\theta}$ in equation (1), we have

$$\frac{dy}{dx} = \frac{\overbrace{f'(\theta)\sin\theta + f(\theta)\cos\theta}^{dy/d\theta}}{\underbrace{f'(\theta)\cos\theta - f(\theta)\sin\theta}_{dx/d\theta}}. \tag{2}$$

If the graph passes through the origin for some angle θ_0, then $f(\theta_0) = 0$, and equation (2) simplifies to

$$\frac{dy}{dx} = \frac{\sin\theta_0}{\cos\theta_0} = \tan\theta_0,$$

provided $f'(\theta_0) \ne 0$. Assuming $\cos\theta_0 \ne 0$, $\tan\theta_0$ is the slope of the line $\theta = \theta_0$, which also passes through the origin. In this case, we conclude that if $f(\theta_0) = 0$, then the tangent

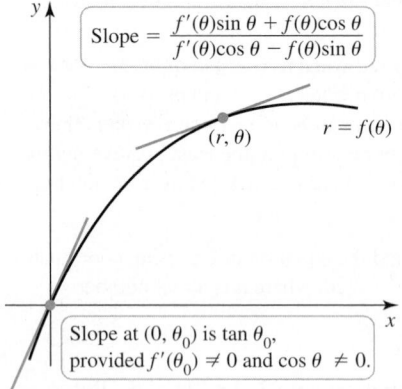

Figure 12.33

QUICK CHECK 1 Verify that if $y = f(\theta) \sin \theta$, then $y'(\theta) = f'(\theta) \sin \theta + f(\theta) \cos \theta$ (which was used earlier to find dy/dx). ◄

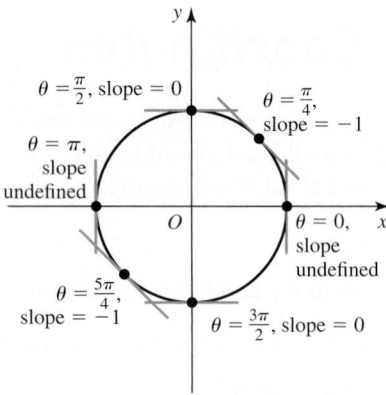

Figure 12.34

line at $(0, \theta_0)$ is simply $\theta = \theta_0$ (Figure 12.33). If $f(\theta_0) = 0$, $f'(\theta_0) \neq 0$, and $\cos \theta_0 = 0$, then the graph has a vertical tangent line at the origin.

> **THEOREM 12.2 Slope of a Tangent Line**
> Let f be a differentiable function at θ_0. The slope of the line tangent to the curve $r = f(\theta)$ at the point $(f(\theta_0), \theta_0)$ is
> $$\frac{dy}{dx} = \frac{f'(\theta_0) \sin \theta_0 + f(\theta_0) \cos \theta_0}{f'(\theta_0) \cos \theta_0 - f(\theta_0) \sin \theta_0},$$
> provided the denominator is nonzero at the point. At angles θ_0 for which $f(\theta_0) = 0$, $f'(\theta_0) \neq 0$, and $\cos \theta_0 \neq 0$, the tangent line is $\theta = \theta_0$ with slope $\tan \theta_0$.

EXAMPLE 1 Slopes on a circle Find the slopes of the lines tangent to the circle $r = f(\theta) = 10$.

SOLUTION In this case, $f(\theta)$ is constant (independent of θ). Therefore, $f'(\theta) = 0$, $f(\theta) \neq 0$, and the slope formula becomes

$$\frac{dy}{dx} = \frac{f'(\theta) \sin \theta + f(\theta) \cos \theta}{f'(\theta) \cos \theta - f(\theta) \sin \theta} = -\frac{\cos \theta}{\sin \theta} = -\cot \theta.$$

We can check a few points to see that this result makes sense. With $\theta = 0$ and $\theta = \pi$, the slope $\dfrac{dy}{dx} = -\cot \theta$ is undefined, which implies the tangent lines are vertical at these points (Figure 12.34). With $\theta = \pi/2$ and $\theta = 3\pi/2$, the slope is zero; with $\theta = 3\pi/4$ and $\theta = 7\pi/4$, the slope is 1; and with $\theta = \pi/4$ and $\theta = 5\pi/4$, the slope is -1. At all points $P(r, \theta)$ on the circle, the slope of the line OP from the origin to P is $\tan \theta$, which is the negative reciprocal of $-\cot \theta$. Therefore, OP is perpendicular to the tangent line at all points P on the circle.

Related Exercises 12–13 ◄

EXAMPLE 2 Vertical and horizontal tangent lines Find the points on the interval $-\pi \leq \theta \leq \pi$ at which the cardioid $r = f(\theta) = 1 - \cos \theta$ has a vertical or horizontal tangent line.

SOLUTION Applying Theorem 12.2, we find that

$$\frac{dy}{dx} = \frac{f'(\theta) \sin \theta + f(\theta) \cos \theta}{f'(\theta) \cos \theta - f(\theta) \sin \theta} \qquad f'(\theta) = \sin \theta$$

$$= \frac{\sin \theta \sin \theta + (1 - \cos \theta) \cos \theta}{\underbrace{\sin \theta \cos \theta - (1 - \cos \theta) \sin \theta}_{\sin \theta\,(2 \cos \theta - 1)}} \qquad \text{Substitute for } f(\theta) \text{ and } f'(\theta).$$

$$\overset{\sin^2 \theta = 1 - \cos^2 \theta}{=} -\frac{(2 \cos^2 \theta - \cos \theta - 1)}{\sin \theta\,(2 \cos \theta - 1)} \qquad \text{Simplify.}$$

$$= -\frac{(2 \cos \theta + 1)(\cos \theta - 1)}{\sin \theta\,(2 \cos \theta - 1)}. \qquad \text{Factor the numerator.}$$

The points with a horizontal tangent line satisfy $\dfrac{dy}{dx} = 0$ and occur where the numerator is zero and the denominator is nonzero. The numerator is zero when $\theta = 0$ and $\pm 2\pi/3$. Because the denominator is *not* zero when $\theta = \pm 2\pi/3$, horizontal tangent lines occur at $\theta = \pm 2\pi/3$ (Figure 12.35).

Vertical tangent lines occur where the numerator of $\dfrac{dy}{dx}$ is nonzero and the denominator is zero. The denominator is zero when $\theta = 0$, $\pm \pi$, and $\pm \pi/3$, and the numerator is not

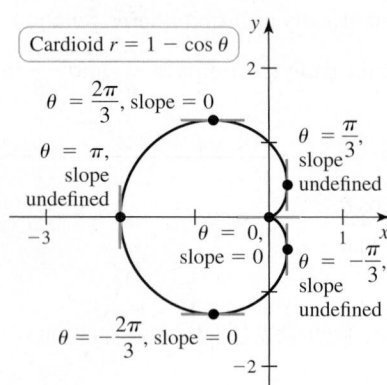

Cardioid $r = 1 - \cos \theta$

$\theta = \dfrac{2\pi}{3}$, slope $= 0$

$\theta = \pi$, slope undefined

$\theta = \dfrac{\pi}{3}$, slope undefined

$\theta = 0$, slope $= 0$

$\theta = -\dfrac{\pi}{3}$, slope undefined

$\theta = -\dfrac{2\pi}{3}$, slope $= 0$

Figure 12.35

zero when $\theta = \pm \pi$ and $\pm \pi/3$. Therefore, vertical tangent lines occur at $\theta = \pm \pi$ and $\pm \pi/3$.

The point $(0, 0)$ on the curve must be handled carefully because both the numerator and denominator of $\dfrac{dy}{dx}$ equal 0 at $\theta = 0$. Notice that with $f(\theta) = 1 - \cos \theta$, we have $f(0) = f'(0) = 0$. Therefore, $\dfrac{dy}{dx}$ may be computed as a limit using l'Hôpital's Rule. As $\theta \to 0^+$, we find that

$$\frac{dy}{dx} = \lim_{\theta \to 0^+} \left(-\frac{(2 \cos \theta + 1)(\cos \theta - 1)}{\sin \theta (2 \cos \theta - 1)} \right)$$

$$= \lim_{\theta \to 0^+} \frac{4 \cos \theta \sin \theta - \sin \theta}{-2 \sin^2 \theta + 2 \cos^2 \theta - \cos \theta} \quad \text{L'Hôpital's Rule}$$

$$= \frac{0}{1} = 0. \quad \text{Evaluate the limit.}$$

QUICK CHECK 2 What is the slope of the line tangent to the cardioid in Example 2 at the point corresponding to $\theta = \pi/4$? ◄

A similar calculation using l'Hôpital's Rule shows that as $\theta \to 0^-$, $\dfrac{dy}{dx} \to 0$. Therefore, the curve has a slope of 0 at $(0, 0)$.

Related Exercises 25–26 ◄

Area of Regions Bounded by Polar Curves

The problem of finding the area of a region bounded by polar curves brings us back to the slice-and-sum strategy used extensively in Chapters 5 and 6. The objective is to find the area of the region R bounded by the graph of $r = f(\theta)$ between the two rays $\theta = \alpha$ and $\theta = \beta$ (Figure 12.36a). We assume f is continuous and nonnegative on $[\alpha, \beta]$.

The area of R is found by slicing the region in the radial direction, creating wedge-shaped slices. The interval $[\alpha, \beta]$ is partitioned into n subintervals by choosing the grid points

$$\alpha = \theta_0 < \theta_1 < \theta_2 < \cdots < \theta_k < \cdots < \theta_n = \beta.$$

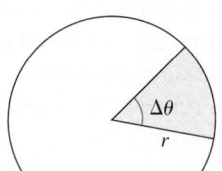

Area of circle $= \pi r^2$
Area of $\Delta\theta/(2\pi)$ of a circle
$= \left(\dfrac{\Delta\theta}{2\pi}\right)\pi r^2 = \dfrac{1}{2} r^2 \Delta\theta$

We let $\Delta\theta_k = \theta_k - \theta_{k-1}$, for $k = 1, 2, \ldots, n$, and we let θ_k^* be any point of the interval $[\theta_{k-1}, \theta_k]$. The kth slice is approximated by the sector of a circle swept out by an angle $\Delta\theta_k$ with radius $f(\theta_k^*)$ (Figure 12.36b). Therefore, the area of the kth slice is approximately $\frac{1}{2} f(\theta_k^*)^2 \Delta\theta_k$, for $k = 1, 2, \ldots, n$ (Figure 12.36c). To find the approximate area of R, we sum the areas of these slices:

$$\text{area} \approx \sum_{k=1}^{n} \frac{1}{2} f(\theta_k^*)^2 \, \Delta\theta_k.$$

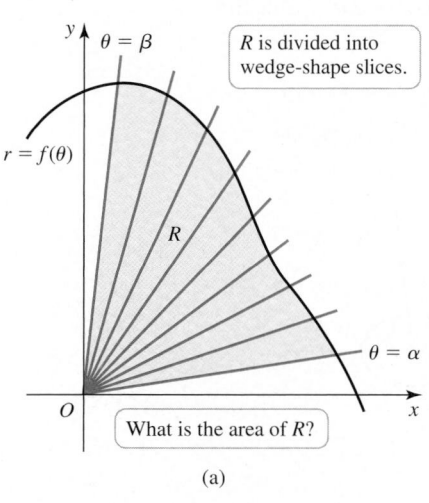
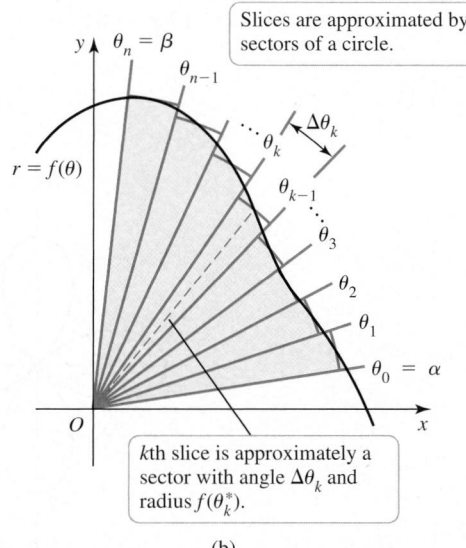
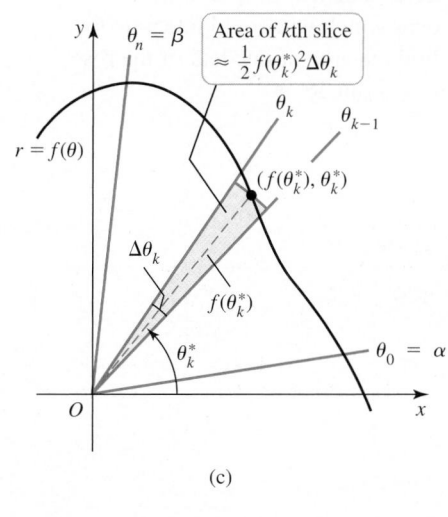

Figure 12.36

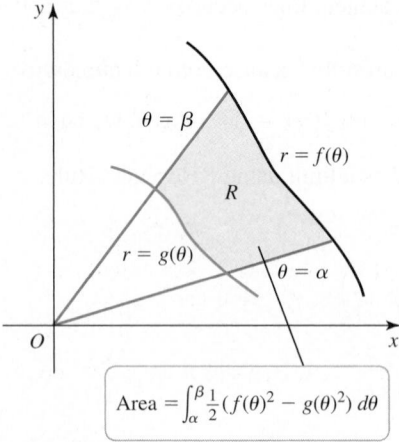

$$\text{Area} = \int_\alpha^\beta \frac{1}{2}(f(\theta)^2 - g(\theta)^2)\,d\theta$$

Figure 12.37

▶ If R is bounded by the graph of $r = f(\theta)$ between $\theta = \alpha$ and $\theta = \beta$, then $g(\theta) = 0$ and the area of R is
$$\int_\alpha^\beta \frac{1}{2} f(\theta)^2 \, d\theta.$$

▶ Though we assume $r = f(\theta) \geq 0$ when deriving the formula for the area of a region bounded by a polar curve, the formula is valid when $r < 0$ (see, for example, Exercise 54).

▶ The equation $r = 2 \cos 2\theta$ is unchanged when θ is replaced with $-\theta$ (symmetry about the x-axis) and when θ is replaced with $\pi - \theta$ (symmetry about the y-axis).

QUICK CHECK 3 Give an interval over which you could integrate to find the area of one leaf of the rose $r = 2 \sin 3\theta$. ◀

This approximation is a Riemann sum, and the approximation improves as we take more sectors ($n \to \infty$) and let $\Delta\theta_k \to 0$, for all k. The exact area is given by $\lim\limits_{n \to \infty} \sum\limits_{k=1}^{n} \frac{1}{2} f(\theta_k^*)^2 \, \Delta\theta_k$, which we identify as the definite integral $\int_\alpha^\beta \frac{1}{2} f(\theta)^2 \, d\theta$.

With a slight modification, a more general result is obtained for the area of a region R bounded by two curves, $r = f(\theta)$ and $r = g(\theta)$, between the rays $\theta = \alpha$ and $\theta = \beta$ (**Figure 12.37**). We assume f and g are continuous and $f(\theta) \geq g(\theta) \geq 0$ on $[\alpha, \beta]$. To find the area of R, we subtract the area of the region bounded by $r = g(\theta)$ from the area of the entire region bounded by $r = f(\theta)$ (all between $\theta = \alpha$ and $\theta = \beta$); that is,

$$\text{area} = \int_\alpha^\beta \frac{1}{2} f(\theta)^2 \, d\theta - \int_\alpha^\beta \frac{1}{2} g(\theta)^2 \, d\theta = \int_\alpha^\beta \frac{1}{2}(f(\theta)^2 - g(\theta)^2) \, d\theta.$$

DEFINITION Area of Regions in Polar Coordinates

Let R be the region bounded by the graphs of $r = f(\theta)$ and $r = g(\theta)$, between $\theta = \alpha$ and $\theta = \beta$, where f and g are continuous and $f(\theta) \geq g(\theta) \geq 0$ on $[\alpha, \beta]$. The area of R is

$$\int_\alpha^\beta \frac{1}{2}(f(\theta)^2 - g(\theta)^2) \, d\theta.$$

EXAMPLE 3 Area of a polar region Find the area of the four-leaf rose $r = f(\theta) = 2 \cos 2\theta$.

SOLUTION The graph of the rose (**Figure 12.38**) *appears* to be symmetric about the x- and y-axes; in fact, these symmetries can be proved. Appealing to this symmetry, we find the area of one-half of a leaf and then multiply the result by 8 to obtain the area of the full rose. The upper half of the rightmost leaf is generated as θ increases from $\theta = 0$ (when $r = 2$) to $\theta = \pi/4$ (when $r = 0$). Therefore, the area of the entire rose is

$$8 \int_0^{\pi/4} \frac{1}{2} f(\theta)^2 \, d\theta = 4 \int_0^{\pi/4} (2 \cos 2\theta)^2 \, d\theta \qquad f(\theta) = 2 \cos 2\theta$$

$$= 16 \int_0^{\pi/4} \cos^2 2\theta \, d\theta \qquad \text{Simplify.}$$

$$= 16 \int_0^{\pi/4} \frac{1 + \cos 4\theta}{2} \, d\theta \qquad \text{Half-angle formula}$$

$$= (8\theta + 2 \sin 4\theta) \Big|_0^{\pi/4} \qquad \text{Fundamental Theorem}$$

$$= (2\pi + 0) - (0 + 0) = 2\pi. \qquad \text{Simplify.}$$

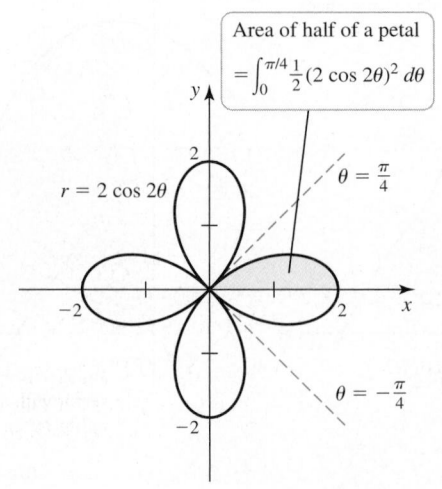

Figure 12.38

Related Exercises 38–39 ◀

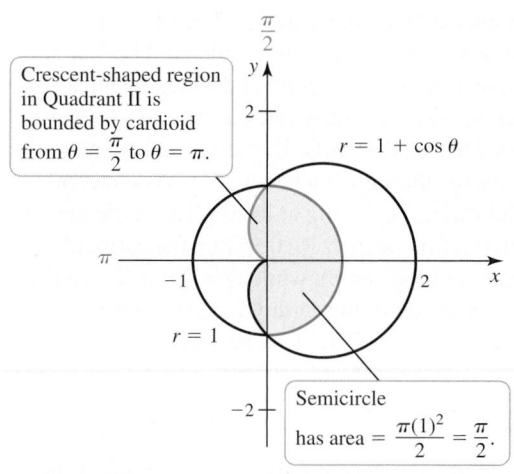

Crescent-shaped region in Quadrant II is bounded by cardioid from $\theta = \dfrac{\pi}{2}$ to $\theta = \pi$.

$r = 1 + \cos\theta$

$r = 1$

Semicircle has area $= \dfrac{\pi(1)^2}{2} = \dfrac{\pi}{2}$.

Figure 12.39

EXAMPLE 4 Areas of polar regions Consider the circle $r = 1$ and the cardioid $r = 1 + \cos\theta$ (Figure 12.39).

a. Find the area of the region inside the circle and inside the cardioid.

b. Find the area of the region inside the circle and outside the cardioid.

SOLUTION

a. The points of intersection of the two curves can be found by solving $1 + \cos\theta = 1$, or $\cos\theta = 0$. The solutions are $\theta = \pm\pi/2$. The region inside the circle and inside the cardioid consists of two subregions (Figure 12.39):

- a semicircle with radius 1 in the first and fourth quadrants bounded by the circle $r = 1$, and

- two crescent-shaped regions in the second and third quadrants bounded by the cardioid $r = 1 + \cos\theta$ and the y-axis.

The area of the semicircle is $\pi/2$. To find the area of the upper crescent-shaped region in the second quadrant, notice that it is bounded by $r = 1 + \cos\theta$, as θ varies from $\pi/2$ to π. Therefore, its area is

$$\int_{\pi/2}^{\pi} \frac{1}{2}(1 + \cos\theta)^2\,d\theta = \int_{\pi/2}^{\pi} \frac{1}{2}(1 + 2\cos\theta + \cos^2\theta)\,d\theta \qquad \text{Expand.}$$

$$= \frac{1}{2}\int_{\pi/2}^{\pi}\left(1 + 2\cos\theta + \frac{1 + \cos 2\theta}{2}\right)d\theta \qquad \text{Half-angle formula}$$

$$= \frac{1}{2}\left(\theta + 2\sin\theta + \frac{\theta}{2} + \frac{\sin 2\theta}{4}\right)\Big|_{\pi/2}^{\pi} \qquad \text{Evaluate integral.}$$

$$= \frac{3\pi}{8} - 1. \qquad \text{Simplify.}$$

The area of the entire region (two crescents and a semicircle) is

$$2\left(\frac{3\pi}{8} - 1\right) + \frac{\pi}{2} = \frac{5\pi}{4} - 2.$$

b. The region inside the circle and outside the cardioid is bounded by the outer curve $r = 1$ and the inner curve $r = 1 + \cos\theta$ on the interval $[\pi/2, 3\pi/2]$ (Figure 12.40). Using the symmetry about the x-axis, the area of the region is

$$2\int_{\pi/2}^{\pi} \frac{1}{2}(1^2 - (1 + \cos\theta)^2)\,d\theta = \int_{\pi/2}^{\pi}(-2\cos\theta - \cos^2\theta)\,d\theta \qquad \text{Simplify integrand.}$$

$$= 2 - \frac{\pi}{4}. \qquad \text{Evaluate integral.}$$

Note that the regions in parts (a) and (b) make up the interior of a circle of radius 1; indeed, their areas have a sum of π.

Related Exercise 45 ◄

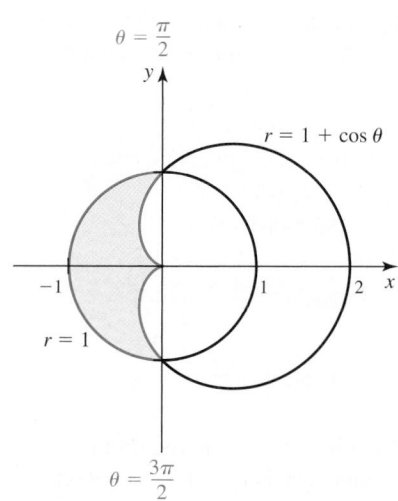

$\theta = \dfrac{\pi}{2}$

$r = 1 + \cos\theta$

$r = 1$

$\theta = \dfrac{3\pi}{2}$

Figure 12.40

Part of the challenge in setting up area integrals in polar coordinates is finding the points of intersection of two polar curves. The following example shows some of the subtleties of this process.

EXAMPLE 5 Points of intersection Find the points of intersection of the circle $r = 3\cos\theta$ and the cardioid $r = 1 + \cos\theta$ (Figure 12.41).

SOLUTION The fact that a point has multiple representations in polar coordinates may lead to subtle difficulties in finding intersection points. We first proceed algebraically. Equating the two expressions for r and solving for θ, we have

$$3\cos\theta = 1 + \cos\theta \quad \text{or} \quad \cos\theta = \frac{1}{2},$$

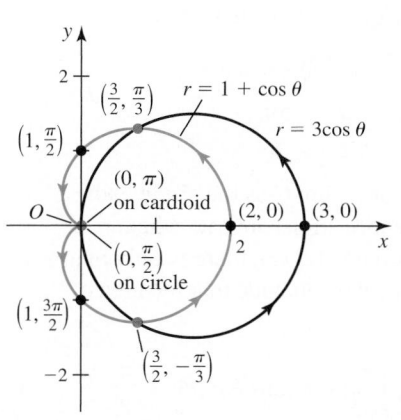

$\left(\dfrac{3}{2}, \dfrac{\pi}{3}\right)$ $r = 1 + \cos\theta$

$\left(1, \dfrac{\pi}{2}\right)$ $r = 3\cos\theta$

$(0, \pi)$ on cardioid

$(2, 0)$ $(3, 0)$

$\left(0, \dfrac{\pi}{2}\right)$ on circle

$\left(1, \dfrac{3\pi}{2}\right)$

$\left(\dfrac{3}{2}, -\dfrac{\pi}{3}\right)$

Figure 12.41

which has roots $\theta = \pm\pi/3$. Therefore, two intersection points are $(3/2, \pi/3)$ and $(3/2, -\pi/3)$ (Figure 12.41). Without examining graphs of the curves, we might be tempted to stop here. Yet the figure shows another intersection point at the origin O that has not been detected. To find this intersection point, we must investigate the way in which the two curves are generated. As θ increases from 0 to 2π, the cardioid is generated counterclockwise, beginning at $(2, 0)$. The cardioid passes through O when $\theta = \pi$. As θ increases from 0 to π, the circle is generated counterclockwise, beginning at $(3, 0)$. The circle passes through O when $\theta = \pi/2$. Therefore, the intersection point O is $(0, \pi)$ on the cardioid (and these coordinates do not satisfy the equation of the circle), while O is $(0, \pi/2)$ on the circle (and these coordinates do not satisfy the equation of the cardioid). There is no foolproof rule for detecting such "hidden" intersection points. Care must be used.

Related Exercise 29 ◄

EXAMPLE 6 Computing areas Example 5 discussed the points of intersection of the curves $r = 3\cos\theta$ (a circle) and $r = 1 + \cos\theta$ (a cardioid). Use those results to compute the areas of the following non-overlapping regions in **Figure 12.42**.

a. region A   **b.** region B   **c.** region C

SOLUTION

a. It is evident that region A is bounded on the inside by the cardioid and on the outside by the circle between the points $Q(\theta = 0)$ and $P(\theta = \pi/3)$. Therefore, the area of region A is

$$\frac{1}{2}\int_0^{\pi/3}\left((3\cos\theta)^2 - (1 + \cos\theta)^2\right)d\theta$$

$$= \frac{1}{2}\int_0^{\pi/3}(8\cos^2\theta - 1 - 2\cos\theta)\,d\theta \qquad \text{Simplify.}$$

$$= \frac{1}{2}\int_0^{\pi/3}(3 + 4\cos 2\theta - 2\cos\theta)\,d\theta \qquad \cos^2\theta = \frac{1 + \cos 2\theta}{2}$$

$$= \frac{1}{2}(3\theta + 2\sin 2\theta - 2\sin\theta)\Big|_0^{\pi/3} = \frac{\pi}{2}. \qquad \text{Evaluate integral.}$$

b. Examining region B, notice that a ray drawn from the origin enters the region immediately. There is no inner boundary, and the outer boundary is $r = 1 + \cos\theta$ on $0 \le \theta \le \pi/3$ and $r = 3\cos\theta$ on $\pi/3 \le \theta \le \pi/2$ (recall from Example 5 that $\theta = \pi/2$ is the angle at which the circle intersects the origin). Therefore, we slice the region into two parts at $\theta = \pi/3$ and write two integrals for its area:

$$\text{area of region B} = \frac{1}{2}\int_0^{\pi/3}(1 + \cos\theta)^2\,d\theta + \frac{1}{2}\int_{\pi/3}^{\pi/2}(3\cos\theta)^2\,d\theta.$$

While these integrals may be evaluated directly, it's easier to notice that

$$\text{area of region B} = \text{area of semicircle } OPQ - \text{area of region A}.$$

Because $r = 3\cos\theta$ is a circle with a radius of $3/2$, we have

$$\text{area of region B} = \frac{1}{2}\cdot\pi\left(\frac{3}{2}\right)^2 - \frac{\pi}{2} = \frac{5\pi}{8}.$$

c. It's easy to *incorrectly* identify the inner boundary of region C as the circle and the outer boundary as the cardioid. While these identifications are true when $\pi/3 \le \theta \le \pi/2$ (notice again the radial lines in Figure 12.42), there is only one boundary curve (the cardioid) when $\pi/2 \le \theta \le \pi$. We conclude that the area of region C is

$$\frac{1}{2}\int_{\pi/3}^{\pi/2}\left((1 + \cos\theta)^2 - (3\cos\theta)^2\right)d\theta + \frac{1}{2}\int_{\pi/2}^{\pi}(1 + \cos\theta)^2\,d\theta = \frac{\pi}{8}.$$

Related Exercise 61 ◄

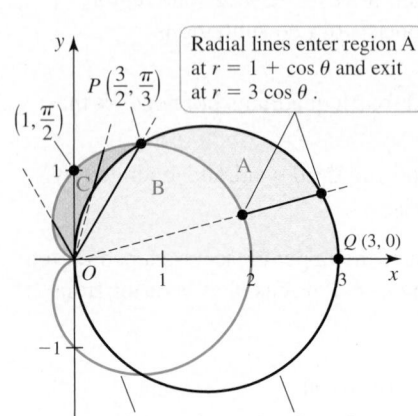

Radial lines enter region A at $r = 1 + \cos\theta$ and exit at $r = 3\cos\theta$.

$P\left(\frac{3}{2}, \frac{\pi}{3}\right)$

$\left(1, \frac{\pi}{2}\right)$

$Q(3, 0)$

$r = 1 + \cos\theta$ $r = 3\cos\theta$

Figure 12.42

➤ One way to verify that the inner and outer boundaries of a region have been correctly identified is to draw a ray from the origin through the region—the ray should enter the region at the inner boundary and exit the region at the outer boundary. In Example 6a, this is the case for every ray through region A, for $0 \le \theta \le \pi/3$.

Arc Length of a Polar Curve

> ▶ Recall from Section 12.2 that to convert from polar to Cartesian coordinates we use the relations
>
> $$x = r \cos \theta \quad \text{and} \quad y = r \sin \theta.$$

We now answer the arc length question for polar curves: Given the polar equation $r = f(\theta)$, what is the length of the corresponding curve for $\alpha \le \theta \le \beta$ (assuming the curve does not retrace itself on this interval)? The key idea is to express the polar equation as a set of parametric equations in Cartesian coordinates and then use the arc length formula for parametric equations derived in Section 12.1. Letting θ play the role of a parameter and using $r = f(\theta)$, parametric equations for the polar curve are

$$x = r \cos \theta = f(\theta) \cos \theta \quad \text{and} \quad y = r \sin \theta = f(\theta) \sin \theta,$$

where $\alpha \le \theta \le \beta$. The arc length formula in terms of the parameter θ is

$$L = \int_\alpha^\beta \sqrt{\left(\frac{dx}{d\theta}\right)^2 + \left(\frac{dy}{d\theta}\right)^2} \, d\theta,$$

where

$$\frac{dx}{d\theta} = f'(\theta) \cos \theta - f(\theta) \sin \theta \quad \text{and} \quad \frac{dy}{d\theta} = f'(\theta) \sin \theta + f(\theta) \cos \theta.$$

When substituted into the arc length formula and simplified, the result is a new arc length integral (Exercise 84).

QUICK CHECK 4 Use the arc length formula to verify that the circumference of the circle $r = f(\theta) = 1$, for $0 \le \theta \le 2\pi$, is 2π. ◀

Arc Length of a Polar Curve

Let f have a continuous derivative on the interval $[\alpha, \beta]$. The **arc length** of the polar curve $r = f(\theta)$ on $[\alpha, \beta]$ is

$$L = \int_\alpha^\beta \sqrt{f(\theta)^2 + f'(\theta)^2} \, d\theta.$$

EXAMPLE 7 Arc length of polar curves

a. Find the arc length of the spiral $r = f(\theta) = \theta$, for $0 \le \theta \le 2\pi$ (Figure 12.43).

b. Find the arc length of the cardioid $r = 1 + \cos \theta$ (Figure 12.44).

SOLUTION

a.
$$L = \int_0^{2\pi} \sqrt{\theta^2 + 1} \, d\theta \qquad \text{\small $f(\theta) = \theta$ and $f'(\theta) = 1$}$$

$$= \left(\frac{\theta}{2}\sqrt{\theta^2 + 1} + \frac{1}{2}\ln\left(\theta + \sqrt{\theta^2 + 1}\right)\right)\Big|_0^{2\pi} \qquad \text{\small Table of integrals or trigonometric substitution}$$

$$= \pi\sqrt{4\pi^2 + 1} + \frac{1}{2}\ln\left(2\pi + \sqrt{4\pi^2 + 1}\right) \qquad \text{\small Substitute limits of integration.}$$

$$\approx 21.26 \qquad \text{\small Evaluate.}$$

b. The cardioid is symmetric about the x-axis, and its upper half is generated for $0 \le \theta \le \pi$. The length of the full curve is twice the length of its upper half:

$$L = 2\int_0^\pi \sqrt{(1 + \cos \theta)^2 + (-\sin \theta)^2} \, d\theta \quad \text{\small $f(\theta) = 1 + \cos \theta$; $f'(\theta) = -\sin \theta$}$$

$$= 2\int_0^\pi \sqrt{2 + 2\cos \theta} \, d\theta \qquad \text{\small Simplify.}$$

$$= 2\int_0^\pi \sqrt{4\cos^2 (\theta/2)} \, d\theta \qquad \text{\small $1 + \cos \theta = 2\cos^2 (\theta/2)$}$$

$$= 4\int_0^\pi \cos (\theta/2) \, d\theta \qquad \text{\small $\cos (\theta/2) \ge 0$, for $0 \le \theta \le \pi$}$$

$$= 8\sin (\theta/2)\Big|_0^\pi = 8. \qquad \text{\small Integrate and simplify.}$$

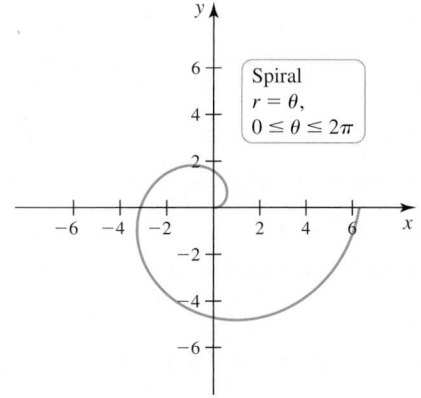

Figure 12.43

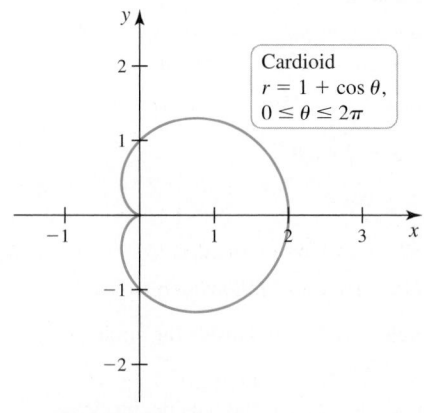

Figure 12.44

Related Exercises 67–68 ◀

SECTION 12.3 EXERCISES

Getting started

1. Express the polar equation $r = f(\theta)$ in parametric form in Cartesian coordinates, where θ is the parameter.

2. Explain why the slope of the line $\theta = \pi/2$ is undefined.

3. Explain why the slope of the line tangent to the polar graph of $r = f(\theta)$ is not $dr/d\theta$.

4. What integral must be evaluated to find the area of the region bounded by the polar graphs of $r = f(\theta)$ and $r = g(\theta)$ on the interval $\alpha \le \theta \le \beta$, where $f(\theta) \ge g(\theta) \ge 0$?

5. What is the slope of the line $\theta = \pi/3$?

6. Find a polar equation of the line with slope -1 that passes through the origin.

7. Find the area of the shaded region.

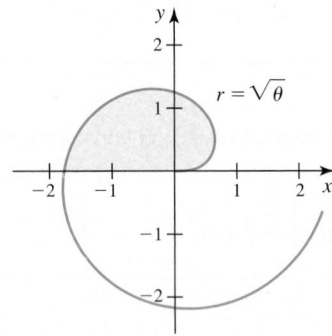

8. Without calculating derivatives, determine the slopes of each of the lines tangent to the curve $r = 8 \cos \theta - 4$ at the origin.

9. Explain why the point with polar coordinates $(0, 0)$ is an intersection point of the curves $r = \cos \theta$ and $r = \sin \theta$ even though $\cos \theta \ne \sin \theta$ when $\theta = 0$.

10. Explain why the point $(-1, 3\pi/2)$ is on the polar graph of $r = 1 + \cos \theta$ even though it does not satisfy the equation $r = 1 + \cos \theta$.

Practice Exercises

11–20. Slopes of tangent lines *Find the slope of the line tangent to the following polar curves at the given points.*

11. $r = 1 - \sin \theta;\ \left(\dfrac{1}{2}, \dfrac{\pi}{6} \right)$ 12. $r = 4 \cos \theta;\ \left(2, \dfrac{\pi}{3} \right)$

13. $r = 8 \sin \theta;\ \left(4, \dfrac{5\pi}{6} \right)$

14. $r = 4 + \sin \theta;\ (4, 0)$ and $\left(3, \dfrac{3\pi}{2} \right)$

15. $r = 6 + 3 \cos \theta;\ (3, \pi)$ and $(9, 0)$

16. $r = 2 \sin 3\theta;$ at the tips of the leaves

17. $r = 4 \cos 2\theta;$ at the tips of the leaves

18. $r = 1 + 2 \sin 2\theta;\ \left(3, \dfrac{\pi}{4} \right)$ 19. $r^2 = 4 \cos 2\theta;\ \left(0, \pm\dfrac{\pi}{4} \right)$

20. $r = 2\theta;\ \left(\dfrac{\pi}{2}, \dfrac{\pi}{4} \right)$

21. **Tangent line at the origin** Find the polar equation of the line tangent to the polar curve $r = \cos \theta + \sin \theta$ at the origin, and then find the slope of this tangent line.

22. **Tangent line at the origin** Find the polar equation of the line tangent to the polar curve $r = 4 \cos \theta$ at the origin. Explain why the slope of this line is undefined.

⊤ 23–24. Multiple tangent lines at a point

a. *Give the smallest interval $[0, P]$ that generates the entire polar curve and use a graphing utility to graph the curve.*
b. *Find a polar equation and the slope of each line tangent to the curve at the origin.*

23. $r = \cos 3\theta + \sin 3\theta$ 24. $r = 1 + 2 \cos 2\theta$

25–28. Horizontal and vertical tangents *Find the points at which the following polar curves have horizontal or vertical tangent lines.*

25. $r = 4 \cos \theta$ 26. $r = 2 + 2 \sin \theta$

⊤ 27. $r = \sin 2\theta$ **⊤ 28.** $r = 3 + 6 \sin \theta$

29–32. Intersection points *Use algebraic methods to find as many intersection points of the following curves as possible. Use graphical methods to identify the remaining intersection points.*

29. $r = 2 \cos \theta$ and $r = 1 + \cos \theta$

30. $r = 1 - \sin \theta$ and $r = 1 + \cos \theta$

31. $r = 1$ and $r = 2 \sin 2\theta$

32. $r = \cos 2\theta$ and $r = \sin 2\theta$

33–40. Areas of regions *Make a sketch of the region and its bounding curves. Find the area of the region.*

33. The region inside the curve $r = \sqrt{\cos \theta}$

34. The region inside the right lobe of $r = \sqrt{\cos 2\theta}$

35. The region inside the circle $r = 8 \sin \theta$

36. The region inside the cardioid $r = 4 + 4 \sin \theta$

37. The region inside the limaçon $r = 2 + \cos \theta$

38. The region inside all the leaves of the rose $r = 3 \sin 2\theta$

39. The region inside one leaf of $r = \cos 3\theta$

40. The region inside the inner loop of $r = \cos \theta - 1/2$

41–44. Intersection points and area

a. *Find all the intersection points of the following curves.*
b. *Find the area of the entire region that lies within both curves.*

41. $r = 3 \sin \theta$ and $r = 3 \cos \theta$

42. $r = 2 + 2 \sin \theta$ and $r = 2 - 2 \sin \theta$

43. $r = 1 + \sin \theta$ and $r = 1 + \cos \theta$

44. $r = 1$ and $r = \sqrt{2} \cos 2\theta$

45–60. Areas of regions *Find the area of the following regions.*

45. The region outside the circle $r = 1/2$ and inside the circle $r = \cos \theta$

46. The region inside the curve $r = \sqrt{\cos \theta}$ and outside the circle $r = 1/\sqrt{2}$

47. The region inside the curve $r = \sqrt{\cos\theta}$ and inside the circle $r = 1/\sqrt{2}$ in the first quadrant

48. The region inside the right lobe of $r = \sqrt{\cos 2\theta}$ and inside the circle $r = 1/\sqrt{2}$ in the first quadrant

49. The region inside one leaf of the rose $r = \cos 5\theta$

50. The region inside the rose $r = 4\cos 2\theta$ and outside the circle $r = 2$

51. The region inside the rose $r = 4\sin 2\theta$ and inside the circle $r = 2$

52. The region inside the lemniscate $r^2 = 2\sin 2\theta$ and outside the circle $r = 1$

53. The region common to the circles $r = 2\sin\theta$ and $r = 1$

54. The region inside the inner loop of the limaçon $r = 2 + 4\cos\theta$

55. The region inside the outer loop but outside the inner loop of the limaçon $r = 3 - 6\sin\theta$

56. The region common to the circle $r = 3\cos\theta$ and the cardioid $r = 1 + \cos\theta$

57. The region inside the lemniscate $r^2 = 6\sin 2\theta$

58. The region inside the limaçon $r = 2 - 4\sin\theta$

59. The region inside the limaçon $r = 4 - 2\cos\theta$

60. The region inside the complete three-leaf rose $r = 2\cos 3\theta$

61. Two curves, three regions Determine the intersection points of the polar curves $r = 4 - 2\cos\theta$ and $r = 2 + 2\cos\theta$, and then find areas of regions A, B, and C (see figure).

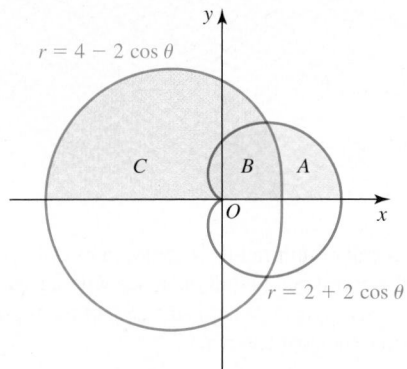

62. Find the area in the first quadrant inside the curve $r = 4 + \cos 2\theta$ but outside the curve $r = 3\cos 2\theta$ (see figure), after first finding the intersection points of the two curves.

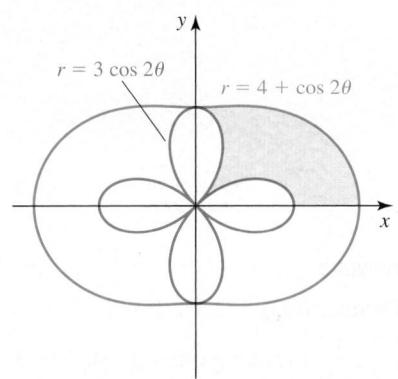

63–74. Arc length of polar curves *Find the length of the following polar curves.*

63. The complete circle $r = a\sin\theta$, where $a > 0$

64. The complete cardioid $r = 2 - 2\sin\theta$

65. The spiral $r = \theta^2$, for $0 \le \theta \le 2\pi$

66. The spiral $r = e^\theta$, for $0 \le \theta \le 2\pi n$, where n is a positive integer

67. The complete cardioid $r = 4 + 4\sin\theta$

68. The spiral $r = 4\theta^2$, for $0 \le \theta \le 6$

69. The spiral $r = 2e^{2\theta}$, for $0 \le \theta \le \ln 8$

70. The parabola $r = \dfrac{\sqrt{2}}{1 + \cos\theta}$, for $0 \le \theta \le \dfrac{\pi}{2}$

71. The curve $r = \sin^3\dfrac{\theta}{3}$, for $0 \le \theta \le \dfrac{\pi}{2}$

▣ 72. The three-leaf rose $r = 2\cos 3\theta$

▣ 73. The complete limaçon $r = 4 - 2\cos\theta$

▣ 74. The complete lemniscate $r^2 = 6\sin 2\theta$

75. Explain why or why not Determine whether the following statements are true and give an explanation or counterexample.

   a. The area of the region bounded by the polar graph of $r = f(\theta)$ on the interval $[\alpha, \beta]$ is $\int_\alpha^\beta f(\theta)\,d\theta$.

   b. The slope of the line tangent to the polar curve $r = f(\theta)$ at a point (r, θ) is $f'(\theta)$.

   c. There may be more than one line that is tangent to a polar curve at some points on the curve.

Explorations and Challenges

76. Area calculation Use polar coordinates to determine the area bounded on the right by the unit circle $x^2 + y^2 = 1$ and bounded on the left by the vertical line $x = \sqrt{2}/2$.

▣ 77. Spiral tangent lines Use a graphing utility to determine the first three points with $\theta \ge 0$ at which the spiral $r = 2\theta$ has a horizontal tangent line. Find the first three points with $\theta \ge 0$ at which the spiral $r = 2\theta$ has a vertical tangent line.

78. Spiral arc length Consider the spiral $r = 4\theta$, for $\theta \ge 0$.

   a. Use a trigonometric substitution to find the length of the spiral, for $0 \le \theta \le \sqrt{8}$.

   b. Find $L(\theta)$, the length of the spiral on the interval $[0, \theta]$, for any $\theta \ge 0$.

   c. Show that $L'(\theta) > 0$. Is $L''(\theta)$ positive or negative? Interpret your answer.

79. Spiral arc length Find the length of the entire spiral $r = e^{-a\theta}$, for $\theta \ge 0$ and $a > 0$.

80. Area of roses Assume m is a positive integer.

   a. *Even number of leaves*: What is the relationship between the total area enclosed by the $4m$-leaf rose $r = \cos(2m\theta)$ and m?

   b. *Odd number of leaves*: What is the relationship between the total area enclosed by the $(2m + 1)$-leaf rose $r = \cos((2m + 1)\theta)$ and m?

81. Regions bounded by a spiral Let R_n be the region bounded by the nth turn and the $(n + 1)$st turn of the spiral $r = e^{-\theta}$ in the first and second quadrants, for $\theta \geq 0$ (see figure).

a. Find the area A_n of R_n.

b. Evaluate $\lim\limits_{n \to \infty} A_n$.

c. Evaluate $\lim\limits_{n \to \infty} \dfrac{A_{n+1}}{A_n}$.

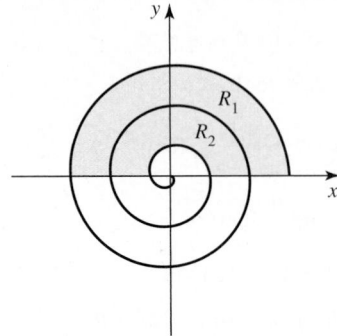

82. Tangents and normals Let a polar curve be described by $r = f(\theta)$, and let ℓ be the line tangent to the curve at the point $P(x, y) = P(r, \theta)$ (see figure).

a. Explain why $\tan \alpha = dy/dx$.

b. Explain why $\tan \theta = y/x$.

c. Let φ be the angle between ℓ and the line through O and P. Prove that $\tan \varphi = f(\theta)/f'(\theta)$.

d. Prove that the values of θ for which ℓ is parallel to the x-axis satisfy $\tan \theta = -f(\theta)/f'(\theta)$.

e. Prove that the values of θ for which ℓ is parallel to the y-axis satisfy $\tan \theta = f(\theta)/f'(\theta)$.

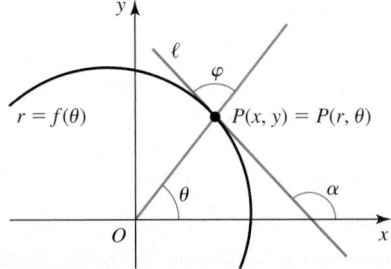

▣ 83. Isogonal curves Let a curve be described by $r = f(\theta)$, where $f(\theta) > 0$ on its domain. Referring to the figure in Exercise 82, a curve is **isogonal** provided the angle φ is constant for all θ.

a. Prove that φ is constant for all θ provided $\cot \varphi = \dfrac{f'(\theta)}{f(\theta)}$ is constant, which implies that $\dfrac{d}{d\theta}(\ln f(\theta)) = k$, where k is a constant.

b. Use part (a) to prove that the family of logarithmic spirals $r = Ce^{k\theta}$ consists of isogonal curves, where C and k are constants.

c. Graph the curve $r = 2e^{2\theta}$ and confirm the result of part (b).

84. Arc length for polar curves Prove that the length of the curve $r = f(\theta)$, for $\alpha \leq \theta \leq \beta$, is

$$L = \int_{\alpha}^{\beta} \sqrt{f(\theta)^2 + f'(\theta)^2} \, d\theta.$$

85–87. Grazing goat problems *Consider the following sequence of problems related to grazing goats tied to a rope. (See the Guided Project* Grazing goat problems.*)*

85. A circular corral of unit radius is enclosed by a fence. A goat inside the corral is tied to the fence with a rope of length $0 \leq a \leq 2$ (see figure). What is the area of the region (inside the corral) that the goat can graze? Check your answer with the special cases $a = 0$ and $a = 2$.

86. A circular concrete slab of unit radius is surrounded by grass. A goat is tied to the edge of the slab with a rope of length $0 \leq a \leq 2$ (see figure). What is the area of the grassy region that the goat can graze? Note that the rope can extend over the concrete slab. Check your answer with the special cases $a = 0$ and $a = 2$.

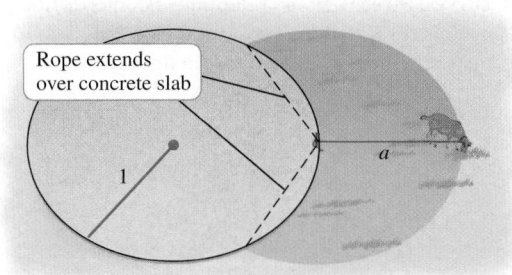

Rope extends over concrete slab

87. A circular corral of unit radius is enclosed by a fence. A goat is outside the corral and tied to the fence with a rope of length $0 \leq a \leq \pi$ (see figure). What is the area of the region (outside the corral) that the goat can reach?

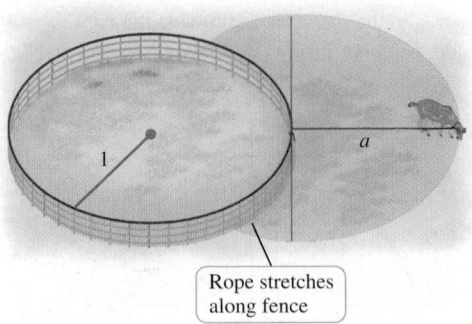

Rope stretches along fence

QUICK CHECK ANSWERS

1. Apply the Product Rule.   **2.** $\sqrt{2} + 1$

3. $\left[0, \dfrac{\pi}{3}\right]$ or $\left[\dfrac{\pi}{3}, \dfrac{2\pi}{3}\right]$ (among others)   **4.** 2π ◄

12.4 Conic Sections

Conic sections are best visualized as the Greeks did over 2000 years ago by slicing a double cone with a plane (**Figure 12.45**). Three of the seven different sets of points that arise in this way are *ellipses*, *parabolas*, and *hyperbolas*. These curves have practical applications and broad theoretical importance. For example, celestial bodies travel in orbits that are modeled by ellipses and hyperbolas. Mirrors for telescopes are designed using the properties of conic sections. And architectural structures, such as domes and arches, are sometimes based on these curves.

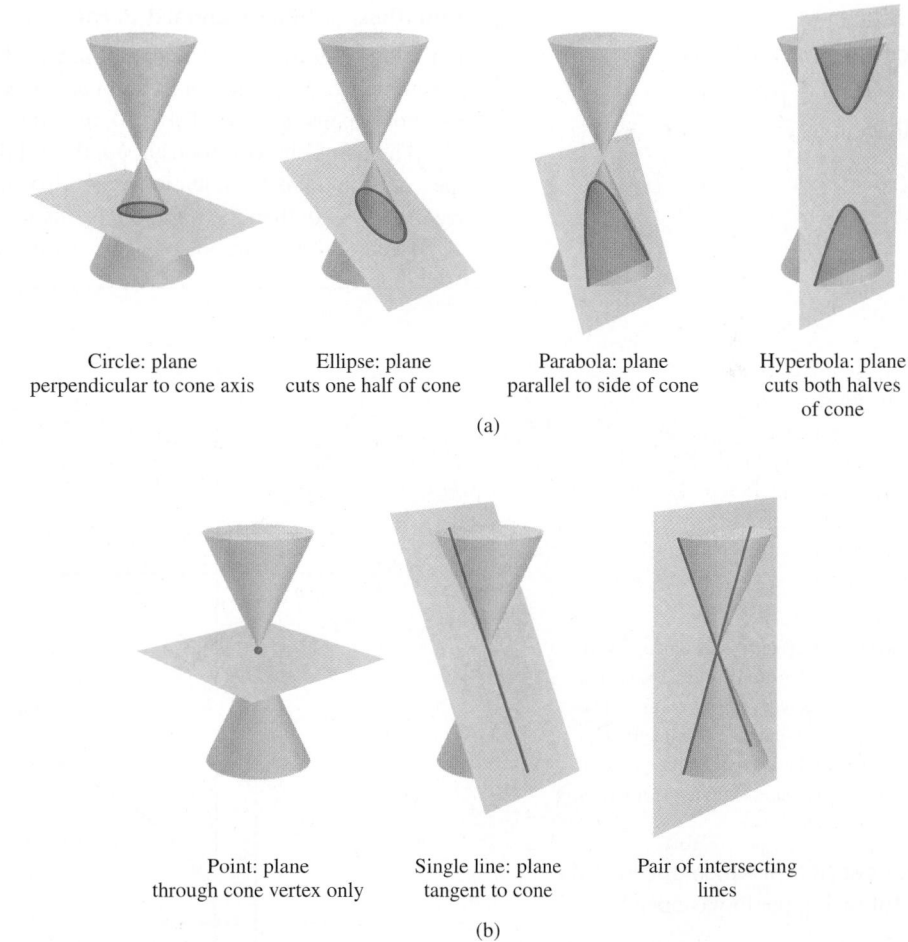

Circle: plane
perpendicular to cone axis

Ellipse: plane
cuts one half of cone

Parabola: plane
parallel to side of cone

Hyperbola: plane
cuts both halves
of cone

(a)

Point: plane
through cone vertex only

Single line: plane
tangent to cone

Pair of intersecting
lines

(b)

Figure 12.45 The standard conic sections (a) are the intersection sets of a double cone and a plane that does not pass through the vertex of the cone. Degenerate conic sections (lines and points) are produced when a plane passes through the vertex of the cone (b).

Parabolas

A **parabola** is the set of points in a plane that are equidistant from a fixed point F (called the **focus**) and a fixed line (called the **directrix**). In the four standard orientations, a parabola may open upward, downward, to the right, or to the left. We derive the equation of the parabola that opens upward.

Suppose the focus F is on the y-axis at $(0, p)$ and the directrix is the horizontal line $y = -p$, where $p > 0$. The parabola is the set of points P that satisfy the defining property $|PF| = |PL|$, where $L(x, -p)$ is the point on the directrix closest to P

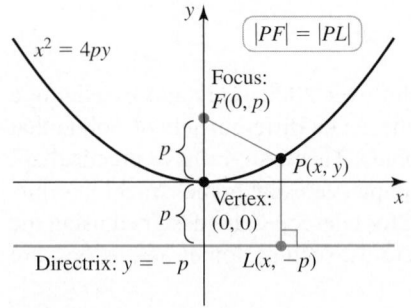

Figure 12.46

QUICK CHECK 1 Verify that $\sqrt{x^2 + (y - p)^2} = y + p$ is equivalent to $x^2 = 4py$. ◄

▶ Recall that a curve is symmetric with respect to the x-axis if $(x, -y)$ is on the curve whenever (x, y) is on the curve. So a y^2-term indicates symmetry with respect to the x-axis. Similarly, an x^2-term indicates symmetry with respect to the y-axis.

QUICK CHECK 2 In which direction do the following parabolas open?

a. $y^2 = -4x$ **b.** $x^2 = 4y$ ◄

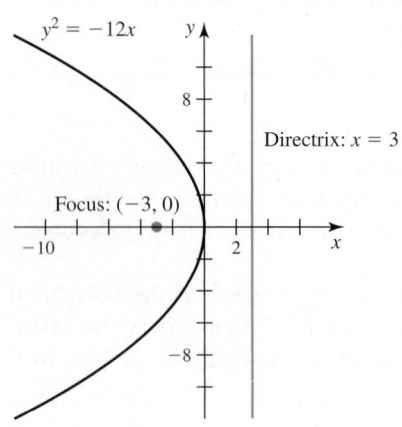

Figure 12.48

(Figure 12.46). Consider an arbitrary point $P(x, y)$ that satisfies this condition. Applying the distance formula, we have

$$\underbrace{\sqrt{x^2 + (y - p)^2}}_{|PF|} = \underbrace{y + p.}_{|PL|}$$

Squaring both sides of this equation and simplifying gives the equation $x^2 = 4py$. This is the equation of a parabola that is symmetric about the y-axis and opens upward. The **vertex** of the parabola is the point closest to the directrix; in this case, it is $(0, 0)$ (which satisfies $|PF| = |PL| = p$).

The equations of the other three standard parabolas are derived in a similar way.

Equations of Four Standard Parabolas

Let p be a real number. The parabola with focus at $(0, p)$ and directrix $y = -p$ is symmetric about the y-axis and has the equation $x^2 = 4py$. If $p > 0$, then the parabola opens *upward*; if $p < 0$, then the parabola opens *downward*.

The parabola with focus at $(p, 0)$ and directrix $x = -p$ is symmetric about the x-axis and has the equation $y^2 = 4px$. If $p > 0$, then the parabola opens *to the right*; if $p < 0$, then the parabola opens *to the left*.

Each of these parabolas has its vertex at the origin (**Figure 12.47**).

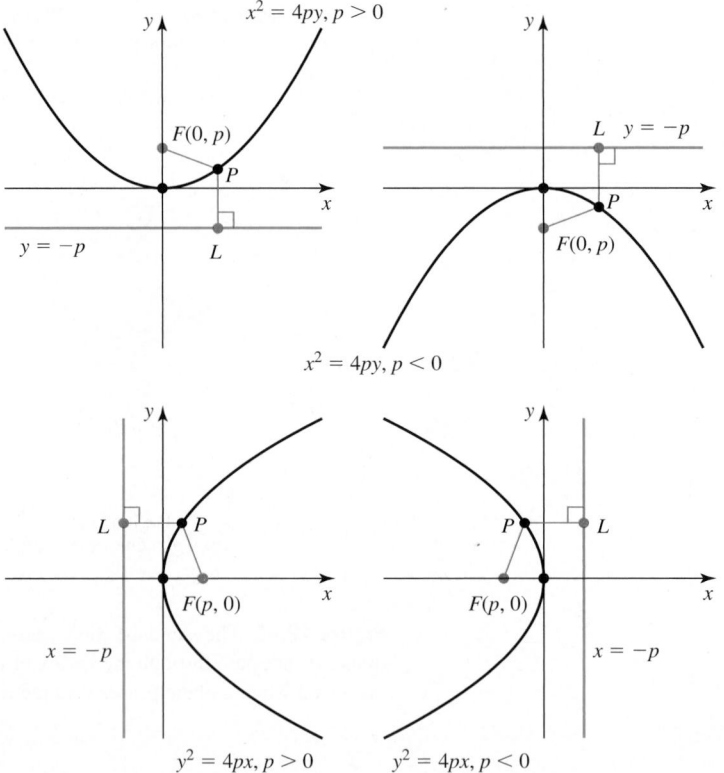

Figure 12.47

EXAMPLE 1   Graphing parabolas Find the focus and directrix of the parabola $y^2 = -12x$. Sketch its graph.

SOLUTION The y^2-term indicates that the parabola is symmetric with respect to the x-axis. Rewriting the equation as $x = -y^2/12$, we see that $x \leq 0$ for all y, implying that the parabola opens to the left. Comparing $y^2 = -12x$ to the standard form $y^2 = 4px$, we see that $p = -3$; therefore, the focus is $(-3, 0)$, and the directrix is $x = 3$ (**Figure 12.48**).

Related Exercises 13–14 ◄

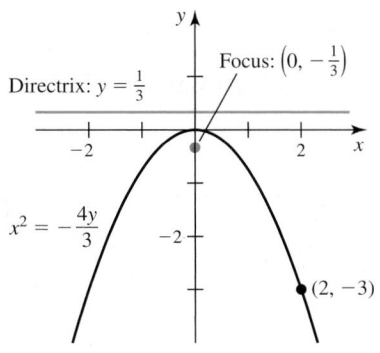

Directrix: $y = \frac{1}{3}$

Focus: $\left(0, -\frac{1}{3}\right)$

$x^2 = -\dfrac{4y}{3}$

$(2, -3)$

Figure 12.49

EXAMPLE 2 Equations of parabolas Find the equation of the parabola with vertex $(0, 0)$ that opens downward and passes through the point $(2, -3)$.

SOLUTION The standard parabola that opens downward has the equation $x^2 = 4py$. The point $(2, -3)$ must satisfy this equation. Substituting $x = 2$ and $y = -3$ into $x^2 = 4py$, we find that $p = -\frac{1}{3}$. Therefore, the focus is at $\left(0, -\frac{1}{3}\right)$, the directrix is $y = \frac{1}{3}$, and the equation of the parabola is $x^2 = -4y/3$, or $y = -3x^2/4$ (**Figure 12.49**).

Related Exercises 35–36 ◄

Reflection Property Parabolas have a property that makes them useful in the design of reflectors and transmitters. A particle approaching a parabola on any line parallel to the axis of the parabola is reflected on a line that passes through the focus (**Figure 12.50**); this property is used to focus incoming light by a parabolic mirror on a telescope. Alternatively, signals emanating from the focus are reflected on lines parallel to the axis, a property used to design radio transmitters and headlights (Exercise 83).

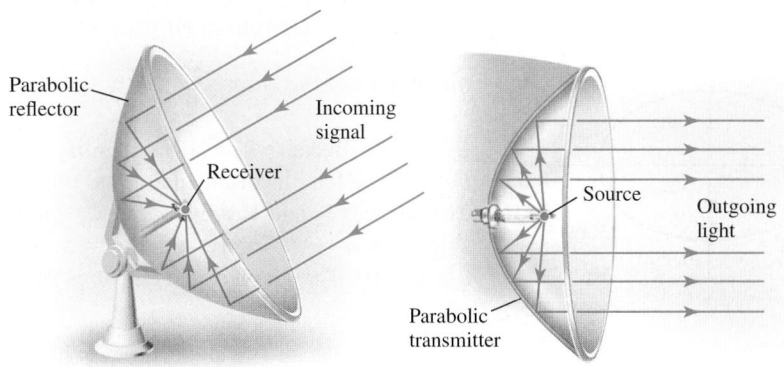

Parabolic reflector

Incoming signal

Receiver

Source

Outgoing light

Parabolic transmitter

Figure 12.50

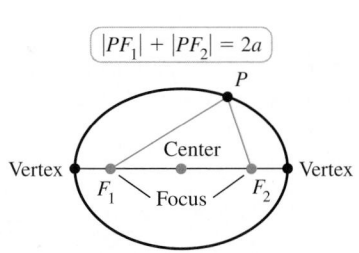

$|PF_1| + |PF_2| = 2a$

P

Center

Vertex

Vertex

F_1

Focus

F_2

Figure 12.51

Ellipses

An **ellipse** is the set of points in a plane whose distances from two fixed points have a constant sum that we denote $2a$ (**Figure 12.51**). Each of the two fixed points is a **focus** (plural **foci**). The equation of an ellipse is simplest if the foci are on the x-axis at $(\pm c, 0)$ or on the y-axis at $(0, \pm c)$. In either case, the **center** of the ellipse is $(0, 0)$. If the foci are on the x-axis, the points $(\pm a, 0)$ lie on the ellipse and are called **vertices**. If the foci are on the y-axis, the vertices are $(0, \pm a)$ (**Figure 12.52**). A short calculation (Exercise 85) using the definition of the ellipse results in the following equations for an ellipse.

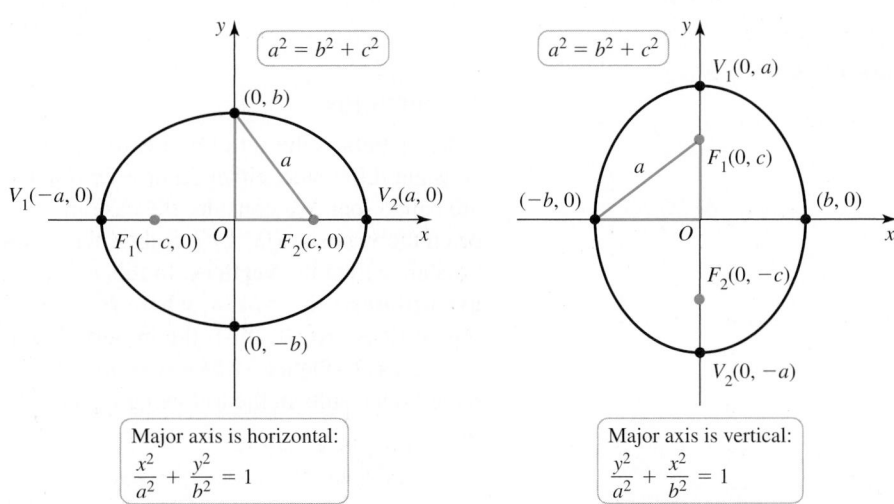

$a^2 = b^2 + c^2$

$(0, b)$

a

$V_1(-a, 0)$

$V_2(a, 0)$

$F_1(-c, 0)$ O $F_2(c, 0)$

$(0, -b)$

Major axis is horizontal:
$$\frac{x^2}{a^2} + \frac{y^2}{b^2} = 1$$

$a^2 = b^2 + c^2$

$V_1(0, a)$

$F_1(0, c)$

a

$(-b, 0)$

$(b, 0)$

O

$F_2(0, -c)$

$V_2(0, -a)$

Major axis is vertical:
$$\frac{y^2}{a^2} + \frac{x^2}{b^2} = 1$$

Figure 12.52

▶ When necessary, we distinguish between the *major-axis vertices* $(\pm a, 0)$ or $(0, \pm a)$, and the *minor-axis vertices* $(\pm b, 0)$ or $(0, \pm b)$. The word *vertices* (without further description) is understood to mean *major-axis vertices*.

> **Equations of Standard Ellipses**
>
> An ellipse centered at the origin with foci F_1 and F_2 at $(\pm c, 0)$ and vertices V_1 and V_2 at $(\pm a, 0)$ has the equation
>
> $$\frac{x^2}{a^2} + \frac{y^2}{b^2} = 1, \quad \text{where } a^2 = b^2 + c^2.$$
>
> An ellipse centered at the origin with foci at $(0, \pm c)$ and vertices at $(0, \pm a)$ has the equation
>
> $$\frac{y^2}{a^2} + \frac{x^2}{b^2} = 1, \quad \text{where } a^2 = b^2 + c^2.$$
>
> In both cases, $a > b > 0$ and $a > c > 0$, the length of the long axis (called the **major axis**) is $2a$, and the length of the short axis (called the **minor axis**) is $2b$.

QUICK CHECK 3 In the case that the vertices and foci are on the x-axis, show that the length of the minor axis of an ellipse is $2b$. ◄

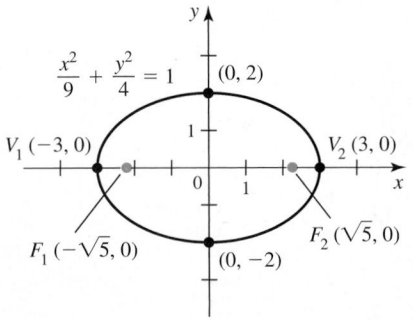

Figure 12.53

EXAMPLE 3 Graphing ellipses Find the vertices, foci, and length of the major and minor axes of the ellipse $\dfrac{x^2}{9} + \dfrac{y^2}{4} = 1$. Graph the ellipse.

SOLUTION Because $9 > 4$, we identify $a^2 = 9$ and $b^2 = 4$. Therefore, $a = 3$ and $b = 2$. The lengths of the major and minor axes are $2a = 6$ and $2b = 4$, respectively. The vertices V_1 and V_2 are at $(\pm 3, 0)$ and lie on the x-axis, as do the foci. The relationship $c^2 = a^2 - b^2$ implies that $c^2 = 5$, or $c = \sqrt{5}$. Therefore, the foci F_1 and F_2 are at $(\pm\sqrt{5}, 0)$. The graph of the ellipse is shown in **Figure 12.53**.

Related Exercises 15–16 ◄

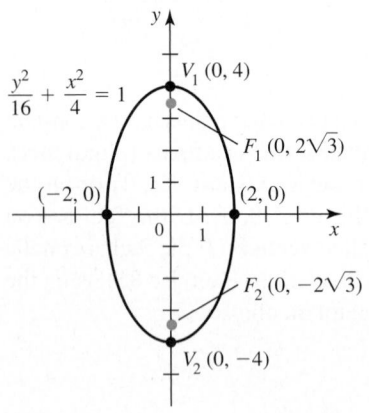

Figure 12.54

EXAMPLE 4 Equation of an ellipse Find the equation of the ellipse centered at the origin with its foci on the y-axis, a major axis of length 8, and a minor axis of length 4. Graph the ellipse.

SOLUTION Because the length of the major axis is 8, the vertices V_1 and V_2 are located at $(0, \pm 4)$, and $a = 4$. Because the length of the minor axis is 4, we have $b = 2$. Therefore, the equation of the ellipse is

$$\frac{y^2}{16} + \frac{x^2}{4} = 1.$$

Using the relation $c^2 = a^2 - b^2$, we find that $c = 2\sqrt{3}$ and the foci F_1 and F_2 are at $(0, \pm 2\sqrt{3})$. The ellipse is shown in **Figure 12.54**.

Related Exercise 39 ◄

Hyperbolas

A **hyperbola** is the set of points in a plane whose distances from two fixed points have a constant difference, either $2a$ or $-2a$ (**Figure 12.55**). As with ellipses, the two fixed points are called **foci**. We consider the case in which the foci are either on the x-axis at $(\pm c, 0)$ or on the y-axis at $(0, \pm c)$. If the foci are on the x-axis, the points $(\pm a, 0)$ on the hyperbola are called the **vertices**. In this case, the hyperbola has no y-intercepts, but it has the **asymptotes** $y = \pm bx/a$, where $b^2 = c^2 - a^2$. Similarly, if the foci are on the y-axis, the vertices are $(0, \pm a)$, the hyperbola has no x-intercepts, and it has the asymptotes $y = \pm ax/b$ (**Figure 12.56**). A short calculation (Exercise 86) using the definition of the hyperbola results in the following equations for standard hyperbolas.

▶ Asymptotes that are not parallel to one of the coordinate axes, as in the case of the standard hyperbolas, are called **oblique**, or **slant**, **asymptotes**.

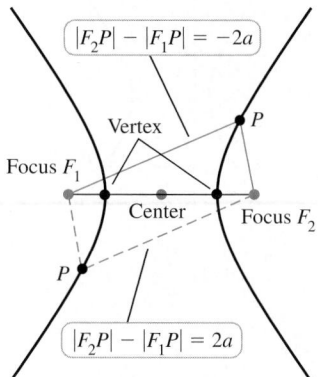

Figure 12.55

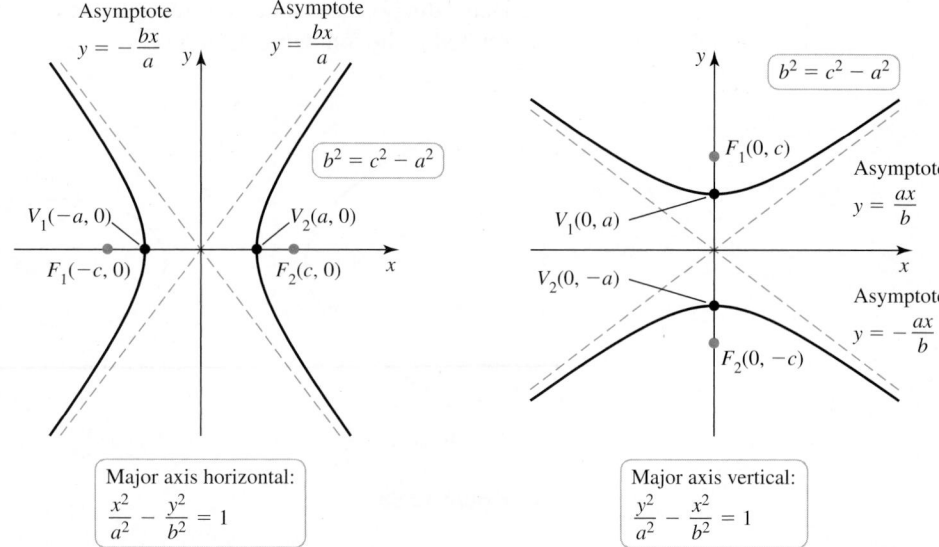

Major axis horizontal:
$$\frac{x^2}{a^2} - \frac{y^2}{b^2} = 1$$

Major axis vertical:
$$\frac{y^2}{a^2} - \frac{x^2}{b^2} = 1$$

Figure 12.56

▶ Notice that the asymptotes for hyperbolas are $y = \pm bx/a$ when the vertices are on the x-axis and $y = \pm ax/b$ when the vertices are on the y-axis (the roles of a and b are reversed).

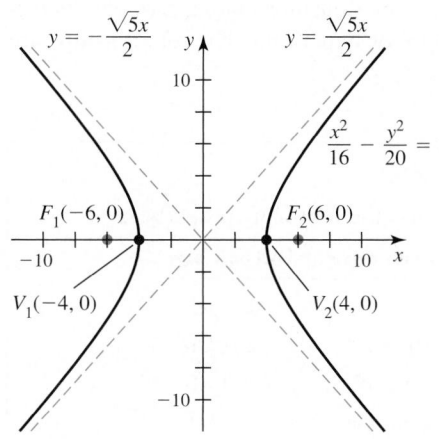

Figure 12.57

QUICK CHECK 4 Identify the vertices and foci of the hyperbola $y^2 - x^2/4 = 1$. ◀

Equations of Standard Hyperbolas

A hyperbola centered at the origin with foci F_1 and F_2 at $(\pm c, 0)$ and vertices V_1 and V_2 at $(\pm a, 0)$ has the equation

$$\frac{x^2}{a^2} - \frac{y^2}{b^2} = 1, \quad \text{where} \quad b^2 = c^2 - a^2.$$

The hyperbola has **asymptotes** $y = \pm bx/a$.

A hyperbola centered at the origin with foci at $(0, \pm c)$ and vertices at $(0, \pm a)$ has the equation

$$\frac{y^2}{a^2} - \frac{x^2}{b^2} = 1, \quad \text{where} \quad b^2 = c^2 - a^2.$$

The hyperbola has **asymptotes** $y = \pm ax/b$.
In both cases, $c > a > 0$ and $c > b > 0$.

EXAMPLE 5 Graphing hyperbolas Find the equation of the hyperbola centered at the origin with vertices V_1 and V_2 at $(\pm 4, 0)$ and foci F_1 and F_2 at $(\pm 6, 0)$. Graph the hyperbola.

SOLUTION Because the foci are on the x-axis, the vertices are also on the x-axis, and there are no y-intercepts. With $a = 4$ and $c = 6$, we have $b^2 = c^2 - a^2 = 20$, or $b = 2\sqrt{5}$. Therefore, the equation of the hyperbola is

$$\frac{x^2}{16} - \frac{y^2}{20} = 1.$$

The asymptotes are $y = \pm bx/a = \pm \sqrt{5}x/2$ (**Figure 12.57**).

Related Exercise 41 ◀

Eccentricity and Directrix

Parabolas, ellipses, and hyperbolas may also be developed in a single unified way called the *eccentricity-directrix* approach. We let ℓ be a line called the **directrix** and F be a point not on ℓ called a **focus**. The **eccentricity** is a real number $e > 0$. Consider the set C of points P in a plane with the property that the distance $|PF|$ equals e multiplied by the perpendicular distance $|PL|$ from P to ℓ (**Figure 12.58**); that is,

$$|PF| = e|PL| \quad \text{or} \quad \frac{|PF|}{|PL|} = e = \text{constant}.$$

▶ The conic section lies in the plane formed by the directrix and the focus.

Depending on the value of e, the set C is one of the three standard conic sections, as described in the following theorem.

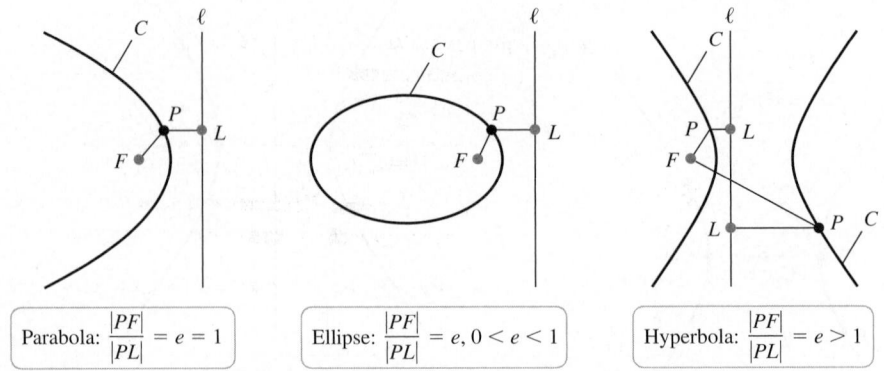

Figure 12.58

▶ Theorem 12.3 for ellipses and hyperbolas describes how the entire curve is generated using just one focus and one directrix. Nevertheless, all ellipses and hyperbolas have two foci and two directrices.

THEOREM 12.3 Eccentricity-Directrix Theorem

Suppose ℓ is a line, F is a point not on ℓ, and e is a positive real number. Let C be the set of points P in a plane with the property that $\dfrac{|PF|}{|PL|} = e$, where $|PL|$ is the perpendicular distance from P to ℓ.

1. If $e = 1$, C is a **parabola**.
2. If $0 < e < 1$, C is an **ellipse**.
3. If $e > 1$, C is a **hyperbola**.

The proof of this theorem is straightforward; it requires an algebraic calculation that is found in Appendix A. The proof establishes relationships among five parameters a, b, c, d, and e that are characteristic of any ellipse or hyperbola. The relationships are given in the following summary.

SUMMARY Properties of Ellipses and Hyperbolas

An ellipse or a hyperbola centered at the origin has the following properties.

	Foci on x-axis	Foci on y-axis
Major-axis vertices:	$(\pm a, 0)$	$(0, \pm a)$
Minor-axis vertices (for ellipses):	$(0, \pm b)$	$(\pm b, 0)$
Foci:	$(\pm c, 0)$	$(0, \pm c)$
Directrices:	$x = \pm d$	$y = \pm d$

Eccentricity: $0 < e < 1$ for ellipses, $e > 1$ for hyperbolas.

Given any two of the five parameters a, b, c, d, and e, the other three are found using the relations

$$c = ae, \quad d = \frac{a}{e},$$

$$b^2 = a^2 - c^2 \text{ (for ellipses)}, \qquad b^2 = c^2 - a^2 \text{ (for hyperbolas)}.$$

QUICK CHECK 5 Given an ellipse with $a = 3$ and $e = \frac{1}{2}$, what are the values of b, c, and d? ◄

EXAMPLE 6   Equation of an ellipse Find the equation of the ellipse centered at the origin with foci F_1 and F_2 at $(0, \pm 4)$ and eccentricity $e = \frac{1}{2}$. Give the length of the major and minor axes, the location of the vertices, and the directrices. Graph the ellipse.

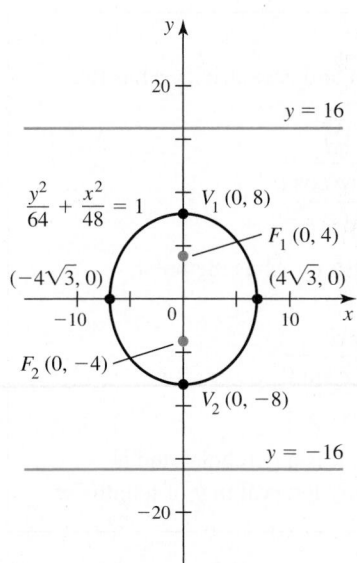

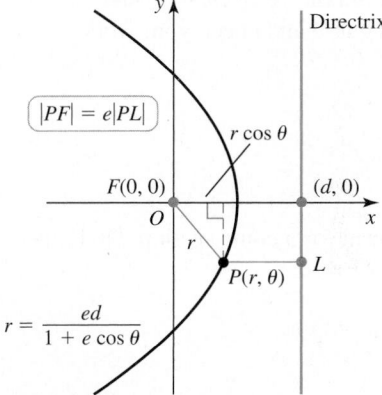

SOLUTION An ellipse with its major axis along the y-axis has the equation

$$\frac{y^2}{a^2} + \frac{x^2}{b^2} = 1,$$

where a and b must be determined (with $a > b$). Because the foci are at $(0, \pm 4)$, we have $c = 4$. Using $e = \frac{1}{2}$ and the relation $c = ae$, it follows that $a = c/e = 8$. So the length of the major axis is $2a = 16$, and the major-axis vertices V_1 and V_2 are $(0, \pm 8)$. Also, $d = a/e = 16$, so the directrices are $y = \pm 16$. Finally, $b^2 = a^2 - c^2 = 48$, or $b = 4\sqrt{3}$. So the length of the minor axis is $2b = 8\sqrt{3}$, and the minor-axis vertices are $(\pm 4\sqrt{3}, 0)$ (Figure 12.59). The equation of the ellipse is

$$\frac{y^2}{64} + \frac{x^2}{48} = 1.$$

Related Exercises 53–54 ◀

Figure 12.59

Polar Equations of Conic Sections

It turns out that conic sections have a natural representation in polar coordinates, provided we use the eccentricity-directrix approach given in Theorem 12.3. Furthermore, a single polar equation covers parabolas, ellipses, and hyperbolas.

When working in polar equations, the key is to place a focus of the conic section at the origin of the coordinate system. We begin by placing one focus F at the origin and taking a directrix perpendicular to the x-axis through $(d, 0)$, where $d > 0$ (Figure 12.60).

We now use the definition $\dfrac{|PF|}{|PL|} = e$, where $P(r, \theta)$ is an arbitrary point on the conic.

As shown in Figure 12.60, $|PF| = r$ and $|PL| = d - r\cos\theta$. The condition $\dfrac{|PF|}{|PL|} = e$

implies that $r = e(d - r\cos\theta)$. Solving for r, we have

$$r = \frac{ed}{1 + e\cos\theta}.$$

A similar derivation (Exercise 84) with the directrix at $x = -d$, where $d > 0$, results in the equation

$$r = \frac{ed}{1 - e\cos\theta}.$$

For horizontal directrices at $y = \pm d$ (Figure 12.61), a similar argument (Exercise 84) leads to the equations

$$r = \frac{ed}{1 \pm e\sin\theta}.$$

Figure 12.60

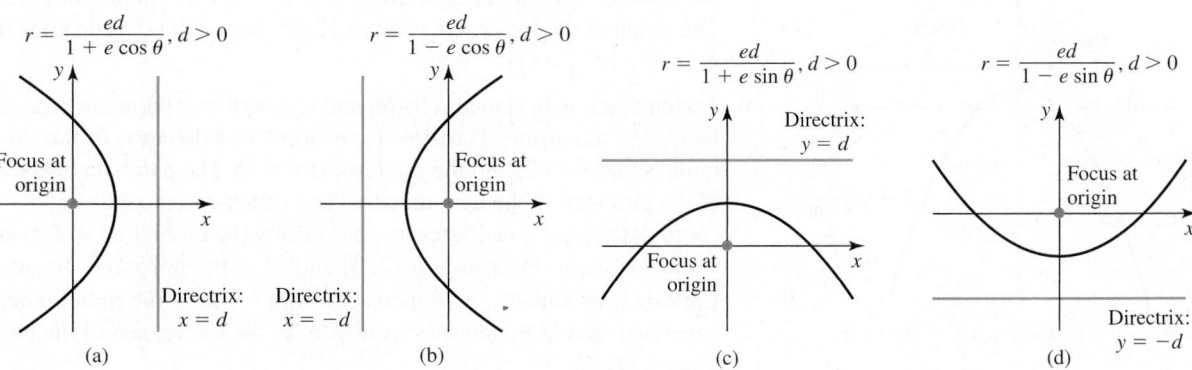

Figure 12.61

> **THEOREM 12.4   Polar Equations of Conic Sections**
> Let $d > 0$. The conic section with a focus at the origin and eccentricity e has the polar equation
>
> $$r = \frac{ed}{1 + e \cos \theta} \quad \text{or} \quad r = \frac{ed}{1 - e \cos \theta}.$$
>
> <u>if one directrix is $x = d$</u>   <u>if one directrix is $x = -d$</u>
>
> The conic section with a focus at the origin and eccentricity e has the polar equation
>
> $$r = \frac{ed}{1 + e \sin \theta} \quad \text{or} \quad r = \frac{ed}{1 - e \sin \theta}.$$
>
> <u>if one directrix is $y = d$</u>   <u>if one directrix is $y = -d$</u>
>
> If $0 < e < 1$, the conic section is an ellipse; if $e = 1$, it is a parabola; and if $e > 1$, it is a hyperbola. The curves are defined over any interval in θ of length 2π.

QUICK CHECK 6 On which axis do the vertices and foci of the conic section $r = 2/(1 - 2 \sin \theta)$ lie? ◄

EXAMPLE 7   Conic sections in polar coordinates Find the vertices, foci, and directrices of the following conic sections. Graph each curve and check your work with a graphing utility.

a. $r = \dfrac{8}{2 + 3 \cos \theta}$ **b.** $r = \dfrac{2}{1 + \sin \theta}$

SOLUTION

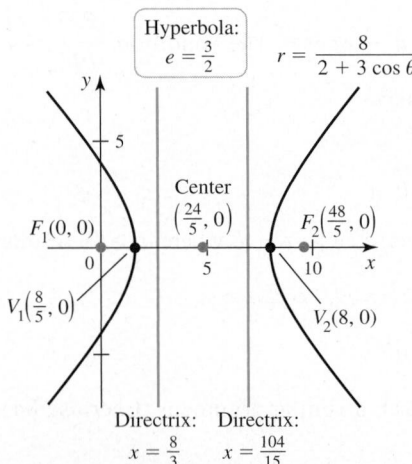

Figure 12.62

a. The equation must be expressed in standard polar form for a conic section. Dividing numerator and denominator by 2, we have

$$r = \frac{4}{1 + \frac{3}{2} \cos \theta},$$

which allows us to identify $e = \frac{3}{2}$. Therefore, the equation describes a hyperbola (because $e > 1$) with one focus at the origin.

    The directrices are vertical (because $\cos \theta$ appears in the equation). Knowing that $ed = 4$, we have $d = 4/e = \frac{8}{3}$, and one directrix is $x = \frac{8}{3}$. Letting $\theta = 0$ and $\theta = \pi$, the polar coordinates of the vertices are $\left(\frac{8}{5}, 0\right)$ and $(-8, \pi)$; equivalently, the vertices are $\left(\frac{8}{5}, 0\right)$ and $(8, 0)$ in Cartesian coordinates (**Figure 12.62**). The center of the hyperbola is halfway between the vertices; therefore, its Cartesian coordinates are $\left(\frac{24}{5}, 0\right)$. The distance between the focus at $(0, 0)$ and the nearest vertex $\left(\frac{8}{5}, 0\right)$ is $\frac{8}{5}$. Therefore, the other focus is $\frac{8}{5}$ units to the right of the vertex $(8, 0)$. So the Cartesian coordinates of the foci are $\left(\frac{48}{5}, 0\right)$ and $(0, 0)$. Because the directrices are symmetric about the center and the left directrix is $x = \frac{8}{3}$, the right directrix is $x = \frac{104}{15} \approx 6.9$. The graph of the hyperbola (Figure 12.62) is generated as θ varies from 0 to 2π (with $\theta \neq \pm \cos^{-1}\left(-\frac{2}{3}\right)$).

b. The equation is in standard form, and it describes a parabola because $e = 1$. The sole focus is at the origin. The directrix is horizontal (because of the $\sin \theta$ term); $ed = 2$ implies that $d = 2$, and the directrix is $y = 2$. The parabola opens downward because of the plus sign in the denominator. The vertex corresponds to $\theta = \frac{\pi}{2}$ and has polar coordinates $\left(1, \frac{\pi}{2}\right)$, or Cartesian coordinates $(0, 1)$. Setting $\theta = 0$ and $\theta = \pi$, the parabola crosses the x-axis at $(2, 0)$ and $(2, \pi)$ in polar coordinates, or $(\pm 2, 0)$ in Cartesian coordinates. As θ increases from $-\frac{\pi}{2}$ to $\frac{\pi}{2}$, the right branch of the parabola is generated, and as θ increases from $\frac{\pi}{2}$ to $\frac{3\pi}{2}$, the left branch of the parabola is generated (**Figure 12.63**).

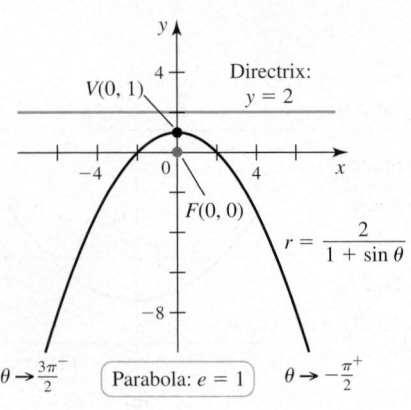

Figure 12.63

Related Exercises 57, 60 ◄

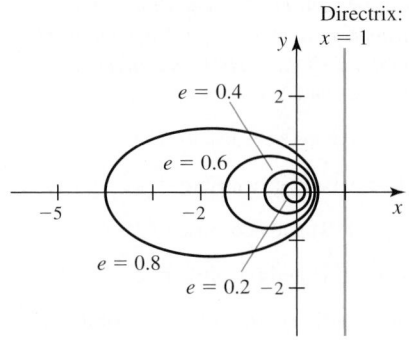

Directrix: $x = 1$

$e = 0.4$

$e = 0.6$

$e = 0.8$

$e = 0.2$

Figure 12.64

EXAMPLE 8   Conics in polar coordinates Use a graphing utility to plot the curves

$$r = \frac{e}{1 + e \cos \theta}, \text{ with } e = 0.2, 0.4, 0.6, \text{ and } 0.8. \text{ Comment on the effect of varying the}$$

eccentricity, e.

SOLUTION Because $0 < e < 1$ in each case, all the curves are ellipses. Notice that the equation is in standard form with $d = 1$; therefore, the curves have the same directrix, $x = d = 1$. As the eccentricity increases, the ellipses becomes more elongated. Small values of e correspond to more circular ellipses (**Figure 12.64**).

Related Exercises 65–66 ◄

SECTION 12.4 EXERCISES

Getting Started

1. Give the property that defines all parabolas.

2. Give the property that defines all ellipses.

3. Give the property that defines all hyperbolas.

4. Sketch the three basic conic sections in standard position with vertices and foci on the x-axis.

5. Sketch the three basic conic sections in standard position with vertices and foci on the y-axis.

6. What is the equation of the standard parabola with its vertex at the origin that opens downward?

7. What is the equation of the standard ellipse with vertices at $(\pm a, 0)$ and foci at $(\pm c, 0)$?

8. What is the equation of the standard hyperbola with vertices at $(0, \pm a)$ and foci at $(0, \pm c)$?

9. Given vertices $(\pm a, 0)$ and the eccentricity e, what are the coordinates of the foci of an ellipse and a hyperbola?

10. Give the equation in polar coordinates of a conic section with a focus at the origin, eccentricity e, and a directrix $x = d$, where $d > 0$.

11. What are the equations of the asymptotes of a standard hyperbola with vertices on the x-axis?

12. How does the eccentricity determine the type of conic section?

Practice Exercises

13–30. Graphing conic sections *Determine whether the following equations describe a parabola, an ellipse, or a hyperbola, and then sketch a graph of the curve. For each parabola, specify the location of the focus and the equation of the directrix; for each ellipse, label the coordinates of the vertices and foci, and find the lengths of the major and minor axes; for each hyperbola, label the coordinates of the vertices and foci, and find the equations of the asymptotes.*

13. $x^2 = 12y$

14. $y^2 = 20x$

15. $\frac{x^2}{4} + y^2 = 1$

16. $\frac{x^2}{9} + \frac{y^2}{4} = 1$

17. $x = -\frac{y^2}{16}$

18. $4x = -y^2$

19. $\frac{x^2}{4} - y^2 = 1$

20. $\frac{y^2}{16} - \frac{x^2}{9} = 1$

21. $4x^2 - y^2 = 16$

22. $25y^2 - 4x^2 = 100$

23. $8y = -3x^2$

24. $12x = 5y^2$

25. $\frac{x^2}{4} + \frac{y^2}{16} = 1$

26. $x^2 + \frac{y^2}{9} = 1$

27. $\frac{x^2}{5} + \frac{y^2}{7} = 1$

28. $12x^2 + 5y^2 = 60$

29. $\frac{x^2}{3} - \frac{y^2}{5} = 1$

30. $10x^2 - 7y^2 = 140$

31–38. Equations of parabolas *Find an equation of the following parabolas. Unless otherwise specified, assume the vertex is at the origin.*

31. A parabola that opens to the right with directrix $x = -4$

32. A parabola that opens downward with directrix $y = 6$

33. A parabola with focus at $(3, 0)$

34. A parabola with focus at $(-4, 0)$

35. A parabola symmetric about the y-axis that passes through the point $(2, -6)$

36. A parabola symmetric about the x-axis that passes through the point $(1, -4)$

37.

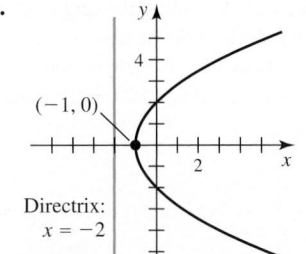

$(-1, 0)$

Directrix: $x = -2$

38.

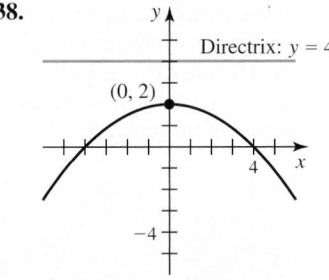

Directrix: $y = 4$

$(0, 2)$

39–50. Equations of ellipses and hyperbolas *Find an equation of the following ellipses and hyperbolas, assuming the center is at the origin.*

39. An ellipse whose major axis is on the x-axis with length 8 and whose minor axis has length 6

40. An ellipse with vertices $(\pm 6, 0)$ and foci $(\pm 4, 0)$

41. A hyperbola with vertices $(\pm 4, 0)$ and foci $(\pm 6, 0)$

42. A hyperbola with vertices $(\pm 1, 0)$ that passes through $(5/3, 8)$

43. An ellipse with vertices $(\pm 5, 0)$, passing through the point $(4, 3/5)$

44. An ellipse with vertices $(0, \pm 10)$, passing through the point $(\sqrt{3}/2, 5)$

45. A hyperbola with vertices $(\pm 2, 0)$ and asymptotes $y = \pm 3x/2$

46. A hyperbola with vertices $(0, \pm 4)$ and asymptotes $y = \pm 2x$

47. 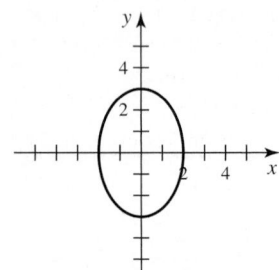   **48.**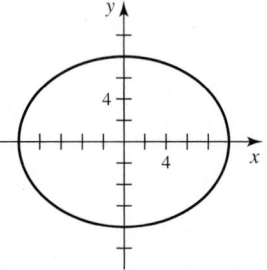

49. 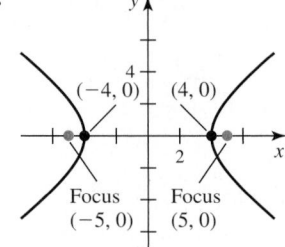   **50.**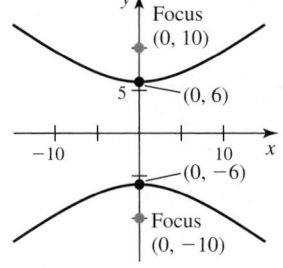

51. Explain why or why not Determine whether the following statements are true and give an explanation or counterexample.

  a. The hyperbola $\dfrac{x^2}{4} - \dfrac{y^2}{9} = 1$ has no y-intercept.

  b. On every ellipse, there are exactly two points at which the curve has slope s, where s is any real number.

  c. Given the directrices and foci of a standard hyperbola, it is possible to find its vertices, eccentricity, and asymptotes.

  d. The point on a parabola closest to the focus is the vertex.

52. Golden Gate Bridge Completed in 1937, San Francisco's Golden Gate Bridge is 2.7 km long and weighs about 890,000 tons. The length of the span between the two central towers is 1280 m; the towers themselves extend 152 m above the roadway. The cables that support the deck of the bridge between the two towers hang in a parabola (see figure). Assuming the origin is midway between the towers on the deck of the bridge, find an equation that describes the cables. How long is a guy wire that hangs vertically from the cables to the roadway 500 m from the center of the bridge?

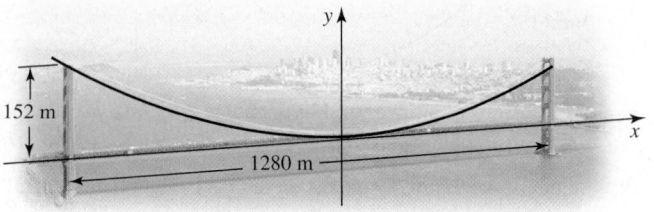

53–56. Eccentricity-directrix approach *Find an equation of the following curves, assuming the center is at the origin. Sketch a graph labeling the vertices, foci, asymptotes (if they exist), and directrices. Use a graphing utility to check your work.*

53. An ellipse with vertices $(\pm 9, 0)$ and eccentricity $1/3$

54. An ellipse with vertices $(0, \pm 9)$ and eccentricity $1/4$

55. A hyperbola with vertices $(\pm 1, 0)$ and eccentricity 3

56. A hyperbola with vertices $(0, \pm 4)$ and eccentricity 2

57–62. Polar equations for conic sections *Graph the following conic sections, labeling the vertices, foci, directrices, and asymptotes (if they exist). Use a graphing utility to check your work.*

57. $r = \dfrac{4}{1 + \cos \theta}$ **58.** $r = \dfrac{3}{2 + \cos \theta}$

59. $r = \dfrac{1}{2 - \cos \theta}$ **60.** $r = \dfrac{6}{3 + 2 \sin \theta}$

61. $r = \dfrac{1}{2 - 2 \sin \theta}$ **62.** $r = \dfrac{12}{3 - \cos \theta}$

63–66. Tracing hyperbolas and parabolas *Graph the following equations. Then use arrows and labeled points to indicate how the curve is generated as θ increases from 0 to 2π.*

63. $r = \dfrac{1}{1 + \sin \theta}$ **64.** $r = \dfrac{1}{1 + 2 \cos \theta}$

65. $r = \dfrac{3}{1 - \cos \theta}$ **66.** $r = \dfrac{1}{1 - 2 \cos \theta}$

T 67. Parabolas with a graphing utility Use a graphing utility to graph the parabolas $r = \dfrac{d}{1 + \cos \theta}$ for $d = 0.25, 0.5, 1, 2, 3$, and 4 on the same set of axes. Explain how the shapes of the curves vary as d changes.

T 68. Hyperbolas with a graphing utility Use a graphing utility to graph the hyperbolas $r = \dfrac{e}{1 + e \cos \theta}$ for $e = 1.1, 1.3, 1.5, 1.7$, and 2 on the same set of axes. Explain how the shapes of the curves vary as e changes.

69–72. Tangent lines *Find an equation of the line tangent to the following curves at the given point.*

69. $x^2 = -6y$; $(-6, -6)$ **70.** $y^2 = 8x$; $(8, -8)$

71. $y^2 - \dfrac{x^2}{64} = 1$; $\left(6, -\dfrac{5}{4}\right)$ **72.** $r = \dfrac{1}{1 + \sin \theta}$; $\left(\dfrac{2}{3}, \dfrac{\pi}{6}\right)$

73. Tangent lines for an ellipse Show that an equation of the line tangent to the ellipse $\dfrac{x^2}{a^2} + \dfrac{y^2}{b^2} = 1$ at the point (x_0, y_0) is

$$\frac{xx_0}{a^2} + \frac{yy_0}{b^2} = 1.$$

74. Tangent lines for a hyperbola Find an equation of the line tangent to the hyperbola $\dfrac{x^2}{a^2} - \dfrac{y^2}{b^2} = 1$ at the point (x_0, y_0).

75–76. Graphs to polar equations *Find a polar equation for each conic section. Assume one focus is at the origin.*

75.

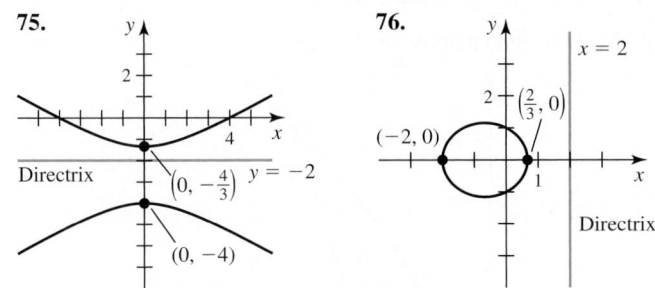

76.

77. Another construction for a hyperbola Suppose two circles, whose centers are at least $2a$ units apart (see figure), are centered at F_1 and F_2, respectively. The radius of one circle is $2a + r$ and the radius of the other circle is r, where $r \geq 0$. Show that as r increases, the intersection points P_1 and P_2 of the two circles describes one branch of a hyperbola with foci at F_1 and F_2.

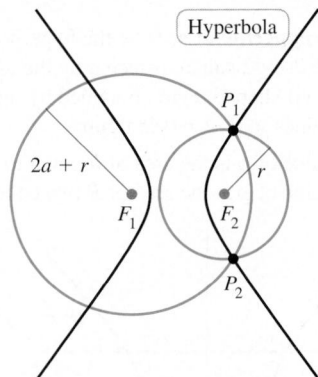

78. The ellipse and the parabola Let R be the region bounded by the upper half of the ellipse $\dfrac{x^2}{2} + y^2 = 1$ and the parabola $y = \dfrac{x^2}{\sqrt{2}}$.

  a. Find the area of R.
  b. Which is greater, the volume of the solid generated when R is revolved about the x-axis or the volume of the solid generated when R is revolved about the y-axis?

79. Volume of an ellipsoid Suppose the ellipse $\dfrac{x^2}{a^2} + \dfrac{y^2}{b^2} = 1$ is revolved about the x-axis. What is the volume of the solid enclosed by the *ellipsoid* that is generated? Is the volume different if the same ellipse is revolved about the y-axis?

80. Area of a sector of a hyperbola Consider the region R bounded by the right branch of the hyperbola $\dfrac{x^2}{a^2} - \dfrac{y^2}{b^2} = 1$ and the vertical line through the right focus.

  a. What is the area of R?
  b. Sketch a graph that shows how the area of R varies with the eccentricity e, for $e > 1$.

81. Volume of a hyperbolic cap Consider the region R bounded by the right branch of the hyperbola $\dfrac{x^2}{a^2} - \dfrac{y^2}{b^2} = 1$ and the vertical line through the right focus.

  a. What is the volume of the solid that is generated when R is revolved about the x-axis?
  b. What is the volume of the solid that is generated when R is revolved about the y-axis?

82. Volume of a paraboloid (Archimedes) The region bounded by the parabola $y = ax^2$ and the horizontal line $y = h$ is revolved about the y-axis to generate a solid bounded by a surface called a *paraboloid* (where $a > 0$ and $h > 0$). Show that the volume of the solid is $3/2$ the volume of the cone with the same base and vertex.

Explorations and Challenges

83. Reflection property of parabolas Consider the parabola $y = \dfrac{x^2}{4p}$ with its focus at $F(0, p)$ (see figure). The goal is to show that the angle of incidence between the ray ℓ and the tangent line L (α in the figure) equals the angle of reflection between the line PF and L (β in the figure). If these two angles are equal, then the reflection property is proved because ℓ is reflected through F.

  a. Let $P(x_0, y_0)$ be a point on the parabola. Show that the slope of the line tangent to the curve at P is $\tan \theta = \dfrac{x_0}{2p}$.
  b. Show that $\tan \varphi = \dfrac{p - y_0}{x_0}$.
  c. Show that $\alpha = \dfrac{\pi}{2} - \theta$; therefore, $\tan \alpha = \cot \theta$.
  d. Note that $\beta = \theta + \varphi$. Use the tangent addition formula $\tan (\theta + \varphi) = \dfrac{\tan \theta + \tan \varphi}{1 - \tan \theta \tan \varphi}$ to show that $\tan \alpha = \tan \beta = \dfrac{2p}{x_0}$.
  e. Conclude that because α and β are acute angles, $\alpha = \beta$.

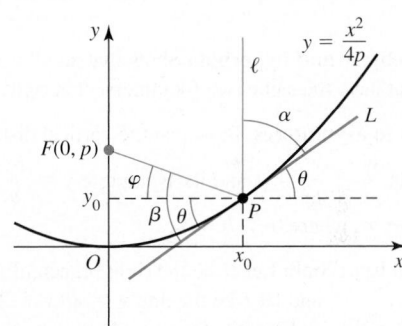

84. Deriving polar equations for conics Use Figures 12.60 and 12.61 to derive the polar equation of a conic section with a focus at the origin in the following three cases.

  a. Vertical directrix at $x = -d$, where $d > 0$
  b. Horizontal directrix at $y = d$, where $d > 0$
  c. Horizontal directrix at $y = -d$, where $d > 0$

85. Equation of an ellipse Consider an ellipse to be the set of points in a plane whose distances from two fixed points have a constant sum $2a$. Derive the equation of an ellipse. Assume the two fixed points are on the x-axis equidistant from the origin.

86. Equation of a hyperbola Consider a hyperbola to be the set of points in a plane whose distances from two fixed points have a constant difference of $2a$ or $-2a$. Derive the equation of a hyperbola. Assume the two fixed points are on the x-axis equidistant from the origin.

87. Equidistant set Show that the set of points equidistant from a circle and a line not passing through the circle is a parabola. Assume the circle, line, and parabola lie in the same plane.

88. Polar equation of a conic Show that the polar equation of an ellipse or hyperbola with one focus at the origin, major axis of length $2a$ on the x-axis, and eccentricity e is

$$r = \frac{a(1 - e^2)}{1 + e \cos \theta}.$$

89. Shared asymptotes Suppose two hyperbolas with eccentricities e and E have perpendicular major axes and share a set of asymptotes. Show that $e^{-2} + E^{-2} = 1$.

90–94. Focal chords *A **focal chord** of a conic section is a line through a focus joining two points of the curve. The **latus rectum** is the focal chord perpendicular to the major axis of the conic. Prove the following properties.*

90. The lines tangent to the endpoints of any focal chord of a parabola $y^2 = 4px$ intersect on the directrix and are perpendicular.

91. Let L be the latus rectum of the parabola $y^2 = 4px$ for $p > 0$. Let F be the focus of the parabola, P be any point on the parabola to the left of L, and D be the (shortest) distance between P and L. Show that for all P, $D + |FP|$ is a constant. Find the constant.

92. The length of the latus rectum of the parabola $y^2 = 4px$ or $x^2 = 4py$ is $4|p|$.

93. The length of the latus rectum of an ellipse centered at the origin is $\dfrac{2b^2}{a} = 2b\sqrt{1 - e^2}$.

94. The length of the latus rectum of a hyperbola centered at the origin is $\dfrac{2b^2}{a} = 2b\sqrt{e^2 - 1}$.

95. Confocal ellipse and hyperbola Show that an ellipse and a hyperbola that have the same two foci intersect at right angles.

96. Approach to asymptotes Show that the vertical distance between a hyperbola $\dfrac{x^2}{a^2} - \dfrac{y^2}{b^2} = 1$ and its asymptote $y = \dfrac{bx}{a}$ approaches zero as $x \to \infty$, where $0 < b < a$.

97. Sector of a hyperbola Let H be the right branch of the hyperbola $x^2 - y^2 = 1$ and let ℓ be the line $y = m(x - 2)$ that passes through the point $(2, 0)$ with slope m, where $-\infty < m < \infty$. Let R be the region in the first quadrant bounded by H and ℓ (see figure). Let $A(m)$ be the area of R. Note that for some values of m, $A(m)$ is not defined.

a. Find the x-coordinates of the intersection points between H and ℓ as functions of m; call them $u(m)$ and $v(m)$, where $v(m) > u(m) > 1$. For what values of m are there two intersection points?

b. Evaluate $\lim\limits_{m \to 1^+} u(m)$ and $\lim\limits_{m \to 1^+} v(m)$.

c. Evaluate $\lim\limits_{m \to \infty} u(m)$ and $\lim\limits_{m \to \infty} v(m)$.

d. Evaluate and interpret $\lim\limits_{m \to \infty} A(m)$.

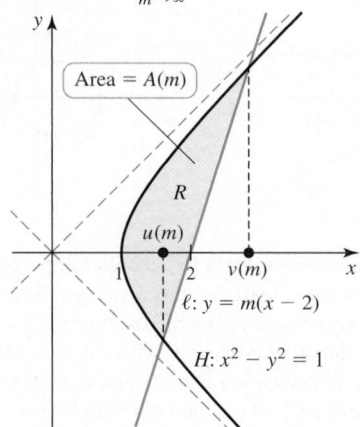

T 98. The anvil of a hyperbola Let H be the hyperbola $x^2 - y^2 = 1$ and let S be the 2-by-2 square bisected by the asymptotes of H. Let R be the anvil-shaped region bounded by the hyperbola and the horizontal lines $y = \pm p$ (see figure).

a. For what value of p is the area of R equal to the area of S?

b. For what value of p is the area of R twice the area of S?

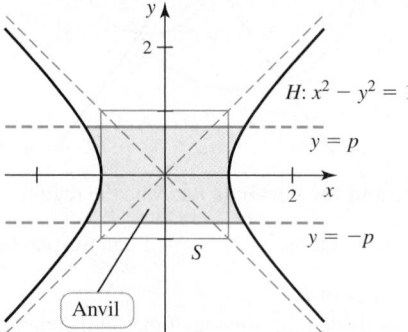

QUICK CHECK ANSWERS

2. a. Left   **b.** Up   **3.** The minor-axis vertices are $(0, \pm b)$. The distance between them is $2b$, which is the length of the minor axis.   **4.** Vertices: $(0, \pm 1)$; foci: $(0, \pm\sqrt{5})$ **5.** $b = 3\sqrt{3}/2, c = 3/2, d = 6$ **6.** y-axis ◄

CHAPTER 12   REVIEW EXERCISES

1. Explain why or why not Determine whether the following statements are true and give an explanation or counterexample.

a. A set of parametric equations for a given curve is always unique.

b. The equations $x = e^t, y = 2e^t$, for $-\infty < t < \infty$, describe a line passing through the origin with slope 2.

c. The polar coordinates $\left(3, -\dfrac{3\pi}{4}\right)$ and $\left(-3, \dfrac{\pi}{4}\right)$ describe the same point in the plane.

d. The area of the region between the inner and outer loops of the limaçon $r = f(\theta) = 1 - 4\cos\theta$ is $\dfrac{1}{2}\displaystyle\int_0^{2\pi} f(\theta)^2\, d\theta$.

e. The hyperbola $\dfrac{y^2}{2} - \dfrac{x^2}{4} = 1$ has no x-intercept.

f. The equation $x^2 + 4y^2 - 2x = 3$ describes an ellipse.

2. Eliminate the parameter in the parametric equations $x = 1 + \sin t$, $y = 3 + 2 \sin t$, for $0 \le t \le \pi/2$, and describe the curve, indicating its positive orientation. How does this curve differ from the curve $x = 1 + \sin t$, $y = 3 + 2 \sin t$, for $\pi/2 \le t \le \pi$?

3–6. Eliminating the parameter *Eliminate the parameter to find a description of the following curves in terms of x and y. Give a geometric description and the positive orientation of the curve.*

3. $x = 4 \cos t$, $y = 3 \sin t$; $0 \le t \le 2\pi$

4. $x = \sin t - 3$, $y = \cos t + 6$; $0 \le t \le \pi$

5. $x = t + 1$, $y = \sqrt{t + 1}$; $3 \le t \le 8$

6. $x = -\cot^2 t$, $y = 1 + \csc^2 t$; $\dfrac{\pi}{4} \le t \le \dfrac{\pi}{2}$

7–8. Parametric curves and tangent lines

a. Eliminate the parameter to obtain an equation in *x* and *y*.
b. Find the slope of the curve at the given value of *t*.
c. Make a sketch of the curve showing the positive orientation of the curve and the tangent line at the point corresponding to the given value of *t*.

7. $x = 8 \cos t + 1$, $y = 8 \sin t + 2$, for $0 \le t \le 2\pi$; $t = \pi/3$

8. $x = 4 \sin 2t$, $y = 3 \cos 2t$, for $0 \le t \le \pi$; $t = \pi/6$

9. Find parametric equations for the curve $x = 5(y - 1)(y - 2) \sin y$ and then plot the curve using a graphing utility.

10–12. Parametric curves

a. Eliminate the parameter to obtain an equation in *x* and *y*.
b. Describe the curve, indicating the positive orientation.
c. Find the slope of the curve at the specified point.

10. $x = t^2 + 4$, $y = -t$, for $-2 < t < 0$; $(5, 1)$

11. $x = 3 \cos (-t)$, $y = 3 \sin (-t) - 1$, for $0 \le t \le \pi$; $(0, -4)$

12. $x = \ln t$, $y = 8 \ln t^2$, for $1 \le t \le e^2$; $(1, 16)$

13. Tangent lines Find an equation of the line tangent to the cycloid $x = t - \sin t$, $y = 1 - \cos t$ at the points corresponding to $t = \pi/6$ and $t = 2\pi/3$.

14–18. Parametric descriptions *Write parametric equations for the following curves. Solutions are not unique.*

14. The segment of the curve $x = y^3 + y + 1$ that starts at $(1, 0)$ and ends at $(11, 2)$.

15. The line segment from $P(-1, 0)$ to $Q(1, 1)$, followed by the line segment from Q to P (two sets of equations are required)

16. The segment of the curve $f(x) = x^3 + 2x$ from $(0, 0)$ to $(2, 12)$

17. The circle $x^2 + y^2 = 9$, generated clockwise

18. The right side of the ellipse $\dfrac{x^2}{9} + \dfrac{y^2}{4} = 1$, generated counterclockwise

19–20. Area bounded by parametric curves *Find the area of the following regions. (Hint: See Exercises 103–105 in Section 12.1.)*

19. The region bounded by the *y*-axis and the parametric curve $x = t^2$, $y = t - t^3$, for $0 \le t \le 1$

20. The region bounded by the *x*-axis and the parametric curve $x = \cos t$, $y = \sin 2t$, for $0 \le t \le \pi/2$

21. Surface of revolution Find the area of the surface obtained by revolving the curve $x = \sin t$, $y = \cos 2t$, for $-\pi/4 \le t \le \pi/4$, about the *x*-axis. (*Hint:* See Exercises 106–111 in Section 12.1.)

22–23. Arc length *Find the length of the following curves.*

22. $x = e^{2t} \sin 3t$, $y = e^{2t} \cos 3t$; $0 \le t \le \pi/3$

23. $x = \cos 2t$, $y = 2t - \sin 2t$; $0 \le t \le \pi/4$

24–26. Sets in polar coordinates *Sketch the following sets of points.*

24. $\theta = \dfrac{\pi}{6}$

25. $0 \le r \le 4$ and $-\dfrac{\pi}{2} \le \theta \le -\dfrac{\pi}{3}$

26. $4 \le r^2 \le 9$

27–32. Polar curves *Graph the following equations.*

27. $r = 2 - 3 \sin \theta$

28. $r = 1 + 2 \cos \theta$

29. $r = 3 \cos 3\theta$

30. $r = e^{-\theta/6}$

31. $r = 3 \sin 4\theta$

32. $r^2 = 4 \cos \theta$

33. Polar valentine Liz wants to show her love for Jake by passing him a valentine on her graphing calculator. Sketch each of the following curves and determine which one Liz should use to get a heart-shaped curve.

a. $r = 5 \cos \theta$ **b.** $r = 1 - \sin \theta$ **c.** $r = \cos 3\theta$

34. Jake's response Jake responds to Liz (Exercise 33) with a graph that shows his love for her is infinite. Sketch each of the following curves. Which one should Jake send to Liz to get an infinity symbol?

a. $r = \theta$, for $\theta \ge 0$ **b.** $r = \dfrac{1}{2} + \sin \theta$ **c.** $r^2 = \cos 2\theta$

35. Polar conversion Write the equation $r^2 + r(2 \sin \theta - 6 \cos \theta) = 0$ in Cartesian coordinates and identify the corresponding curve.

36. Polar conversion Consider the equation $r = \dfrac{4}{\sin \theta + \cos \theta}$.

a. Convert the equation to Cartesian coordinates and identify the curve it describes.
b. Graph the curve and indicate the points that correspond to $\theta = 0$, $\dfrac{\pi}{2}$, and 2π.
c. Give an interval in θ on which the entire curve is generated.

37. Parametric to polar equations Find an equation of the following curve in polar coordinates and describe the curve.

$$x = (1 + \cos t) \cos t, \quad y = (1 + \cos t) \sin t; \quad 0 \le t \le 2\pi$$

38. Cartesian conversion Write the equation $x = y^2$ in polar coordinates and state values of θ that produce the entire graph of the parabola.

39. Cartesian conversion Write the equation $(x - 4)^2 + y^2 = 16$ in polar coordinates and state values of θ that produce the entire graph of the circle.

T 40–41. Slopes of tangent lines

a. *Find all points where the following curves have vertical and horizontal tangent lines.*

b. *Find the slope of the lines tangent to the curve at the origin (when relevant).*

c. *Sketch the curve and all the tangent lines identified in parts (a) and (b).*

40. $r = 3 - 6\cos\theta$ **41.** $r = 1 - \sin\theta$

42–43. Intersection points *Find the intersection points of the following curves.*

42. $r = \sqrt{3}$ and $r = 2\cos 2\theta$ **43.** $r = \sqrt{\cos 3t}$ and $r = \sqrt{\sin 3t}$

44–49. Areas of regions *Find the area of the following regions.*

44. The region inside one leaf of the rose $r = 3\sin 4\theta$

45. The region inside the inner loop of the limaçon $r = 1 + 2\sin\theta$

46. The region inside the limaçon $r = 2 + \cos\theta$ and outside the circle $r = 2$

47. The region inside the lemniscate $r^2 = 4\cos 2\theta$ and outside the circle $r = \sqrt{2}$

48. The region inside $r = \sqrt{2} + \sin\theta$ and inside $r = 3\sin\theta$

49. The region inside the cardioid $r = 1 + \cos\theta$ and outside the cardioid $r = 1 - \cos\theta$

T 50. Intersection points Consider the polar equations $r = 1$ and $r = 2 - 4\cos\theta$.

a. Graph the curves. How many intersection points do you observe?

b. Give approximate polar coordinates of the intersection points.

T 51–52. Arc length of polar curves *Find the approximate length of the following curves.*

51. The limaçon $r = 3 - 6\cos\theta$

52. The limaçon $r = 3 + 2\cos\theta$

53–57. Conic sections

a. *Determine whether the following equations describe a parabola, an ellipse, or a hyperbola.*

b. *Use analytical methods to determine the location of the foci, vertices, and directrices.*

c. *Find the eccentricity of the curve.*

d. *Make an accurate graph of the curve.*

53. $x^2 - \dfrac{y^2}{2} = 1$ **54.** $x = 16y^2$

55. $y^2 - 4x^2 = 16$ **56.** $\dfrac{x^2}{4} + \dfrac{y^2}{25} = 1$

57. $4x^2 + 8y^2 = 16$

58–59. Tangent lines *Find an equation of the line tangent to the following curves at the given point. Check your work with a graphing utility.*

58. $\dfrac{x^2}{16} - \dfrac{y^2}{9} = 1; \left(\dfrac{20}{3}, -4\right)$ **59.** $y^2 = -12x; \left(-\dfrac{4}{3}, -4\right)$

60. A polar conic section Consider the equation $r^2 = \sec 2\theta$.

a. Convert the equation to Cartesian coordinates and identify the curve.

b. Find the vertices, foci, directrices, and eccentricity of the curve.

c. Graph the curve. Explain why the polar equation does not have the form given in the text for conic sections in polar coordinates.

61–64. Polar equations for conic sections *Graph the following conic sections, labeling vertices, foci, directrices, and asymptotes (if they exist). Give the eccentricity of the curve. Use a graphing utility to check your work.*

61. $r = \dfrac{2}{1 + \sin\theta}$ **62.** $r = \dfrac{3}{1 - 2\cos\theta}$

63. $r = \dfrac{4}{2 + \cos\theta}$ **64.** $r = \dfrac{10}{5 + 2\cos\theta}$

65–68. Eccentricity-directrix approach *Find an equation of the following curves, assuming the center is at the origin. Graph the curve, labeling vertices, foci, asymptotes (if they exist), and directrices.*

65. A hyperbola with vertices $(0, \pm 2)$ and directrices $y = \pm 1$

66. An ellipse with foci $(\pm 4, 0)$ and directrices $x = \pm 8$

67. An ellipse with vertices $(0, \pm 4)$ and directrices $y = \pm 10$

68. A hyperbola with vertices $(\pm 4, 0)$ and directrices $x = \pm 2$

69. Conic parameters An ellipse has vertices $(0, \pm 6)$ and foci $(0, \pm 4)$. Find the eccentricity, the directrices, and the minor-axis vertices.

70. Conic parameters A hyperbola has eccentricity $e = 2$ and foci $(0, \pm 2)$. Find the location of the vertices and directrices.

71. Bisecting an ellipse Let R be the region in the first quadrant bounded by the ellipse $\dfrac{x^2}{a^2} + \dfrac{y^2}{b^2} = 1$. Find the value of m (in terms of a and b) such that the line $y = mx$ divides R into two subregions of equal area.

72. Parabola-hyperbola tangency Let P be the parabola $y = px^2$ and H be the right half of the hyperbola $x^2 - y^2 = 1$.

a. For what value of p is P tangent to H?

b. At what point does the tangency occur?

c. Generalize your results for the hyperbola $\dfrac{x^2}{a^2} - \dfrac{y^2}{b^2} = 1$.

73. Another ellipse construction Start with two circles centered at the origin with radii $0 < a < b$ (see figure). Assume the line ℓ through the origin intersects the smaller circle at Q and the larger circle at R. Let $P(x, y)$ have the y-coordinate of Q and the x-coordinate of R. Show that the set of points $P(x, y)$ generated in this way for all lines ℓ through the origin is an ellipse.

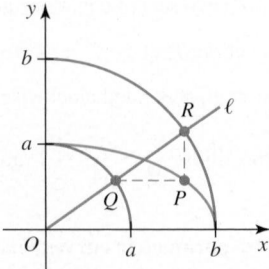

T 74. Hyperbolas The graph of the parametric equations $x = a \tan t$, $y = b \sec t$, for $-\pi < t < \pi$ and $|t| \neq \pi/2$, where a and b are nonzero real numbers, is a hyperbola. Graph the hyperbola with $a = b = 1$. Indicate clearly the direction in which the curve is generated as t increases from $t = -\pi$ to $t = \pi$.

T 75. Lamé curves The *Lamé curve* described by $\left|\dfrac{x}{a}\right|^n + \left|\dfrac{y}{b}\right|^n = 1$,

where a, b, and n are positive real numbers, is a generalization of an ellipse.

a. Express this equation in parametric form (four pairs of equations are needed).

b. Graph the curve for $a = 4$ and $b = 2$, for various values of n.

c. Describe how the curves change as n increases.

76. General equations for a circle Prove that the equations

$$x = a \cos t + b \sin t, \quad y = c \cos t + d \sin t$$

where a, b, c, and d are real numbers, describe a circle of radius R provided $a^2 + c^2 = b^2 + d^2 = R^2$ and $ab + cd = 0$.

Chapter 12 Guided Projects

Applications of the material in this chapter and related topics can be found in the following Guided Projects. For additional information, see the Preface.

- The amazing cycloid
- Parametric art
- Polar art
- Grazing goat problems

- Translations and rotations of axes
- Celestial orbits
- Properties of conic sections

13

Vectors and the Geometry of Space

Chapter Preview   We now make a significant departure from previous chapters by stepping out of the xy-plane ($\mathbb{R}^2$) into three-dimensional space ($\mathbb{R}^3$). The fundamental concept of a *vector*—a quantity with magnitude and direction—is introduced in two and three dimensions. We then develop the algebra associated with vectors (how to add, subtract, and combine them in various ways), and we define two fundamental operations for vectors: the dot product and the cross product. The chapter concludes with a brief survey of basic objects in three-dimensional geometry, namely lines, planes, and elementary surfaces.

13.1 Vectors in the Plane

Imagine a raft drifting down a river, carried by the current. The speed and direction of the raft at a point may be represented by an arrow (**Figure 13.1**). The length of the arrow represents the speed of the raft at that point; longer arrows correspond to greater speeds. The orientation of the arrow gives the direction in which the raft is headed at that point. The arrows at points A and C in Figure 13.1 have the same length and direction, indicating that the raft has the same speed and heading at these locations. The arrow at B is shorter and points to the left of the rock, indicating that the raft slows down as it nears the rock.

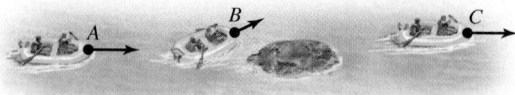

Figure 13.1

Basic Vector Operations

The arrows that describe the raft's motion are examples of *vectors*—quantities that have both *length* (or *magnitude*) and *direction*. Vectors arise naturally in many situations. For example, electric and magnetic fields, the flow of air over an airplane wing, and the velocity and acceleration of elementary particles are described by vectors (**Figure 13.2**). In this section, we examine vectors in the xy-plane; then we extend the concept to three dimensions in Section 13.2.

The vector whose *tail* is at the point P and whose *head* is at the point Q is denoted $\overrightarrow{PQ}$ (**Figure 13.3**). The vector $\overrightarrow{QP}$ has its tail at Q and its head at P. We also label vectors with single boldface characters such as **u** and **v**.

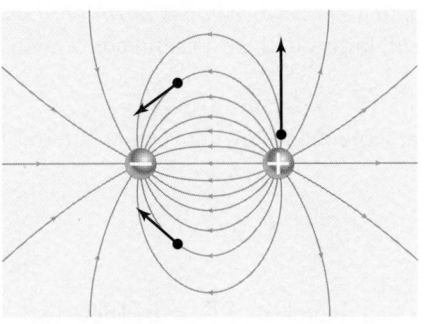

Electric field vectors due to two charges

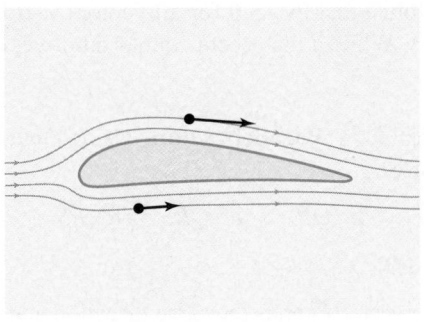

Velocity vectors of air flowing over an airplane wing

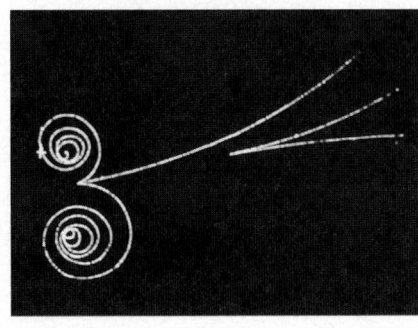

Tracks of elementary particles in a cloud chamber are aligned with the velocity vectors of the particles.

Figure 13.2

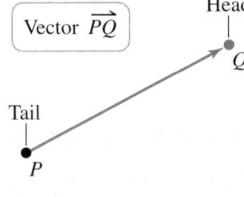

Figure 13.3

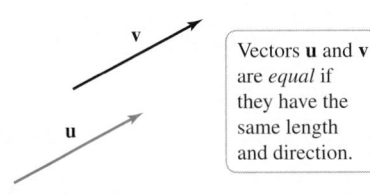

Figure 13.4

Two vectors $\mathbf{u}$ and $\mathbf{v}$ are *equal*, written $\mathbf{u} = \mathbf{v}$, if they have equal length and point in the same direction (**Figure 13.4**). An important fact is that equal vectors do not necessarily have the same location. *Any* two vectors with the same length and direction are equal.

Not all quantities are represented by vectors. For example, mass, temperature, and price have magnitude, but no direction. Such quantities are described by real numbers and are called *scalars*.

> ➤ In this text, *scalar* is another word for *real number*.

> ➤ The vector $\mathbf{v}$ is commonly handwritten as $\vec{v}$. The zero vector is handwritten as $\vec{0}$.

Vectors, Equal Vectors, Scalars, Zero Vector

Vectors are quantities that have both length (or magnitude) and direction. Two vectors are **equal** if they have the same magnitude and direction. Quantities having magnitude but no direction are called **scalars**. One exception is the **zero vector**, denoted $\mathbf{0}$: It has length 0 and no direction.

Scalar Multiplication

A scalar c and a vector $\mathbf{v}$ can be combined using scalar-vector multiplication, or simply *scalar multiplication*. The resulting vector, denoted $c\mathbf{v}$, is called a *scalar multiple* of $\mathbf{v}$. The length of $c\mathbf{v}$ is $|c|$ multiplied by the length of $\mathbf{v}$. The vector $c\mathbf{v}$ has the same direction as $\mathbf{v}$ if $c > 0$. If $c < 0$, then $c\mathbf{v}$ and $\mathbf{v}$ point in opposite directions. If $c = 0$, then the product $0\mathbf{v} = \mathbf{0}$ (the zero vector).

For example, the vector $3\mathbf{v}$ is three times as long as $\mathbf{v}$ and has the same direction as $\mathbf{v}$. The vector $-2\mathbf{v}$ is twice as long as $\mathbf{v}$, but it points in the opposite direction. The vector $\frac{1}{2}\mathbf{v}$ points in the same direction as $\mathbf{v}$ and has half the length of $\mathbf{v}$ (**Figure 13.5**). The vectors $\mathbf{v}$, $3\mathbf{v}$, $-2\mathbf{v}$, and $\frac{1}{2}\mathbf{v}$ are examples of *parallel vectors*: Each one is a scalar multiple of the others.

Same direction as $\mathbf{v}$ and half as long as $\mathbf{v}$

$\frac{1}{2}\mathbf{v}$

$-2\mathbf{v}$

Twice as long as $\mathbf{v}$, pointing in the opposite direction

$3\mathbf{v}$

Same direction as $\mathbf{v}$ and three times as long as $\mathbf{v}$

$\mathbf{v}$

Figure 13.5

DEFINITION **Scalar Multiples and Parallel Vectors**

Given a scalar c and a vector $\mathbf{v}$, the **scalar multiple** $c\mathbf{v}$ is a vector whose length is $|c|$ multiplied by the length of $\mathbf{v}$. If $c > 0$, then $c\mathbf{v}$ has the same direction as $\mathbf{v}$. If $c < 0$, then $c\mathbf{v}$ and $\mathbf{v}$ point in opposite directions. Two vectors are **parallel** if they are scalar multiples of each other.

➤ For convenience, we write $-\mathbf{u}$ for $(-1)\mathbf{u}$, $-c\mathbf{u}$ for $(-c)\mathbf{u}$, and $\mathbf{u}/c$ for $\frac{1}{c}\mathbf{u}$.

QUICK CHECK 1 Describe the length and direction of the vector $-5\mathbf{v}$ relative to $\mathbf{v}$. ◄

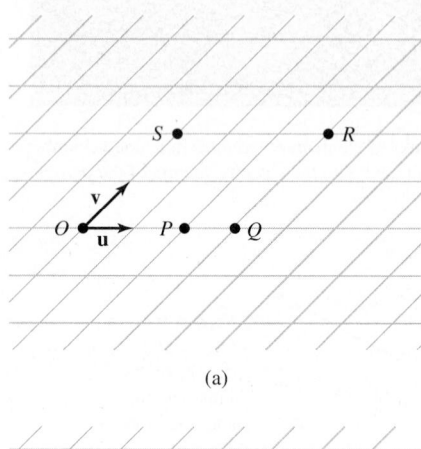

(a)

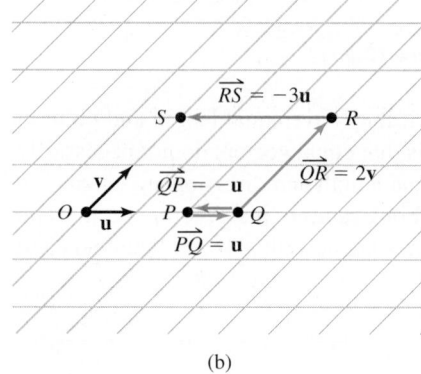

(b)

Figure 13.6

QUICK CHECK 2 Sketch the sum $\mathbf{v}_a + \mathbf{w}$ in Figure 13.7 if the direction of $\mathbf{w}$ is reversed. ◄

Notice that $0\mathbf{v} = \mathbf{0}$ for all vectors $\mathbf{v}$. It follows that *the zero vector is parallel to all vectors*. While it may seem counterintuitive, this result turns out to be a useful convention.

EXAMPLE 1   Parallel vectors Using **Figure 13.6a**, write the following vectors in terms of $\mathbf{u}$ or $\mathbf{v}$.

a. $\overrightarrow{PQ}$ **b.** $\overrightarrow{QP}$ **c.** $\overrightarrow{QR}$ **d.** $\overrightarrow{RS}$

SOLUTION

a. The vector $\overrightarrow{PQ}$ has the same direction and length as $\mathbf{u}$; therefore, $\overrightarrow{PQ} = \mathbf{u}$. These two vectors are equal even though they have different locations (**Figure 13.6b**).

b. Because $\overrightarrow{QP}$ and $\mathbf{u}$ have equal length but opposite directions, $\overrightarrow{QP} = (-1)\mathbf{u} = -\mathbf{u}$.

c. $\overrightarrow{QR}$ points in the same direction as $\mathbf{v}$ and is twice as long as $\mathbf{v}$, so $\overrightarrow{QR} = 2\mathbf{v}$.

d. $\overrightarrow{RS}$ points in the direction opposite that of $\mathbf{u}$ with three times the length of $\mathbf{u}$. Consequently, $\overrightarrow{RS} = -3\mathbf{u}$.

Related Exercise 15 ◄

Vector Addition and Subtraction

To illustrate the idea of vector addition, consider a plane flying horizontally at a constant speed in a crosswind (**Figure 13.7**). The length of vector $\mathbf{v}_a$ represents the plane's *airspeed*, which is the speed the plane would have in still air; $\mathbf{v}_a$ points in the direction of the nose of the plane. The wind vector $\mathbf{w}$ points in the direction of the crosswind and has a length equal to the speed of the crosswind. The combined effect of the motion of the plane and the wind is the *vector sum* $\mathbf{v}_g = \mathbf{v}_a + \mathbf{w}$, which is the velocity of the plane relative to the ground.

Figure 13.8 illustrates two ways to form the vector sum of two nonzero vectors $\mathbf{u}$ and $\mathbf{v}$ geometrically. The first method, called the **Triangle Rule**, places the tail of $\mathbf{v}$ at the head of $\mathbf{u}$. The sum $\mathbf{u} + \mathbf{v}$ is the vector that extends from the tail of $\mathbf{u}$ to the head of $\mathbf{v}$ (**Figure 13.8b**).

When $\mathbf{u}$ and $\mathbf{v}$ are not parallel, another way to form $\mathbf{u} + \mathbf{v}$ is to use the **Parallelogram Rule**. The *tails* of $\mathbf{u}$ and $\mathbf{v}$ are connected to form adjacent sides of a parallelogram; then the remaining two sides of the parallelogram are sketched. The sum $\mathbf{u} + \mathbf{v}$ is the vector that coincides with the diagonal of the parallelogram, beginning at the tails of $\mathbf{u}$ and $\mathbf{v}$ (**Figure 13.8c**). Both the Triangle Rule and the Parallelogram Rule produce the same vector sum $\mathbf{u} + \mathbf{v}$.

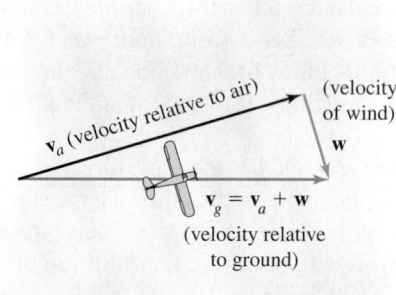

Figure 13.7

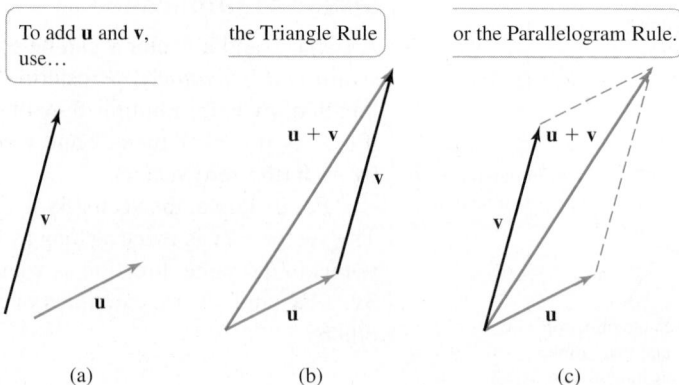

Figure 13.8

QUICK CHECK 3 Use the Triangle Rule to show that the vectors in Figure 13.8 satisfy $\mathbf{u} + \mathbf{v} = \mathbf{v} + \mathbf{u}$. ◄

The difference $\mathbf{u} - \mathbf{v}$ is defined to be the sum $\mathbf{u} + (-\mathbf{v})$. By the Triangle Rule, the tail of $-\mathbf{v}$ is placed at the head of $\mathbf{u}$; then $\mathbf{u} - \mathbf{v}$ extends from the tail of $\mathbf{u}$ to the head of $-\mathbf{v}$ (**Figure 13.9a**). Equivalently, when the tails of $\mathbf{u}$ and $\mathbf{v}$ coincide, $\mathbf{u} - \mathbf{v}$ has its tail at the head of $\mathbf{v}$ and its head at the head of $\mathbf{u}$ (**Figure 13.9b**).

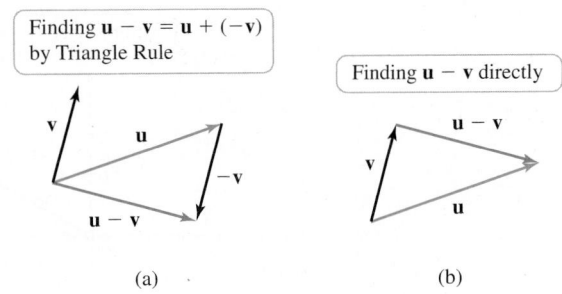

Finding $\mathbf{u} - \mathbf{v} = \mathbf{u} + (-\mathbf{v})$ by Triangle Rule

Finding $\mathbf{u} - \mathbf{v}$ directly

(a)   (b)

Figure 13.9

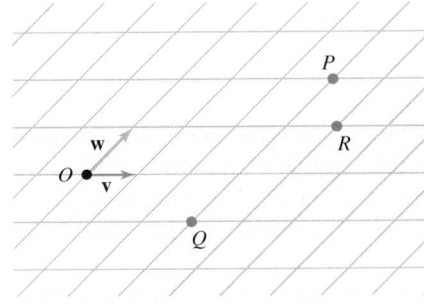

Figure 13.10

EXAMPLE 2   Vector operations Use Figure 13.10 to write the following vectors as sums of scalar multiples of $\mathbf{v}$ and $\mathbf{w}$.

a. $\overrightarrow{OP}$ **b.** $\overrightarrow{OQ}$ **c.** $\overrightarrow{QR}$

SOLUTION

a. Using the Triangle Rule, we start at O, move three lengths of $\mathbf{v}$ in the direction of $\mathbf{v}$ and then two lengths of $\mathbf{w}$ in the direction of $\mathbf{w}$ to reach P. Therefore, $\overrightarrow{OP} = 3\mathbf{v} + 2\mathbf{w}$ (Figure 13.11a).

b. The vector $\overrightarrow{OQ}$ coincides with the diagonal of a parallelogram having adjacent sides equal to $3\mathbf{v}$ and $-\mathbf{w}$. By the Parallelogram Rule, $\overrightarrow{OQ} = 3\mathbf{v} - \mathbf{w}$ (Figure 13.11b).

c. The vector $\overrightarrow{QR}$ lies on the diagonal of a parallelogram having adjacent sides equal to $\mathbf{v}$ and $2\mathbf{w}$. Therefore, $\overrightarrow{QR} = \mathbf{v} + 2\mathbf{w}$ (Figure 13.11c).

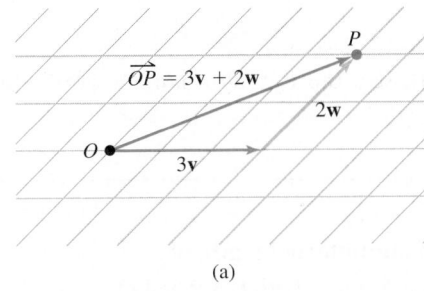

(a)

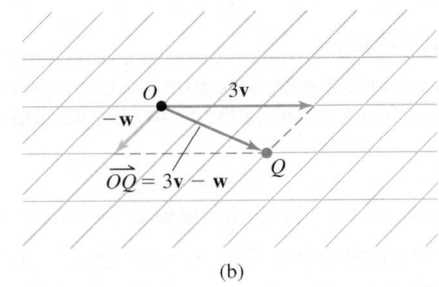

(b)

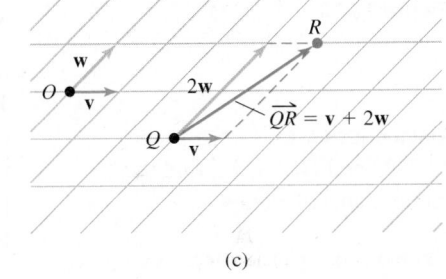

(c)

Figure 13.11

Related Exercises 17–18 ◄

Vector Components

So far, vectors have been examined from a geometric point of view. To do calculations with vectors, it is necessary to introduce a coordinate system. We begin by considering a vector $\mathbf{v}$ whose tail is at the origin in the Cartesian plane and whose head is at the point (v_1, v_2) (Figure 13.12a).

> ➤ Round brackets (a, b) enclose the *coordinates* of a point, while angle brackets $\langle a, b \rangle$ enclose the *components* of a vector. Note that in component form, the zero vector is $\mathbf{0} = \langle 0, 0 \rangle$.

DEFINITION   Position Vectors and Vector Components

A vector $\mathbf{v}$ with its tail at the origin and head at the point (v_1, v_2) is called a **position vector** (or is said to be in **standard position**) and is written $\langle v_1, v_2 \rangle$. The real numbers v_1 and v_2 are the x- and y-**components** of $\mathbf{v}$, respectively. The position vectors $\mathbf{u} = \langle u_1, u_2 \rangle$ and $\mathbf{v} = \langle v_1, v_2 \rangle$ are **equal** if and only if $u_1 = v_1$ and $u_2 = v_2$.

There are infinitely many vectors equal to the position vector $\mathbf{v}$, all with the same length and direction (Figure 13.12b). It is important to abide by the convention that $\mathbf{v} = \langle v_1, v_2 \rangle$ refers to the position vector $\mathbf{v}$ *or to any other vector equal to* $\mathbf{v}$.

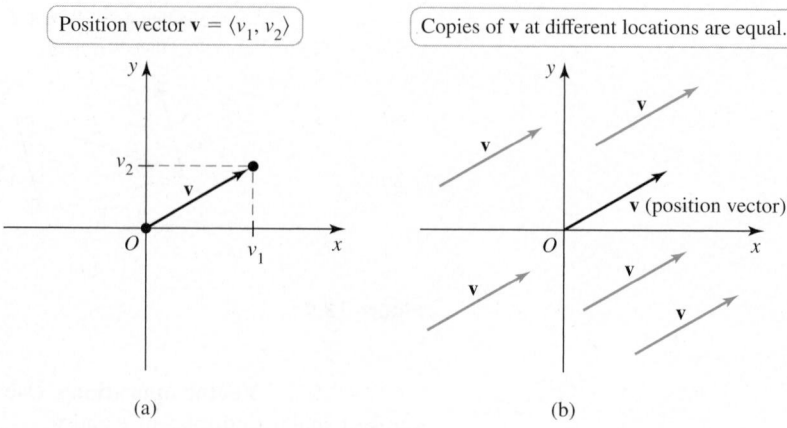

Figure 13.12

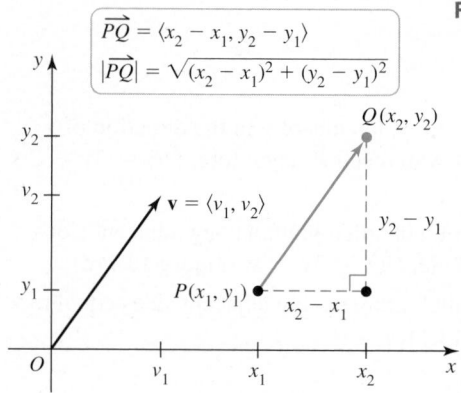

Figure 13.13

Now consider the vector $\overrightarrow{PQ}$ equal to $\mathbf{v} = \langle v_1, v_2 \rangle$, but not in standard position, with its tail at the point $P(x_1, y_1)$ and its head at the point $Q(x_2, y_2)$. The x-component of $\overrightarrow{PQ}$ is the difference in the x-coordinates of Q and P, or $x_2 - x_1$. The y-component of $\overrightarrow{PQ}$ is the difference in the y-coordinates, $y_2 - y_1$ (**Figure 13.13**). Therefore, $\overrightarrow{PQ} = \langle x_2 - x_1, y_2 - y_1 \rangle = \langle v_1, v_2 \rangle = \mathbf{v}$.

As already noted, there are infinitely many vectors equal to a given position vector. All these vectors have the same length and direction; therefore, they are all equal. In other words, two arbitrary vectors are equal if they are equal to the same position vector. For example, the vector $\overrightarrow{PQ}$ from $P(2, 5)$ to $Q(6, 3)$ and the vector $\overrightarrow{AB}$ from $A(7, 12)$ to $B(11, 10)$ are equal because they both equal the position vector $\langle 4, -2 \rangle$.

Magnitude

The magnitude of a vector is simply its length. By the Pythagorean Theorem and Figure 13.13, we have the following definition.

QUICK CHECK 4 Given the points $P(2, 3)$ and $Q(-4, 1)$, find the components of $\overrightarrow{PQ}$. ◄

▶ Just as the absolute value $|p - q|$ gives the distance between the points p and q on the number line, the magnitude $|\overrightarrow{PQ}|$ is the distance between the points P and Q. The magnitude of a vector is also called its **norm**.

DEFINITION   Magnitude of a Vector

Given the points $P(x_1, y_1)$ and $Q(x_2, y_2)$, the **magnitude**, or **length**, of $\overrightarrow{PQ} = \langle x_2 - x_1, y_2 - y_1 \rangle$, denoted $|\overrightarrow{PQ}|$, is the distance between P and Q:

$$|\overrightarrow{PQ}| = \sqrt{(x_2 - x_1)^2 + (y_2 - y_1)^2}.$$

The magnitude of the position vector $\mathbf{v} = \langle v_1, v_2 \rangle$ is $|\mathbf{v}| = \sqrt{v_1^2 + v_2^2}$.

EXAMPLE 3   Calculating components and magnitude   Given the points $O(0, 0)$, $P(-3, 4)$, and $Q(6, 5)$, find the components and magnitude of the following vectors.

a. $\overrightarrow{OP}$ **b.** $\overrightarrow{PQ}$

SOLUTION

a. The vector $\overrightarrow{OP}$ is the position vector whose head is located at $P(-3, 4)$. Therefore, $\overrightarrow{OP} = \langle -3, 4 \rangle$ and its magnitude is $|\overrightarrow{OP}| = \sqrt{(-3)^2 + 4^2} = 5$.

b. $\overrightarrow{PQ} = \langle 6 - (-3), 5 - 4 \rangle = \langle 9, 1 \rangle$ and $|\overrightarrow{PQ}| = \sqrt{9^2 + 1^2} = \sqrt{82}$.

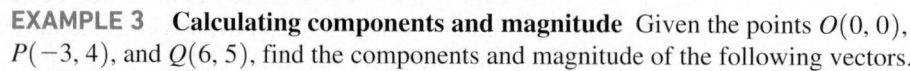

Related Exercise 19 ◄

Vector Operations in Terms of Components

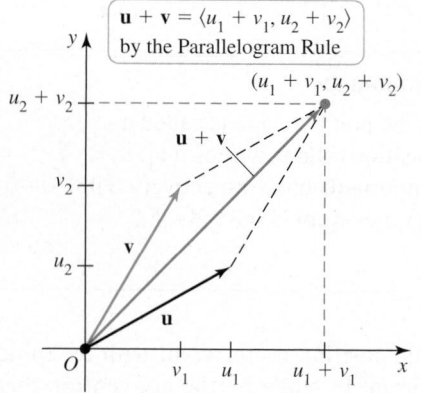

Figure 13.14

We now show how vector addition, vector subtraction, and scalar multiplication are performed using components. Suppose $\mathbf{u} = \langle u_1, u_2 \rangle$ and $\mathbf{v} = \langle v_1, v_2 \rangle$. The vector sum of $\mathbf{u}$ and $\mathbf{v}$ is $\mathbf{u} + \mathbf{v} = \langle u_1 + v_1, u_2 + v_2 \rangle$. This definition of a vector sum is consistent with the Parallelogram Rule given earlier (**Figure 13.14**).

For a scalar c and a vector $\mathbf{u}$, the scalar multiple $c\mathbf{u}$ is $c\mathbf{u} = \langle cu_1, cu_2 \rangle$; that is, the scalar c multiplies each component of $\mathbf{u}$. If $c > 0$, $\mathbf{u}$ and $c\mathbf{u}$ have the same direction (Figure 13.15a). If $c < 0$, $\mathbf{u}$ and $c\mathbf{u}$ have opposite directions (Figure 13.15b). In either case, $|c\mathbf{u}| = |c||\mathbf{u}|$ (Exercise 83).

Notice that $\mathbf{u} - \mathbf{v} = \mathbf{u} + (-\mathbf{v})$, where $-\mathbf{v} = \langle -v_1, -v_2 \rangle$. Therefore, the vector difference of $\mathbf{u}$ and $\mathbf{v}$ is $\mathbf{u} - \mathbf{v} = \langle u_1 - v_1, u_2 - v_2 \rangle$.

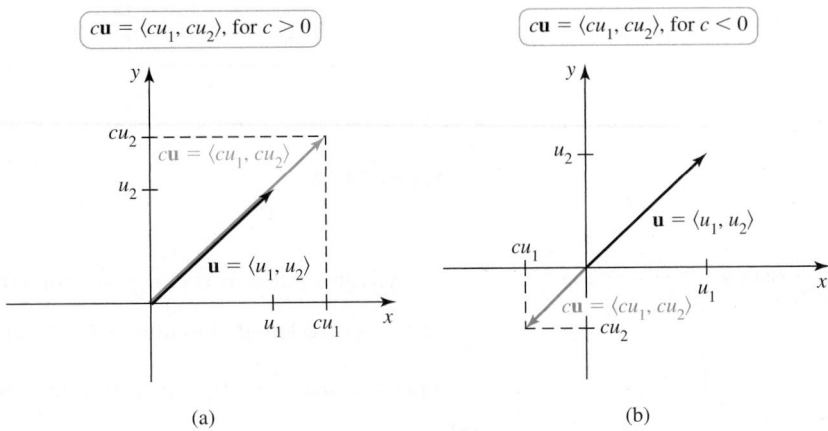

Figure 13.15

> **DEFINITION** **Vector Operations in $\mathbb{R}^2$**
>
> Suppose c is a scalar, $\mathbf{u} = \langle u_1, u_2 \rangle$, and $\mathbf{v} = \langle v_1, v_2 \rangle$.
>
> $\mathbf{u} + \mathbf{v} = \langle u_1 + v_1, u_2 + v_2 \rangle$ Vector addition
>
> $\mathbf{u} - \mathbf{v} = \langle u_1 - v_1, u_2 - v_2 \rangle$ Vector subtraction
>
> $c\mathbf{u} = \langle cu_1, cu_2 \rangle$ Scalar multiplication

▶ Recall that $\mathbb{R}^2$ (pronounced *R-two*) stands for the *xy*-plane or the set of all ordered pairs of real numbers.

EXAMPLE 4 **Vector operations** Let $\mathbf{u} = \langle -1, 2 \rangle$ and $\mathbf{v} = \langle 2, 3 \rangle$.

a. Evaluate $|\mathbf{u} + \mathbf{v}|$.

b. Simplify $2\mathbf{u} - 3\mathbf{v}$.

c. Find two vectors half as long as $\mathbf{u}$ and parallel to $\mathbf{u}$.

SOLUTION

a. Because $\mathbf{u} + \mathbf{v} = \langle -1, 2 \rangle + \langle 2, 3 \rangle = \langle 1, 5 \rangle$, we have
$|\mathbf{u} + \mathbf{v}| = \sqrt{1^2 + 5^2} = \sqrt{26}$.

b. $2\mathbf{u} - 3\mathbf{v} = 2\langle -1, 2 \rangle - 3\langle 2, 3 \rangle = \langle -2, 4 \rangle - \langle 6, 9 \rangle = \langle -8, -5 \rangle$

c. The vectors $\frac{1}{2}\mathbf{u} = \frac{1}{2}\langle -1, 2 \rangle = \langle -\frac{1}{2}, 1 \rangle$ and $-\frac{1}{2}\mathbf{u} = -\frac{1}{2}\langle -1, 2 \rangle = \langle \frac{1}{2}, -1 \rangle$ have half the length of $\mathbf{u}$ and are parallel to $\mathbf{u}$.

Related Exercises 26, 28, 30 ◀

Unit Vectors

A **unit vector** is any vector with length 1. Two useful unit vectors are the **coordinate unit vectors** $\mathbf{i} = \langle 1, 0 \rangle$ and $\mathbf{j} = \langle 0, 1 \rangle$ (Figure 13.16). These vectors are directed along the coordinate axes and enable us to express all vectors in an alternative form. For example, by the Triangle Rule (Figure 13.17a),

$$\langle 3, 4 \rangle = 3\langle 1, 0 \rangle + 4\langle 0, 1 \rangle = 3\mathbf{i} + 4\mathbf{j}.$$

In general, the vector $\mathbf{v} = \langle v_1, v_2 \rangle$ (Figure 13.17b) is also written

$$\mathbf{v} = v_1\langle 1, 0 \rangle + v_2\langle 0, 1 \rangle = v_1\mathbf{i} + v_2\mathbf{j}.$$

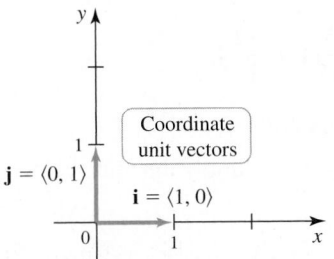

Figure 13.16

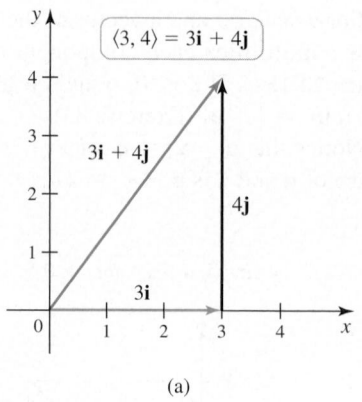

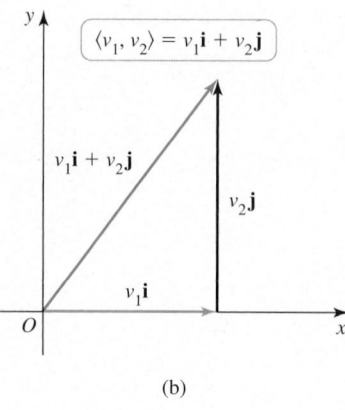

(a)   (b)

Figure 13.17

▶ Coordinate unit vectors are also called **standard basis vectors**.

$\mathbf{u} = \dfrac{\mathbf{v}}{|\mathbf{v}|}$ and $-\mathbf{u} = -\dfrac{\mathbf{v}}{|\mathbf{v}|}$ have length 1.

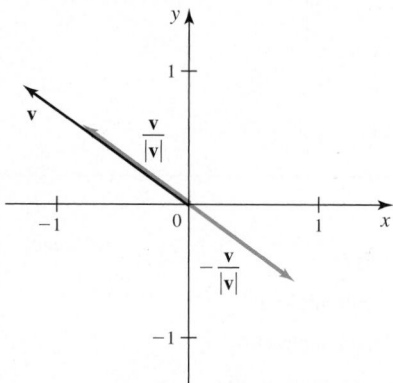

Figure 13.18

QUICK CHECK 5 Find vectors of length 10 parallel to the unit vector $\mathbf{u} = \left\langle \frac{3}{5}, \frac{4}{5} \right\rangle$. ◄

Given a nonzero vector $\mathbf{v}$, we sometimes need to construct a new vector parallel to $\mathbf{v}$ of a specified length. Dividing $\mathbf{v}$ by its length, we obtain the vector $\mathbf{u} = \dfrac{\mathbf{v}}{|\mathbf{v}|}$. Because $\mathbf{u}$ is a positive scalar multiple of $\mathbf{v}$, it follows that $\mathbf{u}$ has the same direction as $\mathbf{v}$. Furthermore, $\mathbf{u}$ is a unit vector because $|\mathbf{u}| = \dfrac{|\mathbf{v}|}{|\mathbf{v}|} = 1$. The vector $-\mathbf{u} = -\dfrac{\mathbf{v}}{|\mathbf{v}|}$ is also a unit vector with a direction opposite that of $\mathbf{v}$ (**Figure 13.18**). Therefore, $\pm\dfrac{\mathbf{v}}{|\mathbf{v}|}$ are unit vectors parallel to $\mathbf{v}$ that point in opposite directions.

To construct a vector that points in the direction of $\mathbf{v}$ and has a specified length $c > 0$, we form the vector $\dfrac{c\mathbf{v}}{|\mathbf{v}|}$. It is a positive scalar multiple of $\mathbf{v}$, so it points in the direction of $\mathbf{v}$, and its length is $\left|\dfrac{c\mathbf{v}}{|\mathbf{v}|}\right| = |c|\dfrac{|\mathbf{v}|}{|\mathbf{v}|} = c$. The vector $-\dfrac{c\mathbf{v}}{|\mathbf{v}|}$ points in the opposite direction and also has length c. With this construction, we can also write $\mathbf{v}$ as the product of its magnitude and a unit vector in the direction of $\mathbf{v}$:

$$\mathbf{v} = \underbrace{|\mathbf{v}|}_{\text{magnitude}} \cdot \underbrace{\frac{\mathbf{v}}{|\mathbf{v}|}}_{\text{direction}}.$$

EXAMPLE 5 Magnitude and unit vectors Consider the points $P(1, -2)$ and $Q(6, 10)$.

a. Find $\overrightarrow{PQ}$ and two unit vectors parallel to $\overrightarrow{PQ}$.

b. Find two vectors of length 2 parallel to $\overrightarrow{PQ}$.

c. Express $\overrightarrow{PQ}$ as the product of its magnitude and a unit vector.

SOLUTION

a. $\overrightarrow{PQ} = \langle 6 - 1, 10 - (-2) \rangle = \langle 5, 12 \rangle$, or $5\mathbf{i} + 12\mathbf{j}$. Because $|\overrightarrow{PQ}| = \sqrt{5^2 + 12^2} = \sqrt{169} = 13$, a unit vector parallel to $\overrightarrow{PQ}$ is

$$\frac{\overrightarrow{PQ}}{|\overrightarrow{PQ}|} = \frac{\langle 5, 12 \rangle}{13} = \left\langle \frac{5}{13}, \frac{12}{13} \right\rangle = \frac{5}{13}\mathbf{i} + \frac{12}{13}\mathbf{j}.$$

The unit vector parallel to $\overrightarrow{PQ}$ with the opposite direction is $\left\langle -\frac{5}{13}, -\frac{12}{13} \right\rangle$.

b. To obtain two vectors of length 2 that are parallel to $\overrightarrow{PQ}$, we multiply the unit vector $\frac{5}{13}\mathbf{i} + \frac{12}{13}\mathbf{j}$ by ± 2:

$$2\left(\frac{5}{13}\mathbf{i} + \frac{12}{13}\mathbf{j}\right) = \frac{10}{13}\mathbf{i} + \frac{24}{13}\mathbf{j} \quad \text{and} \quad -2\left(\frac{5}{13}\mathbf{i} + \frac{12}{13}\mathbf{j}\right) = -\frac{10}{13}\mathbf{i} - \frac{24}{13}\mathbf{j}.$$

QUICK CHECK 6 Verify that the vector $\left\langle \frac{5}{13}, \frac{12}{13} \right\rangle$ has length 1. ◄

c. The unit vector $\left\langle \frac{5}{13}, \frac{12}{13} \right\rangle$ points in the direction of $\overrightarrow{PQ}$, so we have

$$\overrightarrow{PQ} = |\overrightarrow{PQ}| \frac{\overrightarrow{PQ}}{|\overrightarrow{PQ}|} = 13 \left\langle \frac{5}{13}, \frac{12}{13} \right\rangle.$$

Related Exercises 34, 43, 45 ◄

Properties of Vector Operations

When we stand back and look at vector operations, ten general properties emerge. For example, the first property says that vector addition is commutative, which means $\mathbf{u} + \mathbf{v} = \mathbf{v} + \mathbf{u}$. This property is proved by letting $\mathbf{u} = \langle u_1, u_2 \rangle$ and $\mathbf{v} = \langle v_1, v_2 \rangle$. By the commutative property of addition for real numbers,

> ➤ The Parallelogram Rule illustrates the commutative property $\mathbf{u} + \mathbf{v} = \mathbf{v} + \mathbf{u}$.

$$\mathbf{u} + \mathbf{v} = \langle u_1 + v_1, u_2 + v_2 \rangle = \langle v_1 + u_1, v_2 + u_2 \rangle = \mathbf{v} + \mathbf{u}.$$

The proofs of other properties are outlined in Exercises 78–81.

SUMMARY Properties of Vector Operations

Suppose $\mathbf{u}$, $\mathbf{v}$, and $\mathbf{w}$ are vectors and a and c are scalars. Then the following properties hold (for vectors in any number of dimensions).

1. $\mathbf{u} + \mathbf{v} = \mathbf{v} + \mathbf{u}$ Commutative property of addition
2. $(\mathbf{u} + \mathbf{v}) + \mathbf{w} = \mathbf{u} + (\mathbf{v} + \mathbf{w})$ Associative property of addition
3. $\mathbf{v} + \mathbf{0} = \mathbf{v}$ Additive identity
4. $\mathbf{v} + (-\mathbf{v}) = \mathbf{0}$ Additive inverse
5. $c(\mathbf{u} + \mathbf{v}) = c\mathbf{u} + c\mathbf{v}$ Distributive property 1
6. $(a + c)\mathbf{v} = a\mathbf{v} + c\mathbf{v}$ Distributive property 2
7. $0\mathbf{v} = \mathbf{0}$ Multiplication by zero scalar
8. $c\mathbf{0} = \mathbf{0}$ Multiplication by zero vector
9. $1\mathbf{v} = \mathbf{v}$ Multiplicative identity
10. $a(c\mathbf{v}) = (ac)\mathbf{v}$ Associative property of scalar multiplication

---

These properties allow us to solve vector equations. For example, to solve the equation $\mathbf{u} + \mathbf{v} = \mathbf{w}$ for $\mathbf{u}$, we proceed as follows:

$$(\mathbf{u} + \mathbf{v}) + (-\mathbf{v}) = \mathbf{w} + (-\mathbf{v}) \quad \text{Add } -\mathbf{v} \text{ to both sides.}$$
$$\mathbf{u} + \underbrace{(\mathbf{v} + (-\mathbf{v}))}_{0} = \mathbf{w} + (-\mathbf{v}) \quad \text{Property 2}$$
$$\mathbf{u} + \mathbf{0} = \mathbf{w} - \mathbf{v} \quad \text{Property 4}$$
$$\mathbf{u} = \mathbf{w} - \mathbf{v}. \quad \text{Property 3}$$

QUICK CHECK 7 Solve $3\mathbf{u} + 4\mathbf{v} = 12\mathbf{w}$ for $\mathbf{u}$. ◄

Applications of Vectors

Vectors have countless practical applications, particularly in the physical sciences and engineering. These applications are explored throughout the remainder of this text. For now, we present two common uses of vectors: to describe velocities and forces.

> ➤ *Velocity of the boat relative to the water* means the velocity (direction and speed) the boat has relative to someone traveling with the current.

Velocity Vectors Consider a motorboat crossing a river whose current is everywhere represented by the constant vector $\mathbf{w}$ (Figure 13.19); this means that $|\mathbf{w}|$ is the speed of the moving water and $\mathbf{w}$ points in the direction of the moving water. Assume the vector $\mathbf{v}_w$ gives the velocity of the boat relative to the water. The combined effect of $\mathbf{w}$ and $\mathbf{v}_w$ is the sum $\mathbf{v}_g = \mathbf{v}_w + \mathbf{w}$, which is the velocity of the boat that would be observed by someone on the shore (or on the ground).

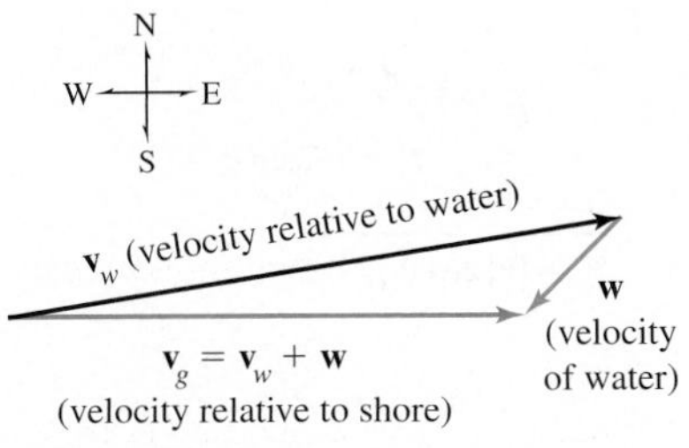

Figure 13.19

EXAMPLE 6 Speed of a boat in a current Suppose the water in a river moves southwest (45° west of south) at 4 mi/hr and a motorboat travels due east at 15 mi/hr relative to the shore. Determine the speed of the boat and its heading relative to the moving water (Figure 13.19).

SOLUTION To solve this problem, the vectors are placed in a coordinate system (Figure 13.20). Because the boat moves east at 15 mi/hr, the velocity relative to the shore is $\mathbf{v}_g = \langle 15, 0 \rangle$. To obtain the components of $\mathbf{w} = \langle w_x, w_y \rangle$, observe that $|\mathbf{w}| = 4$ and that the lengths of the sides of the 45–45–90 triangle in Figure 13.20 are

$$|w_x| = |w_y| = |\mathbf{w}| \cos 45° = \frac{4}{\sqrt{2}} = 2\sqrt{2}.$$

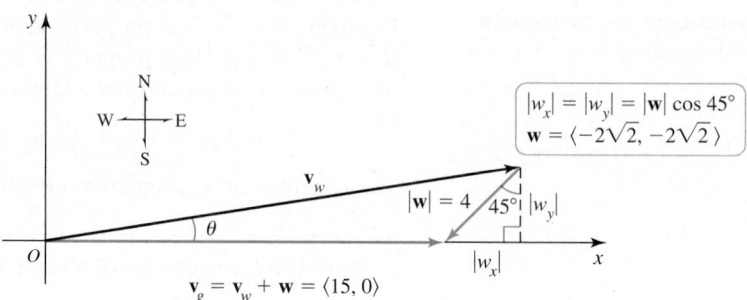

Figure 13.20

Given the orientation of $\mathbf{w}$ (southwest), $\mathbf{w} = \langle -2\sqrt{2}, -2\sqrt{2} \rangle$. Because $\mathbf{v}_g = \mathbf{v}_w + \mathbf{w}$ (Figure 13.20),

$$\mathbf{v}_w = \mathbf{v}_g - \mathbf{w} = \langle 15, 0 \rangle - \langle -2\sqrt{2}, -2\sqrt{2} \rangle$$
$$= \langle 15 + 2\sqrt{2}, 2\sqrt{2} \rangle.$$

The magnitude of $\mathbf{v}_w$ is

$$|\mathbf{v}_w| = \sqrt{(15 + 2\sqrt{2})^2 + (2\sqrt{2})^2} \approx 18.$$

> Recall that the lengths of the legs of a 45–45–90 triangle are equal and are $1/\sqrt{2}$ times the length of the hypotenuse.

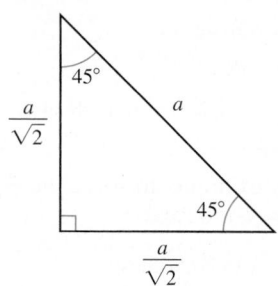

Therefore, the speed of the boat relative to the water is approximately 18 mi/hr.

The heading of the boat is given by the angle θ between $\mathbf{v}_w$ and the positive x-axis. The x-component of $\mathbf{v}_w$ is $15 + 2\sqrt{2}$ and the y-component is $2\sqrt{2}$. Therefore,

$$\theta = \tan^{-1} \left(\frac{2\sqrt{2}}{15 + 2\sqrt{2}} \right) \approx 9°.$$

The heading of the boat is approximately 9° north of east, and its speed relative to the water is approximately 18 mi/hr.

Related Exercises 56–57 ◄

> The magnitude of $\mathbf{F}$ is typically measured in pounds (lb) or newtons (N), where $1 \text{ N} = 1 \text{ kg-m/s}^2$.

> The vector $\langle \cos \theta, \sin \theta \rangle$ is a unit vector. Therefore, any position vector $\mathbf{v}$ may be written $\mathbf{v} = \langle |\mathbf{v}| \cos \theta, |\mathbf{v}| \sin \theta \rangle$, where θ is the angle that $\mathbf{v}$ makes with the positive x-axis.

Force Vectors   Suppose a child pulls on the handle of a wagon at an angle of θ with the horizontal (**Figure 13.21a**). The vector $\mathbf{F}$ represents the force exerted on the wagon; it has a magnitude $|\mathbf{F}|$ and a direction given by θ. We denote the horizontal and vertical components of $\mathbf{F}$ by F_x and F_y, respectively. From **Figure 13.21b**, we see that $F_x = |\mathbf{F}| \cos \theta$, $F_y = |\mathbf{F}| \sin \theta$, and the force vector is $\mathbf{F} = \langle |\mathbf{F}| \cos \theta, |\mathbf{F}| \sin \theta \rangle$.

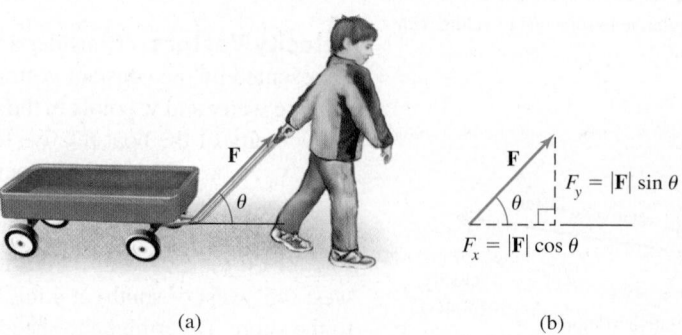

(a)

(b)

Figure 13.21

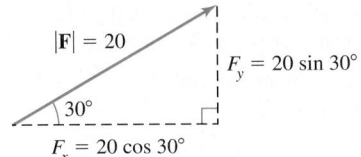

$|\mathbf{F}| = 20$

$F_y = 20 \sin 30°$

$30°$

$F_x = 20 \cos 30°$

Figure 13.22

EXAMPLE 7 **Finding force vectors** A child pulls a wagon (Figure 13.21) with a force of $|\mathbf{F}| = 20$ lb at an angle of $\theta = 30°$ to the horizontal. Find the force vector $\mathbf{F}$.

SOLUTION The force vector (Figure 13.22) is

$$\mathbf{F} = \langle |\mathbf{F}| \cos \theta, |\mathbf{F}| \sin \theta \rangle = \langle 20 \cos 30°, 20 \sin 30° \rangle = \langle 10\sqrt{3}, 10 \rangle.$$

Related Exercise 60 ◄

EXAMPLE 8 **Balancing forces** A 400-lb engine is suspended from two chains that form 60° angles with a horizontal ceiling (Figure 13.23). How much weight does each chain support?

SOLUTION Let $\mathbf{F}_1$ and $\mathbf{F}_2$ denote the forces exerted by the chains on the engine, and let $\mathbf{F}_3$ be the downward force due to the weight of the engine (Figure 13.23). Placing the vectors in a standard coordinate system (Figure 13.24), we find that $\mathbf{F}_1 = \langle |\mathbf{F}_1| \cos 60°, |\mathbf{F}_1| \sin 60° \rangle$, $\mathbf{F}_2 = \langle -|\mathbf{F}_2| \cos 60°, |\mathbf{F}_2| \sin 60° \rangle$, and $\mathbf{F}_3 = \langle 0, -400 \rangle$.

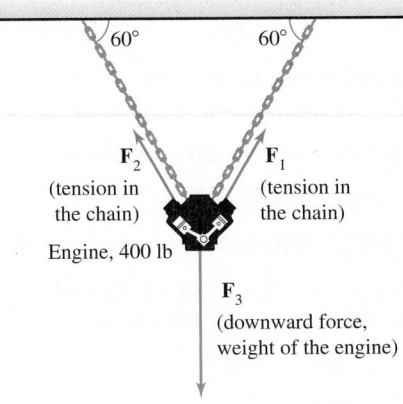

$60°$ $60°$

$\mathbf{F}_2$
(tension in
the chain)

$\mathbf{F}_1$
(tension in
the chain)

Engine, 400 lb

$\mathbf{F}_3$
(downward force,
weight of the engine)

Figure 13.23

➤ The components of $\mathbf{F}_2$ in Example 8 can also be computed using an angle of 120°. That is, $\mathbf{F}_2 = \langle |\mathbf{F}_2| \cos 120°, |\mathbf{F}_2| \sin 120° \rangle$.

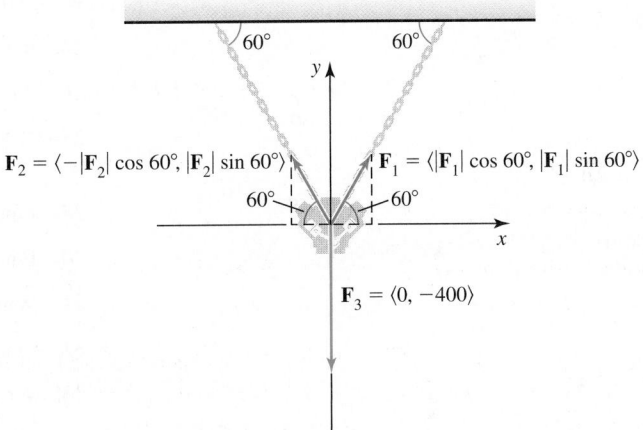

$60°$ $60°$

y

$\mathbf{F}_2 = \langle -|\mathbf{F}_2| \cos 60°, |\mathbf{F}_2| \sin 60° \rangle$ $\mathbf{F}_1 = \langle |\mathbf{F}_1| \cos 60°, |\mathbf{F}_1| \sin 60° \rangle$

$60°$ $60°$

x

$\mathbf{F}_3 = \langle 0, -400 \rangle$

Figure 13.24

Because the engine is in equilibrium (the chains and engine are stationary), the sum of the forces is zero; that is, $\mathbf{F}_1 + \mathbf{F}_2 + \mathbf{F}_3 = \mathbf{0}$ or $\mathbf{F}_1 + \mathbf{F}_2 = -\mathbf{F}_3$. Therefore,

$$\langle |\mathbf{F}_1| \cos 60° - |\mathbf{F}_2| \cos 60°, |\mathbf{F}_1| \sin 60° + |\mathbf{F}_2| \sin 60° \rangle = \langle 0, 400 \rangle.$$

Equating corresponding components, we obtain two equations to be solved for $|\mathbf{F}_1|$ and $|\mathbf{F}_2|$:

$$|\mathbf{F}_1| \cos 60° - |\mathbf{F}_2| \cos 60° = 0 \text{ and}$$
$$|\mathbf{F}_1| \sin 60° + |\mathbf{F}_2| \sin 60° = 400.$$

Factoring the first equation, we find that $(|\mathbf{F}_1| - |\mathbf{F}_2|) \cos 60° = 0$, which implies that $|\mathbf{F}_1| = |\mathbf{F}_2|$. Replacing $|\mathbf{F}_2|$ with $|\mathbf{F}_1|$ in the second equation gives $2|\mathbf{F}_1| \sin 60° = 400$. Noting that $\sin 60° = \sqrt{3}/2$ and solving for $|\mathbf{F}_1|$, we find that $|\mathbf{F}_1| = 400/\sqrt{3} \approx 231$. Each chain must be able to support a weight of approximately 231 lb.

Related Exercise 63 ◄

SECTION 13.1 EXERCISES

Getting Started

1. Interpret the following statement: Points have a location, but no size or direction; nonzero vectors have a size and direction, but no location.

2. What is a position vector?

3. Given a position vector $\mathbf{v}$, why are there infinitely many vectors equal to $\mathbf{v}$?

4. Use the points $P(3, 1)$ and $Q(7, 1)$ to find position vectors equal to $\overrightarrow{PQ}$ and $\overrightarrow{QP}$.

5. If $\mathbf{u} = \langle u_1, u_2 \rangle$ and $\mathbf{v} = \langle v_1, v_2 \rangle$, how do you find $\mathbf{u} + \mathbf{v}$?

6. Find two unit vectors parallel to $\langle 2, 3 \rangle$.

7. Is $\langle 1, 1 \rangle$ a unit vector? Explain.

8. Evaluate $\langle 3, 1 \rangle + \langle 2, 4 \rangle$ and illustrate the sum geometrically using the Parallelogram Rule.

9. How do you compute the magnitude of $\mathbf{v} = \langle v_1, v_2 \rangle$?

10. Write the vector $\mathbf{v} = \langle v_1, v_2 \rangle$ in terms of the unit vectors $\mathbf{i}$ and $\mathbf{j}$.

11. How do you compute $|\overrightarrow{PQ}|$ from the coordinates of the points P and Q?

12. The velocity of a kayak on a lake is $\mathbf{v} = \langle 2\sqrt{2}, 2\sqrt{2} \rangle$. Find the speed and heading of the kayak. Assume the positive x-axis points east and the positive y-axis points north. Assume the coordinates of $\mathbf{v}$ are in feet per second.

Practice Exercises

13–18. Vector operations *Refer to the figure and carry out the following vector operations.*

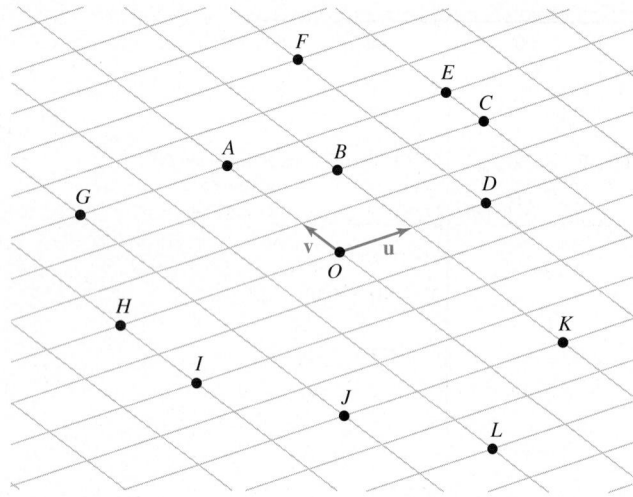

13. Scalar multiples Which of the following vectors equal $\overrightarrow{CE}$? (There may be more than one correct answer.)

   a. $\mathbf{v}$ **b.** $\frac{1}{2}\overrightarrow{HI}$ **c.** $\frac{1}{3}\overrightarrow{OA}$ **d.** $\mathbf{u}$ **e.** $\frac{1}{2}\overrightarrow{IH}$

14. Scalar multiples Which of the following vectors equal $\overrightarrow{BK}$? (There may be more than one correct answer.)

   a. $6\mathbf{v}$ **b.** $-6\mathbf{v}$ **c.** $3\overrightarrow{HI}$ **d.** $3\overrightarrow{IH}$ **e.** $3\overrightarrow{AO}$

15. Scalar multiples Write the following vectors as scalar multiples of $\mathbf{u}$ or $\mathbf{v}$.

   a. $\overrightarrow{OA}$ **b.** $\overrightarrow{OD}$ **c.** $\overrightarrow{OH}$ **d.** $\overrightarrow{AG}$ **e.** $\overrightarrow{CE}$

16. Scalar multiples Write the following vectors as scalar multiples of $\mathbf{u}$ or $\mathbf{v}$.

   a. $\overrightarrow{IH}$ **b.** $\overrightarrow{HI}$ **c.** $\overrightarrow{JK}$ **d.** $\overrightarrow{FD}$ **e.** $\overrightarrow{EA}$

17. Vector addition Write the following vectors as sums of scalar multiples of $\mathbf{u}$ and $\mathbf{v}$.

   a. $\overrightarrow{OE}$ **b.** $\overrightarrow{OB}$ **c.** $\overrightarrow{OF}$ **d.** $\overrightarrow{OG}$ **e.** $\overrightarrow{OC}$
   f. $\overrightarrow{OI}$ **g.** $\overrightarrow{OJ}$ **h.** $\overrightarrow{OK}$ **i.** $\overrightarrow{OL}$

18. Vector addition Write the following vectors as sums of scalar multiples of $\mathbf{u}$ and $\mathbf{v}$.

   a. $\overrightarrow{BF}$ **b.** $\overrightarrow{DE}$ **c.** $\overrightarrow{AF}$ **d.** $\overrightarrow{AD}$ **e.** $\overrightarrow{CD}$
   f. $\overrightarrow{JD}$ **g.** $\overrightarrow{JI}$ **h.** $\overrightarrow{DB}$ **i.** $\overrightarrow{IL}$

19. Components and magnitudes Define the points $O(0, 0)$, $P(3, 2)$, $Q(4, 2)$, and $R(-6, -1)$. For each vector, do the following.

   (i) Sketch the vector in an xy-coordinate system.
   (ii) Compute the magnitude of the vector.

   a. $\overrightarrow{OP}$ **b.** $\overrightarrow{QP}$ **c.** $\overrightarrow{RQ}$

20. Finding vectors from two points Given the points $A(-2, 0)$, $B(6, 16)$, $C(1, 4)$, $D(5, 4)$, $E(\sqrt{2}, \sqrt{2})$, and $F(3\sqrt{2}, -4\sqrt{2})$, find the position vector equal to the following vectors.

   a. $\overrightarrow{AB}$ **b.** $\overrightarrow{AC}$ **c.** $\overrightarrow{EF}$ **d.** $\overrightarrow{CD}$

21–23. Components and equality *Define the points $P(-3, -1)$, $Q(-1, 2)$, $R(1, 2)$, $S(3, 5)$, $T(4, 2)$, and $U(6, 4)$.*

21. Sketch $\overrightarrow{QU}$, $\overrightarrow{PT}$, and $\overrightarrow{RS}$ and their corresponding position vectors.

22. Find the equal vectors among $\overrightarrow{PQ}$, $\overrightarrow{RS}$, and $\overrightarrow{TU}$.

23. Consider the vectors $\overrightarrow{QT}$ and $\overrightarrow{SU}$: Which vector is equal to $\langle 5, 0 \rangle$?

24–27. Vector operations *Let $\mathbf{u} = \langle 4, -2 \rangle$, $\mathbf{v} = \langle -4, 6 \rangle$, and $\mathbf{w} = \langle 0, 8 \rangle$. Express the following vectors in the form $\langle a, b \rangle$.*

24. $\mathbf{u} + \mathbf{v}$ **25.** $\mathbf{w} - \mathbf{u}$

26. $2\mathbf{u} + 3\mathbf{v}$ **27.** $10\mathbf{u} - 3\mathbf{v} + \mathbf{w}$

28–31. Vector operations *Let $\mathbf{u} = \langle 3, -4 \rangle$, $\mathbf{v} = \langle 1, 1 \rangle$, and $\mathbf{w} = \langle -1, 0 \rangle$.*

28. Find $|\mathbf{u} + \mathbf{v} + \mathbf{w}|$.   **29.** Find $|-2\mathbf{v}|$.

30. Find two vectors parallel to $\mathbf{u}$ with four times the magnitude of $\mathbf{u}$.

31. Which has the greater magnitude, $\mathbf{u} - \mathbf{v}$ or $\mathbf{w} - \mathbf{u}$?

32. Find a unit vector in the direction of $\mathbf{v} = \langle -6, 8 \rangle$.

33. Write $\mathbf{v} = \langle -5, 12 \rangle$ as a product of its magnitude and a unit vector in the direction of $\mathbf{v}$.

34. Consider the points $P(2, 7)$ and $Q(6, 4)$. Write $\overrightarrow{PQ}$ as a product of its magnitude and a unit vector in the direction of $\overrightarrow{PQ}$.

35. Find the vector $\mathbf{v}$ of length 6 that has the same direction as the unit vector $\langle 1/2, \sqrt{3}/2 \rangle$.

36. Find the vector $\mathbf{v}$ that has a magnitude of 10 and a direction opposite that of the unit vector $\langle 3/5, -4/5 \rangle$.

37. Find the vector in the direction of $\langle 5, -12 \rangle$ with length 3.

38. Find the vector pointing in the direction opposite that of $\langle 6, -8 \rangle$ with length 20.

39. Find a vector in the same direction as $\langle 3, -2 \rangle$ with length 10.

40. Let $\mathbf{v} = \langle 8, 15 \rangle$.

   a. Find a vector in the direction of $\mathbf{v}$ that is three times as long as $\mathbf{v}$.
   b. Find a vector in the direction of $\mathbf{v}$ that has length 3.

41–46. Unit vectors *Define the points $P(-4, 1)$, $Q(3, -4)$, and $R(2, 6)$.*

41. Express $\overrightarrow{QR}$ in the form $a\mathbf{i} + b\mathbf{j}$.

42. Express $\overrightarrow{PQ}$ in the form $a\mathbf{i} + b\mathbf{j}$.

43. Find two unit vectors parallel to $\overrightarrow{PR}$.

44. Find the unit vector with the same direction as $\overrightarrow{QR}$.

45. Find two vectors parallel to $\overrightarrow{QP}$ with length 4.

46. Find two vectors parallel to $\overrightarrow{RP}$ with length 4.

47. Unit vectors

   a. Find two unit vectors parallel to $\mathbf{v} = 6\mathbf{i} - 8\mathbf{j}$.

   b. Find b if $\mathbf{v} = \langle 1/3, b \rangle$ is a unit vector.

   c. Find all values of a such that $\mathbf{w} = a\mathbf{i} - \dfrac{a}{3}\mathbf{j}$ is a unit vector.

48. Vectors from polar coordinates Suppose O is the origin and P has polar coordinates (r, θ). Show that $\overrightarrow{OP} = \langle r \cos \theta, r \sin \theta \rangle$.

49. Vectors from polar coordinates Find the position vector $\overrightarrow{OP}$ if O is the origin and P has polar coordinates $(8, 5\pi/6)$.

50. Find the velocity $\mathbf{v}$ of an ocean freighter that is traveling northeast (45° east of north) at 40 km/hr.

51. Find the velocity $\mathbf{v}$ of an ocean freighter that is traveling 30° south of east at 30 km/hr.

52. Find a force vector of magnitude 100 that is directed 45° south of east.

▣ 53–55. Airplanes and crosswinds *Assume each plane flies horizontally in a crosswind that blows horizontally.*

53. An airplane flies east to west at 320 mi/hr relative to the air in a crosswind that blows at 40 mi/hr toward the southwest (45° south of west).

   a. Find the velocity of the plane relative to the air $\mathbf{v}_a$, the velocity of the crosswind $\mathbf{w}$, and the velocity of the plane relative to the ground $\mathbf{v}_g$.

   b. Find the ground speed and heading of the plane relative to the ground.

▣ 54. A commercial jet flies west to east at 400 mi/hr relative to the air, and it flies at 420 mi/hr at a heading of 5° north of east relative to the ground.

   a. Find the velocity of the plane relative to the air $\mathbf{v}_a$, the velocity of the plane relative to the ground $\mathbf{v}_g$, and the crosswind $\mathbf{w}$.

   b. Find the speed and heading of the wind.

55. Determine the necessary air speed and heading that a pilot must maintain in order to fly her commercial jet north at a speed of 480 mi/hr relative to the ground in a crosswind that is blowing 60° south of east at 20 mi/hr.

56. A boat in a current The water in a river moves south at 10 mi/hr. A motorboat travels due east at a speed of 20 mi/hr relative to the shore. Determine the speed and direction of the boat relative to the moving water.

57. Another boat in a current The water in a river moves south at 5 km/hr. A motorboat travels due east at a speed of 40 km/hr relative to the water. Determine the speed of the boat relative to the shore.

58. Parachute in the wind In still air, a parachute with a payload falls vertically at a terminal speed of 4 m/s. Find the direction and magnitude of its terminal velocity relative to the ground if it falls in a steady wind blowing horizontally from west to east at 10 m/s.

59. Boat in a wind A sailboat floats in a current that flows due east at 1 m/s. Because of a wind, the boat's actual speed relative to the shore is $\sqrt{3}$ m/s in a direction 30° north of east. Find the speed and direction of the wind.

60. Towing a boat A boat is towed with a force of 150 lb with a rope that makes an angle of 30° to the horizontal. Find the horizontal and vertical components of the force.

61. Pulling a suitcase Suppose you pull a suitcase with a strap that makes a 60° angle with the horizontal. The magnitude of the force you exert on the suitcase is 40 lb.

   a. Find the horizontal and vertical components of the force.

   b. Is the horizontal component of the force greater if the angle of the strap is 45° instead of 60°?

   c. Is the vertical component of the force greater if the angle of the strap is 45° instead of 60°?

62. Which is greater? Which has a greater horizontal component, a 100-N force directed at an angle of 60° above the horizontal or a 60-N force directed at an angle of 30° above the horizontal?

63. Suspended load If a 500-lb load is suspended by two chains (see figure), what is the magnitude of the force each chain must be able to support?

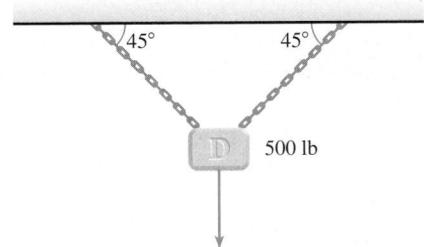

64. Net force Three forces are applied to an object, as shown in the figure. Find the magnitude and direction of the sum of the forces.

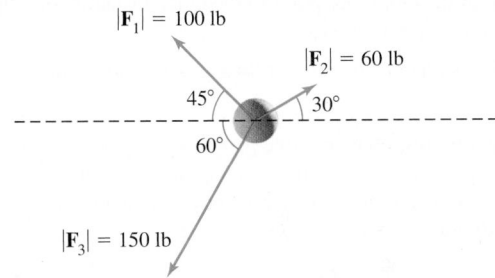

65. Explain why or why not Determine whether the following statements are true and give an explanation or counterexample.

   a. José travels from point A to point B in the plane by following vector $\mathbf{u}$, then vector $\mathbf{v}$, and then vector $\mathbf{w}$. If he starts at A and follows $\mathbf{w}$, then $\mathbf{v}$, and then $\mathbf{u}$, he still arrives at B.

   b. Maria travels from A to B in the plane by following the vector $\mathbf{u}$. By following $-\mathbf{u}$, she returns from B to A.

   c. $|\mathbf{u} + \mathbf{v}| \geq |\mathbf{u}|$, for all vectors $\mathbf{u}$ and $\mathbf{v}$.

   d. $|\mathbf{u} + \mathbf{v}| \geq |\mathbf{u}| + |\mathbf{v}|$, for all vectors $\mathbf{u}$ and $\mathbf{v}$.

   e. Parallel vectors have the same length.

   f. If $\overrightarrow{AB} = \overrightarrow{CD}$, then $A = C$ and $B = D$.

   g. If $\mathbf{u}$ and $\mathbf{v}$ are perpendicular, then $|\mathbf{u} + \mathbf{v}| = |\mathbf{u}| + |\mathbf{v}|$.

   h. If $\mathbf{u}$ and $\mathbf{v}$ are parallel and have the same direction, then $|\mathbf{u} + \mathbf{v}| = |\mathbf{u}| + |\mathbf{v}|$.

66. Equal vectors For the points $A(3, 4)$, $B(6, 10)$, $C(a + 2, b + 5)$, and $D(b + 4, a - 2)$, find the values of a and b such that $\overrightarrow{AB} = \overrightarrow{CD}$.

67–69. Vector equations *Use the properties of vectors to solve the following equations for the unknown vector* $\mathbf{x} = \langle a, b \rangle$. *Let* $\mathbf{u} = \langle 2, -3 \rangle$ *and* $\mathbf{v} = \langle -4, 1 \rangle$.

67. $10\mathbf{x} = \mathbf{u}$ **68.** $2\mathbf{x} + \mathbf{u} = \mathbf{v}$ **69.** $3\mathbf{x} - 4\mathbf{u} = \mathbf{v}$

70. Solve the pair of equations $2\mathbf{u} + 3\mathbf{v} = \mathbf{i}$, $\mathbf{u} - \mathbf{v} = \mathbf{j}$ for the vectors $\mathbf{u}$ and $\mathbf{v}$. Assume $\mathbf{i} = \langle 1, 0 \rangle$ and $\mathbf{j} = \langle 0, 1 \rangle$.

Explorations and Challenges

71–73. Linear combinations *A sum of scalar multiples of two or more vectors (such as $c_1\mathbf{u} + c_2\mathbf{v} + c_3\mathbf{w}$, where c_i are scalars) is called a **linear combination** of the vectors. Let $\mathbf{i} = \langle 1, 0 \rangle$, $\mathbf{j} = \langle 0, 1 \rangle$, $\mathbf{u} = \langle 1, 1 \rangle$, and $\mathbf{v} = \langle -1, 1 \rangle$.*

71. Express $\langle 4, -8 \rangle$ as a linear combination of $\mathbf{i}$ and $\mathbf{j}$ (that is, find scalars c_1 and c_2 such that $\langle 4, -8 \rangle = c_1\mathbf{i} + c_2\mathbf{j}$).

72. Express $\langle 4, -8 \rangle$ as a linear combination of $\mathbf{u}$ and $\mathbf{v}$.

73. For arbitrary real numbers a and b, express $\langle a, b \rangle$ as a linear combination of $\mathbf{u}$ and $\mathbf{v}$.

74. Ant on a page An ant walks due east at a constant speed of 2 mi/hr on a sheet of paper that rests on a table. Suddenly, the sheet of paper starts moving southeast at $\sqrt{2}$ mi/hr. Describe the motion of the ant relative to the table.

75. Clock vectors Consider the 12 vectors that have their tails at the center of a (circular) clock and their heads at the numbers on the edge of the clock.

    a. What is the sum of these 12 vectors?

    b. If the 12:00 vector is removed, what is the sum of the remaining 11 vectors?

    c. By removing one or more of these 12 clock vectors, explain how to make the sum of the remaining vectors as large as possible in magnitude.

    d. Consider the 11 vectors that originate at the number 12 at the top of the clock and point to the other 11 numbers. What is the sum of these vectors?

(*Source: Calculus*, by Gilbert Strang, Wellesley-Cambridge Press, 1991)

76. Three-way tug-of-war Three people located at A, B, and C pull on ropes tied to a ring. Find the magnitude and direction of the force with which the person at C must pull so that no one moves (the system is at equilibrium).

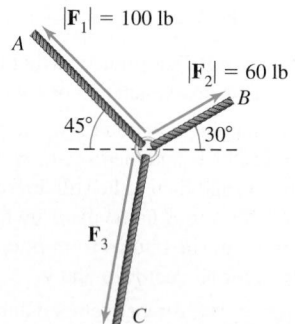

77–81. *Prove the following vector properties using components. Then make a sketch to illustrate the property geometrically. Suppose $\mathbf{u}$, $\mathbf{v}$, and $\mathbf{w}$ are vectors in the xy-plane and a and c are scalars.*

77. $\mathbf{u} + \mathbf{v} = \mathbf{v} + \mathbf{u}$ Commutative property

78. $(\mathbf{u} + \mathbf{v}) + \mathbf{w} = \mathbf{u} + (\mathbf{v} + \mathbf{w})$ Associative property

79. $a(c\mathbf{v}) = (ac)\mathbf{v}$ Associative property

80. $a(\mathbf{u} + \mathbf{v}) = a\mathbf{u} + a\mathbf{v}$ Distributive property 1

81. $(a + c)\mathbf{v} = a\mathbf{v} + c\mathbf{v}$ Distributive property 2

82. Midpoint of a line segment Use vectors to show that the midpoint of the line segment joining $P(x_1, y_1)$ and $Q(x_2, y_2)$ is the point $\left(\dfrac{x_1 + x_2}{2}, \dfrac{y_1 + y_2}{2} \right)$. (*Hint:* Let O be the origin and let M be the midpoint of PQ. Draw a picture and show that $\overrightarrow{OM} = \overrightarrow{OP} + \dfrac{1}{2}\overrightarrow{PQ} = \overrightarrow{OP} + \dfrac{1}{2}(\overrightarrow{OQ} - \overrightarrow{OP})$.)

83. Magnitude of scalar multiple Prove that $|c\mathbf{v}| = |c||\mathbf{v}|$, where c is a scalar and $\mathbf{v}$ is a vector.

84. Equality of vectors Assume $\overrightarrow{PQ}$ equals $\overrightarrow{RS}$. Does it follow that $\overrightarrow{PR}$ is equal to $\overrightarrow{QS}$? Prove your conclusion.

85. Linear independence A pair of nonzero vectors in the plane is *linearly dependent* if one vector is a scalar multiple of the other. Otherwise, the pair is *linearly independent*.

    a. Which pairs of the following vectors are linearly dependent and which are linearly independent: $\mathbf{u} = \langle 2, -3 \rangle$, $\mathbf{v} = \langle -12, 18 \rangle$, and $\mathbf{w} = \langle 4, 6 \rangle$?

    b. Geometrically, what does it mean for a pair of nonzero vectors in the plane to be linearly dependent? Linearly independent?

    c. Prove that if a pair of vectors $\mathbf{u}$ and $\mathbf{v}$ is linearly independent, then given any vector $\mathbf{w}$, there are constants c_1 and c_2 such that $\mathbf{w} = c_1\mathbf{u} + c_2\mathbf{v}$.

86. Perpendicular vectors Show that two nonzero vectors $\mathbf{u} = \langle u_1, u_2 \rangle$ and $\mathbf{v} = \langle v_1, v_2 \rangle$ are perpendicular to each other if $u_1v_1 + u_2v_2 = 0$.

87. Parallel and perpendicular vectors Let $\mathbf{u} = \langle a, 5 \rangle$ and $\mathbf{v} = \langle 2, 6 \rangle$.

    a. Find the value of a such that $\mathbf{u}$ is parallel to $\mathbf{v}$.

    b. Find the value of a such that $\mathbf{u}$ is perpendicular to $\mathbf{v}$.

88. The Triangle Inequality Suppose $\mathbf{u}$ and $\mathbf{v}$ are vectors in the plane.

    a. Use the Triangle Rule for adding vectors to explain why $|\mathbf{u} + \mathbf{v}| \leq |\mathbf{u}| + |\mathbf{v}|$. This result is known as the *Triangle Inequality*.

    b. Under what conditions is $|\mathbf{u} + \mathbf{v}| = |\mathbf{u}| + |\mathbf{v}|$?

QUICK CHECK ANSWERS

1. The vector $-5\mathbf{v}$ is five times as long as $\mathbf{v}$ and points in the opposite direction.   **2.** $\mathbf{v}_a + \mathbf{w}$ points in a northeasterly direction.   **3.** Constructing $\mathbf{u} + \mathbf{v}$ and $\mathbf{v} + \mathbf{u}$ using the Triangle Rule produces vectors having the same length and direction.   **4.** $\overrightarrow{PQ} = \langle -6, -2 \rangle$ **5.** $10\mathbf{u} = \langle 6, 8 \rangle$ and $-10\mathbf{u} = \langle -6, -8 \rangle$

6. $\left| \left\langle \dfrac{5}{13}, \dfrac{12}{13} \right\rangle \right| = \sqrt{\dfrac{25 + 144}{169}} = \sqrt{\dfrac{169}{169}} = 1$

7. $\mathbf{u} = -\dfrac{4}{3}\mathbf{v} + 4\mathbf{w}$ ◄

13.2 Vectors in Three Dimensions

Up to this point, our study of calculus has been limited to functions, curves, and vectors that can be plotted in the two-dimensional xy-plane. However, a two-dimensional coordinate system is insufficient for modeling many physical phenomena. For example, to describe the trajectory of a jet gaining altitude, we need two coordinates, say x and y, to measure east–west and north–south distances. In addition, another coordinate, say z, is needed to measure the altitude of the jet. By adding a third coordinate and creating an ordered triple (x, y, z), the location of the jet can be described. The set of all points described by the triples (x, y, z) is called *three-dimensional space*, *xyz-space*, or $\mathbb{R}^3$. Many of the properties of *xyz*-space are extensions of familiar ideas you have seen in the xy-plane.

The *xyz*-Coordinate System

> ➤ The notation $\mathbb{R}^3$ (pronounced *R-three*) stands for the set of all ordered triples of real numbers.

A three-dimensional coordinate system is created by adding a new axis, called the **z-axis**, to the familiar xy-coordinate system. The new z-axis is inserted through the origin perpendicular to the x- and y-axes (Figure 13.25). The result is a new coordinate system called the **three-dimensional rectangular coordinate system** or the **xyz-coordinate system**.

We use a conventional **right-handed coordinate system**: If the curled fingers of the right hand are rotated from the positive x-axis to the positive y-axis, the thumb points in the direction of the positive z-axis (Figure 13.25).

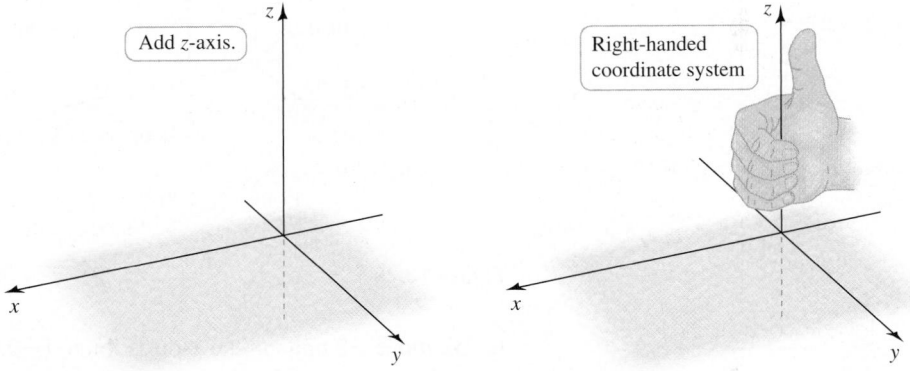

Figure 13.25

The coordinate plane containing the x-axis and y-axis is still called the xy-plane. We now have two new coordinate planes: the **xz-plane** containing the x-axis and the z-axis, and the **yz-plane** containing the y-axis and the z-axis. Taken together, these three coordinate planes divide *xyz*-space into eight regions called **octants** (Figure 13.26).

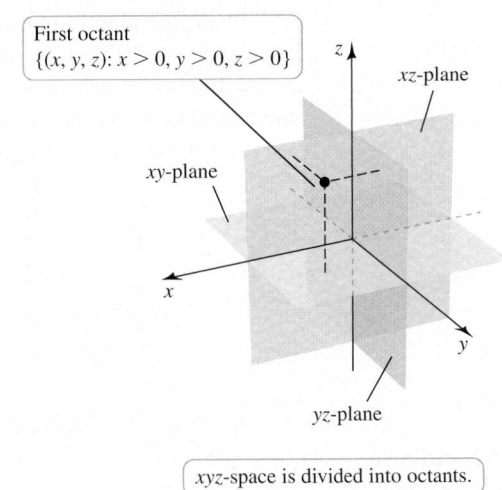

Figure 13.26

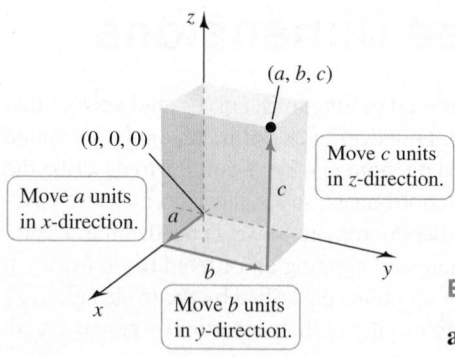

Figure 13.27

The point where all three axes intersect is the **origin**, which has coordinates $(0, 0, 0)$. An ordered triple (a, b, c) refers to the point in xyz-space that is found by starting at the origin, moving a units in the x-direction, b units in the y-direction, and c units in the z-direction. With a negative coordinate, you move in the negative direction along the corresponding coordinate axis. To visualize this point, it's helpful to construct a rectangular box with one vertex at the origin and the opposite vertex at the point (a, b, c) (**Figure 13.27**).

EXAMPLE 1   **Plotting points in xyz-space** Plot the following points.

a. $(3, 4, 5)$ **b.** $(-2, -3, 5)$

SOLUTION

a. Starting at $(0, 0, 0)$, we move 3 units in the x-direction to the point $(3, 0, 0)$, then 4 units in the y-direction to the point $(3, 4, 0)$, and finally 5 units in the z-direction to reach the point $(3, 4, 5)$ (**Figure 13.28**).

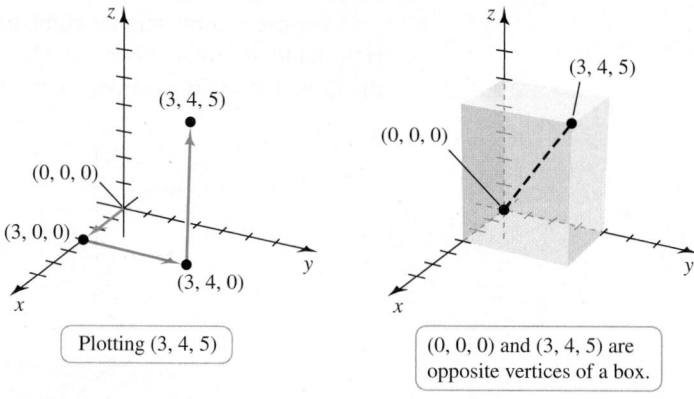

Figure 13.28

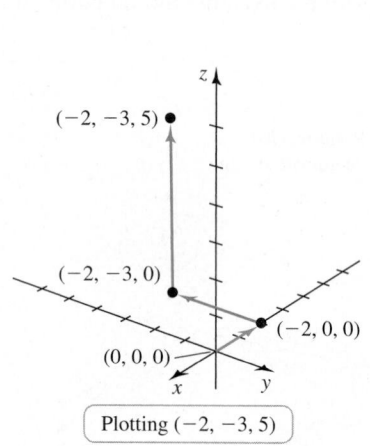

Figure 13.29

QUICK CHECK 1 Suppose the positive x-, y-, and z-axes point east, north, and upward, respectively. Describe the location of the points $(-1, -1, 0)$, $(1, 0, 1)$, and $(-1, -1, -1)$ relative to the origin. ◄

▶ Planes that are not parallel to the coordinate planes are discussed in Section 13.5.

b. We move -2 units in the x-direction to $(-2, 0, 0)$, -3 units in the y-direction to $(-2, -3, 0)$, and 5 units in the z-direction to reach $(-2, -3, 5)$ (**Figure 13.29**).

Related Exercises 13–14 ◄

Equations of Simple Planes

The xy-plane consists of all points in xyz-space that have a z-coordinate of 0. Therefore, the xy-plane is the set $\{(x, y, z): z = 0\}$; it is represented by the equation $z = 0$. Similarly, the xz-plane has the equation $y = 0$, and the yz-plane has the equation $x = 0$.

Planes parallel to one of the coordinate planes are easy to describe. For example, the equation $x = 2$ describes the set of all points whose x-coordinate is 2 and whose y- and z-coordinates are arbitrary; this plane is parallel to and 2 units from the yz-plane. Similarly, the equation $y = a$ describes a plane that is everywhere $|a|$ units from the xz-plane, and $z = a$ is the equation of a horizontal plane $|a|$ units from the xy-plane (**Figure 13.30**).

QUICK CHECK 2 To which coordinate planes are the planes $x = -2$ and $z = 16$ parallel? ◄

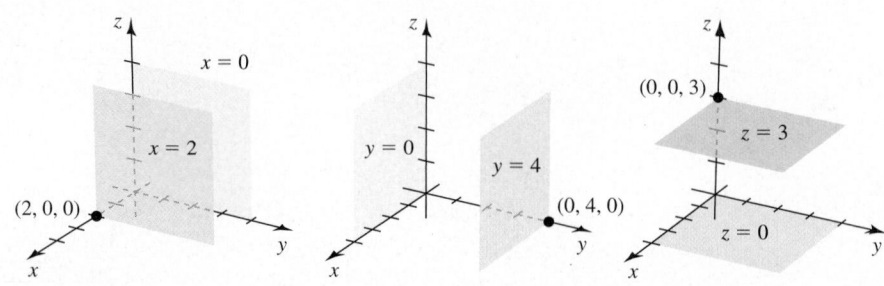

Figure 13.30

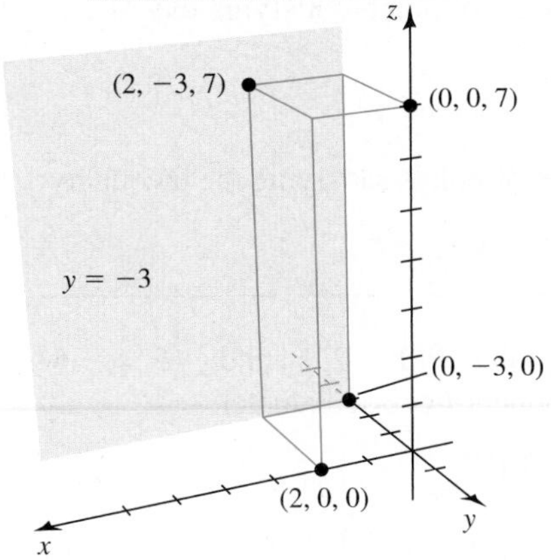

Plane is parallel to the xz-plane
and passes through $(2, -3, 7)$.

Figure 13.31

EXAMPLE 2 Parallel planes

Determine the equation of the plane parallel to the xz-plane passing through the point $(2, -3, 7)$.

SOLUTION Points on a plane parallel to the xz-plane have the same y-coordinate. Therefore, the plane passing through the point $(2, -3, 7)$ with a y-coordinate of -3 has the equation $y = -3$ (Figure 13.31).

Related Exercises 20–22 ◄

Distances in xyz-Space

Recall that the distance between two points (x_1, y_1) and (x_2, y_2) in the xy-plane is $\sqrt{(x_2 - x_1)^2 + (y_2 - y_1)^2}$. This distance formula is useful in deriving a similar formula for the distance between two points $P(x_1, y_1, z_1)$ and $Q(x_2, y_2, z_2)$ in xyz-space.

Figure 13.32 shows the points P and Q, together with the auxiliary point $R(x_2, y_2, z_1)$, which has the same z-coordinate as P and the same x- and y-coordinates as Q. The line segment PR has length $|PR| = \sqrt{(x_2 - x_1)^2 + (y_2 - y_1)^2}$ and is one leg of the right triangle $\triangle PRQ$. The length of the hypotenuse of that triangle is the distance between P and Q:

$$|PQ| = \sqrt{|PR|^2 + |RQ|^2} = \sqrt{\underbrace{(x_2 - x_1)^2 + (y_2 - y_1)^2}_{|PR|^2} + \underbrace{(z_2 - z_1)^2}_{|RQ|^2}}.$$

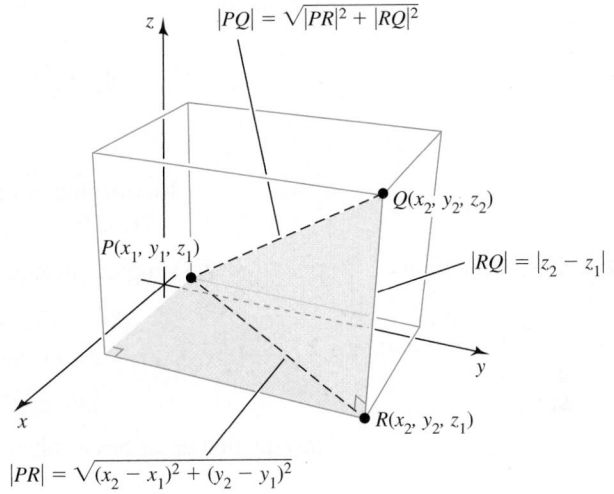

Figure 13.32

$$\text{Midpoint} = \left(\frac{x_1 + x_2}{2}, \frac{y_1 + y_2}{2}, \frac{z_1 + z_2}{2} \right)$$

Figure 13.33

Distance Formula in xyz-Space

The distance between the points $P(x_1, y_1, z_1)$ and $Q(x_2, y_2, z_2)$ is

$$\sqrt{(x_2 - x_1)^2 + (y_2 - y_1)^2 + (z_2 - z_1)^2}.$$

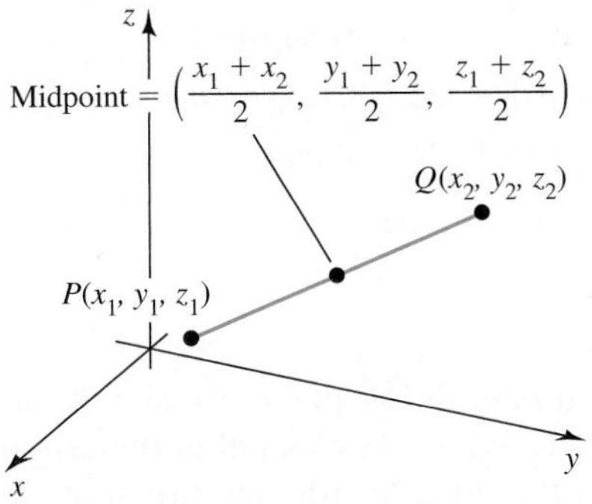

By using the distance formula, we can derive the formula (Exercise 81) for the **midpoint** of the line segment joining $P(x_1, y_1, z_1)$ and $Q(x_2, y_2, z_2)$, which is found by averaging the x-, y-, and z-coordinates (Figure 13.33):

$$\text{Midpoint} = \left(\frac{x_1 + x_2}{2}, \frac{y_1 + y_2}{2}, \frac{z_1 + z_2}{2} \right).$$

Equation of a Sphere

A *sphere* is the set of all points that are a constant distance r from a point (a, b, c); r is the *radius* of the sphere, and (a, b, c) is the *center* of the sphere. A *ball* centered at (a, b, c) with radius r consists of all the points inside and on the sphere centered at (a, b, c) with radius r (Figure 13.34). We now use the distance formula to translate these statements.

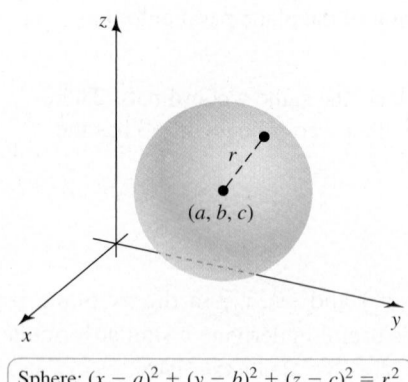

Sphere: $(x - a)^2 + (y - b)^2 + (z - c)^2 = r^2$

Ball: $(x - a)^2 + (y - b)^2 + (z - c)^2 \leq r^2$

Figure 13.34

▶ Just as a circle is the boundary of a disk in two dimensions, a *sphere* is the boundary of a *ball* in three dimensions. We have defined a *closed ball*, which includes its boundary. An *open ball* does not contain its boundary.

QUICK CHECK 3 Describe the solution set of the equation

$(x - 1)^2 + y^2 + (z + 1)^2 + 4 = 0.$ ◀

> **DEFINITION Spheres and Balls**
>
> A **sphere** centered at (a, b, c) with radius r is the set of points satisfying the equation
>
> $$(x - a)^2 + (y - b)^2 + (z - c)^2 = r^2.$$
>
> A **ball** centered at (a, b, c) with radius r is the set of points satisfying the inequality
>
> $$(x - a)^2 + (y - b)^2 + (z - c)^2 \leq r^2.$$

EXAMPLE 3 Equation of a sphere Consider the points $P(1, -2, 5)$ and $Q(3, 4, -6)$. Find an equation of the sphere for which the line segment PQ is a diameter.

SOLUTION The center of the sphere is the midpoint of PQ:

$$\left(\frac{1 + 3}{2}, \frac{-2 + 4}{2}, \frac{5 - 6}{2} \right) = \left(2, 1, -\frac{1}{2} \right).$$

The diameter of the sphere is the distance $|PQ|$, which is

$$\sqrt{(3 - 1)^2 + (4 + 2)^2 + (-6 - 5)^2} = \sqrt{161}.$$

Therefore, the sphere's radius is $\frac{1}{2} \sqrt{161}$, its center is $\left(2, 1, -\frac{1}{2} \right)$, and it is described by the equation

$$(x - 2)^2 + (y - 1)^2 + \left(z + \frac{1}{2} \right)^2 = \left(\frac{1}{2} \sqrt{161} \right)^2 = \frac{161}{4}.$$

Related Exercises 27–28 ◀

EXAMPLE 4 Identifying equations Describe the set of points that satisfy the equation $x^2 + y^2 + z^2 - 2x + 6y - 8z = -1$.

SOLUTION We simplify the equation by completing the square and factoring:

$$(x^2 - 2x) + (y^2 + 6y) + (z^2 - 8z) = -1 \quad \text{Group terms.}$$
$$(x^2 - 2x + 1) + (y^2 + 6y + 9) + (z^2 - 8z + 16) = 25 \quad \text{Complete the square.}$$
$$(x - 1)^2 + (y + 3)^2 + (z - 4)^2 = 25. \quad \text{Factor.}$$

The equation describes a sphere of radius 5 with center $(1, -3, 4)$.

Related Exercises 31–32 ◀

Vectors in $\mathbb{R}^3$

Vectors in $\mathbb{R}^3$ are straightforward extensions of vectors in the xy-plane; we simply include a third component. The position vector $\mathbf{v} = \langle v_1, v_2, v_3 \rangle$ has its tail at the origin and its head at the point (v_1, v_2, v_3). Vectors having the same length and direction are equal. Therefore, the vector from $P(x_1, y_1, z_1)$ to $Q(x_2, y_2, z_2)$ is denoted $\overrightarrow{PQ}$ and is equal to the position vector $\langle x_2 - x_1, y_2 - y_1, z_2 - z_1 \rangle$. It is also equal to all vectors such as $\overrightarrow{RS}$ (Figure 13.35) that have the same length and direction as $\mathbf{v}$.

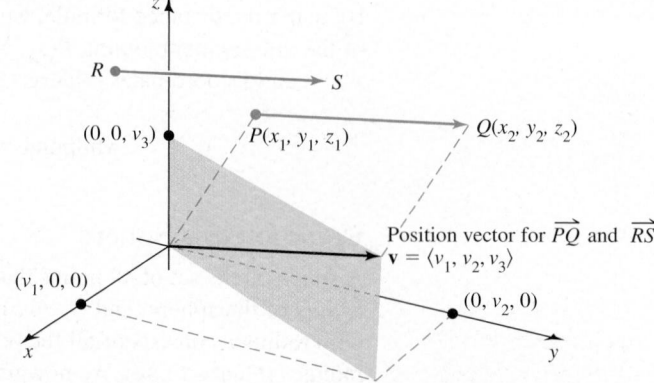

Figure 13.35

QUICK CHECK 4 Which of the following vectors are parallel to each other?

a. $\mathbf{u} = \langle -2, 4, -6 \rangle$

b. $\mathbf{v} = \langle 4, -8, 12 \rangle$

c. $\mathbf{w} = \langle -1, 2, 3 \rangle$ ◄

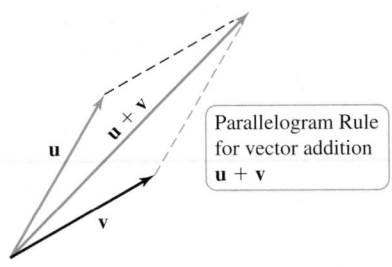

Parallelogram Rule for vector addition $\mathbf{u} + \mathbf{v}$

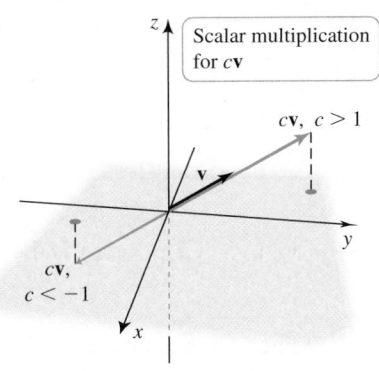

Scalar multiplication for $c\mathbf{v}$

Figure 13.36

The operations of vector addition and scalar multiplication in $\mathbb{R}^2$ generalize in a natural way to three dimensions. For example, the sum of two vectors is found geometrically using the Triangle Rule or the Parallelogram Rule (Section 13.1). The sum is found analytically by adding the respective components of the two vectors. As with two-dimensional vectors, scalar multiplication corresponds to stretching or compressing a vector, possibly with a reversal of direction. Two nonzero vectors are parallel if one is a scalar multiple of the other (Figure 13.36).

DEFINITION Vector Operations in $\mathbb{R}^3$

Let c be a scalar, $\mathbf{u} = \langle u_1, u_2, u_3 \rangle$, and $\mathbf{v} = \langle v_1, v_2, v_3 \rangle$.

$$\mathbf{u} + \mathbf{v} = \langle u_1 + v_1, u_2 + v_2, u_3 + v_3 \rangle \quad \text{Vector addition}$$

$$\mathbf{u} - \mathbf{v} = \langle u_1 - v_1, u_2 - v_2, u_3 - v_3 \rangle \quad \text{Vector subtraction}$$

$$c\mathbf{u} = \langle cu_1, cu_2, cu_3 \rangle \quad \text{Scalar multiplication}$$

EXAMPLE 5   Vectors in $\mathbb{R}^3$ Let $\mathbf{u} = \langle 2, -4, 1 \rangle$ and $\mathbf{v} = \langle 3, 0, -1 \rangle$. Find the components of the following vectors and draw them in $\mathbb{R}^3$.

a. $\dfrac{1}{2}\mathbf{u}$ **b.** $-2\mathbf{v}$ **c.** $\mathbf{u} + 2\mathbf{v}$

SOLUTION

a. Using the definition of scalar multiplication, $\dfrac{1}{2}\mathbf{u} = \dfrac{1}{2}\langle 2, -4, 1 \rangle = \left\langle 1, -2, \dfrac{1}{2} \right\rangle$. The vector $\dfrac{1}{2}\mathbf{u}$ has the same direction as $\mathbf{u}$ with half the length of $\mathbf{u}$ (Figure 13.37).

b. Using scalar multiplication, $-2\mathbf{v} = -2\langle 3, 0, -1 \rangle = \langle -6, 0, 2 \rangle$. The vector $-2\mathbf{v}$ has the direction opposite that of $\mathbf{v}$ and twice the length of $\mathbf{v}$ (Figure 13.38).

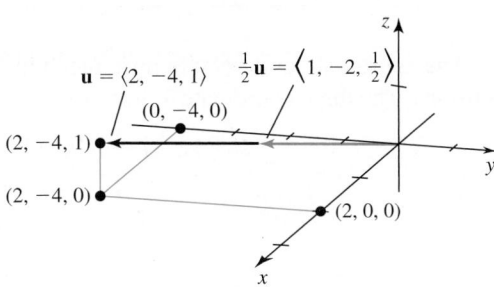

Figure 13.37

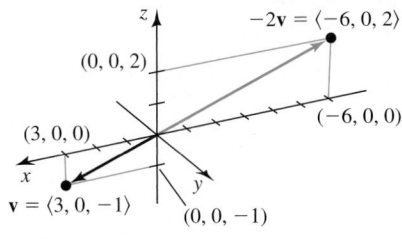

Figure 13.38

c. Using vector addition and scalar multiplication,

$$\mathbf{u} + 2\mathbf{v} = \langle 2, -4, 1 \rangle + 2\langle 3, 0, -1 \rangle = \langle 8, -4, -1 \rangle.$$

The vector $\mathbf{u} + 2\mathbf{v}$ is drawn by applying the Parallelogram Rule to $\mathbf{u}$ and $2\mathbf{v}$ (Figure 13.39).

$\mathbf{u} + 2\mathbf{v}$ by the Parallelogram Rule

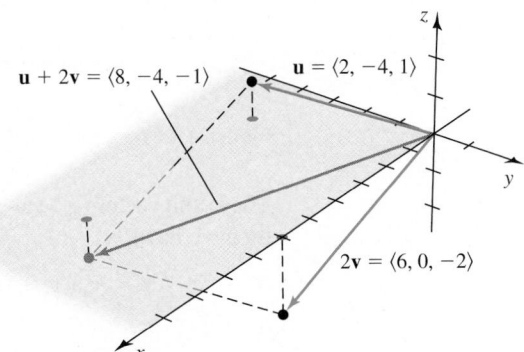

Figure 13.39

Related Exercise 39 ◄

Magnitude and Unit Vectors

The magnitude of the vector $\overrightarrow{PQ}$ from $P(x_1, y_1, z_1)$ to $Q(x_2, y_2, z_2)$ is denoted $|\overrightarrow{PQ}|$; it is the distance between P and Q and is given by the distance formula (Figure 13.40).

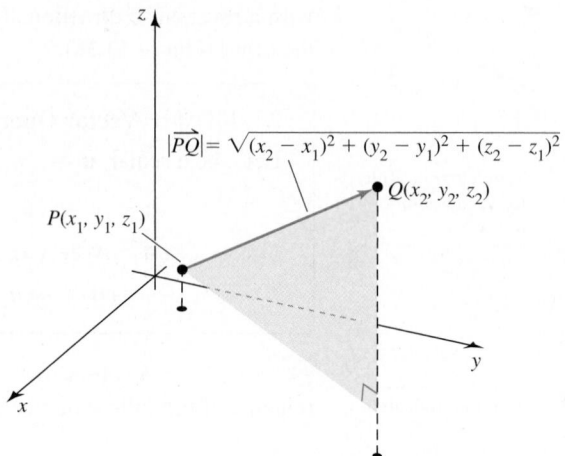

Figure 13.40

DEFINITION Magnitude of a Vector

The **magnitude** (or **length**) of the vector $\overrightarrow{PQ} = \langle x_2 - x_1, y_2 - y_1, z_2 - z_1 \rangle$ is the distance from $P(x_1, y_1, z_1)$ to $Q(x_2, y_2, z_2)$:

$$|\overrightarrow{PQ}| = \sqrt{(x_2 - x_1)^2 + (y_2 - y_1)^2 + (z_2 - z_1)^2}.$$

The coordinate unit vectors introduced in Section 13.1 extend naturally to three dimensions. The three coordinate unit vectors in $\mathbb{R}^3$ (Figure 13.41) are

$$\mathbf{i} = \langle 1, 0, 0 \rangle, \quad \mathbf{j} = \langle 0, 1, 0 \rangle, \quad \text{and} \quad \mathbf{k} = \langle 0, 0, 1 \rangle.$$

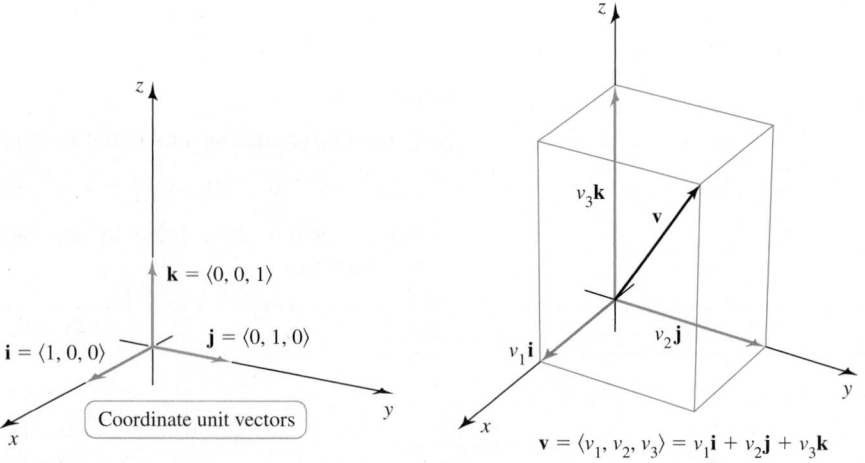

Figure 13.41

These unit vectors give an alternative way of expressing position vectors. If $\mathbf{v} = \langle v_1, v_2, v_3 \rangle$, then we have

$$\mathbf{v} = v_1 \langle 1, 0, 0 \rangle + v_2 \langle 0, 1, 0 \rangle + v_3 \langle 0, 0, 1 \rangle = v_1 \mathbf{i} + v_2 \mathbf{j} + v_3 \mathbf{k}.$$

EXAMPLE 6   **Magnitudes and unit vectors** Consider the points $P(5, 3, 1)$ and $Q(-7, 8, 1)$.

a. Express $\overrightarrow{PQ}$ in terms of the unit vectors **i**, **j**, and **k**.

b. Find the magnitude of $\overrightarrow{PQ}$.

c. Find the position vector of magnitude 10 in the direction of $\overrightarrow{PQ}$.

SOLUTION

a. $\overrightarrow{PQ}$ is equal to the position vector $\langle -7 - 5, 8 - 3, 1 - 1 \rangle = \langle -12, 5, 0 \rangle$. Therefore, $\overrightarrow{PQ} = -12\mathbf{i} + 5\mathbf{j}$.

b. $|\overrightarrow{PQ}| = |-12\mathbf{i} + 5\mathbf{j}| = \sqrt{12^2 + 5^2} = \sqrt{169} = 13$

QUICK CHECK 5 Which vector has the smaller magnitude: $\mathbf{u} = 3\mathbf{i} - \mathbf{j} - \mathbf{k}$ or $\mathbf{v} = 2(\mathbf{i} + \mathbf{j} + \mathbf{k})$? ◄

c. The unit vector in the direction of $\overrightarrow{PQ}$ is $\mathbf{u} = \dfrac{\overrightarrow{PQ}}{|\overrightarrow{PQ}|} = \dfrac{1}{13}\langle -12, 5, 0 \rangle$. Therefore, the vector in the direction of **u** with a magnitude of 10 is $10\mathbf{u} = \dfrac{10}{13}\langle -12, 5, 0 \rangle$.

Related Exercises 45, 68 ◄

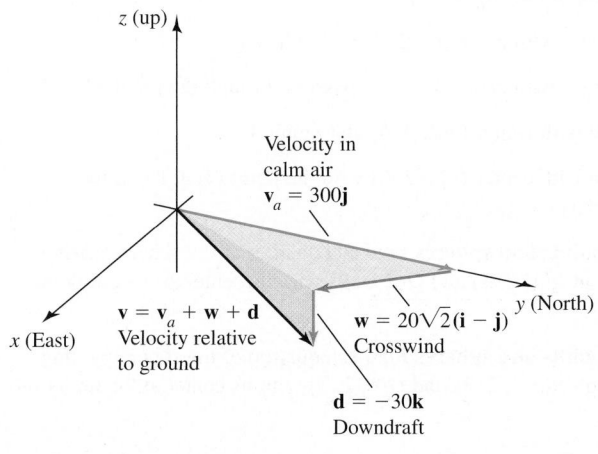

Figure 13.42

EXAMPLE 7   **Flight in crosswinds** A plane is flying horizontally due north in calm air at 300 mi/hr when it encounters a horizontal crosswind blowing southeast at 40 mi/hr and a downdraft blowing vertically downward at 30 mi/hr. What are the resulting speed and direction of the plane relative to the ground?

SOLUTION Let the unit vectors **i**, **j**, and **k** point east, north, and upward, respectively (**Figure 13.42**). The velocity of the plane relative to the air (300 mi/hr due north) is $\mathbf{v}_a = 300\mathbf{j}$. The crosswind blows 45° south of east, so its component to the east is $40 \cos 45° = 20\sqrt{2}$ (in the **i**-direction) and its component to the south is $40 \cos 45° = 20\sqrt{2}$ (in the negative **j**-direction). Therefore, the crosswind may be expressed as $\mathbf{w} = 20\sqrt{2}\mathbf{i} - 20\sqrt{2}\mathbf{j}$. Finally, the downdraft in the negative **k**-direction is $\mathbf{d} = -30\mathbf{k}$. The velocity of the plane relative to the ground is the sum of $\mathbf{v}_a$, **w**, and **d**:

$$\begin{aligned}\mathbf{v} &= \mathbf{v}_a + \mathbf{w} + \mathbf{d} \\ &= 300\mathbf{j} + (20\sqrt{2}\mathbf{i} - 20\sqrt{2}\mathbf{j}) - 30\mathbf{k} \\ &= 20\sqrt{2}\mathbf{i} + (300 - 20\sqrt{2})\mathbf{j} - 30\mathbf{k}.\end{aligned}$$

Figure 13.42 shows the velocity vector of the plane. A quick calculation shows that the speed is $|\mathbf{v}| \approx 275$ mi/hr. The direction of the plane is slightly east of north and downward. In the next section, we present methods for precisely determining the direction of a vector.

Related Exercises 51–52 ◄

SECTION 13.2 EXERCISES

Getting Started

1. Explain how to plot the point $(3, -2, 1)$ in $\mathbb{R}^3$.

2. What is the y-coordinate of all points in the xz-plane?

3. Describe the plane $x = 4$.

4. What position vector is equal to the vector from $(3, 5, -2)$ to $(0, -6, 3)$?

5. Let $\mathbf{u} = \langle 3, 5, -7 \rangle$ and $\mathbf{v} = \langle 6, -5, 1 \rangle$. Evaluate $\mathbf{u} + \mathbf{v}$ and $3\mathbf{u} - \mathbf{v}$.

6. What is the magnitude of a vector joining two points $P(x_1, y_1, z_1)$ and $Q(x_2, y_2, z_2)$?

7. Which point is farther from the origin, $(3, -1, 2)$ or $(0, 0, -4)$?

8. Express the vector from $P(-1, -4, 6)$ to $Q(1, 3, -6)$ as a position vector in terms of **i**, **j**, and **k**.

Practice Exercises

9–12. Points in $\mathbb{R}^3$ *Find the coordinates of the vertices A, B, and C of the following rectangular boxes.*

9.

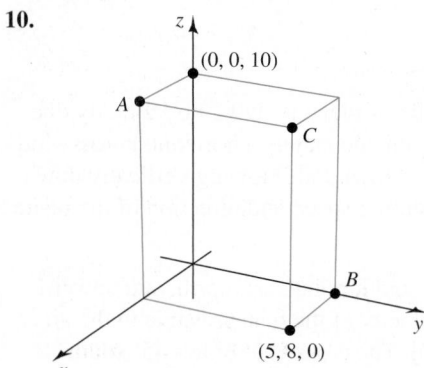

10.

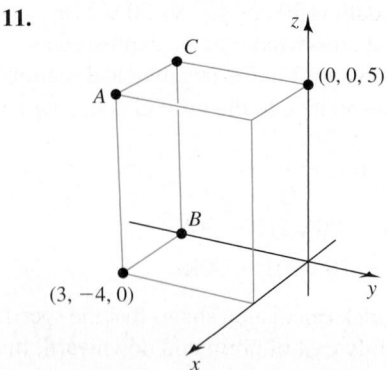

11.

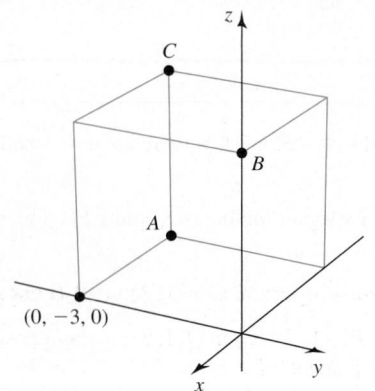

12. Assume all the edges have the same length.

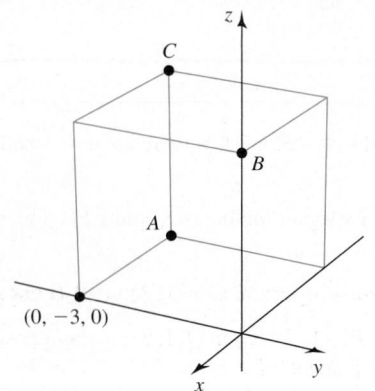

13–14. Plotting points in $\mathbb{R}^3$ *For each point $P(x, y, z)$ given below, let $A(x, y, 0)$, $B(x, 0, z)$, and $C(0, y, z)$ be points in the xy-, xz-, and yz-planes, respectively. Plot and label the points A, B, C, and P in $\mathbb{R}^3$.*

13. a. $P(2, 2, 4)$　　**b.** $P(1, 2, 5)$　　**c.** $P(-2, 0, 5)$

14. a. $P(-3, 2, 4)$　　**b.** $P(4, -2, -3)$　　**c.** $P(-2, -4, -3)$

15–20. Sketching planes *Sketch the following planes in the window* $[0, 5] \times [0, 5] \times [0, 5]$.

15. $x = 2$　　**16.** $z = 3$　　**17.** $y = 2$　　**18.** $z = y$

19. The plane that passes through $(2, 0, 0)$, $(0, 3, 0)$, and $(0, 0, 4)$

20. The plane parallel to the xz-plane containing the point $(1, 2, 3)$

21. Planes Sketch the plane parallel to the xy-plane through $(2, 4, 2)$ and find its equation.

22. Planes Sketch the plane parallel to the yz-plane through $(2, 4, 2)$ and find its equation.

23–26. Spheres and balls *Find an equation or inequality that describes the following objects.*

23. A sphere with center $(1, 2, 3)$ and radius 4

24. A sphere with center $(1, 2, 0)$ passing through the point $(3, 4, 5)$

25. A ball with center $(-2, 0, 4)$ and radius 1

26. A ball with center $(0, -2, 6)$ with the point $(1, 4, 8)$ on its boundary

27. Midpoints and spheres Find an equation of the sphere passing through $P(1, 0, 5)$ and $Q(2, 3, 9)$ with its center at the midpoint of PQ.

28. Midpoints and spheres Find an equation of the sphere passing through $P(-4, 2, 3)$ and $Q(0, 2, 7)$ with its center at the midpoint of PQ.

29–38. Identifying sets *Give a geometric description of the following sets of points.*

29. $(x - 1)^2 + y^2 + z^2 - 9 = 0$

30. $(x + 1)^2 + y^2 + z^2 - 2y - 24 = 0$

31. $x^2 + y^2 + z^2 - 2y - 4z - 4 = 0$

32. $x^2 + y^2 + z^2 - 6x + 6y - 8z - 2 = 0$

33. $x^2 + y^2 - 14y + z^2 \geq -13$

34. $x^2 + y^2 - 14y + z^2 \leq -13$

35. $x^2 + y^2 + z^2 - 8x - 14y - 18z \leq 79$

36. $x^2 + y^2 + z^2 - 8x + 14y - 18z \geq 65$

37. $x^2 - 2x + y^2 + 6y + z^2 + 10 = 0$

38. $x^2 - 4x + y^2 + 6y + z^2 + 14 = 0$

39–44. Vector operations *For the given vectors $\mathbf{u}$ and $\mathbf{v}$, evaluate the following expressions.*

a. $3\mathbf{u} + 2\mathbf{v}$　　*b.* $4\mathbf{u} - \mathbf{v}$　　*c.* $|\mathbf{u} + 3\mathbf{v}|$

39. $\mathbf{u} = \langle 4, -3, 0 \rangle$, $\mathbf{v} = \langle 0, 1, 1 \rangle$

40. $\mathbf{u} = \langle -2, -3, 0 \rangle$, $\mathbf{v} = \langle 1, 2, 1 \rangle$

41. $\mathbf{u} = \langle -2, 1, -2 \rangle$, $\mathbf{v} = \langle 1, 1, 1 \rangle$

42. $\mathbf{u} = \langle -5, 0, 2 \rangle, \mathbf{v} = \langle 3, 1, 1 \rangle$

43. $\mathbf{u} = \langle -7, 11, 8 \rangle, \mathbf{v} = \langle 3, -5, -1 \rangle$

44. $\mathbf{u} = \langle -4, -8\sqrt{3}, 2\sqrt{2} \rangle, \mathbf{v} = \langle 2, 3\sqrt{3}, -\sqrt{2} \rangle$

45–50. Unit vectors and magnitude *Consider the following points P and Q.*

a. *Find $\overrightarrow{PQ}$ and state your answer in two forms: $\langle a, b, c \rangle$ and $a\mathbf{i} + b\mathbf{j} + c\mathbf{k}$.*

b. *Find the magnitude of $\overrightarrow{PQ}$.*

c. *Find two unit vectors parallel to $\overrightarrow{PQ}$.*

45. $P(1, 5, 0), Q(3, 11, 2)$ 46. $P(5, 11, 12), Q(1, 14, 13)$

47. $P(-3, 1, 0), Q(-3, -4, 1)$ 48. $P(3, 8, 12), Q(3, 9, 11)$

49. $P(0, 0, 2), Q(-2, 4, 0)$

50. $P(a, b, c), Q(1, 1, -1)$ (a, b, and c are real numbers)

51. **Flight in crosswinds** A model airplane is flying horizontally due north at 20 mi/hr when it encounters a horizontal crosswind blowing east at 20 mi/hr and a downdraft blowing vertically downward at 10 mi/hr.

    a. Find the position vector that represents the velocity of the plane relative to the ground.

    b. Find the speed of the plane relative to the ground.

52. **Another crosswind flight** A model airplane is flying horizontally due east at 10 mi/hr when it encounters a horizontal crosswind blowing south at 5 mi/hr and an updraft blowing vertically upward at 5 mi/hr.

    a. Find the position vector that represents the velocity of the plane relative to the ground.

    b. Find the speed of the plane relative to the ground.

53. **Crosswinds** A small plane is flying horizontally due east in calm air at 250 mi/hr when it encounters a horizontal crosswind blowing southwest at 50 mi/hr and a 30-mi/hr updraft. Find the resulting speed of the plane, and describe with a sketch the approximate direction of the velocity relative to the ground.

54. **Combined force** An object at the origin is acted on by the forces $\mathbf{F}_1 = 20\mathbf{i} - 10\mathbf{j}$, $\mathbf{F}_2 = 30\mathbf{j} + 10\mathbf{k}$, and $\mathbf{F}_3 = 40\mathbf{j} + 20\mathbf{k}$. Find the magnitude of the combined force, and describe the approximate direction of the force.

55. **Submarine course** A submarine climbs at an angle of 30° above the horizontal with a heading to the northeast. If its speed is 20 knots, find the components of the velocity in the east, north, and vertical directions.

56. **Maintaining equilibrium** An object is acted on by the forces $\mathbf{F}_1 = \langle 10, 6, 3 \rangle$ and $\mathbf{F}_2 = \langle 0, 4, 9 \rangle$. Find the force $\mathbf{F}_3$ that must act on the object so that the sum of the forces is zero.

57. **Explain why or why not** Determine whether the following statements are true and give an explanation or counterexample.

    a. Suppose $\mathbf{u}$ and $\mathbf{v}$ are distinct vectors that both make a 45° angle with $\mathbf{w}$ in $\mathbb{R}^3$. Then $\mathbf{u} + \mathbf{v}$ makes a 45° angle with $\mathbf{w}$.

    b. Suppose $\mathbf{u}$ and $\mathbf{v}$ are distinct vectors that both make a 90° angle with $\mathbf{w}$ in $\mathbb{R}^3$. Then $\mathbf{u} + \mathbf{v}$ can never make a 90° angle with $\mathbf{w}$.

    c. $\mathbf{i} + \mathbf{j} + \mathbf{k} = \mathbf{0}$

    d. The intersection of the planes $x = 1$, $y = 1$, and $z = 1$ is a point.

58–60. Sets of points *Describe with a sketch the sets of points (x, y, z) satisfying the following equations.*

58. $(x + 1)(y - 3) = 0$

59. $x^2 y^2 z^2 > 0$

60. $y - z = 0$

61–64. Sets of points

61. Give a geometric description of the set of points (x, y, z) satisfying the pair of equations $z = 0$ and $x^2 + y^2 = 1$. Sketch a figure of this set of points.

62. Give a geometric description of the set of points (x, y, z) satisfying the pair of equations $z = x^2$ and $y = 0$. Sketch a figure of this set of points.

63. Give a geometric description of the set of points (x, y, z) that lie on the intersection of the sphere $x^2 + y^2 + z^2 = 5$ and the plane $z = 1$.

64. Give a geometric description of the set of points (x, y, z) that lie on the intersection of the sphere $x^2 + y^2 + z^2 = 36$ and the plane $z = 6$.

65. **Describing a circle** Find a pair of equations describing a circle of radius 3 centered at $(2, 4, 1)$ that lies in a plane parallel to the xz-plane.

66. **Describing a line** Find a pair of equations describing a line passing through the point $(-2, -5, 1)$ that is parallel to the x-axis.

67. Write the vector $\mathbf{v} = \langle 2, -4, 4 \rangle$ as a product of its magnitude and a unit vector with the same direction as $\mathbf{v}$.

68. Find the vector of length 10 with the same direction as $\mathbf{w} = \langle 2, \sqrt{2}, \sqrt{3} \rangle$.

69. Find a vector of length 5 in the direction opposite that of $\langle 3, -2, \sqrt{3} \rangle$.

70–73. Parallel vectors of varying lengths *Find vectors parallel to $\mathbf{v}$ of the given length.*

70. $\mathbf{v} = \langle 3, -2, 6 \rangle$; length = 10

71. $\mathbf{v} = \langle 6, -8, 0 \rangle$; length = 20

72. $\mathbf{v} = \overrightarrow{PQ}$ with $P(1, 0, 1)$ and $Q(2, -1, 1)$; length = 3

73. $\mathbf{v} = \overrightarrow{PQ}$ with $P(3, 4, 0)$ and $Q(2, 3, 1)$; length = 3

74. **Collinear points** Determine the values of x and y such that the points $(1, 2, 3)$, $(4, 7, 1)$, and $(x, y, 2)$ are collinear (lie on a line).

75. **Collinear points** Determine whether the points P, Q, and R are collinear (lie on a line) by comparing $\overrightarrow{PQ}$ and $\overrightarrow{PR}$. If the points are collinear, determine which point lies between the other two points.

    a. $P(1, 6, -5), Q(2, 5, -3), R(4, 3, 1)$

    b. $P(1, 5, 7), Q(5, 13, -1), R(0, 3, 9)$

    c. $P(1, 2, 3), Q(2, -3, 6), R(3, -1, 9)$

    d. $P(9, 5, 1), Q(11, 18, 4), R(6, 3, 0)$

76. **Lengths of the diagonals of a box** What is the longest diagonal of a rectangular 2 ft × 3 ft × 4 ft box?

Explorations and Challenges

77. Three-cable load A 500-lb load hangs from three cables of equal length that are anchored at the points $(-2, 0, 0)$, $(1, \sqrt{3}, 0)$, and $(1, -\sqrt{3}, 0)$. The load is located at $(0, 0, -2\sqrt{3})$. Find the vectors describing the forces on the cables due to the load.

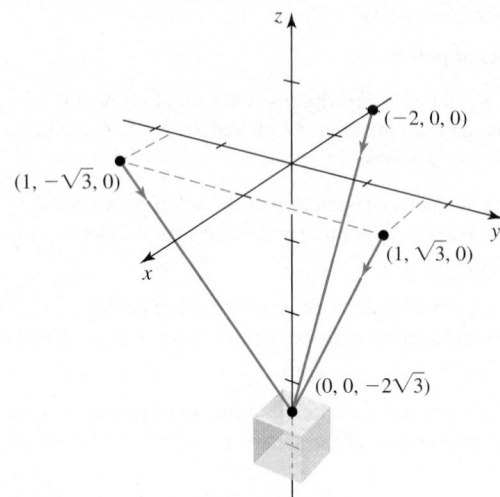

78. Four-cable load A 500-lb load hangs from four cables of equal length that are anchored at the points $(\pm 2, 0, 0)$ and $(0, \pm 2, 0)$. The load is located at $(0, 0, -4)$. Find the vectors describing the forces on the cables due to the load.

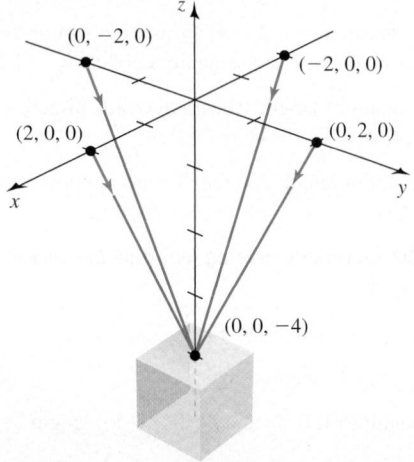

79. Possible parallelograms The points $O(0, 0, 0)$, $P(1, 4, 6)$, and $Q(2, 4, 3)$ lie at three vertices of a parallelogram. Find all possible locations of the fourth vertex.

80. Diagonals of parallelograms Two sides of a parallelogram are formed by the vectors $\mathbf{u}$ and $\mathbf{v}$. Prove that the diagonals of the parallelogram are $\mathbf{u} + \mathbf{v}$ and $\mathbf{u} - \mathbf{v}$.

81. Midpoint formula Prove that the midpoint of the line segment joining $P(x_1, y_1, z_1)$ and $Q(x_2, y_2, z_2)$ is

$$\left(\frac{x_1 + x_2}{2}, \frac{y_1 + y_2}{2}, \frac{z_1 + z_2}{2} \right).$$

82. Equation of a sphere For constants a, b, c, and d, show that the equation

$$x^2 + y^2 + z^2 - 2ax - 2by - 2cz = d$$

describes a sphere centered at (a, b, c) with radius r, where $r^2 = d + a^2 + b^2 + c^2$, provided $d + a^2 + b^2 + c^2 > 0$.

83. Medians of a triangle—without coordinates Assume $\mathbf{u}$, $\mathbf{v}$, and $\mathbf{w}$ are vectors in $\mathbb{R}^3$ that form the sides of a triangle (see figure). Use the following steps to prove that the medians intersect at a point that divides each median in a 2:1 ratio. The proof does not use a coordinate system.

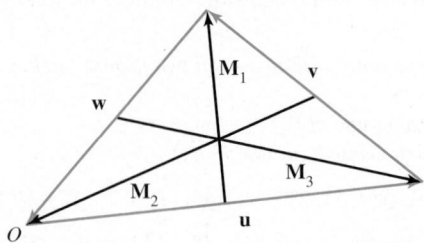

a. Show that $\mathbf{u} + \mathbf{v} + \mathbf{w} = \mathbf{0}$.
b. Let $\mathbf{M}_1$ be the median vector from the midpoint of $\mathbf{u}$ to the opposite vertex. Define $\mathbf{M}_2$ and $\mathbf{M}_3$ similarly. Using the geometry of vector addition, show that $\mathbf{M}_1 = \dfrac{\mathbf{u}}{2} + \mathbf{v}$. Find analogous expressions for $\mathbf{M}_2$ and $\mathbf{M}_3$.
c. Let $\mathbf{a}$, $\mathbf{b}$, and $\mathbf{c}$ be the vectors from O to the points one-third of the way along $\mathbf{M}_1$, $\mathbf{M}_2$, and $\mathbf{M}_3$, respectively. Show that

$$\mathbf{a} = \mathbf{b} = \mathbf{c} = \frac{\mathbf{u} - \mathbf{w}}{3}.$$

d. Conclude that the medians intersect at a point that divides each median in a 2:1 ratio.

84. Medians of a triangle—with coordinates In contrast to the proof in Exercise 83, we now use coordinates and position vectors to prove the same result. Without loss of generality, let $P(x_1, y_1, 0)$ and $Q(x_2, y_2, 0)$ be two points in the xy-plane, and let $R(x_3, y_3, z_3)$ be a third point such that P, Q, and R do not lie on a line. Consider $\triangle PQR$.

a. Let M_1 be the midpoint of the side PQ. Find the coordinates of M_1 and the components of the vector $\overrightarrow{RM_1}$.
b. Find the vector $\overrightarrow{OZ_1}$ from the origin to the point Z_1 two-thirds of the way along $\overrightarrow{RM_1}$.
c. Repeat the calculation of part (b) with the midpoint M_2 of RQ and the vector $\overrightarrow{PM_2}$ to obtain the vector $\overrightarrow{OZ_2}$.
d. Repeat the calculation of part (b) with the midpoint M_3 of PR and the vector $\overrightarrow{QM_3}$ to obtain the vector $\overrightarrow{OZ_3}$.
e. Conclude that the medians of $\triangle PQR$ intersect at a point. Give the coordinates of the point.
f. With $P(2, 4, 0)$, $Q(4, 1, 0)$, and $R(6, 3, 4)$, find the point at which the medians of $\triangle PQR$ intersect.

85. The amazing quadrilateral property—without coordinates The points P, Q, R, and S, joined by the vectors $\mathbf{u}$, $\mathbf{v}$, $\mathbf{w}$, and $\mathbf{x}$, are the vertices of a quadrilateral in $\mathbb{R}^3$. *The four points need not lie in a plane* (see figure). Use the following steps to prove that the line segments joining the midpoints of the sides of the quadrilateral form a parallelogram. The proof does not use a coordinate system.

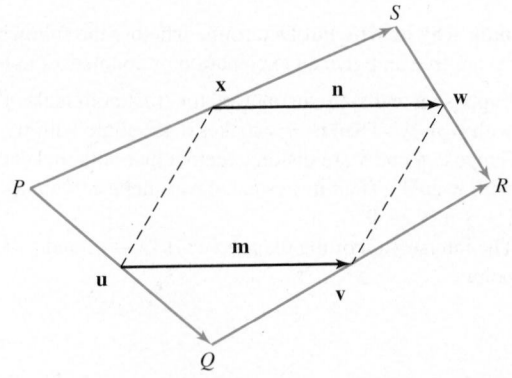

a. Use vector addition to show that $\mathbf{u} + \mathbf{v} = \mathbf{w} + \mathbf{x}$.
b. Let $\mathbf{m}$ be the vector that joins the midpoints of PQ and QR.
   Show that $\mathbf{m} = \dfrac{\mathbf{u} + \mathbf{v}}{2}$.
c. Let $\mathbf{n}$ be the vector that joins the midpoints of PS and SR.
   Show that $\mathbf{n} = \dfrac{\mathbf{x} + \mathbf{w}}{2}$.
d. Combine parts (a), (b), and (c) to conclude that $\mathbf{m} = \mathbf{n}$.
e. Explain why part (d) implies that the line segments joining the midpoints of the sides of the quadrilateral form a parallelogram.

86. **The amazing quadrilateral property—with coordinates** Prove the quadrilateral property in Exercise 85, assuming the coordinates of P, Q, R, and S are $P(x_1, y_1, 0)$, $Q(x_2, y_2, 0)$, $R(x_3, y_3, 0)$, and $S(x_4, y_4, z_4)$, where we assume P, Q, and R lie in the xy-plane without loss of generality.

QUICK CHECK ANSWERS

1. Southwest; due east and upward; southwest and downward 2. yz-plane; xy-plane 3. No solution 4. $\mathbf{u}$ and $\mathbf{v}$ are parallel. 5. $|\mathbf{u}| = \sqrt{11}$ and $|\mathbf{v}| = \sqrt{12} = 2\sqrt{3}$; $\mathbf{u}$ has the smaller magnitude. ◄

13.3 Dot Products

➤ The dot product is also called the *scalar product*, a term we do not use in order to avoid confusion with *scalar multiplication*.

The *dot product* is used to determine the angle between two vectors. It is also a tool for calculating *projections*—the measure of how much of a given vector lies in the direction of another vector.

To see the usefulness of the dot product, consider an example. Recall that the work done by a constant force F in moving an object a distance d is $W = Fd$ (Section 6.7). This rule is valid provided the force acts in the direction of motion (**Figure 13.43a**). Now assume the force is a vector $\mathbf{F}$ applied at an angle θ to the direction of motion; the resulting displacement of the object is a vector $\mathbf{d}$. In this case, the work done by the force is the component of the force in the direction of motion multiplied by the distance moved by the object, which is $W = (|\mathbf{F}| \cos \theta)|\mathbf{d}|$ (**Figure 13.43b**). We call this product of the magnitudes of two vectors and the cosine of the angle between them the dot product.

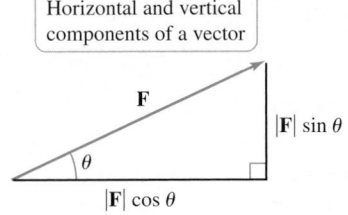

Horizontal and vertical components of a vector

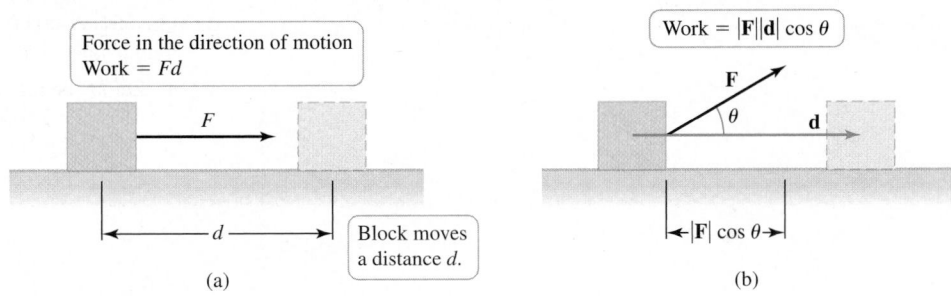

Force in the direction of motion
Work = Fd

Block moves a distance d.

(a)

Work = $|\mathbf{F}||\mathbf{d}| \cos \theta$

(b)

Figure 13.43

Two Forms of the Dot Product

The example of work done by a force leads to our first definition of the dot product. We then give an equivalent formula that is often better suited for computation.

DEFINITION Dot Product

Given two nonzero vectors $\mathbf{u}$ and $\mathbf{v}$ in two or three dimensions, their **dot product** is

$$\mathbf{u} \cdot \mathbf{v} = |\mathbf{u}||\mathbf{v}| \cos \theta,$$

where θ is the angle between $\mathbf{u}$ and $\mathbf{v}$ with $0 \le \theta \le \pi$ (**Figure 13.44**). If $\mathbf{u} = \mathbf{0}$ or $\mathbf{v} = \mathbf{0}$, then $\mathbf{u} \cdot \mathbf{v} = 0$, and θ is undefined.

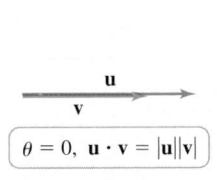

$\theta = 0$, $\mathbf{u} \cdot \mathbf{v} = |\mathbf{u}||\mathbf{v}|$

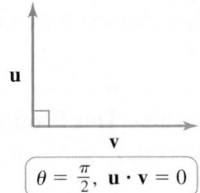

$\mathbf{u} \cdot \mathbf{v} = |\mathbf{u}||\mathbf{v}| \cos \theta > 0$

$\theta = \dfrac{\pi}{2}$, $\mathbf{u} \cdot \mathbf{v} = 0$

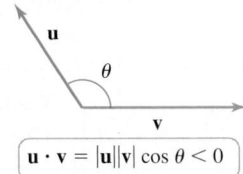

$\mathbf{u} \cdot \mathbf{v} = |\mathbf{u}||\mathbf{v}| \cos \theta < 0$

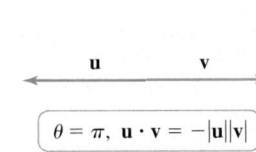

$\theta = \pi$, $\mathbf{u} \cdot \mathbf{v} = -|\mathbf{u}||\mathbf{v}|$

Figure 13.44

The dot product of two vectors is itself a scalar. Two special cases immediately arise:

- **u** and **v** are parallel ($\theta = 0$ or $\theta = \pi$) if and only if $\mathbf{u} \cdot \mathbf{v} = \pm |\mathbf{u}||\mathbf{v}|$.
- **u** and **v** are perpendicular ($\theta = \pi/2$) if and only if $\mathbf{u} \cdot \mathbf{v} = 0$.

The second case gives rise to the important property of *orthogonality*.

> ▶ In two and three dimensions, the terms *orthogonal* and *perpendicular* are used interchangeably. *Orthogonal* is a more general term that also applies in more than three dimensions.

DEFINITION Orthogonal Vectors

Two vectors **u** and **v** are **orthogonal** if and only if $\mathbf{u} \cdot \mathbf{v} = 0$. The zero vector is orthogonal to all vectors. In two or three dimensions, two nonzero orthogonal vectors are perpendicular to each other.

> **QUICK CHECK 1** Sketch two nonzero vectors **u** and **v** with $\theta = 0$. Sketch two nonzero vectors **u** and **v** with $\theta = \pi$. ◀

EXAMPLE 1 Dot products Compute the dot products of the following vectors.

a. $\mathbf{u} = 2\mathbf{i} - 6\mathbf{j}$ and $\mathbf{v} = 12\mathbf{k}$
b. $\mathbf{u} = \langle \sqrt{3}, 1 \rangle$ and $\mathbf{v} = \langle 0, 1 \rangle$

SOLUTION

a. The vector **u** lies in the *xy*-plane and the vector **v** is perpendicular to the *xy*-plane. Therefore, $\theta = \dfrac{\pi}{2}$, **u** and **v** are orthogonal, and $\mathbf{u} \cdot \mathbf{v} = 0$ (Figure 13.45a).

b. As shown in Figure 13.45b, **u** and **v** form two sides of a 30–60–90 triangle in the *xy*-plane, with an angle of $\pi/3$ between them. Because $|\mathbf{u}| = 2$, $|\mathbf{v}| = 1$, and $\cos \pi/3 = 1/2$, the dot product is

$$\mathbf{u} \cdot \mathbf{v} = |\mathbf{u}||\mathbf{v}| \cos \theta = 2 \cdot 1 \cdot \frac{1}{2} = 1.$$

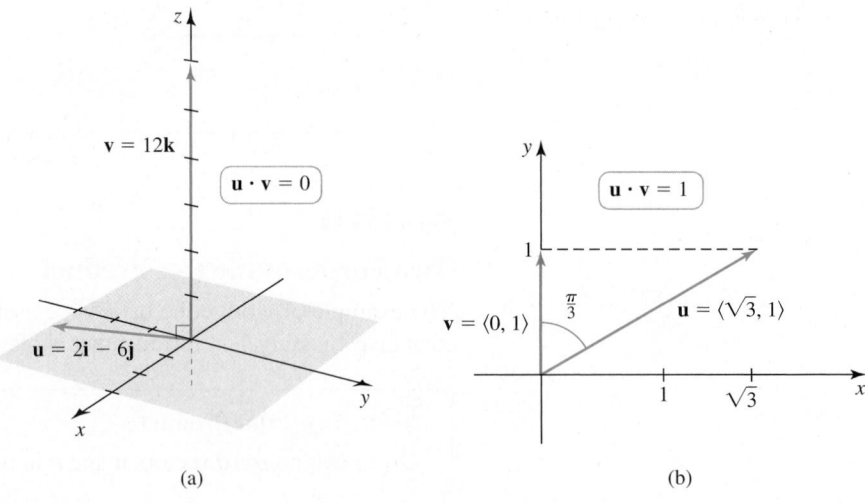

Figure 13.45

(a)   (b)

Related Exercises 16–17 ◀

Computing a dot product in this manner requires knowing the angle θ between the vectors. Often the angle is not known; in fact, it may be exactly what we seek. For this reason, we present another method for computing the dot product that does not require knowing θ.

> ▶ In $\mathbb{R}^2$ with $\mathbf{u} = \langle u_1, u_2 \rangle$ and $\mathbf{v} = \langle v_1, v_2 \rangle$, $\mathbf{u} \cdot \mathbf{v} = u_1 v_1 + u_2 v_2$.

THEOREM 13.1 Dot Product

Given two vectors $\mathbf{u} = \langle u_1, u_2, u_3 \rangle$ and $\mathbf{v} = \langle v_1, v_2, v_3 \rangle$,

$$\mathbf{u} \cdot \mathbf{v} = u_1 v_1 + u_2 v_2 + u_3 v_3.$$

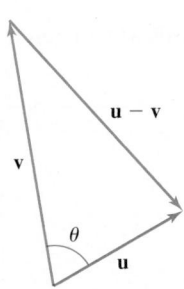

Figure 13.46

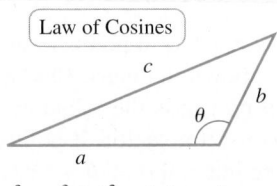

Law of Cosines

$$c^2 = a^2 + b^2 - 2ab\cos\theta$$

Proof: Consider two position vectors $\mathbf{u} = \langle u_1, u_2, u_3 \rangle$ and $\mathbf{v} = \langle v_1, v_2, v_3 \rangle$, and suppose θ is the angle between them. The vector $\mathbf{u} - \mathbf{v}$ forms the third side of a triangle (Figure 13.46). By the Law of Cosines,

$$|\mathbf{u} - \mathbf{v}|^2 = |\mathbf{u}|^2 + |\mathbf{v}|^2 - \underbrace{2|\mathbf{u}||\mathbf{v}|\cos\theta}_{\mathbf{u}\,\cdot\,\mathbf{v}}.$$

The definition of the dot product, $\mathbf{u} \cdot \mathbf{v} = |\mathbf{u}||\mathbf{v}|\cos\theta$, allows us to write

$$\mathbf{u} \cdot \mathbf{v} = |\mathbf{u}||\mathbf{v}|\cos\theta = \frac{1}{2}(|\mathbf{u}|^2 + |\mathbf{v}|^2 - |\mathbf{u} - \mathbf{v}|^2). \tag{1}$$

Using the definition of magnitude, we find that

$$|\mathbf{u}|^2 = u_1^2 + u_2^2 + u_3^2, \qquad |\mathbf{v}|^2 = v_1^2 + v_2^2 + v_3^2,$$

and

$$|\mathbf{u} - \mathbf{v}|^2 = (u_1 - v_1)^2 + (u_2 - v_2)^2 + (u_3 - v_3)^2.$$

Expanding the terms in $|\mathbf{u} - \mathbf{v}|^2$ and simplifying yields

$$|\mathbf{u}|^2 + |\mathbf{v}|^2 - |\mathbf{u} - \mathbf{v}|^2 = 2(u_1v_1 + u_2v_2 + u_3v_3).$$

Substituting into expression (1) gives a compact expression for the dot product:

$$\mathbf{u} \cdot \mathbf{v} = u_1v_1 + u_2v_2 + u_3v_3.$$

This new representation of $\mathbf{u} \cdot \mathbf{v}$ has two immediate consequences.

1. Combining it with the definition of dot product gives

$$\mathbf{u} \cdot \mathbf{v} = u_1v_1 + u_2v_2 + u_3v_3 = |\mathbf{u}||\mathbf{v}|\cos\theta.$$

If $\mathbf{u}$ and $\mathbf{v}$ are both nonzero, then

$$\cos\theta = \frac{u_1v_1 + u_2v_2 + u_3v_3}{|\mathbf{u}||\mathbf{v}|} = \frac{\mathbf{u} \cdot \mathbf{v}}{|\mathbf{u}||\mathbf{v}|},$$

and we have a way to compute θ.

2. Notice that $\mathbf{u} \cdot \mathbf{u} = u_1^2 + u_2^2 + u_3^2 = |\mathbf{u}|^2$. Therefore, we have a relationship between the dot product and the magnitude of a vector: $|\mathbf{u}| = \sqrt{\mathbf{u} \cdot \mathbf{u}}$ or $|\mathbf{u}|^2 = \mathbf{u} \cdot \mathbf{u}$.

QUICK CHECK 2 Use Theorem 13.1 to compute the dot products $\mathbf{i} \cdot \mathbf{j}$, $\mathbf{i} \cdot \mathbf{k}$, and $\mathbf{j} \cdot \mathbf{k}$ for the unit coordinate vectors. What do you conclude about the angles between these vectors? ◄

EXAMPLE 2 Dot products and angles Let $\mathbf{u} = \langle \sqrt{3}, 1, 0 \rangle$, $\mathbf{v} = \langle 1, \sqrt{3}, 0 \rangle$, and $\mathbf{w} = \langle 1, \sqrt{3}, 2\sqrt{3} \rangle$.

a. Compute $\mathbf{u} \cdot \mathbf{v}$.
b. Find the angle between $\mathbf{u}$ and $\mathbf{v}$.
c. Find the angle between $\mathbf{u}$ and $\mathbf{w}$.

SOLUTION

a. $\mathbf{u} \cdot \mathbf{v} = \langle \sqrt{3}, 1, 0 \rangle \cdot \langle 1, \sqrt{3}, 0 \rangle = \sqrt{3} + \sqrt{3} + 0 = 2\sqrt{3}$

b. Note that $|\mathbf{u}| = \sqrt{\mathbf{u} \cdot \mathbf{u}} = \sqrt{\langle \sqrt{3}, 1, 0 \rangle \cdot \langle \sqrt{3}, 1, 0 \rangle} = 2$, and similarly, $|\mathbf{v}| = 2$. Therefore,

$$\cos\theta = \frac{\mathbf{u} \cdot \mathbf{v}}{|\mathbf{u}||\mathbf{v}|} = \frac{2\sqrt{3}}{2 \cdot 2} = \frac{\sqrt{3}}{2}.$$

Because $0 \le \theta \le \pi$, it follows that $\theta = \cos^{-1}\dfrac{\sqrt{3}}{2} = \dfrac{\pi}{6}$.

c. $\cos\theta = \dfrac{\mathbf{u} \cdot \mathbf{w}}{|\mathbf{u}||\mathbf{w}|} = \dfrac{\langle \sqrt{3}, 1, 0 \rangle \cdot \langle 1, \sqrt{3}, 2\sqrt{3} \rangle}{|\langle \sqrt{3}, 1, 0 \rangle||\langle 1, \sqrt{3}, 2\sqrt{3} \rangle|} = \dfrac{2\sqrt{3}}{2 \cdot 4} = \dfrac{\sqrt{3}}{4}$

It follows that

$$\theta = \cos^{-1}\frac{\sqrt{3}}{4} \approx 1.12 \text{ rad} \approx 64.3°.$$

Related Exercises 20, 26 ◄

> Theorem 13.1 extends to vectors
> with any number of components.
> If $\mathbf{u} = \langle u_1, \ldots, u_n \rangle$ and
> $\mathbf{v} = \langle v_1, \ldots, v_n \rangle$, then
>
> $$\mathbf{u} \cdot \mathbf{v} = u_1 v_1 + \cdots + u_n v_n.$$
>
> The properties in Theorem 13.2 also apply
> in two or more dimensions.

Properties of Dot Products   The properties of the dot product in the following theorem are easily proved using vector components (Exercises 79–81).

THEOREM 13.2   Properties of the Dot Product

Suppose $\mathbf{u}$, $\mathbf{v}$, and $\mathbf{w}$ are vectors and let c be a scalar.

1. $\mathbf{u} \cdot \mathbf{v} = \mathbf{v} \cdot \mathbf{u}$ Commutative property
2. $c(\mathbf{u} \cdot \mathbf{v}) = (c\mathbf{u}) \cdot \mathbf{v} = \mathbf{u} \cdot (c\mathbf{v})$ Associative property
3. $\mathbf{u} \cdot (\mathbf{v} + \mathbf{w}) = \mathbf{u} \cdot \mathbf{v} + \mathbf{u} \cdot \mathbf{w}$ Distributive property

---

Orthogonal Projections

Given vectors $\mathbf{u}$ and $\mathbf{v}$, how closely aligned are they? That is, how much of $\mathbf{u}$ points in the direction of $\mathbf{v}$? This question is answered using *projections*. As shown in **Figure 13.47a**, the projection of the vector $\mathbf{u}$ onto a nonzero vector $\mathbf{v}$, denoted $\text{proj}_{\mathbf{v}}\mathbf{u}$, is the "shadow" cast by $\mathbf{u}$ onto the line through $\mathbf{v}$. The projection of $\mathbf{u}$ onto $\mathbf{v}$ is itself a vector; it points in the same direction as $\mathbf{v}$ if the angle between $\mathbf{u}$ and $\mathbf{v}$ lies in the interval $0 \le \theta < \pi/2$ (**Figure 13.47b**); it points in the direction opposite that of $\mathbf{v}$ if the angle between $\mathbf{u}$ and $\mathbf{v}$ lies in the interval $\pi/2 < \theta \le \pi$ (**Figure 13.47c**). If $\theta = \dfrac{\pi}{2}$, $\mathbf{u}$ and $\mathbf{v}$ are orthogonal, and there is no shadow.

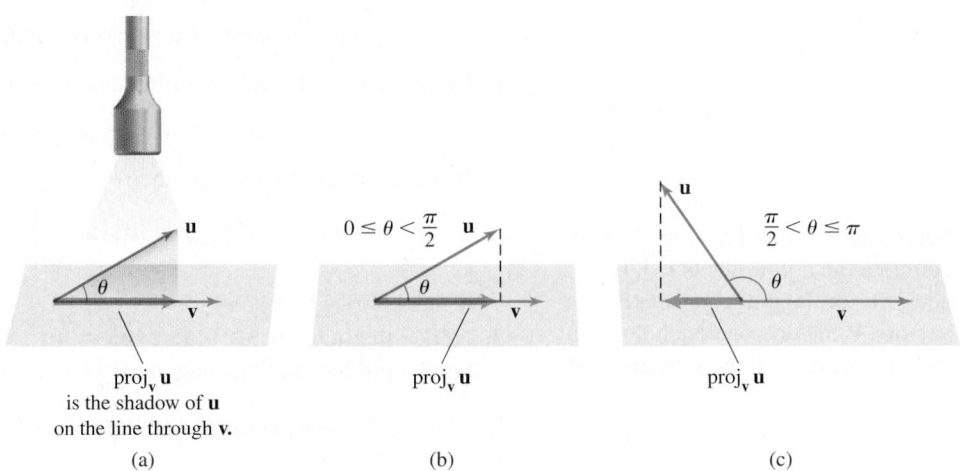

proj$_{\mathbf{v}}$ **u**
is the shadow of **u**
on the line through **v**.

(a)   (b)   (c)

Figure 13.47

To find the projection of $\mathbf{u}$ onto $\mathbf{v}$, we proceed as follows: With the tails of $\mathbf{u}$ and $\mathbf{v}$ together, we drop a perpendicular line segment from the head of $\mathbf{u}$ to the point P on the line through $\mathbf{v}$ (**Figure 13.48**). The vector $\overrightarrow{OP}$ is the *orthogonal projection of $\mathbf{u}$ onto $\mathbf{v}$*. An expression for $\text{proj}_{\mathbf{v}}\mathbf{u}$ is found using two observations.

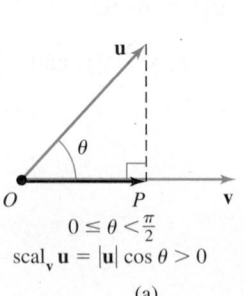

$0 \le \theta < \frac{\pi}{2}$

$\text{scal}_{\mathbf{v}}\,\mathbf{u} = |\mathbf{u}| \cos \theta > 0$

(a)

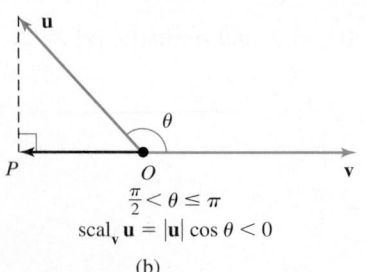

$\frac{\pi}{2} < \theta \le \pi$

$\text{scal}_{\mathbf{v}}\,\mathbf{u} = |\mathbf{u}| \cos \theta < 0$

(b)

Figure 13.48

- If $0 \le \theta < \pi/2$, then $\text{proj}_{\mathbf{v}}\mathbf{u}$ has length $|\mathbf{u}| \cos \theta$ and points in the direction of the unit vector $\mathbf{v}/|\mathbf{v}|$ (**Figure 13.48a**). Therefore,

$$\text{proj}_{\mathbf{v}}\mathbf{u} = \underbrace{|\mathbf{u}| \cos \theta}_{\text{length}} \underbrace{\left(\frac{\mathbf{v}}{|\mathbf{v}|} \right)}_{\text{direction}}.$$

We define the *scalar component of $\mathbf{u}$ in the direction of $\mathbf{v}$* to be $\text{scal}_{\mathbf{v}}\mathbf{u} = |\mathbf{u}| \cos \theta$. In this case, $\text{scal}_{\mathbf{v}}\mathbf{u}$ is the length of $\text{proj}_{\mathbf{v}}\mathbf{u}$.

- If $\pi/2 < \theta \le \pi$, then $\text{proj}_{\mathbf{v}}\mathbf{u}$ has length $-|\mathbf{u}| \cos \theta$ (which is positive) and points in the direction of $-\mathbf{v}/|\mathbf{v}|$ (**Figure 13.48b**). Therefore,

$$\text{proj}_{\mathbf{v}}\mathbf{u} = -\underbrace{|\mathbf{u}| \cos \theta}_{\text{length}} \cdot \underbrace{\left(-\frac{\mathbf{v}}{|\mathbf{v}|} \right)}_{\text{direction}} = |\mathbf{u}| \cos \theta \left(\frac{\mathbf{v}}{|\mathbf{v}|} \right).$$

In this case, $\text{scal}_{\mathbf{v}}\mathbf{u} = |\mathbf{u}| \cos \theta < 0$.

➤ Notice that $\text{scal}_\mathbf{v}\mathbf{u}$ may be positive, negative, or zero. However, $|\text{scal}_\mathbf{v}\mathbf{u}|$ is the length of $\text{proj}_\mathbf{v}\mathbf{u}$. The projection $\text{proj}_\mathbf{v}\mathbf{u}$ is defined for all vectors $\mathbf{u}$, but only for nonzero vectors $\mathbf{v}$.

We see that in both cases, the expression for $\text{proj}_\mathbf{v}\mathbf{u}$ is the same:

$$\text{proj}_\mathbf{v}\mathbf{u} = \underbrace{|\mathbf{u}|\cos\theta}_{\text{scal}_\mathbf{v}\mathbf{u}}\left(\frac{\mathbf{v}}{|\mathbf{v}|}\right) = \text{scal}_\mathbf{v}\mathbf{u}\left(\frac{\mathbf{v}}{|\mathbf{v}|}\right).$$

Note that if $\theta = \dfrac{\pi}{2}$, then $\text{proj}_\mathbf{v}\mathbf{u} = \mathbf{0}$ and $\text{scal}_\mathbf{v}\mathbf{u} = 0$.

Using properties of the dot product, $\text{proj}_\mathbf{v}\mathbf{u}$ may be written in different ways:

$$\begin{aligned}
\text{proj}_\mathbf{v}\mathbf{u} &= |\mathbf{u}|\cos\theta\left(\frac{\mathbf{v}}{|\mathbf{v}|}\right) \\[2mm]
&= \frac{\mathbf{u}\cdot\mathbf{v}}{|\mathbf{v}|}\left(\frac{\mathbf{v}}{|\mathbf{v}|}\right) \qquad |\mathbf{u}|\cos\theta = \frac{|\mathbf{u}||\mathbf{v}|\cos\theta}{|\mathbf{v}|} = \frac{\mathbf{u}\cdot\mathbf{v}}{|\mathbf{v}|} \\[2mm]
&= \underbrace{\left(\frac{\mathbf{u}\cdot\mathbf{v}}{\mathbf{v}\cdot\mathbf{v}}\right)}_{\text{scalar}}\mathbf{v}. \qquad \text{Regroup terms; } |\mathbf{v}|^2 = \mathbf{v}\cdot\mathbf{v}
\end{aligned}$$

The first two expressions show that $\text{proj}_\mathbf{v}\mathbf{u}$ is a scalar multiple of the unit vector $\dfrac{\mathbf{v}}{|\mathbf{v}|}$, whereas the last expression shows that $\text{proj}_\mathbf{v}\mathbf{u}$ is a scalar multiple of $\mathbf{v}$.

QUICK CHECK 3 Let $\mathbf{u} = 4\mathbf{i} - 3\mathbf{j}$. By inspection (not calculations), find the orthogonal projection of $\mathbf{u}$ onto $\mathbf{i}$ and onto $\mathbf{j}$. Find the scalar component of $\mathbf{u}$ in the direction of $\mathbf{i}$ and in the direction of $\mathbf{j}$. ◄

DEFINITION (Orthogonal) Projection of u onto v

The **orthogonal projection of u onto v**, denoted $\text{proj}_\mathbf{v}\mathbf{u}$, where $\mathbf{v} \neq \mathbf{0}$, is

$$\text{proj}_\mathbf{v}\mathbf{u} = |\mathbf{u}|\cos\theta\left(\frac{\mathbf{v}}{|\mathbf{v}|}\right).$$

The orthogonal projection may also be computed with the formulas

$$\text{proj}_\mathbf{v}\mathbf{u} = \text{scal}_\mathbf{v}\mathbf{u}\left(\frac{\mathbf{v}}{|\mathbf{v}|}\right) = \left(\frac{\mathbf{u}\cdot\mathbf{v}}{\mathbf{v}\cdot\mathbf{v}}\right)\mathbf{v},$$

where the **scalar component of u in the direction of v** is

$$\text{scal}_\mathbf{v}\mathbf{u} = |\mathbf{u}|\cos\theta = \frac{\mathbf{u}\cdot\mathbf{v}}{|\mathbf{v}|}.$$

EXAMPLE 3 Orthogonal projections Find $\text{proj}_\mathbf{v}\mathbf{u}$ and $\text{scal}_\mathbf{v}\mathbf{u}$ for the following vectors and illustrate each result.

a. $\mathbf{u} = \langle 4, 1\rangle, \mathbf{v} = \langle 3, 4\rangle$
b. $\mathbf{u} = \langle -4, -3\rangle, \mathbf{v} = \langle 1, -1\rangle$

SOLUTION

a. The scalar component of $\mathbf{u}$ in the direction of $\mathbf{v}$ (Figure 13.49) is

$$\text{scal}_\mathbf{v}\mathbf{u} = \frac{\mathbf{u}\cdot\mathbf{v}}{|\mathbf{v}|} = \frac{\langle 4, 1\rangle\cdot\langle 3, 4\rangle}{|\langle 3, 4\rangle|} = \frac{16}{5}.$$

Because $\dfrac{\mathbf{v}}{|\mathbf{v}|} = \left\langle \dfrac{3}{5}, \dfrac{4}{5}\right\rangle$, we have

$$\text{proj}_\mathbf{v}\mathbf{u} = \text{scal}_\mathbf{v}\mathbf{u}\left(\frac{\mathbf{v}}{|\mathbf{v}|}\right) = \frac{16}{5}\left\langle\frac{3}{5}, \frac{4}{5}\right\rangle = \frac{16}{25}\langle 3, 4\rangle.$$

b. Using another formula for $\text{proj}_\mathbf{v}\mathbf{u}$, we have

$$\text{proj}_\mathbf{v}\mathbf{u} = \left(\frac{\mathbf{u}\cdot\mathbf{v}}{\mathbf{v}\cdot\mathbf{v}}\right)\mathbf{v} = \left(\frac{\langle -4, -3\rangle\cdot\langle 1, -1\rangle}{\langle 1, -1\rangle\cdot\langle 1, -1\rangle}\right)\langle 1, -1\rangle = -\frac{1}{2}\langle 1, -1\rangle.$$

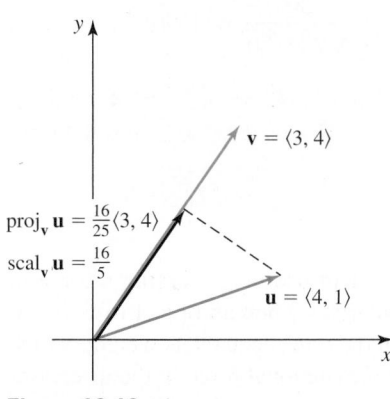

Figure 13.49

$\text{proj}_\mathbf{v}\mathbf{u} = \dfrac{16}{25}\langle 3, 4\rangle$

$\text{scal}_\mathbf{v}\mathbf{u} = \dfrac{16}{5}$

$\mathbf{v} = \langle 3, 4\rangle$

$\mathbf{u} = \langle 4, 1\rangle$

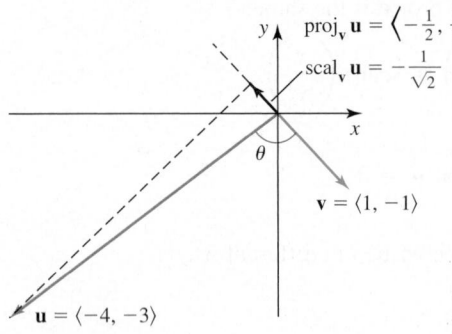

Figure 13.50

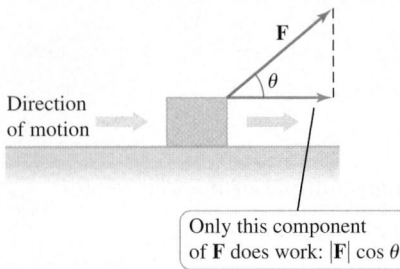

Figure 13.51

➤ If the unit of force is newtons (N) and the distance is measured in meters, then the unit of work is joules (J), where $1 \text{ J} = 1$ N-m. If force is measured in pounds and distance is measured in feet, then work has units of ft-lb.

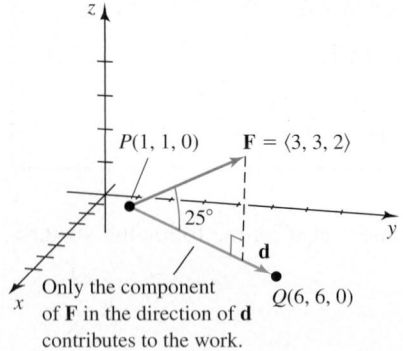

Only the component of **F** in the direction of **d** contributes to the work.

Figure 13.52

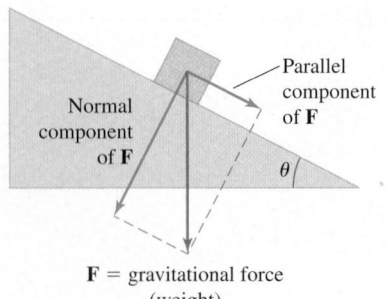

$\mathbf{F}$ = gravitational force (weight)

Figure 13.53

The vectors **v** and $\text{proj}_\mathbf{v}\mathbf{u}$ point in opposite directions because $\pi/2 < \theta \leq \pi$ (Figure 13.50). This fact is reflected in the scalar component of **u** in the direction of **v**, which is negative:

$$\text{scal}_\mathbf{v}\mathbf{u} = \frac{\langle -4, -3 \rangle \cdot \langle 1, -1 \rangle}{|\langle 1, -1 \rangle|} = -\frac{1}{\sqrt{2}}.$$

Related Exercises 35–36 ◄

Applications of Dot Products

Work and Force   In the opening of this section, we observed that if a constant force **F** acts at an angle θ to the direction of motion of an object (Figure 13.51), the work done by the force is

$$W = |\mathbf{F}| \cos \theta \, |\mathbf{d}| = \mathbf{F} \cdot \mathbf{d}.$$

Notice that the work is a scalar, and if the force acts in a direction orthogonal to the motion, then $\theta = \pi/2$, $\mathbf{F} \cdot \mathbf{d} = 0$, and no work is done by the force.

DEFINITION Work

Let a constant force **F** be applied to an object, producing a displacement **d**. If the angle between **F** and **d** is θ, then the **work** done by the force is

$$W = |\mathbf{F}||\mathbf{d}| \cos \theta = \mathbf{F} \cdot \mathbf{d}.$$

EXAMPLE 4   Calculating work   A force $\mathbf{F} = \langle 3, 3, 2 \rangle$ (in newtons) moves an object along a line segment from $P(1, 1, 0)$ to $Q(6, 6, 0)$ (in meters). What is the work done by the force? Interpret the result.

SOLUTION   The displacement of the object is $\mathbf{d} = \overrightarrow{PQ} = \langle 6 - 1, 6 - 1, 0 - 0 \rangle = \langle 5, 5, 0 \rangle$. Therefore, the work done by the force is

$$W = \mathbf{F} \cdot \mathbf{d} = \langle 3, 3, 2 \rangle \cdot \langle 5, 5, 0 \rangle = 30 \text{ J}.$$

To interpret this result, notice that the angle between the force and the displacement vector satisfies

$$\cos \theta = \frac{\mathbf{F} \cdot \mathbf{d}}{|\mathbf{F}||\mathbf{d}|} = \frac{\langle 3, 3, 2 \rangle \cdot \langle 5, 5, 0 \rangle}{|\langle 3, 3, 2 \rangle||\langle 5, 5, 0 \rangle|} = \frac{30}{\sqrt{22}\sqrt{50}} \approx 0.905.$$

Therefore, $\theta \approx 0.44$ rad $\approx 25°$. The magnitude of the force is $|\mathbf{F}| = \sqrt{22} \approx 4.7$ N, but only the component of that force in the direction of motion, $|\mathbf{F}| \cos \theta \approx \sqrt{22} \cos 0.44 \approx 4.2$ N, contributes to the work (Figure 13.52).

Related Exercises 44, 46 ◄

Parallel and Normal Forces   Projections find frequent use in expressing a force in terms of orthogonal components. A common situation arises when an object rests on an inclined plane (Figure 13.53). The gravitational force on the object equals its weight, which is directed vertically downward. The projections of the gravitational force in the directions **parallel** to and **normal** (or perpendicular) to the plane are of interest. Specifically, the projection of the force parallel to the plane determines the tendency of the object to slide down the plane, while the projection of the force normal to the plane determines its tendency to "stick" to the plane.

EXAMPLE 5   Components of a force   A 10-lb block rests on a plane that is inclined at 30° above the horizontal. Find the components of the gravitational force parallel to and normal (perpendicular) to the plane.

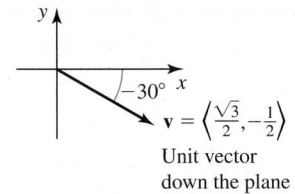

$v = \left\langle \frac{\sqrt{3}}{2}, -\frac{1}{2} \right\rangle$

Unit vector
down the plane

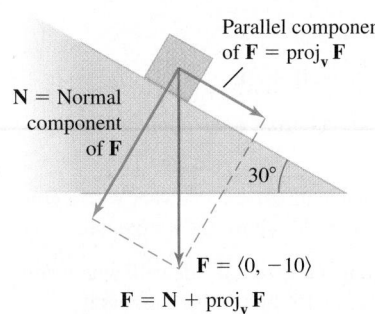

Parallel component
of $\mathbf{F} = \text{proj}_v \mathbf{F}$

$\mathbf{N} = $ Normal
component
of $\mathbf{F}$

$30°$

$\mathbf{F} = \langle 0, -10 \rangle$

$\mathbf{F} = \mathbf{N} + \text{proj}_v \mathbf{F}$

Figure 13.54

SOLUTION The gravitational force $\mathbf{F}$ acting on the block equals the weight of the block (10 lb); we regard the block as a point mass. Using the coordinate system shown in Figure 13.54, the force acts in the negative y-direction; therefore, $\mathbf{F} = \langle 0, -10 \rangle$. The direction *down* the plane is given by the unit vector $\mathbf{v} = \langle \cos(-30°), \sin(-30°) \rangle = \left\langle \frac{\sqrt{3}}{2}, -\frac{1}{2} \right\rangle$ (check that $|\mathbf{v}| = 1$). The component of the gravitational force parallel to the plane is

$$\text{proj}_v \mathbf{F} = \left(\frac{\mathbf{F} \cdot \mathbf{v}}{\underbrace{\mathbf{v} \cdot \mathbf{v}}_{\mathbf{v} \cdot \mathbf{v} = 1}} \right) \mathbf{v} = \left(\underbrace{\langle 0, -10 \rangle}_{\mathbf{F}} \cdot \underbrace{\left\langle \frac{\sqrt{3}}{2}, -\frac{1}{2} \right\rangle}_{\mathbf{v}} \right) \underbrace{\left\langle \frac{\sqrt{3}}{2}, -\frac{1}{2} \right\rangle}_{\mathbf{v}} = 5 \left\langle \frac{\sqrt{3}}{2}, -\frac{1}{2} \right\rangle.$$

Let the component of $\mathbf{F}$ normal to the plane be $\mathbf{N}$. Note that $\mathbf{F} = \text{proj}_v \mathbf{F} + \mathbf{N}$, so

$$\mathbf{N} = \mathbf{F} - \text{proj}_v \mathbf{F} = \langle 0, -10 \rangle - 5 \left\langle \frac{\sqrt{3}}{2}, -\frac{1}{2} \right\rangle = -5 \left\langle \frac{\sqrt{3}}{2}, \frac{3}{2} \right\rangle.$$

Figure 13.54 shows how the components of $\mathbf{F}$ parallel to and normal to the plane combine to form the total force $\mathbf{F}$.

Related Exercises 47, 49 ◄

SECTION 13.3 EXERCISES

Getting Started

1. Express the dot product of $\mathbf{u}$ and $\mathbf{v}$ in terms of their magnitudes and the angle between them.

2. Express the dot product of $\mathbf{u}$ and $\mathbf{v}$ in terms of the components of the vectors.

3. Compute $\langle 2, 3, -6 \rangle \cdot \langle 1, -8, 3 \rangle$.

4. Use the definition of the dot product to explain why $\mathbf{v} \cdot \mathbf{v} = |\mathbf{v}|^2$.

5. Explain how to find the angle between two nonzero vectors.

6. Find the angle θ between $\mathbf{u}$ and $\mathbf{v}$ if $\text{scal}_v \mathbf{u} = -2$ and $|\mathbf{u}| = 4$. Assume $0 \le \theta \le \pi$.

7. Find $\text{proj}_v \mathbf{u}$ if $\text{scal}_v \mathbf{u} = -2$ and $\mathbf{v} = \langle 2, -1, -2 \rangle$.

8. Use a dot product to determine whether the vectors $\mathbf{u} = \langle 1, 2, 3 \rangle$ and $\mathbf{v} = \langle 4, 1, -2 \rangle$ are orthogonal.

9. Find $\mathbf{u} \cdot \mathbf{v}$ if $\mathbf{u}$ and $\mathbf{v}$ are unit vectors and the angle between $\mathbf{u}$ and $\mathbf{v}$ is π.

10. Explain how the work done by a force in moving an object is computed using dot products.

11. Suppose $\mathbf{v}$ is a nonzero position vector in the xy-plane. How many position vectors with length 2 in the xy-plane are orthogonal to $\mathbf{v}$?

12. Suppose $\mathbf{v}$ is a nonzero position vector in xyz-space. How many position vectors with length 2 in xyz-space are orthogonal to $\mathbf{v}$?

Practice Exercises

13–16. Dot product from the definition *Consider the following vectors $\mathbf{u}$ and $\mathbf{v}$. Sketch the vectors, find the angle between the vectors, and compute the dot product using the definition $\mathbf{u} \cdot \mathbf{v} = |\mathbf{u}||\mathbf{v}| \cos \theta$.*

13. $\mathbf{u} = 4\mathbf{i}$ and $\mathbf{v} = 6\mathbf{j}$

14. $\mathbf{u} = \langle -3, 2, 0 \rangle$ and $\mathbf{v} = \langle 0, 0, 6 \rangle$

15. $\mathbf{u} = \langle 10, 0 \rangle$ and $\mathbf{v} = \langle 10, 10 \rangle$

16. $\mathbf{u} = \langle -\sqrt{3}, 1 \rangle$ and $\mathbf{v} = \langle \sqrt{3}, 1 \rangle$

17. **Dot product from the definition** Compute $\mathbf{u} \cdot \mathbf{v}$ if $\mathbf{u}$ and $\mathbf{v}$ are unit vectors and the angle between them is $\pi/3$.

18. **Dot product from the definition** Compute $\mathbf{u} \cdot \mathbf{v}$ if $\mathbf{u}$ is a unit vector, $|\mathbf{v}| = 2$, and the angle between them is $3\pi/4$.

19–28. Dot products and angles *Compute the dot product of the vectors $\mathbf{u}$ and $\mathbf{v}$, and find the angle between the vectors.*

19. $\mathbf{u} = \mathbf{i} + \mathbf{j}$ and $\mathbf{v} = \mathbf{i} - \mathbf{j}$

20. $\mathbf{u} = \langle 10, 0 \rangle$ and $\mathbf{v} = \langle -5, 5 \rangle$

21. $\mathbf{u} = \mathbf{i}$ and $\mathbf{v} = \mathbf{i} + \sqrt{3}\mathbf{j}$

22. $\mathbf{u} = \sqrt{2}\mathbf{i} + \sqrt{2}\mathbf{j}$ and $\mathbf{v} = -\sqrt{2}\mathbf{i} - \sqrt{2}\mathbf{j}$

23. $\mathbf{u} = 4\mathbf{i} + 3\mathbf{j}$ and $\mathbf{v} = 4\mathbf{i} - 6\mathbf{j}$

24. $\mathbf{u} = \langle 3, 4, 0 \rangle$ and $\mathbf{v} = \langle 0, 4, 5 \rangle$

25. $\mathbf{u} = \langle -10, 0, 4 \rangle$ and $\mathbf{v} = \langle 1, 2, 3 \rangle$

26. $\mathbf{u} = \langle 3, -5, 2 \rangle$ and $\mathbf{v} = \langle -9, 5, 1 \rangle$

27. $\mathbf{u} = 2\mathbf{i} - 3\mathbf{k}$ and $\mathbf{v} = \mathbf{i} + 4\mathbf{j} + 2\mathbf{k}$

28. $\mathbf{u} = \mathbf{i} - 4\mathbf{j} - 6\mathbf{k}$ and $\mathbf{v} = 2\mathbf{i} - 4\mathbf{j} + 2\mathbf{k}$

29–30. Angles of a triangle *For the given points P, Q, and R, find the approximate measurements of the angles of $\triangle PQR$.*

29. $P(0, -1, 3), Q(2, 2, 1), R(-2, 2, 4)$

30. $P(1, -4), Q(2, 7), R(-2, 2)$

31–34. Sketching orthogonal projections *Find* $\text{proj}_\mathbf{v}\mathbf{u}$ *and* $\text{scal}_\mathbf{v}\mathbf{u}$ *by inspection without using formulas.*

31.

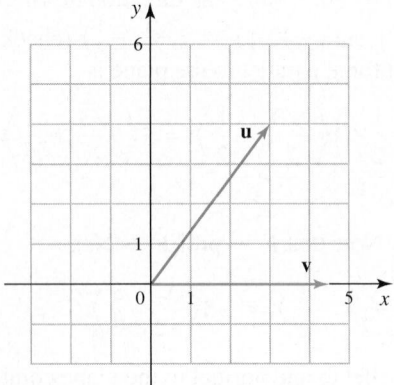

32.

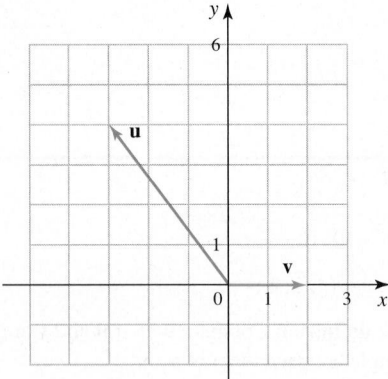

33.

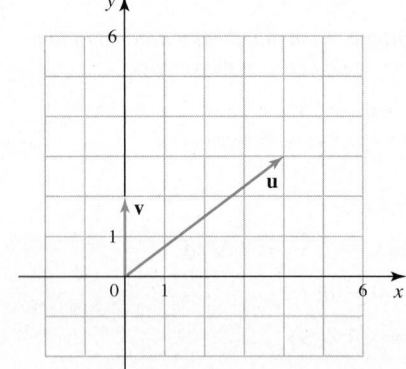

34.

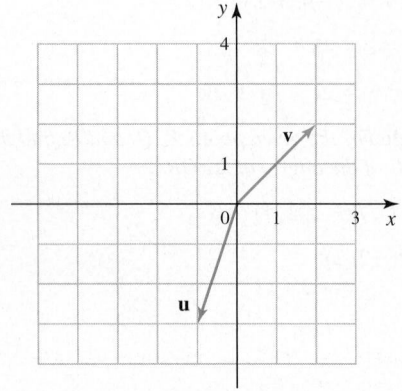

35–40. Calculating orthogonal projections *For the given vectors* $\mathbf{u}$ *and* $\mathbf{v}$, *calculate* $\text{proj}_\mathbf{v}\mathbf{u}$ *and* $\text{scal}_\mathbf{v}\mathbf{u}$.

35. $\mathbf{u} = \langle -1, 4 \rangle$ and $\mathbf{v} = \langle -4, 2 \rangle$

36. $\mathbf{u} = \langle 10, 5 \rangle$ and $\mathbf{v} = \langle 2, 6 \rangle$

37. $\mathbf{u} = \langle -8, 0, 2 \rangle$ and $\mathbf{v} = \langle 1, 3, -3 \rangle$

38. $\mathbf{u} = \langle 5, 0, 15 \rangle$ and $\mathbf{v} = \langle 0, 4, -2 \rangle$

39. $\mathbf{u} = 5\mathbf{i} + \mathbf{j} - 5\mathbf{k}$ and $\mathbf{v} = -\mathbf{i} + \mathbf{j} - 2\mathbf{k}$

40. $\mathbf{u} = \mathbf{i} + 4\mathbf{j} + 7\mathbf{k}$ and $\mathbf{v} = 2\mathbf{i} - 4\mathbf{j} + 2\mathbf{k}$

41–46. Computing work *Calculate the work done in the following situations.*

41. A suitcase is pulled 50 ft along a horizontal sidewalk with a constant force of 30 lb at an angle of 30° above the horizontal.

T 42. A stroller is pushed 20 m along a horizontal sidewalk with a constant force of 10 N at an angle of 15° below the horizontal.

43. A sled is pulled 10 m along horizontal ground with a constant force of 5 N at an angle of 45° above the horizontal.

44. A constant force $\mathbf{F} = \langle 4, 3, 2 \rangle$ (in newtons) moves an object from $(0, 0, 0)$ to $(8, 6, 0)$. (Distance is measured in meters.)

45. A constant force $\mathbf{F} = \langle 40, 30 \rangle$ (in newtons) is used to move a sled horizontally 10 m.

46. A constant force $\mathbf{F} = \langle 2, 4, 1 \rangle$ (in newtons) moves an object from $(0, 0, 0)$ to $(2, 4, 6)$. (Distance is measured in meters.)

47–48. Parallel and normal forces *Find the components of the vertical force* $\mathbf{F} = \langle 0, -10 \rangle$ *in the directions parallel to and normal to the following planes. Show that the total force is the sum of the two component forces.*

47. A plane that makes an angle of $\pi/3$ with the positive x-axis

48. A plane that makes an angle of $\theta = \tan^{-1}\dfrac{4}{5}$ with the positive x-axis

49. Mass on a plane A 100-kg object rests on an inclined plane at an angle of 45° to the floor. Find the components of the force parallel to and perpendicular to the plane.

T 50. Forces on an inclined plane An object on an inclined plane does not slide, provided the component of the object's weight parallel to the plane $|\mathbf{W}_{\text{par}}|$ is less than or equal to the magnitude of the opposing frictional force $|\mathbf{F}_{\text{f}}|$. The magnitude of the frictional force, in turn, is proportional to the component of the object's weight perpendicular to the plane $|\mathbf{W}_{\text{perp}}|$ (see figure). The constant of proportionality is the coefficient of static friction $\mu > 0$. Suppose a 100-lb block rests on a plane that is tilted at an angle of $\theta = 20°$ to the horizontal.

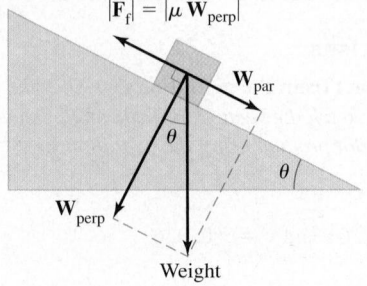

a. Find $|\mathbf{W}_{\text{par}}|$ and $|\mathbf{W}_{\text{perp}}|$. (*Hint:* It is not necessary to find $\mathbf{W}_{\text{par}}$ and $\mathbf{W}_{\text{perp}}$ first.)

b. The condition for the block not sliding is $|\mathbf{W}_{\text{par}}| \leq \mu |\mathbf{W}_{\text{perp}}|$. If $\mu = 0.65$, does the block slide?

c. What is the critical angle above which the block slides?

51. Explain why or why not Determine whether the following statements are true and give an explanation or counterexample.

a. $\text{proj}_{\mathbf{v}}\mathbf{u} = \text{proj}_{\mathbf{u}}\mathbf{v}$.

b. If nonzero vectors $\mathbf{u}$ and $\mathbf{v}$ have the same magnitude, they make equal angles with $\mathbf{u} + \mathbf{v}$.

c. $(\mathbf{u} \cdot \mathbf{i})^2 + (\mathbf{u} \cdot \mathbf{j})^2 + (\mathbf{u} \cdot \mathbf{k})^2 = |\mathbf{u}|^2$.

d. If $\mathbf{u}$ is orthogonal to $\mathbf{v}$ and $\mathbf{v}$ is orthogonal to $\mathbf{w}$, then $\mathbf{u}$ is orthogonal to $\mathbf{w}$.

e. The vectors orthogonal to $\langle 1, 1, 1 \rangle$ lie on the same line.

f. If $\text{proj}_{\mathbf{v}}\mathbf{u} = \mathbf{0}$, then vectors $\mathbf{u}$ and $\mathbf{v}$ (both nonzero) are orthogonal.

52. For what value of a is the vector $\mathbf{v} = \langle 4, -3, 7 \rangle$ orthogonal to $\mathbf{w} = \langle a, 8, 3 \rangle$?

53. For what value of c is the vector $\mathbf{v} = \langle 2, -5, c \rangle$ orthogonal to $\mathbf{w} = \langle 3, 2, 9 \rangle$?

54. Find two vectors that are orthogonal to $\langle 0, 1, 1 \rangle$ and to each other.

55. Let a and b be real numbers. Find all vectors $\langle 1, a, b \rangle$ orthogonal to $\langle 4, -8, 2 \rangle$.

56. Find three mutually orthogonal unit vectors in $\mathbb{R}^3$ besides $\pm\mathbf{i}$, $\pm\mathbf{j}$, and $\pm\mathbf{k}$.

57. Equal angles Consider the set of all unit position vectors $\mathbf{u}$ in $\mathbb{R}^3$ that make a $60°$ angle with the unit vector $\mathbf{k}$ in $\mathbb{R}^3$.

a. Prove that $\text{proj}_{\mathbf{k}}\mathbf{u}$ is the same for all vectors in this set.

b. Is $\text{scal}_{\mathbf{k}}\mathbf{u}$ the same for all vectors in this set?

58–61. Vectors with equal projections *Given a fixed vector $\mathbf{v}$, there is an infinite set of vectors $\mathbf{u}$ with the same value of $\text{proj}_{\mathbf{v}}\mathbf{u}$.*

58. Find another vector that has the same projection onto $\mathbf{v} = \langle 1, 1 \rangle$ as $\mathbf{u} = \langle 1, 2 \rangle$. Draw a picture.

59. Let $\mathbf{v} = \langle 1, 1 \rangle$. Give a description of the position vectors $\mathbf{u}$ such that $\text{proj}_{\mathbf{v}}\mathbf{u} = \text{proj}_{\mathbf{v}}\langle 1, 2 \rangle$.

60. Find another vector that has the same projection onto $\mathbf{v} = \langle 1, 1, 1 \rangle$ as $\mathbf{u} = \langle 1, 2, 3 \rangle$.

61. Let $\mathbf{v} = \langle 0, 0, 1 \rangle$. Give a description of all position vectors $\mathbf{u}$ such that $\text{proj}_{\mathbf{v}}\mathbf{u} = \text{proj}_{\mathbf{v}}\langle 1, 2, 3 \rangle$.

62–65. Decomposing vectors *For the following vectors $\mathbf{u}$ and $\mathbf{v}$, express $\mathbf{u}$ as the sum $\mathbf{u} = \mathbf{p} + \mathbf{n}$, where $\mathbf{p}$ is parallel to $\mathbf{v}$ and $\mathbf{n}$ is orthogonal to $\mathbf{v}$.*

62. $\mathbf{u} = \langle 4, 3 \rangle$, $\mathbf{v} = \langle 1, 1 \rangle$

63. $\mathbf{u} = \langle -2, 2 \rangle$, $\mathbf{v} = \langle 2, 1 \rangle$

64. $\mathbf{u} = \langle 4, 3, 0 \rangle$, $\mathbf{v} = \langle 1, 1, 1 \rangle$

65. $\mathbf{u} = \langle -1, 2, 3 \rangle$, $\mathbf{v} = \langle 2, 1, 1 \rangle$

66–69. An alternative line definition *Given a fixed point $P_0(x_0, y_0)$ and a nonzero vector $\mathbf{n} = \langle a, b \rangle$, the set of points $P(x, y)$ for which $\overline{P_0P}$ is orthogonal to $\mathbf{n}$ is a line ℓ (see figure). The vector $\mathbf{n}$ is called a normal vector or a vector normal to ℓ.*

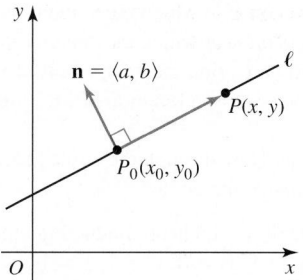

66. Show that the equation of the line passing through $P_0(x_0, y_0)$ with a normal vector $\mathbf{n} = \langle a, b \rangle$ is $a(x - x_0) + b(y - y_0) = 0$. (*Hint:* For a point $P(x, y)$ on ℓ, examine $\mathbf{n} \cdot \overline{P_0P}$.)

67. Use the result of Exercise 66 to find an equation of the line passing through the point $P_0(2, 6)$ with a normal vector $\mathbf{n} = \langle 3, -7 \rangle$. Write the final answer in the form $ax + by = c$.

68. Use the result of Exercise 66 to find an equation of the line passing through the point $P_0(1, -3)$ with a normal vector $\mathbf{n} = \langle 4, 0 \rangle$.

69. Suppose a line is normal to $\mathbf{n} = \langle 5, 3 \rangle$. What is the slope of the line?

Explorations and Challenges

70–72. Orthogonal unit vectors in $\mathbb{R}^2$ *Consider the vectors $\mathbf{I} = \langle 1/\sqrt{2}, 1/\sqrt{2} \rangle$ and $\mathbf{J} = \langle -1/\sqrt{2}, 1/\sqrt{2} \rangle$.*

70. Show that $\mathbf{I}$ and $\mathbf{J}$ are orthogonal unit vectors.

71. Express $\mathbf{I}$ and $\mathbf{J}$ in terms of the usual unit coordinate vectors $\mathbf{i}$ and $\mathbf{j}$. Then write $\mathbf{i}$ and $\mathbf{j}$ in terms of $\mathbf{I}$ and $\mathbf{J}$.

72. Write the vector $\langle 2, -6 \rangle$ in terms of $\mathbf{I}$ and $\mathbf{J}$.

73. Orthogonal unit vectors in $\mathbb{R}^3$ Consider the vectors $\mathbf{I} = \langle 1/2, 1/2, 1/\sqrt{2} \rangle$, $\mathbf{J} = \langle -1/\sqrt{2}, 1/\sqrt{2}, 0 \rangle$, and $\mathbf{K} = \langle 1/2, 1/2, -1/\sqrt{2} \rangle$.

a. Sketch $\mathbf{I}$, $\mathbf{J}$, and $\mathbf{K}$ and show that they are unit vectors.

b. Show that $\mathbf{I}$, $\mathbf{J}$, and $\mathbf{K}$ are pairwise orthogonal.

c. Express the vector $\langle 1, 0, 0 \rangle$ in terms of $\mathbf{I}$, $\mathbf{J}$, and $\mathbf{K}$.

74. Flow through a circle Suppose water flows in a thin sheet over the xy-plane with a uniform velocity given by the vector $\mathbf{v} = \langle 1, 2 \rangle$; this means that at all points of the plane, the velocity of the water has components 1 m/s in the x-direction and 2 m/s in the y-direction (see figure). Let C be an imaginary unit circle (that does not interfere with the flow).

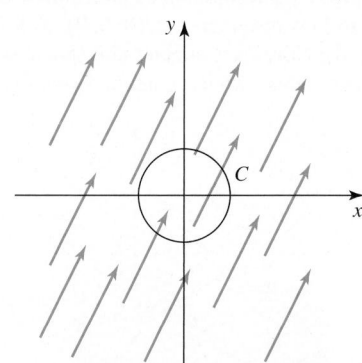

a. Show that at the point (x, y) on the circle C, the outward-pointing unit vector normal to C is $\mathbf{n} = \langle x, y \rangle$.

b. Show that at the point $(\cos \theta, \sin \theta)$ on the circle C, the outward-pointing unit vector normal to C is also $\mathbf{n} = \langle \cos \theta, \sin \theta \rangle$.

c. Find all points on C at which the velocity is normal to C.

d. Find all points on C at which the velocity is tangential to C.

e. At each point on C, find the component of $\mathbf{v}$ normal to C. Express the answer as a function of (x, y) and as a function of θ.

f. What is the net flow through the circle? Does water accumulate inside the circle?

75. Heat flux Let D be a solid heat-conducting cube formed by the planes $x = 0$, $x = 1$, $y = 0$, $y = 1$, $z = 0$, $z = 1$. The heat flow at every point of D is given by the constant vector $\mathbf{Q} = \langle 0, 2, 1 \rangle$.

a. Through which faces of D does $\mathbf{Q}$ point into D?

b. Through which faces of D does $\mathbf{Q}$ point out of D?

c. On which faces of D is $\mathbf{Q}$ tangential to D (pointing neither in nor out of D)?

d. Find the scalar component of $\mathbf{Q}$ normal to the face $x = 0$.

e. Find the scalar component of $\mathbf{Q}$ normal to the face $z = 1$.

f. Find the scalar component of $\mathbf{Q}$ normal to the face $y = 0$.

76. Hexagonal circle packing The German mathematician Carl Friedrich Gauss proved that the densest way to pack circles with the same radius in the plane is to place the centers of the circles on a hexagonal grid (see figure). Some molecular structures use this packing or its three-dimensional analog. Assume all circles have a radius of 1, and let $\mathbf{r}_{ij}$ be the vector that extends from the center of circle i to the center of circle j, for $i, j = 0, 1, \ldots, 6$.

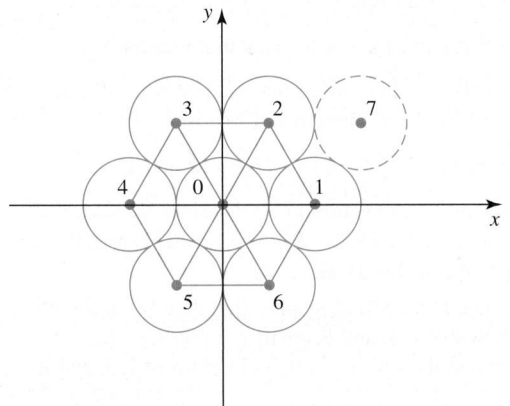

a. Find $\mathbf{r}_{0j}$ for $j = 1, 2, \ldots, 6$.

b. Find $\mathbf{r}_{12}$, $\mathbf{r}_{34}$, and $\mathbf{r}_{61}$.

c. Imagine that circle 7 is added to the arrangement as shown in the figure. Find $\mathbf{r}_{07}$, $\mathbf{r}_{17}$, $\mathbf{r}_{47}$, and $\mathbf{r}_{75}$.

77. Hexagonal sphere packing Imagine three unit spheres (radius equal to 1) with centers at $O(0, 0, 0)$, $P(\sqrt{3}, -1, 0)$, and $Q(\sqrt{3}, 1, 0)$. Now place another unit sphere symmetrically on top of these spheres with its center at R (see figure).

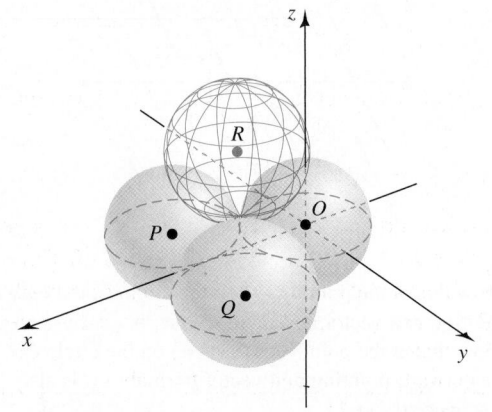

a. Find the coordinates of R. (*Hint:* The distance between the centers of any two spheres is 2.)

b. Let $\mathbf{r}_{IJ}$ be the vector from point I to point J. Find $\mathbf{r}_{OP}$, $\mathbf{r}_{OQ}$, $\mathbf{r}_{PQ}$, $\mathbf{r}_{OR}$, and $\mathbf{r}_{PR}$.

78–81. Properties of dot products *Let* $\mathbf{u} = \langle u_1, u_2, u_3 \rangle$, $\mathbf{v} = \langle v_1, v_2, v_3 \rangle$, *and* $\mathbf{w} = \langle w_1, w_2, w_3 \rangle$. *Prove the following vector properties, where c is a scalar.*

78. $|\mathbf{u} \cdot \mathbf{v}| \le |\mathbf{u}| |\mathbf{v}|$

79. $\mathbf{u} \cdot \mathbf{v} = \mathbf{v} \cdot \mathbf{u}$ Commutative property

80. $c(\mathbf{u} \cdot \mathbf{v}) = (c\mathbf{u}) \cdot \mathbf{v} = \mathbf{u} \cdot (c\mathbf{v})$ Associative property

81. $\mathbf{u} \cdot (\mathbf{v} + \mathbf{w}) = \mathbf{u} \cdot \mathbf{v} + \mathbf{u} \cdot \mathbf{w}$ Distributive property

82. Distributive properties

a. Show that $(\mathbf{u} + \mathbf{v}) \cdot (\mathbf{u} + \mathbf{v}) = |\mathbf{u}|^2 + 2\mathbf{u} \cdot \mathbf{v} + |\mathbf{v}|^2$.

b. Show that $(\mathbf{u} + \mathbf{v}) \cdot (\mathbf{u} + \mathbf{v}) = |\mathbf{u}|^2 + |\mathbf{v}|^2$ if $\mathbf{u}$ is orthogonal to $\mathbf{v}$.

c. Show that $(\mathbf{u} + \mathbf{v}) \cdot (\mathbf{u} - \mathbf{v}) = |\mathbf{u}|^2 - |\mathbf{v}|^2$.

83. Direction angles and cosines Let $\mathbf{v} = \langle a, b, c \rangle$ and let α, β, and γ be the angles between $\mathbf{v}$ and the positive x-axis, the positive y-axis, and the positive z-axis, respectively (see figure).

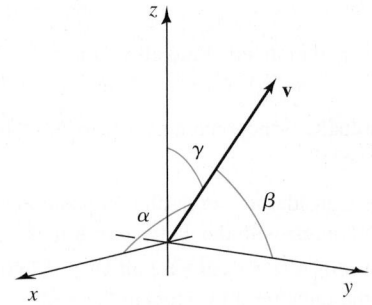

a. Prove that $\cos^2 \alpha + \cos^2 \beta + \cos^2 \gamma = 1$.

b. Find a vector that makes a $45°$ angle with $\mathbf{i}$ and $\mathbf{j}$. What angle does it make with $\mathbf{k}$?

c. Find a vector that makes a $60°$ angle with $\mathbf{i}$ and $\mathbf{j}$. What angle does it make with $\mathbf{k}$?

d. Is there a vector that makes a $30°$ angle with $\mathbf{i}$ and $\mathbf{j}$? Explain.

e. Find a vector $\mathbf{v}$ such that $\alpha = \beta = \gamma$. What is the angle?

84–88. Cauchy-Schwarz Inequality *The definition* $\mathbf{u} \cdot \mathbf{v} = |\mathbf{u}| |\mathbf{v}| \cos \theta$ *implies that* $|\mathbf{u} \cdot \mathbf{v}| \le |\mathbf{u}| |\mathbf{v}|$ *(because* $|\cos \theta| \le 1$*). This inequality, known as the Cauchy-Schwarz Inequality, holds in any number of dimensions and has many consequences.*

84. What conditions on $\mathbf{u}$ and $\mathbf{v}$ lead to equality in the Cauchy-Schwarz Inequality?

85. Verify that the Cauchy-Schwarz Inequality holds for $\mathbf{u} = \langle 3, -5, 6 \rangle$ and $\mathbf{v} = \langle -8, 3, 1 \rangle$.

86. Geometric-arithmetic mean Use the vectors $\mathbf{u} = \langle \sqrt{a}, \sqrt{b} \rangle$ and $\mathbf{v} = \langle \sqrt{b}, \sqrt{a} \rangle$ to show that $\sqrt{ab} \le \dfrac{a + b}{2}$, where $a \ge 0$ and $b \ge 0$.

87. Triangle Inequality Consider the vectors $\mathbf{u}$, $\mathbf{v}$, and $\mathbf{u} + \mathbf{v}$ (in any number of dimensions). Use the following steps to prove that $|\mathbf{u} + \mathbf{v}| \le |\mathbf{u}| + |\mathbf{v}|$.

a. Show that $|\mathbf{u} + \mathbf{v}|^2 = (\mathbf{u} + \mathbf{v}) \cdot (\mathbf{u} + \mathbf{v}) = |\mathbf{u}|^2 + 2\mathbf{u} \cdot \mathbf{v} + |\mathbf{v}|^2$.

b. Use the Cauchy-Schwarz Inequality to show that
$$|\mathbf{u} + \mathbf{v}|^2 \leq (|\mathbf{u}| + |\mathbf{v}|)^2.$$

c. Conclude that $|\mathbf{u} + \mathbf{v}| \leq |\mathbf{u}| + |\mathbf{v}|$.

d. Interpret the Triangle Inequality geometrically in $\mathbb{R}^2$ or $\mathbb{R}^3$.

88. **Algebra inequality** Show that
$$(u_1 + u_2 + u_3)^2 \leq 3(u_1^2 + u_2^2 + u_3^2),$$

for any real numbers u_1, u_2, and u_3. (*Hint:* Use the Cauchy-Schwarz Inequality in three dimensions with $\mathbf{u} = \langle u_1, u_2, u_3 \rangle$ and choose $\mathbf{v}$ in the right way.)

89. **Diagonals of a parallelogram** Consider the parallelogram with adjacent sides $\mathbf{u}$ and $\mathbf{v}$.

　　a. Show that the diagonals of the parallelogram are $\mathbf{u} + \mathbf{v}$ and $\mathbf{u} - \mathbf{v}$.

b. Prove that the diagonals have the same length if and only if $\mathbf{u} \cdot \mathbf{v} = 0$.

c. Show that the sum of the squares of the lengths of the diagonals equals the sum of the squares of the lengths of the sides.

QUICK CHECK ANSWERS

1. If $\theta = 0$, then $\mathbf{u}$ and $\mathbf{v}$ are parallel and point in the same direction. If $\theta = \pi$, then $\mathbf{u}$ and $\mathbf{v}$ are parallel and point in opposite directions.　**2.** All these dot products are zero, and the unit vectors are mutually orthogonal. The angle between two different unit vectors is $\pi/2$.　**3.** $\text{proj}_\mathbf{j}\mathbf{u} = 4\mathbf{i}$, $\text{proj}_\mathbf{j}\mathbf{u} = -3\mathbf{j}$, $\text{scal}_\mathbf{i}\mathbf{u} = 4$, $\text{scal}_\mathbf{j}\mathbf{u} = -3$ ◄

13.4 Cross Products

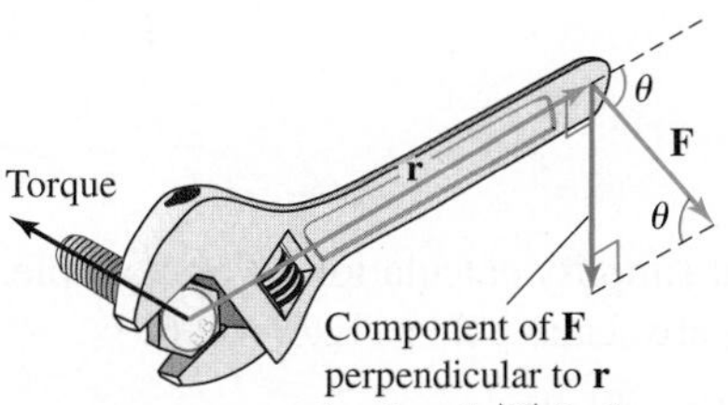

The dot product combines two vectors to produce a *scalar* result. There is an equally fundamental way to combine two vectors in $\mathbb{R}^3$ and obtain a *vector* result. This operation, known as the *cross product* (or *vector product*), may be motivated by a physical application.

　　Suppose you want to loosen a bolt with a wrench. As you apply force to the end of the wrench in the plane perpendicular to the bolt, the "twisting power" you generate depends on three variables:

• the magnitude of the force $\mathbf{F}$ applied to the wrench;

• the length $|\mathbf{r}|$ of the wrench;

• the angle at which the force is applied to the wrench.

The twisting generated by a force acting at a distance from a pivot point is called **torque** (from the Latin *to twist*). The torque is a vector whose magnitude is proportional to $|\mathbf{F}|$, $|\mathbf{r}|$, and $\sin\theta$, where θ is the angle between $\mathbf{F}$ and $\mathbf{r}$ (**Figure 13.55**). If the force is applied parallel to the wrench—for example, if you pull the wrench ($\theta = 0$) or push the wrench ($\theta = \pi$)—there is no twisting effect; if the force is applied perpendicular to the wrench ($\theta = \pi/2$), the twisting effect is maximized. The direction of the torque vector is defined to be orthogonal to both $\mathbf{F}$ and $\mathbf{r}$. As we will see shortly, the torque is expressed in terms of the cross product of $\mathbf{F}$ and $\mathbf{r}$.

Torque

Component of **F** perpendicular to **r** has length $|\mathbf{F}|\sin\theta$.

Figure 13.55

The Cross Product

The preceding physical example leads to the following definition of the cross product.

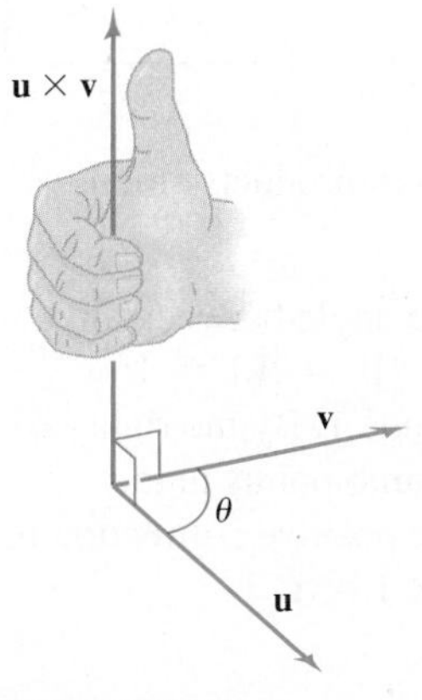

u × v

Figure 13.56

DEFINITION Cross Product

Given two nonzero vectors $\mathbf{u}$ and $\mathbf{v}$ in $\mathbb{R}^3$, the **cross product** $\mathbf{u} \times \mathbf{v}$ is a vector with magnitude

$$|\mathbf{u} \times \mathbf{v}| = |\mathbf{u}||\mathbf{v}| \sin\theta,$$

where $0 \leq \theta \leq \pi$ is the angle between $\mathbf{u}$ and $\mathbf{v}$. The direction of $\mathbf{u} \times \mathbf{v}$ is given by the **right-hand rule**: When you put the vectors tail to tail and let the fingers of your right hand curl from $\mathbf{u}$ to $\mathbf{v}$, the direction of $\mathbf{u} \times \mathbf{v}$ is the direction of your thumb, orthogonal to both $\mathbf{u}$ and $\mathbf{v}$ (**Figure 13.56**). When $\mathbf{u} \times \mathbf{v} = \mathbf{0}$, the direction of $\mathbf{u} \times \mathbf{v}$ is undefined.

QUICK CHECK 1 Sketch the vectors $\mathbf{u} = \langle 1, 2, 0 \rangle$ and $\mathbf{v} = \langle -1, 2, 0 \rangle$. Which way does $\mathbf{u} \times \mathbf{v}$ point? Which way does $\mathbf{v} \times \mathbf{u}$ point? ◄

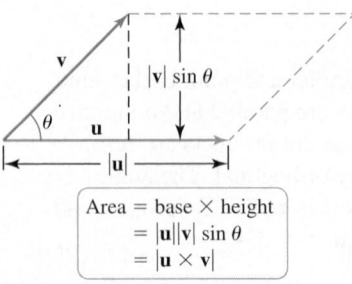

Area = base × height
= $|\mathbf{u}||\mathbf{v}| \sin \theta$
= $|\mathbf{u} \times \mathbf{v}|$

Figure 13.57

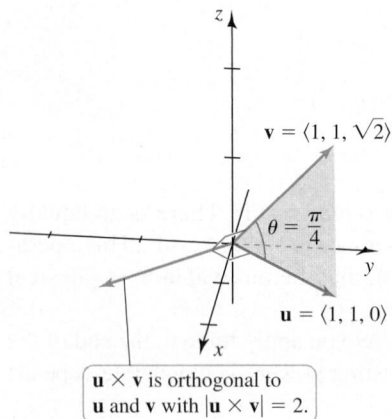

$\mathbf{u} \times \mathbf{v}$ is orthogonal to $\mathbf{u}$ and $\mathbf{v}$ with $|\mathbf{u} \times \mathbf{v}| = 2$.

Figure 13.58

QUICK CHECK 2 Explain why the vector $2\mathbf{u} \times 3\mathbf{v}$ points in the same direction as the vector $\mathbf{u} \times \mathbf{v}$. ◄

The following theorem is an immediate consequence of the definition of the cross product.

THEOREM 13.3   Geometry of the Cross Product
Let $\mathbf{u}$ and $\mathbf{v}$ be two nonzero vectors in $\mathbb{R}^3$.

1. The vectors $\mathbf{u}$ and $\mathbf{v}$ are parallel ($\theta = 0$ or $\theta = \pi$) if and only if $\mathbf{u} \times \mathbf{v} = \mathbf{0}$.

2. If $\mathbf{u}$ and $\mathbf{v}$ are two sides of a parallelogram (**Figure 13.57**), then the area of the parallelogram is

$$|\mathbf{u} \times \mathbf{v}| = |\mathbf{u}||\mathbf{v}| \sin \theta.$$

EXAMPLE 1   A cross product Find the magnitude and direction of $\mathbf{u} \times \mathbf{v}$, where $\mathbf{u} = \langle 1, 1, 0 \rangle$ and $\mathbf{v} = \langle 1, 1, \sqrt{2} \rangle$.

SOLUTION Because $\mathbf{u}$ is one side of a 45–45–90 triangle and $\mathbf{v}$ is the hypotenuse (**Figure 13.58**), we have $\theta = \pi/4$ and $\sin \theta = \frac{1}{\sqrt{2}}$. Also, $|\mathbf{u}| = \sqrt{2}$ and $|\mathbf{v}| = 2$, so the magnitude of $\mathbf{u} \times \mathbf{v}$ is

$$|\mathbf{u} \times \mathbf{v}| = |\mathbf{u}||\mathbf{v}| \sin \theta = \sqrt{2} \cdot 2 \cdot \frac{1}{\sqrt{2}} = 2.$$

The direction of $\mathbf{u} \times \mathbf{v}$ is given by the right-hand rule: $\mathbf{u} \times \mathbf{v}$ is orthogonal to $\mathbf{u}$ and $\mathbf{v}$ (**Figure 13.58**).

Related Exercises 14–15 ◄

Properties of the Cross Product

The cross product has several algebraic properties that simplify calculations. For example, scalars factor out of a cross product; that is, if a and b are scalars, then (Exercise 69)

$$(a\mathbf{u}) \times (b\mathbf{v}) = ab(\mathbf{u} \times \mathbf{v}).$$

The order in which the cross product is performed is important. The magnitudes of $\mathbf{u} \times \mathbf{v}$ and $\mathbf{v} \times \mathbf{u}$ are equal. However, applying the right-hand rule shows that $\mathbf{u} \times \mathbf{v}$ and $\mathbf{v} \times \mathbf{u}$ point in opposite directions. Therefore, $\mathbf{u} \times \mathbf{v} = -(\mathbf{v} \times \mathbf{u})$. There are two distributive properties for the cross product, whose proofs are omitted.

THEOREM 13.4   Properties of the Cross Product
Let $\mathbf{u}$, $\mathbf{v}$, and $\mathbf{w}$ be nonzero vectors in $\mathbb{R}^3$, and let a and b be scalars.

1. $\mathbf{u} \times \mathbf{v} = -(\mathbf{v} \times \mathbf{u})$ Anticommutative property

2. $(a\mathbf{u}) \times (b\mathbf{v}) = ab(\mathbf{u} \times \mathbf{v})$ Associative property

3. $\mathbf{u} \times (\mathbf{v} + \mathbf{w}) = (\mathbf{u} \times \mathbf{v}) + (\mathbf{u} \times \mathbf{w})$ Distributive property

4. $(\mathbf{u} + \mathbf{v}) \times \mathbf{w} = (\mathbf{u} \times \mathbf{w}) + (\mathbf{v} \times \mathbf{w})$ Distributive property

---

EXAMPLE 2   Cross products of unit vectors Evaluate all the cross products among the coordinate unit vectors $\mathbf{i}$, $\mathbf{j}$, and $\mathbf{k}$.

SOLUTION These vectors are mutually orthogonal, which means the angle between any two distinct vectors is $\theta = \pi/2$ and $\sin \theta = 1$. Furthermore, $|\mathbf{i}| = |\mathbf{j}| = |\mathbf{k}| = 1$. Therefore, the cross product of any two distinct vectors has magnitude 1. By the right-hand rule, when the fingers of the right hand curl from $\mathbf{i}$ to $\mathbf{j}$, the thumb points in the direction of the positive z-axis (**Figure 13.59**). The unit vector in the positive z-direction is $\mathbf{k}$, so $\mathbf{i} \times \mathbf{j} = \mathbf{k}$. Similar calculations show that $\mathbf{j} \times \mathbf{k} = \mathbf{i}$ and $\mathbf{k} \times \mathbf{i} = \mathbf{j}$.

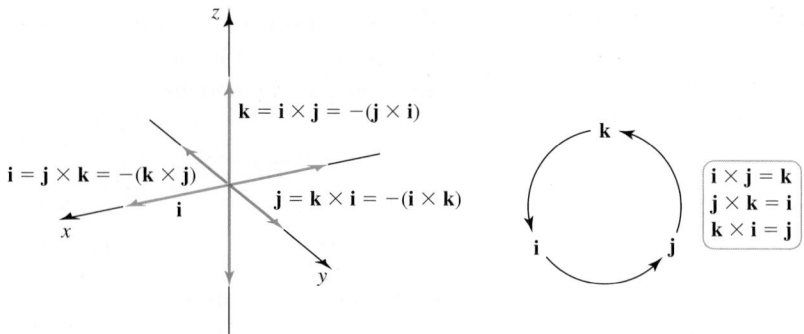

Figure 13.59

By property 1 of Theorem 13.4, $\mathbf{j} \times \mathbf{i} = -(\mathbf{i} \times \mathbf{j}) = -\mathbf{k}$, so $\mathbf{j} \times \mathbf{i}$ and $\mathbf{i} \times \mathbf{j}$ point in opposite directions. Similarly, $\mathbf{k} \times \mathbf{j} = -\mathbf{i}$ and $\mathbf{i} \times \mathbf{k} = -\mathbf{j}$. These relationships are easily remembered with the circle diagram in Figure 13.59. Finally, the angle between any unit vector and itself is $\theta = 0$. Therefore, $\mathbf{i} \times \mathbf{i} = \mathbf{j} \times \mathbf{j} = \mathbf{k} \times \mathbf{k} = \mathbf{0}$.

Related Exercises 17–19 ◀

THEOREM 13.5   Cross Products of Coordinate Unit Vectors

$$\mathbf{i} \times \mathbf{j} = -(\mathbf{j} \times \mathbf{i}) = \mathbf{k} \qquad \mathbf{j} \times \mathbf{k} = -(\mathbf{k} \times \mathbf{j}) = \mathbf{i}$$
$$\mathbf{k} \times \mathbf{i} = -(\mathbf{i} \times \mathbf{k}) = \mathbf{j} \qquad \mathbf{i} \times \mathbf{i} = \mathbf{j} \times \mathbf{j} = \mathbf{k} \times \mathbf{k} = \mathbf{0}$$

What is missing so far is an efficient method for finding the components of the cross product of two vectors in $\mathbb{R}^3$. Let $\mathbf{u} = u_1\mathbf{i} + u_2\mathbf{j} + u_3\mathbf{k}$ and $\mathbf{v} = v_1\mathbf{i} + v_2\mathbf{j} + v_3\mathbf{k}$. Using the distributive properties of the cross product (Theorem 13.4), we have

$$\mathbf{u} \times \mathbf{v} = (u_1\mathbf{i} + u_2\mathbf{j} + u_3\mathbf{k}) \times (v_1\mathbf{i} + v_2\mathbf{j} + v_3\mathbf{k})$$
$$= u_1v_1\underbrace{(\mathbf{i} \times \mathbf{i})}_{0} + u_1v_2\underbrace{(\mathbf{i} \times \mathbf{j})}_{\mathbf{k}} + u_1v_3\underbrace{(\mathbf{i} \times \mathbf{k})}_{-\mathbf{j}}$$
$$+ u_2v_1\underbrace{(\mathbf{j} \times \mathbf{i})}_{-\mathbf{k}} + u_2v_2\underbrace{(\mathbf{j} \times \mathbf{j})}_{0} + u_2v_3\underbrace{(\mathbf{j} \times \mathbf{k})}_{\mathbf{i}}$$
$$+ u_3v_1\underbrace{(\mathbf{k} \times \mathbf{i})}_{\mathbf{j}} + u_3v_2\underbrace{(\mathbf{k} \times \mathbf{j})}_{-\mathbf{i}} + u_3v_3\underbrace{(\mathbf{k} \times \mathbf{k})}_{0}.$$

> The determinant of the matrix A is denoted both $|A|$ and det A. The formula for the determinant of a 3×3 matrix A is
> $$\begin{vmatrix} a_1 & a_2 & a_3 \\ b_1 & b_2 & b_3 \\ c_1 & c_2 & c_3 \end{vmatrix} = a_1\begin{vmatrix} b_2 & b_3 \\ c_2 & c_3 \end{vmatrix} - a_2\begin{vmatrix} b_1 & b_3 \\ c_1 & c_3 \end{vmatrix} + a_3\begin{vmatrix} b_1 & b_2 \\ c_1 & c_2 \end{vmatrix},$$
> where
> $$\begin{vmatrix} a & b \\ c & d \end{vmatrix} = ad - bc.$$

This formula looks impossible to remember until we see that it fits the pattern used to evaluate 3×3 determinants. Specifically, if we compute the determinant of the matrix

$$\begin{array}{l} \text{Unit vectors} \quad \rightarrow \\ \text{Components of } \mathbf{u} \quad \rightarrow \\ \text{Components of } \mathbf{v} \quad \rightarrow \end{array} \begin{pmatrix} \mathbf{i} & \mathbf{j} & \mathbf{k} \\ u_1 & u_2 & u_3 \\ v_1 & v_2 & v_3 \end{pmatrix}$$

(expanding about the first row), the following formula for the cross product emerges (see margin note).

THEOREM 13.6   Evaluating the Cross Product

Let $\mathbf{u} = u_1\mathbf{i} + u_2\mathbf{j} + u_3\mathbf{k}$ and $\mathbf{v} = v_1\mathbf{i} + v_2\mathbf{j} + v_3\mathbf{k}$. Then

$$\mathbf{u} \times \mathbf{v} = \begin{vmatrix} \mathbf{i} & \mathbf{j} & \mathbf{k} \\ u_1 & u_2 & u_3 \\ v_1 & v_2 & v_3 \end{vmatrix} = \begin{vmatrix} u_2 & u_3 \\ v_2 & v_3 \end{vmatrix}\mathbf{i} - \begin{vmatrix} u_1 & u_3 \\ v_1 & v_3 \end{vmatrix}\mathbf{j} + \begin{vmatrix} u_1 & u_2 \\ v_1 & v_2 \end{vmatrix}\mathbf{k}.$$

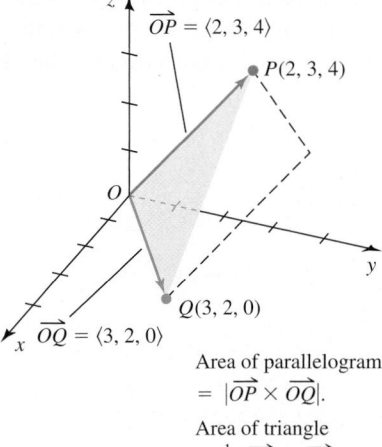

$\overrightarrow{OP} = \langle 2, 3, 4 \rangle$
$P(2, 3, 4)$
$\overrightarrow{OQ} = \langle 3, 2, 0 \rangle$
$Q(3, 2, 0)$

Area of parallelogram
$= |\overrightarrow{OP} \times \overrightarrow{OQ}|.$

Area of triangle
$= \frac{1}{2}|\overrightarrow{OP} \times \overrightarrow{OQ}|.$

Figure 13.60

EXAMPLE 3   Area of a triangle Find the area of the triangle with vertices $O(0, 0, 0)$, $P(2, 3, 4)$, and $Q(3, 2, 0)$ (**Figure 13.60**).

SOLUTION First consider the parallelogram, two of whose sides are the vectors $\overrightarrow{OP}$ and $\overrightarrow{OQ}$. By Theorem 13.3, the area of this parallelogram is $|\overrightarrow{OP} \times \overrightarrow{OQ}|$. Computing the cross product, we find that

$$\overrightarrow{OP} \times \overrightarrow{OQ} = \begin{vmatrix} \mathbf{i} & \mathbf{j} & \mathbf{k} \\ 2 & 3 & 4 \\ 3 & 2 & 0 \end{vmatrix} = \begin{vmatrix} 3 & 4 \\ 2 & 0 \end{vmatrix} \mathbf{i} - \begin{vmatrix} 2 & 4 \\ 3 & 0 \end{vmatrix} \mathbf{j} + \begin{vmatrix} 2 & 3 \\ 3 & 2 \end{vmatrix} \mathbf{k}$$

$$= -8\mathbf{i} + 12\mathbf{j} - 5\mathbf{k}.$$

Therefore, the area of the parallelogram is

$$|\overrightarrow{OP} \times \overrightarrow{OQ}| = |-8\mathbf{i} + 12\mathbf{j} - 5\mathbf{k}| = \sqrt{233} \approx 15.26.$$

The triangle with vertices O, P, and Q forms half of the parallelogram, so its area is $\sqrt{233}/2 \approx 7.63$.

Related Exercises 34–36 ◀

EXAMPLE 4   **Vector orthogonal to two vectors**   Find a vector orthogonal to the two vectors $\mathbf{u} = -\mathbf{i} + 6\mathbf{k}$ and $\mathbf{v} = 2\mathbf{i} - 5\mathbf{j} - 3\mathbf{k}$.

SOLUTION A vector orthogonal to $\mathbf{u}$ and $\mathbf{v}$ is parallel to $\mathbf{u} \times \mathbf{v}$ (Figure 13.61). One such orthogonal vector is

$$\mathbf{u} \times \mathbf{v} = \begin{vmatrix} \mathbf{i} & \mathbf{j} & \mathbf{k} \\ -1 & 0 & 6 \\ 2 & -5 & -3 \end{vmatrix}$$

$$= (0 + 30)\mathbf{i} - (3 - 12)\mathbf{j} + (5 - 0)\mathbf{k}$$

$$= 30\mathbf{i} + 9\mathbf{j} + 5\mathbf{k}.$$

Any scalar multiple of this vector is also orthogonal to $\mathbf{u}$ and $\mathbf{v}$.

Related Exercises 42–44 ◀

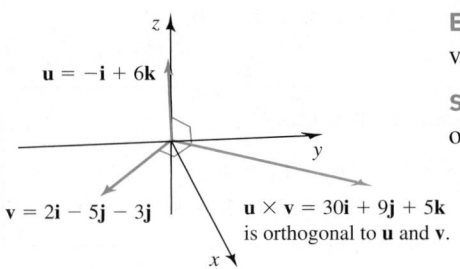

Figure 13.61

QUICK CHECK 3 A good check on a cross product calculation is to verify that $\mathbf{u}$ and $\mathbf{v}$ are orthogonal to the computed $\mathbf{u} \times \mathbf{v}$. In Example 4, verify that $\mathbf{u} \cdot (\mathbf{u} \times \mathbf{v}) = 0$ and $\mathbf{v} \cdot (\mathbf{u} \times \mathbf{v}) = 0$. ◀

Applications of the Cross Product

We now investigate two physical applications of the cross product.

Torque   Returning to the example of applying a force to a wrench, suppose a force $\mathbf{F}$ is applied to the point P at the head of a vector $\mathbf{r} = \overrightarrow{OP}$ (Figure 13.62). The **torque**, or twisting effect, produced by the force about the point O is given by $\boldsymbol{\tau} = \mathbf{r} \times \mathbf{F}$. The torque vector has a magnitude of

$$|\boldsymbol{\tau}| = |\mathbf{r} \times \mathbf{F}| = |\mathbf{r}||\mathbf{F}| \sin \theta,$$

where θ is the angle between $\mathbf{r}$ and $\mathbf{F}$. The direction of the torque is given by the right-hand rule; it is orthogonal to both $\mathbf{r}$ and $\mathbf{F}$. As noted earlier, if $\mathbf{r}$ and $\mathbf{F}$ are parallel, then $\sin \theta = 0$ and the torque is zero. For a given $\mathbf{r}$ and $\mathbf{F}$, the maximum torque occurs when $\mathbf{F}$ is applied in a direction orthogonal to $\mathbf{r}$ ($\theta = \pi/2$).

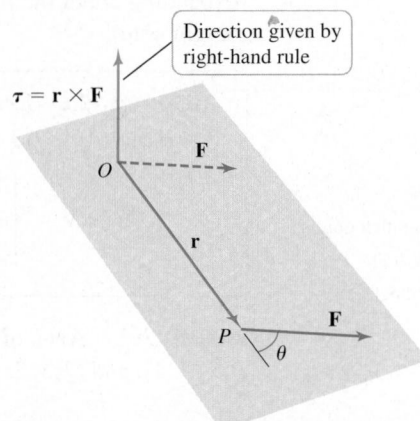

Figure 13.62

EXAMPLE 5 **Tightening a bolt** A force of 20 N is applied to a wrench attached to a bolt in a direction perpendicular to the bolt (Figure 13.63). Which produces more torque: applying the force at an angle of 60° on a wrench that is 0.15 m long or applying the force at an angle of 135° on a wrench that is 0.25 m long? In each case, what is the direction of the torque?

▶ When standard threads are added to the bolt in Figure 13.63, the forces used in Example 5 cause the bolt to move upward into a nut—in the direction of the torque.

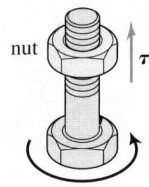

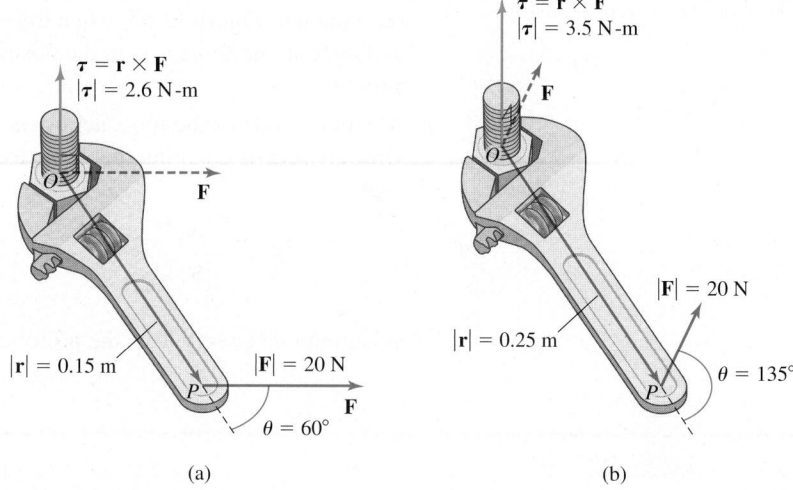

Figure 13.63

SOLUTION The magnitude of the torque in the first case is

$$|\boldsymbol{\tau}| = |\mathbf{r}||\mathbf{F}| \sin\theta = (0.15 \text{ m})(20 \text{ N}) \sin 60° \approx 2.6 \text{ N-m.}$$

In the second case, the magnitude of the torque is

$$|\boldsymbol{\tau}| = |\mathbf{r}||\mathbf{F}| \sin\theta = (0.25 \text{ m})(20 \text{ N}) \sin 135° \approx 3.5 \text{ N-m.}$$

The second instance gives the greater torque. In both cases, the torque is orthogonal to $\mathbf{r}$ and $\mathbf{F}$, parallel to the shaft of the bolt (Figure 13.63).

Related Exercises 47, 49 ◀

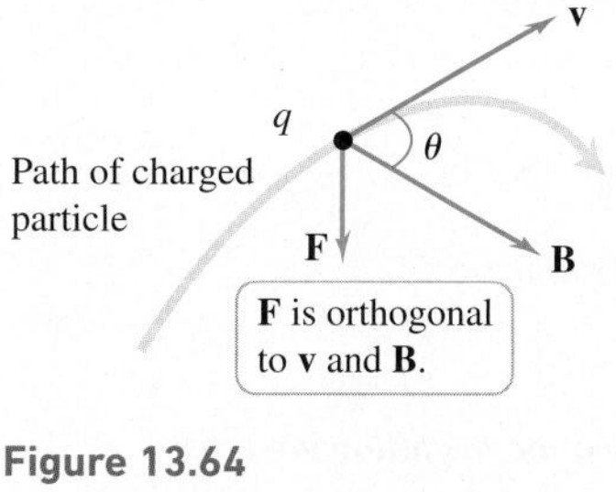

Magnetic Force on a Moving Charge Moving electric charges (either an isolated charge or a current in a wire) experience a force when they pass through a magnetic field. For an isolated charge q, the force is given by $\mathbf{F} = q(\mathbf{v} \times \mathbf{B})$, where $\mathbf{v}$ is the velocity of the charge and $\mathbf{B}$ is the magnetic field. The magnitude of the force is

$$|\mathbf{F}| = |q||\mathbf{v} \times \mathbf{B}| = |q||\mathbf{v}||\mathbf{B}| \sin\theta,$$

where θ is the angle between $\mathbf{v}$ and $\mathbf{B}$ (Figure 13.64). Note that the sign of the charge also determines the direction of the force. If the velocity vector is parallel to the magnetic field, the charge experiences no force. The maximum force occurs when the velocity is orthogonal to the magnetic field.

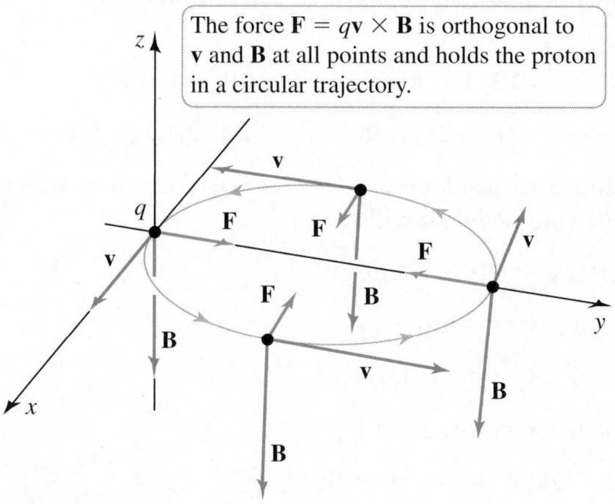

Path of charged particle

F is orthogonal to **v** and **B**.

Figure 13.64

The force $\mathbf{F} = q\mathbf{v} \times \mathbf{B}$ is orthogonal to $\mathbf{v}$ and $\mathbf{B}$ at all points and holds the proton in a circular trajectory.

EXAMPLE 6 **Force on a proton** A proton with a mass of 1.7×10^{-27} kg and a charge of $q = +1.6 \times 10^{-19}$ coulombs (C) moves along the x-axis with a speed of $|\mathbf{v}| = 9 \times 10^5$ m/s. When it reaches $(0, 0, 0)$, a uniform magnetic field is turned on. The field has a constant strength of 1 tesla (1 T) and is directed along the negative z-axis (Figure 13.65).

a. Find the magnitude and direction of the force on the proton at the instant it enters the magnetic field.

b. Assume the proton loses no energy and the force in part (a) acts as a *centripetal force* with magnitude $|\mathbf{F}| = m|\mathbf{v}|^2/R$ that keeps the proton in a circular orbit of radius R. Find the radius of the orbit.

Figure 13.65

➤ The standard unit of magnetic field strength is the tesla (T, named after Nicola Tesla). A typical strong bar magnet has a strength of about 1 T. In terms of other units, 1 T = 1 kg/(C-s), where C is the unit of charge called the *coulomb*. Therefore, the units of force in Example 6a are kg-m/s², or newtons.

SOLUTION

a. Expressed as vectors, we have $\mathbf{v} = 9 \times 10^5\,\mathbf{i}$ and $\mathbf{B} = -\mathbf{k}$. Therefore, the force on the proton in newtons is

$$\mathbf{F} = q(\mathbf{v} \times \mathbf{B}) = 1.6 \times 10^{-19}\big((9 \times 10^5\,\mathbf{i}) \times (-\mathbf{k})\big)$$
$$= 1.44 \times 10^{-13}\mathbf{j}.$$

As shown in Figure 13.65, when the proton enters the magnetic field in the positive *x*-direction, the force acts in the positive *y*-direction, which changes the path of the proton.

b. The magnitude of the force acting on the proton remains 1.44×10^{-13} N at all times (from part (a)). Equating this force to the centripetal force $|\mathbf{F}| = m|\mathbf{v}|^2/R$, we find that

$$R = \frac{m|\mathbf{v}|^2}{|\mathbf{F}|} = \frac{(1.7 \times 10^{-27}\text{ kg})\,(9 \times 10^5\text{ m/s})^2}{1.44 \times 10^{-13}\text{ N}} \approx 0.01\text{ m}.$$

Assuming no energy loss, the proton moves in a circular orbit of radius 0.01 m.

Related Exercises 55, 67 ◀

SECTION 13.4 EXERCISES

Getting Started

1. What is the magnitude of the cross product of two parallel vectors?

2. If **u** and **v** are orthogonal, what is the magnitude of $\mathbf{u} \times \mathbf{v}$?

3. Suppose **u** and **v** are nonzero vectors. What is the geometric relationship between **u** and **v** under each of the following conditions?

    a. $\mathbf{u} \cdot \mathbf{v} = 0$ **b.** $\mathbf{u} \times \mathbf{v} = \mathbf{0}$

4. Use a geometric argument to explain why $\mathbf{u} \cdot (\mathbf{u} \times \mathbf{v}) = 0$.

5. Compute $|\mathbf{u} \times \mathbf{v}|$ if **u** and **v** are unit vectors and the angle between them is $\pi/4$.

6. Compute $|\mathbf{u} \times \mathbf{v}|$ if $|\mathbf{u}| = 3$ and $|\mathbf{v}| = 4$ and the angle between **u** and **v** is $2\pi/3$.

7. Find $\mathbf{v} \times \mathbf{u}$ if $\mathbf{u} \times \mathbf{v} = 3\mathbf{i} + 2\mathbf{j} - 7\mathbf{k}$.

8. For any vector **v** in $\mathbb{R}^3$, explain why $\mathbf{v} \times \mathbf{v} = \mathbf{0}$.

9. Explain how to use a determinant to compute $\mathbf{u} \times \mathbf{v}$.

10. Explain how to find the torque produced by a force using cross products.

Practice Exercises

11–12. Cross products from the definition *Find the cross product* $\mathbf{u} \times \mathbf{v}$ *in each figure.*

11.

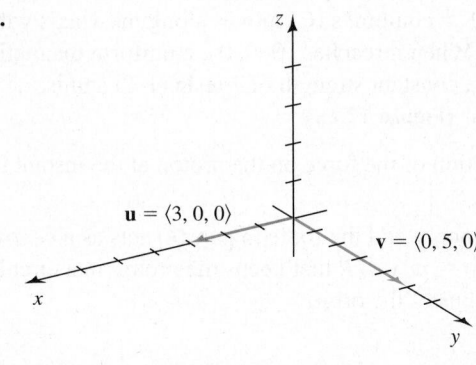

$\mathbf{u} = \langle 3, 0, 0 \rangle$

$\mathbf{v} = \langle 0, 5, 0 \rangle$

12.

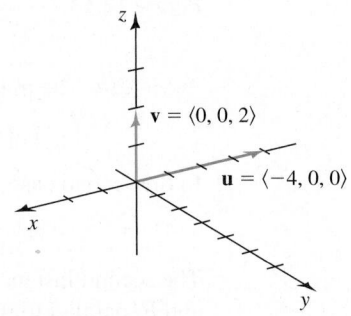

$\mathbf{v} = \langle 0, 0, 2 \rangle$

$\mathbf{u} = \langle -4, 0, 0 \rangle$

13–16. Cross products from the definition *Sketch the following vectors* **u** *and* **v**. *Then compute* $|\mathbf{u} \times \mathbf{v}|$ *and show the cross product on your sketch.*

13. $\mathbf{u} = \langle 0, -2, 0 \rangle, \mathbf{v} = \langle 0, 1, 0 \rangle$

14. $\mathbf{u} = \langle 0, 4, 0 \rangle, \mathbf{v} = \langle 0, 0, -8 \rangle$

15. $\mathbf{u} = \langle 3, 3, 0 \rangle, \mathbf{v} = \langle 3, 3, 3\sqrt{2} \rangle$

16. $\mathbf{u} = \langle 0, -2, -2 \rangle, \mathbf{v} = \langle 0, 2, -2 \rangle$

17–22. Coordinate unit vectors *Compute the following cross products. Then make a sketch showing the two vectors and their cross product.*

17. $\mathbf{j} \times \mathbf{k}$ 18. $\mathbf{i} \times \mathbf{k}$ 19. $-\mathbf{j} \times \mathbf{k}$

20. $3\mathbf{j} \times \mathbf{i}$ 21. $-2\mathbf{i} \times 3\mathbf{k}$ 22. $2\mathbf{j} \times (-5)\mathbf{i}$

23–28. Computing cross products *Find the cross products* $\mathbf{u} \times \mathbf{v}$ *and* $\mathbf{v} \times \mathbf{u}$ *for the following vectors* **u** *and* **v**.

23. $\mathbf{u} = \langle 3, 5, 0 \rangle, \mathbf{v} = \langle 0, 3, -6 \rangle$

24. $\mathbf{u} = \langle -4, 1, 1 \rangle, \mathbf{v} = \langle 0, 1, -1 \rangle$

25. $\mathbf{u} = \langle 2, 3, -9 \rangle, \mathbf{v} = \langle -1, 1, -1 \rangle$

26. $\mathbf{u} = \langle 3, -4, 6 \rangle, \mathbf{v} = \langle 1, 2, -1 \rangle$

27. $\mathbf{u} = 3\mathbf{i} - \mathbf{j} - 2\mathbf{k}, \mathbf{v} = \mathbf{i} + 3\mathbf{j} - 2\mathbf{k}$

28. $\mathbf{u} = 2\mathbf{i} - 10\mathbf{j} + 15\mathbf{k}, \mathbf{v} = 0.5\mathbf{i} + \mathbf{j} - 0.6\mathbf{k}$

29–32. Area of a parallelogram *Find the area of the parallelogram that has two adjacent sides* **u** *and* **v**.

29. $\mathbf{u} = 3\mathbf{i} - \mathbf{j}, \mathbf{v} = 3\mathbf{j} + 2\mathbf{k}$

30. $\mathbf{u} = -3\mathbf{i} + 2\mathbf{k}, \mathbf{v} = \mathbf{i} + \mathbf{j} + \mathbf{k}$

31. $\mathbf{u} = 2\mathbf{i} - \mathbf{j} - 2\mathbf{k}, \mathbf{v} = 3\mathbf{i} + 2\mathbf{j} - \mathbf{k}$

32. $\mathbf{u} = 8\mathbf{i} + 2\mathbf{j} - 3\mathbf{k}, \mathbf{v} = 2\mathbf{i} + 4\mathbf{j} - 4\mathbf{k}$

33–38. Areas of triangles *Find the area of the following triangles T.*

33. The vertices of T are $A(0, 0, 0)$, $B(3, 0, 1)$, and $C(1, 1, 0)$.

34. The vertices of T are $O(0, 0, 0)$, $P(1, 2, 3)$, and $Q(6, 5, 4)$.

35. The vertices of T are $A(5, 6, 2)$, $B(7, 16, 4)$, and $C(6, 7, 3)$.

36. The vertices of T are $A(-1, -5, -3)$, $B(-3, -2, -1)$, and $C(0, -5, -1)$.

37. The sides of T are $\mathbf{u} = \langle 3, 3, 3 \rangle$, $\mathbf{v} = \langle 6, 0, 6 \rangle$, and $\mathbf{u} - \mathbf{v}$.

38. The sides of T are $\mathbf{u} = \langle 0, 6, 0 \rangle$, $\mathbf{v} = \langle 4, 4, 4 \rangle$, and $\mathbf{u} - \mathbf{v}$.

39. Collinear points and cross products Explain why the points A, B, and C in $\mathbb{R}^3$ are collinear if and only if $\overrightarrow{AB} \times \overrightarrow{AC} = \mathbf{0}$.

40–41. Collinear points *Use cross products to determine whether the points A, B, and C are collinear.*

40. $A(3, 2, 1)$, $B(5, 4, 7)$, and $C(9, 8, 19)$

41. $A(-3, -2, 1)$, $B(1, 4, 7)$, and $C(4, 10, 14)$

42–44. Orthogonal vectors *Find a vector orthogonal to the given vectors.*

42. $\langle 1, 2, 3 \rangle$ and $\langle -2, 4, -1 \rangle$

43. $\langle 0, 1, 2 \rangle$ and $\langle -2, 0, 3 \rangle$

44. $\langle 8, 0, 4 \rangle$ and $\langle -8, 2, 1 \rangle$

45–48. Computing torque *Answer the following questions about torque.*

45. Let $\mathbf{r} = \overrightarrow{OP} = \mathbf{i} + \mathbf{j} + \mathbf{k}$. A force $\mathbf{F} = \langle 20, 0, 0 \rangle$ is applied at P. Find the torque about O that is produced.

46. Let $\mathbf{r} = \overrightarrow{OP} = \mathbf{i} - \mathbf{j} + 2\mathbf{k}$. A force $\mathbf{F} = \langle 10, 10, 0 \rangle$ is applied at P. Find the torque about O that is produced.

47. Let $\mathbf{r} = \overrightarrow{OP} = 10\mathbf{i}$. Which is greater (in magnitude): the torque about O when a force $\mathbf{F} = 5\mathbf{i} - 5\mathbf{k}$ is applied at P or the torque about O when a force $\mathbf{F} = 4\mathbf{i} - 3\mathbf{j}$ is applied at P?

48. A pump handle has a pivot at $(0, 0, 0)$ and extends to $P(5, 0, -5)$. A force $\mathbf{F} = \langle 1, 0, -10 \rangle$ is applied at P. Find the magnitude and direction of the torque about the pivot.

49. Tightening a bolt Suppose you apply a force of 20 N to a 0.25-meter-long wrench attached to a bolt in a direction perpendicular to the bolt. Determine the magnitude of the torque when the force is applied at an angle of 45° to the wrench.

50. Opening a laptop A force of 1.5 lb is applied in a direction perpendicular to the screen of a laptop at a distance of 10 in from the hinge of the screen. Find the magnitude of the torque (in ft-lb) that you apply.

51. Bicycle brakes A set of caliper brakes exerts a force on the rim of a bicycle wheel that creates a frictional force **F** of 40 N perpendicular to the radius of the wheel (see figure). Assuming the wheel

has a radius of 33 cm, find the magnitude and direction of the torque about the axle of the wheel.

52. Arm torque A horizontally outstretched arm supports a weight of 20 lb in a hand (see figure). If the distance from the shoulder to the elbow is 1 ft and the distance from the elbow to the hand is 1 ft, find the magnitude and describe the direction of the torque about (a) the shoulder and (b) the elbow. (The units of torque in this case are ft-lb.)

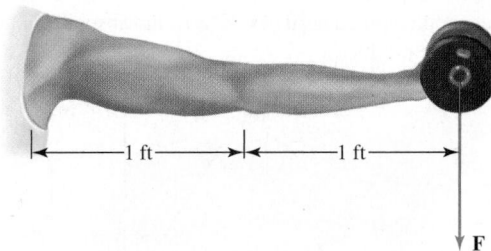

53–56. Force on a moving charge *Answer the following questions about force on a moving charge.*

53. A particle with a unit positive charge ($q = 1$) enters a constant magnetic field $\mathbf{B} = \mathbf{i} + \mathbf{j}$ with a velocity $\mathbf{v} = 20\mathbf{k}$. Find the magnitude and direction of the force on the particle. Make a sketch of the magnetic field, the velocity, and the force.

54. A particle with a unit negative charge ($q = -1$) enters a constant magnetic field $\mathbf{B} = 5\mathbf{k}$ with a velocity $\mathbf{v} = \mathbf{i} + 2\mathbf{j}$. Find the magnitude and direction of the force on the particle. Make a sketch of the magnetic field, the velocity, and the force.

55. An electron ($q = -1.6 \times 10^{-19}$ C) enters a constant 2-T magnetic field at an angle of 45° to the field with a speed of 2×10^5 m/s. Find the magnitude of the force on the electron.

56. A proton ($q = 1.6 \times 10^{-19}$ C) with velocity $2 \times 10^6 \mathbf{j}$ m/s experiences a force in newtons of $\mathbf{F} = 5 \times 10^{-12} \mathbf{k}$ as it passes through the origin. Find the magnitude and direction of the magnetic field at that instant.

57. Explain why or why not Determine whether the following statements are true and give an explanation or counterexample.

a. The cross product of two nonzero vectors is a nonzero vector.

b. $|\mathbf{u} \times \mathbf{v}|$ is less than both $|\mathbf{u}|$ and $|\mathbf{v}|$.

c. If **u** points east and **v** points south, then $\mathbf{u} \times \mathbf{v}$ points west.

d. If $\mathbf{u} \times \mathbf{v} = \mathbf{0}$ and $\mathbf{u} \cdot \mathbf{v} = 0$, then either $\mathbf{u} = \mathbf{0}$ or $\mathbf{v} = \mathbf{0}$.

e. Law of Cancellation? If $\mathbf{u} \times \mathbf{v} = \mathbf{u} \times \mathbf{w}$, then $\mathbf{v} = \mathbf{w}$.

58. Finding an unknown Find the value of a such that $\langle a, a, 2 \rangle \times \langle 1, a, 3 \rangle = \langle 2, -4, 2 \rangle$.

59. Vector equation Find all vectors **u** that satisfy the equation
$$\langle 1, 1, 1 \rangle \times \mathbf{u} = \langle -1, -1, 2 \rangle.$$

60. Vector equation Find all vectors **u** that satisfy the equation
$$\langle 1, 1, 1 \rangle \times \mathbf{u} = \langle 0, 0, 1 \rangle.$$

61. Area of a triangle Find the area of the triangle with vertices on the coordinate axes at the points $(a, 0, 0)$, $(0, b, 0)$, and $(0, 0, c)$, in terms of a, b, and c.

Explorations and Challenges

62–66. Scalar triple product *Another operation with vectors is the scalar triple product, defined to be* $\mathbf{u} \cdot (\mathbf{v} \times \mathbf{w})$, *for nonzero vectors* **u**, **v**, *and* **w** *in* $\mathbb{R}^3$.

62. Express **u**, **v**, and **w** in terms of their components, and show that $\mathbf{u} \cdot (\mathbf{v} \times \mathbf{w})$ equals the determinant

$$\begin{vmatrix} u_1 & u_2 & u_3 \\ v_1 & v_2 & v_3 \\ w_1 & w_2 & w_3 \end{vmatrix}.$$

63. Consider the *parallelepiped* (slanted box) determined by the position vectors **u**, **v**, and **w** (see figure). Show that the volume of the parallelepiped is $|\mathbf{u} \cdot (\mathbf{v} \times \mathbf{w})|$, the absolute value of the scalar triple product.

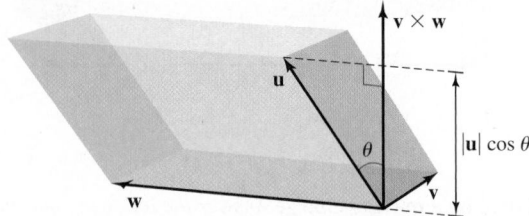

64. Find the volume of the parallelepiped determined by the position vectors $\mathbf{u} = \langle 3, 1, 0 \rangle$, $\mathbf{v} = \langle 2, 4, 1 \rangle$, and $\mathbf{w} = \langle 1, 1, 5 \rangle$ (see Exercise 63).

65. Explain why the position vectors **u**, **v**, and **w** are coplanar if and only if $|\mathbf{u} \cdot (\mathbf{v} \times \mathbf{w})| = 0$. (*Hint:* See Exercise 63).

66. Prove that $\mathbf{u} \cdot (\mathbf{v} \times \mathbf{w}) = (\mathbf{u} \times \mathbf{v}) \cdot \mathbf{w}$.

67. Electron speed An electron with a mass of 9.1×10^{-31} kg and a charge of -1.6×10^{-19} C travels in a circular path with no loss of energy in a magnetic field of 0.05 T that is orthogonal to the path of the electron (see figure). If the radius of the path is 0.002 m, what is the speed of the electron?

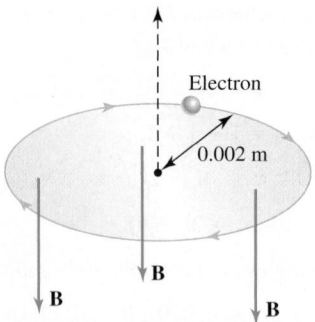

68. Three proofs Prove that $\mathbf{u} \times \mathbf{u} = \mathbf{0}$ in three ways.
   a. Use the definition of the cross product.
   b. Use the determinant formulation of the cross product.
   c. Use the property that $\mathbf{u} \times \mathbf{v} = -(\mathbf{v} \times \mathbf{u})$.

69. Associative property Prove in two ways that for scalars a and b, $(a\mathbf{u}) \times (b\mathbf{v}) = ab(\mathbf{u} \times \mathbf{v})$. Use the definition of the cross product and the determinant formula.

70–72. Possible identities *Determine whether the following statements are true using a proof or counterexample. Assume* **u**, **v**, *and* **w** *are nonzero vectors in* $\mathbb{R}^3$.

70. $\mathbf{u} \times (\mathbf{u} \times \mathbf{v}) = \mathbf{0}$

71. $(\mathbf{u} - \mathbf{v}) \times (\mathbf{u} + \mathbf{v}) = 2\mathbf{u} \times \mathbf{v}$

72. $\mathbf{u} \cdot (\mathbf{v} \times \mathbf{w}) = \mathbf{w} \cdot (\mathbf{u} \times \mathbf{v})$

73–74. Identities *Prove the following identities. Assume* **u**, **v**, **w**, *and* **x** *are nonzero vectors in* $\mathbb{R}^3$.

73. $\mathbf{u} \times (\mathbf{v} \times \mathbf{w}) = (\mathbf{u} \cdot \mathbf{w})\mathbf{v} - (\mathbf{u} \cdot \mathbf{v})\mathbf{w}$ Vector triple product

74. $(\mathbf{u} \times \mathbf{v}) \cdot (\mathbf{w} \times \mathbf{x}) = (\mathbf{u} \cdot \mathbf{w})(\mathbf{v} \cdot \mathbf{x}) - (\mathbf{u} \cdot \mathbf{x})(\mathbf{v} \cdot \mathbf{w})$

75. Cross product equations Suppose **u** and **v** are nonzero vectors in $\mathbb{R}^3$.
   a. Prove that the equation $\mathbf{u} \times \mathbf{z} = \mathbf{v}$ has a nonzero solution **z** if and only if $\mathbf{u} \cdot \mathbf{v} = 0$. (*Hint:* Take the dot product of both sides with **v**.)
   b. Explain this result geometrically.

QUICK CHECK ANSWERS

1. $\mathbf{u} \times \mathbf{v}$ points in the positive z-direction; $\mathbf{v} \times \mathbf{u}$ points in the negative z-direction.   **2.** The vector $2\mathbf{u}$ points in the same direction as **u**, and the vector $3\mathbf{v}$ points in the same direction as **v**. So the right-hand rule gives the same direction for $2\mathbf{u} \times 3\mathbf{v}$ as it does for $\mathbf{u} \times \mathbf{v}$.   **3.** $\mathbf{u} \cdot (\mathbf{u} \times \mathbf{v}) = \langle -1, 0, 6 \rangle \cdot \langle 30, 9, 5 \rangle = -30 + 0 + 30 = 0$. A similar calculation shows that $\mathbf{v} \cdot (\mathbf{u} \times \mathbf{v}) = 0$. ◄

13.5 Lines and Planes in Space

In Chapter 1, we reviewed the catalog of standard functions and their associated graphs. For example, the graph of a linear equation in two variables ($y = mx + b$) is a line, the graph of a quadratic equation ($y = ax^2 + bx + c$) is a parabola, and both of these graphs lie in the xy-plane (two-dimensional space). Our immediate aim is to begin a similar journey through three-dimensional space. What are the basic geometrical objects in three dimensions, and how do we describe them with equations?

Certainly among the most fundamental objects in three dimensions are the line and plane. In this section, we develop equations for both lines and planes and explore their properties and uses.

Lines in Space

Two distinct points in $\mathbb{R}^3$ determine a unique line. Alternatively, one point and a direction also determine a unique line. We use both of these properties to derive two different descriptions of lines in space: one using parametric equations, and one using vector equations.

Let ℓ be the line passing through the point $P_0(x_0, y_0, z_0)$ parallel to the nonzero vector $\mathbf{v} = \langle a, b, c \rangle$, where P_0 and $\mathbf{v}$ are given. The fixed point P_0 is associated with the position vector $\mathbf{r}_0 = \overrightarrow{OP_0} = \langle x_0, y_0, z_0 \rangle$. We let $P(x, y, z)$ be a variable point on ℓ and let $\mathbf{r} = \overrightarrow{OP} = \langle x, y, z \rangle$ be the position vector associated with P (Figure 13.66). Because ℓ is parallel to $\mathbf{v}$, the vector $\overrightarrow{P_0P}$ is also parallel to $\mathbf{v}$; therefore, $\overrightarrow{P_0P} = t\mathbf{v}$, where t is a real number. By vector addition, we see that $\overrightarrow{OP} = \overrightarrow{OP_0} + \overrightarrow{P_0P}$, or $\overrightarrow{OP} = \overrightarrow{OP_0} + t\mathbf{v}$. Expressing these vectors in terms of their components, we obtain a *vector equation* for a line:

$$\text{Component form} \quad \underbrace{\langle x, y, z \rangle}_{\mathbf{r} = \overrightarrow{OP}} = \underbrace{\langle x_0, y_0, z_0 \rangle}_{\mathbf{r}_0 = \overrightarrow{OP_0}} + t\underbrace{\langle a, b, c \rangle}_{\mathbf{v}} \quad \text{or}$$

$$\text{Vector form} \quad \mathbf{r} = \mathbf{r}_0 + t\mathbf{v}.$$

Equating components, we obtain *parametric equations* for a line:

$$x = x_0 + at, \qquad y = y_0 + bt, \qquad z = z_0 + ct, \quad \text{for } -\infty < t < \infty.$$

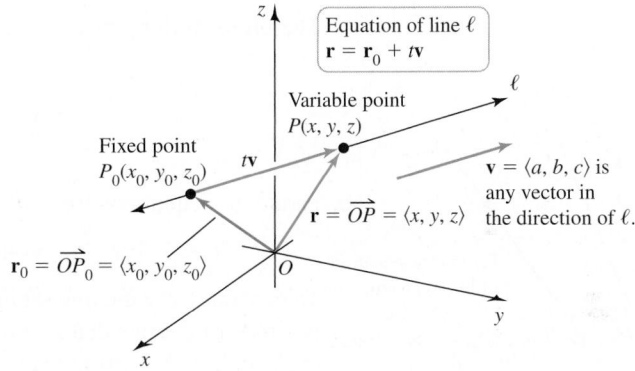

Figure 13.66

QUICK CHECK 1 Describe the line $\mathbf{r} = t\mathbf{k}$, for $-\infty < t < \infty$. Describe the line $\mathbf{r} = t(\mathbf{i} + \mathbf{j} + 0\mathbf{k})$, for $-\infty < t < \infty$. ◄

The parameter t determines the location of points on the line, where $t = 0$ corresponds to P_0. If t increases from 0, we move along the line in the direction of $\mathbf{v}$, and if t decreases from 0, we move along the line in the direction of $-\mathbf{v}$. As t varies over all real numbers $(-\infty < t < \infty)$, the entire line ℓ is generated. If, instead of knowing the direction $\mathbf{v}$ of the line, we are given two points $P_0(x_0, y_0, z_0)$ and $P_1(x_1, y_1, z_1)$, then the direction of the line is $\mathbf{v} = \overrightarrow{P_0P_1} = \langle x_1 - x_0, y_1 - y_0, z_1 - z_0 \rangle$.

▶ There are infinitely many equations for the same line. The direction vector is determined only up to a scalar multiple.

> **Equation of a Line**
>
> A **vector equation of the line** passing through the point $P_0(x_0, y_0, z_0)$ in the direction of the vector $\mathbf{v} = \langle a, b, c \rangle$ is $\mathbf{r} = \mathbf{r}_0 + t\mathbf{v}$, or
>
> $$\langle x, y, z \rangle = \langle x_0, y_0, z_0 \rangle + t\langle a, b, c \rangle, \quad \text{for } -\infty < t < \infty.$$
>
> Equivalently, the corresponding **parametric equations of the line** are
>
> $$x = x_0 + at, \quad y = y_0 + bt, \quad z = z_0 + ct, \quad \text{for } -\infty < t < \infty.$$

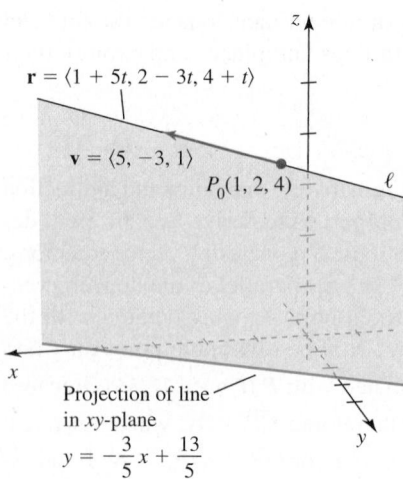

$\mathbf{r} = \langle 1 + 5t, 2 - 3t, 4 + t \rangle$

$\mathbf{v} = \langle 5, -3, 1 \rangle$

$P_0(1, 2, 4)$

ℓ

Projection of line
in xy-plane
$y = -\dfrac{3}{5}x + \dfrac{13}{5}$

Figure 13.67

EXAMPLE 1   Equations of lines Find a vector equation of the line ℓ that passes through the point $P_0(1, 2, 4)$ in the direction of $\mathbf{v} = \langle 5, -3, 1 \rangle$, and then find the corresponding parametric equations of ℓ.

SOLUTION We are given $\mathbf{r}_0 = \langle 1, 2, 4 \rangle$. Therefore, an equation of the line is

$$\mathbf{r} = \mathbf{r}_0 + t\mathbf{v} = \langle 1, 2, 4 \rangle + t\langle 5, -3, 1 \rangle = \langle 1 + 5t, 2 - 3t, 4 + t \rangle,$$

for $-\infty < t < \infty$ (Figure 13.67). The corresponding parametric equations are

$$x = 1 + 5t, \quad y = 2 - 3t, \quad z = 4 + t, \quad \text{for} \quad -\infty < t < \infty.$$

The line is easier to visualize if it is plotted together with its projection in the xy-plane. Setting $z = 0$ (the equation of the xy-plane), parametric equations of the projection line are $x = 1 + 5t, y = 2 - 3t$, and $z = 0$. Eliminating t from these equations, an equation of the projection line is $y = -\frac{3}{5}x + \frac{13}{5}$ (Figure 13.67).

Related Exercises 11–12 ◄

EXAMPLE 2   Equation of a line and a line segment Let ℓ be the line that passes through the points $P_0(-3, 5, 8)$ and $P_1(4, 2, -1)$.

a. Find an equation of ℓ.

b. Find parametric equations of the line segment that extends from P_0 to P_1.

SOLUTION

a. The direction of the line is

$$\mathbf{v} = \overrightarrow{P_0P_1} = \langle 4 - (-3), 2 - 5, -1 - 8 \rangle = \langle 7, -3, -9 \rangle.$$

Therefore, with $\mathbf{r}_0 = \langle -3, 5, 8 \rangle$, a vector equation of ℓ is

$$\mathbf{r} = \mathbf{r}_0 + t\mathbf{v}, \quad \text{or}$$
$$\langle x, y, z \rangle = \langle -3, 5, 8 \rangle + t\langle 7, -3, -9 \rangle$$
$$= \langle -3 + 7t, 5 - 3t, 8 - 9t \rangle.$$

b. Parametric equations for ℓ are

$$x = -3 + 7t, \quad y = 5 - 3t, \quad z = 8 - 9t, \quad \text{for} \quad -\infty < t < \infty.$$

To generate only the line segment from P_0 to P_1, we simply restrict the values of the parameter t. Notice that $t = 0$ corresponds to $P_0(-3, 5, 8)$, and $t = 1$ corresponds to $P_1(4, 2, -1)$. Letting t vary from 0 to 1 generates the line segment from P_0 to P_1. Therefore, parametric equations of the line segment are

$$x = -3 + 7t, \quad y = 5 - 3t, \quad z = 8 - 9t, \quad \text{for } 0 \le t \le 1.$$

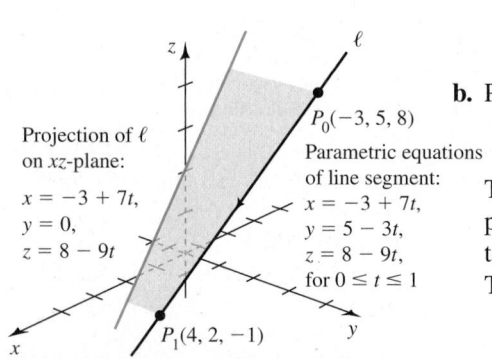

Projection of ℓ
on xz-plane:
$x = -3 + 7t,$
$y = 0,$
$z = 8 - 9t$

$P_0(-3, 5, 8)$

Parametric equations
of line segment:
$x = -3 + 7t,$
$y = 5 - 3t,$
$z = 8 - 9t,$
for $0 \le t \le 1$

ℓ

$P_1(4, 2, -1)$

Figure 13.68

The graph of ℓ, which includes the line segment from P_0 to P_1, is shown in Figure 13.68, along with the projection of ℓ in the xz-plane. The parametric equations of the projection line are found by setting $y = 0$, which is the equation of the xz-plane.

Related Exercises 16, 29–30 ◄

QUICK CHECK 2 In the equation of the line

$$\langle x, y, z \rangle = \langle x_0, y_0, z_0 \rangle + t\langle x_1 - x_0, y_1 - y_0, z_1 - z_0 \rangle,$$

what value of t corresponds to the point $P_0(x_0, y_0, z_0)$? What value of t corresponds to the point $P_1(x_1, y_1, z_1)$? ◄

Distance from a Point to a Line

Three-dimensional geometry has practical applications in such diverse fields as orbital mechanics, ballistics, computer graphics, and regression analysis. For example, determining the distance from a point to a line is an important calculation in problems such as programming video games. We first derive a formula for this distance and then illustrate how the formula is used to determine whether (virtual) billiard balls collide in a video game.

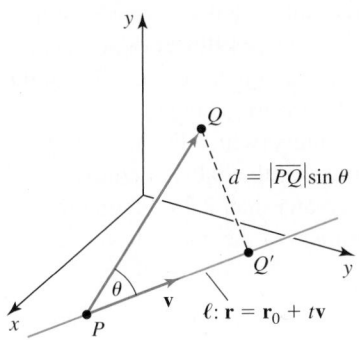

Figure 13.69

Consider a point Q and a line ℓ containing the point P, where ℓ is given by the vector equation $\mathbf{r} = \mathbf{r}_0 + t\mathbf{v}$. Our goal is to find the distance d between Q and ℓ. The geometry of the problem is shown in Figure 13.69, where we have placed the tail of $\mathbf{v}$, which is a vector parallel to ℓ, at the point P. We drop a perpendicular from Q to the point Q' on ℓ to form the right triangle PQQ'; the shortest distance d from Q to ℓ is the distance from Q to Q'. From trigonometry, we know that $d = |\overrightarrow{PQ}|\sin\theta$, where θ is the angle between $\mathbf{v}$ and $\overrightarrow{PQ}$. Using the definition of the magnitude of the cross product, we can also write

$$|\mathbf{v} \times \overrightarrow{PQ}| = |\mathbf{v}|\underbrace{|\overrightarrow{PQ}|\sin\theta}_{d} = |\mathbf{v}|d. \quad \text{Cross product definition}$$

Dividing both sides of this equation by $|\mathbf{v}|$ leads to the desired result,

$$d = \frac{|\mathbf{v} \times \overrightarrow{PQ}|}{|\mathbf{v}|}.$$

QUICK CHECK 3 Find the distance between the point $Q(1, 0, 3)$ and the line $\langle x, y, z \rangle = t\langle 2, 1, 2 \rangle$. Note that $P(0, 0, 0)$ lies on the line and $\mathbf{v} = \langle 2, 1, 2 \rangle$ is parallel to the line. ◄

> **Distance Between a Point and a Line**
>
> The distance d between the point Q and the line $\mathbf{r} = \mathbf{r}_0 + t\mathbf{v}$ is
>
> $$d = \frac{|\mathbf{v} \times \overrightarrow{PQ}|}{|\mathbf{v}|},$$
>
> where P is any point on the line and $\mathbf{v}$ is a vector parallel to the line.

The computer program for a billiards video game must keep track of the locations of all the balls on a two-dimensional screen. Although it is possible to write code that tracks each pixel in every ball, it is much simpler to track only the center pixel of each ball. As explained in the next example, the question of whether two balls collide during the game is answered by computing the distance between a point and a line.

EXAMPLE 3   Video game calculation Arria is playing a billiards video game on her iPad. The playing surface is represented in the video game by the rectangle in the first quadrant with opposite corners at the origin and the point $(100, 50)$ (Figure 13.70). Suppose the cue ball is located at $P(25, 16)$, and Arria shoots the ball with an angle of 30° above the x-axis, aiming for a target ball located at $Q(75, 46)$. If the cue ball is struck with sufficient force, will it collide with the target ball? Assume the diameter of each ball is 2.25.

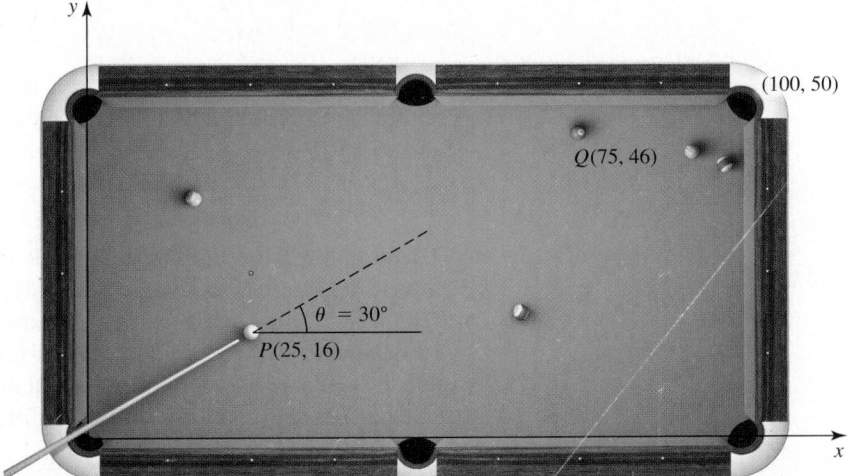

Figure 13.70

SOLUTION We assume the video game stores the locations of all the balls by representing each ball with the coordinates of its center, and that the path of a pool shot is represented by the equation of a line. To determine whether two balls collide, it helps to look first at the situation where the cue ball barely touches the target ball. As illustrated in **Figure 13.71**, the balls will *not* collide when the distance between their centers (which lie on a line perpendicular to the line of the shot) is greater than the diameter of the balls, or equivalently, when the distance d between Q and the line of the shot is greater than 2.25. Stating this result in another way leads to a useful test for the programmers of the game: If $d < 2.25$, the balls *will* collide.

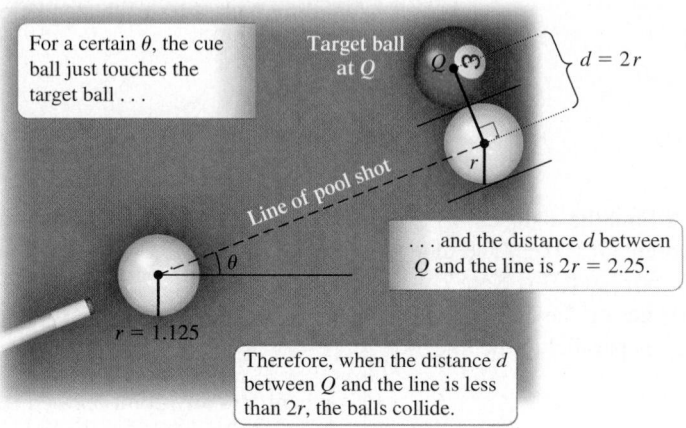

For a certain θ, the cue ball just touches the target ball . . .

Target ball at Q

$d = 2r$

Line of pool shot

r

. . . and the distance d between Q and the line is $2r = 2.25$.

$r = 1.125$

Therefore, when the distance d between Q and the line is less than $2r$, the balls collide.

Figure 13.71

To find d for Arria's attempt, we need a vector parallel to the line of the shot, and we need the vector from the cue ball at $P(25, 16)$ to the target ball at $Q(75, 46)$. Because the cue ball is aimed at an angle 30° above the x-axis, a vector parallel to the line of the shot is $\mathbf{v} = \langle \cos 30°, \sin 30° \rangle = \left\langle \dfrac{\sqrt{3}}{2}, \dfrac{1}{2} \right\rangle$. Notice that $\overrightarrow{PQ} = \langle 75, 46 \rangle - \langle 25, 16 \rangle = \langle 50, 30 \rangle$. Therefore, the distance between $Q(75, 46)$ and the line of the pool shot is

> Example 3 is a two-dimensional problem while the cross product is defined for three-dimensional vectors. Therefore, we must embed the 2D vectors of the example into three dimensions by adding a z-component of 0.

$$d = \frac{|\mathbf{v} \times \overrightarrow{PQ}|}{|\mathbf{v}|} = \frac{\left| \left\langle \dfrac{\sqrt{3}}{2}, \dfrac{1}{2}, 0 \right\rangle \times \langle 50, 30, 0 \rangle \right|}{\left| \left\langle \dfrac{\sqrt{3}}{2}, \dfrac{1}{2}, 0 \right\rangle \right|}.$$

The cross product in the numerator is

$$\begin{vmatrix} \mathbf{i} & \mathbf{j} & \mathbf{k} \\ \sqrt{3}/2 & 1/2 & 0 \\ 50 & 30 & 0 \end{vmatrix} = \left(\frac{\sqrt{3}}{2} 30 - \frac{1}{2} 50 \right) \mathbf{k} \approx 0.98\mathbf{k}.$$

Because $\mathbf{v}$ is a unit vector, its length is 1, and we find that $d = |0.98\mathbf{k}| = 0.98 < 2.25$; therefore, the balls collide. Additional calculations enable the programmers to determine the directions in which the two balls travel after the collision.

Related Exercise 41 ◄

Determining whether two balls collide in a video game may not seem very important. However, the same mathematical methods used to create realistic video games are employed in flight simulators that train military and civilian pilots. The principles used to design video games are also used to train surgeons and to assist them in performing surgery. Example 4 looks at another crucial calculation used in designing the virtual-world tools that are becoming part of our daily lives: finding points of intersection.

EXAMPLE 4 **Points of intersection** Determine whether the lines ℓ_1 and ℓ_2 intersect. If they do, find the point of intersection.

a. $\ell_1: x = 2 + 3t, y = 3t, z = 1 - t$ $\qquad$ $\ell_2: x = 4 + 2s, y = -3 + 3s, z = -2s$

b. $\ell_1: x = 3 + t, y = 4 - t, z = 5 + 3t$ $\quad$ $\ell_2: x = 2s, y = -1 + 2s, z = 4s$

SOLUTION

a. Let's first check whether the lines are parallel; if they are, there is no point of intersection (unless the lines are identical). Reading the coefficients in the parametric equations for each line, we find that $\mathbf{v}_1 = \langle 3, 3, -1 \rangle$ is parallel to ℓ_1 and that $\mathbf{v}_2 = \langle 2, 3, -2 \rangle$ is parallel to ℓ_2. Because $\mathbf{v}_1$ is not a constant multiple of $\mathbf{v}_2$, the lines are not parallel. In $\mathbb{R}^3$, this fact alone does not guarantee that the lines intersect. Two lines that are neither parallel nor intersecting are said to be **skew**.

Determining whether two lines intersect amounts to solving a system of three linear equations in two variables. We set the x-, y-, and z-components of each line equal to one another, which results in the following system:

> Given two skew lines in $\mathbb{R}^3$, one can always find two parallel planes in which the lines lie: one line in one plane, and the other line in a plane parallel to the first.

$$2 + 3t = 4 + 2s \quad (1) \quad \text{Equate the } x\text{-components.}$$
$$3t = -3 + 3s \quad (2) \quad \text{Equate the } y\text{-components.}$$
$$1 - t = -2s. \quad (3) \quad \text{Equate the } z\text{-components.}$$

When equation (2) is subtracted from equation (1), the result is $2 = 7 - s$, or $s = 5$. Substituting $s = 5$ into equation (1) or (2) yields $t = 4$. However, when these values are substituted into (3), a false statement results, which implies that the system of equations is *inconsistent* and the lines do not intersect. We conclude that ℓ_1 and ℓ_2 are skew.

b. Proceeding as we did in part (a), we note that the lines are not parallel and solve the system

$$3 + t = 2s$$
$$4 - t = -1 + 2s$$
$$5 + 3t = 4s.$$

When the first two equations are added to eliminate the variable t, we find that $s = 2$, which implies that $t = 1$. When these values are substituted into the last equation, a true statement results, which means we have a solution to the system and the lines intersect. To find the point of intersection, we substitute $t = 1$ into the parametric equations for ℓ_1 and arrive at $(4, 3, 8)$. You can verify that when $s = 2$ is substituted into the equations for ℓ_2, the same point of intersection results.

Related Exercises 32–33 ◄

Equations of Planes

Intuitively, a plane is a flat surface with infinite extent in all directions. Three noncollinear points (not all on the same line) determine a unique plane in $\mathbb{R}^3$. A plane in $\mathbb{R}^3$ is also uniquely determined by one point in the plane and any nonzero vector orthogonal (perpendicular) to the plane. Such a vector, called a *normal vector*, specifies the orientation of the plane.

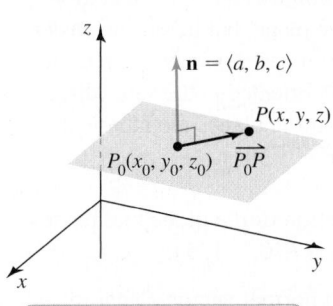

The orientation of a plane is specified by a normal vector $\mathbf{n}$. All vectors $\overrightarrow{P_0P}$ in the plane are orthogonal to $\mathbf{n}$.

Figure 13.72

DEFINITION **Plane in $\mathbb{R}^3$**
Given a fixed point P_0 and a nonzero **normal vector $\mathbf{n}$**, the set of points P in $\mathbb{R}^3$ for which $\overrightarrow{P_0P}$ is orthogonal to $\mathbf{n}$ is called a **plane** (Figure 13.72).

We now derive an equation of the plane passing through the point $P_0(x_0, y_0, z_0)$ with nonzero normal vector $\mathbf{n} = \langle a, b, c \rangle$. Notice that for any point $P(x, y, z)$ in the plane,

▶ Just as the slope determines the orientation of a line in $\mathbb{R}^2$, a normal vector determines the orientation of a plane in $\mathbb{R}^3$.

the vector $\overrightarrow{P_0P} = \langle x - x_0, y - y_0, z - z_0 \rangle$ lies in the plane and is orthogonal to **n**. This orthogonality relationship is written and simplified as follows:

$$\mathbf{n} \cdot \overrightarrow{P_0P} = 0 \qquad \text{Dot product of orthogonal vectors}$$

$$\langle a, b, c \rangle \cdot \langle x - x_0, y - y_0, z - z_0 \rangle = 0 \qquad \text{Substitute vector components.}$$

$$a(x - x_0) + b(y - y_0) + c(z - z_0) = 0 \qquad \text{Expand the dot product.}$$

$$ax + by + cz = d. \qquad d = ax_0 + by_0 + cz_0$$

This important result states that the most general linear equation in three variables, $ax + by + cz = d$, describes a plane in $\mathbb{R}^3$.

▶ A vector $\mathbf{n} = \langle a, b, c \rangle$ is used to describe a *plane* by specifying a direction *orthogonal* to the plane. By contrast, a vector $\mathbf{v} = \langle a, b, c \rangle$ is used to describe a *line* by specifying a direction *parallel* to the line.

> **General Equation of a Plane in $\mathbb{R}^3$**
>
> The plane passing through the point $P_0(x_0, y_0, z_0)$ with a nonzero normal vector $\mathbf{n} = \langle a, b, c \rangle$ is described by the equation
>
> $$a(x - x_0) + b(y - y_0) + c(z - z_0) = 0 \quad \text{or} \quad ax + by + cz = d,$$
>
> where $d = ax_0 + by_0 + cz_0$.

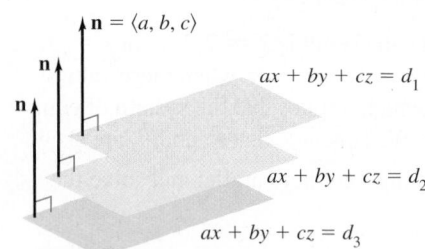

The normal vectors of parallel planes have the same direction.

Figure 13.73

The coefficients a, b, and c in the equation of a plane determine the *orientation* of the plane, while the constant term d determines the *location* of the plane. If a, b, and c are held constant and d is varied, a family of parallel planes is generated, all with the same orientation (**Figure 13.73**).

QUICK CHECK 4 Consider the equation of a plane in the form $\mathbf{n} \cdot \overrightarrow{P_0P} = 0$. Explain why the equation of the plane depends only on the direction, not on the length, of the normal vector **n**. ◄

EXAMPLE 5   Equation of a plane

a. Find an equation of the plane passing through $P_0(2, -3, 4)$ with a normal vector $\mathbf{n} = \langle -1, 2, 3 \rangle$.

b. Find an equation of the plane passing through $P_0(2, -3, 4)$ that is perpendicular to the line $x = 3 + 2t, y = -4t, z = 1 - 6t$.

SOLUTION

a. Substituting the components of **n** ($a = -1$, $b = 2$, and $c = 3$) and the coordinates of P_0 ($x_0 = 2$, $y_0 = -3$, and $z_0 = 4$) into the equation of a plane, we have

$$a(x - x_0) + b(y - y_0) + c(z - z_0) = 0 \qquad \text{General equation of a plane}$$

$$(-1)(x - 2) + 2(y - (-3)) + 3(z - 4) = 0 \qquad \text{Substitute.}$$

$$-x + 2y + 3z = 4. \qquad \text{Simplify.}$$

The plane is shown in **Figure 13.74**.

b. Note that $\mathbf{v} = \langle 2, -4, -6 \rangle$ is parallel to the given line and therefore perpendicular to the plane, so we have a vector normal to the plane. We could carry out calculations similar to those found in part (a) to find the equation of the plane, but here is an easier solution. Observe that **v** is a multiple of the normal vector in part (a) ($\mathbf{v} = -2\mathbf{n}$), and therefore **v** and **n** are parallel, which implies both planes are oriented in the same direction. Because both planes pass through P_0, we conclude that the planes are identical.

Related Exercises 43–44 ◄

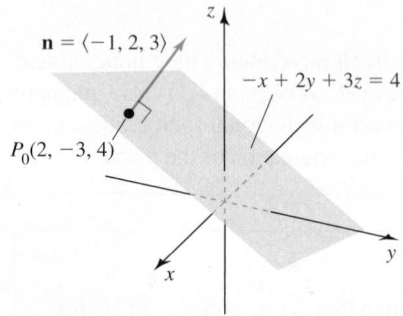

Figure 13.74

EXAMPLE 6   A plane through three points Find an equation of the plane that passes through the (noncollinear) points $P(2, -1, 3)$, $Q(1, 4, 0)$, and $R(0, -1, 5)$.

SOLUTION To write an equation of the plane, we must find a normal vector. Because P, Q, and R lie in the plane, the vectors $\overrightarrow{PQ} = \langle -1, 5, -3 \rangle$ and $\overrightarrow{PR} = \langle -2, 0, 2 \rangle$ also lie in the plane. The cross product $\overrightarrow{PQ} \times \overrightarrow{PR}$ is perpendicular to both $\overrightarrow{PQ}$ and $\overrightarrow{PR}$; therefore, a vector normal to the plane is

▶ Three points P, Q, and R determine a plane provided they are not collinear. If P, Q, and R *are* collinear, then the vectors $\overrightarrow{PQ}$ and $\overrightarrow{PR}$ are parallel, which implies that $\overrightarrow{PQ} \times \overrightarrow{PR} = \mathbf{0}$.

$$\mathbf{n} = \overrightarrow{PQ} \times \overrightarrow{PR} = \begin{vmatrix} \mathbf{i} & \mathbf{j} & \mathbf{k} \\ -1 & 5 & -3 \\ -2 & 0 & 2 \end{vmatrix} = 10\mathbf{i} + 8\mathbf{j} + 10\mathbf{k}.$$

QUICK CHECK 5 Verify that in Example 6, the same equation for the plane results if either Q or R is used as the fixed point in the plane. ◄

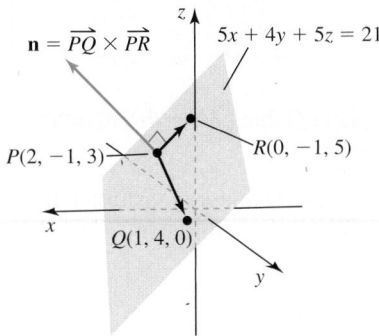

$\mathbf{n} = \overrightarrow{PQ} \times \overrightarrow{PR}$

$5x + 4y + 5z = 21$

$P(2, -1, 3)$

$R(0, -1, 5)$

$Q(1, 4, 0)$

$\overrightarrow{PQ}$ and $\overrightarrow{PR}$ lie in the same plane.
$\overrightarrow{PQ} \times \overrightarrow{PR}$ is orthogonal to the plane.

Figure 13.75

► There is a possibility for confusion here. Working in $\mathbb{R}^3$ with no other restrictions, the equation $-3y - z = 6$ describes a plane that is parallel to the x-axis (because x is unspecified). To make it clear that $-3y - z = 6$ is a line in the yz-plane, the condition $x = 0$ is included.

Any nonzero scalar multiple of $\mathbf{n}$ may be used as the normal vector. Choosing $\mathbf{n} = \langle 5, 4, 5 \rangle$ and $P_0(2, -1, 3)$ as the fixed point in the plane (**Figure 13.75**), an equation of the plane is

$$5(x - 2) + 4(y - (-1)) + 5(z - 3) = 0 \quad \text{or} \quad 5x + 4y + 5z = 21.$$

Using either Q or R as the fixed point in the plane leads to an equivalent equation of the plane.

Related Exercises 49–50 ◄

EXAMPLE 7 Properties of a plane Let Q be the plane described by the equation $2x - 3y - z = 6$.

a. Find a vector normal to Q.

b. Find the points at which Q intersects the coordinate axes and plot Q.

c. Describe the sets of points at which Q intersects the yz-plane, the xz-plane, and the xy-plane.

SOLUTION

a. The coefficients of x, y, and z in the equation of Q are the components of a vector normal to Q. Therefore, a normal vector is $\mathbf{n} = \langle 2, -3, -1 \rangle$ (or any nonzero multiple of $\mathbf{n}$).

b. The point (x, y, z) at which Q intersects the x-axis must have $y = z = 0$. Substituting $y = z = 0$ into the equation of Q gives $x = 3$, so Q intersects the x-axis at $(3, 0, 0)$. Similarly, Q intersects the y-axis at $(0, -2, 0)$, and Q intersects the z-axis at $(0, 0, -6)$. Connecting the three intercepts with straight lines allows us to visualize the plane (**Figure 13.76**).

c. All points in the yz-plane have $x = 0$. Setting $x = 0$ in the equation of Q gives the equation $-3y - z = 6$, which, with the condition $x = 0$, describes a line in the yz-plane. If we set $y = 0$, then Q intersects the xz-plane in the line $2x - z = 6$, where $y = 0$. If $z = 0$, then Q intersects the xy-plane in the line $2x - 3y = 6$, where $z = 0$ (**Figure 13.76**).

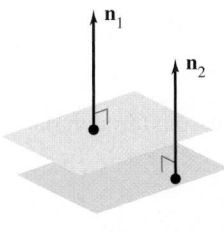

Two distinct planes are parallel if $\mathbf{n}_1$ and $\mathbf{n}_2$ are parallel.

$\mathbf{n}_1$

$\mathbf{n}_2$

(a)

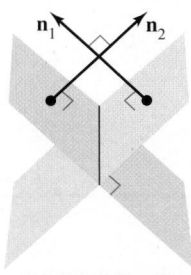

$\mathbf{n}_1$ $\mathbf{n}_2$

Two planes are orthogonal if $\mathbf{n}_1 \cdot \mathbf{n}_2 = 0$.

(b)

Figure 13.77

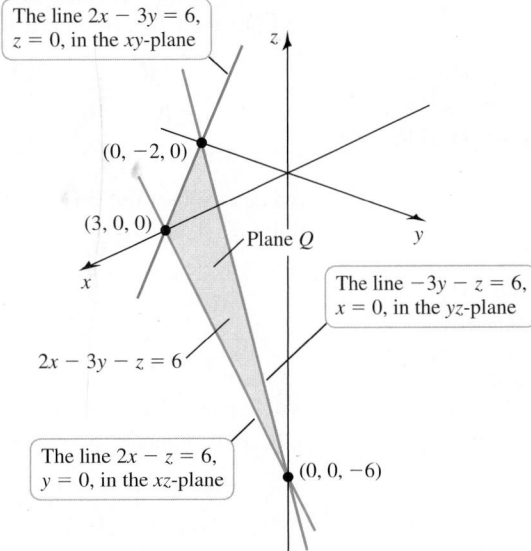

The line $2x - 3y = 6$, $z = 0$, in the xy-plane

$(0, -2, 0)$

$(3, 0, 0)$

Plane Q

The line $-3y - z = 6$, $x = 0$, in the yz-plane

$2x - 3y - z = 6$

The line $2x - z = 6$, $y = 0$, in the xz-plane

$(0, 0, -6)$

Figure 13.76

Related Exercise 61 ◄

Parallel and Orthogonal Planes

The normal vectors of distinct planes tell us about the relative orientation of the planes. Two cases are of particular interest: Two distinct planes may be *parallel* (**Figure 13.77a**) and two intersecting planes may be *orthogonal* (**Figure 13.77b**).

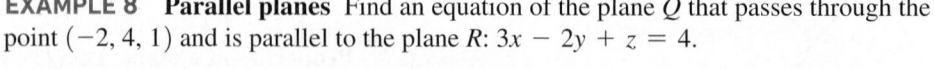

> **DEFINITION Parallel and Orthogonal Planes**
>
> Two distinct planes are **parallel** if their respective normal vectors are parallel (that is, the normal vectors are scalar multiples of each other). Two planes are **orthogonal** if their respective normal vectors are orthogonal (that is, the dot product of the normal vectors is zero).

QUICK CHECK 6 Determine whether the planes $2x - 3y + 6z = 12$ and $6x + 8y + 2z = 1$ are parallel, orthogonal, or neither. ◄

EXAMPLE 8 Parallel planes Find an equation of the plane Q that passes through the point $(-2, 4, 1)$ and is parallel to the plane R: $3x - 2y + z = 4$.

SOLUTION The vector $\mathbf{n} = \langle 3, -2, 1 \rangle$ is normal to R. Because Q and R are parallel, $\mathbf{n}$ is also normal to Q. We conclude that an equation of Q (which passes through $(-2, 4, 1)$ and has a normal vector $\langle 3, -2, 1 \rangle$) is given by

$$3(x + 2) - 2(y - 4) + (z - 1) = 0 \quad \text{or} \quad 3x - 2y + z = -13.$$

Related Exercises 51–52 ◄

EXAMPLE 9 Intersecting planes Find an equation of the line of intersection of the planes Q: $x + 2y + z = 5$ and R: $2x + y - z = 7$.

SOLUTION First note that the vectors normal to the planes, $\mathbf{n}_Q = \langle 1, 2, 1 \rangle$ and $\mathbf{n}_R = \langle 2, 1, -1 \rangle$, are *not* multiples of each other. Therefore, the planes are not parallel and they must intersect in a line; call it ℓ. To find an equation of ℓ, we need two pieces of information: a point on ℓ and a vector pointing in the direction of ℓ. Here is one of several ways to find a point on ℓ. Setting $z = 0$ in the equations of the planes gives equations of the lines in which the planes intersect the xy-plane:

$$x + 2y = 5$$
$$2x + y = 7.$$

Solving these equations simultaneously, we find that $x = 3$ and $y = 1$. Combining this result with $z = 0$, we see that $(3, 1, 0)$ is a point on ℓ (**Figure 13.78**).

We next find a vector parallel to ℓ. Because ℓ lies in Q and R, it is orthogonal to the normal vectors $\mathbf{n}_Q$ and $\mathbf{n}_R$. Therefore, the cross product of $\mathbf{n}_Q$ and $\mathbf{n}_R$ is a vector parallel to ℓ (**Figure 13.78**). In this case, the cross product is

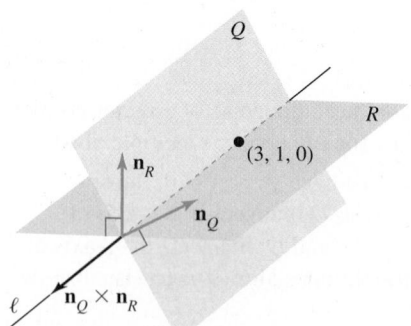

$\mathbf{n}_Q \times \mathbf{n}_R$ is a vector perpendicular to $\mathbf{n}_Q$ and $\mathbf{n}_R$.
Line ℓ is perpendicular to $\mathbf{n}_Q$ and $\mathbf{n}_R$.
Therefore, ℓ and $\mathbf{n}_Q \times \mathbf{n}_R$ are parallel to each other.

Figure 13.78

$$\mathbf{n}_Q \times \mathbf{n}_R = \begin{vmatrix} \mathbf{i} & \mathbf{j} & \mathbf{k} \\ 1 & 2 & 1 \\ 2 & 1 & -1 \end{vmatrix} = -3\mathbf{i} + 3\mathbf{j} - 3\mathbf{k} = \langle -3, 3, -3 \rangle.$$

➤ By setting $z = 0$ and solving the two resulting equations, we find the point that lies on both planes *and* lies in the xy-plane ($z = 0$).

An equation of the line ℓ in the direction of the vector $\langle -3, 3, -3 \rangle$ passing through the point $(3, 1, 0)$ is

$$\begin{aligned} \mathbf{r} &= \langle x_0, y_0, z_0 \rangle + t\langle a, b, c \rangle && \text{Equation of a line} \\ &= \langle 3, 1, 0 \rangle + t\langle -3, 3, -3 \rangle && \text{Substitute.} \\ &= \langle 3 - 3t, 1 + 3t, -3t \rangle, && \text{Simplify.} \end{aligned}$$

➤ Any nonzero scalar multiple of $\langle -3, 3, -3 \rangle$ can be used for the direction of ℓ. For example, another equation of ℓ is $\mathbf{r} = \langle 3 + t, 1 - t, t \rangle$.

where $-\infty < t < \infty$. You can check that any point (x, y, z) with $x = 3 - 3t$, $y = 1 + 3t$, and $z = -3t$ satisfies the equations of both planes.

Related Exercise 74 ◄

SECTION 13.5 EXERCISES

Getting Started

1. Find a position vector that is parallel to the line $x = 2 + 4t, y = 5 - 8t, z = 9t$.

2. Find the parametric equations of the line $\mathbf{r} = \langle 1, 2, 3 \rangle + t\langle 4, 0, -6 \rangle$.

3. Explain how to find a vector in the direction of the line segment from $P_0(x_0, y_0, z_0)$ to $P_1(x_1, y_1, z_1)$.

4. Find the vector equation of the line through the points $P_0(x_0, y_0, z_0)$ and $P_1(x_1, y_1, z_1)$.

5. Determine whether the plane $x + y + z = 9$ and the line $x = t, y = t + 1, z = t + 2$ are parallel, perpendicular, or neither. Be careful.

6. Determine whether the plane $x + y + z = 9$ and the line $x = t, y = -2t + 1, z = t + 2$ are parallel, perpendicular, or neither.

7. Give two pieces of information that, taken together, uniquely determine a plane.

8. Find a vector normal to the plane $-2x - 3y = 12 - 4z$.

9. Where does the plane $-2x - 3y + 4z = 12$ intersect the coordinate axes?

10. Give an equation of the plane with a normal vector $\mathbf{n} = \langle 1, 1, 1 \rangle$ that passes through the point $(1, 0, 0)$.

Practice Exercises

11–26. Equations of lines *Find both the parametric and the vector equations of the following lines.*

11. The line through $(0, 0, 1)$ in the direction of the vector $\mathbf{v} = \langle 4, 7, 0 \rangle$

12. The line through $(-3, 2, -1)$ in the direction of the vector $\mathbf{v} = \langle 1, -2, 0 \rangle$

13. The line through $(0, 0, 1)$ parallel to the y-axis

14. The line through $(-2, 4, 3)$ parallel to the x-axis

15. The line through $(0, 0, 0)$ and $(1, 2, 3)$

16. The line through $(-3, 4, 6)$ and $(5, -1, 0)$

17. The line through $(0, 0, 0)$ that is parallel to the line $\mathbf{r} = \langle 3 - 2t, 5 + 8t, 7 - 4t \rangle$

18. The line through $(1, -3, 4)$ that is parallel to the line $x = 3 + 4t, y = 5 - t, z = 7$

19. The line through $(0, 0, 0)$ that is perpendicular to both $\mathbf{u} = \langle 1, 0, 2 \rangle$ and $\mathbf{v} = \langle 0, 1, 1 \rangle$

20. The line through $(-3, 4, 2)$ that is perpendicular to both $\mathbf{u} = \langle 1, 1, -5 \rangle$ and $\mathbf{v} = \langle 0, 4, 0 \rangle$

21. The line through $(-2, 5, 3)$ that is perpendicular to both $\mathbf{u} = \mathbf{i} + \mathbf{j} - 2\mathbf{k}$ and the x-axis

22. The line through $(0, 2, 1)$ that is perpendicular to both $\mathbf{u} = 4\mathbf{i} + 3\mathbf{j} - 5\mathbf{k}$ and the z-axis

23. The line through $(1, 2, 3)$ that is perpendicular to the lines $x = 3 - 2t, y = 5 + 8t, z = 7 - 4t$ and $x = -2t, y = 5 + t, z = 7 - t$

24. The line through $(1, 0, -1)$ that is perpendicular to the lines $x = 3 + 2t, y = 3t, z = -4t$ and $x = t, y = t, z = -t$

25. The line that is perpendicular to the lines $\mathbf{r} = \langle 4t, 1 + 2t, 3t \rangle$ and $\mathbf{R} = \langle -1 + s, -7 + 2s, -12 + 3s \rangle$, and passes through the point of intersection of the lines $\mathbf{r}$ and $\mathbf{R}$

26. The line that is perpendicular to the lines $\mathbf{r} = \langle -2 + 3t, 2t, 3t \rangle$ and $\mathbf{R} = \langle -6 + s, -8 + 2s, -12 + 3s \rangle$, and passes through the point of intersection of the lines $\mathbf{r}$ and $\mathbf{R}$

27–30. Line segments *Find parametric equations for the line segment joining the first point to the second point.*

27. $(0, 0, 0)$ and $(1, 2, 3)$ **28.** $(1, 0, 1)$ and $(0, -2, 1)$

29. $(2, 4, 8)$ and $(7, 5, 3)$ **30.** $(-1, -8, 4)$ and $(-9, 5, -3)$

31–37. Parallel, intersecting, or skew lines *Determine whether the following pairs of lines are parallel, intersect at a single point, or are skew. If the lines are parallel, determine whether they are the same line (and thus intersect at all points). If the lines intersect at a single point, determine the point of intersection.*

31. $\mathbf{r} = \langle 1, 3, 2 \rangle + t\langle 6, -7, 1 \rangle$; $\mathbf{R} = \langle 10, 6, 14 \rangle + s\langle 3, 1, 4 \rangle$

32. $x = 2t, y = t + 2, z = 3t - 1$ and $x = 5s - 2, y = s + 4$, $z = 5s + 1$

33. $x = 4, y = 6 - t, z = 1 + t$ and $x = -3 - 7s, y = 1 + 4s$, $z = 4 - s$

34. $x = 4 + 5t, y = -2t, z = 1 + 3t$ and $x = 10s, y = 6 + 4s$, $z = 4 + 6s$

35. $x = 1 + 2t, y = 7 - 3t, z = 6 + t$ and $x = -9 + 6t$, $y = 22 - 9t, z = 1 + 3t$

36. $\mathbf{r} = \langle 3, 1, 0 \rangle + t\langle 4, -6, 4 \rangle$; $\mathbf{R} = \langle 0, 5, 4 \rangle + s\langle -2, 3, -2 \rangle$

37. $\mathbf{r} = \langle 4 + t, -2t, 1 + 3t \rangle$; $\mathbf{R} = \langle 1 - 7s, 6 + 14s, 4 - 21s \rangle$

38. Intersecting lines and colliding particles Consider the lines

$$\mathbf{r} = \langle 2 + 2t, 8 + t, 10 + 3t \rangle \quad \text{and}$$
$$\mathbf{R} = \langle 6 + s, 10 - 2s, 16 - s \rangle.$$

  a. Determine whether the lines intersect (have a common point), and if so, find the coordinates of that point.
  b. If $\mathbf{r}$ and $\mathbf{R}$ describe the paths of two particles, do the particles collide? Assume $t \geq 0$ and $s \geq 0$ measure time in seconds, and that motion starts at $s = t = 0$.

39–40. Distance from a point to a line *Find the distance between the given point Q and the given line.*

39. $Q(-5, 2, 9)$; $x = 5t + 7, y = 2 - t, z = 12t + 4$

40. $Q(5, 6, 1)$; $x = 1 + 3t, y = 3 - 4t, z = t + 1$

T 41. Billiards shot A cue ball in a billiards video game lies at $P(25, 16)$ (see figure). Refer to Example 3, where we assume the diameter of each ball is 2.25 screen units, and pool balls are represented by the point at their center.

  a. The cue ball is aimed at an angle of 58° above the negative x-axis toward a target ball at $A(5, 45)$. Do the balls collide?
  b. The cue ball is aimed at the point $(50, 25)$ in an attempt to hit a target ball at $B(76, 40)$. Do the balls collide?
  c. The cue ball is aimed at an angle θ above the x-axis in the general direction of a target ball at $C(75, 30)$. What range of angles (for $0 \leq \theta \leq \pi/2$) will result in a collision? Express your answer in degrees.

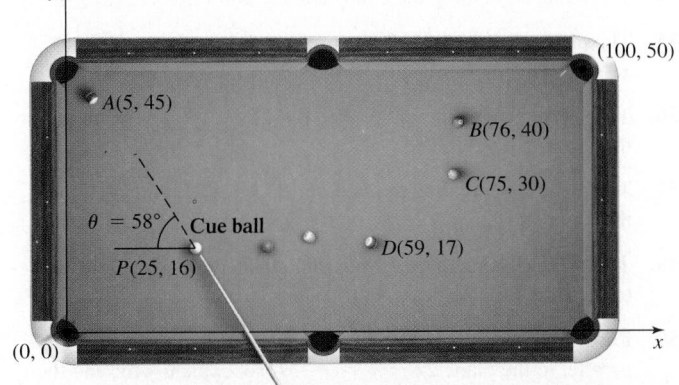

T 42. Bank shot Refer to the figure in Exercise 41. The cue ball lies at $P(25, 16)$ and Jerrod hopes to hit ball $D(59, 17)$; a direct shot isn't an option with other balls blocking the path. Instead, he attempts a bank shot, aiming the cue ball at an angle 45° below the x-axis. Will the balls collide? Assume the angles at which the cue ball meets and leaves the bumper are equal and that the diameter of each ball is 2 screen units. (*Hint:* The cue ball will bounce off the bumper when its center hits an "imaginary bumper" one unit above the bumper; see following figure.)

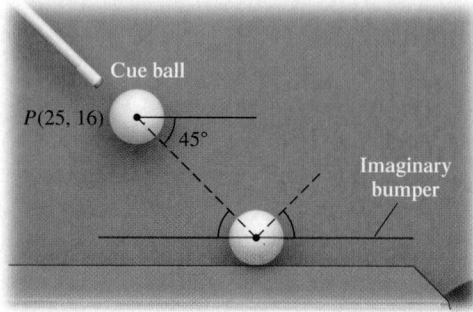

43–58. Equations of planes *Find an equation of the following planes.*

43. The plane passing through the point $P_0(0, 2, -2)$ with a normal vector $\mathbf{n} = \langle 1, 1, -1 \rangle$

44. The plane passing through the point $P_0(2, 3, 0)$ with a normal vector $\mathbf{n} = \langle -1, 2, -3 \rangle$

45. The plane that is parallel to the vectors $\langle 1, 0, 1 \rangle$ and $\langle 0, 2, 1 \rangle$, passing through the point $(1, 2, 3)$

46. The plane that is parallel to the vectors $\langle 1, -3, 1 \rangle$ and $\langle 4, 2, 0 \rangle$, passing through the point $(3, 0, -2)$

47. The plane passing through the origin that is perpendicular to the line $x = t, y = 1 + 4t, z = 7t$

48. The plane passing through the point $(2, -3, 5)$ that is perpendicular to the line $x = 2t, y = 1 + 3t, z = 5 + 4t$

49. The plane passing through the points $(1, 0, 3)$, $(0, 4, 2)$, and $(1, 1, 1)$

50. The plane passing through the points $(2, -1, 4)$, $(1, 1, -1)$, and $(-4, 1, 1)$

51. The plane passing through the point $P_0(1, 0, 4)$ that is parallel to the plane $-x + 2y - 4z = 1$

52. The plane passing through the point $P_0(0, 2, -2)$ that is parallel to the plane $2x + y - z = 1$

53. The plane containing the x-axis and the point $P_0(1, 2, 3)$

54. The plane containing the z-axis and the point $P_0(3, -1, 2)$

55. The plane passing through the origin and containing the line $x = t - 1, y = 2t, z = 3t + 4$

56. The plane passing though the point $P_0(1, -2, 3)$ and containing the line $\mathbf{r} = \langle t, -t, 2t \rangle$

57. The plane passing though the point $P_0(-4, 1, 2)$ and containing the line $\mathbf{r} = \langle 2t - 2, -2t, -4t + 1 \rangle$

58. The plane passing through the origin that contains the line of intersection of the planes $x + y + 2z = 0$ and $x - y = 4$

59. **Parallel planes** Is the line $x = t + 1, y = 2t + 3, z = 4t + 5$ parallel to the plane $2x - y = -2$? If so, explain why, and then find an equation of the plane containing the line that is parallel to the plane $2x - y = -2$.

60. Do the lines $x = t, y = 2t + 1, z = 3t + 4$ and $x = 2s - 2, y = 2s - 1, z = 3s + 1$ intersect each other at only one point? If so, find a plane that contains both lines.

61–64. Properties of planes *Find the points at which the following planes intersect the coordinate axes, and find equations of the lines where the planes intersect the coordinate planes. Sketch a graph of the plane.*

61. $3x - 2y + z = 6$

62. $-4x + 8z = 16$

63. $x + 3y - 5z - 30 = 0$

64. $12x - 9y + 4z + 72 = 0$

65–68. Pairs of planes *Determine whether the following pairs of planes are parallel, orthogonal, or neither.*

65. $x + y + 4z = 10$ and $-x - 3y + z = 10$

66. $2x + 2y - 3z = 10$ and $-10x - 10y + 15z = 10$

67. $3x + 2y - 3z = 10$ and $-6x - 10y + z = 10$

68. $3x + 2y + 2z = 10$ and $-6x - 10y + 19z = 10$

69–70. Equations of planes *For the following sets of planes, determine which pairs of planes in the set are parallel, which pairs are orthogonal, and which pairs are identical.*

69. $Q: 3x - 2y + z = 12$; $R: -x + 2y/3 - z/3 = 0$; $S: -x + 2y + 7z = 1$; $T: 3x/2 - y + z/2 = 6$

70. $Q: x + y - z = 0$; $R: y + z = 0$; $S: x - y = 0$; $T: x + y + z = 0$

71–72. Lines normal to planes *Find an equation of the following lines.*

71. The line passing through the point $P_0(2, 1, 3)$ that is normal to the plane $2x - 4y + z = 10$

72. The line passing through the point $P_0(0, -10, -3)$ that is normal to the plane $x + 4z = 2$

73–76. Intersecting planes *Find an equation of the line of intersection of the planes Q and R.*

73. $Q: -x + 2y + z = 1$; $R: x + y + z = 0$

74. $Q: x + 2y - z = 1$; $R: x + y + z = 1$

75. $Q: 2x - y + 3z - 1 = 0$; $R: -x + 3y + z - 4 = 0$

76. $Q: x - y - 2z = 1$; $R: x + y + z = -1$

77–80. Line-plane intersections *Find the point (if it exists) at which the following planes and lines intersect.*

77. $x = 3$ and $\mathbf{r} = \langle t, t, t \rangle$

78. $y = -2$ and $\mathbf{r} = \langle 2t + 1, -t + 4, t - 6 \rangle$

79. $3x + 2y - 4z = -3$ and $x = -2t + 5, y = 3t - 5, z = 4t - 6$

80. $2x - 3y + 3z = 2$ and $x = 3t, y = t, z = -t$

81. **Explain why or why not** Determine whether the following statements are true and give an explanation or counterexample.

   a. The line $\mathbf{r} = \langle 3, -1, 4 \rangle + t\langle 6, -2, 8 \rangle$ passes through the origin.

   b. Any two nonparallel lines in $\mathbb{R}^3$ intersect.

   c. The plane $x + y + z = 0$ and the line $x = t, y = t, z = t$ are parallel.

d. The vector equations $\mathbf{r} = \langle 1, 2, 3 \rangle + t\langle 1, 1, 1 \rangle$ and $\mathbf{R} = \langle 1, 2, 3 \rangle + t\langle -2, -2, -2 \rangle$ describe the same line.

e. The equations $x + y - z = 1$ and $-x - y + z = 1$ describe the same plane.

f. Any two distinct lines in $\mathbb{R}^3$ determine a unique plane.

g. The vector $\langle -1, -5, 7 \rangle$ is perpendicular to both the line $x = 1 + 5t, y = 3 - t, z = 1$ and the line $x = 7t, y = 3, z = 3 + t$.

82. Distance from a point to a plane Suppose P is a point in the plane $ax + by + cz = d$. The distance from any point Q to the plane equals the length of the orthogonal projection of $\vec{PQ}$ onto a vector $\mathbf{n} = \langle a, b, c \rangle$ normal to the plane. Use this information to show that the distance from Q to the plane is $\left| \vec{PQ} \cdot \mathbf{n} \right| / |\mathbf{n}|$.

83. Find the distance from the point $Q(6, -2, 4)$ to the plane $2x - y + 2z = 4$.

84. Find the distance from the point $Q(1, 2, -4)$ to the plane $2x - 5z = 5$.

Explorations and Challenges

85–86. Symmetric equations for a line *If we solve for t in the parametric equations of the line* $x = x_0 + at, y = y_0 + bt, z = z_0 + ct$, *we obtain the symmetric equations*

$$\frac{x - x_0}{a} = \frac{y - y_0}{b} = \frac{z - z_0}{c},$$

provided a, b, and c do not equal 0.

85. Find symmetric equations of the line $\mathbf{r} = \langle 1, 2, 0 \rangle + t\langle 4, 7, 2 \rangle$.

86. Find parametric and symmetric equations of the line passing through the points $P(1, -2, 3)$ and $Q(2, 3, -1)$.

T 87. Angle between planes The angle between two planes is the smallest angle θ between the normal vectors of the planes, where the directions of the normal vectors are chosen so that $0 \le \theta \le \pi/2$. Find the angle between the planes $5x + 2y - z = 0$ and $-3x + y + 2z = 0$.

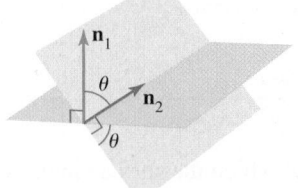

88. Intercepts Let a, b, c, and d be constants. Find the points at which the plane $ax + by + cz = d$ intersects the x-, y-, and z-axes.

89. A family of orthogonal planes Find an equation for a family of planes that are orthogonal to the planes $2x + 3y = 4$ and $-x - y + 2z = 8$.

90. Orthogonal plane Find an equation of the plane passing through $(0, -2, 4)$ that is orthogonal to the planes $2x + 5y - 3z = 0$ and $-x + 5y + 2z = 8$.

91. Three intersecting planes Describe the set of all points (if any) at which all three planes $x + 3z = 3, y + 4z = 6$, and $x + y + 6z = 9$ intersect.

92. Three intersecting planes Describe the set of all points (if any) at which all three planes $x + 2y + 2z = 3, y + 4z = 6$, and $x + 2y + 8z = 9$ intersect.

93. T-shirt profits A clothing company makes a profit of $10 on its long-sleeved T-shirts and a profit of $5 on its short-sleeved T-shirts. Assuming there is a $200 setup cost, the profit on T-shirt sales is $z = 10x + 5y - 200$, where x is the number of long-sleeved T-shirts sold and y is the number of short-sleeved T-shirts sold. Assume x and y are nonnegative.

a. Graph the plane that gives the profit using the window $[0, 40] \times [0, 40] \times [-400, 400]$.

b. If $x = 20$ and $y = 10$, is the profit positive or negative?

c. Describe the values of x and y for which the company breaks even (for which the profit is zero). Mark this set on your graph.

QUICK CHECK ANSWERS

1. The z-axis; the line $y = x$ in the xy-plane   **2.** When $t = 0$, the point on the line is P_0; when $t = 1$, the point on the line is P_1.   **3.** $d = \sqrt{26}/3$ **4.** Because the right side of the equation is 0, the equation can be multiplied by any nonzero constant (changing the length of $\mathbf{n}$) without changing the graph.   **6.** The planes are orthogonal because $\langle 2, -3, 6 \rangle \cdot \langle 6, 8, 2 \rangle = 0$. ◄

13.6 Cylinders and Quadric Surfaces

In Section 13.5, we discovered that lines in three-dimensional space are described by parametric equations (or vector equations) that are linear in the variable. We also saw that planes are described with linear equations in three variables. In this section, we take this progression one step further and investigate the geometry of three-dimensional objects described by quadratic equations in three variables. The result is a collection of *quadric surfaces* that you will encounter frequently throughout the remainder of the text. You saw one such surface in Section 13.2: A sphere with radius a centered at the origin with an equation of $x^2 + y^2 + z^2 = a^2$ is an example of a quadric surface. We also introduce a family of surfaces called *cylinders*, some of which are quadric surfaces.

Cylinders and Traces

In everyday language, we use the word *cylinder* to describe the surface that forms, say, the curved wall of a paint can. In the context of three-dimensional surfaces, the term *cylinder* refers to a surface that is parallel to a line. In this text, we focus on cylinders that are parallel to one of the coordinate axes. Equations for such cylinders are easy to identify: The variable corresponding to the coordinate axis parallel to the cylinder is missing from the equation.

For example, working in $\mathbb{R}^3$, the equation $y = x^2$ does not include z, which means that z is arbitrary and can take on all values. Therefore, $y = x^2$ describes the cylinder consisting of all lines parallel to the z-axis that pass through the parabola $y = x^2$ in the xy-plane (Figure 13.79a). In a similar way, the equation $y = z^2$ in $\mathbb{R}^3$ is missing the variable x, so it describes a cylinder parallel to the x-axis. The cylinder consists of lines parallel to the x-axis that pass through the parabola $y = z^2$ in the yz-plane (Figure 13.79b).

> ➤ The parabolic cylinder in Figure 13.79a can also be described as the surface swept out by translating the plane curve $y = x^2$ up and down (or parallel to) the z-axis.

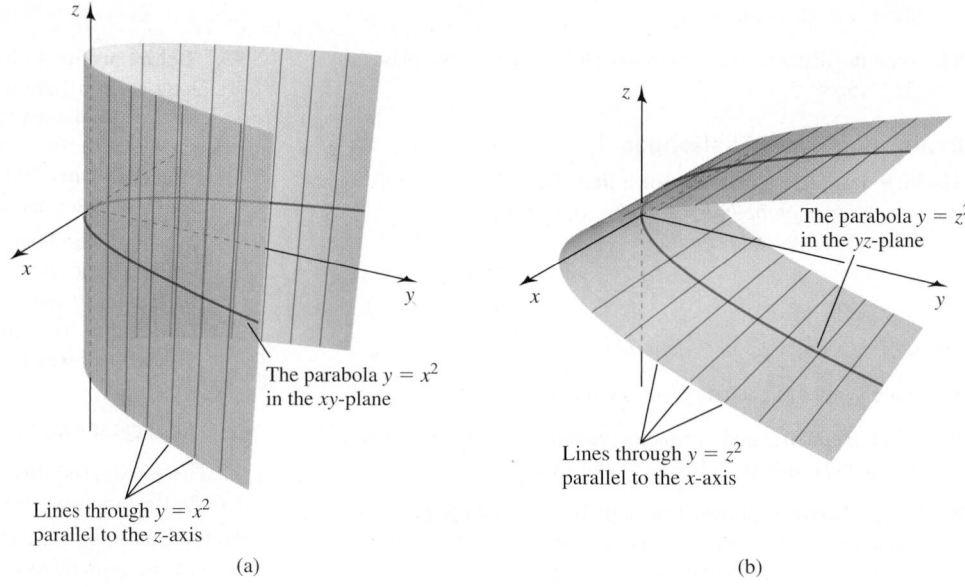

The parabola $y = z^2$ in the yz-plane

The parabola $y = x^2$ in the xy-plane

Lines through $y = z^2$ parallel to the x-axis

Lines through $y = x^2$ parallel to the z-axis

(a)   (b)

Figure 13.79

QUICK CHECK 1 To which coordinate axis in $\mathbb{R}^3$ is the cylinder $z - 2 \ln x = 0$ parallel? To which coordinate axis in $\mathbb{R}^3$ is the cylinder $y = 4z^2 - 1$ parallel? ◄

Graphing surfaces—and cylinders in particular—is facilitated by identifying the *traces* of the surface.

DEFINITION Trace

A **trace** of a surface is the set of points at which the surface intersects a plane that is parallel to one of the coordinate planes. The traces in the coordinate planes are called the ***xy*-trace**, the ***yz*-trace**, and the ***xz*-trace** (Figure 13.80).

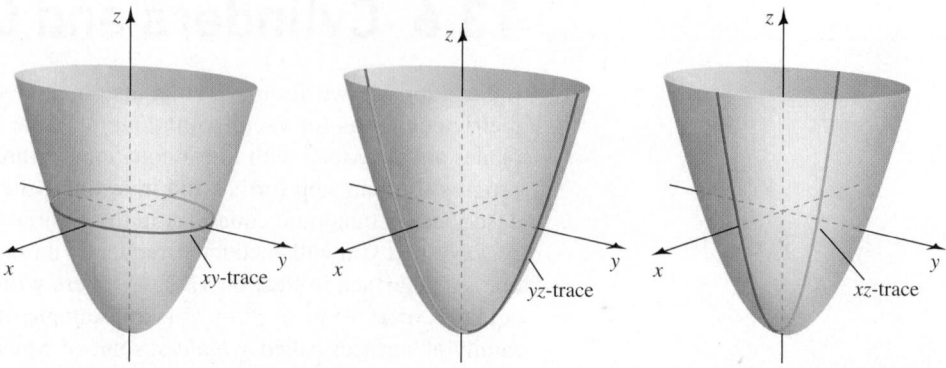

xy-trace yz-trace xz-trace

Figure 13.80

EXAMPLE 1   Graphing cylinders Sketch the graphs of the following cylinders in $\mathbb{R}^3$. Identify the axis to which each cylinder is parallel.

a. $x^2 + 4y^2 = 16$ **b.** $x - \sin z = 0$

SOLUTION

a. As an equation in $\mathbb{R}^3$, the variable z is absent. Therefore, z assumes all real values and the graph is a cylinder consisting of lines parallel to the z-axis passing through the curve $x^2 + 4y^2 = 16$ in the xy-plane. You can sketch the cylinder in the following steps.

   1. Rewriting the given equation as $\dfrac{x^2}{4^2} + \dfrac{y^2}{2^2} = 1$, we see that the trace of the cylinder in the xy-plane (the xy-trace) is an ellipse. We begin by drawing this ellipse.

   2. Next draw a second trace (a copy of the ellipse in Step 1) in a plane parallel to the xy-plane.

   3. Now draw lines parallel to the z-axis through the two traces to fill out the cylinder (**Figure 13.81a**).

   The resulting surface, called an *elliptic cylinder*, runs parallel to the z-axis (**Figure 13.81b**).

b. As an equation in $\mathbb{R}^3$, $x - \sin z = 0$ is missing the variable y. Therefore, y assumes all real values and the graph is a cylinder consisting of lines parallel to the y-axis passing through the curve $x = \sin z$ in the xz-plane. You can sketch the cylinder in the following steps.

   1. Graph the curve $x = \sin z$ in the xz-plane, which is the xz-trace of the surface.

   2. Draw a second trace (a copy of the curve in Step 1) in a plane parallel to the xz-plane.

   3. Draw lines parallel to the y-axis passing through the two traces (**Figure 13.82a**).

   The result is a cylinder, running parallel to the y-axis, consisting of copies of the curve $x = \sin z$ (**Figure 13.82b**).

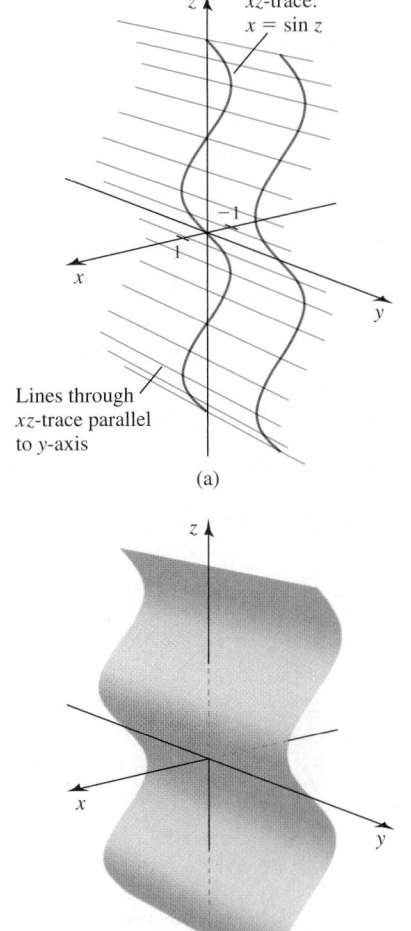

Figure (a): xz-trace: $x = \sin z$. Lines through xz-trace parallel to y-axis. (a)

Figure (b)

Figure 13.82

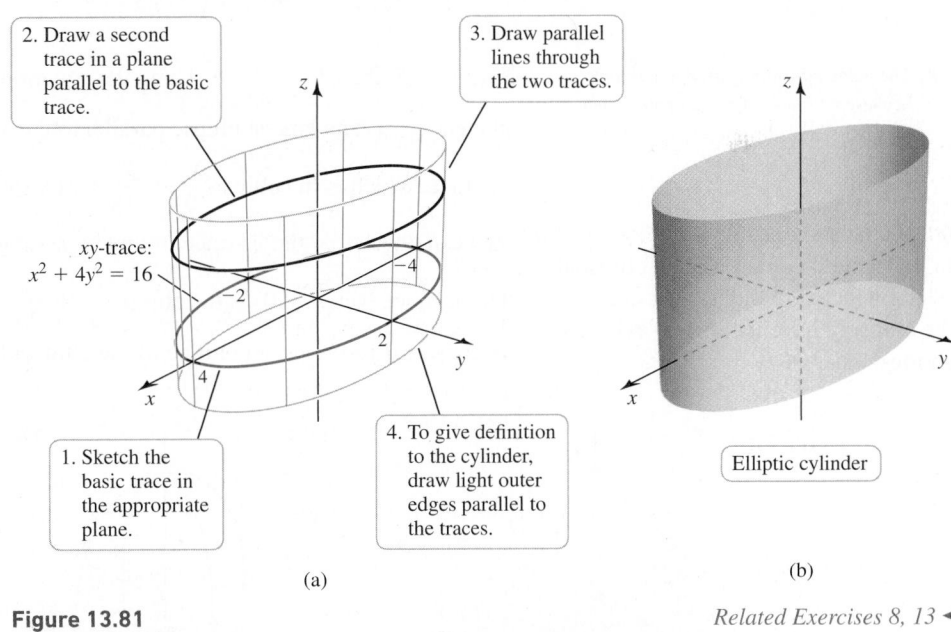

2. Draw a second trace in a plane parallel to the basic trace.

3. Draw parallel lines through the two traces.

xy-trace: $x^2 + 4y^2 = 16$

1. Sketch the basic trace in the appropriate plane.

4. To give definition to the cylinder, draw light outer edges parallel to the traces.

Elliptic cylinder

(a)   (b)

Figure 13.81   *Related Exercises 8, 13* ◄

Quadric Surfaces

Quadric surfaces are described by the general quadratic (second-degree) equation in three variables,

$$Ax^2 + By^2 + Cz^2 + Dxy + Exz + Fyz + Gx + Hy + Iz + J = 0,$$

where the coefficients $A, \ldots, J$ are constants and not all of $A, B, C, D, E,$ and F are zero. We do not attempt a detailed study of this large family of surfaces. However, a few standard surfaces are worth investigating.

Apart from their mathematical interest, quadric surfaces have a variety of practical uses. Paraboloids (defined in Example 3) share the reflective properties of their two-dimensional counterparts (Section 12.4) and are used to design satellite dishes, headlamps, and mirrors in telescopes. Cooling towers for nuclear power plants have the shape of hyperboloids of one sheet. Ellipsoids appear in the design of water tanks and gears.

Making hand sketches of quadric surfaces can be challenging. Here are a few general ideas to keep in mind as you sketch their graphs.

1. **Intercepts** Determine the points, if any, where the surface intersects the coordinate axes. To find these intercepts, set x, y, and z equal to zero in pairs in the equation of the surface, and solve for the third coordinate.

2. **Traces** As illustrated in the following examples, finding traces of the surface helps visualize the surface. For example, setting $z = 0$ or $z = z_0$ (a constant) gives the traces in planes parallel to the xy-plane.

3. **Completing the figure** Sketch at least two traces in parallel planes (for example, traces with $z = 0$ and $z = \pm 1$). Then draw smooth curves that pass through the traces to fill out the surface.

> ➤ Working with quadric surfaces requires familiarity with conic sections (Section 12.4).

QUICK CHECK 2 Explain why the elliptic cylinder discussed in Example 1a is a quadric surface. ◄

EXAMPLE 2   An ellipsoid The surface defined by the equation $\dfrac{x^2}{a^2} + \dfrac{y^2}{b^2} + \dfrac{z^2}{c^2} = 1$ is an *ellipsoid*. Graph the ellipsoid with $a = 3$, $b = 4$, and $c = 5$.

SOLUTION Setting x, y, and z equal to zero in pairs gives the intercepts $(\pm 3, 0, 0)$, $(0, \pm 4, 0)$, and $(0, 0, \pm 5)$. Note that points in $\mathbb{R}^3$ with $|x| > 3$ or $|y| > 4$ or $|z| > 5$ do not satisfy the equation of the surface (because the left side of the equation is a sum of nonnegative terms that cannot exceed 1). Therefore, the entire surface is contained in the rectangular box defined by $|x| \le 3$, $|y| \le 4$, and $|z| \le 5$.

The trace in the horizontal plane $z = z_0$ is found by substituting $z = z_0$ into the equation of the ellipsoid, which gives

$$\frac{x^2}{9} + \frac{y^2}{16} + \frac{z_0^2}{25} = 1 \quad \text{or} \quad \frac{x^2}{9} + \frac{y^2}{16} = 1 - \frac{z_0^2}{25}.$$

> ➤ The name *ellipsoid* is used in Example 2 because all traces of this surface, when they exist, are ellipses.

If $|z_0| < 5$, then $1 - \dfrac{z_0^2}{25} > 0$, and the equation describes an ellipse in the horizontal plane $z = z_0$. The largest ellipse parallel to the xy-plane occurs with $z_0 = 0$; it is the xy-trace, which is the ellipse $\dfrac{x^2}{9} + \dfrac{y^2}{16} = 1$ with axes of length 6 and 8 (**Figure 13.83a**). You can check that the yz-trace, found by setting $x = 0$, is the ellipse $\dfrac{y^2}{16} + \dfrac{z^2}{25} = 1$.

QUICK CHECK 3 Assume $0 < c < b < a$ in the general equation of an ellipsoid. Along which coordinate axis does the ellipsoid have its longest axis? Its shortest axis? ◄

The xz-trace (set $y = 0$) is the ellipse $\dfrac{x^2}{9} + \dfrac{z^2}{25} = 1$ (**Figure 13.83b**). When we sketch the xy-, xz-, and yz-traces, an outline of the ellipsoid emerges (**Figure 13.83c**).

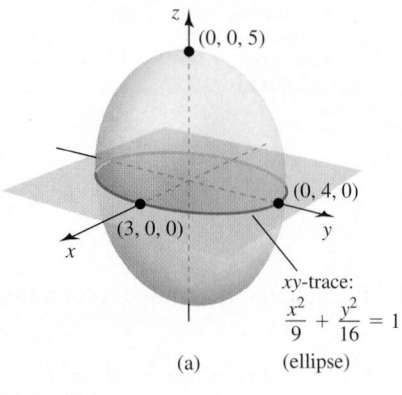

(a)

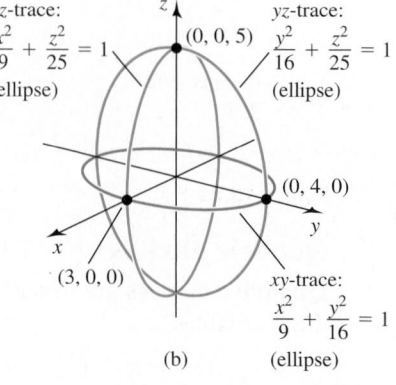

(b)

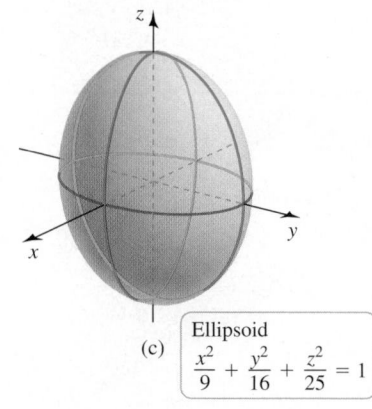

(c)

Figure 13.83

Related Exercise 29 ◄

EXAMPLE 3   An elliptic paraboloid The surface defined by the equation $z = \dfrac{x^2}{a^2} + \dfrac{y^2}{b^2}$ is an *elliptic paraboloid*. Graph the elliptic paraboloid with $a = 4$ and $b = 2$.

SOLUTION Note that the only intercept of the coordinate axes is $(0, 0, 0)$, which is the *vertex* of the paraboloid. The trace in the horizontal plane $z = z_0$, where $z_0 > 0$, satisfies the equation $\dfrac{x^2}{16} + \dfrac{y^2}{4} = z_0$, which describes an ellipse; there is no horizontal trace when $z_0 < 0$ (**Figure 13.84a**). The trace in the vertical plane $x = x_0$ is the parabola $z = \dfrac{x_0^2}{16} + \dfrac{y^2}{4}$ (**Figure 13.84b**); the trace in the vertical plane $y = y_0$ is the parabola $z = \dfrac{x^2}{16} + \dfrac{y_0^2}{4}$ (**Figure 13.84c**).

> The name *elliptic paraboloid* reflects the fact that the traces of this surface are parabolas and ellipses. Two of the three traces in the coordinate planes are parabolas, so this surface is called a paraboloid rather than an ellipsoid.

To graph the surface, we sketch the xz-trace $z = \dfrac{x^2}{16}$ (setting $y = 0$) and the yz-trace $z = \dfrac{y^2}{4}$ (setting $x = 0$). When these traces are combined with an elliptical trace $\dfrac{x^2}{16} + \dfrac{y^2}{4} = z_0$ in a plane $z = z_0$, an outline of the surface appears (**Figure 13.84d**).

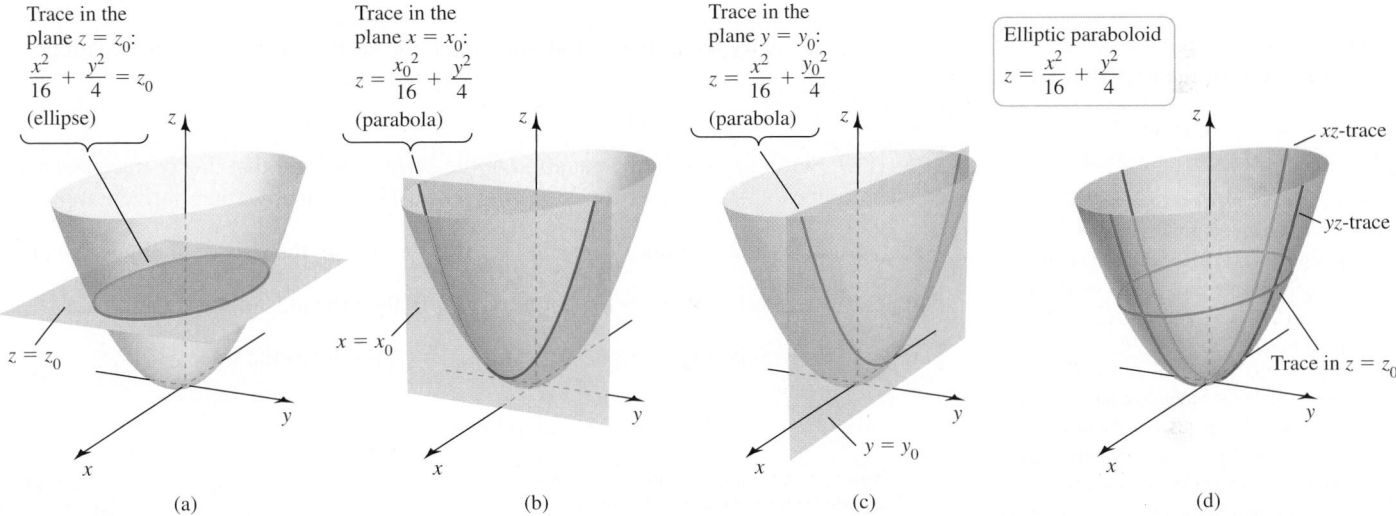

Figure 13.84

Related Exercise 32 ◄

QUICK CHECK 4 The elliptic paraboloid $x = \dfrac{y^2}{3} + \dfrac{z^2}{7}$ is a bowl-shaped surface. Along which axis does the bowl open? ◄

> To be completely accurate, this surface should be called an *elliptic hyperboloid of one sheet* because the traces are ellipses and hyperbolas.

EXAMPLE 4   A hyperboloid of one sheet Graph the surface defined by the equation $\dfrac{x^2}{4} + \dfrac{y^2}{9} - z^2 = 1$.

SOLUTION The intercepts of the coordinate axes are $(0, \pm 3, 0)$ and $(\pm 2, 0, 0)$. Setting $z = z_0$, the traces in horizontal planes are ellipses of the form $\dfrac{x^2}{4} + \dfrac{y^2}{9} = 1 + z_0^2$. This equation has solutions for all choices of z_0, so the surface has traces in all horizontal planes. These elliptical traces increase in size as $|z_0|$ increases (**Figure 13.85a**), with the smallest trace being the ellipse $\dfrac{x^2}{4} + \dfrac{y^2}{9} = 1$ in the xy-plane. Setting $y = 0$, the xz-trace is the hyperbola $\dfrac{x^2}{4} - z^2 = 1$; with $x = 0$, the yz-trace is the hyperbola $\dfrac{y^2}{9} - z^2 = 1$ (**Figure 13.85b, c**). In fact, the intersection of the surface with any vertical plane is a hyperbola. The resulting surface is a *hyperboloid of one sheet* (**Figure 13.85d**).

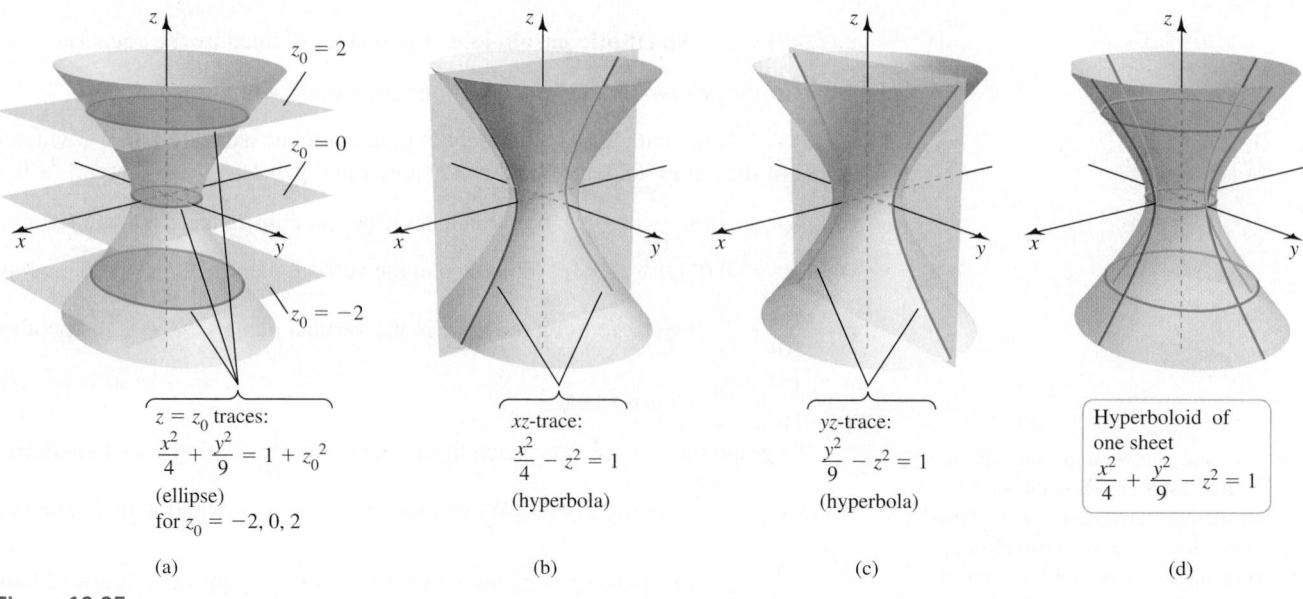

$z = z_0$ traces:
$$\frac{x^2}{4} + \frac{y^2}{9} = 1 + z_0^2$$
(ellipse)
for $z_0 = -2, 0, 2$

(a)

xz-trace:
$$\frac{x^2}{4} - z^2 = 1$$
(hyperbola)

(b)

yz-trace:
$$\frac{y^2}{9} - z^2 = 1$$
(hyperbola)

(c)

Hyperboloid of one sheet
$$\frac{x^2}{4} + \frac{y^2}{9} - z^2 = 1$$

(d)

Figure 13.85

Related Exercise 33 ◄

QUICK CHECK 5 Which coordinate axis is the axis of the hyperboloid
$$\frac{y^2}{a^2} + \frac{z^2}{b^2} - \frac{x^2}{c^2} = 1?$$ ◄

➤ The name *hyperbolic paraboloid* tells us that the traces are hyperbolas and parabolas. Two of the three traces in the coordinate planes are parabolas, so this surface is a paraboloid rather than a hyperboloid.

➤ The hyperbolic paraboloid has a feature called a *saddle point*. For the surface in Example 5, if you walk from the saddle point at the origin in the direction of the x-axis, you move uphill. If you walk from the saddle point in the direction of the y-axis, you move downhill. Saddle points are examined in detail in Section 15.7.

EXAMPLE 5   A hyperbolic paraboloid   Graph the surface defined by the equation
$$z = x^2 - \frac{y^2}{4}.$$

SOLUTION Setting $z = 0$ in the equation of the surface, we see that the xy-trace consists of the two lines $y = \pm 2x$. However, slicing the surface with any other horizontal plane $z = z_0$ produces a hyperbola $x^2 - \frac{y^2}{4} = z_0$. If $z_0 > 0$, then the axis of the hyperbola is parallel to the x-axis. On the other hand, if $z_0 < 0$, then the axis of the hyperbola is parallel to the y-axis (**Figure 13.86a**). Setting $x = x_0$ produces the trace $z = x_0^2 - \frac{y^2}{4}$,

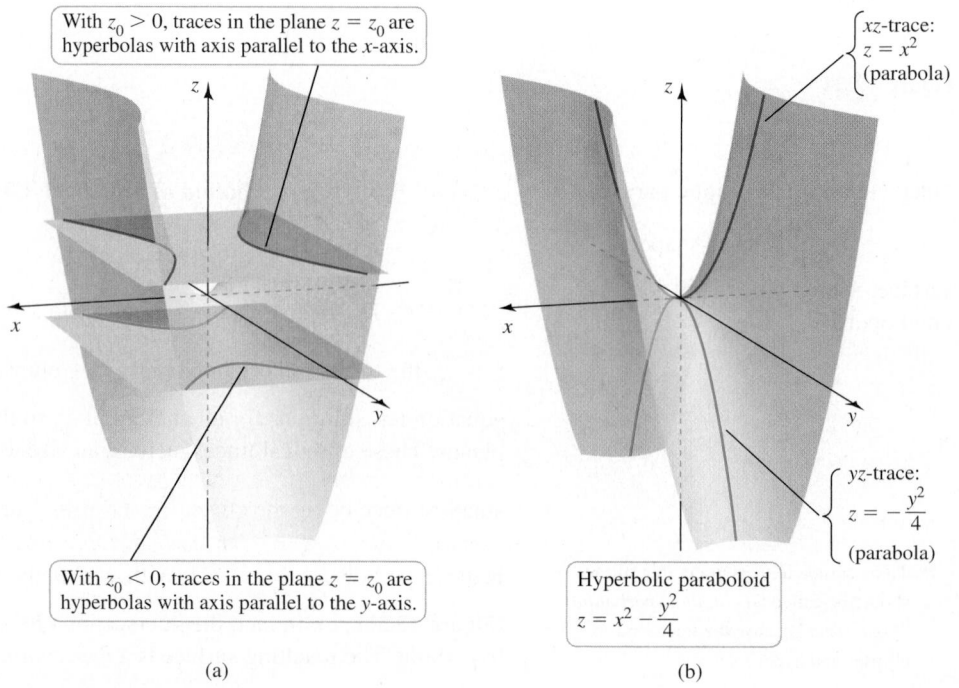

With $z_0 > 0$, traces in the plane $z = z_0$ are hyperbolas with axis parallel to the x-axis.

With $z_0 < 0$, traces in the plane $z = z_0$ are hyperbolas with axis parallel to the y-axis.

(a)

xz-trace:
$z = x^2$
(parabola)

yz-trace:
$z = -\dfrac{y^2}{4}$
(parabola)

Hyperbolic paraboloid
$$z = x^2 - \frac{y^2}{4}$$

(b)

Figure 13.86

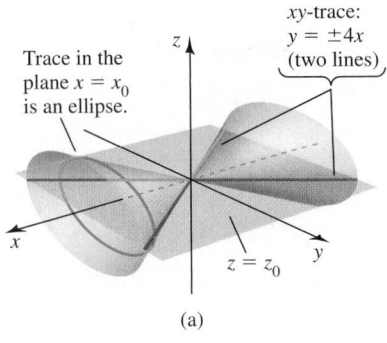

Trace in the plane $x = x_0$ is an ellipse.

xy-trace: $y = \pm 4x$ (two lines)

$z = z_0$

(a)

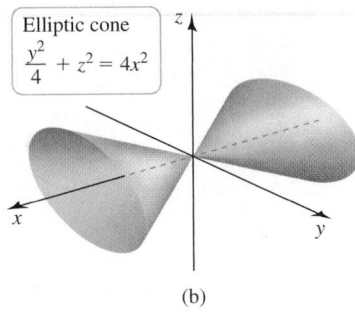

Elliptic cone
$$\frac{y^2}{4} + z^2 = 4x^2$$

(b)

Figure 13.87

which is the equation of a parabola that opens downward in a plane parallel to the yz-plane. You can check that traces in planes parallel to the xz-plane are parabolas that open upward. The resulting surface is a *hyperbolic paraboloid* (Figure 13.86b).

Related Exercise 35 ◄

EXAMPLE 6 **Elliptic cones** Graph the surface defined by the equation $\dfrac{y^2}{4} + z^2 = 4x^2$.

SOLUTION The only point at which the surface intersects the coordinate axes is $(0, 0, 0)$. Traces in the planes $x = x_0$ are ellipses of the form $\dfrac{y^2}{4} + z^2 = 4x_0^2$ that shrink in size as x_0 approaches 0. Setting $y = 0$, the xz-trace satisfies the equation $z^2 = 4x^2$ or $z = \pm 2x$, which are equations of two lines in the xz-plane that intersect at the origin. Setting $z = 0$, the xy-trace satisfies $y^2 = 16x^2$ or $y = \pm 4x$, which describes two lines in the xy-plane that intersect at the origin (Figure 13.87a). The complete surface consists of two *cones* opening in opposite directions along the x-axis with a common vertex at the origin (Figure 13.87b).

Related Exercise 38 ◄

EXAMPLE 7 **A hyperboloid of two sheets** Graph the surface defined by the equation
$$-16x^2 - 4y^2 + z^2 + 64x - 80 = 0.$$

SOLUTION We first regroup terms, which yields
$$-16\underbrace{(x^2 - 4x)}_{\text{complete the square}} - 4y^2 + z^2 - 80 = 0,$$

and then complete the square in x:
$$-16(\underbrace{x^2 - 4x + 4}_{(x-2)^2} - 4) - 4y^2 + z^2 - 80 = 0.$$

Collecting terms and dividing by 16 gives the equation
$$-(x - 2)^2 - \frac{y^2}{4} + \frac{z^2}{16} = 1.$$

Notice that if $z = 0$, the equation has no solution, so the surface does not intersect the xy-plane. The traces in planes parallel to the xz- and yz-planes are hyperbolas. If $|z_0| \geq 4$, the trace in the plane $z = z_0$ is an ellipse. This equation describes a *hyperboloid of two sheets*, with its axis parallel to the z-axis and shifted 2 units in the positive x-direction (Figure 13.88).

Related Exercise 56 ◄

> The equation $-x^2 - \dfrac{y^2}{4} + \dfrac{z^2}{16} = 1$ describes a hyperboloid of two sheets with its axis on the z-axis. Therefore, the equation in Example 7 describes the same surface shifted 2 units in the positive x-direction.

Hyperboloid of two sheets
$$-(x - 2)^2 - \frac{y^2}{4} + \frac{z^2}{16} = 1$$

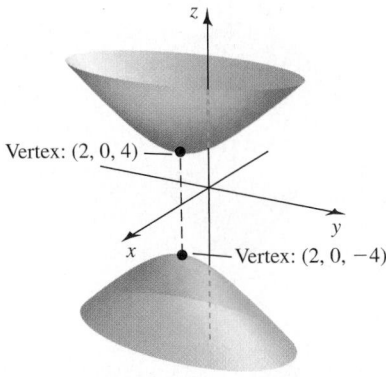

Vertex: $(2, 0, 4)$

Vertex: $(2, 0, -4)$

Figure 13.88

QUICK CHECK 6 In which variable(s) should you complete the square to identify the surface $x = y^2 + 2y + z^2 - 4z + 16$? Name and describe the surface. ◄

Table 13.1 (where a, b, and c are nonzero real numbers) summarizes the standard quadric surfaces. It is important to note that the same surfaces with different orientations are obtained when the roles of the variables are interchanged. For this reason, Table 13.1 summarizes many more surfaces than those listed.

Table 13.1

Name	Standard Equation	Features	Graph
Ellipsoid	$\dfrac{x^2}{a^2} + \dfrac{y^2}{b^2} + \dfrac{z^2}{c^2} = 1$	All traces are ellipses.	
Elliptic paraboloid	$z = \dfrac{x^2}{a^2} + \dfrac{y^2}{b^2}$	Traces with $z = z_0 > 0$ are ellipses. Traces with $x = x_0$ or $y = y_0$ are parabolas.	
Hyperboloid of one sheet	$\dfrac{x^2}{a^2} + \dfrac{y^2}{b^2} - \dfrac{z^2}{c^2} = 1$	Traces with $z = z_0$ are ellipses for all z_0. Traces with $x = x_0$ or $y = y_0$ are hyperbolas.	
Hyperboloid of two sheets	$-\dfrac{x^2}{a^2} - \dfrac{y^2}{b^2} + \dfrac{z^2}{c^2} = 1$	Traces with $z = z_0$ with $|z_0| > |c|$ are ellipses. Traces with $x = x_0$ and $y = y_0$ are hyperbolas.	
Elliptic cone	$\dfrac{x^2}{a^2} + \dfrac{y^2}{b^2} = \dfrac{z^2}{c^2}$	Traces with $z = z_0 \neq 0$ are ellipses. Traces with $x = x_0$ or $y = y_0$ are hyperbolas or intersecting lines.	
Hyperbolic paraboloid	$z = \dfrac{x^2}{a^2} - \dfrac{y^2}{b^2}$	Traces with $z = z_0 \neq 0$ are hyperbolas. Traces with $x = x_0$ or $y = y_0$ are parabolas.	

SECTION 13.6 EXERCISES

Getting Started

1. To which coordinate axes are the following cylinders in $\mathbb{R}^3$ parallel: $x^2 + 2y^2 = 8$, $z^2 + 2y^2 = 8$, and $x^2 + 2z^2 = 8$?

2. Describe the graph of $x = z^2$ in $\mathbb{R}^3$.

3. What is a trace of a surface?

4. What is the name of the surface defined by the equation $y = \dfrac{x^2}{4} + \dfrac{z^2}{8}$?

5. What is the name of the surface defined by the equation $x^2 + \dfrac{y^2}{3} + 2z^2 = 1$?

6. What is the name of the surface defined by the equation $-y^2 - \dfrac{z^2}{2} + x^2 = 1$?

Practice Exercises

7–14. Cylinders in $\mathbb{R}^3$ *Consider the following cylinders in $\mathbb{R}^3$.*

a. Identify the coordinate axis to which the cylinder is parallel.
b. Sketch the cylinder.

7. $z = y^2$

8. $x^2 + 4y^2 = 4$

9. $x^2 + z^2 = 4$

10. $x = z^2 - 4$

11. $y - x^3 = 0$

12. $x - 2z^2 = 0$

13. $z - \ln y = 0$

14. $x - \dfrac{1}{y} = 0$

15–20. Identifying quadric surfaces *Identify the following quadric surfaces by name. Find and describe the xy-, xz-, and yz-traces, when they exist.*

15. $25x^2 + 25y^2 + z^2 = 25$

16. $25x^2 + 25y^2 - z^2 = 25$

17. $25x^2 + 25y^2 - z = 0$

18. $25x^2 - 25y^2 - z = 0$

19. $-25x^2 - 25y^2 + z^2 = 25$

20. $-25x^2 - 25y^2 + z^2 = 0$

21–28. Identifying surfaces *Identify the following surfaces by name.*

21. $y = 4z^2 - x^2$

22. $-y^2 - 9z^2 + \dfrac{x^2}{4} = 1$

23. $y = \dfrac{x^2}{6} + \dfrac{z^2}{16}$

24. $z^2 + 4y^2 - x^2 = 1$

25. $y^2 - z^2 = 2$

26. $x^2 + 4y^2 = 1$

27. $9x^2 + 4z^2 - 36y = 0$

28. $9y^2 + 4z^2 - 36x^2 = 0$

29–51. Quadric surfaces *Consider the following equations of quadric surfaces.*

a. Find the intercepts with the three coordinate axes, when they exist.
b. Find the equations of the xy-, xz-, and yz-traces, when they exist.
c. Identify and sketch a graph of the surface.

29. $x^2 + \dfrac{y^2}{4} + \dfrac{z^2}{9} = 1$

30. $4x^2 + y^2 + \dfrac{z^2}{2} = 1$

31. $x = y^2 + z^2$

32. $z = \dfrac{x^2}{4} + \dfrac{y^2}{9}$

33. $\dfrac{x^2}{25} + \dfrac{y^2}{9} - z^2 = 1$

34. $\dfrac{y^2}{4} + \dfrac{z^2}{9} - \dfrac{x^2}{16} = 1$

35. $z = \dfrac{x^2}{9} - y^2$

36. $y = \dfrac{x^2}{16} - 4z^2$

37. $x^2 + \dfrac{y^2}{4} = z^2$

38. $4y^2 + z^2 = x^2$

39. $\dfrac{x^2}{3} + 3y^2 + \dfrac{z^2}{12} = 3$

40. $\dfrac{x^2}{6} + 24y^2 + \dfrac{z^2}{24} - 6 = 0$

41. $9x - 81y^2 - \dfrac{z^2}{4} = 0$

42. $2y - \dfrac{x^2}{8} - \dfrac{z^2}{18} = 0$

43. $\dfrac{y^2}{16} + 36z^2 - \dfrac{x^2}{4} - 9 = 0$

44. $9z^2 + x^2 - \dfrac{y^2}{3} - 1 = 0$

45. $5x - \dfrac{y^2}{5} + \dfrac{z^2}{20} = 0$

46. $6y + \dfrac{x^2}{6} - \dfrac{z^2}{24} = 0$

47. $\dfrac{z^2}{32} + \dfrac{y^2}{18} = 2x^2$

48. $\dfrac{x^2}{3} + \dfrac{z^2}{12} = 3y^2$

49. $-x^2 + \dfrac{y^2}{4} - \dfrac{z^2}{9} = 1$

50. $-\dfrac{x^2}{6} - 24y^2 + \dfrac{z^2}{24} - 6 = 0$

51. $-\dfrac{x^2}{3} + 3y^2 - \dfrac{z^2}{12} = 1$

52. Describe the relationship between the graphs of the quadric surfaces $x^2 + y^2 - z^2 + 2z = 1$ and $x^2 + y^2 - z^2 = 0$, and state the names of the surfaces.

53. Describe the relationship between the graphs of $x^2 + 4y^2 + 9z^2 = 100$ and $x^2 + 4y^2 + 9z^2 + 54z = 19$, and state the names of the surfaces.

54–58. Identifying surfaces *Identify and briefly describe the surfaces defined by the following equations.*

54. $x^2 + y^2 + 4z^2 + 2x = 0$

55. $9x^2 + y^2 - 4z^2 + 2y = 0$

56. $-x^2 - y^2 + \dfrac{z^2}{9} + 6x - 8y = 26$

57. $\dfrac{x^2}{4} + y^2 - 2x - 10y - z^2 + 41 = 0$

58. $z = -x^2 - y^2$

59. **Explain why or why not** Determine whether the following statements are true and give an explanation or counterexample.

   a. The graph of the equation $y = z^2$ in $\mathbb{R}^3$ is both a cylinder and a quadric surface.
   b. The xy-traces of the ellipsoid $x^2 + 2y^2 + 3z^2 = 16$ and the cylinder $x^2 + 2y^2 = 16$ are identical.
   c. Traces of the surface $y = 3x^2 - z^2$ in planes parallel to the xy-plane are parabolas.
   d. Traces of the surface $y = 3x^2 - z^2$ in planes parallel to the xz-plane are parabolas.
   e. The graph of the ellipsoid $x^2 + 2y^2 + 3(z - 4)^2 = 25$ is obtained by shifting the graph of the ellipsoid $x^2 + 2y^2 + 3z^2 = 25$ down 4 units.

60. Matching graphs with equations Match equations a–f with surfaces A–F.

a. $y - z^2 = 0$

b. $2x + 3y - z = 5$

c. $4x^2 + \dfrac{y^2}{9} + z^2 = 1$

d. $x^2 + \dfrac{y^2}{9} - z^2 = 1$

e. $x^2 + \dfrac{y^2}{9} = z^2$

f. $y = |x|$

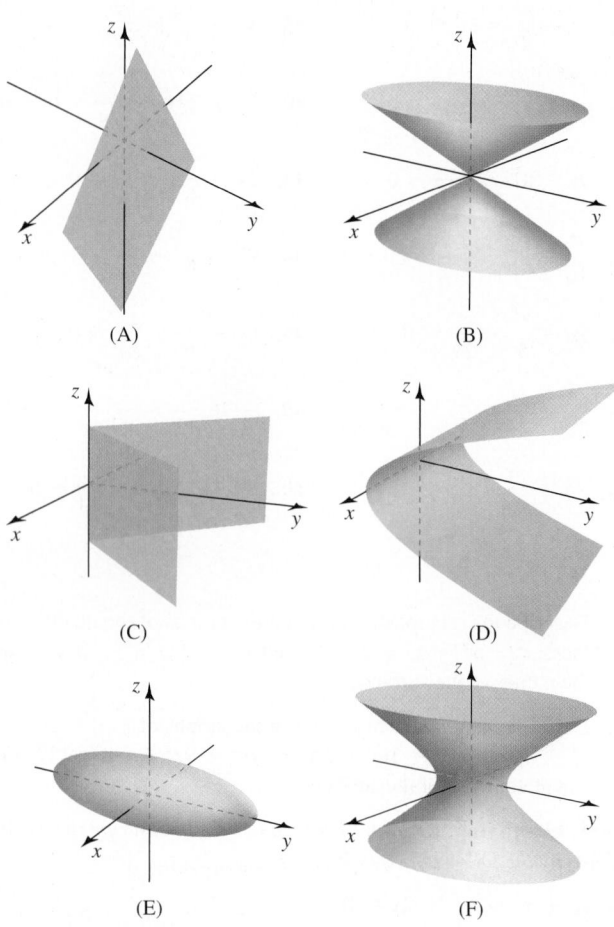

(A)   (B)

(C)   (D)

(E)   (F)

Explorations and Challenges

61. Solids of revolution Which of the quadric surfaces in Table 13.1 can be generated by revolving a curve in one of the coordinate planes about a coordinate axis, assuming $a = b = c \neq 0$?

62. Solids of revolution Consider the ellipse $x^2 + 4y^2 = 1$ in the xy-plane.

a. If this ellipse is revolved about the x-axis, what is the equation of the resulting ellipsoid?

b. If this ellipse is revolved about the y-axis, what is the equation of the resulting ellipsoid?

63. Volume Find the volume of the solid that is bounded between the planes $z = 0$ and $z = 3$ and the cylinders $y = x^2$ and $y = 2 - x^2$.

64. Light cones The idea of a *light cone* appears in the Special Theory of Relativity. The xy-plane (see figure) represents all of three-dimensional space, and the z-axis is the time axis (t-axis). If an event E occurs at the origin, the interior of the future light cone ($t > 0$) represents all events in the future that could be affected by E, assuming no signal travels faster than the speed of light. The interior of the past light cone ($t < 0$) represents all events in the

past that could have affected E, again assuming no signal travels faster than the speed of light.

a. If time is measured in seconds and distance (x and y) is measured in light-seconds (the distance light travels in 1 s), the light cone makes a 45° angle with the xy-plane. Write the equation of the light cone in this case.

b. Suppose distance is measured in meters and time is measured in seconds. Write the equation of the light cone in this case, given that the speed of light is 3×10^8 m/s.

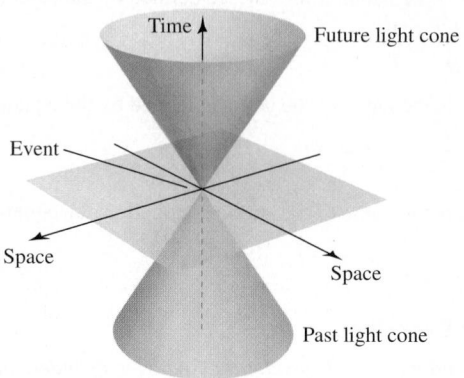

65. Designing an NFL football A *prolate spheroid* is a surface of revolution obtained by rotating an ellipse about its major axis.

a. Explain why one possible equation for a prolate spheroid is
$$\frac{x^2 + z^2}{a^2} + \frac{y^2}{b^2} = 1, \text{ where } b > a > 0.$$

b. According to the National Football League (NFL) rulebook, the shape of an NFL football is required to be a prolate spheroid with a long axis between 11 and 11.25 inches long and a short circumference (the circumference of the xz-trace) between 21 and 21.25 inches. Find an equation for the shape of the football if the long axis is 11.1 inches and the short circumference is 21.1 inches.

66. Hand tracking Researchers are developing hand tracking software that will allow computers to track and recognize detailed hand movements for better human-computer interaction. One three-dimensional hand model under investigation is constructed from a set of truncated quadrics (see figure). For example, the palm of the hand consists of a truncated elliptic cylinder, capped off by the upper half of an ellipsoid. Suppose the palm of the hand is modeled by the truncated cylinder $4x^2/9 + 4y^2 = 1$, for $0 \leq z \leq 3$. Find an equation of the upper half of an ellipsoid, whose bottom corresponds with the top of the cylinder, if the distance from the top of the truncated cylinder to the top of the ellipsoid is $1/2$.

(*Source: Computer Vision and Pattern Recognition*, 2, Dec 2001)

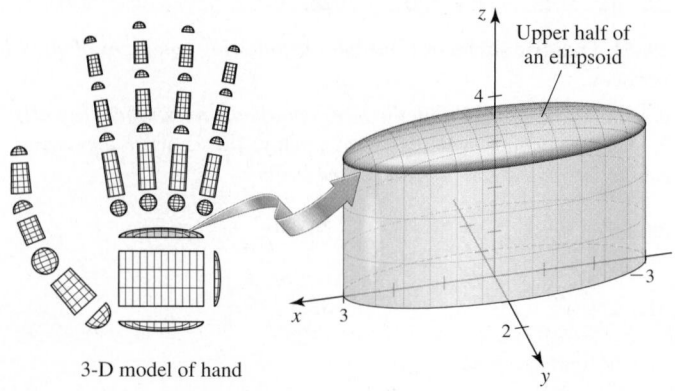

3-D model of hand

67. Designing a snow cone A surface, having the shape of an oblong snow cone, consists of a truncated cone, $\dfrac{x^2}{2} + y^2 = \dfrac{z^2}{8}$, for $0 \le z \le 3$, capped off by the upper half of an ellipsoid. Find an equation for the upper half of the ellipsoid so that the bottom edge of the truncated ellipsoid and the top edge of the cone coincide, and the distance from the top of the cone to the top of the ellipsoid is $3/2$.

68. Designing a glass The outer, lateral side of a 6-inch-tall glass has the shape of the truncated hyperboloid of one sheet $\dfrac{x^2}{a^2} + \dfrac{y^2}{b^2} - \dfrac{z^2}{c^2} = 1$, for $0 \le z \le 6$. If the base of the glass has a radius of 1 inch and the top of the glass has a radius of 2 inches, find the values of a^2, b^2, and c^2 that satisfy these conditions. Assume horizontal traces of the glass are circular.

QUICK CHECK ANSWERS

1. y-axis; x-axis   **2.** The equation $x^2 + 4y^2 = 16$ is a special case of the general equation for quadric surfaces; all the coefficients except A, B, and J are zero.   **3.** x-axis; z-axis   **4.** Positive x-axis   **5.** x-axis   **6.** Complete the square in y and z; elliptic paraboloid with its axis parallel to the x-axis ◄

CHAPTER 13   REVIEW EXERCISES

1. Explain why or why not Determine whether the following statements are true and give an explanation or counterexample.

a. Given two vectors $\mathbf{u}$ and $\mathbf{v}$, it is always true that $2\mathbf{u} + \mathbf{v} = \mathbf{v} + 2\mathbf{u}$.

b. The vector in the direction of $\mathbf{u}$ with the length of $\mathbf{v}$ equals the vector in the direction of $\mathbf{v}$ with the length of $\mathbf{u}$.

c. If $\mathbf{u} \ne \mathbf{0}$ and $\mathbf{u} + \mathbf{v} = \mathbf{0}$, then $\mathbf{u}$ and $\mathbf{v}$ are parallel.

d. The lines $x = 3 + t, y = 4 + 2t, z = 2 - t$ and $x = 2t, y = 4t, z = t$ are parallel.

e. The lines $x = 3 + t, y = 4 + 2t, z = 2 - t$ and the plane $x + 2y + 5z = 3$ are parallel.

f. There is always a plane orthogonal to both of two distinct intersecting planes.

2–5. Working with vectors Let $\mathbf{u} = \langle 3, -4 \rangle$ and $\mathbf{v} = \langle -1, 2 \rangle$. Evaluate each of the following.

2. $\mathbf{u} - \mathbf{v}$ **3.** $-3\mathbf{v}$

4. $\mathbf{u} + 2\mathbf{v}$ **5.** $2\mathbf{v} - \mathbf{u}$

6–15. Working with vectors Let $\mathbf{u} = \langle 2, 4, -5 \rangle$, $\mathbf{v} = \langle -6, 10, 2 \rangle$, and $\mathbf{w} = \langle 4, -8, 8 \rangle$.

6. Compute $\mathbf{u} - 3\mathbf{v}$.

7. Compute $|\mathbf{u} + \mathbf{v}|$.

8. Find the unit vector with the same direction as $\mathbf{u}$.

9. Write the vector $\mathbf{w}$ as a product of its magnitude and a unit vector in the direction of $\mathbf{w}$.

10. Find a vector in the direction of $\mathbf{w}$ that is 10 times as long as $\mathbf{w}$.

11. Find a vector in the direction of $\mathbf{w}$ with a length of 10.

12. Compute $\mathbf{u} \cdot \mathbf{v}$

13. Compute $\mathbf{u} \times \mathbf{v}$

14. For what value of a is the vector $\mathbf{v}$ orthogonal to $\mathbf{y} = \langle a, 1, -3 \rangle$?

15. For what value of a is the vector $\mathbf{w}$ parallel to $\mathbf{y} = \langle a, 6, -6 \rangle$?

16. Scalar multiples Find scalars a, b, and c such that
$$\langle 2, 2, 2 \rangle = a\langle 1, 1, 0 \rangle + b\langle 0, 1, 1 \rangle + c\langle 1, 0, 1 \rangle.$$

17. Velocity vectors Assume the positive x-axis points east and the positive y-axis points north.

a. An airliner flies northwest at a constant altitude at 550 mi/hr in calm air. Find a and b such that its velocity may be expressed in the form $\mathbf{v} = a\mathbf{i} + b\mathbf{j}$.

b. An airliner flies northwest at a constant altitude at 550 mi/hr relative to the air in a southerly crosswind $\mathbf{w} = \langle 0, 40 \rangle$. Find the velocity of the airliner relative to the ground.

18. Position vectors Let $\overrightarrow{PQ}$ extend from $P(2, 0, 6)$ to $Q(2, -8, 5)$.

a. Find the position vector equal to $\overrightarrow{PQ}$.

b. Find the midpoint M of the line segment PQ. Then find the magnitude of $\overrightarrow{PM}$.

c. Find a vector of length 8 with direction opposite to that of $\overrightarrow{PQ}$.

19–21. Spheres and balls Use set notation to describe the following sets.

19. The sphere of radius 4 centered at $(1, 0, -1)$

20. The points inside the sphere of radius 10 centered at $(2, 4, -3)$

21. The points outside the sphere of radius 2 centered at $(0, 1, 0)$

22–25. Identifying sets Give a geometric description of the following sets of points.

22. $x^2 - 6x + y^2 + 8y + z^2 - 2z - 23 = 0$

23. $x^2 - x + y^2 + 4y + z^2 - 6z + 11 \le 0$

24. $x^2 + y^2 - 10y + z^2 - 6z = -34$

25. $x^2 - 6x + y^2 + z^2 - 20z + 9 > 0$

26. Combined force An object at the origin is acted on by the forces $\mathbf{F}_1 = -10\mathbf{i} + 20\mathbf{k}$, $\mathbf{F}_2 = 40\mathbf{j} + 10\mathbf{k}$, and $\mathbf{F}_3 = -50\mathbf{i} + 20\mathbf{j}$. Find the magnitude of the combined force, and use a sketch to illustrate the direction of the combined force.

27. Falling probe A remote sensing probe falls vertically with a terminal velocity of 60 m/s when it encounters a horizontal crosswind blowing north at 4 m/s and an updraft blowing vertically at 10 m/s. Find the magnitude and direction of the resulting velocity relative to the ground.

28. Crosswinds A small plane is flying north in calm air at 250 mi/hr when it is hit by a horizontal crosswind blowing northeast at 40 mi/hr and a 25 mi/hr downdraft. Find the resulting velocity and speed of the plane.

29. Net force Jack pulls east on a rope attached to a camel with a force of 40 lb. Jill pulls north on a rope attached to the same camel with a force of 30 lb. What is the magnitude and direction of the force on the camel? Assume the vectors lie in a horizontal plane.

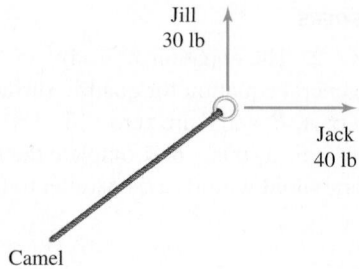

30. Canoe in a current A woman in a canoe paddles due west at 4 mi/hr relative to the water in a current that flows northwest at 2 mi/hr. Find the speed and direction of the canoe relative to the shore.

31. Set of points Describe the set of points satisfying both the equations $x^2 + z^2 = 1$ and $y = 2$.

32–33. Angles and projections

a. Find the angle between **u** and **v**.
b. Compute $\text{proj}_v \mathbf{u}$ and $\text{scal}_v \mathbf{u}$.
c. Compute $\text{proj}_u \mathbf{v}$ and $\text{scal}_u \mathbf{v}$.

32. $\mathbf{u} = -3\mathbf{j} + 4\mathbf{k}, \mathbf{v} = -4\mathbf{i} + \mathbf{j} + 5\mathbf{k}$

33. $\mathbf{u} = -\mathbf{i} + 2\mathbf{j} + 2\mathbf{k}, \mathbf{v} = 3\mathbf{i} + 6\mathbf{j} + 6\mathbf{k}$

34. Parallelepiped Find the volume of a parallelepiped determined by the position vectors $\mathbf{u} = \langle 2, 4, -5 \rangle, \mathbf{v} = \langle -6, 10, 2 \rangle$, and $\mathbf{w} = \langle 4, -8, 8 \rangle$ (see Exercise 63 in Section 13.4).

35–36. Computing work *Calculate the work done in the following situations.*

35. A suitcase is pulled 25 ft along a horizontal sidewalk with a constant force of 20 lb at an angle of 45° above the horizontal.

36. A constant force $\mathbf{F} = \langle 2, 3, 4 \rangle$ (in newtons) moves an object from $(0, 0, 0)$ to $(2, 1, 6)$. (Distance is measured in meters.)

37–38. Inclined plane *A 180-lb man stands on a hillside that makes an angle of 30° with the horizontal, producing a force of* $\mathbf{W} = \langle 0, -180 \rangle$.

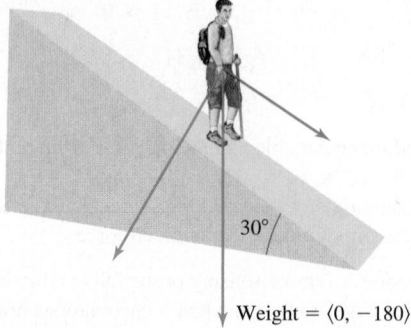

37. Find the component of his weight in the downward direction perpendicular to the hillside and in the downward direction parallel to the hillside.

38. How much work is done when the man moves 10 ft up the hillside?

39. Area of a parallelogram Find the area of the parallelogram with the vertices $(1, 2, 3), (1, 0, 6)$, and $(4, 2, 4)$.

40. Area of a triangle Find the area of the triangle with the vertices $(1, 0, 3), (5, 0, -1)$, and $(0, 2, -2)$.

41. Unit normal vector Find unit vectors normal to the vectors $\langle 2, -6, 9 \rangle$ and $\langle -1, 0, 6 \rangle$.

42. Angle in two ways Find the angle between $\langle 2, 0, -2 \rangle$ and $\langle 2, 2, 0 \rangle$ using (a) the dot product and (b) the cross product.

43. Let $\mathbf{r} = \overrightarrow{OP} = 3\mathbf{i} + 2\mathbf{j} + \mathbf{k}$. A force $\mathbf{F} = \langle 10, 10, 0 \rangle$ is applied at P. Find the torque about O that is produced.

44. Suppose you apply a force of $|\mathbf{F}| = 50$ N near the end of a wrench attached to a bolt (see figure). Determine the magnitude of the torque when the force is applied at an angle of 60° to the wrench. Assume the distance along the wrench from the center of the bolt to the point where the force is applied is $|\mathbf{r}| = 0.25$ m.

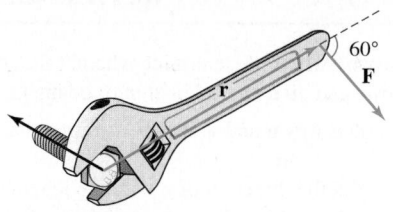

45. Knee torque Jan does leg lifts with a 10-kg weight attached to her foot, so the resulting force is $mg \approx 98$ N directed vertically downward (see figure). If the distance from her knee to the weight is 0.4 m and her lower leg makes an angle of θ to the vertical, find the magnitude of the torque about her knee as her leg is lifted (as a function of θ). What are the minimum and maximum magnitudes of the torque? Does the direction of the torque change as her leg is lifted?

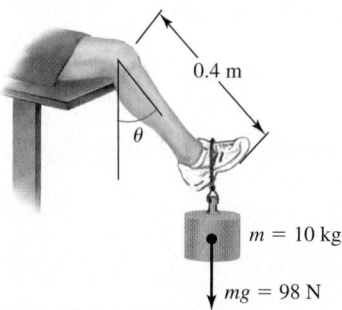

46–50. Lines in space *Find an equation of the following lines or line segments.*

46. The line that passes through the points $(2, 6, -1)$ and $(-6, 4, 0)$

47. The line segment that joins the points $(0, -3, 9)$ and $(2, -8, 1)$

48. The line through the point $(0, 1, 1)$ and parallel to the line $\mathbf{R} = \langle 1 + 2t, 3 - 5t, 7 + 6t \rangle$.

49. The line through the point $(0, 1, 1)$ that is orthogonal to both $\langle 0, -1, 3 \rangle$ and $\langle 2, -1, 2 \rangle$.

50. The line through the point $(0, 1, 4)$ and orthogonal to the vector $\langle -2, 1, 7 \rangle$ and the y-axis.

51. Equations of planes Consider the plane passing through the points $(0, 0, 3)$, $(1, 0, -6)$, and $(1, 2, 3)$.

    a. Find an equation of the plane.
    b. Find the intercepts of the plane with the three coordinate axes.
    c. Make a sketch of the plane.

52–53. Intersecting planes *Find an equation of the line of intersection of the planes Q and R.*

52. $Q: 2x + y - z = 0$, $R: -x + y + z = 1$

53. $Q: -3x + y + 2z = 0$, $R: 3x + 3y + 4z - 12 = 0$

54–57. Equations of planes *Find an equation of the following planes.*

54. The plane passing through $(5, 0, 2)$ that is parallel to the plane $2x + y - z = 0$

55. The plane containing the lines $x = 5 + t, y = 3 - 2t, z = 1$ and $x = 4s, y = 5s, z = 3 - 2s$, if possible

56. The plane passing through $(2, -3, 1)$ normal to the line $\langle x, y, z \rangle = \langle 2 + t, 3t, 2 - 3t \rangle$

57. The plane passing through $(-2, 3, 1)$, $(1, 1, 0)$, and $(-1, 0, 1)$

58. Distance from a point to a line Find the distance from the point $(1, 2, 3)$ to the line $x = 2 + t, y = 3, z = 1 - 3t$.

59. Distance from a point to a plane Find the distance from the point $(2, 2, 2)$ to the plane $x + 2y + 2z = 1$.

60–74. Identifying surfaces *Consider the surfaces defined by the following equations.*

a. Identify and briefly describe the surface.
b. Find the xy-, xz-, and yz-traces, when they exist.
c. Find the intercepts with the three coordinate axes, when they exist.
d. Sketch the surface.

60. $z - \sqrt{x} = 0$

61. $3z = \dfrac{x^2}{12} - \dfrac{y^2}{48}$

62. $\dfrac{x^2}{100} + 4y^2 + \dfrac{z^2}{16} = 1$

63. $y^2 = 4x^2 + \dfrac{z^2}{25}$

64. $\dfrac{4x^2}{9} + \dfrac{9z^2}{4} = y^2$

65. $4z = \dfrac{x^2}{4} + \dfrac{y^2}{9}$

66. $\dfrac{x^2}{16} + \dfrac{z^2}{36} - \dfrac{y^2}{100} = 1$

67. $y^2 + 4z^2 - 2x^2 = 1$

68. $-\dfrac{x^2}{16} + \dfrac{z^2}{36} - \dfrac{y^2}{25} = 4$

69. $\dfrac{x^2}{4} + \dfrac{y^2}{16} - z^2 = 4$

70. $x = \dfrac{y^2}{64} - \dfrac{z^2}{9}$

71. $\dfrac{x^2}{4} + \dfrac{y^2}{16} + z^2 = 4$

72. $y - e^{-x} = 0$

73. $\dfrac{y^2}{49} + \dfrac{x^2}{9} = \dfrac{z^2}{64}$

74. $y = 4x^2 + \dfrac{z^2}{9}$

75. Matching surfaces Match equations a–d with surfaces A–D.

    a. $z = \sqrt{2x^2 + 3y^2 + 1} - 1$ **b.** $z = -3y^2$
    c. $z = 2x^2 - 3y^2 + 1$ **d.** $z = \sqrt{2x^2 + 3y^2 - 1}$

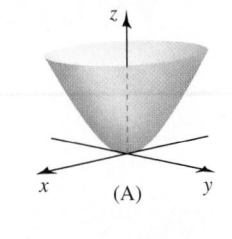
(A)

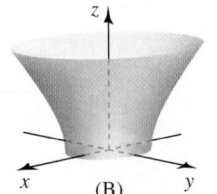
(B)

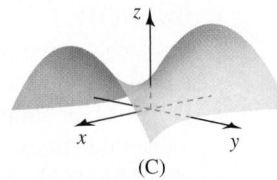
(C)

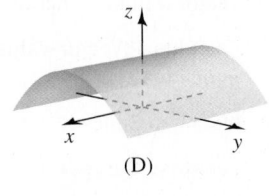
(D)

76. Designing a water bottle The lateral surface of a water bottle consists of a circular cylinder of radius 2 and height 6, topped off by a truncated hyperboloid of one sheet of height 2 (see figure). Assume the top of the truncated hyperboloid has a radius of 1/2. Find two equations that, when graphed together, form the lateral surface of the bottle. Answers may vary.

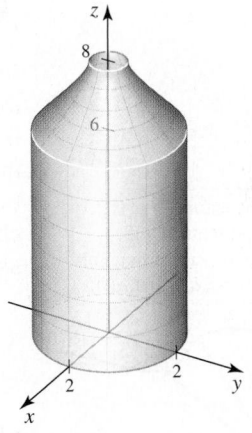

Chapter 13 Guided Projects

Applications of the material in this chapter and related topics can be found in the following Guided Projects. For additional information, see the Preface.

- Intercepting a UFO
- CORDIC algorithms: How your calculator works

14

Vector-Valued Functions

Chapter Preview In Chapter 13, we used vectors to represent static quantities, such as the constant force applied to the end of a wrench or the constant velocity of a boat in a current. In this chapter, we put vectors in motion by introducing *vector-valued functions*, or simply *vector functions*. Our first task is to investigate the graphs of vector-valued functions and to study them in the setting of calculus. Everything you already know about limits, derivatives, and integrals applies to this new family of functions. Also, with the calculus of vector functions, we can solve a wealth of practical problems involving the motion of objects in space. The chapter closes with an exploration of arc length, curvature, and tangent and normal vectors, all important features of space curves.

14.1 Vector-Valued Functions

Imagine a projectile moving along a path in three-dimensional space; it could be an electron or a comet, a soccer ball or a rocket. If you take a snapshot of the object, its position is described by a static position vector $\mathbf{r} = \langle x, y, z \rangle$. However, if you want to describe the full trajectory of the object as it unfolds in time, you must represent the object's position with a *vector-valued function* such as $\mathbf{r}(t) = \langle x(t), y(t), z(t) \rangle$ whose components change in time (**Figure 14.1**). The goal of this section is to describe continuous motion using vector-valued functions.

Vector-Valued Functions

A function of the form $\mathbf{r}(t) = \langle x(t), y(t), z(t) \rangle$ may be viewed in two ways.

- It is a set of three parametric equations that describe a curve in space.

- It is also a **vector-valued function**, which means that the three dependent variables (x, y, and z) are the components of $\mathbf{r}$, and each component varies with respect to a single independent variable t (that often represents time).

Here is the connection between these perspectives: As t varies, a point $(x(t), y(t), z(t))$ on a parametric curve is also the head of the position vector $\mathbf{r}(t) = \langle x(t), y(t), z(t) \rangle$. In other words, a vector-valued function is a set of parametric equations written in vector form. It is useful to keep both of these interpretations in mind as you work with vector-valued functions.

Although our focus is on vector functions whose graphs lie in three-dimensional space, vector functions can be given in any number of dimensions. In fact, you became acquainted with the essential ideas behind two-dimensional vector functions

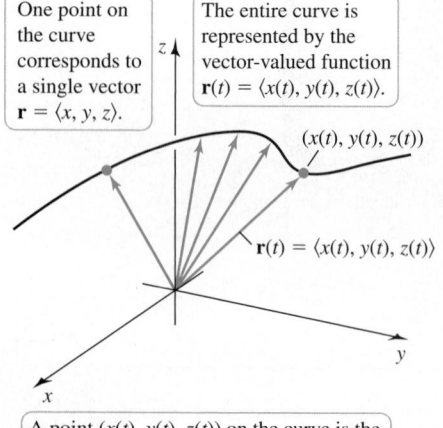

One point on the curve corresponds to a single vector $\mathbf{r} = \langle x, y, z \rangle$.

The entire curve is represented by the vector-valued function $\mathbf{r}(t) = \langle x(t), y(t), z(t) \rangle$.

A point $(x(t), y(t), z(t))$ on the curve is the head of the vector $\mathbf{r}(t) = \langle x(t), y(t), z(t) \rangle$.

Figure 14.1

in Section 12.1 when you studied parametric equations. For example, recall that the parametric equations

$$x = a \cos t, \ y = a \sin t, \quad \text{for } 0 \le t \le 2\pi$$

describe a circle of radius a centered at the origin. The corresponding vector function is

$$\mathbf{r}(t) = \langle a \cos t, a \sin t \rangle, \quad \text{for } 0 \le t \le 2\pi.$$

All the plane curves described by parametric equations in Section 12.1 are easily converted to vector functions in the same manner.

Curves in Space

We now explore general vector-valued functions of the form

$$\mathbf{r}(t) = \langle f(t), g(t), h(t) \rangle = f(t)\mathbf{i} + g(t)\mathbf{j} + h(t)\mathbf{k},$$

where f, g, and h are defined on an interval $a \le t \le b$. The **domain** of $\mathbf{r}$ is the largest set of values of t on which all of f, g, and h are defined.

Figure 14.2 illustrates how a parameterized curve is generated by such a function. As the parameter t varies over the interval $a \le t \le b$, each value of t produces a position vector that corresponds to a point on the curve, starting at the initial vector $\mathbf{r}(a)$ and ending at the terminal vector $\mathbf{r}(b)$. The resulting parameterized curve can either have finite length or extend indefinitely. The curve may also cross itself or close and retrace itself. As shown in Example 1, when f, g, and h are linear functions of t, the resulting curve is a line or line segment.

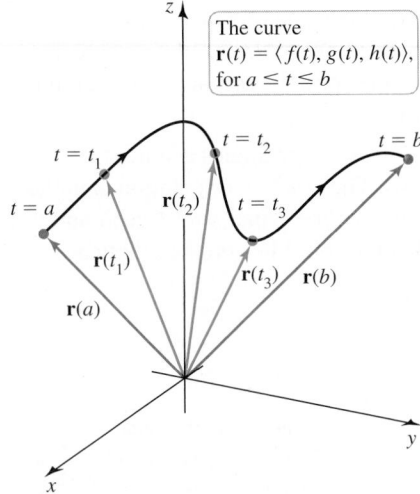

The curve
$\mathbf{r}(t) = \langle f(t), g(t), h(t) \rangle$,
for $a \le t \le b$

Figure 14.2

EXAMPLE 1   Lines as vector-valued functions Find a vector function for the line that passes through the points $P(2, -1, 4)$ and $Q(3, 0, 6)$.

SOLUTION Recall from Section 13.5 that parametric equations of the line parallel to the vector $\mathbf{v} = \langle a, b, c \rangle$ and passing through the point $P_0(x_0, y_0, z_0)$ are

$$x = x_0 + at, \quad y = y_0 + bt, \quad z = z_0 + ct.$$

The vector $\mathbf{v} = \overrightarrow{PQ} = \langle 3 - 2, 0 - (-1), 6 - 4 \rangle = \langle 1, 1, 2 \rangle$ is parallel to the line, and we let $P_0 = P(2, -1, 4)$. Therefore, parametric equations for the line are

$$x = 2 + t, \quad y = -1 + t, \quad z = 4 + 2t,$$

and the corresponding vector function for the line is

$$\mathbf{r}(t) = \langle 2 + t, -1 + t, 4 + 2t \rangle,$$

with a domain of all real numbers. As t increases, the line is generated in the direction of $\overrightarrow{PQ}$. Just as we did with parametric equations, we can restrict the domain to a finite interval to produce a vector function for a line segment (see Quick Check 1).

Related Exercises 9, 13 ◄

QUICK CHECK 1 Restrict the domain of the vector function in Example 1 to produce a line segment that goes from $P(2, -1, 4)$ to $R(5, 1, 8)$. ◄

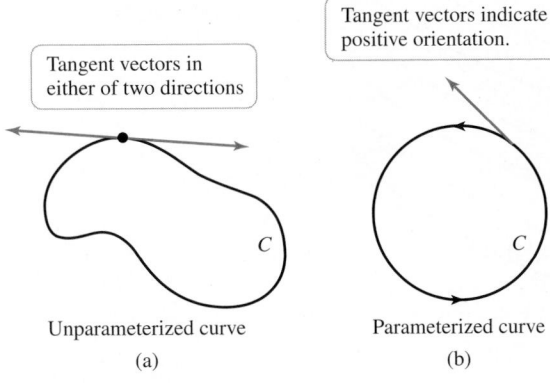

Tangent vectors in either of two directions

Tangent vectors indicate positive orientation.

Unparameterized curve
(a)

Parameterized curve
(b)

Figure 14.3

Orientation of Curves If a smooth curve C is viewed only as a set of points, then at any point of C, it is possible to draw tangent vectors in two directions (**Figure 14.3a**). On the other hand, a parameterized curve described by the function $\mathbf{r}(t)$, where $a \le t \le b$, has a natural direction, or **orientation**. The *positive* orientation is the direction in which the curve is generated as the parameter increases from a to b. For example, the positive orientation of the circle $\mathbf{r}(t) = \langle \cos t, \sin t \rangle$, for $0 \le t \le 2\pi$, is counterclockwise (**Figure 14.3b**), and the positive orientation of the line in Example 1 is in the direction of the vector $\overrightarrow{PQ}$. An important property of all parameterized curves is the relationship between the orientation of a given curve and its tangent vectors (to be defined precisely in Section 14.2): At all points, the tangent vectors point in the direction of the positive orientation of the curve.

EXAMPLE 2   A spiral Graph the curve described by the equation

$$\mathbf{r}(t) = 4 \cos t \, \mathbf{i} + \sin t \, \mathbf{j} + \frac{t}{2\pi} \, \mathbf{k},$$

where (a) $0 \le t \le 2\pi$ and (b) $-\infty < t < \infty$.

SOLUTION

a. We begin by setting $z = 0$ to determine the projection of the curve in the xy-plane. The resulting function $\mathbf{r}(t) = 4 \cos t \, \mathbf{i} + \sin t \, \mathbf{j}$ implies that $x = 4 \cos t$ and $y = \sin t$; these equations describe an ellipse in the xy-plane whose positive direction is counterclockwise (**Figure 14.4a**). Because $z = \frac{t}{2\pi}$, the value of z increases from 0 to 1 as t increases from 0 to 2π. Therefore, the curve rises out of the xy-plane to create an elliptical spiral (or coil). Over the interval $[0, 2\pi]$, the spiral begins at $(4, 0, 0)$, circles the z-axis once, and ends at $(4, 0, 1)$ (**Figure 14.4b**).

b. Letting the parameter vary over the interval $-\infty < t < \infty$ generates a spiral that winds around the z-axis endlessly in both directions. The positive orientation is in the upward direction (increasing z-direction). Noticing once more that $x = 4 \cos t$ and $y = \sin t$ are x- and y-components of $\mathbf{r}$, we see that the spiral lies on the elliptical cylinder $\left(\dfrac{x}{4}\right)^2 + y^2 = \cos^2 t + \sin^2 t = 1$ (**Figure 14.4c**).

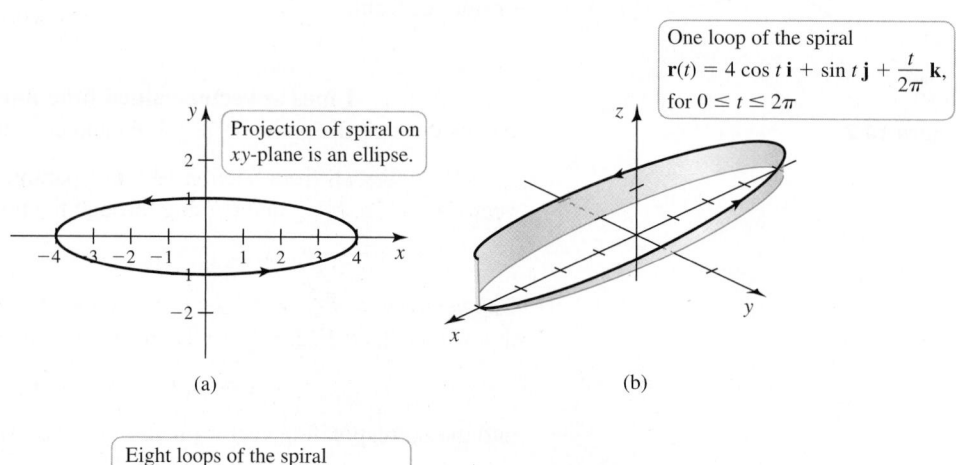

Projection of spiral on xy-plane is an ellipse.

(a)

One loop of the spiral
$\mathbf{r}(t) = 4 \cos t \, \mathbf{i} + \sin t \, \mathbf{j} + \dfrac{t}{2\pi} \, \mathbf{k}$,
for $0 \le t \le 2\pi$

(b)

> Recall that the functions $\sin at$ and $\cos at$ oscillate a times over the interval $[0, 2\pi]$. Therefore, their period is $2\pi/a$.

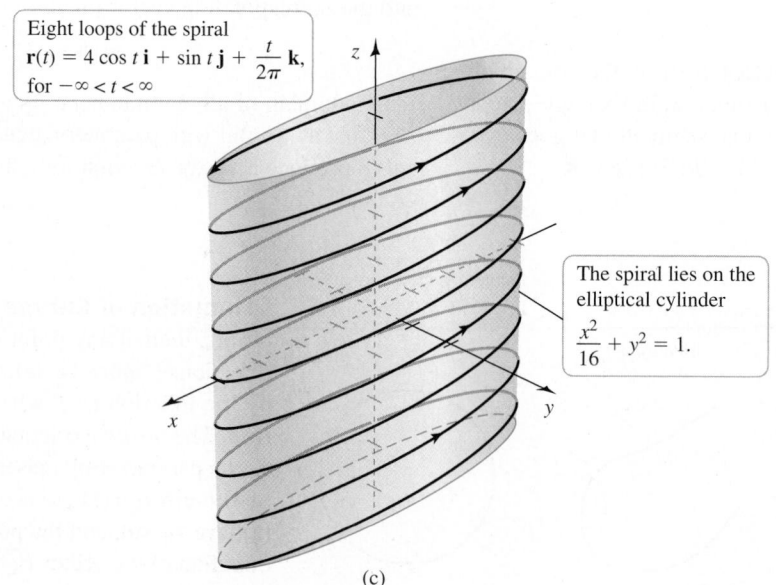

Eight loops of the spiral
$\mathbf{r}(t) = 4 \cos t \, \mathbf{i} + \sin t \, \mathbf{j} + \dfrac{t}{2\pi} \, \mathbf{k}$,
for $-\infty < t < \infty$

The spiral lies on the elliptical cylinder
$\dfrac{x^2}{16} + y^2 = 1.$

(c)

Figure 14.4

Related Exercise 24 ◄

EXAMPLE 3   Roller coaster curve Graph the curve

$$\mathbf{r}(t) = \cos t\,\mathbf{i} + \sin t\,\mathbf{j} + 0.4 \sin 2t\,\mathbf{k}, \quad \text{for } 0 \le t \le 2\pi.$$

SOLUTION Without the z-component, the resulting function $\mathbf{r}(t) = \cos t\,\mathbf{i} + \sin t\,\mathbf{j}$ describes a circle of radius 1 in the xy-plane. The z-component of the function varies between -0.4 and 0.4 with a period of π units. Therefore, on the interval $[0, 2\pi]$, the z-coordinates of points on the curve oscillate twice between -0.4 and 0.4, while the x- and y-coordinates describe a circle. The result is a curve that circles the z-axis once in the counterclockwise direction with two peaks and two valleys (**Figure 14.5a**).

The space curve in this example is not particularly complicated, but visualizing a given curve is easier when we determine the surface(s) on which it lies. Writing the vector function $\mathbf{r}(t) = \cos t\,\mathbf{i} + \sin t\,\mathbf{j} + 0.4 \sin 2t\,\mathbf{k}$ in parametric form, we have

$$x = \cos t, \quad y = \sin t, \quad z = 0.4 \sin 2t, \quad \text{for } 0 \le t \le 2\pi.$$

Noting that $x^2 + y^2 = \cos^2 t + \sin^2 t = 1$, we conclude that the curve lies on the cylinder $x^2 + y^2 = 1$. In this case, we can also eliminate the parameter by writing

$$\begin{aligned} z &= 0.4 \sin 2t \\ &= 0.4(2\,\underbrace{\sin t}_{y}\,\underbrace{\cos t}_{x}) \qquad \text{Double angle identity: } \sin 2t = 2 \sin t \cos t \\ &= 0.8xy, \qquad\qquad x = \cos t,\ y = \sin t \end{aligned}$$

which implies that the curve also lies on the hyperbolic paraboloid $z = 0.8xy$ (see margin note). In fact, the roller coaster curve is the curve in which the surfaces $x^2 + y^2 = 1$ and $z = 0.8xy$ intersect, as shown in **Figure 14.5b**.

▶ The graph of $z = 0.8xy$ is a rotation of the quadric surface $z = 0.4(x^2 - y^2)$, which we recognize as a hyperbolic paraboloid. See the Guided Project *Translation and rotation of axes.*

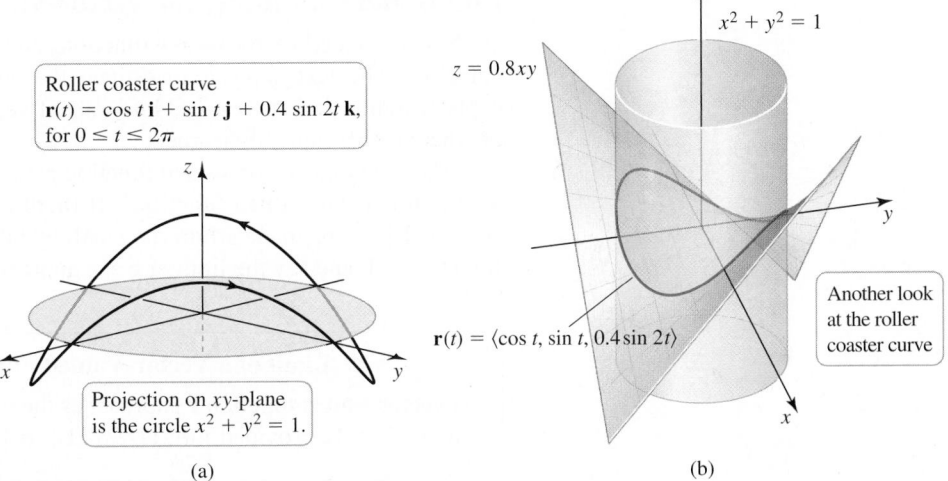

Roller coaster curve
$\mathbf{r}(t) = \cos t\,\mathbf{i} + \sin t\,\mathbf{j} + 0.4 \sin 2t\,\mathbf{k},$
for $0 \le t \le 2\pi$

Projection on xy-plane
is the circle $x^2 + y^2 = 1$.

(a)

$z = 0.8xy$

$x^2 + y^2 = 1$

$\mathbf{r}(t) = \langle \cos t, \sin t, 0.4 \sin 2t \rangle$

Another look at the roller coaster curve

(b)

Figure 14.5

Related Exercises 28, 55 ◀

EXAMPLE 4   Slinky curve Use a graphing utility to graph the curve

$$\mathbf{r}(t) = (3 + \cos 15t) \cos t\,\mathbf{i} + (3 + \cos 15t) \sin t\,\mathbf{j} + \sin 15t\,\mathbf{k},$$

for $0 \le t \le 2\pi$, and discuss its properties.

SOLUTION The factor $A(t) = 3 + \cos 15t$ that appears in the x- and y-components is a varying amplitude for $\cos t\,\mathbf{i}$ and $\sin t\,\mathbf{j}$. Its effect is seen in the graph of the x-component $A(t) \cos t$ (**Figure 14.6a**). For $0 \le t \le 2\pi$, the curve consists of one period of $3 \cos t$ with 15 small oscillations superimposed on it. As a result, the x-component of $\mathbf{r}$ varies from -4 to 4 with 15 small oscillations along the way. A similar behavior is seen in the y-component of $\mathbf{r}$. Finally, the z-component of $\mathbf{r}$, which is $\sin 15t$, oscillates between -1 and 1 fifteen times over $[0, 2\pi]$. Combining these effects, we discover a coil-shaped curve that circles the z-axis in the counterclockwise direction and closes on itself. **Figures 14.6b** and **14.6c** show two views, one looking along the xy-plane and the other from overhead on the z-axis. It can be shown (Exercise 56) that eliminating the parameter from the parametric equations defining $\mathbf{r}$ leads to a standard equation of a torus in Cartesian coordinates—in this case, $(3 - \sqrt{x^2 + y^2})^2 + z^2 = 1$—and therefore, the curve lies on this torus, as seen in Figure 14.6b.

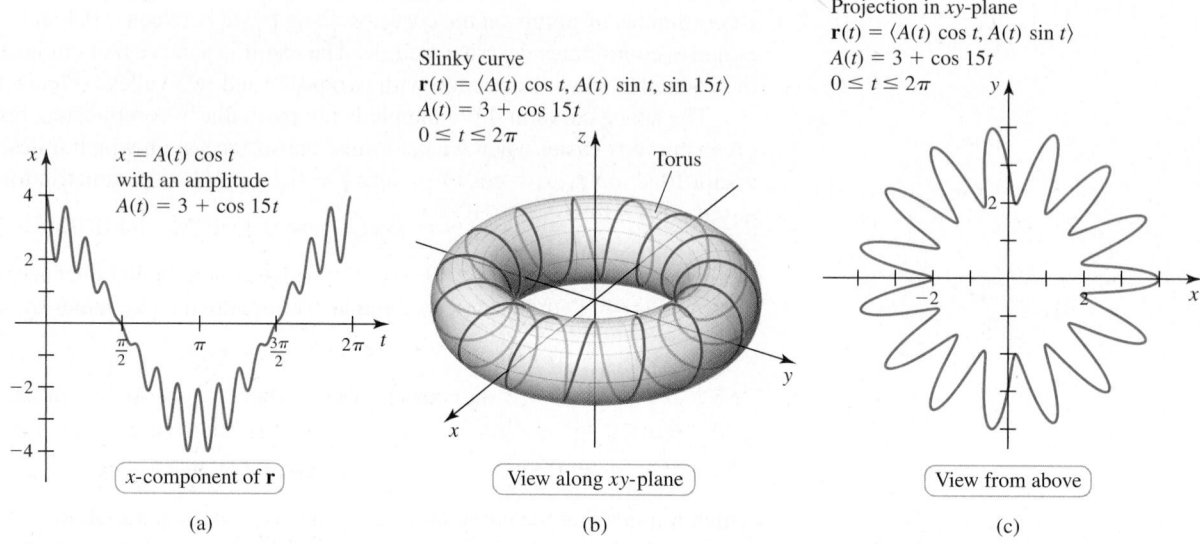

Slinky curve
$\mathbf{r}(t) = \langle A(t) \cos t, A(t) \sin t, \sin 15t \rangle$
$A(t) = 3 + \cos 15t$
$0 \le t \le 2\pi$

Projection in xy-plane
$\mathbf{r}(t) = \langle A(t) \cos t, A(t) \sin t \rangle$
$A(t) = 3 + \cos 15t$
$0 \le t \le 2\pi$

$x = A(t) \cos t$
with an amplitude
$A(t) = 3 + \cos 15t$

| x-component of **r** | View along xy-plane | View from above |

(a)   (b)   (c)

Figure 14.6

Related Exercises 27, 56 ◀

Limits and Continuity for Vector-Valued Functions

We have presented vector valued functions and established their relationship to parametric equations. The next step is to investigate the calculus of vector-valued functions. The concepts of limits, derivatives, and integrals of vector-valued functions are direct extensions of what you have already learned.

The limit of a vector-valued function $\mathbf{r}(t) = f(t)\mathbf{i} + g(t)\mathbf{j} + h(t)\mathbf{k}$ is defined much as it is for scalar-valued functions. If there is a vector $\mathbf{L}$ such that the scalar function $|\mathbf{r}(t) - \mathbf{L}|$ can be made arbitrarily small by taking t sufficiently close to a, then we write $\lim_{t \to a} \mathbf{r}(t) = \mathbf{L}$ and say the limit of $\mathbf{r}$ as t approaches a is $\mathbf{L}$.

> **DEFINITION Limit of a Vector-Valued Function**
>
> A vector-valued function $\mathbf{r}$ approaches the limit $\mathbf{L}$ as t approaches a, written $\lim_{t \to a} \mathbf{r}(t) = \mathbf{L}$, provided $\lim_{t \to a} |\mathbf{r}(t) - \mathbf{L}| = 0$.

Notice that while $\mathbf{r}$ is vector valued, $|\mathbf{r}(t) - \mathbf{L}|$ is a function of the single variable t, to which our familiar limit theorems apply. Therefore, this definition and a short calculation (Exercise 66) lead to a straightforward method for computing limits of the vector-valued function $\mathbf{r} = \langle f, g, h \rangle$. Suppose

$$\lim_{t \to a} f(t) = L_1, \qquad \lim_{t \to a} g(t) = L_2, \qquad \text{and} \qquad \lim_{t \to a} h(t) = L_3.$$

Then

$$\lim_{t \to a} \mathbf{r}(t) = \left\langle \lim_{t \to a} f(t), \lim_{t \to a} g(t), \lim_{t \to a} h(t) \right\rangle = \langle L_1, L_2, L_3 \rangle.$$

In other words, the limit of $\mathbf{r}$ is determined by computing the limits of its components.

The limits laws in Chapter 2 have analogs for vector-valued functions. For example, if $\lim_{t \to a} \mathbf{r}(t)$ and $\lim_{t \to a} \mathbf{s}(t)$ exist and c is a scalar, then

$$\lim_{t \to a} (\mathbf{r}(t) + \mathbf{s}(t)) = \lim_{t \to a} \mathbf{r}(t) + \lim_{t \to a} \mathbf{s}(t) \quad \text{and} \quad \lim_{t \to a} c\mathbf{r}(t) = c \lim_{t \to a} \mathbf{r}(t).$$

The idea of continuity also extends directly to vector-valued functions. A function $\mathbf{r}(t) = f(t)\mathbf{i} + g(t)\mathbf{j} + h(t)\mathbf{k}$ is **continuous at** a provided $\lim_{t \to a} \mathbf{r}(t) = \mathbf{r}(a)$. Specifically,

▶ Continuity is often taken as part of the definition of a parameterized curve.

if the component functions f, g, and h are continuous at a, then $\mathbf{r}$ is also continuous at a, and vice versa. The function $\mathbf{r}$ is **continuous on an interval** I if it is continuous for all t in I.

Continuity has the same intuitive meaning in this setting as it does for scalar-valued functions. If $\mathbf{r}$ is continuous on an interval, the curve it describes has no breaks or gaps, which is an important property when $\mathbf{r}$ describes the trajectory of an object.

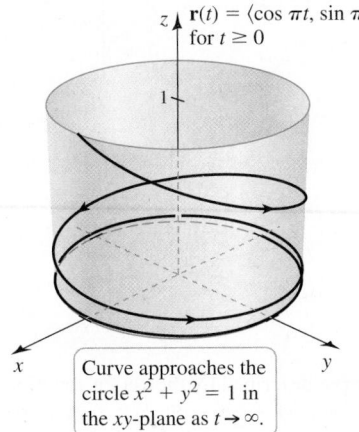

$\mathbf{r}(t) = \langle \cos \pi t, \sin \pi t, e^{-t} \rangle$, for $t \geq 0$

Curve approaches the circle $x^2 + y^2 = 1$ in the xy-plane as $t \to \infty$.

Figure 14.7

QUICK CHECK 2 Explain why the curve in Example 5 lies on the cylinder $x^2 + y^2 = 1$, as shown in Figure 14.7. ◀

EXAMPLE 5   Limits and continuity Consider the function

$$\mathbf{r}(t) = \cos \pi t \, \mathbf{i} + \sin \pi t \, \mathbf{j} + e^{-t} \mathbf{k}, \quad \text{for } t \geq 0.$$

a. Evaluate $\lim\limits_{t \to 2} \mathbf{r}(t)$.

b. Evaluate $\lim\limits_{t \to \infty} \mathbf{r}(t)$.

c. At what points is $\mathbf{r}$ continuous?

SOLUTION

a. We evaluate the limit of each component of $\mathbf{r}$:

$$\lim_{t \to 2} \mathbf{r}(t) = \lim_{t \to 2} \left(\underbrace{\cos \pi t}_{\to 1} \, \mathbf{i} + \underbrace{\sin \pi t}_{\to 0} \, \mathbf{j} + \underbrace{e^{-t}}_{\to e^{-2}} \mathbf{k} \right) = \mathbf{i} + e^{-2} \mathbf{k}.$$

b. Note that although $\lim\limits_{t \to \infty} e^{-t} = 0$, $\lim\limits_{t \to \infty} \cos t$ and $\lim\limits_{t \to \infty} \sin t$ do not exist. Therefore, $\lim\limits_{t \to \infty} \mathbf{r}(t)$ does not exist. As shown in **Figure 14.7**, the curve is a coil that approaches the unit circle in the xy-plane.

c. Because the components of $\mathbf{r}$ are continuous for all t, $\mathbf{r}$ is also continuous for all t.

Related Exercise 31 ◀

SECTION 14.1 EXERCISES

Getting Started

1. How many independent variables does the function $\mathbf{r}(t) = \langle f(t), g(t), h(t) \rangle$ have?

2. How many dependent scalar variables does the function $\mathbf{r}(t) = \langle f(t), g(t), h(t) \rangle$ have?

3. Why is $\mathbf{r}(t) = \langle f(t), g(t), h(t) \rangle$ called a vector-valued function?

4. In what plane does the curve $\mathbf{r}(t) = t\,\mathbf{i} + t^2\,\mathbf{k}$ lie?

5. How do you evaluate $\lim\limits_{t \to a} \mathbf{r}(t)$, where $\mathbf{r}(t) = \langle f(t), g(t), h(t) \rangle$?

6. How do you determine whether $\mathbf{r}(t) = f(t)\mathbf{i} + g(t)\mathbf{j} + h(t)\mathbf{k}$ is continuous at $t = a$?

7. Find a function $\mathbf{r}(t)$ for the line passing through the points $P(0, 0, 0)$ and $Q(1, 2, 3)$. Express your answer in terms of $\mathbf{i}$, $\mathbf{j}$, and $\mathbf{k}$.

8. Find a function $\mathbf{r}(t)$ whose graph is a circle of radius 1 parallel to the xy-plane and centered at $(0, 0, 10)$.

Practice Exercises

9–14. Lines and line segments *Find a function $\mathbf{r}(t)$ that describes the given line or line segment.*

9. The line through $P(2, 3, 7)$ and $Q(4, 6, 3)$

10. The line through $P(0, -3, 2)$ that is parallel to the line $\mathbf{r}(t) = \langle 4, 6 - t, 1 + t \rangle$

11. The line through $P(3, 4, 5)$ that is orthogonal to the plane $2x - z = 4$

12. The line of intersection of the planes $2x + 3y + 4z = 7$ and $2x + 3y + 5z = 8$

13. The line segment from $P(1, 2, 1)$ to $Q(0, 2, 3)$

14. The line segment from $P(-4, -2, 1)$ to $Q(-2, -2, 3)$

15–26. Graphing curves *Graph the curves described by the following functions, indicating the positive orientation.*

15. $\mathbf{r}(t) = \langle 2 \cos t, 2 \sin t \rangle$, for $0 \leq t \leq 2\pi$

16. $\mathbf{r}(t) = \langle 1 + \cos t, 2 + \sin t \rangle$, for $0 \leq t \leq 2\pi$

17. $\mathbf{r}(t) = \langle t, 2t \rangle$, for $0 \leq t \leq 1$

18. $\mathbf{r}(t) = \langle 3 \cos t, 2 \sin t \rangle$, for $0 \leq t \leq 2\pi$

19. $\mathbf{r}(t) = \langle \cos t, 0, \sin t \rangle$, for $0 \leq t \leq 2\pi$

20. $\mathbf{r}(t) = \langle 0, 4 \cos t, 16 \sin t \rangle$, for $0 \leq t \leq 2\pi$

21. $\mathbf{r}(t) = \cos t\,\mathbf{i} + \mathbf{j} + \sin t\,\mathbf{k}$, for $0 \leq t \leq 2\pi$

22. $\mathbf{r}(t) = 2 \cos t\,\mathbf{i} + 2 \sin t\,\mathbf{j} + 2\,\mathbf{k}$, for $0 \leq t \leq 2\pi$

T 23. $\mathbf{r}(t) = t \cos t\,\mathbf{i} + t \sin t\,\mathbf{j} + t\,\mathbf{k}$, for $0 \leq t \leq 6\pi$

T 24. $\mathbf{r}(t) = 4 \sin t\,\mathbf{i} + 4 \cos t\,\mathbf{j} + e^{-t/10}\,\mathbf{k}$, for $0 \leq t < \infty$

T 25. $\mathbf{r}(t) = e^{-t/20} \sin t\,\mathbf{i} + e^{-t/20} \cos t\,\mathbf{j} + t\,\mathbf{k}$, for $0 \leq t < \infty$

T 26. $\mathbf{r}(t) = e^{-t/10}\,\mathbf{i} + 3 \cos t\,\mathbf{j} + 3 \sin t\,\mathbf{k}$, for $0 \leq t < \infty$

T **27–30. Exotic curves** *Graph the curves described by the following functions. Use analysis to anticipate the shape of the curve before using a graphing utility.*

27. $\mathbf{r}(t) = \cos 15t\,\mathbf{i} + (4 + \sin 15t) \cos t\,\mathbf{j} + (4 + \sin 15t) \sin t\,\mathbf{k}$, for $0 \leq t \leq 2\pi$

28. $\mathbf{r}(t) = 2 \cos t\,\mathbf{i} + 4 \sin t\,\mathbf{j} + \cos 10t\,\mathbf{k}$, for $0 \leq t \leq 2\pi$

29. $\mathbf{r}(t) = \sin t\,\mathbf{i} + \sin^2 t\,\mathbf{j} + \dfrac{t}{5\pi}\,\mathbf{k}$, for $0 \le t \le 10\pi$

30. $\mathbf{r}(t) = \cos t \sin 3t\,\mathbf{i} + \sin t \sin 3t\,\mathbf{j} + \sqrt{t}\,\mathbf{k}$, for $0 \le t \le 9$

31–36. Limits *Evaluate the following limits.*

31. $\displaystyle\lim_{t \to \pi/2}\left(\cos 2t\,\mathbf{i} - 4\sin t\,\mathbf{j} + \dfrac{2t}{\pi}\,\mathbf{k}\right)$

32. $\displaystyle\lim_{t \to \ln 2}\left(2e^t\,\mathbf{i} + 6e^{-t}\,\mathbf{j} - 4e^{-2t}\,\mathbf{k}\right)$

33. $\displaystyle\lim_{t \to \infty}\left(e^{-t}\,\mathbf{i} - \dfrac{2t}{t+1}\,\mathbf{j} + \tan^{-1} t\,\mathbf{k}\right)$

34. $\displaystyle\lim_{t \to 2}\left(\dfrac{t}{t^2 + 1}\,\mathbf{i} - 4e^{-t}\sin \pi t\,\mathbf{j} + \dfrac{1}{\sqrt{4t + 1}}\,\mathbf{k}\right)$

35. $\displaystyle\lim_{t \to 0}\left(\dfrac{\sin t}{t}\,\mathbf{i} - \dfrac{e^t - t - 1}{t}\,\mathbf{j} + \dfrac{\cos t + t^2/2 - 1}{t^2}\,\mathbf{k}\right)$

36. $\displaystyle\lim_{t \to 0}\left(\dfrac{\tan t}{t}\,\mathbf{i} - \dfrac{3t}{\sin t}\,\mathbf{j} + \sqrt{t + 1}\,\mathbf{k}\right)$

37. Explain why or why not Determine whether the following statements are true and give an explanation or counterexample.

   a. The projection of the curve $\mathbf{r}(t) = \langle t, \cos t, t^2\rangle$ in the xz-plane is a parabola.
   b. The curve $\mathbf{r}(t) = \langle \sin t, \cos t, \sin t\rangle$ lies on a unit sphere.
   c. The curve $\mathbf{r}(t) = \langle e^{-t}, \sin t, -\cos t\rangle$ approaches a circle as $t \to \infty$.
   d. If $\mathbf{r}(t) = e^{-t^2}\langle 1, 1, 1\rangle$, then $\displaystyle\lim_{t \to \infty} \mathbf{r}(t) = \lim_{t \to -\infty} \mathbf{r}(t)$.

38–41. Domains *Find the domain of the following vector-valued functions.*

38. $\mathbf{r}(t) = \dfrac{2}{t - 1}\,\mathbf{i} + \dfrac{3}{t + 2}\,\mathbf{j}$

39. $\mathbf{r}(t) = \sqrt{t + 2}\,\mathbf{i} + \sqrt{2 - t}\,\mathbf{j}$

40. $\mathbf{r}(t) = \cos 2t\,\mathbf{i} + e^{\sqrt{t}}\,\mathbf{j} + \dfrac{12}{t}\,\mathbf{k}$

41. $\mathbf{r}(t) = \sqrt{4 - t^2}\,\mathbf{i} + \sqrt{t}\,\mathbf{j} - \dfrac{2}{\sqrt{1 + t}}\,\mathbf{k}$

42–44. Curve-plane intersections *Find the points (if they exist) at which the following planes and curves intersect.*

42. $y = 1$; $\mathbf{r}(t) = \langle 10\cos t, 2\sin t, 1\rangle$, for $0 \le t \le 2\pi$

43. $z = 16$; $\mathbf{r}(t) = \langle t, 2t, 4 + 3t\rangle$, for $-\infty < t < \infty$

44. $y + x = 0$; $\mathbf{r}(t) = \langle \cos t, \sin t, t\rangle$, for $0 \le t \le 4\pi$

45. Matching functions with graphs Match functions a–f with the appropriate graphs A–F.

   a. $\mathbf{r}(t) = \langle t, -t, t\rangle$ **b.** $\mathbf{r}(t) = \langle t^2, t, t\rangle$
   c. $\mathbf{r}(t) = \langle 4\cos t, 4\sin t, 2\rangle$ **d.** $\mathbf{r}(t) = \langle 2t, \sin t, \cos t\rangle$
   e. $\mathbf{r}(t) = \langle \sin t, \cos t, \sin 2t\rangle$ **f.** $\mathbf{r}(t) = \langle \sin t, 2t, \cos t\rangle$

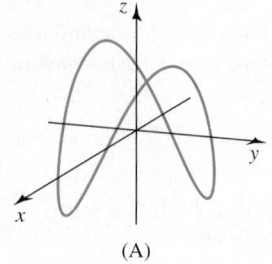

(A)

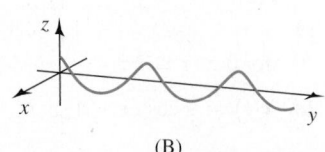

(B)

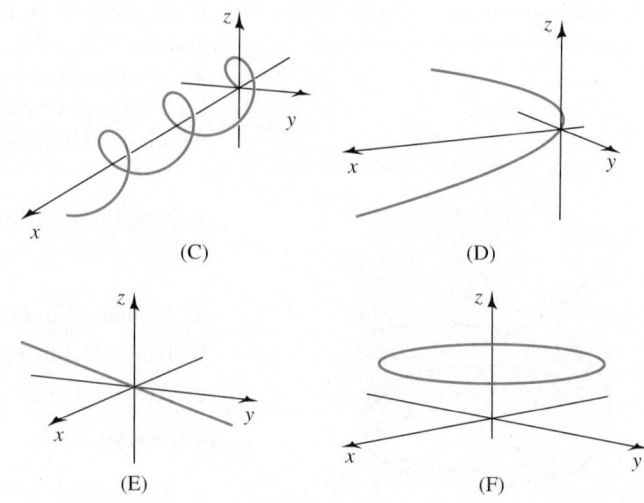

(C)   (D)

(E)   (F)

46. Upward path Consider the curve described by the vector function $\mathbf{r}(t) = (50e^{-t}\cos t)\,\mathbf{i} + (50e^{-t}\sin t)\,\mathbf{j} + (5 - 5e^{-t})\,\mathbf{k}$, for $t \ge 0$.

   a. What is the initial point of the path corresponding to $\mathbf{r}(0)$?
   b. What is $\displaystyle\lim_{t \to \infty} \mathbf{r}(t)$?
   c. Eliminate the parameter t to show that the curve $\mathbf{r}(t)$ lies on the surface $z = 5 - r/10$, where $r^2 = x^2 + y^2$.

47–50. Curve of intersection *Find a function $\mathbf{r}(t)$ that describes the curve where the following surfaces intersect. Answers are not unique.*

47. $z = 4$; $z = x^2 + y^2$

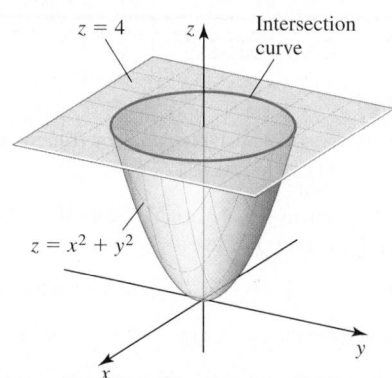

48. $z = 3x^2 + y^2 + 1$; $z = 5 - x^2 - 3y^2$

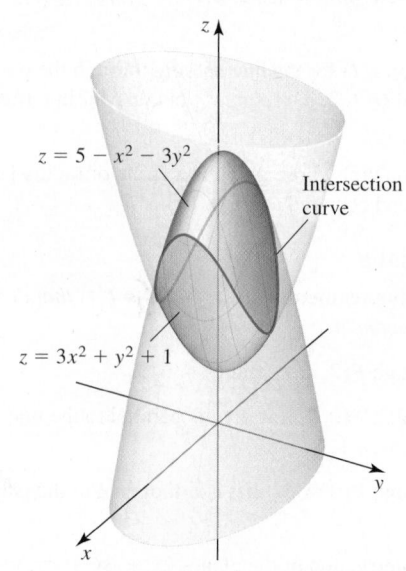

49. $x^2 + y^2 = 25; z = 2x + 2y$

50. $z = y + 1; z = x^2 + 1$

T 51. Golf slice A golfer launches a tee shot down a horizontal fairway; it follows a path given by $\mathbf{r}(t) = \langle at, (75 - 0.1a)t, -5t^2 + 80t \rangle$, where $t \geq 0$ measures time in seconds and $\mathbf{r}$ has units of feet. The y-axis points straight down the fairway and the z-axis points vertically upward. The parameter a is the slice factor that determines how much the shot deviates from a straight path down the fairway.

　a. With no slice ($a = 0$), describe the shot. How far does the ball travel horizontally (the distance between the point where the ball leaves the ground and the point where it first strikes the ground)?

　b. With a slice ($a = 0.2$), how far does the ball travel horizontally?

　c. How far does the ball travel horizontally with $a = 2.5$?

52–56. Curves on surfaces *Verify that the curve* $\mathbf{r}(t)$ *lies on the given surface. Give the name of the surface.*

52. $\mathbf{r}(t) = (t \cos t)\mathbf{i} + (t \sin t)\mathbf{j} + t\mathbf{k}; x^2 + y^2 = z^2$

53. $\mathbf{r}(t) = (\sqrt{t^2 + 1} \cos t)\mathbf{i} + (\sqrt{t^2 + 1} \sin t)\mathbf{j} + t\mathbf{k};$ $x^2 + y^2 - z^2 = 1$

54. $\mathbf{r}(t) = \langle \sqrt{t} \cos t, \sqrt{t} \sin t, t \rangle; z = x^2 + y^2$

55. $\mathbf{r}(t) = \langle 0, 2 \cos t, 3 \sin t \rangle; x^2 + \dfrac{y^2}{4} + \dfrac{z^2}{9} = 1$

56. $\mathbf{r}(t) = \langle (3 + \cos 15t) \cos t, (3 + \cos 15t) \sin t, \sin 15t \rangle;$ $(3 - \sqrt{x^2 + y^2})^2 + z^2 = 1$ (*Hint:* See Example 4.)

57–58. Closest point on a curve *Find the point P on the curve* $\mathbf{r}(t)$ *that lies closest to* P_0 *and state the distance between* P_0 *and P.*

57. $\mathbf{r}(t) = t^2\mathbf{i} + t\mathbf{j} + t\mathbf{k}; P_0(1, 1, 15)$

T 58. $\mathbf{r}(t) = \cos t\,\mathbf{i} + \sin t\,\mathbf{j} + t\mathbf{k}; P_0(1, 1, 3)$

Explorations and Challenges

59–61. Curves on spheres

T 59. Graph the curve $\mathbf{r}(t) = \left\langle \dfrac{1}{2} \sin 2t, \dfrac{1}{2}(1 - \cos 2t), \cos t \right\rangle$ and prove that it lies on the surface of a sphere centered at the origin.

60. Prove that for integers m and n, the curve

$$\mathbf{r}(t) = \langle a \sin mt \cos nt, b \sin mt \sin nt, c \cos mt \rangle$$

lies on the surface of a sphere provided $a^2 = b^2 = c^2$.

61. Find the period of the function in Exercise 60; that is, in terms of m and n, find the smallest positive real number T such that $\mathbf{r}(t + T) = \mathbf{r}(t)$ for all t.

62–65. Closed plane curves *Consider the curve*

$$\mathbf{r}(t) = \langle a \cos t + b \sin t, c \cos t + d \sin t, e \cos t + f \sin t \rangle,$$

where a, b, c, d, e, and f are real numbers. It can be shown that this curve lies in a plane.

62. Assuming the curve lies in a plane, show that it is a circle centered at the origin with radius R provided $a^2 + c^2 + e^2 = b^2 + d^2 + f^2 = R^2$ and $ab + cd + ef = 0$.

T 63. Graph the following curve and describe it.

$$\mathbf{r}(t) = \left(\frac{1}{\sqrt{2}} \cos t + \frac{1}{\sqrt{3}} \sin t \right)\mathbf{i}$$
$$+ \left(-\frac{1}{\sqrt{2}} \cos t + \frac{1}{\sqrt{3}} \sin t \right)\mathbf{j} + \left(\frac{1}{\sqrt{3}} \sin t \right)\mathbf{k}$$

T 64. Graph the following curve and describe it.

$$\mathbf{r}(t) = (2 \cos t + 2 \sin t)\mathbf{i} + (-\cos t + 2 \sin t)\mathbf{j} + (\cos t - 2 \sin t)\mathbf{k}$$

65. Find a general expression for a nonzero vector orthogonal to the plane containing the curve

$$\mathbf{r}(t) = \langle a \cos t + b \sin t, c \cos t + d \sin t, e \cos t + f \sin t \rangle,$$

where $\langle a, c, e \rangle \times \langle b, d, f \rangle \neq \mathbf{0}$.

66. Limits of vector functions Let $\mathbf{r}(t) = \langle f(t), g(t), h(t) \rangle$.

　a. Assume $\lim_{t \to a} \mathbf{r}(t) = \mathbf{L} = \langle L_1, L_2, L_3 \rangle$, which means that $\lim_{t \to a} |\mathbf{r}(t) - \mathbf{L}| = 0$. Prove that

$$\lim_{t \to a} f(t) = L_1, \quad \lim_{t \to a} g(t) = L_2, \quad \text{and} \quad \lim_{t \to a} h(t) = L_3.$$

　b. Assume $\lim_{t \to a} f(t) = L_1, \lim_{t \to a} g(t) = L_2$, and $\lim_{t \to a} h(t) = L_3$. Prove that $\lim_{t \to a} \mathbf{r}(t) = \mathbf{L} = \langle L_1, L_2, L_3 \rangle$, which means that $\lim_{t \to a} |\mathbf{r}(t) - \mathbf{L}| = 0$.

QUICK CHECK ANSWERS

1. $0 \leq t \leq 2$　**2.** The x- and y-components of the curve are $x = \cos \pi t$ and $y = \sin \pi t$, and $x^2 + y^2 = \cos^2 \pi t + \sin^2 \pi t = 1$. ◄

14.2 Calculus of Vector-Valued Functions

We now turn to the topic of ultimate interest in this chapter: the calculus of vector-valued functions. Everything you learned about differentiating and integrating functions of the form $y = f(x)$ carries over to vector-valued functions $\mathbf{r}(t)$; you simply apply the rules of differentiation and integration to the individual components of $\mathbf{r}$.

The Derivative and Tangent Vector

Consider the function $\mathbf{r}(t) = f(t)\mathbf{i} + g(t)\mathbf{j} + h(t)\mathbf{k}$, where f, g, and h are differentiable functions on an interval $a < t < b$. The first task is to explain the meaning of the *derivative* of a vector-valued function and to show how to compute it. We begin with the definition of the derivative—now from a vector perspective:

$$\mathbf{r}'(t) = \lim_{\Delta t \to 0} \frac{\Delta \mathbf{r}}{\Delta t} = \lim_{\Delta t \to 0} \frac{\mathbf{r}(t + \Delta t) - \mathbf{r}(t)}{\Delta t}.$$

Let's first look at the geometry of this limit. The function $\mathbf{r}(t) = f(t)\,\mathbf{i} + g(t)\,\mathbf{j} + h(t)\,\mathbf{k}$ describes a parameterized curve in space. Let P be a point on that curve associated with the position vector $\mathbf{r}(t)$, and let Q be a nearby point associated with the position vector $\mathbf{r}(t + \Delta t)$, where $\Delta t > 0$ is a small increment in t **(Figure 14.8a)**. The difference $\Delta \mathbf{r} = \mathbf{r}(t + \Delta t) - \mathbf{r}(t)$ is the vector $\overrightarrow{PQ}$, where we assume $\Delta \mathbf{r} \neq \mathbf{0}$. Because Δt is a scalar, the direction of $\Delta \mathbf{r}/\Delta t$ is the same as the direction of $\overrightarrow{PQ}$.

As Δt approaches 0, Q approaches P and the vector $\Delta \mathbf{r}/\Delta t$ approaches a limiting vector that we denote $\mathbf{r}'(t)$ **(Figure 14.8b)**. This new vector $\mathbf{r}'(t)$ has two important interpretations.

> ➤ An analogous argument can be given for $\Delta t < 0$, with the same result. Figure 14.8 illustrates the tangent vector $\mathbf{r}'$ for $\Delta t > 0$.

- The vector $\mathbf{r}'(t)$ points in the direction of the curve at P. For this reason, $\mathbf{r}'(t)$ is a *tangent vector* at P (provided it is not the zero vector).

> ➤ Section 14.3 is devoted to problems of motion in two and three dimensions.

- The vector $\mathbf{r}'(t)$ is the *derivative* of $\mathbf{r}$ with respect to t; it gives the rate of change of the function $\mathbf{r}(t)$ at the point P. In fact, if $\mathbf{r}(t)$ is the position function of a moving object, then $\mathbf{r}'(t)$ is the velocity vector of the object, which always points in the direction of motion, and $|\mathbf{r}'(t)|$ is the speed of the object.

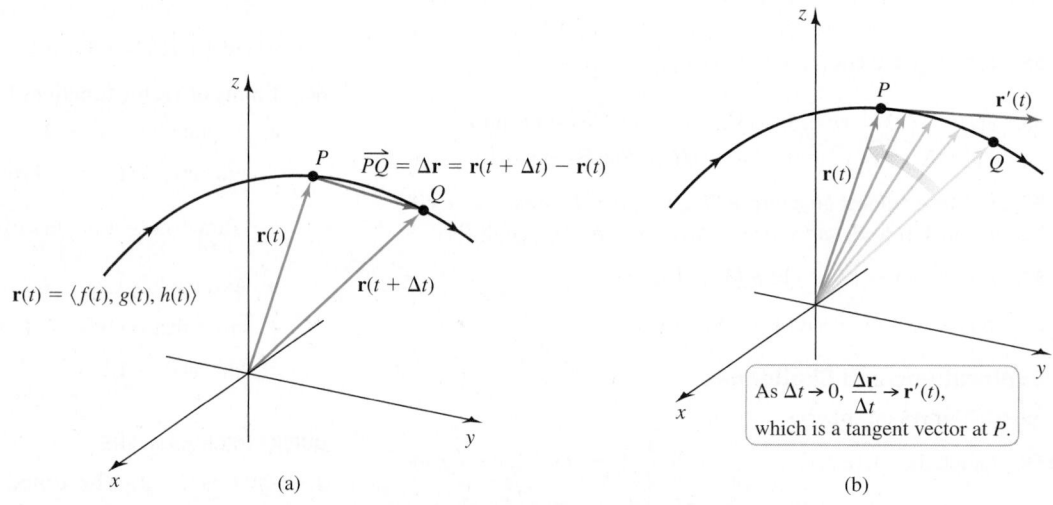

Figure 14.8

We now evaluate the limit that defines $\mathbf{r}'(t)$ by expressing $\mathbf{r}$ in terms of its components and using the properties of limits.

$$\mathbf{r}'(t) = \lim_{\Delta t \to 0} \frac{\mathbf{r}(t + \Delta t) - \mathbf{r}(t)}{\Delta t}$$

$$= \lim_{\Delta t \to 0} \frac{(f(t + \Delta t)\,\mathbf{i} + g(t + \Delta t)\,\mathbf{j} + h(t + \Delta t)\,\mathbf{k}) - (f(t)\,\mathbf{i} + g(t)\,\mathbf{j} + h(t)\,\mathbf{k})}{\Delta t}$$
Substitute components of r.

$$= \lim_{\Delta t \to 0} \left(\frac{f(t + \Delta t) - f(t)}{\Delta t}\,\mathbf{i} + \frac{g(t + \Delta t) - g(t)}{\Delta t}\,\mathbf{j} + \frac{h(t + \Delta t) - h(t)}{\Delta t}\,\mathbf{k} \right)$$
Rearrange terms inside of limit.

$$= \underbrace{\lim_{\Delta t \to 0} \frac{f(t + \Delta t) - f(t)}{\Delta t}}_{f'(t)}\,\mathbf{i} + \underbrace{\lim_{\Delta t \to 0} \frac{g(t + \Delta t) - g(t)}{\Delta t}}_{g'(t)}\,\mathbf{j} + \underbrace{\lim_{\Delta t \to 0} \frac{h(t + \Delta t) - h(t)}{\Delta t}}_{h'(t)}\,\mathbf{k}$$
Limit of sum equals sum of limits.

Because f, g, and h are differentiable scalar-valued functions of the variable t, the three limits in the last step are identified as the derivatives of f, g, and h, respectively. Therefore, there are no surprises:

$$\mathbf{r}'(t) = f'(t)\,\mathbf{i} + g'(t)\,\mathbf{j} + h'(t)\,\mathbf{k}.$$

In other words, to differentiate the vector-valued function $\mathbf{r}(t)$, we simply differentiate each of its components with respect to t.

DEFINITION Derivative and Tangent Vector

Let $\mathbf{r}(t) = f(t)\,\mathbf{i} + g(t)\,\mathbf{j} + h(t)\,\mathbf{k}$, where f, g, and h are differentiable functions on (a, b). Then $\mathbf{r}$ has a **derivative** (or is **differentiable**) on (a, b) and

$$\mathbf{r}'(t) = f'(t)\,\mathbf{i} + g'(t)\,\mathbf{j} + h'(t)\,\mathbf{k}.$$

Provided $\mathbf{r}'(t) \neq \mathbf{0}$, $\mathbf{r}'(t)$ is a **tangent vector** at the point corresponding to $\mathbf{r}(t)$.

EXAMPLE 1 Derivative of vector functions Compute the derivative of the following functions.

a. $\mathbf{r}(t) = \langle t^3, 3t^2, t^3/6 \rangle$ **b.** $\mathbf{r}(t) = e^{-t}\,\mathbf{i} + 10\sqrt{t}\,\mathbf{j} + 2\cos 3t\,\mathbf{k}$

SOLUTION

a. $\mathbf{r}'(t) = \langle 3t^2, 6t, t^2/2 \rangle$; note that $\mathbf{r}$ is differentiable for all t and $\mathbf{r}'(0) = \mathbf{0}$.

b. $\mathbf{r}'(t) = -e^{-t}\,\mathbf{i} + \dfrac{5}{\sqrt{t}}\,\mathbf{j} - 6\sin 3t\,\mathbf{k}$; the function $\mathbf{r}$ is differentiable for $t > 0$.

Related Exercises 11–12 ◄

QUICK CHECK 1 Let $\mathbf{r}(t) = \langle t, t, t \rangle$. Compute $\mathbf{r}'(t)$ and interpret the result. ◄

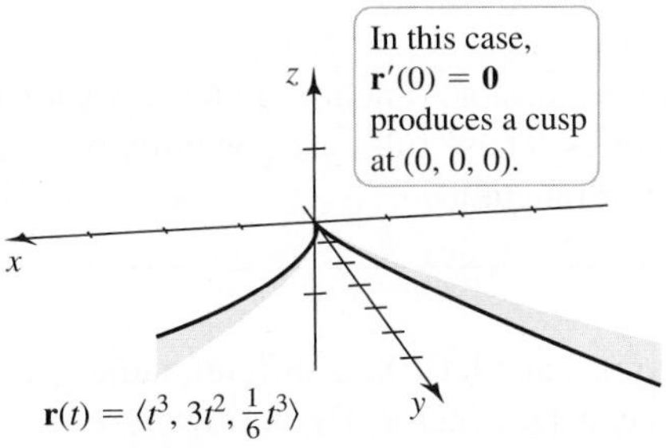

In this case, $\mathbf{r}'(0) = \mathbf{0}$ produces a cusp at $(0, 0, 0)$.

$\mathbf{r}(t) = \langle t^3, 3t^2, \frac{1}{6}t^3 \rangle$

Figure 14.9

➤ If a curve has a cusp at a point, then $\mathbf{r}'(t) = \mathbf{0}$ at that point. However, the converse is not true; it may happen that $\mathbf{r}'(t) = \mathbf{0}$ at a point that is not a cusp (Exercise 95).

The condition that $\mathbf{r}'(t) \neq \mathbf{0}$ in order for the tangent vector to be defined requires explanation. Consider the function $\mathbf{r}(t) = \langle t^3, 3t^2, t^3/6 \rangle$. As shown in Example 1a, $\mathbf{r}'(0) = \mathbf{0}$; that is, all three components of $\mathbf{r}'(t)$ are zero simultaneously when $t = 0$. We see in **Figure 14.9** that this otherwise smooth curve has a *cusp*, or a sharp point, at the origin. If $\mathbf{r}$ describes the motion of an object, then $\mathbf{r}'(t) = \mathbf{0}$ means that the velocity (and speed) of the object is zero at a point. At such a stationary point, the object *may* change direction abruptly, creating a cusp in its trajectory. For this reason, we say a function $\mathbf{r}(t) = \langle f(t), g(t), h(t) \rangle$ is **smooth** on an interval if f, g, and h are differentiable *and* $\mathbf{r}'(t) \neq \mathbf{0}$ on that interval. Smooth curves have no cusps or corners.

Unit Tangent Vector In situations in which only the direction (but not the length) of the tangent vector is of interest, we work with the *unit tangent vector*. It is the vector with magnitude 1, formed by dividing $\mathbf{r}'(t)$ by its length.

QUICK CHECK 2 Suppose $\mathbf{r}'(t)$ has units of m/s. Explain why $\mathbf{T}(t) = \mathbf{r}'(t)/|\mathbf{r}'(t)|$ is dimensionless (has no units) and carries information only about direction. ◄

DEFINITION Unit Tangent Vector

Let $\mathbf{r}(t) = f(t)\,\mathbf{i} + g(t)\,\mathbf{j} + h(t)\,\mathbf{k}$ be a smooth parameterized curve, for $a \leq t \leq b$. The **unit tangent vector** for a particular value of t is

$$\mathbf{T}(t) = \frac{\mathbf{r}'(t)}{|\mathbf{r}'(t)|}.$$

EXAMPLE 2 Unit tangent vectors Find the unit tangent vectors for the following parameterized curves.

a. $\mathbf{r}(t) = \langle t^2, 4t, 4\ln t \rangle$,   for $t > 0$

b. $\mathbf{r}(t) = \langle 10, 3\cos t, 3\sin t \rangle$,   for $0 \leq t \leq 2\pi$

SOLUTION

a. A tangent vector is $\mathbf{r}'(t) = \langle 2t, 4, 4/t \rangle$, which has a magnitude of

$$|\mathbf{r}'(t)| = \sqrt{(2t)^2 + 4^2 + \left(\frac{4}{t}\right)^2} \quad \text{Definition of magnitude}$$

$$= \sqrt{4t^2 + 16 + \frac{16}{t^2}} \quad \text{Expand.}$$

$$= \sqrt{\left(2t + \frac{4}{t}\right)^2} \quad \text{Factor.}$$

$$= 2t + \frac{4}{t}. \quad \text{Simplify.}$$

Therefore, the unit tangent vector for a particular value of t is

$$\mathbf{T}(t) = \frac{\langle 2t, 4, 4/t \rangle}{2t + 4/t}.$$

As shown in **Figure 14.10**, the unit tangent vectors change direction along the curve but maintain unit length.

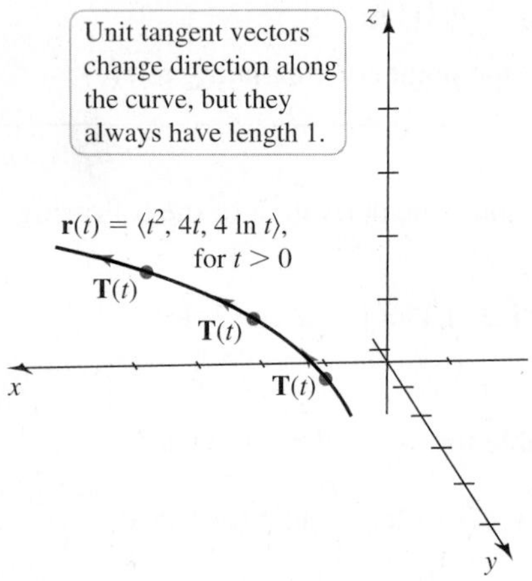

b. In this case, $\mathbf{r}'(t) = \langle 0, -3 \sin t, 3 \cos t \rangle$ and

$$|\mathbf{r}'(t)| = \sqrt{0^2 + (-3 \sin t)^2 + (3 \cos t)^2} = \sqrt{9\underbrace{(\sin^2 t + \cos^2 t)}_{1}} = 3.$$

Therefore, the unit tangent vector for a particular value of t is

$$\mathbf{T}(t) = \frac{1}{3}\langle 0, -3 \sin t, 3 \cos t \rangle = \langle 0, -\sin t, \cos t \rangle.$$

The direction of $\mathbf{T}$ changes along the curve, but its length remains 1.

Related Exercises 25, 27 ◄

Unit tangent vectors change direction along the curve, but they always have length 1.

$\mathbf{r}(t) = \langle t^2, 4t, 4 \ln t \rangle$, for $t > 0$

$\mathbf{T}(t)$

$\mathbf{T}(t)$

$\mathbf{T}(t)$

Figure 14.10

Derivative Rules The rules for derivatives for single-variable functions either carry over directly to vector-valued functions or have close analogs. These rules are generally proved by working on the individual components of the vector function.

THEOREM 14.1   Derivative Rules

Let $\mathbf{u}$ and $\mathbf{v}$ be differentiable vector-valued functions, and let f be a differentiable scalar-valued function, all at a point t. Let $\mathbf{c}$ be a constant vector. The following rules apply.

1. $\dfrac{d}{dt}(\mathbf{c}) = \mathbf{0}$ Constant Rule

2. $\dfrac{d}{dt}(\mathbf{u}(t) + \mathbf{v}(t)) = \mathbf{u}'(t) + \mathbf{v}'(t)$ Sum Rule

3. $\dfrac{d}{dt}(f(t)\mathbf{u}(t)) = f'(t)\mathbf{u}(t) + f(t)\mathbf{u}'(t)$ Product Rule

4. $\dfrac{d}{dt}(\mathbf{u}(f(t))) = \mathbf{u}'(f(t))f'(t)$ Chain Rule

5. $\dfrac{d}{dt}(\mathbf{u}(t) \cdot \mathbf{v}(t)) = \mathbf{u}'(t) \cdot \mathbf{v}(t) + \mathbf{u}(t) \cdot \mathbf{v}'(t)$ Dot Product Rule

6. $\dfrac{d}{dt}(\mathbf{u}(t) \times \mathbf{v}(t)) = \mathbf{u}'(t) \times \mathbf{v}(t) + \mathbf{u}(t) \times \mathbf{v}'(t)$ Cross Product Rule

---

➤ With the exception of the Cross Product Rule, these rules apply to vector-valued functions with any number of components. Notice that we have three new product rules, all of which mimic the original Product Rule. In Rule 4, $\mathbf{u}$ must be differentiable at $f(t)$.

QUICK CHECK 3 Let $\mathbf{u}(t) = \langle t, t, t \rangle$ and $\mathbf{v}(t) = \langle 1, 1, 1 \rangle$. Compute $\dfrac{d}{dt}(\mathbf{u}(t) \cdot \mathbf{v}(t))$ using Derivative Rule 5, and show that it agrees with the result obtained by first computing the dot product and differentiating directly. ◄

The proofs of these rules are assigned in Exercises 92–94 with the exception of the following representative proofs.

Proof of the Chain Rule: Let $\mathbf{u}(t) = \langle u_1(t), u_2(t), u_3(t) \rangle$, which implies that

$$\mathbf{u}(f(t)) = u_1(f(t))\,\mathbf{i} + u_2(f(t))\,\mathbf{j} + u_3(f(t))\,\mathbf{k}.$$

We now apply the ordinary Chain Rule componentwise:

$$\frac{d}{dt}\left(\mathbf{u}(f(t))\right) = \frac{d}{dt}\left(u_1(f(t))\,\mathbf{i} + u_2(f(t))\,\mathbf{j} + u_3(f(t))\,\mathbf{k}\right) \qquad \text{Components of } \mathbf{u}$$

$$= \frac{d}{dt}\left(u_1(f(t))\right)\mathbf{i} + \frac{d}{dt}\left(u_2(f(t))\right)\mathbf{j} + \frac{d}{dt}\left(u_3(f(t))\right)\mathbf{k} \qquad \begin{array}{l}\text{Differentiate each}\\\text{component.}\end{array}$$

$$= u_1{}'(f(t))\,f'(t)\,\mathbf{i} + u_2{}'(f(t))\,f'(t)\,\mathbf{j} + u_3{}'(f(t))\,f'(t)\,\mathbf{k} \qquad \text{Chain Rule}$$

$$= \left(u_1{}'(f(t))\,\mathbf{i} + u_2{}'(f(t))\,\mathbf{j} + u_3{}'(f(t))\,\mathbf{k}\right)f'(t) \qquad \text{Factor } f'(t).$$

$$= \mathbf{u}'(f(t))\,f'(t). \qquad \text{Definition of } \mathbf{u}'$$

◄

Proof of the Dot Product Rule: We use the standard Product Rule on each component. Let $\mathbf{u}(t) = \langle u_1(t), u_2(t), u_3(t)\rangle$ and $\mathbf{v}(t) = \langle v_1(t), v_2(t), v_3(t)\rangle$. Then

$$\frac{d}{dt}(\mathbf{u}\cdot\mathbf{v}) = \frac{d}{dt}(u_1 v_1 + u_2 v_2 + u_3 v_3) \qquad \text{Definition of dot product}$$

$$= u_1{}'v_1 + u_1 v_1{}' + u_2{}'v_2 + u_2 v_2{}' + u_3{}'v_3 + u_3 v_3{}' \qquad \text{Product Rule}$$

$$= \underbrace{u_1{}'v_1 + u_2{}'v_2 + u_3{}'v_3}_{\mathbf{u}'\cdot\mathbf{v}} + \underbrace{u_1 v_1{}' + u_2 v_2{}' + u_3 v_3{}'}_{\mathbf{u}\cdot\mathbf{v}'} \qquad \text{Rearrange.}$$

$$= \mathbf{u}'\cdot\mathbf{v} + \mathbf{u}\cdot\mathbf{v}'.$$

◄

EXAMPLE 3 **Derivative rules** Compute the following derivatives, where

$$\mathbf{u}(t) = t\,\mathbf{i} + t^2\,\mathbf{j} - t^3\,\mathbf{k} \quad \text{and} \quad \mathbf{v}(t) = \sin t\,\mathbf{i} + 2\cos t\,\mathbf{j} + \cos t\,\mathbf{k}.$$

a. $\dfrac{d}{dt}(\mathbf{v}(t^2))$ **b.** $\dfrac{d}{dt}(t^2\,\mathbf{v}(t))$ **c.** $\dfrac{d}{dt}(\mathbf{u}(t)\cdot\mathbf{v}(t))$

SOLUTION

a. Note that $\mathbf{v}'(t) = \cos t\,\mathbf{i} - 2\sin t\,\mathbf{j} - \sin t\,\mathbf{k}$. Using the Chain Rule, we have

$$\frac{d}{dt}(\mathbf{v}(t^2)) = \mathbf{v}'(t^2)\frac{d}{dt}(t^2) = \underbrace{(\cos t^2\,\mathbf{i} - 2\sin t^2\,\mathbf{j} - \sin t^2\,\mathbf{k})}_{\mathbf{v}'(t^2)}(2t).$$

b. $\dfrac{d}{dt}(t^2\,\mathbf{v}(t)) = \dfrac{d}{dt}(t^2)\mathbf{v}(t) + t^2\dfrac{d}{dt}(\mathbf{v}(t)) \qquad \text{Product Rule}$

$$= 2t\,\mathbf{v}(t) + t^2\,\mathbf{v}'(t)$$

$$= 2t\underbrace{(\sin t\,\mathbf{i} + 2\cos t\,\mathbf{j} + \cos t\,\mathbf{k})}_{\mathbf{v}(t)} + t^2\underbrace{(\cos t\,\mathbf{i} - 2\sin t\,\mathbf{j} - \sin t\,\mathbf{k})}_{\mathbf{v}'(t)}$$

Differentiate.

$$= (2t\sin t + t^2\cos t)\mathbf{i} + (4t\cos t - 2t^2\sin t)\mathbf{j} + (2t\cos t - t^2\sin t)\mathbf{k}$$

Collect terms.

c. $\dfrac{d}{dt}(\mathbf{u}(t)\cdot\mathbf{v}(t)) = \mathbf{u}'(t)\cdot\mathbf{v}(t) + \mathbf{u}(t)\cdot\mathbf{v}'(t) \qquad \text{Dot Product Rule}$

$$= (\mathbf{i} + 2t\,\mathbf{j} - 3t^2\,\mathbf{k})\cdot(\sin t\,\mathbf{i} + 2\cos t\,\mathbf{j} + \cos t\,\mathbf{k})$$
$$+ (t\,\mathbf{i} + t^2\,\mathbf{j} - t^3\,\mathbf{k})\cdot(\cos t\,\mathbf{i} - 2\sin t\,\mathbf{j} - \sin t\,\mathbf{k}) \qquad \text{Differentiate.}$$

$$= (\sin t + 4t\cos t - 3t^2\cos t) + (t\cos t - 2t^2\sin t + t^3\sin t) \qquad \text{Dot products}$$

$$= (1 - 2t^2 + t^3)\sin t + (5t - 3t^2)\cos t \qquad \text{Simplify.}$$

Note that the result is a scalar. The same result is obtained if you first compute $\mathbf{u}\cdot\mathbf{v}$ and then differentiate.

Related Exercises 33, 36, 37 ◄

Higher-Order Derivatives Higher-order derivatives of vector-valued functions are computed in the expected way: We simply differentiate each component multiple times. Second derivatives feature prominently in the next section, playing the role of acceleration.

EXAMPLE 4 Higher-order derivatives Compute the first, second, and third derivative of $\mathbf{r}(t) = \langle t^2, 8 \ln t, 3e^{-2t} \rangle$.

SOLUTION Differentiating once, we have $\mathbf{r}'(t) = \langle 2t, 8/t, -6e^{-2t} \rangle$. Differentiating again produces $\mathbf{r}''(t) = \langle 2, -8/t^2, 12e^{-2t} \rangle$. Differentiating once more, we have $\mathbf{r}'''(t) = \langle 0, 16/t^3, -24e^{-2t} \rangle$.

Related Exercise 58 ◄

Integrals of Vector-Valued Functions

An **antiderivative** of the vector function $\mathbf{r}$ is a function $\mathbf{R}$ such that $\mathbf{R}' = \mathbf{r}$. If

$$\mathbf{r}(t) = f(t)\,\mathbf{i} + g(t)\,\mathbf{j} + h(t)\,\mathbf{k},$$

then an antiderivative of $\mathbf{r}$ is

$$\mathbf{R}(t) = F(t)\,\mathbf{i} + G(t)\,\mathbf{j} + H(t)\,\mathbf{k},$$

where F, G, and H are antiderivatives of f, g, and h, respectively. This fact follows by differentiating the components of $\mathbf{R}$ and verifying that $\mathbf{R}' = \mathbf{r}$. The collection of all antiderivatives of $\mathbf{r}$ is the *indefinite integral* of $\mathbf{r}$.

DEFINITION Indefinite Integral of a Vector-Valued Function

Let $\mathbf{r}(t) = f(t)\,\mathbf{i} + g(t)\,\mathbf{j} + h(t)\,\mathbf{k}$ be a vector function, and let $\mathbf{R}(t) = F(t)\,\mathbf{i} + G(t)\,\mathbf{j} + H(t)\,\mathbf{k}$, where F, G, and H are antiderivatives of f, g, and h, respectively. The **indefinite integral** of $\mathbf{r}$ is

$$\int \mathbf{r}(t)\,dt = \mathbf{R}(t) + \mathbf{C},$$

where $\mathbf{C}$ is an arbitrary constant vector. Alternatively, in component form,

$$\int \langle f(t), g(t), h(t) \rangle\,dt = \langle F(t), G(t), H(t) \rangle + \langle C_1, C_2, C_3 \rangle.$$

EXAMPLE 5 Indefinite integrals Compute

$$\int \left(\frac{t}{\sqrt{t^2 + 2}}\,\mathbf{i} + e^{-3t}\,\mathbf{j} + (\sin 4t + 1)\,\mathbf{k} \right) dt.$$

> The substitution $u = t^2 + 2$ is used to evaluate the $\mathbf{i}$-component of the integral.

SOLUTION We compute the indefinite integral of each component:

$$\int \left(\frac{t}{\sqrt{t^2 + 2}}\,\mathbf{i} + e^{-3t}\,\mathbf{j} + (\sin 4t + 1)\,\mathbf{k} \right) dt$$

$$= \left(\sqrt{t^2 + 2} + C_1 \right)\mathbf{i} + \left(-\frac{1}{3}e^{-3t} + C_2 \right)\mathbf{j} + \left(-\frac{1}{4}\cos 4t + t + C_3 \right)\mathbf{k}$$

$$= \sqrt{t^2 + 2}\,\mathbf{i} - \frac{1}{3}e^{-3t}\,\mathbf{j} + \left(t - \frac{1}{4}\cos 4t \right)\mathbf{k} + \mathbf{C}. \quad \text{Let } \mathbf{C} = C_1\,\mathbf{i} + C_2\,\mathbf{j} + C_3\,\mathbf{k}.$$

QUICK CHECK 4 Let $\mathbf{r}(t) = \langle 1, 2t, 3t^2 \rangle$. Compute $\int \mathbf{r}(t)\,dt$. ◄

The constants C_1, C_2, and C_3 are combined to form one vector constant $\mathbf{C}$ at the end of the calculation.

Related Exercise 63 ◄

EXAMPLE 6 Finding one antiderivative Find $\mathbf{r}(t)$ such that $\mathbf{r}'(t) = \langle 10, \sin t, t \rangle$ and $\mathbf{r}(0) = \mathbf{j}$.

SOLUTION The required function $\mathbf{r}$ is an antiderivative of $\langle 10, \sin t, t \rangle$:

$$\mathbf{r}(t) = \int \langle 10, \sin t, t \rangle\,dt = \left\langle 10t, -\cos t, \frac{t^2}{2} \right\rangle + \mathbf{C},$$

where $\mathbf{C}$ is an arbitrary constant vector. The condition $\mathbf{r}(0) = \mathbf{j}$ allows us to determine $\mathbf{C}$; substituting $t = 0$ implies that $\mathbf{r}(0) = \langle 0, -1, 0 \rangle + \mathbf{C} = \mathbf{j}$, where $\mathbf{j} = \langle 0, 1, 0 \rangle$. Solving for $\mathbf{C}$, we have $\mathbf{C} = \langle 0, 1, 0 \rangle - \langle 0, -1, 0 \rangle = \langle 0, 2, 0 \rangle$. Therefore,

$$\mathbf{r}(t) = \left\langle 10t, 2 - \cos t, \frac{t^2}{2} \right\rangle.$$

Related Exercise 65 ◄

Definite integrals are evaluated by applying the Fundamental Theorem of Calculus to each component of a vector-valued function.

DEFINITION Definite Integral of a Vector-Valued Function

Let $\mathbf{r}(t) = f(t)\,\mathbf{i} + g(t)\,\mathbf{j} + h(t)\,\mathbf{k}$, where f, g, and h are integrable on the interval $[a, b]$. The **definite integral** of $\mathbf{r}$ on $[a, b]$ is

$$\int_a^b \mathbf{r}(t)\,dt = \left(\int_a^b f(t)\,dt \right)\mathbf{i} + \left(\int_a^b g(t)\,dt \right)\mathbf{j} + \left(\int_a^b h(t)\,dt \right)\mathbf{k}.$$

EXAMPLE 7 Definite integrals Evaluate

$$\int_0^\pi \left(\mathbf{i} + 3 \cos \frac{t}{2}\,\mathbf{j} - 4t\,\mathbf{k} \right) dt.$$

SOLUTION

$$\int_0^\pi \left(\mathbf{i} + 3 \cos \frac{t}{2}\,\mathbf{j} - 4t\,\mathbf{k} \right) dt = t\,\mathbf{i}\Big|_0^\pi + 6 \sin \frac{t}{2}\,\mathbf{j}\Big|_0^\pi - 2t^2\,\mathbf{k}\Big|_0^\pi \quad \text{Evaluate integrals for each component.}$$

$$= \pi\,\mathbf{i} + 6\mathbf{j} - 2\pi^2\,\mathbf{k} \qquad \text{Simplify.}$$

Related Exercise 75 ◄

With the tools of differentiation and integration in hand, we are prepared to tackle some practical problems, notably the motion of objects in space.

SECTION 14.2 EXERCISES

Getting Started

1. What is the derivative of $\mathbf{r}(t) = \langle f(t), g(t), h(t) \rangle$?

2. Explain the geometric meaning of $\mathbf{r}'(t)$.

3. Given a tangent vector on an oriented curve, how do you find the unit tangent vector?

4. Compute $\mathbf{r}''(t)$ when $\mathbf{r}(t) = \langle t^{10}, 8t, \cos t \rangle$.

5. How do you find the indefinite integral of $\mathbf{r}(t) = \langle f(t), g(t), h(t) \rangle$?

6. How do you evaluate $\int_a^b \mathbf{r}(t)\,dt$?

7. Find $\mathbf{C}$ if $\mathbf{r}(t) = \langle e^t, 3 \cos t, t + 10 \rangle + \mathbf{C}$ and $\mathbf{r}(0) = \langle 0, 0, 0 \rangle$.

8. Find the unit tangent vector at $t = 0$ for the parameterized curve $\mathbf{r}(t)$ if $\mathbf{r}'(t) = \langle e^t + 5, \sin t + 2, \cos t + 2 \rangle$.

Practice Exercises

9–16. Derivatives of vector-valued functions *Differentiate the following functions.*

9. $\mathbf{r}(t) = \langle \cos t, t^2, \sin t \rangle$

10. $\mathbf{r}(t) = 4e^t\,\mathbf{i} + 5\,\mathbf{j} + \ln t\,\mathbf{k}$

11. $\mathbf{r}(t) = \left\langle 2t^3, 6\sqrt{t}, \dfrac{3}{t} \right\rangle$

12. $\mathbf{r}(t) = \langle 4, 3 \cos 2t, 2 \sin 3t \rangle$

13. $\mathbf{r}(t) = e^t\,\mathbf{i} + 2e^{-t}\,\mathbf{j} - 4e^{2t}\,\mathbf{k}$

14. $\mathbf{r}(t) = \tan t\,\mathbf{i} + \sec t\,\mathbf{j} + \cos^2 t\,\mathbf{k}$

15. $\mathbf{r}(t) = \langle te^{-t}, t \ln t, t \cos t \rangle$

16. $\mathbf{r}(t) = \langle (t + 1)^{-1}, \tan^{-1} t, \ln(t + 1) \rangle$

17–22. Tangent vectors *Find a tangent vector at the given value of t for the following parameterized curves.*

17. $\mathbf{r}(t) = \langle t, 3t^2, t^3 \rangle, t = 1$ 18. $\mathbf{r}(t) = \langle e^t, e^{3t}, e^{5t} \rangle, t = 0$

19. $\mathbf{r}(t) = \langle t, \cos 2t, 2 \sin t \rangle, t = \dfrac{\pi}{2}$

20. $\mathbf{r}(t) = \left\langle 2 \sin t, 3 \cos t, \sin \dfrac{t}{2} \right\rangle, t = \pi$

21. $\mathbf{r}(t) = 2t^4\,\mathbf{i} + 6t^{3/2}\,\mathbf{j} + \dfrac{10}{t}\,\mathbf{k}, t = 1$

22. $\mathbf{r}(t) = 2e^t\,\mathbf{i} + e^{-2t}\,\mathbf{j} + 4e^{2t}\,\mathbf{k}, t = \ln 3$

23–28. Unit tangent vectors *Find the unit tangent vector for the following parameterized curves.*

23. $\mathbf{r}(t) = \langle 2t, 2t, t \rangle$, for $0 \le t \le 1$

24. $\mathbf{r}(t) = \langle \cos t, \sin t, 2 \rangle$, for $0 \le t \le 2\pi$

25. $\mathbf{r}(t) = \langle 8, \cos 2t, 2 \sin 2t \rangle$, for $0 \le t \le 2\pi$

26. $\mathbf{r}(t) = \langle \sin t, \cos t, \cos t \rangle$, for $0 \le t \le 2\pi$

27. $\mathbf{r}(t) = \left\langle t, 2, \dfrac{2}{t} \right\rangle$, for $t \ge 1$

28. $\mathbf{r}(t) = \langle e^{2t}, 2e^{2t}, 2e^{-3t} \rangle$, for $t \ge 0$

29–32. Unit tangent vectors at a point *Find the unit tangent vector at the given value of t for the following parameterized curves.*

29. $\mathbf{r}(t) = \langle \cos 2t, 4, 3 \sin 2t \rangle; t = \dfrac{\pi}{2}$

30. $\mathbf{r}(t) = \langle \sin t, \cos t, e^{-t} \rangle; t = 0$

31. $\mathbf{r}(t) = \left\langle 6t, 6, \dfrac{3}{t} \right\rangle; t = 1$

32. $\mathbf{r}(t) = \langle \sqrt{7}e^t, 3e^t, 3e^t \rangle; t = \ln 2$

33–38. Derivative rules *Let* $\mathbf{u}(t) = 2t^3 \mathbf{i} + (t^2 - 1)\mathbf{j} - 8\mathbf{k}$ *and* $\mathbf{v}(t) = e^t \mathbf{i} + 2e^{-t}\mathbf{j} - e^{2t}\mathbf{k}$. *Compute the derivative of the following functions.*

33. $(t^{12} + 3t)\mathbf{u}(t)$

34. $(4t^8 - 6t^3)\mathbf{v}(t)$

35. $\mathbf{u}(t^4 - 2t)$

36. $\mathbf{v}(\sqrt{t})$

37. $\mathbf{u}(t) \cdot \mathbf{v}(t)$

38. $\mathbf{u}(t) \times \mathbf{v}(t)$

39–42. Derivative rules *Suppose* $\mathbf{u}$ *and* $\mathbf{v}$ *are differentiable functions at* $t = 0$ *with* $\mathbf{u}(0) = \langle 0, 1, 1 \rangle$, $\mathbf{u}'(0) = \langle 0, 7, 1 \rangle$, $\mathbf{v}(0) = \langle 0, 1, 1 \rangle$, *and* $\mathbf{v}'(0) = \langle 1, 1, 2 \rangle$. *Evaluate the following expressions.*

39. $\dfrac{d}{dt}(\mathbf{u} \cdot \mathbf{v})\Big|_{t=0}$

40. $\dfrac{d}{dt}(\mathbf{u} \times \mathbf{v})\Big|_{t=0}$

41. $\dfrac{d}{dt}(\cos t\,\mathbf{u}(t))\Big|_{t=0}$

42. $\dfrac{d}{dt}(\mathbf{u}(\sin t))\Big|_{t=0}$

43–48. Derivative rules *Let* $\mathbf{u}(t) = \langle 1, t, t^2 \rangle$, $\mathbf{v}(t) = \langle t^2, -2t, 1 \rangle$, *and* $g(t) = 2\sqrt{t}$. *Compute the derivatives of the following functions.*

43. $\mathbf{v}(e^t)$ **44.** $\mathbf{u}(t^3)$ **45.** $\mathbf{v}(g(t))$

46. $g(t)\mathbf{v}(t)$ **47.** $\mathbf{u}(t) \times \mathbf{v}(t)$ **48.** $\mathbf{u}(t) \cdot \mathbf{v}(t)$

49–52. Derivative rules *Compute the following derivatives.*

49. $\dfrac{d}{dt}(t^2(\mathbf{i} + 2\mathbf{i} - 2t\,\mathbf{k}) \cdot (e^t \mathbf{i} + 2e^t \mathbf{j} - 3e^{-t}\mathbf{k}))$

50. $\dfrac{d}{dt}((t^3 \mathbf{i} - 2t\,\mathbf{j} - 2\,\mathbf{k}) \times (t\,\mathbf{i} - t^2\mathbf{j} - t^3\mathbf{k}))$

51. $\dfrac{d}{dt}((3t^2\mathbf{i} + \sqrt{t}\,\mathbf{j} - 2t^{-1}\mathbf{k}) \cdot (\cos t\,\mathbf{i} + \sin 2t\,\mathbf{j} - 3t\,\mathbf{k}))$

52. $\dfrac{d}{dt}((t^3 \mathbf{i} + 6\mathbf{j} - 2\sqrt{t}\,\mathbf{k}) \times (3t\,\mathbf{i} - 12t^2\mathbf{j} - 6t^{-2}\mathbf{k}))$

53–58. Higher-order derivatives *Compute* $\mathbf{r}''(t)$ *and* $\mathbf{r}'''(t)$ *for the following functions.*

53. $\mathbf{r}(t) = \langle t^2 + 1, t + 1, 1 \rangle$

54. $\mathbf{r}(t) = \langle 3t^{12} - t^2, t^8 + t^3, t^{-4} - 2 \rangle$

55. $\mathbf{r}(t) = \langle \cos 3t, \sin 4t, \cos 6t \rangle$

56. $\mathbf{r}(t) = \langle e^{4t}, 2e^{-4t} + 1, 2e^{-t} \rangle$

57. $\mathbf{r}(t) = \sqrt{t+4}\,\mathbf{i} + \dfrac{t}{t+1}\mathbf{j} - e^{-t^2}\mathbf{k}$

58. $\mathbf{r}(t) = \tan t\,\mathbf{i} + \left(t + \dfrac{1}{t} \right)\mathbf{j} - \ln(t+1)\mathbf{k}$

59–64. Indefinite integrals *Compute the indefinite integral of the following functions.*

59. $\mathbf{r}(t) = \langle t^4 - 3t, 2t - 1, 10 \rangle$

60. $\mathbf{r}(t) = \left\langle 5t^{-4} - t^2, t^6 - 4t^3, \dfrac{2}{t} \right\rangle$

61. $\mathbf{r}(t) = \langle 2 \cos t, 2 \sin 3t, 4 \cos 8t \rangle$

62. $\mathbf{r}(t) = te^t \mathbf{i} + t \sin t^2 \mathbf{j} - \dfrac{2t}{\sqrt{t^2 + 4}}\mathbf{k}$

63. $\mathbf{r}(t) = e^{3t}\mathbf{i} + \dfrac{1}{1+t^2}\mathbf{j} - \dfrac{1}{\sqrt{2t}}\mathbf{k}$

64. $\mathbf{r}(t) = 2^t \mathbf{i} + \dfrac{1}{1+2t}\mathbf{j} + \ln t\,\mathbf{k}$

65–70. Finding r from r′ *Find the function* $\mathbf{r}$ *that satisfies the given conditions.*

65. $\mathbf{r}'(t) = \langle e^t, \sin t, \sec^2 t \rangle;\quad \mathbf{r}(0) = \langle 2, 2, 2 \rangle$

66. $\mathbf{r}'(t) = \langle 0, 2, 2t \rangle;\quad \mathbf{r}(1) = \langle 4, 3, -5 \rangle$

67. $\mathbf{r}'(t) = \langle 1, 2t, 3t^2 \rangle;\quad \mathbf{r}(1) = \langle 4, 3, -5 \rangle$

68. $\mathbf{r}'(t) = \left\langle \sqrt{t}, \cos \pi t, \dfrac{4}{t} \right\rangle;\quad \mathbf{r}(1) = \langle 2, 3, 4 \rangle$

69. $\mathbf{r}'(t) = \langle e^{2t}, 1 - 2e^{-t}, 1 - 2e^t \rangle;\quad \mathbf{r}(0) = \langle 1, 1, 1 \rangle$

70. $\mathbf{r}'(t) = \dfrac{t}{t^2 + 1}\mathbf{i} + te^{-t^2}\mathbf{j} - \dfrac{2t}{\sqrt{t^2 + 4}}\mathbf{k};\quad \mathbf{r}(0) = \mathbf{i} + \dfrac{3}{2}\mathbf{j} - 3\mathbf{k}$

71–78. Definite integrals *Evaluate the following definite integrals.*

71. $\displaystyle\int_{-1}^{1}(\mathbf{i} + t\,\mathbf{j} + 3t^2\,\mathbf{k})\,dt$

72. $\displaystyle\int_{1}^{4}(6t^2\mathbf{i} + 8t^3\mathbf{j} + 9t^2\,\mathbf{k})\,dt$

73. $\displaystyle\int_{0}^{\ln 2}(e^t\mathbf{i} + e^t \cos(\pi e^t)\,\mathbf{j})\,dt$

74. $\displaystyle\int_{1/2}^{1}\left(\dfrac{3}{1+2t}\mathbf{i} - \pi \csc^2\left(\dfrac{\pi}{2}t\right)\mathbf{k} \right)dt$

75. $\displaystyle\int_{-\pi}^{\pi}(\sin t\,\mathbf{i} + \cos t\,\mathbf{j} + 2t\,\mathbf{k})\,dt$

76. $\displaystyle\int_{0}^{\ln 2}(e^{-t}\mathbf{i} + 2e^{2t}\mathbf{j} - 4e^t\,\mathbf{k})\,dt$

77. $\displaystyle\int_{0}^{2} te^t(\mathbf{i} + 2\mathbf{j} - \mathbf{k})\,dt$

78. $\displaystyle\int_{0}^{\pi/4}(\sec^2 t\,\mathbf{i} - 2 \cos t\,\mathbf{j} - \mathbf{k})\,dt$

79. Explain why or why not Determine whether the following statements are true and give an explanation or counterexample.

a. The vectors $\mathbf{r}(t)$ and $\mathbf{r}'(t)$ are parallel for all values of t in the domain.

b. The curve described by the function $\mathbf{r}(t) = \langle t, t^2 - 2t, \cos \pi t \rangle$ is smooth, for $-\infty < t < \infty$.

c. If f, g, and h are odd integrable functions and a is a real number, then

$$\int_{-a}^{a}(f(t)\,\mathbf{i} + g(t)\,\mathbf{j} + h(t)\,\mathbf{k})\,dt = \mathbf{0}.$$

80–83. Tangent lines *Suppose the vector-valued function* $\mathbf{r}(t) = \langle f(t), g(t), h(t) \rangle$ *is smooth on an interval containing the point* t_0. *The line tangent to* $\mathbf{r}(t)$ *at* $t = t_0$ *is the line parallel to the tangent vector* $\mathbf{r}'(t_0)$ *that passes through* $(f(t_0), g(t_0), h(t_0))$. *For each of the following functions, find an equation of the line tangent to the curve at* $t = t_0$. *Choose an orientation for the line that is the same as the direction of* $\mathbf{r}'$.

80. $\mathbf{r}(t) = \langle e^t, e^{2t}, e^{3t} \rangle; t_0 = 0$

81. $\mathbf{r}(t) = \langle 2 + \cos t, 3 + \sin 2t, t \rangle; t_0 = \dfrac{\pi}{2}$

82. $\mathbf{r}(t) = \langle \sqrt{2t + 1}, \sin \pi t, 4 \rangle; t_0 = 4$

83. $\mathbf{r}(t) = \langle 3t - 1, 7t + 2, t^2 \rangle; t_0 = 1$

Explorations and Challenges

84–89. Relationship between r and r′

84. Consider the circle $\mathbf{r}(t) = \langle a \cos t, a \sin t \rangle$, for $0 \le t \le 2\pi$, where a is a positive real number. Compute $\mathbf{r}'$ and show that it is orthogonal to $\mathbf{r}$ for all t.

85. Consider the parabola $\mathbf{r}(t) = \langle at^2 + 1, t \rangle$, for $-\infty < t < \infty$, where a is a positive real number. Find all points on the parabola at which $\mathbf{r}$ and $\mathbf{r}'$ are orthogonal.

86. Consider the curve $\mathbf{r}(t) = \langle \sqrt{t}, 1, t \rangle$, for $t > 0$. Find all points on the curve at which $\mathbf{r}$ and $\mathbf{r}'$ are orthogonal.

87. Consider the helix $\mathbf{r}(t) = \langle \cos t, \sin t, t \rangle$, for $-\infty < t < \infty$. Find all points on the helix at which $\mathbf{r}$ and $\mathbf{r}'$ are orthogonal.

88. Consider the ellipse $\mathbf{r}(t) = \langle 2 \cos t, 8 \sin t, 0 \rangle$, for $0 \le t \le 2\pi$. Find all points on the ellipse at which $\mathbf{r}$ and $\mathbf{r}'$ are orthogonal.

89. Give two families of curves in $\mathbb{R}^3$ for which $\mathbf{r}$ and $\mathbf{r}'$ are parallel for all t in the domain.

90. **Motion on a sphere** Prove that $\mathbf{r}$ describes a curve that lies on the surface of a sphere centered at the origin ($x^2 + y^2 + z^2 = a^2$ with $a \ge 0$) if and only if $\mathbf{r}$ and $\mathbf{r}'$ are orthogonal at all points of the curve.

91. **Vectors r and r′ for lines**
   a. If $\mathbf{r}(t) = \langle at, bt, ct \rangle$ with $\langle a, b, c \rangle \ne \langle 0, 0, 0 \rangle$, show that the angle between $\mathbf{r}$ and $\mathbf{r}'$ is constant for all $t > 0$.
   b. If $\mathbf{r}(t) = \langle x_0 + at, y_0 + bt, z_0 + ct \rangle$, where x_0, y_0, and z_0 are not all zero, show that the angle between $\mathbf{r}$ and $\mathbf{r}'$ varies with t.
   c. Explain the results of parts (a) and (b) geometrically.

92. **Proof of Sum Rule** By expressing $\mathbf{u}$ and $\mathbf{v}$ in terms of their components, prove that
$$\frac{d}{dt}(\mathbf{u}(t) + \mathbf{v}(t)) = \mathbf{u}'(t) + \mathbf{v}'(t).$$

93. **Proof of Product Rule** By expressing $\mathbf{u}$ in terms of its components, prove that
$$\frac{d}{dt}(f(t)\mathbf{u}(t)) = f'(t)\mathbf{u}(t) + f(t)\mathbf{u}'(t).$$

94. **Proof of Cross Product Rule** Prove that
$$\frac{d}{dt}(\mathbf{u}(t) \times \mathbf{v}(t)) = \mathbf{u}'(t) \times \mathbf{v}(t) + \mathbf{u}(t) \times \mathbf{v}'(t).$$

There are two ways to proceed: Either express $\mathbf{u}$ and $\mathbf{v}$ in terms of their three components or use the definition of the derivative.

T 95. Cusps and noncusps
   a. Graph the curve $\mathbf{r}(t) = \langle t^3, t^3 \rangle$. Show that $\mathbf{r}'(0) = \mathbf{0}$ and the curve does not have a cusp at $t = 0$. Explain.
   b. Graph the curve $\mathbf{r}(t) = \langle t^3, t^2 \rangle$. Show that $\mathbf{r}'(0) = \mathbf{0}$ and the curve has a cusp at $t = 0$. Explain.
   c. The functions $\mathbf{r}(t) = \langle t, t^2 \rangle$ and $\mathbf{p}(t) = \langle t^2, t^4 \rangle$ both satisfy $y = x^2$. Explain how the curves they parameterize are different.
   d. Consider the curve $\mathbf{r}(t) = \langle t^m, t^n \rangle$, where $m > 1$ and $n > 1$ are integers with no common factors. Is it true that the curve has a cusp at $t = 0$ if one (not both) of m and n is even? Explain.

QUICK CHECK ANSWERS

1. $\mathbf{r}(t)$ describes a line, so its tangent vector $\mathbf{r}'(t) = \langle 1, 1, 1 \rangle$ has constant direction and magnitude.   **2.** Both $\mathbf{r}'$ and $|\mathbf{r}'|$ have units of m/s. In forming $\mathbf{r}'/|\mathbf{r}'|$, the units cancel and $\mathbf{T}(t)$ is without units.   **3.** $\dfrac{d}{dt}(\mathbf{u}(t) \cdot \mathbf{v}(t)) =$ $\langle 1, 1, 1 \rangle \cdot \langle 1, 1, 1 \rangle + \langle t, t, t \rangle \cdot \langle 0, 0, 0 \rangle = 3$. $\dfrac{d}{dt}(\langle t, t, t \rangle \cdot \langle 1, 1, 1 \rangle) = \dfrac{d}{dt}(3t) = 3$.   **4.** $\langle t, t^2, t^3 \rangle + \mathbf{C}$, where $\mathbf{C} = \langle a, b, c \rangle$, and a, b, and c are real numbers ◄

14.3 Motion in Space

It is a remarkable fact that given the forces acting on an object and its initial position and velocity, the motion of the object in three-dimensional space can be modeled for all future times. To be sure, the accuracy of the results depends on how well the various forces on the object are described. For example, it may be more difficult to predict the trajectory of a spinning soccer ball than the path of a space station orbiting Earth. Nevertheless, as shown in this section, by combining Newton's Second Law of Motion with everything we have learned about vectors, it is possible to solve a variety of moving-body problems.

Position, Velocity, Speed, Acceleration

Until now, we have studied objects that move in one dimension (along a line). The next step is to consider the motion of objects in two dimensions (in a plane) and three

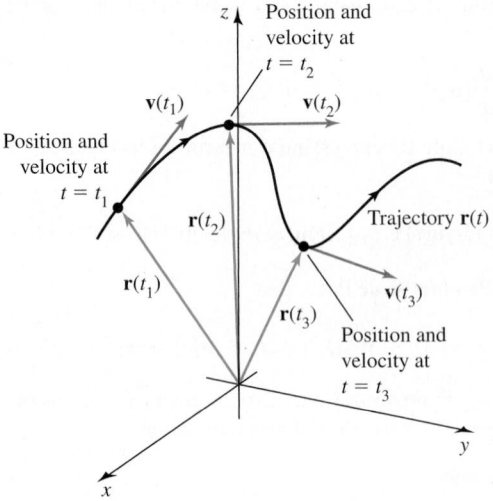

Figure 14.11

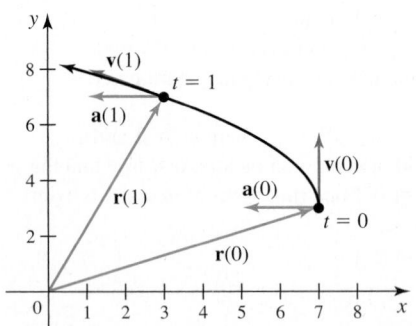

Figure 14.12

➤ In the case of two-dimensional motion, $\mathbf{r}(t) = \langle x(t), y(t) \rangle$, $\mathbf{v}(t) = \mathbf{r}'(t)$, and $\mathbf{a}(t) = \mathbf{r}''(t)$.

QUICK CHECK 1 Given $\mathbf{r}(t) = \langle t, t^2, t^3 \rangle$, find $\mathbf{v}(t)$ and $\mathbf{a}(t)$. ◄

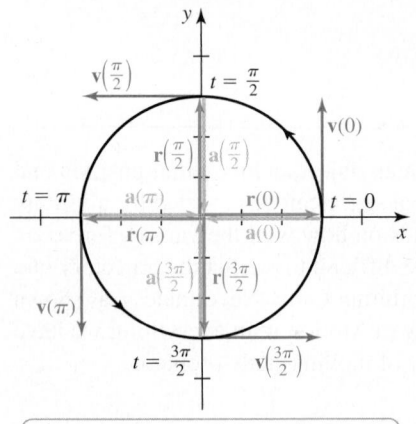

Circular motion: At all times $\mathbf{a}(t) = -\mathbf{r}(t)$ and $\mathbf{v}(t)$ is orthogonal to $\mathbf{r}(t)$ and $\mathbf{a}(t)$.

Figure 14.13

dimensions (in space). We work in a three-dimensional coordinate system and let the vector-valued function $\mathbf{r}(t) = \langle x(t), y(t), z(t) \rangle$ describe the *position* of a moving object at times $t \geq 0$. The curve described by $\mathbf{r}$ is the *path* or *trajectory* of the object (**Figure 14.11**). Just as with one-dimensional motion, the rate of change of the position function with respect to time is the *instantaneous velocity* of the object—a vector with three components corresponding to the velocity in the x-, y-, and z-directions:

$$\mathbf{v}(t) = \mathbf{r}'(t) = \langle x'(t), y'(t), z'(t) \rangle.$$

This expression should look familiar. The velocity vectors of a moving object are simply tangent vectors; that is, at any point, the velocity vector is tangent to the trajectory (Figure 14.11).

As with one-dimensional motion, the *speed* of an object moving in three dimensions is the magnitude of its velocity vector:

$$|\mathbf{v}(t)| = |\langle x'(t), y'(t), z'(t) \rangle| = \sqrt{x'(t)^2 + y'(t)^2 + z'(t)^2}.$$

The speed is a nonnegative scalar-valued function.

Finally, the *acceleration* of a moving object is the rate of change of the velocity:

$$\mathbf{a}(t) = \mathbf{v}'(t) = \mathbf{r}''(t).$$

Although the position vector gives the path of a moving object and the velocity vector is always tangent to the path, the acceleration vector is more difficult to visualize. **Figure 14.12** shows one particular instance of two-dimensional motion. The trajectory is a segment of a parabola and is traced out by the position vectors (shown at $t = 0$ and $t = 1$). As expected, the velocity vectors are tangent to the trajectory. In this case, the acceleration is $\mathbf{a} = \langle -2, 0 \rangle$; it is constant in magnitude and direction for all times. The relationships among $\mathbf{r}$, $\mathbf{v}$, and $\mathbf{a}$ are explored in the coming examples.

DEFINITION Position, Velocity, Speed, Acceleration

Let the **position** of an object moving in three-dimensional space be given by $\mathbf{r}(t) = \langle x(t), y(t), z(t) \rangle$, for $t \geq 0$. The **velocity** of the object is

$$\mathbf{v}(t) = \mathbf{r}'(t) = \langle x'(t), y'(t), z'(t) \rangle.$$

The **speed** of the object is the scalar function

$$|\mathbf{v}(t)| = \sqrt{x'(t)^2 + y'(t)^2 + z'(t)^2}.$$

The **acceleration** of the object is $\mathbf{a}(t) = \mathbf{v}'(t) = \mathbf{r}''(t)$.

EXAMPLE 1   Velocity and acceleration for circular motion Consider the two-dimensional motion given by the position vector

$$\mathbf{r}(t) = \langle x(t), y(t) \rangle = \langle 3 \cos t, 3 \sin t \rangle, \quad \text{for } 0 \leq t \leq 2\pi.$$

a. Sketch the trajectory of the object.

b. Find the velocity and speed of the object.

c. Find the acceleration of the object.

d. Sketch the position, velocity, and acceleration vectors, for $t = 0$, $\pi/2$, π, and $3\pi/2$.

SOLUTION

a. Notice that

$$x(t)^2 + y(t)^2 = 9(\cos^2 t + \sin^2 t) = 9,$$

which is an equation of a circle centered at the origin with radius 3. The object moves on this circle in the counterclockwise direction (**Figure 14.13**).

b. $\mathbf{v}(t) = \langle x'(t), y'(t) \rangle = \langle -3 \sin t, 3 \cos t \rangle$ Velocity vector

$$|\mathbf{v}(t)| = \sqrt{x'(t)^2 + y'(t)^2}$$ Definition of speed

$$= \sqrt{(-3 \sin t)^2 + (3 \cos t)^2}$$

$$= \sqrt{9\underbrace{(\sin^2 t + \cos^2 t)}_{1}} = 3$$

The velocity vector has a constant magnitude and a continuously changing direction.

c. Differentiating the velocity, we find that $\mathbf{a}(t) = \mathbf{v}'(t) = \langle -3 \cos t, -3 \sin t \rangle = -\mathbf{r}(t)$. In this case, the acceleration vector is the negative of the position vector at all times.

d. The relationships among $\mathbf{r}$, $\mathbf{v}$, and $\mathbf{a}$ at four points in time are shown in Figure 14.13. The velocity vector is always tangent to the trajectory and has length 3, while the acceleration vector and position vector each have length 3 and point in opposite directions. At all times, $\mathbf{v}$ is orthogonal to $\mathbf{r}$ and $\mathbf{a}$.

Related Exercise 13 ◄

EXAMPLE 2 **Comparing trajectories** Consider the trajectories described by the position functions

$$\mathbf{r}(t) = \left\langle t, t^2 - 4, \frac{t^3}{4} - 8 \right\rangle, \quad \text{for } t \geq 0, \text{ and}$$

$$\mathbf{R}(t) = \left\langle t^2, t^4 - 4, \frac{t^6}{4} - 8 \right\rangle, \quad \text{for } t \geq 0,$$

where t is measured in the same time units for both functions.

a. Graph and compare the trajectories using a graphing utility.

b. Find the velocity vectors associated with the position functions.

SOLUTION

a. Plotting the position functions at selected values of t results in the trajectories shown in Figure 14.14. Because $\mathbf{r}(0) = \mathbf{R}(0) = \langle 0, -4, -8 \rangle$, both curves have the same initial point. For $t \geq 0$, the two curves consist of the same points, but they are traced out differently. For example, both curves pass through the point $(4, 12, 8)$, but that point corresponds to $\mathbf{r}(4)$ on the first curve and $\mathbf{R}(2)$ on the second curve. In general, $\mathbf{r}(t^2) = \mathbf{R}(t)$, for $t \geq 0$.

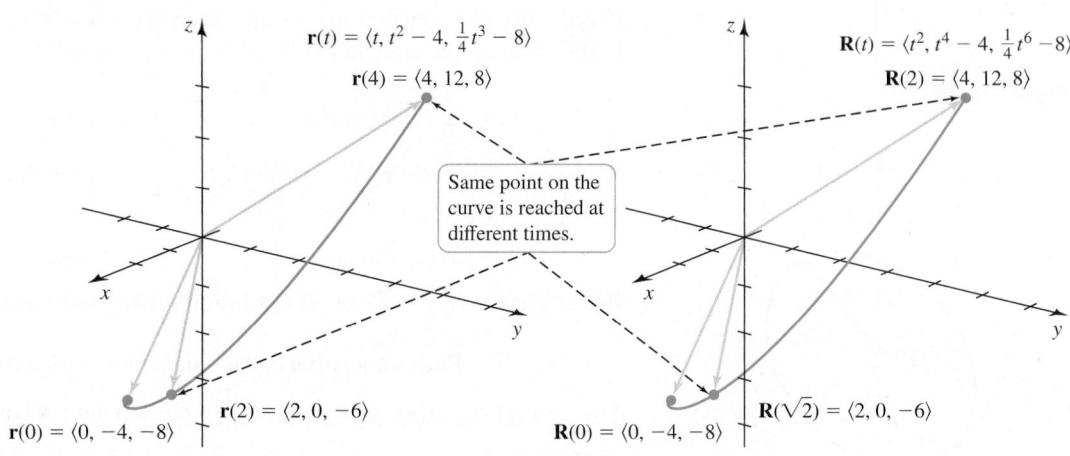

Figure 14.14

b. The velocity vectors are

$$\mathbf{r}'(t) = \left\langle 1, 2t, \frac{3t^2}{4} \right\rangle \quad \text{and} \quad \mathbf{R}'(t) = \left\langle 2t, 4t^3, \frac{3}{2}t^5 \right\rangle.$$

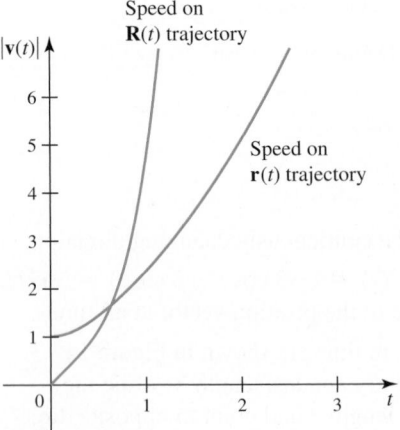

Figure 14.15

QUICK CHECK 2 Find the functions that give the speed of the two objects in Example 2, for $t \geq 0$ (corresponding to the graphs in Figure 14.15). ◄

➤ See Exercise 83 for a discussion of nonuniform straight-line motion.

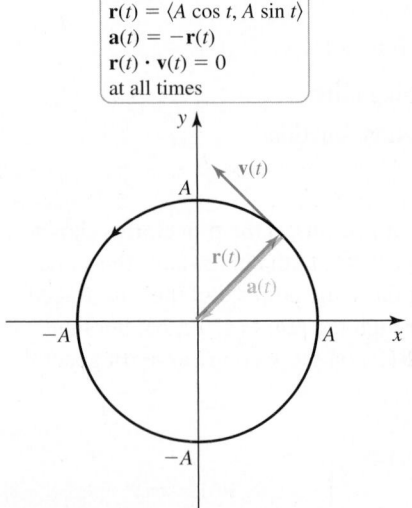

Figure 14.16

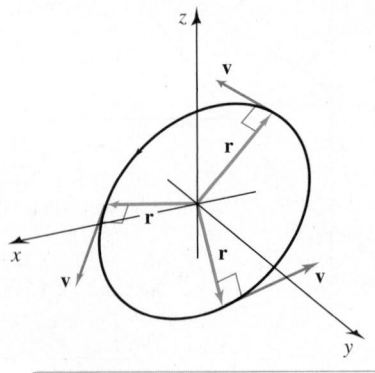

On a trajectory on which $|\mathbf{r}|$ is constant, $\mathbf{v}$ is orthogonal to $\mathbf{r}$ at all points.

Figure 14.17

The difference in the motion on the two curves is revealed by the graphs of the speeds associated with the trajectories (**Figure 14.15**). The object on the first trajectory reaches the point $(4, 12, 8)$ at $t = 4$, where its speed is $|\mathbf{r}'(4)| = |\langle 1, 8, 12 \rangle| \approx 14.5$. The object on the second trajectory reaches the same point $(4, 12, 8)$ at $t = 2$, where its speed is $|\mathbf{R}'(2)| = |\langle 4, 32, 48 \rangle| \approx 57.8$.

Related Exercise 21 ◄

Straight-Line and Circular Motion

Two types of motion in space arise frequently and deserve to be singled out. First consider a trajectory described by the vector function

$$\mathbf{r}(t) = \langle x_0 + at, y_0 + bt, z_0 + ct \rangle, \quad \text{for } t \geq 0,$$

where $x_0, y_0, z_0, a, b,$ and c are constants. This function describes a straight-line trajectory with an initial position $\langle x_0, y_0, z_0 \rangle$ and a direction given by the vector $\langle a, b, c \rangle$ (Section 13.5). The velocity on this trajectory is the constant $\mathbf{v}(t) = \mathbf{r}'(t) = \langle a, b, c \rangle$ in the direction of the trajectory, and the acceleration is $\mathbf{a} = \langle 0, 0, 0 \rangle$. The motion associated with this function is **uniform** (constant velocity) **straight-line motion**.

A different situation is **circular motion** (Example 1). Consider the two-dimensional circular path

$$\mathbf{r}(t) = \langle A \cos t, A \sin t \rangle, \quad \text{for } 0 \leq t \leq 2\pi,$$

where A is a nonzero constant (**Figure 14.16**). The velocity and acceleration vectors are

$$\mathbf{v}(t) = \langle -A \sin t, A \cos t \rangle \quad \text{and}$$
$$\mathbf{a}(t) = \langle -A \cos t, -A \sin t \rangle = -\mathbf{r}(t).$$

Notice that $\mathbf{r}$ and $\mathbf{a}$ are parallel but point in opposite directions. Furthermore, $\mathbf{r} \cdot \mathbf{v} = \mathbf{a} \cdot \mathbf{v} = 0$; therefore, the position and acceleration vectors are both orthogonal to the velocity vectors at any given point (Figure 14.16). Finally, $\mathbf{r}, \mathbf{v},$ and $\mathbf{a}$ have constant magnitude A and variable directions. The conclusion that $\mathbf{r} \cdot \mathbf{v} = 0$ applies to any motion for which $|\mathbf{r}|$ is constant; that is, to any motion on a circle or a sphere (**Figure 14.17**).

THEOREM 14.2 Motion with Constant $|\mathbf{r}|$
Let $\mathbf{r}$ describe a path on which $|\mathbf{r}|$ is constant (motion on a circle or sphere centered at the origin). Then $\mathbf{r} \cdot \mathbf{v} = 0$, which means the position vector and the velocity vector are orthogonal at all times for which the functions are defined.

Proof: If $\mathbf{r}$ has constant magnitude, then $|\mathbf{r}(t)|^2 = \mathbf{r}(t) \cdot \mathbf{r}(t) = c$ for some constant c. Differentiating the equation $\mathbf{r}(t) \cdot \mathbf{r}(t) = c$, we have

$$\begin{aligned} 0 &= \frac{d}{dt} \left(\mathbf{r}(t) \cdot \mathbf{r}(t) \right) & \text{Differentiate both sides of } |\mathbf{r}(t)|^2 = c. \\ &= \mathbf{r}'(t) \cdot \mathbf{r}(t) + \mathbf{r}(t) \cdot \mathbf{r}'(t) & \text{Derivative of dot product (Theorem 14.1)} \\ &= 2\mathbf{r}'(t) \cdot \mathbf{r}(t) & \text{Simplify.} \\ &= 2\mathbf{v}(t) \cdot \mathbf{r}(t). & \mathbf{r}'(t) = \mathbf{v}(t) \end{aligned}$$

Because $\mathbf{r}(t) \cdot \mathbf{v}(t) = 0$ for all t, it follows that $\mathbf{r}$ and $\mathbf{v}$ are orthogonal for all t. ◄

EXAMPLE 3 Path on a sphere An object moves on a trajectory described by

$$\mathbf{r}(t) = \langle x(t), y(t), z(t) \rangle = \langle 3 \cos t, 5 \sin t, 4 \cos t \rangle, \quad \text{for } 0 \leq t \leq 2\pi.$$

a. Show that the object moves on a sphere and find the radius of the sphere.

b. Find the velocity and speed of the object.

c. Consider the curve $\mathbf{r}(t) = \langle 5 \cos t, 5 \sin t, 5 \sin 2t \rangle$, which is the roller coaster curve from Example 3 of Section 14.1, with different coefficients. Show that this curve does not lie on a sphere. How could $\mathbf{r}$ be modified so that it describes a curve that lies on a sphere of radius 1, centered at the origin?

▶ For generalizations of this example and explorations of trajectories that lie on spheres and ellipses, see Exercises 79 and 82.

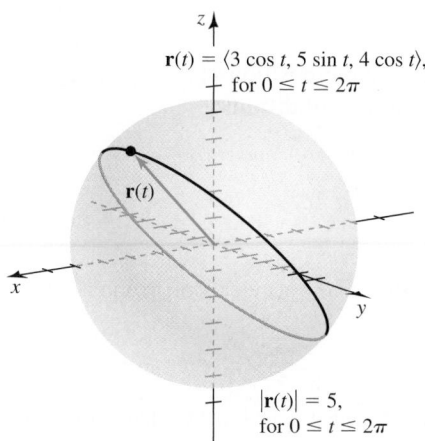

$\mathbf{r}(t) = \langle 3 \cos t, 5 \sin t, 4 \cos t \rangle$, for $0 \le t \le 2\pi$

$|\mathbf{r}(t)| = 5$, for $0 \le t \le 2\pi$

Figure 14.18

$\mathbf{u}(t) = \mathbf{r}(t)/|\mathbf{r}(t)|$, where
$\mathbf{r}(t) = \langle 5 \cos t, 5 \sin t, 5 \sin 2t \rangle$
$0 \le t \le 2\pi$

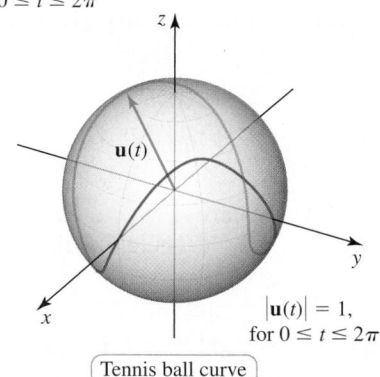

$|\mathbf{u}(t)| = 1$, for $0 \le t \le 2\pi$

Tennis ball curve

Figure 14.19

QUICK CHECK 3 Explain how to modify the curve $\mathbf{r}(t)$ given in Example 3c so that it lies on a sphere of radius 5 centered at the origin. ◀

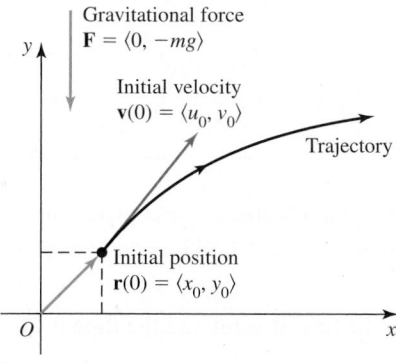

Gravitational force
$\mathbf{F} = \langle 0, -mg \rangle$

Initial velocity
$\mathbf{v}(0) = \langle u_0, v_0 \rangle$

Trajectory

Initial position
$\mathbf{r}(0) = \langle x_0, y_0 \rangle$

Figure 14.20

SOLUTION

a. $|\mathbf{r}(t)|^2 = x(t)^2 + y(t)^2 + z(t)^2$ Square of the distance from the origin

$= (3 \cos t)^2 + (5 \sin t)^2 + (4 \cos t)^2$ Substitute.

$= 25 \cos^2 t + 25 \sin^2 t$ Simplify.

$= 25 \underbrace{(\cos^2 t + \sin^2 t)}_{1} = 25$ Factor.

Therefore, $|\mathbf{r}(t)| = 5$, for $0 \le t \le 2\pi$, and the trajectory lies on a sphere of radius 5 centered at the origin (**Figure 14.18**).

b. $\mathbf{v}(t) = \mathbf{r}'(t) = \langle -3 \sin t, 5 \cos t, -4 \sin t \rangle$ Velocity vector

$|\mathbf{v}(t)| = \sqrt{\mathbf{v}(t) \cdot \mathbf{v}(t)}$ Speed of the object

$= \sqrt{9 \sin^2 t + 25 \cos^2 t + 16 \sin^2 t}$ Evaluate the dot product.

$= \sqrt{25 \underbrace{(\sin^2 t + \cos^2 t)}_{1}}$ Simplify.

$= 5$ Simplify.

The speed of the object is always 5. You should verify that $\mathbf{r}(t) \cdot \mathbf{v}(t) = 0$, for all t, implying that $\mathbf{r}$ and $\mathbf{v}$ are always orthogonal.

c. We first compute the distance from the origin to the curve:

$|\mathbf{r}(t)| = \sqrt{(5 \cos t)^2 + (5 \sin t)^2 + (5 \sin 2t)^2}$ Distance from origin to curve

$= \sqrt{25(\cos^2 t + \sin^2 t + \sin^2 2t)}$ Simplify.

$= 5\sqrt{1 + \sin^2 2t}.$ $\cos^2 t + \sin^2 t = 1$

It is clear that $|\mathbf{r}(t)|$ is not constant, and therefore the curve does not lie on a sphere.

One way to modify the curve so that it does lie on a sphere is to divide each output vector $\mathbf{r}(t)$ by its length. In fact, as long as $|\mathbf{r}(t)| \ne 0$ on the interval of interest, we can force any path onto a sphere (centered at the origin) with this modification. The function

$$\mathbf{u}(t) = \frac{\mathbf{r}(t)}{|\mathbf{r}(t)|} = \left\langle \frac{\cos t}{\sqrt{1 + \sin^2 2t}}, \frac{\sin t}{\sqrt{1 + \sin^2 2t}}, \frac{\sin 2t}{\sqrt{1 + \sin^2 2t}} \right\rangle$$

describes a curve on which $|\mathbf{u}(t)|$ is constant because $\dfrac{\mathbf{r}(t)}{|\mathbf{r}(t)|}$ is a unit vector. We conclude that the new curve, which is reminiscent of the seam on a tennis ball (**Figure 14.19**), lies on the unit sphere centered at the origin.

Related Exercises 32–33 ◀

Two-Dimensional Motion in a Gravitational Field

Newton's Second Law of Motion, which is used to model the motion of most objects, states that

$$\underbrace{\text{mass}}_{m} \cdot \underbrace{\text{acceleration}}_{\mathbf{a}(t) = \mathbf{r}''(t)} = \underbrace{\text{sum of all forces.}}_{\sum \mathbf{F}_k}$$

The governing law says something about the *acceleration* of an object, and in order to describe the motion fully, we must find the velocity and position from the acceleration.

Finding Velocity and Position from Acceleration   We begin with the case of two-dimensional projectile motion in which the only force acting on the object is the gravitational force; for the moment, air resistance and other possible external forces are neglected.

A convenient coordinate system uses a y-axis that points vertically upward and an x-axis that points in the direction of horizontal motion. The gravitational force is in the negative y-direction and is given by $\mathbf{F} = \langle 0, -mg \rangle$, where m is the mass of the object and $g = 9.8 \text{ m/s}^2 = 32 \text{ ft/s}^2$ is the acceleration due to gravity (**Figure 14.20**).

With these observations, Newton's Second Law takes the form

$$m\mathbf{a}(t) = \mathbf{F} = \langle 0, -mg \rangle.$$

Significantly, the mass of the object cancels, leaving the vector equation

$$\mathbf{a}(t) = \langle 0, -g \rangle. \tag{1}$$

In order to find the velocity $\mathbf{v}(t) = \langle x'(t), y'(t) \rangle$ and the position $\mathbf{r}(t) = \langle x(t), y(t) \rangle$ from this equation, we must be given the following **initial conditions**:

Initial velocity at $t = 0$: $\mathbf{v}(0) = \langle u_0, v_0 \rangle$ and

Initial position at $t = 0$: $\mathbf{r}(0) = \langle x_0, y_0 \rangle.$

We proceed in two steps.

▶ Recall that an antiderivative of 0 is a constant C and an antiderivative of $-g$ is $-gt + C$.

1. **Solve for the velocity** The velocity is an antiderivative of the acceleration in equation (1). Integrating the acceleration, we have

$$\mathbf{v}(t) = \int \mathbf{a}(t) \, dt = \int \langle 0, -g \rangle \, dt = \langle 0, -gt \rangle + \mathbf{C},$$

where $\mathbf{C}$ is an arbitrary constant vector. The arbitrary constant is determined by substituting $t = 0$ and using the initial condition $\mathbf{v}(0) = \langle u_0, v_0 \rangle$. We find that $\mathbf{v}(0) = \langle 0, 0 \rangle + \mathbf{C} = \langle u_0, v_0 \rangle$, or $\mathbf{C} = \langle u_0, v_0 \rangle$. Therefore, the velocity is

▶ You have a choice. You may do these calculations in vector notation as we have done here, or you may work with individual components.

$$\mathbf{v}(t) = \langle 0, -gt \rangle + \langle u_0, v_0 \rangle = \langle u_0, -gt + v_0 \rangle. \tag{2}$$

Notice that the horizontal component of velocity is simply the initial horizontal velocity u_0 for all time. The vertical component of velocity decreases linearly from its initial value of v_0.

2. **Solve for the position** The position is an antiderivative of the velocity given by equation (2):

$$\mathbf{r}(t) = \int \mathbf{v}(t) \, dt = \int \langle u_0, -gt + v_0 \rangle \, dt = \left\langle u_0 t, -\frac{1}{2} gt^2 + v_0 t \right\rangle + \mathbf{C},$$

where $\mathbf{C}$ is an arbitrary constant vector. Substituting $t = 0$, we have $\mathbf{r}(0) = \langle 0, 0 \rangle + \mathbf{C} = \langle x_0, y_0 \rangle$, which implies that $\mathbf{C} = \langle x_0, y_0 \rangle$. Therefore, the position of the object, for $t \geq 0$, is

$$\mathbf{r}(t) = \left\langle u_0 t, -\frac{1}{2} gt^2 + v_0 t \right\rangle + \langle x_0, y_0 \rangle = \left\langle \underbrace{u_0 t + x_0}_{x(t)}, \underbrace{-\frac{1}{2} gt^2 + v_0 t + y_0}_{y(t)} \right\rangle.$$

SUMMARY   Two-Dimensional Motion in a Gravitational Field

Consider an object moving in a plane with a horizontal x-axis and a vertical y-axis, subject only to the force of gravity. Given the initial velocity $\mathbf{v}(0) = \langle u_0, v_0 \rangle$ and the initial position $\mathbf{r}(0) = \langle x_0, y_0 \rangle$, the velocity of the object, for $t \geq 0$, is

$$\mathbf{v}(t) = \langle x'(t), y'(t) \rangle = \langle u_0, -gt + v_0 \rangle$$

and the position is

$$\mathbf{r}(t) = \langle x(t), y(t) \rangle = \left\langle u_0 t + x_0, -\frac{1}{2} gt^2 + v_0 t + y_0 \right\rangle.$$

EXAMPLE 4   Flight of a baseball A baseball is hit from 3 ft above home plate with an initial velocity in ft/s of $\mathbf{v}(0) = \langle u_0, v_0 \rangle = \langle 80, 80 \rangle$. Neglect all forces other than gravity.

a. Find the position and velocity of the ball between the time it is hit and the time it first hits the ground.

b. Show that the trajectory of the ball is a segment of a parabola.

c. Assuming a flat playing field, how far does the ball travel horizontally? Plot the trajectory of the ball.

d. What is the maximum height of the ball?

e. Does the ball clear a 20-ft fence that is 380 ft from home plate (directly under the path of the ball)?

SOLUTION Assume the origin is located at home plate. Because distances are measured in feet, we use $g = 32 \text{ ft/s}^2$.

a. Substituting $x_0 = 0$ and $y_0 = 3$ into the equation for $\mathbf{r}$, the position of the ball is

$$\mathbf{r}(t) = \langle x(t), y(t) \rangle = \langle 80t, -16t^2 + 80t + 3 \rangle, \quad \text{for } t \geq 0. \tag{3}$$

We then compute $\mathbf{v}(t) = \mathbf{r}'(t) = \langle 80, -32t + 80 \rangle$.

b. Equation (3) says that the horizontal position is $x = 80t$ and the vertical position is $y = -16t^2 + 80t + 3$. Substituting $t = x/80$ into the equation for y gives

$$y = -16 \left(\frac{x}{80} \right)^2 + x + 3 = -\frac{x^2}{400} + x + 3,$$

which is an equation of a parabola.

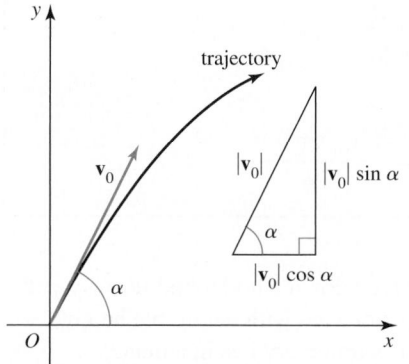

Figure 14.21

> The equation in Example 4c can be solved using the quadratic formula or a root-finder on a calculator.

QUICK CHECK 4 Write the functions $x(t)$ and $y(t)$ in Example 4 in the case that $x_0 = 0$, $y_0 = 2$, $u_0 = 100$, and $v_0 = 60$. ◀

c. The ball lands on the ground at the value of $t > 0$ at which $y = 0$. Solving $y(t) = -16t^2 + 80t + 3 = 0$, we find that $t \approx -0.04$ and $t \approx 5.04$ s. The first root is not relevant for the problem at hand, so we conclude that the ball lands when $t \approx 5.04$ s. The horizontal distance traveled by the ball is $x(5.04) \approx 403$ ft. The path of the ball in the xy-coordinate system on the time interval $[0, 5.04]$ is shown in Figure 14.21.

d. The ball reaches its maximum height at the time its vertical velocity is zero. Solving $y'(t) = -32t + 80 = 0$, we find that $t = 2.5$ s. The height at that time is $y(2.5) = 103$ ft.

e. The ball reaches a horizontal distance of 380 ft (the distance to the fence) when $x(t) = 80t = 380$. Solving for t, we find that $t = 4.75$ s. The height of the ball at that time is $y(4.75) = 22$ ft. So, indeed, the ball clears a 20-ft fence.

Related Exercises 41, 43 ◀

Range, Time of Flight, Maximum Height   Having solved one specific motion problem, we can make some general observations about two-dimensional projectile motion in a gravitational field. Assume the motion of an object begins at the origin; that is, $x_0 = y_0 = 0$. Also assume the object is launched at an angle of α ($0 \leq \alpha \leq \pi/2$) above the horizontal with an initial speed $|\mathbf{v}_0|$ (Figure 14.22). This means that the initial velocity is

$$\langle u_0, v_0 \rangle = \langle |\mathbf{v}_0| \cos \alpha, |\mathbf{v}_0| \sin \alpha \rangle.$$

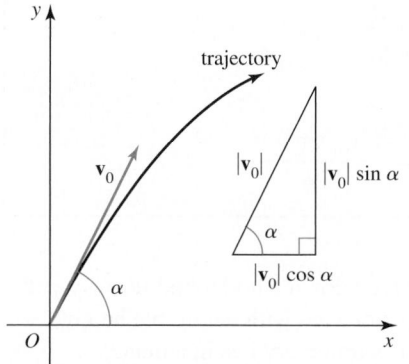

Figure 14.22

Substituting these values into the general expressions for the velocity and position, we find that the velocity of the object is

$$\mathbf{v}(t) = \langle u_0, -gt + v_0 \rangle = \langle |\mathbf{v}_0| \cos \alpha, -gt + |\mathbf{v}_0| \sin \alpha \rangle.$$

The position of the object (with $x_0 = y_0 = 0$) is

$$\mathbf{r}(t) = \langle x(t), y(t) \rangle = \langle (|\mathbf{v}_0| \cos \alpha) t, -gt^2/2 + (|\mathbf{v}_0| \sin \alpha) t \rangle.$$

Notice that the motion is determined entirely by the parameters $|\mathbf{v}_0|$ and α. Several general conclusions now follow.

> The other root of the equation $y(t) = 0$ is $t = 0$, the time the object leaves the ground.

1. Assuming the object is launched from the origin over horizontal ground, it returns to the ground when $y(t) = -gt^2/2 + (|\mathbf{v}_0| \sin \alpha)t = 0$. Solving for t, the **time of flight** is $T = 2|\mathbf{v}_0|(\sin \alpha)/g$.

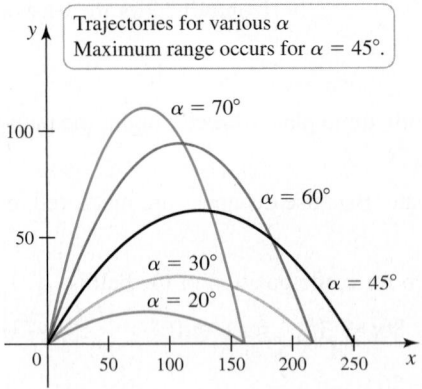

Figure 14.23

QUICK CHECK 5 Show that the range attained with an angle α equals the range attained with the angle $\pi/2 - \alpha$. ◄

2. The **range** of the object, which is the horizontal distance it travels, is the x-coordinate of the trajectory when $t = T$:

$$
\begin{aligned}
x(T) &= (|\mathbf{v}_0| \cos \alpha)T \\
&= (|\mathbf{v}_0| \cos \alpha)\frac{2|\mathbf{v}_0| \sin \alpha}{g} \qquad \text{Substitute for } T. \\
&= \frac{2|\mathbf{v}_0|^2 \sin \alpha \cos \alpha}{g} \qquad \text{Simplify.} \\
&= \frac{|\mathbf{v}_0|^2 \sin 2\alpha}{g}. \qquad 2 \sin \alpha \cos \alpha = \sin 2\alpha
\end{aligned}
$$

Note that on the interval $0 \le \alpha \le \pi/2$, $\sin 2\alpha$ has a maximum value of 1 when $\alpha = \pi/4$, so the maximum range is $|\mathbf{v}_0|^2/g$. In other words, in an ideal world, firing an object from the ground at an angle of $\pi/4$ (45°) maximizes its range. Notice that the ranges obtained with the angles α and $\pi/2 - \alpha$ are equal (**Figure 14.23**).

3. The maximum height of the object is reached when the vertical velocity is zero, or when $y'(t) = -gt + |\mathbf{v}_0| \sin \alpha = 0$. Solving for t, the maximum height is reached at $t = |\mathbf{v}_0|(\sin \alpha)/g = T/2$, which is half the time of flight. The object spends equal amounts of time ascending and descending. The maximum height is

$$
y\left(\frac{T}{2}\right) = \frac{(|\mathbf{v}_0| \sin \alpha)^2}{2g}.
$$

4. Finally, by eliminating t from the equations for $x(t)$ and $y(t)$, it can be shown (Exercise 72) that the trajectory of the object is a segment of a parabola.

> ▶ Use caution with the formulas in the summary box: They are applicable only when the initial position of the object is the origin. Exercise 73 addresses the case where the initial position of the object is $\langle 0, y_0 \rangle$.

SUMMARY   Two-Dimensional Motion

Assume an object traveling over horizontal ground, acted on only by the gravitational force, has an initial position $\langle x_0, y_0 \rangle = \langle 0, 0 \rangle$ and initial velocity $\langle u_0, v_0 \rangle = \langle |\mathbf{v}_0| \cos \alpha, |\mathbf{v}_0| \sin \alpha \rangle$. The trajectory, which is a segment of a parabola, has the following properties.

$$
\text{time of flight} = T = \frac{2|\mathbf{v}_0| \sin \alpha}{g}
$$

$$
\text{range} = \frac{|\mathbf{v}_0|^2 \sin 2\alpha}{g}
$$

$$
\text{maximum height} = y\left(\frac{T}{2}\right) = \frac{(|\mathbf{v}_0| \sin \alpha)^2}{2g}
$$

EXAMPLE 5   Flight of a golf ball A golf ball is driven down a horizontal fairway with an initial speed of 55 m/s at an initial angle of 25° (from a tee with negligible height). Neglect all forces except gravity, and assume the ball's trajectory lies in a plane.

a. When the ball first touches the ground, how far has it traveled horizontally and how long has it been in the air?

b. What is the maximum height of the ball?

c. At what angles should the ball be hit to reach a green that is 300 m from the tee?

SOLUTION

a. Using the range formula with $\alpha = 25°$ and $|\mathbf{v}_0| = 55$ m/s, the ball travels

$$
\frac{|\mathbf{v}_0|^2 \sin 2\alpha}{g} = \frac{(55 \text{ m/s})^2 \sin 50°}{9.8 \text{ m/s}^2} \approx 236 \text{ m}.
$$

The time of the flight is

$$T = \frac{2|\mathbf{v}_0| \sin \alpha}{g} = \frac{2(55 \text{ m/s}) \sin 25°}{9.8 \text{ m/s}^2} \approx 4.7 \text{ s}.$$

b. The maximum height of the ball is

$$\frac{(|\mathbf{v}_0| \sin \alpha)^2}{2g} = \frac{((55 \text{ m/s})(\sin 25°))^2}{2(9.8 \text{ m/s}^2)} \approx 27.6 \text{ m}.$$

c. Letting R denote the range and solving the range formula for $\sin 2\alpha$, we find that $\sin 2\alpha = Rg/|\mathbf{v}_0|^2$. For a range of $R = 300$ m and an initial speed of $|\mathbf{v}_0| = 55$ m/s, the required angle satisfies

$$\sin 2\alpha = \frac{Rg}{|\mathbf{v}_0|^2} = \frac{(300 \text{ m})(9.8 \text{ m/s}^2)}{(55 \text{ m/s})^2} \approx 0.972.$$

For the ball to travel a horizontal distance of exactly 300 m, the required angles are $\alpha = \frac{1}{2} \sin^{-1} 0.972 \approx 38.2°$ or $51.8°$.

Related Exercises 42, 45 ◄

Three-Dimensional Motion

To solve three-dimensional motion problems, we adopt a coordinate system in which the x- and y-axes point in two perpendicular horizontal directions (for example, east and north), while the positive z-axis points vertically upward (**Figure 14.24**). Newton's Second Law now has three components and appears in the form

$$m\mathbf{a}(t) = \langle mx''(t), my''(t), mz''(t) \rangle = \mathbf{F}.$$

If only the gravitational force is present (now in the negative z-direction), then the force vector is $\mathbf{F} = \langle 0, 0, -mg \rangle$; the equation of motion is then $\mathbf{a}(t) = \langle 0, 0, -g \rangle$. Other effects, such as crosswinds, spins, or slices, can be modeled by including other force components.

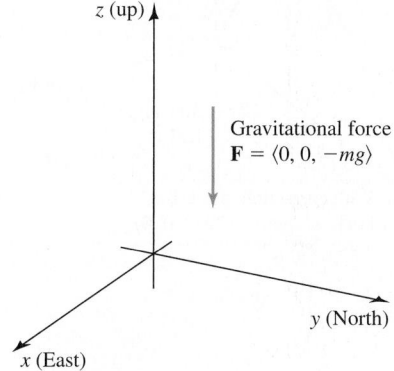

z (up)

Gravitational force
$\mathbf{F} = \langle 0, 0, -mg \rangle$

y (North)

x (East)

Figure 14.24

EXAMPLE 6   Projectile motion A small projectile is fired to the east over horizontal ground with an initial speed of $|\mathbf{v}_0| = 300$ m/s at an angle of $\alpha = 30°$ above the horizontal. A crosswind blows from south to north, producing an acceleration of the projectile of 0.36 m/s^2 to the north.

a. Where does the projectile land? How far does it land from its launch site?

b. In order to correct for the crosswind and make the projectile land due east of the launch site, at what angle from due east must the projectile be fired? Assume the initial speed $|\mathbf{v}_0| = 300$ m/s and the angle of elevation $\alpha = 30°$ are the same as in part (a).

SOLUTION

a. Letting $g = 9.8 \text{ m/s}^2$, the equations of motion are $\mathbf{a}(t) = \mathbf{v}'(t) = \langle 0, 0.36, -9.8 \rangle$. Proceeding as in the two-dimensional case, the indefinite integral of the acceleration is the velocity function

$$\mathbf{v}(t) = \langle 0, 0.36t, -9.8t \rangle + \mathbf{C},$$

where $\mathbf{C}$ is an arbitrary constant vector. With an initial speed $|\mathbf{v}_0| = 300$ m/s and an angle of elevation of $\alpha = 30°$ (**Figure 14.25a**), the initial velocity is

$$\mathbf{v}(0) = \langle 300 \cos 30°, 0, 300 \sin 30° \rangle = \langle 150\sqrt{3}, 0, 150 \rangle.$$

Substituting $t = 0$ and using the initial condition, we find that $\mathbf{C} = \langle 150\sqrt{3}, 0, 150 \rangle$. Therefore, the velocity function is

$$\mathbf{v}(t) = \langle 150\sqrt{3}, 0.36t, -9.8t + 150 \rangle.$$

Integrating the velocity function produces the position function

$$\mathbf{r}(t) = \langle 150\sqrt{3}t, 0.18t^2, -4.9t^2 + 150t \rangle + \mathbf{C}.$$

Using the initial condition $\mathbf{r}(0) = \langle 0, 0, 0 \rangle$, we find that $\mathbf{C} = \langle 0, 0, 0 \rangle$, and the position function is

$$\mathbf{r}(t) = \langle x(t), y(t), z(t) \rangle = \langle 150\sqrt{3}t, 0.18t^2, -4.9t^2 + 150t \rangle.$$

The projectile lands when $z(t) = -4.9t^2 + 150t = 0$. Solving for t, the positive root, which gives the time of flight, is $T = 150/4.9 \approx 30.6$ s. The x- and y-coordinates at that time are

$$x(T) \approx 7953 \text{ m} \quad \text{and} \quad y(T) \approx 169 \text{ m}.$$

Therefore, the projectile lands approximately 7953 m east and 169 m north of the firing site. Because the projectile started at $(0, 0, 0)$, it traveled a horizontal distance of $\sqrt{7953^2 + 169^2} \approx 7955$ m (Figure 14.25a).

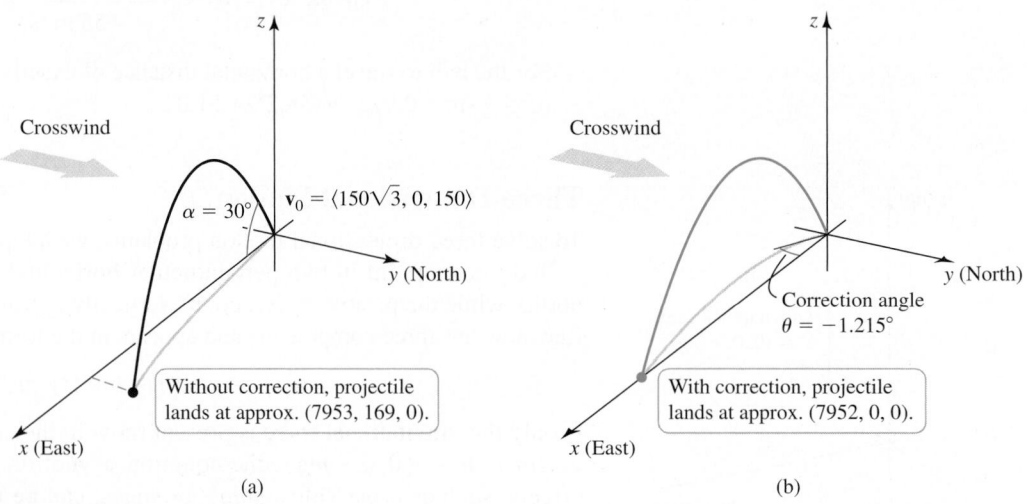

Figure 14.25

b. Keeping the initial speed of the projectile equal to $|\mathbf{v}_0| = 300$ m/s, we decompose the horizontal component of the speed, $150\sqrt{3}$ m/s, into an east component, $u_0 = 150\sqrt{3} \cos \theta$, and a north component, $v_0 = 150\sqrt{3} \sin \theta$, where θ is the angle relative to due east; we must determine the correction angle θ (Figure 14.25b). The x- and y-components of the position are

$$x(t) = (150\sqrt{3} \cos \theta)t \quad \text{and} \quad y(t) = 0.18t^2 + (150\sqrt{3} \sin \theta)t.$$

These changes in the initial velocity affect the x- and y-equations, but not the z-equation. Therefore, the time of flight is still $T = 150/4.9 \approx 30.6$ s. The aim is to choose θ so that the projectile lands on the x-axis (due east from the launch site), which means $y(T) = 0$. Solving

$$y(T) = 0.18T^2 + (150\sqrt{3} \sin \theta)T = 0,$$

with $T = 150/4.9$, we find that $\sin \theta \approx -0.0212$; therefore, $\theta \approx -0.0212$ rad $\approx -1.215°$. In other words, the projectile must be fired at a horizontal angle of $1.215°$ to the *south* of east to correct for the northerly crosswind (Figure 14.25b). The landing location of the projectile is $x(T) \approx 7952$ m and $y(T) = 0$.

Related Exercises 52–53 ◄

SECTION 14.3 EXERCISES

Getting Started

1. Given the position function $\mathbf{r}$ of a moving object, explain how to find the velocity, speed, and acceleration of the object.

2. What is the relationship between the position and velocity vectors for motion on a circle?

3. Write Newton's Second Law of Motion in vector form.

4. Write Newton's Second Law of Motion for three-dimensional motion with only the gravitational force (acting in the z-direction).

5. Given the acceleration of an object and its initial velocity, how do you find the velocity of the object, for $t \geq 0$?

6. Given the velocity of an object and its initial position, how do you find the position of the object, for $t \geq 0$?

7. The velocity of a moving object, for $t \geq 0$, is
$\mathbf{r}'(t) = \langle 60, 96 - 32t \rangle$ ft/s.

   a. When is the vertical component of velocity of the object equal to 0?

   b. Find $\mathbf{r}(t)$ if $\mathbf{r}(0) = \langle 0, 3 \rangle$.

8. A baseball is hit 2 feet above home plate, and the position of the ball t seconds later is $\mathbf{r}(t) = \langle 40t, -16t^2 + 31t + 2 \rangle$ ft. Find each of the following values.

   a. The time of flight of the baseball

   b. The range of the baseball

Practice Exercises

9–20. Velocity and acceleration from position *Consider the following position functions.*

a. Find the velocity and speed of the object.

b. Find the acceleration of the object.

9. $\mathbf{r}(t) = \langle 3t^2 + 1, 4t^2 + 3 \rangle$, for $t \geq 0$

10. $\mathbf{r}(t) = \left\langle \frac{5}{2}t^2 + 3, 6t^2 + 10 \right\rangle$, for $t \geq 0$

11. $\mathbf{r}(t) = \langle 2 + 2t, 1 - 4t \rangle$, for $t \geq 0$

12. $\mathbf{r}(t) = \langle 1 - t^2, 3 + 2t^3 \rangle$, for $t \geq 0$

13. $\mathbf{r}(t) = \langle 8 \sin t, 8 \cos t \rangle$, for $0 \leq t \leq 2\pi$

14. $\mathbf{r}(t) = \langle 3 \cos t, 4 \sin t \rangle$, for $0 \leq t \leq 2\pi$

15. $\mathbf{r}(t) = \left\langle t^2 + 3, t^2 + 10, \frac{1}{2}t^2 \right\rangle$, for $t \geq 0$

16. $\mathbf{r}(t) = \langle 2e^{2t} + 1, e^{2t} - 1, 2e^{2t} - 10 \rangle$, for $t \geq 0$

17. $\mathbf{r}(t) = \langle 3 + t, 2 - 4t, 1 + 6t \rangle$, for $t \geq 0$

18. $\mathbf{r}(t) = \langle 3 \sin t, 5 \cos t, 4 \sin t \rangle$, for $0 \leq t \leq 2\pi$

19. $\mathbf{r}(t) = \langle 1, t^2, e^{-t} \rangle$, for $t \geq 0$

20. $\mathbf{r}(t) = \langle 13 \cos 2t, 12 \sin t\, 2t, 5 \sin 2t \rangle$, for $0 \leq t \leq \pi$

T 21–26. Comparing trajectories *Consider the following position functions $\mathbf{r}$ and $\mathbf{R}$ for two objects.*

a. Find the interval $[c, d]$ over which the $\mathbf{R}$ trajectory is the same as the $\mathbf{r}$ trajectory over $[a, b]$.

b. Find the velocity for both objects.

c. Graph the speed of the two objects over the intervals $[a, b]$ and $[c, d]$, respectively.

21. $\mathbf{r}(t) = \langle t, t^2 \rangle$, $[a, b] = [0, 2]$; $\mathbf{R}(t) = \langle 2t, 4t^2 \rangle$ on $[c, d]$

22. $\mathbf{r}(t) = \langle 1 + 3t, 2 + 4t \rangle$, $[a, b] = [0, 6]$; $\mathbf{R}(t) = \langle 1 + 9t, 2 + 12t \rangle$ on $[c, d]$

23. $\mathbf{r}(t) = \langle \cos t, 4 \sin t \rangle$, $[a, b] = [0, 2\pi]$; $\mathbf{R}(t) = \langle \cos 3t, 4 \sin 3t \rangle$ on $[c, d]$

24. $\mathbf{r}(t) = \langle 2 - e^t, 4 - e^{-t} \rangle$, $[a, b] = [0, \ln 10]$; $\mathbf{R}(t) = \langle 2 - t, 4 - 1/t \rangle$ on $[c, d]$

25. $\mathbf{r}(t) = \langle 4 + t^2, 3 - 2t^4, 1 + 3t^6 \rangle$, $[a, b] = [0, 6]$; $\mathbf{R}(t) = \langle 4 + \ln t, 3 - 2 \ln^2 t, 1 + 3 \ln^3 t \rangle$ on $[c, d]$
For graphing, let $c = 1$ and $d = 20$.

26. $\mathbf{r}(t) = \langle 2 \cos 2t, \sqrt{2} \sin 2t, \sqrt{2} \sin 2t \rangle$, $[a, b] = [0, \pi]$; $\mathbf{R}(t) = \langle 2 \cos 4t, \sqrt{2} \sin 4t, \sqrt{2} \sin 4t \rangle$ on $[c, d]$

T 27–28. Carnival rides

27. Consider a carnival ride where Andrea is at point P that moves counterclockwise around a circle centered at C while the arm, represented by the line segment from the origin O to point C, moves counterclockwise about the origin (see figure). Andrea's position (in feet) at time t (in seconds) is

$$\mathbf{r}(t) = \langle 20 \cos t + 10 \cos 5t, 20 \sin t + 10 \sin 5t \rangle.$$

   a. Plot a graph of $\mathbf{r}(t)$, for $0 \leq t \leq 2\pi$.

   b. Find the velocity $\mathbf{v}(t)$.

   c. Show that the speed $|\mathbf{v}(t)| = v(t) = 10\sqrt{29 + 20 \cos 4t}$ and plot the speed, for $0 \leq t \leq 2\pi$. (*Hint:* Use the identity $\sin mx \sin nx + \cos mx \cos nx = \cos((m - n)x)$.)

   d. Determine Andrea's maximum and minimum speeds.

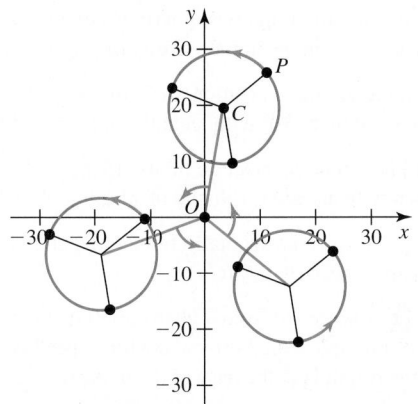

T 28. Suppose the carnival ride in Exercise 27 is modified so that Andrea's position P (in ft) at time t (in s) is

$$\mathbf{r}(t) = \langle 20 \cos t + 10 \cos 5t, 20 \sin t + 10 \sin 5t, 5 \sin 2t \rangle.$$

   a. Describe how this carnival ride differs from the ride in Exercise 27.

   b. Find the speed function $|\mathbf{v}(t)| = v(t)$ and plot its graph.

   c. Find Andrea's maximum and minimum speeds.

29–32. Trajectories on circles and spheres *Determine whether the following trajectories lie on either a circle in $\mathbb{R}^2$ or a sphere in $\mathbb{R}^3$ centered at the origin. If so, find the radius of the circle or sphere, and show that the position vector and the velocity vector are everywhere orthogonal.*

29. $\mathbf{r}(t) = \langle 8 \cos 2t, 8 \sin 2t \rangle$, for $0 \leq t \leq \pi$

30. $\mathbf{r}(t) = \langle 4 \sin t, 2 \cos t \rangle$, for $0 \leq t \leq 2\pi$

31. $\mathbf{r}(t) = \langle \sin t + \sqrt{3} \cos t, \sqrt{3} \sin t - \cos t \rangle$, for $0 \leq t \leq 2\pi$

32. $\mathbf{r}(t) = \langle 3 \sin t, 5 \cos t, 4 \sin t \rangle$, for $0 \leq t \leq 2\pi$

33–34. Path on a sphere *Show that the following trajectories lie on a sphere centered at the origin, and find the radius of the sphere.*

33. $\mathbf{r}(t) = \left\langle \dfrac{5 \sin t}{\sqrt{1 + \sin^2 2t}}, \dfrac{5 \cos t}{\sqrt{1 + \sin^2 2t}}, \dfrac{5 \sin 2t}{\sqrt{1 + \sin^2 2t}} \right\rangle$,
for $0 \leq t \leq 2\pi$

34. $\mathbf{r}(t) = \left\langle \dfrac{4 \cos t}{\sqrt{4 + t^2}}, \dfrac{2t}{\sqrt{4 + t^2}}, \dfrac{4 \sin t}{\sqrt{4 + t^2}} \right\rangle$, for $0 \leq t \leq 4\pi$

35–40. Solving equations of motion *Given an acceleration vector, initial velocity $\langle u_0, v_0 \rangle$, and initial position $\langle x_0, y_0 \rangle$, find the velocity and position vectors for $t \geq 0$.*

35. $\mathbf{a}(t) = \langle 0, 1 \rangle$, $\langle u_0, v_0 \rangle = \langle 2, 3 \rangle$, $\langle x_0, y_0 \rangle = \langle 0, 0 \rangle$

36. $\mathbf{a}(t) = \langle 1, 2 \rangle$, $\langle u_0, v_0 \rangle = \langle 1, 1 \rangle$, $\langle x_0, y_0 \rangle = \langle 2, 3 \rangle$

37. $\mathbf{a}(t) = \langle 0, 10 \rangle$, $\langle u_0, v_0 \rangle = \langle 0, 5 \rangle$, $\langle x_0, y_0 \rangle = \langle 1, -1 \rangle$

38. $\mathbf{a}(t) = \langle 1, t \rangle$, $\langle u_0, v_0 \rangle = \langle 2, -1 \rangle$, $\langle x_0, y_0 \rangle = \langle 0, 8 \rangle$

39. $\mathbf{a}(t) = \langle \cos t, 2 \sin t \rangle$, $\langle u_0, v_0 \rangle = \langle 0, 1 \rangle$, $\langle x_0, y_0 \rangle = \langle 1, 0 \rangle$

40. $\mathbf{a}(t) = \langle e^{-t}, 1 \rangle$, $\langle u_0, v_0 \rangle = \langle 1, 0 \rangle$, $\langle x_0, y_0 \rangle = \langle 0, 0 \rangle$

■ 41–46. Two-dimensional motion *Consider the motion of the following objects. Assume the x-axis is horizontal, the positive y-axis is vertical, the ground is horizontal, and only the gravitational force acts on the object.*

a. *Find the velocity and position vectors, for $t \geq 0$.*
b. *Graph the trajectory.*
c. *Determine the time of flight and range of the object.*
d. *Determine the maximum height of the object.*

41. A soccer ball has an initial position (in m) of $\langle x_0, y_0 \rangle = \langle 0, 0 \rangle$ when it is kicked with an initial velocity of $\langle u_0, v_0 \rangle = \langle 30, 6 \rangle$ m/s.

42. A golf ball has an initial position (in ft) of $\langle x_0, y_0 \rangle = \langle 0, 0 \rangle$ when it is hit at an angle of $30°$ with an initial speed of 150 ft/s.

43. A baseball has an initial position (in ft) of $\langle x_0, y_0 \rangle = \langle 0, 6 \rangle$ when it is thrown with an initial velocity of $\langle u_0, v_0 \rangle = \langle 80, 10 \rangle$ ft/s.

44. A baseball is thrown horizontally from a height of 10 ft above the ground with a speed of 132 ft/s.

45. A projectile is launched from a platform 20 ft above the ground at an angle of $60°$ above the horizontal with a speed of 250 ft/s. Assume the origin is at the base of the platform.

46. A rock is thrown from the edge of a vertical cliff 40 m above the ground at an angle of $45°$ above the horizontal with a speed of $10\sqrt{2}$ m/s. Assume the origin is at the foot of the cliff.

47–50. Solving equations of motion *Given an acceleration vector, initial velocity $\langle u_0, v_0, w_0 \rangle$, and initial position $\langle x_0, y_0, z_0 \rangle$, find the velocity and position vectors, for $t \geq 0$.*

47. $\mathbf{a}(t) = \langle 0, 0, 10 \rangle$, $\langle u_0, v_0, w_0 \rangle = \langle 1, 5, 0 \rangle$, $\langle x_0, y_0, z_0 \rangle = \langle 0, 5, 0 \rangle$

48. $\mathbf{a}(t) = \langle 1, t, 4t \rangle$, $\langle u_0, v_0, w_0 \rangle = \langle 20, 0, 0 \rangle$, $\langle x_0, y_0, z_0 \rangle = \langle 0, 0, 0 \rangle$

49. $\mathbf{a}(t) = \langle \sin t, \cos t, 1 \rangle$, $\langle u_0, v_0, w_0 \rangle = \langle 0, 2, 0 \rangle$, $\langle x_0, y_0, z_0 \rangle = \langle 0, 0, 0 \rangle$

50. $\mathbf{a}(t) = \langle t, e^{-t}, 1 \rangle$, $\langle u_0, v_0, w_0 \rangle = \langle 0, 0, 1 \rangle$, $\langle x_0, y_0, z_0 \rangle = \langle 4, 0, 0 \rangle$

■ 51–56. Three-dimensional motion *Consider the motion of the following objects. Assume the x-axis points east, the y-axis points north, the positive z-axis is vertical and opposite g, the ground is horizontal, and only the gravitational force acts on the object unless otherwise stated.*

a. *Find the velocity and position vectors, for $t \geq 0$.*
b. *Make a sketch of the trajectory.*
c. *Determine the time of flight and range of the object.*
d. *Determine the maximum height of the object.*

51. A bullet is fired from a rifle 1 m above the ground in a northeast direction. The initial velocity of the bullet is $\langle 200, 200, 0 \rangle$ m/s.

52. A golf ball is hit east down a fairway with an initial velocity of $\langle 50, 0, 30 \rangle$ m/s. A crosswind blowing to the south produces an acceleration of the ball of -0.8 m/s².

53. A baseball is hit 3 ft above home plate with an initial velocity of $\langle 60, 80, 80 \rangle$ ft/s. The spin on the baseball produces a horizontal acceleration of the ball of 10 ft/s² in the eastward direction.

54. A baseball is hit 3 ft above home plate with an initial velocity of $\langle 30, 30, 80 \rangle$ ft/s. The spin on the baseball produces a horizontal acceleration of the ball of 5 ft/s² in the northward direction.

55. A small rocket is fired from a launch pad 10 m above the ground with an initial velocity, in m/s, of $\langle 300, 400, 500 \rangle$. A crosswind blowing to the north produces an acceleration of the rocket of 2.5 m/s².

56. A soccer ball is kicked from the point $\langle 0, 0, 0 \rangle$ with an initial velocity of $\langle 0, 80, 80 \rangle$ ft/s. The spin on the ball produces an acceleration of $\langle 1.2, 0, 0 \rangle$ ft/s².

57. Explain why or why not Determine whether the following statements are true and give an explanation or counterexample.

a. If the speed of an object is constant, then its velocity components are constant.

b. The functions $\mathbf{r}(t) = \langle \cos t, \sin t \rangle$ and $\mathbf{R}(t) = \langle \sin t^2, \cos t^2 \rangle$ generate the same set of points, for $t \geq 0$.

c. A velocity vector of variable magnitude cannot have a constant direction.

d. If the acceleration of an object is $\mathbf{a}(t) = \mathbf{0}$, for all $t \geq 0$, then the velocity of the object is constant.

e. If you double the initial speed of a projectile, its range also doubles (assume no forces other than gravity act on the projectile).

f. If you double the initial speed of a projectile, its time of flight also doubles (assume no forces other than gravity).

g. A trajectory with $\mathbf{v}(t) = \mathbf{a}(t) \neq \mathbf{0}$, for all t, is possible.

■ 58–61. Trajectory properties *Find the time of flight, range, and maximum height of the following two-dimensional trajectories, assuming no forces other than gravity. In each case, the initial position is $\langle 0, 0 \rangle$ and the initial velocity is $\mathbf{v}_0 = \langle u_0, v_0 \rangle$.*

58. $\langle u_0, v_0 \rangle = \langle 10, 20 \rangle$ ft/s

59. Initial speed $|\mathbf{v}_0| = 150$ m/s, launch angle $\alpha = 30°$

60. $\langle u_0, v_0 \rangle = \langle 40, 80 \rangle$ m/s

61. Initial speed $|\mathbf{v}_0| = 400$ ft/s, launch angle $\alpha = 60°$

62. Motion on the moon The acceleration due to gravity on the moon is approximately $g/6$ (one-sixth its value on Earth). Compare the time of flight, range, and maximum height of a projectile on the moon with the corresponding values on Earth.

63. Firing angles A projectile is fired over horizontal ground from the origin with an initial speed of 60 m/s. What firing angles produce a range of 300 m?

■ 64. Firing strategies Suppose you wish to fire a projectile over horizontal ground from the origin and attain a range of 1000 m.

a. Sketch a graph of the initial speed required for all firing angles $0 < \alpha < \pi/2$.

b. What firing angle requires the least initial speed?

65. Speed on an ellipse An object moves along an ellipse given by the function $\mathbf{r}(t) = \langle a \cos t, b \sin t \rangle$, for $0 \leq t \leq 2\pi$, where $a > 0$ and $b > 0$.

a. Find the velocity and speed of the object in terms of a and b, for $0 \leq t \leq 2\pi$.

b. With $a = 1$ and $b = 6$, graph the speed function, for $0 \le t \le 2\pi$. Mark the points on the trajectory at which the speed is a minimum and a maximum.

c. Is it true that the object speeds up along the flattest (straightest) parts of the trajectory and slows down where the curves are sharpest?

d. For general a and b, find the ratio of the maximum speed to the minimum speed on the ellipse (in terms of a and b).

Explorations and Challenges

66. Golf shot A golfer stands 390 ft (130 yd) horizontally from the hole and 40 ft below the hole (see figure). Assuming the ball is hit with an initial speed of 150 ft/s, at what angle(s) should it be hit to land in the hole? Assume the path of the ball lies in a plane.

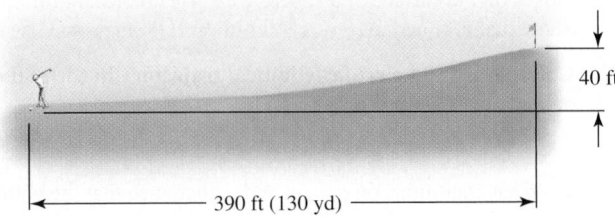

67. Another golf shot A golfer stands 420 ft (140 yd) horizontally from the hole and 50 ft above the hole (see figure). Assuming the ball is hit with an initial speed of 120 ft/s, at what angle(s) should it be hit to land in the hole? Assume the path of the ball lies in a plane.

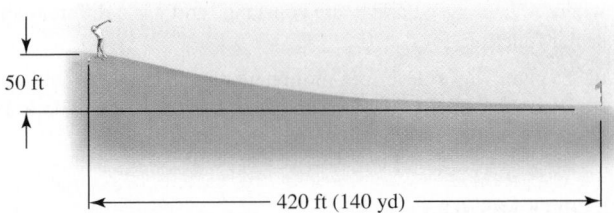

T 68. Initial speed of a golf shot A golfer stands 390 ft horizontally from the hole and 40 ft below the hole (see figure for Exercise 66). If the ball leaves the ground at an initial angle of $45°$ with the horizontal, with what initial speed should it be hit to land in the hole?

T 69. Initial speed of a golf shot A golfer stands 420 ft horizontally from the hole and 50 ft above the hole (see figure for Exercise 67). If the ball leaves the ground at an initial angle of $30°$ with the horizontal, with what initial speed should it be hit to land in the hole?

T 70. Ski jump The lip of a ski jump is 8 m above the outrun that is sloped at an angle of $30°$ to the horizontal (see figure).

a. If the initial velocity of a ski jumper at the lip of the jump is $\langle 40, 0 \rangle$ m/s, what is the length of the jump (distance from the origin to the landing point)? Assume only gravity affects the motion.

b. Assume air resistance produces a constant horizontal acceleration of 0.15 m/s² opposing the motion. What is the length of the jump?

c. Suppose the takeoff ramp is tilted upward at an angle of $\theta°$, so that the skier's initial velocity is $40\langle \cos\theta, \sin\theta \rangle$ m/s. What value of θ maximizes the length of the jump? Express your answer in degrees and neglect air resistance.

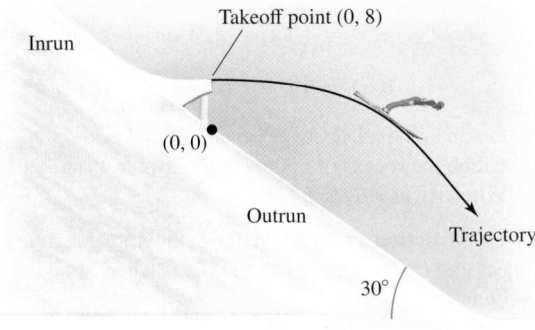

71. Designing a baseball pitch A baseball leaves the hand of a pitcher 6 vertical feet above and 60 horizontal feet from home plate. Assume the coordinate axes are oriented as shown in the figure.

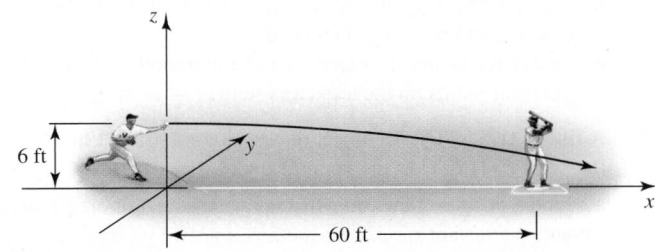

a. Suppose a pitch is thrown with an initial velocity of $\langle 130, 0, -3 \rangle$ ft/s (about 90 mi/hr). In the absence of all forces except gravity, how far above the ground is the ball when it crosses home plate and how long does it take the pitch to arrive?

b. What vertical velocity component should the pitcher use so that the pitch crosses home plate exactly 3 ft above the ground?

c. A simple model to describe the curve of a baseball assumes the spin of the ball produces a constant sideways acceleration (in the y-direction) of c ft/s². Suppose a pitcher throws a curve ball with $c = 8$ ft/s² (one fourth the acceleration of gravity). How far does the ball move in the y-direction by the time it reaches home plate, assuming an initial velocity of $\langle 130, 0, -3 \rangle$ ft/s?

d. In part (c), does the ball curve more in the first half of its trip to the plate or in the second half? How does this fact affect the batter?

e. Suppose the pitcher releases the ball from an initial position of $\langle 0, -3, 6 \rangle$ with initial velocity $\langle 130, 0, -3 \rangle$. What value of the spin parameter c is needed to put the ball over home plate passing through the point $(60, 0, 3)$?

72. Parabolic trajectories Show that the two-dimensional trajectory

$$x(t) = u_0 t + x_0 \text{ and } y(t) = -\frac{gt^2}{2} + v_0 t + y_0, \text{ for } 0 \le t \le T,$$

of an object moving in a gravitational field is a segment of a parabola for some value of $T > 0$. Find T such that $y(T) = 0$.

73. Time of flight, range, height Derive the formulas for time of flight, range, and maximum height in the case that an object is launched from the initial position $\langle 0, y_0 \rangle$ above the horizontal ground with initial velocity $|\mathbf{v}_0|\langle \cos\alpha, \sin\alpha \rangle$.

74. A race Two people travel from $P(4, 0)$ to $Q(-4, 0)$ along the paths given by

$$\mathbf{r}(t) = \left\langle 4 \cos \frac{\pi t}{8}, 4 \sin \frac{\pi t}{8} \right\rangle \quad \text{and}$$

$$\mathbf{R}(t) = \langle 4 - t, (4 - t)^2 - 16 \rangle.$$

a. Graph both paths between P and Q.
b. Graph the speeds of both people between P and Q.
c. Who arrives at Q first?

75. Circular motion Consider an object moving along the circular trajectory $\mathbf{r}(t) = \langle A \cos \omega t, A \sin \omega t \rangle$, where A and ω are constants.

a. Over what time interval $[0, T]$ does the object traverse the circle once?
b. Find the velocity and speed of the object. Is the velocity constant in either direction or magnitude? Is the speed constant?
c. Find the acceleration of the object.
d. How are the position and velocity related? How are the position and acceleration related?
e. Sketch the position, velocity, and acceleration vectors at four different points on the trajectory with $A = \omega = 1$.

76. A linear trajectory An object moves along a straight line from the point $P(1, 2, 4)$ to the point $Q(-6, 8, 10)$.

a. Find a position function $\mathbf{r}$ that describes the motion if it occurs with a constant speed over the time interval $[0, 5]$.
b. Find a position function $\mathbf{r}$ that describes the motion if it occurs with speed e^t.

77. A circular trajectory An object moves clockwise around a circle centered at the origin with radius 5 m beginning at the point $(0, 5)$.

a. Find a position function $\mathbf{r}$ that describes the motion if the object moves with a constant speed, completing 1 lap every 12 s.
b. Find a position function $\mathbf{r}$ that describes the motion if it occurs with speed e^{-t}.

78. A helical trajectory An object moves on the helix $\langle \cos t, \sin t, t \rangle$, for $t \geq 0$.

a. Find a position function $\mathbf{r}$ that describes the motion if it occurs with a constant speed of 10.
b. Find a position function $\mathbf{r}$ that describes the motion if it occurs with speed t.

79. Tilted ellipse Consider the curve $\mathbf{r}(t) = \langle \cos t, \sin t, c \sin t \rangle$, for $0 \leq t \leq 2\pi$, where c is a real number. Assuming the curve lies in a plane, prove that the curve is an ellipse in that plane.

80. Equal area property Consider the ellipse $\mathbf{r}(t) = \langle a \cos t, b \sin t \rangle$, for $0 \leq t \leq 2\pi$, where a and b are real numbers. Let θ be the angle between the position vector and the x-axis.

a. Show that $\tan \theta = \dfrac{b}{a} \tan t$.
b. Find $\theta'(t)$.
c. Recall that the area bounded by the polar curve $r = f(\theta)$ on the interval $[0, \theta]$ is $A(\theta) = \dfrac{1}{2} \int_0^\theta (f(u))^2 \, du$. Letting $f(\theta(t)) = |\mathbf{r}(\theta(t))|$, show that $A'(t) = \dfrac{1}{2} ab$.
d. Conclude that as an object moves around the ellipse, it sweeps out equal areas in equal times.

81. Another property of constant $|\mathbf{r}|$ motion Suppose an object moves on the surface of a sphere with $|\mathbf{r}(t)|$ constant for all t. Show that $\mathbf{r}(t)$ and $\mathbf{a}(t) = \mathbf{r}''(t)$ satisfy $\mathbf{r}(t) \cdot \mathbf{a}(t) = -|\mathbf{v}(t)|^2$.

82. Conditions for a circular/elliptical trajectory in the plane An object moves along a path given by

$$\mathbf{r}(t) = \langle a \cos t + b \sin t, c \cos t + d \sin t \rangle, \text{ for } 0 \leq t \leq 2\pi.$$

a. What conditions on a, b, c, and d guarantee that the path is a circle?
b. What conditions on a, b, c, and d guarantee that the path is an ellipse?

83. Nonuniform straight-line motion Consider the motion of an object given by the position function

$$\mathbf{r}(t) = f(t) \langle a, b, c \rangle + \langle x_0, y_0, z_0 \rangle, \quad \text{for } t \geq 0,$$

where a, b, c, x_0, y_0, and z_0 are constants, and f is a differentiable scalar function, for $t \geq 0$.

a. Explain why $\mathbf{r}$ describes motion along a line.
b. Find the velocity function. In general, is the velocity constant in magnitude or direction along the path?

QUICK CHECK ANSWERS

1. $\mathbf{v}(t) = \langle 1, 2t, 3t^2 \rangle$, $\mathbf{a}(t) = \langle 0, 2, 6t \rangle$
2. $|\mathbf{r}'(t)| = \sqrt{1 + 4t^2 + 9t^4/16}$; $|\mathbf{R}'(t)| = \sqrt{4t^2 + 16t^6 + 9t^{10}/4}$ 3. The curve $\mathbf{R}(t) = 5\mathbf{r}(t)/|\mathbf{r}(t)|$ lies on a sphere of radius 5.
4. $x(t) = 100t$, $y(t) = -16t^2 + 60t + 2$
5. $\sin(2(\pi/2 - \alpha)) = \sin(\pi - 2\alpha) = \sin 2\alpha$ ◄

14.4 Length of Curves

We return now to a recurring theme: determining the arc length of a curve. In Section 6.5, we learned how to find the arc length of curves of the form $y = f(x)$, and in Sections 12.1 and 12.3, we discovered formulas for the arc length of a plane curve described parametrically or described in polar coordinates. In this section, we extend these ideas to handle the arc length of a three-dimensional curve described by a vector function. We also discover how to formulate a parametric description of a curve using arc length as a parameter.

Arc Length

Suppose a curve is described by the vector-valued function $\mathbf{r}(t) = \langle f(t), g(t), h(t) \rangle$, for $a \le t \le b$, where f', g', and h' are continuous on $[a, b]$. In Section 12.1, we showed that the arc length L of the two-dimensional curve $\mathbf{r}(t) = \langle f(t), g(t) \rangle$, for $a \le t \le b$ is given by

$$L = \int_a^b \sqrt{f'(t)^2 + g'(t)^2}\, dt.$$

An analogous arc length formula for three-dimensional curves follows using a similar argument. The length of the curve $\mathbf{r}(t) = \langle f(t), g(t), h(t) \rangle$ on the interval $[a, b]$ is

$$L = \int_a^b \sqrt{f'(t)^2 + g'(t)^2 + h'(t)^2}\, dt.$$

Noting that $\mathbf{r}'(t) = \langle f'(t), g'(t), h'(t) \rangle$, we state the following definition.

> ▶ An important fact is that the arc length of a smooth parameterized curve is independent of the choice of parameter (Exercise 54).
>
> ▶ For curves in the *xy*-plane, we set $h(t) = 0$ in the definition of arc length.

DEFINITION Arc Length for Vector Functions

Consider the parameterized curve $\mathbf{r}(t) = \langle f(t), g(t), h(t) \rangle$, where f', g', and h' are continuous, and the curve is traversed once for $a \le t \le b$. The **arc length** of the curve between $(f(a), g(a), h(a))$ and $(f(b), g(b), h(b))$ is

$$L = \int_a^b \sqrt{f'(t)^2 + g'(t)^2 + h'(t)^2}\, dt = \int_a^b |\mathbf{r}'(t)|\, dt.$$

QUICK CHECK 1 What does the arc length formula give for the length of the line $\mathbf{r}(t) = \langle 2t, t, -2t \rangle$, for $0 \le t \le 3$? ◄

The following application of arc length leads to an integral that is difficult to evaluate exactly.

EXAMPLE 1 Lengths of planetary orbits According to Kepler's first law, the planets revolve about the sun in elliptical orbits. A vector function that describes an ellipse in the *xy*-plane is

$$\mathbf{r}(t) = \langle a \cos t, b \sin t \rangle, \qquad \text{where } 0 \le t \le 2\pi.$$

If $a > b > 0$, then $2a$ is the length of the major axis and $2b$ is the length of the minor axis (**Figure 14.26**). Verify the lengths of the planetary orbits given in Table 14.1. Distances are given in terms of the astronomical unit (AU), which is the length of the semimajor axis of Earth's orbit, or about 93 million miles.

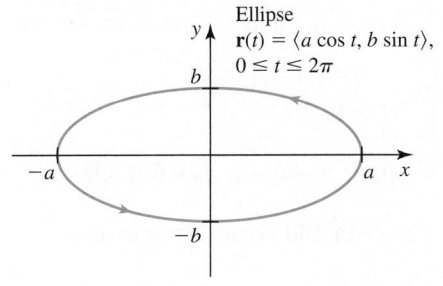

Ellipse
$\mathbf{r}(t) = \langle a \cos t, b \sin t \rangle,$
$0 \le t \le 2\pi$

Figure 14.26

> ▶ The German astronomer and mathematician Johannes Kepler (1571–1630) worked with the meticulously gathered data of Tycho Brahe to formulate three empirical laws obeyed by planets and comets orbiting the sun. The work of Kepler formed the foundation for Newton's laws of gravitation developed 50 years later.
>
> ▶ In September 2006, Pluto joined the ranks of Ceres, Haumea, Makemake, and Eris as one of five dwarf planets in our solar system.

Table 14.1

Planet	Semimajor axis, a (AU)	Semiminor axis, b (AU)	$\alpha = b/a$	Orbit length (AU)
Mercury	0.387	0.379	0.979	2.407
Venus	0.723	0.723	1.000	4.543
Earth	1.000	0.999	0.999	6.280
Mars	1.524	1.517	0.995	9.554
Jupiter	5.203	5.179	0.995	32.616
Saturn	9.539	9.524	0.998	59.888
Uranus	19.182	19.161	0.999	120.458
Neptune	30.058	30.057	1.000	188.857

SOLUTION Using the arc length formula, the length of a general elliptical orbit is

$$L = \int_0^{2\pi} \sqrt{x'(t)^2 + y'(t)^2}\, dt$$

$$= \int_0^{2\pi} \sqrt{(-a \sin t)^2 + (b \cos t)^2}\, dt \quad \text{Substitute for } x'(t) \text{ and } y'(t).$$

$$= \int_0^{2\pi} \sqrt{a^2 \sin^2 t + b^2 \cos^2 t}\, dt. \quad \text{Simplify.}$$

Factoring a^2 out of the square root and letting $\alpha = b/a$, we have

$$L = \int_0^{2\pi} \sqrt{a^2 \left(\sin^2 t + (b/a)^2 \cos^2 t\right)} \, dt \qquad \text{Factor out } a^2.$$

$$= a \int_0^{2\pi} \sqrt{\sin^2 t + \alpha^2 \cos^2 t} \, dt \qquad \text{Let } \alpha = b/a.$$

$$= 4a \int_0^{\pi/2} \sqrt{\sin^2 t + \alpha^2 \cos^2 t} \, dt. \qquad \text{Use symmetry; quarter orbit on } [0, \pi/2].$$

▶ The integral that gives the length of an ellipse is a *complete elliptic integral of the second kind*. Many reference books and software packages provide approximate values of this integral.

Unfortunately, an antiderivative for this integrand cannot be found in terms of elementary functions, so we have two options: This integral is well known and values have been tabulated for various values of α. Alternatively, we may use a calculator to approximate the integral numerically (see Section 8.8). Using numerical integration, the orbit lengths in Table 14.1 are obtained. For example, the length of Mercury's orbit with $a = 0.387$ and $\alpha = 0.979$ is

▶ Although rounded values for α appear in Table 14.1, the calculations in Example 1 were done in full precision and were rounded to three decimal places only in the final step.

$$L = 4a \int_0^{\pi/2} \sqrt{\sin^2 t + \alpha^2 \cos^2 t} \, dt$$

$$= 1.548 \int_0^{\pi/2} \sqrt{\sin^2 t + 0.959 \cos^2 t} \, dt \qquad \text{Simplify.}$$

$$\approx 2.407. \qquad \text{Approximate using calculator.}$$

The fact that α is close to 1 for all the planets means that their orbits are nearly circular. For this reason, the lengths of the orbits shown in the table are nearly equal to $2\pi a$, which is the length of a circular orbit with radius a.

Related Exercises 27–28 ◀

▶ Recall from Chapter 6 that the distance traveled by an object in one dimension is $\int_a^b |\mathbf{v}(t)| \, dt$. The arc length formula generalizes this formula to three dimensions.

If the function $\mathbf{r}(t) = \langle x(t), y(t), z(t) \rangle$ is the position function for a moving object, then the arc length formula has a natural interpretation. Recall that $\mathbf{v}(t) = \mathbf{r}'(t)$ is the velocity of the object and $|\mathbf{v}(t)| = |\mathbf{r}'(t)|$ is the speed of the object. Therefore, the arc length formula becomes

$$L = \int_a^b |\mathbf{r}'(t)| \, dt = \int_a^b |\mathbf{v}(t)| \, dt.$$

This formula is an analog of the familiar *distance = speed × elapsed time* formula.

EXAMPLE 2 Flight of an eagle An eagle rises at a rate of 100 vertical ft/min on a helical path given by

$$\mathbf{r}(t) = \langle 250 \cos t, 250 \sin t, 100t \rangle$$

(Figure 14.27), where $\mathbf{r}$ is measured in feet and t is measured in minutes. How far does it travel in 10 min?

SOLUTION The speed of the eagle is

$$|\mathbf{v}(t)| = \sqrt{x'(t)^2 + y'(t)^2 + z'(t)^2}$$

$$= \sqrt{(-250 \sin t)^2 + (250 \cos t)^2 + 100^2} \qquad \text{Substitute derivatives.}$$

$$= \sqrt{250^2 (\sin^2 t + \cos^2 t) + 100^2} \qquad \text{Combine terms.}$$

$$= \sqrt{250^2 + 100^2} \approx 269. \qquad \sin^2 t + \cos^2 t = 1$$

The constant speed makes the arc length integral easy to evaluate:

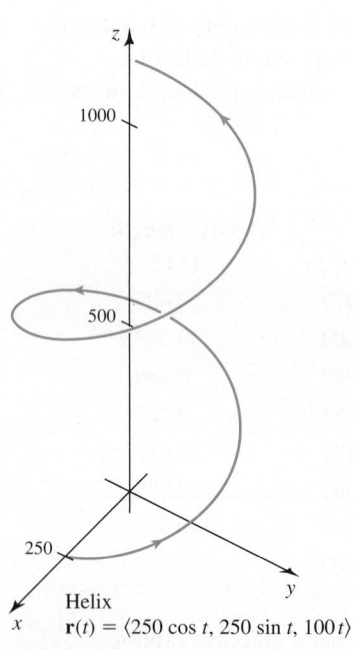

Helix
$\mathbf{r}(t) = \langle 250 \cos t, 250 \sin t, 100t \rangle$

Figure 14.27

$$L = \int_0^{10} |\mathbf{v}(t)| \, dt \approx \int_0^{10} 269 \, dt = 2690.$$

The eagle travels approximately 2690 ft in 10 min.

Related Exercise 25 ◀

▶ The standard parameterization for a helix winding counterclockwise around the z-axis is $\mathbf{r}(t) = \langle a \cos t, a \sin t, bt \rangle$. A helix has the property that its tangent vector makes a constant angle with the axis around which it winds.

Arc Length as a Parameter

Until now, the parameter t used to describe a curve $\mathbf{r}(t) = \langle f(t), g(t), h(t) \rangle$ has been chosen either for convenience or because it represents time in some specified unit. We now introduce the most natural parameter for describing curves; that parameter is *arc length*. Let's see what it means for a curve to be *parameterized by arc length*.

Consider the following two characterizations of the unit circle centered at the origin:

$$\langle \cos t, \sin t \rangle, \text{ for } 0 \le t \le 2\pi \quad \text{and} \quad \langle \cos 2t, \sin 2t \rangle, \text{ for } 0 \le t \le \pi.$$

In the first description, as the parameter t increases from $t = 0$ to $t = 2\pi$, the full circle is generated and the arc length s of the curve also increases from $s = 0$ to $s = 2\pi$. In other words, as the parameter t increases, it measures the arc length of the curve that is generated (**Figure 14.28a**).

In the second description, as t varies from $t = 0$ to $t = \pi$, the full circle is generated and the arc length increases from $s = 0$ to $s = 2\pi$. In this case, the length of the interval in t does not equal the length of the curve generated; therefore, the parameter t does not correspond to arc length (**Figure 14.28b**). In general, there are infinitely many ways to parameterize a given curve; however, for a given initial point and orientation, arc length is the parameter for only one of them.

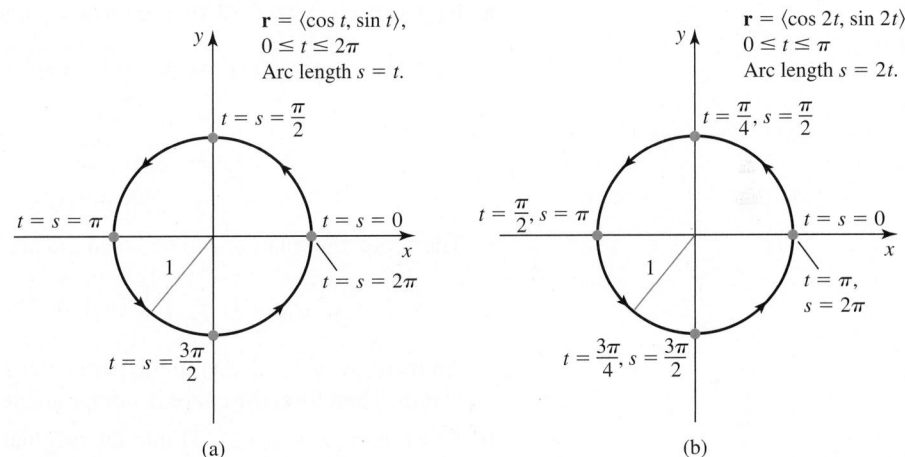

QUICK CHECK 2 Consider the portion of a circle $\mathbf{r}(t) = \langle \cos t, \sin t \rangle$, for $a \le t \le b$. Show that the arc length of the curve is $b - a$. ◄

Figure 14.28

The Arc Length Function   Suppose a smooth curve is represented by the function $\mathbf{r}(t) = \langle f(t), g(t), h(t) \rangle$, for $t \ge a$, where t is a parameter. Notice that as t increases, the length of the curve also increases. Using the arc length formula, the length of the curve from $\mathbf{r}(a)$ to $\mathbf{r}(t)$ is

$$s(t) = \int_a^t \sqrt{f'(u)^2 + g'(u)^2 + h'(u)^2}\, du = \int_a^t |\mathbf{v}(u)|\, du.$$

> ► Notice that t is the independent variable of the function $s(t)$, so a different symbol u is used for the variable of integration. It is common to use s as the arc length function.

This equation gives the relationship between the arc length of a curve and any parameter t used to describe the curve.

An important consequence of this relationship arises if we differentiate both sides with respect to t using the Fundamental Theorem of Calculus:

$$\frac{ds}{dt} = \frac{d}{dt}\left(\int_a^t |\mathbf{v}(u)|\, du \right) = |\mathbf{v}(t)|.$$

Specifically, if t represents time and $\mathbf{r}$ is the position of an object moving on the curve, then the rate of change of the arc length with respect to time is the speed of the object. Notice that if $\mathbf{r}(t)$ describes a smooth curve, then $|\mathbf{v}(t)| \ne 0$; hence $ds/dt > 0$, and s is an increasing function of t—as t increases, the arc length also increases. If $\mathbf{r}(t)$ is a curve on which $|\mathbf{v}(t)| = 1$, then

$$s(t) = \int_a^t |\mathbf{v}(u)|\, du = \int_a^t 1\, du = t - a,$$

which means the parameter t corresponds to arc length. These observations are summarized in the following theorem.

> **THEOREM 14.3 Arc Length as a Function of a Parameter**
> Let $\mathbf{r}(t)$ describe a smooth curve, for $t \geq a$. The arc length is given by
>
> $$s(t) = \int_a^t |\mathbf{v}(u)|\, du,$$
>
> where $|\mathbf{v}| = |\mathbf{r}'|$. Equivalently, $\dfrac{ds}{dt} = |\mathbf{v}(t)|$. If $|\mathbf{v}(t)| = 1$, for all $t \geq a$, then the parameter t corresponds to arc length.

EXAMPLE 3 Arc length parameterization Consider the helix
$\mathbf{r}(t) = \langle 2 \cos t, 2 \sin t, 4t \rangle$, for $t \geq 0$.

a. Find the arc length function $s(t)$.

b. Find another description of the helix that uses arc length as the parameter.

SOLUTION

a. Note that $\mathbf{r}'(t) = \langle -2 \sin t, 2 \cos t, 4 \rangle$ and

$$
\begin{aligned}
|\mathbf{v}(t)| = |\mathbf{r}'(t)| &= \sqrt{(-2 \sin t)^2 + (2 \cos t)^2 + 4^2} \\
&= \sqrt{4(\sin^2 t + \cos^2 t) + 4^2} \qquad \text{Simplify.} \\
&= \sqrt{4 + 4^2} \qquad\qquad\qquad \sin^2 t + \cos^2 t = 1 \\
&= \sqrt{20} = 2\sqrt{5}. \qquad\qquad \text{Simplify.}
\end{aligned}
$$

Therefore, the relationship between the arc length s and the parameter t is

$$s(t) = \int_a^t |\mathbf{v}(u)|\, du = \int_0^t 2\sqrt{5}\, du = 2\sqrt{5}\, t.$$

An increase of $1/(2\sqrt{5})$ in the parameter t corresponds to an increase of 1 in the arc length. Therefore, the curve is not parameterized by arc length.

b. Substituting $t = s/(2\sqrt{5})$ into the original parametric description of the helix, we find that the description with arc length as a parameter is (using a different function name)

$$\mathbf{r}_1(s) = \left\langle 2 \cos\left(\frac{s}{2\sqrt{5}}\right), 2 \sin\left(\frac{s}{2\sqrt{5}}\right), \frac{2s}{\sqrt{5}} \right\rangle, \quad \text{for } s \geq 0.$$

This description has the property that an increment of Δs in the parameter corresponds to an increment of exactly Δs in the arc length.

Related Exercises 37–39 ◄

QUICK CHECK 3 Does the line $\mathbf{r}(t) = \langle t, t, t \rangle$ have arc length as a parameter? Explain. ◄

As you will see in Section 14.5, using arc length as a parameter—when it can be done—generally leads to simplified calculations.

SECTION 14.4 EXERCISES

Getting Started

1. Find the length of the line given by $\mathbf{r}(t) = \langle t, 2t \rangle$, for $a \leq t \leq b$.

2. Explain how to find the length of the curve
$$\mathbf{r}(t) = \langle f(t), g(t), h(t) \rangle, \text{ for } a \leq t \leq b.$$

3. Express the arc length of a curve in terms of the speed of an object moving along the curve.

4. Suppose an object moves in space with the position function $\mathbf{r}(t) = \langle x(t), y(t), z(t) \rangle$. Write the integral that gives the distance it travels between $t = a$ and $t = b$.

5. An object moves on a trajectory given by
$$\mathbf{r}(t) = \langle 10 \cos 2t, 10 \sin 2t \rangle, \text{ for } 0 \leq t \leq \pi.$$
How far does it travel?

6. Use calculus to find the length of the line segment $\mathbf{r}(t) = \langle t, -8t, 4t \rangle$, for $0 \leq t \leq 2$. Verify your answer without using calculus.

7. Explain what it means for a curve to be parameterized by its arc length.

8. Is the curve $\mathbf{r}(t) = \langle \cos t, \sin t \rangle$ parameterized by its arc length? Explain.

Practice Exercises

9–22. Arc length calculations *Find the length of the following two- and three-dimensional curves.*

9. $\mathbf{r}(t) = \langle 3t^2 - 1, 4t^2 + 5 \rangle$, for $0 \le t \le 1$

10. $\mathbf{r}(t) = \langle 3t - 1, 4t + 5, t \rangle$, for $0 \le t \le 1$

11. $\mathbf{r}(t) = \langle 3 \cos t, 3 \sin t \rangle$, for $0 \le t \le \pi$

12. $\mathbf{r}(t) = \langle 4 \cos 3t, 4 \sin 3t \rangle$, for $0 \le t \le 2\pi/3$

13. $\mathbf{r}(t) = \langle \cos t + t \sin t, \sin t - t \cos t \rangle$, for $0 \le t \le \pi/2$

14. $\mathbf{r}(t) = \langle \cos t + \sin t, \cos t - \sin t \rangle$, for $0 \le t \le 2\pi$

15. $\mathbf{r}(t) = \langle 2 + 3t, 1 - 4t, -4 + 3t \rangle$, for $1 \le t \le 6$

16. $\mathbf{r}(t) = \langle 4 \cos t, 4 \sin t, 3t \rangle$, for $0 \le t \le 6\pi$

17. $\mathbf{r}(t) = \langle t, 8 \sin t, 8 \cos t \rangle$, for $0 \le t \le 4\pi$

18. $\mathbf{r}(t) = \left\langle \dfrac{t^2}{2}, \dfrac{(2t+1)^{3/2}}{3} \right\rangle$, for $0 \le t \le 2$

19. $\mathbf{r}(t) = \langle e^{2t}, 2e^{2t} + 5, 2e^{2t} - 20 \rangle$, for $0 \le t \le \ln 2$

20. $\mathbf{r}(t) = \langle t^2, t^3 \rangle$, for $0 \le t \le 4$

21. $\mathbf{r}(t) = \langle \cos^3 t, \sin^3 t \rangle$, for $0 \le t \le \pi/2$

22. $\mathbf{r}(t) = \langle 2 \cos t, 2\sqrt{3} \cos t, 4 \sin t \rangle$, for $0 \le t \le 2\pi$

23–26. Speed and arc length *For the following trajectories, find the speed associated with the trajectory, and then find the length of the trajectory on the given interval.*

23. $\mathbf{r}(t) = \langle 2t^3, -t^3, 5t^3 \rangle$, for $0 \le t \le 4$

24. $\mathbf{r}(t) = \langle 5 \cos t^2, 5 \sin t^2, 12t^2 \rangle$, for $0 \le t \le 2$

25. $\mathbf{r}(t) = \langle 13 \sin 2t, 12 \cos 2t, 5 \cos 2t \rangle$, for $0 \le t \le \pi$

26. $\mathbf{r}(t) = \langle e^t \sin t, e^t \cos t, e^t \rangle$, for $0 \le t \le \ln 2$

T 27. Speed of Earth Verify that the length of one orbit of Earth is approximately 6.280 AU (see Table 14.1). Then determine the average speed of Earth relative to the sun in miles per hour. (*Hint:* It takes Earth 365.25 days to orbit the sun.)

T 28. Speed of Jupiter Verify that the length of one orbit of Jupiter is approximately 32.616 AU (see Table 14.1). Then determine the average speed of Jupiter relative to the sun in miles per hour. (*Hint:* It takes Jupiter 11.8618 Earth years to orbit the sun.)

T 29–32. Arc length approximations *Use a calculator to approximate the length of the following curves. In each case, simplify the arc length integral as much as possible before finding an approximation.*

29. $\mathbf{r}(t) = \langle 2 \cos t, 4 \sin t \rangle$, for $0 \le t \le 2\pi$

30. $\mathbf{r}(t) = \langle 2 \cos t, 4 \sin t, 6 \cos t \rangle$, for $0 \le t \le 2\pi$

31. $\mathbf{r}(t) = \langle t, 4t^2, 10 \rangle$, for $-2 \le t \le 2$

32. $\mathbf{r}(t) = \langle e^t, 2e^{-t}, t \rangle$, for $0 \le t \le \ln 3$

33–42. Arc length parametrization *Determine whether the following curves use arc length as a parameter. If not, find a description that uses arc length as a parameter.*

33. $\mathbf{r}(t) = \langle 1, \sin t, \cos t \rangle$, for $t \ge 1$

34. $\mathbf{r}(t) = \left\langle \dfrac{t}{\sqrt{3}}, \dfrac{t}{\sqrt{3}}, \dfrac{t}{\sqrt{3}} \right\rangle$, for $0 \le t \le 10$

35. $\mathbf{r}(t) = \langle t, 2t \rangle$, for $0 \le t \le 3$

36. $\mathbf{r}(t) = \langle t + 1, 2t - 3, 6t \rangle$, for $0 \le t \le 10$

37. $\mathbf{r}(t) = \langle 2 \cos t, 2 \sin t \rangle$, for $0 \le t \le 2\pi$

38. $\mathbf{r}(t) = \langle 17 \cos t, 15 \sin t, 8 \sin t \rangle$, for $0 \le t \le \pi$

39. $\mathbf{r}(t) = \langle \cos t^2, \sin t^2 \rangle$, for $0 \le t \le \sqrt{\pi}$

40. $\mathbf{r}(t) = \langle t^2, 2t^2, 4t^2 \rangle$, for $1 \le t \le 4$

41. $\mathbf{r}(t) = \langle e^t, e^t, e^t \rangle$, for $t \ge 0$

42. $\mathbf{r}(t) = \left\langle \dfrac{\cos t}{\sqrt{2}}, \dfrac{\cos t}{\sqrt{2}}, \sin t \right\rangle$, for $0 \le t \le 10$

43. **Explain why or why not** Determine whether the following statements are true and give an explanation or counterexample.

   a. If an object moves on a trajectory with constant speed S over a time interval $a \le t \le b$, then the length of the trajectory is $S(b - a)$.

   b. The curves defined by
   $$\mathbf{r}(t) = \langle f(t), g(t) \rangle \quad \text{and} \quad \mathbf{R}(t) = \langle g(t), f(t) \rangle$$
   have the same length over the interval $[a, b]$.

   c. The curve $\mathbf{r}(t) = \langle f(t), g(t) \rangle$, for $0 \le a \le t \le b$, and the curve $\mathbf{R}(t) = \langle f(t^2), g(t^2) \rangle$, for $\sqrt{a} \le t \le \sqrt{b}$, have the same length.

   d. The curve $\mathbf{r}(t) = \langle t, t^2, 3t^2 \rangle$, for $1 \le t \le 4$, is parameterized by arc length.

44. **Length of a line segment** Consider the line segment joining the points $P(x_0, y_0, z_0)$ and $Q(x_1, y_1, z_1)$.

   a. Find a function $\mathbf{r}(t)$ for the segment PQ.

   b. Use the arc length formula to find the length of PQ.

   c. Use geometry (distance formula) to verify the result of part (b).

45. **Tilted circles** Let the curve C be described by
   $$\mathbf{r}(t) = \langle a \cos t, b \sin t, c \sin t \rangle,$$
   where a, b, and c are real positive numbers.

   a. Assume C lies in a plane. Show that C is a circle centered at the origin, provided $a^2 = b^2 + c^2$.

   b. Find the arc length of the circle in part (a).

   c. Assuming C lies in a plane, find the conditions for which $\mathbf{r}(t) = \langle a \cos t + b \sin t, c \cos t + d \sin t, e \cos t + f \sin t \rangle$ describes a circle. Then find its arc length.

46. **A family of arc length integrals** Find the length of the curve $\mathbf{r}(t) = \langle t^m, t^m, t^{3m/2} \rangle$, for $0 \le a \le t \le b$, where m is a real number. Express the result in terms of m, a, and b.

47. **A special case** Suppose a curve is described by
   $$\mathbf{r}(t) = \langle Ah(t), Bh(t) \rangle, \text{ for } a \le t \le b,$$
   where A and B are constants and h has a continuous derivative.

   a. Show that the length of the curve is
   $$\sqrt{A^2 + B^2} \int_a^b |h'(t)| \, dt.$$

   b. Use part (a) to find the length of the curve $x = 2t^3$, $y = 5t^3$, for $0 \le t \le 4$.

   c. Use part (a) to find the length of the curve $x = 4/t$, $y = 10/t$, for $1 \le t \le 8$.

Explorations and Challenges

T 48. Toroidal magnetic field A circle of radius a that is centered at $(A, 0)$ is revolved about the y-axis to create a torus (assume $a < A$). When current flows through a copper wire that is wrapped around this torus, a magnetic field is created and the strength of this field depends on the amount of copper wire used. If the wire is wrapped evenly around the torus a total of k times, the shape of the wire is modeled by the function

$$\mathbf{r}(t) = \langle (A + a\cos kt)\cos t, (A + a\cos kt)\sin t, a\sin kt\rangle,$$

for $0 \le t \le 2\pi$. Determine the amount of copper required if $A = 4$ in, $a = 1$ in, and $k = 35$.

T 49. Projectile trajectories A projectile (such as a baseball or a cannonball) launched from the origin with an initial horizontal velocity u_0 and an initial vertical velocity v_0 moves in a parabolic trajectory given by

$$\mathbf{r}(t) = \left\langle u_0 t, -\frac{1}{2}gt^2 + v_0 t\right\rangle, \quad \text{for } t \ge 0,$$

where air resistance is neglected and $g = 9.8$ m/s^2 is the acceleration due to gravity (see Section 14.3).

a. Let $u_0 = 20$ m/s and $v_0 = 25$ m/s. Assuming the projectile is launched over horizontal ground, at what time does it return to Earth?

b. Find the integral that gives the length of the trajectory from launch to landing.

c. Evaluate the integral in part (b) by first making the change of variables $u = -gt + v_0$. The resulting integral is evaluated either by making a second change of variables or by using a calculator. What is the length of the trajectory?

d. How far does the projectile land from its launch site?

50. Variable speed on a circle Consider a particle that moves in a plane according to the function $\mathbf{r}(t) = \langle \sin t^2, \cos t^2\rangle$ with an initial position $(0, 1)$ at $t = 0$.

a. Describe the path of the particle, including the time required to return to the initial position.

b. What is the length of the path in part (a)?

c. Describe how the motion of this particle differs from the motion described by the equations $x = \sin t$ and $y = \cos t$.

d. Consider the motion described by $x = \sin t^n$ and $y = \cos t^n$, where n is a positive integer. Describe the path of the particle, including the time required to return to the initial position.

e. What is the length of the path in part (d) for any positive integer n?

f. If you were watching a race on a circular path between two runners, one moving according to $x = \sin t$ and $y = \cos t$ and one according to $x = \sin t^2$ and $y = \cos t^2$, who would win and when would one runner pass the other?

51. Arc length parameterization Prove that the line $\mathbf{r}(t) = \langle x_0 + at, y_0 + bt, z_0 + ct\rangle$ is parameterized by arc length, provided $a^2 + b^2 + c^2 = 1$.

52. Arc length parameterization Prove that the curve $\mathbf{r}(t) = \langle a\cos t, b\sin t, c\sin t\rangle$ is parameterized by arc length, provided $a^2 = b^2 + c^2 = 1$.

53. Lengths of related curves Suppose a curve is given by $\mathbf{r}(t) = \langle f(t), g(t)\rangle$, where f' and g' are continuous, for $a \le t \le b$. Assume the curve is traversed once, for $a \le t \le b$, and the length of the curve between $(f(a), g(a))$ and $(f(b), g(b))$ is L. Prove that for any nonzero constant c, the length of the curve defined by $\mathbf{r}(t) = \langle cf(t), cg(t)\rangle$, for $a \le t \le b$, is $|c|L$.

54. Change of variables Consider the parameterized curves $\mathbf{r}(t) = \langle f(t), g(t), h(t)\rangle$ and $\mathbf{R}(t) = \langle f(u(t)), g(u(t)), h(u(t))\rangle$, where f, g, h, and u are continuously differentiable functions and u has an inverse on $[a, b]$.

a. Show that the curve generated by $\mathbf{r}$ on the interval $a \le t \le b$ is the same as the curve generated by $\mathbf{R}$ on $u^{-1}(a) \le t \le u^{-1}(b)$ (or $u^{-1}(b) \le t \le u^{-1}(a)$).

b. Show that the lengths of the two curves are equal. (*Hint:* Use the Chain Rule and a change of variables in the arc length integral for the curve generated by $\mathbf{R}$.)

QUICK CHECK ANSWERS

1. 9 **2.** For $a \le t \le b$, the curve C generated is $(b - a)/2\pi$ of a full circle. Because the full circle has a length of 2π, the curve C has a length of $b - a$. **3.** No. If t increases by 1 unit, the length of the curve increases by $\sqrt{3}$ units. ◄

14.5 Curvature and Normal Vectors

We know how to find tangent vectors and lengths of curves in space, but much more can be said about the shape of such curves. In this section, we introduce several new concepts. *Curvature* measures how *fast* a curve turns at a point, the *normal vector* gives the *direction* in which a curve turns, and the *binormal vector* and the *torsion* describe the twisting of a curve.

Curvature

Imagine driving a car along a winding mountain road. There are two ways to change the velocity of the car (that is, to accelerate). You can change the *speed* of the car or you can change the *direction* of the car. A change of speed is relatively easy to describe, so we postpone that discussion and focus on the change of direction. The rate at which the car changes direction is related to the notion of *curvature*.

Unit Tangent Vector   Recall from Section 14.2 that if $\mathbf{r}(t) = \langle x(t), y(t), z(t) \rangle$ is a smooth oriented curve, then the unit tangent vector at a point is the unit vector that points in the direction of the tangent vector $\mathbf{r}'(t)$; that is,

$$\mathbf{T}(t) = \frac{\mathbf{r}'(t)}{|\mathbf{r}'(t)|} = \frac{\mathbf{v}(t)}{|\mathbf{v}(t)|}.$$

Because $\mathbf{T}$ is a unit vector, its length does not change along the curve. The only way $\mathbf{T}$ can change is through a change in direction.

How quickly does $\mathbf{T}$ change (in direction) as we move along the curve? If a small increment in arc length Δs along the curve results in a large change in the direction of $\mathbf{T}$, the curve is turning quickly over that interval and we say it has a large *curvature* (Figure 14.29a). If a small increment Δs in arc length results in a small change in the direction of $\mathbf{T}$, the curve is turning slowly over that interval and it has a small curvature (Figure 14.29b). The magnitude of the rate at which the direction of $\mathbf{T}$ changes with respect to arc length is the curvature of the curve.

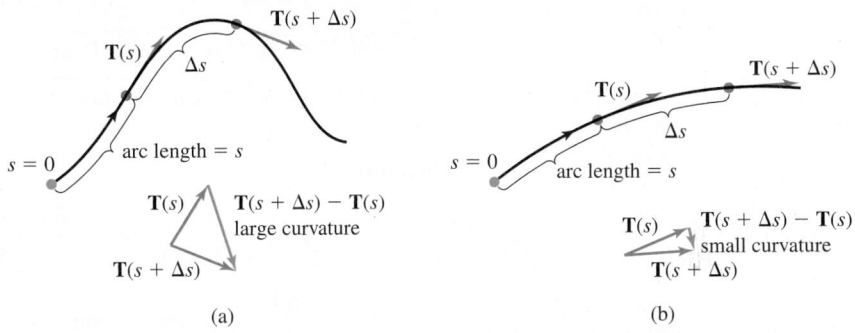

(a)   **(b)**

Figure 14.29

> Recall that the unit tangent vector at a point depends on the orientation of the curve. The curvature does not depend on the orientation of the curve, but it does depend on the shape of the curve. The Greek letter κ (kappa) is used to denote curvature.

DEFINITION   Curvature

Let $\mathbf{r}$ describe a smooth parameterized curve. If s denotes arc length and $\mathbf{T} = \mathbf{r}'/|\mathbf{r}'|$ is the unit tangent vector, the **curvature** is $\kappa(s) = \left| \dfrac{d\mathbf{T}}{ds} \right|$.

Note that κ is a nonnegative scalar-valued function. A large value of κ at a point indicates a tight curve that changes direction quickly. If κ is small, then the curve is relatively flat and its direction changes slowly. The minimum curvature (zero) occurs on a straight line, where the tangent vector never changes direction along the curve.

In order to evaluate $d\mathbf{T}/ds$, a description of the curve in terms of the arc length appears to be needed, but it may be difficult to obtain. A short calculation leads to the first of two practical curvature formulas.

Letting t be an arbitrary parameter, we begin with the Chain Rule and write $\dfrac{d\mathbf{T}}{dt} = \dfrac{d\mathbf{T}}{ds} \cdot \dfrac{ds}{dt}$. After dividing both sides of this equation by $ds/dt = |\mathbf{v}|$, we take absolute values and arrive at

$$\kappa = \left| \frac{d\mathbf{T}}{ds} \right| = \frac{|d\mathbf{T}/dt|}{|ds/dt|} = \frac{1}{|\mathbf{v}|} \left| \frac{d\mathbf{T}}{dt} \right|.$$

This derivation is a proof of the following theorem.

THEOREM 14.4   Curvature Formula

Let $\mathbf{r}(t)$ describe a smooth parameterized curve, where t is any parameter. If $\mathbf{v} = \mathbf{r}'$ is the velocity and $\mathbf{T}$ is the unit tangent vector, then the curvature is

$$\kappa(t) = \frac{1}{|\mathbf{v}|} \left| \frac{d\mathbf{T}}{dt} \right| = \frac{|\mathbf{T}'(t)|}{|\mathbf{r}'(t)|}.$$

EXAMPLE 1   **Lines have zero curvature** Consider the line

$$\mathbf{r}(t) = \langle x_0 + at, y_0 + bt, z_0 + ct \rangle, \text{ for } -\infty < t < \infty.$$

Show that $\kappa = 0$ at all points on the line.

SOLUTION Note that $\mathbf{r}'(t) = \langle a, b, c \rangle$ and $|\mathbf{r}'(t)| = |\mathbf{v}(t)| = \sqrt{a^2 + b^2 + c^2}$. Therefore,

$$\mathbf{T}(t) = \frac{\mathbf{r}'(t)}{|\mathbf{r}'(t)|} = \frac{\langle a, b, c \rangle}{\sqrt{a^2 + b^2 + c^2}}.$$

Because $\mathbf{T}$ is a constant, $\dfrac{d\mathbf{T}}{dt} = \mathbf{0}$; therefore, $\kappa = 0$ at all points of the line.

Related Exercise 11 ◄

EXAMPLE 2   **Circles have constant curvature** Consider the circle $\mathbf{r}(t) = \langle R \cos t, R \sin t \rangle$, for $0 \le t \le 2\pi$, where $R > 0$. Show that $\kappa = 1/R$.

SOLUTION We compute $\mathbf{r}'(t) = \langle -R \sin t, R \cos t \rangle$ and

$$
\begin{aligned}
|\mathbf{v}(t)| = |\mathbf{r}'(t)| &= \sqrt{(-R \sin t)^2 + (R \cos t)^2} \\
&= \sqrt{R^2 (\sin^2 t + \cos^2 t)} \qquad \text{Simplify.} \\
&= R. \qquad\qquad\qquad \sin^2 t + \cos^2 t = 1, R > 0
\end{aligned}
$$

Therefore,

$$\mathbf{T}(t) = \frac{\mathbf{r}'(t)}{|\mathbf{r}'(t)|} = \frac{\langle -R \sin t, R \cos t \rangle}{R} = \langle -\sin t, \cos t \rangle, \text{ and}$$

$$\frac{d\mathbf{T}}{dt} = \langle -\cos t, -\sin t \rangle.$$

> ► The curvature of a curve at a point can also be visualized in terms of a **circle of curvature**, which is a circle of radius R that is tangent to the curve at that point. The curvature at the point is $\kappa = 1/R$. See Exercises 70–73.

Combining these observations, the curvature is

$$\kappa = \frac{1}{|\mathbf{v}|}\left|\frac{d\mathbf{T}}{dt}\right| = \frac{1}{R}|\langle -\cos t, -\sin t \rangle| = \frac{1}{R}\underbrace{\sqrt{\cos^2 t + \sin^2 t}}_{1} = \frac{1}{R}.$$

The curvature of a circle is constant; a circle with a small radius has a large curvature, and vice versa.

Related Exercise 12 ◄

> **QUICK CHECK 1** What is the curvature of the circle $\mathbf{r}(t) = \langle 3 \sin t, 3 \cos t \rangle$? ◄

An Alternative Curvature Formula A second curvature formula, which pertains specifically to trajectories of moving objects, is easier to use in some cases. The calculation is instructive because it relies on many properties of vector functions. In the end, a remarkably simple formula emerges.

Again consider a smooth curve $\mathbf{r}(t) = \langle x(t), y(t), z(t) \rangle$, where $\mathbf{v}(t) = \mathbf{r}'(t)$ and $\mathbf{a}(t) = \mathbf{v}'(t)$ are the velocity and acceleration of an object moving along that curve, respectively. We assume $\mathbf{v}(t) \ne \mathbf{0}$ and $\mathbf{a}(t) \ne \mathbf{0}$. Because $\mathbf{T} = \mathbf{v}/|\mathbf{v}|$, we begin by writing $\mathbf{v} = |\mathbf{v}|\,\mathbf{T}$ and differentiating both sides with respect to t:

$$\mathbf{a} = \frac{d\mathbf{v}}{dt} = \frac{d}{dt}(|\mathbf{v}(t)|\,\mathbf{T}(t)) = \frac{d}{dt}(|\mathbf{v}(t)|)\mathbf{T}(t) + |\mathbf{v}(t)|\frac{d\mathbf{T}}{dt}. \quad \text{Product Rule}$$

We now form $\mathbf{v} \times \mathbf{a}$:

> ► Distributive law for cross products:
> $$\mathbf{w} \times (\mathbf{u} + \mathbf{v}) = (\mathbf{w} \times \mathbf{u}) + (\mathbf{w} \times \mathbf{v})$$
> $$(\mathbf{u} + \mathbf{v}) \times \mathbf{w} = (\mathbf{u} \times \mathbf{w}) + (\mathbf{v} \times \mathbf{w})$$

$$
\begin{aligned}
\mathbf{v} \times \mathbf{a} &= \underbrace{|\mathbf{v}|\mathbf{T}}_{\mathbf{v}} \times \underbrace{\left(\frac{d}{dt}(|\mathbf{v}|)\mathbf{T} + |\mathbf{v}|\frac{d\mathbf{T}}{dt}\right)}_{\mathbf{a}} \\
&= \underbrace{|\mathbf{v}|\mathbf{T} \times \left(\frac{d}{dt}(|\mathbf{v}|)\right)\mathbf{T}}_{\mathbf{0}} + |\mathbf{v}|\mathbf{T} \times |\mathbf{v}|\frac{d\mathbf{T}}{dt} \quad \text{Distributive law for cross products}
\end{aligned}
$$

The first term in this expression has the form $a\mathbf{T} \times b\mathbf{T}$, where a and b are scalars. Therefore, $a\mathbf{T}$ and $b\mathbf{T}$ are parallel vectors and $a\mathbf{T} \times b\mathbf{T} = \mathbf{0}$. To simplify the second term, recall that a vector $\mathbf{u}(t)$ of constant length has the property that $\mathbf{u}$ and $d\mathbf{u}/dt$ are orthogonal (Section 14.3). Because $\mathbf{T}$ is a unit vector, it has constant length, and $\mathbf{T}$ and $d\mathbf{T}/dt$ are orthogonal. Furthermore, scalar multiples of $\mathbf{T}$ and $d\mathbf{T}/dt$ are also orthogonal. Therefore, the magnitude of the second term simplifies as follows:

$$\left| |\mathbf{v}|\mathbf{T} \times |\mathbf{v}|\frac{d\mathbf{T}}{dt} \right| = |\mathbf{v}||\mathbf{T}|\left| |\mathbf{v}|\frac{d\mathbf{T}}{dt} \right| \underbrace{\sin\theta}_{1} \qquad |\mathbf{u} \times \mathbf{v}| = |\mathbf{u}||\mathbf{v}|\sin\theta$$

$$= |\mathbf{v}|^2 \left|\frac{d\mathbf{T}}{dt}\right| \underbrace{|\mathbf{T}|}_{1} \qquad \text{Simplify, } \theta = \pi/2.$$

$$= |\mathbf{v}|^2 \left|\frac{d\mathbf{T}}{dt}\right|. \qquad |\mathbf{T}| = 1$$

> Recall that the magnitude of the cross product of nonzero vectors is $|\mathbf{u} \times \mathbf{v}| = |\mathbf{u}||\mathbf{v}|\sin\theta$, where θ is the angle between the vectors. If the vectors are orthogonal, $\sin\theta = 1$ and $|\mathbf{u} \times \mathbf{v}| = |\mathbf{u}||\mathbf{v}|$.

The final step is to use Theorem 14.4 and substitute $\left|\dfrac{d\mathbf{T}}{dt}\right| = \kappa|\mathbf{v}|$. Putting these results together, we find that

$$|\mathbf{v} \times \mathbf{a}| = |\mathbf{v}|^2 \left|\frac{d\mathbf{T}}{dt}\right| = |\mathbf{v}|^2 \kappa|\mathbf{v}| = \kappa|\mathbf{v}|^3.$$

> Note that $\mathbf{a}(t) = \mathbf{0}$ corresponds to straight-line motion and $\kappa = 0$. If $\mathbf{v}(t) = \mathbf{0}$, the object is at rest and κ is undefined.

Solving for the curvature gives $\kappa = \dfrac{|\mathbf{v} \times \mathbf{a}|}{|\mathbf{v}|^3}$.

THEOREM 14.5 Alternative Curvature Formula
Let $\mathbf{r}$ be the position of an object moving on a smooth curve. The **curvature** at a point on the curve is

$$\kappa = \frac{|\mathbf{v} \times \mathbf{a}|}{|\mathbf{v}|^3},$$

where $\mathbf{v} = \mathbf{r}'$ is the velocity and $\mathbf{a} = \mathbf{v}'$ is the acceleration.

QUICK CHECK 2 Use the alternative curvature formula to compute the curvature of the curve $\mathbf{r}(t) = \langle t^2, 10, -10 \rangle$. ◄

EXAMPLE 3 Curvature of a parabola Find the curvature of the parabola $\mathbf{r}(t) = \langle t, at^2 \rangle$, for $-\infty < t < \infty$, where $a > 0$ is a real number.

SOLUTION The alternative formula works well in this case. We find that $\mathbf{v}(t) = \mathbf{r}'(t) = \langle 1, 2at \rangle$ and $\mathbf{a}(t) = \mathbf{v}'(t) = \langle 0, 2a \rangle$. To compute the cross product $\mathbf{v} \times \mathbf{a}$, we append a third component of 0 to each vector:

$$\mathbf{v} \times \mathbf{a} = \begin{vmatrix} \mathbf{i} & \mathbf{j} & \mathbf{k} \\ 1 & 2at & 0 \\ 0 & 2a & 0 \end{vmatrix} = 2a\,\mathbf{k}.$$

Therefore, the curvature is

$$\kappa(t) = \frac{|\mathbf{v} \times \mathbf{a}|}{|\mathbf{v}|^3} = \frac{|2a\,\mathbf{k}|}{|\langle 1, 2at \rangle|^3} = \frac{2a}{(1 + 4a^2 t^2)^{3/2}}.$$

The curvature is a maximum at the vertex of the parabola where $t = 0$ and $\kappa = 2a$. The curvature decreases as one moves along the curve away from the vertex, as shown in Figure 14.30 with $a = 1$.

Related Exercise 23 ◄

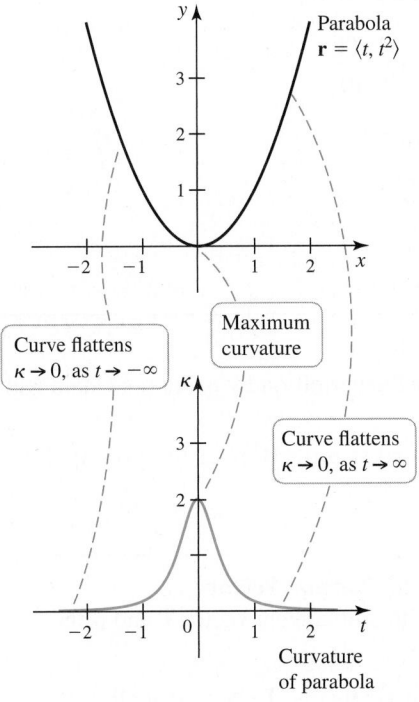

Curve flattens $\kappa \to 0$, as $t \to -\infty$

Maximum curvature

Curve flattens $\kappa \to 0$, as $t \to \infty$

Parabola $\mathbf{r} = \langle t, t^2 \rangle$

Curvature of parabola

Figure 14.30

EXAMPLE 4 Curvature of a helix Find the curvature of the helix $\mathbf{r}(t) = \langle a\cos t, a\sin t, bt \rangle$, for $-\infty < t < \infty$, where $a > 0$ and $b > 0$ are real numbers.

SOLUTION We use the alternative curvature formula, with

$$\mathbf{v}(t) = \mathbf{r}'(t) = \langle -a\sin t, a\cos t, b \rangle \quad \text{and}$$
$$\mathbf{a}(t) = \mathbf{v}'(t) = \langle -a\cos t, -a\sin t, 0 \rangle.$$

The cross product $\mathbf{v} \times \mathbf{a}$ is

$$\mathbf{v} \times \mathbf{a} = \begin{vmatrix} \mathbf{i} & \mathbf{j} & \mathbf{k} \\ -a \sin t & a \cos t & b \\ -a \cos t & -a \sin t & 0 \end{vmatrix} = ab \sin t \, \mathbf{i} - ab \cos t \, \mathbf{j} + a^2 \, \mathbf{k}.$$

Therefore,

$$\begin{aligned} |\mathbf{v} \times \mathbf{a}| &= |ab \sin t \, \mathbf{i} - ab \cos t \, \mathbf{j} + a^2 \, \mathbf{k}| \\ &= \sqrt{a^2 b^2 \underbrace{(\sin^2 t + \cos^2 t)}_{1} + a^4} \\ &= a\sqrt{a^2 + b^2}. \end{aligned}$$

> In the curvature formula for the helix, if $b = 0$, the helix becomes a circle of radius a with $\kappa = \dfrac{1}{a}$. At the other extreme, holding a fixed and letting $b \to \infty$ stretches and straightens the helix so that $\kappa \to 0$.

By a familiar calculation, $|\mathbf{v}| = |\langle -a \sin t, a \cos t, b \rangle| = \sqrt{a^2 + b^2}$. Therefore,

$$\kappa = \frac{|\mathbf{v} \times \mathbf{a}|}{|\mathbf{v}|^3} = \frac{a\sqrt{a^2 + b^2}}{(\sqrt{a^2 + b^2})^3} = \frac{a}{a^2 + b^2}.$$

A similar calculation shows that all helices of this form have constant curvature.

Related Exercise 22 ◄

Principal Unit Normal Vector

The curvature answers the question of how *fast* a curve turns. The principal unit normal vector determines the *direction* in which a curve turns. Specifically, the magnitude of $d\mathbf{T}/ds$ is the curvature: $\kappa = |d\mathbf{T}/ds|$. What about the direction of $d\mathbf{T}/ds$? If only the direction, but not the magnitude, of a vector is of interest, it is convenient to work with a unit vector that has the same direction as the original vector. We apply this idea to $d\mathbf{T}/ds$. The unit vector that points in the direction of $d\mathbf{T}/ds$ is the *principal unit normal vector*.

> The principal unit normal vector depends on the shape of the curve but not on the orientation of the curve.

DEFINITION Principal Unit Normal Vector

Let $\mathbf{r}$ describe a smooth curve parameterized by arc length. The **principal unit normal vector** at a point P on the curve at which $\kappa \neq 0$ is

$$\mathbf{N}(s) = \frac{d\mathbf{T}/ds}{|d\mathbf{T}/ds|} = \frac{1}{\kappa} \frac{d\mathbf{T}}{ds}.$$

For other parameters, we use the equivalent formula

$$\mathbf{N}(t) = \frac{d\mathbf{T}/dt}{|d\mathbf{T}/dt|},$$

evaluated at the value of t corresponding to P.

The practical formula $\mathbf{N} = \dfrac{d\mathbf{T}/dt}{|d\mathbf{T}/dt|}$ follows from the definition by using the Chain Rule to write $\dfrac{d\mathbf{T}}{ds} = \dfrac{d\mathbf{T}}{dt} \cdot \dfrac{dt}{ds}$ (Exercise 80). Two important properties of the principal unit normal vector follow from the definition.

THEOREM 14.6 Properties of the Principal Unit Normal Vector

Let $\mathbf{r}$ describe a smooth parameterized curve with unit tangent vector $\mathbf{T}$ and principal unit normal vector $\mathbf{N}$.

1. $\mathbf{T}$ and $\mathbf{N}$ are orthogonal at all points of the curve; that is, $\mathbf{T} \cdot \mathbf{N} = 0$ at all points where $\mathbf{N}$ is defined.

2. The principal unit normal vector points to the inside of the curve—in the direction that the curve is turning.

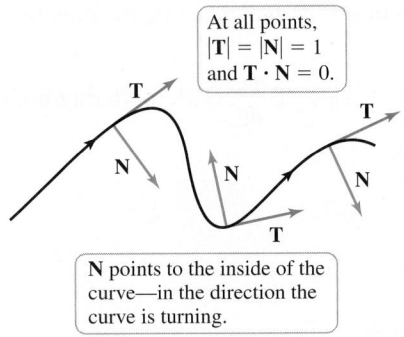

At all points,
$|\mathbf{T}| = |\mathbf{N}| = 1$
and $\mathbf{T} \cdot \mathbf{N} = 0$.

N points to the inside of the curve—in the direction the curve is turning.

Figure 14.31

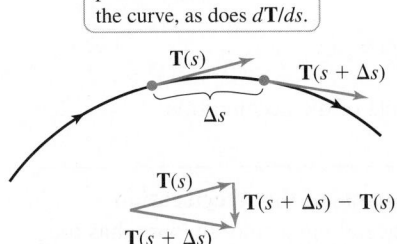

For small Δs,
$\mathbf{T}(s + \Delta s) - \mathbf{T}(s)$
points to the inside of the curve, as does $d\mathbf{T}/ds$.

Figure 14.32

QUICK CHECK 3 Consider the parabola $\mathbf{r}(t) = \langle t, -t^2 \rangle$. Does the principal unit normal vector point in the positive y-direction or the negative y-direction along the curve? ◀

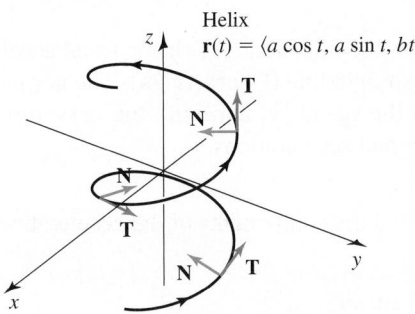

$\mathbf{T} \cdot \mathbf{N} = 0$ at all points of the curve.
$\mathbf{T}$ points in the direction of the curve.
$\mathbf{N}$ points to the inside of the curve.

Figure 14.33

QUICK CHECK 4 Why is the principal unit normal vector for a straight line undefined? ◀

Proof:

1. As a unit vector, $\mathbf{T}$ has constant length. Therefore, by Theorem 14.2, $\mathbf{T}$ and $d\mathbf{T}/dt$ (or $\mathbf{T}$ and $d\mathbf{T}/ds$) are orthogonal. Because $\mathbf{N}$ is a scalar multiple of $d\mathbf{T}/ds$, $\mathbf{T}$ and $\mathbf{N}$ are orthogonal (**Figure 14.31**).

2. We motivate—but do not prove—this fact by recalling that

$$\frac{d\mathbf{T}}{ds} = \lim_{\Delta s \to 0} \frac{\mathbf{T}(s + \Delta s) - \mathbf{T}(s)}{\Delta s}.$$

Therefore, $d\mathbf{T}/ds$ points in the approximate direction of $\mathbf{T}(s + \Delta s) - \mathbf{T}(s)$ when Δs is small. As shown in **Figure 14.32**, this difference points in the direction in which the curve is turning. Because $\mathbf{N}$ is a positive scalar multiple of $d\mathbf{T}/ds$, it points in the same direction. ◀

EXAMPLE 5 Principal unit normal vector for a helix Find the principal unit normal vector for the helix $\mathbf{r}(t) = \langle a \cos t, a \sin t, bt \rangle$, for $-\infty < t < \infty$, where $a > 0$ and $b > 0$ are real numbers.

SOLUTION Several preliminary calculations are needed. First, we have $\mathbf{v}(t) = \mathbf{r}'(t) = \langle -a \sin t, a \cos t, b \rangle$. Therefore,

$$
\begin{aligned}
|\mathbf{v}(t)| = |\mathbf{r}'(t)| &= \sqrt{(-a \sin t)^2 + (a \cos t)^2 + b^2} \\
&= \sqrt{a^2 (\sin^2 t + \cos^2 t) + b^2} && \text{Simplify.} \\
&= \sqrt{a^2 + b^2}. && \sin^2 t + \cos^2 t = 1
\end{aligned}
$$

The unit tangent vector is

$$\mathbf{T}(t) = \frac{\mathbf{r}'(t)}{|\mathbf{r}'(t)|} = \frac{\langle -a \sin t, a \cos t, b \rangle}{\sqrt{a^2 + b^2}}.$$

Notice that $\mathbf{T}$ points along the curve in an upward direction (at an angle to the horizontal that satisfies the equation $\tan \theta = b/a$; **Figure 14.33**). We can now calculate the principal unit normal vector. First, we determine that

$$\frac{d\mathbf{T}}{dt} = \frac{d}{dt}\left(\frac{\langle -a \sin t, a \cos t, b \rangle}{\sqrt{a^2 + b^2}} \right) = \frac{\langle -a \cos t, -a \sin t, 0 \rangle}{\sqrt{a^2 + b^2}}$$

and

$$\left| \frac{d\mathbf{T}}{dt} \right| = \frac{a}{\sqrt{a^2 + b^2}}.$$

The principal unit normal vector now follows:

$$\mathbf{N} = \frac{d\mathbf{T}/dt}{|d\mathbf{T}/dt|} = \frac{\dfrac{\langle -a \cos t, -a \sin t, 0 \rangle}{\sqrt{a^2 + b^2}}}{\dfrac{a}{\sqrt{a^2 + b^2}}} = \langle -\cos t, -\sin t, 0 \rangle.$$

Several important checks should be made. First note that $\mathbf{N}$ is a unit vector; that is, $|\mathbf{N}| = 1$. It should also be confirmed that $\mathbf{T} \cdot \mathbf{N} = 0$; that is, the unit tangent vector and the principal unit normal vector are everywhere orthogonal. Finally, $\mathbf{N}$ is parallel to the xy-plane and points inward toward the z-axis, in the direction the curve turns (Figure 14.33). Notice that in the special case $b = 0$, the trajectory is a circle, but the normal vector is still $\mathbf{N} = \langle -\cos t, -\sin t, 0 \rangle$.

Related Exercise 28 ◀

Components of the Acceleration

The vectors $\mathbf{T}$ and $\mathbf{N}$ may be used to gain insight into how moving objects accelerate. Recall the observation made earlier that the two ways to change the velocity of an object (accelerate) are to change its *speed* and to change its *direction* of motion. We show that changing the speed produces acceleration in the direction of **T** and changing the direction produces acceleration in the direction of **N**.

We begin with the fact that $\mathbf{T} = \dfrac{\mathbf{v}}{|\mathbf{v}|}$ or $\mathbf{v} = \mathbf{T}|\mathbf{v}| = \mathbf{T}\dfrac{ds}{dt}$. Differentiating both

▶ Recall that the speed is $|\mathbf{v}| = ds/dt$, where s is arc length.

sides of $\mathbf{v} = \mathbf{T}\dfrac{ds}{dt}$ with respect to t gives

$$
\begin{aligned}
\mathbf{a} = \frac{d\mathbf{v}}{dt} &= \frac{d}{dt}\left(\mathbf{T}\frac{ds}{dt}\right) \\[2mm]
&= \frac{d\mathbf{T}}{dt}\frac{ds}{dt} + \mathbf{T}\frac{d^2s}{dt^2} \qquad \text{Product Rule} \\[2mm]
&= \underbrace{\frac{d\mathbf{T}}{ds}}_{\kappa\mathbf{N}}\underbrace{\frac{ds}{dt}}_{|\mathbf{v}|}\frac{ds}{dt} + \mathbf{T}\frac{d^2s}{dt^2} \qquad \text{Chain Rule: } \frac{d\mathbf{T}}{dt} = \frac{d\mathbf{T}}{ds}\frac{ds}{dt} \\[2mm]
&= \kappa\mathbf{N}|\mathbf{v}|^2 + \mathbf{T}\frac{d^2s}{dt^2}. \qquad \text{Substitute.}
\end{aligned}
$$

We now identify the normal and tangential components of the acceleration.

▶ Note that a_N and a_T are defined even at points where $\kappa = 0$ and **N** is undefined.

THEOREM 14.7   Tangential and Normal Components of the Acceleration
The acceleration vector of an object moving in space along a smooth curve has the following representation in terms of its **tangential component** a_T (in the direction of **T**) and its **normal component** a_N (in the direction of **N**):

$$\mathbf{a} = a_N\mathbf{N} + a_T\mathbf{T},$$

where $a_N = \kappa|\mathbf{v}|^2 = \dfrac{|\mathbf{v}\times\mathbf{a}|}{|\mathbf{v}|}$ and $a_T = \dfrac{d^2s}{dt^2}$.

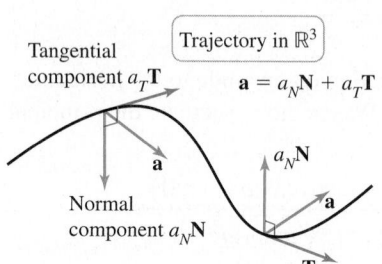

Tangential component $a_T\mathbf{T}$

Trajectory in $\mathbb{R}^3$

$\mathbf{a} = a_N\mathbf{N} + a_T\mathbf{T}$

Normal component $a_N\mathbf{N}$

Figure 14.34

The tangential component of the acceleration, in the direction of **T**, is the usual acceleration $a_T = d^2s/dt^2$ of an object moving along a straight line (**Figure 14.34**). The normal component, in the direction of **N**, increases with the speed $|\mathbf{v}|$ and with the curvature. Higher speeds on tighter curves produce greater normal accelerations.

EXAMPLE 6   Acceleration on a circular path Find the components of the acceleration on the circular trajectory

$$\mathbf{r}(t) = \langle R\cos\omega t, R\sin\omega t\rangle,$$

where R and ω are positive real numbers.

SOLUTION We find that $\mathbf{r}'(t) = \langle -R\omega\sin\omega t, R\omega\cos\omega t\rangle$, $|\mathbf{v}(t)| = |\mathbf{r}'(t)| = R\omega$, and, by Example 2, $\kappa = 1/R$. Recall that $ds/dt = |\mathbf{v}(t)|$, which is constant; therefore, $d^2s/dt^2 = 0$ and the tangential component of the acceleration is zero. The acceleration is

$$\mathbf{a} = \kappa|\mathbf{v}|^2\mathbf{N} + \underbrace{\frac{d^2s}{dt^2}}_{0}\mathbf{T} = \frac{1}{R}(R\omega)^2\mathbf{N} = R\omega^2\mathbf{N}.$$

On a circular path (traversed at constant speed), the acceleration is entirely in the normal direction, orthogonal to the tangent vectors. The acceleration increases with the radius of the circle R and with the frequency of the motion ω.

Related Exercise 36 ◀

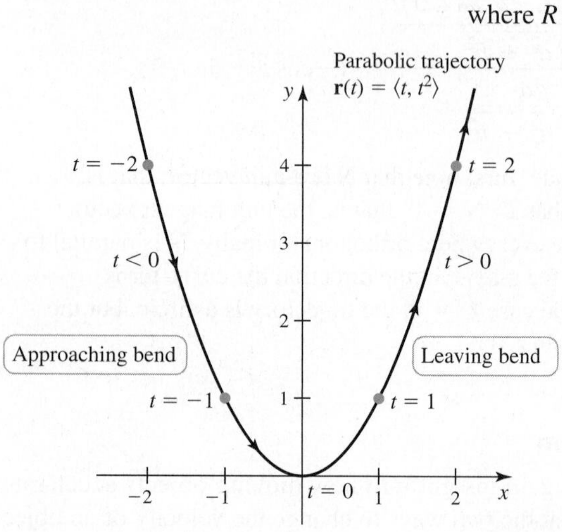

EXAMPLE 7   A bend in the road The driver of a car follows the parabolic trajectory $\mathbf{r}(t) = \langle t, t^2\rangle$, for $-2 \le t \le 2$, through a sharp bend (**Figure 14.35**). Find the tangential and normal components of the acceleration of the car.

Parabolic trajectory $\mathbf{r}(t) = \langle t, t^2\rangle$

$t = -2$

$t < 0$

$t > 0$

$t = 2$

Approaching bend

Leaving bend

$t = -1$

$t = 1$

$t = 0$

Figure 14.35

SOLUTION The velocity and acceleration vectors are easily computed:
$\mathbf{v}(t) = \mathbf{r}'(t) = \langle 1, 2t \rangle$ and $\mathbf{a}(t) = \mathbf{r}''(t) = \langle 0, 2 \rangle$. The goal is to express $\mathbf{a} = \langle 0, 2 \rangle$ in terms of $\mathbf{T}$ and $\mathbf{N}$. A short calculation reveals that

$$\mathbf{T} = \frac{\mathbf{v}}{|\mathbf{v}|} = \frac{\langle 1, 2t \rangle}{\sqrt{1 + 4t^2}} \quad \text{and} \quad \mathbf{N} = \frac{d\mathbf{T}/dt}{|d\mathbf{T}/dt|} = \frac{\langle -2t, 1 \rangle}{\sqrt{1 + 4t^2}}.$$

We now have two ways to proceed. One is to compute the normal and tangential components of the acceleration directly using the definitions. More efficient is to note that $\mathbf{T}$ and $\mathbf{N}$ are orthogonal unit vectors, and then to compute the scalar projections of $\mathbf{a} = \langle 0, 2 \rangle$ in the directions of $\mathbf{T}$ and $\mathbf{N}$. We find that

$$a_N = \mathbf{a} \cdot \mathbf{N} = \langle 0, 2 \rangle \cdot \frac{\langle -2t, 1 \rangle}{\sqrt{1 + 4t^2}} = \frac{2}{\sqrt{1 + 4t^2}}$$

and

$$a_T = \mathbf{a} \cdot \mathbf{T} = \langle 0, 2 \rangle \cdot \frac{\langle 1, 2t \rangle}{\sqrt{1 + 4t^2}} = \frac{4t}{\sqrt{1 + 4t^2}}.$$

You should verify that at all times (Exercise 76),

$$\mathbf{a} = a_N \mathbf{N} + a_T \mathbf{T} = \frac{2}{\sqrt{1 + 4t^2}} (\mathbf{N} + 2t\,\mathbf{T}) = \langle 0, 2 \rangle.$$

> Using the fact that $|\mathbf{T}| = |\mathbf{N}| = 1$, we have, from Section 13.3,
>
> $$a_N = \text{scal}_\mathbf{N} \mathbf{a} = \frac{\mathbf{a} \cdot \mathbf{N}}{|\mathbf{N}|} = \mathbf{a} \cdot \mathbf{N}$$
>
> and
>
> $$a_T = \text{scal}_\mathbf{T} \mathbf{a} = \frac{\mathbf{a} \cdot \mathbf{T}}{|\mathbf{T}|} = \mathbf{a} \cdot \mathbf{T} = \frac{\mathbf{v} \cdot \mathbf{a}}{|\mathbf{v}|}.$$

Let's interpret these results. First notice that the driver negotiates the curve in a sensible way: The speed $|\mathbf{v}| = \sqrt{1 + 4t^2}$ decreases as the car approaches the origin (the tightest part of the curve) and increases as it moves away from the origin (Figure 14.36). As the car approaches the origin ($t < 0$), $\mathbf{T}$ points in the direction of the trajectory and $\mathbf{N}$ points to the inside of the curve. However, $a_T = \dfrac{d^2s}{dt^2} < 0$ when $t < 0$, so $a_T \mathbf{T}$ points in the direction opposite that of $\mathbf{T}$ (corresponding to a deceleration). As the car leaves the origin ($t > 0$), $a_T > 0$ (corresponding to an acceleration) and $a_T \mathbf{T}$ and $\mathbf{T}$ point in the direction of the trajectory. At all times, $\mathbf{N}$ points to the inside of the curve (Figure 14.36; Exercise 78).

Related Exercise 38 ◄

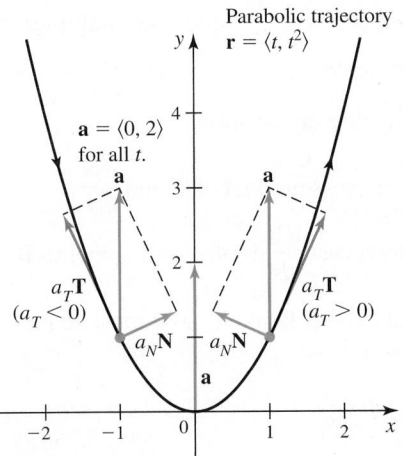

Figure 14.36

QUICK CHECK 5 Verify that $\mathbf{T}$ and $\mathbf{N}$ given in Example 7 satisfy $|\mathbf{T}| = |\mathbf{N}| = 1$ and that $\mathbf{T} \cdot \mathbf{N} = 0$. ◄

The Binormal Vector and Torsion

We have seen that the curvature function and the principal unit normal vector tell us how quickly and in what direction a curve turns. For curves in two dimensions, these quantities give a fairly complete description of motion along the curve. However, in three dimensions, a curve has more "room" in which to change its course, and another descriptive function is often useful. Figure 14.37 shows a smooth parameterized curve C with its unit tangent vector $\mathbf{T}$ and its principal unit normal vector $\mathbf{N}$ at two different points. These two vectors determine a plane called the *osculating plane* (Figure 14.37b). The question we now ask is, How quickly does the curve C move out of the plane determined by $\mathbf{T}$ and $\mathbf{N}$?

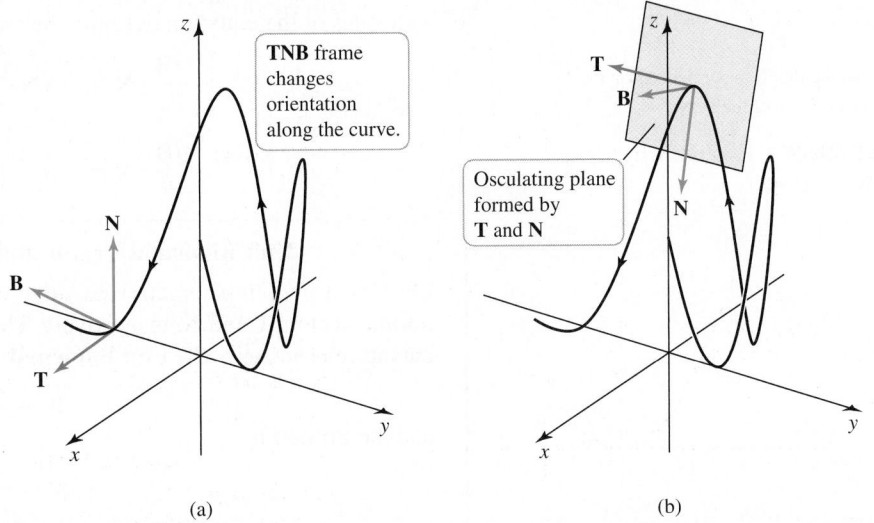

(a)　　　　　(b)

Figure 14.37

To answer this question, we begin by defining the *unit binormal vector* $\mathbf{B} = \mathbf{T} \times \mathbf{N}$. By the definition of the cross product, $\mathbf{B}$ is orthogonal to $\mathbf{T}$ and $\mathbf{N}$. Because $\mathbf{T}$ and $\mathbf{N}$ are unit vectors, $\mathbf{B}$ is also a unit vector. Notice that $\mathbf{T}$, $\mathbf{N}$, and $\mathbf{B}$ form a right-handed coordinate system (like the *xyz*-coordinate system) that changes its orientation as we move along the curve. This coordinate system is often called the **TNB frame** (Figure 14.37).

The rate at which the curve C twists out of the plane determined by $\mathbf{T}$ and $\mathbf{N}$ is the rate at which $\mathbf{B}$ changes as we move along C, which is $\dfrac{d\mathbf{B}}{ds}$. A short calculation leads to a practical formula for the twisting of the curve. Differentiating the cross product $\mathbf{T} \times \mathbf{N}$, we find that

> ► The **TNB** frame is also called the Frenet-Serret frame, after two 19th-century French mathematicians, Jean Frenet and Joseph Serret.

$$\frac{d\mathbf{B}}{ds} = \frac{d}{ds}(\mathbf{T} \times \mathbf{N})$$

$$= \underbrace{\frac{d\mathbf{T}}{ds} \times \mathbf{N}}_{\text{parallel vectors}} + \mathbf{T} \times \frac{d\mathbf{N}}{ds} \qquad \text{Product Rule for cross products}$$

$$= \mathbf{T} \times \frac{d\mathbf{N}}{ds}. \qquad \frac{d\mathbf{T}}{ds} \text{ and } \mathbf{N} \text{ are parallel; } \frac{d\mathbf{T}}{ds} \times \mathbf{N} = \mathbf{0}.$$

QUICK CHECK 6 Explain why $\mathbf{B} = \mathbf{T} \times \mathbf{N}$ is a unit vector. ◄

Notice that by definition, $\mathbf{N} = \dfrac{1}{\kappa}\dfrac{d\mathbf{T}}{ds}$, which implies that $\mathbf{N}$ and $\dfrac{d\mathbf{T}}{ds}$ are scalar multiples of each other. Therefore, their cross product is the zero vector.

The properties of $\dfrac{d\mathbf{B}}{ds}$ become clear with the following observations.

- $\dfrac{d\mathbf{B}}{ds}$ is orthogonal to both $\mathbf{T}$ and $\dfrac{d\mathbf{N}}{ds}$, because it is the cross product of $\mathbf{T}$ and $\dfrac{d\mathbf{N}}{ds}$.

- Applying Theorem 14.2 to the unit vector $\mathbf{B}$, it follows that $\dfrac{d\mathbf{B}}{ds}$ is also orthogonal to $\mathbf{B}$.

> ► Note that $\mathbf{B}$ is a unit vector (of constant length). Therefore, by Theorem 14.2, $\mathbf{B}$ and $\mathbf{B}'(t)$ are orthogonal. Because $\mathbf{B}'(t)$ and $\mathbf{B}'(s)$ are parallel, it follows that $\mathbf{B}$ and $\mathbf{B}'(s)$ are orthogonal.

- By the previous two observations, $\dfrac{d\mathbf{B}}{ds}$ is orthogonal to both $\mathbf{B}$ and $\mathbf{T}$, so it must be parallel to $\mathbf{N}$.

Because $\dfrac{d\mathbf{B}}{ds}$ is parallel to (a scalar multiple of) $\mathbf{N}$, we write

$$\frac{d\mathbf{B}}{ds} = -\tau\mathbf{N},$$

> ► The negative sign in the definition of the torsion is conventional. However, τ may be positive or negative (or zero), and in general, it varies along the curve.

where the scalar τ is the *torsion*. Notice that $\left|\dfrac{d\mathbf{B}}{ds}\right| = |-\tau\mathbf{N}| = |-\tau|$, so the magnitude of the torsion equals the magnitude of $\dfrac{d\mathbf{B}}{ds}$, which is the rate at which the curve twists out of the **TN**-plane.

A short calculation gives a method for computing the torsion. We take the dot product of both sides of the equation defining the torsion with $\mathbf{N}$:

> ► Notice that $\mathbf{B}$ and τ depend on the orientation of the curve.

$$\frac{d\mathbf{B}}{ds} \cdot \mathbf{N} = -\tau\underbrace{\mathbf{N} \cdot \mathbf{N}}_{1}$$

QUICK CHECK 7 Explain why $\mathbf{N} \cdot \mathbf{N} = 1$. ◄

$$\frac{d\mathbf{B}}{ds} \cdot \mathbf{N} = -\tau. \qquad \mathbf{N} \text{ is a unit vector.}$$

DEFINITION Unit Binormal Vector and Torsion

Let C be a smooth parameterized curve with unit tangent and principal unit normal vectors $\mathbf{T}$ and $\mathbf{N}$, respectively. Then at each point of the curve at which the curvature is nonzero, the **unit binormal vector** is

$$\mathbf{B} = \mathbf{T} \times \mathbf{N},$$

and the **torsion** is

$$\tau = -\frac{d\mathbf{B}}{ds} \cdot \mathbf{N}.$$

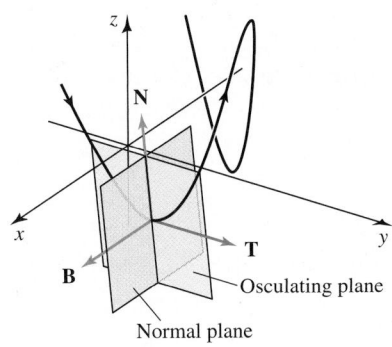

Figure 14.38

▶ The third plane formed by the vectors **T** and **B** is called the *rectifying plane*.

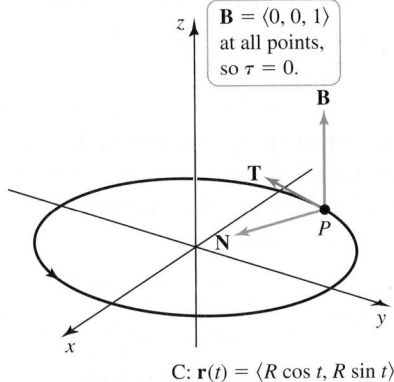

$C: \mathbf{r}(t) = \langle R \cos t, R \sin t \rangle$

Figure 14.39

Figure 14.38 provides some interpretation of the curvature and the torsion. First, we see a smooth curve C passing through a point where the mutually orthogonal vectors **T**, **N**, and **B** are defined. The **osculating plane** is defined by the vectors **T** and **N**. The plane orthogonal to the osculating plane containing **N** is called the **normal plane**. Because **N** and $\dfrac{d\mathbf{B}}{ds}$ are parallel, $\dfrac{d\mathbf{B}}{ds}$ also lies in the normal plane. The torsion, which is equal in magnitude to $\left|\dfrac{d\mathbf{B}}{ds}\right|$, gives the rate at which the curve moves *out of* the osculating plane. In a complementary way, the curvature, which is equal to $\left|\dfrac{d\mathbf{T}}{ds}\right|$, gives the rate at which the curve turns *within* the osculating plane. Two examples will clarify these concepts.

EXAMPLE 8 Unit binormal vector Consider the circle C defined by

$$\mathbf{r}(t) = \langle R \cos t, R \sin t \rangle, \text{ for } 0 \le t \le 2\pi, \text{ with } R > 0.$$

a. Without doing any calculations, find the unit binormal vector **B** and determine the torsion.

b. Use the definition of **B** to calculate **B** and confirm your answer in part (a).

SOLUTION

a. The circle C lies in the xy-plane, so at all points on the circle, **T** and **N** are in the xy-plane. Therefore, at all points of the circle, $\mathbf{B} = \mathbf{T} \times \mathbf{N}$ is the unit vector in the positive z-direction (by the right-hand rule); that is, $\mathbf{B} = \mathbf{k}$. Because **B** changes in neither length nor direction, $\dfrac{d\mathbf{B}}{ds} = \mathbf{0}$ and $\tau = 0$ (Figure 14.39).

b. Building on the calculations of Example 2, we find that

$$\mathbf{T} = \langle -\sin t, \cos t \rangle \quad \text{and} \quad \mathbf{N} = \langle -\cos t, -\sin t \rangle.$$

Therefore, the unit binormal vector is

$$\mathbf{B} = \mathbf{T} \times \mathbf{N} = \begin{vmatrix} \mathbf{i} & \mathbf{j} & \mathbf{k} \\ -\sin t & \cos t & 0 \\ -\cos t & -\sin t & 0 \end{vmatrix} = 0 \cdot \mathbf{i} - 0 \cdot \mathbf{j} + 1 \cdot \mathbf{k} = \mathbf{k}.$$

As in part (a), it follows that the torsion is zero.

Related Exercise 41 ◀

Generalizing Example 8, it can be shown that the binormal vector of any curve that lies in the xy-plane is always parallel to the z-axis; therefore, the torsion of the curve is everywhere zero.

EXAMPLE 9 Torsion of a helix Compute the torsion of the helix
$\mathbf{r}(t) = \langle a \cos t, a \sin t, bt \rangle$, for $t \ge 0$, $a > 0$, and $b > 0$.

SOLUTION In Example 5, we found that

$$\mathbf{T} = \frac{\langle -a \sin t, a \cos t, b \rangle}{\sqrt{a^2 + b^2}} \quad \text{and} \quad \mathbf{N} = \langle -\cos t, -\sin t, 0 \rangle.$$

Therefore,

$$\mathbf{B} = \mathbf{T} \times \mathbf{N} = \frac{1}{\sqrt{a^2 + b^2}} \begin{vmatrix} \mathbf{i} & \mathbf{j} & \mathbf{k} \\ -a \sin t & a \cos t & b \\ -\cos t & -\sin t & 0 \end{vmatrix} = \frac{\langle b \sin t, -b \cos t, a \rangle}{\sqrt{a^2 + b^2}}.$$

The next step is to determine $\dfrac{d\mathbf{B}}{ds}$, which we do in the same way we computed $\dfrac{d\mathbf{T}}{ds}$, by writing

$$\frac{d\mathbf{B}}{dt} = \frac{d\mathbf{B}}{ds} \cdot \frac{ds}{dt} \quad \text{or} \quad \frac{d\mathbf{B}}{ds} = \frac{d\mathbf{B}/dt}{ds/dt}.$$

In this case,

$$\frac{ds}{dt} = |\mathbf{r}'(t)| = \sqrt{a^2 \sin^2 t + a^2 \cos^2 t + b^2} = \sqrt{a^2 + b^2}.$$

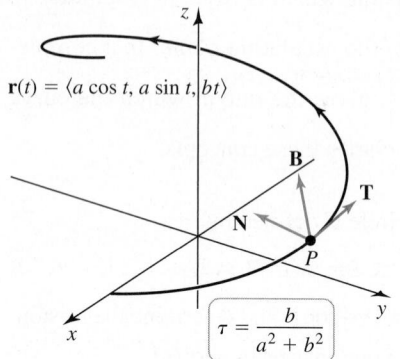

$\mathbf{r}(t) = \langle a \cos t, a \sin t, bt \rangle$

$\tau = \dfrac{b}{a^2 + b^2}$

Figure 14.40

Computing $\dfrac{d\mathbf{B}}{dt}$, we have

$$\frac{d\mathbf{B}}{ds} = \frac{d\mathbf{B}/dt}{ds/dt} = \frac{\langle b \cos t, b \sin t, 0 \rangle}{a^2 + b^2}.$$

The final step is to compute the torsion:

$$\tau = -\frac{d\mathbf{B}}{ds} \cdot \mathbf{N} = -\frac{\langle b \cos t, b \sin t, 0 \rangle}{a^2 + b^2} \cdot \langle -\cos t, -\sin t, 0 \rangle = \frac{b}{a^2 + b^2}.$$

We see that the torsion is constant over the helix. In Example 4, we found that the curvature of a helix is also constant. This special property of circular helices means that the curve turns about its axis at a constant rate and rises vertically at a constant rate (Figure 14.40).

Related Exercise 47 ◄

Example 9 suggests that the computation of the binormal vector and the torsion can be involved. We close by stating some alternative formulas for $\mathbf{B}$ and τ that may simplify calculations in some cases. Letting $\mathbf{v} = \mathbf{r}'(t)$ and $\mathbf{a} = \mathbf{v}'(t) = \mathbf{r}''(t)$, the binormal vector can be written compactly as (Exercise 83)

$$\mathbf{B} = \mathbf{T} \times \mathbf{N} = \frac{\mathbf{v} \times \mathbf{a}}{|\mathbf{v} \times \mathbf{a}|}.$$

We also state without proof that the torsion may be expressed in either of the forms

$$\tau = \frac{(\mathbf{v} \times \mathbf{a}) \cdot \mathbf{a}'}{|\mathbf{v} \times \mathbf{a}|^2} \quad \text{or} \quad \tau = \frac{(\mathbf{r}' \times \mathbf{r}'') \cdot \mathbf{r}'''}{|\mathbf{r}' \times \mathbf{r}''|^2}.$$

SUMMARY Formulas for Curves in Space

Position function: $\mathbf{r}(t) = \langle x(t), y(t), z(t) \rangle$

Velocity: $\mathbf{v} = \mathbf{r}'$

Acceleration: $\mathbf{a} = \mathbf{v}'$

Unit tangent vector: $\mathbf{T} = \dfrac{\mathbf{v}}{|\mathbf{v}|}$

Principal unit normal vector: $\mathbf{N} = \dfrac{d\mathbf{T}/dt}{|d\mathbf{T}/dt|}$ (provided $d\mathbf{T}/dt \neq \mathbf{0}$)

Curvature: $\kappa = \left| \dfrac{d\mathbf{T}}{ds} \right| = \dfrac{1}{|\mathbf{v}|} \left| \dfrac{d\mathbf{T}}{dt} \right| = \dfrac{|\mathbf{v} \times \mathbf{a}|}{|\mathbf{v}|^3}$

Components of acceleration: $\mathbf{a} = a_N \mathbf{N} + a_T \mathbf{T}$, where $a_N = \kappa |\mathbf{v}|^2 = \dfrac{|\mathbf{v} \times \mathbf{a}|}{|\mathbf{v}|}$
and $a_T = \dfrac{d^2 s}{dt^2} = \dfrac{\mathbf{v} \cdot \mathbf{a}}{|\mathbf{v}|}$

Unit binormal vector: $\mathbf{B} = \mathbf{T} \times \mathbf{N} = \dfrac{\mathbf{v} \times \mathbf{a}}{|\mathbf{v} \times \mathbf{a}|}$

Torsion: $\tau = -\dfrac{d\mathbf{B}}{ds} \cdot \mathbf{N} = \dfrac{(\mathbf{v} \times \mathbf{a}) \cdot \mathbf{a}'}{|\mathbf{v} \times \mathbf{a}|^2} = \dfrac{(\mathbf{r}' \times \mathbf{r}'') \cdot \mathbf{r}'''}{|\mathbf{r}' \times \mathbf{r}''|^2}$

SECTION 14.5 EXERCISES

Getting Started

1. What is the curvature of a straight line?

2. Explain in words the meaning of *the curvature of a curve*. Is it a scalar function or a vector function?

3. Give a practical formula for computing curvature.

4. Interpret *the principal unit normal vector of a curve*. Is it a scalar function or a vector function?

5. Give a practical formula for computing the principal unit normal vector.

6. Explain how to decompose the acceleration vector of a moving object into its tangential and normal components.

7. Explain how the vectors **T**, **N**, and **B** are related geometrically.

8. How do you compute **B**?

9. Give a geometrical interpretation of torsion.

10. How do you compute torsion?

Practice Exercises

11–20. Curvature *Find the unit tangent vector* **T** *and the curvature* κ *for the following parameterized curves.*

11. $\mathbf{r}(t) = \langle 2t + 1, 4t - 5, 6t + 12 \rangle$

12. $\mathbf{r}(t) = \langle 2 \cos t, -2 \sin t \rangle$

13. $\mathbf{r}(t) = \langle 2t, 4 \sin t, 4 \cos t \rangle$

14. $\mathbf{r}(t) = \langle \cos t^2, \sin t^2 \rangle$

15. $\mathbf{r}(t) = \langle \sqrt{3} \sin t, \sin t, 2 \cos t \rangle$

16. $\mathbf{r}(t) = \langle t, \ln \cos t \rangle$

17. $\mathbf{r}(t) = \langle t, 2t^2 \rangle$

18. $\mathbf{r}(t) = \langle \cos^3 t, \sin^3 t \rangle$

19. $\mathbf{r}(t) = \left\langle \int_0^t \cos \frac{\pi u^2}{2} \, du, \int_0^t \sin \frac{\pi u^2}{2} \, du \right\rangle, t > 0$

20. $\mathbf{r}(t) = \left\langle \int_0^t \cos u^2 \, du, \int_0^t \sin u^2 \, du \right\rangle, t > 0$

21–26. Alternative curvature formula *Use the alternative curvature formula* $\kappa = \dfrac{|\mathbf{v} \times \mathbf{a}|}{|\mathbf{v}|^3}$ *to find the curvature of the following parameterized curves.*

21. $\mathbf{r}(t) = \langle -3 \cos t, 3 \sin t, 0 \rangle$

22. $\mathbf{r}(t) = \langle 4t, 3 \sin t, 3 \cos t \rangle$

23. $\mathbf{r}(t) = \langle 4 + t^2, t, 0 \rangle$

24. $\mathbf{r}(t) = \langle \sqrt{3} \sin t, \sin t, 2 \cos t \rangle$

25. $\mathbf{r}(t) = \langle 4 \cos t, \sin t, 2 \cos t \rangle$

26. $\mathbf{r}(t) = \langle e^t \cos t, e^t \sin t, e^t \rangle$

27–34. Principal unit normal vector *Find the unit tangent vector* **T** *and the principal unit normal vector* **N** *for the following parameterized curves. In each case, verify that* $|\mathbf{T}| = |\mathbf{N}| = 1$ *and* $\mathbf{T} \cdot \mathbf{N} = 0$.

27. $\mathbf{r}(t) = \langle 2 \sin t, 2 \cos t \rangle$ 28. $\mathbf{r}(t) = \langle 4 \sin t, 4 \cos t, 10t \rangle$

29. $\mathbf{r}(t) = \left\langle \dfrac{t^2}{2}, 4 - 3t, 1 \right\rangle$ 30. $\mathbf{r}(t) = \left\langle \dfrac{t^2}{2}, \dfrac{t^3}{3} \right\rangle, t > 0$

31. $\mathbf{r}(t) = \langle \cos t^2, \sin t^2 \rangle$ 32. $\mathbf{r}(t) = \langle \cos^3 t, \sin^3 t \rangle$

33. $\mathbf{r}(t) = \langle t^2, t \rangle$ 34. $\mathbf{r}(t) = \langle t, \ln \cos t \rangle$

35–40. Components of the acceleration *Consider the following trajectories of moving objects. Find the tangential and normal components of the acceleration.*

35. $\mathbf{r}(t) = \langle t, 1 + 4t, 2 - 6t \rangle$ 36. $\mathbf{r}(t) = \langle 10 \cos t, -10 \sin t \rangle$

37. $\mathbf{r}(t) = \langle e^t \cos t, e^t \sin t, e^t \rangle$ 38. $\mathbf{r}(t) = \langle t, t^2 + 1 \rangle$

39. $\mathbf{r}(t) = \langle t^3, t^2 \rangle$ 40. $\mathbf{r}(t) = \langle 20 \cos t, 20 \sin t, 30t \rangle$

41–44. Computing the binormal vector and torsion *In Exercises 27–30, the unit tangent vector* **T** *and the principal unit normal vector* **N** *were computed for the following parameterized curves. Use the definitions to compute their unit binormal vector and torsion.*

41. $\mathbf{r}(t) = \langle 2 \sin t, 2 \cos t \rangle$ 42. $\mathbf{r}(t) = \langle 4 \sin t, 4 \cos t, 10t \rangle$

43. $\mathbf{r}(t) = \left\langle \dfrac{t^2}{2}, 4 - 3t, 1 \right\rangle$ 44. $\mathbf{r}(t) = \left\langle \dfrac{t^2}{2}, \dfrac{t^3}{3} \right\rangle, t > 0$

45–48. Computing the binormal vector and torsion *Use the definitions to compute the unit binormal vector and torsion of the following curves.*

45. $\mathbf{r}(t) = \langle 2 \cos t, 2 \sin t, -t \rangle$

46. $\mathbf{r}(t) = \langle t, \cosh t, -\sinh t \rangle$

47. $\mathbf{r}(t) = \langle 12t, 5 \cos t, 5 \sin t \rangle$

48. $\mathbf{r}(t) = \langle \sin t - t \cos t, \cos t + t \sin t, t \rangle$

49. **Explain why or why not** Determine whether the following statements are true and give an explanation or counterexample.

   a. The position, unit tangent, and principal unit normal vectors (**r**, **T**, and **N**) at a point lie in the same plane.

   b. The vectors **T** and **N** at a point depend on the orientation of a curve.

   c. The curvature at a point depends on the orientation of a curve.

   d. An object with unit speed ($|\mathbf{v}| = 1$) on a circle of radius R has an acceleration of $\mathbf{a} = \mathbf{N}/R$.

   e. If the speedometer of a car reads a constant 60 mi/hr, the car is not accelerating.

   f. A curve in the xy-plane that is concave up at all points has positive torsion.

   g. A curve with large curvature also has large torsion.

50. **Special formula: Curvature for** $y = f(x)$ Assume f is twice differentiable. Prove that the curve $y = f(x)$ has curvature

$$\kappa(x) = \frac{|f''(x)|}{(1 + f'(x)^2)^{3/2}}.$$

   (*Hint:* Use the parametric description $x = t, y = f(t)$.)

51–54. Curvature for $y = f(x)$ *Use the result of Exercise 50 to find the curvature function of the following curves.*

51. $f(x) = x^2$ 52. $f(x) = \sqrt{a^2 - x^2}$

53. $f(x) = \ln x$ 54. $f(x) = \ln \cos x$

55. Special formula: Curvature for plane curves Show that the parametric curve $\mathbf{r}(t) = \langle f(t), g(t) \rangle$, where f and g are twice differentiable, has curvature

$$\kappa(t) = \frac{|f'g'' - f''g'|}{((f')^2 + (g')^2)^{3/2}},$$

where all derivatives are taken with respect to t.

56–59. Curvature for plane curves *Use the result of Exercise 55 to find the curvature function of the following curves.*

56. $\mathbf{r}(t) = \langle a \sin t, a \cos t \rangle$ (circle)

57. $\mathbf{r}(t) = \langle a \sin t, b \cos t \rangle$ (ellipse)

58. $\mathbf{r}(t) = \langle a \cos^3 t, a \sin^3 t \rangle$ (astroid)

59. $\mathbf{r}(t) = \langle t, at^2 \rangle$ (parabola)

When appropriate, consider using the special formulas derived in Exercises 50 and 55 in the remaining exercises.

60–63. Same paths, different velocity *The position functions of objects A and B describe different motion along the same path for $t \geq 0$.*

a. *Sketch the path followed by both A and B.*
b. *Find the velocity and acceleration of A and B and discuss the differences.*
c. *Express the acceleration of A and B in terms of the tangential and normal components and discuss the differences.*

60. $A: \mathbf{r}(t) = \langle 1 + 2t, 2 - 3t, 4t \rangle, B: \mathbf{r}(t) = \langle 1 + 6t, 2 - 9t, 12t \rangle$

61. $A: \mathbf{r}(t) = \langle t, 2t, 3t \rangle, B: \mathbf{r}(t) = \langle t^2, 2t^2, 3t^2 \rangle$

62. $A: \mathbf{r}(t) = \langle \cos t, \sin t \rangle, B: \mathbf{r}(t) = \langle \cos 3t, \sin 3t \rangle$

63. $A: \mathbf{r}(t) = \langle \cos t, \sin t \rangle, B: \mathbf{r}(t) = \langle \cos t^2, \sin t^2 \rangle$

T 64–67. Graphs of the curvature *Consider the following curves.*

a. *Graph the curve.*
b. *Compute the curvature.*
c. *Graph the curvature as a function of the parameter.*
d. *Identify the points (if any) at which the curve has a maximum or minimum curvature.*
e. *Verify that the graph of the curvature is consistent with the graph of the curve.*

64. $\mathbf{r}(t) = \langle t, t^2 \rangle$, for $-2 \leq t \leq 2$ (parabola)

65. $\mathbf{r}(t) = \langle t - \sin t, 1 - \cos t \rangle$, for $0 \leq t \leq 2\pi$ (cycloid)

66. $\mathbf{r}(t) = \langle t, \sin t \rangle$, for $0 \leq t \leq \pi$ (sine curve)

67. $\mathbf{r}(t) = \left\langle \dfrac{t^2}{2}, \dfrac{t^3}{3} \right\rangle$, for $t > 0$

68. Curvature of ln x Find the curvature of $f(x) = \ln x$, for $x > 0$, and find the point at which it is a maximum. What is the value of the maximum curvature?

69. Curvature of e^x Find the curvature of $f(x) = e^x$ and find the point at which it is a maximum. What is the value of the maximum curvature?

70. Circle and radius of curvature Choose a point P on a smooth curve C in the plane. The **circle of curvature** (or **osculating circle**) at P is the circle that (a) is tangent to C at P, (b) has the same curvature as C at P, and (c) lies on the same side of C as the principal unit normal $\mathbf{N}$ (see figure). The **radius of curvature** is

the radius of the circle of curvature. Show that the radius of curvature is $1/\kappa$, where κ is the curvature of C at P.

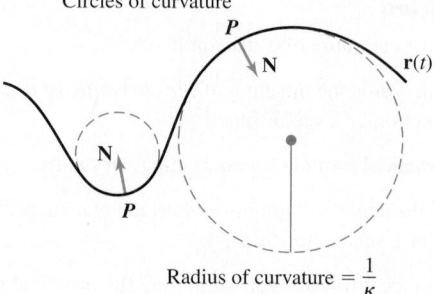

Circles of curvature

Radius of curvature $= \dfrac{1}{\kappa}$

71–73. Finding the radius of curvature *Find the radius of curvature (see Exercise 70) of the following curves at the given point. Then write an equation of the circle of curvature at the point.*

71. $\mathbf{r}(t) = \langle t, t^2 \rangle$ (parabola) at $t = 0$

72. $y = \ln x$ at $x = 1$

73. $\mathbf{r}(t) = \langle t - \sin t, 1 - \cos t \rangle$ (cycloid) at $t = \pi$

74. Designing a highway curve The function

$$\mathbf{r}(t) = \left\langle \int_0^t \cos \frac{u^2}{2} du, \int_0^t \sin \frac{u^2}{2} du \right\rangle,$$

whose graph is called a **clothoid** or **Euler spiral** (Figure A), has applications in the design of railroad tracks, roller coasters, and highways.

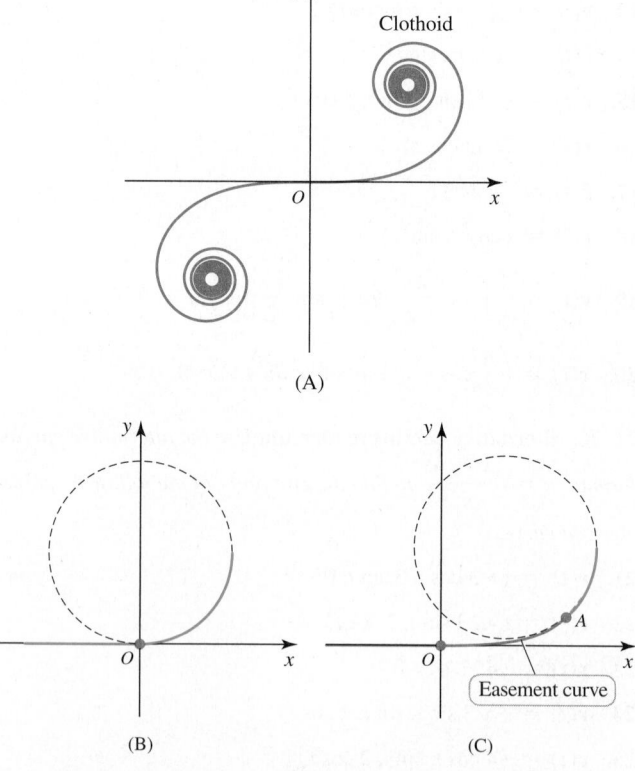

(A)

(B)   (C)

a. A car moves from left to right on a straight highway, approaching a curve at the origin (Figure B). Sudden changes in curvature at the start of the curve may cause the driver to jerk the steering wheel. Suppose the curve starting at the origin is a segment of a circle of radius a. Explain why there is a sudden change in the curvature of the road at the origin. (*Hint:* See Exercise 70.)

b. A better approach is to use a segment of a clothoid as an easement curve, in between the straight highway and a circle, to avoid sudden changes in curvature (Figure C). Assume the easement curve corresponds to the clothoid $\mathbf{r}(t)$, for $0 \le t \le 1.2$. Find the curvature of the easement curve as a function of t, and explain why this curve eliminates the sudden change in curvature at the origin.

c. Find the radius of a circle connected to the easement curve at point A (that corresponds to $t = 1.2$ on the curve $\mathbf{r}(t)$) so that the curvature of the circle matches the curvature of the easement curve at point A.

75. Curvature of the sine curve The function $f(x) = \sin nx$, where n is a positive real number, has a local maximum at $x = \dfrac{\pi}{2n}$. Compute the curvature κ of f at this point. How does κ vary (if at all) as n varies?

76. Parabolic trajectory In Example 7 it was shown that for the parabolic trajectory $\mathbf{r}(t) = \langle t, t^2 \rangle$, $\mathbf{a} = \langle 0, 2 \rangle$ and
$$\mathbf{a} = \frac{2}{\sqrt{1 + 4t^2}}\,(\mathbf{N} + 2t\,\mathbf{T}).$$
Show that the second expression for $\mathbf{a}$ reduces to the first expression.

T 77. Parabolic trajectory Consider the parabolic trajectory
$$x = (V_0 \cos \alpha)t, \quad y = (V_0 \sin \alpha)t - \frac{1}{2}gt^2,$$
where V_0 is the initial speed, α is the angle of launch, and g is the acceleration due to gravity. Consider all times $[0, T]$ for which $y \ge 0$.

a. Find and graph the speed, for $0 \le t \le T$.
b. Find and graph the curvature, for $0 \le t \le T$.
c. At what times (if any) do the speed and curvature have maximum and minimum values?

Explorations and Challenges

78. Relationship between T, N, and a Show that if an object accelerates in the sense that $\dfrac{d^2s}{dt^2} > 0$ and $\kappa \ne 0$, then the acceleration vector lies between $\mathbf{T}$ and $\mathbf{N}$ in the plane of $\mathbf{T}$ and $\mathbf{N}$. Show that if an object decelerates in the sense that $\dfrac{d^2s}{dt^2} < 0$, then the acceleration vector lies in the plane of $\mathbf{T}$ and $\mathbf{N}$, but not between $\mathbf{T}$ and $\mathbf{N}$.

79. Zero curvature Prove that the curve
$$\mathbf{r}(t) = \langle a + bt^p, c + dt^p, e + ft^p \rangle,$$
where a, b, c, d, e, and f are real numbers and p is a positive integer, has zero curvature. Give an explanation.

80. Practical formula for N Show that the definition of the principal unit normal vector $\mathbf{N} = \dfrac{d\mathbf{T}/ds}{|d\mathbf{T}/ds|}$ implies the practical formula $\mathbf{N} = \dfrac{d\mathbf{T}/dt}{|d\mathbf{T}/dt|}$. Use the Chain Rule and recall that $|\mathbf{v}| = \dfrac{ds}{dt} > 0$.

T 81. Maximum curvature Consider the "superparabolas" $f_n(x) = x^{2n}$, where n is a positive integer.

a. Find the curvature function of f_n, for $n = 1, 2$, and 3.
b. Plot f_n and their curvature functions, for $n = 1, 2$, and 3, and check for consistency.
c. At what points does the maximum curvature occur, for $n = 1, 2$, and 3?
d. Let the maximum curvature for f_n occur at $x = \pm z_n$. Using either analytical methods or a calculator, determine $\displaystyle\lim_{n\to\infty} z_n$. Interpret your result.

82. Alternative derivation of curvature Derive the computational formula for curvature using the following steps.

a. Use the tangential and normal components of the acceleration to show that $\mathbf{v} \times \mathbf{a} = \kappa |\mathbf{v}|^3 \mathbf{B}$. (Note that $\mathbf{T} \times \mathbf{T} = \mathbf{0}$.)
b. Solve the equation in part (a) for κ and conclude that
$$\kappa = \frac{|\mathbf{v} \times \mathbf{a}|}{|\mathbf{v}|^3},$$
as shown in the text.

83. Computational formula for B Use the result of part (a) of Exercise 82 and the formula for κ to show that
$$\mathbf{B} = \frac{\mathbf{v} \times \mathbf{a}}{|\mathbf{v} \times \mathbf{a}|}.$$

84. Torsion formula Show that the formula defining torsion, $\tau = -\dfrac{d\mathbf{B}}{ds} \cdot \mathbf{N}$, is equivalent to $\tau = -\dfrac{1}{|\mathbf{v}|}\dfrac{d\mathbf{B}}{dt} \cdot \mathbf{N}$. The second formula is generally easier to use.

85. Descartes' four-circle solution Consider the four mutually tangent circles shown in the figure that have radii a, b, c, and d, and curvatures $A = \dfrac{1}{a}, B = \dfrac{1}{b}, C = \dfrac{1}{c}$, and $D = \dfrac{1}{d}$. Prove Descartes' result (1643) that
$$(A + B + C + D)^2 = 2(A^2 + B^2 + C^2 + D^2).$$

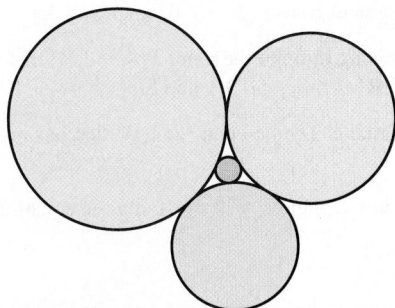

CHAPTER 14 REVIEW EXERCISES

1. **Explain why or why not** Determine whether the following statements are true and give an explanation or counterexample.

a. If $\mathbf{r}(t) = \langle \cos t, e^t, t \rangle + \mathbf{C}$ and $\mathbf{r}(0) = \langle 0, 0, 0 \rangle$, then $\mathbf{C} = \langle 0, 0, 0 \rangle$.

b. The curvature of a circle of radius 5 is $\kappa = 1/5$.

c. The graph of $\mathbf{r}(t) = \langle 3 \cos t, 0, 6 \sin t \rangle$ is an ellipse in the xz-plane.

d. If $\mathbf{r}'(t) = \mathbf{0}$, then $\mathbf{r}(t) = \langle a, b, c \rangle$, where a, b, and c are real numbers.

e. The parameterized curve $\mathbf{r}(t) = \langle 5 \cos t, 12 \cos t, 13 \sin t \rangle$ has arc length as a parameter.

f. The position vector and the principal unit normal are always parallel on a smooth curve.

2. **Sets of points** Describe the set of points satisfying the equations $x^2 + z^2 = 1$ and $y = 2$.

T **3–6. Graphing curves** *Sketch the curves described by the following functions, indicating the orientation of the curve. Use analysis and describe the shape of the curve before using a graphing utility.*

3. $\mathbf{r}(t) = (2t + 1)\mathbf{i} + t\mathbf{j}$

4. $\mathbf{r}(t) = \langle \cos t, 1 + \cos^2 t \rangle$, for $0 \le t \le \pi/2$

5. $\mathbf{r}(t) = 4 \cos t\,\mathbf{i} + \mathbf{j} + 4 \sin t\,\mathbf{k}$, for $0 \le t \le 2\pi$

6. $\mathbf{r}(t) = e^t\mathbf{i} + 2e^t\mathbf{j} + \mathbf{k}$, for $t \ge 0$

7. **Intersection curve** A sphere S and a plane P intersect along the curve $\mathbf{r}(t) = \sin t\,\mathbf{i} + \sqrt{2} \cos t\,\mathbf{j} + \sin t\,\mathbf{k}$, for $0 \le t \le 2\pi$. Find equations for S and P and describe the curve $\mathbf{r}$.

8–13. Vector-valued functions *Find a function $\mathbf{r}(t)$ that describes each of the following curves.*

8. The line segment from $P(2, -3, 0)$ to $Q(1, 4, 9)$

9. The line passing through the point $P(4, -2, 3)$ that is orthogonal to the lines $\mathbf{R}(t) = \langle t, 5t, 2t \rangle$ and $\mathbf{S}(t) = \langle t + 1, -1, 3t - 1 \rangle$

10. A circle of radius 3 centered at $(2, 1, 0)$ that lies in the plane $y = 1$

11. An ellipse in the plane $x = 2$ satisfying the equation $\dfrac{y^2}{9} + \dfrac{z^2}{16} = 1$

12. The projection of the curve onto the xy-plane is the parabola $y = x^2$, and the projection of the curve onto the xz-plane is the line $z = x$.

13. The projection of the curve onto the xy-plane is the unit circle $x^2 + y^2 = 1$, and the projection of the curve onto the yz-plane is the line segment $z = y$, for $-1 \le y \le 1$.

14–15. Intersection curve *Find the curve $\mathbf{r}(t)$ where the following surfaces intersect.*

14. $z = x^2 - 5y^2$; $z = 10x^2 + 4y^2 - 36$

15. $x^2 + 7y^2 + 2z^2 = 9$; $z = y$

16–19. Working with vector-valued functions *For each vector-valued function $\mathbf{r}$, carry out the following steps.*

a. Evaluate $\lim_{t \to 0} \mathbf{r}(t)$ and $\lim_{t \to \infty} \mathbf{r}(t)$, if each exists.

b. Find $\mathbf{r}'(t)$ and evaluate $\mathbf{r}'(0)$.

c. Find $\mathbf{r}''(t)$.

d. Evaluate $\int \mathbf{r}(t)\,dt$.

16. $\mathbf{r}(t) = \langle t + 1, t^2 - 3 \rangle$ **17.** $\mathbf{r}(t) = \left\langle \dfrac{1}{2t + 1}, \dfrac{t}{t + 1} \right\rangle$

18. $\mathbf{r}(t) = \langle e^{-2t}, te^{-t}, \tan^{-1}t \rangle$ **19.** $\mathbf{r}(t) = \langle \sin 2t, 3 \cos 4t, t \rangle$

20–21. Definite integrals *Evaluate the following definite integrals.*

20. $\displaystyle\int_1^3 \left(6t^2\mathbf{i} + 4t\mathbf{j} + \dfrac{1}{t}\mathbf{k} \right) dt$

21. $\displaystyle\int_{-1}^1 \left(\sin \pi t\,\mathbf{i} + \mathbf{j} + \dfrac{2}{1 + t^2}\mathbf{k} \right) dt$

22–24. Derivative rules *Suppose $\mathbf{u}$ and $\mathbf{v}$ are differentiable functions at $t = 0$ with $\mathbf{u}(0) = \langle 2, 7, 0 \rangle$, $\mathbf{u}'(0) = \langle 3, 1, 2 \rangle$, $\mathbf{v}(0) = \langle 3, -1, 0 \rangle$, and $\mathbf{v}'(0) = \langle 5, 0, 3 \rangle$. Evaluate the following expressions.*

22. $\dfrac{d}{dt}(\mathbf{u} \cdot \mathbf{v})\Big|_{t=0}$ **23.** $\dfrac{d}{dt}(\mathbf{u} \times \mathbf{v})\Big|_{t=0}$

24. $\dfrac{d}{dt}(\mathbf{u}(e^{5t} - 1))\Big|_{t=0}$

25–27. Finding r from r′ *Find the function $\mathbf{r}$ that satisfies the given conditions.*

25. $\mathbf{r}'(t) = \langle 1, \sin 2t, \sec^2 t \rangle$; $\mathbf{r}(0) = \langle 2, 2, 2 \rangle$

26. $\mathbf{r}'(t) = \langle e^t, 2e^{2t}, 6e^{3t} \rangle$; $\mathbf{r}(0) = \langle 1, 3, -1 \rangle$

27. $\mathbf{r}'(t) = \left\langle \dfrac{4}{1 + t^2}, 2t + 1, 3t^2 \right\rangle$; $\mathbf{r}(1) = \langle 0, 0, 0 \rangle$

28–29. Unit tangent vectors *Find the unit tangent vector $\mathbf{T}(t)$ for the following parameterized curves. Then determine the unit tangent vector at the given value of t.*

28. $\mathbf{r}(t) = \langle 8, 3 \sin 2t, 3 \cos 2t \rangle$, for $0 \le t \le \pi$; $t = \pi/4$

29. $\mathbf{r}(t) = \langle 2e^t, e^{2t}, t \rangle$, for $0 \le t \le 2\pi$; $t = 0$

30–31. Velocity and acceleration from position *Consider the following position functions.*

a. Find the velocity and speed of the object.

b. Find the acceleration of the object.

30. $\mathbf{r}(t) = \left\langle \dfrac{5}{3}t^3 + 1, t^2 + 10t \right\rangle$, for $t \ge 0$

31. $\mathbf{r}(t) = \left\langle e^{4t} + 1, e^{4t}, \dfrac{1}{2}e^{4t} + 1 \right\rangle$, for $t \ge 0$

32–33. Solving equations of motion *Given an acceleration vector, initial velocity $\langle u_0, v_0 \rangle$, and initial position $\langle x_0, y_0 \rangle$, find the velocity and position vectors for $t \ge 0$.*

32. $\mathbf{a}(t) = \langle 1, 4 \rangle$, $\langle u_0, v_0 \rangle = \langle 4, 3 \rangle$, $\langle x_0, y_0 \rangle = \langle 0, 2 \rangle$

33. $\mathbf{a}(t) = \langle \cos t, 2 \sin t \rangle$, $\langle u_0, v_0 \rangle = \langle 2, 1 \rangle$, $\langle x_0, y_0 \rangle = \langle 1, 2 \rangle$

T **34.** **Orthogonal r and r′** Find all points on the ellipse $\mathbf{r}(t) = \langle 1, 8 \sin t, \cos t \rangle$, for $0 \le t \le 2\pi$, at which $\mathbf{r}(t)$ and $\mathbf{r}'(t)$ are orthogonal. Sketch the curve and the tangent vectors to verify your conclusion.

T 35–36. Modeling motion *Consider the motion of the following objects. Assume the x-axis is horizontal, the positive y-axis is vertical, the ground is horizontal, and only the gravitational force acts on the object.*

   a. *Find the velocity and position vectors, for $t \geq 0$.*

   b. *Determine the time of flight and range of the object.*

   c. *Determine the maximum height of the object.*

35. A baseball has an initial position $\langle x_0, y_0 \rangle = \langle 0, 3 \rangle$ ft when it is hit at an angle of $60°$ with an initial speed of 80 ft/s.

36. A rock is thrown from 2 m above horizontal ground at an angle of $30°$ above the horizontal with a speed of 6 m/s. Assume the initial position of the rock is $\langle x_0, y_0 \rangle = \langle 0, 2 \rangle$.

T 37. A baseball is hit 2 ft above home plate with an initial velocity of $\langle 40, 20, 40 \rangle$ ft/s. The spin on the baseball produces a horizontal acceleration of the ball of 4 ft/s² in the eastward direction. Assume the positive x-axis points east and the positive y-axis points north.

   a. Find the velocity and position vectors, for $t \geq 0$. Assume the origin is at home plate.

   b. When does the ball hit the ground? Round your answer to three digits to the right of the decimal place.

   c. How far does it land from home plate? Round your answer to the nearest whole number.

38. Firing angles A projectile is fired over horizontal ground from the ground with an initial speed of 40 m/s. What firing angles produce a range of 100 m?

T 39. Projectile motion A projectile is launched from the origin, which is a point 50 ft from a 30-ft vertical cliff (see figure). It is launched at a speed of $50\sqrt{2}$ ft/s at an angle of $45°$ to the horizontal. Assume the ground is horizontal on top of the cliff and that only the gravitational force affects the motion of the object.

   a. Give the coordinates of the landing spot of the projectile on the top of the cliff.

   b. What is the maximum height reached by the projectile?

   c. What is the time of flight?

   d. Write the integral that gives the length of the trajectory.

   e. Approximate the length of the trajectory.

   f. What is the range of launch angles needed to clear the edge of the cliff?

40. Baseball motion A toddler on level ground throws a baseball into the air at an angle of $30°$ with the ground from a height of 2 ft. If the ball lands 10 ft from the child, determine the initial speed of the ball.

T 41. Closest point Find the approximate location of the point on the curve $\mathbf{r}(t) = \langle t, t^2 + 1, 3t \rangle$ that lies closest to the point $P(3, 1, 6)$.

T 42. Basketball shot A basketball is shot at an angle of $45°$ to the horizontal. The center of the basketball is at the point $A(0, 8)$ at the moment it is released, and it passes through the center of the basketball hoop that is located at the point $B(18, 10)$. Assume the basketball does not hit the front of the hoop (otherwise it might not pass through the basket). The validity of this assumption is investigated in parts (d), (e), and (f).

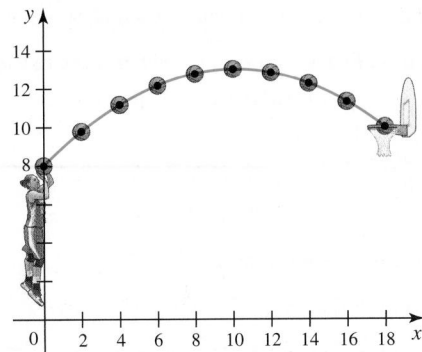

   a. Determine the initial speed of the basketball.

   b. Find the initial velocity $\mathbf{v}(0)$ at the moment it is released.

   c. Find the position function $\mathbf{r}(t)$ of the center of the basketball t seconds after the ball is released. Assume $\mathbf{r}(0) = \langle 0, 8 \rangle$.

   d. Find the distance $s(t)$ between the center of the basketball and the front of the basketball hoop t seconds after the ball is released. Assume the diameter of the basketball hoop is 18 inches.

   e. Determine the closest distance (in inches) between the center of the basketball and the front of the basketball hoop.

   f. Is the assumption that the basketball does not hit the front of the hoop valid? Use the fact that the diameter of a women's basketball is about 9.23 inches. (*Hint:* The ball hits the front of the hoop if, during its flight, the distance from the center of the ball to the front of the hoop is less than the radius of the basketball.)

43–46. Arc length *Find the arc length of the following curves.*

43. $\mathbf{r}(t) = \left\langle t^2, \dfrac{4\sqrt{2}}{3} t^{3/2}, 2t \right\rangle$, for $1 \leq t \leq 3$

44. $\mathbf{r}(t) = \langle 2t^{9/2}, t^3 \rangle$, for $0 \leq t \leq 2$

T 45. $\mathbf{r}(t) = \langle \sin t, t + \cos t, 4t \rangle$, for $0 \leq t \leq \dfrac{\pi}{2}$

46. $\mathbf{r}(t) = \langle t, \ln \sec t, \ln (\sec t + \tan t) \rangle$, for $0 \leq t \leq \dfrac{\pi}{4}$

47. Velocity and trajectory length The acceleration of a wayward firework is given by $\mathbf{a}(t) = \sqrt{2}\,\mathbf{j} + 2t\mathbf{k}$, for $0 \leq t \leq 3$. Suppose the initial velocity of the firework is $\mathbf{v}(0) = \mathbf{i}$.

   a. Find the velocity of the firework, for $0 \leq t \leq 3$.

   b. Find the length of the trajectory of the firework over the interval $0 \leq t \leq 3$.

48–49. Arc length parameterization *Find a description of the following curves that uses arc length as a parameter.*

48. $\mathbf{r}(t) = (1 + 4t)\mathbf{i} - 3t\mathbf{j}$, for $t \geq 1$

49. $\mathbf{r}(t) = \left\langle t^2, \dfrac{4\sqrt{2}}{3} t^{3/2}, 2t \right\rangle$, for $t \geq 0$

50. Tangents and normals for an ellipse Consider the ellipse $\mathbf{r}(t) = \langle 3 \cos t, 4 \sin t \rangle$, for $0 \le t \le 2\pi$.

a. Find the tangent vector $\mathbf{r}'$, the unit tangent vector $\mathbf{T}$, and the principal unit normal vector $\mathbf{N}$ at all points on the curve.

b. At what points does $|\mathbf{r}'|$ have maximum and minimum values?

c. At what points does the curvature have maximum and minimum values? Interpret this result in light of part (b).

d. Find the points (if any) at which $\mathbf{r}$ and $\mathbf{N}$ are parallel.

■ 51–54. Properties of space curves *Do the following calculations.*

a. *Find the tangent vector and the unit tangent vector.*

b. *Find the curvature.*

c. *Find the principal unit normal vector.*

d. *Verify that $|\mathbf{N}| = 1$ and $\mathbf{T} \cdot \mathbf{N} = 0$.*

e. *Graph the curve and sketch $\mathbf{T}$ and $\mathbf{N}$ at two points.*

51. $\mathbf{r}(t) = \langle 6 \cos t, 3 \sin t \rangle$, for $0 \le t \le 2\pi$

52. $\mathbf{r}(t) = \cos t\, \mathbf{i} + 2 \sin t\, \mathbf{j} + \mathbf{k}$, for $0 \le t \le 2\pi$

53. $\mathbf{r}(t) = \cos t\, \mathbf{i} + 2 \cos t\, \mathbf{j} + \sqrt{5} \sin t\, \mathbf{k}$, for $0 \le t \le 2\pi$

54. $\mathbf{r}(t) = t\, \mathbf{i} + 2 \cos t\, \mathbf{j} + 2 \sin t\, \mathbf{k}$, for $0 \le t \le 2\pi$

55–58. Analyzing motion *Consider the position vector of the following moving objects.*

a. *Find the normal and tangential components of the acceleration.*

b. *Graph the trajectory and sketch the normal and tangential components of the acceleration at two points on the trajectory. Show that their sum gives the total acceleration.*

55. $\mathbf{r}(t) = 2 \cos t\, \mathbf{i} + 2 \sin t\, \mathbf{j}$, for $0 \le t \le 2\pi$

56. $\mathbf{r}(t) = 3t\, \mathbf{i} + (4 - t)\, \mathbf{j} + t\, \mathbf{k}$, for $t \ge 0$

57. $\mathbf{r}(t) = (t^2 + 1)\, \mathbf{i} + 2t\, \mathbf{j}$, for $t \ge 0$

58. $\mathbf{r}(t) = 2 \cos t\, \mathbf{i} + 2 \sin t\, \mathbf{j} + 10t\, \mathbf{k}$, for $0 \le t \le 2\pi$

59. Computing the binormal vector and torsion Compute the unit binormal vector $\mathbf{B}$ and the torsion of the curve $\mathbf{r}(t) = \langle t, t^2, t^3 \rangle$ at $t = 1$.

60–61. Curve analysis *Carry out the following steps for the given curves C.*

a. *Find $\mathbf{T}(t)$ at all points of C.*

b. *Find $\mathbf{N}(t)$ and the curvature at all points of C.*

c. *Sketch the curve and show $\mathbf{T}(t)$ and $\mathbf{N}(t)$ at the points of C corresponding to $t = 0$ and $t = \pi/2$.*

d. *Are the results of parts (a) and (b) consistent with the graph?*

e. *Find $\mathbf{B}(t)$ at all points of C.*

f. *On the graph of part (c), plot $\mathbf{B}(t)$ at the points of C corresponding to $t = 0$ and $t = \pi/2$.*

g. *Describe three calculations that serve to check the accuracy of your results in parts (a)–(f).*

h. *Compute the torsion at all points of C. Interpret this result.*

60. $C: \mathbf{r}(t) = \langle 3 \sin t, 4 \sin t, 5 \cos t \rangle$, for $0 \le t \le 2\pi$

61. $C: \mathbf{r}(t) = \langle 3 \sin t, 3 \cos t, 4t \rangle$, for $0 \le t \le 2\pi$

62. Torsion of a plane curve Suppose $\mathbf{r}(t) = \langle f(t), g(t), h(t) \rangle$, where f, g, and h are the quadratic functions $f(t) = a_1 t^2 + b_1 t + c_1$, $g(t) = a_2 t^2 + b_2 t + c_2$, and $h(t) = a_3 t^2 + b_3 t + c_3$, and where at least one of the leading coefficients a_1, a_2, and a_3 is nonzero. Apart from a set of degenerate cases (for example, $\mathbf{r}(t) = \langle t^2, t^2, t^2 \rangle$, whose graph is a line), it can be shown that the graph of $\mathbf{r}(t)$ is a parabola that lies in a plane (Exercise 63).

a. Show by direct computation that $\mathbf{v} \times \mathbf{a}$ is constant. Then explain why the unit binormal vector is constant at all points on the curve. What does this result say about the torsion of the curve?

b. Compute $\mathbf{a}'(t)$ and explain why the torsion is zero at all points on the curve for which the torsion is defined.

63. Families of plane curves Let f and g be continuous on an interval I. Consider the curve

$$C: \mathbf{r}(t) = \langle a_1 f(t) + a_2 g(t) + a_3, b_1 f(t) + b_2 g(t) + b_3,$$
$$c_1 f(t) + c_2 g(t) + c_3 \rangle,$$

for t in I, and where a_i, b_i, and c_i, for $i = 1, 2$, and 3, are real numbers.

a. Show that, in general, C lies in a plane.

b. Explain why the torsion is zero at all points of C for which the torsion is defined.

64. Length of a DVD groove The capacity of a single-sided, single-layer digital versatile disc (DVD) is approximately 4.7 billion bytes—enough to store a two-hour movie. (Newer double-sided, double-layer DVDs have about four times that capacity, and Blu-ray discs are in the range of 50 gigabytes.) A DVD consists of a single "groove" that spirals outward from the inner edge to the outer edge of the storage region.

a. First consider the spiral given in polar coordinates by $r = \dfrac{t\theta}{2\pi}$, where $0 \le \theta \le 2\pi N$ and successive loops of the spiral are t units apart. Explain why this spiral has N loops and why the entire spiral has a radius of $R = Nt$ units. Sketch three loops of the spiral.

b. Write an integral for the length L of the spiral with N loops.

c. The integral in part (b) can be evaluated exactly, but a good approximation can also be made. Assuming N is large, explain why $\theta^2 + 1 \approx \theta^2$. Use this approximation to simplify the integral in part (b) and show that $L \approx t\pi N^2 = \dfrac{\pi R^2}{t}$.

d. Now consider a DVD with an inner radius of $r = 2.5$ cm and an outer radius of $R = 5.9$ cm. Model the groove by a spiral with a thickness of $t = 1.5$ microns $= 1.5 \times 10^{-6}$ m. Because of the hole in the DVD, the lower limit in the arc length integral is not $\theta = 0$. What are the limits of integration?

e. Use the approximation in part (c) to find the length of the DVD groove. Express your answer in centimeters and miles.

Chapter 14 Guided Projects

Applications of the material in this chapter and related topics can be found in the following Guided Projects. For additional information, see the Preface.

• Designing a trajectory

• Bezier curves for graphic design

• Kepler's laws

15

Functions of Several Variables

Chapter Preview   The vectors of Chapter 13 and the vector-valued functions of Chapter 14 took us into three-dimensional space for the first time. In this chapter, we step into three-dimensional space along a different path by considering functions with several independent variables and one dependent variable. All the familiar properties of single-variable functions—domains, graphs, limits, continuity, and derivatives—have generalizations for multivariable functions, although you will also see subtle differences and new features. With functions of several independent variables, we work with *partial derivatives*, which, in turn, give rise to directional derivatives and the *gradient*, a fundamental concept in calculus. Partial derivatives allow us to find maximum and minimum values of multivariable functions. We define tangent planes, rather than tangent lines, that enable us to make linear approximations. The chapter ends with a survey of optimization problems in several variables.

15.1 Graphs and Level Curves

In Chapter 14, we discussed vector-valued functions with one independent variable and several dependent variables. We now reverse the situation and consider functions with several independent variables and one dependent variable. Such functions are aptly called *functions of several variables* or *multivariable functions*.

To set the stage, consider the following practical questions that illustrate a few of the many applications of functions of several variables.

- What is the probability that one man selected randomly from a large group of men weighs more than 200 pounds and is over 6 feet tall? (The answer depends on two variables, weight and height.)

- Where on the wing of an airliner flying at a speed of 550 mi/hr is the pressure greatest? (Pressure depends on the x-, y-, and z-coordinates of various points on the wing.)

- A physician knows the optimal blood concentration of an antibiotic needed by a patient. What dose of antibiotic is needed and how often should it be given to reach this optimal level? (The concentration depends (at least) on the amount of the dose, the frequency with which it is administered, and the weight of the patient.)

Although we don't answer these questions immediately, they clearly suggest the scope and importance of the topic. First, we must introduce the idea of a function of several variables.

Functions of Two Variables

The key concepts related to functions of several variables are most easily presented in the case of two independent variables; the extension to three or more variables is then straight-forward. In general, functions of two variables are written *explicitly* in the form

$$z = f(x, y)$$

or *implicitly* in the form

$$F(x, y, z) = 0.$$

Both forms are important, but for now, we consider explicitly defined functions.

The concepts of domain and range carry over directly from functions of a single variable.

DEFINITION Function, Domain, and Range with Two Independent Variables

A **function** $z = f(x, y)$ assigns to each point (x, y) in a set D in $\mathbb{R}^2$ a unique real number z in a subset of $\mathbb{R}$. The set D is the **domain** of f. The **range** of f is the set of real numbers z that are assumed as the points (x, y) vary over the domain (Figure 15.1).

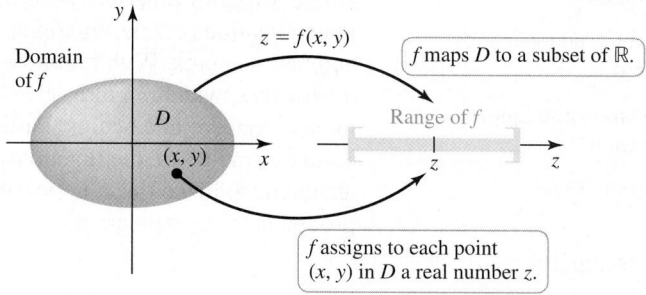

Figure 15.1

As with functions of one variable, a function of several variables may have a domain that is restricted by the context of the problem. For example, if the independent variables correspond to price or length or population, they take only nonnegative values, even though the associated function may be defined for negative values of the variables. If not stated otherwise, D is the set of all points for which the function is defined.

A polynomial in x and y consists of sums and products of polynomials in x and polynomials in y; for example, $f(x, y) = x^2y - 2xy - xy^2$. Such polynomials are defined for all values of x and y, so their domain is $\mathbb{R}^2$. A quotient of two polynomials in x and y, such as $h(x, y) = \dfrac{xy}{x - y}$, is a rational function in x and y. The domain of a rational function excludes points at which the denominator is zero, so the domain of h is $\{(x, y): x \neq y\}$.

EXAMPLE 1   Finding domains Find the domain of the function $g(x, y) = \sqrt{4 - x^2 - y^2}$.

SOLUTION Because g involves a square root, its domain consists of ordered pairs (x, y) for which $4 - x^2 - y^2 \geq 0$ or $x^2 + y^2 \leq 4$. Therefore, the domain of g is $\{(x, y): x^2 + y^2 \leq 4\}$, which is the set of points on or within the circle of radius 2 centered at the origin in the xy-plane (a *disk* of radius 2) (Figure 15.2).

Related Exercises 17–18 ◄

Graphs of Functions of Two Variables

The **graph** of a function f of two variables is the set of points (x, y, z) that satisfy the equation $z = f(x, y)$. More specifically, for each point (x, y) in the domain of f, the point $(x, y, f(x, y))$ lies on the graph of f (Figure 15.3). A similar definition applies to relations of the form $F(x, y, z) = 0$.

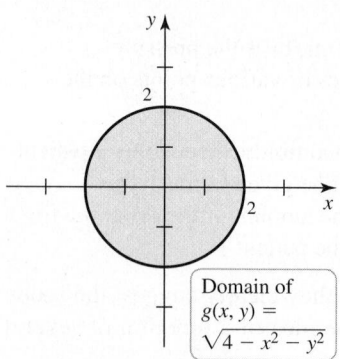

Figure 15.2

QUICK CHECK 1 Find the domains of $f(x, y) = \sin xy$ and $g(x, y) = \sqrt{(x^2 + 1)y}$. ◄

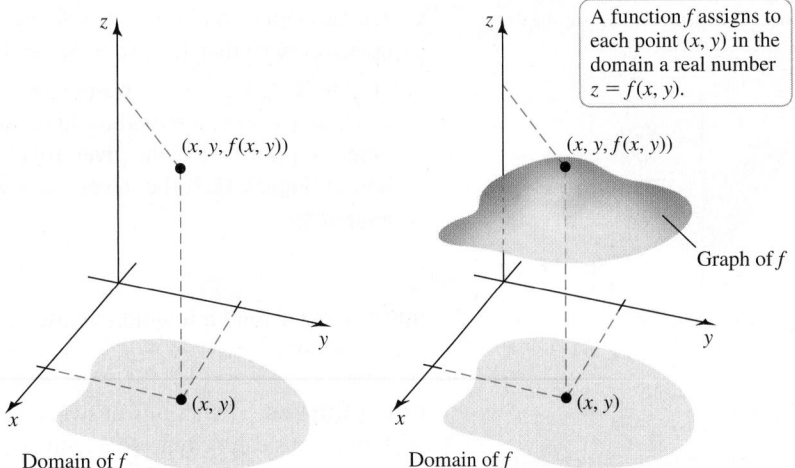

Figure 15.3

QUICK CHECK 2 Does the graph of a hyperboloid of one sheet represent a function? Does the graph of a cone with its axis parallel to the x-axis represent a function? ◄

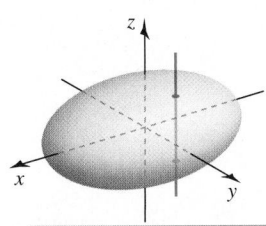

An ellipsoid does not pass the vertical line test: not the graph of a function.

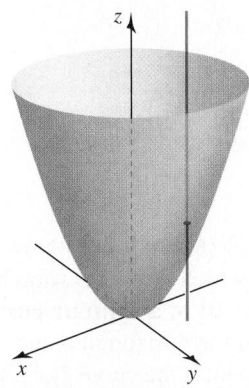

This elliptic paraboloid passes the vertical line test: graph of a function.

Figure 15.4

Like functions of one variable, functions of two variables must pass a **vertical line test**. A relation of the form $F(x, y, z) = 0$ is a function provided every line parallel to the z-axis intersects the graph of the relation at most once. For example, an ellipsoid (discussed in Section 13.6) is not the graph of a function because some vertical lines intersect the surface twice. On the other hand, an elliptic paraboloid of the form $z = ax^2 + by^2$ does represent a function (**Figure 15.4**).

EXAMPLE 2 **Graphing two-variable functions** Find the domain and range of the following functions. Then sketch a graph.

a. $f(x, y) = 2x + 3y - 12$ **b.** $g(x, y) = x^2 + y^2$ **c.** $h(x, y) = \sqrt{1 + x^2 + y^2}$

SOLUTION

a. Letting $z = f(x, y)$, we have the equation $z = 2x + 3y - 12$, or $2x + 3y - z = 12$, which describes a plane with a normal vector $\langle 2, 3, -1 \rangle$ (Section 13.5). The domain consists of all points in $\mathbb{R}^2$, and the range is $\mathbb{R}$. We sketch the surface by noting that the x-intercept is $(6, 0, 0)$ (setting $y = z = 0$); the y-intercept is $(0, 4, 0)$ and the z-intercept is $(0, 0, -12)$ (**Figure 15.5**).

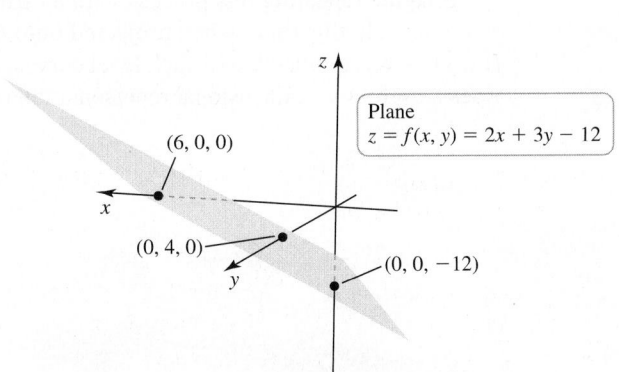

Plane
$z = f(x, y) = 2x + 3y - 12$

$(6, 0, 0)$

$(0, 4, 0)$

$(0, 0, -12)$

Figure 15.5

Paraboloid
$z = f(x, y) = x^2 + y^2$

Figure 15.6

b. Letting $z = g(x, y)$, we have the equation $z = x^2 + y^2$, which describes an elliptic paraboloid that opens upward with vertex $(0, 0, 0)$. The domain is $\mathbb{R}^2$ and the range consists of all nonnegative real numbers (**Figure 15.6**).

Upper sheet of hyperboloid of two sheets
$z = \sqrt{1 + x^2 + y^2}$

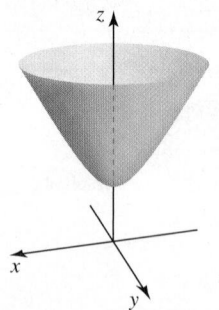

Figure 15.7

▶ To anticipate results that appear later in the chapter, notice how the streams in the topographic map—which flow downhill—cross the level curves roughly at right angles.

c. The domain of the function is $\mathbb{R}^2$ because the quantity under the square root is always positive. Note that $1 + x^2 + y^2 \geq 1$, so the range is $\{z: z \geq 1\}$. Squaring both sides of $z = \sqrt{1 + x^2 + y^2}$, we obtain $z^2 = 1 + x^2 + y^2$, or $-x^2 - y^2 + z^2 = 1$. This is the equation of a hyperboloid of two sheets that opens along the z-axis. Because the range is $\{z: z \geq 1\}$, the given function represents only the upper sheet of the hyperboloid (**Figure 15.7**; the lower sheet was introduced when we squared the original equation).

Related Exercises 25, 27, 29 ◀

QUICK CHECK 3 Find a function whose graph is the lower half of the hyperboloid $-x^2 - y^2 + z^2 = 1$. ◀

Level Curves   Functions of two variables are represented by surfaces in $\mathbb{R}^3$. However, such functions can be represented in another illuminating way, which is used to make topographic maps (**Figure 15.8**).

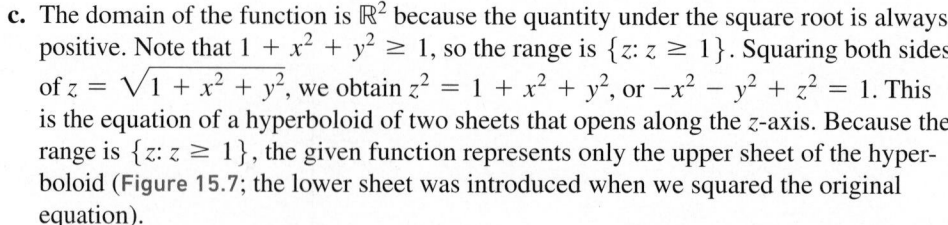

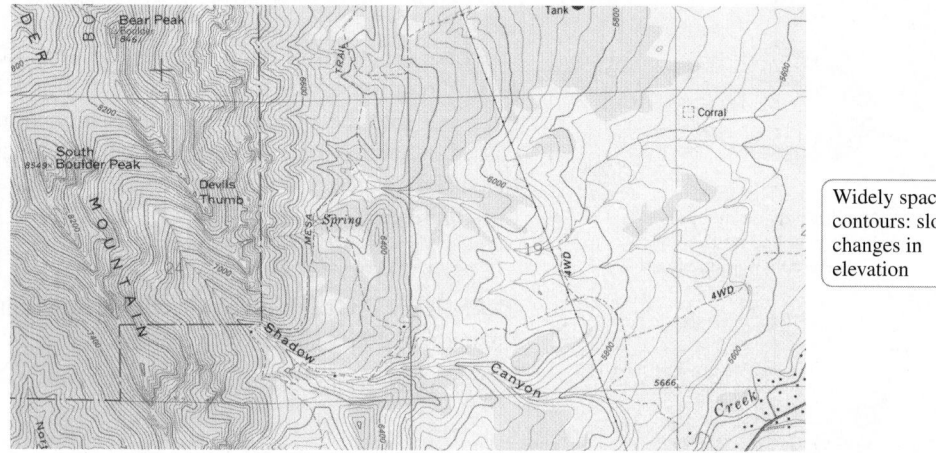

Closely spaced contours: rapid changes in elevation

Widely spaced contours: slow changes in elevation

Figure 15.8

▶ A contour curve is a trace (Section 13.6) in the plane $z = z_0$.

▶ A level curve may not always be a single curve. It might consist of a point ($x^2 + y^2 = 0$) or it might consist of several lines or curves ($xy = 0$).

Consider a surface defined by the function $z = f(x, y)$ (**Figure 15.9**). Now imagine stepping onto the surface and walking along a path on which your elevation has the constant value $z = z_0$. The path you walk on the surface is part of a **contour curve**; the complete contour curve is the intersection of the surface and the horizontal plane $z = z_0$. When the contour curve is projected onto the xy-plane, the result is the curve $f(x, y) = z_0$. This curve in the xy-plane is called a **level curve**.

Imagine repeating this process with a different constant value of z, say, $z = z_1$. The path you walk this time, when projected onto the xy-plane, is part of another level curve $f(x, y) = z_1$. A collection of such level curves, corresponding to different values of z, provides a useful two-dimensional representation of the surface (**Figure 15.10**).

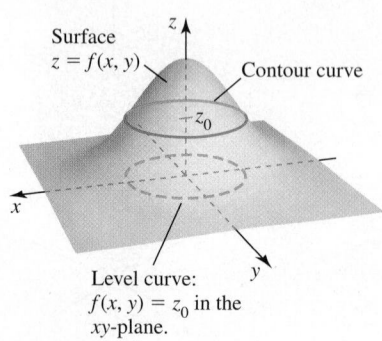

Surface
$z = f(x, y)$
Contour curve
z_0

Level curve:
$f(x, y) = z_0$ in the
xy-plane.

Figure 15.9

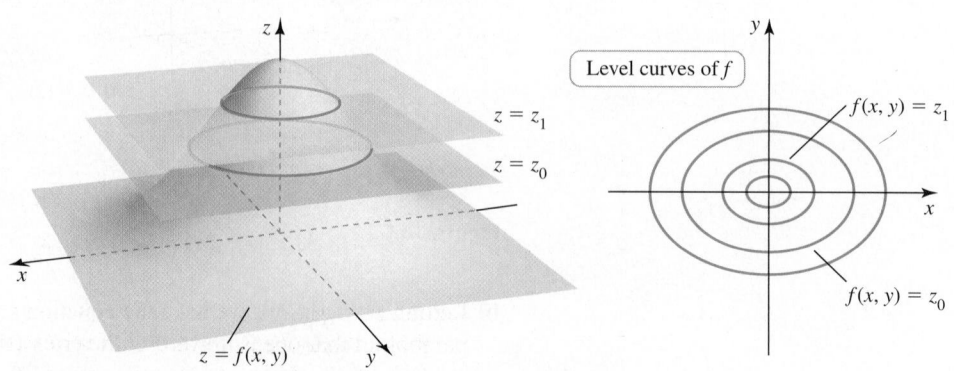

$z = z_1$

$z = z_0$

Level curves of f

$f(x, y) = z_1$

$f(x, y) = z_0$

$z = f(x, y)$

QUICK CHECK 4 Can two level curves of a function intersect? Explain. ◀

Figure 15.10

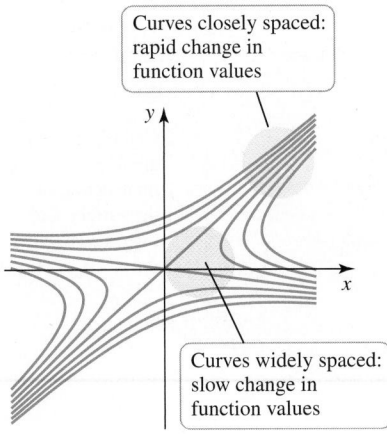

Curves closely spaced: rapid change in function values

Curves widely spaced: slow change in function values

Figure 15.11

QUICK CHECK 5 Describe in words the level curves of the top half of the sphere $x^2 + y^2 + z^2 = 1$. ◄

Assuming two adjacent level curves always correspond to the same change in z, widely spaced level curves indicate gradual changes in z-values, while closely spaced level curves indicate rapid changes in some directions (**Figure 15.11**). Concentric closed level curves generally indicate either a peak or a depression on the surface.

EXAMPLE 3 **Level curves** Find and sketch the level curves of the following surfaces.

a. $f(x, y) = y - x^2 - 1$ **b.** $f(x, y) = e^{-x^2 - y^2}$

SOLUTION

a. The level curves are described by the equation $y - x^2 - 1 = z_0$, where z_0 is a constant in the range of f. For all values of z_0, these curves are parabolas in the xy-plane, as seen by writing the equation in the form $y = x^2 + z_0 + 1$. For example:

- With $z_0 = 0$, the level curve is the parabola $y = x^2 + 1$; along this curve, the surface has an elevation (z-coordinate) of 0.
- With $z_0 = -1$, the level curve is $y = x^2$; along this curve, the surface has an elevation of -1.
- With $z_0 = 1$, the level curve is $y = x^2 + 2$, along which the surface has an elevation of 1.

As shown in **Figure 15.12a**, the level curves form a family of shifted parabolas. When these level curves are labeled with their z-coordinates, the graph of the surface $z = f(x, y)$ can be visualized (**Figure 15.12b**).

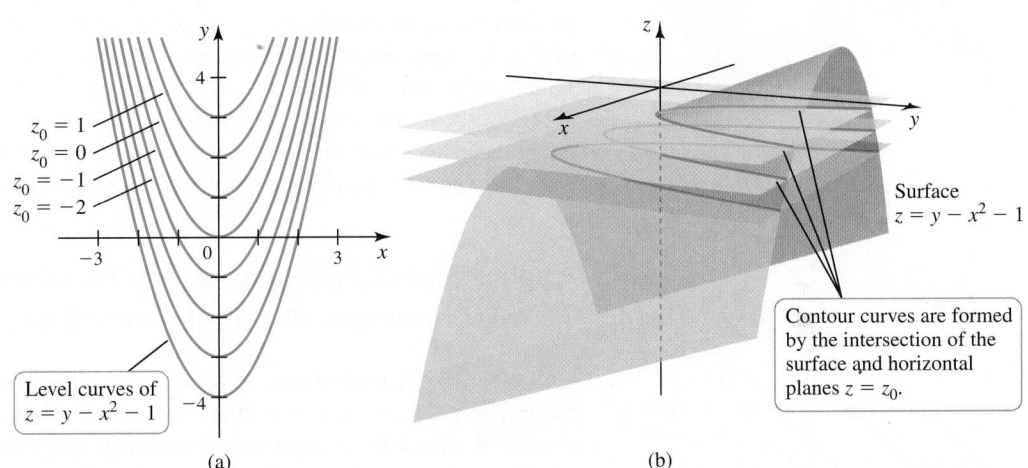

$z_0 = 1$
$z_0 = 0$
$z_0 = -1$
$z_0 = -2$

Level curves of $z = y - x^2 - 1$

(a)

Surface $z = y - x^2 - 1$

Contour curves are formed by the intersection of the surface and horizontal planes $z = z_0$.

(b)

Figure 15.12

b. The level curves satisfy the equation $e^{-x^2 - y^2} = z_0$, where z_0 is a positive constant. Taking the natural logarithm of both sides gives the equation $x^2 + y^2 = -\ln z_0$, which describes circular level curves. These curves can be sketched for all values of z_0 with $0 < z_0 \leq 1$ (because the right side of $x^2 + y^2 = -\ln z_0$ must be nonnegative). For example:

- With $z_0 = 1$, the level curve satisfies the equation $x^2 + y^2 = 0$, whose solution is the single point $(0, 0)$; at this point, the surface has an elevation of 1.
- With $z_0 = e^{-1}$, the level curve is $x^2 + y^2 = -\ln e^{-1} = 1$, which is a circle centered at $(0, 0)$ with a radius of 1; along this curve the surface has an elevation of $e^{-1} \approx 0.37$.

In general, the level curves are circles centered at $(0, 0)$; as the radii of the circles increase, the corresponding z-values decrease. **Figure 15.13a** shows the level curves, with larger z-values corresponding to darker shades. From these labeled level curves, we can reconstruct the graph of the surface (**Figure 15.13b**).

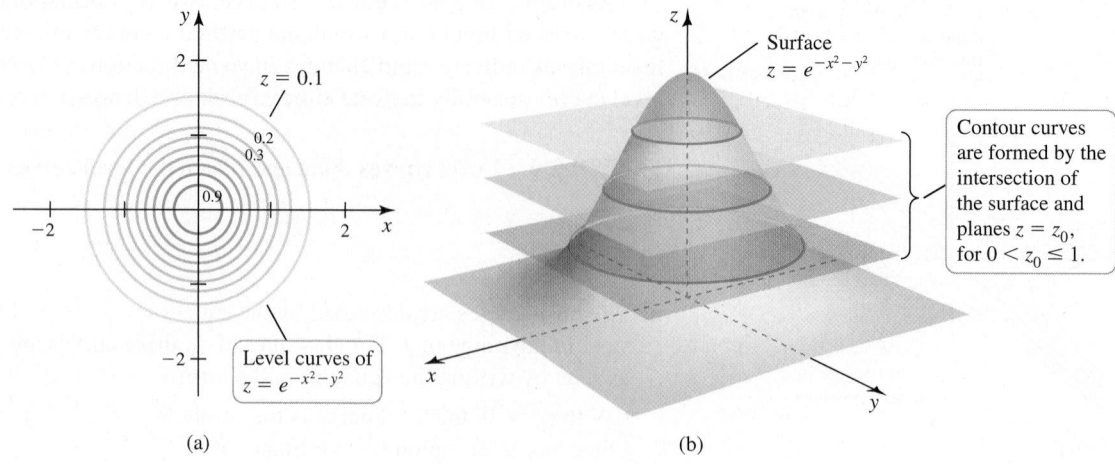

(a)

(b)

Figure 15.13

Related Exercises 37, 40 ◄

QUICK CHECK 6 Does the surface in Example 3b have a level curve for $z_0 = 0$? Explain. ◄

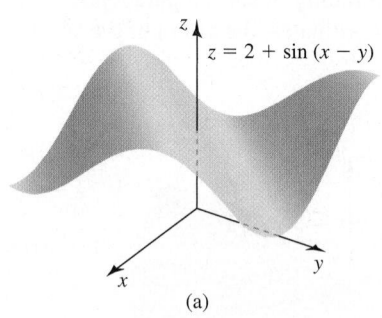

(a)

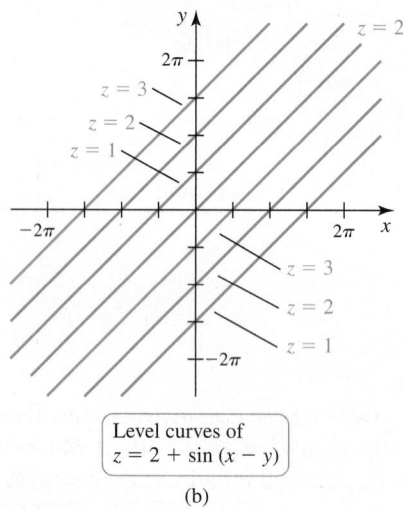

Level curves of
$z = 2 + \sin(x - y)$

(b)

Figure 15.14

EXAMPLE 4   Level curves The graph of the function

$$f(x, y) = 2 + \sin(x - y)$$

is shown in **Figure 15.14a**. Sketch several level curves of the function.

SOLUTION The level curves are $f(x, y) = 2 + \sin(x - y) = z_0$, or $\sin(x - y) = z_0 - 2$. Because $-1 \le \sin(x - y) \le 1$, the admissible values of z_0 satisfy $-1 \le z_0 - 2 \le 1$, or, equivalently, $1 \le z_0 \le 3$. For example, when $z_0 = 2$, the level curves satisfy $\sin(x - y) = 0$. The solutions of this equation are $x - y = k\pi$, or $y = x - k\pi$, where k is an integer. Therefore, the surface has an elevation of 2 on this set of lines. With $z_0 = 1$ (the minimum value of z), the level curves satisfy $\sin(x - y) = -1$. The solutions are $x - y = -\pi/2 + 2k\pi$, where k is an integer; along these lines, the surface has an elevation of 1. Here we have an example in which each level curve is an infinite collection of lines of slope 1 (**Figure 15.14b**).

Related Exercise 43 ◄

Applications of Functions of Two Variables

The following examples offer two of many applications of functions of two variables.

EXAMPLE 5   A probability function of two variables Suppose on a particular day, the fraction of students on campus infected with flu is r, where $0 \le r \le 1$. If you have n random (possibly repeated) encounters with students during the day, the probability of meeting *at least* one infected person is $p(n, r) = 1 - (1 - r)^n$ (**Figure 15.15a**). Discuss this probability function.

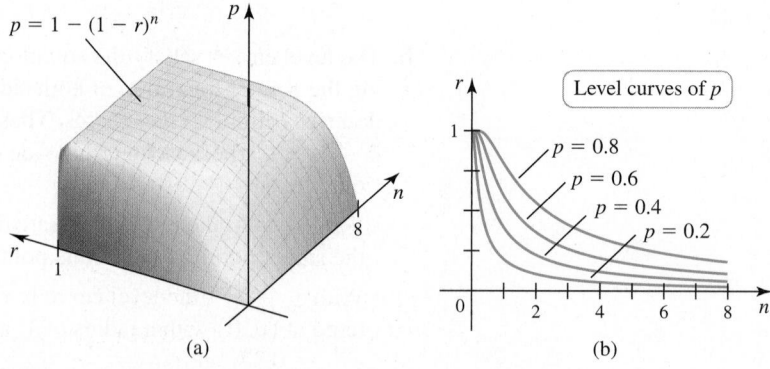

(a)

(b)

Figure 15.15

SOLUTION The independent variable r is restricted to the interval $[0, 1]$ because it is a fraction of the population. The other independent variable n is any nonnegative integer; for the purposes of graphing, we treat n as a real number in the interval $[0, 8]$. With $0 \le r \le 1$, note that $0 \le 1 - r \le 1$. If n is nonnegative, then $0 \le (1 - r)^n \le 1$, and

QUICK CHECK 7 In Example 5, if 50% of the population is infected, what is the probability of meeting at least one infected person in five encounters? ◄

Table 15.1

			n		
	2	**5**	**10**	**15**	**20**
0.05	0.10	0.23	0.40	0.54	0.64
0.1	0.19	0.41	0.65	0.79	0.88
0.3	0.51	0.83	0.97	1	1
0.5	0.75	0.97	1	1	1
0.7	0.91	1	1	1	1

r labels the rows.

it follows that $0 \le p(n, r) \le 1$. Therefore, the range of the function is $[0, 1]$, which is consistent with the fact that p is a probability.

The level curves (Figure 15.15b) show that for a fixed value of n, the probability of at least one encounter increases with r; and for a fixed value of r, the probability increases with n. Therefore, as r increases or as n increases, the probability approaches 1 (at a surprising rate). If 10% of the population is infected ($r = 0.1$) and you have $n = 10$ encounters, then the probability of at least one encounter with an infected person is $p(10, 0.1) \approx 0.651$, which is about 2 in 3.

A numerical view of this function is given in Table 15.1, where we see probabilities tabulated for various values of n and r (rounded to two digits). The numerical values confirm the preceding observations.

Related Exercise 44 ◄

EXAMPLE 6 **Electric potential function in two variables** The electric field at points in the xy-plane due to two point charges located at $(0, 0)$ and $(1, 0)$ is related to the electric potential function

$$\varphi(x, y) = \frac{2}{\sqrt{x^2 + y^2}} + \frac{2}{\sqrt{(x - 1)^2 + y^2}}.$$

Discuss the electric potential function.

> The electric potential function, often denoted φ (pronounced *fee* or *fie*), is a scalar-valued function from which the electric field can be computed. Potential functions are discussed in detail in Chapter 17.

> A function that grows without bound near a point, as in the case of the electric potential function, is said to have a *singularity* at that point. A singularity is analogous to a vertical asymptote in a function of one variable.

SOLUTION The domain of the function contains all points of $\mathbb{R}^2$ except $(0, 0)$ and $(1, 0)$ where the charges are located. As these points are approached, the potential function becomes arbitrarily large (Figure 15.16a). The potential approaches zero as x or y increases in magnitude. These observations imply that the range of the potential function is all positive real numbers. The level curves of φ are closed curves, encircling either a single charge (at small distances) or both charges (at larger distances; Figure 15.16b).

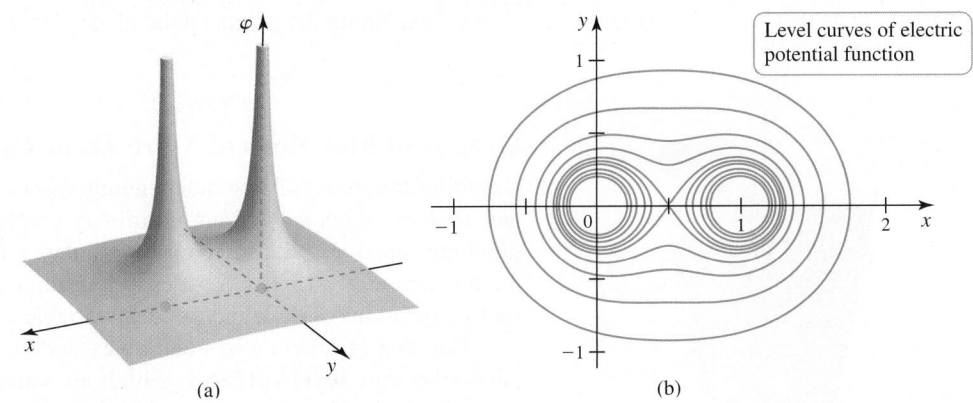

Level curves of electric potential function

QUICK CHECK 8 In Example 6, what is the electric potential at the point $\left(\frac{1}{2}, 0\right)$? ◄

Figure 15.16

(a)   (b)

Related Exercise 45 ◄

Functions of More Than Two Variables

Many properties of functions of two independent variables extend naturally to functions of three or more variables. A function of three variables is defined explicitly in the form $w = f(x, y, z)$ and implicitly in the form $F(x, y, z, w) = 0$. With more than three independent variables, the variables are usually written $x_1, \ldots, x_n$. Table 15.2 shows the progression of functions of several variables.

Table 15.2

Number of Independent Variables	Explicit Form	Implicit Form	Graph Resides In . . .
1	$y = f(x)$	$F(x, y) = 0$	$\mathbb{R}^2$ (xy-plane)
2	$z = f(x, y)$	$F(x, y, z) = 0$	$\mathbb{R}^3$ (xyz-space)
3	$w = f(x, y, z)$	$F(x, y, z, w) = 0$	$\mathbb{R}^4$
n	$x_{n+1} = f(x_1, x_2, \ldots, x_n)$	$F(x_1, x_2, \ldots, x_n, x_{n+1}) = 0$	$\mathbb{R}^{n+1}$

The concepts of domain and range extend from the one- and two-variable cases in an obvious way.

> **DEFINITION Function, Domain, and Range with n Independent Variables**
>
> The **function** $x_{n+1} = f(x_1, x_2, \ldots, x_n)$ assigns a unique real number x_{n+1} to each point $(x_1, x_2, \ldots, x_n)$ in a set D in $\mathbb{R}^n$. The set D is the **domain** of f. The **range** is the set of real numbers x_{n+1} that are assumed as the points $(x_1, x_2, \ldots, x_n)$ vary over the domain.

EXAMPLE 7   Finding domains Find the domain of the following functions.

a. $g(x, y, z) = \sqrt{16 - x^2 - y^2 - z^2}$ **b.** $h(x, y, z) = \dfrac{12y^2}{z - y}$

SOLUTION

a. Values of the variables that make the argument of a square root negative must be excluded from the domain. In this case, the quantity under the square root is nonnegative provided

$$16 - x^2 - y^2 - z^2 \geq 0, \quad \text{or} \quad x^2 + y^2 + z^2 \leq 16.$$

> ➤ Recall that a closed ball of radius r is the set of all points on or within a sphere of radius r.

Therefore, the domain of g is a closed ball in $\mathbb{R}^3$ of radius 4 centered at the origin.

b. Values of the variables that make a denominator zero must be excluded from the domain. In this case, the denominator vanishes for all points in $\mathbb{R}^3$ that satisfy $z - y = 0$, or $y = z$. Therefore, the domain of h is the set $\{(x, y, z): y \neq z\}$. This set is $\mathbb{R}^3$ excluding the points on the plane $y = z$.

> **QUICK CHECK 9** What is the domain of the function $w = f(x, y, z) = 1/xyz$? ◄

Related Exercises 51–52 ◄

Graphs of Functions of More Than Two Variables

Graphing functions of *two* independent variables requires a three-dimensional coordinate system, which is the limit of ordinary graphing methods. Clearly, difficulties arise in graphing functions with three or more independent variables. For example, the graph of the function $w = f(x, y, z)$ resides in four dimensions. Here are two approaches to representing functions of three independent variables.

The idea of level curves can be extended. With the function $w = f(x, y, z)$, level curves become **level surfaces**, which are surfaces in $\mathbb{R}^3$ on which w is constant. For example, the level surfaces of the function

$$w = f(x, y, z) = \sqrt{z - x^2 - 2y^2}$$

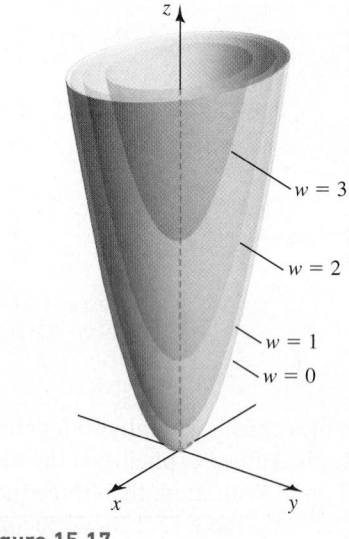

Figure 15.17

satisfy $w = \sqrt{z - x^2 - 2y^2} = C$, where C is a nonnegative constant. This equation is satisfied when $z = x^2 + 2y^2 + C^2$. Therefore, the level surfaces are elliptic paraboloids, stacked one inside another (**Figure 15.17**).

Another approach to displaying functions of three variables is to use colors to represent the fourth dimension. **Figure 15.18a** shows the electrical activity of the heart at one snapshot in time. The three independent variables correspond to locations in the heart. At each point, the value of the electrical activity, which is the dependent variable, is coded by colors.

In **Figure 15.18b**, the dependent variable is the switching speed in an integrated circuit, again represented by colors, as it varies over points of the domain. Software to produce such images, once expensive and inefficient, has become much more accessible.

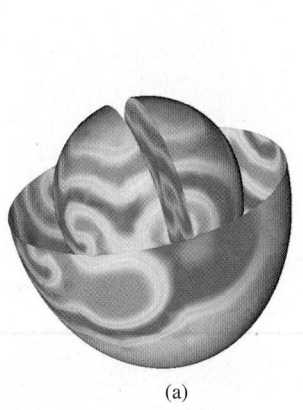

(a)   (b)

Figure 15.18

SECTION 15.1 EXERCISES

Getting Started

1. A function is defined by $z = x^2y - xy^2$. Identify the independent and dependent variables.

2. What is the domain of $f(x, y) = x^2y - xy^2$?

3. What is the domain of $g(x, y) = \dfrac{1}{xy}$?

4. What is the domain of $h(x, y) = \sqrt{x - y}$?

5. How many axes (or how many dimensions) are needed to graph the function $z = f(x, y)$? Explain.

6. Explain how to graph the level curves of a surface $z = f(x, y)$.

7. Given the function $f(x, y) = \sqrt{10 - x + y}$, evaluate $f(2, 1)$ and $f(-9, -3)$.

8. Given the function $g(x, y, z) = \dfrac{x + y}{z}$, evaluate $g(1, 5, 3)$ and $g(3, 7, 2)$.

9–10. *The function $z = f(x, y)$ gives the elevation z (in hundreds of feet) of a hillside above the point (x, y). Use the level curves of f to answer the following questions (see figure).*

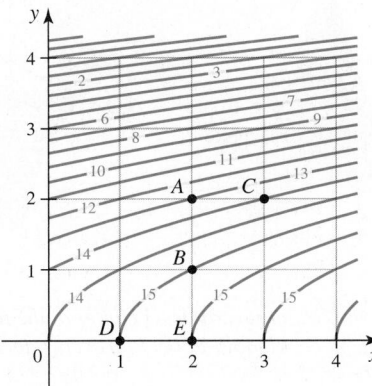

9. Katie and Zeke are standing on the surface above the point $A(2, 2)$.

   a. At what elevation are Katie and Zeke standing?
   b. Katie hikes south to the point on the surface above $B(2, 1)$ and Zeke hikes east to the point on the surface above $C(3, 2)$. Who experienced the greater elevation change and what is the difference in their elevations?

10. Katie and Zeke are standing on the surface above $D(1, 0)$. Katie hikes on the surface above the level curve containing $D(1, 0)$ to $B(2, 1)$ and Zeke walks east along the surface to $E(2, 0)$. What can be said about the elevations of Katie and Zeke during their hikes?

11. Describe in words the level curves of the paraboloid $z = x^2 + y^2$.

12. How many axes (or how many dimensions) are needed to graph the level surfaces of $w = f(x, y, z)$? Explain.

13. The domain of $Q = f(u, v, w, x, y, z)$ lies in $\mathbb{R}^n$ for what value of n? Explain.

14. Give two methods for graphically representing a function with three independent variables.

Practice Exercises

15–24. Domains *Find the domain of the following functions.*

15. $f(x, y) = 2xy - 3x + 4y$ 16. $f(x, y) = \cos(x^2 - y^2)$

17. $f(x, y) = \sqrt{25 - x^2 - y^2}$ 18. $f(x, y) = \dfrac{1}{\sqrt{x^2 + y^2 - 25}}$

19. $f(x, y) = \sin\dfrac{x}{y}$ 20. $f(x, y) = \dfrac{12}{y^2 - x^2}$

21. $g(x, y) = \ln(x^2 - y)$ 22. $f(x, y) = \sin^{-1}(y - x^2)$

23. $g(x, y) = \sqrt{\dfrac{xy}{x^2 + y^2}}$ 24. $h(x, y) = \sqrt{x - 2y + 4}$

25–33. Graphs of familiar functions *Use what you learned about surfaces in Sections 13.5 and 13.6 to sketch a graph of the following functions. In each case, identify the surface and state the domain and range of the function.*

25. $f(x, y) = 6 - x - 2y$ 26. $g(x, y) = 4$

27. $p(x, y) = x^2 - y^2$ 28. $h(x, y) = 2x^2 + 3y^2$

29. $G(x, y) = -\sqrt{1 + x^2 + y^2}$

30. $F(x, y) = \sqrt{1 - x^2 - y^2}$

31. $P(x, y) = \sqrt{x^2 + y^2 - 1}$

32. $H(x, y) = \sqrt{x^2 + y^2}$ 33. $g(x, y) = \sqrt{16 - 4x^2}$

34. Matching level curves with surfaces Match surfaces a–f in the figure with level curves A–F.

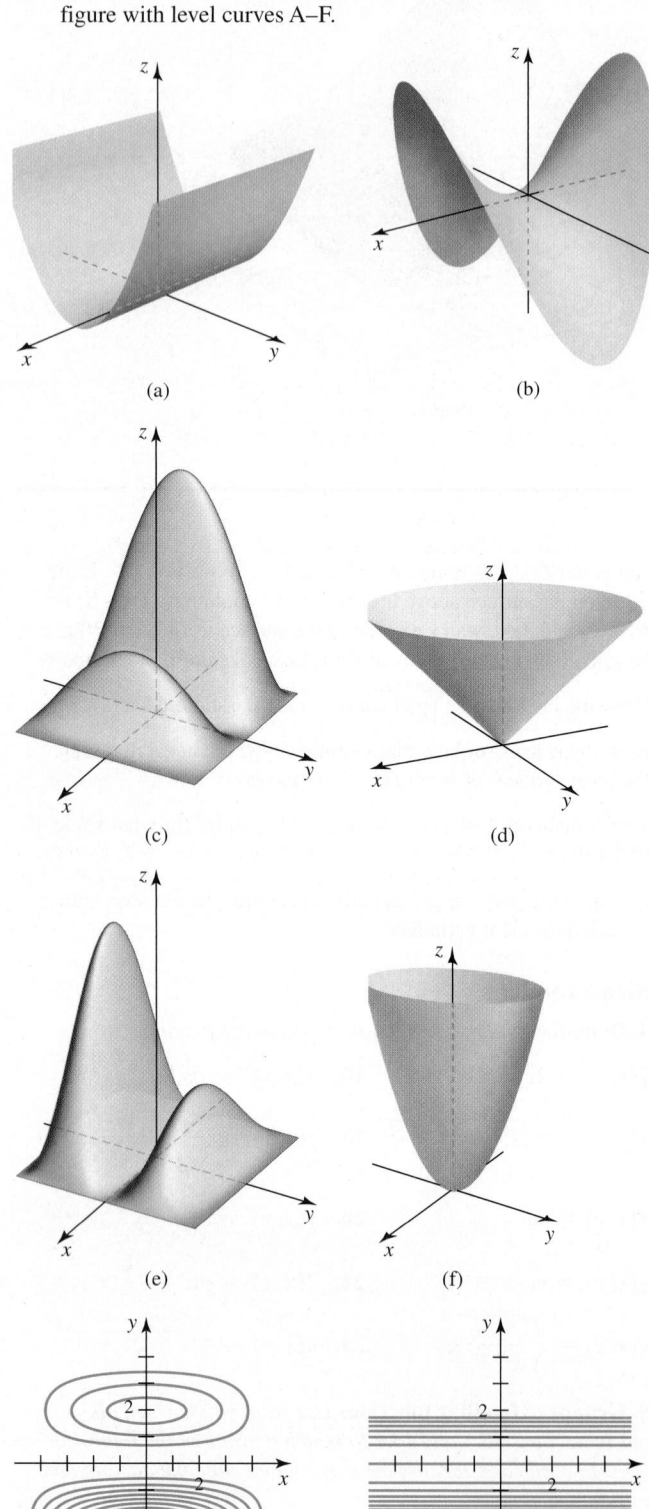

(a)

(b)

(c)

(d)

(e)

(f)

(A)

(B)

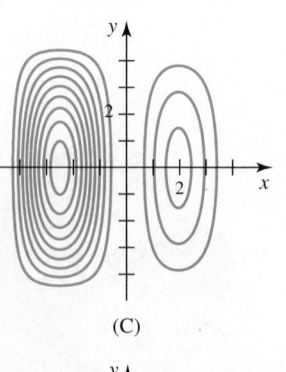

(C)

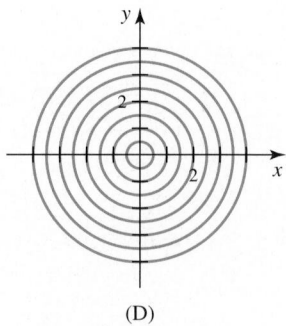

(D)

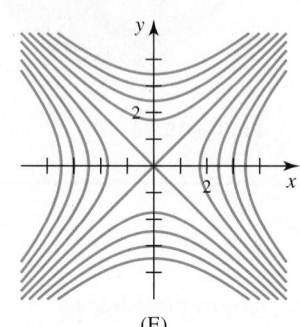

(E)

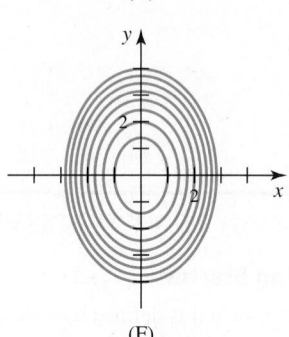

(F)

35. Matching surfaces Match functions a–d with surfaces A–D in the figure.

a. $f(x, y) = \cos xy$ **b.** $g(x, y) = \ln (x^2 + y^2)$

c. $h(x, y) = \dfrac{1}{x - y}$ **d.** $p(x, y) = \dfrac{1}{1 + x^2 + y^2}$

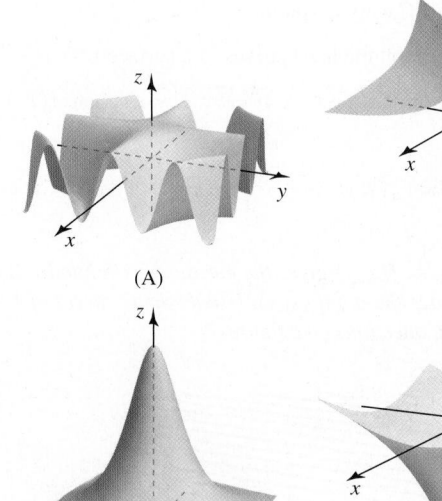

(A)

(B)

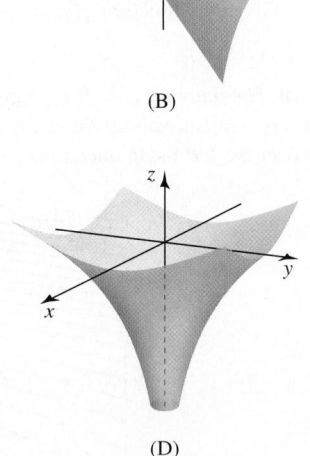

(C)

(D)

T 36–43. Level curves *Graph several level curves of the following functions using the given window. Label at least two level curves with their z-values.*

36. $z = x^2 + y^2$; $[-4, 4] \times [-4, 4]$

37. $z = x - y^2$; $[0, 4] \times [-2, 2]$

38. $z = 2x - y$; $[-2, 2] \times [-2, 2]$

39. $z = \sqrt{x^2 + 4y^2}$; $[-8, 8] \times [-8, 8]$

40. $z = e^{-x^2 - 2y^2}$; $[-2, 2] \times [-2, 2]$

41. $z = \sqrt{25 - x^2 - y^2}$; $[-6, 6] \times [-6, 6]$

42. $z = \sqrt{y - x^2 - 1}$; $[-5, 5] \times [-5, 5]$

43. $z = 3 \cos(2x + y)$; $[-2, 2] \times [-2, 2]$

44. **Earned run average** A baseball pitcher's earned run average (ERA) is $A(e, i) = 9e/i$, where e is the number of earned runs given up by the pitcher and i is the number of innings pitched. Good pitchers have low ERAs. Assume $e \geq 0$ and $i > 0$ are real numbers.

    a. The single-season major league record for the lowest ERA was set by Dutch Leonard of the Detroit Tigers in 1914. During that season, Dutch pitched a total of 224 innings and gave up just 24 earned runs. What was his ERA?

    b. Determine the ERA of a relief pitcher who gives up 4 earned runs in one-third of an inning.

    c. Graph the level curve $A(e, i) = 3$ and describe the relationship between e and i in this case.

T 45. **Electric potential function** The electric potential function for two positive charges, one at $(0, 1)$ with twice the strength of the charge at $(0, -1)$, is given by

$$\varphi(x, y) = \frac{2}{\sqrt{x^2 + (y - 1)^2}} + \frac{1}{\sqrt{x^2 + (y + 1)^2}}.$$

    a. Graph the electric potential using the window $[-5, 5] \times [-5, 5] \times [0, 10]$.

    b. For what values of x and y is the potential φ defined?

    c. Is the electric potential greater at $(3, 2)$ or $(2, 3)$?

    d. Describe how the electric potential varies along the line $y = x$.

T 46. **Cobb-Douglas production function** The output Q of an economic system subject to two inputs, such as labor L and capital K, is often modeled by the Cobb-Douglas production function $Q(L, K) = cL^a K^b$, where a, b, and c are positive real numbers. When $a + b = 1$, the case is called *constant returns to scale*. Suppose $a = 1/3$, $b = 2/3$, and $c = 40$.

    a. Graph the output function using the window $[0, 20] \times [0, 20] \times [0, 500]$.

    b. If L is held constant at $L = 10$, write the function that gives the dependence of Q on K.

    c. If K is held constant at $K = 15$, write the function that gives the dependence of Q on L.

T 47. **Resistors in parallel** Two resistors wired in parallel in an electrical circuit give an effective resistance of $R(x, y) = \dfrac{xy}{x + y}$, where x and y are the positive resistances of the individual resistors (typically measured in ohms).

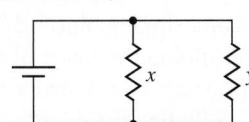

    a. Graph the resistance function using the window $[0, 10] \times [0, 10] \times [0, 5]$.

    b. Estimate the maximum value of R, for $0 < x \leq 10$ and $0 < y \leq 10$.

    c. Explain what it means to say that the resistance function is symmetric in x and y.

T 48. **Level curves of a savings account** Suppose you make a one-time deposit of P dollars into a savings account that earns interest at an annual rate of $p\%$ compounded continuously. The balance in the account after t years is $B(P, r, t) = Pe^{rt}$, where $r = p/100$ (for example, if the annual interest rate is 4%, then $r = 0.04$). Let the interest rate be fixed at $r = 0.04$.

    a. With a target balance of $2000, find the set of all points (P, t) that satisfy $B = 2000$. This curve gives all deposits P and times t that result in a balance of $2000.

    b. Repeat part (a) with $B = \$500, \$1000, \$1500$, and $2500, and draw the resulting level curves of the balance function.

    c. In general, on one level curve, if t increases, does P increase or decrease?

T 49. **Level curves of a savings plan** Suppose you make monthly deposits of P dollars into an account that earns interest at a *monthly* rate of $p\%$. The balance in the account after t years is

$$B(P, r, t) = P\left(\frac{(1 + r)^{12t} - 1}{r}\right), \text{ where } r = \frac{p}{100} \text{ (for example,}$$

if the annual interest rate is 9%, then $p = \dfrac{9}{12} = 0.75$ and $r = 0.0075$). Let the time of investment be fixed at $t = 20$ years.

    a. With a target balance of $20,000, find the set of all points (P, r) that satisfy $B = 20,000$. This curve gives all deposits P and monthly interest rates r that result in a balance of $20,000 after 20 years.

    b. Repeat part (a) with $B = \$5000, \$10,000, \$15,000$, and $25,000, and draw the resulting level curves of the balance function.

50–56. Domains of functions of three or more variables *Find the domain of the following functions. If possible, give a description of the domains (for example, all points outside a sphere of radius 1 centered at the origin).*

50. $f(x, y, z) = 2xyz - 3xz + 4yz$

51. $g(x, y, z) = \dfrac{1}{x - z}$

52. $p(x, y, z) = \sqrt{x^2 + y^2 + z^2 - 9}$

53. $f(x, y, z) = \sqrt{y - z}$

54. $Q(x, y, z) = \dfrac{10}{1 + x^2 + y^2 + 4z^2}$

55. $F(x, y, z) = \sqrt{y - x^2}$

56. $f(w, x, y, z) = \sqrt{1 - w^2 - x^2 - y^2 - z^2}$

57. **Explain why or why not** Determine whether the following statements are true and give an explanation or counterexample.

    a. The domain of the function $f(x, y) = 1 - |x - y|$ is $\{(x, y) : x \geq y\}$.

    b. The domain of the function $Q = g(w, x, y, z)$ is a region in $\mathbb{R}^3$.

    c. All level curves of the plane $z = 2x - 3y$ are lines.

58. **Quarterback passer ratings** One measurement of the quality of a quarterback in the National Football League is known as the *quarterback passer rating*. The rating formula is

$$R(c, t, i, y) = \frac{50 + 20c + 80t - 100i + 100y}{24}, \text{ where } c\% \text{ of}$$

a quarterback's passes were completed, $t\%$ of his passes were thrown for touchdowns, $i\%$ of his passes were intercepted, and an average of y yards were gained per attempted pass.

a. In the 2016/17 NFL playoffs, Atlanta Falcons quarterback Matt Ryan completed 71.43% of his passes, 9.18% of his passes were thrown for touchdowns, none of his passes were intercepted, and he gained an average of 10.35 yards per passing attempt. What was his passer rating in the 2016 playoffs?

b. In the 2016 regular season, New England Patriots quarterback Tom Brady completed 67.36% of his passes, 6.48% of his passes were thrown for touchdowns, 0.46% of his passes were intercepted, and he gained an average of 8.23 yards per passing attempt. What was his passer rating in the 2016 regular season?

c. If c, t, and y remain fixed, what happens to the quarterback passer rating as i increases? Explain your answer with and without mathematics.

(*Source:* www.nfl.com)

T 59. Ideal Gas Law Many gases can be modeled by the Ideal Gas Law, $PV = nRT$, which relates the temperature (T, measured in kelvins (K)), pressure (P, measured in pascals (Pa)), and volume (V, measured in m^3) of a gas. Assume the quantity of gas in question is $n = 1$ mole (mol). The gas constant has a value of $R = 8.3 \text{ m}^3 \text{Pa/mol-K}$.

a. Consider T to be the dependent variable, and plot several level curves (called *isotherms*) of the temperature surface in the region $0 \le P \le 100{,}000$ and $0 \le V \le 0.5$.

b. Consider P to be the dependent variable, and plot several level curves (called *isobars*) of the pressure surface in the region $0 \le T \le 900$ and $0 < V \le 0.5$.

c. Consider V to be the dependent variable, and plot several level curves of the volume surface in the region $0 \le T \le 900$ and $0 < P \le 100{,}000$.

Explorations and Challenges

T 60. Water waves A snapshot of a water wave moving toward shore is described by the function $z = 10 \sin(2x - 3y)$, where z is the height of the water surface above (or below) the xy-plane, which is the level of undisturbed water.

a. Graph the height function using the window $[-5, 5] \times [-5, 5] \times [-15, 15]$.

b. For what values of x and y is z defined?

c. What are the maximum and minimum values of the water height?

d. Give a vector in the xy-plane that is orthogonal to the level curves of the crests and troughs of the wave (which is parallel to the direction of wave propagation).

T 61. Approximate mountains Suppose the elevation of Earth's surface over a 16-mi by 16-mi region is approximated by the function
$$z = 10e^{-(x^2+y^2)} + 5e^{-((x+5)^2+(y-3)^2)/10} + 4e^{-2((x-4)^2+(y+1)^2)}.$$

a. Graph the height function using the window $[-8, 8] \times [-8, 8] \times [0, 15]$.

b. Approximate the points (x, y) where the peaks in the landscape appear.

c. What are the approximate elevations of the peaks?

T 62–68. Graphing functions

a. *Determine the domain and range of the following functions.*

b. *Graph each function using a graphing utility. Be sure to experiment with the window and orientation to give the best perspective on the surface.*

62. $g(x, y) = e^{-xy}$

63. $f(x, y) = |xy|$

64. $p(x, y) = 1 - |x - 1| + |y + 1|$

65. $h(x, y) = \dfrac{x + y}{x - y}$

66. $G(x, y) = \ln(2 + \sin(x + y))$

67. $F(x, y) = \tan^2(x - y)$ **68.** $P(x, y) = \cos x \sin 2y$

T 69–72. Peaks and valleys *The following functions have exactly one isolated peak or one isolated depression (one local maximum or minimum). Use a graphing utility to approximate the coordinates of the peak or depression.*

69. $f(x, y) = x^2 y^2 - 8x^2 - y^2 + 6$

70. $g(x, y) = (x^2 - x - 2)(y^2 + 2y)$

71. $h(x, y) = 1 - e^{-(x^2 + y^2 - 2x)}$

72. $p(x, y) = 2 + |x - 1| + |y - 1|$

73. Level curves of planes Prove that the level curves of the plane $ax + by + cz = d$ are parallel lines in the xy-plane, provided $a^2 + b^2 \ne 0$ and $c \ne 0$.

74–77. Level surfaces *Find an equation for the family of level surfaces corresponding to f. Describe the level surfaces.*

74. $f(x, y, z) = \dfrac{1}{x^2 + y^2 + z^2}$ **75.** $f(x, y, z) = x^2 + y^2 - z$

76. $f(x, y, z) = x^2 - y^2 - z$ **77.** $f(x, y, z) = \sqrt{x^2 + 2z^2}$

78–81. Challenge domains *Find the domain of the following functions. Specify the domain mathematically, and then describe it in words or with a sketch.*

78. $g(x, y, z) = \dfrac{10}{x^2 - (y + z)x + yz}$

79. $f(x, y) = \sin^{-1}(x - y)^2$

80. $f(x, y, z) = \ln(z - x^2 - y^2 + 2x + 3)$

81. $h(x, y, z) = \sqrt[4]{z^2 - xz + yz - xy}$

82. Other balls The closed unit ball in $\mathbb{R}^3$ centered at the origin is the set $\{(x, y, z): x^2 + y^2 + z^2 \le 1\}$. Describe the following alternative unit balls.

a. $\{(x, y, z): |x| + |y| + |z| \le 1\}$

b. $\{(x, y, z): \max\{|x|, |y|, |z|\} \le 1\}$, where $\max\{a, b, c\}$ is the maximum value of a, b, and c

QUICK CHECK ANSWERS

1. $\mathbb{R}^2$; $\{(x, y): y \ge 0\}$ **2.** No; no
3. $z = -\sqrt{1 + x^2 + y^2}$ **4.** No; otherwise the function would have two values at a single point.   **5.** Concentric circles   **6.** No; $z = 0$ is not in the range of the function.
7. 0.97   **8.** 8   **9.** $\{(x, y, z): x \ne 0 \text{ and } y \ne 0 \text{ and } z \ne 0\}$ (which is $\mathbb{R}^3$, excluding the coordinate planes) ◄

15.2 Limits and Continuity

You have now seen examples of functions of several variables, but calculus has not yet entered the picture. In this section, we revisit topics encountered in single-variable calculus and see how they apply to functions of several variables. We begin with the fundamental concepts of limits and continuity.

Limit of a Function of Two Variables

A function f of two variables has a limit L as $P(x, y)$ approaches a fixed point $P_0(a, b)$ if $|f(x, y) - L|$ can be made arbitrarily small for all P in the domain that are sufficiently close to P_0. If such a limit exists, we write

$$\lim_{(x, y) \to (a, b)} f(x, y) = \lim_{P \to P_0} f(x, y) = L.$$

To make this definition more precise, *close to* must be defined carefully.

A point x on the number line is close to another point a provided the distance $|x - a|$ is small (**Figure 15.19a**). In $\mathbb{R}^2$, a point $P(x, y)$ is close to another point $P_0(a, b)$ if the distance between them $|PP_0| = \sqrt{(x - a)^2 + (y - b)^2}$ is small (**Figure 15.19b**). When we say *for all P close to P_0*, it means that $|PP_0|$ is small for points P *on all sides of P_0*.

With this understanding of closeness, we can give a formal definition of a limit with two independent variables. This definition parallels the formal definition of a limit given in Section 2.7 (**Figure 15.20**).

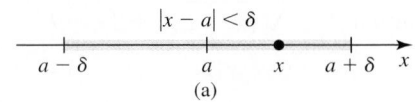

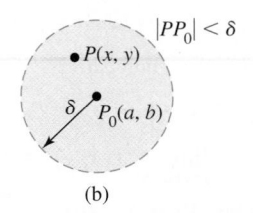

Figure 15.19

> ➤ The formal definition extends naturally to any number of variables. With n variables, the limit point is $P_0(a_1, \ldots, a_n)$, the variable point is $P(x_1, \ldots, x_n)$, and $|PP_0| = \sqrt{(x_1 - a_1)^2 + \cdots + (x_n - a_n)^2}$.

DEFINITION Limit of a Function of Two Variables

The function f has the **limit** L as $P(x, y)$ approaches $P_0(a, b)$, written

$$\lim_{(x, y) \to (a, b)} f(x, y) = \lim_{P \to P_0} f(x, y) = L,$$

if, given any $\varepsilon > 0$, there exists a $\delta > 0$ such that

$$|f(x, y) - L| < \varepsilon$$

whenever (x, y) is in the domain of f and

$$0 < |PP_0| = \sqrt{(x - a)^2 + (y - b)^2} < \delta.$$

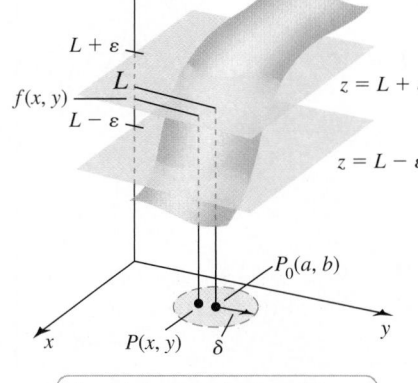

$f(x, y)$ is between $L - \varepsilon$ and $L + \varepsilon$ whenever $P(x, y)$ is within δ of P_0.

Figure 15.20

The condition $|PP_0| < \delta$ means that the distance between $P(x, y)$ and $P_0(a, b)$ is less than δ as P approaches P_0 from all possible directions (**Figure 15.21**). Therefore, the limit exists only if $f(x, y)$ approaches L as P approaches P_0 *along all possible paths* in the domain of f. As shown in upcoming examples, this interpretation is critical in determining whether a limit exists.

As with functions of one variable, we first establish limits of the simplest functions.

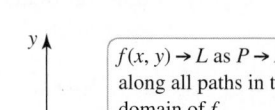

$f(x, y) \to L$ as $P \to P_0$ along all paths in the domain of f.

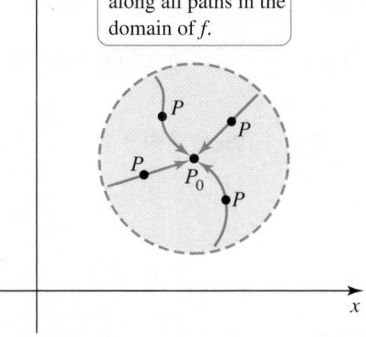

Figure 15.21

THEOREM 15.1   Limits of Constant and Linear Functions

Let a, b, and c be real numbers.

1. Constant function $f(x, y) = c$: $\displaystyle\lim_{(x, y) \to (a, b)} c = c$

2. Linear function $f(x, y) = x$: $\displaystyle\lim_{(x, y) \to (a, b)} x = a$

3. Linear function $f(x, y) = y$: $\displaystyle\lim_{(x, y) \to (a, b)} y = b$

Proof:

1. Consider the constant function $f(x, y) = c$ and assume $\varepsilon > 0$ is given. To prove that the value of the limit is $L = c$, we must produce a $\delta > 0$ such that $|f(x, y) - L| < \varepsilon$

whenever $0 < \sqrt{(x - a)^2 + (y - b)^2} < \delta$. For constant functions, we may use *any* $\delta > 0$. Then, for every (x, y) in the domain of f,

$$|f(x, y) - L| = |f(x, y) - c| = |c - c| = 0 < \varepsilon$$

whenever $0 < \sqrt{(x - a)^2 + (y - b)^2} < \delta$.

2. Assume $\varepsilon > 0$ is given and take $\delta = \varepsilon$. The condition $0 < \sqrt{(x - a)^2 + (y - b)^2} < \delta$ implies that

$$
\begin{aligned}
0 < \sqrt{(x - a)^2 + (y - b)^2} &< \varepsilon \qquad \delta = \varepsilon \\
\sqrt{(x - a)^2} &< \varepsilon \qquad (x - a)^2 \le (x - a)^2 + (y - b)^2 \\
|x - a| &< \varepsilon. \qquad \sqrt{x^2} = |x| \text{ for real numbers } x
\end{aligned}
$$

Because $f(x, y) = x$ and $L = a$, we have shown that $|f(x, y) - L| < \varepsilon$ whenever $0 < \sqrt{(x - a)^2 + (y - b)^2} < \delta$. Therefore, $\lim\limits_{(x, y) \to (a, b)} f(x, y) = L$, or $\lim\limits_{(x, y) \to (a, b)} x = a$. The proof that $\lim\limits_{(x, y) \to (a, b)} y = b$ is similar (Exercise 86).   ◄

Using the three basic limits in Theorem 15.1, we can compute limits of more complicated functions. The only tools needed are limit laws analogous to those given in Theorem 2.3. The proofs of these laws are examined in Exercises 88–89.

THEOREM 15.2   Limit Laws for Functions of Two Variables

Let L and M be real numbers and suppose $\lim\limits_{(x, y) \to (a, b)} f(x, y) = L$ and $\lim\limits_{(x, y) \to (a, b)} g(x, y) = M$. Assume c is a constant, and $n > 0$ is an integer.

1. **Sum** $\lim\limits_{(x, y) \to (a, b)} (f(x, y) + g(x, y)) = L + M$

2. **Difference** $\lim\limits_{(x, y) \to (a, b)} (f(x, y) - g(x, y)) = L - M$

3. **Constant multiple** $\lim\limits_{(x, y) \to (a, b)} cf(x, y) = cL$

4. **Product** $\lim\limits_{(x, y) \to (a, b)} f(x, y)g(x, y) = LM$

5. **Quotient** $\lim\limits_{(x, y) \to (a, b)} \dfrac{f(x, y)}{g(x, y)} = \dfrac{L}{M}$, provided $M \ne 0$

6. **Power** $\lim\limits_{(x, y) \to (a, b)} (f(x, y))^n = L^n$

7. **Root** $\lim\limits_{(x, y) \to (a, b)} (f(x, y))^{1/n} = L^{1/n}$, where we assume $L > 0$ if n is even.

➤ Recall that a polynomial in two variables consists of sums and products of polynomials in x and polynomials in y. A rational function is the quotient of two polynomials.

Combining Theorems 15.1 and 15.2 allows us to find limits of polynomial, rational, and algebraic functions in two variables.

EXAMPLE 1   Limits of two-variable functions Evaluate $\lim\limits_{(x, y) \to (2, 8)} (3x^2 y + \sqrt{xy})$.

SOLUTION All the operations in this function appear in Theorem 15.2. Therefore, we can apply the limit laws directly.

$$
\begin{aligned}
\lim_{(x, y) \to (2, 8)} (3x^2 y + \sqrt{xy}) &= \lim_{(x, y) \to (2, 8)} 3x^2 y + \lim_{(x, y) \to (2, 8)} \sqrt{xy} \qquad \text{Law 1} \\
&= 3 \lim_{(x, y) \to (2, 8)} x^2 \cdot \lim_{(x, y) \to (2, 8)} y \\
&\quad + \sqrt{\lim_{(x, y) \to (2, 8)} x \cdot \lim_{(x, y) \to (2, 8)} y} \qquad \text{Laws 3, 4, 7} \\
&= 3 \cdot 2^2 \cdot 8 + \sqrt{2 \cdot 8} = 100 \qquad \begin{array}{l}\text{Law 6 and} \\ \text{Theorem 15.1}\end{array}
\end{aligned}
$$

Related Exercise 16 ◄

QUICK CHECK 1 Which of the following limits exist?

a. $\lim\limits_{(x,\,y)\to(1,\,1)} 3x^{12}y^2$

b. $\lim\limits_{(x,\,y)\to(0,\,0)} 3x^{-2}y^2$

c. $\lim\limits_{(x,\,y)\to(1,\,2)} \sqrt{x-y^2}$ ◀

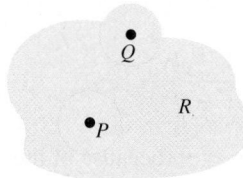

> Q is a boundary point: Every disk centered at Q contains points in R and points not in R.

> P is an interior point: There is a disk centered at P that lies entirely in R.

Figure 15.22

➤ The definitions of *interior point* and *boundary point* apply to regions in $\mathbb{R}^3$ if we replace *disk* by *ball*.

➤ Many sets, such as the annulus $\{(x, y): 2 \le x^2 + y^2 < 5\}$, are neither open nor closed.

QUICK CHECK 2 Give an example of a set that contains none of its boundary points. ◀

➤ Recall that this same method was used with functions of one variable. For example, after the common factor $x - 2$ is canceled, the function

$$g(x) = \frac{x^2 - 4}{x - 2}$$

becomes $g(x) = x + 2$, provided $x \ne 2$. In this case, 2 plays the role of a boundary point.

In Example 1, the value of the limit equals the value of the function at (a, b); in other words, $\lim\limits_{(x,\,y)\to(a,\,b)} f(x, y) = f(a, b)$ and the limit can be evaluated by substitution. This is a property of *continuous* functions, discussed later in this section.

Limits at Boundary Points

This is an appropriate place to make some definitions that are used in the remainder of the text.

DEFINITION Interior and Boundary Points

Let R be a region in $\mathbb{R}^2$. An **interior point** P of R lies entirely within R, which means it is possible to find a disk centered at P that contains only points of R (**Figure 15.22**).

A **boundary point** Q of R lies on the edge of R in the sense that *every* disk centered at Q contains at least one point in R and at least one point not in R.

For example, let R be the points in $\mathbb{R}^2$ satisfying $x^2 + y^2 < 9$. The boundary points of R lie on the circle $x^2 + y^2 = 9$. The interior points lie inside that circle and satisfy $x^2 + y^2 < 9$. Notice that the boundary points of a set need not lie in the set.

DEFINITION Open and Closed Sets

A region is **open** if it consists entirely of interior points. A region is **closed** if it contains all its boundary points.

An example of an open region in $\mathbb{R}^2$ is the open disk $\{(x, y): x^2 + y^2 < 9\}$. An example of a closed region in $\mathbb{R}^2$ is the square $\{(x, y): |x| \le 1, |y| \le 1\}$. Later in the text, we encounter interior and boundary points of three-dimensional sets such as balls, boxes, and pyramids.

Suppose $P_0(a, b)$ is a boundary point of the domain of f. The limit $\lim\limits_{(x,\,y)\to(a,\,b)} f(x, y)$ exists, even if P_0 is not in the domain of f, provided $f(x, y)$ approaches the same value as (x, y) approaches (a, b) *along all paths that lie in the domain* (**Figure 15.23**).

Consider the function $f(x, y) = \dfrac{x^2 - y^2}{x - y}$ whose domain is $\{(x, y): x \ne y\}$. Provided $x \ne y$, we may cancel the factor $(x - y)$ from the numerator and denominator and write

$$f(x, y) = \frac{x^2 - y^2}{x - y} = \frac{(x - y)(x + y)}{x - y} = x + y.$$

The graph of f (**Figure 15.24**) is the plane $z = x + y$, with points corresponding to the line $x = y$ removed.

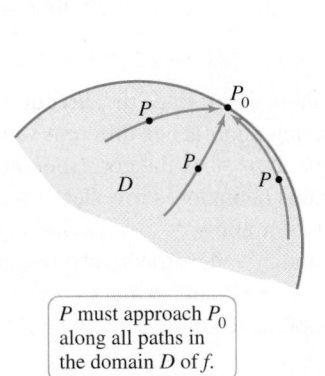

> P must approach P_0 along all paths in the domain D of f.

Figure 15.23

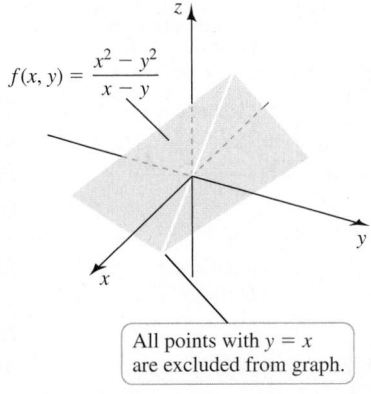

$$f(x, y) = \frac{x^2 - y^2}{x - y}$$

> All points with $y = x$ are excluded from graph.

Figure 15.24

Now we examine $\lim\limits_{(x,y)\to(4,4)} \dfrac{x^2 - y^2}{x - y}$, where $(4, 4)$ is a boundary point of the domain of f but does not lie in the domain. For this limit to exist, $f(x, y)$ must approach the same value along all paths to $(4, 4)$ that lie in the domain of f—that is, all paths approaching $(4, 4)$ that do not intersect $x = y$. To evaluate the limit, we proceed as follows:

$$\lim_{(x,y)\to(4,4)} \frac{x^2 - y^2}{x - y} = \lim_{(x,y)\to(4,4)} (x + y) \qquad \text{Assume } x \neq y, \text{ cancel } x - y.$$

$$= 4 + 4 = 8. \qquad \text{Same limit along all paths in the domain}$$

To emphasize, we let $(x, y) \to (4, 4)$ along all paths that do not intersect $x = y$, which lies outside the domain of f. Along all admissible paths, the function approaches 8.

QUICK CHECK 3 Can the limit
$$\lim_{(x,y)\to(0,0)} \frac{x^2 - xy}{x} \text{ be evaluated by}$$
direct substitution? ◄

EXAMPLE 2 Limits at boundary points Evaluate $\lim\limits_{(x,y)\to(4,1)} \dfrac{xy - 4y^2}{\sqrt{x} - 2\sqrt{y}}$.

SOLUTION Points in the domain of this function satisfy $x \geq 0$ and $y \geq 0$ (because of the square roots) and $x \neq 4y$ (to ensure the denominator is nonzero). We see that the point $(4, 1)$ lies on the boundary of the domain. Multiplying the numerator and denominator by the algebraic conjugate of the denominator, the limit is computed as follows:

$$\lim_{(x,y)\to(4,1)} \frac{xy - 4y^2}{\sqrt{x} - 2\sqrt{y}} = \lim_{(x,y)\to(4,1)} \frac{(xy - 4y^2)(\sqrt{x} + 2\sqrt{y})}{(\sqrt{x} - 2\sqrt{y})(\sqrt{x} + 2\sqrt{y})} \qquad \text{Multiply by conjugate.}$$

$$= \lim_{(x,y)\to(4,1)} \frac{y(x - 4y)(\sqrt{x} + 2\sqrt{y})}{x - 4y} \qquad \text{Simplify.}$$

$$= \lim_{(x,y)\to(4,1)} y(\sqrt{x} + 2\sqrt{y}) \qquad \begin{array}{l}\text{Cancel } x - 4y, \\ \text{assumed to be nonzero.}\end{array}$$

$$= 4. \qquad \text{Evaluate limit.}$$

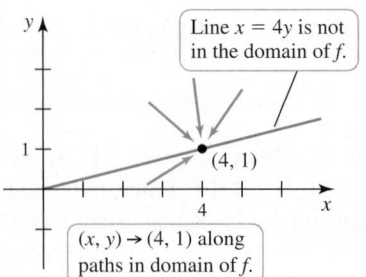

Line $x = 4y$ is not in the domain of f.

$(4, 1)$

$(x, y) \to (4, 1)$ along paths in domain of f.

Figure 15.25

Because points on the line $x = 4y$ are outside the domain of the function, we assume $x - 4y \neq 0$. Along all other paths to $(4, 1)$, the function values approach 4 (Figure 15.25).

Related Exercises 26–27 ◄

► Notice that if we choose any path of the form $y = mx$, then $y \to 0$ as $x \to 0$. Therefore, $\lim\limits_{(x,y)\to(0,0)}$ can be replaced by $\lim\limits_{x\to 0}$ along this path. A similar argument applies to paths of the form $y = mx^p$, for $p > 0$.

EXAMPLE 3 Nonexistence of a limit Investigate the limit $\lim\limits_{(x,y)\to(0,0)} \dfrac{(x + y)^2}{x^2 + y^2}$.

SOLUTION The domain of the function is $\{(x, y): (x, y) \neq (0, 0)\}$; therefore, the limit is at a boundary point outside the domain. Suppose we let (x, y) approach $(0, 0)$ along the line $y = mx$ for a fixed constant m. Substituting $y = mx$ and noting that $y \to 0$ as $x \to 0$, we have

$$\lim_{\substack{(x,y)\to(0,0) \\ (\text{along } y = mx)}} \frac{(x + y)^2}{x^2 + y^2} = \lim_{x\to 0} \frac{(x + mx)^2}{x^2 + m^2 x^2} = \lim_{x\to 0} \frac{x^2(1 + m)^2}{x^2(1 + m^2)} = \frac{(1 + m)^2}{1 + m^2}.$$

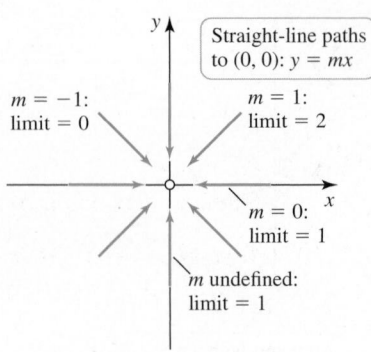

Straight-line paths to $(0, 0)$: $y = mx$

$m = -1$: limit $= 0$

$m = 1$: limit $= 2$

$m = 0$: limit $= 1$

m undefined: limit $= 1$

Figure 15.26

The constant m determines the direction of approach to $(0, 0)$. Therefore, depending on m, the function approaches different values as (x, y) approaches $(0, 0)$ (Figure 15.26). For example, if $m = 0$, the corresponding limit is 1, and if $m = -1$, the limit is 0. The reason for this behavior is revealed if we plot the surface and look at two level curves. The lines $y = x$ and $y = -x$ (excluding the origin) are level curves of the function for $z = 2$ and $z = 0$, respectively. (Figure 15.27). Therefore, as $(x, y) \to (0, 0)$ along $y = x$, $f(x, y) \to 2$, and as $(x, y) \to (0, 0)$ along $y = -x$, $f(x, y) \to 0$. Because the function approaches different values along different paths, we conclude that the *limit does not exist*.

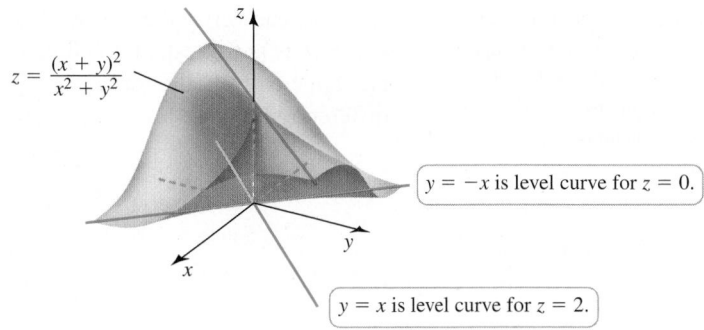

$z = \dfrac{(x + y)^2}{x^2 + y^2}$

$y = -x$ is level curve for $z = 0$.

$y = x$ is level curve for $z = 2$.

Figure 15.27   *Related Exercise 30* ◄

The strategy used in Example 3 is an effective way to prove the nonexistence of a limit.

QUICK CHECK 4 What is the analog of the Two-Path Test for functions of a single variable? ◄

PROCEDURE   Two-Path Test for Nonexistence of Limits

If $f(x, y)$ approaches two different values as (x, y) approaches (a, b) along two different paths in the domain of f, then $\displaystyle\lim_{(x, y) \to (a, b)} f(x, y)$ does not exist.

Continuity of Functions of Two Variables

The following definition of continuity for functions of two variables is analogous to the continuity definition for functions of one variable.

DEFINITION   Continuity

The function f is continuous at the point (a, b) provided

1. f is defined at (a, b),

2. $\displaystyle\lim_{(x, y) \to (a, b)} f(x, y)$ exists, and

3. $\displaystyle\lim_{(x, y) \to (a, b)} f(x, y) = f(a, b)$.

A function of two (or more) variables is continuous at a point, provided its limit equals its value at that point (which implies the limit and the value both exist). The definition of continuity applies at boundary points of the domain of f, provided the limits in the definition are taken along all paths that lie in the domain. Because limits of polynomials and rational functions can be evaluated by substitution at points of their domains (that is, $\displaystyle\lim_{(x, y) \to (a, b)} f(x, y) = f(a, b)$), it follows that polynomials and rational functions are continuous at all points of their domains.

EXAMPLE 4   Checking continuity Determine the points at which the following function is continuous.

$$f(x, y) = \begin{cases} \dfrac{3xy^2}{x^2 + y^4} & \text{if } (x, y) \neq (0, 0) \\ 0 & \text{if } (x, y) = (0, 0) \end{cases}$$

SOLUTION The function $\dfrac{3xy^2}{x^2 + y^4}$ is a rational function, so it is continuous at all points of its domain, which consists of all points of $\mathbb{R}^2$ except $(0, 0)$. To determine whether f is continuous at $(0, 0)$, we must show that

$$\lim_{(x, y) \to (0, 0)} \frac{3xy^2}{x^2 + y^4}$$

exists and equals $f(0, 0) = 0$ along all paths that approach $(0, 0)$.

▶ The choice of $x = my^2$ for paths to $(0, 0)$ is not obvious. Notice that if x is replaced with my^2 in f, the result involves the same power of y (in this case, y^4) in the numerator and denominator, which may be canceled.

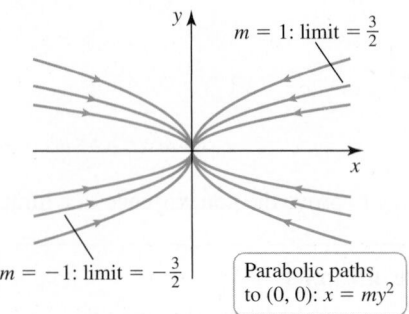

$m = 1$: limit $= \frac{3}{2}$

$m = -1$: limit $= -\frac{3}{2}$

Parabolic paths to $(0, 0)$: $x = my^2$

Figure 15.28

You can verify that as (x, y) approaches $(0, 0)$ along paths of the form $y = mx$, where m is any constant, the function values approach $f(0, 0) = 0$. However, along parabolic paths of the form $x = my^2$ (where m is a nonzero constant), the limit behaves differently (**Figure 15.28**). This time we substitute $x = my^2$ and note that $x \to 0$ as $y \to 0$:

$$\lim_{\substack{(x, y) \to (0, 0) \\ (\text{along } x = my^2)}} \frac{3xy^2}{x^2 + y^4} = \lim_{y \to 0} \frac{3(my^2)y^2}{(my^2)^2 + y^4} \quad \text{Substitute } x = my^2.$$

$$= \lim_{y \to 0} \frac{3my^4}{m^2y^4 + y^4} \quad \text{Simplify.}$$

$$= \lim_{y \to 0} \frac{3m}{m^2 + 1} \quad \text{Cancel } y^4.$$

$$= \frac{3m}{m^2 + 1}.$$

We see that along parabolic paths, the limit depends on the approach path. For example, with $m = 1$, along the path $x = y^2$ the function values approach $\frac{3}{2}$; with $m = -1$, along the path $x = -y^2$ the function values approach $-\frac{3}{2}$ (**Figure 15.29**). Because $f(x, y)$ approaches two different numbers along two different paths, the limit at $(0, 0)$ does not exist, and f is not continuous at $(0, 0)$.

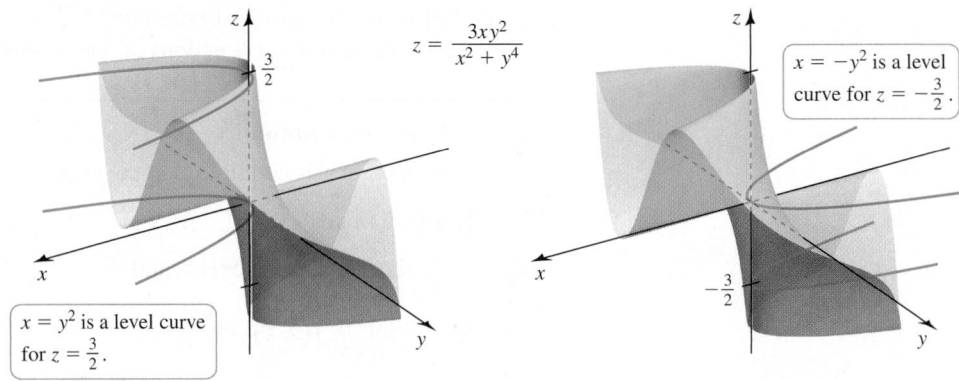

$z = \dfrac{3xy^2}{x^2 + y^4}$

$x = -y^2$ is a level curve for $z = -\frac{3}{2}$.

$x = y^2$ is a level curve for $z = \frac{3}{2}$.

Figure 15.29   *Related Exercises 41–42* ◀

QUICK CHECK 5 Which of the following functions are continuous at $(0, 0)$?

a. $f(x, y) = 2x^2y^5$

b. $f(x, y) = \dfrac{2x^2y^5}{x - 1}$

c. $f(x, y) = 2x^{-2}y^5$ ◀

Composite Functions   Recall that for functions of a single variable, compositions of continuous functions are also continuous. The following theorem gives the analogous result for functions of two variables; it is proved in Appendix A.

THEOREM 15.3   **Continuity of Composite Functions**

If $u = g(x, y)$ is continuous at (a, b) and $z = f(u)$ is continuous at $g(a, b)$, then the composite function $z = f(g(x, y))$ is continuous at (a, b).

With Theorem 15.3, we can easily analyze the continuity of many functions. For example, $\sin x$, $\cos x$, and e^x are continuous functions of a single variable, for all real values of x. Therefore, compositions of these functions with polynomials in x and y (for example, $\sin(x^2y)$ and $e^{x^4-y^2}$) are continuous for all real numbers x and y. Similarly, $\sqrt{x}$ is a continuous function of a single variable, for $x \geq 0$. Therefore, $\sqrt{u(x, y)}$ is continuous at (x, y) provided u is continuous at (x, y) and $u(x, y) \geq 0$. As long as we observe restrictions on domains, then compositions of continuous functions are also continuous.

EXAMPLE 5   **Continuity of composite functions.** Determine the points at which the following functions are continuous.

a. $h(x, y) = \ln(x^2 + y^2 + 4)$ **b.** $h(x, y) = e^{x/y}$

SOLUTION

a. This function is the composition $f(g(x, y))$, where

$$f(u) = \ln u \quad \text{and} \quad u = g(x, y) = x^2 + y^2 + 4.$$

As a polynomial, g is continuous for all (x, y) in $\mathbb{R}^2$. The function f is continuous for $u > 0$. Because $u = x^2 + y^2 + 4 > 0$ for all (x, y), it follows that h is continuous at all points of $\mathbb{R}^2$.

b. Letting $f(u) = e^u$ and $u = g(x, y) = x/y$, we have $h(x, y) = f(g(x, y))$. Note that f is continuous at all points of $\mathbb{R}$ and g is continuous at all points of $\mathbb{R}^2$ provided $y \neq 0$. Therefore, h is continuous on the set $\{(x, y): y \neq 0\}$.

Related Exercises 48–49 ◄

Functions of Three Variables

The work we have done with limits and continuity of functions of two variables extends to functions of three or more variables. Specifically, the limit laws of Theorem 15.2 apply to functions of the form $w = f(x, y, z)$. Polynomials and rational functions are continuous at all points of their domains, and limits of these functions may be evaluated by direct substitution at all points of their domains. Compositions of continuous functions of the form $f(g(x, y, z))$ are also continuous at points at which $g(x, y, z)$ is in the domain of f.

EXAMPLE 6   Functions of three variables

a. Evaluate $\displaystyle\lim_{(x, y, z) \to (2, \pi/2, 0)} \frac{x^2 \sin y}{z^2 + 4}$.

b. Find the points at which $h(x, y, z) = \sqrt{x^2 + y^2 + z^2 - 1}$ is continuous.

SOLUTION

a. This function consists of products and quotients of functions that are continuous at $(2, \pi/2, 0)$. Therefore, the limit is evaluated by direct substitution:

$$\lim_{(x, y, z) \to (2, \pi/2, 0)} \frac{x^2 \sin y}{z^2 + 4} = \frac{2^2 \sin (\pi/2)}{0^2 + 4} = 1.$$

b. This function is a composition in which the outer function $f(u) = \sqrt{u}$ is continuous for $u \geq 0$. The inner function

$$g(x, y, z) = x^2 + y^2 + z^2 - 1$$

is nonnegative provided $x^2 + y^2 + z^2 \geq 1$. Therefore, h is continuous at all points on or outside the unit sphere in $\mathbb{R}^3$.

Related Exercise 55 ◄

SECTION 15.2 EXERCISES

Getting Started

1. Explain what $\displaystyle\lim_{(x, y) \to (a, b)} f(x, y) = L$ means.

2. Explain why $f(x, y)$ must approach a unique number L as (x, y) approaches (a, b) along *all* paths in the domain in order for $\displaystyle\lim_{(x, y) \to (a, b)} f(x, y)$ to exist.

3. What does it means to say that limits of polynomials may be evaluated by direct substitution?

4. Suppose (a, b) is on the boundary of the domain of f. Explain how you would determine whether $\displaystyle\lim_{(x, y) \to (a, b)} f(x, y)$ exists.

5. Explain how examining limits along multiple paths may prove the nonexistence of a limit.

6. Explain why evaluating a limit along a finite number of paths does not prove the existence of a limit of a function of several variables.

7. What three conditions must be met for a function f to be continuous at the point (a, b)?

8. Let R be the unit disk $\{(x, y): x^2 + y^2 \leq 1\}$ with $(0, 0)$ removed. Is $(0, 0)$ a boundary point of R? Is R open or closed?

9. At what points of $\mathbb{R}^2$ is a rational function of two variables continuous?

10. Evaluate $\displaystyle\lim_{(x, y, z) \to (1, 1, -1)} xy^2 z^3$.

11. Evaluate $\displaystyle\lim_{(x, y) \to (5, -5)} \frac{x^2 - y^2}{x + y}$.

12. Let $f(x) = \dfrac{x^2 - 2x - y^2 + 1}{x^2 - 2x + y^2 + 1}$. Use the Two-Path Test to show that $\lim\limits_{(x,\,y)\to(1,\,0)} f(x)$ does not exist. (*Hint:* Examine $\lim\limits_{\substack{(x,\,y)\to(1,\,0)\\ (\text{along } y = 0)}} f(x)$ and $\lim\limits_{\substack{(x,\,y)\to(1,\,0)\\ (\text{along } x = 1)}} f(x)$ first.)

Practice Exercises

13–28. Limits of functions *Evaluate the following limits.*

13. $\lim\limits_{(x,\,y)\to(2,\,9)} 101$

14. $\lim\limits_{(x,\,y)\to(1,\,-3)} (3x + 4y - 2)$

15. $\lim\limits_{(x,\,y)\to(-3,\,3)} (4x^2 - y^2)$

16. $\lim\limits_{(x,\,y)\to(2,\,-1)} (xy^8 - 3x^2y^3)$

17. $\lim\limits_{(x,\,y)\to(0,\,\pi)} \dfrac{\cos xy + \sin xy}{2y}$

18. $\lim\limits_{(x,\,y)\to(e^2,\,4)} \ln\sqrt{xy}$

19. $\lim\limits_{(x,\,y)\to(2,\,0)} \dfrac{x^2 - 3xy^2}{x + y}$

20. $\lim\limits_{(u,\,v)\to(1,\,-1)} \dfrac{10uv - 2v^2}{u^2 + v^2}$

21. $\lim\limits_{(x,\,y)\to(6,\,2)} \dfrac{x^2 - 3xy}{x - 3y}$

22. $\lim\limits_{(x,\,y)\to(1,\,-2)} \dfrac{y^2 + 2xy}{y + 2x}$

23. $\lim\limits_{(x,\,y)\to(3,\,1)} \dfrac{x^2 - 7xy + 12y^2}{x - 3y}$

24. $\lim\limits_{(x,\,y)\to(-1,\,1)} \dfrac{2x^2 - xy - 3y^2}{x + y}$

25. $\lim\limits_{(x,\,y)\to(2,\,2)} \dfrac{y^2 - 4}{xy - 2x}$

26. $\lim\limits_{(x,\,y)\to(4,\,5)} \dfrac{\sqrt{x + y} - 3}{x + y - 9}$

27. $\lim\limits_{(x,\,y)\to(1,\,2)} \dfrac{\sqrt{y} - \sqrt{x + 1}}{y - x - 1}$

28. $\lim\limits_{(u,\,v)\to(8,\,8)} \dfrac{u^{1/3} - v^{1/3}}{u^{2/3} - v^{2/3}}$

29–34. Nonexistence of limits *Use the Two-Path Test to prove that the following limits do not exist.*

29. $\lim\limits_{(x,\,y)\to(0,\,0)} \dfrac{x + 2y}{x - 2y}$

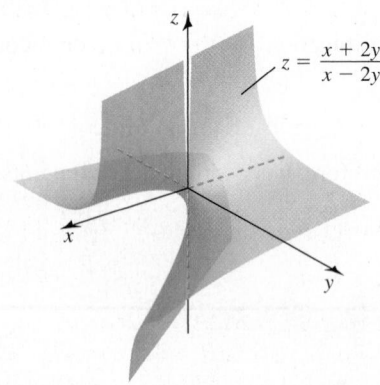

$z = \dfrac{x + 2y}{x - 2y}$

30. $\lim\limits_{(x,\,y)\to(0,\,0)} \dfrac{4xy}{3x^2 + y^2}$

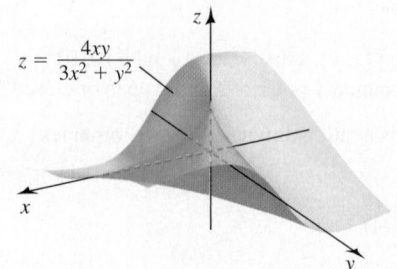

$z = \dfrac{4xy}{3x^2 + y^2}$

31. $\lim\limits_{(x,\,y)\to(0,\,0)} \dfrac{y^4 - 2x^2}{y^4 + x^2}$

32. $\lim\limits_{(x,\,y)\to(0,\,0)} \dfrac{x^3 - y^2}{x^3 + y^2}$

33. $\lim\limits_{(x,\,y)\to(0,\,0)} \dfrac{y^3 + x^3}{xy^2}$

34. $\lim\limits_{(x,\,y)\to(0,\,0)} \dfrac{y}{\sqrt{x^2 - y^2}}$

35–54. Continuity *At what points of $\mathbb{R}^2$ are the following functions continuous?*

35. $f(x, y) = x^2 + 2xy - y^3$

36. $f(x, y) = \dfrac{xy}{x^2y^2 + 1}$

37. $p(x, y) = \dfrac{4x^2y^2}{x^4 + y^2}$

38. $S(x, y) = \dfrac{2xy}{x^2 - y^2}$

39. $f(x, y) = \dfrac{2}{x(y^2 + 1)}$

40. $f(x, y) = \dfrac{x^2 + y^2}{x(y^2 - 1)}$

41. $f(x, y) = \begin{cases} \dfrac{xy}{x^2 + y^2} & \text{if } (x, y) \neq (0, 0) \\ 0 & \text{if } (x, y) = (0, 0) \end{cases}$

42. $f(x, y) = \begin{cases} \dfrac{y^4 - 2x^2}{y^4 + x^2} & \text{if } (x, y) \neq (0, 0) \\ 0 & \text{if } (x, y) = (0, 0) \end{cases}$

43. $f(x, y) = \sqrt{x^2 + y^2}$

44. $f(x, y) = e^{x^2 + y^2}$

45. $f(x, y) = \sin xy$

46. $g(x, y) = \ln(x - y)$

47. $h(x, y) = \cos(x + y)$

48. $p(x, y) = e^{x - y}$

49. $f(x, y) = \ln(x^2 + y^2)$

50. $f(x, y) = \sqrt{4 - x^2 - y^2}$

51. $g(x, y) = \sqrt[3]{x^2 + y^2 - 9}$

52. $h(x, y) = \dfrac{\sqrt{x - y}}{4}$

53. $f(x, y) = \begin{cases} \dfrac{\sin(x^2 + y^2)}{x^2 + y^2} & \text{if } (x, y) \neq (0, 0) \\ 1 & \text{if } (x, y) = (0, 0) \end{cases}$

54. $f(x, y) = \begin{cases} \dfrac{1 - \cos(x^2 + y^2)}{x^2 + y^2} & \text{if } (x, y) \neq (0, 0) \\ 0 & \text{if } (x, y) = (0, 0) \end{cases}$

55–60. Limits of functions of three variables *Evaluate the following limits.*

55. $\lim\limits_{(x,\,y,\,z)\to(1,\,\ln 2,\,3)} ze^{xy}$

56. $\lim\limits_{(x,\,y,\,z)\to(0,\,1,\,0)} \ln(1 + y)e^{xz}$

57. $\lim\limits_{(x,\,y,\,z)\to(1,\,1,\,1)} \dfrac{yz - xy - xz - x^2}{yz + xy + xz - y^2}$

58. $\lim\limits_{(x,\,y,\,z)\to(1,\,1,\,1)} \dfrac{x - \sqrt{xz} - \sqrt{xy} + \sqrt{yz}}{x - \sqrt{xz} + \sqrt{xy} - \sqrt{yz}}$

59. $\lim\limits_{(x,\,y,\,z)\to(1,\,1,\,1)} \dfrac{x^2 + xy - xz - yz}{x - z}$

60. $\lim\limits_{(x,\,y,\,z)\to(1,\,-1,\,1)} \dfrac{xz + 5x + yz + 5y}{x + y}$

61. Explain why or why not Determine whether the following statements are true and give an explanation or counterexample.

a. If the limits $\lim\limits_{(x,0)\to(0,0)} f(x,0)$ and $\lim\limits_{(0,y)\to(0,0)} f(0,y)$ exist and equal L, then $\lim\limits_{(x,y)\to(0,0)} f(x,y) = L$.

b. If $\lim\limits_{(x,y)\to(a,b)} f(x,y)$ equals a finite number L, then f is continuous at (a,b).

c. If f is continuous at (a,b), then $\lim\limits_{(x,y)\to(a,b)} f(x,y)$ exists.

d. If P is a boundary point of the domain of f, then P is in the domain of f.

62–76. Miscellaneous limits *Use the method of your choice to evaluate the following limits.*

62. $\lim\limits_{(x,y)\to(0,0)} \dfrac{y^2}{x^8+y^2}$

63. $\lim\limits_{(x,y)\to(0,1)} \dfrac{y\sin x}{x(y+1)}$

64. $\lim\limits_{(x,y)\to(1,1)} \dfrac{x^2+xy-2y^2}{2x^2-xy-y^2}$

65. $\lim\limits_{(x,y)\to(1,0)} \dfrac{y\ln y}{x}$

66. $\lim\limits_{(x,y)\to(0,0)} \dfrac{|xy|}{xy}$

67. $\lim\limits_{(x,y)\to(0,0)} \dfrac{|x-y|}{|x+y|}$

68. $\lim\limits_{(u,v)\to(-1,0)} \dfrac{uve^{-v}}{u^2+v^2}$

69. $\lim\limits_{(x,y)\to(2,0)} \dfrac{1-\cos y}{xy^2}$

70. $\lim\limits_{(x,y)\to(4,0)} x^2 y \ln xy$

71. $\lim\limits_{(x,y)\to(1,0)} \dfrac{\sin xy}{xy}$

72. $\lim\limits_{(x,y)\to(0,\pi/2)} \dfrac{1-\cos xy}{4x^2y^3}$

73. $\lim\limits_{(x,y)\to(0,2)} (2xy)^{xy}$

74. $\lim\limits_{(x,y)\to(3,3)} \dfrac{x^2+2xy-6x+y^2-6y}{x+y-6}$

75. $\lim\limits_{(x,y)\to(1,2)} \dfrac{x^2+2xy-x+y^2-y-6}{x+y-3}$

76. $\lim\limits_{(x,y)\to(0,0)} \tan^{-1}\dfrac{(2+(x+y)^2+(x-y)^2)}{2e^{x^2+y^2}}$

77. Piecewise function Let

$$f(x,y) = \begin{cases} \dfrac{\sin(x^2+y^2-1)}{x^2+y^2-1} & \text{if } x^2+y^2 \neq 1 \\ b & \text{if } x^2+y^2 = 1. \end{cases}$$

Find the value of b for which f is continuous at all points in $\mathbb{R}^2$.

78. Piecewise function Let

$$f(x,y) = \begin{cases} \dfrac{1+2xy-\cos xy}{xy} & \text{if } xy \neq 0 \\ a & \text{if } xy = 0. \end{cases}$$

Find the value of a for which f is continuous at all points in $\mathbb{R}^2$.

79–81. Limits using polar coordinates *Limits at $(0,0)$ may be easier to evaluate by converting to polar coordinates. Remember that the same limit must be obtained as $r \to 0$ along all paths in the domain to $(0,0)$. Evaluate the following limits or state that they do not exist.*

79. $\lim\limits_{(x,y)\to(0,0)} \dfrac{x^2y}{x^2+y^2}$

80. $\lim\limits_{(x,y)\to(0,0)} \dfrac{x-y}{\sqrt{x^2+y^2}}$

81. $\lim\limits_{(x,y)\to(0,0)} \dfrac{x^2+y^2+x^2y^2}{x^2+y^2}$

Explorations and Challenges

82. Sine limits Verify that $\lim\limits_{(x,y)\to(0,0)} \dfrac{\sin x+\sin y}{x+y} = 1$.

83. Nonexistence of limits Show that $\lim\limits_{(x,y)\to(0,0)} \dfrac{ax^m y^n}{bx^{m+n}+cy^{m+n}}$ does not exist when a, b, and c are nonzero real numbers and m and n are positive integers.

84. Nonexistence of limits Show that $\lim\limits_{(x,y)\to(0,0)} \dfrac{ax^{2(p-n)}y^n}{bx^{2p}+cy^p}$ does not exist when a, b, and c are nonzero real numbers and n and p are positive integers with $p \geq n$.

85. Filling in a function value The domain of $f(x,y) = e^{-1/(x^2+y^2)}$ excludes $(0,0)$. How should f be defined at $(0,0)$ to make it continuous there?

86. Limit proof Use the formal definition of a limit to prove that $\lim\limits_{(x,y)\to(a,b)} y = b$. (*Hint:* Take $\delta = \varepsilon$.)

87. Limit proof Use the formal definition of a limit to prove that $\lim\limits_{(x,y)\to(a,b)} (x+y) = a+b$. (*Hint:* Take $\delta = \varepsilon/2$.)

88. Proof of Limit Law 1 Use the formal definition of a limit to prove that
$$\lim\limits_{(x,y)\to(a,b)} (f(x,y)+g(x,y)) = \lim\limits_{(x,y)\to(a,b)} f(x,y) + \lim\limits_{(x,y)\to(a,b)} g(x,y).$$

89. Proof of Limit Law 3 Use the formal definition of a limit to prove that $\lim\limits_{(x,y)\to(a,b)} cf(x,y) = c\lim\limits_{(x,y)\to(a,b)} f(x,y)$.

QUICK CHECK ANSWERS

1. The limit exists only for (a).   **2.** $\{(x,y): x^2+y^2 < 2\}$
3. If a factor of x is first canceled, then the limit may be evaluated by substitution.   **4.** If the left and right limits at a point are not equal, then the two-sided limit does not exist.
5. (a) and (b) are continuous at $(0,0)$. ◄

15.3 Partial Derivatives

The derivative of a function of one variable, $y = f(x)$, measures the rate of change of y with respect to x, and it gives slopes of tangent lines. The analogous idea for functions of several variables presents a new twist: Derivatives may be defined with respect to any of the independent variables. For example, we can compute the derivative of $f(x, y)$ with respect to x or y. The resulting derivatives are called *partial derivatives*; they still represent rates of change and they are associated with slopes of tangents. Therefore, much of what you have learned about derivatives applies to functions of several variables. However, much is also different.

Derivatives with Two Variables

Consider a function f defined on a domain D in the xy-plane. Suppose f represents the elevation of the land (above sea level) over D. Imagine that you are on the surface $z = f(x, y)$ at the point $(a, b, f(a, b))$ and you are asked to determine the slope of the surface where you are standing. Your answer should be, *it depends*!

Figure 15.30a shows a function that resembles the landscape in Figure 15.30b. Suppose you are standing at the point $P(0, 0, f(0, 0))$, which lies on the pass or the saddle. The surface behaves differently depending on the direction in which you walk. If you walk east (positive x-direction), the elevation increases and your path takes you upward on the surface. If you walk north (positive y-direction), the elevation decreases and your path takes you downward on the surface. In fact, in every direction you walk from the point P, the function values change at different rates. So how should the slope or the rate of change at a given point be defined?

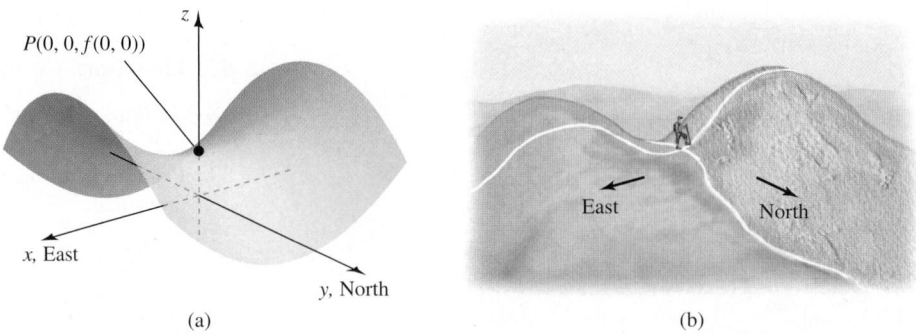

Figure 15.30

The answer to this question involves *partial derivatives*, which arise when we hold all but one independent variable fixed and then compute an ordinary derivative with respect to the remaining variable. Suppose we move along the surface $z = f(x, y)$, starting at the point $(a, b, f(a, b))$ in such a way that $y = b$ is fixed and only x varies. The resulting path is a curve (a trace) on the surface that varies in the x-direction (Figure 15.31). This curve is the intersection of the surface with the vertical plane $y = b$; it is described by $z = f(x, b)$, which is a function of the single variable x. We know how to compute the slope of this curve: It is the ordinary derivative of $f(x, b)$ with respect to x. This derivative is called the *partial derivative of f with respect to x*, denoted $\partial f / \partial x$ or f_x. When evaluated at (a, b), its value is defined by the limit

$$f_x(a, b) = \lim_{h \to 0} \frac{f(a + h, b) - f(a, b)}{h},$$

provided this limit exists. Notice that the y-coordinate is fixed at $y = b$ in this limit. If we replace (a, b) with the variable point (x, y), then f_x becomes a function of x and y.

In a similar way, we can move along the surface $z = f(x, y)$ from the point $(a, b, f(a, b))$ in such a way that $x = a$ is fixed and only y varies. Now the result is

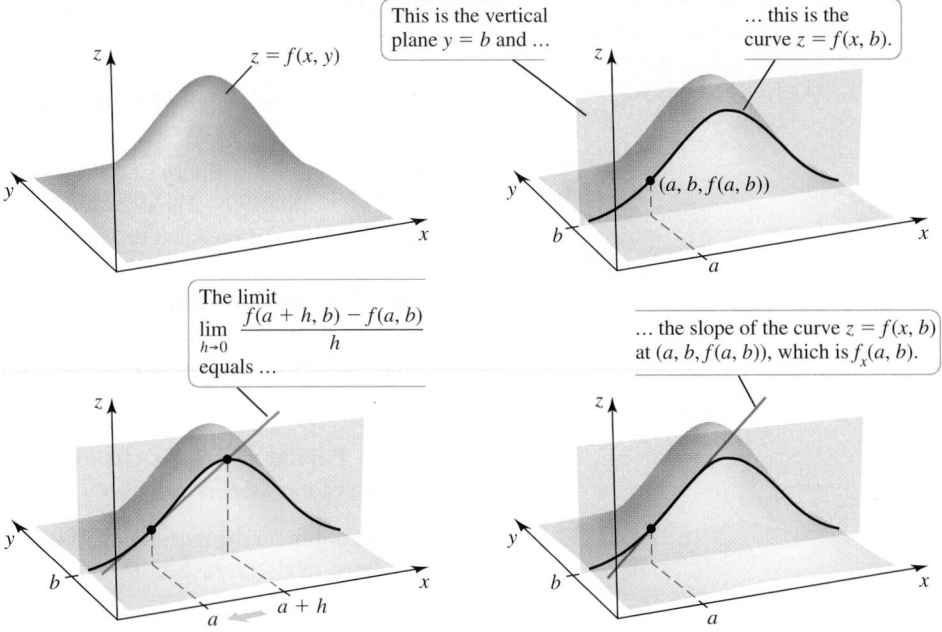

Figure 15.31

a trace described by $z = f(a, y)$, which is the intersection of the surface and the plane $x = a$ (**Figure 15.32**). The slope of this curve at (a, b) is given by the ordinary derivative of $f(a, y)$ with respect to y. This derivative is called the *partial derivative of f with respect to y*, denoted $\partial f / \partial y$ or f_y. When evaluated at (a, b), it is defined by the limit

$$f_y(a, b) = \lim_{h \to 0} \frac{f(a, b + h) - f(a, b)}{h},$$

provided this limit exists. If we replace (a, b) with the variable point (x, y), then f_y becomes a function of x and y.

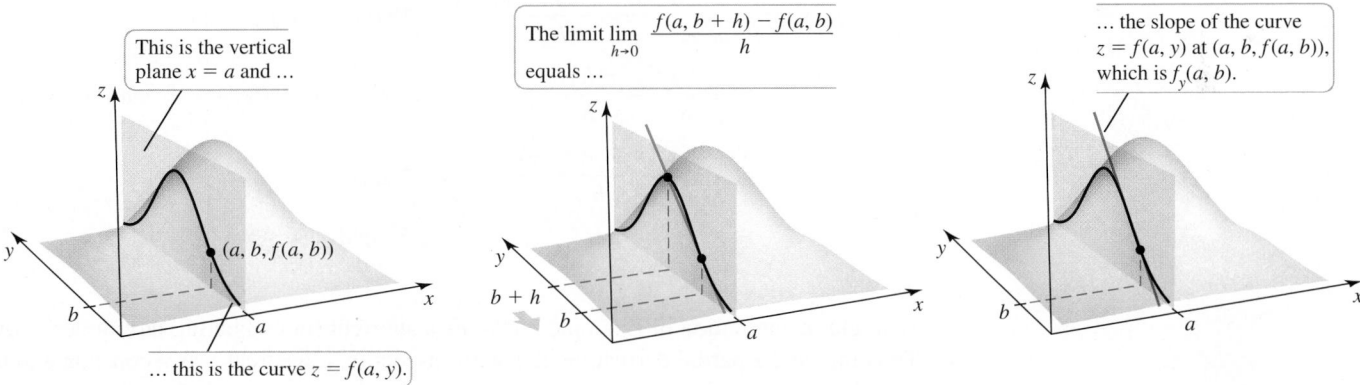

Figure 15.32

DEFINITION Partial Derivatives

The **partial derivative of f with respect to x at the point (a, b)** is

$$f_x(a, b) = \lim_{h \to 0} \frac{f(a + h, b) - f(a, b)}{h}.$$

The **partial derivative of f with respect to y at the point (a, b)** is

$$f_y(a, b) = \lim_{h \to 0} \frac{f(a, b + h) - f(a, b)}{h},$$

provided these limits exist.

▶ Recall that f' is a function, while $f'(a)$ is the value of the derivative at $x = a$. In the same way, f_x and f_y are functions of x and y, while $f_x(a, b)$ and $f_y(a, b)$ are their values at (a, b).

Notation   The partial derivatives evaluated at a point (a, b) are denoted in any of the following ways:

$$\frac{\partial f}{\partial x}(a, b) = \frac{\partial f}{\partial x}\bigg|_{(a, b)} = f_x(a, b) \quad \text{and} \quad \frac{\partial f}{\partial y}(a, b) = \frac{\partial f}{\partial y}\bigg|_{(a, b)} = f_y(a, b).$$

Notice that the d in the ordinary derivative df/dx has been replaced with ∂ in the partial derivatives $\partial f/\partial x$ and $\partial f/\partial y$. The notation $\partial/\partial x$ is an instruction or operator: It says, "Take the partial derivative with respect to x of the function that follows."

Calculating Partial Derivatives   We begin by calculating partial derivatives using the limit definition. The procedure in Example 1 should look familiar. It echoes the method used in Chapter 3 when we first introduced ordinary derivatives.

EXAMPLE 1   Partial derivatives from the definition Suppose $f(x, y) = x^2 y$. Use the limit definition of partial derivatives to compute $f_x(x, y)$ and $f_y(x, y)$.

SOLUTION We compute the partial derivatives at an arbitrary point (x, y) in the domain. The partial derivative with respect to x is

$$\begin{aligned}
f_x(x, y) &= \lim_{h \to 0} \frac{f(x + h, y) - f(x, y)}{h} && \text{Definition of } f_x \text{ at } (x, y) \\
&= \lim_{h \to 0} \frac{(x + h)^2 y - x^2 y}{h} && \text{Substitute for } f(x + h, y) \text{ and } f(x, y). \\
&= \lim_{h \to 0} \frac{(x^2 + 2xh + h^2 - x^2)y}{h} && \text{Factor and expand.} \\
&= \lim_{h \to 0} (2x + h)y && \text{Simplify and cancel } h. \\
&= 2xy. && \text{Evaluate limit.}
\end{aligned}$$

In a similar way, the partial derivative with respect to y is

$$\begin{aligned}
f_y(x, y) &= \lim_{h \to 0} \frac{f(x, y + h) - f(x, y)}{h} && \text{Definition of } f_y \text{ at } (x, y) \\
&= \lim_{h \to 0} \frac{x^2(y + h) - x^2 y}{h} && \text{Substitute for } f(x, y + h) \text{ and } f(x, y). \\
&= \lim_{h \to 0} \frac{x^2(y + h - y)}{h} && \text{Factor.} \\
&= x^2. && \text{Simplify and evaluate limit.}
\end{aligned}$$

Related Exercise 11 ◀

A careful examination of Example 1 reveals a shortcut for evaluating partial derivatives. To compute the partial derivative of f with respect to x, we treat y as a constant and take an ordinary derivative with respect to x:

$$\frac{\partial}{\partial x}(x^2 y) = y \underbrace{\frac{\partial}{\partial x}(x^2)}_{2x} = 2xy. \quad \text{Treat } y \text{ as a constant.}$$

Similarly, we treat x (and therefore x^2) as a constant to evaluate the partial derivative of f with respect to y:

$$\frac{\partial}{\partial y}(x^2 y) = x^2 \underbrace{\frac{\partial}{\partial y}(y)}_{1} = x^2. \quad \text{Treat } x \text{ as a constant.}$$

The next two examples illustrate the process.

EXAMPLE 2   Partial derivatives Let $f(x, y) = x^3 - y^2 + 4$.

a. Compute $\dfrac{\partial f}{\partial x}$ and $\dfrac{\partial f}{\partial y}$.

b. Evaluate each derivative at $(2, -4)$.

SOLUTION

a. We compute the partial derivative with respect to x assuming y is a constant; the Power Rule gives

$$\frac{\partial f}{\partial x} = \frac{\partial}{\partial x}(\underbrace{x^3}_{\text{variable}} - \underbrace{y^2 + 4}_{\substack{\text{constant with} \\ \text{respect to } x}}) = 3x^2 + 0 = 3x^2.$$

The partial derivative with respect to y is computed by treating x as a constant; using the Power Rule gives

$$\frac{\partial f}{\partial y} = \frac{\partial}{\partial y}(\underbrace{x^3}_{\substack{\text{constant} \\ \text{with respect} \\ \text{to } y}} - \underbrace{y^2}_{\text{variable}} + \underbrace{4}_{\text{constant}}) = -2y.$$

QUICK CHECK 1 Compute f_x and f_y for $f(x, y) = 2xy$. ◄

b. It follows that $f_x(2, -4) = (3x^2)\big|_{(2,-4)} = 12$ and $f_y(2, -4) = (-2y)\big|_{(2,-4)} = 8$.

Related Exercise 16 ◄

EXAMPLE 3   Partial derivatives Compute the partial derivatives of the following functions.

a. $f(x, y) = \sin xy$ **b.** $g(x, y) = x^2 e^{xy}$

SOLUTION

a. Treating y as a constant and differentiating with respect to x, we have

$$\frac{\partial f}{\partial x} = \frac{\partial}{\partial x}(\sin xy) = y \cos xy.$$

Holding x fixed and differentiating with respect to y, we have

$$\frac{\partial f}{\partial y} = \frac{\partial}{\partial y}(\sin xy) = x \cos xy.$$

> Recall that
> $$\frac{d}{dx}(\sin 2x) = 2\cos 2x.$$
> Replacing 2 with the constant y, we have
> $$\frac{\partial}{\partial x}(\sin xy) = y \cos xy.$$

b. To compute the partial derivative with respect to x, we call on the Product Rule. Holding y fixed, we have

$$\frac{\partial g}{\partial x} = \frac{\partial}{\partial x}(x^2 e^{xy})$$

$$= \frac{\partial}{\partial x}(x^2)e^{xy} + x^2 \frac{\partial}{\partial x}(e^{xy}) \quad \text{Product Rule}$$

$$= 2xe^{xy} + x^2 y e^{xy} \qquad\qquad \text{Evaluate partial derivatives.}$$

$$= xe^{xy}(2 + xy). \qquad\qquad \text{Simplify.}$$

> Because x and y are *independent* variables,
> $$\frac{\partial}{\partial x}(y) = 0 \quad \text{and} \quad \frac{\partial}{\partial y}(x) = 0.$$

Treating x as a constant, the partial derivative with respect to y is

$$\frac{\partial g}{\partial y} = \frac{\partial}{\partial y}(x^2 e^{xy}) = x^2 \underbrace{\frac{\partial}{\partial y}(e^{xy})}_{xe^{xy}} = x^3 e^{xy}.$$

Related Exercises 17, 21 ◄

Higher-Order Partial Derivatives

Just as we have higher-order derivatives of functions of one variable, we also have higher-order partial derivatives. For example, given a function f and its partial derivative f_x, we can take the derivative of f_x with respect to x or with respect to y, which accounts for two of the four possible *second-order partial derivatives*. Table 15.3 summarizes the notation for second partial derivatives.

Table 15.3

Notation 1	Notation 2	What we say . . .
$\dfrac{\partial}{\partial x}\left(\dfrac{\partial f}{\partial x}\right) = \dfrac{\partial^2 f}{\partial x^2}$	$(f_x)_x = f_{xx}$	*d squared f dx squared or f-x-x*
$\dfrac{\partial}{\partial y}\left(\dfrac{\partial f}{\partial y}\right) = \dfrac{\partial^2 f}{\partial y^2}$	$(f_y)_y = f_{yy}$	*d squared f dy squared or f-y-y*
$\dfrac{\partial}{\partial x}\left(\dfrac{\partial f}{\partial y}\right) = \dfrac{\partial^2 f}{\partial x \partial y}$	$(f_y)_x = f_{yx}$	*f-y-x*
$\dfrac{\partial}{\partial y}\left(\dfrac{\partial f}{\partial x}\right) = \dfrac{\partial^2 f}{\partial y \partial x}$	$(f_x)_y = f_{xy}$	*f-x-y*

QUICK CHECK 2 Which of the following expressions are equivalent to each other: (a) f_{xy}, (b) f_{yx}, or (c) $\dfrac{\partial^2 f}{\partial y \partial x}$?

Write $\dfrac{\partial^2 f}{\partial p \partial q}$ in subscript notation. ◄

The order of differentiation can make a difference in the **mixed partial derivatives** f_{xy} and f_{yx}. So it is important to use the correct notation to reflect the order in which derivatives are taken. For example, the notations $\dfrac{\partial^2 f}{\partial x \partial y}$ and f_{yx} both mean $\dfrac{\partial}{\partial x}\left(\dfrac{\partial f}{\partial y}\right)$; that is, differentiate first with respect to y, then with respect to x.

EXAMPLE 4 **Second partial derivatives** Find the four second partial derivatives of $f(x, y) = 3x^4 y - 2xy + 5xy^3$.

SOLUTION First, we compute

$$\frac{\partial f}{\partial x} = \frac{\partial}{\partial x}\left(3x^4 y - 2xy + 5xy^3\right) = 12x^3 y - 2y + 5y^3$$

and

$$\frac{\partial f}{\partial y} = \frac{\partial}{\partial y}\left(3x^4 y - 2xy + 5xy^3\right) = 3x^4 - 2x + 15xy^2.$$

For the second partial derivatives, we have

$$\frac{\partial^2 f}{\partial x^2} = \frac{\partial}{\partial x}\left(\frac{\partial f}{\partial x}\right) = \frac{\partial}{\partial x}\left(12x^3 y - 2y + 5y^3\right) = 36x^2 y,$$

$$\frac{\partial^2 f}{\partial y^2} = \frac{\partial}{\partial y}\left(\frac{\partial f}{\partial y}\right) = \frac{\partial}{\partial y}\left(3x^4 - 2x + 15xy^2\right) = 30xy,$$

$$\frac{\partial^2 f}{\partial x \partial y} = \frac{\partial}{\partial x}\left(\frac{\partial f}{\partial y}\right) = \frac{\partial}{\partial x}\left(3x^4 - 2x + 15xy^2\right) = 12x^3 - 2 + 15y^2, \text{ and}$$

$$\frac{\partial^2 f}{\partial y \partial x} = \frac{\partial}{\partial y}\left(\frac{\partial f}{\partial x}\right) = \frac{\partial}{\partial y}\left(12x^3 y - 2y + 5y^3\right) = 12x^3 - 2 + 15y^2.$$

Related Exercises 39–40 ◄

QUICK CHECK 3 Compute f_{xxx} and f_{xxy} for $f(x, y) = x^3 y$. ◄

Equality of Mixed Partial Derivatives Notice that the two mixed partial derivatives in Example 4 are equal; that is, $f_{xy} = f_{yx}$. It turns out that most of the functions we encounter in this text have this property. Sufficient conditions for equality of mixed partial derivatives are given in a theorem attributed to the French mathematician Alexis Clairaut (1713–1765). The proof is found in advanced texts.

> **THEOREM 15.4 (Clairaut) Equality of Mixed Partial Derivatives**
> Assume f is defined on an open set D of $\mathbb{R}^2$, and that f_{xy} and f_{yx} are continuous throughout D. Then $f_{xy} = f_{yx}$ at all points of D.

Assuming sufficient continuity, Theorem 15.4 can be extended to higher derivatives of f. For example, $f_{xyx} = f_{xxy} = f_{yxx}$.

Functions of Three Variables

Everything we learned about partial derivatives of functions with two variables carries over to functions of three or more variables, as illustrated in Example 5.

EXAMPLE 5   Partial derivatives with more than two variables Find f_x, f_y, and f_z when $f(x, y, z) = e^{-xy} \cos z$.

SOLUTION To find f_x, we treat y and z as constants and differentiate with respect to x:

$$\frac{\partial f}{\partial x} = \frac{\partial}{\partial x} (\underbrace{e^{-xy}}_{\substack{y \text{ is} \\ \text{constant}}} \cdot \underbrace{\cos z}_{\text{constant}}) = -ye^{-xy} \cos z.$$

Holding x and z constant and differentiating with respect to y, we have

$$\frac{\partial f}{\partial y} = \frac{\partial}{\partial y} (\underbrace{e^{-xy}}_{\substack{x \text{ is} \\ \text{constant}}} \cdot \underbrace{\cos z}_{\text{constant}}) = -xe^{-xy} \cos z.$$

To find f_z, we hold x and y constant and differentiate with respect to z:

$$\frac{\partial f}{\partial z} = \frac{\partial}{\partial z} (\underbrace{e^{-xy}}_{\text{constant}} \cos z) = -e^{-xy} \sin z.$$

Related Exercises 55–56 ◄

QUICK CHECK 4 Compute f_{xz} and f_{zz} for $f(x, y, z) = xyz - x^2 z + yz^2$. ◄

Applications of Partial Derivatives When functions are used in realistic applications (for example, to describe velocity, pressure, investment fund balance, or population), they often involve more than one independent variable. For this reason, partial derivatives appear frequently in mathematical modeling.

EXAMPLE 6   Ideal Gas Law The pressure P, volume V, and temperature T of an ideal gas are related by the equation $PV = kT$, where $k > 0$ is a constant depending on the amount of gas.

a. Determine the rate of change of the pressure with respect to the volume at constant temperature. Interpret the result.

b. Determine the rate of change of the pressure with respect to the temperature at constant volume. Interpret the result.

c. Explain these results using level curves.

> ➤ Implicit differentiation can also be used with partial derivatives. Instead of solving for P, we could differentiate both sides of $PV = kT$ with respect to V holding T fixed. Using the Product Rule, $P_V V + P = 0$, which implies that $P_V = -P/V$. Substituting $P = kT/V$, we have $P_V = -kT/V^2$.

> ➤ In the Ideal Gas Law, temperature is a positive variable because it is measured in kelvins.

SOLUTION Expressing the pressure as a function of volume and temperature, we have

$$P = k\frac{T}{V}.$$

a. We find the partial derivative $\partial P/\partial V$ by holding T constant and differentiating P with respect to V:

$$\frac{\partial P}{\partial V} = \frac{\partial}{\partial V}\left(k\frac{T}{V}\right) = kT\frac{\partial}{\partial V}(V^{-1}) = -\frac{kT}{V^2}.$$

Recognizing that P, V, and T are always positive, we see that $\dfrac{\partial P}{\partial V} < 0$, which means that the pressure is a decreasing function of volume at a constant temperature.

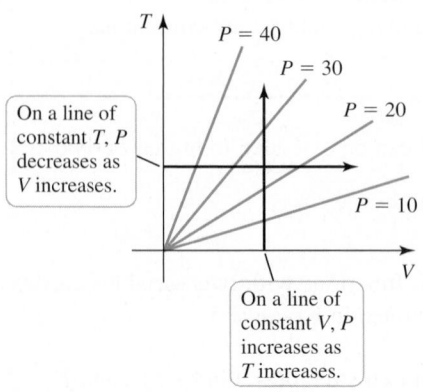

On a line of constant T, P decreases as V increases.

On a line of constant V, P increases as T increases.

Figure 15.33

QUICK CHECK 5 Explain why, in Figure 15.33, the slopes of the level curves increase as the pressure increases. ◄

b. The partial derivative $\partial P/\partial T$ is found by holding V constant and differentiating P with respect to T:

$$\frac{\partial P}{\partial T} = \frac{\partial}{\partial T}\left(k\,\frac{T}{V}\right) = \frac{k}{V}.$$

In this case, $\partial P/\partial T > 0$, which says that the pressure is an increasing function of temperature at constant volume.

c. The level curves (Section 15.1) of the pressure function are curves in the VT-plane that satisfy $k\,\dfrac{T}{V} = P_0$, where P_0 is a constant. Solving for T, the level curves are given by $T = \dfrac{1}{k}\,P_0 V$. Because $\dfrac{P_0}{k}$ is a positive constant, the level curves are lines in the first quadrant (Figure 15.33) with slope P_0/k. The fact that $\dfrac{\partial P}{\partial V} < 0$ (from part (a)) means that if we hold $T > 0$ fixed and move in the direction of increasing V on a *horizontal* line, we cross level curves corresponding to decreasing pressures. Similarly, $\dfrac{\partial P}{\partial T} > 0$ (from part (b)) means that if we hold $V > 0$ fixed and move in the direction of increasing T on a *vertical* line, we cross level curves corresponding to increasing pressures.

Related Exercise 69 ◄

Differentiability

We close this section with a technical matter that bears on the remainder of the chapter. Although we know how to compute partial derivatives of a function of several variables, we have not said what it means for such a function to be *differentiable* at a point. It is tempting to conclude that if the partial derivatives f_x and f_y exist at a point, then f is differentiable there. However, it is not so simple.

Recall that a function f of one variable is differentiable at $x = a$ provided the limit

$$f'(a) = \lim_{\Delta x \to 0}\frac{f(a + \Delta x) - f(a)}{\Delta x}$$

exists. If f is differentiable at a, it means that the curve is smooth at the point $(a, f(a))$ (no jumps, corners, or cusps); furthermore, the curve has a unique tangent line at that point with slope $f'(a)$. Differentiability for a function of several variables should carry the same properties: The surface should be smooth at the point in question, and something analogous to a unique tangent line should exist at the point.

Staying with the one-variable case, we define the quantity

$$\varepsilon = \underbrace{\frac{f(a + \Delta x) - f(a)}{\Delta x}}_{\text{slope of secant line}} - \underbrace{f'(a)}_{\substack{\text{slope of} \\ \text{tangent line}}},$$

where ε is viewed as a function of Δx. Notice that ε is the difference between the slopes of secant lines and the slope of the tangent line at the point $(a, f(a))$. If f is differentiable at a, then this difference approaches zero as $\Delta x \to 0$; therefore, $\lim_{\Delta x \to 0}\varepsilon = 0$. Multiplying both sides of the definition of ε by Δx gives

$$\varepsilon\,\Delta x = f(a + \Delta x) - f(a) - f'(a)\,\Delta x.$$

Rearranging, we have the change in the function $y = f(x)$:

$$\Delta y = f(a + \Delta x) - f(a) = f'(a)\,\Delta x + \underbrace{\varepsilon\,\Delta x}_{\varepsilon \to 0 \text{ as } \Delta x \to 0}.$$

➤ Notice that $f'(a)\,\Delta x$ is the approximate change in the function given by a linear approximation.

This expression says that in the one-variable case, if f is differentiable at a, then the change in f between a and a nearby point $a + \Delta x$ is represented by $f'(a)\,\Delta x$ plus a quantity $\varepsilon\,\Delta x$, where $\lim_{\Delta x \to 0}\varepsilon = 0$.

The analogous requirement with several variables is the definition of differentiability for functions of two (or more) variables.

DEFINITION Differentiability

The function $z = f(x, y)$ is **differentiable at (a, b)** provided $f_x(a, b)$ and $f_y(a, b)$ exist and the change $\Delta z = f(a + \Delta x, b + \Delta y) - f(a, b)$ equals

$$\Delta z = f_x(a, b)\,\Delta x + f_y(a, b)\,\Delta y + \varepsilon_1 \Delta x + \varepsilon_2 \Delta y,$$

where for fixed a and b, ε_1 and ε_2 are functions that depend only on Δx and Δy, with $(\varepsilon_1, \varepsilon_2) \to (0, 0)$ as $(\Delta x, \Delta y) \to (0, 0)$. A function is **differentiable** on an open set R if it is differentiable at every point of R.

Several observations are needed here. First, the definition extends to functions of more than two variables. Second, we show how differentiability is related to linear approximation and the existence of a *tangent plane* in Section 15.6. Finally, the conditions of the definition are generally difficult to verify. The following theorem may be useful in checking differentiability.

THEOREM 15.5   Conditions for Differentiability

Suppose the function f has partial derivatives f_x and f_y defined on an open set containing (a, b), with f_x and f_y continuous at (a, b). Then f is differentiable at (a, b).

As shown in Example 7, the existence of f_x and f_y at (a, b) is not enough to ensure differentiability of f at (a, b). However, by Theorem 15.5, if f_x and f_y are continuous at (a, b) (and defined in an open set containing (a, b)), then we can conclude f is differentiable there. Polynomials and rational functions are differentiable at all points of their domains, as are compositions of exponential, logarithmic, and trigonometric functions with other differentiable functions. The proof of this theorem is given in Appendix A.

We close with the analog of Theorem 3.1, which states that differentiability implies continuity.

THEOREM 15.6   Differentiable Implies Continuous

If a function f is differentiable at (a, b), then it is continuous at (a, b).

Proof: By the definition of differentiability,

$$\Delta z = f_x(a, b)\,\Delta x + f_y(a, b)\,\Delta y + \varepsilon_1 \Delta x + \varepsilon_2 \Delta y,$$

where $(\varepsilon_1, \varepsilon_2) \to (0, 0)$ as $(\Delta x, \Delta y) \to (0, 0)$. Because f is assumed to be differentiable, we see that as Δx and Δy approach 0,

$$\lim_{(\Delta x, \Delta y) \to (0, 0)} \Delta z = 0.$$

Also, because $\Delta z = f(a + \Delta x, b + \Delta y) - f(a, b)$, it follows that

$$\lim_{(\Delta x, \Delta y) \to (0, 0)} f(a + \Delta x, b + \Delta y) = f(a, b),$$

which implies continuity of f at (a, b). ◄

> Recall that continuity requires that
> $$\lim_{(x, y) \to (a, b)} f(x, y) = f(a, b),$$
> which is equivalent to
> $$\lim_{(\Delta x, \Delta y) \to (0, 0)} f(a + \Delta x, b + \Delta y) = f(a, b).$$

EXAMPLE 7   A nondifferentiable function Discuss the differentiability and continuity of the function

$$f(x, y) = \begin{cases} \dfrac{3xy}{x^2 + y^2} & \text{if } (x, y) \neq (0, 0) \\ 0 & \text{if } (x, y) = (0, 0). \end{cases}$$

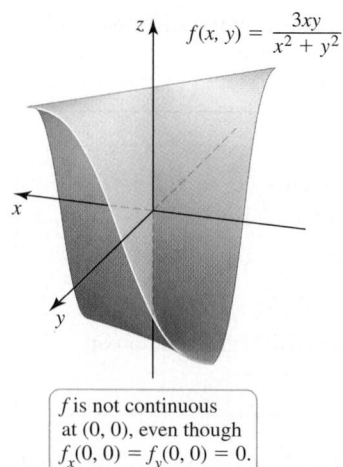

z

$f(x, y) = \dfrac{3xy}{x^2 + y^2}$

x

y

f is not continuous
at $(0, 0)$, even though
$f_x(0, 0) = f_y(0, 0) = 0$.

Figure 15.34

▶ The relationships between the existence
and continuity of partial derivatives and
whether a function is differentiable are
further explored in Exercises 96–97.

SOLUTION As a rational function, f is continuous and differentiable at all points $(x, y) \neq (0, 0)$. The interesting behavior occurs at the origin. Using calculations similar to those in Example 4 in Section 15.2, it can be shown that if the origin is approached along the line $y = mx$, then

$$\lim_{\substack{(x, y) \to (0, 0) \\ (\text{along } y = mx)}} \frac{3xy}{x^2 + y^2} = \frac{3m}{m^2 + 1}.$$

Therefore, the value of the limit depends on the direction of approach, which implies that the limit does not exist, and f is not continuous at $(0, 0)$. By Theorem 15.6, it follows that f is not differentiable at $(0, 0)$. **Figure 15.34** shows the discontinuity of f at the origin. Let's look at the first partial derivatives of f at $(0, 0)$. A short calculation shows that

$$f_x(0, 0) = \lim_{h \to 0} \frac{f(0 + h, 0) - f(0, 0)}{h} = \lim_{h \to 0} \frac{0 - 0}{h} = 0,$$

$$f_y(0, 0) = \lim_{h \to 0} \frac{f(0, 0 + h) - f(0, 0)}{h} = \lim_{h \to 0} \frac{0 - 0}{h} = 0.$$

Despite the fact that its first partial derivatives exist at $(0, 0)$, f is not differentiable at $(0, 0)$. As noted earlier, the existence of first partial derivatives at a point is not enough to ensure differentiability at that point.

Related Exercises 77–78 ◀

SECTION 15.3 EXERCISES

Getting Started

1. Suppose you are standing on the surface $z = f(x, y)$ at the point $(a, b, f(a, b))$. Interpret the meaning of $f_x(a, b)$ and $f_y(a, b)$ in terms of slopes or rates of change.

2. Let $f(x, y) = 3x^2 + y^3$.
   a. Compute f_x and f_y.
   b. Evaluate each derivative at $(1, 3)$.
   c. Find the four second partial derivatives of f.

3. Given the graph of a function $z = f(x, y)$ and its traces in the planes $x = 4$, $y = 1$, and $y = 3$ (see figure), determine whether the following partial derivatives are positive or negative.

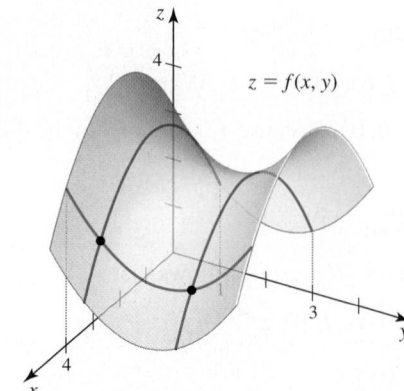

z

4

$z = f(x, y)$

3

y

4

x

   a. $f_x(4, 1)$ b. $f_y(4, 1)$ c. $f_x(4, 3)$ d. $f_y(4, 3)$

4. Find f_x and f_y when $f(x, y) = y^8 + 2x^6 + 2xy$.

5. Find f_x and f_y when $f(x, y) = 3x^2y + 2$.

6. Find the four second partial derivatives of $f(x, y) = x^2y^3$.

7. Verify that $f_{xy} = f_{yx}$, for $f(x, y) = 2x^3 + 3y^2 + 1$.

8. Verify that $f_{xy} = f_{yx}$, for $f(x, y) = xe^y$.

9. Find f_x, f_y, and f_z, for $f(x, y, z) = xy + xz + yz$.

10. The volume of a right circular cylinder with radius r and height h is $V = \pi r^2 h$. Is the volume an increasing or decreasing function of the radius at a fixed height (assume $r > 0$ and $h > 0$)?

Practice Exercises

11–14. Evaluating partial derivatives using limits *Use the limit definition of partial derivatives to evaluate $f_x(x, y)$ and $f_y(x, y)$ for the following functions.*

11. $f(x, y) = 5xy$

12. $f(x, y) = x + y^2 + 4$

13. $f(x, y) = \dfrac{x}{y}$

14. $f(x, y) = \sqrt{xy}$

15–37. Partial derivatives *Find the first partial derivatives of the following functions.*

15. $f(x, y) = xe^y$

16. $f(x, y) = 4x^3y^2 + 3x^2y^3 + 10$

17. $f(x, y) = e^{x^2y}$

18. $f(x, y) = (3xy + 4y^2 + 1)^5$

19. $f(w, z) = \dfrac{w}{w^2 + z^2}$

20. $f(s, t) = \dfrac{s - t}{s + t}$

21. $f(x, y) = x \cos xy$

22. $f(x, y) = \tan^{-1} \dfrac{x^2}{y^2}$

23. $s(y, z) = z^2 \tan yz$

24. $g(x, z) = x \ln(z^2 + x^2)$

25. $G(s, t) = \dfrac{\sqrt{st}}{s + t}$

26. $F(p, q) = \sqrt{p^2 + pq + q^2}$

27. $f(x, y) = x^{2y}$

28. $g(x, y) = \cos^5(x^2 y^3)$

29. $h(x, y) = x - \sqrt{x^2 - 4y}$

30. $h(u, v) = \sqrt{\dfrac{uv}{u - v}}$

31. $f(x, y) = \displaystyle\int_x^{y^3} e^{t^2} dt$

32. $g(x, y) = y \sin^{-1}\sqrt{xy}$

33. $f(x, y) = 1 - \tan^{-1}(x^2 + y^2)$

34. $f(x, y) = \ln(1 + e^{-xy})$

35. $h(x, y) = (1 + 2y)^x$

36. $f(x, y) = 1 - \cos(2(x + y)) + \cos^2(x + y)$

37. $f(x, y) = \displaystyle\int_x^y h(s)\, ds$, where h is continuous for all real numbers

38–47. Second partial derivatives *Find the four second partial derivatives of the following functions.*

38. $f(x, y) = x^2 \sin y$

39. $h(x, y) = x^3 + xy^2 + 1$

40. $f(x, y) = 2x^5 y^2 + x^2 y$

41. $f(x, y) = y^3 \sin 4x$

42. $f(x, y) = \sin^2(x^3 y)$

43. $p(u, v) = \ln(u^2 + v^2 + 4)$

44. $Q(r, s) = \dfrac{e^{r^3 s}}{s}$

45. $F(r, s) = re^s$

46. $H(x, y) = \sqrt{4 + x^2 + y^2}$

47. $f(x, y) = \tan^{-1}(x^3 y^2)$

48–53. Equality of mixed partial derivatives *Verify that $f_{xy} = f_{yx}$ for the following functions.*

48. $f(x, y) = 3x^2 y^{-1} - 2x^{-1} y^2$

49. $f(x, y) = e^{x+y}$

50. $f(x, y) = \sqrt{xy}$

51. $f(x, y) = \cos xy$

52. $f(x, y) = e^{\sin xy}$

53. $f(x, y) = (2x - y^3)^4$

54–62. Partial derivatives with more than two variables *Find the first partial derivatives of the following functions.*

54. $G(r, s, t) = \sqrt{rs + rt + st}$

55. $h(x, y, z) = \cos(x + y + z)$

56. $g(x, y, z) = 2x^2 y - 3xz^4 + 10y^2 z^2$

57. $F(u, v, w) = \dfrac{u}{v + w}$

58. $Q(x, y, z) = \tan xyz$

59. $G(r, s, t) = \sqrt{rs^3 t^5}$

60. $g(w, x, y, z) = \cos(w + x) \sin(y - z)$

61. $h(w, x, y, z) = \dfrac{wz}{xy}$

62. $F(w, x, y, z) = w\sqrt{x + 2y + 3z}$

63. Exploiting patterns Let $R(t) = \dfrac{at + b}{ct + d}$ and

$g(x, y, z) = \dfrac{4x - 2y - 2z}{-6x + 3y - 3z}$.

a. Verify that $R'(t) = \dfrac{ad - bc}{(ct + d)^2}$.

b. Use the derivative $R'(t)$ to find the first partial derivatives of g.

64–67. Estimating partial derivatives from a table *The following table shows values of a function $f(x, y)$ for values of x from 2 to 2.5 and values of y from 3 to 3.5. Use this table to estimate the values of the following partial derivatives.*

y \ x	2	2.1	2.2	2.3	2.4	2.5
3	4.243	4.347	4.450	4.550	4.648	4.743
3.1	4.384	4.492	4.598	4.701	4.802	4.902
3.2	4.525	4.637	4.746	4.853	4.957	5.060
3.3	4.667	4.782	4.895	5.005	5.112	5.218
3.4	4.808	4.930	5.043	5.156	5.267	5.376
3.5	4.950	5.072	5.191	5.308	5.422	5.534

64. $f_x(2, 3)$

65. $f_y(2, 3)$

66. $f_x(2.2, 3.4)$

67. $f_y(2.4, 3.3)$

68. Estimating partial derivatives from a graph Use the level curves of f (see figure) to estimate the values of f_x and f_y at $A(0.42, 0.5)$.

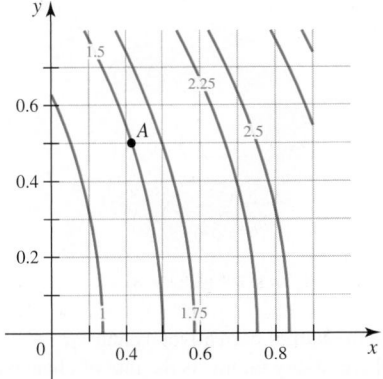

69. Gas law calculations Consider the Ideal Gas Law $PV = kT$, where $k > 0$ is a constant. Solve this equation for V in terms of P and T.

a. Determine the rate of change of the volume with respect to the pressure at constant temperature. Interpret the result.

b. Determine the rate of change of the volume with respect to the temperature at constant pressure. Interpret the result.

c. Assuming $k = 1$, draw several level curves of the volume function, and interpret the results as in Example 6.

70. Body mass index The body mass index (BMI) for an adult human is given by the function $B = \dfrac{w}{h^2}$, where w is the weight measured in kilograms and h is the height measured in meters.

a. Find the rate of change of the BMI with respect to weight at a constant height.

b. For fixed h, is the BMI an increasing or decreasing function of w? Explain.

c. Find the rate of change of the BMI with respect to height at a constant weight.

d. For fixed w, is the BMI an increasing or decreasing function of h? Explain.

71. Resistors in parallel Two resistors in an electrical circuit with resistance R_1 and R_2 wired in parallel with a constant voltage give an effective resistance of R, where $\dfrac{1}{R} = \dfrac{1}{R_1} + \dfrac{1}{R_2}$.

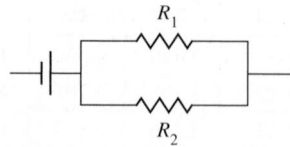

a. Find $\dfrac{\partial R}{\partial R_1}$ and $\dfrac{\partial R}{\partial R_2}$ by solving for R and differentiating.

b. Find $\dfrac{\partial R}{\partial R_1}$ and $\dfrac{\partial R}{\partial R_2}$ by differentiating implicitly.

c. Describe how an increase in R_1 with R_2 constant affects R.

d. Describe how a decrease in R_2 with R_1 constant affects R.

72. Spherical caps The volume of the cap of a sphere of radius r and thickness h is $V = \dfrac{\pi}{3} h^2 (3r - h)$, for $0 \le h \le 2r$.

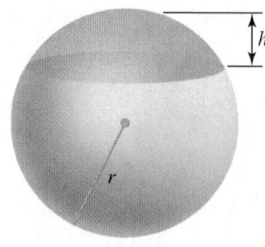

$$V = \tfrac{\pi}{3} h^2 (3r - h)$$

a. Compute the partial derivatives V_h and V_r.

b. For a sphere of any radius, is the rate of change of volume with respect to r greater when $h = 0.2r$ or when $h = 0.8r$?

c. For a sphere of any radius, for what value of h is the rate of change of volume with respect to r equal to 1?

d. For a fixed radius r, for what value of h ($0 \le h \le 2r$) is the rate of change of volume with respect to h the greatest?

73–76. Heat equation *The flow of heat along a thin conducting bar is governed by the one-dimensional heat equation (with analogs for thin plates in two dimensions and for solids in three dimensions):*

$$\frac{\partial u}{\partial t} = k \frac{\partial^2 u}{\partial x^2},$$

where u is a measure of the temperature at a location x on the bar at time t and the positive constant k is related to the conductivity of the material. Show that the following functions satisfy the heat equation with k = 1.

73. $u(x, t) = 4e^{-4t} \cos 2x$ **74.** $u(x, t) = 10e^{-t} \sin x$

75. $u(x, t) = Ae^{-a^2 t} \cos ax$, for any real numbers a and A

76. $u(x, t) = e^{-t}(2 \sin x + 3 \cos x)$

77–78. Nondifferentiability? *Consider the following functions f.*

a. *Is f continuous at $(0, 0)$?*

b. *Is f differentiable at $(0, 0)$?*

c. *If possible, evaluate $f_x(0, 0)$ and $f_y(0, 0)$.*

d. *Determine whether f_x and f_y are continuous at $(0, 0)$.*

e. *Explain why Theorems 15.5 and 15.6 are consistent with the results in parts (a)–(d).*

77. $f(x, y) = \begin{cases} -\dfrac{xy}{x^2 + y^2} & \text{if } (x, y) \ne (0, 0) \\ 0 & \text{if } (x, y) = (0, 0) \end{cases}$

78. $f(x, y) = \begin{cases} \dfrac{2xy^2}{x^2 + y^4} & \text{if } (x, y) \ne (0, 0) \\ 0 & \text{if } (x, y) = (0, 0) \end{cases}$

79. Explain why or why not Determine whether the following statements are true and give an explanation or counterexample.

a. $\dfrac{\partial}{\partial x}(y^{10}) = 10y^9$.

b. $\dfrac{\partial^2}{\partial x \partial y}(\sqrt{xy}) = \dfrac{1}{\sqrt{xy}}$.

c. If f has continuous partial derivatives of all orders, then $f_{xxy} = f_{yxx}$.

80. Mixed partial derivatives

a. Consider the function $w = f(x, y, z)$. List all possible second partial derivatives that could be computed.

b. Let $f(x, y, z) = x^2 y + 2xz^2 - 3y^2 z$ and determine which second partial derivatives are equal.

c. How many second partial derivatives does $p = g(w, x, y, z)$ have?

Explorations and Challenges

81. Partial derivatives and level curves Consider the function $z = x/y^2$.

a. Compute z_x and z_y.

b. Sketch the level curves for $z = 1, 2, 3$, and 4.

c. Move along the horizontal line $y = 1$ in the xy-plane and describe how the corresponding z-values change. Explain how this observation is consistent with z_x as computed in part (a).

d. Move along the vertical line $x = 1$ in the xy-plane and describe how the corresponding z-values change. Explain how this observation is consistent with z_y as computed in part (a).

82. Volume of a box A box with a square base of length x and height h has a volume $V = x^2 h$.

a. Compute the partial derivatives V_x and V_h.

b. For a box with $h = 1.5$ m, use linear approximation to estimate the change in volume if x increases from $x = 0.5$ m to $x = 0.51$ m.

c. For a box with $x = 0.5$ m, use linear approximation to estimate the change in volume if h decreases from $h = 1.5$ m to $h = 1.49$ m.

d. For a fixed height, does a 10% change in x always produce (approximately) a 10% change in V? Explain.

e. For a fixed base length, does a 10% change in h always produce (approximately) a 10% change in V? Explain.

83. Electric potential function The electric potential in the xy-plane associated with two positive charges, one at $(0, 1)$ with twice the magnitude of the charge at $(0, -1)$, is

$$\varphi(x, y) = \frac{2}{\sqrt{x^2 + (y - 1)^2}} + \frac{1}{\sqrt{x^2 + (y + 1)^2}}.$$

a. Compute φ_x and φ_y.

b. Describe how φ_x and φ_y behave as $x, y \to \pm\infty$.

c. Evaluate $\varphi_x(0, y)$, for all $y \ne \pm 1$. Interpret this result.

d. Evaluate $\varphi_y(x, 0)$, for all x. Interpret this result.

T 84. Cobb-Douglas production function The output Q of an economic system subject to two inputs, such as labor L and capital K, is often modeled by the Cobb-Douglas production function $Q(L, K) = cL^a K^b$. Suppose $a = \dfrac{1}{3}$, $b = \dfrac{2}{3}$, and $c = 1$.

a. Evaluate the partial derivatives Q_L and Q_K.
b. Suppose $L = 10$ is fixed and K increases from $K = 20$ to $K = 20.5$. Use linear approximation to estimate the change in Q.
c. Suppose $K = 20$ is fixed and L decreases from $L = 10$ to $L = 9.5$. Use linear approximation to estimate the change in Q.
d. Graph the level curves of the production function in the first quadrant of the LK-plane for $Q = 1, 2,$ and 3.
e. Use the graph of part (d). If you move along the vertical line $L = 2$ in the positive K-direction, how does Q change? Is this consistent with Q_K computed in part (a)?
f. Use the graph of part (d). If you move along the horizontal line $K = 2$ in the positive L-direction, how does Q change? Is this consistent with Q_L computed in part (a)?

85. An identity Show that if $f(x, y) = \dfrac{ax + by}{cx + dy}$, where $a, b, c,$ and d are real numbers with $ad - bc = 0$, then $f_x = f_y = 0$, for all x and y in the domain of f. Give an explanation.

86. Wave on a string Imagine a string that is fixed at both ends (for example, a guitar string). When plucked, the string forms a standing wave. The displacement u of the string varies with position x and with time t. Suppose it is given by $u = f(x, t) = 2 \sin(\pi x) \sin(\pi t/2)$, for $0 \le x \le 1$ and $t \ge 0$ (see figure). At a fixed point in time, the string forms a wave on $[0, 1]$. Alternatively, if you focus on a point on the string (fix a value of x), that point oscillates up and down in time.

a. What is the period of the motion in time?
b. Find the rate of change of the displacement with respect to time at a constant position (which is the vertical velocity of a point on the string).
c. At a fixed time, what point on the string is moving fastest?
d. At a fixed position on the string, when is the string moving fastest?
e. Find the rate of change of the displacement with respect to position at a constant time (which is the slope of the string).
f. At a fixed time, where is the slope of the string greatest?

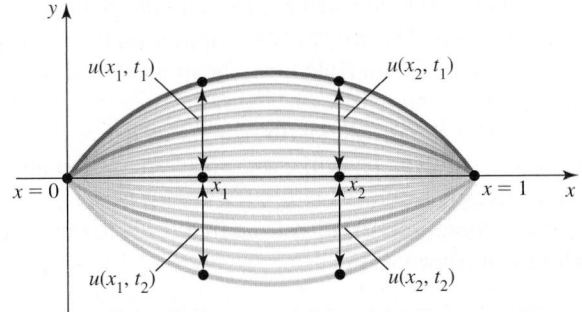

87–89. Wave equation *Traveling waves (for example, water waves or electromagnetic waves) exhibit periodic motion in both time and position. In one dimension, some types of wave motion are governed by the one-dimensional wave equation*

$$\frac{\partial^2 u}{\partial t^2} = c^2 \frac{\partial^2 u}{\partial x^2},$$

where $u(x, t)$ is the height or displacement of the wave surface at position x and time t, and c is the constant speed of the wave. Show that the following functions are solutions of the wave equation.

87. $u(x, t) = \cos(2(x + ct))$

88. $u(x, t) = 5 \cos(2(x + ct)) + 3 \sin(x - ct)$

89. $u(x, t) = Af(x + ct) + Bg(x - ct)$, where A and B are constants, and f and g are twice differentiable functions of one variable.

90–93. Laplace's equation *A classical equation of mathematics is Laplace's equation, which arises in both theory and applications. It governs ideal fluid flow, electrostatic potentials, and the steady-state distribution of heat in a conducting medium. In two dimensions, Laplace's equation is*

$$\frac{\partial^2 u}{\partial x^2} + \frac{\partial^2 u}{\partial y^2} = 0.$$

*Show that the following functions are **harmonic**; that is, they satisfy Laplace's equation.*

90. $u(x, y) = e^{-x} \sin y$

91. $u(x, y) = x(x^2 - 3y^2)$

92. $u(x, y) = e^{ax} \cos ay$, for any real number a

93. $u(x, y) = \tan^{-1}\left(\dfrac{y}{x - 1}\right) - \tan^{-1}\left(\dfrac{y}{x + 1}\right)$

94–95. Differentiability *Use the definition of differentiability to prove that the following functions are differentiable at $(0, 0)$. You must produce functions ε_1 and ε_2 with the required properties.*

94. $f(x, y) = x + y$ **95.** $f(x, y) = xy$

96–97. Nondifferentiability? *Consider the following functions f.*

a. Is f continuous at $(0, 0)$?
b. Is f differentiable at $(0, 0)$?
c. If possible, evaluate $f_x(0, 0)$ and $f_y(0, 0)$.
d. Determine whether f_x and f_y are continuous at $(0, 0)$.
e. Explain why Theorems 15.5 and 15.6 are consistent with the results in parts (a)–(d).

96. $f(x, y) = 1 - |xy|$ **97.** $f(x, y) = \sqrt{|xy|}$

98. Cauchy-Riemann equations In the advanced subject of complex variables, a function typically has the form $f(x, y) = u(x, y) + iv(x, y)$, where u and v are real-valued functions and $i = \sqrt{-1}$ is the imaginary unit. A function $f = u + iv$ is said to be *analytic* (analogous to differentiable) if it satisfies the Cauchy-Riemann equations: $u_x = v_y$ and $u_y = -v_x$.

a. Show that $f(x, y) = (x^2 - y^2) + i(2xy)$ is analytic.
b. Show that $f(x, y) = x(x^2 - 3y^2) + iy(3x^2 - y^2)$ is analytic.
c. Show that if $f = u + iv$ is analytic, then $u_{xx} + u_{yy} = 0$ and $v_{xx} + v_{yy} = 0$. Assume u and v satisfy the conditions in Theorem 15.4.

99. Derivatives of an integral Let h be continuous for all real numbers. Find f_x and f_y when $f(x, y) = \int_1^{xy} h(s)\,ds$.

QUICK CHECK ANSWERS

1. $f_x = 2y$; $f_y = 2x$ **2.** (a) and (c) are the same; f_{qp}
3. $f_{xxx} = 6y$; $f_{xxy} = 6x$ **4.** $f_{xz} = y - 2x$; $f_{zz} = 2y$
5. The equations of the level curves are $T = \dfrac{1}{k}P_0 V$. As the pressure P_0 increases, the slopes of these lines increase. ◄

15.4 The Chain Rule

In this section, we combine ideas based on the Chain Rule (Section 3.7) with what we know about partial derivatives (Section 15.3) to develop new methods for finding derivatives of functions of several variables. To illustrate the importance of these methods, consider the following situation.

Economists modeling manufacturing systems often work with *production functions* that relate the productivity (output) of the system to all the variables on which it depends (input). A simplified production function might take the form $P = F(L, K, R)$, where L, K, and R represent the availability of labor, capital, and natural resources, respectively. However, the variables L, K, and R may be intermediate variables that depend on other variables. For example, it might be that L is a function of the unemployment rate u, K is a function of the prime interest rate i, and R is a function of time t (seasonal availability of resources). Even in this simplified model, we see that productivity, which is the dependent variable, is ultimately related to many other variables (**Figure 15.35**). Of critical interest to an economist is how changes in one variable determine changes in other variables. For instance, if the unemployment rate increases by 0.1% and the interest rate decreases by 0.2%, what is the effect on productivity? In this section, we develop the tools needed to answer such questions.

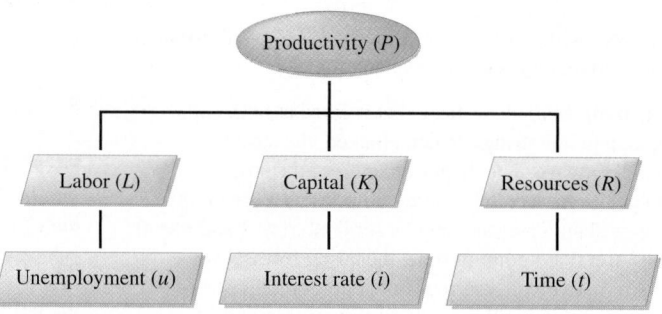

Figure 15.35

The Chain Rule with One Independent Variable

Recall the basic Chain Rule: If y is a function of u and u is a function of t, then $\dfrac{dy}{dt} = \dfrac{dy}{du}\dfrac{du}{dt}$. We first extend the Chain Rule to composite functions of the form $z = f(x, y)$, where x and y are functions of t. What is $\dfrac{dz}{dt}$?

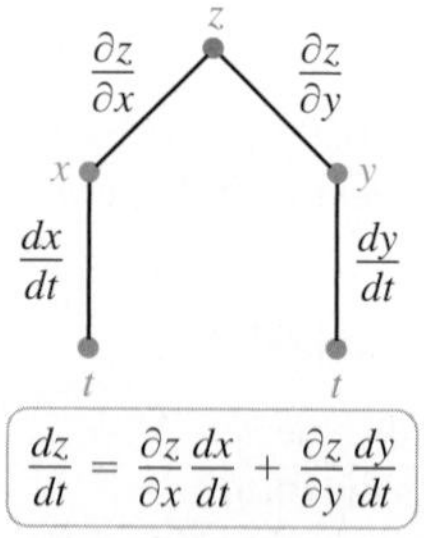

We illustrate the relationships among the variables t, x, y, and z using a *tree diagram* (**Figure 15.36**). To find dz/dt, first notice that z depends on x, which in turn depends on t. The change in z with respect to x is the partial derivative $\partial z/\partial x$, and the change in x with respect to t is the ordinary derivative dx/dt. These derivatives appear on the corresponding branches of the tree diagram. Using the Chain Rule idea, the product of these derivatives gives the change in z with respect to t through x.

Similarly, z also depends on y. The change in z with respect to y is $\partial z/\partial y$, and the change in y with respect to t is dy/dt. The product of these derivatives, which appear on the corresponding branches of the tree, gives the change in z with respect to t through y. Summing the contributions to dz/dt along each branch of the tree leads to the following theorem, the proof of which is found in Appendix A.

THEOREM 15.7 Chain Rule (One Independent Variable)
Let z be a differentiable function of x and y on its domain, where x and y are differentiable functions of t on an interval I. Then

$$\frac{dz}{dt} = \frac{\partial z}{\partial x}\frac{dx}{dt} + \frac{\partial z}{\partial y}\frac{dy}{dt}.$$

Figure 15.36

$$\frac{dz}{dt} = \frac{\partial z}{\partial x}\frac{dx}{dt} + \frac{\partial z}{\partial y}\frac{dy}{dt}$$

▶ A subtle observation about notation should be made. If $z = f(x, y)$, where x and y are functions of another variable t, it is common to write $z = f(t)$ to show that z ultimately depends on t. However, these two functions denoted f are actually different. We *should* write (or at least remember) that in fact $z = F(t)$, where F is a function other than f. This distinction is often overlooked for the sake of convenience.

QUICK CHECK 1 Explain why Theorem 15.7 reduces to the Chain Rule for a function of one variable in the case that $z = f(x)$ and $x = g(t)$. ◀

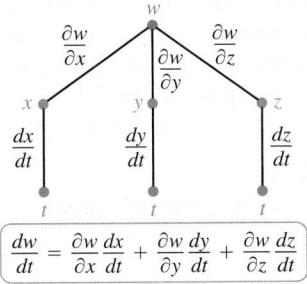

$$\frac{dw}{dt} = \frac{\partial w}{\partial x}\frac{dx}{dt} + \frac{\partial w}{\partial y}\frac{dy}{dt} + \frac{\partial w}{\partial z}\frac{dz}{dt}$$

Figure 15.37

Before presenting examples, several comments are in order.

- With $z = f(x(t), y(t))$, the dependent variable is z and the sole independent variable is t. The variables x and y are **intermediate variables**.

- The choice of notation for partial and ordinary derivatives in the Chain Rule is important. We write the ordinary derivatives dx/dt and dy/dt because x and y depend only on t. We write the partial derivatives $\partial z/\partial x$ and $\partial z/\partial y$ because z is a function of both x and y. Finally, we write dz/dt as an ordinary derivative because z ultimately depends only on t.

- Theorem 15.7 generalizes directly to functions of more than two intermediate variables (Figure 15.37). For example, if $w = f(x, y, z)$, where x, y, and z are functions of the single independent variable t, then

$$\frac{dw}{dt} = \frac{\partial w}{\partial x}\frac{dx}{dt} + \frac{\partial w}{\partial y}\frac{dy}{dt} + \frac{\partial w}{\partial z}\frac{dz}{dt}.$$

> ➤ If f, x, and y are simple, as in Example 1, it is possible to substitute $x(t)$ and $y(t)$ into f, producing a function of t only, and then differentiate with respect to t. But this approach quickly becomes impractical with more complicated functions, and the Chain Rule offers a great advantage.

EXAMPLE 1   Chain Rule with one independent variable Let $z = x^2 - 3y^2 + 20$, where $x = 2 \cos t$ and $y = 2 \sin t$.

a. Find $\dfrac{dz}{dt}$ and evaluate it at $t = \pi/4$.

b. Interpret the result geometrically.

SOLUTION

a. Computing the intermediate derivatives and applying the Chain Rule (Theorem 15.7), we find that

$$\frac{dz}{dt} = \frac{\partial z}{\partial x}\frac{dx}{dt} + \frac{\partial z}{\partial y}\frac{dy}{dt}$$

$$= \underbrace{(2x)}_{\frac{\partial z}{\partial x}}\underbrace{(-2\sin t)}_{\frac{dx}{dt}} + \underbrace{(-6y)}_{\frac{\partial z}{\partial y}}\underbrace{(2\cos t)}_{\frac{dy}{dt}} \quad \text{Evaluate derivatives.}$$

$$= -4x\sin t - 12y\cos t \qquad\qquad \text{Simplify.}$$

$$= -8\cos t\sin t - 24\sin t\cos t \qquad \text{Substitute } x = 2\cos t, y = 2\sin t.$$

$$= -16\sin 2t. \qquad\qquad\qquad \text{Simplify; } \sin 2t = 2\sin t\cos t.$$

Substituting $t = \pi/4$ gives $\left.\dfrac{dz}{dt}\right|_{t=\pi/4} = -16$.

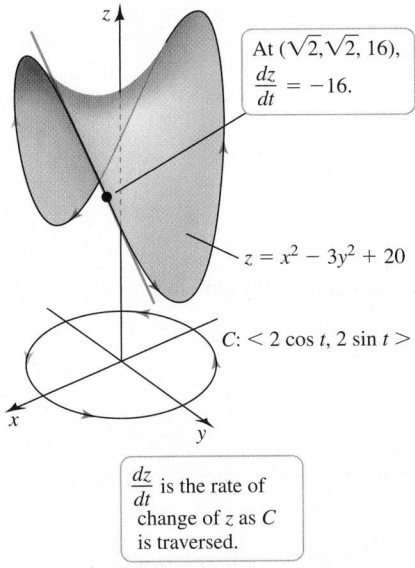

At $(\sqrt{2}, \sqrt{2}, 16)$, $\dfrac{dz}{dt} = -16$.

$z = x^2 - 3y^2 + 20$

$C: \langle 2\cos t, 2\sin t \rangle$

$\dfrac{dz}{dt}$ is the rate of change of z as C is traversed.

Figure 15.38

b. The parametric equations $x = 2\cos t$, $y = 2\sin t$, for $0 \le t \le 2\pi$, describe a circle C of radius 2 in the xy-plane. Imagine walking on the surface $z = x^2 - 3y^2 + 20$ directly above the circle C consistent with positive (counterclockwise) orientation of C. Your path rises and falls as you walk (Figure 15.38); the rate of change of your elevation z with respect to t is given by dz/dt. For example, when $t = \pi/4$, the corresponding point on the surface is $(\sqrt{2}, \sqrt{2}, 16)$. At that point, z decreases at a rate of -16 (by part (a)) as you walk on the surface above C.

Related Exercises 10, 12 ◄

The Chain Rule with Several Independent Variables

The ideas behind the Chain Rule of Theorem 15.7 can be modified to cover a variety of situations in which functions of several variables are composed with one another. For example, suppose z depends on two intermediate variables x and y, each of which depends on the independent variables s and t. Once again, a tree diagram (Figure 15.39) helps organize the relationships among variables. The dependent variable z now ultimately depends on the two independent variables s and t, so it makes sense to ask about the rates of change of z with respect to either s or t, which are $\partial z/\partial s$ and $\partial z/\partial t$, respectively.

To compute $\partial z/\partial s$, we note that there are two paths in the tree (in red in Figure 15.39) that connect z to s and contribute to $\partial z/\partial s$. Along one path, z changes with respect to x

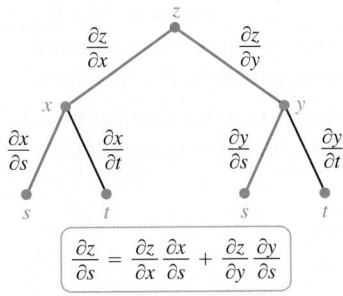

$$\frac{\partial z}{\partial s} = \frac{\partial z}{\partial x}\frac{\partial x}{\partial s} + \frac{\partial z}{\partial y}\frac{\partial y}{\partial s}$$

Figure 15.39

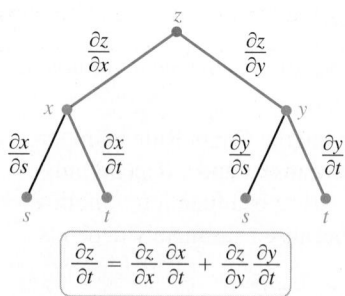

Figure 15.40

QUICK CHECK 2 Suppose $w = f(x, y, z)$, where $x = g(s, t)$, $y = h(s, t)$, and $z = p(s, t)$. Extend Theorem 15.8 to write a formula for $\partial w/\partial t$. ◄

(with rate of change $\partial z/\partial x$) and x changes with respect to s (with rate of change $\partial x/\partial s$). Along the other path, z changes with respect to y (with rate of change $\partial z/\partial y$) and y changes with respect to s (with rate of change $\partial y/\partial s$). We use a Chain Rule calculation along each path and combine the results. A similar argument leads to $\partial z/\partial t$ (Figure 15.40).

THEOREM 15.8 Chain Rule (Two Independent Variables)

Let z be a differentiable function of x and y, where x and y are differentiable functions of s and t. Then

$$\frac{\partial z}{\partial s} = \frac{\partial z}{\partial x}\frac{\partial x}{\partial s} + \frac{\partial z}{\partial y}\frac{\partial y}{\partial s} \quad \text{and} \quad \frac{\partial z}{\partial t} = \frac{\partial z}{\partial x}\frac{\partial x}{\partial t} + \frac{\partial z}{\partial y}\frac{\partial y}{\partial t}.$$

EXAMPLE 2 Chain Rule with two independent variables Let $z = \sin 2x \cos 3y$, where $x = s + t$ and $y = s - t$. Evaluate $\partial z/\partial s$ and $\partial z/\partial t$.

SOLUTION The tree diagram in Figure 15.39 gives the Chain Rule formula for $\partial z/\partial s$: We form products of the derivatives along the red branches connecting z to s and add the results. The partial derivative is

$$\frac{\partial z}{\partial s} = \frac{\partial z}{\partial x}\frac{\partial x}{\partial s} + \frac{\partial z}{\partial y}\frac{\partial y}{\partial s}$$

$$= \underbrace{2 \cos 2x \cos 3y}_{\frac{\partial z}{\partial x}} \cdot \underbrace{1}_{\frac{\partial x}{\partial s}} + \underbrace{(-3 \sin 2x \sin 3y)}_{\frac{\partial z}{\partial y}} \cdot \underbrace{1}_{\frac{\partial y}{\partial s}}$$

$$= 2 \cos (2\underbrace{(s + t)}_{x}) \cos (3\underbrace{(s - t)}_{y}) - 3 \sin (2\underbrace{(s + t)}_{x}) \sin (3\underbrace{(s - t)}_{y}).$$

Following the branches of Figure 15.40 connecting z to t, we have

$$\frac{\partial z}{\partial t} = \frac{\partial z}{\partial x}\frac{\partial x}{\partial t} + \frac{\partial z}{\partial y}\frac{\partial y}{\partial t}$$

$$= \underbrace{2 \cos 2x \cos 3y}_{\frac{\partial z}{\partial x}} \cdot \underbrace{1}_{\frac{\partial x}{\partial t}} + \underbrace{(-3 \sin 2x \sin 3y)}_{\frac{\partial z}{\partial y}} \cdot \underbrace{-1}_{\frac{\partial y}{\partial t}}$$

$$= 2 \cos (2\underbrace{(s + t)}_{x}) \cos (3\underbrace{(s - t)}_{y}) + 3 \sin (2\underbrace{(s + t)}_{x}) \sin (3\underbrace{(s - t)}_{y}).$$

Related Exercise 22 ◄

EXAMPLE 3 More variables Let w be a function of x, y, and z, each of which is a function of s and t.

a. Draw a labeled tree diagram showing the relationships among the variables.

b. Write the Chain Rule formula for $\dfrac{\partial w}{\partial s}$.

SOLUTION

a. Because w is a function of x, y, and z, the upper branches of the tree (**Figure 15.41**) are labeled with the partial derivatives w_x, w_y, and w_z. Each of x, y, and z is a function of two variables, so the lower branches of the tree also require partial derivative labels.

b. Extending Theorem 15.8, we take the three paths through the tree that connect w to s (red branches in Figure 15.41). Multiplying the derivatives that appear on each path and adding gives the result

$$\frac{\partial w}{\partial s} = \frac{\partial w}{\partial x}\frac{\partial x}{\partial s} + \frac{\partial w}{\partial y}\frac{\partial y}{\partial s} + \frac{\partial w}{\partial z}\frac{\partial z}{\partial s}.$$

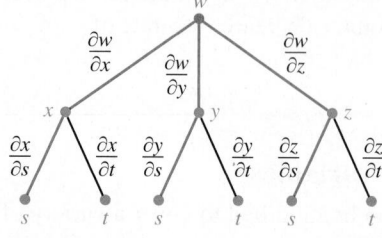

Figure 15.41

QUICK CHECK 3 If Q is a function of w, x, y, and z, each of which is a function of r, s, and t, how many dependent variables, intermediate variables, and independent variables are there? ◄

Related Exercises 25–26 ◄

It is probably clear by now that we can create a Chain Rule for any set of relationships among variables. The key is to draw an accurate tree diagram and label the branches of the tree with the appropriate derivatives.

EXAMPLE 4   A different kind of tree   Let w be a function of z, where z is a function of x and y, and each of x and y is a function of t. Draw a labeled tree diagram and write the Chain Rule formula for dw/dt.

SOLUTION   The dependent variable w is related to the independent variable t through two paths in the tree: $w \to z \to x \to t$ and $w \to z \to y \to t$ (**Figure 15.42**). At the top of the tree, w is a function of the single variable z, so the rate of change is the ordinary derivative dw/dz. The tree below z looks like Figure 15.36. Multiplying the derivatives on each of the two branches connecting w to t and adding the results, we have

$$\frac{dw}{dt} = \frac{dw}{dz}\frac{\partial z}{\partial x}\frac{dx}{dt} + \frac{dw}{dz}\frac{\partial z}{\partial y}\frac{dy}{dt} = \frac{dw}{dz}\left(\frac{\partial z}{\partial x}\frac{dx}{dt} + \frac{\partial z}{\partial y}\frac{dy}{dt}\right).$$

Related Exercise 31 ◄

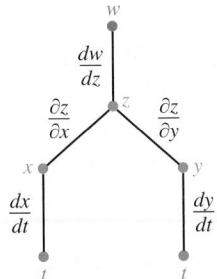

Figure 15.42

Implicit Differentiation

Using the Chain Rule for partial derivatives, the technique of implicit differentiation can be put in a larger perspective. Recall that if x and y are related through an implicit relationship, such as $\sin xy + \pi y^2 = x$, then dy/dx is computed using implicit differentiation (Section 3.8). Another way to compute dy/dx is to define the function $F(x, y) = \sin xy + \pi y^2 - x$. Notice that the original relationship $\sin xy + \pi y^2 = x$ is $F(x, y) = 0$.

To find dy/dx, we treat x as the independent variable and differentiate both sides of $F(x, y(x)) = 0$ with respect to x. The derivative of the right side is 0. On the left side, we use the Chain Rule of Theorem 15.7:

$$\frac{\partial F}{\partial x}\underbrace{\frac{dx}{dx}}_{1} + \frac{\partial F}{\partial y}\frac{dy}{dx} = 0.$$

Noting that $dx/dx = 1$ and solving for dy/dx, we obtain the following theorem.

> ► The question of whether a relationship of the form $F(x, y) = 0$ or $F(x, y, z) = 0$ determines one or more functions is addressed by a theorem of advanced calculus called the Implicit Function Theorem.

THEOREM 15.9   Implicit Differentiation
Let F be differentiable on its domain and suppose $F(x, y) = 0$ defines y as a differentiable function of x. Provided $F_y \neq 0$,

$$\frac{dy}{dx} = -\frac{F_x}{F_y}.$$

> ► The method of Theorem 15.9 generalizes to computing $\dfrac{\partial z}{\partial x}$ and $\dfrac{\partial z}{\partial y}$ with functions of the form $F(x, y, z) = 0$ (Exercise 56).

EXAMPLE 5   Implicit differentiation   Find dy/dx when $F(x, y) = \sin xy + \pi y^2 - x = 0$.

SOLUTION   Computing the partial derivatives of F with respect to x and y, we find that

$$F_x = y \cos xy - 1 \quad \text{and} \quad F_y = x \cos xy + 2\pi y.$$

Therefore,

$$\frac{dy}{dx} = -\frac{F_x}{F_y} = -\frac{y \cos xy - 1}{x \cos xy + 2\pi y}.$$

QUICK CHECK 4   Use the method of Example 5 to find dy/dx when $F(x, y) = x^2 + xy - y^3 - 7 = 0$. Compare your solution to Example 3 in Section 3.8. Which method is easier? ◄

As with many implicit differentiation calculations, the result is left in terms of both x and y. The same result is obtained using the methods of Section 3.8.

Related Exercises 37 ◄

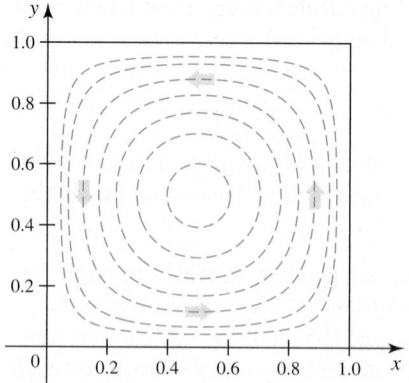

Figure 15.43

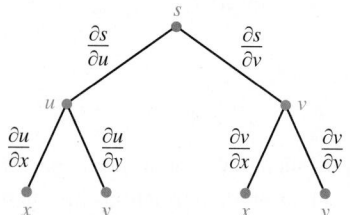

Figure 15.44

EXAMPLE 6   Fluid flow A basin of circulating water is represented by the square region $\{(x, y): 0 \le x \le 1, 0 \le y \le 1\}$, where x is positive in the eastward direction and y is positive in the northward direction. The velocity components of the water are

the east-west velocity $u(x, y) = 2 \sin \pi x \cos \pi y$ and

the north-south velocity $v(x, y) = -2 \cos \pi x \sin \pi y$;

these velocity components produce the flow pattern shown in **Figure 15.43**. The *streamlines* shown in the figure are the paths followed by small parcels of water. The speed of the water at a point (x, y) is given by the function $s(x, y) = \sqrt{u(x, y)^2 + v(x, y)^2}$. Find $\partial s/\partial x$ and $\partial s/\partial y$, the rates of change of the water speed in the x- and y-directions, respectively.

SOLUTION The dependent variable s depends on the independent variables x and y through the intermediate variables u and v (**Figure 15.44**). Theorem 15.8 applies here in the form

$$\frac{\partial s}{\partial x} = \frac{\partial s}{\partial u}\frac{\partial u}{\partial x} + \frac{\partial s}{\partial v}\frac{\partial v}{\partial x} \quad \text{and} \quad \frac{\partial s}{\partial y} = \frac{\partial s}{\partial u}\frac{\partial u}{\partial y} + \frac{\partial s}{\partial v}\frac{\partial v}{\partial y}.$$

The derivatives $\partial s/\partial u$ and $\partial s/\partial v$ are easier to find if we square the speed function to obtain $s^2 = u^2 + v^2$ and then use implicit differentiation. To compute $\partial s/\partial u$, we differentiate both sides of $s^2 = u^2 + v^2$ with respect to u:

$$2s\frac{\partial s}{\partial u} = 2u, \quad \text{which implies that} \quad \frac{\partial s}{\partial u} = \frac{u}{s}.$$

Similarly, differentiating $s^2 = u^2 + v^2$ with respect to v gives

$$2s\frac{\partial s}{\partial v} = 2v, \quad \text{which implies that} \quad \frac{\partial s}{\partial v} = \frac{v}{s}.$$

Now the Chain Rule leads to $\dfrac{\partial s}{\partial x}$:

$$\frac{\partial s}{\partial x} = \frac{\partial s}{\partial u}\frac{\partial u}{\partial x} + \frac{\partial s}{\partial v}\frac{\partial v}{\partial x}$$

$$= \underbrace{\frac{u}{s}}_{\frac{\partial s}{\partial u}}\underbrace{(2\pi \cos \pi x \cos \pi y)}_{\frac{\partial u}{\partial x}} + \underbrace{\frac{v}{s}}_{\frac{\partial s}{\partial v}}\underbrace{(2\pi \sin \pi x \sin \pi y)}_{\frac{\partial v}{\partial x}}$$

$$= \frac{2\pi}{s}(u \cos \pi x \cos \pi y + v \sin \pi x \sin \pi y).$$

A similar calculation shows that

$$\frac{\partial s}{\partial y} = -\frac{2\pi}{s}(u \sin \pi x \sin \pi y + v \cos \pi x \cos \pi y).$$

As a final step, you could replace s, u, and v with their definitions in terms of x and y.

Related Exercises 41–42 ◄

EXAMPLE 7   Second derivatives Let $z = f(x, y) = \dfrac{x}{y}$, where $x = s + t^2$ and

$y = s^2 - t$. Compute $\dfrac{\partial^2 z}{\partial s^2} = z_{ss}$, $\dfrac{\partial^2 z}{\partial t \partial s} = z_{st}$, and $\dfrac{\partial^2 z}{\partial t^2} = z_{tt}$, and express the results in terms of s and t. We use subscripts for partial derivatives in this example to simplify the notation.

SOLUTION First, we need some ground rules. In this example, it is possible to express f in terms of s and t by substituting, after which the result could be differentiated directly to find the required partial derivatives. Unfortunately, this maneuver is not always possible in practice (see Exercises 72 and 73). Therefore, to make this example as useful as possible, we develop general formulas for the second partial derivatives and make substitutions only in the last step.

Figures 15.39 and 15.40 show the relationships among the variables, and Example 2 demonstrates the calculation of the first partial derivatives. Throughout these calculations, it is important to remember the meaning of differentiation with respect to s and t:

$$(\ )_s = (\ )_x \, x_s + (\ )_y \, y_s \text{ and } (\ )_t = (\ )_x \, x_t + (\ )_y \, y_t.$$

Let's compute the first partial derivatives:

$$z_s = z_x x_s + z_y y_s \text{ and } z_t = z_x x_t + z_y y_t.$$

Differentiating z_s with respect to s, we have

$$
\begin{aligned}
z_{ss} &= (z_x x_s + z_y y_s)_s \\
&= (z_x)_s \, x_s + z_x x_{ss} + (z_y)_s \, y_s + z_y y_{ss} && \text{Product Rule (twice)} \\
&= \underbrace{(z_{xx} x_s + z_{xy} y_s)}_{(z_x)_s} x_s + z_x x_{ss} && \text{Differentiate } z_x \text{ and } z_y \text{ with respect to } s. \\
&\quad + \underbrace{(z_{yx} x_s + z_{yy} y_s)}_{(z_y)_s} y_s + z_y y_{ss} \\
&= z_{xx} x_s^2 + 2 z_{xy} x_s y_s + z_{yy} y_s^2 + z_x x_{ss} + z_y y_{ss}. && \text{Simplify with } z_{xy} = z_{yx}.
\end{aligned}
$$

At this point, we substitute

$$z_x = \frac{1}{y}, \, z_y = -\frac{x}{y^2}, \, z_{xx} = 0, \, z_{xy} = -\frac{1}{y^2}, \, z_{yy} = \frac{2x}{y^3}, \, x_s = 1, \, x_{ss} = 0, \, y_s = 2s, \text{ and } y_{ss} = 2$$

and simplify to find that

$$z_{ss} = \frac{2(s^3 + 3st + 3s^2 t^2 + t^3)}{(s^2 - t)^3}.$$

Differentiating z_s with respect to t, a similar procedure produces z_{st}:

$$
\begin{aligned}
z_{st} &= (z_x x_s + z_y y_s)_t \\
&= (z_x)_t \, x_s + z_x x_{st} + (z_y)_t \, y_s + z_y y_{st} && \text{Product Rule (twice)} \\
&= \underbrace{(z_{xx} x_t + z_{xy} y_t)}_{(z_x)_t} x_s + z_x x_{st} && \text{Differentiate } z_x \text{ and } z_y \\
&&& \text{with respect to } t. \\
&\quad + \underbrace{(z_{yx} x_t + z_{yy} y_t)}_{(z_y)_t} y_s + z_y y_{st} \\
&= z_{xx} x_s x_t + z_{xy} x_s y_t + z_{xy} x_t y_s + z_{yy} y_s y_t + z_x x_{st} + z_y y_{st}. && \text{Simplify with } z_{xy} = z_{yx}.
\end{aligned}
$$

Substituting in terms of s and t with $x_{st} = 0$ and $y_{st} = 0$, we have

$$z_{st} = -\frac{3s^2 + t + 4s^3 t}{(s^2 - t)^3}.$$

An analogous calculation gives

$$z_{tt} = \frac{2s(1 + s^3)}{(s^2 - t)^3}.$$

Related Exercise 45 ◄

SECTION 15.4 EXERCISES

Getting Started

1. Suppose $z = f(x, y)$, where x and y are functions of t. How many dependent, intermediate, and independent variables are there?

2. Let z be a function of x and y, while x and y are functions of t. Explain how to find dz/dt.

3. Suppose w is a function of x, y, and z, which are each functions of t. Explain how to find dw/dt.

4. Let $z = f(x, y)$, $x = g(s, t)$, and $y = h(s, t)$. Explain how to find $\partial z/\partial t$.

5. Given that $w = F(x, y, z)$, and x, y, and z are functions of r and s, sketch a Chain Rule tree diagram with branches labeled with the appropriate derivatives.

6. Suppose $F(x, y) = 0$ and y is a differentiable function of x. Explain how to find dy/dx.

7. Evaluate dz/dt, where $z = x^2 + y^3$, $x = t^2$, and $y = t$, using Theorem 15.7. Check your work by writing z as a function of t and evaluating dz/dt.

8. Evaluate dz/dt, where $z = xy^2$, $x = t^2$, and $y = t$, using Theorem 15.7. Check your work by writing z as a function of t and evaluating dz/dt.

Practice Exercises

9–18. Chain Rule with one independent variable *Use Theorem 15.7 to find the following derivatives.*

9. dz/dt, where $z = x \sin y$, $x = t^2$, and $y = 4t^3$

10. dz/dt, where $z = x^2 y - xy^3$, $x = t^2$, and $y = t^{-2}$

11. dw/dt, where $w = \cos 2x \sin 3y$, $x = t/2$, and $y = t^4$

12. dz/dt, where $z = \sqrt{r^2 + s^2}$, $r = \cos 2t$, and $s = \sin 2t$

13. dz/dt, where $z = (x + 2y)^{10}$, $x = \sin^2 t$, $y = (3t + 4)^5$

14. $\dfrac{dz}{dt}$, where $z = \dfrac{x^{20}}{y^{10}}$, $x = \tan^{-1} t$, $y = \ln(t^2 + 1)$

15. dw/dt, where $w = xy \sin z$, $x = t^2$, $y = 4t^3$, and $z = t + 1$

16. dQ/dt, where $Q = \sqrt{x^2 + y^2 + z^2}$, $x = \sin t$, $y = \cos t$, and $z = \cos t$

17. dV/dt, where $V = xyz$, $x = e^t$, $y = 2t + 3$, and $z = \sin t$

18. $\dfrac{dU}{dt}$, where $U = \dfrac{xy^2}{z^8}$, $x = e^t$, $y = \sin 3t$, and $z = 4t + 1$

19–26. Chain Rule with several independent variables *Find the following derivatives.*

19. z_s and z_t, where $z = x^2 \sin y$, $x = s - t$, and $y = t^2$

20. z_s and z_t, where $z = \sin(2x + y)$, $x = s^2 - t^2$, and $y = s^2 + t^2$

21. z_s and z_t, where $z = xy - x^2 y$, $x = s + t$, and $y = s - t$

22. z_s and z_t, where $z = \sin x \cos 2y$, $x = s + t$, and $y = s - t$

23. z_s and z_t, where $z = e^{x+y}$, $x = st$, and $y = s + t$

24. z_s and z_t, where $z = \sin xy$, $x = s^2 t$, and $y = (s + t)^{10}$

25. w_s and w_t, where $w = \dfrac{x - z}{y + z}$, $x = s + t$, $y = st$, and $z = s - t$

26. w_r, w_s, and w_t, where $w = \sqrt{x^2 + y^2 + z^2}$, $x = st$, $y = rs$, and $z = rt$

27. Changing cylinder The volume of a right circular cylinder with radius r and height h is $V = \pi r^2 h$.

  a. Assume r and h are functions of t. Find $V'(t)$.
  b. Suppose $r = e^t$ and $h = e^{-2t}$, for $t \geq 0$. Use part (a) to find $V'(t)$.
  c. Does the volume of the cylinder in part (b) increase or decrease as t increases?

28. Changing pyramid The volume of a pyramid with a square base x units on a side and a height of h is $V = \dfrac{1}{3}x^2 h$.

  a. Assume x and h are functions of t. Find $V'(t)$.
  b. Suppose $x = \dfrac{t}{t + 1}$ and $h = \dfrac{1}{t + 1}$, for $t \geq 0$. Use part (a) to find $V'(t)$.
  c. Does the volume of the pyramid in part (b) increase or decrease as t increases?

29–30. Derivative practice two ways *Find the indicated derivative in two ways:*

a. *Replace x and y to write z as a function of t, and differentiate.*
b. *Use the Chain Rule.*

29. $z'(t)$, where $z = \dfrac{1}{x} + \dfrac{1}{y}$, $x = t^2 + 2t$, and $y = t^3 - 2$

30. $z'(t)$, where $z = \ln(x + y)$, $x = te^t$, and $y = e^t$

31–34. Making trees *Use a tree diagram to write the required Chain Rule formula.*

31. w is a function of z, where z is a function of p, q, and r, each of which is a function of t. Find dw/dt.

32. $w = f(x, y, z)$, where $x = g(t)$, $y = h(s, t)$, and $z = p(r, s, t)$. Find $\partial w/\partial t$.

33. $u = f(v)$, where $v = g(w, x, y)$, $w = h(z)$, $x = p(t, z)$, and $y = q(t, z)$. Find $\partial u/\partial z$.

34. $u = f(v, w, x)$, where $v = g(r, s, t)$, $w = h(r, s, t)$, $x = p(r, s, t)$, and $r = F(z)$. Find $\partial u/\partial z$.

35–40. Implicit differentiation *Use Theorem 15.9 to evaluate dy/dx. Assume each equation implicitly defines y as a differentiable function of x.*

35. $x^2 - 2y^2 - 1 = 0$

36. $x^3 + 3xy^2 - y^5 = 0$

37. $2 \sin xy = 1$

38. $ye^{xy} - 2 = 0$

39. $\sqrt{x^2 + 2xy + y^4} = 3$

40. $y \ln(x^2 + y^2 + 4) = 3$

41–42. Fluid flow *The x- and y-components of a fluid moving in two dimensions are given by the following functions u and v. The speed of the fluid at (x, y) is $s(x, y) = \sqrt{u(x, y)^2 + v(x, y)^2}$. Use the Chain Rule to find $\partial s/\partial x$ and $\partial s/\partial y$.*

41. $u(x, y) = 2y$ and $v(x, y) = -2x$; $x \geq 0$ and $y \geq 0$

42. $u(x, y) = x(1 - x)(1 - 2y)$ and $v(x, y) = y(y - 1)(1 - 2x)$; $0 \leq x \leq 1, 0 \leq y \leq 1$

43–48. Second derivatives *For the following sets of variables, find all the relevant second derivatives. In all cases, first find general expressions for the second derivatives and then substitute variables at the last step.*

43. $f(x, y) = x^2 y$, where $x = s + t$ and $y = s - t$

44. $f(x, y) = x^2 y - xy^2$, where $x = st$ and $y = s/t$

45. $f(x, y) = y/x$, where $x = s^2 + t^2$ and $y = s^2 - t^2$

46. $f(x, y) = e^{x-y}$, where $x = s^2$ and $y = 3t^2$

47. $f(x, y, z) = xy + xz - yz$, where $x = s^2 - 2s$, $y = 2/s^2$, and $z = 3s^2 - 2$

48. $f(x, y) = xy$, where $x = s + 2t - u$ and $y = s + 2t + u$

49. Explain why or why not Determine whether the following statements are true and give an explanation or counterexample. Assume all partial derivatives exist.

a. If $z = (x + y) \sin xy$, where x and y are functions of s, then
$$\frac{\partial z}{\partial s} = \frac{dz}{dx} \frac{dx}{ds}.$$

b. Given that $w = f(x(s, t), y(s, t), z(s, t))$, the rate of change of w with respect to t is dw/dt.

50–54. Derivative practice *Find the indicated derivative for the following functions.*

50. $\partial z/\partial p$, where $z = x/y$, $x = p + q$, and $y = p - q$

51. dw/dt, where $w = xyz$, $x = 2t^4$, $y = 3t^{-1}$, and $z = 4t^{-3}$

52. $\partial w/\partial x$, where $w = \cos z - \cos x \cos y + \sin x \sin y$, and $z = x + y$

53. $\dfrac{\partial z}{\partial x}$, where $\dfrac{1}{x} + \dfrac{1}{y} + \dfrac{1}{z} = 1$

54. $\partial z/\partial x$, where $xy - z = 1$

55. Change on a line Suppose $w = f(x, y, z)$ and ℓ is the line $\mathbf{r}(t) = \langle at, bt, ct \rangle$, for $-\infty < t < \infty$.

a. Find $w'(t)$ on ℓ (in terms of a, b, c, w_x, w_y, and w_z).
b. Apply part (a) to find $w'(t)$ when $f(x, y, z) = xyz$.
c. Apply part (a) to find $w'(t)$ when $f(x, y, z) = \sqrt{x^2 + y^2 + z^2}$.
d. For a general twice differentiable function $w = f(x, y, z)$, find $w''(t)$.

56. Implicit differentiation rule with three variables Assume $F(x, y, z(x, y)) = 0$ implicitly defines z as a differentiable function of x and y. Extend Theorem 15.9 to show that
$$\frac{\partial z}{\partial x} = -\frac{F_x}{F_z} \quad \text{and} \quad \frac{\partial z}{\partial y} = -\frac{F_y}{F_z}.$$

57–59. Implicit differentiation with three variables *Use the result of Exercise 56 to evaluate $\partial z/\partial x$ and $\partial z/\partial y$ for the following relations.*

57. $xy + xz + yz = 3$ **58.** $x^2 + 2y^2 - 3z^2 = 1$

59. $xyz + x + y - z = 0$

60. More than one way Let $e^{xyz} = 2$. Find z_x and z_y in three ways (and check for agreement).

a. Use the result of Exercise 56.
b. Take logarithms of both sides and differentiate $xyz = \ln 2$.
c. Solve for z and differentiate $z = \dfrac{\ln 2}{xy}$.

61–64. Walking on a surface *Consider the following surfaces specified in the form $z = f(x, y)$ and the oriented curve C in the xy-plane.*

a. In each case, find $z'(t)$.
b. Imagine that you are walking on the surface directly above the curve C in the direction of positive orientation. Find the values of t for which you are walking uphill (that is, z is increasing).

61. $z = x^2 + 4y^2 + 1$, $C: x = \cos t$, $y = \sin t$; $0 \le t \le 2\pi$

62. $z = 4x^2 - y^2 + 1$, $C: x = \cos t$, $y = \sin t$; $0 \le t \le 2\pi$

63. $z = \sqrt{1 - x^2 - y^2}$, $C: x = e^{-t}$, $y = e^{-t}$; $t \ge \dfrac{1}{2} \ln 2$

64. $z = 2x^2 + y^2 + 1$, $C: x = 1 + \cos t$, $y = \sin t$; $0 \le t \le 2\pi$

65. Conservation of energy A projectile with mass m is launched into the air on a parabolic trajectory. For $t \ge 0$, its horizontal and vertical coordinates are $x(t) = u_0 t$ and $y(t) = -\dfrac{1}{2} g t^2 + v_0 t$, respectively, where u_0 is the initial horizontal velocity, v_0 is the initial vertical velocity, and g is the acceleration due to gravity. Recalling that $u(t) = x'(t)$ and $v(t) = y'(t)$ are the components of the velocity, the energy of the projectile (kinetic plus potential) is

$$E(t) = \frac{1}{2} m(u^2 + v^2) + mgy.$$

Use the Chain Rule to compute $E'(t)$ and show that $E'(t) = 0$, for all $t \ge 0$. Interpret the result.

66. Utility functions in economics Economists use *utility functions* to describe consumers' relative preference for two or more commodities (for example, vanilla vs. chocolate ice cream or leisure time vs. material goods). The Cobb-Douglas family of utility functions has the form $U(x, y) = x^a y^{1-a}$, where x and y are the amounts of two commodities and $0 < a < 1$ is a parameter. Level curves on which the utility function is constant are called *indifference curves*; the preference is the same for all combinations of x and y along an indifference curve (see figure).

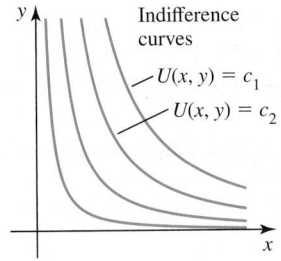

a. The marginal utilities of the commodities x and y are defined to be $\partial U/\partial x$ and $\partial U/\partial y$, respectively. Compute the marginal utilities for the utility function $U(x, y) = x^a y^{1-a}$.
b. The marginal rate of substitution (MRS) is the slope of the indifference curve at the point (x, y). Use the Chain Rule to show that for $U(x, y) = x^a y^{1-a}$, the MRS is $-\dfrac{a}{1-a} \dfrac{y}{x}$.
c. Find the MRS for the utility function $U(x, y) = x^{0.4} y^{0.6}$ at $(x, y) = (8, 12)$.

67. Constant volume tori The volume of a solid torus is given by $V = \dfrac{\pi^2}{4}(R + r)(R - r)^2$, where r and R are the inner and outer radii and $R > r$ (see figure).

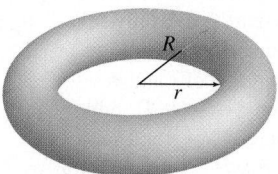

a. If R and r increase at the same rate, does the volume of the torus increase, decrease, or remain constant?
b. If R and r decrease at the same rate, does the volume of the torus increase, decrease, or remain constant?

68. Body surface area One of several empirical formulas that relates the surface area S of a human body to the height h and weight w of the body is the Mosteller formula $S(h, w) = \dfrac{1}{60}\sqrt{hw}$, where h is measured in cm, w is measured in kg, and S is measured in square meters. Suppose h and w are functions of t.

   a. Find $S'(t)$.

   b. Show that the condition under which the surface area remains constant as h and w change is $wh'(t) + hw'(t) = 0$.

   c. Show that part (b) implies that for constant surface area, h and w must be inversely related; that is, $h = C/w$, where C is a constant.

69. The Ideal Gas Law The pressure, temperature, and volume of an ideal gas are related by $PV = kT$, where $k > 0$ is a constant. Any two of the variables may be considered independent, which determines the dependent variable.

   a. Use implicit differentiation to compute the partial derivatives $\dfrac{\partial P}{\partial V}, \dfrac{\partial T}{\partial P},$ and $\dfrac{\partial V}{\partial T}$.

   b. Show that $\dfrac{\partial P}{\partial V}\dfrac{\partial T}{\partial P}\dfrac{\partial V}{\partial T} = -1$. (See Exercise 75 for a generalization.)

70. Variable density The density of a thin circular plate of radius 2 is given by $\rho(x, y) = 4 + xy$. The edge of the plate is described by the parametric equations $x = 2\cos t$, $y = 2\sin t$, for $0 \le t \le 2\pi$.

   a. Find the rate of change of the density with respect to t on the edge of the plate.

   b. At what point(s) on the edge of the plate is the density a maximum?

T 71. Spiral through a domain Suppose you follow the helical path $C: x = \cos t$, $y = \sin t$, $z = t$, for $t \ge 0$, through the domain of the function $w = f(x, y, z) = \dfrac{xyz}{z^2 + 1}$.

   a. Find $w'(t)$ along C.

   b. Estimate the point (x, y, z) on C at which w has its maximum value.

Explorations and Challenges

72. Change of coordinates Recall that Cartesian and polar coordinates are related through the transformation equations

$$\begin{cases} x = r\cos\theta \\ y = r\sin\theta \end{cases} \quad\text{or}\quad \begin{cases} r^2 = x^2 + y^2 \\ \tan\theta = y/x. \end{cases}$$

   a. Evaluate the partial derivatives x_r, y_r, x_θ, and y_θ.

   b. Evaluate the partial derivatives r_x, r_y, θ_x, and θ_y.

   c. For a function $z = f(x, y)$, find z_r and z_θ, where x and y are expressed in terms of r and θ.

   d. For a function $z = g(r, \theta)$, find z_x and z_y, where r and θ are expressed in terms of x and y.

   e. Show that $\left(\dfrac{\partial z}{\partial x}\right)^2 + \left(\dfrac{\partial z}{\partial y}\right)^2 = \left(\dfrac{\partial z}{\partial r}\right)^2 + \dfrac{1}{r^2}\left(\dfrac{\partial z}{\partial \theta}\right)^2$.

73. Change of coordinates continued An important derivative operation in many applications is called the Laplacian; in Cartesian coordinates, for $z = f(x, y)$, the Laplacian is $z_{xx} + z_{yy}$. Determine the Laplacian in polar coordinates using the following steps.

   a. Begin with $z = g(r, \theta)$ and write z_x and z_y in terms of polar coordinates (see Exercise 72).

   b. Use the Chain Rule to find $z_{xx} = \dfrac{\partial}{\partial x}(z_x)$. There should be two major terms, which, when expanded and simplified, result in five terms.

   c. Use the Chain Rule to find $z_{yy} = \dfrac{\partial}{\partial y}(z_y)$. There should be two major terms, which, when expanded and simplified, result in five terms.

   d. Combine parts (b) and (c) to show that

$$z_{xx} + z_{yy} = z_{rr} + \frac{1}{r}z_r + \frac{1}{r^2}z_{\theta\theta}.$$

74. Geometry of implicit differentiation Suppose x and y are related by the equation $F(x, y) = 0$. Interpret the solution of this equation as the set of points (x, y) that lie on the intersection of the surface $z = F(x, y)$ with the xy-plane ($z = 0$).

   a. Make a sketch of a surface and its intersection with the xy-plane. Give a geometric interpretation of the result that $\dfrac{dy}{dx} = -\dfrac{F_x}{F_y}$.

   b. Explain geometrically what happens at points where $F_y = 0$.

75. General three-variable relationship In the implicit relationship $F(x, y, z) = 0$, any two of the variables may be considered independent, which then determines the dependent variable. To avoid confusion, we may use a subscript to indicate which variable is held fixed in a derivative calculation; for example, $\left(\dfrac{\partial z}{\partial x}\right)_y$ means that y is held fixed in taking the partial derivative of z with respect to x. (In this context, the subscript does *not* mean a derivative.)

   a. Differentiate $F(x, y, z) = 0$ with respect to x, holding y fixed, to show that $\left(\dfrac{\partial z}{\partial x}\right)_y = -\dfrac{F_x}{F_z}$.

   b. As in part (a), find $\left(\dfrac{\partial y}{\partial z}\right)_x$ and $\left(\dfrac{\partial x}{\partial y}\right)_z$.

   c. Show that $\left(\dfrac{\partial z}{\partial x}\right)_y\left(\dfrac{\partial y}{\partial z}\right)_x\left(\dfrac{\partial x}{\partial y}\right)_z = -1$.

   d. Find the relationship analogous to part (c) for the case $F(w, x, y, z) = 0$.

76. Second derivative Let $f(x, y) = 0$ define y as a twice differentiable function of x.

   a. Show that $y''(x) = -\dfrac{f_{xx}f_y^2 - 2f_xf_yf_{xy} + f_{yy}f_x^2}{f_y^3}$.

   b. Verify part (a) using the function $f(x, y) = xy - 1$.

77. Subtleties of the Chain Rule Let $w = f(x, y, z) = 2x + 3y + 4z$, which is defined for all (x, y, z) in $\mathbb{R}^3$. Suppose we are interested in the partial derivative w_x on a subset of $\mathbb{R}^3$, such as the plane P given by $z = 4x - 2y$. The point to be made is that the result is not unique unless we specify which variables are considered independent.

   a. We could proceed as follows. On the plane P, consider x and y as the independent variables, which means z depends on x and y, so we write $w = f(x, y, z(x, y))$. Differentiate with respect to x, holding y fixed, to show that $\left(\dfrac{\partial w}{\partial x}\right)_y = 18$, where the subscript y indicates that y is held fixed.

b. Alternatively, on the plane P, we could consider x and z as the independent variables, which means y depends on x and z, so we write $w = f(x, y(x, z), z)$ and differentiate with respect to x, holding z fixed. Show that $\left(\dfrac{\partial w}{\partial x}\right)_z = 8$, where the subscript z indicates that z is held fixed.

c. Make a sketch of the plane $z = 4x - 2y$ and interpret the results of parts (a) and (b) geometrically.

d. Repeat the arguments of parts (a) and (b) to find $\left(\dfrac{\partial w}{\partial y}\right)_x$, $\left(\dfrac{\partial w}{\partial y}\right)_z$, $\left(\dfrac{\partial w}{\partial z}\right)_x$, and $\left(\dfrac{\partial w}{\partial z}\right)_y$.

QUICK CHECK ANSWERS

1. If $z = f(x(t))$, then $\dfrac{\partial z}{\partial y} = 0$, and the original Chain Rule results. **2.** $\dfrac{\partial w}{\partial t} = \dfrac{\partial w}{\partial x}\dfrac{\partial x}{\partial t} + \dfrac{\partial w}{\partial y}\dfrac{\partial y}{\partial t} + \dfrac{\partial w}{\partial z}\dfrac{\partial z}{\partial t}$ **3.** One dependent variable, four intermediate variables, and three independent variables **4.** $\dfrac{dy}{dx} = \dfrac{2x + y}{3y^2 - x}$; in this case, using $\dfrac{dy}{dx} = -\dfrac{F_x}{F_y}$ is more efficient. ◄

15.5 Directional Derivatives and the Gradient

Partial derivatives tell us a lot about the rate of change of a function on its domain. However, they do not *directly* answer some important questions. For example, suppose you are standing at a point $(a, b, f(a, b))$ on the surface $z = f(x, y)$. The partial derivatives f_x and f_y tell you the rate of change (or slope) of the surface at that point in the directions parallel to the x-axis and y-axis, respectively. But you could walk in an infinite number of directions from that point and find a different rate of change in every direction. With this observation in mind, we pose several questions.

- Suppose you are standing on a surface and you walk in a direction *other* than a coordinate direction—say, northwest or south-southeast. What is the rate of change of the function in such a direction?

- Suppose you are standing on a surface and you release a ball at your feet and let it roll. In which direction will it roll?

- If you are hiking up a mountain, in what direction should you walk after each step if you want to follow the steepest path?

These questions are answered in this section by introducing the *directional derivative*, followed by one of the central concepts of calculus—the *gradient*.

Directional Derivatives

Let $(a, b, f(a, b))$ be a point on the surface $z = f(x, y)$ and let $\mathbf{u}$ be a unit vector in the xy-plane (Figure 15.45). Our aim is to find the rate of change of f in the direction $\mathbf{u}$ at $P_0(a, b)$. In general, this rate of change is neither $f_x(a, b)$ nor $f_y(a, b)$ (unless $\mathbf{u} = \langle 1, 0 \rangle$ or $\mathbf{u} = \langle 0, 1 \rangle$), but it turns out to be a combination of $f_x(a, b)$ and $f_y(a, b)$.

Figure 15.46a shows the unit vector $\mathbf{u} = \langle u_1, u_2 \rangle$; its x- and y-components are u_1 and u_2, respectively. The derivative we seek must be computed along the line ℓ in the xy-plane through P_0 in the direction of $\mathbf{u}$. A neighboring point P, which is h units from P_0 along ℓ, has coordinates $P(a + hu_1, b + hu_2)$ (Figure 15.46b).

Now imagine the plane Q perpendicular to the xy-plane, containing ℓ. This plane cuts the surface $z = f(x, y)$ in a curve C. Consider two points on C corresponding to P_0 and P; they have z-coordinates $f(a, b)$ and $f(a + hu_1, b + hu_2)$ (Figure 15.47). The slope of the secant line between these points is

$$\frac{f(a + hu_1, b + hu_2) - f(a, b)}{h}.$$

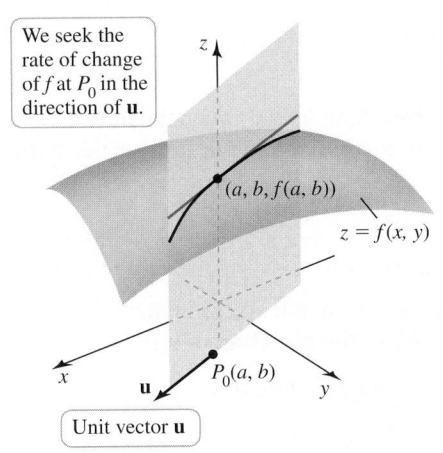

We seek the rate of change of f at P_0 in the direction of $\mathbf{u}$.

Unit vector $\mathbf{u}$

Figure 15.45

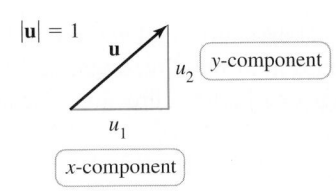

$|\mathbf{u}| = 1$

u_2 — *y*-component

u_1

x-component

(a)

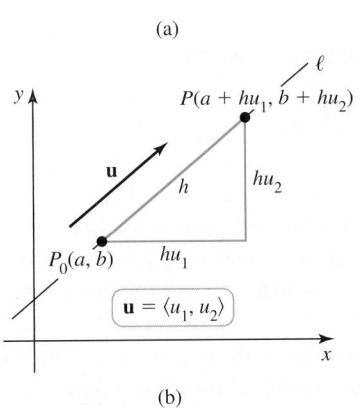

$P(a + hu_1, b + hu_2)$

$\mathbf{u}$

h

hu_2

$P_0(a, b)$ hu_1

$\mathbf{u} = \langle u_1, u_2 \rangle$

(b)

Figure 15.46

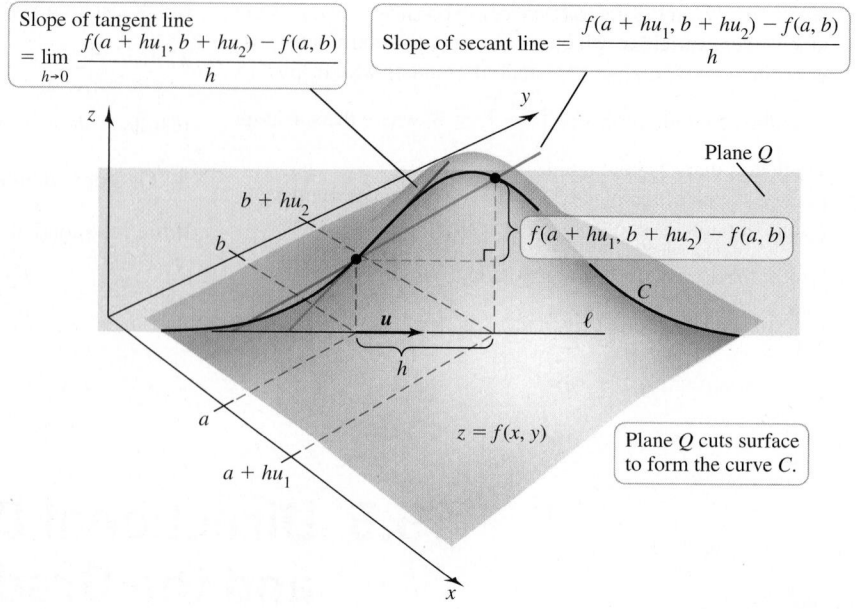

Figure 15.47

The derivative of f in the direction of $\mathbf{u}$ is obtained by letting $h \to 0$; when the limit exists, it is called the *directional derivative of f at (a, b) in the direction of $\mathbf{u}$*. It gives the slope of the line tangent to the curve C in the plane Q.

> ▶ The definition of the directional derivative looks like the definition of the ordinary derivative if we write it as
>
> $$\lim_{P \to P_0} \frac{f(P) - f(P_0)}{|P - P_0|},$$
>
> where P approaches P_0 along the line ℓ.

DEFINITION Directional Derivative

Let f be differentiable at (a, b) and let $\mathbf{u} = \langle u_1, u_2 \rangle$ be a unit vector in the xy-plane. The **directional derivative of f at (a, b) in the direction of $\mathbf{u}$** is

$$D_{\mathbf{u}} f(a, b) = \lim_{h \to 0} \frac{f(a + hu_1, b + hu_2) - f(a, b)}{h},$$

provided the limit exists.

As motivation, it is instructive to see how the directional derivative includes the ordinary derivative in one variable. Setting $u_2 = 0$ in the definition of the directional derivative and ignoring the second variable gives the rate of change of f in the x-direction. The directional derivative then becomes

$$\lim_{h \to 0} \frac{f(a + hu_1) - f(a)}{h}.$$

Multiplying the numerator and denominator of this quotient by u_1, we have

$$u_1 \underbrace{\lim_{h \to 0} \frac{f(a + hu_1) - f(a)}{hu_1}}_{f'(a)} = u_1 f'(a).$$

Only because $\mathbf{u}$ is a unit vector and $u_1 = 1$ does the directional derivative reduce to the ordinary derivative $f'(a)$ in the x-direction. A similar argument may be used in the y-direction. Choosing $\mathbf{u}$ to be a unit vector gives the simplest formulas for the directional derivative.

As with ordinary derivatives, we would prefer to evaluate directional derivatives without taking limits. Fortunately, there is an easy way to express a directional derivative in terms of partial derivatives.

The key is to define a function that is equal to f along the line ℓ through (a, b) in the direction of the unit vector $\mathbf{u} = \langle u_1, u_2 \rangle$. The points on ℓ satisfy the parametric equations

$$x = a + su_1 \quad \text{and} \quad y = b + su_2,$$

QUICK CHECK 1 Explain why, when $\mathbf{u} = \langle 1, 0 \rangle$ in the definition of the directional derivative, the result is $f_x(a, b)$, and when $\mathbf{u} = \langle 0, 1 \rangle$, the result is $f_y(a, b)$. ◀

➤ To see that *s* is an arc length parameter, note that the line ℓ may be written in the form

$$\mathbf{r}(s) = \langle a + su_1, b + su_2 \rangle.$$

Therefore, $\mathbf{r}'(s) = \langle u_1, u_2 \rangle$ and $|\mathbf{r}'(s)| = 1$. It follows by the discussion in Section 14.4 that *s* is an arc length parameter.

➤ Note that $g'(s)$ does not correctly measure the slope of *f* along ℓ unless **u** is a unit vector.

where $-\infty < s < \infty$. Because **u** is a unit vector, the parameter *s* corresponds to arc length. As *s* increases, the points (x, y) move along ℓ in the direction of **u** with $s = 0$ corresponding to (a, b). Now we define the function

$$g(s) = f(\underbrace{a + su_1}_{x}, \underbrace{b + su_2}_{y}),$$

which gives the values of *f* along ℓ. The derivative of *f* along ℓ is $g'(s)$ (see margin note), and when evaluated at $s = 0$, it is the directional derivative of *f* at (a, b); that is, $g'(0) = D_{\mathbf{u}} f(a, b)$.

Noting that $\dfrac{dx}{ds} = u_1$ and $\dfrac{dy}{ds} = u_2$, we apply the Chain Rule to find that

$$D_{\mathbf{u}} f(a, b) = g'(0) = \left(\frac{\partial f}{\partial x} \underbrace{\frac{dx}{ds}}_{u_1} + \frac{\partial f}{\partial y} \underbrace{\frac{dy}{ds}}_{u_2} \right)\Bigg|_{s=0} \quad \text{Chain Rule}$$

$$= f_x(a, b)u_1 + f_y(a, b)u_2 \qquad s = 0 \text{ corresponds to } (a, b).$$

$$= \langle f_x(a, b), f_y(a, b) \rangle \cdot \langle u_1, u_2 \rangle. \quad \text{Identify dot product.}$$

We see that the directional derivative is a weighted average of the partial derivatives $f_x(a, b)$ and $f_y(a, b)$, with the components of **u** serving as the weights. In other words, knowing the slope of the surface in the *x*- and *y*-directions allows us to find the slope in any direction. Notice that the directional derivative can be written as a dot product, which provides a practical formula for computing directional derivatives.

QUICK CHECK 2 In the parametric description $x = a + su_1$ and $y = b + su_2$, where $\mathbf{u} = \langle u_1, u_2 \rangle$ is a unit vector, show that any positive change Δs in *s* produces a line segment of length Δs. ◄

> **THEOREM 15.10 Directional Derivative**
> Let *f* be differentiable at (a, b) and let $\mathbf{u} = \langle u_1, u_2 \rangle$ be a unit vector in the *xy*-plane. The **directional derivative of *f* at (a, b) in the direction of u** is
> $$D_{\mathbf{u}} f(a, b) = \langle f_x(a, b), f_y(a, b) \rangle \cdot \langle u_1, u_2 \rangle.$$

EXAMPLE 1 Computing directional derivatives Consider the paraboloid $z = f(x, y) = \frac{1}{4}(x^2 + 2y^2) + 2$. Let P_0 be the point $(3, 2)$ and consider the unit vectors

$$\mathbf{u} = \left\langle \frac{1}{\sqrt{2}}, \frac{1}{\sqrt{2}} \right\rangle \quad \text{and} \quad \mathbf{v} = \left\langle \frac{1}{2}, -\frac{\sqrt{3}}{2} \right\rangle.$$

a. Find the directional derivative of *f* at P_0 in the directions of **u** and **v**.

b. Graph the surface and interpret the directional derivatives.

SOLUTION

a. We see that $f_x = x/2$ and $f_y = y$; evaluated at $(3, 2)$, we have $f_x(3, 2) = 3/2$ and $f_y(3, 2) = 2$. The directional derivatives in the directions **u** and **v** are

$$D_{\mathbf{u}} f(3, 2) = \langle f_x(3, 2), f_y(3, 2) \rangle \cdot \langle u_1, u_2 \rangle$$
$$= \frac{3}{2} \cdot \frac{1}{\sqrt{2}} + 2 \cdot \frac{1}{\sqrt{2}} = \frac{7}{2\sqrt{2}} \approx 2.47 \text{ and}$$
$$D_{\mathbf{v}} f(3, 2) = \langle f_x(3, 2), f_y(3, 2) \rangle \cdot \langle v_1, v_2 \rangle$$
$$= \frac{3}{2} \cdot \frac{1}{2} + 2 \left(-\frac{\sqrt{3}}{2} \right) = \frac{3}{4} - \sqrt{3} \approx -0.98.$$

➤ It is understood that the line tangent to *C* in the direction of **u** lies in the vertical plane *Q* containing **u**.

b. In the direction of **u**, the directional derivative is approximately 2.47. Because it is positive, the function is increasing at $(3, 2)$ in this direction. Equivalently, if *Q* is the vertical plane containing **u**, and *C* is the curve along which the surface intersects *Q*, then the slope of the line tangent to *C* is approximately 2.47 (**Figure 15.48a**). In the direction of **v**, the directional derivative is approximately −0.98. Because it is negative, the function is decreasing in this direction. In this case, the vertical plane *Q* contains **v** and again *C* is the curve along which the surface intersects *Q*; the slope of the line tangent to *C* is approximately −0.98 (**Figure 15.48b**).

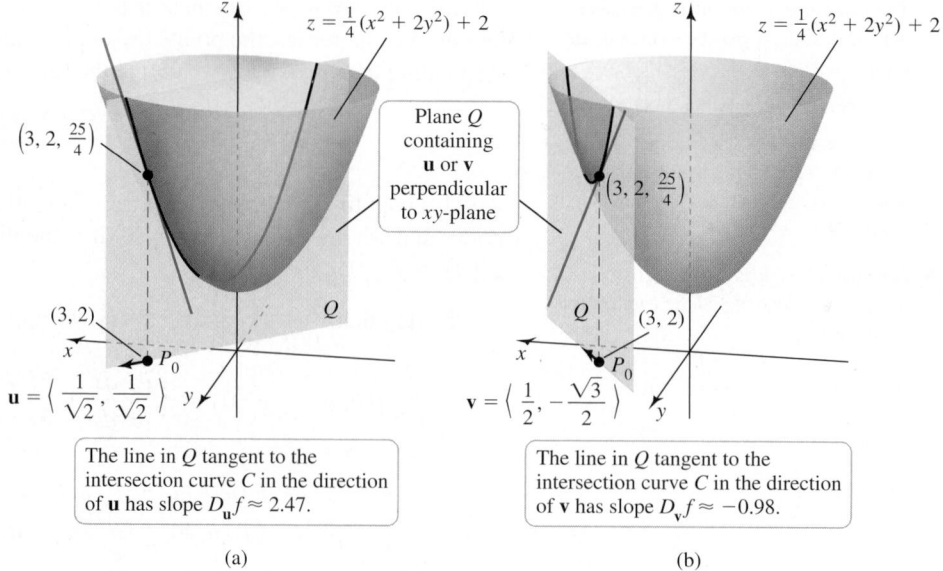

(a)

(b)

The line in Q tangent to the intersection curve C in the direction of $\mathbf{u}$ has slope $D_{\mathbf{u}}f \approx 2.47$.

The line in Q tangent to the intersection curve C in the direction of $\mathbf{v}$ has slope $D_{\mathbf{v}}f \approx -0.98$.

QUICK CHECK 3 In Example 1, evaluate $D_{-\mathbf{u}}f(3, 2)$ and $D_{-\mathbf{v}}f(3, 2)$. ◄

Figure 15.48

Related Exercises 11–12 ◄

The Gradient Vector

We have seen that the directional derivative can be written as a dot product: $D_{\mathbf{u}} f(a, b) = \langle f_x(a, b), f_y(a, b) \rangle \cdot \langle u_1, u_2 \rangle$. The vector $\langle f_x(a, b), f_y(a, b) \rangle$ that appears in the dot product is important in its own right and is called the *gradient* of f.

> ► Recall that the unit coordinate vectors in $\mathbb{R}^2$ are $\mathbf{i} = \langle 1, 0 \rangle$ and $\mathbf{j} = \langle 0, 1 \rangle$. The gradient of f is also written grad f, read *grad f*.

DEFINITION Gradient (Two Dimensions)

Let f be differentiable at the point (x, y). The **gradient** of f at (x, y) is the vector-valued function

$$\nabla f(x, y) = \langle f_x(x, y), f_y(x, y) \rangle = f_x(x, y)\, \mathbf{i} + f_y(x, y)\, \mathbf{j}.$$

With the definition of the gradient, the directional derivative of f at (a, b) in the direction of the unit vector $\mathbf{u}$ can be written

$$D_{\mathbf{u}} f(a, b) = \nabla f(a, b) \cdot \mathbf{u}.$$

The gradient satisfies sum, product, and quotient rules analogous to those for ordinary derivatives (Exercise 85).

EXAMPLE 2 Computing gradients Find $\nabla f(3, 2)$ for $f(x, y) = x^2 + 2xy - y^3$.

SOLUTION Computing $f_x = 2x + 2y$ and $f_y = 2x - 3y^2$, we have

$$\nabla f(x, y) = \langle 2(x + y), 2x - 3y^2 \rangle = 2(x + y)\, \mathbf{i} + (2x - 3y^2)\, \mathbf{j}.$$

Substituting $x = 3$ and $y = 2$ gives

$$\nabla f(3, 2) = \langle 10, -6 \rangle = 10\mathbf{i} - 6\mathbf{j}.$$

Related Exercises 13–15 ◄

EXAMPLE 3 Computing directional derivatives with gradients Let

$$f(x, y) = 3 - \frac{x^2}{10} + \frac{xy^2}{10}.$$

a. Compute $\nabla f(3, -1)$.

b. Compute $D_{\mathbf{u}} f(3, -1)$, where $\mathbf{u} = \left\langle \dfrac{1}{\sqrt{2}}, -\dfrac{1}{\sqrt{2}} \right\rangle$.

c. Compute the directional derivative of f at $(3, -1)$ in the direction of the vector $\langle 3, 4 \rangle$.

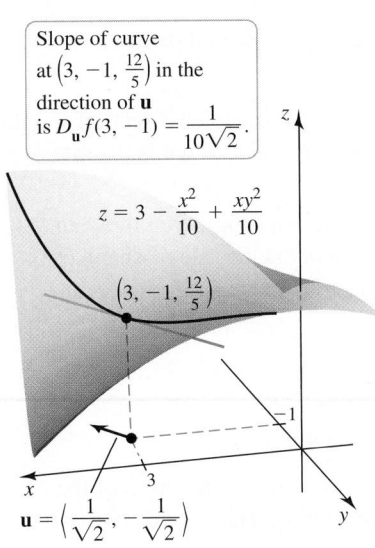

Slope of curve at $\left(3, -1, \frac{12}{5}\right)$ in the direction of **u** is $D_{\mathbf{u}} f(3, -1) = \frac{1}{10\sqrt{2}}$.

$z = 3 - \frac{x^2}{10} + \frac{xy^2}{10}$

$\left(3, -1, \frac{12}{5}\right)$

$\mathbf{u} = \left\langle \frac{1}{\sqrt{2}}, -\frac{1}{\sqrt{2}} \right\rangle$

Figure 15.49

SOLUTION

a. Note that $f_x = -x/5 + y^2/10$ and $f_y = xy/5$. Therefore,

$$\nabla f(3, -1) = \left\langle -\frac{x}{5} + \frac{y^2}{10}, \frac{xy}{5} \right\rangle \bigg|_{(3, -1)} = \left\langle -\frac{1}{2}, -\frac{3}{5} \right\rangle.$$

b. Before computing the directional derivative, it is important to verify that **u** is a unit vector (in this case, it is). The required directional derivative is

$$D_{\mathbf{u}} f(3, -1) = \nabla f(3, -1) \cdot \mathbf{u} = \left\langle -\frac{1}{2}, -\frac{3}{5} \right\rangle \cdot \left\langle \frac{1}{\sqrt{2}}, -\frac{1}{\sqrt{2}} \right\rangle = \frac{1}{10\sqrt{2}}.$$

Figure 15.49 shows the line tangent to the intersection curve in the plane corresponding to **u**; its slope is $D_{\mathbf{u}} f(3, -1)$.

c. In this case, the direction is given in terms of a nonunit vector. The vector $\langle 3, 4 \rangle$ has length 5, so the unit vector in the direction of $\langle 3, 4 \rangle$ is $\mathbf{u} = \left\langle \frac{3}{5}, \frac{4}{5} \right\rangle$. The directional derivative at $(3, -1)$ in the direction of **u** is

$$D_{\mathbf{u}} f(3, -1) = \nabla f(3, -1) \cdot \mathbf{u} = \left\langle -\frac{1}{2}, -\frac{3}{5} \right\rangle \cdot \left\langle \frac{3}{5}, \frac{4}{5} \right\rangle = -\frac{39}{50},$$

which gives the slope of the surface in the direction of $\langle 3, 4 \rangle$ at $(3, -1)$.

Related Exercises 22, 27 ◄

Interpretations of the Gradient

The gradient is important not only in calculating directional derivatives; it plays many other roles in multivariable calculus. Our present goal is to develop some intuition about the meaning of the gradient.

➤ Recall that $\mathbf{u} \cdot \mathbf{v} = |\mathbf{u}||\mathbf{v}| \cos \theta$, where θ is the angle between **u** and **v**.

We have seen that the directional derivative of f at (a, b) in the direction of the unit vector **u** is $D_{\mathbf{u}} f(a, b) = \nabla f(a, b) \cdot \mathbf{u}$. Using properties of the dot product, we have

$$D_{\mathbf{u}} f(a, b) = \nabla f(a, b) \cdot \mathbf{u}$$
$$= |\nabla f(a, b)||\mathbf{u}| \cos \theta$$
$$= |\nabla f(a, b)| \cos \theta, \quad |\mathbf{u}| = 1$$

where θ is the angle between $\nabla f(a, b)$ and **u**. It follows that $D_{\mathbf{u}} f(a, b)$ has its maximum value when $\cos \theta = 1$, which corresponds to $\theta = 0$. Therefore, $D_{\mathbf{u}} f(a, b)$ has its maximum value and f has its greatest rate of *increase* when $\nabla f(a, b)$ and **u** point in the same direction. Note that when $\cos \theta = 1$, the actual rate of increase is $D_{\mathbf{u}} f(a, b) = |\nabla f(a, b)|$ (Figure 15.50).

➤ It is important to remember (but easy to forget) that $\nabla f(a, b)$ lies in the same plane as the domain of f.

Similarly, when $\theta = \pi$, we have $\cos \theta = -1$, and f has its greatest rate of *decrease* when $\nabla f(a, b)$ and **u** point in opposite directions. The actual rate of decrease is $D_{\mathbf{u}} f(a, b) = -|\nabla f(a, b)|$. These observations are summarized as follows: The gradient $\nabla f(a, b)$ points in the *direction of steepest ascent* at (a, b), while $-\nabla f(a, b)$ points in the *direction of steepest descent*.

Notice that $D_{\mathbf{u}} f(a, b) = 0$ when the angle between $\nabla f(a, b)$ and **u** is $\pi/2$, which means $\nabla f(a, b)$ and **u** are orthogonal (Figure 15.50). These observations justify the following theorem.

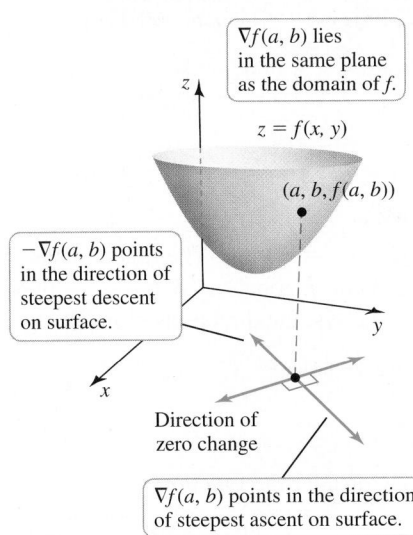

$\nabla f(a, b)$ lies in the same plane as the domain of f.

$z = f(x, y)$

$(a, b, f(a, b))$

$-\nabla f(a, b)$ points in the direction of steepest descent on surface.

Direction of zero change

$\nabla f(a, b)$ points in the direction of steepest ascent on surface.

Figure 15.50

THEOREM 15.11   Directions of Change

Let f be differentiable at (a, b) with $\nabla f(a, b) \neq \mathbf{0}$.

1. f has its maximum rate of increase at (a, b) in the direction of the gradient $\nabla f(a, b)$. The rate of change in this direction is $|\nabla f(a, b)|$.

2. f has its maximum rate of decrease at (a, b) in the direction of $-\nabla f(a, b)$. The rate of change in this direction is $-|\nabla f(a, b)|$.

3. The directional derivative is zero in any direction orthogonal to $\nabla f(a, b)$.

EXAMPLE 4   Steepest ascent and descent   Consider the bowl-shaped paraboloid $z = f(x, y) = 4 + x^2 + 3y^2$.

a. If you are located on the paraboloid at the point $\left(2, -\frac{1}{2}, \frac{35}{4}\right)$, in which direction should you move in order to *ascend* on the surface at the maximum rate? What is the rate of change?

b. If you are located at the point $\left(2, -\frac{1}{2}, \frac{35}{4}\right)$, in which direction should you move in order to *descend* on the surface at the maximum rate? What is the rate of change?

c. At the point $(3, 1, 16)$, in what direction(s) is there no change in the function values?

SOLUTION

a. At the point $\left(2, -\frac{1}{2}\right)$, the value of the gradient is

$$\nabla f\left(2, -\tfrac{1}{2}\right) = \langle 2x, 6y \rangle \big|_{(2, -1/2)} = \langle 4, -3 \rangle.$$

Therefore, the direction of steepest ascent in the xy-plane is in the direction of the gradient vector $\langle 4, -3 \rangle$ (or $\mathbf{u} = \frac{1}{5}\langle 4, -3 \rangle$, as a unit vector). The rate of change is $\left|\nabla f\left(2, -\tfrac{1}{2}\right)\right| = \left|\langle 4, -3 \rangle\right| = 5$ **(Figure 15.51a)**.

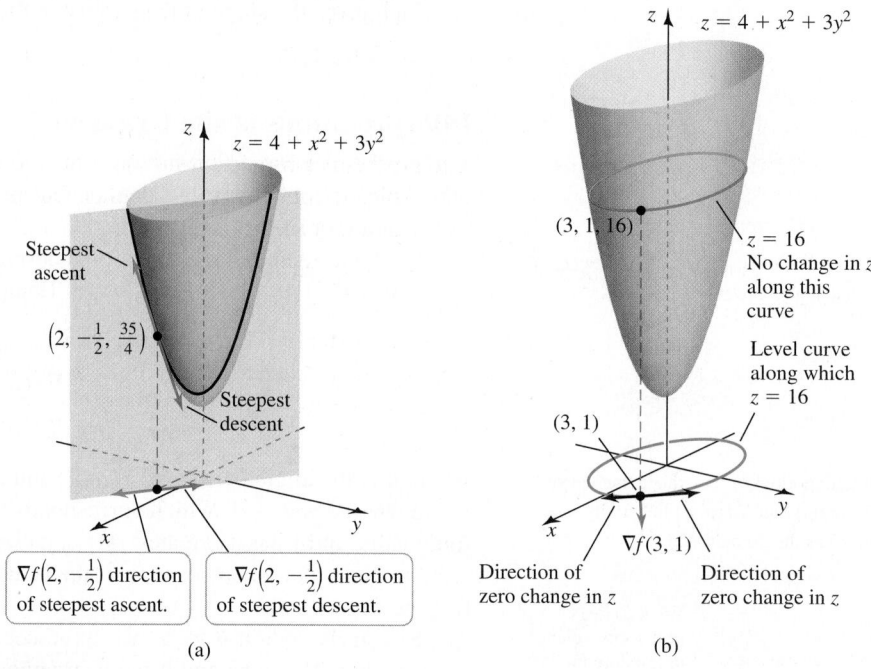

Steepest ascent

$\left(2, -\tfrac{1}{2}, \tfrac{35}{4}\right)$

Steepest descent

$z = 4 + x^2 + 3y^2$

$\nabla f\left(2, -\tfrac{1}{2}\right)$ direction of steepest ascent.	$-\nabla f\left(2, -\tfrac{1}{2}\right)$ direction of steepest descent.

$(3, 1, 16)$

$z = 16$
No change in z along this curve

Level curve along which $z = 16$

$(3, 1)$

$\nabla f(3, 1)$

Direction of zero change in z	Direction of zero change in z

$z = 4 + x^2 + 3y^2$

(a)   (b)

Figure 15.51

b. The direction of steepest *descent* is the direction of $-\nabla f\left(2, -\tfrac{1}{2}\right) = \langle -4, 3 \rangle$ (or $\mathbf{u} = \frac{1}{5}\langle -4, 3 \rangle$, as a unit vector). The rate of change is $-\left|\nabla f\left(2, -\tfrac{1}{2}\right)\right| = -5$.

> Note that $\langle 6, 6 \rangle$ and $\langle 6, -6 \rangle$ are orthogonal because $\langle 6, 6 \rangle \cdot \langle 6, -6 \rangle = 0$.

c. At the point $(3, 1)$, the value of the gradient is $\nabla f(3, 1) = \langle 6, 6 \rangle$. The function has zero change if we move in either of the two directions orthogonal to $\langle 6, 6 \rangle$; these two directions are parallel to $\langle 6, -6 \rangle$. In terms of unit vectors, the directions of no change are $\mathbf{u} = \dfrac{1}{\sqrt{2}}\langle -1, 1 \rangle$ and $\mathbf{u} = \dfrac{1}{\sqrt{2}}\langle 1, -1 \rangle$ **(Figure 15.51b)**.

Related Exercises 31–32 ◀

EXAMPLE 5   Interpreting directional derivatives   Consider the function $f(x, y) = 3x^2 - 2y^2$.

a. Compute $\nabla f(x, y)$ and $\nabla f(2, 3)$.

b. Let $\mathbf{u} = \langle \cos\theta, \sin\theta \rangle$ be a unit vector. At $(2, 3)$, for what values of θ (measured relative to the positive x-axis), with $0 \le \theta < 2\pi$, does the directional derivative have its maximum and minimum values? What are those values?

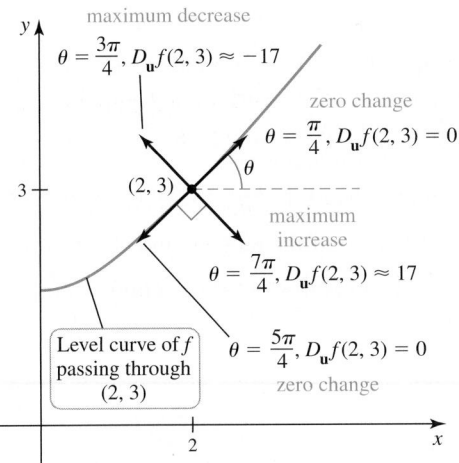

Figure 15.52

SOLUTION

a. The gradient is $\nabla f(x, y) = \langle f_x, f_y \rangle = \langle 6x, -4y \rangle$, and at $(2, 3)$ we have $\nabla f(2, 3) = \langle 12, -12 \rangle$.

b. The gradient $\nabla f(2, 3) = \langle 12, -12 \rangle$ makes an angle of $7\pi/4$ with the positive x-axis. So the maximum rate of change of f occurs in this direction, and that rate of change is $|\nabla f(2, 3)| = |\langle 12, -12 \rangle| = 12\sqrt{2} \approx 17$. The direction of maximum decrease is opposite the direction of the gradient, which corresponds to $\theta = 3\pi/4$. The maximum rate of decrease is the negative of the maximum rate of increase, or $-12\sqrt{2} \approx -17$. The function has zero change in the directions orthogonal to the gradient, which correspond to $\theta = \pi/4$ and $\theta = 5\pi/4$.

Figure 15.52 summarizes these conclusions. Notice that the gradient at $(2, 3)$ appears to be orthogonal to the level curve of f passing through $(2, 3)$. We next see that this is always the case.

Related Exercises 37–38 ◄

The Gradient and Level Curves

Theorem 15.11 states that in any direction orthogonal to the gradient $\nabla f(a, b)$, the function f does not change at (a, b). Recall from Section 15.1 that the curve $f(x, y) = z_0$, where z_0 is a constant, is a *level curve*, on which function values are constant. Combining these two observations, we conclude that the gradient $\nabla f(a, b)$ is orthogonal to the line tangent to the level curve through (a, b).

> **THEOREM 15.12 The Gradient and Level Curves**
> Given a function f differentiable at (a, b), the line tangent to the level curve of f at (a, b) is orthogonal to the gradient $\nabla f(a, b)$, provided $\nabla f(a, b) \neq \mathbf{0}$.

Proof: Consider the function $z = f(x, y)$ and its level curve $f(x, y) = z_0$, where the constant z_0 is chosen so that the curve passes through the point (a, b). Let $\mathbf{r}(t) = \langle x(t), y(t) \rangle$ be a parameterization for the level curve near (a, b) (where it is smooth), and let $\mathbf{r}(t_0)$ correspond to the point (a, b). We now differentiate $f(x, y) = z_0$ with respect to t. The derivative of the right side is 0. Applying the Chain Rule to the left side results in

$$\frac{d}{dt}(f(x, y)) = \frac{\partial f}{\partial x}\frac{dx}{dt} + \frac{\partial f}{\partial y}\frac{dy}{dt}$$

$$= \underbrace{\left\langle \frac{\partial f}{\partial x}, \frac{\partial f}{\partial y} \right\rangle}_{\nabla f(x, y)} \cdot \underbrace{\left\langle \frac{dx}{dt}, \frac{dy}{dt} \right\rangle}_{\mathbf{r}'(t)}$$

$$= \nabla f(x, y) \cdot \mathbf{r}'(t).$$

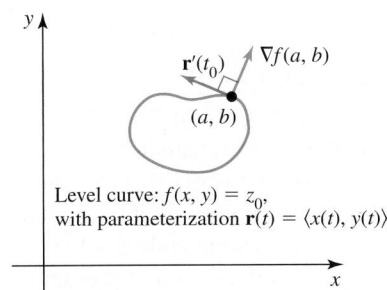

Figure 15.53

QUICK CHECK 4 Draw a circle in the xy-plane centered at the origin and regard it is as a level curve of the surface $z = x^2 + y^2$. At the point (a, a) of the level curve in the xy-plane, the slope of the tangent line is -1. Show that the gradient at (a, a) is orthogonal to the tangent line. ◄

Substituting $t = t_0$, we have $\nabla f(a, b) \cdot \mathbf{r}'(t_0) = 0$, which implies that $\mathbf{r}'(t_0)$ (the tangent vector at (a, b)) is orthogonal to $\nabla f(a, b)$. Figure 15.53 illustrates the geometry of the theorem. ◄

An immediate consequence of Theorem 15.12 is an alternative equation of the tangent line. The curve described by $f(x, y) = z_0$ can be viewed as a level curve in the xy-plane for the surface $z = f(x, y)$. By Theorem 15.12, the line tangent to the curve at (a, b) is orthogonal to $\nabla f(a, b)$. Therefore, if (x, y) is a point on the tangent line, then $\nabla f(a, b) \cdot \langle x - a, y - b \rangle = 0$, which, when simplified, gives an equation of the line tangent to the curve $f(x, y) = z_0$ at (a, b):

$$f_x(a, b)(x - a) + f_y(a, b)(y - b) = 0.$$

EXAMPLE 6 Gradients and level curves Consider the upper sheet $z = f(x, y) = \sqrt{1 + 2x^2 + y^2}$ of a hyperboloid of two sheets.

a. Verify that the gradient at $(1, 1)$ is orthogonal to the corresponding level curve at that point.

b. Find an equation of the line tangent to the level curve at $(1, 1)$.

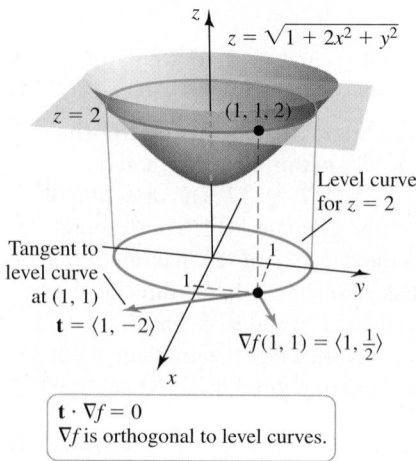

Figure 15.54

▶ The fact that $y' = -2x/y$ may also be obtained using Theorem 15.9: If $F(x, y) = 0$, then $y'(x) = -F_x/F_y$.

SOLUTION

a. You can verify that $(1, 1, 2)$ is on the surface; therefore, $(1, 1)$ is on the level curve corresponding to $z = 2$. Setting $z = 2$ in the equation of the surface and squaring both sides, the equation of the level curve is $4 = 1 + 2x^2 + y^2$, or $2x^2 + y^2 = 3$, which is the equation of an ellipse (**Figure 15.54**). Differentiating $2x^2 + y^2 = 3$ with respect to x gives $4x + 2yy'(x) = 0$, which implies that the slope of the level curve is $y'(x) = -\dfrac{2x}{y}$. Therefore, at the point $(1, 1)$, the slope of the tangent line is -2. Any vector proportional to $\mathbf{t} = \langle 1, -2 \rangle$ has slope -2 and points in the direction of the tangent line.

We now compute the gradient:

$$\nabla f(x, y) = \langle f_x, f_y \rangle = \left\langle \frac{2x}{\sqrt{1 + 2x^2 + y^2}}, \frac{y}{\sqrt{1 + 2x^2 + y^2}} \right\rangle.$$

It follows that $\nabla f(1, 1) = \langle 1, \frac{1}{2} \rangle$ (**Figure 15.54**). The tangent vector $\mathbf{t}$ and the gradient are orthogonal because

$$\mathbf{t} \cdot \nabla f(1, 1) = \langle 1, -2 \rangle \cdot \langle 1, \tfrac{1}{2} \rangle = 0.$$

b. An equation of the line tangent to the level curve at $(1, 1)$ is

$$\underbrace{f_x(1, 1)}_{1}(x - 1) + \underbrace{f_y(1, 1)}_{\frac{1}{2}}(y - 1) = 0,$$

or $y = -2x + 3$.

Related Exercises 49, 52 ◀

EXAMPLE 7 Path of steepest descent The paraboloid $z = f(x, y) = 4 + x^2 + 3y^2$ is shown in **Figure 15.55**. A ball is released at the point $(3, 4, 61)$ on the surface, and it follows the path of steepest descent C to the vertex of the paraboloid.

a. Find an equation of the projection of C in the xy-plane.

b. Find an equation of C on the paraboloid.

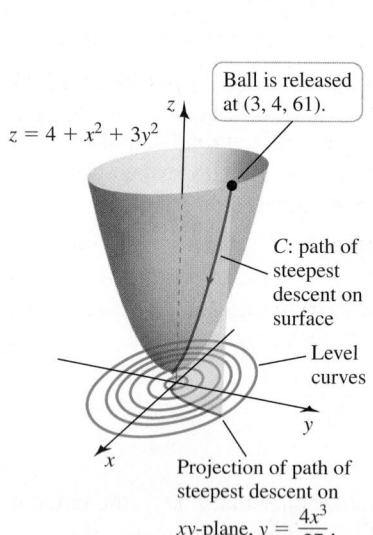

Figure 15.55

SOLUTION

a. The projection of C in the xy-plane points in the direction of $-\nabla f(x, y) = \langle -2x, -6y \rangle$, which means that at the point (x, y), the line tangent to the path has slope $y'(x) = (-6y)/(-2x) = 3y/x$. Therefore, the path in the xy-plane satisfies $y'(x) = 3y/x$ and passes through the initial point $(3, 4)$. You can verify that the solution to this differential equation is $y = 4x^3/27$. Therefore, the projection of the path of steepest descent in the xy-plane is the curve $y = 4x^3/27$. The descent ends at $(0, 0)$, which corresponds to the vertex of the paraboloid (Figure 15.55). At all points of the descent, the projection curve in the xy-plane is orthogonal to the level curves of the paraboloid.

b. To find a parametric description of C, it is easiest to define the parameter $t = x$. Using part (a), we find that

$$y = \frac{4x^3}{27} = \frac{4t^3}{27} \text{ and } z = 4 + x^2 + 3y^2 = 4 + t^2 + \frac{16}{243} t^6.$$

Because $0 \le x \le 3$, the parameter t varies over the interval $0 \le t \le 3$. A parametric description of C is

$$C: \mathbf{r}(t) = \left\langle t, \frac{4t^3}{27}, 4 + t^2 + \frac{16}{243} t^6 \right\rangle, \text{ for } 0 \le t \le 3.$$

QUICK CHECK 5 Verify that $y = 4x^3/27$ satisfies the equation $y'(x) = 3y/x$, with $y(3) = 4$. ◀

With this parameterization, C is traced from $\mathbf{r}(0) = \langle 0, 0, 4 \rangle$ to $\mathbf{r}(3) = \langle 3, 4, 61 \rangle$— in the direction opposite to that of the ball's descent.

Related Exercise 57 ◀

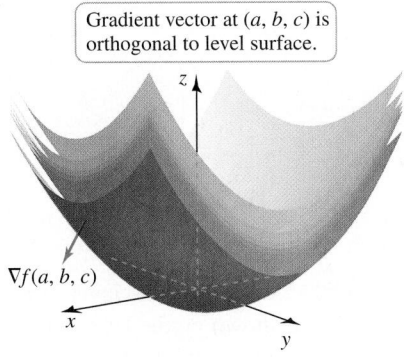

Gradient vector at (a, b, c) is orthogonal to level surface.

$\nabla f(a, b, c)$

Figure 15.56

The Gradient in Three Dimensions

The directional derivative, the gradient, and the idea of a level curve extend immediately to functions of three variables of the form $w = f(x, y, z)$. The main differences are that the gradient is a vector in $\mathbb{R}^3$, and level curves become *level surfaces* (Section 15.1). Here is how the gradient looks when we step up one dimension.

The easiest way to visualize the surface $w = f(x, y, z)$ is to picture its level surfaces— the surfaces in $\mathbb{R}^3$ on which f has a constant value. The level surfaces are given by the equation $f(x, y, z) = C$, where C is a constant (Figure 15.56). The level surfaces *can* be graphed, and they may be viewed as layers of the full four-dimensional surface (like layers of an onion). With this image in mind, we now extend the concepts of directional derivative and gradient to three dimensions.

Given the function $w = f(x, y, z)$, we begin just as we did in the two-variable case and define the directional derivative and the gradient.

DEFINITION Directional Derivative and Gradient in Three Dimensions

Let f be differentiable at (a, b, c) and let $\mathbf{u} = \langle u_1, u_2, u_3 \rangle$ be a unit vector. The **directional derivative of f at (a, b, c) in the direction of u** is

$$D_{\mathbf{u}}f(a, b, c) = \lim_{h \to 0} \frac{f(a + hu_1, b + hu_2, c + hu_3) - f(a, b, c)}{h},$$

provided this limit exists.

The **gradient** of f at the point (x, y, z) is the vector-valued function

$$\nabla f(x, y, z) = \langle f_x(x, y, z), f_y(x, y, z), f_z(x, y, z) \rangle$$
$$= f_x(x, y, z)\,\mathbf{i} + f_y(x, y, z)\,\mathbf{j} + f_z(x, y, z)\,\mathbf{k}.$$

An argument similar to that given in two dimensions leads from the definition of the directional derivative to a computational formula. Given a unit vector $\mathbf{u} = \langle u_1, u_2, u_3 \rangle$, the directional derivative of f in the direction of $\mathbf{u}$ at the point (a, b, c) is

$$D_{\mathbf{u}}\,f(a, b, c) = f_x(a, b, c)\,u_1 + f_y(a, b, c)\,u_2 + f_z(a, b, c)\,u_3.$$

As before, we recognize this expression as a dot product of the vector $\mathbf{u}$ and the vector $\nabla f(a, b, c) = \langle f_x(a, b, c), f_y(a, b, c), f_z(a, b, c) \rangle$, which is the gradient evaluated at (a, b, c). These observations lead to Theorem 15.13, which mirrors Theorems 15.10 and 15.11.

> When we introduce the tangent plane in Section 15.6, we can also claim that $\nabla f(a, b, c)$ is orthogonal to the level surface that passes through (a, b, c).

THEOREM 15.13 Directional Derivative and Interpreting the Gradient

Let f be differentiable at (a, b, c) and let $\mathbf{u} = \langle u_1, u_2, u_3 \rangle$ be a unit vector. The directional derivative of f at (a, b, c) in the direction of $\mathbf{u}$ is

$$D_{\mathbf{u}}f(a, b, c) = \nabla f(a, b, c) \cdot \mathbf{u}$$
$$= \langle f_x(a, b, c), f_y(a, b, c), f_z(a, b, c) \rangle \cdot \langle u_1, u_2, u_3 \rangle.$$

Assuming $\nabla f(a, b, c) \neq \mathbf{0}$, the gradient in three dimensions has the following properties.

1. f has its maximum rate of increase at (a, b, c) in the direction of the gradient $\nabla f(a, b, c)$, and the rate of change in this direction is $|\nabla f(a, b, c)|$.

2. f has its maximum rate of decrease at (a, b, c) in the direction of $-\nabla f(a, b, c)$, and the rate of change in this direction is $-|\nabla f(a, b, c)|$.

3. The directional derivative is zero in any direction orthogonal to $\nabla f(a, b, c)$.

QUICK CHECK 6 Compute $\nabla f(-1, 2, 1)$, where $f(x, y, z) = xy/z$. ◄

EXAMPLE 8 Gradients in three dimensions Consider the function $f(x, y, z) = x^2 + 2y^2 + 4z^2 - 1$ and its level surface $f(x, y, z) = 3$.

a. Find and interpret the gradient at the points $P(2, 0, 0)$, $Q(0, \sqrt{2}, 0)$, $R(0, 0, 1)$, and $S\left(1, 1, \frac{1}{2}\right)$ on the level surface.

b. What are the actual rates of change of f in the directions of the gradients in part (a)?

Level surface of $f(x, y, z) = x^2 + 2y^2 + 4z^2 - 1$
$f(x, y, z) = 3$

$\nabla f(0, 0, 1) = \langle 0, 0, 8 \rangle$

$\nabla f(1, 1, \tfrac{1}{2}) = \langle 2, 4, 4 \rangle$

$\nabla f(2, 0, 0) = \langle 4, 0, 0 \rangle$

$\nabla f(0, \sqrt{2}, 0) = \langle 0, 4\sqrt{2}, 0 \rangle$

Figure 15.57

SOLUTION

a. The gradient is

$$\nabla f = \langle f_x, f_y, f_z \rangle = \langle 2x, 4y, 8z \rangle.$$

Evaluating the gradient at the four points, we find that

$$\nabla f(2, 0, 0) = \langle 4, 0, 0 \rangle, \qquad \nabla f(0, \sqrt{2}, 0) = \langle 0, 4\sqrt{2}, 0 \rangle,$$
$$\nabla f(0, 0, 1) = \langle 0, 0, 8 \rangle, \quad \text{and} \quad \nabla f(1, 1, \tfrac{1}{2}) = \langle 2, 4, 4 \rangle.$$

The level surface $f(x, y, z) = 3$ is an ellipsoid (**Figure 15.57**), which is one layer of a four-dimensional surface. The four points $P, Q, R,$ and S are shown on the level surface with the respective gradient vectors. In each case, the gradient points in the direction that f has its maximum rate of increase. Of particular importance is the fact—to be made clear in the next section—that at each point, the gradient is orthogonal to the level surface.

b. The actual rate of increase of f at (a, b, c) in the direction of the gradient is $|\nabla f(a, b, c)|$. At P, the rate of increase of f in the direction of the gradient is $|\langle 4, 0, 0 \rangle| = 4$; at Q, the rate of increase is $|\langle 0, 4\sqrt{2}, 0 \rangle| = 4\sqrt{2}$; at R, the rate of increase is $|\langle 0, 0, 8 \rangle| = 8$; and at S, the rate of increase is $|\langle 2, 4, 4 \rangle| = 6$.

Related Exercises 59–60 ◄

SECTION 15.5 EXERCISES

Getting Started

1. Explain how a directional derivative is formed from the two partial derivatives f_x and f_y.

2. How do you compute the gradient of the functions $f(x, y)$ and $f(x, y, z)$?

3. Interpret the direction of the gradient vector at a point.

4. Interpret the magnitude of the gradient vector at a point.

5. Given a function f, explain the relationship between the gradient and the level curves of f.

6. The level curves of the surface $z = x^2 + y^2$ are circles in the xy-plane centered at the origin. Without computing the gradient, what is the direction of the gradient at $(1, 1)$ and $(-1, -1)$ (determined up to a scalar multiple)?

7. Suppose f is differentiable at $(3, 4)$, $\nabla f(3, 4) = \langle -\sqrt{3}, 1 \rangle$, and $\mathbf{u} = \left\langle \dfrac{\sqrt{3}}{2}, -\dfrac{1}{2} \right\rangle$. Compute $D_{\mathbf{u}} f(3, 4)$.

8. Suppose f is differentiable at $(9, 9)$, $\nabla f(9, 9) = \langle 3, 1 \rangle$, and $\mathbf{w} = \langle 1, -1 \rangle$. Compute the directional derivative of f at $(9, 9)$ in the direction of the vector $\mathbf{w}$.

9. Suppose f is differentiable at $(3, 4)$. Assume $\mathbf{u}, \mathbf{v},$ and $\mathbf{w}$ are unit vectors, $\mathbf{v} = -\mathbf{u}$, $\mathbf{w} \cdot \nabla f(3, 4) = 0$, and $D_{\mathbf{u}} f(3, 4) = 7$. Find $D_{\mathbf{v}} f(3, 4)$ and $D_{\mathbf{w}} f(3, 4)$.

10. Suppose f is differentiable at $(1, 2)$ and $\nabla f(1, 2) = \langle 3, 4 \rangle$. Find the slope of the line tangent to the level curve of f at $(1, 2)$.

Practice Exercises

11. Directional derivatives Consider the function

$$f(x, y) = 8 - \frac{x^2}{2} - y^2,$$ whose graph is a paraboloid (see figure).

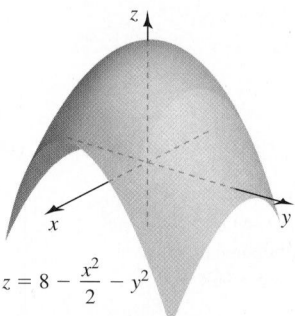

$z = 8 - \dfrac{x^2}{2} - y^2$

	$(a, b) = (2, 0)$	$(a, b) = (0, 2)$	$(a, b) = (1, 1)$
$\mathbf{u} = \left\langle \frac{\sqrt{2}}{2}, \frac{\sqrt{2}}{2} \right\rangle$			
$\mathbf{v} = \left\langle -\frac{\sqrt{2}}{2}, \frac{\sqrt{2}}{2} \right\rangle$			
$\mathbf{w} = \left\langle -\frac{\sqrt{2}}{2}, -\frac{\sqrt{2}}{2} \right\rangle$			

a. Fill in the table with the values of the directional derivative at the points (a, b) in the directions given by the unit vectors $\mathbf{u}, \mathbf{v},$ and $\mathbf{w}$.

b. Interpret each of the directional derivatives computed in part (a) at the point $(2, 0)$.

12. Directional derivatives Consider the function $f(x, y) = 2x^2 + y^2$, whose graph is a paraboloid (see figure).

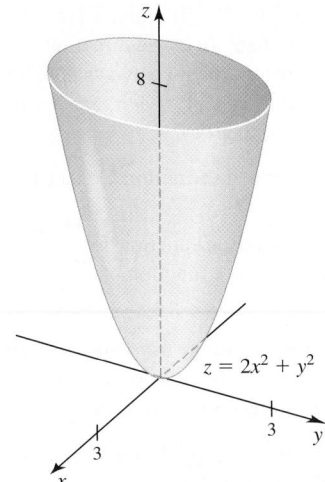

$z = 2x^2 + y^2$

	$(a, b) = (1, 0)$	$(a, b) = (1, 1)$	$(a, b) = (1, 2)$
$\mathbf{u} = \langle 1, 0 \rangle$			
$\mathbf{v} = \langle \frac{\sqrt{2}}{2}, \frac{\sqrt{2}}{2} \rangle$			
$\mathbf{w} = \langle 0, 1 \rangle$			

a. Fill in the table with the values of the directional derivative at the points (a, b) in the directions given by the unit vectors $\mathbf{u}$, $\mathbf{v}$, and $\mathbf{w}$.

b. Interpret each of the directional derivatives computed in part (a) at the point $(1, 0)$.

13–20. Computing gradients *Compute the gradient of the following functions and evaluate it at the given point P.*

13. $f(x, y) = 2 + 3x^2 - 5y^2$; $P(2, -1)$

14. $f(x, y) = 4x^2 - 2xy + y^2$; $P(-1, -5)$

15. $g(x, y) = x^2 - 4x^2y - 8xy^2$; $P(-1, 2)$

16. $p(x, y) = \sqrt{12 - 4x^2 - y^2}$; $P(-1, -1)$

17. $f(x, y) = xe^{2xy}$; $P(1, 0)$

18. $f(x, y) = \sin(3x + 2y)$; $P(\pi, 3\pi/2)$

19. $F(x, y) = e^{-x^2 - 2y^2}$; $P(-1, 2)$

20. $h(x, y) = \ln(1 + x^2 + 2y^2)$; $P(2, -3)$

21–30. Computing directional derivatives with the gradient *Compute the directional derivative of the following functions at the given point P in the direction of the given vector. Be sure to use a unit vector for the direction vector.*

21. $f(x, y) = x^2 - y^2$; $P(-1, -3)$; $\langle \frac{3}{5}, -\frac{4}{5} \rangle$

22. $f(x, y) = 3x^2 + y^3$; $P(3, 2)$; $\langle \frac{5}{13}, \frac{12}{13} \rangle$

23. $f(x, y) = 10 - 3x^2 + \frac{y^4}{4}$; $P(2, -3)$; $\langle \frac{\sqrt{3}}{2}, -\frac{1}{2} \rangle$

24. $g(x, y) = \sin \pi(2x - y)$; $P(-1, -1)$; $\langle \frac{5}{13}, -\frac{12}{13} \rangle$

25. $f(x, y) = \sqrt{4 - x^2 - 2y}$; $P(2, -2)$; $\langle \frac{1}{\sqrt{5}}, \frac{2}{\sqrt{5}} \rangle$

26. $f(x, y) = 13e^{xy}$; $P(1, 0)$; $\langle 5, 12 \rangle$

27. $f(x, y) = 3x^2 + 2y + 5$; $P(1, 2)$; $\langle -3, 4 \rangle$

28. $h(x, y) = e^{-x-y}$; $P(\ln 2, \ln 3)$; $\langle 1, 1 \rangle$

29. $g(x, y) = \ln(4 + x^2 + y^2)$; $g(-1, 2)$; $\langle 2, 1 \rangle$

30. $f(x, y) = \dfrac{x}{x - y}$; $P(4, 1)$; $\langle -1, 2 \rangle$

31–36. Direction of steepest ascent and descent *Consider the following functions and points P.*

a. *Find the unit vectors that give the direction of steepest ascent and steepest descent at P.*

b. *Find a vector that points in a direction of no change in the function at P.*

31. $f(x, y) = x^2 - 4y^2 - 9$; $P(1, -2)$

32. $f(x, y) = x^2 + 4xy - y^2$; $P(2, 1)$

33. $f(x, y) = x^4 - x^2y + y^2 + 6$; $P(-1, 1)$

34. $p(x, y) = \sqrt{20 + x^2 + 2xy - y^2}$; $P(1, 2)$

35. $F(x, y) = e^{-x^2/2 - y^2/2}$; $P(-1, 1)$

36. $f(x, y) = 2\sin(2x - 3y)$; $P(0, \pi)$

37–42. Interpreting directional derivatives *A function f and a point P are given. Let θ correspond to the direction of the directional derivative.*

a. *Find the gradient and evaluate it at P.*

b. *Find the angles θ (with respect to the positive x-axis) associated with the directions of maximum increase, maximum decrease, and zero change.*

c. *Write the directional derivative at P as a function of θ; call this function g.*

d. *Find the value of θ that maximizes g(θ) and find the maximum value.*

e. *Verify that the value of θ that maximizes g corresponds to the direction of the gradient. Verify that the maximum value of g equals the magnitude of the gradient.*

37. $f(x, y) = 10 - 2x^2 - 3y^2$; $P(3, 2)$

38. $f(x, y) = 8 + x^2 + 3y^2$; $P(-3, -1)$

39. $f(x, y) = \sqrt{2 + x^2 + y^2}$; $P(\sqrt{3}, 1)$

40. $f(x, y) = \sqrt{12 - x^2 - y^2}$; $P(-1, -1/\sqrt{3})$

41. $f(x, y) = e^{-x^2 - 2y^2}$; $P(-1, 0)$

42. $f(x, y) = \ln(1 + 2x^2 + 3y^2)$; $P(3/4, -\sqrt{3})$

43–46. Directions of change *Consider the following functions f and points P. Sketch the xy-plane showing P and the level curve through P. Indicate (as in Figure 15.52) the directions of maximum increase, maximum decrease, and no change for f.*

43. $f(x, y) = 8 + 4x^2 + 2y^2$; $P(2, -4)$

44. $f(x, y) = -4 + 6x^2 + 3y^2$; $P(-1, -2)$

45. $f(x, y) = x^2 + xy + y^2 + 7$; $P(-3, 3)$

46. $f(x, y) = \tan(2x + 2y)$; $P(\pi/16, \pi/16)$

47–50. Level curves *Consider the paraboloid* $f(x, y) = 16 - \dfrac{x^2}{4} - \dfrac{y^2}{16}$ *and the point P on the given level curve of f. Compute the slope of the line tangent to the level curve at P, and verify that the tangent line is orthogonal to the gradient at that point.*

47. $f(x, y) = 0; P(0, 16)$ **48.** $f(x, y) = 0; P(8, 0)$

49. $f(x, y) = 12; P(4, 0)$ **50.** $f(x, y) = 12; P(2\sqrt{3}, 4)$

51–54. Level curves *Consider the upper half of the ellipsoid*

$f(x, y) = \sqrt{1 - \dfrac{x^2}{4} - \dfrac{y^2}{16}}$ *and the point P on the given level curve*

of f. Compute the slope of the line tangent to the level curve at P, and verify that the tangent line is orthogonal to the gradient at that point.

51. $f(x, y) = \dfrac{\sqrt{3}}{2}; P\left(\dfrac{1}{2}, \sqrt{3}\right)$ **52.** $f(x, y) = \dfrac{1}{\sqrt{2}}; P(0, \sqrt{8})$

53. $f(x, y) = \dfrac{1}{\sqrt{2}}; P(\sqrt{2}, 0)$ **54.** $f(x, y) = \dfrac{1}{\sqrt{2}}; P(1, 2)$

55–58. Path of steepest descent *Consider each of the following surfaces and the point P on the surface.*

a. *Find the gradient of f.*

b. *Let C' be the path of steepest descent on the surface beginning at P, and let C be the projection of C' on the xy-plane. Find an equation of C in the xy-plane.*

c. *Find parametric equations for the path C' on the surface.*

55. $f(x, y) = 4 + x$ (a plane); $P(4, 4, 8)$

56. $f(x, y) = y + x$ (a plane); $P(2, 2, 4)$

57. $f(x, y) = 4 - x^2 - 2y^2$ (a paraboloid); $P(1, 1, 1)$

58. $f(x, y) = y + x^{-1}; P(1, 2, 3)$

59–66. Gradients in three dimensions *Consider the following functions f, points P, and unit vectors* **u**.

a. *Compute the gradient of f and evaluate it at P.*

b. *Find the unit vector in the direction of maximum increase of f at P.*

c. *Find the rate of change of the function in the direction of maximum increase at P.*

d. *Find the directional derivative at P in the direction of the given vector.*

59. $f(x, y, z) = x^2 + 2y^2 + 4z^2 + 10; P(1, 0, 4); \left\langle \dfrac{1}{\sqrt{2}}, 0, \dfrac{1}{\sqrt{2}} \right\rangle$

60. $f(x, y, z) = 4 - x^2 + 3y^2 + \dfrac{z^2}{2}; P(0, 2, -1); \left\langle 0, \dfrac{1}{\sqrt{2}}, -\dfrac{1}{\sqrt{2}} \right\rangle$

61. $f(x, y, z) = 1 + 4xyz; P(1, -1, -1); \left\langle \dfrac{1}{\sqrt{3}}, \dfrac{1}{\sqrt{3}}, -\dfrac{1}{\sqrt{3}} \right\rangle$

62. $f(x, y, z) = xy + yz + xz + 4; P(2, -2, 1); \left\langle 0, -\dfrac{1}{\sqrt{2}}, -\dfrac{1}{\sqrt{2}} \right\rangle$

63. $f(x, y, z) = 1 + \sin(x + 2y - z); P\left(\dfrac{\pi}{6}, \dfrac{\pi}{6}, -\dfrac{\pi}{6}\right); \left\langle \dfrac{1}{3}, \dfrac{2}{3}, \dfrac{2}{3} \right\rangle$

64. $f(x, y, z) = e^{xyz-1}; P(0, 1, -1); \left\langle -\dfrac{2}{3}, \dfrac{2}{3}, -\dfrac{1}{3} \right\rangle$

65. $f(x, y, z) = \ln(1 + x^2 + y^2 + z^2); P(1, 1, -1); \left\langle \dfrac{2}{3}, \dfrac{2}{3}, -\dfrac{1}{3} \right\rangle$

66. $f(x, y, z) = \dfrac{x - z}{y - z}; P(3, 2, -1); \left\langle \dfrac{1}{3}, \dfrac{2}{3}, -\dfrac{2}{3} \right\rangle$

67. Explain why or why not Determine whether the following statements are true and give an explanation or counterexample.

a. If $f(x, y) = x^2 + y^2 - 10$, then $\nabla f(x, y) = 2x + 2y$.

b. Because the gradient gives the direction of maximum increase of a function, the gradient is always positive.

c. The gradient of $f(x, y, z) = 1 + xyz$ has four components.

d. If $f(x, y, z) = 4$, then $\nabla f = \mathbf{0}$.

68. Gradient of a composite function Consider the function $F(x, y, z) = e^{xyz}$.

a. Write F as a composite function $f \circ g$, where f is a function of one variable and g is a function of three variables.

b. Relate ∇F to ∇g.

69–72. Directions of zero change *Find the directions in the xy-plane in which the following functions have zero change at the given point. Express the directions in terms of unit vectors.*

69. $f(x, y) = 12 - 4x^2 - y^2; P(1, 2, 4)$

70. $f(x, y) = x^2 - 4y^2 - 8; P(4, 1, 4)$

71. $f(x, y) = \sqrt{3 + 2x^2 + y^2}; P(1, -2, 3)$

72. $f(x, y) = e^{1-xy}; P(1, 0, e)$

73. Steepest ascent on a plane Suppose a long sloping hillside is described by the plane $z = ax + by + c$, where a, b, and c are constants. Find the path in the xy-plane, beginning at (x_0, y_0), that corresponds to the path of steepest ascent on the hillside.

74. Gradient of a distance function Let (a, b) be a given point in $\mathbb{R}^2$, and let $d = f(x, y)$ be the distance between (a, b) and the variable point (x, y).

a. Show that the graph of f is a cone.

b. Show that the gradient of f at any point other than (a, b) is a unit vector.

c. Interpret the direction and magnitude of ∇f.

75–78. Looking ahead—tangent planes *Consider the following surfaces* $f(x, y, z) = 0$, *which may be regarded as a level surface of the function* $w = f(x, y, z)$. *A point* $P(a, b, c)$ *on the surface is also given.*

a. *Find the (three-dimensional) gradient of f and evaluate it at P.*

b. *The set of all vectors orthogonal to the gradient with their tails at P form a plane. Find an equation of that plane (soon to be called the tangent plane).*

75. $f(x, y, z) = x^2 + y^2 + z^2 - 3 = 0; P(1, 1, 1)$

76. $f(x, y, z) = 8 - xyz = 0; P(2, 2, 2)$

77. $f(x, y, z) = e^{x+y-z} - 1 = 0; P(1, 1, 2)$

78. $f(x, y, z) = xy + xz - yz - 1 = 0; P(1, 1, 1)$

▣ 79. A traveling wave A snapshot (frozen in time) of a set of water waves is described by the function $z = 1 + \sin(x - y)$, where z gives the height of the waves and (x, y) are coordinates in the horizontal plane $z = 0$.

a. Use a graphing utility to graph $z = 1 + \sin(x - y)$.

b. The crests and the troughs of the waves are aligned in the direction in which the height function has zero change. Find the direction in which the crests and troughs are aligned.

c. If you were surfing on one of these waves and wanted the steepest descent from the crest to the trough, in which direction would you point your surfboard (given in terms of a unit vector in the xy-plane)?

d. Check that your answers to parts (b) and (c) are consistent with the graph of part (a).

80. Traveling waves in general Generalize Exercise 79 by considering a set of waves described by the function $z = A + \sin(ax - by)$, where a, b, and A are real numbers.

   a. Find the direction in which the crests and troughs of the waves are aligned. Express your answer as a unit vector in terms of a and b.

   b. Find the surfer's direction—that is, the direction of steepest descent from a crest to a trough. Express your answer as a unit vector in terms of a and b.

Explorations and Challenges

81–83. Potential functions *Potential functions arise frequently in physics and engineering. A potential function has the property that a field of interest (for example, an electric field, a gravitational field, or a velocity field) is the gradient of the potential (or sometimes the negative of the gradient of the potential). (Potential functions are considered in depth in Chapter 17.)*

81. Electric potential due to a point charge The electric field due to a point charge of strength Q at the origin has a potential function $\varphi = kQ/r$, where $r^2 = x^2 + y^2 + z^2$ is the square of the distance between a variable point $P(x, y, z)$ and the charge, and $k > 0$ is a physical constant. The electric field is given by $\mathbf{E} = -\nabla\varphi$, where $\nabla\varphi$ is the gradient in three dimensions.

   a. Show that the three-dimensional electric field due to a point charge is given by

$$\mathbf{E}(x, y, z) = kQ\left\langle \frac{x}{r^3}, \frac{y}{r^3}, \frac{z}{r^3} \right\rangle.$$

   b. Show that the electric field at a point has a magnitude $|\mathbf{E}| = \dfrac{kQ}{r^2}$. Explain why this relationship is called an inverse square law.

82. Gravitational potential The gravitational potential associated with two objects of mass M and m is $\varphi = -GMm/r$, where G is the gravitational constant. If one of the objects is at the origin and the other object is at $P(x, y, z)$, then $r^2 = x^2 + y^2 + z^2$ is the square of the distance between the objects. The gravitational field at P is given by $\mathbf{F} = -\nabla\varphi$, where $\nabla\varphi$ is the gradient in three dimensions. Show that the force has a magnitude $|\mathbf{F}| = GMm/r^2$. Explain why this relationship is called an inverse square law.

83. Velocity potential In two dimensions, the motion of an ideal fluid (an incompressible and irrotational fluid) is governed by a velocity potential φ. The velocity components of the fluid, u in the x-direction and v in the y-direction, are given by $\langle u, v \rangle = \nabla\varphi$. Find the velocity components associated with the velocity potential $\varphi(x, y) = \sin \pi x \sin 2\pi y$.

84. Gradients for planes Prove that for the plane described by $f(x, y) = Ax + By$, where A and B are nonzero constants, the gradient is constant (independent of (x, y)). Interpret this result.

85. Rules for gradients Use the definition of the gradient (in two or three dimensions), assume f and g are differentiable functions on $\mathbb{R}^2$ or $\mathbb{R}^3$, and let c be a constant. Prove the following gradient rules.

   a. Constants Rule: $\nabla(cf) = c\nabla f$
   b. Sum Rule: $\nabla(f + g) = \nabla f + \nabla g$
   c. Product Rule: $\nabla(fg) = (\nabla f)g + f\nabla g$
   d. Quotient Rule: $\nabla\left(\dfrac{f}{g}\right) = \dfrac{g\nabla f - f\nabla g}{g^2}$
   e. Chain Rule: $\nabla(f \circ g) = f'(g)\nabla g$, where f is a function of one variable

86–91. Using gradient rules *Use the gradient rules of Exercise 85 to find the gradient of the following functions.*

86. $f(x, y) = xy \cos(xy)$ **87.** $f(x, y) = \dfrac{x + y}{x^2 + y^2}$

88. $f(x, y) = \ln(1 + x^2 + y^2)$

89. $f(x, y, z) = \sqrt{25 - x^2 - y^2 - z^2}$

90. $f(x, y, z) = (x + y + z)e^{xyz}$

91. $f(x, y, z) = \dfrac{x + yz}{y + xz}$

QUICK CHECK ANSWERS

1. If $\mathbf{u} = \langle u_1, u_2 \rangle = \langle 1, 0 \rangle$ then

$$D_\mathbf{u} f(a, b) = \lim_{h \to 0} \frac{f(a + hu_1, b + hu_2) - f(a, b)}{h}$$
$$= \lim_{h \to 0} \frac{f(a + h, b) - f(a, b)}{h} = f_x(a, b).$$

Similarly, when $\mathbf{u} = \langle 0, 1 \rangle$, the partial derivative $f_y(a, b)$ results. **2.** The vector from (a, b) to $(a + \Delta su_1, b + \Delta su_2)$ is $\langle \Delta su_1, \Delta su_2 \rangle = \Delta s \langle u_1, u_2 \rangle = \Delta s\mathbf{u}$. Its length is $|\Delta s\mathbf{u}| = \Delta s|\mathbf{u}| = \Delta s$. Therefore, s measures arc length. **3.** Reversing (negating) the direction vector negates the directional derivative, so the respective values are approximately -2.47 and 0.98. **4.** The gradient is $\langle 2x, 2y \rangle$, which, evaluated at (a, a), is $\langle 2a, 2a \rangle$. Taking the dot product of the gradient and the vector $\langle -1, 1 \rangle$ (a vector parallel to a line of slope -1), we see that $\langle 2a, 2a \rangle \cdot \langle -1, 1 \rangle = 0$. **6.** $\langle 2, -1, 2 \rangle$ ◄

15.6 Tangent Planes and Linear Approximation

In Section 4.6, we saw that if we zoom in on a point on a smooth curve (one described by a differentiable function), the curve looks more and more like the tangent line at that point. Once we have the tangent line at a point, it can be used to approximate function values and to estimate changes in the dependent variable. In this section, the analogous story is developed in three dimensions. Now we see that differentiability at a point (as discussed in Section 15.3) implies the existence of a tangent *plane* at that point (**Figure 15.58**).

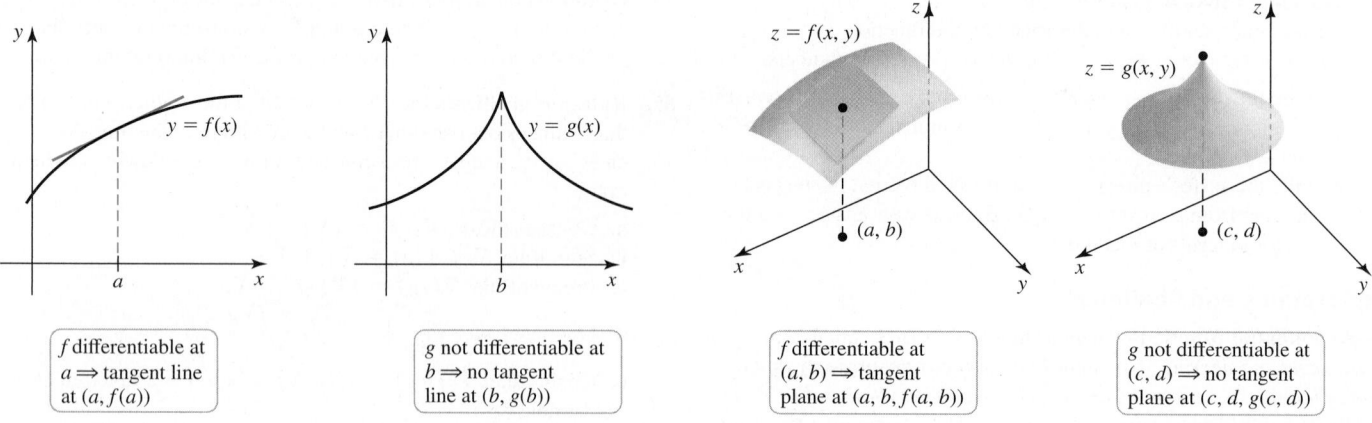

f differentiable at $a \Rightarrow$ tangent line at $(a, f(a))$	g not differentiable at $b \Rightarrow$ no tangent line at $(b, g(b))$	f differentiable at $(a, b) \Rightarrow$ tangent plane at $(a, b, f(a, b))$	g not differentiable at $(c, d) \Rightarrow$ no tangent plane at $(c, d, g(c, d))$

Figure 15.58

Consider a smooth surface described by a differentiable function f, and focus on a single point on the surface. As we zoom in on that point (**Figure 15.59**), the surface appears more and more like a plane. The first step is to define this plane carefully; it is called the *tangent plane*. Once we have the tangent plane, we can use it to approximate function values and to estimate changes in the dependent variable.

Tangent Planes

Recall that a surface in $\mathbb{R}^3$ may be defined in at least two different ways:

• **Explicitly** in the form $z = f(x, y)$ or
• **Implicitly** in the form $F(x, y, z) = 0$.

It is easiest to begin by considering a surface defined implicitly by $F(x, y, z) = 0$, where F is differentiable at a particular point. Such a surface may be viewed as a level surface of a function $w = F(x, y, z)$; it is the level surface for $w = 0$.

QUICK CHECK 1 Write the function $z = xy + x - y$ in the form $F(x, y, z) = 0$. ◄

Tangent Planes for $F(x, y, z) = 0$ To find an equation of the tangent plane, consider a smooth curve $C: \mathbf{r}(t) = \langle x(t), y(t), z(t) \rangle$ that lies on the surface $F(x, y, z) = 0$ (**Figure 15.60a**). Because the points of C lie on the surface, we have $F(x(t), y(t), z(t)) = 0$.

As we zoom in on a smooth surface ...

... it appears more like a plane.

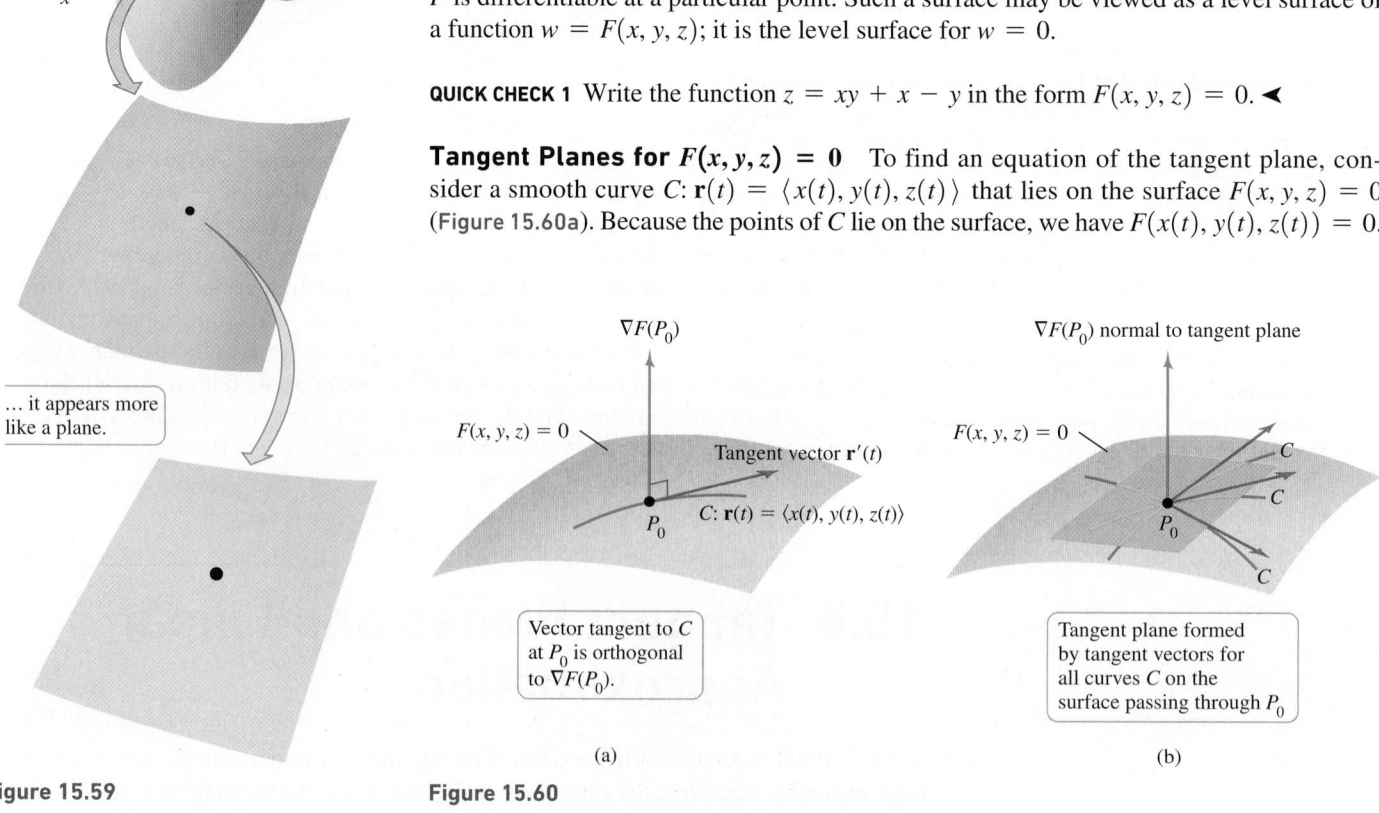

Vector tangent to C at P_0 is orthogonal to $\nabla F(P_0)$.	Tangent plane formed by tangent vectors for all curves C on the surface passing through P_0

(a)

(b)

Figure 15.59   **Figure 15.60**

Differentiating both sides of this equation with respect to t reveals a useful relationship. The derivative of the right side is 0. The Chain Rule applied to the left side yields

$$\frac{d}{dt}\big(F(x(t), y(t), z(t))\big) = \frac{\partial F}{\partial x}\frac{dx}{dt} + \frac{\partial F}{\partial y}\frac{dy}{dt} + \frac{\partial F}{\partial z}\frac{dz}{dt}$$

$$= \underbrace{\left\langle \frac{\partial F}{\partial x}, \frac{\partial F}{\partial y}, \frac{\partial F}{\partial z} \right\rangle}_{\nabla F(x,\,y,\,z)} \cdot \underbrace{\left\langle \frac{dx}{dt}, \frac{dy}{dt}, \frac{dz}{dt} \right\rangle}_{\mathbf{r}'(t)}$$

$$= \nabla F(x, y, z) \cdot \mathbf{r}'(t).$$

Therefore, $\nabla F(x, y, z) \cdot \mathbf{r}'(t) = 0$ and at any point on the curve, the tangent vector $\mathbf{r}'(t)$ is orthogonal to the gradient.

Now fix a point $P_0(a, b, c)$ on the surface, assume $\nabla F(a, b, c) \neq \mathbf{0}$, and let C be any smooth curve on the surface passing through P_0. We have shown that any vector tangent to C is orthogonal to $\nabla F(a, b, c)$ at P_0. Because this argument applies to *all* smooth curves on the surface passing through P_0, the tangent vectors for all these curves (with their tails at P_0) are orthogonal to $\nabla F(a, b, c)$; therefore, they all lie in the same plane (Figure 15.60b). This plane is called the *tangent plane* at P_0. We can easily find an equation of the tangent plane because we know both a point on the plane $P_0(a, b, c)$ and a normal vector $\nabla F(a, b, c)$; an equation is

$$\nabla F(a, b, c) \cdot \langle x - a, y - b, z - c \rangle = 0.$$

> Recall that an equation of the plane passing though (a, b, c) with a normal vector $\mathbf{n} = \langle n_1, n_2, n_3 \rangle$ is $n_1(x - a) + n_2(y - b) + n_3(z - c) = 0$.

> If $\mathbf{r}$ is a position vector corresponding to an arbitrary point on the tangent plane, and $\mathbf{r_0}$ is a position vector corresponding to a fixed point (a, b, c) on the plane, then an equation of the tangent plane may be written concisely as
> $$\nabla F(a, b, c) \cdot (\mathbf{r} - \mathbf{r_0}) = 0.$$
> Notice the analogy with tangent lines and level curves (Section 15.5). An equation of the line tangent to $f(x, y) = 0$ at (a, b) is
> $$\nabla f(a, b) \cdot \langle x - a, y - b \rangle = 0.$$

DEFINITION   **Equation of the Tangent Plane for** $F(x, y, z) = 0$

Let F be differentiable at the point $P_0(a, b, c)$ with $\nabla F(a, b, c) \neq \mathbf{0}$. The plane tangent to the surface $F(x, y, z) = 0$ at P_0, called the **tangent plane**, is the plane passing through P_0 orthogonal to $\nabla F(a, b, c)$. An equation of the tangent plane is

$$F_x(a, b, c)(x - a) + F_y(a, b, c)(y - b) + F_z(a, b, c)(z - c) = 0.$$

EXAMPLE 1   **Equation of a tangent plane**   Consider the ellipsoid

$$F(x, y, z) = \frac{x^2}{9} + \frac{y^2}{25} + z^2 - 1 = 0.$$

a. Find an equation of the plane tangent to the ellipsoid at $\left(0, 4, \frac{3}{5}\right)$.

b. At what points on the ellipsoid is the tangent plane horizontal?

SOLUTION

a. Notice that we have written the equation of the ellipsoid in the implicit form $F(x, y, z) = 0$. The gradient of F is $\nabla F(x, y, z) = \left\langle \frac{2x}{9}, \frac{2y}{25}, 2z \right\rangle$. Evaluating at $\left(0, 4, \frac{3}{5}\right)$, we have

$$\nabla F\left(0, 4, \frac{3}{5}\right) = \left\langle 0, \frac{8}{25}, \frac{6}{5} \right\rangle.$$

An equation of the tangent plane at this point is

$$0 \cdot (x - 0) + \frac{8}{25}(y - 4) + \frac{6}{5}\left(z - \frac{3}{5}\right) = 0,$$

or $4y + 15z = 25$. The equation does not involve x, so the tangent plane is parallel to (does not intersect) the x-axis (Figure 15.61).

b. A horizontal plane has a normal vector of the form $\langle 0, 0, c \rangle$, where $c \neq 0$. A plane tangent to the ellipsoid has a normal vector $\nabla F(x, y, z) = \left\langle \frac{2x}{9}, \frac{2y}{25}, 2z \right\rangle$. Therefore, the ellipsoid has a horizontal tangent plane when $F_x = \frac{2x}{9} = 0$ and $F_y = \frac{2y}{25} = 0$, or

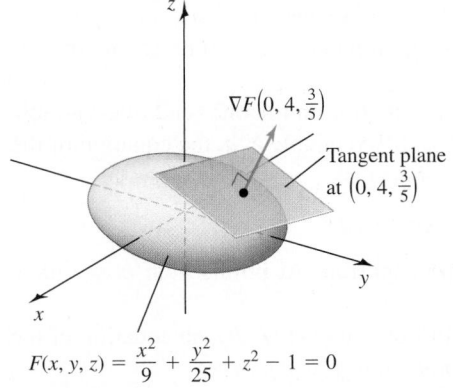

$\nabla F\left(0, 4, \frac{3}{5}\right)$

Tangent plane at $\left(0, 4, \frac{3}{5}\right)$

$F(x, y, z) = \dfrac{x^2}{9} + \dfrac{y^2}{25} + z^2 - 1 = 0$

Figure 15.61

when $x = 0$ and $y = 0$. Substituting these values into the original equation for the ellipsoid, we find that horizontal planes occur at $(0, 0, 1)$ and $(0, 0, -1)$.

Related Exercises 14, 16 ◀

▶ This result extends Theorem 15.12, which states that for functions $f(x, y) = 0$, the gradient at a point is orthogonal to the level curve that passes through that point.

The preceding discussion allows us to confirm a claim made in Section 15.5. The surface $F(x, y, z) = 0$ is a level surface of the function $w = F(x, y, z)$ (corresponding to $w = 0$). At any point on that surface, the tangent plane has a normal vector $\nabla F(x, y, z)$. Therefore, the gradient $\nabla F(x, y, z)$ is orthogonal to the level surface $F(x, y, z) = 0$ at all points of the domain at which F is differentiable.

Tangent Planes for $z = f(x, y)$ Surfaces in $\mathbb{R}^3$ are often defined explicitly in the form $z = f(x, y)$. In this situation, the equation of the tangent plane is a special case of the general equation just derived. The equation $z = f(x, y)$ is written as $F(x, y, z) = z - f(x, y) = 0$, and the gradient of F at the point $(a, b, f(a, b))$ is

▶ To be clear, when $F(x, y, z) = z - f(x, y)$, we have $F_x = -f_x$, $F_y = -f_y$, and $F_z = 1$.

$$\nabla F(a, b, f(a, b)) = \langle F_x(a, b, f(a, b)), F_y(a, b, f(a, b)), F_z(a, b, f(a, b)) \rangle$$
$$= \langle -f_x(a, b), -f_y(a, b), 1 \rangle.$$

Using the tangent plane definition, an equation of the plane tangent to the surface $z = f(x, y)$ at the point $(a, b, f(a, b))$ is

$$-f_x(a, b)(x - a) - f_y(a, b)(y - b) + 1(z - f(a, b)) = 0.$$

After some rearranging, we obtain an equation of the tangent plane.

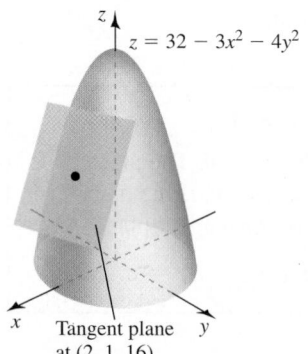

$z = 32 - 3x^2 - 4y^2$

Tangent plane at $(2, 1, 16)$

Figure 15.62

> **Tangent Plane for $z = f(x, y)$**
>
> Let f be differentiable at the point (a, b). An equation of the plane tangent to the surface $z = f(x, y)$ at the point $(a, b, f(a, b))$ is
>
> $$z = f_x(a, b)(x - a) + f_y(a, b)(y - b) + f(a, b).$$

EXAMPLE 2   **Tangent plane for $z = f(x, y)$** Find an equation of the plane tangent to the paraboloid $z = f(x, y) = 32 - 3x^2 - 4y^2$ at $(2, 1, 16)$.

SOLUTION The partial derivatives are $f_x = -6x$ and $f_y = -8y$. Evaluating the partial derivatives at $(2, 1)$, we have $f_x(2, 1) = -12$ and $f_y(2, 1) = -8$. Therefore, an equation of the tangent plane (**Figure 15.62**) is

$$z = f_x(a, b)(x - a) + f_y(a, b)(y - b) + f(a, b)$$
$$= -12(x - 2) - 8(y - 1) + 16$$
$$= -12x - 8y + 48.$$

Related Exercises 17–18 ◀

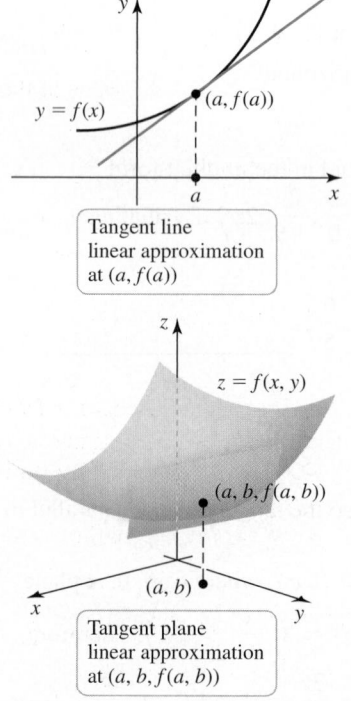

$y = f(x)$

$(a, f(a))$

Tangent line linear approximation at $(a, f(a))$

$z = f(x, y)$

$(a, b, f(a, b))$

(a, b)

Tangent plane linear approximation at $(a, b, f(a, b))$

Figure 15.63

Linear Approximation

With a function of the form $y = f(x)$, the tangent line at a point often gives good approximations to the function near that point. A straightforward extension of this idea applies to approximating functions of two variables with tangent planes. As before, the method is called *linear approximation*.

Figure 15.63 shows the details of linear approximation in the one- and two-variable cases. In the one-variable case (Section 4.6), if f is differentiable at a, the equation of the line tangent to the curve $y = f(x)$ at the point $(a, f(a))$ is

$$L(x) = f(a) + f'(a)(x - a).$$

The tangent line gives an approximation to the function. At points near a, we have $f(x) \approx L(x)$.

The two-variable case is analogous. If f is differentiable at (a, b), an equation of the plane tangent to the surface $z = f(x, y)$ at the point $(a, b, f(a, b))$ is

$$L(x, y) = f_x(a, b)(x - a) + f_y(a, b)(y - b) + f(a, b).$$

This tangent plane is the linear approximation to f at (a, b). At points near (a, b), we have $f(x, y) \approx L(x, y)$. The pattern established here continues for linear approximations in higher dimensions: For each additional variable, a new term is added to the approximation formula.

> ▶ The term *linear approximation* applies in $\mathbb{R}^2$, in $\mathbb{R}^3$, and in higher dimensions. Recall that lines in $\mathbb{R}^2$ and planes in $\mathbb{R}^3$ are described by linear functions of the independent variables. In both cases, we call the linear approximation L.

DEFINITION Linear Approximation

Let f be differentiable at (a, b). The linear approximation to the surface $z = f(x, y)$ at the point $(a, b, f(a, b))$ is the tangent plane at that point, given by the equation

$$L(x, y) = f_x(a, b)(x - a) + f_y(a, b)(y - b) + f(a, b).$$

For a function of three variables, the linear approximation to $w = f(x, y, z)$ at the point $(a, b, c, f(a, b, c))$ is given by

$$L(x, y, z) = f_x(a, b, c)(x - a) + f_y(a, b, c)(y - b)$$
$$+ f_z(a, b, c)(z - c) + f(a, b, c).$$

EXAMPLE 3 Linear approximation Let $f(x, y) = \dfrac{5}{x^2 + y^2}$.

a. Find the linear approximation to the function at the point $(-1, 2, 1)$.

b. Use the linear approximation to estimate the value of $f(-1.05, 2.1)$.

SOLUTION

a. The partial derivatives of f are

$$f_x = -\frac{10x}{(x^2 + y^2)^2} \quad \text{and} \quad f_y = -\frac{10y}{(x^2 + y^2)^2}.$$

Evaluated at $(-1, 2)$, we have $f_x(-1, 2) = \frac{2}{5} = 0.4$ and $f_y(-1, 2) = -\frac{4}{5} = -0.8$. Therefore, the linear approximation to the function at $(-1, 2, 1)$ is

$$\begin{aligned} L(x, y) &= f_x(-1, 2)(x - (-1)) + f_y(-1, 2)(y - 2) + f(-1, 2) \\ &= 0.4(x + 1) - 0.8(y - 2) + 1 \\ &= 0.4x - 0.8y + 3. \end{aligned}$$

The surface and the tangent plane are shown in **Figure 15.64**.

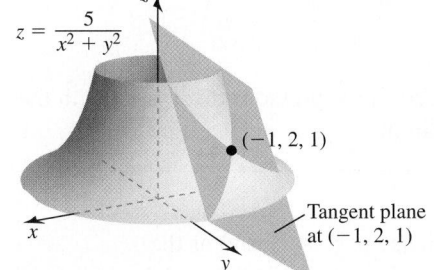

$z = \dfrac{5}{x^2 + y^2}$

$(-1, 2, 1)$

Tangent plane at $(-1, 2, 1)$

Figure 15.64

> ▶ Relative error = $\dfrac{|\text{approximation} - \text{exact value}|}{|\text{exact value}|}$

b. The value of the function at the point $(-1.05, 2.1)$ is approximated by the value of the linear approximation at that point, which is

$$L(-1.05, 2.1) = 0.4(-1.05) - 0.8(2.1) + 3 = 0.90.$$

In this case, we can easily evaluate $f(-1.05, 2.1) \approx 0.907$ and compare the linear approximation with the exact value; the approximation has a relative error of about 0.8%.

Related Exercise 36 ◀

QUICK CHECK 2 Look at the graph of the surface in Example 3 (Figure 15.64) and explain why $f_x(-1, 2) > 0$ and $f_y(-1, 2) < 0$. ◀

Differentials and Change

Recall that for a function of the form $y = f(x)$, if the independent variable changes from x to $x + dx$, the corresponding change Δy in the dependent variable is approximated by the differential $dy = f'(x)\,dx$, which is the change in the linear approximation. Therefore, $\Delta y \approx dy$, with the approximation improving as dx approaches 0.

For functions of the form $z = f(x, y)$, we start with the linear approximation to the surface

$$f(x, y) \approx L(x, y) = f_x(a, b)(x - a) + f_y(a, b)(y - b) + f(a, b).$$

The exact change in the function between the points (a, b) and (x, y) is

$$\Delta z = f(x, y) - f(a, b).$$

Replacing $f(x, y)$ with its linear approximation, the change Δz is approximated by

$$\Delta z \approx \underbrace{L(x, y) - f(a, b)}_{dz} = f_x(a, b)\underbrace{(x - a)}_{dx} + f_y(a, b)\underbrace{(y - b)}_{dy}.$$

> ▶ Alternative notation for the differential at (a, b) is $dz|_{(a, b)}$ or $df|_{(a, b)}$.

The change in the x-coordinate is $dx = x - a$ and the change in the y-coordinate is $dy = y - b$ (**Figure 15.65**). As before, we let the differential dz denote the change in the linear approximation. Therefore, the approximate change in the z-coordinate is

$$\Delta z \approx dz = \underbrace{f_x(a, b)\, dx}_{\substack{\text{change in } z \text{ due} \\ \text{to change in } x}} + \underbrace{f_y(a, b)\, dy}_{\substack{\text{change in } z \text{ due} \\ \text{to change in } y}}.$$

This expression says that if we move the independent variables from (a, b) to $(x, y) = (a + dx, b + dy)$, the corresponding change in the dependent variable Δz has two contributions—one due to the change in x and one due to the change in y. If dx and dy are small in magnitude, then so is Δz. The approximation $\Delta z \approx dz$ improves as dx and dy approach 0. The relationships among the differentials are illustrated in Figure 15.65.

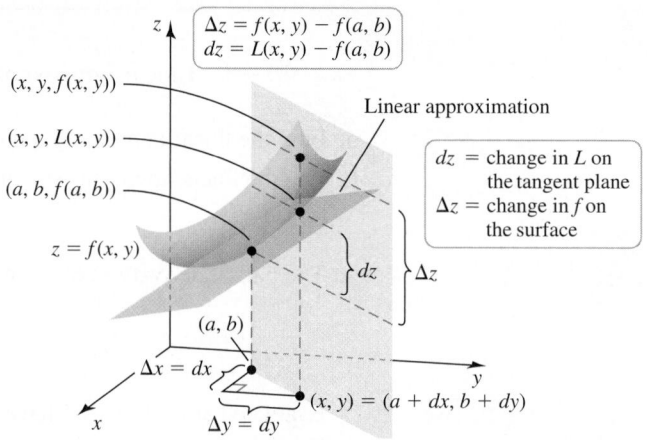

Figure 15.65

> **QUICK CHECK 3** Explain why, if $dx = 0$ or $dy = 0$ in the change formula for Δz, the result is the change formula for one variable. ◀

More generally, we replace the fixed point (a, b) in the previous discussion with the variable point (x, y) to arrive at the following definition.

DEFINITION The differential dz

Let f be differentiable at the point (x, y). The change in $z = f(x, y)$ as the independent variables change from (x, y) to $(x + dx, y + dy)$ is denoted Δz and is approximated by the differential dz:

$$\Delta z \approx dz = f_x(x, y)\, dx + f_y(x, y)\, dy.$$

EXAMPLE 4 Approximating function change Let $z = f(x, y) = \dfrac{5}{x^2 + y^2}$.

Approximate the change in z when the independent variables change from $(-1, 2)$ to $(-0.93, 1.94)$.

SOLUTION If the independent variables change from $(-1, 2)$ to $(-0.93, 1.94)$, then $dx = 0.07$ (an increase) and $dy = -0.06$ (a decrease). Using the values of the partial derivatives evaluated in Example 3, the corresponding change in z is approximately

$$\begin{aligned}
dz &= f_x(-1, 2)\, dx + f_y(-1, 2)\, dy \\
&= 0.4(0.07) + (-0.8)(-0.06) \\
&= 0.076.
\end{aligned}$$

Again, we can check the accuracy of the approximation. The actual change is $f(-0.93, 1.94) - f(-1, 2) \approx 0.080$, so the approximation has a 5% error.

Related Exercise 40 ◀

EXAMPLE 5   **Body mass index**   The body mass index (BMI) for an adult human is given by the function $B(w, h) = w/h^2$, where w is weight measured in kilograms and h is height measured in meters.

a. Use differentials to approximate the change in the BMI when weight increases from 55 to 56.5 kg and height increases from 1.65 to 1.66 m.

b. Which produces a greater *percentage* change in the BMI, a 1% change in the weight (at a constant height) or a 1% change in the height (at a constant weight)?

SOLUTION

a. The approximate change in the BMI is $dB = B_w\, dw + B_h\, dh$, where the derivatives are evaluated at $w = 55$ and $h = 1.65$, and the changes in the independent variables are $dw = 1.5$ and $dh = 0.01$. Evaluating the partial derivatives, we find that

$$B_w(w, h) = \frac{1}{h^2}, \qquad\qquad B_w(55, 1.65) \approx 0.37,$$

$$B_h(w, h) = -\frac{2w}{h^3}, \quad \text{and} \quad B_h(55, 1.65) \approx -24.49.$$

Therefore, the approximate change in the BMI is

$$\begin{aligned} dB &= B_w(55, 1.65)\, dw + B_h(55, 1.65)\, dh \\ &\approx (0.37)(1.5) + (-24.49)(0.01) \\ &\approx 0.56 - 0.25 \\ &= 0.31. \end{aligned}$$

As expected, an increase in weight *increases* the BMI, while an increase in height *decreases* the BMI. In this case, the two contributions combine for a net increase in the BMI.

b. The changes dw, dh, and dB that appear in the differential change formula in part (a) are *absolute changes*. The corresponding *relative*, or *percentage*, changes are $\dfrac{dw}{w}, \dfrac{dh}{h}$, and $\dfrac{dB}{B}$. To introduce relative changes into the change formula, we divide both sides of $dB = B_w\, dw + B_h\, dh$ by $B = w/h^2 = wh^{-2}$. The result is

$$\begin{aligned} \frac{dB}{B} &= B_w \frac{dw}{wh^{-2}} + B_h \frac{dh}{wh^{-2}} \\ &= \frac{1}{h^2}\frac{dw}{wh^{-2}} - \frac{2w}{h^3}\frac{dh}{wh^{-2}} \quad \text{Substitute for } B_w \text{ and } B_h. \\ &= \underbrace{\frac{dw}{w}}_{\substack{\text{relative} \\ \text{change} \\ \text{in } w}} - 2\underbrace{\frac{dh}{h}}_{\substack{\text{relative} \\ \text{change} \\ \text{in } h}}. \qquad \text{Simplify.} \end{aligned}$$

> ► See Exercises 68–69 for general results about relative or percentage changes in functions.

QUICK CHECK 4 In Example 5, interpret the facts that $B_w > 0$ and $B_h < 0$, for $w, h > 0$. ◄

This expression relates the relative changes in w, h, and B. With h constant ($dh = 0$), a 1% change in w ($dw/w = 0.01$) produces approximately a 1% change of the same sign in B. With w constant ($dw = 0$), a 1% change in h ($dh/h = 0.01$) produces approximately a 2% change in B of the opposite sign. We see that the BMI formula is more sensitive to small changes in h than in w.

Related Exercise 44 ◄

The differential for functions of two variables extends naturally to more variables. For example, if f is differentiable at (x, y, z) with $w = f(x, y, z)$, then

$$dw = f_x(x, y, z)\, dx + f_y(x, y, z)\, dy + f_z(x, y, z)\, dz.$$

The differential dw (or df) gives the approximate change in f at the point (x, y, z) due to changes of dx, dy, and dz in the independent variables.

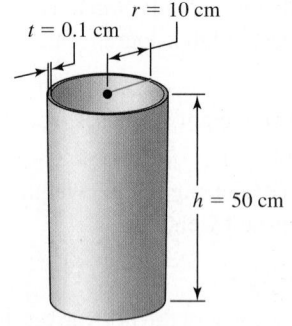

Figure 15.66

EXAMPLE 6   Manufacturing errors   A company manufactures cylindrical aluminum tubes to rigid specifications. The tubes are designed to have an outside radius of $r = 10$ cm, a height of $h = 50$ cm, and a thickness of $t = 0.1$ cm (Figure 15.66). The manufacturing process produces tubes with a maximum error of ± 0.05 cm in the radius and height, and a maximum error of ± 0.0005 cm in the thickness. The volume of the cylindrical tube is $V(r, h, t) = \pi h t (2r - t)$. Use differentials to estimate the maximum error in the volume of a tube.

SOLUTION   The approximate change in the volume of a tube due to changes dr, dh, and dt in the radius, height, and thickness, respectively, is

$$dV = V_r\, dr + V_h\, dh + V_t\, dt.$$

The partial derivatives evaluated at $r = 10$, $h = 50$, and $t = 0.1$ are

$$\begin{aligned}
V_r(r, h, t) &= 2\pi h t, & V_r(10, 50, 0.1) &= 10\pi, \\
V_h(r, h, t) &= \pi t (2r - t), & V_h(10, 50, 0.1) &= 1.99\pi, \\
V_t(r, h, t) &= 2\pi h (r - t), & V_t(10, 50, 0.1) &= 990\pi.
\end{aligned}$$

We let $dr = dh = 0.05$ and $dt = 0.0005$ be the maximum errors in the radius, height, and thickness, respectively. The maximum error in the volume is approximately

$$\begin{aligned}
dV &= V_r(10, 50, 0.1)\, dr + V_h(10, 50, 0.1)\, dh + V_t(10, 50, 0.1)\, dt \\
&= 10\pi(0.05) + 1.99\pi(0.05) + 990\pi(0.0005) \\
&\approx 1.57 + 0.31 + 1.56 \\
&= 3.44.
\end{aligned}$$

The maximum error in the volume is approximately 3.44 cm³. Notice that the "magnification factor" for the thickness (990π) is roughly 100 and 500 times greater than the magnification factors for the radius and height, respectively. This means that for the same errors in r, h, and t, the volume is far more sensitive to errors in the thickness. The partial derivatives allow us to do a sensitivity analysis to determine which independent (input) variables are most critical in producing change in the dependent (output) variable.

Related Exercise 52 ◄

SECTION 15.6 EXERCISES

Getting Started

1. Suppose **n** is a vector normal to the tangent plane of the surface $F(x, y, z) = 0$ at a point. How is **n** related to the gradient of F at that point?

2. Write the explicit function $z = xy^2 + x^2y - 10$ in the implicit form $F(x, y, z) = 0$.

3. Write an equation for the plane tangent to the surface $F(x, y, z) = 0$ at the point (a, b, c).

4. Write an equation for the plane tangent to the surface $z = f(x, y)$ at the point $(a, b, f(a, b))$.

5. Explain how to approximate a function f at a point near (a, b), where the values of f, f_x, and f_y are known at (a, b).

6. Explain how to approximate the change in a function f when the independent variables change from (a, b) to $(a + \Delta x, b + \Delta y)$.

7. Write the approximate change formula for a function $z = f(x, y)$ at the point (x, y) in terms of differentials.

8. Write the differential dw for the function $w = f(x, y, z)$.

9–10. Suppose $f(1, 2) = 4$, $f_x(1, 2) = 5$, and $f_y(1, 2) = -3$.

9. Find an equation of the plane tangent to the surface $z = f(x, y)$ at the point $P_0(1, 2, 4)$.

10. Find the linear approximation to f at $P_0(1, 2, 4)$, and use it to estimate $f(1.01, 1.99)$.

11–12. Suppose $F(0, 2, 1) = 0$, $F_x(0, 2, 1) = 3$, $F_y(0, 2, 1) = -1$, and $F_z(0, 2, 1) = 6$.

11. Find an equation of the plane tangent to the surface $F(x, y, z) = 0$ at the point $P_0(0, 2, 1)$.

12. Find the linear approximation to the function $w = F(x, y, z)$ at the point $P_0(0, 2, 1)$ and use it to estimate $F(0.1, 2, 0.99)$.

Practice Exercises

13–28. Tangent planes *Find an equation of the plane tangent to the following surfaces at the given points (two planes and two equations).*

13. $x^2 + y + z = 3$; $(1, 1, 1)$ and $(2, 0, -1)$

14. $x^2 + y^3 + z^4 = 2$; $(1, 0, 1)$ and $(-1, 0, 1)$

15. $xy + xz + yz - 12 = 0$; $(2, 2, 2)$ and $(2, 0, 6)$

16. $x^2 + y^2 - z^2 = 0$; $(3, 4, 5)$ and $(-4, -3, 5)$

17. $z = 4 - 2x^2 - y^2$; $(2, 2, -8)$ and $(-1, -1, 1)$

18. $z = 2 + 2x^2 + \dfrac{y^2}{2}$; $\left(-\dfrac{1}{2}, 1, 3\right)$ and $(3, -2, 22)$

19. $z = e^{xy}$; $(1, 0, 1)$ and $(0, 1, 1)$

20. $z = \sin xy + 2$; $(1, 0, 2)$ and $(0, 5, 2)$

21. $xy \sin z = 1$; $(1, 2, \pi/6)$ and $(-2, -1, 5\pi/6)$

22. $yze^{xz} - 8 = 0$; $(0, 2, 4)$ and $(0, -8, -1)$

23. $z^2 - x^2/16 - y^2/9 - 1 = 0$; $(4, 3, -\sqrt{3})$ and $(-8, 9, \sqrt{14})$

24. $2x + y^2 - z^2 = 0$; $(0, 1, 1)$ and $(4, 1, -3)$

25. $z = x^2 e^{x-y}$; $(2, 2, 4)$ and $(-1, -1, 1)$

26. $z = \ln(1 + xy)$; $(1, 2, \ln 3)$ and $(-2, -1, \ln 3)$

27. $z = \dfrac{x - y}{x^2 + y^2}$; $\left(1, 2, -\dfrac{1}{5}\right)$ and $\left(2, -1, \dfrac{3}{5}\right)$

28. $z = 2\cos(x - y) + 2$; $\left(\dfrac{\pi}{6}, -\dfrac{\pi}{6}, 3\right)$ and $\left(\dfrac{\pi}{3}, \dfrac{\pi}{3}, 4\right)$

29–32. Tangent planes *Find an equation of the plane tangent to the following surfaces at the given point.*

29. $z = \tan^{-1}(xy)$; $\left(1, 1, \dfrac{\pi}{4}\right)$ **30.** $z = \tan^{-1}(x + y)$; $(0, 0, 0)$

31. $\sin xyz = \dfrac{1}{2}$; $\left(\pi, 1, \dfrac{1}{6}\right)$ **32.** $\dfrac{x + z}{y - z} = 2$; $(4, 2, 0)$

33–38. Linear approximation

a. *Find the linear approximation to the function f at the given point.*
b. *Use part (a) to estimate the given function value.*

33. $f(x, y) = xy + x - y$; $(2, 3)$; estimate $f(2.1, 2.99)$.

34. $f(x, y) = 12 - 4x^2 - 8y^2$; $(-1, 4)$; estimate $f(-1.05, 3.95)$.

35. $f(x, y) = -x^2 + 2y^2$; $(3, -1)$; estimate $f(3.1, -1.04)$.

36. $f(x, y) = \sqrt{x^2 + y^2}$; $(3, -4)$; estimate $f(3.06, -3.92)$.

37. $f(x, y, z) = \ln(1 + x + y + 2z)$; $(0, 0, 0)$; estimate $f(0.1, -0.2, 0.2)$.

38. $f(x, y, z) = \dfrac{x + y}{x - z}$; $(3, 2, 4)$; estimate $f(2.95, 2.05, 4.02)$.

39–42. Approximate function change *Use differentials to approximate the change in z for the given changes in the independent variables.*

39. $z = 2x - 3y - 2xy$ when (x, y) changes from $(1, 4)$ to $(1.1, 3.9)$

40. $z = -x^2 + 3y^2 + 2$ when (x, y) changes from $(-1, 2)$ to $(-1.05, 1.9)$

41. $z = e^{x+y}$ when (x, y) changes from $(0, 0)$ to $(0.1, -0.05)$

42. $z = \ln(1 + x + y)$ when (x, y) changes from $(0, 0)$ to $(-0.1, 0.03)$

43. Changes in torus surface area The surface area of a torus with an inner radius r and an outer radius $R > r$ is $S = 4\pi^2(R^2 - r^2)$.

   a. If r increases and R decreases, does S increase or decrease, or is it impossible to say?

   b. If r increases and R increases, does S increase or decrease, or is it impossible to say?

   c. Estimate the change in the surface area of the torus when r changes from $r = 3.00$ to $r = 3.05$ and R changes from $R = 5.50$ to $R = 5.65$.

   d. Estimate the change in the surface area of the torus when r changes from $r = 3.00$ to $r = 2.95$ and R changes from $R = 7.00$ to $R = 7.04$.

   e. Find the relationship between the changes in r and R that leaves the surface area (approximately) unchanged.

44. Changes in cone volume The volume of a right circular cone with radius r and height h is $V = \pi r^2 h/3$.

   a. Approximate the change in the volume of the cone when the radius changes from $r = 6.5$ to $r = 6.6$ and the height changes from $h = 4.20$ to $h = 4.15$.

   b. Approximate the change in the volume of the cone when the radius changes from $r = 5.40$ to $r = 5.37$ and the height changes from $h = 12.0$ to $h = 11.96$.

45. Area of an ellipse The area of an ellipse with axes of length $2a$ and $2b$ is $A = \pi ab$. Approximate the percent change in the area when a increases by 2% and b increases by 1.5%.

46. Volume of a paraboloid The volume of a segment of a circular paraboloid (see figure) with radius r and height h is $V = \pi r^2 h/2$. Approximate the percent change in the volume when the radius decreases by 1.5% and the height increases by 2.2%.

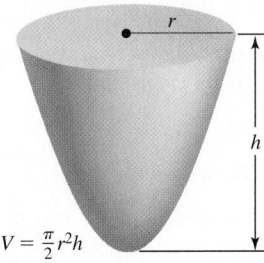

$$V = \frac{\pi}{2} r^2 h$$

47–50. Differentials with more than two variables *Write the differential dw in terms of the differentials of the independent variables.*

47. $w = f(x, y, z) = xy^2 + x^2 z + yz^2$

48. $w = f(x, y, z) = \sin(x + y - z)$

49. $w = f(u, x, y, z) = \dfrac{u + x}{y + z}$

50. $w = f(p, q, r, s) = \dfrac{pq}{rs}$

T 51. Law of Cosines The side lengths of any triangle are related by the Law of Cosines,
$$c^2 = a^2 + b^2 - 2ab\cos\theta.$$

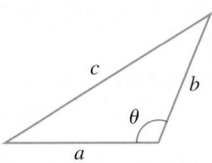

   a. Estimate the change in the side length c when a changes from $a = 2$ to $a = 2.03$, b changes from $b = 4.00$ to $b = 3.96$, and θ changes from $\theta = \dfrac{\pi}{3}$ to $\theta = \dfrac{\pi}{3} + \dfrac{\pi}{90}$.

   b. If a changes from $a = 2$ to $a = 2.03$ and b changes from $b = 4.00$ to $b = 3.96$, is the resulting change in c greater in magnitude when $\theta = \dfrac{\pi}{20}$ (small angle) or when $\theta = \dfrac{9\pi}{20}$ (close to a right angle)?

52. Travel cost The cost of a trip that is L miles long, driving a car that gets m miles per gallon, with gas costs of $\$p/\text{gal}$ is $C = Lp/m$ dollars. Suppose you plan a trip of $L = 1500$ mi in a car that gets $m = 32$ mi/gal, with gas costs of $p = \$3.80/\text{gal}$.

a. Explain how the cost function is derived.
b. Compute the partial derivatives C_L, C_m, and C_p. Explain the meaning of the signs of the derivatives in the context of this problem.
c. Estimate the change in the total cost of the trip if L changes from $L = 1500$ to $L = 1520$, m changes from $m = 32$ to $m = 31$, and p changes from $p = \$3.80$ to $p = \$3.85$.
d. Is the total cost of the trip (with $L = 1500$ mi, $m = 32$ mi/gal, and $p = \$3.80$) more sensitive to a 1% change in L, in m, or in p (assuming the other two variables are fixed)? Explain.

53. Explain why or why not Determine whether the following statements are true and give an explanation or counterexample.

a. The planes tangent to the cylinder $x^2 + y^2 = 1$ in $\mathbb{R}^3$ all have the form $ax + bz + c = 0$.
b. Suppose $w = xy/z$, for $x > 0$, $y > 0$, and $z > 0$. A decrease in z with x and y fixed results in an increase in w.
c. The gradient $\nabla F(a, b, c)$ lies in the plane tangent to the surface $F(x, y, z) = 0$ at (a, b, c).

54–57. Horizontal tangent planes *Find the points at which the following surfaces have horizontal tangent planes.*

54. $x^2 + 2y^2 + z^2 - 2x - 2z - 2 = 0$

55. $x^2 + y^2 - z^2 - 2x + 2y + 3 = 0$

56. $z = \sin(x - y)$ in the region $-2\pi \leq x \leq 2\pi, -2\pi \leq y \leq 2\pi$

57. $z = \cos 2x \sin y$ in the region $-\pi \leq x \leq \pi, -\pi \leq y \leq \pi$

58. Heron's formula The area of a triangle with sides of length a, b, and c is given by a formula from antiquity called Heron's formula:

$$A = \sqrt{s(s - a)(s - b)(s - c)},$$

where $s = \dfrac{1}{2}(a + b + c)$ is the *semiperimeter* of the triangle.

a. Find the partial derivatives A_a, A_b, and A_c.
b. A triangle has sides of length $a = 2$, $b = 4$, $c = 5$. Estimate the change in the area when a increases by 0.03, b decreases by 0.08, and c increases by 0.6.
c. For an equilateral triangle with $a = b = c$, estimate the percent change in the area when all sides increase in length by $p\%$.

59. Surface area of a cone A cone with height h and radius r has a lateral surface area (the curved surface only, excluding the base) of $S = \pi r\sqrt{r^2 + h^2}$.

a. Estimate the change in the surface area when r increases from $r = 2.50$ to $r = 2.55$ and h decreases from $h = 0.60$ to $h = 0.58$.
b. When $r = 100$ and $h = 200$, is the surface area more sensitive to a small change in r or a small change in h? Explain.

60. Line tangent to an intersection curve Consider the paraboloid $z = x^2 + 3y^2$ and the plane $z = x + y + 4$, which intersects the paraboloid in a curve C at $(2, 1, 7)$ (see figure). Find the equation of the line tangent to C at the point $(2, 1, 7)$. Proceed as follows.

a. Find a vector normal to the plane at $(2, 1, 7)$.
b. Find a vector normal to the plane tangent to the paraboloid at $(2, 1, 7)$.
c. Argue that the line tangent to C at $(2, 1, 7)$ is orthogonal to both normal vectors found in parts (a) and (b). Use this fact to find a direction vector for the tangent line.

d. Knowing a point on the tangent line and the direction of the tangent line, write an equation of the tangent line in parametric form.

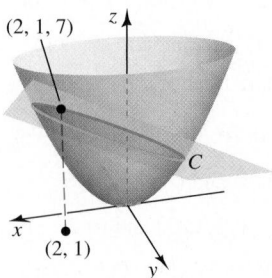

61. Batting averages Batting averages in baseball are defined by $A = x/y$, where $x \geq 0$ is the total number of hits and $y > 0$ is the total number of at-bats. Treat x and y as positive real numbers and note that $0 \leq A \leq 1$.

a. Use differentials to estimate the change in the batting average if the number of hits increases from 60 to 62 and the number of at-bats increases from 175 to 180.
b. If a batter currently has a batting average of $A = 0.350$, does the average decrease more if the batter fails to get a hit than it increases if the batter gets a hit?
c. Does the answer to part (b) depend on the current batting average? Explain.

62. Water-level changes A conical tank with radius 0.50 m and height 2.00 m is filled with water (see figure). Water is released from the tank, and the water level drops by 0.05 m (from 2.00 m to 1.95 m). Approximate the change in the volume of water in the tank. (*Hint:* When the water level drops, both the radius and height of the cone of water change.)

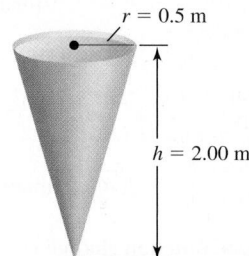

63. Flow in a cylinder Poiseuille's Law is a fundamental law of fluid dynamics that describes the flow velocity of a viscous incompressible fluid in a cylinder (it is used to model blood flow through veins and arteries). It says that in a cylinder of radius R and length L, the velocity of the fluid $r \leq R$ units from the centerline of the cylinder is $V = \dfrac{P}{4\,Lv}(R^2 - r^2)$, where P is the difference in the pressure between the ends of the cylinder, and v is the viscosity of the fluid (see figure). Assuming P and v are constant, the velocity V along the centerline of the cylinder $(r = 0)$ is $V = \dfrac{kR^2}{L}$, where k is a constant that we will take to be $k = 1$.

a. Estimate the change in the centerline velocity $(r = 0)$ if the radius of the flow cylinder increases from $R = 3$ cm to $R = 3.05$ cm and the length increases from $L = 50$ cm to $L = 50.5$ cm.
b. Estimate the percent change in the centerline velocity if the radius of the flow cylinder R decreases by 1% and its length L increases by 2%.

c. Complete the following sentence: If the radius of the cylinder increases by $p\%$, then the length of the cylinder must increase by approximately ___% in order for the velocity to remain constant.

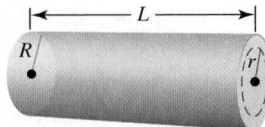

Explorations and Challenges

64. Floating-point operations In general, real numbers (with infinite decimal expansions) cannot be represented exactly in a computer by floating-point numbers (with finite decimal expansions). Suppose floating-point numbers on a particular computer carry an error of at most 10^{-16}. Estimate the maximum error that is committed in evaluating the following functions. Express the error in absolute and relative (percent) terms.

a. $f(x, y) = xy$
b. $f(x, y) = \dfrac{x}{y}$

c. $F(x, y, z) = xyz$
d. $F(x, y, z) = \dfrac{x/y}{z}$

65. Probability of at least one encounter Suppose in a large group of people, a fraction $0 \le r \le 1$ of the people have flu. The probability that in n random encounters you will meet at least one person with flu is $P = f(n, r) = 1 - (1 - r)^n$. Although n is a positive integer, regard it as a positive real number.

a. Compute f_r and f_n.
b. How sensitive is the probability P to the flu rate r? Suppose you meet $n = 20$ people. Approximately how much does the probability P increase if the flu rate increases from $r = 0.1$ to $r = 0.11$ (with n fixed)?
c. Approximately how much does the probability P increase if the flu rate increases from $r = 0.9$ to $r = 0.91$ with $n = 20$?
d. Interpret the results of parts (b) and (c).

66. Two electrical resistors When two electrical resistors with resistance $R_1 > 0$ and $R_2 > 0$ are wired in parallel in a circuit (see figure), the combined resistance R, measured in ohms (Ω), is given by $\dfrac{1}{R} = \dfrac{1}{R_1} + \dfrac{1}{R_2}$.

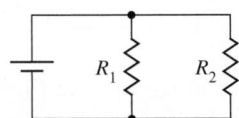

a. Estimate the change in R if R_1 increases from 2 Ω to 2.05 Ω and R_2 decreases from 3 Ω to 2.95 Ω.
b. Is it true that if $R_1 = R_2$ and R_1 increases by the same small amount as R_2 decreases, then R is approximately unchanged? Explain.
c. Is it true that if R_1 and R_2 increase, then R increases? Explain.
d. Suppose $R_1 > R_2$ and R_1 increases by the same small amount as R_2 decreases. Does R increase or decrease?

67. Three electrical resistors Extending Exercise 66, when three electrical resistors with resistances $R_1 > 0$, $R_2 > 0$, and $R_3 > 0$ are wired in parallel in a circuit (see figure), the combined resistance R, measured in ohms (Ω), is given by $\dfrac{1}{R} = \dfrac{1}{R_1} + \dfrac{1}{R_2} + \dfrac{1}{R_3}$.

Estimate the change in R if R_1 increases from 2 Ω to 2.05 Ω, R_2 decreases from 3 Ω to 2.95 Ω, and R_3 increases from 1.5 Ω to 1.55 Ω.

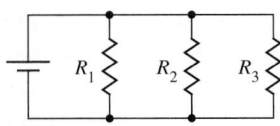

68. Power functions and percent change Suppose $z = f(x, y) = x^a y^b$, where a and b are real numbers. Let $\dfrac{dx}{x}, \dfrac{dy}{y}$, and $\dfrac{dz}{z}$ be the approximate relative (percent) changes in x, y, and z, respectively. Show that $\dfrac{dz}{z} = \dfrac{a(dx)}{x} + \dfrac{b(dy)}{y}$; that is, the relative changes are additive when weighted by the exponents a and b.

69. Logarithmic differentials Let f be a differentiable function of one or more variables that is positive on its domain.

a. Show that $d(\ln f) = \dfrac{df}{f}$.
b. Use part (a) to explain the statement that the absolute change in $\ln f$ is approximately equal to the relative change in f.
c. Let $f(x, y) = xy$, note that $\ln f = \ln x + \ln y$, and show that relative changes add; that is, $\dfrac{df}{f} = \dfrac{dx}{x} + \dfrac{dy}{y}$.
d. Let $f(x, y) = \dfrac{x}{y}$, note that $\ln f = \ln x - \ln y$, and show that relative changes subtract; that is, $\dfrac{df}{f} = \dfrac{dx}{x} - \dfrac{dy}{y}$.
e. Show that in a product of n numbers, $f = x_1 x_2 \cdots x_n$, the relative change in f is approximately equal to the sum of the relative changes in the variables.

70. Distance from a plane to an ellipsoid (Adapted from 1938 Putnam Exam) Consider the ellipsoid $\dfrac{x^2}{a^2} + \dfrac{y^2}{b^2} + \dfrac{z^2}{c^2} = 1$ and the plane P given by $Ax + By + Cz + 1 = 0$. Let $h = (A^2 + B^2 + C^2)^{-1/2}$ and $m = (a^2 A^2 + b^2 B^2 + c^2 C^2)^{1/2}$.

a. Find the equation of the plane tangent to the ellipsoid at the point (p, q, r).
b. Find the two points on the ellipsoid at which the tangent plane is parallel to P, and find equations of the tangent planes.
c. Show that the distance between the origin and the plane P is h.
d. Show that the distance between the origin and the tangent planes is hm.
e. Find a condition that guarantees the plane P does not intersect the ellipsoid.

QUICK CHECK ANSWERS

1. $F(x, y, z) = z - xy - x + y = 0$ **2.** If you walk in the positive x-direction from $(-1, 2, 1)$, then you walk uphill. If you walk in the positive y-direction from $(-1, 2, 1)$, then you walk downhill.   **3.** If $\Delta x = 0$, then the change formula becomes $\Delta z \approx f_y(a, b)\,\Delta y$, which is the change formula for the single variable y. If $\Delta y = 0$, then the change formula becomes $\Delta z \approx f_x(a, b)\,\Delta x$, which is the change formula for the single variable x.   **4.** The BMI increases with weight w and decreases with height h. ◄

15.7 Maximum/Minimum Problems

In Chapter 4, we showed how to use derivatives to find maximum and minimum values of functions of a single variable. When those techniques are extended to functions of two variables, we discover both similarities and differences. The landscape of a surface is far more complicated than the profile of a curve in the plane, so we see more interesting features when working with several variables. In addition to peaks (maximum values) and hollows (minimum values), we encounter winding ridges, long valleys, and mountain passes. Yet despite these complications, many of the ideas used for single-variable functions reappear in higher dimensions. For example, the Second Derivative Test, suitably adapted for two variables, plays a central role. As with single-variable functions, the techniques developed here are useful for solving practical optimization problems.

Local Maximum/Minimum Values

The concepts of local maximum and minimum values encountered in Chapter 4 extend readily to functions of two variables of the form $z = f(x, y)$. Figure 15.67 shows a general surface defined on a domain D, which is a subset of $\mathbb{R}^2$. The surface has peaks (local high points) and hollows (local low points) at points in the interior of D. The goal is to locate and classify these extreme points.

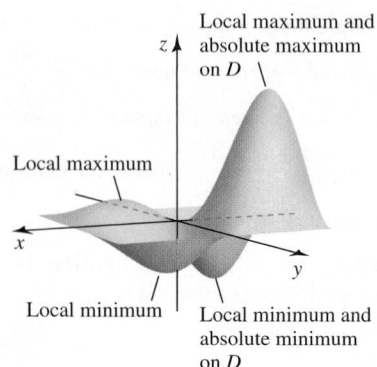

Local maximum and absolute maximum on D

Local maximum

Local minimum

Local minimum and absolute minimum on D

Figure 15.67

> We maintain the convention adopted in Chapter 4 that local maxima or minima occur at interior points of the domain. Recall that an open disk centered at (a, b) is the set of points within a circle centered at (a, b).

DEFINITION Local Maximum/Minimum Values

Suppose (a, b) is a point in a region R on which f is defined. If $f(x, y) \leq f(a, b)$ for all (x, y) in the domain of f and in some open disk centered at (a, b), then $f(a, b)$ is a **local maximum value** of f. If $f(x, y) \geq f(a, b)$ for all (x, y) in the domain of f and in some open disk centered at (a, b), then $f(a, b)$ is a **local minimum value** of f. Local maximum and local minimum values are also called **local extreme values** or **local extrema**.

In familiar terms, a local maximum is a point on a surface from which you cannot walk uphill. A local minimum is a point from which you cannot walk downhill. The following theorem is the analog of Theorem 4.2.

THEOREM 15.14 Derivatives and Local Maximum/Minimum Values

If f has a local maximum or minimum value at (a, b) and the partial derivatives f_x and f_y exist at (a, b), then $f_x(a, b) = f_y(a, b) = 0$.

Proof: Suppose f has a local maximum value at (a, b). The function of one variable $g(x) = f(x, b)$, obtained by holding $y = b$ fixed, also has a local maximum at (a, b). By Theorem 4.2, $g'(a) = 0$. However, $g'(a) = f_x(a, b)$; therefore, $f_x(a, b) = 0$. Similarly, the function $h(y) = f(a, y)$, obtained by holding $x = a$ fixed, has a local maximum at (a, b), which implies that $f_y(a, b) = h'(b) = 0$. An analogous argument is used for the local minimum case.   ◄

QUICK CHECK 1 The paraboloid $z = x^2 + y^2 - 4x + 2y + 5$ has a local minimum at $(2, -1)$. Verify the conclusion of Theorem 15.14 for this function. ◄

Suppose f is differentiable at (a, b) (ensuring the existence of a tangent plane) and f has a local extremum at (a, b). Then $f_x(a, b) = f_y(a, b) = 0$, which, when substituted into the equation of the tangent plane, gives the equation $z = f(a, b)$ (a constant). Therefore, if the tangent plane exists at a local extremum, then it is horizontal there.

Recall that for a function of one variable, the condition $f'(a) = 0$ does not guarantee a local extremum at a. A similar precaution must be taken with Theorem 15.14. The conditions $f_x(a, b) = f_y(a, b) = 0$ do not imply that f has a local extremum at (a, b), as we show momentarily. Theorem 15.14 provides *candidates* for local extrema. We call these candidates *critical points*, as we did for functions of one variable. Therefore, the procedure for locating local maximum and minimum values is to find the critical points and then determine whether these candidates correspond to genuine local maximum and minimum values.

DEFINITION Critical Point

An interior point (a, b) in the domain of f is a **critical point** of f if either

1. $f_x(a, b) = f_y(a, b) = 0$, or
2. at least one of the partial derivatives f_x and f_y does not exist at (a, b).

EXAMPLE 1   Finding critical points Find the critical points of $f(x, y) = xy(x - 2)(y + 3)$.

SOLUTION This function is differentiable at all points of $\mathbb{R}^2$, so the critical points occur only at points where $f_x(x, y) = f_y(x, y) = 0$. Computing and simplifying the partial derivatives, these conditions become

$$f_x(x, y) = 2y(x - 1)(y + 3) = 0$$
$$f_y(x, y) = x(x - 2)(2y + 3) = 0.$$

We must now identify all (x, y) pairs that satisfy both equations. The first equation is satisfied if and only if $y = 0, x = 1$, or $y = -3$. We consider each of these cases.

- Substituting $y = 0$, the second equation is $3x(x - 2) = 0$, which has solutions $x = 0$ and $x = 2$. So $(0, 0)$ and $(2, 0)$ are critical points.
- Substituting $x = 1$, the second equation is $-(2y + 3) = 0$, which has the solution $y = -\frac{3}{2}$. So $\left(1, -\frac{3}{2}\right)$ is a critical point.
- Substituting $y = -3$, the second equation is $-3x(x - 2) = 0$, which has roots $x = 0$ and $x = 2$. So $(0, -3)$ and $(2, -3)$ are critical points.

We find that there are five critical points: $(0, 0)$, $(2, 0)$, $\left(1, -\frac{3}{2}\right)$, $(0, -3)$, and $(2, -3)$. Some of these critical points may correspond to local maximum or minimum values. We will return to this example and a complete analysis shortly.

Related Exercises 15, 18 ◄

Second Derivative Test

Critical points are candidates for local extreme values. With functions of one variable, the Second Derivative Test is used to determine whether critical points correspond to local maxima or minima (the test can also be inconclusive). The analogous test for functions of two variables not only detects local maxima and minima, but also identifies another type of point known as a *saddle point*.

DEFINITION Saddle Point

Consider a function f that is differentiable at a critical point (a, b). Then f has a **saddle point** at (a, b) if, in every open disk centered at (a, b), there are points (x, y) for which $f(x, y) > f(a, b)$ and points for which $f(x, y) < f(a, b)$.

If (a, b) is a critical point of f and f has a saddle point at (a, b), then from the point $(a, b, f(a, b))$, it is possible to walk uphill in some directions and downhill in other directions. The function $f(x, y) = x^2 - y^2$ (a hyperbolic paraboloid) is a good example to remember. The surface *rises* from the critical point $(0, 0)$ along the x-axis and *falls* from $(0, 0)$ along the y-axis (**Figure 15.68**). We can easily check that $f_x(0, 0) = f_y(0, 0) = 0$, demonstrating that critical points do not necessarily correspond to local maxima or minima.

QUICK CHECK 2 Consider the plane tangent to a surface at a saddle point. In what direction does the normal to the plane point? ◄

> ► The usual image of a saddle point is that of a mountain pass (or a horse saddle), where you can walk upward in some directions and downward in other directions. The definition of a saddle point given here includes other less common situations. For example, with this definition, the cylinder $z = x^3$ has a line of saddle points along the y-axis.

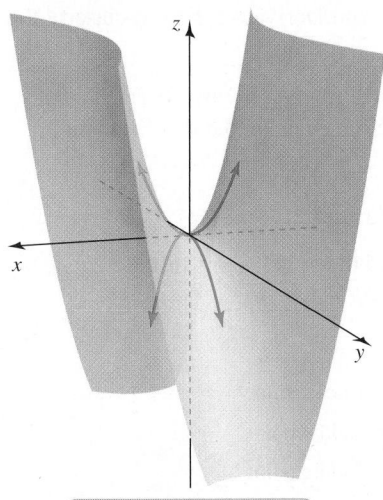

The hyperbolic paraboloid $z = x^2 - y^2$ has a saddle point at $(0, 0)$.

Figure 15.68

> ▶ The Second Derivative Test for functions of a single variable states that if a is a critical point with $f'(a) = 0$, then $f''(a) > 0$ implies that f has a local minimum at a and $f''(a) < 0$ implies that f has a local maximum at a; if $f''(a) = 0$, the test is inconclusive. Theorem 15.15 is easier to remember if you notice the parallels between the two second derivative tests.

THEOREM 15.15   Second Derivative Test

Suppose the second partial derivatives of f are continuous throughout an open disk centered at the point (a, b), where $f_x(a, b) = f_y(a, b) = 0$. Let $D(x, y) = f_{xx}(x, y)\, f_{yy}(x, y) - (f_{xy}(x, y))^2$.

1. If $D(a, b) > 0$ and $f_{xx}(a, b) < 0$, then f has a local maximum value at (a, b).
2. If $D(a, b) > 0$ and $f_{xx}(a, b) > 0$, then f has a local minimum value at (a, b).
3. If $D(a, b) < 0$, then f has a saddle point at (a, b).
4. If $D(a, b) = 0$, then the test is inconclusive.

The proof of this theorem is given in Appendix A, but a few comments are in order. The test relies on the quantity $D(x, y) = f_{xx} f_{yy} - (f_{xy})^2$, which is called the **discriminant** of f. It can be remembered as the 2×2 determinant of the **Hessian** matrix $\begin{pmatrix} f_{xx} & f_{xy} \\ f_{yx} & f_{yy} \end{pmatrix}$, where $f_{xy} = f_{yx}$, provided these derivatives are continuous (Theorem 15.4). The condition $D(x, y) > 0$ means that the surface has the same general behavior in all directions near (a, b); either the surface rises in all directions or it falls in all directions. In the case that $D(a, b) = 0$, the test is inconclusive: (a, b) could correspond to a local maximum, a local minimum, or a saddle point.

Finally, another useful characterization of a saddle point can be derived from Theorem 15.15: The tangent plane at a saddle point lies both above and below the surface.

QUICK CHECK 3 Compute the discriminant $D(x, y)$ of $f(x, y) = x^2 y^2$. ◄

EXAMPLE 2   Analyzing critical points Use the Second Derivative Test to classify the critical points of $f(x, y) = x^2 + 2y^2 - 4x + 4y + 6$.

SOLUTION We begin with the following derivative calculations:

$$f_x = 2x - 4, \qquad f_y = 4y + 4,$$
$$f_{xx} = 2, \qquad f_{xy} = f_{yx} = 0, \quad \text{and} \quad f_{yy} = 4.$$

Setting both f_x and f_y equal to zero yields the single critical point $(2, -1)$. The value of the discriminant at the critical point is $D(2, -1) = f_{xx} f_{yy} - (f_{xy})^2 = 8 > 0$. Furthermore, $f_{xx}(2, -1) = 2 > 0$. By the Second Derivative Test, f has a local minimum at $(2, -1)$; the value of the function at that point is $f(2, -1) = 0$ (Figure 15.69).

Related Exercise 24 ◄

EXAMPLE 3   Analyzing critical points Use the Second Derivative Test to classify the critical points of $f(x, y) = xy(x - 2)(y + 3)$.

SOLUTION In Example 1, we determined that the critical points of f are $(0, 0)$, $(2, 0)$, $\left(1, -\frac{3}{2}\right)$, $(0, -3)$, and $(2, -3)$. The derivatives needed to evaluate the discriminant are

$$f_x = 2y(x - 1)(y + 3), \qquad f_y = x(x - 2)(2y + 3),$$
$$f_{xx} = 2y(y + 3), \qquad f_{xy} = 2(2y + 3)(x - 1), \quad \text{and} \quad f_{yy} = 2x(x - 2).$$

The values of the discriminant at the critical points and the conclusions of the Second Derivative Test are shown in Table 15.4.

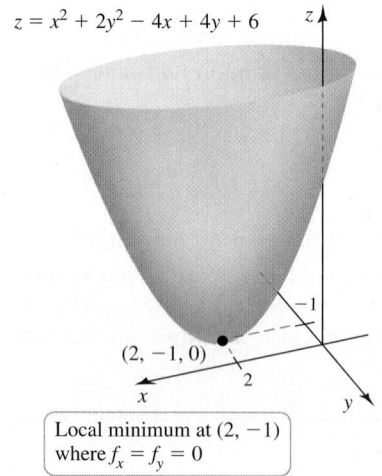

$z = x^2 + 2y^2 - 4x + 4y + 6$

$(2, -1, 0)$

Local minimum at $(2, -1)$ where $f_x = f_y = 0$

Figure 15.69

Table 15.4

(x, y)	$D(x, y)$	f_{xx}	Conclusion
$(0, 0)$	-36	0	Saddle point
$(2, 0)$	-36	0	Saddle point
$\left(1, -\frac{3}{2}\right)$	9	$-\frac{9}{2}$	Local maximum
$(0, -3)$	-36	0	Saddle point
$(2, -3)$	-36	0	Saddle point

The surface described by f has one local maximum at $\left(1, -\frac{3}{2}\right)$, surrounded by four saddle points (**Figure 15.70a**). The structure of the surface may also be visualized by plotting the level curves of f (**Figure 15.70b**).

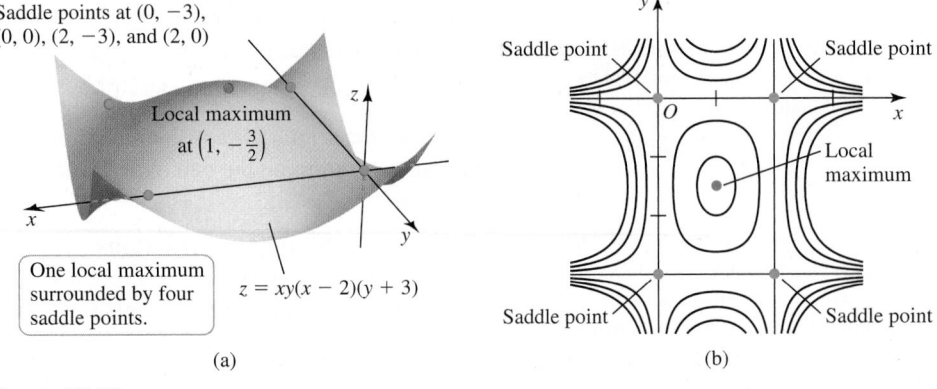

Saddle points at $(0, -3)$, $(0, 0)$, $(2, -3)$, and $(2, 0)$

Local maximum at $\left(1, -\frac{3}{2}\right)$

One local maximum surrounded by four saddle points.

$z = xy(x - 2)(y + 3)$

(a)

Saddle point · Saddle point · Local maximum · Saddle point · Saddle point

(b)

Figure 15.70

Related Exercise 27 ◄

EXAMPLE 4 Inconclusive tests Apply the Second Derivative Test to the following functions and interpret the results.

a. $f(x, y) = 2x^4 + y^4$ **b.** $f(x, y) = 2 - xy^2$

SOLUTION

a. The critical points of f satisfy the conditions

$$f_x = 8x^3 = 0 \quad \text{and} \quad f_y = 4y^3 = 0,$$

so the sole critical point is $(0, 0)$. The second partial derivatives evaluated at $(0, 0)$ are

$$f_{xx}(0, 0) = f_{xy}(0, 0) = f_{yy}(0, 0) = 0.$$

We see that $D(0, 0) = 0$, and the Second Derivative Test is inconclusive. While the bowl-shaped surface (**Figure 15.71**) described by f has a local minimum at $(0, 0)$, the surface also has a broad flat bottom, which makes the local minimum "invisible" to the Second Derivative Test.

b. The critical points of this function satisfy

$$f_x(x, y) = -y^2 = 0 \quad \text{and} \quad f_y(x, y) = -2xy = 0.$$

The solutions of these equations have the form $(a, 0)$, where a is a real number. It is easy to check that the second partial derivatives evaluated at $(a, 0)$ are

$$f_{xx}(a, 0) = f_{xy}(a, 0) = 0 \quad \text{and} \quad f_{yy}(a, 0) = -2a.$$

Therefore, the discriminant is $D(a, 0) = 0$, and the Second Derivative Test is inconclusive. **Figure 15.72** shows that f has a flat ridge above the x-axis that the Second Derivative Test is unable to classify.

Related Exercises 29–30 ◄

Absolute Maximum and Minimum Values

As in the one-variable case, we are often interested in knowing where a function of two or more variables attains its extreme values over its domain (or a subset of its domain).

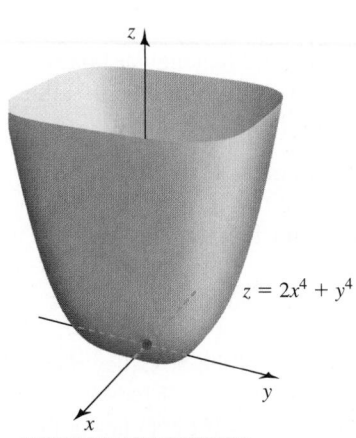

$z = 2x^4 + y^4$

Local minimum at $(0, 0)$, but the Second Derivative Test is inconclusive.

Figure 15.71

▶ The same "flat" behavior occurs with functions of one variable, such as $f(x) = x^4$. Although f has a local minimum at $x = 0$, the Second Derivative Test is inconclusive.

▶ It is not surprising that the Second Derivative Test is inconclusive in Example 4b. The function has a line of local maxima at $(a, 0)$ for $a > 0$, a line of local minima at $(a, 0)$ for $a < 0$, and a saddle point at $(0, 0)$.

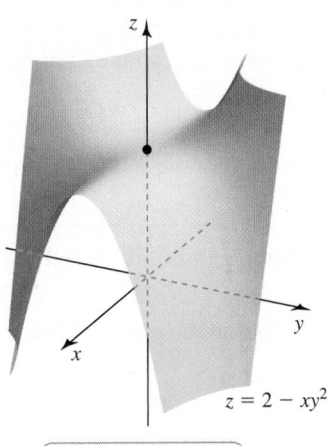

$z = 2 - xy^2$

Second derivative test fails to detect saddle point at $(0, 0)$.

Figure 15.72

DEFINITION Absolute Maximum/Minimum Values

Let f be defined on a set R in $\mathbb{R}^2$ containing the point (a, b). If $f(a, b) \geq f(x, y)$ for every (x, y) in R, then $f(a, b)$ is an **absolute maximum value** of f on R. If $f(a, b) \leq f(x, y)$ for every (x, y) in R, then $f(a, b)$ is an **absolute minimum value** of f on R.

It should be noted that the Extreme Value Theorem of Chapter 4 has an analog in $\mathbb{R}^2$ (or in higher dimensions): A function that is continuous on a closed bounded set in $\mathbb{R}^2$ attains its absolute maximum and absolute minimum values on that set. Absolute maximum and minimum values on a closed bounded set R occur in two ways.

▶ Recall that a *closed set* in $\mathbb{R}^2$ is a set that includes its boundary. A *bounded set* in $\mathbb{R}^2$ is a set that may be enclosed by a circle of finite radius.

• They may be local maximum or minimum values at interior points of R, where they are associated with critical points.

• They may occur on the boundary of R.

Therefore, the search for absolute maximum and minimum values on a closed bounded set amounts to examining the behavior of the function on the boundary of R and at the interior points of R.

▶ Example 5 is a *constrained optimization problem*, in which the goal is to maximize the volume subject to an additional condition called a *constraint*. We return to such problems in the next section and present another method of solution.

EXAMPLE 5 Shipping regulations A shipping company handles rectangular boxes provided the sum of the length, width, and height of the box does not exceed 96 in. Find the dimensions of the box that meets this condition and has the largest volume.

SOLUTION Let x, y, and z be the dimensions of the box; its volume is $V = xyz$. The box with the maximum volume must also satisfy the condition $x + y + z = 96$, which is used to eliminate any one of the variables from the volume function. Noting that $z = 96 - x - y$, the volume function becomes

$$V(x, y) = xy(96 - x - y).$$

Notice that because x, y, and $96 - x - y$ are dimensions of the box, they must be nonnegative. The condition $96 - x - y \geq 0$ implies that $x + y \leq 96$. Therefore, among points in the xy-plane, the constraint is met only if (x, y) lies in the triangle bounded by the lines $x = 0$, $y = 0$, and $x + y = 96$ (Figure 15.73).

At this stage, we have reduced the original problem to a related problem: Find the absolute maximum value of $V(x, y) = xy(96 - x - y)$ over the triangular region

$$R = \{(x, y): 0 \leq x \leq 96, 0 \leq y \leq 96 - x\}.$$

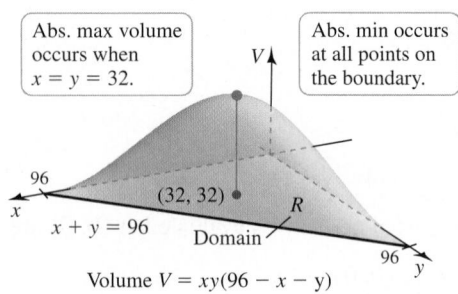

Abs. max volume occurs when $x = y = 32$.

Abs. min occurs at all points on the boundary.

$(32, 32)$

$x + y = 96$ Domain

Volume $V = xy(96 - x - y)$

Figure 15.73

The boundaries of R consist of the line segments $x = 0, 0 \leq y \leq 96$; $y = 0, 0 \leq x \leq 96$; and $x + y = 96, 0 \leq x \leq 96$. We find that on these boundary segments, $V = 0$. To determine the behavior of V at interior points of R, we need to find critical points. The critical points of V satisfy

$$V_x = 96y - 2xy - y^2 = y(96 - 2x - y) = 0$$
$$V_y = 96x - 2xy - x^2 = x(96 - 2y - x) = 0.$$

You can check that these two equations have four solutions: $(0, 0)$, $(96, 0)$, $(0, 96)$, and $(32, 32)$. The first three solutions lie on the boundary of the domain, where $V = 0$. At the fourth point, we have $V(32, 32) = 32{,}768 \text{ in}^3$, which is the absolute maximum volume of the box. The dimensions of the box with maximum volume are $x = 32$, $y = 32$, and $z = 96 - x - y = 32$ (it is a cube). We also found that V has an absolute minimum of 0 at every point on the boundary of R.

Related Exercise 43 ◀

We summarize the method of solution given in Example 5 in the following procedure box.

PROCEDURE Finding Absolute Maximum/Minimum Values on Closed Bounded Sets

Let f be continuous on a closed bounded set R in $\mathbb{R}^2$. To find the absolute maximum and minimum values of f on R:

1. Determine the values of f at all critical points in R.

2. Find the maximum and minimum values of f on the boundary of R.

3. The greatest function value found in Steps 1 and 2 is the absolute maximum value of f on R, and the least function value found in Steps 1 and 2 is the absolute minimum value of f on R.

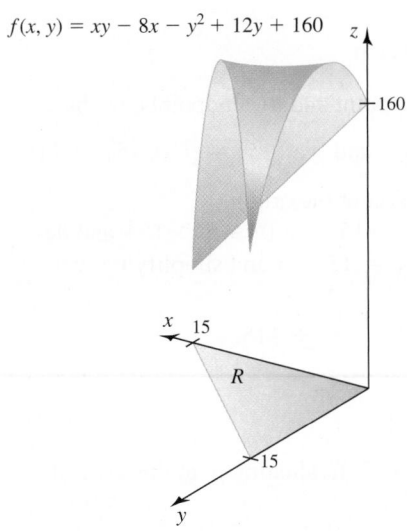

$f(x, y) = xy - 8x - y^2 + 12y + 160$

Figure 15.74

The techniques for carrying out Step 1 of this process have been presented. The challenge often lies in locating extreme values on the boundary. Examples 6 and 7 illustrate two approaches to handling the boundary of R. The first expresses the boundary using functions of a single variable, and the second describes the boundary parametrically. In both cases, finding extreme values on the boundary becomes a one-variable problem. In the next section, we discuss an alternative method for finding extreme values on boundaries.

EXAMPLE 6 Extreme values over a region Find the absolute maximum and minimum values of $f(x, y) = xy - 8x - y^2 + 12y + 160$ over the triangular region $R = \{(x, y): 0 \le x \le 15, 0 \le y \le 15 - x\}$.

SOLUTION Figure 15.74 shows the graph of f over the region R. The goal is to determine the absolute maximum and minimum values of f over R—including the boundary of R. We begin by finding the critical points of f on the interior of R. The partial derivatives of f are

$$f_x(x, y) = y - 8 \quad \text{and} \quad f_y(x, y) = x - 2y + 12.$$

The conditions $f_x(x, y) = f_y(x, y) = 0$ are satisfied only when $(x, y) = (4, 8)$, which is a point in the interior of R. This critical point is a candidate for the location of an extreme value of f, and the value of the function at this point is $f(4, 8) = 192$.

To search for extrema on the boundary of R, we consider each edge of R separately. Let C_1 be the line segment $\{(x, y): y = 0, \text{for } 0 \le x \le 15\}$ on the x-axis, and define the single-variable function g_1 to equal f at all points along C_1 (Figure 15.75). We substitute $y = 0$ and find that g_1 has the form

$$g_1(x) = f(x, 0) = 160 - 8x.$$

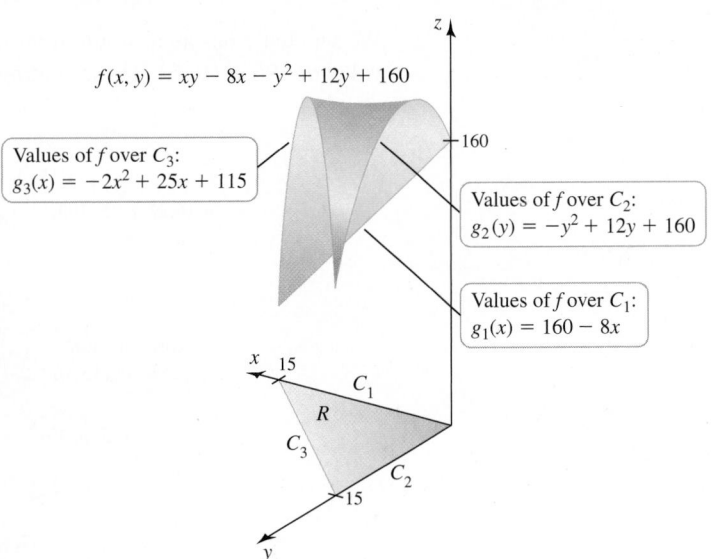

$f(x, y) = xy - 8x - y^2 + 12y + 160$

Values of f over C_3:
$g_3(x) = -2x^2 + 25x + 115$

Values of f over C_2:
$g_2(y) = -y^2 + 12y + 160$

Values of f over C_1:
$g_1(x) = 160 - 8x$

Figure 15.75

Using what we learned in Chapter 4, the candidates for absolute extreme values of g_1 on $0 \le x \le 15$ occur at critical points and endpoints. Specifically, the critical points of g_1 correspond to values where its derivative is zero, but in this case $g_1'(x) = -8$. So there is no critical point, which implies that the extreme values of g_1 occur at the endpoints of the interval $[0, 15]$. At the endpoints, we find that

$$g_1(0) = f(0, 0) = 160 \quad \text{and} \quad g_1(15) = f(15, 0) = 40.$$

Let's set aside this information while we do a similar analysis on the other two edges of the boundary of R.

Let C_2 be the line segment $\{(x, y): x = 0, \text{for } 0 \le y \le 15\}$ and define g_2 to equal f on C_2 (Figure 15.75). Substituting $x = 0$, we see that

$$g_2(y) = f(0, y) = -y^2 + 12y + 160.$$

The critical points of g_2 satisfy

$$g_2'(y) = -2y + 12 = 0,$$

which has the single root $y = 6$. Evaluating g_2 at this point and the endpoints, we have

$$g_2(6) = f(0, 6) = 196, \quad g_2(0) = f(0, 0) = 160, \quad \text{and} \quad g_2(15) = f(0, 15) = 115.$$

Observe that $g_1(0) = g_2(0)$ because C_1 and C_2 intersect at the origin.

Finally, we let C_3 be the line segment $\{(x, y): y = 15 - x, 0 \le x \le 15\}$ and define g_3 to equal f on C_3 (Figure 15.75). Substituting $y = 15 - x$ and simplifying, we find that

$$g_3(x) = f(x, 15 - x) = -2x^2 + 25x + 115.$$

The critical points of g_3 satisfy

$$g_3'(x) = -4x + 25,$$

whose only root on the interval $0 \le x \le 15$ is $x = 6.25$. Evaluating g_3 at this critical point and the endpoints, we have

$$g_3(6.25) = f(6.25, 8.75) = 193.125, \quad g_3(15) = f(15, 0) = 40, \quad \text{and}$$
$$g_3(0) = f(0, 15) = 115.$$

Observe that $g_3(15) = g_1(15)$ and $g_3(0) = g_2(15)$; the only new candidate for the location of an extreme value is the point $(6.25, 8.75)$.

Collecting and summarizing our work, we have 6 candidates for absolute extreme values:

$$f(4, 8) = 192, \quad f(0, 0) = 160, \quad f(15, 0) = 40, \quad f(0, 6) = 196,$$
$$f(0, 15) = 115, \quad \text{and} \quad f(6.25, 8.75) = 193.125.$$

We see that f has an absolute minimum value of 40 at $(15, 0)$ and an absolute maximum value of 196 at $(0, 6)$. These findings are illustrated in **Figure 15.76**.

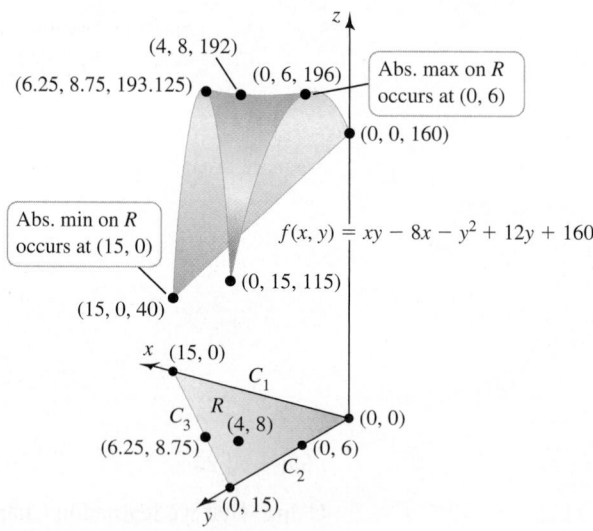

Figure 15.76

Related Exercise 52 ◄

> Finding absolute extrema on a closed set does not require using the Second Derivative Test to classify the critical points. In Example 7, the test *could* be used to show that $\left(\frac{1}{\sqrt{3}}, 0\right)$ and $\left(-\frac{1}{\sqrt{3}}, 0\right)$ correspond to a saddle point and a local maximum, respectively, but that information isn't needed.

EXAMPLE 7   Absolute maximum and minimum values Find the absolute maximum and minimum values of $f(x, y) = \frac{1}{2}(x^3 - x - y^2) + 3$ on the region $R = \{(x, y) : x^2 + y^2 \le 1\}$ (the closed disk centered at $(0, 0)$ with radius 1).

SOLUTION We begin by locating the critical points of f on the interior of R. The critical points satisfy the equations

$$f_x(x, y) = \frac{1}{2}(3x^2 - 1) = 0 \quad \text{and} \quad f_y(x, y) = -y = 0,$$

> Recall that a parametric description of a circle of radius $|a|$ centered at the origin is $x = a \cos \theta$, $y = a \sin \theta$, for $0 \le \theta \le 2\pi$.

which have the solutions $x = \pm \dfrac{1}{\sqrt{3}}$ and $y = 0$. The values of the function at these

points are $f\left(\dfrac{1}{\sqrt{3}}, 0\right) = 3 - \dfrac{1}{3\sqrt{3}}$ and $f\left(-\dfrac{1}{\sqrt{3}}, 0\right) = 3 + \dfrac{1}{3\sqrt{3}}$.

We now determine the maximum and minimum values of f on the boundary of R, which is a circle of radius 1 described by the parametric equations

$$x = \cos \theta, \quad y = \sin \theta, \quad \text{for} \quad 0 \le \theta \le 2\pi.$$

Substituting x and y in terms of θ into the function f, we obtain a new function $g(\theta)$ that gives the values of f on the boundary of R:

$$g(\theta) = \frac{1}{2}(\cos^3 \theta - \cos \theta - \sin^2 \theta) + 3.$$

Finding the maximum and minimum boundary values is now a one-variable problem. The critical points of g satisfy

$$g'(\theta) = \frac{1}{2}(-3\cos^2 \theta \sin \theta + \sin \theta - 2 \sin \theta \cos \theta)$$

$$= -\frac{1}{2}\sin \theta (3 \cos^2 \theta + 2 \cos \theta - 1)$$

$$= -\frac{1}{2}\sin \theta (3 \cos \theta - 1)(\cos \theta + 1) = 0.$$

> There are two solutions to the equation $\cos \theta = \frac{1}{3}$ on the interval $(0, 2\pi)$. Recall, however, that using the inverse cosine to solve the equation reveals only the solution $\theta = \cos^{-1} \frac{1}{3}$ because the range of $\cos^{-1} x$ is $[0, \pi]$. The other solution, $\theta = 2\pi - \cos^{-1} \frac{1}{3}$, is found using symmetry.

This condition is satisfied when $\sin \theta = 0$, $\cos \theta = \dfrac{1}{3}$, or $\cos \theta = -1$. The solutions of

these equations on the interval $(0, 2\pi)$ are $\theta = \pi$, $\theta = \cos^{-1}\dfrac{1}{3}$, and $\theta = 2\pi - \cos^{-1}\dfrac{1}{3}$,

which correspond to the points $(-1, 0)$, $\left(\dfrac{1}{3}, \dfrac{2\sqrt{2}}{3}\right)$, and $\left(\dfrac{1}{3}, -\dfrac{2\sqrt{2}}{3}\right)$ in the xy-plane,

respectively. Notice that the endpoints of the interval ($\theta = 0$ and $\theta = 2\pi$) correspond to the same point on the boundary of R, namely $(1, 0)$.

Having completed the first two steps of the procedure, we have six function values to consider:

- $f\left(\dfrac{1}{\sqrt{3}}, 0\right) = 3 - \dfrac{1}{3\sqrt{3}} \approx 2.81$ and $f\left(-\dfrac{1}{\sqrt{3}}, 0\right) = 3 + \dfrac{1}{3\sqrt{3}} \approx 3.19$ (critical points),

- $f(-1, 0) = 3$ (boundary point),

- $f\left(\dfrac{1}{3}, \dfrac{2\sqrt{2}}{3}\right) = f\left(\dfrac{1}{3}, -\dfrac{2\sqrt{2}}{3}\right) = \dfrac{65}{27}$ (boundary points), and

- $f(1, 0) = 3$ (boundary point).

> Observe that the level curves of f in Figure 15.77b appear to be tangent to the blue curve $x^2 + y^2 = 1$ (the boundary of the region R) at the points corresponding to the maximum and minimum values of f on this boundary. The significance of this observation is explained in Section 15.8.

The greatest value of f on R, $f\left(-\dfrac{1}{\sqrt{3}}, 0\right) = 3 + \dfrac{1}{3\sqrt{3}}$, is the absolute maxi-

mum value, and it occurs at an interior point (**Figure 15.77a**). The least value,

$f\left(\dfrac{1}{3}, \dfrac{2\sqrt{2}}{3}\right) = f\left(\dfrac{1}{3}, -\dfrac{2\sqrt{2}}{3}\right) = \dfrac{65}{27}$, is the absolute minimum value, and it occurs at

two symmetric boundary points. Also revealing is the plot of the level curves of the surface with the boundary of R superimposed (**Figure 15.77b**). As the boundary of R is traversed, the values of f vary, reaching a maximum value of 3 at $(1, 0)$ and $(-1, 0)$, and

a minimum value of $\dfrac{65}{27}$ at $\left(\dfrac{1}{3}, \dfrac{2\sqrt{2}}{3}\right)$ and $\left(\dfrac{1}{3}, -\dfrac{2\sqrt{2}}{3}\right)$.

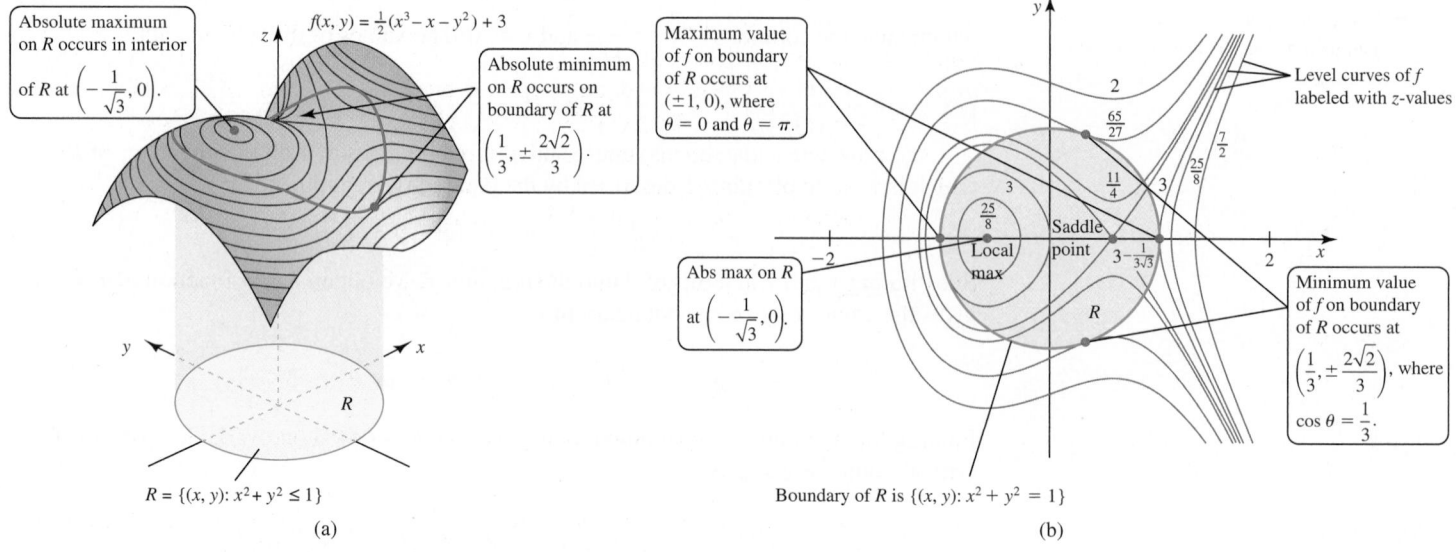

$f(x, y) = \frac{1}{2}(x^3 - x - y^2) + 3$

Absolute maximum on R occurs in interior of R at $\left(-\frac{1}{\sqrt{3}}, 0\right)$.

Absolute minimum on R occurs on boundary of R at $\left(\frac{1}{3}, \pm\frac{2\sqrt{2}}{3}\right)$.

Maximum value of f on boundary of R occurs at $(\pm 1, 0)$, where $\theta = 0$ and $\theta = \pi$.

Level curves of f labeled with z-values

Abs max on R at $\left(-\frac{1}{\sqrt{3}}, 0\right)$.

Local max

Saddle point

Minimum value of f on boundary of R occurs at $\left(\frac{1}{3}, \pm\frac{2\sqrt{2}}{3}\right)$, where $\cos\theta = \frac{1}{3}$.

$R = \{(x, y): x^2 + y^2 \leq 1\}$

(a)

Boundary of R is $\{(x, y): x^2 + y^2 = 1\}$

(b)

Figure 15.77

Related Exercises 47–48 ◄

Open and/or Unbounded Regions   Finding absolute maximum and minimum values of a function on an open region (for example, $R = \{(x, y) = x^2 + y^2 < 9\}$) or an unbounded domain (for example, $R = \{(x, y): x > 0, y > 0\}$) presents additional challenges. Because there is no systematic procedure for dealing with such problems, some ingenuity is generally needed. Notice that absolute extrema may not exist on such regions.

EXAMPLE 8   Absolute extreme values on an open region   Find the absolute maximum and minimum values of $f(x, y) = 4 - x^2 - y^2$ on the open disk $R = \{(x, y): x^2 + y^2 < 1\}$ (if they exist).

SOLUTION   You should verify that f has a critical point at $(0, 0)$ and it corresponds to a local maximum (on an inverted paraboloid). Moving away from $(0, 0)$ in all directions, the function values decrease, so f also has an absolute maximum value of 4 at $(0, 0)$. The boundary of R is the unit circle $\{(x, y): x^2 + y^2 = 1\}$, which is not contained in R. As (x, y) approaches any point on the unit circle along any path in R, the function values $f(x, y) = 4 - (x^2 + y^2)$ decrease and approach 3 but never reach 3. Therefore, f does not have an absolute minimum on R.

Related Exercise 59 ◄

QUICK CHECK 4   Does the linear function $f(x, y) = 2x + 3y$ have an absolute maximum or minimum value on the open unit square $\{(x, y): 0 < x < 1, 0 < y < 1\}$? ◄

▶ Notice that $\frac{\partial}{\partial x}(d^2) = 2d\frac{\partial d}{\partial x}$ and $\frac{\partial}{\partial y}(d^2) = 2d\frac{\partial d}{\partial y}$. Because $d \geq 0$, d^2 and d have the same critical points.

EXAMPLE 9   Absolute extreme values on an open region   Find the point(s) on the plane $x + 2y + z = 2$ closest to the point $P(2, 0, 4)$.

SOLUTION   Suppose (x, y, z) is a point on the plane, which means that $z = 2 - x - 2y$. The distance between $P(2, 0, 4)$ and (x, y, z) that we seek to minimize is

$$d(x, y, z) = \sqrt{(x - 2)^2 + y^2 + (z - 4)^2}.$$

It is easier to minimize d^2, which has the same critical points as d. Squaring d and eliminating z using $z = 2 - x - 2y$, we have

$$f(x, y) = (d(x, y, z))^2 = (x - 2)^2 + y^2 + \underbrace{(-x - 2y - 2)^2}_{z - 4}$$

$$= 2x^2 + 5y^2 + 4xy + 8y + 8.$$

The critical points of f satisfy the equations

$$f_x = 4x + 4y = 0 \quad \text{and} \quad f_y = 4x + 10y + 8 = 0,$$

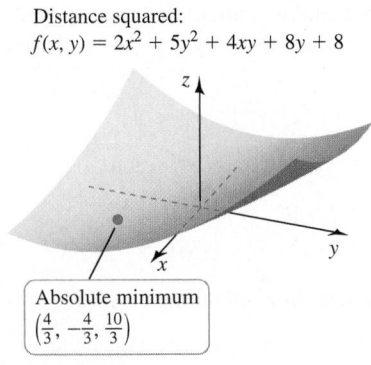

Distance squared:
$$f(x, y) = 2x^2 + 5y^2 + 4xy + 8y + 8$$

Absolute minimum
$\left(\frac{4}{3}, -\frac{4}{3}, \frac{10}{3}\right)$

Figure 15.78

whose only solution is $x = \frac{4}{3}$, $y = -\frac{4}{3}$. The Second Derivative Test confirms that this point corresponds to a local minimum of f. We now ask: Does $\left(\frac{4}{3}, -\frac{4}{3}\right)$ correspond to the *absolute* minimum value of f over the entire domain of the problem, which is $\mathbb{R}^2$? Because the domain has no boundary, we cannot check values of f on the boundary. Instead, we argue geometrically that there is exactly one point on the plane that is closest to P. We have found a point that is closest to P among nearby points on the plane. As we move away from this point, the values of f increase without bound. Therefore, $\left(\frac{4}{3}, -\frac{4}{3}\right)$ corresponds to the absolute minimum value of f. A graph of f (Figure 15.78) confirms this reasoning, and we conclude that the point $\left(\frac{4}{3}, -\frac{4}{3}, \frac{10}{3}\right)$ is the point on the plane nearest P.

Related Exercises 62–63 ◄

SECTION 15.7 EXERCISES

Getting Started

1. Describe the appearance of a smooth surface with a local maximum at a point.

2. Describe the usual appearance of a smooth surface at a saddle point.

3. What are the conditions for a critical point of a function f?

4. If $f_x(a, b) = f_y(a, b) = 0$, does it follow that f has a local maximum or local minimum at (a, b)? Explain.

5. Consider the function $z = f(x, y)$. What is the discriminant of f, and how do you compute it?

6. Explain how the Second Derivative Test is used.

7. What is an absolute minimum value of a function f on a set R in $\mathbb{R}^2$?

8. What is the procedure for locating absolute maximum and minimum values on a closed bounded domain?

9–12. *Assume the second derivatives of f are continuous throughout the xy-plane and $f_x(0, 0) = f_y(0, 0) = 0$. Use the given information and the Second Derivative Test to determine whether f has a local minimum, a local maximum, or a saddle point at $(0, 0)$, or state that the test is inconclusive.*

9. $f_{xx}(0, 0) = 5$, $f_{yy}(0, 0) = 3$, and $f_{xy}(0, 0) = -4$

10. $f_{xx}(0, 0) = -6$, $f_{yy}(0, 0) = -3$, and $f_{xy}(0, 0) = 4$

11. $f_{xx}(0, 0) = 8$, $f_{yy}(0, 0) = 5$, and $f_{xy}(0, 0) = -6$

12. $f_{xx}(0, 0) = -9$, $f_{yy}(0, 0) = -4$, and $f_{xy}(0, 0) = -6$

Practice Exercises

13–22. Critical points *Find all critical points of the following functions.*

13. $f(x, y) = 3x^2 - 4y^2$

14. $f(x, y) = x^2 - 6x + y^2 + 8y$

15. $f(x, y) = 3x^2 + 3y - y^3$

16. $f(x, y) = x^3 - 12x + 6y^2$

17. $f(x, y) = x^4 + y^4 - 16xy$

18. $f(x, y) = \frac{x^3}{3} - \frac{y^3}{3} + 3xy$

19. $f(x, y) = x^4 - 2x^2 + y^2 - 4y + 5$

20. $f(x, y) = x^3 + 6xy - 6x + y^2 - 2y$

21. $f(x, y) = y^3 + 6xy + x^2 - 18y - 6x$

22. $f(x, y) = e^{8x^2y^2 - 24x^2 - 8xy^4}$

23–40. Analyzing critical points *Find the critical points of the following functions. Use the Second Derivative Test to determine (if possible) whether each critical point corresponds to a local maximum, a local minimum, or a saddle point. If the Second Derivative Test is inconclusive, determine the behavior of the function at the critical points.*

23. $f(x, y) = -4x^2 + 8y^2 - 3$

24. $f(x, y) = x^4 + y^4 - 4x - 32y + 10$

25. $f(x, y) = 4 + 2x^2 + 3y^2$

26. $f(x, y) = xye^{-x-y}$

27. $f(x, y) = x^4 + 2y^2 - 4xy$

28. $f(x, y) = (4x - 1)^2 + (2y + 4)^2 + 1$

29. $f(x, y) = 4 + x^4 + 3y^4$

30. $f(x, y) = x^4y^2$

31. $f(x, y) = \sqrt{x^2 + y^2 - 4x + 5}$

32. $f(x, y) = \tan^{-1}xy$

33. $f(x, y) = 2xye^{-x^2-y^2}$ 34. $f(x, y) = x^2 + xy^2 - 2x + 1$

35. $f(x, y) = \dfrac{x}{1 + x^2 + y^2}$ 36. $f(x, y) = \dfrac{x - 1}{x^2 + y^2}$

37. $f(x, y) = x^4 + 4x^2(y - 2) + 8(y - 1)^2$

38. $f(x, y) = xe^{-x-y}\sin y$, for $|x| \le 2, 0 \le y \le \pi$

39. $f(x, y) = ye^x - e^y$

40. $f(x, y) = \sin(2\pi x)\cos(\pi y)$, for $|x| \le \dfrac{1}{2}$ and $|y| \le \dfrac{1}{2}$

41–42. Inconclusive tests *Show that the Second Derivative Test is inconclusive when applied to the following functions at $(0, 0)$. Describe the behavior of the function at $(0, 0)$.*

41. $f(x, y) = x^2y - 3$ 42. $f(x, y) = \sin(x^2y^2)$

43. **Shipping regulations** A shipping company handles rectangular boxes provided the sum of the height and the girth of the box does not exceed 96 in. (The girth is the perimeter of the smallest side of the box.) Find the dimensions of the box that meets this condition and has the largest volume.

44. **Cardboard boxes** A lidless box is to be made using 2 m² of cardboard. Find the dimensions of the box with the largest possible volume.

45. Cardboard boxes A lidless cardboard box is to be made with a volume of 4 m³. Find the dimensions of the box that requires the least amount of cardboard.

46. Optimal box Find the dimensions of the largest rectangular box in the first octant of the *xyz*-coordinate system that has one vertex at the origin and the opposite vertex on the plane $x + 2y + 3z = 6$.

47–56. Absolute maxima and minima *Find the absolute maximum and minimum values of the following functions on the given region R.*

47. $f(x, y) = x^2 + y^2 - 2y + 1; R = \{(x, y): x^2 + y^2 \le 4\}$

48. $f(x, y) = 2x^2 + y^2; R = \{(x, y): x^2 + y^2 \le 16\}$

49. $f(x, y) = 4 + 2x^2 + y^2;$
$R = \{(x, y): -1 \le x \le 1, -1 \le y \le 1\}$

50. $f(x, y) = 6 - x^2 - 4y^2;$
$R = \{(x, y): -2 \le x \le 2, -1 \le y \le 1\}$

51. $f(x, y) = 2x^2 - 4x + 3y^2 + 2;$
$R = \{(x, y): (x - 1)^2 + y^2 \le 1\}$

52. $f(x, y) = x^2 + y^2 - 2x - 2y; R$ is the closed region bounded by the triangle with vertices $(0, 0)$, $(2, 0)$, and $(0, 2)$.

53. $f(x, y) = -2x^2 + 4x - 3y^2 - 6y - 1;$
$R = \{(x, y): (x - 1)^2 + (y + 1)^2 \le 1\}$

54. $f(x, y) = \sqrt{x^2 + y^2 - 2x + 2};$
$R = \{(x, y): x^2 + y^2 \le 4, y \ge 0\}$

55. $f(x, y) = \dfrac{2y^2 - x^2}{2 + 2x^2y^2}; R$ is the closed region bounded by the lines $y = x, y = 2x$, and $y = 2$.

56. $f(x, y) = \sqrt{x^2 + y^2}; R$ is the closed region bounded by the ellipse $\dfrac{x^2}{4} + y^2 = 1$.

T 57. Pectin Extraction An increase in world production of processed fruit has led to an increase in fruit waste. One way of reducing this waste is to find useful waste byproducts. For example, waste from pineapples is reduced by extracting pectin from pineapple peels (pectin is commonly used as a thickening agent in jam and jellies, and it is also widely used in the pharmaceutical industry). Pectin extraction involves heating and drying the peels, then grinding the peels into a fine powder. The powder is next placed in a solution with a particular pH level H, for $1.5 \le H \le 2.5$, and heated to a temperature T (in degrees Celsius), for $70 \le T \le 90$. The percentage of the powder $F(H, T)$ that becomes extracted pectin is

$$F(H, T) = -0.042T^2 - 0.213TH - 11.219H^2 + 7.327T$$
$$+ 58.729H - 342.684.$$

a. It can be shown that F attains its absolute maximum in the interior of the domain $D = \{(H, T): 1.5 \le H \le 2.5, 70 \le T \le 90\}$. Find the pH level H and temperature T that together maximize the amount of pectin extracted from the powder.

b. What is the maximum percentage of pectin that can be extracted from the powder? Round your answer to the nearest whole number. (*Source: Carpathian Journal of Food Science and Technology*, Dec 2014)

58–61. Absolute extrema on open and/or unbounded regions *If possible, find the absolute maximum and minimum values of the following functions on the region R.*

58. $f(x, y) = x + 3y; R = \{(x, y): |x| < 1, |y| < 2\}$

59. $f(x, y) = x^2 + y^2 - 4; R = \{(x, y): x^2 + y^2 < 4\}$

60. $f(x, y) = x^2 - y^2; R = \{(x, y): |x| < 1, |y| < 1\}$

61. $f(x, y) = 2e^{-x-y}; R = \{(x, y): x \ge 0, y \ge 0\}$

62–66. Absolute extrema on open and/or unbounded regions

62. Find the point on the plane $x + y + z = 4$ nearest the point $P(5, 4, 4)$.

63. Find the point on the plane $x - y + z = 2$ nearest the point $P(1, 1, 1)$.

64. Find the point on the paraboloid $z = x^2 + y^2$ nearest the point $P(3, 3, 1)$.

65. Find the points on the cone $z^2 = x^2 + y^2$ nearest the point $P(6, 8, 0)$.

66. Rectangular boxes with a volume of 10 m³ are made of two materials. The material for the top and bottom of the box costs $10/m² and the material for the sides of the box costs $1/m². What are the dimensions of the box that minimize the cost of the box?

67. Explain why or why not Determine whether the following statements are true and give an explanation or counterexample. Assume f is differentiable at the points in question.

a. The fact that $f_x(2, 2) = f_y(2, 2) = 0$ implies that f has a local maximum, local minimum, or saddle point at $(2, 2)$.

b. The function f could have a local maximum at (a, b) where $f_y(a, b) \ne 0$.

c. The function f could have both an absolute maximum and an absolute minimum at two different points that are not critical points.

d. The tangent plane is horizontal at a point on a smooth surface corresponding to a critical point.

68–69. Extreme points from contour plots *Based on the level curves that are visible in the following graphs, identify the approximate locations of the local maxima, local minima, and saddle points.*

68.

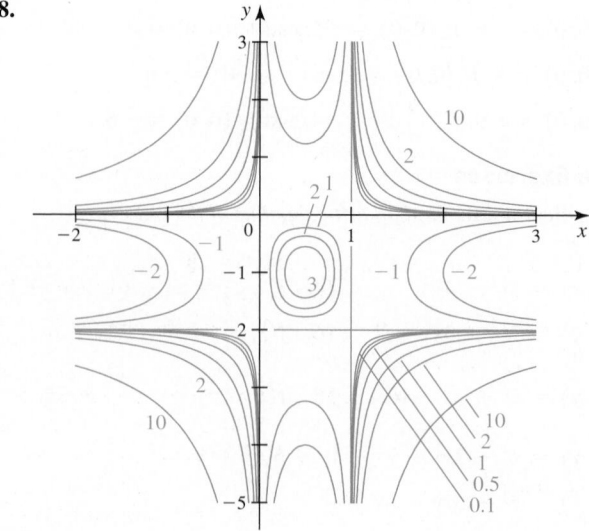

69.

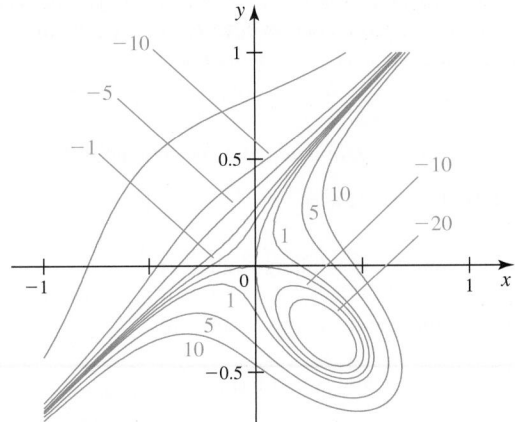

70. Optimal box Find the dimensions of the rectangular box with maximum volume in the first octant with one vertex at the origin and the opposite vertex on the ellipsoid $36x^2 + 4y^2 + 9z^2 = 36$.

Explorations and Challenges

71. Magic triples Let x, y, and z be nonnegative numbers with $x + y + z = 200$.

  a. Find the values of x, y, and z that minimize $x^2 + y^2 + z^2$.

  b. Find the values of x, y, and z that minimize $\sqrt{x^2 + y^2 + z^2}$.

  c. Find the values of x, y, and z that maximize xyz.

  d. Find the values of x, y, and z that maximize $x^2y^2z^2$.

72. Maximum/minimum of linear functions Let R be a closed bounded region in $\mathbb{R}^2$ and let $f(x, y) = ax + by + c$, where a, b, and c are real numbers, with a and b not both zero. Give a geometric argument explaining why the absolute maximum and minimum values of f over R occur on the boundaries of R.

T 73. Optimal locations Suppose n houses are located at the distinct points (x_1, y_1), (x_2, y_2), . . . , (x_n, y_n). A power substation must be located at a point such that the *sum of the squares* of the distances between the houses and the substation is minimized.

  a. Find the optimal location of the substation in the case that $n = 3$ and the houses are located at $(0, 0)$, $(2, 0)$, and $(1, 1)$.

  b. Find the optimal location of the substation in the case that $n = 3$ and the houses are located at distinct points (x_1, y_1), (x_2, y_2), and (x_3, y_3).

  c. Find the optimal location of the substation in the general case of n houses located at distinct points (x_1, y_1), (x_2, y_2), . . . , (x_n, y_n).

  d. You might argue that the locations found in parts (a), (b) and (c) are not optimal because they result from minimizing the sum of the *squares* of the distances, not the sum of the distances themselves. Use the locations in part (a) and write the function that gives the sum of the distances. Note that minimizing this function is much more difficult than in part (a). Then use a graphing utility to determine whether the optimal location is the same in the two cases. (Also see Exercise 81 about Steiner's problem.)

74–75. Least squares approximation *In its many guises, least squares approximation arises in numerous areas of mathematics and statistics. Suppose you collect data for two variables (for example, height and shoe size) in the form of pairs (x_1, y_1), (x_2, y_2), . . . , (x_n, y_n). The data may be plotted as a scatterplot in the xy-plane, as shown in the figure. The technique known as* linear regression *asks the question: What is the equation of the line that "best fits" the data? The least squares*

criterion for best fit requires that the sum of the squares of the vertical distances between the line and the data points be a minimum.

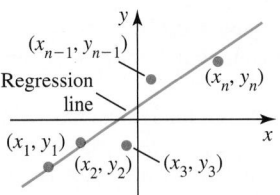

74. Let the equation of the best-fit line be $y = mx + b$, where the slope m and the y-intercept b must be determined using the least squares condition. First assume there are three data points $(1, 2)$, $(3, 5)$, and $(4, 6)$. Show that the function of m and b that gives the sum of the squares of the vertical distances between the line and the three data points is

$$E(m, b) = ((m + b) - 2)^2 + ((3m + b) - 5)^2 + ((4m + b) - 6)^2.$$

Find the critical points of E and find the values of m and b that minimize E. Graph the three data points and the best-fit line.

T 75. Generalize the procedure in Exercise 74 by assuming n data points (x_1, y_1), (x_2, y_2), . . . , (x_n, y_n) are given. Write the function $E(m, b)$ (summation notation allows for a more compact calculation). Show that the coefficients of the best-fit line are

$$m = \frac{\left(\sum x_k\right)\left(\sum y_k\right) - n\sum x_k y_k}{\left(\sum x_k\right)^2 - n\sum x_k^2},$$

$$b = \frac{1}{n}\left(\sum y_k - m\sum x_k\right),$$

where all sums run from $k = 1$ to $k = n$.

T 76–77. Least squares practice *Use the results of Exercise 75 to find the best-fit line for the following data sets. Plot the points and the best-fit line.*

76. $(0, 0)$, $(2, 3)$, $(4, 5)$ **77.** $(-1, 0)$, $(0, 6)$, $(3, 8)$

78. Second Derivative Test Suppose the conditions of the Second Derivative Test are satisfied on an open disk containing the point (a, b). Use the test to prove that if (a, b) is a critical point of f at which $f_x(a, b) = f_y(a, b) = 0$ and $f_{xx}(a, b) < 0 < f_{yy}(a, b)$ or $f_{yy}(a, b) < 0 < f_{xx}(a, b)$, then f has a saddle point at (a, b).

79. Maximum area triangle Among all triangles with a perimeter of 9 units, find the dimensions of the triangle with the maximum area. It may be easiest to use Heron's formula, which states that the area of a triangle with side length a, b, and c is $A = \sqrt{s(s - a)(s - b)(s - c)}$, where $2s$ is the perimeter of the triangle.

80. Slicing plane Find an equation of the plane passing through the point $(3, 2, 1)$ that slices off the solid in the first octant with the least volume.

T 81. Steiner's problem for three points Given three distinct noncollinear points A, B, and C in the plane, find the point P in the plane such that the sum of the distances $|AP| + |BP| + |CP|$ is a minimum. Here is how to proceed with three points, assuming the triangle formed by the three points has no angle greater than $2\pi/3$ ($120°$).

  a. Assume the coordinates of the three given points are $A(x_1, y_1)$, $B(x_2, y_2)$, and $C(x_3, y_3)$. Let $d_1(x, y)$ be the distance between $A(x_1, y_1)$ and a variable point $P(x, y)$. Compute the gradient of d_1 and show that it is a unit vector pointing along the line between the two points.

b. Define d_2 and d_3 in a similar way and show that ∇d_2 and ∇d_3 are also unit vectors in the direction of the line between the two points.

c. The goal is to minimize $f(x, y) = d_1 + d_2 + d_3$. Show that the condition $f_x = f_y = 0$ implies that $\nabla d_1 + \nabla d_2 + \nabla d_3 = 0$.

d. Explain why part (c) implies that the optimal point P has the property that the three line segments AP, BP, and CP all intersect symmetrically in angles of $2\pi/3$.

e. What is the optimal solution if one of the angles in the triangle is greater than $2\pi/3$ (just draw a picture)?

f. Estimate the Steiner point for the three points $(0, 0)$, $(0, 1)$, and $(2, 0)$.

▣ 82. Solitary critical points A function of *one* variable has the property that a local maximum (or minimum) occurring at the only critical point is also the absolute maximum (or minimum) (for example, $f(x) = x^2$). Does the same result hold for a function of *two* variables? Show that the following functions have the property that they have a single local maximum (or minimum), occurring at the only critical point, but the local maximum (or minimum) is not an absolute maximum (or minimum) on $\mathbb{R}^2$.

a. $f(x, y) = 3xe^y - x^3 - e^{3y}$

b. $f(x, y) = (2y^2 - y^4)\left(e^x + \dfrac{1}{1 + x^2}\right) - \dfrac{1}{1 + x^2}$

This property has the following interpretation. Suppose a surface has a single local minimum that is not the absolute minimum. Then water can be poured into the basin around the local minimum and the surface never overflows, even though there are points on the surface below the local minimum.
(*Source: Mathematics Magazine*, May 1985, and *Calculus and Analytical Geometry*, 2nd ed., Philip Gillett, 1984)

▣ 83. Two mountains without a saddle Show that the following functions have two local maxima but no other extreme points (therefore, no saddle or basin between the mountains).

a. $f(x, y) = -(x^2 - 1)^2 - (x^2 - e^y)^2$

b. $f(x, y) = 4x^2e^y - 2x^4 - e^{4y}$

(*Source:* Ira Rosenholtz, *Mathematics Magazine*, Feb 1987)

84. Powers and roots Assume $x + y + z = 1$ with $x \geq 0$, $y \geq 0$, and $z \geq 0$.

a. Find the maximum and minimum values of $(1 + x^2)(1 + y^2)(1 + z^2)$.

b. Find the maximum and minimum values of $(1 + \sqrt{x})(1 + \sqrt{y})(1 + \sqrt{z})$.

(*Source: Math Horizons*, Apr 2004)

85. Ellipsoid inside a tetrahedron (1946 Putnam Exam) Let P be a plane tangent to the ellipsoid $\dfrac{x^2}{a^2} + \dfrac{y^2}{b^2} + \dfrac{z^2}{c^2} = 1$ at a point in the first octant. Let T be the tetrahedron in the first octant bounded by P and the coordinate planes $x = 0$, $y = 0$, and $z = 0$. Find the minimum volume of T. (The volume of a tetrahedron is one-third the area of the base times the height.)

QUICK CHECK ANSWERS

1. $f_x(2, -1) = f_y(2, -1) = 0$ **2.** Vertically, in the directions $\langle 0, 0, \pm 1 \rangle$ **3.** $D(x, y) = -12x^2y^2$
4. It has neither an absolute maximum nor an absolute minimum value on this set. ◄

15.8 Lagrange Multipliers

One of many challenges in economics and marketing is predicting the behavior of consumers. Basic models of consumer behavior often involve a *utility function* that expresses consumers' combined preference for several different amenities. For example, a simple utility function might have the form $U = f(\ell, g)$, where ℓ represents the amount of leisure time and g represents the number of consumable goods. The model assumes consumers try to maximize their utility function, but they do so under certain constraints on the variables of the problem. For example, increasing leisure time may increase utility, but leisure time produces no income for consumable goods. Similarly, consumable goods may also increase utility, but they require income, which reduces leisure time. We first develop a general method for solving such constrained optimization problems and then return to economics problems later in the section.

The Basic Idea

We start with a typical constrained optimization problem with two independent variables and give its method of solution; a generalization to more variables then follows. We seek maximum and/or minimum values of a differentiable **objective function** f with the restriction that x and y must lie on a **constraint** curve C in the xy-plane given by $g(x, y) = 0$ (Figure 15.79).

Figure 15.80 shows the details of a typical situation in which we assume the (green) level curves of f have increasing z-values moving away from the origin. Now imagine moving along the (red) constraint curve C: $g(x, y) = 0$ toward the point $P(a, b)$. As we approach P (from either side), the values of f evaluated on C increase, and as we move past P along C, the values of f decrease.

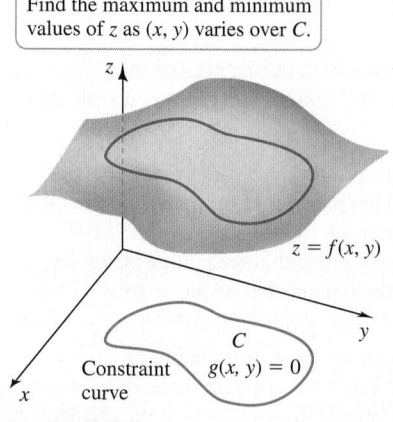

Find the maximum and minimum values of z as (x, y) varies over C.

$z = f(x, y)$

Constraint curve C $g(x, y) = 0$

Figure 15.79

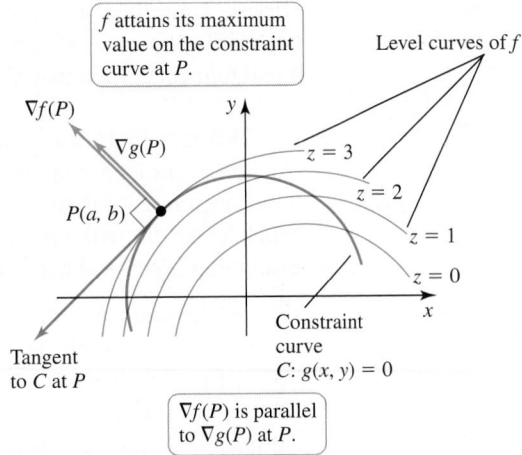

Figure 15.80

What is special about the point P at which f appears to have a local maximum value on C? From Theorem 15.12, we know that at any point $P(a, b)$ on a level curve of f, the line tangent to the level curve at P is orthogonal to $\nabla f(a, b)$. Figure 15.80 also suggests that the line tangent to the level curve of f at P is tangent to the constraint curve C at P. We prove this fact shortly. This observation implies that $\nabla f(a, b)$ is also orthogonal to the line tangent to C at $P(a, b)$.

We need one more observation. The constraint curve C is just one level curve of the function $z = g(x, y)$. Using Theorem 15.12 again, the line tangent to C at $P(a, b)$ is orthogonal to $\nabla g(a, b)$. We have now found two vectors $\nabla f(a, b)$ and $\nabla g(a, b)$ that are both orthogonal to the line tangent to the level curve C at $P(a, b)$. Therefore, these two gradient vectors are parallel. These properties characterize the point P at which f has a local extremum on the constraint curve. They are the basis of the method of *Lagrange multipliers* that we now formalize.

Lagrange Multipliers with Two Independent Variables

The major step in establishing the method of Lagrange multipliers is to prove that Figure 15.80 is drawn correctly; that is, at the point on the constraint curve C where f has a local extreme value, the line tangent to C is orthogonal to $\nabla f(a, b)$ and $\nabla g(a, b)$.

▶ The Greek lowercase ℓ is λ; it is read *lambda*.

> **THEOREM 15.16   Parallel Gradients**
> Let f be a differentiable function in a region of $\mathbb{R}^2$ that contains the smooth curve C given by $g(x, y) = 0$. Assume f has a local extreme value on C at a point $P(a, b)$. Then $\nabla f(a, b)$ is orthogonal to the line tangent to C at P. Assuming $\nabla g(a, b) \neq \mathbf{0}$, it follows that there is a real number λ (called a **Lagrange multiplier**) such that $\nabla f(a, b) = \lambda \nabla g(a, b)$.

Proof: Because C is smooth, it can be expressed parametrically in the form $C: \mathbf{r}(t) = \langle x(t), y(t) \rangle$, where x and y are differentiable functions on an interval in t that contains t_0 with $P(a, b) = (x(t_0), y(t_0))$. As we vary t and follow C, the rate of change of f is given by the Chain Rule:

$$\frac{df}{dt} = \frac{\partial f}{\partial x}\frac{dx}{dt} + \frac{\partial f}{\partial y}\frac{dy}{dt} = \nabla f \cdot \mathbf{r}'(t).$$

At the point $(x(t_0), y(t_0)) = (a, b)$ at which f has a local extreme value, we have $\left. \dfrac{df}{dt} \right|_{t=t_0} = 0$, which implies that $\nabla f(a, b) \cdot \mathbf{r}'(t_0) = 0$. Because $\mathbf{r}'(t)$ is tangent to C, the gradient $\nabla f(a, b)$ is orthogonal to the line tangent to C at P.

To prove the second assertion, note that the constraint curve C given by $g(x, y) = 0$ is also a level curve of the surface $z = g(x, y)$. Recall that gradients are orthogonal to level curves. Therefore, at the point $P(a, b)$, $\nabla g(a, b)$ is orthogonal to C at (a, b). Because both $\nabla f(a, b)$ and $\nabla g(a, b)$ are orthogonal to C, the two gradients are parallel, so there is a real number λ such that $\nabla f(a, b) = \lambda \nabla g(a, b)$. ◄

Theorem 15.16 leads directly to the method of Lagrange Multipliers, which produces candidates for *local* maxima and minima of f on the constraint curve. In many problems, however, the goal is to find *absolute* maxima and minima of f on the constraint curve. Much as we did with optimization problems in one variable, we find absolute extrema by examining both local extrema and endpoints. Several different cases arise:

- If the constraint curve is bounded (it lies within a circle of finite radius) and it closes on itself (for example, an ellipse), then we know that the absolute extrema of f exist. In this case, there are no endpoints to consider, and the absolute extrema are found among the local extrema.

- If the constraint curve is bounded and includes its endpoints but does not close on itself (for example, a closed line segment), then the absolute extrema of f exist, and we find them by examining the local extrema and the endpoints.

- In the case that the constraint curve is unbounded (for example, a line or a parabola) or the curve excludes one or both of its endpoints, we have no guarantee that absolute extrema exist. We can find local extrema, but they must be examined carefully to determine whether they are, in fact, absolute extrema (see Example 2 and Exercise 65).

We deal first with the case of finding absolute extrema on closed and bounded constraint curves.

QUICK CHECK 1 It can be shown that the function $f(x, y) = x^2 + y^2$ attains its absolute minimum value on the curve

$$C: g(x, y) = \frac{1}{4}(x - 3)^2 - y = 0$$

at the point $(1, 1)$. Verify that $\nabla f(1, 1)$ and $\nabla g(1, 1)$ are parallel, and that both vectors are orthogonal to the line tangent to C at $(1, 1)$, thereby confirming Theorem 15.16. ◄

PROCEDURE   Lagrange Multipliers: Absolute Extrema on Closed and Bounded Constraint Curves

Let the objective function f and the constraint function g be differentiable on a region of $\mathbb{R}^2$ with $\nabla g(x, y) \neq \mathbf{0}$ on the curve $g(x, y) = 0$. To locate the absolute maximum and minimum values of f subject to the constraint $g(x, y) = 0$, carry out the following steps.

1. Find the values of x, y, and λ (if they exist) that satisfy the equations

$$\nabla f(x, y) = \lambda \nabla g(x, y) \quad \text{and} \quad g(x, y) = 0.$$

2. Evaluate f at the values (x, y) found in Step 1 and at the endpoints of the constraint curve (if they exist). Select the largest and smallest corresponding function values. These values are the absolute maximum and minimum values of f subject to the constraint.

▶ *In principle*, it is possible to solve a constrained optimization problem by solving the constraint equation for one of the variables and eliminating that variable in the objective function. In practice, this method is often impractical, particularly with three or more variables or two or more constraints.

Notice that $\nabla f = \lambda \nabla g$ is a vector equation: $\langle f_x, f_y \rangle = \lambda \langle g_x, g_y \rangle$. It is satisfied provided $f_x = \lambda g_x$ and $f_y = \lambda g_y$. Therefore, the crux of the method is solving the three equations

$$f_x = \lambda g_x, \qquad f_y = \lambda g_y, \qquad \text{and} \qquad g(x, y) = 0,$$

for the three variables x, y, and λ.

EXAMPLE 1   Lagrange multipliers with two variables Find the absolute maximum and minimum values of the objective function $f(x, y) = x^2 + y^2 + 2$, where x and y lie on the ellipse C given by $g(x, y) = x^2 + xy + y^2 - 4 = 0$.

SOLUTION Because C is closed and bounded, the absolute maximum and minimum values of f exist. Figure 15.81a shows the paraboloid $z = f(x, y)$ above the ellipse C in the xy-plane. As the ellipse is traversed, the corresponding function values on the surface vary. The goal is to find the maximum and minimum of these function values. An alternative view is given in Figure 15.81b, where we see the level curves of f and the constraint curve C. As the ellipse is traversed, the values of f vary, reaching maximum and minimum values along the way.

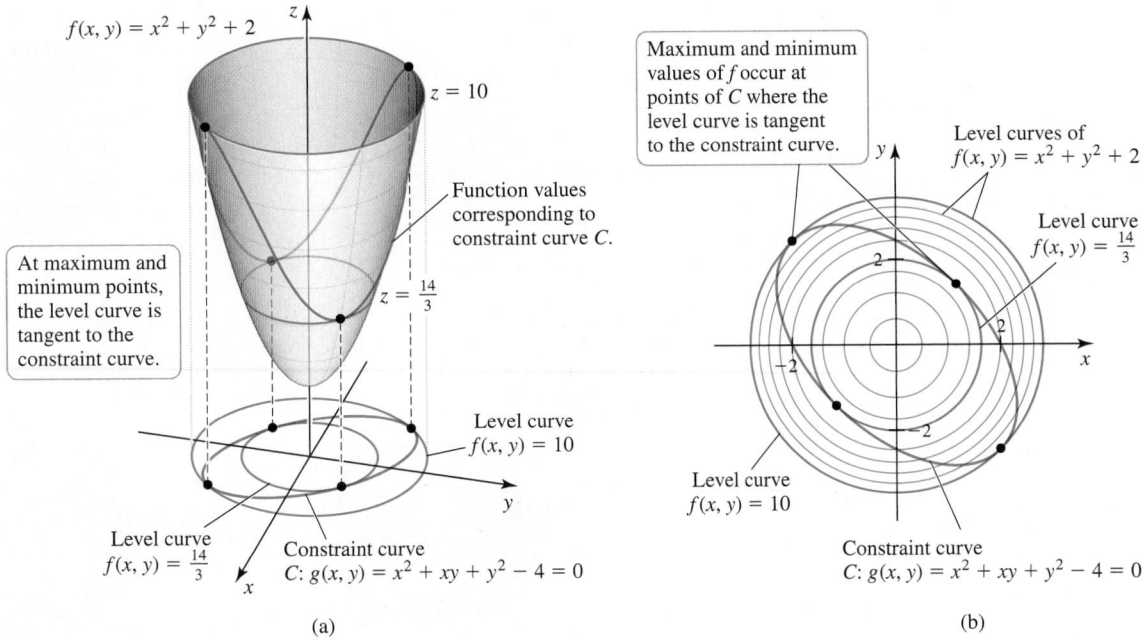

$f(x, y) = x^2 + y^2 + 2$

$z = 10$

Function values corresponding to constraint curve C.

Maximum and minimum values of f occur at points of C where the level curve is tangent to the constraint curve.

Level curves of $f(x, y) = x^2 + y^2 + 2$

Level curve $f(x, y) = \frac{14}{3}$

At maximum and minimum points, the level curve is tangent to the constraint curve.

$z = \frac{14}{3}$

Level curve $f(x, y) = 10$

Level curve $f(x, y) = 10$

Level curve $f(x, y) = \frac{14}{3}$

Constraint curve $C: g(x, y) = x^2 + xy + y^2 - 4 = 0$

Constraint curve $C: g(x, y) = x^2 + xy + y^2 - 4 = 0$

(a)   (b)

Figure 15.81

Noting that $\nabla f(x, y) = \langle 2x, 2y \rangle$ and $\nabla g(x, y) = \langle 2x + y, x + 2y \rangle$, the equations that result from $\nabla f = \lambda \nabla g$ and the constraint are

$$\underbrace{2x = \lambda(2x + y),}_{f_x = \lambda g_x} \qquad \underbrace{2y = \lambda(x + 2y),}_{f_y = \lambda g_y} \qquad \text{and} \qquad \underbrace{x^2 + xy + y^2 - 4 = 0.}_{\text{constraint } g(x, y) = 0}$$

Subtracting the second equation from the first leads to

$$(x - y)(2 - \lambda) = 0,$$

which implies that $y = x$, or $\lambda = 2$. In the case that $y = x$, the constraint equation simplifies to $3x^2 - 4 = 0$, or $x = \pm\frac{2}{\sqrt{3}}$. Therefore, two candidates for locations of extreme values are $\left(\frac{2}{\sqrt{3}}, \frac{2}{\sqrt{3}}\right)$ and $\left(-\frac{2}{\sqrt{3}}, -\frac{2}{\sqrt{3}}\right)$.

Substituting $\lambda = 2$ into the first equation leads to $y = -x$, and then the constraint equation simplifies to $x^2 - 4 = 0$, or $x = \pm 2$. These values give two additional points of interest, $(2, -2)$ and $(-2, 2)$. Evaluating f at each of these points, we find that $f\left(\frac{2}{\sqrt{3}}, \frac{2}{\sqrt{3}}\right) = f\left(-\frac{2}{\sqrt{3}}, -\frac{2}{\sqrt{3}}\right) = \frac{14}{3}$ and $f(2, -2) = f(-2, 2) = 10$. Therefore, the absolute maximum value of f on C is 10, which occurs at $(2, -2)$ and $(-2, 2)$, and the absolute minimum value of f on C is 14/3, which occurs at $\left(\frac{2}{\sqrt{3}}, \frac{2}{\sqrt{3}}\right)$ and $\left(-\frac{2}{\sqrt{3}}, -\frac{2}{\sqrt{3}}\right)$. Notice that the value of λ is not used in the final result.

Related Exercises 9–10 ◀

QUICK CHECK 2 Choose any point on the constraint curve in Figure 15.81b other than a solution point. Draw ∇f and ∇g at that point and show that they are not parallel. ◀

Lagrange Multipliers with Three Independent Variables

The technique just outlined extends to three or more independent variables. With three variables, suppose an objective function $w = f(x, y, z)$ is given; its level surfaces are surfaces in $\mathbb{R}^3$ (**Figure 15.82a**). The constraint equation takes the form $g(x, y, z) = 0$, which is another surface S in $\mathbb{R}^3$ (**Figure 15.82b**). To find the local maximum and minimum values of f on S (assuming they exist), we must find the points (a, b, c) on S at which $\nabla f(a, b, c)$ is parallel to $\nabla g(a, b, c)$, assuming $\nabla g(a, b, c) \neq \mathbf{0}$ (**Figure 15.82c, d**). In the case where the surface $g(x, y, z) = 0$ is closed and bounded, the procedure for finding the absolute maximum and minimum values of $f(x, y, z)$, where the point (x, y, z) is constrained to lie on S, is similar to the procedure for two variables.

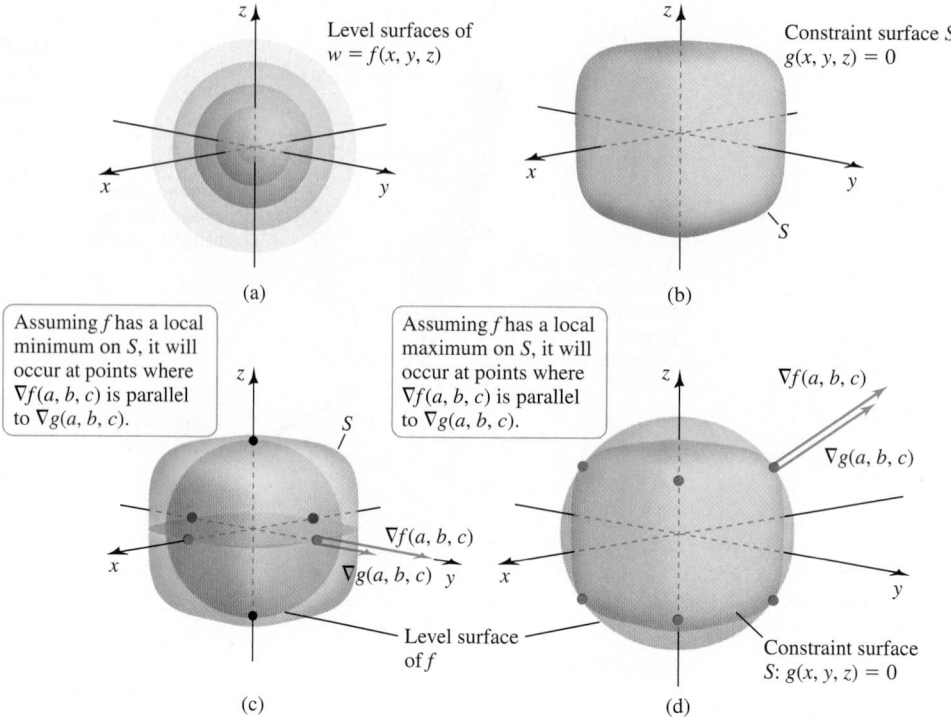

Figure 15.82

▶ If the constraint surface S: $g(x, y, z) = 0$ has a boundary curve C (see figure), then each point on C is a candidate for the location of an absolute maximum or minimum value of f, and these points must be analyzed in Step 2 of the procedure. We avoid this case in the exercise set.

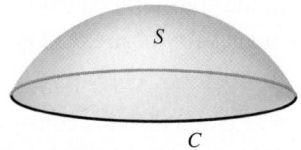

▶ Problems similar to Example 2 were solved in Section 15.7 using ordinary optimization techniques. These methods may or may not be easier to apply than Lagrange multipliers.

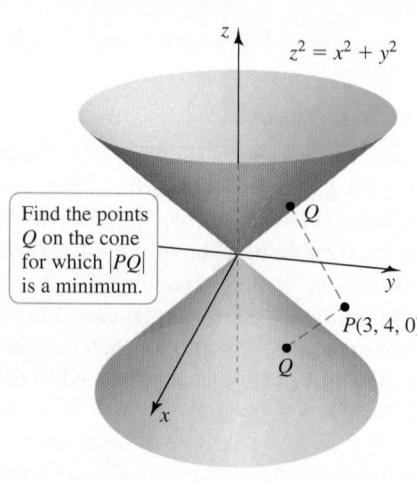

Figure 15.83

PROCEDURE   Lagrange Multipliers: Absolute Extrema on Closed and Bounded Constraint Surfaces

Let f and g be differentiable on a region of $\mathbb{R}^3$ with $\nabla g(x, y, z) \ne \mathbf{0}$ on the surface $g(x, y, z) = 0$. To locate the absolute maximum and minimum values of f subject to the constraint $g(x, y, z) = 0$, carry out the following steps.

1. Find the values of x, y, z, and λ that satisfy the equations

$$\nabla f(x, y, z) = \lambda \nabla g(x, y, z) \quad \text{and} \quad g(x, y, z) = 0.$$

2. Among the points (x, y, z) found in Step 1, select the largest and smallest corresponding function values. These values are the absolute maximum and minimum values of f subject to the constraint.

Now there are four equations to be solved for x, y, z, and λ:

$$f_x(x, y, z) = \lambda g_x(x, y, z), \qquad f_y(x, y, z) = \lambda g_y(x, y, z),$$
$$f_z(x, y, z) = \lambda g_z(x, y, z), \quad \text{and} \quad g(x, y, z) = 0.$$

As in the two-variable case, special care must be given to constraint surfaces that are not closed and bounded. We examine one such case in Example 2.

EXAMPLE 2   A geometry problem Find the least distance between the point $P(3, 4, 0)$ and the surface of the cone $z^2 = x^2 + y^2$.

SOLUTION The cone is not bounded, so we begin our calculations recognizing that solutions are only candidates for local extrema. Figure 15.83 shows both sheets of the cone and the point $P(3, 4, 0)$. Because P is in the xy-plane, we anticipate two solutions, one for each sheet of the cone. The distance between P and any point $Q(x, y, z)$ on the cone is

$$d(x, y, z) = \sqrt{(x - 3)^2 + (y - 4)^2 + z^2}.$$

In many distance problems, it is easier to work with the *square* of the distance to avoid dealing with square roots. This maneuver is allowable because if a point minimizes $(d(x, y, z))^2$, it also minimizes $d(x, y, z)$. Therefore, we define

$$f(x, y, z) = (d(x, y, z))^2 = (x - 3)^2 + (y - 4)^2 + z^2.$$

The constraint is the condition that the point (x, y, z) must lie on the cone, which implies $z^2 = x^2 + y^2$, or $g(x, y, z) = z^2 - x^2 - y^2 = 0$.

Now we proceed with Lagrange multipliers; the conditions are

$$f_x(x, y, z) = \lambda g_x(x, y, z), \quad \text{or } 2(x - 3) = \lambda(-2x), \quad \text{or } x(1 + \lambda) = 3, \tag{1}$$

$$f_y(x, y, z) = \lambda g_y(x, y, z), \quad \text{or } 2(y - 4) = \lambda(-2y), \quad \text{or } y(1 + \lambda) = 4, \tag{2}$$

$$f_z(x, y, z) = \lambda g_z(x, y, z), \quad \text{or } 2z = \lambda(2z), \quad \text{or } z = \lambda z, \text{ and} \tag{3}$$

$$g(x, y, z) = z^2 - x^2 - y^2 = 0. \tag{4}$$

The solutions of equation (3) (the simplest of the four equations) are either $z = 0$, or $\lambda = 1$ and $z \neq 0$. In the first case, if $z = 0$, then by equation (4), $x = y = 0$; however, $x = 0$ and $y = 0$ do not satisfy (1) and (2). So no solution results from this case.

On the other hand, if $\lambda = 1$ in equation (3), then by (1) and (2), we find that $x = \frac{3}{2}$ and $y = 2$. Using (4), the corresponding values of z are $\pm\frac{5}{2}$. Therefore, the two solutions and the values of f are

$$x = \tfrac{3}{2}, \quad y = 2, \quad z = \tfrac{5}{2}, \quad \text{with } f\left(\tfrac{3}{2}, 2, \tfrac{5}{2}\right) = \tfrac{25}{2}, \text{ and}$$

$$x = \tfrac{3}{2}, \quad y = 2, \quad z = -\tfrac{5}{2}, \quad \text{with } f\left(\tfrac{3}{2}, 2, -\tfrac{5}{2}\right) = \tfrac{25}{2}.$$

You can check that moving away from $\left(\frac{3}{2}, 2, \pm\frac{5}{2}\right)$ in any direction on the cone has the effect of increasing the values of f. Therefore, the points correspond to *local* minima of f. Do these points also correspond to *absolute* minima? The domain of this problem is unbounded; however, one can argue geometrically that f increases without bound moving away from $\left(\frac{3}{2}, 2, \pm\frac{5}{2}\right)$ on the cone with $|x| \to \infty$ and $|y| \to \infty$. Therefore, these points correspond to absolute minimum values and the points on the cone nearest to $(3, 4, 0)$ are $\left(\frac{3}{2}, 2, \pm\frac{5}{2}\right)$, at a distance of $\sqrt{\dfrac{25}{2}} = \dfrac{5}{\sqrt{2}}$. (Recall that $f = d^2$.)

Related Exercises 32–34 ◄

Economic Models In the opening of this section, we briefly described how utility functions are used to model consumer behavior. We now look in more detail at some specific—admittedly simple—utility functions and the constraints that are imposed on them.

As described earlier, a prototype model for consumer behavior uses two independent variables: leisure time ℓ and consumable goods g. A utility function $U = f(\ell, g)$ measures consumer preferences for various combinations of leisure time and consumable goods. The following assumptions about utility functions are commonly made.

1. Utility increases if any variable increases (essentially, *more is better*).

2. Various combinations of leisure time and consumable goods have the same utility; that is, giving up some leisure time for additional consumable goods (or vice versa) results in the same utility.

The level curves of a typical utility function are shown in **Figure 15.84**. Assumption 1 is reflected by the fact that the utility values on the level curves increase as either ℓ or g increases. Consistent with Assumption 2, a single level curve shows the combinations of ℓ and g that have the same utility; for this reason, economists call the level curves *indifference curves*. Notice that if ℓ increases, then g must decrease on a level curve to maintain the same utility, and vice versa.

Economic models assert that consumers maximize utility subject to constraints on leisure time and consumable goods. One assumption that leads to a reasonable constraint is that an increase in leisure time implies a linear decrease in consumable goods. Therefore, the constraint curve is a line with negative slope (**Figure 15.85**). When such a constraint is superimposed on the level curves of the utility function, the optimization problem becomes evident. Among all points on the constraint line, which one maximizes utility? A solution is marked in the figure; at this point, the utility has a maximum value (between 2.5 and 3.0).

► With three independent variables, it is possible to impose two constraints. These problems are explored in Exercises 61–64.

QUICK CHECK 3 In Example 2, is there a point that *maximizes* the distance between $(3, 4, 0)$ and the cone? If the point $(3, 4, 0)$ were replaced by $(3, 4, 1)$, how many minimizing solutions would there be? ◄

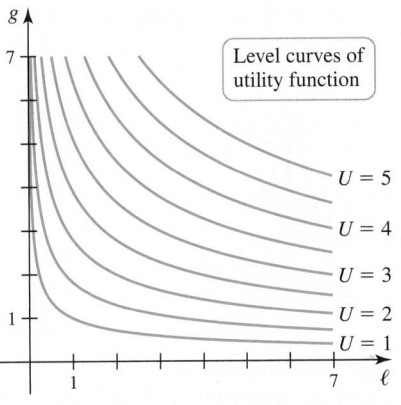

Figure 15.84

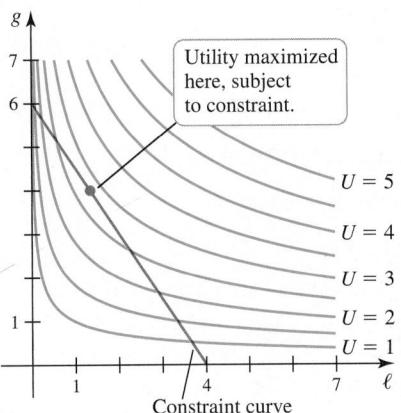

Figure 15.85

EXAMPLE 3   Constrained optimization of utility Find the absolute maximum value of the utility function $U = f(\ell, g) = \ell^{1/3}g^{2/3}$, subject to the constraint $G(\ell, g) = 3\ell + 2g - 12 = 0$, where $\ell \geq 0$ and $g \geq 0$.

SOLUTION The constraint is closed and bounded, so we expect to find an absolute maximum value of f. The level curves of the utility function and the linear constraint are shown in Figure 15.85. The solution follows the Lagrange multiplier method with two variables. The gradient of the utility function is

$$\nabla f(\ell, g) = \left\langle \frac{\ell^{-2/3}g^{2/3}}{3}, \frac{2\ell^{1/3}g^{-1/3}}{3} \right\rangle = \frac{1}{3}\left\langle \left(\frac{g}{\ell}\right)^{2/3}, 2\left(\frac{\ell}{g}\right)^{1/3} \right\rangle.$$

The gradient of the constraint function is $\nabla G(\ell, g) = \langle 3, 2 \rangle$. Therefore, the equations that must be solved are

$$\frac{1}{3}\left(\frac{g}{\ell}\right)^{2/3} = 3\lambda, \quad \frac{2}{3}\left(\frac{\ell}{g}\right)^{1/3} = 2\lambda, \quad \text{and} \quad G(\ell, g) = 3\ell + 2g - 12 = 0.$$

Eliminating λ from the first two equations leads to the condition $g = 3\ell$, which, when substituted into the constraint equation, gives the solution $\ell = \frac{4}{3}$ and $g = 4$. This point is a candidate for the location of the absolute maximum; the other candidates are the endpoints of the constraint curve, $(4, 0)$ and $(0, 6)$. The actual values of the utility function at these points are $U = f\left(\frac{4}{3}, 4\right) = 4/\sqrt[3]{3} \approx 2.8$ and $f(4, 0) = f(0, 6) = 0$. We conclude that the maximum value of f is 2.8; this solution occurs at $\ell = \frac{4}{3}$ and $g = 4$, and it is consistent with Figure 15.85.

Related Exercise 38 ◄

QUICK CHECK 4 In Figure 15.85, explain why, if you move away from the optimal point along the constraint line, the utility decreases. ◄

SECTION 15.8 EXERCISES

Getting Started

1. Explain why, at a point that maximizes or minimizes f subject to a constraint $g(x, y) = 0$, the gradient of f is parallel to the gradient of g. Use a diagram.

2. Describe the steps used to find the absolute maximum value and absolute minimum value of a differentiable function on a circle centered at the origin of the xy-plane.

3. For functions $f(x, y) = x + 4y$ and $g(x, y) = x^2 + y^2 - 1$, write the Lagrange multiplier conditions that must be satisfied by a point that maximizes or minimizes f subject to the constraint $g(x, y) = 0$.

4. For functions $f(x, y, z) = xyz$ and $g(x, y, z) = x^2 + 2y^2 + 3z^2 - 1$, write the Lagrange multiplier conditions that must be satisfied by a point that maximizes or minimizes f subject to the constraint $g(x, y) = 0$.

5–6. The following figures show the level curves of f and the constraint curve g(x, y) = 0. Estimate the maximum and minimum values of f subject to the constraint. At each point where an extreme value occurs, indicate the direction of ∇f and a possible direction of ∇g.

5.

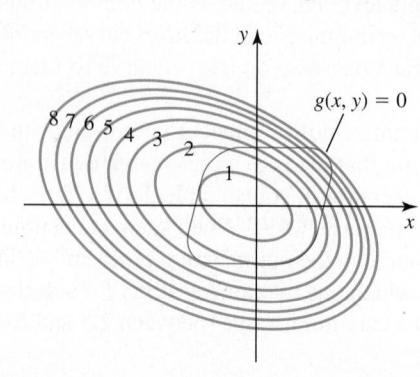

6.

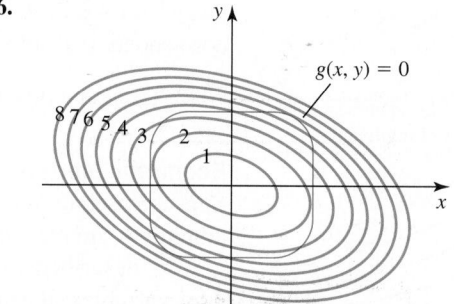

Practice Exercises

7–26. Lagrange multipliers *Each function f has an absolute maximum value and absolute minimum value subject to the given constraint. Use Lagrange multipliers to find these values.*

7. $f(x, y) = x + 2y$ subject to $x^2 + y^2 = 4$

8. $f(x, y) = xy^2$ subject to $x^2 + y^2 = 1$

9. $f(x, y) = x + y$ subject to $x^2 - xy + y^2 = 1$

10. $f(x, y) = x^2 + y^2$ subject to $2x^2 + 3xy + 2y^2 = 7$

11. $f(x, y) = xy$ subject to $x^2 + y^2 - xy = 9$

12. $f(x, y) = x - y$ subject to $x^2 + y^2 - 3xy = 20$

13. $f(x, y) = e^{xy}$ subject to $x^2 + xy + y^2 = 9$

14. $f(x, y) = x^2y$ subject to $x^2 + y^2 = 9$

15. $f(x, y) = 2x^2 + y^2$ subject to $x^2 + 2y + y^2 = 15$

16. $f(x, y) = x^2$ subject to $x^2 + xy + y^2 = 3$

17. $f(x, y, z) = x + 3y - z$ subject to $x^2 + y^2 + z^2 = 4$

18. $f(x, y, z) = xyz$ subject to $x^2 + 2y^2 + 4z^2 = 9$

19. $f(x, y, z) = x$ subject to $x^2 + y^2 + z^2 - z = 1$

20. $f(x, y, z) = x - z$ subject to $x^2 + y^2 + z^2 - y = 2$

21. $f(x, y, z) = x + y + z$ subject to $x^2 + y^2 + z^2 - xy = 5$

22. $f(x, y, z) = x + y + z$ subject to $x^2 + y^2 + z^2 - 2x - 2y = 1$

23. $f(x, y, z) = 2x + z^2$ subject to $x^2 + y^2 + 2z^2 = 25$

24. $f(x, y, z) = xy - z$ subject to $x^2 + y^2 + z^2 - xy = 1$

25. $f(x, y, z) = x^2 + y + z$ subject to $2x^2 + 2y^2 + z^2 = 2$

26. $f(x, y, z) = (xyz)^{1/2}$ subject to $x + y + z = 1$ with $x \geq 0$, $y \geq 0, z \geq 0$

27–36. Applications of Lagrange multipliers *Use Lagrange multipliers in the following problems. When the constraint curve is unbounded, explain why you have found an absolute maximum or minimum value.*

27. Shipping regulations A shipping company requires that the sum of length plus girth of rectangular boxes not exceed 108 in. Find the dimensions of the box with maximum volume that meets this condition. (The girth is the perimeter of the smallest side of the box.)

28. Box with minimum surface area Find the dimensions of the rectangular box with a volume of 16 ft³ that has minimum surface area.

T 29. Extreme distances to an ellipse Find the minimum and maximum distances between the ellipse $x^2 + xy + 2y^2 = 1$ and the origin.

30. Maximum area rectangle in an ellipse Find the dimensions of the rectangle of maximum area with sides parallel to the coordinate axes that can be inscribed in the ellipse $4x^2 + 16y^2 = 16$.

31. Maximum perimeter rectangle in an ellipse Find the dimensions of the rectangle of maximum perimeter with sides parallel to the coordinate axes that can be inscribed in the ellipse $2x^2 + 4y^2 = 3$.

32. Minimum distance to a plane Find the point on the plane $2x + 3y + 6z - 10 = 0$ closest to the point $(-2, 5, 1)$.

33. Minimum distance to a surface Find the point on the surface $4x + y - 1 = 0$ closest to the point $(1, 2, -3)$.

34. Minimum distance to a cone Find the points on the cone $z^2 = x^2 + y^2$ closest to the point $(1, 2, 0)$.

35. Extreme distances to a sphere Find the minimum and maximum distances between the sphere $x^2 + y^2 + z^2 = 9$ and the point $(2, 3, 4)$.

36. Maximum volume cylinder in a sphere Find the dimensions of the right circular cylinder of maximum volume that can be inscribed in a sphere of radius 16.

37–40. Maximizing utility functions *Find the values of ℓ and g with $\ell \geq 0$ and $g \geq 0$ that maximize the following utility functions subject to the given constraints. Give the value of the utility function at the optimal point.*

37. $U = f(\ell, g) = 10\ell^{1/2}g^{1/2}$ subject to $3\ell + 6g = 18$

38. $U = f(\ell, g) = 32\ell^{2/3}g^{1/3}$ subject to $4\ell + 2g = 12$

39. $U = f(\ell, g) = 8\ell^{4/5}g^{1/5}$ subject to $10\ell + 8g = 40$

40. $U = f(\ell, g) = \ell^{1/6}g^{5/6}$ subject to $4\ell + 5g = 20$

41. Explain why or why not Determine whether the following statements are true and give an explanation or counterexample.

   a. Suppose you are standing at the center of a sphere looking at a point P on the surface of the sphere. Your line of sight to P is orthogonal to the plane tangent to the sphere at P.

   b. At a point that maximizes f on the curve $g(x, y) = 0$, the dot product $\nabla f \cdot \nabla g$ is zero.

42–47. Alternative method *Solve the following problems from Section 15.7 using Lagrange multipliers.*

42. Exercise 43   **43.** Exercise 44   **44.** Exercise 45

45. Exercise 46   **46.** Exercise 70   **47.** Exercise 63

48–51. Absolute maximum and minimum values *Find the absolute maximum and minimum values of the following functions over the given regions R. Use Lagrange multipliers to check for extreme points on the boundary.*

48. $f(x, y) = x^2 + 4y^2 + 1; R = \{(x, y): x^2 + 4y^2 \leq 1\}$

49. $f(x, y) = x^2 + y^2 - 2y + 1; R = \{(x, y): x^2 + y^2 \leq 4\}$ (This is Exercise 47, Section 15.7.)

50. $f(x, y) = 2x^2 + y^2; R = \{(x, y): x^2 + y^2 \leq 16\}$ (This is Exercise 48, Section 15.7.)

51. $f(x, y) = 2x^2 - 4x + 3y^2 + 2;$
$R = \{(x, y): (x - 1)^2 + y^2 \leq 1\}$ (This is Exercise 51, Section 15.7.)

52. Extreme points on flattened spheres The equation $x^{2n} + y^{2n} + z^{2n} = 1$, where n is a positive integer, describes a flattened sphere. Define the extreme points to be the points on the flattened sphere with a maximum distance from the origin.

   a. Find all the extreme points on the flattened sphere with $n = 2$. What is the distance between the extreme points and the origin?

   b. Find all the extreme points on the flattened sphere for integers $n > 2$. What is the distance between the extreme points and the origin?

   c. Give the location of the extreme points in the limit as $n \to \infty$. What is the limiting distance between the extreme points and the origin as $n \to \infty$?

53–55. Production functions *Economists model the output of manufacturing systems using production functions that have many of the same properties as utility functions. The family of Cobb-Douglas production functions has the form $P = f(K, L) = CK^aL^{1-a}$, where K represents capital, L represents labor, and C and a are positive real numbers with $0 < a < 1$. If the cost of capital is p dollars per unit, the cost of labor is q dollars per unit, and the total available budget is B, then the constraint takes the form $pK + qL = B$. Find the values of K and L that maximize the following production functions subject to the given constraint, assuming $K \geq 0$ and $L \geq 0$.*

53. $P = f(K, L) = K^{1/2}L^{1/2}$ for $20K + 30L = 300$

54. $P = f(K, L) = 10K^{1/3}L^{2/3}$ for $30K + 60L = 360$

55. Given the production function $P = f(K, L) = K^aL^{1-a}$ and the budget constraint $pK + qL = B$, where $a, p, q,$ and B are given,

   show that P is maximized when $K = \dfrac{aB}{p}$ and $L = \dfrac{(1 - a)B}{q}$.

56. Temperature of an elliptical plate The temperature of points on an elliptical plate $x^2 + y^2 + xy \leq 1$ is given by $T(x, y) = 25(x^2 + y^2)$. Find the hottest and coldest temperatures on the edge of the plate.

Explorations and Challenges

57–59. Maximizing a sum

57. Find the maximum value of $x_1 + x_2 + x_3 + x_4$ subject to the condition that $x_1^2 + x_2^2 + x_3^2 + x_4^2 = 16$.

58. Generalize Exercise 57 and find the maximum value of $x_1 + x_2 + \cdots + x_n$ subject to the condition that $x_1^2 + x_2^2 + \cdots + x_n^2 = c^2$ for a real number c and a positive integer n.

59. Generalize Exercise 57 and find the maximum value of $a_1 x_1 + a_2 x_2 + \cdots + a_n x_n$ subject to the condition that $x_1^2 + x_2^2 + \cdots + x_n^2 = 1$, for given positive real numbers $a_1, \ldots, a_n$ and a positive integer n.

60. Geometric and arithmetic means Given positive numbers $x_1, \ldots, x_n$, prove that the geometric mean $(x_1 x_2 \cdots x_n)^{1/n}$ is no greater than the arithmetic mean $\dfrac{x_1 + \cdots + x_n}{n}$ in the following cases.

   a. Find the maximum value of xyz, subject to $x + y + z = k$, where k is a positive real number and $x > 0$, $y > 0$, and $z > 0$. Use the result to prove that
$$(xyz)^{1/3} \leq \frac{x + y + z}{3}.$$

   b. Generalize part (a) and show that
$$(x_1 x_2 \cdots x_n)^{1/n} \leq \frac{x_1 + \cdots + x_n}{n}.$$

61. Problems with two constraints Given a differentiable function $w = f(x, y, z)$, the goal is to find its absolute maximum and minimum values (assuming they exist) subject to the constraints $g(x, y, z) = 0$ and $h(x, y, z) = 0$, where g and h are also differentiable.

   a. Imagine a level surface of the function f and the constraint surfaces $g(x, y, z) = 0$ and $h(x, y, z) = 0$. Note that g and h intersect (in general) in a curve C on which maximum and minimum values of f must be found. Explain why ∇g and ∇h are orthogonal to their respective surfaces.

   b. Explain why ∇f lies in the plane formed by ∇g and ∇h at a point of C where f has a maximum or minimum value.

   c. Explain why part (b) implies that $\nabla f = \lambda \nabla g + \mu \nabla h$ at a point of C where f has a maximum or minimum value, where λ and μ (the Lagrange multipliers) are real numbers.

   d. Conclude from part (c) that the equations that must be solved for maximum or minimum values of f subject to two constraints are $\nabla f = \lambda \nabla g + \mu \nabla h$, $g(x, y, z) = 0$, and $h(x, y, z) = 0$.

62–64. Two-constraint problems *Use the result of Exercise 61 to solve the following problems.*

62. The planes $x + 2z = 12$ and $x + y = 6$ intersect in a line L. Find the point on L nearest the origin.

63. Find the maximum and minimum values of $f(x, y, z) = xyz$ subject to the conditions that $x^2 + y^2 = 4$ and $x + y + z = 1$.

64. Find the maximum and minimum values of $f(x, y, z) = x^2 + y^2 + z^2$ on the curve on which the cone $z^2 = 4x^2 + 4y^2$ and the plane $2x + 4z = 5$ intersect.

65. Check assumptions Consider the function $f(x, y) = xy + x + y + 100$ subject to the constraint $xy = 4$.

   a. Use the method of Lagrange multipliers to write a system of three equations with three variables x, y, and λ.

   b. Solve the system in part (a) to verify that $(x, y) = (-2, -2)$ and $(x, y) = (2, 2)$ are solutions.

   c. Let the curve C_1 be the branch of the constraint curve corresponding to $x > 0$. Calculate $f(2, 2)$ and determine whether this value is an absolute maximum or minimum value of f over C_1. (*Hint:* Let $h_1(x)$, for $x > 0$, equal the values of f over the curve C_1 and determine whether h_1 attains an absolute maximum or minimum value at $x = 2$.)

   d. Let the curve C_2 be the branch of the constraint curve corresponding to $x < 0$. Calculate $f(-2, -2)$ and determine whether this value is an absolute maximum or minimum value of f over C_2. (*Hint:* Let $h_2(x)$, for $x < 0$, equal the values of f over the curve C_2 and determine whether h_2 attains an absolute maximum or minimum value at $x = -2$.)

   e. Show that the method of Lagrange multipliers fails to find the absolute maximum and minimum values of f over the constraint curve $xy = 4$. Reconcile your explanation with the method of Lagrange multipliers.

QUICK CHECK ANSWERS

1. Note that $\nabla f(1, 1) = \langle 2x, 2y \rangle \big|_{(1, 1)} = \langle 2, 2 \rangle$ and $\nabla g(1, 1) = \langle \frac{1}{2}(x - 3), -1 \rangle \big|_{(1, 1)} = \langle -1, -1 \rangle$, which implies the gradients are multiples of one another, and therefore parallel. The equation of the line tangent to C at $(1, 1)$ is $y = -x + 2$; therefore, the vector $\mathbf{v} = \langle 1, -1 \rangle$ is parallel to this tangent line. Because $\nabla f(1, 1) \cdot \mathbf{v} = 0$ and $\nabla g(1, 1) \cdot \mathbf{v} = 0$, both gradients are orthogonal to the tangent line.   **3.** The distance between $(3, 4, 0)$ and the cone can be arbitrarily large, so there is no maximizing solution. If the point of interest is not in the xy-plane, there is one minimizing solution.   **4.** If you move along the constraint line away from the optimal solution in either direction, you cross level curves of the utility function with decreasing values. ◄

CHAPTER 15   REVIEW EXERCISES

1. Explain why or why not Determine whether the following statements are true and give an explanation or counterexample.

  a. The level curves of $g(x, y) = e^{x+y}$ are lines.
  b. The equation $z^2 = 2x^2 - 6y^2$ determines z as a single function of x and y.
  c. If f has continuous partial derivatives of all orders, then $f_{xxy} = f_{yyx}$.
  d. Given the surface $z = f(x, y)$, the gradient $\nabla f(a, b)$ lies in the plane tangent to the surface at $(a, b, f(a, b))$.

2–5. Domains *Find the domain of the following functions. Make a sketch of the domain in the xy-plane.*

2. $f(x, y) = \sin^{-1}\left(\dfrac{x^2 + y^2}{4}\right)$ **3.** $f(x, y) = \sqrt{x - y^2}$

4. $f(x, y) = \ln xy$ **5.** $f(x, y) = \sqrt{9x^2 + 4y^2 - 36}$

6–7. Graphs *Describe the graph of the following functions, and state the domain and range of the function.*

6. $f(x, y) = -\sqrt{x^2 + y^2}$ **7.** $g(x, y) = -\sqrt{x^2 + y^2 - 1}$

8–9. Level curves *Make a sketch of several level curves of the following functions. Label at least two level curves with their z-values.*

8. $f(x, y) = x^2 - y$ **9.** $f(x, y) = x^2 + 4y^2$

10. Matching level curves with surfaces Match level curve plots a–d with surfaces A–D.

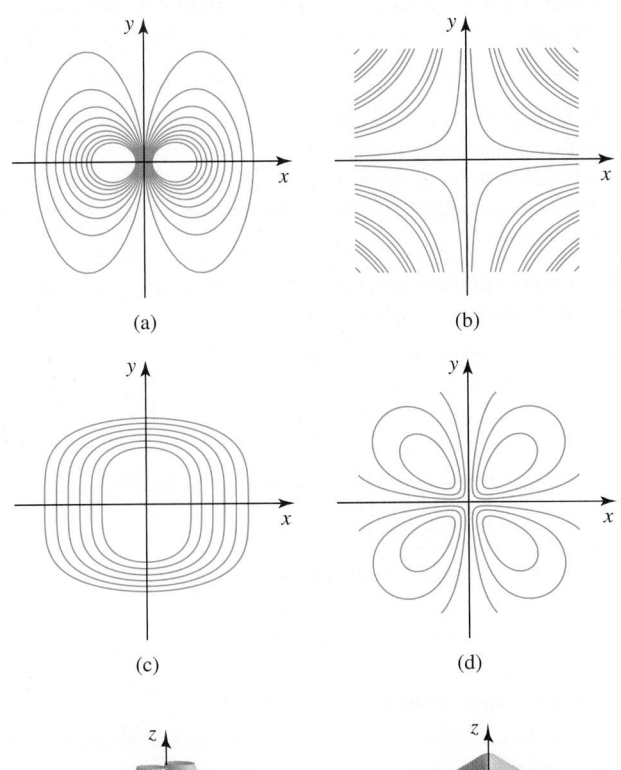

(a)   (b)   (c)   (d)   (A)   (B)

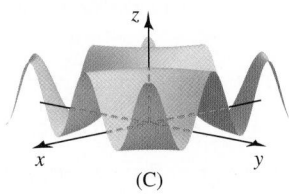
(C)

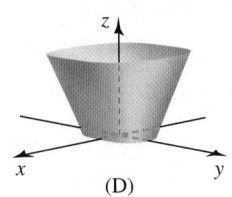
(D)

11–18. Limits *Evaluate the following limits or determine that they do not exist.*

11. $\displaystyle\lim_{(x, y)\to(4, -2)} (10x - 5y + 6xy)$

12. $\displaystyle\lim_{(x, y)\to(1, 1)} \frac{xy}{x + y}$ **13.** $\displaystyle\lim_{(x, y)\to(0, 0)} \frac{x + y}{xy}$

14. $\displaystyle\lim_{(x, y)\to(0, 0)} \frac{\sin xy}{x^2 + y^2}$ **15.** $\displaystyle\lim_{(x, y)\to(-1, 1)} \frac{x^2 - y^2}{x^2 - xy - 2y^2}$

16. $\displaystyle\lim_{(x, y)\to(0, 0)} \frac{25x^6 y^4}{\sin^2(x^3 y^2)}$

17. $\displaystyle\lim_{(x, y, z)\to(2, 2, 3)} \frac{x^2 z - 3x^2 - y^2 z + 3y^2}{xz - 3x - yz + 3y}$

18. $\displaystyle\lim_{(x, y, z)\to(3, 4, 7)} \frac{\sqrt{x + y} - \sqrt{z}}{x + y - z}$

19–20. Continuity *At what points of $\mathbb{R}^2$ are the following functions continuous?*

19. $f(x, y) = \ln(y - x^2 - 1)$ **20.** $g(x, y) = \dfrac{1}{x^2 + y^2}$

21–26. Partial derivatives *Find the first partial derivatives of the following functions.*

21. $f(x, y) = 3x^2 y^5$ **22.** $g(x, y, z) = 4xyz^2 - \dfrac{3x}{y}$

23. $f(x, y) = \dfrac{x^2}{x^2 + y^2}$ **24.** $g(x, y, z) = \dfrac{xyz}{x + y}$

25. $f(x, y) = xye^{xy}$ **26.** $g(u, v) = u\cos v - v\sin u$

27–28. Second partial derivatives *Find the four second partial derivatives of the following functions.*

27. $f(x, y) = e^{2xy}$ **28.** $H(p, r) = p^2\sqrt{p + 2r}$

29–30. Laplace's equation *Verify that the following functions satisfy Laplace's equation,* $\dfrac{\partial^2 u}{\partial x^2} + \dfrac{\partial^2 u}{\partial y^2} = 0.$

29. $u(x, y) = y(3x^2 - y^2)$ **30.** $u(x, y) = \ln(x^2 + y^2)$

31. Region between spheres Two spheres have the same center and radii r and R, where $0 < r < R$. The volume of the region between the spheres is $V(r, R) = \dfrac{4\pi}{3}(R^3 - r^3)$.

  a. First use your intuition. If r is held fixed, how does V change as R increases? What is the sign of V_R? If R is held fixed, how does V change as r increases (up to the value of R)? What is the sign of V_r?
  b. Compute V_r and V_R. Are the results consistent with part (a)?

c. Consider spheres with $R = 3$ and $r = 1$. Does the volume change more if R is increased by $\Delta R = 0.1$ (with r fixed) or if r is decreased by $\Delta r = 0.1$ (with R fixed)?

32–35. Chain Rule *Use the Chain Rule to evaluate the following derivatives.*

32. $w'(t)$, where $w = x \cos yz$, $x = t^2 + 1$, $y = t$, and $z = t^3$

33. $w'(t)$, where $w = z \ln(x^2 + y^2)$, $x = 3e^t$, $y = 4e^t$, and $z = t$

34. w_s, w_t, w_{ss}, w_{tt}, and w_{st}, where $w = xyz$, $x = 2st$, $y = st^2$, and $z = s^2t$

35. w_r, w_s and w_t, where $w = \ln(xy^2)$, $x = rst$, and $y = r + s$

36–37. Implicit differentiation *Find dy/dx for the following implicit relations using Theorem 15.9.*

36. $2x^2 + 3xy - 3y^4 = 2$ **37.** $y \ln(x^2 + y^2) = 4$

38–39. Walking on a surface *Consider the following surfaces and parameterized curves C in the xy-plane.*

a. *In each case, find $z'(t)$ on C.*
b. *Imagine that you are walking on the surface directly above C consistent with the positive orientation of C. Find the values of t for which you are walking uphill.*

38. $z = 4x^2 + y^2 - 2$; $C: x = \cos t$, $y = \sin t$, for $0 \le t \le 2\pi$

39. $z = x^2 - 2y^2 + 4$; $C: x = 2 \cos t$, $y = 2 \sin t$, for $0 \le t \le 2\pi$

40. Constant volume cones Suppose the radius of a right circular cone increases as $r(t) = t^a$ and the height decreases as $h(t) = t^{-b}$, for $t \ge 1$, where a and b are positive constants. What is the relationship between a and b such that the volume of the cone remains constant (that is, $V'(t) = 0$, where $V = (\pi/3)r^2h$)?

41. Directional derivatives Consider the function $f(x, y) = 2x^2 - 4y^2 + 10$, whose graph is shown in the figure.

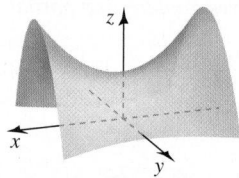

a. Fill in the table showing the values of the directional derivative at points (a, b) in the directions given by the unit vectors $\mathbf{u}$, $\mathbf{v}$, and $\mathbf{w}$.

	$(a, b) = (0, 0)$	$(a, b) = (2, 0)$	$(a, b) = (1, 1)$
$\mathbf{u} = \left\langle \frac{\sqrt{2}}{2}, \frac{\sqrt{2}}{2} \right\rangle$			
$\mathbf{v} = \left\langle -\frac{\sqrt{2}}{2}, \frac{\sqrt{2}}{2} \right\rangle$			
$\mathbf{w} = \left\langle -\frac{\sqrt{2}}{2}, -\frac{\sqrt{2}}{2} \right\rangle$			

b. Interpret each of the directional derivatives computed in part (a) at the point $(2, 0)$.

42–47. Computing directional derivatives *Compute the gradient of the following functions, evaluate it at the given point P, and evaluate the directional derivative at that point in the direction of the given vector.*

42. $f(x, y) = x^2$; $P(1, 2)$; $\mathbf{u} = \left\langle \frac{1}{\sqrt{2}}, -\frac{1}{\sqrt{2}} \right\rangle$

43. $g(x, y) = x^2y^3$; $P(-1, 1)$; $\mathbf{u} = \left\langle \frac{5}{13}, \frac{12}{13} \right\rangle$

44. $f(x, y) = \dfrac{x}{y^2}$; $P(0, 3)$; $\mathbf{u} = \left\langle \frac{\sqrt{3}}{2}, \frac{1}{2} \right\rangle$

45. $h(x, y) = \sqrt{2 + x^2 + 2y^2}$; $P(2, 1)$; $\mathbf{u} = \left\langle \frac{3}{5}, \frac{4}{5} \right\rangle$

46. $f(x, y, z) = xe^{1+y^2+z^2}$; $P(0, 1, -2)$; $\mathbf{u} = \left\langle \frac{4}{9}, \frac{1}{9}, \frac{8}{9} \right\rangle$

47. $f(x, y, z) = \sin xy + \cos z$; $P(1, \pi, 0)$; $\mathbf{u} = \left\langle \frac{2}{7}, \frac{3}{7}, -\frac{6}{7} \right\rangle$

48–49. Direction of steepest ascent and descent

a. *Find the unit vectors that give the direction of steepest ascent and steepest descent at P.*
b. *Find a unit vector that points in a direction of no change.*

48. $f(x, y) = \ln(1 + xy)$; $P(2, 3)$

49. $f(x, y) = \sqrt{4 - x^2 - y^2}$; $P(-1, 1)$

50–51. Level curves *Consider the paraboloid $f(x, y) = 8 - 2x^2 - y^2$. For the following level curves $f(x, y) = C$ and points (a, b), compute the slope of the line tangent to the level curve at (a, b) and verify that the tangent line is orthogonal to the gradient at that point.*

50. $f(x, y) = 5$; $(a, b) = (1, 1)$

51. $f(x, y) = 0$; $(a, b) = (2, 0)$

52. Directions of zero change Find the directions in which the function $f(x, y) = 4x^2 - y^2$ has zero change at the point $(1, 1, 3)$. Express the directions in terms of unit vectors.

53. Electric potential due to a charged cylinder An infinitely long charged cylinder of radius R with its axis along the z-axis has an electric potential $V = k \ln \dfrac{R}{r}$, where r is the distance between a variable point $P(x, y)$ and the axis of the cylinder ($r^2 = x^2 + y^2$) and k is a physical constant. The electric field at a point (x, y) in the xy-plane is given by $\mathbf{E} = -\nabla V$, where ∇V is the two-dimensional gradient. Compute the electric field at a point (x, y) with $r > R$.

54–59. Tangent planes *Find an equation of the plane tangent to the following surfaces at the given points.*

54. $z = 2x^2 + y^2$; $(1, 1, 3)$ and $(0, 2, 4)$

55. $x^2 + \dfrac{y^2}{4} - \dfrac{z^2}{9} = 1$; $(0, 2, 0)$ and $\left(1, 1, \frac{3}{2}\right)$

56. $x^2 - 2x + y^2 + 4y + 3z^2 = 2$; $(3, -2, 1)$ and $(-1, -2, 1)$

57. $e^{xy^2z^3-1} = 1$; $(1, 1, 1)$ and $(1, -1, 1)$

58. $z - \tan^{-1} xy = 0$; $(1, 1, \pi/4)$ and $(1, \sqrt{3}, \pi/3)$

59. $\sqrt{\dfrac{x + y}{z}} = 1$; $(2, 2, 4)$ and $(10, -1, 9)$

T 60–61. Linear approximation

a. *Find the linear approximation to the function f at the point (a, b).*
b. *Use part (a) to estimate the given function value.*

60. $f(x, y) = 4 \cos(2x - y)$; $(a, b) = \left(\frac{\pi}{4}, \frac{\pi}{4}\right)$; estimate $f(0.8, 0.8)$.

61. $f(x, y) = (x + y)e^{xy}$; $(a, b) = (2, 0)$; estimate $f(1.95, 0.05)$.

62. Changes in a function Estimate the change in the function $f(x, y) = -2y^2 + 3x^2 + xy$ when (x, y) changes from $(1, -2)$ to $(1.05, -1.9)$.

63. Volume of a cylinder The volume of a cylinder with radius r and height h is $V = \pi r^2 h$. Find the approximate percentage change in the volume when the radius decreases by 3% and the height increases by 2%.

64. Volume of an ellipsoid The volume of an ellipsoid with axes of length $2a$, $2b$, and $2c$ is $V = \pi abc$. Find the percentage change in the volume when a increases by 2%, b increases by 1.5%, and c decreases by 2.5%.

65. Water level changes A hemispherical tank with a radius of 1.50 m is filled with water to a depth of 1.00 m. Water is released from the tank, and the water level drops by 0.05 m (from 1.00 m to 0.95 m).

   a. Approximate the change in the volume of water in the tank.

      The volume of a spherical cap is $V = \dfrac{1}{3}\pi h^2(3r - h)$, where r is the radius of the sphere and h is the thickness of the cap (in this case, the depth of the water).

   b. Approximate the change in the surface area of the water in the tank.

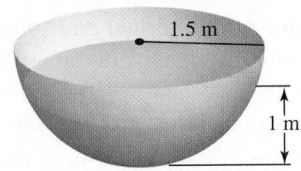

66–69. Analyzing critical points *Identify the critical points of the following functions. Then determine whether each critical point corresponds to a local maximum, local minimum, or saddle point. State when your analysis is inconclusive. Confirm your results using a graphing utility.*

66. $f(x, y) = x^4 + y^4 - 16xy$ **67.** $f(x, y) = \dfrac{x^3}{3} - \dfrac{y^3}{3} + 2xy$

68. $f(x, y) = xy(2 + x)(y - 3)$

69. $f(x, y) = 10 - x^3 - y^3 - 3x^2 + 3y^2$

70–73. Absolute maxima and minima *Find the absolute maximum and minimum values of the following functions on the specified region R.*

70. $f(x, y) = \dfrac{x^3}{3} - \dfrac{y^3}{3} + 2xy$ on the rectangle
$R = \{(x, y): 0 \le x \le 3, -1 \le y \le 1\}$

71. $f(x, y) = x^4 + y^4 - 4xy + 1$ on the square
$R = \{(x, y): -2 \le x \le 2, -2 \le y \le 2\}$

72. $f(x, y) = x^2 y - y^3$ on the triangle
$R = \{(x, y): 0 \le x \le 2, 0 \le y \le 2 - x\}$

73. $f(x, y) = xy$ on the semicircular disk
$R = \left\{(x, y): -1 \le x \le 1, 0 \le y \le \sqrt{1 - x^2}\right\}$

74. Least distance What point on the plane $x + y + 4z = 8$ is closest to the origin? Give an argument showing you have found an absolute minimum of the distance function.

75–78. Lagrange multipliers *Use Lagrange multipliers to find the absolute maximum and minimum values of f (if they exist) subject to the given constraint.*

75. $f(x, y) = 2x + y + 10$ subject to $2(x - 1)^2 + 4(y - 1)^2 = 1$

76. $f(x, y) = xy$ subject to $3x^2 - 2xy + 3y^2 = 4$

77. $f(x, y, z) = x + 2y - z$ subject to $x^2 + y^2 + z^2 = 1$

78. $f(x, y, z) = x^2 y^2 z$ subject to $2x^2 + y^2 + z^2 = 25$

79. Maximum perimeter rectangle Use Lagrange multipliers to find the dimensions of the rectangle with the maximum perimeter that can be inscribed with sides parallel to the coordinate axes in the ellipse $\dfrac{x^2}{a^2} + \dfrac{y^2}{b^2} = 1$.

80. Minimum surface area cylinder Use Lagrange multipliers to find the dimensions of the right circular cylinder of minimum surface area (including the circular ends) with a volume of 32π in³. Give an argument showing you have found an absolute minimum.

81. Minimum distance to a cone Find the point(s) on the cone $z^2 - x^2 - y^2 = 0$ that are closest to the point $(1, 3, 1)$. Give an argument showing you have found an absolute minimum of the distance function.

82. Gradient of a distance function Let $P_0(a, b, c)$ be a fixed point in $\mathbb{R}^3$, and let $d(x, y, z)$ be the distance between P_0 and a variable point $P(x, y, z)$.

   a. Compute $\nabla d(x, y, z)$.

   b. Show that $\nabla d(x, y, z)$ points in the direction from P_0 to P and has magnitude 1 for all (x, y, z).

   c. Describe the level surfaces of d and give the direction of $\nabla d(x, y, z)$ relative to the level surfaces of d.

   d. Discuss $\lim\limits_{P \to P_0} \nabla d(x, y, z)$.

83. Minimum distance to a paraboloid Use Lagrange multipliers to find the point on the paraboloid $z = x^2 + y^2$ that lies closest to the point $(5, 10, 3)$. Give an argument showing you have found an absolute minimum of the distance function.

Chapter 15 Guided Projects

Applications of the material in this chapter and related topics can be found in the following Guided Projects. For additional information, see the Preface.

- Traveling waves
- Ecological diversity
- Economic production functions

16

Multiple Integration

Chapter Preview We have now generalized limits and derivatives to functions of several variables. The next step is to carry out a similar process with respect to integration. As you know, single (one-variable) integrals are developed from Riemann sums and are used to compute areas of regions in $\mathbb{R}^2$. In an analogous way, we use Riemann sums to develop double (two-variable) and triple (three-variable) integrals, which are used to compute volumes of solid regions in $\mathbb{R}^3$. These multiple integrals have many applications in statistics, science, and engineering, including calculating the mass, the center of mass, and moments of inertia of solids with a variable density. Another significant development in this chapter is the appearance of cylindrical and spherical coordinates. These alternative coordinate systems often simplify the evaluation of integrals in three-dimensional space. The chapter closes with the two- and three-dimensional versions of the substitution (change of variables) rule. The overall lesson of the chapter is that we can integrate functions over most geometrical objects, from intervals on the x-axis to regions in the plane bounded by curves to complicated three-dimensional solids.

16.1 Double Integrals over Rectangular Regions

In Chapter 15 the concept of differentiation was extended to functions of several variables. In this chapter, we extend integration to multivariable functions. By the close of the chapter, we will have completed Table 16.1, which is a basic road map for calculus.

Table 16.1

	Derivatives	Integrals
Single variable: $f(x)$	$f'(x)$	$\int_a^b f(x)\,dx$
Several variables: $f(x, y)$ and $f(x, y, z)$	$\dfrac{\partial f}{\partial x}, \dfrac{\partial f}{\partial y}, \dfrac{\partial f}{\partial z}$	$\iint\limits_R f(x, y)\,dA, \iiint\limits_D f(x, y, z)\,dV$

Volumes of Solids

The problem of finding the net area of a region bounded by a curve led to the definite integral in Chapter 5. Recall that we began that discussion by approximating the region with a collection of rectangles and then formed a Riemann sum of the areas of the rectangles. Under appropriate conditions, as the number of rectangles increases, the sum approaches the value of the definite integral, which is the net area of the region.

We now carry out an analogous procedure with surfaces defined by functions of the form $z = f(x, y)$, where, for the moment, we assume $f(x, y) \geq 0$ on a region R in the xy-plane (Figure 16.1a). The goal is to determine the volume of the solid bounded by the surface and R. In general terms, the solid is first approximated by *boxes* (Figure 16.1b). The sum of the volumes of these boxes, which is a Riemann sum, approximates the volume of the solid. Under appropriate conditions, as the number of boxes increases, the approximations converge to the value of a *double integral*, which is the volume of the solid.

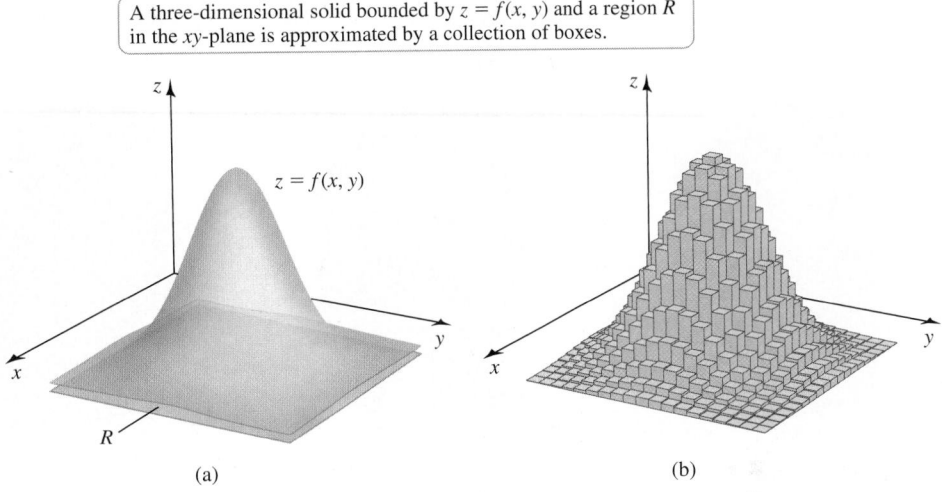

A three-dimensional solid bounded by $z = f(x, y)$ and a region R in the xy-plane is approximated by a collection of boxes.

(a)   (b)

Figure 16.1

> We adopt the convention that Δx_k and Δy_k are the side lengths of the kth rectangle, for $k = 1, \ldots, n$, even though there are generally fewer than n different values of Δx_k and Δy_k. This convention is used throughout the chapter.

We assume $z = f(x, y)$ is a nonnegative function defined on a *rectangular* region $R = \{(x, y): a \leq x \leq b, c \leq y \leq d\}$. A **partition** of R is formed by dividing R into n rectangular subregions using lines parallel to the x- and y-axes (not necessarily uniformly spaced). The rectangles may be numbered in any systematic way; for example, left to right and then bottom to top. The side lengths of the kth rectangle are denoted Δx_k and Δy_k, so the area of the kth rectangle is $\Delta A_k = \Delta x_k \Delta y_k$. We also let (x_k^*, y_k^*) be any point in the kth rectangle, for $1 \leq k \leq n$ (Figure 16.2).

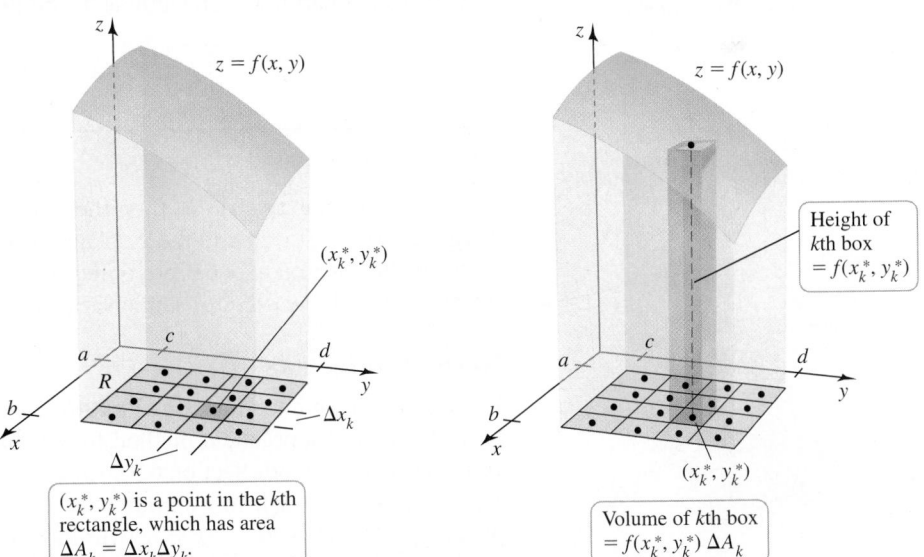

(x_k^*, y_k^*) is a point in the kth rectangle, which has area $\Delta A_k = \Delta x_k \Delta y_k$.

Figure 16.2

Height of kth box $= f(x_k^*, y_k^*)$

Volume of kth box $= f(x_k^*, y_k^*) \Delta A_k$

Figure 16.3

To approximate the volume of the solid bounded by the surface $z = f(x, y)$ and the region R, we construct boxes on each of the n rectangles; each box has a height of $f(x_k^*, y_k^*)$ and a base with area ΔA_k, for $1 \leq k \leq n$ (Figure 16.3). Therefore, the volume of the kth box is

$$f(x_k^*, y_k^*) \Delta A_k = f(x_k^*, y_k^*) \Delta x_k \Delta y_k.$$

QUICK CHECK 1 Explain why the sum for the volume is an approximation. How can the approximation be improved? ◄

The sum of the volumes of the n boxes gives an approximation to the volume of the solid:

$$V \approx \sum_{k=1}^{n} f(x_k^*, y_k^*)\, \Delta A_k.$$

We now let Δ be the maximum length of the diagonals of the rectangles in the partition. As $\Delta \to 0$, the areas of *all* the rectangles approach zero ($\Delta A_k \to 0$) and the number of rectangles increases ($n \to \infty$). If the approximations given by these Riemann sums have a limit as $\Delta \to 0$, then we define the volume of the solid to be that limit (**Figure 16.4**).

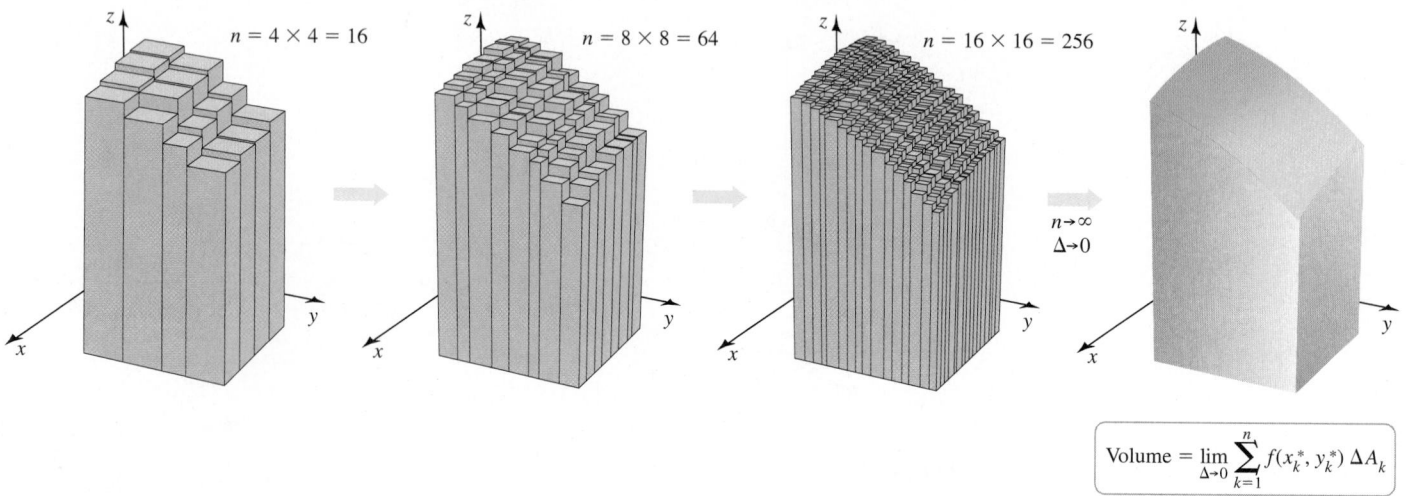

$$\text{Volume} = \lim_{\Delta \to 0} \sum_{k=1}^{n} f(x_k^*, y_k^*)\, \Delta A_k$$

Figure 16.4

► The functions that we encounter in this text are integrable. Advanced methods are needed to prove that continuous functions and many functions with finite discontinuities are also integrable.

DEFINITION Double Integrals

A function f defined on a rectangular region R in the xy-plane is **integrable** on R if $\lim_{\Delta \to 0} \sum_{k=1}^{n} f(x_k^*, y_k^*) \Delta A_k$ exists for all partitions of R and for all choices of (x_k^*, y_k^*) within those partitions. The limit is the **double integral of f over R**, which we write

$$\iint\limits_{R} f(x, y)\, dA = \lim_{\Delta \to 0} \sum_{k=1}^{n} f(x_k^*, y_k^*) \Delta A_k.$$

If f is nonnegative on R, then the double integral equals the volume of the solid bounded by $z = f(x, y)$ and the xy-plane over R. If f is negative on parts of R, the value of the double integral may be zero or negative, and the result is interpreted as a *net volume* (by analogy with *net area* for single variable integrals).

Iterated Integrals

Evaluating double integrals using limits of Riemann sums is tedious and rarely done. Fortunately, there is a practical method for evaluating double integrals that is based on the general slicing method (Section 6.3). An example illustrates the technique.

Suppose we wish to compute the volume of the solid region bounded by the plane $z = f(x, y) = 6 - 2x - y$ over the rectangular region $R = \{(x, y): 0 \le x \le 1, 0 \le y \le 2\}$ (**Figure 16.5**). By definition, the volume is given by the double integral

$$V = \iint\limits_{R} f(x, y)\, dA = \iint\limits_{R} (6 - 2x - y)\, dA.$$

According to the general slicing method (see margin note and figure), we can compute this volume by taking vertical slices through the solid parallel to the yz-plane (Figure 16.5). The slice at the point x has a cross-sectional area denoted $A(x)$. In general, as x varies, the

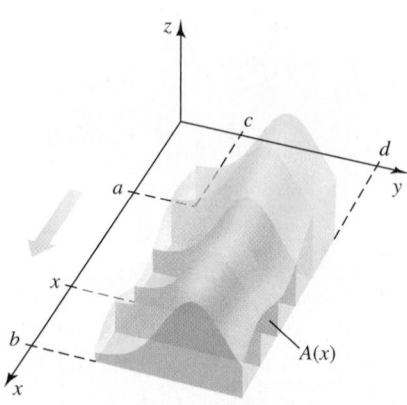

If a solid is sliced parallel to the y-axis and perpendicular to the xy-plane, and the cross-sectional area of the slice at the point x is $A(x)$, then the volume of the solid region is

$$V = \int_a^b A(x)\, dx.$$

area $A(x)$ also changes, so we integrate these cross-sectional areas from $x = 0$ to $x = 1$ to obtain the volume

$$V = \int_0^1 A(x)\,dx.$$

The important observation is that for a fixed value of x, $A(x)$ is the area of the plane region under the curve $z = 6 - 2x - y$. This area is computed by integrating f with respect to y from $y = 0$ to $y = 2$, holding x fixed; that is,

$$A(x) = \int_0^2 (6 - 2x - y)\,dy,$$

where $0 \le x \le 1$, and x is treated as a constant in the integration. Substituting for $A(x)$, we have

$$V = \int_0^1 A(x)\,dx = \int_0^1 \underbrace{\left(\int_0^2 (6 - 2x - y)\,dy \right)}_{A(x)} dx.$$

The expression that appears on the right side of this equation is called an **iterated integral** (meaning repeated integral). We first evaluate the inner integral with respect to y holding x fixed; the result is a function of x. Then the outer integral is evaluated with respect to x; the result is a real number, which is the volume of the solid in Figure 16.5. Both of these integrals are ordinary one-variable integrals.

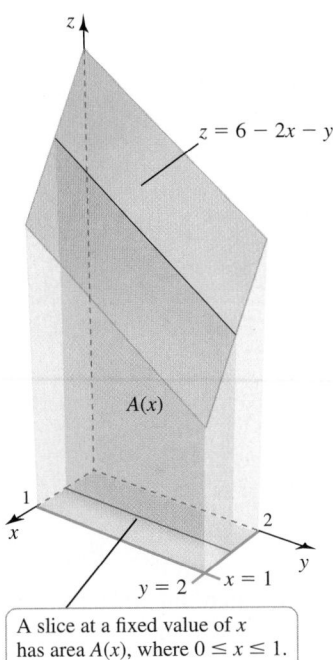

A slice at a fixed value of x has area $A(x)$, where $0 \le x \le 1$.

Figure 16.5

EXAMPLE 1   Evaluating an iterated integral Evaluate $V = \int_0^1 A(x)\,dx$, where $A(x) = \int_0^2 (6 - 2x - y)\,dy$.

SOLUTION Using the Fundamental Theorem of Calculus, holding x constant, we have

$$
\begin{aligned}
A(x) &= \int_0^2 (6 - 2x - y)\,dy \\
&= \left(6y - 2xy - \frac{y^2}{2} \right)\Big|_0^2 \quad \text{Evaluate integral with respect to } y; \\
&\qquad\qquad\qquad\qquad\qquad\quad x \text{ is constant.} \\
&= (12 - 4x - 2) - 0 \quad \text{Simplify; limits are in } y. \\
&= 10 - 4x. \quad\quad\quad\quad\quad\ \text{Simplify.}
\end{aligned}
$$

Substituting $A(x) = 10 - 4x$ into the volume integral, we have

$$
\begin{aligned}
V &= \int_0^1 A(x)\,dx \\
&= \int_0^1 (10 - 4x)\,dx \quad \text{Substitute for } A(x). \\
&= (10x - 2x^2)\Big|_0^1 \quad \text{Evaluate integral with respect to } x. \\
&= 8. \quad\quad\quad\quad\quad\quad\ \text{Simplify.}
\end{aligned}
$$

Related Exercises 10, 25 ◄

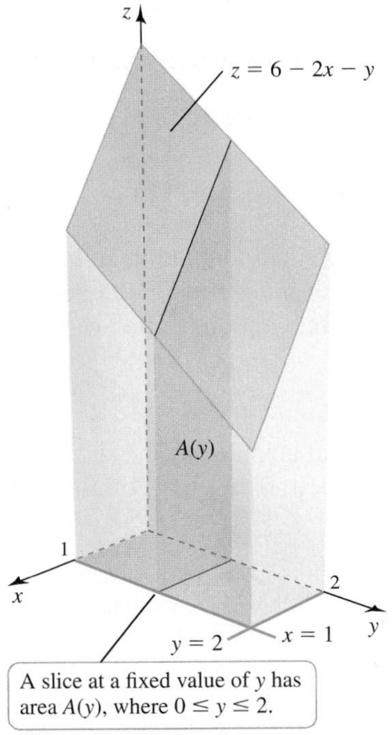

EXAMPLE 2   Same double integral, different order Example 1 used slices through the solid parallel to the yz-plane. Compute the volume of the same solid using vertical slices through the solid parallel to the xz-plane, for $0 \le y \le 2$ (Figure 16.6).

SOLUTION In this case, $A(y)$ is the area of a slice through the solid for a fixed value of y in the interval $0 \le y \le 2$. This area is computed by integrating $z = 6 - 2x - y$ from $x = 0$ to $x = 1$, holding y fixed; that is,

$$A(y) = \int_0^1 (6 - 2x - y)\,dx,$$

where $0 \le y \le 2$.

A slice at a fixed value of y has area $A(y)$, where $0 \le y \le 2$.

Figure 16.6

Using the general slicing method again, the volume is

$$V = \int_0^2 A(y)\,dy \qquad \text{General slicing method}$$

$$= \int_0^2 \left(\underbrace{\int_0^1 (6 - 2x - y)\,dx}_{A(y)} \right) dy \qquad \text{Substitute for } A(y).$$

$$= \int_0^2 \left((6x - x^2 - yx)\Big|_0^1 \right) dy \qquad \begin{array}{l}\text{Evaluate inner integral with respect}\\ \text{to } x;\ y \text{ is constant.}\end{array}$$

$$= \int_0^2 (5 - y)\,dy \qquad \text{Simplify; limits are in } x.$$

$$= \left(5y - \frac{y^2}{2} \right)\Big|_0^2 \qquad \text{Evaluate outer integral with respect to } y.$$

$$= 8. \qquad \text{Simplify.}$$

Related Exercise 37 ◄

Several important comments are in order. First, the two iterated integrals give the same value for the double integral. Second, the notation of the iterated integral must be used carefully. When we write $\int_c^d \int_a^b f(x, y)\,dx\,dy$, it means $\int_c^d \left(\int_a^b f(x, y)\,dx \right) dy$. The *inner* integral with respect to x is evaluated first, holding y fixed, and the variable runs from $x = a$ to $x = b$. The result of that integration is a constant or a function of y, which is then integrated in the *outer* integral, with the variable running from $y = c$ to $y = d$. The order of integration is signified by the order of dx and dy.

Similarly, $\int_a^b \int_c^d f(x, y)\,dy\,dx$ means $\int_a^b \left(\int_c^d f(x, y)\,dy \right) dx$. The inner integral with respect to y is evaluated first, holding x fixed. The result is then integrated with respect to x. In both cases, the limits of integration in the iterated integrals determine the boundaries of the rectangular *region of integration*.

Examples 1 and 2 illustrate one version of *Fubini's Theorem*, a deep result that relates double integrals to iterated integrals. The first version of the theorem applies to double integrals over rectangular regions.

> **QUICK CHECK 2** Consider the integral $\int_3^4 \int_1^2 f(x, y)\,dx\,dy$. Give the limits of integration and the variable of integration for the first (inner) integral and the second (outer) integral. Sketch the region of integration. ◄

> ➤ The area of the kth rectangle in the partition is $\Delta A_k = \Delta x_k \Delta y_k$, where Δx_k and Δy_k are the lengths of the sides of that rectangle. Accordingly, the *element of area dA* in the double integral becomes $dx\,dy$ or $dy\,dx$ in the iterated integral.

THEOREM 16.1   (Fubini) Double Integrals over Rectangular Regions
Let f be continuous on the rectangular region $R = \{(x, y): a \le x \le b, c \le y \le d\}$. The double integral of f over R may be evaluated by either of two iterated integrals:

$$\iint\limits_R f(x, y)\,dA = \int_c^d \int_a^b f(x, y)\,dx\,dy = \int_a^b \int_c^d f(x, y)\,dy\,dx.$$

The importance of Fubini's Theorem is twofold: It says that double integrals may be evaluated by iterated integrals. It also says that the order of integration in the iterated integrals does not matter (although in practice, one order of integration is often easier to use than the other).

EXAMPLE 3   A double integral Find the volume of the solid bounded by the surface $f(x, y) = 4 + 9x^2y^2$ over the region $R = \{(x, y): -1 \le x \le 1, 0 \le y \le 2\}$. Use both possible orders of integration.

SOLUTION Because $f(x, y) > 0$ on R, the volume of the region is given by the double integral $\iint_R (4 + 9x^2y^2)\,dA$. By Fubini's Theorem, the double integral is evaluated as an iterated integral. If we first integrate with respect to x, the area of a cross section of the solid for a fixed value of y is given by $A(y)$ (Figure 16.7a). The volume of the region is

$$\iint\limits_R (4 + 9x^2y^2)\,dA = \int_0^2 \underbrace{\int_{-1}^1 (4 + 9x^2y^2)\,dx}_{A(y)}\,dy \qquad \text{Convert to an iterated integral.}$$

$$= \int_0^2 (4x + 3x^3y^2)\Big|_{-1}^1\,dy \qquad \begin{array}{l}\text{Evaluate inner integral} \\ \text{with respect to } x.\end{array}$$

$$= \int_0^2 (8 + 6y^2)\,dy \qquad \text{Simplify.}$$

$$= (8y + 2y^3)\Big|_0^2 \qquad \begin{array}{l}\text{Evaluate outer integral} \\ \text{with respect to } y.\end{array}$$

$$= 32. \qquad \text{Simplify.}$$

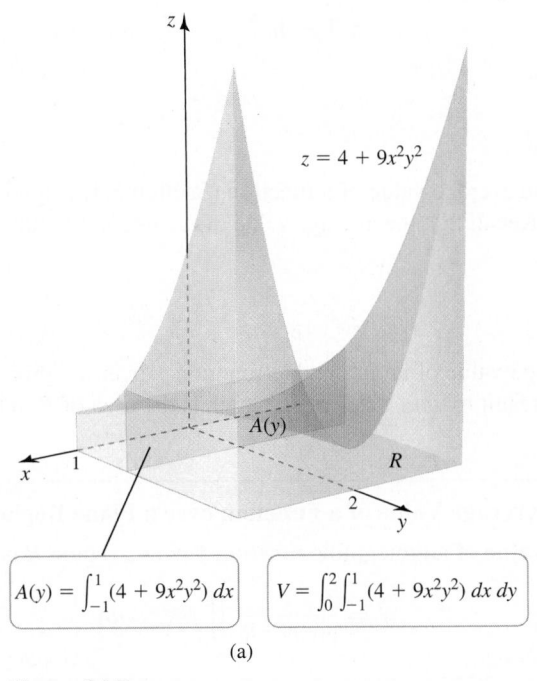

$$A(y) = \int_{-1}^1 (4 + 9x^2y^2)\,dx \qquad V = \int_0^2 \int_{-1}^1 (4 + 9x^2y^2)\,dx\,dy$$

(a)

$$A(x) = \int_0^2 (4 + 9x^2y^2)\,dy \qquad V = \int_{-1}^1 \int_0^2 (4 + 9x^2y^2)\,dy\,dx$$

(b)

Figure 16.7

Alternatively, if we integrate first with respect to y, the area of a cross section of the solid for a fixed value of x is given by $A(x)$ (Figure 16.7b). The volume of the region is

$$\iint\limits_R (4 + 9x^2y^2)\,dA = \int_{-1}^1 \underbrace{\int_0^2 (4 + 9x^2y^2)\,dy}_{A(x)}\,dx \qquad \text{Convert to an iterated integral.}$$

$$= \int_{-1}^1 (4y + 3x^2y^3)\Big|_0^2\,dx \qquad \begin{array}{l}\text{Evaluate inner integral with} \\ \text{respect to } y.\end{array}$$

$$= \int_{-1}^1 (8 + 24x^2)\,dx \qquad \text{Simplify.}$$

$$= (8x + 8x^3)\Big|_{-1}^1 = 32. \qquad \begin{array}{l}\text{Evaluate outer integral with} \\ \text{respect to } x.\end{array}$$

QUICK CHECK 3 Write the iterated integral $\int_{-10}^{10} \int_0^{20} (x^2y + 2xy^3)\,dy\,dx$ with the order of integration reversed. ◄

As guaranteed by Fubini's Theorem, the two iterated integrals are equal, both giving the value of the double integral and the volume of the solid.

Related Exercises 26, 39 ◄

The following example shows that sometimes the order of integration must be chosen carefully either to save work or to make the integration possible.

EXAMPLE 4   Choosing a convenient order of integration Evaluate $\iint_R ye^{xy}\, dA$, where $R = \{(x, y): 0 \le x \le 1, 0 \le y \le \ln 2\}$.

SOLUTION The iterated integral $\int_0^1 \int_0^{\ln 2} ye^{xy}\, dy\, dx$ requires first integrating ye^{xy} with respect to y, which entails integration by parts. An easier approach is to integrate first with respect to x:

$$
\begin{aligned}
\int_0^{\ln 2} \int_0^1 ye^{xy}\, dx\, dy &= \int_0^{\ln 2} e^{xy} \Big|_0^1\, dy && \text{Evaluate inner integral with respect to } x. \\
&= \int_0^{\ln 2} (e^y - 1)\, dy && \text{Simplify.} \\
&= (e^y - y) \Big|_0^{\ln 2} && \text{Evaluate outer integral with respect to } y. \\
&= 1 - \ln 2. && \text{Simplify.}
\end{aligned}
$$

Related Exercises 41, 43 ◄

Average Value

The concept of the average value of a function (Section 5.4) extends naturally to functions of two variables. Recall that the average value of the integrable function f over the interval $[a, b]$ is

$$
\bar{f} = \frac{1}{b - a} \int_a^b f(x)\, dx.
$$

To find the average value of an integrable function f over a region R, we integrate f over R and divide the result by the "size" of R, which is the area of R in the two-variable case.

> ➤ The same definition of average value applies to more general regions in the plane.

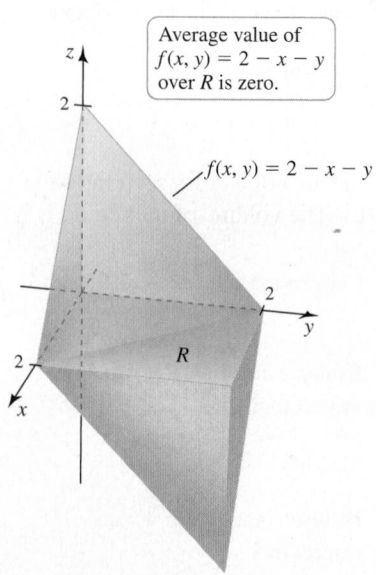

Average value of $f(x, y) = 2 - x - y$ over R is zero.

$f(x, y) = 2 - x - y$

R

Figure 16.8

> ➤ An average value of 0 means that over the region R, the volume of the solid above the xy-plane and below the surface equals the volume of the solid below the xy-plane and above the surface.

DEFINITION   Average Value of a Function over a Plane Region

The **average value** of an integrable function f over a region R is

$$
\bar{f} = \frac{1}{\text{area of } R} \iint_R f(x, y)\, dA.
$$

EXAMPLE 5   Average value Find the average value of the quantity $2 - x - y$ over the square $R = \{(x, y): 0 \le x \le 2, 0 \le y \le 2\}$ (Figure 16.8).

SOLUTION The area of the region R is 4. Letting $f(x, y) = 2 - x - y$, the average value of f is

$$
\begin{aligned}
\frac{1}{\text{area of } R} \iint_R f(x, y)\, dA &= \frac{1}{4} \iint_R (2 - x - y)\, dA \\
&= \frac{1}{4} \int_0^2 \int_0^2 (2 - x - y)\, dx\, dy && \text{Convert to an iterated integral.} \\
&= \frac{1}{4} \int_0^2 \left(2x - \frac{x^2}{2} - xy \right) \Big|_0^2\, dy && \text{Evaluate inner integral with respect to } x. \\
&= \frac{1}{4} \int_0^2 (2 - 2y)\, dy && \text{Simplify.} \\
&= 0. && \text{Evaluate outer integral with respect to } y.
\end{aligned}
$$

Related Exercise 46 ◄

SECTION 16.1 EXERCISES

Getting Started

1. Write an iterated integral that gives the volume of the solid bounded by the surface $f(x, y) = xy$ over the square $R = \{(x, y): 0 \le x \le 2, 1 \le y \le 3\}$.

2. Write an iterated integral that gives the volume of a box with height 10 and base $R = \{(x, y): 0 \le x \le 5, -2 \le y \le 4\}$.

3. Write two iterated integrals that equal $\iint_R f(x, y)\, dA$, where $R = \{(x, y): -2 \le x \le 4, 1 \le y \le 5\}$.

4. Consider the integral $\int_1^3 \int_{-1}^1 (2y^2 + xy)\, dy\, dx$. State the variable of integration in the first (inner) integral and the limits of integration. State the variable of integration in the second (outer) integral and the limits of integration.

5. Region $R = \{(x, y): 0 \le x \le 4, 0 \le y \le 6\}$ is partitioned into six equal subregions (see figure, which also shows the level curves of a function f continuous on the region R). Estimate the value of $\iint_R f(x, y)\, dA$ by evaluating the Riemann sum $\sum_{k=1}^{6} f(x_k^*, y_k^*)\Delta A_k$, where (x_k^*, y_k^*) is the center of the kth subregion, for $k = 1, \ldots, 6$.

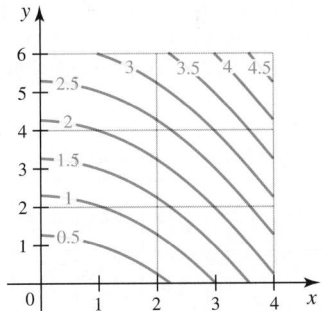

6. Draw a solid whose volume is given by the double integral $\int_0^6 \int_1^2 10\, dy\, dx$. Then evaluate the integral using geometry.

Practice Exercises

7–24. Iterated integrals *Evaluate the following iterated integrals.*

7. $\int_0^2 \int_0^1 4xy\, dx\, dy$

8. $\int_1^2 \int_0^1 (3x^2 + 4y^3)\, dy\, dx$

9. $\int_1^3 \int_0^2 x^2 y\, dx\, dy$

10. $\int_0^3 \int_{-2}^1 (2x + 3y)\, dx\, dy$

11. $\int_1^3 \int_0^{\pi/2} x \sin y\, dy\, dx$

12. $\int_1^3 \int_1^2 (y^2 + y)\, dx\, dy$

13. $\int_1^4 \int_0^4 \sqrt{uv}\, du\, dv$

14. $\int_0^{\pi/4} \int_0^3 r \sec \theta\, dr\, d\theta$

15. $\int_1^{\ln 5} \int_0^{\ln 3} e^{x+y}\, dx\, dy$

16. $\int_0^{\pi/2} \int_0^1 uv \cos (u^2 v)\, du\, dv$

17. $\int_0^1 \int_0^1 t^2 e^{st}\, ds\, dt$

18. $\int_0^2 \int_0^1 \frac{8xy}{1 + x^4}\, dx\, dy$

19. $\int_1^e \int_0^1 4(p + q) \ln q\, dp\, dq$

20. $\int_0^1 \int_0^{\pi} y^2 \cos xy\, dx\, dy$

21. $\int_1^2 \int_1^2 \frac{x}{x + y}\, dy\, dx$

22. $\int_0^2 \int_0^1 x^5 y^2 e^{x^3 y^3}\, dy\, dx$

23. $\int_0^1 \int_1^4 \frac{3y}{\sqrt{x + y^2}}\, dx\, dy$

24. $\int_0^1 \int_0^1 x^2 y^2 e^{x^3 y}\, dx\, dy$

25–35. Double integrals *Evaluate each double integral over the region R by converting it to an iterated integral.*

25. $\iint_R (x + 2y)\, dA; R = \{(x, y): 0 \le x \le 3, 1 \le y \le 4\}$

26. $\iint_R (x^2 + xy)\, dA; R = \{(x, y): 1 \le x \le 2, -1 \le y \le 1\}$

27. $\iint_R s^2 t \sin (st^2)\, dA; R = \{(s, t): 0 \le s \le \pi, 0 \le t \le 1\}$

28. $\iint_R \frac{x}{1 + xy}\, dA; R = \{(x, y): 0 \le x \le 1, 0 \le y \le 1\}$

29. $\iint_R \sqrt{\frac{x}{y}}\, dA; R = \{(x, y): 0 \le x \le 1, 1 \le y \le 4\}$

30. $\iint_R xy \sin x^2\, dA; R = \{(x, y): 0 \le x \le \sqrt{\pi/2}, 0 \le y \le 1\}$

31. $\iint_R e^{x+2y}\, dA; R = \{(x, y): 0 \le x \le \ln 2, 1 \le y \le \ln 3\}$

32. $\iint_R (x^2 - y^2)^2\, dA; R = \{(x, y): -1 \le x \le 2, 0 \le y \le 1\}$

33. $\iint_R (x^5 - y^5)^2\, dA; R = \{(x, y): 0 \le x \le 1, -1 \le y \le 1\}$

34. $\iint_R \cos (x\sqrt{y})\, dA; R = \{(x, y): 0 \le x \le 1, 0 \le y \le \pi^2/4\}$

35. $\iint_R x^3 y \cos (x^2 y^2)\, dA; R = \{(x, y): 0 \le x \le \sqrt{\pi/2}, 0 \le y \le 1\}$

36–39. Volumes of solids *Find the volume of the following solids.*

36. The solid beneath the cylinder $f(x, y) = e^{-x}$ and above the region $R = \{(x, y): 0 \le x \le \ln 4, -2 \le y \le 2\}$

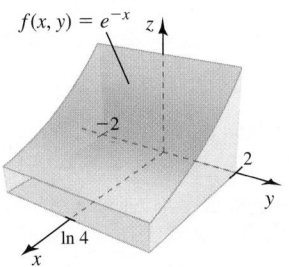

37. The solid beneath the plane $f(x, y) = 24 - 3x - 4y$ and above the region $R = \{(x, y): -1 \le x \le 3, 0 \le y \le 2\}$

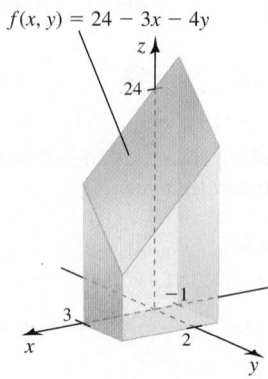

$f(x, y) = 24 - 3x - 4y$

38. The solid in the first octant bounded above by the surface $z = 9xy\sqrt{1 - x^2}\sqrt{4 - y^2}$ and below by the xy-plane

39. The solid in the first octant bounded by the surface $z = xy^2\sqrt{1 - x^2}$ and the planes $z = 0$ and $y = 3$

40–45. Choose a convenient order *When converted to an iterated integral, the following double integrals are easier to evaluate in one order than the other. Find the best order and evaluate the integral.*

40. $\displaystyle\iint\limits_{R} y \cos xy \, dA; R = \{(x, y): 0 \le x \le 1, 0 \le y \le \pi/3\}$

41. $\displaystyle\iint\limits_{R} (y + 1)e^{x(y+1)} \, dA; R = \{(x, y): 0 \le x \le 1, -1 \le y \le 1\}$

42. $\displaystyle\iint\limits_{R} x \sec^2 xy \, dA; R = \{(x, y): 0 \le x \le \pi/3, 0 \le y \le 1\}$

43. $\displaystyle\iint\limits_{R} 6x^5 e^{x^3 y} \, dA; R = \{(x, y): 0 \le x \le 2, 0 \le y \le 2\}$

44. $\displaystyle\iint\limits_{R} y^3 \sin(xy^2) \, dA; R = \{(x, y): 0 \le x \le 2, 0 \le y \le \sqrt{\pi/2}\}$

45. $\displaystyle\iint\limits_{R} \frac{x}{(1 + xy)^2} \, dA; R = \{(x, y): 0 \le x \le 4, 1 \le y \le 2\}$

46–48. Average value *Compute the average value of the following functions over the region R.*

46. $f(x, y) = 4 - x - y; R = \{(x, y): 0 \le x \le 2, 0 \le y \le 2\}$

47. $f(x, y) = e^{-y}; R = \{(x, y): 0 \le x \le 6, 0 \le y \le \ln 2\}$

48. $f(x, y) = \sin x \sin y; R = \{(x, y): 0 \le x \le \pi, 0 \le y \le \pi\}$

49. Average value Find the average squared distance between the points of $R = \{(x, y): -2 \le x \le 2, 0 \le y \le 2\}$ and the origin.

50. Average value Find the average squared distance between the points of $R = \{(x, y): 0 \le x \le 3, 0 \le y \le 3\}$ and the point $(3, 3)$.

51. Explain why or why not Determine whether the following statements are true and give an explanation or counterexample.

a. The region of integration for $\int_4^6 \int_1^3 4 \, dx \, dy$ is a square.

b. If f is continuous on $\mathbb{R}^2$, then
$$\int_4^6 \int_1^3 f(x, y) \, dx \, dy = \int_4^6 \int_1^3 f(x, y) \, dy \, dx.$$

c. If f is continuous on $\mathbb{R}^2$, then
$$\int_4^6 \int_1^3 f(x, y) \, dx \, dy = \int_1^3 \int_4^6 f(x, y) \, dy \, dx.$$

52. Symmetry Evaluate the following integrals using symmetry arguments. Let $R = \{(x, y): -a \le x \le a, -b \le y \le b\}$, where a and b are positive real numbers.

a. $\displaystyle\iint\limits_{R} xye^{-(x^2+y^2)} \, dA$

b. $\displaystyle\iint\limits_{R} \frac{\sin(x - y)}{x^2 + y^2 + 1} \, dA$

53. Computing populations The population densities in nine districts of a rectangular county are shown in the figure.

a. Use the fact that population = (population density) × (area) to estimate the population of the county.

b. Explain how the calculation of part (a) is related to Riemann sums and double integrals.

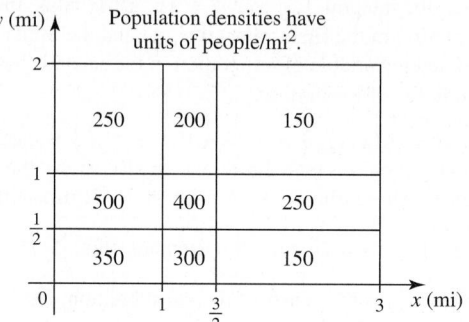

Population densities have units of people/mi².

	250	200	150
	500	400	250
	350	300	150

T 54. Approximating water volume The varying depth of an 18 m × 25 m swimming pool is measured in 15 different rectangles of equal area (see figure). Approximate the volume of water in the pool.

Depth readings have units of m.

0.75	1.25	1.75	2.25	2.75
1	1.5	2.0	2.5	3.0
1	1.5	2.0	2.5	3.0

Explorations and Challenges

55. Cylinders Let S be the solid in $\mathbb{R}^3$ between the cylinder $z = f(x)$ and the region $R = \{(x, y): a \le x \le b, c \le y \le d\}$, where $f(x) \ge 0$ on R. Explain why $\int_c^d \int_a^b f(x) \, dx \, dy$ equals the area of the constant cross section of S multiplied by $(d - c)$, which is the volume of S.

56. Product of integrals Suppose $f(x, y) = g(x)h(y)$, where g and h are continuous functions for all real values of x and y.

a. Show that $\int_c^d \int_a^b f(x, y) \, dx \, dy = \left(\int_a^b g(x) \, dx\right)\left(\int_c^d h(y) \, dy\right)$. Interpret this result geometrically.

b. Write $\left(\int_a^b g(x) \, dx\right)^2$ as an iterated integral.

c. Use the result of part (a) to evaluate $\int_0^{2\pi} \int_{10}^{30} e^{-4y^2} \cos x \, dy \, dx$.

57. Solving for a parameter Let $R = \{(x, y): 0 \le x \le \pi, 0 \le y \le a\}$. For what values of a, with $0 \le a \le \pi$, is $\iint_R \sin(x + y) \, dA$ equal to 1?

58–59. Zero average value *Find the value of $a > 0$ such that the average value of the following functions over $R = \{(x, y): 0 \le x \le a, 0 \le y \le a\}$ is zero.*

58. $f(x, y) = x + y - 8$ **59.** $f(x, y) = 4 - x^2 - y^2$

60. Maximum integral Consider the plane $x + 3y + z = 6$ over the rectangle R with vertices at $(0, 0)$, $(a, 0)$, $(0, b)$, and (a, b), where the vertex (a, b) lies on the line where the plane intersects the xy-plane (so $a + 3b = 6$). Find the point (a, b) for which the volume of the solid between the plane and R is a maximum.

61. Density and mass Suppose a thin rectangular plate, represented by a region R in the xy-plane, has a density given by the function $\rho(x, y)$; this function gives the *area density* in units such as grams per square centimeter (g/cm^2). The mass of the plate is $\iint_R \rho(x, y)\, dA$. Assume $R = \{(x, y): 0 \le x \le \pi/2, 0 \le y \le \pi\}$ and find the mass of the plates with the following density functions.

a. $\rho(x, y) = 1 + \sin x$ **b.** $\rho(x, y) = 1 + \sin y$
c. $\rho(x, y) = 1 + \sin x \sin y$

62. Approximating volume Propose a method based on Riemann sums to approximate the volume of the shed shown in the figure (the peak of the roof is directly above the rear corner of the shed). Carry out the method and provide an estimate of the volume.

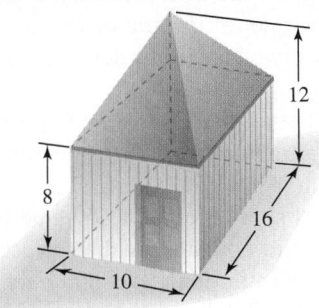

63. An identity Suppose the second partial derivatives of f are continuous on $R = \{(x, y): 0 \le x \le a, 0 \le y \le b\}$. Simplify

$$\iint_R \frac{\partial^2 f}{\partial x\, \partial y}\, dA.$$

QUICK CHECK ANSWERS

1. The sum gives the volume of a collection of rectangular boxes, and these boxes do not exactly fill the solid region under the surface. The approximation is improved by using more boxes.   **2.** Inner integral: x runs from $x = 1$ to $x = 2$; outer integral: y runs from $y = 3$ to $y = 4$. The region is the rectangle $\{(x, y): 1 \le x \le 2, 3 \le y \le 4\}$.
3. $\int_0^{20} \int_{-10}^{10} (x^2 y + 2xy^3)\, dx\, dy$ ◄

16.2 Double Integrals over General Regions

Evaluating double integrals over rectangular regions is a useful place to begin our study of multiple integrals. Problems of practical interest, however, usually involve nonrectangular regions of integration. The goal of this section is to extend the methods presented in Section 16.1 so that they apply to more general regions of integration.

General Regions of Integration

Consider a function f defined over a closed, bounded *nonrectangular* region R in the xy-plane. As with rectangular regions, we use a partition consisting of rectangles, but now, such a partition does not cover R exactly. In this case, only the n rectangles that lie entirely within R are considered to be in the partition (**Figure 16.9**). When f is nonnegative on R, the volume of the solid bounded by the surface $z = f(x, y)$ and the xy-plane over R is approximated by the Riemann sum

$$V \approx \sum_{k=1}^{n} f(x_k^*, y_k^*)\Delta A_k,$$

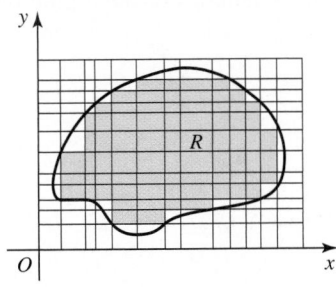

Figure 16.9

where $\Delta A_k = \Delta x_k \Delta y_k$ is the area of the kth rectangle and (x_k^*, y_k^*) is any point in the kth rectangle, for $1 \le k \le n$. As before, we define Δ to be the maximum length of the diagonals of the rectangles in the partition.

Under the assumptions that f is continuous on R and that the boundary of R consists of a finite number of smooth curves, two things occur as $\Delta \to 0$ and the number of rectangles increases $(n \to \infty)$.

- The rectangles in the partition fill R more and more completely; that is, the union of the rectangles approaches R.
- Over all partitions and all choices of (x_k^*, y_k^*) within a partition, the Riemann sums approach a (unique) limit.

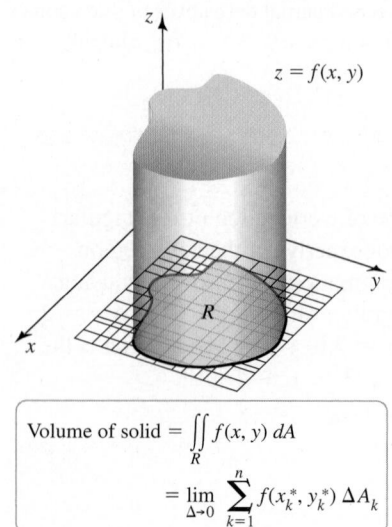

Volume of solid $= \iint\limits_{R} f(x, y)\, dA$

$= \lim\limits_{\Delta \to 0} \sum\limits_{k=1}^{n} f(x_k^*, y_k^*)\, \Delta A_k$

Figure 16.10

The limit approached by the Riemann sums is the **double integral of f over R**; that is,

$$\iint\limits_{R} f(x, y)\, dA = \lim\limits_{\Delta \to 0} \sum\limits_{k=1}^{n} f(x_k^*, y_k^*)\, \Delta A_k.$$

When this limit exists, f is **integrable** over R. If f is nonnegative on R, then the double integral equals the volume of the solid bounded by the surface $z = f(x, y)$ and the xy-plane over R (Figure 16.10).

The double integral $\iint_{R} f(x, y)\, dA$ has another common interpretation. Suppose R represents a thin plate whose density at the point (x, y) is $f(x, y)$. The units of density are mass per unit area, so the product $f(x_k^*, y_k^*)\, \Delta A_k$ approximates the mass of the kth rectangle in R. Summing the masses of the rectangles gives an approximation to the total mass of R. In the limit as $n \to \infty$ and $\Delta \to 0$, the double integral equals the mass of the plate.

Iterated Integrals

Double integrals over nonrectangular regions are also evaluated using iterated integrals. However, in this more general setting, the order of integration is critical. Most of the double integrals we encounter fall into one of two categories determined by the shape of the region R.

The first type of region has the property that its lower and upper boundaries are the graphs of continuous functions $y = g(x)$ and $y = h(x)$, respectively, for $a \le x \le b$. Such regions have any of the forms shown in Figure 16.11.

Once again, we appeal to the general slicing method. Assume for the moment that f is nonnegative on R and consider the solid bounded by the surface $z = f(x, y)$ and R (Figure 16.12). Imagine taking vertical slices through the solid parallel to the yz-plane. The cross section through the solid at a fixed value of x extends from the lower curve $y = g(x)$ to the upper curve $y = h(x)$. The area of that cross section is

$$A(x) = \int_{g(x)}^{h(x)} f(x, y)\, dy, \qquad \text{for } a \le x \le b.$$

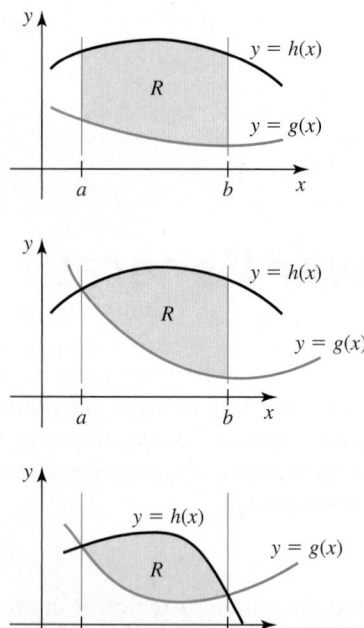

Figure 16.11

The volume of the solid is given by a double integral; it is evaluated by integrating the cross-sectional areas $A(x)$ from $x = a$ to $x = b$:

$$\iint\limits_{R} f(x, y)\, dA = \int_{a}^{b} \underbrace{\int_{g(x)}^{h(x)} f(x, y)\, dy}_{A(x)}\, dx.$$

The limits of integration in the iterated integral describe the boundaries of the region of integration R.

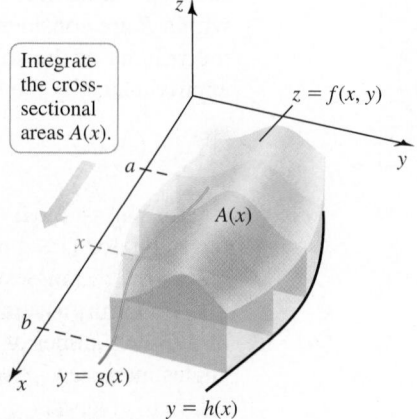

Figure 16.12

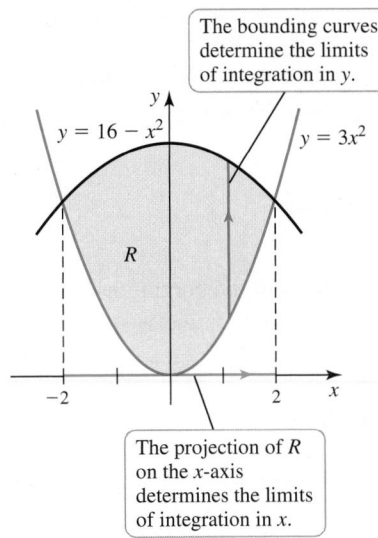

The bounding curves determine the limits of integration in y.

$y = 16 - x^2$ $y = 3x^2$

R

-2 2 x

The projection of R on the x-axis determines the limits of integration in x.

Figure 16.13

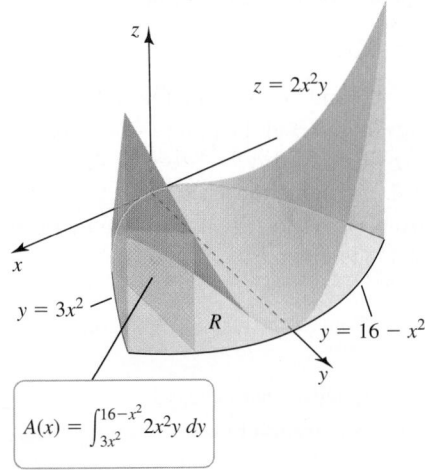

$z = 2x^2y$

x

$y = 3x^2$ R $y = 16 - x^2$

y

$A(x) = \int_{3x^2}^{16-x^2} 2x^2y \, dy$

Figure 16.14

QUICK CHECK 1 A region R is bounded by the x- and y-axes and the line $x + y = 2$. Suppose you integrate first with respect to y. Give the limits of the iterated integral over R. ◄

EXAMPLE 1   Evaluating a double integral Express the integral $\iint_R 2x^2y \, dA$ as an iterated integral, where R is the region bounded by the parabolas $y = 3x^2$ and $y = 16 - x^2$. Then evaluate the integral.

SOLUTION The region R is bounded below and above by the graphs of $g(x) = 3x^2$ and $h(x) = 16 - x^2$, respectively. Solving $3x^2 = 16 - x^2$, we find that these curves intersect at $x = -2$ and $x = 2$, which are the limits of integration in the x-direction (Figure 16.13).

Figure 16.14 shows the solid bounded by the surface $z = 2x^2y$ and the region R. A typical vertical cross section through the solid parallel to the yz-plane at a fixed value of x has area

$$A(x) = \int_{3x^2}^{16-x^2} 2x^2y \, dy.$$

Integrating these cross-sectional areas between $x = -2$ and $x = 2$, the iterated integral becomes

$$\iint_R 2x^2y \, dA = \int_{-2}^{2} \underbrace{\int_{3x^2}^{16-x^2} 2x^2y \, dy}_{A(x)} \, dx \qquad \text{Convert to an iterated integral.}$$

$$= \int_{-2}^{2} x^2 y^2 \Big|_{3x^2}^{16-x^2} \, dx \qquad \begin{array}{l}\text{Evaluate inner integral} \\ \text{with respect to } y.\end{array}$$

$$= \int_{-2}^{2} x^2 ((16 - x^2)^2 - (3x^2)^2) \, dx \qquad \text{Simplify.}$$

$$= \int_{-2}^{2} (-8x^6 - 32x^4 + 256x^2) \, dx \qquad \text{Simplify.}$$

$$\approx 663.2. \qquad \begin{array}{l}\text{Evaluate outer integral} \\ \text{with respect to } x.\end{array}$$

Because $z = 2x^2y \geq 0$ on R, the value of the integral is the volume of the solid shown in Figure 16.14.

Related Exercises 12, 46 ◄

Change of Perspective   Suppose the region of integration R is bounded on the left and right by the graphs of continuous functions $x = g(y)$ and $x = h(y)$, respectively, on the interval $c \leq y \leq d$. Such regions may take any of the forms shown in Figure 16.15.

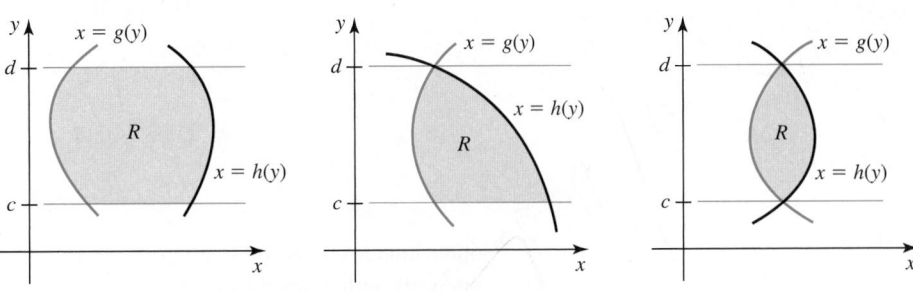

Figure 16.15

To find the volume of the solid bounded by the surface $z = f(x, y)$ and R, we now take vertical slices parallel to the xz-plane. The double integral $\iint_R f(x, y) \, dA$ is then converted to an iterated integral in which the inner integration is with respect to x over the interval $g(y) \leq x \leq h(y)$ and the outer integration is with respect to y over the interval $c \leq y \leq d$. The evaluation of double integrals in these two cases is summarized in the following theorem.

► Theorem 16.2 is another version of Fubini's Theorem. With integrals over nonrectangular regions, the order of integration cannot be simply switched; that is,

$$\int_a^b \int_{g(x)}^{h(x)} f(x, y)\, dy\, dx$$

$$\neq \int_{g(x)}^{h(x)} \int_a^b f(x, y)\, dx\, dy.$$

The *element of area dA* corresponds to the area of a small rectangle in the partition. Comparing the double integral to the iterated integral, we see that the element of area is $dA = dy\, dx$ or $dA = dx\, dy$, which is consistent with the area formula for rectangles.

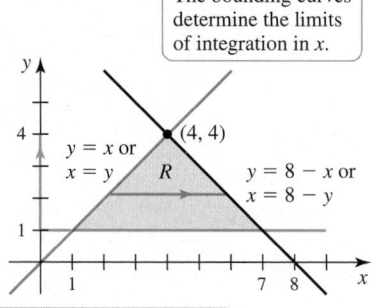

The bounding curves determine the limits of integration in x.

The projection of R on the y-axis determines the limits of integration in y.

Figure 16.16

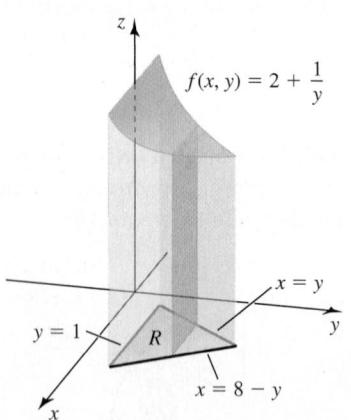

Figure 16.17

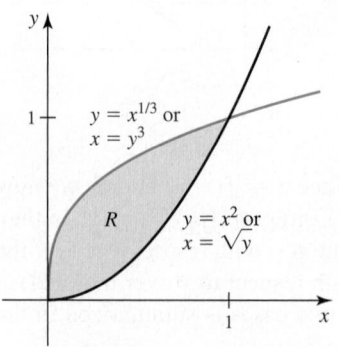

R is bounded above and below, and on the right and left, by curves.

Figure 16.18

> **THEOREM 16.2   Double Integrals over Nonrectangular Regions**
> Let R be a region bounded below and above by the graphs of the continuous functions $y = g(x)$ and $y = h(x)$, respectively, and by the lines $x = a$ and $x = b$ (Figure 16.11). If f is continuous on R, then
>
> $$\iint\limits_R f(x, y)\, dA = \int_a^b \int_{g(x)}^{h(x)} f(x, y)\, dy\, dx.$$
>
> Let R be a region bounded on the left and right by the graphs of the continuous functions $x = g(y)$ and $x = h(y)$, respectively, and the lines $y = c$ and $y = d$ (Figure 16.15). If f is continuous on R, then
>
> $$\iint\limits_R f(x, y)\, dA = \int_c^d \int_{g(y)}^{h(y)} f(x, y)\, dx\, dy.$$

EXAMPLE 2   Computing a volume Find the volume of the solid below the surface $f(x, y) = 2 + \dfrac{1}{y}$ and above the region R in the xy-plane bounded by the lines $y = x, y = 8 - x$, and $y = 1$. Notice that $f(x, y) > 0$ on R.

SOLUTION The region R is bounded on the left by $x = y$ and bounded on the right by $y = 8 - x$, or $x = 8 - y$ (**Figure 16.16**). These lines intersect at the point $(4, 4)$. We take vertical slices through the solid parallel to the xz-plane from $y = 1$ to $y = 4$. To visualize these slices, it helps to draw lines through R parallel to the x-axis.

Integrating the cross-sectional areas of slices from $y = 1$ to $y = 4$, the volume of the solid beneath the graph of f and above R (**Figure 16.17**) is given by

$$\iint\limits_R \left(2 + \frac{1}{y}\right) dA = \int_1^4 \int_y^{8-y} \left(2 + \frac{1}{y}\right) dx\, dy \qquad \text{Convert to an iterated integral.}$$

$$= \int_1^4 \left(2 + \frac{1}{y}\right) x \Big|_y^{8-y}\, dy \qquad \begin{array}{l}\text{Evaluate inner integral} \\ \text{with respect to } x.\end{array}$$

$$= \int_1^4 \left(2 + \frac{1}{y}\right)(8 - 2y)\, dy \qquad \text{Simplify.}$$

$$= \int_1^4 \left(14 - 4y + \frac{8}{y}\right) dy \qquad \text{Simplify.}$$

$$= \left(14y - 2y^2 + 8 \ln |y|\right)\Big|_1^4 \qquad \begin{array}{l}\text{Evaluate outer integral} \\ \text{with respect to } y.\end{array}$$

$$= 12 + 8 \ln 4 \approx 23.09. \qquad \text{Simplify.}$$

Related Exercise 74 ◄

QUICK CHECK 2 Could the integral in Example 2 be evaluated by integrating first (inner integral) with respect to y? ◄

Choosing and Changing the Order of Integration

Occasionally, a region of integration is bounded above and below by a pair of curves *and* the region is bounded on the right and left by a pair of curves. For example, the region R in **Figure 16.18** is bounded above by $y = x^{1/3}$ and below by $y = x^2$, and it is bounded on the right by $x = \sqrt{y}$ and on the left by $x = y^3$. In these cases, we can choose either of two orders of integration; however, one order of integration may be preferable. The following examples illustrate the valuable techniques of choosing and changing the order of integration.

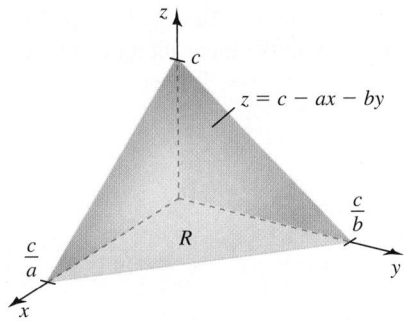

Figure 16.19

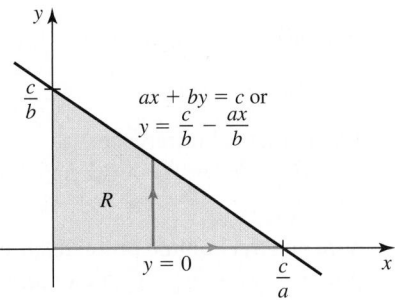

Figure 16.20

▶ In Example 3, it is just as easy to view R as being bounded on the left and the right by the lines $x = 0$ and $x = c/a - by/a$, respectively, and integrating first with respect to x.

EXAMPLE 3   **Volume of a tetrahedron** Find the volume of the tetrahedron (pyramid with four triangular faces) in the first octant bounded by the plane $z = c - ax - by$ and the coordinate planes ($x = 0$, $y = 0$, and $z = 0$). Assume a, b, and c are positive real numbers (Figure 16.19).

SOLUTION Let R be the triangular base of the tetrahedron in the xy-plane; it is bounded by the x- and y-axes and the line $ax + by = c$ (found by setting $z = 0$ in the equation of the plane; Figure 16.20). We can view R as being bounded below and above by the lines $y = 0$ and $y = c/b - ax/b$, respectively. The boundaries on the left and right are then $x = 0$ and $x = c/a$, respectively. Therefore, the volume of the solid region between the plane and R is

$$\iint_R (c - ax - by)\, dA = \int_0^{c/a} \int_0^{c/b - ax/b} (c - ax - by)\, dy\, dx \qquad \text{Convert to an iterated integral.}$$

$$= \int_0^{c/a} \left(cy - axy - \frac{by^2}{2} \right) \Big|_0^{c/b - ax/b} dx \qquad \text{Evaluate inner integral with respect to } y.$$

$$= \int_0^{c/a} \frac{(ax - c)^2}{2b}\, dx \qquad \text{Simplify and factor.}$$

$$= \frac{c^3}{6ab}. \qquad \text{Evaluate outer integral with respect to } x.$$

This result illustrates the volume formula for a tetrahedron. The lengths of the legs of the triangular base are c/a and c/b, which means the area of the base is $c^2/(2ab)$. The height of the tetrahedron is c. The general volume formula is

$$V = \frac{c^3}{6ab} = \frac{1}{3}\underbrace{\frac{c^2}{2ab}}_{\substack{\text{area of} \\ \text{base}}} \cdot \underbrace{c}_{\text{height}} = \frac{1}{3}\,(\text{area of base})(\text{height}).$$

Related Exercise 73 ◀

EXAMPLE 4   **Changing the order of integration** Consider the iterated integral $\int_0^{\sqrt{\pi}} \int_y^{\sqrt{\pi}} \sin x^2 \, dx \, dy$. Sketch the region of integration determined by the limits of integration and then evaluate the iterated integral.

SOLUTION The region of integration is $R = \{(x, y): y \le x \le \sqrt{\pi}, 0 \le y \le \sqrt{\pi}\}$, which is a triangle (Figure 16.21a). Evaluating the iterated integral as given (integrating first with respect to x) requires integrating $\sin x^2$, a function whose antiderivative is not expressible in terms of elementary functions. Therefore, this order of integration is not feasible.

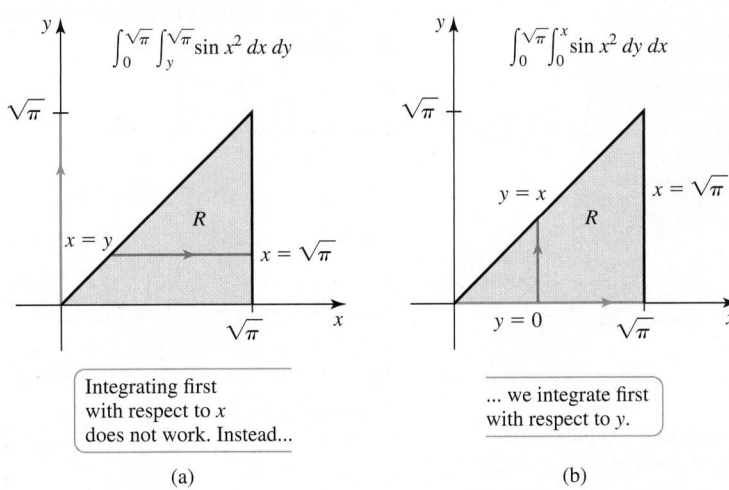

Figure 16.21

Instead, we change our perspective (Figure 16.21b) and integrate first with respect to y. With this order of integration, y runs from $y = 0$ to $y = x$ in the inner integral, and x runs from $x = 0$ to $x = \sqrt{\pi}$ in the outer integral:

$$\iint\limits_{R} \sin x^2 \, dA = \int_0^{\sqrt{\pi}} \int_0^x \sin x^2 \, dy \, dx$$

$$= \int_0^{\sqrt{\pi}} y \sin x^2 \Big|_0^x \, dx \qquad \text{Evaluate inner integral with respect to } y; \sin x^2 \text{ is constant.}$$

$$= \int_0^{\sqrt{\pi}} x \sin x^2 \, dx \qquad \text{Simplify.}$$

$$= -\frac{1}{2} \cos x^2 \Big|_0^{\sqrt{\pi}} \qquad \text{Evaluate outer integral with respect to } x.$$

$$= 1. \qquad \text{Simplify.}$$

QUICK CHECK 3 Change the order of integration of the integral $\int_0^1 \int_0^y f(x, y) \, dx \, dy$. ◄

This example shows that the order of integration can make a practical difference.

Related Exercises 58, 64 ◄

Regions Between Two Surfaces

An extension of the preceding ideas allows us to solve more general volume problems. Let $z = f(x, y)$ and $z = g(x, y)$ be continuous functions with $f(x, y) \geq g(x, y)$ on a region R in the xy-plane. Suppose we wish to compute the volume of the solid between the two surfaces over the region R (Figure 16.22). Forming a Riemann sum for the volume, the height of a typical box within the solid is the vertical distance $f(x, y) - g(x, y)$ between the upper and lower surfaces. Therefore, the volume of the solid between the surfaces is

$$V = \iint\limits_{R} (f(x, y) - g(x, y)) \, dA.$$

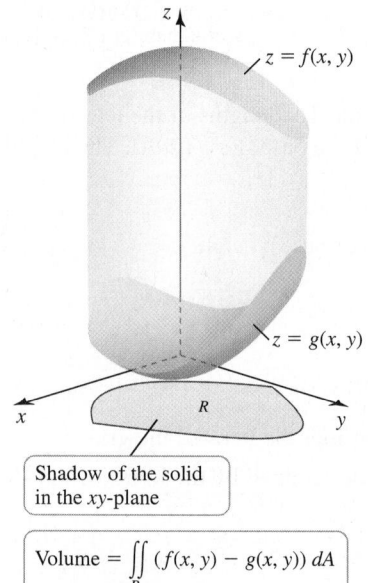

Shadow of the solid in the xy-plane

Volume $= \iint\limits_{R} (f(x, y) - g(x, y)) \, dA$

Figure 16.22

EXAMPLE 5 Region bounded by two surfaces Find the volume of the solid bounded by the parabolic cylinder $z = 1 + x^2$ and the planes $z = 5 - y$ and $y = 0$ (Figure 16.23).

SOLUTION The upper surface bounding the solid is $z = 5 - y$ and the lower surface is $z = 1 + x^2$; these two surfaces intersect along a curve C. Solving $5 - y = 1 + x^2$, we find that $y = 4 - x^2$, which is the projection of C onto the xy-plane. The back wall of the solid is the plane $y = 0$, and its projection onto the xy-plane is the x-axis. This line $(y = 0)$ intersects the parabola $y = 4 - x^2$ at $x = \pm 2$. Therefore, the region of integration (Figure 16.23) is

$$R = \{(x, y): 0 \leq y \leq 4 - x^2, -2 \leq x \leq 2\}.$$

Notice that both R and the solid are symmetric about the yz-plane. Therefore, the volume of the entire solid is twice the volume of that part of the solid that lies in the first octant. The volume of the solid is

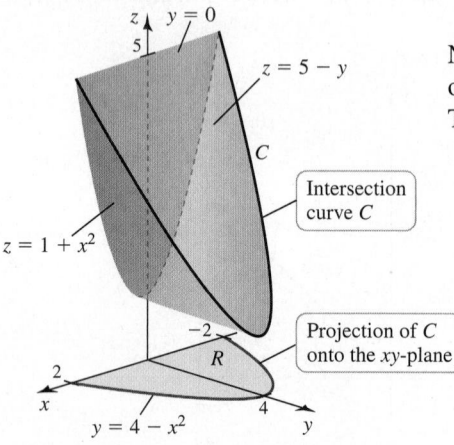

Figure 16.23

▶ To use symmetry to simplify a double integral, you must check that both the region of integration and the integrand have the same symmetry.

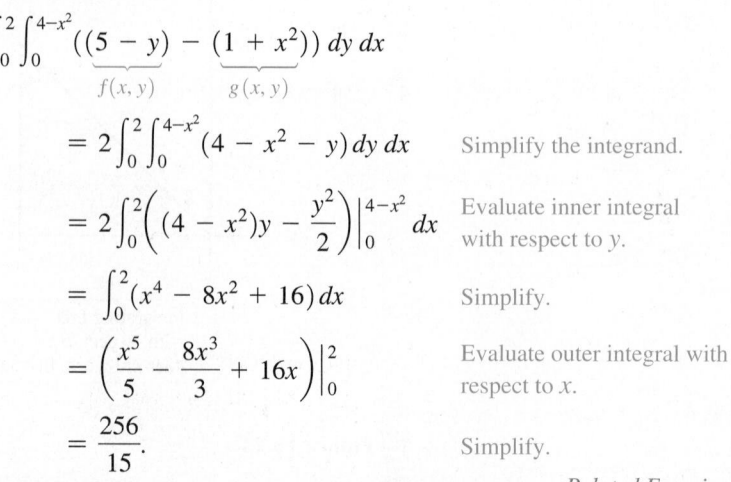

$$2\int_0^2 \int_0^{4-x^2} (\underbrace{(5 - y)}_{f(x, y)} - \underbrace{(1 + x^2)}_{g(x, y)}) \, dy \, dx$$

$$= 2\int_0^2 \int_0^{4-x^2} (4 - x^2 - y) \, dy \, dx \qquad \text{Simplify the integrand.}$$

$$= 2\int_0^2 \left((4 - x^2)y - \frac{y^2}{2} \right) \Big|_0^{4-x^2} \, dx \qquad \text{Evaluate inner integral with respect to } y.$$

$$= \int_0^2 (x^4 - 8x^2 + 16) \, dx \qquad \text{Simplify.}$$

$$= \left(\frac{x^5}{5} - \frac{8x^3}{3} + 16x \right) \Big|_0^2 \qquad \text{Evaluate outer integral with respect to } x.$$

$$= \frac{256}{15}. \qquad \text{Simplify.}$$

Related Exercises 78–79 ◄

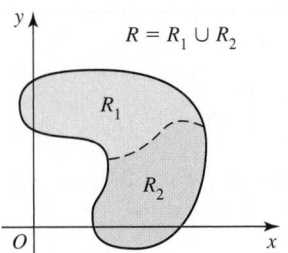

Figure 16.24

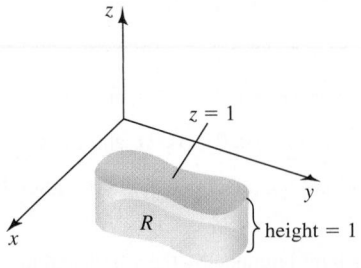

Volume of solid = (area of R) × (height)
= area of R = $\iint\limits_R 1\, dA$

Figure 16.25

➤ We are solving a familiar area problem first encountered in Section 6.2. Suppose R is bounded above by $y = h(x)$ and below by $y = g(x)$, for $a \le x \le b$. Using a double integral, the area of R is

$$\iint\limits_R dA = \int_a^b \int_{g(x)}^{h(x)} dy\, dx$$

$$= \int_a^b (h(x) - g(x))\, dx,$$

which is a result obtained in Section 6.2.

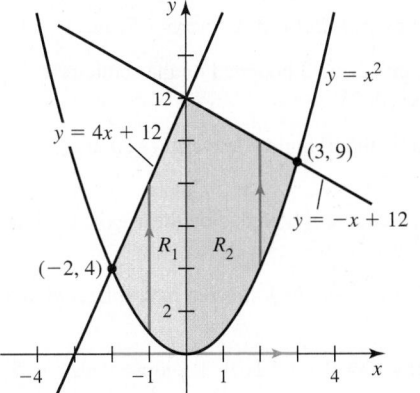

Figure 16.26

Decomposition of Regions

We occasionally encounter regions that are more complicated than those considered so far. A technique called *decomposition* allows us to subdivide a region of integration into two (or more) subregions. If the integrals over the subregions can be evaluated separately, the results are added to obtain the value of the original integral. For example, the region R in Figure 16.24 is divided into two nonoverlapping subregions R_1 and R_2. By partitioning these regions and using Riemann sums, it can be shown that

$$\iint\limits_R f(x, y)\, dA = \iint\limits_{R_1} f(x, y)\, dA + \iint\limits_{R_2} f(x, y)\, dA.$$

This method is illustrated in Example 6. The analog of decomposition with single variable integrals is the property $\int_a^b f(x)\, dx = \int_a^p f(x)\, dx + \int_p^b f(x)\, dx$.

Finding Area by Double Integrals

An interesting application of double integrals arises when the integrand is $f(x, y) = 1$. The integral $\iint\limits_R 1\, dA$ gives the volume of the solid between the horizontal plane $z = 1$ and the region R. Because the height of this solid is 1, its volume equals (numerically) the area of R (Figure 16.25). Therefore, we have a way to compute areas of regions in the xy-plane using double integrals.

Areas of Regions by Double Integrals

Let R be a region in the xy-plane. Then

$$\text{area of } R = \iint\limits_R dA.$$

EXAMPLE 6 **Area of a plane region** Find the area of the region R bounded by $y = x^2$, $y = -x + 12$, and $y = 4x + 12$ (Figure 16.26).

SOLUTION The region R in its entirety is bounded neither above and below by two curves, nor on the left and right by two curves. However, when decomposed along the y-axis, R may be viewed as two regions R_1 and R_2, each of which is bounded above and below by a pair of curves. Notice that the parabola $y = x^2$ and the line $y = -x + 12$ intersect in the first quadrant at the point $(3, 9)$, while the parabola and the line $y = 4x + 12$ intersect in the second quadrant at the point $(-2, 4)$.

To find the area of R, we integrate the function $f(x, y) = 1$ over R_1 and R_2; the area is

$$\iint\limits_{R_1} 1\, dA + \iint\limits_{R_2} 1\, dA \qquad \text{Decompose region.}$$

$$= \int_{-2}^{0} \int_{x^2}^{4x+12} 1\, dy\, dx + \int_{0}^{3} \int_{x^2}^{-x+12} 1\, dy\, dx \qquad \text{Convert to iterated integrals.}$$

$$= \int_{-2}^{0} (4x + 12 - x^2)\, dx + \int_{0}^{3} (-x + 12 - x^2)\, dx \qquad \begin{array}{l}\text{Evaluate inner integrals}\\\text{with respect to } y.\end{array}$$

$$= \left(2x^2 + 12x - \frac{x^3}{3}\right)\Big|_{-2}^{0} + \left(-\frac{x^2}{2} + 12x - \frac{x^3}{3}\right)\Big|_{0}^{3} \qquad \begin{array}{l}\text{Evaluate outer integrals}\\\text{with respect to } x.\end{array}$$

$$= \frac{40}{3} + \frac{45}{2} = \frac{215}{6}. \qquad \text{Simplify.}$$

Related Exercise 86 ◄

QUICK CHECK 4 Consider the triangle R with vertices $(-1, 0)$, $(1, 0)$, and $(0, 1)$ as a region of integration. If we integrate first with respect to x, does R need to be decomposed? If we integrate first with respect to y, does R need to be decomposed? ◄

SECTION 16.2 EXERCISES

Getting Started

1. Describe and sketch a region that is bounded above and below by two curves.

2. Describe and sketch a region that is bounded on the left and on the right by two curves.

3. Which order of integration is preferable to integrate $f(x, y) = xy$ over $R = \{(x, y): y - 1 \le x \le 1 - y, 0 \le y \le 1\}$?

4. Which order of integration would you use to find the area of the region bounded by the x-axis and the lines $y = 2x + 3$ and $y = 3x - 4$ using a double integral?

5. Change the order of integration in the integral $\int_0^1 \int_{y^2}^{\sqrt{y}} f(x, y)\, dx\, dy$.

6. Sketch the region of integration for $\int_{-2}^{2} \int_{x^2}^{4} e^{xy}\, dy\, dx$.

7. Sketch the region of integration for $\int_0^2 \int_0^{2x} dy\, dx$ and use geometry to evaluate the iterated integral.

8. Describe a solid whose volume equals $\int_{-4}^{4} \int_{-\sqrt{16-x^2}}^{\sqrt{16-x^2}} 10\, dy\, dx$ and evaluate this iterated integral using geometry.

9–10. Consider the region R shown in the figure and write an iterated integral of a continuous function f over R.

9.
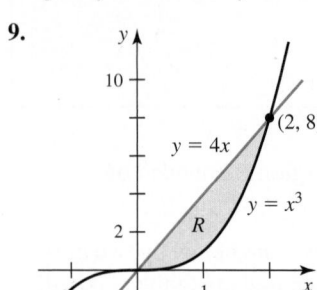

10.
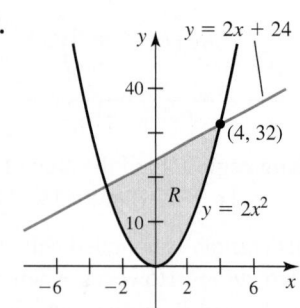

Practice Exercises

11–27. Evaluating integrals Evaluate the following integrals.

11. $\int_0^1 \int_x^1 6y\, dy\, dx$

12. $\int_0^1 \int_0^{2x} 15xy^2\, dy\, dx$

13. $\int_0^2 \int_{x^2}^{2x} xy\, dy\, dx$

14. $\int_{-\pi/4}^{\pi/4} \int_{\sin x}^{\cos x} dy\, dx$

15. $\int_{-2}^2 \int_{x^2}^{8-x^2} x\, dy\, dx$

16. $\int_0^{\ln 2} \int_{e^x}^2 dy\, dx$

17. $\int_0^1 \int_0^x 2e^{x^2}\, dy\, dx$

18. $\int_0^{\sqrt[3]{\pi/2}} \int_0^x y\cos x^3\, dy\, dx$

19. $\int_0^{\ln 2} \int_{e^y}^2 \frac{y}{x}\, dx\, dy$

20. $\int_0^4 \int_y^{2y} xy\, dx\, dy$

21. $\int_0^{\pi/2} \int_y^{\pi/2} 6\sin(2x - 3y)\, dx\, dy$

22. $\int_0^{\pi/2} \int_0^{\cos y} e^{\sin y}\, dx\, dy$

23. $\int_0^{\pi/2} \int_0^{y\cos y} dx\, dy$

24. $\int_0^1 \int_{\tan^{-1} x}^{\pi/4} 2x\, dy\, dx$

25. $\int_0^4 \int_{-\sqrt{16-y^2}}^{\sqrt{16-y^2}} 2xy\, dx\, dy$

26. $\int_0^1 \int_0^x 2e^x\, dy\, dx$

27. $\int_{\pi/2}^{\pi} \int_0^{y^2} \cos\frac{x}{y}\, dx\, dy$

28–34. Regions of integration Sketch each region R and write an iterated integral of a continuous function f over R. Use the order dy dx.

28. $R = \{(x, y): 0 \le x \le 2, 3x^2 \le y \le -6x + 24\}$

29. $R = \{(x, y): 1 \le x \le 2, x + 1 \le y \le 2x + 4\}$

30. $R = \{(x, y): 0 \le x \le 4, x^2 \le y \le 8\sqrt{x}\}$

31. R is the triangular region with vertices $(0, 0)$, $(0, 2)$, and $(1, 0)$.

32. R is the triangular region with vertices $(0, 0)$, $(0, 2)$, and $(1, 1)$.

33. R is the region in the first quadrant bounded by a circle of radius 1 centered at the origin.

34. R is the region in the first quadrant bounded by the y-axis and the parabolas $y = x^2$ and $y = 1 - x^2$.

35–42. Regions of integration Write an iterated integral of a continuous function f over the region R. Use the order dy dx. Start by sketching the region of integration if it is not supplied.

35.
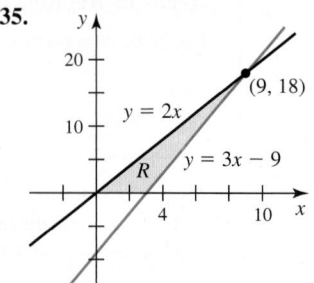

36.
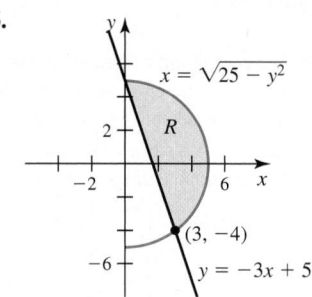

37. R is the region bounded by $y = 4 - x$, $y = 1$, and $x = 0$.

38. $R = \{(x, y): 0 \le x \le y(1 - y)\}$.

39. R is the region bounded by $y = 2x + 3$, $y = 3x - 7$, and $y = 0$.

40. R is the region in quadrants 2 and 3 bounded by the semicircle with radius 3 centered at $(0, 0)$.

41. R is the region bounded by the triangle with vertices $(0, 0)$, $(2, 0)$, and $(1, 1)$.

42. R is the region in the first quadrant bounded by the x-axis, the line $x = 6 - y$, and the curve $y = \sqrt{x}$.

43–56. Evaluating integrals Evaluate the following integrals. A sketch is helpful.

43. $\iint_R xy\, dA$; R is bounded by $x = 0$, $y = 2x + 1$, and $y = -2x + 5$.

44. $\iint_R (x + y)\, dA$; R is the region in the first quadrant bounded by $x = 0$, $y = x^2$, and $y = 8 - x^2$.

45. $\iint_R y^2\, dA$; R is bounded by $x = 1$, $y = 2x + 2$, and $y = -x - 1$.

46. $\iint_R x^2 y\, dA$; R is the region in quadrants 1 and 4 bounded by the semicircle of radius 4 centered at $(0, 0)$.

47. $\iint_R 12y\, dA$; R is bounded by $y = 2 - x$, $y = \sqrt{x}$, and $y = 0$.

48. $\iint_R y^2\, dA$; R is bounded by $y = 1$, $y = 1 - x$, and $y = x - 1$.

49. $\iint_R 3xy \, dA$; R is the region in the first quadrant bounded by $y = 2 - x$, $y = 0$, and $x = 4 - y^2$.

50. $\iint_R (x + y) \, dA$; R is bounded by $y = |x|$ and $y = 4$.

51. $\iint_R 3x^2 \, dA$; R is bounded by $y = 0$, $y = 2x + 4$, and $y = x^3$.

52. $\iint_R 8xy \, dA$; $R = \{(x, y): 0 \le y \le \sec x, 0 \le x \le \pi/4\}$

53. $\iint_R (x + y) \, dA$; R is the region bounded by $y = 1/x$ and $y = 5/2 - x$.

54. $\iint_R \dfrac{y}{1 + x + y^2} \, dA$; $R = \{(x, y): 0 \le \sqrt{x} \le y, 0 \le y \le 1\}$

55. $\iint_R x \sec^2 y \, dA$; $R = \{(x, y): 0 \le y \le x^2, 0 \le x \le \sqrt{\pi}/2\}$

56. $\iint_R \dfrac{8xy}{1 + x^2 + y^2} \, dA$; $R = \{(x, y): 0 \le y \le x, 0 \le x \le 2\}$

57–62. Changing order of integration *Reverse the order of integration in the following integrals.*

57. $\displaystyle\int_0^2 \int_{x^2}^{2x} f(x, y) \, dy \, dx$

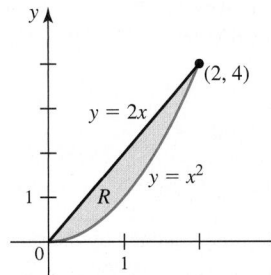

58. $\displaystyle\int_0^3 \int_0^{6-2x} f(x, y) \, dy \, dx$

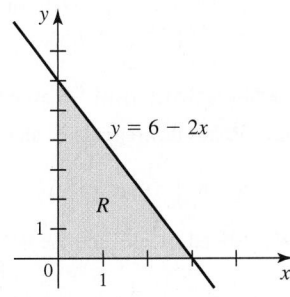

59. $\displaystyle\int_{1/2}^1 \int_0^{-\ln y} f(x, y) \, dx \, dy$

60. $\displaystyle\int_0^1 \int_1^{e^y} f(x, y) \, dx \, dy$

61. $\displaystyle\int_0^1 \int_0^{\cos^{-1} y} f(x, y) \, dx \, dy$

62. $\displaystyle\int_1^e \int_0^{\ln x} f(x, y) \, dy \, dx$

63–68. Changing order of integration *Reverse the order of integration and evaluate the integral.*

63. $\displaystyle\int_0^1 \int_y^1 e^{x^2} \, dx \, dy$

64. $\displaystyle\int_0^\pi \int_x^\pi \sin y^2 \, dy \, dx$

65. $\displaystyle\int_0^{1/2} \int_{y^2}^{1/4} y \cos(16\pi x^2) \, dx \, dy$

66. $\displaystyle\int_0^4 \int_{\sqrt{x}}^2 \dfrac{x}{y^5 + 1} \, dy \, dx$

67. $\displaystyle\int_0^{\sqrt[3]{\pi}} \int_y^{\sqrt[3]{\pi}} x^4 \cos(x^2 y) \, dx \, dy$

68. $\displaystyle\int_0^2 \int_0^{4-x^2} \dfrac{xe^{2y}}{4 - y} \, dy \, dx$

69–70. Two integrals to one *Draw the regions of integration and write the following integrals as a single iterated integral.*

69. $\displaystyle\int_0^1 \int_{e^y}^e f(x, y) \, dx \, dy + \int_{-1}^0 \int_{e^{-y}}^e f(x, y) \, dx \, dy$

70. $\displaystyle\int_{-4}^0 \int_0^{\sqrt{16-x^2}} f(x, y) \, dy \, dx + \int_0^4 \int_0^{4-x} f(x, y) \, dy \, dx$

71–80. Volumes *Find the volume of the following solids.*

71. The solid bounded by the cylinder $z = 2 - y^2$, the xy-plane, the xz-plane, and the planes $y = x$ and $x = 1$

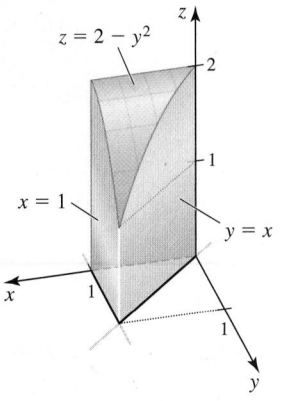

72. The solid bounded between the cylinder $z = 2 \sin^2 x$ and the xy-plane over the region $R = \{(x, y): 0 \le x \le y \le \pi\}$

73. The tetrahedron bounded by the coordinate planes ($x = 0$, $y = 0$, and $z = 0$) and the plane $z = 8 - 2x - 4y$

74. The solid in the first octant bounded by the coordinate planes and the surface $z = 1 - y - x^2$

75. The segment of the cylinder $x^2 + y^2 = 1$ bounded above by the plane $z = 12 + x + y$ and below by $z = 0$

76. The solid S between the surfaces $z = e^{x-y}$ and $z = -e^{x-y}$, where S intersects the xy-plane in the region $R = \{(x, y): 0 \le x \le y, 0 \le y \le 1\}$

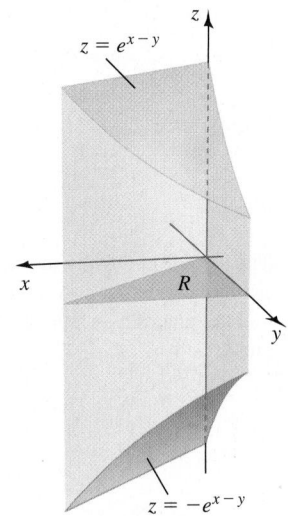

77. The solid above the region $R = \{(x, y): 0 \le x \le 1, 0 \le y \le 2 - x\}$ and between the planes $-4x - 4y + z = 0$ and $-2x - y + z = 8$

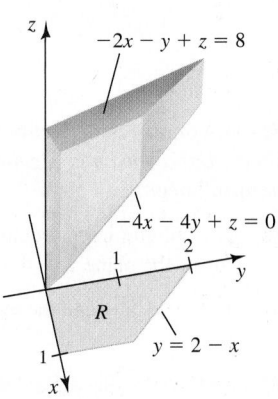

78. The solid in the first octant bounded by the planes $x = 0$, $y = 0$, $z = 1$, and $z = 2 - y$, and the cylinder $y = 1 - x^2$

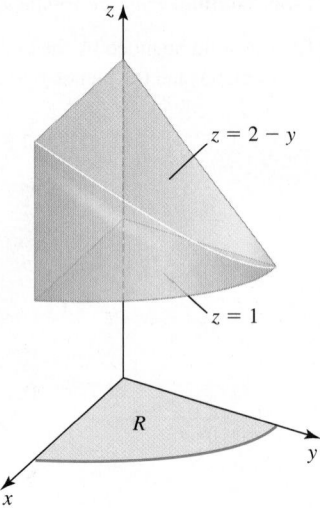

79. The solid above the region $R = \{(x, y): 0 \le x \le 1,$ $0 \le y \le 1 - x\}$ bounded by the paraboloids $z = x^2 + y^2$ and $z = 2 - x^2 - y^2$ and the coordinate planes in the first octant

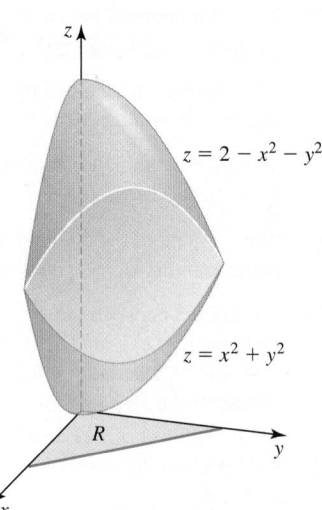

80. The solid bounded by the parabolic cylinder $z = x^2 + 1$, and the planes $z = y + 1$ and $y = 1$

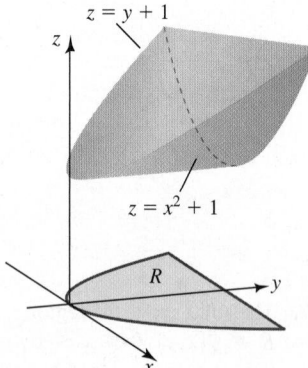

T 81–84. Volume using technology *Find the volume of the following solids. Use a computer algebra system to evaluate an appropriate iterated integral.*

81. The column with a square base $R = \{(x, y): |x| \le 1, |y| \le 1\}$ cut by the plane $z = 4 - x - y$

82. The solid between the paraboloid $z = x^2 + y^2$ and the plane $z = 1 - 2y$

83. The wedge sliced from the cylinder $x^2 + y^2 = 1$ by the planes $z = a(2 - x)$ and $z = a(x - 2)$, where $a > 0$

84. The solid bounded by the elliptical cylinder $x^2 + 3y^2 = 12$, the plane $z = 0$, and the paraboloid $z = 3x^2 + y^2 + 1$

85–90. Area of plane regions *Use double integrals to compute the area of the following regions.*

85. The region bounded by the parabola $y = x^2$ and the line $y = 4$

86. The region bounded by the parabola $y = x^2$ and the line $y = x + 2$

87. The region in the first quadrant bounded by $y = e^x$ and $x = \ln 2$

88. The region bounded by $y = 1 + \sin x$ and $y = 1 - \sin x$ on the interval $[0, \pi]$

89. The region in the first quadrant bounded by $y = x^2$, $y = 5x + 6$, and $y = 6 - x$

90. The region bounded by the lines $x = 0$, $x = 4$, $y = x$, and $y = 2x + 1$

91. Explain why or why not Determine whether the following statements are true and give an explanation or counterexample.

    a. In the iterated integral $\int_c^d \int_a^b f(x, y)\, dx\, dy$, the limits a and b must be constants or functions of x.

    b. In the iterated integral $\int_c^d \int_a^b f(x, y)\, dx\, dy$, the limits c and d must be functions of y.

    c. Changing the order of integration gives $\int_0^2 \int_1^y f(x, y)\, dx\, dy = \int_1^y \int_0^2 f(x, y)\, dy\, dx$.

Explorations and Challenges

92. Related integrals Evaluate each integral.

    a. $\displaystyle\int_0^4 \int_0^4 (4 - x - y)\, dx\, dy$ **b.** $\displaystyle\int_0^4 \int_0^4 |4 - x - y|\, dx\, dy$

93. Sliced block Find the volume of the solid bounded by the planes $x = 0$, $x = 5$, $z = y - 1$, $z = -2y - 1$, $z = 0$, and $z = 2$.

94. Square region Consider the region $R = \{(x, y): |x| + |y| \le 1\}$ shown in the figure.

    a. Use a double integral to show that the area of R is 2.

    b. Find the volume of the square column whose base is R and whose upper surface is $z = 12 - 3x - 4y$.

    c. Find the volume of the solid above R and beneath the cylinder $x^2 + z^2 = 1$.

    d. Find the volume of the pyramid whose base is R and whose vertex is on the z-axis at $(0, 0, 6)$.

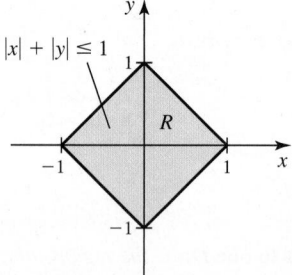

95–96. Average value *Use the definition for the average value of a function over a region R (Section 16.1),* $\overline{f} = \dfrac{1}{\text{area of } R} \iint_R f(x, y)\, dA$.

95. Find the average value of $a - x - y$ over the region $R = \{(x, y): x + y \le a, x \ge 0, y \ge 0\}$, where $a > 0$.

96. Find the average value of $z = a^2 - x^2 - y^2$ over the region $R = \{(x, y): x^2 + y^2 \le a^2\}$, where $a > 0$.

T 97–98. Area integrals *Consider the following regions R. Use a computer algebra system to evaluate the integrals.*

a. *Sketch the region R.*

b. *Evaluate $\iint_R dA$ to determine the area of the region.*

c. *Evaluate $\iint_R xy\,dA$.*

97. *R is the region between both branches of $y = 1/x$ and the lines $y = x + 3/2$ and $y = x - 3/2$.*

98. *R is the region bounded by the ellipse $x^2/18 + y^2/36 = 1$ with $y \le 4x/3$.*

99–102. Improper integrals *Many improper double integrals may be handled using the techniques for improper integrals in one variable (Section 8.9). For example, under suitable conditions on f,*

$$\int_a^\infty \int_{g(x)}^{h(x)} f(x, y)\,dy\,dx = \lim_{b \to \infty} \int_a^b \int_{g(x)}^{h(x)} f(x, y)\,dy\,dx.$$

Use or extend the one-variable methods for improper integrals to evaluate the following integrals.

99. $\displaystyle\int_1^\infty \int_0^{e^{-x}} xy\,dy\,dx$

100. $\displaystyle\int_1^\infty \int_0^{1/x^2} \frac{2y}{x}\,dy\,dx$

101. $\displaystyle\int_0^\infty \int_0^\infty e^{-x-y}\,dy\,dx$

102. $\displaystyle\int_{-\infty}^\infty \int_{-\infty}^\infty \frac{1}{(x^2+1)(y^2+1)}\,dy\,dx$

QUICK CHECK ANSWERS

1. Inner integral: $0 \le y \le 2 - x$; outer integral: $0 \le x \le 2$ **2.** Yes; however, two separate iterated integrals would be required.   **3.** $\int_0^1 \int_x^1 f(x, y)\,dy\,dx$ **4.** No; yes ◀

16.3 Double Integrals in Polar Coordinates

▶ Recall the conversions between Cartesian and polar coordinates (Section 12.2):

$x = r \cos\theta, y = r \sin\theta$, or

$r^2 = x^2 + y^2, \tan\theta = y/x$.

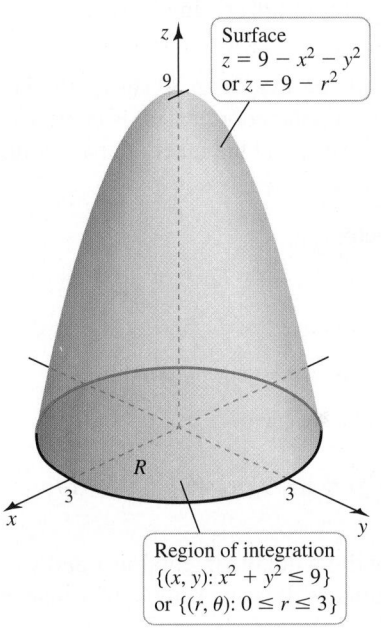

Figure 16.27

In Chapter 12, we explored polar coordinates and saw that in certain situations, they simplify problems considerably. The same is true when it comes to integration over plane regions. In this section, we learn how to formulate double integrals in polar coordinates and how to change double integrals from Cartesian coordinates to polar coordinates.

Moving from Rectangular to Polar Coordinates

Suppose we want to find the volume of the solid bounded by the paraboloid $z = 9 - x^2 - y^2$ and the xy-plane (**Figure 16.27**). The intersection of the paraboloid and the xy-plane $(z = 0)$ is the curve $9 - x^2 - y^2 = 0$, or $x^2 + y^2 = 9$. Therefore, the region of integration R is the disk of radius 3 in the xy-plane, centered at the origin, which, when expressed in Cartesian coordinates, is $R = \{(x, y): -\sqrt{9 - x^2} \le y \le \sqrt{9 - x^2}, -3 \le x \le 3\}$. Using the relationship $r^2 = x^2 + y^2$ for converting Cartesian to polar coordinates, the region of integration expressed in polar coordinates is simply $R = \{(r, \theta): 0 \le r \le 3, 0 \le \theta \le 2\pi\}$. Furthermore, the paraboloid expressed in polar coordinates is $z = 9 - r^2$. This problem (which is solved in Example 1) illustrates how both the integrand and the region of integration in a double integral can be simplified by working in polar coordinates.

The region of integration in this problem is an example of a **polar rectangle**. In polar coordinates, it has the form $R = \{(r, \theta): 0 \le a \le r \le b, \alpha \le \theta \le \beta\}$, where $\beta - \alpha \le 2\pi$ and a, b, α, and β are constants (**Figure 16.28**). Polar rectangles are the analogs of rectangles in Cartesian coordinates. For this reason, the methods used in Section 16.1 for evaluating double integrals over rectangles can be extended to polar rectangles. The goal is to evaluate integrals of the form $\iint_R f(x, y)\,dA$, where f is a continuous function on the polar rectangle R. If f is nonnegative on R, this integral equals the volume of the solid bounded by the surface $z = f(x, y)$ and the region R in the xy-plane.

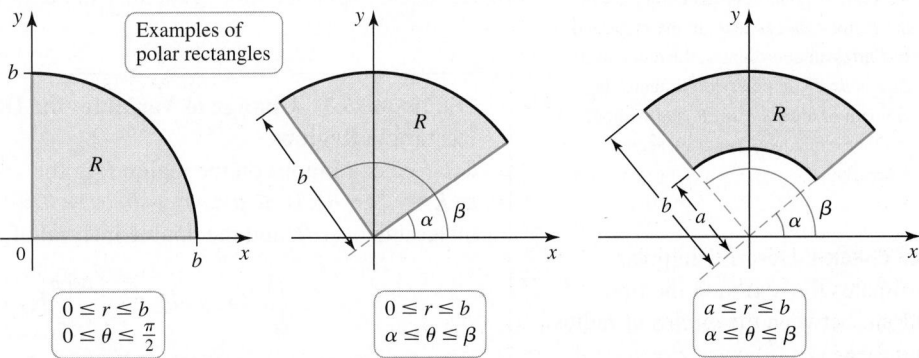

Figure 16.28

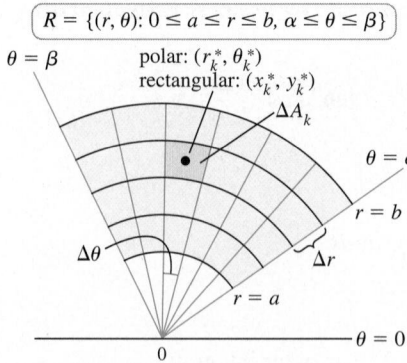

Figure 16.29

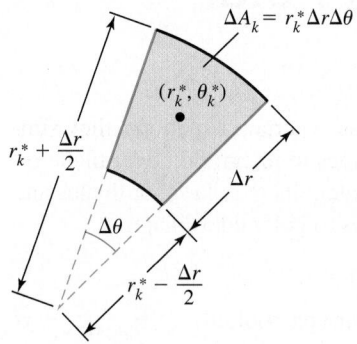

Figure 16.30

▶ Recall that the area of a sector of a circle of radius r subtended by an angle θ is $\frac{1}{2}r^2\theta$.

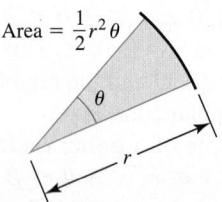

Area $= \frac{1}{2}r^2\theta$

▶ The most common error in evaluating integrals in polar coordinates is to omit the factor r that appears in the integrand. In Cartesian coordinates, the element of area is $dx\,dy$; in polar coordinates, the element of area is $r\,dr\,d\theta$, and without the factor of r, area is not measured correctly.

QUICK CHECK 1 Describe in polar coordinates the region in the first quadrant between the circles of radius 1 and 2. ◀

Our approach is to divide $[a, b]$ into M subintervals of equal length $\Delta r = (b - a)/M$. We similarly divide $[\alpha, \beta]$ into m subintervals of equal length $\Delta\theta = (\beta - \alpha)/m$. Now look at the arcs of the circles centered at the origin with radii

$$r = a, r = a + \Delta r, r = a + 2\Delta r, \ldots, r = b$$

and the rays

$$\theta = \alpha, \theta = \alpha + \Delta\theta, \theta = a + 2\Delta\theta, \ldots, \theta = \beta$$

emanating from the origin (**Figure 16.29**). The arcs and rays divide the region R into $n = Mm$ polar rectangles that we number in a convenient way from $k = 1$ to $k = n$. The area of the kth rectangle is denoted ΔA_k, and we let (r_k^*, θ_k^*) be the polar coordinates of an arbitrary point in that rectangle. Note that this point also has the Cartesian coordinates $(x_k^*, y_k^*) = (r_k^* \cos \theta_k^*, r_k^* \sin \theta_k^*)$. If f is continuous on R, the volume of the solid region beneath the surface $z = f(x, y)$ and above R may be computed with Riemann sums using either ordinary rectangles (as in Sections 16.1 and 16.2) or polar rectangles. Here, we now use polar rectangles.

Consider the "box" whose base is the kth polar rectangle and whose height is $f(x_k^*, y_k^*)$; its volume is $f(x_k^*, y_k^*)\Delta A_k$, for $k = 1, \ldots, n$. Therefore, the volume of the solid region beneath the surface $z = f(x, y)$ with a base R is approximately

$$V = \sum_{k=1}^{n} f(x_k^*, y_k^*)\Delta A_k.$$

This approximation to the volume is a Riemann sum. We let Δ be the maximum value of Δr and $\Delta\theta$. If f is continuous on R, then as $n \to \infty$ and $\Delta \to 0$, the sum approaches a double integral; that is,

$$\iint\limits_{R} f(x, y)\,dA = \lim_{\Delta \to 0} \sum_{k=1}^{n} f(x_k^*, y_k^*)\Delta A_k = \lim_{\Delta \to 0} \sum_{k=1}^{n} f(r_k^* \cos \theta_k^*, r_k^* \sin \theta_k^*)\Delta A_k. \quad (1)$$

The next step is to express ΔA_k in terms of Δr and $\Delta\theta$. **Figure 16.30** shows the kth polar rectangle, with an area of ΔA_k. The point (r_k^*, θ_k^*) (in polar coordinates) is chosen so that the outer arc of the polar rectangle has radius $r_k^* + \Delta r/2$ and the inner arc has radius $r_k^* - \Delta r/2$. The area of the polar rectangle is

$$\Delta A_k = (\text{area of outer sector}) - (\text{area of inner sector})$$

$$= \frac{1}{2}\left(r_k^* + \frac{\Delta r}{2}\right)^2 \Delta\theta - \frac{1}{2}\left(r_k^* - \frac{\Delta r}{2}\right)^2 \Delta\theta \qquad \text{Area of sector} = \frac{1}{2}r^2\Delta\theta$$

$$= r_k^* \Delta r \Delta\theta. \qquad\qquad\qquad\qquad \text{Expand and simplify.}$$

Substituting this expression for ΔA_k into equation (1), we have

$$\iint\limits_{R} f(x, y)\,dA = \lim_{\Delta \to 0} \sum_{k=1}^{n} f(x_k^*, y_k^*)\Delta A_k = \lim_{\Delta \to 0} \sum_{k=1}^{n} f(r_k^* \cos \theta_k^*, r_k^* \sin \theta_k^*)r_k^* \Delta r \Delta\theta.$$

This observation leads to a theorem that allows us to write a double integral in x and y as an iterated integral of $f(r \cos \theta, r \sin \theta)r$ in polar coordinates. It is an example of a change of variables, explained more generally in Section 16.7.

THEOREM 16.3   Change of Variables for Double Integrals over Polar Rectangle Regions

Let f be continuous on the region R in the xy-plane expressed in polar coordinates as $R = \{(r, \theta): 0 \le a \le r \le b, \alpha \le \theta \le \beta\}$, where $\beta - \alpha \le 2\pi$. Then f is integrable over R, and the double integral of f over R is

$$\iint\limits_{R} f(x, y)\,dA = \int_{\alpha}^{\beta}\int_{a}^{b} f(r \cos \theta, r \sin \theta)\,r\,dr\,d\theta.$$

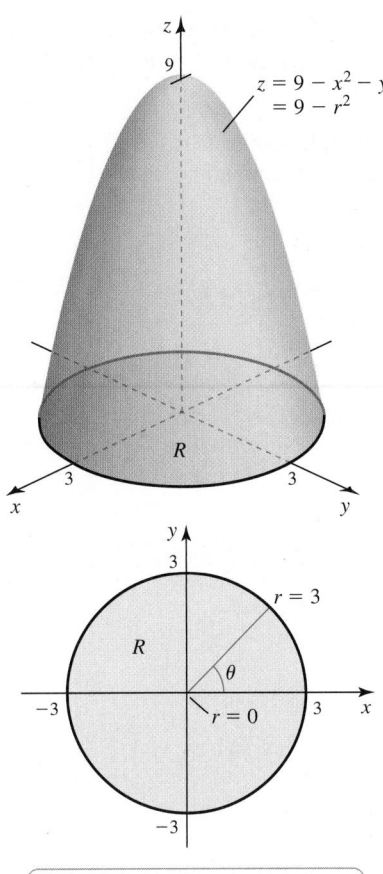

$z = 9 - x^2 - y^2$
$= 9 - r^2$

R

$R = \{(r, \theta): 0 \le r \le 3, 0 \le \theta \le 2\pi\}$

Figure 16.31

QUICK CHECK 2 Express the functions $f(x, y) = (x^2 + y^2)^{5/2}$ and $h(x, y) = x^2 - y^2$ in polar coordinates. ◄

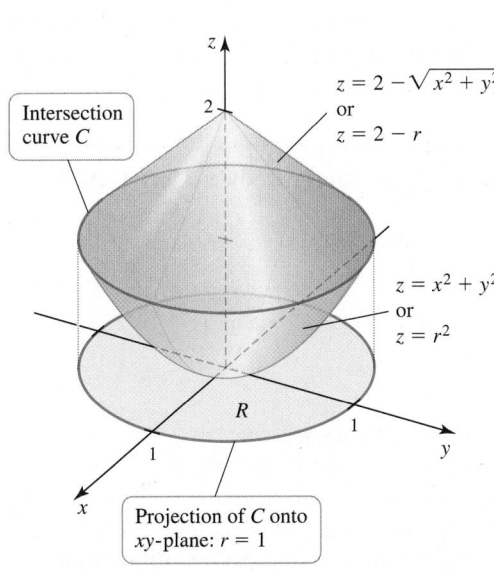

Intersection curve C

$z = 2 - \sqrt{x^2 + y^2}$
or
$z = 2 - r$

$z = x^2 + y^2$
or
$z = r^2$

R

Projection of C onto xy-plane: $r = 1$

Figure 16.32

QUICK CHECK 3 Give a geometric explanation for the extraneous root $z = 4$ found in Example 2. ◄

EXAMPLE 1 **Volume of a paraboloid cap** Find the volume of the solid bounded by the paraboloid $z = 9 - x^2 - y^2$ and the xy-plane.

SOLUTION Using $x^2 + y^2 = r^2$, the surface is described in polar coordinates by $z = 9 - r^2$. The paraboloid intersects the xy-plane $(z = 0)$ when $z = 9 - r^2 = 0$, or $r = 3$. Therefore, the intersection curve is the circle of radius 3 centered at the origin. The resulting region of integration is the disk $R = \{(r, \theta): 0 \le r \le 3, 0 \le \theta \le 2\pi\}$ (**Figure 16.31**). Integrating over R in polar coordinates, the volume is

$$V = \int_0^{2\pi} \int_0^3 \underbrace{(9 - r^2)}_{z} r \, dr \, d\theta \quad \text{Iterated integral for volume}$$

$$= \int_0^{2\pi} \left(\frac{9r^2}{2} - \frac{r^4}{4} \right) \Big|_0^3 d\theta \quad \text{Evaluate inner integral with respect to } r.$$

$$= \int_0^{2\pi} \frac{81}{4} \, d\theta = \frac{81\pi}{2}. \quad \text{Evaluate outer integral with respect to } \theta.$$

Related Exercises 12, 16 ◄

EXAMPLE 2 **Region bounded by two surfaces** Find the volume of the region bounded by the paraboloid $z = x^2 + y^2$ and the cone $z = 2 - \sqrt{x^2 + y^2}$.

SOLUTION As discussed in Section 16.2, the volume of a solid bounded by two surfaces $z = f(x, y)$ and $z = g(x, y)$ over a region R in the xy-plane is given by $\iint_R (f(x, y) - g(x, y)) \, dA$, where $f(x, y) \ge g(x, y)$ over R. Because the paraboloid $z = x^2 + y^2$ lies below the cone $z = 2 - \sqrt{x^2 + y^2}$ (**Figure 16.32**), the volume of the solid bounded by the surfaces is

$$V = \iint_R \left((2 - \sqrt{x^2 + y^2}) - (x^2 + y^2) \right) dA,$$

where the boundary of R is the curve of intersection C of the surfaces projected onto the xy-plane. To find C, we set the equations of the surfaces equal to one another. Writing $x^2 + y^2 = 2 - \sqrt{x^2 + y^2}$ seems like a good start, but it leads to algebraic difficulties. Instead, we write the equation of the cone as $\sqrt{x^2 + y^2} = 2 - z$ and then substitute this equation into the equation for the paraboloid:

$$z = (2 - z)^2 \qquad \begin{array}{l} z = x^2 + y^2 \text{ (paraboloid) and} \\ \sqrt{x^2 + y^2} = 2 - z \text{ (cone)} \end{array}$$

$$z^2 - 5z + 4 = 0 \qquad \text{Simplify.}$$

$$(z - 1)(z - 4) = 0 \qquad \text{Factor.}$$

$$z = 1 \text{ or } z = 4. \qquad \text{Solve for } z.$$

The solution $z = 4$ is an extraneous root (see Quick Check 3). Setting $z = 1$ in the equation of either the paraboloid or the cone leads to $x^2 + y^2 = 1$, which is an equation of the curve C in the plane $z = 1$. Projecting C onto the xy-plane, we conclude that the region of integration (written in polar coordinates) is $R = \{(r, \theta): 0 \le r \le 1, 0 \le \theta \le 2\pi\}$.

Converting the original volume integral into polar coordinates and evaluating it over R, we have

$$V = \iint_R \left((2 - \sqrt{x^2 + y^2}) - (x^2 + y^2) \right) dA \quad \text{Double integral for volume}$$

$$= \int_0^{2\pi} \int_0^1 (2 - r - r^2) r \, dr \, d\theta \quad \begin{array}{l} \text{Convert to polar coordinates;} \\ x^2 + y^2 = r^2. \end{array}$$

$$= \int_0^{2\pi} \left(r^2 - \frac{1}{3} r^3 - \frac{1}{4} r^4 \right) \Big|_0^1 d\theta \quad \text{Evaluate the inner integral.}$$

$$= \int_0^{2\pi} \frac{5}{12} \, d\theta = \frac{5\pi}{6}. \quad \text{Evaluate the outer integral.}$$

Related Exercises 33, 40 ◄

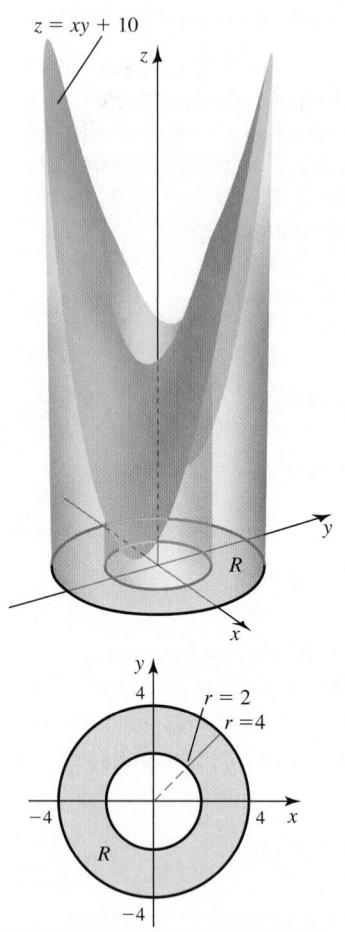

$z = xy + 10$

$R = \{(r, \theta): 2 \le r \le 4, 0 \le \theta \le 2\pi\}$

Figure 16.33

EXAMPLE 3 **Annular region** Find the volume of the region beneath the surface $z = xy + 10$ and above the annular region $R = \{(r, \theta): 2 \le r \le 4, 0 \le \theta \le 2\pi\}$. (An *annulus* is the region between two concentric circles.)

SOLUTION The region of integration suggests working in polar coordinates (Figure 16.33). Substituting $x = r \cos \theta$ and $y = r \sin \theta$, the integrand becomes

$$
\begin{aligned}
xy + 10 &= (r \cos \theta)(r \sin \theta) + 10 && \text{Substitute for } x \text{ and } y. \\
&= r^2 \sin \theta \cos \theta + 10 && \text{Simplify.} \\
&= \tfrac{1}{2} r^2 \sin 2\theta + 10. && \sin 2\theta = 2 \sin \theta \cos \theta
\end{aligned}
$$

Substituting the integrand into the volume integral, we have

$$
\begin{aligned}
V &= \int_0^{2\pi} \int_2^4 \left(\tfrac{1}{2} r^2 \sin 2\theta + 10 \right) r \, dr \, d\theta && \text{Iterated integral for volume} \\
&= \int_0^{2\pi} \int_2^4 \left(\tfrac{1}{2} r^3 \sin 2\theta + 10r \right) dr \, d\theta && \text{Simplify.} \\
&= \int_0^{2\pi} \left(\frac{r^4}{8} \sin 2\theta + 5r^2 \right) \Big|_2^4 d\theta && \text{Evaluate inner integral with respect to } r. \\
&= \int_0^{2\pi} (30 \sin 2\theta + 60) \, d\theta && \text{Simplify.} \\
&= (15(-\cos 2\theta) + 60\theta) \Big|_0^{2\pi} = 120\pi. && \text{Evaluate outer integral with respect to } \theta.
\end{aligned}
$$

Related Exercises 22, 38 ◄

More General Polar Regions

In Section 16.2 we generalized double integrals over rectangular regions to double integrals over nonrectangular regions. In an analogous way, the method for integrating over a polar rectangle may be extended to more general regions. Consider a region (described in polar coordinates) bounded by two rays $\theta = \alpha$ and $\theta = \beta$, where $\beta - \alpha \le 2\pi$, and two curves $r = g(\theta)$ and $r = h(\theta)$ (Figure 16.34):

$$ R = \{(r, \theta): 0 \le g(\theta) \le r \le h(\theta), \alpha \le \theta \le \beta\}. $$

The double integral $\iint_R f(x, y) \, dA$ is expressed as an iterated integral in which the inner integral has limits $r = g(\theta)$ and $r = h(\theta)$, and the outer integral runs from $\theta = \alpha$ to $\theta = \beta$; the integrand is transformed into polar coordinates as before. If f is nonnegative on R, the double integral gives the volume of the solid bounded by the surface $z = f(x, y)$ and R.

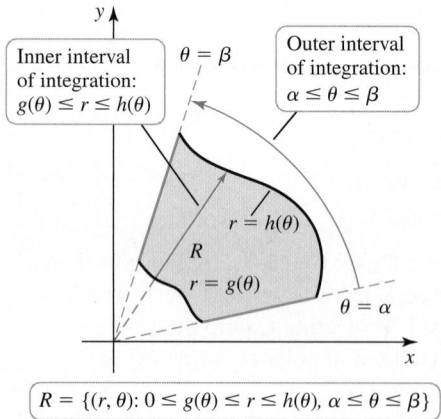

$R = \{(r, \theta): 0 \le g(\theta) \le r \le h(\theta), \alpha \le \theta \le \beta\}$

Figure 16.34

➤ For the type of region described in Theorem 16.4, with the boundaries in the radial direction expressed as functions of θ, the inner integral is always with respect to r.

➤ Recall from Section 12.2 that the polar equation $r = 2a \sin \theta$ describes a circle of radius $|a|$ with center $(0, a)$. The polar equation $r = 2a \cos \theta$ describes a circle of radius $|a|$ with center $(a, 0)$.

THEOREM 16.4 **Change of Variables for Double Integrals over More General Polar Regions**

Let f be continuous on the region R in the xy-plane expressed in polar coordinates as

$$ R = \{(r, \theta): 0 \le g(\theta) \le r \le h(\theta), \alpha \le \theta \le \beta\}, $$

where $0 < \beta - \alpha \le 2\pi$. Then

$$ \iint_R f(x, y) \, dA = \int_\alpha^\beta \int_{g(\theta)}^{h(\theta)} f(r \cos \theta, r \sin \theta) \, r \, dr \, d\theta. $$

EXAMPLE 4 **Specifying regions** Write an iterated integral in polar coordinates for $\iint_R g(r, \theta) \, dA$ for the following regions R in the xy-plane.

a. The region outside the circle $r = 2$ (with radius 2 centered at $(0, 0)$) and inside the circle $r = 4 \cos \theta$ (with radius 2 centered at $(2, 0)$)

b. The region inside both circles of part (a)

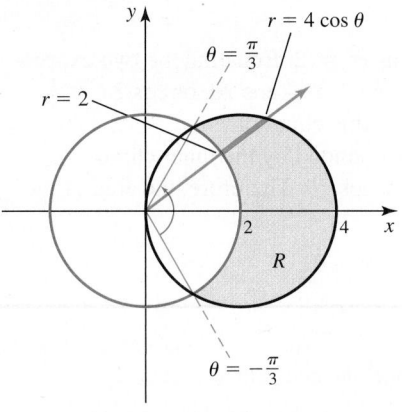

Figure 16.35

SOLUTION

a. Equating the two expressions for r, we have $4\cos\theta = 2$ or $\cos\theta = \frac{1}{2}$, so the circles intersect when $\theta = \pm\pi/3$ (**Figure 16.35**). The inner boundary of R is the circle $r = 2$, and the outer boundary is the circle $r = 4\cos\theta$. Therefore, the region of integration is $R = \{(r, \theta): 2 \le r \le 4\cos\theta, -\pi/3 \le \theta \le \pi/3\}$ and the iterated integral is

$$\iint\limits_R g(r, \theta)\, dA = \int_{-\pi/3}^{\pi/3}\int_2^{4\cos\theta} g(r, \theta)\, r\, dr\, d\theta.$$

b. From part (a), we know that the circles intersect when $\theta = \pm\pi/3$. The region R consists of three subregions R_1, R_2, and R_3 (**Figure 16.36a**).

- For $-\pi/2 \le \theta \le -\pi/3$, R_1 is bounded by $r = 0$ (inner curve) and $r = 4\cos\theta$ (outer curve) (**Figure 16.36b**).
- For $-\pi/3 \le \theta \le \pi/3$, R_2 is bounded by $r = 0$ (inner curve) and $r = 2$ (outer curve) (**Figure 16.36c**).
- For $\pi/3 \le \theta \le \pi/2$, R_3 is bounded by $r = 0$ (inner curve) and $r = 4\cos\theta$ (outer curve) (**Figure 16.36d**).

Therefore, the double integral is expressed in three parts:

$$\iint\limits_R g(r, \theta)\, dA = \int_{-\pi/2}^{-\pi/3}\int_0^{4\cos\theta} g(r, \theta)\, r\, dr\, d\theta + \int_{-\pi/3}^{\pi/3}\int_0^2 g(r, \theta)\, r\, dr\, d\theta$$

$$+ \int_{\pi/3}^{\pi/2}\int_0^{4\cos\theta} g(r, \theta)\, r\, dr\, d\theta.$$

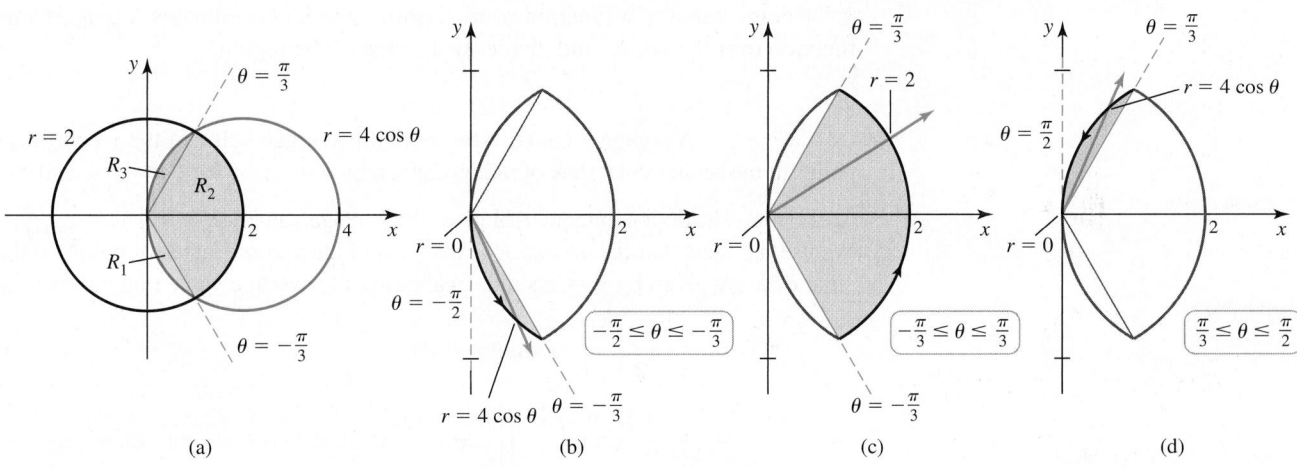

In R_1, radial lines begin at the origin and exit at $r = 4\cos\theta$.

In R_2, radial lines begin at the origin and exit at $r = 2$.

In R_3, radial lines begin at the origin and exit at $r = 4\cos\theta$.

(a)   (b)   (c)   (d)

Figure 16.36

Related Exercise 44 ◄

Areas of Regions

In Cartesian coordinates, the area of a region R in the xy-plane is computed by integrating the function $f(x, y) = 1$ over R; that is, $A = \iint_R dA$. This fact extends to polar coordinates.

▶ Do not forget the factor of r in the area integral!

Area of Polar Regions

The area of the polar region $R = \{(r, \theta): 0 \le g(\theta) \le r \le h(\theta), \alpha \le \theta \le \beta\}$, where $0 < \beta - \alpha \le 2\pi$, is

$$A = \iint\limits_R dA = \int_\alpha^\beta \int_{g(\theta)}^{h(\theta)} r\, dr\, d\theta.$$

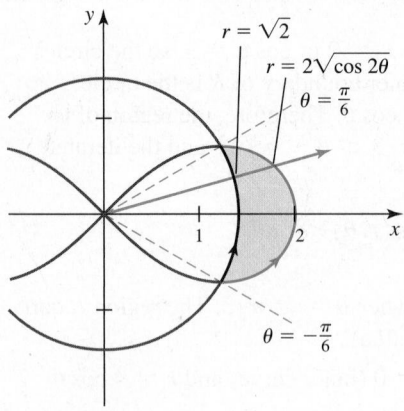

Figure 16.37

QUICK CHECK 4 Express the area of the disk $R = \{(r, \theta): 0 \le r \le a, 0 \le \theta \le 2\pi\}$ in terms of a double integral in polar coordinates. ◄

EXAMPLE 5   Area within a lemniscate Compute the area of the region in the first and fourth quadrants outside the circle $r = \sqrt{2}$ and inside the lemniscate $r^2 = 4 \cos 2\theta$ (Figure 16.37).

SOLUTION The equation of the circle can be written as $r^2 = 2$. Equating the two expressions for r^2, the circle and the lemniscate intersect when $2 = 4 \cos 2\theta$, or $\cos 2\theta = \frac{1}{2}$. The angles in the first and fourth quadrants that satisfy this equation are $\theta = \pm \pi/6$ (Figure 16.37). The region between the two curves is bounded by the inner curve $r = g(\theta) = \sqrt{2}$ and the outer curve $r = h(\theta) = 2\sqrt{\cos 2\theta}$. Therefore, the area of the region is

$$
\begin{aligned}
A &= \int_{-\pi/6}^{\pi/6} \int_{\sqrt{2}}^{2\sqrt{\cos 2\theta}} r \, dr \, d\theta \\
&= \int_{-\pi/6}^{\pi/6} \left(\frac{r^2}{2} \right) \Bigg|_{\sqrt{2}}^{2\sqrt{\cos 2\theta}} d\theta \quad \text{Evaluate inner integral with respect to } r. \\
&= \int_{-\pi/6}^{\pi/6} (2 \cos 2\theta - 1) \, d\theta \quad \text{Simplify.} \\
&= (\sin 2\theta - \theta) \Bigg|_{-\pi/6}^{\pi/6} \quad \text{Evaluate outer integral with respect to } \theta. \\
&= \sqrt{3} - \frac{\pi}{3}. \quad \text{Simplify.}
\end{aligned}
$$

Related Exercises 50–51 ◄

Average Value over a Planar Polar Region

We have encountered the average value of a function in several different settings. To find the average value of a function over a region in polar coordinates, we again integrate the function over the region and divide by the area of the region.

EXAMPLE 6   Average y-coordinate Find the average value of the y-coordinates of the points in the semicircular disk of radius a given by $R = \{(r, \theta): 0 \le r \le a, 0 \le \theta \le \pi\}$.

SOLUTION The double integral that gives the average value we seek is $\bar{y} = \frac{1}{\text{area of } R} \iint_R y \, dA$. We use the facts that the area of R is $\pi a^2/2$ and the y-coordinates of points in the semicircular disk are given by $y = r \sin \theta$. Evaluating the average value integral, we find that

$$
\begin{aligned}
\bar{y} &= \frac{1}{\pi a^2/2} \int_0^\pi \int_0^a r \sin \theta \, r \, dr \, d\theta \\
&= \frac{2}{\pi a^2} \int_0^\pi \sin \theta \left(\frac{r^3}{3} \right) \Bigg|_0^a d\theta \quad \text{Evaluate inner integral with respect to } r. \\
&= \frac{2}{\pi a^2} \frac{a^3}{3} \int_0^\pi \sin \theta \, d\theta \quad \text{Simplify.} \\
&= \frac{2a}{3\pi} (-\cos \theta) \Bigg|_0^\pi \quad \text{Evaluate outer integral with respect to } \theta. \\
&= \frac{4a}{3\pi}. \quad \text{Simplify.}
\end{aligned}
$$

Note that $4/(3\pi) \approx 0.42$, so the average value of the y-coordinates is less than half the radius of the disk.

Related Exercise 53 ◄

SECTION 16.3 EXERCISES

Getting Started

1. Draw the polar region $\{(r, \theta): 1 \le r \le 2, 0 \le \theta \le \pi/2\}$. Why is it called a polar rectangle?

2. Write the double integral $\iint_R f(x, y)\, dA$ as an iterated integral in polar coordinates when $R = \{(r, \theta): a \le r \le b, \alpha \le \theta \le \beta\}$.

3. Sketch in the xy-plane the region of integration for the integral
$$\int_{-\pi/6}^{\pi/6} \int_{1/2}^{\cos 2\theta} g(r, \theta)\, r\, dr\, d\theta.$$

4. Explain why the element of area in Cartesian coordinates $dx\, dy$ becomes $r\, dr\, d\theta$ in polar coordinates.

5. How do you find the area of a polar region $R = \{(r, \theta): g(\theta) \le r \le h(\theta), \alpha \le \theta \le \beta\}$?

6. How do you find the average value of a function over a region that is expressed in polar coordinates?

7–10. Polar rectangles *Sketch the following polar rectangles.*

7. $R = \{(r, \theta): 0 \le r \le 5, 0 \le \theta \le \pi/2\}$

8. $R = \{(r, \theta): 2 \le r \le 3, \pi/4 \le \theta \le 5\pi/4\}$

9. $R = \{(r, \theta): 1 \le r \le 4, -\pi/4 \le \theta \le 2\pi/3\}$

10. $R = \{(r, \theta): 4 \le r \le 5, -\pi/3 \le \theta \le \pi/2\}$

Practice Exercises

11–14. Volume of solids *Find the volume of the solid bounded by the surface $z = f(x, y)$ and the xy-plane.*

11. $f(x, y) = 4 - \sqrt{x^2 + y^2}$

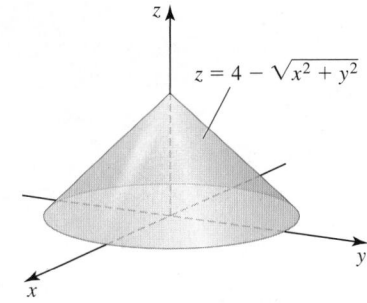

12. $f(x, y) = 16 - 4(x^2 + y^2)$ 13. $f(x, y) = e^{-(x^2+y^2)/8} - e^{-2}$

14. $f(x, y) = \dfrac{20}{1 + x^2 + y^2} - 2$

15–18. Solids bounded by paraboloids *Find the volume of the solid below the paraboloid $z = 4 - x^2 - y^2$ and above the following polar rectangles.*

15. $R = \{(r, \theta): 0 \le r \le 1,$
$0 \le \theta \le 2\pi\}$

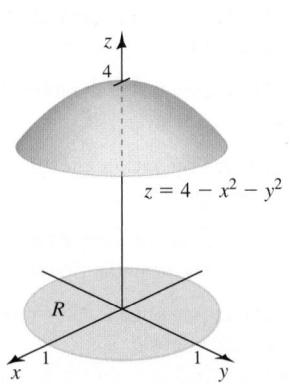

16. $R = \{(r, \theta): 0 \le r \le 2, 0 \le \theta \le 2\pi\}$

17. $R = \{(r, \theta): 1 \le r \le 2, 0 \le \theta \le 2\pi\}$

18. $R = \{(r, \theta): 1 \le r \le 2, -\pi/2 \le \theta \le \pi/2\}$

19–20. Solids bounded by hyperboloids *Find the volume of the solid below the hyperboloid $z = 5 - \sqrt{1 + x^2 + y^2}$ and above the following polar rectangles.*

19. $R = \{(r, \theta): \sqrt{3} \le r \le 2\sqrt{2}, 0 \le \theta \le 2\pi\}$

20. $R = \{(r, \theta): \sqrt{3} \le r \le \sqrt{15}, -\pi/2 \le \theta \le \pi\}$

21–30. Cartesian to polar coordinates *Evaluate the following integrals using polar coordinates. Assume (r, θ) are polar coordinates. A sketch is helpful.*

21. $\displaystyle\iint_R (x^2 + y^2)\, dA; R = \{(r, \theta): 0 \le r \le 4, 0 \le \theta \le 2\pi\}$

22. $\displaystyle\iint_R 2xy\, dA; R = \{(r, \theta): 1 \le r \le 3, 0 \le \theta \le \pi/2\}$

23. $\displaystyle\iint_R 2xy\, dA; R = \{(x, y): x^2 + y^2 \le 9, y \ge 0\}$

24. $\displaystyle\iint_R \frac{dA}{1 + x^2 + y^2}; R = \{(r, \theta): 1 \le r \le 2, 0 \le \theta \le \pi\}$

25. $\displaystyle\iint_R \frac{dA}{\sqrt{16 - x^2 - y^2}}; R = \{(x, y): x^2 + y^2 \le 4, x \ge 0, y \ge 0\}$

26. $\displaystyle\iint_R e^{-x^2 - y^2}\, dA; R = \{(x, y): x^2 + y^2 \le 9\}$

27. $\displaystyle\int_{-1}^{1} \int_{-\sqrt{1-x^2}}^{\sqrt{1-x^2}} (x^2 + y^2)^{3/2}\, dy\, dx$

28. $\displaystyle\int_{0}^{3} \int_{0}^{\sqrt{9-x^2}} \sqrt{x^2 + y^2}\, dy\, dx$

29. $\displaystyle\iint_R \sqrt{x^2 + y^2}\, dA; R = \{(x, y): 1 \le x^2 + y^2 \le 4\}$

30. $\displaystyle\int_{-4}^{4} \int_{0}^{\sqrt{16-y^2}} (16 - x^2 - y^2)\, dx\, dy$

31–40. Volume between surfaces *Find the volume of the following solids.*

31. The solid bounded by the paraboloid $z = x^2 + y^2$ and the plane $z = 9$

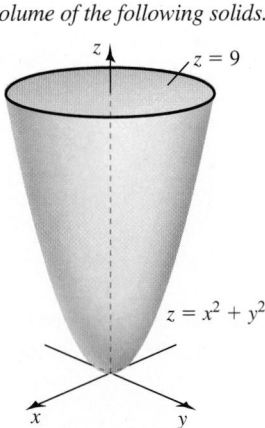

32. The solid bounded by the paraboloid $z = 2 - x^2 - y^2$ and the plane $z = 1$

33. The solid bounded by the paraboloids $z = x^2 + y^2$ and $z = 2 - x^2 - y^2$

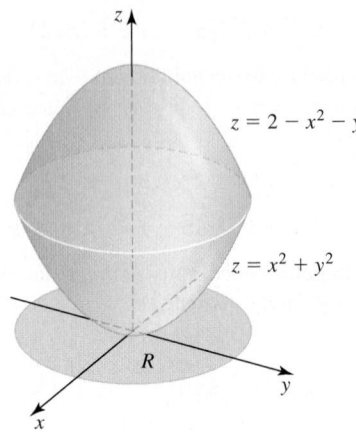

34. The solid bounded by the paraboloids $z = 2x^2 + y^2$ and $z = 27 - x^2 - 2y^2$

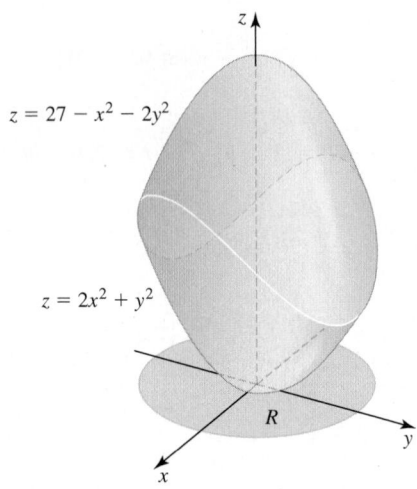

35. The solid bounded below by the paraboloid $z = x^2 + y^2 - x - y$ and above by the plane $x + y + z = 4$

36. The solid bounded by the cylinder $x^2 + y^2 = 4$ and the planes $z = 3 - x$ and $z = x - 3$

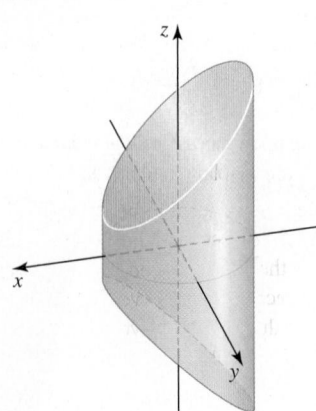

37. The solid bounded by the paraboloid $z = 18 - x^2 - 3y^2$ and the hyperbolic paraboloid $z = x^2 - y^2$

38. The solid outside the cylinder $x^2 + y^2 = 1$ that is bounded above by the hyperbolic paraboloid $z = -x^2 + y^2 + 8$ and below by the paraboloid $z = x^2 + 3y^2$

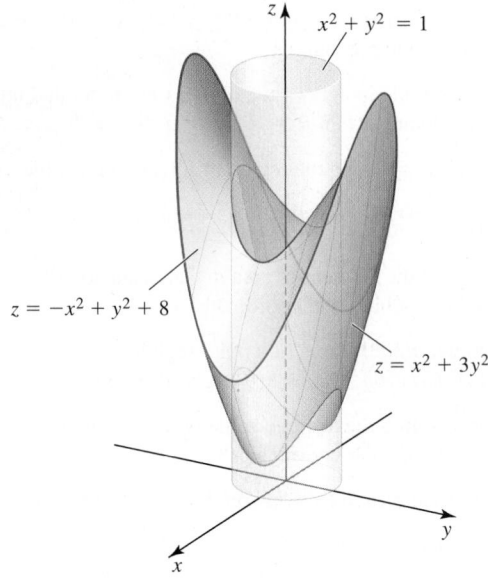

39. The solid outside the cylinder $x^2 + y^2 = 1$ that is bounded above by the sphere $x^2 + y^2 + z^2 = 8$ and below by the cone $z = \sqrt{x^2 + y^2}$

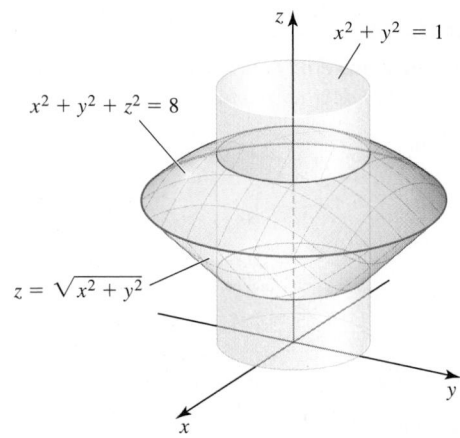

40. The solid bounded by the cone $z = 2 - \sqrt{x^2 + y^2}$ and the upper half of a hyperboloid of two sheets $z = \sqrt{1 + x^2 + y^2}$

41–46. Describing general regions *Sketch the following regions R. Then express* $\iint_R g(r, \theta)\, dA$ *as an iterated integral over R in polar coordinates.*

41. The region inside the limaçon $r = 1 + \frac{1}{2}\cos\theta$

42. The region inside the leaf of the rose $r = 2\sin 2\theta$ in the first quadrant

43. The region inside the lobe of the lemniscate $r^2 = 2\sin 2\theta$ in the first quadrant

44. The region outside the circle $r = 2$ and inside the circle $r = 4\sin\theta$

45. The region outside the circle $r = 1$ and inside the rose $r = 2\sin 3\theta$ in the first quadrant

46. The region outside the circle $r = 1/2$ and inside the cardioid $r = 1 + \cos\theta$

47–52. Computing areas *Use a double integral to find the area of the following regions.*

47. The annular region $\{(r, \theta): 1 \leq r \leq 2, 0 \leq \theta \leq \pi\}$

48. The region bounded by the cardioid $r = 2(1 - \sin\theta)$

49. The region bounded by all leaves of the rose $r = 2\cos 3\theta$

50. The region inside both the cardioid $r = 1 - \cos\theta$ and the circle $r = 1$

51. The region inside both the cardioid $r = 1 + \sin\theta$ and the cardioid $r = 1 + \cos\theta$

52. The region bounded by the spiral $r = 2\theta$, for $0 \leq \theta \leq \pi$, and the x-axis

53–54. Average values *Find the following average values.*

53. The average distance between points of the disk $\{(r, \theta): 0 \leq r \leq a\}$ and the origin

54. The average value of $1/r^2$ over the annulus $\{(r, \theta): 2 \leq r \leq 4\}$

55. Explain why or why not Determine whether the following statements are true and give an explanation or counterexample.

    a. Let R be the unit disk centered at $(0, 0)$. Then $\iint_R (x^2 + y^2)\, dA = \int_0^{2\pi} \int_0^1 r^2\, dr\, d\theta$.

    b. The average distance between the points of the hemisphere $z = \sqrt{4 - x^2 - y^2}$ and the origin is 2 (calculus not required).

    c. The integral $\int_0^1 \int_0^{\sqrt{1-y^2}} e^{x^2+y^2}\, dx\, dy$ is easier to evaluate in polar coordinates than in Cartesian coordinates.

56. Areas of circles Use integration to show that the circles $r = 2a\cos\theta$ and $r = 2a\sin\theta$ have the same area, which is πa^2.

57. Filling bowls with water Which bowl holds the most water when all the bowls are filled to a depth of 4 units?

    • The paraboloid $z = x^2 + y^2$, for $0 \leq z \leq 4$

    • The cone $z = \sqrt{x^2 + y^2}$, for $0 \leq z \leq 4$

    • The hyperboloid $z = \sqrt{1 + x^2 + y^2}$, for $1 \leq z \leq 5$

58. Equal volumes To what height (above the bottom of the bowl) must the cone and paraboloid bowls of Exercise 57 be filled to hold the same volume of water as the hyperboloid bowl filled to a depth of 4 units $(1 \leq z \leq 5)$?

59. Volume of a hyperbolic paraboloid Consider the surface $z = x^2 - y^2$.

    a. Find the region in the xy-plane in polar coordinates for which $z \geq 0$.

    b. Let $R = \{(r, \theta): 0 \leq r \leq a, -\pi/4 \leq \theta \leq \pi/4\}$, which is a sector of a circle of radius a. Find the volume of the region below the hyperbolic paraboloid and above the region R.

60. Volume of a sphere Use double integrals in polar coordinates to verify that the volume of a sphere of radius a is $\frac{4}{3}\pi a^3$.

61. Volume Find the volume of the solid bounded by the cylinder $(x - 1)^2 + y^2 = 1$, the plane $z = 0$, and the cone $z = \sqrt{x^2 + y^2}$ (see figure). (*Hint:* Use symmetry.)

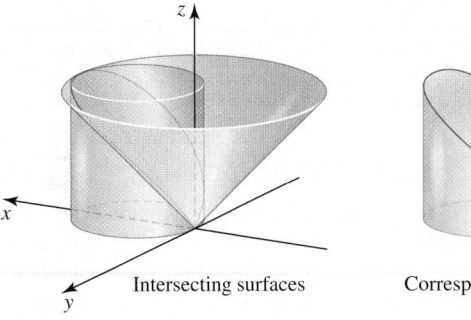

Intersecting surfaces   Corresponding solid

62. Volume Find the volume of the solid bounded by the paraboloid $z = 2x^2 + 2y^2$, the plane $z = 0$, and the cylinder $x^2 + (y - 1)^2 = 1$. (*Hint:* Use symmetry.)

Explorations and Challenges

63–64. Miscellaneous integrals *Evaluate the following integrals using the method of your choice. A sketch is helpful.*

63. $\displaystyle\iint_R \frac{dA}{4 + \sqrt{x^2 + y^2}}; R = \left\{(r, \theta): 0 \leq r \leq 2, \frac{\pi}{2} \leq \theta \leq \frac{3\pi}{2}\right\}$

64. $\displaystyle\iint_R \frac{x - y}{x^2 + y^2 + 1}\, dA; R$ is the region bounded by the unit circle centered at the origin.

65–68. Improper integrals *Improper integrals arise in polar coordinates when the radial coordinate r becomes arbitrarily large. Under certain conditions, these integrals are treated in the usual way:*

$$\int_\alpha^\beta \int_a^\infty g(r, \theta)\, r\, dr\, d\theta = \lim_{b \to \infty} \int_\alpha^\beta \int_a^b g(r, \theta)\, r\, dr\, d\theta.$$

Use this technique to evaluate the following integrals.

65. $\displaystyle\int_0^{\pi/2} \int_1^\infty \frac{\cos\theta}{r^3}\, r\, dr\, d\theta$

66. $\displaystyle\iint_R \frac{dA}{(x^2 + y^2)^{5/2}}; R = \{(r, \theta): 1 \leq r < \infty, 0 \leq \theta \leq 2\pi\}$

67. $\displaystyle\iint_R e^{-x^2-y^2}\, dA; R = \left\{(r, \theta): 0 \leq r < \infty, 0 \leq \theta \leq \frac{\pi}{2}\right\}$

68. $\displaystyle\iint_R \frac{dA}{(1 + x^2 + y^2)^2}; R$ is the first quadrant.

⊤ 69. Slicing a hemispherical cake A cake is shaped like a hemisphere of radius 4 with its base on the xy-plane. A wedge of the cake is removed by making two slices from the center of the cake outward, perpendicular to the xy-plane and separated by an angle of φ.

    a. Use a double integral to find the volume of the slice for $\varphi = \pi/4$. Use geometry to check your answer.

    b. Now suppose the cake is sliced horizontally at $z = a > 0$ and let D be the piece of cake above the plane $z = a$. For what approximate value of a is the volume of D equal to the volume in part (a)?

T 70. Mass from density data The following table gives the density (in units of g/cm^2) at selected points (in polar coordinates) of a thin semicircular plate of radius 3. Estimate the mass of the plate and explain your method.

	$\theta = 0$	$\theta = \pi/4$	$\theta = \pi/2$	$\theta = 3\pi/4$	$\theta = \pi$
$r = 1$	2.0	2.1	2.2	2.3	2.4
$r = 2$	2.5	2.7	2.9	3.1	3.3
$r = 3$	3.2	3.4	3.5	3.6	3.7

71. A mass calculation Suppose the density of a thin plate represented by the polar region R is $\rho(r, \theta)$ (in units of mass per area). The mass of the plate is $\iint_R \rho(r, \theta)\, dA$. Find the mass of the thin half annulus $R = \{(r, \theta): 1 \le r \le 4, 0 \le \theta \le \pi\}$ with a density $\rho(r, \theta) = 4 + r \sin \theta$.

72. Area formula In Section 12.3 it was shown that the area of a region enclosed by the polar curve $r = g(\theta)$ and the rays $\theta = \alpha$ and $\theta = \beta$, where $\beta - \alpha \le 2\pi$, is $A = \frac{1}{2} \int_\alpha^\beta r^2\, d\theta$. Prove this result using the area formula with double integrals.

73. Normal distribution An important integral in statistics associated with the normal distribution is $I = \int_{-\infty}^{\infty} e^{-x^2}\, dx$. It is evaluated in the following steps.

a. In Section 8.9, it is shown that $\int_0^{\infty} e^{-x^2}\, dx$ converges (in the narrative following Example 7). Use this result to explain why $\int_{-\infty}^{\infty} e^{-x^2}\, dx$ converges.

b. Assume

$$I^2 = \left(\int_{-\infty}^{\infty} e^{-x^2}\, dx \right)\left(\int_{-\infty}^{\infty} e^{-y^2}\, dy \right) = \int_{-\infty}^{\infty} \int_{-\infty}^{\infty} e^{-x^2 - y^2}\, dx\, dy,$$

where we have chosen the variables of integration to be x and y and then written the product as an iterated integral. Evaluate

this integral in polar coordinates and show that $I = \sqrt{\pi}$. Why is the solution $I = -\sqrt{\pi}$ rejected?

c. Evaluate $\int_0^{\infty} e^{-x^2}\, dx$, $\int_0^{\infty} x e^{-x^2}\, dx$, and $\int_0^{\infty} x^2 e^{-x^2}\, dx$ (using part (a) if needed).

74. Existence of integrals For what values of p does the integral

$$\iint_R \frac{dA}{(x^2 + y^2)^p}$$ exist in the following cases? Assume (r, θ) are

polar coordinates.

a. $R = \{(r, \theta): 1 \le r < \infty, 0 \le \theta \le 2\pi\}$
b. $R = \{(r, \theta): 0 \le r \le 1, 0 \le \theta \le 2\pi\}$

75. Integrals in strips Consider the integral

$$I = \iint_R \frac{dA}{(1 + x^2 + y^2)^2}$$

where $R = \{(x, y): 0 \le x \le 1, 0 \le y \le a\}$.

a. Evaluate I for $a = 1$. (*Hint:* Use polar coordinates.)
b. Evaluate I for arbitrary $a > 0$.
c. Let $a \to \infty$ in part (b) to find I over the infinite strip $R = \{(x, y): 0 \le x \le 1, 0 \le y < \infty\}$.

QUICK CHECK ANSWERS

1. $R = \{(r, \theta): 1 \le r \le 2, 0 \le \theta \le \pi/2\}$
2. r^5, $r^2 (\cos^2 \theta - \sin^2 \theta) = r^2 \cos 2\theta$
3. $z = 2 - \sqrt{x^2 + y^2}$ is the lower half of the double-napped cone $(2 - z)^2 = x^2 + y^2$. Imagine both halves of this cone in Figure 16.32: It is apparent that the paraboloid $z = x^2 + y^2$ intersects the cone twice, once when $z = 1$ and once when $z = 4$.   4. $\int_0^{2\pi} \int_0^a r\, dr\, d\theta = \pi a^2$ ◄

16.4 Triple Integrals

At this point, you may see a pattern that is developing with respect to integration. In Chapter 5, we introduced integrals of single-variable functions. In the first three sections of this chapter, we moved up one dimension to double integrals of two-variable functions. In this section, we take another step and investigate triple integrals of three-variable functions. There is no end to the progression of multiple integrals. It is possible to define integrals with respect to any number of variables. For example, problems in statistics and statistical mechanics involve integration over regions of many dimensions.

Triple Integrals in Rectangular Coordinates

Consider a function $w = f(x, y, z)$ that is defined on a closed and bounded region D of $\mathbb{R}^3$. The graph of f lies in four-dimensional space and is the set of points $(x, y, z, f(x, y, z))$, where (x, y, z) is in D. Despite the difficulty in representing f in $\mathbb{R}^3$, we may still define the integral of f over D. We first create a partition of D by slicing the region with three sets of planes that run parallel to the xz-, yz-, and xy-planes (**Figure 16.38**). This partition subdivides D into small boxes that are ordered in a convenient way from $k = 1$ to $k = n$. The partition includes all boxes that are wholly contained in D. The kth box has side lengths Δx_k, Δy_k, and Δz_k, and volume $\Delta V_k = \Delta x_k \Delta y_k \Delta z_k$. We let (x_k^*, y_k^*, z_k^*) be an arbitrary point in the kth box, for $k = 1, \ldots, n$.

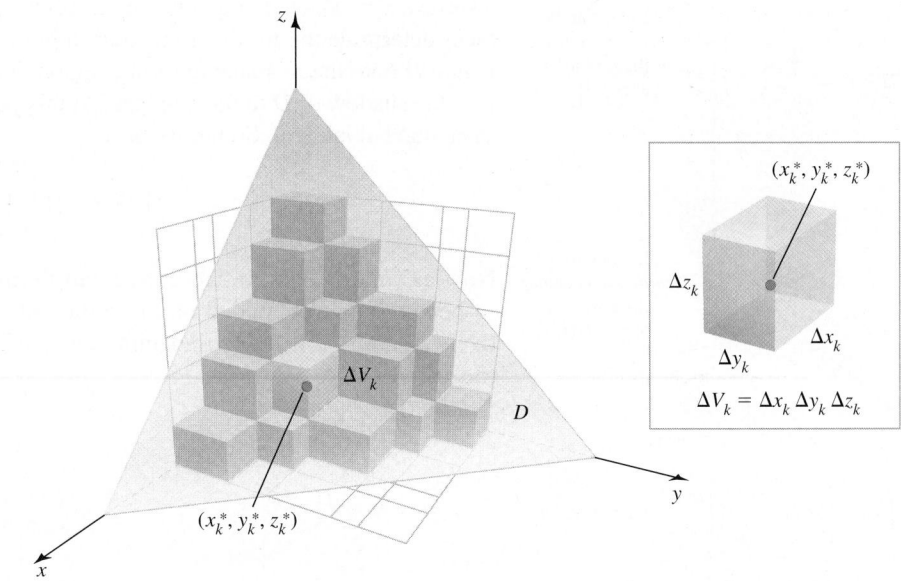

Figure 16.38

A Riemann sum is now formed, in which the kth term is the function value $f(x_k^*, y_k^*, z_k^*)$ multiplied by the volume of the kth box:

$$\sum_{k=1}^{n} f(x_k^*, y_k^*, z_k^*) \Delta V_k.$$

We let Δ denote the maximum length of the diagonals of the boxes. As the number of boxes n increases, while Δ approaches zero, two things happen.

- For commonly encountered regions, the region formed by the collection of boxes approaches the region D.
- If f is continuous, the Riemann sum approaches a limit.

The limit of the Riemann sum is the **triple integral of f over D**, and we write

$$\iiint_D f(x, y, z)\, dV = \lim_{\Delta \to 0} \sum_{k=1}^{n} f(x_k^*, y_k^*, z_k^*) \Delta V_k.$$

The kth box in the partition has volume $\Delta V_k = \Delta x_k \Delta y_k \Delta z_k$, where Δx_k, Δy_k, and Δz_k are the side lengths of the box. Accordingly, the *element of volume* in the triple integral, which we denote dV, becomes $dx\, dy\, dz$ (or some rearrangement of dx, dy, and dz) in an iterated integral.

We give two immediate interpretations of a triple integral. First, if $f(x, y, z) = 1$, then the Riemann sum simply adds up the volumes of the boxes in the partition. In the limit as $\Delta \to 0$, the triple integral $\iiint_D dV$ gives the volume of the region D. Second, suppose D is a solid three-dimensional object and its density varies from point to point according to the function $f(x, y, z)$. The units of density are mass per unit volume, so the product $f(x_k^*, y_k^*, z_k^*) \Delta V_k$ approximates the mass of the kth box in D. Summing the masses of the boxes gives an approximation to the total mass of D. In the limit as $\Delta \to 0$, the triple integral gives the mass of the object.

As with double integrals, a version of Fubini's Theorem expresses a triple integral in terms of an iterated integral in x, y, and z. The situation becomes interesting because with three variables, there are *six* possible orders of integration.

> ➤ Notice the analogy between double and triple integrals:
>
> $$\text{area of } R = \iint_R dA \quad \text{and}$$
>
> $$\text{volume of } D = \iiint_D dV.$$
>
> The use of triple integrals to compute the mass of an object is discussed in detail in Section 16.6.

QUICK CHECK 1 List the six orders in which the three differentials dx, dy, and dz may be written. ◄

Finding Limits of Integration We discuss one of the six orders of integration in detail; the others are examined in the examples. Suppose a region D in $\mathbb{R}^3$ is bounded above by

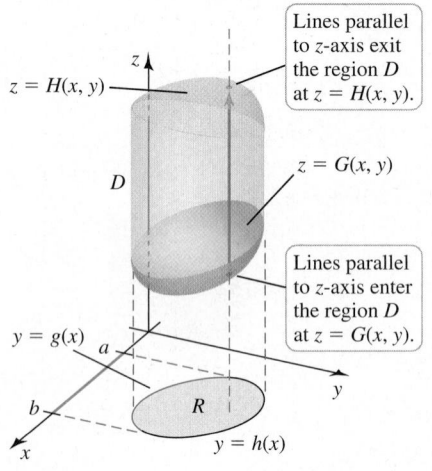

$$\iiint\limits_{D} f(x, y, z) \, dV = \iint\limits_{R} \left(\int_{G(x, y)}^{H(x, y)} f(x, y, z) \, dz \right) dA$$

Figure 16.39

a surface $z = H(x, y)$ and below by a surface $z = G(x, y)$ (**Figure 16.39**). These two surfaces determine the limits of integration in the z-direction. The next step is to project the region D onto the xy-plane to form a region that we call R (**Figure 16.40**). You can think of R as the shadow of D in the xy-plane. At this point, we can begin to write the triple integral as an iterated integral. So far, we have

$$\iiint\limits_{D} f(x, y, z) \, dV = \iint\limits_{R} \left(\int_{G(x, y)}^{H(x, y)} f(x, y, z) \, dz \right) dA.$$

Now assume R is bounded above and below by the curves $y = h(x)$ and $y = g(x)$, respectively, and bounded on the right and left by the lines $x = a$ and $x = b$, respectively (**Figure 16.40**). The remaining integration over R is carried out as a double integral (Section 16.2).

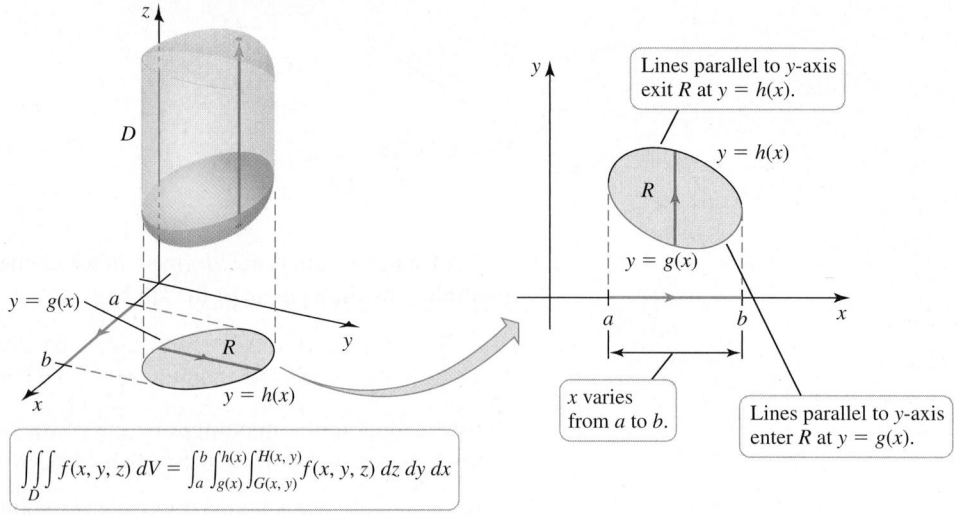

$$\iiint\limits_{D} f(x, y, z) \, dV = \int_a^b \int_{g(x)}^{h(x)} \int_{G(x, y)}^{H(x, y)} f(x, y, z) \, dz \, dy \, dx$$

Figure 16.40

Table 16.2

Integral	Variable	Interval
Inner	z	$G(x, y) \le z \le H(x, y)$
Middle	y	$g(x) \le y \le h(x)$
Outer	x	$a \le x \le b$

The intervals that describe D are summarized in Table 16.2, which can then be used to formulate the limits of integration. To integrate over all points of D, we carry out the following steps.

1. Integrate with respect to z from $z = G(x, y)$ to $z = H(x, y)$; the result (in general) is a function of x and y.

2. Integrate with respect to y from $y = g(x)$ to $y = h(x)$; the result (in general) is a function of x.

3. Integrate with respect to x from $x = a$ to $x = b$; the result is (always) a real number.

▶ Theorem 16.5 is a version of Fubini's Theorem. Five other versions could be written for the other orders of integration.

THEOREM 16.5 Triple Integrals

Let f be continuous over the region

$$D = \{ (x, y, z) : a \le x \le b, g(x) \le y \le h(x), G(x, y) \le z \le H(x, y) \},$$

where g, h, G, and H are continuous functions. Then f is integrable over D and the triple integral is evaluated as the iterated integral

$$\iiint\limits_{D} f(x, y, z) \, dV = \int_a^b \int_{g(x)}^{h(x)} \int_{G(x, y)}^{H(x, y)} f(x, y, z) \, dz \, dy \, dx.$$

We now illustrate this procedure with several examples.

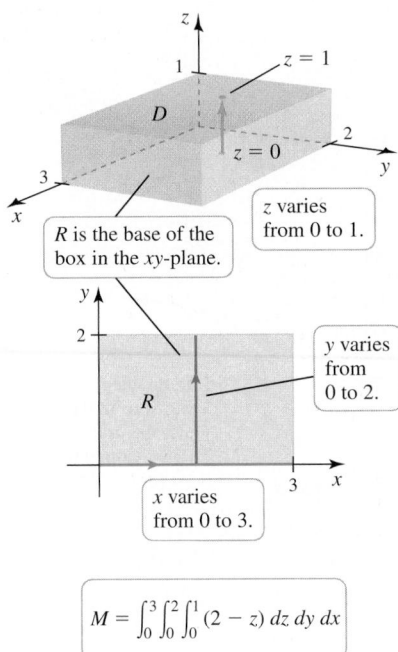

R is the base of the box in the xy-plane.

z varies from 0 to 1.

y varies from 0 to 2.

x varies from 0 to 3.

$$M = \int_0^3 \int_0^2 \int_0^1 (2 - z)\, dz\, dy\, dx$$

Figure 16.41

Table 16.3

Integral	Variable	Interval
Inner	z	$0 \le z \le 1$
Middle	y	$0 \le y \le 2$
Outer	x	$0 \le x \le 3$

QUICK CHECK 2 Write the integral in Example 1 in the orders *dx dy dz* and *dx dz dy*. ◄

EXAMPLE 1   Mass of a box A solid box D is bounded by the planes $x = 0$, $x = 3$, $y = 0$, $y = 2$, $z = 0$, and $z = 1$. The density of the box decreases linearly in the positive z-direction and is given by $f(x, y, z) = 2 - z$. Find the mass of the box.

SOLUTION The mass of the box is found by integrating the density $f(x, y, z) = 2 - z$ over the box. Because the limits of integration for all three variables are constant, the iterated integral may be written in any order. Using the order of integration $dz\, dy\, dx$ (**Figure 16.41**), the limits of integration are shown in Table 16.3.

The mass of the box is

$$M = \iiint_D (2 - z)\, dV$$

$$= \int_0^3 \int_0^2 \int_0^1 (2 - z)\, dz\, dy\, dx \qquad \text{Convert to an iterated integral.}$$

$$= \int_0^3 \int_0^2 \left(2z - \frac{z^2}{2}\right)\Big|_0^1 dy\, dx \qquad \text{Evaluate inner integral with respect to } z.$$

$$= \int_0^3 \int_0^2 \frac{3}{2}\, dy\, dx \qquad \text{Simplify.}$$

$$= \int_0^3 \left(\frac{3y}{2}\right)\Big|_0^2 dx \qquad \text{Evaluate middle integral with respect to } y.$$

$$= \int_0^3 3\, dx = 9. \qquad \text{Evaluate outer integral with respect to } x \text{ and simplify.}$$

The result makes sense: The density of the box varies linearly from 1 (at the top of the box) to 2 (at the bottom); if the box had a constant density of 1, its mass would be (volume) × (density) = 6; if the box had a constant density of 2, its mass would be 12. The actual mass is the average of 6 and 12, as you might expect.

Any other order of integration produces the same result. For example, with the order *dy dx dz*, the iterated integral is

$$M = \iiint_D (2 - z)\, dV = \int_0^1 \int_0^3 \int_0^2 (2 - z)\, dy\, dx\, dz = 9.$$

Related Exercises 8–9 ◄

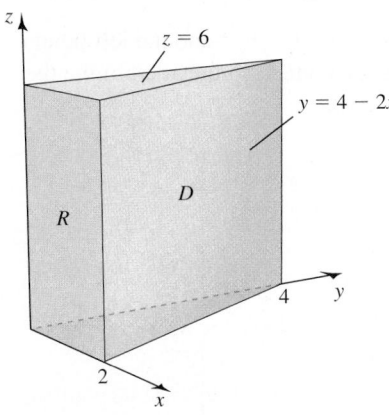

Figure 16.42

EXAMPLE 2   Volume of a prism Find the volume of the prism D in the first octant bounded by the planes $y = 4 - 2x$ and $z = 6$ (**Figure 16.42**).

SOLUTION The prism may be viewed in several different ways. Letting the base of the prism be in the xz-plane, the upper surface of the prism is the plane $y = 4 - 2x$, and the lower surface is $y = 0$. The projection of the prism onto the xz-plane is the rectangle $R = \{(x, z): 0 \le x \le 2, 0 \le z \le 6\}$. One possible order of integration in this case is *dy dx dz*.

Inner integral with respect to y: A line through the prism parallel to the y-axis enters the prism through the rectangle R at $y = 0$ and exits the prism at the plane $y = 4 - 2x$. Therefore, we first integrate with respect to y over the interval $0 \le y \le 4 - 2x$ (**Figure 16.43a**).

Middle integral with respect to x: The limits of integration for the middle and outer integrals must cover the region R in the xz-plane. A line parallel to the x-axis enters R at $x = 0$ and exits R at $x = 2$. So we integrate with respect to x over the interval $0 \le x \le 2$ (**Figure 16.43b**).

Outer integral with respect to z: To cover all of R, the line segments from $x = 0$ to $x = 2$ must run from $z = 0$ to $z = 6$. So we integrate with respect to z over the interval $0 \le z \le 6$ (**Figure 16.43b**).

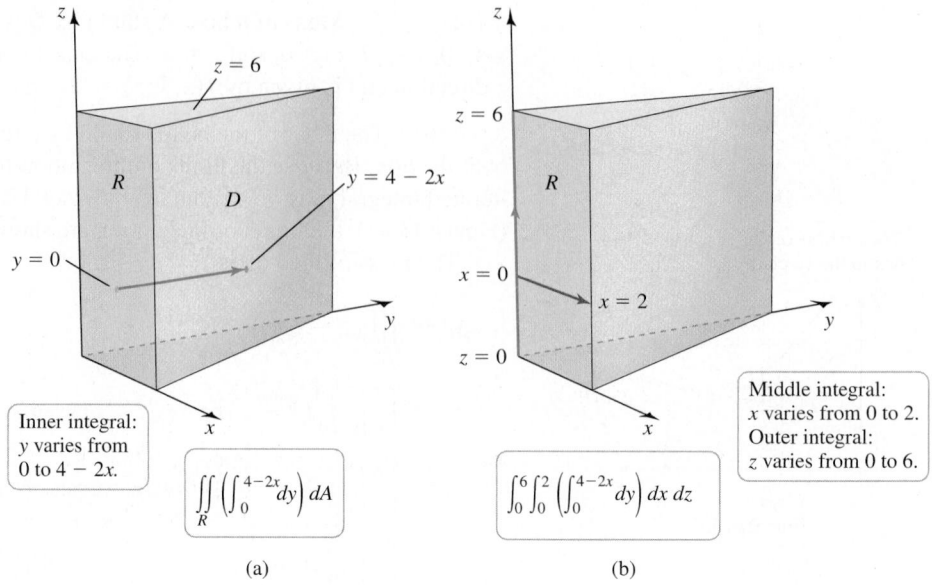

Figure 16.43

Integrating $f(x, y, z) = 1$, the volume of the prism is

$$V = \iiint_D dV = \int_0^6 \int_0^2 \int_0^{4-2x} dy \, dx \, dz$$

$$= \int_0^6 \int_0^2 (4 - 2x) \, dx \, dz \qquad \text{Evaluate inner integral with respect to } y.$$

$$= \int_0^6 (4x - x^2) \Big|_0^2 \, dz \qquad \text{Evaluate middle integral with respect to } x.$$

$$= \int_0^6 4 \, dz \qquad \text{Simplify.}$$

$$= 24. \qquad \text{Evaluate outer integral with respect to } z.$$

Related Exercises 15, 18 ◄

▶ The volume of the prism could also be found using geometry: The area of the triangular base in the *xy*-plane is 4 and the height of the prism is 6. Therefore, the volume is $6 \cdot 4 = 24$.

QUICK CHECK 3 Write the integral in Example 2 in the orders $dz \, dy \, dx$ and $dx \, dy \, dz$. ◄

EXAMPLE 3   A volume integral Find the volume of the solid D bounded by the paraboloids $y = x^2 + 3z^2 + 1$ and $y = 5 - 3x^2 - z^2$ (**Figure 16.44a**).

SOLUTION The right boundary of D is the surface $y = 5 - 3x^2 - z^2$ and the left boundary is $y = x^2 + 3z^2 + 1$. These surfaces are functions of x and z, so they determine the limits of integration for the inner integral in the y-direction.

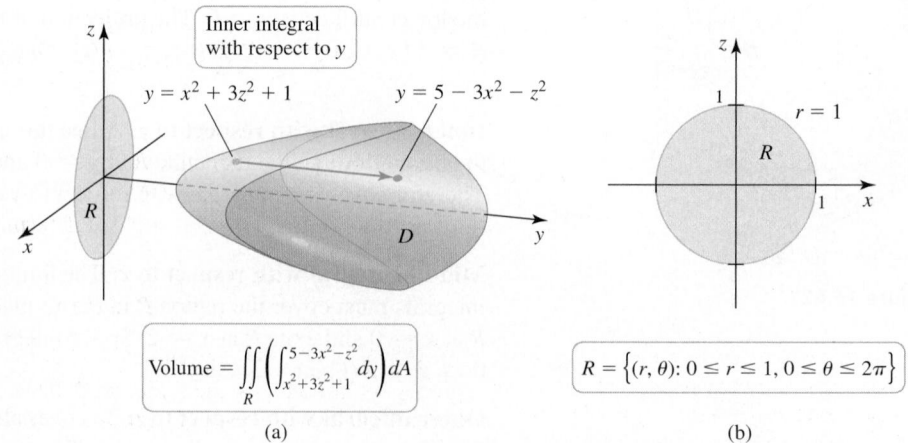

Figure 16.44

A key step in the calculation is finding the curve of intersection between the two surfaces and projecting it onto the xz-plane to form the boundary of the region R, where R is the projection of D onto the xz-plane. Equating the y-coordinates of the surfaces, we have $x^2 + 3z^2 + 1 = 5 - 3x^2 - z^2$, which, when simplified, is the equation of a unit circle centered at the origin in the xz-plane:

$$x^2 + z^2 = 1.$$

Observe that a line through the solid parallel to the y-axis enters the solid at $y = x^2 + 3z^2 + 1$ and exits at $y = 5 - 3x^2 - z^2$. Therefore, for fixed values of x and z, we integrate in y over the interval $x^2 + 3z^2 + 1 \le y \le 5 - 3x^2 - z^2$ (Figure 16.44a). After evaluating the inner integral with respect to y, we have

$$V = \iint\limits_{R} \left(\int_{x^2+3z^2+1}^{5-3x^2-z^2} dy \right) dA = \iint\limits_{R} \left(y \Big|_{x^2+3z^2+1}^{5-3x^2-z^2} \right) dA = \iint\limits_{R} 4(1 - x^2 - z^2)\, dA.$$

The region R is bounded by a circle, so it is advantageous to evaluate the remaining double integral in polar coordinates, where θ and r have the same meaning in the xz-plane as they do in the xy-plane. Note that R is expressed in polar coordinates as $R = \{(r, \theta): 0 \le r \le 1, 0 \le \theta \le 2\pi\}$ (Figure 16.44b). Using the relationship $x^2 + z^2 = r^2$, we change variables and evaluate the double integral:

$$
\begin{aligned}
V &= \iint\limits_{R} 4(1 - x^2 - z^2)\, dA \\
&= \int_{0}^{2\pi} \int_{0}^{1} 4(1 - r^2)\, r\, dr\, d\theta &&\text{Convert to polar coordinates: } x^2 + z^2 = r^2 \\
& &&\text{and } dA = r\, dr\, d\theta. \\
&= \int_{0}^{2\pi} \int_{0}^{1} 2u\, du\, d\theta &&\text{Let } u = 1 - r^2 \Rightarrow du = -2r\, dr. \\
&= \int_{0}^{2\pi} \left(u^2 \Big|_{0}^{1} \right) d\theta &&\text{Evaluate inner integral.} \\
&= \int_{0}^{2\pi} d\theta = 2\pi. &&\text{Evaluate outer integral.}
\end{aligned}
$$

Related Exercises 23, 25 ◀

Changing the Order of Integration

As with double integrals, choosing an appropriate order of integration may simplify the evaluation of a triple integral. Therefore, it is important to become proficient at changing the order of integration.

EXAMPLE 4   Changing the order of integration Consider the integral

$$\int_{0}^{\sqrt[4]{\pi}} \int_{0}^{z} \int_{y}^{z} 12y^2 z^3 \sin x^4\, dx\, dy\, dz.$$

a. Sketch the region of integration D.

b. Evaluate the integral by changing the order of integration.

SOLUTION

a. We begin by finding the projection of the region of integration D on the appropriate coordinate plane; call the projection R. Because the inner integration is with respect to x, R lies in the yz-plane, and it is determined by the limits on the middle and outer integrals. We see that

$$R = \{(y, z): 0 \le y \le z, 0 \le z \le \sqrt[4]{\pi}\},$$

which is a triangular region in the yz-plane bounded by the z-axis and the lines $y = z$ and $z = \sqrt[4]{\pi}$. Using the limits on the inner integral, for each point in R we let x vary from the plane $x = y$ to the plane $x = z$. In so doing, the points fill an inverted tetrahedron in the first octant with its vertex at the origin, which is D (Figure 16.45).

b. It is difficult to evaluate the integral in the given order $(dx\, dy\, dz)$ because the antiderivative of $\sin x^4$ is not expressible in terms of elementary functions. If we integrate first with respect to y, we introduce a factor in the integrand that enables us to use a substitution

> ➤ How do we know to switch the order of integration so the inner integral is with respect to y? Often we do not know in advance whether a new order of integration will work, and some trial and error is needed. In this case, either y^2 or z^3 is easier to integrate than $\sin x^4$, so either y or z is a likely variable for the inner integral.

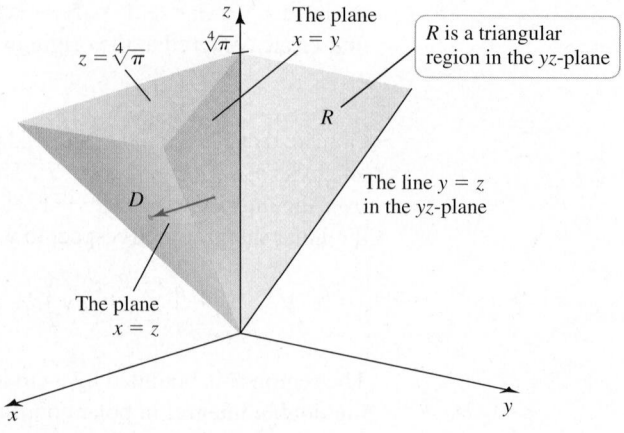

Figure 16.45

to integrate $\sin x^4$. With the order of integration $dy\,dx\,dz$, the bounds of integration for the inner integral extend from the plane $y = 0$ to the plane $y = x$ (Figure 16.46a). Furthermore, the projection of D onto the xz-plane is the region R, which must be covered by the middle and outer integrals (Figure 16.46b). In this case, we draw a line segment parallel to the x-axis to see that the limits of the middle integral run from $x = 0$ to $x = z$. Then we include all these segments from $z = 0$ to $z = \sqrt[4]{\pi}$ to obtain the outer limits of integration in z. The integration proceeds as follows:

$$\int_0^{\sqrt[4]{\pi}} \int_0^z \int_0^x 12y^2z^3 \sin x^4 \, dy \, dx \, dz = \int_0^{\sqrt[4]{\pi}} \int_0^z (4y^3z^3 \sin x^4)\Big|_0^x dx \, dz \qquad \text{Evaluate inner integral with respect to } y.$$

$$= \int_0^{\sqrt[4]{\pi}} \int_0^z 4x^3z^3 \sin x^4 \, dx \, dz \qquad \text{Simplify.}$$

$$= \int_0^{\sqrt[4]{\pi}} z^3(-\cos x^4)\Big|_0^z dz \qquad \text{Evaluate middle integral with respect to } x;\ u = x^4.$$

$$= \int_0^{\sqrt[4]{\pi}} z^3(1 - \cos z^4)\, dz \qquad \text{Simplify.}$$

$$= \left(\frac{z^4}{4} - \frac{\sin z^4}{4}\right)\Big|_0^{\sqrt[4]{\pi}} \qquad \text{Evaluate outer integral with respect to } z;\ u = z^4.$$

$$= \frac{\pi}{4}. \qquad \text{Simplify.}$$

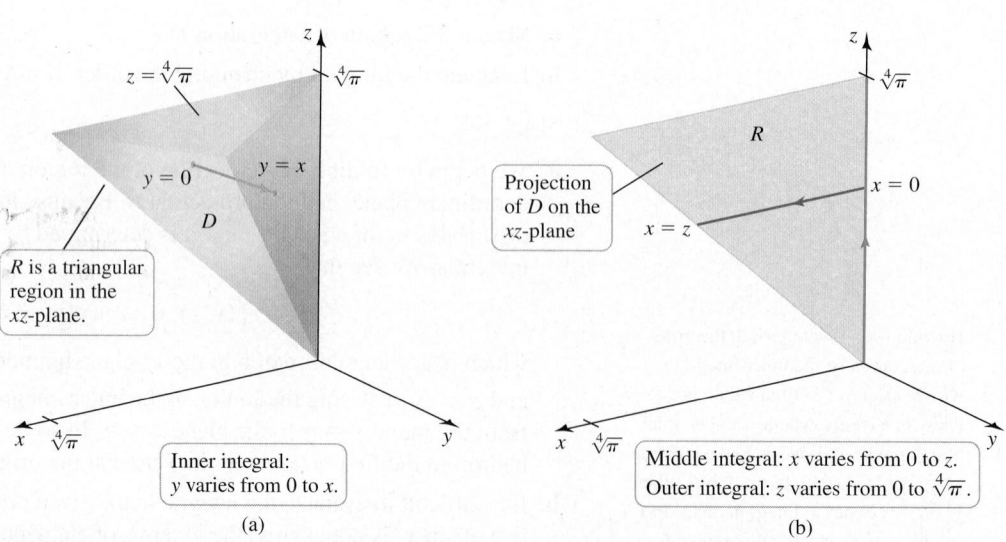

Figure 16.46

Related Exercises 47, 56 ◄

Average Value of a Function of Three Variables

The idea of the average value of a function extends naturally from the one- and two-variable cases. The average value of a function of three variables is found by integrating the function over the region of interest and dividing by the volume of the region.

DEFINITION Average Value of a Function of Three Variables

If f is continuous on a region D of $\mathbb{R}^3$, then the **average value** of f over D is

$$\bar{f} = \frac{1}{\text{volume of } D} \iiint_D f(x, y, z)\, dV.$$

EXAMPLE 5 Average temperature Consider a block of a conducting material occupying the region

$$D = \{(x, y, z): 0 \le x \le 2, 0 \le y \le 2, 0 \le z \le 1\}.$$

Due to heat sources on its boundaries, the temperature in the block is given by $T(x, y, z) = 250xy \sin \pi z$. Find the average temperature of the block.

SOLUTION We must integrate the temperature function over the block and divide by the volume of the block, which is 4. One way to evaluate the temperature integral is as follows:

$$\iiint_D 250xy \sin \pi z\, dV = 250 \int_0^2 \int_0^2 \int_0^1 xy \sin \pi z\, dz\, dy\, dx \qquad \text{Convert to an iterated integral.}$$

$$= 250 \int_0^2 \int_0^2 xy \frac{1}{\pi}(-\cos \pi z)\Big|_0^1 dy\, dx \qquad \text{Evaluate inner integral with respect to } z.$$

$$= \frac{500}{\pi} \int_0^2 \int_0^2 xy\, dy\, dx \qquad \text{Simplify.}$$

$$= \frac{500}{\pi} \int_0^2 x \left(\frac{y^2}{2}\right)\Big|_0^2 dx \qquad \text{Evaluate middle integral with respect to } y.$$

$$= \frac{1000}{\pi} \int_0^2 x\, dx \qquad \text{Simplify.}$$

$$= \frac{1000}{\pi} \left(\frac{x^2}{2}\right)\Big|_0^2 = \frac{2000}{\pi}. \qquad \text{Evaluate outer integral with respect to } x.$$

Dividing by the volume of the region, we find that the average temperature is $(2000/\pi)/4 = 500/\pi \approx 159.2$.

Related Exercises 51–52 ◄

QUICK CHECK 4 Without integrating, what is the average value of $f(x, y, z) = \sin x \sin y \sin z$ on the cube $\{(x, y, z): -1 \le x \le 1, -1 \le y \le 1, -1 \le z \le 1\}$? Use symmetry arguments. ◄

SECTION 16.4 EXERCISES

Getting Started

1. Sketch the region $D = \{(x, y, z): x^2 + y^2 \le 4, 0 \le z \le 4\}$.

2. Write an iterated integral for $\iiint_D f(x, y, z)\, dV$, where D is the box $\{(x, y, z): 0 \le x \le 3, 0 \le y \le 6, 0 \le z \le 4\}$.

3. Write an iterated integral for $\iiint_D f(x, y, z)\, dV$, where D is a sphere of radius 9 centered at $(0, 0, 0)$. Use the order $dz\, dy\, dx$.

4. Sketch the region of integration for the integral
$$\int_0^1 \int_0^{\sqrt{1-z^2}} \int_0^{\sqrt{1-y^2-z^2}} f(x, y, z)\, dx\, dy\, dz.$$

5. Write the integral in Exercise 4 in the order $dy\, dx\, dz$.

6. Write an integral for the average value of $f(x, y, z) = xyz$ over the region bounded by the paraboloid $z = 9 - x^2 - y^2$ and the xy-plane (assuming the volume of the region is known).

Practice Exercises

7–14. Integrals over boxes *Evaluate the following integrals. A sketch of the region of integration may be useful.*

7. $\int_{-2}^2 \int_3^6 \int_0^2 dx\, dy\, dz$

8. $\int_{-1}^1 \int_{-1}^2 \int_0^1 6xyz\, dy\, dx\, dz$

9. $\int_{-2}^2 \int_1^2 \int_1^e \frac{xy^2}{z}\, dz\, dx\, dy$

10. $\int_0^{\ln 4} \int_0^{\ln 3} \int_0^{\ln 2} e^{-x+y+z}\, dx\, dy\, dz$

11. $\int_0^{\pi} \int_0^{\pi} \int_0^{\pi/2} \sin \pi x \cos y \sin 2z \, dy \, dx \, dz$

$\int_0^2 \int_1^2 \int_0^1 yze^x \, dx \, dz \, dy$

13. $\iiint_D (xy + xz + yz) \, dV; D = \{(x, y, z): -1 \le x \le 1,$
$-2 \le y \le 2, -3 \le z \le 3\}$

14. $\iiint_D xyze^{-x^2-y^2} \, dV; D = \{(x, y, z): 0 \le x \le \sqrt{\ln 2},$
$0 \le y \le \sqrt{\ln 4}, 0 \le z \le 1\}$

15–29. Volumes of solids *Use a triple integral to find the volume of the following solids.*

15. The solid in the first octant bounded by the plane $2x + 3y + 6z = 12$ and the coordinate planes

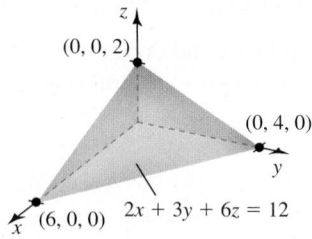

$(0, 0, 2)$ $(0, 4, 0)$
y
$(6, 0, 0)$ $2x + 3y + 6z = 12$
x

16. The solid in the first octant formed when the cylinder $z = \sin y$, for $0 \le y \le \pi$, is sliced by the planes $y = x$ and $x = 0$

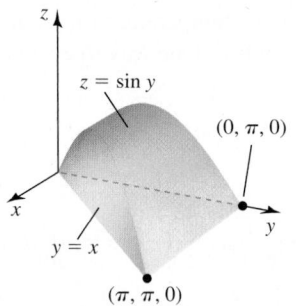

$z = \sin y$ $(0, \pi, 0)$
x y
$y = x$
$(\pi, \pi, 0)$

17. The wedge above the xy-plane formed when the cylinder $x^2 + y^2 = 4$ is cut by the planes $z = 0$ and $y = -z$

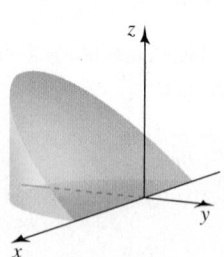

18. The prism in the first octant bounded by $z = 2 - 4x$ and $y = 8$

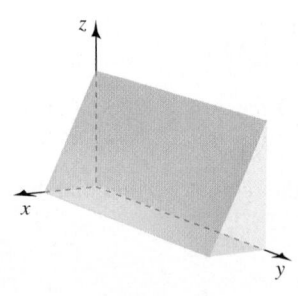

19. The solid bounded by the surfaces $z = e^y$ and $z = 1$ over the rectangle $\{(x, y): 0 \le x \le 1, 0 \le y \le \ln 2\}$

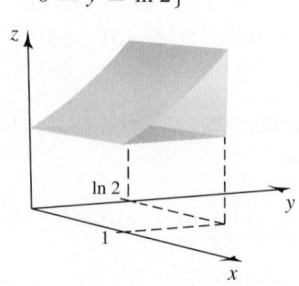

$\ln 2$

20. The wedge bounded by the parabolic cylinder $y = x^2$ and the planes $z = 3 - y$ and $z = 0$

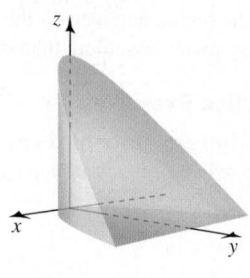

21. The solid between the sphere $x^2 + y^2 + z^2 = 19$ and the hyperboloid $z^2 - x^2 - y^2 = 1$, for $z > 0$

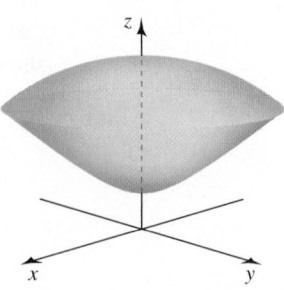

22. The solid bounded below by the cone $z = \sqrt{x^2 + y^2}$ and bounded above by the sphere $x^2 + y^2 + z^2 = 8$

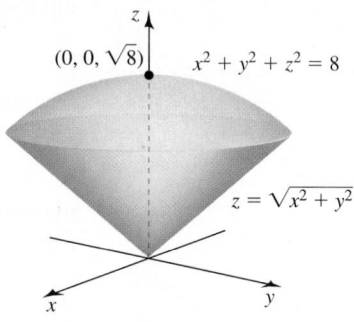

$(0, 0, \sqrt{8})$ $x^2 + y^2 + z^2 = 8$
$z = \sqrt{x^2 + y^2}$
x y

23. The solid bounded by the cylinder $y = 9 - x^2$ and the paraboloid $y = 2x^2 + 3z^2$

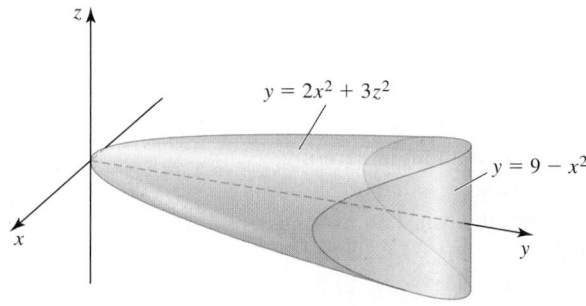

$y = 2x^2 + 3z^2$
x $y = 9 - x^2$
y

24. The wedge in the first octant bounded by the cylinder $x = z^2$ and the planes $z = 2 - x$, $y = 2$, $y = 0$, and $z = 0$

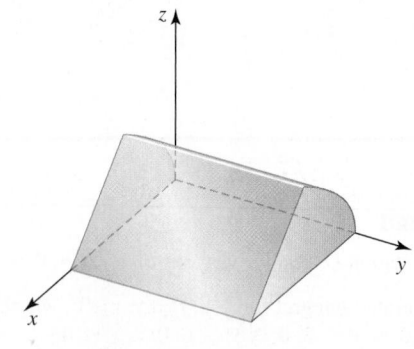

25. The wedge of the cylinder $x^2 + 4z^2 = 4$ created by the planes $y = 3 - x$ and $y = x - 3$

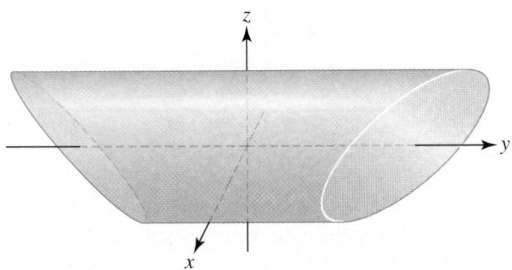

26. The solid bounded by $x = 0$, $x = 2$, $y = 0$, $y = e^{-z}$, $z = 0$, and $z = 1$

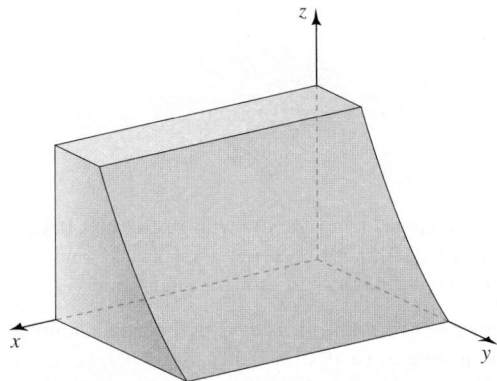

27. The solid bounded by $x = 0$, $x = 1 - z^2$, $y = 0$, $z = 0$, and $z = 1 - y$

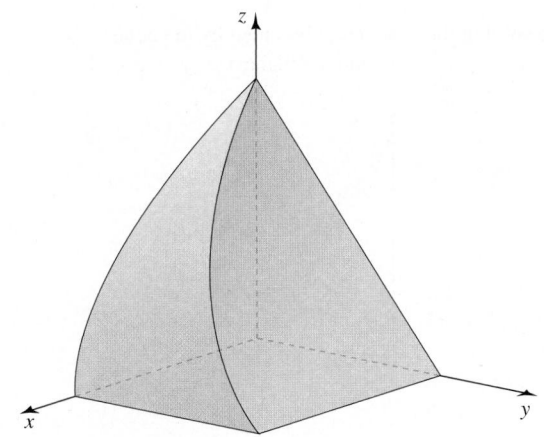

28. The solid bounded by $x = 0$, $y = z^2$, $z = 0$, and $z = 2 - x - y$

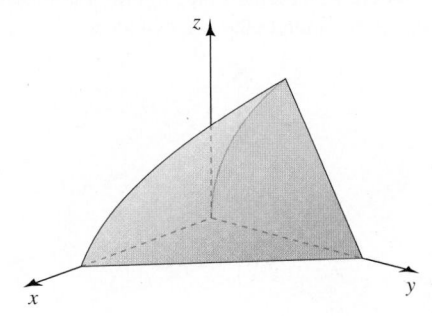

29. The solid bounded by $x = 0$, $x = 2$, $y = z$, $y = z + 1$, $z = 0$, and $z = 4$

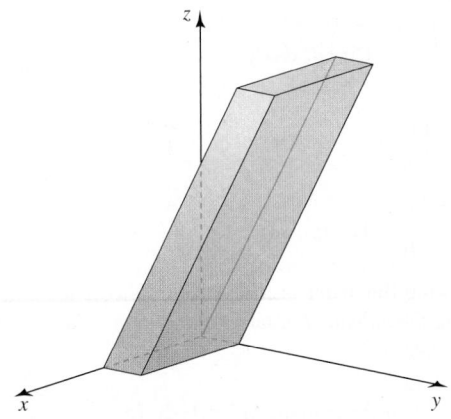

30–35. Six orderings *Let D be the solid in the first octant bounded by the planes $y = 0$, $z = 0$, and $y = x$, and the cylinder $4x^2 + z^2 = 4$. Write the triple integral of $f(x, y, z)$ over D in the given order of integration.*

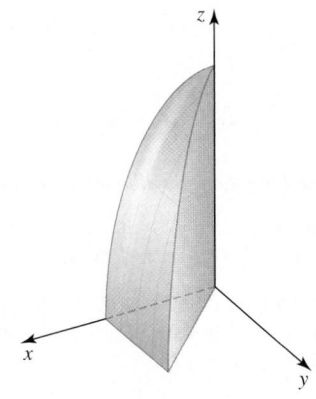

30. $dz\,dy\,dx$ **31.** $dz\,dx\,dy$ **32.** $dy\,dx\,dz$

33. $dy\,dz\,dx$ **34.** $dx\,dy\,dz$ **35.** $dx\,dz\,dy$

36. **All six orders** Let D be the solid bounded by $y = x$, $z = 1 - y^2$, $x = 0$, and $z = 0$. Write triple integrals over D in all six possible orders of integration.

37. **Changing order of integration** Write the integral $\int_0^2 \int_0^1 \int_0^{1-y} dz\,dy\,dx$ in the five other possible orders of integration.

38–46. Triple integrals *Evaluate the following integrals.*

38. $\displaystyle\int_0^\pi \int_0^\pi \int_0^{\sin x} \sin y\,dz\,dx\,dy$

39. $\displaystyle\int_0^2 \int_0^4 \int_{y^2}^4 \sqrt{x}\,dz\,dx\,dy$

40. $\displaystyle\int_0^1 \int_0^{\sqrt{1-x^2}} \int_0^{\sqrt{1-x^2-y^2}} 2xz\,dz\,dy\,dx$

41. $\displaystyle\int_0^1 \int_0^{\sqrt{1-x^2}} \int_0^{\sqrt{1-x^2}} dz\,dy\,dx$

42. $\displaystyle\int_1^6 \int_0^{4-2y/3} \int_0^{12-2y-3z} \frac{1}{y}\,dx\,dz\,dy$

43. $\displaystyle\int_0^3 \int_0^{\sqrt{9-z^2}} \int_0^{\sqrt{1+x^2+z^2}} dy \, dx \, dz$

44. $\displaystyle\int_0^1 \int_y^{2-y} \int_0^{2-x-y} 15xy \, dz \, dx \, dy$

45. $\displaystyle\int_0^{\ln 8} \int_1^{\sqrt{z}} \int_{\ln y}^{\ln 2y} e^{x+y^2-z} \, dx \, dy \, dz$

46. $\displaystyle\int_0^1 \int_0^{\sqrt{1-x^2}} \int_0^{2-x} 4yz \, dz \, dy \, dx$

47–50. Changing the order of integration *Rewrite the following integrals using the indicated order of integration, and then evaluate the resulting integral.*

47. $\displaystyle\int_0^5 \int_{-1}^0 \int_0^{4x+4} dy \, dx \, dz$ in the order $dz \, dx \, dy$

48. $\displaystyle\int_0^1 \int_{-2}^2 \int_0^{\sqrt{4-y^2}} dz \, dy \, dx$ in the order $dy \, dz \, dx$

49. $\displaystyle\int_0^1 \int_0^{\sqrt{1-x^2}} \int_0^{\sqrt{1-x^2}} dy \, dz \, dx$ in the order $dz \, dy \, dx$

50. $\displaystyle\int_0^4 \int_0^{\sqrt{16-x^2}} \int_0^{\sqrt{16-x^2-z^2}} dy \, dz \, dx$ in the order $dx \, dy \, dz$

51–54. Average value *Find the following average values.*

51. The average value of $f(x, y, z) = 8xy \cos z$ over the points inside the box $D = \{(x, y, z) : 0 \le x \le 1, 0 \le y \le 2, 0 \le z \le \pi/2\}$

T 52. The average temperature in the box $D = \{(x, y, z) :$
$0 \le x \le \ln 2, 0 \le y \le \ln 4, 0 \le z \le \ln 8\}$ with a temperature distribution of $T(x, y, z) = 128e^{-x-y-z}$

T 53. The average of the *squared* distance between the origin and points in the solid cylinder $D = \{(x, y, z) : x^2 + y^2 \le 4, 0 \le z \le 2\}$

54. The average z-coordinate of points on and within a hemisphere of radius 4 centered at the origin with its base in the xy-plane

55. **Explain why or why not** Determine whether the following statements are true and give an explanation or counterexample.

  a. An iterated integral of a function over the box
  $D = \{(x, y, z) : 0 \le x \le a, 0 \le y \le b, 0 \le z \le c\}$ can be expressed in eight different ways.
  b. One possible iterated integral of f over the prism
  $D = \{(x, y, z) : 0 \le x \le 1, 0 \le y \le 3x - 3, 0 \le z \le 5\}$ is
  $\int_0^{3x-3} \int_0^1 \int_0^5 f(x, y, z) \, dz \, dx \, dy$.
  c. The region $D = \{(x, y, z) : 0 \le x \le 1, 0 \le y \le \sqrt{1-x^2},$
  $0 \le z \le \sqrt{1-x^2}\}$ is a sphere.

Explorations and Challenges

56. **Changing the order of integration** Use another order of integration to evaluate $\displaystyle\int_1^4 \int_z^{4z} \int_0^{\pi^2} \frac{\sin\sqrt{yz}}{x^{3/2}} \, dy \, dx \, dz$.

57–62. Miscellaneous volumes *Use a triple integral to compute the volume of the following regions.*

57. The wedge of the square column $|x| + |y| = 1$ created by the planes $z = 0$ and $x + y + z = 1$

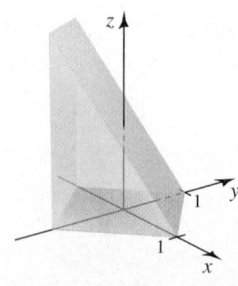

58. The solid common to the cylinders $z = \sin x$ and $z = \sin y$ over the square $R = \{(x, y) : 0 \le x \le \pi, 0 \le y \le \pi\}$ (The figure shows the cylinders, but not the common region.)

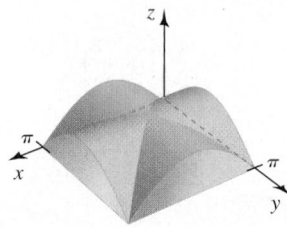

59. The parallelepiped (slanted box) with vertices $(0, 0, 0)$, $(1, 0, 0)$, $(0, 1, 0)$, $(1, 1, 0)$, $(0, 1, 1)$, $(1, 1, 1)$, $(0, 2, 1)$, and $(1, 2, 1)$ (Use integration and find the best order of integration.)

60. The larger of two solids formed when the parallelepiped (slanted box) with vertices $(0, 0, 0)$, $(2, 0, 0)$, $(0, 2, 0)$, $(2, 2, 0)$, $(0, 1, 1)$, $(2, 1, 1)$, $(0, 3, 1)$, and $(2, 3, 1)$ is sliced by the plane $y = 2$

61. The pyramid with vertices $(0, 0, 0)$, $(2, 0, 0)$, $(2, 2, 0)$, $(0, 2, 0)$, and $(0, 0, 4)$

62. The solid in the first octant bounded by the cone $z = 1 - \sqrt{x^2 + y^2}$ and the plane $x + y + z = 1$

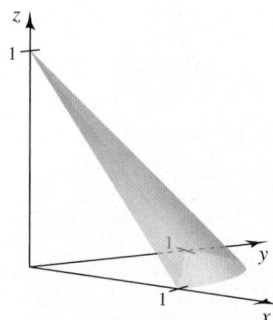

63. **Two cylinders** The x- and y-axes form the axes of two right circular cylinders with radius 1 (see figure). Find the volume of the solid that is common to the two cylinders.

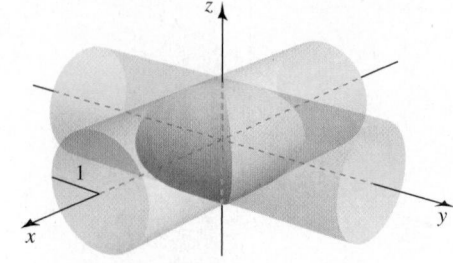

64. Three cylinders The coordinate axes form the axes of three right circular cylinders with radius 1 (see figure). Find the volume of the solid that is common to the three cylinders.

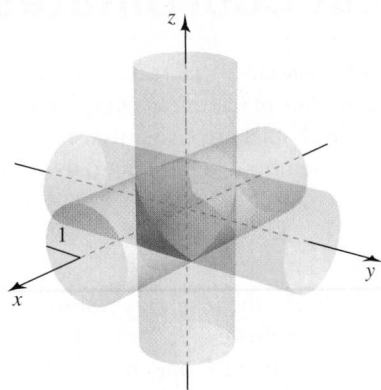

65. Dividing the cheese Suppose a wedge of cheese fills the region in the first octant bounded by the planes $y = z$, $y = 4$, and $x = 4$. You could divide the wedge into two pieces of equal volume by slicing the wedge with the plane $x = 2$. Instead find a with $0 < a < 4$ such that slicing the wedge with the plane $y = a$ divides the wedge into two pieces of equal volume.

66. Partitioning a cube Consider the region
$$D_1 = \{(x, y, z): 0 \le x \le y \le z \le 1\}.$$
   a. Find the volume of D_1.
   b. Let $D_2, \ldots, D_6$ be the "cousins" of D_1 formed by rearranging x, y, and z in the inequality $0 \le x \le y \le z \le 1$. Show that the volumes of $D_1, \ldots, D_6$ are equal.
   c. Show that the union of $D_1, \ldots, D_6$ is a unit cube.

67–71. General volume formulas *Find equations for the bounding surfaces, set up a volume integral, and evaluate the integral to obtain a volume formula for each region. Assume a, b, c, r, R, and h are positive constants.*

67. Cone Find the volume of a right circular cone with height h and base radius r.

68. Tetrahedron Find the volume of a tetrahedron whose vertices are located at $(0, 0, 0)$, $(a, 0, 0)$, $(0, b, 0)$, and $(0, 0, c)$.

69. Spherical cap Find the volume of the cap of a sphere of radius R with height h.

70. Frustum of a cone Find the volume of a truncated cone of height h whose ends have radii r and R.

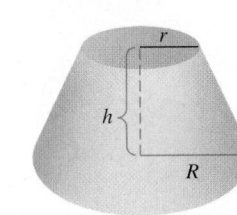

71. Ellipsoid Find the volume of an ellipsoid with axes of lengths $2a$, $2b$, and $2c$.

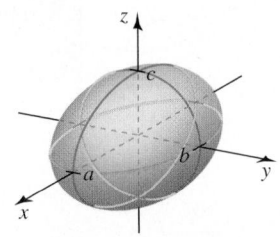

72. Exponential distribution The occurrence of random events (such as phone calls or e-mail messages) is often idealized using an exponential distribution. If λ is the average rate of occurrence of such an event, assumed to be constant over time, then the average time between occurrences is λ^{-1} (for example, if phone calls arrive at a rate of $\lambda = 2/\text{min}$, then the mean time between phone calls is $\lambda^{-1} = 1/2$ min). The exponential distribution is given by $f(t) = \lambda e^{-\lambda t}$, for $0 \le t < \infty$.

   a. Suppose you work at a customer service desk and phone calls arrive at an average rate of $\lambda_1 = 0.8/\text{min}$ (meaning the average time between phone calls is $1/0.8 = 1.25$ min). The probability that a phone call arrives during the interval $[0, T]$ is $p(T) = \int_0^T \lambda_1 e^{-\lambda_1 t} \, dt$. Find the probability that a phone call arrives during the first 45 s (0.75 min) that you work at the desk.

   b. Now suppose walk-in customers also arrive at your desk at an average rate of $\lambda_2 = 0.1/\text{min}$. The probability that a phone call *and* a customer arrive during the interval $[0, T]$ is
   $$p(T) = \int_0^T \int_0^T \lambda_1 e^{-\lambda_1 t} \lambda_2 e^{-\lambda_2 s} \, dt \, ds.$$
   Find the probability that a phone call and a customer arrive during the first 45 s that you work at the desk.

   c. E-mail messages also arrive at your desk at an average rate of $\lambda_3 = 0.05/\text{min}$. The probability that a phone call *and* a customer *and* an e-mail message arrive during the interval $[0, T]$ is
   $$p(T) = \int_0^T \int_0^T \int_0^T \lambda_1 e^{-\lambda_1 t} \lambda_2 e^{-\lambda_2 s} \lambda_3 e^{-\lambda_3 u} \, dt \, ds \, du.$$
   Find the probability that a phone call and a customer and an e-mail message arrive during the first 45 s that you work at the desk.

73. Hypervolume Find the "volume" of the four-dimensional pyramid bounded by $w + x + y + z + 1 = 0$ and the coordinate planes $w = 0$, $x = 0$, $y = 0$, $z = 0$.

74. An identity (Putnam Exam 1941) Let f be a continuous function on $[0, 1]$. Prove that
$$\int_0^1 \int_x^1 \int_x^y f(x)f(y)f(z) \, dz \, dy \, dx = \frac{1}{6}\left(\int_0^1 f(x) \, dx\right)^3.$$

QUICK CHECK ANSWERS

1. $dx \, dy \, dz$, $dx \, dz \, dy$, $dy \, dx \, dz$, $dy \, dz \, dx$, $dz \, dx \, dy$, $dz \, dy \, dx$

2. $\int_0^1 \int_0^2 \int_0^3 (2 - z) \, dx \, dy \, dz$, $\int_0^2 \int_0^1 \int_0^3 (2 - z) \, dx \, dz \, dy$

3. $\int_0^2 \int_0^{4-2x} \int_0^6 dz \, dy \, dx$, $\int_0^6 \int_0^4 \int_0^{2-y/2} dx \, dy \, dz$

4. 0 ($\sin x$, $\sin y$, and $\sin z$ are odd functions.) ◄

16.5 Triple Integrals in Cylindrical and Spherical Coordinates

When evaluating triple integrals, you may have noticed that some regions (such as spheres, cones, and cylinders) have awkward descriptions in Cartesian coordinates. In this section, we examine two other coordinate systems in $\mathbb{R}^3$ that are easier to use when working with certain types of regions. These coordinate systems are helpful not only for integration, but also for general problem solving.

Cylindrical Coordinates

▶ In cylindrical coordinates, r and θ are the usual polar coordinates, with the additional restriction that $r \geq 0$. Adding the z-coordinate lifts points in the polar plane into $\mathbb{R}^3$.

When we extend polar coordinates from $\mathbb{R}^2$ to $\mathbb{R}^3$, the result is *cylindrical coordinates*. In this coordinate system, a point P in $\mathbb{R}^3$ has coordinates (r, θ, z), where r and θ are polar coordinates for the point P^*, which is the projection of P onto the xy-plane (**Figure 16.47**). As in Cartesian coordinates, the z-coordinate is the signed vertical distance between P and the xy-plane. Any point in $\mathbb{R}^3$ can be represented by cylindrical coordinates using the intervals $0 \leq r < \infty$, $0 \leq \theta \leq 2\pi$, and $-\infty < z < \infty$.

Many sets of points have simple representations in cylindrical coordinates. For example, the set $\{(r, \theta, z): r = a\}$ is the set of points whose distance from the z-axis is a, which is a right circular cylinder of radius a. The set $\{(r, \theta, z): \theta = \theta_0\}$ is the set of points with a constant θ coordinate; it is a vertical half-plane emanating from the z-axis in the direction $\theta = \theta_0$. Table 16.4 summarizes these and other sets that are ideal for integration in cylindrical coordinates.

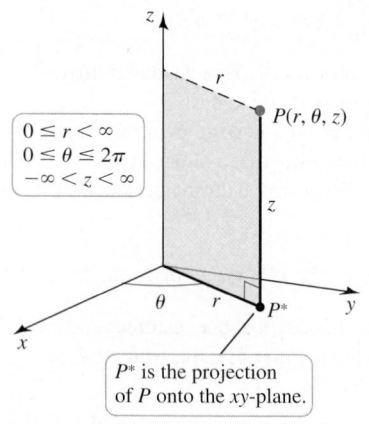

$0 \leq r < \infty$
$0 \leq \theta \leq 2\pi$
$-\infty < z < \infty$

$P(r, \theta, z)$

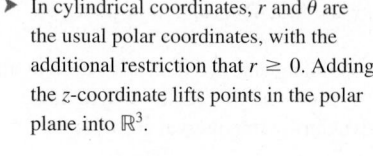

P^* is the projection of P onto the xy-plane.

Figure 16.47

Table 16.4

Name	Description	Example
Cylinder	$\{(r, \theta, z): r = a\}, a > 0$	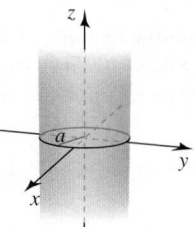
Cylindrical shell	$\{(r, \theta, z): 0 < a \leq r \leq b\}$	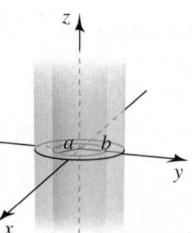
Vertical half-plane	$\{(r, \theta, z): \theta = \theta_0\}$	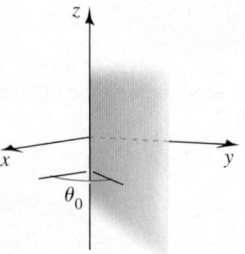

Table 16.4   **(Continued)**

Name	Description	Example
Horizontal plane	$\{(r, \theta, z): z = a\}$	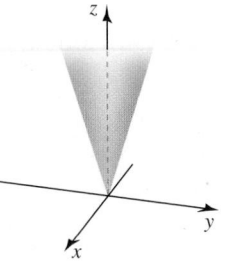
Cone	$\{(r, \theta, z): z = ar\}, a \neq 0$	

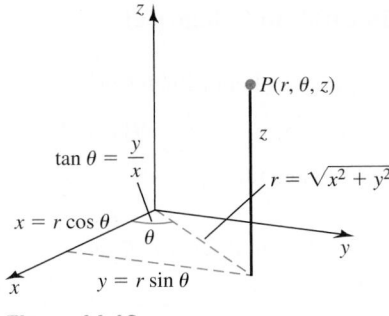

(a)

(b)

Figure 16.48

EXAMPLE 1   Sets in cylindrical coordinates Identify and sketch the following sets in cylindrical coordinates.

a. $Q = \{(r, \theta, z): 1 \leq r \leq 3, z \geq 0\}$

b. $S = \{(r, \theta, z): z = 1 - r, 0 \leq r \leq 1\}$

SOLUTION

a. The set Q is a cylindrical shell with inner radius 1 and outer radius 3 that extends indefinitely along the positive z-axis (**Figure 16.48a**). Because θ is unspecified, it takes on all values.

b. To identify this surface, it helps to work in steps. The set $S_1 = \{(r, \theta, z): z = r\}$ is a cone that opens *upward* with its vertex at the origin. Similarly, the set $S_2 = \{(r, \theta, z): z = -r\}$ is a cone that opens *downward* with its vertex at the origin. Therefore, S is S_2 shifted vertically upward by 1 unit; it is a cone that opens downward with its vertex at $(0, 0, 1)$. Because $0 \leq r \leq 1$, the base of the cone is on the xy-plane (**Figure 16.48b**).

Related Exercise 11 ◄

Equations for transforming Cartesian coordinates to cylindrical coordinates, and vice versa, are often needed for integration. We simply use the rules for polar coordinates (Section 12.2) with no change in the z-coordinate (**Figure 16.49**).

Figure 16.49

QUICK CHECK 1 Find the cylindrical coordinates of the point with rectangular coordinates $(1, -1, 5)$. Find the rectangular coordinates of the point with cylindrical coordinates $(2, \pi/3, 5)$. ◄

Transformations Between Cylindrical and Rectangular Coordinates	
Rectangular → Cylindrical	**Cylindrical → Rectangular**
$r^2 = x^2 + y^2$	$x = r \cos \theta$
$\tan \theta = y/x$	$y = r \sin \theta$
$z = z$	$z = z$

Integration in Cylindrical Coordinates

Among the uses of cylindrical coordinates is the evaluation of triple integrals of the form $\iiint_D f(x, y, z) \, dV$. We begin with a region D in $\mathbb{R}^3$ and partition it into cylindrical wedges formed by changes of Δr, $\Delta \theta$, and Δz in the coordinate directions (**Figure 16.50**). Those wedges that lie entirely within D are labeled from $k = 1$ to $k = n$ in some convenient order. We let $(r_k^*, \theta_k^*, z_k^*)$ be the cylindrical coordinates of an arbitrary point in the kth wedge. This point also has Cartesian coordinates $(x_k^*, y_k^*, z_k^*) = (r_k^* \cos \theta_k^*, r_k^* \sin \theta_k^*, z_k^*)$.

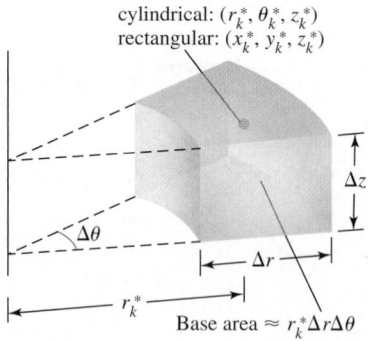

cylindrical: $(r_k^*, \theta_k^*, z_k^*)$
rectangular: (x_k^*, y_k^*, z_k^*)

Base area $\approx r_k^* \Delta r \Delta \theta$

Approximate volume $\Delta V_k = r_k^* \Delta r \Delta \theta \Delta z$

Figure 16.50

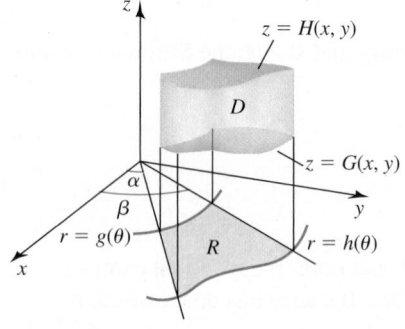

Figure 16.51

▶ The order of the differentials specifies
the order in which the integrals are
evaluated, so we write the volume
element dV as $dz\, r\, dr\, d\theta$. Do not lose
sight of the factor of r in the integrand.
It plays the same role as it does in the
area element $dA = r\, dr\, d\theta$ in polar
coordinates.

As shown in Figure 16.50, the base of the kth wedge is a polar rectangle with an approximate area of $r_k^* \Delta r \Delta \theta$ (Section 16.3). The height of the wedge is Δz. Multiplying these dimensions together, the approximate volume of the wedge is $\Delta V_k = r_k^* \Delta r \Delta \theta \Delta z$, for $k = 1, \ldots, n$.

We now assume $f(x, y, z)$ is continuous on D and form a Riemann sum over the region by adding function values multiplied by the corresponding approximate volumes:

$$\sum_{k=1}^{n} f(x_k^*, y_k^*, z_k^*) \Delta V_k = \sum_{k=1}^{n} f(r_k^* \cos \theta_k^*, r_k^* \sin \theta_k^*, z_k^*) \Delta V_k.$$

Let Δ be the maximum value of Δr, $\Delta \theta$, and Δz, for $k = 1, \ldots, n$. As $n \to \infty$ and $\Delta \to 0$, the Riemann sums approach a limit called the **triple integral of f over D in cylindrical coordinates**:

$$\iiint_D f(x, y, z)\, dV = \lim_{\Delta \to 0} \sum_{k=1}^{n} f(r_k^* \cos \theta_k^*, r_k^* \sin \theta_k^*, z_k^*) \underbrace{r_k^* \Delta r \Delta \theta \Delta z}_{\Delta V_k}.$$

The rightmost sum tells us how to write a triple integral in x, y, and z as an iterated integral of $f(r \cos \theta, r \sin \theta, z)\, r$ in cylindrical coordinates.

Finding Limits of Integration   We show how to find the limits of integration in one common situation involving cylindrical coordinates. Suppose D is a region in $\mathbb{R}^3$ consisting of points between the surfaces $z = G(x, y)$ and $z = H(x, y)$, where x and y belong to a region R in the xy-plane and $G(x, y) \le H(x, y)$ on R (**Figure 16.51**). Assuming f is continuous on D, the triple integral of f over D may be expressed as the iterated integral

$$\iiint_D f(x, y, z)\, dV = \iint_R \left(\int_{G(x, y)}^{H(x, y)} f(x, y, z)\, dz \right) dA.$$

The inner integral with respect to z runs from the lower surface $z = G(x, y)$ to the upper surface $z = H(x, y)$, leaving an outer double integral over R.

If the region R is described in polar coordinates by

$$\{(r, \theta): g(\theta) \le r \le h(\theta), \alpha \le \theta \le \beta\},$$

then we evaluate the double integral over R in polar coordinates (Section 16.3). The effect is a change of variables from rectangular to cylindrical coordinates. Letting $x = r \cos \theta$ and $y = r \sin \theta$, we have the following result, which is another change of variables formula.

> **THEOREM 16.6   Change of Variables for Triple Integrals in Cylindrical Coordinates**
>
> Let f be continuous over the region D, expressed in cylindrical coordinates as
>
> $$D = \{(r, \theta, z): 0 \le g(\theta) \le r \le h(\theta), \alpha \le \theta \le \beta, G(x, y) \le z \le H(x, y)\}.$$
>
> Then f is integrable over D, and the triple integral of f over D is
>
> $$\iiint_D f(x, y, z)\, dV = \int_{\alpha}^{\beta} \int_{g(\theta)}^{h(\theta)} \int_{G(r \cos \theta, r \sin \theta)}^{H(r \cos \theta, r \sin \theta)} f(r \cos \theta, r \sin \theta, z)\, dz\, r\, dr\, d\theta.$$

Notice that the integrand and the limits of integration are converted from Cartesian to cylindrical coordinates. As with triple integrals in Cartesian coordinates, there are two immediate interpretations of this integral. If $f = 1$, then the triple integral $\iiint_D dV$ equals the volume of the region D. Also, if f describes the density of an object occupying the region D, the triple integral equals the mass of the object.

EXAMPLE 2 **Switching coordinate systems** Evaluate the integral

$$I = \int_0^{2\sqrt{2}} \int_{-\sqrt{8-x^2}}^{\sqrt{8-x^2}} \int_{-1}^{2} \sqrt{1 + x^2 + y^2} \, dz \, dy \, dx.$$

SOLUTION Evaluating this integral as it is given in Cartesian coordinates requires a tricky trigonometric substitution in the middle integral, followed by an even more difficult integral. Notice that z varies between the planes $z = -1$ and $z = 2$, while x and y vary over half of a disk in the xy-plane. Therefore, D is half of a solid cylinder (**Figure 16.52a**), which suggests a change to cylindrical coordinates.

The limits of integration in cylindrical coordinates are determined as follows:

Inner integral with respect to z A line through the half cylinder parallel to the z-axis enters at $z = -1$ and leaves at $z = 2$, so we integrate over the interval $-1 \le z \le 2$ (**Figure 16.52b**).

Middle integral with respect to r The projection of the half cylinder onto the xy-plane is the half disk R of radius $2\sqrt{2}$ centered at the origin, so r varies over the interval $0 \le r \le 2\sqrt{2}$ (**Figure 16.52c**).

Outer integral with respect to θ The half disk R is swept out by letting θ vary over the interval $-\pi/2 \le \theta \le \pi/2$ (Figure 16.52c).

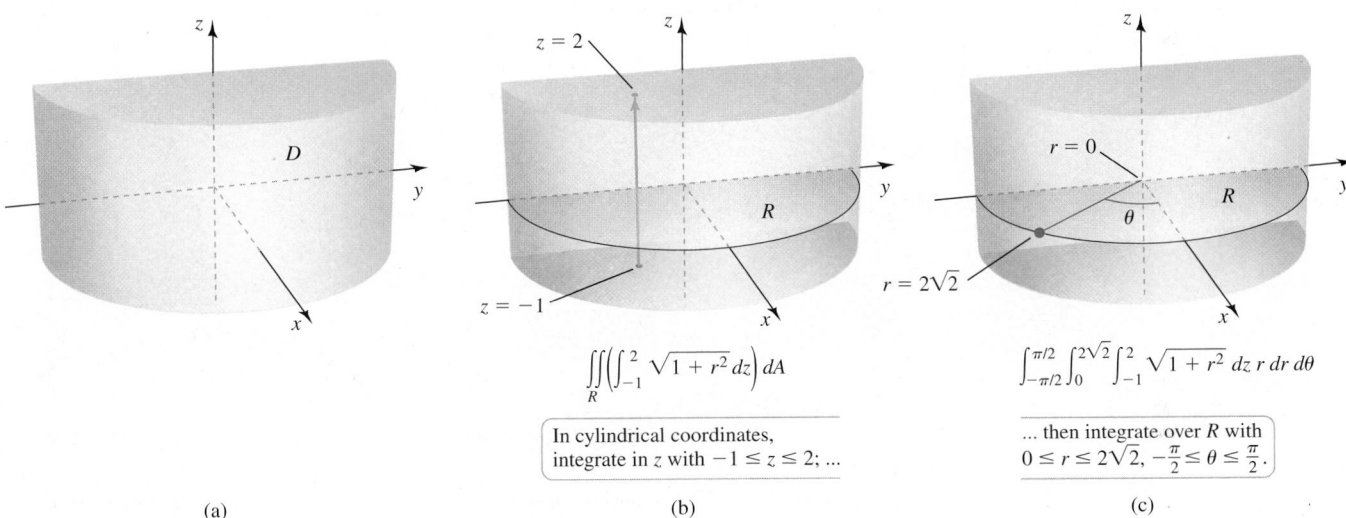

(a)

Figure 16.52

$$\iint_R \left(\int_{-1}^{2} \sqrt{1 + r^2} \, dz \right) dA$$

In cylindrical coordinates, integrate in z with $-1 \le z \le 2$; ...

(b)

$$\int_{-\pi/2}^{\pi/2} \int_0^{2\sqrt{2}} \int_{-1}^{2} \sqrt{1 + r^2} \, dz \, r \, dr \, d\theta$$

... then integrate over R with $0 \le r \le 2\sqrt{2}, -\frac{\pi}{2} \le \theta \le \frac{\pi}{2}$.

(c)

We also convert the integrand to cylindrical coordinates:

$$f(x, y, z) = \sqrt{1 + \underbrace{x^2 + y^2}_{r^2}} = \sqrt{1 + r^2}.$$

The evaluation of the integral in cylindrical coordinates now follows:

$$I = \int_{-\pi/2}^{\pi/2} \int_0^{2\sqrt{2}} \int_{-1}^{2} \sqrt{1 + r^2} \, dz \, r \, dr \, d\theta \qquad \text{Convert to cylindrical coordinates.}$$

$$= 3 \int_{-\pi/2}^{\pi/2} \int_0^{2\sqrt{2}} \sqrt{1 + r^2} \, r \, dr \, d\theta \qquad \text{Evaluate inner integral with respect to } z.$$

$$= \int_{-\pi/2}^{\pi/2} (1 + r^2)^{3/2} \Big|_0^{2\sqrt{2}} \, d\theta \qquad \text{Evaluate middle integral with respect to } r.$$

$$= \int_{-\pi/2}^{\pi/2} 26 \, d\theta = 26\pi. \qquad \text{Evaluate outer integral with respect to } \theta.$$

Related Exercises 17–18 ◄

QUICK CHECK 2 Find the limits of integration for a triple integral in cylindrical coordinates that gives the volume of a cylinder with height 20 and a circular base of radius 10 centered at the origin in the xy-plane. ◄

As illustrated in Example 2, triple integrals given in rectangular coordinates may be more easily evaluated after converting to cylindrical coordinates. Answering the following questions may help you choose the best coordinate system for a particular integral.

• In which coordinate system is the region of integration most easily described?
• In which coordinate system is the integrand most easily expressed?
• In which coordinate system is the triple integral most easily evaluated?

In general, if an integral in one coordinate system is difficult to evaluate, consider using a different coordinate system.

EXAMPLE 3   Mass of a solid paraboloid Find the mass of the solid D bounded by the paraboloid $z = 4 - r^2$ and the plane $z = 0$ (**Figure 16.53a**), where the density of the solid, given in cylindrical coordinates, is $f(r, \theta, z) = 5 - z$ (heavy near the base and light near the vertex).

SOLUTION The z-coordinate runs from the base $z = 0$ to the surface $z = 4 - r^2$ (**Figure 16.53b**). The projection R of the region D onto the xy-plane is found by setting $z = 0$ in the equation of the surface, $z = 4 - r^2$. The positive value of r satisfying the equation $4 - r^2 = 0$ is $r = 2$, so in polar coordinates $R = \{(r, \theta): 0 \le r \le 2, 0 \le \theta \le 2\pi\}$, which is a disk of radius 2 (**Figure 16.53c**).

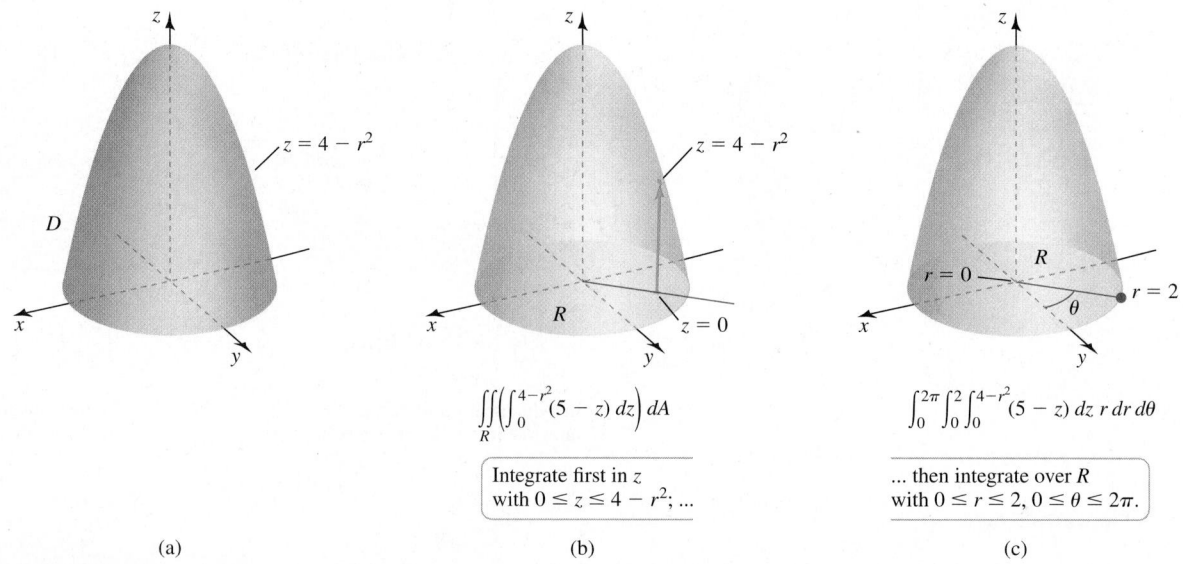

$$\iint_R \left(\int_0^{4-r^2} (5 - z)\, dz \right) dA$$

Integrate first in z
with $0 \le z \le 4 - r^2$; ...

$$\int_0^{2\pi} \int_0^2 \int_0^{4-r^2} (5 - z)\, dz\, r\, dr\, d\theta$$

... then integrate over R
with $0 \le r \le 2, 0 \le \theta \le 2\pi$.

(a)   (b)   (c)

Figure 16.53

➤ In Example 3, the integrand is independent of θ, so the integral with respect to θ could have been done first, producing a factor of 2π.

The mass is computed by integrating the density function over D:

$$\iiint_D f(r, \theta, z)\, dV = \int_0^{2\pi} \int_0^2 \int_0^{4-r^2} (5 - z)\, dz\, r\, dr\, d\theta \qquad \text{Integrate density.}$$

$$= \int_0^{2\pi} \int_0^2 \left(5z - \frac{z^2}{2} \right) \Big|_0^{4-r^2} r\, dr\, d\theta \qquad \begin{array}{l}\text{Evaluate inner integral with} \\ \text{respect to } z.\end{array}$$

$$= \frac{1}{2} \int_0^{2\pi} \int_0^2 (24r - 2r^3 - r^5)\, dr\, d\theta \qquad \text{Simplify.}$$

$$= \int_0^{2\pi} \frac{44}{3}\, d\theta \qquad \begin{array}{l}\text{Evaluate middle integral} \\ \text{with respect to } r.\end{array}$$

$$= \frac{88\pi}{3}. \qquad \begin{array}{l}\text{Evaluate outer integral} \\ \text{with respect to } \theta.\end{array}$$

Related Exercises 25–26 ◄

> Recall that to find the volume of a region D using a triple integral, we set $f = 1$ and evaluate
> $$V = \iiint\limits_{D} dV.$$

EXAMPLE 4   Volume between two surfaces Find the volume of the solid D between the cone $z = \sqrt{x^2 + y^2}$ and the inverted paraboloid $z = 12 - x^2 - y^2$ (**Figure 16.54a**).

SOLUTION Because $x^2 + y^2 = r^2$, the equation of the cone in cylindrical coordinates becomes $z = r$, and the equation of the paraboloid becomes $z = 12 - r^2$. The inner integral in z runs from the cone $z = r$ (the lower surface) to the paraboloid $z = 12 - r^2$ (the upper surface) (**Figure 16.54b**). We project D onto the xy-plane to produce the region R, whose boundary is determined by the intersection of the two surfaces. Equating the z-coordinates in the equations of the two surfaces, we have $12 - r^2 = r$, or $(r - 3)(r + 4) = 0$. Because $r \geq 0$, the relevant root is $r = 3$. Therefore, the projection of D onto the xy-plane is the polar region $R = \{(r, \theta): 0 \leq r \leq 3, 0 \leq \theta \leq 2\pi\}$, which is a disk of radius 3 centered at $(0, 0)$ (**Figure 16.54c**).

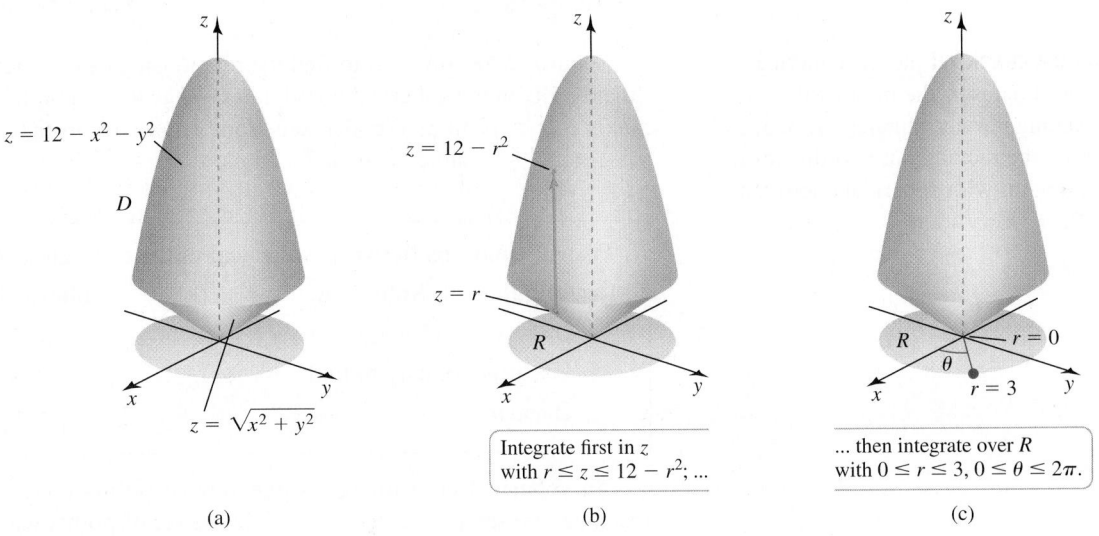

(a)

Integrate first in z with $r \leq z \leq 12 - r^2$; ...

(b)

... then integrate over R with $0 \leq r \leq 3, 0 \leq \theta \leq 2\pi$.

(c)

Figure 16.54

The volume of the region is

$$\iiint\limits_{D} dV = \int_0^{2\pi} \int_0^3 \int_r^{12-r^2} dz \, r \, dr \, d\theta$$

$$= \int_0^{2\pi} \int_0^3 (12 - r^2 - r) \, r \, dr \, d\theta \qquad \text{Evaluate inner integral with respect to } z.$$

$$= \int_0^{2\pi} \frac{99}{4} \, d\theta \qquad \text{Evaluate middle integral with respect to } r.$$

$$= \frac{99\pi}{2}. \qquad \text{Evaluate outer integral with respect to } \theta.$$

Related Exercises 30–31 ◄

> The coordinate ρ (pronounced "rho") in spherical coordinates should not be confused with r in cylindrical coordinates, which is the distance from P to the z-axis.

> The coordinate φ is called the *colatitude* because it is $\pi/2$ minus the latitude of points in the Northern Hemisphere. Physicists may reverse the roles of θ and φ; that is, θ is the colatitude and φ is the polar angle.

Spherical Coordinates

In spherical coordinates, a point P in $\mathbb{R}^3$ is represented by three coordinates (ρ, φ, θ) (**Figure 16.55**).

• ρ is the distance from the origin to P.

• φ is the angle between the positive z-axis and the line OP.

• θ is the same angle as in cylindrical coordinates; it measures rotation about the z-axis relative to the positive x-axis.

All points in $\mathbb{R}^3$ can be represented by spherical coordinates using the intervals $0 \leq \rho < \infty, 0 \leq \varphi \leq \pi$, and $0 \leq \theta \leq 2\pi$.

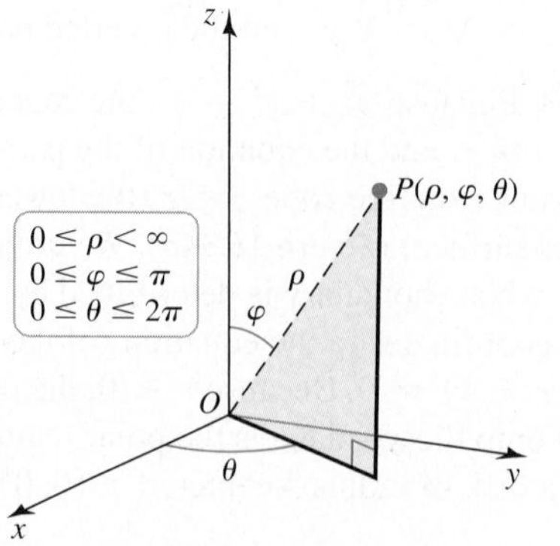

$$0 \le \rho < \infty$$
$$0 \le \varphi \le \pi$$
$$0 \le \theta \le 2\pi$$

$P(\rho, \varphi, \theta)$

Figure 16.55

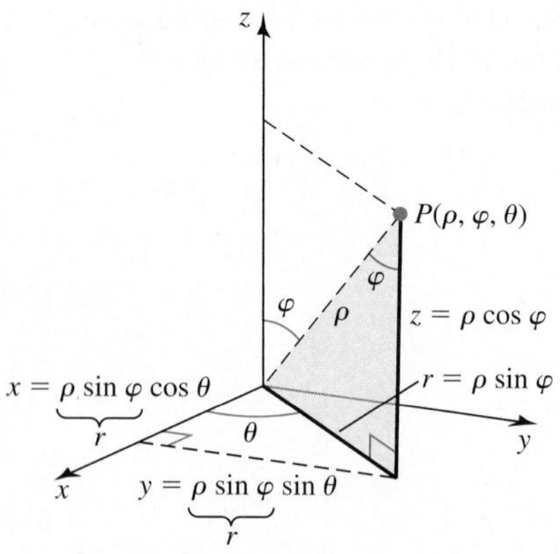

$P(\rho, \varphi, \theta)$

$z = \rho \cos \varphi$

$x = \rho \sin \varphi \cos \theta$

$r = \rho \sin \varphi$

$y = \rho \sin \varphi \sin \theta$

Figure 16.56

QUICK CHECK 3 Find the spherical coordinates of the point with rectangular coordinates $(1, \sqrt{3}, 2)$. Find the rectangular coordinates of the point with spherical coordinates $(2, \pi/4, \pi/4)$. ◄

Figure 16.56 allows us to find the relationships among rectangular and spherical coordinates. Given the spherical coordinates (ρ, φ, θ) of a point P, the distance from P to the z-axis is $r = \rho \sin \varphi$. We also see from Figure 16.56 that $x = r \cos \theta = \rho \sin \varphi \cos \theta$, $y = r \sin \theta = \rho \sin \varphi \sin \theta$, and $z = \rho \cos \varphi$.

Transformations Between Spherical and Rectangular Coordinates

Rectangular → Spherical	**Spherical → Rectangular**
$\rho^2 = x^2 + y^2 + z^2$	$x = \rho \sin \varphi \cos \theta$
Use trigonometry to find	$y = \rho \sin \varphi \sin \theta$
φ and θ.	$z = \rho \cos \varphi$

In spherical coordinates, some sets of points have simple representations. For instance, the set $\{(\rho, \varphi, \theta): \rho = a\}$ is the set of points whose ρ-coordinate is constant, which is a sphere of radius a centered at the origin. The set $\{(\rho, \varphi, \theta): \varphi = \varphi_0\}$ is the set of points with a constant φ-coordinate; it is a cone with its vertex at the origin and whose sides make an angle φ_0 with the positive z-axis.

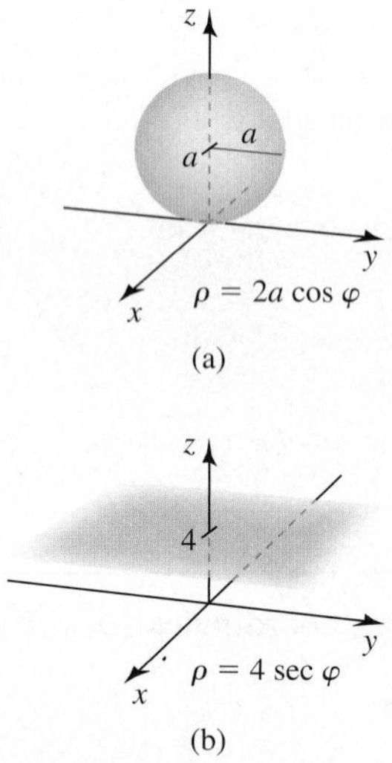

EXAMPLE 5 **Sets in spherical coordinates** Express the following sets in rectangular coordinates and identify the set. Assume a is a positive real number.

a. $\{(\rho, \varphi, \theta): \rho = 2a \cos \varphi, 0 \le \varphi \le \pi/2, 0 \le \theta \le 2\pi\}$

b. $\{(\rho, \varphi, \theta): \rho = 4 \sec \varphi, 0 \le \varphi < \pi/2, 0 \le \theta \le 2\pi\}$

SOLUTION

a. To avoid working with square roots, we multiply both sides of $\rho = 2a \cos \varphi$ by ρ to obtain $\rho^2 = 2a \rho \cos \varphi$. Substituting rectangular coordinates, we have $x^2 + y^2 + z^2 = 2az$. Completing the square results in the equation

$$x^2 + y^2 + (z - a)^2 = a^2.$$

This is the equation of a sphere centered at $(0, 0, a)$ with radius a (Figure 16.57a). With the limits $0 \le \varphi \le \pi/2$ and $0 \le \theta \le 2\pi$, the set describes a full sphere.

b. The equation $\rho = 4 \sec \varphi$ is first written $\rho \cos \varphi = 4$. Noting that $z = \rho \cos \varphi$, the set consists of all points with $z = 4$, which is a horizontal plane (Figure 16.57b).

Related Exercises 37–38 ◄

a a

$\rho = 2a \cos \varphi$

(a)

4

$\rho = 4 \sec \varphi$

(b)

Figure 16.57

Table 16.5 summarizes some sets that have simple descriptions in spherical coordinates.

Table 16.5

Name	Description	Example
Sphere, radius a, center $(0, 0, 0)$	$\{(\rho, \varphi, \theta): \rho = a\}, a > 0$	
Cone	$\{(\rho, \varphi, \theta): \varphi = \varphi_0\}, \varphi_0 \neq 0, \pi/2, \pi$	
Vertical half-plane	$\{(\rho, \varphi, \theta): \theta = \theta_0\}$	
Horizontal plane, $z = a$	$a > 0: \{(\rho, \varphi, \theta): \rho = a \sec \varphi, 0 \leq \varphi < \pi/2\}$ $a < 0: \{(\rho, \varphi, \theta): \rho = a \sec \varphi, \pi/2 < \varphi \leq \pi\}$	
Cylinder, radius $a > 0$	$\{(\rho, \varphi, \theta): \rho = a \csc \varphi, 0 < \varphi < \pi\}$	
Sphere, radius $a > 0$, center $(0, 0, a)$	$\{(\rho, \varphi, \theta): \rho = 2a \cos \varphi, 0 \leq \varphi \leq \pi/2\}$	

➤ Notice that the set of points (ρ, φ, θ) with $\varphi = \pi/2$ is the xy-plane, and if $\pi/2 < \varphi_0 < \pi$, the set of points with $\varphi = \varphi_0$ is a cone that opens downward.

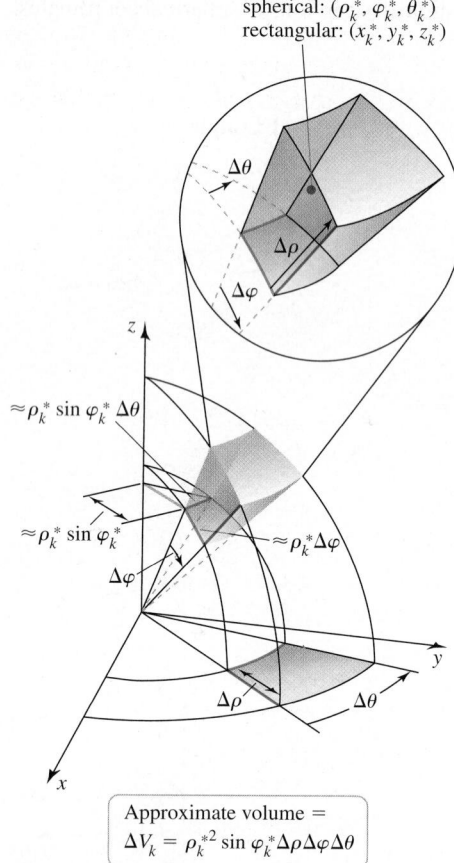

spherical: $(\rho_k^*, \varphi_k^*, \theta_k^*)$
rectangular: (x_k^*, y_k^*, z_k^*)

$\Delta\theta$

$\Delta\rho$

$\Delta\varphi$

z

$\approx \rho_k^* \sin \varphi_k^* \Delta\theta$

$\approx \rho_k^* \sin \varphi_k^*$

$\approx \rho_k^* \Delta\varphi$

$\Delta\varphi$

$\Delta\rho$

$\Delta\theta$

y

x

> Approximate volume =
> $\Delta V_k = \rho_k^{*2} \sin \varphi_k^* \Delta\rho \Delta\varphi \Delta\theta$

Figure 16.58

➤ Recall that the length s of a circular arc
of radius r subtended by an angle θ is
$s = r\theta$.

Integration in Spherical Coordinates

We now investigate triple integrals in spherical coordinates over a region D in $\mathbb{R}^3$. The region D is partitioned into "spherical boxes" that are formed by changes of $\Delta\rho$, $\Delta\varphi$, and $\Delta\theta$ in the coordinate directions (**Figure 16.58**). Those boxes that lie entirely within D are labeled from $k = 1$ to $k = n$. We let $(\rho_k^*, \varphi_k^*, \theta_k^*)$ be the spherical coordinates for an arbitrary point in the kth box. This point also has Cartesian coordinates

$$(x_k^*, y_k^*, z_k^*) = (\rho_k^* \sin \varphi_k^* \cos \theta_k^*, \rho_k^* \sin \varphi_k^* \sin \theta_k^*, \rho_k^* \cos \varphi_k^*).$$

To approximate the volume of a typical box, note that the length of the box in the ρ-direction is $\Delta\rho$ (Figure 16.58). The approximate length of the kth box in the θ-direction is the length of an arc of a circle of radius $\rho_k^* \sin \varphi_k^*$ subtended by an angle $\Delta\theta$; this length is $\rho_k^* \sin \varphi_k^* \Delta\theta$. The approximate length of the box in the φ-direction is the length of an arc of radius ρ_k^* subtended by an angle $\Delta\varphi$; this length is $\rho_k^* \Delta\varphi$. Multiplying these dimensions together, the approximate volume of the kth spherical box is $\Delta V_k = \rho_k^{*2} \sin \varphi_k^* \Delta\rho \Delta\varphi \Delta\theta$, for $k = 1, \ldots, n$.

We now assume $f(x, y, z)$ is continuous on D and form a Riemann sum over the region by adding function values multiplied by the corresponding approximate volumes:

$$\sum_{k=1}^{n} f(x_k^*, y_k^*, z_k^*) \Delta V_k = \sum_{k=1}^{n} f(\rho_k^* \sin \varphi_k^* \cos \theta_k^*, \rho_k^* \sin \varphi_k^* \sin \theta_k^*, \rho_k^* \cos \varphi_k^*) \Delta V_k.$$

We let Δ denote the maximum value of $\Delta\rho$, $\Delta\varphi$, and $\Delta\theta$. As $n \to \infty$ and $\Delta \to 0$, the Riemann sums approach a limit called the **triple integral of f over D in spherical coordinates**:

$$\iiint_D f(x, y, z) \, dV$$

$$= \lim_{\Delta \to 0} \sum_{k=1}^{n} f(\rho_k^* \sin \varphi_k^* \cos \theta_k^*, \rho_k^* \sin \varphi_k^* \sin \theta_k^*, \rho_k^* \cos \varphi_k^*) \underbrace{\rho_k^{*2} \sin \varphi_k^* \Delta\rho \Delta\varphi \Delta\theta}_{\Delta V_k}.$$

The rightmost sum tells us how to write a triple integral in x, y, and z as an iterated integral of $f(\rho \sin \varphi \cos \theta, \rho \sin \varphi \sin \theta, \rho \cos \varphi) \rho^2 \sin \varphi$ in spherical coordinates.

Finding Limits of Integration   We consider a common situation in which the region of integration D, expressed in spherical coordinates, has the form

$$D = \{(\rho, \varphi, \theta): 0 \leq g(\varphi, \theta) \leq \rho \leq h(\varphi, \theta), a \leq \varphi \leq b, \alpha \leq \theta \leq \beta\}.$$

In other words, D is bounded in the ρ-direction by two surfaces given by g and h. In the angular directions, the region lies between two cones ($a \leq \varphi \leq b$) and two half-planes ($\alpha \leq \theta \leq \beta$) (**Figure 16.59**).

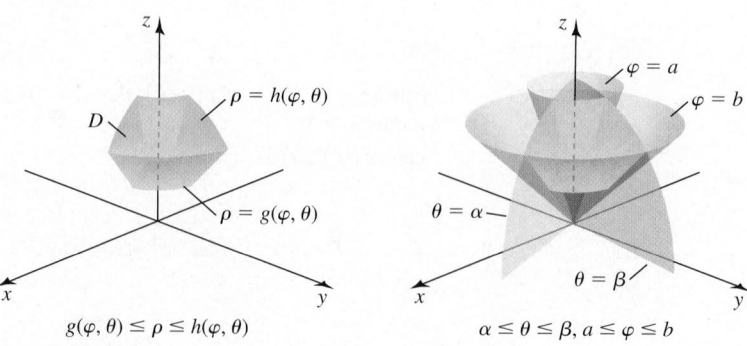

$$g(\varphi, \theta) \leq \rho \leq h(\varphi, \theta) \qquad \alpha \leq \theta \leq \beta, a \leq \varphi \leq b$$

Figure 16.59

For this type of region, the inner integral is with respect to ρ, which varies from $\rho = g(\varphi, \theta)$ to $\rho = h(\varphi, \theta)$. As ρ varies between these limits, imagine letting θ and φ vary over the intervals $a \leq \varphi \leq b$ and $\alpha \leq \theta \leq \beta$. The effect is to sweep out all points of D. Notice that the middle and outer integrals, with respect to θ and φ, may be done in either order (Figure 16.60).

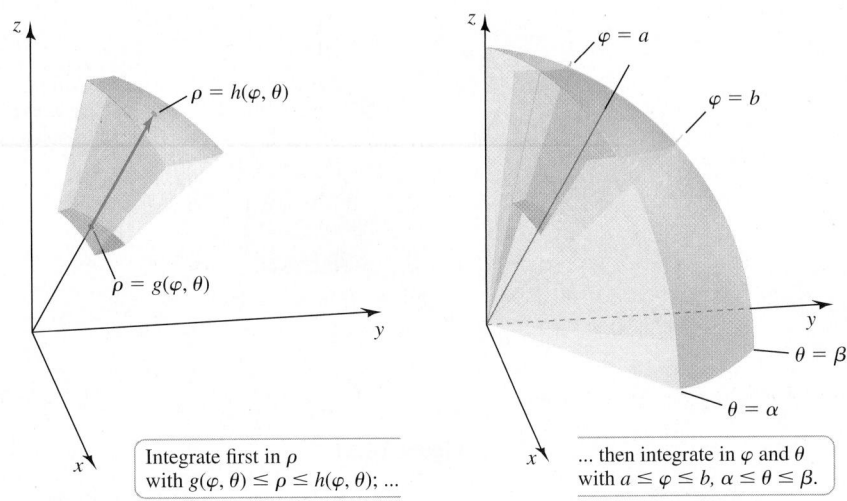

Figure 16.60

In summary, to integrate over all points of D, we carry out the following steps.

1. Integrate with respect to ρ from $\rho = g(\varphi, \theta)$ to $\rho = h(\varphi, \theta)$; the result (in general) is a function of φ and θ.

2. Integrate with respect to φ from $\varphi = a$ to $\varphi = b$; the result (in general) is a function of θ.

3. Integrate with respect to θ from $\theta = \alpha$ to $\theta = \beta$; the result is (always) a real number.

Another change of variables expresses the triple integral as an iterated integral in spherical coordinates.

> ▶ The element of volume in spherical coordinates is $dV = \rho^2 \sin \varphi \, d\rho \, d\varphi \, d\theta$.

THEOREM 16.7   Change of Variables for Triple Integrals in Spherical Coordinates

Let f be continuous over the region D, expressed in spherical coordinates as

$$D = \{(\rho, \varphi, \theta): 0 \leq g(\varphi, \theta) \leq \rho \leq h(\varphi, \theta), a \leq \varphi \leq b, \alpha \leq \theta \leq \beta\}.$$

Then f is integrable over D, and the triple integral of f over D is

$$\iiint\limits_{D} f(x, y, z) \, dV$$

$$= \int_{\alpha}^{\beta} \int_{a}^{b} \int_{g(\varphi, \theta)}^{h(\varphi, \theta)} f(\rho \sin \varphi \cos \theta, \rho \sin \varphi \sin \theta, \rho \cos \varphi) \rho^2 \sin \varphi \, d\rho \, d\varphi \, d\theta.$$

If the integrand is given in terms of Cartesian coordinates x, y, and z, it must be expressed in spherical coordinates before integrating. As with other triple integrals, if $f = 1$, then the triple integral equals the volume of D. If f is a density function for an object occupying the region D, then the triple integral equals the mass of the object.

EXAMPLE 6   A triple integral Evaluate $\iiint_{D} (x^2 + y^2 + z^2)^{-3/2} \, dV$, where D is the region in the first octant between two spheres of radius 1 and 2 centered at the origin.

SOLUTION Both the integrand f and the region D are greatly simplified when expressed in spherical coordinates. The integrand becomes

$$(x^2 + y^2 + z^2)^{-3/2} = (\rho^2)^{-3/2} = \rho^{-3},$$

while the region of integration (**Figure 16.61**) is

$$D = \{(\rho, \varphi, \theta): 1 \le \rho \le 2, 0 \le \varphi \le \pi/2, 0 \le \theta \le \pi/2\}.$$

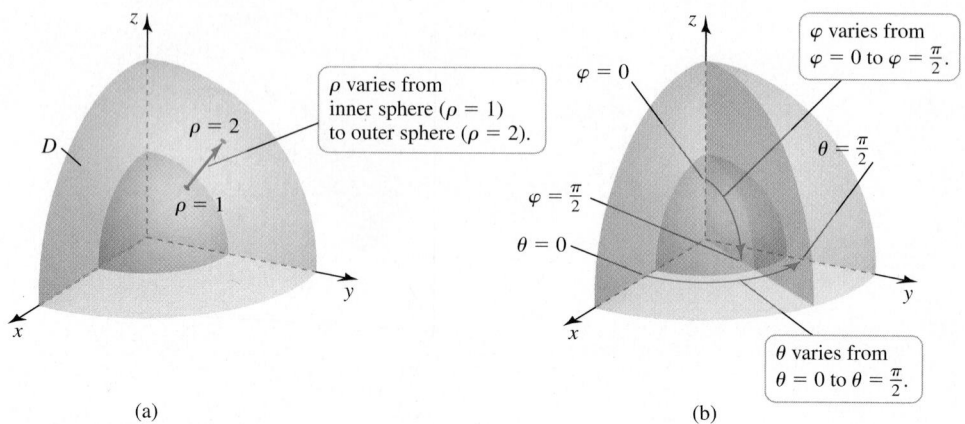

(a)   (b)

Figure 16.61

The integral is evaluated as follows:

$$\iiint\limits_{D} f(x, y, z)\, dV = \int_0^{\pi/2} \int_0^{\pi/2} \int_1^2 \rho^{-3}\, \rho^2 \sin\varphi\, d\rho\, d\varphi\, d\theta \qquad \text{Convert to spherical coordinates.}$$

$$= \int_0^{\pi/2} \int_0^{\pi/2} \int_1^2 \rho^{-1} \sin\varphi\, d\rho\, d\varphi\, d\theta \qquad \text{Simplify.}$$

$$= \int_0^{\pi/2} \int_0^{\pi/2} \ln|\rho|\,\Big|_1^2 \sin\varphi\, d\varphi\, d\theta \qquad \text{Evaluate inner integral with respect to } \rho.$$

$$= \ln 2 \int_0^{\pi/2} \int_0^{\pi/2} \sin\varphi\, d\varphi\, d\theta \qquad \text{Simplify.}$$

$$= \ln 2 \int_0^{\pi/2} (-\cos\varphi)\,\Big|_0^{\pi/2} d\theta \qquad \text{Evaluate middle integral with respect to } \varphi.$$

$$= \ln 2 \int_0^{\pi/2} d\theta = \frac{\pi \ln 2}{2}. \qquad \text{Evaluate outer integral with respect to } \theta.$$

Related Exercises 41, 43 ◄

EXAMPLE 7   Ice cream cone Find the volume of the solid region D that lies inside the cone $\varphi = \pi/6$ and inside the sphere $\rho = 4$ (**Figure 16.62a**).

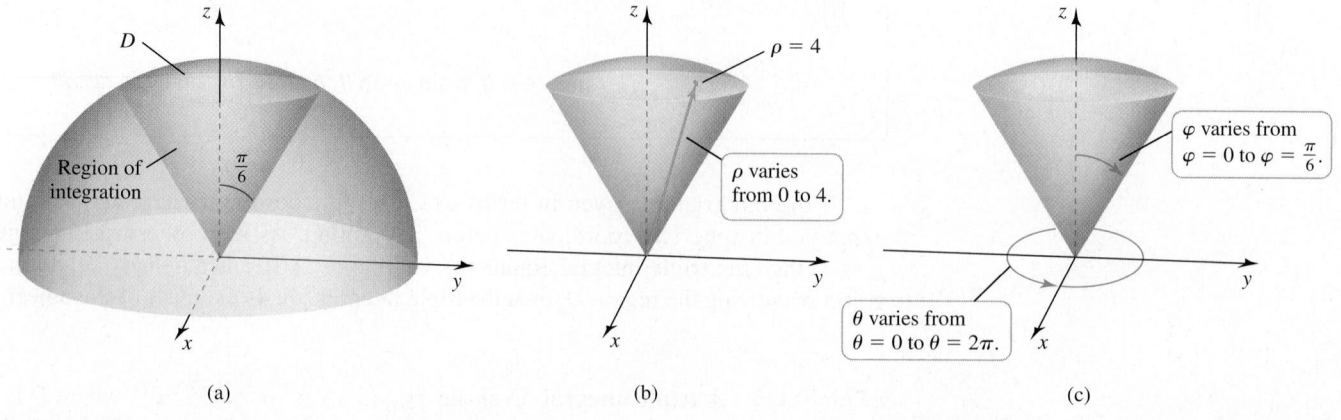

(a)   (b)   (c)

Figure 16.62

SOLUTION To find the volume, we evaluate a triple integral with $f = 1$. In the radial direction, the region extends from the origin $\rho = 0$ to the sphere $\rho = 4$ (**Figure 16.62b**). To sweep out all points of D, φ varies from 0 to $\pi/6$, and θ varies from 0 to 2π (**Figure 16.62c**). The volume of the region is

$$\iiint_D dV = \int_0^{2\pi} \int_0^{\pi/6} \int_0^4 \rho^2 \sin\varphi \, d\rho \, d\varphi \, d\theta \quad \text{Convert to an iterated integral.}$$

$$= \int_0^{2\pi} \int_0^{\pi/6} \frac{\rho^3}{3}\Big|_0^4 \sin\varphi \, d\varphi \, d\theta \quad \text{Evaluate inner integral with respect to } \rho.$$

$$= \frac{64}{3} \int_0^{2\pi} \int_0^{\pi/6} \sin\varphi \, d\varphi \, d\theta \quad \text{Simplify.}$$

$$= \frac{64}{3} \int_0^{2\pi} \underbrace{(-\cos\varphi)\Big|_0^{\pi/6}}_{1 - \sqrt{3}/2} d\theta \quad \text{Evaluate middle integral with respect to } \varphi.$$

$$= \frac{32}{3}(2 - \sqrt{3}) \int_0^{2\pi} d\theta \quad \text{Simplify.}$$

$$= \frac{64\pi}{3}(2 - \sqrt{3}). \quad \text{Evaluate outer integral with respect to } \theta.$$

Related Exercises 49, 51 ◀

SECTION 16.5 EXERCISES

Getting Started

1. Explain how cylindrical coordinates are used to describe a point in $\mathbb{R}^3$.

2. Explain how spherical coordinates are used to describe a point in $\mathbb{R}^3$.

3. Describe the set $\{(r, \theta, z): r = 4z\}$ in cylindrical coordinates.

4. Describe the set $\{(\rho, \varphi, \theta): \varphi = \pi/4\}$ in spherical coordinates.

5. Explain why $dz \, r \, dr \, d\theta$ is the volume of a small "box" in cylindrical coordinates.

6. Explain why $\rho^2 \sin\varphi \, d\rho \, d\varphi \, d\theta$ is the volume of a small "box" in spherical coordinates.

7. Write the integral $\iiint_D w(r, \theta, z) \, dV$ as an iterated integral, where the region D, expressed in cylindrical coordinates, is $D = \{(r, \theta, z): G(r, \theta) \leq z \leq H(r, \theta), g(\theta) \leq r \leq h(\theta), \alpha \leq \theta \leq \beta\}$.

8. Write the integral $\iiint_D w(\rho, \varphi, \theta) \, dV$ as an iterated integral, where the region D, expressed in spherical coordinates, is $D = \{(\rho, \varphi, \theta): g(\varphi, \theta) \leq \rho \leq h(\varphi, \theta), a \leq \varphi \leq b, \alpha \leq \theta \leq \beta\}$.

9. What coordinate system is *suggested* if the integrand of a triple integral involves $x^2 + y^2$?

10. What coordinate system is *suggested* if the integrand of a triple integral involves $x^2 + y^2 + z^2$?

Practice Exercises

11–14. Sets in cylindrical coordinates *Identify and sketch the following sets in cylindrical coordinates.*

11. $\{(r, \theta, z): 0 \leq r \leq 3, 0 \leq \theta \leq \pi/3, 1 \leq z \leq 4\}$

12. $\{(r, \theta, z): 0 \leq \theta \leq \pi/2, z = 1\}$

13. $\{(r, \theta, z): 2r \leq z \leq 4\}$

14. $\{(r, \theta, z): 0 \leq z \leq 8 - 2r\}$

15–22. Integrals in cylindrical coordinates *Evaluate the following integrals in cylindrical coordinates. The figures, if given, illustrate the region of integration.*

15. $\displaystyle\int_0^{2\pi} \int_0^1 \int_{-1}^1 dz \, r \, dr \, d\theta$

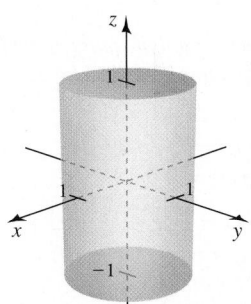

16. $\displaystyle\int_0^3 \int_{-\sqrt{9-y^2}}^{\sqrt{9-y^2}} \int_0^{9-3\sqrt{x^2+y^2}} dz \, dx \, dy$

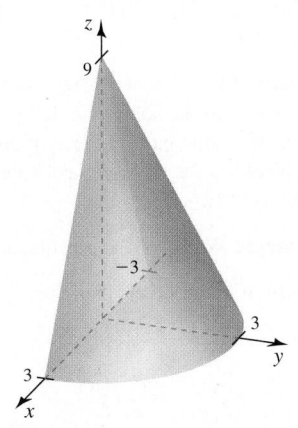

17. $\int_{-1}^{1} \int_{-\sqrt{1-y^2}}^{\sqrt{1-y^2}} \int_{-1}^{1} (x^2 + y^2)^{3/2} \, dz \, dx \, dy$

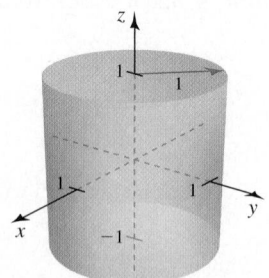

18. $\int_{-3}^{3} \int_{0}^{\sqrt{9-x^2}} \int_{0}^{2} \frac{1}{1 + x^2 + y^2} \, dz \, dy \, dx$

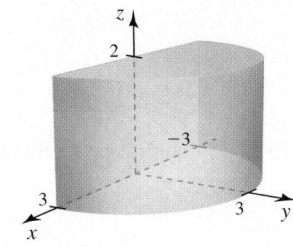

19. $\int_{0}^{4} \int_{0}^{\sqrt{2}/2} \int_{x}^{\sqrt{1-x^2}} e^{-x^2-y^2} \, dy \, dx \, dz$

20. $\int_{-4}^{4} \int_{-\sqrt{16-x^2}}^{\sqrt{16-x^2}} \int_{\sqrt{x^2+y^2}}^{4} dz \, dy \, dx$

21. $\int_{0}^{3} \int_{0}^{\sqrt{9-x^2}} \int_{0}^{\sqrt{x^2+y^2}} (x^2 + y^2)^{-1/2} \, dz \, dy \, dx$

22. $\int_{-1}^{1} \int_{0}^{1/2} \int_{\sqrt{3}y}^{\sqrt{1-y^2}} (x^2 + y^2)^{1/2} \, dx \, dy \, dz$

23–26. Mass from density *Find the mass of the following objects with the given density functions. Assume (r, θ, z) are cylindrical coordinates.*

23. The solid cylinder $D = \{(r, \theta, z): 0 \le r \le 4, 0 \le z \le 10\}$ with density $\rho(r, \theta, z) = 1 + z/2$

24. The solid cylinder $D = \{(r, \theta, z): 0 \le r \le 3, 0 \le z \le 2\}$ with density $\rho(r, \theta, z) = 5e^{-r^2}$

25. The solid cone $D = \{(r, \theta, z): 0 \le z \le 6 - r, 0 \le r \le 6\}$ with density $\rho(r, \theta, z) = 7 - z$

26. The solid paraboloid $D = \{(r, \theta, z): 0 \le z \le 9 - r^2, 0 \le r \le 3\}$ with density $\rho(r, \theta, z) = 1 + z/9$

27. Which weighs more? For $0 \le r \le 1$, the solid bounded by the cone $z = 4 - 4r$ and the solid bounded by the paraboloid $z = 4 - 4r^2$ have the same base in the xy-plane and the same height. Which object has the greater mass if the density of both objects is $\rho(r, \theta, z) = 10 - 2z$?

28. Which weighs more? Which of the objects in Exercise 27 weighs more if the density of both objects is $\rho(r, \theta, z) = \dfrac{8}{\pi} e^{-z}$?

29–34. Volumes in cylindrical coordinates *Use cylindrical coordinates to find the volume of the following solids.*

29. The solid bounded by the plane $z = 0$ and the hyperboloid $z = 3 - \sqrt{1 + x^2 + y^2}$

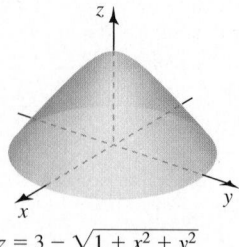

$z = 3 - \sqrt{1 + x^2 + y^2}$

30. The solid bounded by the plane $z = 25$ and the paraboloid $z = x^2 + y^2$

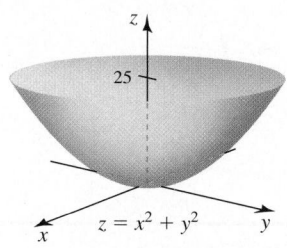

$z = x^2 + y^2$

31. The solid bounded by the plane $z = \sqrt{29}$ and the hyperboloid $z = \sqrt{4 + x^2 + y^2}$

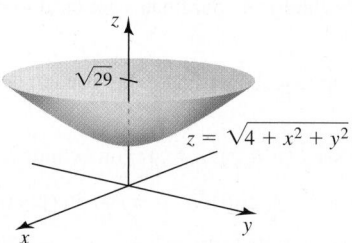

$z = \sqrt{4 + x^2 + y^2}$

32. The solid cylinder whose height is 4 and whose base is the disk $\{(r, \theta): 0 \le r \le 2 \cos \theta\}$

33. The solid in the first octant bounded by the cylinder $r = 1$, and the planes $z = x$ and $z = 0$

34. The solid bounded by the cylinders $r = 1$ and $r = 2$, and the planes $z = 4 - x - y$ and $z = 0$

35–38. Sets in spherical coordinates *Identify and sketch the following sets in spherical coordinates.*

35. $\{(\rho, \varphi, \theta): 1 \le \rho \le 3\}$

36. $\{(\rho, \varphi, \theta): \rho = 2 \csc \varphi, 0 < \varphi < \pi\}$

37. $\{(\rho, \varphi, \theta): \rho = 4 \cos \varphi, 0 \le \varphi \le \pi/2\}$

38. $\{(\rho, \varphi, \theta): \rho = 2 \sec \varphi, 0 \le \varphi < \pi/2\}$

T 39–40. Latitude, longitude, and distances *Assume Earth is a sphere with radius $r = 3960$ miles, oriented in xyz-space so that its center passes through the origin O, the positive z-axis passes through the North Pole, and the xz-plane passes through Greenwich, England (the intersection of Earth's surface and the xz-plane is called the prime meridian). The location of a point on Earth is given by its latitude* *(degrees north or south of the equator) and its longitude (degrees east or west of the prime meridian).*

39. Seattle has a latitude of 47.6° North and a longitude of 122.3° West; Rome, Italy, has a latitude of 41.9° North and a longitude of 12.5° East.

    a. Find the approximate spherical and rectangular coordinates of Seattle. Express the angular coordinates in radians.

    b. Find the approximate spherical and rectangular coordinates of Rome.

    c. Consider the intersection curve of a sphere, and a plane passing through the center of the sphere and two points A and B on the sphere. It can be shown that the arc length of the segment of the intersection curve from A to B is the shortest distance on the sphere from A to B. Find the approximate shortest distance from Seattle to Rome. (*Hint:* Recall that $\cos\theta = \dfrac{\mathbf{u}\cdot\mathbf{v}}{|\mathbf{u}||\mathbf{v}|}$, where $0 \le \theta \le \pi$ is the angle between $\mathbf{u}$ and $\mathbf{v}$; use the arc length formula $s = r\theta$ to find the distance.)

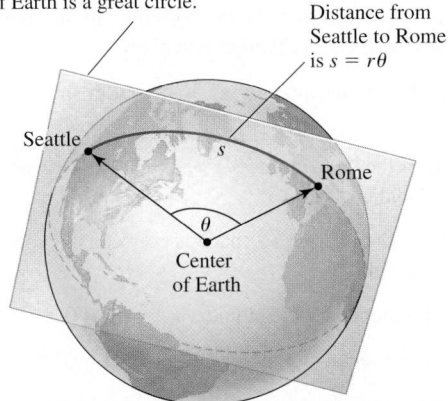

40. Los Angeles has a latitude of 34.05° North and a longitude of 118.24° West, and New York City has a latitude of 40.71° North and a longitude of 74.01° West. Find the approximate shortest distance from Los Angeles to New York City.

41–47. Integrals in spherical coordinates *Evaluate the following integrals in spherical coordinates.*

41. $\displaystyle\iiint_D (x^2 + y^2 + z^2)^{5/2}\, dV$; D is the unit ball.

42. $\displaystyle\iiint_D e^{-(x^2+y^2+z^2)^{3/2}}\, dV$; D is the unit ball.

43. $\displaystyle\iiint_D \frac{dV}{(x^2 + y^2 + z^2)^{3/2}}$; D is the solid between the spheres of radius 1 and 2 centered at the origin.

44. $\displaystyle\int_0^{2\pi}\int_0^{\pi/3}\int_0^{4\sec\varphi} \rho^2 \sin\varphi\, d\rho\, d\varphi\, d\theta$

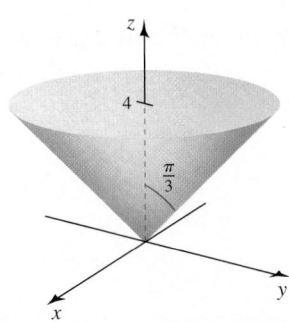

45. $\displaystyle\int_0^{\pi}\int_0^{\pi/6}\int_{2\sec\varphi}^{4} \rho^2 \sin\varphi\, d\rho\, d\varphi\, d\theta$

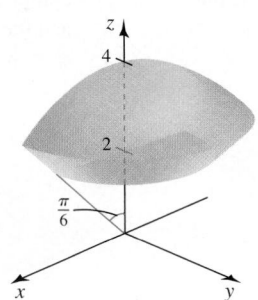

46. $\displaystyle\int_0^{2\pi}\int_0^{\pi/4}\int_1^{2\sec\varphi} (\rho^{-3})\, \rho^2 \sin\varphi\, d\rho\, d\varphi\, d\theta$

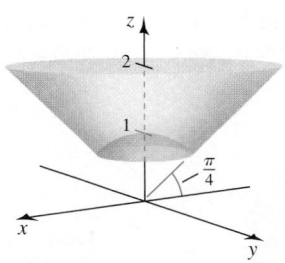

47. $\displaystyle\int_0^{2\pi}\int_{\pi/6}^{\pi/3}\int_0^{2\csc\varphi} \rho^2 \sin\varphi\, d\rho\, d\varphi\, d\theta$

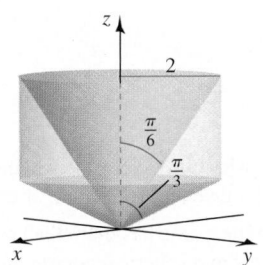

48–54. Volumes in spherical coordinates *Use spherical coordinates to find the volume of the following solids.*

48. A ball of radius $a > 0$

49. The solid bounded by the sphere $\rho = 2\cos\varphi$ and the hemisphere $\rho = 1$, $z \ge 0$

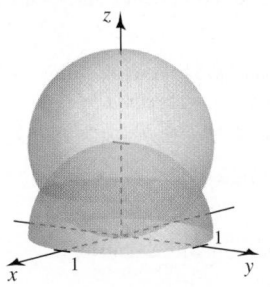

50. The solid cardioid of revolution $D = \{(\rho, \varphi, \theta): 0 \le \rho \le 1 + \cos\varphi, 0 \le \varphi \le \pi, 0 \le \theta \le 2\pi\}$

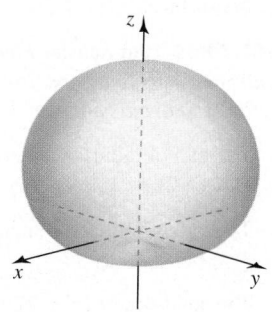

51. The solid outside the cone $\varphi = \pi/4$ and inside the sphere $\rho = 4 \cos \varphi$

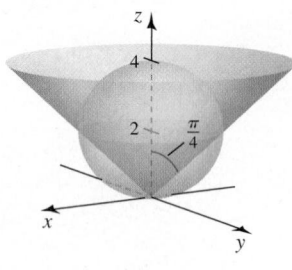

52. The solid bounded by the cylinders $r = 1$ and $r = 2$, and the cones $\varphi = \pi/6$ and $\varphi = \pi/3$

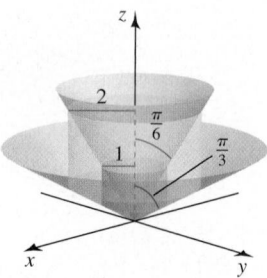

53. That part of the ball $\rho \le 4$ that lies between the planes $z = 2$ and $z = 2\sqrt{3}$

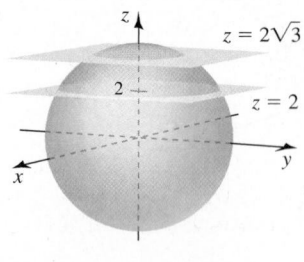

54. The solid lying between the planes $z = 1$ and $z = 2$ that is bounded by the cone $z = (x^2 + y^2)^{1/2}$

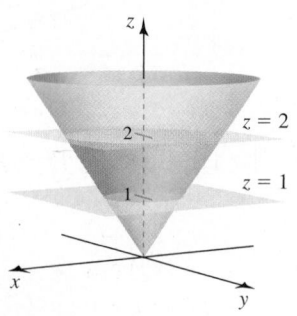

55. Explain why or why not Determine whether the following statements are true and give an explanation or counterexample.

   a. Any point on the z-axis has more than one representation in both cylindrical and spherical coordinates.

   b. The sets $\{(r, \theta, z): r = z\}$ in cylindrical coordinates and the set $\{(\rho, \varphi, \theta): \varphi = \pi/4\}$ in spherical coordinates describe the same set of points.

56. Spherical to rectangular Convert the equation $\rho^2 = \sec 2\varphi$, where $0 \le \varphi < \pi/4$, to rectangular coordinates and identify the surface.

57. Spherical to rectangular Convert the equation $\rho^2 = -\sec 2\varphi$, where $\pi/4 < \varphi \le \pi/2$, to rectangular coordinates and identify the surface.

58–61. Mass from density *Find the mass of the following solids with the given density functions. Note that density is described by the function f to avoid confusion with the radial spherical coordinate ρ.*

58. The ball of radius 4 centered at the origin with a density $f(\rho, \varphi, \theta) = 1 + \rho$

59. The ball of radius 8 centered at the origin with a density $f(\rho, \varphi, \theta) = 2e^{-\rho^3}$

60. The solid cone $\{(r, \theta, z): 0 \le z \le 4, 0 \le r \le \sqrt{3}z, 0 \le \theta \le 2\pi\}$ with a density $f(r, \theta, z) = 5 - z$

61. The solid cylinder $\{(r, \theta, z): 0 \le r \le 2, 0 \le \theta \le 2\pi, -1 \le z \le 1\}$ with a density $f(r, \theta, z) = (2 - |z|)(4 - r)$

62–63. Changing order of integration *If possible, write an iterated integral in cylindrical coordinates of a function $g(r, \theta, z)$ for the following regions in the specified orders. Sketch the region of integration.*

62. The solid outside the cylinder $r = 1$ and inside the sphere $\rho = 5$, for $z \ge 0$, in the orders $dz\,dr\,d\theta$, $dr\,dz\,d\theta$, and $d\theta\,dz\,dr$

63. The solid above the cone $z = r$ and below the sphere $\rho = 2$, for $z \ge 0$, in the orders $dz\,dr\,d\theta$, $dr\,dz\,d\theta$, and $d\theta\,dz\,dr$

64–65. Changing order of integration *If possible, write iterated integrals in spherical coordinates for the following regions in the specified orders. Sketch the region of integration. Assume g is continuous on the region.*

64. $\int_0^{2\pi} \int_0^{\pi/2} \int_0^{4 \sec \varphi} g(\rho, \varphi, \theta)\rho^2 \sin \varphi\, d\rho\, d\varphi\, d\theta$ in the orders $d\rho\, d\theta\, d\varphi$ and $d\theta\, d\rho\, d\varphi$

65. $\int_0^{2\pi} \int_{\pi/6}^{\pi/2} \int_{\csc \varphi}^{2} g(\rho, \varphi, \theta)\rho^2 \sin \varphi\, d\rho\, d\varphi\, d\theta$ in the orders $d\rho\, d\theta\, d\varphi$ and $d\theta\, d\rho\, d\varphi$

66–71. Miscellaneous volumes *Choose the best coordinate system and find the volume of the following solids. Surfaces are specified using the coordinates that give the simplest description, but the simplest integration may be with respect to different variables.*

66. The solid inside the sphere $\rho = 1$ and below the cone $\varphi = \pi/4$, for $z \ge 0$

67. That part of the solid cylinder $r \le 2$ that lies between the cones $\varphi = \pi/3$ and $\varphi = 2\pi/3$

68. That part of the ball $\rho \le 2$ that lies between the cones $\varphi = \pi/3$ and $\varphi = 2\pi/3$

69. The solid bounded by the cylinder $r = 1$, for $0 \le z \le x + y$

70. The solid inside the cylinder $r = 2 \cos \theta$, for $0 \le z \le 4 - x$

71. The wedge cut from the cardioid cylinder $r = 1 + \cos \theta$ by the planes $z = 2 - x$ and $z = x - 2$

72. Volume of a drilled hemisphere Find the volume of material remaining in a hemisphere of radius 2 after a cylindrical hole of radius 1 is drilled through the center of the hemisphere perpendicular to its base.

73. Density distribution A right circular cylinder with height 8 cm and radius 2 cm is filled with water. A heated filament running along its axis produces a variable density in the water given by $\rho(r) = 1 - 0.05e^{-0.01r^2} \text{g/cm}^3$ (ρ stands for density here, not for the radial spherical coordinate). Find the mass of the water in the cylinder. Neglect the volume of the filament.

74. Charge distribution A spherical cloud of electric charge has a known charge density $Q(\rho)$, where ρ is the spherical coordinate. Find the total charge in the cloud in the following cases.

   a. $Q(\rho) = \dfrac{2 \times 10^{-4}}{\rho^4}, 1 \le \rho < \infty$

   b. $Q(\rho) = (2 \times 10^{-4})e^{-0.01\rho^3}, 0 \le \rho < \infty$

Explorations and Challenges

75. Gravitational field due to spherical shell A point mass m is a distance d from the center of a thin spherical shell of mass M and radius R. The magnitude of the gravitational force on the point mass is given by the integral

$$F(d) = \frac{GMm}{4\pi} \int_0^{2\pi} \int_0^{\pi} \frac{(d - R\cos\varphi)\sin\varphi}{(R^2 + d^2 - 2Rd\cos\varphi)^{3/2}} \, d\varphi \, d\theta,$$

where G is the gravitational constant.

a. Use the change of variable $x = \cos\varphi$ to evaluate the integral and show that if $d > R$, then $F(d) = GMm/d^2$, which means the force is the same as it would be if the mass of the shell were concentrated at its center.

b. Show that if $d < R$ (the point mass is inside the shell), then $F = 0$.

76. Water in a gas tank Before a gasoline-powered engine is started, water must be drained from the bottom of the fuel tank. Suppose the tank is a right circular cylinder on its side with a length of 2 ft and a radius of 1 ft. If the water level is 6 in above the lowest part of the tank, determine how much water must be drained from the tank.

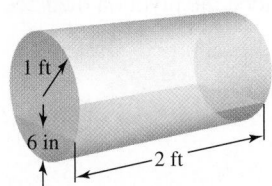

77–80. General volume formulas *Use integration to find the volume of the following solids. In each case, choose a convenient coordinate system, find equations for the bounding surfaces, set up a triple integral, and evaluate the integral. Assume a, b, c, r, R, and h are positive constants.*

77. Cone Find the volume of a solid right circular cone with height h and base radius r.

78. Spherical cap Find the volume of the cap of a sphere of radius R with thickness h.

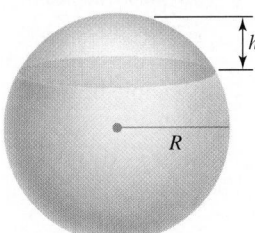

79. Frustum of a cone Find the volume of a truncated solid cone of height h whose ends have radii r and R.

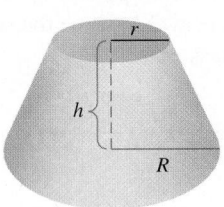

80. Ellipsoid Find the volume of a solid ellipsoid with axes of lengths $2a$, $2b$, and $2c$.

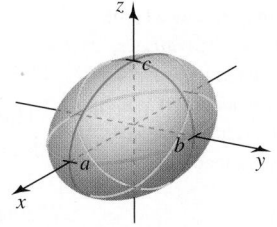

81. Intersecting spheres One sphere is centered at the origin and has a radius of R. Another sphere is centered at $(0, 0, r)$ and has a radius of r, where $r > R/2$. What is the volume of the region common to the two spheres?

QUICK CHECK ANSWERS

1. $\left(\sqrt{2}, 7\pi/4, 5\right), \left(1, \sqrt{3}, 5\right)$
2. $0 \le r \le 10, 0 \le \theta \le 2\pi, 0 \le z \le 20$
3. $\left(2\sqrt{2}, \pi/4, \pi/3\right), \left(1, 1, \sqrt{2}\right)$ ◄

16.6 Integrals for Mass Calculations

Intuition says that a thin circular disk (such as a DVD without a hole) should balance on a pencil placed at the center of the disk (Figure 16.63). If, however, you were given a thin plate with an irregular shape, at what point would it balance? This question asks about the *center of mass* of a thin object (thin enough that it can be treated as a two-dimensional region). Similarly, given a solid object with an irregular shape and variable density, where is the point at which all of the mass of the object would be located if it were treated as a point mass? In this section, we use integration to compute the center of mass of one-, two-, and three-dimensional objects.

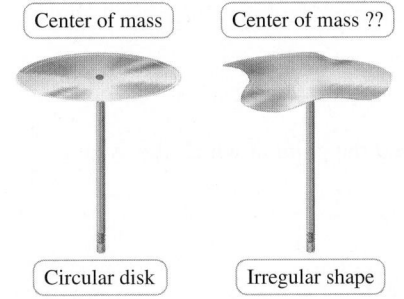

Figure 16.63

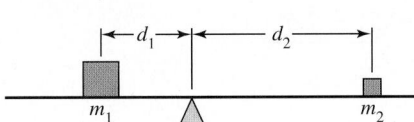

Figure 16.64

Sets of Individual Objects

Methods for finding the center of mass of an object are ultimately based on a well-known playground principle: If two people with masses m_1 and m_2 sit at distances d_1 and d_2 from the pivot point of a seesaw (with no mass), then the seesaw balances, provided $m_1 d_1 = m_2 d_2$ (Figure 16.64).

QUICK CHECK 1 A 90-kg person sits 2 m from the balance point of a seesaw. How far from that point must a 60-kg person sit to balance the seesaw? Assume the seesaw has no mass. ◀

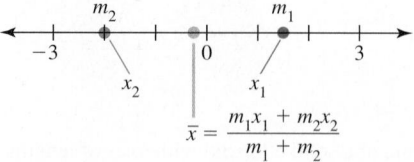

Figure 16.65

▶ The center of mass may be viewed as the weighted average of the x-coordinates, with the masses serving as the weights. Notice how the units work out: If x_1 and x_2 have units of meters and m_1 and m_2 have units of kilograms, then $\bar{x}$ has units of meters.

QUICK CHECK 2 Solve the equation $m_1(x_1 - \bar{x}) + m_2(x_2 - \bar{x}) = 0$ for $\bar{x}$ to verify the preceding expression for the center of mass. ◀

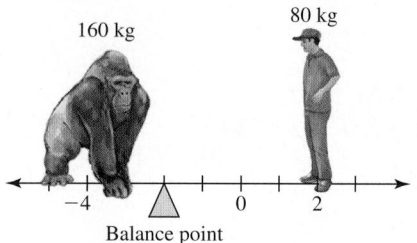

Figure 16.66

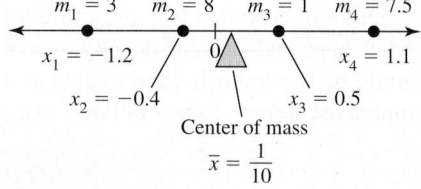

Figure 16.67

To generalize the problem, we introduce a coordinate system with the origin at $x = 0$ (Figure 16.65). Suppose the location of the balance point $\bar{x}$ is unknown. The coordinates of the two masses m_1 and m_2 are denoted x_1 and x_2, respectively, with $x_1 > x_2$. The mass m_1 is a distance $x_1 - \bar{x}$ from the balance point (because distance is positive and $x_1 > \bar{x}$). The mass m_2 is a distance $\bar{x} - x_2$ from the balance point (because $\bar{x} > x_2$). The playground principle becomes

$$m_1\underbrace{(x_1 - \bar{x})}_{\substack{\text{distance from} \\ \text{balance point} \\ \text{to } m_1}} = m_2\underbrace{(\bar{x} - x_2)}_{\substack{\text{distance from} \\ \text{balance point} \\ \text{to } m_2}},$$

or $m_1(x_1 - \bar{x}) + m_2(x_2 - \bar{x}) = 0$.

Solving this equation for $\bar{x}$, we find that the balance point or *center of mass* of the two-mass system is located at

$$\bar{x} = \frac{m_1 x_1 + m_2 x_2}{m_1 + m_2}.$$

The quantities $m_1 x_1$ and $m_2 x_2$ are called *moments about the origin* (or just *moments*). The location of the center of mass is the sum of the moments divided by the sum of the masses.

For example, an 80-kg man standing 2 m to the right of the origin will balance a 160-kg gorilla sitting 4 m to the left of the origin, provided the pivot on their seesaw is placed at

$$\bar{x} = \frac{80(2) + 160(-4)}{80 + 160} = -2,$$

or 2 m to the left of the origin (Figure 16.66).

Several Objects on a Line Generalizing the preceding argument to n objects having masses $m_1, m_2, \ldots,$ and m_n with coordinates $x_1, x_2, \ldots,$ and x_n, respectively, the balance condition becomes

$$m_1(x_1 - \bar{x}) + m_2(x_2 - \bar{x}) + \cdots + m_n(x_n - \bar{x}) = \sum_{k=1}^{n} m_k(x_k - \bar{x}) = 0.$$

Solving this equation for the location of the center of mass, we find that

$$\bar{x} = \frac{m_1 x_1 + m_2 x_2 + \cdots + m_n x_n}{m_1 + m_2 + \cdots + m_n} = \frac{\displaystyle\sum_{k=1}^{n} m_k x_k}{\displaystyle\sum_{k=1}^{n} m_k}.$$

Again, the location of the center of mass is the sum of the moments $m_1 x_1, m_2 x_2, \ldots,$ and $m_n x_n$ divided by the sum of the masses.

EXAMPLE 1 **Center of mass for four objects** Find the point at which the system shown in Figure 16.67 balances.

SOLUTION The center of mass is

$$\begin{aligned}
\bar{x} &= \frac{m_1 x_1 + m_2 x_2 + m_3 x_3 + m_4 x_4}{m_1 + m_2 + m_3 + m_4} \\
&= \frac{3(-1.2) + 8(-0.4) + 1(0.5) + 7.5(1.1)}{3 + 8 + 1 + 7.5} \\
&= \frac{1}{10}.
\end{aligned}$$

The balancing point is slightly to the right of the origin.

Related Exercises 7–8 ◀

Continuous Objects in One Dimension

Now consider a thin rod or wire with density ρ that varies along the length of the rod (Figure 16.68). The density in this case has units of mass per length (for example, g/cm). As before, we want to determine the location $\bar{x}$ at which the rod balances on a pivot.

Using the slice-and-sum strategy, we divide the rod, which corresponds to the interval $a \leq x \leq b$, into n subintervals, each with a width of $\Delta x = \dfrac{b - a}{n}$ (Figure 16.69). The corresponding grid points are

$$x_0 = a, \; x_1 = a + \Delta x, \ldots, x_k = a + k\,\Delta x, \ldots, \text{ and } x_n = b.$$

The mass of the kth segment of the rod is approximately the density at x_k multiplied by the length of the interval, or $m_k \approx \rho(x_k)\Delta x$.

> ▶ Density is usually measured in units of *mass per volume*. However, for thin, narrow objects such as rods and wires, linear density with units of *mass per length* is used. For thin, flat objects such as plates and sheets, area density with units of *mass per area* is used.

QUICK CHECK 3 In Figure 16.68, suppose $a = 0$, $b = 3$, and the density of the rod in g/cm is $\rho(x) = 4 - x$. Where is the rod lightest? Heaviest? ◀

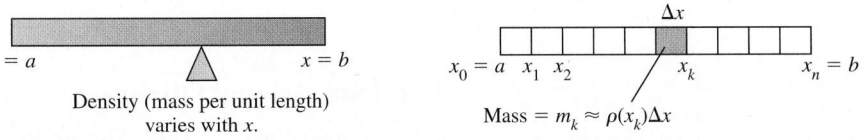

Figure 16.68

Figure 16.69

We now use the center-of-mass formula for several distinct objects to write the approximate center of mass of the rod as

$$\bar{x} = \frac{\displaystyle\sum_{k=1}^{n} m_k x_k}{\displaystyle\sum_{k=1}^{n} m_k} \approx \frac{\displaystyle\sum_{k=1}^{n} \left(\rho(x_k)\Delta x\right) x_k}{\displaystyle\sum_{k=1}^{n} \rho(x_k)\Delta x}.$$

> ▶ An object consisting of two different materials that meet at an interface has a discontinuous density function. Physical density functions either are continuous or have a finite number of discontinuities.

To model a rod with a continuous density, we let $\Delta x \to 0$ and $n \to \infty$; the center of mass of the rod is

> ▶ We assume the rod has positive mass and the limits in the numerator and denominator exist, so the limit of the quotient is the quotient of the limits.

$$\bar{x} = \lim_{\Delta x \to 0} \frac{\displaystyle\sum_{k=1}^{n} \left(\rho(x_k)\Delta x\right) x_k}{\displaystyle\sum_{k=1}^{n} \rho(x_k)\Delta x} = \frac{\displaystyle\lim_{\Delta x \to 0}\sum_{k=1}^{n} x_k\,\rho(x_k)\Delta x}{\displaystyle\lim_{\Delta x \to 0}\sum_{k=1}^{n} \rho(x_k)\Delta x} = \frac{\displaystyle\int_a^b x\rho(x)\,dx}{\displaystyle\int_a^b \rho(x)\,dx}.$$

As discussed in Section 6.7, the denominator of the last fraction, $\int_a^b \rho(x)\,dx$, is the mass of the rod. The numerator is the "sum" of the moments of the individual pieces of the rod, which is called the *total moment*.

> ▶ The units of a moment are mass × length. The center of mass is a moment divided by a mass, which has units of length. Notice that if the density is constant, then ρ effectively does not enter the calculation of $\bar{x}$.

DEFINITION Center of Mass in One Dimension

Let ρ be an integrable density function on the interval $[a, b]$ (which represents a thin rod or wire). The **center of mass** is located at the point $\bar{x} = \dfrac{M}{m}$, where the **total moment** M and mass m are

$$M = \int_a^b x\rho(x)\,dx \quad \text{and} \quad m = \int_a^b \rho(x)\,dx.$$

Observe the parallels between the discrete and continuous cases:

$$n \text{ individual objects: } \quad \bar{x} = \frac{\displaystyle\sum_{k=1}^{n} x_k m_k}{\displaystyle\sum_{k=1}^{n} m_k}; \quad \text{continuous object: } \quad \bar{x} = \frac{\displaystyle\int_a^b x\rho(x)\,dx}{\displaystyle\int_a^b \rho(x)\,dx}.$$

EXAMPLE 2   Center of mass of a one-dimensional object Suppose a thin 2-m bar is made of an alloy whose density in kg/m is $\rho(x) = 1 + x^2$, where $0 \le x \le 2$. Find the center of mass of the bar.

SOLUTION The total mass of the bar in kilograms is

$$m = \int_a^b \rho(x)\, dx = \int_0^2 (1 + x^2)\, dx = \left(x + \frac{x^3}{3} \right) \Big|_0^2 = \frac{14}{3}.$$

The total moment of the bar, with units kg-m, is

$$M = \int_a^b x\rho(x)\, dx = \int_0^2 x(1 + x^2)\, dx = \left(\frac{x^2}{2} + \frac{x^4}{4} \right) \Big|_0^2 = 6.$$

> Notice that the density of the bar increases with x. As a consistency check, our calculation must result in a center of mass to the right of the midpoint of the bar.

Therefore, the center of mass is located at $\bar{x} = \dfrac{M}{m} = \dfrac{9}{7} \approx 1.29$ m.

Related Exercises 10–11 ◄

Two-Dimensional Objects

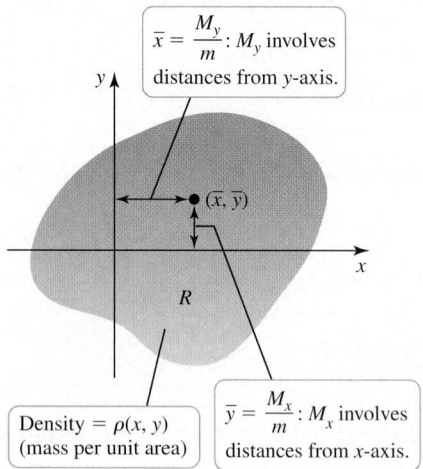

$\bar{x} = \dfrac{M_y}{m}$: M_y involves distances from y-axis.

Density $= \rho(x, y)$ (mass per unit area)

$\bar{y} = \dfrac{M_x}{m}$: M_x involves distances from x-axis.

Figure 16.70

> The moment with respect to the y-axis M_y is a weighted average of distances from the y-axis, so it has x in the integrand (the distance between a point and the y-axis). Similarly, the moment with respect to the x-axis M_x is a weighted average of distances from the x-axis, so it has y in the integrand.

QUICK CHECK 4 Explain why the integral for M_y has x in the integrand. Explain why the density drops out of the center-of-mass calculation if it is constant. ◄

In two dimensions, we start with an integrable density function $\rho(x, y)$ defined over a closed bounded region R in the xy-plane. The density is now an *area density* with units of mass per area (for example, kg/m²). The region represents a thin plate (or *lamina*). The center of mass is the point at which a pivot must be located to balance the plate. If the density is constant, the location of the center of mass depends only on the shape of the plate, in which case the center of mass is called the *centroid*.

For a two- or three-dimensional object, the coordinates for the center of mass are computed independently by applying the one-dimensional argument in each coordinate direction (**Figure 16.70**). The mass of the plate is the integral of the density function over R:

$$m = \iint_R \rho(x, y)\, dA.$$

In analogy with the moment calculation in the one-dimensional case, we now define two moments.

DEFINITION Center of Mass in Two Dimensions

Let ρ be an integrable area density function defined over a closed bounded region R in $\mathbb{R}^2$. The coordinates of the center of mass of the object represented by R are

$$\bar{x} = \frac{M_y}{m} = \frac{1}{m} \iint_R x\rho(x, y)\, dA \quad \text{and} \quad \bar{y} = \frac{M_x}{m} = \frac{1}{m} \iint_R y\rho(x, y)\, dA,$$

where $m = \iint_R \rho(x, y)\, dA$ is the mass, and M_y and M_x are the moments with respect to the y-axis and x-axis, respectively. If ρ is constant, the center of mass is called the **centroid** and is independent of the density.

As before, the center-of-mass coordinates are weighted averages of the distances from the coordinate axes. For two- and three-dimensional objects, the center of mass need not lie within the object (Exercises 51, 61, and 62).

EXAMPLE 3   Centroid calculation Find the centroid (center of mass) of the unit-density, dart-shaped region bounded by the y-axis and the curves $y = e^{-x} - \frac{1}{2}$ and $y = \frac{1}{2} - e^{-x}$ (**Figure 16.71**).

SOLUTION Because the region is symmetric about the x-axis and the density is constant, the y-coordinate of the center of mass is $\bar{y} = 0$. This leaves the integrals for m and M_y to evaluate.

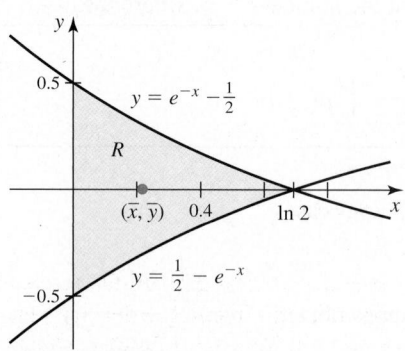

$y = e^{-x} - \frac{1}{2}$

R

$(\bar{x}, \bar{y})$ 0.4 $\ln 2$

$y = \frac{1}{2} - e^{-x}$

Figure 16.71

The first task is to find the point at which the curves intersect. Solving $e^{-x} - \frac{1}{2} = \frac{1}{2} - e^{-x}$, we find that $x = \ln 2$, from which it follows that $y = 0$. Therefore, the intersection point is $(\ln 2, 0)$. The moment M_y (with $\rho = 1$) is given by

$$M_y = \int_0^{\ln 2} \int_{1/2-e^{-x}}^{e^{-x}-1/2} x \, dy \, dx \qquad \text{Definition of } M_y$$

$$= \int_0^{\ln 2} x\left(\left(e^{-x} - \frac{1}{2}\right) - \left(\frac{1}{2} - e^{-x}\right)\right) dx \quad \text{Evaluate inner integral.}$$

$$= \int_0^{\ln 2} x(2e^{-x} - 1) \, dx. \qquad \text{Simplify.}$$

Using integration by parts for this integral, we find that

$$M_y = \int_0^{\ln 2} \underbrace{x}_{u} \underbrace{(2e^{-x} - 1)}_{dv} \, dx$$

$$= -x(2e^{-x} + x)\Big|_0^{\ln 2} + \int_0^{\ln 2} (2e^{-x} + x) \, dx \quad \text{Integration by parts}$$

$$= 1 - \ln 2 - \frac{1}{2}\ln^2 2 \approx 0.067. \qquad \text{Evaluate and simplify.}$$

With $\rho = 1$, the mass of the region is given by

$$m = \int_0^{\ln 2} \int_{1/2-e^{-x}}^{e^{-x}-1/2} dy \, dx \qquad \text{Definition of } m$$

$$= \int_0^{\ln 2} (2e^{-x} - 1) \, dx \qquad \text{Evaluate inner integral.}$$

$$= (-2e^{-x} - x)\Big|_0^{\ln 2} \qquad \text{Evaluate outer integral.}$$

$$= 1 - \ln 2 \approx 0.307. \quad \text{Simplify.}$$

Therefore, the x-coordinate of the center of mass is $\bar{x} = \dfrac{M_y}{m} \approx 0.217$. The center of mass is located approximately at $(0.217, 0)$.

Related Exercise 18 ◀

EXAMPLE 4 Variable-density plate Find the center of mass of the rectangular plate $R = \{(x, y): -1 \leq x \leq 1, 0 \leq y \leq 1\}$ with a density of $\rho(x, y) = 2 - y$ (heavy at the lower edge and light at the top edge; **Figure 16.72**).

SOLUTION Because the plate is symmetric with respect to the y-axis and because the density is independent of x, we have $\bar{x} = 0$. We must still compute m and M_x.

$$m = \iint_R \rho(x, y) \, dA = \int_{-1}^{1} \int_0^1 (2 - y) \, dy \, dx = \frac{3}{2}\int_{-1}^{1} dx = 3$$

$$M_x = \iint_R y\rho(x, y) \, dA = \int_{-1}^{1} \int_0^1 y(2 - y) \, dy \, dx = \frac{2}{3}\int_{-1}^{1} dx = \frac{4}{3}$$

Therefore, the center-of-mass coordinates are

$$\bar{x} = \frac{M_y}{m} = 0 \quad \text{and} \quad \bar{y} = \frac{M_x}{m} = \frac{4/3}{3} = \frac{4}{9}.$$

Related Exercise 21 ◀

> The density does not enter the center-of-mass calculation when the density is constant. So it is easiest to set $\rho = 1$.

> If possible, try to arrange the coordinate system so that at least one of the integrations in the center-of-mass calculation can be avoided by using symmetry. Often the mass (or area) can be found using geometry if the density is constant.

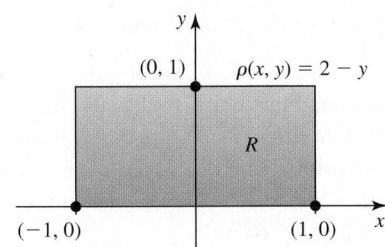

Figure 16.72

> To verify that $\bar{x} = 0$, notice that to find M_y, we integrate an odd function in x over $-1 \leq x \leq 1$; the result is zero.

Three-Dimensional Objects

We now extend the preceding arguments to compute the center of mass of three-dimensional solids. Assume D is a closed bounded region in $\mathbb{R}^3$, on which an integrable density function ρ is defined. The units of the density are mass per volume (for example, g/cm^3). The coordinates of the center of mass depend on the mass of the region, which by

Section 16.4 is the integral of the density function over D. Three moments enter the picture: M_{yz} involves distances from the yz-plane; therefore, it has an x in the integrand. Similarly, M_{xz} involves distances from the xz-plane, so it has a y in the integrand, and M_{xy} involves distances from the xy-plane, so it has a z in the integrand. As before, the coordinates of the center of mass are the total moments divided by the total mass (**Figure 16.73**).

QUICK CHECK 5 Explain why the integral for the moment M_{xy} has z in the integrand. ◄

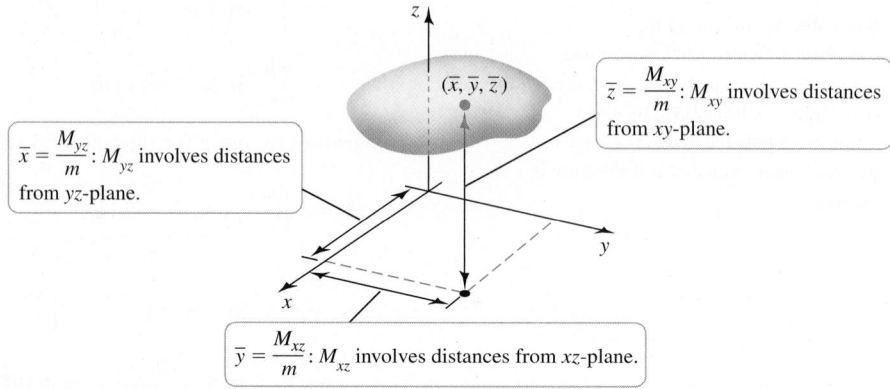

$\bar{x} = \dfrac{M_{yz}}{m}$: M_{yz} involves distances from yz-plane.

$\bar{z} = \dfrac{M_{xy}}{m}$: M_{xy} involves distances from xy-plane.

$\bar{y} = \dfrac{M_{xz}}{m}$: M_{xz} involves distances from xz-plane.

Figure 16.73

DEFINITION Center of Mass in Three Dimensions

Let ρ be an integrable density function on a closed bounded region D in $\mathbb{R}^3$. The coordinates of the center of mass of the region are

$$\bar{x} = \frac{M_{yz}}{m} = \frac{1}{m} \iiint_D x\rho(x, y, z)\, dV, \quad \bar{y} = \frac{M_{xz}}{m} = \frac{1}{m} \iiint_D y\rho(x, y, z)\, dV, \text{ and}$$

$$\bar{z} = \frac{M_{xy}}{m} = \frac{1}{m} \iiint_D z\rho(x, y, z)\, dV,$$

where $m = \iiint_D \rho(x, y, z)\, dV$ is the mass, and M_{yz}, M_{xz}, and M_{xy} are the moments with respect to the coordinate planes.

EXAMPLE 5 Center of mass with constant density Find the center of mass of the constant-density solid cone D bounded by the surface $z = 4 - \sqrt{x^2 + y^2}$ and $z = 0$ (**Figure 16.74**).

SOLUTION Because the cone is symmetric about the z-axis and has uniform density, the center of mass lies on the z-axis; that is, $\bar{x} = 0$ and $\bar{y} = 0$. Setting $z = 0$, the base of the cone in the xy-plane is the disk of radius 4 centered at the origin. Therefore, the cone has height 4 and radius 4; by the volume formula, its volume is $\pi r^2 h/3 = 64\pi/3$. The cone has a constant density, so we assume $\rho = 1$ and its mass is $m = 64\pi/3$.

To obtain the value of $\bar{z}$, only M_{xy} needs to be calculated, which is most easily done in cylindrical coordinates. The cone is described by the equation $z = 4 - \sqrt{x^2 + y^2} = 4 - r$. The projection of the cone onto the xy-plane, which is the region of integration in the xy-plane, is the disk $R = \{(r, \theta): 0 \le r \le 4, 0 \le \theta \le 2\pi\}$. The integration for M_{xy} now follows:

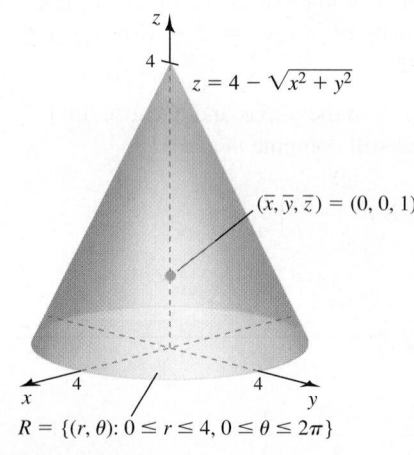

$z = 4 - \sqrt{x^2 + y^2}$

$(\bar{x}, \bar{y}, \bar{z}) = (0, 0, 1)$

$R = \{(r, \theta): 0 \le r \le 4, 0 \le \theta \le 2\pi\}$

Figure 16.74

$$M_{xy} = \iiint_D z\, dV \qquad \text{Definition of } M_{xy} \text{ with } \rho = 1$$

$$= \int_0^{2\pi} \int_0^4 \int_0^{4-r} z\, dz\, r\, dr\, d\theta \qquad \text{Convert to an iterated integral.}$$

$$= \int_0^{2\pi} \int_0^4 \frac{z^2}{2} \Big|_0^{4-r} r\, dr\, d\theta \qquad \text{Evaluate inner integral with respect to } z.$$

$$= \frac{1}{2} \int_0^{2\pi} \int_0^4 r(4 - r)^2\, dr\, d\theta \qquad \text{Simplify.}$$

$$= \frac{1}{2}\int_0^{2\pi}\frac{64}{3}\,d\theta \qquad \text{Evaluate middle integral with respect to } r.$$

$$= \frac{64\pi}{3}. \qquad \text{Evaluate outer integral with respect to } \theta.$$

The z-coordinate of the center of mass is $\bar{z} = \dfrac{M_{xy}}{m} = \dfrac{64\pi/3}{64\pi/3} = 1$, and the center of mass is located at $(0, 0, 1)$. It can be shown (Exercise 55) that the center of mass of a constant-density cone of height h is located $h/4$ units from the base on the axis of the cone, independent of the radius.

Related Exercise 28 ◄

EXAMPLE 6   Center of mass with variable density Find the center of mass of the interior of the hemisphere D of radius a with its base on the xy-plane. The density of the object given in spherical coordinates is $f(\rho, \varphi, \theta) = 2 - \rho/a$ (heavy near the center and light near the outer surface; Figure 16.75).

SOLUTION The center of mass lies on the z-axis because of the symmetry of both the solid and the density function; therefore, $\bar{x} = \bar{y} = 0$. Only the integrals for m and M_{xy} need to be evaluated, and they should be done in spherical coordinates.

The integral for the mass is

$$m = \iiint_D f(\rho, \varphi, \theta)\,dV \qquad \text{Definition of } m$$

$$= \int_0^{2\pi}\int_0^{\pi/2}\int_0^a \left(2 - \frac{\rho}{a}\right)\rho^2 \sin\varphi\,d\rho\,d\varphi\,d\theta \qquad \text{Convert to an iterated integral.}$$

$$= \int_0^{2\pi}\int_0^{\pi/2}\left(\frac{2\rho^3}{3} - \frac{\rho^4}{4a}\right)\bigg|_0^a \sin\varphi\,d\varphi\,d\theta \qquad \text{Evaluate inner integral with respect to } \rho.$$

$$= \int_0^{2\pi}\int_0^{\pi/2}\frac{5a^3}{12}\sin\varphi\,d\varphi\,d\theta \qquad \text{Simplify.}$$

$$= \frac{5a^3}{12}\int_0^{2\pi}\underbrace{(-\cos\varphi)\bigg|_0^{\pi/2}}_{1}\,d\theta \qquad \text{Evaluate middle integral with respect to } \varphi.$$

$$= \frac{5a^3}{12}\int_0^{2\pi}d\theta \qquad \text{Simplify.}$$

$$= \frac{5\pi a^3}{6}. \qquad \text{Evaluate outer integral with respect to } \theta.$$

In spherical coordinates, $z = \rho\cos\varphi$, so the integral for the moment M_{xy} is

$$M_{xy} = \iiint_D z\,f(\rho, \varphi, \theta)\,dV \qquad \text{Definition of } M_{xy}$$

$$= \int_0^{2\pi}\int_0^{\pi/2}\int_0^a \underbrace{\rho\cos\varphi}_{z}\left(2 - \frac{\rho}{a}\right)\rho^2 \sin\varphi\,d\rho\,d\varphi\,d\theta \qquad \text{Convert to an iterated integral.}$$

$$= \int_0^{2\pi}\int_0^{\pi/2}\left(\frac{\rho^4}{2} - \frac{\rho^5}{5a}\right)\bigg|_0^a \sin\varphi\cos\varphi\,d\varphi\,d\theta \qquad \text{Evaluate inner integral with respect to } \rho.$$

$$= \int_0^{2\pi}\int_0^{\pi/2}\frac{3a^4}{10}\underbrace{\sin\varphi\cos\varphi}_{(\sin 2\varphi)/2}\,d\varphi\,d\theta \qquad \text{Simplify.}$$

$$= \frac{3a^4}{10}\int_0^{2\pi}\underbrace{\left(-\frac{\cos 2\varphi}{4}\right)\bigg|_0^{\pi/2}}_{1/2}\,d\theta \qquad \text{Evaluate middle integral with respect to } \varphi.$$

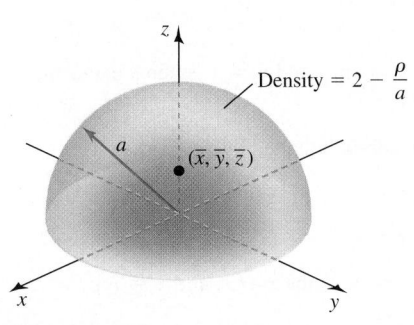

Density $= 2 - \dfrac{\rho}{a}$

$(\bar{x}, \bar{y}, \bar{z})$

Figure 16.75

$$= \frac{3a^4}{20} \int_0^{2\pi} d\theta \qquad \text{Simplify.}$$

$$= \frac{3\pi a^4}{10}. \qquad \text{Evaluate outer integral with respect to } \theta.$$

The z-coordinate of the center of mass is $\bar{z} = \dfrac{M_{xy}}{m} = \dfrac{3\pi a^4/10}{5\pi a^3/6} = \dfrac{9a}{25} = 0.36a$. It can be shown (Exercise 56) that the center of mass of a uniform-density hemispherical solid of radius a is $3a/8 = 0.375a$ units above the base. In this case, the variable density lowers the center of mass toward the base.

Related Exercise 35 ◄

SECTION 16.6 EXERCISES

Getting Started

1. Explain how to find the balance point for two people on opposite ends of a (massless) plank that rests on a pivot.

2. If a thin 1-m cylindrical rod has a density of $\rho = 1$ g/cm for its left half and a density of $\rho = 2$ g/cm for its right half, what is its mass and where is its center of mass?

3. Explain how to find the center of mass of a thin plate with a variable density.

4. In the integral for the moment M_x of a thin plate, why does y appear in the integrand?

5. Explain how to find the center of mass of a three-dimensional object with a variable density.

6. In the integral for the moment M_{xz} with respect to the xz-plane of a solid, why does y appear in the integrand?

Practice Exercises

7–8. Individual masses on a line *Sketch the following systems on a number line and find the location of the center of mass.*

7. $m_1 = 10$ kg located at $x = 3$ m; $m_2 = 3$ kg located at $x = -1$ m

8. $m_1 = 8$ kg located at $x = 2$ m; $m_2 = 4$ kg located at $x = -4$ m; $m_3 = 1$ kg located at $x = 0$ m

9–14. One-dimensional objects *Find the mass and center of mass of the thin rods with the following density functions.*

9. $\rho(x) = 1 + \sin x$, for $0 \le x \le \pi$

10. $\rho(x) = 1 + x^3$, for $0 \le x \le 1$

11. $\rho(x) = 2 - \dfrac{x^2}{16}$, for $0 \le x \le 4$

12. $\rho(x) = 2 + \cos x$, for $0 \le x \le \pi$

13. $\rho(x) = \begin{cases} 1 & \text{if } 0 \le x \le 2 \\ 1 + x & \text{if } 2 < x \le 4 \end{cases}$

14. $\rho(x) = \begin{cases} x^2 & \text{if } 0 \le x \le 1 \\ x(2 - x) & \text{if } 1 < x \le 2 \end{cases}$

15–20. Centroid calculations *Find the mass and centroid (center of mass) of the following thin plates, assuming constant density. Sketch the region corresponding to the plate and indicate the location of the center of mass. Use symmetry when possible to simplify your work.*

15. The region bounded by $y = \sin x$ and $y = 1 - \sin x$ between $x = \pi/4$ and $x = 3\pi/4$

16. The region in the first quadrant bounded by $x^2 + y^2 = 16$

17. The region bounded by $y = 1 - |x|$ and the x-axis

18. The region bounded by $y = e^x$, $y = e^{-x}$, $x = 0$, and $x = \ln 2$

19. The region bounded by $y = \ln x$, the x-axis, and $x = e$

20. The region bounded by $x^2 + y^2 = 1$ and $x^2 + y^2 = 9$, for $y \ge 0$

21–26. Variable-density plates *Find the center of mass of the following plane regions with variable density. Describe the distribution of mass in the region.*

21. $R = \{(x, y): 0 \le x \le 4, 0 \le y \le 2\}$; $\rho(x, y) = 1 + x/2$

22. $R = \{(x, y): 0 \le x \le 1, 0 \le y \le 5\}$; $\rho(x, y) = 2e^{-y/2}$

23. The triangular plate in the first quadrant bounded by $x + y = 4$ with $\rho(x, y) = 1 + x + y$

24. The upper half ($y \ge 0$) of the disk bounded by the circle $x^2 + y^2 = 4$ with $\rho(x, y) = 1 + y/2$

25. The upper half ($y \ge 0$) of the plate bounded by the ellipse $x^2 + 9y^2 = 9$ with $\rho(x, y) = 1 + y$

26. The quarter disk in the first quadrant bounded by $x^2 + y^2 = 4$ with $\rho(x, y) = 1 + x^2 + y^2$

27–32. Center of mass of constant-density solids *Find the center of mass of the following solids, assuming a constant density of 1. Sketch the region and indicate the location of the centroid. Use symmetry when possible and choose a convenient coordinate system.*

27. The upper half of the ball $x^2 + y^2 + z^2 \le 16$ (for $z \ge 0$)

28. The solid bounded by the paraboloid $z = x^2 + y^2$ and the plane $z = 25$

29. The tetrahedron in the first octant bounded by $z = 1 - x - y$ and the coordinate planes

30. The solid bounded by the cone $z = 16 - r$ and the plane $z = 0$

31. The sliced solid cylinder bounded by $x^2 + y^2 = 1$, $z = 0$, and $y + z = 1$

32. The solid bounded by the upper half ($z \ge 0$) of the ellipsoid $4x^2 + 4y^2 + z^2 = 16$

33–38. Variable-density solids *Find the coordinates of the center of mass of the following solids with variable density.*

33. $R = \{(x, y, z): 0 \le x \le 4, 0 \le y \le 1, 0 \le z \le 1\}$;
$\rho(x, y, z) = 1 + x/2$

34. The region bounded by the paraboloid $z = 4 - x^2 - y^2$ and $z = 0$ with $\rho(x, y, z) = 5 - z$

35. The solid bounded by the upper half of the sphere $\rho = 6$ and $z = 0$ with density $f(\rho, \varphi, \theta) = 1 + \rho/4$

36. The interior of the cube in the first octant formed by the planes $x = 1$, $y = 1$, and $z = 1$, with $\rho(x, y, z) = 2 + x + y + z$

37. The interior of the prism formed by the planes $z = x$, $x = 1$, and $y = 4$, and the coordinate planes, with $\rho(x, y, z) = 2 + y$

38. The solid bounded by the cone $z = 9 - r$ and $z = 0$ with $\rho(r, \theta, z) = 1 + z$

39. Explain why or why not Determine whether the following statements are true and give an explanation or counterexample.

   a. A thin plate of constant density that is symmetric about the x-axis has a center of mass with an x-coordinate of zero.
   b. A thin plate of constant density that is symmetric about both the x-axis and the y-axis has its center of mass at the origin.
   c. The center of mass of a thin plate must lie on the plate.
   d. The center of mass of a connected solid region (all in one piece) must lie within the region.

40. Limiting center of mass A thin rod of length L has a linear density given by $\rho(x) = 2e^{-x/3}$ on the interval $0 \le x \le L$. Find the mass and center of mass of the rod. How does the center of mass change as $L \to \infty$?

41. Limiting center of mass A thin rod of length L has a linear density given by $\rho(x) = \dfrac{10}{1 + x^2}$ on the interval $0 \le x \le L$. Find the mass and center of mass of the rod. How does the center of mass change as $L \to \infty$?

42. Limiting center of mass A thin plate is bounded by the graphs of $y = e^{-x}$, $y = -e^{-x}$, $x = 0$, and $x = L$. Find its center of mass. How does the center of mass change as $L \to \infty$?

43–44. Two-dimensional plates *Find the mass and center of mass of the thin constant-density plates shown in the figure.*

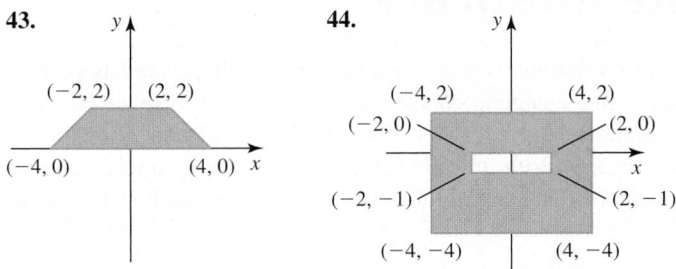

43. (−2, 2) (2, 2) (−4, 0) (4, 0)

44. (−4, 2) (4, 2) (−2, 0) (2, 0) (−2, −1) (2, −1) (−4, −4) (4, −4)

45–50. Centroids *Use polar coordinates to find the centroid of the following constant-density plane regions.*

45. The semicircular disk $R = \{(r, \theta): 0 \le r \le 2, 0 \le \theta \le \pi\}$

46. The quarter-circular disk $R = \{(r, \theta): 0 \le r \le 2, 0 \le \theta \le \pi/2\}$

47. The region bounded by the cardioid $r = 1 + \cos \theta$

48. The region bounded by the cardioid $r = 3 - 3 \cos \theta$

49. The region bounded by one leaf of the rose $r = \sin 2\theta$, for $0 \le \theta \le \pi/2$

50. The region bounded by the limaçon $r = 2 + \cos \theta$

51. Semicircular wire A thin (one-dimensional) wire of constant density is bent into the shape of a semicircle of radius r. Find the location of its center of mass. (*Hint:* Treat the wire as a thin half-annulus with width Δa, and then let $\Delta a \to 0$.)

52. Parabolic region A thin plate of unit density occupies the region between the parabola $y = ax^2$ and the horizontal line $y = b$, where $a > 0$ and $b > 0$. Show that the center of mass is $\left(0, \dfrac{3b}{5}\right)$, independent of a.

53. Circular crescent Find the center of mass of the region in the first quadrant bounded by the circle $x^2 + y^2 = a^2$ and the lines $x = a$ and $y = a$, where $a > 0$.

54–57. Centers of mass for general objects *Consider the following two- and three-dimensional regions with variable dimensions. Specify the surfaces and curves that bound the region, choose a convenient coordinate system, and compute the center of mass assuming constant density. All parameters are positive real numbers.*

54. A solid rectangular box has sides of lengths a, b, and c. Where is the center of mass relative to the faces of the box?

55. A solid cone has a base with a radius of a and a height of h. How far from the base is the center of mass?

56. A solid is enclosed by a hemisphere of radius a. How far from the base is the center of mass?

57. A region is enclosed by an isosceles triangle with two sides of length s and a base of length b. How far from the base is the center of mass?

Explorations and Challenges

58. A tetrahedron is bounded by the coordinate planes and the plane $x/a + y/a + z/a = 1$. What are the coordinates of the center of mass?

59. A solid is enclosed by the upper half of an ellipsoid with a circular base of radius r and a height of a. How far from the base is the center of mass?

60. Geographic vs. population center Geographers measure the *geographical center* of a country (which is the centroid) and the *population center* of a country (which is the center of mass computed with the population density). A hypothetical country is shown in the figure with the location and population of five towns. Assuming no one lives outside the towns, find the geographical center of the country and the population center of the country.

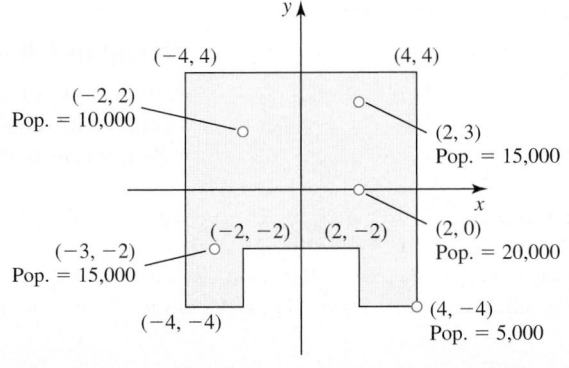

(−4, 4) (4, 4) (−2, 2) Pop. = 10,000 (2, 3) Pop. = 15,000 (−2, −2) (2, −2) (2, 0) Pop. = 20,000 (−3, −2) Pop. = 15,000 (−4, −4) (4, −4) Pop. = 5,000

61. Center of mass on the edge Consider the thin constant-density plate $\{(r, \theta): 0 < a \le r \le 1, 0 \le \theta \le \pi\}$ bounded by two semicircles and the x-axis.

    a. Find and graph the y-coordinate of the center of mass of the plate as a function of a.

    b. For what value of a is the center of mass on the edge of the plate?

62. Center of mass on the edge Consider the constant-density solid $\{(\rho, \varphi, \theta): 0 < a \le \rho \le 1, 0 \le \varphi \le \pi/2, 0 \le \theta \le 2\pi\}$ bounded by two hemispheres and the xy-plane.

    a. Find and graph the z-coordinate of the center of mass of the plate as a function of a.

    b. For what value of a is the center of mass on the edge of the solid?

63. Draining a soda can A cylindrical soda can has a radius of 4 cm and a height of 12 cm. When the can is full of soda, the center of mass of the contents of the can is 6 cm above the base on the axis of the can (halfway along the axis of the can). As the can is drained, the center of mass descends for a while. However, when the can is empty (filled only with air), the center of mass is once again 6 cm above the base on the axis of the can. Find the depth of soda in the can for which the center of mass is at its lowest point. Neglect the mass of the can, and assume the density of the soda is 1 g/cm^3 and the density of air is 0.001 g/cm^3.

64. Triangle medians A triangular region has a base that connects the vertices $(0, 0)$ and $(b, 0)$, and a third vertex at (a, h), where $a > 0, b > 0$, and $h > 0$.

    a. Show that the centroid of the triangle is $\left(\dfrac{a + b}{3}, \dfrac{h}{3}\right)$.

    b. Recall that the three medians of a triangle extend from each vertex to the midpoint of the opposite side. Knowing that the medians of a triangle intersect in a point M and that each median bisects the triangle, conclude that the centroid of the triangle is M.

65. The golden earring A disk of radius r is removed from a larger disk of radius R to form an earring (see figure). Assume the earring is a thin plate of uniform density.

    a. Find the center of mass of the earring in terms of r and R. (*Hint:* Place the origin of a coordinate system either at the center of the large disk or at Q; either way, the earring is symmetric about the x-axis.)

    b. Show that the ratio $\dfrac{R}{r}$ such that the center of mass lies at the point P (on the edge of the inner disk) is the golden mean $\dfrac{1 + \sqrt{5}}{2} \approx 1.618$.

    (*Source:* P. Glaister, *Golden Earrings*, *Mathematical Gazette*, 80, 1996)

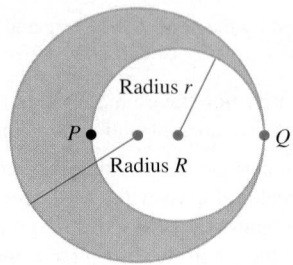

QUICK CHECK ANSWERS

1. 3 m **3.** It is heaviest at $x = 0$ and lightest at $x = 3$.
4. The distance from the point (x, y) to the y-axis is x. The constant density appears in the integral for the moment, and it appears in the integral for the mass. Therefore, the density cancels when we divide the two integrals. **5.** The distance from the xy-plane to a point (x, y, z) is z. ◄

16.7 Change of Variables in Multiple Integrals

Converting double integrals from rectangular coordinates to polar coordinates (Section 16.3) and converting triple integrals from rectangular coordinates to cylindrical or spherical coordinates (Section 16.5) are examples of a general procedure known as a *change of variables*. The idea is not new: The Substitution Rule introduced in Chapter 5 with single-variable integrals is also a change of variables. The aim of this section is to show you how to change variables in double and triple integrals.

Recap of Change of Variables

Recall how a change of variables is used to simplify a single-variable integral. For example, to simplify the integral $\int_0^1 2\sqrt{2x + 1}\, dx$, we choose a new variable $u = 2x + 1$, which means that $du = 2\, dx$. Therefore,

$$\int_0^1 2\sqrt{2x + 1}\, dx = \int_1^3 \sqrt{u}\, du.$$

This equality means that the area under the curve $y = 2\sqrt{2x + 1}$ from $x = 0$ to $x = 1$ equals the area under the curve $y = \sqrt{u}$ from $u = 1$ to $u = 3$ (Figure 16.76). The relation $du = 2\,dx$ relates the length of a small interval on the u-axis to the length of the corresponding interval on the x-axis.

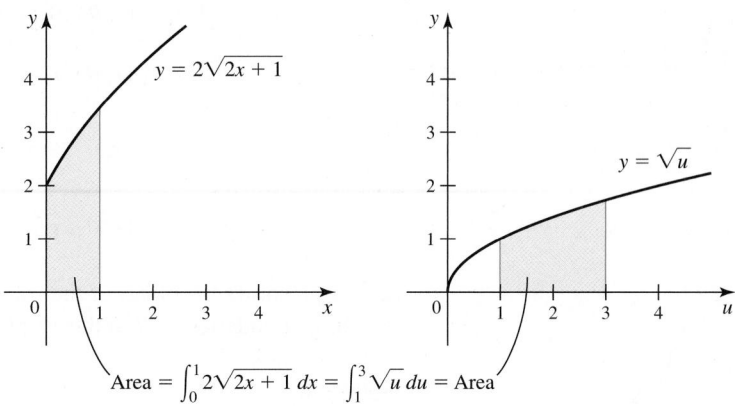

$$\text{Area} = \int_0^1 2\sqrt{2x + 1}\,dx = \int_1^3 \sqrt{u}\,du = \text{Area}$$

Figure 16.76

Similarly, some double and triple integrals can be simplified through a change of variables. For example, the region of integration for

$$\int_0^1 \int_0^{\sqrt{1-x^2}} e^{1-x^2-y^2}\,dy\,dx$$

is the quarter disk $R = \{(x, y): x \ge 0, y \ge 0, x^2 + y^2 \le 1\}$. Changing variables to polar coordinates with $x = r \cos \theta$, $y = r \sin \theta$, and $dy\,dx = r\,dr\,d\theta$, we have

$$\int_0^1 \int_0^{\sqrt{1-x^2}} e^{1-x^2-y^2}\,dy\,dx \;\overset{\substack{x = r\cos\theta \\ y = r\sin\theta}}{=}\; \int_0^{\pi/2} \int_0^1 e^{1-r^2} r\,dr\,d\theta.$$

In this case, the original region of integration R is transformed into a new region $S = \{(r, \theta): 0 \le r \le 1, 0 \le \theta \le \pi/2\}$, which is a rectangle in the $r\theta$-plane.

Transformations in the Plane

A change of variables in a double integral is a *transformation* that relates two sets of variables, (u, v) and (x, y). It is written compactly as $(x, y) = T(u, v)$. Because it relates pairs of variables, T has two components,

$$T: x = g(u, v) \quad \text{and} \quad y = h(u, v).$$

Geometrically, T takes a region S in the uv-plane and "maps" it point by point to a region R in the xy-plane (Figure 16.77). We write the outcome of this process as $R = T(S)$ and call R the **image** of S under T.

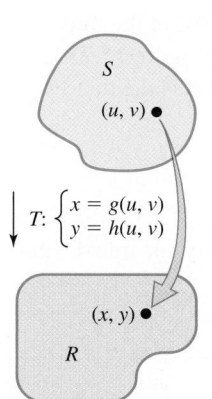

Figure 16.77

➤ In Example 1, we have replaced the coordinates u and v with the familiar polar coordinates r and θ.

EXAMPLE 1   Image of a transformation Consider the transformation from polar to rectangular coordinates given by

$$T: \qquad x = g(r, \theta) = r \cos \theta \quad \text{and} \quad y = h(r, \theta) = r \sin \theta.$$

Find the image under this transformation of the rectangle

$$S = \{(r, \theta): 0 \le r \le 1, 0 \le \theta \le \pi/2\}.$$

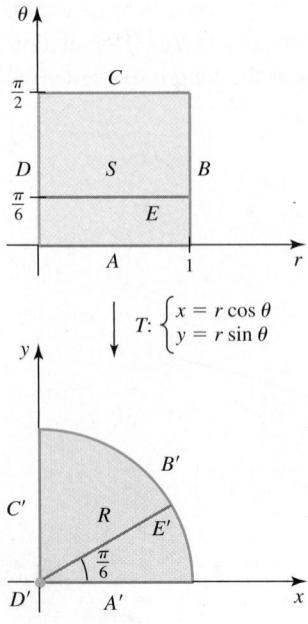

Figure 16.78

SOLUTION If we apply T to every point of S (Figure 16.78), what is the resulting set R in the xy-plane? One way to answer this question is to walk around the boundary of S, let's say counterclockwise, and determine the corresponding path in the xy-plane. In the $r\theta$-plane, we let the horizontal axis be the r-axis and the vertical axis be the θ-axis. Starting at the origin, we denote the edges of the rectangle S as follows.

$$A = \{(r, \theta): 0 \le r \le 1, \theta = 0\} \quad \text{Lower boundary}$$

$$B = \left\{(r, \theta): r = 1, 0 \le \theta \le \frac{\pi}{2}\right\} \quad \text{Right boundary}$$

$$C = \left\{(r, \theta): 0 \le r \le 1, \theta = \frac{\pi}{2}\right\} \quad \text{Upper boundary}$$

$$D = \left\{(r, \theta): r = 0, 0 \le \theta \le \frac{\pi}{2}\right\} \quad \text{Left boundary}$$

Table 16.6 shows the effect of the transformation on the four boundaries of S; the corresponding boundaries of R in the xy-plane are denoted A', B', C', and D' (Figure 16.78).

Table 16.6

Boundary of S in $r\theta$-plane	Transformation equations	Boundary of R in xy-plane
A: $0 \le r \le 1, \theta = 0$	$x = r \cos \theta = r$, $y = r \sin \theta = 0$	A': $0 \le x \le 1, y = 0$
B: $r = 1, 0 \le \theta \le \pi/2$	$x = r \cos \theta = \cos \theta$, $y = r \sin \theta = \sin \theta$	B': quarter unit circle
C: $0 \le r \le 1, \theta = \pi/2$	$x = r \cos \theta = 0$, $y = r \sin \theta = r$	C': $x = 0, 0 \le y \le 1$
D: $r = 0, 0 \le \theta \le \pi/2$	$x = r \cos \theta = 0$, $y = r \sin \theta = 0$	D': single point $(0, 0)$

The image of the rectangular boundary of S is the boundary of R. Furthermore, it can be shown that every point in the interior of R is the image of one point in the interior of S. (For example, the horizontal line segment E in the $r\theta$-plane in Figure 16.78 is mapped to the line segment E' in the xy-plane.) Therefore, the image of S is the quarter disk R in the xy-plane.

Related Exercises 11–12 ◀

Recall that a function f is *one-to-one* on an interval I if $f(x_1) = f(x_2)$ only when $x_1 = x_2$, where x_1 and x_2 are points of I. We need an analogous property for transformations when changing variables.

QUICK CHECK 1 How would the image of S change in Example 1 if $S = \{(r, \theta): 0 \le r \le 1, 0 \le \theta \le \pi\}$? ◀

DEFINITION One-to-One Transformation

A transformation T from a region S to a region R is one-to-one on S if $T(P) = T(Q)$ only when $P = Q$, where P and Q are points in S.

Notice that the polar coordinate transformation in Example 1 is not one-to-one on the rectangle $S = \{(r, \theta): 0 \le r \le 1, 0 \le \theta \le \pi/2\}$ (because all points with $r = 0$ map to the point $(0, 0)$). However, this transformation *is* one-to-one on the interior of S.

We can now anticipate how a transformation (change of variables) is used to simplify a double integral. Suppose we have the integral $\iint_R f(x, y)\, dA$. The goal is to find a transformation to a new set of coordinates (u, v) such that the new equivalent integral $\iint_S f(x(u, v), y(u, v))\, dA$ involves a simple region S (such as a rectangle), a simple integrand, or both. The next theorem allows us to do exactly that, but it first requires a new concept.

▶ The Jacobian is named after the German mathematician Carl Gustav Jacob Jacobi (1804–1851). In some books, the Jacobian is the matrix of partial derivatives. In others, as here, the Jacobian is the determinant of the matrix of partial derivatives. Both $J(u, v)$ and $\dfrac{\partial(x, y)}{\partial(u, v)}$ are used to refer to the Jacobian.

DEFINITION Jacobian Determinant of a Transformation of Two Variables

Given a transformation $T: x = g(u, v)$, $y = h(u, v)$, where g and h are differentiable on a region of the uv-plane, the **Jacobian determinant** (or **Jacobian**) of T is

$$J(u, v) = \frac{\partial(x, y)}{\partial(u, v)} = \begin{vmatrix} \dfrac{\partial x}{\partial u} & \dfrac{\partial x}{\partial v} \\ \dfrac{\partial y}{\partial u} & \dfrac{\partial y}{\partial v} \end{vmatrix} = \frac{\partial x}{\partial u}\frac{\partial y}{\partial v} - \frac{\partial x}{\partial v}\frac{\partial y}{\partial u}.$$

QUICK CHECK 2 Find $J(u, v)$ if $x = u + v$, $y = 2v$. ◄

The Jacobian is easiest to remember as the determinant of a 2×2 matrix of partial derivatives. With the Jacobian in hand, we can state the change-of-variables rule for double integrals.

▶ The condition that g and h have continuous first partial derivatives ensures that the new integrand is integrable.

THEOREM 16.8   Change of Variables for Double Integrals

Let $T: x = g(u, v)$, $y = h(u, v)$ be a transformation that maps a closed bounded region S in the uv-plane to a region R in the xy-plane. Assume T is one-to-one on the interior of S and g and h have continuous first partial derivatives there. If f is continuous on R, then

$$\iint\limits_{R} f(x, y)\, dA = \iint\limits_{S} f(g(u, v), h(u, v))\, |J(u, v)|\, dA.$$

▶ In the integral over R, dA corresponds to $dx\, dy$. In the integral over S, dA corresponds to $du\, dv$. The relation $dx\, dy = |J|\, du\, dv$ is the analog of $du = g'(x)\, dx$ in a change of variables with one variable.

The proof of this result is technical and is found in advanced texts. The factor $|J(u, v)|$ that appears in the second integral is the absolute value of the Jacobian. Matching the area elements in the two integrals of Theorem 16.8, we see that $dx\, dy = |J(u, v)|\, du\, dv$. This expression shows that the Jacobian is a magnification (or reduction) factor: It relates the area of a small region $dx\, dy$ in the xy-plane to the area of the corresponding region $du\, dv$ in the uv-plane. If the transformation equations are linear, then this relationship is exact in the sense that area$(T(S)) = |J(u, v)| \cdot$ area of S (see Exercise 60). The way in which the Jacobian arises is explored in Exercise 61.

EXAMPLE 2   Jacobian of the polar-to-rectangular transformation Compute the Jacobian of the transformation

$$T: \qquad x = g(r, \theta) = r \cos \theta \quad \text{and} \quad y = h(r, \theta) = r \sin \theta.$$

SOLUTION The necessary partial derivatives are

$$\frac{\partial x}{\partial r} = \cos \theta, \qquad \frac{\partial x}{\partial \theta} = -r \sin \theta, \qquad \frac{\partial y}{\partial r} = \sin \theta, \quad \text{and} \quad \frac{\partial y}{\partial \theta} = r \cos \theta.$$

Therefore,

$$J(r, \theta) = \frac{\partial(x, y)}{\partial(r, \theta)} = \begin{vmatrix} \dfrac{\partial x}{\partial r} & \dfrac{\partial x}{\partial \theta} \\ \dfrac{\partial y}{\partial r} & \dfrac{\partial y}{\partial \theta} \end{vmatrix} = \begin{vmatrix} \cos \theta & -r \sin \theta \\ \sin \theta & r \cos \theta \end{vmatrix} = r(\cos^2 \theta + \sin^2 \theta) = r.$$

This determinant calculation confirms the change-of-variables formula for polar coordinates: $dx\, dy$ becomes $r\, dr\, d\theta$.

Related Exercise 20 ◄

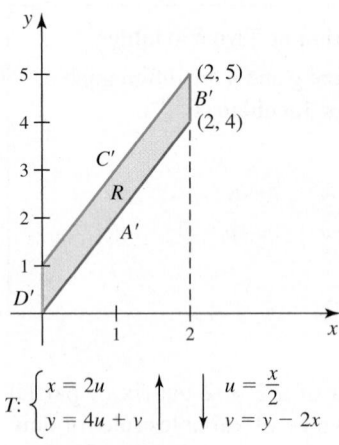

$$T: \begin{cases} x = 2u \\ y = 4u + v \end{cases} \qquad \begin{cases} u = \dfrac{x}{2} \\ v = y - 2x \end{cases}$$

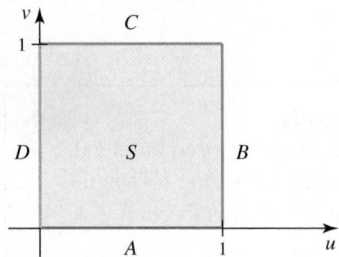

Figure 16.79

➤ The relations that "go the other direction" make up the inverse transformation, usually denoted T^{-1}.

Table 16.7

(x, y)	(u, v)
$(0, 0)$	$(0, 0)$
$(0, 1)$	$(0, 1)$
$(2, 5)$	$(1, 1)$
$(2, 4)$	$(1, 0)$

➤ T is an example of a *shearing transformation*. The greater the u-coordinate of a point, the more that point is displaced in the v-direction. It also involves a uniform stretch in the u-direction.

We are now ready for a change of variables. To transform the integral $\iint_R f(x, y)\, dA$ into $\iint_S f(x(u, v), y(u, v))|J(u, v)|\, dA$, we must find the transformation $x = g(u, v)$ and $y = h(u, v)$, and then use it to find the new region of integration S. The next example illustrates how the region S is found, assuming the transformation is given.

EXAMPLE 3   Double integral with a change of variables given Evaluate the integral $\iint_R \sqrt{2x(y - 2x)}\, dA$, where R is the parallelogram in the xy-plane with vertices $(0, 0)$, $(0, 1)$, $(2, 4)$, and $(2, 5)$ (Figure 16.79). Use the transformation

$$T: x = 2u \quad \text{and} \quad y = 4u + v.$$

SOLUTION To what region S in the uv-plane is R mapped? Because T takes points in the uv-plane and assigns them to points in the xy-plane, we must reverse the process by solving $x = 2u$, $y = 4u + v$ for u and v.

$$\text{First equation: } x = 2u \;\Rightarrow\; u = \frac{x}{2}$$

$$\text{Second equation: } y = 4u + v \;\Rightarrow\; v = y - 4u = y - 2x$$

Rather than walk around the boundary of R in the xy-plane to determine the resulting region S in the uv-plane, it suffices to find the images of the vertices of R. You should confirm that the vertices map as shown in Table 16.7.

Connecting the points in the uv-plane in order, we see that S is the unit square $\{(u, v): 0 \le u \le 1, 0 \le v \le 1\}$ (Figure 16.79). These inequalities determine the limits of integration in the uv-plane.

Replacing $2x$ with $4u$ and $y - 2x$ with v, the original integrand becomes $\sqrt{2x(y - 2x)} = \sqrt{4uv}$. The Jacobian is

$$J(u, v) = \begin{vmatrix} \dfrac{\partial x}{\partial u} & \dfrac{\partial x}{\partial v} \\[2mm] \dfrac{\partial y}{\partial u} & \dfrac{\partial y}{\partial v} \end{vmatrix} = \begin{vmatrix} 2 & 0 \\ 4 & 1 \end{vmatrix} = 2.$$

The integration now follows:

$$\iint_R \sqrt{2x(y - 2x)}\, dA = \iint_S \sqrt{4uv}\; \underbrace{|J(u, v)|}_{2}\, dA \qquad \text{Change variables.}$$

$$= \int_0^1 \int_0^1 \sqrt{4uv}\; 2\, du\, dv \qquad \text{Convert to an iterated integral.}$$

$$= 4 \int_0^1 \frac{2}{3} \sqrt{v}\; (u^{3/2}) \Big|_0^1 dv \qquad \text{Evaluate inner integral.}$$

$$= \frac{8}{3} \cdot \frac{2}{3} (v^{3/2}) \Big|_0^1 = \frac{16}{9}. \qquad \text{Evaluate outer integral.}$$

The effect of the change of variables is illustrated in **Figure 16.80**, where we see the surface $z = \sqrt{2x(y - 2x)}$ over the region R and the surface $w = 2\sqrt{4uv}$ over the region S. The volumes of the solids beneath the two surfaces are equal, but the integral over S is easier to evaluate.

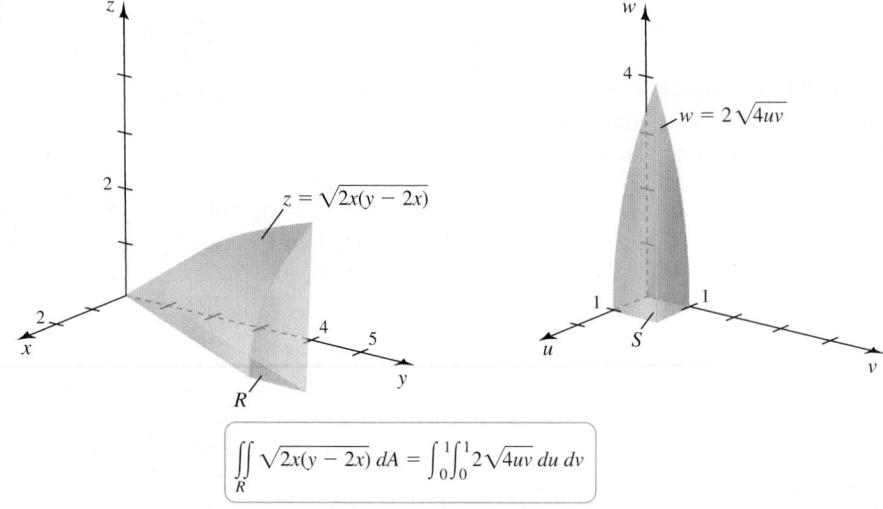

$$\iint_R \sqrt{2x(y-2x)}\,dA = \int_0^1\int_0^1 2\sqrt{4uv}\,du\,dv$$

Figure 16.80

Related Exercise 29 ◄

QUICK CHECK 3 Solve the equations $u = x + y$, $v = -x + 2y$ for x and y. ◄

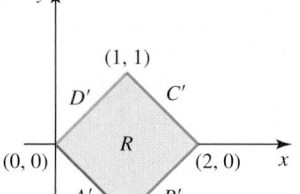

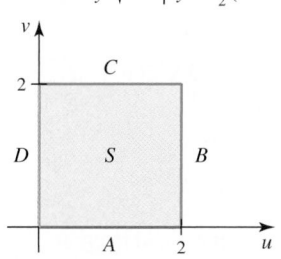

Figure 16.81

> ► The transformation in Example 4 is a *rotation*. It rotates the points of R about the origin 45° in the counterclockwise direction (it also increases lengths by a factor of $\sqrt{2}$). In this example, the change of variables $u = x + y$ and $v = x - y$ would work just as well.

> ► An appropriate change of variables for a double integral is not always obvious. Some trial and error is often needed to come up with a transformation that simplifies the integrand and/or the region of integration. Strategies are discussed at the end of this section.

In Example 3, the required transformation was given. More practically, we must deduce an appropriate transformation from the form of either the integrand or the region of integration.

EXAMPLE 4   Change of variables determined by the integrand Evaluate $\iint_R \sqrt{\dfrac{x - y}{x + y + 1}}\,dA$, where R is the square with vertices $(0, 0)$, $(1, -1)$, $(2, 0)$, and $(1, 1)$ (Figure 16.81).

SOLUTION Evaluating the integral as it stands requires splitting the region R into two subregions; furthermore, the integrand presents difficulties. The terms $x + y$ and $x - y$ in the integrand suggest the new variables

$$u = x - y \quad \text{and} \quad v = x + y.$$

To determine the region S in the uv-plane that corresponds to R under this transformation, we find the images of the vertices of R in the uv-plane and connect them in order. The result is the square $S = \{(u, v): 0 \le u \le 2, 0 \le v \le 2\}$ (Figure 16.81). Before computing the Jacobian, we express x and y in terms of u and v. Adding the two equations and solving for x, we have $x = (u + v)/2$. Subtracting the two equations and solving for y gives $y = (v - u)/2$. The Jacobian now follows:

$$J(u, v) = \begin{vmatrix} \dfrac{\partial x}{\partial u} & \dfrac{\partial x}{\partial v} \\[2mm] \dfrac{\partial y}{\partial u} & \dfrac{\partial y}{\partial v} \end{vmatrix} = \begin{vmatrix} \dfrac{1}{2} & \dfrac{1}{2} \\[2mm] -\dfrac{1}{2} & \dfrac{1}{2} \end{vmatrix} = \dfrac{1}{2}.$$

With the choice of new variables, the original integrand $\sqrt{\dfrac{x - y}{x + y + 1}}$ becomes $\sqrt{\dfrac{u}{v + 1}}$.

The integration in the uv-plane may now be done:

$$\iint_R \sqrt{\dfrac{x - y}{x + y + 1}}\,dA = \iint_S \sqrt{\dfrac{u}{v + 1}}\,|J(u, v)|\,dA \qquad \text{Change of variables}$$

$$= \int_0^2\int_0^2 \sqrt{\dfrac{u}{v + 1}}\,\dfrac{1}{2}\,du\,dv \qquad \text{Convert to an iterated integral.}$$

$$= \dfrac{1}{2}\int_0^2 (v + 1)^{-1/2}\dfrac{2}{3}\,(u^{3/2})\Big|_0^2\,dv \qquad \text{Evaluate inner integral.}$$

$$= \frac{2^{3/2}}{3} 2(v+1)^{1/2}\Big|_0^2 \qquad \text{Evaluate outer integral.}$$

$$= \frac{4\sqrt{2}}{3}(\sqrt{3}-1). \qquad \text{Simplify.}$$

Related Exercises 32, 36 ◄

QUICK CHECK 4 In Example 4, what is the ratio of the area of S to the area of R? How is this ratio related to J? ◄

EXAMPLE 5   Change of variables determined by the region Let R be the region in the first quadrant bounded by the parabolas $x = y^2$, $x = y^2 - 4$, $x = 9 - y^2$, and $x = 16 - y^2$ (Figure 16.82). Evaluate $\iint_R y^2 \, dA$.

SOLUTION Notice that the bounding curves may be written as $x - y^2 = 0$, $x - y^2 = -4$, $x + y^2 = 9$, and $x + y^2 = 16$. The first two parabolas have the form $x - y^2 = C$, where C is a constant, which suggests the new variable $u = x - y^2$. The last two parabolas have the form $x + y^2 = C$, which suggests the new variable $v = x + y^2$. Therefore, the new variables are

$$u = x - y^2 \quad \text{and} \quad v = x + y^2.$$

The boundary curves of S are $u = -4$, $u = 0$, $v = 9$, and $v = 16$. Therefore, the new region is $S = \{(u, v): -4 \leq u \leq 0, 9 \leq v \leq 16\}$ (Figure 16.82). To compute the Jacobian, we must find the transformation T by writing x and y in terms of u and v. Solving for x and y, and observing that $y \geq 0$ for all points in R, we find that

$$T: \quad x = \frac{u+v}{2} \quad \text{and} \quad y = \sqrt{\frac{v-u}{2}}.$$

The points of S satisfy $v > u$, so $\sqrt{v - u}$ is defined. Now the Jacobian may be computed:

$$J(u, v) = \begin{vmatrix} \dfrac{\partial x}{\partial u} & \dfrac{\partial x}{\partial v} \\[2mm] \dfrac{\partial y}{\partial u} & \dfrac{\partial y}{\partial v} \end{vmatrix} = \begin{vmatrix} \dfrac{1}{2} & \dfrac{1}{2} \\[2mm] -\dfrac{1}{2\sqrt{2(v-u)}} & \dfrac{1}{2\sqrt{2(v-u)}} \end{vmatrix} = \frac{1}{2\sqrt{2(v-u)}}.$$

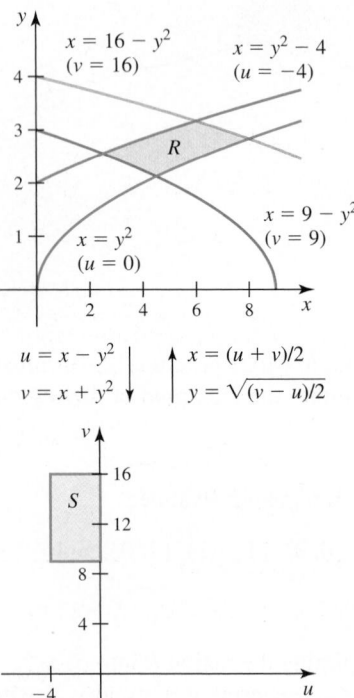

Figure 16.82

The change of variables proceeds as follows:

$$\iint_R y^2 \, dA = \int_9^{16}\int_{-4}^0 \underbrace{\frac{v-u}{2}}_{y^2} \underbrace{\frac{1}{2\sqrt{2(v-u)}}}_{|J(u,v)|} \, du \, dv \qquad \text{Convert to an iterated integral.}$$

$$= \frac{1}{4\sqrt{2}}\int_9^{16}\int_{-4}^0 \sqrt{v-u}\, du \, dv \qquad \text{Simplify.}$$

$$= \frac{1}{4\sqrt{2}}\frac{2}{3}\int_9^{16}\left(-(v-u)^{3/2}\right)\Big|_{-4}^0 dv \qquad \text{Evaluate inner integral.}$$

$$= \frac{1}{6\sqrt{2}}\int_9^{16}\left((v+4)^{3/2} - v^{3/2}\right) dv \qquad \text{Simplify.}$$

$$= \frac{1}{6\sqrt{2}}\frac{2}{5}\left((v+4)^{5/2} - v^{5/2}\right)\Big|_9^{16} \qquad \text{Evaluate outer integral.}$$

$$= \frac{\sqrt{2}}{30}\left(32 \cdot 5^{5/2} - 13^{5/2} - 781\right) \qquad \text{Simplify.}$$

$$\approx 18.79.$$

Related Exercises 33–34 ◄

Change of Variables in Triple Integrals

With triple integrals, we work with a transformation T of the form

$$T: \quad x = g(u, v, w), \quad y = h(u, v, w), \quad \text{and} \quad z = p(u, v, w).$$

In this case, T maps a region S in uvw-space to a region D in xyz-space. As before, the goal is to transform the integral $\iiint_D f(x, y, z) \, dV$ into a new integral over the region S that is easier to evaluate. First, we need a Jacobian.

▶ Recall that expanding about the first row yields

$$\begin{vmatrix} a_{11} & a_{12} & a_{13} \\ a_{21} & a_{22} & a_{23} \\ a_{31} & a_{32} & a_{33} \end{vmatrix}$$

$$= a_{11}(a_{22}a_{33} - a_{23}a_{32})$$
$$- a_{12}(a_{21}a_{33} - a_{23}a_{31})$$
$$+ a_{13}(a_{21}a_{32} - a_{22}a_{31}).$$

> **DEFINITION** **Jacobian Determinant of a Transformation of Three Variables**
>
> Given a transformation $T: x = g(u, v, w)$, $y = h(u, v, w)$, and $z = p(u, v, w)$, where g, h, and p are differentiable on a region of uvw-space, the **Jacobian determinant** (or **Jacobian**) of T is
>
> $$J(u, v, w) = \frac{\partial(x, y, z)}{\partial(u, v, w)} = \begin{vmatrix} \dfrac{\partial x}{\partial u} & \dfrac{\partial x}{\partial v} & \dfrac{\partial x}{\partial w} \\ \dfrac{\partial y}{\partial u} & \dfrac{\partial y}{\partial v} & \dfrac{\partial y}{\partial w} \\ \dfrac{\partial z}{\partial u} & \dfrac{\partial z}{\partial v} & \dfrac{\partial z}{\partial w} \end{vmatrix}.$$

The Jacobian is evaluated as a 3×3 determinant and is a function of u, v, and w. A change of variables with respect to three variables proceeds in analogy to the two-variable case.

▶ If we match the elements of volume in both integrals, then $dx\,dy\,dz = |J(u, v, w)|\,du\,dv\,dw$. As before, the Jacobian is a magnification (or reduction) factor, now relating the volume of a small region in xyz-space to the volume of the corresponding region in uvw-space.

▶ To see that triple integrals in cylindrical and spherical coordinates as derived in Section 16.5 are consistent with this change-of-variables formulation, see Exercises 46 and 47.

> **THEOREM 16.9** **Change of Variables for Triple Integrals**
>
> Let $T: x = g(u, v, w)$, $y = h(u, v, w)$, and $z = p(u, v, w)$ be a transformation that maps a closed bounded region S in uvw-space to a region $D = T(S)$ in xyz-space. Assume T is one-to-one on the interior of S and g, h, and p have continuous first partial derivatives there. If f is continuous on D, then
>
> $$\iiint_D f(x, y, z)\,dV = \iiint_S f(g(u, v, w), h(u, v, w), p(u, v, w))\,|J(u, v, w)|\,dV.$$

EXAMPLE 6 **A triple integral** Use a change of variables to evaluate $\iiint_D xz\,dV$, where D is a parallelepiped bounded by the planes

$$y = x, \qquad y = x + 2, \qquad z = x, \qquad z = x + 3, \qquad z = 0, \quad \text{and} \quad z = 4$$

(Figure 16.83a).

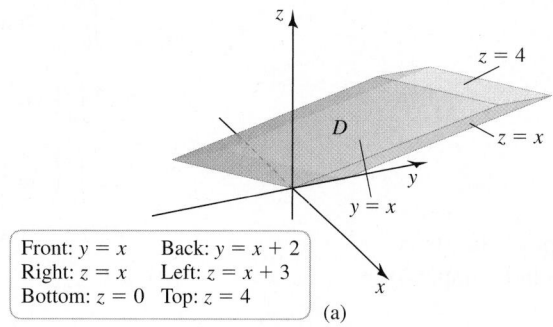

Front: $y = x$ Back: $y = x + 2$
Right: $z = x$ Left: $z = x + 3$
Bottom: $z = 0$ Top: $z = 4$

(a)

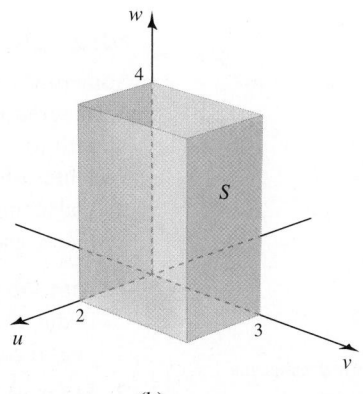

(b)

Figure 16.83

SOLUTION The key is to note that D is bounded by three pairs of parallel planes.

- $y - x = 0$ and $y - x = 2$
- $z - x = 0$ and $z - x = 3$
- $z = 0$ and $z = 4$

These combinations of variables suggest the new variables

$$u = y - x, \qquad v = z - x, \qquad \text{and} \quad w = z.$$

With this choice, the new region of integration (**Figure 16.83b**) is the rectangular box

$$S = \{(u, v, w) : 0 \le u \le 2, 0 \le v \le 3, 0 \le w \le 4\}.$$

To compute the Jacobian, we must express x, y, and z in terms of u, v, and w. A few steps of algebra lead to the transformation

$$T: \qquad x = w - v, \qquad y = u - v + w, \qquad \text{and} \quad z = w.$$

The resulting Jacobian is

> It is easiest to expand the Jacobian determinant in Example 6 about the third row.

$$J(u, v, w) = \begin{vmatrix} \dfrac{\partial x}{\partial u} & \dfrac{\partial x}{\partial v} & \dfrac{\partial x}{\partial w} \\[6pt] \dfrac{\partial y}{\partial u} & \dfrac{\partial y}{\partial v} & \dfrac{\partial y}{\partial w} \\[6pt] \dfrac{\partial z}{\partial u} & \dfrac{\partial z}{\partial v} & \dfrac{\partial z}{\partial w} \end{vmatrix} = \begin{vmatrix} 0 & -1 & 1 \\ 1 & -1 & 1 \\ 0 & 0 & 1 \end{vmatrix} = 1.$$

Noting that the integrand is $xz = (w - v)w = w^2 - vw$, the integral may now be evaluated:

$$\iiint_D xz \, dV = \iiint_S (w^2 - vw) \, |J(u, v, w)| \, dV \qquad \text{Change variables.}$$

$$= \int_0^4 \int_0^3 \int_0^2 (w^2 - vw) \underbrace{1}_{|J(u,v,w)|} du \, dv \, dw \qquad \text{Convert to an iterated integral.}$$

$$= \int_0^4 \int_0^3 2(w^2 - vw) \, dv \, dw \qquad \text{Evaluate inner integral.}$$

$$= 2 \int_0^4 \left(vw^2 - \frac{v^2 w}{2} \right) \Big|_0^3 dw \qquad \text{Evaluate middle integral.}$$

$$= 2 \int_0^4 \left(3w^2 - \frac{9w}{2} \right) dw \qquad \text{Simplify.}$$

QUICK CHECK 5 Interpret a Jacobian with a value of 1 (as in Example 6). ◄

$$= 2 \left(w^3 - \frac{9w^2}{4} \right) \Big|_0^4 = 56. \qquad \text{Evaluate outer integral.}$$

Related Exercises 40–41 ◄

Strategies for Choosing New Variables

Sometimes a change of variables simplifies the integrand but leads to an awkward region of integration. Conversely, the new region of integration may be simplified at the expense of additional complications in the integrand. Here are a few suggestions for finding new variables of integration. The observations are made with respect to double integrals, but they also apply to triple integrals. As before, R is the original region of integration in the xy-plane, and S is the new region in the uv-plane.

1. **Aim for simple regions of integration in the uv-plane** The new region of integration in the uv-plane should be as simple as possible. Double integrals are easiest to evaluate over rectangular regions with sides parallel to the coordinate axes.

> Inverting the transformation means solving for x and y in terms of u and v, or vice versa.

2. **Is $(x, y) \to (u, v)$ or $(u, v) \to (x, y)$ better?** For some problems it is easier to write (x, y) as functions of (u, v); in other cases, the opposite is true. Depending on the

problem, inverting the transformation (finding relations that go in the opposite direction) may be easy, difficult, or impossible.

- If you know (x, y) in terms of (u, v) (that is, $x = g(u, v)$ and $y = h(u, v)$), then computing the Jacobian is straightforward, as is sketching the region R given the region S. However, the transformation must be inverted to determine the shape of S.
- If you know (u, v) in terms of (x, y) (that is, $u = G(x, y)$ and $v = H(x, y)$), then sketching the region S is straightforward. However, the transformation must be inverted to compute the Jacobian.

3. **Let the integrand suggest new variables** New variables are often chosen to simplify the integrand. For example, the integrand $\sqrt{\dfrac{x - y}{x + y}}$ calls for new variables $u = x - y$ and $v = x + y$ (or $u = x + y, v = x - y$). There is, however, no guarantee that this change of variables will simplify the region of integration. In cases in which only one combination of variables appears, let one new variable be that combination and let the other new variable be unchanged. For example, if the integrand is $(x + 4y)^{3/2}$, try letting $u = x + 4y$ and $v = y$.

4. **Let the region suggest new variables** Example 5 illustrates an ideal situation. It occurs when the region R is bounded by two pairs of "parallel" curves in the families $g(x, y) = C_1$ and $h(x, y) = C_2$ (Figure 16.84). In this case, the new region of integration is a rectangle $S = \{(u, v): a_1 \le u \le a_2, b_1 \le v \le b_2\}$, where $u = g(x, y)$ and $v = h(x, y)$.

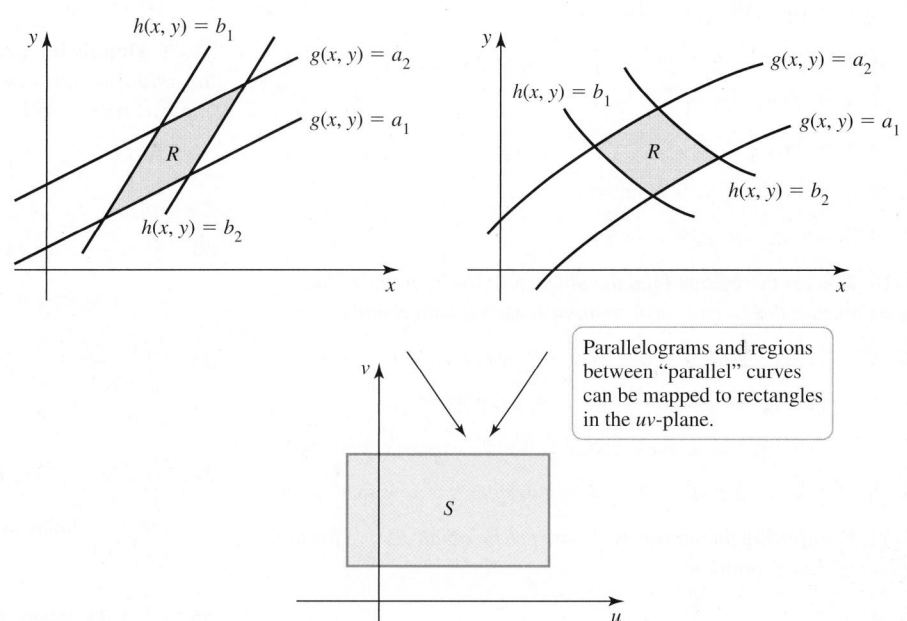

Figure 16.84

As another example, suppose the region is bounded by the lines $y = x$ (or $y/x = 1$) and $y = 2x$ (or $y/x = 2$) and by the hyperbolas $xy = 1$ and $xy = 3$. Then the new variables should be $u = xy$ and $v = y/x$ (or vice versa). The new region of integration is the rectangle $S = \{(u, v): 1 \le u \le 3, 1 \le v \le 2\}$.

SECTION 16.7 EXERCISES

Getting Started

1. Suppose S is the unit square in the first quadrant of the uv-plane. Describe the image of the transformation $T: x = 2u, y = 2v$.

2. Explain how to compute the Jacobian of the transformation $T: x = g(u, v), y = h(u, v)$.

3. Using the transformation $T: x = u + v, y = u - v$, the image of the unit square $S = \{(u, v): 0 \le u \le 1, 0 \le v \le 1\}$ is a region R in the xy-plane. Explain how to change variables in the integral $\iint_R f(x, y)\, dA$ to find a new integral over S.

4. Suppose S is the unit cube in the first octant of uvw-space with one vertex at the origin. What is the image of the transformation $T: x = u/2, y = v/2, z = w/2$?

Practice Exercises

5–12. Transforming a square *Let $S = \{(u, v): 0 \le u \le 1, 0 \le v \le 1\}$ be a unit square in the uv-plane. Find the image of S in the xy-plane under the following transformations.*

5. $T: x = 2u, y = v/2$

6. $T: x = -u, y = -v$

7. $T: x = (u + v)/2, y = (u - v)/2$

8. $T: x = 2u + v, y = 2u$

9. $T: x = u^2 - v^2, y = 2uv$

10. $T: x = 2uv, y = u^2 - v^2$

11. $T: x = u \cos \pi v, y = u \sin \pi v$

12. $T: x = v \sin \pi u, y = v \cos \pi u$

13–16. Images of regions *Find the image R in the xy-plane of the region S using the given transformation T. Sketch both R and S.*

13. $S = \{(u, v): v \le 1 - u, u \ge 0, v \ge 0\}; T: x = u, y = v^2$

14. $S = \{(u, v): u^2 + v^2 \le 1\}; T: x = 2u, y = 4v$

15. $S = \{(u, v): 1 \le u \le 3, 2 \le v \le 4\}; T: x = u/v, y = v$

16. $S = \{(u, v): 2 \le u \le 3, 3 \le v \le 6\}; T: x = u, y = v/u$

17–22. Computing Jacobians *Compute the Jacobian $J(u, v)$ for the following transformations.*

17. $T: x = 3u, y = -3v$

18. $T: x = 4v, y = -2u$

19. $T: x = 2uv, y = u^2 - v^2$

20. $T: x = u \cos \pi v, y = u \sin \pi v$

21. $T: x = (u + v)/\sqrt{2}, y = (u - v)/\sqrt{2}$

22. $T: x = u/v, y = v$

23–26. Solve and compute Jacobians *Solve the following relations for x and y, and compute the Jacobian $J(u, v)$.*

23. $u = x + y, v = 2x - y$

24. $u = xy, v = x$

25. $u = 2x - 3y, v = y - x$

26. $u = x + 4y, v = 3x + 2y$

27–30. Double integrals—transformation given *To evaluate the following integrals, carry out these steps.*

a. *Sketch the original region of integration R in the xy-plane and the new region S in the uv-plane using the given change of variables.*

b. *Find the limits of integration for the new integral with respect to u and v.*

c. *Compute the Jacobian.*

d. *Change variables and evaluate the new integral.*

27. $\iint_R xy\, dA$, where R is the square with vertices $(0, 0), (1, 1), (2, 0)$, and $(1, -1)$; use $x = u + v, y = u - v$.

28. $\iint_R x^2 y\, dA$, where $R = \{(x, y): 0 \le x \le 2, x \le y \le x + 4\}$; use $x = 2u, y = 4v + 2u$.

29. $\iint_R x^2\sqrt{x + 2y}\, dA$, where $R = \{(x, y): 0 \le x \le 2, -x/2 \le y \le 1 - x\}$; use $x = 2u, y = v - u$.

30. $\iint_R xy\, dA$, where R is bounded by the ellipse $9x^2 + 4y^2 = 36$; use $x = 2u, y = 3v$.

31–36. Double integrals—your choice of transformation *Evaluate the following integrals using a change of variables. Sketch the original and new regions of integration, R and S.*

31. $\int_0^1 \int_y^{y+2} \sqrt{x - y}\, dx\, dy$

32. $\iint_R \sqrt{y^2 - x^2}\, dA$, where R is the diamond bounded by $y - x = 0$, $y - x = 2, y + x = 0$, and $y + x = 2$

33. $\iint_R \left(\frac{y - x}{y + 2x + 1}\right)^4 dA$, where R is the parallelogram bounded by $y - x = 1, y - x = 2, y + 2x = 0$, and $y + 2x = 4$

34. $\iint_R e^{xy}\, dA$, where R is the region in the first quadrant bounded by the hyperbolas $xy = 1$ and $xy = 4$, and the lines $y/x = 1$ and $y/x = 3$

35. $\iint_R xy\, dA$, where R is the region bounded by the hyperbolas $xy = 1$ and $xy = 4$, and the lines $y = 1$ and $y = 3$

36. $\iint_R (x - y)\sqrt{x - 2y}\, dA$, where R is the triangular region bounded by $y = 0, x - 2y = 0$, and $x - y = 1$

37–40. Jacobians in three variables *Evaluate the Jacobians $J(u, v, w)$ for the following transformations.*

37. $x = v + w, y = u + w, z = u + v$

38. $x = u + v - w, y = u - v + w, z = -u + v + w$

39. $x = vw, y = uw, z = u^2 - v^2$

40. $u = x - y, v = x - z, w = y + z$ (*Hint:* Solve for x, y, and z first.)

41–44. Triple integrals *Use a change of variables to evaluate the following integrals.*

41. $\iiint\limits_{D} xy\, dV$; D is bounded by the planes $y - x = 0$, $y - x = 2$, $z - y = 0$, $z - y = 1$, $z = 0$, and $z = 3$.

42. $\iiint\limits_{D} dV$; D is bounded by the planes $y - 2x = 0$, $y - 2x = 1$, $z - 3y = 0$, $z - 3y = 1$, $z - 4x = 0$, and $z - 4x = 3$.

43. $\iiint\limits_{D} z\, dV$; D is bounded by the paraboloid $z = 16 - x^2 - 4y^2$ and the xy-plane. Use $x = 4u\cos v$, $y = 2u\sin v$, $z = w$.

44. $\iiint\limits_{D} dV$; D is bounded by the upper half of the ellipsoid $x^2/9 + y^2/4 + z^2 = 1$ and the xy-plane. Use $x = 3u$, $y = 2v$, $z = w$.

45. Explain why or why not Determine whether the following statements are true and give an explanation or counterexample.

   a. If the transformation $T: x = g(u, v)$, $y = h(u, v)$ is linear in u and v, then the Jacobian is a constant.

   b. The transformation $x = au + bv$, $y = cu + dv$ generally maps triangular regions to triangular regions.

   c. The transformation $x = 2v$, $y = -2u$ maps circles to circles.

46. Cylindrical coordinates Evaluate the Jacobian for the transformation from cylindrical coordinates (r, θ, Z) to rectangular coordinates (x, y, z): $x = r\cos\theta$, $y = r\sin\theta$, $z = Z$. Show that $J(r, \theta, Z) = r$.

47. Spherical coordinates Evaluate the Jacobian for the transformation from spherical to rectangular coordinates: $x = \rho\sin\varphi\cos\theta$, $y = \rho\sin\varphi\sin\theta$, $z = \rho\cos\varphi$. Show that $J(\rho, \varphi, \theta) = \rho^2\sin\varphi$.

48–52. Ellipse problems *Let R be the region bounded by the ellipse $x^2/a^2 + y^2/b^2 = 1$, where $a > 0$ and $b > 0$ are real numbers. Let T be the transformation $x = au$, $y = bv$.*

48. Find the area of R.

49. Evaluate $\iint\limits_{R} |xy|\, dA$.

50. Find the center of mass of the upper half of R ($y \ge 0$) assuming it has a constant density.

51. Find the average square of the distance between points of R and the origin.

52. Find the average distance between points in the upper half of R and the x-axis.

53–56. Ellipsoid problems *Let D be the solid bounded by the ellipsoid $x^2/a^2 + y^2/b^2 + z^2/c^2 = 1$, where $a > 0$, $b > 0$, and $c > 0$ are real numbers. Let T be the transformation $x = au$, $y = bv$, $z = cw$.*

53. Find the volume of D.

54. Evaluate $\iiint\limits_{D} |xyz|\, dV$.

55. Find the center of mass of the upper half of D ($z \ge 0$) assuming it has a constant density.

56. Find the average square of the distance between points of D and the origin.

Explorations and Challenges

57. Parabolic coordinates Let T be the transformation $x = u^2 - v^2$, $y = 2uv$.

   a. Show that the lines $u = a$ in the uv-plane map to parabolas in the xy-plane that open in the negative x-direction with vertices on the positive x-axis.

   b. Show that the lines $v = b$ in the uv-plane map to parabolas in the xy-plane that open in the positive x-direction with vertices on the negative x-axis.

   c. Evaluate $J(u, v)$.

   d. Use a change of variables to find the area of the region bounded by $x = 4 - y^2/16$ and $x = y^2/4 - 1$.

   e. Use a change of variables to find the area of the curved rectangle above the x-axis bounded by $x = 4 - y^2/16$, $x = 9 - y^2/36$, $x = y^2/4 - 1$, and $x = y^2/64 - 16$.

   f. Describe the effect of the transformation $x = 2uv$, $y = u^2 - v^2$ on horizontal and vertical lines in the uv-plane.

58. Shear transformations in $\mathbb{R}^2$ The transformation T in $\mathbb{R}^2$ given by $x = au + bv$, $y = cv$, where a, b, and c are positive real numbers, is a *shear transformation*. Let S be the unit square $\{(u, v): 0 \le u \le 1, 0 \le v \le 1\}$. Let $R = T(S)$ be the image of S.

   a. Explain with pictures the effect of T on S.

   b. Compute the Jacobian of T.

   c. Find the area of R and compare it to the area of S (which is 1).

   d. Assuming a constant density, find the center of mass of R (in terms of a, b, and c) and compare it to the center of mass of S, which is $\left(\frac{1}{2}, \frac{1}{2}\right)$.

   e. Find an analogous transformation that gives a shear in the y-direction.

59. Shear transformations in $\mathbb{R}^3$ The transformation T in $\mathbb{R}^3$ given by

$$x = au + bv + cw, \qquad y = dv + ew, \qquad z = w,$$

where a, b, c, d, and e are positive real numbers, is one of many possible shear transformations in $\mathbb{R}^3$. Let S be the unit cube $\{(u, v, w): 0 \le u \le 1, 0 \le v \le 1, 0 \le w \le 1\}$. Let $D = T(S)$ be the image of S.

   a. Explain with pictures and words the effect of T on S.

   b. Compute the Jacobian of T.

   c. Find the volume of D and compare it to the volume of S (which is 1).

   d. Assuming a constant density, find the center of mass of D and compare it to the center of mass of S, which is $\left(\frac{1}{2}, \frac{1}{2}, \frac{1}{2}\right)$.

60. Linear transformations Consider the linear transformation T in $\mathbb{R}^2$ given by $x = au + bv$, $y = cu + dv$, where a, b, c, and d are real numbers, with $ad \ne bc$.

   a. Find the Jacobian of T.

   b. Let S be the square in the uv-plane with vertices $(0, 0)$, $(1, 0)$, $(0, 1)$, and $(1, 1)$, and let $R = T(S)$. Show that $\text{area}(R) = |J(u, v)|$.

   c. Let ℓ be the line segment joining the points P and Q in the uv-plane. Show that $T(\ell)$ (the image of ℓ under T) is the line segment joining $T(P)$ and $T(Q)$ in the xy-plane. (*Hint:* Use vectors.)

   d. Show that if S is a parallelogram in the uv-plane and $R = T(S)$, then $\text{area}(R) = |J(u, v)| \cdot \text{area of } S$. (*Hint:* Without loss of generality, assume the vertices of S are $(0, 0)$, $(A, 0)$, (B, C), and $(A + B, C)$, where A, B, and C are positive, and use vectors.)

61. Meaning of the Jacobian The Jacobian is a magnification (or reduction) factor that relates the area of a small region near the point (u, v) to the area of the image of that region near the point (x, y).

   a. Suppose S is a rectangle in the uv-plane with vertices $O(0, 0)$, $P(\Delta u, 0)$, $(\Delta u, \Delta v)$, and $Q(0, \Delta v)$ (see figure). The image of S under the transformation $x = g(u, v)$, $y = h(u, v)$ is a region R in the xy-plane. Let O', P', and Q' be the images of O, P, and Q, respectively, in the xy-plane, where O', P', and Q' do not all lie on the same line. Explain why the coordinates of O', P', and Q' are $(g(0, 0), h(0, 0))$, $(g(\Delta u, 0), h(\Delta u, 0))$, and $(g(0, \Delta v), h(0, \Delta v))$, respectively.

   b. Use a Taylor series in both variables to show that
   $$g(\Delta u, 0) \approx g(0, 0) + g_u(0, 0)\Delta u,$$
   $$g(0, \Delta v) \approx g(0, 0) + g_v(0, 0)\Delta v,$$
   $$h(\Delta u, 0) \approx h(0, 0) + h_u(0, 0)\Delta u, \text{ and}$$
   $$h(0, \Delta v) \approx h(0, 0) + h_v(0, 0)\Delta v,$$

   where $g_u(0, 0)$ is $\dfrac{\partial x}{\partial u}$ evaluated at $(0, 0)$, with similar meanings for g_v, h_u, and h_v.

   c. Consider the parallelogram determined by the vectors $\overrightarrow{O'P'}$ and $\overrightarrow{O'Q'}$. Use the cross product to show that the area of the parallelogram is approximately $|J(u, v)|\,\Delta u\,\Delta v$.

   d. Explain why the ratio of the area of R to the area of S is approximately $|J(u, v)|$.

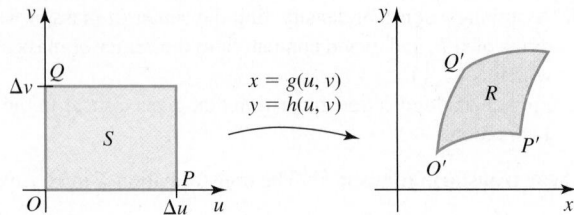

62. Open and closed boxes Consider the region R bounded by three pairs of parallel planes: $ax + by = 0$, $ax + by = 1$; $cx + dz = 0$, $cx + dz = 1$; and $ey + fz = 0$, $ey + fz = 1$, where $a, b, c, d, e,$ and f are real numbers. For the purposes of evaluating triple integrals, when do these six planes bound a finite region? Carry out the following steps.

   a. Find three vectors $\mathbf{n}_1$, $\mathbf{n}_2$, and $\mathbf{n}_3$ each of which is normal to one of the three pairs of planes.

   b. Show that the three normal vectors lie in a plane if their triple scalar product $\mathbf{n}_1 \cdot (\mathbf{n}_2 \times \mathbf{n}_3)$ is zero.

   c. Show that the three normal vectors lie in a plane if $ade + bcf = 0$.

   d. Assuming $\mathbf{n}_1$, $\mathbf{n}_2$, and $\mathbf{n}_3$ lie in a plane P, find a vector $\mathbf{N}$ that is normal to P. Explain why a line in the direction of $\mathbf{N}$ does not intersect any of the six planes, and therefore the six planes do not form a bounded region.

   e. Consider the change of variables $u = ax + by$, $v = cx + dz$, $w = ey + fz$. Show that
   $$J(x, y, z) = \frac{\partial(u, v, w)}{\partial(x, y, z)} = -ade - bcf.$$

   What is the value of the Jacobian if R is unbounded?

QUICK CHECK ANSWERS

1. The image is a semicircular disk of radius 1.
2. $J(u, v) = 2$ 3. $x = 2u/3 - v/3$, $y = u/3 + v/3$
4. The ratio is 2, which is $1/J(u, v)$.   5. It means that the volume of a small region in xyz-space is unchanged when it is transformed by T into a small region in uvw-space. ◄

CHAPTER 16   REVIEW EXERCISES

1. Explain why or why not Determine whether the following statements are true and give an explanation or counterexample.

   a. Assuming g is integrable and $a, b, c,$ and d are constants,
   $$\int_c^d \int_a^b g(x, y)\,dx\,dy = \left(\int_a^b g(x, y)\,dx\right)\left(\int_c^d g(x, y)\,dy\right).$$

   b. The spherical equation $\varphi = \pi/2$, the cylindrical equation $z = 0$, and the rectangular equation $z = 0$ all describe the same set of points.

   c. Changing the order of integration in $\iiint_D f(x, y, z)\,dx\,dy\,dz$ from $dx\,dy\,dz$ to $dy\,dz\,dx$ requires also changing the integrand from $f(x, y, z)$ to $f(y, z, x)$.

   d. The transformation $T: x = v$, $y = -u$ maps a square in the uv-plane to a triangle in the xy-plane.

2–4. Evaluating integrals *Evaluate the following integrals as they are written.*

2. $\displaystyle\int_1^2 \int_1^4 \frac{xy}{(x^2 + y^2)^2}\,dx\,dy$ 3. $\displaystyle\int_1^3 \int_1^{e^x} \frac{x}{y}\,dy\,dx$

4. $\displaystyle\int_1^2 \int_0^{\ln x} x^3 e^y\,dy\,dx$

5–7. Changing the order of integration *Assuming f is integrable, change the order of integration in the following integrals.*

5. $\displaystyle\int_{-1}^1 \int_{x^2}^1 f(x, y)\,dy\,dx$ 6. $\displaystyle\int_0^2 \int_{y-1}^1 f(x, y)\,dx\,dy$

7. $\displaystyle\int_0^1 \int_0^{\sqrt{1-y^2}} f(x, y)\,dx\,dy$

8–10. Area of plane regions *Use double integrals to compute the area of the following regions. Make a sketch of the region.*

8. The region bounded by the lines $y = -x - 4$, $y = x$, and $y = 2x - 4$

9. The region bounded by $y = |x|$ and $y = 20 - x^2$

10. The region between the curves $y = x^2$ and $y = 1 + x - x^2$

11–16. Miscellaneous double integrals *Choose a convenient method for evaluating the following integrals.*

11. $\displaystyle\iint_R \frac{2y}{\sqrt{x^4 + 1}}\, dA$; R is the region bounded by $x = 1$, $x = 2$, $y = x^{3/2}$, and $y = 0$.

12. $\displaystyle\iint_R x^{-1/2}\, e^y\, dA$; R is the region bounded by $x = 1$, $x = 4$, $y = \sqrt{x}$, and $y = 0$.

13. $\displaystyle\iint_R (x + y)\, dA$; R is the disk bounded by the circle $r = 4 \sin \theta$.

14. $\displaystyle\iint_R (x^2 + y^2)\, dA$; R is the region $\{(x, y): 0 \le x \le 2, 0 \le y \le x\}$.

15. $\displaystyle\int_0^1 \int_{y^{1/3}}^1 x^{10} \cos(\pi x^4 y)\, dx\, dy$

16. $\displaystyle\int_0^2 \int_{y^2}^4 x^8 y \sqrt{1 + x^4 y^2}\, dx\, dy$

17–18. Cartesian to polar coordinates *Evaluate the following integrals over the specified region. Assume (r, θ) are polar coordinates.*

17. $\displaystyle\iint_R 3x^2 y\, dA$; $R = \{(r, \theta): 0 \le r \le 1, 0 \le \theta \le \pi/2\}$

18. $\displaystyle\iint_R \frac{dA}{(1 + x^2 + y^2)^2}$; $R = \{(r, \theta): 1 \le r \le 4, 0 \le \theta \le \pi\}$

19–21. Computing areas *Sketch the following regions and use a double integral to find their areas.*

19. The region bounded by all leaves of the rose $r = 3 \cos 2\theta$

20. The region inside both of the circles $r = 2$ and $r = 4 \cos \theta$

21. The region that lies inside both of the cardioids $r = 2 - 2 \cos \theta$ and $r = 2 + 2 \cos \theta$

22–23. Average values

22. Find the average value of $z = \sqrt{16 - x^2 - y^2}$ over the disk in the xy-plane centered at the origin with radius 4.

23. Find the average distance from the points in the solid cone bounded by $z = 2\sqrt{x^2 + y^2}$ to the z-axis, for $0 \le z \le 8$.

24–26. Changing order of integration *Rewrite the following integrals using the indicated order of integration.*

24. $\displaystyle\int_0^1 \int_0^{\sqrt{1-z^2}} \int_0^{\sqrt{1-x^2}} f(x, y, z)\, dy\, dx\, dz$ in the order $dz\, dy\, dx$

25. $\displaystyle\int_0^1 \int_0^{\sqrt{1-x^2}} \int_{2\sqrt{x^2+y^2}}^2 f(x, y, z)\, dz\, dy\, dx$ in the order $dx\, dz\, dy$

26. $\displaystyle\int_0^2 \int_0^{9-x^2} \int_0^x f(x, y, z)\, dy\, dz\, dx$ in the order $dz\, dx\, dy$

27–31. Triple integrals *Evaluate the following integrals, changing the order of integration if needed.*

27. $\displaystyle\int_0^1 \int_{-z}^z \int_{-\sqrt{1-x^2}}^{\sqrt{1-x^2}} dy\, dx\, dz$

28. $\displaystyle\int_0^\pi \int_0^y \int_0^{\sin x} dz\, dx\, dy$

29. $\displaystyle\int_1^9 \int_0^1 \int_{2y}^2 \frac{4 \sin x^2}{\sqrt{z}}\, dx\, dy\, dz$

30. $\displaystyle\int_0^2 \int_{-\sqrt{2-x^2/2}}^{\sqrt{2-x^2/2}} \int_{x^2+3y^2}^{8-x^2-y^2} dz\, dy\, dx$

31. $\displaystyle\int_0^2 \int_0^{y^{1/3}} \int_0^{y^2} yz^5(1 + x + y^2 + z^6)^2\, dx\, dz\, dy$

32–38. Volumes of solids *Find the volume of the following solids.*

32. The solid beneath the paraboloid $f(x, y) = 12 - x^2 - 2y^2$ and above the region $R = \{(x, y): 1 \le x \le 2, 0 \le y \le 1\}$

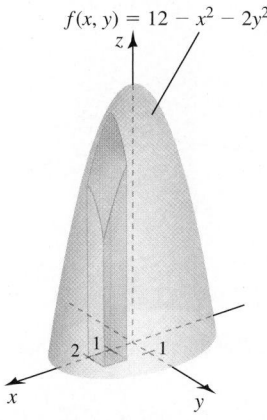

$f(x, y) = 12 - x^2 - 2y^2$

33. The solid bounded by the surfaces $x = 0$, $y = 0$, $z = 3 - 2y$, and $z = 2x^2 + 1$

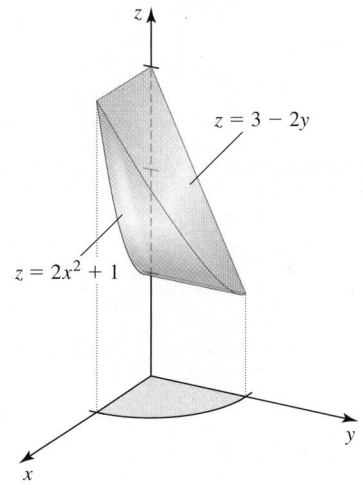

$z = 3 - 2y$

$z = 2x^2 + 1$

34. The prism in the first octant bounded by the planes $y = 3 - 3x$ and $z = 2$

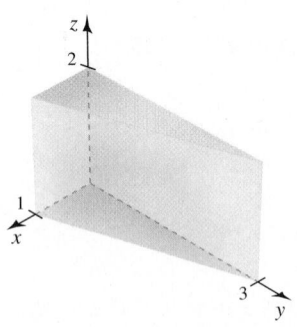

35. One of the wedges formed when the cylinder $x^2 + y^2 = 4$ is cut by the planes $z = 0$ and $y = z$

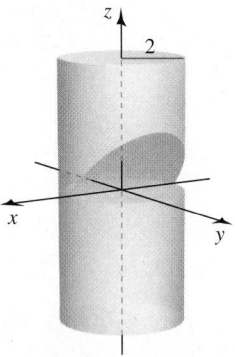

36. The solid bounded by the parabolic cylinders $z = y^2 + 1$ and $z = 2 - x^2$

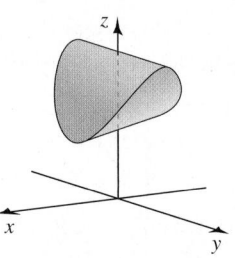

37. The solid common to the two cylinders $x^2 + y^2 = 4$ and $x^2 + z^2 = 4$

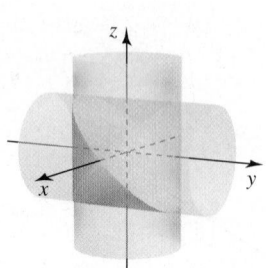

38. The tetrahedron with vertices $(0, 0, 0)$, $(1, 0, 0)$, $(1, 1, 0)$, and $(1, 1, 1)$

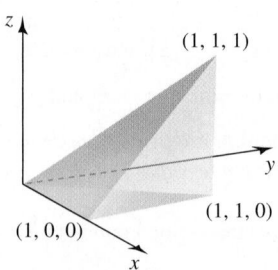

39. Single to double integral Evaluate $\int_0^{1/2}(\sin^{-1} 2x - \sin^{-1} x)\, dx$ by converting it to a double integral.

40. Tetrahedron limits Let D be the tetrahedron with vertices at $(0, 0, 0)$, $(1, 0, 0)$, $(0, 2, 0)$, and $(0, 0, 3)$. Suppose the volume of D is to be found using a triple integral. Give the limits of integration for the six possible orderings of the variables.

41. Polar to Cartesian Evaluate $\int_0^{\pi/4}\int_0^{\sec\theta} r^3\, dr\, d\theta$ using rectangular coordinates, where (r, θ) are polar coordinates.

42–43. Average value

42. Find the average of the *square* of the distance between the origin and the points in the solid paraboloid $D = \{(x, y, z): 0 \le z \le 4 - x^2 - y^2\}$.

43. Find the average x-coordinate of the points in the prism $D = \{(x, y, z): 0 \le x \le 1, 0 \le y \le 3 - 3x, 0 \le z \le 2\}$.

44–45. Integrals in cylindrical coordinates *Evaluate the following integrals in cylindrical coordinates.*

44. $\int_0^3\int_0^{\sqrt{9-x^2}}\int_0^3 (x^2 + y^2)^{3/2}\, dz\, dy\, dx$

45. $\int_{-1}^1\int_{-2}^2\int_0^{\sqrt{1-y^2}} \dfrac{1}{(1 + x^2 + y^2)^2}\, dx\, dz\, dy$

46–47. Volumes in cylindrical coordinates *Use integration in cylindrical coordinates to find the volume of the following solids.*

46. The solid bounded by the hemisphere $z = \sqrt{9 - x^2 - y^2}$ and the hyperboloid $z = \sqrt{1 + x^2 + y^2}$.

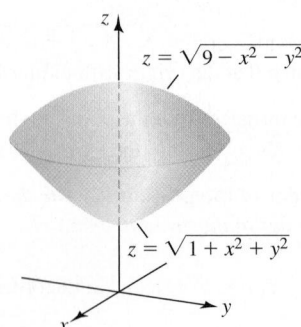

47. The solid cylinder whose height is 4 and whose base is the disk $\{(r, \theta): 0 \le r \le 2 \cos \theta\}$

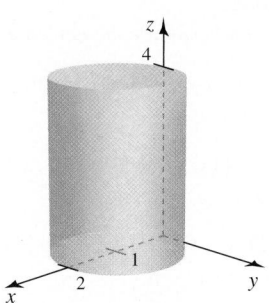

48–49. Integrals in spherical coordinates *Evaluate the following integrals in spherical coordinates.*

48. $\displaystyle\int_0^{2\pi} \int_0^{\pi/2} \int_0^{2 \cos \varphi} \rho^2 \sin \varphi \, d\rho \, d\varphi \, d\theta$

49. $\displaystyle\int_0^{\pi} \int_0^{\pi/4} \int_{2 \sec \varphi}^{4 \sec \varphi} \rho^2 \sin \varphi \, d\rho \, d\varphi \, d\theta$

50–52. Volumes in spherical coordinates *Use integration in spherical coordinates to find the volume of the following solids.*

50. The solid cardioid of revolution $D = \{(\rho, \varphi, \theta): 0 \le \rho \le (1 - \cos \varphi)/2, 0 \le \varphi \le \pi, 0 \le \theta \le 2\pi\}$

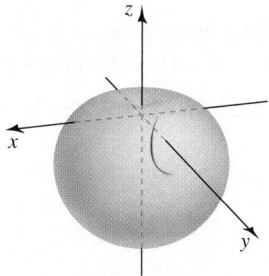

51. The solid rose petal of revolution
$D = \{(\rho, \varphi, \theta): 0 \le \rho \le 4 \sin 2\varphi,$
$0 \le \varphi \le \pi/2, 0 \le \theta \le 2\pi\}$

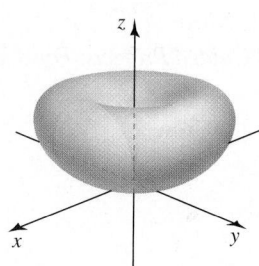

52. The solid above the cone $\varphi = \pi/4$ and inside the sphere $\rho = 4 \cos \varphi$

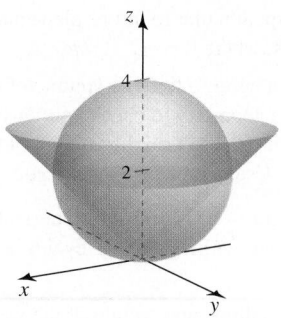

53–56. Center of mass of constant-density plates *Find the center of mass (centroid) of the following thin, constant-density plates. Sketch the region corresponding to the plate and indicate the location of the center of mass. Use symmetry whenever possible to simplify your work.*

53. The region bounded by $y = \sin x$ and $y = 0$ between $x = 0$ and $x = \pi$

54. The region bounded by $y = x^3$ and $y = x^2$ between $x = 0$ and $x = 1$

55. The half-annulus $\{(r, \theta): 2 \le r \le 4, 0 \le \theta \le \pi\}$

56. The region bounded by $y = x^2$ and $y = a^2 - x^2$, where $a > 0$

57–58. Center of mass of constant-density solids *Find the center of mass of the following solids, assuming a constant density. Use symmetry whenever possible and choose a convenient coordinate system.*

57. The paraboloid bowl bounded by $z = x^2 + y^2$ and $z = 36$

58. The tetrahedron bounded by $z = 4 - x - 2y$ and the coordinate planes

59–60. Variable-density solids *Find the coordinates of the center of mass of the following solids with the given density.*

59. The upper half of the ball $\left\{(\rho, \varphi, \theta): 0 \le \rho \le 16, 0 \le \varphi \le \dfrac{\pi}{2},\right.$ $\left. 0 \le \theta \le 2\pi\right\}$ with density $f(\rho, \varphi, \theta) = 1 + \rho/4$

60. The cube in the first octant bounded by the planes $x = 2$, $y = 2$, and $z = 2$, with $\rho(x, y, z) = 1 + x + y + z$

61–64. Center of mass for general objects *Consider the following two- and three-dimensional regions. Compute the center of mass, assuming constant density. All parameters are positive real numbers.*

61. A solid is bounded by a paraboloid with a circular base of radius R and height h. How far from the base is the center of mass?

62. Let R be the region enclosed by an equilateral triangle with sides of length s. What is the perpendicular distance between the center of mass of R and the edges of R?

63. A sector of a circle in the first quadrant is bounded between the x-axis, the line $y = x$, and the circle $x^2 + y^2 = a^2$. What are the coordinates of the center of mass?

64. An ice cream cone is bounded above by the sphere $x^2 + y^2 + z^2 = a^2$ and below by the upper half of the cone $z^2 = x^2 + y^2$. What are the coordinates of the center of mass?

65. Slicing a conical cake A cake is shaped like a solid cone with radius 4 and height 2, with its base on the *xy*-plane. A wedge of the cake is removed by making two slices from the axis of the cone outward, perpendicular to the *xy*-plane and separated by an angle of *Q* radians, where $0 < Q < 2\pi$.

    a. Use a double integral to find the volume of the slice for $Q = \pi/4$. Use geometry to check your answer.

    b. Use a double integral to find the volume of the slice for any $0 < Q < 2\pi$. Use geometry to check your answer.

66. Volume and weight of a fish tank A spherical fish tank with a radius of 1 ft is filled with water to a level 6 in below the top of the tank.

    a. Determine the volume and weight of the water in the fish tank. (The weight density of water is about 62.5 lb/ft^3.)

    b. How much additional water must be added to completely fill the tank?

67–70. Transforming a square *Let* $S = \{(u, v): 0 \le u \le 1,$ $0 \le v \le 1\}$ *be a unit square in the uv-plane. Find the image of S in the xy-plane under the following transformations.*

67. $T: x = v, y = u$ **68.** $T: x = -v, y = u$

69. $T: x = 3u + v, y = u + 3v$

70. $T: x = u, y = 2v + 2$

71–74. Computing Jacobians *Compute the Jacobian J(u, v) of the following transformations.*

71. $T: x = 4u - v, y = -2u + 3v$

72. $T: x = u + v, y = u - v$

73. $T: x = 3u, y = 2v + 2$ **74.** $T: x = u^2 - v^2, y = 2uv$

75–78. Double integrals—transformation given *To evaluate the following integrals, carry out these steps.*

a. Sketch the original region of integration R and the new region S using the given change of variables.

b. Find the limits of integration for the new integral with respect to u and v.

c. Compute the Jacobian.

d. Change variables and evaluate the new integral.

75. $\iint_R xy^2 \, dA; R = \{(x, y): y/3 \le x \le (y + 6)/3, 0 \le y \le 3\};$ use $x = u + v/3, y = v.$

76. $\iint_R 3xy^2 \, dA; R = \{(x, y): 0 \le x \le 2, x \le y \le x + 4\};$ use $x = 2u, y = 4v + 2u.$

77. $\iint_R (x - y + 1)(x - y)^9 \, dy \, dx, R = \{(x, y): 0 \le y \le x \le 1\};$ use $x = u + v, y = v - u.$

78. $\iint_R xy^2 \, dA; R$ is the region between the hyperbolas $xy = 1$ and $xy = 4$ and the lines $y = 1$ and $y = 4$; use $x = u/v, y = v.$

79–80. Double integrals *Evaluate the following integrals using a change of variables. Sketch the original and new regions of integration, R and S.*

79. $\iint_R y^4 \, dA; R$ is the region bounded by the hyperbolas $xy = 1$ and $xy = 4$ and the lines $y/x = 1$ and $y/x = 3.$

80. $\iint_R (y^2 + xy - 2x^2) \, dA; R$ is the region bounded by the lines $y = x, y = x - 3, y = -2x + 3,$ and $y = -2x - 3.$

81–82. Triple integrals *Use a change of variables to evaluate the following integrals.*

81. $\iiint_D yz \, dV; D$ is bounded by the planes $x + 2y = 1, x + 2y = 2,$ $x - z = 0, x - z = 2, 2y - z = 0,$ and $2y - z = 3.$

82. $\iiint_D x \, dV; D$ is bounded by the planes $y - 2x = 0, y - 2x = 1,$ $z - 3y = 0, z - 3y = 1, z - 4x = 0,$ and $z - 4x = 3.$

Chapter 16 Guided Projects

Applications of the material in this chapter and related topics can be found in the following Guided Projects. For additional information, see the Preface.

- How big are *n*-balls?
- Electrical field integrals
- The tilted cylinder problem
- The exponential Eiffel Tower
- Moments of inertia
- Gravitational fields

17

Vector Calculus

Chapter Preview This culminating chapter of the text provides a beautiful, unifying conclusion to our study of calculus. Many ideas and themes that have appeared throughout the text come together in these final pages. First, we combine vector-valued functions (Chapter 14) and functions of several variables (Chapter 15) to form *vector fields*. Once vector fields have been introduced and illustrated through their many applications, we explore the calculus of vector fields. Concepts such as limits and continuity carry over directly. The extension of derivatives to vector fields leads to two new operations that underlie this chapter: the *curl* and the *divergence*. When integration is extended to vector fields, we discover new versions of the Fundamental Theorem of Calculus. The chapter ends with a final look at the Fundamental Theorem of Calculus and the several related forms in which it has appeared throughout the text.

17.1 Vector Fields

We live in a world filled with phenomena that can be represented by vector fields. Imagine sitting in a window seat looking out at the wing of an airliner. Although you can't see it, air is rushing over and under the wing. Focus on a point near the wing and visualize the motion of the air at that point at a single instant of time. The motion is described by a velocity vector with three components—for example, east-west, north-south, and up-down. At another point near the wing at the same time, the air is moving at a different speed and direction, and a different velocity vector is associated with that point. In general, at one instant in time, every point around the wing has a velocity vector associated with it (**Figure 17.1**). This collection of velocity vectors—a unique vector for each point in space—is a function called a *vector field*.

Other examples of vector fields include the wind patterns in a hurricane (**Figure 17.2a**) and the circulation of water in a heat exchanger (**Figure 17.2b**). Gravitational, magnetic, and electric force fields are also represented by vector fields (**Figure 17.2c**), as are the stresses and strains in buildings and bridges. Beyond physics and engineering, the transport of a chemical pollutant in a lake and human migration patterns can be modeled by vector fields.

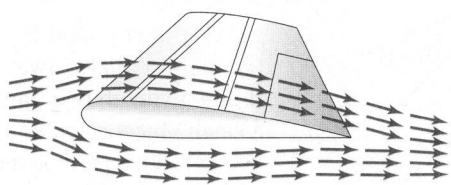

Figure 17.1

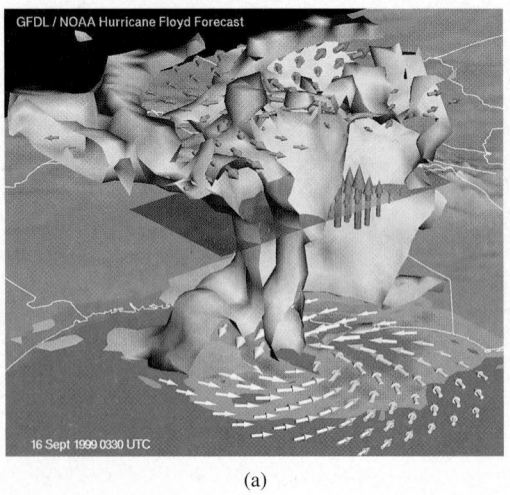

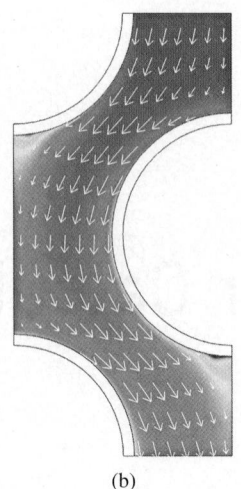

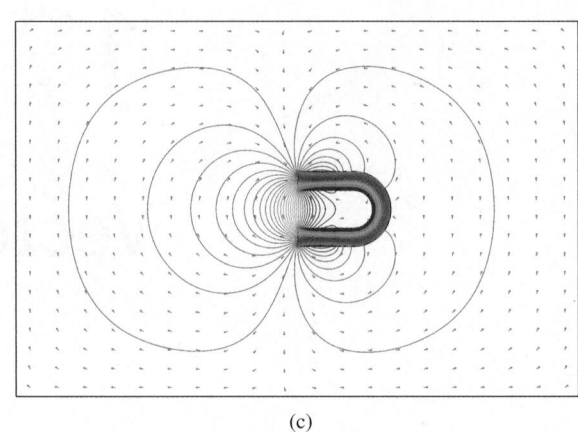

(a)   (b)   (c)

Figure 17.2

Vector Fields in Two Dimensions

To solidify the idea of a vector field, we begin by exploring vector fields in $\mathbb{R}^2$. From there, it is a short step to vector fields in $\mathbb{R}^3$.

> ➤ Notice that a vector field is both a vector-valued function (Chapter 14) and a function of several (Chapter 15).

DEFINITION **Vector Fields in Two Dimensions**

Let f and g be defined on a region R of $\mathbb{R}^2$. A **vector field** in $\mathbb{R}^2$ is a function $\mathbf{F}$ that assigns to each point in R a vector $\langle f(x, y), g(x, y) \rangle$. The vector field is written as

$$\mathbf{F}(x, y) = \langle f(x, y), g(x, y) \rangle \quad \text{or}$$
$$\mathbf{F}(x, y) = f(x, y)\, \mathbf{i} + g(x, y)\, \mathbf{j}.$$

A vector field $\mathbf{F} = \langle f, g \rangle$ is continuous or differentiable on a region R of $\mathbb{R}^2$ if f and g are continuous or differentiable on R, respectively.

A vector field cannot be represented graphically in its entirety. Instead, we plot a representative sample of vectors that illustrates the general appearance of the vector field. Consider the vector field defined by

$$\mathbf{F}(x, y) = \langle 2x, 2y \rangle = 2x\, \mathbf{i} + 2y\, \mathbf{j}.$$

At selected points $P(x, y)$, we plot a vector with its tail at P equal to the value of $\mathbf{F}(x, y)$. For example, $\mathbf{F}(1, 1) = \langle 2, 2 \rangle$, so we draw a vector equal to $\langle 2, 2 \rangle$ with its tail at the point $(1, 1)$. Similarly, $\mathbf{F}(-2, -3) = \langle -4, -6 \rangle$, so at the point $(-2, -3)$, we draw a vector equal to $\langle -4, -6 \rangle$. We can make the following general observations about the vector field $\mathbf{F}(x, y) = \langle 2x, 2y \rangle$.

• For every (x, y) except $(0, 0)$, the vector $\mathbf{F}(x, y)$ points in the direction of $\langle 2x, 2y \rangle$, which is directly outward from the origin.

• The length of $\mathbf{F}(x, y)$ is $|\mathbf{F}| = |\langle 2x, 2y \rangle| = 2\sqrt{x^2 + y^2}$, which increases with distance from the origin.

The vector field $\mathbf{F} = \langle 2x, 2y \rangle$ is an example of a *radial vector field* because its vectors point radially away from the origin (**Figure 17.3**). If $\mathbf{F}$ represents the velocity of a fluid moving in two dimensions, the graph of the vector field gives a vivid image of how a small object, such as a cork, moves through the fluid. In this case, at every point of the vector field, a particle moves in the direction of the arrow at that point with a speed equal to the length of the arrow. For this reason, vector fields are sometimes called *flows*. When sketching vector fields, it is often useful to draw continuous curves that are aligned with the vector field. Such curves are called *flow curves* or *streamlines*; we examine their properties in greater detail later in this section.

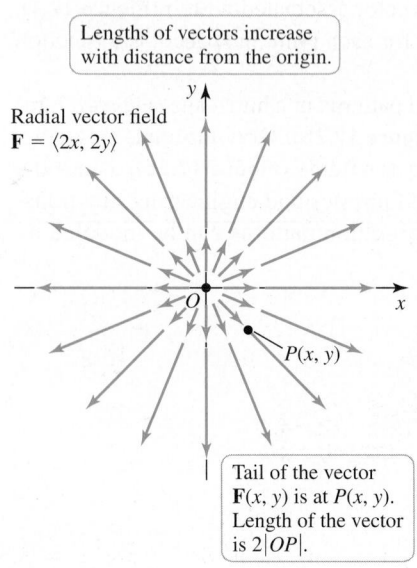

Lengths of vectors increase with distance from the origin.

Radial vector field
$\mathbf{F} = \langle 2x, 2y \rangle$

Tail of the vector $\mathbf{F}(x, y)$ is at $P(x, y)$. Length of the vector is $2|OP|$.

Figure 17.3

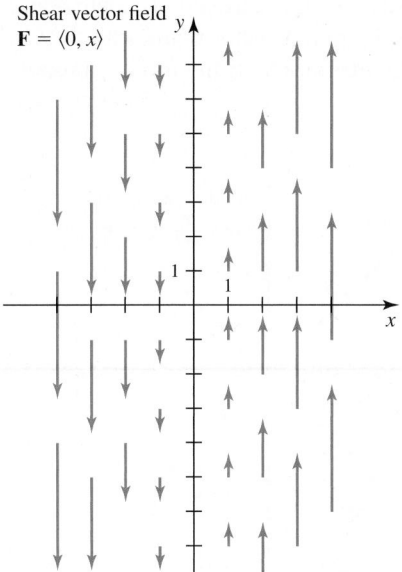

Shear vector field
$\mathbf{F} = \langle 0, x \rangle$

Figure 17.4

➤ Drawing vectors with their actual length often leads to cluttered pictures of vector fields. For this reason, most of the vector fields in this chapter are illustrated with proportional scaling: All vectors are multiplied by a scalar chosen to make the vector field as understandable as possible.

➤ A useful observation for two-dimensional vector fields $\mathbf{F} = \langle f, g \rangle$ is that the slope of the vector at (x, y) is $g(x, y)/f(x, y)$. In Example 1a, the slopes are everywhere undefined; in part (b), the slopes are everywhere 0, and in part (c), the slopes are $-x/y$.

QUICK CHECK 1 If the vector field in Example 1c describes the velocity of a fluid and you place a small cork in the plane at $(2, 0)$, what path will it follow? ◄

EXAMPLE 1   **Vector fields**   Sketch representative vectors of the following vector fields.

a. $\mathbf{F}(x, y) = \langle 0, x \rangle = x\,\mathbf{j}$ (a shear field)

b. $\mathbf{F}(x, y) = \langle 1 - y^2, 0 \rangle = (1 - y^2)\,\mathbf{i}$, for $|y| \le 1$ (channel flow)

c. $\mathbf{F}(x, y) = \langle -y, x \rangle = -y\,\mathbf{i} + x\,\mathbf{j}$ (a rotation field)

SOLUTION

a. This vector field is independent of y. Furthermore, because the x-component of $\mathbf{F}$ is zero, all vectors in the field (for $x \ne 0$) point in the y-direction: upward for $x > 0$ and downward for $x < 0$. The magnitudes of the vectors in the field increase with distance from the y-axis (**Figure 17.4**). The flow curves for this field are vertical lines. If $\mathbf{F}$ represents a velocity field, a particle right of the y-axis moves upward, a particle left of the y-axis moves downward, and a particle on the y-axis is stationary.

b. In this case, the vector field is independent of x and the y-component of $\mathbf{F}$ is zero. Because $1 - y^2 > 0$ for $|y| < 1$, vectors in this region point in the positive x-direction. The x-component of the vector field is zero at the boundaries $y = \pm 1$ and increases to 1 along the center of the strip, $y = 0$. This vector field might model the flow of water in a straight shallow channel (**Figure 17.5**); its flow curves are horizontal lines, indicating motion in the direction of the positive x-axis.

c. It often helps to determine the vector field along the coordinate axes.

- When $y = 0$ (along the x-axis), we have $\mathbf{F}(x, 0) = \langle 0, x \rangle$. With $x > 0$, this vector field consists of vectors pointing upward, increasing in length as x increases. With $x < 0$, the vectors point downward, increasing in length as $|x|$ increases.

- When $x = 0$ (along the y-axis), we have $\mathbf{F}(0, y) = \langle -y, 0 \rangle$. If $y > 0$, the vectors point in the negative x-direction, increasing in length as y increases. If $y < 0$, the vectors point in the positive x-direction, increasing in length as $|y|$ increases.

A few more representative vectors show that this vector field has a counterclockwise rotation about the origin; the magnitudes of the vectors increase with distance from the origin (**Figure 17.6**).

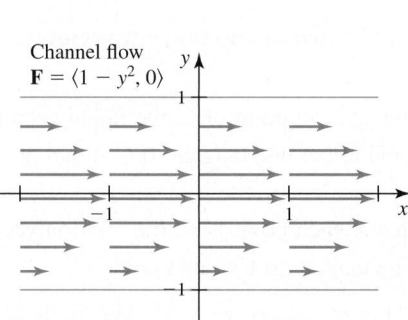

Channel flow
$\mathbf{F} = \langle 1 - y^2, 0 \rangle$

Figure 17.5

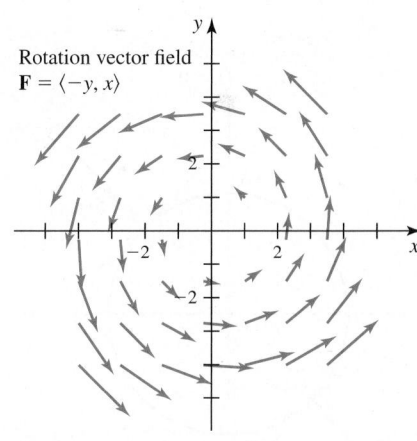

Rotation vector field
$\mathbf{F} = \langle -y, x \rangle$

Figure 17.6

Related Exercises 10, 11, 13 ◄

Radial Vector Fields in $\mathbb{R}^2$ Radial vector fields in $\mathbb{R}^2$ have the property that their vectors point directly toward or away from the origin at all points (except the origin), parallel to the position vectors $\mathbf{r} = \langle x, y \rangle$. We will work with radial vector fields of the form

$$\mathbf{F}(x, y) = \frac{\mathbf{r}}{|\mathbf{r}|^p} = \frac{\langle x, y \rangle}{|\mathbf{r}|^p} = \underbrace{\frac{\mathbf{r}}{|\mathbf{r}|}}_{\substack{\text{unit} \\ \text{vector}}} \underbrace{\frac{1}{|\mathbf{r}|^{p-1}}}_{\text{magnitude}},$$

where p is a real number. **Figure 17.7** illustrates radial fields with $p = 1$ and $p = 3$. These vector fields (and their three-dimensional counterparts) play an important role in many applications. For example, central forces, such as gravitational or electrostatic forces between point masses or charges, are described by radial vector fields with $p = 3$. These forces obey an inverse square law in which the magnitude of the force is proportional to $1/|\mathbf{r}|^2$.

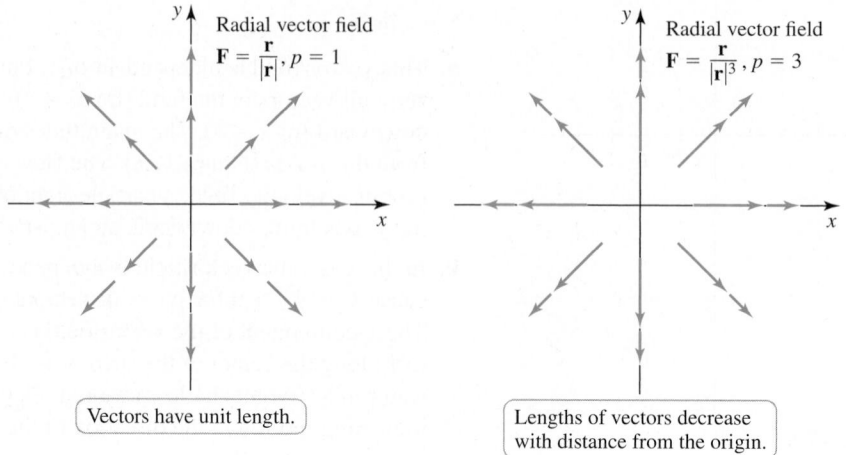

Radial vector field $\mathbf{F} = \dfrac{\mathbf{r}}{|\mathbf{r}|}, p = 1$

Vectors have unit length.

Radial vector field $\mathbf{F} = \dfrac{\mathbf{r}}{|\mathbf{r}|^3}, p = 3$

Lengths of vectors decrease with distance from the origin.

Figure 17.7

> **DEFINITION Radial Vector Fields in $\mathbb{R}^2$**
>
> Let $\mathbf{r} = \langle x, y \rangle$. A vector field of the form $\mathbf{F} = f(x, y)\,\mathbf{r}$, where f is a scalar-valued function, is a **radial vector field**. Of specific interest are the radial vector fields
>
> $$\mathbf{F}(x, y) = \frac{\mathbf{r}}{|\mathbf{r}|^p} = \frac{\langle x, y \rangle}{|\mathbf{r}|^p},$$
>
> where p is a real number. At every point (except the origin), the vectors of this field are directed outward from the origin with a magnitude of $|\mathbf{F}| = \dfrac{1}{|\mathbf{r}|^{p-1}}$.

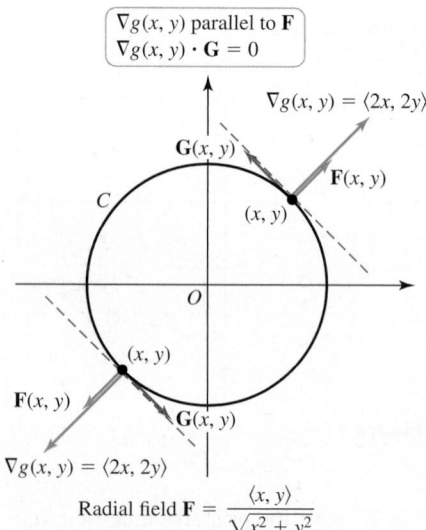

$\nabla g(x, y)$ parallel to $\mathbf{F}$
$\nabla g(x, y) \cdot \mathbf{G} = 0$

$\nabla g(x, y) = \langle 2x, 2y \rangle$

$G(x, y)$

C

$\mathbf{F}(x, y)$

(x, y)

O

(x, y)

$F(x, y)$

$G(x, y)$

$\nabla g(x, y) = \langle 2x, 2y \rangle$

Radial field $\mathbf{F} = \dfrac{\langle x, y \rangle}{\sqrt{x^2 + y^2}}$

Rotation field $\mathbf{G} = \dfrac{\langle -y, x \rangle}{\sqrt{x^2 + y^2}}$

Figure 17.8

QUICK CHECK 2 In Example 2, verify that $\nabla g(x, y) \cdot \mathbf{G}(x, y) = 0$. In parts (a) and (b) of Example 2, verify that $|\mathbf{F}| = 1$ and $|\mathbf{G}| = 1$ at all points excluding the origin. ◀

EXAMPLE 2   Normal and tangent vectors Let C be the circle $x^2 + y^2 = a^2$, where $a > 0$.

a. Show that at each point of C, the radial vector field $\mathbf{F}(x, y) = \dfrac{\mathbf{r}}{|\mathbf{r}|} = \dfrac{\langle x, y \rangle}{\sqrt{x^2 + y^2}}$ is orthogonal to the line tangent to C at that point.

b. Show that at each point of C, the rotation vector field $\mathbf{G}(x, y) = \dfrac{\langle -y, x \rangle}{\sqrt{x^2 + y^2}}$ is parallel to the line tangent to C at that point.

SOLUTION Let $g(x, y) = x^2 + y^2$. The circle C described by the equation $g(x, y) = a^2$ may be viewed as a level curve of the surface $z = x^2 + y^2$. As shown in Theorem 15.12 (Section 15.5), the gradient $\nabla g(x, y) = \langle 2x, 2y \rangle$ is orthogonal to the line tangent to C at (x, y) (Figure 17.8).

a. Notice that $\nabla g(x, y)$ is parallel to $\mathbf{F} = \langle x, y \rangle / |\mathbf{r}|$ at the point (x, y). It follows that $\mathbf{F}$ is also orthogonal the line tangent to C at (x, y).

b. Notice that

$$\nabla g(x, y) \cdot \mathbf{G}(x, y) = \langle 2x, 2y \rangle \cdot \frac{\langle -y, x \rangle}{|\mathbf{r}|} = 0.$$

Therefore, $\nabla g(x, y)$ is orthogonal to the vector field $\mathbf{G}$ at (x, y), which implies that $\mathbf{G}$ is parallel to the tangent line at (x, y).

Related Exercises 27–28 ◀

Vector Fields in Three Dimensions

Vector fields in three dimensions are conceptually the same as vector fields in two dimensions. The vector $\mathbf{F}$ now has three components, each of which depends on three variables.

DEFINITION **Vector Fields and Radial Vector Fields in $\mathbb{R}^3$**

Let f, g, and h be defined on a region D of $\mathbb{R}^3$. A **vector field** in $\mathbb{R}^3$ is a function $\mathbf{F}$ that assigns to each point in D a vector $\langle f(x, y, z), g(x, y, z), h(x, y, z) \rangle$. The vector field is written as

$$\mathbf{F}(x, y, z) = \langle f(x, y, z), g(x, y, z), h(x, y, z) \rangle \quad \text{or}$$
$$\mathbf{F}(x, y, z) = f(x, y, z)\,\mathbf{i} + g(x, y, z)\,\mathbf{j} + h(x, y, z)\,\mathbf{k}.$$

A vector field $\mathbf{F} = \langle f, g, h \rangle$ is continuous or differentiable on a region D of $\mathbb{R}^3$ if f, g, and h are continuous or differentiable on D, respectively. Of particular importance are the **radial vector fields**

$$\mathbf{F}(x, y, z) = \frac{\mathbf{r}}{|\mathbf{r}|^p} = \frac{\langle x, y, z \rangle}{|\mathbf{r}|^p},$$

where p is a real number.

EXAMPLE 3 **Vector fields in $\mathbb{R}^3$** Sketch and discuss the following vector fields.

a. $\mathbf{F}(x, y, z) = \langle x, y, e^{-z} \rangle$, for $z \geq 0$

b. $\mathbf{F}(x, y, z) = \langle 0, 0, 1 - x^2 - y^2 \rangle$, for $x^2 + y^2 \leq 1$

SOLUTION

a. First consider the x- and y-components of $\mathbf{F}$ in the xy-plane ($z = 0$), where $\mathbf{F} = \langle x, y, 1 \rangle$. This vector field looks like a radial field in the first two components, increasing in magnitude with distance from the z-axis. However, each vector also has a constant vertical component of 1. In horizontal planes $z = z_0 > 0$, the radial pattern remains the same, but the vertical component decreases as z increases. As $z \to \infty$, $e^{-z} \to 0$ and the vector field approaches a horizontal radial field (**Figure 17.9**).

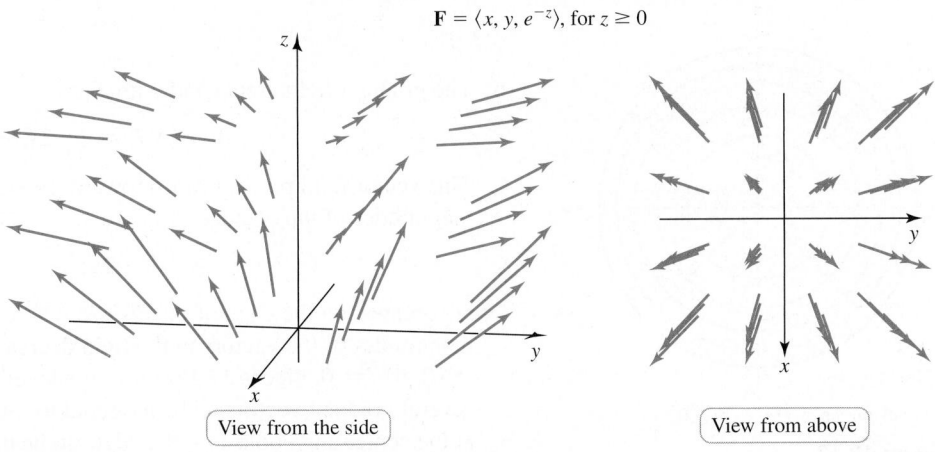

$\mathbf{F} = \langle x, y, e^{-z} \rangle$, for $z \geq 0$

View from the side View from above

Figure 17.9

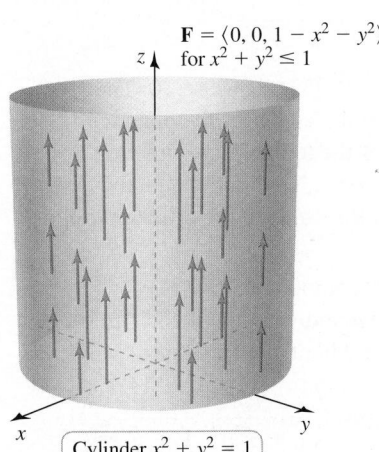

$\mathbf{F} = \langle 0, 0, 1 - x^2 - y^2 \rangle$, for $x^2 + y^2 \leq 1$

Cylinder $x^2 + y^2 = 1$

Figure 17.10

b. Regarding $\mathbf{F}$ as a velocity field for points in and on the cylinder $x^2 + y^2 = 1$, there is no motion in the x- or y-direction. The z-component of the vector field may be written $1 - r^2$, where $r^2 = x^2 + y^2$ is the square of the distance from the z-axis. We see that the z-component increases from 0 on the boundary of the cylinder ($r = 1$) to a maximum value of 1 along the centerline of the cylinder ($r = 0$) (**Figure 17.10**). This vector field models the flow of a fluid inside a tube (such as a blood vessel).

Related Exercises 20, 22 ◄

➤ Physicists often use the convention that a gradient field and its potential function are related by $\mathbf{F} = -\nabla\varphi$ (with a negative sign).

The vector field $\mathbf{F} = \nabla\varphi$ is orthogonal to the level curves of φ at (x, y).

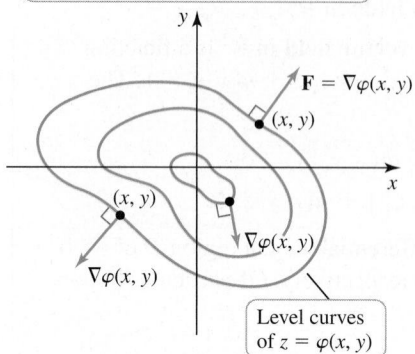

Figure 17.11

Gradient Fields and Potential Functions   One way to generate a vector field is to start with a differentiable scalar-valued function φ, take its gradient, and let $\mathbf{F} = \nabla\varphi$. A vector field defined as the gradient of a scalar-valued function φ is called a *gradient field*, and φ is called a *potential function*.

Suppose φ is a differentiable function on a region R of $\mathbb{R}^2$ and consider the surface $z = \varphi(x, y)$. Recall from Chapter 15 that this function may also be represented by level curves in the xy-plane. At each point (a, b) on a level curve, the gradient $\nabla\varphi(a, b) = \langle \varphi_x(a, b), \varphi_y(a, b) \rangle$ is orthogonal to the level curve at (a, b) (**Figure 17.11**). Therefore, the vectors of $\mathbf{F} = \nabla\varphi$ point in a direction orthogonal to the level curves of φ.

The idea extends to gradients of functions of three variables. If φ is differentiable on a region D of $\mathbb{R}^3$, then $\mathbf{F} = \nabla\varphi = \langle \varphi_x, \varphi_y, \varphi_z \rangle$ is a vector field that points in a direction orthogonal to the level *surfaces* of φ.

Gradient fields are useful because of the physical meaning of the gradient. For example, if φ represents the temperature in a conducting material, then the gradient field $\mathbf{F} = \nabla\varphi$ evaluated at a point indicates the direction in which the temperature increases most rapidly at that point. According to a basic physical law, heat diffuses in the direction of the vector field $-\mathbf{F} = -\nabla\varphi$, the direction in which the temperature *decreases* most rapidly; that is, heat flows "down the gradient" from relatively hot regions to cooler regions. Similarly, water on a smooth surface tends to flow down the elevation gradient.

QUICK CHECK 3 Find the gradient field associated with the function $\varphi(x, y, z) = xyz$. ◄

DEFINITION Gradient Fields and Potential Functions

Let φ be differentiable on a region of $\mathbb{R}^2$ or $\mathbb{R}^3$. The vector field $\mathbf{F} = \nabla\varphi$ is a **gradient field** and the function φ is a **potential function** for $\mathbf{F}$.

➤ A potential function plays the role of an antiderivative of a vector field: Derivatives of the potential function produce the vector field. If φ is a potential function for a gradient field, then $\varphi + C$ is also a potential function for that gradient field, for any constant C.

Gradient vectors ∇T (not drawn to scale) are orthogonal to the level curves.

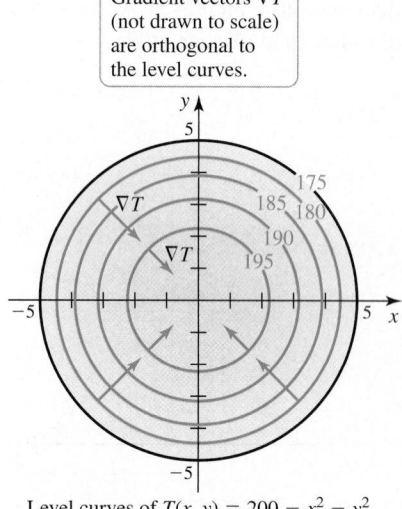

Level curves of $T(x, y) = 200 - x^2 - y^2$

Figure 17.12

EXAMPLE 4   Gradient fields

a. Sketch and interpret the gradient field associated with the temperature function $T = 200 - x^2 - y^2$ on the circular plate $R = \{(x, y): x^2 + y^2 \le 25\}$.

b. Sketch and interpret the gradient field associated with the velocity potential $\varphi = \tan^{-1} xy$.

SOLUTION

a. The gradient field associated with T is

$$\mathbf{F} = \nabla T = \langle -2x, -2y \rangle = -2\langle x, y \rangle.$$

This vector field points inward toward the origin at all points of R except $(0, 0)$. The magnitudes of the vectors,

$$|\mathbf{F}| = \sqrt{(-2x)^2 + (-2y)^2} = 2\sqrt{x^2 + y^2},$$

are greatest on the edge of the disk R, where $x^2 + y^2 = 25$ and $|\mathbf{F}| = 10$. The magnitudes of the vectors in the field decrease toward the center of the plate with $|\mathbf{F}(0, 0)| = 0$. **Figure 17.12** shows the level curves of the temperature function with several gradient vectors, all orthogonal to the level curves. Note that the plate is hottest at the center and coolest on the edge, so heat diffuses *outward*, in the direction opposite that of the gradient.

b. The gradient of a velocity potential gives the velocity components of a two-dimensional flow; that is, $\mathbf{F} = \langle u, v \rangle = \nabla\varphi$, where u and v are the velocities in the x- and y-directions, respectively. Computing the gradient, we find that

$$\mathbf{F} = \langle \varphi_x, \varphi_y \rangle = \left\langle \frac{1}{1 + (xy)^2} \cdot y, \frac{1}{1 + (xy)^2} \cdot x \right\rangle = \left\langle \frac{y}{1 + x^2y^2}, \frac{x}{1 + x^2y^2} \right\rangle.$$

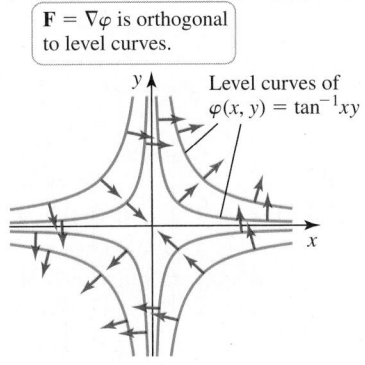

F = ∇φ is orthogonal to level curves.

Level curves of $\varphi(x, y) = \tan^{-1}xy$

Figure 17.13

Notice that the level curves of φ are the hyperbolas $xy = C$ or $y = C/x$. At all points, the vector field is orthogonal to the level curves (**Figure 17.13**).

Related Exercises 38, 47 ◄

Equipotential Curves and Surfaces The preceding example illustrates a beautiful geometric connection between a gradient field and its associated potential function. Let φ be a potential function for the vector field **F** in $\mathbb{R}^2$; that is, $\mathbf{F} = \nabla\varphi$. The level curves of a potential function are called **equipotential curves** (curves on which the potential function is constant).

Because the equipotential curves are level curves of φ, the vector field $\mathbf{F} = \nabla\varphi$ is everywhere orthogonal to the equipotential curves (**Figure 17.14**). The vector field may be visualized by drawing continuous **flow curves** or **streamlines** that are everywhere orthogonal to the equipotential curves. These ideas also apply to vector fields in $\mathbb{R}^3$, in which case the vector field is orthogonal to the **equipotential surfaces**.

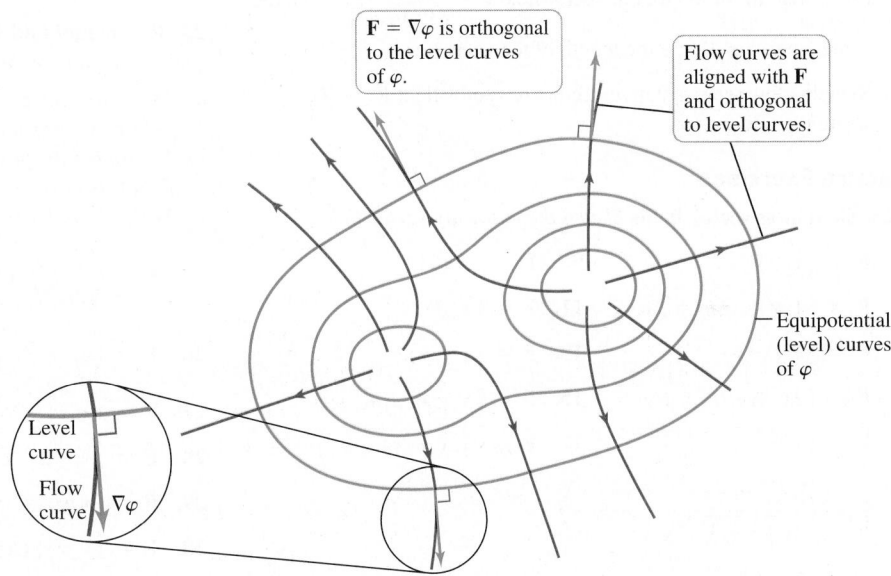

F = ∇φ is orthogonal to the level curves of φ.

Flow curves are aligned with **F** and orthogonal to level curves.

Equipotential (level) curves of φ

Level curve

Flow curve ∇φ

Figure 17.14

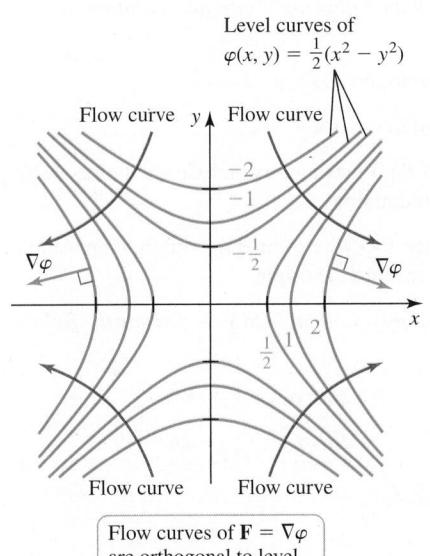

Level curves of $\varphi(x, y) = \frac{1}{2}(x^2 - y^2)$

Flow curve   Flow curve

∇φ   ∇φ

Flow curve   Flow curve

Flow curves of **F** = ∇φ are orthogonal to level curves of φ everywhere.

Figure 17.15

▶ We use the fact that a line with slope a/b points in the direction of the vectors $\langle 1, a/b \rangle$ or $\langle b, a \rangle$.

EXAMPLE 5 **Equipotential curves** The equipotential curves for the potential function $\varphi(x, y) = (x^2 - y^2)/2$ are shown in green in **Figure 17.15**.

a. Find the gradient field associated with φ and verify that the gradient field is orthogonal to the equipotential curve at $(2, 1)$.

b. Verify that the vector field $\mathbf{F} = \nabla\varphi$ is orthogonal to the equipotential curves at all points (x, y).

SOLUTION

a. The level (or equipotential) curves are the hyperbolas $(x^2 - y^2)/2 = C$, where C is a constant. The slope at any point on a level curve $\varphi(x, y) = C$ (Section 15.4) is

$$\frac{dy}{dx} = -\frac{\varphi_x}{\varphi_y} = \frac{x}{y}.$$

At the point $(2, 1)$, the slope of the level curve is $dy/dx = 2$, so the vector tangent to the curve points in the direction $\langle 1, 2 \rangle$. The gradient field is given by $\mathbf{F} = \nabla\varphi = \langle x, -y \rangle$, so $\mathbf{F}(2, 1) = \nabla\varphi(2, 1) = \langle 2, -1 \rangle$. The dot product of the tangent vector $\langle 1, 2 \rangle$ and the gradient is $\langle 1, 2 \rangle \cdot \langle 2, -1 \rangle = 0$; therefore, the two vectors are orthogonal.

b. In general, the line tangent to the equipotential curve at (x, y) is parallel to the vector $\langle y, x \rangle$, and the vector field at that point is $\mathbf{F} = \langle x, -y \rangle$. The vector field and the tangent vectors are orthogonal because $\langle y, x \rangle \cdot \langle x, -y \rangle = 0$.

Related Exercise 52 ◄

SECTION 17.1 EXERCISES

Getting Started

1. How is a vector field $\mathbf{F} = \langle f, g, h \rangle$ used to describe the motion of air at one instant in time?

2. Sketch the vector field $\mathbf{F} = \langle x, y \rangle$.

3. How do you graph the vector field $\mathbf{F} = \langle f(x, y), g(x, y) \rangle$?

4. Given a differentiable, scalar-valued function φ, why is the gradient of φ a vector field?

5. Interpret the gradient field of the temperature function $T = f(x, y)$.

6. Show that all the vectors in vector field $\mathbf{F} = \dfrac{\sqrt{2}\langle x, y \rangle}{\sqrt{x^2 + y^2}}$ have the same length, and state the length of the vectors.

7. Sketch a few representative vectors of vector field $\mathbf{F} = \langle 0, 1 \rangle$ along the line $y = 2$.

Practice Exercises

8–23. Sketching vector fields *Sketch the following vector fields.*

8. $\mathbf{F} = \langle 1, 0 \rangle$

9. $\mathbf{F} = \langle -1, 1 \rangle$

10. $\mathbf{F} = \langle 1, y \rangle$

11. $\mathbf{F} = \langle x, 0 \rangle$

12. $\mathbf{F} = \langle -x, -y \rangle$

13. $\mathbf{F} = \langle x, -y \rangle$

14. $\mathbf{F} = \langle 2x, 3y \rangle$

15. $\mathbf{F} = \langle y, -x \rangle$

16. $\mathbf{F} = \langle x + y, y \rangle$

17. $\mathbf{F} = \langle x, y - x \rangle$

18. $\mathbf{F} = \left\langle \dfrac{x}{\sqrt{x^2 + y^2}}, \dfrac{y}{\sqrt{x^2 + y^2}} \right\rangle$

19. $\mathbf{F} = \langle e^{-x}, 0 \rangle$

20. $\mathbf{F} = \langle 0, 0, 1 \rangle$

T 21. $\mathbf{F} = \langle 1, 0, z \rangle$

T 22. $\mathbf{F} = \dfrac{\langle x, y, z \rangle}{\sqrt{x^2 + y^2 + z^2}}$

T 23. $\mathbf{F} = \langle y, -x, 0 \rangle$

24. **Matching vector fields with graphs** Match vector fields a–d with graphs A–D.

a. $\mathbf{F} = \langle 0, x^2 \rangle$
b. $\mathbf{F} = \langle x - y, x \rangle$
c. $\mathbf{F} = \langle 2x, -y \rangle$
d. $\mathbf{F} = \langle y, x \rangle$

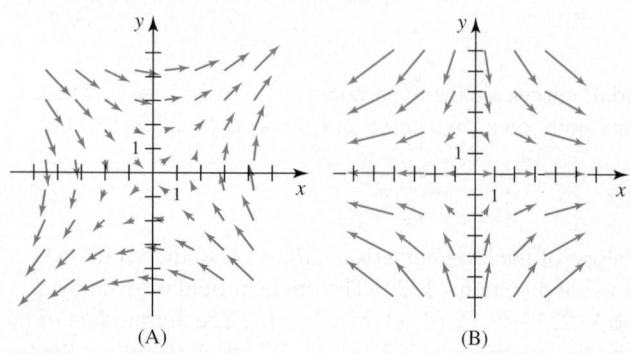

(A)   (B)

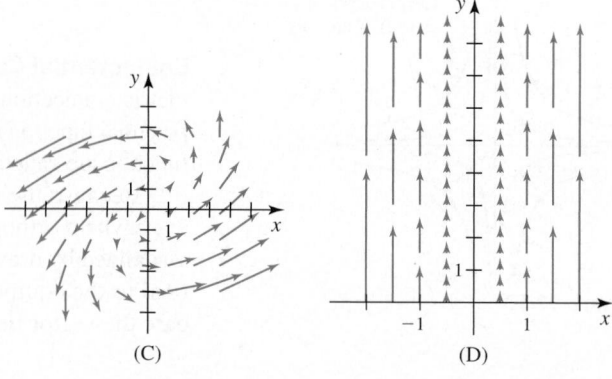

(C)   (D)

25–30. Normal and tangential components *For the vector field $\mathbf{F}$ and curve C, complete the following:*

a. *Determine the points (if any) along the curve C at which the vector field $\mathbf{F}$ is tangent to C.*

b. *Determine the points (if any) along the curve C at which the vector field $\mathbf{F}$ is normal to C.*

c. *Sketch C and a few representative vectors of $\mathbf{F}$ on C.*

25. $\mathbf{F} = \left\langle \dfrac{1}{2}, 0 \right\rangle$; $C = \{(x, y): y - x^2 = 1\}$

26. $\mathbf{F} = \left\langle \dfrac{y}{2}, -\dfrac{x}{2} \right\rangle$; $C = \{(x, y): y - x^2 = 1\}$

27. $\mathbf{F} = \langle x, y \rangle$; $C = \{(x, y): x^2 + y^2 = 4\}$

28. $\mathbf{F} = \langle y, -x \rangle$; $C = \{(x, y): x^2 + y^2 = 1\}$

29. $\mathbf{F} = \langle x, y \rangle$; $C = \{(x, y): x = 1\}$

30. $\mathbf{F} = \langle y, x \rangle$; $C = \{(x, y): x^2 + y^2 = 1\}$

31–34. Design your own vector field *Specify the component functions of a vector field $\mathbf{F}$ in $\mathbb{R}^2$ with the following properties. Solutions are not unique.*

31. $\mathbf{F}$ is everywhere normal to the line $y = x$.

32. $\mathbf{F}$ is everywhere normal to the line $x = 2$.

33. At all points except $(0, 0)$, $\mathbf{F}$ has unit magnitude and points away from the origin along radial lines.

34. The flow of $\mathbf{F}$ is counterclockwise around the origin, increasing in magnitude with distance from the origin.

35–42. Gradient fields *Find the gradient field $\mathbf{F} = \nabla \varphi$ for the following potential functions φ.*

35. $\varphi(x, y) = x^2 y - y^2 x$

36. $\varphi(x, y) = \sqrt{xy}$

37. $\varphi(x, y) = x/y$

38. $\varphi(x, y) = \tan^{-1}(x/y)$

39. $\varphi(x, y, z) = \dfrac{x^2 + y^2 + z^2}{2}$

40. $\varphi(x, y, z) = \ln(1 + x^2 + y^2 + z^2)$

41. $\varphi(x, y, z) = (x^2 + y^2 + z^2)^{-1/2}$

42. $\varphi(x, y, z) = e^{-z} \sin(x + y)$

43–46. Gradient fields on curves *For the potential function φ and points A, B, C, and D on the level curve $\varphi(x, y) = 0$, complete the following steps.*

a. *Find the gradient field $\mathbf{F} = \nabla\varphi$.*
b. *Evaluate $\mathbf{F}$ at the points A, B, C, and D.*
c. *Plot the level curve $\varphi(x, y) = 0$ and the vectors $\mathbf{F}$ at the points A, B, C, and D.*

43. $\varphi(x, y) = y - 2x$; $A(-1, -2)$, $B(0, 0)$, $C(1, 2)$, and $D(2, 4)$

44. $\varphi(x, y) = \dfrac{1}{2}x^2 - y$; $A(-2, 2)$, $B(-1, 1/2)$, $C(1, 1/2)$, and $D(2, 2)$

45. $\varphi(x, y) = -y + \sin x$; $A(\pi/2, 1)$, $B(\pi, 0)$, $C(3\pi/2, -1)$, and $D(2\pi, 0)$

T 46. $\varphi(x, y) = \dfrac{32 - x^4 - y^4}{32}$; $A(2, 2)$, $B(-2, 2)$, $C(-2, -2)$, and $D(2, -2)$

T 47–48. Gradient fields *Find the gradient field $\mathbf{F} = \nabla\varphi$ for the potential function φ. Sketch a few level curves of φ and a few vectors of $\mathbf{F}$.*

47. $\varphi(x, y) = x^2 + y^2$, for $x^2 + y^2 \le 16$

48. $\varphi(x, y) = x + y$, for $|x| \le 2$, $|y| \le 2$

49–52. Equipotential curves *Consider the following potential functions and the graphs of their equipotential curves.*

a. *Find the associated gradient field $\mathbf{F} = \nabla\varphi$.*
b. *Show that the vector field is orthogonal to the equipotential curve at the point $(1, 1)$. Illustrate this result on the figure.*
c. *Show that the vector field is orthogonal to the equipotential curve at all points (x, y).*
d. *Sketch two flow curves representing $\mathbf{F}$ that are everywhere orthogonal to the equipotential curves.*

49. $\varphi(x, y) = 2x + 3y$

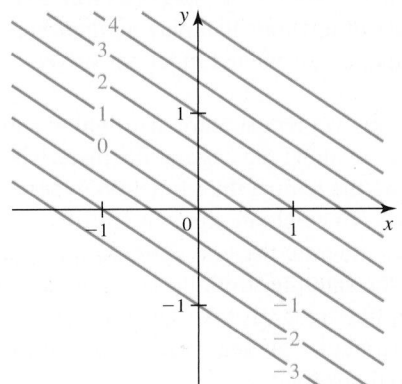

50. $\varphi(x, y) = x + y^2$

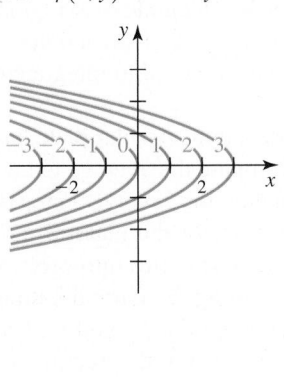

51. $\varphi(x, y) = e^{x-y}$

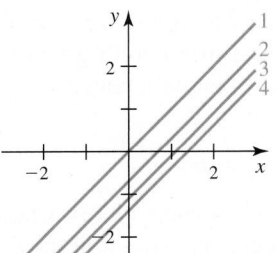

52. $\varphi(x, y) = x^2 + 2y^2$

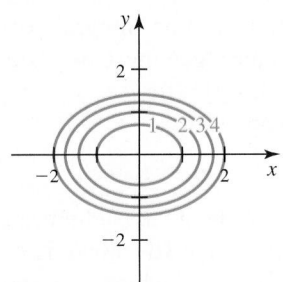

53. Explain why or why not Determine whether the following statements are true and give an explanation or counterexample.

a. The vector field $\mathbf{F} = \langle 3x^2, 1 \rangle$ is a gradient field for both $\varphi_1(x, y) = x^3 + y$ and $\varphi_2(x, y) = y + x^3 + 100$.

b. The vector field $\mathbf{F} = \dfrac{\langle y, x \rangle}{\sqrt{x^2 + y^2}}$ is constant in direction and magnitude on the unit circle.

c. The vector field $\mathbf{F} = \dfrac{\langle y, x \rangle}{\sqrt{x^2 + y^2}}$ is neither a radial field nor a rotation field.

Explorations and Challenges

54. Electric field due to a point charge The electric field in the xy-plane due to a point charge at $(0, 0)$ is a gradient field with a potential function $V(x, y) = \dfrac{k}{\sqrt{x^2 + y^2}}$, where $k > 0$ is a physical constant.

a. Find the components of the electric field in the x- and y-directions, where $\mathbf{E}(x, y) = -\nabla V(x, y)$.

b. Show that the vectors of the electric field point in the radial direction (outward from the origin) and the radial component of $\mathbf{E}$ can be expressed as $E_r = \dfrac{k}{r^2}$, where $r = \sqrt{x^2 + y^2}$.

c. Show that the vector field is orthogonal to the equipotential curves at all points in the domain of V.

55. Electric field due to a line of charge The electric field in the xy-plane due to an infinite line of charge along the z-axis is a gradient field with a potential function $V(x, y) = c \ln\left(\dfrac{r_0}{\sqrt{x^2 + y^2}}\right)$, where $c > 0$ is a constant and r_0 is a reference distance at which the potential is assumed to be 0 (see figure).

a. Find the components of the electric field in the x- and y-directions, where $\mathbf{E}(x, y) = -\nabla V(x, y)$.

b. Show that the electric field at a point in the xy-plane is directed outward from the origin and has magnitude $|\mathbf{E}| = c/r$, where $r = \sqrt{x^2 + y^2}$.

c. Show that the vector field is orthogonal to the equipotential curves at all points in the domain of V.

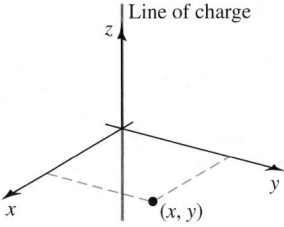

56. Gravitational force due to a mass The gravitational force on a point mass m due to a point mass M is a gradient field with potential $U(r) = GMm/r$, where G is the gravitational constant and $r = \sqrt{x^2 + y^2 + z^2}$ is the distance between the masses.

a. Find the components of the gravitational force in the x-, y-, and z-directions, where $\mathbf{F}(x, y, z) = -\nabla U(x, y, z)$.

b. Show that the gravitational force points in the radial direction (outward from point mass M) and the radial component is $F(r) = GMm/r^2$.

c. Show that the vector field is orthogonal to the equipotential surfaces at all points in the domain of U.

57–61. Flow curves in the plane *Let* $\mathbf{F}(x, y) = \langle f(x, y), g(x, y) \rangle$ *be defined on* $\mathbb{R}^2$.

57. Explain why the flow curves or streamlines of $\mathbf{F}$ satisfy
$$y' = \frac{g(x, y)}{f(x, y)}$$
and are everywhere tangent to the vector field.

58. Find and graph the flow curves for the vector field $\mathbf{F} = \langle 1, x \rangle$.

59. Find and graph the flow curves for the vector field $\mathbf{F} = \langle x, x \rangle$.

60. Find and graph the flow curves for the vector field $\mathbf{F} = \langle y, x \rangle$.
Note that $\dfrac{d}{dx}(y^2) = 2yy'(x)$.

61. Find and graph the flow curves for the vector field $\mathbf{F} = \langle -y, x \rangle$.

62–63. Unit vectors in polar coordinates

62. Vectors in $\mathbb{R}^2$ may also be expressed in terms of polar coordinates. The standard coordinate unit vectors in polar coordinates are denoted $\mathbf{u}_r$ and $\mathbf{u}_\theta$ (see figure). Unlike the coordinate unit vectors in Cartesian coordinates, $\mathbf{u}_r$ and $\mathbf{u}_\theta$ change their direction depending on the point (r, θ). Use the figure to show that for $r > 0$, the following relationships among the unit vectors in Cartesian and polar coordinates hold:

$$\mathbf{u}_r = \cos\theta\,\mathbf{i} + \sin\theta\,\mathbf{j} \qquad \mathbf{i} = \mathbf{u}_r\cos\theta - \mathbf{u}_\theta\sin\theta$$
$$\mathbf{u}_\theta = -\sin\theta\,\mathbf{i} + \cos\theta\,\mathbf{j} \qquad \mathbf{j} = \mathbf{u}_r\sin\theta + \mathbf{u}_\theta\cos\theta$$

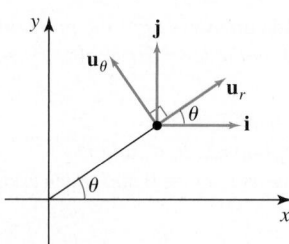

63. Verify that the relationships in Exercise 62 are consistent when $\theta = 0, \pi/2, \pi$, and $3\pi/2$.

64–66. Vector fields in polar coordinates *A vector field in polar coordinates has the form* $\mathbf{F}(r, \theta) = f(r, \theta)\,\mathbf{u}_r + g(r, \theta)\,\mathbf{u}_\theta$, *where the unit vectors are defined in Exercise 62. Sketch the following vector fields and express them in Cartesian coordinates.*

64. $\mathbf{F} = \mathbf{u}_r$ **65.** $\mathbf{F} = \mathbf{u}_\theta$ **66.** $\mathbf{F} = r\,\mathbf{u}_\theta$

67. **Cartesian vector field to polar vector field** Write the vector field $\mathbf{F} = \langle -y, x \rangle$ in polar coordinates and sketch the field.

QUICK CHECK ANSWERS

1. The particle follows a circular path around the origin.
3. $\nabla\varphi = \langle yz, xz, xy \rangle$ ◄

17.2 Line Integrals

With integrals of a single variable, we integrate over intervals in $\mathbb{R}$ (the real line). With double and triple integrals, we integrate over regions in $\mathbb{R}^2$ or $\mathbb{R}^3$. *Line integrals* (which really should be called *curve integrals*) are another class of integrals that play an important role in vector calculus. They are used to integrate either scalar-valued functions or vector fields along curves.

Suppose a thin, circular plate has a known temperature distribution and you must compute the average temperature along the edge of the plate. The required calculation involves integrating the temperature function over the *curved* boundary of the plate. Similarly, to calculate the amount of work needed to put a satellite into orbit, we integrate the gravitational force (a vector field) along the curved path of the satellite. Both these calculations require line integrals. As you will see, line integrals take several different forms. It is the goal of this section to distinguish among these various forms and show how and when each form should be used.

Scalar Line Integrals in the Plane

We focus first on line integrals of scalar-valued functions over curves in the xy-plane. Assume C is a smooth curve of finite length given by $\mathbf{r}(t) = \langle x(t), y(t) \rangle$, for $a \le t \le b$. We divide $[a, b]$ into n subintervals using the grid points

$$a = t_0 < t_1 < \cdots < t_{n-1} < t_n = b.$$

This partition of $[a, b]$ divides C into n subarcs (**Figure 17.16**), where the arc length of the kth subarc is denoted Δs_k. Let t_k^* be a point in the kth subinterval $[t_{k-1}, t_k]$, which corresponds to a point $(x(t_k^*), y(t_k^*))$ on the kth subarc of C, for $k = 1, 2, \ldots, n$.

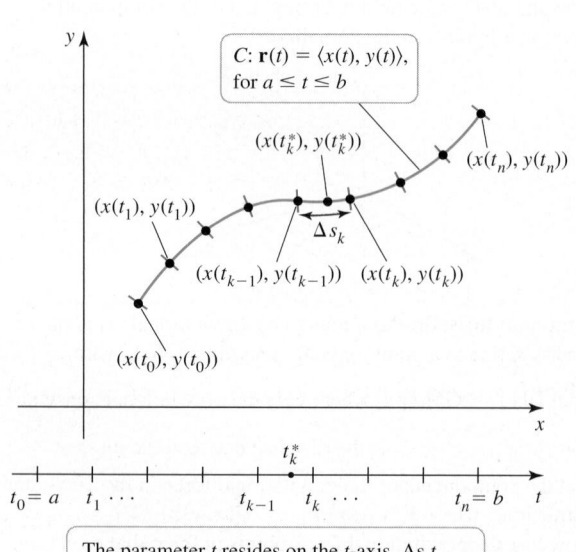

$C: \mathbf{r}(t) = \langle x(t), y(t) \rangle$, for $a \le t \le b$

The parameter t resides on the t-axis. As t varies from a to b, the curve C in the xy-plane is generated from $(x(a), y(a))$ to $(x(b), y(b))$.

Figure 17.16

Now consider a scalar-valued function $z = f(x, y)$ defined on a region containing C. Evaluating f at $(x(t_k^*), y(t_k^*))$ and multiplying this value by Δs_k, we form the sum

$$S_n = \sum_{k=1}^{n} f(x(t_k^*), y(t_k^*)) \Delta s_k,$$

which is similar to a Riemann sum. We now let Δ be the maximum value of $\{\Delta s_1, \ldots, \Delta s_n\}$. If the limit of the sum as $n \to \infty$ and $\Delta \to 0$ exists over all partitions, then the limit is called *the line integral of f over C.*

DEFINITION Scalar Line Integral in the Plane

Suppose the scalar-valued function f is defined on a region containing the smooth curve C given by $\mathbf{r}(t) = \langle x(t), y(t) \rangle$, for $a \le t \le b$. The **line integral of f over C** is

$$\int_C f(x(t), y(t)) \, ds = \lim_{\Delta \to 0} \sum_{k=1}^{n} f(x(t_k^*), y(t_k^*)) \Delta s_k,$$

provided this limit exists over all partitions of $[a, b]$. When the limit exists, f is said to be **integrable** on C.

The more compact notations $\int_C f(\mathbf{r}(t)) \, ds$, $\int_C f(x, y) \, ds$, and $\int_C f \, ds$ are also used for the line integral of f over C. It can be shown that if f is continuous on a region containing C, then f is integrable over C.

There are several useful interpretations of the line integral of a scalar function. If $f(x, y) = 1$, the line integral $\int_C ds$ gives the length of the curve C, just as the ordinary integral $\int_a^b dx$ gives the length of the interval $[a, b]$, which is $b - a$. If $f(x, y) \ge 0$ on C, then $\int_C f(x, y) \, ds$ can be viewed as the area of one side of the vertical, curtain-like surface that lies between the graphs of f and C (Figure 17.17). This interpretation results from regarding the product $f(x(t_k^*), y(t_k^*)) \Delta s_k$ as an approximation to the area of the kth panel of the curtain. Similarly, if f is a density function for a thin wire represented by the curve C, then $\int_C f(x, y) \, ds$ gives the mass of the wire—the product $f(x(t_k^*), y(t_k^*)) \Delta s_k$ is an approximation to the mass of the kth piece of the wire (Exercises 35–36).

Evaluating Line Integrals

The line integral of f over C given in the definition is not an ordinary Riemann integral, because the integrand is expressed as a function of t while the variable of integration is the arc length parameter s. We need a practical way to evaluate such integrals; the key is to use a change of variables to convert a line integral into an ordinary integral. Let C be given by $\mathbf{r}(t) = \langle x(t), y(t) \rangle$, for $a \le t \le b$. Recall from Section 14.4 that the length of C over the interval $[a, t]$ is

$$s(t) = \int_a^t |\mathbf{r}'(u)| \, du.$$

Differentiating both sides of this equation and using the Fundamental Theorem of Calculus yields $s'(t) = |\mathbf{r}'(t)|$. We now make a standard change of variables using the relationship

$$ds = s'(t) \, dt = |\mathbf{r}'(t)| \, dt.$$

Relying on a result from advanced calculus, the original line integral with respect to s can be converted into an ordinary integral with respect to t:

$$\int_C f \, ds = \int_a^b f(x(t), y(t)) \underbrace{|\mathbf{r}'(t)| \, dt}_{ds}.$$

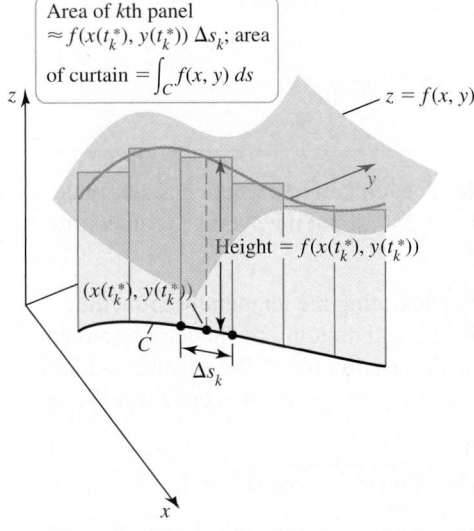

Area of kth panel $\approx f(x(t_k^*), y(t_k^*)) \, \Delta s_k$; area of curtain $= \int_C f(x, y) \, ds$

$z = f(x, y)$

Height $= f(x(t_k^*), y(t_k^*))$

$(x(t_k^*), y(t_k^*))$

C

Δs_k

Figure 17.17

QUICK CHECK 1 Explain mathematically why differentiating the arc length integral leads to $s'(t) = |\mathbf{r}'(t)|$. ◀

▶ If t represents time, then the relationship $ds = |\mathbf{r}'(t)| \, dt$ is a generalization of the familiar formula

distance = speed · time.

▶ The value of a line integral of a scalar-valued function is independent of the parameterization of C and independent of the direction in which C is traversed (Exercises 64–65).

THEOREM 17.1   Evaluating Scalar Line Integrals in $\mathbb{R}^2$

Let f be continuous on a region containing a smooth curve C: $\mathbf{r}(t) = \langle x(t), y(t) \rangle$, for $a \le t \le b$. Then

$$\int_C f \, ds = \int_a^b f(x(t), y(t)) |\mathbf{r}'(t)| \, dt$$

$$= \int_a^b f(x(t), y(t)) \sqrt{x'(t)^2 + y'(t)^2} \, dt.$$

If t represents time and C is the path of a moving object, then $|\mathbf{r}'(t)|$ is the speed of the object. The *speed factor* $|\mathbf{r}'(t)|$ that appears in the integral relates distance traveled along the curve as measured by s to the elapsed time as measured by the parameter t.

Notice that if $f(x, y) = 1$, then the line integral is $\int_a^b \sqrt{x'(t)^2 + y'(t)^2} \, dt$, which is the arc length formula for C. Theorem 17.1 leads to the following procedure for evaluating line integrals.

PROCEDURE   Evaluating the Line Integral $\int_C f \, ds$

1. Find a parametric description of C in the form $\mathbf{r}(t) = \langle x(t), y(t) \rangle$, for $a \le t \le b$.

2. Compute $|\mathbf{r}'(t)| = \sqrt{x'(t)^2 + y'(t)^2}$.

3. Make substitutions for x and y in the integrand and evaluate an ordinary integral:

$$\int_C f \, ds = \int_a^b f(x(t), y(t)) |\mathbf{r}'(t)| \, dt.$$

▶ When we compute the average value by an ordinary integral, we divide by the length of the interval of integration. Analogously, when we compute the average value by a line integral, we divide by the length of the curve L:

$$\bar{f} = \frac{1}{L} \int_C f \, ds.$$

EXAMPLE 1   Average temperature on a circle The temperature of the circular plate $R = \{(x, y): x^2 + y^2 \le 1\}$ is $T(x, y) = 100(x^2 + 2y^2)$. Find the average temperature along the edge of the plate.

SOLUTION Calculating the average value requires integrating the temperature function over the boundary circle $C = \{(x, y): x^2 + y^2 = 1\}$ and dividing by the length (circumference) of C. The first step is to find a parametric description for C. We use the standard parameterization for a unit circle centered at the origin, $\mathbf{r} = \langle x, y \rangle = \langle \cos t, \sin t \rangle$, for $0 \le t \le 2\pi$. Next, we compute the speed factor

$$|\mathbf{r}'(t)| = \sqrt{x'(t)^2 + y'(t)^2} = \sqrt{(-\sin t)^2 + (\cos t)^2} = 1.$$

We substitute $x = \cos t$ and $y = \sin t$ into the temperature function and express the line integral as an ordinary integral with respect to t:

$$\int_C T(x, y) \, ds = \int_0^{2\pi} \underbrace{100(x(t)^2 + 2y(t)^2)}_{T(t)} \underbrace{|\mathbf{r}'(t)|}_{1} \, dt \qquad \text{Write the line integral as an ordinary integral with respect to } t; \, ds = |\mathbf{r}'(t)| \, dt.$$

$$= 100 \int_0^{2\pi} (\cos^2 t + 2 \sin^2 t) \, dt \qquad \text{Substitute for } x \text{ and } y.$$

$$= 100 \underbrace{\int_0^{2\pi} (1 + \sin^2 t) \, dt}_{3\pi} \qquad \cos^2 t + \sin^2 t = 1$$

$$= 300\pi. \qquad \text{Use } \sin^2 t = \frac{1 - \cos 2t}{2} \text{ and integrate.}$$

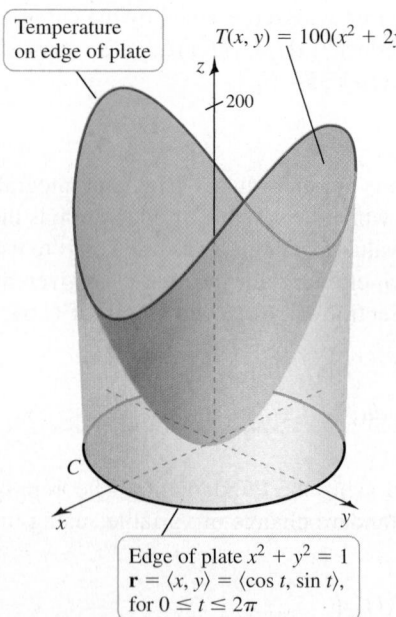

Temperature on edge of plate

$T(x, y) = 100(x^2 + 2y^2)$

Edge of plate $x^2 + y^2 = 1$
$\mathbf{r} = \langle x, y \rangle = \langle \cos t, \sin t \rangle$,
for $0 \le t \le 2\pi$

Figure 17.18

▶ The line integral in Example 1 also gives the area of the vertical cylindrical curtain that hangs between the surface and C in Figure 17.18.

The geometry of this line integral is shown in **Figure 17.18**. The temperature function on the boundary of C is a function of t. The line integral is an ordinary integral with respect to t over the interval $[0, 2\pi]$. To find the average value, we divide the line integral of the temperature by the length of the curve, which is 2π. Therefore, the average temperature on the boundary of the plate is $300\pi/(2\pi) = 150$.

Related Exercise 17 ◀

QUICK CHECK 2 Suppose $\mathbf{r}(t) = \langle t, 0 \rangle$, for $a \le t \le b$, is a parametric description of C; note that C is the interval $[a, b]$ on the x-axis. Show that $\int_C f(x, y)\, ds = \int_a^b f(t, 0)\, dt$, which is an ordinary, single-variable integral introduced in Chapter 5. ◄

> ► If $f(x, y, z) = 1$, then the line integral gives the length of C.

> ► Recall that a parametric equation of a line is
>
> $\mathbf{r}(t) = \langle x_0, y_0, z_0 \rangle + t \langle a, b, c \rangle,$
>
> where $\langle x_0, y_0, z_0 \rangle$ is a position vector associated with a fixed point on the line and $\langle a, b, c \rangle$ is a vector parallel to the line.

Line Integrals in $\mathbb{R}^3$

The argument that leads to line integrals on plane curves extends immediately to three or more dimensions. Here is the corresponding evaluation theorem for line integrals in $\mathbb{R}^3$.

THEOREM 17.2   Evaluating Scalar Line Integrals in $\mathbb{R}^3$

Let f be continuous on a region containing a smooth curve $C: \mathbf{r}(t) = \langle x(t), y(t), z(t) \rangle$, for $a \le t \le b$. Then

$$\int_C f\, ds = \int_a^b f(x(t), y(t), z(t)) |\mathbf{r}'(t)|\, dt$$

$$= \int_a^b f(x(t), y(t), z(t)) \sqrt{x'(t)^2 + y'(t)^2 + z'(t)^2}\, dt.$$

EXAMPLE 2   Line integrals in $\mathbb{R}^3$ Evaluate $\int_C (xy + 2z)\, ds$ on the following line segments.

a. The line segment from $P(1, 0, 0)$ to $Q(0, 1, 1)$
b. The line segment from $Q(0, 1, 1)$ to $P(1, 0, 0)$

SOLUTION

a. A parametric description of the line segment from $P(1, 0, 0)$ to $Q(0, 1, 1)$ is

$$\mathbf{r}(t) = \langle 1, 0, 0 \rangle + t\langle -1, 1, 1 \rangle = \langle 1 - t, t, t \rangle, \text{ for } 0 \le t \le 1.$$

The speed factor is

$$|\mathbf{r}'(t)| = \sqrt{x'(t)^2 + y'(t)^2 + z'(t)^2} = \sqrt{(-1)^2 + 1^2 + 1^2} = \sqrt{3}.$$

Substituting $x = 1 - t$, $y = t$, and $z = t$, the value of the line integral is

$$\int_C (xy + 2z)\, ds = \int_0^1 (\underbrace{(1-t)}_x \underbrace{t}_y + \underbrace{2t}_{2z})\sqrt{3}\, dt \quad \text{Substitute for } x, y, \text{ and } z.$$

$$= \sqrt{3} \int_0^1 (3t - t^2)\, dt \qquad \text{Simplify.}$$

$$= \sqrt{3}\left(\frac{3t^2}{2} - \frac{t^3}{3} \right)\Big|_0^1 \qquad \text{Integrate.}$$

$$= \frac{7\sqrt{3}}{6}. \qquad \text{Evaluate.}$$

b. The line segment from $Q(0, 1, 1)$ to $P(1, 0, 0)$ may be described parametrically by

$$\mathbf{r}(t) = \langle 0, 1, 1 \rangle + t\langle 1, -1, -1 \rangle = \langle t, 1 - t, 1 - t \rangle, \text{ for } 0 \le t \le 1.$$

The speed factor is

$$|\mathbf{r}'(t)| = \sqrt{x'(t)^2 + y'(t)^2 + z'(t)^2} = \sqrt{1^2 + (-1)^2 + (-1)^2} = \sqrt{3}.$$

We substitute $x = t$, $y = 1 - t$, and $z = 1 - t$ and do a calculation similar to that in part (a). The value of the line integral is again $\dfrac{7\sqrt{3}}{6}$, emphasizing the fact that a scalar line integral is independent of the orientation and parameterization of the curve.

Related Exercises 32–33 ◄

EXAMPLE 3   Flight of an eagle An eagle soars on the ascending spiral path

$$C: \mathbf{r}(t) = \langle x(t), y(t), z(t) \rangle = \left\langle 2400 \cos \frac{t}{2}, 2400 \sin \frac{t}{2}, 500t \right\rangle,$$

where x, y, and z are measured in feet and t is measured in minutes. How far does the eagle fly over the time interval $0 \le t \le 10$?

> ▶ Because we are finding the length of a curve, the integrand in this line integral is $f(x, y, z) = 1$.

SOLUTION The distance traveled is found by integrating the element of arc length ds along C, that is, $L = \int_C ds$. We now make a change of variables to the parameter t using

$$|\mathbf{r}'(t)| = \sqrt{x'(t)^2 + y'(t)^2 + z'(t)^2}$$

$$= \sqrt{\left(-1200 \sin \frac{t}{2}\right)^2 + \left(1200 \cos \frac{t}{2}\right)^2 + 500^2} \quad \text{Substitute derivatives.}$$

$$= \sqrt{1200^2 + 500^2} = 1300. \qquad \sin^2 \frac{t}{2} + \cos^2 \frac{t}{2} = 1$$

It follows that the distance traveled is

QUICK CHECK 3 What is the speed of the eagle in Example 3? ◀

$$L = \int_C ds = \int_0^{10} |\mathbf{r}'(t)| \, dt = \int_0^{10} 1300 \, dt = 13{,}000 \text{ ft.}$$

Related Exercise 39 ◀

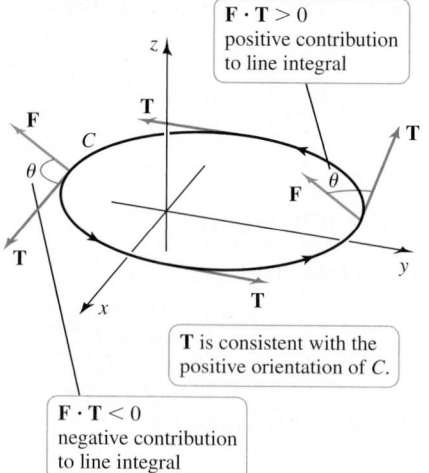

F · T > 0
positive contribution to line integral

T is consistent with the positive orientation of C.

F · T < 0
negative contribution to line integral

Figure 17.19

Line Integrals of Vector Fields

Line integrals along curves in $\mathbb{R}^2$ or $\mathbb{R}^3$ may also have integrands that involve vector fields. Such line integrals are different from scalar line integrals in two respects.

- Recall that an *oriented curve* is a parameterized curve for which a direction is specified. The *positive* orientation is the direction in which the curve is generated as the parameter increases. For example, the positive orientation of the circle $\mathbf{r}(t) = \langle \cos t, \sin t \rangle$, for $0 \le t \le 2\pi$, is counterclockwise. As we will see, vector line integrals must be evaluated on oriented curves, and the value of a line integral depends on the orientation.

- The line integral of a vector field $\mathbf{F}$ along an oriented curve involves a specific component of $\mathbf{F}$ relative to the curve. We begin by defining vector line integrals for the *tangential* component of $\mathbf{F}$, a situation that has many physical applications.

Let C: $\mathbf{r}(s) = \langle x(s), y(s), z(s) \rangle$ be a smooth oriented curve in $\mathbb{R}^3$ parameterized by arc length and let $\mathbf{F}$ be a vector field that is continuous on a region containing C. At each point of C, the unit tangent vector $\mathbf{T}$ points in the positive direction on C (**Figure 17.19**). The component of $\mathbf{F}$ in the direction of $\mathbf{T}$ at a point of C is $|\mathbf{F}| \cos \theta$, where θ is the angle between $\mathbf{F}$ and $\mathbf{T}$. Because $\mathbf{T}$ is a unit vector,

$$|\mathbf{F}| \cos \theta = |\mathbf{F}| |\mathbf{T}| \cos \theta = \mathbf{F} \cdot \mathbf{T}.$$

The first line integral of a vector field $\mathbf{F}$ that we introduce is the line integral of the scalar $\mathbf{F} \cdot \mathbf{T}$ along the curve C. When we integrate $\mathbf{F} \cdot \mathbf{T}$ along C, the effect is to add up the components of $\mathbf{F}$ in the direction of C at each point of C.

> ▶ The component of $\mathbf{F}$ in the direction of $\mathbf{T}$ is the scalar component of $\mathbf{F}$ in the direction of $\mathbf{T}$, $\text{scal}_\mathbf{T}\, \mathbf{F}$, as defined in Section 13.3. Note that $|\mathbf{T}| = 1$.

DEFINITION   **Line Integral of a Vector Field**

Let $\mathbf{F}$ be a vector field that is continuous on a region containing a smooth oriented curve C parameterized by arc length. Let $\mathbf{T}$ be the unit tangent vector at each point of C consistent with the orientation. The line integral of $\mathbf{F}$ over C is $\int_C \mathbf{F} \cdot \mathbf{T} \, ds$.

> ▶ Some texts let $d\mathbf{s}$ stand for $\mathbf{T}\, ds$. Then the line integral $\int_C \mathbf{F} \cdot \mathbf{T} \, ds$ is written $\int_C \mathbf{F} \cdot d\mathbf{s}$.

Just as we did for line integrals of scalar-valued functions, we need a method for evaluating vector line integrals when the parameter is not the arc length. Suppose C has a parameterization $\mathbf{r}(t) = \langle x(t), y(t), z(t) \rangle$, for $a \le t \le b$. Recall from Section 14.2 that the unit tangent vector at a point on the curve is $\mathbf{T} = \dfrac{\mathbf{r}'(t)}{|\mathbf{r}'(t)|}$. Using the fact that $ds = |\mathbf{r}'(t)| \, dt$, the line integral becomes

$$\int_C \mathbf{F} \cdot \mathbf{T} \, ds = \int_a^b \mathbf{F} \cdot \underbrace{\frac{\mathbf{r}'(t)}{|\mathbf{r}'(t)|}}_{\mathbf{T}} \underbrace{|\mathbf{r}'(t)| \, dt}_{ds} = \int_a^b \mathbf{F} \cdot \mathbf{r}'(t) \, dt.$$

▶ Keep in mind that $f(t)$ stands for $f(x(t), y(t), z(t))$ with analogous expressions for $g(t)$ and $h(t)$.

This integral may be written in several equivalent forms. If $\mathbf{F} = \langle f, g, h \rangle$, then the line integral is expressed in component form as

$$\int_C \mathbf{F} \cdot \mathbf{T} \, ds = \int_a^b \mathbf{F} \cdot \mathbf{r}'(t) \, dt = \int_a^b (f(t)x'(t) + g(t)y'(t) + h(t)z'(t)) \, dt.$$

Another useful form is obtained by noting that

$$dx = x'(t) \, dt, \qquad dy = y'(t) \, dt, \qquad dz = z'(t) \, dt.$$

Making these replacements in the previous integral results in the form

$$\int_C \mathbf{F} \cdot \mathbf{T} \, ds = \int_C f \, dx + g \, dy + h \, dz.$$

Finally, if we let $d\mathbf{r} = \langle dx, dy, dz \rangle$, then $f \, dx + g \, dy + h \, dz = \mathbf{F} \cdot d\mathbf{r}$, and we have

$$\int_C \mathbf{F} \cdot \mathbf{T} \, ds = \int_C \mathbf{F} \cdot d\mathbf{r}.$$

It is helpful to become familiar with these various forms of the line integral.

Different Forms of Line Integrals of Vector Fields

The line integral $\int_C \mathbf{F} \cdot \mathbf{T} \, ds$ may be expressed in the following forms, where $\mathbf{F} = \langle f, g, h \rangle$ and C has a parameterization $\mathbf{r}(t) = \langle x(t), y(t), z(t) \rangle$, for $a \le t \le b$:

$$\int_a^b \mathbf{F} \cdot \mathbf{r}'(t) \, dt = \int_a^b (f(t)x'(t) + g(t)y'(t) + h(t)z'(t)) \, dt$$

$$= \int_C f \, dx + g \, dy + h \, dz$$

$$= \int_C \mathbf{F} \cdot d\mathbf{r}.$$

For line integrals in the plane, we let $\mathbf{F} = \langle f, g \rangle$ and assume C is parameterized in the form $\mathbf{r}(t) = \langle x(t), y(t) \rangle$, for $a \le t \le b$. Then

$$\int_a^b \mathbf{F} \cdot \mathbf{r}'(t) \, dt = \int_a^b (f(t)x'(t) + g(t)y'(t)) \, dt = \int_C f \, dx + g \, dy = \int_C \mathbf{F} \cdot d\mathbf{r}.$$

▶ We use the convention that $-C$ is the curve C with the opposite orientation.

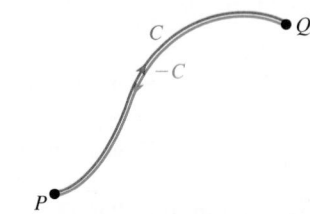

EXAMPLE 4   Different paths Evaluate $\int_C \mathbf{F} \cdot \mathbf{T} \, ds$ with $\mathbf{F} = \langle y - x, x \rangle$ on the following oriented paths in $\mathbb{R}^2$ (**Figure 17.20**).

a. The quarter-circle C_1 from $P(0, 1)$ to $Q(1, 0)$
b. The quarter-circle $-C_1$ from $Q(1, 0)$ to $P(0, 1)$
c. The path C_2 from $P(0, 1)$ to $Q(1, 0)$ via two line segments through $O(0, 0)$

SOLUTION

a. Working in $\mathbb{R}^2$, a parametric description of the curve C_1 with the required (clockwise) orientation is $\mathbf{r}(t) = \langle \sin t, \cos t \rangle$, for $0 \le t \le \pi/2$. Along C_1, the vector field is

$$\mathbf{F} = \langle y - x, x \rangle = \langle \cos t - \sin t, \sin t \rangle.$$

The velocity vector is $\mathbf{r}'(t) = \langle \cos t, -\sin t \rangle$, so the integrand of the line integral is

$$\mathbf{F} \cdot \mathbf{r}'(t) = \langle \cos t - \sin t, \sin t \rangle \cdot \langle \cos t, -\sin t \rangle = \underbrace{\cos^2 t - \sin^2 t}_{\cos 2t} - \underbrace{\sin t \cos t}_{\frac{1}{2} \sin 2t}.$$

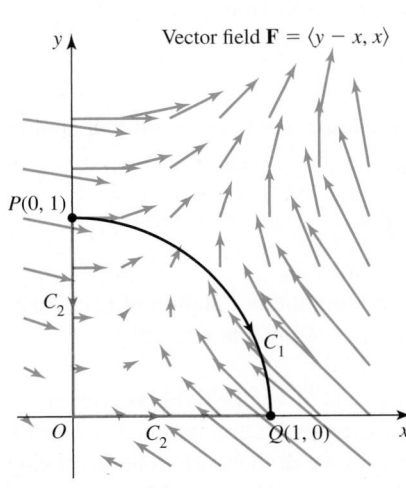

Figure 17.20

The value of the line integral of **F** over C_1 is

$$\int_0^{\pi/2} \mathbf{F} \cdot \mathbf{r}'(t)\, dt = \int_0^{\pi/2} \left(\cos 2t - \frac{1}{2} \sin 2t \right) dt \qquad \text{Substitute for } \mathbf{F} \cdot \mathbf{r}'(t).$$

$$= \left(\frac{1}{2} \sin 2t + \frac{1}{4} \cos 2t \right) \Big|_0^{\pi/2} \qquad \text{Evaluate integral.}$$

$$= -\frac{1}{2}. \qquad \text{Simplify.}$$

b. A parameterization of the curve $-C_1$ from Q to P is $\mathbf{r}(t) = \langle \cos t, \sin t \rangle$, for $0 \le t \le \pi/2$. The vector field along the curve is

$$\mathbf{F} = \langle y - x, x \rangle = \langle \sin t - \cos t, \cos t \rangle,$$

and the velocity vector is $\mathbf{r}'(t) = \langle -\sin t, \cos t \rangle$. A calculation similar to that in part (a) results in

$$\int_{-C_1} \mathbf{F} \cdot \mathbf{T}\, ds = \int_0^{\pi/2} \mathbf{F} \cdot \mathbf{r}'(t)\, dt = \frac{1}{2}.$$

Comparing the results of parts (a) and (b), we see that reversing the orientation of C_1 reverses the sign of the line integral of the vector field.

c. The path C_2 consists of two line segments.

- The segment from P to O is parameterized by $\mathbf{r}(t) = \langle 0, 1 - t \rangle$, for $0 \le t \le 1$. Therefore, $\mathbf{r}'(t) = \langle 0, -1 \rangle$ and $\mathbf{F} = \langle y - x, x \rangle = \langle 1 - t, 0 \rangle$. On this segment, $\mathbf{T} = \langle 0, -1 \rangle$.
- The line segment from O to Q is parameterized by $\mathbf{r}(t) = \langle t, 0 \rangle$, for $0 \le t \le 1$. Therefore, $\mathbf{r}'(t) = \langle 1, 0 \rangle$ and $\mathbf{F} = \langle y - x, x \rangle = \langle -t, t \rangle$. On this segment, $\mathbf{T} = \langle 1, 0 \rangle$.

> Line integrals of vector fields satisfy properties similar to those of ordinary integrals. Suppose C is a smooth curve from A to B, C_1 is the curve from A to P, and C_2 is the curve from P to B, where P is a point on C between A and B. Then
>
> $$\int_C \mathbf{F} \cdot d\mathbf{r} = \int_{C_1} \mathbf{F} \cdot d\mathbf{r} + \int_{C_2} \mathbf{F} \cdot d\mathbf{r}.$$

The line integral is split into two parts and evaluated as follows:

$$\int_{C_2} \mathbf{F} \cdot \mathbf{T}\, ds = \int_{PO} \mathbf{F} \cdot \mathbf{T}\, ds + \int_{OQ} \mathbf{F} \cdot \mathbf{T}\, ds$$

$$= \int_0^1 \langle 1 - t, 0 \rangle \cdot \langle 0, -1 \rangle\, dt + \int_0^1 \langle -t, t \rangle \cdot \langle 1, 0 \rangle\, dt \qquad \begin{array}{l}\text{Substitute} \\ \text{for } x, y, \mathbf{r}'.\end{array}$$

$$= \int_0^1 0\, dt + \int_0^1 (-t)\, dt \qquad \text{Simplify.}$$

$$= -\frac{1}{2}. \qquad \text{Evaluate integrals.}$$

The line integrals in parts (a) and (c) have the same value and run from P to Q, but along different paths. We might ask: For what vector fields are the values of a line integral independent of path? We return to this question in Section 17.3.

Related Exercises 42–43 ◄

The solutions to parts (a) and (b) of Example 4 illustrate a general result that applies to line integrals of vector fields:

$$\int_{-C} \mathbf{F} \cdot \mathbf{T}\, ds = -\int_C \mathbf{F} \cdot \mathbf{T}\, ds.$$

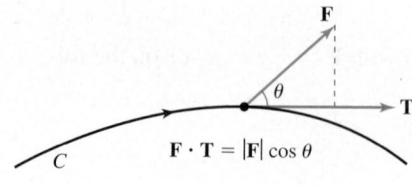

$$\mathbf{F} \cdot \mathbf{T} = |\mathbf{F}| \cos \theta$$

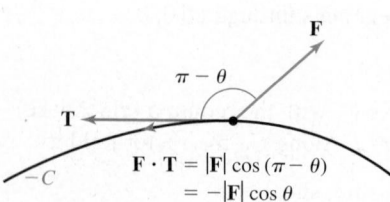

$$\mathbf{F} \cdot \mathbf{T} = |\mathbf{F}| \cos (\pi - \theta)$$
$$= -|\mathbf{F}| \cos \theta$$

Reversing the orientation of C changes the sign of $\mathbf{F} \cdot \mathbf{T}$ at each point on C.

Figure 17.21

Figure 17.21 provides the justification of this fact: Reversing the orientation of C changes the sign of $\mathbf{F} \cdot \mathbf{T}$ at each point of C, which changes the sign of the line integral.

Work Integrals   A common application of line integrals of vector fields is computing the work done in moving an object in a force field (for example, a gravitational or electric field). First recall (Section 6.7) that if **F** is a *constant* force field, the work done in moving an object a distance d along the x-axis is $W = F_x d$, where $F_x = |\mathbf{F}| \cos \theta$ is the component

➤ Remember that the value of $\int_C f \, ds$ (the line integral of a *scalar* function) does not depend on the orientation of C.

QUICK CHECK 4 Suppose a two-dimensional force field is everywhere directed outward from the origin, and C is a circle centered at the origin. What is the angle between the field and the unit vectors tangent to C? ◄

of the force along the x-axis (**Figure 17.22a**). Only the component of **F** in the direction of motion contributes to the work. More generally, if **F** is a *variable* force field, the work done in moving an object from $x = a$ to $x = b$ is $W = \int_a^b F_x(x) \, dx$, where again F_x is the component of the force **F** in the direction of motion (parallel to the x-axis, **Figure 17.22b**).

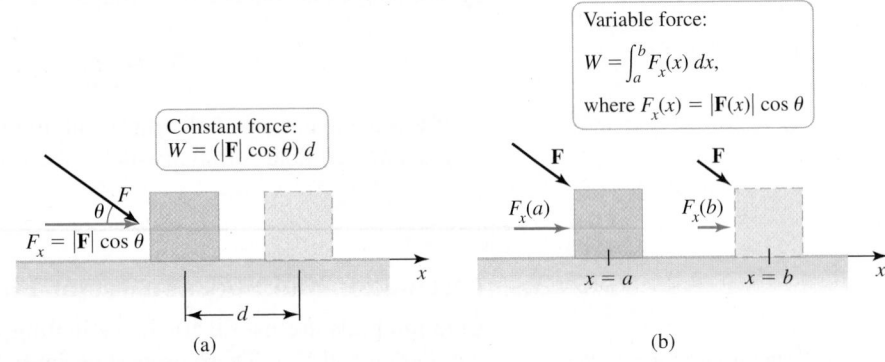

Figure 17.22

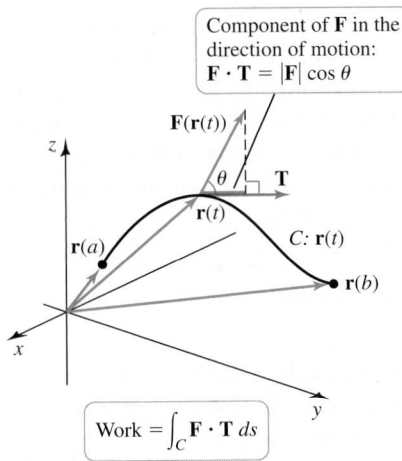

$$\text{Work} = \int_C \mathbf{F} \cdot \mathbf{T} \, ds$$

Figure 17.23

➤ Just to be clear, a work integral is nothing more than a line integral of the tangential component of a force field.

We now take this progression one step further. Let **F** be a variable force field defined in a region D of $\mathbb{R}^3$ and suppose C is a smooth oriented curve in D along which an object moves. The direction of motion at each point of C is given by the unit tangent vector **T**. Therefore, the component of **F** in the direction of motion is $\mathbf{F} \cdot \mathbf{T}$, which is the tangential component of **F** along C. Summing the contributions to the work at each point of C, the work done in moving an object along C in the presence of the force is the line integral of $\mathbf{F} \cdot \mathbf{T}$ (**Figure 17.23**).

DEFINITION Work Done in a Force Field

Let **F** be a continuous force field in a region D of $\mathbb{R}^3$. Let

$$C: \mathbf{r}(t) = \langle x(t), y(t), z(t) \rangle, \text{ for } a \le t \le b,$$

be a smooth curve in D with a unit tangent vector **T** consistent with the orientation. The work done in moving an object along C in the positive direction is

$$W = \int_C \mathbf{F} \cdot \mathbf{T} \, ds = \int_a^b \mathbf{F} \cdot \mathbf{r}'(t) \, dt.$$

EXAMPLE 5 An inverse square force Gravitational and electrical forces between point masses and point charges obey inverse square laws: They act along the line joining the centers and they vary as $1/r^2$, where r is the distance between the centers. The force of attraction (or repulsion) of an inverse square force field is given by the vector field $\mathbf{F} = \dfrac{k \langle x, y, z \rangle}{(x^2 + y^2 + z^2)^{3/2}}$, where k is a physical constant. Because $\mathbf{r} = \langle x, y, z \rangle$, this force may also be written $\mathbf{F} = \dfrac{k\mathbf{r}}{|\mathbf{r}|^3}$. Find the work done in moving an object along the following paths.

a. C_1 is the line segment from $(1, 1, 1)$ to (a, a, a), where $a > 1$.

b. C_2 is the extension of C_1 produced by letting $a \to \infty$.

SOLUTION

a. A parametric description of C_1 consistent with the orientation is $\mathbf{r}(t) = \langle t, t, t \rangle$, for $1 \le t \le a$, with $\mathbf{r}'(t) = \langle 1, 1, 1 \rangle$. In terms of the parameter t, the force field is

$$\mathbf{F} = \frac{k \langle x, y, z \rangle}{(x^2 + y^2 + z^2)^{3/2}} = \frac{k \langle t, t, t \rangle}{(3t^2)^{3/2}}.$$

The dot product that appears in the work integral is

$$\mathbf{F} \cdot \mathbf{r}'(t) = \frac{k \langle t, t, t \rangle}{(3t^2)^{3/2}} \cdot \langle 1, 1, 1 \rangle = \frac{3kt}{3\sqrt{3}\, t^3} = \frac{k}{\sqrt{3}\, t^2}.$$

Therefore, the work done is

$$W = \int_1^a \mathbf{F} \cdot \mathbf{r}'(t)\, dt = \frac{k}{\sqrt{3}} \int_1^a t^{-2}\, dt = \frac{k}{\sqrt{3}}\left(1 - \frac{1}{a}\right).$$

b. The path C_2 is obtained by letting $a \to \infty$ in part (a). The required work is

$$W = \lim_{a \to \infty} \frac{k}{\sqrt{3}}\left(1 - \frac{1}{a}\right) = \frac{k}{\sqrt{3}}.$$

If $\mathbf{F}$ is a gravitational field, this result implies that the work required to escape Earth's gravitational field is finite (which makes space flight possible).

Related Exercise 55 ◄

Circulation and Flux of a Vector Field

Line integrals are useful for investigating two important properties of vector fields: *circulation* and *flux*. These properties apply to any vector field, but they are particularly relevant and easy to visualize if you think of $\mathbf{F}$ as the velocity field for a moving fluid.

> ➤ In the definition of circulation, a *closed curve* is a curve whose initial and terminal points are the same, as defined formally in Section 17.3.

> ➤ Although we define circulation integrals for smooth curves, these integrals may be computed on piecewise-smooth curves. We adopt the convention that *piecewise* refers to a curve with finitely many pieces.

Circulation   We assume $\mathbf{F} = \langle f, g, h \rangle$ is a continuous vector field on a region D of $\mathbb{R}^3$, and we take C to be a *closed* smooth oriented curve in D. The *circulation* of $\mathbf{F}$ along C is a measure of how much of the vector field points in the direction of C. More simply, as you travel along C in the positive direction, how much of the vector field is at your back and how much of it is in your face? To determine the circulation, we simply "add up" the components of $\mathbf{F}$ in the direction of the unit tangent vector $\mathbf{T}$ at each point. Therefore, circulation integrals are another example of line integrals of vector fields.

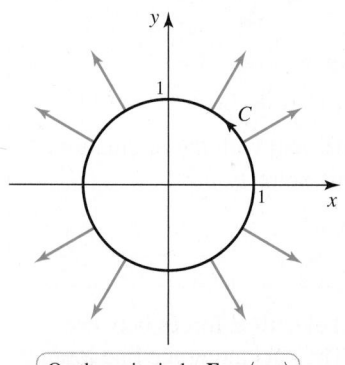

On the unit circle, $\mathbf{F} = \langle x, y \rangle$ is orthogonal to C and has zero circulation on C.

(a)

> **DEFINITION** **Circulation**
>
> Let $\mathbf{F}$ be a continuous vector field on a region D of $\mathbb{R}^3$, and let C be a closed smooth oriented curve in D. The **circulation** of $\mathbf{F}$ on C is $\int_C \mathbf{F} \cdot \mathbf{T}\, ds$, where $\mathbf{T}$ is the unit vector tangent to C consistent with the orientation.

EXAMPLE 6   Circulation of two-dimensional flows Let C be the unit circle with counterclockwise orientation. Find the circulation on C of the following vector fields.

a. The radial vector field $\mathbf{F} = \langle x, y \rangle$

b. The rotation vector field $\mathbf{F} = \langle -y, x \rangle$

SOLUTION

a. The unit circle with the specified orientation is described parametrically by $\mathbf{r}(t) = \langle \cos t, \sin t \rangle$, for $0 \le t \le 2\pi$. Therefore, $\mathbf{r}'(t) = \langle -\sin t, \cos t \rangle$ and the circulation of the radial field $\mathbf{F} = \langle x, y \rangle$ is

$$\int_C \mathbf{F} \cdot \mathbf{T}\, ds = \int_0^{2\pi} \mathbf{F} \cdot \mathbf{r}'(t)\, dt \qquad \text{Evaluation of a line integral}$$

$$= \int_0^{2\pi} \underbrace{\langle \cos t, \sin t \rangle}_{\mathbf{F}\, =\, \langle x, y \rangle} \cdot \underbrace{\langle -\sin t, \cos t \rangle}_{\mathbf{r}'(t)}\, dt \qquad \text{Substitute for } \mathbf{F} \text{ and } \mathbf{r}'.$$

$$= \int_0^{2\pi} 0\, dt = 0. \qquad \text{Simplify.}$$

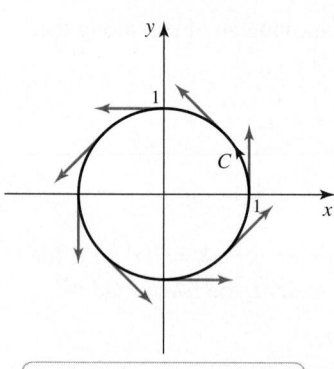

On the unit circle, $\mathbf{F} = \langle -y, x \rangle$ is tangent to C and has positive circulation on C.

(b)

Figure 17.24

The tangential component of the radial field is zero everywhere on C, so the circulation is zero (**Figure 17.24a**).

b. The circulation for the rotation field $\mathbf{F} = \langle -y, x \rangle$ is

$$\int_C \mathbf{F} \cdot \mathbf{T} \, ds = \int_0^{2\pi} \mathbf{F} \cdot \mathbf{r}'(t) \, dt \qquad \text{Evaluation of a line integral}$$

$$= \int_0^{2\pi} \underbrace{\langle -\sin t, \cos t \rangle}_{\mathbf{F} = \langle -y, x \rangle} \cdot \underbrace{\langle -\sin t, \cos t \rangle}_{\mathbf{r}'(t)} \, dt \qquad \text{Substitute for } \mathbf{F} \text{ and } \mathbf{r}'.$$

$$= \int_0^{2\pi} \underbrace{(\sin^2 t + \cos^2 t)}_{1} \, dt \qquad \text{Simplify.}$$

$$= 2\pi.$$

In this case, at every point of C, the rotation field is in the direction of the tangent vector; the result is a positive circulation (**Figure 17.24b**).

Related Exercise 57 ◄

EXAMPLE 7 **Circulation of a three-dimensional flow** Find the circulation of the vector field $\mathbf{F} = \langle z, x, -y \rangle$ on the tilted ellipse $C: \mathbf{r}(t) = \langle \cos t, \sin t, \cos t \rangle$, for $0 \le t \le 2\pi$ (**Figure 17.25a**).

SOLUTION We first determine that

$$\mathbf{r}'(t) = \langle x'(t), y'(t), z'(t) \rangle = \langle -\sin t, \cos t, -\sin t \rangle.$$

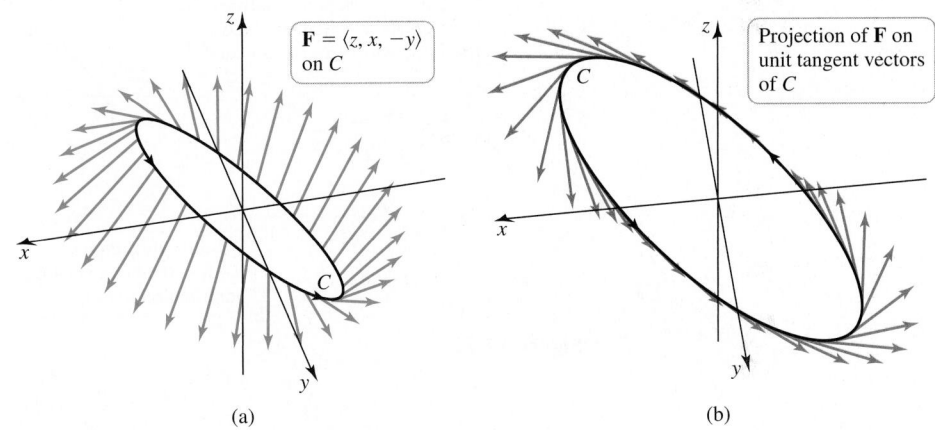

$\mathbf{F} = \langle z, x, -y \rangle$ on C

Projection of $\mathbf{F}$ on unit tangent vectors of C

(a)   (b)

Figure 17.25

Substituting $x = \cos t$, $y = \sin t$, and $z = \cos t$ into $\mathbf{F} = \langle z, x, -y \rangle$, the circulation is

$$\int_C \mathbf{F} \cdot \mathbf{T} \, ds = \int_0^{2\pi} \mathbf{F} \cdot \mathbf{r}'(t) \, dt \qquad \text{Evaluation of a line integral}$$

$$= \int_0^{2\pi} \langle \cos t, \cos t, -\sin t \rangle \cdot \langle -\sin t, \cos t, -\sin t \rangle \, dt \qquad \text{Substitute for } \mathbf{F} \text{ and } \mathbf{r}'.$$

$$= \int_0^{2\pi} (-\sin t \cos t + 1) \, dt \qquad \begin{array}{l}\text{Simplify;} \\ \sin^2 t + \cos^2 t = 1.\end{array}$$

$$= 2\pi. \qquad \text{Evaluate integral.}$$

Figure 17.25b shows the projection of the vector field on the unit tangent vectors at various points on C. The circulation is the "sum" of the scalar components associated with these projections, which, in this case, is positive.

Related Exercise 53 ◄

► In the definition of flux, the non-self-intersecting property of C means that C is a *simple* curve, as defined formally in Section 17.3.

Flux of Two-Dimensional Vector Fields Assume $\mathbf{F} = \langle f, g \rangle$ is a continuous vector field on a region R of $\mathbb{R}^2$. We let C be a smooth oriented curve in R that does not intersect itself; C may or may not be closed. To compute the *flux* of the vector field across C, we "add up" the components of $\mathbf{F}$ *orthogonal* or *normal* to C at each point of C. Notice that every point on C has two unit vectors normal to C. Therefore, we let $\mathbf{n}$ denote the unit vector in the xy-plane normal to C in a direction to be defined momentarily. Once the direction of $\mathbf{n}$ is defined, the component of $\mathbf{F}$ normal to C is $\mathbf{F} \cdot \mathbf{n}$, and the flux is the line integral of $\mathbf{F} \cdot \mathbf{n}$ along C, which we denote $\int_C \mathbf{F} \cdot \mathbf{n}\, ds$.

The first step is to define the unit normal vector at a point P of C. Because C lies in the xy-plane, the unit vector $\mathbf{T}$ tangent at P also lies in the xy-plane. Therefore, its z-component is 0, and we let $\mathbf{T} = \langle T_x, T_y, 0 \rangle$. As always, $\mathbf{k} = \langle 0, 0, 1 \rangle$ is the unit vector in the z-direction. Because a unit vector $\mathbf{n}$ in the xy-plane normal to C is orthogonal to both $\mathbf{T}$ and $\mathbf{k}$, we determine the direction of $\mathbf{n}$ by letting $\mathbf{n} = \mathbf{T} \times \mathbf{k}$. This choice has two implications.

> ➤ Recall that $\mathbf{a} \times \mathbf{b}$ is orthogonal to both $\mathbf{a}$ and $\mathbf{b}$.

- If C is a closed curve oriented counterclockwise (when viewed from above), the unit normal vector points *outward* along the curve (**Figure 17.26a**). When $\mathbf{F}$ also points outward at a point on C, the angle θ between $\mathbf{F}$ and $\mathbf{n}$ satisfies $0 \le \theta < \frac{\pi}{2}$ (**Figure 17.26b**). At all such points, $\mathbf{F} \cdot \mathbf{n} > 0$ and there is a positive contribution to the flux across C. When $\mathbf{F}$ points inward at a point on C, $\frac{\pi}{2} < \theta \le \pi$ and $\mathbf{F} \cdot \mathbf{n} < 0$, which means there is a negative contribution to the flux at that point.

- If C is not a closed curve, the unit normal vector points to the right (when viewed from above) as the curve is traversed in the positive direction.

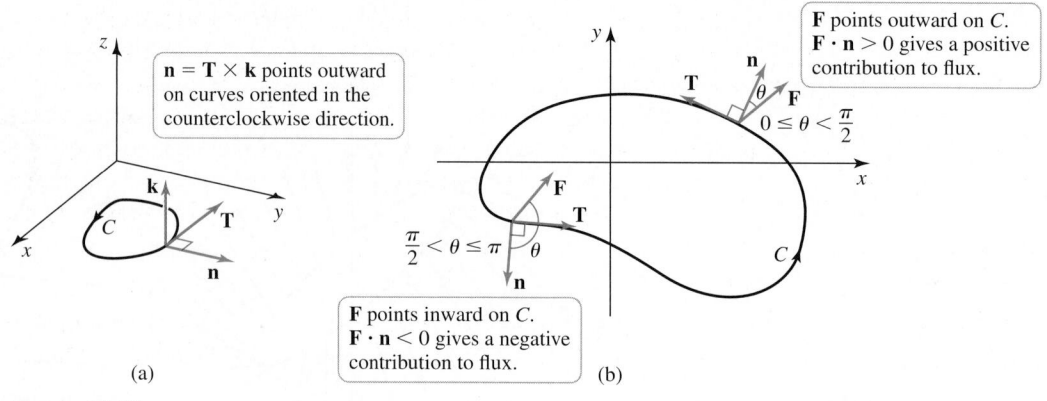

F points outward on C.
$\mathbf{F} \cdot \mathbf{n} > 0$ gives a positive contribution to flux.

$\mathbf{n} = \mathbf{T} \times \mathbf{k}$ points outward on curves oriented in the counterclockwise direction.

$0 \le \theta < \frac{\pi}{2}$

$\frac{\pi}{2} < \theta \le \pi$

F points inward on C.
$\mathbf{F} \cdot \mathbf{n} < 0$ gives a negative contribution to flux.

(a)   (b)

Figure 17.26

QUICK CHECK 5 Sketch a closed curve on a sheet of paper and draw a unit tangent vector $\mathbf{T}$ on the curve pointing in the counterclockwise direction. Explain why $\mathbf{n} = \mathbf{T} \times \mathbf{k}$ is an *outward* unit normal vector. ◄

Calculating the cross product that defines the unit normal vector $\mathbf{n}$, we find that

$$\mathbf{n} = \mathbf{T} \times \mathbf{k} = \begin{vmatrix} \mathbf{i} & \mathbf{j} & \mathbf{k} \\ T_x & T_y & 0 \\ 0 & 0 & 1 \end{vmatrix} = T_y \mathbf{i} - T_x \mathbf{j}.$$

Because $\mathbf{T} = \dfrac{\mathbf{r}'(t)}{|\mathbf{r}'(t)|}$, the components of $\mathbf{T}$ are

$$\mathbf{T} = \langle T_x, T_y, 0 \rangle = \frac{\langle x'(t), y'(t), 0 \rangle}{|\mathbf{r}'(t)|}.$$

We now have an expression for the unit normal vector:

$$\mathbf{n} = T_y \mathbf{i} - T_x \mathbf{j} = \frac{y'(t)}{|\mathbf{r}'(t)|} \mathbf{i} - \frac{x'(t)}{|\mathbf{r}'(t)|} \mathbf{j} = \frac{\langle y'(t), -x'(t) \rangle}{|\mathbf{r}'(t)|}.$$

To evaluate the flux integral $\int_C \mathbf{F} \cdot \mathbf{n}\, ds$, we make a familiar change of variables by letting $ds = |\mathbf{r}'(t)|\, dt$. The flux of $\mathbf{F} = \langle f, g \rangle$ across C is then

$$\int_C \mathbf{F} \cdot \mathbf{n}\, ds = \int_a^b \mathbf{F} \cdot \underbrace{\frac{\langle y'(t), -x'(t) \rangle}{|\mathbf{r}'(t)|}}_{\mathbf{n}} \underbrace{|\mathbf{r}'(t)|\, dt}_{ds} = \int_a^b (f(t)y'(t) - g(t)x'(t))\, dt.$$

This is one useful form of the flux integral. Alternatively, we can note that $dx = x'(t)\, dt$ and $dy = y'(t)\, dt$ and write

$$\int_C \mathbf{F} \cdot \mathbf{n}\, ds = \int_C f\, dy - g\, dx.$$

▶ Like circulation integrals, flux integrals may be computed on piecewise-smooth curves by finding the flux on each piece and adding the results.

DEFINITION Flux

Let $\mathbf{F} = \langle f, g \rangle$ be a continuous vector field on a region R of $\mathbb{R}^2$. Let $C: \mathbf{r}(t) = \langle x(t), y(t) \rangle$, for $a \le t \le b$, be a smooth oriented curve in R that does not intersect itself. The **flux** of the vector field $\mathbf{F}$ across C is

$$\int_C \mathbf{F} \cdot \mathbf{n}\, ds = \int_a^b (f(t)y'(t) - g(t)x'(t))\, dt,$$

where $\mathbf{n} = \mathbf{T} \times \mathbf{k}$ is the unit normal vector and $\mathbf{T}$ is the unit tangent vector consistent with the orientation. If C is a closed curve with counterclockwise orientation, $\mathbf{n}$ is the outward normal vector, and the flux integral gives the **outward flux** across C.

The concepts of circulation and flux can be visualized in terms of headwinds and crosswinds. Suppose the wind patterns in your neighborhood can be modeled with a vector field $\mathbf{F}$ (that doesn't change with time). Now imagine taking a walk around the block in a counterclockwise direction along a closed path. At different points along your walk, you encounter winds from various directions and with various speeds. The circulation of the wind field $\mathbf{F}$ along your path is the net amount of headwind (negative contribution) and tailwind (positive contribution) that you encounter during your walk. The flux of $\mathbf{F}$ across your path is the net amount of crosswind (positive from your left and negative from your right) encountered on your walk.

EXAMPLE 8 Flux of two-dimensional flows Find the outward flux across the unit circle with counterclockwise orientation for the following vector fields.

a. The radial vector field $\mathbf{F} = \langle x, y \rangle$

b. The rotation vector field $\mathbf{F} = \langle -y, x \rangle$

SOLUTION

a. The unit circle with counterclockwise orientation has a description $\mathbf{r}(t) = \langle x(t), y(t) \rangle = \langle \cos t, \sin t \rangle$, for $0 \le t \le 2\pi$. Therefore, $x'(t) = -\sin t$ and $y'(t) = \cos t$. The components of $\mathbf{F}$ are $f = x(t) = \cos t$ and $g = y(t) = \sin t$. It follows that the outward flux is

$$\int_a^b (f(t)y'(t) - g(t)x'(t))\, dt = \int_0^{2\pi} (\underbrace{\cos t}_{f(t)}\underbrace{\cos t}_{y'(t)} - \underbrace{\sin t}_{g(t)}\underbrace{(-\sin t)}_{x'(t)})\, dt$$

$$= \int_0^{2\pi} 1\, dt = 2\pi. \qquad \cos^2 t + \sin^2 t = 1$$

Because the radial field points outward and is aligned with the unit normal vectors on C, the outward flux is positive (**Figure 17.27a**).

b. For the rotation field, $f = -y(t) = -\sin t$ and $g = x(t) = \cos t$. The outward flux is

$$\int_a^b (f(t)y'(t) - g(t)x'(t))\, dt = \int_0^{2\pi} (\underbrace{-\sin t}_{f(t)}\underbrace{\cos t}_{y'(t)} - \underbrace{\cos t}_{g(t)}\underbrace{(-\sin t)}_{x'(t)})\, dt$$

$$= \int_0^{2\pi} 0\, dt = 0.$$

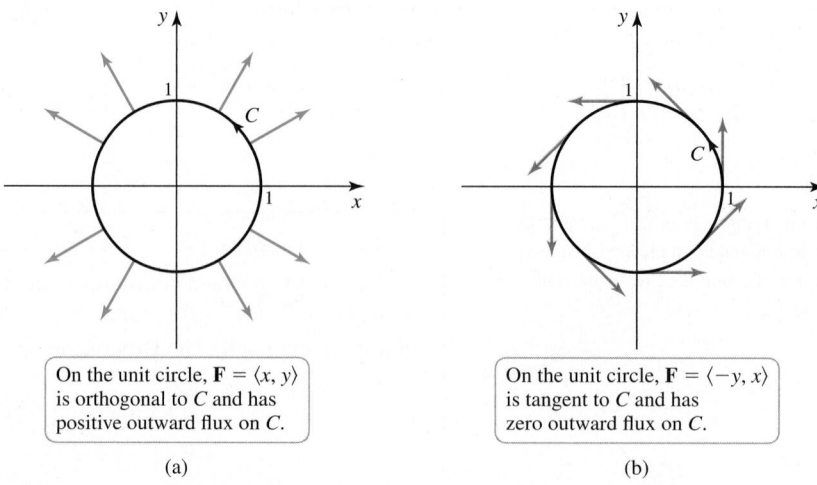

On the unit circle, $\mathbf{F} = \langle x, y \rangle$ is orthogonal to C and has positive outward flux on C.

(a)

On the unit circle, $\mathbf{F} = \langle -y, x \rangle$ is tangent to C and has zero outward flux on C.

(b)

Figure 17.27

Because the rotation field is orthogonal to $\mathbf{n}$ at all points of C, the outward flux across C is zero (**Figure 17.27b**). The results of Examples 6 and 8 are worth remembering: On a unit circle centered at the origin, the *radial* vector field $\langle x, y \rangle$ has outward flux 2π and zero circulation. The *rotation* vector field $\langle -y, x \rangle$ has zero outward flux and circulation 2π.

Related Exercises 59–60 ◄

SECTION 17.2 EXERCISES

Getting Started

1. How does a line integral differ from the single-variable integral $\int_a^b f(x)\,dx$?

2. If a curve C is given by $\mathbf{r}(t) = \langle t, t^2 \rangle$, what is $|\mathbf{r}'(t)|$?

3. Given that C is the curve $\mathbf{r}(t) = \langle \cos t, t \rangle$, for $\pi/2 \le t \le \pi$, convert the line integral $\int_C \dfrac{x}{y}\,ds$ to an ordinary integral. Do not evaluate the integral.

4–7. *Find a parametric description $\mathbf{r}(t)$ for the following curves.*

4. The segment of the curve $x = \sin \pi y$ from $(0, 0)$ to $(0, 3)$

5. The line segment from $(1, 2, 3)$ to $(5, 4, 0)$

6. The quarter-circle from $(1, 0)$ to $(0, 1)$ with its center at the origin

7. The segment of the parabola $x = y^2 + 1$ from $(5, 2)$ to $(17, 4)$

8. Find an expression for the vector field $\mathbf{F} = \langle x - y, y - x \rangle$ (in terms of t) along the unit circle $\mathbf{r}(t) = \langle \cos t, \sin t \rangle$.

9. Suppose C is the curve $\mathbf{r}(t) = \langle t, t^3 \rangle$, for $0 \le t \le 2$, and $\mathbf{F} = \langle x, 2y \rangle$. Evaluate $\int_C \mathbf{F} \cdot \mathbf{T}\,ds$ using the following steps.
   a. Convert the line integral $\int_C \mathbf{F} \cdot \mathbf{T}\,ds$ to an ordinary integral.
   b. Evaluate the integral in part (a).

10. Suppose C is the circle $\mathbf{r}(t) = \langle \cos t, \sin t \rangle$, for $0 \le t \le 2\pi$, and $\mathbf{F} = \langle 1, x \rangle$. Evaluate $\int_C \mathbf{F} \cdot \mathbf{n}\,ds$ using the following steps.
   a. Convert the line integral $\int_C \mathbf{F} \cdot \mathbf{n}\,ds$ to an ordinary integral.
   b. Evaluate the integral in part (a).

11. State two other forms for the line integral $\int_C \mathbf{F} \cdot \mathbf{T}\,ds$ given that $\mathbf{F} = \langle f, g, h \rangle$.

12–13. *Assume f is continuous on a region containing the smooth curve C from point A to point B and suppose $\int_C f\,ds = 10$.*

12. Explain the meaning of the curve $-C$ and state the value of $\int_{-C} f\,ds$.

13. Suppose P is a point on the curve C between A and B, where C_1 is the part of the curve from A to P, and C_2 is the part of the curve from P to B. Assuming $\int_{C_1} f\,ds = 3$, find the value of $\int_{C_2} f\,ds$.

14. Consider the graph of an ellipse C, oriented counterclockwise. Graphs of the vector fields $\mathbf{F}_1$, $\mathbf{F}_2$, $\mathbf{F}_3$, and $\mathbf{F}_4$ along the curve C are given (see figures). Along C, $\mathbf{F}_1$ and $\mathbf{F}_4$ are tangent to C, and $\mathbf{F}_2$ and $\mathbf{F}_3$ are normal to C. Determine whether the following integrals are positive, negative, or equal to 0.

   a. $\displaystyle\int_C \mathbf{F}_1 \cdot \mathbf{T}\,ds$ b. $\displaystyle\int_C \mathbf{F}_2 \cdot \mathbf{T}\,ds$ c. $\displaystyle\int_C \mathbf{F}_3 \cdot \mathbf{T}\,ds$

   d. $\displaystyle\int_C \mathbf{F}_4 \cdot \mathbf{T}\,ds$ e. $\displaystyle\int_C \mathbf{F}_1 \cdot \mathbf{n}\,ds$ f. $\displaystyle\int_C \mathbf{F}_2 \cdot \mathbf{n}\,ds$

   g. $\displaystyle\int_C \mathbf{F}_3 \cdot \mathbf{n}\,ds$ h. $\displaystyle\int_C \mathbf{F}_4 \cdot \mathbf{n}\,ds$

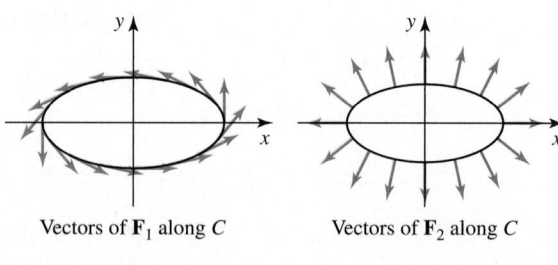

Vectors of $\mathbf{F}_1$ along C Vectors of $\mathbf{F}_2$ along C

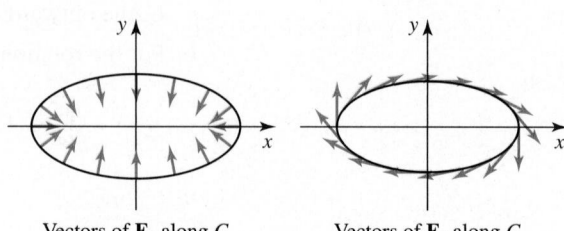

Vectors of $\mathbf{F}_3$ along C Vectors of $\mathbf{F}_4$ along C

15. How is the circulation of a vector field on a closed smooth oriented curve calculated?

16. Given a two-dimensional vector field **F** and a smooth oriented curve C, what is the meaning of the flux of **F** across C?

Practice Exercises

17–34. Scalar line integrals *Evaluate the following line integrals along the curve C.*

17. $\int_C xy \, ds$; C is the unit circle $\mathbf{r}(t) = \langle \cos t, \sin t \rangle$, for $0 \le t \le 2\pi$.

18. $\int_C (x^2 - 2y^2) \, ds$; C is the line segment $\mathbf{r}(t) = \left\langle \dfrac{t}{\sqrt{2}}, \dfrac{t}{\sqrt{2}} \right\rangle$, for $0 \le t \le 4$.

19. $\int_C (2x + y) \, ds$; C is the line segment $\mathbf{r}(t) = \langle 3t, 4t \rangle$, for $0 \le t \le 2$.

20. $\int_C x \, ds$; C is the curve $\mathbf{r}(t) = \langle t^3, 4t \rangle$, for $0 \le t \le 1$.

21. $\int_C xy^3 \, ds$; C is the quarter-circle $\mathbf{r}(t) = \langle 2\cos t, 2\sin t \rangle$, for $0 \le t \le \pi/2$.

22. $\int_C 3x \cos y \, ds$; C is the curve $\mathbf{r}(t) = \langle \sin t, t \rangle$, for $0 \le t \le \pi/2$.

23. $\int_C (y - z) \, ds$; C is the helix $\mathbf{r}(t) = \langle 3\cos t, 3\sin t, 4t \rangle$, for $0 \le t \le 2\pi$.

24. $\int_C (x - y + 2z) \, ds$; C is the circle $\mathbf{r}(t) = \langle 1, 3\cos t, 3\sin t \rangle$, for $0 \le t \le 2\pi$.

25. $\int_C (x^2 + y^2) \, ds$; C is the circle of radius 4 centered at $(0, 0)$.

26. $\int_C (x^2 + y^2) \, ds$; C is the line segment from $(0, 0)$ to $(5, 5)$.

27. $\int_C \dfrac{x}{x^2 + y^2} \, ds$; C is the line segment from $(1, 1)$ to $(10, 10)$.

28. $\int_C (xy)^{1/3} \, ds$; C is the curve $y = x^2$, for $0 \le x \le 1$.

29. $\int_C xy \, ds$; C is a portion of the ellipse $\dfrac{x^2}{4} + \dfrac{y^2}{16} = 1$ in the first quadrant, oriented counterclockwise.

30. $\int_C (2x - 3y) \, ds$; C is the line segment from $(-1, 0)$ to $(0, 1)$ followed by the line segment from $(0, 1)$ to $(1, 0)$.

31. $\int_C (x + y + z) \, ds$; C is the semicircle $\mathbf{r}(t) = \langle 2\cos t, 0, 2\sin t \rangle$, for $0 \le t \le \pi$.

32. $\int_C \dfrac{xy}{z} \, ds$; C is the line segment from $(1, 4, 1)$ to $(3, 6, 3)$.

33. $\int_C xz \, ds$; C is the line segment from $(0, 0, 0)$ to $(3, 2, 6)$ followed by the line segment from $(3, 2, 6)$ to $(7, 9, 10)$.

34. $\int_C xe^{yz} \, ds$; C is $\mathbf{r}(t) = \langle t, 2t, -2t \rangle$, for $0 \le t \le 2$.

35–36. Mass and density *A thin wire represented by the smooth curve C with a density ρ (mass per unit length) has a mass $M = \int_C \rho \, ds$. Find the mass of the following wires with the given density.*

35. $C: \{(x, y): y = 2x^2, 0 \le x \le 3\}$; $\rho(x, y) = 1 + xy$

36. $C: \mathbf{r}(\theta) = \langle \cos\theta, \sin\theta \rangle$, for $0 \le \theta \le \pi$; $\rho(\theta) = 2\theta/\pi + 1$

37–38. Average values *Find the average value of the following functions on the given curves.*

37. $f(x, y) = x + 2y$ on the line segment from $(1, 1)$ to $(2, 5)$

38. $f(x, y) = xe^y$ on the unit circle centered at the origin

39–40. Length of curves *Use a scalar line integral to find the length of the following curves.*

39. $\mathbf{r}(t) = \left\langle 20\sin\dfrac{t}{4}, 20\cos\dfrac{t}{4}, \dfrac{t}{2} \right\rangle$, for $0 \le t \le 2$

40. $\mathbf{r}(t) = \langle 30\sin t, 40\sin t, 50\cos t \rangle$, for $0 \le t \le 2\pi$

41–46. Line integrals of vector fields in the plane *Given the following vector fields and oriented curves C, evaluate $\int_C \mathbf{F} \cdot \mathbf{T} \, ds$.*

41. $\mathbf{F} = \langle x, y \rangle$ on the parabola $\mathbf{r}(t) = \langle 4t, t^2 \rangle$, for $0 \le t \le 1$

42. $\mathbf{F} = \langle -y, x \rangle$ on the semicircle $\mathbf{r}(t) = \langle 4\cos t, 4\sin t \rangle$, for $0 \le t \le \pi$

43. $\mathbf{F} = \langle y, x \rangle$ on the line segment from $(1, 1)$ to $(5, 10)$

44. $\mathbf{F} = \langle -y, x \rangle$ on the parabola $y = x^2$ from $(0, 0)$ to $(1, 1)$

45. $\mathbf{F} = \dfrac{\langle x, y \rangle}{(x^2 + y^2)^{3/2}}$ on the curve $\mathbf{r}(t) = \langle t^2, 3t^2 \rangle$, for $1 \le t \le 2$

46. $\mathbf{F} = \dfrac{\langle x, y \rangle}{x^2 + y^2}$ on the line segment $\mathbf{r}(t) = \langle t, 4t \rangle$, for $1 \le t \le 10$

47–48. Line integrals from graphs *Determine whether $\int_C \mathbf{F} \cdot d\mathbf{r}$ along the paths C_1 and C_2 shown in the following vector fields is positive or negative. Explain your reasoning.*

a. $\int_{C_1} \mathbf{F} \cdot d\mathbf{r}$ **b.** $\int_{C_2} \mathbf{F} \cdot d\mathbf{r}$

47.

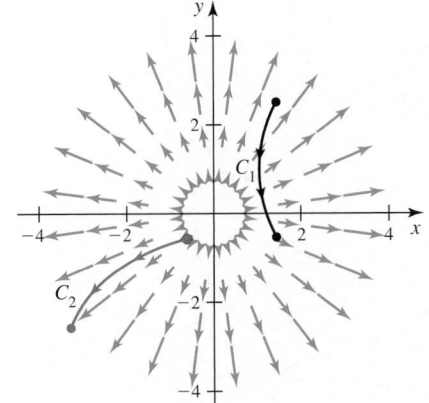

48.

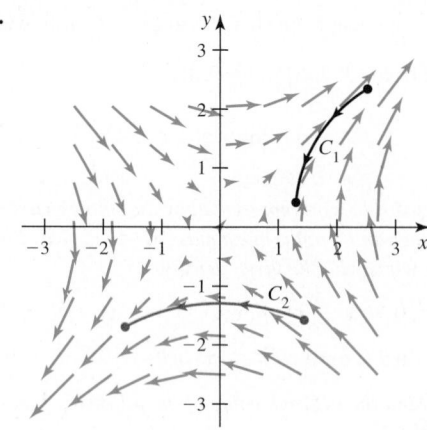

49–56. Work integrals *Given the force field* **F**, *find the work required to move an object on the given oriented curve.*

49. $\mathbf{F} = \langle y, -x \rangle$ on the line segment from $(1, 2)$ to $(0, 0)$ followed by the line segment from $(0, 0)$ to $(0, 4)$

50. $\mathbf{F} = \langle x, y \rangle$ on the line segment from $(-1, 0)$ to $(0, 8)$ followed by the line segment from $(0, 8)$ to $(2, 8)$

51. $\mathbf{F} = \langle y, x \rangle$ on the parabola $y = 2x^2$ from $(0, 0)$ to $(2, 8)$

52. $\mathbf{F} = \langle y, -x \rangle$ on the line segment $y = 10 - 2x$ from $(1, 8)$ to $(3, 4)$

53. $\mathbf{F} = \langle x, y, z \rangle$ on the tilted ellipse $\mathbf{r}(t) = \langle 4 \cos t, 4 \sin t, 4 \cos t \rangle$, for $0 \le t \le 2\pi$

54. $\mathbf{F} = \langle -y, x, z \rangle$ on the helix $\mathbf{r}(t) = \left\langle 2 \cos t, 2 \sin t, \dfrac{t}{2\pi} \right\rangle$, for $0 \le t \le 2\pi$

55. $\mathbf{F} = \dfrac{\langle x, y, z \rangle}{(x^2 + y^2 + z^2)^{3/2}}$ on the line segment from $(1, 1, 1)$ to $(10, 10, 10)$

Ⓣ 56. $\mathbf{F} = \dfrac{\langle x, y, z \rangle}{x^2 + y^2 + z^2}$ on the line segment from $(1, 1, 1)$ to $(8, 4, 2)$

57–58. Circulation *Consider the following vector fields* **F** *and closed oriented curves* C *in the plane (see figures).*

a. *Based on the picture, make a conjecture about whether the circulation of* **F** *on* C *is positive, negative, or zero.*

b. *Compute the circulation and interpret the result.*

57. $\mathbf{F} = \langle y - x, x \rangle$; $C: \mathbf{r}(t) = \langle 2 \cos t, 2 \sin t \rangle$, for $0 \le t \le 2\pi$

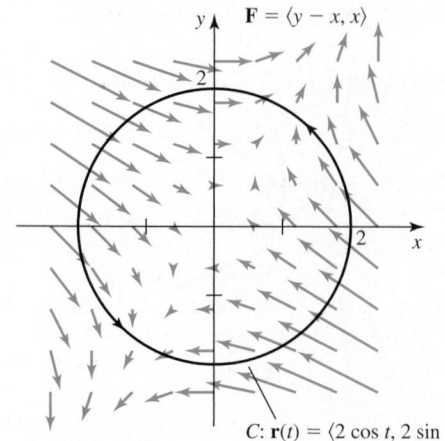

58. $\mathbf{F} = \dfrac{\langle y, -2x \rangle}{\sqrt{4x^2 + y^2}}$; $C: \mathbf{r}(t) = \langle 2 \cos t, 4 \sin t \rangle$, for $0 \le t \le 2\pi$

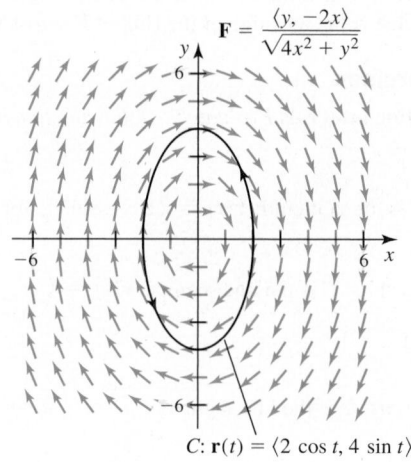

$C: \mathbf{r}(t) = \langle 2 \cos t, 4 \sin t \rangle$

59–60. Flux *Consider the vector fields and curves in Exercises 57–58.*

a. *Based on the picture, make a conjecture about whether the outward flux of* **F** *across* C *is positive, negative, or zero.*

b. *Compute the flux for the vector fields and curves.*

59. **F** and C given in Exercise 57

60. **F** and C given in Exercise 58

61. Explain why or why not Determine whether the following statements are true and give an explanation or counterexample.

a. If a curve has a parametric description $\mathbf{r}(t) = \langle x(t), y(t), z(t) \rangle$, where t is the arc length, then $|\mathbf{r}'(t)| = 1$.

b. The vector field $\mathbf{F} = \langle y, x \rangle$ has both zero circulation along and zero flux across the unit circle centered at the origin.

c. If at all points of a path a force acts in a direction orthogonal to the path, then no work is done in moving an object along the path.

d. The flux of a vector field across a curve in $\mathbb{R}^2$ can be computed using a line integral.

62. Flying into a headwind An airplane flies in the xz-plane, where x increases in the eastward direction and $z \ge 0$ represents vertical distance above the ground. A wind blows horizontally out of the west, producing a force $\mathbf{F} = \langle 150, 0 \rangle$. On which path between the points $(100, 50)$ and $(-100, 50)$ is more work done overcoming the wind?

a. The line segment $\mathbf{r}(t) = \langle x(t), z(t) \rangle = \langle -t, 50 \rangle$, for $-100 \le t \le 100$

b. The arc of the circle $\mathbf{r}(t) = \langle 100 \cos t, 50 + 100 \sin t \rangle$, for $0 \le t \le \pi$

63. Flying into a headwind

a. How does the result of Exercise 62 change if the force due to the wind is $\mathbf{F} = \langle 141, 50 \rangle$ (approximately the same magnitude, but a different direction)?

b. How does the result of Exercise 62 change if the force due to the wind is $\mathbf{F} = \langle 141, -50 \rangle$ (approximately the same magnitude, but a different direction)?

64. Changing orientation Let $f(x, y) = x + 2y$ and let C be the unit circle.

a. Find a parameterization of C with a counterclockwise orientation and evaluate $\int_C f \, ds$.

b. Find a parameterization of C with a clockwise orientation and evaluate $\int_C f\, ds$.

c. Compare the results of parts (a) and (b).

65. Changing orientation Let $f(x, y) = x$ and let C be the segment of the parabola $y = x^2$ joining $O(0, 0)$ and $P(1, 1)$.

a. Find a parameterization of C in the direction from O to P. Evaluate $\int_C f\, ds$.

b. Find a parameterization of C in the direction from P to O. Evaluate $\int_C f\, ds$.

c. Compare the results of parts (a) and (b).

66. Work in a rotation field Consider the rotation field $\mathbf{F} = \langle -y, x \rangle$ and the three paths shown in the figure. Compute the work done on each of the three paths. Does it appear that the line integral $\int_C \mathbf{F} \cdot \mathbf{T}\, ds$ is independent of the path, where C is any path from $(1, 0)$ to $(0, 1)$?

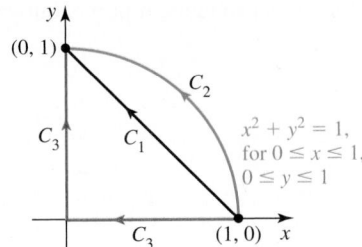

67. Work in a hyperbolic field Consider the hyperbolic force field $\mathbf{F} = \langle y, x \rangle$ (the streamlines are hyperbolas) and the three paths shown in the figure for Exercise 66. Compute the work done in the presence of $\mathbf{F}$ on each of the three paths. Does it appear that the line integral $\int_C \mathbf{F} \cdot \mathbf{T}\, ds$ is independent of the path, where C is any path from $(1, 0)$ to $(0, 1)$?

68–72. Assorted line integrals *Evaluate each line integral using the given curve C.*

68. $\displaystyle\int_C x^2\, dx + dy + y\, dz$; C is the curve $\mathbf{r}(t) = \langle t, 2t, t^2 \rangle$, for $0 \le t \le 3$.

⊤ 69. $\displaystyle\int_C x^3 y\, dx + xz\, dy + (x + y)^2\, dz$; C is the helix $\mathbf{r}(t) = \langle 2t, \sin t, \cos t \rangle$, for $0 \le t \le 4\pi$.

⊤ 70. $\displaystyle\int_C \frac{x^2}{y^4}\, ds$; C is the segment of the parabola $x = 3y^2$ from $(3, 1)$ to $(27, 3)$.

71. $\displaystyle\int_C \frac{y}{\sqrt{x^2 + y^2}}\, dx - \frac{x}{\sqrt{x^2 + y^2}}\, dy$; C is a quarter-circle from $(0, 4)$ to $(4, 0)$.

72. $\displaystyle\int_C (x + y)\, dx + (x - y)\, dy + x\, dz$; C is the line segment from $(1, 2, 4)$ to $(3, 8, 13)$.

73. Flux across curves in a vector field Consider the vector field $\mathbf{F} = \langle y, x \rangle$ shown in the figure.

a. Compute the outward flux across the quarter-circle $C: \mathbf{r}(t) = \langle 2 \cos t, 2 \sin t \rangle$, for $0 \le t \le \pi/2$.

b. Compute the outward flux across the quarter-circle $C: \mathbf{r}(t) = \langle 2 \cos t, 2 \sin t \rangle$, for $\pi/2 \le t \le \pi$.

c. Explain why the flux across the quarter-circle in the third quadrant equals the flux computed in part (a).

d. Explain why the flux across the quarter-circle in the fourth quadrant equals the flux computed in part (b).

e. What is the outward flux across the full circle?

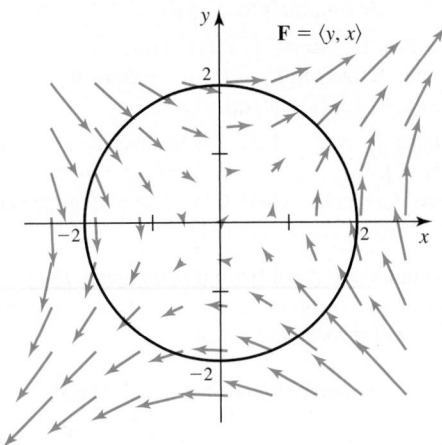

Explorations and Challenges

74–75. Zero circulation fields

74. For what values of b and c does the vector field $\mathbf{F} = \langle by, cx \rangle$ have zero circulation on the unit circle centered at the origin and oriented counterclockwise?

75. Consider the vector field $\mathbf{F} = \langle ax + by, cx + dy \rangle$. Show that $\mathbf{F}$ has zero circulation on any oriented circle centered at the origin, for any a, b, c, and d, provided $b = c$.

76–77. Zero flux fields

76. For what values of a and d does the vector field $\mathbf{F} = \langle ax, dy \rangle$ have zero flux across the unit circle centered at the origin and oriented counterclockwise?

77. Consider the vector field $\mathbf{F} = \langle ax + by, cx + dy \rangle$. Show that $\mathbf{F}$ has zero flux across any oriented circle centered at the origin, for any a, b, c, and d, provided $a = -d$.

78. Heat flux in a plate A square plate $R = \{(x, y): 0 \le x \le 1, 0 \le y \le 1\}$ has a temperature distribution $T(x, y) = 100 - 50x - 25y$.

a. Sketch two level curves of the temperature in the plate.

b. Find the gradient of the temperature $\nabla T(x, y)$.

c. Assume the flow of heat is given by the vector field $\mathbf{F} = -\nabla T(x, y)$. Compute $\mathbf{F}$.

d. Find the outward heat flux across the boundary $\{(x, y): x = 1, 0 \le y \le 1\}$.

e. Find the outward heat flux across the boundary $\{(x, y): 0 \le x \le 1, y = 1\}$.

79. Inverse force fields Consider the radial field $$\mathbf{F} = \frac{\mathbf{r}}{|\mathbf{r}|^p} = \frac{\langle x, y, z \rangle}{|\mathbf{r}|^p}, \text{ where } p > 1 \text{ (the inverse square law}$$ corresponds to $p = 3$). Let C be the line segment from $(1, 1, 1)$ to (a, a, a), where $a > 1$, given by $\mathbf{r}(t) = \langle t, t, t \rangle$, for $1 \le t \le a$.

a. Find the work done in moving an object along C with $p = 2$.

b. If $a \to \infty$ in part (a), is the work finite?

c. Find the work done in moving an object along C with $p = 4$.

d. If $a \to \infty$ in part (c), is the work finite?

e. Find the work done in moving an object along C for any $p > 1$.

f. If $a \to \infty$ in part (e), for what values of p is the work finite?

80. Line integrals with respect to dx and dy Given a vector field $\mathbf{F} = \langle f, 0 \rangle$ and curve C with parameterization $\mathbf{r}(t) = \langle x(t), y(t) \rangle$, for $a \le t \le b$, we see that the line integral $\int_C f \, dx + g \, dy$ simplifies to $\int_C f \, dx$.

   a. Show that $\int_C f \, dx = \int_a^b f(t) x'(t) \, dt$.
   b. Use the vector field $\mathbf{F} = \langle 0, g \rangle$ to show that $\int_C g \, dy = \int_a^b g(t) y'(t) \, dt$.
   c. Evaluate $\int_C xy \, dx$, where C is the line segment from $(0, 0)$ to $(5, 12)$.
   d. Evaluate $\int_C xy \, dy$, where C is a segment of the parabola $x = y^2$ from $(1, -1)$ to $(1, 1)$.

81–82. Looking ahead: Area from line integrals *The area of a region R in the plane, whose boundary is the curve C, may be computed using line integrals with the formula*

$$\text{area of } R = \int_C x \, dy = -\int_C y \, dx.$$

81. Let R be the rectangle with vertices $(0, 0)$, $(a, 0)$, $(0, b)$, and (a, b), and let C be the boundary of R oriented counterclockwise. Use the formula $A = \int_C x \, dy$ to verify that the area of the rectangle is ab.

82. Let $R = \{(r, \theta): 0 \le r \le a, 0 \le \theta \le 2\pi\}$ be the disk of radius a centered at the origin, and let C be the boundary of R oriented counterclockwise. Use the formula $A = -\int_C y \, dx$ to verify that the area of the disk is πa^2.

QUICK CHECK ANSWERS

1. The Fundamental Theorem of Calculus says that $\dfrac{d}{dt} \displaystyle\int_a^t f(u) \, du = f(t)$, which applies to differentiating the arc length integral.   **2.** Note that $x = t, y = 0$, and $|\mathbf{r}'(t)| = \sqrt{1^2 + 0^2} = 1$. Therefore, $\displaystyle\int_C f(x, y) \, ds = \int_a^b f(t, 0) \, dt$.   **3.** 1300 ft/min   **4.** $\pi/2$

5. $\mathbf{T}$ and $\mathbf{k}$ are unit vectors, so $\mathbf{n}$ is a unit vector. By the right-hand rule for cross products, $\mathbf{n}$ points outward from the curve. ◄

17.3 Conservative Vector Fields

This is an action-packed section in which several fundamental ideas come together. At the heart of the matter are two questions:

- When can a vector field be expressed as the gradient of a potential function? A vector field with this property will be defined as a *conservative* vector field.
- What special properties do conservative vector fields have?

After some preliminary definitions, we present a test to determine whether a vector field in $\mathbb{R}^2$ or $\mathbb{R}^3$ is conservative. This test is followed by a procedure to find a potential function for a conservative field. We then develop several equivalent properties shared by all conservative vector fields.

Types of Curves and Regions

Many of the results in the remainder of this text rely on special properties of regions and curves. It's best to collect these definitions in one place for easy reference.

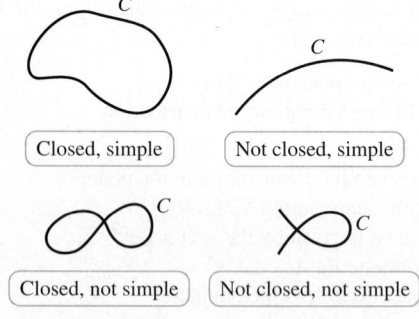

Closed, simple   Not closed, simple

Closed, not simple   Not closed, not simple

Figure 17.28

➤ Recall that all points of an open set are interior points. An open set does not contain its boundary points.

➤ Roughly speaking, connected means that R is all in one piece and simply connected in $\mathbb{R}^2$ means that R has no holes. $\mathbb{R}^2$ and $\mathbb{R}^3$ are themselves connected and simply connected.

> **DEFINITION Simple and Closed Curves**
>
> Suppose a curve C (in $\mathbb{R}^2$ or $\mathbb{R}^3$) is described parametrically by $\mathbf{r}(t)$, where $a \le t \le b$. Then C is a **simple curve** if $\mathbf{r}(t_1) \ne \mathbf{r}(t_2)$ for all t_1 and t_2, with $a < t_1 < t_2 < b$; that is, C never intersects itself between its endpoints. The curve C is **closed** if $\mathbf{r}(a) = \mathbf{r}(b)$; that is, the initial and terminal points of C are the same (Figure 17.28).

In all that follows, we generally assume that R in $\mathbb{R}^2$ (or D in $\mathbb{R}^3$) is an open region. Open regions are further classified according to whether they are *connected* and whether they are *simply connected*.

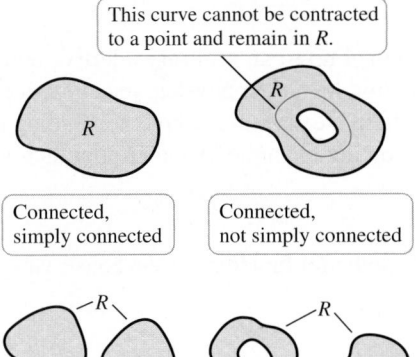

This curve cannot be contracted to a point and remain in R.

Connected, simply connected

Connected, not simply connected

Not connected, simply connected

Not connected, not simply connected

Figure 17.29

QUICK CHECK 1 Is a figure-8 curve simple? Closed? Is a torus connected? Simply connected? ◄

▸ The term *conservative* refers to conservation of energy. See Exercise 66 for an example of conservation of energy in a conservative force field.

▸ Depending on the context and the interpretation of the vector field, the potential function φ may be defined such that $\mathbf{F} = -\nabla\varphi$ (with a negative sign).

> **DEFINITION** **Connected and Simply Connected Regions**
>
> An open region R in $\mathbb{R}^2$ (or D in $\mathbb{R}^3$) is **connected** if it is possible to connect any two points of R by a continuous curve lying in R. An open region R is **simply connected** if every closed simple curve in R can be deformed and contracted to a point in R (Figure 17.29).

Test for Conservative Vector Fields

We begin with the central definition of this section.

> **DEFINITION** **Conservative Vector Field**
>
> A vector field $\mathbf{F}$ is said to be **conservative** on a region (in $\mathbb{R}^2$ or $\mathbb{R}^3$) if there exists a scalar function φ such that $\mathbf{F} = \nabla\varphi$ on that region.

Suppose the components of $\mathbf{F} = \langle f, g, h \rangle$ have continuous first partial derivatives on a region D in $\mathbb{R}^3$. Also assume $\mathbf{F}$ is conservative, which means by definition that there is a potential function φ such that $\mathbf{F} = \nabla\varphi$. Matching the components of $\mathbf{F}$ and $\nabla\varphi$, we see that $f = \varphi_x$, $g = \varphi_y$, and $h = \varphi_z$. Recall from Theorem 15.4 that if a function has continuous second partial derivatives, the order of differentiation in the second partial derivatives does not matter. Under these conditions on φ, we conclude the following:

- $\varphi_{xy} = \varphi_{yx}$, which implies that $f_y = g_x$,
- $\varphi_{xz} = \varphi_{zx}$, which implies that $f_z = h_x$, and
- $\varphi_{yz} = \varphi_{zy}$, which implies that $g_z = h_y$.

These observations constitute half of the proof of the following theorem. The remainder of the proof is given in Section 17.4.

> **THEOREM 17.3** **Test for Conservative Vector Fields**
> Let $\mathbf{F} = \langle f, g, h \rangle$ be a vector field defined on a connected and simply connected region D of $\mathbb{R}^3$, where f, g, and h have continuous first partial derivatives on D. Then $\mathbf{F}$ is a conservative vector field on D (there is a potential function φ such that $\mathbf{F} = \nabla\varphi$) if and only if
> $$\frac{\partial f}{\partial y} = \frac{\partial g}{\partial x}, \qquad \frac{\partial f}{\partial z} = \frac{\partial h}{\partial x}, \quad \text{and} \quad \frac{\partial g}{\partial z} = \frac{\partial h}{\partial y}.$$
>
> For vector fields in $\mathbb{R}^2$, we have the single condition $\dfrac{\partial f}{\partial y} = \dfrac{\partial g}{\partial x}$.

EXAMPLE 1 **Testing for conservative fields** Determine whether the following vector fields are conservative on $\mathbb{R}^2$ and $\mathbb{R}^3$, respectively.

a. $\mathbf{F} = \langle e^x \cos y, -e^x \sin y \rangle$ **b.** $\mathbf{F} = \langle 2xy - z^2, x^2 + 2z, 2y - 2xz \rangle$

SOLUTION

a. Letting $f(x, y) = e^x \cos y$ and $g(x, y) = -e^x \sin y$, we see that
$$\frac{\partial f}{\partial y} = -e^x \sin y = \frac{\partial g}{\partial x}.$$

The conditions of Theorem 17.3 are met and $\mathbf{F}$ is conservative.

b. Letting $f(x, y, z) = 2xy - z^2$, $g(x, y, z) = x^2 + 2z$, and $h(x, y, z) = 2y - 2xz$, we have
$$\frac{\partial f}{\partial y} = 2x = \frac{\partial g}{\partial x}, \qquad \frac{\partial f}{\partial z} = -2z = \frac{\partial h}{\partial x}, \qquad \frac{\partial g}{\partial z} = 2 = \frac{\partial h}{\partial y}.$$

By Theorem 17.3, $\mathbf{F}$ is conservative.   *Related Exercises 13–14* ◄

Finding Potential Functions

QUICK CHECK 2 Explain why a potential function for a conservative vector field is determined up to an additive constant. ◄

Like antiderivatives, potential functions are determined up to an arbitrary additive constant. Unless an additive constant in a potential function has some physical meaning, it is usually omitted. Given a conservative vector field, there are several methods for finding a potential function. One method is shown in the following example. Another approach is illustrated in Exercise 71.

EXAMPLE 2   Finding potential functions Find a potential function for the conservative vector fields in Example 1.

a. $\mathbf{F} = \langle e^x \cos y, -e^x \sin y \rangle$

b. $\mathbf{F} = \langle 2xy - z^2, x^2 + 2z, 2y - 2xz \rangle$

SOLUTION

a. A potential function φ for $\mathbf{F} = \langle f, g \rangle$ has the property that $\mathbf{F} = \nabla \varphi$ and satisfies the conditions

$$\varphi_x = f(x, y) = e^x \cos y \quad \text{and} \quad \varphi_y = g(x, y) = -e^x \sin y.$$

The first equation is integrated with respect to x (holding y fixed) to obtain

$$\int \varphi_x \, dx = \int e^x \cos y \, dx,$$

which implies that

$$\varphi(x, y) = e^x \cos y + c(y).$$

> ▶ This procedure may begin with either of the two conditions, $\varphi_x = f$ or $\varphi_y = g$.

In this case, the "constant of integration" $c(y)$ is an arbitrary function of y. You can check the preceding calculation by noting that

$$\frac{\partial \varphi}{\partial x} = \frac{\partial}{\partial x} (e^x \cos y + c(y)) = e^x \cos y = f(x, y).$$

To find the arbitrary function $c(y)$, we differentiate $\varphi(x, y) = e^x \cos y + c(y)$ with respect to y and equate the result to g (recall that $\varphi_y = g$):

$$\varphi_y = -e^x \sin y + c'(y) \quad \text{and} \quad g = -e^x \sin y.$$

We conclude that $c'(y) = 0$, which implies that $c(y)$ is any real number, which we typically take to be zero. So a potential function is $\varphi(x, y) = e^x \cos y$, a result that may be checked by differentiation.

> ▶ This procedure may begin with any of the three conditions.

b. The method of part (a) is more elaborate with three variables. A potential function φ must now satisfy these conditions:

$$\varphi_x = f = 2xy - z^2, \qquad \varphi_y = g = x^2 + 2z, \quad \text{and} \quad \varphi_z = h = 2y - 2xz.$$

Integrating the first condition with respect to x (holding y and z fixed), we have

$$\varphi = \int (2xy - z^2) \, dx = x^2 y - xz^2 + c(y, z).$$

Because the integration is with respect to x, the arbitrary "constant" is a function of y and z. To find $c(y, z)$, we differentiate φ with respect to y, which results in

$$\varphi_y = x^2 + c_y(y, z).$$

Equating φ_y and $g = x^2 + 2z$, we see that $c_y(y, z) = 2z$. To obtain $c(y, z)$, we integrate $c_y(y, z) = 2z$ with respect to y (holding z fixed), which results in $c(y, z) = 2yz + d(z)$. The "constant" of integration is now a function of z, which we call $d(z)$. At this point, a potential function looks like

$$\varphi(x, y, z) = x^2 y - xz^2 + 2yz + d(z).$$

To determine $d(z)$, we differentiate φ with respect to z:

$$\varphi_z = -2xz + 2y + d'(z).$$

QUICK CHECK 3 Verify by differentiation that the potential functions found in Example 2 produce the corresponding vector fields. ◄

Equating φ_z and $h = 2y - 2xz$, we see that $d'(z) = 0$, or $d(z)$ is a real number, which we generally take to be zero. Putting it all together, a potential function is

$$\varphi = x^2 y - xz^2 + 2yz.$$

Related Exercises 19, 24 ◄

PROCEDURE **Finding Potential Functions in $\mathbb{R}^3$**

Suppose $\mathbf{F} = \langle f, g, h \rangle$ is a conservative vector field. To find φ such that $\mathbf{F} = \nabla\varphi$, use the following steps:

1. Integrate $\varphi_x = f$ with respect to x to obtain φ, which includes an arbitrary function $c(y, z)$.

2. Compute φ_y and equate it to g to obtain an expression for $c_y(y, z)$.

3. Integrate $c_y(y, z)$ with respect to y to obtain $c(y, z)$, including an arbitrary function $d(z)$.

4. Compute φ_z and equate it to h to get $d(z)$.

A similar procedure beginning with $\varphi_y = g$ or $\varphi_z = h$ may be easier in some cases.

► Compare the two versions of the Fundamental Theorem.

$$\int_a^b F'(x)\, dx = F(b) - F(a)$$

$$\int_C \nabla\varphi \cdot d\mathbf{r} = \varphi(B) - \varphi(A)$$

Fundamental Theorem for Line Integrals and Path Independence

Knowing how to find potential functions, we now investigate their properties. The first property is one of several beautiful parallels to the Fundamental Theorem of Calculus.

THEOREM 17.4 **Fundamental Theorem for Line Integrals**

Let R be a region in $\mathbb{R}^2$ or $\mathbb{R}^3$ and let φ be a differentiable potential function defined on R. If $\mathbf{F} = \nabla\varphi$ (which means that $\mathbf{F}$ is conservative), then

$$\int_C \mathbf{F} \cdot \mathbf{T}\, ds = \int_C \mathbf{F} \cdot d\mathbf{r} = \varphi(B) - \varphi(A),$$

for all points A and B in R and all piecewise-smooth oriented curves C in R from A to B.

Proof: Let the curve C in $\mathbb{R}^3$ be given by $\mathbf{r}(t) = \langle x(t), y(t), z(t) \rangle$, for $a \le t \le b$, where $\mathbf{r}(a)$ and $\mathbf{r}(b)$ are the position vectors for the points A and B, respectively. By the Chain Rule, the rate of change of φ with respect to t along C is

$$\frac{d\varphi}{dt} = \frac{\partial\varphi}{\partial x}\frac{dx}{dt} + \frac{\partial\varphi}{\partial y}\frac{dy}{dt} + \frac{\partial\varphi}{\partial z}\frac{dz}{dt} \qquad \text{Chain Rule}$$

$$= \left\langle \frac{\partial\varphi}{\partial x}, \frac{\partial\varphi}{\partial y}, \frac{\partial\varphi}{\partial z} \right\rangle \cdot \left\langle \frac{dx}{dt}, \frac{dy}{dt}, \frac{dz}{dt} \right\rangle \qquad \text{Identify the dot product.}$$

$$= \nabla\varphi \cdot \mathbf{r}'(t) \qquad\qquad \mathbf{r} = \langle x, y, z \rangle$$

$$= \mathbf{F} \cdot \mathbf{r}'(t). \qquad\qquad \mathbf{F} = \nabla\varphi$$

Evaluating the line integral and using the Fundamental Theorem of Calculus, it follows that

$$\int_C \mathbf{F} \cdot d\mathbf{r} = \int_a^b \mathbf{F} \cdot \mathbf{r}'(t)\, dt$$

$$= \int_a^b \frac{d\varphi}{dt}\, dt \qquad \mathbf{F} \cdot \mathbf{r}'(t) = \frac{d\varphi}{dt}$$

$$= \varphi(B) - \varphi(A). \qquad \begin{array}{l}\text{Fundamental Theorem of Calculus; } t = b \text{ corresponds} \\ \text{to } B \text{ and } t = a \text{ corresponds to } A.\end{array} \qquad ◄$$

Here is the meaning of Theorem 17.4: If $\mathbf{F}$ is a conservative vector field, then the value of a line integral of $\mathbf{F}$ depends only on the endpoints of the path. For this reason, we say the line integral is *independent of path*, which means that to evaluate line integrals of conservative vector fields, we do not need a parameterization of the path.

If we think of φ as an antiderivative of the vector field $\mathbf{F}$, then the parallel to the Fundamental Theorem of Calculus is clear. The line integral of $\mathbf{F}$ is the difference of the values of φ evaluated at the endpoints. Theorem 17.4 motivates the following definition.

DEFINITION Independence of Path

Let $\mathbf{F}$ be a continuous vector field with domain R. If $\displaystyle\int_{C_1} \mathbf{F} \cdot d\mathbf{r} = \int_{C_2} \mathbf{F} \cdot d\mathbf{r}$ for all piecewise-smooth curves C_1 and C_2 in R with the same initial and terminal points, then the line integral is **independent of path**.

An important question concerns the converse of Theorem 17.4. With additional conditions on the domain R, the converse turns out to be true.

THEOREM 17.5

Let $\mathbf{F}$ be a continuous vector field on an open connected region R in $\mathbb{R}^2$. If

$$\int_C \mathbf{F} \cdot d\mathbf{r} \text{ is independent of path, then } \mathbf{F} \text{ is conservative; that is, there exists a}$$

potential function φ such that $\mathbf{F} = \nabla\varphi$ on R.

▶ We state and prove Theorem 17.5 in two variables. It is easily extended to three or more variables.

Proof: Let $P(a, b)$ and $Q(x, y)$ be interior points of R and define $\displaystyle\varphi(x, y) = \int_C \mathbf{F} \cdot d\mathbf{r}$, where C is a piecewise-smooth path from P to Q, and $\mathbf{F} = \langle f, g \rangle$. Because the integral defining φ is independent of path, any piecewise-smooth path in R from P to Q can be used. The goal is to compute the directional derivative $D_{\mathbf{u}}\varphi(x, y)$, where $\mathbf{u} = \langle u_1, u_2 \rangle$ is an arbitrary unit vector, and then show that $\mathbf{F} = \nabla\varphi$. We let $S(x + tu_1, y + tu_2)$ be a point in R near Q and then apply the definition of the directional derivative at Q:

$$D_{\mathbf{u}}\varphi(x, y) = \lim_{t \to 0} \frac{\varphi(x + tu_1, y + tu_2) - \varphi(x, y)}{t}$$

$$= \lim_{t \to 0} \frac{1}{t} \left(\int_P^S \mathbf{F} \cdot d\mathbf{r} - \int_P^Q \mathbf{F} \cdot d\mathbf{r} \right)$$

$$= \lim_{t \to 0} \frac{1}{t} \int_Q^S \mathbf{F} \cdot d\mathbf{r}.$$

Using path independence, we choose the path from Q to S to be a line parameterized by $\mathbf{r}(s) = \langle x + su_1, y + su_2 \rangle$, for $0 \le s \le t$. Noting that $\mathbf{r}'(s) = \mathbf{u}$, the directional derivative is

$$D_{\mathbf{u}}\varphi(x, y) = \lim_{t \to 0} \frac{1}{t} \int_Q^S \mathbf{F} \cdot d\mathbf{r}$$

$$= \lim_{t \to 0} \frac{1}{t} \int_0^t \mathbf{F}(x + su_1, y + su_2) \cdot \mathbf{r}'(s) \, ds \qquad \text{Change line integral to ordinary integral.}$$

$$= \lim_{t \to 0} \frac{\displaystyle\int_0^t \mathbf{F}(x + su_1, y + su_2) \cdot \mathbf{r}'(s) \, ds - \int_0^0 \mathbf{F}(x + su_1, y + su_2) \cdot \mathbf{r}'(s) \, ds}{t}$$

$$\text{Second integral equals 0.}$$

$$= \frac{d}{dt} \int_0^t \mathbf{F}(x + su_1, y + su_2) \cdot \mathbf{u} \, ds \bigg|_{t=0} \qquad \text{Identify difference quotient;} \quad \mathbf{r}'(s) = \mathbf{u}$$

$$= \mathbf{F}(x, y) \cdot \mathbf{u}. \qquad \text{Fundamental Theorem of Calculus}$$

Choosing $\mathbf{u} = \mathbf{i} = \langle 1, 0 \rangle$, we see that $D_{\mathbf{i}}\varphi(x, y) = \varphi_x(x, y) = \mathbf{F}(x, y) \cdot \mathbf{i} = f(x, y)$. Similarly, choosing $\mathbf{u} = \mathbf{j} = \langle 0, 1 \rangle$, we have $D_{\mathbf{j}}\varphi(x, y) = \varphi_y(x, y) = \mathbf{F}(x, y) \cdot \mathbf{j} = g(x, y)$. Therefore, $\mathbf{F} = \langle f, g \rangle = \langle \varphi_x, \varphi_y \rangle = \nabla\varphi$, and $\mathbf{F}$ is conservative. ◄

EXAMPLE 3 **Verifying path independence** Consider the potential function $\varphi(x, y) = (x^2 - y^2)/2$ and its gradient field $\mathbf{F} = \langle x, -y \rangle$.

- Let C_1 be the quarter-circle $\mathbf{r}(t) = \langle \cos t, \sin t \rangle$, for $0 \le t \le \pi/2$, from $A(1, 0)$ to $B(0, 1)$.
- Let C_2 be the line $\mathbf{r}(t) = \langle 1 - t, t \rangle$, for $0 \le t \le 1$, also from A to B.

Evaluate the line integrals of $\mathbf{F}$ on C_1 and C_2, and show that both are equal to $\varphi(B) - \varphi(A)$.

SOLUTION On C_1, we have $\mathbf{r}'(t) = \langle -\sin t, \cos t \rangle$ and $\mathbf{F} = \langle x, -y \rangle = \langle \cos t, -\sin t \rangle$. The line integral on C_1 is

$$\int_{C_1} \mathbf{F} \cdot d\mathbf{r} = \int_a^b \mathbf{F} \cdot \mathbf{r}'(t) \, dt$$

$$= \int_0^{\pi/2} \underbrace{\langle \cos t, -\sin t \rangle}_{\mathbf{F}} \cdot \underbrace{\langle -\sin t, \cos t \rangle \, dt}_{\mathbf{r}'(t) \, dt} \quad \text{Substitute for } \mathbf{F} \text{ and } \mathbf{r}'.$$

$$= \int_0^{\pi/2} (-\sin 2t) \, dt \quad\quad\quad 2 \sin t \cos t = \sin 2t$$

$$= \left(\frac{1}{2} \cos 2t \right) \Big|_0^{\pi/2} = -1. \quad\quad \text{Evaluate the integral.}$$

On C_2, we have $\mathbf{r}'(t) = \langle -1, 1 \rangle$ and $\mathbf{F} = \langle x, -y \rangle = \langle 1 - t, -t \rangle$; therefore,

$$\int_{C_2} \mathbf{F} \cdot d\mathbf{r} = \int_0^1 \underbrace{\langle 1 - t, -t \rangle}_{\mathbf{F}} \cdot \underbrace{\langle -1, 1 \rangle \, dt}_{d\mathbf{r}} \quad \text{Substitute for } \mathbf{F} \text{ and } d\mathbf{r}.$$

$$= \int_0^1 (-1) \, dt = -1. \quad\quad \text{Simplify.}$$

The two line integrals have the same value, which is

$$\varphi(B) - \varphi(A) = \varphi(0, 1) - \varphi(1, 0) = -\frac{1}{2} - \frac{1}{2} = -1.$$

Related Exercises 31–32 ◄

EXAMPLE 4 **Line integral of a conservative vector field** Evaluate

$$\int_C ((2xy - z^2) \mathbf{i} + (x^2 + 2z) \mathbf{j} + (2y - 2xz) \mathbf{k}) \cdot d\mathbf{r},$$

where C is a simple curve from $A(-3, -2, -1)$ to $B(1, 2, 3)$.

SOLUTION This vector field is conservative and has a potential function $\varphi = x^2 y - xz^2 + 2yz$ (Example 2). By the Fundamental Theorem for line integrals,

$$\int_C ((2xy - z^2) \mathbf{i} + (x^2 + 2z) \mathbf{j} + (2y - 2xz) \mathbf{k}) \cdot d\mathbf{r}$$

$$= \int_C \nabla \underbrace{(x^2 y - xz^2 + 2yz)}_{\varphi} \cdot d\mathbf{r}$$

$$= \varphi(1, 2, 3) - \varphi(-3, -2, -1) = 16.$$

Related Exercise 34 ◄

QUICK CHECK 4 Explain why the vector field $\nabla(xy + xz - yz)$ is conservative. ◄

Line Integrals on Closed Curves

It is a short step to another characterization of conservative vector fields. Suppose C is a simple *closed* piecewise-smooth oriented curve in $\mathbb{R}^2$ or $\mathbb{R}^3$. To distinguish line integrals on closed curves, we adopt the notation $\oint_C \mathbf{F} \cdot d\mathbf{r}$, where the small circle on the integral sign indicates that C is a closed curve. Let A be any point on C and think of A as both the initial point and the final point of C. Assuming $\mathbf{F}$ is a conservative vector field on an open connected region R containing C, it follows by Theorem 17.4 that

▶ Notice the analogy to $\int_a^a f(x)\,dx = 0$, which is true of all integrable functions.

$$\oint_C \mathbf{F} \cdot d\mathbf{r} = \varphi(A) - \varphi(A) = 0.$$

Because A is an arbitrary point on C, we see that the line integral of a conservative vector field on a closed curve is zero.

An argument can be made in the opposite direction as well: Suppose $\oint_C \mathbf{F} \cdot d\mathbf{r} = 0$ on all simple closed piecewise-smooth oriented curves in a region R, and let A and B be distinct points in R. Let C_1 denote any curve from A to B, let C_2 be any curve from B to A (distinct from and not intersecting C_1), and let C be the closed curve consisting of C_1 followed by C_2 (**Figure 17.30**). Then

$$0 = \oint_C \mathbf{F} \cdot d\mathbf{r} = \int_{C_1} \mathbf{F} \cdot d\mathbf{r} + \int_{C_2} \mathbf{F} \cdot d\mathbf{r}.$$

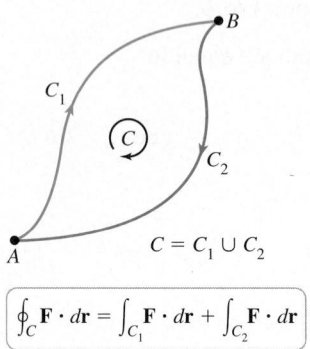

$$\oint_C \mathbf{F} \cdot d\mathbf{r} = \int_{C_1} \mathbf{F} \cdot d\mathbf{r} + \int_{C_2} \mathbf{F} \cdot d\mathbf{r}$$

Figure 17.30

Therefore, $\int_{C_1} \mathbf{F} \cdot d\mathbf{r} = -\int_{C_2} \mathbf{F} \cdot d\mathbf{r} = \int_{-C_2} \mathbf{F} \cdot d\mathbf{r}$, where $-C_2$ is the curve C_2 traversed in the opposite direction (from A to B). We see that the line integral has the same value on two arbitrary paths between A and B. It follows that the value of the line integral is independent of path, and by Theorem 17.5, $\mathbf{F}$ is conservative. This argument is a proof of the following theorem.

> **THEOREM 17.6 Line Integrals on Closed Curves**
> Let R be an open connected region in $\mathbb{R}^2$ or $\mathbb{R}^3$. Then $\mathbf{F}$ is a conservative vector field on R if and only if $\oint_C \mathbf{F} \cdot d\mathbf{r} = 0$ on all simple closed piecewise-smooth oriented curves C in R.

EXAMPLE 5   A closed curve line integral in $\mathbb{R}^3$ Evaluate $\int_C \nabla(-xy + xz + yz) \cdot d\mathbf{r}$ on the curve C: $\mathbf{r}(t) = \langle \sin t, \cos t, \sin t \rangle$, for $0 \le t \le 2\pi$, without using Theorem 17.4 or Theorem 17.6.

SOLUTION The components of the vector field are

$$\mathbf{F} = \nabla(-xy + xz + yz) = \langle -y + z, -x + z, x + y \rangle.$$

Note that $\mathbf{r}'(t) = \langle \cos t, -\sin t, \cos t \rangle$ and $d\mathbf{r} = \mathbf{r}'(t)\,dt$. Substituting values of x, y, and z, the value of the line integral is

$$\oint_C \mathbf{F} \cdot d\mathbf{r} = \oint_C \langle -y + z, -x + z, x + y \rangle \cdot d\mathbf{r} \qquad \text{Substitute for } \mathbf{F}.$$

$$= \int_0^{2\pi} \sin 2t\,dt \qquad \text{Substitute for } x, y, z, d\mathbf{r}.$$

$$= -\frac{1}{2} \cos 2t \Big|_0^{2\pi} = 0. \qquad \text{Evaluate integral.}$$

The line integral of this conservative vector field on the closed curve C is zero. In fact, by Theorem 17.6, the line integral vanishes on any simple closed piecewise-smooth oriented curve.

Related Exercise 50 ◀

Summary of the Properties of Conservative Vector Fields

We have established three equivalent properties of conservative vector fields $\mathbf{F}$ defined on an open connected region R in $\mathbb{R}^2$ (or D in $\mathbb{R}^3$).

- There exists a potential function φ such that $\mathbf{F} = \nabla\varphi$ (definition).
- $\int_C \mathbf{F} \cdot d\mathbf{r} = \varphi(B) - \varphi(A)$ for all points A and B in R and all piecewise-smooth oriented curves C in R from A to B (path independence).
- $\oint_C \mathbf{F} \cdot d\mathbf{r} = 0$ on all simple piecewise-smooth closed oriented curves C in R.

The connections between these properties were established by Theorems 17.4, 17.5, and 17.6 in the following way:

$$\overbrace{\text{Theorems 17.4 and 17.5}}\qquad\qquad \overbrace{\text{Theorem 17.6}}$$

$$\text{Path independence} \iff \mathbf{F} \text{ is conservative } (\nabla\varphi = \mathbf{F}) \iff \oint_C \mathbf{F} \cdot d\mathbf{r} = \mathbf{0}.$$

SECTION 17.3 EXERCISES

Getting Started

1. Explain with pictures what is meant by a simple curve and a closed curve.

2. Explain with pictures what is meant by a connected region and a simply connected region.

3. How do you determine whether a vector field in $\mathbb{R}^2$ is conservative (has a potential function φ such that $\mathbf{F} = \nabla\varphi$)?

4. How do you determine whether a vector field in $\mathbb{R}^3$ is conservative?

5. Briefly describe how to find a potential function φ for a conservative vector field $\mathbf{F} = \langle f, g \rangle$.

6. If $\mathbf{F}$ is a conservative vector field on a region R, how do you evaluate $\int_C \mathbf{F} \cdot d\mathbf{r}$, where C is a path between two points A and B in R?

7. If $\mathbf{F}$ is a conservative vector field on a region R, what is the value of $\oint_C \mathbf{F} \cdot d\mathbf{r}$, where C is a simple closed piecewise-smooth oriented curve in R?

8. Give three equivalent properties of conservative vector fields.

Practice Exercises

9–16. Testing for conservative vector fields *Determine whether the following vector fields are conservative (in $\mathbb{R}^2$ or $\mathbb{R}^3$).*

9. $\mathbf{F} = \langle 1, 1 \rangle$

10. $\mathbf{F} = \langle x, y \rangle$

11. $\mathbf{F} = \langle -y, x \rangle$

12. $\mathbf{F} = \langle -y, x + y \rangle$

13. $\mathbf{F} = \langle e^{-x} \cos y, e^{-x} \sin y \rangle$

14. $\mathbf{F} = \langle 2x^3 + xy^2, 2y^3 - x^2 y \rangle$

15. $\mathbf{F} = \langle yz \cos xz, \sin xz, xy \cos xz \rangle$

16. $\mathbf{F} = \langle ye^{x-z}, e^{x-z}, ye^{x-z} \rangle$

17–30. Finding potential functions *Determine whether the following vector fields are conservative on the specified region. If so, determine a potential function. Let R^* and D^* be open regions of $\mathbb{R}^2$ and $\mathbb{R}^3$, respectively, that do not include the origin.*

17. $\mathbf{F} = \langle x, y \rangle$ on $\mathbb{R}^2$

18. $\mathbf{F} = \langle -y, -x \rangle$ on $\mathbb{R}^2$

19. $\mathbf{F} = \left\langle x^3 - xy, \dfrac{x^2}{2} + y \right\rangle$ on $\mathbb{R}^2$

20. $\mathbf{F} = \dfrac{\langle x, y \rangle}{x^2 + y^2}$ on R^*

21. $\mathbf{F} = \dfrac{\langle x, y \rangle}{\sqrt{x^2 + y^2}}$ on R^*

22. $\mathbf{F} = \langle y, x, x - y \rangle$ on $\mathbb{R}^3$

23. $\mathbf{F} = \langle z, 1, x \rangle$ on $\mathbb{R}^3$

24. $\mathbf{F} = \langle yz, xz, xy \rangle$ on $\mathbb{R}^3$

25. $\mathbf{F} = \langle e^z, e^z, e^z(x - y) \rangle$ on $\mathbb{R}^3$

26. $\mathbf{F} = \langle 1, -z, y \rangle$ on $\mathbb{R}^3$

27. $\mathbf{F} = \langle y + z, x + z, x + y \rangle$ on $\mathbb{R}^3$

28. $\mathbf{F} = \dfrac{\langle x, y, z \rangle}{x^2 + y^2 + z^2}$ on D^*

29. $\mathbf{F} = \dfrac{\langle x, y, z \rangle}{\sqrt{x^2 + y^2 + z^2}}$ on D^*

30. $\mathbf{F} = \langle x^3, 2y, -z^3 \rangle$ on $\mathbb{R}^3$

31–34. Evaluating line integrals *Use the given potential function φ of the gradient field $\mathbf{F}$ and the curve C to evaluate the line integral $\int_C \mathbf{F} \cdot d\mathbf{r}$ in two ways.*

a. *Use a parametric description of C and evaluate the integral directly.*
b. *Use the Fundamental Theorem for line integrals.*

31. $\varphi(x, y) = xy$; $C: \mathbf{r}(t) = \langle \cos t, \sin t \rangle$, for $0 \le t \le \pi$

32. $\varphi(x, y) = x + 3y$; $C: \mathbf{r}(t) = \langle 2 - t, t \rangle$, for $0 \le t \le 2$

33. $\varphi(x, y, z) = (x^2 + y^2 + z^2)/2$; $C: \mathbf{r}(t) = \langle \cos t, \sin t, t/\pi \rangle$, for $0 \le t \le 2\pi$

34. $\varphi(x, y, z) = xy + xz + yz$; $C: \mathbf{r}(t) = \langle t, 2t, 3t \rangle$, for $0 \le t \le 4$

35–38. Applying the Fundamental Theorem of Line Integrals *Suppose the vector field $\mathbf{F}$ is continuous on $\mathbb{R}^2$, $\mathbf{F} = \langle f, g \rangle = \nabla\varphi$, $\varphi(1, 2) = 7$, $\varphi(3, 6) = 10$, and $\varphi(6, 4) = 20$. Evaluate the following integrals for the given curve C, if possible.*

35. $\displaystyle\int_C \mathbf{F} \cdot d\mathbf{r}$; $C: \mathbf{r}(t) = \langle 2t - 1, t^2 + t \rangle$, for $1 \le t \le 2$

36. $\displaystyle\int_C \mathbf{F} \cdot \mathbf{T} \, ds$; C is a smooth curve from $(1, 2)$ to $(6, 4)$.

37. $\displaystyle\int_C f \, dx + g \, dy$; C is the path consisting of the line segment from $A(6, 4)$ to $B(1, 2)$ followed by the line segment from $B(1, 2)$ to $C(3, 6)$.

38. $\displaystyle\oint_C \mathbf{F} \cdot d\mathbf{r}$; C is a circle, oriented clockwise, starting and ending at the point $A(6, 4)$.

39–44. Using the Fundamental Theorem for line integrals *Verify that the Fundamental Theorem for line integrals can be used to evaluate the given integral, and then evaluate the integral.*

39. $\int_C \langle 2x, 2y \rangle \cdot d\mathbf{r}$, where C is a smooth curve from $(0, 1)$ to $(3, 4)$

40. $\int_C \langle 1, 1, 1 \rangle \cdot d\mathbf{r}$, where C is a smooth curve from $(1, -1, 2)$ to $(3, 0, 7)$

41. $\int_C \nabla(e^{-x} \cos y) \cdot d\mathbf{r}$, where C is the line segment from $(0, 0)$ to $(\ln 2, 2\pi)$

42. $\int_C \nabla(1 + x^2 yz) \cdot d\mathbf{r}$, where C is the helix $\mathbf{r}(t) = \langle \cos 2t, \sin 2t, t \rangle$, for $0 \le t \le 4\pi$

43. $\int_C \cos(2y - z) \, dx - 2x \sin(2y - z) \, dy + x \sin(2y - z) \, dz$, where C is the curve $\mathbf{r}(t) = \langle t^2, t, t \rangle$, for $0 \le t \le \pi$

44. $\int_C e^x y \, dx + e^x \, dy$, where C is the parabola $\mathbf{r}(t) = \langle t + 1, t^2 \rangle$, for $-1 \le t \le 3$

45–50. Line integrals of vector fields on closed curves *Evaluate $\oint_C \mathbf{F} \cdot d\mathbf{r}$ for the following vector fields and closed oriented curves C by parameterizing C. If the integral is not zero, give an explanation.*

45. $\mathbf{F} = \langle x, y \rangle$; C is the circle of radius 4 centered at the origin oriented counterclockwise.

46. $\mathbf{F} = \langle y, x \rangle$; C is the circle of radius 8 centered at the origin oriented counterclockwise.

47. $\mathbf{F} = \langle x, y \rangle$; C is the triangle with vertices $(0, \pm 1)$ and $(1, 0)$ oriented counterclockwise.

48. $\mathbf{F} = \langle y, -x \rangle$; C is the circle of radius 3 centered at the origin oriented counterclockwise.

49. $\mathbf{F} = \langle x, y, z \rangle$; C: $\mathbf{r}(t) = \langle \cos t, \sin t, 2 \rangle$, for $0 \le t \le 2\pi$

50. $\mathbf{F} = \langle y - z, z - x, x - y \rangle$; C: $\mathbf{r}(t) = \langle \cos t, \sin t, \cos t \rangle$, for $0 \le t \le 2\pi$

51–52. Evaluating line integrals using level curves *Suppose the vector field F, whose potential function is φ, is continuous on $\mathbb{R}^2$. Use the curves C_1 and C_2 and level curves of φ (see figure) to evaluate the following line integrals.*

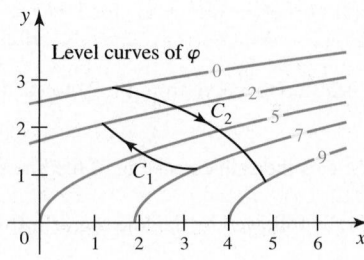

51. $\int_{C_1} \mathbf{F} \cdot d\mathbf{r}$

52. $\int_{C_2} \mathbf{F} \cdot d\mathbf{r}$

53–56. Line integrals *Evaluate the following line integrals using a method of your choice.*

53. $\oint_C \mathbf{F} \cdot d\mathbf{r}$, where $\mathbf{F} = \langle 2xy + z^2, x^2, 2xz \rangle$ and C is the circle $\mathbf{r}(t) = \langle 3 \cos t, 4 \cos t, 5 \sin t \rangle$, for $0 \le t \le 2\pi$

54. $\oint_C e^{-x}(\cos y \, dx + \sin y \, dy)$, where C is the square with vertices $(\pm 1, \pm 1)$ oriented counterclockwise

55. $\int_C \nabla(\sin xy) \cdot d\mathbf{r}$, where C is the line segment from $(0, 0)$ to $(2, \pi/4)$

56. $\int_C x^3 \, dx + y^3 \, dy$, where C is the curve $\mathbf{r}(t) = \langle 1 + \sin t, \cos^2 t \rangle$, for $0 \le t \le \pi/2$

57. **Explain why or why not** Determine whether the following statements are true and give an explanation or counterexample.

   a. If $\mathbf{F} = \langle -y, x \rangle$ and C is the circle of radius 4 centered at $(1, 0)$ oriented counterclockwise, then $\oint_C \mathbf{F} \cdot d\mathbf{r} = 0$.

   b. If $\mathbf{F} = \langle x, -y \rangle$ and C is the circle of radius 4 centered at $(1, 0)$ oriented counterclockwise, then $\oint_C \mathbf{F} \cdot d\mathbf{r} = 0$.

   c. A constant vector field is conservative on $\mathbb{R}^2$.

   d. The vector field $\mathbf{F} = \langle f(x), g(y) \rangle$ is conservative on $\mathbb{R}^2$ (assume f and g are defined for all real numbers).

   e. Gradient fields are conservative.

58. **Closed-curve integrals** Evaluate $\oint_C ds$, $\oint_C dx$, and $\oint_C dy$, where C is the unit circle oriented counterclockwise.

59–62. Work in force fields *Find the work required to move an object in the following force fields along a line segment between the given points. Check to see whether the force is conservative.*

59. $\mathbf{F} = \langle x, 2 \rangle$ from $A(0, 0)$ to $B(2, 4)$

60. $\mathbf{F} = \langle x, y \rangle$ from $A(1, 1)$ to $B(3, -6)$

61. $\mathbf{F} = \langle x, y, z \rangle$ from $A(1, 2, 1)$ to $B(2, 4, 6)$

62. $\mathbf{F} = e^{x+y}\langle 1, 1, z \rangle$ from $A(0, 0, 0)$ to $B(-1, 2, -4)$

63. Suppose C is a circle centered at the origin in a vector field $\mathbf{F}$ (see figure).

   a. If C is oriented counterclockwise, is $\oint_C \mathbf{F} \cdot d\mathbf{r}$ positive, negative, or zero?

   b. If C is oriented clockwise, is $\oint_C \mathbf{F} \cdot d\mathbf{r}$ positive, negative, or zero?

   c. Is $\mathbf{F}$ conservative in $\mathbb{R}^2$? Explain.

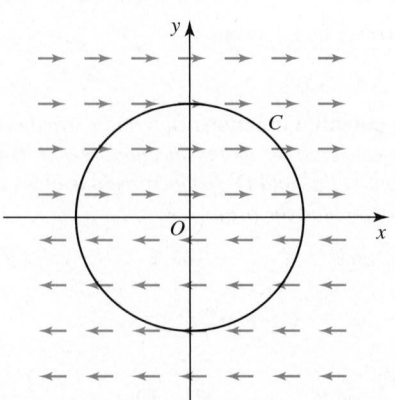

64. A vector field that is continuous in $\mathbb{R}^2$ is given (see figure). Is it conservative?

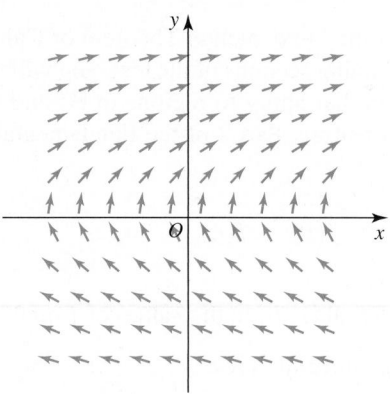

65. Work by a constant force Evaluate a line integral to show that the work done in moving an object from point A to point B in the presence of a constant force $\mathbf{F} = \langle a, b, c \rangle$ is $\mathbf{F} \cdot \overrightarrow{AB}$.

Explorations and Challenges

66. Conservation of energy Suppose an object with mass m moves in a region R in a conservative force field given by $\mathbf{F} = -\nabla\varphi$, where φ is a potential function in a region R. The motion of the object is governed by Newton's Second Law of Motion, $\mathbf{F} = m\mathbf{a}$, where $\mathbf{a}$ is the acceleration. Suppose the object moves from point A to point B in R.

a. Show that the equation of motion is $m\dfrac{d\mathbf{v}}{dt} = -\nabla\varphi$.

b. Show that $\dfrac{d\mathbf{v}}{dt} \cdot \mathbf{v} = \dfrac{1}{2}\dfrac{d}{dt}(\mathbf{v} \cdot \mathbf{v})$.

c. Take the dot product of both sides of the equation in part (a) with $\mathbf{v}(t) = \mathbf{r}'(t)$ and integrate along a curve between A and B. Use part (b) and the fact that $\mathbf{F}$ is conservative to show that the total energy (kinetic plus potential) $\dfrac{1}{2}m|\mathbf{v}|^2 + \varphi$ is the same at A and B. Conclude that because A and B are arbitrary, energy is conserved in R.

67. Gravitational potential The gravitational force between two point masses M and m is
$$\mathbf{F} = GMm\frac{\mathbf{r}}{|\mathbf{r}|^3} = GMm\frac{\langle x, y, z \rangle}{(x^2 + y^2 + z^2)^{3/2}},$$
where G is the gravitational constant.

a. Verify that this force field is conservative on any region excluding the origin.

b. Find a potential function φ for this force field such that $\mathbf{F} = -\nabla\varphi$.

c. Suppose the object with mass m is moved from a point A to a point B, where A is a distance r_1 from M, and B is a distance r_2 from M. Show that the work done in moving the object is
$$GMm\left(\frac{1}{r_2} - \frac{1}{r_1}\right).$$

d. Does the work depend on the path between A and B? Explain.

68. Radial fields in $\mathbb{R}^3$ are conservative Prove that the radial field
$$\mathbf{F} = \frac{\mathbf{r}}{|\mathbf{r}|^p},$$
where $\mathbf{r} = \langle x, y, z \rangle$ and p is a real number, is conservative on any region not containing the origin. For what values of p is $\mathbf{F}$ conservative on a region that contains the origin?

69. Rotation fields are usually not conservative

a. Prove that the rotation field $\mathbf{F} = \dfrac{\langle -y, x \rangle}{|\mathbf{r}|^p}$, where $\mathbf{r} = \langle x, y \rangle$, is not conservative for $p \neq 2$.

b. For $p = 2$, show that $\mathbf{F}$ is conservative on any region not containing the origin.

c. Find a potential function for $\mathbf{F}$ when $p = 2$.

70. Linear and quadratic vector fields

a. For what values of a, b, c, and d is the field $\mathbf{F} = \langle ax + by, cx + dy \rangle$ conservative?

b. For what values of a, b, and c is the field $\mathbf{F} = \langle ax^2 - by^2, cxy \rangle$ conservative?

71. Alternative construction of potential functions in $\mathbb{R}^2$ Assume the vector field $\mathbf{F}$ is conservative on $\mathbb{R}^2$, so that the line integral $\int_C \mathbf{F} \cdot d\mathbf{r}$ is independent of path. Use the following procedure to construct a potential function φ for the vector field $\mathbf{F} = \langle f, g \rangle = \langle 2x - y, -x + 2y \rangle$.

a. Let A be $(0, 0)$ and let B be an arbitrary point (x, y). Define $\varphi(x, y)$ to be the work required to move an object from A to B, where $\varphi(A) = 0$. Let C_1 be the path from A to $(x, 0)$ to B, and let C_2 be the path from A to $(0, y)$ to B. Draw a picture.

b. Evaluate $\int_{C_1} \mathbf{F} \cdot d\mathbf{r} = \int_{C_1} f\,dx + g\,dy$ and conclude that $\varphi(x, y) = x^2 - xy + y^2$.

c. Verify that the same potential function is obtained by evaluating the line integral over C_2.

72–75. Alternative construction of potential functions *Use the procedure in Exercise 71 to construct potential functions for the following fields.*

72. $\mathbf{F} = \langle -y, -x \rangle$

73. $\mathbf{F} = \langle x, y \rangle$

74. $\mathbf{F} = \dfrac{\mathbf{r}}{|\mathbf{r}|}$, where $\mathbf{r} = \langle x, y \rangle$

75. $\mathbf{F} = \langle 2x^3 + xy^2, 2y^3 + x^2y \rangle$

QUICK CHECK ANSWERS

1. A figure-8 is closed but not simple; a torus is connected but not simply connected. **2.** The vector field is obtained by differentiating the potential function. So additive constants in the potential give the same vector field: $\nabla(\varphi + C) = \nabla\varphi$, where C is a constant. **3.** Show that $\nabla(e^x \cos y) = \langle e^x \cos y, -e^x \sin y \rangle$, which is the original vector field. A similar calculation may be done for part (b). **4.** The vector field $\nabla(xy + xz - yz)$ is the gradient of $xy + xz - yz$, so the vector field is conservative. ◄

17.4 Green's Theorem

The preceding section gave a version of the Fundamental Theorem of Calculus that applies to line integrals. In this and the remaining sections of the text, you will see additional extensions of the Fundamental Theorem that apply to regions in $\mathbb{R}^2$ and $\mathbb{R}^3$. All these fundamental theorems share a common feature. Part 2 of the Fundamental Theorem of Calculus (Chapter 5) says

$$\int_a^b \frac{df}{dx}\, dx = f(b) - f(a),$$

which relates the integral of $\dfrac{df}{dx}$ on an interval $[a, b]$ to the values of f on the boundary of $[a, b]$. The Fundamental Theorem for line integrals says

$$\int_C \nabla\varphi \cdot d\mathbf{r} = \varphi(B) - \varphi(A),$$

which relates the integral of $\nabla\varphi$ on a piecewise-smooth oriented curve C to the boundary values of φ. (The boundary consists of the two endpoints A and B.)

The subject of this section is Green's Theorem, which is another step in this progression. It relates the double integral of derivatives of a function over a region in $\mathbb{R}^2$ to function values on the boundary of that region.

Circulation Form of Green's Theorem

Throughout this section, unless otherwise stated, we assume curves in the plane are simple closed piecewise-smooth oriented curves. By a result called the *Jordan Curve Theorem*, such curves have a well-defined interior such that when the curve is traversed in the counterclockwise direction (viewed from above), the interior is on the left. With this orientation, there is a unique outward unit normal vector that points to the right (at points where the curve is smooth). We also assume curves in the plane lie in regions that are both connected and simply connected.

Suppose the vector field $\mathbf{F}$ is defined on a region R whose boundary is the closed curve C. As we have seen, the circulation $\oint_C \mathbf{F} \cdot d\mathbf{r}$ (Section 17.2) measures the net component of $\mathbf{F}$ in the direction tangent to C. It is easiest to visualize the circulation when $\mathbf{F}$ represents the velocity of a fluid moving in two dimensions. For example, let C be the unit circle with a counterclockwise orientation. The vector field $\mathbf{F} = \langle -y, x \rangle$ has a positive circulation of 2π on C (Section 17.2) because the vector field is everywhere tangent to C (Figure 17.31). A nonzero circulation on a closed curve says that the vector field must have some property *inside* the curve that produces the circulation. You can think of this property as a *net rotation*.

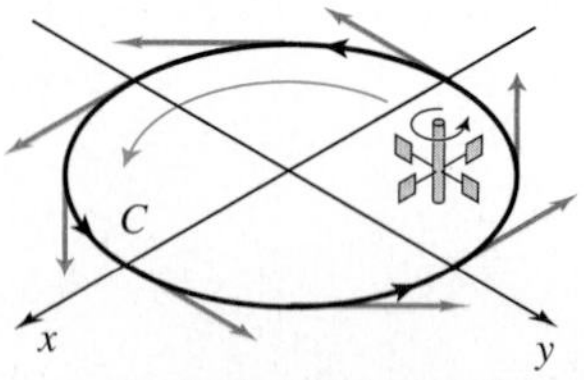

To visualize the rotation of a vector field, imagine a small paddle wheel, fixed at a point in the vector field, with its axis perpendicular to the xy-plane (Figure 17.31). The strength of the rotation at that point is seen in the speed at which the paddle wheel spins, and the direction of the rotation is the direction in which the paddle wheel spins. At a different point in the vector field, the paddle wheel will, in general, have a different speed and direction of rotation.

The first form of Green's Theorem relates the circulation on C to the double integral, over the region R, of a quantity that measures rotation at each point of R.

Paddle wheel at one point of vector field.

$\mathbf{F} = \langle -y, x \rangle$ has positive (counterclockwise) circulation on C.

Figure 17.31

> The circulation form of Green's Theorem is also called the *tangential*, or *curl*, form.

THEOREM 17.7 Green's Theorem—Circulation Form

Let C be a simple closed piecewise-smooth curve, oriented counterclockwise, that encloses a connected and simply connected region R in the plane. Assume $\mathbf{F} = \langle f, g \rangle$, where f and g have continuous first partial derivatives in R. Then

$$\underbrace{\oint_C \mathbf{F} \cdot d\mathbf{r}}_{\text{circulation}} = \underbrace{\oint_C f\, dx + g\, dy}_{\text{circulation}} = \iint_R \left(\frac{\partial g}{\partial x} - \frac{\partial f}{\partial y} \right) dA.$$

The proof of a special case of the theorem is given at the end of this section. Notice that the two line integrals on the left side of Green's Theorem give the circulation of the vector field on C. The double integral on the right side involves the quantity $\dfrac{\partial g}{\partial x} - \dfrac{\partial f}{\partial y}$, which describes the rotation of the vector field *within* C that produces the circulation *on* C. This quantity is called the *two-dimensional curl* of the vector field.

Figure 17.32 illustrates how the curl measures the rotation of a particular vector field at a point P. If the horizontal component of the field decreases in the y-direction at P ($f_y < 0$) and the vertical component increases in the x-direction at P ($g_x > 0$), then $\dfrac{\partial g}{\partial x} - \dfrac{\partial f}{\partial y} > 0$, and the field has a counterclockwise rotation at P. The double integral in Green's Theorem computes the net rotation of the field throughout R. The theorem says that the net rotation throughout R equals the circulation on the boundary of R.

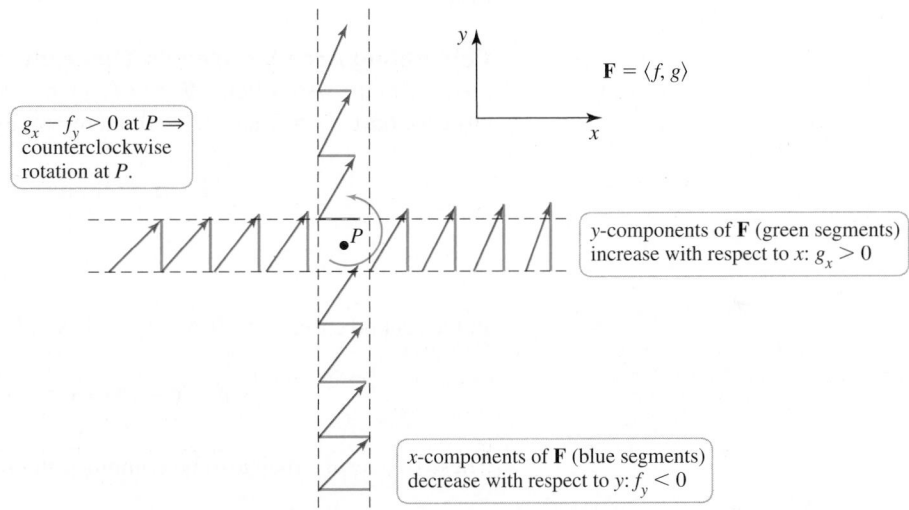

QUICK CHECK 1 Compute $\dfrac{\partial g}{\partial x} - \dfrac{\partial f}{\partial y}$ for the radial vector field $\mathbf{F} = \langle x, y \rangle$. What does this tell you about the circulation on a simple closed curve? ◄

Figure 17.32

Green's Theorem has an important consequence when applied to a conservative vector field. Recall from Theorem 17.3 that if $\mathbf{F} = \langle f, g \rangle$ is conservative, then its components satisfy the condition $f_y = g_x$. If R is a region of $\mathbb{R}^2$ on which the conditions of Green's Theorem are satisfied, then for a conservative field, we have

$$\oint_C \mathbf{F} \cdot d\mathbf{r} = \iint_R \underbrace{\left(\frac{\partial g}{\partial x} - \frac{\partial f}{\partial y} \right)}_{0} dA = 0.$$

> In some cases, the rotation of a vector field is not obvious. For example, the parallel flow in a channel $\mathbf{F} = \langle 0, 1 - x^2 \rangle$, for $|x| \le 1$, has a nonzero curl for $x \ne 0$. See Exercise 72.

Green's Theorem confirms the fact (Theorem 17.6) that if $\mathbf{F}$ is a conservative vector field in a region, then the circulation $\oint_C \mathbf{F} \cdot d\mathbf{r}$ is zero on any simple closed curve in the region. A two-dimensional vector field $\mathbf{F} = \langle f, g \rangle$ for which $\dfrac{\partial g}{\partial x} - \dfrac{\partial f}{\partial y} = 0$ at all points of a region is said to be *irrotational*, because it produces zero circulation on closed curves in the region. Irrotational vector fields on simply connected regions in $\mathbb{R}^2$ are conservative.

DEFINITION Two-Dimensional Curl

The **two-dimensional curl** of the vector field $\mathbf{F} = \langle f, g \rangle$ is $\dfrac{\partial g}{\partial x} - \dfrac{\partial f}{\partial y}$. If the curl is zero throughout a region, the vector field is **irrotational** on that region.

Evaluating circulation integrals of conservative vector fields on closed curves is easy. The integral is always zero. Green's Theorem provides a way to evaluate circulation integrals for nonconservative vector fields.

EXAMPLE 1 Circulation of a rotation field Consider the rotation vector field $\mathbf{F} = \langle -y, x \rangle$ on the unit disk $R = \{(x, y): x^2 + y^2 \le 1\}$ (Figure 17.31). In Example 6 of Section 17.2, we showed that $\oint_C \mathbf{F} \cdot d\mathbf{r} = 2\pi$, where C is the boundary of R oriented counterclockwise. Confirm this result using Green's Theorem.

SOLUTION Note that $f(x, y) = -y$ and $g(x, y) = x$; therefore, the curl of $\mathbf{F}$ is $\dfrac{\partial g}{\partial x} - \dfrac{\partial f}{\partial y} = 2$. By Green's Theorem,

$$\oint_C \mathbf{F} \cdot d\mathbf{r} = \iint_R \underbrace{\left(\frac{\partial g}{\partial x} - \frac{\partial f}{\partial y} \right)}_{2} dA = \iint_R 2 \, dA = 2 \times \text{area of } R = 2\pi.$$

The curl $\dfrac{\partial g}{\partial x} - \dfrac{\partial f}{\partial y}$ is nonzero on R, which results in a nonzero circulation on the boundary of R.

Related Exercises 18, 20 ◄

Calculating Area by Green's Theorem A useful consequence of Green's Theorem arises with the vector fields $\mathbf{F} = \langle f, g \rangle = \langle 0, x \rangle$ and $\mathbf{F} = \langle f, g \rangle = \langle y, 0 \rangle$. In the first case, we have $g_x = 1$ and $f_y = 0$; therefore, by Green's Theorem,

$$\oint_C \mathbf{F} \cdot d\mathbf{r} = \oint_C \underset{\mathbf{F} \cdot d\mathbf{r}}{x \, dy} = \iint_R \underset{\frac{\partial g}{\partial x} - \frac{\partial f}{\partial y} = 1}{dA} = \text{area of } R.$$

In the second case, $g_x = 0$ and $f_y = 1$, and Green's Theorem says

$$\oint_C \mathbf{F} \cdot d\mathbf{r} = \oint_C y \, dx = -\iint_R dA = -\text{area of } R.$$

These two results may also be combined in one statement to give the following theorem.

THEOREM 17.8 Area of a Plane Region by Line Integrals
Under the conditions of Green's Theorem, the area of a region R enclosed by a curve C is

$$\oint_C x \, dy = -\oint_C y \, dx = \frac{1}{2} \oint_C (x \, dy - y \, dx).$$

A remarkably simple calculation of the area of an ellipse follows from this result.

EXAMPLE 2 Area of an ellipse Find the area of the ellipse $\dfrac{x^2}{a^2} + \dfrac{y^2}{b^2} = 1$.

SOLUTION An ellipse with counterclockwise orientation is described parametrically by $\mathbf{r}(t) = \langle x, y \rangle = \langle a \cos t, b \sin t \rangle$, for $0 \le t \le 2\pi$. Noting that $dx = -a \sin t \, dt$ and $dy = b \cos t \, dt$, we have

$$x \, dy - y \, dx = (a \cos t)(b \cos t) \, dt - (b \sin t)(-a \sin t) \, dt$$
$$= ab (\cos^2 t + \sin^2 t) \, dt$$
$$= ab \, dt.$$

Expressing the line integral as an ordinary integral with respect to t, the area of the ellipse is

$$\frac{1}{2} \oint_C \underbrace{(x \, dy - y \, dx)}_{ab \, dt} = \frac{ab}{2} \int_0^{2\pi} dt = \pi ab.$$

Related Exercises 22–23 ◄

Flux Form of Green's Theorem

Let C be a closed curve enclosing a region R in $\mathbb{R}^2$ and let $\mathbf{F}$ be a vector field defined on R. We assume C and R have the previously stated properties; specifically, C is oriented counterclockwise with an outward normal vector $\mathbf{n}$. Recall that the outward flux of $\mathbf{F}$ across C is $\oint_C \mathbf{F} \cdot \mathbf{n}\, ds$ (Section 17.2). The second form of Green's Theorem relates the flux across C to a property of the vector field within R that produces the flux.

> ▶ The flux form of Green's Theorem is also called the *normal*, or *divergence*, form.

THEOREM 17.9   Green's Theorem—Flux Form

Let C be a simple closed piecewise-smooth curve, oriented counterclockwise, that encloses a connected and simply connected region R in the plane. Assume $\mathbf{F} = \langle f, g \rangle$, where f and g have continuous first partial derivatives in R. Then

$$\underbrace{\oint_C \mathbf{F} \cdot \mathbf{n}\, ds}_{\text{outward flux}} = \underbrace{\oint_C f\, dy - g\, dx}_{\text{outward flux}} = \iint_R \left(\frac{\partial f}{\partial x} + \frac{\partial g}{\partial y} \right) dA,$$

where $\mathbf{n}$ is the outward unit normal vector on the curve.

> ▶ The two forms of Green's Theorem are related in the following way: Applying the circulation form of the theorem to $\mathbf{F} = \langle -g, f \rangle$ results in the flux form, and applying the flux form of the theorem to $\mathbf{F} = \langle g, -f \rangle$ results in the circulation form.

The two line integrals on the left side of Theorem 17.9 give the outward flux of the vector field across C. The double integral on the right side involves the quantity $\dfrac{\partial f}{\partial x} + \dfrac{\partial g}{\partial y}$, which is the property of the vector field that produces the flux across C. This quantity is called the *two-dimensional divergence*.

Figure 17.33 illustrates how the divergence measures the flux of a particular vector field at a point P. If $f_x > 0$ at P, it indicates an expansion of the vector field in the x-direction (if f_x is negative, it indicates a contraction). Similarly, if $g_y > 0$ at P, it indicates an expansion of the vector field in the y-direction. The combined effect of $f_x + g_y > 0$ at a point is a net outward flux across a small circle enclosing P.

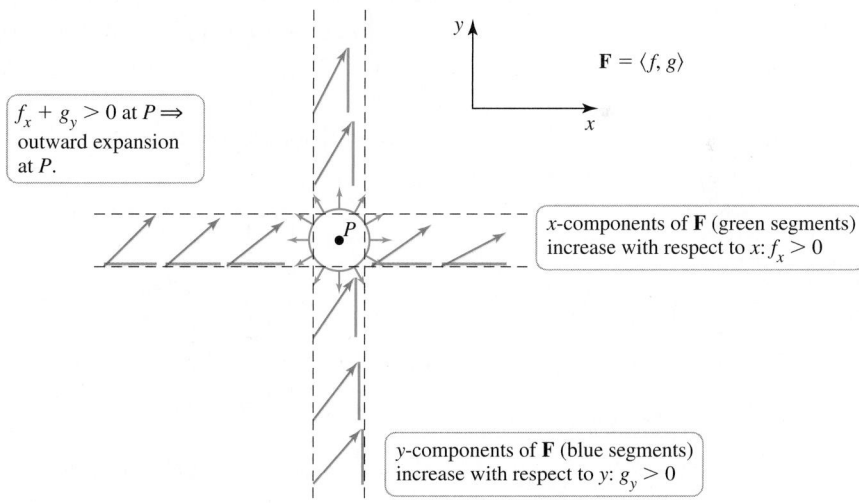

Figure 17.33

If the divergence of $\mathbf{F}$ is zero throughout a region on which $\mathbf{F}$ satisfies the conditions of Theorem 17.9, then the outward flux across the boundary is zero. Vector fields with a zero divergence are said to be *source free*. If the divergence is positive throughout R, the outward flux across C is positive, meaning that the vector field acts as a *source* in R. If the divergence is negative throughout R, the outward flux across C is negative, meaning that the vector field acts as a *sink* in R.

> **QUICK CHECK 2** Compute $\dfrac{\partial f}{\partial x} + \dfrac{\partial g}{\partial y}$ for the rotation field $\mathbf{F} = \langle -y, x \rangle$. What does this tell you about the outward flux of $\mathbf{F}$ across a simple closed curve? ◀

DEFINITION   Two-Dimensional Divergence

The **two-dimensional divergence** of the vector field $\mathbf{F} = \langle f, g \rangle$ is $\dfrac{\partial f}{\partial x} + \dfrac{\partial g}{\partial y}$. If the divergence is zero throughout a region, the vector field is **source free** on that region.

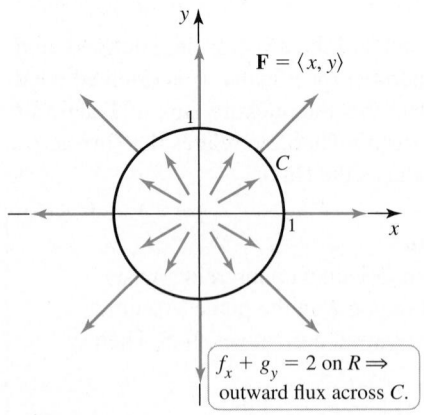

$$\mathbf{F} = \langle x, y \rangle$$

$f_x + g_y = 2$ on $R \Rightarrow$
outward flux across C.

Figure 17.34

EXAMPLE 3 Outward flux of a radial field Use Green's Theorem to compute the outward flux of the radial field $\mathbf{F} = \langle x, y \rangle$ across the unit circle $C = \{(x, y): x^2 + y^2 = 1\}$ (Figure 17.34). Interpret the result.

SOLUTION We have already calculated the outward flux of the radial field across C as a line integral and found it to be 2π (Example 8, Section 17.2). Computing the outward flux using Green's Theorem, note that $f(x, y) = x$ and $g(x, y) = y$; therefore, the divergence of $\mathbf{F}$ is $\dfrac{\partial f}{\partial x} + \dfrac{\partial g}{\partial y} = 2$. By Green's Theorem, we have

$$\oint_C \mathbf{F} \cdot \mathbf{n}\, ds = \iint_R \left(\frac{\partial f}{\partial x} + \frac{\partial g}{\partial y} \right) dA = \iint_R \underbrace{2}_{2}\, dA = 2 \times \text{area of } R = 2\pi.$$

The positive divergence on R results in an outward flux of the vector field across the boundary of R.

Related Exercise 27 ◄

As with the circulation form, the flux form of Green's Theorem can be used in either direction: to simplify line integrals or to simplify double integrals.

EXAMPLE 4 Line integral as a double integral Evaluate

$$\oint_C (4x^3 + \sin y^2)\, dy - (4y^3 + \cos x^2)\, dx,$$

where C is the boundary of the disk $R = \{(x, y): x^2 + y^2 \le 4\}$ oriented counterclockwise.

SOLUTION Letting $f(x, y) = 4x^3 + \sin y^2$ and $g(x, y) = 4y^3 + \cos x^2$, Green's Theorem takes the form

$$\oint_C \underbrace{(4x^3 + \sin y^2)}_{f}\, dy - \underbrace{(4y^3 + \cos x^2)}_{g}\, dx$$

$$= \iint_R \left(\underbrace{12x^2}_{f_x} + \underbrace{12y^2}_{g_y} \right) dA \qquad \text{Green's Theorem, flux form}$$

$$= 12 \int_0^{2\pi} \int_0^2 r^2 \underbrace{r\, dr\, d\theta}_{dA} \qquad \text{Polar coordinates; } x^2 + y^2 = r^2$$

$$= 12 \int_0^{2\pi} \frac{r^4}{4} \bigg|_0^2\, d\theta \qquad \text{Evaluate inner integral.}$$

$$= 48 \int_0^{2\pi} d\theta = 96\pi. \qquad \text{Evaluate outer integral.}$$

Related Exercises 35–36 ◄

Circulation and Flux on More General Regions

Some ingenuity is required to extend both forms of Green's Theorem to more complicated regions. The next two examples illustrate Green's Theorem on two such regions: a half-annulus and a full annulus.

EXAMPLE 5 Circulation on a half-annulus Consider the vector field $\mathbf{F} = \langle y^2, x^2 \rangle$ on the half-annulus $R = \{(x, y): 1 \le x^2 + y^2 \le 9, y \ge 0\}$, whose boundary is C. Find the circulation on C, assuming it has the orientation shown in Figure 17.35.

SOLUTION The circulation on C is

$$\oint_C f\, dx + g\, dy = \oint_C y^2\, dx + x^2\, dy.$$

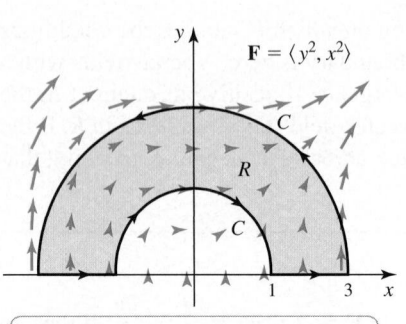

$$\mathbf{F} = \langle y^2, x^2 \rangle$$

Circulation on boundary of R is negative.

Figure 17.35

With the given orientation, the curve runs counterclockwise on the outer semicircle and clockwise on the inner semicircle. Identifying $f(x, y) = y^2$ and $g(x, y) = x^2$, the circulation form of Green's Theorem converts the line integral into a double integral. The double integral is most easily evaluated in polar coordinates using $x = r \cos \theta$ and $y = r \sin \theta$:

$$\oint_C \underbrace{y^2}_{f} \, dx + \underbrace{x^2}_{g} \, dy = \iint_R (\underbrace{2x}_{g_x} - \underbrace{2y}_{f_y}) \, dA \qquad \text{Green's Theorem, circulation form}$$

$$= 2 \int_0^\pi \int_1^3 (r \cos \theta - r \sin \theta) \underbrace{r \, dr \, d\theta}_{dA} \qquad \text{Convert to polar coordinates.}$$

$$= 2 \int_0^\pi (\cos \theta - \sin \theta) \frac{r^3}{3} \Big|_1^3 \, d\theta \qquad \text{Evaluate inner integral.}$$

$$= \frac{52}{3} \int_0^\pi (\cos \theta - \sin \theta) \, d\theta \qquad \text{Simplify.}$$

$$= -\frac{104}{3}. \qquad \text{Evaluate outer integral.}$$

The vector field (Figure 17.35) suggests why the circulation is negative. The field is roughly *opposed* to the direction of C on the outer semicircle but roughly aligned with the direction of C on the inner semicircle. Because the outer semicircle is longer and the field has greater magnitudes on the outer curve than on the inner curve, the greater contribution to the circulation is negative.

Related Exercise 41 ◄

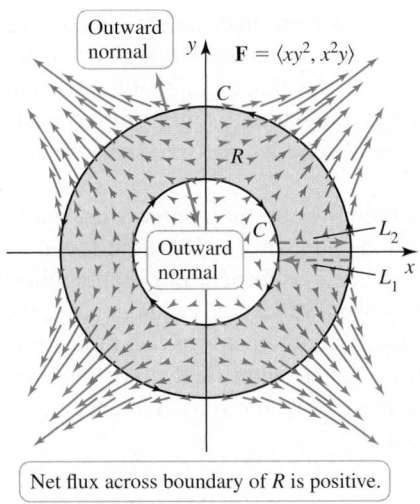

Net flux across boundary of R is positive.

Figure 17.36

➤ Another way to deal with the flux across the annulus is to apply Green's Theorem to the entire disk $|r| \le 2$ and compute the flux across the outer circle. Then apply Green's Theorem to the disk $|r| \le 1$ and compute the flux across the inner circle. Note that the flux *out of* the inner disk is a flux *into* the annulus. Therefore, the difference of the two fluxes gives the net flux for the annulus.

EXAMPLE 6   Flux across the boundary of an annulus Find the outward flux of the vector field $\mathbf{F} = \langle xy^2, x^2y \rangle$ across the boundary of the annulus $R = \{(x, y): 1 \le x^2 + y^2 \le 4\}$, which, when expressed in polar coordinates, is the set $\{(r, \theta): 1 \le r \le 2, 0 \le \theta \le 2\pi\}$ (Figure 17.36).

SOLUTION Because the annulus R is not simply connected, Green's Theorem does not apply as stated in Theorem 17.9. This difficulty is overcome by defining the curve C shown in Figure 17.36, which is simple, closed, and piecewise smooth. The connecting links L_1 and L_2 below and above the x-axis are traversed in opposite directions. Letting L_1 and L_2 approach the x-axis, the contributions to the line integral cancel on L_1 and L_2. Because of this cancellation, we take C to be the curve that runs counterclockwise on the outer boundary and clockwise on the inner boundary.

Using the flux form of Green's Theorem and converting to polar coordinates, we have

$$\oint_C \mathbf{F} \cdot \mathbf{n} \, ds = \oint_C f \, dy - g \, dx = \oint_C xy^2 \, dy - x^2y \, dx \qquad \text{Substitute for } f \text{ and } g.$$

$$= \iint_R (\underbrace{y^2}_{f_x} + \underbrace{x^2}_{g_y}) \, dA \qquad \text{Green's Theorem, flux form}$$

$$= \int_0^{2\pi} \int_1^2 (r^2) \, r \, dr \, d\theta \qquad \text{Polar coordinates; } x^2 + y^2 = r^2$$

$$= \int_0^{2\pi} \frac{r^4}{4} \Big|_1^2 \, d\theta \qquad \text{Evaluate inner integral.}$$

$$= \frac{15}{4} \int_0^{2\pi} d\theta \qquad \text{Simplify.}$$

$$= \frac{15\pi}{2}. \qquad \text{Evaluate outer integral.}$$

> ➤ Notice that the divergence of the vector field in Example 6 ($x^2 + y^2$) is positive on R, so we expect an outward flux across C.

> ➤ Potential function for $\mathbf{F} = \langle f, g \rangle$:
> $$\varphi_x = f \quad \text{and} \quad \varphi_y = g$$
> Stream function for $\mathbf{F} = \langle f, g \rangle$:
> $$\psi_x = -g \quad \text{and} \quad \psi_y = f$$

QUICK CHECK 3 Show that $\psi = \dfrac{1}{2}(y^2 - x^2)$ is a stream function for the vector field $\mathbf{F} = \langle y, x \rangle$. Show that $\mathbf{F}$ has zero divergence. ◄

> ➤ In fluid dynamics, velocity fields that are both conservative and source free are called *ideal flows*. They model fluids that are irrotational and incompressible.

> ➤ Methods for finding solutions of Laplace's equation are discussed in advanced mathematics courses.

Figure 17.36 shows the vector field and explains why the flux across C is positive. Because the field increases in magnitude moving away from the origin, the outward flux across the outer boundary is greater than the inward flux across the inner boundary. Hence, the net outward flux across C is positive.

Related Exercise 42 ◄

Stream Functions

We can now see a wonderful parallel between circulation properties (and conservative vector fields) and flux properties (and source-free fields). We need one more piece to complete the picture; it is the *stream function*, which plays the same role for source-free fields that the potential function plays for conservative fields.

Consider a two-dimensional vector field $\mathbf{F} = \langle f, g \rangle$ that is differentiable on a region R. A **stream function** for the vector field—if it exists—is a function ψ (pronounced *psigh* or *psee*) that satisfies

$$\frac{\partial \psi}{\partial y} = f, \qquad \frac{\partial \psi}{\partial x} = -g.$$

If we compute the divergence of a vector field $\mathbf{F} = \langle f, g \rangle$ that has a stream function and use the fact that $\psi_{xy} = \psi_{yx}$, then

$$\frac{\partial f}{\partial x} + \frac{\partial g}{\partial y} = \frac{\partial}{\partial x}\left(\frac{\partial \psi}{\partial y}\right) + \frac{\partial}{\partial y}\left(-\frac{\partial \psi}{\partial x}\right) = 0.$$
$$\underbrace{\qquad\qquad\qquad\qquad}_{\psi_{yx} = \psi_{xy}}$$

We see that the existence of a stream function guarantees that the vector field has zero divergence or, equivalently, is source free. The converse is also true on simply connected regions of $\mathbb{R}^2$.

As discussed in Section 17.1, the level curves of a stream function are called flow curves or streamlines—and for good reason. It can be shown (Exercise 70) that the vector field $\mathbf{F}$ is everywhere tangent to the streamlines, which means that a graph of the streamlines shows the flow of the vector field. Finally, just as circulation integrals of a conservative vector field are independent of path, flux integrals of a source-free field are also independent of path (Exercise 69).

Vector fields that are both conservative and source free are quite interesting mathematically because they have both a potential function and a stream function. It can be shown that the level curves of the potential and stream functions form orthogonal families; that is, at each point of intersection, the line tangent to one level curve is orthogonal to the line tangent to the other level curve (equivalently, the gradient vector of one function is orthogonal to the gradient vector of the other function). Such vector fields have zero curl ($g_x - f_y = 0$) and zero divergence ($f_x + g_y = 0$). If we write the zero divergence condition in terms of the potential function φ, we find that

$$0 = f_x + g_y = \varphi_{xx} + \varphi_{yy}.$$

Writing the zero curl condition in terms of the stream function ψ, we find that

$$0 = g_x - f_y = -\psi_{xx} - \psi_{yy}.$$

We see that the potential function and the stream function both satisfy an important equation known as **Laplace's equation**:

$$\varphi_{xx} + \varphi_{yy} = 0 \quad \text{and} \quad \psi_{xx} + \psi_{yy} = 0.$$

Any function satisfying Laplace's equation can be used as a potential function or stream function for a conservative, source-free vector field. These vector fields are used in fluid dynamics, electrostatics, and other modeling applications.

Table 17.1 shows the parallel properties of conservative and source-free vector fields in two dimensions. We assume C is a simple piecewise-smooth oriented curve and either is closed or has endpoints A and B.

Table 17.1

Conservative Fields $\mathbf{F} = \langle f, g \rangle$	Source-Free Fields $\mathbf{F} = \langle f, g \rangle$
• curl $= \dfrac{\partial g}{\partial x} - \dfrac{\partial f}{\partial y} = 0$	• divergence $= \dfrac{\partial f}{\partial x} + \dfrac{\partial g}{\partial y} = 0$
• Potential function φ with $\mathbf{F} = \nabla\varphi$ or $f = \dfrac{\partial\varphi}{\partial x}$, $g = \dfrac{\partial\varphi}{\partial y}$	• Stream function ψ with $f = \dfrac{\partial\psi}{\partial y}$, $g = -\dfrac{\partial\psi}{\partial x}$
• Circulation $= \oint_C \mathbf{F} \cdot d\mathbf{r} = 0$ on all closed curves C.	• Flux $= \oint_C \mathbf{F} \cdot \mathbf{n} \, ds = 0$ on all closed curves C.
• Evaluation of line integral $\displaystyle\int_C \mathbf{F} \cdot d\mathbf{r} = \varphi(B) - \varphi(A)$	• Evaluation of line integral $\displaystyle\int_C \mathbf{F} \cdot \mathbf{n} \, ds = \psi(B) - \psi(A)$

With Green's Theorem in the picture, we may also give a concise summary of the various cases that arise with line integrals of both the circulation and flux types (Table 17.2).

Table 17.2

Circulation/work integrals: $\displaystyle\int_C \mathbf{F} \cdot \mathbf{T} \, ds = \int_C \mathbf{F} \cdot d\mathbf{r} = \int_C f \, dx + g \, dy$

	C closed	C not closed
F conservative ($\mathbf{F} = \nabla\varphi$)	$\displaystyle\oint_C \mathbf{F} \cdot d\mathbf{r} = 0$	$\displaystyle\int_C \mathbf{F} \cdot d\mathbf{r} = \varphi(B) - \varphi(A)$
F not conservative	Green's Theorem $\displaystyle\oint_C \mathbf{F} \cdot d\mathbf{r} = \iint_R (g_x - f_y) \, dA$	Direct evaluation $\displaystyle\int_C \mathbf{F} \cdot d\mathbf{r} = \int_a^b (fx' + gy') \, dt$

Flux integrals: $\displaystyle\int_C \mathbf{F} \cdot \mathbf{n} \, ds = \int_C f \, dy - g \, dx$

	C closed	C not closed
F source free ($f = \psi_y$, $g = -\psi_x$)	$\displaystyle\oint_C \mathbf{F} \cdot \mathbf{n} \, ds = 0$	$\displaystyle\int_C \mathbf{F} \cdot \mathbf{n} \, ds = \psi(B) - \psi(A)$
F not source free	Green's Theorem $\displaystyle\oint_C \mathbf{F} \cdot \mathbf{n} \, ds = \iint_R (f_x + g_y) \, dA$	Direct evaluation $\displaystyle\int_C \mathbf{F} \cdot \mathbf{n} \, ds = \int_a^b (fy' - gx') \, dt$

Proof of Green's Theorem on Special Regions

The proof of Green's Theorem is straightforward when restricted to special regions. We consider regions R enclosed by a simple closed smooth curve C oriented in the counterclockwise direction, such that the region can be expressed in two ways (**Figure 17.37**):

▶ This restriction on R means that lines parallel to the coordinate axes intersect the boundary of R at most twice.

• $R = \{(x, y): a \leq x \leq b, G_1(x) \leq y \leq G_2(x)\}$ or
• $R = \{(x, y): H_1(y) \leq x \leq H_2(y), c \leq y \leq d\}$.

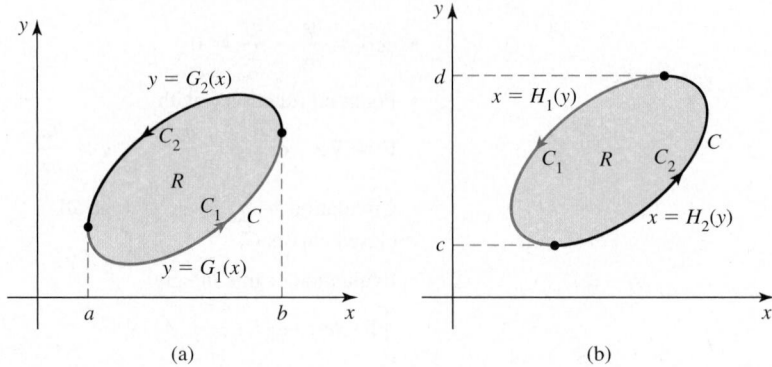

(a)   (b)

Figure 17.37

Under these conditions, we prove the circulation form of Green's Theorem:

$$\oint_C f \, dx + g \, dy = \iint_R \left(\frac{\partial g}{\partial x} - \frac{\partial f}{\partial y} \right) dA.$$

Beginning with the term $\displaystyle\iint_R \frac{\partial f}{\partial y} \, dA$, we write this double integral as an iterated integral, where $G_1(x) \leq y \leq G_2(x)$ in the inner integral and $a \leq x \leq b$ in the outer integral (**Figure 17.37a**). The upper curve is labeled C_2 and the lower curve is labeled C_1. Notice that the inner integral of $\dfrac{\partial f}{\partial y}$ with respect to y gives $f(x, y)$. Therefore, the first step of the double integration is

$$\iint_R \frac{\partial f}{\partial y} \, dA = \int_a^b \int_{G_1(x)}^{G_2(x)} \frac{\partial f}{\partial y} \, dy \, dx \qquad \text{Convert to an iterated integral.}$$

$$= \int_a^b \Big(\underbrace{f(x, G_2(x))}_{\text{on } C_2} - \underbrace{f(x, G_1(x))}_{\text{on } C_1} \Big) \, dx.$$

Over the interval $a \leq x \leq b$, the points $(x, G_2(x))$ trace out the upper part of C (labeled C_2) in the *negative* (clockwise) direction. Similarly, over the interval $a \leq x \leq b$, the points $(x, G_1(x))$ trace out the lower part of C (labeled C_1) in the *positive* (counterclockwise) direction.

Therefore,

$$\iint_R \frac{\partial f}{\partial y} \, dA = \int_a^b \big(f(x, G_2(x)) - f(x, G_1(x)) \big) \, dx$$

$$= \int_{-C_2} f \, dx - \int_{C_1} f \, dx$$

$$= -\int_{C_2} f \, dx - \int_{C_1} f \, dx \qquad\qquad \int_{-C_2} f \, dx = -\int_{C_2} f \, dx$$

$$= -\oint_C f \, dx. \qquad\qquad\qquad\qquad \int_C f \, dx = \int_{C_1} f \, dx + \int_{C_2} f \, dx$$

A similar argument applies to the double integral of $\dfrac{\partial g}{\partial x}$, except we use the bounding curves $x = H_1(y)$ and $x = H_2(y)$, where C_1 is now the left curve and C_2 is the right curve (Figure 17.37b). We have

$$\iint\limits_R \frac{\partial g}{\partial x}\, dA = \int_c^d \int_{H_1(y)}^{H_2(y)} \frac{\partial g}{\partial x}\, dx\, dy \qquad \text{Convert to an iterated integral.}$$

$$= \int_c^d \left(\underbrace{g(H_2(y), y)}_{C_2} - \underbrace{g(H_1(y), y)}_{-C_1} \right) dy \qquad \int \frac{\partial g}{\partial x}\, dx = g$$

$$= \int_{C_2} g\, dy - \int_{-C_1} g\, dy$$

$$= \int_{C_2} g\, dy + \int_{C_1} g\, dy \qquad\qquad \int_{-C_1} g\, dy = -\int_{C_1} g\, dy$$

$$= \oint_C g\, dy. \qquad\qquad\qquad \int_C g\, dy = \int_{C_1} g\, dy + \int_{C_2} g\, dy$$

Combining these two calculations results in

$$\iint\limits_R \left(\frac{\partial g}{\partial x} - \frac{\partial f}{\partial y} \right) dA = \oint_C f\, dx + g\, dy.$$

QUICK CHECK 4 Explain why Green's Theorem proves that if $g_x = f_y$, then the vector field $\mathbf{F} = \langle f, g \rangle$ is conservative. ◄

As mentioned earlier, with a change of notation (replace g with f and f with $-g$), the flux form of Green's Theorem is obtained. This proof also completes the list of equivalent properties of conservative fields given in Section 17.3: From Green's Theorem, it follows that if $\dfrac{\partial g}{\partial x} = \dfrac{\partial f}{\partial y}$ on a simply connected region R, then the vector field $\mathbf{F} = \langle f, g \rangle$ is conservative on R.

SECTION 17.4 EXERCISES

Getting Started

1. Explain why the two forms of Green's Theorem are analogs of the Fundamental Theorem of Calculus.

2. Referring to both forms of Green's Theorem, match each idea in Column 1 to an idea in Column 2:

Line integral for flux	Double integral of the curl
Line integral for circulation	Double integral of the divergence

3. How do you use a line integral to compute the area of a plane region?

4. Why does a two-dimensional vector field with zero curl on a region have zero circulation on a closed curve that bounds the region?

5. Why does a two-dimensional vector field with zero divergence on a region have zero outward flux across a closed curve that bounds the region?

6. Sketch a two-dimensional vector field that has zero curl everywhere in the plane.

7. Sketch a two-dimensional vector field that has zero divergence everywhere in the plane.

8. Discuss one of the parallels between a conservative vector field and a source-free vector field.

9–14. *Assume C is a circle centered at the origin, oriented counterclockwise, that encloses disk R in the plane. Complete the following steps for each vector field $\mathbf{F}$.*

a. *Calculate the two-dimensional curl of $\mathbf{F}$.*
b. *Calculate the two-dimensional divergence of $\mathbf{F}$.*
c. *Is $\mathbf{F}$ irrotational on R?*
d. *Is $\mathbf{F}$ source free on R?*

9. $\mathbf{F} = \langle x, y \rangle$ 10. $\mathbf{F} = \langle y, -x \rangle$

11. $\mathbf{F} = \langle y, -3x \rangle$ 12. $\mathbf{F} = \langle x^2 + 2xy, -2xy - y^2 \rangle$

13. $\mathbf{F} = \langle 4x^3 y, xy^2 + x^4 \rangle$ 14. $\mathbf{F} = \langle 4x^3 + y, 12xy \rangle$

15. Suppose C is the boundary of region $R = \{(x, y): x^2 \le y \le x \le 1\}$, oriented counterclockwise (see figure); let $\mathbf{F} = \langle 1, x \rangle$.

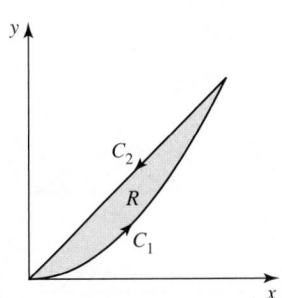

a. Compute the two-dimensional curl of **F** and determine whether **F** is irrotational.

b. Find parameterizations $\mathbf{r}_1(t)$ and $\mathbf{r}_2(t)$ for C_1 and C_2, respectively.

c. Evaluate both the line integral and the double integral in the circulation form of Green's Theorem and check for consistency.

d. Compute the two-dimensional divergence of **F** and use the flux form of Green's Theorem to explain why the outward flux is 0.

16. Suppose C is the boundary of region $R = \{(x, y): 2x^2 - 2x \le y \le 0\}$, oriented counterclockwise (see figure); let $\mathbf{F} = \langle x, 1 \rangle$.

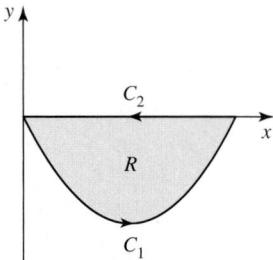

a. Compute the two-dimensional curl of **F** and use the circulation form of Green's Theorem to explain why the circulation is 0.

b. Compute the two-dimensional divergence of **F** and determine whether **F** is source free.

c. Find parameterizations $\mathbf{r}_1(t)$ and $\mathbf{r}_2(t)$ for C_1 and C_2, respectively.

d. Evaluate both the line integral and the double integral in the flux form of Green's Theorem and check for consistency.

Practice Exercises

17–20. Green's Theorem, circulation form *Consider the following regions R and vector fields **F**.*

a. *Compute the two-dimensional curl of the vector field.*

b. *Evaluate both integrals in Green's Theorem and check for consistency.*

17. $\mathbf{F} = \langle 2y, -2x \rangle$; R is the region bounded by $y = \sin x$ and $y = 0$, for $0 \le x \le \pi$.

18. $\mathbf{F} = \langle -3y, 3x \rangle$; R is the triangle with vertices $(0, 0)$, $(1, 0)$, and $(0, 2)$.

19. $\mathbf{F} = \langle -2xy, x^2 \rangle$; R is the region bounded by $y = x(2 - x)$ and $y = 0$.

20. $\mathbf{F} = \langle 0, x^2 + y^2 \rangle$; $R = \{(x, y): x^2 + y^2 \le 1\}$

21–26. Area of regions *Use a line integral on the boundary to find the area of the following regions.*

21. A disk of radius 5

22. A region bounded by an ellipse with major and minor axes of lengths 12 and 8, respectively

23. $\{(x, y): x^2 + y^2 \le 16\}$

24. The region shown in the figure

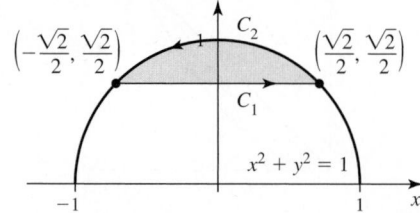

25. The region bounded by the parabolas $\mathbf{r}(t) = \langle t, 2t^2 \rangle$ and $\mathbf{r}(t) = \langle t, 12 - t^2 \rangle$, for $-2 \le t \le 2$

26. The region bounded by the curve $\mathbf{r}(t) = \langle t(1 - t^2), 1 - t^2 \rangle$, for $-1 \le t \le 1$ (*Hint:* Plot the curve.)

27–30. Green's Theorem, flux form *Consider the following regions R and vector fields **F**.*

a. *Compute the two-dimensional divergence of the vector field.*

b. *Evaluate both integrals in Green's Theorem and check for consistency.*

27. $\mathbf{F} = \langle x, y \rangle$; $R = \{(x, y): x^2 + y^2 \le 4\}$

28. $\mathbf{F} = \langle x, -3y \rangle$; R is the triangle with vertices $(0, 0)$, $(1, 2)$, and $(0, 2)$.

29. $\mathbf{F} = \langle 2xy, x^2 \rangle$; $R = \{(x, y): 0 \le y \le x(2 - x)\}$

30. $\mathbf{F} = \langle x^2 + y^2, 0 \rangle$; $R = \{(x, y): x^2 + y^2 \le 1\}$

31–40. Line integrals *Use Green's Theorem to evaluate the following line integrals. Assume all curves are oriented counterclockwise. A sketch is helpful.*

31. $\oint_C \langle 3y + 1, 4x^2 + 3 \rangle \cdot d\mathbf{r}$, where C is the boundary of the rectangle with vertices $(0, 0)$, $(4, 0)$, $(4, 2)$, and $(0, 2)$

32. $\oint_C \langle \sin y, x \rangle \cdot d\mathbf{r}$, where C is the boundary of the triangle with vertices $(0, 0)$, $\left(\frac{\pi}{2}, 0\right)$, and $\left(\frac{\pi}{2}, \frac{\pi}{2}\right)$

33. $\oint_C xe^y \, dx + x \, dy$, where C is the boundary of the region bounded by the curves $y = x^2$, $x = 2$, and the x-axis

34. $\oint_C \frac{1}{1 + y^2} \, dx + y \, dy$, where C is the boundary of the triangle with vertices $(0, 0)$, $(1, 0)$, and $(1, 1)$

35. $\oint_C (2x + e^{y^2}) \, dy - (4y^2 + e^{x^2}) \, dx$, where C is the boundary of the square with vertices $(0, 0)$, $(1, 0)$, $(1, 1)$, and $(0, 1)$

36. $\oint_C (2x - 3y) \, dy - (3x + 4y) \, dx$, where C is the unit circle

37. $\oint_C f \, dy - g \, dx$, where $\langle f, g \rangle = \langle 0, xy \rangle$ and C is the triangle with vertices $(0, 0)$, $(2, 0)$, and $(0, 4)$

38. $\oint_C f \, dy - g \, dx$, where $\langle f, g \rangle = \langle x^2, 2y^2 \rangle$ and C is the upper half of the unit circle and the line segment $-1 \le x \le 1$, oriented clockwise

39. The circulation line integral of $\mathbf{F} = \langle x^2 + y^2, 4x + y^3 \rangle$, where C is the boundary of $\{(x, y): 0 \le y \le \sin x, 0 \le x \le \pi\}$

40. The flux line integral of $\mathbf{F} = \langle e^{x-y}, e^{y-x} \rangle$, where C is the boundary of $\{(x, y): 0 \le y \le x, 0 \le x \le 1\}$

41–48. Circulation and flux *For the following vector fields, compute (a) the circulation on, and (b) the outward flux across, the boundary of the given region. Assume boundary curves are oriented counterclockwise.*

41. $\mathbf{F} = \langle x, y \rangle$; R is the half-annulus $\{(r, \theta); 1 \le r \le 2,$ $0 \le \theta \le \pi\}$.

42. $\mathbf{F} = \langle -y, x \rangle$; R is the annulus $\{(r, \theta): 1 \le r \le 3,$ $0 \le \theta \le 2\pi\}$.

43. $\mathbf{F} = \langle 2x + y, x - 4y \rangle$; R is the quarter-annulus $\{(r, \theta): 1 \le r \le 4, 0 \le \theta \le \pi/2\}$.

44. $\mathbf{F} = \langle x - y, -x + 2y \rangle$; R is the parallelogram $\{(x, y): 1 - x \le y \le 3 - x, 0 \le x \le 1\}$.

45. $\mathbf{F} = \nabla\left(\sqrt{x^2 + y^2}\right)$; R is the half-annulus $\{(r, \theta): 1 \le r \le 3, 0 \le \theta \le \pi\}$.

46. $\mathbf{F} = \left\langle \ln(x^2 + y^2), \tan^{-1}\dfrac{y}{x} \right\rangle$; R is the eighth-annulus $\{(r, \theta): 1 \le r \le 2, 0 \le \theta \le \pi/4\}$.

47. $\mathbf{F} = \langle x + y^2, x^2 - y \rangle$; $R = \{(x, y): y^2 \le x \le 2 - y^2\}$.

48. $\mathbf{F} = \langle y \cos x, -\sin x \rangle$; R is the square $\{(x, y): 0 \le x \le \pi/2, 0 \le y \le \pi/2\}$.

49. Explain why or why not Determine whether the following statements are true and give an explanation or counterexample.

 a. The work required to move an object around a closed curve C in the presence of a vector force field is the circulation of the force field on the curve.

 b. If a vector field has zero divergence throughout a region (on which the conditions of Green's Theorem are met), then the circulation on the boundary of that region is zero.

 c. If the two-dimensional curl of a vector field is positive throughout a region (on which the conditions of Green's Theorem are met), then the circulation on the boundary of that region is positive (assuming counterclockwise orientation).

50–51. Special line integrals *Prove the following identities, where C is a simple closed smooth oriented curve.*

50. $\displaystyle\oint_C dx = \oint_C dy = 0$

51. $\displaystyle\oint_C f(x)\,dx + g(y)\,dy = 0$, where f and g have continuous derivatives on the region enclosed by C

52. Double integral to line integral Use the flux form of Green's Theorem to evaluate $\iint_R (2xy + 4y^3)\,dA$, where R is the triangle with vertices $(0, 0)$, $(1, 0)$, and $(0, 1)$.

53. Area line integral Show that the value of

$$\oint_C xy^2\,dx + (x^2 y + 2x)\,dy$$

depends only on the area of the region enclosed by C.

54. Area line integral In terms of the parameters a and b, how is the value of $\oint_C ay\,dx + bx\,dy$ related to the area of the region enclosed by C, assuming counterclockwise orientation of C?

55–58. Stream function *Recall that if the vector field $\mathbf{F} = \langle f, g \rangle$ is source free (zero divergence), then a stream function ψ exists such that $f = \psi_y$ and $g = -\psi_x$.*

 a. *Verify that the given vector field has zero divergence.*

 b. *Integrate the relations $f = \psi_y$ and $g = -\psi_x$ to find a stream function for the field.*

55. $\mathbf{F} = \langle 4, 2 \rangle$ **56.** $\mathbf{F} = \langle y^2, x^2 \rangle$

57. $\mathbf{F} = \langle -e^{-x}\sin y, e^{-x}\cos y \rangle$ **58.** $\mathbf{F} = \langle x^2, -2xy \rangle$

Explorations and Challenges

59–62. Ideal flow *A two-dimensional vector field describes **ideal flow** if it has both zero curl and zero divergence on a simply connected region.*

 a. *Verify that both the curl and the divergence of the given field are zero.*

 b. *Find a potential function φ and a stream function ψ for the field.*

 c. *Verify that φ and ψ satisfy Laplace's equation*
$$\varphi_{xx} + \varphi_{yy} = \psi_{xx} + \psi_{yy} = 0.$$

59. $\mathbf{F} = \langle e^x \cos y, -e^x \sin y \rangle$

60. $\mathbf{F} = \langle x^3 - 3xy^2, y^3 - 3x^2 y \rangle$

61. $\mathbf{F} = \left\langle \tan^{-1}\dfrac{y}{x}, \dfrac{1}{2}\ln(x^2 + y^2) \right\rangle$, for $x > 0$

62. $\mathbf{F} = \dfrac{\langle x, y \rangle}{x^2 + y^2}$, for $x > 0, y > 0$

63. Flow in an ocean basin An idealized two-dimensional ocean is modeled by the square region $R = \left[-\dfrac{\pi}{2}, \dfrac{\pi}{2}\right] \times \left[-\dfrac{\pi}{2}, \dfrac{\pi}{2}\right]$ with boundary C. Consider the stream function $\psi(x, y) = 4\cos x \cos y$ defined on R (see figure).

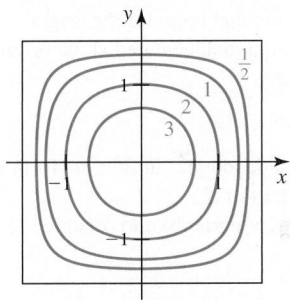

 a. The horizontal (east-west) component of the velocity is $u = \psi_y$ and the vertical (north-south) component of the velocity is $v = -\psi_x$. Sketch a few representative velocity vectors and show that the flow is counterclockwise around the region.

 b. Is the velocity field source free? Explain.

 c. Is the velocity field irrotational? Explain.

 d. Let C be the boundary of R. Find the total outward flux across C.

 e. Find the circulation on C assuming counterclockwise orientation.

64. Green's Theorem as a Fundamental Theorem of Calculus Show that if the circulation form of Green's Theorem is applied to the vector field $\left\langle 0, \dfrac{f(x)}{c} \right\rangle$, where $c > 0$ and $R = \{(x, y): a \le x \le b, 0 \le y \le c\}$, then the result is the Fundamental Theorem of Calculus,

$$\int_a^b \frac{df}{dx}\,dx = f(b) - f(a).$$

65. Green's Theorem as a Fundamental Theorem of Calculus Show that if the flux form of Green's Theorem is applied to the vector field $\left\langle \dfrac{f(x)}{c}, 0 \right\rangle$, where $c > 0$ and $R = \{(x, y): a \le x \le b, 0 \le y \le c\}$, then the result is the Fundamental Theorem of Calculus,

$$\int_a^b \frac{df}{dx}\, dx = f(b) - f(a).$$

66. What's wrong? Consider the rotation field $\mathbf{F} = \dfrac{\langle -y, x \rangle}{x^2 + y^2}$.

a. Verify that the two-dimensional curl of $\mathbf{F}$ is zero, which suggests that the double integral in the circulation form of Green's Theorem is zero.

b. Use a line integral to verify that the circulation on the unit circle of the vector field is 2π.

c. Explain why the results of parts (a) and (b) do not agree.

67. What's wrong? Consider the radial field $\mathbf{F} = \dfrac{\langle x, y \rangle}{x^2 + y^2}$.

a. Verify that the divergence of $\mathbf{F}$ is zero, which suggests that the double integral in the flux form of Green's Theorem is zero.

b. Use a line integral to verify that the outward flux across the unit circle of the vector field is 2π.

c. Explain why the results of parts (a) and (b) do not agree.

68. Conditions for Green's Theorem Consider the radial field

$$\mathbf{F} = \langle f, g \rangle = \frac{\langle x, y \rangle}{\sqrt{x^2 + y^2}} = \frac{\mathbf{r}}{|\mathbf{r}|}.$$

a. Explain why the conditions of Green's Theorem do not apply to $\mathbf{F}$ on a region that includes the origin.

b. Let R be the unit disk centered at the origin and compute

$$\iint\limits_R \left(\frac{\partial f}{\partial x} + \frac{\partial g}{\partial y} \right) dA.$$

c. Evaluate the line integral in the flux form of Green's Theorem on the boundary of R.

d. Do the results of parts (b) and (c) agree? Explain.

69. Flux integrals Assume the vector field $\mathbf{F} = \langle f, g \rangle$ is source free (zero divergence) with stream function ψ. Let C be any smooth simple curve from A to the distinct point B. Show that the flux integral $\int_C \mathbf{F} \cdot \mathbf{n}\, ds$ is independent of path; that is, $\int_C \mathbf{F} \cdot \mathbf{n}\, ds = \psi(B) - \psi(A)$.

70. Streamlines are tangent to the vector field Assume the vector field $\mathbf{F} = \langle f, g \rangle$ is related to the stream function ψ by $\psi_y = f$ and $\psi_x = -g$ on a region R. Prove that at all points of R, the vector field is tangent to the streamlines (the level curves of the stream function).

71. Streamlines and equipotential lines Assume that on $\mathbb{R}^2$, the vector field $\mathbf{F} = \langle f, g \rangle$ has a potential function φ such that $f = \varphi_x$ and $g = \varphi_y$, and it has a stream function ψ such that $f = \psi_y$ and $g = -\psi_x$. Show that the equipotential curves (level curves of φ) and the streamlines (level curves of ψ) are everywhere orthogonal.

72. Channel flow The flow in a long shallow channel is modeled by the velocity field $\mathbf{F} = \langle 0, 1 - x^2 \rangle$, where $R = \{(x, y): |x| \le 1 \text{ and } |y| < 5\}$.

a. Sketch R and several streamlines of $\mathbf{F}$.

b. Evaluate the curl of $\mathbf{F}$ on the lines $x = 0$, $x = 1/4$, $x = 1/2$, and $x = 1$.

c. Compute the circulation on the boundary of the region R.

d. How do you explain the fact that the curl of $\mathbf{F}$ is nonzero at points of R, but the circulation is zero?

QUICK CHECK ANSWERS

1. $g_x - f_y = 0$, which implies zero circulation on a closed curve. **2.** $f_x + g_y = 0$, which implies zero flux across a closed curve. **3.** $\psi_y = y$ is the x-component of $\mathbf{F} = \langle y, x \rangle$, and $-\psi_x = x$ is the y-component of $\mathbf{F}$. Also, the divergence of $\mathbf{F}$ is $y_x + x_y = 0$. **4.** If the curl is zero on a region, then all closed-path integrals are zero, which is a condition (Section 17.3) for a conservative field. ◄

17.5 Divergence and Curl

Green's Theorem sets the stage for the final act in our exploration of calculus. The last four sections of this chapter have the following goal: to lift both forms of Green's Theorem out of the plane ($\mathbb{R}^2$) and into space ($\mathbb{R}^3$). It is done as follows.

• The circulation form of Green's Theorem relates a line integral over a simple closed oriented curve in the plane to a double integral over the enclosed region. In an analogous manner, we will see that *Stokes' Theorem* (Section 17.7) relates a line integral over a simple closed oriented curve in $\mathbb{R}^3$ to a double integral over a surface whose boundary is that curve.

• The flux form of Green's Theorem relates a line integral over a simple closed oriented curve in the plane to a double integral over the enclosed region. Similarly, the *Divergence Theorem* (Section 17.8) relates an integral over a closed oriented surface in $\mathbb{R}^3$ to a triple integral over the region enclosed by that surface.

In order to make these extensions, we need a few more tools.

• The two-dimensional divergence and two-dimensional curl must be extended to three dimensions (this section).

• The idea of a *surface integral* must be introduced (Section 17.6).

▶ Review: The divergence measures the
expansion or contraction of a vector
field at each point. The flux form of
Green's Theorem implies that if the two-
dimensional divergence of a vector field
is zero throughout a simply connected
plane region, then the outward flux
across the boundary of the region is zero.
If the divergence is nonzero, Green's
Theorem gives the outward flux across
the boundary.

The Divergence

Recall that in two dimensions, the divergence of the vector field $\mathbf{F} = \langle f, g \rangle$ is $\dfrac{\partial f}{\partial x} + \dfrac{\partial g}{\partial y}$.
The extension to three dimensions is straightforward. If $\mathbf{F} = \langle f, g, h \rangle$ is a differentiable vector field defined on a region of $\mathbb{R}^3$, the divergence of $\mathbf{F}$ is $\dfrac{\partial f}{\partial x} + \dfrac{\partial g}{\partial y} + \dfrac{\partial h}{\partial z}$. The interpretation of the three-dimensional divergence is much the same as it is in two dimensions. It measures the expansion or contraction of the vector field at each point. If the divergence is zero at all points of a region, the vector field is *source free* on that region.

Recall the *del operator* ∇ that was introduced in Section 15.5 to define the gradient:

$$\nabla = \mathbf{i}\frac{\partial}{\partial x} + \mathbf{j}\frac{\partial}{\partial y} + \mathbf{k}\frac{\partial}{\partial z} = \left\langle \frac{\partial}{\partial x}, \frac{\partial}{\partial y}, \frac{\partial}{\partial z} \right\rangle.$$

This object is not really a vector; it is an operation that is applied to a function or a vector field. Applying it directly to a scalar function f results in the gradient of f:

$$\nabla f = \frac{\partial f}{\partial x}\mathbf{i} + \frac{\partial f}{\partial y}\mathbf{j} + \frac{\partial f}{\partial z}\mathbf{k} = \langle f_x, f_y, f_z \rangle.$$

▶ In evaluating $\nabla \cdot \mathbf{F}$ as a dot product,
each component of ∇ is applied to
the corresponding component of $\mathbf{F}$,
producing $f_x + g_y + h_z$.

However, if we form the *dot product* of ∇ and a vector field $\mathbf{F} = \langle f, g, h \rangle$, the result is

$$\nabla \cdot \mathbf{F} = \left\langle \frac{\partial}{\partial x}, \frac{\partial}{\partial y}, \frac{\partial}{\partial z} \right\rangle \cdot \langle f, g, h \rangle = \frac{\partial f}{\partial x} + \frac{\partial g}{\partial y} + \frac{\partial h}{\partial z},$$

which is the divergence of $\mathbf{F}$, also denoted div $\mathbf{F}$. Like all dot products, the divergence is a scalar; in this case, it is a scalar-valued function.

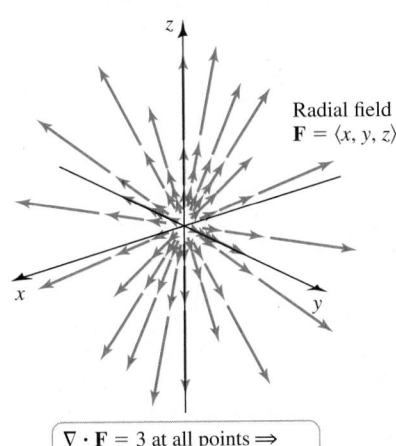

Radial field
$\mathbf{F} = \langle x, y, z \rangle$

$\nabla \cdot \mathbf{F} = 3$ at all points $\Rightarrow$
vector field expands outward
at all points.

(a)

> **DEFINITION Divergence of a Vector Field**
>
> The **divergence** of a vector field $\mathbf{F} = \langle f, g, h \rangle$ that is differentiable on a region of $\mathbb{R}^3$ is
>
> $$\text{div } \mathbf{F} = \nabla \cdot \mathbf{F} = \frac{\partial f}{\partial x} + \frac{\partial g}{\partial y} + \frac{\partial h}{\partial z}.$$
>
> If $\nabla \cdot \mathbf{F} = 0$, the vector field is **source free**.

EXAMPLE 1 Computing the divergence Compute the divergence of the following vector fields.

a. $\mathbf{F} = \langle x, y, z \rangle$ (a radial field)

b. $\mathbf{F} = \langle -y, x - z, y \rangle$ (a rotation field)

c. $\mathbf{F} = \langle -y, x, z \rangle$ (a spiral flow)

SOLUTION

a. The divergence is $\nabla \cdot \mathbf{F} = \nabla \cdot \langle x, y, z \rangle = \dfrac{\partial x}{\partial x} + \dfrac{\partial y}{\partial y} + \dfrac{\partial z}{\partial z} = 1 + 1 + 1 = 3.$

Because the divergence is positive, the flow expands outward at all points (Figure 17.38a).

b. The divergence is

$$\nabla \cdot \mathbf{F} = \nabla \cdot \langle -y, x - z, y \rangle = \frac{\partial(-y)}{\partial x} + \frac{\partial(x - z)}{\partial y} + \frac{\partial y}{\partial z} = 0 + 0 + 0 = 0,$$

so the field is source free.

c. This field is a combination of the two-dimensional rotation field $\mathbf{F} = \langle -y, x \rangle$ and a vertical flow in the z-direction; the net effect is a field that spirals upward for $z > 0$ and spirals downward for $z < 0$ (Figure 17.38b). The divergence is

$$\nabla \cdot \mathbf{F} = \nabla \cdot \langle -y, x, z \rangle = \frac{\partial(-y)}{\partial x} + \frac{\partial x}{\partial y} + \frac{\partial z}{\partial z} = 0 + 0 + 1 = 1.$$

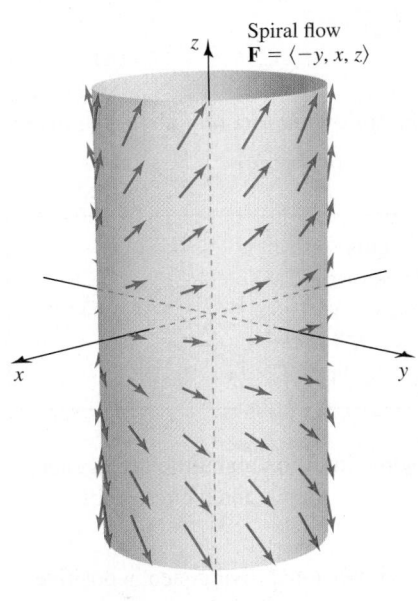

Spiral flow
$\mathbf{F} = \langle -y, x, z \rangle$

(b)

Figure 17.38

QUICK CHECK 1 Show that if a vector field has the form $\mathbf{F} = \langle f(y, z), g(x, z), h(x, y) \rangle$, then div $\mathbf{F} = 0$. ◄

The rotational part of the field in x and y does not contribute to the divergence. However, the z-component of the field produces a nonzero divergence.

Related Exercises 10–11 ◄

Divergence of a Radial Vector Field   The vector field considered in Example 1a is just one of many radial fields that have important applications (for example, the inverse square laws of gravitation and electrostatics). The following example leads to a general result for the divergence of radial vector fields.

EXAMPLE 2   Divergence of a radial field Compute the divergence of the radial vector field

$$\mathbf{F} = \frac{\mathbf{r}}{|\mathbf{r}|} = \frac{\langle x, y, z \rangle}{\sqrt{x^2 + y^2 + z^2}}.$$

SOLUTION This radial field has the property that it is directed outward from the origin and all vectors have unit length ($|\mathbf{F}| = 1$). Let's compute one piece of the divergence; the others follow the same pattern. Using the Quotient Rule, the derivative with respect to x of the first component of $\mathbf{F}$ is

$$\frac{\partial}{\partial x} \left(\frac{x}{(x^2 + y^2 + z^2)^{1/2}} \right) = \frac{(x^2 + y^2 + z^2)^{1/2} - x^2 (x^2 + y^2 + z^2)^{-1/2}}{x^2 + y^2 + z^2} \qquad \text{Quotient Rule}$$

$$= \frac{|\mathbf{r}| - x^2 |\mathbf{r}|^{-1}}{|\mathbf{r}|^2} \qquad \sqrt{x^2 + y^2 + z^2} = |\mathbf{r}|$$

$$= \frac{|\mathbf{r}|^2 - x^2}{|\mathbf{r}|^3}. \qquad \text{Simplify.}$$

A similar calculation of the y- and z-derivatives yields $\dfrac{|\mathbf{r}|^2 - y^2}{|\mathbf{r}|^3}$ and $\dfrac{|\mathbf{r}|^2 - z^2}{|\mathbf{r}|^3}$, respectively. Adding the three terms, we find that

$$\nabla \cdot \mathbf{F} = \frac{|\mathbf{r}|^2 - x^2}{|\mathbf{r}|^3} + \frac{|\mathbf{r}|^2 - y^2}{|\mathbf{r}|^3} + \frac{|\mathbf{r}|^2 - z^2}{|\mathbf{r}|^3}$$

$$= 3 \frac{|\mathbf{r}|^2}{|\mathbf{r}|^3} - \frac{x^2 + y^2 + z^2}{|\mathbf{r}|^3} \qquad \text{Collect terms.}$$

$$= \frac{2}{|\mathbf{r}|}. \qquad x^2 + y^2 + z^2 = |\mathbf{r}|^2$$

Related Exercise 18 ◄

Examples 1a and 2 give two special cases of the following theorem about the divergence of radial vector fields (Exercise 73).

> **THEOREM 17.10   Divergence of Radial Vector Fields**
> For a real number p, the divergence of the radial vector field
> $$\mathbf{F} = \frac{\mathbf{r}}{|\mathbf{r}|^p} = \frac{\langle x, y, z \rangle}{(x^2 + y^2 + z^2)^{p/2}} \quad \text{is} \quad \nabla \cdot \mathbf{F} = \frac{3 - p}{|\mathbf{r}|^p}.$$

EXAMPLE 3   Divergence from a graph To gain some intuition about the divergence, consider the two-dimensional vector field $\mathbf{F} = \langle f, g \rangle = \langle x^2, y \rangle$ and a circle C of radius 2 centered at the origin (**Figure 17.39**).

a. Without computing it, determine whether the two-dimensional divergence is positive or negative at the point $Q(1, 1)$. Why?

b. Confirm your conjecture in part (a) by computing the two-dimensional divergence at Q.

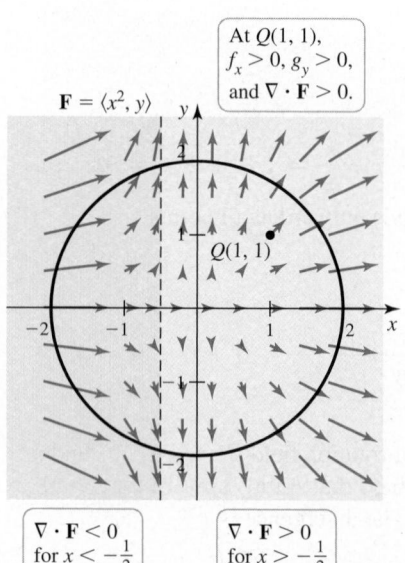

At $Q(1, 1)$, $f_x > 0, g_y > 0$, and $\nabla \cdot \mathbf{F} > 0$.

$\mathbf{F} = \langle x^2, y \rangle$

$\nabla \cdot \mathbf{F} < 0$ for $x < -\frac{1}{2}$

$\nabla \cdot \mathbf{F} > 0$ for $x > -\frac{1}{2}$

Figure 17.39

c. Based on part (b), over what regions within the circle is the divergence positive and over what regions within the circle is the divergence negative?

d. By inspection of the figure, on what part of the circle is the flux across the boundary outward? Is the net flux out of the circle positive or negative?

> To understand the conclusion of Example 3a, note that as you move through the point Q from left to right, the horizontal components of the vectors increase in length ($f_x > 0$). As you move through the point Q in the upward direction, the vertical components of the vectors also increase in length ($g_y > 0$).

SOLUTION

a. At $Q(1, 1)$ the x-component and the y-component of the field are increasing ($f_x > 0$ and $g_y > 0$), so the field is expanding at that point and the two-dimensional divergence is positive.

b. Calculating the two-dimensional divergence, we find that

$$\nabla \cdot \mathbf{F} = \frac{\partial}{\partial x}(x^2) + \frac{\partial}{\partial y}(y) = 2x + 1.$$

At $Q(1, 1)$ the divergence is 3, confirming part (a).

QUICK CHECK 2 Verify the claim made in Example 3d by showing that the net outward flux of $\mathbf{F}$ across C is positive. (*Hint:* If you use Green's Theorem to evaluate the integral $\int_C f\,dy - g\,dx$, convert to polar coordinates.) ◄

c. From part (b), we see that $\nabla \cdot \mathbf{F} = 2x + 1 > 0$, for $x > -\frac{1}{2}$, and $\nabla \cdot \mathbf{F} < 0$, for $x < -\frac{1}{2}$. To the left of the line $x = -\frac{1}{2}$ the field is contracting, and to the right of the line the field is expanding.

d. Using Figure 17.39, it appears that the field is tangent to the circle at two points with $x \approx -1$. For points on the circle with $x < -1$, the flow is into the circle; for points on the circle with $x > -1$, the flow is out of the circle. It appears that the net outward flux across C is positive. The points where the field changes from inward to outward may be determined exactly (Exercise 46).   *Related Exercises 21–22* ◄

The Curl

> Review: The *two-dimensional curl* $g_x - f_y$ measures the rotation of a vector field at a point. The circulation form of Green's Theorem implies that if the two-dimensional curl of a vector field is zero throughout a simply connected region, then the circulation on the boundary of the region is also zero. If the curl is nonzero, Green's Theorem gives the circulation along the curve.

Just as the divergence $\nabla \cdot \mathbf{F}$ is the dot product of the *del operator* and $\mathbf{F}$, the three-dimensional curl is the cross product $\nabla \times \mathbf{F}$. If we formally use the notation for the cross product in terms of a 3×3 determinant, we obtain the definition of the curl:

$$\nabla \times \mathbf{F} = \begin{vmatrix} \mathbf{i} & \mathbf{j} & \mathbf{k} \\ \dfrac{\partial}{\partial x} & \dfrac{\partial}{\partial y} & \dfrac{\partial}{\partial z} \\ f & g & h \end{vmatrix} \begin{matrix} \leftarrow \text{Unit Vectors} \\ \\ \leftarrow \text{Components of } \nabla \\ \\ \leftarrow \text{Components of } \mathbf{F} \end{matrix}$$

$$= \left(\frac{\partial h}{\partial y} - \frac{\partial g}{\partial z}\right)\mathbf{i} + \left(\frac{\partial f}{\partial z} - \frac{\partial h}{\partial x}\right)\mathbf{j} + \left(\frac{\partial g}{\partial x} - \frac{\partial f}{\partial y}\right)\mathbf{k}.$$

The curl of a vector field, also denoted curl $\mathbf{F}$, is a vector with three components. Notice that the $\mathbf{k}$-component of the curl ($g_x - f_y$) is the two-dimensional curl, which gives the rotation in the xy-plane at a point. The $\mathbf{i}$- and $\mathbf{j}$-components of the curl correspond to the rotation of the vector field in planes parallel to the yz-plane (orthogonal to $\mathbf{i}$) and in planes parallel to the xz-plane (orthogonal to $\mathbf{j}$) (Figure 17.40).

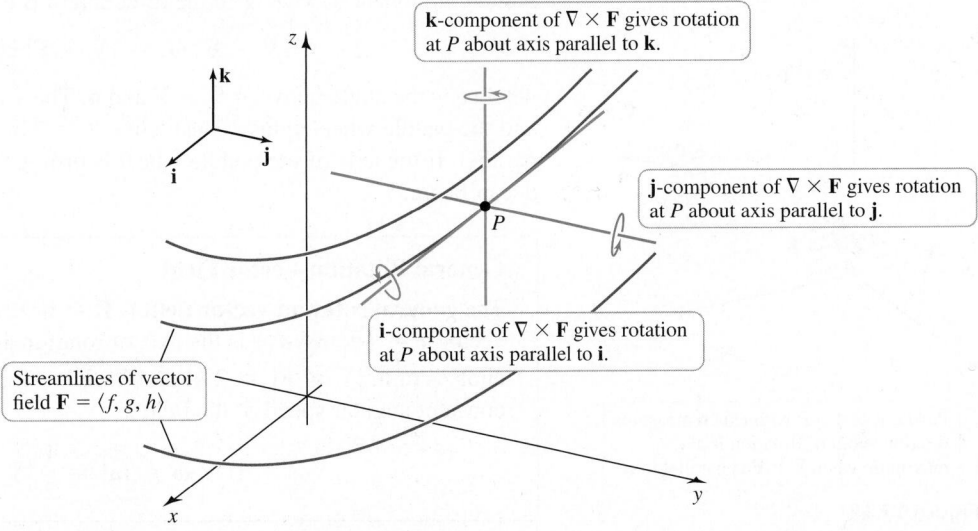

k-component of $\nabla \times \mathbf{F}$ gives rotation at P about axis parallel to $\mathbf{k}$.

j-component of $\nabla \times \mathbf{F}$ gives rotation at P about axis parallel to $\mathbf{j}$.

i-component of $\nabla \times \mathbf{F}$ gives rotation at P about axis parallel to $\mathbf{i}$.

Streamlines of vector field $\mathbf{F} = \langle f, g, h \rangle$

Figure 17.40

> **DEFINITION Curl of a Vector Field**
>
> The **curl** of a vector field $\mathbf{F} = \langle f, g, h \rangle$ that is differentiable on a region of $\mathbb{R}^3$ is
>
> $$\nabla \times \mathbf{F} = \operatorname{curl} \mathbf{F}$$
>
> $$= \left(\frac{\partial h}{\partial y} - \frac{\partial g}{\partial z} \right) \mathbf{i} + \left(\frac{\partial f}{\partial z} - \frac{\partial h}{\partial x} \right) \mathbf{j} + \left(\frac{\partial g}{\partial x} - \frac{\partial f}{\partial y} \right) \mathbf{k}.$$
>
> If $\nabla \times \mathbf{F} = \mathbf{0}$, the vector field is **irrotational**.

Curl of a General Rotation Vector Field   We can clarify the physical meaning of the curl by considering the vector field $\mathbf{F} = \mathbf{a} \times \mathbf{r}$, where $\mathbf{a} = \langle a_1, a_2, a_3 \rangle$ is a nonzero constant vector and $\mathbf{r} = \langle x, y, z \rangle$. Writing out its components, we see that

$$\mathbf{F} = \mathbf{a} \times \mathbf{r} = \begin{vmatrix} \mathbf{i} & \mathbf{j} & \mathbf{k} \\ a_1 & a_2 & a_3 \\ x & y & z \end{vmatrix} = (a_2 z - a_3 y)\, \mathbf{i} + (a_3 x - a_1 z)\, \mathbf{j} + (a_1 y - a_2 x)\, \mathbf{k}.$$

This vector field is a *general rotation field* in three dimensions. With $a_1 = a_2 = 0$ and $a_3 = 1$, we have the familiar two-dimensional rotation field $\langle -y, x \rangle$ with its axis in the **k**-direction. More generally, **F** is the superposition of three rotation fields with axes in the **i**-, **j**-, and **k**-directions. The result is a single rotation field with an axis in the direction of **a** (Figure 17.41).

Three calculations tell us a lot about the general rotation field. The first calculation confirms that $\nabla \cdot \mathbf{F} = 0$ (Exercise 42). Just as with rotation fields in two dimensions, the divergence of a general rotation field is zero.

The second calculation (Exercises 43–44) uses the right-hand rule for cross products to show that the vector field $\mathbf{F} = \mathbf{a} \times \mathbf{r}$ is indeed a rotation field that circles the vector **a** in a counterclockwise direction looking along the length of **a** from head to tail (Figure 17.41).

The third calculation (Exercise 45) says that $\nabla \times \mathbf{F} = 2\mathbf{a}$. Therefore, the curl of the general rotation field is in the direction of the axis of rotation **a** (Figure 17.41). The magnitude of the curl is $|\nabla \times \mathbf{F}| = 2|\mathbf{a}|$. It can be shown (Exercise 52) that if **F** is a velocity field, then $|\mathbf{a}|$ is the constant angular speed of rotation of the field, denoted ω. The angular speed is the rate (radians per unit time) at which a small particle in the vector field rotates about the axis of the field. Therefore, the angular speed is half the magnitude of the curl, or

$$\omega = |\mathbf{a}| = \frac{1}{2} |\nabla \times \mathbf{F}|.$$

The rotation field $\mathbf{F} = \mathbf{a} \times \mathbf{r}$ suggests a related question. Suppose a paddle wheel is placed in the vector field **F** at a point P with the axis of the wheel in the direction of a unit vector **n** (Figure 17.42). How should **n** be chosen so the paddle wheel spins fastest? The scalar component of $\nabla \times \mathbf{F}$ in the direction of **n** is

$$(\nabla \times \mathbf{F}) \cdot \mathbf{n} = |\nabla \times \mathbf{F}| \cos \theta, \quad (|n| = 1)$$

where θ is the angle between $\nabla \times \mathbf{F}$ and **n**. The scalar component is greatest in magnitude and the paddle wheel spins fastest when $\theta = 0$ or $\theta = \pi$; that is, when **n** and $\nabla \times \mathbf{F}$ are parallel. If the axis of the paddle wheel is orthogonal to $\nabla \times \mathbf{F}$ ($\theta = \pm\pi/2$), the wheel doesn't spin.

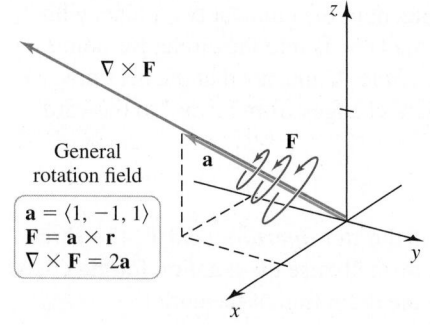

Figure 17.41

General rotation field

$\mathbf{a} = \langle 1, -1, 1 \rangle$
$\mathbf{F} = \mathbf{a} \times \mathbf{r}$
$\nabla \times \mathbf{F} = 2\mathbf{a}$

➤ Just as $\nabla f \cdot \mathbf{n}$ is the directional derivative in the direction **n**, $(\nabla \times \mathbf{F}) \cdot \mathbf{n}$ is the directional spin in the direction **n**.

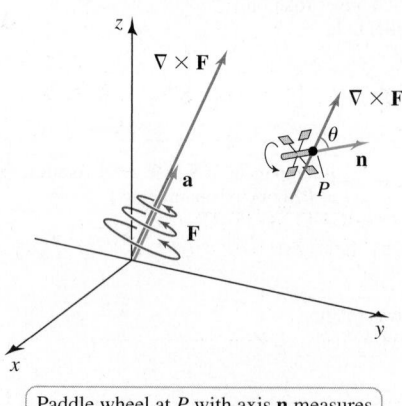

Paddle wheel at P with axis **n** measures rotation about **n**. Rotation is a maximum when $\nabla \times \mathbf{F}$ is parallel to **n**.

Figure 17.42

> **General Rotation Vector Field**
>
> The **general rotation vector field** is $\mathbf{F} = \mathbf{a} \times \mathbf{r}$, where the nonzero constant vector $\mathbf{a} = \langle a_1, a_2, a_3 \rangle$ is the axis of rotation and $\mathbf{r} = \langle x, y, z \rangle$. For all nonzero choices of **a**, $|\nabla \times \mathbf{F}| = 2|\mathbf{a}|$ and $\nabla \cdot \mathbf{F} = 0$. If **F** is a velocity field, then the constant angular speed of the field is
>
> $$\omega = |\mathbf{a}| = \frac{1}{2} |\nabla \times \mathbf{F}|.$$

QUICK CHECK 3 Show that if a vector field has the form $\mathbf{F} = \langle f(x), g(y), h(z) \rangle$, then $\nabla \times \mathbf{F} = \mathbf{0}$. ◄

EXAMPLE 4 **Curl of a rotation field** Compute the curl of the rotation field $\mathbf{F} = \mathbf{a} \times \mathbf{r}$, where $\mathbf{a} = \langle 2, -1, 1 \rangle$ and $\mathbf{r} = \langle x, y, z \rangle$ (Figure 17.41). What are the direction and the magnitude of the curl?

SOLUTION A quick calculation shows that

$$\mathbf{F} = \mathbf{a} \times \mathbf{r} = (-y - z)\,\mathbf{i} + (x - 2z)\,\mathbf{j} + (x + 2y)\,\mathbf{k}.$$

The curl of the vector field is

$$\nabla \times \mathbf{F} = \begin{vmatrix} \mathbf{i} & \mathbf{j} & \mathbf{k} \\ \dfrac{\partial}{\partial x} & \dfrac{\partial}{\partial y} & \dfrac{\partial}{\partial z} \\ -y - z & x - 2z & x + 2y \end{vmatrix} = 4\mathbf{i} - 2\mathbf{j} + 2\mathbf{k} = 2\mathbf{a}.$$

We have confirmed that $\nabla \times \mathbf{F} = 2\mathbf{a}$ and that the direction of the curl is the direction of $\mathbf{a}$, which is the axis of rotation. The magnitude of $\nabla \times \mathbf{F}$ is $|2\mathbf{a}| = 2\sqrt{6}$, which is twice the angular speed of rotation.

Related Exercises 25–26 ◄

Working with Divergence and Curl

The divergence and curl satisfy some of the same properties that ordinary derivatives satisfy. For example, given a real number c and differentiable vector fields $\mathbf{F}$ and $\mathbf{G}$, we have the following properties.

Divergence Properties

$$\nabla \cdot (\mathbf{F} + \mathbf{G}) = \nabla \cdot \mathbf{F} + \nabla \cdot \mathbf{G}$$

$$\nabla \cdot (c\mathbf{F}) = c(\nabla \cdot \mathbf{F})$$

Curl Properties

$$\nabla \times (\mathbf{F} + \mathbf{G}) = (\nabla \times \mathbf{F}) + (\nabla \times \mathbf{G})$$

$$\nabla \times (c\mathbf{F}) = c(\nabla \times \mathbf{F})$$

These and other properties are explored in Exercises 65–72.

Additional properties that have importance in theory and applications are presented in the following theorems and examples.

THEOREM 17.11 **Curl of a Conservative Vector Field**

Suppose $\mathbf{F}$ is a conservative vector field on an open region D of $\mathbb{R}^3$. Let $\mathbf{F} = \nabla\varphi$, where φ is a potential function with continuous second partial derivatives on D. Then $\nabla \times \mathbf{F} = \nabla \times \nabla\varphi = \mathbf{0}$: The curl of the gradient is the zero vector and $\mathbf{F}$ is irrotational.

Proof: We must calculate $\nabla \times \nabla\varphi$:

$$\nabla \times \nabla\varphi = \begin{vmatrix} \mathbf{i} & \mathbf{j} & \mathbf{k} \\ \dfrac{\partial}{\partial x} & \dfrac{\partial}{\partial y} & \dfrac{\partial}{\partial z} \\ \varphi_x & \varphi_y & \varphi_z \end{vmatrix} = \underbrace{(\varphi_{zy} - \varphi_{yz})}_{0}\mathbf{i} + \underbrace{(\varphi_{xz} - \varphi_{zx})}_{0}\mathbf{j} + \underbrace{(\varphi_{yx} - \varphi_{xy})}_{0}\mathbf{k} = \mathbf{0}.$$

The mixed partial derivatives are equal by Clairaut's Theorem (Theorem 15.4).

The converse of this theorem (if $\nabla \times \mathbf{F} = \mathbf{0}$, then $\mathbf{F}$ is a conservative field) is handled in Section 17.7 by means of Stokes' Theorem. ◄

> First note that $\nabla \times \mathbf{F}$ is a vector, so it makes sense to take the divergence of the curl.

THEOREM 17.12 **Divergence of the Curl**

Suppose $\mathbf{F} = \langle f, g, h \rangle$, where f, g, and h have continuous second partial derivatives. Then $\nabla \cdot (\nabla \times \mathbf{F}) = 0$: The divergence of the curl is zero.

Proof: Again, a calculation is needed:

$$\nabla \cdot (\nabla \times \mathbf{F})$$

$$= \frac{\partial}{\partial x}\left(\frac{\partial h}{\partial y} - \frac{\partial g}{\partial z}\right) + \frac{\partial}{\partial y}\left(\frac{\partial f}{\partial z} - \frac{\partial h}{\partial x}\right) + \frac{\partial}{\partial z}\left(\frac{\partial g}{\partial x} - \frac{\partial f}{\partial y}\right)$$

$$= \underbrace{(h_{yx} - h_{xy})}_{0} + \underbrace{(g_{xz} - g_{zx})}_{0} + \underbrace{(f_{zy} - f_{yz})}_{0} = 0.$$

Clairaut's Theorem (Theorem 15.4) ensures that the mixed partial derivatives are equal. ◄

The gradient, the divergence, and the curl may be combined in many ways—some of which are undefined. For example, the gradient of the curl $(\nabla(\nabla \times \mathbf{F}))$ and the curl of the divergence $(\nabla \times (\nabla \cdot \mathbf{F}))$ are undefined. However, a combination that *is* defined and is important is the divergence of the gradient $\nabla \cdot \nabla u$, where u is a scalar-valued function. This combination is denoted $\nabla^2 u$ and is called the **Laplacian** of u; it arises in many physical situations (Exercises 56–58, 62). Carrying out the calculation, we find that

$$\nabla \cdot \nabla u = \frac{\partial}{\partial x}\frac{\partial u}{\partial x} + \frac{\partial}{\partial y}\frac{\partial u}{\partial y} + \frac{\partial}{\partial z}\frac{\partial u}{\partial z} = \frac{\partial^2 u}{\partial x^2} + \frac{\partial^2 u}{\partial y^2} + \frac{\partial^2 u}{\partial z^2}.$$

We close with a result that is useful in its own right but is also intriguing because it parallels the Product Rule from single-variable calculus.

THEOREM 17.13 Product Rule for the Divergence

Let u be a scalar-valued function that is differentiable on a region D and let $\mathbf{F}$ be a vector field that is differentiable on D. Then

$$\nabla \cdot (u\mathbf{F}) = \nabla u \cdot \mathbf{F} + u(\nabla \cdot \mathbf{F}).$$

QUICK CHECK 4 Is $\nabla \cdot (u\mathbf{F})$ a vector function or a scalar function? ◄

The rule says that the "derivative" of the product is the "derivative" of the first function multiplied by the second function plus the first function multiplied by the "derivative" of the second function. However, in each instance, "derivative" must be interpreted correctly for the operations to make sense. The proof of the theorem requires a direct calculation (Exercise 67). Other similar vector calculus identities are presented in Exercises 68–72.

EXAMPLE 5 More properties of radial fields Let $\mathbf{r} = \langle x, y, z \rangle$ and let
$\varphi = \dfrac{1}{|\mathbf{r}|} = (x^2 + y^2 + z^2)^{-1/2}$ be a potential function.

a. Find the associated gradient field $\mathbf{F} = \nabla\left(\dfrac{1}{|\mathbf{r}|}\right)$.

b. Compute $\nabla \cdot \mathbf{F}$.

SOLUTION

a. The gradient has three components. Computing the first component reveals a pattern:

$$\frac{\partial \varphi}{\partial x} = \frac{\partial}{\partial x}(x^2 + y^2 + z^2)^{-1/2} = -\frac{1}{2}(x^2 + y^2 + z^2)^{-3/2}\, 2x = -\frac{x}{|\mathbf{r}|^3}.$$

Making a similar calculation for the y- and z-derivatives, the gradient is

$$\mathbf{F} = \nabla\left(\frac{1}{|\mathbf{r}|}\right) = -\frac{\langle x, y, z \rangle}{|\mathbf{r}|^3} = -\frac{\mathbf{r}}{|\mathbf{r}|^3}.$$

This result reveals that $\mathbf{F}$ is an inverse square vector field (for example, a gravitational or electric field), and its potential function is $\varphi = \dfrac{1}{|\mathbf{r}|}$.

b. The divergence $\nabla \cdot \mathbf{F} = \nabla \cdot \left(-\dfrac{\mathbf{r}}{|\mathbf{r}|^3} \right)$ involves a product of the vector function $\mathbf{r} = \langle x, y, z \rangle$ and the scalar function $|\mathbf{r}|^{-3}$. Applying Theorem 17.13, we find that

$$\nabla \cdot \mathbf{F} = \nabla \cdot \left(-\dfrac{\mathbf{r}}{|\mathbf{r}|^3} \right) = -\nabla \dfrac{1}{|\mathbf{r}|^3} \cdot \mathbf{r} - \dfrac{1}{|\mathbf{r}|^3} \nabla \cdot \mathbf{r}.$$

A calculation similar to part (a) shows that $\nabla \dfrac{1}{|\mathbf{r}|^3} = -\dfrac{3\mathbf{r}}{|\mathbf{r}|^5}$ (Exercise 35). Therefore,

$$\nabla \cdot \mathbf{F} = \nabla \cdot \left(-\dfrac{\mathbf{r}}{|\mathbf{r}|^3} \right) = -\underbrace{\nabla \dfrac{1}{|\mathbf{r}|^3}}_{-3\mathbf{r}/|\mathbf{r}|^5} \cdot \mathbf{r} - \dfrac{1}{|\mathbf{r}|^3} \underbrace{\nabla \cdot \mathbf{r}}_{3}$$

$$= \dfrac{3\mathbf{r}}{|\mathbf{r}|^5} \cdot \mathbf{r} - \dfrac{3}{|\mathbf{r}|^3} \qquad \text{Substitute for } \nabla \dfrac{1}{|\mathbf{r}|^3}.$$

$$= \dfrac{3|\mathbf{r}|^2}{|\mathbf{r}|^5} - \dfrac{3}{|\mathbf{r}|^3} \qquad \mathbf{r} \cdot \mathbf{r} = |\mathbf{r}|^2$$

$$= 0.$$

The result is consistent with Theorem 17.10 (with $p = 3$): The divergence of an inverse square vector field in $\mathbb{R}^3$ is zero. It does not happen for any other radial fields of this form.

Related Exercises 35–36 ◄

Summary of Properties of Conservative Vector Fields

We can now extend the list of equivalent properties of conservative vector fields $\mathbf{F}$ defined on an open connected region. Theorem 17.11 is added to the list given at the end of Section 17.3.

Properties of a Conservative Vector Field

Let $\mathbf{F}$ be a conservative vector field whose components have continuous second partial derivatives on an open connected region D in $\mathbb{R}^3$. Then $\mathbf{F}$ has the following equivalent properties.

1. There exists a potential function φ such that $\mathbf{F} = \nabla \varphi$ (definition).
2. $\int_C \mathbf{F} \cdot d\mathbf{r} = \varphi(B) - \varphi(A)$ for all points A and B in D and all piecewise-smooth oriented curves C in D from A to B.
3. $\oint_C \mathbf{F} \cdot d\mathbf{r} = 0$ on all simple piecewise-smooth closed oriented curves C in D.
4. $\nabla \times \mathbf{F} = \mathbf{0}$ at all points of D.

SECTION 17.5 EXERCISES

Getting Started

1. Explain how to compute the divergence of the vector field $\mathbf{F} = \langle f, g, h \rangle$.

2. Interpret the divergence of a vector field.

3. What does it mean if the divergence of a vector field is zero throughout a region?

4. Explain how to compute the curl of the vector field $\mathbf{F} = \langle f, g, h \rangle$.

5. Interpret the curl of a general rotation vector field.

6. What does it mean if the curl of a vector field is zero throughout a region?

7. What is the value of $\nabla \cdot (\nabla \times \mathbf{F})$?

8. What is the value of $\nabla \times \nabla u$?

Practice Exercises

9–16. Divergence of vectors fields *Find the divergence of the following vector fields.*

9. $\mathbf{F} = \langle 2x, 4y, -3z \rangle$

10. $\mathbf{F} = \langle -2y, 3x, z \rangle$

11. $\mathbf{F} = \langle 12x, -6y, -6z \rangle$ **12.** $\mathbf{F} = \langle x^2yz, -xy^2z, -xyz^2 \rangle$

13. $\mathbf{F} = \langle x^2 - y^2, y^2 - z^2, z^2 - x^2 \rangle$

14. $\mathbf{F} = \langle e^{-x+y}, e^{-y+z}, e^{-z+x} \rangle$

15. $\mathbf{F} = \dfrac{\langle x, y, z \rangle}{1 + x^2 + y^2}$

16. $\mathbf{F} = \langle yz \sin x, xz \cos y, xy \cos z \rangle$

17–20. Divergence of radial fields *Calculate the divergence of the following radial fields. Express the result in terms of the position vector* $\mathbf{r}$ *and its length* $|\mathbf{r}|$. *Check for agreement with Theorem 17.10.*

17. $\mathbf{F} = \dfrac{\langle x, y, z \rangle}{x^2 + y^2 + z^2} = \dfrac{\mathbf{r}}{|\mathbf{r}|^2}$

18. $\mathbf{F} = \dfrac{\langle x, y, z \rangle}{(x^2 + y^2 + z^2)^{3/2}} = \dfrac{\mathbf{r}}{|\mathbf{r}|^3}$

19. $\mathbf{F} = \dfrac{\langle x, y, z \rangle}{(x^2 + y^2 + z^2)^2} = \dfrac{\mathbf{r}}{|\mathbf{r}|^4}$

20. $\mathbf{F} = \langle x, y, z \rangle (x^2 + y^2 + z^2) = \mathbf{r}|\mathbf{r}|^2$

21–22. Divergence and flux from graphs *Consider the following vector fields, the circle C, and two points P and Q.*

a. *Without computing the divergence, does the graph suggest that the divergence is positive or negative at P and Q? Justify your answer.*

b. *Compute the divergence and confirm your conjecture in part (a).*

c. *On what part of C is the flux outward? Inward?*

d. *Is the net outward flux across C positive or negative?*

21. $\mathbf{F} = \langle x, x + y \rangle$

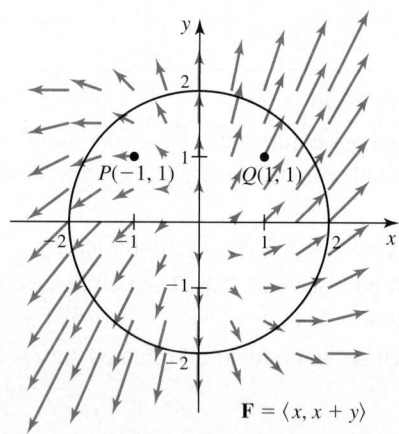

$\mathbf{F} = \langle x, x + y \rangle$

22. $\mathbf{F} = \langle x, y^2 \rangle$

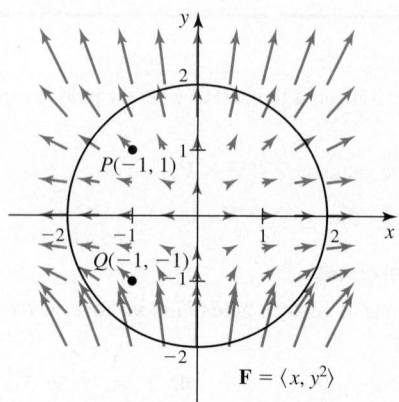

$\mathbf{F} = \langle x, y^2 \rangle$

23–26. Curl of a rotation field *Consider the following vector fields, where* $\mathbf{r} = \langle x, y, z \rangle$.

a. *Compute the curl of the field and verify that it has the same direction as the axis of rotation.*

b. *Compute the magnitude of the curl of the field.*

23. $\mathbf{F} = \langle 1, 0, 0 \rangle \times \mathbf{r}$ **24.** $\mathbf{F} = \langle 1, -1, 0 \rangle \times \mathbf{r}$

25. $\mathbf{F} = \langle 1, -1, 1 \rangle \times \mathbf{r}$ **26.** $\mathbf{F} = \langle 1, -2, -3 \rangle \times \mathbf{r}$

27–34. Curl of a vector field *Compute the curl of the following vector fields.*

27. $\mathbf{F} = \langle x^2 - y^2, xy, z \rangle$ **28.** $\mathbf{F} = \langle 0, z^2 - y^2, -yz \rangle$

29. $\mathbf{F} = \langle x^2 - z^2, 1, 2xz \rangle$ **30.** $\mathbf{F} = \mathbf{r} = \langle x, y, z \rangle$

31. $\mathbf{F} = \dfrac{\langle x, y, z \rangle}{(x^2 + y^2 + z^2)^{3/2}} = \dfrac{\mathbf{r}}{|\mathbf{r}|^3}$

32. $\mathbf{F} = \dfrac{\langle x, y, z \rangle}{(x^2 + y^2 + z^2)^{1/2}} = \dfrac{\mathbf{r}}{|\mathbf{r}|}$

33. $\mathbf{F} = \langle z^2 \sin y, xz^2 \cos y, 2xz \sin y \rangle$

34. $\mathbf{F} = \langle 3xz^3 e^{y^2}, 2xz^3 e^{y^2}, 3xz^2 e^{y^2} \rangle$

35–38. Derivative rules *Prove the following identities. Use Theorem 17.13 (Product Rule) whenever possible.*

35. $\nabla\left(\dfrac{1}{|\mathbf{r}|^3}\right) = -\dfrac{3\mathbf{r}}{|\mathbf{r}|^5}$ (used in Example 5)

36. $\nabla\left(\dfrac{1}{|\mathbf{r}|^2}\right) = -\dfrac{2\mathbf{r}}{|\mathbf{r}|^4}$

37. $\nabla \cdot \nabla\left(\dfrac{1}{|\mathbf{r}|^2}\right) = \dfrac{2}{|\mathbf{r}|^4}$ (*Hint:* Use Exercise 36.)

38. $\nabla(\ln |\mathbf{r}|) = \dfrac{\mathbf{r}}{|\mathbf{r}|^2}$

39. Explain why or why not Determine whether the following statements are true and give an explanation or counterexample.

a. For a function f of a single variable, if $f'(x) = 0$ for all x in the domain, then f is a constant function. If $\nabla \cdot \mathbf{F} = 0$ for all points in the domain, then $\mathbf{F}$ is constant.

b. If $\nabla \times \mathbf{F} = \mathbf{0}$, then $\mathbf{F}$ is constant.

c. A vector field consisting of parallel vectors has zero curl.

d. A vector field consisting of parallel vectors has zero divergence.

e. curl $\mathbf{F}$ is orthogonal to $\mathbf{F}$.

40. Another derivative combination Let $\mathbf{F} = \langle f, g, h \rangle$ and let u be a differentiable scalar-valued function.

a. Take the dot product of $\mathbf{F}$ and the del operator; then apply the result to u to show that

$$(\mathbf{F} \cdot \nabla)u = \left(f\frac{\partial}{\partial x} + g\frac{\partial}{\partial y} + h\frac{\partial}{\partial z}\right)u$$
$$= f\frac{\partial u}{\partial x} + g\frac{\partial u}{\partial y} + h\frac{\partial u}{\partial z}.$$

b. Evaluate $(\mathbf{F} \cdot \nabla)(xy^2z^3)$ at $(1, 1, 1)$, where $\mathbf{F} = \langle 1, 1, 1 \rangle$.

41. Does it make sense? Are the following expressions defined? If so, state whether the result is a scalar or a vector. Assume $\mathbf{F}$ is a sufficiently differentiable vector field and φ is a sufficiently differentiable scalar-valued function.

a. $\nabla \cdot \varphi$ b. $\nabla \mathbf{F}$ c. $\nabla \cdot \nabla \varphi$
d. $\nabla(\nabla \cdot \varphi)$ e. $\nabla(\nabla \times \varphi)$ f. $\nabla \cdot (\nabla \cdot \mathbf{F})$
g. $\nabla \times \nabla \varphi$ h. $\nabla \times (\nabla \cdot \mathbf{F})$ i. $\nabla \times (\nabla \times \mathbf{F})$

42. Zero divergence of the rotation field Show that the general rotation field $\mathbf{F} = \mathbf{a} \times \mathbf{r}$, where $\mathbf{a}$ is a nonzero constant vector and $\mathbf{r} = \langle x, y, z \rangle$, has zero divergence.

43. General rotation fields

a. Let $\mathbf{a} = \langle 0, 1, 0 \rangle$, let $\mathbf{r} = \langle x, y, z \rangle$, and consider the rotation field $\mathbf{F} = \mathbf{a} \times \mathbf{r}$. Use the right-hand rule for cross products to find the direction of $\mathbf{F}$ at the points $(0, 1, 1)$, $(1, 1, 0)$, $(0, 1, -1)$, and $(-1, 1, 0)$.

b. With $\mathbf{a} = \langle 0, 1, 0 \rangle$, explain why the rotation field $\mathbf{F} = \mathbf{a} \times \mathbf{r}$ circles the y-axis in the counterclockwise direction looking along $\mathbf{a}$ from head to tail (that is, in the negative y-direction).

44. General rotation fields Generalize Exercise 43 to show that the rotation field $\mathbf{F} = \mathbf{a} \times \mathbf{r}$ circles the vector $\mathbf{a}$ in the counterclockwise direction looking along $\mathbf{a}$ from head to tail.

45. Curl of the rotation field For the general rotation field $\mathbf{F} = \mathbf{a} \times \mathbf{r}$, where $\mathbf{a}$ is a nonzero constant vector and $\mathbf{r} = \langle x, y, z \rangle$, show that curl $\mathbf{F} = 2\mathbf{a}$.

46. Inward to outward Find the exact points on the circle $x^2 + y^2 = 2$ at which the field $\mathbf{F} = \langle f, g \rangle = \langle x^2, y \rangle$ switches from pointing inward to pointing outward on the circle, or vice versa.

47. Maximum divergence Within the cube $\{(x, y, z): |x| \leq 1, |y| \leq 1, |z| \leq 1\}$, where does div $\mathbf{F}$ have the greatest magnitude when $\mathbf{F} = \langle x^2 - y^2, xy^2 z, 2xz \rangle$?

48. Maximum curl Let $\mathbf{F} = \langle z, 0, -y \rangle$.

a. Find the scalar component of curl $\mathbf{F}$ in the direction of the unit vector $\mathbf{n} = \langle 1, 0, 0 \rangle$.

b. Find the scalar component of curl $\mathbf{F}$ in the direction of the unit vector $\mathbf{n} = \left\langle \frac{1}{\sqrt{3}}, -\frac{1}{\sqrt{3}}, \frac{1}{\sqrt{3}} \right\rangle$.

c. Find the unit vector $\mathbf{n}$ that maximizes $\mathrm{scal}_\mathbf{n} \langle -1, 1, 0 \rangle$ and state the value of $\mathrm{scal}_\mathbf{n} \langle -1, 1, 0 \rangle$ in this direction.

49. Zero component of the curl For what vectors $\mathbf{n}$ is $(\mathrm{curl}\ \mathbf{F}) \cdot \mathbf{n} = 0$ when $\mathbf{F} = \langle y, -2z, -x \rangle$?

50–51. Find a vector field Find a vector field $\mathbf{F}$ with the given curl. In each case, is the vector field you found unique?

50. curl $\mathbf{F} = \langle 0, 1, 0 \rangle$ **51.** curl $\mathbf{F} = \langle 0, z, -y \rangle$

Explorations and Challenges

52. Curl and angular speed Consider the rotational velocity field $\mathbf{v} = \mathbf{a} \times \mathbf{r}$, where $\mathbf{a}$ is a nonzero constant vector and $\mathbf{r} = \langle x, y, z \rangle$. Use the fact that an object moving in a circular path of radius R with speed $|\mathbf{v}|$ has an angular speed of $\omega = |\mathbf{v}|/R$.

a. Sketch a position vector $\mathbf{a}$, which is the axis of rotation for the vector field, and a position vector $\mathbf{r}$ of a point P in $\mathbb{R}^3$. Let θ be the angle between the two vectors. Show that the perpendicular distance from P to the axis of rotation is $R = |\mathbf{r}| \sin \theta$.

b. Show that the speed of a particle in the velocity field is $|\mathbf{a} \times \mathbf{r}|$ and that the angular speed of the object is $|\mathbf{a}|$.

c. Conclude that $\omega = \frac{1}{2} |\nabla \times \mathbf{v}|$.

53. Paddle wheel in a vector field Let $\mathbf{F} = \langle z, 0, 0 \rangle$ and let $\mathbf{n}$ be a unit vector aligned with the axis of a paddle wheel located on the x-axis (see figure).

a. If the paddle wheel is oriented with $\mathbf{n} = \langle 1, 0, 0 \rangle$, in what direction (if any) does the wheel spin?

b. If the paddle wheel is oriented with $\mathbf{n} = \langle 0, 1, 0 \rangle$, in what direction (if any) does the wheel spin?

c. If the paddle wheel is oriented with $\mathbf{n} = \langle 0, 0, 1 \rangle$, in what direction (if any) does the wheel spin?

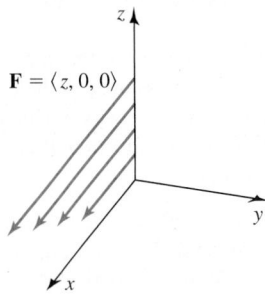

54. Angular speed Consider the rotational velocity field $\mathbf{v} = \langle -2y, 2z, 0 \rangle$.

a. If a paddle wheel is placed in the xy-plane with its axis normal to this plane, what is its angular speed?

b. If a paddle wheel is placed in the xz-plane with its axis normal to this plane, what is its angular speed?

c. If a paddle wheel is placed in the yz-plane with its axis normal to this plane, what is its angular speed?

55. Angular speed Consider the rotational velocity field $\mathbf{v} = \langle 0, 10z, -10y \rangle$. If a paddle wheel is placed in the plane $x + y + z = 1$ with its axis normal to this plane, how fast does the paddle wheel spin (in revolutions per unit time)?

56–58. Heat flux *Suppose a solid object in $\mathbb{R}^3$ has a temperature distribution given by $T(x, y, z)$. The heat flow vector field in the object is $\mathbf{F} = -k\nabla T$, where the conductivity $k > 0$ is a property of the material. Note that the heat flow vector points in the direction opposite to that of the gradient, which is the direction of greatest temperature decrease. The divergence of the heat flow vector is $\nabla \cdot \mathbf{F} = -k\nabla \cdot \nabla T = -k\nabla^2 T$ (the Laplacian of T). Compute the heat flow vector field and its divergence for the following temperature distributions.*

56. $T(x, y, z) = 100e^{-\sqrt{x^2+y^2+z^2}}$

57. $T(x, y, z) = 100e^{-x^2+y^2+z^2}$

58. $T(x, y, z) = 100\left(1 + \sqrt{x^2 + y^2 + z^2}\right)$

59. Gravitational potential The potential function for the gravitational force field due to a mass M at the origin acting on a mass m is $\varphi = GMm/|\mathbf{r}|$, where $\mathbf{r} = \langle x, y, z \rangle$ is the position vector of the mass m, and G is the gravitational constant.

a. Compute the gravitational force field $\mathbf{F} = -\nabla \varphi$.

b. Show that the field is irrotational; that is, show that $\nabla \times \mathbf{F} = \mathbf{0}$.

60. Electric potential The potential function for the force field due to a charge q at the origin is $\varphi = \dfrac{1}{4\pi\varepsilon_0}\dfrac{q}{|\mathbf{r}|}$, where $\mathbf{r} = \langle x, y, z\rangle$ is the position vector of a point in the field, and ε_0 is the permittivity of free space.

  a. Compute the force field $\mathbf{F} = -\nabla\varphi$.
  b. Show that the field is irrotational; that is, show that $\nabla \times \mathbf{F} = \mathbf{0}$.

61. Navier-Stokes equation The Navier-Stokes equation is the fundamental equation of fluid dynamics that models the flow in everything from bathtubs to oceans. In one of its many forms (incompressible, viscous flow), the equation is

$$\rho\left(\frac{\partial \mathbf{V}}{\partial t} + (\mathbf{V}\cdot\nabla)\mathbf{V}\right) = -\nabla p + \mu(\nabla\cdot\nabla)\mathbf{V}.$$

In this notation, $\mathbf{V} = \langle u, v, w\rangle$ is the three-dimensional velocity field, p is the (scalar) pressure, ρ is the constant density of the fluid, and μ is the constant viscosity. Write out the three component equations of this vector equation. (See Exercise 40 for an interpretation of the operations.)

T 62. Stream function and vorticity The rotation of a three-dimensional velocity field $\mathbf{V} = \langle u, v, w\rangle$ is measured by the **vorticity** $\omega = \nabla \times \mathbf{V}$. If $\omega = \mathbf{0}$ at all points in the domain, the flow is irrotational.

  a. Which of the following velocity fields is irrotational: $\mathbf{V} = \langle 2, -3y, 5z\rangle$ or $\mathbf{V} = \langle y, x - z, -y\rangle$?
  b. Recall that for a two-dimensional source-free flow $\mathbf{V} = \langle u, v, 0\rangle$, a stream function $\psi(x, y)$ may be defined such that $u = \psi_y$ and $v = -\psi_x$. For such a two-dimensional flow, let $\zeta = \mathbf{k}\cdot\nabla\times\mathbf{V}$ be the $\mathbf{k}$-component of the vorticity. Show that $\nabla^2\psi = \nabla\cdot\nabla\psi = -\zeta$.
  c. Consider the stream function $\psi(x, y) = \sin x \sin y$ on the square region $R = \{(x, y): 0 \le x \le \pi, 0 \le y \le \pi\}$. Find the velocity components u and v; then sketch the velocity field.
  d. For the stream function in part (c), find the vorticity function ζ as defined in part (b). Plot several level curves of the vorticity function. Where on R is it a maximum? A minimum?

63. Ampère's Law One of Maxwell's equations for electromagnetic waves is $\nabla \times \mathbf{B} = C\dfrac{\partial \mathbf{E}}{\partial t}$, where $\mathbf{E}$ is the electric field, $\mathbf{B}$ is the magnetic field, and C is a constant.

  a. Show that the fields
  $$\mathbf{E}(z, t) = A\sin(kz - \omega t)\mathbf{i} \text{ and } \mathbf{B}(z, t) = A\sin(kz - \omega t)\mathbf{j}$$
  satisfy the equation for constants A, k, and ω, provided $\omega = k/C$.
  b. Make a rough sketch showing the directions of $\mathbf{E}$ and $\mathbf{B}$.

64. Splitting a vector field Express the vector field $\mathbf{F} = \langle xy, 0, 0\rangle$ in the form $\mathbf{V} + \mathbf{W}$, where $\nabla\cdot\mathbf{V} = 0$ and $\nabla \times \mathbf{W} = \mathbf{0}$.

65. Properties of div and curl Prove the following properties of the divergence and curl. Assume $\mathbf{F}$ and $\mathbf{G}$ are differentiable vector fields and c is a real number.

  a. $\nabla\cdot(\mathbf{F} + \mathbf{G}) = \nabla\cdot\mathbf{F} + \nabla\cdot\mathbf{G}$
  b. $\nabla\times(\mathbf{F} + \mathbf{G}) = (\nabla\times\mathbf{F}) + (\nabla\times\mathbf{G})$
  c. $\nabla\cdot(c\mathbf{F}) = c(\nabla\cdot\mathbf{F})$
  d. $\nabla\times(c\mathbf{F}) = c(\nabla\times\mathbf{F})$

66. Equal curls If two functions of one variable, f and g, have the property that $f' = g'$, then f and g differ by a constant. Prove or disprove: If $\mathbf{F}$ and $\mathbf{G}$ are nonconstant vector fields in $\mathbb{R}^2$ with curl $\mathbf{F}$ = curl $\mathbf{G}$ and div $\mathbf{F}$ = div $\mathbf{G}$ at all points of $\mathbb{R}^2$, then $\mathbf{F}$ and $\mathbf{G}$ differ by a constant vector.

67–72. Identities *Prove the following identities. Assume φ is a differentiable scalar-valued function and $\mathbf{F}$ and $\mathbf{G}$ are differentiable vector fields, all defined on a region of $\mathbb{R}^3$.*

67. $\nabla\cdot(\varphi\mathbf{F}) = \nabla\varphi\cdot\mathbf{F} + \varphi\nabla\cdot\mathbf{F}$ (Product Rule)

68. $\nabla\times(\varphi\mathbf{F}) = (\nabla\varphi\times\mathbf{F}) + (\varphi\nabla\times\mathbf{F})$ (Product Rule)

69. $\nabla\cdot(\mathbf{F}\times\mathbf{G}) = \mathbf{G}\cdot(\nabla\times\mathbf{F}) - \mathbf{F}\cdot(\nabla\times\mathbf{G})$

70. $\nabla\times(\mathbf{F}\times\mathbf{G}) = (\mathbf{G}\cdot\nabla)\mathbf{F} - \mathbf{G}(\nabla\cdot\mathbf{F}) - (\mathbf{F}\cdot\nabla)\mathbf{G} + \mathbf{F}(\nabla\cdot\mathbf{G})$

71. $\nabla(\mathbf{F}\cdot\mathbf{G}) = (\mathbf{G}\cdot\nabla)\mathbf{F} + (\mathbf{F}\cdot\nabla)\mathbf{G} + \mathbf{G}\times(\nabla\times\mathbf{F}) + \mathbf{F}\times(\nabla\times\mathbf{G})$

72. $\nabla\times(\nabla\times\mathbf{F}) = \nabla(\nabla\cdot\mathbf{F}) - (\nabla\cdot\nabla)\mathbf{F}$

73. Divergence of radial fields Prove that for a real number p, with $\mathbf{r} = \langle x, y, z\rangle$, $\nabla\cdot\dfrac{\langle x, y, z\rangle}{|\mathbf{r}|^p} = \dfrac{3 - p}{|\mathbf{r}|^p}$.

74. Gradients and radial fields Prove that for a real number p, with $\mathbf{r} = \langle x, y, z\rangle$, $\nabla\left(\dfrac{1}{|\mathbf{r}|^p}\right) = \dfrac{-p\mathbf{r}}{|\mathbf{r}|^{p+2}}$.

75. Divergence of gradient fields Prove that for a real number p, with $\mathbf{r} = \langle x, y, z\rangle$, $\nabla\cdot\nabla\left(\dfrac{1}{|\mathbf{r}|^p}\right) = \dfrac{p(p - 1)}{|\mathbf{r}|^{p+2}}$.

QUICK CHECK ANSWERS

1. The x-derivative of the divergence is applied to $f(y, z)$, which gives zero. Similarly, the y- and z-derivatives are zero. **2.** The net outward flux is 4π.   **3.** In the curl, the first component of $\mathbf{F}$ is differentiated only with respect to y and z, so the contribution from the first component is zero. Similarly, the second and third components of $\mathbf{F}$ make no contribution to the curl.   **4.** The divergence is a scalar-valued function. ◄

17.6 Surface Integrals

We have studied integrals on the real line, on regions in the plane, on solid regions in space, and along curves in space. One situation is still unexplored. Suppose a sphere has a known temperature distribution; perhaps it is cold near the poles and warm near the equator. How do you find the average temperature over the entire sphere? In analogy with other average value calculations, we should expect to "add up" the temperature values over the sphere and divide by the surface area of the sphere. Because the temperature varies continuously over the sphere, adding up means integrating. How do you integrate a function over a surface? This question leads to *surface integrals*.

It helps to keep curves, arc length, and line integrals in mind as we discuss surfaces, surface area, and surface integrals. What we discover about surfaces parallels what we already know about curves—all "lifted" up one dimension.

Parameterized Surfaces

A curve in $\mathbb{R}^2$ is defined parametrically by $\mathbf{r}(t) = \langle x(t), y(t) \rangle$, for $a \leq t \leq b$; it requires one parameter and two dependent variables. Stepping up one dimension to define a surface in $\mathbb{R}^3$, we need *two* parameters and *three* dependent variables. Letting u and v be parameters, the general parametric description of a surface has the form

$$\mathbf{r}(u, v) = \langle x(u, v), y(u, v), z(u, v) \rangle.$$

We make the assumption that the parameters vary over a rectangle $R = \{(u, v): a \leq u \leq b, c \leq v \leq d\}$ (Figure 17.43). As the parameters (u, v) vary over R, the vector $\mathbf{r}(u, v) = \langle x(u, v), y(u, v), z(u, v) \rangle$ sweeps out a surface S in $\mathbb{R}^3$.

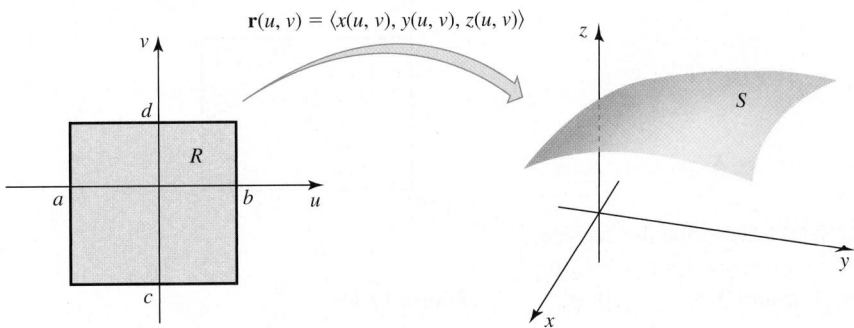

A rectangle in the *uv*-plane is mapped to a surface in *xyz*-space.

Figure 17.43

We work extensively with three surfaces that are easily described in parametric form. As with parameterized curves, a parametric description of a surface is not unique.

Cylinders   In Cartesian coordinates, the set

$$\{(x, y, z): x = a \cos \theta, y = a \sin \theta, 0 \leq \theta \leq 2\pi, 0 \leq z \leq h\},$$

where $a > 0$, is a cylindrical surface of radius a and height h with its axis along the z-axis. Using the parameters $u = \theta$ and $v = z$, a parametric description of the cylinder is

$$\mathbf{r}(u, v) = \langle x(u, v), y(u, v), z(u, v) \rangle = \langle a \cos u, a \sin u, v \rangle,$$

where $0 \leq u \leq 2\pi$ and $0 \leq v \leq h$ (Figure 17.44).

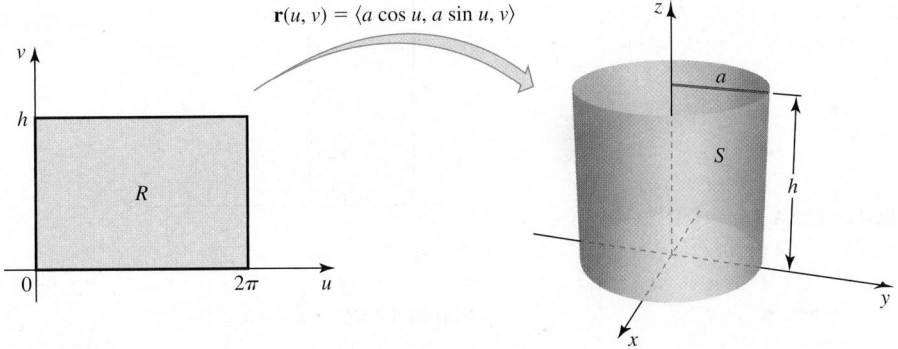

Figure 17.44

Parallel Concepts

Curves	Surfaces
Arc length	Surface area
Line integrals	Surface integrals
One-parameter description	Two-parameter description

QUICK CHECK 1 Describe the surface $\mathbf{r}(u, v) = \langle 2 \cos u, 2 \sin u, v \rangle$, for $0 \leq u \leq \pi$ and $0 \leq v \leq 1$. ◀

Cones   The surface of a cone of height h and radius a with its vertex at the origin is described in cylindrical coordinates by

$$\{(r, \theta, z): 0 \le r \le a, 0 \le \theta \le 2\pi, z = rh/a\}.$$

▶ Note that when $r = 0$, $z = 0$ and when $r = a$, $z = h$.

For a fixed value of z, we have $r = az/h$; therefore, on the surface of the cone,

$$x = r \cos \theta = \frac{az}{h} \cos \theta \quad \text{and} \quad y = r \sin \theta = \frac{az}{h} \sin \theta.$$

▶ Recall the relationships between polar and rectangular coordinates:

$x = r \cos \theta, y = r \sin \theta$, and $x^2 + y^2 = r^2$.

Using the parameters $u = \theta$ and $v = z$, the parametric description of the conical surface is

$$\mathbf{r}(u, v) = \langle x(u, v), y(u, v), z(u, v) \rangle = \left\langle \frac{av}{h} \cos u, \frac{av}{h} \sin u, v \right\rangle,$$

where $0 \le u \le 2\pi$ and $0 \le v \le h$ (Figure 17.45).

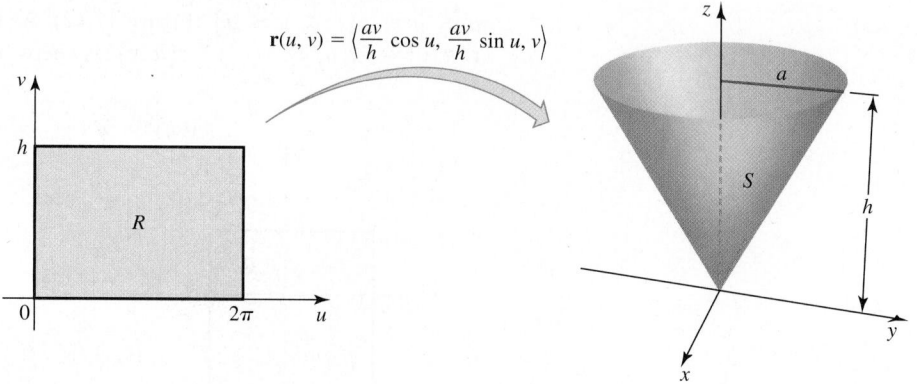

QUICK CHECK 2 Describe the surface $\mathbf{r}(u, v) = \langle v \cos u, v \sin u, v \rangle$, for $0 \le u \le \pi$ and $0 \le v \le 10$. ◀

Figure 17.45

▶ The complete cylinder, cone, and sphere are generated as the angle variable θ varies over the half-open interval $[0, 2\pi)$. As in previous chapters, we will use the closed interval $[0, 2\pi]$.

Spheres   The parametric description of a sphere of radius a centered at the origin comes directly from spherical coordinates:

$$\{(\rho, \varphi, \theta): \rho = a, 0 \le \varphi \le \pi, 0 \le \theta \le 2\pi\}.$$

Recall the following relationships among spherical and rectangular coordinates (Section 16.5):

$$x = a \sin \varphi \cos \theta, \qquad y = a \sin \varphi \sin \theta, \qquad z = a \cos \varphi.$$

When we define the parameters $u = \varphi$ and $v = \theta$, a parametric description of the sphere is

$$\mathbf{r}(u, v) = \langle a \sin u \cos v, a \sin u \sin v, a \cos u \rangle,$$

where $0 \le u \le \pi$ and $0 \le v \le 2\pi$ (Figure 17.46).

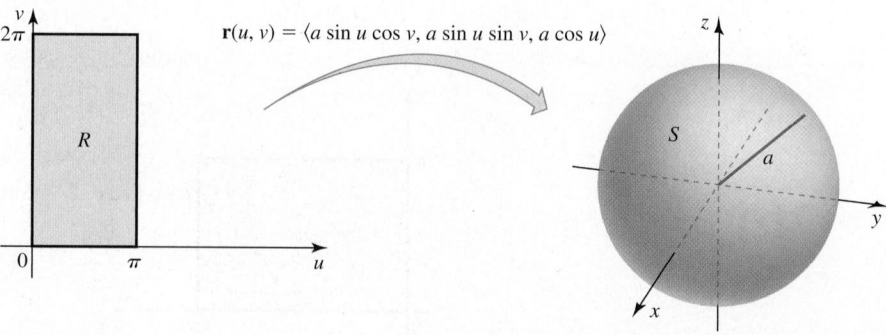

QUICK CHECK 3 Describe the surface $\mathbf{r}(u, v) = \langle 4 \sin u \cos v, 4 \sin u \sin v, 4 \cos u \rangle$, for $0 \le u \le \pi/2$ and $0 \le v \le \pi$. ◀

Figure 17.46

EXAMPLE 1 **Parametric surfaces** Find parametric descriptions for the following surfaces.

a. The plane $3x - 2y + z = 2$

b. The paraboloid $z = x^2 + y^2$, for $0 \le z \le 9$

SOLUTION

a. Defining the parameters $u = x$ and $v = y$, we find that

$$z = 2 - 3x + 2y = 2 - 3u + 2v.$$

Therefore, a parametric description of the plane is

$$\mathbf{r}(u, v) = \langle u, v, 2 - 3u + 2v \rangle,$$

for $-\infty < u < \infty$ and $-\infty < v < \infty$.

b. Thinking in terms of polar coordinates, we let $u = \theta$ and $v = \sqrt{z}$, which means that $z = v^2$. The equation of the paraboloid is $x^2 + y^2 = z = v^2$, so v plays the role of the polar coordinate r. Therefore, $x = v \cos \theta = v \cos u$ and $y = v \sin \theta = v \sin u$. A parametric description for the paraboloid is

$$\mathbf{r}(u, v) = \langle v \cos u, v \sin u, v^2 \rangle,$$

where $0 \le u \le 2\pi$ and $0 \le v \le 3$.

Alternatively, we could choose $u = \theta$ and $v = z$. The resulting description is

$$\mathbf{r}(u, v) = \langle \sqrt{v} \cos u, \sqrt{v} \sin u, v \rangle,$$

where $0 \le u \le 2\pi$ and $0 \le v \le 9$.   *Related Exercises 9, 12* ◀

Surface Integrals of Scalar-Valued Functions

We now develop the surface integral of a scalar-valued function f defined on a smooth parameterized surface S described by the equation

$$\mathbf{r}(u, v) = \langle x(u, v), y(u, v), z(u, v) \rangle,$$

where the parameters vary over a rectangle $R = \{(u, v): a \le u \le b, c \le v \le d\}$. The functions x, y, and z are assumed to have continuous partial derivatives with respect to u and v. The rectangular region R in the uv-plane is partitioned into rectangles, with sides of length Δu and Δv, that are ordered in some convenient way, for $k = 1, \ldots, n$. The kth rectangle R_k, which has area $\Delta A = \Delta u \Delta v$, corresponds to a curved patch S_k on the surface S (**Figure 17.47**), which has area ΔS_k. We let (u_k, v_k) be the lower-left corner point of R_k. The parameterization then assigns (u_k, v_k) to a point $P(x(u_k, v_k), y(u_k, v_k), z(u_k, v_k))$, or more simply, $P(x_k, y_k, z_k)$, on S_k. To construct the surface integral, we define a Riemann sum, which adds up function values multiplied by areas of the respective patches:

▶ A more general approach allows (u_k, v_k) to be an arbitrary point in the kth rectangle. The outcome of the two approaches is the same.

$$\sum_{k=1}^{n} f(x(u_k, v_k), y(u_k, v_k), z(u_k, v_k)) \Delta S_k.$$

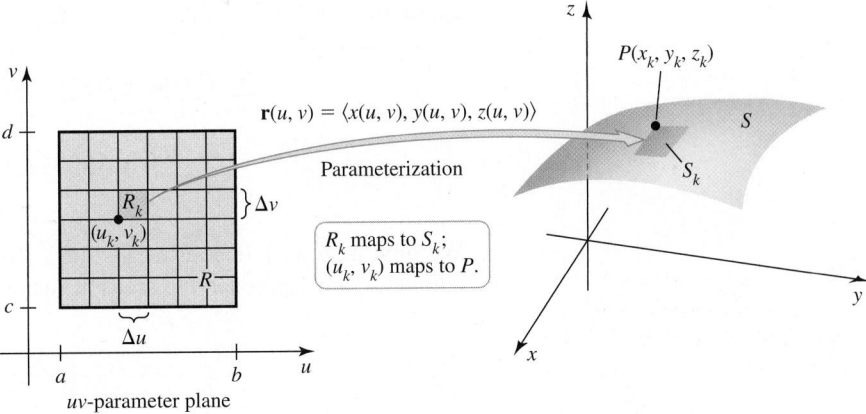

Figure 17.47

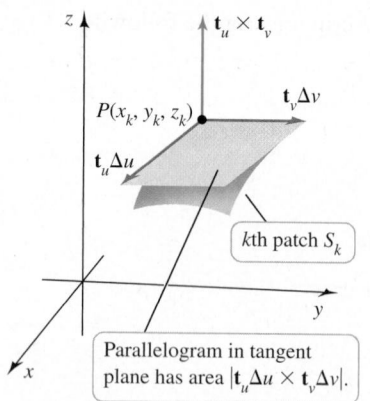

Parallelogram in tangent plane has area $|\mathbf{t}_u\Delta u \times \mathbf{t}_v\Delta v|$.

Figure 17.48

▶ In general, the vectors $\mathbf{t}_u$ and $\mathbf{t}_v$ are different for each patch, so they should carry a subscript k. To keep the notation as simple as possible, we have suppressed the subscripts on these vectors with the understanding that they change with k. These tangent vectors are given by partial derivatives because in each case, either u or v is held constant, while the other variable changes.

The crucial step is computing ΔS_k, the area of the kth patch S_k.

Figure 17.48 shows the patch S_k and the point $P(x_k, y_k, z_k)$. Two special vectors are tangent to the surface at P; these vectors lie in the plane tangent to S at P.

• $\mathbf{t}_u$ is a vector tangent to the surface corresponding to a change in u with v constant in the uv-plane.

• $\mathbf{t}_v$ is a vector tangent to the surface corresponding to a change in v with u constant in the uv-plane.

Because the surface S may be written $\mathbf{r}(u, v) = \langle x(u, v), y(u, v), z(u, v) \rangle$, a tangent vector corresponding to a change in u with v fixed is

$$\mathbf{t}_u = \frac{\partial \mathbf{r}}{\partial u} = \left\langle \frac{\partial x}{\partial u}, \frac{\partial y}{\partial u}, \frac{\partial z}{\partial u} \right\rangle.$$

Similarly, a tangent vector corresponding to a change in v with u fixed is

$$\mathbf{t}_v = \frac{\partial \mathbf{r}}{\partial v} = \left\langle \frac{\partial x}{\partial v}, \frac{\partial y}{\partial v}, \frac{\partial z}{\partial v} \right\rangle.$$

Now consider an increment Δu in u with v fixed; the corresponding change in $\mathbf{r}$, which is $\mathbf{r}(u + \Delta u, v) - \mathbf{r}(u, v)$, can be approximated using the definition of the partial derivative of $\mathbf{r}$ with respect to u. Specifically, when Δu is small, we have

$$\frac{\partial \mathbf{r}}{\partial u} \approx \frac{1}{\Delta u}\big(\mathbf{r}(u + \Delta u, v) - \mathbf{r}(u, v)\big).$$

Multiplying both sides of this equation by Δu and recalling that $\mathbf{t}_u = \dfrac{\partial \mathbf{r}}{\partial u}$, we see that the change in $\mathbf{r}$ corresponding to the increment Δu is approximated by the vector

$$\mathbf{t}_u\Delta u \approx \underbrace{\mathbf{r}(u + \Delta u, v) - \mathbf{r}(u, v)}_{\text{change in } \mathbf{r} \text{ corresponding to } \Delta u}.$$

Using a similar line of reasoning, the change in $\mathbf{r}$ corresponding to the increment Δv (with u fixed) is approximated by the vector

$$\mathbf{t}_v\Delta v \approx \underbrace{\mathbf{r}(u, v + \Delta v) - \mathbf{r}(u, v)}_{\text{change in } \mathbf{r} \text{ corresponding to } \Delta v}.$$

As nonzero scalar multiples of $\mathbf{t}_u$ and $\mathbf{t}_v$, the vectors $\mathbf{t}_u\Delta u$ and $\mathbf{t}_v\Delta v$ are also tangent to the surface. They determine a parallelogram that lies in the plane tangent to S at P (Figure 17.48); the area of this parallelogram approximates the area of the kth patch S_k, which is ΔS_k.

Appealing to the cross product (Section 13.4), the area of the parallelogram is

$$|\mathbf{t}_u\Delta u \times \mathbf{t}_v\Delta v| = |\mathbf{t}_u \times \mathbf{t}_v|\, \Delta u\, \Delta v \approx \Delta S_k.$$

Note that $\mathbf{t}_u \times \mathbf{t}_v$ is evaluated at (u_k, v_k) and is a vector normal to the surface at P, which we assume to be nonzero at all points of S.

We write the Riemann sum with the observation that the areas of the parallelograms approximate the areas of the patches S_k:

$$\sum_{k=1}^{n} f(x(u_k, v_k), y(u_k, v_k), z(u_k, v_k))\,\Delta S_k$$

$$\approx \sum_{k=1}^{n} f(x(u_k, v_k), y(u_k, v_k), z(u_k, v_k))\underbrace{|\mathbf{t}_u \times \mathbf{t}_v|\,\Delta u\,\Delta v}_{\approx \Delta S_k}.$$

We now assume f is continuous on S. As Δu and Δv approach zero, the areas of the parallelograms approach the areas of the corresponding patches on S. We define the limit

▶ The role that the factor $|\mathbf{t}_u \times \mathbf{t}_v|\, dA$ plays in surface integrals is analogous to the role played by $|\mathbf{r}'(t)|\, dt$ in line integrals.

of this Riemann sum to be the surface integral of f over S, which we write $\iint_S f(x, y, z)\, dS$. The surface integral is evaluated as an ordinary double integral over the region R in the uv-plane:

$$\iint_S f(x, y, z)\, dS = \lim_{\Delta u, \Delta v \to 0} \sum_{k=1}^{n} f(x(u_k, v_k), y(u_k, v_k), z(u_k, v_k))|\mathbf{t}_u \times \mathbf{t}_v|\,\Delta u\, \Delta v$$

$$= \iint_R f(x(u, v), y(u, v), z(u, v))|\mathbf{t}_u \times \mathbf{t}_v|\, dA.$$

If R is a rectangular region, as we have assumed, the double integral becomes an iterated integral with respect to u and v with constant limits. In the special case that $f(x, y, z) = 1$, the integral gives the surface area of S.

▶ The condition that $\mathbf{t}_u \times \mathbf{t}_v$ be nonzero means $\mathbf{t}_u$ and $\mathbf{t}_v$ are nonzero and not parallel. If $\mathbf{t}_u \times \mathbf{t}_v \neq \mathbf{0}$ at all points, then the surface is *smooth*. The value of the integral is independent of the parameterization of S.

DEFINITION Surface Integral of Scalar-Valued Functions on Parameterized Surfaces

Let f be a continuous scalar-valued function on a smooth surface S given parametrically by $\mathbf{r}(u, v) = \langle x(u, v), y(u, v), z(u, v) \rangle$, where u and v vary over $R = \{(u, v): a \le u \le b, c \le v \le d\}$. Assume also that the tangent vectors

$$\mathbf{t}_u = \frac{\partial \mathbf{r}}{\partial u} = \left\langle \frac{\partial x}{\partial u}, \frac{\partial y}{\partial u}, \frac{\partial z}{\partial u} \right\rangle \text{ and } \mathbf{t}_v = \frac{\partial \mathbf{r}}{\partial v} = \left\langle \frac{\partial x}{\partial v}, \frac{\partial y}{\partial v}, \frac{\partial z}{\partial v} \right\rangle \text{ are continuous on } R \text{ and}$$

the normal vector $\mathbf{t}_u \times \mathbf{t}_v$ is nonzero on R. Then the **surface integral of f over S** is

$$\iint_S f(x, y, z)\, dS = \iint_R f(x(u, v), y(u, v), z(u, v))|\mathbf{t}_u \times \mathbf{t}_v|\, dA.$$

If $f(x, y, z) = 1$, this integral equals the surface area of S.

EXAMPLE 2 Surface area of a cylinder and sphere Find the surface area of the following surfaces.

a. A cylinder with radius $a > 0$ and height h (excluding the circular ends)

b. A sphere of radius a

SOLUTION The critical step is evaluating the normal vector $\mathbf{t}_u \times \mathbf{t}_v$. It needs to be done only once for any given surface.

a. As shown before, a parametric description of the cylinder is

$$\mathbf{r}(u, v) = \langle x(u, v), y(u, v), z(u, v) \rangle = \langle a \cos u, a \sin u, v \rangle,$$

where $0 \le u \le 2\pi$ and $0 \le v \le h$. The required normal vector is

$$\mathbf{t}_u \times \mathbf{t}_v = \begin{vmatrix} \mathbf{i} & \mathbf{j} & \mathbf{k} \\ \dfrac{\partial x}{\partial u} & \dfrac{\partial y}{\partial u} & \dfrac{\partial z}{\partial u} \\ \dfrac{\partial x}{\partial v} & \dfrac{\partial y}{\partial v} & \dfrac{\partial z}{\partial v} \end{vmatrix} \qquad \text{Definition of cross product}$$

$$= \begin{vmatrix} \mathbf{i} & \mathbf{j} & \mathbf{k} \\ -a \sin u & a \cos u & 0 \\ 0 & 0 & 1 \end{vmatrix} \qquad \text{Evaluate derivatives.}$$

$$= \langle a \cos u, a \sin u, 0 \rangle. \qquad \text{Compute cross product.}$$

Notice that this normal vector points outward from the cylinder, away from the z-axis (Figure 17.49). It follows that

$$|\mathbf{t}_u \times \mathbf{t}_v| = \sqrt{a^2 \cos^2 u + a^2 \sin^2 u} = a.$$

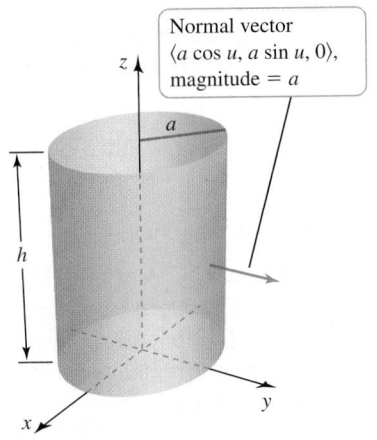

Normal vector $\langle a \cos u, a \sin u, 0 \rangle$, magnitude $= a$

Cylinder: $\mathbf{r}(u, v) = \langle a \cos u, a \sin u, v \rangle$, $0 \le u \le 2\pi$ and $0 \le v \le h$

Figure 17.49

Setting $f(x, y, z) = 1$, the surface area of the cylinder is

$$\iint_S 1 \, dS = \iint_R \underbrace{|\mathbf{t}_u \times \mathbf{t}_v|}_{a} \, dA = \int_0^{2\pi} \int_0^h a \, dv \, du = 2\pi a h,$$

confirming the formula for the surface area of a cylinder (excluding the ends).

b. A parametric description of the sphere is

$$\mathbf{r}(u, v) = \langle a \sin u \cos v, a \sin u \sin v, a \cos u \rangle,$$

where $0 \le u \le \pi$ and $0 \le v \le 2\pi$. The required normal vector is

> ➤ Recall that for the sphere, $u = \varphi$ and $v = \theta$, where φ and θ are spherical coordinates. The element of surface area in spherical coordinates is $dS = a^2 \sin \varphi \, d\varphi \, d\theta$.

$$\mathbf{t}_u \times \mathbf{t}_v = \begin{vmatrix} \mathbf{i} & \mathbf{j} & \mathbf{k} \\ a \cos u \cos v & a \cos u \sin v & -a \sin u \\ -a \sin u \sin v & a \sin u \cos v & 0 \end{vmatrix}$$

$$= \langle a^2 \sin^2 u \cos v, a^2 \sin^2 u \sin v, a^2 \sin u \cos u \rangle.$$

Computing $|\mathbf{t}_u \times \mathbf{t}_v|$ requires several steps (Exercise 70). However, the needed result is quite simple: $|\mathbf{t}_u \times \mathbf{t}_v| = a^2 \sin u$ and the normal vector $\mathbf{t}_u \times \mathbf{t}_v$ points outward from the surface of the sphere (**Figure 17.50**). With $f(x, y, z) = 1$, the surface area of the sphere is

$$\iint_S 1 \, dS = \iint_R \underbrace{|\mathbf{t}_u \times \mathbf{t}_v|}_{a^2 \sin u} \, dA = \int_0^{2\pi} \int_0^\pi a^2 \sin u \, du \, dv = 4\pi a^2,$$

confirming the formula for the surface area of a sphere.

Sphere:
$\mathbf{r}(u, v) = \langle a \sin u \cos v, a \sin u \sin v, a \cos u \rangle,$
$0 \le u \le \pi$ and $0 \le v \le 2\pi$

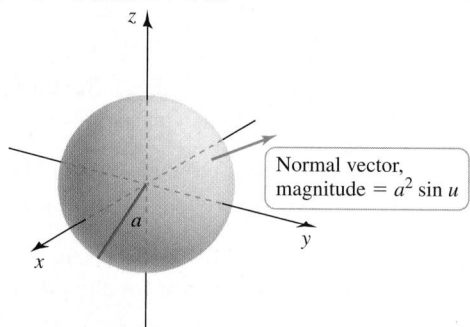

Normal vector, magnitude $= a^2 \sin u$

Figure 17.50

Related Exercises 19, 22 ◄

EXAMPLE 3   Surface area of a partial cylinder Find the surface area of the cylinder $\{(r, \theta): r = 4, 0 \le \theta \le 2\pi\}$ between the planes $z = 0$ and $z = 16 - 2x$ (excluding the top and bottom surfaces).

SOLUTION Figure 17.51 shows the cylinder bounded by the two planes. With $u = \theta$ and $v = z$, a parametric description of the cylinder is

$$\mathbf{r}(u, v) = \langle x(u, v), y(u, v), z(u, v) \rangle = \langle 4 \cos u, 4 \sin u, v \rangle.$$

The challenge is finding the limits on v, which is the z-coordinate. The plane $z = 16 - 2x$ intersects the cylinder in an ellipse; along this ellipse, as u varies between 0 and 2π, the parameter v also changes. To find the relationship between u and v along this intersection curve, notice that at any point on the cylinder, we have $x = 4 \cos u$ (remember that $u = \theta$). Making this substitution in the equation of the plane, we have

$$z = 16 - 2x = 16 - 2(4 \cos u) = 16 - 8 \cos u.$$

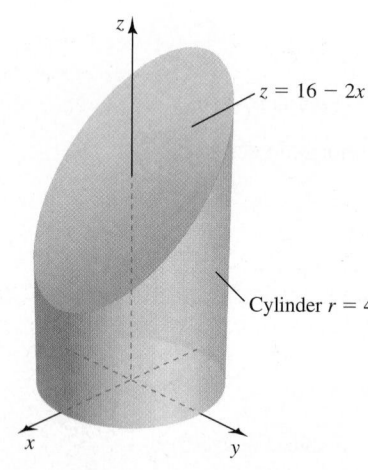

$z = 16 - 2x$

Cylinder $r = 4$

Sliced cylinder is generated by $\mathbf{r}(u, v) = \langle 4 \cos u, 4 \sin u, v \rangle$, where $0 \le u \le 2\pi, 0 \le v \le 16 - 8 \cos u$.

Figure 17.51

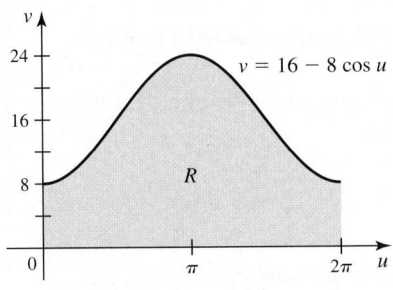

Region of integration in the uv-plane is
$R = \{(u, v): 0 \le u \le 2\pi,$
$0 \le v \le 16 - 8 \cos u\}.$

Figure 17.52

Substituting $v = z$, the relationship between u and v is $v = 16 - 8 \cos u$ (Figure 17.52). Therefore, the region of integration in the uv-plane is

$$R = \{(u, v): 0 \le u \le 2\pi, 0 \le v \le 16 - 8 \cos u\}.$$

Recall from Example 2a that for the cylinder, $|\mathbf{t}_u \times \mathbf{t}_v| = a = 4$. Setting $f(x, y, z) = 1$, the surface integral for the area is

$$\iint_S 1 \, dS = \iint_R \underbrace{|\mathbf{t}_u \times \mathbf{t}_v|}_{4} \, dA$$

$$= \int_0^{2\pi} \int_0^{16-8\cos u} 4 \, dv \, du$$

$$= 4 \int_0^{2\pi} (16 - 8 \cos u) \, du \quad \text{Evaluate inner integral.}$$

$$= 4(16u - 8 \sin u)\Big|_0^{2\pi} \quad \text{Evaluate outer integral.}$$

$$= 128\pi. \quad \text{Simplify.}$$

Related Exercise 24 ◄

EXAMPLE 4 **Average temperature on a sphere** The temperature on the surface of a sphere of radius a varies with latitude according to the function $T(\varphi, \theta) = 10 + 50 \sin \varphi$, for $0 \le \varphi \le \pi$ and $0 \le \theta \le 2\pi$ (φ and θ are spherical coordinates, so the temperature is $10°$ at the poles, increasing to $60°$ at the equator). Find the average temperature over the sphere.

SOLUTION We use the parametric description of a sphere. With $u = \varphi$ and $v = \theta$, the temperature function becomes $f(u, v) = 10 + 50 \sin u$. Integrating the temperature over the sphere using the fact that $|\mathbf{t}_u \times \mathbf{t}_v| = a^2 \sin u$ (Example 2b), we have

$$\iint_S (10 + 50 \sin u) \, dS = \iint_R (10 + 50 \sin u) \underbrace{|\mathbf{t}_u \times \mathbf{t}_v|}_{a^2 \sin u} \, dA$$

$$= \int_0^{\pi} \int_0^{2\pi} (10 + 50 \sin u) a^2 \sin u \, dv \, du$$

$$= 2\pi a^2 \int_0^{\pi} (10 + 50 \sin u) \sin u \, du \quad \text{Evaluate inner integral.}$$

$$= 10\pi a^2 (4 + 5\pi). \quad \text{Evaluate outer integral.}$$

The average temperature is the integrated temperature $10\pi a^2(4 + 5\pi)$ divided by the surface area of the sphere $4\pi a^2$, so the average temperature is $(20 + 25\pi)/2 \approx 49.3°$.

Related Exercise 42 ◄

Surface Integrals on Explicitly Defined Surfaces Suppose a smooth surface S is defined not parametrically, but explicitly, in the form $z = g(x, y)$ over a region R in the xy-plane. Such a surface may be treated as a parameterized surface. We simply define the parameters to be $u = x$ and $v = y$. Making these substitutions into the expression for $\mathbf{t}_u$ and $\mathbf{t}_v$, a short calculation (Exercise 71) reveals that $\mathbf{t}_u = \mathbf{t}_x = \langle 1, 0, z_x \rangle$, $\mathbf{t}_v = \mathbf{t}_y = \langle 0, 1, z_y \rangle$, and the required normal vector is

$$\mathbf{t}_x \times \mathbf{t}_y = \langle -z_x, -z_y, 1 \rangle.$$

> This is a familiar result: A normal to the surface $z = g(x, y)$ at a point is a constant multiple of the gradient of $z - g(x, y)$, which is $\langle -g_x, -g_y, 1 \rangle = \langle -z_x, -z_y, 1 \rangle$. The factor $\sqrt{z_x^2 + z_y^2 + 1}$ is analogous to the factor $\sqrt{f'(x)^2 + 1}$ that appears in arc length integrals.

It follows that

$$|\mathbf{t}_x \times \mathbf{t}_y| = |\langle -z_x, -z_y, 1 \rangle| = \sqrt{z_x^2 + z_y^2 + 1}.$$

With these observations, the surface integral over S can be expressed as a double integral over a region R in the xy-plane.

> If the surface S in Theorem 17.14 is generated by revolving a curve in the xy-plane about the x-axis, the theorem gives the standard surface area formula for surfaces of revolution (Exercise 75).

THEOREM 17.14   Evaluation of Surface Integrals of Scalar-Valued Functions on Explicitly Defined Surfaces
Let f be a continuous function on a smooth surface S given by $z = g(x, y)$, for (x, y) in a region R. The surface integral of f over S is

$$\iint_S f(x, y, z)\, dS = \iint_R f(x, y, g(x, y))\sqrt{z_x^2 + z_y^2 + 1}\, dA.$$

If $f(x, y, z) = 1$, the surface integral equals the area of the surface.

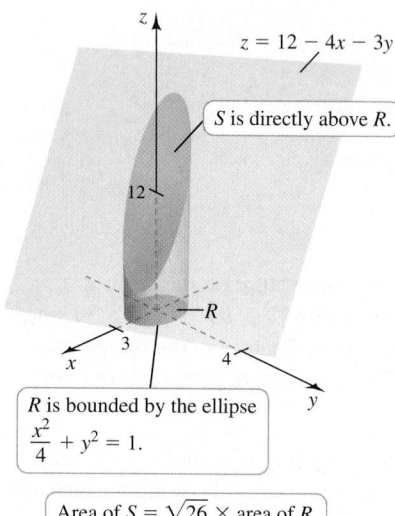

R is bounded by the ellipse $\dfrac{x^2}{4} + y^2 = 1$.

Area of $S = \sqrt{26} \times$ area of R.

Figure 17.53

QUICK CHECK 4 The plane $z = y$ forms a 45° angle with the xy-plane. Suppose the plane is the roof of a room and the xy-plane is the floor of the room. Then 1 ft² on the floor becomes how many square feet when projected on the roof? ◄

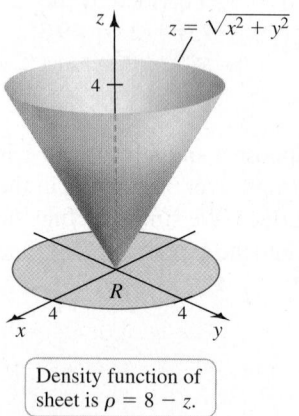

Density function of sheet is $\rho = 8 - z$.

Figure 17.54

EXAMPLE 5   Area of a roof over an ellipse Find the area of the surface S that lies in the plane $z = 12 - 4x - 3y$ directly above the region R bounded by the ellipse $x^2/4 + y^2 = 1$ (Figure 17.53).

SOLUTION Because we are computing the area of the surface, we take $f(x, y, z) = 1$. Note that $z_x = -4$ and $z_y = -3$, so the factor $\sqrt{z_x^2 + z_y^2 + 1}$ has the value $\sqrt{(-4)^2 + (-3)^2 + 1} = \sqrt{26}$ (a constant because the surface is a plane). The relevant surface integral is

$$\iint_S 1\, dS = \iint_R \underbrace{\sqrt{z_x^2 + z_y^2 + 1}}_{\sqrt{26}}\, dA = \sqrt{26}\iint_R dA.$$

The double integral that remains is simply the area of the region R bounded by the ellipse. Because the ellipse has semiaxes of length $a = 2$ and $b = 1$, its area is $\pi ab = 2\pi$. Therefore, the area of S is $2\pi\sqrt{26}$.

This result has a useful interpretation. The plane surface S is not horizontal, so it has a greater area than the horizontal region R beneath it. The factor that converts the area of R to the area of S is $\sqrt{26}$. Notice that if the roof *were* horizontal, then the surface would be $z = c$, the area conversion factor would be 1, and the area of the roof would equal the area of the floor beneath it.

Related Exercises 29–30 ◄

EXAMPLE 6   Mass of a conical sheet A thin conical sheet is described by the surface $z = (x^2 + y^2)^{1/2}$, for $0 \le z \le 4$. The density of the sheet in g/cm² is $\rho = f(x, y, z) = (8 - z)$ (decreasing from 8 g/cm² at the vertex to 4 g/cm² at the top of the cone; Figure 17.54). What is the mass of the cone?

SOLUTION We find the mass by integrating the density function over the surface of the cone. The projection of the cone on the xy-plane is found by setting $z = 4$ (the top of the cone) in the equation of the cone. We find that $(x^2 + y^2)^{1/2} = 4$; therefore, the region of integration is the disk $R = \{(x, y): x^2 + y^2 \le 16\}$. The next step is to compute z_x and z_y in order to evaluate $\sqrt{z_x^2 + z_y^2 + 1}$. Differentiating $z^2 = x^2 + y^2$ implicitly gives $2zz_x = 2x$, or $z_x = x/z$. Similarly, $z_y = y/z$. Using the fact that $z^2 = x^2 + y^2$, we have

$$\sqrt{z_x^2 + z_y^2 + 1} = \sqrt{(x/z)^2 + (y/z)^2 + 1} = \sqrt{\underbrace{\frac{x^2 + y^2}{z^2}}_{1} + 1} = \sqrt{2}.$$

To integrate the density over the conical surface, we set $f(x, y, z) = 8 - z$. Replacing z in the integrand by $r = (x^2 + y^2)^{1/2}$ and using polar coordinates, the mass in grams is given by

$$\iint_S f(x, y, z)\, dS = \iint_R f(x, y, z)\underbrace{\sqrt{z_x^2 + z_y^2 + 1}}_{\sqrt{2}}\, dA$$

$$= \sqrt{2}\iint_R (8 - z)\, dA \qquad \text{Substitute.}$$

$$= \sqrt{2} \iint\limits_{R} (8 - \sqrt{x^2 + y^2}) \, dA \qquad z = \sqrt{x^2 + y^2}$$

$$= \sqrt{2} \int_0^{2\pi} \int_0^4 (8 - r) \, r \, dr \, d\theta \qquad \text{Polar coordinates}$$

$$= \sqrt{2} \int_0^{2\pi} \left(4r^2 - \frac{r^3}{3}\right)\Big|_0^4 \, d\theta \qquad \text{Evaluate inner integral.}$$

$$= \frac{128\sqrt{2}}{3} \int_0^{2\pi} d\theta \qquad \text{Simplify.}$$

$$= \frac{256\pi\sqrt{2}}{3} \approx 379. \qquad \text{Evaluate outer integral.}$$

As a check, note that the surface area of the cone is $\pi r \sqrt{r^2 + h^2} \approx 71 \text{ cm}^2$. If the entire cone had the maximum density $\rho = 8 \text{ g/cm}^2$, its mass would be approximately 568 g. If the entire cone had the minimum density $\rho = 4 \text{ g/cm}^2$, its mass would be approximately 284 g. The actual mass is between these extremes and closer to the low value because the cone is lighter at the top, where the surface area is greater.

Related Exercise 36 ◄

Table 17.3 summarizes the essential relationships for the explicit and parametric descriptions of cylinders, cones, spheres, and paraboloids. The listed normal vectors are chosen to point away from the z-axis.

Table 17.3

Surface	Explicit Description $z = g(x, y)$		Parametric Description	
	Equation	Normal vector; magnitude $\pm\langle -z_x, -z_y, 1\rangle$; $|\langle -z_x, -z_y, 1\rangle|$	Equation	Normal vector; magnitude $t_u \times t_v$; $|t_u \times t_v|$
Cylinder	$x^2 + y^2 = a^2$, $0 \le z \le h$	$\langle x, y, 0\rangle$; a	$\mathbf{r} = \langle a\cos u, a\sin u, v\rangle$, $0 \le u \le 2\pi, 0 \le v \le h$	$\langle a\cos u, a\sin u, 0\rangle$; a
Cone	$z^2 = x^2 + y^2$, $0 \le z \le h$	$\langle x/z, y/z, -1\rangle$; $\sqrt{2}$	$\mathbf{r} = \langle v\cos u, v\sin u, v\rangle$, $0 \le u \le 2\pi, 0 \le v \le h$	$\langle v\cos u, v\sin u, -v\rangle$; $\sqrt{2}v$
Sphere	$x^2 + y^2 + z^2 = a^2$	$\langle x/z, y/z, 1\rangle$; a/z	$\mathbf{r} = \langle a\sin u\cos v,$ $a\sin u\sin v, a\cos u\rangle$, $0 \le u \le \pi, 0 \le v \le 2\pi$	$\langle a^2\sin^2 u\cos v, a^2\sin^2 u\sin v,$ $a^2\sin u\cos u\rangle$; $a^2\sin u$
Paraboloid	$z = x^2 + y^2$, $0 \le z \le h$	$\langle 2x, 2y, -1\rangle$; $\sqrt{1 + 4(x^2 + y^2)}$	$\mathbf{r} = \langle v\cos u, v\sin u, v^2\rangle$, $0 \le u \le 2\pi, 0 \le v \le \sqrt{h}$	$\langle 2v^2\cos u, 2v^2\sin u, -v\rangle$; $v\sqrt{1 + 4v^2}$

QUICK CHECK 5 Explain why the explicit description for a cylinder $x^2 + y^2 = a^2$ cannot be used for a surface integral over a cylinder, and a parametric description must be used. ◄

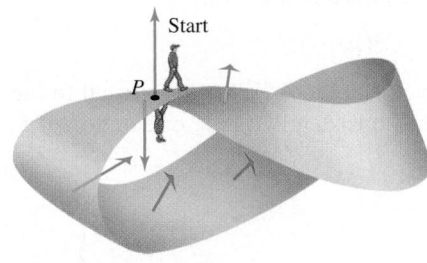

Figure 17.55

Surface Integrals of Vector Fields

Before beginning a discussion of surface integrals of vector fields, we must address two technical issues about surfaces and normal vectors.

The surfaces we consider in this text are called **two-sided**, or **orientable**, surfaces. To be orientable, a surface must have the property that the normal vectors vary continuously over the surface. In other words, when you walk on any closed path on an orientable surface and return to your starting point, your head must point in the same direction it did when you started. A well-known example of a *nonorientable* surface is the Möbius strip (Figure 17.55). Suppose you start walking the length of the Möbius strip at a point P with your head pointing upward. When you return to P, your head points in the opposite direction, or downward. Therefore, the Möbius strip is not orientable.

At any point of a parameterized orientable surface, there are two unit normal vectors. Therefore, the second point concerns the orientation of the surface or, equivalently, the direction of the normal vector. Once the direction of the normal vector is determined, the surface becomes **oriented**.

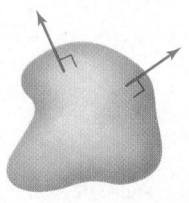

Surfaces that enclose a region are oriented so normal vectors point in the outward direction.

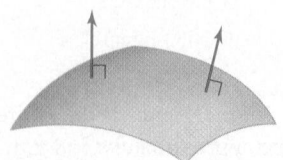

For other surfaces, the orientation of the surface must be specified.

Figure 17.56

We make the common assumption that—unless specified otherwise—a closed orientable surface that fully encloses a region (such as a sphere) is oriented so that the normal vectors point in the *outward direction*. For a surface that does not enclose a region in $\mathbb{R}^3$, the orientation must be specified in some way. For example, we might specify that the normal vectors for a particular surface point in the general direction of the positive z-axis; that is, in an upward direction (**Figure 17.56**).

Now recall that the parameterization of a surface defines a normal vector $\mathbf{t}_u \times \mathbf{t}_v$ at each point. In many cases, the normal vectors are consistent with the specified orientation, in which case no adjustments need to be made. If the direction of $\mathbf{t}_u \times \mathbf{t}_v$ is not consistent with the specified orientation, then the sign of $\mathbf{t}_u \times \mathbf{t}_v$ must be reversed before doing calculations. This process is demonstrated in the following examples.

Flux Integrals   It turns out that the most common surface integral of a vector field is a *flux integral*. Consider a vector field $\mathbf{F} = \langle f, g, h \rangle$, continuous on a region in $\mathbb{R}^3$, that represents the flow of a fluid or the transport of a substance. Given a smooth oriented surface S, we aim to compute the net flux of the vector field across the surface. In a small region containing a point P, the flux across the surface is proportional to the component of $\mathbf{F}$ in the direction of the unit normal vector $\mathbf{n}$ at P. If θ is the angle between $\mathbf{F}$ and $\mathbf{n}$, then this component is $\mathbf{F} \cdot \mathbf{n} = |\mathbf{F}||\mathbf{n}| \cos \theta = |\mathbf{F}| \cos \theta$ (because $|\mathbf{n}| = 1$; **Figure 17.57a**). We have the following special cases.

- If $\mathbf{F}$ and the unit normal vector are aligned at P ($\theta = 0$), then the component of $\mathbf{F}$ in the direction $\mathbf{n}$ is $\mathbf{F} \cdot \mathbf{n} = |\mathbf{F}|$; that is, all of $\mathbf{F}$ flows across the surface in the direction of $\mathbf{n}$ (**Figure 17.57b**).

- If $\mathbf{F}$ and the unit normal vector point in opposite directions at P ($\theta = \pi$), then the component of $\mathbf{F}$ in the direction $\mathbf{n}$ is $\mathbf{F} \cdot \mathbf{n} = -|\mathbf{F}|$; that is, all of $\mathbf{F}$ flows across the surface in the direction opposite that of $\mathbf{n}$ (**Figure 17.57c**).

- If $\mathbf{F}$ and the unit normal vector are orthogonal at P ($\theta = \pi/2$), then the component of $\mathbf{F}$ in the direction $\mathbf{n}$ is $\mathbf{F} \cdot \mathbf{n} = 0$; that is, none of $\mathbf{F}$ flows across the surface at that point (**Figure 17.57d**).

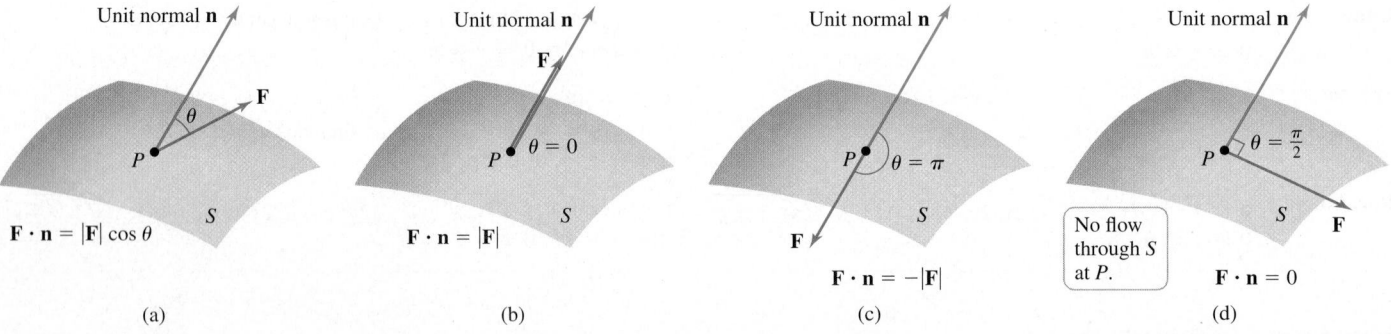

Figure 17.57

The flux integral, denoted $\iint_S \mathbf{F} \cdot \mathbf{n} \, dS$ or $\iint_S \mathbf{F} \cdot d\mathbf{S}$, simply adds up the components of $\mathbf{F}$ normal to the surface at all points of the surface. Notice that $\mathbf{F} \cdot \mathbf{n}$ is a scalar-valued function. Here is how the flux integral is computed.

Suppose the smooth oriented surface S is parameterized in the form

$$\mathbf{r}(u, v) = \langle x(u, v), y(u, v), z(u, v) \rangle,$$

where u and v vary over a region R in the uv-plane. The required vector normal to the surface at a point is $\mathbf{t}_u \times \mathbf{t}_v$, which we assume to be consistent with the orientation of S.

> If $\mathbf{t}_u \times \mathbf{t}_v$ is not consistent with the specified orientation, its sign must be reversed.

Therefore, the *unit* normal vector consistent with the orientation is $\mathbf{n} = \dfrac{\mathbf{t}_u \times \mathbf{t}_v}{|\mathbf{t}_u \times \mathbf{t}_v|}$.

Appealing to the definition of the surface integral for parameterized surfaces, the flux integral is

$$\iint_S \mathbf{F} \cdot \mathbf{n} \, dS = \iint_R \mathbf{F} \cdot \mathbf{n} |\mathbf{t}_u \times \mathbf{t}_v| \, dA \qquad \text{Definition of surface integral}$$

$$= \iint_R \mathbf{F} \cdot \underbrace{\frac{\mathbf{t}_u \times \mathbf{t}_v}{|\mathbf{t}_u \times \mathbf{t}_v|}}_{\mathbf{n}} |\mathbf{t}_u \times \mathbf{t}_v| \, dA \qquad \text{Substitute for } \mathbf{n}.$$

$$= \iint_R \mathbf{F} \cdot (\mathbf{t}_u \times \mathbf{t}_v) \, dA. \qquad \text{Convenient cancellation}$$

The remarkable occurrence in the flux integral is the cancellation of the factor $|\mathbf{t}_u \times \mathbf{t}_v|$.

The special case in which the surface S is specified in the form $z = s(x, y)$ follows directly by recalling that the required normal vector is $\mathbf{t}_u \times \mathbf{t}_v = \langle -z_x, -z_y, 1 \rangle$. In this case, with $\mathbf{F} = \langle f, g, h \rangle$, the integrand of the surface integral is $\mathbf{F} \cdot (\mathbf{t}_u \times \mathbf{t}_v) = -fz_x - gz_y + h$.

> The value of the surface integral is independent of the parameterization. However, in contrast to a surface integral of a scalar-valued function, the value of a surface integral of a vector field depends on the orientation of the surface. Changing the orientation changes the sign of the result.

DEFINITION Surface Integral of a Vector Field

Suppose $\mathbf{F} = \langle f, g, h \rangle$ is a continuous vector field on a region of $\mathbb{R}^3$ containing a smooth oriented surface S. If S is defined parametrically as $\mathbf{r}(u, v) = \langle x(u, v), y(u, v), z(u, v) \rangle$, for (u, v) in a region R, then

$$\iint_S \mathbf{F} \cdot \mathbf{n} \, dS = \iint_R \mathbf{F} \cdot (\mathbf{t}_u \times \mathbf{t}_v) \, dA,$$

where $\mathbf{t}_u = \dfrac{\partial \mathbf{r}}{\partial u} = \left\langle \dfrac{\partial x}{\partial u}, \dfrac{\partial y}{\partial u}, \dfrac{\partial z}{\partial u} \right\rangle$ and $\mathbf{t}_v = \dfrac{\partial \mathbf{r}}{\partial v} = \left\langle \dfrac{\partial x}{\partial v}, \dfrac{\partial y}{\partial v}, \dfrac{\partial z}{\partial v} \right\rangle$ are continuous on R, the normal vector $\mathbf{t}_u \times \mathbf{t}_v$ is nonzero on R, and the direction of the normal vector is consistent with the orientation of S. If S is defined in the form $z = s(x, y)$, for (x, y) in a region R, then

$$\iint_S \mathbf{F} \cdot \mathbf{n} \, dS = \iint_R (-fz_x - gz_y + h) \, dA.$$

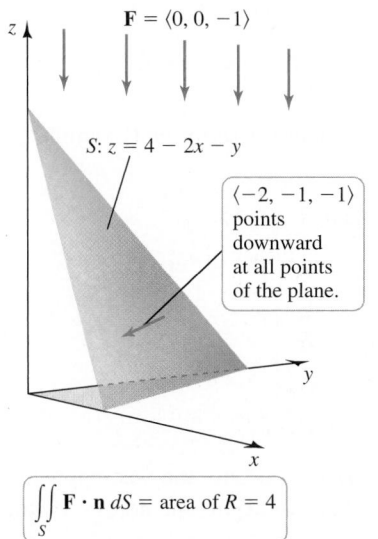

$$\iint_S \mathbf{F} \cdot \mathbf{n} \, dS = \text{area of } R = 4$$

Figure 17.58

EXAMPLE 7 Rain on a roof Consider the vertical vector field $\mathbf{F} = \langle 0, 0, -1 \rangle$, corresponding to a constant downward flow. Find the flux in the downward direction across the surface S, which is the plane $z = 4 - 2x - y$ in the first octant.

SOLUTION In this case, the surface is given explicitly. With $z = 4 - 2x - y$, we have $z_x = -2$ and $z_y = -1$. Therefore, the required normal vector is $\langle -z_x, -z_y, 1 \rangle = \langle 2, 1, 1 \rangle$, which points *upward* (the z-component of the vector is positive). Because we are interested in the *downward* flux of $\mathbf{F}$ across S, the surface must be oriented such that the normal vectors point downward. So we take the normal vector to be $\langle -2, -1, -1 \rangle$ (Figure 17.58). Letting R be the region in the xy-plane beneath S and noting that $\mathbf{F} = \langle f, g, h \rangle = \langle 0, 0, -1 \rangle$, the flux integral is

$$\iint_S \mathbf{F} \cdot \mathbf{n} \, dS = \iint_R \langle 0, 0, -1 \rangle \cdot \langle -2, -1, -1 \rangle \, dA = \iint_R dA = \text{area of } R.$$

The base R is a triangle in the xy-plane with vertices $(0, 0)$, $(2, 0)$, and $(0, 4)$, so its area is 4. Therefore, the *downward* flux across S is 4. This flux integral has an interesting interpretation. If the vector field $\mathbf{F}$ represents the rate of rainfall with units of, say, g/m² per unit time, then the flux integral gives the mass of rain (in grams) that falls on the surface in a unit of time. This result says that (because the vector field is vertical) the mass of rain that falls on the roof equals the mass that would fall on the floor beneath the roof if the roof were not there. This property is explored further in Exercise 73.

Related Exercises 43–44 ◀

EXAMPLE 8   Flux of the radial field Consider the radial vector field
$\mathbf{F} = \langle f, g, h \rangle = \langle x, y, z \rangle$. Is the upward flux of the field greater across the hemisphere $x^2 + y^2 + z^2 = 1$, for $z \geq 0$, or across the paraboloid $z = 1 - x^2 - y^2$, for $z \geq 0$? Note that the two surfaces have the same base in the xy-plane and the same high point $(0, 0, 1)$. Use the explicit description for the hemisphere and a parametric description for the paraboloid.

SOLUTION The base of both surfaces in the xy-plane is the unit disk $R = \{(x, y): x^2 + y^2 \leq 1\}$, which, when expressed in polar coordinates, is the set $\{(r, \theta): 0 \leq r \leq 1, 0 \leq \theta \leq 2\pi\}$. To use the explicit description for the hemisphere, we must compute z_x and z_y. Differentiating $x^2 + y^2 + z^2 = 1$ implicitly, we find that $z_x = -x/z$ and $z_y = -y/z$. Therefore, the required normal vector is $\langle x/z, y/z, 1 \rangle$, which points upward on the surface. The flux integral is evaluated by substituting for f, g, h, z_x, and z_y; eliminating z from the integrand; and converting the integral in x and y to an integral in polar coordinates:

> ▶ Recall that the required normal vector for an explicitly defined surface $z = s(x, y)$ is $\langle -z_x, -z_y, 1 \rangle$.

$$
\iint_S \mathbf{F} \cdot \mathbf{n} \, dS = \iint_R (-fz_x - gz_y + h) \, dA
$$

$$
= \iint_R \left(x\frac{x}{z} + y\frac{y}{z} + z \right) dA \qquad \text{Substitute.}
$$

$$
= \iint_R \left(\frac{x^2 + y^2 + z^2}{z} \right) dA \qquad \text{Simplify.}
$$

$$
= \iint_R \left(\frac{1}{z} \right) dA \qquad x^2 + y^2 + z^2 = 1
$$

$$
= \iint_R \left(\frac{1}{\sqrt{1 - x^2 - y^2}} \right) dA \qquad z = \sqrt{1 - x^2 - y^2}
$$

$$
= \int_0^{2\pi} \int_0^1 \left(\frac{1}{\sqrt{1 - r^2}} \right) r \, dr \, d\theta \qquad \text{Polar coordinates}
$$

$$
= \int_0^{2\pi} \left(-\sqrt{1 - r^2} \right) \Big|_0^1 \, d\theta \qquad \begin{array}{l}\text{Evaluate inner integral} \\ \text{as an improper integral.}\end{array}
$$

$$
= \int_0^{2\pi} d\theta = 2\pi. \qquad \text{Evaluate outer integral.}
$$

For the paraboloid $z = 1 - x^2 - y^2$, we use the parametric description (Example 1b or Table 17.3)

$$
\mathbf{r}(u, v) = \langle x, y, z \rangle = \langle v \cos u, v \sin u, 1 - v^2 \rangle,
$$

for $0 \leq u \leq 2\pi$ and $0 \leq v \leq 1$. The required vector normal to the surface is

$$
\mathbf{t}_u \times \mathbf{t}_v = \begin{vmatrix} \mathbf{i} & \mathbf{j} & \mathbf{k} \\ -v \sin u & v \cos u & 0 \\ \cos u & \sin u & -2v \end{vmatrix}
$$

$$
= \langle -2v^2 \cos u, -2v^2 \sin u, -v \rangle.
$$

Notice that the normal vectors point downward on the surface (because the z-component is negative for $0 < v \leq 1$). In order to find the upward flux, we negate the normal vector and use the upward normal vector

$$
-(\mathbf{t}_u \times \mathbf{t}_v) = \langle 2v^2 \cos u, 2v^2 \sin u, v \rangle.
$$

The flux integral is evaluated by substituting for $\mathbf{F} = \langle x, y, z \rangle$ and $-(\mathbf{t}_u \times \mathbf{t}_v)$ and then evaluating an iterated integral in u and v:

$$\iint\limits_S \mathbf{F} \cdot \mathbf{n} \, dS = \int_0^1 \int_0^{2\pi} \langle v \cos u, v \sin u, 1 - v^2 \rangle \cdot \langle 2v^2 \cos u, 2v^2 \sin u, v \rangle \, du \, dv$$

Substitute for F and $-(\mathbf{t}_u \times \mathbf{t}_v)$.

$$= \int_0^1 \int_0^{2\pi} (v^3 + v) \, du \, dv \qquad \text{Simplify.}$$

QUICK CHECK 6 Explain why the upward flux for the radial field in Example 8 is greater for the hemisphere than for the paraboloid. ◄

$$= 2\pi \left(\frac{v^4}{4} + \frac{v^2}{2} \right) \Big|_0^1 = \frac{3\pi}{2}. \qquad \text{Evaluate integrals.}$$

We see that the upward flux is greater for the hemisphere than for the paraboloid.

Related Exercises 45, 47 ◄

SECTION 17.6 EXERCISES

Getting Started

1. Give a parametric description for a cylinder with radius a and height h, including the intervals for the parameters.

2. Give a parametric description for a cone with radius a and height h, including the intervals for the parameters.

3. Give a parametric description for a sphere with radius a, including the intervals for the parameters.

4. Explain how to compute the surface integral of a scalar-valued function f over a cone using an explicit description of the cone.

5. Explain how to compute the surface integral of a scalar-valued function f over a sphere using a parametric description of the sphere.

6. Explain what it means for a surface to be orientable.

7. Describe the usual orientation of a closed surface such as a sphere.

8. Why is the upward flux of a vertical vector field $\mathbf{F} = \langle 0, 0, 1 \rangle$ across a surface equal to the area of the projection of the surface in the xy-plane?

Practice Exercises

9–14. Parametric descriptions *Give a parametric description of the form $\mathbf{r}(u, v) = \langle x(u, v), y(u, v), z(u, v) \rangle$ for the following surfaces. The descriptions are not unique. Specify the required rectangle in the uv-plane.*

9. The plane $2x - 4y + 3z = 16$

10. The cap of the sphere $x^2 + y^2 + z^2 = 16$, for $2\sqrt{2} \le z \le 4$

11. The frustum of the cone $z^2 = x^2 + y^2$, for $2 \le z \le 8$

12. The cone $z^2 = 4(x^2 + y^2)$, for $0 \le z \le 4$

13. The portion of the cylinder $x^2 + y^2 = 9$ in the first octant, for $0 \le z \le 3$

14. The cylinder $y^2 + z^2 = 36$, for $0 \le x \le 9$

15–18. Identify the surface *Describe the surface with the given parametric representation.*

15. $\mathbf{r}(u, v) = \langle u, v, 2u + 3v - 1 \rangle$, for $1 \le u \le 3, 2 \le v \le 4$

16. $\mathbf{r}(u, v) = \langle u, u + v, 2 - u - v \rangle$, for $0 \le u \le 2, 0 \le v \le 2$

17. $\mathbf{r}(u, v) = \langle v \cos u, v \sin u, 4v \rangle$, for $0 \le u \le \pi, 0 \le v \le 3$

18. $\mathbf{r}(u, v) = \langle v, 6 \cos u, 6 \sin u \rangle$, for $0 \le u \le 2\pi, 0 \le v \le 2$

19–24. Surface area using a parametric description *Find the area of the following surfaces using a parametric description of the surface.*

19. The half-cylinder $\{(r, \theta, z): r = 4, 0 \le \theta \le \pi, 0 \le z \le 7\}$

20. The plane $z = 3 - x - 3y$ in the first octant

21. The plane $z = 10 - x - y$ above the square $|x| \le 2, |y| \le 2$

22. The hemisphere $x^2 + y^2 + z^2 = 100$, for $z \ge 0$

23. A cone with base radius r and height h, where r and h are positive constants

24. The cap of the sphere $x^2 + y^2 + z^2 = 4$, for $1 \le z \le 2$

25–28. Surface integrals using a parametric description *Evaluate the surface integral $\iint_S f \, dS$ using a parametric description of the surface.*

25. $f(x, y, z) = x^2 + y^2$, where S is the hemisphere $x^2 + y^2 + z^2 = 36$, for $z \ge 0$

26. $f(x, y, z) = y$, where S is the cylinder $x^2 + y^2 = 9, 0 \le z \le 3$

27. $f(x, y, z) = x$, where S is the cylinder $x^2 + z^2 = 1, 0 \le y \le 3$

28. $f(\rho, \varphi, \theta) = \cos \varphi$, where S is the part of the unit sphere in the first octant

29–34. Surface area using an explicit description *Find the area of the following surfaces using an explicit description of the surface.*

29. The part of the plane $z = 2x + 2y + 4$ over the region R bounded by the triangle with vertices $(0, 0), (2, 0)$, and $(2, 4)$

30. The part of the plane $z = x + 3y + 5$ over the region $R = \{(x, y): 1 \le x^2 + y^2 \le 4\}$

31. The cone $z^2 = 4(x^2 + y^2)$, for $0 \le z \le 4$

T 32. The trough $z = \dfrac{1}{2}x^2$, for $-1 \le x \le 1, 0 \le y \le 4$

33. The paraboloid $z = 2(x^2 + y^2)$, for $0 \le z \le 8$

34. The part of the hyperbolic paraboloid $z = 3 + x^2 - y^2$ above the sector $R = \{(r, \theta): 0 \le r \le \sqrt{2}, 0 \le \theta \le \pi/2\}$

35–38. Surface integrals using an explicit description *Evaluate the surface integral $\iint_S f(x, y, z)\, dS$ using an explicit representation of the surface.*

35. $f(x, y, z) = xy$; S is the plane $z = 2 - x - y$ in the first octant.

36. $f(x, y, z) = x^2 + y^2$; S is the paraboloid $z = x^2 + y^2$, for $0 \le z \le 1$.

37. $f(x, y, z) = 25 - x^2 - y^2$; S is the hemisphere centered at the origin with radius 5, for $z \ge 0$.

38. $f(x, y, z) = e^z$; S is the plane $z = 8 - x - 2y$ in the first octant.

39–42. Average values

39. Find the average temperature on that part of the plane $2x + 2y + z = 4$ over the square $0 \le x \le 1, 0 \le y \le 1$, where the temperature is given by $T(x, y, z) = e^{2x+y+z-3}$.

T 40. Find the average squared distance between the origin and the points on the paraboloid $z = 4 - x^2 - y^2$, for $z \ge 0$.

41. Find the average value of the function $f(x, y, z) = xyz$ on the unit sphere in the first octant.

42. Find the average value of the temperature function $T(x, y, z) = 100 - 25z$ on the cone $z^2 = x^2 + y^2$, for $0 \le z \le 2$.

43–48. Surface integrals of vector fields *Find the flux of the following vector fields across the given surface with the specified orientation. You may use either an explicit or a parametric description of the surface.*

43. $\mathbf{F} = \langle 0, 0, -1 \rangle$ across the slanted face of the tetrahedron $z = 4 - x - y$ in the first octant; normal vectors point upward.

44. $\mathbf{F} = \langle x, y, z \rangle$ across the slanted face of the tetrahedron $z = 10 - 2x - 5y$ in the first octant; normal vectors point upward.

45. $\mathbf{F} = \langle x, y, z \rangle$ across the slanted surface of the cone $z^2 = x^2 + y^2$, for $0 \le z \le 1$; normal vectors point upward.

46. $\mathbf{F} = \langle e^{-y}, 2z, xy \rangle$ across the curved sides of the surface $S = \{(x, y, z): z = \cos y, |y| \le \pi, 0 \le x \le 4\}$; normal vectors point upward.

47. $\mathbf{F} = \mathbf{r}/|\mathbf{r}|^3$ across the sphere of radius a centered at the origin, where $\mathbf{r} = \langle x, y, z \rangle$; normal vectors point outward.

48. $\mathbf{F} = \langle -y, x, 1 \rangle$ across the cylinder $y = x^2$, for $0 \le x \le 1$, $0 \le z \le 4$; normal vectors point in the general direction of the positive y-axis.

49. Explain why or why not Determine whether the following statements are true and give an explanation or counterexample.

a. If the surface S is given by $\{(x, y, z): 0 \le x \le 1, 0 \le y \le 1, z = 10\}$, then $\iint_S f(x, y, z)\, dS = \int_0^1 \int_0^1 f(x, y, 10)\, dx\, dy$.

b. If the surface S is given by $\{(x, y, z): 0 \le x \le 1, 0 \le y \le 1, z = x\}$, then $\iint_S f(x, y, z)\, dS = \int_0^1 \int_0^1 f(x, y, x)\, dx\, dy$.

c. The surface $\mathbf{r} = \langle v \cos u, v \sin u, v^2 \rangle$, for $0 \le u \le \pi$, $0 \le v \le 2$, is the same as the surface $\mathbf{r} = \langle \sqrt{v} \cos 2u, \sqrt{v} \sin 2u, v \rangle$, for $0 \le u \le \pi/2, 0 \le v \le 4$.

d. Given the standard parameterization of a sphere, the normal vectors $\mathbf{t}_u \times \mathbf{t}_v$ are outward normal vectors.

50–53. Miscellaneous surface integrals *Evaluate the following integrals using the method of your choice. Assume normal vectors point either outward or upward.*

50. $\iint_S \nabla \ln |\mathbf{r}| \cdot \mathbf{n}\, dS$, where S is the hemisphere $x^2 + y^2 + z^2 = a^2$, for $z \ge 0$, and where $\mathbf{r} = \langle x, y, z \rangle$

51. $\iint_S |\mathbf{r}|\, dS$, where S is the cylinder $x^2 + y^2 = 4$, for $0 \le z \le 8$, where $\mathbf{r} = \langle x, y, z \rangle$

52. $\iint_S xyz\, dS$, where S is that part of the plane $z = 6 - y$ that lies in the cylinder $x^2 + y^2 = 4$

53. $\iint_S \dfrac{\langle x, 0, z \rangle}{\sqrt{x^2 + z^2}} \cdot \mathbf{n}\, dS$, where S is the cylinder $x^2 + z^2 = a^2$, $|y| \le 2$

54. Cone and sphere The cone $z^2 = x^2 + y^2$, for $z \ge 0$, cuts the sphere $x^2 + y^2 + z^2 = 16$ along a curve C.

a. Find the surface area of the sphere below C, for $z \ge 0$.
b. Find the surface area of the sphere above C.
c. Find the surface area of the cone below C, for $z \ge 0$.

T 55. Cylinder and sphere Consider the sphere $x^2 + y^2 + z^2 = 4$ and the cylinder $(x - 1)^2 + y^2 = 1$, for $z \ge 0$. Find the surface area of the cylinder inside the sphere.

56. Flux on a tetrahedron Find the upward flux of the field $\mathbf{F} = \langle x, y, z \rangle$ across the plane $\dfrac{x}{a} + \dfrac{y}{b} + \dfrac{z}{c} = 1$ in the first octant, where a, b, and c are positive real numbers. Show that the flux equals c times the area of the base of the region. Interpret the result physically.

57. Flux across a cone Consider the field $\mathbf{F} = \langle x, y, z \rangle$ and the cone $z^2 = \dfrac{x^2 + y^2}{a^2}$, for $0 \le z \le 1$.

a. Show that when $a = 1$, the outward flux across the cone is zero. Interpret the result.
b. Find the outward flux (away from the z-axis), for any $a > 0$. Interpret the result.

58. Surface area formula for cones Find the general formula for the surface area of a cone with height h and base radius a (excluding the base).

59. Surface area formula for spherical cap A sphere of radius a is sliced parallel to the equatorial plane at a distance $a - h$ from the equatorial plane (see figure). Find the general formula for the surface area of the resulting spherical cap (excluding the base) with thickness h.

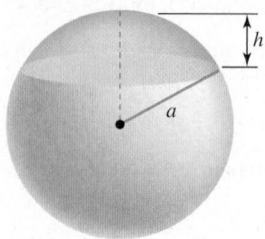

Explorations and Challenges

60. Radial fields and spheres Consider the radial field $\mathbf{F} = \mathbf{r}/|\mathbf{r}|^p$, where $\mathbf{r} = \langle x, y, z \rangle$ and p is a real number. Let S be the sphere of radius a centered at the origin. Show that the outward flux of $\mathbf{F}$ across the sphere is $4\pi/a^{p-3}$. It is instructive to do the calculation using both an explicit and a parametric description of the sphere.

61–63. Heat flux *The heat flow vector field for conducting objects is $\mathbf{F} = -k\nabla T$, where $T(x, y, z)$ is the temperature in the object and $k > 0$ is a constant that depends on the material. Compute the outward flux of $\mathbf{F}$ across the following surfaces S for the given temperature distributions. Assume $k = 1$.*

61. $T(x, y, z) = 100e^{-x-y}$; S consists of the faces of the cube $|x| \le 1$, $|y| \le 1$, $|z| \le 1$.

62. $T(x, y, z) = 100e^{-x^2-y^2-z^2}$; S is the sphere $x^2 + y^2 + z^2 = a^2$.

63. $T(x, y, z) = -\ln(x^2 + y^2 + z^2)$; S is the sphere $x^2 + y^2 + z^2 = a^2$.

64. Flux across a cylinder Let S be the cylinder $x^2 + y^2 = a^2$, for $-L \le z \le L$.

  a. Find the outward flux of the field $\mathbf{F} = \langle x, y, 0 \rangle$ across S.

  b. Find the outward flux of the field $\mathbf{F} = \dfrac{\langle x, y, 0 \rangle}{(x^2 + y^2)^{p/2}} = \dfrac{\mathbf{r}}{|\mathbf{r}|^p}$ across S, where $|\mathbf{r}|$ is the distance from the z-axis and p is a real number.

  c. In part (b), for what values of p is the outward flux finite as $a \to \infty$ (with L fixed)?

  d. In part (b), for what values of p is the outward flux finite as $L \to \infty$ (with a fixed)?

65. Flux across concentric spheres Consider the radial fields
$$\mathbf{F} = \dfrac{\langle x, y, z \rangle}{(x^2 + y^2 + z^2)^{p/2}} = \dfrac{\mathbf{r}}{|\mathbf{r}|^p},$$
where p is a real number. Let S consist of the spheres A and B centered at the origin with radii $0 < a < b$, respectively. The total outward flux across S consists of the flux out of S across the outer sphere B minus the flux into S across the inner sphere A.

  a. Find the total flux across S with $p = 0$. Interpret the result.

  b. Show that for $p = 3$ (an inverse square law), the flux across S is independent of a and b.

66–69. Mass and center of mass *Let S be a surface that represents a thin shell with density ρ. The moments about the coordinate planes (see Section 16.6) are $M_{yz} = \iint_S x\rho(x, y, z)\, dS$, $M_{xz} = \iint_S y\rho(x, y, z)\, dS$, and $M_{xy} = \iint_S z\rho(x, y, z)\, dS$. The coordinates of the center of mass of the shell are $\bar{x} = \dfrac{M_{yz}}{m}$, $\bar{y} = \dfrac{M_{xz}}{m}$, and $\bar{z} = \dfrac{M_{xy}}{m}$, where m is the mass of the shell. Find the mass and center of mass of the following shells. Use symmetry whenever possible.*

66. The constant-density hemispherical shell $x^2 + y^2 + z^2 = a^2$, $z \ge 0$

67. The constant-density cone with radius a, height h, and base in the xy-plane

68. The constant-density half-cylinder $x^2 + z^2 = a^2$, $-\dfrac{h}{2} \le y \le \dfrac{h}{2}$, $z \ge 0$

69. The cylinder $x^2 + y^2 = a^2$, $0 \le z \le 2$, with density $\rho(x, y, z) = 1 + z$

70. Outward normal to a sphere Show that $|\mathbf{t}_u \times \mathbf{t}_v| = a^2 \sin u$ for a sphere of radius a defined parametrically by $\mathbf{r}(u, v) = \langle a \sin u \cos v, a \sin u \sin v, a \cos u \rangle$, where $0 \le u \le \pi$ and $0 \le v \le 2\pi$.

71. Special case of surface integrals of scalar-valued functions Suppose a surface S is defined as $z = g(x, y)$ on a region R. Show that $\mathbf{t}_x \times \mathbf{t}_y = \langle -z_x, -z_y, 1 \rangle$ and that
$$\iint_S f(x, y, z)\, dS = \iint_R f(x, y, g(x, y))\sqrt{z_x^2 + z_y^2 + 1}\, dA.$$

72. Surfaces of revolution Suppose $y = f(x)$ is a continuous and positive function on $[a, b]$. Let S be the surface generated when the graph of f on $[a, b]$ is revolved about the x-axis.

  a. Show that S is described parametrically by $\mathbf{r}(u, v) = \langle u, f(u) \cos v, f(u) \sin v \rangle$, for $a \le u \le b$, $0 \le v \le 2\pi$.

  b. Find an integral that gives the surface area of S.

  c. Apply the result of part (b) to the surface generated with $f(x) = x^3$, for $1 \le x \le 2$.

  d. Apply the result of part (b) to the surface generated with $f(x) = (25 - x^2)^{1/2}$, for $3 \le x \le 4$.

73. Rain on roofs Let $z = s(x, y)$ define a surface over a region R in the xy-plane, where $z \ge 0$ on R. Show that the downward flux of the vertical vector field $\mathbf{F} = \langle 0, 0, -1 \rangle$ across S equals the area of R. Interpret the result physically.

74. Surface area of a torus

  a. Show that a torus with radii $R > r$ (see figure) may be described parametrically by $\mathbf{r}(u, v) = \langle (R + r \cos u) \cos v, (R + r \cos u) \sin v, r \sin u \rangle$, for $0 \le u \le 2\pi$, $0 \le v \le 2\pi$.

  b. Show that the surface area of the torus is $4\pi^2 Rr$.

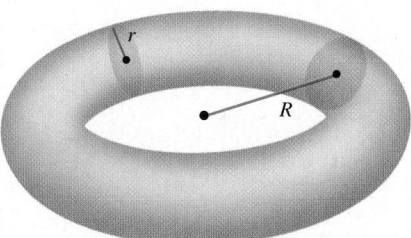

75. Surfaces of revolution—single variable Let f be differentiable and positive on the interval $[a, b]$. Let S be the surface generated when the graph of f on $[a, b]$ is revolved about the x-axis. Use Theorem 17.14 to show that the area of S (as given in Section 6.6) is
$$\int_a^b 2\pi f(x)\sqrt{1 + f'(x)^2}\, dx.$$

QUICK CHECK ANSWERS

1. A half-cylinder with height 1 and radius 2 with its axis along the z-axis **2.** A half-cone with height 10 and radius 10 **3.** A quarter-sphere with radius 4 **4.** $\sqrt{2}$ **5.** The cylinder $x^2 + y^2 = a^2$ does not represent a function, so z_x and z_y cannot be computed. **6.** The vector field is everywhere orthogonal to the hemisphere, so the hemisphere has maximum flux at every point. ◄

17.7 Stokes' Theorem

With the divergence, the curl, and surface integrals in hand, we are ready to present two of the crowning results of calculus. Fortunately, all the heavy lifting has been done. In this section, you will see Stokes' Theorem, and in the next section, we present the Divergence Theorem.

Stokes' Theorem

Stokes' Theorem is the three-dimensional version of the circulation form of Green's Theorem. Recall that if C is a closed simple piecewise-smooth oriented curve in the xy-plane enclosing a simply connected region R, and $\mathbf{F} = \langle f, g \rangle$ is a differentiable vector field on R, then Green's Theorem says that

$$\underbrace{\oint_C \mathbf{F} \cdot d\mathbf{r}}_{\text{circulation}} = \iint_R \underbrace{(g_x - f_y)}_{\text{curl or rotation}} dA.$$

The line integral on the left gives the circulation along the boundary of R. The double integral on the right sums the curl of the vector field over all points of R. If $\mathbf{F}$ represents a fluid flow, the theorem says that the cumulative rotation of the flow within R equals the circulation along the boundary.

In Stokes' Theorem, the plane region R in Green's Theorem becomes an oriented surface S in $\mathbb{R}^3$. The circulation integral in Green's Theorem remains a circulation integral, but now over the closed simple piecewise-smooth oriented curve C that forms the boundary of S. The double integral of the curl in Green's Theorem becomes a surface integral of the three-dimensional curl (**Figure 17.59**).

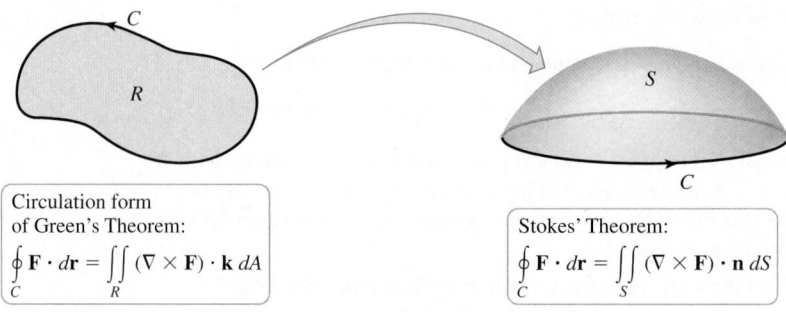

Circulation form
of Green's Theorem:
$$\oint_C \mathbf{F} \cdot d\mathbf{r} = \iint_R (\nabla \times \mathbf{F}) \cdot \mathbf{k} \, dA$$

Stokes' Theorem:
$$\oint_C \mathbf{F} \cdot d\mathbf{r} = \iint_S (\nabla \times \mathbf{F}) \cdot \mathbf{n} \, dS$$

Figure 17.59

Stokes' Theorem involves an oriented curve C and an oriented surface S on which there are two unit normal vectors at every point. These orientations must be consistent and the normal vectors must be chosen correctly. Here is the right-hand rule that relates the orientations of S and C and determines the choice of the normal vectors:

> If the fingers of your right hand curl in the positive direction around C, then your right thumb points in the (general) direction of the vectors normal to S (**Figure 17.60**).

A common situation occurs when C has a counterclockwise orientation when viewed from above; then the vectors normal to S point upward.

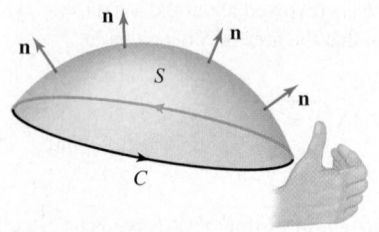

Figure 17.60

▶ The right-hand rule tells you which of two normal vectors at a point of S to use. Remember that the direction of normal vectors changes continuously on an oriented surface.

THEOREM 17.15   Stokes' Theorem
Let S be an oriented surface in $\mathbb{R}^3$ with a piecewise-smooth closed boundary C whose orientation is consistent with that of S. Assume $\mathbf{F} = \langle f, g, h \rangle$ is a vector field whose components have continuous first partial derivatives on S. Then

$$\oint_C \mathbf{F} \cdot d\mathbf{r} = \iint_S (\nabla \times \mathbf{F}) \cdot \mathbf{n} \, dS,$$

where $\mathbf{n}$ is the unit vector normal to S determined by the orientation of S.

QUICK CHECK 1 Suppose S is a region in the xy-plane with a boundary oriented counterclockwise. What is the normal to S? Explain why Stokes' Theorem becomes the circulation form of Green's Theorem. ◄

The meaning of Stokes' Theorem is much the same as for the circulation form of Green's Theorem: Under the proper conditions, the accumulated rotation of the vector field over the surface S (as given by the normal component of the curl) equals the net circulation on the boundary of S. An outline of the proof of Stokes' Theorem is given at the end of this section. First, we look at some special cases that give further insight into the theorem.

If $\mathbf{F}$ is a conservative vector field on a domain D, then it has a potential function φ such that $\mathbf{F} = \nabla\varphi$. Because $\nabla \times \nabla\varphi = \mathbf{0}$, it follows that $\nabla \times \mathbf{F} = \mathbf{0}$ (Theorem 17.11); therefore, the circulation integral is zero on all closed curves in D. Recall that the circulation integral is also a work integral for the force field $\mathbf{F}$, which emphasizes the fact that no work is done in moving an object on a closed path in a conservative force field. Among the important conservative vector fields are the radial fields $\mathbf{F} = \mathbf{r}/|\mathbf{r}|^p$, which generally have zero curl and zero circulation on closed curves.

> ➤ Recall that for a constant nonzero vector $\mathbf{a}$ and the position vector $\mathbf{r} = \langle x, y, z \rangle$, the field $\mathbf{F} = \mathbf{a} \times \mathbf{r}$ is a rotation field. In Example 1,
> $$\mathbf{F} = \langle 0, 1, 1 \rangle \times \langle x, y, z \rangle.$$

EXAMPLE 1   Verifying Stokes' Theorem Confirm that Stokes' Theorem holds for the vector field $\mathbf{F} = \langle z - y, x, -x \rangle$, where S is the hemisphere $x^2 + y^2 + z^2 = 4$, for $z \geq 0$, and C is the circle $x^2 + y^2 = 4$ oriented counterclockwise.

SOLUTION The orientation of C implies that vectors normal to S should point in the outward direction. The vector field is a rotation field $\mathbf{a} \times \mathbf{r}$, where $\mathbf{a} = \langle 0, 1, 1 \rangle$ and $\mathbf{r} = \langle x, y, z \rangle$; so the axis of rotation points in the direction of the vector $\langle 0, 1, 1 \rangle$ (Figure 17.61). We first compute the circulation integral in Stokes' Theorem. The curve C with the given orientation is parameterized as $\mathbf{r}(t) = \langle 2 \cos t, 2 \sin t, 0 \rangle$, for $0 \leq t \leq 2\pi$; therefore, $\mathbf{r}'(t) = \langle -2 \sin t, 2 \cos t, 0 \rangle$. The circulation integral is

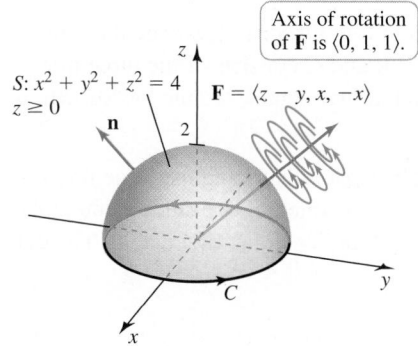

Figure 17.61

$$\oint_C \mathbf{F} \cdot d\mathbf{r} = \int_0^{2\pi} \mathbf{F} \cdot \mathbf{r}'(t)\, dt \qquad \text{Definition of line integral}$$

$$= \int_0^{2\pi} \langle \underbrace{z - y}_{-2\sin t}, x, -x \rangle \cdot \langle -2 \sin t, 2 \cos t, 0 \rangle\, dt \quad \text{Substitute.}$$

$$= \int_0^{2\pi} 4(\sin^2 t + \cos^2 t)\, dt \qquad \text{Simplify.}$$

$$= 4 \int_0^{2\pi} dt \qquad \sin^2 t + \cos^2 t = 1$$

$$= 8\pi. \qquad \text{Evaluate integral.}$$

The surface integral requires computing the curl of the vector field:

$$\nabla \times \mathbf{F} = \nabla \times \langle z - y, x, -x \rangle = \begin{vmatrix} \mathbf{i} & \mathbf{j} & \mathbf{k} \\ \dfrac{\partial}{\partial x} & \dfrac{\partial}{\partial y} & \dfrac{\partial}{\partial z} \\ z - y & x & -x \end{vmatrix} = \langle 0, 2, 2 \rangle.$$

Recall from Section 17.6 (Table 17.3) that the required outward normal to the hemisphere is $\langle x/z, y/z, 1 \rangle$. The region of integration is the base of the hemisphere in the xy-plane, which is

$$R = \{(x, y): x^2 + y^2 \leq 4\}, \text{ or, in polar coordinates,}$$
$$\{(r, \theta): 0 \leq r \leq 2, 0 \leq \theta \leq 2\pi\}.$$

Combining these results, the surface integral in Stokes' Theorem is

$$\iint_S \underbrace{(\nabla \times \mathbf{F})}_{\langle 0, 2, 2 \rangle} \cdot \mathbf{n}\, dS = \iint_R \langle 0, 2, 2 \rangle \cdot \left\langle \frac{x}{z}, \frac{y}{z}, 1 \right\rangle dA \qquad \begin{array}{l}\text{Substitute and convert to a}\\ \text{double integral over } R.\end{array}$$

$$= \iint_R \left(\frac{2y}{\sqrt{4 - x^2 - y^2}} + 2 \right) dA \qquad \begin{array}{l}\text{Simplify and use}\\ z = \sqrt{4 - x^2 - y^2}.\end{array}$$

$$= \int_0^{2\pi} \int_0^2 \left(\frac{2r \sin \theta}{\sqrt{4 - r^2}} + 2 \right) r\, dr\, d\theta. \quad \text{Convert to polar coordinates.}$$

We integrate first with respect to θ because the integral of $\sin\theta$ from 0 to 2π is zero and the first term in the integral is eliminated. Therefore, the surface integral reduces to

▶ In eliminating the first term of this double integral, we note that the improper integral $\int_0^2 \dfrac{r^2}{\sqrt{4-r^2}}\,dr$ has a finite value.

$$\iint\limits_S (\nabla\times\mathbf{F})\cdot\mathbf{n}\,dS = \int_0^2\int_0^{2\pi}\left(\frac{2r^2\sin\theta}{\sqrt{4-r^2}}+2r\right)d\theta\,dr$$

$$= \int_0^2\int_0^{2\pi} 2r\,d\theta\,dr \qquad \int_0^{2\pi}\sin\theta\,d\theta = 0$$

$$= 4\pi\int_0^2 r\,dr \qquad\qquad \text{Evaluate inner integral.}$$

$$= 8\pi. \qquad\qquad\qquad \text{Evaluate outer integral.}$$

Computed either as a line integral or as a surface integral, the vector field has a positive circulation along the boundary of S, which is produced by the net rotation of the field over the surface S.

Related Exercises 5–6 ◀

In Example 1, it was possible to evaluate both the line integral and the surface integral that appear in Stokes' Theorem. Often the theorem provides an easier way to evaluate difficult line integrals.

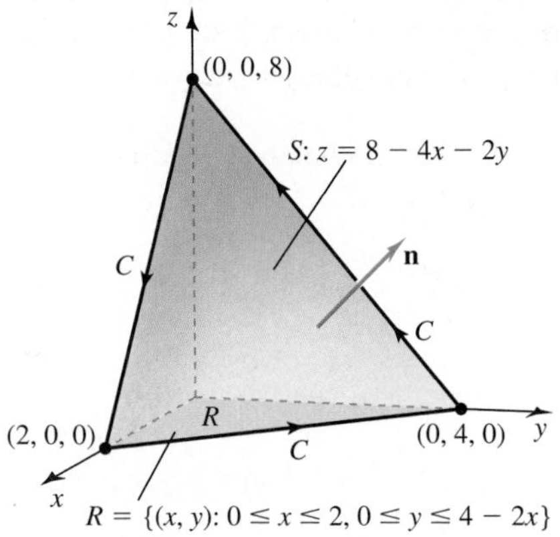

Figure 17.62

▶ Recall that for an explicitly defined surface S given by $z = s(x,y)$ over a region R with $\mathbf{F} = \langle f, g, h\rangle$,

$$\iint\limits_S \mathbf{F}\cdot\mathbf{n}\,dS = \iint\limits_R (-fz_x - gz_y + h)\,dA.$$

In Example 2, $\mathbf{F}$ is replaced with $\nabla\times\mathbf{F}$.

EXAMPLE 2 Using Stokes' Theorem to evaluate a line integral Evaluate the line integral $\oint_C \mathbf{F}\cdot d\mathbf{r}$, where $\mathbf{F} = z\,\mathbf{i} - z\,\mathbf{j} + (x^2-y^2)\,\mathbf{k}$ and C consists of the three line segments that bound the plane $z = 8 - 4x - 2y$ in the first octant, oriented as shown in Figure 17.62.

SOLUTION Evaluating the line integral directly involves parameterizing the three line segments. Instead, we use Stokes' Theorem to convert the line integral to a surface integral, where S is that portion of the plane $z = 8 - 4x - 2y$ that lies in the first octant. The curl of the vector field is

$$\nabla\times\mathbf{F} = \nabla\times\langle z, -z, x^2-y^2\rangle = \begin{vmatrix} \mathbf{i} & \mathbf{j} & \mathbf{k} \\ \dfrac{\partial}{\partial x} & \dfrac{\partial}{\partial y} & \dfrac{\partial}{\partial z} \\ z & -z & x^2-y^2 \end{vmatrix} = \langle 1-2y, 1-2x, 0\rangle.$$

The appropriate vector normal to the plane $z = 8 - 4x - 2y$ is $\langle -z_x, -z_y, 1\rangle = \langle 4, 2, 1\rangle$, which points upward, consistent with the orientation of C. The triangular region R in the xy-plane beneath S is found by setting $z = 0$ in the equation of the plane; we find that $R = \{(x,y): 0\le x\le 2, 0\le y\le 4-2x\}$. The surface integral in Stokes' Theorem may now be evaluated:

$$\iint\limits_S (\nabla\times\mathbf{F})\cdot\mathbf{n}\,dS = \iint\limits_R \underbrace{\langle 1-2y, 1-2x, 0\rangle}_{\langle 1-2y,\,1-2x,\,0\rangle}\cdot\langle 4,2,1\rangle\,dA \quad \text{Substitute and convert to a double integral over } R.$$

$$= \int_0^2\int_0^{4-2x}(6-4x-8y)\,dy\,dx \quad \text{Simplify.}$$

$$= -\frac{88}{3}. \quad \text{Evaluate integrals.}$$

The circulation around the boundary of R is negative, indicating a net circulation in the clockwise direction on C (looking from above).

Related Exercises 13, 16 ◀

In other situations, Stokes' Theorem may be used to convert a difficult surface integral into a relatively easy line integral, as illustrated in the next example.

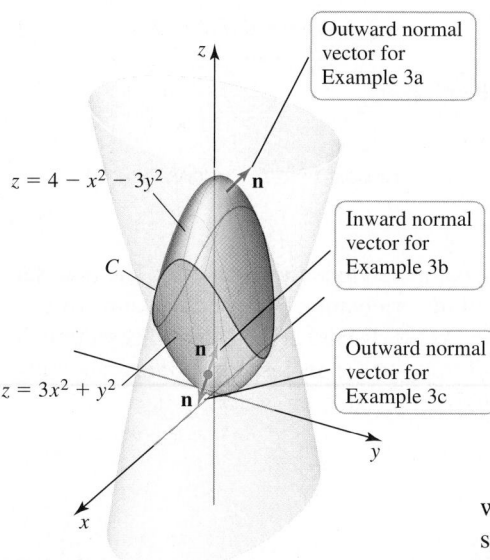

z = 4 − x² − 3y²

Outward normal vector for Example 3a

Inward normal vector for Example 3b

Outward normal vector for Example 3c

z = 3x² + y²

Figure 17.63

EXAMPLE 3   Using Stokes' Theorem to evaluate a surface integral Evaluate $\iint_S (\nabla \times \mathbf{F}) \cdot \mathbf{n}\, dS$, where $\mathbf{F} = -y\,\mathbf{i} + x\,\mathbf{j} + z\,\mathbf{k}$, in the following cases.

a. S is the part of the paraboloid $z = 4 - x^2 - 3y^2$ that lies within the paraboloid $z = 3x^2 + y^2$ (the blue surface in **Figure 17.63**). Assume $\mathbf{n}$ points in the upward direction on S.

b. S is the part of the paraboloid $z = 3x^2 + y^2$ that lies within the paraboloid $z = 4 - x^2 - 3y^2$, with $\mathbf{n}$ pointing in the upward direction on S.

c. S is the surface in part (b), but $\mathbf{n}$ pointing in the downward direction on S.

SOLUTION

a. Finding a parametric description for S is challenging, so we use Stokes' Theorem to convert the surface integral into a line integral along the curve C that bounds S. Note that C is the intersection between the paraboloids $z = 4 - x^2 - 3y^2$ and $z = 3x^2 + y^2$. Eliminating z from these equations, we find that the projection of C onto the xy-plane is the circle $x^2 + y^2 = 1$, which suggests that we choose $x = \cos t$ and $y = \sin t$ for the x- and y-components of the equations for C. To find the z-component, we substitute x and y into the equation of either paraboloid. Choosing $z = 3x^2 + y^2$, we find that a parametric description of C is $\mathbf{r}(t) = \langle \cos t, \sin t, 3\cos^2 t + \sin^2 t \rangle$; note that C is oriented in the counterclockwise direction, consistent with the orientation of S.

To evaluate the line integral in Stokes' Theorem, it is helpful to first compute $\mathbf{F} \cdot \mathbf{r}'(t)$. Along C, the vector field is $\mathbf{F} = \langle -y, x, z \rangle = \langle -\sin t, \cos t, 3\cos^2 t + \sin^2 t \rangle$. Differentiating $\mathbf{r}$ yields $\mathbf{r}'(t) = \langle -\sin t, \cos t, -4\cos t \sin t \rangle$, which leads to

$$\mathbf{F} \cdot \mathbf{r}'(t) = \langle -\sin t, \cos t, 3\cos^2 t + \sin^2 t \rangle \cdot \langle -\sin t, \cos t, -4\cos t \sin t \rangle$$
$$= \underbrace{\sin^2 t + \cos^2 t}_{1} - 12\cos^3 t \sin t - 4\sin^3 t \cos t.$$

Noting that $\sin^2 t + \cos^2 t = 1$, we are ready to evaluate the integral:

$$\iint_S (\nabla \times \mathbf{F}) \cdot \mathbf{n}\, dS = \oint_C \mathbf{F} \cdot d\mathbf{r} \qquad \text{Stokes' Theorem}$$

$$= \int_0^{2\pi} \mathbf{F} \cdot \mathbf{r}'(t)\, dt \qquad \text{Definition of line integral}$$

$$= \int_0^{2\pi} (1 - 12\cos^3 t \sin t - 4\cos t \sin^3 t)\, dt \quad \text{Substitute.}$$

$$= \int_0^{2\pi} 1\, dt - \underbrace{\int_0^{2\pi} 12\cos^3 t \sin t\, dt}_{0} - \underbrace{\int_0^{2\pi} 4\cos t \sin^3 t\, dt}_{0}$$

Split integral.

$$= 2\pi. \qquad \text{Evaluate integrals.}$$

A standard substitution in the last two integrals of the final step shows that both integrals equal 0.

b. Because the lower surface ($z = 3x^2 + y^2$) shares the same boundary C with the upper surface ($z = 4 - x^2 - 3y^2$), and because both surfaces have an upward-pointing normal vector, the line integral resulting from an application of Stokes' Theorem is identical to the integral in part (a). For this surface S with its associated normal vector, we conclude that $\iint_S (\nabla \times \mathbf{F}) \cdot \mathbf{n}\, dS = \oint_C \mathbf{F} \cdot d\mathbf{r} = 2\pi$. In fact, the value of this integral is 2π for *any* surface whose boundary is C and whose normal vectors point in the upward direction.

▶ Recall that $x = \cos t, y = \sin t$ is a standard parameterization for the unit circle centered at the origin with counterclockwise orientation. The parameterization $x = \sin t, y = \cos t$ reverses the orientation.

c. In this case, $\mathbf{n}$ points downward. We use the parameterization $\mathbf{r}(t) = \langle \sin t, \cos t, 3\cos^2 t + \sin^2 t \rangle$ for C so that C is oriented in the clockwise direction, consistent with the orientation of S. You should verify that, when duplicating the calculations in part (a) with a new description for C, we have

$$\mathbf{F} \cdot \mathbf{r}'(t) = \underbrace{-\sin^2 t - \cos^2 t}_{-1} - 12\cos^3 t \sin t - 4\sin^3 t \cos t.$$

Therefore, the required integral is

$$\iint_S (\nabla \times \mathbf{F}) \cdot \mathbf{n} \, dS = \oint_C \mathbf{F} \cdot d\mathbf{r} = \int_0^{2\pi} \mathbf{F} \cdot \mathbf{r}'(t) \, dt$$

$$= \int_0^{2\pi} (-1 - 12 \cos^3 t \sin t - 4 \cos t \sin^3 t) \, dt$$

$$= -2\pi.$$

QUICK CHECK 2 In Example 3a, we used the parameterization $\mathbf{r}(t) = \langle \cos t, \sin t, 3 \cos^2 t + \sin^2 t \rangle$ for C. Confirm that the parameterization $C: \mathbf{r}(t) = \langle \cos t, \sin t, 4 - \cos^2 t - 3 \sin^2 t \rangle$ also results in an answer of 2π. ◄

This result is perhaps not surprising when compared to parts (a) and (b): The reversal of the orientation of S requires a reversal of the orientation of C, and we know from Section 17.2 that $\int_C \mathbf{F} \cdot d\mathbf{r} = -\int_{-C} \mathbf{F} \cdot d\mathbf{r}$. As we discuss at the end of this section, it follows that the surface integral over the closed surface enclosed by *both* paraboloids (with normal vectors everywhere outward) has the value $2\pi - 2\pi = 0$.

Related Exercises 21–22 ◄

Interpreting the Curl

Stokes' Theorem leads to another interpretation of the curl at a point in a vector field. We need the idea of the **average circulation**. If C is the boundary of an oriented surface S, we define the average circulation of $\mathbf{F}$ over S as

$$\frac{1}{\text{area of } S} \oint_C \mathbf{F} \cdot d\mathbf{r} = \frac{1}{\text{area of } S} \iint_S (\nabla \times \mathbf{F}) \cdot \mathbf{n} \, dS,$$

where Stokes' Theorem is used to convert the circulation integral to a surface integral.

First consider a general rotation field $\mathbf{F} = \mathbf{a} \times \mathbf{r}$, where $\mathbf{a} = \langle a_1, a_2, a_3 \rangle$ is a constant nonzero vector and $\mathbf{r} = \langle x, y, z \rangle$. Recall that $\mathbf{F}$ describes the rotation about an axis in the direction of $\mathbf{a}$ with angular speed $\omega = |\mathbf{a}|$. We also showed that $\mathbf{F}$ has a constant curl, $\nabla \times \mathbf{F} = \nabla \times (\mathbf{a} \times \mathbf{r}) = 2\mathbf{a}$. We now take S to be a small circular disk centered at a point P, whose normal vector $\mathbf{n}$ makes an angle θ with the axis $\mathbf{a}$ (Figure 17.64). Let C be the boundary of S with a counterclockwise orientation.

The average circulation of this vector field on S is

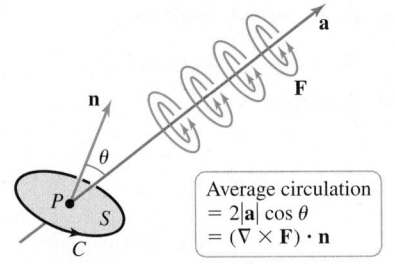

Figure 17.64

Average circulation
$= 2|\mathbf{a}| \cos \theta$
$= (\nabla \times \mathbf{F}) \cdot \mathbf{n}$

$$\frac{1}{\text{area of } S} \iint_S \underbrace{(\nabla \times \mathbf{F}) \cdot \mathbf{n}}_{\text{constant}} \, dS \qquad \text{Definition}$$

$$= \frac{1}{\text{area of } S} (\nabla \times \mathbf{F}) \cdot \mathbf{n} \cdot \text{area of } S \qquad \iint_S dS = \text{area of } S$$

$$= \underbrace{(\nabla \times \mathbf{F})}_{2\mathbf{a}} \cdot \mathbf{n} \qquad \text{Simplify.}$$

$$= 2|\mathbf{a}| \cos \theta. \qquad |\mathbf{n}| = 1, |\nabla \times \mathbf{F}| = 2|\mathbf{a}|$$

> Recall that $\mathbf{n}$ is a unit normal vector with $|\mathbf{n}| = 1$. By definition, the dot product gives $\mathbf{a} \cdot \mathbf{n} = |\mathbf{a}| \cos \theta$.

If the normal vector $\mathbf{n}$ is aligned with $\nabla \times \mathbf{F}$ (which is parallel to $\mathbf{a}$), then $\theta = 0$ and the average circulation on S has its maximum value of $2|\mathbf{a}|$. However, if the vector normal to the surface S is orthogonal to the axis of rotation ($\theta = \pi/2$), the average circulation is zero.

We see that for a general rotation field $\mathbf{F} = \mathbf{a} \times \mathbf{r}$, the curl of $\mathbf{F}$ has the following interpretations, where S is a small disk centered at a point P with a normal vector $\mathbf{n}$.

- The scalar component of $\nabla \times \mathbf{F}$ at P in the direction of $\mathbf{n}$, which is $(\nabla \times \mathbf{F}) \cdot \mathbf{n} = 2|\mathbf{a}| \cos \theta$, is the average circulation of $\mathbf{F}$ on S.

- The direction of $\nabla \times \mathbf{F}$ at P is the direction that maximizes the average circulation of $\mathbf{F}$ on S. Equivalently, it is the direction in which the axis of a paddle wheel should be oriented to obtain the maximum angular speed.

A similar argument may be applied to a general vector field (with a variable curl) to give an analogous interpretation of the curl at a point (Exercise 48).

EXAMPLE 4 **Horizontal channel flow** Consider the velocity field
$\mathbf{v} = \langle 0, 1 - x^2, 0 \rangle$, for $|x| \le 1$ and $|z| \le 1$, which represents a horizontal flow in the
y-direction (**Figure 17.65a**).

a. Suppose you place a paddle wheel at the point $P\left(\frac{1}{2}, 0, 0\right)$. Using physical arguments,
in which of the coordinate directions should the axis of the wheel point in order for the
wheel to spin? In which direction does it spin? What happens if you place the wheel at
$Q\left(-\frac{1}{2}, 0, 0\right)$?

b. Compute and graph the curl of $\mathbf{v}$ and provide an interpretation.

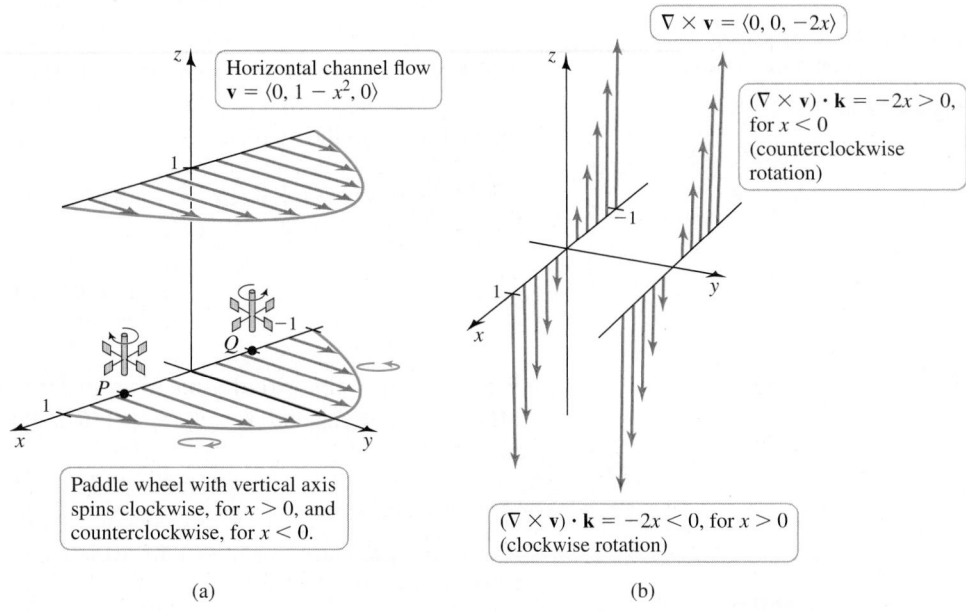

$\nabla \times \mathbf{v} = \langle 0, 0, -2x \rangle$

Horizontal channel flow
$\mathbf{v} = \langle 0, 1 - x^2, 0 \rangle$

$(\nabla \times \mathbf{v}) \cdot \mathbf{k} = -2x > 0$,
for $x < 0$
(counterclockwise
rotation)

Paddle wheel with vertical axis
spins clockwise, for $x > 0$, and
counterclockwise, for $x < 0$.

$(\nabla \times \mathbf{v}) \cdot \mathbf{k} = -2x < 0$, for $x > 0$
(clockwise rotation)

(a)   (b)

Figure 17.65

SOLUTION

a. If the axis of the wheel is aligned with the x-axis at P, the flow strikes the upper and
lower halves of the wheel symmetrically and the wheel does not spin. If the axis of the
wheel is aligned with the y-axis, the flow is parallel to the axis of the wheel and the
wheel does not spin. If the axis of the wheel is aligned with the z-axis at P, the flow in
the y-direction is greater for $x < \frac{1}{2}$ than it is for $x > \frac{1}{2}$. Therefore, a wheel located at
$P\left(\frac{1}{2}, 0, 0\right)$ spins in the clockwise direction, looking from above (Figure 17.65a). Using a similar argument, we conclude that a vertically oriented paddle wheel placed at
$Q\left(-\frac{1}{2}, 0, 0\right)$ spins in the counterclockwise direction (when viewed from above).

b. A short calculation shows that

$$\nabla \times \mathbf{v} = \begin{vmatrix} \mathbf{i} & \mathbf{j} & \mathbf{k} \\ \dfrac{\partial}{\partial x} & \dfrac{\partial}{\partial y} & \dfrac{\partial}{\partial z} \\ 0 & 1 - x^2 & 0 \end{vmatrix} = -2x\,\mathbf{k}.$$

QUICK CHECK 3 In Example 4, explain
why a paddle wheel with its axis
aligned with the z-axis does not spin
when placed on the y-axis. ◄

As shown in **Figure 17.65b**, the curl points in the z-direction, which is the direction of
the paddle wheel axis that gives the maximum angular speed of the wheel. Consider
the z-component of the curl, which is $(\nabla \times \mathbf{v}) \cdot \mathbf{k} = -2x$. At $x = 0$, this component
is zero, meaning the wheel does not spin at any point along the y-axis when its axis is
aligned with the z-axis. For $x > 0$, we see that $(\nabla \times \mathbf{v}) \cdot \mathbf{k} < 0$, which corresponds
to clockwise rotation of the vector field. For $x < 0$, we have $(\nabla \times \mathbf{v}) \cdot \mathbf{k} > 0$, corresponding to counterclockwise rotation.

Related Exercise 26 ◄

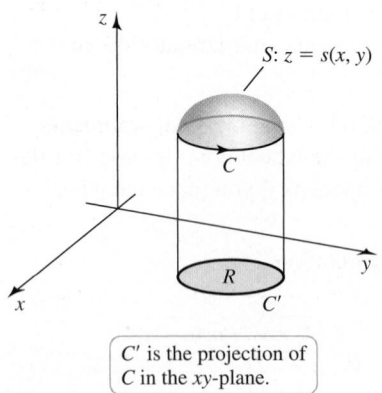

C' is the projection of C in the xy-plane.

Figure 17.66

Proof of Stokes' Theorem

The proof of the most general case of Stokes' Theorem is intricate. However, a proof of a special case is instructive and relies on several previous results.

Consider the case in which the surface S is the graph of the function $z = s(x, y)$, defined on a region in the xy-plane. Let C be the curve that bounds S with a counterclockwise orientation, let R be the projection of S in the xy-plane, and let C' be the projection of C in the xy-plane (Figure 17.66).

Letting $\mathbf{F} = \langle f, g, h \rangle$, the line integral in Stokes' Theorem is

$$\oint_C \mathbf{F} \cdot d\mathbf{r} = \oint_C f \, dx + g \, dy + h \, dz.$$

The key observation for this integral is that along C (which is the boundary of S), $dz = z_x \, dx + z_y \, dy$. Making this substitution, we convert the line integral on C to a line integral on C' in the xy-plane:

$$\oint_C \mathbf{F} \cdot d\mathbf{r} = \oint_{C'} f \, dx + g \, dy + \underbrace{h(z_x \, dx + z_y \, dy)}_{dz}$$

$$= \oint_{C'} \underbrace{(f + hz_x)}_{M(x, y)} dx + \underbrace{(g + hz_y)}_{N(x, y)} dy.$$

We now apply the circulation form of Green's Theorem to this line integral with $M(x, y) = f + hz_x$ and $N(x, y) = g + hz_y$; the result is

$$\oint_{C'} M \, dx + N \, dy = \iint_R (N_x - M_y) \, dA.$$

A careful application of the Chain Rule (remembering that z is a function of x and y, Exercise 49) reveals that

$$M_y = f_y + f_z z_y + h z_{xy} + z_x (h_y + h_z z_y) \quad \text{and}$$
$$N_x = g_x + g_z z_x + h z_{yx} + z_y (h_x + h_z z_x).$$

Making these substitutions in the line integral and simplifying (note that $z_{xy} = z_{yx}$ is needed), we have

$$\oint_C \mathbf{F} \cdot d\mathbf{r} = \iint_R (z_x(g_z - h_y) + z_y(h_x - f_z) + (g_x - f_y)) \, dA. \tag{1}$$

Now let's look at the surface integral in Stokes' Theorem. The upward vector normal to the surface is $\langle -z_x, -z_y, 1 \rangle$. Substituting the components of $\nabla \times \mathbf{F}$, the surface integral takes the form

$$\iint_S (\nabla \times \mathbf{F}) \cdot \mathbf{n} \, dS = \iint_R ((h_y - g_z)(-z_x) + (f_z - h_x)(-z_y) + (g_x - f_y)) \, dA,$$

which upon rearrangement becomes the integral in (1).

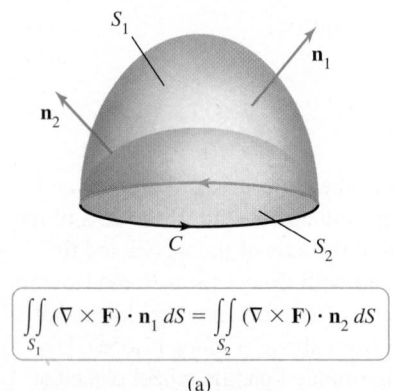

$$\iint_{S_1} (\nabla \times \mathbf{F}) \cdot \mathbf{n}_1 \, dS = \iint_{S_2} (\nabla \times \mathbf{F}) \cdot \mathbf{n}_2 \, dS$$

(a)

Two Final Notes on Stokes' Theorem

1. Stokes' Theorem allows a surface integral $\iint_S (\nabla \times \mathbf{F}) \cdot \mathbf{n} \, dS$ to be evaluated using only the values of the vector field on the boundary C. This means that if a closed curve C is the boundary of two different smooth oriented surfaces S_1 and S_2, which both have an orientation consistent with that of C, then the integrals of $(\nabla \times \mathbf{F}) \cdot \mathbf{n}$ on the two surfaces are equal; that is,

$$\iint_{S_1} (\nabla \times \mathbf{F}) \cdot \mathbf{n}_1 \, dS = \iint_{S_2} (\nabla \times \mathbf{F}) \cdot \mathbf{n}_2 \, dS,$$

where $\mathbf{n}_1$ and $\mathbf{n}_2$ are the respective unit normal vectors consistent with the orientation of the surfaces (Figure 17.67a; see Example 3).

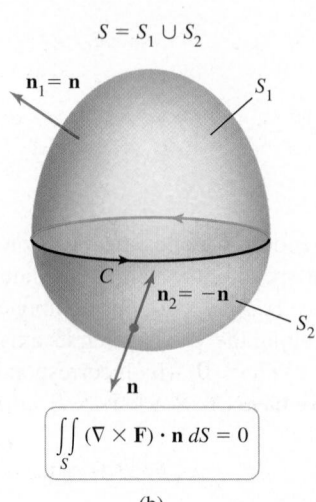

$$\iint_S (\nabla \times \mathbf{F}) \cdot \mathbf{n} \, dS = 0$$

(b)

Figure 17.67

Now let's take a different perspective. Suppose S is a *closed* surface consisting of S_1 and S_2 with a common boundary curve C (**Figure 17.67b**). Let $\mathbf{n}$ represent the outward unit normal vector for the entire surface S. It follows that $\mathbf{n}$ points in the same direction as $\mathbf{n}_1$ and in the direction opposite to that of $\mathbf{n}_2$ (Figure 17.67b). Therefore, $\iint_{S_1} (\nabla \times \mathbf{F}) \cdot \mathbf{n} \, dS$ and $\iint_{S_2} (\nabla \times \mathbf{F}) \cdot \mathbf{n} \, dS$ are equal in magnitude and of opposite sign, from which we conclude that

$$\iint_S (\nabla \times \mathbf{F}) \cdot \mathbf{n} \, dS = \iint_{S_1} (\nabla \times \mathbf{F}) \cdot \mathbf{n} \, dS + \iint_{S_2} (\nabla \times \mathbf{F}) \cdot \mathbf{n} \, dS = 0.$$

This argument can be adapted to show that $\iint_S (\nabla \times \mathbf{F}) \cdot \mathbf{n} \, dS = 0$ over any closed oriented surface S (Exercise 50).

2. We can now resolve an assertion made in Section 17.5. There we proved (Theorem 17.11) that if $\mathbf{F}$ is a conservative vector field, then $\nabla \times \mathbf{F} = \mathbf{0}$; we claimed, but did not prove, that the converse is true. The converse follows directly from Stokes' Theorem.

THEOREM 17.16   Curl F $=$ 0 Implies F Is Conservative

Suppose $\nabla \times \mathbf{F} = \mathbf{0}$ throughout an open simply connected region D of $\mathbb{R}^3$. Then $\oint_C \mathbf{F} \cdot d\mathbf{r} = 0$ on all closed simple smooth curves C in D, and $\mathbf{F}$ is a conservative vector field on D.

Proof: Given a closed simple smooth curve C, an advanced result states that C is the boundary of at least one smooth oriented surface S in D. By Stokes' Theorem,

$$\oint_C \mathbf{F} \cdot d\mathbf{r} = \iint_S \underbrace{(\nabla \times \mathbf{F}) \cdot \mathbf{n} \, dS}_{0} = 0.$$

Because the line integral equals zero over all such curves in D, the vector field is conservative on D by Theorem 17.6. ◀

SECTION 17.7 EXERCISES

Getting Started

1. Explain the meaning of the integral $\oint_C \mathbf{F} \cdot d\mathbf{r}$ in Stokes' Theorem.

2. Explain the meaning of the integral $\iint_S (\nabla \times \mathbf{F}) \cdot \mathbf{n} \, dS$ in Stokes' Theorem.

3. Explain the meaning of Stokes' Theorem.

4. Why does a conservative vector field produce zero circulation around a closed curve?

Practice Exercises

5–10. Verifying Stokes' Theorem *Verify that the line integral and the surface integral of Stokes' Theorem are equal for the following vector fields, surfaces S, and closed curves C. Assume C has counterclockwise orientation and S has a consistent orientation.*

5. $\mathbf{F} = \langle y, -x, 10 \rangle$; S is the upper half of the sphere $x^2 + y^2 + z^2 = 1$ and C is the circle $x^2 + y^2 = 1$ in the xy-plane.

6. $\mathbf{F} = \langle 0, -x, y \rangle$; S is the upper half of the sphere $x^2 + y^2 + z^2 = 4$ and C is the circle $x^2 + y^2 = 4$ in the xy-plane.

7. $\mathbf{F} = \langle x, y, z \rangle$; S is the paraboloid $z = 8 - x^2 - y^2$, for $0 \leq z \leq 8$, and C is the circle $x^2 + y^2 = 8$ in the xy-plane.

8. $\mathbf{F} = \langle 2z, -4x, 3y \rangle$; S is the cap of the sphere $x^2 + y^2 + z^2 = 169$ above the plane $z = 12$ and C is the boundary of S.

9. $\mathbf{F} = \langle y - z, z - x, x - y \rangle$; S is the cap of the sphere $x^2 + y^2 + z^2 = 16$ above the plane $z = \sqrt{7}$ and C is the boundary of S.

10. $\mathbf{F} = \langle -y, -x - z, y - x \rangle$; S is the part of the plane $z = 6 - y$ that lies in the cylinder $x^2 + y^2 = 16$ and C is the boundary of S.

11–16. Stokes' Theorem for evaluating line integrals *Evaluate the line integral $\oint_C \mathbf{F} \cdot d\mathbf{r}$ by evaluating the surface integral in Stokes' Theorem with an appropriate choice of S. Assume C has a counterclockwise orientation.*

11. $\mathbf{F} = \langle 2y, -z, x \rangle$; C is the circle $x^2 + y^2 = 12$ in the plane $z = 0$.

12. $\mathbf{F} = \langle y, xz, -y \rangle$; C is the ellipse $x^2 + y^2/4 = 1$ in the plane $z = 1$.

13. $\mathbf{F} = \langle x^2 - z^2, y, 2xz \rangle$; C is the boundary of the plane $z = 4 - x - y$ in the first octant.

14. $\mathbf{F} = \langle x^2 - y^2, z^2 - x^2, y^2 - z^2 \rangle$; C is the boundary of the square $|x| \leq 1, |y| \leq 1$ in the plane $z = 0$.

15. $\mathbf{F} = \langle y^2, -z^2, x \rangle$; C is the circle $\mathbf{r}(t) = \langle 3 \cos t, 4 \cos t, 5 \sin t \rangle$, for $0 \le t \le 2\pi$.

16. $\mathbf{F} = \langle 2xy \sin z, x^2 \sin z, x^2 y \cos z \rangle$; C is the boundary of the plane $z = 8 - 2x - 4y$ in the first octant.

17–24. Stokes' Theorem for evaluating surface integrals *Evaluate the line integral in Stokes' Theorem to determine the value of the surface integral $\iint_S (\nabla \times \mathbf{F}) \cdot \mathbf{n} \, dS$. Assume $\mathbf{n}$ points in an upward direction.*

17. $\mathbf{F} = \langle x, y, z \rangle$; S is the upper half of the ellipsoid $x^2/4 + y^2/9 + z^2 = 1$.

18. $\mathbf{F} = \mathbf{r}/|\mathbf{r}|$; S is the paraboloid $x = 9 - y^2 - z^2$, for $0 \le x \le 9$ (excluding its base), and $\mathbf{r} = \langle x, y, z \rangle$.

19. $\mathbf{F} = \langle 2y, -z, x - y - z \rangle$; S is the cap of the sphere $x^2 + y^2 + z^2 = 25$, for $3 \le x \le 5$ (excluding its base).

20. $\mathbf{F} = \langle x + y, y + z, z + x \rangle$; S is the tilted disk enclosed by $\mathbf{r}(t) = \langle \cos t, 2 \sin t, \sqrt{3} \cos t \rangle$.

21. $\mathbf{F} = \langle y, z - x, -y \rangle$; S is the part of the paraboloid $z = 2 - x^2 - 2y^2$ that lies within the cylinder $x^2 + y^2 = 1$.

22. $\mathbf{F} = \langle 4x, -8z, 4y \rangle$; S is the part of the paraboloid $z = 1 - 2x^2 - 3y^2$ that lies within the paraboloid $z = 2x^2 + y^2$.

23. $\mathbf{F} = \langle y, 1, z \rangle$; S is the part of the surface $z = 2\sqrt{x}$ that lies within the cone $z = \sqrt{x^2 + y^2}$.

24. $\mathbf{F} = \langle e^x, 1/z, y \rangle$; S is the part of the surface $z = 4 - 3y^2$ that lies within the paraboloid $z = x^2 + y^2$.

25–28. Interpreting and graphing the curl *For the following velocity fields, compute the curl, make a sketch of the curl, and interpret the curl.*

25. $\mathbf{v} = \langle 0, 0, y \rangle$ **26.** $\mathbf{v} = \langle 1 - z^2, 0, 0 \rangle$

27. $\mathbf{v} = \langle -2z, 0, 1 \rangle$ **28.** $\mathbf{v} = \langle 0, -z, y \rangle$

29. Explain why or why not Determine whether the following statements are true and give an explanation or counterexample.

    a. A paddle wheel with its axis in the direction $\langle 0, 1, -1 \rangle$ would not spin when put in the vector field $\mathbf{F} = \langle 1, 1, 2 \rangle \times \langle x, y, z \rangle$.

    b. Stokes' Theorem relates the flux of a vector field $\mathbf{F}$ across a surface to values of $\mathbf{F}$ on the boundary of the surface.

    c. A vector field of the form $\mathbf{F} = \langle a + f(x), b + g(y), c + h(z) \rangle$, where a, b, and c are constants, has zero circulation on a closed curve.

    d. If a vector field has zero circulation on all simple closed smooth curves C in a region D, then $\mathbf{F}$ is conservative on D.

30–33. Conservative fields *Use Stokes' Theorem to find the circulation of the following vector fields around any simple closed smooth curve C.*

30. $\mathbf{F} = \langle 2x, -2y, 2z \rangle$ **31.** $\mathbf{F} = \nabla(x \sin y e^z)$

32. $\mathbf{F} = \langle 3x^2 y, x^3 + 2yz^2, 2y^2 z \rangle$

33. $\mathbf{F} = \langle y^2 z^3, 2xyz^3, 3xy^2 z^2 \rangle$

34–38. Tilted disks *Let S be the disk enclosed by the curve $C: \mathbf{r}(t) = \langle \cos \varphi \cos t, \sin t, \sin \varphi \cos t \rangle$, for $0 \le t \le 2\pi$, where $0 \le \varphi \le \pi/2$ is a fixed angle.*

34. What is the area of S? Find a vector normal to S.

35. What is the length of C?

36. Use Stokes' Theorem and a surface integral to find the circulation on C of the vector field $\mathbf{F} = \langle -y, x, 0 \rangle$ as a function of φ. For what value of φ is the circulation a maximum?

37. What is the circulation on C of the vector field $\mathbf{F} = \langle -y, -z, x \rangle$ as a function of φ? For what value of φ is the circulation a maximum?

38. Consider the vector field $\mathbf{F} = \mathbf{a} \times \mathbf{r}$, where $\mathbf{a} = \langle a_1, a_2, a_3 \rangle$ is a constant nonzero vector and $\mathbf{r} = \langle x, y, z \rangle$. Show that the circulation is a maximum when $\mathbf{a}$ points in the direction of the normal to S.

39. Circulation in a plane A circle C in the plane $x + y + z = 8$ has a radius of 4 and center $(2, 3, 3)$. Evaluate $\oint_C \mathbf{F} \cdot d\mathbf{r}$ for $\mathbf{F} = \langle 0, -z, 2y \rangle$, where C has a counterclockwise orientation when viewed from above. Does the circulation depend on the radius of the circle? Does it depend on the location of the center of the circle?

40. No integrals Let $\mathbf{F} = \langle 2z, z, 2y + x \rangle$, and let S be the hemisphere of radius a with its base in the xy-plane and center at the origin.

    a. Evaluate $\iint_S (\nabla \times \mathbf{F}) \cdot \mathbf{n} \, dS$ by computing $\nabla \times \mathbf{F}$ and appealing to symmetry.

    b. Evaluate the line integral using Stokes' Theorem to check part (a).

41. Compound surface and boundary Begin with the paraboloid $z = x^2 + y^2$, for $0 \le z \le 4$, and slice it with the plane $y = 0$. Let S be the surface that remains for $y \ge 0$ (including the planar surface in the xz-plane) (see figure). Let C be the semicircle and line segment that bound the cap of S in the plane $z = 4$ with counterclockwise orientation. Let $\mathbf{F} = \langle 2z + y, 2x + z, 2y + x \rangle$.

    a. Describe the direction of the vectors normal to the surface that are consistent with the orientation of C.

    b. Evaluate $\iint_S (\nabla \times \mathbf{F}) \cdot \mathbf{n} \, dS$.

    c. Evaluate $\oint_C \mathbf{F} \cdot d\mathbf{r}$ and check for agreement with part (b).

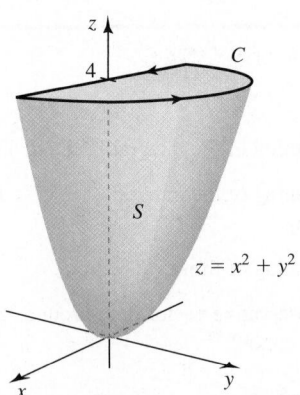

42. Ampère's Law The French physicist André-Marie Ampère (1775–1836) discovered that an electrical current I in a wire produces a magnetic field $\mathbf{B}$. A special case of Ampère's Law relates the current to the magnetic field through the equation $\oint_C \mathbf{B} \cdot d\mathbf{r} = \mu I$, where C is any closed curve through which the wire passes and μ is a physical constant. Assume the current I is given in terms of the current density $\mathbf{J}$ as $I = \iint_S \mathbf{J} \cdot \mathbf{n} \, dS$, where S is an oriented surface with C as a boundary. Use Stokes' Theorem to show that an equivalent form of Ampère's Law is $\nabla \times \mathbf{B} = \mu \mathbf{J}$.

43. Maximum surface integral Let S be the paraboloid $z = a(1 - x^2 - y^2)$, for $z \ge 0$, where $a > 0$ is a real number. Let $\mathbf{F} = \langle x - y, y + z, z - x \rangle$. For what value(s) of a (if any) does $\iint_S (\nabla \times \mathbf{F}) \cdot \mathbf{n} \, dS$ have its maximum value?

Explorations and Challenges

44. **Area of a region in a plane** Let R be a region in a plane that has a unit normal vector $\mathbf{n} = \langle a, b, c \rangle$ and boundary C. Let $\mathbf{F} = \langle bz, cx, ay \rangle$.

    a. Show that $\nabla \times \mathbf{F} = \mathbf{n}$.
    b. Use Stokes' Theorem to show that

    $$\text{area of } R = \oint_C \mathbf{F} \cdot d\mathbf{r}.$$

    c. Consider the curve C given by $\mathbf{r} = \langle 5 \sin t, 13 \cos t, 12 \sin t \rangle$, for $0 \le t \le 2\pi$. Prove that C lies in a plane by showing that $\mathbf{r} \times \mathbf{r}'$ is constant for all t.
    d. Use part (b) to find the area of the region enclosed by C in part (c). (*Hint:* Find the unit normal vector that is consistent with the orientation of C.)

45. **Choosing a more convenient surface** The goal is to evaluate $A = \iint_S (\nabla \times \mathbf{F}) \cdot \mathbf{n} \, dS$, where $\mathbf{F} = \langle yz, -xz, xy \rangle$ and S is the surface of the upper half of the ellipsoid $x^2 + y^2 + 8z^2 = 1$ $(z \ge 0)$.

    a. Evaluate a surface integral over a more convenient surface to find the value of A.
    b. Evaluate A using a line integral.

46. **Radial fields and zero circulation** Consider the radial vector fields $\mathbf{F} = \mathbf{r}/|\mathbf{r}|^p$, where p is a real number and $\mathbf{r} = \langle x, y, z \rangle$. Let C be any circle in the xy-plane centered at the origin.

    a. Evaluate a line integral to show that the field has zero circulation on C.
    b. For what values of p does Stokes' Theorem apply? For those values of p, use the surface integral in Stokes' Theorem to show that the field has zero circulation on C.

47. **Zero curl** Consider the vector field

    $$\mathbf{F} = -\frac{y}{x^2 + y^2}\mathbf{i} + \frac{x}{x^2 + y^2}\mathbf{j} + z\,\mathbf{k}.$$

    a. Show that $\nabla \times \mathbf{F} = \mathbf{0}$.
    b. Show that $\oint_C \mathbf{F} \cdot d\mathbf{r}$ is not zero on a circle C in the xy-plane enclosing the origin.
    c. Explain why Stokes' Theorem does not apply in this case.

48. **Average circulation** Let S be a small circular disk of radius R centered at the point P with a unit normal vector $\mathbf{n}$. Let C be the boundary of S.

    a. Express the average circulation of the vector field $\mathbf{F}$ on S as a surface integral of $\nabla \times \mathbf{F}$.
    b. Argue that for small R, the average circulation approaches $(\nabla \times \mathbf{F})|_P \cdot \mathbf{n}$ (the component of $\nabla \times \mathbf{F}$ in the direction of $\mathbf{n}$ evaluated at P) with the approximation improving as $R \to 0$.

49. **Proof of Stokes' Theorem** Confirm the following step in the proof of Stokes' Theorem. If $z = s(x, y)$ and f, g, and h are functions of x, y, and z, with $M = f + hz_x$ and $N = g + hz_y$, then

    $$M_y = f_y + f_z z_y + h z_{xy} + z_x(h_y + h_z z_y) \quad \text{and}$$
    $$N_x = g_x + g_z z_x + h z_{yx} + z_y(h_x + h_z z_x).$$

50. **Stokes' Theorem on closed surfaces** Prove that if $\mathbf{F}$ satisfies the conditions of Stokes' Theorem, then $\iint_S (\nabla \times \mathbf{F}) \cdot \mathbf{n} \, dS = 0$, where S is a smooth surface that encloses a region.

51. **Rotated Green's Theorem** Use Stokes' Theorem to write the circulation form of Green's Theorem in the yz-plane.

QUICK CHECK ANSWERS

1. If S is a region in the xy-plane, $\mathbf{n} = \mathbf{k}$ and $(\nabla \times \mathbf{F}) \cdot \mathbf{n}$ becomes $g_x - f_y$. **3.** The vector field is symmetric about the y-axis. ◄

17.8 Divergence Theorem

Vector fields can represent electric or magnetic fields, air velocities in hurricanes, or blood flow in an artery. These and other vector phenomena suggest movement of a "substance." A frequent question concerns the amount of a substance that flows across a surface—for example, the amount of water that passes across the membrane of a cell per unit time. Such flux calculations may be done using flux integrals as in Section 17.6. The Divergence Theorem offers an alternative method. In effect, it says that instead of integrating the flow into and out of a region across its boundary, you may also add up all the sources (or sinks) of the flow throughout the region.

Divergence Theorem

> ➤ Circulation form of Green's Theorem → Stokes' Theorem
>
> Flux form of Green's Theorem → Divergence Theorem

The Divergence Theorem is the three-dimensional version of the flux form of Green's Theorem. Recall that if R is a region in the xy-plane, C is the simple closed piecewise-smooth oriented boundary of R, and $\mathbf{F} = \langle f, g \rangle$ is a vector field, then Green's Theorem says that

$$\underbrace{\oint_C \mathbf{F} \cdot \mathbf{n} \, ds}_{\text{flux across } C} = \iint_R \underbrace{(f_x + g_y)}_{\text{divergence}} \, dA.$$

The line integral on the left gives the flux across the boundary of R. The double integral on the right measures the net expansion or contraction of the vector field within R. If $\mathbf{F}$ represents a fluid flow or the transport of a material, the theorem says that the cumulative effect of the sources (or sinks) of the flow within R equals the net flow across its boundary.

The Divergence Theorem is a direct extension of Green's Theorem. The plane region in Green's Theorem becomes a solid region D in $\mathbb{R}^3$, and the closed curve in Green's Theorem becomes the oriented surface S that encloses D. The flux integral in Green's Theorem becomes a surface integral over S, and the double integral in Green's Theorem becomes a triple integral over D of the three-dimensional divergence (**Figure 17.68**).

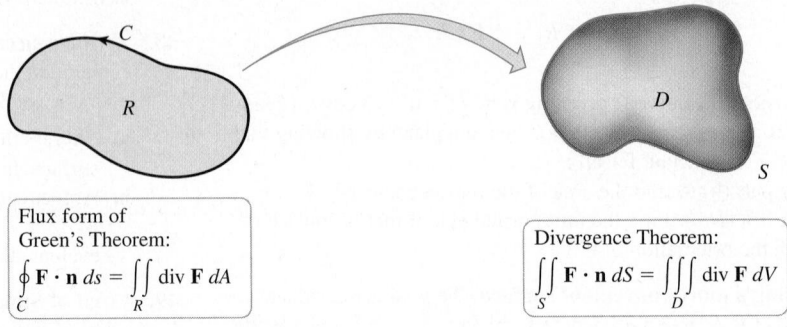

Flux form of
Green's Theorem:

$$\oint_C \mathbf{F} \cdot \mathbf{n}\, ds = \iint_R \operatorname{div} \mathbf{F}\, dA$$

Divergence Theorem:

$$\iint_S \mathbf{F} \cdot \mathbf{n}\, dS = \iiint_D \operatorname{div} \mathbf{F}\, dV$$

Figure 17.68

THEOREM 17.17   Divergence Theorem
Let $\mathbf{F}$ be a vector field whose components have continuous first partial derivatives in a connected and simply connected region D in $\mathbb{R}^3$ enclosed by an oriented surface S. Then

$$\iint_S \mathbf{F} \cdot \mathbf{n}\, dS = \iiint_D \nabla \cdot \mathbf{F}\, dV,$$

where $\mathbf{n}$ is the outward unit normal vector on S.

QUICK CHECK 1 Interpret the Divergence Theorem in the case that $\mathbf{F} = \langle a, b, c \rangle$ is a constant vector field and D is a ball. ◄

The surface integral on the left gives the flux of the vector field across the boundary; a positive flux integral means there is a net flow of the field out of the region. The triple integral on the right is the cumulative expansion or contraction of the field over the region D. The proof of a special case of the theorem is given later in this section.

EXAMPLE 1   Verifying the Divergence Theorem Consider the radial field $\mathbf{F} = \langle x, y, z \rangle$ and let S be the sphere $x^2 + y^2 + z^2 = a^2$ that encloses the region D. Assume $\mathbf{n}$ is the outward unit normal vector on the sphere. Evaluate both integrals of the Divergence Theorem.

SOLUTION The divergence of $\mathbf{F}$ is

$$\nabla \cdot \mathbf{F} = \frac{\partial}{\partial x}(x) + \frac{\partial}{\partial y}(y) + \frac{\partial}{\partial z}(z) = 3.$$

Integrating over D, we have

$$\iiint_D \nabla \cdot \mathbf{F}\, dV = \iiint_D 3\, dV = 3 \times \text{volume of } D = 3 \cdot \frac{4}{3}\pi a^3 = 4\pi a^3.$$

To evaluate the surface integral, we parameterize the sphere (Section 17.6, Table 17.3) in the form

$$\mathbf{r} = \langle x, y, z \rangle = \langle a \sin u \cos v, a \sin u \sin v, a \cos u \rangle,$$

where $R = \{(u, v): 0 \le u \le \pi, 0 \le v \le 2\pi\}$ (u and v are the spherical coordinates φ and θ, respectively). The surface integral is

$$\iint_S \mathbf{F} \cdot \mathbf{n}\, dS = \iint_R \mathbf{F} \cdot (\mathbf{t}_u \times \mathbf{t}_v)\, dA,$$

where the required vector normal to the surface is

$$\mathbf{t}_u \times \mathbf{t}_v = \langle a^2 \sin^2 u \cos v, a^2 \sin^2 u \sin v, a^2 \sin u \cos u \rangle.$$

▶ See Exercise 32 for an alternative
evaluation of the surface integral.

Substituting for $\mathbf{F} = \langle x, y, z \rangle$ and $\mathbf{t}_u \times \mathbf{t}_v$, we find after simplifying that
$\mathbf{F} \cdot (\mathbf{t}_u \times \mathbf{t}_v) = a^3 \sin u$. Therefore, the surface integral becomes

$$\iint_S \mathbf{F} \cdot \mathbf{n} \, dS = \iint_R \underbrace{\mathbf{F} \cdot (\mathbf{t}_u \times \mathbf{t}_v)}_{a^3 \sin u} \, dA$$

$$= \int_0^{2\pi} \int_0^{\pi} a^3 \sin u \, du \, dv \quad \text{Substitute for } \mathbf{F} \text{ and } \mathbf{t}_u \times \mathbf{t}_v.$$

$$= 4\pi a^3. \quad \text{Evaluate integrals.}$$

The two integrals of the Divergence Theorem are equal.

Related Exercise 9 ◀

EXAMPLE 2 Divergence Theorem with a rotation field Consider the rotation field

$$\mathbf{F} = \mathbf{a} \times \mathbf{r} = \langle 1, 0, 1 \rangle \times \langle x, y, z \rangle = \langle -y, x - z, y \rangle.$$

Let S be the hemisphere $x^2 + y^2 + z^2 = a^2$, for $z \geq 0$, together with its base in the
xy-plane. Find the net outward flux across S.

SOLUTION To find the flux using surface integrals, two surfaces must be considered (the
hemisphere and its base). The Divergence Theorem gives a simpler solution. Note that

$$\nabla \cdot \mathbf{F} = \frac{\partial}{\partial x}(-y) + \frac{\partial}{\partial y}(x - z) + \frac{\partial}{\partial z}(y) = 0.$$

We see that the flux across the hemisphere is zero.

Related Exercise 13 ◀

With Stokes' Theorem, rotation fields are noteworthy because they have a nonzero curl.
With the Divergence Theorem, the situation is reversed. As suggested by Example 2, pure
rotation fields of the form $\mathbf{F} = \mathbf{a} \times \mathbf{r}$ have zero divergence (Exercise 16). However, with
the Divergence Theorem, radial fields are interesting and have many physical applications.

EXAMPLE 3 Computing flux with the Divergence Theorem Find the net
outward flux of the field $\mathbf{F} = xyz\langle 1, 1, 1 \rangle$ across the boundaries of the cube
$D = \{(x, y, z) : 0 \leq x \leq 1, 0 \leq y \leq 1, 0 \leq z \leq 1\}$.

SOLUTION Computing a surface integral involves the six faces of the cube. The Divergence Theorem gives the outward flux with a single integral over D. The divergence of
the field is

$$\nabla \cdot \mathbf{F} = \frac{\partial}{\partial x}(xyz) + \frac{\partial}{\partial y}(xyz) + \frac{\partial}{\partial z}(xyz) = yz + xz + xy.$$

The integral over D is a standard triple integral:

$$\iiint_D \nabla \cdot \mathbf{F} \, dV = \iiint_D (yz + xz + xy) \, dV$$

$$= \int_0^1 \int_0^1 \int_0^1 (yz + xz + xy) \, dx \, dy \, dz \quad \text{Convert to a triple integral.}$$

$$= \frac{3}{4}. \quad \text{Evaluate integrals.}$$

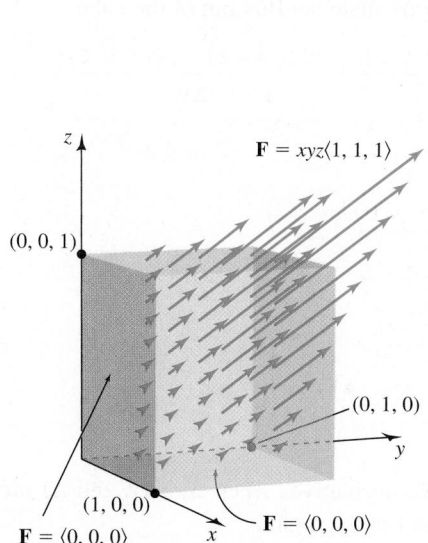

$\mathbf{F} = xyz\langle 1, 1, 1 \rangle$

$(0, 0, 1)$

$(0, 1, 0)$

$(1, 0, 0)$

$\mathbf{F} = \langle 0, 0, 0 \rangle$

$\mathbf{F} = \langle 0, 0, 0 \rangle$

Figure 17.69

On three faces of the cube (those that lie in the coordinate planes), we see that
$\mathbf{F}(0, y, z) = \mathbf{F}(x, 0, z) = \mathbf{F}(x, y, 0) = \mathbf{0}$, so there is no contribution to the flux on these
faces (Figure 17.69). On the other three faces, the vector field has components out of the
cube. Therefore, the net outward flux is positive, as calculated.

Related Exercises 18–19 ◀

QUICK CHECK 2 In Example 3, does the vector field have negative components anywhere in the cube D? Is the divergence negative anywhere in D? ◄

▶ The mass transport is also called the *flux density*; when multiplied by an area, it gives the flux. We use the convention that flux has units of mass per unit time.

▶ Check the units: If **F** has units of mass/(area · time), then the flux has units of mass/time (**n** has no units).

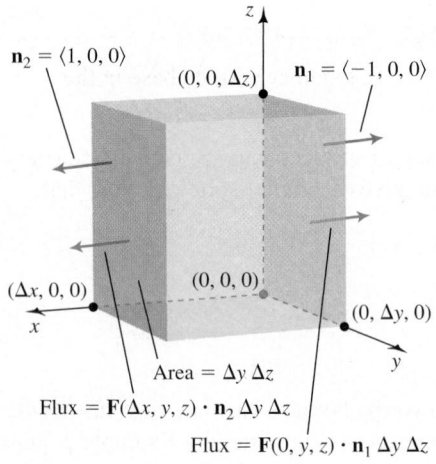

Figure 17.70

Interpretation of the Divergence Theorem Using Mass Transport   Suppose **v** is the velocity field of a material, such as water or molasses, and ρ is its constant density. The vector field $\mathbf{F} = \rho\mathbf{v} = \langle f, g, h \rangle$ describes the **mass transport** of the material, with units of $(\text{mass/vol}) \times (\text{length/time}) = \text{mass}/(\text{area} \cdot \text{time})$; typical units of mass transport are $\text{g/m}^2/\text{s}$. This means that **F** gives the mass of material flowing past a point (in each of the three coordinate directions) per unit of surface area per unit of time. When **F** is multiplied by an area, the result is the **flux**, with units of mass/unit time.

Now consider a small cube located in the vector field with its faces parallel to the coordinate planes. One vertex is located at $(0, 0, 0)$, the opposite vertex is at $(\Delta x, \Delta y, \Delta z)$, and (x, y, z) is an arbitrary point in the cube (**Figure 17.70**). The goal is to compute the approximate flux of material across the faces of the cube. We begin with the flux across the two parallel faces $x = 0$ and $x = \Delta x$.

The outward unit vectors normal to the faces $x = 0$ and $x = \Delta x$ are $\mathbf{n}_1 = \langle -1, 0, 0 \rangle$ and $\mathbf{n}_2 = \langle 1, 0, 0 \rangle$, respectively. Each face has area $\Delta y \, \Delta z$, so the approximate net flux across these faces is

$$\underbrace{\mathbf{F}(\Delta x, y, z)}_{x\,=\,\Delta x\text{ face}} \cdot \underbrace{\mathbf{n}_2}_{\langle 1,0,0\rangle} \Delta y \, \Delta z + \underbrace{\mathbf{F}(0, y, z)}_{x\,=\,0\text{ face}} \cdot \underbrace{\mathbf{n}_1}_{\langle -1,0,0\rangle} \Delta y \, \Delta z$$

$$= (f(\Delta x, y, z) - f(0, y, z)) \, \Delta y \, \Delta z.$$

Note that if $f(\Delta x, y, z) > f(0, y, z)$, the net flux across these two faces of the cube is positive, which means the net flow is *out* of the cube. Letting $\Delta V = \Delta x \, \Delta y \, \Delta z$ be the volume of the cube, we rewrite the net flux as

$$(f(\Delta x, y, z) - f(0, y, z)) \, \Delta y \, \Delta z$$

$$= \frac{f(\Delta x, y, z) - f(0, y, z)}{\Delta x} \Delta x \, \Delta y \, \Delta z \quad \text{Multiply by } \frac{\Delta x}{\Delta x}.$$

$$= \frac{f(\Delta x, y, z) - f(0, y, z)}{\Delta x} \Delta V. \qquad \Delta V = \Delta x \, \Delta y \, \Delta z$$

A similar argument can be applied to the other two pairs of faces. The approximate net flux across the faces $y = 0$ and $y = \Delta y$ is

$$\frac{g(x, \Delta y, z) - g(x, 0, z)}{\Delta y} \Delta V,$$

and the approximate net flux across the faces $z = 0$ and $z = \Delta z$ is

$$\frac{h(x, y, \Delta z) - h(x, y, 0)}{\Delta z} \Delta V.$$

Adding these three individual fluxes gives the approximate net flux out of the cube:

$$\text{net flux out of cube} \approx \left(\underbrace{\frac{f(\Delta x, y, z) - f(0, y, z)}{\Delta x}}_{\approx \frac{\partial f}{\partial x}(0,0,0)} + \underbrace{\frac{g(x, \Delta y, z) - g(x, 0, z)}{\Delta y}}_{\approx \frac{\partial g}{\partial y}(0,0,0)} \right.$$

$$\left. + \underbrace{\frac{h(x, y, \Delta z) - h(x, y, 0)}{\Delta z}}_{\approx \frac{\partial h}{\partial z}(0,0,0)} \right) \Delta V$$

$$\approx \left. \left(\frac{\partial f}{\partial x} + \frac{\partial g}{\partial y} + \frac{\partial h}{\partial z} \right) \right|_{(0,0,0)} \Delta V$$

$$= (\nabla \cdot \mathbf{F})(0, 0, 0) \, \Delta V.$$

Notice how the three quotients approximate partial derivatives when Δx, Δy, and Δz are small. A similar argument may be made at any point in the region.

Taking one more step, we show informally how the Divergence Theorem arises. Suppose the small cube we just analyzed is one of many small cubes of volume ΔV that fill a region D. We label the cubes $k = 1, \ldots, n$ and apply the preceding argument to each cube, letting $(\nabla \cdot \mathbf{F})_k$ be the divergence evaluated at a point in the kth cube. Adding the individual contributions to the net flux from each cube, we obtain the approximate net flux across the boundary of D:

$$\text{net flux out of } D \approx \sum_{k=1}^{n} (\nabla \cdot \mathbf{F})_k \, \Delta V.$$

Letting the volume of the cubes ΔV approach 0, and letting the number of cubes n increase, we obtain an integral over D:

$$\text{net flux out of } D = \lim_{n \to \infty} \sum_{k=1}^{n} (\nabla \cdot \mathbf{F})_k \, \Delta V = \iiint_D \nabla \cdot \mathbf{F} \, dV.$$

The net flux across the boundary of D is also given by $\iint_S \mathbf{F} \cdot \mathbf{n} \, dS$. Equating the surface integral and the volume integral gives the Divergence Theorem. Now we look at a formal proof.

Proof of the Divergence Theorem

We prove the Divergence Theorem under special conditions on the region D. Let R be the projection of D in the xy-plane (**Figure 17.71**); that is,

$$R = \{(x, y): (x, y, z) \text{ is in } D\}.$$

Assume the boundary of D is S and let $\mathbf{n}$ be the unit vector normal to S that points outward.

Letting $\mathbf{F} = \langle f, g, h \rangle = f\,\mathbf{i} + g\,\mathbf{j} + h\,\mathbf{k}$, the surface integral in the Divergence Theorem is

$$\iint_S \mathbf{F} \cdot \mathbf{n} \, dS = \iint_S (f\,\mathbf{i} + g\,\mathbf{j} + h\,\mathbf{k}) \cdot \mathbf{n} \, dS$$

$$= \iint_S f\,\mathbf{i} \cdot \mathbf{n} \, dS + \iint_S g\,\mathbf{j} \cdot \mathbf{n} \, dS + \iint_S h\,\mathbf{k} \cdot \mathbf{n} \, dS.$$

The volume integral in the Divergence Theorem is

$$\iiint_D \nabla \cdot \mathbf{F} \, dV = \iiint_D \left(\frac{\partial f}{\partial x} + \frac{\partial g}{\partial y} + \frac{\partial h}{\partial z} \right) dV.$$

Matching terms of the surface and volume integrals, the theorem is proved by showing that

$$\iint_S f\,\mathbf{i} \cdot \mathbf{n} \, dS = \iiint_D \frac{\partial f}{\partial x} \, dV, \tag{1}$$

$$\iint_S g\,\mathbf{j} \cdot \mathbf{n} \, dS = \iiint_D \frac{\partial g}{\partial y} \, dV, \text{ and} \tag{2}$$

$$\iint_S h\,\mathbf{k} \cdot \mathbf{n} \, dS = \iiint_D \frac{\partial h}{\partial z} \, dV. \tag{3}$$

We work on equation (3) assuming special properties for D. Suppose D is bounded by two surfaces $S_1: z = p(x, y)$ and $S_2: z = q(x, y)$, where $p(x, y) \leq q(x, y)$ on R (Figure 17.71). The Fundamental Theorem of Calculus is used in the triple integral to show that

$$\iiint_D \frac{\partial h}{\partial z} \, dV = \iint_R \int_{p(x, y)}^{q(x, y)} \frac{\partial h}{\partial z} \, dz \, dx \, dy$$

$$= \iint_R \big(h(x, y, q(x, y)) - h(x, y, p(x, y)) \big) \, dx \, dy. \quad \text{Evaluate inner integral.}$$

Now let's turn to the surface integral in equation (3), $\iint_S h\,\mathbf{k} \cdot \mathbf{n} \, dS$, and note that S consists of three pieces: the lower surface S_1, the upper surface S_2, and the vertical sides S_3 of the surface (if they exist). The normal to S_3 is everywhere orthogonal to $\mathbf{k}$, so $\mathbf{k} \cdot \mathbf{n} = 0$

> In making this argument, notice that for two adjacent cubes, the flux into one cube equals the flux out of the other cube across the common face. Therefore, there is a cancellation of fluxes throughout the interior of D.

QUICK CHECK 3 Draw the unit cube $D = \{(x, y, z): 0 \leq x \leq 1, 0 \leq y \leq 1, 0 \leq z \leq 1\}$ and sketch the vector field $\mathbf{F} = \langle x, -y, 2z \rangle$ on the six faces of the cube. Compute and interpret div $\mathbf{F}$. ◄

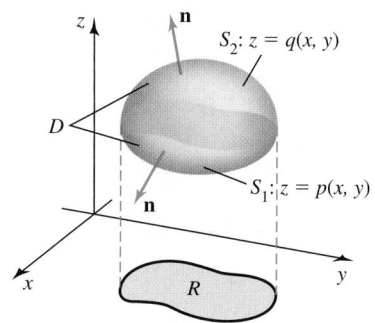

Figure 17.71

and the S_3 integral makes no contribution. What remains is to compute the surface integrals over S_1 and S_2.

The required outward normal to S_2 (which is the graph of $z = q(x, y)$) is $\langle -q_x, -q_y, 1 \rangle$. The outward normal to S_1 (which is the graph of $z = p(x, y)$) points *downward*, so it is given by $\langle p_x, p_y, -1 \rangle$. The surface integral of (3) becomes

$$\iint_S h\, \mathbf{k} \cdot \mathbf{n}\, dS = \iint_{S_2} h(x, y, z)\, \mathbf{k} \cdot \mathbf{n}\, dS + \iint_{S_1} h(x, y, z)\, \mathbf{k} \cdot \mathbf{n}\, dS$$

$$= \iint_R h(x, y, q(x, y))\, \underbrace{\mathbf{k} \cdot \langle -q_x, -q_y, 1 \rangle}_{1}\, dx\, dy$$

$$+ \iint_R h(x, y, p(x, y))\, \underbrace{\mathbf{k} \cdot \langle p_x, p_y, -1 \rangle}_{-1}\, dx\, dy$$

$$= \iint_R h(x, y, q(x, y))\, dx\, dy - \iint_R h(x, y, p(x, y))\, dx\, dy. \quad \text{Simplify.}$$

Observe that both the volume integral and the surface integral of (3) reduce to the same integral over R. Therefore, $\iint_S h\, \mathbf{k} \cdot \mathbf{n}\, dS = \iiint_D \dfrac{\partial h}{\partial z}\, dV$.

Equations (1) and (2) are handled in a similar way.

- To prove (1), we make the special assumption that D is also bounded by two surfaces, $S_1 : x = s(y, z)$ and $S_2 : x = t(y, z)$, where $s(y, z) \leq t(y, z)$.
- To prove (2), we assume D is bounded by two surfaces, $S_1 : y = u(x, z)$ and $S_2 : y = v(x, z)$, where $u(x, z) \leq v(x, z)$.

When combined, equations (1), (2), and (3) yield the Divergence Theorem.   ◄

Divergence Theorem for Hollow Regions

The Divergence Theorem may be extended to more general solid regions. Here we consider the important case of hollow regions. Suppose D is a region consisting of all points inside a closed oriented surface S_2 and outside a closed oriented surface S_1, where S_1 lies within S_2 (**Figure 17.72**). Therefore, the boundary of D consists of S_1 and S_2, which we denote S. (Note that D is simply connected.)

We let $\mathbf{n}_1$ and $\mathbf{n}_2$ be the outward unit normal vectors for S_1 and S_2, respectively. Note that $\mathbf{n}_1$ points into D, so the outward normal to S on S_1 is $-\mathbf{n}_1$. With this observation, the Divergence Theorem takes the following form.

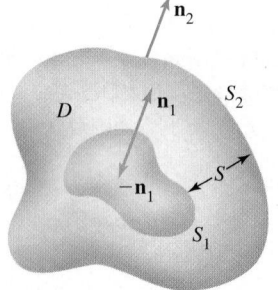

$\mathbf{n}_1$ is the outward unit normal to S_1 and points into D. The outward unit normal to S on S_1 is $-\mathbf{n}_1$.

Figure 17.72

▶ It's important to point out again that $\mathbf{n}_1$ is the unit normal that we would use for S_1 alone, independent of S. It is the outward unit normal to S_1, but it points into D.

> **THEOREM 17.18 Divergence Theorem for Hollow Regions**
> Suppose the vector field $\mathbf{F}$ satisfies the conditions of the Divergence Theorem on a region D bounded by two oriented surfaces S_1 and S_2, where S_1 lies within S_2. Let S be the entire boundary of D ($S = S_1 \cup S_2$) and let $\mathbf{n}_1$ and $\mathbf{n}_2$ be the outward unit normal vectors for S_1 and S_2, respectively. Then
> $$\iiint_D \nabla \cdot \mathbf{F}\, dV = \iint_S \mathbf{F} \cdot \mathbf{n}\, dS = \iint_{S_2} \mathbf{F} \cdot \mathbf{n}_2\, dS - \iint_{S_1} \mathbf{F} \cdot \mathbf{n}_1\, dS.$$

This form of the Divergence Theorem is applicable to vector fields that are not differentiable at the origin, as is the case with some important radial vector fields.

EXAMPLE 4 Flux for an inverse square field Consider the inverse square vector field

$$\mathbf{F} = \frac{\mathbf{r}}{|\mathbf{r}|^3} = \frac{\langle x, y, z \rangle}{(x^2 + y^2 + z^2)^{3/2}}.$$

► Recall that an inverse square force is proportional to $1/|\mathbf{r}|^2$ multiplied by a unit vector in the radial direction, which is $\mathbf{r}/|\mathbf{r}|$. Combining these two factors gives $\mathbf{F} = \mathbf{r}/|\mathbf{r}|^3$.

a. Find the net outward flux of $\mathbf{F}$ across the surface of the region $D = \{(x, y, z): a^2 \leq x^2 + y^2 + z^2 \leq b^2\}$ that lies between concentric spheres with radii a and b.

b. Find the outward flux of $\mathbf{F}$ across any sphere that encloses the origin.

SOLUTION

a. Although the vector field is undefined at the origin, it is defined and differentiable in D, which excludes the origin. In Section 17.5 (Exercise 73) it was shown that the divergence of the radial field $\mathbf{F} = \dfrac{\mathbf{r}}{|\mathbf{r}|^p}$ with $p = 3$ is 0. We let S be the union of S_2, the larger sphere of radius b, and S_1, the smaller sphere of radius a. Because $\iiint_D \nabla \cdot \mathbf{F} \, dV = 0$, the Divergence Theorem implies that

$$\iint_S \mathbf{F} \cdot \mathbf{n} \, dS = \iint_{S_2} \mathbf{F} \cdot \mathbf{n}_2 \, dS - \iint_{S_1} \mathbf{F} \cdot \mathbf{n}_1 \, dS = 0.$$

Therefore, the next flux across S is zero.

b. Part (a) implies that

$$\underbrace{\iint_{S_2} \mathbf{F} \cdot \mathbf{n}_2 \, dS}_{\text{out of } D} = \underbrace{\iint_{S_1} \mathbf{F} \cdot \mathbf{n}_1 \, dS}_{\text{into } D}.$$

We see that the flux out of D across S_2 equals the flux into D across S_1. To find that flux, we evaluate the surface integral over S_1 on which $|\mathbf{r}| = a$. (Because the fluxes are equal, S_2 could also be used.)

The easiest way to evaluate the surface integral is to note that on the sphere S_1, the unit outward normal vector is $\mathbf{n}_1 = \mathbf{r}/|\mathbf{r}|$. Therefore, the surface integral is

$$
\begin{aligned}
\iint_{S_1} \mathbf{F} \cdot \mathbf{n}_1 \, dS &= \iint_{S_1} \frac{\mathbf{r}}{|\mathbf{r}|^3} \cdot \frac{\mathbf{r}}{|\mathbf{r}|} \, dS && \text{Substitute for } \mathbf{F} \text{ and } \mathbf{n}_1. \\
&= \iint_{S_1} \frac{|\mathbf{r}|^2}{|\mathbf{r}|^4} \, dS && \mathbf{r} \cdot \mathbf{r} = |\mathbf{r}|^2 \\
&= \iint_{S_1} \frac{1}{a^2} \, dS && |\mathbf{r}| = a \\
&= \frac{4\pi a^2}{a^2} && \text{Surface area} = 4\pi a^2 \\
&= 4\pi.
\end{aligned}
$$

The same result is obtained using S_2 or any smooth surface enclosing the origin. The flux of the inverse square field across *any* surface enclosing the origin is 4π. As shown in Exercise 46, among radial fields, this property holds only for the inverse square field $(p = 3)$.

Related Exercises 26–27 ◄

Gauss' Law

Applying the Divergence Theorem to electric fields leads to one of the fundamental laws of physics. The electric field due to a point charge Q located at the origin is given by the inverse square law,

$$\mathbf{E}(x, y, z) = \frac{Q}{4\pi\varepsilon_0} \frac{\mathbf{r}}{|\mathbf{r}|^3},$$

where $\mathbf{r} = \langle x, y, z \rangle$ and ε_0 is a physical constant called the *permittivity of free space*.

According to the calculation of Example 4, the flux of the field $\dfrac{\mathbf{r}}{|\mathbf{r}|^3}$ across any surface that encloses the origin is 4π. Therefore, the flux of the electric field across any surface enclosing the origin is $\dfrac{Q}{4\pi\varepsilon_0}\cdot 4\pi = \dfrac{Q}{\varepsilon_0}$ (Figure 17.73a). This is one statement of Gauss' Law: If S is a surface that encloses a point charge Q, then the flux of the electric field across S is

$$\iint\limits_S \mathbf{E}\cdot\mathbf{n}\,dS = \frac{Q}{\varepsilon_0}.$$

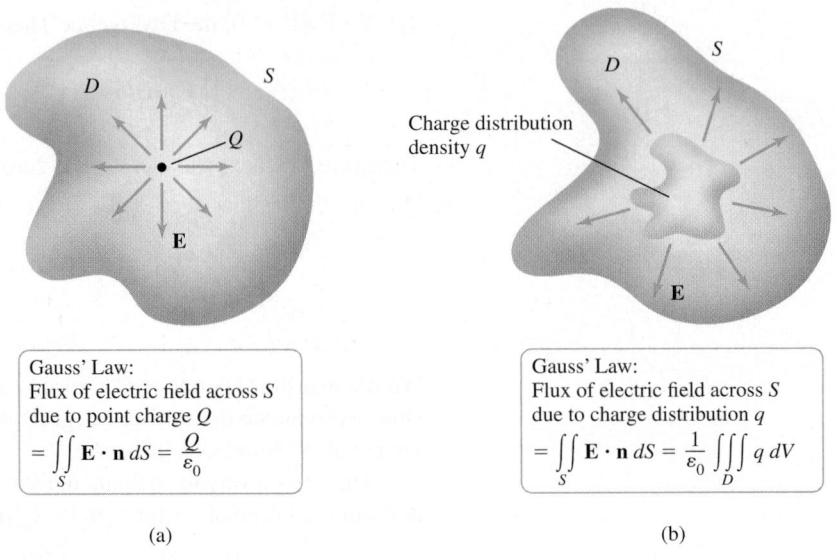

(a)

Gauss' Law:
Flux of electric field across S
due to point charge Q

$= \displaystyle\iint\limits_S \mathbf{E}\cdot\mathbf{n}\,dS = \dfrac{Q}{\varepsilon_0}$

(b)

Charge distribution
density q

Gauss' Law:
Flux of electric field across S
due to charge distribution q

$= \displaystyle\iint\limits_S \mathbf{E}\cdot\mathbf{n}\,dS = \dfrac{1}{\varepsilon_0}\iiint\limits_D q\,dV$

Figure 17.73

In fact, Gauss' Law applies to more general charge distributions (Exercise 39). If $q(x, y, z)$ is a charge density (charge per unit volume) defined on a region D enclosed by S, then the total charge within D is $Q = \iiint_D q(x, y, z)\,dV$ (Figure 17.73b). Replacing Q with this triple integral, Gauss' Law takes the form

$$\iint\limits_S \mathbf{E}\cdot\mathbf{n}\,dS = \frac{1}{\varepsilon_0}\underbrace{\iiint\limits_D q(x, y, z)\,dV}_{Q}.$$

Gauss' Law applies to other inverse square fields. In a slightly different form, it also governs heat transfer. If T is the temperature distribution in a solid body D, then the heat flow vector field is $\mathbf{F} = -k\nabla T$. (Heat flows down the temperature gradient.) If $q(x, y, z)$ represents the sources of heat within D, Gauss' Law says

$$\iint\limits_S \mathbf{F}\cdot\mathbf{n}\,dS = -k\iint\limits_S \nabla T\cdot\mathbf{n}\,dS = \iiint\limits_D q(x, y, z)\,dV.$$

We see that, in general, the flux of material (fluid, heat, electric field lines) across the boundary of a region is the cumulative effect of the sources within the region.

A Final Perspective

Table 17.4 offers a look at the progression of fundamental theorems of calculus that have appeared throughout this text. Each theorem builds on its predecessors, extending the same basic idea to a different situation or to higher dimensions.

In all cases, the statement is effectively the same: The cumulative (integrated) effect of the *derivatives* of a function throughout a region is determined by the values of the function on the boundary of that region. This principle underlies much of our understanding of the world around us.

Table 17.4

Fundamental Theorem of Calculus	$\int_a^b f'(x)\,dx = f(b) - f(a)$	

Fundamental Theorem for Line Integrals $\int_C \nabla f \cdot d\mathbf{r} = f(B) - f(A)$

Green's Theorem (Circulation form) $\iint_R (g_x - f_y)\,dA = \oint_C f\,dx + g\,dy$

Stokes' Theorem $\iint_S (\nabla \times \mathbf{F}) \cdot \mathbf{n}\,dS = \oint_C \mathbf{F} \cdot d\mathbf{r}$

Divergence Theorem $\iiint_D \nabla \cdot \mathbf{F}\,dV = \iint_S \mathbf{F} \cdot \mathbf{n}\,dS$

SECTION 17.8 EXERCISES

Getting Started

1. Explain the meaning of the surface integral in the Divergence Theorem.

2. Interpret the volume integral in the Divergence Theorem.

3. Explain the meaning of the Divergence Theorem.

4. What is the net outward flux of the rotation field $\mathbf{F} = \langle 2z + y, -x, -2x \rangle$ across the surface that encloses any region?

5. What is the net outward flux of the radial field $\mathbf{F} = \langle x, y, z \rangle$ across the sphere of radius 2 centered at the origin?

6. What is the divergence of an inverse square vector field?

7. Suppose div $\mathbf{F} = 0$ in a region enclosed by two concentric spheres. What is the relationship between the outward fluxes across the two spheres?

8. If div $\mathbf{F} > 0$ in a region enclosed by a small cube, is the net flux of the field into or out of the cube?

Practice Exercises

9–12. Verifying the Divergence Theorem *Evaluate both integrals of the Divergence Theorem for the following vector fields and regions. Check for agreement.*

9. $\mathbf{F} = \langle 2x, 3y, 4z \rangle$; $D = \{(x, y, z): x^2 + y^2 + z^2 \le 4\}$

10. $\mathbf{F} = \langle -x, -y, -z \rangle$; $D = \{(x, y, z): |x| \le 1, |y| \le 1, |z| \le 1\}$

11. $\mathbf{F} = \langle z - y, x, -x \rangle$; $D = \{(x, y, z): x^2/4 + y^2/8 + z^2/12 \le 1\}$

12. $\mathbf{F} = \langle x^2, y^2, z^2 \rangle$; $D = \{(x, y, z): |x| \le 1, |y| \le 2, |z| \le 3\}$

13–16. Rotation fields

13. Find the net outward flux of the field $\mathbf{F} = \langle 2z - y, x, -2x \rangle$ across the sphere of radius 1 centered at the origin.

14. Find the net outward flux of the field $\mathbf{F} = \langle z - y, x - z, y - x \rangle$ across the boundary of the cube $\{(x, y, z): |x| \le 1, |y| \le 1, |z| \le 1\}$.

15. Find the net outward flux of the field $\mathbf{F} = \langle bz - cy, cx - az, ay - bx \rangle$ across any smooth closed surface in $\mathbb{R}^3$, where a, b, and c are constants.

16. Find the net outward flux of $\mathbf{F} = \mathbf{a} \times \mathbf{r}$ across any smooth closed surface in $\mathbb{R}^3$, where $\mathbf{a}$ is a constant nonzero vector and $\mathbf{r} = \langle x, y, z \rangle$.

17–24. Computing flux *Use the Divergence Theorem to compute the net outward flux of the following fields across the given surface S.*

17. $\mathbf{F} = \langle x, -2y, 3z \rangle$; S is the sphere $\{(x, y, z): x^2 + y^2 + z^2 = 6\}$.

18. $\mathbf{F} = \langle x^2, 2xz, y^2 \rangle$; S is the surface of the cube cut from the first octant by the planes $x = 1$, $y = 1$, and $z = 1$.

19. $\mathbf{F} = \langle x, 2y, z \rangle$; S is the boundary of the tetrahedron in the first octant formed by the plane $x + y + z = 1$.

20. $\mathbf{F} = \langle x^2, y^2, z^2 \rangle$; S is the sphere $\{(x, y, z): x^2 + y^2 + z^2 = 25\}$.

21. $\mathbf{F} = \langle y - 2x, x^3 - y, y^2 - z \rangle$; S is the sphere
$\{(x, y, z): x^2 + y^2 + z^2 = 4\}$.

22. $\mathbf{F} = \langle y + z, x + z, x + y \rangle$; S consists of the faces of the cube
$\{(x, y, z): |x| \le 1, |y| \le 1, |z| \le 1\}$.

23. $\mathbf{F} = \langle x, y, z \rangle$; S is the surface of the paraboloid
$z = 4 - x^2 - y^2$, for $z \ge 0$, plus its base in the xy-plane.

24. $\mathbf{F} = \langle x, y, z \rangle$; S is the surface of the cone $z^2 = x^2 + y^2$, for
$0 \le z \le 4$, plus its top surface in the plane $z = 4$.

25–30. Divergence Theorem for more general regions *Use the Divergence Theorem to compute the net outward flux of the following vector fields across the boundary of the given regions D.*

25. $\mathbf{F} = \langle z - x, x - y, 2y - z \rangle$; D is the region between the spheres of radius 2 and 4 centered at the origin.

26. $\mathbf{F} = \mathbf{r}|\mathbf{r}| = \langle x, y, z \rangle \sqrt{x^2 + y^2 + z^2}$; D is the region between the spheres of radius 1 and 2 centered at the origin.

27. $\mathbf{F} = \dfrac{\mathbf{r}}{|\mathbf{r}|} = \dfrac{\langle x, y, z \rangle}{\sqrt{x^2 + y^2 + z^2}}$; D is the region between the spheres of radius 1 and 2 centered at the origin.

28. $\mathbf{F} = \langle z - y, x - z, 2y - x \rangle$; D is the region between two cubes:
$\{(x, y, z): 1 \le |x| \le 3, 1 \le |y| \le 3, 1 \le |z| \le 3\}$.

29. $\mathbf{F} = \langle x^2, -y^2, z^2 \rangle$; D is the region in the first octant between the planes $z = 4 - x - y$ and $z = 2 - x - y$.

30. $\mathbf{F} = \langle x, 2y, 3z \rangle$; D is the region between the cylinders
$x^2 + y^2 = 1$ and $x^2 + y^2 = 4$, for $0 \le z \le 8$.

31. Explain why or why not Determine whether the following statements are true and give an explanation or counterexample.

a. If $\nabla \cdot \mathbf{F} = 0$ at all points of a region D, then $\mathbf{F} \cdot \mathbf{n} = 0$ at all points of the boundary of D.

b. If $\iint_S \mathbf{F} \cdot \mathbf{n} \, dS = 0$ on all closed surfaces in $\mathbb{R}^3$, then $\mathbf{F}$ is constant.

c. If $|\mathbf{F}| < 1$, then $\left| \iiint_D \nabla \cdot \mathbf{F} \, dV \right|$ is less than the area of the surface of D.

32. Flux across a sphere Consider the radial field $\mathbf{F} = \langle x, y, z \rangle$ and let S be the sphere of radius a centered at the origin. Compute the outward flux of $\mathbf{F}$ across S using the representation $z = \pm \sqrt{a^2 - x^2 - y^2}$ for the sphere (either symmetry or two surfaces must be used).

33–35. Flux integrals *Compute the outward flux of the following vector fields across the given surfaces S. You should decide which integral of the Divergence Theorem to use.*

33. $\mathbf{F} = \langle x^2 e^y \cos z, -4xe^y \cos z, 2xe^y \sin z \rangle$; S is the boundary of the ellipsoid $x^2/4 + y^2 + z^2 = 1$.

34. $\mathbf{F} = \langle -yz, xz, 1 \rangle$; S is the boundary of the ellipsoid
$x^2/4 + y^2/4 + z^2 = 1$.

35. $\mathbf{F} = \langle x \sin y, -\cos y, z \sin y \rangle$; S is the boundary of the region bounded by the planes $x = 1$, $y = 0$, $y = \pi/2$, $z = 0$, and $z = x$.

36. Radial fields Consider the radial vector field
$$\mathbf{F} = \frac{\mathbf{r}}{|\mathbf{r}|^p} = \frac{\langle x, y, z \rangle}{(x^2 + y^2 + z^2)^{p/2}}.$$ Let S be the sphere of radius a
centered at the origin.

a. Use a surface integral to show that the outward flux of $\mathbf{F}$ across S is $4\pi a^{3-p}$. Recall that the unit normal to the sphere is $\mathbf{r}/|\mathbf{r}|$.

b. For what values of p does $\mathbf{F}$ satisfy the conditions of the Divergence Theorem? For these values of p, use the fact (Theorem 17.10) that $\nabla \cdot \mathbf{F} = \dfrac{3 - p}{|\mathbf{r}|^p}$ to compute the flux across S using the Divergence Theorem.

37. Singular radial field Consider the radial field
$$\mathbf{F} = \frac{\mathbf{r}}{|\mathbf{r}|} = \frac{\langle x, y, z \rangle}{(x^2 + y^2 + z^2)^{1/2}}.$$

a. Evaluate a surface integral to show that $\iint_S \mathbf{F} \cdot \mathbf{n} \, dS = 4\pi a^2$, where S is the surface of a sphere of radius a centered at the origin.

b. Note that the first partial derivatives of the components of $\mathbf{F}$ are undefined at the origin, so the Divergence Theorem does not apply directly. Nevertheless, the flux across the sphere as computed in part (a) is finite. Evaluate the triple integral of the Divergence Theorem as an improper integral as follows. Integrate div $\mathbf{F}$ over the region between two spheres of radius a and $0 < \varepsilon < a$. Then let $\varepsilon \to 0^+$ to obtain the flux computed in part (a).

38. Logarithmic potential Consider the potential function
$$\varphi(x, y, z) = \frac{1}{2} \ln (x^2 + y^2 + z^2) = \ln |\mathbf{r}|, \text{ where } \mathbf{r} = \langle x, y, z \rangle.$$

a. Show that the gradient field associated with φ is
$$\mathbf{F} = \frac{\mathbf{r}}{|\mathbf{r}|^2} = \frac{\langle x, y, z \rangle}{x^2 + y^2 + z^2}.$$

b. Show that $\iint_S \mathbf{F} \cdot \mathbf{n} \, dS = 4\pi a$, where S is the surface of a sphere of radius a centered at the origin.

c. Compute div $\mathbf{F}$.

d. Note that $\mathbf{F}$ is undefined at the origin, so the Divergence Theorem does not apply directly. Evaluate the volume integral as described in Exercise 37.

39. Gauss' Law for electric fields The electric field due to a point charge Q is $\mathbf{E} = \dfrac{Q}{4\pi\varepsilon_0} \dfrac{\mathbf{r}}{|\mathbf{r}|^3}$, where $\mathbf{r} = \langle x, y, z \rangle$, and ε_0 is a constant.

a. Show that the flux of the field across a sphere of radius a centered at the origin is $\iint_S \mathbf{E} \cdot \mathbf{n} \, dS = \dfrac{Q}{\varepsilon_0}$.

b. Let S be the boundary of the region between two spheres centered at the origin of radius a and b, respectively, with $a < b$. Use the Divergence Theorem to show that the net outward flux across S is zero.

c. Suppose there is a distribution of charge within a region D. Let $q(x, y, z)$ be the charge density (charge per unit volume). Interpret the statement that
$$\iint_S \mathbf{E} \cdot \mathbf{n} \, dS = \frac{1}{\varepsilon_0} \iiint_D q(x, y, z) \, dV.$$

d. Assuming $\mathbf{E}$ satisfies the conditions of the Divergence Theorem on D, conclude from part (c) that $\nabla \cdot \mathbf{E} = \dfrac{q}{\varepsilon_0}$.

e. Because the electric force is conservative, it has a potential function φ. From part (d), conclude that $\nabla^2 \varphi = \nabla \cdot \nabla \varphi = \dfrac{q}{\varepsilon_0}$.

40. Gauss' Law for gravitation The gravitational force due to a point mass M at the origin is proportional to $\mathbf{F} = GM\mathbf{r}/|\mathbf{r}|^3$, where $\mathbf{r} = \langle x, y, z \rangle$ and G is the gravitational constant.

a. Show that the flux of the force field across a sphere of radius a centered at the origin is $\iint_S \mathbf{F} \cdot \mathbf{n} \, dS = 4\pi GM$.

b. Let S be the boundary of the region between two spheres centered at the origin of radius a and b, respectively, with $a < b$. Use the Divergence Theorem to show that the net outward flux across S is zero.

c. Suppose there is a distribution of mass within a region D. Let $\rho(x, y, z)$ be the mass density (mass per unit volume). Interpret the statement that

$$\iint_S \mathbf{F} \cdot \mathbf{n} \, dS = 4\pi G \iiint_D \rho(x, y, z) \, dV.$$

d. Assuming $\mathbf{F}$ satisfies the conditions of the Divergence Theorem on D, conclude from part (c) that $\nabla \cdot \mathbf{F} = 4\pi G\rho$.

e. Because the gravitational force is conservative, it has a potential function φ. From part (d), conclude that $\nabla^2 \varphi = 4\pi G\rho$.

41–45. Heat transfer *Fourier's Law of heat transfer (or heat conduction) states that the heat flow vector* $\mathbf{F}$ *at a point is proportional to the negative gradient of the temperature; that is,* $\mathbf{F} = -k\nabla T$, *which means that heat energy flows from hot regions to cold regions. The constant* $k > 0$ *is called the conductivity, which has metric units of* $J/(m\text{-}s\text{-}K)$. *A temperature function for a region* D *is given. Find the net outward heat flux* $\iint_S \mathbf{F} \cdot \mathbf{n} \, dS = -k \iint_S \nabla T \cdot \mathbf{n} \, dS$ *across the boundary* S *of* D. *In some cases, it may be easier to use the Divergence Theorem and evaluate a triple integral. Assume* $k = 1$.

41. $T(x, y, z) = 100 + x + 2y + z$;
$D = \{(x, y, z): 0 \le x \le 1, 0 \le y \le 1, 0 \le z \le 1\}$

42. $T(x, y, z) = 100 + x^2 + y^2 + z^2$;
$D = \{(x, y, z): 0 \le x \le 1, 0 \le y \le 1, 0 \le z \le 1\}$

43. $T(x, y, z) = 100 + e^{-z}$;
$D = \{(x, y, z): 0 \le x \le 1, 0 \le y \le 1, 0 \le z \le 1\}$

44. $T(x, y, z) = 100 + x^2 + y^2 + z^2$; D is the unit sphere centered at the origin.

T 45. $T(x, y, z) = 100e^{-x^2 - y^2 - z^2}$; D is the sphere of radius a centered at the origin.

Explorations and Challenges

46. Inverse square fields are special Let $\mathbf{F}$ be a radial field $\mathbf{F} = \mathbf{r}/|\mathbf{r}|^p$, where p is a real number and $\mathbf{r} = \langle x, y, z \rangle$. With $p = 3$, $\mathbf{F}$ is an inverse square field.

a. Show that the net flux across a sphere centered at the origin is independent of the radius of the sphere only for $p = 3$.

b. Explain the observation in part (a) by finding the flux of $\mathbf{F} = \mathbf{r}/|\mathbf{r}|^p$ across the boundaries of a spherical box $\{(\rho, \varphi, \theta): a \le \rho \le b, \varphi_1 \le \varphi \le \varphi_2, \theta_1 \le \theta \le \theta_2\}$ for various values of p.

47. A beautiful flux integral Consider the potential function $\varphi(x, y, z) = G(\rho)$, where G is any twice differentiable function and $\rho = \sqrt{x^2 + y^2 + z^2}$; therefore, G depends only on the distance from the origin.

a. Show that the gradient vector field associated with φ is

$$\mathbf{F} = \nabla\varphi = G'(\rho)\frac{\mathbf{r}}{\rho}, \text{ where } \mathbf{r} = \langle x, y, z \rangle \text{ and } \rho = |\mathbf{r}|.$$

b. Let S be the sphere of radius a centered at the origin and let D be the region enclosed by S. Show that the flux of $\mathbf{F}$ across S is $\iint_S \mathbf{F} \cdot \mathbf{n} \, dS = 4\pi a^2 G'(a)$.

c. Show that $\nabla \cdot \mathbf{F} = \nabla \cdot \nabla\varphi = \dfrac{2G'(\rho)}{\rho} + G''(\rho)$.

d. Use part (c) to show that the flux across S (as given in part (b)) is also obtained by the volume integral $\iiint_D \nabla \cdot \mathbf{F} \, dV$. (*Hint:* Use spherical coordinates and integrate by parts.)

48. Integration by parts (Gauss' Formula) Recall the Product Rule of Theorem 17.13: $\nabla \cdot (u\mathbf{F}) = \nabla u \cdot \mathbf{F} + u(\nabla \cdot \mathbf{F})$.

a. Integrate both sides of this identity over a solid region D with a closed boundary S, and use the Divergence Theorem to prove an integration by parts rule:

$$\iiint_D u(\nabla \cdot \mathbf{F}) \, dV = \iint_S u\mathbf{F} \cdot \mathbf{n} \, dS - \iiint_D \nabla u \cdot \mathbf{F} \, dV.$$

b. Explain the correspondence between this rule and the integration by parts rule for single-variable functions.

c. Use integration by parts to evaluate $\iiint_D (x^2y + y^2z + z^2x) \, dV$, where D is the cube in the first octant cut by the planes $x = 1$, $y = 1$, and $z = 1$.

49. Green's Formula Write Gauss' Formula of Exercise 48 in two dimensions—that is, where $\mathbf{F} = \langle f, g \rangle$, D is a plane region R and C is the boundary of R. Show that the result is Green's Formula:

$$\iint_R u(f_x + g_y) \, dA = \oint_C u(\mathbf{F} \cdot \mathbf{n}) \, ds - \iint_R (fu_x + gu_y) \, dA.$$

Show that with $u = 1$, one form of Green's Theorem appears. Which form of Green's Theorem is it?

50. Green's First Identity Prove Green's First Identity for twice differentiable scalar-valued functions u and v defined on a region D:

$$\iiint_D (u\nabla^2 v + \nabla u \cdot \nabla v) \, dV = \iint_S u\nabla v \cdot \mathbf{n} \, dS,$$

where $\nabla^2 v = \nabla \cdot \nabla v$. You may apply Gauss' Formula in Exercise 48 to $\mathbf{F} = \nabla v$ or apply the Divergence Theorem to $\mathbf{F} = u\nabla v$.

51. Green's Second Identity Prove Green's Second Identity for scalar-valued functions u and v defined on a region D:

$$\iiint_D (u\nabla^2 v - v\nabla^2 u) \, dV = \iint_S (u\nabla v - v\nabla u) \cdot \mathbf{n} \, dS.$$

(*Hint:* Reverse the roles of u and v in Green's First Identity.)

52–54. Harmonic functions *A scalar-valued function* φ *is harmonic on a region* D *if* $\nabla^2 \varphi = \nabla \cdot \nabla\varphi = 0$ *at all points of* D.

52. Show that the potential function $\varphi(x, y, z) = |\mathbf{r}|^{-p}$ is harmonic provided $p = 0$ or $p = 1$, where $\mathbf{r} = \langle x, y, z \rangle$. To what vector fields do these potentials correspond?

53. Show that if φ is harmonic on a region D enclosed by a surface S, then $\iint_S \nabla\varphi \cdot \mathbf{n} \, dS = 0$.

54. Show that if u is harmonic on a region D enclosed by a surface S, then $\iint_S u\nabla u \cdot \mathbf{n} \, dS = \iiint_D |\nabla u|^2 \, dV$.

55. Miscellaneous integral identities Prove the following identities.

a. $\iiint_D \nabla \times \mathbf{F} \, dV = \iint_S (\mathbf{n} \times \mathbf{F}) \, dS$ (*Hint:* Apply the Divergence Theorem to each component of the identity.)

b. $\iint_S (\mathbf{n} \times \nabla\varphi) \, dS = \oint_C \varphi \, d\mathbf{r}$ (*Hint:* Apply Stokes' Theorem to each component of the identity.)

1. If **F** is constant, then div **F** $= 0$, so $\iiint_D \nabla \cdot \mathbf{F}\, dV = \iint_S \mathbf{F} \cdot \mathbf{n}\, dS = 0$. This means that all the "material" that flows into one side of D flows out of the other side of D.
2. The vector field and the divergence are positive throughout D. **3.** The vector field has no flow into or out of the cube on the faces $x = 0$, $y = 0$, and $z = 0$ because the vectors of **F** on these faces are parallel to the faces. The vector field points out of the cube on the $x = 1$ and $z = 1$ faces and into the cube on the $y = 1$ face. div **F** $= 2$, so there is a net flow out of the cube. ◄

CHAPTER 17   REVIEW EXERCISES

1. **Explain why or why not** Determine whether the following statements are true and give an explanation or counterexample.

    a. The rotation field $\mathbf{F} = \langle -y, x \rangle$ has zero curl and zero divergence.
    b. $\nabla \times \nabla \varphi = \mathbf{0}$.
    c. Two vector fields with the same curl differ by a constant vector field.
    d. Two vector fields with the same divergence differ by a constant vector field.
    e. If $\mathbf{F} = \langle x, y, z \rangle$ and S encloses a region D, then $\iint_S \mathbf{F} \cdot \mathbf{n}\, dS$ is three times the volume of D.

2. **Matching vector fields** Match vector fields a–f with the graphs A–F. Let $\mathbf{r} = \langle x, y \rangle$.

    a. $\mathbf{F} = \langle x, y \rangle$ **b.** $\mathbf{F} = \langle -2y, 2x \rangle$
    c. $\mathbf{F} = \mathbf{r}/|\mathbf{r}|$ **d.** $\mathbf{F} = \langle y - x, x \rangle$
    e. $\mathbf{F} = \langle e^{-y}, e^{-x} \rangle$ **f.** $\mathbf{F} = \langle \sin \pi x, \sin \pi y \rangle$

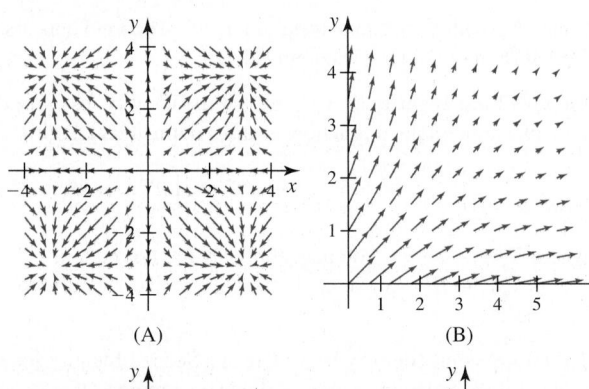

               (A)   (B)

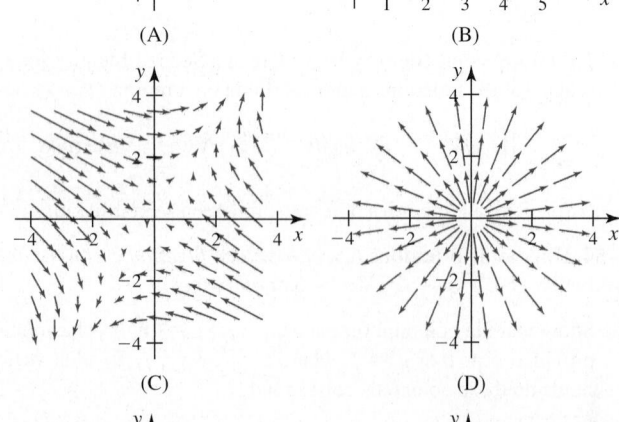

               (C)   (D)

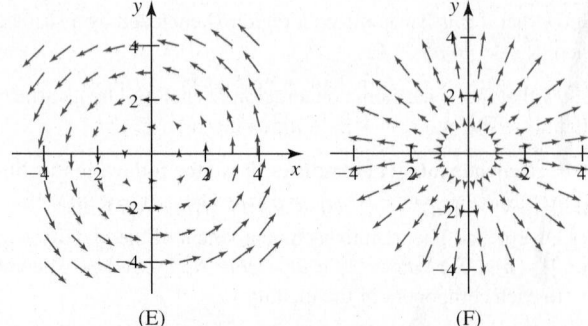

               (E)   (F)

3–4. Gradient fields in $\mathbb{R}^2$ *Find the vector field $\mathbf{F} = \nabla \varphi$ for the following potential functions. Sketch a few level curves of φ and a few vectors of $\mathbf{F}$ along the level curves.*

3. $\varphi(x, y) = x^2 + 4y^2$, for $|x| \leq 5$, $|y| \leq 5$

4. $\varphi(x, y) = (x^2 - y^2)/2$, for $|x| \leq 2$, $|y| \leq 2$

5–6. Gradient fields in $\mathbb{R}^3$ *Find the vector field $\mathbf{F} = \nabla \varphi$ for the following potential functions.*

5. $\varphi(x, y, z) = 1/|\mathbf{r}|$, where $\mathbf{r} = \langle x, y, z \rangle$

6. $\varphi(x, y, z) = \dfrac{1}{2} e^{-x^2 - y^2 - z^2}$

7. **Normal component** Let C be the circle of radius 2 centered at the origin with counterclockwise orientation. Give the unit outward normal vector at any point (x, y) on C.

8–10. Line integrals *Evaluate the following line integrals.*

8. $\displaystyle\int_C (x^2 - 2xy + y^2)\, ds$; C is the upper half of a circle
    $\mathbf{r}(t) = \langle 5 \cos t, 5 \sin t \rangle$, for $0 \leq t \leq \pi$.

9. $\displaystyle\int_C y e^{-xz}\, ds$; C is the path $\mathbf{r}(t) = \langle 2t, 3t, -6t \rangle$, for $0 \leq t \leq 2$.

10. $\displaystyle\int_C (xz - y^2)\, ds$; C is the line segment from $(0, 1, 2)$ to $(-3, 7, -1)$.

11. **Two parameterizations** Verify that $\oint_C (x - 2y + 3z)\, ds$ has the same value when C is given by $\mathbf{r}(t) = \langle 2 \cos t, 2 \sin t, 0 \rangle$, for $0 \leq t \leq 2\pi$, and by $\mathbf{r}(t) = \langle 2 \cos t^2, 2 \sin t^2, 0 \rangle$, for $0 \leq t \leq \sqrt{2\pi}$.

12. **Work integral** Find the work done in moving an object from $P(1, 0, 0)$ to $Q(0, 1, 0)$ in the presence of the force $\mathbf{F} = \langle 1, 2y, -4z \rangle$ along the following paths.

    a. The line segment from P to Q
    b. The line segment from P to $O(0, 0, 0)$ followed by the line segment from O to Q
    c. The arc of the quarter circle from P to Q
    d. Is the work independent of the path?

13–14. Work integrals in $\mathbb{R}^3$ *Given the force field $\mathbf{F}$, find the work required to move an object on the given curve.*

13. $\mathbf{F} = \langle -y, z, x \rangle$ on the path consisting of the line segment from $(0, 0, 0)$ to $(0, 1, 0)$ followed by the line segment from $(0, 1, 0)$ to $(0, 1, 4)$

14. $\mathbf{F} = \dfrac{\langle x, y, z \rangle}{(x^2 + y^2 + z^2)^{3/2}}$ on the path $\mathbf{r}(t) = \langle t^2, 3t^2, -t^2 \rangle$, for $1 \leq t \leq 2$

15–18. Circulation and flux *Find the circulation and the outward flux of the following vector fields for the curve* $\mathbf{r}(t) = \langle 2\cos t, 2\sin t \rangle$, *for* $0 \le t \le 2\pi$.

15. $\mathbf{F} = \langle y - x, y \rangle$ **16.** $\mathbf{F} = \langle x, y \rangle$

17. $\mathbf{F} = \mathbf{r}/|\mathbf{r}|^2$, where $\mathbf{r} = \langle x, y \rangle$

18. $\mathbf{F} = \langle x - y, x \rangle$

19. Flux in channel flow Consider the flow of water in a channel whose boundaries are the planes $y = \pm L$ and $z = \pm 1/2$. The velocity field in the channel is $\mathbf{v} = \langle v_0(L^2 - y^2), 0, 0 \rangle$. Find the flux across the cross section of the channel at $x = 0$ in terms of v_0 and L.

20–23. Conservative vector fields and potentials *Determine whether the following vector fields are conservative on their domains. If so, find a potential function.*

20. $\mathbf{F} = \langle y^2, 2xy \rangle$ **21.** $\mathbf{F} = \langle y, x + z^2, 2yz \rangle$

22. $\mathbf{F} = \langle e^x \cos y, -e^x \sin y \rangle$ **23.** $\mathbf{F} = e^z \langle y, x, xy \rangle$

24–27. Evaluating line integrals *Evaluate the line integral* $\int_C \mathbf{F} \cdot d\mathbf{r}$ *for the following vector fields* $\mathbf{F}$ *and curves* C *in two ways.*

a. *By parameterizing* C
b. *By using the Fundamental Theorem for line integrals, if possible*

24. $\mathbf{F} = \nabla(x^2 y)$; C: $\mathbf{r}(t) = \langle 9 - t^2, t \rangle$, for $0 \le t \le 3$

25. $\mathbf{F} = \nabla(xyz)$; C: $\mathbf{r}(t) = \langle \cos t, \sin t, t/\pi \rangle$, for $0 \le t \le \pi$

26. $\mathbf{F} = \langle x, -y \rangle$; C is the square with vertices $(\pm 1, \pm 1)$ with counterclockwise orientation.

27. $\mathbf{F} = \langle y, z, -x \rangle$; C: $\mathbf{r}(t) = \langle \cos t, \sin t, 4 \rangle$, for $0 \le t \le 2\pi$

28. Radial fields in $\mathbb{R}^2$ are conservative Prove that the radial field $\mathbf{F} = \mathbf{r}/|\mathbf{r}|^p$, where $\mathbf{r} = \langle x, y \rangle$ and p is a real number, is conservative on $\mathbb{R}^2$ with the origin removed. For what value of p is $\mathbf{F}$ conservative on $\mathbb{R}^2$ (including the origin)?

29–32. Green's Theorem for line integrals *Use either form of Green's Theorem to evaluate the following line integrals.*

29. $\oint_C xy^2\, dx + x^2 y\, dy$; C is the triangle with vertices $(0, 0)$, $(2, 0)$, and $(0, 2)$ with counterclockwise orientation.

30. $\oint_C (-3y + x^{3/2})\, dx + (x - y^{2/3})\, dy$; C is the boundary of the half-disk $\{(x, y): x^2 + y^2 \le 2, y \ge 0\}$ with counterclockwise orientation.

31. $\oint_C (x^3 + xy)\, dy + (2y^2 - 2x^2 y)\, dx$; C is the square with vertices $(\pm 1, \pm 1)$ with counterclockwise orientation.

32. $\oint_C 3x^3\, dy - 3y^3\, dx$; C is the circle of radius 4 centered at the origin with *clockwise* orientation.

33–34. Areas of plane regions *Find the area of the following regions using a line integral.*

33. The region enclosed by the ellipse $x^2 + 4y^2 = 16$

34. The region bounded by the hypocycloid $\mathbf{r}(t) = \langle \cos^3 t, \sin^3 t \rangle$, for $0 \le t \le 2\pi$

35–36. Circulation and flux *Consider the following vector fields.*

a. *Compute the circulation on the boundary of the region R (with counterclockwise orientation).*
b. *Compute the outward flux across the boundary of R.*

35. $\mathbf{F} = \mathbf{r}/|\mathbf{r}|$, where $\mathbf{r} = \langle x, y \rangle$ and R is the half-annulus $\{(r, \theta): 1 \le r \le 3, 0 \le \theta \le \pi\}$

36. $\mathbf{F} = \langle -\sin y, x \cos y \rangle$, where R is the square $\{(x, y): 0 \le x \le \pi/2, 0 \le y \le \pi/2\}$

37. Parameters Let $\mathbf{F} = \langle ax + by, cx + dy \rangle$, where a, b, c, and d are constants.

a. For what values of a, b, c, and d is $\mathbf{F}$ conservative?
b. For what values of a, b, c, and d is $\mathbf{F}$ source free?
c. For what values of a, b, c, and d is $\mathbf{F}$ conservative and source free?

38–41. Divergence and curl *Compute the divergence and curl of the following vector fields. State whether the field is source free or irrotational.*

38. $\mathbf{F} = \langle yz, xz, xy \rangle$

39. $\mathbf{F} = \mathbf{r}|\mathbf{r}| = \langle x, y, z \rangle \sqrt{x^2 + y^2 + z^2}$

40. $\mathbf{F} = \langle \sin xy, \cos yz, \sin xz \rangle$

41. $\mathbf{F} = \langle 2xy + z^4, x^2, 4xz^3 \rangle$

42. Identities Prove that $\nabla\left(\dfrac{1}{|\mathbf{r}|^4}\right) = -\dfrac{4\mathbf{r}}{|\mathbf{r}|^6}$ and use the result to prove that $\nabla \cdot \nabla\left(\dfrac{1}{|\mathbf{r}|^4}\right) = \dfrac{12}{|\mathbf{r}|^6}$.

43. Maximum curl Let $\mathbf{F} = \langle z, x, -y \rangle$.

a. What is the scalar component of curl $\mathbf{F}$ in the direction of $\mathbf{n} = \langle 1, 0, 0 \rangle$?
b. What is the scalar component of curl $\mathbf{F}$ in the direction of $\mathbf{n} = \langle 0, -1/\sqrt{2}, 1/\sqrt{2} \rangle$?
c. In the direction of what unit vector $\mathbf{n}$ is the scalar component of curl $\mathbf{F}$ a maximum?

44. Paddle wheel in a vector field Let $\mathbf{F} = \langle 0, 2x, 0 \rangle$ and let $\mathbf{n}$ be a unit vector aligned with the axis of a paddle wheel located on the y-axis.

a. If the axis of the paddle wheel is aligned with $\mathbf{n} = \langle 1, 0, 0 \rangle$, how fast does it spin?
b. If the axis of the paddle wheel is aligned with $\mathbf{n} = \langle 0, 0, 1 \rangle$, how fast does it spin?
c. For what direction $\mathbf{n}$ does the paddle wheel spin fastest?

45–48. Surface areas *Use a surface integral to find the area of the following surfaces.*

45. The hemisphere $x^2 + y^2 + z^2 = 9$, for $z \ge 0$

46. The frustum of the cone $z^2 = x^2 + y^2$, for $2 \le z \le 4$ (excluding the bases)

47. The plane $z = 6 - x - y$ above the square $|x| \le 1$, $|y| \le 1$

48. The surface $f(x, y) = \sqrt{2}\, xy$ above the polar region $\{(r, \theta): 0 \le r \le 2, 0 \le \theta \le 2\pi\}$

49–51. Surface integrals *Evaluate the following surface integrals.*

49. $\iint_S (1 + yz)\, dS$; S is the plane $x + y + z = 2$ in the first octant.

50. $\iint_S \langle 0, y, z \rangle \cdot \mathbf{n} \, dS$; S is the curved surface of the cylinder $y^2 + z^2 = a^2$, for $|x| \le 8$, with outward normal vectors.

51. $\iint_S (x - y + z) \, dS$; S is the entire surface, including the base, of the hemisphere $x^2 + y^2 + z^2 = 4$, for $z \ge 0$.

52–53. Flux integrals *Find the flux of the following vector fields across the given surface. Assume the vectors normal to the surface point outward.*

52. $\mathbf{F} = \langle x, y, z \rangle$ across the curved surface of the cylinder $x^2 + y^2 = 1$, for $|z| \le 8$

53. $\mathbf{F} = \mathbf{r}/|\mathbf{r}|$ across the sphere of radius a centered at the origin, where $\mathbf{r} = \langle x, y, z \rangle$

54. Three methods Find the surface area of the paraboloid $z = x^2 + y^2$, for $0 \le z \le 4$, in three ways.

   a. Use an explicit description of the surface.

   b. Use the parametric description $\mathbf{r} = \langle v \cos u, v \sin u, v^2 \rangle$.

   c. Use the parametric description $\mathbf{r} = \langle \sqrt{v} \cos u, \sqrt{v} \sin u, v \rangle$.

55. Flux across hemispheres and paraboloids Let S be the hemisphere $x^2 + y^2 + z^2 = a^2$, for $z \ge 0$, and let T be the paraboloid $z = a - (x^2 + y^2)/a$, for $z \ge 0$, where $a > 0$. Assume the surfaces have outward normal vectors.

   a. Verify that S and T have the same base ($x^2 + y^2 \le a^2$) and the same high point $(0, 0, a)$.

   b. Which surface has the greater area?

   c. Show that the flux of the radial field $\mathbf{F} = \langle x, y, z \rangle$ across S is $2\pi a^3$.

   d. Show that the flux of the radial field $\mathbf{F} = \langle x, y, z \rangle$ across T is $3\pi a^3/2$.

56. Surface area of an ellipsoid Consider the ellipsoid $x^2/a^2 + y^2/b^2 + z^2/c^2 = 1$, where a, b, and c are positive real numbers.

   a. Show that the surface is described by the parametric equations

$$\mathbf{r}(u, v) = \langle a \cos u \sin v, b \sin u \sin v, c \cos v \rangle$$

   for $0 \le u \le 2\pi$, $0 \le v \le \pi$.

   b. Write an integral for the surface area of the ellipsoid.

57–58. Stokes' Theorem for line integrals *Evaluate the line integral $\oint_C \mathbf{F} \cdot d\mathbf{r}$ using Stokes' Theorem. Assume C has counterclockwise orientation.*

57. $\mathbf{F} = \langle xz, yz, xy \rangle$; C is the circle $x^2 + y^2 = 4$ in the xy-plane.

58. $\mathbf{F} = \langle x^2 - y^2, x, 2yz \rangle$; C is the boundary of the plane $z = 6 - 2x - y$ in the first octant.

59–60. Stokes' Theorem for surface integrals *Use Stokes' Theorem to evaluate the surface integral $\iint_S (\nabla \times \mathbf{F}) \cdot \mathbf{n} \, dS$. Assume $\mathbf{n}$ is the outward normal.*

59. $\mathbf{F} = \langle -z, x, y \rangle$, where S is the hyperboloid $z = 10 - \sqrt{1 + x^2 + y^2}$, for $z \ge 0$

60. $\mathbf{F} = \langle x^2 - z^2, y^2, xz \rangle$, where S is the hemisphere $x^2 + y^2 + z^2 = 4$, for $y \ge 0$

61. Conservative fields Use Stokes' Theorem to find the circulation of the vector field $\mathbf{F} = \nabla(10 - x^2 + y^2 + z^2)$ around any smooth closed curve C with counterclockwise orientation.

62–64. Computing fluxes *Use the Divergence Theorem to compute the outward flux of the following vector fields across the given surfaces S.*

62. $\mathbf{F} = \langle -x, x - y, x - z \rangle$; S is the surface of the cube cut from the first octant by the planes $x = 1$, $y = 1$, and $z = 1$.

63. $\mathbf{F} = \langle x^3, y^3, z^3 \rangle /3$; S is the sphere $\{(x, y, z): x^2 + y^2 + z^2 = 9\}$.

64. $\mathbf{F} = \langle x^2, y^2, z^2 \rangle$; S is the cylinder $\{(x, y, z): x^2 + y^2 = 4, 0 \le z \le 8\}$.

65–66. General regions *Use the Divergence Theorem to compute the outward flux of the following vector fields across the boundary of the given regions D.*

65. $\mathbf{F} = \langle x^3, y^3, 10 \rangle$; D is the region between the hemispheres of radius 1 and 2 centered at the origin with bases in the xy-plane.

66. $\mathbf{F} = \dfrac{\mathbf{r}}{|\mathbf{r}|^3} = \dfrac{\langle x, y, z \rangle}{(x^2 + y^2 + z^2)^{3/2}}$; D is the region between two spheres with radii 1 and 2 centered at $(5, 5, 5)$.

67. Flux integrals Compute the outward flux of the field $\mathbf{F} = \langle x^2 + x \sin y, y^2 + 2 \cos y, z^2 + z \sin y \rangle$ across the surface S that is the boundary of the prism bounded by the planes $y = 1 - x$, $x = 0$, $y = 0$, $z = 0$, and $z = 4$.

68. Stokes' Theorem on a compound surface Consider the surface S consisting of the quarter-sphere $x^2 + y^2 + z^2 = a^2$, for $z \ge 0$ and $x \ge 0$, and the half-disk in the yz-plane $y^2 + z^2 \le a^2$, for $z \ge 0$. The boundary of S in the xy-plane is C, which consists of the semicircle $x^2 + y^2 = a^2$, for $x \ge 0$, and the line segment $[-a, a]$ on the y-axis, with a counterclockwise orientation. Let $\mathbf{F} = \langle 2z - y, x - z, y - 2x \rangle$.

   a. Describe the direction in which the normal vectors point on S.

   b. Evaluate $\oint_C \mathbf{F} \cdot d\mathbf{r}$

   c. Evaluate $\iint_S (\nabla \times \mathbf{F}) \cdot \mathbf{n} \, dS$ and check for agreement with part (b).

Chapter 17 Guided Projects

Applications of the material in this chapter and related topics can be found in the following Guided Projects. For additional information, see the Preface.

• Ideal fluid flow

• Maxwell's equations

• Planimeters and vector fields

• Vector calculus in other coordinate systems

A

Appendix

Proofs of Selected Theorems

THEOREM 2.3 **Limit Laws**

Assume $\lim\limits_{x \to a} f(x)$ and $\lim\limits_{x \to a} g(x)$ exist. The following properties hold, where c is a real number, and $n > 0$ is an integer.

1. **Sum** $\lim\limits_{x \to a} (f(x) + g(x)) = \lim\limits_{x \to a} f(x) + \lim\limits_{x \to a} g(x)$

2. **Difference** $\lim\limits_{x \to a} (f(x) - g(x)) = \lim\limits_{x \to a} f(x) - \lim\limits_{x \to a} g(x)$

3. **Constant multiple** $\lim\limits_{x \to a} (cf(x)) = c \lim\limits_{x \to a} f(x)$

4. **Product** $\lim\limits_{x \to a} (f(x)g(x)) = \left(\lim\limits_{x \to a} f(x)\right)\left(\lim\limits_{x \to a} g(x)\right)$

5. **Quotient** $\lim\limits_{x \to a} \dfrac{f(x)}{g(x)} = \dfrac{\lim\limits_{x \to a} f(x)}{\lim\limits_{x \to a} g(x)}$, provided $\lim\limits_{x \to a} g(x) \neq 0$

6. **Power** $\lim\limits_{x \to a} (f(x))^n = \left(\lim\limits_{x \to a} f(x)\right)^n$

7. **Root** $\lim\limits_{x \to a} (f(x))^{1/n} = \left(\lim\limits_{x \to a} f(x)\right)^{1/n}$, provided $f(x) > 0$, for x near a, if n is even

---

Proof: The proof of Law 1 is given in Example 6 of Section 2.7. The proof of Law 2 is analogous to that of Law 1; the triangle inequality in the form $|x - y| \leq |x| + |y|$ is used (Exercise 43, Section 2.7). The proof of Law 3 is outlined in Exercise 44 of Section 2.7. The proofs of Laws 4 and 5 are given below. The proof of Law 6 involves the repeated use of Law 4. Law 7 is a special case of Theorem 2.12, whose proof is given in this appendix (p. AP-2). ◄

Proof of Product Law: Let $L = \lim\limits_{x \to a} f(x)$ and $M = \lim\limits_{x \to a} g(x)$. Using the definition of a limit, the goal is to show that given any $\varepsilon > 0$, it is possible to specify a $\delta > 0$ such that $|f(x)g(x) - LM| < \varepsilon$ whenever $0 < |x - a| < \delta$. Notice that

$$\begin{aligned}
|f(x)g(x) - LM| &= |f(x)g(x) - Lg(x) + Lg(x) - LM| && \text{Add and subtract } Lg(x). \\
&= |(f(x) - L)g(x) + (g(x) - M)L| && \text{Group terms.} \\
&\leq |(f(x) - L)g(x)| + |(g(x) - M)L| && \text{Triangle inequality} \\
&= |f(x) - L||g(x)| + |g(x) - M||L|. && |xy| = |x||y|
\end{aligned}$$

> Real numbers x and y obey the triangle inequality $|x + y| \le |x| + |y|$.

We now use the definition of the limits of f and g, and we note that L and M are fixed real numbers. Given $\varepsilon > 0$, there exist $\delta_1 > 0$ and $\delta_2 > 0$ such that

$$|f(x) - L| < \frac{\varepsilon}{2(|M| + 1)} \quad \text{and} \quad |g(x) - M| < \frac{\varepsilon}{2(|L| + 1)}$$

> $|g(x) - M| < 1$ implies that $g(x)$ is less than 1 unit from M. Therefore, whether $g(x)$ and M are positive or negative, $|g(x)| < |M| + 1$.

whenever $0 < |x - a| < \delta_1$ and $0 < |x - a| < \delta_2$, respectively. Furthermore, by the definition of the limit of g, there exits a $\delta_3 > 0$ such that $|g(x) - M| < 1$ whenever $0 < |x - a| < \delta_3$. It follows that $|g(x)| < |M| + 1$ whenever $0 < |x - a| < \delta_3$. Now take δ to be the minimum of δ_1, δ_2, and δ_3. Then for $0 < |x - a| < \delta$, we have

$$|f(x)g(x) - LM| \le \underbrace{|f(x) - L|}_{<\frac{\varepsilon}{2(|M| + 1)}} \underbrace{|g(x)|}_{<(|M| + 1)} + \underbrace{|g(x) - M|}_{<\frac{\varepsilon}{2(|L| + 1)}} |L|$$

$$< \frac{\varepsilon}{2} + \frac{\varepsilon}{2} \underbrace{\frac{|L|}{|L| + 1}}_{<1} < \frac{\varepsilon}{2} + \frac{\varepsilon}{2} = \varepsilon.$$

It follows that $\lim\limits_{x \to a} (f(x)g(x)) = LM$.   ◄

Proof of Quotient Law: We first prove that if $\lim\limits_{x \to a} g(x) = M$ exists, where $M \ne 0$, then $\lim\limits_{x \to a} \dfrac{1}{g(x)} = \dfrac{1}{M}$. The Quotient Law then follows when we replace g with $1/g$ in the Product Law. Therefore, the goal is to show that given any $\varepsilon > 0$, it is possible to specify a $\delta > 0$ such that $\left| \dfrac{1}{g(x)} - \dfrac{1}{M} \right| < \varepsilon$ whenever $0 < |x - a| < \delta$. First note that $M \ne 0$ and $g(x)$ can be made arbitrarily close to M. For this reason, there exists a $\delta_1 > 0$ such that $|g(x)| > |M|/2$, or equivalently, $1/|g(x)| < 2/|M|$, whenever $0 < |x - a| < \delta_1$. Furthermore, using the definition of the limit of g, given any $\varepsilon > 0$, there exists a $\delta_2 > 0$ such that $|g(x) - M| < \dfrac{\varepsilon|M|^2}{2}$ whenever $0 < |x - a| < \delta_2$. Now take δ to be the minimum of δ_1 and δ_2. Then for $0 < |x - a| < \delta$, we have

> Note that if $|g(x)| > |M|/2$, then $1/|g(x)| < 2/|M|$.

$$\left| \frac{1}{g(x)} - \frac{1}{M} \right| = \left| \frac{M - g(x)}{Mg(x)} \right| \qquad \text{Common denominator}$$

$$= \frac{1}{|M|} \underbrace{\frac{1}{|g(x)|}}_{<\frac{2}{|M|}} \underbrace{|g(x) - M|}_{<\frac{\varepsilon|M|^2}{2}} \qquad \text{Rewrite.}$$

$$< \frac{1}{|M|} \frac{2}{|M|} \frac{\varepsilon|M|^2}{2} = \varepsilon. \qquad \text{Simplify.}$$

By the definition of a limit, we have $\lim\limits_{x \to a} \dfrac{1}{g(x)} = \dfrac{1}{M}$. The proof can be completed by applying the Product Law with g replaced with $1/g$.

THEOREM 2.12   Limits of Composite Functions

1. If g is continuous at a and f is continuous at $g(a)$, then

$$\lim_{x \to a} f(g(x)) = f\left(\lim_{x \to a} g(x) \right).$$

2. If $\lim\limits_{x \to a} g(x) = L$ and f is continuous at L, then

$$\lim_{x \to a} f(g(x)) = f\left(\lim_{x \to a} g(x) \right).$$

Proof: Part 1 was proved in Section 2.6 (it is also a direct consequence of Part 2). Let $\varepsilon > 0$ be given. We need to show that there exists $\delta > 0$ such that $|f(g(x)) - f(L)| < \varepsilon$ whenever $0 < |x - a| < \delta$. Because f is continuous at L, there exists $\delta_1 > 0$ such that

$$|f(y) - f(L)| < \varepsilon \text{ whenever } |y - L| < \delta_1.$$

We also know that $\lim\limits_{x \to a} g(x) = L$, so there exists $\delta > 0$ such that

$$|g(x) - L| < \delta_1 \text{ whenever } 0 < |x - a| < \delta.$$

Combining these results, if $0 < |x - a| < \delta$, then $|g(x) - L| < \delta_1$, and the inequality $|g(x) - L| < \delta_1$ implies that $|f(g(x)) - f(L)| < \varepsilon$. Therefore, $|f(g(x)) - f(L)| < \varepsilon$ whenever $0 < |x - a| < \delta$.   ◀

THEOREM 10.20   Ratio Test

Let $\sum a_k$ be an infinite series, and let $r = \lim\limits_{k \to \infty} \left| \dfrac{a_{k+1}}{a_k} \right|$.

1. If $r < 1$, the series converges absolutely, and therefore it converges (Theorem 10.19).
2. If $r > 1$ (including $r = \infty$), the series diverges.
3. If $r = 1$, the test is inconclusive.

Proof: We consider three cases: $r < 1$, $r > 1$, and $r = 1$.

1. Assume $r < 1$ and choose a number R such that $r < R < 1$. Because the sequence $\left\{ \left| \dfrac{a_{k+1}}{a_k} \right| \right\}$ converges to a number r less than R, eventually all the terms in the tail of the sequence $\left\{ \left| \dfrac{a_{k+1}}{a_k} \right| \right\}$ are less than R. That is, there is a positive integer N such that $\left| \dfrac{a_{k+1}}{a_k} \right| < R$, for all $k > N$. Multiplying both sides of this inequality by $\dfrac{|a_k|}{R^{k+1}}$, we have $\dfrac{|a_{k+1}|}{R^{k+1}} < \dfrac{|a_k|}{R^k}$, for all $k > N$. So the sequence $\left\{ \dfrac{|a_k|}{R^k} \right\}_{k=N+1}^{\infty}$ is decreasing and it follows that $\dfrac{|a_k|}{R^k} < \dfrac{|a_{N+1}|}{R^{N+1}}$, for all $k \geq N + 1$. By letting $c = \dfrac{|a_{N+1}|}{R^{N+1}}$, we have

$$0 < |a_k| \leq cR^k, \text{ for all } k \geq N + 1. \text{ Let } S_n \text{ represent the } n\text{th partial sum of } \sum_{k=N+1}^{\infty} |a_k|;$$

note that the partial sums of this series are bounded by a convergent geometric series:

$$\begin{aligned}
S_n &= |a_{N+1}| + |a_{N+2}| + \cdots + |a_{N+n}| \\
&\leq cR^{N+1} + cR^{N+2} + \cdots + cR^{N+n} \\
&< cR^{N+1} + cR^{N+2} + \cdots + cR^{N+n} + \cdots \\
&= \frac{cR^{N+1}}{1 - R}.
\end{aligned}$$

Because the sequence $\{S_n\}$ is increasing (each partial sum in the sequence consists of positive terms) and is bounded above by $\dfrac{cR^{N+1}}{1 - R}$, it converges by the Bounded Monotonic Sequences Theorem (Theorem 10.5). Therefore, $\sum\limits_{k=N+1}^{\infty} |a_k|$ converges and we conclude that $\sum\limits_{k=1}^{\infty} |a_k|$ converges (Theorem 10.8), which in turn implies that $\sum\limits_{k=1}^{\infty} a_k$ converges (Theorem 10.19).

2. If $r > 1$, there is a positive integer N for which $\left|\dfrac{a_{k+1}}{a_k}\right| > 1$, or equivalently $|a_{k+1}| > |a_k|$, for all $k > N$. So every term in the sequence $\{|a_k|\}_{k=N+1}^{\infty}$ is greater than or equal to the positive number $|a_{N+1}|$, which implies that $\lim\limits_{n\to\infty} a_n \neq 0$. Therefore, the series $\sum\limits_{k=N+1}^{\infty} a_k$ diverges by the Divergence Test, and we conclude that $\sum\limits_{k=1}^{\infty} a_k$ diverges (Theorem 10.8).

3. In the case that $r = 1$, the series $\sum\limits_{k=1}^{\infty} a_k$ may or may not converge. For example, both the divergent harmonic series $\sum\limits_{k=1}^{\infty}\dfrac{1}{k}$ and the convergent p-series $\sum\limits_{k=1}^{\infty}\dfrac{1}{k^2}$ produce a value of $r = 1$. ◄

THEOREM 10.21 Root Test

Let $\sum a_k$ be an infinite series, and let $\rho = \lim\limits_{k\to\infty}\sqrt[k]{|a_k|}$.

1. If $\rho < 1$, the series converges absolutely, and therefore it converges (Theorem 10.19).

2. If $\rho > 1$ (including $\rho = \infty$), the series diverges.

3. If $\rho = 1$, the test is inconclusive.

Proof: We consider three cases: $\rho < 1$, $\rho > 1$, and $\rho = 1$.

1. Assume $\rho < 1$ and choose a number R such that $\rho < R < 1$. Because the sequence $\left\{\sqrt[k]{|a_k|}\right\}$ converges to a number less than R, there is a positive integer N such that $\sqrt[k]{|a_k|} < R$, or equivalently $|a_k| < R^k$, for all $k > N$. Let S_n represent the nth partial sum of $\sum\limits_{k=N+1}^{\infty} |a_k|$; note that the partial sums of this series are bounded by a convergent geometric series:

$$
\begin{aligned}
S_n &= |a_{N+1}| + |a_{N+2}| + \cdots + |a_{N+n}| \\
&\leq R^{N+1} + R^{N+2} + \cdots + R^{N+n} \\
&< R^{N+1} + R^{N+2} + \cdots + R^{N+n} + \cdots \\
&= \frac{R^{N+1}}{1-R}.
\end{aligned}
$$

Because the sequence $\{S_n\}$ is nondecreasing (each partial sum in the sequence consists of nonnegative terms) and is bounded above by $\dfrac{R^{N+1}}{1-R}$, it converges by the Bounded Monotonic Sequences Theorem (Theorem 10.5). Therefore, $\sum\limits_{k=N+1}^{\infty} |a_k|$ converges, and we conclude that $\sum\limits_{k=1}^{\infty} |a_k|$ converges (Theorem 10.8), which in turn implies that $\sum\limits_{k=1}^{\infty} a_k$ converges (Theorem 10.19).

2. If $\rho > 1$, there is an integer N for which $\sqrt[k]{|a_k|} > 1$, or equivalently $|a_k| > 1$, for all $k > N$. So every term in the sequence $\{|a_k|\}_{k=N+1}^{\infty}$ is greater than or equal to 1, which implies that $\lim\limits_{n\to\infty} a_n \neq 0$. Therefore, the series $\sum\limits_{k=N+1}^{\infty} a_k$ diverges by the Divergence Test, and we conclude that $\sum\limits_{k=1}^{\infty} a_k$ diverges (Theorem 10.8).

3. If $\rho = 1$, the series $\sum\limits_{k=1}^{\infty} a_k$ may or may not converge. For example, both the divergent harmonic series $\sum\limits_{k=1}^{\infty}\dfrac{1}{k}$ and the convergent p-series $\sum\limits_{k=1}^{\infty}\dfrac{1}{k^2}$ produce a value of $\rho = 1$. ◄

THEOREM 11.3   Convergence of Power Series

A power series $\sum\limits_{k=0}^{\infty} c_k(x-a)^k$ centered at a converges in one of three ways:

1. The series converges absolutely for all x. It follows, by Theorem 10.19, that the series converges, in which case the interval of convergence is $(-\infty, \infty)$ and the radius of convergence is $R = \infty$.

2. There is a real number $R > 0$ such that the series converges absolutely (and therefore converges) for $|x-a| < R$ and diverges for $|x-a| > R$, in which case the radius of convergence is R.

3. The series converges only at a, in which case the radius of convergence is $R = 0$.

Proof: Without loss of generality, we take $a = 0$. (If $a \neq 0$, the following argument may be shifted so it is centered at $x = a$.) The proof hinges on a preliminary result:

If $\sum\limits_{k=0}^{\infty} c_k x^k$ converges for $x = b \neq 0$, then it converges absolutely, for

$|x| < |b|$. If $\sum\limits_{k=0}^{\infty} c_k x^k$ diverges for $x = d$, then it diverges, for $|x| > |d|$.

To prove these facts, assume that $\sum\limits_{k=0}^{\infty} c_k b^k$ converges, which implies that $\lim\limits_{k\to\infty} c_k b^k = 0$.
Then there exists a real number $M > 0$ such that $\left|c_k b^k\right| < M$, for $k = 0, 1, 2, 3, \ldots$. It follows that

$$\sum_{k=0}^{\infty} |c_k x^k| = \sum_{k=0}^{\infty} \underbrace{|c_k b^k|}_{<M} \left|\frac{x}{b}\right|^k < M \sum_{k=0}^{\infty} \left|\frac{x}{b}\right|^k .$$

If $|x| < |b|$, then $|x/b| < 1$ and $\sum\limits_{k=0}^{\infty} \left|\dfrac{x}{b}\right|^k$ is a convergent geometric series. Therefore, $\sum\limits_{k=0}^{\infty} |c_k x^k|$ converges by the comparison test, which implies that $\sum\limits_{k=0}^{\infty} c_k x^k$ converges absolutely for $|x| < |b|$. The second half of the preliminary result is proved by supposing the series diverges at $x = d$. The series cannot converge at a point x_0 with $|x_0| > |d|$ because by the preceding argument, it would converge for $|x| < |x_0|$, which includes $x = d$. Therefore, the series diverges for $|x| > |d|$.

Now we may deal with the three cases in the theorem. Let S be the set of real numbers for which the series converges, which always includes 0. If $S = \{0\}$, then we have Case 3. If S consists of all real numbers, then we have Case 1. For Case 2, assume that $d \neq 0$ is a point at which the series diverges. By the preliminary result, the series diverges for $|x| > |d|$. Therefore, if x is in S, then $|x| < |d|$, which implies that S is bounded. By the Least Upper Bound Property for real numbers, S has a least upper bound R, such that $x \leq R$, for all x in S. If $|x| > R$, then x is not in S and the series diverges. If $|x| < R$, then x is not the least upper bound of S and there exists a number b in S with $|x| < b \leq R$.

Because the series converges at $x = b$, by the preliminary result, $\sum\limits_{k=0}^{\infty} |c_k x^k|$ converges for $|x| < |b|$. Therefore, the series $\sum\limits_{k=0}^{\infty} c_k x^k$ converges absolutely (which, by Theorem 10.19, implies the series converges) for $|x| < R$ and diverges for $|x| > R$. ◀

> ▶ The Least Upper Bound Property for real numbers states that if a nonempty set S is bounded (that is, there exists a number M, called an *upper bound*, such that $x \leq M$ for all x in S), then S has a *least upper bound* L, which is the smallest of the upper bounds.

THEOREM 12.3   Eccentricity-Directrix Theorem

Suppose ℓ is a line, F is a point not on ℓ, and e is a positive real number. Let C be

the set of points P in a plane with the property that $\dfrac{|PF|}{|PL|} = e$, where $|PL|$ is the
perpendicular distance from P to ℓ.

1. If $e = 1$, C is a **parabola**.
2. If $0 < e < 1$, C is an **ellipse**.
3. If $e > 1$, C is a **hyperbola**.

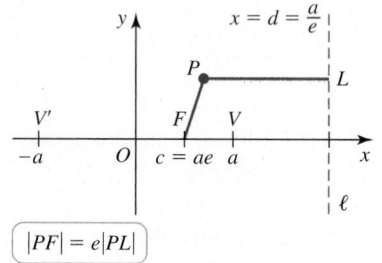

$|PF| = e|PL|$

Figure A.1

Proof: If $e = 1$, then the defining property becomes $|PF| = |PL|$, which is the standard definition of a parabola (Section 12.4). We prove the result for ellipses $(0 < e < 1)$; a small modification handles the case of hyperbolas $(e > 1)$.

Let E be the curve whose points satisfy $|PF| = e\,|PL|$ and suppose $0 < e < 1$; the goal is to show that E is an ellipse. There are two points on the x-axis (the *vertices*), call them V and V', that satisfy $|VF| = e|VL|$ and $|V'F| = e|V'L|$. (Note that $|VL|$ is the perpendicular distance from V to the line ℓ, not the distance from the points labeled V and L in **Figure A.1**. $|V'L|$ should also be interpreted in this way.) We choose the origin such that V and V' have coordinates $(a, 0)$ and $(-a, 0)$, respectively (Figure A.1). We locate the point F (a *focus*) at $(c, 0)$ and let ℓ (a *directrix*) be the line $x = d$, where $c > 0$ and $d > 0$. These choices place the center of E at the origin. Notice that we have four parameters (a, c, d, and e) that must be related.

Because the vertex $V(a, 0)$ is on E, it satisfies the defining property $|PF| = e\,|PL|$, with $P = V$. This condition implies that $a - c = e(d - a)$. Because the vertex $V'(-a, 0)$ is on the curve E, it also satisfies the defining property $|PF| = e\,|PL|$, with $P = V'$. This condition implies that $a + c = e(d + a)$. Solving these two equations for c and d, we find that $c = ae$ and $d = a/e$. To summarize, the parameters a, c, d, and e are related by the equations

$$c = ae \quad \text{and} \quad a = de.$$

Because $e < 1$, it follows that $c < a < d$.

We now use the property $|PF| = e\,|PL|$ with an arbitrary point $P(x, y)$ on the curve E. Figure A.1 shows the geometry with the focus $(c, 0) = (ae, 0)$ and the directrix $x = d = a/e$. The condition $|PF| = e\,|PL|$ becomes

$$\sqrt{(x - ae)^2 + y^2} = e\left(\frac{a}{e} - x\right).$$

The goal is to find the simplest possible relationship between x and y. Squaring both sides and collecting terms, we have

$$(1 - e^2)x^2 + y^2 = a^2(1 - e^2).$$

Dividing through by $a^2(1 - e^2)$ gives the equation of the standard ellipse:

$$\frac{x^2}{a^2} + \frac{y^2}{a^2(1 - e^2)} = \frac{x^2}{a^2} + \frac{y^2}{b^2} = 1, \quad \text{where} \quad b^2 = a^2(1 - e^2).$$

This is the equation of an ellipse centered at the origin with vertices and foci on the x-axis.

The preceding proof is now applied with $e > 1$. The argument for ellipses with $0 < e < 1$ led to the equation

$$\frac{x^2}{a^2} + \frac{y^2}{a^2(1 - e^2)} = 1.$$

With $e > 1$, we have $1 - e^2 < 0$, so we write $(1 - e^2) = -(e^2 - 1)$. The resulting equation describes a hyperbola centered at the origin with the foci on the x-axis:

$$\frac{x^2}{a^2} - \frac{y^2}{b^2} = 1, \quad \text{where} \quad b^2 = a^2(e^2 - 1).$$

◄

> **THEOREM 15.3**   **Continuity of Composite Functions**
> If $u = g(x, y)$ is continuous at (a, b) and $z = f(u)$ is continuous at $g(a, b)$, then the composite function $z = f(g(x, y))$ is continuous at (a, b).

Proof: Let P and P_0 represent the points (x, y) and (a, b), respectively. Let $u = g(P)$ and $u_0 = g(P_0)$. The continuity of f at u_0 means that $\lim_{u \to u_0} f(u) = f(u_0)$. This limit implies that given any $\varepsilon > 0$, there exists a $\delta^* > 0$ such that

$$|f(u) - f(u_0)| < \varepsilon \quad \text{whenever} \quad 0 < |u - u_0| < \delta^*.$$

The continuity of g at P_0 means that $\lim_{P \to P_0} g(P) = g(P_0)$. Letting $|P - P_0|$ denote the distance between P and P_0, this limit implies that given any $\delta^* > 0$, there exists a $\delta > 0$ such that

$$|g(P) - g(P_0)| = |u - u_0| < \delta^* \quad \text{whenever} \quad 0 < |P - P_0| < \delta.$$

We now combine these two statements. Given any $\varepsilon > 0$, there exists a $\delta > 0$ such that

$$|f(g(P)) - f(g(P_0))| = |f(u) - f(u_0)| < \varepsilon \quad \text{whenever} \quad 0 < |P - P_0| < \delta.$$

Therefore, $\lim_{(x,y) \to (a,b)} f(g(x, y)) = f(g(a, b))$ and $z = f(g(x, y))$ is continuous at (a, b). ◄

> **THEOREM 15.5**   **Conditions for Differentiability**
> Suppose the function f has partial derivatives f_x and f_y defined on an open set containing (a, b), with f_x and f_y continuous at (a, b). Then f is differentiable at (a, b).

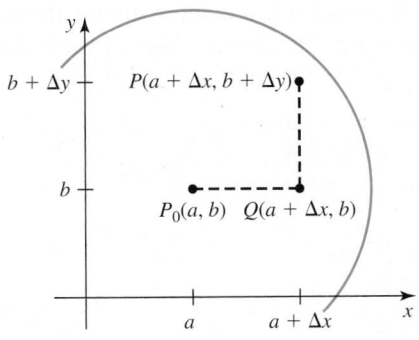

Figure A.2

Proof: Figure A.2 shows a region on which the conditions of the theorem are satisfied containing the points $P_0(a, b)$, $Q(a + \Delta x, b)$, and $P(a + \Delta x, b + \Delta y)$. By the definition of differentiability of f at P_0, we must show that

$$\Delta z = f(P) - f(P_0) = f_x(a, b)\Delta x + f_y(a, b)\Delta y + \varepsilon_1 \Delta x + \varepsilon_2 \Delta y,$$

where ε_1 and ε_2 depend only on $a, b, \Delta x,$ and Δy, with $(\varepsilon_1, \varepsilon_2) \to (0, 0)$ as $(\Delta x, \Delta y) \to (0, 0)$. We can view the change Δz taking place in two stages:

- $\Delta z_1 = f(a + \Delta x, b) - f(a, b)$ is the change in z as (x, y) moves from P_0 to Q.
- $\Delta z_2 = f(a + \Delta x, b + \Delta y) - f(a + \Delta x, b)$ is the change in z as (x, y) moves from Q to P.

Applying the Mean Value Theorem to the first variable and noting that f is differentiable with respect to x, we have

$$\Delta z_1 = f(a + \Delta x, b) - f(a, b) = f_x(c, b)\Delta x,$$

where c lies in the interval $(a, a + \Delta x)$. Similarly, applying the Mean Value Theorem to the second variable and noting that f is differentiable with respect to y, we have

$$\Delta z_2 = f(a + \Delta x, b + \Delta y) - f(a + \Delta x, b) = f_y(a + \Delta x, d)\Delta y,$$

where d lies in the interval $(b, b + \Delta y)$. We now express Δz as the sum of Δz_1 and Δz_2:

$$
\begin{aligned}
\Delta z &= \Delta z_1 + \Delta z_2 \\
&= f_x(c, b)\Delta x + f_y(a + \Delta x, d)\Delta y \\
&= (\underbrace{f_x(c, b) - f_x(a, b)}_{\varepsilon_1} + f_x(a, b))\Delta x \qquad \text{Add and subtract } f_x(a, b). \\
&\quad + (\underbrace{f_y(a + \Delta x, d) - f_y(a, b)}_{\varepsilon_2} + f_y(a, b))\Delta y \quad \text{Add and subtract } f_y(a, b). \\
&= (f_x(a, b) + \varepsilon_1)\Delta x + (f_y(a, b) + \varepsilon_2)\Delta y.
\end{aligned}
$$

Note that as $\Delta x \to 0$ and $\Delta y \to 0$, we have $c \to a$ and $d \to b$. Because f_x and f_y are continuous at (a, b), it follows that as $\Delta x \to 0$ and $\Delta y \to 0$,

$$\varepsilon_1 = f_x(c, b) - f_x(a, b) \to 0 \quad \text{and} \quad \varepsilon_2 = f_y(a + \Delta x, d) - f_y(a, b) \to 0.$$

Therefore, the condition for differentiability of f at (a, b) has been proved.   ◄

THEOREM 15.7   Chain Rule (One Independent Variable)

Let $z = f(x, y)$ be a differentiable function of x and y on its domain, where x and y are differentiable functions of t on an interval I. Then

$$\frac{dz}{dt} = \frac{\partial f}{\partial x}\frac{dx}{dt} + \frac{\partial f}{\partial y}\frac{dy}{dt}.$$

Proof: Assume $(a, b) = (x(t), y(t))$ is in the domain of f, where t is in I. Let $\Delta x = x(t + \Delta t) - x(t)$ and $\Delta y = y(t + \Delta t) - y(t)$. Because f is differentiable at (a, b), we know (Section 15.3) that

$$\Delta z = \frac{\partial f}{\partial x}(a, b)\Delta x + \frac{\partial f}{\partial y}(a, b)\Delta y + \varepsilon_1 \Delta x + \varepsilon_2 \Delta y,$$

where $(\varepsilon_1, \varepsilon_2) \to (0, 0)$ as $(\Delta x, \Delta y) \to (0, 0)$. Dividing this equation by Δt gives

$$\frac{\Delta z}{\Delta t} = \frac{\partial f}{\partial x}\frac{\Delta x}{\Delta t} + \frac{\partial f}{\partial y}\frac{\Delta y}{\Delta t} + \varepsilon_1\frac{\Delta x}{\Delta t} + \varepsilon_2\frac{\Delta y}{\Delta t}.$$

As $\Delta t \to 0$, several things occur. First, because $x = g(t)$ and $y = h(t)$ are differentiable on I, $\dfrac{\Delta x}{\Delta t}$ and $\dfrac{\Delta y}{\Delta t}$ approach $\dfrac{dx}{dt}$ and $\dfrac{dy}{dt}$, respectively. Similarly, $\dfrac{\Delta z}{\Delta t}$ approaches $\dfrac{dz}{dt}$ as $\Delta t \to 0$. The fact that x and y are continuous on I (because they are differentiable there) means that $\Delta x \to 0$ and $\Delta y \to 0$ as $\Delta t \to 0$. Therefore, because $(\varepsilon_1, \varepsilon_2) \to (0, 0)$ as $(\Delta x, \Delta y) \to (0, 0)$, it follows that $(\varepsilon_1, \varepsilon_2) \to (0, 0)$ as $\Delta t \to 0$. Letting $\Delta t \to 0$, we have

$$\underbrace{\lim_{\Delta t \to 0}\frac{\Delta z}{\Delta t}}_{\frac{dz}{dt}} = \frac{\partial f}{\partial x}\underbrace{\lim_{\Delta t \to 0}\frac{\Delta x}{\Delta t}}_{\frac{dx}{dt}} + \frac{\partial f}{\partial y}\underbrace{\lim_{\Delta t \to 0}\frac{\Delta y}{\Delta t}}_{\frac{dy}{dt}} + \underbrace{\lim_{\Delta t \to 0}\underbrace{\varepsilon_1}_{\to 0}\underbrace{\frac{\Delta x}{\Delta t}}_{\frac{dx}{dt}}} + \underbrace{\lim_{\Delta t \to 0}\underbrace{\varepsilon_2}_{\to 0}\underbrace{\frac{\Delta y}{\Delta t}}_{\frac{dy}{dt}}}$$

or

$$\frac{dz}{dt} = \frac{\partial f}{\partial x}\frac{dx}{dt} + \frac{\partial f}{\partial y}\frac{dy}{dt}.$$ ◄

THEOREM 15.15   Second Derivative Test

Suppose the second partial derivatives of f are continuous throughout an open disk centered at the point (a, b), where $f_x(a, b) = f_y(a, b) = 0$. Let $D(x, y) = f_{xx}(x, y) f_{yy}(x, y) - (f_{xy}(x, y))^2$.

1. If $D(a, b) > 0$ and $f_{xx}(a, b) < 0$, then f has a local maximum value at (a, b).

2. If $D(a, b) > 0$ and $f_{xx}(a, b) > 0$, then f has a local minimum value at (a, b).

3. If $D(a, b) < 0$, then f has a saddle point at (a, b).

4. If $D(a, b) = 0$, then the test is inconclusive.

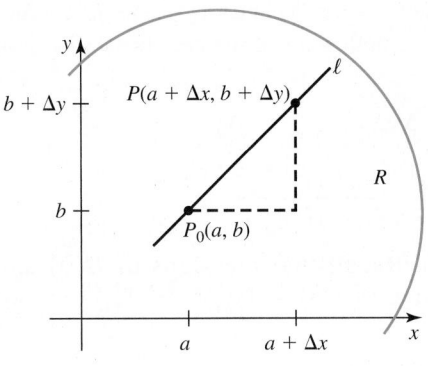

Figure A.3

Proof: The proof relies on a two-variable version of Taylor's Theorem, which we prove first. **Figure A.3** shows the open disk R on which the conditions of Theorem 15.15 are satisfied; it contains the points $P_0(a, b)$ and $P(a + \Delta x, b + \Delta y)$. The line ℓ through $P_0 P$ has a parametric description

$$\langle x(t), y(t) \rangle = \langle a + t\Delta x, b + t\Delta y \rangle,$$

where $t = 0$ corresponds to P_0 and $t = 1$ corresponds to P.

We now let $F(t) = f(a + t\Delta x, b + t\Delta y)$ be the value of f along that part of ℓ that lies in R. By the Chain Rule, we have

$$F'(t) = f_x \underbrace{x'(t)}_{\Delta x} + f_y \underbrace{y'(t)}_{\Delta y} = f_x \Delta x + f_y \Delta y.$$

Differentiating again with respect to t (f_x and f_y are differentiable), we use $f_{xy} = f_{yx}$ to obtain

$$F''(t) = \frac{\partial F'}{\partial x} \underbrace{x'(t)}_{\Delta x} + \frac{\partial F'}{\partial y} \underbrace{y'(t)}_{\Delta y}$$

$$= \frac{\partial}{\partial x}(f_x \Delta x + f_y \Delta y)\Delta x + \frac{\partial}{\partial y}(f_x \Delta x + f_y \Delta y)\Delta y$$

$$= f_{xx} \Delta x^2 + 2f_{xy} \Delta x\Delta y + f_{yy} \Delta y^2.$$

Noting that F meets the conditions of Taylor's Theorem for one variable with $n = 1$, we write

$$F(t) = F(0) + F'(0)(t - 0) + \frac{1}{2}F''(c)(t - 0)^2,$$

where c is between 0 and t. Setting $t = 1$, it follows that

$$\bullet \quad F(1) = F(0) + F'(0) + \frac{1}{2}F''(c),$$

where $0 < c < 1$. Recalling that $F(t) = f(a + t\Delta x, b + t\Delta y)$ and invoking the condition $f_x(a, b) = f_y(a, b) = 0$, we have

$$F(1) = f(a + \Delta x, b + \Delta y)$$

$$= f(a, b) + \underbrace{f_x(a, b)\Delta x + f_y(a, b)\Delta y}_{F'(0) = 0}$$

$$+ \frac{1}{2}(f_{xx} \Delta x^2 + 2f_{xy} \Delta x\Delta y + f_{yy} \Delta y^2)\Big|_{(a+c\Delta x,\, b+c\Delta y)}$$

$$= f(a, b) + \underbrace{\frac{1}{2}(f_{xx} \Delta x^2 + 2f_{xy} \Delta x\Delta y + f_{yy} \Delta y^2)\Big|_{(a+c\Delta x,\, b+c\Delta y)}}_{H(c)}$$

$$= f(a, b) + \frac{1}{2}H(c).$$

The existence and type of extreme point at (a, b) are determined by the sign of $f(a + \Delta x, b + \Delta y) - f(a, b)$ (for example, if $f(a + \Delta x, b + \Delta y) - f(a, b) \geq 0$ for all Δx and Δy near 0, then f has a local minimum at (a, b)). Note that $f(a + \Delta x, b + \Delta y) - f(a, b)$ has the same sign as the quantity we have denoted $H(c)$. Assuming $H(0) \neq 0$, for Δx and Δy sufficiently small and nonzero, the sign of $H(c)$ is the same as the sign of

$$H(0) = f_{xx}(a, b)\Delta x^2 + 2f_{xy}(a, b)\Delta x\Delta y + f_{yy}(a, b)\Delta y^2$$

(because the second partial derivatives are continuous at (a, b) and $(a + c\Delta x, b + c\Delta y)$ can be made arbitrarily close to (a, b)). Multiplying both sides of the previous expression by f_{xx} and rearranging terms leads to

$$f_{xx} H(0) = f_{xx}^2 \, \Delta x^2 + 2f_{xy} f_{xx} \, \Delta x \Delta y + f_{yy} f_{xx} \, \Delta y^2$$
$$= \underbrace{(f_{xx} \Delta x + f_{xy} \Delta y)^2}_{\geq 0} + \underbrace{(f_{xx} f_{yy} - f_{xy}^2)}_{D} \Delta y^2,$$

where all derivatives are evaluated at (a, b). Recall that the signs of $H(0)$ and $f(a + \Delta x, b + \Delta y) - f(a, b)$ are the same. Letting $D(a, b) = (f_{xx} f_{yy} - f_{xy}^2)|_{(a, b)}$, we reach the following conclusions:

- If $D(a, b) > 0$ and $f_{xx}(a, b) < 0$, then $H(0) < 0$ (for Δx and Δy sufficiently close to 0) and $f(a + \Delta x, b + \Delta y) - f(a, b) < 0$. Therefore, f has a local maximum value at (a, b).

- If $D(a, b) > 0$ and $f_{xx}(a, b) > 0$, then $H(0) > 0$ (for Δx and Δy sufficiently close to 0) and $f(a + \Delta x, b + \Delta y) - f(a, b) > 0$. Therefore, f has a local minimum value at (a, b).

- If $D(a, b) < 0$, then $H(0) > 0$ for some small nonzero values of Δx and Δy (implying $f(a + \Delta x, b + \Delta y) > f(a, b)$), and $H(0) < 0$ for other small nonzero values of Δx and Δy (implying $f(a + \Delta x, b + \Delta y) < f(a, b)$). (The relative sizes of $(f_{xx} \Delta x + f_{xy} \Delta y)^2$ and $(f_{xx} f_{yy} - f_{xy}^2) \Delta y^2$ can be adjusted by varying Δx and Δy.) Therefore, f has a saddle point at (a, b).

- If $D(a, b) = 0$, then $H(0)$ may be zero, in which case the sign of $H(c)$ cannot be determined. Therefore, the test is inconclusive.   ◄

Answers

CHAPTER 1

Section 1.1 Exercises, pp. 9–13

1. A function is a rule that assigns to each value of the independent variable in the domain a unique value of the dependent variable in the range. **3.** B **5.** The first statement **7.** $D = \mathbb{R}, R = [-10, \infty)$
9. The independent variable is h; the dependent variable is V; $D = [0, 50]$. **11.** $-3; 1/8; 1/(2x)$ **13.** The domain of $f \circ g$ consists of all x in the domain of g such that $g(x)$ is in the domain of f.
15. a. 4 **b.** 1 **c.** 3 **d.** 3 **e.** 8 **f.** 1 **17.** 15.4 ft/s; radiosonde rises at an average rate of 15.4 ft/s during the first 5 seconds of its flight. **19.** 2; 2; 2; -2 **21.** A is even, B is odd, and C is even.
23. $D = \{x: x \neq 2\}; R = \{y: y \neq -1\}$ **25.** $D = [-\sqrt{7}, \sqrt{7}]$; $R = [0, \sqrt{7}]$ **27.** $D = \mathbb{R}$ **29.** $D = [-3, 3]$
31. a. $[0, 3 + \sqrt{14}]$

b.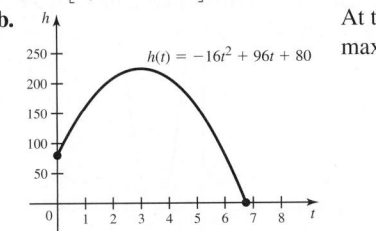
At time $t = 3$, the maximum height is 224 ft.

33. $1/z^3$ **35.** $1/(y^3 - 3)$ **37.** $(u^2 - 4)^3$ **39.** $\dfrac{x - 3}{10 - 3x}$ **41.** x
43. $g(x) = x^3 - 5, f(x) = x^{10}$ **45.** $g(x) = x^4 + 2, f(x) = \sqrt{x}$
47. $|x^2 - 4|; D = \mathbb{R}$ **49.** $\dfrac{1}{|x - 2|}; D = \{x: x \neq 2\}$
51. $\dfrac{1}{x^2 - 6}; D = \{x: x \neq \sqrt{6}, -\sqrt{6}\}$
53. $x^4 - 8x^2 + 12; D = \mathbb{R}$ **55.** $f(x) = x - 3$
57. $f(x) = x^2$ **59.** $f(x) = x^2$ **61. a.** True **b.** False **c.** True
d. False **e.** False **f.** True **g.** True **h.** False **i.** True
63. 3 **65.** $2x + h$ **67.** $-\dfrac{2}{x(x + h)}$ **69.** $x + a + 1$
71. $x^2 + ax + a^2 - 2$ **73.** $\dfrac{4(x + a)}{a^2 x^2}$ **75. a.** 864 ft/hr;
the hiker's elevation increases at an average rate of 864 ft/hr.
b. -487 ft/hr; the hiker's elevation decreases at an average rate of 487 ft/hr. **c.** The hiker might have stopped to rest during this interval of time and/or the trail was level during this portion of the hike.
77. a.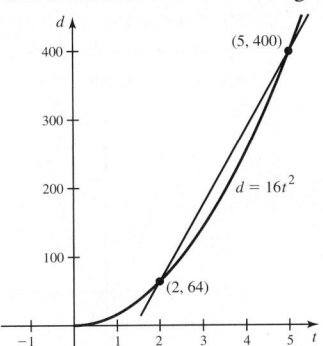

b. $m_{\text{sec}} = 112$ ft/s; the object falls at an average rate of 112 ft/s.

79. y-axis **81.** No symmetry **83.** x-axis, y-axis, origin
85. Origin **87. a.** 4 **b.** 1 **c.** 3 **d.** -2 **e.** -1 **f.** 7
89.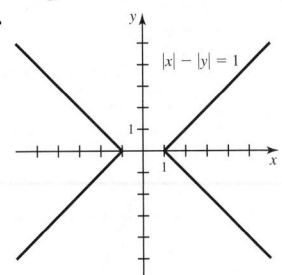

91. The equation $y = 2 - \sqrt{-x^2 + 6x + 16}$ can be rewritten as $(x - 3)^2 + (y - 2)^2 = 5^2$. Because $y \leq 2$, the function is the lower half of a circle of radius 5 centered at $(3, 2)$.
$D = [-2, 8]; R = [-3, 2]$ **93.** $f(x) = 3x - 2$ or $f(x) = -3x + 4$
95. $f(x) = x^2 - 6$ **97.** $\dfrac{1}{\sqrt{x + h} + \sqrt{x}}; \dfrac{1}{\sqrt{x} + \sqrt{a}}$
99. $\dfrac{3}{\sqrt{x}(x + h) + x\sqrt{x + h}}; \dfrac{3}{x\sqrt{a} + a\sqrt{x}}$ **101.** None **103.** y-axis

Section 1.2 Exercises, pp. 22–27

1. A formula, a graph, a table, words **3.** $y = -\frac{2}{3}x - 1$ **5.** The set of all real numbers for which the denominator does not equal 0
7. $y = \begin{cases} x + 3 & \text{if } x < 0 \\ -\frac{1}{2}x + 3 & \text{if } x \geq 0 \end{cases}$ **9.** Shift the graph to the left 2 units.
11. Compress the graph horizontally by a factor of $\frac{1}{3}$.
13. $f(x) = |x - 2| + 3; g(x) = -|x + 2| - 1$
15. $f(x) = 2x + 1$

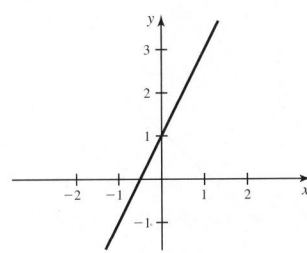

17. $f(x) = 3x - 7$ **19.** $C_s = 5.71$; 856.5 million
21. $d = -3p/50 + 27; D = [0, 450]$

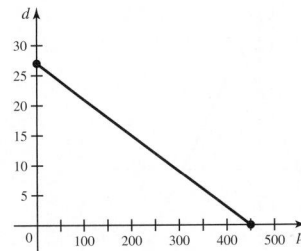

23. a. $p(t) = 328.3t + 1875$ **b.** 4830
25. $f(x) = \begin{cases} 3 & \text{if } x \leq 3 \\ 2x - 3 & \text{if } x > 3 \end{cases}$

27. $c(t) = \begin{cases} 0.05t & \text{if } 0 \le t \le 60 \\ 1.2 + 0.03t & \text{if } 60 < t \le 120 \end{cases}$

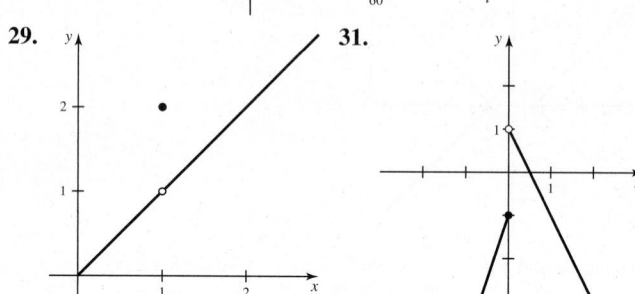

29. 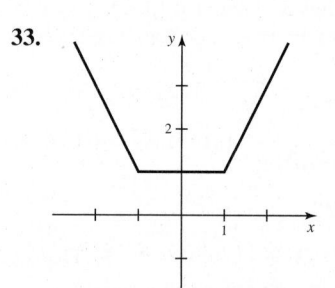 **31.**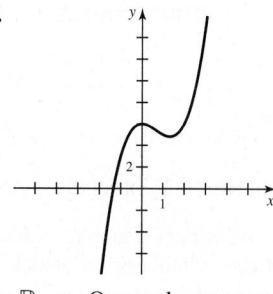

33. 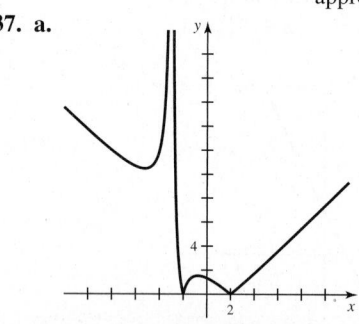 **35. a.**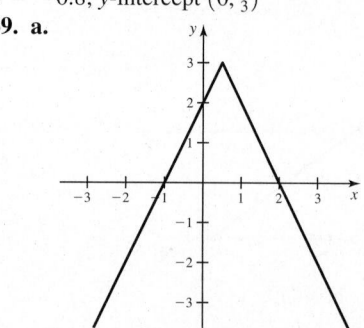

b. $D = \mathbb{R}$ **c.** One peak near $x = 0$; one valley near $x = 4/3$; x-intercept approx. $(-1.3, 0)$, y-intercept $(0, 6)$

37. a.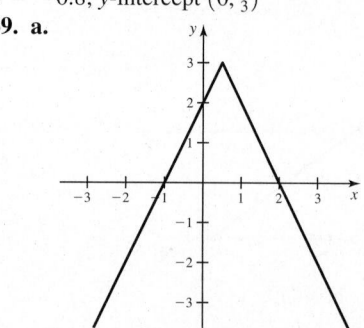

b. $D = \{x : x \ne -3\}$ **c.** Undefined at $x = -3$; a valley near $x = -5.2$; x-intercepts (and valleys) at $(-2, 0)$ and $(2, 0)$; a peak near $x = -0.8$; y-intercept $(0, \frac{4}{3})$

39. a.

b. $D = \mathbb{R}$ **c.** One peak at $x = \frac{1}{2}$; x-intercepts $(-1, 0)$ and $(2, 0)$; y-intercept $(0, 2)$ **41. a.** A, D, F, I
b. E **c.** B, H **d.** I **e.** A

43. a.

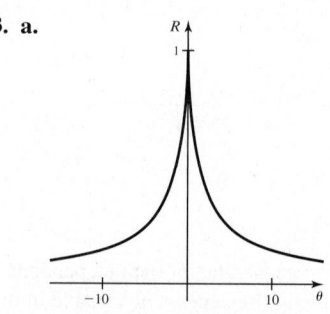

b. $\theta = 0$; vision is sharpest when we look straight ahead.
c. $|\theta| \le 0.19°$ (less than $\frac{1}{5}$ of a degree) **45.** $S(x) = 2$
47. $S(x) = \begin{cases} 1 & \text{if } x < 0 \\ -\frac{1}{2} & \text{if } x > 0 \end{cases}$
49. a. 12 **b.** 36 **c.** $A(x) = 6x$ **51. a.** 12 **b.** 21
c. $A(x) = \begin{cases} 8x - x^2 & \text{if } 0 \le x \le 3 \\ 2x + 9 & \text{if } x > 3 \end{cases}$
53. a. True **b.** False **c.** True **d.** False
55. a. Shift 3 units to the right. **b.** Horizontal compression by a factor of $\frac{1}{2}$, then shift 2 units to the right.

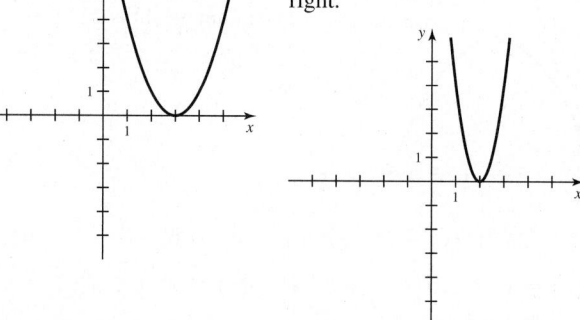

c. Shift to the right 2 units, vertically stretch by a factor of 3, reflect across the x-axis, and shift up 4 units. **d.** Horizontal stretch by a factor of 3, horizontal shift right 2 units, vertical stretch by a factor of 6, and vertical shift up 1 unit.

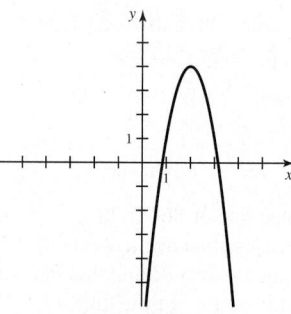 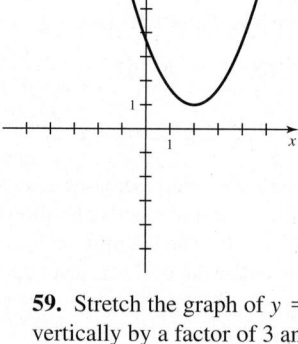

57. Shift the graph of $y = x^2$ right 2 units and up 1 unit. **59.** Stretch the graph of $y = x^2$ vertically by a factor of 3 and reflect across the x-axis.

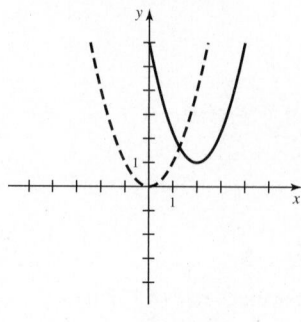 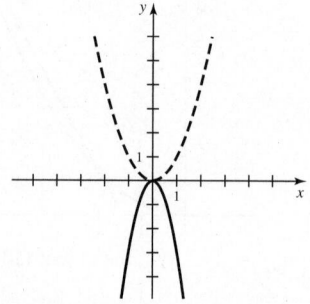

61. Shift the graph of $y = x^2$ left 3 units and stretch vertically by a factor of 2.

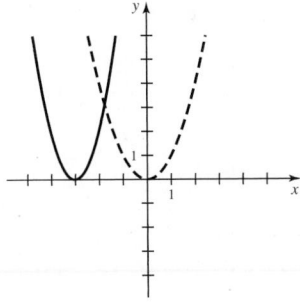

63. Shift the graph of $y = x^2$ to the left $\frac{1}{2}$ unit, stretch vertically by a factor of 4, reflect across the x-axis, and then shift up 13 units.

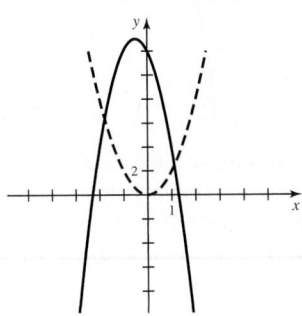

b.

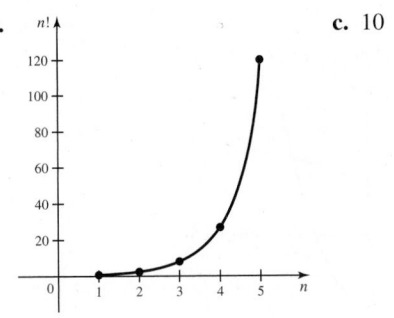

c. 10

Section 1.3 Exercises, pp. 35–39

1. $D = \mathbb{R}; R = (0, \infty)$

3.

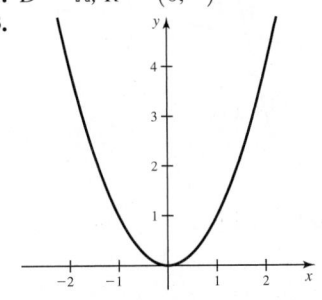

65. $(0, 0); (2, 8)$ **67.** $(0, 0); (4, 16)$

69.

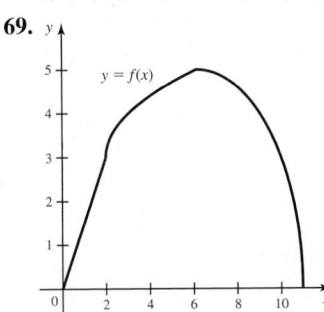

71.

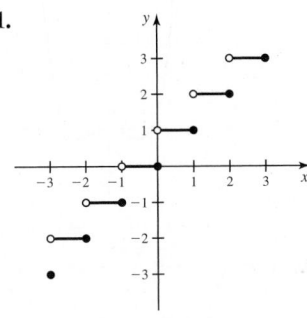

5. $(-\infty, -1], [-1, 1], [1, \infty)$ **7.** If a function f is not one-to-one, then there are domain values, x_1 and x_2, such that $x_1 \neq x_2$ but $f(x_1) = f(x_2)$. If f^{-1} exists, then by definition, $f^{-1}(f(x_1)) = x_1$ and $f^{-1}(f(x_2)) = x_2$, so f^{-1} assigns two different range values to the single domain value of $f(x_1)$.

9. $f^{-1}(x) = \frac{1}{2}x$ **11.**

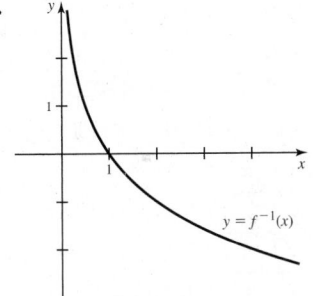

73.

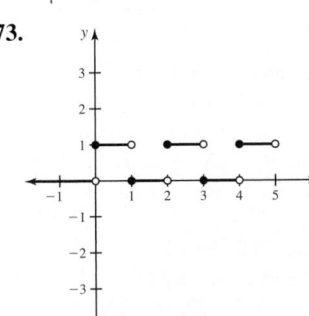

75.

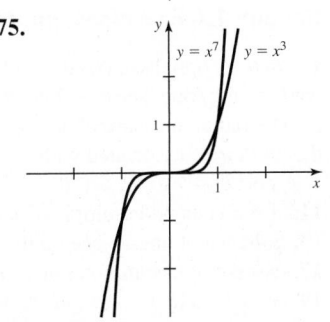

13. $g_1(x) = x^2 + 1; D = [0, \infty); R = [1, \infty);$ $g_1^{-1}(x) = \sqrt{x - 1}; D = [1, \infty); R = [0, \infty)$

15. The expression $\log_b x$ represents the power to which b must be raised to obtain x.   **17.** $D = (0, \infty); R = \mathbb{R}$

19. a. 3   **b.** 4   **c.** -2 **d.** 3   **e.** $1/2$ **21.** $(-\infty, \infty)$

23. $(-\infty, 5) \cup (5, \infty)$ **25.** $(-\infty, 0), (0, \infty)$

27. $f^{-1}(x) = -\frac{1}{4}x + 2$

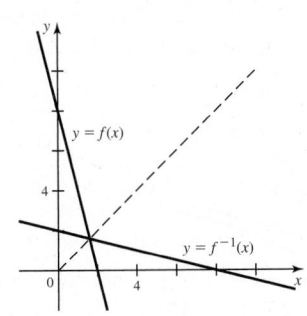

77. a. 0.9; 90% chance that server will win from deuce given that such servers win 75% of their service points   **b.** 0.1; 10% chance that server will win from deuce given that such servers win 25% of their service points   **79. a.** $f(m) = 350m + 1200$ **b.** Buy

81. a. $S(x) = x^2 + \frac{500}{x}$ **b.** Approximately 6.3 ft

85. a.

n	1	2	3	4	5
$n!$	1	2	6	24	120

29. $f^{-1}(x) = x^2 - 2$

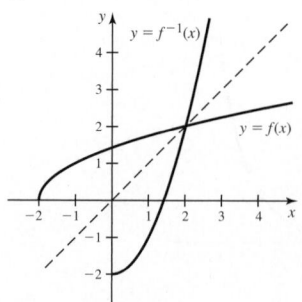

31. $f^{-1}(x) = 2 + \sqrt{x+1}$

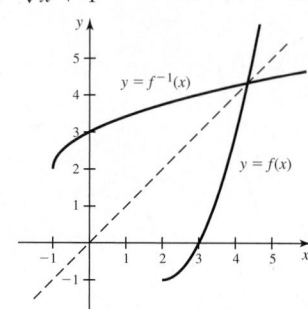

33. $f^{-1}(x) = \sqrt{\dfrac{2}{x} - 1}$ **35.** $f^{-1}(x) = \dfrac{1}{2}\ln x - 3$

37. $f^{-1}(x) = \dfrac{e^x - 1}{3}$ **39.** $f^{-1}(x) = -\dfrac{1}{2}\log_{10} x$

41. $f^{-1}(x) = \ln\left(\dfrac{2x}{1-x}\right)$

43. a. $f_1(x) = \sqrt{1-x^2};\ 0 \le x \le 1$
$f_2(x) = \sqrt{1-x^2};\ -1 \le x \le 0$
$f_3(x) = -\sqrt{1-x^2};\ -1 \le x \le 0$
$f_4(x) = -\sqrt{1-x^2};\ 0 \le x \le 1$

b. $f_1^{-1}(x) = \sqrt{1-x^2};\ 0 \le x \le 1$
$f_2^{-1}(x) = -\sqrt{1-x^2};\ 0 \le x \le 1$
$f_3^{-1}(x) = -\sqrt{1-x^2};\ -1 \le x \le 0$
$f_4^{-1}(x) = \sqrt{1-x^2};\ -1 \le x \le 0$

45. -0.2 **47.** 1.19 **49.** $-0.09\overline{6}$ **51.** 1000 **53.** 2 **55.** $1/e$
57. $\ln 21/\ln 7$ **59.** $\ln 5/(3\ln 3) + 5/3$ **61.** 451 years
63. 9.53 years   **65. a.** No   **b.** $f^{-1}(h) = 2 - \dfrac{1}{4}\sqrt{64 - h}$

c. $f^{-1}(h) = 2 + \dfrac{1}{4}\sqrt{64 - h}$ **d.** 0.542 s   **e.** 3.837 s

67. $\dfrac{\ln 15}{\ln 2} \approx 3.9069$ **69.** $\dfrac{\ln 40}{\ln 4} \approx 2.6610$ **71.** $e^{x \ln 2}$

73. $\log_5 |x|/\log_5 e$ **75.** e **77. a.** False   **b.** False   **c.** False
d. True   **e.** False   **f.** False   **g.** True   **79.** A is $y = \log_2 x$; B is
$y = \log_4 x$; C is $y = \log_{10} x$.

81.

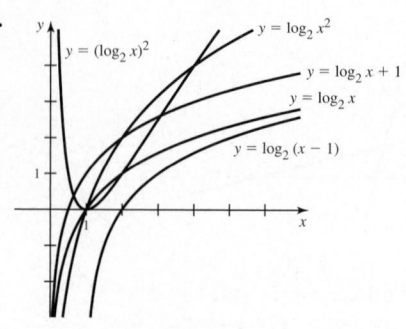

83. a.

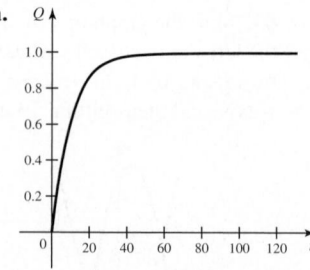

b. Vertical scaling; steady state equals a.   **c.** Horizontal scaling;
steady state remains constant.   **d.** a
85. $f^{-1}(x) = \sqrt{x - 5} + 1, x \ge 5$ **87.** $f^{-1}(x) = \sqrt[3]{x} - 1, D = \mathbb{R}$
89. $f_1^{-1}(x) = \sqrt{2/x - 2}, D_1 = (0, 1]; f_2^{-1}(x) = -\sqrt{2/x - 2},$
$D_2 = (0, 1]$
95. a.

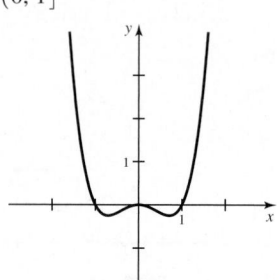

f is one-to-one on the intervals
$(-\infty, -1/\sqrt{2}], [-1/\sqrt{2}, 0],$
$[0, 1/\sqrt{2}],$ and $[1/\sqrt{2}, \infty)$.

b. $x = \sqrt{\dfrac{1 \pm \sqrt{4y + 1}}{2}}, -\sqrt{\dfrac{1 \pm \sqrt{4y + 1}}{2}}$

Section 1.4 Exercises, pp. 48–51

1. $\sin\theta = \text{opp/hyp}; \cos\theta = \text{adj/hyp}; \tan\theta = \text{opp/adj};$
$\cot\theta = \text{adj/opp}; \sec\theta = \text{hyp/adj}; \csc\theta = \text{hyp/opp}$ **3.** 3 s
5. The radian measure of an angle θ is the length s of an arc on
the unit circle associated with θ.   **7.** $\sin^2\theta + \cos^2\theta = 1,$
$1 + \cot^2\theta = \csc^2\theta, \tan^2\theta + 1 = \sec^2\theta$ **9.** $\theta = 3\pi/2$
11. $\{x: x \text{ is an odd multiple of } \pi/2\}$
13. Sine is not one-to-one on its domain.   **15.** $3\pi/4$
17. Horizontal asymptotes at $y = \pi/2$ and $y = -\pi/2$
19. $-1/2$ **21.** 1   **23.** $-1/\sqrt{3}$ **25.** $1/\sqrt{3}$ **27.** 1   **29.** -1
31. Undefined   **33.** $\dfrac{\sqrt{2 + \sqrt{3}}}{2}$ or $\dfrac{\sqrt{6} + \sqrt{2}}{4}$
35. $\pi/4 + n\pi, n = 0, \pm1, \pm2, \ldots$
37. $\pi/6, 5\pi/6, 7\pi/6, 11\pi/6$
39. $\pi/4 + 2n\pi, 3\pi/4 + 2n\pi, n = 0, \pm1, \pm2, \ldots$
41. $0, \pi/2, \pi, 3\pi/2$
43. $\pi/12, 5\pi/12, 3\pi/4, 13\pi/12, 17\pi/12, 7\pi/4$
45. $0.1007; 1.4701$ **47.** $17.3°; 72.7°$ **49.** $\pi/2$ **51.** $-\pi/6$
53. $\pi/3$ **55.** $2\pi/3$ **57.** -1 **59.** $\sin\theta = \dfrac{12}{13}; \tan\theta = \dfrac{12}{5}$
61. $\sqrt{1 - x^2}$ **63.** $\dfrac{\sqrt{4 - x^2}}{2}$ **65.** $2x\sqrt{1 - x^2}$
67. $\sec\theta = \dfrac{r}{x} = \dfrac{1}{x/r} = \dfrac{1}{\cos\theta}$
69. Dividing both sides of $\cos^2\theta + \sin^2\theta = 1$ by $\cos^2\theta$ gives
$1 + \tan^2\theta = \sec^2\theta$.   **71.** Because $\cos(\pi/2 - \theta) = \sin\theta,$
for all θ, $1/\cos(\pi/2 - \theta) = 1/\sin\theta$, excluding integer
multiples of π, and $\sec(\pi/2 - \theta) = \csc\theta$.

73. $\cos^{-1} x + \cos^{-1}(-x) = \theta + (\pi - \theta) = \pi$

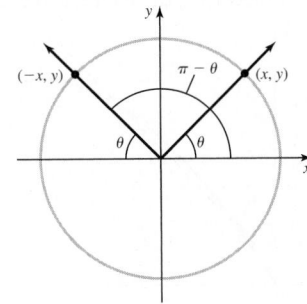

75. $\pi/3$ **77.** $\pi/3$ **79.** $\pi/4$ **81.** $\pi/2 - 2$

83. $\dfrac{1}{\sqrt{x^2 + 1}}$ **85.** $1/x$ **87.** $x/\sqrt{x^2 + 16}$

89. $\theta = \sin^{-1}\dfrac{x}{6} = \tan^{-1}\left(\dfrac{x}{\sqrt{36 - x^2}}\right) = \sec^{-1}\left(\dfrac{6}{\sqrt{36 - x^2}}\right)$

91. a. False   **b.** False   **c.** False   **d.** False   **e.** True   **f.** False
g. True   **h.** False   **93.** $\sin\theta = \frac{12}{13}$; $\tan\theta = \frac{12}{5}$; $\sec\theta = \frac{13}{5}$;
$\csc\theta = \frac{13}{12}$; $\cot\theta = \frac{5}{12}$ **95.** $\sin\theta = \frac{12}{13}$; $\cos\theta = \frac{5}{13}$; $\tan\theta = \frac{12}{5}$;
$\sec\theta = \frac{13}{5}$; $\cot\theta = \frac{5}{12}$ **97.** Amp $= 3$; period $= 6\pi$
99. Amp $= 3.6$; period $= 48$ **103.** Area of circle is πr^2; $\theta/(2\pi)$
represents the proportion of area swept out by a central angle θ.
Therefore, the area of such a sector is $(\theta/2\pi)\pi r^2 = r^2\theta/2$.
105. Stretch the graph of $y = \cos x$ horizontally by a factor of 3,
stretch vertically by a factor of 2, and reflect across the x-axis.

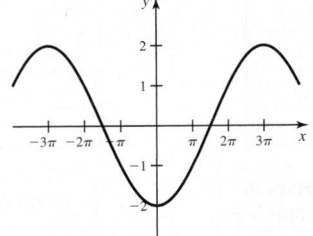

107. Stretch the graph of $y = \cos x$ horizontally by a factor of $24/\pi$;
then stretch it vertically by a factor of 3.6 and shift it up 2 units.

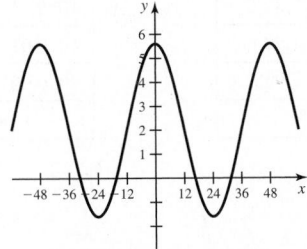

109. $y = 3\sin(\pi t/12 - 3\pi/4) + 13$ **111.** About 6 ft
113. $d(t) = 10\cos(4\pi t/3)$ **115.** h

Chapter 1 Review Exercises, pp. 51–55

1. a. True   **b.** False   **c.** False   **d.** True   **e.** False   **f.** False
g. True   **3.** f is one-to-one but not g.
5. $D = \{w: w \neq 2\}$; $R = \{y: y \neq 5\}$
7. $D = (-\infty, -1] \cup [3, \infty)$; $R = [0, \infty)$ **9.** Yes; no   **11.** 8
13. 7   **15.** 8   **17.** -2 **19. a.** 1   **b.** $\sqrt{x^3}$ **c.** $\sin^3\sqrt{x}$
d. $\mathbb{R}$ **e.** $[-1, 1]$ **21.** $2x + h - 2$; $x + a - 2$
23. $3x^2 + 3xh + h^2$; $x^2 + ax + a^2$ **25. a.** $y = \dfrac{5}{2}x - 8$
b. $y = \dfrac{3}{4}x + 3$ **c.** $y = \dfrac{1}{2}x - 2$ **27.** $B = -\dfrac{1}{500}a + 212$

29. a.

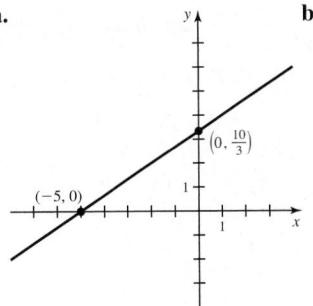

b.

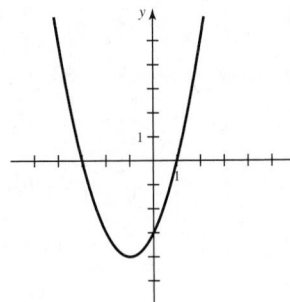

c.

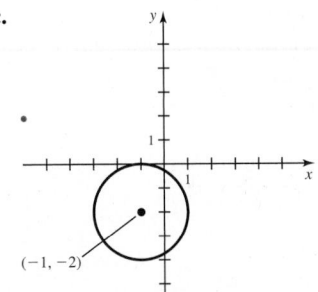

d

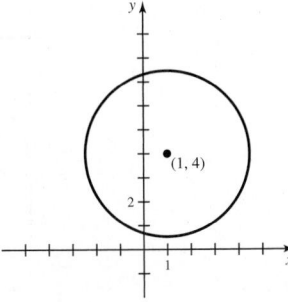

31.

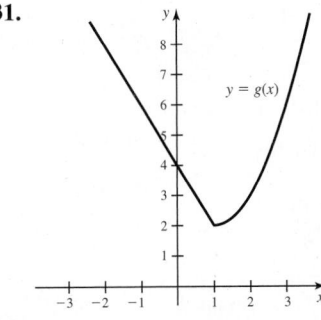

33. a.

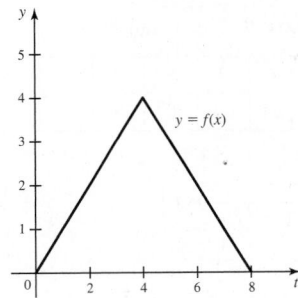

b. 2; 14   **c.** $A(x) = \begin{cases} x^2/2 & \text{if } 0 \leq x \leq 4 \\ -x^2/2 + 8x - 16 & \text{if } 4 < x \leq 8 \end{cases}$

35. $f(x) = \begin{cases} 4x & \text{if } x < 0 \\ 0 & \text{if } x \geq 0 \end{cases}$

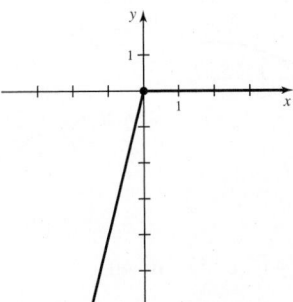

37. $D_f = \mathbb{R}$, $R_f = \mathbb{R}$; $D_g = [0, \infty)$, $R_g = [0, \infty)$
39. Shift $y = x^2$ left 3 units and down 12 units.
41. a. y-axis   **b.** y-axis   **c.** x-axis, y-axis, origin   **43.** $x = 2$

45. $t = \dfrac{e^4 - 4}{5}$ **47.** $\theta = \dfrac{\pi}{4}, \dfrac{3\pi}{4}, \dfrac{5\pi}{4}, \dfrac{7\pi}{4}$ **49.** $\theta = \pm\dfrac{\pi}{12}, \pm\dfrac{5\pi}{12}$

51. Approx. 35 years   **53.** $(-\infty, 0], [0, 2],$ and $[2, \infty)$

55. $f^{-1}(x) = -\dfrac{1}{4}x + \dfrac{3}{2}$ **57.** $f^{-1}(x) = 2 + \sqrt{x - 1}$

59. $f^{-1}(x) = -\sqrt{\dfrac{x - 1}{3}}$ **61.** $f^{-1}(x) = \sqrt{\ln x - 1}$

63. $f^{-1}(x) = \dfrac{4x^2}{(6 - x)^2}$, for $0 \le x < 6$

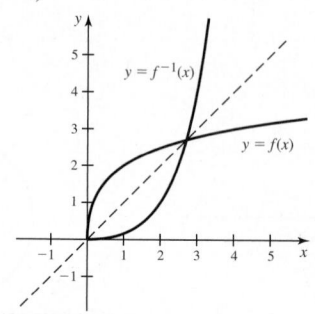

65. a. $f(t) = -2\cos\dfrac{\pi t}{3}$ **b.** $f(t) = 5\sin\dfrac{\pi t}{12} + 15$

67. a. F   **b.** E   **c.** D   **d.** B   **e.** C   **f.** A

69. $(7\pi/6, -1/2); (11\pi/6, -1/2)$ **71.** $-\dfrac{\sqrt{2 + \sqrt{2}}}{2}$

73. $\pi/6$ **75.** $-\pi/2$ **77.** x, provided $-1 \le x \le 1$

79. $\cos\theta = \frac{5}{13}; \tan\theta = \frac{12}{5}; \cot\theta = \frac{5}{12}; \sec\theta = \frac{13}{5}; \csc\theta = \frac{13}{12}$

81. $\dfrac{\sqrt{16 - x^2}}{4}$ **83.** $\pi/2 - \theta$ **85.** 0

87. $\tan 2\theta = \dfrac{\sin 2\theta}{\cos 2\theta} = \dfrac{2\sin\theta\cos\theta}{\cos^2\theta - \sin^2\theta}$

$= \dfrac{2\sin\theta\cos\theta/\cos^2\theta}{(\cos^2\theta - \sin^2\theta)/\cos^2\theta} = \dfrac{2\tan\theta}{1 - \tan^2\theta}$

89. a.

n	1	2	3	4	5	6	7	8	9	10
T(n)	1	5	14	30	55	91	140	204	285	385

b. $D = \{n: n \text{ is a positive integer}\}$ **c.** 14

91. $s(t) = 117.5 - 87.5\sin\left(\dfrac{\pi}{182.5}(t - 95)\right)$

$S(t) = 844.5 + 87.5\sin\left(\dfrac{\pi}{182.5}(t - 67)\right)$

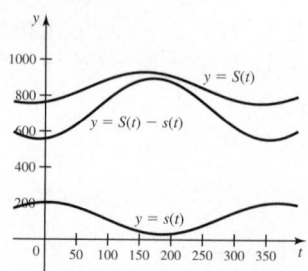

CHAPTER 2

Section 2.1 Exercises, pp. 61–62

1. $\dfrac{s(b) - s(a)}{b - a}$ **3.** 20   **5. a.** 36   **b.** 44   **c.** 52   **d.** 60

7. 47.84, 47.984, 47.9984; instantaneous velocity appears to be 48

9. $\dfrac{f(b) - f(a)}{b - a}$ **11.** The instantaneous velocity at $t = a$ is the slope of the line tangent to the position curve at $t = a$.   **13. a.** 48

b. 64   **c.** 80   **d.** $16(6 - h)$ **15.** $m_{\text{sec}} = 60$; the slope is the average velocity of the object over the interval $[0.5, 2]$.

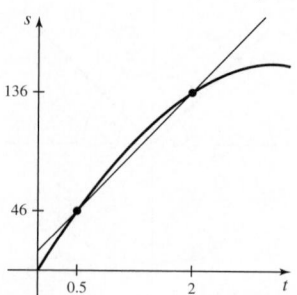

17.

Time interval	Average velocity
$[1, 2]$	80
$[1, 1.5]$	88
$[1, 1.1]$	94.4
$[1, 1.01]$	95.84
$[1, 1.001]$	95.984
$v_{\text{inst}} = 96$	

19.

Time interval	Average velocity
$[2, 3]$	20
$[2.9, 3]$	5.60
$[2.99, 3]$	4.16
$[2.999, 3]$	4.016
$[2.9999, 3]$	4.002
$v_{\text{inst}} = 4$	

21.

Time interval	Average velocity
$[3, 3.5]$	-24
$[3, 3.1]$	-17.6
$[3, 3.01]$	-16.16
$[3, 3.001]$	-16.016
$[3, 3.0001]$	-16.002
$v_{\text{inst}} = -16$	

23.

Time interval	Average velocity
$[0, 1]$	36.372
$[0, 0.5]$	67.318
$[0, 0.1]$	79.468
$[0, 0.01]$	79.995
$[0, 0.001]$	80.000
$v_{\text{inst}} = 80$	

25.

Interval	Slope of secant line
$[1, 2]$	6
$[1.5, 2]$	7
$[1.9, 2]$	7.8
$[1.99, 2]$	7.98
$[1.999, 2]$	7.998
$m_{\text{tan}} = 8$	

27.

Interval	Slope of secant line
$[0, 1]$	1.718
$[0, 0.5]$	1.297
$[0, 0.1]$	1.052
$[0, 0.01]$	1.005
$[0, 0.001]$	1.001
$m_{\text{tan}} = 1$	

29. a.

(graph)

b. $(2, -1)$

c.

Interval	Slope of secant line
$[2, 2.5]$	0.5
$[2, 2.1]$	0.1
$[2, 2.01]$	0.01
$[2, 2.001]$	0.001
$[2, 2.0001]$	0.0001
$m_{\text{tan}} = 0$	

31. a.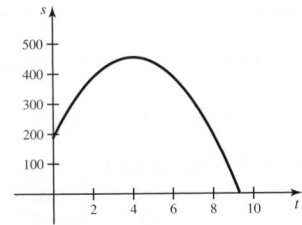

b. $t = 4$

c.

Interval	Average velocity
$[4, 4.5]$	-8
$[4, 4.1]$	-1.6
$[4, 4.01]$	-0.16
$[4, 4.001]$	-0.016
$[4, 4.0001]$	-0.0016
$v_{\text{inst}} = 0$	

d. $0 \le t < 4$ **e.** $4 < t \le 9$ **33.** 0.6366, 0.9589, 0.9996, 1

Section 2.2 Exercises, pp. 67–71

1. As x approaches a from either side, the values of $f(x)$ approach L.
3. a. 5 **b.** 3 **c.** Does not exist **d.** 1 **e.** 2
5. a. -1 **b.** 1 **c.** 2 **d.** 2
7. a.

x	$f(x)$	x	$f(x)$
1.9	3.9	2.1	4.1
1.99	3.99	2.01	4.01
1.999	3.999	2.001	4.001
1.9999	3.9999	2.0001	4.0001

b. 4

9. a.

t	$g(t)$	t	$g(t)$
8.9	5.983287	9.1	6.016621
8.99	5.998333	9.01	6.001666
8.999	5.999833	9.001	6.000167

b. 6

11. As x approaches a from the right, the values of $f(x)$ approach L.
13. $L = M$ **15. a.** 0 **b.** 1 **c.** 0
d. Does not exist; $\lim_{x \to 1^-} f(x) \ne \lim_{x \to 1^+} f(x)$ **17. a.** 3 **b.** 2
c. 2 **d.** 2 **e.** 2 **f.** 4 **g.** 1 **h.** Does not exist
i. 3 **j.** 3 **k.** 3 **l.** 3

19.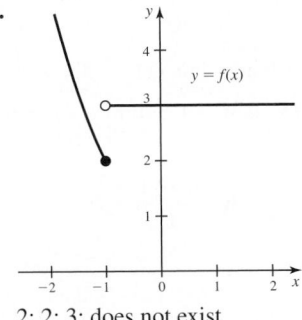

2; 2; 3; does not exist

21.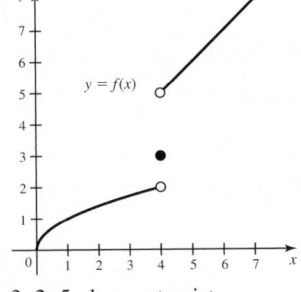

3; 2; 5; does not exist

23.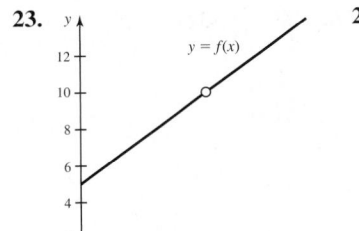

undefined; 10; 10; 10

25.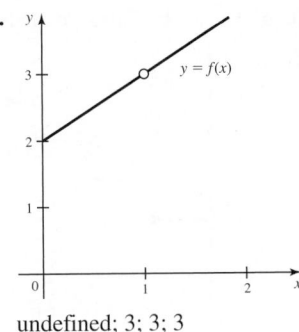

undefined; 3; 3; 3

27. From the graph and table, the limit appears to be 0.

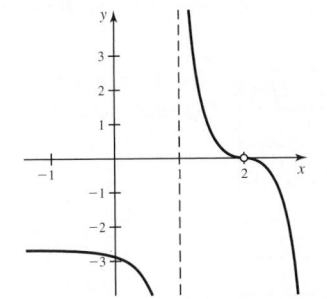

x	1.99	1.999	1.9999
$f(x)$	0.0021715	0.00014476	0.000010857
x	2.0001	2.001	2.01
$f(x)$	-0.000010857	-0.00014476	-0.0021715

29. From the graph and table, the limit appears to be 2.

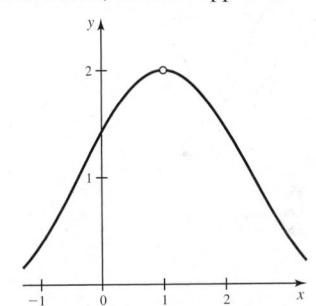

x	0.9	0.99	0.999
$f(x)$	1.993342	1.999933	1.999999
x	1.001	1.01	1.1
$f(x)$	1.999999	1.999933	1.993342

31.

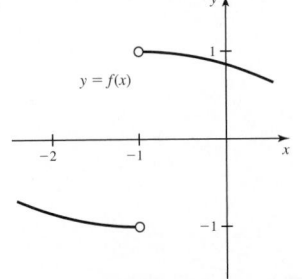

x	-1.1	-1.01	-1.001
$g(x)$	-0.9983342	-0.9999833	-0.9999998
x	-0.999	-0.99	-0.9
$g(x)$	0.9999998	0.9999833	0.9983342

From the table and the graph, it appears that the limit does not exist.

33. a. False   **b.** False   **c.** False   **d.** False   **e.** True

35. a.

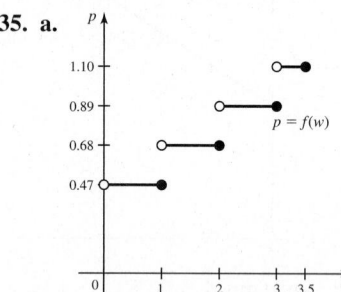

b. 0.89   **c.** Because $\lim\limits_{w\to3^-} f(w) = 0.89$ and $\lim\limits_{w\to3^+} f(w) = 1.1$, we know that $\lim\limits_{w\to3^-} f(w) \neq \lim\limits_{w\to3^+} f(w)$. So $\lim\limits_{w\to3} f(w)$ does not exist.

37. 3   **39.** 16   **41.** 1

43. a.

x	$\sin(1/x)$
$2/\pi$	1
$2/(3\pi)$	-1
$2/(5\pi)$	1
$2/(7\pi)$	-1
$2/(9\pi)$	1
$2/(11\pi)$	-1

The function values alternate between 1 and -1.

b. The function values alternate between 1 and -1 infinitely many times on the interval $(0, h)$ no matter how small $h > 0$ becomes.   **c.** Does not exist

45.

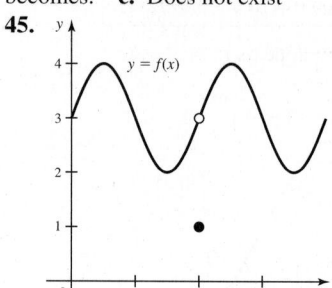

47.

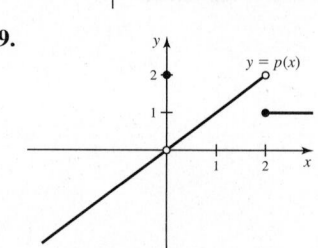

49.

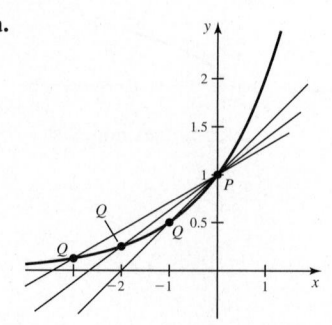

51. a. $-2, -1, 1, 2$ **b.** $2, 2, 2$
c. $\lim\limits_{x\to a^-} \lfloor x \rfloor = a - 1$ and $\lim\limits_{x\to a^+} \lfloor x \rfloor = a$, if a is an integer
d. $\lim\limits_{x\to a^-} \lfloor x \rfloor = \lfloor a \rfloor$ and $\lim\limits_{x\to a^+} \lfloor x \rfloor = \lfloor a \rfloor$, if a is not an integer
e. Limit exists provided a is not an integer
53. a. 8   **b.** 5   **55. a.** $2; 3; 4$ **b.** p **57.** p/q

Section 2.3 Exercises, pp. 79–82

1. $\lim\limits_{x\to a} p(x) = p(a)$ **3.** Those values of a for which the denominator is not zero   **5.** $\dfrac{x^2 - 7x + 12}{x - 3} = x - 4$, for $x \neq 3; -1$
7. 32; Constant Multiple Law   **9.** 5; Difference Law
11. 8; Quotient and Difference Laws   **13.** 4; Root and Power
Laws   **15.** 5; 1   **17.** 20   **19.** 5   **21.** -45 **23.** 8   **25.** 3
27. 3   **29.** -5 **31.** 8   **33.** 2   **35.** -8 **37.** -1 **39.** -12 **41.** $\frac{1}{6}$
43. $-\frac{1}{36}$ **45.** $2\sqrt{a}$ **47.** $\frac{1}{8}$ **49.** $-\frac{1}{16}$ **51.** 5   **53.** 10   **55.** 2
57. -54 **59.** 0   **61.** 1   **63.** 1   **65.** $1/2$ **67.** Does not exist
69. Does not exist   **71. a.** False   **b.** False   **c.** False   **d.** False
e. False   **73. a.** 2   **b.** 0   **c.** Does not exist   **75. a.** 0
b. $\sqrt{x - 2}$ is undefined for $x < 2$.   **77.** 0.0435 N/C
79. $\lim\limits_{S\to0^+} r(S) = 0$; the radius of the cylinder approaches 0 as the surface area of the cylinder approaches 0.

81. a. Because $\left| \sin\dfrac{1}{x} \right| \leq 1$, for all $x \neq 0$, we have that

$$|x| \left| \sin\dfrac{1}{x} \right| \leq |x|.$$

That is, $\left| x\sin\dfrac{1}{x} \right| \leq |x|$, so $-|x| \leq x\sin\dfrac{1}{x} \leq |x|$, for all $x \neq 0$.

b.

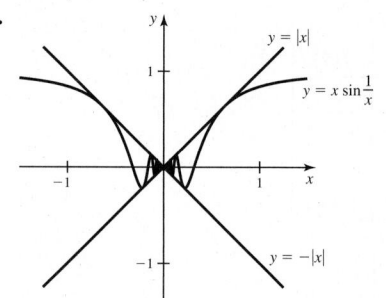

c. $\lim\limits_{x\to0} -|x| = 0$ and $\lim\limits_{x\to0} |x| = 0$; by part (a) and the Squeeze Theorem, $\lim\limits_{x\to0} x\sin\dfrac{1}{x} = 0$.

83. a.

b. $\lim\limits_{x\to0} \dfrac{\sin x}{x} = 1$

85. Because $\lim\limits_{x\to0^-} |x| = \lim\limits_{x\to0^-} (-x) = 0$ and $\lim\limits_{x\to0^+} |x| = \lim\limits_{x\to0^+} x = 0$, we know that $\lim\limits_{x\to0} |x| = 0$.   **87.** 1   **89.** $a = -13$; $\lim\limits_{x\to-1} g(x) = 6$
91. 6   **93.** $5a^4$

95. a.

b. $\dfrac{2^x - 1}{x}$

c.

x	$\dfrac{2^x - 1}{x}$
-1	0.5
-0.1	0.6697
-0.01	0.6908
-0.001	0.6929
-0.0001	0.6931
-0.00001	0.6931
Limit $\approx$ 0.693	

97. 6; 5   **99.** $\dfrac{1}{3}$ **101.** $f(x) = x - 1, g(x) = \dfrac{5}{x - 1}$

103. $b = 2$ and $c = -8$; yes   **105.** 6; 4

Section 2.4 Exercises, pp. 88–91

1. As x approaches a from the right, the values of $f(x)$ are negative and become arbitrarily large in magnitude.

3. A vertical asymptote for a function f is a vertical line $x = a$, where one (or more) of the following is true:
$$\lim_{x \to a^-} f(x) = \pm\infty; \ \lim_{x \to a^+} f(x) = \pm\infty.$$

5. ∞ **7. a.** ∞ **b.** ∞ **c.** ∞ **d.** ∞ **e.** $-\infty$ **f.** Does not exist
9. a. $-\infty$ **b.** $-\infty$ **c.** $-\infty$ **d.** ∞ **e.** $-\infty$ **f.** Does not exist
11. a. ∞ **b.** $-\infty$ **c.** $-\infty$ **d.** ∞ **13.** $-\infty$ **15.** No; there is a
vertical asymptote at $x = 2$ but not at $x = 1$.

17.

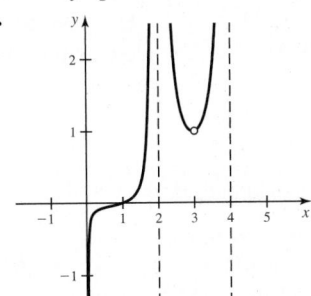

19. a and b are correct.   **21. a.** ∞ **b.** $-\infty$ **c.** Does not exist
23. a. $-\infty$ **b.** $-\infty$ **c.** $-\infty$ **25. a.** ∞ **b.** $-\infty$ **c.** Does not
exist   **27. a.** $-\infty$ **b.** $-\infty$ **c.** $-\infty$ **29. a.** ∞ **b.** Does not
exist   **c.** Does not exist   **31. a.** ∞ **b.** $1/54$ **c.** Does not exist
33. -5 **35.** ∞ **37.** $-\infty$ **39.** ∞ **41.** $-\infty$ **43.** ∞
45. a. $1/10$ **b.** $-\infty$ **c.** ∞; vertical asymptote: $x = -5$
47. $x = 3$; $\lim\limits_{x \to 3^+} f(x) = -\infty$; $\lim\limits_{x \to 3^-} f(x) = \infty$; $\lim\limits_{x \to 3} f(x)$
does not exist   **49.** $x = 0$ and $x = 2$; $\lim\limits_{x \to 0^+} f(x) = \infty$;
$\lim\limits_{x \to 0^-} f(x) = -\infty$; $\lim\limits_{x \to 0} f(x)$ does not exist; $\lim\limits_{x \to 2^+} f(x) = \infty$;
$\lim\limits_{x \to 2^-} f(x) = \infty$; $\lim\limits_{x \to 2} f(x) = \infty$
51. a. $-\infty$ **b.** ∞ **c.** $-\infty$ **d.** ∞ **53. a.** False   **b.** True
c. False   **55.** $r(x) = \dfrac{(x - 1)^2}{(x - 1)(x - 2)^2}$ **57.** $f(x) = \dfrac{1}{x - 6}$

59. $x = 0$ **61.** $x = -1$ **63.** $\theta = 10k + 5$, for any integer k
65. $x = 0$ **67. a.** $a = 4$ or $a = 3$ **b.** Either $a > 4$ or $a < 3$

c. $3 < a < 4$ **69. a.** $\dfrac{1}{\sqrt[3]{h}}$ regardless of the sign of h

b. $\lim\limits_{h \to 0^+} \dfrac{1}{\sqrt[3]{h}} = \infty$; $\lim\limits_{h \to 0^-} \dfrac{1}{\sqrt[3]{h}} = -\infty$; the tangent line
at $(0, 0)$ is vertical.

Section 2.5 Exercises, pp. 100–102

1. As $x < 0$ becomes arbitrarily large in magnitude, the corresponding values of f approach 10.   **3.** ∞ **5.** 0   **7.** 0   **9.** 3   **11.** 0
13. ∞; 0; 0   **15.** 3; 3   **17.** 0   **19.** 0   **21.** ∞ **23.** $-\infty$
25. $2/3$ **27.** $-\infty$ **29.** 5   **31.** -4 **33.** $-3/2$ **35.** 0

37. $\lim\limits_{x \to \infty} f(x) = \lim\limits_{x \to -\infty} f(x) = \frac{1}{5}; y = \frac{1}{5}$

39. $\lim\limits_{x \to \infty} f(x) = \lim\limits_{x \to -\infty} f(x) = 2; y = 2$

41. $\lim\limits_{x \to \infty} f(x) = \lim\limits_{x \to -\infty} f(x) = 0; y = 0$

43. $\lim\limits_{x \to \infty} f(x) = \infty$; $\lim\limits_{x \to -\infty} f(x) = -\infty$; none

45. $\lim\limits_{x \to \infty} f(x) = \lim\limits_{x \to -\infty} f(x) = \frac{4}{9}; y = \frac{4}{9}$

47. $\lim\limits_{x \to \infty} f(x) = \frac{2}{3}$; $\lim\limits_{x \to -\infty} f(x) = -2; y = \frac{2}{3}; y = -2$

49. $\lim\limits_{x \to \infty} f(x) = \lim\limits_{x \to -\infty} f(x) = \dfrac{1}{4 + \sqrt{3}}; y = \dfrac{1}{4 + \sqrt{3}}$

51. a. $y = x - 6$ **b.** $x = -6$ **53. a.** $y = \frac{1}{3}x - \frac{4}{9}$ **b.** $x = \frac{2}{3}$
c.

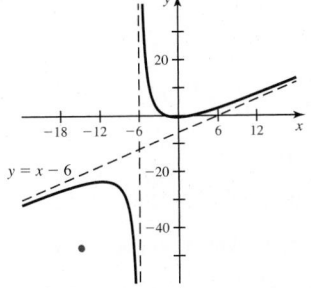

c.

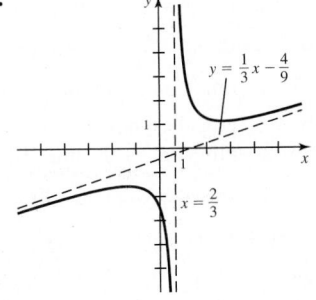

55. a. $y = 4x + 4$ **b.** No vertical asymptote
c.

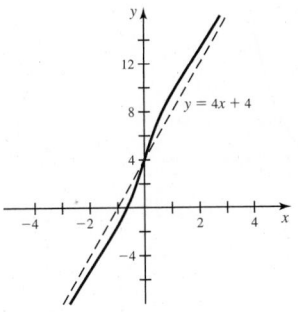

57. $\lim\limits_{x \to \infty} (-3e^{-x}) = 0$;
$\lim\limits_{x \to -\infty} (-3e^{-x}) = -\infty$

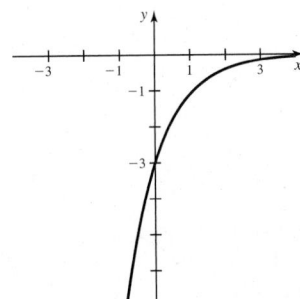

59. $\lim\limits_{x \to \infty} (1 - \ln x) = -\infty$;
$\lim\limits_{x \to 0^+} (1 - \ln x) = \infty$

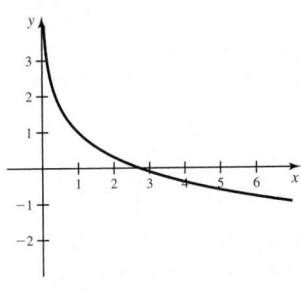

61. $\lim\limits_{x\to\infty} \sin x$ does not exist; $\lim\limits_{x\to-\infty} \sin x$ does not exist

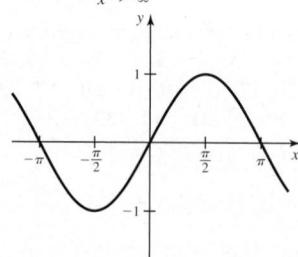

63. a. False   **b.** False   **c.** True   **d.** False   **65.** 3500
67. No steady state   **69.** 2
71. a. $\lim\limits_{x\to\infty} f(x) = \lim\limits_{x\to-\infty} f(x) = 2; y = 2$
b. $x = 0; \lim\limits_{x\to0^+} f(x) = \infty; \lim\limits_{x\to0^-} f(x) = -\infty$
73. a. $\lim\limits_{x\to\infty} f(x) = \lim\limits_{x\to-\infty} f(x) = 3; y = 3$
b. $x = -3$ and $x = 4; \lim\limits_{x\to-3^-} f(x) = \infty; \lim\limits_{x\to-3^+} f(x) = -\infty;$
$\lim\limits_{x\to4^-} f(x) = -\infty; \lim\limits_{x\to4^+} f(x) = \infty$
75. a. $\lim\limits_{x\to\infty} f(x) = \lim\limits_{x\to-\infty} f(x) = 1; y = 1$
b. $x = 0; \lim\limits_{x\to0^+} f(x) = \infty; \lim\limits_{x\to0^-} f(x) = -\infty$
77. a. $\lim\limits_{x\to\infty} f(x) = 1; \lim\limits_{x\to-\infty} f(x) = -1; y = 1$ and $y = -1$
b. No vertical asymptote
79. a. $\lim\limits_{x\to\infty} f(x) = 0; \lim\limits_{x\to-\infty} f(x) = 0; y = 0$
b. No vertical asymptote   **81. a.** $\lim\limits_{x\to\infty} f(x) = 2; \lim\limits_{x\to-\infty} f(x)$ does not
exist; $y = 2$ **b.** $x = 0; \lim\limits_{x\to0^+} f(x) = \infty; \lim\limits_{x\to0^-} f(x)$ does not exist
83. a. $\dfrac{\pi}{2}$ **b.** $\dfrac{\pi}{2}$ **85. a.** $\lim\limits_{x\to\infty} \sinh x = \infty; \lim\limits_{x\to-\infty} \sinh x = -\infty$
b. $\sinh 0 = 0$

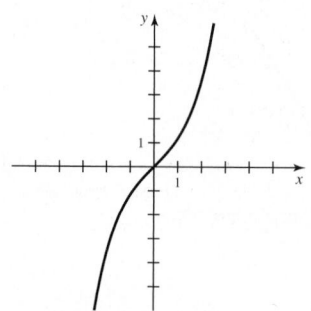

87.

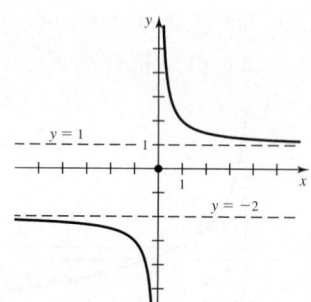

89. 1   **91.** 0   **93. a.** No; f has a horizontal asymptote if $m = n$,
and it has a slant asymptote if $m = n + 1$.   **b.** Yes;
$f(x) = x^4/\sqrt{x^6 + 1}$ **95.** $y = 3$ and $y = 2$

97. $y = 2; x = \sqrt[3]{e}$

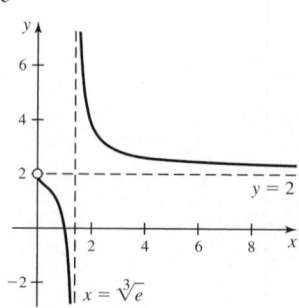

Section 2.6 Exercises, pp. 112–115

1. a, c   **3.** A function is continuous on an interval if it is continuous
at each point of the interval. If the interval contains endpoints, then the
function must be right- or left-continuous at those points.   **5.** $a = 1$,
item 1; $a = 2$, item 3; $a = 3$, item 2   **7.** $a = 1$, item 1; $a = 2$, item 2;
$a = 3$, item 1   **9. a.** $\lim\limits_{x\to a^-} f(x) = f(a)$ **b.** $\lim\limits_{x\to a^+} f(x) = f(a)$
11. $(0, 1), (1, 2), (2, 3], (3, 5)$; left-continuous at 3
13. $(0, 1), (1, 2), [2, 3), (3, 5)$; right-continuous at 2
15. $\{x: x \neq 0\}, \{x: x \neq 0\}$ **17.** No; $f(-5)$ is undefined.
19. No; $f(1)$ is undefined.   **21.** No; $\lim\limits_{x\to1} f(x) = 2$ but $f(1) = 3$.
23. No; $f(4)$ is undefined.   **25.** $(-\infty, \infty)$
27. $(-\infty, -3), (-3, 3), (3, \infty)$ **29.** $(-\infty, -2), (-2, 2), (2, \infty)$
31. 1   **33.** $2\sqrt{6}$ **35.** 16   **37.** $\ln 2$ **39. a.** $\lim\limits_{x\to1} f(x)$ does not exist.
b. Continuous from the right   **c.** $(-\infty, 1), [1, \infty)$ **41.** $(-\infty, 5]$;
left-continuous at 5   **43.** $(-\infty, -2\sqrt{2}], [2\sqrt{2}, \infty)$; left-continuous at
$-2\sqrt{2}$; right-continuous at $2\sqrt{2}$ **45.** $(-\infty, \infty)$ **47.** $(-\infty, \infty)$
49. 3   **51.** 1   **53.** 4   **55.** 2   **57.** $-\frac{1}{2}$ **59.** 4
61. $(n\pi, (n + 1)\pi)$, where n is an integer; $\sqrt{2}; -\infty$
63. $\left(\dfrac{n\pi}{2}, \left(\dfrac{n}{2} + 1\right)\dfrac{\pi}{2}\right)$, where n is an odd integer; $\infty; \sqrt{3} - 2$
65. $(-\infty, 0), (0, \infty); \infty; -\infty$ **67. b.** $x \approx 0.835$
69. b. $x \approx -0.285; x \approx 0.778; x \approx 4.507$ **71. b.** $x \approx -0.567$
73. a. True   **b.** True   **c.** False   **d.** False   **75. a.** $A(r)$ is
continuous on $[0, 0.08]$, and 7000 is between $A(0) = 5000$ and
$A(0.08) = 11,098.20$. By the Intermediate Value Theorem, there is at
least one c in $(0, 0.08)$ such that $A(c) = 7000$.
b. $c \approx 0.034$ or 3.4%

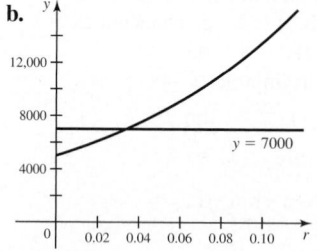

77. $[0, \pi/2]; 0.45$ **79.** $(-\infty, \infty)$ **81.** $[0, 16), (16, \infty)$
83. The vertical line segments should not appear.

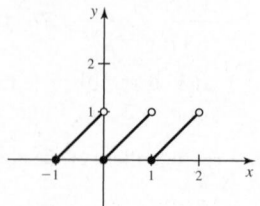

85. a, b.

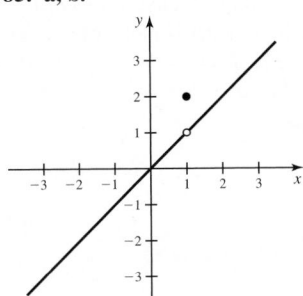

87. a. 2 **b.** 8 **c.** No; $\lim_{x\to 1^-} g(x) = 2$ and $\lim_{x\to 1^+} g(x) = 8$.

89. $\lim_{x\to 0} f(x) = 6$, $\lim_{x\to -\infty} f(x) = 10$, and $\lim_{x\to \infty} f(x) = 2$; no vertical asymptote; $y = 2$ and $y = 10$ are the horizontal asymptotes.

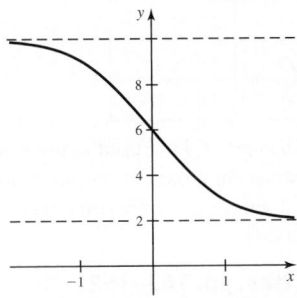

91. $x_1 = \frac{1}{7}$; $x_2 = \frac{1}{2}$; $x_3 = \frac{3}{5}$ **93.** Yes. Imagine there is a clone of the monk who walks down the path at the same time the monk walks up the path. The monk and his clone must cross paths at some time between dawn and dusk. **95.** No; f cannot be made continuous at $x = a$ by redefining $f(a)$. **97.** Removable discontinuity

99. $x = 0$ removable discontinuity; $x = 1$ infinite discontinuity
101. a. Yes **b.** No **103. a.** For example, $f(x) = 1/(x - 1)$, $g(x) = x + 1$ **b.** For continuity, g must be continuous at 0, and f must be continuous at $g(0)$.

Section 2.7 Exercises, pp. 124–128

1. 1 **3.** c **5.** Given any $\varepsilon > 0$ there exists a $\delta > 0$ such that $|f(x) - L| < \varepsilon$ whenever $0 < |x - a| < \delta$. **7.** $0 < \delta \le 2$
9. a. $\delta = 1$ **b.** $\delta = \frac{1}{2}$ **11. a.** $\delta = 2$ **b.** $\delta = \frac{1}{2}$
13. a. $0 < \delta \le 1$ **b.** $0 < \delta \le 0.79$ **15. a.** $0 < \delta \le 1$
b. $0 < \delta \le \frac{1}{2}$ **c.** $0 < \delta \le \varepsilon$ **17. a.** $0 < \delta \le 1$ **b.** $0 < \delta \le \frac{1}{2}$
c. $0 < \delta \le \frac{\varepsilon}{2}$ **19.** $\delta = \varepsilon/8$ **21.** $\delta = \varepsilon$ **23.** $\delta = \varepsilon$
25. $\delta = \varepsilon/3$ **27.** $\delta = \sqrt{\varepsilon}$ **29.** $\delta = \min\{1, \varepsilon/8\}$ **31.** $\delta = \varepsilon/2$
33. $\delta = \min\{1, 6\varepsilon\}$ **35.** $\delta = \min\{1/20, \varepsilon/200\}$
37. $\delta = \min\{1, \sqrt{\varepsilon/2}\}$ **39.** $\delta = \varepsilon/|m|$ if $m \ne 0$; use any $\delta > 0$
if $m = 0$ **41.** $\delta = \min\{1, 8\varepsilon/15\}$ **45.** $\delta = 1/\sqrt{N}$
47. $\delta = 1/\sqrt{N - 1}$ **49. a.** False **b.** False **c.** True **d.** True
51. For $x > a$, $|x - a| = x - a$. **53. a.** $\delta = \varepsilon/2$ **b.** $\delta = \varepsilon/3$
c. Because $\lim_{x\to 0^+} f(x) = \lim_{x\to 0^-} f(x) = -4$, $\lim_{x\to 0} f(x) = -4$.
55. $\delta = \varepsilon^2$ **57. a.** For each $N > 0$ there exists $\delta > 0$ such that $f(x) > N$ whenever $0 < x - a < \delta$. **b.** For each $N < 0$ there exists $\delta > 0$ such that $f(x) < N$ whenever $0 < a - x < \delta$.
c. For each $N > 0$ there exists $\delta > 0$ such that $f(x) > N$ whenever $0 < a - x < \delta$. **59.** $\delta = 1/N$ **61.** $\delta = (-10/M)^{1/4}$
65. $N = 1/\varepsilon$ **67.** $N = M - 1$

Chapter 2 Review Exercises, pp. 128–130

1. a. False **b.** False **c.** False **d.** True **e.** False **f.** False
g. False **h.** True **3.** 12 ft/s **5.** $x = -1$; $\lim_{x\to -1} f(x)$ does not exist; $x = 1$; $\lim_{x\to 1} f(x) \ne f(1)$; $x = 3$; $f(3)$ is undefined.

7. a. 1.414 **b.** $\sqrt{2}$
9.

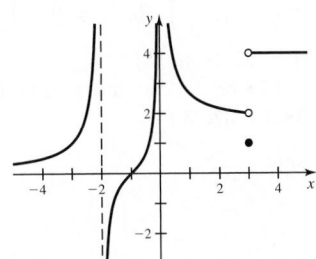

11. $\sqrt{11}$ **13.** 13 **15.** 2 **17.** $\frac{1}{3}$ **19.** $-\frac{1}{16}$ **21.** 108 **23.** $\frac{1}{108}$
25. 0 **27.** $-\infty$ **29.** ∞ **31.** 4 **33.** $-\infty$ **35.** $\frac{1}{2}$ **37.** $-3/\sqrt{a}$
39. $2/(1 - a)$ **41.** $3\pi/2 + 2$ **43.** 1; ∞ **45.** $2/3$
47. $-1/3$; $2/7$ **49.** 5 **51.** $-\infty$
53. a.

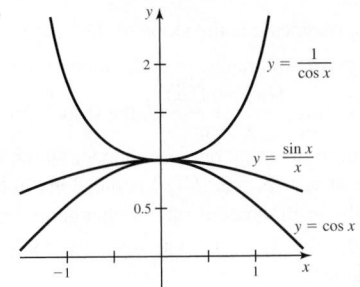

b. $\lim_{x\to 0} \cos x \le \lim_{x\to 0} \frac{\sin x}{x} \le \lim_{x\to 0} \frac{1}{\cos x}$;
$$1 \le \lim_{x\to 0} \frac{\sin x}{x} \le 1;$$
$$\lim_{x\to 0} \frac{\sin x}{x} = 1$$
55. $\lim_{x\to \infty} f(x) = -4$; $\lim_{x\to -\infty} f(x) = -4$
57. $\lim_{x\to \infty} f(x) = 1$; $\lim_{x\to -\infty} f(x) = -\infty$
59. $\lim_{x\to \infty} f(x) = 2$; $\lim_{x\to -\infty} f(x) = 5$ **61. a.** ∞; $-\infty$
b. $y = 3x + 2$ is the slant asymptote.
63. a. $-\infty$; ∞ **b.** $y = -x - 2$ is the slant asymptote.
65. a. ∞, $-\infty$ **b.** $y = 4x + 5$ is the slant asymptote.
67. Horizontal asymptotes at $y = 2/\pi$ and $y = -2/\pi$; vertical asymptote at $x = 0$
69.

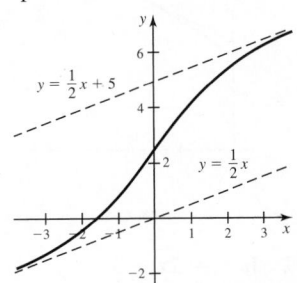

71. No; $f(5)$ does not exist. **73.** Yes; $h(5) = \lim_{x\to 5} h(x) = 4$
75. $(-\infty, -\sqrt{5}]$ and $[\sqrt{5}, \infty)$; left-continuous at $-\sqrt{5}$ and right-continuous at $\sqrt{5}$ **77.** $(-\infty, -5)$, $(-5, 0)$, $(0, 5)$, and $(5, \infty)$
79. $a = 3, b = 0$
81.

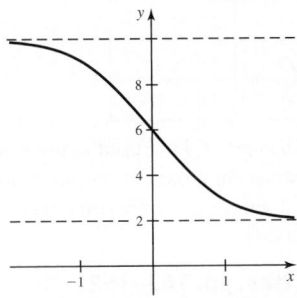

83. a. Let $f(x) = x - \cos x$; $f(0) < 0 < f\left(\dfrac{\pi}{2}\right)$ **b.** $x \approx 0.739$

85. a. $m(0) < 30 < m(5)$ and $m(5) > 30 > m(15)$
b. $m = 30$ when $t \approx 2.4$ hr and $t \approx 10.8$ hr   **c.** No; the maximum amount is approximately $m(5.5) \approx 38.5$.   **87.** $\delta = \varepsilon$

89. $\delta = \min\left\{1, \dfrac{\varepsilon}{15}\right\}$ **91.** $\delta = 1/\sqrt[4]{N}$

CHAPTER 3

Section 3.1 Exercises, pp. 137–140

1. Given the point $(a, f(a))$ and any point $(x, f(x))$ near $(a, f(a))$, the slope of the secant line joining these points is $\dfrac{f(x) - f(a)}{x - a}$. The limit of this quotient as x approaches a is the slope of the tangent line at the point.   **3.** The average rate of change over the interval $[a, x]$ is $\dfrac{f(x) - f(a)}{x - a}$. The value of $\displaystyle\lim_{x \to a} \dfrac{f(x) - f(a)}{x - a}$ is the slope of the tangent line; it is also the limit of average rates of change, which is the instantaneous rate of change at $x = a$.   **5.** $f'(a)$ is the slope of the tangent line at $(a, f(a))$ or the instantaneous rate of change in f at a.
7. $f(2) = 7$; $f'(2) = 4$ **9.** $y = 3x - 1$ **11.** -5 **13.** 68 ft/s
15. a. 6   **b.** $y = 6x - 14$
c.

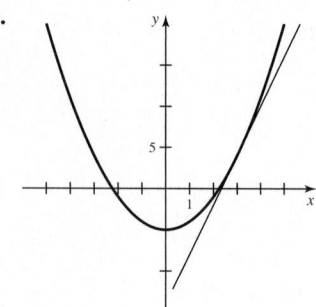

17. a. -1 **b.** $y = -x - 2$
c.

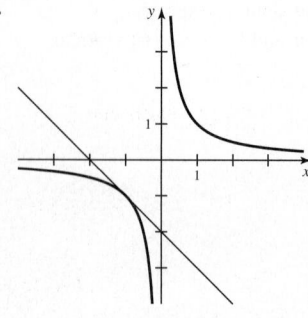

19. a. $\frac{1}{2}$ **b.** $y = \frac{1}{2}x + 2$
c.

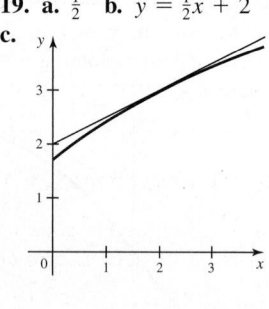

21. a. 2   **b.** $y = 2x + 1$ **23. a.** 2   **b.** $y = 2x - 3$
25. a. 4   **b.** $y = 4x - 8$ **27. a.** 3   **b.** $y = 3x - 2$
29. a. $\frac{2}{25}$ **b.** $y = \frac{2}{25}x + \frac{7}{25}$ **31. a.** $\frac{1}{4}$ **b.** $y = \frac{1}{4}x + \frac{7}{4}$
33. a. 8   **b.** $y = 8x$ **35. a.** -14 **b.** $y = -14x - 16$
37. a. -4 **b.** $y = -4x + 3$ **39. a.** $\frac{1}{3}$ **b.** $y = \frac{1}{3}x + \frac{5}{3}$
41. a. $-\frac{1}{100}$ **b.** $y = -\frac{x}{100} + \frac{3}{20}$ **43.** $-\frac{1}{4}$ **45.** $\frac{1}{5}$ **47. a.** True
b. False   **c.** True   **49.** $d'(4) = 128$ ft/s; the object falls with an instantaneous speed of 128 ft/s four seconds after being dropped.
51. $v'(3) = -4$ m/s per second; the instantaneous rate of change in the car's speed is -4 m/s^2 at $t = 3$ s.
53. a. $L'(1.5) \approx 4.3$ mm/week; the talon is growing at a rate of approximately 4.3 mm/week at $t = 1.5$ weeks (answers will vary).   **b.** $L'(a) \approx 0$, for $a \geq 4$; the talon stops growing at $t = 4$ weeks.   **55.** $D'(60) \approx 0.05$ hr/day; the number of

daylight hours is increasing at about 0.05 hr/day, 60 days after Jan 1. $D'(170) \approx 0$ hr/day; the number of daylight hours is neither increasing nor decreasing 170 days after Jan 1.   **57.** $f(x) = 5x^2$; $a = 2$; 20
59. $f(x) = x^4$; $a = 2$; 32   **61.** $f(x) = |x|$; $a = -1$; -1
63. a.

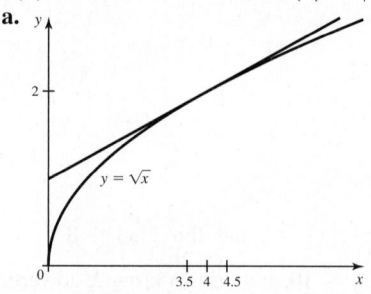

b.

h	Approximation	Error
0.1	0.25002	2.0×10^{-5}
0.01	0.25000	2.0×10^{-7}
0.001	0.25000	2.0×10^{-9}

c. Values of x on both sides of 4 are used in the formula.
d. The centered difference approximations are more accurate than the forward and backward difference approximations.   **65. a.** 0.39470, 0.41545   **b.** 0.02, 0.0003

Section 3.2 Exercises, pp. 148–152

1. f' is the slope function of f.   **3.** $\dfrac{dy}{dx}$ is the limit of $\dfrac{\Delta y}{\Delta x}$ as $\Delta x \to 0$.

5.

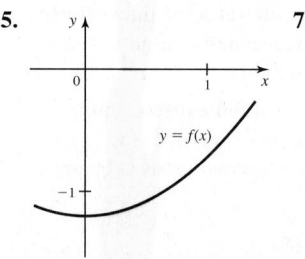

7. Yes

9. A line with a y-intercept of 1 and a slope of 3
11. $f'(x) = 7$ **13.** $\dfrac{dy}{dx} = 2x$; $\dfrac{dy}{dx}\Big|_{x=3} = 6$; $\dfrac{dy}{dx}\Big|_{x=-2} = -4$
15. a–C; b–C; c–A; d–B
17.

19. a. Not continuous at $x = 1$ **b.** Not differentiable at $x = 0, 1$
21. a. $f'(x) = 5$ **b.** $f'(1) = 5$; $f'(2) = 5$
23. a. $f'(x) = 8x$ **b.** $f'(2) = 16$; $f'(4) = 32$
25. a. $f'(x) = -\dfrac{1}{(x + 1)^2}$ **b.** $f'\left(-\dfrac{1}{2}\right) = -4$; $f'(5) = -\dfrac{1}{36}$
27. a. $f'(t) = -\dfrac{1}{2t^{3/2}}$ **b.** $f'(9) = -\dfrac{1}{54}$; $f'\left(\dfrac{1}{4}\right) = -4$
29. a. $f'(s) = 12s^2 + 3$ **b.** $f'(-3) = 111$; $f'(-1) = 15$
31. a. $v(t) = -32t + 100$ **b.** $v(1) = 68$ ft/s; $v(2) = 36$ ft/s
33. $\dfrac{dy}{dx} = \dfrac{1}{(x + 2)^2}$; $\dfrac{dy}{dx}\Big|_{x=2} = \dfrac{1}{16}$ **35. a.** $6x + 2$
b. $y = 8x - 13$ **37. a.** $\dfrac{3}{2\sqrt{3x + 1}}$ **b.** $y = 3x/10 + 13/5$

39. a. $\dfrac{6}{(3x+1)^2}$ **b.** $y = -3x/2 - 5/2$

41. a. Approximately 10 kW; approximately -5 kW
b. $t = 6, 18$ **c.** $t = 12$ **43. a.** $2ax + b$ **b.** $8x - 3$ **c.** 5
45. a. C, D **b.** A, B, E **c.** A, B, E, D, C **47.** a–D; b–C; c–B; d–A

49.
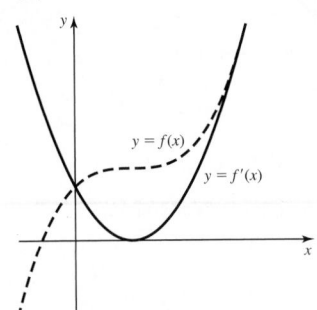

51.
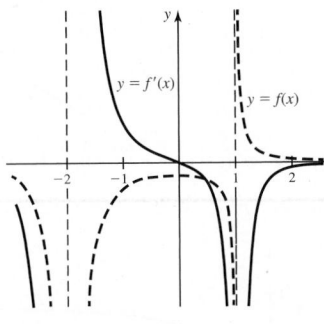

53. a. $x = 1$ **b.** $x = 1, x = 2$ **c.**
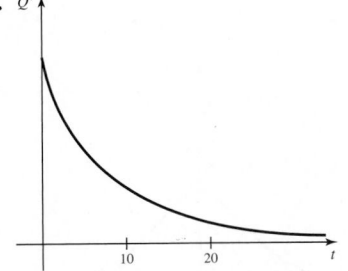

55. a. $t = 0$ **b.** Positive **c.** Decreasing
d. Q'
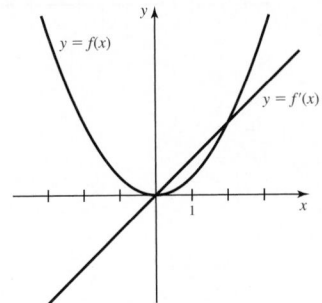

57. a. True **b.** True **c.** False **59.** $a = 4$
61. Yes

63. $y = -\dfrac{x}{3} - \dfrac{2}{3}$ **65.** $y = \dfrac{x}{2} + \dfrac{3}{2}$ **67.** $(1, 2), (5, 26)$

69. $(1, 1), \left(-\dfrac{1}{2}, -2\right)$ **71. b.** $f'_+(2) = 1; f'_-(2) = -1$

c. f is continuous but not differentiable at $x = 2$.

73. a.
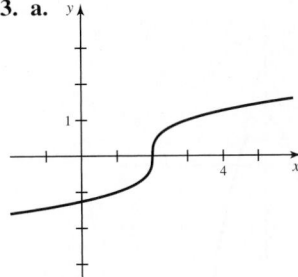
Vertical tangent line $x = 2$

b.
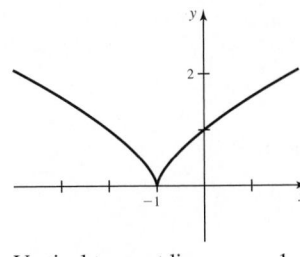
Vertical tangent line $x = -1$

c.
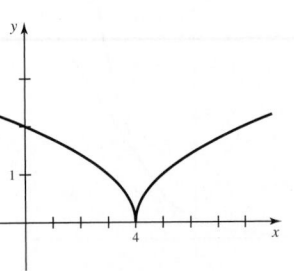
Vertical tangent line $x = 4$

d.
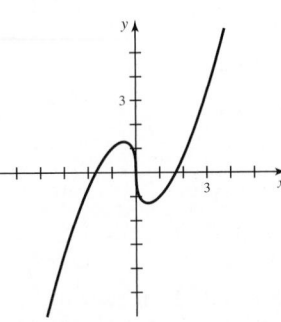
Vertical tangent line $x = 0$

75. $f'(x) = \dfrac{1}{3}x^{-2/3}$ and $\displaystyle\lim_{x \to 0^-} |f'(x)| = \lim_{x \to 0^+} |f'(x)| = \infty$

77. a.
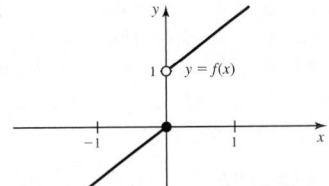

b. 1 **c.** 1 **d.**
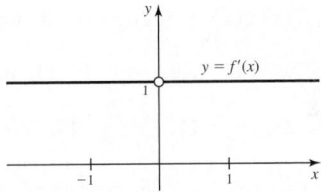

e. f is not differentiable at 0 because it is not continuous at 0.

Section 3.3 Exercises, pp. 159–162

1. Using the definition can be tedious. **3.** $f(x) = e^x$ **5.** Take the
product of the constant and the derivative of the function. **7.** 4

9. $-\dfrac{1}{2}$ **11.** -2 **13.** 7.5 **15.** $10t^9; 90t^8; 720t^7$ **17.** $\dfrac{2}{5}$ **19.** $5x^4$

21. 0 **23.** $15x^2$ **25.** t **27.** 8 **29.** $200t$ **31.** $12x^3 + 7$
33. $40x^3 - 32$ **35.** $6w^2 + 6w + 10$ **37.** $3e^x + 5$

39. $\begin{cases} 2x & \text{if } x < 0 \\ 4x + 1 & \text{if } x > 0 \end{cases}$ **41. a.** $d'(t) = 32t$; ft/s; the velocity of

the stone **b.** 576 ft; approx. 131 mi/hr **43. a.** $A'(t) = -\frac{1}{25}t + 2$
measures the rate at which the city grows in mi^2/yr. **b.** 1.6 mi^2/yr
c. 1200 people/yr

45. $w'(x) = \begin{cases} 0.4 & \text{if } 19 < x < 21 \\ 0.8 & \text{if } 21 < x < 32 \\ 1.5 & \text{if } \phantom{21 < } x > 32 \end{cases}$ **47.** $18x^2 + 6x + 4$

49. $2w$, for $w \neq 0$ **51.** $4x^3 + 4x$ **53.** 1, for $x \neq 1$

55. $\dfrac{1}{2\sqrt{x}}$, for $x \neq a$ **57.** e^w

59. a. $y = -6x + 5$　**b.**

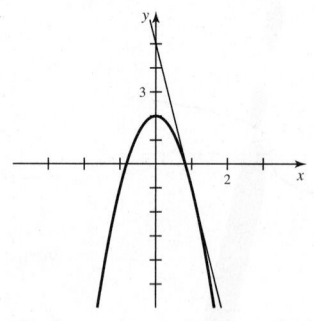

61. a. $y = 3x + 3 - 3 \ln 3$　**b.**

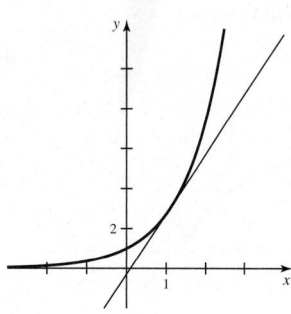

63. a. $x = 3$　**b.** $x = 4$
65. a. $(-1, 11), (2, -16)$　**b.** $(-3, -41), (4, 36)$
67. a. $(4, 4)$　**b.** $(16, 0)$　**69.** $f'(x) = 20x^3 + 30x^2 + 3$;
$f''(x) = 60x^2 + 60x$; $f'''(x) = 120x + 60$
71. $f'(x) = 1$; $f''(x) = f'''(x) = 0$, for $x \neq -1$
73. a. False　**b.** True　**c.** False　**d.** False　**e.** False
75. a. $y = 7x - 1$　**b.** $y = -2x + 5$　**c.** $y = 16x + 4$
77. $b = 2, c = 3$　**79.** -10　**81.** 4　**83. a.** $f(x) = x + e^x$; $a = 0$
b. 2　**85. a.** $f(x) = \sqrt{x}$; $a = 9$　**b.** $\frac{1}{6}$　**87. a.** $f(x) = e^x$; $a = 3$
b. e^3　**89.** 3　**91.** 1　**95. d.** $\frac{n}{2} x^{n/2 - 1}$　**97. c.** $2e^{2x}$

Section 3.4 Exercises, pp. 168–170

1. $\dfrac{d}{dx}(f(x)\, g(x)) = f'(x)\, g(x) + f(x)\, g'(x)$　**3.** $6x + 5$

5. $\dfrac{5}{(3x + 2)^2}$　**7. a.** $2x - 1$　**9. a.** $6x + 1$　**11. a.** $2w$, for $w \neq 0$

13. 1, for $x \neq a$　**15.** $23; -\dfrac{7}{4}$　**17.** $\dfrac{2}{27}; \dfrac{3}{8}$　**19.** $36x^5 - 12x^3$

21. $\dfrac{1}{(x + 1)^2}$　**23.** $e^t t^{2/3}\left(t + \dfrac{5}{3}\right)$　**25.** $\dfrac{e^x}{(e^x + 1)^2}$　**27.** $e^{-x}(1 - x)$

29. $-\dfrac{1}{(t - 1)^2}$　**31.** $4x^3$　**33.** $e^w(w^3 + 3w^2 - 1)$　**35.** $t^2 e^t$

37. $\dfrac{e^x(x^2 - 2x - 1)}{(x^2 - 1)^2}$　**39.** $-27x^{-10}$　**41.** $6t - 42/t^8$

43. $-3/t^2 - 2/t^3$　**45.** $\dfrac{e^x(x^2 - x - 5)}{(x - 2)^2}$

47. $\dfrac{e^x(x^2 + x + 1)}{(x + 1)^2}$　**49.** $\dfrac{\sqrt{w}}{(\sqrt{w} - w)^2}$　**51.** $\dfrac{5w^{2/3}}{3(w^{5/3} + 1)^2}$

53. $8x - \dfrac{2}{(5x + 1)^2}$　**55.** $\dfrac{r - 6\sqrt{r} - 1}{2\sqrt{r}(r + 1)^2}$

57. $300x^9 + 135x^8 + 105x^6 + 120x^3 + 45x^2 + 15$　**59.** $e^x + 8x$
61. a. $y = -3x/2 + 17/2$　**b.**

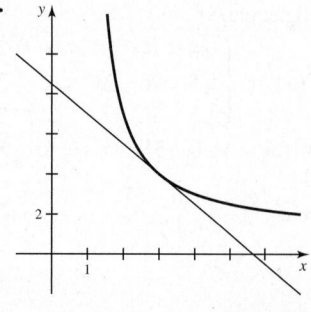

63. a. $y = 3x + 1$　**b.**

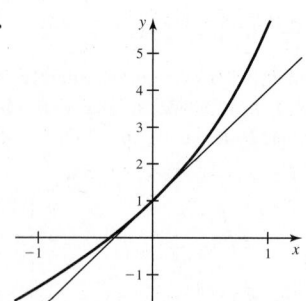

65. a. $p'(t) = \left(\dfrac{20}{t + 2}\right)^2$　**b.** $p'(5) \approx 8.16$　**c.** $t = 0$
d. $\displaystyle\lim_{t \to \infty} p(t) = 200$; the population approaches a steady state.
e.

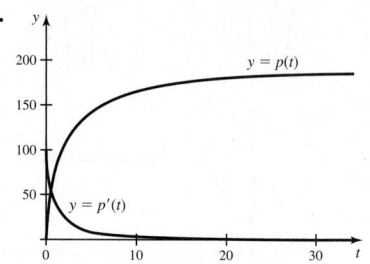

67. a. $F'(x) = -\dfrac{1.8 \times 10^{10}\, Qq}{x^3}$ N/m　**b.** -1.8×10^{19} N/m
c. $|F'(x)|$ decreases as x increases.　**69. a.** False　**b.** False
c. False　**d.** False　**71.** $4x - \dfrac{1}{x^2}; 2\left(\dfrac{1}{x^3} + 2\right); -\dfrac{6}{x^4}$

73. $\dfrac{x^2 + 2x - 7}{(x + 1)^2}; \dfrac{16}{(x + 1)^3}$

75. a. $y = -\dfrac{108}{169}x + \dfrac{567}{169}$　**b.**

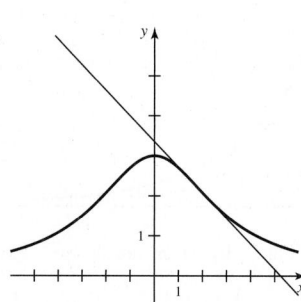

77. $-\dfrac{3}{2}$　**79.** $\dfrac{1}{9}$　**81.** $\dfrac{7}{8}$

83. a.

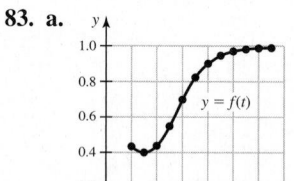

b. $t \approx 3$

c. $f'(3) \approx 0.28\, \dfrac{\text{mm/g}}{\text{week}}$; at a young age, the bird's wings
are growing quickly relative to its weight.
d. $f'(6.5) \approx 0.003\, \dfrac{\text{mm/g}}{\text{week}}$; the rate of change of the ratio of wing
chord length to mass is nearly 0.　**85.** $\dfrac{15}{2}$　**87.** $-\dfrac{5}{2}$　**89.** 1

91. a. $y = -2x + 16$　**b.** $y = -\dfrac{5}{9}x + \dfrac{23}{9}$

93. -90　**97.** $f''g + 2f'g' + fg''$　**99. a.** $f'gh + fg'h + fgh'$
b. $e^x(x^2 + 4x - 1)$

Section 3.5 Exercises, pp. 175–178

1. $\dfrac{\sin x}{x}$ is undefined at $x = 0$.   **3.** The tangent and
cotangent functions are defined as ratios of the sine and cosine
functions.   **5.** -1 **7.** $y = x$ **9.** $-\sin x - \cos x$ **11.** 3 **13.** $\frac{7}{3}$
15. 5 **17.** 7 **19.** $\frac{1}{4}$ **21.** a/b **23.** $\cos x - \sin x$
25. $e^{-x}(\cos x - \sin x)$ **27.** $\sin x + x \cos x$ **29.** $-\dfrac{1}{1 + \sin x}$
31. $\cos^2 x - \sin^2 x = \cos 2x$ **33.** $-2 \sin x \cos x = -\sin 2x$
35. $w^2 \cos w$ **37.** $x \cos 2x + \dfrac{1}{2} \sin 2x$ **39.** $\dfrac{1}{1 + \cos x}$
41. $\dfrac{2 \sin x}{(1 + \cos x)^2}$ **43.** $\sec x \tan x - \csc x \cot x$
45. $e^x \csc x(1 - \cot x)$ **47.** $-\dfrac{\csc x}{1 + \csc x}$
49. $\cos^2 z - \sin^2 z = \cos 2z$ **51.** $2 \sin^2 x$
55. a.

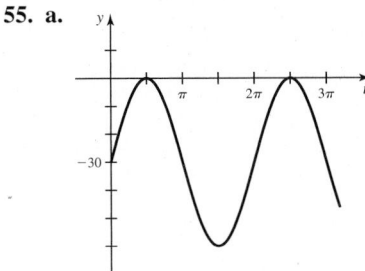

b. $v(t) = 30 \cos t$

c.

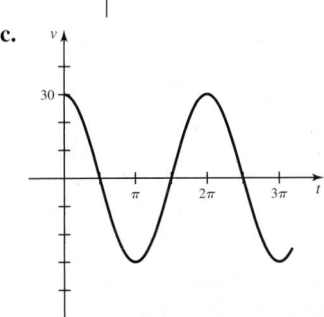

d. $v(t) = 0$, for $t = (2k + 1)\dfrac{\pi}{2}$, where k is any nonnegative
integer; the position is $y\left((2k + 1)\dfrac{\pi}{2}\right) = 0$ if k is even or
$y\left((2k + 1)\dfrac{\pi}{2}\right) = -60$ if k is odd. **e.** $v(t)$ has a maximum
at $t = 2k\pi$, where k is a nonnegative integer; the position is
$y(2k\pi) = -30$. **f.** $a(t) = -30 \sin t$

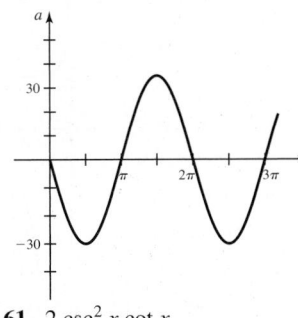

57. $2 \cos x - x \sin x$ **59.** $2e^x \cos x$ **61.** $2 \csc^2 x \cot x$
63. $2 (\sec^2 x \tan x + \csc^2 x \cot x)$ **65. a.** False **b.** False
c. True **d.** True **67.** 2 **69.** $-\frac{1}{2}$ **71.** $\frac{4}{3}$

73. a. $y = \sqrt{3}x + 2 - \dfrac{\pi\sqrt{3}}{6}$ **75. a.** $y = -2\sqrt{3}x + \dfrac{2\sqrt{3}\pi}{3} + 1$

b.

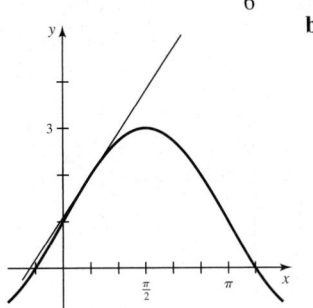

b.

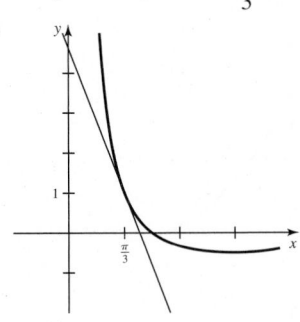

77. $x = 7\pi/6 + 2k\pi$ and $x = 11\pi/6 + 2k\pi$, where k is an integer
85. $a = 0$ **87. a.** $2 \sin x \cos x$ **b.** $3 \sin^2 x \cos x$ **c.** $4 \sin^3 x \cos x$
d. $n \sin^{n-1} x \cos x$; the conjecture is true for $n = 1$. If it holds
for $n = k$, then when $n = k + 1$, we have $\dfrac{d}{dx}(\sin^{k+1} x) =$
$\dfrac{d}{dx}(\sin^k x \cdot \sin x) = \sin^k x \cos x + \sin x \cdot k \sin^{k-1} x \cos x =$
$(k + 1) \sin^k x \cos x$.

89. Because D is a difference quotient for f (and $h = 0.01$ is small),
D is a good approximation to f'. Therefore, the graph of D is nearly
indistinguishable from the graph of $f'(x) = \cos x$.

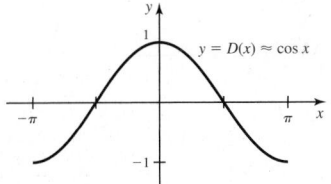

Section 3.6 Exercises, pp. 186–191

1. The average rate of change is $\dfrac{f(x + \Delta x) - f(x)}{\Delta x}$, whereas
the instantaneous rate of change is the limit as Δx goes to zero in
this quotient. **3.** Small **5.** At 15 weeks, the puppy grows at a rate
of 1.75 lb/week. **7.** If the position of the object at time t is $s(t)$, then
the acceleration at time t is $a(t) = d^2s/dt^2$. **9.** $v'(T) = 0.6$; the
speed of sound increases approximately 0.6 m/s for each increase of
1°C. **11. a.** 40 mi/hr **b.** 40 mi/hr; yes **c.** -60 mi/hr;
-60 mi/hr; south **d.** The patrol car drives away from the station
going north until about 10:08, when it turns around and heads south,
toward the station. It continues south until it passes the station at about
11:02 and keeps going south until about 11:40, when it turns around
and heads north. **13.** The first 200 stoves cost, on average, $70 to
produce. When 200 stoves have already been produced, the 201st stove
costs $65 to produce.

15. a.

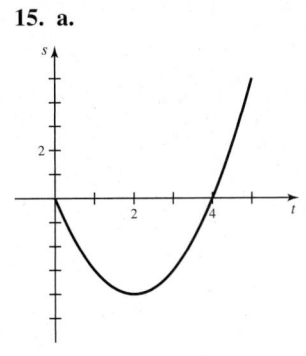

b. $v(t) = 2t - 4$; stationary at
$t = 2$, to the right on $(2, 5]$, to the
left on $[0, 2)$

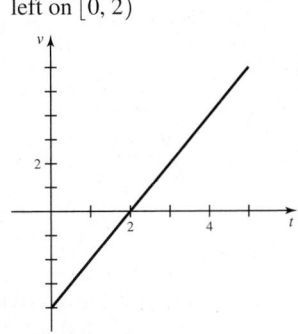

c. $v(1) = -2$ ft/s; $a(1) = 2$ ft/s² **d.** $a(2) = 2$ ft/s² **e.** $(2, 5]$

17. a.

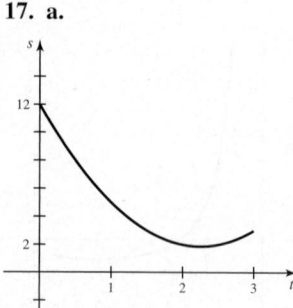

b. $v(t) = 4t - 9$; stationary at $t = \frac{9}{4}$, to the right on $\left(\frac{9}{4}, 3\right]$, to the left on $\left[0, \frac{9}{4}\right)$

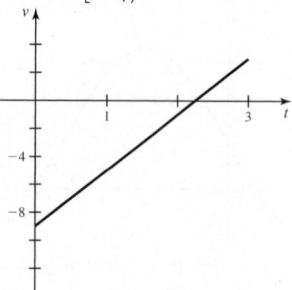

c. $v(1) = -5$ ft/s; $a(1) = 4$ ft/s²
d. $a\left(\frac{9}{4}\right) = 4$ ft/s² **e.** $\left(\frac{9}{4}, 3\right]$

19. a.

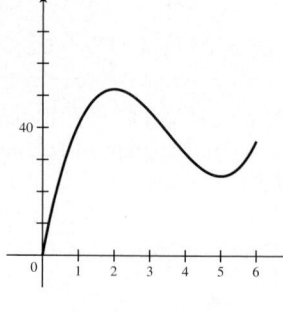

b. $v(t) = 6t^2 - 42t + 60$; stationary at $t = 2$ and $t = 5$, to the right on $[0, 2)$ and $(5, 6]$, to the left on $(2, 5)$

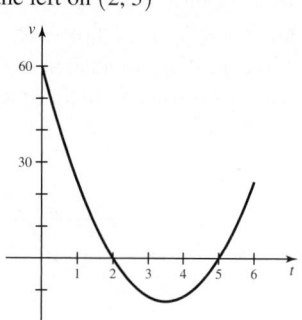

c. $v(1) = 24$ ft/s; $a(1) = -30$ ft/s² **d.** $a(2) = -18$ ft/s²; $a(5) = 18$ ft/s² **e.** $\left(2, \frac{7}{2}\right), (5, 6]$ **21.** -64 ft/s; 64 ft/s
23. a. $v(t) = -32t + 32$ **b.** At $t = 1$ s **c.** 64 ft **d.** At $t = 3$ s
e. -64 ft/s **f.** $(1, 3)$ **25. a.** $v(t) = -32t + 64$ **b.** At $t = 2$ s
c. 96 ft **d.** At $2 + \sqrt{6}$ s **e.** $-32\sqrt{6}$ ft/s **f.** $(2, 2 + \sqrt{6})$
27. Approx. 90.5 ft/s **29. a.** $\overline{C}(x) = \dfrac{1000}{x} + 0.1$; $C'(x) = 0.1$
b. $\overline{C}(2000) = \$0.60$/item; $C'(2000) = \$0.10$/item
c. The average cost per item when 2000 items are produced is $\$0.60$/item. The cost of producing the 2001st item is $\$0.10$.
31. a. $\overline{C}(x) = -0.01x + 40 + 100/x$; $C'(x) = -0.02x + 40$
b. $\overline{C}(1000) = \$30.10$/item; $C'(1000) = \$20$/item
c. The average cost per item is about $\$30.10$ when 1000 items are produced. The cost of producing the 1001st item is $\$20$. **33. a.** 20
b. $\$20$ **c.** $E(p) = \dfrac{p}{p - 20}$ **d.** Elastic for $p > 10$; inelastic for $0 < p < 10$ **e.** 2.5% **f.** 2.5% **35. a.** False **b.** True **c.** False
d. True **37.** 240 ft **39.** 64 ft/s **41. a.** $t = 1, 2, 3$ **b.** It is moving in the positive direction for t in $(0, 1)$ and $(2, 3)$; it is moving in the negative direction for t in $(1, 2)$ and $t > 3$.

c.

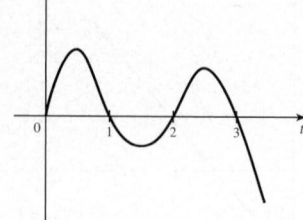

d. $\left(0, \frac{1}{2}\right), \left(1, \frac{3}{2}\right), \left(2, \frac{5}{2}\right), (3, \infty)$

43. a. $P(x) = 0.02x^2 + 50x - 100$
b. $\dfrac{P(x)}{x} = 0.02x + 50 - \dfrac{100}{x}$; $\dfrac{dP}{dx} = 0.04x + 50$

c. $\dfrac{P(500)}{500} = 59.8$; $\dfrac{dp}{dx}(500) = 70$
d. The profit, on average, for each of the first 500 items produced is 59.8; the profit for the 501st item produced is 70.
45. a. $P(x) = 0.04x^2 + 100x - 800$
b. $\dfrac{P(x)}{x} = 0.04x + 100 - \dfrac{800}{x}$; $\dfrac{dp}{dx} = 0.08x + 100$
c. $\dfrac{P(1000)}{1000} = 139.2$; $p'(1000) = 180$
d. The average profit per item for each of the first 1000 items produced is $\$139.20$. The profit for the 1001st item produced is $\$180$.
47. About 1935; approximately 890,000 people/yr (answers will vary)

49. a.

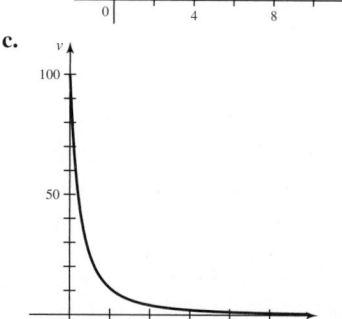

b. $v = \dfrac{100}{(t + 1)^2}$

c.

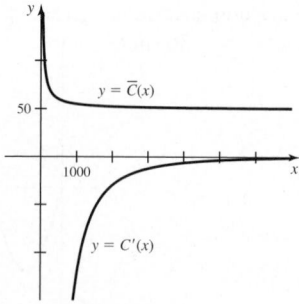

The marble moves fastest at the beginning and slows considerably over the first 5 s. It continues to slow but never actually stops.
d. $t = 4$ s **e.** $t = -1 + \sqrt{2} \approx 0.414$ s
51. a. $C'(x) = -\dfrac{125,000,000}{x^2} + 1.5$;

$$\overline{C}(x) = \dfrac{C(x)}{25,000} = 50 + \dfrac{5000}{x} + 0.00006x$$

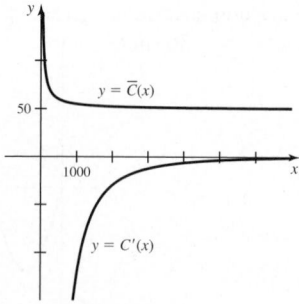

b. $C'(5000) = -3.5$; $\overline{C}(5000) = 51.3$ **c.** Marginal cost: If the batch size is increased from 5000 to 5001, then the cost of producing 25,000 gadgets will *decrease* by about $\$3.50$. Average cost: When batch size is 5000, it costs $\$51.30$ *per item* to produce all 25,000 gadgets.

53. a. $R(p) = \dfrac{100p}{p^2 + 1}$ **b.** $R'(p) = \dfrac{100(1 - p^2)}{(p^2 + 1)^2}$

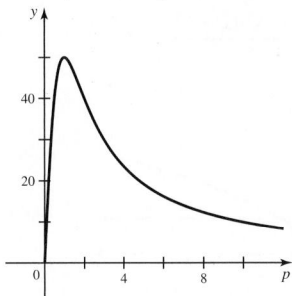 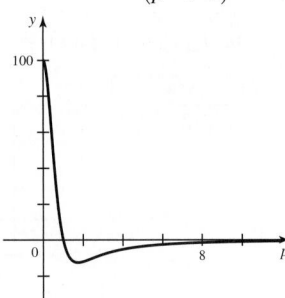

c. $p = 1$

55. a.

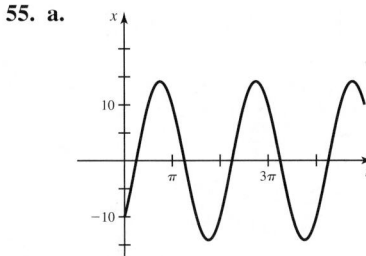

b. $dx/dt = 10 \cos t + 10 \sin t$

c. $t = 3\pi/4 + k\pi$, where k is any positive integer

d. The graph implies that the spring never stops oscillating. In reality, the weight would eventually come to rest.

57. a. Juan starts faster than Jean and opens up a big lead. Then Juan slows down while Jean speeds up. Jean catches up, and the race finishes in a tie. **b.** Same average velocity **c.** Tie **d.** At $t = 2$, $\theta'(2) = \pi/2$ rad/min; $\theta'(4) = \pi =$ Jean's greatest velocity **e.** At $t = 2$, $\varphi'(2) = \pi/2$ rad/min; $\varphi'(0) = \pi =$ Juan's greatest velocity **59. a.** $v(t) = -15e^{-t}(\sin t + \cos t)$; $v(1) \approx -7.6$ m/s, $v(3) \approx 0.63$ m/s **b.** Down $(0, 2.4)$ and $(5.5, 8.6)$; up $(2.4, 5.5)$ and $(8.6, 10)$ **c.** ≈ 0.65 m/s **61. a.** $-T'(1) = -80, -T'(3) = 80$ **b.** $-T'(x) < 0$ for $0 \le x < 2$; $-T'(x) > 0$ for $2 < x \le 4$ **c.** Near $x = 0$, with $x > 0$, $-T'(x) < 0$, so heat flows toward the end of the rod. Similarly, near $x = 4$, with $x < 4$, $-T'(x) > 0$.

Section 3.7 Exercises, pp. 196–200

1. $\dfrac{dy}{dx} = \dfrac{dy}{du} \cdot \dfrac{du}{dx}$; $\dfrac{d}{dx}(f(g(x))) = f'(g(x)) \cdot g'(x)$

3. $u = x^3 + x + 1$; $y = u^4$; $4(x^3 + x + 1)^3(3x^2 + 1)$

5. $u = \cos x, y = u^3, dy/dx = -3 \cos^2 x \sin x$;

$u = x^3, y = \cos u, dy/dx = -3x^2 \sin x^3$ **7.** $g(x), x$ **9.** $\dfrac{2}{\sqrt{4x + 1}}$

11. 50 **13.** ke^{kx} **15.** $u = 3x + 7$; $f(u) = u^{10}$; $30(3x + 7)^9$

17. $u = \sin x$; $f(u) = u^5$; $5 \sin^4 x \cos x$

19. $u = x^2 + 1$; $f(u) = \sqrt{u}$; $\dfrac{x}{\sqrt{x^2 + 1}}$

21. $u = 4x^2 + 1$; $f(u) = e^u$; $8xe^{4x^2 + 1}$

23. $u = 5x^2$; $f(u) = \tan u$; $10x \sec^2 5x^2$ **25. a.** 100 **b.** -100 **c.** -16 **d.** 40 **e.** 40 **27.** $10(6x + 7)(3x^2 + 7x)^9$

29. $\dfrac{5}{\sqrt{10x + 1}}$ **31.** $-\dfrac{315x^2}{(7x^3 + 1)^4}$ **33.** $3 \sec(3x + 1) \tan(3x + 1)$

35. $e^x \sec^2 e^x$ **37.** $(12x^2 + 3)\cos(4x^3 + 3x + 1)$

39. $\dfrac{10}{3(5x + 1)^{1/3}}$ **41.** $-\dfrac{3}{2^{7/4}x^{3/4}(4x - 3)^{5/4}}$

43. $5 \sec x (\sec x + \tan x)^5$ **45.** $25(12x^5 - 9x^2)(2x^6 - 3x^3 + 3)^{24}$

47. $9(1 + 2 \tan u)^{3.5} \sec^2 u$ **49.** $-\dfrac{\cot x \csc^2 x}{\sqrt{1 + \cot^2 x}}$

51. $\dfrac{2}{3}e^x - e^{-x}$ **53.** $e^x \cos(\sin e^x) \cos e^x$

55. $-15 \sin^4(\cos 3x)(\sin 3x)(\cos(\cos 3x))$

57. $\dfrac{2e^{2t}}{(1 + e^{2t})^2}$ **59.** $\dfrac{1}{2\sqrt{x + \sqrt{x}}}\left(1 + \dfrac{1}{2\sqrt{x}}\right)$

61. $f'(g(x^2))g'(x^2) \, 2x$ **63.** $\dfrac{5x^4}{(x + 1)^6}$

65. $xe^{x^2 + 1}(2 \sin x^3 + 3 x \cos x^3)$ **67.** $\theta(2 + 5\theta \tan 5\theta) \sec 5\theta$

69. $4((x + 2)(x^2 + 1))^3(3x + 1)(x + 1)$ **71.** $\dfrac{4x^3 - 2 \sin 2x}{5(x^4 + \cos 2x)^{4/5}}$

73. $2(p + 3)(\sin p^2 + p(p + 3) \cos p^2)$

75. $f'(x)/(2\sqrt{f(x)})$ **77. a.** True **b.** True **c.** True **d.** False **79.** -0.297 hPa/min **81.** Approx. 0.33 g/day; mass is increasing by 0.33 g/day 65 days after the diet switch.

83. a. \$297.77 **b.** \$11.85/yr **c.** $y = 11.85t + 179.27$

85. a. $x = -\dfrac{1}{2}$ **b.** The line tangent to the graph of $f(x)$ at $x = -\dfrac{1}{2}$ is horizontal. **87.** $2 \cos x^2 - 4x^2 \sin x^2$ **89.** $4e^{-2x^2}(4x^2 - 1)$

91. $y = 6x - 1$ **93. a.** $h(4) = 9, h'(4) = -6$ **b.** $y = -6x + 33$ **95.** $y = 6x + 3 - 3 \ln 3$ **97. a.** -3π **b.** -5π

99. a. $\dfrac{d^2 y}{dt^2} = -\dfrac{y_0 k}{m} \cos\left(t\sqrt{\dfrac{k}{m}}\right)$

101. a.

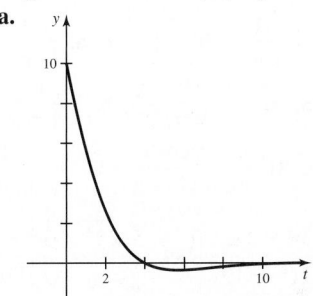

b. $v(t) = -5e^{-t/2}\left(\dfrac{\pi}{4} \sin \dfrac{\pi t}{8} + \cos \dfrac{\pi t}{8}\right)$

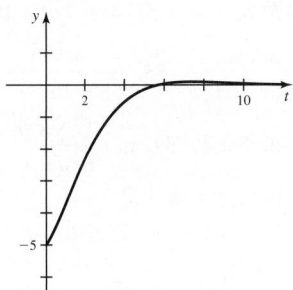

103. a. 10.88 hr **b.** $D'(t) = \dfrac{6\pi}{365} \sin\left(\dfrac{2\pi(t + 10)}{365}\right)$

c. 2.87 min/day; on March 1, the length of day is increasing at a rate of about 2.87 min/day.

d.

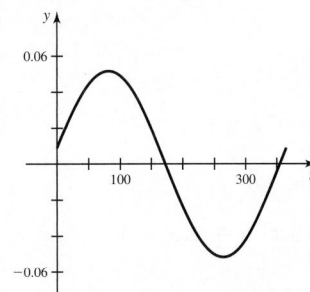

e. Most rapidly: approximately March 22 and September 22; least rapidly: approximately December 21 and June 21

105. a. $E'(t) = 400 + 200 \cos \dfrac{\pi t}{12}$ MW

b. At noon; $E'(0) = 600$ MW   **c.** At midnight; $E'(12) = 200$ MW

d.

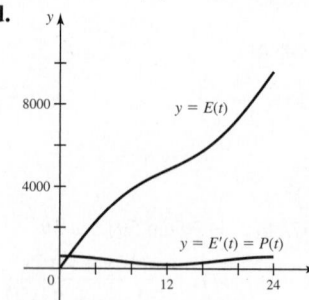

109. a. $g(x) = (x^2 - 3)^5$; $a = 2$ **b.** 20

111. a. $g(x) = \sin x^2$; $a = \pi/2$ **b.** $\pi \cos(\pi^2/4)$ **113.** $10 f'(25)$

Section 3.8 Exercises, pp. 205–208

1. There may be more than one expression for y or y'.

3. When derived implicitly, dy/dx is usually given in terms

of both x and y.   **5.** $\dfrac{1}{2y}$ **7.** $\dfrac{1}{\cos y}$ **9. a.** $(0, 0)$, $(0, -1)$, $(0, 1)$

c. Slope at $(0, 0)$ is 2; slope at $(0, -1)$ and $(1, 0)$ is -1.

11. $\dfrac{d^2 y}{dx^2} = -\dfrac{2}{9y^5}$ **13. a.** $-\dfrac{x^3}{y^3}$ **b.** 1   **15. a.** $\dfrac{2}{y}$ **b.** 1

17. a. $\dfrac{20x^3}{\cos y}$ **b.** -20 **19. a.** $-\dfrac{1}{\sin y}$ **b.** -1 **21. a.** $-\dfrac{y}{x}$ **b.** -7

23. a. $-\dfrac{1}{4x^{2/3}y^{1/3}}$ **b.** $-\dfrac{1}{4}$ **25. a.** $-\dfrac{3y}{x + 3y^{2/3}}$ **b.** $-\dfrac{24}{13}$

27. $\dfrac{\cos x}{1 - \cos y}$ **29.** $-\dfrac{1}{1 + \sin y}$ **31.** $\dfrac{1 - y \cos xy}{x \cos xy - 1}$ **33.** $\dfrac{1}{2y \sin y^2 + e^y}$

35. $\dfrac{3x^2(x - y)^2 + 2y}{2x}$ **37.** $\dfrac{13y - 18x^2}{21y^2 - 13x}$ **39.** $\dfrac{5\sqrt{x^4 + y^2} - 2x^3}{y - 6y^2\sqrt{x^4 + y^2}}$

41. a. $\dfrac{dK}{dL} = -\dfrac{K}{2L}$ **b.** -4 **43.** $\dfrac{dr}{dh} = \dfrac{h - 2r}{h}$; -3

45. b. $y = -5x$ **47. b.** $y = -5x/4 + 7/2$ **49. b.** $y = \dfrac{x}{2}$

51. $-\dfrac{1}{4y^3}$ **53.** $\dfrac{\sin y}{(\cos y - 1)^3}$ **55.** $\dfrac{4e^{2y}}{(1 - 2e^{2y})^3}$ **57. a.** False

b. True   **c.** False   **d.** False   **59. a.** $\dfrac{y(3\sqrt{x} + 2y^{3/2})}{x(\sqrt{x} - 2y^{3/2})}$ **b.** -5

61. **a.** $y = x - 1$ and $y = -x + 2$

b.

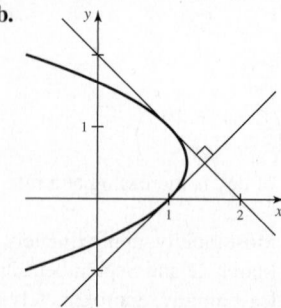

63. a. $y' = -\dfrac{2xy}{x^2 + 4}$ **b.** $y = \dfrac{1}{2}x + 2$, $y = -\dfrac{1}{2}x + 2$

c. $-\dfrac{16x}{(x^2 + 4)^2}$ **65. a.** $\left(\dfrac{5}{4}, \dfrac{1}{2}\right)$ **b.** No

67. Horizontal: $y = -6$, $y = 0$; vertical: $x = 1$, $x = 3$

69. a. $\dfrac{dy}{dx} = 0$ on the $y = 1$ branch; $\dfrac{dy}{dx} = \dfrac{1}{2y + 1}$ on the other two

branches.   **b.** $f_1(x) = 1$, $f_2(x) = \dfrac{-1 + \sqrt{4x - 3}}{2}$,

$f_3(x) = \dfrac{-1 - \sqrt{4x - 3}}{2}$ **c.**

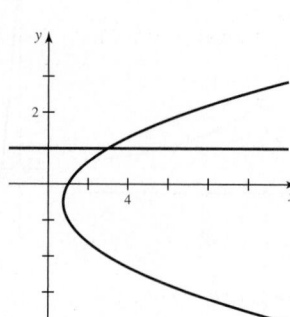

71. a. $\dfrac{dy}{dx} = \dfrac{x - x^3}{y}$ **b.** $f_1(x) = \sqrt{x^2 - \dfrac{x^4}{2}}$; $f_2(x) = -\sqrt{x^2 - \dfrac{x^4}{2}}$

c.

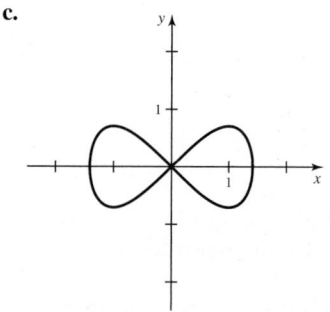

73. $y = \dfrac{x}{5}$

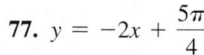

75. $y = \dfrac{4x}{5} - \dfrac{3}{5}$

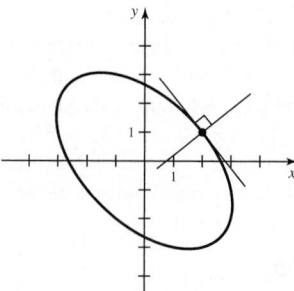

77. $y = -2x + \dfrac{5\pi}{4}$

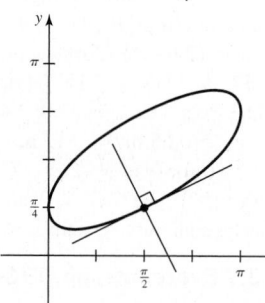

79. a. Tangent line $y = -\dfrac{9x}{11} + \dfrac{20}{11}$; normal line $y = \dfrac{11x}{9} - \dfrac{2}{9}$

b.

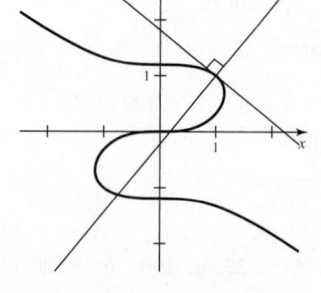

81. a. Tangent line $y = -\dfrac{x}{3} + \dfrac{8}{3}$;

normal line $y = 3x - 4$

b.

83. For $y = mx$, $dy/dx = m$; for $x^2 + y^2 = a^2$, $dy/dx = -x/y$.

85. For $xy = a$, $dy/dx = -y/x$; for $x^2 - y^2 = b$, $dy/dx = x/y$.

Because $(-y/x) \cdot (x/y) = -1$, the families of curves

form orthogonal trajectories.   **87.** $\dfrac{7y^2 - 3x^2 - 4xy^2 - 4x^3}{2y(2x^2 + 2y^2 - 7x)}$

89. $\dfrac{2y^2(5 + 8x\sqrt{y})}{(1 + 2x\sqrt{y})^3}$ **91.** No horizontal tangent line; vertical tangent

lines at $(2, 1), (-2, 1)$ **93.** No horizontal tangent line;

vertical tangent lines at $(0, 0), \left(\frac{3\sqrt{3}}{2}, \sqrt{3}\right), \left(-\frac{3\sqrt{3}}{2}, -\sqrt{3}\right)$

Section 3.9 Exercises, pp. 215–218

1. $x = e^y \Rightarrow 1 = e^y y'(x) \Rightarrow y'(x) = 1/e^y = 1/x$

3. $\dfrac{d}{dx}(\ln kx) = \dfrac{d}{dx}(\ln k + \ln x) = \dfrac{d}{dx}(\ln x)$ **5.** $f'(x) = \dfrac{1}{x \ln b}$;

if $b = e$, then $f'(x) = \dfrac{1}{x}$.   **7.** $(x^2 + 1)^x$ **9.** $\dfrac{x}{x^2 + 1}$

11. $f(x) = e^{h(x) \ln g(x)}$ **13.** $\dfrac{1 + x}{x}$ **15.** $\dfrac{1}{x}$ **17.** $2/x$ **19.** $\cot x$

21. $\dfrac{4x^3}{x^4 + 1}$ **23.** $2/(1 - x^2)$ **25.** $(x^2 + 1)/x + 2x \ln x$

27. $-2x \ln x^2$ or $-4x \ln x$ **29.** $1/(x \ln x)$ **31.** $\dfrac{1}{x(\ln x + 1)^2}$

33. ex^{e-1} **35.** $\pi(2^x + 1)^{\pi-1}2^x \ln 2$ **37.** $8^x \ln 8$ **39.** $5 \cdot 4^x \ln 4$

41. $2^{3+\sin x}(\ln 2)\cos x$ **43.** $3^x \cdot x^2 (x \ln 3 + 3)$

45. $1000(1.045)^{4t} \ln 1.045$ **47.** $\dfrac{2^x \ln 2}{(2^x + 1)^2}$

49. $x^{\cos x-1}(\cos x - x \ln x \sin x); -\ln(\pi/2)$

51. $x^{\sqrt{x}}\left(\dfrac{2 + \ln x}{2\sqrt{x}}\right); 4(2 + \ln 4)$

53. $\dfrac{(\sin x)^{\ln x}(\ln(\sin x) + x(\ln x)\cot x)}{x}; 0$

55. $(4 \sin x + 2)^{\cos x}\left(\dfrac{2\cos^2 x}{2 \sin x + 1} - \sin x \ln (4 \sin x + 2)\right); 1$

57. a. Approx. 28.7 s   **b.** -46.512 s/1000 ft

c. $dT/da = -2.74 \cdot 2^{-0.274a} \ln 2$

At $a = 8, \dfrac{dT}{da} = -0.4156$ min/1000 ft

$= -24.938$ s/1000 ft.

If a plane travels at 30,000 feet and increases its altitude by 1000 feet, the time of useful consciousness decreases by about 25 seconds.

59. $y = x \sin 1 + 1 - \sin 1$ **61.** $y = e^{2/e}$ and $y = e^{-2/e}$

63. $\dfrac{8x}{(x^2 - 1) \ln 3}$ **65.** $-\sin x (\ln (\cos^2 x) + 2)$

67. $-\dfrac{\ln 4}{x \ln^2 x}$ **69.** $\dfrac{12}{3x + 1}$ **71.** $\dfrac{1}{2x}$

73. $\dfrac{2}{2x - 1} + \dfrac{3}{x + 2} + \dfrac{8}{1 - 4x}$ **75.** $10x^{10x}(1 + \ln x)$

77. $\dfrac{(x + 1)^{10}}{(2x - 4)^8}\left(\dfrac{10}{x + 1} - \dfrac{8}{x - 2}\right)$ **79.** $2x^{\ln x-1} \ln x$

81. $\dfrac{(x + 1)^{3/2}(x - 4)^{5/2}}{(5x + 3)^{2/3}}\left(\dfrac{3}{2(x + 1)} + \dfrac{5}{2(x - 4)} - \dfrac{10}{3(5x + 3)}\right)$

83. $(\sin x)^{\tan x}(1 + (\sec^2 x) \ln \sin x)$

85. $\left(1 + \dfrac{1}{x}\right)^x\left(\ln\left(1 + \dfrac{1}{x}\right) - \dfrac{1}{x + 1}\right)$

87. a. False **b.** False **c.** False **d.** False **e.** True **f.** True

89. $-\dfrac{1}{x^2 \ln 10}$ **91.** $\dfrac{2}{x}$ **93.** $3^x \ln 3$

95. $y = 2$

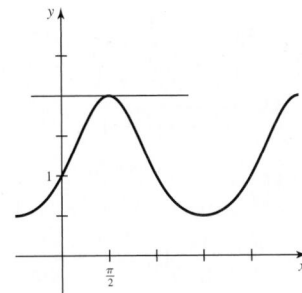

97. a.

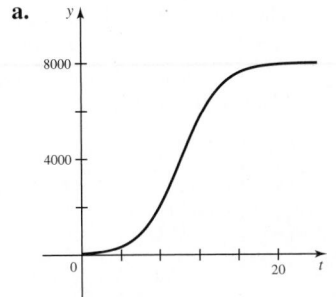

b. $t = 2 \ln 265 \approx 11.2$ years; approx. 14.5 years

c. $P'(0) \approx 25$ fish/year; $P'(5) \approx 264$ fish/year

d.

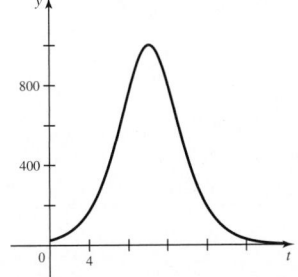

The population is growing fastest after about 10 years.

99. b. $r(11) \approx 0.0133; r(21) \approx 0.0118$; the relative growth rate is decreasing.   **c.** $\lim_{t\to\infty} r(t) = 0$; as the population gets close to carrying capacity, the relative growth rate approaches zero.

101. a. $A(5) = \$17{,}443$

$A(15) = \$72{,}705$

$A(25) = \$173{,}248$

$A(35) = \$356{,}178$

$\$5526.20$/year, $\$10{,}054.30$/year, $\$18{,}293$/year

b. $A(40) = \$497{,}873$

c. $\dfrac{dA}{dt} = 600{,}000 \ln (1.005)((1.005)^{12t})$

$\approx (2992.5)(1.005)^{12t}$

A increases at an increasing rate.

103. $p = e^{1/e}; (e, e)$ **105.** $1/e$ **107.** $27(1 + \ln 3)$

Section 3.10 Exercises, pp. 225–227

1. $\dfrac{d}{dx}(\sin^{-1} x) = \dfrac{1}{\sqrt{1 - x^2}}; \dfrac{d}{dx}(\tan^{-1} x) = \dfrac{1}{1 + x^2}$;

$\dfrac{d}{dx}(\sec^{-1} x) = \dfrac{1}{|x|\sqrt{x^2 - 1}}$ **3.** $\dfrac{1}{5}$ **5.** $\dfrac{1}{4}$ **7. a.** $\dfrac{1}{2}$ **b.** $\dfrac{2}{3}$

c. Cannot be determined   **d.** $\dfrac{3}{2}$ **9.** $y = \dfrac{1}{7}x + \dfrac{13}{7}$ **11.** $\dfrac{2}{\sqrt{3}}$

13. $\dfrac{2}{\sqrt{1 - 4x^2}}$ **15.** $-\dfrac{4w}{\sqrt{1 - 4w^2}}$ **17.** $-\dfrac{2e^{-2x}}{\sqrt{1 - e^{-4x}}}$

19. $\dfrac{10}{100x^2 + 1}$ **21.** $\dfrac{4y}{1 + (2y^2 - 4)^2}$ **23.** $-\dfrac{1}{2\sqrt{z}(1 + z)}$

25. $6x^2 \cot^{-1} x$ **27.** $\dfrac{2w^5}{1 + w^4}$ **29.** $\dfrac{1}{|x|\sqrt{x^2 - 1}}$

31. $-\dfrac{1}{|2u + 1|\sqrt{u^2 + u}}$ **33.** $\dfrac{2y}{(y^2 + 1)^2 + 1}$

35. $\dfrac{1}{x|\ln x|\sqrt{(\ln x)^2 - 1}}$ **37.** $-\dfrac{e^x \sec^2 e^x}{|\tan e^x|\sqrt{\tan^2 e^x - 1}}$

39. $-\dfrac{e^s}{1 + e^{2s}}$ **41.** $y = x + \dfrac{\pi}{4} - \dfrac{1}{2}$ **43.** $y = -\dfrac{4}{\sqrt{6}}x + \dfrac{\pi}{3} + \dfrac{2}{\sqrt{3}}$

45. a. Approx. -0.00055 rad/m

b.

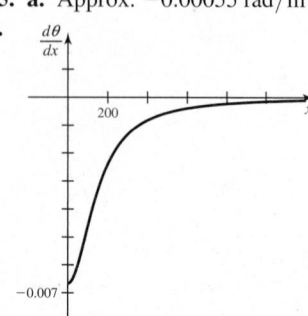

The magnitude of the change in angular size, $|d\theta/dx|$, is greatest when the boat is at the skyscraper (that is, at $x = 0$).

47. $\frac{1}{3}$ **49.** $\frac{e}{5}$ **51.** $\frac{1}{2}$ **53.** 4 **55.** $\frac{1}{12}$ **57.** $\frac{1}{4}$ **59.** $\frac{5}{4}$ **61. a.** True
b. False **c.** True **d.** True **e.** True
63. a.

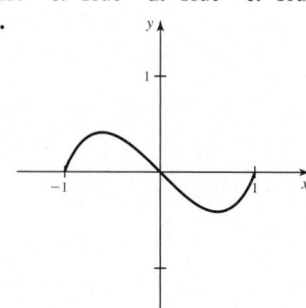

b. $f'(x) = 2x \sin^{-1} x + \dfrac{x^2 - 1}{\sqrt{1 - x^2}}$

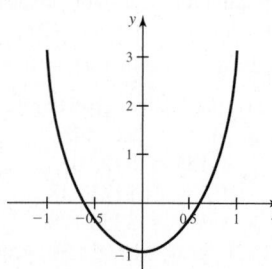

65. a.

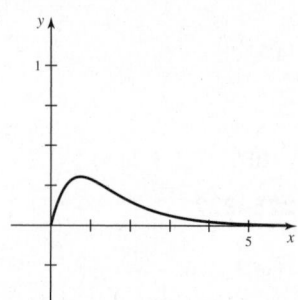

b. $f'(x) = \dfrac{e^{-x}}{1 + x^2} - e^{-x} \tan^{-1} x$

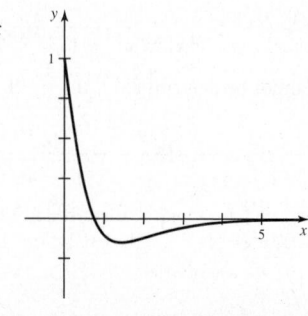

67. $\dfrac{1}{3}$ **69.** $1/(2\sqrt{x + 4})$ **71.** $\dfrac{1}{3x}$ **73.** $\dfrac{1}{12x \ln 10}$ **75.** $2x$

77. $-2/x^3$ **79. b.** $-0.0041, -0.0289,$ and -0.1984
c. $\lim\limits_{\ell \to 10^+} d\theta/d\ell = -\infty$ **d.** The length ℓ is decreasing.
81. a. $1/\sqrt{D^2 - c^2}$ **b.** $1/D$ **85.** Use the identity
$\cot^{-1} x + \tan^{-1} x = \pi/2$.

Section 3.11 Exercises, pp. 231–236

1. As the side length s of a cube changes, the surface area $6s^2$ changes as well. **3.** The other two opposite sides decrease in length.

5. a. $V = 200h$; $\dfrac{dV}{dt} = 200 \dfrac{dh}{dt}$ **b.** 50 ft^3/min

c. $\dfrac{1}{20}$ ft/min **7. a.** $\dfrac{dV}{dt} = 4\pi r^2 \dfrac{dr}{dt}$ **b.** 128π in^3/min

c. $\dfrac{1}{10\pi}$ in/min **9.** 59 **11. a.** 40 m^2/s **b.** 80 m^2/s

13. a. 4 m^2/s **b.** $\sqrt{2}$ m^2/s **c.** $2\sqrt{2}$ m/s **15. a.** $\dfrac{1}{4\pi}$ cm/s

b. $\dfrac{1}{2}$ cm/s **17.** -40π ft^2/min **19.** $\dfrac{3}{80\pi}$ in/min

23. 720.3 mi/hr **25.** $\dfrac{3\sqrt{5}}{2}$ ft/s **27.** 57.9 ft/s **29.** 4.66 in/s

31. $\dfrac{\pi}{2}$ ft^3/min **33.** -75π cm^3/s **35.** 2592π cm^3/s

37. 9π ft^3/min **39.** $\dfrac{1}{25\pi}$ m/min **41.** $\dfrac{5}{24}$ ft/s

43. $-\dfrac{8}{3}$ ft/s, $-\dfrac{32}{3}$ ft/s **45.** $\dfrac{d\theta}{dt} = \dfrac{1}{5}$ rad/s, $\dfrac{d\theta}{dt} = \dfrac{1}{8}$ rad/s
47. -0.0201 rad/s **49.** $10 \tan 20°$ km/hr $\approx$ 3.6 km/hr

51. a. 187.5 ft/s **b.** 0.938 rad/s **53. a.** $P = \dfrac{1}{2} v^2 \dfrac{dm}{dt}$

c. $17{,}388.7$ W **d.** 4347.2 W **55.** 11.06 m/hr

57. $\dfrac{1}{500}$ m/min; 2000 min **59.** 0.543 rad/hr

61. $\dfrac{d\theta}{dt} = 0$ rad/s, for all $t \geq 0$ **63. a.** $-\dfrac{\sqrt{3}}{10}$ m/hr **b.** -1 m^2/hr

Chapter 3 Review Exercises, pp. 236–240

1. a. False **b.** False **c.** False **d.** False **e.** True

3. $-\dfrac{2x}{(x^2 + 5)^2}$ **9.** $2x^2 + 2\pi x + 7$ **11.** $2^x \ln 2$

13. $2e^{2\theta}$ **15.** $6x^3\sqrt{1 + x^4}$ **17.** $5t^2 \cos t + 10t \sin t$

19. $-x^2 e^{-x}$ **21.** $\dfrac{2 \sec 2w \tan 2w}{(\sec 2w + 1)^2}$ **23.** $3 \tan 3x$

25. $1000t(5t^2 + 10)^{99}$ **27.** $3x^2 \cot x^3$ **29.** $\dfrac{1}{t\sqrt{t^2 - 1}}$

31. $(8\theta + 12) \sec^2 (\theta^2 + 3\theta + 2)$ **33.** $\dfrac{1 - 5 \ln w}{w^6}$

35. $\dfrac{32u^2 + 8u + 1}{(8u + 1)^2}$ **37.** $(\sec^2 \sin \theta) \cos \theta$

39. $-\dfrac{\cos \sqrt{\cos^2 x + 1} \cos x \sin x}{\sqrt{\cos^2 x + 1}}$ **41.** $\dfrac{e^t}{2(e^t + 1)}$

43. $2 \tan^{-1}(\cot x)$ **45.** $(2 + \ln x) \ln x$ **47.** $(2x - 1) 2^{x^2 - x} \ln 2$

49. $(x^2 + 1)^{\ln x}\left(\dfrac{\ln (x^2 + 1)}{x} + \dfrac{2x \ln x}{x^2 + 1}\right)$ **51.** $\dfrac{1}{|x|\sqrt{x^2 - 1}}$

53. $6 \cot^{-1} 3x$ **55.** $1 + \csc(x - y)$

57. $\dfrac{y \cos x}{e^y - 1 - \sin x}$ **59.** $-\dfrac{xy}{x^2 + 2y^2}$

61. $\dfrac{(3x+5)^{10}\sqrt{x^2+5}}{(x^3+1)^{50}}\left(\dfrac{30}{3x+5}+\dfrac{x}{x^2+5}-\dfrac{150x^2}{x^3+1}\right)$

63. $\sqrt{3}+\pi/6$ **65.** 1 **67.** $2^x \ln 2(x\ln 2+2)$ **69.** $\dfrac{6\ln x-5}{x^4}$

71. $\dfrac{2(xy+y^2)}{(x+2y)^3}=\dfrac{2}{(x+2y)^3}$ **73.** $y=x$ **75.** $y=-\dfrac{4x}{5}+\dfrac{24}{5}$

77. $x^2 f'(x)+2xf(x)$ **79.** $\dfrac{g(x)(xf'(x)+f(x))-xf(x)g'(x)}{g^2(x)}$

81. a–D; b–C; c–B; d–A

83.
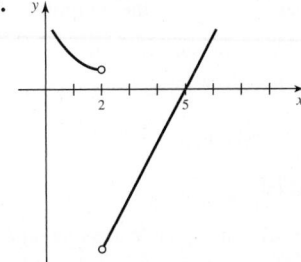

85. a. 27 **b.** $\frac{16}{27}$ **c.** 72 **d.** 1215 **e.** $\frac{1}{9}$ **87.** $\frac{6}{13}$
89. $(f^{-1})'(x)=-3/x^4$ **91. a.** $\frac{1}{4}$ **b.** 1 **c.** $\frac{1}{3}$
93. $y=24x-118$ **95. a.** 84 ft/s **b.** 7 s **c.** 384 ft
d. 96 ft/s **97. a.** \$200,366; \$21,552/yr
b. 14 yr; \$12,551/yr **99. a.** 2.70 million people/yr
b. The slope of the secant line through the two points is
approximately equal to the slope of that tangent line at $t=55$.
c. 2.217 million people/yr **101. a.** 40 m/s **b.** 20/3 m/s
c. 15 m/s **d.**

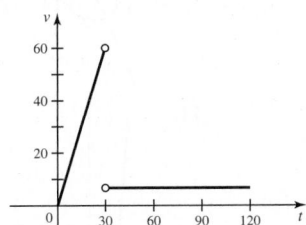

e. The skydiver deployed the parachute. **103.** $x=4; x=6$
105. $f(x)=\tan\left(\pi\sqrt{3x-11}\right)$, $a=5$; $f'(5)=3\pi/4$
107. a. $\overline{C}(3000)=\$341.67$; $C'(3000)=\$280$ **b.** The average
cost of producing the first 3000 lawn mowers is \$341.67 per mower.
The cost of producing the 3001st lawn mower is \$280.
109. a. 6550 people/yr **b.** $p'(40)=4800$ people/yr
111. 50 mi/hr **113.** $-5\sin 65° $ ft/s ≈ -4.5 ft/s
115. -0.166 rad/s **117.** 1.5 ft/s **119. a.** $(f^{-1})'(1/\sqrt{2})=\sqrt{2}$

CHAPTER 4

Section 4.1 Exercises, pp. 247–250

1. f has an absolute maximum at c in $[a,b]$ if $f(x)\le f(c)$ for all x in
$[a,b]$. f has an absolute minimum at c in $[a,b]$ if $f(x)\ge f(c)$ for all
x in $[a,b]$. **3.** The function must be continuous on a closed interval.
5.

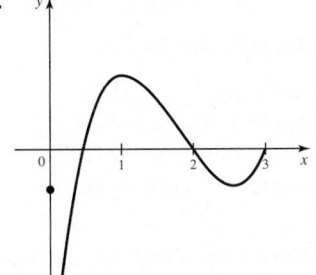

7.

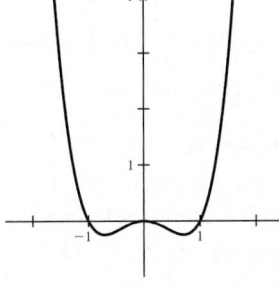

9. Evaluate the function at the critical points and at the endpoints of
the interval. **11.** Abs. min at $x=c_2$; abs. max at $x=b$ **13.** Abs.
min at $x=a$; no abs. max **15.** Local min at $x=q, s$; local max at
$x=p, r$; abs. min at $x=a$; abs. max at $x=b$ **17.** Local max at
$x=p, r$; local min at $x=q$; abs. max at $x=p$; abs. min at $x=b$
19.

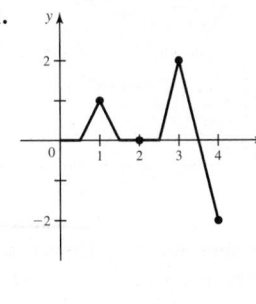

21.

23. $x=\frac{2}{3}$ **25.** $x=\pm 3$ **27.** $x=-\frac{2}{3},\frac{1}{3}$ **29.** $x=\pm\dfrac{2a}{\sqrt{3}}$
31. $t=\pm 1$ **33.** $x=0$ **35.** $x=1$ **37.** $x=-4,0$
39. If $a\ge 0$, there is no critical point. If $a<0$, $x=2a/3$ is
the only critical point. **41.** $t=\pm a$ **43.** Abs. max: -1 at
$x=3$; abs. min: -10 at $x=0$ **45.** Abs. max: 0 at $x=0, 3$;
abs. min: -4 at $x=-1, 2$ **47.** Abs. max: 234 at $x=3$;
abs. min: -38 at $x=-1$ **49.** Abs. max: 1 at $x=0, \pi$; abs. min: 0 at
$x=\pi/2$ **51.** Abs. max: 1 at $x=\pi/6$; abs. min: -1 at $x=-\pi/6$
53. Abs. min: $(\sqrt{1/e})^{1/e}$ at $x=1/(2e)$; abs. max: 2 at $x=1$
55. Abs. max: $1+\pi$ at $x=-1$; abs. min: 1 at $x=1$
57. Abs. max: 11 at $x=1$; abs. min: -16 at $x=4$
59. Abs. max: 27 at $x=-3$; abs. min: $-\frac{19}{12}$ at $x=\frac{1}{2}$
61. Abs. max: $\dfrac{1}{100,000}$ at $x=1$; abs. min: $-\dfrac{1}{100,000}$ at $x=-1$
63. Abs. max: $\sqrt{2}$ at $x=\pm\pi/4$; abs. min: 1 at $x=0$
65. Abs. max: $27/e^3$ at $x=3$; abs. min: $-e$ at $x=-1$
67. Abs. max: 3 at $x=\pm 1$; abs. min: 0 at $x=-2, 0, 2$
69. a. The velocity of the downstream wind v_2 is less than or equal to
the velocity of the upstream wind, so $0\le v_2\le v_1$, or $0\le\dfrac{v_2}{v_1}\le 1$.
b. $R(1)=0$ **c.** $R(0)=\frac{1}{2}$ **d.** 0.593 is the maximum fraction of
power that can be extracted from a wind stream by a wind turbine.
71. $t=2$ s **73.** $t=2$ s **75. a.** 50 **b.** 45 **77. a.** False
b. False **c.** False **d.** True **79. a.** $x=-0.96, 2.18, 5.32$
b. Abs. max: 3.72 at $x=2.18$; abs. min: -32.80 at $x=5.32$
c.

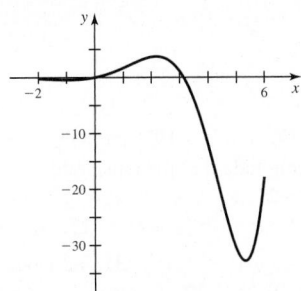

81. a. $x=\tan^{-1}2+k\pi$, for $k=-2,-1,0,1$
b. $x=\tan^{-1}2+k\pi$, for $k=-2,0$, correspond to local max;
$x=\tan^{-1}2+k\pi$, for $k=-1,1$, correspond to local min.
c. Abs. max: 2.24; abs. min: -2.24 **83. a.** $x=5-4\sqrt{2}$
b. $x=5-4\sqrt{2}$ corresponds to a local max. **c.** No abs. max or min
85. Abs. max: 4 at $x=-1$; abs. min: -8 at $x=3$

87. a. $T(x)=\dfrac{\sqrt{2500+x^2}}{2}+\dfrac{50-x}{4}$ **b.** $x=50/\sqrt{3}$

c. $T(50/\sqrt{3}) \approx 34.15$, $T(0) = 37.50$, $T(50) \approx 35.36$
d.

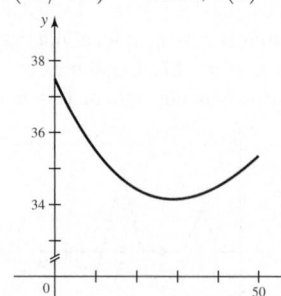

89. a. 1, 3, 0, 1   **b.** Because $h'(2) \neq 0$, h does not have a local extreme value at $x = 2$. However, g may have a local extremum at $x = 2$ (because $g'(2) = 0$).   **91. a.** $f(x) - f(c) \le 0$ for all x near c
b. $\lim\limits_{x \to c^+} \dfrac{f(x) - f(c)}{x - c} \le 0$ **c.** $\lim\limits_{x \to c^-} \dfrac{f(x) - f(c)}{x - c} \ge 0$
d. Because $f'(c)$ exists, $\lim\limits_{x \to c^+} \dfrac{f(x) - f(c)}{x - c} = \lim\limits_{x \to c^-} \dfrac{f(x) - f(c)}{x - c}$. By parts (b) and (c), we must have that $f'(c) = 0$.

Section 4.2 Exercises, pp. 254–257

1. If f is a continuous function on the closed interval $[a, b]$ and is differentiable on (a, b), and the slope of the secant line that joins $(a, f(a))$ to $(b, f(b))$ is zero, then there is at least one value c in (a, b) at which the slope of the line tangent to f at $(c, f(c))$ is also zero.

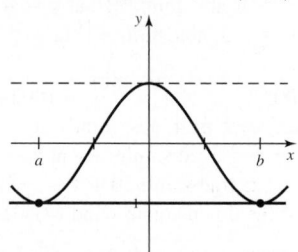

3. $f(x) = |x|$ is not differentiable at 0.   **5. b.** $c = 1$
7. b. $c = \pm 2/\sqrt[4]{5} \approx \pm 1.34$
9.

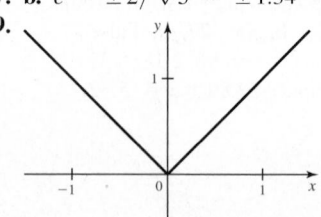

11. $x = \frac{1}{3}$

13. $x = \pi/4$ **15.** Does not apply   **17.** $x = \frac{5}{3}$ **19.** Average lapse rate $= -6.3°$/km. You cannot conclude that the lapse rate at a point exceeds the threshold value.   **21. a.** Yes   **b.** $c = \frac{1}{2}$
23. a. Does not apply   **25. a.** Yes   **b.** $c = \ln(e - 1)$ **27. a.** Yes
b. $c \approx \pm 0.881$ **29. a.** Yes   **b.** $c = \sqrt{1 - 9/\pi^2}$ **31. a.** Does not apply   **33. a.** False   **b.** True   **c.** False   **37.** h and p
39.

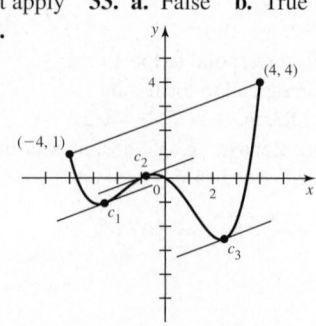

41. No such point exists; function is not continuous at 2.
43. The car's average velocity is $(30 - 0)/(28/60) = 64.3$ mi/hr. By the MVT, the car's instantaneous velocity was 64.3 mi/hr at some time.
45. Average speed $= 11.625$ mi/hr. By the MVT, the speed was exactly 11.625 mi/hr at least once. By the Intermediate Value Theorem, all speeds between 0 and 11.625 mi/hr were reached. Because the initial and final speeds were 0 mi/hr, the speed of 11 mi/hr was reached at least twice.
47. $\dfrac{f(b) - f(a)}{b - a} = A(a + b) + B$ and $f'(x) = 2Ax + B$;
$2Ax + B = A(a + b) + B$ implies that $x = \dfrac{a + b}{2}$, the midpoint of $[a, b]$.   **49.** $\tan^2 x$ and $\sec^2 x$ differ by a constant; in fact, $\tan^2 x - \sec^2 x = -1$.   **53.** *Hint:* By the MVT, there is a value of c in $(2, 4)$ for which $\dfrac{f(4) - f(2)}{4 - 2} = f'(c)$.   **57. b.** $c = \frac{1}{2}$

Section 4.3 Exercises, pp. 267–270

1. f is increasing on I if $f'(x) > 0$ for all x in I; f is decreasing on I if $f'(x) < 0$ for all x in I.   **3. a.** $x = 3$ **b.** Increasing on $(3, \infty)$; decreasing on $(-\infty, 3)$
5.   **7.**

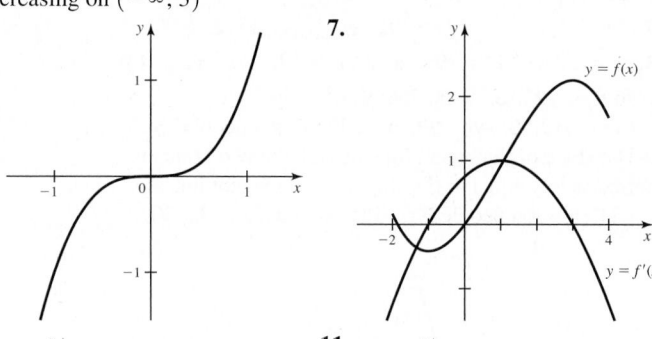

9.   **11.**

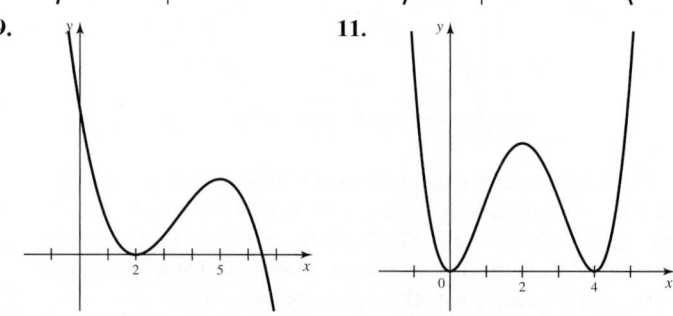

13. a. Concave up on $(-\infty, 2)$; concave down on $(2, \infty)$
b. Inflection point at $x = 2$ **15.** Yes; consider the graph of $y = \sqrt{x}$ on $(0, \infty)$.   **17.** $f(x) = x^4$ **19.** Increasing on $(-\infty, 0)$; decreasing on $(0, \infty)$ **21.** Decreasing on $(-\infty, 1)$; increasing on $(1, \infty)$
23. Increasing on $(-\infty, 1)$ and $(4, \infty)$; decreasing on $(1, 4)$
25. Increasing on $(-\infty, 1/2)$; decreasing on $(1/2, \infty)$
27. Increasing on $(-\infty, 0)$, $(1, 2)$; decreasing on $(0, 1)$, $(2, \infty)$
29. Increasing on $\left(-\dfrac{1}{\sqrt{e}}, 0\right)$, $\left(\dfrac{1}{\sqrt{e}}, \infty\right)$; decreasing on $\left(-\infty, -\dfrac{1}{\sqrt{e}}\right)$, $\left(0, \dfrac{1}{\sqrt{e}}\right)$ **31.** Increasing on $\left(\dfrac{\pi}{6}, \dfrac{5\pi}{6}\right)$; decreasing on $\left(0, \dfrac{\pi}{6}\right)$, $\left(\dfrac{5\pi}{6}, 2\pi\right)$ **33.** Increasing on $(-\pi, -2\pi/3)$, $(-\pi/3, 0)$, $(\pi/3, 2\pi/3)$; decreasing on $(-2\pi/3, -\pi/3)$, $(0, \pi/3)$, $(2\pi/3, \pi)$ **35.** Increasing on $(-1, 0)$, $(1, \infty)$; decreasing on $(-\infty, -1)$, $(0, 1)$ **37.** Increasing on $(-3, 1)$; decreasing on $(1, 3)$ **39.** Increasing on $(1, 4)$; decreasing on $(-\infty, 1)$, $(4, \infty)$ **41.** Increasing on $\left(-\infty, -\frac{1}{2}\right)$, $\left(0, \frac{1}{2}\right)$; decreasing on $\left(-\frac{1}{2}, 0\right)$, $\left(\frac{1}{2}, \infty\right)$ **43.** Increasing on $(-1, 1)$; decreasing on $(-\infty, -1)$, $(1, \infty)$ **45. a.** $x = 0$ **b.** Local min at $x = 0$ **c.** Abs. min: 3 at $x = 0$; abs. max: 12 at $x = -3$

47. a. $x = \pm\sqrt{2}$ **b.** Local min at $x = -\sqrt{2}$; local max at $x = \sqrt{2}$
c. Abs. max: 2 at $x = \sqrt{2}$; abs. min: -2 at $x = -\sqrt{2}$
49. a. $x = \pm\sqrt{3}$ **b.** Local min at $x = -\sqrt{3}$; local max
at $x = \sqrt{3}$ **c.** Abs. max: 28 at $x = -4$; abs. min: $-6\sqrt{3}$ at
$x = -\sqrt{3}$ **51. a.** $x = 2$ and $x = 0$ **b.** Local max at $x = 0$;
local min at $x = 2$ **c.** Abs. min: $-10\sqrt[3]{25}$ at $x = -5$; abs. max: 0
at $x = 0, 5$ **53. a.** $x = e^{-2}$ **b.** Local min at $x = e^{-2}$
c. Abs. min: $-2/e$ at $x = e^{-2}$; no abs. max **55.** Abs. max: $1/e$ at
$x = 1$ **57.** Abs. min: $36\sqrt[3]{\pi/6}$ at $r = \sqrt[3]{6/\pi}$
59. **61.**

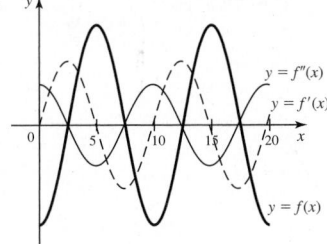

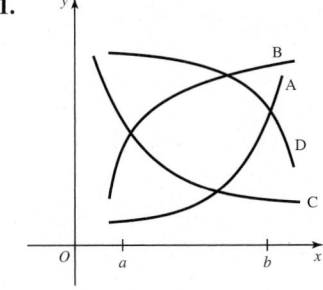
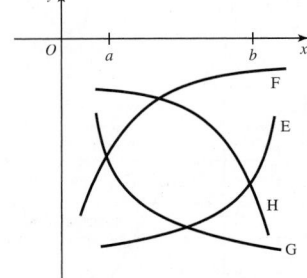

63. Concave up on $(-\infty, 0)$, $(1, \infty)$; concave down on $(0, 1)$;
inflection points at $x = 0$ and $x = 1$ **65.** Concave up on $(-\infty, 0)$,
$(2, \infty)$; concave down on $(0, 2)$; inflection points at $x = 0$ and
$x = 2$ **67.** Concave down on $(-\infty, 1)$; concave up on $(1, \infty)$;
inflection point at $x = 1$ **69.** Concave up on $(-1/\sqrt{3}, 1/\sqrt{3})$;
concave down on $(-\infty, -1/\sqrt{3})$, $(1/\sqrt{3}, \infty)$; inflection points at
$t = \pm 1/\sqrt{3}$ **71.** Concave up on $(-\infty, -1)$, $(1, \infty)$; concave down
on $(-1, 1)$; inflection points at $x = \pm 1$ **73.** Concave up on $(0, 1)$;
concave down on $(1, \infty)$; inflection point at $x = 1$ **75.** Concave
up on $(0, 2)$, $(4, \infty)$; concave down on $(-\infty, 0)$, $(2, 4)$; inflection
points at $x = 0, 2, 4$ **77.** Critical pts. $x = 0$ and $x = 2$; local max
at $x = 0$, local min at $x = 2$ **79.** Critical pt. $x = 0$; local max at
$x = 0$ **81.** Critical pt. $x = 6$; local min at $x = 6$ **83.** Critical pts.
$x = 0$ and $x = 1$; local max at $x = 0$; local min at $x = 1$
85. Critical pts. $x = 0$ and $x = 2$; local min at $x = 0$; local max at
$x = 2$ **87.** Critical pt. $x = e^5$; local min at $x = e^5$ **89.** Critical pts.
$t = -3$ and $t = 2$; local max at $t = -3$; local min at $t = 2$
91. Critical pt. $x = -a$; inconclusive at $x = -a$
93. Critical pts. $x = \pm\sqrt[3]{e}$; local min at $x = \pm\sqrt[3]{e}$
95. a. True **b.** False **c.** True **d.** False **e.** False
97. **99.** a–C–i, b–B–iii, c–A–ii

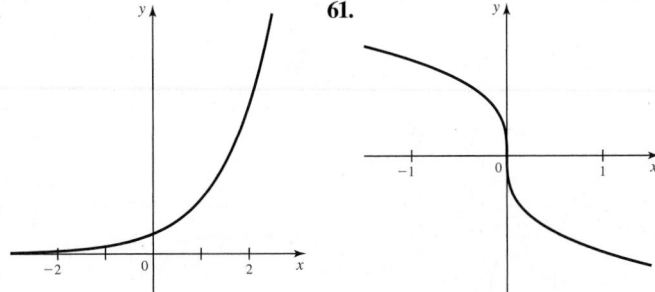

101.

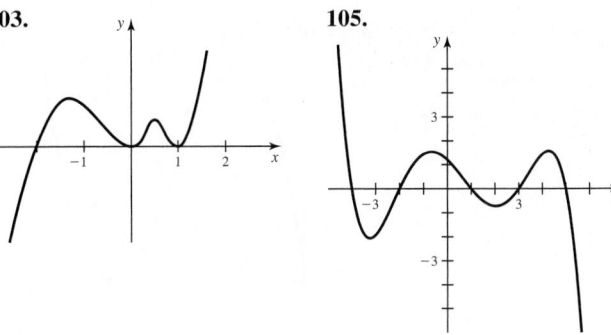

103.

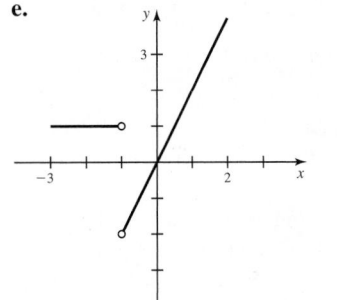

105.

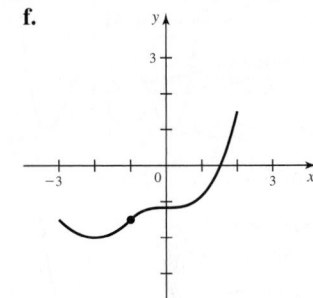

107. a. Increasing on $(-2, 2)$; decreasing on $(-3, -2)$
b. Critical pts. $x = -2$ and $x = 0$; local min at $x = -2$; neither a
local max nor min at $x = 0$ **c.** Inflection pts. at $x = -1$ and $x = 0$
d. Concave up on $(-3, -1)$, $(0, 2)$; concave down on $(-1, 0)$
e. **f.**

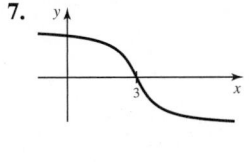
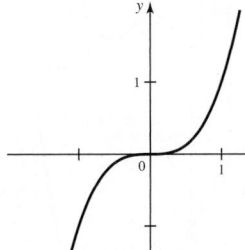

109. a. $E = \dfrac{p}{p - 50}$ **b.** -1.4% **c.** $E'(p) = -\dfrac{ab}{(a - bp)^2} < 0$,
for $p \geq 0, p \neq a/b$ **d.** $E(p) = -b$, for $p \geq 0$
111. a. 300 **b.** $t = \sqrt{10}$ **c.** $t = \sqrt{b/3}$

Section 4.4 Exercises, pp. 277–280

1. We need to know on which interval(s) to graph f. **3.** No; the
domain of any polynomial is $(-\infty, \infty)$; there is no vertical asymptote.
Also, $\lim\limits_{x \to \pm\infty} p(x) = \pm\infty$, where p is any polynomial; there is no horizontal asymptote. **5.** Evaluate the function at the critical points and
at the endpoints. Then find the largest and smallest values among those
candidates.
7. **9.**

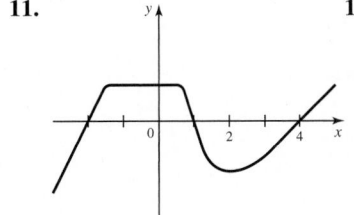
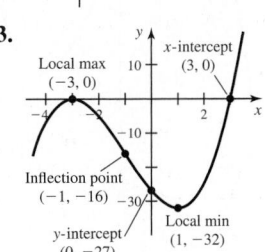

15.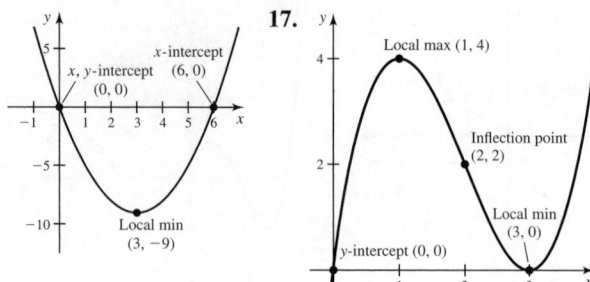

17.

19.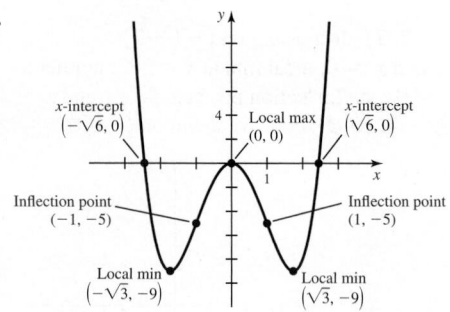

21.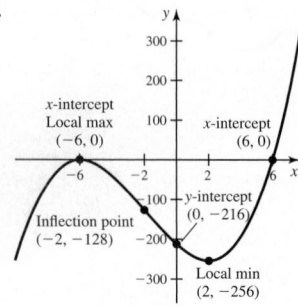

23.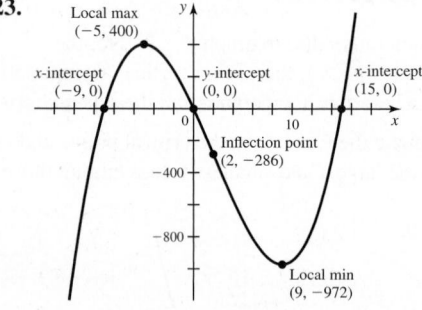

25.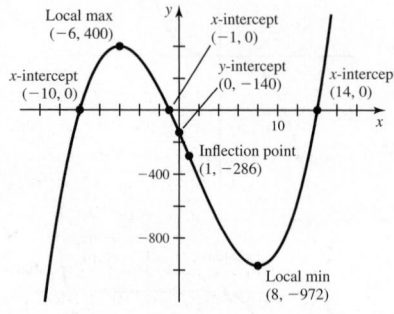

27.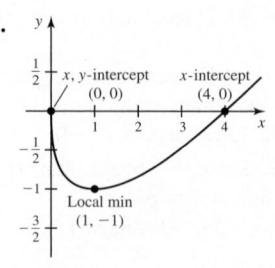

29.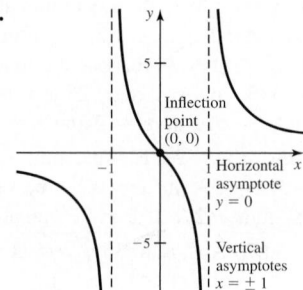

31.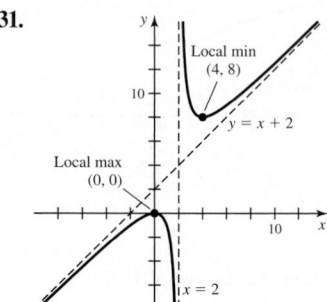

33.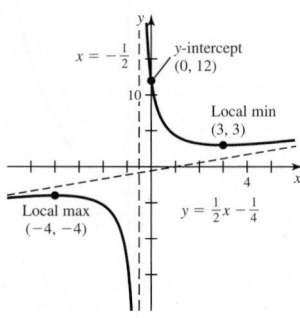

35.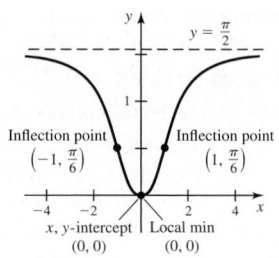

37.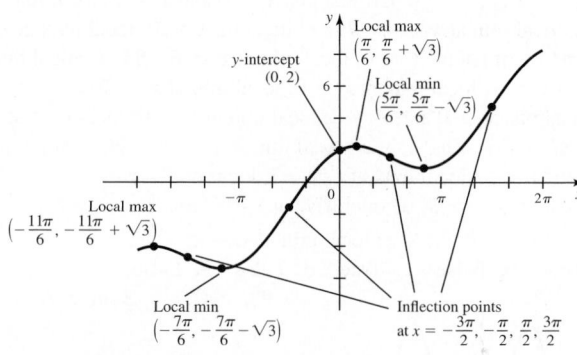

39.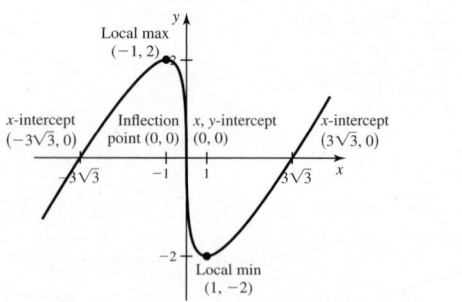

41.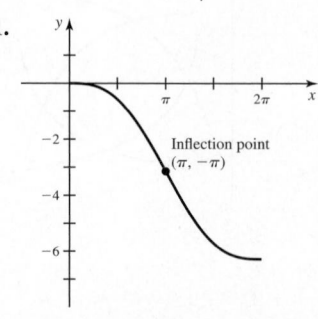

43.

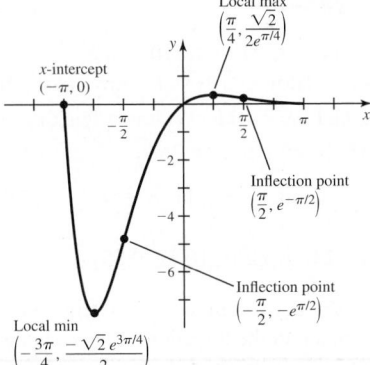

45.

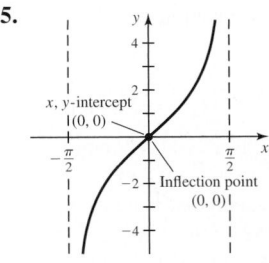

47. Critical pts. at $x = 1, 3$; local max at $x = 1$; local min at $x = 3$; inflection pt. at $x = 2$; increasing on $(0, 1), (3, 4)$; decreasing on $(1, 3)$; concave up on $(2, 4)$; concave down on $(0, 2)$

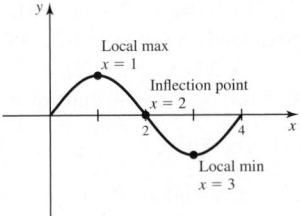

49.

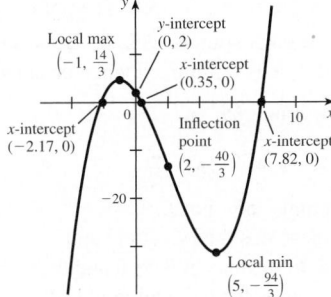

51.

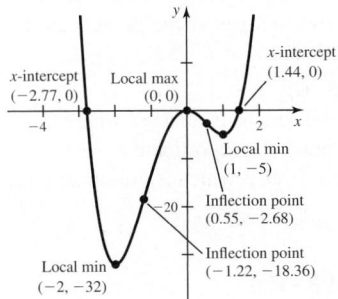

53.

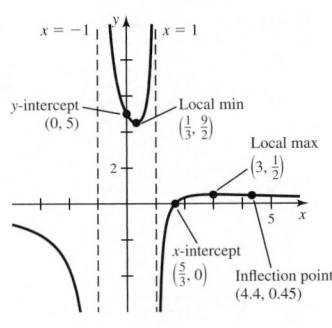

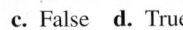

55. a. False   **b.** False
c. False   **d.** True

57.

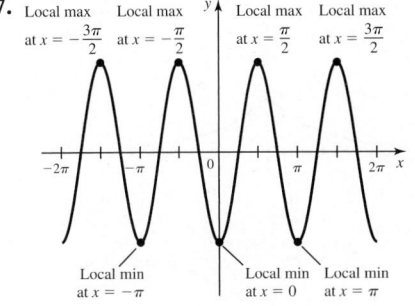

59.

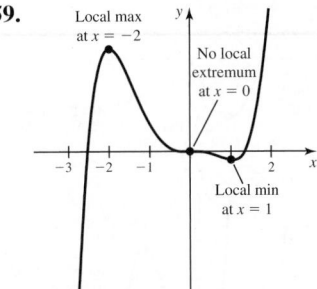

61. (A) a.

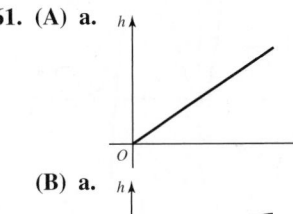

b. Water is being added at all times.   **c.** No concavity
d. h' has an abs. max at all points of $[0, 10]$.

(B) a.

c. Concave down   **d.** h' has abs. max at $t = 0$.

(C) a.

c. Concave up   **d.** h' has abs. max at $t = 10$.

(D) a.

c. Concave down on $(0, 5)$, then concave up on $(5, 10)$; inflection pt. at $t = 5$ **d.** h' has abs. max at $t = 0$ and $t = 10$.

(E) a.

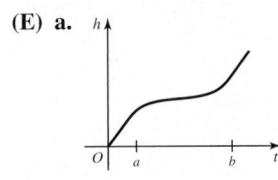

c. First, no concavity; then concave down, no concavity, concave up, and, finally, no concavity
d. h' has abs. max at all points of an interval $[0, a]$ and $[b, 10]$.

(F) a.

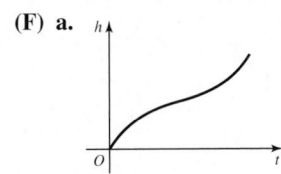

c. Concave down on $(0, 5)$; concave up on $(5, 10)$; inflection pt. at $t = 5$ **d.** h' has abs. max at $t = 0$ and $t = 10$.

63. Local max of $e^{1/e}$ at $x = e$

65. $f'(0)$ does not exist.

67.

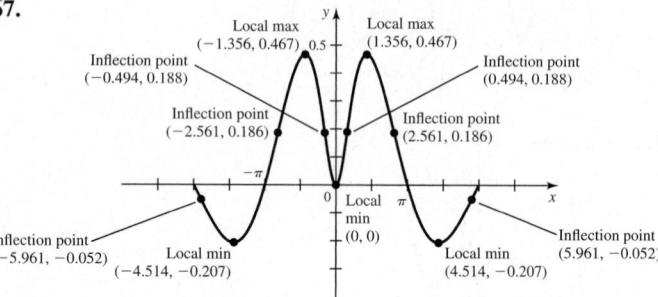

69.

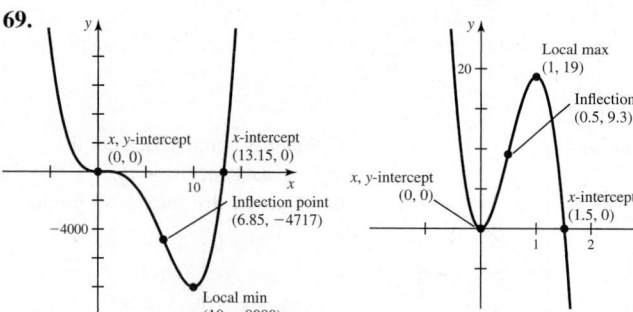

71.

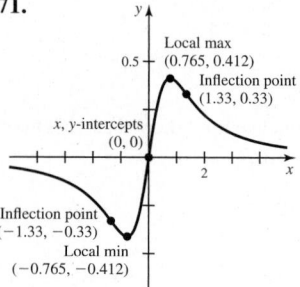

73.

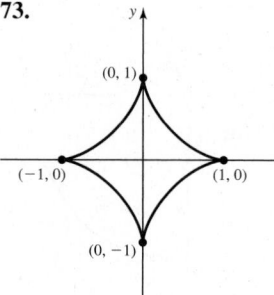

75.

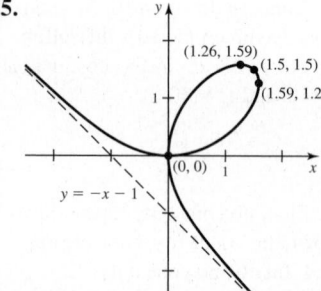

77.

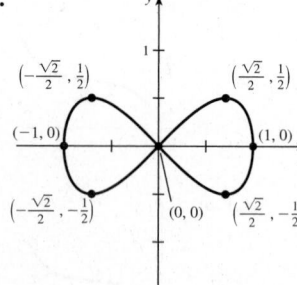

79.

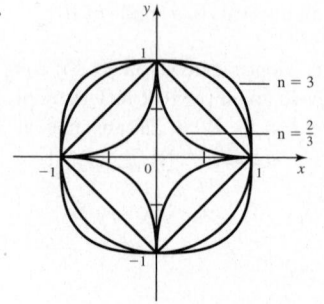

Section 4.5 Exercises, pp. 284–291

1. Objective function, constraint(s)   **3.** $Q = x^2(10 - x)$; $Q = (10 - y)^2 y$ **5. a.** $P(x) = 100x - 10x^2$ **b.** Abs. max: 250 at $x = 5$ **7.** $\frac{23}{2}$ and $\frac{23}{2}$ **9.** $5\sqrt{2}$ and $5\sqrt{2}$ **11.** Width = length = $\frac{5}{2}$
13. Width = length = 10   **15.** $x = \sqrt{6}, y = 2\sqrt{6}$
17. $\frac{10}{\sqrt{2}}$ by $\frac{5}{\sqrt{2}}$ **19.** Length = width = height = 2
21. $\frac{4}{\sqrt[3]{5}}$ ft by $\frac{4}{\sqrt[3]{5}}$ ft by $5^{2/3}$ ft   **23.** Approx. $(0.59, 0.65)$
25. $(5, 15)$, distance ≈ 47.4 **27. a.** A point $8/\sqrt{5}$ mi from the point on the shore nearest the woman in the direction of the restaurant **b.** $9/\sqrt{13}$ mi/hr   **29.** A point $7\sqrt{3}/6$ mi from the point on shore nearest the island, in the direction of the power station
31. 18.2 ft   **33.** $h = \dfrac{20}{\sqrt{3}}; r = 20\sqrt{\frac{2}{3}}$
35. a. $r = \sqrt[3]{177/\pi} \approx 3.83$ cm; $h = 2\sqrt[3]{177/\pi} \approx 7.67$ cm
b. $r = \sqrt[3]{177/2\pi} \approx 3.04$ cm; $h = 2\sqrt[3]{708/\pi} \approx 12.17$ cm; part (b) is closer to the real can.   **37.** $\sqrt{15}$ m by $2\sqrt{15}$ m
39. $12''$ by $6''$ by $3''$; 216 in³   **41.** Lower rectangular pane is approximately 5.6 ft wide by 2.8 ft high.   **43.** $r/h = \sqrt{2}$
45. $r = \sqrt{2}R/\sqrt{3}; h = 2R/\sqrt{3}$ **47.** 3:1   **49. a.** 0, 30, 25
b. 42.5 mi/hr   **c.** The units of $p/g(v)$ are \$/mi and so are the units of w/v. Therefore, $L\left(\dfrac{p}{g(v)} + \dfrac{w}{v}\right)$ gives the total cost of a trip of L miles.   **d.** Approx. 62.9 mi/hr   **e.** Neither; the zeros of $C'(v)$ are independent of L.   **f.** Decreased slightly, to 62.5 mi/hr
g. Decreased to 60.8 mi/hr   **51.** $\sqrt{30} \approx 5.5$ ft   **53.** The point $12/\left(\sqrt[3]{2} + 1\right) \approx 5.3$ m from the weaker source   **55. b.** Because the speed of light is constant, travel time is minimized when distance is minimized.   **57.** $r = h = \sqrt[3]{450/\pi}$ m   **59. a.** $\dfrac{a_1 + a_2}{2}$
b. $\dfrac{a_1 + a_2 + a_3}{3}$ **c.** $\dfrac{a_1 + a_2 + \cdots + a_n}{n}$ **61.** $\dfrac{\pi}{3}$
63. When the seat is at its lowest point   **65.** For $L \le 4r$, max at $\theta = 0$ and $\theta = 2\pi$; min at $\theta = \cos^{-1}(-L/(4r))$ and $\theta = 2\pi - \cos^{-1}(-L/(4r))$. For $L > 4r$, max at $\theta = 0$ and $\theta = 2\pi$; min at $\theta = \pi$.   **67. a.** $P = 2/\sqrt{3}$ units from the midpoint of the base   **69.** You can run 12 mi/hr if you run toward the point $3/16$ mi ahead of the locomotive (when it passes the point nearest you).   **71. a.** $r = 2R/3; h = \frac{1}{3}H$ **b.** $r = R/2; h = H/2$
73. $(1 + \sqrt{3})$ mi ≈ 2.7 mi   **75.** (i) $\left(p - \frac{1}{2}, \sqrt{p - \frac{1}{2}}\right)$
(ii) $(0, 0)$ **77.** Let the angle of the cuts be φ_1 and φ_2, where $\varphi_1 + \varphi_2 = \theta$. The volume of the notch is proportional to $\tan \varphi_1 + \tan \varphi_2 = \tan \varphi_1 + \tan(\theta - \varphi_1)$, which is minimized when $\varphi_1 = \varphi_2 = \dfrac{\theta}{2}$.   **79.** $x \approx 38.81, y \approx 55.03$

Section 4.6 Exercises, pp. 298–300

1.

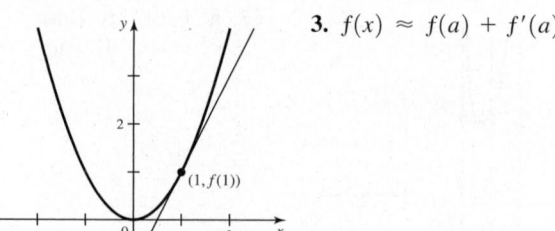

3. $f(x) \approx f(a) + f'(a)(x - a)$

5. $L(x) = 3x - 1; 2.3$ **7.** 2.7   **9.** $dy = f'(x)\, dx$

11. Approx. 25   **13.** 61 mi/hr; 61.02 mi/hr
15. $L(x) = T(0) + T'(0)(x - 0) = D - (D/60)x = D(1 - x/60)$
17. 84 min; 84.21 min   **19.** $L(x) = 9x - 4$ **21.** $L(t) = t + 5$
23. $L(x) = 3x - 5$ **25. a.** $L(x) = -4x + 16$ **b.** 7.6
c. 0.13% error   **27. a.** $L(x) = x$ **b.** 0.9   **c.** 40% error

29. a. $L(x) = 1$ **b.** 1   **c.** 0.005% error   **31. a.** $L(x) = \dfrac{1}{2} - \dfrac{x}{48}$

b. 0.5   **c.** 0.003% error   **33. a.** $L(x) = 1 - x$;   **b.** $1/1.1 \approx 0.9$
c. 1% error   **35. a.** $L(x) = 1 - x$ **b.** $e^{-0.03} \approx 0.97$
c. 0.05% error   **37.** $a = 200; \frac{1}{203} \approx 0.004925$

39. $a = 144; \sqrt{146} \approx \frac{145}{12}$ **41.** $a = 1; \ln 1.05 \approx 0.05$

43. $a = 0; e^{0.06} \approx 1.06$ **45.** $a = 512; \dfrac{1}{\sqrt[3]{510}} \approx 0.125$

47. a. $L(x) = -2x + 4$ **b.**

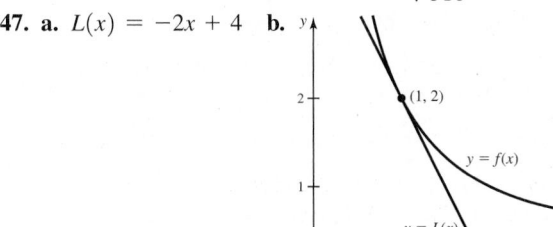

c. Underestimates   **d.** $f''(1) = 4 > 0$

49. a. $L(x) = -\dfrac{1}{2}x + \dfrac{1}{2}(1 + \ln 2)$

b.

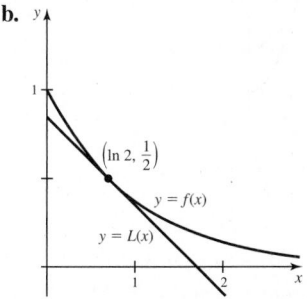

c. Underestimates   **d.** $f''(\ln 2) = \dfrac{1}{2} > 0$

51. $E(x) \leq 1$ when $-7.26 \leq x \leq 8.26$, which corresponds to driving times for 1 mi from about 53 s to 68 s. Therefore, $L(x)$ gives approximations to $s(x)$ that are within 1 mi/hr of the true value when you drive 1 mi in t seconds, where $53 < t < 68$.
53. a. True   **b.** False   **c.** True   **d.** True
55. $\Delta V \approx 10\pi$ ft^3 **57.** $\Delta V \approx -40\pi$ cm^3

59. $\Delta S \approx -\dfrac{59\pi}{5\sqrt{34}}$ m^2 **61.** $dy = 2\,dx$ **63.** $dy = -\dfrac{3}{x^4}\,dx$

65. $dy = a\sin x\,dx$ **67.** $dy = (9x^2 - 4)\,dx$
69. $dy = \sec^2 x\,dx$
71. $L(x) = 2 + (x - 8)/12$

x	Linear approx.	Exact value	Percent error
8.1	$2.008\overline{3}$	2.00829885	1.7×10^{-3}
8.01	$2.0008\overline{3}$	2.000832986	1.7×10^{-5}
8.001	$2.00008\overline{3}$	2.00008333	1.7×10^{-7}
8.0001	$2.0000083\overline{3}$	2.000008333	1.7×10^{-9}
7.9999	$1.9999916\overline{6}$	1.999991667	1.7×10^{-9}
7.999	$1.999991\overline{6}$	1.999991663	1.7×10^{-7}
7.99	$1.99991\overline{6}$	1.999166319	1.7×10^{-5}
7.9	$1.991\overline{6}$	1.991631701	1.8×10^{-3}

73. a. f; the rate at which f' is changing at 1 is smaller than the rate at which g' is changing at 1. The graph of f bends away from the linear function more slowly than the graph of g.   **b.** The larger the value of $|f''(a)|$, the greater the deviation of the curve $y = f(x)$ from the tangent line at points near $x = a$.

Section 4.7 Exercises, pp. 310–312

1. If $\lim\limits_{x \to a} f(x) = 0$ and $\lim\limits_{x \to a} g(x) = 0$, then we say $\lim\limits_{x \to a} f(x)/g(x)$ is an indeterminate form $0/0$.   **3.** Take the limit of the quotient of the derivatives of the functions.

5. a. $\lim\limits_{x \to 0}\left(x^2 \cdot \dfrac{1}{x^2}\right) = 1$ **b.** $\lim\limits_{x \to 0}\left(2x^2 \cdot \dfrac{1}{x^2}\right) = 2$

7. If $\lim\limits_{x \to a} f(x)g(x)$ has the indeterminate form $0 \cdot \infty$, then
$\lim\limits_{x \to a}\left(\dfrac{f(x)}{1/g(x)}\right)$ has the indeterminate form $0/0$ or ∞/∞.

9. $\dfrac{1}{5}$ **11.** If $\lim\limits_{x \to a} f(x) = 1$ and $\lim\limits_{x \to a} g(x) = \infty$, then $f(x)^{g(x)}$ has the form 1^∞ as $x \to a$, which is meaningless; so direct substitution does not work.   **13.** $\lim\limits_{x \to \infty} \dfrac{g(x)}{f(x)} = 0$ **15.** $\ln x, x^3, 2^x, x^x$

17. -1 **19.** $\frac{3}{4}$ **21.** $\frac{1}{2}$ **23.** $\frac{1}{2}$ **25.** $\frac{1}{e}$ **27.** -1 **29.** $\frac{12}{5}$
31. 4   **33.** $\frac{9}{16}$ **35.** $\frac{1}{2}$ **37.** $-\frac{2}{3}$ **39.** $\frac{1}{24}$ **41.** 1   **43.** 4
45. 1   **47.** $-\frac{1}{2}$ **49.** $\frac{1}{\pi^2}$ **51.** $\frac{1}{3}$ **53.** 1   **55.** $\frac{7}{6}$ **57.** 1
59. $-\frac{1}{2}$ **61.** 0   **63.** -8 **65.** 0   **67.** $\frac{1}{2}$ **69.** $\frac{\ln 3}{\ln 2}$
71. $-\frac{9}{4}$ **73.** $\frac{1}{6}$ **75.** 1   **77.** 1   **79.** e **81.** e^{a+1} **83.** e

85. b. $\lim\limits_{m \to \infty}(1 + r/m)^m = \lim\limits_{m \to \infty}\left(1 + \dfrac{1}{(m/r)}\right)^{(m/r)r} = e^r$

87. 3   **89.** $\frac{1}{2}$ **91.** e **93.** $\ln a - \ln b$ **95.** $e^{0.01x}$ **97.** Comparable growth rates   **99.** x^x **101.** 1.00001^x **103.** e^{x^2} **105. a.** False
b. False   **c.** False   **d.** False   **e.** True   **f.** True
107.

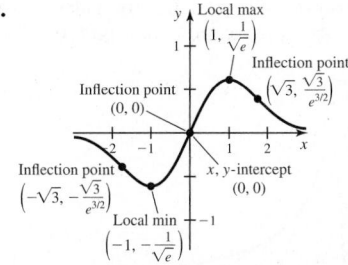

109.

111. $\sqrt{a/c}$ **113.** $\lim\limits_{x \to \infty} \dfrac{x^p}{b^x} = \lim\limits_{t \to \infty} \dfrac{\ln^p t}{t \ln^p b} = 0$, where $t = b^x$

115. Show $\lim\limits_{x \to \infty} \dfrac{\log_a x}{\log_b x} = \dfrac{\ln b}{\ln a} \neq 0$.   **117.** $1/3$ **121. a.** $b > e$
b. e^{ax} grows faster than e^x as $x \to \infty$, for $a > 1$; e^{ax} grows slower than e^x as $x \to \infty$, for $0 < a < 1$.

Section 4.8 Exercises, pp. 318–321

1. Newton's method generates a sequence of x-intercepts of lines tangent to the graph of f to approximate the roots of f.
3. $x_1 = 2, x_2 = 1, x_3 = 0$ **5.** $x_1 = 0.75$ **7.** Generally, if two successive Newton approximations agree in their first p digits, then those approximations have p digits of accuracy. The method is terminated when the desired accuracy is reached.

9. $x_{n+1} = x_n - \dfrac{x_n^2 - 6}{2x_n} = \dfrac{x_n^2 + 6}{2x_n}; x_1 = 2.5, x_2 = 2.45$

11. $x_{n+1} = x_n - \dfrac{e^{-x_n} - x_n}{e^{-x_n} - 1}; x_1 = 0.564382, x_2 = 0.567142$

13.

n	x_n
0	3
1	3.1667
2	3.16228
3	3.16228

$r \approx 3.16228$

15.

n	x_n
0	0.5
1	0.51096
2	0.51097
3	0.51097

$r \approx 0.51097$

17.

n	x_n
0	1.2
1	1.16935
2	1.16561
3	1.16556
4	1.16556

$r \approx 1.16556$

19.

n	x_n
0	0.75
1	0.73915
2	0.73909
3	0.73909

$r \approx 0.73909$

21. $x \approx -0.335408, 1.333057$ **23.** $x \approx 0.179295$
25. $x \approx 0.620723, 3.03645$ **27.** $x \approx 0, 1.895494, -1.895494$
29. $x \approx -2.114908, 0.254102, 1.860806$
31. $x \approx 0.062997, 2.230120$

33.

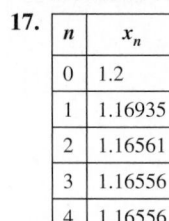

The tumor decreases in size and then starts growing again. It decreases to half its size after about 6.4 days.

35. b. $r \approx 7.3\%$ **37. a.** $t = \pi/4$ **b.** $t \approx 1.33897$
c. $t \approx 2.35619$ **d.** $t \approx 2.90977$
39. $p(x) = x^4 - 7; r \approx 1.62658$
41. $p(x) = x^3 + 9; r \approx -2.08008$ **43.** $x \approx 2.798386$
45. $x \approx -0.666667, 1.5, 1.666667$ **47. a.** True   **b.** False
c. False   **49.** $x \approx 0.739085$ **51.** $x = 0$ and $x \approx 1.047198$

53. a.

n	x_n
0	2
1	5.33333
2	11.0553
3	22.2931

b.

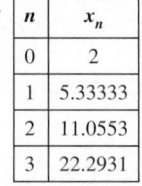

c. The tangent lines intersect the x-axis farther and farther away from the root r.   **55. b.** $x \approx 0.14285714$ is approximately $\frac{1}{7}$.
57. $\lambda = 1.29011, 2.37305, 3.40918$

Section 4.9 Exercises, pp. 331–334

1. The derivative, an antiderivative   **3.** $x + C$, where C is an arbitrary constant   **5.** $\dfrac{x^{p+1}}{p + 1} + C$, where $p \neq -1$ **7.** $\ln|x| + C$

9. 0   **11.** $x^5 + C$ **13.** $-2\cos x + x + C$ **15.** $3\tan x + C$
17. $y^{-2} + C$ **19.** $e^x + C$ **21.** $\tan^{-1}s + C$ **23.** $\frac{1}{2}x^6 - \frac{1}{2}x^{10} + C$
25. $\frac{8}{3}x^{3/2} - 8x^{1/2} + C$ **27.** $\frac{25}{3}s^3 + 15s^2 + 9s + C$
29. $\frac{9}{4}x^{4/3} + 6x^{2/3} + 6x + C$ **31.** $-x^3 + \frac{11}{2}x^2 + 4x + C$
33. $-x^{-3} + 2x + 3x^{-1} + C$ **35.** $x^4 - 3x^2 + C$
37. $\dfrac{1}{2}x^2 + 6x + C$ **39.** $-\cot\theta + 2\theta^3/3 - 3\theta^2/2 + C$
41. $-2\cot y - 3\csc y + C$ **43.** $\tan x - x + C$
45. $\tan\theta + \sec\theta + C$ **47.** $t^3 - 2\cot t + C$
49. $\sec\theta + \tan\theta + \theta + C$ **51.** $\frac{1}{2}\ln|y| + C$ **53.** $3\sin^{-1}x + C$
55. $4\sec^{-1}|x| + C$ **57.** $\frac{1}{6}\sec^{-1}|x| + C$ **59.** $t + \ln|t| + C$
61. $e^{x+2} + C$ **63.** $e^w - 4w + C$ **65.** $\ln|x| + 2\sqrt{x} + C$
67. $\frac{4}{15}x^{15/2} - \frac{24}{11}x^{11/6} + C$ **69.** $\dfrac{1}{6}x^6 - \dfrac{2}{3}x^3 + x + 1$
71. $2x^4 - \cos x + 3$ **73.** $\sec v + 1, -\pi/2 < v < \pi/2$
75. $y^3 + 5\ln y + 2, y > 0$ **77.** $f(x) = x^2 - 3x + 4$
79. $g(x) = \dfrac{7}{8}x^8 - \dfrac{x^2}{2} + \dfrac{13}{8}$ **81.** $f(u) = 4(\sin u + \cos u) - 4$
83. $y(t) = 3\ln t + 6t + 2, t > 0$
85. $y(\theta) = \sqrt{2}\sin\theta + \tan\theta + 1, -\pi/2 < \theta < \pi/2$
87. $f(x) = x^2 - 5x + 4$ **89.** $f(x) = \dfrac{3}{2}x^2 - \cos x + 4$

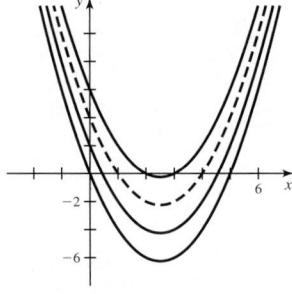 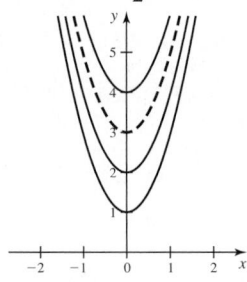

91. $s(t) = t^2 + 4t$ **93.** $s(t) = \frac{4}{3}t^{3/2} + 1$
95. $s(t) = 2t^3 + 2t^2 - 10t$ **97.** $s(t) = -16t^2 + 20t$
99. $s(t) = \frac{1}{30}t^3 + 1$ **101.** $s(t) = t^2 + 4t - 3\sin t + 10$
103. 200 ft   **105.** Runner A overtakes runner B at $t = \pi/2$
107. a. $v(t) = -9.8t + 30$ **b.** $s(t) = -4.9t^2 + 30t$
c. Approx. 45.92 m at time $t \approx 3.06$ **d.** $t \approx 6.12$
109. a. $v(t) = -9.8t + 10$ **b.** $s(t) = -4.9t^2 + 10t + 400$
c. Approx. 405.10 m at time $t \approx 1.02$ **d.** $t \approx 10.11$
111. a. True   **b.** False   **c.** True   **d.** False
e. False   **113.** $F(x) = -\cos x + 3x + 3 - 3\pi$
115. $F(x) = 2x^8 + x^4 + 2x + 1$
117. a. $Q(t) = 10t - t^3/30$ gal

b.

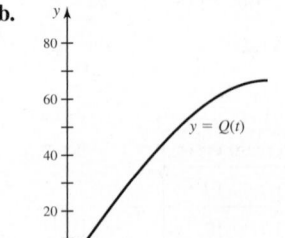

$y = Q(t)$

c. $\dfrac{200}{3}$ gal

Chapter 4 Review Exercises, pp. 334–337

1. a. False **b.** False **c.** True **d.** True **e.** True **f.** False
3.

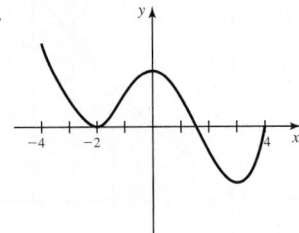

5. a. $x = 0, \pm 10$; increasing on $(-10, 0)$ and $(10, \infty)$, decreasing on $(-\infty, -10)$ and $(0, 10)$ **b.** $x = \pm 6$; concave up on $(-\infty, -6)$ and $(6, \infty)$, concave down on $(-6, 6)$ **c.** Local min at $x = -10, 10$; local max at $x = 0$ **d.**

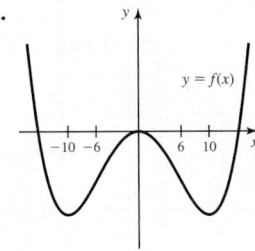

7. Critical pts. $x = 0, \pm 1$; abs. max: 33 at $x = \pm 2$; abs. min: 6 at $x = \pm 1$ **9.** $x = 3$ and $x = -2$; no abs. max or min
11. Critical pt. $x = 1$; abs. max: $\ln 2$ at $x = 0, 2$; abs. min: 0 at $x = 1$
13. Critical pts. $x = \dfrac{2\pi}{3}, \dfrac{4\pi}{3}$; abs. max: $\dfrac{3\sqrt{3}}{8}$ at $x = \dfrac{4\pi}{3}$; abs. min: $-\dfrac{3\sqrt{3}}{8}$ at $x = \dfrac{2\pi}{3}$ **15.** Critical pt. $x = 1/e$; abs. min: $10 - 2/e$ at $x = 1/e$

17.

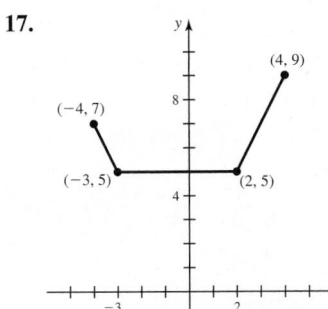

Critical pts.: all x in the interval $[-3, 2]$; abs. max: 9 at $x = 4$; abs. and local min: 5 for x in $[-3, 2]$; local max: 5 for x in $(-3, 2)$

19. a. Increasing on $(-\infty, -1)$ and $(1, \infty)$; decreasing on $(-1, 1)$ **b.** Concave up on $(0, \infty)$ and concave down on $(-\infty, 0)$
21. Inflection pt. $x = 0$ **23.** Critical pts. $x = -a, a/2$; inflection pts. $x = 0, -a$
25.

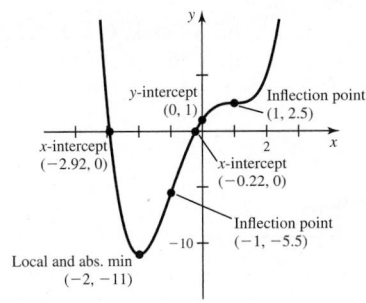

27.

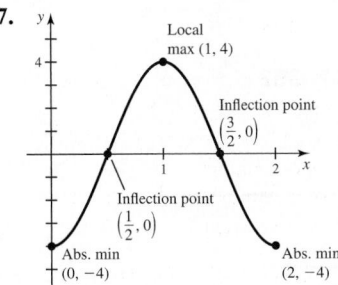

29.

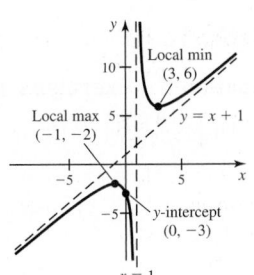

31.

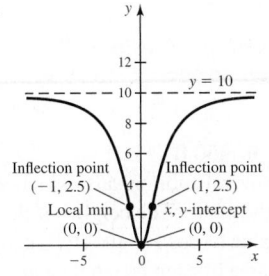

33.

35. Approx. 2.5″ by 3.5″ by 9.5″ **37.** Approx. 59 m from the loudest speaker **39.** $r = 4\sqrt{6}/3$; $h = 4\sqrt{3}/3$ **41.** $x = 7, y = 14$
43. $p = q = 5\sqrt{2}$ **45. a.** $L(x) = \frac{2}{9}x + 3$ **b.** $\frac{85}{9} \approx 9.44$; overestimate **47.** $f(x) = 1/x^2$; $a = 4$; $1/4.2^2 \approx 0.05625$
49. $\Delta h \approx -112$ ft **51.** $c = 2.5$ **53. a.** $\frac{100}{9}$ cells/week **b.** $t = 2$ weeks **55.** $-0.434259, 0.767592, 1$ **57.** $0, \pm 0.948683$
59. 0 **61.** 0 **63.** 12 **65.** $\frac{2}{3}$ **67.** ∞ **69.** 0 **71.** 1 **73.** 0
75. 1 **77.** 1 **79.** $1/e^3$ **81.** 1 **83.** $x^{1/2}$ **85.** $\sqrt{x}$ **87.** 3^x
89. Comparable growth rates **91.** $\frac{4}{3}x^3 + 2x^2 + x + C$
93. $-\dfrac{1}{x} + \dfrac{4}{3}x^{-3/2} + C$ **95.** $\theta + 3\sin\theta + C$ **97.** $\tan x + C$
99. $12\ln|x| + C$ **101.** $\tan^{-1}x + C$ **103.** $\frac{4}{7}x^{7/4} + \frac{2}{7}x^{7/2} + C$
105. $f(t) = -\cos t + t^2 + 6$ **107.** $h(x) = \dfrac{1}{3}x^3 - x - \tan^{-1}x + \dfrac{\pi}{4}$
109. $v(t) = -9.8t + 120$; $s(t) = -4.9t^2 + 120t + 125$. The rocket reaches a height of 859.69 m at time $t \approx 12.24$ s and then falls to the ground, hitting at time $t \approx 25.49$ s. **111. a.** $v(t) = 64 - 32t$
b. $s(t) = -16t^2 + 64t + 128$ **c.** $t = 2$; 192 ft
d. $-64\sqrt{3}$ ft/s ≈ -110.9 ft/s **113.** 1; 1
115. $\lim\limits_{x \to 0^+} f(x) = 1$; $\lim\limits_{x \to 0^+} g(x) = 0$

CHAPTER 5

Section 5.1 Exercises, pp. 347–352

1. Displacement $= 105$ m

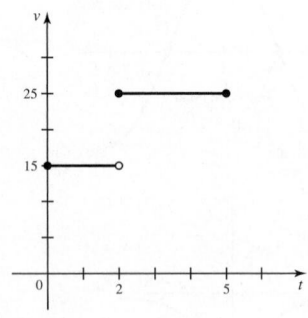

3. a. 440 ft   **b.** 400 ft   **5. a.** 340 ft   **b.** 330 ft
7. Subdivide the interval $[0, \pi/2]$ into several subintervals, which will be the bases of rectangles that fit under the curve. The heights of the rectangles are computed by taking the value of $\cos x$ at the right-hand endpoint of each base. We calculate the area of each rectangle and add them to get a lower bound on the area.   **9.** Left sum: 34; right sum: 24   **11.** 0.5; 1, 1.5, 2, 2.5, 3; 1, 1.5, 2, 2.5; 1.5, 2, 2.5, 3; 1.25, 1.75, 2.25, 2.75   **13.** Underestimate; the rectangles all fit under the curve.   **15. a.** 67 ft   **b.** 67.75 ft   **17.** 40 m
19. 2.78 m   **21.** 148.96 mi   **23.** 20; 25
25. a. c.

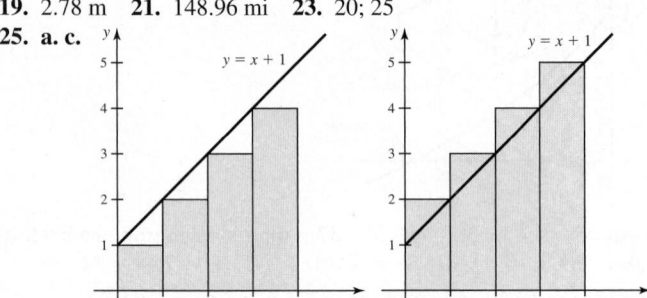

Left Riemann sum underestimates area.   Right Riemann sum overestimates area.

b. $\Delta x = 1; x_0 = 0, x_1 = 1, x_2 = 2, x_3 = 3, x_4 = 4$ **d.** 10, 14
27. a. c.

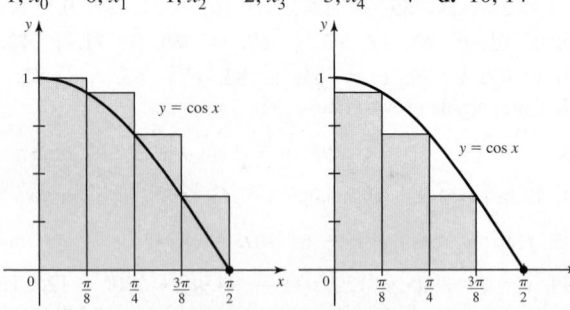

Left Riemann sum overestimates area.   Right Riemann sum underestimates area.

b. $\Delta x = \pi/8; 0, \pi/8, \pi/4, 3\pi/8, \pi/2$ **d.** 1.18; 0.79

29. a. c.

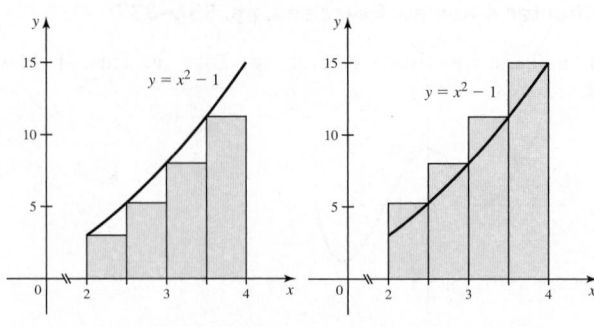

Left Riemann sum underestimates area.   Right Riemann sum overestimates area.

b. $\Delta x = 0.5; 2, 2.5, 3, 3.5, 4$ **d.** 13.75; 19.75
31. a. c.

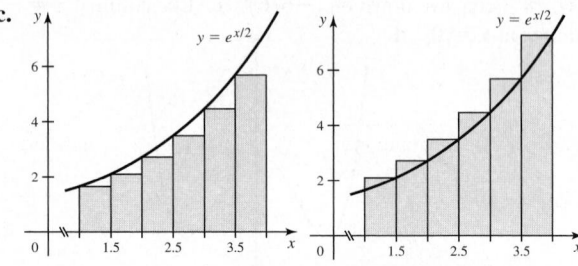

Left Riemann sum underestimates area.   Right Riemann sum overestimates area.

b. $\Delta x = 0.5; x_0 = 1, x_1 = 1.5, x_2 = 2, x_3 = 2.5, x_4 = 3, x_5 = 3.5,$ $x_6 = 4$ **d.** 10.11, 12.98   **33.** 670   **35. a.** 10,500 m; 10,350 m
b. Left Riemann sum   **c.** Increase the number of subintervals in the partition.
37. a. c.

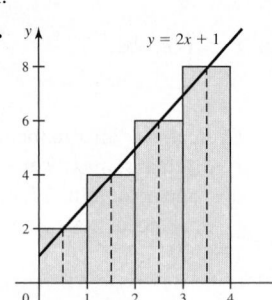

b. $\Delta x = 1; 0, 1, 2, 3, 4$
d. 20

39. a. c.

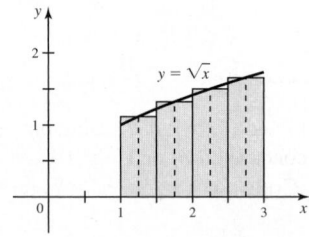

b. $\Delta x = \frac{1}{2}; 1, \frac{3}{2}, 2, \frac{5}{2}, 3$
d. 2.80

41. a. c.

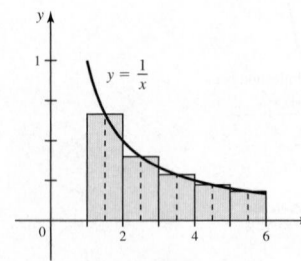

b. $\Delta x = 1; 1, 2, 3, 4, 5, 6$
d. 1.76

43. 5.5, 3.5   **45. b.** 110, 117.5   **47. a.** $\displaystyle\sum_{k=1}^{5} k$ **b.** $\displaystyle\sum_{k=1}^{6} (k+3)$

c. $\displaystyle\sum_{k=1}^{4} k^2$ **d.** $\displaystyle\sum_{k=1}^{4}\frac{1}{k}$ **49. a.** 55   **b.** 48   **c.** 30   **d.** 60   **e.** 6

f. 6   **g.** 85   **h.** 0   **51. a.** Left: $\dfrac{3}{10}\displaystyle\sum_{k=1}^{40}\sqrt{\dfrac{k-1}{10}} \approx 15.6809;$

right: $\dfrac{3}{10}\displaystyle\sum_{k=1}^{40}\sqrt{\dfrac{k}{10}} \approx 16.2809;$ midpoint: $\dfrac{3}{10}\displaystyle\sum_{k=1}^{40}\sqrt{\dfrac{k-0.5}{10}} \approx 16.0055$

b. 16   **53. a.** Left: $\dfrac{1}{25}\displaystyle\sum_{k=1}^{75}\left(\left(2+\dfrac{k-1}{25}\right)^2 - 1\right) \approx 35.5808;$

right: $\dfrac{1}{25}\displaystyle\sum_{k=1}^{75}\left(\left(2+\dfrac{k}{25}\right)^2 - 1\right) \approx 36.4208;$

midpoint: $\dfrac{1}{25}\displaystyle\sum_{k=1}^{75}\left(\left(2+\dfrac{k-0.5}{25}\right)^2 - 1\right) \approx 35.9996$ **b.** 36

55.

n	Right Riemann sum
10	21.96
30	21.9956
60	21.9989
80	21.9994

The sums approach 22.

57.

n	Right Riemann sum
10	3.14159
30	3.14159
60	3.14159
80	3.14159

The sums approach π.

59. a. True   **b.** False   **c.** True   **61.** $\displaystyle\sum_{k=1}^{50}\left(\dfrac{4k}{50}+1\right)\cdot\dfrac{4}{50} = 12.16$

63. $\displaystyle\sum_{k=1}^{32}\left(3+\dfrac{2k-1}{8}\right)^3\cdot\dfrac{1}{4} \approx 3639.1$ **65.** $[1,5]; 4$ **67.** $[2,6]; 4$

69. a.

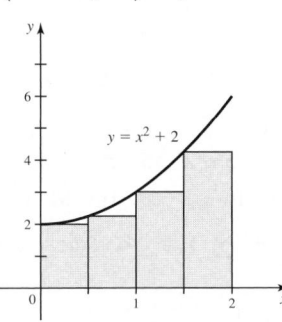

Left Riemann sum is $\dfrac{23}{4} = 5.75.$

b.

Midpoint Riemann sum is $\dfrac{53}{8} = 6.625.$

c.

Right Riemann sum is $\dfrac{31}{4} = 7.75.$

71. a. The object is speeding up on the interval $(0, 1)$, moving at a constant rate on $(1, 3)$, slowing down on $(3, 5)$, and moving at a constant rate on $(5, 6)$.   **b.** 30 m   **c.** 50 m   **d.** $s(t) = 80 + 10t$

73. a. 14.5 g   **b.** 29.5 g   **c.** 44 g   **d.** $x = \frac{19}{3}$ cm

75. $s(t) = \begin{cases} 30t & \text{if } 0 \le t \le 2 \\ 50t - 40 & \text{if } 2 < t \le 2.5 \\ 44t - 25 & \text{if } 2.5 < t \le 3 \end{cases}$

77.

n	Midpoint Riemann sum
16	4.7257
32	4.7437
64	4.7485

The sums approach 4.75.

81. Underestimates for decreasing functions, independent of concavity; overestimates for increasing functions, independent of concavity

Section 5.2 Exercises, pp. 364–367

1. The difference between the area bounded by the curve above the x-axis and the area bounded by the curve below the x-axis   **3.** 60; 0

5. $-12; -18; -16$

7.

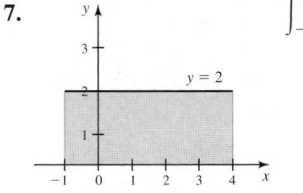

$\displaystyle\int_{-1}^{4} 2\, dx = 10$

9.

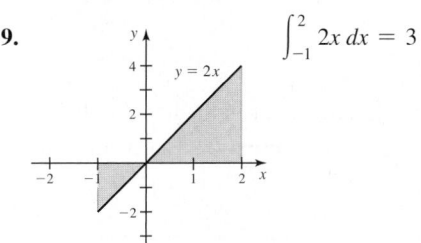

$\displaystyle\int_{-1}^{2} 2x\, dx = 3$

11. Both integrals equal 0.   **13.** The length of the interval $[a, a]$ is $a - a = 0$, so the net area is 0.   **15.** $\dfrac{a^2}{2}$

17. a.

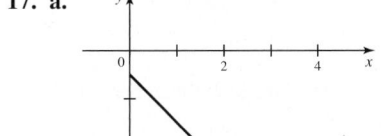

b. $-16, -24, -20$

19. a.

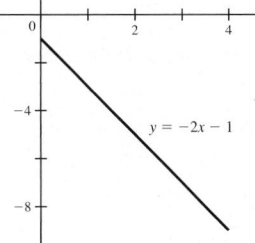

b. $-0.948, -0.948, -1.026$

21. a.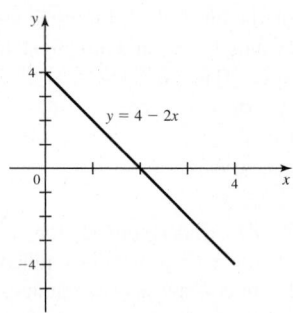

b. 4, −4, 0   **c.** Positive contributions on $[0, 2)$; negative contributions on $(2, 4]$

23. a.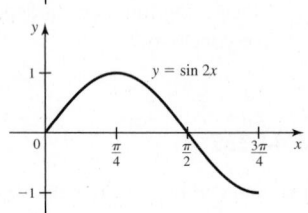

b. 0.735, 0.146, 0.530
c. Positive contributions on $(0, \pi/2)$; negative contributions on $(\pi/2, 3\pi/4]$

25. a.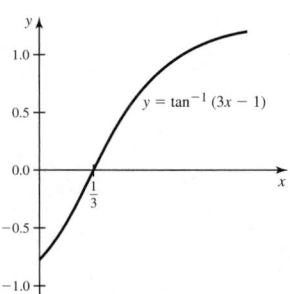

b. 0.082; 0.555; 0.326

c. Positive contributions on $(\frac{1}{3}, 1]$; negative contributions on $[0, \frac{1}{3})$

27. 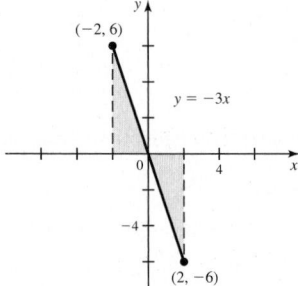   The area is 12; the net area is 0.

29. 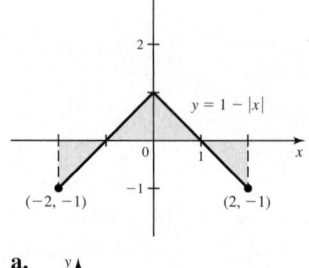   The area is 2; the net area is 0.

31. a.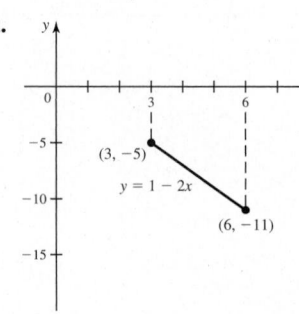

b. $\Delta x = \frac{1}{2}$; 3, 3.5, 4, 4.5, 5, 5.5, 6   **c.** −22.5; −25.5
d. The left Riemann sum overestimates the integral; the right Riemann sum underestimates the integral.

33. a.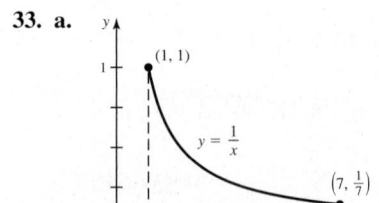

b. $\Delta x = 1$; 1, 2, 3, 4, 5, 6, 7
c. $\frac{49}{20}; \frac{223}{140}$ **d.** The left Riemann sum overestimates the integral; the right Riemann sum underestimates the integral.

35. $\int_0^2 (x^2 + 1)\, dx$ **37.** $\int_1^2 x \ln x\, dx$

39. 16 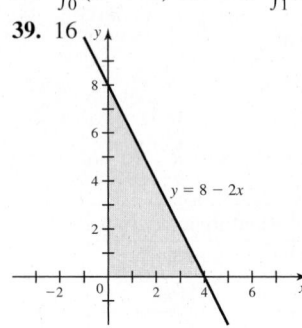   **41.** $-\frac{5}{2}$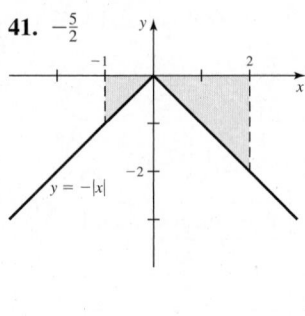

43. 4π 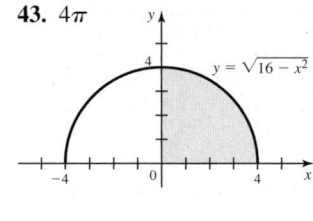   **45.** 26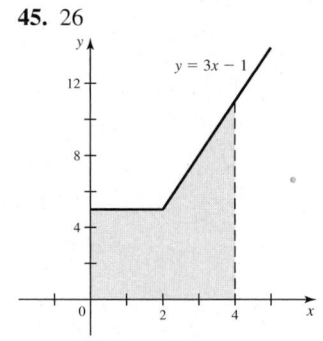

47. π **49.** -2π **51. a.** −32   **b.** $-\frac{32}{3}$ **c.** −64   **d.** Not possible
53. a. 10   **b.** −3   **c.** −16   **d.** 3   **55. a.** 15   **b.** 5   **c.** 3
d. −2   **e.** 24   **f.** −10   **57. a.** $\frac{3}{2}$ **b.** $-\frac{3}{4}$ **59.** 16   **61.** 6

63. 32   **65.** −16   **67.** $\frac{\pi}{4} + 2$ **69. a.** True   **b.** True   **c.** True
d. False   **e.** False

71. a. Left: $\displaystyle\sum_{k=1}^{n}\left(\left(\frac{k-1}{n}\right)^2 + 1\right)\cdot\frac{1}{n}$;

right: $\displaystyle\sum_{k=1}^{n}\left(\left(\frac{k}{n}\right)^2 + 1\right)\cdot\frac{1}{n}$

b.

n	Left Riemann sum	Right Riemann sum
20	1.30875	1.35875
50	1.3234	1.3434
100	1.32835	1.33835

Estimate: $\frac{4}{3}$

73. a. Left: $\displaystyle\sum_{k=1}^{n}\cos^{-1}\left(\frac{k-1}{n}\right)\frac{1}{n}$;

right: $\displaystyle\sum_{k=1}^{n}\cos^{-1}\left(\frac{k}{n}\right)\frac{1}{n}$

b.

n	Left Riemann sum	Right Riemann sum
20	1.03619	0.95765
50	1.01491	0.983494
100	1.00757	0.99186

Estimate: 1

75. a. $\displaystyle\sum_{k=1}^{n} 2\sqrt{1 + \left(k - \frac{1}{2}\right)\frac{3}{n}} \cdot \frac{3}{n}$

b.

n	Midpoint Riemann sum
20	9.33380
50	9.33341
100	9.33335

Estimate: $\dfrac{28}{3}$

77. a. $\displaystyle\sum_{k=1}^{n} \left(4\left(k - \frac{1}{2}\right)\frac{4}{n} - \left(\left(k - \frac{1}{2}\right)\frac{4}{n}\right)^2\right) \cdot \frac{4}{n}$

b.

n	Midpoint Riemann sum
20	10.6800
50	10.6688
100	10.6672

Estimate: $\dfrac{32}{3}$

79. 6 **81.** 104 **83.** 18 **85.** 2 **87.** $25\pi/2$ **89.** 25 **91.** 35
95. For any such partition on $[0, 1]$, the grid points are $x_k = k/n$, for $k = 0, 1, \ldots, n$. That is, x_k is rational for each k so that $f(x_k) = 1$, for $k = 0, 1, \ldots, n$. Therefore, the left, right, and midpoint Riemann sums are $\displaystyle\sum_{k=1}^{n} 1 \cdot (1/n) = 1$.

Section 5.3 Exercises, pp. 377–381

1. A is an antiderivative of f; $A'(x) = f(x)$.

3. $\displaystyle\int_a^b f(x)\, dx = F(b) - F(a)$, where F is any antiderivative of f.

5. Increasing **7.** The derivative of the integral of f is f, or
$\dfrac{d}{dx}\left(\displaystyle\int_a^x f(t)\, dt\right) = f(x)$. **9.** $f(x), 0$ **11.** 16 **13. a.** 0 **b.** -9
c. 25 **d.** 0 **e.** 16
15. a. $A(x) = 5x$ **b.** $A'(x) = 5$

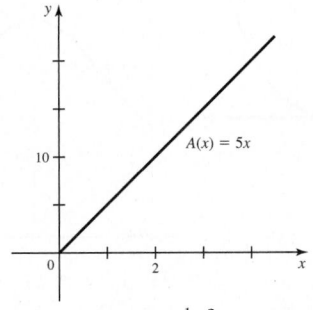

17. a. $A(2) = 2, A(4) = 8; A(x) = \frac{1}{2}x^2$ **b.** $F(4) = 6, F(6) = 16$;
$F(x) = \frac{1}{2}x^2 - 2$ **c.** $A(x) - F(x) = \frac{1}{2}x^2 - \left(\frac{1}{2}x^2 - 2\right) = 2$
19. a.

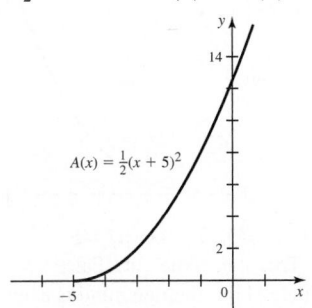

b. $A'(x) = \left(\frac{1}{2}(x + 5)^2\right)' = x + 5 = f(x)$

21. a.

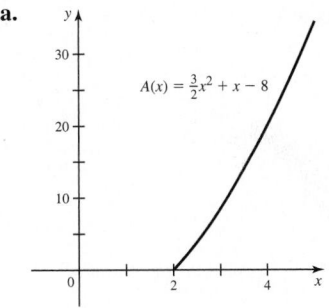

b. $A'(x) = \left(\frac{3}{2}x^2 + x - 8\right)' = 3x + 1 = f(x)$ **23.** $\frac{7}{3}$
25. $-\frac{125}{6}$ **27.** $-\frac{10}{3}$

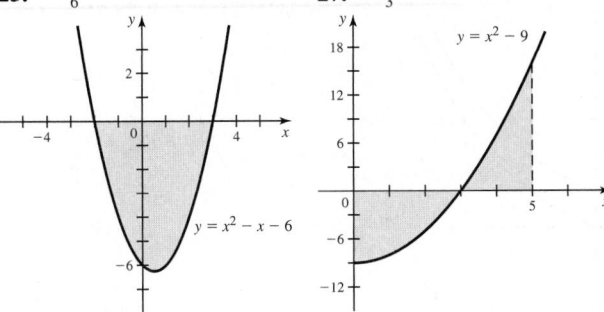

29. 16 **31.** 90 **33.** $\frac{7}{6}$ **35.** 8 **37.** $-\frac{32}{3}$ **39.** $-\frac{5}{2}$ **41.** $\frac{9}{2}$ **43.** $-\frac{3}{8}$
45. 1 **47.** $3 \ln 2$ **49.** $\frac{45}{4}$ **51.** $\frac{2}{3}$ **53.** 1 **55.** 2 **57.** $\frac{\pi}{12}$
59. $\frac{3}{2} + 4 \ln 2$ **61.** $\frac{3\pi}{2} - 1$

63. (i) $\frac{14}{3}$ **(ii)** $\frac{14}{3}$ **65. (i)** -51.2 **(ii)** 51.2

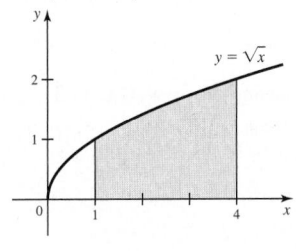
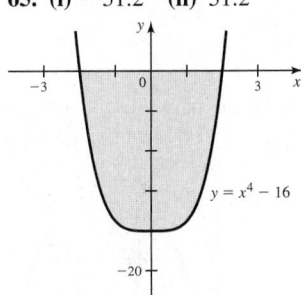

67. $\frac{94}{3}$ **69.** $\ln 2$ **71.** 2 **73.** $x^2 + x + 1$ **75.** $-\sqrt{x^4 + 1}$
77. $3/x^4$ **79.** $-(\cos^4 x + 6) \sin x$ **81.** $-\dfrac{\cos z}{\sin^4 z + 1}$
83. $\dfrac{9}{t}$ **85.** $2\sqrt{1 + x^2}$ **87.** a–C, b–B, c–D, d–A
89. a. $x = 0, x \approx 3.5$ **b.** Local min at $x \approx 1.5$; local max at
$x \approx 8.5$ **c.**

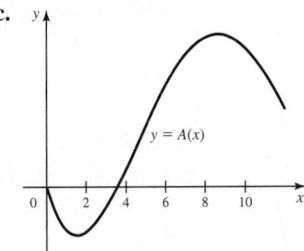

91. a. $x = 0, 10$ **b.** Local max at $x = 5$

c.

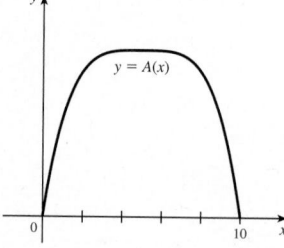

93. $-\pi, -\pi + \frac{9}{2}, -\pi + 9, 5 - \pi$

95. a. $A(x) = e^x - 1$

b.

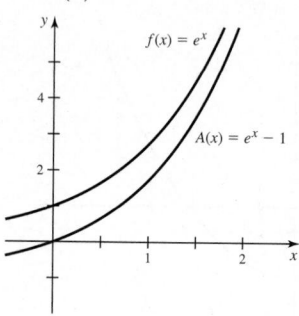

c. $A(\ln 2) = 1; A(\ln 4) = 3$

97. a. $A(x) = \sin x$

b.

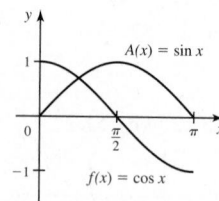

c. $A\left(\frac{\pi}{2}\right) = 1; A(\pi) = 0$

99. Critical pts. $x = 0, 3$, and 4; increasing on $(-\infty, 0), (0, 3)$, and $(4, \infty)$; decreasing on $(3, 4)$

101. a.

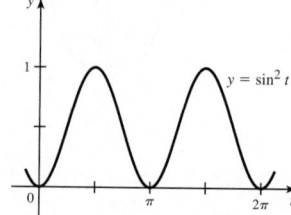

b. $g'(x) = \sin^2 x$

c.

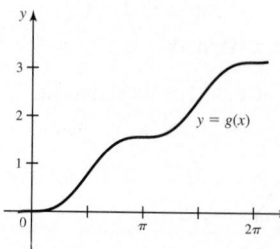

103.

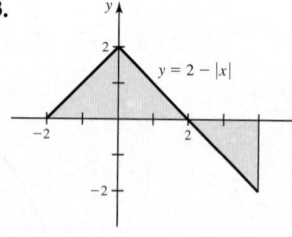

Area $= 6$

105.

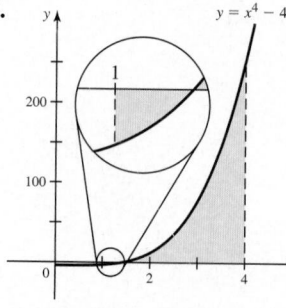

Area ≈ 194.05

107. a. True   **b.** True
c. False   **d.** True
109. 3

111. a.

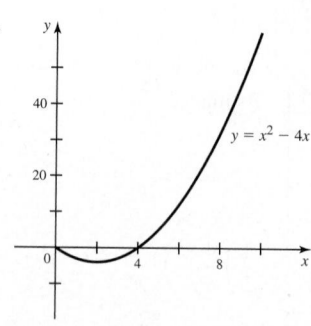

b. $b = 6$ **c.** $b = \dfrac{3a}{2}$

113. $f(x) = -2 \sin x + 3$ **115.** $\pi/2 \approx 1.57$

117. $(S'(x))^2 + \left(\dfrac{S''(x)}{2x}\right)^2 = (\sin x^2)^2 + \left(\dfrac{2x \cos x^2}{2x}\right)^2$

$$= \sin^2 x^2 + \cos^2 x^2 = 1$$

119. c. The summation relationship is a discrete analog of the Fundamental Theorem. Summing the difference quotient and integrating the derivative over the relevant interval give the difference of the function values at the endpoints.

Section 5.4 Exercises, pp. 385–387

1. If f is odd, the regions between f and the positive x-axis and between f and the negative x-axis are reflections of each other through the origin. Therefore, on $[-a, a]$, the areas cancel each other.
3. a. 9   **b.** 0   **5.** $3x^3$ and x are odd functions.   **7.** Even; even
9. If f is continuous on $[a, b]$, then there is a c in (a, b) such that

$f(c) = \dfrac{1}{b - a} \int_a^b f(x)\, dx$.   **11.** 0   **13.** $\dfrac{1000}{3}$ **15.** $\dfrac{16}{3}$ **17.** $-\dfrac{88}{3}$

19. 0   **21.** 2   **23.** 0

25. 0

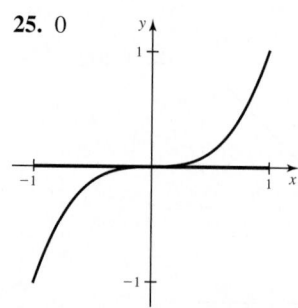

27. $\dfrac{\pi}{4}$

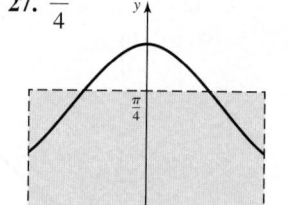

29. $2/\pi$

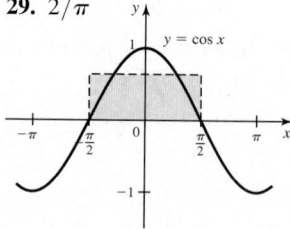

31. $1/(n + 1)$

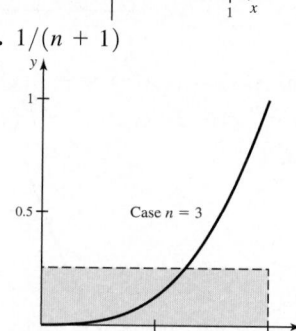

33. 2000   **35.** 21 m/s   **37.** $20/\pi$ **39.** 2   **41.** $a/\sqrt{3}$
43. $c = \pm\frac{1}{2}$ **45. a.** True   **b.** True   **c.** True   **d.** False
47. 420 ft   **49.** $f(g(-x)) = f(g(x)) \Rightarrow$ the integrand is even;

$\int_{-a}^a f(g(x))\, dx = 2 \int_0^a f(g(x))\, dx$ **51.** $p(g(-x)) = p(g(x)) \Rightarrow$

the integrand is even; $\int_{-a}^a p(g(x))\, dx = 2 \int_0^a p(g(x))\, dx$

53. a. $a/6$ **b.** $(3 \pm \sqrt{3})/6$, independent of a

57.

Even	Even
Even	Odd

Section 5.5 Exercises, pp. 395–398

1. The Chain Rule   **3.** $u = g(x)$ **5.** The lower bound a becomes $g(a)$ and the upper bound b becomes $g(b)$.   **7.** $\dfrac{(x^2 + 1)^5}{5} + C$

9. $\frac{1}{4}\sin^4 x + C$ **11.** $\dfrac{(x + 1)^{13}}{13} + C$ **13.** $\dfrac{(2x + 1)^{3/2}}{3} + C$

15. a. $\frac{1}{10}e^{10x} + C$ **b.** $\frac{1}{5}\sec 5x + C$ **c.** $-\frac{1}{7}\cos 7x + C$

d. $7\sin\frac{x}{7} + C$ **e.** $\frac{1}{27}\tan^{-1}\frac{x}{3} + C$ **f.** $\sin^{-1}\frac{x}{6} + C$

17. $\dfrac{(x^2 - 1)^{100}}{100} + C$ **19.** $-\dfrac{(1 - 4x^3)^{1/2}}{3} + C$ **21.** $\dfrac{(x^2 + x)^{11}}{11} + C$

23. $\dfrac{(x^4 + 16)^7}{28} + C$ **25.** $\frac{1}{2}\sin^{-1}\frac{x}{3} + C$ **27.** $\dfrac{4^{x^3}}{\ln 2} + C$

29. $\dfrac{(x^6 - 3x^2)^5}{30} + C$ **31.** $\frac{3}{5}\sin^{-1}5x + C$ **33.** $\frac{1}{6}\tan^{-1}\dfrac{e^w}{6} + C$

35. $-\frac{1}{2}\csc x^2 + C$ **37.** $\frac{1}{10}\tan(10x + 7) + C$ **39.** $\dfrac{10^{4t+1}}{4\ln 10} + C$

41. $\frac{1}{2}\tan^2 x + C$ **43.** $\frac{1}{7}\sec^7 x + C$ **45.** $\frac{\sqrt{2}}{4}$ **47.** $\frac{7}{2}$ **49.** 1   **51.** $\frac{1}{3}$

53. $\dfrac{2 - \sqrt{2}}{2}$ **55.** $(e^9 - 1)/3$ **57.** $\sqrt{2} - 1$ **59.** $\frac{\pi}{6}$ **61.** $\frac{1}{2}\ln 17$

63. $\frac{\pi}{9}$ **65.** $\frac{1}{3}$ **67.** $\frac{3}{4}(4 - 3^{2/3})$ **69.** $\frac{32}{3}$ **71.** $-\ln 3$ **73.** $\frac{1}{7}$

75. 10 m/s   **77. a.** 160   **b.** $\dfrac{4800}{49} \approx 98$ **c.** $\Delta p = \displaystyle\int_0^T \dfrac{200}{(t + 1)^r}\,dt$;

decreases as r increases   **d.** $r \approx 1.28$ **e.** As $t \to \infty$, the population approaches 100.   **79.** $\frac{2}{3}(x - 4)^{1/2}(x + 8) + C$

81. $\frac{3}{5}(x + 4)^{2/3}(x - 6) + C$ **83.** $\frac{3}{112}(2x + 1)^{4/3}(8x - 3) + C$

85. $\dfrac{(x + 10)^{10}(x - 1)}{11} + C$ **87.** π

89. $\dfrac{\theta}{2} - \frac{1}{4}\sin\left(\dfrac{6\theta + \pi}{3}\right) + C$ **91.** $\frac{\pi}{4}$ **93.** $\ln\frac{9}{8}$ **95. a.** True

b. True   **c.** False   **d.** False   **e.** False   **97.** 1   **99.** $\frac{2}{3}$; constant

101. a. π/p **b.** 0   **103.** $2/\pi$ **105.** One area is $\displaystyle\int_4^9 \dfrac{(\sqrt{x} - 1)^2}{2\sqrt{x}}\,dx$.

Changing variables by letting $u = \sqrt{x} - 1$ yields $\displaystyle\int_1^2 u^2\,du$, which is the other area.   **107.** 7297/12   **109.** $\frac{2}{15}(3 - 2a)(1 + a)^{3/2} + \frac{4}{15}a^{5/2}$

111. $\frac{1}{3}\sec^3 \theta + C$ **113. a.** $I = \frac{1}{8}x - \frac{1}{32}\sin 4x + C$

b. $I = \frac{1}{8}x - \frac{1}{32}\sin 4x + C$

117. $\frac{4}{3}(-2 + \sqrt{1 + x})\sqrt{1 + \sqrt{1 + x}} + C$ **119.** $-4 + \sqrt{17}$

Chapter 5 Review Exercises, pp. 398–402

1. a. True   **b.** False   **c.** True   **d.** True   **e.** False
f. True   **g.** True

3. a. 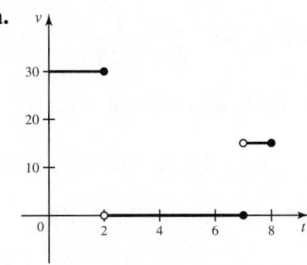   **b.** 75   **c.** The area is the distance the diver ascends.

5. 9.34; 10.28; 9.82

7.

n	Midpoint Riemann sum
10	114.167
30	114.022
60	114.006

$\displaystyle\int_1^{25}\sqrt{2x - 1}\,dx = 114$

9. a. $1((3\cdot2 - 2) + (3\cdot3 - 2) + (3\cdot4 - 2)) = 21$

b. $\displaystyle\sum_{k=1}^n \frac{3}{n}\left(3\left(1 + \dfrac{3k}{n}\right) - 2\right)$ **c.** $\frac{33}{2}$ **11.** $-\frac{16}{3}$ **13.** 56

15. a. 20   **b.** 0   **c.** 80   **d.** 10   **e.** 0   **17.** 18   **19.** 10
21. Not enough information   **23. a.** 8.5   **b.** −4.5   **c.** 0   **d.** 11.5

25. 4π **27.** A: $\displaystyle\int_0^x f(t)\,dt$; B: $f(x)$; C: $f'(x)$ **29.** $\sqrt{1 + x^4 + x^6}$

31. $-\sin x^6$ **33.** $\dfrac{2}{x^{10} + 1}$ **35.** Increasing on $(3, 6)$; decreasing on $(-\infty, 3)$ and $(6, \infty)$ **39.** $\frac{212}{5}$ **41.** $x^9 - x^7 + C$

43. $\frac{7}{6}$ **45.** $\dfrac{4}{\sqrt{3}}$ **47.** $\dfrac{\pi}{12}$ **49.** $-\dfrac{4}{3\sin^{3/4}x} + C$

51. $\frac{1}{3}\sin x^3 + C$ **53.** $\frac{1}{28}\tan^{-1}\left(\dfrac{\sin 7w}{4}\right) + C$ **55.** $\dfrac{1}{\ln 2}$

57. 78   **59.** $\frac{5}{6}e^2(e^3 - 1)$ **61.** $e^{e^x} + C$ **63.** $\frac{1}{2}\sin^{-1}2x + C$

65. $\pi + \dfrac{3\sqrt{3}}{4}$ **67.** $\dfrac{\pi}{2}$ **69.** $\frac{1}{3}\ln\frac{9}{2}$ **71.** 0   **73.** $\cos\dfrac{1}{x} + C$

75. $\ln|\tan^{-1}x| + C$ **77.** $(x + 3)^{11}\left(\dfrac{11x - 3}{132}\right) + C$

79. 1   **81.** $\dfrac{\pi}{12}$ **83.** 0   **85.** 48   **87.** $\dfrac{256}{3}$ **89.** 8   **91.** $-\frac{4}{15}; \frac{4}{15}$

93. Approx. 431.5 ft   **95.** Displacement = 0; distance = $20/\pi$

97. $\dfrac{3}{2\ln 2}$ **99. a.** $5/2, c = 3.5$ **b.** $3, c = 3$ and $c = 5$ **101.** 24

103. $f(1) = 0$; $f'(x) > 0$ on $[1, \infty)$; $f''(x) < 0$ on $[1, \infty)$

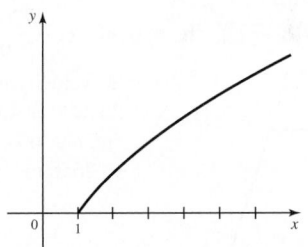

105. a. $\dfrac{3}{2}, \dfrac{5}{6}$ **b.** x **c.** $\frac{1}{2}x^2$ **d.** $-1, \frac{1}{2}$ **e.** 1, 1   **f.** $\dfrac{3}{2}$ **107.** e^4

113. a. Increasing on $(-\infty, 1)$ and $(2, \infty)$; decreasing on $(1, 2)$

b. Concave up on $\left(\frac{13}{8}, \infty\right)$; concave down on $\left(-\infty, \frac{13}{8}\right)$

c. Local max at $x = 1$; local min at $x = 2$ **d.** Inflection point at $x = \frac{13}{8}$ **115.** Differentiating the first equation gives the second equation; no.   **117.** $\sqrt[4]{12}$

CHAPTER 6

Section 6.1 Exercises, pp. 410–416

1. The position $s(t)$ is the location of the object relative to the origin. The displacement is the change in position between time $t = a$ and $t = b$. The distance traveled between $t = a$ and $t = b$ is $\int_a^b |v(t)|\, dt$, where $v(t)$ is the velocity at time t.   **3.** The displacement between $t = a$ and $t = b$ is $\int_a^b v(t)\, dt$.   **5.** $Q(t) = Q(0) + \int_0^t Q'(x)\, dx$

7. a. $[0, 1), (3, 5)$ **b.** -4 mi   **c.** 26 mi   **d.** 6 mi   **e.** 6 mi on the positive side of the initial position   **9. a.** 3   **b.** $\frac{13}{3}$ **c.** 3

d. $s(t) = \begin{cases} -\dfrac{t^2}{2} + 2t & \text{if } 0 \le t \le 3 \\[2mm] \dfrac{3t^2}{2} - 10t + 18 & \text{if } 3 < t \le 4 \\[2mm] -t^2 + 10t - 22 & \text{if } 4 < t \le 5 \end{cases}$

11. a. 3 m   **b.** 3 m; 0 m; 3 m; 0 m   **c.** 12 m
13. a. Positive direction for $2 < t \le 3$; negative direction for $0 < t < 2$ **b.** 0 m   **c.** 8 m   **15. a.** Positive direction for $0 \le t < 2$ and $4 < t \le 5$; negative direction for $2 < t < 4$
b. 20 m   **c.** 28 m   **17. a.** $s(t) = 2 - \cos t$ **19. a.** $s(t) = 6t - t^2$

21. a. $s(t) = 9t - \dfrac{t^3}{3} - 2$ **23. a.** $s(t) = 2 \sin \pi t$

b.

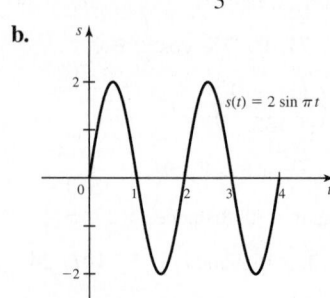

c. $\frac{3}{2}, \frac{7}{2}, \frac{11}{2}$
d. $\frac{1}{2}, \frac{5}{2}, \frac{9}{2}$

$s(t) = 2 \sin \pi t$

25. a. $s(t) = 10t(48 - t^2)$ **b.** 880 mi   **c.** $\dfrac{2720\sqrt{6}}{9} \approx 740.29$ mi

27.

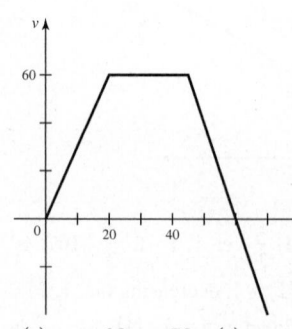

a. Velocity is a maximum for $20 \le t \le 45$; $v = 0$ at $t = 0$ and $t = 60$ **b.** 1200 m
c. 2550 m   **d.** 2100 m

29. $v(t) = -32t + 70$; $s(t) = -16t^2 + 70t + 10$
31. $v(t) = -9.8t + 20$; $s(t) = -4.9t^2 + 20t$
33. $v(t) = -\frac{1}{200} t^2 + 10$; $s(t) = -\frac{1}{600} t^3 + 10t$
35. $v(t) = \frac{1}{2} \sin 2t + 5$; $s(t) = -\frac{1}{4} \cos 2t + 5t + \frac{29}{4}$
37. a. $s(t) = 44t^2$ **b.** 704 ft   **c.** $\sqrt{30} \approx 5.477$ s

d. $\dfrac{5\sqrt{33}}{11} \approx 2.611$ s   **e.** Approx. 180.023 ft

39. 6.154 mi; 1.465 mi   **41. a.** 2639 people
b. $P(t) = 250 + 20t^{3/2} + 30t$ people   **43. a.** 1897 cells;
1900 cells   **b.** $N(t) = 1900 - 400e^{-0.25t}$ **45. a.** 27,250 barrels

b. 31,000 barrels   **c.** 4000 barrels   **47. a.** $\dfrac{10^7(1 - e^{-kt})}{k}$

b. $\dfrac{10^7}{k}$ = total number of barrels of oil extracted if the nation extracts

the oil indefinitely, and it has at least $\dfrac{10^7}{k}$ barrels of oil in reserve.

c. $k = \dfrac{1}{200} = 0.005$ **d.** Approx. 138.6 yr

49. a. $\dfrac{120}{\pi} + 40 \approx 78.20$ m^3

b. $Q(t) = 20\left(t + \dfrac{12}{\pi} \sin\left(\dfrac{\pi}{12}t\right)\right)$ **c.** Approx. 122.6 hr

51. a. $V(t) = 5 + \cos\dfrac{\pi t}{2}$ **b.** 15 breaths/min   **c.** 2 L, 6 L

53. a. 7200 MWh or 2.592×10^{13} J   **b.** 16,000 kg; 5,840,000 kg
c. 450 g; 164,250 g   **d.** About 1500 turbines   **55. a.** \$96,875
b. \$86,875   **57. a.** \$69,583.33   **b.** \$139,583.33   **59. a.** False
b. True   **c.** True   **d.** True   **61.** $\frac{2}{3}$ **63.** $\frac{25}{3}$

65. a.

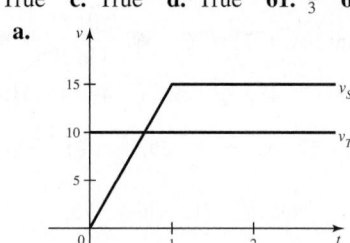

b. Theo   **c.** Sasha
d. Theo hits the 10-mi mark before Sasha; Sasha and Theo hit the 15-mi mark at the same time; Sasha hits the 20-mi mark before Theo.   **e.** Sasha
f. Theo

67. Approx. 11:23 A.M.

69. $\int_a^b f'(x)\, dx = f(b) - f(a) = g(b) - g(a) = \int_a^b g'(x)\, dx$

Section 6.2 Exercises, pp. 420–425

1. $\int_a^b (f(x) - g(x))\, dx + \int_b^c (g(x) - f(x))\, dx$

3.

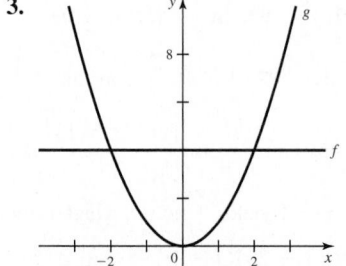

5. a. 1

7. $\int_0^1 y\, dy + \int_1^2 (2 - y)\, dy$

9. $\dfrac{9}{2}$ **11.** $3\pi - 2$

13. $\dfrac{5}{2} - \dfrac{1}{\ln 2}$

15. $2 - \sqrt{2}$

17. $\dfrac{4}{3}$ **19.** $\dfrac{32}{3}$

21. 9   **23.** $\dfrac{81}{2}$ **25.** 2   **27.** $\dfrac{8\sqrt{2} - 7}{6}$ **29.** $\dfrac{125}{2}$

31. a. $\int_0^1 (\sqrt{x} - x^3)\, dx$ **b.** $\int_0^1 (\sqrt[3]{y} - y^2)\, dy$

33. $\frac{19}{6} \approx 3.17$ km; the faster runner jogged approximately 3.17 km farther than the slower runner.   **35. a.** 7   **b.** 4   **37.** 25   **39.** $\frac{81}{32}$
41. $\pi - 2$ **43.** $\frac{1}{2} + \ln 2$ **45.** $\frac{7}{3}$ **47.** 3   **49.** $\frac{64}{5}$ **51.** $\ln 2$
53. $\frac{5}{24}$ **55.** $\frac{63}{4}$ **57.** $\frac{9}{2}$ **59.** $\frac{32}{3}$ **61.** $\frac{15}{8} - 2\ln 2$ **63.** $\frac{17}{3}$

65. a. False   **b.** False   **c.** True   **67.** $\frac{4}{9}$ **69.** $\dfrac{n-1}{2(n+1)}$

71. $A_n = \dfrac{n-1}{n+1}$; $\lim\limits_{n \to \infty} A_n = 1$; the region approximates a square

with side length of 1.   **73.** $k = 1 - \dfrac{1}{\sqrt{2}}$ **75. a.** The lowest $p\%$
of households own exactly $p\%$ of the wealth for $0 \le p \le 100$.

b. The function must be one-to-one and its graph must lie below $y = x$ because the poorest $p\%$ cannot own more than $p\%$ of the wealth.   **c.** $p = 1.1$ is most equitable; $p = 4$ is least equitable.   **e.** $G(p) = \dfrac{p - 1}{p + 1}$ **f.** $0 \le G \le 1$ for $p \ge 1$ **g.** $\frac{5}{18}$

77. a.

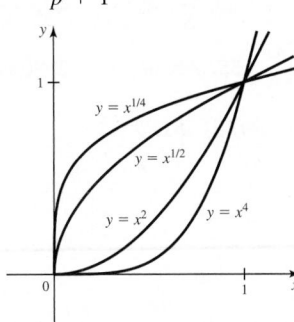

b. $A_n(x)$ is the net area of the region between the graphs of f and g from 0 to x.   **c.** $x = n^{n/(n^2 - 1)}$; the root decreases with n.

Section 6.3 Exercises, pp. 434–439

1. $A(x)$ is the area of the cross section through the solid at the point x.

3. a. $3 - x$ **b.** $\int_0^2 (3 - x)\, dx$ **5. a.** $\sqrt{\cos x}$ **b.** $\pi \cos x$

c. $\int_0^{\pi/2} \pi \cos x\, dx$ **7. a.** $\sqrt{x} + 1$ **b.** 1   **c.** $\pi((\sqrt{x} + 1)^2 - 1)$

d. $\int_0^4 \pi((\sqrt{x} + 1)^2 - 1)\, dx$ **9. a.** $\sqrt{x}$ **b.** πx **c.** $\int_0^4 \pi x\, dx$

11. $\frac{4}{3}$ **13.** 1   **15.** $\frac{\pi}{3}$ **17.** 36π **19.** $\frac{15\pi}{32}$ **21.** $\frac{\pi^2}{2}$ **23.** $\frac{32\pi}{3}$

25. $\frac{5\pi}{6}$ **27.** $\frac{2\pi}{5}$ **29.** $\frac{\pi^2}{4}$ **31.** $\frac{\pi^2}{2}$ **33.** $\frac{\pi(\pi - 2)}{8}$

35. $\frac{4\pi - \pi^2}{4}$ **37.** $\frac{128\pi}{5}$ **39.** $\pi \ln 3$ **41.** $\frac{\pi}{2}(e^4 - 1)$

43. $\frac{49\pi}{2}$ **45.** Volumes are equal.   **47.** x-axis   **49.** $\frac{\pi}{2}$

51. $\pi\sqrt{3}$ **53.** $\frac{\pi}{6}$ **55.** $2\pi(8 + \pi)$ **57.** $(6\sqrt{3} - 2\pi)\pi$

59. 4π **61. a.** False   **b.** True   **c.** True

63. Volume $(S) = 8\pi a^{5/2}/15$; volume $(T) = \pi a^{5/2}/3$

65. a.   **b.**

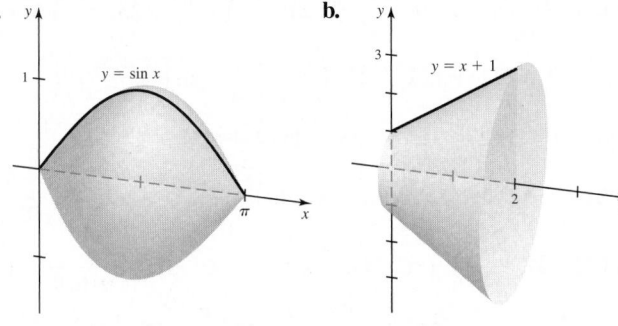

67. Left: 166π; right: 309π; midpoint: 219π **69. a.** $\frac{1}{3} V_C$ **b.** $\frac{2}{3} V_C$

71. $24\pi^2$ **73. b.** $V = \pi r^2 h$

Section 6.4 Exercises, pp. 447–451

1. $\int_a^b 2\pi x(f(x) - g(x))\, dx$ **3.** $x; y$ **5. a.** x **b.** $2 - x^2 - x$

c. $\int_0^1 2\pi x(2 - x^2 - x)\, dx$ **7. a.** $2 - y$ **b.** $4 - (2 - y)^2 = 4y - y^2$

c. $\int_0^2 2\pi(2 - y)(4y - y^2)\, dy$ **9.** $\frac{\pi}{6}$ **11.** π **13.** 8π **15.** $\frac{32\pi}{3}$

17. π **19.** $\frac{\pi}{2}$ **21.** $\frac{81\pi}{2}$ **23.** $\frac{2\pi}{3}$ **25.** $\frac{3\pi}{10}$ **27.** 90π

29. $2\pi e(e - 1)$ **31.** π **33.** $\frac{\pi}{5}$ **35.** $\frac{4\pi}{15}$ **37.** 500π

39. $\frac{11\pi}{6}$ **41.** $\frac{5\pi}{6}$ **43.** $\frac{23\pi}{15}$ **45.** $\frac{52\pi}{15}$ **47.** $\frac{36\pi}{5}$

51. a. $4\pi \int_1^5 x\sqrt{4 - (x - 3)^2}\, dx$ **b.** $12\pi \int_{-2}^2 \sqrt{4 - y^2}\, dy$

c. $24\pi^2$ **53.** $\frac{\pi}{9}$ **55.** $\frac{16\pi}{3}$ **57.** $\frac{608\pi}{3}$ **59.** $\pi(\sqrt{e} - 1)^2$

61. $\frac{5\pi}{6}$ **63. a.** True   **b.** False   **c.** True   **65.** 24π **67.** 54π

69. a. $V_1 = \dfrac{\pi}{15}(3a^2 + 10a + 15); V_2 = \dfrac{\pi}{2}(a + 2)$

b. $V(S_1) = V(S_2)$, for $a = 0$ and $a = -\frac{5}{6}$ **71.** 10π

73. a. $27\sqrt{3}\pi r^3/8$ **b.** $54\sqrt{2}/(3 + \sqrt{2})^3$ **c.** $500\pi/3$

Section 6.5 Exercises, pp. 455–457

1. Determine whether f has a continuous derivative on $[a, b]$. If so, calculate $f'(x)$ and evaluate the integral $\int_a^b \sqrt{1 + f'(x)^2}\, dx$.

3. $\int_{-2}^5 \sqrt{1 + 9x^4}\, dx$ **5.** $\int_0^2 \sqrt{1 + 4e^{-4x}}\, dx$ **7.** $4\sqrt{5}$

9. $8\sqrt{65}$ **11.** 168   **13.** $\frac{4}{3}$ **15.** $\frac{123}{32}$ **17.** $\frac{123}{32}$ **19.** $7\sqrt{5}$

21. a. $\int_{-1}^1 \sqrt{1 + 4x^2}\, dx$ **b.** 2.96   **23. a.** $\int_1^4 \sqrt{1 + \dfrac{1}{x^2}}\, dx$ **b.** 3.34

25. a. $\int_3^4 \sqrt{\dfrac{4y - 7}{4y - 8}}\, dy$ **b.** 1.08   **27. a.** $\int_0^\pi \sqrt{1 + 4 \sin^2 2x}\, dx$

b. 5.27   **29. a.** $\int_1^{10} \sqrt{1 + 1/x^4}\, dx$ **b.** 9.15

31. Approx. 1326 m   **33. a.** False   **b.** True   **c.** False

35. a. $f(x) = \pm 4x^3/3 + C$ **b.** $f(x) = \pm 3 \sin 2x + C$

37. $y = 1 - x^2$ **39. a.** $L/2$ **b.** L/c

Section 6.6 Exercises, pp. 463–465

1. 15π **3.** Evaluate $\int_a^b 2\pi f(x)\sqrt{1 + f'(x)^2}\, dx$ **5. a.** $4\sqrt{2}\pi$

7. $156\sqrt{10}\pi$ **9.** $\frac{2912\pi}{3}$ **11.** $\frac{\pi}{9}(17^{3/2} - 1)$ **13.** 2π

15. $15\sqrt{17}\,\pi$ **17.** $\frac{\pi}{8}(16 + e^8 - e^{-8})$ **19.** 96π

21. $\frac{9\pi}{125}\, \text{m}^3$ **23. a.** False   **b.** False   **c.** True   **d.** False

25. a. $\int_0^{\pi/2} 2\pi(\cos x)\sqrt{1 + \sin^2 x}\, dx$ **b.** Approx. 7.21

27. a. $\int_0^{\pi/4} 2\pi(\tan x)\sqrt{1 + \sec^4 x}\, dx$ **b.** Approx. 3.84

29. $\frac{12\pi a^2}{5}$ **31.** $\frac{53\pi}{9}$ **33.** $\frac{275\pi}{32}$ **35.** $\frac{48,143\pi}{48}$ **39. a.** $\frac{6}{R}$ **b.** $\frac{3}{R}$

c. $\frac{9 + 4\pi\sqrt{3}}{6R\sqrt[3]{4}}$ **d.** The sphere   **e.** A sphere

41. a. $c^2 A$ **b.** A

Section 6.7 Exercises, pp. 473–477

1. 150 g   **3.** 25 J   **5.** Horizontal cross sections of water at various locations in the tank are lifted different distances.   **7.** 39,200 N/m²

9. $\int_5^{10} 25\pi \rho g(15 - y)\, dy$ **11.** $\int_0^{10} 25\pi \rho g(10 - y)\, dy$ **13.** $\pi + 2$

15. 3   **17.** $(2\sqrt{2} - 1)/3$ **19.** 10   **21.** 9 J   **23. a.** $k = 150$

b. 12 J   **c.** 6.75 J   **d.** 9 J   **25. a.** 112.5 J   **b.** 12.5 J

27. a. 31.25 J   **b.** 312.5 J   **29. a.** 625 J   **b.** 391 J
31. a. 22,050 J   **b.** 36,750 J   **33.** 3675 J   **35.** 1.15×10^7 J
37 3.94×10^6 J   **39. a.** 66,150π J   **b.** No   **41. a.** 2.10×10^8 J
b. 3.78×10^8 J   **43. a.** 32,667 J   **b.** Yes   **45.** 7.70×10^3 J
47. 1.47×10^7 N   **49.** 2.94×10^7 N   **51.** 6533 N   **53.** 6737.5 N
55. 8×10^5 N   **57. a.** True   **b.** True   **c.** True   **d.** False
59. a. Compared to a linear spring, $F(x) = 16x$, the restoring force is
less for large displacements.   **b.** 17.87 J   **c.** 31.6 J   **61.** 1,381,800 J
63. 0.28 J   **65. a.** Yes   **b.** 4.296 m   **67.** Left: 16,730 N;
right: 14,700 N   **69. a.** 8.87×10^9 J
b. 500 $GMx/(R(x + R)) = (2 \times 10^{17})x/(R(x + R))$ J
c. GMm/R **d.** $v = \sqrt{2GM/R}$

Chapter 6 Review Exercises, pp. 478–482

1. a. True   **b.** True   **c.** True   **3. a.** Positive direction for
$0 \le t < \frac{1}{2}$ and $2 < t \le 3$; negative direction for $\frac{1}{2} < t < 2$
b. 9 m   **c.** 22.5 m   **d.** $s(t) = 4t^3 - 15t^2 + 12t + 1$
5. $s(t) = 20t - 5t^2$; displacement $= 20t - 5t^2$;
$$D(t) = \begin{cases} 20t - 5t^2 & \text{if } 0 \le t < 2 \\ 5t^2 - 20t + 40 & \text{if } 2 \le t \le 4 \end{cases}$$
7. a. $v(t) = -\dfrac{8}{\pi}\cos\dfrac{\pi t}{4}; s(t) = -\dfrac{32}{\pi^2}\sin\dfrac{\pi t}{4}$ **b.** Min value $= -\dfrac{32}{\pi^2}$;
max value $= \dfrac{32}{\pi^2}$ **c.** 0; 0   **9. a.** $R(t) = 3t^{4/3}$
b. $R(t) = \begin{cases} 3t^{4/3} & \text{if } 0 \le t \le 8 \\ 2t + 32 & \text{if } t > 8 \end{cases}$ **c.** $t = 59$ min

11. a.   **b.** $10 \ln 4 \approx 13.86$ s

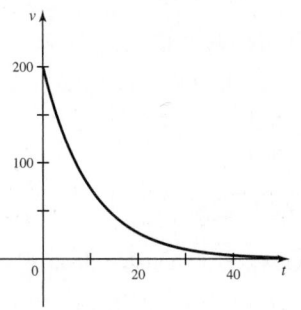

c. $s(t) = 2000(1 - e^{-t/10})$ **d.** No

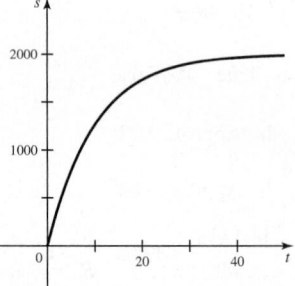

13. a. $s_{\text{Tom}}(t) = -10e^{-2t} + 10$
$s_{\text{Sue}}(t) = -15e^{-t} + 15$

b. $t = 0$ and $t = \ln 2$ **c.** Sue   **15.** $1 - \dfrac{\pi}{4}$ **17.** $e - 2$ **19.** $\dfrac{7}{3}$

---

21. 8   **23.** 1   **25.** $\dfrac{1}{3}$ **27.** $R_1: \dfrac{7}{6}; R_2: \dfrac{10}{3}; R_3: 4\sqrt{3} - \dfrac{10}{3}$ **29.** $\dfrac{11\pi}{15}$
31. $\dfrac{14\pi}{3}$ **33.** $\int_1^3 2\pi(3 - x)(2\sqrt{x} - 3 + x)\,dx$ **35.** $\dfrac{7}{3}$ **37.** $\dfrac{31\pi}{5}$
39. $R_1: \sqrt{3}; R_2: \dfrac{4\pi}{3} - \sqrt{3}$ **41.** $\dfrac{1}{3}$ **43.** $\dfrac{5}{6}$ **45.** $\dfrac{8}{15}$ **47.** $\dfrac{8\pi}{5}$
49. $\pi(e - 1)^2$ **51.** π **53.** $\dfrac{512\pi}{15}$ **55.** About $y = -2: 80\pi$;
about $x = -2: 112\pi$ **57.** $c = 5$ **59.** 1   **61.** $2\sqrt{3} - \dfrac{4}{3}$
63. $\int_2^4 \sqrt{4x^2 + 8x + 5}\,dx \approx 16.127$
65. $\sqrt{b^2 + 1} - \sqrt{2} + \ln\left(\dfrac{(\sqrt{b^2 + 1} - 1)(1 + \sqrt{2})}{b}\right); b \approx 2.715$
67. a. 9π **b.** $\dfrac{9\pi}{2}$ **69. a.** $\dfrac{263,439\pi}{4096}$ **b.** $\dfrac{483}{64}$ **c.** $\dfrac{\pi}{8}(84 + \ln 2)$
d. $\dfrac{264,341\pi}{18,432}$ **71.** $\left(450 - \dfrac{450}{e}\right)$ g   **73. a.** 562.5 J   **b.** 56.25 J
75. a. 980 J   **b.** 627.2 J   **77. a.** 1,411,200 J   **b.** 940,800 J
79. a. 1,477,805 J   **b.** The work required to pump out the top 3 m
of water is 1,015,991 J, and the work required to pump out the bottom
3 m of water is 461,814 J. More work is required to pump out the top
3 m of water.   **81.** 4,987,592 J   **83.** 5716.7 N   **85.** 5.2×10^7 N

CHAPTER 7

Section 7.1 Exercises, pp. 490–492

1. $D = (0, \infty), R = (-\infty, \infty)$ **3.** $\dfrac{4^x}{\ln 4} + C$
5. $e^{x\ln 3}, e^{\pi\ln x}, e^{(\sin x)(\ln x)}$ **7.** $3(\ln x + 1)$ **9.** $\dfrac{\cos(\ln x)}{x}, x > 0$
11. $-\dfrac{5}{x(\ln 2x)^6}$ **13.** $4^{2x+1}x^{4x}(1 + \ln 2x)$ **15.** $(\ln 2)\,2^{x^2+1}\,x$
17. $2(x + 1)^{2x}\left(\dfrac{x}{x + 1} + \ln(x + 1)\right)$
19. $y^{\sin y}\left(\cos y \ln y + \dfrac{\sin y}{y}\right)$ **21.** $-20xe^{-10x^2}$ **23.** $x^{2x}(2\ln x + 2)$
25. $-(1/x)^x(1 + \ln x)$ **27.** $\left(-\dfrac{4}{x + 4} + \ln\left(\dfrac{x + 4}{x}\right)\right)\left(1 + \dfrac{4}{x}\right)^x$
29. $6(1 - \ln 2)$ **31.** $\dfrac{3}{8}$ **33.** $\dfrac{1}{2}\ln(4 + e^{2x}) + C$ **35.** $\dfrac{1}{\ln 2} - \dfrac{1}{\ln 3}$
37. $4 - \dfrac{4}{e^2}$ **39.** $2e^{\sqrt{x}} + C$ **41.** $\ln|e^x - e^{-x}| + C$ **43.** $\dfrac{99}{10\ln 10}$
45. 3   **47.** $\dfrac{6^{x^3+8}}{3\ln 6} + C$ **49.** $\dfrac{1}{6}e^{3x^2+1} + C$ **51.** $-\dfrac{1}{9^x\ln 9} + C$
53. $\dfrac{10^{x^3}}{3\ln 10} + C$ **55.** $\dfrac{3 \cdot 3^{\ln 2} - 1}{\ln 3}$ **57.** $\dfrac{32}{3}$ **59.** $\dfrac{1}{3}\ln\dfrac{65}{16}$
61. $2e^{5+\sqrt{x}} + C$ **63.**

h	$(1 + 2h)^{1/h}$	h	$(1 + 2h)^{1/h}$
10^{-1}	6.1917	-10^{-1}	9.3132
10^{-2}	7.2446	-10^{-2}	7.5404
10^{-3}	7.3743	-10^{-3}	7.4039
10^{-4}	7.3876	-10^{-4}	7.3905
10^{-5}	7.3889	-10^{-5}	7.3892
10^{-6}	7.3890	-10^{-6}	7.3891

$\lim_{h \to 0}(1 + 2h)^{1/h} = e^2$

65.

x	$\dfrac{2^x - 1}{x}$	x	$\dfrac{2^x - 1}{x}$
10^{-1}	0.71773	-10^{-1}	0.66967
10^{-2}	0.69556	-10^{-2}	0.69075
10^{-3}	0.69339	-10^{-3}	0.69291
10^{-4}	0.69317	-10^{-4}	0.69312
10^{-5}	0.69315	-10^{-5}	0.69314
10^{-6}	0.69315	-10^{-6}	0.69315

$$\lim_{x \to 0} \frac{2^x - 1}{x} = \ln 2$$

67. a. True **b.** False **c.** False **d.** False **e.** True

69. $\dfrac{\ln p}{p - 1}, 0$ **71. a.** No **b.** No

75. $\ln 2 = \displaystyle\int_1^2 \frac{dt}{t} < L_2 = \frac{5}{6} < 1$

$\ln 3 = \displaystyle\int_1^3 \frac{dt}{t} > R_7$

$= 2\left(\dfrac{1}{9} + \dfrac{1}{11} + \dfrac{1}{13} + \dfrac{1}{15} + \dfrac{1}{17} + \dfrac{1}{19} + \dfrac{1}{21}\right) > 1$

Section 7.2 Exercises, pp. 499–501

1. The relative growth is constant. **3.** The time it takes a function to double in value **5.** $T_2 = (\ln 2)/k$ **7.** $\dfrac{\ln 2}{20} \approx 0.03466$

9. Compound interest, world population **11.** $\ln 1.11 \approx 0.1044$.

13. $\dfrac{df}{dt} = 10.5$; $\dfrac{dg}{dt} \cdot \dfrac{1}{g} = \dfrac{1}{10}$

15. a. $\ln 1.024 \approx 0.02372$; $y(t) = 90{,}000\, e^{t \ln 1.024}$ **b.** 2028

17. a. $\dfrac{\ln 1.1}{10} \approx 0.009531$; $y(t) = 50{,}000\, e^{t \ln 1.1/10}$ **b.** 60,500

19. a. $\ln 1.016 \approx 0.01587$; $y(t) = 100\, e^{t \ln 1.016}$ **b.** \$126.88

21. 3.71% **23. a.** 88.1 years; 423.4 million

b. 99.4 years; 412.2 million **25.** 28.7 million **27.** 2026

29. $a(t) = 20 e^{(t/36) \ln 0.5}$ mg with $t = 0$ at midnight; 15.87 mg; 119.6 hr $\approx$ 5 days **31.** 1.798 million; the downward turn in the population size may be temporary. **33.** 18,928 ft; 125,754 ft

35. 1.055 billion yr **37.** 6.2 hours **39.** 2 dollars **41.** 1044 days

43. a. False **b.** False **c.** True **d.** True **e.** True

45. a. $V_1(t) = 0.495 e^{-0.1216t}$ **b.** $V_2(t) = 0.005 e^{0.239t}$

c. $V(t) = 0.495 e^{-0.1216t} + 0.005 e^{0.239t}$

d.

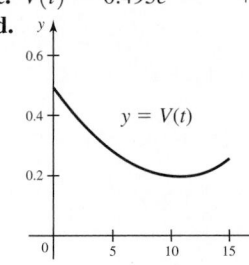

The tumor initially shrinks significantly in size but eventually starts growing again. **e.** 10.9 days; give a second treatment just before the end of the 10th day after the first treatment.

47. a. Bob; Abe **b.** $y = 4 \ln(t + 1)$ and $y = 8 - 8e^{-t/2}$; Bob

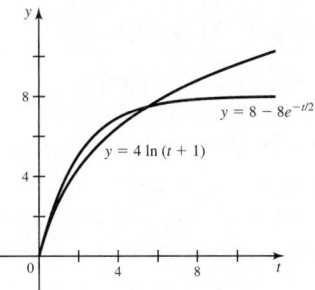

49. 10.034%; no **51.** 1.3 s

53. $k = \ln(1 + r)$; $r = 2^{1/T_2} - 1$; $T_2 = (\ln 2)/k$

Section 7.3 Exercises, pp. 513–517

1. $\cosh x = \dfrac{e^x + e^{-x}}{2}$; $\sinh x = \dfrac{e^x - e^{-x}}{2}$ **3.** $\cosh^2 x - \sinh^2 x = 1$

5. $\sinh^{-1} x = \ln(x + \sqrt{x^2 + 1})$ **7.** Evaluate $\sinh^{-1}\frac{1}{5}$.

9. $\displaystyle\int \frac{dx}{16 - x^2} = \frac{1}{4} \coth^{-1}\frac{x}{4} + C$ when $|x| > 4$; the values in the interval of integration $6 \le x \le 8$ satisfy $|x| > 4$.

23. $2 \cosh x \sinh x$ **25.** $2 \tanh x \operatorname{sech}^2 x$ **27.** $-2 \tanh 2x$

29. $2x(3x \sinh 3x + \cosh 3x)\cosh 3x$ **31.** $4/\sqrt{16x^2 - 1}$

33. $2v/\sqrt{v^4 + 1}$ **35.** $\sinh^{-1} x$ **37.** $(\sinh 2x)/2 + C$

39. $\ln(1 + \cosh x) + C$ **41.** $x - \tanh x + C$

43. $(\cosh^4 3 - 1)/12 \approx 856$ **45.** $\ln(5/4)$

47. $\dfrac{1}{2\sqrt{2}} \coth^{-1}\left(\dfrac{x}{2\sqrt{2}}\right) + C$ **49.** $\tanh^{-1}(e^x/6)/6 + C$

51. $-\operatorname{sech}^{-1}(x^4/2)/8 + C$ **53.** $-\operatorname{csch} z + C$

55. $\ln \sqrt{3} \cdot \ln(4/3) \approx 0.158$ **57.** $\dfrac{x^2 + 1}{2x} + C$

59. a. The values of $y = \coth x$ are close to 1 on $[5, 10]$.

b. $\ln(\sinh 10) - \ln(\sinh 5) \approx 5.0000454$; $|\text{error}| \approx 0.0000454$

61.

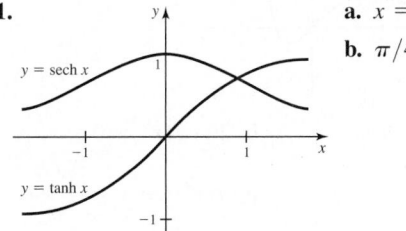

a. $x = \sinh^{-1} 1 = \ln(1 + \sqrt{2})$

b. $\pi/4 - \ln \sqrt{2} \approx 0.44$

63. $\sinh^{-1} 2 = \ln(2 + \sqrt{5})$ **65.** $-(\ln 5)/3 \approx -0.54$

67. $3 \ln\left(\dfrac{\sqrt{5} + 2}{\sqrt{2} + 1}\right) = 3(\sinh^{-1} 2 - \sinh^{-1} 1)$

69. $\dfrac{1}{15}\left(17 - \dfrac{8}{\ln(5/3)}\right) \approx 0.09$

71. a. Sag $= f(50) - f(0) = a(\cosh(50/a) - 1) = 10$; now divide by a. **b.** $t \approx 0.08$ **c.** $a = 10/t \approx 125$; $L = 250 \sinh(2/5) \approx 102.7$ ft **73.** $\lambda \approx 32.81$ m

75. b. When $d/\lambda < 0.05$, $2\pi d/\lambda$ is small. Because $\tanh x \approx x$ for small values of x, $\tanh(2\pi d/\lambda) \approx 2\pi d/\lambda$; therefore,

$$v = \sqrt{\frac{g\lambda}{2\pi} \tanh\left(\frac{2\pi d}{\lambda}\right)} \approx \sqrt{\frac{g\lambda}{2\pi} \cdot \frac{2\pi d}{\lambda}} = \sqrt{gd}.$$

c. $v = \sqrt{gd}$ is a function of depth alone; when depth d decreases, v also decreases. **77. a.** False **b.** False **c.** True **d.** False

79. a. 1 **b.** 0 **c.** Undefined **d.** 1 **e.** 13/12 **f.** 40/9

g. $\left(\dfrac{e^2 + 1}{2e}\right)^2$ **h.** Undefined **i.** ln 4 **j.** 1 **81.** $x = 0$

83. $x = \pm\tanh^{-1}(1/\sqrt{3}) = \pm\ln(2 + \sqrt{3})/2 \approx \pm0.658$
85. $\tan^{-1}(\sinh 1) - \pi/4 \approx 0.08$ **87.** Applying l'Hôpital's Rule twice brings you back to the initial limit; $\lim_{x\to\infty} \tanh x = 1$.

89. $2/\pi$ **91.** 1 **93.** $12(3\ln(3 + \sqrt{8}) - \sqrt{8}) \approx 29.5$
95. a. Approx. 360.8 m **b.** First 100 m: $t \approx 4.72$ s, $v_{av} \approx 21.2$ m/s; second 100 m: $t \approx 2.25$ s, $v_{av} \approx 44.5$ m/s **97. a.** $\sqrt{mg/k}$

b. $35\sqrt{3} \approx 60.6$ m/s **c.** $t = \sqrt{\dfrac{m}{kg}}\tanh^{-1} 0.95 = \dfrac{\ln 39}{2}\sqrt{\dfrac{m}{kg}}$

d. Approx. 736.5 m **109.** $\ln(21/4) \approx 1.66$

Chapter 7 Review Exercises, pp. 518–519

1. a. False **b.** False **c.** False **d.** True **3.** ln 4
5. $\frac{1}{2}\ln(x^2 + 8x + 25) + C$
7. $\cosh^{-1}(x/3) + C = \ln(x + \sqrt{x^2 - 9}) + C$
9. $\tanh^{-1}(1/3)/9 = (\ln 2)/18 \approx 0.0385$

11. $x^{3x^2+1}\left(6x\ln x + 3x + \dfrac{1}{x}\right)$ **13.** $\sinh^2 t + \cosh^2 t$

15. $3\sinh(6x - 2)$ **17.** $-\csc x$ **19.** $\dfrac{2x}{\sqrt{x^4 - 1}}$

21. Approx. 7.3 hours **23. a.** $y(t) = 29{,}000e^{(t\ln 2)/2}$
b. Approx. 41,996,486 transistors (which closely approximates the actual number of transistors) **25.** 48.37 yr
27. Local max at $x = -\frac{1}{2}(\sqrt{5} + 1)$; local min at $x = \frac{1}{2}(\sqrt{5} - 1)$; inflection points at $x = -3$ and $x = 0$; $\lim_{x\to-\infty} f(x) = 0$;

$\lim_{x\to\infty} f(x) = \infty$

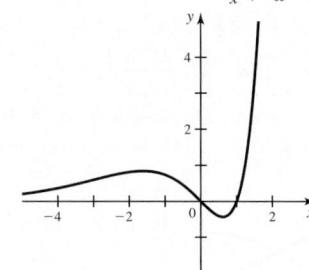

29. a.

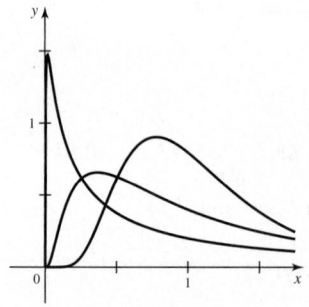

b. $\lim_{x\to 0} f(x) = 0$

d. $f(x^*) = \dfrac{1}{\sqrt{2\pi}}\dfrac{e^{\sigma^2/2}}{\sigma}$

e. $\sigma = 1$

31. $L(x) = \frac{5}{3} + \frac{4}{3}(x - \ln 3)$; $\cosh 1 \approx 1.535$
33. a. $\cosh x$ **b.** $(1 - x\tanh x)\operatorname{sech} x$

CHAPTER 8

Section 8.1 Exercises, pp. 523–525

1. $u = 4 - 7x$ **3.** $\sin^2 x = \dfrac{1 - \cos 2x}{2}$ **5.** Complete the square in

$x^2 - 4x - 9$. **7.** $\dfrac{1}{15(3 - 5x)^3} + C$ **9.** $\dfrac{\sqrt{2}}{4}$ **11.** $\frac{1}{2}\ln^2 2x + C$

13. $\ln(e^x + 1) + C$ **15.** $\dfrac{32}{3}$ **17.** $\dfrac{21}{110}$

19. $\dfrac{(\ln w - 1)^9}{9} + \dfrac{(\ln w - 1)^8}{8} + C$

21. $\dfrac{1}{2}\ln(x^2 + 4) + \tan^{-1}\dfrac{x}{2} + C$

23. $-\dfrac{1}{3}\ln|\csc(3e^x + 4) + \cot(3e^x + 4)| + C$ **25.** 1

27. $3\sqrt{1 - x^2} + 2\sin^{-1} x + C$ **29.** $\ln(\sqrt{2} + 1)$

31. $\dfrac{1}{3}\tan^{-1}\left(\dfrac{x - 1}{3}\right) + C$ **33.** $\dfrac{x^2}{2} + x + \ln(x^2 + x + 2) + C$

35. $\dfrac{3\pi + 10}{12}$ **37.** $\sin^{-1}\left(\dfrac{\theta + 3}{6}\right) + C$ **39.** $\tan\theta - \sec\theta + C$

41. $-x - \cot x - \csc x + C$ **43.** $\dfrac{1}{3}\ln(1 + \sinh 3x) + C$

45. $\frac{1}{2}\ln|e^{2x} - 2| + C$ **47.** $x - \ln|x + 1| + C$

49. $\dfrac{4}{5}(9 + \sqrt{t + 1})^{3/2}(\sqrt{t + 1} - 6) + C$ **51.** $\dfrac{\ln 4 - \pi}{4}$

53. $\ln|\sec(e^x + 1) + \tan(e^x + 1)| + C$

55. $\dfrac{2\sin^3 x}{3} + C$ **57.** $2\tan^{-1}\sqrt{x} + C$

59. $\dfrac{1}{2}\ln(x^2 + 6x + 13) - \dfrac{5}{2}\tan^{-1}\left(\dfrac{x + 3}{2}\right) + C$

61. $-\dfrac{1}{e^x + 1} + C$ **63.** $\frac{1}{2}$ **65. a.** False **b.** False **c.** False

d. False **69. a.** $\dfrac{\tan^2 x}{2} + C$ **b.** $\dfrac{\sec^2 x}{2} + C$ **c.** The antiderivatives

differ by a constant. **71. a.** $\frac{1}{2}(x + 1)^2 - 2(x + 1) + \ln|x + 1| + C$

b. $\dfrac{x^2}{2} - x + \ln|x + 1| + C$ **c.** The antiderivatives differ by a

constant. **73.** $\dfrac{\ln 26}{3}$ **75.** $\dfrac{2}{3}(5\sqrt{5} - 1)\pi$

77. $\pi\left(\dfrac{9}{2} - \dfrac{5\sqrt{5}}{6}\right)$ **79.** $\dfrac{2048 + 1763\sqrt{41}}{9375}$

Section 8.2 Exercises, pp. 529–532

1. Product Rule **3.** $\dfrac{x^2(2\ln x - 1)}{4} + C$ **5.** Products for which the choice for dv is easily integrated and when the resulting new integral is no more difficult than the original integral
7. $(\tan x + 2)\ln(\tan x + 2) - \tan x + C$

9. $\dfrac{1}{5}x\sin 5x + \dfrac{1}{25}\cos 5x + C$ **11.** $\dfrac{e^{6t}}{36}(6t - 1) + C$

13. $\dfrac{x^2}{4}(2\ln 10x - 1) + C$ **15.** $(w + 2)\sin 2w + \dfrac{1}{2}\cos 2w + C$

17. $\dfrac{3^x}{\ln 3}\left(x - \dfrac{1}{\ln 3}\right) + C$ **19.** $-\dfrac{1}{9x^9}\left(\ln x + \dfrac{1}{9}\right) + C$

21. $\dfrac{1}{8}\sin 2x - \dfrac{x}{4}\cos 2x + C$ **23.** $\dfrac{1}{4}(1 - 2x^2)\cos 2x + \dfrac{x}{2}\sin 2x + C$

25. $-e^{-t}(t^2 + 2t + 2) + C$ **27.** $\dfrac{e^x}{2}(\sin x + \cos x) + C$

29. $-\dfrac{e^{-x}}{17}(\sin 4x + 4\cos 4x) + C$

31. $-e^{2x}\cos e^x + 2e^x\sin e^x + 2\cos e^x + C$ **33.** π **35.** $-\frac{1}{2}$

37. $\dfrac{1}{9}(5e^6 + 1)$ **39.** $\dfrac{\pi - 2}{2}$ **41. a.** $x\tan^{-1} x - \dfrac{1}{2}\ln(1 + x^2) + C$

b. $\frac{1}{2}x^2 \tan^{-1} x^2 - \frac{1}{4} \ln (1 + x^4) + C$ **43.** $\pi(1 - \ln 2)$ **45.** π

47. $\frac{2\pi}{27} (13e^6 - 1)$ **49. a.** False   **b.** True   **c.** True

51. Let $u = x^n$ and $dv = \cos ax \, dx$.   **53.** Let $u = \ln^n x$ and $dv = dx$.

55. $\frac{x^2 \sin 5x}{5} + \frac{2x \cos 5x}{25} - \frac{2 \sin 5x}{125} + C$

57. $6 - 2e$ **59. a.** $\frac{2}{3}(x - 2)\sqrt{x + 1} + C$

61. $\int \log_b x \, dx = \int \frac{\ln x}{\ln b} \, dx = \frac{1}{\ln b}(x \ln x - x) + C$

63. $2\sqrt{x} \sin \sqrt{x} + 2 \cos \sqrt{x} + C$

65. Let $u = x$ and $dv = f''(x) \, dx$.

67. $2e^3$ **69.** x-axis: $\pi^2/2$; y-axis: $2\pi^2$ **71.** $\pi(\pi - 2)$

75. a. $t = k\pi$, for $k = 0, 1, 2, \ldots$ **b.** $\dfrac{e^{-\pi} + 1}{2\pi}$

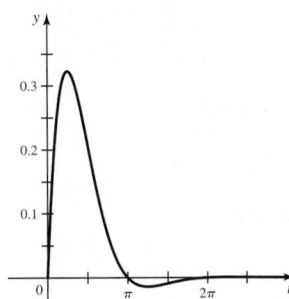

c. $(-1)^n \left(\dfrac{e^{\pi} + 1}{2\pi e^{(n+1)\pi}} \right)$

d. $a_n = a_{n-1} \cdot \dfrac{1}{e^{\pi}}$

77. c. $\int f(x)g(x)dx = f(x)G_1(x) - f'(x)G_2(x) + f''(x)G_3(x) - \int f'''(x)G_3(x)dx$

f and its derivatives	g and its integrals
$f(x)$ +	$g(x)$
$f'(x)$ −	$G_1(x)$
$f''(x)$ +	$G_2(x)$
$f'''(x)$ −	$G_3(x)$

d. $\int x^2 e^{x/2} dx = 2x^2 e^{x/2} - 8xe^{x/2} + 16e^{x/2} + C$

f and its derivatives	g and its integrals
x^2 +	$e^{x/2}$
$2x$ −	$2e^{x/2}$
2 +	$4e^{x/2}$
0 −	$8e^{x/2}$

$\dfrac{d^n}{dx^n}(x^2) = 0$, for $n \geq 3$, so all entries in the left column of the table beyond row three are 0, which results in no additional contribution to the antiderivative.   **e.** $x^3 \sin x + 3x^2 \cos x - 6x \sin x - 6 \cos x + C$; five rows are needed because $\dfrac{d^n}{dx^n}(x^3) = 0$, for $n \geq 4$.

f. $\dfrac{d^k}{dx^k}(p_n(x)) = 0$, for $k \geq n + 1$

79. a. $\int e^x \cos x \, dx = e^x \sin x + e^x \cos x - \int e^x \cos x \, dx$

b. $\frac{1}{2}(e^x \sin x + e^x \cos x) + C$ **c.** $-\frac{3}{13}e^{-2x} \cos 3x - \frac{2}{13}e^{-2x} \sin 3x + C$

81. a. $I_1 = -\frac{1}{2}e^{-x^2} + C$ **b.** $I_3 = -\frac{1}{2}e^{-x^2}(x^2 + 1) + C$

c. $I_5 = -\frac{1}{2}e^{-x^2}(x^4 + 2x^2 + 2) + C$

Section 8.3 Exercises, pp. 536–538

1. $\sin^2 x = \frac{1}{2}(1 - \cos 2x); \cos^2 x = \frac{1}{2}(1 + \cos 2x)$ **3.** Rewrite $\sin^3 x$ as $(1 - \cos^2 x) \sin x$.   **5.** A reduction formula expresses an integral with a power in the integrand in terms of another integral with a smaller power in the integrand.   **7.** Let $u = \tan x$.

9. $\sin x - \frac{1}{3}\sin^3 x + C$ **11.** $\frac{x}{2} - \frac{1}{12}\sin 6x + C$

13. $-\cos x + \frac{2}{3}\cos^3 x - \frac{1}{5}\cos^5 x + C$ **15.** $\frac{1}{5}\cos^5 x - \frac{1}{3}\cos^3 x + C$

17. $\frac{2}{3}\sin^{3/2} x - \frac{2}{7}\sin^{7/2} x + C$ **19.** $\frac{7}{24}$ **21.** $\frac{8}{45}$

23. $\frac{1}{8}x - \frac{1}{32}\sin 4x + C$ **25.** $\frac{1}{48}\sin^3 2x + \frac{1}{16}x - \frac{1}{64}\sin 4x + C$

27. $\tan x - x + C$ **29.** $-\frac{1}{3}\cot^3 x + \cot x + x + C$

31. $4 \tan^5 x - \frac{20}{3}\tan^3 x + 20 \tan x - 20x + C$ **33.** $\tan^{10} x + C$

35. $\frac{1}{3}\sec^3 x + C$ **37.** $\frac{1}{3}\tan^3 (\ln \theta) + \tan (\ln \theta) + C$ **39.** $\ln 4$

41. $\frac{7}{6}$ **43.** $\frac{1}{8}\tan^2 4x + \frac{1}{4}\ln |\cos 4x| + C$ **45.** $\frac{2}{3}\tan^{3/2} x + C$

47. $\tan x - \cot x + C$ **49.** $\frac{1}{25}$ **51.** $-2 \cot x - \frac{\cot^3 x}{3} + C$

53. $\frac{4}{3}$ **55.** $\frac{4}{3} - \ln \sqrt{3}$ **57.** $8\sqrt{2}/3$ **59.** $\sqrt{2}$ **61.** $2\sqrt{2}/3$

63. a. True   **b.** False   **65.** $\frac{2\pi}{35}$ **67.** $\frac{1}{8}\cos 4x - \frac{1}{20}\cos 10x + C$

69. $\frac{1}{2}\sin x - \frac{1}{10}\sin 5x + C$ **73.** $\frac{1}{2} - \ln \sqrt{2}$ **75. a.** $\frac{\pi}{2}; \frac{\pi}{2}$

b. $\frac{\pi}{2}$, for all n **d.** Yes   **e.** $\frac{3\pi}{8}$, for all n

Section 8.4 Exercises, pp. 543–546

1. $x = 3 \sec \theta$ **3.** $x = 10 \sin \theta$ **5.** $\sqrt{4 - x^2}/x$ **7.** $\pi/6$

9. $\frac{25\pi}{3}$ **11.** $\frac{\pi}{12}$ **13.** $\sin^{-1}\frac{x}{4} + C$ **15.** $-\dfrac{\sqrt{x^2 + 9}}{9x} + C$

17. $2 - \frac{\pi}{2}$ **19.** $\ln (\sqrt{x^2 - 81} + x) + C$

21. $\frac{x}{2}\sqrt{64 - x^2} + 32 \sin^{-1}\frac{x}{8} + C$ **23.** $\dfrac{x}{25\sqrt{25 - x^2}} + C$

25. $-3 \ln \left| \dfrac{\sqrt{9 - x^2} + 3}{x} \right| + \sqrt{9 - x^2} + C$ **27.** $\sqrt{2}/6$

29. $\frac{1}{16}\left(\tan^{-1}\frac{x}{2} + \frac{2x}{x^2 + 4} \right) + C$

31. $8 \sin^{-1}(x/4) - x\sqrt{16 - x^2}/2 + C$

33. $\sqrt{x^2 - 9} - 3 \sec^{-1}(x/3) + C$

35. $-1/\sqrt{x^2 - 1} - \sec^{-1} x + C$ **37.** $2 - \sqrt{2}$

39. $x/\sqrt{100 - x^2} - \sin^{-1}(x/10) + C$ **41.** $x/\sqrt{1 + 4x^2} + C$

43. $\frac{\ln 3}{2}$ **45.** $81/(2(81 - x^2)) + \ln \sqrt{81 - x^2} + C$

47. $\frac{1}{16}(1 - \sqrt{3} - \ln (21 - 12\sqrt{3}))$ **49.** $\frac{1}{3} + \frac{\ln 3}{4}$

51. $\frac{x}{2}\sqrt{4 + x^2} - 2 \ln (x + \sqrt{4 + x^2}) + C$

53. $\frac{9}{10}\cos^{-1}\frac{5}{3x} - \dfrac{\sqrt{9x^2 - 25}}{2x^2} + C$

55. $\dfrac{\sec^{-1}\frac{x}{10}}{2000} + \dfrac{\sqrt{x^2 - 100}}{200 x^2} + C$

57. a. False   **b.** True   **c.** False   **d.** False

61. $\sin^{-1}\left(\dfrac{x + 1}{2} \right) + C$ **63.** $\frac{1}{3}\tan^{-1}\left(\dfrac{x + 3}{3} \right) + C$

65. $\dfrac{\pi\sqrt{2}}{48}$ **67.** $\dfrac{x - 4}{\sqrt{9 + 8x - x^2}} - \sin^{-1}\left(\dfrac{x - 4}{5} \right) + C$

69. $\ln((2 + \sqrt{3})(\sqrt{2} - 1))$

71. $\dfrac{1}{81} + \dfrac{\ln 3}{108}$

73. $3\sqrt{3} - \pi$

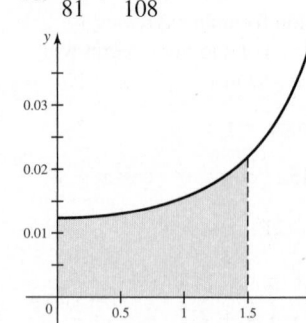

75. $\dfrac{3}{80}$ **77.** $\dfrac{1}{4a}\left(20a\sqrt{1 + 400a^2} + \ln(20a + \sqrt{1 + 400a^2})\right)$

81. b. $\displaystyle\lim_{L\to\infty} \dfrac{kQ}{a\sqrt{a^2 + L^2}} = \lim_{L\to\infty} 2\rho k \dfrac{1}{a\sqrt{\left(\dfrac{a}{L}\right)^2 + 1}} = \dfrac{2\rho k}{a}$

85. a. $\dfrac{1}{\sqrt{g}}\left(\dfrac{\pi}{2} - \sin^{-1}\left(\dfrac{2\cos b - \cos a + 1}{\cos a + 1}\right)\right)$

b. For $b = \pi$, the descent time is $\dfrac{\pi}{\sqrt{g}}$, a constant.

Section 8.5 Exercises, pp. 554–556

1. Rational functions   **3. a.** $\dfrac{A}{x - 3}$ **b.** $\dfrac{A}{x - 4}, \dfrac{B}{(x - 4)^2}, \dfrac{C}{(x - 4)^3}$

c. $\dfrac{Ax + B}{x^2 + 2x + 6}$ **5.** $\dfrac{A}{x - 4} + \dfrac{B}{x - 5}$ **7.** $\dfrac{A}{x - 5} + \dfrac{B}{(x - 5)^2}$

9. $\dfrac{A}{x} + \dfrac{B}{x + 1} + \dfrac{C}{x - 1} + \dfrac{D}{x + 2} + \dfrac{E}{x - 2}$ **11.** $\dfrac{A}{x} + \dfrac{Bx + C}{x^2 + 1}$

13. $\dfrac{A}{x - 2} + \dfrac{B}{(x - 2)^2} + \dfrac{Cx + D}{x^2 + x + 2} + \dfrac{Ex + F}{(x^2 + x + 2)^2}$

15. $\dfrac{A}{x - 2} + \dfrac{B}{(x - 2)^2} + \dfrac{C}{x + 2} + \dfrac{D}{(x + 2)^2} + \dfrac{Ex + F}{x^2 + 4} + \dfrac{Gx + H}{(x^2 + 4)^2}$

17. $\dfrac{2}{x - 1} + \dfrac{3}{x - 2}$ **19.** $\dfrac{1}{x - 4} - \dfrac{1}{x + 2}$ **21.** $2 + \dfrac{3}{x + 1} - \dfrac{4}{x + 2}$

23. $\ln\left|\dfrac{x - 1}{x + 2}\right| + C$ **25.** $3\ln\left|\dfrac{x - 1}{x + 1}\right| + C$

27. $3\ln|x - 1| - \dfrac{1}{3}\ln|3x - 2| + C$ **29.** $-\ln 4$

31. $\ln\left|\dfrac{(x - 2)^2(x + 1)}{(x + 2)^2(x - 1)}\right| + C$ **33.** $3x + \ln\dfrac{(x - 2)^{14}}{|x - 1|} + C$

35. $\ln\left|\dfrac{x(x - 2)^3}{(x + 2)^3}\right| + C$ **37.** $\ln\left|\dfrac{(x - 3)^{1/3}(x + 1)}{(x + 3)^{1/3}(x - 1)}\right|^{1/16} + C$

39. $\dfrac{9}{x} + \ln\left|\dfrac{x - 9}{x}\right| + C$ **41.** $\ln 2 - \dfrac{3}{4}$ **43.** $-\dfrac{2}{x} + \ln\left|\dfrac{x + 1}{x}\right|^2 + C$

45. $\dfrac{5}{x} + \ln\left|\dfrac{x}{x + 1}\right|^6 + C$ **47.** $\dfrac{x^2}{2} + 2\ln|x - 5| - \dfrac{10}{x - 5} + C$

49. $\dfrac{3}{x - 1} + \ln\left|\dfrac{(x - 1)^5}{x^4}\right| + C$ **51.** $\ln|x + 1| + \tan^{-1}x + C$

53. $\ln(x + 1)^2 + \tan^{-1}(x + 1) + C$

55. $\ln\left|\dfrac{(x - 1)^2}{x^2 + 4x + 5}\right| + 14\tan^{-1}(x + 2) + C$

57. $\dfrac{1}{2}\ln|x^2 + 3| - \dfrac{1}{x^2 + 3} + C$

59. $\dfrac{1}{2}\ln(x^2 + 6x + 10) - 3\tan^{-1}(x + 3) - \dfrac{1}{x^2 + 6x + 10} + C$

61. $\ln\left(\dfrac{x^2}{x^2 + 1}\right) + \dfrac{1}{x^2 + 1} + C$

63. $\sqrt{\dfrac{3}{7}}\tan^{-1}\left(\sqrt{\dfrac{3}{7}}x\right) - \dfrac{1}{6(3x^2 + 7)} + C$

65. a. False   **b.** False   **c.** False   **d.** True   **67.** $\ln 6$

69. $\left(\dfrac{24}{5} - 2\ln 5\right)\pi$ **71.** $\dfrac{2}{3}\pi\ln 2$ **73.** $\ln\sqrt{\left|\dfrac{x - 1}{x + 1}\right|} + C$

75. $\dfrac{A}{x} + \dfrac{Bx + C}{x^2 + 1} + \dfrac{Dx + E}{(x^2 + 1)^2} + \dfrac{Fx + G}{x^2 + x + 4} + \dfrac{Hx + I}{(x^2 + x + 4)^2}$;

$\dfrac{1}{16x} - \dfrac{x + 10}{100(x^2 + 1)} + \dfrac{4x + 3}{50(x^2 + 1)^2} - \dfrac{21x - 19}{400(x^2 + x + 4)} -$

$\dfrac{4x + 1}{20(x^2 + x + 4)^2}$ **77.** $\ln\left|\dfrac{e^x - 1}{e^x + 2}\right|^{1/3} + C$

79. $\dfrac{1}{4}\ln\left(\dfrac{1 + \sin t}{1 - \sin t} - \dfrac{2}{1 + \sin t}\right) + C$

81. $\tan^{-1}e^x - \dfrac{1}{2(e^{2x} + 1)} + C$ **83.** $x - \ln(1 + e^x) + C$

89. $-\cot x - \csc x + C = -\cot(x/2) + C$

91. $\dfrac{1}{\sqrt{2}}\ln\dfrac{\sqrt{2} + 1}{\sqrt{2} - 1}$ **93. a.** Car A **b.** Car C

c. $S_A(t) = 88t - 88\ln|t + 1|$;

$S_B(t) = 88\left(t - \ln(t + 1)^2 - \dfrac{1}{t + 1} + 1\right)$;

$S_C(t) = 88(t - \tan^{-1}t)$

d. Car C **95.** Because $\dfrac{x^4(1 - x)^4}{1 + x^2} > 0$ on $(0, 1)$,

$\displaystyle\int_0^1 \dfrac{x^4(1 - x^4)}{1 + x^2}\,dx > 0$; therefore, $\dfrac{22}{7} > \pi$.

Section 8.6 Exercises, pp. 560–562

1. Integrate by parts.   **3.** Let $x = 8\sin\theta$.   **5.** Use the method of partial fractions.   **7.** $\dfrac{\pi}{4}$ **9.** $\dfrac{\pi}{6}$ **11.** $\dfrac{5}{4} - \dfrac{3\pi}{8}$ **13.** $-\dfrac{\sqrt{1 - e^{2x}}}{e^x} + C$

15. $\dfrac{4}{\ln 2}$ **17.** $\dfrac{3e^4}{2}$ **19.** $\dfrac{16}{35}$ **21.** $\dfrac{x^{10}}{10}\ln 3x - \dfrac{x^{10}}{100} + C$

23. $\ln|\cos x + 1| - \ln|\cos x| + C$

25. $\ln\left|\dfrac{x}{1 + \sqrt{1 - x^2}}\right| + C$ **27.** $\dfrac{3x}{8} - \dfrac{1}{2}\sin x + \dfrac{1}{16}\sin 2x + C$

29. $\ln|\sin x + \sin^2 x| + C$ **31.** $6\sin^{-1}\dfrac{x}{2} + \dfrac{3}{2}x\sqrt{4 - x^2} + C$

33. $\dfrac{1}{a}\tan^{-1}\dfrac{e^x}{a} + C$ **35.** $\dfrac{11}{6}$ **37.** $\dfrac{\sqrt{3} + 1}{2}$

39. $-\cos x\ln(\sin x) - \ln|\csc x + \cot x| + \cos x + C$

41. $-\dfrac{2}{5}\cot^{5/2}x - \dfrac{2}{9}\cot^{9/2}x + C$ **43.** $\dfrac{\sin^{-1}x^{10}}{10} + C$ **45.** $\ln\dfrac{4}{3} - \dfrac{1}{6}$

47. $x^2 + 3x + 4\ln|x - 2| + \ln|x + 1| + C$

49. $\dfrac{\sec^{11}x}{11} - \dfrac{\sec^9 x}{9} + C$ **51.** $\dfrac{4}{7}(2^{7/4} - 1)$

53. $-\dfrac{\cot^2 e^x}{2} - \ln|\sin e^x| + C$ **55.** $\ln|x^3 + x| + 3\tan^{-1}x + C$

57. $-2\sqrt{x}\cos\sqrt{x} + 2\sin\sqrt{x} + C$ **59.** $-\dfrac{1}{x} - \tan^{-1}x + C$

61. $\dfrac{\sqrt{2}e^{\pi/4} - 1}{2}$ 63. $\dfrac{x^{a+1}}{a+1}\left(\ln x - \dfrac{1}{a+1}\right) + C$

65. $\dfrac{\pi}{18}$ 67. $\dfrac{1}{54}(\sin^{-1}3x - 3x\sqrt{1-9x^2}) + C$

69. $\dfrac{1-\sqrt{1-x^2}}{x} + C$ 71. $-2\cot x + 2\csc x - x + C$

73. $\dfrac{40\sqrt{5}}{3} - \dfrac{224}{15}$ 75. $\dfrac{7\pi^2}{144}$ 77. $x\cos^{-1}x - \sqrt{1-x^2} + C$

79. $-\dfrac{\sin^{-1}x}{x} + \ln\left|\dfrac{x}{1+\sqrt{1-x^2}}\right| + C$

81. $\ln|x| + 2\tan^{-1}x - \dfrac{3}{2(x^2+1)} + C$

83. $\dfrac{\sin^{999}e^x}{999} - \dfrac{\sin^{1001}e^x}{1001} + C$ 85. a. True b. True c. False

d. False 87. $\pi(\sqrt{2} + \ln(1+\sqrt{2})) \approx 7.212$

89. $\dfrac{\pi(4\sqrt{2}+3)}{3} \approx 9.065$ 91. $9800\pi\ln 2 \approx 21{,}340.3$ J

93. $4x - 2\ln(e^{2x} + 2e^x + 17) - \tan^{-1}\left(\dfrac{e^x+1}{4}\right) + C$

95. $\dfrac{1}{4}\ln|\tan x + 1| - \dfrac{1}{4}\ln|\tan x - 1| + \dfrac{x}{2} + C$

97. $x\tan^{-1}\sqrt[3]{x} - \dfrac{x^{2/3}}{2} + \dfrac{1}{2}\ln(1+x^{2/3}) + C$

99. $\pi\left(\sqrt{5} - \sqrt{2} + \dfrac{1}{2}\ln\left(\dfrac{\sqrt{5}-1}{\sqrt{5}+1}\right) - \dfrac{1}{2}\ln\left(\dfrac{\sqrt{2}-1}{\sqrt{2}+1}\right)\right) \approx 3.839$

Section 8.7 Exercises, pp. 565–567

1. Substitutions, integration by parts, partial fractions 3. The CAS may not include the constant of integration, and it may use a trigonometric identity or other algebraic simplification.

5. $-\dfrac{1}{3}\sin^3 e^x + \sin e^x + C$ 7. $x\cos^{-1}x - \sqrt{1-x^2} + C$

9. $\ln(x + \sqrt{16+x^2}) + C$ 11. $\frac{3}{4}(2u - 7\ln|7+2u|) + C$

13. $-\dfrac{1}{4}\cot 2x + C$ 15. $\frac{1}{12}(2x-1)\sqrt{4x+1} + C$

17. $\frac{1}{3}\ln\left|x + \sqrt{x^2 - \left(\frac{10}{3}\right)^2}\right| + C$ 19. $\ln(e^x + \sqrt{4+e^{2x}}) + C$

21. $-\dfrac{1}{2}\ln\left|\dfrac{2+\sin x}{\sin x}\right| + C$

23. $\dfrac{2\ln^2 x - 1}{4}\sin^{-1}(\ln x) + \dfrac{\ln x\sqrt{1-\ln^2 x}}{4} + C$

25. $\dfrac{x}{16\sqrt{16+9x^2}} + C$ 27. $-\dfrac{1}{12}\ln\left|\dfrac{12 + \sqrt{144-x^2}}{x}\right| + C$

29. $2x + x\ln^2 x - 2x\ln x + C$

31. $\dfrac{x+5}{2}\sqrt{x^2+10x} - \dfrac{25}{2}\ln|x+5+\sqrt{x^2+10x}| + C$

33. $\dfrac{1}{3}\tan^{-1}\left(\dfrac{x+1}{3}\right) + C$ 35. $\ln x - \dfrac{1}{10}\ln(x^{10}+1) + C$

37. $2\ln(\sqrt{x-6} + \sqrt{x}) + C$

39. $-\dfrac{\tan^{-1}x^3}{3x^3} + \ln\left|\dfrac{x}{(x^6+1)^{1/6}}\right| + C$ 41. $4\sqrt{17} + \ln(4+\sqrt{17})$

43. $\sqrt{5} - \sqrt{2} + \ln\left(\dfrac{2+2\sqrt{2}}{1+\sqrt{5}}\right)$ 45. $\dfrac{128\pi}{3}$ 47. $\dfrac{\pi^2}{4}$

49. $\dfrac{(x-3)\sqrt{3+2x}}{3} + C$ 51. $\dfrac{1}{3}\tan 3x - x + C$

53. $\dfrac{1540 + 243\ln 3}{8}$

55. $\dfrac{(x^2-a^2)^{3/2}}{3} - a^2\sqrt{x^2-a^2} + a^3\cos^{-1}\dfrac{a}{x} + C$ 57. $\dfrac{\pi}{4}$

59. $-\dfrac{x}{8}(2x^2 - 5a^2)\sqrt{a^2-x^2} + \dfrac{3a^4}{8}\sin^{-1}\dfrac{x}{a} + C$

61. $2 - \dfrac{\pi^2}{12} - \ln 4$ 63. $\dfrac{27{,}456\sqrt{15}}{7} \approx 15{,}190.9$

65. $\dfrac{1}{8}e^{2x}(4x^3 - 6x^2 + 6x - 3) + C$ 67. $\dfrac{\tan^3 3y}{9} - \dfrac{\tan 3y}{3} + y + C$

69. $\dfrac{1}{24}(128 - 78\sqrt{2} - 3\ln(3+2\sqrt{2}))$

71. $\dfrac{1}{a^2}(ax - b\ln|b+ax|) + C$

73. $\dfrac{1}{a^2}\left(\dfrac{(ax+b)^{n+2}}{n+2} - \dfrac{b(ax+b)^{n+1}}{n+1}\right) + C$

75. a. True b. True

79. $\dfrac{1}{16}((8x^2-1)\sin^{-1}2x + 2x\sqrt{1-4x^2}) + C$

81. $-\dfrac{\tan^{-1}x}{x} + \ln\left(\dfrac{|x|}{\sqrt{x^2+1}}\right) + C$ 83. b. $\dfrac{\pi}{8}\ln 2$

85. a.

θ_0	T
0.10	6.27927
0.20	6.26762
0.30	6.24854
0.40	6.22253
0.50	6.19021
0.60	6.15236
0.70	6.10979
0.80	6.06338
0.90	6.01399
1.00	5.96247

b. All are within 10%.

87. b. $\dfrac{63\pi}{512}$ c. Decrease

Section 8.8 Exercises, pp. 578–582

1. $\frac{1}{2}$ 3. The Trapezoid Rule approximates areas under curves using trapezoids. 5. 42 7. $\dfrac{112}{3}$ 9. $-1, 1, 3, 5, 7, 9$

11. $1.59 \times 10^{-3}; 5.04 \times 10^{-4}$ 13. $1.72 \times 10^{-3}; 6.32 \times 10^{-4}$

15. 576; 640; 656 17. 0.643950551 19. 704; 672; 664

21. 0.622 23. 2.28476811; 2.33512377 25. 1.76798499

27. $M(25) \approx 0.63703884$, $T(25) \approx 0.63578179$; 6.58×10^{-4}, 1.32×10^{-3}

29.

n	$M(n)$	$T(n)$	Error in $M(n)$	Error in $T(n)$
4	99	102	1.00	2.00
8	99.75	100.5	0.250	0.500
16	99.9375	100.125	6.3×10^{-2}	0.125
32	99.984375	100.03125	1.6×10^{-2}	3.1×10^{-2}

31.

n	$M(n)$	$T(n)$	Error in $M(n)$	Error in $T(n)$
4	1.50968181	1.48067370	9.7×10^{-3}	1.9×10^{-2}
8	1.50241228	1.49517776	2.4×10^{-3}	4.8×10^{-3}
16	1.50060256	1.49879502	6.0×10^{-4}	1.2×10^{-3}
32	1.50015061	1.49969879	1.5×10^{-4}	3.0×10^{-4}

33.

n	$M(n)$	$T(n)$	Error in $M(n)$	Error in $T(n)$
4	-1.96×10^{-16}	0	2.0×10^{-16}	0
8	7.63×10^{-17}	-1.41×10^{-16}	7.6×10^{-17}	1.4×10^{-16}
16	1.61×10^{-16}	1.09×10^{-17}	1.6×10^{-16}	1.1×10^{-17}
32	6.27×10^{-17}	-4.77×10^{-17}	6.3×10^{-17}	4.8×10^{-17}

35. $T(4) \approx 690.3$ million ft³; $S(4) \approx 692.2$ million ft³ (answers may vary)   **37.** 54.5°F, Trapezoid Rule   **39.** 35.0°F, Trapezoid Rule   **41. a.** Left sum: 204.917; right sum: 261.375; Trapezoid Rule: 233.146; the approximations measure the average temperature of the curling iron on $[0, 120]$.   **b.** Left sum: underestimate; right sum: overestimate; Trapezoid Rule: underestimate   **c.** 305°F is the change in temperature over $[0, 120]$.   **43. a.** 5907.5   **b.** 5965   **c.** 5917
45. a. $T(25) \approx 3.19623162$
$T(50) \approx 3.19495398$
b. $S(50) \approx 3.19452809$
c. $e_T(50) \approx 4.3 \times 10^{-4}$
$e_S(50) \approx 4.5 \times 10^{-8}$
47. a. $T(50) \approx 1.00008509$
$T(100) \approx 1.00002127$
b. $S(100) \approx 1.00000000$
c. $e_T(100) \approx 2.1 \times 10^{-5}$
$e_S(100) \approx 4.6 \times 10^{-9}$

49.

n	$T(n)$	$S(n)$	Error in $T(n)$	Error in $S(n)$
4	1820.0000	—	284	—
8	1607.7500	1537.0000	71.8	1
16	1553.9844	1536.0625	18.0	6.3×10^{-2}
32	1540.4990	1536.0039	4.50	3.9×10^{-3}

51.

n	$T(n)$	$S(n)$	Error in $T(n)$	Error in $S(n)$
4	0.46911538	—	5.3×10^{-2}	—
8	0.50826998	0.52132152	1.3×10^{-2}	2.9×10^{-4}
16	0.51825968	0.52158957	3.4×10^{-3}	1.7×10^{-5}
32	0.52076933	0.52160588	8.4×10^{-4}	1.1×10^{-6}

53. a. True   **b.** False   **c.** True

55.

n	$M(n)$	$T(n)$	Error in $M(n)$	Error in $T(n)$
4	0.40635058	0.40634782	1.4×10^{-6}	1.4×10^{-6}
8	0.40634920	0.40634920	7.6×10^{-10}	7.6×10^{-10}
16	0.40634920	0.40634920	6.6×10^{-13}	6.6×10^{-13}
32	0.40634920	0.40634920	8.9×10^{-16}	7.8×10^{-16}

57.

n	$M(n)$	$T(n)$	Error in $M(n)$	Error in $T(n)$
4	4.72531819	4.72507878	1.2×10^{-4}	1.2×10^{-4}
8	4.72519850	4.72519849	9.1×10^{-9}	9.1×10^{-9}
16	4.72519850	4.72519850	0	8.9×10^{-16}
32	4.72519850	4.72519850	0	8.9×10^{-16}

63. Approximations will vary; exact value is 68.26894921 . . . .
65. a. Approx. 1.6×10^{11} barrels   **b.** Approx. 6.8×10^{10} barrels
67. a. $M(50) \approx 34.4345566$
b. $f''(x) = \dfrac{3(x^4 + 4x)}{4(x^3 + 1)^{3/2}}$ **d.** $E_M \le 0.0028$
69. a. $T(40) = 0.874799972\dots$ **b.** $f''(x) = e^x \cos e^x - e^{2x} \sin e^x$
d. $E_T \le \dfrac{1}{3200}$ **71. a.** $S(20) \approx 0.97774576$
b. $E_S \le 3.5 \times 10^{-8}$ **73.** Approximations will vary; exact value is 38.753792 . . . .   **77.** Overestimate   **79.** $S(20) \approx 1.00000175$

Section 8.9 Exercises, pp. 590–593

1. The interval of integration is infinite or the integrand is unbounded on the interval of integration.   **3.** $\displaystyle\lim_{b \to \infty} \int_2^b \dfrac{dx}{x^{1/5}}$ **5.** $\displaystyle\int_{-\infty}^{\infty} f(x)\,dx$
7. $\dfrac{1}{3}$ **9.** Diverges   **11.** $\dfrac{1}{a}$ **13.** Diverges   **15.** $\dfrac{\pi}{10}$
17. Diverges   **19.** Diverges   **21.** $\dfrac{1}{\pi}$ **23.** $\dfrac{\pi}{4}$ **25.** $\dfrac{\pi}{6}$ **27.** 0
29. $\dfrac{\pi^3}{12}$ **31.** $\ln 2$ **33.** Diverges   **35.** Diverges   **37.** 6
39. Diverges   **41.** Diverges   **43.** $2(e - 1)$ **45.** Diverges
47. $4 \cdot 10^{3/4}/3$ **49.** Diverges   **51.** π **53.** -1
55. $\ln(2 + \sqrt{3})$ **57.** 2   **59.** \$41,666.67   **61.** 0.76   **63.** 20,000 hr
65. $\dfrac{\pi}{3}$ **67.** $3\pi/2$ **69.** $\pi/\ln 2$ **71.** 2π **73.** Does not exist
75. $\dfrac{72 \cdot 2^{1/3}\,\pi}{5}$ **77.** Converges   **79.** Diverges   **81.** Converges
83. Diverges   **85.** Converges   **87. a.** True   **b.** False   **c.** False
d. True   **e.** True   **89.** $1/b - 1/a$ **91. a.** $A(a, b) = \dfrac{e^{-ab}}{a}$, for $a > 0$
b. $b = g(a) = -\dfrac{1}{a} \ln 2a$ **c.** $b^* = -2/e$ **93.** π **107. a.** π
b. $\pi/(4e^2)$ **109. a.** $6.28 \times 10^7\,m\,J$ **b.** 11.2 km/s   **c.** ≤ 9 mm

Chapter 8 Review Exercises, pp. 593–596

1. a. True   **b.** False   **c.** False   **d.** True   **e.** False
3. $2(x - 8)\sqrt{x + 4} + C$ **5.** $\dfrac{1}{3}\sqrt{x + 2}\,(x - 4) + C$ **7.** $\dfrac{\pi}{4}$
9. $\dfrac{4}{105}$ **11.** $\sqrt{t - 1} - \tan^{-1}\sqrt{t - 1} + C$ **13.** $\dfrac{2}{15}(1 - e^{3\pi})$
15. $7 + \ln 40 - \ln 17$ **17.** $2 \ln |x| + 3 \tan^{-1}(x + 1) + C$
19. $\dfrac{2}{x + 3} - \dfrac{2}{(x + 3)^2} + \ln |x + 3| + C$ **21.** $\sqrt{3} - 1 - \dfrac{\pi}{12}$
23. $\dfrac{1}{5} \tan^5 t + C$ **25.** $\dfrac{\pi}{8}$ **27.** $\dfrac{\sqrt{w^2 + 2w - 8}}{9(w + 1)} + C$ **29.** $-\dfrac{\cot^5 x}{5} + C$
31. $\dfrac{x \cosh 2x}{2} - \dfrac{\sinh 2x}{4} + C$ **33.** $\dfrac{1}{15} \sec^5 3\theta - \dfrac{1}{9} \sec^3 3\theta + C$

35. $\frac{1}{6}(x^2 - 8)\sqrt{x^2 + 4} + C$ **37.** $\frac{1}{x + 1} + \ln|(x + 1)(x^2 + 4)| + C$

39. $\frac{t - \ln(2 + e^t)}{2} + C$ **41.** $\frac{1}{4}(\csc 4\theta - \cot 4\theta) + C$

43. $\frac{e^x}{2}(\sin x - \cos x) + C$

45. $\ln|x| - \frac{1}{x} + \frac{1}{2}\ln(x^2 + 4x + 9) - \frac{2}{\sqrt{5}}\tan^{-1}\left(\frac{x + 2}{\sqrt{5}}\right) + C$

47. $\frac{\theta}{2} + \frac{1}{16}\sin 8\theta + C$ **49.** $\frac{\sec^{49} 2z}{98} + C$ **51.** $\frac{4}{15}$

53. $2\sqrt{x} - 3\sqrt[3]{x} + 6\sqrt[6]{x} - 6\ln(\sqrt[6]{x} + 1) + C$

55. $-\frac{\sqrt{9 - y^2}}{9\sqrt{2}\, y} + C$ **57.** $\frac{\pi}{9}$ **59.** $-\mathrm{sech}\, x + C$ **61.** $\frac{\pi}{3}$

63. $\frac{1}{8}\ln\left|\frac{x - 5}{x + 3}\right| + C$ **65.** $\frac{\ln 2}{4} + \frac{\pi}{8}$ **67.** 3   **69.** $\frac{1}{3}\ln\left|\frac{x - 2}{x + 1}\right| + C$

71. $2(x - 2\ln|x + 2|) + C$ **73.** $e^{2t}/(2\sqrt{1 + e^{4t}}) + C$

75. a. $\sec e^x + C$ **b.** $e^x \sec e^x - \ln|\sec e^x + \tan e^x| + C$

77. $\frac{\sqrt{6}}{3}\tan^{-1}\sqrt{\frac{2x - 3}{3}} + C$

79. $\frac{1}{4}\sec^3 x \tan x + \frac{3}{8}\sec x \tan x + \frac{3}{8}\ln|\sec x + \tan x| + C$

81. $2(\ln^3 2 - 3\ln^2 2 + \ln 64 - 3)$ **83.** 1   **85.** $\frac{\pi}{2}$

87. $\frac{2\pi}{\sqrt{3}}$ **89.** Converges   **91.** Diverges   **93.** 1.196288

95. $M(4) = 44; T(4) = 42; S(4) = \frac{124}{3}$

97. $M(40) \approx 0.398236; T(40) \approx 0.398771; S(40) \approx 0.398416$

99. 0.886227   **101.** y-axis   **103.** $\pi(e - 2)$ **105.** $\frac{\pi}{2}(e^2 - 3)$

107. a. 1.603   **b.** 1.870   **c.** $b \ln b - b = a \ln a - a$
d. Decreasing   **109.** $20/(3\pi)$ **111.** 1901 cars

113. a. $I(p) = \frac{1}{(p - 1)^2}(1 - pe^{1-p})$ if $p \neq 1, I(1) = \frac{1}{2}$ **b.** $0, \infty$

c. $I(0) = 1$ **115.** 0.4054651   **117.** $n = 2$

119. a. $V_1(a) = \pi(a \ln^2 a - 2a \ln a + 2(a - 1))$

b. $V_2(a) = \frac{\pi}{2}(2a^2 \ln a - a^2 + 1)$

c. $V_2(a) > V_1(a)$ for all $a > 1$

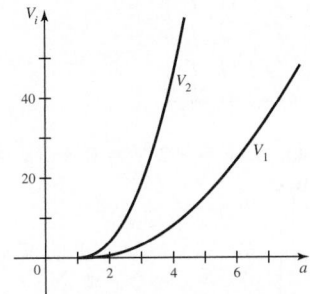

121. $a = \ln 2/(2b)$ **123.** $\ln(1 + \sqrt{2}/2)$

CHAPTER 9

Section 9.1 Exercises, pp. 604–606

1. a. 1   **b.** Linear   **3.** Yes   **5.** $\frac{\pi}{2} < t < \frac{3\pi}{2}$

21. $y = 3t - \frac{e^{-2t}}{2} + C$ **23.** $y = 2\ln|\sec 2x| - 3\sin x + C$

25. $y = 2t^6 + 6t^{-1} - 2t^2 + C_1 t + C_2$

27. $u = \frac{x^{11}}{2} + \frac{x^9}{2} - \frac{x^7}{2} + \frac{5}{x} + C_1 x + C_2$

29. $u = \ln(x^2 + 4) - \tan^{-1}\frac{x}{2} + C$ **31.** $y = \sin^{-1} x + C_1 x + C_2$

33. $y = e^t + t + 3$ **35.** $y = x^3 + x^{-3} - 2, x > 0$

37. $y = -t^5 + 2t^3 + 1$ **39.** $y = e^t(t - 2) + 2(t + 1)$

41. $u = \frac{1}{4}\tan^{-1}\frac{x}{4} - 4x + 2$ **43. a.** $v(t) = -9.8t + 29.4$;

$s(t) = -4.9t^2 + 29.4t + 30$; the object is above the ground for
approximately $0 \leq t \leq 6.89$.   **b.** The highest point of 74.1 m is
reached at $t = 3$ s.   **45.** The amount of resource is increasing for
$H < 75$ and is constant if $H = 75$. If $H = 100$, the resource
vanishes at approximately 28 time units.

47. $h = (\sqrt{1.96} - 0.1t\sqrt{2g})^2 \approx (1.4 - 0.44t)^2, 0 \leq t \leq 3.16$;
the tank is empty after approximately 3.16 s.

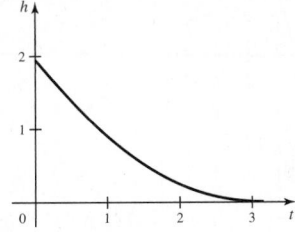

49. a. False   **b.** False   **c.** True   **51. c.** $y = C_1 \sin kt + C_2 \cos kt$

53. b. $C = \frac{K - 50}{50}$

c.

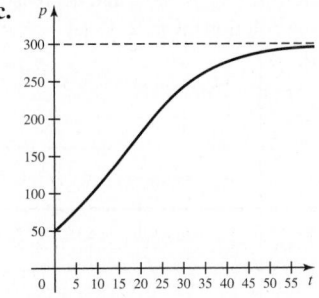

55. c. The decay rate is greater
for the $n = 1$ model.

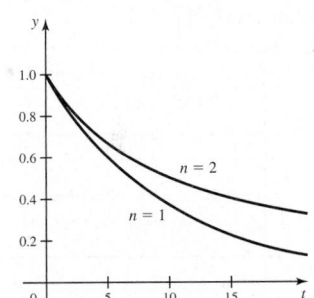

d. 300

Section 9.2 Exercises, pp. 611–614

1. At selected points (t_0, y_0) in the region of interest draw a short line
segment with slope $f(t_0, y_0)$.   **3.** $y(3.1) \approx 1.6$
5. a. D   **b.** B   **c.** A   **d.** C
7.

9. An initial condition of $y(0) = -1$ leads to a constant solution. For any other initial condition, the solutions are increasing over time.

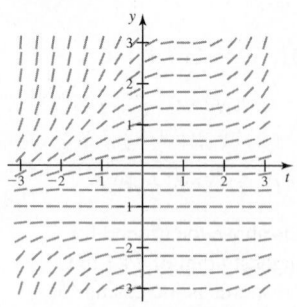

11. An initial condition of $y(0) = 1$ leads to a constant solution. Initial conditions $y(0) = A$ lead to solutions that are increasing over time if $A > 1$.

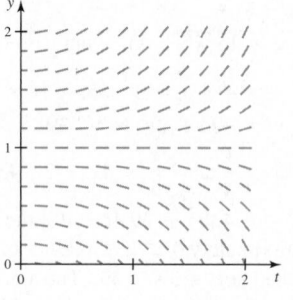

13.

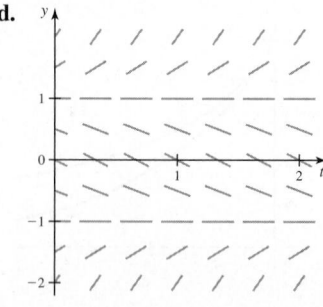

15.

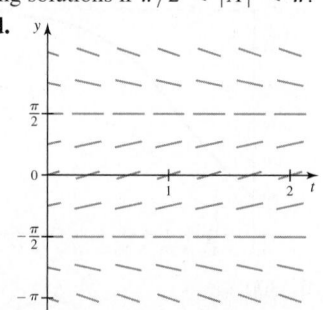

17. a. $y = 1, y = -1$
b. Solutions are increasing for $|y| > 1$ and decreasing for $|y| < 1$. **c.** Initial conditions $y(0) = A$ lead to increasing solutions if $|A| > 1$ and decreasing solutions if $|A| < 1$.
d.

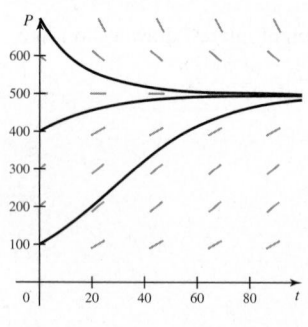

19. a. $y = \pi/2, y = -\pi/2$
b. Solutions are increasing for $|y| < \pi/2$ and decreasing for $|y| > \pi/2$. **c.** Initial conditions $y(0) = A$ lead to increasing solutions if $|A| < \pi/2$ and decreasing solutions if $\pi/2 < |A| < \pi$.
d.

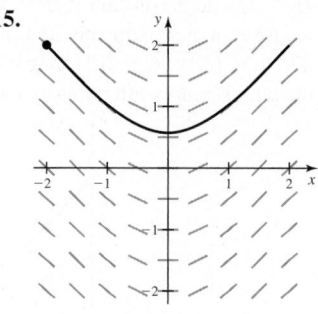

21. The equilibrium solutions are $P = 0$ and $P = 500$.

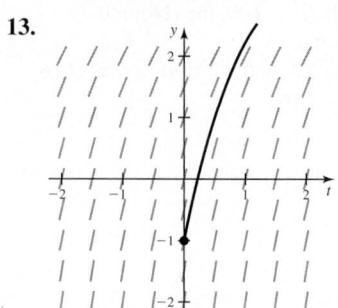

23. The equilibrium solutions are $P = 0$ and $P = 3200$.

25. $y(0.5) \approx u_1 = 4; y(1) \approx u_2 = 8$
27. $y(0.1) \approx u_1 = 1.1; y(0.2) \approx u_2 = 1.19$

29. a.

Δt	approximation to $y(0.2)$	approximation to $y(0.4)$
0.20000	0.80000	0.64000
0.10000	0.81000	0.65610
0.05000	0.81451	0.66342
0.02500	0.81665	0.66692

b.

Δt	errors for $y(0.2)$	errors for $y(0.4)$
0.20000	0.01873	0.03032
0.10000	0.00873	0.01422
0.05000	0.00422	0.00690
0.02500	0.00208	0.00340

c. Time step $\Delta t = 0.025$; smaller time steps generally produce more accurate results. **d.** Halving the time steps results in approximately halving the error.

31. a.

Δt	approximation to $y(0.2)$	approximation to $y(0.4)$
0.20000	3.20000	3.36000
0.10000	3.19000	3.34390
0.05000	3.18549	3.33658
0.02500	3.18335	3.33308

b.

Δt	errors for $y(0.2)$	errors for $y(0.4)$
0.20000	0.01873	0.03032
0.10000	0.00873	0.01422
0.05000	0.00422	0.00690
0.02500	0.00208	0.00340

c. Time step $\Delta t = 0.025$; smaller time steps generally produce more accurate results. **d.** Halving the time steps results in approximately halving the error. **33. a.** $y(2) \approx 0.00604662$ **b.** 0.012269 **c.** $y(2) \approx 0.0115292$ **d.** Error in part (c) is approximately half of the error in part (b). **35. a.** $y(4) \approx 3.05765$ **b.** 0.0339321 **c.** $y(4) \approx 3.0739$ **d.** Error in part (c) is approximately half of the error in part (b). **37. a.** True **b.** False
39. a. $y = 3$ **b, c.**

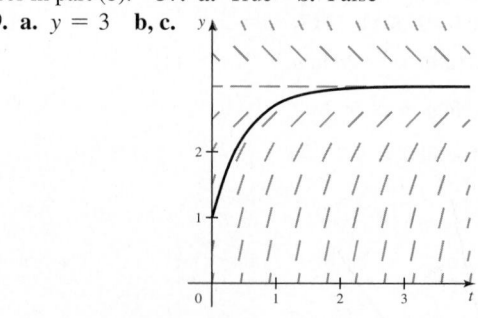

41. a. $y = 0$ and $y = 3$
b, c.

43. a. $y = -2, y = 0$, and $y = 3$
b, c.

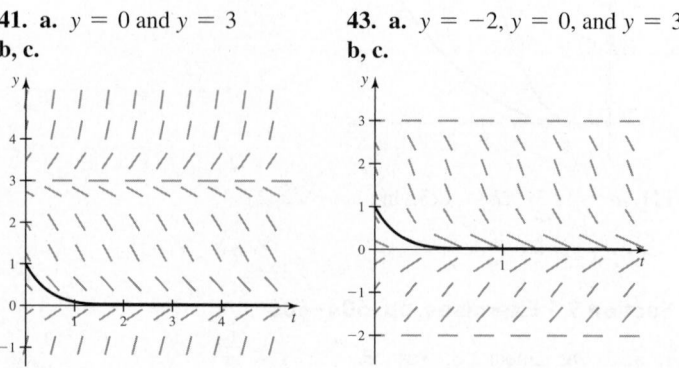

45. a. $\Delta t = \dfrac{b-a}{N}$ **b.** $u_1 = A + f(a, A)\dfrac{b-a}{N}$

c. $u_{k+1} = u_k + f(t_k, u_k)\dfrac{b-a}{N}$, where $u_0 = A$ and $t_k = a + k(b-a)/N$, for $k = 0, 1, 2, \ldots, N-1$.

47. a.

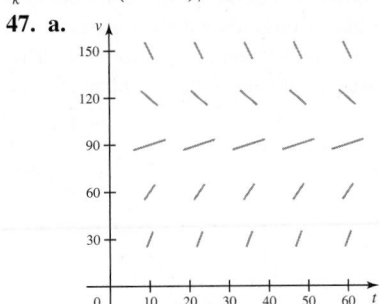

b. Increasing for $A < 98$ and decreasing for $A > 98$
c. $v(t) = 98$

c.

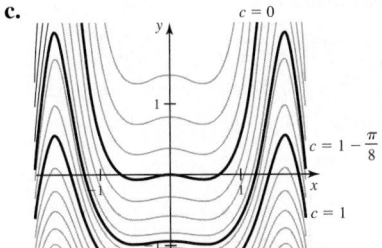

45. $y = kx$ **47. b.** $\sqrt{gm/k}$

c. $v = \sqrt{\dfrac{g}{a}}\dfrac{Ce^{2\sqrt{ag}\,t} - 1}{Ce^{2\sqrt{ag}\,t} + 1}$, $t \geq 0$, where $a = \dfrac{k}{m}$

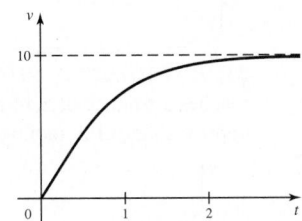

Section 9.3 Exercises, pp. 618–620

1. A first-order separable differential equation has the form $g(y)\,y'(t) = h(t)$, where the factor $g(y)$ is a function of y and $h(t)$ is a function of t. **3.** No **5.** $y = \dfrac{t^4}{4} + C$

7. $y = \pm\sqrt{2t^3 + C}$ **9.** $y = -2\ln\left(\dfrac{1}{2}\cos t + C\right)$

11. $y = \dfrac{x}{1 + Cx}$ **13.** $y = \pm\dfrac{1}{\sqrt{C - \cos t}}$ **15.** $u = \ln\left(\dfrac{e^{2x}}{2} + C\right)$

17. $y = \sqrt{t^3 + 81}$ **19.** Not separable **21.** $y(t) = -e^{e^t - 1}$

23. $y = \ln(e^x + 2)$ **25.** $y = \ln\left(\dfrac{\ln^4 t}{4} + 1\right)$

27. $y = \sqrt{\tan t}$, $0 < t < \pi/2$ **29.** $y = \sqrt{t^2 + 3}$ **31.** $y = \ln t + 2$

33. $y^3 - 3y = 2t^3$, $-1 < t < 1$

35. $\cos u = 2 - 2\sin\dfrac{x}{2}$, $\dfrac{\pi}{3} < x < \dfrac{5\pi}{3}$

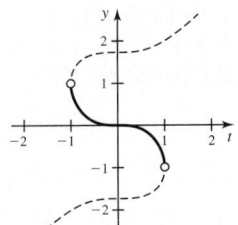

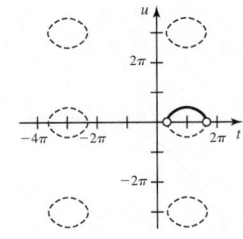

37.

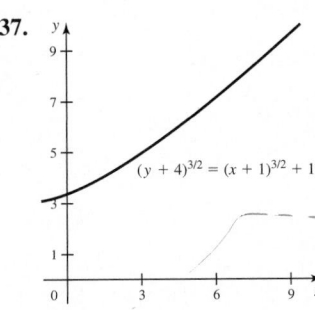

$(y+4)^{3/2} = (x+1)^{3/2} + 19$

39. a.

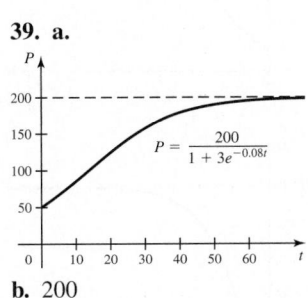

$P = \dfrac{200}{1 + 3e^{-0.08t}}$

b. 200

41. a. True **b.** False **c.** True

43. a. $y = -2\ln\left(\dfrac{x^2}{4} + \cos x^2 + C\right)$ **b.** $C = 0, 1, 1 - \dfrac{\pi}{8}$

49. a. $h = \left(\sqrt{H} - \dfrac{kt}{2}\right)^2$, $0 \leq t \leq \dfrac{2\sqrt{H}}{k}$

b. $h = (\sqrt{0.5} - 0.05t)^2$, $0 \leq t \leq 14.1$

c. Approx. 14.1 s

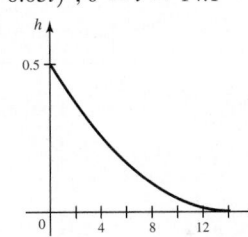

51. a.

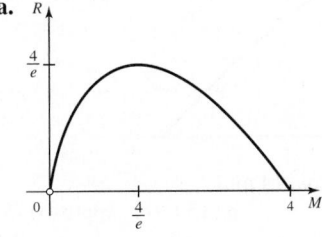

R is positive if $0 < M < 4$; R has a maximum value when $M = \dfrac{4}{e}$; $\lim_{M \to 0} R(M) = 0$.

b. $M(t) = 4^{1 - e^{-t}}$, $t \geq 0$; the tumor grows quickly at first and then the rate of growth slows down; the limiting size of the tumor is 4.

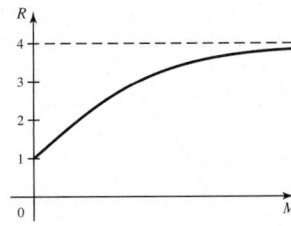

53. a. $y = \dfrac{1}{1-t}$, $t < 1$ **b.** $y = \dfrac{1}{\sqrt{2}\sqrt{1-t}}$, $t < 1$

c. $y = \dfrac{1}{(n(1-t))^{1/n}}$, $t < 1$; as $t \to 1^-$, $y \to \infty$

Section 9.4 Exercises, pp. 625–627

1. $y = 17e^{-10t} - 13$ **3.** $y = Ce^{-4t} + \frac{3}{2}$ **5.** $y = Ce^{3t} + \frac{4}{3}$

7. $y = Ce^{-2x} - 2$ **9.** $u = Ce^{-12t} + \frac{5}{4}$ **11.** $y = 7e^{3t} + 2$

13. $y = 4(e^{2t} - 1)$ **15.** $y = 4(2e^{3t-3} - 1)$

17. $y = \frac{3}{2}$; unstable

19. $y = -3$; stable

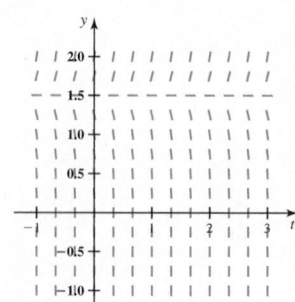

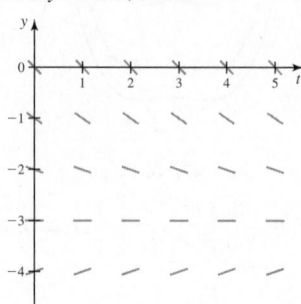

21. $u = -3$; stable

23. $B = 100{,}000 - 50{,}000e^{0.005t}$; reaches a balance of zero after approximately 139 months

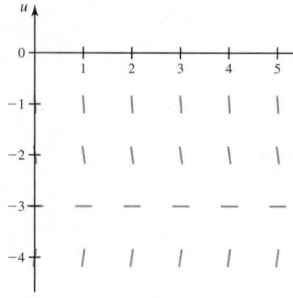

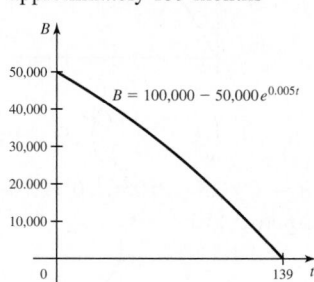

25. $B = 200{,}000 - 100{,}000e^{0.0075t}$; reaches a balance of zero after approximately 93 months

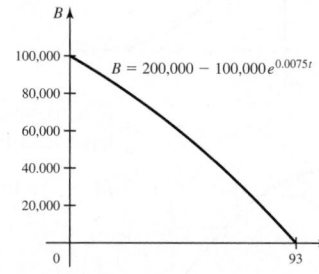

27. Approx. 32 min   **29.** Approx. 14 min

31. a.   **b.** 150   **c.** Approx. 115.1 hr

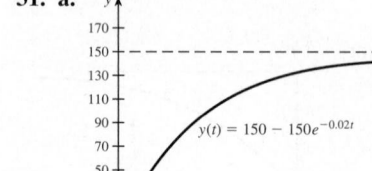
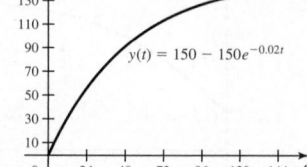

33. a. $h = 16 \text{ yr}^{-1}$ **b.** 25,000   **35. a.** False   **b.** True   **c.** False
d. False   **37. a.** $B = 20{,}000 + 20{,}000e^{0.03t}$; the unpaid balance is growing because the monthly payment of $600 is less than the interest on the unpaid balance.   **b.** $20,000   **c.** $\dfrac{m}{r}$

39. $y = 1 + \dfrac{t}{2} + \dfrac{5}{2t}, t > 0$ **41.** $y = \dfrac{1}{2}e^{3t} + \dfrac{7}{2}e^{t}$

45. $y(t) = \dfrac{6}{t}, t > 0$ **47.** $y = \dfrac{9t^5 + 20t^3 + 15t + 76}{15(t^2 + 1)}$

Section 9.5 Exercises, pp. 634–636

1. The growth rate function specifies the rate of growth of the population. The population is increasing when the growth rate function is positive, and the population is decreasing when the growth rate function is negative.   **3.** If the growth rate function is positive (it does not matter whether it is increasing or decreasing), then the population is increasing.   **5.** It is a linear, first-order differential equation.   **7.** The solution curves in the *FH*-plane are closed curves that circulate around the equilibrium point.

9.

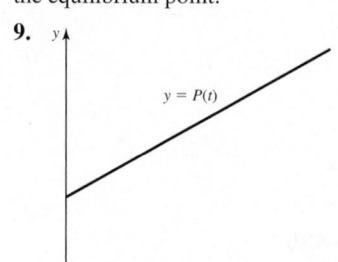

11.

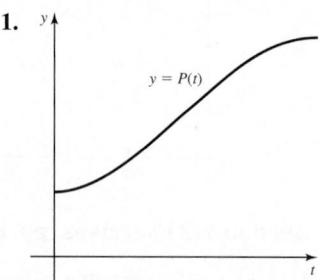

13.

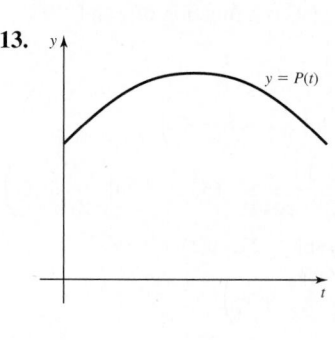

15. $P' = 0.2\, P\left(1 - \dfrac{P}{300}\right)$;
$P = \dfrac{300}{5e^{-0.2t} + 1}, t \geq 0$

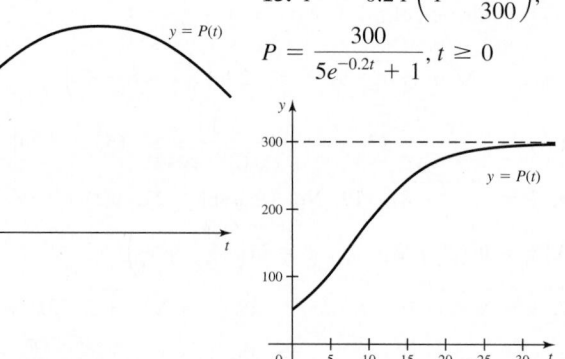

17. $P = \dfrac{2000}{9e^{-\ln(27/7)t} + 1}, t \geq 0$ **19.** $M = K\left(\dfrac{M_0}{K}\right)^{e^{-rt}}, t \geq 0$

21. $M = 1200 \cdot 0.075^{\exp(-0.05t)}, t \geq 0$

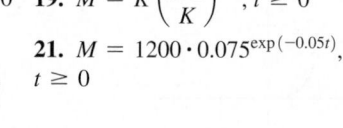

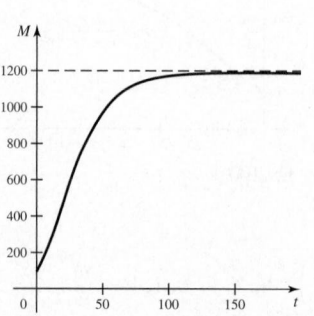

23. a. $m'(t) = -0.008t + 80, m(0) = 0$
b. $m = 10{,}000 - 10{,}000e^{-0.008t}, t \geq 0$
25. a. $m'(t) = -0.005\, t + 100, m(0) = 80{,}000$
b. $m = 60{,}000e^{-0.005t} + 20{,}000, t \geq 0$

27. a. x is the predator population; y is the prey population.
b. $x' = 0$ on the lines $x = 0$ and $y = \frac{1}{2}$; $y' = 0$ on the lines $y = 0$
and $x = \frac{1}{4}$.   **c.** $(0,0)$, $\left(\frac{1}{4}, \frac{1}{2}\right)$
d. $x' > 0$ and $y' > 0$ for $0 < x < \frac{1}{4}, y > \frac{1}{2}$
$\qquad x' > 0$ and $y' < 0$ for $x > \frac{1}{4}, y > \frac{1}{2}$
$\qquad x' < 0$ and $y' < 0$ for $x > \frac{1}{4}, 0 < y < \frac{1}{2}$
$\qquad x' < 0$ and $y' > 0$ for $0 < x < \frac{1}{4}, 0 < y < \frac{1}{2}$
e. The solution evolves in the clockwise direction.

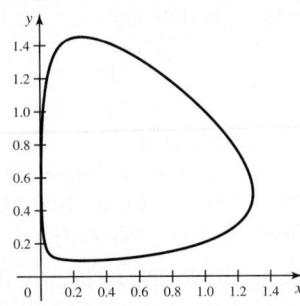

29. a. x is the predator population; y is the prey population.
b. $x' = 0$ on the lines $x = 0$ and $y = 3$; $y' = 0$ on the lines $y = 0$
and $x = 2$.   **c.** $(0,0)$, $(2,3)$
d. $x' > 0$ and $y' > 0$ for $0 < x < 2, y > 3$
$\qquad x' > 0$ and $y' < 0$ for $x > 2, y > 3$
$\qquad x' < 0$ and $y' < 0$ for $x > 2, 0 < y < 3$
$\qquad x' < 0$ and $y' > 0$ for $0 < x < 2, 0 < y < 3$
e. The solution evolves in the clockwise direction.

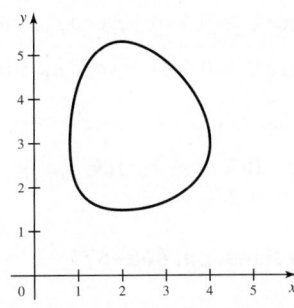

31. a. True   **b.** True   **c.** True   **35. c.** $\lim\limits_{t \to \infty} m(t) = C_i V$, which is
the amount of substance in the tank when the tank is filled with the
inflow solution.   **d.** Increasing R increases the rate at which the
solution in the tank reaches the steady-state concentration.

37. a. $I = \dfrac{V}{R} e^{-t/(RC)}$ **b.** $Q = VC(1 - e^{-t/(RC)})$

39. a. $y'(x) = \dfrac{y(c - dx)}{x(-a + by)}$ **c.**

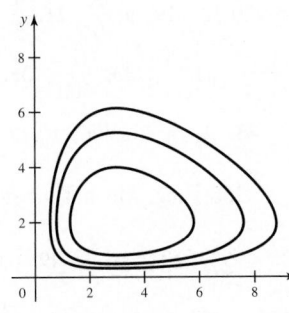

Chapter 9 Review Exercises, pp. 636–638

1. a. False   **b.** False   **c.** True   **d.** True   **e.** False
3. $y = Ce^{-2t} + 3$ **5.** $y = Ce^{t^2}$ **7.** $y = Ce^{\tan^{-1}t}$
9. $y = \tan(t^2 + t + C)$ **11.** $y = \sin t + t^2 + 1$
13. $Q = 8(1 - e^{t-1})$ **15.** $u = (3 + t^{2/3})^{3/2}, t > 0$

17. $s = \dfrac{t\sqrt{2}}{\sqrt{t^2 + 1}}$

19. a, b.

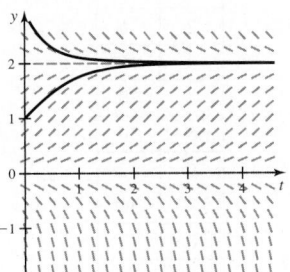

c. $0 < A < 2$
d. $A > 2$ or $A < 0$
e. $y = 0$ and $y = 2$

21. a. 1.05, 1.09762   **b.** 1.04939, 1.09651   **c.** 0.00217, 0.00106;
the error in part (b) is smaller.   **23.** $y = -3$ (unstable), $y = 0$
(stable), $y = 5$ (unstable)   **25.** $y = -1$ (unstable), $y = 0$ (stable),
$y = 2$ (unstable)   **27. a.** 0.0713   **b.** $P = \dfrac{1600}{79e^{-0.0713t} + 1}, t \geq 0$

c. Approx. 61 hours   **29. a.** $m = 2000(1 - e^{-0.005t})$
b. 2000 g   **c.** Approx. 599 minutes   **31. a.** x represents the
predator.   **b.** $x'(t) = 0$ when $x = 0$ and $y = 2$. $y'(t) = 0$ when
$y = 0$ and $x = 5$.   **c.** $(0,0)$ and $(5,2)$ **d.** $x' > 0, y' > 0$ when
$0 < x < 5$ and $y > 2$; $x' > 0, y' < 0$ when $x > 5$ and $y > 2$;
$x' < 0, y' < 0$ when $x > 5$ and $0 < y < 2$; $x' < 0, y' > 0$ when
$0 < x < 5$ and $0 < y < 2$
e. Clockwise direction

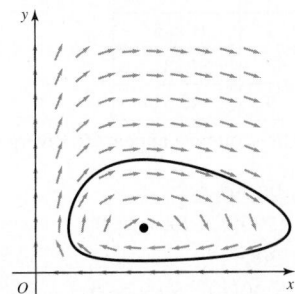

33. a. $p_1 = 3, p_2 = -4$ **b.** $y(t) = t^3 - t^{-4}, t > 0$

CHAPTER 10

Section 10.1 Exercises, pp. 647–649

1. A sequence is an ordered list of numbers. Example: $1, \frac{1}{3}, \frac{1}{9}, \frac{1}{27}, \ldots$
3. 1, 1, 2, 6, 24   **5.** $a_n = (-1)^{n+1} n$, for $n = 1, 2, 3, \ldots$;
$a_n = (-1)^n (n + 1)$, for $n = 0, 1, 2, \ldots$ (Answers may vary.)

7. e **9.** 1, 5, 14, 30   **11.** $\sum\limits_{k=1}^{\infty} 10$ (Answer is not unique.)

13. $\frac{1}{10}, \frac{1}{100}, \frac{1}{1000}, \frac{1}{10,000}$ **15.** $-\frac{1}{2}, \frac{1}{4}, -\frac{1}{8}, \frac{1}{16}$ **17.** $\frac{4}{3}, \frac{8}{5}, \frac{16}{9}, \frac{32}{17}$

19. 2, 1, 0, 1   **21.** 2, 4, 8, 16   **23.** 10, 18, 42, 114   **25.** $1, \dfrac{1}{2}, \dfrac{2}{3}, \dfrac{3}{5}$

27. a. $\dfrac{1}{32}, \dfrac{1}{64}$ **b.** $a_1 = 1, a_{n+1} = \dfrac{1}{2} a_n$, for $n \geq 1$ **c.** $a_n = \dfrac{1}{2^{n-1}}$,
for $n \geq 1$ **29. a.** 32, 64   **b.** $a_1 = 1, a_{n+1} = 2a_n$, for $n \geq 1$
c. $a_n = 2^{n-1}$, for $n \geq 1$ **31. a.** 243, 729   **b.** $a_1 = 1, a_{n+1} = 3a_n$,
for $n \geq 1$ **c.** $a_n = 3^{n-1}$, for $n \geq 1$ **33. a.** $-5, 5$ **b.** $a_1 = -5$,
$a_{n+1} = -a_n$, for $n \geq 1$ **c.** $a_n = (-1)^n \cdot 5$, for $n \geq 1$
35. 9, 99, 999, 9999; diverges   **37.** $\frac{1}{10}, \frac{1}{100}, \frac{1}{1000}, \frac{1}{10,000}$;
converges to 0   **39.** 2, 4, 2, 4; diverges   **41.** 2, 2, 2, 2; converges to 2
43. 54.545, 54.959, 54.996, 55.000; converges to 55

45.

n	a_n
1	0.83333333
2	0.96153846
3	0.99206349
4	0.99840256
5	0.99968010
6	0.99993600
7	0.99998720
8	0.99999744
9	0.99999949
10	0.99999990

The limit appears to be 1.

47.

n	a_n
1	2
2	6
3	12
4	20
5	30
6	42
7	56
8	72
9	90
10	110

The sequence appears to diverge.

49. a. $\frac{5}{2}, \frac{9}{4}, \frac{17}{8}, \frac{33}{16}$ **b.** 2

51.

n	a_n
1	3.00000000
2	3.50000000
3	3.75000000
4	3.87500000
5	3.93750000
6	3.96875000
7	3.98437500
8	3.99218750
9	3.99609375
10	3.99804688

The limit appears to be 4.

53.

n	a_n
0	1
1	5
2	21
3	85
4	341
5	1365
6	5461
7	21,845
8	87,381
9	349,525

The sequence appears to diverge.

55.

n	a_n
1	8.00000000
2	4.41421356
3	4.05050150
4	4.00629289
5	4.00078630
6	4.00009828
7	4.00001229
8	4.00000154
9	4.00000019
10	4.00000002

The limit appears to be 4.

57. a. $20, 10, 5, \frac{5}{2}$
b. $h_n = 20\left(\frac{1}{2}\right)^n$, for $n \geq 0$
59. a. $30, \frac{15}{2}, \frac{15}{8}, \frac{15}{32}$
b. $h_n = 30\left(\frac{1}{4}\right)^n$, for $n \geq 0$
61. $S_1 = 0.3, S_2 = 0.33, S_3 = 0.333,$
$S_4 = 0.3333; \frac{1}{3}$
63. $S_1 = 4, S_2 = 4.9,$
$S_3 = 4.99, S_4 = 4.999; 5$
65. $S_1 = 0.6, S_2 = 0.66, S_3 = 0.666,$
$S_4 = 0.6666; \frac{2}{3}$ **67. a.** $\frac{2}{3}, \frac{4}{5}, \frac{6}{7}, \frac{8}{9}$
b. $S_n = \frac{2n}{2n+1}; \frac{10}{11}, \frac{12}{13}, \frac{14}{15}, \frac{16}{17}$ **c.** 1

69. a. $9, 9.9, 9.99, 9.999$ **b.** $S_n = 10 - (0.1)^{n-1}$; $9.9999, 9.99999,$
$9.999999, 9.9999999$ **c.** 10 **71. a.** True **b.** False **c.** True
73. a. $20, 10, 5, \frac{5}{2}, \frac{5}{4}$ **b.** $M_n = 20\left(\frac{1}{2}\right)^n$, for $n \geq 0$ **c.** $M_0 = 20,$
$M_{n+1} = \frac{1}{2}M_n$, for $n \geq 0$ **d.** $\lim\limits_{n \to \infty} M_n = 0$ **75. a.** $200, 190, 180.5,$
$171.475, 162.90125$ **b.** $d_n = 200(0.95)^n$, for $n \geq 0$
c. $d_0 = 200, d_{n+1} = (0.95)d_n$, for $n \geq 0$ **d.** $\lim\limits_{n \to \infty} d_n = 0$
77. a. $40, 70, 92.5, 109.375$ **b.** 160 **79.** 0.739

Section 10.2 Exercises, pp. 659–662

1. $a_n = \frac{1}{n}, n \geq 1$ **3.** $a_n = \frac{n}{n+1}, n \geq 1$ **5.** Converges for
$-1 < r \leq 1$, diverges otherwise **7.** Diverges monotonically
9. Converges, oscillates; 0 **11.** $\{e^{n/100}\}$ grows faster then $\{n^{100}\}$.

13. 0 **15.** $\frac{3}{2}$ **17.** $\frac{\pi}{4}$ **19.** 2 **21.** 0 **23.** $\frac{1}{4}$ **25.** 2 **27.** 0
29. 0 **31.** 3 **33.** Diverges **35.** $\frac{\pi}{2}$ **37.** 0 **39.** e^2 **41.** e^3
43. $e^{1/4}$ **45.** 0 **47.** 1 **49.** 0 **51.** 6
53.

Diverges

55. 0 **57.** Diverges
59. Diverges **61.** 0 **63.** 0
65. 0 **67.** 0 **69.** 0
71. a. $d_{n+1} = \frac{1}{2}d_n + 80$, for $n \geq 1$
b. 160 mg

73. a. $\$0, \$100, \$200.75, \$302.26, \$404.53$
b. $B_{n+1} = 1.0075B_n + 100$, for $n \geq 0$ **c.** Approx. 43 months
75. 0 **77.** Diverges **79.** 0 **81.** 1 **83. a.** True **b.** False
c. True **d.** True **e.** False **f.** True **85. a.** Nondecreasing
b. $\frac{1}{2}$ **87. a.** Nonincreasing **b.** 2 **89. a.** $d_{n+1} = 0.4d_n + 75;$
$d_1 = 75$ **c.** 125; in the long run there will be approximately 125 mg
of medication in the blood. **91.** 0.607 **93. b.** 9
95. a. $\sqrt{2}, \sqrt{2 + \sqrt{2}}, \sqrt{2 + \sqrt{2 + \sqrt{2}}},$
$\sqrt{2 + \sqrt{2 + \sqrt{2 + \sqrt{2}}}}$, or $1.41421, 1.84776, 1.96157, 1.99037$
c. 2 **97. a.** $1, 1, 2, 3, 5, 8, 13, 21, 34, 55$ **b.** No **99. b.** $1, 2,$
$1.5, 1.6667, 1.6$ **c.** Approx. 1.618 **e.** $\dfrac{a + \sqrt{a^2 + 4b}}{2}$
101. Given a tolerance $\varepsilon > 0$, look beyond a_N, where $N > 1/\varepsilon$.
103. Given a tolerance $\varepsilon > 0$, look beyond a_N, where $N > \frac{1}{4}\sqrt{3/\varepsilon}$,
provided $\varepsilon < \frac{3}{4}$. **105.** Given a tolerance $\varepsilon > 0$, look beyond a_N,
where $N > c/(\varepsilon b^2)$. **107.** $a < 1$ **109.** $\{n^2 + 2n - 17\}_{n=3}^{\infty}$
111. $n = 4, n = 6, n = 25$

Section 10.3 Exercises, pp. 668–671

1. The constant r in the series $\sum\limits_{k=0}^{\infty} ar^k$ **3.** No **5. a.** $a = \frac{2}{3}; r = \frac{1}{5}$
b. $a = \frac{1}{27}; r = -\frac{1}{3}$ **7.** $S_n = \frac{1}{4} - \frac{1}{n+4}; S_{36} = \frac{9}{40}$ **9.** 9841
11. Approx. 1.1905 **13.** Approx. 0.5392 **15.** $\dfrac{1093}{2916}$
17. $\$15,920.22$ **19. a.** $\frac{7}{9}$ **21.** $\frac{4}{3}$ **23.** $\frac{10}{19}$ **25.** 10 **27.** Diverges
29. $\dfrac{1}{e^2 - 1}$ **31.** $\dfrac{1}{7}$ **33.** $\dfrac{1}{500}$ **35.** $\dfrac{3\pi}{\pi + 1}$ **37.** $\dfrac{\pi}{\pi - e}$ **39.** $\dfrac{9}{460}$
41. $\dfrac{4}{11}$ **43.** $A_5 = 266.406; A_{10} = 266.666; A_{30} = 266.667;$
$\lim\limits_{n \to \infty} A_n = 266\frac{2}{3}$ mg, which is the steady-state level. **45.** 400 mg
47. $0.\overline{3} = \sum\limits_{k=1}^{\infty} 3(0.1)^k = \frac{1}{3}$ **49.** $0.\overline{037} = \sum\limits_{k=1}^{\infty} 37(0.001)^k = \frac{1}{27}$
51. $0.\overline{456} = \sum\limits_{k=0}^{\infty} 0.456(0.001)^k = \frac{152}{333}$
53. $0.00\overline{952} = \sum\limits_{k=0}^{\infty} 0.00952(0.001)^k = \frac{238}{24,975}$
55. $S_n = \frac{n}{2n + 4}; \frac{1}{2}$ **57.** $S_n = \frac{1}{7} - \frac{1}{n+7}; \frac{1}{7}$
59. $S_n = \frac{1}{9} - \frac{1}{4n+9}; \frac{1}{9}$ **61.** $S_n = \ln(n + 1)$; diverges

63. $S_n = \dfrac{1}{p+1} - \dfrac{1}{n+p+1}; \dfrac{1}{p+1}$

65. $S_n = \left(\dfrac{1}{\sqrt{2}} + \dfrac{1}{\sqrt{3}}\right) - \left(\dfrac{1}{\sqrt{n+2}} + \dfrac{1}{\sqrt{n+3}}\right); \dfrac{1}{\sqrt{2}} + \dfrac{1}{\sqrt{3}}$

67. $S_n = \dfrac{13}{12} - \dfrac{1}{n+2} - \dfrac{3}{n+3} - \dfrac{1}{n+4}; \dfrac{13}{12}$

69. $S_n = \tan^{-1}(n+1) - \tan^{-1}1; \dfrac{\pi}{4}$ **71. a, b.** $\dfrac{4}{3}$ **73.** $-\dfrac{1}{4}$

75. $\dfrac{2500}{19}$ **77.** $-\dfrac{2}{15}$ **79.** $\dfrac{1}{\ln 2}$ **81.** -2 **83.** $\dfrac{113}{30}$ **85.** $\dfrac{17}{10}$

87. a. True   **b.** True   **c.** False   **d.** False   **e.** True   **f.** False
g. True   **89. a.** $\dfrac{1}{5}$ **b.** Approx. 0.19999695 **91.** Approx. 0.96

95. 462 months   **99.** $\displaystyle\sum_{k=0}^{\infty}\left(\dfrac{1}{4}\right)^k A_1 = \dfrac{A_1}{1 - 1/4} = \dfrac{4}{3}A_1$

101. a. $L_n = 3\left(\dfrac{4}{3}\right)^n$, so $\lim_{n\to\infty} L_n = \infty$ **b.** $\lim_{n\to\infty} A_n = \dfrac{2\sqrt{3}}{5}$

103. $R_n = S - S_n = \dfrac{1}{1-r} - \left(\dfrac{1-r^n}{1-r}\right) = \dfrac{r^n}{1-r}$ **105. a.** 60

b. 9   **107. a.** 13   **b.** 15   **109. a.** $1, \dfrac{5}{6}, \dfrac{2}{3}$, undefined, undefined

b. $(-1, 1)$ **111.** Converges for x in $(-\infty, -2)$ or $(0, \infty)$; $x = \dfrac{1}{2}$

Section 10.4 Exercises, pp. 680–683

1. The series diverges.   **3.** $\lim_{k\to\infty} a_k = 0$ **5.** Converges for $p > 1$ and
diverges for $p \le 1$ **7.** $R_n = S - S_n$ **9.** Diverges
11. Inconclusive   **13.** Diverges   **15.** Diverges   **17.** Diverges
19. Diverges   **21.** Converges   **23.** Diverges   **25.** Converges
27. Diverges   **29.** Converges   **31.** Converges   **33.** Converges
35. Diverges   **37.** Diverges   **39. a.** $S \approx S_2 = 1.0078125$
b. $R_2 < 0.0026042$ **c.** $L_2 = 1.0080411; U_2 = 1.0104167$
41. a. $\dfrac{1}{5n^5}$ **b.** 3   **c.** $L_n = S_n + \dfrac{1}{5(n+1)^5}; U_n = S_n + \dfrac{1}{5n^5}$
d. $(1.017342754, 1.017343512)$ **43. a.** $\dfrac{3^{-n}}{\ln 3}$ **b.** 7
c. $L_n = S_n + \dfrac{3^{-n-1}}{\ln 3}; U_n = S_n + \dfrac{3^{-n}}{\ln 3}$
d. $(0.499996671, 0.500006947)$ **45.** 1.0083   **47. a.** False
b. True   **c.** False   **d.** True   **e.** False   **f.** False   **49.** Converges
51. Converges   **53.** Diverges   **55.** Diverges   **57.** Diverges
59. Converges   **61.** Converges   **63.** Converges   **65. a.** $p > 1$
b. $\displaystyle\sum_{k=2}^{\infty} \dfrac{1}{k(\ln k)^2}$ converges faster.   **67.** $\zeta(3) \approx 1.202, \zeta(5) \approx 1.037$
69. $\dfrac{\pi^2}{8}$ **73. a.** $\dfrac{1}{2}, \dfrac{7}{12}, \dfrac{37}{60}$

Section 10.5 Exercises, pp. 687–688

1. Find an appropriate comparison series. Then take the limit of the
ratio of the terms of the given series and the comparison series as
$n \to \infty$. The value of the limit determines whether the series converges.
3. $\displaystyle\sum_{k=1}^{\infty} \dfrac{1}{k^2}$ **5.** $\displaystyle\sum_{k=1}^{\infty}\left(\dfrac{2}{3}\right)^k$ **7.** $\displaystyle\sum_{k=1}^{\infty} \dfrac{1}{k}$ **9.** Converges
11. Diverges   **13.** Converges   **15.** Converges   **17.** Converges
19. Diverges   **21.** Diverges   **23.** Converges   **25.** Diverges
27. Converges   **29.** Diverges   **31.** Diverges   **33.** Diverges
35. Converges   **37. a.** False   **b.** True   **c.** True   **d.** True

39. Converges   **41.** Diverges   **43.** Diverges   **45.** Diverges
47. Converges   **49.** Diverges   **51.** Converges   **53.** Converges
55. Diverges   **57.** Converges   **59.** Diverges   **61.** Converges

Section 10.6 Exercises, pp. 694–696

1. Because $S_{n+1} - S_n = (-1)^n a_{n+1}$ alternates sign
3. Because the remainder $R_n = S - S_n$ alternates sign
5. $|R_n| = |S - S_n| \le |S_{n+1} - S_n| = a_{n+1}$ **7.** No; if a series
of positive terms converges, it does so absolutely and not conditionally.
9. Yes, $\displaystyle\sum_{k=1}^{\infty} \dfrac{(-1)^k}{k^2}$ has this property.   **11.** Converges   **13.** Diverges
15. Converges   **17.** Converges   **19.** Diverges   **21.** Diverges
23. Diverges   **25.** Diverges   **27.** Converges   **29.** $S_4 = -0.945939$;
$|S - S_4| \le 0.0016$ **31.** $S_5 = 0.70696; |S - S_5| \le 0.001536$
33. 10,000   **35.** 5000   **37.** 10   **39.** -0.973 **41.** -0.269
43. -0.783 **45.** Converges conditionally   **47.** Converges absolutely
49. Converges absolutely   **51.** Converges absolutely   **53.** Diverges
55. Converges conditionally   **57.** Diverges   **59.** Converges absolutely   **61.** Converges conditionally   **63.** Converges absolutely
65. a. False   **b.** True   **c.** True   **d.** True   **e.** False   **f.** True
g. True   **69.** x and y are divergent series.
71. b. $S_{2n} = \displaystyle\sum_{k=1}^{n}\left(\dfrac{1}{k^2} - \dfrac{1}{k}\right)$

Section 10.7 Exercises, pp. 699–700

1. Take the limit of the magnitude of the ratio of consecutive terms
of the series as $k \to \infty$. The value of the limit determines whether
the series converges absolutely or diverges.   **3.** 999,000
5. $\dfrac{1}{(k+2)(k+1)}$ **7.** Ratio Test   **9.** Converges absolutely
11. Diverges   **13.** Converges absolutely   **15.** Converges absolutely
17. Diverges   **19.** Diverges   **21.** Converges absolutely
23. Converges absolutely   **25.** Diverges   **27.** Converges absolutely
29. Diverges   **31. a.** False   **b.** True   **c.** True   **d.** True
33. Converges absolutely   **35.** Diverges   **37.** Converges absolutely
39. Converges conditionally   **41.** Converges absolutely
43. Converges absolutely   **45.** Converges conditionally
47. Converges absolutely   **49.** Converges conditionally
51. $p > 1$ **53.** $p > 1$ **55.** $p < 1$ **57.** Diverges for all p
59. $-1 < x < 1$ **61.** $-1 \le x \le 1$ **63.** $-2 < x < 2$

Section 10.8 Exercises, pp. 703–704

1. Root Test   **3.** Divergence Test   **5.** p-series Test or Limit Comparison Test   **7.** Comparison Test or Limit Comparison Test
9. Alternating Series Test   **11.** Diverges   **13.** Diverges
15. Converges   **17.** Diverges   **19.** Converges   **21.** Converges
23. Converges   **25.** Converges   **27.** Converges   **29.** Diverges
31. Converges   **33.** Diverges   **35.** Converges   **37.** Diverges
39. Diverges   **41.** Converges   **43.** Diverges   **45.** Converges
47. Diverges   **49.** Converges   **51.** Diverges   **53.** Converges
55. Diverges   **57.** Converges   **59.** Converges   **61.** Diverges
63. Diverges   **65.** Converges   **67.** Converges   **69.** Diverges
71. Converges   **73.** Converges   **75.** Converges   **77.** Diverges
79. Diverges   **81.** Converges   **83.** Converges   **85.** Converges
87. a. False   **b.** True   **c.** True   **d.** False   **89.** Diverges
91. Diverges   **93.** Diverges

Chapter 10 Review Exercises, pp. 704–707

1. a. False **b.** False **c.** True **d.** False **e.** True **f.** False
g. False **h.** True **3.** Approx. 1.25; approx. 0.05 **5.** $\lim_{k \to \infty} a_k = 0$,

$\lim_{n \to \infty} S_n = 8$ **7.** $a_k = \dfrac{1}{k}$ **9. a.** 0 **b.** $\dfrac{5}{9}$ **11. a.** Yes; $\lim_{k \to \infty} a_k = 1$

b. No; $\lim_{k \to \infty} a_k \neq 0$ **13.** Diverges **15.** 5 **17.** 0 **19.** 0 **21.** $1/e$

23. Diverges **25. a.** 80, 48, 32, 24, 20 **b.** 16 **27.** Diverges

29. Diverges **31.** Diverges **33.** $\dfrac{3\pi}{4}$ **35.** 3 **37.** 2/9

39. $\dfrac{311}{990}$ **41.** 200 mg **43.** Diverges **45.** Diverges **47.** Converges
49. Converges **51.** Converges **53.** Converges **55.** Converges
57. Diverges **59.** Converges **61.** Converges **63.** Converges
65. Converges **67.** Converges **69.** Converges **71.** Converges
73. Diverges **75.** Diverges **77.** Converges conditionally
79. Converges absolutely **81.** Diverges **83.** Converges absolutely
85. Converges absolutely **87.** Diverges **89. a.** Approx. 1.03666
b. 0.0004 **c.** $L_5 = 1.03685$; $U_5 = 1.03706$ **91.** 0.0067
93. 100 **95. a.** 803 m, 1283 m, $2000(1 - 0.95^N)$ m **b.** 2000 m

97. a. $\dfrac{\pi}{2^{n-1}}$ **b.** 2π **99. a.** $T_1 = \dfrac{\sqrt{3}}{16}, T_2 = \dfrac{7\sqrt{3}}{64}$

b. $T_n = \dfrac{\sqrt{3}}{4}\left(1 - \left(\dfrac{3}{4}\right)^n\right)$ **c.** $\lim_{n \to \infty} T_n = \dfrac{\sqrt{3}}{4}$ **d.** 0

101. $\sqrt{\dfrac{20}{g}\left(\dfrac{1 + \sqrt{p}}{1 - \sqrt{p}}\right)}$ s

CHAPTER 11

Section 11.1 Exercises, pp. 718–721

1. $f(0) = p_2(0)$, $f'(0) = p_2'(0)$, and $f''(0) = p_2''(0)$
3. 1, 1.05, 1.04875 **5.** $p_3(x) = 1 + x^2 + x^3$; 1.048
7. $p_3(x) = 1 + (x - 2) + 2(x - 2)^3$; 0.898
9. a. $p_1(x) = 8 + 12(x - 1)$
b. $p_2(x) = 8 + 12(x - 1) + 3(x - 1)^2$ **c.** 9.2; 9.23
11. a. $p_1(x) = 1 - 2x$ **b.** $p_2(x) = 1 - 2x + 2x^2$ **c.** 0.8, 0.82
13. a. $p_1(x) = 1 - x$ **b.** $p_2(x) = 1 - x + x^2$ **c.** 0.95, 0.9525
15. a. $p_1(x) = 2 + \frac{1}{12}(x - 8)$
b. $p_2(x) = 2 + \frac{1}{12}(x - 8) - \frac{1}{288}(x - 8)^2$ **c.** $1.958\overline{3}$, 1.95747
17. $p_1(x) = 1, p_2(x) = p_3(x) = 1 - 18x^2, p_4(x) = 1 - 18x^2 + 54x^4$
19. $p_3(x) = 1 - 3x + 6x^2 - 10x^3$,
$p_4(x) = 1 - 3x + 6x^2 - 10x^3 + 15x^4$
21. $p_1(x) = 1 + 3(x - 1), p_2(x) = 1 + 3(x - 1) + 3(x - 1)^2$,
$p_3(x) = 1 + 3(x - 1) + 3(x - 1)^2 + (x - 1)^3$

23. $p_3(x) = 1 + \dfrac{1}{e}(x - e) - \dfrac{1}{2e^2}(x - e)^2 + \dfrac{1}{3e^3}(x - e)^3$

25. a. $p_1(x) = -x, p_2(x) = -x - \dfrac{x^2}{2}$

b.

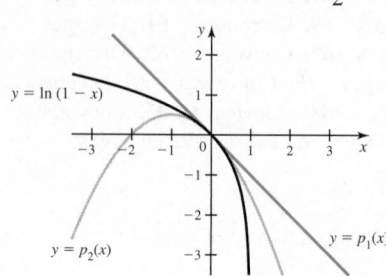

$y = \ln(1 - x)$
$y = p_2(x)$
$y = p_1(x)$

27. a. $p_1(x) = \dfrac{\sqrt{2}}{2} + \dfrac{\sqrt{2}}{2}\left(x - \dfrac{\pi}{4}\right)$,

$p_2(x) = \dfrac{\sqrt{2}}{2} + \dfrac{\sqrt{2}}{2}\left(x - \dfrac{\pi}{4}\right) - \dfrac{\sqrt{2}}{4}\left(x - \dfrac{\pi}{4}\right)^2$

b.

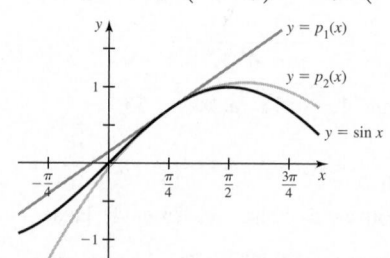

$y = p_1(x)$
$y = p_2(x)$
$y = \sin x$

29. a. 1.0247 **b.** 7.6×10^{-6} **31. a.** 0.8613 **b.** 5.4×10^{-4}
33. a. 1.1274988 **b.** Approx. 8.85×10^{-6} (Answers may vary if
intermediate calculations are rounded.) **35. a.** Approx. -0.10033333
b. Approx. 1.34×10^{-6} (Answers may vary if intermediate calculations are rounded.) **37. a.** 1.0295635 **b.** Approx. 4.86×10^{-7}
(Answers may vary if intermediate calculations are rounded.)
39. a. Approx. 0.52083333 **b.** Approx. 2.62×10^{-4} (Answers
may vary if intermediate calculations are rounded.)

41. $R_n(x) = \dfrac{\sin^{(n+1)}(c)}{(n + 1)!} x^{n+1}$, for c between x and 0

43. $R_n(x) = \dfrac{(-1)^{n+1}e^{-c}}{(n + 1)!} x^{n+1}$, for c between x and 0

45. $R_n(x) = \dfrac{\sin^{(n+1)}(c)}{(n + 1)!}\left(x - \dfrac{\pi}{2}\right)^{n+1}$, for c between x and $\dfrac{\pi}{2}$

47. 2.0×10^{-5} **49.** 1.6×10^{-5} $(e^{0.25} < 2)$ **51.** 2.6×10^{-4}
53. With $n = 4$, $|\text{error}| \leq 2.5 \times 10^{-3}$
55. With $n = 2$, $|\text{error}| \leq 4.2 \times 10^{-2}$ $(e^{0.5} < 2)$
57. With $n = 2$, $|\text{error}| \leq 5.4 \times 10^{-3}$ **59.** 4 **61.** 3 **63.** 1
65. a. False **b.** True **c.** True **d.** True **67. a.** C **b.** E
c. A **d.** D **e.** B **f.** F **69. a.** 0.1; 1.7×10^{-4} **b.** 0.2;
1.3×10^{-3} **71. a.** 0.995; 4.2×10^{-6} **b.** 0.98; 6.7×10^{-5}
73. a. 1.05; 1.3×10^{-3} **b.** 1.1; 5×10^{-3} **75. a.** 1.1; 10^{-2}
b. 1.2; 4×10^{-2}

77. a.

x	$|\sec x - p_2(x)|$	$|\sec x - p_4(x)|$
-0.2	3.39×10^{-4}	5.51×10^{-6}
-0.1	2.09×10^{-5}	8.51×10^{-8}
0.0	0	0
0.1	2.09×10^{-5}	8.51×10^{-8}
0.2	3.39×10^{-4}	5.51×10^{-6}

b. The errors decrease as $|x|$ decreases.

79. a.

x	$|e^{-x} - p_1(x)|$	$|e^{-x} - p_2(x)|$
-0.2	2.14×10^{-2}	1.40×10^{-3}
-0.1	5.17×10^{-3}	1.71×10^{-4}
0.0	0	0
0.1	4.84×10^{-3}	1.63×10^{-4}
0.2	1.87×10^{-2}	1.27×10^{-3}

b. The errors decrease as $|x|$ decreases.

81. Centered at $x = 0$, for all n

83. a. $y = f(a) + f'(a)(x - a)$ **85. a.** $p_5(x) = x - \dfrac{x^3}{6} + \dfrac{x^5}{120}$;

$q_5(x) = -(x - \pi) + \dfrac{1}{6}(x - \pi)^3 - \dfrac{1}{120}(x - \pi)^5$

b.

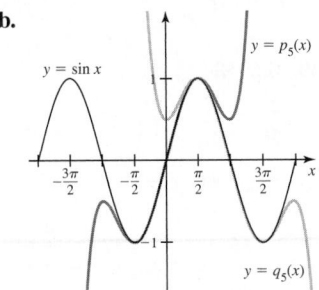

p_5 is a better approximation on $[-\pi, \pi/2]$; q_5 is a better approximation on $(\pi/2, 2\pi]$.

c.

x	$|\sin x - p_5(x)|$	$|\sin x - q_5(x)|$
$\pi/4$	3.6×10^{-5}	7.4×10^{-2}
$\pi/2$	4.5×10^{-3}	4.5×10^{-3}
$3\pi/4$	7.4×10^{-2}	3.6×10^{-5}
$5\pi/4$	2.3	3.6×10^{-5}
$7\pi/4$	20	7.4×10^{-2}

d. p_5 is a better approximation at $x = \pi/4$; at $x = \pi/2$ the errors are equal.

87. a. $p_1(x) = 6 + \dfrac{1}{12}(x - 36)$; $q_1(x) = 7 + \dfrac{1}{14}(x - 49)$

b.

x	$|\sqrt{x} - p_1(x)|$	$|\sqrt{x} - q_1(x)|$
37	5.7×10^{-4}	6.0×10^{-2}
39	5.0×10^{-3}	4.1×10^{-2}
41	1.4×10^{-2}	2.5×10^{-2}
43	2.6×10^{-2}	1.4×10^{-2}
45	4.2×10^{-2}	6.1×10^{-3}
47	6.1×10^{-2}	1.5×10^{-3}

c. p_1 is a better approximation at $x = 37, 39,$ and 41.

Section 11.2 Exercises, pp. 729–730

1. $c_0 + c_1 x + c_2 x^2 + c_3 x^3$ **3.** Ratio and Root Tests   **5.** The radius of convergence does not change. The interval of convergence may change.   **7.** $R = 10$; $[-8, 12]$ **9.** $R = \frac{1}{2}$; $\left(-\frac{1}{2}, \frac{1}{2}\right)$

11. $R = 0$; $\{x: x = 0\}$ **13.** $R = \infty$; $(-\infty, \infty)$ **15.** $R = 3$; $(-3, 3)$
17. $R = \infty$; $(-\infty, \infty)$ **19.** $R = 2$; $(-2, 2)$ **21.** $R = \infty$; $(-\infty, \infty)$
23. $R = 1$; $(0, 2]$ **25.** $R = \frac{1}{4}$; $\left[0, \frac{1}{2}\right]$ **27.** $R = 5$; $(-3, 7)$
29. $R = \infty$; $(-\infty, \infty)$ **31.** $R = \sqrt{3}$; $(-\sqrt{3}, \sqrt{3})$ **33.** $R = 1$; $(0, 2)$
35. $R = \infty$; $(-\infty, \infty)$ **37.** $R = e$ **39.** $R = e^4$

41. $\displaystyle\sum_{k=0}^{\infty} (3x)^k$; $\left(-\frac{1}{3}, \frac{1}{3}\right)$ **43.** $2\displaystyle\sum_{k=0}^{\infty} x^{k+3}$; $(-1, 1)$

45. $4\displaystyle\sum_{k=0}^{\infty} x^{k+12}$; $(-1, 1)$ **47.** $-\displaystyle\sum_{k=1}^{\infty} \dfrac{(3x)^k}{k}$; $\left[-\frac{1}{3}, \frac{1}{3}\right)$

49. $-2\displaystyle\sum_{k=1}^{\infty} \dfrac{x^{k+6}}{k}$; $[-1, 1)$ **51.** $g(x) = 2\displaystyle\sum_{k=1}^{\infty} k(2x)^{k-1}$; $\left(-\frac{1}{2}, \frac{1}{2}\right)$

53. $g(x) = \displaystyle\sum_{k=1}^{\infty} (-1)^k kx^{k-1}$; $(-1, 1)$

55. $g(x) = -\displaystyle\sum_{k=1}^{\infty} \dfrac{3^k x^k}{k}$; $\left[-\frac{1}{3}, \frac{1}{3}\right)$

57. $\displaystyle\sum_{k=1}^{\infty} (-1)^{k+1} 2kx^{2k-1}$; $(-1, 1)$ **59.** $\displaystyle\sum_{k=0}^{\infty} \left(-\dfrac{x}{3}\right)^k$; $(-3, 3)$

61. $\ln 2 - \dfrac{1}{2}\displaystyle\sum_{k=1}^{\infty} \dfrac{x^{2k}}{k \cdot 4^k}$; $(-2, 2)$ **63. a.** True   **b.** True   **c.** True

d. True   **65.** $|x - a| < R$ **67.** $f(x) = \dfrac{1}{3 - \sqrt{x}}$; $1 < x < 9$

69. $f(x) = \dfrac{e^x}{e^x - 1}$; $0 < x < \infty$ **71.** $f(x) = \dfrac{3}{4 - x^2}$; $-2 < x < 2$

73. $\displaystyle\sum_{k=0}^{\infty} \dfrac{(-3x)^k}{k!}$; $-\infty < x < \infty$

75. $\displaystyle\lim_{k \to \infty} \left|\dfrac{c_{k+1} x^{k+1}}{c_k x^k}\right| = \lim_{k \to \infty} \left|\dfrac{c_{k+1} x^{k+m+1}}{c_k x^{k+m}}\right|$, so by the Ratio Test, the two series have the same radius of convergence.
77. a. $f(x) g(x) = c_0 d_0 + (c_0 d_1 + c_1 d_0)x +$

$(c_0 d_2 + c_1 d_1 + c_2 d_0)x^2$ **b.** $\displaystyle\sum_{k=0}^{n} c_k d_{n-k}$

Section 11.3 Exercises, pp. 740–742

1. The nth Taylor polynomial is the nth partial sum of the corresponding Taylor series.   **3.** $\displaystyle\sum_{k=0}^{\infty} \dfrac{(x - 2)^k}{k!}$ **5.** Replace x with x^2 in the Taylor series for $f(x)$; $|x| < 1$.   **7.** The Taylor series for a function f converges to f on an interval if, for all x in the interval, $\displaystyle\lim_{n \to \infty} R_n(x) = 0$, where $R_n(x)$ is the remainder at x.
9. a. $1 - 2(x - 1) + 3(x - 1)^2 - 4(x - 1)^2$

b. $\displaystyle\sum_{k=0}^{\infty} (-1)^k (k + 1)(x - 1)^k$ **c.** $(0, 2)$ **11. a.** $1 - x + \dfrac{x^2}{2!} - \dfrac{x^3}{3!}$

b. $\displaystyle\sum_{k=0}^{\infty} \dfrac{(-1)^k x^k}{k!}$ **c.** $(-\infty, \infty)$ **13. a.** $2 + 6x + 12x^2 + 20x^3$

b. $\displaystyle\sum_{k=0}^{\infty} (k + 1)(k + 2)x^k$ **c.** $(-1, 1)$ **15. a.** $1 - x^2 + x^4 - x^6$

b. $\displaystyle\sum_{k=0}^{\infty} (-1)^k x^{2k}$ **c.** $(-1, 1)$ **17. a.** $1 + 2x + \dfrac{(2x)^2}{2!} + \dfrac{(2x)^3}{3!}$

b. $\displaystyle\sum_{k=0}^{\infty} \dfrac{(2x)^k}{k!}$ **c.** $(-\infty, \infty)$ **19. a.** $\dfrac{x}{2} - \dfrac{x^3}{3 \cdot 2^3} + \dfrac{x^5}{5 \cdot 2^5} - \dfrac{x^7}{7 \cdot 2^7}$

b. $\displaystyle\sum_{k=0}^{\infty} \dfrac{(-1)^k x^{2k+1}}{(2k + 1)2^{2k+1}}$ **c.** $[-2, 2]$

21. a. $1 + (\ln 3)x + \dfrac{\ln^2 3}{2}x^2 + \dfrac{\ln^3 3}{6}x^3$ **b.** $\displaystyle\sum_{k=0}^{\infty} \dfrac{\ln^k 3}{k!}x^k$ **c.** $(-\infty, \infty)$

23. a. $1 + \dfrac{(3x)^2}{2} + \dfrac{(3x)^4}{24} + \dfrac{(3x)^6}{720}$ **b.** $\displaystyle\sum_{k=0}^{\infty} \dfrac{(3x)^{2k}}{(2k)!}$ **c.** $(-\infty, \infty)$

25. a. $(x - 3) - \dfrac{1}{2}(x - 3)^2 + \dfrac{1}{3}(x - 3)^3 - \dfrac{1}{4}(x - 3)^4$

b. $\displaystyle\sum_{k=1}^{\infty} \dfrac{(-1)^{k+1}(x - 3)^k}{k}$ **c.** $(2, 4]$

27. a. $1 - \dfrac{(x - \pi/2)^2}{2!} + \dfrac{(x - \pi/2)^4}{4!} - \dfrac{(x - \pi/2)^6}{6!}$

b. $\displaystyle\sum_{k=0}^{\infty} \dfrac{(-1)^k}{(2k)!}(x - \pi/2)^{2k}$

29. a. $1 - (x - 1) + (x - 1)^2 - (x - 1)^3$ **b.** $\displaystyle\sum_{k=0}^{\infty} (-1)^k (x - 1)^k$

31. a. $\ln 3 + \dfrac{(x - 3)}{3} - \dfrac{(x - 3)^2}{3^2 \cdot 2} + \dfrac{(x - 3)^3}{3^3 \cdot 3}$

b. $\ln 3 + \displaystyle\sum_{k=1}^{\infty} \dfrac{(-1)^{k+1}(x - 3)^k}{k3^k}$

33. a. $2 + 2 (\ln 2)(x - 1) + (\ln^2 2)(x - 1)^2 + \dfrac{\ln^3 2}{3}(x - 1)^3$

b. $\displaystyle\sum_{k=0}^{\infty} \dfrac{2(x - 1)^k \ln^k 2}{k!}$ **35.** $x^2 - \dfrac{x^4}{2} + \dfrac{x^6}{3} - \dfrac{x^8}{4}$

37. $1 + 2x + 4x^2 + 8x^3$ **39.** $1 + \dfrac{x}{2} + \dfrac{x^2}{6} + \dfrac{x^3}{24}$

41. $1 - x^4 + x^8 - x^{12}$ **43.** $x^2 + \dfrac{x^6}{6} + \dfrac{x^{10}}{120} + \dfrac{x^{14}}{5040}$

45. a. $1 - 2x + 3x^2 - 4x^3$ **b.** 0.826

47. a. $1 + \frac{1}{4}x - \frac{3}{32}x^2 + \frac{7}{128}x^3$ **b.** 1.029

49. a. $1 - \frac{2}{3}x + \frac{5}{9}x^2 - \frac{40}{81}x^3$ **b.** 0.895 **51.** $1 + \dfrac{x^2}{2} - \dfrac{x^4}{8} + \dfrac{x^6}{16}$;

$[-1, 1]$ **53.** $3 - \dfrac{3x}{2} - \dfrac{3x^2}{8} - \dfrac{3x^3}{16}$; $[-1, 1]$

55. $a + \dfrac{x^2}{2a} - \dfrac{x^4}{8a^3} + \dfrac{x^6}{16a^5}$; $|x| \le a$

57. $1 - 8x + 48x^2 - 256x^3$ **59.** $\dfrac{1}{16} - \dfrac{x^2}{32} + \dfrac{3x^4}{256} - \dfrac{x^6}{256}$

61. $\dfrac{1}{9} - \dfrac{2}{9}\left(\dfrac{4x}{3}\right) + \dfrac{3}{9}\left(\dfrac{4x}{3}\right)^2 - \dfrac{4}{9}\left(\dfrac{4x}{3}\right)^3$

63. $R_n(x) = \dfrac{f^{(n+1)}(c)}{(n + 1)!} x^{n+1}$, where c is between 0 and x and

$f^{(n+1)}(c) = \pm\sin c$ or $\pm\cos c$. Therefore, $|R_n(x)| \le \dfrac{|x|^{n+1}}{(n + 1)!} \to 0$

as $n \to \infty$, for $-\infty < x < \infty$.   **65.** $R_n(x) = \dfrac{f^{(n+1)}(c)}{(n + 1)!} x^{n+1}$,

where c is between 0 and x and $f^{(n+1)}(c) = (-1)^n e^{-c}$. Therefore,

$|R_n(x)| \le \dfrac{|x|^{n+1}}{e^c(n + 1)!} \to 0$ as $n \to \infty$, for $-\infty < x < \infty$.

67. a. False   **b.** True   **c.** False   **d.** False   **e.** True

69. a. $1 + \dfrac{x^2}{2!} + \dfrac{x^4}{4!} + \dfrac{x^6}{6!}$ **b.** $R = \infty$

71. a. $1 - \frac{2}{3}x^2 + \frac{5}{9}x^4 - \frac{40}{81}x^6$ **b.** $R = 1$

73. a. $1 - \frac{1}{2}x^2 - \frac{1}{8}x^4 - \frac{1}{16}x^6$ **b.** $R = 1$

75. a. $1 - 2x^2 + 3x^4 - 4x^6$ **b.** $R = 1$ **77.** Approx. 3.9149

79. Approx. 1.8989 **85.** $\displaystyle\sum_{k=0}^{\infty} \left(\dfrac{x - 4}{2}\right)^k$ **87.** Use three terms of the

Taylor series for $\cos x$ centered at $a = \pi/4$; 0.766 **89. a.** Use

three terms of the Taylor series for $\sqrt[3]{125 + x}$ centered at $a = 0$;

5.03968 **b.** Use three terms of the Taylor series for $\sqrt[3]{x}$ centered at

$a = 125$; 5.03968 **c.** Yes

Section 11.4 Exercises, pp. 748–750

1. Replace f and g with their Taylor series centered at a and evaluate
the limit.   **3.** Substitute $x = -0.6$ into the Taylor series for e^x
centered at 0. Because the resulting series is an alternating series, the
error can be estimated easily.   **7.** 1 **9.** $\frac{1}{2}$ **11.** 2 **13.** $\frac{2}{3}$ **15.** $\frac{1}{3}$

17. $\frac{3}{5}$ **19.** $-\frac{8}{5}$ **21.** 1 **23.** $\frac{3}{4}$ **25. a.** $1 + x + \dfrac{x^2}{2!} + \cdots + \dfrac{x^n}{n!} + \cdots$

b. e^x **c.** $-\infty < x < \infty$

27. a. $1 - x + x^2 - \cdots (-1)^{n-1}x^{n-1} + \cdots$ **b.** $\dfrac{1}{1 + x}$ **c.** $|x| < 1$

29. a. $-2 + 4x - 8 \cdot \dfrac{x^2}{2!} + \cdots + (-2)^n \dfrac{x^{n-1}}{(n - 1)!} + \cdots$

b. $-2e^{-2x}$ **c.** $-\infty < x < \infty$ **31. a.** $1 - x^2 + x^4 - \cdots$

b. $\dfrac{1}{1 + x^2}$ **c.** $-1 < x < 1$

33. a. $2 + 2t + \dfrac{2t^2}{2!} + \cdots + \dfrac{2t^n}{n!} + \cdots$ **b.** $y(t) = 2e^t$

35. a. $2 + 16t + 24t^2 + 24t^3 + \cdots + \dfrac{3^{n-1} \cdot 16}{n!} t^n + \cdots$

b. $y(t) = \frac{16}{3} e^{3t} - \frac{10}{3}$ **37.** 0.2448 **39.** 0.6958

41. 0.0600 **43.** 0.4994 **45.** $1 + 2 + \dfrac{2^2}{2!} + \dfrac{2^3}{3!}$

47. $1 - 2 + \dfrac{2}{3} - \dfrac{4}{45}$ **49.** $\frac{1}{2} - \frac{1}{8} + \frac{1}{24} - \frac{1}{64}$ **51.** $e - 1$

53. $\displaystyle\sum_{k=1}^{\infty} \dfrac{(-1)^{k+1}x^k}{k}$, for $-1 < x \le 1$; $\ln 2$ **55.** $\dfrac{2}{2 - x}$ **57.** $\dfrac{4}{4 + x^2}$

59. $-\ln (1 - x)$ **61.** $-\dfrac{3x^2}{(3 + x)^2}$ **63.** $\dfrac{6x^2}{(3 - x)^3}$

65. a. False   **b.** False   **c.** True   **67.** $\frac{a}{b}$ **69.** $e^{-1/6}$

71. $f^{(3)}(0) = 0$; $f^{(4)}(0) = 4e$ **73.** $f^{(3)}(0) = 2$; $f^{(4)}(0) = 0$

75. 2 **77.** 1.575 using four terms   **79. a.** $S'(x) = \sin x^2$;

$C'(x) = \cos x^2$ **b.** $\dfrac{x^3}{3} - \dfrac{x^7}{7 \cdot 3!} + \dfrac{x^{11}}{11 \cdot 5!} - \dfrac{x^{15}}{15 \cdot 7!}$;

$x - \dfrac{x^5}{5 \cdot 2!} + \dfrac{x^9}{9 \cdot 4!} - \dfrac{x^{13}}{13 \cdot 6!}$ **c.** $S(0.05) \approx 0.00004166664807$;

$C(-0.25) \approx -0.2499023614$ **d.** 1 **e.** 2

81. a. $1 - \dfrac{x^2}{4} + \dfrac{x^4}{64} - \dfrac{x^6}{2304}$ **b.** $R = \infty$, $-\infty < x < \infty$

Chapter 11 Review Exercises, pp. 750–752

1. a. True   **b.** False   **c.** True   **d.** True   **e.** True

3. $p_2(x) = 1 - \dfrac{3}{2}x^2$ **5.** $p_2(x) = 1 - (x - 1) + \dfrac{3}{2}(x - 1)^2$

7. $p_2(x) = 1 - \dfrac{1}{2}(x - 1)^2$

9. $p_3(x) = \dfrac{5}{4} + \dfrac{3}{4}(x - \ln 2) + \dfrac{5}{8}(x - \ln 2)^2 + \dfrac{1}{8}(x - \ln 2)^3$

11. a. $p_1(x) = 1 + x$; $p_2(x) = 1 + x + \dfrac{x^2}{2}$

b.

n	$p_n(x)$	Error
1	0.92	3.1×10^{-3}
2	0.9232	8.4×10^{-5}

13. a. $p_1(x) = \dfrac{\sqrt{2}}{2} + \dfrac{\sqrt{2}}{2}\left(x - \dfrac{\pi}{4}\right)$;

$p_2(x) = \dfrac{\sqrt{2}}{2} + \dfrac{\sqrt{2}}{2}\left(x - \dfrac{\pi}{4}\right) - \dfrac{\sqrt{2}}{4}\left(x - \dfrac{\pi}{4}\right)^2$

b.

n	$p_n(x)$	Error
1	0.5960	8.2×10^{-3}
2	0.5873	4.7×10^{-4}

15. $|R_3| < \dfrac{\pi^4}{4!}$ **17.** $R = \infty$, $(-\infty, \infty)$ **19.** $R = \infty$, $(-\infty, \infty)$

21. $R = 9$, $(-9, 9)$ **23.** $R = 2$, $[-4, 0)$ **25.** $R = \dfrac{3}{2}$, $[-2, 1]$

27. $R = \dfrac{1}{27}$ **29.** $\displaystyle\sum_{k=0}^{\infty} x^{2k}$; $(-1, 1)$ **31.** $\displaystyle\sum_{k=0}^{\infty} (-5x)^k$; $\left(-\frac{1}{5}, \frac{1}{5}\right)$

33. $\sum_{k=1}^{\infty} k(10x)^{k-1}; \left(-\frac{1}{10}, \frac{1}{10}\right)$ **35.** $1 + 3x + \frac{9x^2}{2!}; \sum_{k=0}^{\infty} \frac{(3x)^k}{k!}$

37. $-(x - \pi/2) + \frac{(x - \pi/2)^3}{3!} - \frac{(x - \pi/2)^5}{5!};$

$\sum_{k=0}^{\infty} (-1)^{k+1} \frac{(x - \pi/2)^{2k+1}}{(2k + 1)!}$

39. $4x - \frac{(4x)^3}{3} + \frac{(4x)^5}{5}; \sum_{k=0}^{\infty} \frac{(-1)^k(4x)^{2k+1}}{2k + 1}$

41. $1 + 2(x - 1)^2 + \frac{2}{3}(x - 1)^4; \sum_{k=0}^{\infty} \frac{4^k(x - 1)^{2k}}{(2k)!}$

43. $1 + \frac{x}{3} - \frac{x^2}{9} + \cdots$ **45.** $1 - \frac{3}{2}x + \frac{3}{2}x^2 - \cdots$

47. $R_n(x) = \frac{(\sinh c + \cosh c)\, x^{n+1}}{(n + 1)!}$, where c is between 0 and x;

$\lim_{n\to\infty} |R_n(x)| = |\sinh c + \cosh c| \lim_{n\to\infty} \frac{|x|^{n+1}}{(n + 1)!} = 0$ because

$|x|^{n+1} \ll (n + 1)!$ for any fixed value of x.

49. $\frac{1}{24}$ **51.** $\frac{1}{8}$ **53.** $\frac{1}{6}$ **55.** Approx. 0.4615   **57.** Approx. 0.3819

59. $11 - \frac{1}{11} - \frac{1}{2 \cdot 11^3} - \frac{1}{2 \cdot 11^5}$ **61.** $-\frac{1}{3} + \frac{1}{3 \cdot 3^3} - \frac{1}{5 \cdot 3^5} + \frac{1}{7 \cdot 3^7}$

63. $y = 4 + 4x + \frac{4^2}{2!}x^2 + \frac{4^3}{3!}x^3 + \cdots + \frac{4^n}{n!}x^n + \cdots = 3 + e^{4x}$

65. a. $\sum_{k=1}^{\infty} \frac{(-1)^{k+1}}{k}$ **b.** $\sum_{k=1}^{\infty} \frac{1}{k2^k}$ **c.** $2\sum_{k=0}^{\infty} \frac{x^{2k+1}}{2k + 1}$

d. $x = \frac{1}{3}; 2\sum_{k=0}^{\infty} \frac{1}{3^{2k+1}(2k + 1)}$ **e.** Series in part (d)

CHAPTER 12

Section 12.1 Exercises, pp. 763–767

1. Plotting $\{(f(t), g(t)): a \le t \le b\}$ generates a curve in the xy-plane.
3. $x = R\cos(\pi t/5), y = -R\sin(\pi t/5)$ **5.** $x = t^2, y = t,$

$-\infty < t < \infty$ **7.** $-\frac{1}{2}$ **9.** $x = t, y = t, 0 \le t \le 6; x = 2t, y = 2t,$

$0 \le t \le 3; x = 3t, y = 3t, 0 \le t \le 2$ (answers will vary)

11. a.

t	-10	-4	0	4	10
x	-20	-8	0	8	20
y	-34	-16	-4	8	26

b.

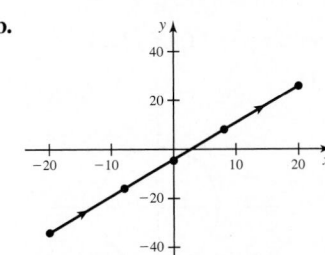

c. $y = \frac{3}{2}x - 4$
d. A line segment rising to the right as t increases

13. a.

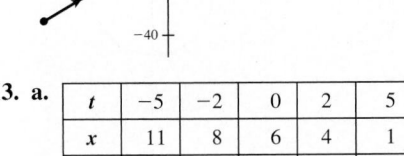

t	-5	-2	0	2	5
x	11	8	6	4	1
y	-18	-9	-3	3	12

b.

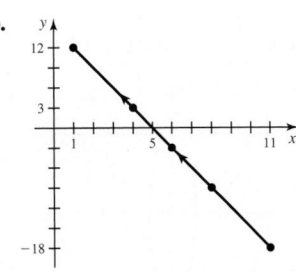

c. $y = -3x + 15$
d. A line segment rising to the left as t increases

15. a. $y = -x + 4$ **b.** A line segment starting at $(3, 1)$ and ending at $(4, 0)$ **17. a.** $y = 3x - 12$ **b.** A line segment starting at $(4, 0)$ and ending at $(8, 12)$ **19. a.** $x^2 + y^2 = 9$ **b.** Lower half of a circle of radius 3 centered at $(0, 0)$; starts at $(-3, 0)$ and ends at $(3, 0)$
21. a. $y = 1 - x^2, -1 \le x \le 1$ **b.** A parabola opening downward with a vertex at $(0, 1)$ starting at $(1, 0)$ and ending at $(-1, 0)$
23. a. $x^2 + (y - 1)^2 = 1$ **b.** A circle of radius 1 centered at $(0, 1)$; generated counterclockwise, starting and ending at $(1, 1)$
25. a. $y = (x + 1)^3$ **b.** A cubic curve rising to the right as r increases   **27. a.** $x^2 + y^2 = 49$ **b.** A circle of radius 7 centered at $(0, 0)$; generated counterclockwise, starting and ending at $(-7, 0)$
29. a. $y = 1, -\infty < x < \infty$ **b.** A horizontal line with y-intercept 1, generated from left to right   **31.** $x^2 + y^2 = 4$ **33.** $y = \sqrt{4 - x^2}$
35. $y = x^2$ **37.** $x = 4\cos t, y = 4\sin t, 0 \le t \le 2\pi$
39. $x = \cos t + 2, y = \sin t + 3, 0 \le t \le 2\pi$
41. $x = 2t, y = 8t; 0 \le t \le 1$
43. $x = t, y = 2t^2 - 4; -1 \le t \le 5$ **45.** $x = 2, y = t; 3 \le t \le 9$
47. $x = 4t - 2, y = -6t + 3; 0 \le t \le 1$ and
$x = t + 1, y = 8t - 11; 1 \le t \le 2$
49. $x = 1 + 2t, y = 1 + 4t; -\infty < t < \infty$
51. $x = t^2, y = t; t \ge 0$
53. $x = 400\cos\left(\frac{4\pi t}{3}\right), y = 400\sin\left(\frac{4\pi t}{3}\right); 0 \le t \le 1.5$
55. $x = 50\cos\left(\frac{\pi t}{12}\right), y(t) = 50\sin\left(\frac{\pi t}{12}\right); 0 \le t \le 24$

57.

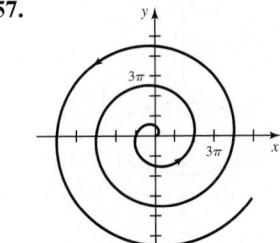

59.

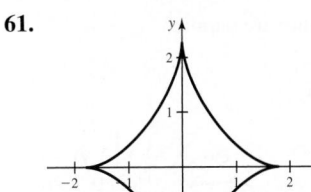

61.

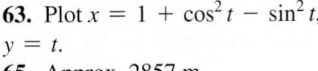

63. Plot $x = 1 + \cos^2 t - \sin^2 t$, $y = t$.
65. Approx. 2857 m

67. a. $\dfrac{dy}{dx} = -2; -2$

b.
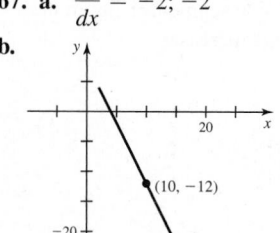

69. a. $\dfrac{dy}{dx} = -8 \cot t; 0$

b.
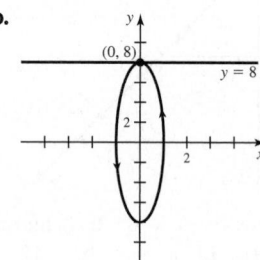

71. a. $\dfrac{dy}{dx} = \dfrac{t^2 + 1}{t^2 - 1}, t \neq 0$; undefined

b.
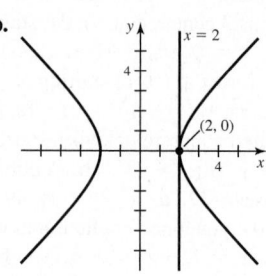

73. $y = \dfrac{13}{4}x + \dfrac{1}{4}$ **75.** $y = x - \dfrac{\pi\sqrt{2}}{4}$ **77.** $\left(-\dfrac{4}{\sqrt{5}}, \dfrac{8}{\sqrt{5}}\right)$ and $\left(\dfrac{4}{\sqrt{5}}, -\dfrac{8}{\sqrt{5}}\right)$ **79.** There is no such point.   **81.** 10   **83.** $\pi\sqrt{2}$

85. $\dfrac{1}{3}(5\sqrt{5} - 8)$ **87.** $\dfrac{3}{2}$ **89. a.** False   **b.** True   **c.** False   **d.** True   **e.** True

91. $0 \leq t \leq 2\pi$

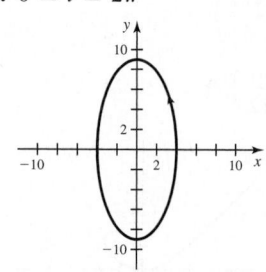

93. $x = 3\cos t, y = \dfrac{3}{2}\sin t$, $0 \leq t \leq 2\pi; \left(\dfrac{x}{3}\right)^2 + \left(\dfrac{2y}{3}\right)^2 = 1$; in the counterclockwise direction

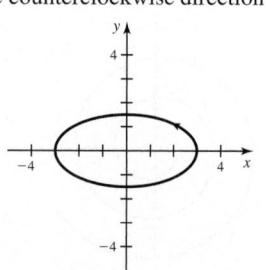

95. a. Lines intersect at $(1, 0)$.   **b.** Lines are parallel.   **c.** Lines intersect at $(4, 6)$.

97.
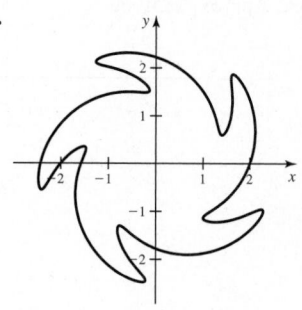

99.
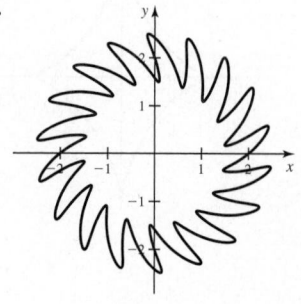

101.
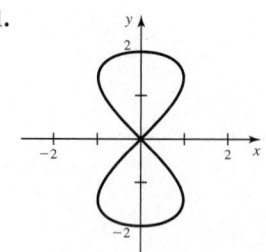

a. $(0, 2)$ and $(0, -2)$
b. $(1, \sqrt{2}), (1, -\sqrt{2})$, $(-1, \sqrt{2}), (-1, -\sqrt{2})$

103. 27π **105.** $\dfrac{3\pi}{8}$ **107. a.** A circle of radius 3 centered at $(0, 4)$

b. A torus (doughnut); $48\pi^2$ **109.** $\dfrac{64\pi}{3}$

111. $\displaystyle\int_0^1 2\pi(e^{3t} + 1)\sqrt{4e^{4t} + 9e^{6t}}\, dt \approx 1445.9$

113. a.
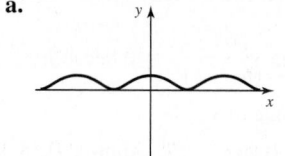

b.
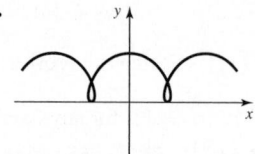

Section 12.2 Exercises, pp. 775–779

1.
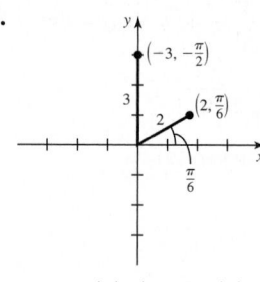

$(-2, -5\pi/6), (2, 13\pi/6)$; $(3, \pi/2), (3, 5\pi/2)$

3. $r^2 = x^2 + y^2, \tan\theta = \dfrac{y}{x}$
5. $r\cos\theta = 5$ or $r = 5\sec\theta$
7. x-axis symmetry occurs if (r, θ) on the graph implies $(r, -\theta)$ is on the graph. y-axis symmetry occurs if (r, θ) on the graph implies $(r, \pi - \theta) = (-r, -\theta)$ is on the graph. Symmetry about the origin occurs if (r, θ) on the graph implies $(-r, \theta) = (r, \theta + \pi)$ is on the graph.

9.
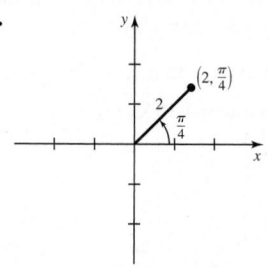

$(-2, -3\pi/4), (2, 9\pi/4)$

11.
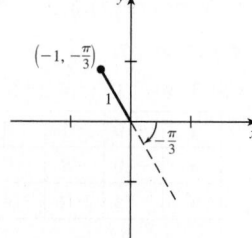

$(1, 2\pi/3), (1, 8\pi/3)$

13.
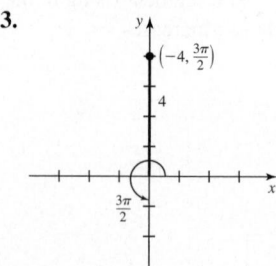

$(4, \pi/2), (4, 5\pi/2)$

15.
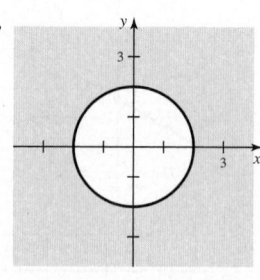

17.

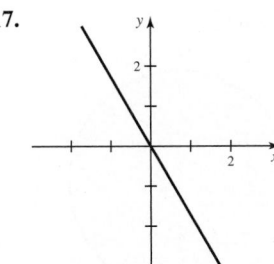

19.

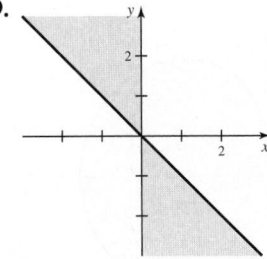

21.

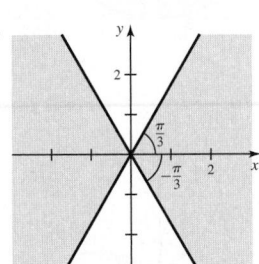

23. $\left(100, -\dfrac{\pi}{4}\right)$

25. $(3\sqrt{2}/2, 3\sqrt{2}/2)$

27. $(1/2, -\sqrt{3}/2)$

29. $(2\sqrt{2}, -2\sqrt{2})$

31. $(2\sqrt{2}, \pi/4), (-2\sqrt{2}, 5\pi/4)$

33. $(2, \pi/3), (-2, 4\pi/3)$

35. $(8, 2\pi/3), (-8, -\pi/3)$

37. $x = -4$; vertical line passing through $(-4, 0)$

39. $x^2 + y^2 = 4$; circle of radius 2 centered at $(0, 0)$

41. $(x - 1)^2 + (y - 1)^2 = 2$; circle of radius $\sqrt{2}$ centered at $(1, 1)$

43. $(x - 3)^2 + (y - 4)^2 = 25$; circle of radius 5 centered at $(3, 4)$

45. $x^2 + (y - 1)^2 = 1$; circle of radius 1 centered at $(0, 1)$ and $x = 0$ **47.** $x^2 + (y - 4)^2 = 16$; circle of radius 4 centered at $(0, 4)$

49. $r = \tan \theta \sec \theta$ **51.** $r^2 = \sec \theta \csc \theta$ or $r^2 = 2 \csc 2\theta$

53.

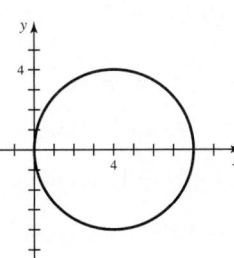

55.

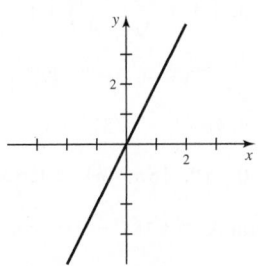

57.

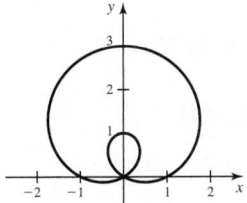

59.

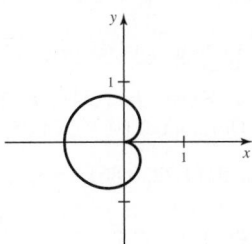

61.

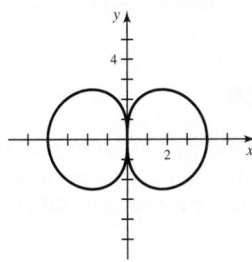

63.

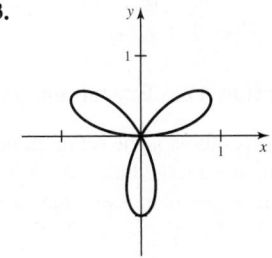

65.

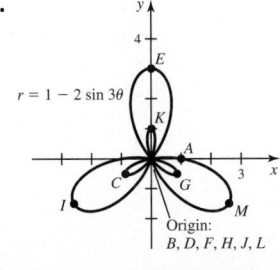

67.

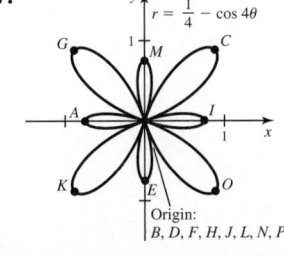

69. $[0, 8\pi]$

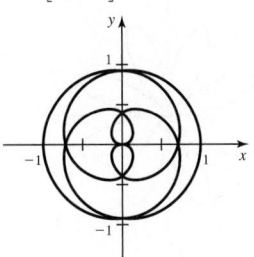

71. $[0, 2\pi]$

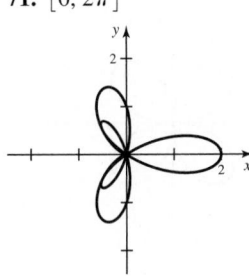

73. $[0, 5\pi]$

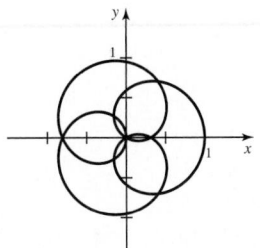

75. $[0, 2\pi]$

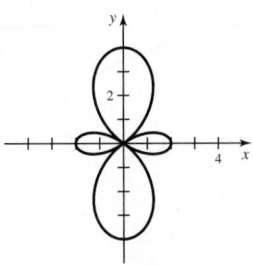

77. a. True   **b.** True   **c.** False   **d.** True   **e.** True

81. A circle of radius 4 and center $(2, \pi/3)$ (polar coordinates)

83. A circle of radius 4 centered at $(2, 3)$ (Cartesian coordinates)

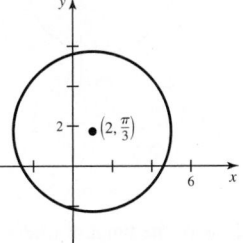

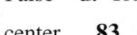

85. a. 132.3 miles   **b.** 264.6 mi/hr

87.

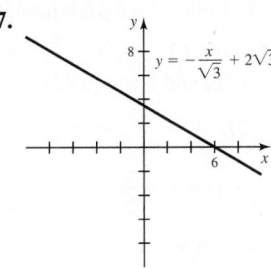

$y = -\dfrac{x}{\sqrt{3}} + 2\sqrt{3}$

89.

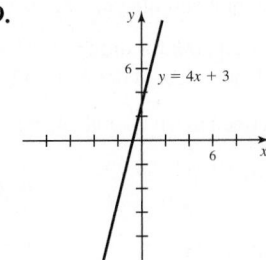

$y = 4x + 3$

91. a. A   **b.** C   **c.** B   **d.** D   **e.** E   **f.** F

93.

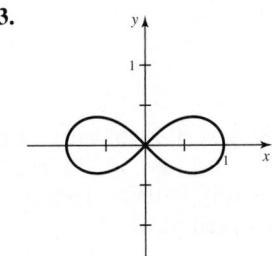

95.

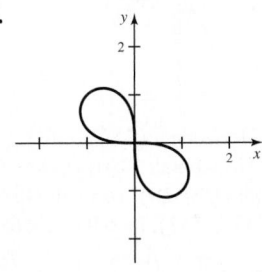

97.

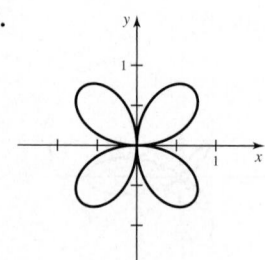

99.

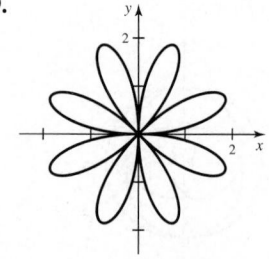

103.

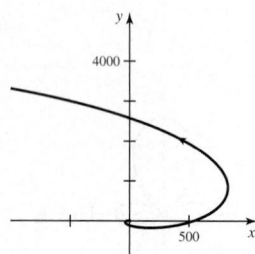

For $a = -1$, the spiral winds inward toward the origin.

105. a.

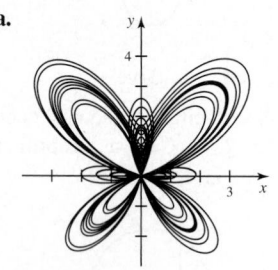

107. a.

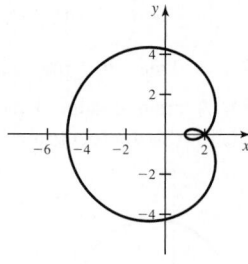

109. Symmetry about the x-axis   **111.** $(x^2 + y^2)^2 = a^2(x^2 - y^2)$

Section 12.3 Exercises, pp. 786–788

1. $x = f(\theta)\cos\theta, y = f(\theta)\sin\theta$ **3.** The slope of the tangent line is the rate of change of the vertical coordinate with respect to the horizontal coordinate.   **5.** $\sqrt{3}$ **7.** $\dfrac{\pi^2}{4}$ **9.** Both curves pass through the origin, but for different values of θ.   **11.** 0   **13.** $-\sqrt{3}$
15. Undefined, undefined   **17.** 0 at $(-4, \pi/2)$ and $(-4, 3\pi/2)$, undefined at $(4, 0)$ and $(4, \pi)$ **19.** ± 1 **21.** $\theta = \dfrac{3\pi}{4}; m = -1$
23. a. $[0, \pi]$

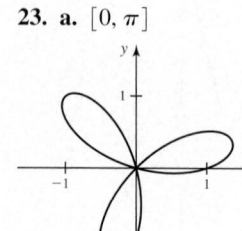

b. $\theta = \dfrac{\pi}{4}, m = 1; \theta = \dfrac{7\pi}{12}, m \approx -3.73;$
$\theta = \dfrac{11\pi}{12}, m \approx -0.27$

25. Horizontal: $(2\sqrt{2}, \pi/4), (-2\sqrt{2}, 3\pi/4)$; vertical: $(0, \pi/2), (4, 0)$
27. Horizontal: $(0, 0)$ $(0.943, 0.955), (-0.943, 2.186), (0.943, 4.097),$
$(-0.943, 5.328)$; vertical: $(0, 0), (0.943, 0.615), (-0.943, 2.526),$
$(0.943, 3.757), (-0.943, 5.668)$ **29.** $(2, 0)$ and $(0, 0)$

31. $\left(1, \dfrac{\pi}{12}\right), \left(1, \dfrac{5\pi}{12}\right), \left(1, \dfrac{7\pi}{12}\right), \left(1, \dfrac{11\pi}{12}\right),$
$\left(1, \dfrac{13\pi}{12}\right), \left(1, \dfrac{17\pi}{12}\right), \left(1, \dfrac{19\pi}{12}\right),$ and $\left(1, \dfrac{23\pi}{12}\right)$

33. 1

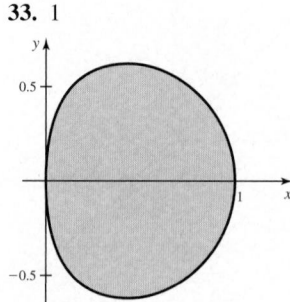

35. 16π

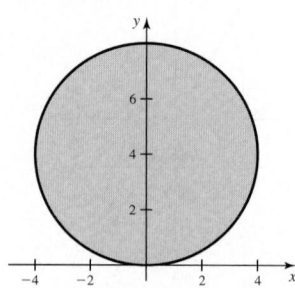

37. $9\pi/2$

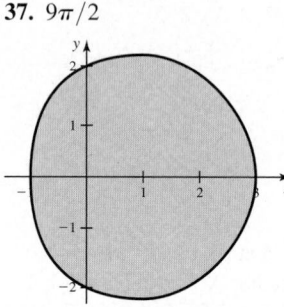

39. $\dfrac{\pi}{12}$

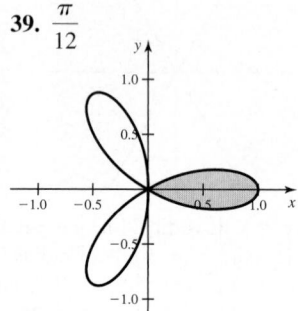

41. a. $(0, 0), \left(\dfrac{3}{\sqrt{2}}, \dfrac{\pi}{4}\right)$ **b.** $\dfrac{9}{8}(\pi - 2)$

43. a. $\left(1 + \dfrac{1}{\sqrt{2}}, \dfrac{\pi}{4}\right), \left(1 - \dfrac{1}{\sqrt{2}}, \dfrac{5\pi}{4}\right), (0, 0)$ **b.** $\dfrac{3\pi}{2} - 2\sqrt{2}$

45. $\dfrac{1}{24}(3\sqrt{3} + 2\pi)$ **47.** $\dfrac{1}{4}(2 - \sqrt{3}) + \dfrac{\pi}{12}$ **49.** $\pi/20$
51. $4(4\pi/3 - \sqrt{3})$ **53.** $2\pi/3 - \sqrt{3}/2$ **55.** $9\pi + 27\sqrt{3}$
57. 6   **59.** 18π **61.** Intersection points: $\left(3, \pm\dfrac{\pi}{3}\right)$; area of region A $= 6\sqrt{3} - 2\pi$; area of region B $= 5\pi - 6\sqrt{3}$; area of region C $= 4\pi + 6\sqrt{3}$ **63.** πa **65.** $\dfrac{8}{3}\left((1 + \pi^2)^{3/2} - 1\right)$
67. 32   **69.** $63\sqrt{5}$ **71.** $\dfrac{2\pi - 3\sqrt{3}}{8}$ **73.** 26.73
75. a. False   **b.** False   **c.** True
77. Horizontal: $(0, 0), (4.05, 2.03), (9.83, 4.91)$;
vertical: $(1.72, 0.86), (6.85, 3.43), (12.87, 6.44)$ **79.** $\dfrac{\sqrt{1 + a^2}}{a}$
81. a. $A_n = \dfrac{1}{4e^{(4n+2)\pi}} - \dfrac{1}{4e^{4n\pi}} - \dfrac{1}{4e^{(4n-2)\pi}} + \dfrac{1}{4e^{(4n-4)\pi}}$ **b.** 0
c. $e^{-4\pi}$ **85.** $(a^2 - 2)\theta^* + \pi - \sin 2\theta^*$, where $\theta^* = \cos^{-1}(a/2)$
87. $a^2(\pi/2 + a/3)$

Section 12.4 Exercises, pp. 797–800

1. A parabola is the set of all points in a plane equidistant from a fixed point and a fixed line.   **3.** A hyperbola is the set of all points in a plane whose distances from two fixed points have a constant difference.

5. Parabola:

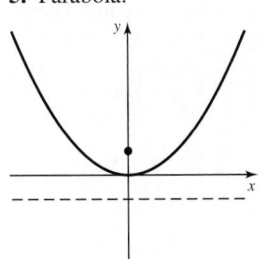

Hyperbola:

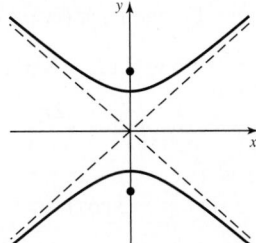

Ellipse:

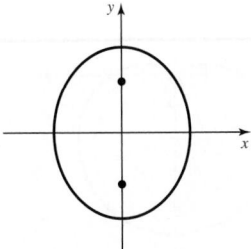

7. $\left(\dfrac{x}{a}\right)^2 + \dfrac{y^2}{a^2 - c^2} = 1$ **9.** $(\pm ae, 0)$ **11.** $y = \pm\dfrac{b}{a}x$

13. Parabola

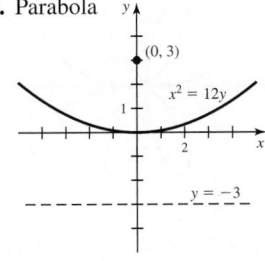

15. Ellipse

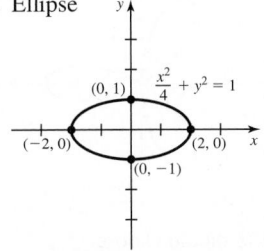

Vertices: $(\pm 2, 0)$; foci: $(\pm \sqrt{3}, 0)$; major axis has length 4; minor axis has length 2.

17. Parabola

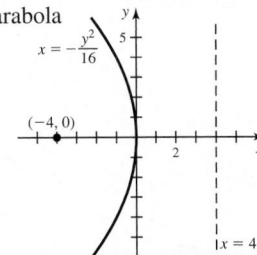

19. Hyperbola

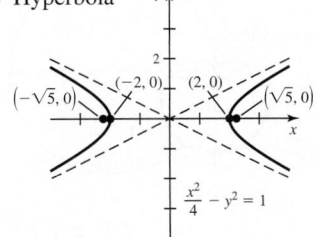

Vertices: $(\pm 2, 0)$; foci: $(\pm \sqrt{5}, 0)$; asymptotes: $y = \pm\dfrac{1}{2}x$

21. Hyperbola

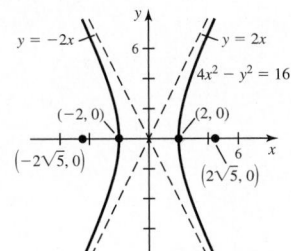

23. Parabola

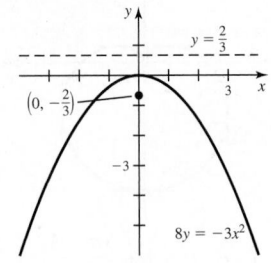

25. Ellipse

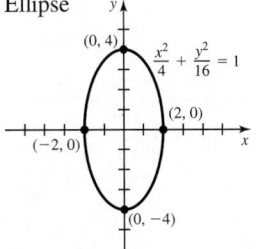

Vertices: $(0, \pm 4)$; foci: $(0, \pm 2\sqrt{3})$; major axis has length 8; minor axis has length 4.

27. Ellipse

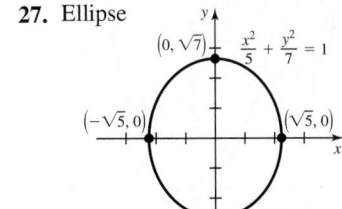

Vertices: $(0, \pm \sqrt{7})$; foci: $(0, \pm \sqrt{2})$; major axis has length $2\sqrt{7}$; minor axis has length $2\sqrt{5}$.

29. Hyperbola

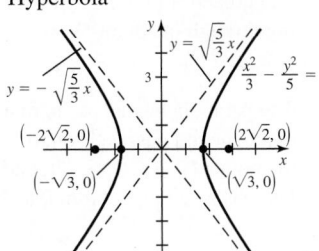

31. $y^2 = 16x$ **33.** $y^2 = 12x$

35. $x^2 = -\dfrac{2}{3}y$

37. $y^2 = 4(x + 1)$

39. $\dfrac{x^2}{16} + \dfrac{y^2}{9} = 1$

41. $\dfrac{x^2}{16} - \dfrac{y^2}{20} = 1$

43. $\dfrac{x^2}{25} + y^2 = 1$ **45.** $\dfrac{x^2}{4} - \dfrac{y^2}{9} = 1$ **47.** $\dfrac{x^2}{4} + \dfrac{y^2}{9} = 1$

49. $\dfrac{x^2}{16} - \dfrac{y^2}{9} = 1$ **51. a.** True   **b.** True   **c.** True   **d.** True

53. $\dfrac{x^2}{81} + \dfrac{y^2}{72} = 1$

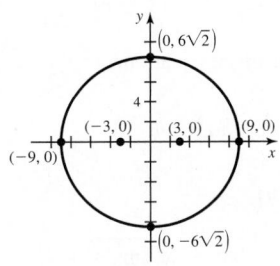

Directrices: $x = \pm 27$

55. $x^2 - \dfrac{y^2}{8} = 1$

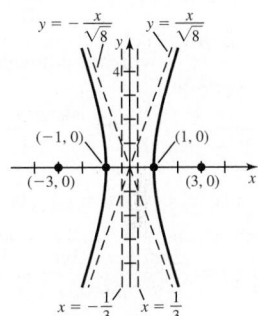

57.

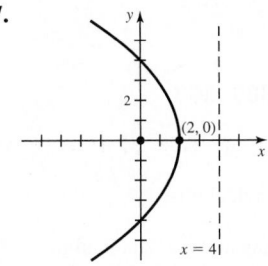

Vertex: $(2, 0)$; focus: $(0, 0)$; directrix: $x = 4$

59.

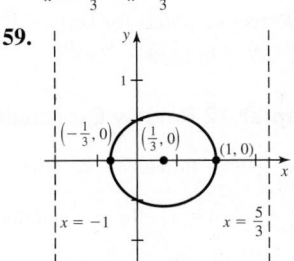

Vertices: $(1, 0)$, $\left(-\dfrac{1}{3}, 0\right)$; center: $\left(\dfrac{1}{3}, 0\right)$; foci: $(0, 0)$, $\left(\dfrac{2}{3}, 0\right)$; directrices: $x = -1$, $x = \dfrac{5}{3}$

61.

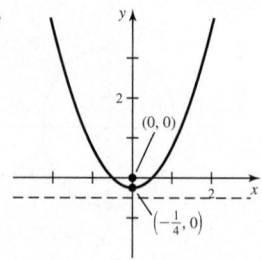

Vertex: $\left(0, -\frac{1}{4}\right)$; focus: $(0, 0)$;
directrix: $y = -\frac{1}{2}$

63.

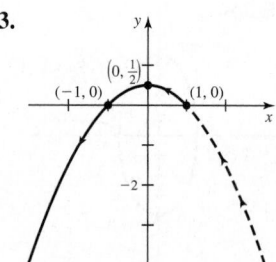

The parabola starts at $(1, 0)$ and goes through quadrants I, II, and III for θ in $[0, 3\pi/2]$; then it approaches $(1, 0)$ by traveling through quadrant IV on $(3\pi/2, 2\pi)$.

65.

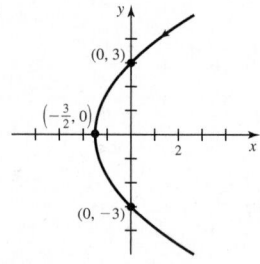

The parabola begins in the first quadrant and passes through the points $(0, 3)$, $\left(-\frac{3}{2}, 0\right)$, and $(0, -3)$ as θ ranges from 0 to 2π.

67. The parabolas open to the left due to the presence of a positive $\cos \theta$ term in the denominator. As d increases, the directrix $x = d$ moves to the right, resulting in wider parabolas.

69. $y = 2x + 6$ **71.** $y = -\frac{3}{40}x - \frac{4}{5}$ **73.** $\frac{dy}{dx} = -\frac{b^2 x}{a^2 y}$, so $\frac{y - y_0}{x - x_0} = -\frac{b^2 x_0}{a^2 y_0}$, which is equivalent to the given equation.

75. $r = \frac{4}{1 - 2\sin\theta}$ **79.** $\frac{4\pi b^2 a}{3}$; $\frac{4\pi a^2 b}{3}$; yes, if $a \neq b$

81. a. $\frac{\pi b^2}{3a^2}(a - c)^2(2a + c)$ **b.** $\frac{4\pi b^4}{3a}$ **91.** $2p$

97. a. $u(m) = \frac{2m^2 - \sqrt{3m^2 + 1}}{m^2 - 1}$; $v(m) = \frac{2m^2 + \sqrt{3m^2 + 1}}{m^2 - 1}$; 2 intersection points for $|m| > 1$ **b.** $\frac{5}{4}, \infty$ **c.** 2, 2
d. $2\sqrt{3} - \ln(\sqrt{3} + 2)$

Chapter 12 Review Exercises, pp. 800–803

1. a. False   **b.** False   **c.** True   **d.** False   **e.** True   **f.** True

3. $\frac{x^2}{16} + \frac{y^2}{9} = 1$; ellipse generated counterclockwise

5. Segment of the parabola $y = \sqrt{x}$ starting at $(4, 2)$ and ending at $(9, 3)$

7. a. $(x - 1)^2 + (y - 2)^2 = 64$
b. $-\frac{1}{\sqrt{3}}$
c.

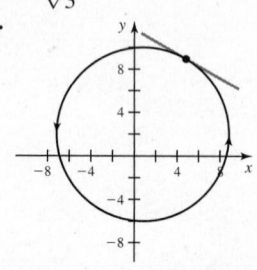

9. $x = 5(t - 1)(t - 2)\sin t$, $y = t$

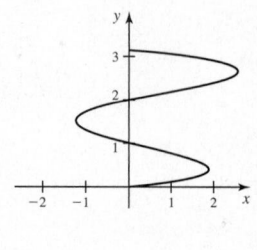

11. a. $x^2 + (y + 1)^2 = 9$ **b.** Lower half of a circle of radius 3 centered at $(0, -1)$, starting at $(3, -1)$ and ending at $(-3, -1)$ **c.** 0

13. At $t = \pi/6$: $y = (2 + \sqrt{3})x + \left(2 - \frac{\pi}{3} - \frac{\pi\sqrt{3}}{6}\right)$; at $t = \frac{2\pi}{3}$: $y = \frac{x}{\sqrt{3}} + 2 - \frac{2\pi}{3\sqrt{3}}$ **15.** $x = -1 + 2t, y = t$, for $0 \leq t \leq 1$; $x = 1 - 2t, y = 1 - t$, for $0 \leq t \leq 1$
17. $x = 3\sin t, y = 3\cos t$, for $0 \leq t \leq 2\pi$
19. $\frac{4}{15}$ **21.** 9.1   **23.** $4 - 2\sqrt{2}$

25.

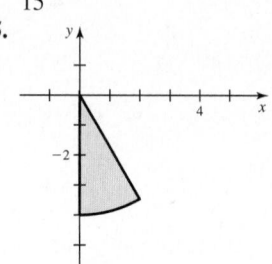

27.

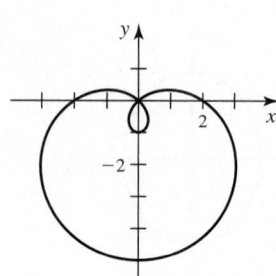

29.

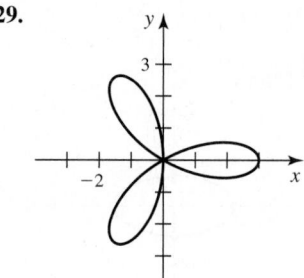

31.

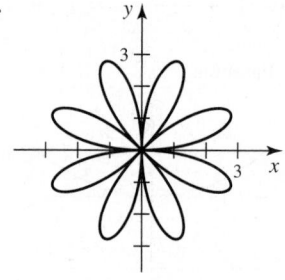

33. Liz should choose $r = 1 - \sin\theta$.

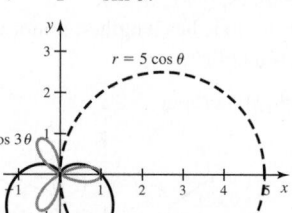

35. $(x - 3)^2 + (y + 1)^2 = 10$; a circle of radius $\sqrt{10}$ centered at $(3, -1)$
37. $r = 1 + \cos\theta$; a cardioid
39. $r = 8\cos\theta, 0 \leq \theta \leq \pi$

41. a. Horizontal: $\left(\frac{1}{2}, \frac{\pi}{6}\right), \left(\frac{1}{2}, \frac{5\pi}{6}\right), \left(2, \frac{3\pi}{2}\right)$;
vertical: $\left(\frac{3}{2}, \frac{7\pi}{6}\right), \left(\frac{3}{2}, \frac{11\pi}{6}\right), \left(0, \frac{\pi}{2}\right)$ **b.** Undefined
c.

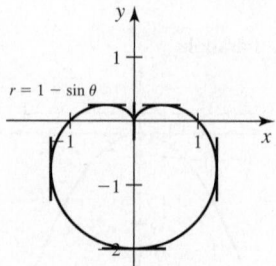

43. $\left(\frac{\pi}{12}, \frac{1}{2^{1/4}}\right), \left(\frac{3\pi}{4}, \frac{1}{2^{1/4}}\right), \left(\frac{17\pi}{12}, \frac{1}{2^{1/4}}\right), (0, 0)$

45. $\pi - \frac{3\sqrt{3}}{2}$ **47.** $2\sqrt{3} - \frac{2\pi}{3}$ **49.** 4   **51.** 40.09

53. a. Hyperbola
b. Foci $(\pm\sqrt{3}, 0)$, vertices $(\pm 1, 0)$, directrices $x = \pm\dfrac{1}{\sqrt{3}}$
c. $e = \sqrt{3}$

d.

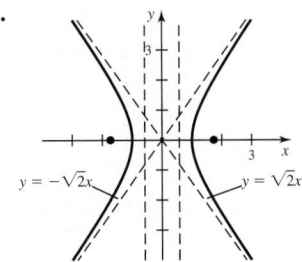

55. a. Hyperbola
b. Foci $(0, \pm 2\sqrt{5})$, vertices $(0, \pm 4)$, directrices $y = \pm\dfrac{8}{\sqrt{5}}$ **c.** $e = \dfrac{\sqrt{5}}{2}$

d.

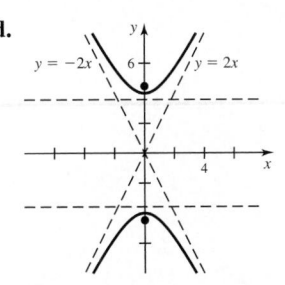

57. a. Ellipse
b. Foci $(\pm\sqrt{2}, 0)$, vertices $(\pm 2, 0)$, directrices $x = \pm 2\sqrt{2}$ **c.** $e = \dfrac{\sqrt{2}}{2}$

d.

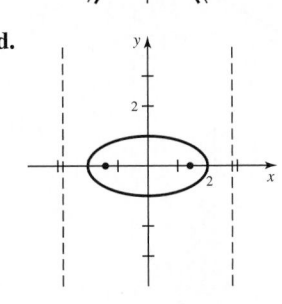

59. $y = \dfrac{3}{2}x - 2$

61. $e = 1$

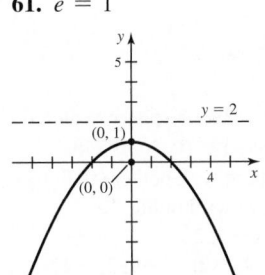

63. $e = \dfrac{1}{2}$

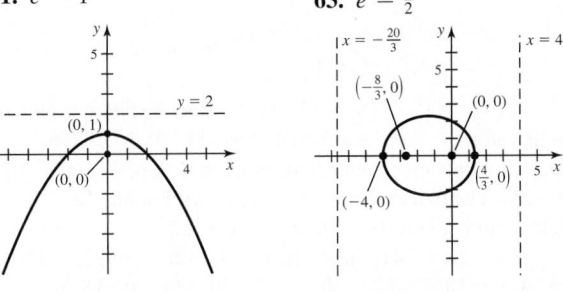

65. $\dfrac{y^2}{4} - \dfrac{x^2}{12} = 1$

67. $\dfrac{y^2}{16} + \dfrac{25x^2}{336} = 1$; foci: $\left(0, \pm\dfrac{8}{5}\right)$

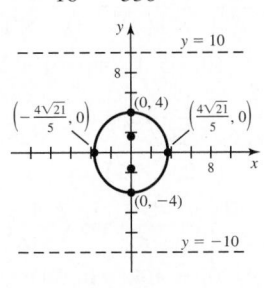

69. $e = 2/3$, $y = \pm 9$, $(\pm 2\sqrt{5}, 0)$ **71.** $m = \dfrac{b}{a}$

75. a. $x = \pm a \cos^{2/n} t$, $y = \pm b \sin^{2/n} t$
c. The curve becomes more rectangular as n increases.

CHAPTER 13

Section 13.1 Exercises, pp. 813–816

3. There are infinitely many vectors with the same direction and length as **v**. **5.** $\mathbf{u} + \mathbf{v} = \langle u_1 + v_1, u_2 + v_2 \rangle$ **7.** No
9. $|\langle v_1, v_2 \rangle| = \sqrt{v_1^2 + v_2^2}$ **11.** If P has coordinates (x_1, y_1) and Q has coordinates (x_2, y_2), then the magnitude of $\overrightarrow{PQ}$ is given by $\sqrt{(x_2 - x_1)^2 + (y_2 - y_1)^2}$. **13.** a, c, e **15. a.** $3\mathbf{v}$ **b.** $2\mathbf{u}$
c. $-3\mathbf{u}$ **d.** $-2\mathbf{u}$ **e.** $\mathbf{v}$ **17. a.** $3\mathbf{u} + 3\mathbf{v}$ **b.** $\mathbf{u} + 2\mathbf{v}$ **c.** $2\mathbf{u} + 5\mathbf{v}$
d. $-2\mathbf{u} + 3\mathbf{v}$ **e.** $3\mathbf{u} + 2\mathbf{v}$ **f.** $-3\mathbf{u} - 2\mathbf{v}$ **g.** $-2\mathbf{u} - 4\mathbf{v}$
h. $\mathbf{u} - 4\mathbf{v}$ **i.** $-\mathbf{u} - 6\mathbf{v}$

19. a.

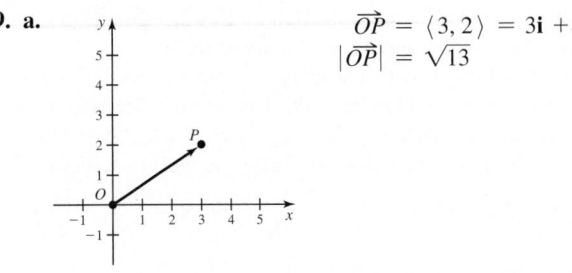

$\overrightarrow{OP} = \langle 3, 2 \rangle = 3\mathbf{i} + 2\mathbf{j}$
$|\overrightarrow{OP}| = \sqrt{13}$

b.

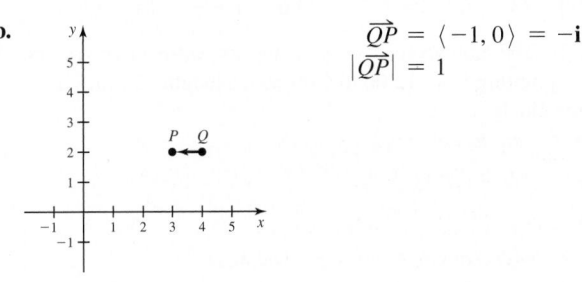

$\overrightarrow{QP} = \langle -1, 0 \rangle = -\mathbf{i}$
$|\overrightarrow{QP}| = 1$

c.

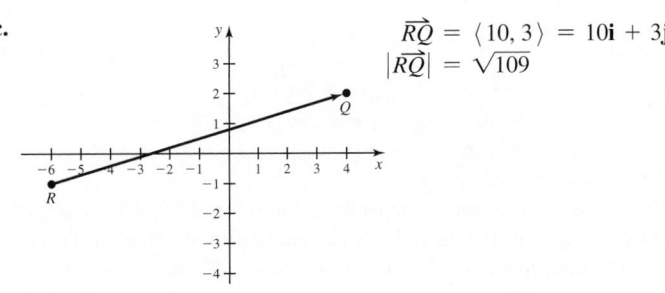

$\overrightarrow{RQ} = \langle 10, 3 \rangle = 10\mathbf{i} + 3\mathbf{j}$
$|\overrightarrow{RQ}| = \sqrt{109}$

21. $\overrightarrow{QU} = \langle 7, 2 \rangle$, $\overrightarrow{PT} = \langle 7, 3 \rangle$, $\overrightarrow{RS} = \langle 2, 3 \rangle$

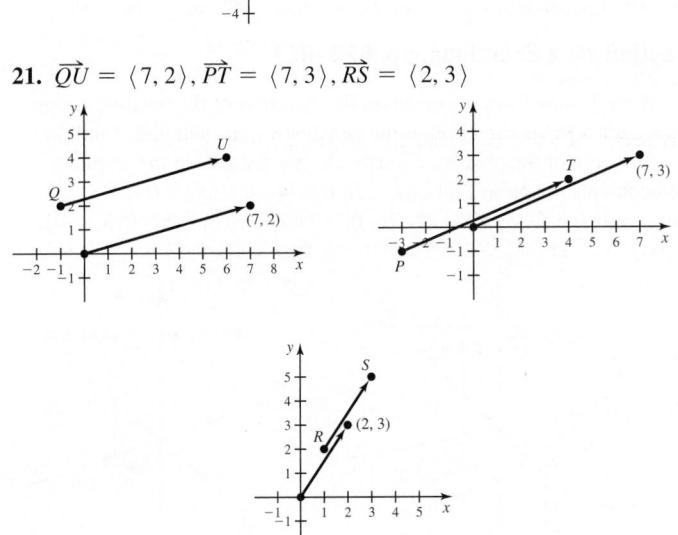

23. $\overrightarrow{QT}$ **25.** $\langle -4, 10 \rangle$ **27.** $\langle 52, -30 \rangle$ **29.** $2\sqrt{2}$

31. $\mathbf{w} - \mathbf{u}$ **33.** $13\left\langle -\dfrac{5}{13}, \dfrac{12}{13} \right\rangle$ **35.** $\langle 3, 3\sqrt{3} \rangle$

37. $\left\langle \dfrac{15}{13}, -\dfrac{36}{13} \right\rangle$ **39.** $\left\langle \dfrac{30}{\sqrt{13}}, -\dfrac{20}{\sqrt{13}} \right\rangle$ **41.** $-\mathbf{i} + 10\mathbf{j}$

43. $\pm\dfrac{1}{\sqrt{61}} \langle 6, 5 \rangle$ **45.** $\left\langle -\dfrac{28}{\sqrt{74}}, \dfrac{20}{\sqrt{74}} \right\rangle, \left\langle \dfrac{28}{\sqrt{74}}, -\dfrac{20}{\sqrt{74}} \right\rangle$

47. a. $\left\langle \dfrac{3}{5}, -\dfrac{4}{5} \right\rangle, \left\langle -\dfrac{3}{5}, \dfrac{4}{5} \right\rangle$ **b.** $b = \pm\dfrac{2\sqrt{2}}{3}$ **c.** $a = \pm\dfrac{3}{\sqrt{10}}$

49. $\langle -4\sqrt{3}, 4 \rangle$ **51.** $\langle 15\sqrt{3}, -15 \rangle$ **53. a.** $\mathbf{v}_a = \langle -320, 0 \rangle$;
$\mathbf{w} = \langle -20\sqrt{2}, -20\sqrt{2} \rangle$; $\mathbf{v}_g = \langle -320 - 20\sqrt{2}, -20\sqrt{2} \rangle$
b. Approx. 349.4 mi/hr; approx. 4.6° south of west
55. Approx. 490.3 mi/hr with a heading of about 1.2° west of north
57. $5\sqrt{65}$ km/hr ≈ 40.3 km/hr   **59.** 1 m/s in the direction 30° east
of north   **61. a.** $\langle 20, 20\sqrt{3} \rangle$ **b.** Yes   **c.** No   **63.** $250\sqrt{2}$ lb
65. a. True   **b.** True   **c.** False   **d.** False   **e.** False   **f.** False
g. False   **h.** True   **67.** $\mathbf{x} = \left\langle \dfrac{1}{5}, -\dfrac{3}{10} \right\rangle$ **69.** $\mathbf{x} = \left\langle \dfrac{4}{3}, -\dfrac{11}{3} \right\rangle$

71. $4\mathbf{i} - 8\mathbf{j}$ **73.** $\langle a, b \rangle = \left(\dfrac{a+b}{2}\right)\mathbf{u} + \left(\dfrac{b-a}{2}\right)\mathbf{v}$

75. a. $\mathbf{0}$ **b.** The 6:00 vector   **c.** Sum any six consecutive vectors.
d. A vector pointing from 12:00 to 6:00 with a length 12 times the
radius of the clock
77. $\mathbf{u} + \mathbf{v} = \langle u_1, u_2 \rangle + \langle v_1, v_2 \rangle = \langle u_1 + v_1, u_2 + v_2 \rangle$
$\qquad\qquad = \langle v_1 + u_1, v_2 + u_2 \rangle = \langle v_1, v_2 \rangle + \langle u_1, u_2 \rangle$
$\qquad\qquad = \mathbf{v} + \mathbf{u}$
79. $a(c\mathbf{v}) = a(c\langle v_1, v_2 \rangle) = a\langle cv_1, cv_2 \rangle$
$\qquad\qquad = \langle acv_1, acv_2 \rangle = \langle (ac)v_1, (ac)v_2 \rangle$
$\qquad\qquad = ac\langle v_1, v_2 \rangle = (ac)\mathbf{v}$
81. $(a + c)\mathbf{v} = (a + c)\langle v_1, v_2 \rangle$
$\qquad\qquad = \langle (a + c)v_1, (a + c)v_2 \rangle$
$\qquad\qquad = \langle av_1 + cv_1, av_2 + cv_2 \rangle$
$\qquad\qquad = \langle av_1, av_2 \rangle + \langle cv_1, cv_2 \rangle$
$\qquad\qquad = a\langle v_1, v_2 \rangle + c\langle v_1, v_2 \rangle$
$\qquad\qquad = a\mathbf{v} + c\mathbf{v}$
85. a. $\{\mathbf{u}, \mathbf{v}\}$ are linearly dependent. $\{\mathbf{u}, \mathbf{w}\}$ and $\{\mathbf{v}, \mathbf{w}\}$ are linearly
independent.   **b.** Two linearly dependent vectors are parallel. Two
linearly independent vectors are not parallel.   **87. a.** $\frac{5}{3}$ **b.** -15

Section 13.2 Exercises, pp. 823–827

1. Move 3 units from the origin in the direction of the positive x-axis,
then 2 units in the direction of the negative y-axis, and then 1 unit in
the direction of the positive z-axis.   **3.** It is parallel to the yz-plane
and contains the point $(4, 0, 0)$.   **5.** $\mathbf{u} + \mathbf{v} = \langle 9, 0, -6 \rangle$;
$3\mathbf{u} - \mathbf{v} = \langle 3, 20, -22 \rangle$ **7.** $(0, 0, -4)$ **9.** $A(3, 0, 5), B(3, 4, 0),$
$C(0, 4, 5)$ **11.** $A(3, -4, 5), B(0, -4, 0), C(0, -4, 5)$
13. a.

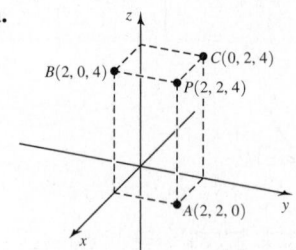

b.

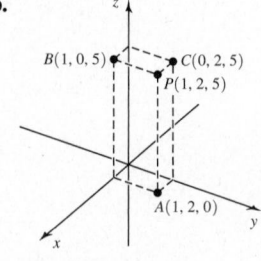

c.

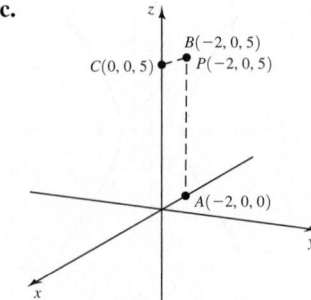

15.

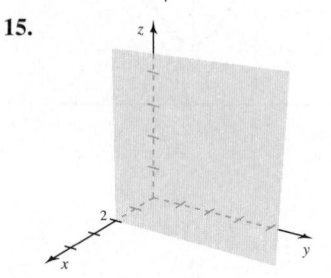

17.

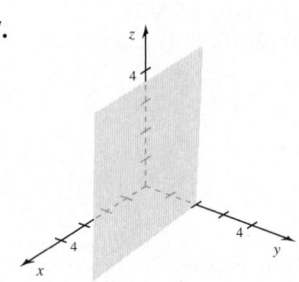

19.

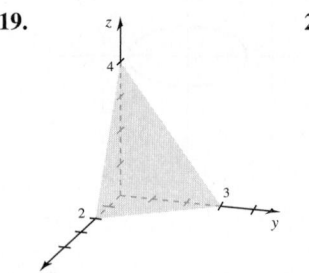

21.

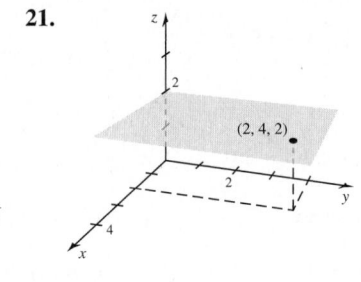

23. $(x - 1)^2 + (y - 2)^2 + (z - 3)^2 = 16$
25. $(x + 2)^2 + y^2 + (z - 4)^2 \le 1$
27. $\left(x - \frac{3}{2}\right)^2 + \left(y - \frac{3}{2}\right)^2 + (z - 7)^2 = \frac{13}{2}$ **29.** A sphere centered
at $(1, 0, 0)$ with radius 3   **31.** A sphere centered at $(0, 1, 2)$ with
radius 3   **33.** All points on or outside the sphere with center $(0, 7, 0)$
and radius 6   **35.** The ball centered at $(4, 7, 9)$ with radius 15
37. The single point $(1, -3, 0)$ **39. a.** $\langle 12, -7, 2 \rangle$
b. $\langle 16, -13, -1 \rangle$ **c.** 5   **41. a.** $\langle -4, 5, -4 \rangle$ **b.** $\langle -9, 3, -9 \rangle$
c. $3\sqrt{2}$ **43. a.** $\langle -15, 23, 22 \rangle$ **b.** $\langle -31, 49, 33 \rangle$ **c.** $3\sqrt{5}$
45. a. $\overrightarrow{PQ} = \langle 2, 6, 2 \rangle = 2\mathbf{i} + 6\mathbf{j} + 2\mathbf{k}$ **b.** $|\overrightarrow{PQ}| = 2\sqrt{11}$
c. $\left\langle \dfrac{1}{\sqrt{11}}, \dfrac{3}{\sqrt{11}}, \dfrac{1}{\sqrt{11}} \right\rangle$ and $\left\langle -\dfrac{1}{\sqrt{11}}, -\dfrac{3}{\sqrt{11}}, -\dfrac{1}{\sqrt{11}} \right\rangle$
47. a. $\overrightarrow{PQ} = \langle 0, -5, 1 \rangle = -5\mathbf{j} + \mathbf{k}$ **b.** $|\overrightarrow{PQ}| = \sqrt{26}$
c. $\left\langle 0, -\dfrac{5}{\sqrt{26}}, \dfrac{1}{\sqrt{26}} \right\rangle$ and $\left\langle 0, \dfrac{5}{\sqrt{26}}, -\dfrac{1}{\sqrt{26}} \right\rangle$
49. a. $\overrightarrow{PQ} = \langle -2, 4, -2 \rangle = -2\mathbf{i} + 4\mathbf{j} - 2\mathbf{k}$ **b.** $|\overrightarrow{PQ}| = 2\sqrt{6}$
c. $\left\langle -\dfrac{1}{\sqrt{6}}, \dfrac{2}{\sqrt{6}}, -\dfrac{1}{\sqrt{6}} \right\rangle$ and $\left\langle \dfrac{1}{\sqrt{6}}, -\dfrac{2}{\sqrt{6}}, \dfrac{1}{\sqrt{6}} \right\rangle$
51. a. $20\mathbf{i} + 20\mathbf{j} - 10\mathbf{k}$;   **b.** 30 mi/hr
53. The speed of the plane is approximately 220 mi/hr; the direction
is slightly south of east and upward.

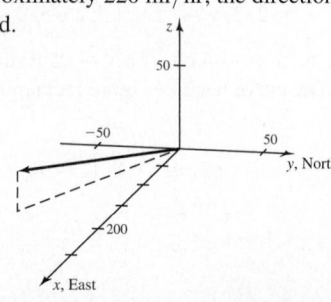

55. $5\sqrt{6}$ knots to the east, $5\sqrt{6}$ knots to the north, 10 knots upward
57. a. False **b.** False **c.** False **d.** True **59.** All points in $\mathbb{R}^3$ except those on the coordinate axes **61.** A circle of radius 1 centered at $(0, 0, 0)$ in the xy-plane

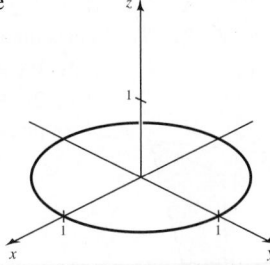

63. A circle of radius 2 centered at $(0, 0, 1)$ in the horizontal plane $z = 1$ **65.** $(x - 2)^2 + (z - 1)^2 = 9, y = 4$ **67.** $6\langle \frac{1}{3}, -\frac{2}{3}, \frac{2}{3} \rangle$

69. $\langle -\frac{15}{4}, \frac{5}{2}, -\frac{5\sqrt{3}}{4} \rangle$ **71.** $\langle 12, -16, 0 \rangle, \langle -12, 16, 0 \rangle$

73. $\langle -\sqrt{3}, -\sqrt{3}, \sqrt{3} \rangle, \langle \sqrt{3}, \sqrt{3}, -\sqrt{3} \rangle$ **75. a.** Collinear; Q is between P and R. **b.** Collinear; P is between Q and R.

c. Noncollinear **d.** Noncollinear **77.** $\langle \frac{500\sqrt{3}}{9}, 0, -\frac{500}{3} \rangle,$
$\langle -\frac{250\sqrt{3}}{9}, -\frac{250}{3}, -\frac{500}{3} \rangle, \langle -\frac{250\sqrt{3}}{9}, \frac{250}{3}, -\frac{500}{3} \rangle$
79. $(3, 8, 9), (-1, 0, 3), (1, 0, -3)$

Section 13.3 Exercises, pp. 833–837

1. $\mathbf{u} \cdot \mathbf{v} = |\mathbf{u}||\mathbf{v}| \cos \theta$ **3.** -40 **5.** $\cos \theta = \frac{\mathbf{u} \cdot \mathbf{v}}{|\mathbf{u}||\mathbf{v}|}$, so

$\theta = \cos^{-1}\left(\frac{\mathbf{u} \cdot \mathbf{v}}{|\mathbf{u}||\mathbf{v}|}\right)$ **7.** $\langle -\frac{4}{3}, \frac{2}{3}, \frac{4}{3} \rangle$ **9.** -1 **11.** 2

13. $\frac{\pi}{2}; 0$ **15.** $100; \frac{\pi}{4}$ **17.** $\frac{1}{2}$ **19.** $0; \frac{\pi}{2}$ **21.** $1; \pi/3$

23. $-2; 93.2°$ **25.** $2; 87.2°$ **27.** $-4; 104°$ **29.** $\angle P = 78.8°,$
$\angle Q = 47.2°, \angle R = 54.0°$ **31.** $\langle 3, 0 \rangle; 3$ **33.** $\langle 0, 3 \rangle; 3$

35. $\frac{6}{5}\langle -2, 1 \rangle; \frac{6}{\sqrt{5}}$ **37.** $\frac{14}{19}\langle -1, -3, 3 \rangle; -\frac{14}{\sqrt{19}}$

39. $-\mathbf{i} + \mathbf{j} - 2\mathbf{k}; \sqrt{6}$ **41.** $750\sqrt{3}$ ft-lb **43.** $25\sqrt{2}$ J

45. 400 J **47.** $\frac{1}{2}\langle 5\sqrt{3}, -15 \rangle, \frac{1}{2}\langle -5\sqrt{3}, -5 \rangle$ **49.** $\langle 490, -490 \rangle,$
$\langle -490, -490 \rangle$ **51. a.** False **b.** True **c.** True **d.** False

e. False **f.** True **53.** $c = \frac{4}{9}$ **55.** $\langle 1, a, 4a - 2 \rangle, a$ real

57. a. $\text{proj}_\mathbf{k}\mathbf{u} = |\mathbf{u}| \cos 60°\left(\frac{\mathbf{k}}{|\mathbf{k}|}\right) = \frac{1}{2}\mathbf{k}$, for all such $\mathbf{u}$ **b.** Yes

59. The heads of the vectors lie on the line $y = 3 - x$.
61. The heads of the vectors lie on the plane $z = 3$.

63. $\mathbf{u} = \langle -\frac{4}{5}, -\frac{2}{5} \rangle + \langle -\frac{6}{5}, \frac{12}{5} \rangle$

65. $\mathbf{u} = \langle 1, \frac{1}{2}, \frac{1}{2} \rangle + \langle -2, \frac{3}{2}, \frac{5}{2} \rangle$ **67.** $3x - 7y = -36$

69. $-\frac{5}{3}$ **71.** $\mathbf{I} = \frac{1}{\sqrt{2}}\mathbf{i} + \frac{1}{\sqrt{2}}\mathbf{j}, \mathbf{J} = -\frac{1}{\sqrt{2}}\mathbf{i} + \frac{1}{\sqrt{2}}\mathbf{j};$

$\mathbf{i} = \frac{1}{\sqrt{2}}(\mathbf{I} - \mathbf{J}), \mathbf{j} = \frac{1}{\sqrt{2}}(\mathbf{I} + \mathbf{J})$ **73. a.** $|\mathbf{I}| = |\mathbf{J}| = |\mathbf{K}| = 1$

b. $\mathbf{I} \cdot \mathbf{J} = 0, \mathbf{I} \cdot \mathbf{K} = 0, \mathbf{J} \cdot \mathbf{K} = 0$ **c.** $\langle 1, 0, 0 \rangle = \frac{1}{2}\mathbf{I} - \frac{1}{\sqrt{2}}\mathbf{J} + \frac{1}{2}\mathbf{K}$

75. a. The faces on $y = 0$ and $z = 0$ **b.** The faces on $y = 1$ and $z = 1$ **c.** The faces on $x = 0$ and $x = 1$ **d.** 0 **e.** 1 **f.** 2

77. a. $\left(\frac{2}{\sqrt{3}}, 0, \frac{2\sqrt{2}}{\sqrt{3}}\right)$ **b.** $\mathbf{r}_{OP} = \langle \sqrt{3}, -1, 0 \rangle, \mathbf{r}_{OQ} = \langle \sqrt{3}, 1, 0 \rangle,$
$\mathbf{r}_{PQ} = \langle 0, 2, 0 \rangle, \mathbf{r}_{OR} = \langle \frac{2}{\sqrt{3}}, 0, \frac{2\sqrt{2}}{\sqrt{3}} \rangle, \mathbf{r}_{PR} = \langle -\frac{\sqrt{3}}{3}, 1, \frac{2\sqrt{2}}{\sqrt{3}} \rangle$
83. a. $\cos^2 \alpha + \cos^2 \beta + \cos^2 \gamma$

$$= \left(\frac{\mathbf{v} \cdot \mathbf{i}}{|\mathbf{v}||\mathbf{i}|}\right)^2 + \left(\frac{\mathbf{v} \cdot \mathbf{j}}{|\mathbf{v}||\mathbf{j}|}\right)^2 + \left(\frac{\mathbf{v} \cdot \mathbf{k}}{|\mathbf{v}||\mathbf{k}|}\right)^2$$

$$= \frac{a^2}{a^2 + b^2 + c^2} + \frac{b^2}{a^2 + b^2 + c^2} + \frac{c^2}{a^2 + b^2 + c^2} = 1$$

b. $\langle 1, 1, 0 \rangle, 90°$ **c.** $\langle \frac{1}{\sqrt{2}}, \frac{1}{\sqrt{2}}, 1 \rangle, 45°$ **d.** No. If so,

$\left(\frac{\sqrt{3}}{2}\right)^2 + \left(\frac{\sqrt{3}}{2}\right)^2 + \cos^2 \gamma = 1$, which has no solution. **e.** 54.7°
85. $|\mathbf{u} \cdot \mathbf{v}| = 33 = \sqrt{33} \cdot \sqrt{33} < \sqrt{70} \cdot \sqrt{74} = |\mathbf{u}||\mathbf{v}|$

Section 13.4 Exercises, pp. 842–844

1. 0 **3. a.** $\mathbf{u}$ is orthogonal to $\mathbf{v}$. **b.** $\mathbf{u}$ is parallel to $\mathbf{v}$. **5.** $\sqrt{2}/2$

7. $-3\mathbf{i} - 2\mathbf{j} + 7\mathbf{k}$ **9.** $\mathbf{u} \times \mathbf{v} = \begin{vmatrix} \mathbf{i} & \mathbf{j} & \mathbf{k} \\ u_1 & u_2 & u_3 \\ v_1 & v_2 & v_3 \end{vmatrix}$ **11.** $15\mathbf{k}$

13. 0

15. 18

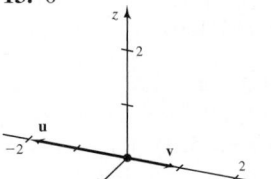

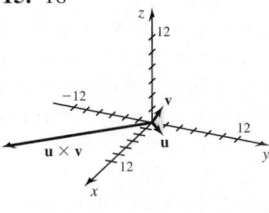

17. $\mathbf{i}$

19. $-\mathbf{i}$

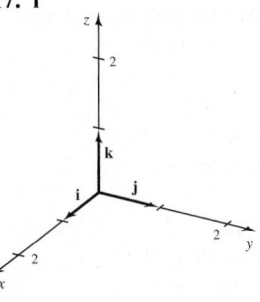

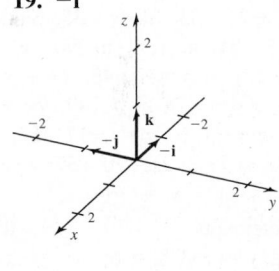

21. $6\mathbf{j}$

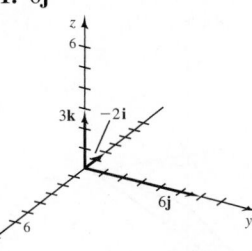

23. $\mathbf{u} \times \mathbf{v} = \langle -30, 18, 9 \rangle, \mathbf{v} \times \mathbf{u} = \langle 30, -18, -9 \rangle$
25. $\mathbf{u} \times \mathbf{v} = \langle 6, 11, 5 \rangle, \mathbf{v} \times \mathbf{u} = \langle -6, -11, -5 \rangle$
27. $\mathbf{u} \times \mathbf{v} = \langle 8, 4, 10 \rangle, \mathbf{v} \times \mathbf{u} = \langle -8, -4, -10 \rangle$ **29.** 11
31. $3\sqrt{10}$ **33.** $\sqrt{11}/2$ **35.** $4\sqrt{2}$ **37.** $9\sqrt{2}$ **41.** Not collinear
43. $\langle 3, -4, 2 \rangle$ **45.** $\langle 0, 20, -20 \rangle$ **47.** The force $\mathbf{F} = 5\mathbf{i} - 5\mathbf{k}$
produces the greater torque. **49.** $5/\sqrt{2}$ N-m **51.** $|\tau| = 13.2$ N-m; direction: into the page

53. The magnitude is $20\sqrt{2}$ at a $135°$ angle with the positive x-axis in the xy-plane.

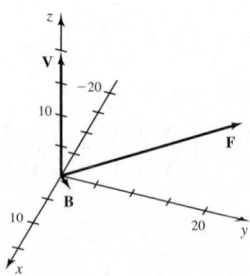

55. 4.53×10^{-14} kg-m/s² **57. a.** False **b.** False **c.** False
d. True **e.** False **59.** $\langle u_1, u_1 + 2, u_1 + 1 \rangle$, u_1 real
61. $\dfrac{\sqrt{(ab)^2 + (ac)^2 + (bc)^2}}{2}$
63. $|\mathbf{u} \cdot (\mathbf{v} \times \mathbf{w})| = |\mathbf{u}||\mathbf{v} \times \mathbf{w}||\cos\theta|$, where $|\mathbf{v} \times \mathbf{w}|$ is the area of the base of the parallelepiped and $|\mathbf{u}||\cos\theta|$ is its height.
67. 1.76×10^7 m/s

Section 13.5 Exercises, pp. 852–855

1. $\langle 4, -8, 9 \rangle$ **3.** $\langle x_1 - x_0, y_1 - y_0, z_1 - z_0 \rangle$ **5.** Perpendicular
7. A point and a normal vector **9.** $(-6, 0, 0), (0, -4, 0), (0, 0, 3)$
11. $x = 4t, y = 7t, z = 1; \mathbf{r} = \langle 0, 0, 1 \rangle + t\langle 4, 7, 0 \rangle$
13. $x = 0, y = t, z = 1; \mathbf{r} = \langle 0, 0, 1 \rangle + t\langle 0, 1, 0 \rangle$
15. $x = t, y = 2t, z = 3t; \mathbf{r} = t\langle 1, 2, 3 \rangle$
17. $x = -2t, y = 8t, z = -4t; \mathbf{r} = t\langle -2, 8, -4 \rangle$
19. $x = -2t, y = -t, z = t; \mathbf{r} = t\langle -2, -1, 1 \rangle$
21. $x = -2, y = 5 - 2t, z = 3 - t; \mathbf{r} = \langle -2, 5, 3 \rangle + t\langle 0, -2, -1 \rangle$ **23.** $x = 1 - 4t, y = 2 + 6t, z = 3 + 14t;$
$\mathbf{r} = \langle 1, 2, 3 \rangle + t\langle -4, 6, 14 \rangle$ **25.** $x = 4, y = 3 - 9t, z = 3 + 6t;$
$\mathbf{r} = \langle 4, 3, 3 \rangle + t\langle 0, -9, 6 \rangle$ **27.** $x = t, y = 2t, z = 3t, 0 \le t \le 1$
29. $x = 2 + 5t, y = 4 + t, z = 8 - 5t, 0 \le t \le 1$ **31.** Intersect
at $(1, 3, 2)$ **33.** Skew **35.** Same line **37.** Parallel, distinct lines
39. 13 **41. a.** Yes **b.** No **c.** $13.16° < \theta < 18.12°$
43. $x + y - z = 4$ **45.** $2x + y - 2z = -2$
47. $x + 4y + 7z = 0$ **49.** $7x + 2y + z = 10$
51. $-x + 2y - 4z = -17$ **53.** $3y - 2z = 0$
55. $8x - 7y + 2z = 0$ **57.** $x + 3y - z = -3$
59. Yes; $2x - y = -1$

61. Intercepts
$x = 2, y = -3, z = 6;$
$3x - 2y = 6, z = 0;$
$-2y + z = 6, x = 0;$ and
$3x + z = 6, y = 0$

63. Intercepts
$x = 30, y = 10, z = -6;$
$x + 3y = 30, z = 0;$
$x - 5z = 30, y = 0;$ and
$3y - 5z = 30, x = 0$

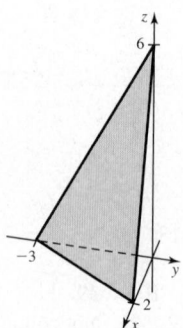
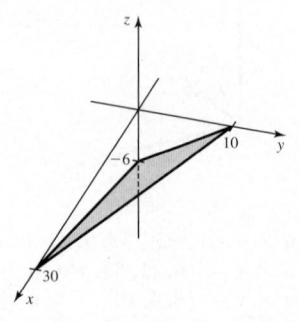

65. Orthogonal **67.** Neither **69.** Q and T are identical; Q, R, and T are parallel; S is orthogonal to Q, R, and T.
71. $\mathbf{r} = \langle 2 + 2t, 1 - 4t, 3 + t \rangle$
73. $x = t, y = 1 + 2t, z = -1 - 3t$

75. $x = \frac{7}{5} + 2t, y = \frac{9}{5} + t, z = -t$ **77.** $(3, 3, 3)$ **79.** $(1, 1, 2)$
81. a. True **b.** False **c.** False **d.** True **e.** False **f.** False
g. True **83.** 6 **85.** $\dfrac{x - 1}{4} = \dfrac{y - 2}{7} = \dfrac{z}{2}$ **87.** Approx. $43°$
89. $6x - 4y + z = d$ **91.** The planes intersect in the point $(3, 6, 0)$.
93. a.

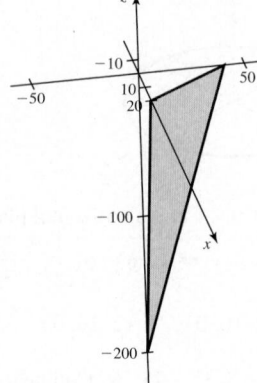

b. Positive
c. $2x + y = 40$, line in the xy-plane

Section 13.6 Exercises, pp. 863–865

1. z-axis; x-axis; y-axis **3.** Intersection of the surface with a plane parallel to one of the coordinate planes **5.** Ellipsoid

7. a. x-axis
b.

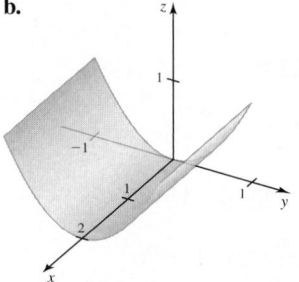

9. a. y-axis
b.

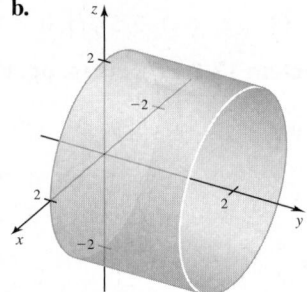

11. a. z-axis
b.

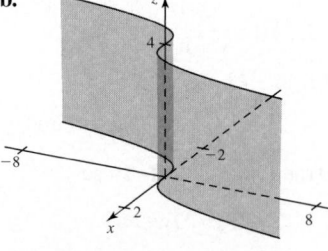

13. a. x-axis
b.

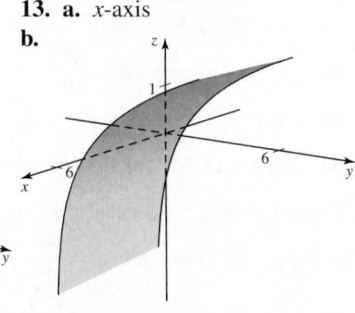

15. Ellipsoid; xy-trace: $x^2 + y^2 = 1$ (circle); xz-trace:
$x^2 + \dfrac{z^2}{25} = 1$ (ellipse); yz-trace: $y^2 + \dfrac{z^2}{25} = 1$ (ellipse)
17. Paraboloid; xy-trace: $(0, 0, 0)$ (a single point); xz-trace:
$z = 25x^2$ (parabola); yz-trace: $z = 25y^2$ (parabola) **19.** Hyperboloid
of two sheets; xz-trace: $z^2 - 25x^2 = 25$ (hyperbola);
yz-trace: $z^2 - 25y^2 = 25$ (hyperbola) **21.** Hyperbolic paraboloid
23. Elliptic paraboloid **25.** Hyperbolic cylinder
27. Elliptic paraboloid

29. a. $x = \pm 1, y = \pm 2,$ $z = \pm 3$

b. $x^2 + \dfrac{y^2}{4} = 1, x^2 + \dfrac{z^2}{9} = 1,$ $\dfrac{y^2}{4} + \dfrac{z^2}{9} = 1$

c. Ellipsoid

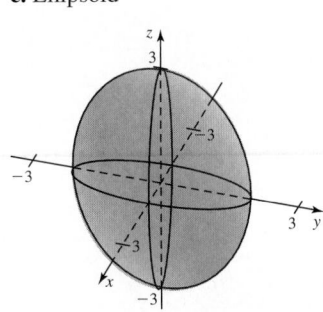

31. a. $x = y = z = 0$
b. $x = y^2, x = z^2,$ origin
c. Elliptic paraboloid

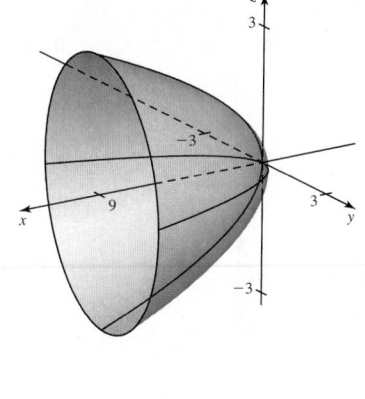

33. a. $x = \pm 5, y = \pm 3,$ no z-intercept

b. $\dfrac{x^2}{25} + \dfrac{y^2}{9} = 1, \dfrac{x^2}{25} - z^2 = 1, \dfrac{y^2}{9} - z^2 = 1$

c. Hyperboloid of one sheet

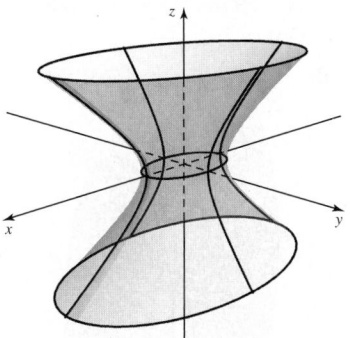

35. a. $x = y = z = 0$ **b.** $\dfrac{x^2}{9} - y^2 = 0, z = \dfrac{x^2}{9}, z = -y^2$

c. Hyperbolic paraboloid

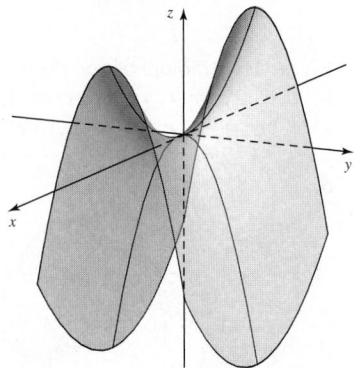

37. a. $x = y = z = 0$

b. Origin, $\dfrac{y^2}{4} = z^2, x^2 = z^2$

c. Elliptic cone

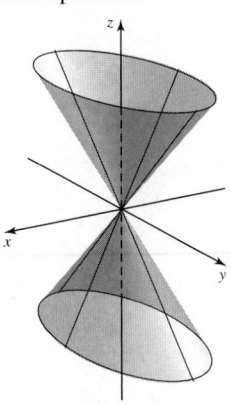

41. a. $x = y = z = 0$
b. Origin,

$x - 9y^2 = 0, 9x - \dfrac{z^2}{4} = 0$

c. Elliptic paraboloid

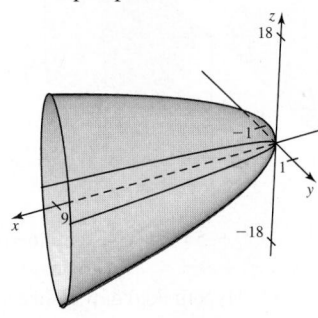

45. a. $x = y = z = 0$

b. $5x - \dfrac{y^2}{5} = 0, 5x + \dfrac{z^2}{20} = 0,$

$-\dfrac{y^2}{5} + \dfrac{z^2}{20} = 0$

c. Hyperbolic paraboloid

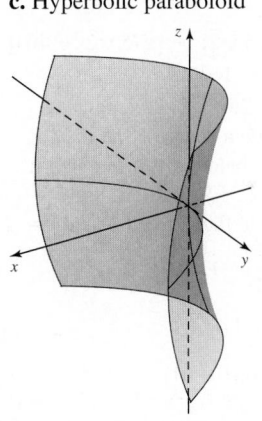

39. a. $x = \pm 3, y = \pm 1, z = \pm 6$

b. $\dfrac{x^2}{3} + 3y^2 = 3, \dfrac{x^2}{3} + \dfrac{z^2}{12} = 3,$

$3y^2 + \dfrac{z^2}{12} = 3$

c. Ellipsoid

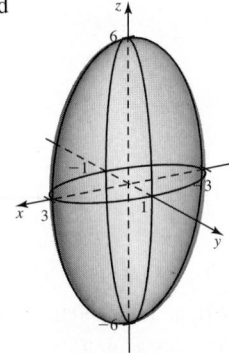

43. a. No x-intercept,

$y = \pm 12, z = \pm\dfrac{1}{2}$

b. $-\dfrac{x^2}{4} + \dfrac{y^2}{16} = 9,$

$-\dfrac{x^2}{4} + 36z^2 = 9, \dfrac{y^2}{16} + 36z^2 = 9$

c. Hyperboloid of one sheet

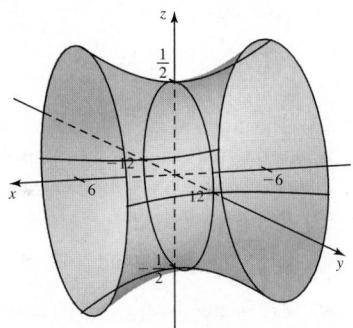

47. a. $x = y = z = 0$

b. $\dfrac{y^2}{18} = 2x^2, \dfrac{z^2}{32} = 2x^2,$ origin

c. Elliptic cone

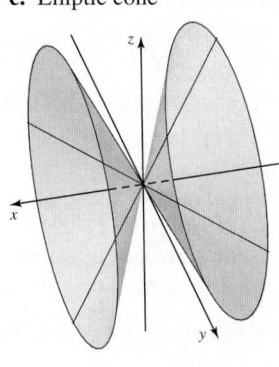

49. a. No x-intercept, $y = \pm 2$, no z-intercept

b. $-x^2 + \dfrac{y^2}{4} = 1$, no xz-trace, $\dfrac{y^2}{4} - \dfrac{z^2}{9} = 1$

c. Hyperboloid of two sheets

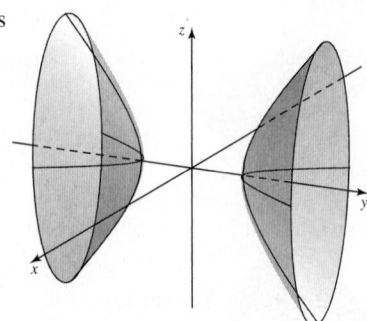

51. a. No x-intercept, $y = \pm \dfrac{\sqrt{3}}{3}$, no z-intercept

b. $-\dfrac{x^2}{3} + 3y^2 = 1$, no xz-trace, $3y^2 - \dfrac{z^2}{12} = 1$

c. Hyperboloid of two sheets

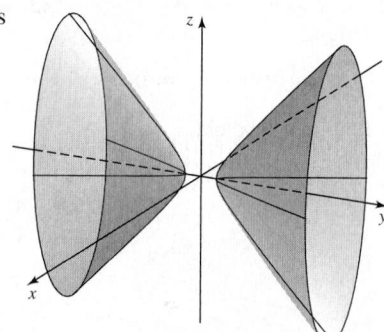

53. The graph of the ellipsoid $x^2 + 4y^2 + 9z^2 + 54z = 19$ is obtained by shifting the graph of the ellipsoid $x^2 + 4y^2 + 9z^2 = 100$ down 3 units.   **55.** Hyperboloid of one sheet   **57.** Hyperboloid of two sheets
59. a. True   **b.** True   **c.** True   **d.** False   **e.** False   **61.** All except the hyperbolic paraboloid   **63.** 8   **65. b.** $\dfrac{x^2 + z^2}{(10.55/\pi)^2} + \dfrac{y^2}{(5.55)^2} = 1$
67. $4x^2 + 8y^2 + 4(z - 3)^2 = 9, 3 \le z \le 4.5$

Chapter 13 Review Exercises, pp. 865–867

1. a. True   **b.** False   **c.** True   **d.** False   **e.** True   **f.** True

3. $\langle 3, -6 \rangle$ **5.** $\langle -5, 8 \rangle$ **7.** $\sqrt{221}$ **9.** $12 \left\langle \dfrac{1}{3}, -\dfrac{2}{3}, \dfrac{2}{3} \right\rangle$

11. $\left\langle \dfrac{10}{3}, -\dfrac{20}{3}, \dfrac{20}{3} \right\rangle$ **13.** $\langle 58, 26, 44 \rangle$ **15.** $a = -3$

17. a. $\mathbf{v} = -275\sqrt{2}\mathbf{i} + 275\sqrt{2}\mathbf{j}$ **b.** $-275\sqrt{2}\mathbf{i} + (275\sqrt{2} + 40)\mathbf{j}$
19. $\{(x, y, z): (x - 1)^2 + y^2 + (z + 1)^2 = 16\}$
21. $\{(x, y, z): x^2 + (y - 1)^2 + z^2 > 4\}$ **23.** A ball centered at $\left(\frac{1}{2}, -2, 3\right)$ of radius $\frac{3}{2}$ **25.** All points outside a sphere of radius 10 centered at $(3, 0, 10)$ **27.** 50.15 m/s; 85.4° below the horizontal in the northerly horizontal direction   **29.** 50 lb; 36.9° north of east
31. A circle of radius 1 centered at $(0, 2, 0)$ in the vertical plane $y = 2$

33. a. 0.68 radian   **b.** $\dfrac{7}{9}\langle 1, 2, 2 \rangle; \dfrac{7}{3}$ **c.** $\dfrac{7}{3}\langle -1, 2, 2 \rangle; 7$

35. $250\sqrt{2}$ ft-lb   **37.** $90\sqrt{3}$ lb; 90 lb   **39.** 11

41. $\pm \left\langle \dfrac{12}{\sqrt{197}}, \dfrac{7}{\sqrt{197}}, \dfrac{2}{\sqrt{197}} \right\rangle$ **43.** $\langle -10, 10, 10 \rangle$

45. $|\tau|(\theta) = 39.2 \sin \theta$ has a maximum value of 39.2 N-m (when $\theta = \pi/2$) and a minimum value of 0 N-m (when $\theta = 0$). Direction does *not* change.   **47.** $\mathbf{r} = \langle 0, -3, 9 \rangle + t\langle 2, -5, -8 \rangle, 0 \le t \le 1$
49. $\mathbf{r} = \langle t, 1 + 6t, 1 + 2t \rangle$
51. a. $18x - 9y + 2z = 6$ **b.** $x = \frac{1}{3}, y = -\frac{2}{3}, z = 3$
c.

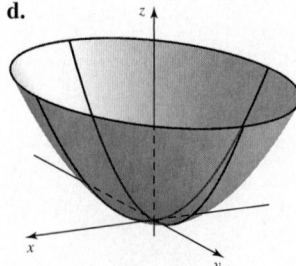

53. $x = t, y = 12 - 9t, z = -6 + 6t$ **55.** $4x + 2y + 13z = 39$
57. $3x + y + 7z = 4$ **59.** 3

61. a. Hyperbolic paraboloid

b. $y^2 = 4x^2, z = \dfrac{x^2}{36}, z = -\dfrac{y^2}{144}$

c. $x = y = z = 0$

d.

63. a. Elliptic cone

b. $y^2 = 4x^2$, origin, $y^2 = \dfrac{z^2}{25}$

c. Origin

d.

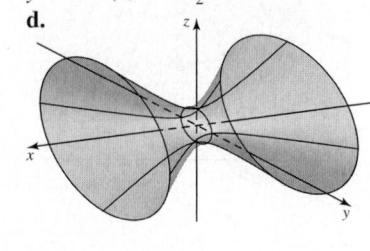

65. a. Elliptic paraboloid

b. Origin, $z = \dfrac{x^2}{16}, z = \dfrac{y^2}{36}$

c. Origin

d.

67. a. Hyperboloid of one sheet
b. $y^2 - 2x^2 = 1, 4z^2 - 2x^2 = 1$, $y^2 + 4z^2 = 1$ **c.** No x-intercept, $y = \pm 1, z = \pm \frac{1}{2}$
d.

69. a. Hyperboloid of one sheet

b. $\dfrac{x^2}{4} + \dfrac{y^2}{16} = 4, \dfrac{x^2}{4} - z^2 = 4, \dfrac{y^2}{16} - z^2 = 4$

c. $x = \pm 4, y = \pm 8$, no z-intercept

d.

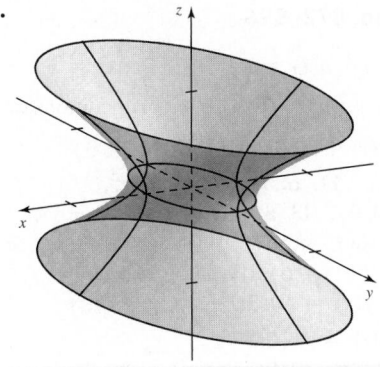

71. a. Ellipsoid   **b.** $\dfrac{x^2}{4} + \dfrac{y^2}{16} = 4, \dfrac{x^2}{4} + z^2 = 4, \dfrac{y^2}{16} + z^2 = 4$

c. $x = \pm 4, y = \pm 8, z = \pm 2$

d.

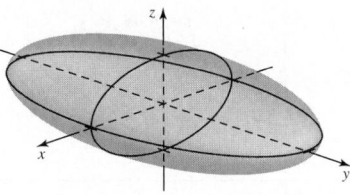

73. a. Elliptic cone   **b.** Origin, $\dfrac{x^2}{9} = \dfrac{z^2}{64}, \dfrac{y^2}{49} = \dfrac{z^2}{64}$ **c.** Origin

d.

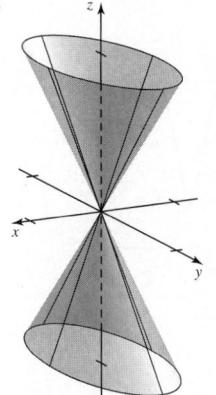

75. a. A   **b.** D   **c.** C   **d.** B

CHAPTER 14

Section 14.1 Exercises, pp. 873–875

1. One   **3.** Its output is a vector.

5. $\lim\limits_{t \to a} \mathbf{r}(t) = \left\langle \lim\limits_{t \to a} f(t), \lim\limits_{t \to a} g(t), \lim\limits_{t \to a} h(t) \right\rangle$

7. $\mathbf{r}(t) = t\,\mathbf{i} + 2t\,\mathbf{j} + 3t\,\mathbf{k}$

9. $\mathbf{r}(t) = \langle 2 + 2t, 3 + 3t, 7 - 4t \rangle$

11. $\mathbf{r}(t) = \langle 3 + 2t, 4, 5 - t \rangle$

13. $\mathbf{r}(t) = \langle 1 - t, 2, 1 + 2t \rangle$, for $0 \le t \le 1$

15.

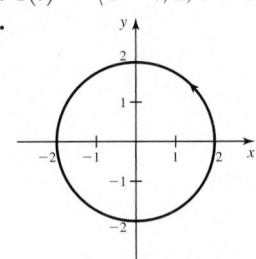

17.

19.

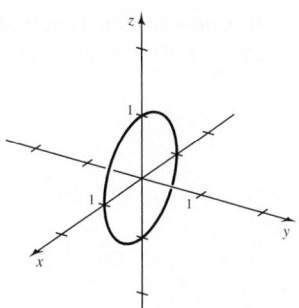

21.

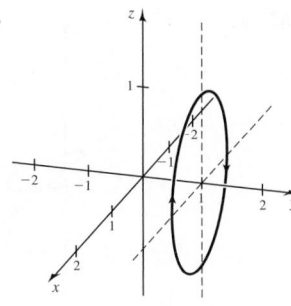

23.

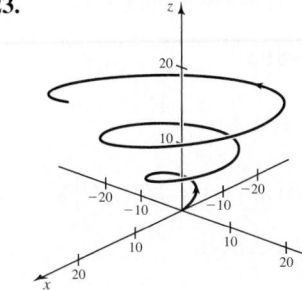

25.

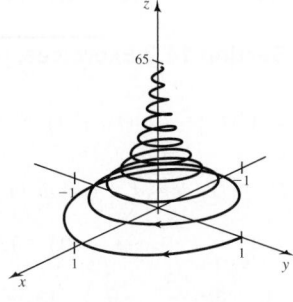

27.

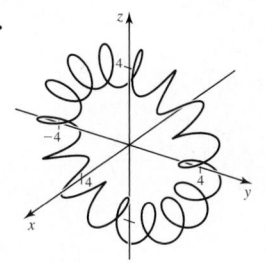

29. When viewed from above, the curve is a portion of the parabola $y = x^2$.

31. $-\mathbf{i} - 4\mathbf{j} + \mathbf{k}$ **33.** $-2\mathbf{j} + \dfrac{\pi}{2}\mathbf{k}$ **35.** $\mathbf{i}$ **37. a.** True   **b.** False

c. True   **d.** True   **39.** $\{t : |t| \le 2\}$ **41.** $\{t : 0 \le t \le 2\}$

43. $(4, 8, 16)$ **45. a.** E   **b.** D   **c.** F   **d.** C   **e.** A   **f.** B

47. $\mathbf{r}(t) = \langle 2 \cos t, 2 \sin t, 4 \rangle$

49. $\mathbf{r}(t) = \langle 5 \cos t, 5 \sin t, 10 \cos t + 10 \sin t \rangle$

51. a. Ball has a parabolic trajectory in the yz-plane; 1200 ft

b. Approx. 1199.7 ft   **c.** 1196 ft   **53.** Hyperboloid of one sheet

55. Ellipsoid   **57.** $(4, 2, 2)$; $\sqrt{179}$

59.

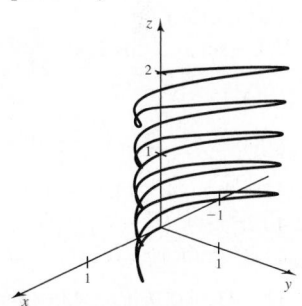

The curve lies on the sphere $x^2 + y^2 + z^2 = 1$.

61. $\dfrac{2\pi}{(m, n)}$, where (m, n) = greatest common factor of m and n

63. a.

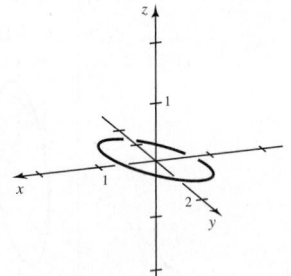

b. Curve is a tilted circle of radius 1 centered at the origin.

65. $\langle cf - ed, be - af, ad - bc \rangle$ or any scalar multiple

Section 14.2 Exercises, pp. 881–883

1. $\mathbf{r}'(t) = \langle f'(t), g'(t), h'(t) \rangle$ **3.** $\mathbf{T}(t) = \dfrac{\mathbf{r}'(t)}{|\mathbf{r}'(t)|}$

5. $\displaystyle \int \mathbf{r}(t)\, dt = \left(\int f(t)\, dt \right)\mathbf{i} + \left(\int g(t)\, dt \right)\mathbf{j} + \left(\int h(t)\, dt \right)\mathbf{k}$

7. $\mathbf{C} = \langle -1, -3, -10 \rangle$ **9.** $\langle -\sin t, 2t, \cos t \rangle$

11. $\left\langle 6t^2, \dfrac{3}{\sqrt{t}}, -\dfrac{3}{t^2} \right\rangle$ **13.** $e^t \mathbf{i} - 2e^{-t}\mathbf{j} - 8e^{2t}\mathbf{k}$

15. $\langle e^{-t}(1 - t), 1 + \ln t, \cos t - t \sin t \rangle$

17. $\langle 1, 6, 3 \rangle$ **19.** $\langle 1, 0, 0 \rangle$ **21.** $8\mathbf{i} + 9\mathbf{j} - 10\mathbf{k}$

23. $\langle 2/3, 2/3, 1/3 \rangle$

25. $\dfrac{\langle 0, -\sin 2t, 2 \cos 2t \rangle}{\sqrt{1 + 3 \cos^2 2t}}$ **27.** $\dfrac{t^2}{\sqrt{t^4 + 4}} \left\langle 1, 0, -\dfrac{2}{t^2} \right\rangle$

29. $\langle 0, 0, -1 \rangle$ **31.** $\left\langle \dfrac{2}{\sqrt{5}}, 0, -\dfrac{1}{\sqrt{5}} \right\rangle$

33. $\langle 30t^{14} + 24t^3, 14t^{13} - 12t^{11} + 9t^2 - 3, -96t^{11} - 24 \rangle$

35. $4t(2t^3 - 1)(t^3 - 2)\langle 3t(t^3 - 2), 1, 0 \rangle$

37. $e^t(2t^3 + 6t^2) - 2e^{-t}(t^2 - 2t - 1) - 16e^{-2t}$

39. 11   **41.** $\langle 0, 7, 1 \rangle$ **43.** $\langle 2e^{2t}, -2e^t, 0 \rangle$ **45.** $\left\langle 4, -\dfrac{2}{\sqrt{t}}, 0 \right\rangle$

47. $\langle 1 + 6t^2, 4t^3, -2 - 3t^2 \rangle$ **49.** $5te^t(t + 2) - 6t^2 e^{-t}(t - 3)$

51. $-3t^2 \sin t + 6t \cos t + 2\sqrt{t} \cos 2t + \dfrac{1}{2\sqrt{t}} \sin 2t$

53. $\langle 2, 0, 0 \rangle, \langle 0, 0, 0 \rangle$ **55.** $\langle -9 \cos 3t, -16 \sin 4t, -36 \cos 6t \rangle,$ $\langle 27 \sin 3t, -64 \cos 4t, 216 \sin 6t \rangle$

57. $\left\langle -\dfrac{1}{4}(t + 4)^{-3/2}, -2(t + 1)^{-3}, 2e^{-t^2}(1 - 2t^2) \right\rangle,$

$\left\langle \dfrac{3}{8}(t + 4)^{-5/2}, 6(t + 1)^{-4}, -4te^{-t^2}(3 - 2t^2) \right\rangle$

59. $\left\langle \dfrac{t^5}{5} - \dfrac{3t^2}{2}, t^2 - t, 10t \right\rangle + \mathbf{C}$

61. $\left\langle 2 \sin t, -\dfrac{2}{3} \cos 3t, \dfrac{1}{2} \sin 8t \right\rangle + \mathbf{C}$

63. $\frac{1}{3}e^{3t}\mathbf{i} + \tan^{-1}t\,\mathbf{j} - \sqrt{2t}\,\mathbf{k} + \mathbf{C}$

65. $\mathbf{r}(t) = \langle e^t + 1, 3 - \cos t, \tan t + 2 \rangle$

67. $\mathbf{r}(t) = \langle t + 3, t^2 + 2, t^3 - 6 \rangle$

69. $\mathbf{r}(t) = \left\langle \frac{1}{2}e^{2t} + \frac{1}{2}, 2e^{-t} + t - 1, t - 2e^t + 3 \right\rangle$

71. $\langle 2, 0, 2 \rangle$ **73.** $\mathbf{i}$ **75.** $\langle 0, 0, 0 \rangle$

77. $(e^2 + 1)\langle 1, 2, -1 \rangle$ **79. a.** False   **b.** True   **c.** True

81. $\langle 2 - t, 3 - 2t, \pi/2 + t \rangle$ **83.** $\langle 2 + 3t, 9 + 7t, 1 + 2t \rangle$

85. $(1, 0)$ **87.** $(1, 0, 0)$ **89.** $\mathbf{r}(t) = \langle a_1 t, a_2 t, a_3 t \rangle$ or $\mathbf{r}(t) = \langle a_1 e^{kt}, a_2 e^{kt}, a_3 e^{kt} \rangle$, where a_i and k are real numbers

Section 14.3 Exercises, pp. 892–896

1. $\mathbf{v}(t) = \mathbf{r}'(t)$, speed $= |\mathbf{r}'(t)|$, $\mathbf{a}(t) = \mathbf{r}''(t)$ **3.** $m\mathbf{a}(t) = \mathbf{F}$

5. $\mathbf{v}(t) = \displaystyle \int \mathbf{a}(t)\, dt = \langle v_1(t), v_2(t) \rangle + \mathbf{C}$. Use initial conditions to find $\mathbf{C}$.   **7. a.** $t = 3$ s   **b.** $\mathbf{r}(t) = \langle 60t, -16t^2 + 96t + 3 \rangle$

9. a. $\langle 6t, 8t \rangle$, $10t$ **b.** $\langle 6, 8 \rangle$ **11. a.** $\mathbf{v}(t) = \langle 2, -4 \rangle$, $|\mathbf{v}(t)| = 2\sqrt{5}$ **b.** $\mathbf{a}(t) = \langle 0, 0 \rangle$ **13. a.** $\mathbf{v}(t) = \langle 8 \cos t, -8 \sin t \rangle$, $|\mathbf{v}(t)| = 8$ **b.** $\mathbf{a}(t) = \langle -8 \sin t, -8 \cos t \rangle$ **15. a.** $\langle 2t, 2t, t \rangle$, $3t$

b. $\langle 2, 2, 1 \rangle$ **17. a.** $\mathbf{v}(t) = \langle 1, -4, 6 \rangle$, $|\mathbf{v}(t)| = \sqrt{53}$

b. $\mathbf{a}(t) = \langle 0, 0, 0 \rangle$ **19. a.** $\mathbf{v}(t) = \langle 0, 2t, -e^{-t} \rangle$, $|\mathbf{v}(t)| = \sqrt{4t^2 + e^{-2t}}$ **b.** $\mathbf{a}(t) = \langle 0, 2, e^{-t} \rangle$

21. a. $[c, d] = [0, 1]$ **b.** $\langle 1, 2t \rangle, \langle 2, 8t \rangle$

c.

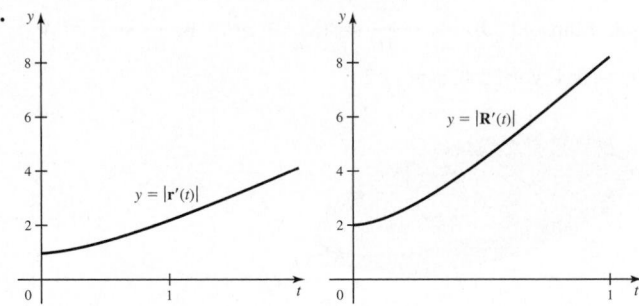

23. a. $\left[0, \frac{2\pi}{3} \right]$ **b.** $\mathbf{V}_\mathbf{r}(t) = \langle -\sin t, 4 \cos t \rangle$, $\mathbf{V}_\mathbf{R}(t) = \langle -3 \sin 3t, 12 \cos 3t \rangle$

c.

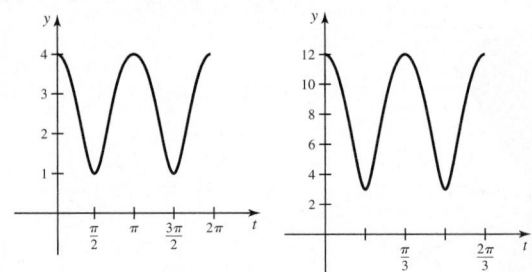

25. a. $[1, e^{36}]$ **b.** $\mathbf{V}_\mathbf{r}(t) = \langle 2t, -8t^3, 18t^5 \rangle$, $\mathbf{V}_\mathbf{R}(t) = \left\langle \dfrac{1}{t}, -\dfrac{4}{t} \ln t, \dfrac{9}{t} \ln^2 t \right\rangle$

c.

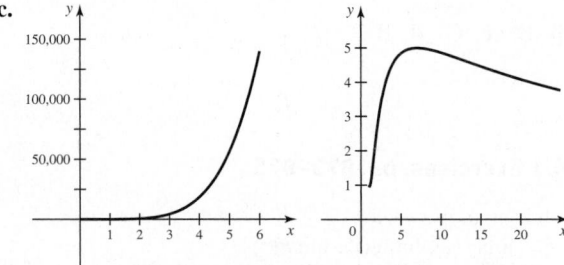

27. a.

b. $\langle -20 \sin t - 50 \sin 5t, 20 \cos t + 50 \cos 5t \rangle$

c. 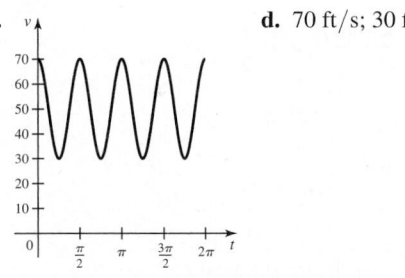   **d.** 70 ft/s; 30 ft/s

29. $\mathbf{r}(t)$ lies on a circle of radius 8; $\langle -16\sin 2t, 16\cos 2t\rangle \cdot \langle 8\cos 2t, 8\sin 2t\rangle = 0$.   **31.** $\mathbf{r}(t)$ lies on a sphere of radius 2; $\langle \cos t - \sqrt{3}\sin t, \sqrt{3}\cos t + \sin t\rangle \cdot \langle \sin t + \sqrt{3}\cos t, \sqrt{3}\sin t - \cos t\rangle = 0$.   **33.** 5

35. $\mathbf{v}(t) = \langle 2, t+3\rangle, \mathbf{r}(t) = \langle 2t, \frac{t^2}{2}+3t\rangle$

37. $\mathbf{v}(t) = \langle 0, 10t+5\rangle, \mathbf{r}(t) = \langle 1, 5t^2+5t-1\rangle$

39. $\mathbf{v}(t) = \langle \sin t, -2\cos t+3\rangle$, $\mathbf{r}(t) = \langle -\cos t+2, -2\sin t+3t\rangle$

41. a. $\mathbf{v}(t) = \langle 30, -9.8t+6\rangle, \mathbf{r}(t) = \langle 30t, -4.9t^2+6t\rangle$
b. 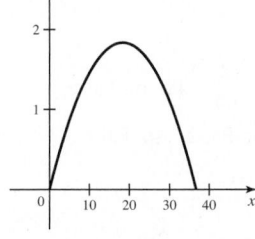   **c.** $T \approx 1.22$ s, range ≈ 36.7 m   **d.** 1.84 m

43. a. $\mathbf{v}(t) = \langle 80, 10-32t\rangle, \mathbf{r}(t) = \langle 80t, -16t^2+10t+6\rangle$
b. 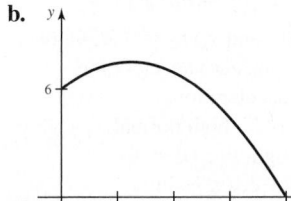   **c.** 1 s, 80 ft   **d.** Max height ≈ 7.56 ft

45. a. $\mathbf{v}(t) = \langle 125, -32t+125\sqrt{3}\rangle$, $\mathbf{r}(t) = \langle 125t, -16t^2+125\sqrt{3}t+20\rangle$
b. 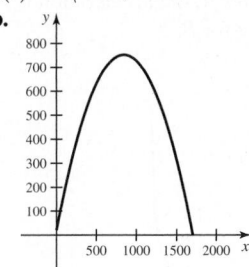   **c.** 13.6 s, 1702.5 ft   **d.** 752.4 ft

47. $\mathbf{v}(t) = \langle 1, 5, 10t\rangle, \mathbf{r}(t) = \langle t, 5t+5, 5t^2\rangle$
49. $\mathbf{v}(t) = \langle -\cos t+1, \sin t+2, t\rangle$, $\mathbf{r}(t) = \langle -\sin t+t, -\cos t+2t+1, \frac{t^2}{2}\rangle$

51. a. $\mathbf{v}(t) = \langle 200, 200, -9.8t\rangle, \mathbf{r}(t) = \langle 200t, 200t, -4.9t^2+1\rangle$
b. 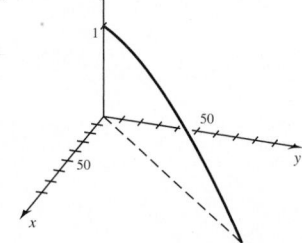   **c.** 0.452 s, 127.8 m   **d.** 1 m

53. a. $\mathbf{v}(t) = \langle 60+10t, 80, 80-32t\rangle$, $\mathbf{r}(t) = \langle 60t+5t^2, 80t, 80t-16t^2+3\rangle$
b. 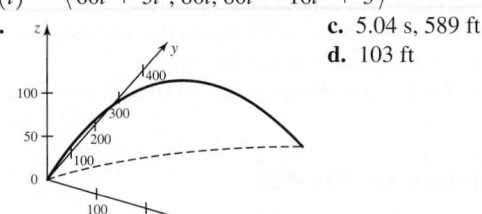   **c.** 5.04 s, 589 ft   **d.** 103 ft

55. a. $\mathbf{v}(t) = \langle 300, 2.5t+400, -9.8t+500\rangle$, $\mathbf{r}(t) = \langle 300t, 1.25t^2+400t, -4.9t^2+500t+10\rangle$
b. 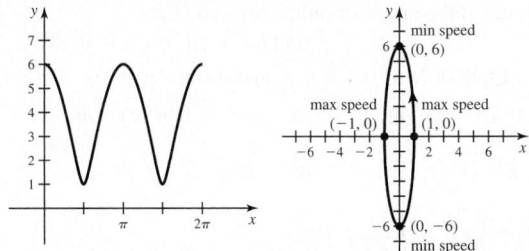   **c.** 102.1 s, 61,941.5 m   **d.** 12,765.1 m

57. a. False **b.** True **c.** False **d.** True **e.** False **f.** True **g.** True   **59.** 15.3 s, 1988.3 m, 287.0 m   **61.** 21.7 s, 4330.1 ft, 1875 ft
63. Approx. 27.4° and 62.6°
65. a. $\mathbf{v}(t) = \langle -a\sin t, b\cos t\rangle; |\mathbf{v}(t)| = \sqrt{a^2\sin^2 t + b^2\cos^2 t}$
b.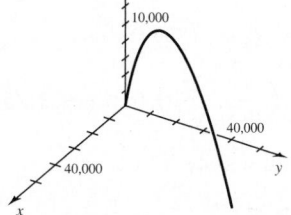
c. Yes **d.** Max $\{\frac{a}{b}, \frac{b}{a}\}$ **67.** Approx. 23.5° or 59.6°
69. 113.4 ft/s   **71. a.** 1.2 ft, 0.46 s **b.** 0.88 ft/s **c.** 0.85 ft
d. More curve in the second half **e.** $c = 28.17$ ft/s²
73. $T = \dfrac{|\mathbf{v}_0|\sin\alpha + \sqrt{|\mathbf{v}_0|^2\sin^2\alpha + 2gy_0}}{g}$,
range $= |\mathbf{v}_0|(\cos\alpha)T$, max height $= y_0 + \dfrac{|\mathbf{v}_0|^2\sin^2\alpha}{2g}$
75. a. $[0, \frac{2\pi}{\omega}]$ **b.** $\mathbf{v}(t) = \langle -A\omega\sin\omega t, A\omega\cos\omega t\rangle$ is not constant; $|\mathbf{v}(t)| = |A\omega|$ is constant. **c.** $\mathbf{a}(t) = \langle -A\omega^2\cos\omega t, -A\omega^2\sin\omega t\rangle$
d. $\mathbf{r}$ and $\mathbf{v}$ are orthogonal; $\mathbf{r}$ and $\mathbf{a}$ are in opposite directions.

e.

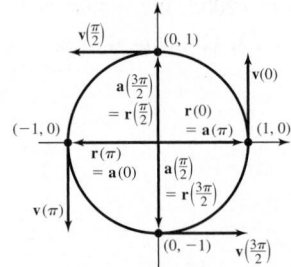

77. a. $r(t) = \langle 5 \sin (\pi t/6), 5 \cos (\pi t/6) \rangle$
b. $r(t) = \langle 5 \sin \left(\frac{1 - e^{-t}}{5}\right), 5 \cos \left(\frac{1 - e^{-t}}{5}\right) \rangle$
79. $\{(\cos t, \sin t, c \sin t): t \in \mathbb{R}\}$ satisfies the equations $x^2 + y^2 = 1$ and $z - cy = 0$ so that $\langle \cos t, \sin t, c \sin t \rangle$ lies on the intersection of a right circular cylinder and a plane, which is an ellipse.
83. a. The direction of r does not change.   **b.** Constant in direction, not in magnitude

Section 14.4 Exercises, pp. 900–902

1. $\sqrt{5}(b - a)$ **3.** $\int_a^b |v(t)| \, dt$ **5.** 20π **7.** If the parameter t used to describe a trajectory also measures the arc length s of the curve that is generated, we say the curve has been parameterized by its arc length.
9. 5   **11.** 3π **13.** $\frac{\pi^2}{8}$ **15.** $5\sqrt{34}$ **17.** $4\pi\sqrt{65}$ **19.** 9   **21.** $\frac{3}{2}$
23. $3t^2\sqrt{30}$; $64\sqrt{30}$ **25.** 26; 26π
27. Approx. 66,626 mi/hr   **29.** 19.38
31. 32.50   **33.** Yes   **35.** No; $r(s) = \left\langle \frac{s}{\sqrt{5}}, \frac{2s}{\sqrt{5}} \right\rangle$, $0 \le s \le 3\sqrt{5}$
37. No; $r(s) = \left\langle 2 \cos \frac{s}{2}, 2 \sin \frac{s}{2} \right\rangle$, $0 \le s \le 4\pi$
39. No; $r(s) = \langle \cos s, \sin s \rangle$, $0 \le s \le \pi$
41. No; $r(s) = \left\langle \frac{s}{\sqrt{3}} + 1, \frac{s}{\sqrt{3}} + 1, \frac{s}{\sqrt{3}} + 1 \right\rangle$, $s \ge 0$
43. a. True   **b.** True   **c.** True   **d.** False   **45. a.** If $a^2 = b^2 + c^2$, then $|r(t)|^2 = (a \cos t)^2 + (b \sin t)^2 + (c \sin t)^2 = a^2$ so that $r(t)$ is a circle centered at the origin of radius $|a|$.   **b.** $2\pi a$
c. If $a^2 + c^2 + e^2 = b^2 + d^2 + f^2$ and $ab + cd + ef = 0$, then $r(t)$ is a circle of radius $\sqrt{a^2 + c^2 + e^2}$ and its arc length is $2\pi\sqrt{a^2 + c^2 + e^2}$.   **47. a.** $\int_a^b \sqrt{(Ah'(t))^2 + (Bh'(t))^2} \, dt$
$= \int_a^b \sqrt{(A^2 + B^2)(h'(t))^2} \, dt = \sqrt{A^2 + B^2} \int_a^b |h'(t)| \, dt$
b. $64\sqrt{29}$ **c.** $\frac{7\sqrt{29}}{4}$ **49. a.** 5.102 s
b. $\int_0^{5.102} \sqrt{400 + (25 - 9.8t)^2} \, dt$ **c.** 124.43 m   **d.** 102.04 m
51. $|v(t)| = \sqrt{a^2 + b^2 + c^2} = 1$, if $a^2 + b^2 + c^2 = 1$
53. $\int_a^b |r'(t)| \, dt = \int_a^b \sqrt{(cf'(t))^2 + (cg'(t))^2} \, dt$
$= |c| \int_a^b \sqrt{(f'(t))^2 + (g'(t))^2} \, dt = |c|L$

Section 14.5 Exercises, pp. 913–915

1. 0   **3.** $\kappa = \frac{1}{|v|}\left|\frac{dT}{dt}\right|$ or $\kappa = \frac{|v \times a|}{|v|^3}$ **5.** $N = \frac{dT/dt}{|dT/dt|}$
7. These three unit vectors are mutually orthogonal at all points of the curve.   **9.** The torsion measures the rate at which the curve rises or twists out of the **TN**-plane at a point.   **11.** $T = \frac{\langle 1, 2, 3 \rangle}{\sqrt{14}}$, $\kappa = 0$
13. $T = \frac{\langle 1, 2 \cos t, -2 \sin t \rangle}{\sqrt{5}}$, $\kappa = \frac{1}{5}$

15. $T = \frac{\langle \sqrt{3} \cos t, \cos t, -2 \sin t \rangle}{2}$, $\kappa = \frac{1}{2}$
17. $T = \frac{\langle 1, 4t \rangle}{\sqrt{1 + 16t^2}}$, $\kappa = \frac{4}{(1 + 16t^2)^{3/2}}$
19. $T = \left\langle \cos \left(\frac{\pi t^2}{2}\right), \sin \left(\frac{\pi t^2}{2}\right) \right\rangle$, $\kappa = \pi t$
21. $\frac{1}{3}$ **23.** $\frac{2}{(4t^2 + 1)^{3/2}}$ **25.** $\frac{2\sqrt{5}}{(20 \sin^2 t + \cos^2 t)^{3/2}}$
27. $T = \langle \cos t, -\sin t \rangle$, $N = \langle -\sin t, -\cos t \rangle$
29. $T = \frac{\langle t, -3, 0 \rangle}{\sqrt{t^2 + 9}}$, $N = \frac{\langle 3, t, 0 \rangle}{\sqrt{t^2 + 9}}$
31. $T = \langle -\sin t^2, \cos t^2 \rangle$, $N = \langle -\cos t^2, -\sin t^2 \rangle$
33. $T = \frac{\langle 2t, 1 \rangle}{\sqrt{4t^2 + 1}}$, $N = \frac{\langle 1, -2t \rangle}{\sqrt{4t^2 + 1}}$ **35.** $a_N = a_T = 0$
37. $a_T = \sqrt{3}e^t$; $a_N = \sqrt{2}e^t$ **39.** $a = \frac{6t}{\sqrt{9t^2 + 4}}N + \frac{18t^2 + 4}{\sqrt{9t^2 + 4}}T$
41. $B(t) = \langle 0, 0, -1 \rangle$, $\tau = 0$ **43.** $B(t) = \langle 0, 0, 1 \rangle$, $\tau = 0$
45. $B(t) = \frac{\langle -\sin t, \cos t, 2 \rangle}{\sqrt{5}}$, $\tau = -\frac{1}{5}$
47. $B(t) = \frac{\langle 5, 12 \sin t, -12 \cos t \rangle}{13}$, $\tau = \frac{12}{169}$ **49. a.** False
b. False   **c.** False   **d.** True   **e.** False   **f.** False   **g.** False
51. $\kappa = \frac{2}{(1 + 4x^2)^{3/2}}$ **53.** $\kappa = \frac{x}{(x^2 + 1)^{3/2}}$
57. $\kappa = \frac{|ab|}{(a^2 \cos^2 t + b^2 \sin^2 t)^{3/2}}$ **59.** $\kappa = \frac{2|a|}{(1 + 4a^2 t^2)^{3/2}}$
61. b. $v_A(t) = \langle 1, 2, 3 \rangle$, $a_A(t) = \langle 0, 0, 0 \rangle$ and $v_B(t) = \langle 2t, 4t, 6t \rangle$, $a_B(t) = \langle 2, 4, 6 \rangle$; A has constant velocity and zero acceleration, while B has increasing speed and constant acceleration.
c. $a_A(t) = 0N + 0T$, $a_B(t) = 0N + 2\sqrt{14}\,T$; both normal components are zero since the path is a straight line ($\kappa = 0$).
63. b. $v_A(t) = \langle -\sin t, \cos t \rangle$, $a_A(t) = \langle -\cos t, -\sin t \rangle$
$v_B(t) = \langle -2t \sin t^2, 2t \cos t^2 \rangle$
$a_B(t) = \langle -4t^2 \cos t^2 - 2 \sin t^2, -4t^2 \sin t^2 + 2 \cos t^2 \rangle$
c. $a_A(t) = N + 0T$, $a_B(t) = 4t^2 N + 2T$; for A, the acceleration is always normal to the curve, but this is not true for B.
65. b. $\kappa = \frac{1}{2\sqrt{2(1 - \cos t)}}$ **c.**

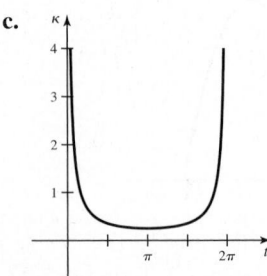

d. Minimum curvature at $t = \pi$
67. b. $\kappa = \frac{1}{t(1 + t^2)^{3/2}}$ **c.**

d. No maximum or minimum curvature

69. $\kappa = \dfrac{e^x}{(1 + e^{2x})^{3/2}}, \left(-\dfrac{\ln 2}{2}, \dfrac{1}{\sqrt{2}}\right), \dfrac{2\sqrt{3}}{9}$

71. $\dfrac{1}{\kappa} = \dfrac{1}{2}; x^2 + \left(y - \dfrac{1}{2}\right)^2 = \dfrac{1}{4}$

73. $\dfrac{1}{\kappa} = 4; (x - \pi)^2 + (y + 2)^2 = 16$

75. $\kappa\left(\dfrac{\pi}{2n}\right) = n^2; \kappa$ increases as n increases.

77. a. Speed $= \sqrt{V_0^2 - 2V_0\, gt \sin \alpha + g^2 t^2}$

b. $\kappa(t) = \dfrac{gV_0 \cos \alpha}{(V_0^2 - 2V_0\, gt \sin \alpha + g^2 t^2)^{3/2}}$

c. Speed has a minimum at $t = \dfrac{V_0 \sin \alpha}{g}$ and $\kappa(t)$ has a maximum at

$t = \dfrac{V_0 \sin \alpha}{g}$. **79.** $\kappa = \dfrac{1}{|\mathbf{v}|} \cdot \left|\dfrac{d\mathbf{T}}{dt}\right|$, where $\mathbf{T} = \dfrac{\langle b, d, f \rangle}{\sqrt{b^2 + d^2 + f^2}}$

and b, d, and f are constant. Therefore, $\dfrac{d\mathbf{T}}{dt} = 0$ so $\kappa = 0$.

81. a. $\kappa_1(x) = \dfrac{2}{(1 + 4x^2)^{3/2}}$

$\kappa_2(x) = \dfrac{12x^2}{(1 + 16x^6)^{3/2}}$

$\kappa_3(x) = \dfrac{30x^4}{(1 + 36x^{10})^{3/2}}$

b.

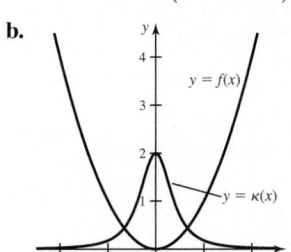
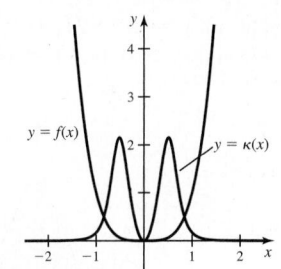
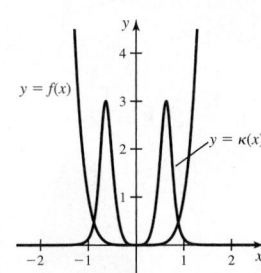

c. κ_1 has its maximum at $x = 0$, κ_2 has its maxima at $x = \pm\sqrt[6]{\frac{1}{56}}$, and κ_3 has its maxima at $x = \pm\sqrt[10]{\frac{1}{99}}$. **d.** $\lim\limits_{n \to \infty} z_n = 1$; the graphs of $y = f_n(x)$ show that as $n \to \infty$, the point corresponding to maximum curvature gets arbitrarily close to the point $(1, 0)$.

Chapter 14 Review Exercises, pp. 916–918

1. a. False **b.** True **c.** True **d.** True **e.** False **f.** False

3.

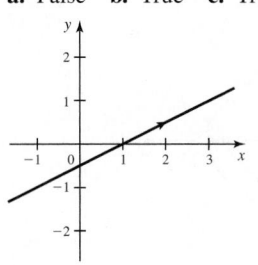

5.

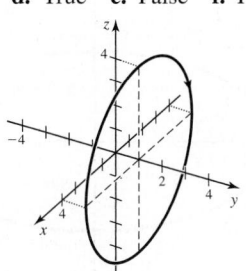

7. $x^2 + y^2 + z^2 = 2; y = z$; a tilted circle of radius $\sqrt{2}$ centered at $(0, 0, 0)$ **9.** $\mathbf{r}(t) = \langle 4 + 15t, -2 - t, 3 - 5t \rangle$

11. $\mathbf{r}(t) = \langle 2, 3 \cos t, 4 \sin t \rangle$, for $0 \leq t \leq 2\pi$

13. $\mathbf{r}(t) = \langle \cos t, \sin t, \sin t \rangle$, for $0 \leq t \leq 2\pi$

15. $\mathbf{r}(t) = \langle 3 \cos t, \sin t, \sin t \rangle$, for $0 \leq t \leq 2\pi$

17. a. $\langle 1, 0 \rangle; \langle 0, 1 \rangle$ **b.** $\left\langle -\dfrac{2}{(2t + 1)^2}, \dfrac{1}{(t + 1)^2} \right\rangle; \langle -2, 1 \rangle$

c. $\left\langle \dfrac{8}{(2t + 1)^3}, -\dfrac{2}{(t + 1)^3} \right\rangle$

d. $\left\langle \dfrac{1}{2} \ln|2t + 1|, t - \ln|t + 1| \right\rangle + \mathbf{C}$

19. a. $\langle 0, 3, 0 \rangle$; does not exist

b. $\langle 2 \cos 2t, -12 \sin 4t, 1 \rangle; \langle 2, 0, 1 \rangle$ **c.** $\langle -4 \sin 2t, -48 \cos 4t, 0 \rangle$

d. $\left\langle -\dfrac{1}{2} \cos 2t, \dfrac{3}{4} \sin 4t, \dfrac{1}{2}t^2 \right\rangle + \mathbf{C}$ **21.** $2\mathbf{j} + \pi\mathbf{k}$

23. $23\mathbf{i} - 41\mathbf{k}$ **25.** $\mathbf{r}(t) = \left\langle t + 2, -\dfrac{1}{2} \cos 2t + \dfrac{5}{2}, \tan t + 2 \right\rangle$

27. $\mathbf{r}(t) = \langle 4 \tan^{-1} t - \pi, t^2 + t - 2, t^3 - 1 \rangle$

29. $\mathbf{T}(t) = \left\langle \dfrac{2e^t}{2e^{2t} + 1}, \dfrac{2e^{2t}}{2e^{2t} + 1}, \dfrac{1}{2e^{2t} + 1} \right\rangle; \left\langle \dfrac{2}{3}, \dfrac{2}{3}, \dfrac{1}{3} \right\rangle$

31. a. $\langle 4e^{4t}, 4e^{4t}, 2e^{4t} \rangle; 6e^{4t}$ **b.** $\langle 16e^{4t}, 16e^{4t}, 8e^{4t} \rangle$

33. $\mathbf{v}(t) = \langle 2 + \sin t, 3 - 2 \cos t \rangle$;

$\mathbf{r}(t) = \langle 2t + 2 - \cos t, 3t + 2 - 2 \sin t \rangle$

35. a. $\mathbf{v}(t) = \langle 40, -32t + 40\sqrt{3} \rangle$;

$\mathbf{r}(t) = \langle 40t, -16t^2 + 40\sqrt{3}t + 3 \rangle$

b. Approx. 4.37 s; approx. 174.9 ft **c.** 78 ft

37. a. $\mathbf{v}(t) = \langle 4t + 40, 20, 40 - 32t \rangle$;

$\mathbf{r}(t) = \langle 2t^2 + 40t, 20t, -16t^2 + 40t + 2 \rangle$ **b.** 2.549 s

c. 126 ft **39. a.** $(116, 30)$ **b.** 39.1 ft **c.** 2.315 s

d. $\displaystyle\int_0^{2.315} \sqrt{50^2 + (-32t + 50)^2}\, dt$ **e.** 129 ft **f.** 41.4° to 79.4°

41. $(1.47, 3.15, 4.4)$ **43.** 12 **45.** Approx. 6.42

47. a. $\mathbf{v}(t) = \mathbf{i} + t\sqrt{2}\,\mathbf{j} + t^2\mathbf{k}$ **b.** 12

49. $\mathbf{r}(s) = \left\langle (\sqrt{1 + s} - 1)^2, \dfrac{4\sqrt{2}}{3}(\sqrt{1 + s} - 1)^{3/2}, \right.$

$\left. 2(\sqrt{1 + s} - 1) \right\rangle$, for $s \geq 0$ **51. a.** $\mathbf{v} = \langle -6 \sin t, 3 \cos t \rangle$,

$\mathbf{T} = \dfrac{\langle -2 \sin t, \cos t \rangle}{\sqrt{1 + 3 \sin^2 t}}$ **b.** $\kappa(t) = \dfrac{2}{3(1 + 3 \sin^2 t)^{3/2}}$

c. $\mathbf{N} = \left\langle -\dfrac{\cos t}{\sqrt{1 + 3 \sin^2 t}}, -\dfrac{2 \sin t}{\sqrt{1 + 3 \sin^2 t}} \right\rangle$

d. $|\mathbf{N}| = \sqrt{\dfrac{\cos^2 t + 4 \sin^2 t}{1 + 3 \sin^2 t}} = \sqrt{\dfrac{(\cos^2 t + \sin^2 t) + 3 \sin^2 t}{1 + 3 \sin^2 t}} = 1$;

$\mathbf{T} \cdot \mathbf{N} = \dfrac{2 \sin t \cos t - 2 \sin t \cos t}{1 + 3 \sin^2 t} = 0$

e.

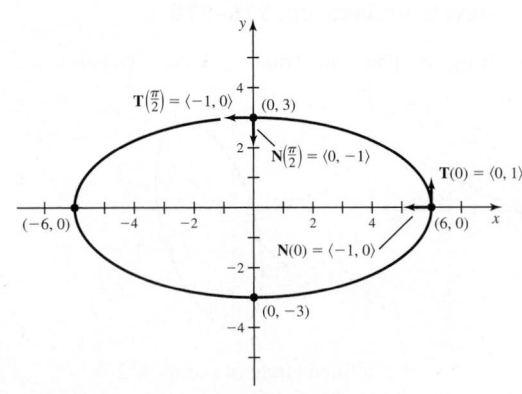

53. a. $\mathbf{v}(t) = \langle -\sin t, -2\sin t, \sqrt{5}\cos t \rangle,$

$\mathbf{T}(t) = \left\langle -\dfrac{1}{\sqrt{5}}\sin t, -\dfrac{2}{\sqrt{5}}\sin t, \cos t \right\rangle$ **b.** $\kappa(t) = \dfrac{1}{\sqrt{5}}$

c. $\mathbf{N}(t) = \left\langle -\dfrac{1}{\sqrt{5}}\cos t, -\dfrac{2}{\sqrt{5}}\cos t, -\sin t \right\rangle$

d. $|\mathbf{N}(t)| = \sqrt{\dfrac{1}{5}\cos^2 t + \dfrac{4}{5}\cos^2 t + \sin^2 t} = 1;$

$\mathbf{T}\cdot\mathbf{N} = \left(\dfrac{1}{5}\cos t\sin t + \dfrac{4}{5}\cos t\sin t\right) - \sin t\cos t = 0$

e.

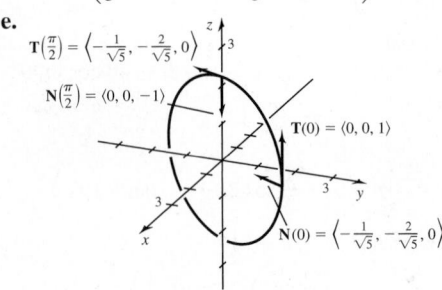

55. a. $\mathbf{a}(t) = 2\mathbf{N} + 0\mathbf{T} = 2\langle -\cos t, -\sin t \rangle$

b.

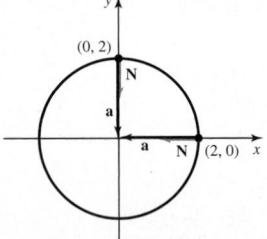

57. a. $a_T = \dfrac{2t}{\sqrt{t^2+1}}$ and $a_N = \dfrac{2}{\sqrt{t^2+1}}$

b.

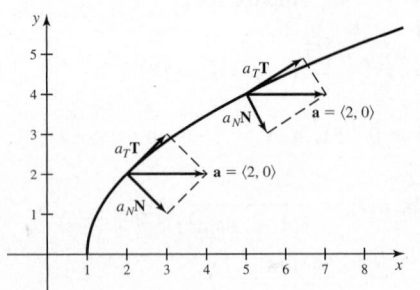

59. $\mathbf{B}(1) = \dfrac{\langle 3, -3, 1 \rangle}{\sqrt{19}};\ \tau = \dfrac{3}{19}$

61. a. $\mathbf{T}(t) = \dfrac{1}{5}\langle 3\cos t, -3\sin t, 4 \rangle$

b. $\mathbf{N}(t) = \langle -\sin t, -\cos t, 0 \rangle;\ \kappa = \dfrac{3}{25}$

c.

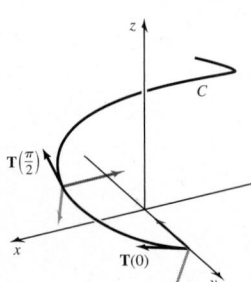

d. Yes

e. $\mathbf{B}(t) = \dfrac{1}{5}\langle 4\cos t, -4\sin t, -3 \rangle$

f. See graph in part (c).

g. Check that $\mathbf{T}$, $\mathbf{N}$, and $\mathbf{B}$ have unit length and are mutually orthogonal.

h. $\tau = -\dfrac{4}{25}$

63. a. Consider first the case where $a_3 = b_3 = c_3 = 0,$ and show that for all $s \neq t$ in I, $\mathbf{r}(t) \times \mathbf{r}(s)$ is a multiple of the constant vector $\langle b_1 c_2 - b_2 c_1, a_2 c_1 - a_1 c_2, a_1 b_2 - a_2 b_1 \rangle$, which implies $\mathbf{r}(t) \times \mathbf{r}(s)$ is always orthogonal to the same vector, and therefore the vectors $\mathbf{r}(t)$ must all lie in the same plane. When a_3, b_3, and c_3 are not necessarily 0, the curve still lies in a plane because these constants represent a simple translation of the curve to a different location in $\mathbb{R}^3$.

b. Because the curve lies in a plane, $\mathbf{B}$ is always normal to the plane and has length 1. Therefore, $\dfrac{d\mathbf{B}}{ds} = \mathbf{0}$ and $\tau = 0$.

CHAPTER 15

Section 15.1 Exercises, pp. 927–930

1. Independent: x and y; dependent: z
3. $D = \{(x, y): x \neq 0 \text{ and } y \neq 0\}$ **5.** Three   **7.** 3; 4
9. a. 1300 ft   **b.** Katie; Katie is 100 ft higher than Zeke.   **11.** Circles
13. $n = 6$ **15.** $D = \mathbb{R}^2$ **17.** $D = \{(x, y): x^2 + y^2 \le 25\}$
19. $D = \{(x, y): y \neq 0\}$ **21.** $D = \{(x, y): y < x^2\}$
23. $D = \{(x, y): xy \ge 0, (x, y) \neq (0, 0)\}$
25. Plane; $D = \mathbb{R}^2, R = \mathbb{R}$

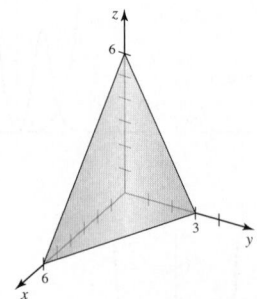

27. Hyperbolic paraboloid; $D = \mathbb{R}^2, R = \mathbb{R}$

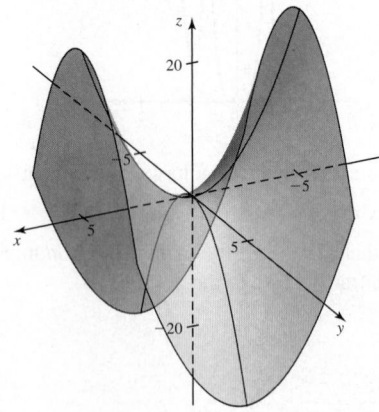

29. Lower part of a hyperboloid of two sheets;
$D = \mathbb{R}^2, R = (-\infty, -1]$

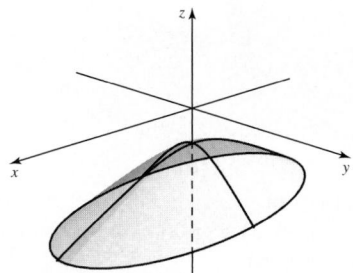

31. Upper half of a hyperboloid of one sheet;
$D = \{(x, y): x^2 + y^2 \geq 1\}, R = [0, \infty)$

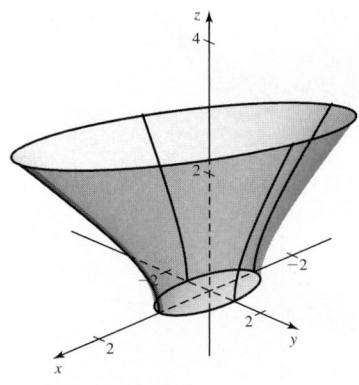

33. Upper half of an elliptical cylinder;
$D = \{(x, y): -2 \leq x \leq 2, -\infty < y < \infty\}, R = [0, 4]$

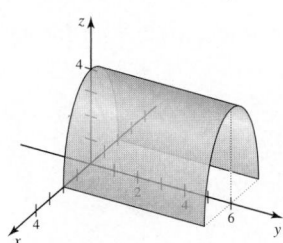

35. a. A **b.** D **c.** B **d.** C

37.

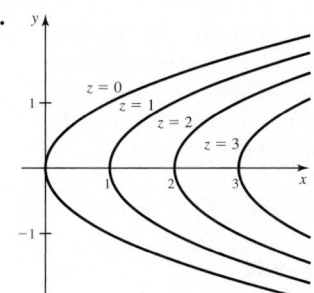

39.

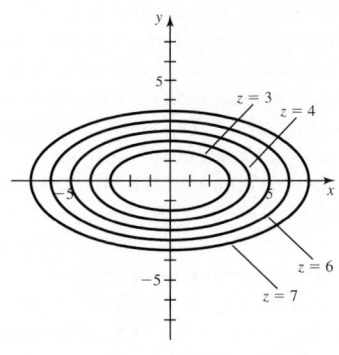

41.

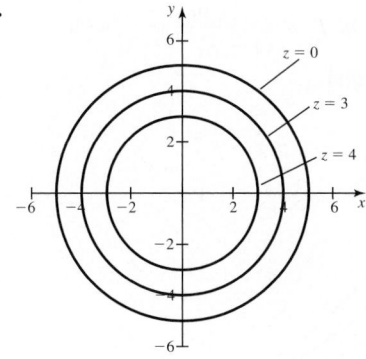

43.

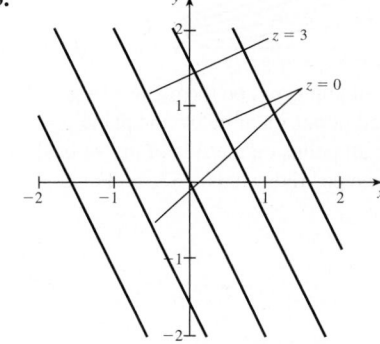

45. a.

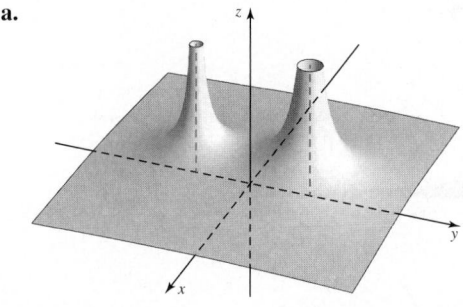

b. $\mathbb{R}^2$ excluding the points $(0, 1)$ and $(0, -1)$

c. $\varphi(2, 3)$ is greater. **d.**

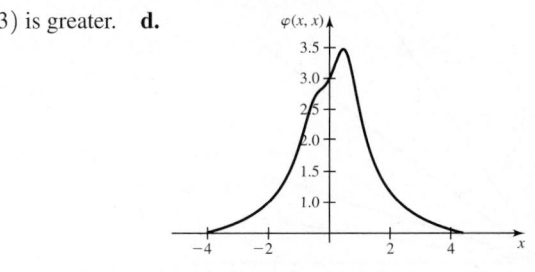

47. a.

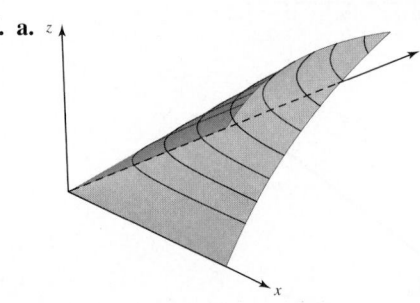

b. $R(10, 10) = 5$
c. $R(x, y) = R(y, x)$

49. a. $P = \dfrac{20{,}000r}{(1 + r)^{240} - 1}$ **b.** $P = \dfrac{Br}{(1 + r)^{240} - 1}$, with

$B = 5000, 10{,}000, 15{,}000, 25{,}000$

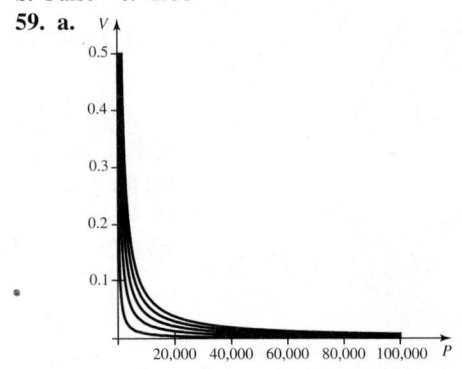

51. $D = \{(x, y, z): x \neq z\}$; all points not on the plane $x = z$
53. $D = \{(x, y, z): y \geq z\}$; all points on or below the plane $y = z$
55. $D = \{(x, y, z): x^2 \leq y\}$; all points on the side of the vertical
cylinder $y = x^2$ that contains the positive y-axis   **57. a.** False
b. False   **c.** True
59. a.

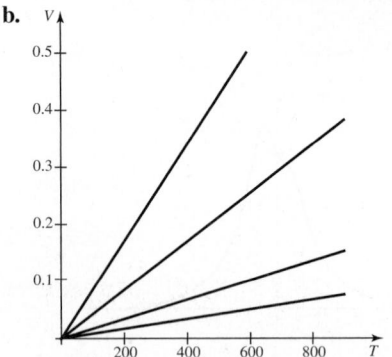

b.

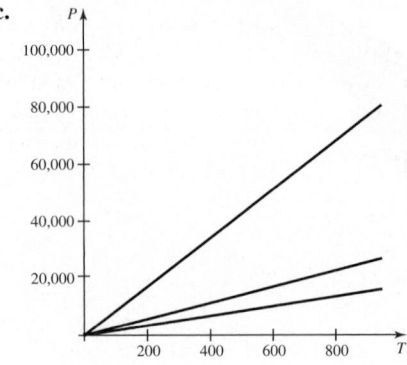

c.

61. a.

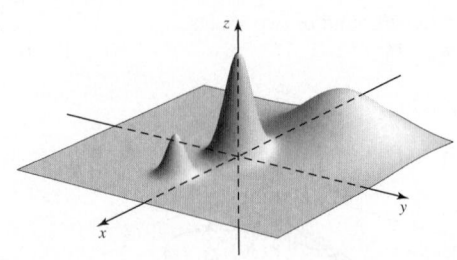

b. $(0, 0), (-5, 3), (4, -1)$
c. $f(0, 0) = 10.17, f(-5, 3) = 5.00, f(4, -1) = 4.00$
63. a. $D = \mathbb{R}^2, R = [0, \infty)$
b.

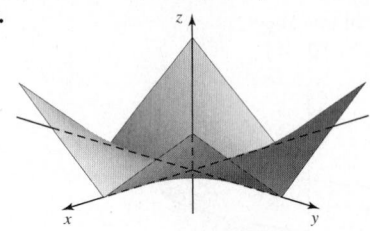

65. a. $D = \{(x, y): x \neq y\}, R = \mathbb{R}$
b.

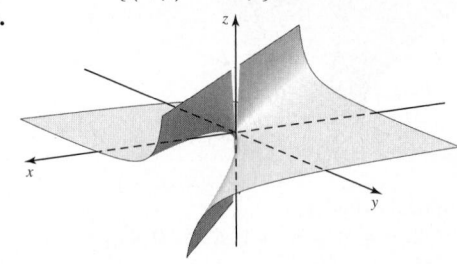

67. a. $D = \{(x, y): y \neq x + \pi/2 + n\pi \text{ for any integer } n\}$,
$R = [0, \infty)$ **b.**

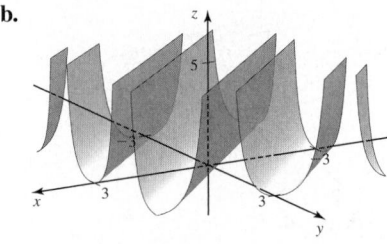

69. Peak at the origin   **71.** Depression at $(1, 0)$ **73.** The level
curves are $ax + by = d - cz_0$, where z_0 is a constant, which are
lines with slope $-a/b$ if $b \neq 0$ or vertical lines if $b = 0$.
75. $z = x^2 + y^2 - C$; paraboloids with vertices at $(0, 0, -C)$
77. $x^2 + 2z^2 = C$; elliptic cylinders parallel to the y-axis
79. $D = \{(x, y): x - 1 \leq y \leq x + 1\}$
81. $D = \{(x, y, z): (x \leq z \text{ and } y \geq -z) \text{ or } (x \geq z \text{ and } y \leq -z)\}$

Section 15.2 Exercises, pp. 937–939

1. The values of $f(x, y)$ are arbitrarily close to L for all (x, y) sufficiently close to (a, b).   **3.** Because polynomials of n variables are
continuous on all of $\mathbb{R}^n$, limits of polynomials can be evaluated with
direct substitution.   **5.** If the function approaches different values
along different paths, the limit does not exist.   **7.** f must be defined,
the limit must exist, and the limit must equal the function value.
9. At any point where the denominator is nonzero   **11.** 10   **13.** 101
15. 27   **17.** $1/(2\pi)$ **19.** 2   **21.** 6   **23.** -1 **25.** 2
27. $1/(2\sqrt{2}) = \sqrt{2}/4$ **29.** $L = 1$ along $y = 0$, and $L = -1$ along
$x = 0$ **31.** $L = 1$ along $x = 0$, and $L = -2$ along $y = 0$
33. $L = 2$ along $y = x$, and $L = 0$ along $y = -x$ **35.** $\mathbb{R}^2$
37. All points except $(0, 0)$ **39.** $\{(x, y): x \neq 0\}$ **41.** All points
except $(0, 0)$ **43.** $\mathbb{R}^2$ **45.** $\mathbb{R}^2$ **47.** $\mathbb{R}^2$ **49.** All points except
$(0, 0)$ **51.** $\mathbb{R}^2$ **53.** $\mathbb{R}^2$ **55.** 6   **57.** -1 **59.** 2   **61. a.** False

b. False   **c.** True   **d.** False   **63.** $\frac{1}{2}$ **65.** 0   **67.** Does not exist
69. $\frac{1}{4}$ **71.** 1   **73.** 1   **75.** 5   **77.** $b = 1$ **79.** 0   **81.** 1   **85.** 0

Section 15.3 Exercises, pp. 948–951

1. $f_x(a, b)$ is the slope of the surface in the direction parallel to the positive x-axis, $f_y(a, b)$ is the slope of the surface in the direction parallel to the positive y-axis, both taken at (a, b).   **3. a.** Negative
b. Negative   **c.** Negative   **d.** Positive   **5.** $f_x(x, y) = 6xy$;
$f_y(x, y) = 3x^2$ **7.** $f_{xy} = 0 = f_{yx}$ **9.** $f_x(x, y, z) = y + z$;
$f_y(x, y, z) = x + z$; $f_z(x, y, z) = x + y$

11. $f_x(x, y) = 5y$; $f_y(x, y) = 5x$ **13.** $f_x(x, y) = \frac{1}{y}$; $f_y(x, y) = -\frac{x}{y^2}$

15. $f_x(x, y) = e^y$; $f_y(x, y) = xe^y$

17. $f_x(x, y) = 2xye^{x^2y}$; $f_y(x, y) = x^2 e^{x^2y}$

19. $f_w(w, z) = \frac{z^2 - w^2}{(w^2 + z^2)^2}$; $f_z(w, z) = -\frac{2wz}{(w^2 + z^2)^2}$

21. $f_x(x, y) = \cos xy - xy \sin xy$; $f_y(x, y) = -x^2 \sin xy$

23. $s_y(y, z) = z^3 \sec^2 yz$; $s_z(y, z) = 2z \tan yz + yz^2 \sec^2 yz$

25. $G_s(s, t) = \frac{\sqrt{st}(t - s)}{2s(s + t)^2}$; $G_t(s, t) = \frac{\sqrt{st}(s - t)}{2t(s + t)^2}$

27. $f_x(x, y) = 2yx^{2y-1}$; $f_y(x, y) = 2x^{2y} \ln x$

29. $h_x(x, y) = \frac{\sqrt{x^2 - 4y} - x}{\sqrt{x^2 - 4y}}$; $h_y(x, y) = \frac{2}{\sqrt{x^2 - 4y}}$

31. $f_x(x, y) = -e^{x^2}$; $f_y(x, y) = 3y^2 e^{y^6}$

33. $f_x(x, y) = -\frac{2x}{1 + (x^2 + y^2)^2}$; $f_y(x, y) = -\frac{2y}{1 + (x^2 + y^2)^2}$

35. $h_x(x, y) = (1 + 2y)^x \ln (1 + 2y)$; $h_y(x, y) = 2x(1 + 2y)^{x-1}$
37. $f_x(x, y) = -h(x)$; $f_y(x, y) = h(y)$
39. $h_{xx}(x, y) = 6x$; $h_{xy}(x, y) = 2y = h_{yx}(x, y)$; $h_{yy}(x, y) = 2x$
41. $f_{xx}(x, y) = -16y^3 \sin 4x$; $f_{xy}(x, y) = 12y^2 \cos 4x = f_{yx}(x, y)$;
$f_{yy}(x, y) = 6y \sin 4x$

43. $p_{uu}(u, v) = \frac{-2u^2 + 2v^2 + 8}{(u^2 + v^2 + 4)^2}$;

$p_{uv}(u, v) = -\frac{4uv}{(u^2 + v^2 + 4)^2} = p_{vu}(u, v)$;

$p_{vv}(u, v) = \frac{2u^2 - 2v^2 + 8}{(u^2 + v^2 + 4)^2}$

45. $F_{rr}(r, s) = 0$; $F_{rs}(r, s) = e^s = F_{sr}(r, s)$; $F_{ss}(r, s) = re^s$

47. $f_{xx}(x, y) = \frac{6xy^2(1 - 2x^6 y^4)}{(1 + x^6 y^4)^2}$;

$f_{xy}(x, y) = \frac{6x^2 y(1 - x^6 y^4)}{(1 + x^6 y^4)^2} = f_{yx}(x, y)$;

$f_{yy}(x, y) = \frac{2x^3(1 - 3x^6 y^4)}{(1 + x^6 y^4)^2}$

49. $f_{xy} = e^{x+y} = f_{yx}$ **51.** $f_{xy} = -(xy \cos xy + \sin xy) = f_{yx}$
53. $f_{xy}(x, y) = -72y^2(2x - y^3)^2 = f_{yx}(x, y)$
55. $h_x(x, y, z) = h_y(x, y, z) = h_z(x, y, z) = -\sin (x + y + z)$

57. $F_u(u, v, w) = \frac{1}{v + w}$; $F_v(u, v, w) = F_w(u, v, w) = -\frac{u}{(v + w)^2}$

59. $G_r(r, s, t) = \frac{s^3 t^5}{2\sqrt{rs^3 t^5}}$;

$G_s(r, s, t) = \frac{3rs^2 t^5}{2\sqrt{rs^3 t^5}}$;

$G_t(r, s, t) = \frac{5rs^3 t^4}{2\sqrt{rs^3 t^5}}$

61. $h_w(w, x, y, z) = \frac{z}{xy}$; $h_x(w, x, y, z) = -\frac{wz}{x^2 y}$;

$h_y(w, x, y, z) = -\frac{wz}{xy^2}$; $h_z(w, x, y, z) = \frac{w}{xy}$

63. b. $g_x(x, y, z) = -\frac{8z}{3(2x - y + z)^2}$;

$g_y(x, y, z) = \frac{4z}{3(2x - y + z)^2}$;

$g_z(x, y, z) = \frac{4(2x - y)}{3(2x - y + z)^2}$

65. 1.41   **67.** 1.55 (answer will vary)   **69. a.** $\frac{\partial V}{\partial P} = -\frac{kT}{P^2}$; volume

decreases with pressure at fixed temperature.   **b.** $\frac{\partial V}{\partial T} = \frac{k}{P}$; volume

increases with temperature at fixed pressure.
c.

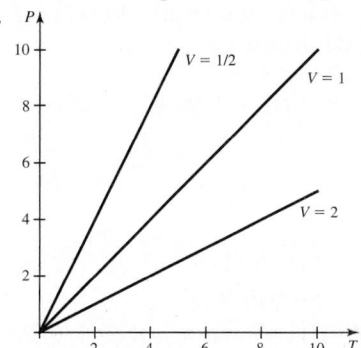

71. a. $\frac{\partial R}{\partial R_1} = \frac{R_2^2}{(R_1 + R_2)^2}$; $\frac{\partial R}{\partial R_2} = \frac{R_1^2}{(R_1 + R_2)^2}$

b. $\frac{\partial R}{\partial R_1} = \frac{R^2}{R_1^2}$; $\frac{\partial R}{\partial R_2} = \frac{R^2}{R_2^2}$ **c.** Increase   **d.** Decrease

73. $u_t = -16e^{-4t} \cos 2x = u_{xx}$ **75.** $u_t = -a^2 A e^{-a^2 t} \cos ax = u_{xx}$
77. a. No   **b.** No   **c.** $f_x(0, 0) = f_y(0, 0) = 0$ **d.** f_x and f_y are
not continuous at $(0, 0)$.   **79. a.** False   **b.** False   **c.** True

81. a. $z_x(x, y) = \frac{1}{y^2}$; $z_y(x, y) = -\frac{2x}{y^3}$

b.

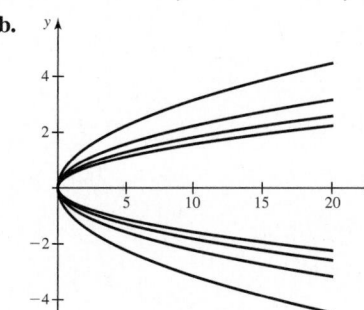

c. z increases as x increases.
d. z increases as y increases
when $y < 0$, z is undefined
for $y = 0$, and z decreases
as y increases for $y > 0$.

83. a. $\varphi_x(x, y) = -\frac{2x}{(x^2 + (y - 1)^2)^{3/2}} - \frac{x}{(x^2 + (y + 1)^2)^{3/2}}$;

$\varphi_y(x, y) = -\frac{2(y - 1)}{(x^2 + (y - 1)^2)^{3/2}} - \frac{y + 1}{(x^2 + (y + 1)^2)^{3/2}}$

b. They both approach zero.   **c.** $\varphi_x(0, y) = 0$

d. $\varphi_y(x, 0) = \dfrac{1}{(x^2 + 1)^{3/2}}$

87. $\dfrac{\partial^2 u}{\partial t^2} = -4c^2 \cos(2(x + ct)) = c^2 \dfrac{\partial^2 u}{\partial x^2}$

89. $\dfrac{\partial^2 u}{\partial t^2} = c^2 A f''(x + ct) + c^2 B g''(x - ct) = c^2 \dfrac{\partial^2 u}{\partial x^2}$

91. $u_{xx} = 6x;\ u_{yy} = -6x$

93. $u_{xx} = \dfrac{2(x - 1)y}{((x - 1)^2 + y^2)^2} - \dfrac{2(x + 1)y}{((x + 1)^2 + y^2)^2};$

$u_{yy} = -\dfrac{2(x - 1)y}{((x - 1)^2 + y^2)^2} + \dfrac{2(x + 1)y}{((x + 1)^2 + y^2)^2}$

95. $\varepsilon_1 = \Delta y,\ \varepsilon_2 = 0$ or $\varepsilon_1 = 0,\ \varepsilon_2 = \Delta x$ **97. a.** f is continuous at $(0, 0)$.   **b.** f is not differentiable at $(0, 0)$.
c. $f_x(0, 0) = f_y(0, 0) = 0$ **d.** f_x and f_y are not continuous at $(0, 0)$.
e. Theorem 15.5 does not apply because f_x and f_y are not continuous at $(0, 0)$; Theorem 15.6 does not apply because f is not differentiable at $(0, 0)$.   **99.** $f_x(x, y) = yh(xy);\ f_y(x, y) = xh(xy)$

Section 15.4 Exercises, pp. 957–961

1. One dependent, two intermediate, and one independent variable
3. Multiply each of the partial derivatives of w by the t-derivative of the corresponding function and add all these expressions.

5.

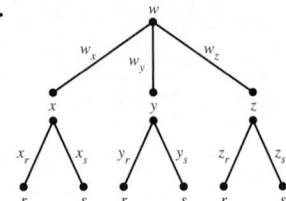

7. $4t^3 + 3t^2$

9. $z'(t) = 2t \sin 4t^3 + 12t^4 \cos 4t^3$

11. $w'(t) = -\sin t \sin 3t^4 + 12t^3 \cos t \cos 3t^4$
13. $z'(t) = 20(\sin^2 t + 2(3t + 4)^5)^9 (\sin t \cos t + 15(3t + 4)^4)$
15. $w'(t) = 20t^4 \sin(t + 1) + 4t^5 \cos(t + 1)$
17. $V'(t) = e^t((2t + 5) \sin t + (2t + 3)\cos t)$
19. $z_s = 2(s - t) \sin t^2;\ z_t = 2(s - t)(t(s - t) \cos t^2 - \sin t^2)$
21. $z_s = 2s - 3s^2 - 2st + t^2;\ z_t = -s^2 - 2t + 2st + 3t^2$
23. $z_s = (t + 1)e^{st + s + t};\ z_t = (s + 1)e^{st + s + t}$
25. $w_s = -\dfrac{2t(t + 1)}{(st + s - t)^2};\ w_t = \dfrac{2s}{(st + s - t)^2}$
27. a. $V'(t) = 2\pi r(t)h(t)r'(t) + \pi r(t)^2 h'(t)$ **b.** $V'(t) = 0$
c. The volume remains constant.
29. $z'(t) = -\dfrac{2t + 2}{(t + 2t)} - \dfrac{3t^2}{(t^3 - 2)}$

31.

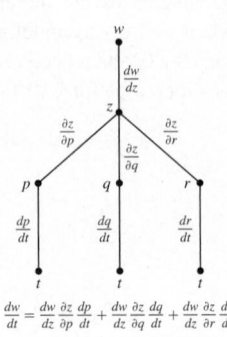

33.

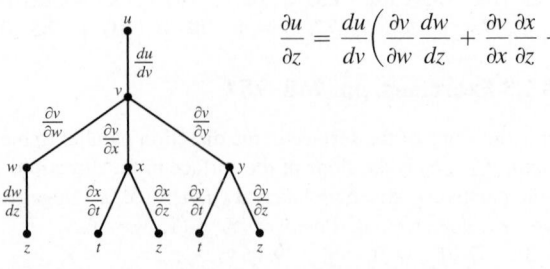

$\dfrac{\partial u}{\partial z} = \dfrac{du}{dv}\left(\dfrac{\partial v}{\partial w}\dfrac{dw}{dz} + \dfrac{\partial v}{\partial x}\dfrac{\partial x}{\partial z} + \dfrac{\partial v}{\partial y}\dfrac{\partial y}{\partial z}\right)$

35. $\dfrac{dy}{dx} = \dfrac{x}{2y}$ **37.** $\dfrac{dy}{dx} = -\dfrac{y}{x}$ **39.** $\dfrac{dy}{dx} = -\dfrac{x + y}{2y^3 + x}$

41. $\dfrac{\partial s}{\partial x} = \dfrac{2x}{\sqrt{x^2 + y^2}};\ \dfrac{\partial s}{\partial y} = \dfrac{2y}{\sqrt{x^2 + y^2}}$

43. $f_{ss} = 2(3s + t);\ f_{st} = 2(s - t);\ f_{tt} = -2(s + 3t)$

45. $f_{ss} = \dfrac{4t^2(-3s^2 + t^2)}{(s^2 + t^2)^3};\ f_{st} = \dfrac{8st(s^2 - t^2)}{(s^2 + t^2)^3};\ f_{tt} = -\dfrac{4(s^4 - 3s^2t^2)}{(s^2 + t^2)^3}$

47. $f''(s) = 4\left(\dfrac{6}{s^4} - \dfrac{2}{s^3} - 1 - 9s + 9s^2\right)$ **49. a.** False   **b.** False

51. $w'(t) = 0$ **53.** $\dfrac{\partial z}{\partial x} = -\dfrac{z^2}{x^2}$ **55. a.** $w'(t) = af_x + bf_y + cf_z$

b. $w'(t) = ayz + bxz + cxy = 3abct^2$

c. $w'(t) = \sqrt{a^2 + b^2 + c^2}\ \dfrac{t}{|t|}$

d. $w''(t) = a^2 f_{xx} + b^2 f_{yy} + c^2 f_{zz} + 2abf_{xy} + 2acf_{xz} + 2bcf_{yz}$

57. $\dfrac{\partial z}{\partial x} = -\dfrac{y + z}{x + y};\ \dfrac{\partial z}{\partial y} = -\dfrac{x + z}{x + y}$ **59.** $\dfrac{\partial z}{\partial x} = -\dfrac{yz + 1}{xy - 1};$

$\dfrac{\partial z}{\partial y} = -\dfrac{xz + 1}{xy - 1}$ **61. a.** $z'(t) = -2x \sin t + 8y \cos t = 3 \sin 2t$

b. $0 < t < \pi/2$ and $\pi < t < 3\pi/2$

63. a. $z'(t) = \dfrac{(x + y)e^{-t}}{\sqrt{1 - x^2 - y^2}} = \dfrac{2e^{-2t}}{\sqrt{1 - 2e^{-2t}}}$ **b.** All $t \geq \frac{1}{2} \ln 2$

65. $E'(t) = mx'x'' + my'y'' + mgy' = 0$
67. a. The volume increases.   **b.** The volume decreases.

69. a. $\dfrac{\partial P}{\partial V} = -\dfrac{P}{V};\ \dfrac{\partial T}{\partial P} = \dfrac{V}{k};\ \dfrac{\partial V}{\partial T} = \dfrac{k}{P}$ **b.** Follows directly from part (a)

71. a. $w'(t) = \dfrac{2t(t^2 + 1) \cos 2t - (t^2 - 1) \sin 2t}{2(t^2 + 1)^2}$

b. Max value of $t \approx 0.838$, $(x, y, z) \approx (0.669, 0.743, 0.838)$

73. a. $z_x = \dfrac{x}{r}z_r - \dfrac{y}{r^2}z_\theta;\ z_y = \dfrac{y}{r}z_r + \dfrac{x}{r^2}z_\theta$

b. $z_{xx} = \dfrac{x^2}{r^2}z_{rr} + \dfrac{y^2}{r^4}z_{\theta\theta} - \dfrac{2xy}{r^3}z_{r\theta} + \dfrac{y^2}{r^3}z_r + \dfrac{2xy}{r^4}z_\theta$

c. $z_{yy} = \dfrac{y^2}{r^2}z_{rr} + \dfrac{x^2}{r^4}z_{\theta\theta} + \dfrac{2xy}{r^3}z_{r\theta} + \dfrac{x^2}{r^3}z_r - \dfrac{2xy}{r^4}z_\theta$

75. a. $\left(\dfrac{\partial z}{\partial x}\right)_y = -\dfrac{F_x}{F_z}$ **b.** $\left(\dfrac{\partial y}{\partial z}\right)_x = -\dfrac{F_z}{F_y};\ \left(\dfrac{\partial x}{\partial y}\right)_z = -\dfrac{F_y}{F_x}$

d. $\left(\dfrac{\partial w}{\partial x}\right)_{y,z}\left(\dfrac{\partial z}{\partial w}\right)_{x,y}\left(\dfrac{\partial y}{\partial z}\right)_{x,w}\left(\dfrac{\partial x}{\partial y}\right)_{z,w} = 1$

77. a. $\left(\dfrac{\partial w}{\partial x}\right)_y = f_x + f_z\dfrac{dz}{dx} = 18$ **b.** $\left(\dfrac{\partial w}{\partial x}\right)_z = f_x + f_y\dfrac{dy}{dx} = 8$

d. $\left(\dfrac{\partial w}{\partial y}\right)_x = -5;\ \left(\dfrac{\partial w}{\partial y}\right)_z = 4;\ \left(\dfrac{\partial w}{\partial z}\right)_x = \dfrac{5}{2};\ \left(\dfrac{\partial w}{\partial z}\right)_y = \dfrac{9}{2}$

Section 15.5 Exercises, pp. 970–973

1. Form the dot product between the unit direction vector **u** and the gradient of the function.   **3.** Direction of steepest ascent
5. The gradient is orthogonal to the level curves of f.
7. -2 **9.** $-7; 0$
11. a.

	$(a,b) = (2,0)$	$(a,b) = (0,2)$	$(a,b) = (1,1)$
$\mathbf{u} = \left\langle \frac{\sqrt{2}}{2}, \frac{\sqrt{2}}{2} \right\rangle$	$-\sqrt{2}$	$-2\sqrt{2}$	$-3\sqrt{2}/2$
$\mathbf{v} = \left\langle -\frac{\sqrt{2}}{2}, \frac{\sqrt{2}}{2} \right\rangle$	$\sqrt{2}$	$-2\sqrt{2}$	$-\sqrt{2}/2$
$\mathbf{w} = \left\langle -\frac{\sqrt{2}}{2}, -\frac{\sqrt{2}}{2} \right\rangle$	$\sqrt{2}$	$2\sqrt{2}$	$3\sqrt{2}/2$

b. The function is decreasing at $(2, 0)$ in the direction of **u** and increasing at $(2, 0)$ in the directions of **v** and **w**.
13. $\nabla f(x, y) = \langle 6x, -10y \rangle$, $\nabla f(2, -1) = \langle 12, 10 \rangle$
15. $\nabla g(x, y) = \langle 2(x - 4xy - 4y^2), -4x(x + 4y) \rangle$, $\nabla g(-1, 2) = \langle -18, 28 \rangle$ **17.** $\nabla f(x, y) = e^{2xy} \langle 1 + 2xy, 2x^2 \rangle$, $\nabla f(1, 0) = \langle 1, 2 \rangle$ **19.** $\nabla F(x, y) = -2e^{-x^2 - 2y^2} \langle x, 2y \rangle$, $\nabla F(-1, 2) = 2e^{-9} \langle 1, -4 \rangle$ **21.** -6 **23.** $\frac{27}{2} - 6\sqrt{3}$
25. $-\dfrac{2}{\sqrt{5}}$ **27.** -2 **29.** 0 **31. a.** Direction of steepest ascent:
$\dfrac{1}{\sqrt{65}} \langle 1, 8 \rangle$; direction of steepest descent: $-\dfrac{1}{\sqrt{65}} \langle 1, 8 \rangle$
b. $\langle -8, 1 \rangle$ **33. a.** Direction of steepest ascent: $\dfrac{1}{\sqrt{5}} \langle -2, 1 \rangle$;
direction of steepest descent: $\dfrac{1}{\sqrt{5}} \langle 2, -1 \rangle$ **b.** $\langle 1, 2 \rangle$
35. a. Direction of steepest ascent: $\dfrac{1}{\sqrt{2}} \langle 1, -1 \rangle$;
direction of steepest descent: $\dfrac{1}{\sqrt{2}} \langle -1, 1 \rangle$ **b.** $\langle 1, 1 \rangle$
37. a. $\nabla f(3, 2) = -12\mathbf{i} - 12\mathbf{j}$
b. Direction of max increase: $\theta = \dfrac{5\pi}{4}$; direction of max decrease:
$\theta = \dfrac{\pi}{4}$; directions of no change: $\theta = \dfrac{3\pi}{4}, \dfrac{7\pi}{4}$
c. $g(\theta) = -12 \cos \theta - 12 \sin \theta$ **d.** $\theta = \dfrac{5\pi}{4}, g\left(\dfrac{5\pi}{4}\right) = 12\sqrt{2}$
e. $\nabla f(3, 2) = 12\sqrt{2} \left\langle \cos \dfrac{5\pi}{4}, \sin \dfrac{5\pi}{4} \right\rangle$, $|\nabla f(3, 2)| = 12\sqrt{2}$
39. a. $\nabla f(\sqrt{3}, 1) = \dfrac{\sqrt{6}}{6} \langle \sqrt{3}, 1 \rangle$ **b.** Direction of max increase:
$\theta = \dfrac{\pi}{6}$; direction of max decrease: $\theta = \dfrac{7\pi}{6}$; directions of no change:
$\theta = \dfrac{2\pi}{3}, \dfrac{5\pi}{3}$ **c.** $g(\theta) = \dfrac{\sqrt{2}}{2} \cos \theta + \dfrac{\sqrt{6}}{6} \sin \theta$ **d.** $\theta = \dfrac{\pi}{6}, g\left(\dfrac{\pi}{6}\right) = \dfrac{\sqrt{6}}{3}$
e. $\nabla f(\sqrt{3}, 1) = \dfrac{\sqrt{6}}{3} \left\langle \cos \dfrac{\pi}{6}, \sin \dfrac{\pi}{6} \right\rangle$, $|\nabla f(\sqrt{3}, 1)| = \dfrac{\sqrt{6}}{3}$
41. a. $\nabla F(-1, 0) = \dfrac{2}{e} \mathbf{i}$ **b.** Direction of max increase: $\theta = 0$; direction of max decrease: $\theta = \pi$; directions of no change: $\theta = \pm \dfrac{\pi}{2}$
c. $g(\theta) = \dfrac{2}{e} \cos \theta$ **d.** $\theta = 0, g(0) = \dfrac{2}{e}$
e. $\nabla F(-1, 0) = \dfrac{2}{e} \langle \cos 0, \sin 0 \rangle$, $|\nabla F(-1, 0)| = \dfrac{2}{e}$
43.

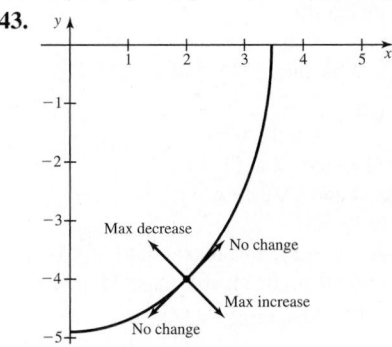

45.

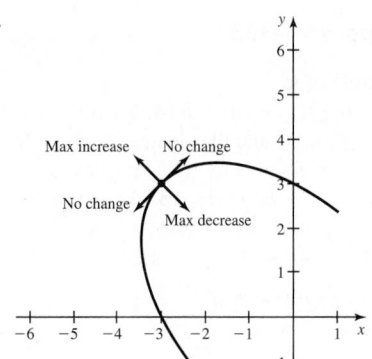

47. $y' = 0$
49. Vertical tangent
51. $y' = -2/\sqrt{3}$
53. Vertical tangent

55. a. $\nabla f = \langle 1, 0 \rangle$ **b.** $x = 4 - t, y = 4, t \geq 0$
c. $\mathbf{r}(t) = \langle 4 - t, 4, 8 - t \rangle$, for $t \geq 0$
57. a. $\nabla f = \langle -2x, -4y \rangle$ **b.** $y = x^2, x \geq 1$
c. $\mathbf{r}(t) = \langle t, t^2, 4 - t^2 - 2t^4 \rangle$, for $t \geq 1$
59. a. $\nabla f(x, y, z) = 2x\mathbf{i} + 4y\mathbf{j} + 8z\mathbf{k}$, $\nabla f(1, 0, 4) = 2\mathbf{i} + 32\mathbf{k}$
b. $\dfrac{1}{\sqrt{257}} (\mathbf{i} + 16\mathbf{k})$ **c.** $2\sqrt{257}$ **d.** $17\sqrt{2}$
61. a. $\nabla f(x, y, z) = 4yz\mathbf{i} + 4xz\mathbf{j} + 4xy\mathbf{k}$,
$\nabla f(1, -1, -1) = 4\mathbf{i} - 4\mathbf{j} - 4\mathbf{k}$ **b.** $\dfrac{1}{\sqrt{3}} (\mathbf{i} - \mathbf{j} - \mathbf{k})$ **c.** $4\sqrt{3}$
d. $\dfrac{4}{\sqrt{3}}$ **63. a.** $\nabla f(x, y, z) = \cos (x + 2y - z)(\mathbf{i} + 2\mathbf{j} - \mathbf{k})$,
$\nabla f\left(\dfrac{\pi}{6}, \dfrac{\pi}{6}, -\dfrac{\pi}{6}\right) = -\dfrac{1}{2}\mathbf{i} - \mathbf{j} + \dfrac{1}{2}\mathbf{k}$ **b.** $\dfrac{1}{\sqrt{6}} (-\mathbf{i} - 2\mathbf{j} + \mathbf{k})$
c. $\sqrt{6}/2$ **d.** $-\dfrac{1}{2}$
65. a. $\nabla f(x, y, z) = \dfrac{2}{1 + x^2 + y^2 + z^2}(x\mathbf{i} + y\mathbf{j} + z\mathbf{k})$,
$\nabla f(1, 1, -1) = \dfrac{1}{2}\mathbf{i} + \dfrac{1}{2}\mathbf{j} - \dfrac{1}{2}\mathbf{k}$ **b.** $\dfrac{1}{\sqrt{3}} (\mathbf{i} + \mathbf{j} - \mathbf{k})$
c. $\dfrac{\sqrt{3}}{2}$ **d.** $\dfrac{5}{6}$ **67. a.** False   **b.** False   **c.** False   **d.** True
69. $\pm \dfrac{1}{\sqrt{5}} (\mathbf{i} - 2\mathbf{j})$ **71.** $\pm \dfrac{1}{\sqrt{2}} (\mathbf{i} + \mathbf{j})$
73. $x = x_0 + at, y = y_0 + bt$ **75. a.** $\nabla f(x, y, z) = \langle 2x, 2y, 2z \rangle$,
$\nabla f(1, 1, 1) = \langle 2, 2, 2 \rangle$ **b.** $x + y + z = 3$
77. a. $\nabla f(x, y, z) = e^{x+y-z} \langle 1, 1, -1 \rangle$, $\nabla f(1, 1, 2) = \langle 1, 1, -1 \rangle$
b. $x + y - z = 0$
79. a.

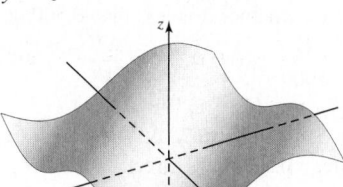

b. $\mathbf{v} = \pm \langle 1, 1 \rangle$
c. $\mathbf{v} = \pm \langle 1, -1 \rangle$

83. $\langle u, v \rangle = \langle \pi \cos \pi x \sin 2\pi y, 2\pi \sin \pi x \cos 2\pi y \rangle$
87. $\nabla f(x, y) = \dfrac{1}{(x^2 + y^2)^2} \langle y^2 - x^2 - 2xy, x^2 - y^2 - 2xy \rangle$
89. $\nabla f(x, y, z) = -\dfrac{1}{\sqrt{25 - x^2 - y^2 - z^2}} \langle x, y, z \rangle$
91. $\nabla f(x, y, z) = \dfrac{(y + xz) \langle 1, z, y \rangle - (x + yz) \langle z, 1, x \rangle}{(y + xz)^2}$
$= \dfrac{1}{(y + xz)^2} \langle y(1 - z^2), x(z^2 - 1), y^2 - x^2 \rangle$

Section 15.6 Exercises, pp. 980–983

1. The gradient of f is a multiple of $\mathbf{n}$.
3. $F_x(a, b, c)(x - a) + F_y(a, b, c)(y - b) + F_z(a, b, c)(z - c) = 0$
5. Multiply the change in x by $f_x(a, b)$ and the change in y by $f_y(a, b)$, and add both terms to f.   **7.** $dz = f_x(x, y)\, dx + f_y(x, y)\, dy$
9. $z = 5x - 3y + 5$ **11.** $3x - y + 6z = 4$ **13.** $2x + y + z = 4$;
$4x + y + z = 7$ **15.** $x + y + z = 6$; $3x + 4y + z = 12$
17. $z = -8x - 4y + 16$ and $z = 4x + 2y + 7$ **19.** $z = y + 1$
and $z = x + 1$ **21.** $x + \dfrac{1}{2}y + \sqrt{3}z = 2 + \dfrac{\sqrt{3}\pi}{6}$ and
$\dfrac{1}{2}x + y + \sqrt{3}z = \dfrac{5\sqrt{3}\pi}{6} - 2$ **23.** $\dfrac{1}{2}x + \dfrac{2}{3}y + 2\sqrt{3}z = -2$ and
$x - 2y + 2\sqrt{14}z = 2$ **25.** $z = 8x - 4y - 4$ and $z = -x - y - 1$
27. $z = \frac{7}{25}x - \frac{1}{25}y - \frac{2}{5}$ and $z = -\frac{7}{25}x + \frac{1}{25}y + \frac{6}{5}$
29. $z = \dfrac{1}{2}x + \dfrac{1}{2}y + \dfrac{\pi}{4} - 1$
31. $\dfrac{1}{6}(x - \pi) + \dfrac{\pi}{6}(y - 1) + \pi\left(z - \dfrac{1}{6}\right) = 0$
33. a. $L(x, y) = 4x + y - 6$ **b.** $L(2.1, 2.99) = 5.39$
35. a. $L(x, y) = -6x - 4y + 7$ **b.** $L(3.1, -1.04) = -7.44$
37. a. $L(x, y, z) = x + y + 2z$ **b.** $L(0.1, -0.2, 0.2) = 0.3$
39. $dz = -6dx - 5dy = -0.1$ **41.** $dz = dx + dy = 0.05$
43. a. The surface area decreases.   **b.** Impossible to say
c. $\Delta S \approx 53.3$ **d.** $\Delta S \approx 33.95$ **e.** $R\, dR = r\, dr$ **45.** $\dfrac{\Delta A}{A} \approx 3.5\%$
47. $dw = (y^2 + 2xz)\, dx + (2xy + z^2)\, dy + (x^2 + 2yz)\, dz$
49. $dw = \dfrac{dx}{y + z} - \dfrac{u + x}{(y + z)^2}\, dy - \dfrac{u + x}{(y + z)^2}\, dz + \dfrac{du}{y + z}$
51. a. $\Delta c \approx 0.035$ **b.** When $\theta = \dfrac{\pi}{20}$ **53. a.** True   **b.** True
c. False   **55.** $(1, -1, 1)$ and $(1, -1, -1)$
57. Points with $x = 0, \pm\dfrac{\pi}{2}, \pm\pi$ and $y = \pm\dfrac{\pi}{2}$, or points with
$x = \pm\dfrac{\pi}{4}, \pm\dfrac{3\pi}{4}$ and $y = 0, \pm\pi$
59. a. $\Delta S \approx 0.749$ **b.** More sensitive to changes in r
61. a. $\Delta A \approx \dfrac{2}{1225} = 0.00163$ **b.** No. The batting average increases
more if the batter gets a hit than it decreases if he fails to get a hit.
c. Yes. The answer depends on whether A is less than 0.500 or greater
than 0.500.   **63. a.** $\Delta V \approx \dfrac{21}{5000} = 0.0042$ **b.** $\dfrac{\Delta V}{V} \approx -4\%$ **c.** $2p$
65. a. $f_r = n(1 - r)^{n-1}, f_n = -(1 - r)^n \ln(1 - r)$
b. $\Delta P \approx 0.027$ **c.** $\Delta P \approx 2 \times 10^{-20}$ **67.** $\Delta R \approx 7/540 \approx 0.013\Omega$
69. a. Apply the Chain Rule.   **b.** Follows directly from (a)
c. $d(\ln(xy)) = \dfrac{dx}{x} + \dfrac{dy}{y}$ **d.** $d(\ln(x/y)) = \dfrac{dx}{x} - \dfrac{dy}{y}$
e. $\dfrac{df}{f} = \dfrac{dx_1}{x_1} + \dfrac{dx_2}{x_2} + \cdots + \dfrac{dx_n}{x_n}$

Section 15.7 Exercises, pp. 993–996

1. The local maximum occurs at the highest point on the surface; you
cannot get to a higher point in any direction.   **3.** The partial derivatives are both zero or do not exist.   **5.** The discriminant is a determinant; it is defined as $D(a, b) = f_{xx}(a, b)\, f_{yy}(a, b) - f_{xy}^2(a, b)$.
7. f has an absolute minimum value on R at (a, b) if $f(a, b) \le f(x, y)$
for all (x, y) in R.   **9.** Saddle point   **11.** Local min   **13.** $(0, 0)$
15. $(0, 1), (0, -1)$ **17.** $(0, 0), (2, 2),$ and $(-2, -2)$
19. $(0, 2), (\pm1, 2)$ **21.** $(3, 0), (-15, 6)$ **23.** Saddle point at $(0, 0)$

25. Local min at $(0, 0)$ **27.** Saddle point at $(0, 0)$; local min at $(1, 1)$
and $(-1, -1)$ **29.** $(0, 0)$; Second Derivative Test is inconclusive;
absolute min of 4 at $(0, 0)$ **31.** Local min at $(2, 0)$
33. Saddle point at $(0, 0)$; local max at $\left(\dfrac{1}{\sqrt{2}}, \dfrac{1}{\sqrt{2}}\right)$ and
$\left(-\dfrac{1}{\sqrt{2}}, -\dfrac{1}{\sqrt{2}}\right)$; local min at $\left(\dfrac{1}{\sqrt{2}}, -\dfrac{1}{\sqrt{2}}\right)$ and $\left(-\dfrac{1}{\sqrt{2}}, \dfrac{1}{\sqrt{2}}\right)$
35. Local min at $(-1, 0)$; local max at $(1, 0)$ **37.** Saddle point at
$(0, 1)$; local min at $(\pm2, 0)$ **39.** Saddle point at $(0, 0)$ **41.** Saddle
point   **43.** Height $= 32$ in, base is 16 in $\times$ 16 in; volume is 8192 in^3
45. 2 m $\times$ 2 m $\times$ 1 m   **47.** Absolute min: $0 = f(0, 1)$; absolute
max: $9 = f(0, -2)$ **49.** Absolute min: $4 = f(0, 0)$; absolute max:
$7 = f(\pm1, \pm1)$ **51.** Absolute min: $0 = f(1, 0)$; absolute max:
$3 = f(1, 1) = f(1, -1)$ **53.** Absolute min:
$1 = f(1, -2) = f(1, 0)$; absolute max: $4 = f(1, -1)$
55. Absolute min: $0 = f(0, 0)$; absolute max: $\dfrac{7}{8} = f\left(\dfrac{1}{\sqrt{2}}, \sqrt{2}\right)$
57. a. 1.83; 82.6°   **b.** 14%   **59.** Absolute min: $-4 = f(0, 0)$; no
absolute max on R **61.** Absolute max: $2 = f(0, 0)$; no absolute min
on R **63.** $P\left(\frac{4}{3}, \frac{2}{3}, \frac{4}{3}\right)$ **65.** $(3, 4, 5), (3, 4, -5)$
67. a. True   **b.** False   **c.** True   **d.** True
69. Local min at $(0.3, -0.3)$; saddle point at $(0, 0)$
71. a.–d. $x = y = z = \dfrac{200}{3}$
73. a. $P\left(1, \frac{1}{3}\right)$ **b.** $P\left(\frac{1}{3}(x_1 + x_2 + x_3), \frac{1}{3}(y_1 + y_2 + y_3)\right)$
c. $P(\bar{x}, \bar{y})$, where $\bar{x} = \dfrac{1}{n}\sum_{k=1}^{n} x_k$ and $\bar{y} = \dfrac{1}{n}\sum_{k=1}^{n} y_k$
d. $d(x, y) = \sqrt{x^2 + y^2} + \sqrt{(x - 2)^2 + y^2} + $
$\sqrt{(x - 1)^2 + (y - 1)^2}$. Absolute max: $1 + \sqrt{3} = f\left(1, \dfrac{1}{\sqrt{3}}\right)$
77. $y = \dfrac{22}{13}x + \dfrac{46}{13}$ **79.** $a = b = c = 3$
81. a. $\nabla d_1(x, y) = \dfrac{x - x_1}{d_1(x, y)}\mathbf{i} + \dfrac{y - y_1}{d_1(x, y)}\mathbf{j}$
b. $\nabla d_2(x, y) = \dfrac{x - x_2}{d_2(x, y)}\mathbf{i} + \dfrac{y - y_2}{d_2(x, y)}\mathbf{j}$;
$\nabla d_3(x, y) = \dfrac{x - x_3}{d_3(x, y)}\mathbf{i} + \dfrac{y - y_3}{d_3(x, y)}\mathbf{j}$
c. Follows from $\nabla f = \nabla d_1 + \nabla d_2 + \nabla d_3$ **d.** Three unit vectors
add to zero.   **e.** P is the vertex at the large angle.
f. $P(0.255457, 0.304504)$ **83. a.** Local max at $(1, 0), (-1, 0)$
b. Local max at $(1, 0)$ and $(-1, 0)$ **85.** $\dfrac{abc\sqrt{3}}{2}$

Section 15.8 Exercises, pp. 1002–1004

1. The level curve of f is tangent to the curve $g = 0$ at the optimal
point; therefore, the gradients are parallel.
3. $1 = 2\lambda x, 4 = 2\lambda y, x^2 + y^2 - 1 = 0$
5. Abs. min: 1; abs. max: 8   **7.** Abs. min: $-2\sqrt{5}$ at
$\left(-\dfrac{2}{\sqrt{5}}, -\dfrac{4}{\sqrt{5}}\right)$; abs. max: $2\sqrt{5}$ at $\left(\dfrac{2}{\sqrt{5}}, \dfrac{4}{\sqrt{5}}\right)$
9. Abs. min: -2 at $(-1, -1)$; abs. max: 2 at $(1, 1)$
11. Abs. min: -3 at $(-\sqrt{3}, \sqrt{3})$ and $(\sqrt{3}, -\sqrt{3})$;
abs. max: 9 at $(3, 3)$ and $(-3, -3)$
13. Abs. min: e^{-9} at $(-3, 3)$ and $(3, -3)$; abs. max: e^3 at $(\sqrt{3}, \sqrt{3})$
and $(-\sqrt{3}, -\sqrt{3})$ **15.** Abs. min: 9 at $(0, 3)$; abs. max: 34 at
$(-\sqrt{15}, -2)$ and $(\sqrt{15}, -2)$ **17.** Abs. min: $-2\sqrt{11}$ at
$\left(-\dfrac{2}{\sqrt{11}}, -\dfrac{6}{\sqrt{11}}, \dfrac{2}{\sqrt{11}}\right)$; abs. max: $2\sqrt{11}$ at $\left(\dfrac{2}{\sqrt{11}}, \dfrac{6}{\sqrt{11}}, -\dfrac{2}{\sqrt{11}}\right)$

19. Abs. min: $-\dfrac{\sqrt{5}}{2}$ at $\left(-\dfrac{\sqrt{5}}{2}, 0, \dfrac{1}{2}\right)$; abs. max: $\dfrac{\sqrt{5}}{2}$ at $\left(\dfrac{\sqrt{5}}{2}, 0, \dfrac{1}{2}\right)$

21. Abs. min: -5 at $(-2, -2, -1)$; abs. max: 5 at $(2, 2, 1)$

23. Abs. min: -10 at $(-5, 0, 0)$; abs. max: $\dfrac{29}{2}$ at $\left(2, 0, \pm\sqrt{\dfrac{21}{2}}\right)$

25. Abs. min: $-\sqrt{3}$ at $\left(0, -\dfrac{1}{\sqrt{3}}, -\dfrac{2}{\sqrt{3}}\right)$; abs. max: $\dfrac{7}{4}$ at $\left(\dfrac{1}{2}, \dfrac{1}{2}, 1\right)$ and $\left(-\dfrac{1}{2}, \dfrac{1}{2}, 1\right)$ **27.** 18 in $\times$ 18 in $\times$ 36 in   **29.** Abs. min: 0.6731; abs. max: 1.1230   **31.** 2×1 **33.** $\left(-\dfrac{3}{17}, \dfrac{29}{17}, -3\right)$

35. Abs. min: $\sqrt{38 - 6\sqrt{29}}$ (or $\sqrt{29} - 3$); abs. max: $\sqrt{38 + 6\sqrt{29}}$ (or $\sqrt{29} + 3$)   **37.** $\ell = 3$ and $g = \dfrac{3}{2}$; $U = 15\sqrt{2}$

39. $\ell = \dfrac{16}{5}$ and $g = 1$; $U = 20.287$ **41. a.** True   **b.** False

43. $\dfrac{\sqrt{6}}{3}$ m $\times \dfrac{\sqrt{6}}{3}$ m $\times \dfrac{\sqrt{6}}{6}$ m   **45.** $2 \times 1 \times \dfrac{2}{3}$ **47.** $P\left(\dfrac{4}{3}, \dfrac{2}{3}, \dfrac{4}{3}\right)$

49. Abs. min: 1; abs. max: 9   **51.** Abs. min: 0; abs. max: 3

53. $K = 7.5$ and $L = 5$ **57.** Abs. max: 8   **59.** Abs. max: $\sqrt{a_1^2 + a_2^2 + a_3^2 + \cdots + a_n^2}$ **61. a.** Gradients are perpendicular to level surfaces.   **b.** If the gradient were not in the plane spanned by ∇g and ∇h, f could be increased (decreased) by moving the point slightly.   **c.** ∇f is a linear combination of ∇g and ∇h, since it belongs to the plane spanned by these two vectors.   **d.** The gradient condition from part (c), as well as the constraints, must be satisfied.

63. Abs. min: $2 - 4\sqrt{2}$; abs. max: $2 + 4\sqrt{2}$

65. a. $y + 1 = \lambda y$, $x + 1 = \lambda x$, $xy = 4$ **c.** Abs. min of 108 over the curve C_1 **d.** Abs. max of 100 over the curve C_2 **e.** The constraint curve is unbounded, so there is no guarantee that an abs. min or max occurs over the curve $xy = 4$.

Chapter 15 Review Exercises, pp. 1005–1007

1. a. True   **b.** False   **c.** False   **d.** False

3. $D = \{(x, y): x \geq y^2\}$

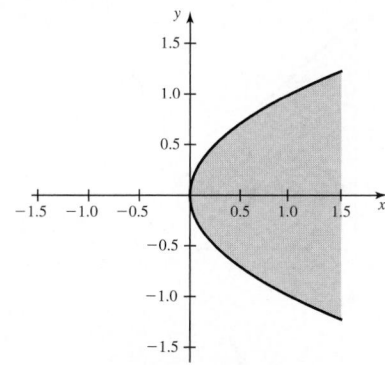

5. $D = \left\{(x, y): \dfrac{x^2}{4} + \dfrac{y^2}{9} \geq 1\right\}$

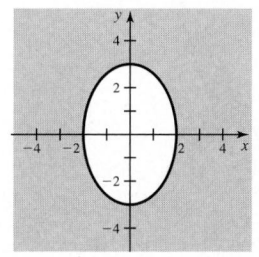

7. $D = \{(x, y): x^2 + y^2 \geq 1\}$; $R = (-\infty, 0]$; lower half of the hyperboloid of one sheet $x^2 + y^2 - z^2 = 1$

9.

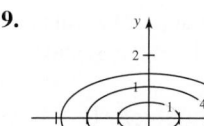

11. 2   **13.** Does not exist
15. $\dfrac{2}{3}$ **17.** 4
19. $\{(x, y): y > x^2 + 1\}$
21. $f_x = 6xy^5$; $f_y = 15x^2y^4$

23. $f_x = \dfrac{2xy^2}{(x^2 + y^2)^2}$; $f_y = -\dfrac{2x^2y}{(x^2 + y^2)^2}$

25. $f_x = y(1 + xy)e^{xy}$; $f_y = x(1 + xy)e^{xy}$ **27.** $f_{xx} = 4y^2e^{2xy}$; $f_{xy} = 2e^{2xy}(2xy + 1) = f_{yx}$; $f_{yy} = 4x^2e^{2xy}$

29. $\dfrac{\partial^2 u}{\partial x^2} = 6y = -\dfrac{\partial^2 u}{\partial y^2}$ **31. a.** V increases with R if r is fixed, $V_R > 0$; V decreases if r increases and R is fixed, $V_r < 0$.
b. $V_r = -4\pi r^2$; $V_R = 4\pi R^2$ **c.** The volume increases more if R is increased.   **33.** $4t + 2 \ln 5$

35. $w_r = \dfrac{3r + s}{r(r + s)}$; $w_s = \dfrac{r + 3s}{s(r + s)}$; $w_t = \dfrac{1}{t}$

37. $\dfrac{dy}{dx} = -\dfrac{2xy}{2y^2 + (x^2 + y^2)\ln(x^2 + y^2)}$

39. a. $z'(t) = -24 \sin t \cos t = -12 \sin 2t$

b. $z'(t) > 0$ for $\dfrac{\pi}{2} < t < \pi$ and $\dfrac{3\pi}{2} < t < 2\pi$

41. a.

	$(a, b) = (0, 0)$	$(a, b) = (2, 0)$	$(a, b) = (1, 1)$
$\mathbf{u} = \left\langle \dfrac{\sqrt{2}}{2}, \dfrac{\sqrt{2}}{2}\right\rangle$	0	$4\sqrt{2}$	$-2\sqrt{2}$
$\mathbf{v} = \left\langle -\dfrac{\sqrt{2}}{2}, \dfrac{\sqrt{2}}{2}\right\rangle$	0	$-4\sqrt{2}$	$-6\sqrt{2}$
$\mathbf{w} = \left\langle -\dfrac{\sqrt{2}}{2}, -\dfrac{\sqrt{2}}{2}\right\rangle$	0	$-4\sqrt{2}$	$2\sqrt{2}$

b. The function is increasing at $(2, 0)$ in the direction of $\mathbf{u}$ and decreasing at $(2, 0)$ in the directions of $\mathbf{v}$ and $\mathbf{w}$.
43. $\nabla g = \langle 2xy^3, 3x^2y^2\rangle$; $\nabla g(-1, 1) = \langle -2, 3\rangle$; $D_{\mathbf{u}} g(-1, 1) = 2$

45. $\nabla h = \left\langle \dfrac{x}{\sqrt{2 + x^2 + 2y^2}}, \dfrac{2y}{\sqrt{2 + x^2 + 2y^2}}\right\rangle$; $\nabla h(2, 1) = \left\langle \dfrac{\sqrt{2}}{2}, \dfrac{\sqrt{2}}{2}\right\rangle$; $D_{\mathbf{u}} h(2, 1) = \dfrac{7\sqrt{2}}{10}$

47. $\nabla f = \langle y \cos xy, x \cos xy, -\sin z\rangle$; $\nabla f(1, \pi, 0) = \langle -\pi, -1, 0\rangle$; $D_{\mathbf{u}} f(1, \pi, 0) = -\dfrac{1}{7}(3 + 2\pi)$

49. a. Direction of steepest ascent: $\mathbf{u} = \dfrac{\sqrt{2}}{2}\mathbf{i} - \dfrac{\sqrt{2}}{2}\mathbf{j}$; direction of steepest descent: $\mathbf{u} = -\dfrac{\sqrt{2}}{2}\mathbf{i} + \dfrac{\sqrt{2}}{2}\mathbf{j}$

b. No change: $\mathbf{u} = \pm\left(\dfrac{\sqrt{2}}{2}\mathbf{i} + \dfrac{\sqrt{2}}{2}\mathbf{j}\right)$

51. Tangent line is vertical; $\nabla f(2, 0) = -8\mathbf{i}$

53. $E = \dfrac{kx}{x^2 + y^2}\mathbf{i} + \dfrac{ky}{x^2 + y^2}\mathbf{j}$

55. $y = 2$ and $12x + 3y - 2z = 12$
57. $x + 2y + 3z = 6$ and $x - 2y + 3z = 6$
59. $x + y - z = 0$ and $x + y - z = 0$
61. a. $L(x, y) = x + 5y$ **b.** $L(1.95, 0.05) = 2.2$ **63.** Approx. -4%
65. a. $\Delta V \approx -0.1\pi$ m^3 **b.** $\Delta S \approx -0.05\pi$ m^2
67. Saddle point at $(0, 0)$; local min at $(2, -2)$

69. Saddle points at $(0, 0)$ and $(-2, 2)$; local max at $(0, 2)$; local min at $(-2, 0)$ **71.** Abs. min: $-1 = f(1, 1) = f(-1, -1)$; abs. max: $49 = f(2, -2) = f(-2, 2)$

73. Abs. min: $-\dfrac{1}{2} = f\left(-\dfrac{1}{\sqrt{2}}, \dfrac{1}{\sqrt{2}}\right)$; abs. max: $\dfrac{1}{2} = f\left(\dfrac{1}{\sqrt{2}}, \dfrac{1}{\sqrt{2}}\right)$

75. Abs. min: $\dfrac{23}{2} = f\left(\dfrac{1}{3}, \dfrac{5}{6}\right)$ abs. max: $\dfrac{29}{2} = f\left(\dfrac{5}{3}, \dfrac{7}{6}\right)$;

77. Abs. min: $-\sqrt{6} = f\left(-\dfrac{\sqrt{6}}{6}, -\dfrac{\sqrt{6}}{3}, \dfrac{\sqrt{6}}{6}\right)$;

abs. max: $\sqrt{6} = f\left(\dfrac{\sqrt{6}}{6}, \dfrac{\sqrt{6}}{3}, -\dfrac{\sqrt{6}}{6}\right)$

79. $\dfrac{2a^2}{\sqrt{a^2 + b^2}}$ by $\dfrac{2b^2}{\sqrt{a^2 + b^2}}$

81. $x = \dfrac{1}{2} + \dfrac{\sqrt{10}}{20}, y = \dfrac{3}{2} + \dfrac{3\sqrt{10}}{20} = 3x, z = \dfrac{1}{2} + \dfrac{\sqrt{10}}{2} = \sqrt{10}x$

83. $(1, 2, 5)$

CHAPTER 16

Section 16.1 Exercises, pp. 1015–1017

1. $\int_0^2 \int_1^3 xy \, dy \, dx$ or $\int_1^3 \int_0^2 xy \, dx \, dy$ **3.** $\int_{-2}^4 \int_1^5 f(x, y) \, dy \, dx$ or

$\int_1^5 \int_{-2}^4 f(x, y) \, dx \, dy$ **5.** 48   **7.** 4   **9.** $\dfrac{32}{3}$ **11.** 4   **13.** $\dfrac{224}{9}$

15. $10 - 2e$ **17.** $\dfrac{1}{2}$ **19.** $e^2 + 3$ **21.** $\dfrac{1}{2}$ **23.** $10\sqrt{5} - 4\sqrt{2} - 14$

25. $\dfrac{117}{2}$ **27.** $\dfrac{\pi^2}{4} + 1$ **29.** $\dfrac{4}{3}$ **31.** $\dfrac{9 - e^2}{2}$ **33.** $\dfrac{4}{11}$ **35.** $\dfrac{1}{4}$

37. 136   **39.** 3   **41.** $e^2 - 3$ **43.** $e^{16} - 17$ **45.** $\ln \dfrac{5}{3}$ **47.** $\dfrac{1}{2 \ln 2}$

49. $\dfrac{8}{3}$ **51. a.** True   **b.** False   **c.** True   **53. a.** 1475   **b.** The sum of products of population densities and areas is a Riemann sum.

55. $\int_c^d \int_a^b f(x) \, dy \, dx = (c - d) \int_a^b f(x) \, dx$. The integral is the area of the cross section of S.   **57.** $a = \pi/6, 5\pi/6$ **59.** $a = \sqrt{6}$

61. a. $\frac{1}{2}\pi^2 + \pi$ **b.** $\frac{1}{2}\pi^2 + \pi$ **c.** $\frac{1}{2}\pi^2 + 2$

63. $f(a, b) - f(a, 0) - f(0, b) + f(0, 0)$

Section 16.2 Exercises, pp. 1024–1027

1.

3. $dx \, dy$ **5.** $\int_0^1 \int_{x^2}^{\sqrt{x}} f(x, y) \, dy \, dx$ **7.** 4   **9.** $\int_0^2 \int_{x^3}^{4x} f(x, y) \, dy \, dx$

11. 2   **13.** $\dfrac{8}{3}$ **15.** 0   **17.** $e - 1$ **19.** $\dfrac{\ln^3 2}{6}$

21. 2   **23.** $\dfrac{\pi}{2} - 1$ **25.** 0   **27.** $\pi - 1$

29.

$\int_1^2 \int_{x+1}^{2x+4} f(x, y) \, dy \, dx$

31.

$\int_0^1 \int_0^{-2x+2} f(x, y) \, dy \, dx$

33.

$\int_0^1 \int_0^{\sqrt{1-x^2}} f(x, y) \, dy \, dx$

35. $\int_0^{18} \int_{y/2}^{(y+9)/3} f(x, y) \, dx \, dy$

37.

$\int_1^4 \int_0^{4-y} f(x, y) \, dx \, dy$

39.

$\int_0^{23} \int_{(y-3)/2}^{(y+7)/3} f(x, y) \, dx \, dy$

41.

$\int_0^1 \int_y^{2-y} f(x, y) \, dx \, dy$

43. 2   **45.** 12   **47.** 5   **49.** 14   **51.** 32   **53.** $\frac{9}{8}$ **55.** $\frac{1}{4}\ln 2$

57. $\int_0^4 \int_{y/2}^{\sqrt{y}} f(x, y)\, dx\, dy$ **59.** $\int_0^{\ln 2} \int_{1/2}^{e^{-x}} f(x, y)\, dy\, dx$

61. $\int_0^{\pi/2} \int_0^{\cos x} f(x, y)\, dy\, dx$ **63.** $\frac{1}{2}(e - 1)$ **65.** 0   **67.** $\frac{2}{3}$

69.

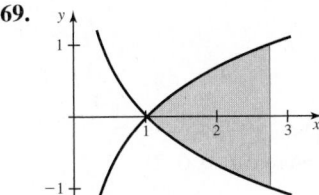

$\int_1^e \int_{-\ln x}^{\ln x} f(x, y)\, dy\, dx$

71. $\frac{11}{12}$ **73.** $\frac{32}{3}$ **75.** 12π **77.** $\frac{43}{6}$ **79.** $\frac{2}{3}$ **81.** 16   **83.** $4a\pi$

85. $\frac{32}{3}$ **87.** 1   **89.** $\frac{140}{3}$ **91. a.** False   **b.** False   **c.** False

93. 30   **95.** $\frac{a}{3}$ **97. a.**

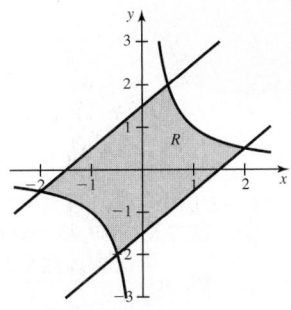

b. $\frac{15}{4} + 4\ln 2$ **c.** $2\ln 2 - \frac{5}{64}$ **99.** $\frac{3}{8e^2}$ **101.** 1

Section 16.3 Exercises, pp. 1033–1036

1. It is called a polar rectangle because r and θ vary between two constants.

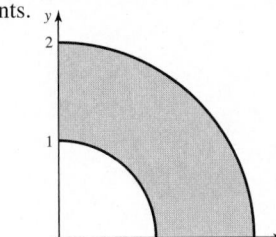

3.

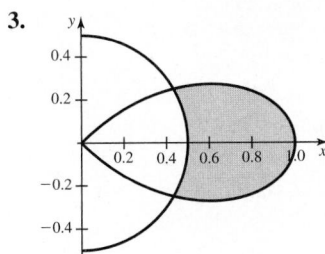

5. Evaluate the integral $\int_\alpha^\beta \int_{g(\theta)}^{h(\theta)} r\, dr\, d\theta$.

7.

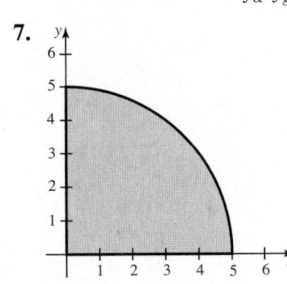

9.

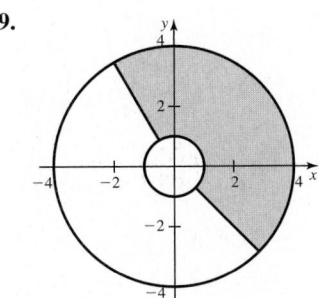

11. $\frac{64\pi}{3}$ **13.** $(8 - 24e^{-2})\pi$ **15.** $\frac{7\pi}{2}$ **17.** $\frac{9\pi}{2}$ **19.** $\frac{37\pi}{3}$

21. 128π **23.** 0   **25.** $(2 - \sqrt{3})\pi$ **27.** $2\pi/5$ **29.** $\frac{14\pi}{3}$

31. $\frac{81\pi}{2}$ **33.** π **35.** 8π **37.** 81π **39.** $\frac{2\pi}{3}(7\sqrt{7} - 15)$

41.

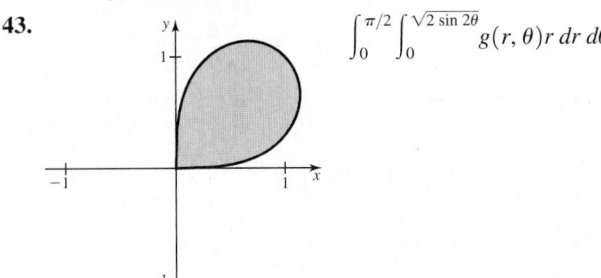

$\int_0^{2\pi} \int_0^{1+\frac{1}{2}\cos\theta} g(r, \theta)r\, dr\, d\theta$

43.

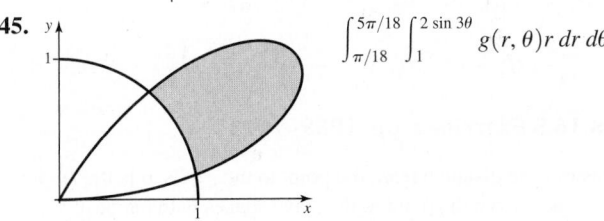

$\int_0^{\pi/2} \int_0^{\sqrt{2\sin 2\theta}} g(r, \theta)r\, dr\, d\theta$

45.

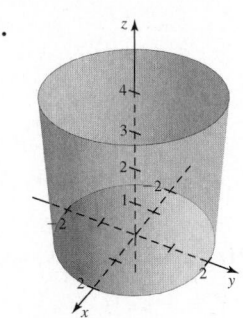

$\int_{\pi/18}^{5\pi/18} \int_1^{2\sin 3\theta} g(r, \theta)r\, dr\, d\theta$

47. $3\pi/2$ **49.** π **51.** $\frac{3\pi}{2} - 2\sqrt{2}$ **53.** $2a/3$

55. a. False   **b.** True   **c.** True   **57.** The hyperboloid $\left(V = \frac{112\pi}{3}\right)$

59. a. $R = \{(r, \theta): -\pi/4 \le \theta \le \pi/4 \text{ or } 3\pi/4 \le \theta \le 5\pi/4\}$

b. $\frac{a^4}{4}$ **61.** $\frac{32}{9}$ **63.** $2\pi\left(1 - 2\ln\frac{3}{2}\right)$ **65.** 1

67. $\pi/4$ **69. a.** $\frac{16\pi}{3}$ **b.** 2.78   **71.** $30\pi + 42$

73. c. $\sqrt{\pi}/2, 1/2$, and $\sqrt{\pi}/4$ **75. a.** $I = \frac{\sqrt{2}}{2}\tan^{-1}\frac{\sqrt{2}}{2}$

b. $I = \frac{\sqrt{2}}{4}\tan^{-1}\frac{\sqrt{2}a}{2} + \frac{a}{2\sqrt{a^2 + 1}}\tan^{-1}\frac{1}{\sqrt{a^2 + 1}}$ **c.** $\frac{\sqrt{2}\pi}{8}$

Section 16.4 Exercises, pp. 1043–1047

1.

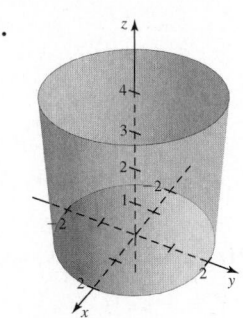

3. $\int_{-9}^9 \int_{-\sqrt{81-x^2}}^{\sqrt{81-x^2}} \int_{-\sqrt{81-x^2-y^2}}^{\sqrt{81-x^2-y^2}} f(x, y, z)\, dz\, dy\, dx$

5. $\int_0^1 \int_0^{\sqrt{1-z^2}} \int_0^{\sqrt{1-z^2-x^2}} f(x, y, z)\, dy\, dx\, dz$ **7.** 24   **9.** 8   **11.** $\dfrac{2}{\pi}$

13. 0   **15.** 8   **17.** $\dfrac{16}{3}$ **19.** $1 - \ln 2$

21. $\dfrac{2\pi(1 + 19\sqrt{19} - 20\sqrt{10})}{3}$ **23.** $\dfrac{27\pi}{2}$ **25.** 12π

27. $\dfrac{5}{12}$ **29.** 8   **31.** $\int_0^1 \int_y^1 \int_0^{2\sqrt{1-x^2}} f(x, y, z)\, dz\, dx\, dy$

33. $\int_0^1 \int_0^{2\sqrt{1-x^2}} \int_0^x f(x, y, z)\, dy\, dz\, dx$

35. $\int_0^1 \int_0^{2\sqrt{1-y^2}} \int_{y^2}^{\frac{1}{2}\sqrt{4-z^2}} f(x, y, z)\, dx\, dz\, dy$

37. $\int_0^1 \int_0^2 \int_0^{1-y} dz\, dx\, dy$, $\int_0^2 \int_0^1 \int_0^{1-z} dy\, dz\, dx$, $\int_0^1 \int_0^2 \int_0^{1-z} dy\, dx\, dz$,

$\int_0^1 \int_0^{1-y} \int_0^2 dx\, dz\, dy$, $\int_0^1 \int_0^{1-z} \int_0^2 dx\, dy\, dz$ **39.** $\dfrac{256}{9}$ **41.** $\dfrac{2}{3}$

43. $(10\sqrt{10} - 1)\dfrac{\pi}{6}$ **45.** $\dfrac{3\ln 2}{2} + \dfrac{e}{16} - 1$

47. $\int_0^4 \int_{y/4-1}^0 \int_0^5 dz\, dx\, dy = 10$ **49.** $\int_0^1 \int_0^{\sqrt{1-x^2}} \int_0^{\sqrt{1-x^2}} dz\, dy\, dx = \dfrac{2}{3}$

51. $\dfrac{8}{\pi}$ **53.** $\dfrac{10}{3}$ **55. a.** False   **b.** False   **c.** False   **57.** 2

59. 1   **61.** $\dfrac{16}{3}$ **63.** $\dfrac{16}{3}$ **65.** $a = 2\sqrt{2}$ **67.** $V = \dfrac{\pi r^2 h}{3}$

69. $V = \dfrac{\pi h^2}{3}(3R - h)$ **71.** $V = \dfrac{4\pi abc}{3}$ **73.** $\dfrac{1}{24}$

Section 16.5 Exercises, pp. 1059–1063

1. r measures the distance from the point to the z axis, θ is the angle that the segment from the point to the z-axis makes with the positive xz-plane, and z is the directed distance from the point to the xy-plane.
3. A cone   **5.** It approximates the volume of the cylindrical wedge formed by the changes Δr, $\Delta \theta$, and Δz.
7. $\int_\alpha^\beta \int_{g(\theta)}^{h(\theta)} \int_{G(r, \theta)}^{H(r, \theta)} w(r, \theta, z)\, r\, dz\, dr\, d\theta$ **9.** Cylindrical coordinates

11.

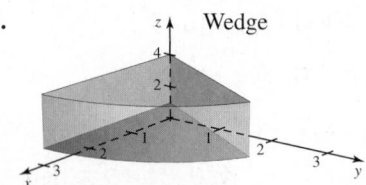

Wedge

13.

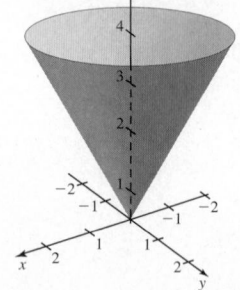

Solid bounded by cone and plane

15. 2π **17.** $4\pi/5$ **19.** $\pi(1 - e^{-1})/2$ **21.** $9\pi/4$

23. 560π **25.** 396π **27.** The paraboloid $(V = 44\pi/3)$

29. $\dfrac{20\pi}{3}$ **31.** $\dfrac{(16 + 17\sqrt{29})\pi}{3}$ **33.** $\dfrac{1}{3}$

35.

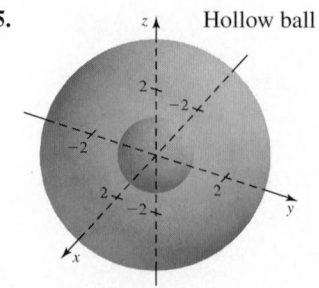

Hollow ball

37.

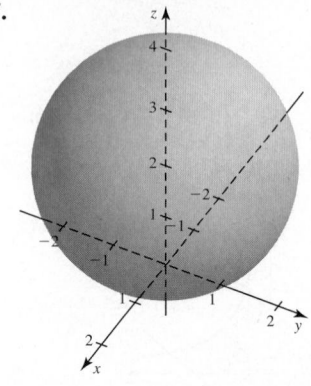

Sphere of radius $r = 2$, centered at $(0, 0, 2)$

39. a. $(3960, 0.74, -2.13)$, $(-1426.85, -2257.05, 2924.28)$
b. $(3960, 0.84, 0.22)$, $(2877.61, 637.95, 2644.62)$ **c.** 5666 mi

41. $\dfrac{\pi}{2}$ **43.** $4\pi \ln 2$ **45.** $\pi\left(\dfrac{188}{9} - \dfrac{32\sqrt{3}}{3}\right)$ **47.** $\dfrac{32\pi\sqrt{3}}{9}$

49. $\dfrac{5\pi}{12}$ **51.** $\dfrac{8\pi}{3}$ **53.** $\dfrac{8\pi}{3}(9\sqrt{3} - 11)$ **55. a.** True   **b.** True

57. $z = \sqrt{x^2 + y^2 - 1}$; upper half of a hyperboloid of one sheet

59. $\dfrac{8\pi}{3}(1 - e^{-512}) \approx \dfrac{8\pi}{3}$ **61.** 32π

63.

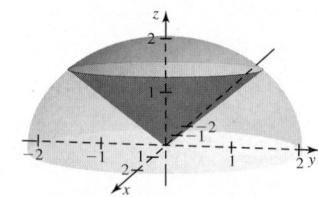

$\int_0^{2\pi} \int_0^{\sqrt{2}} \int_r^{\sqrt{4-r^2}} g(r, \theta, z)\, r\, dz\, dr\, d\theta$,

$\int_0^{2\pi} \int_0^{\sqrt{2}} \int_0^z g(r, \theta, z)\, r\, dr\, dz\, d\theta$

$+ \int_0^{2\pi} \int_{\sqrt{2}}^2 \int_0^{\sqrt{4-z^2}} g(r, \theta, z)\, r\, dr\, dz\, d\theta$,

$\int_0^{\sqrt{2}} \int_r^{\sqrt{4-r^2}} \int_0^{2\pi} g(r, \theta, z)\, r\, d\theta\, dz\, dr$

65.

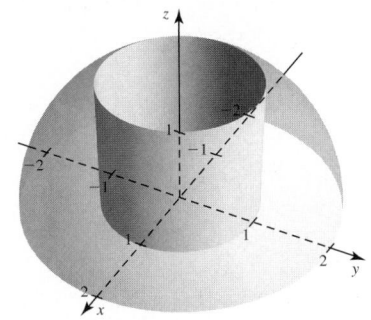

$$\int_{\pi/6}^{\pi/2}\int_0^{2\pi}\int_{\csc\varphi}^2 g(\rho,\varphi,\theta)\rho^2 \sin\varphi \; d\rho \; d\theta \; d\varphi,$$

$$\int_{\pi/6}^{\pi/2}\int_{\csc\varphi}^2\int_0^{2\pi} g(\rho,\varphi,\theta)\rho^2 \sin\varphi \; d\theta \; d\rho \; d\varphi$$

67. $32\sqrt{3}\pi/9$ **69.** $2\sqrt{2}/3$ **71.** $7\pi/2$ **73.** 95.6036

77. $V = \dfrac{\pi r^2 h}{3}$ **79.** $V = \dfrac{\pi}{3}(R^2 + rR + r^2)h$

81. $V = \dfrac{\pi R^3(8r - 3R)}{12r}$

Section 16.6 Exercises, pp. 1070–1072

1. The pivot should be located at the center of mass of the system.
3. Use a double integral. Integrate the density function over the region occupied by the plate.   **5.** Use a triple integral to find the mass of the object and the three moments.

7.

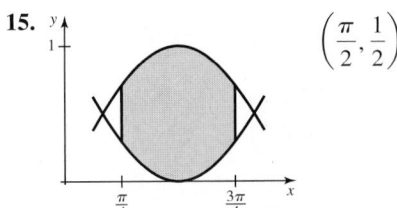

$\frac{27}{13}$ **9.** Mass is $2 + \pi$; $\bar{x} = \frac{\pi}{2}$

11. Mass is $\frac{20}{3}$; $\bar{x} = \frac{9}{5}$ **13.** Mass is 10; $\bar{x} = \frac{8}{3}$

15.

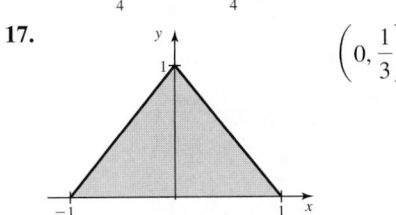

$\left(\dfrac{\pi}{2}, \dfrac{1}{2}\right)$

17.

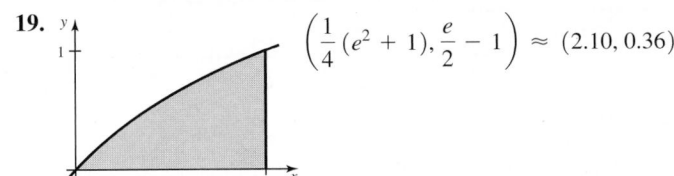

$\left(0, \dfrac{1}{3}\right)$

19.

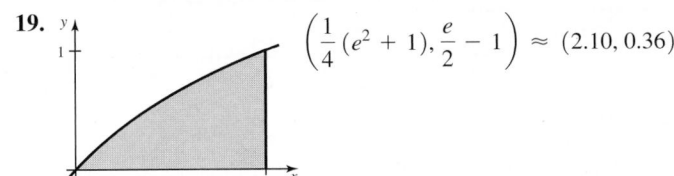

$\left(\dfrac{1}{4}(e^2 + 1), \dfrac{e}{2} - 1\right) \approx (2.10, 0.36)$

21. $\left(\frac{7}{3}, 1\right)$; density increases to the right.   **23.** $\left(\frac{16}{11}, \frac{16}{11}\right)$; density increases toward the hypotenuse of the triangle.

25. $\left(0, \frac{16 + 3\pi}{16 + 12\pi}\right) \approx (0, 0.4735)$; density increases away from the x-axis.

27.

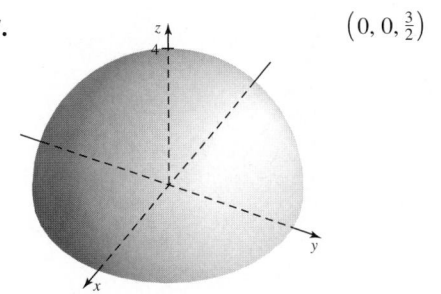

$\left(0, 0, \dfrac{3}{2}\right)$

29.

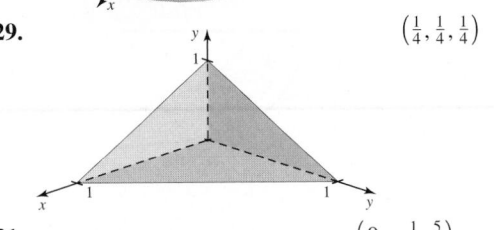

$\left(\dfrac{1}{4}, \dfrac{1}{4}, \dfrac{1}{4}\right)$

31.

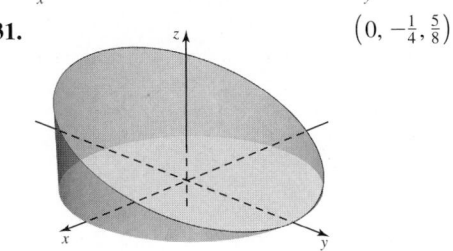

$\left(0, -\dfrac{1}{4}, \dfrac{5}{8}\right)$

33. $\left(\frac{7}{3}, \frac{1}{2}, \frac{1}{2}\right)$ **35.** $\left(0, 0, \frac{198}{85}\right)$ **37.** $\left(\frac{2}{3}, \frac{7}{3}, \frac{1}{3}\right)$ **39. a.** False
b. True   **c.** False   **d.** False   **41.** $\bar{x} = \dfrac{\ln(1 + L^2)}{2\tan^{-1}L}$, $\displaystyle\lim_{L\to\infty} \bar{x} = \infty$

43. $\left(0, \dfrac{8}{9}\right)$ **45.** $\left(0, \dfrac{8}{3\pi}\right)$ **47.** $\left(\dfrac{5}{6}, 0\right)$ **49.** $\left(\dfrac{128}{105\pi}, \dfrac{128}{105\pi}\right)$

51. On the line of symmetry, $2a/\pi$ units above the diameter

53. $\left(\dfrac{2a}{3(4 - \pi)}, \dfrac{2a}{3(4 - \pi)}\right)$ **55.** $h/4$ units

57. $h/3$ units, where h is the height of the triangle   **59.** $3a/8$ units

61. a. $\left(0, \dfrac{4(1 + a + a^2)}{3(1 + a)\pi}\right)$

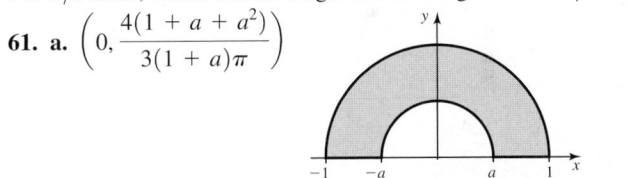

b. $a = \dfrac{1}{2}\left(-1 + \sqrt{1 + \dfrac{16}{3\pi - 4}}\right) \approx 0.4937$

63. Depth $= \dfrac{40\sqrt{10} - 4}{333}$ cm ≈ 0.3678 cm

65. a. $(\bar{x}, \bar{y}) = \left(\dfrac{-r^2}{R + r}, 0\right)$ (origin at center of large circle);

$(\bar{x}, \bar{y}) = \left(\dfrac{R^2 + Rr + r^2}{R + r}, 0\right)$ (origin at common point of the circles)
b. *Hint:* Solve $\bar{x} = R - 2r$.

Section 16.7 Exercises, pp. 1082–1084

1. The image of S is the 2×2 square with vertices at $(0, 0)$, $(2, 0)$, $(2, 2)$, and $(0, 2)$.   **3.** $\int_0^1\int_0^1 f(u + v, u - v)\, 2\, du\, dv$
5. The rectangle with vertices at $(0, 0)$, $(2, 0)$, $\left(2, \frac{1}{2}\right)$, and $\left(0, \frac{1}{2}\right)$
7. The square with vertices at $(0, 0)$, $\left(\frac{1}{2}, \frac{1}{2}\right)$, $(1, 0)$, and $\left(\frac{1}{2}, -\frac{1}{2}\right)$
9. The region above the x-axis and bounded by the curves $y^2 = 4 \pm 4x$ **11.** The upper half of the unit circle

13.

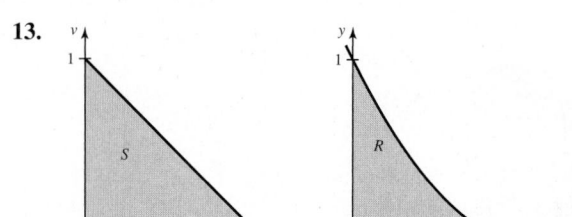

b. $0 \le u \le 1, 0 \le v \le 1 - u$ **c.** $J(u, v) = 2$ **d.** $256\sqrt{2}/945$

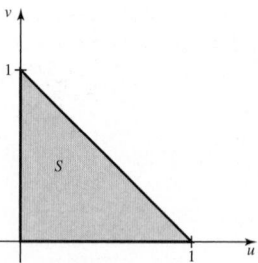

31. $4\sqrt{2}/3$

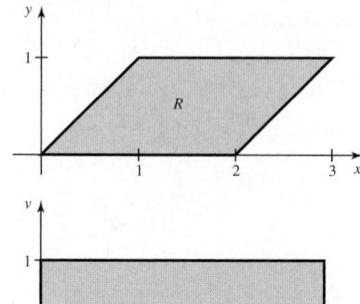

15.

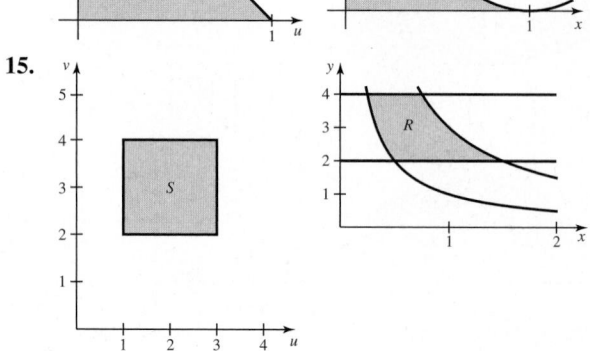

17. -9 **19.** $-4(u^2 + v^2)$ **21.** -1
23. $x = (u + v)/3, y = (2u - v)/3; -1/3$
25. $x = -(u + 3v), y = -(u + 2v); -1$
27. a.

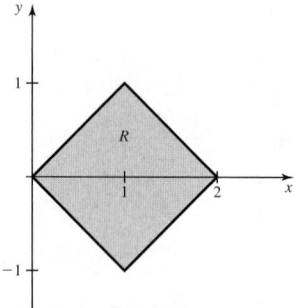

b. $0 \le u \le 1, 0 \le v \le 1$ **c.** $J(u, v) = -2$ **d.** 0

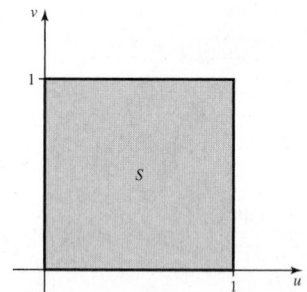

29. a.

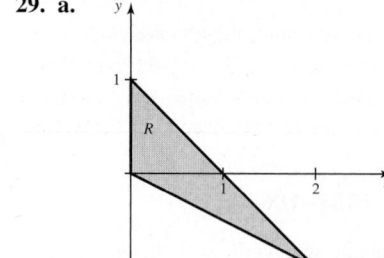

33. $3844/5625$

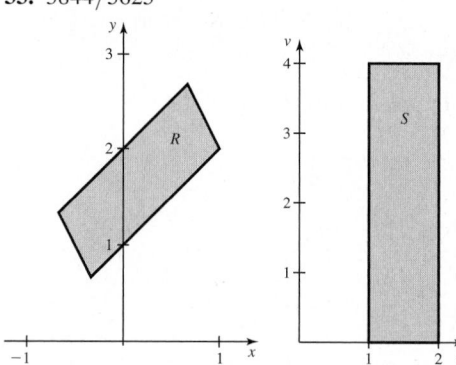

35. $\dfrac{15 \ln 3}{2}$

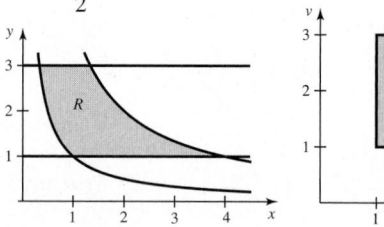

37. 2 **39.** $2w(u^2 - v^2)$ **41.** 5 **43.** $1024\pi/3$

45. a. True **b.** True **c.** True

47. *Hint:* $J(\rho, \varphi, \theta) = \begin{vmatrix} \sin \varphi \cos \theta & \rho \cos \varphi \cos \theta & -\rho \sin \varphi \sin \theta \\ \sin \varphi \sin \theta & \rho \cos \varphi \sin \theta & \rho \sin \varphi \cos \theta \\ \cos \varphi & -\rho \sin \varphi & 0 \end{vmatrix}$

49. $a^2b^2/2$ **51.** $(a^2 + b^2)/4$ **53.** $4\pi abc/3$

55. $(\bar{x}, \bar{y}, \bar{z}) = \left(0, 0, \dfrac{3c}{8}\right)$ **57. a.** $x = a^2 - \dfrac{y^2}{4a^2}$

b. $x = \dfrac{y^2}{4b^2} - b^2$ **c.** $J(u, v) = 4(u^2 + v^2)$ **d.** $\dfrac{80}{3}$ **e.** 160

f. Vertical lines become parabolas opening downward with vertices on the positive *y*-axis, and horizontal lines become parabolas opening upward with vertices on the negative *y*-axis.   **59. a.** *S* is stretched in the positive *u*- and *v*-directions but not in the *w*-direction. The amount of stretching increases with *u* and *v*.   **b.** $J(u, v, w) = ad$

c. Volume = ad **d.** $\left(\dfrac{a + b + c}{2}, \dfrac{d + e}{2}, \dfrac{1}{2}\right)$

Chapter 16 Review Exercises, pp. 1084–1088

1. a. False   **b.** True   **c.** False   **d.** False   **3.** $\frac{26}{3}$

5. $\displaystyle\int_0^1 \int_{-\sqrt{y}}^{\sqrt{y}} f(x, y)\, dx\, dy$ **7.** $\displaystyle\int_0^1 \int_0^{\sqrt{1-x^2}} f(x, y)\, dy\, dx$

9. $\dfrac{304}{3}$ **11.** $\dfrac{\sqrt{17} - \sqrt{2}}{2}$

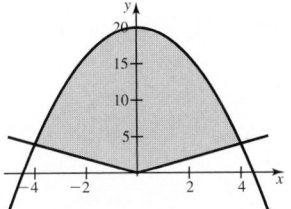

13. 8π **15.** $\dfrac{2}{7\pi^2}$ **17.** $\dfrac{1}{5}$

19. $\dfrac{9\pi}{2}$

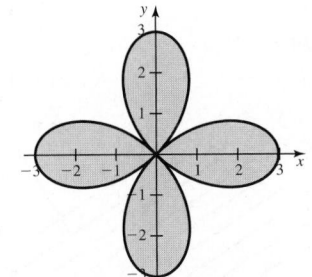

21. $6\pi - 16$

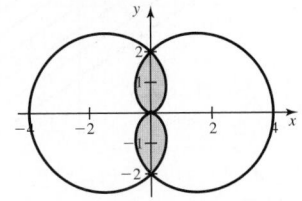

23. 2   **25.** $\displaystyle\int_0^1 \int_{2y}^2 \int_0^{\sqrt{z^2-4y^2}/2} f(x, y, z)\, dx\, dz\, dy$ **27.** $\pi - \dfrac{4}{3}$

29. $8\sin^2 2 = 4(1 - \cos 4)$ **31.** $\dfrac{848}{9}$ **33.** $\dfrac{8}{15}$ **35.** $\dfrac{16}{3}$

37. $\dfrac{128}{3}$ **39.** $\dfrac{\pi}{6} - \dfrac{\sqrt{3}}{2} + \dfrac{1}{2}$ **41.** $\dfrac{1}{3}$ **43.** $\dfrac{1}{3}$ **45.** π

47. 4π **49.** $\dfrac{28\pi}{3}$ **51.** $\dfrac{2048\pi}{105}$

53.

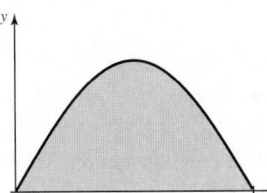

$(\bar{x}, \bar{y}) = \left(\dfrac{\pi}{2}, \dfrac{\pi}{8}\right)$

55.

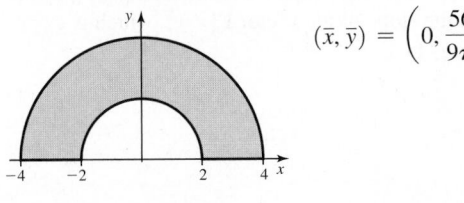

$(\bar{x}, \bar{y}) = \left(0, \dfrac{56}{9\pi}\right)$

57. $(\bar{x}, \bar{y}, \bar{z}) = (0, 0, 24)$ **59.** $(\bar{x}, \bar{y}, \bar{z}) = \left(0, 0, \dfrac{63}{10}\right)$

61. $\dfrac{h}{3}$ **63.** $\left(\dfrac{4\sqrt{2}a}{3\pi}, \dfrac{4(2 - \sqrt{2})a}{3\pi}\right)$ **65. a.** $\dfrac{4\pi}{3}$ **b.** $\dfrac{16Q}{3}$

67. $R = \{(x, y): 0 \le x \le 1, 0 \le y \le 1\}$

69. The parallelogram with vertices $(0, 0)$, $(3, 1)$, $(4, 4)$, and $(1, 3)$

71. 10   **73.** 6

75. a.

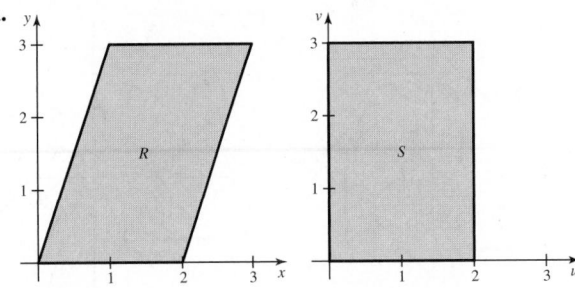

b. $0 \le u \le 2, 0 \le v \le 3$ **c.** $J(u, v) = 1$ **d.** $\frac{63}{2}$

77. a.

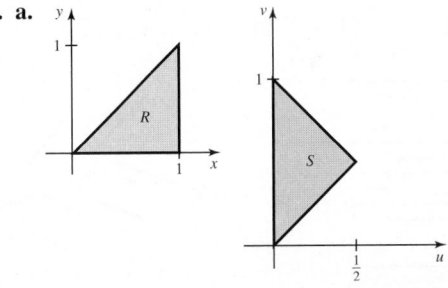

b. $u \le v \le 1 - u, 0 \le u \le \dfrac{1}{2}$ **c.** $J(u, v) = 2$ **d.** $\dfrac{1}{60}$

79. 42

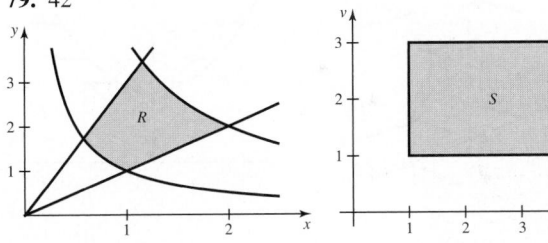

81. $-\dfrac{7}{16}$

CHAPTER 17

Section 17.1 Exercises, pp. 1096–1098

1. $\mathbf{F} = \langle f, g, h \rangle$ evaluated at (x, y, z) is the velocity vector of an air particle at (x, y, z) at a fixed point in time.   **3.** At selected points (a, b), plot the vector $\langle f(a, b), g(a, b) \rangle$.   **5.** It shows the direction in which the temperature increases the fastest and the amount of increase.

7.

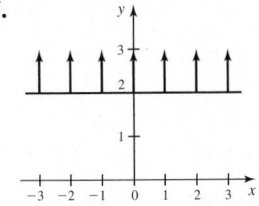

9.

11.

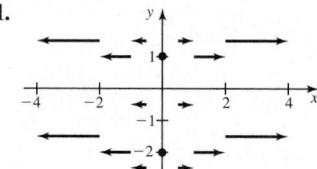

13.

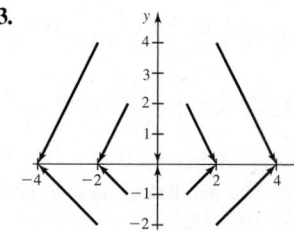

15.

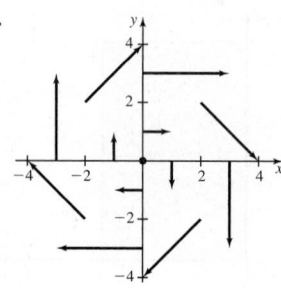

17.

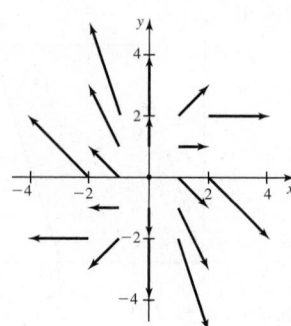

19.

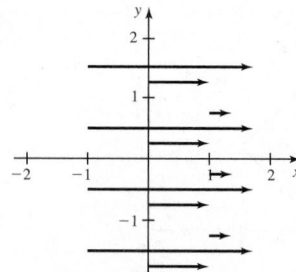

21.

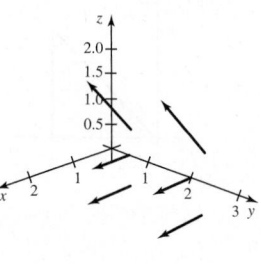

23.

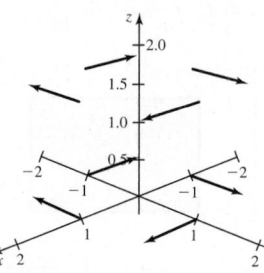

25. a. $(0, 1)$ **b.** None

c.

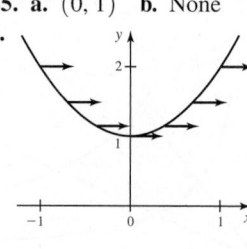

27. a. None
b. At all points on C

c.

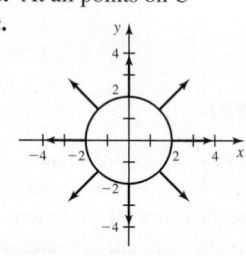

29. a. None
b. $(1, 0)$

c.

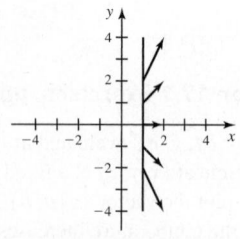

31. $\mathbf{F} = \langle -y, x \rangle$ or $\mathbf{F} = \langle -1, 1 \rangle$

33. $\mathbf{F}(x, y) = \dfrac{\langle x, y \rangle}{\sqrt{x^2 + y^2}} = \dfrac{\mathbf{r}}{|\mathbf{r}|}, \; \mathbf{F}(0, 0) = \mathbf{0}$

35. $\nabla \varphi(x, y) = \langle 2xy - y^2, x^2 - 2xy \rangle$

37. $\nabla \varphi(x, y) = \langle 1/y, -x/y^2 \rangle$ **39.** $\nabla \varphi(x, y, z) = \langle x, y, z \rangle = \mathbf{r}$

41. $\nabla \varphi(x, y, z) = -(x^2 + y^2 + z^2)^{-3/2} \langle x, y, z \rangle = -\dfrac{\mathbf{r}}{|\mathbf{r}|^3}$

43. a. $\mathbf{F} = \langle -2, 1 \rangle$
b. $\mathbf{F}(-1, -2) = \mathbf{F}(0, 0) = \mathbf{F}(1, 2) = \mathbf{F}(2, 4) = \langle -2, 1 \rangle$
c.

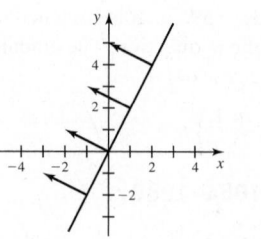

45. a. $\mathbf{F} = \langle \cos x, -1 \rangle$

b. $\mathbf{F}\left(\dfrac{\pi}{2}, 1\right) = \langle 0, -1 \rangle; \; \mathbf{F}(\pi, 0) = \langle -1, -1 \rangle;$

$\mathbf{F}\left(\dfrac{3\pi}{2}, -1\right) = \langle 0, -1 \rangle; \; \mathbf{F}(2\pi, 0) = \langle 1, -1 \rangle$

c.

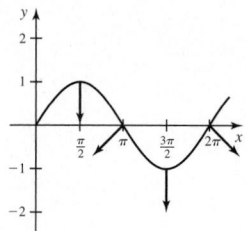

47. $\nabla \varphi(x, y) = 2 \langle x, y \rangle$

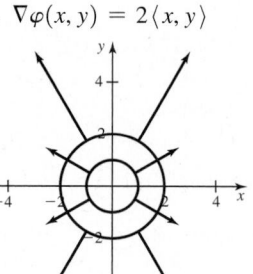

49. a. $\nabla \varphi(x, y) = \langle 2, 3 \rangle$
b. $y' = -2/3, \; \langle 1, -\tfrac{2}{3} \rangle \cdot \nabla \varphi(1, 1) = 0$
c. $y' = -2/3, \; \langle 1, -\tfrac{2}{3} \rangle \cdot \nabla \varphi(x, y) = 0$
d.

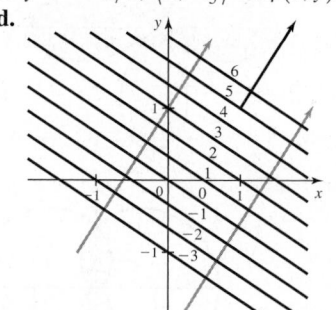

51. a. $\nabla \varphi(x, y) = \langle e^{x-y}, -e^{x-y} \rangle = e^{x-y} \langle 1, -1 \rangle$
b. $y' = 1, \; \langle 1, 1 \rangle \cdot \nabla \varphi(1, 1) = 0$
c. $y' = 1, \; \langle 1, 1 \rangle \cdot \nabla \varphi(x, y) = 0$
d.

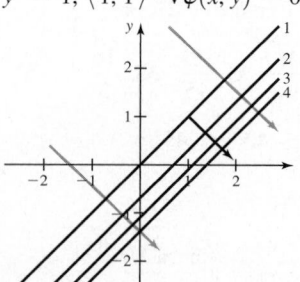

53. a. True   **b.** False   **c.** True   **55. a.** $\mathbf{E} = \dfrac{c}{x^2 + y^2} \langle x, y \rangle$

b. $|\mathbf{E}| = \left| \dfrac{c}{|\mathbf{r}|^2} \mathbf{r} \right| = \dfrac{c}{r}$ **c.** *Hint:* The equipotential curves are circles centered at the origin.   **57.** The slope of the streamline at (x, y) is $y'(x)$, which equals the slope of the vector $\mathbf{F}(x, y)$, which is g/f. Therefore, $y'(x) = g/f$.

59.

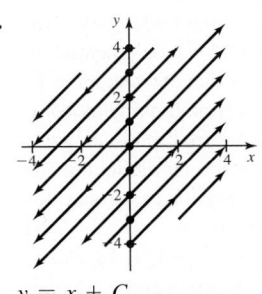

$y = x + C$

61.

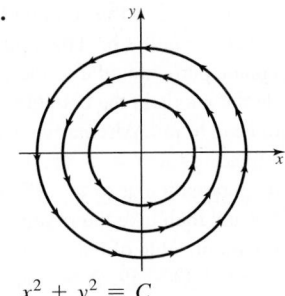

$x^2 + y^2 = C$

63. For $\theta = 0$: $\mathbf{u}_r = \mathbf{i}$ and $\mathbf{u}_\theta = \mathbf{j}$
for $\theta = \frac{\pi}{2}$: $\mathbf{u}_r = \mathbf{j}$ and $\mathbf{u}_\theta = -\mathbf{i}$
for $\theta = \pi$: $\mathbf{u}_r = -\mathbf{i}$ and $\mathbf{u}_\theta = -\mathbf{j}$
for $\theta = \frac{3\pi}{2}$: $\mathbf{u}_r = -\mathbf{j}$ and $\mathbf{u}_\theta = \mathbf{i}$

65.

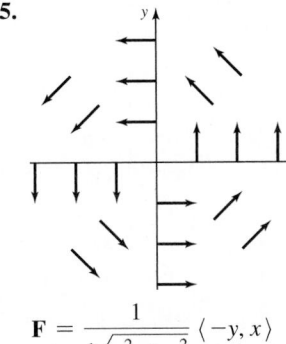

$\mathbf{F} = \dfrac{1}{\sqrt{x^2 + y^2}} \langle -y, x \rangle$

67.

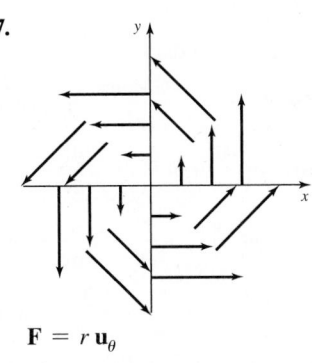

$\mathbf{F} = r\,\mathbf{u}_\theta$

Section 17.2 Exercises, pp. 1110–1114

1. A line integral is taken along a curve; an ordinary single-variable
integral is taken along an interval.   **3.** $\int_{\pi/2}^{\pi} \frac{1}{t} \cos t \sqrt{\sin^2 t + 1}\, dt$
5. $\mathbf{r}(t) = \langle 1 + 4t, 2 + 2t, 3 - 3t \rangle$, for $0 \le t \le 1$
7. $\mathbf{r}(t) = \langle t^2 + 1, t \rangle$, for $2 \le t \le 4$
9. a. $\int_0^2 (t + 6t^5)\,dt$ **b.** 66   **11.** $\int_C \mathbf{F} \cdot d\mathbf{r}$ and $\int_C f\,dx + g\,dy + h\,dz$
13. 7   **15.** Take the line integral of $\mathbf{F} \cdot \mathbf{T}$ along the curve with arc
length as the parameter.   **17.** 0   **19.** 100   **21.** 8   **23.** $-40\pi^2$
25. 128π **27.** $\dfrac{\sqrt{2}}{2} \ln 10$ **29.** $\dfrac{112}{9}$ **31.** 8   **33.** 414   **35.** 409.5
37. $\dfrac{15}{2}$ **39.** $\sqrt{101}$ **41.** $\dfrac{17}{2}$ **43.** 49   **45.** $\dfrac{3}{4\sqrt{10}}$ **47. a.** Negative
b. Positive   **49.** 0   **51.** 16   **53.** 0   **55.** $\dfrac{3\sqrt{3}}{10}$ **57. b.** 0
59. a. Negative   **b.** -4π **61. a.** True   **b.** True   **c.** True   **d.** True
63. a. Both paths require the same work: $W = 28{,}200$.
b. Both paths require the same work: $W = 28{,}200$.
65. a. $\dfrac{5\sqrt{5} - 1}{12}$ **b.** $\dfrac{5\sqrt{5} - 1}{12}$ **c.** The results are identical.
67. The work equals zero for all three paths.
69. $8\pi(48 + 7\pi - 128\pi^2) \approx -29{,}991.4$ **71.** 2π **73. a.** 4   **b.** -4
e. 0   **75.** *Hint:* Show that $\displaystyle\int_C \mathbf{F} \cdot \mathbf{T}\, ds = \pi r^2(c - b)$.
77. *Hint:* Show that $\displaystyle\int_C \mathbf{F} \cdot \mathbf{n}\, ds = \pi r^2(a + d)$.   **79. a.** $\ln a$ **b.** No
c. $\dfrac{1}{6}\left(1 - \dfrac{1}{a^2}\right)$ **d.** Yes   **e.** $W = \dfrac{3^{1-p/2}}{2 - p}(a^{2-p} - 1)$, for $p \ne 2$;
otherwise, $W = \ln a$.   **f.** $p > 2$ **81.** ab

Section 17.3 Exercises, pp. 1121–1123

1. A simple curve has no self-intersections; the initial and terminal
points of a closed curve are identical.   **3.** Test for equality of partial
derivatives as given in Theorem 17.3.   **5.** Integrate f with respect
to x and make the constant of integration a function of y to obtain
$\varphi = \int f\,dx + h(y)$; finally, set $\frac{\partial \varphi}{\partial y} = g$ to determine h.   **7.** 0
9. Conservative   **11.** Not conservative   **13.** Conservative
15. Conservative   **17.** $\varphi(x, y) = \frac{1}{2}(x^2 + y^2)$ **19.** Not conservative
21. $\varphi(x, y) = \sqrt{x^2 + y^2}$ **23.** $\varphi(x, y, z) = xz + y$
25. Not conservative   **27.** $\varphi(x, y, z) = xy + yz + zx$
29. $\varphi(x, y) = \sqrt{x^2 + y^2 + z^2}$ **31. a, b.** 0   **33. a, b.** 2   **35.** 3
37. -10 **39.** 24   **41.** $-\frac{1}{2}$ **43.** $-\pi^2$ **45.** 0   **47.** 0   **49.** 0
51. -5 **53.** 0   **55.** 1   **57. a.** False   **b.** True   **c.** True   **d.** True
e. True   **59.** 10   **61.** 25   **63. a.** Negative   **b.** Positive   **c.** No
67. a. Compare partial derivatives.
b. $\varphi(x, y, z) = \dfrac{GMm}{\sqrt{x^2 + y^2 + z^2}} = \dfrac{GMm}{|\mathbf{r}|}$
c. $\varphi(B) - \varphi(A) = GMm\left(\dfrac{1}{r_2} - \dfrac{1}{r_1}\right)$ **d.** No
69. a. $\dfrac{\partial}{\partial y}\left(\dfrac{-y}{(x^2 + y^2)^{p/2}}\right) = \dfrac{-x^2 + (p - 1)y^2}{(x^2 + y^2)^{1+p/2}}$ and
$\dfrac{\partial}{\partial x}\left(\dfrac{x}{(x^2 + y^2)^{p/2}}\right) = \dfrac{(1 - p)x^2 + y^2}{(x^2 + y^2)^{1+p/2}}$
b. The two partial derivatives in (a) are equal if $p = 2$.
c. $\varphi(x, y) = \tan^{-1}(y/x)$ **73.** $\varphi(x, y) = \frac{1}{2}(x^2 + y^2)$
75. $\varphi(x, y) = \frac{1}{2}(x^4 + x^2y^2 + y^4)$

Section 17.4 Exercises, pp. 1133–1136

1. In both forms, the integral of a *derivative* is computed from
boundary data.   **3.** Area $= \frac{1}{2}\oint_C x\,dy - y\,dx$, where C encloses the
region   **5.** The integral in the flux form of Green's Theorem vanishes.
7. $\mathbf{F} = \langle y, x \rangle$

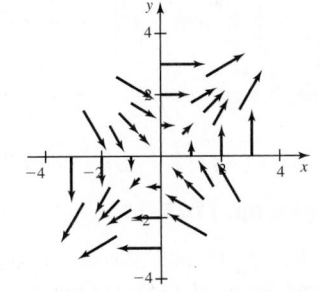

9. a. 0   **b.** 2   **c.** Yes   **d.** No   **11. a.** -4 **b.** 0   **c.** No   **d.** Yes
13. a. y^2 **b.** $12x^2y + 2xy$ **c.** No   **d.** No   **15. a.** 1; no
b. $\mathbf{r}_1(t) = \langle t, t^2 \rangle$, for $0 \le t \le 1$, and $\mathbf{r}_2(t) = \langle 1 - t, 1 - t \rangle$, for
$0 \le t \le 1$ (answers may vary)   **c.** Both integrals equal $\dfrac{1}{6}$.   **d.** 0
17. a. -4 **b.** Both integrals equal -8.   **19. a.** $4x$ **b.** $\dfrac{16}{3}$
21. 25π **23.** 16π **25.** 32   **27. a.** 2   **b.** Both integrals equal 8π.
29. a. $2y$ **b.** $\dfrac{16}{15}$ **31.** 104   **33.** $\dfrac{31 - 3e^4}{6}$ **35.** 6   **37.** $\dfrac{8}{3}$
39. $8 - \dfrac{\pi}{2}$ **41. a.** 0   **b.** 3π **43. a.** 0   **b.** $-\dfrac{15\pi}{2}$ **45. a.** 0
b. 2π **47. a.** $\dfrac{16}{3}$ **b.** 0   **49. a.** True   **b.** False   **c.** True

51. Note: $\dfrac{\partial f}{\partial y} = 0 = \dfrac{\partial g}{\partial x}$ **53.** The integral becomes $\iint_R 2\, dA$.

55. a. $f_x = g_y = 0$ **b.** $\psi(x, y) = -2x + 4y$

57. a. $f_x = e^{-x} \sin y = -g_y$ **b.** $\psi(x, y) = e^{-x} \cos y$

59. a. *Hint:* $f_x = e^x \cos y, f_y = -e^x \sin y,$
$g_x = -e^x \sin y, g_y = -e^x \cos y$
b. $\varphi(x, y) = e^x \cos y,\ \psi(x, y) = e^x \sin y$

61. a. *Hint:* $f_x = -\dfrac{y}{x^2 + y^2}, f_y = \dfrac{x}{x^2 + y^2},$
$g_x = \dfrac{x}{x^2 + y^2}, g_y = \dfrac{y}{x^2 + y^2}$
b. $\varphi(x, y) = x \tan^{-1}\dfrac{y}{x} + \dfrac{y}{2} \ln(x^2 + y^2) - y,$
$\psi(x, y) = y \tan^{-1}\dfrac{y}{x} - \dfrac{x}{2} \ln(x^2 + y^2) + x$

63. a.

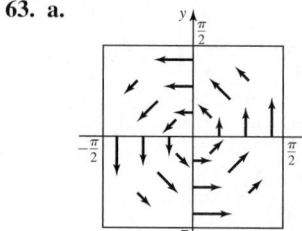

$F = \langle -4 \cos x \sin y, 4 \sin x \cos y \rangle$ **b.** Yes, the divergence equals zero.
c. No, the two-dimensional curl equals $8 \cos x \cos y$.   **d.** 0   **e.** 32
67. c. The vector field is undefined at the origin.
69.

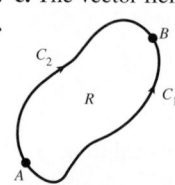

Basic ideas: Let C_1 and C_2 be two smooth simple curves from A to B.
$$\int_{C_1} \mathbf{F} \cdot \mathbf{n}\, ds - \int_{C_2} \mathbf{F} \cdot \mathbf{n}\, ds = \oint_C \mathbf{F} \cdot \mathbf{n}\, ds = \iint_R (f_x + g_y)\, dA = 0$$
and $\displaystyle\int_{C_1} \mathbf{F} \cdot \mathbf{n}\, ds = \int_{C_1} \psi_x\, dx + \psi_y\, dy = \int_{C_1} d\psi = \psi(B) - \psi(A)$

71. Use $\nabla \varphi \cdot \nabla \psi = \langle f, g \rangle \cdot \langle -g, f \rangle = 0$.

Section 17.5 Exercises, pp. 1143–1146

1. Compute $f_x + g_y + h_z$.   **3.** There is no source or sink.
5. It indicates the axis and the angular speed of the circulation at a
point.   **7.** 0   **9.** 3   **11.** 0   **13.** $2(x + y + z)$
15. $\dfrac{x^2 + y^2 + 3}{(1 + x^2 + y^2)^2}$ **17.** $\dfrac{1}{|\mathbf{r}|^2}$ **19.** $-\dfrac{1}{|\mathbf{r}|^4}$ **21. a.** Positive for
both points   **b.** div $\mathbf{F} = 2$ **c.** Outward everywhere   **d.** Positive
23. a. curl $\mathbf{F} = 2\mathbf{i}$ **b.** $|\text{curl } \mathbf{F}| = 2$
25. a. curl $\mathbf{F} = 2\mathbf{i} - 2\mathbf{j} + 2\mathbf{k}$ **b.** $|\text{curl } \mathbf{F}| = 2\sqrt{3}$ **27.** $3y\,\mathbf{k}$
29. $-4z\,\mathbf{j}$ **31.** 0   **33.** 0   **35.** Follows from partial differentiation of
$\dfrac{1}{(x^2 + y^2 + z^2)^{3/2}}$ **37.** Combine Exercise 36 with Theorem 17.10.
39. a. False   **b.** False   **c.** False   **d.** False   **e.** False   **41. a.** No
b. No   **c.** Yes, scalar function   **d.** No   **e.** No   **f.** No   **g.** Yes,
vector field   **h.** No   **i.** Yes, vector field   **43. a.** At $(0, 1, 1)$,
$\mathbf{F}$ points in the positive x-direction; at $(1, 1, 0)$, $\mathbf{F}$ points in the negative z-direction; at $(0, 1, -1)$, $\mathbf{F}$ points in the negative x-direction;
and at $(-1, 1, 0)$, $\mathbf{F}$ points in the positive z-direction. These vectors

circle the y-axis in the counterclockwise direction looking along $\mathbf{a}$
from head to tail.   **b.** The argument in part (a) can be repeated in
any plane perpendicular to the y-axis to show that the vectors of $\mathbf{F}$
circle the y-axis in the counterclockwise direction looking along $\mathbf{a}$
from head to tail. Alternatively, computing the cross product, we
find that $\mathbf{F} = \mathbf{a} \times \mathbf{r} = \langle z, 0, -x \rangle$, which is a rotation field in any
plane perpendicular to $\mathbf{a}$.   **45.** Compute an explicit expression for
$\mathbf{a} \times \mathbf{r}$ and then take the required partial derivatives.   **47.** div $\mathbf{F}$ has
a maximum value of 6 at $(1, 1, 1)$, $(1, -1, -1)$, $(-1, 1, 1)$, and
$(-1, -1, -1)$.   **49.** $\mathbf{n} = \langle a, b, 2a + b \rangle$, where a and b are real
numbers   **51.** $\mathbf{F} = \frac{1}{2}(y^2 + z^2)\,\mathbf{i}$; no   **53. a.** The wheel does not spin.
b. Clockwise, looking in the positive y-direction   **c.** The wheel does
not spin.   **55.** $\omega = \dfrac{10}{\sqrt{3}}$, or $\dfrac{5}{\sqrt{3}\pi} \approx 0.9189$ revolution per unit time
57. $\mathbf{F} = -200ke^{-x^2 + y^2 + z^2}(-x\,\mathbf{i} + y\,\mathbf{j} + z\,\mathbf{k})$
$\nabla \cdot \mathbf{F} = -200k(1 + 2(x^2 + y^2 + z^2))e^{-x^2 + y^2 + z^2}$
59. a. $\mathbf{F} = -\dfrac{GMm\mathbf{r}}{|\mathbf{r}|^3}$ **b.** See Theorem 17.11.
61. $\rho\left(\dfrac{\partial u}{\partial t} + u\dfrac{\partial u}{\partial x} + v\dfrac{\partial u}{\partial y} + w\dfrac{\partial u}{\partial z}\right) = -\dfrac{\partial p}{\partial x} + \mu\left(\dfrac{\partial^2 u}{\partial x^2} + \dfrac{\partial^2 u}{\partial y^2} + \dfrac{\partial^2 u}{\partial z^2}\right)$
$\rho\left(\dfrac{\partial v}{\partial t} + u\dfrac{\partial v}{\partial x} + v\dfrac{\partial v}{\partial y} + w\dfrac{\partial v}{\partial z}\right) = -\dfrac{\partial p}{\partial y} + \mu\left(\dfrac{\partial^2 v}{\partial x^2} + \dfrac{\partial^2 v}{\partial y^2} + \dfrac{\partial^2 v}{\partial z^2}\right)$
$\rho\left(\dfrac{\partial w}{\partial t} + u\dfrac{\partial w}{\partial x} + v\dfrac{\partial w}{\partial y} + w\dfrac{\partial w}{\partial z}\right) = -\dfrac{\partial p}{\partial z} + \mu\left(\dfrac{\partial^2 w}{\partial x^2} + \dfrac{\partial^2 w}{\partial y^2} + \dfrac{\partial^2 w}{\partial z^2}\right)$
63. a. Use $\nabla \times \mathbf{B} = -Ak \cos(kz - \omega t)\,\mathbf{i}$ and
$\dfrac{\partial \mathbf{E}}{\partial t} = -A\omega \cos(kz - \omega t)\,\mathbf{i}$.   **b.**

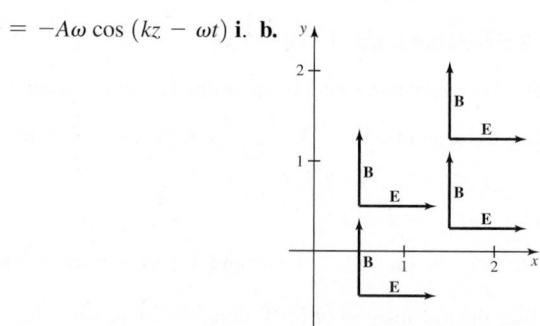

Section 17.6 Exercises, pp. 1159–1161

1. $\mathbf{r}(u, v) = \langle a \cos u, a \sin u, v \rangle, 0 \le u \le 2\pi, 0 \le v \le h$
3. $\mathbf{r}(u, v) = \langle a \sin u \cos v, a \sin u \sin v, a \cos u \rangle, 0 \le u \le \pi,$
$0 \le v \le 2\pi$ **5.** Use the parameterization from Exercise 3 and
compute $\displaystyle\int_0^\pi \int_0^{2\pi} f(a \sin u \cos v, a \sin u \sin v, a \cos u)\, a^2 \sin u\, dv\, du$.
7. The normal vectors point outward.   **9.** $\langle u, v, \frac{1}{3}(16 - 2u + 4v) \rangle$,
$|u| < \infty, |v| < \infty$ **11.** $\langle v \cos u, v \sin u, v \rangle, 0 \le u \le 2\pi,$
$2 \le v \le 8$ **13.** $\langle 3 \cos u, 3 \sin u, v \rangle, 0 \le u \le \dfrac{\pi}{2}, 0 \le v \le 3$
15. The plane $z = 2x + 3y - 1$ **17.** Part of the upper half of the
cone $z^2 = 16x^2 + 16y^2$ of height 12 and radius 3 (with $y \ge 0$)
19. 28π **21.** $16\sqrt{3}$ **23.** $\pi r\sqrt{r^2 + h^2}$ **25.** 1728π **27.** 0
29. 12   **31.** $4\pi\sqrt{5}$ **33.** $\dfrac{(65\sqrt{65} - 1)\pi}{24}$ **35.** $\dfrac{2\sqrt{3}}{3}$ **37.** $\dfrac{1250\pi}{3}$
39. $e - 1$ **41.** $\dfrac{1}{4\pi}$ **43.** -8 **45.** 0   **47.** 4π **49. a.** True
b. False   **c.** True   **d.** True   **51.** $8\pi(4\sqrt{17} + \ln(\sqrt{17} + 4))$
53. $8\pi a$ **55.** 8   **57. a.** 0   **b.** 0; the flow is tangent to the surface
(radial flow).   **59.** $2\pi ah$ **61.** $-400\left(e - \dfrac{1}{e}\right)^2$ **63.** $8\pi a$

65. a. $4\pi(b^3 - a^3)$ **b.** The net flux is zero. **67.** $\left(0, 0, \frac{2}{3}h\right)$

69. $\left(0, 0, \frac{7}{6}\right)$ **73.** Flux $= \displaystyle\iint_S \mathbf{F} \cdot \mathbf{n} \, dS = \iint_R dA$

Section 17.7 Exercises, pp. 1169–1171

1. The integral measures the circulation along the closed curve C.
3. Under certain conditions, the accumulated rotation of the vector field over the surface S equals the net circulation on the boundary of S.
5. Both integrals equal -2π. **7.** Both integrals equal zero. **9.** Both integrals equal -18π. **11.** -24π **13.** $-\frac{128}{3}$ **15.** 15π **17.** 0 **19.** 0 **21.** -2π **23.** -4π **25.** $\nabla \times \mathbf{v} = \langle 1, 0, 0 \rangle$; a paddle wheel with its axis aligned with the x-axis will spin with maximum angular speed counterclockwise (looking in the negative x-direction) at all points. **27.** $\nabla \times \mathbf{v} = \langle 0, -2, 0 \rangle$; a paddle wheel with its axis aligned with the y-axis will spin with maximum angular speed clockwise (looking in the negative y-direction) at all points. **29. a.** False **b.** False **c.** True **d.** True **31.** 0 **33.** 0 **35.** 2π **37.** $\pi(\cos\varphi - \sin\varphi)$; maximum for $\varphi = 0$ **39.** The circulation is 48π; it depends on the radius of the circle but not on the center. **41. a.** The normal vectors point toward the z-axis on the curved surface of S and in the direction of $\langle 0, 1, 0 \rangle$ on the flat surface of S. **b.** 2π **c.** 2π **43.** The integral is π for all a. **45. a.** 0 **b.** 0 **47. b.** 2π for any circle of radius r centered at the origin **c.** $\mathbf{F}$ is not differentiable along the z-axis.

49. Apply the Chain Rule. **51.** $\displaystyle\int_C \mathbf{F} \cdot d\mathbf{r} = \iint_R \left(\frac{\partial h}{\partial y} - \frac{\partial g}{\partial z} \right) dA$

Section 17.8 Exercises, pp. 1179–1182

1. The surface integral measures the flow across the boundary.
3. The flux across the boundary equals the cumulative expansion or contraction of the vector field inside the region. **5.** 32π
7. The outward fluxes are equal. **9.** Both integrals equal 96π.
11. Both integrals equal zero. **13.** 0 **15.** 0 **17.** $16\sqrt{6}\pi$ **19.** $\frac{2}{3}$
21. $-\frac{128}{3}\pi$ **23.** 24π **25.** -224π **27.** 12π **29.** 20
31. a. False **b.** False **c.** True **33.** 0 **35.** $\frac{3}{2}$ **37. b.** The net flux between the two spheres is $4\pi(a^2 - \varepsilon^2)$. **39. b.** Use $\nabla \cdot \mathbf{E} = 0$.

c. The flux across S is the sum of the contributions from the individual charges. **d.** For an arbitrary volume, we find

$$\frac{1}{\varepsilon_0} \iiint_D q(x, y, z) \, dV = \iint_S \mathbf{E} \cdot \mathbf{n} \, dS = \iiint_D \nabla \cdot \mathbf{E} \, dV.$$

e. Use $\nabla^2 \varphi = \nabla \cdot \nabla \varphi$. **41.** 0 **43.** $e^{-1} - 1$ **45.** $800\pi a^3 e^{-a^2}$

Chapter 17 Review Exercises, pp. 1182–1184

1. a. False **b.** True **c.** False **d.** False **e.** True
3. $\nabla \varphi = \langle 2x, 8y \rangle$

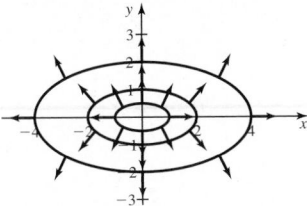

5. $-\dfrac{\mathbf{r}}{|\mathbf{r}|^3}$ **7.** $\mathbf{n} = \frac{1}{2}\langle x, y \rangle$ **9.** $\frac{7}{8}(e^{48} - 1)$ **11.** Both integrals equal zero. **13.** 0 **15.** The circulation is -4π; the outward flux is zero. **17.** The circulation is zero; the outward flux is 2π.

19. $\dfrac{4v_0 L^3}{3}$ **21.** $\varphi(x, y, z) = xy + yz^2$ **23.** $\varphi(x, y, z) = xye^z$

25. 0 for both methods **27. a.** $-\pi$ **b.** $\mathbf{F}$ is not conservative.
29. 0 **31.** $\frac{20}{3}$ **33.** 8π **35.** The circulation is zero; the outward flux equals 2π. **37. a.** $b = c$ **b.** $a = -d$ **c.** $a = -d$ and $b = c$
39. $\nabla \cdot \mathbf{F} = 4\sqrt{x^2 + y^2 + z^2} = 4|\mathbf{r}|$, $\nabla \times \mathbf{F} = \mathbf{0}$, $\nabla \cdot \mathbf{F} \neq 0$; irrotational but not source free **41.** $\nabla \cdot \mathbf{F} = 2y + 12xz^2$, $\nabla \times \mathbf{F} = \mathbf{0}$, $\nabla \cdot \mathbf{F} \neq 0$; irrotational but not source free

43. a. -1 **b.** 0 **c.** $\mathbf{n} = \dfrac{1}{\sqrt{3}}\langle -1, 1, 1 \rangle$ **45.** 18π **47.** $4\sqrt{3}$

49. $\dfrac{8\sqrt{3}}{3}$ **51.** 8π **53.** $4\pi a^2$ **55. a.** Use $x = y = 0$ to confirm the highest point; use $z = 0$ to confirm the base. **b.** The hemisphere S has the greater surface area—$2\pi a^2$ for S versus $\dfrac{5\sqrt{5} - 1}{6}\pi a^2$ for T.

57. 0 **59.** 99π **61.** 0 **63.** $\dfrac{972}{5}\pi$ **65.** $\dfrac{124}{5}\pi$ **67.** $\dfrac{32}{3}$

Index

Index of Applications

TABLE OF INTEGRALS

Substitution Rule	Integration by Parts
$\int f(g(x))g'(x)\,dx = \int f(u)\,du \quad (u = g(x))$	$\int u\,dv = uv - \int v\,du$
$\int_a^b f(g(x))g'(x)\,dx = \int_{g(a)}^{g(b)} f(u)\,du$	$\int_a^b uv'\,dx = uv\Big|_a^b - \int_a^b vu'\,dx$

Basic Integrals

1. $\int x^n\,dx = \dfrac{1}{n+1}x^{n+1} + C;\ n \neq -1$

2. $\int \dfrac{dx}{x} = \ln|x| + C$

3. $\int \cos ax\,dx = \dfrac{1}{a}\sin ax + C$

4. $\int \sin ax\,dx = -\dfrac{1}{a}\cos ax + C$

5. $\int \tan x\,dx = \ln|\sec x| + C$

6. $\int \cot x\,dx = \ln|\sin x| + C$

7. $\int \sec x\,dx = \ln|\sec x + \tan x| + C$

8. $\int \csc x\,dx = -\ln|\csc x + \cot x| + C$

9. $\int e^{ax}\,dx = \dfrac{1}{a}e^{ax} + C$

10. $\int b^{ax}\,dx = \dfrac{1}{a\ln b}b^{ax} + C;\ b > 0, b \neq 1$

11. $\int \ln x\,dx = x\ln x - x + C$

12. $\int \log_b x\,dx = \dfrac{1}{\ln b}(x\ln x - x) + C$

13. $\int \dfrac{dx}{x^2 + a^2} = \dfrac{1}{a}\tan^{-1}\dfrac{x}{a} + C$

14. $\int \dfrac{dx}{\sqrt{a^2 - x^2}} = \sin^{-1}\dfrac{x}{a} + C, a > 0$

15. $\int \dfrac{dx}{x\sqrt{x^2 - a^2}} = \dfrac{1}{a}\sec^{-1}\left|\dfrac{x}{a}\right| + C, a > 0$

16. $\int \sin^{-1} x\,dx = x\sin^{-1}x + \sqrt{1 - x^2} + C$

17. $\int \cos^{-1} x\,dx = x\cos^{-1}x - \sqrt{1 - x^2} + C$

18. $\int \tan^{-1} x\,dx = x\tan^{-1}x - \dfrac{1}{2}\ln(1 + x^2) + C$

19. $\int \sec^{-1} x\,dx = x\sec^{-1}x - \ln(x + \sqrt{x^2 - 1}) + C$

20. $\int \sinh x\,dx = \cosh x + C$

21. $\int \cosh x\,dx = \sinh x + C$

22. $\int \operatorname{sech}^2 x\,dx = \tanh x + C$

23. $\int \operatorname{csch}^2 x\,dx = -\coth x + C$

24. $\int \operatorname{sech} x \tanh x\,dx = -\operatorname{sech} x + C$

25. $\int \operatorname{csch} x \coth x\,dx = -\operatorname{csch} x + C$

26. $\int \tanh x\,dx = \ln \cosh x + C$

27. $\int \coth x\,dx = \ln|\sinh x| + C$

28. $\int \operatorname{sech} x\,dx = \tan^{-1}\sinh x + C = \sin^{-1}\tanh x + C$

29. $\int \operatorname{csch} x\,dx = \ln|\tanh(x/2)| + C$

Trigonometric Integrals

30. $\int \cos^2 x\,dx = \dfrac{x}{2} + \dfrac{\sin 2x}{4} + C$

31. $\int \sin^2 x\,dx = \dfrac{x}{2} - \dfrac{\sin 2x}{4} + C$

32. $\int \sec^2 ax\,dx = \dfrac{1}{a}\tan ax + C$

33. $\int \csc^2 ax\,dx = -\dfrac{1}{a}\cot ax + C$

34. $\int \tan^2 x\,dx = \tan x - x + C$

35. $\int \cot^2 x\,dx = -\cot x - x + C$

36. $\int \cos^3 x\,dx = -\dfrac{1}{3}\sin^3 x + \sin x + C$

37. $\int \sin^3 x\,dx = \dfrac{1}{3}\cos^3 x - \cos x + C$

38. $\displaystyle\int \sec^3 x\, dx = \frac{1}{2}\sec x \tan x + \frac{1}{2}\ln\left|\sec x + \tan x\right| + C$

39. $\displaystyle\int \csc^3 x\, dx = -\frac{1}{2}\csc x \cot x - \frac{1}{2}\ln\left|\csc x + \cot x\right| + C$

40. $\displaystyle\int \tan^3 x\, dx = \frac{1}{2}\tan^2 x - \ln\left|\sec x\right| + C$

41. $\displaystyle\int \cot^3 x\, dx = -\frac{1}{2}\cot^2 x - \ln\left|\sin x\right| + C$

42. $\displaystyle\int \sec^n ax \tan ax\, dx = \frac{1}{na}\sec^n ax + C;\ n \neq 0$

43. $\displaystyle\int \csc^n ax \cot ax\, dx = -\frac{1}{na}\csc^n ax + C;\ n \neq 0$

44. $\displaystyle\int \frac{dx}{1 + \sin ax} = -\frac{1}{a}\tan\left(\frac{\pi}{4} - \frac{ax}{2}\right) + C$

45. $\displaystyle\int \frac{dx}{1 - \sin ax} = \frac{1}{a}\tan\left(\frac{\pi}{4} + \frac{ax}{2}\right) + C$

46. $\displaystyle\int \frac{dx}{1 + \cos ax} = \frac{1}{a}\tan\frac{ax}{2} + C$

47. $\displaystyle\int \frac{dx}{1 - \cos ax} = -\frac{1}{a}\cot\frac{ax}{2} + C$

48. $\displaystyle\int \sin mx \cos nx\, dx = -\frac{\cos (m + n)x}{2(m + n)} - \frac{\cos (m - n)x}{2(m - n)} + C;\ m^2 \neq n^2$

49. $\displaystyle\int \sin mx \sin nx\, dx = \frac{\sin (m - n)x}{2(m - n)} - \frac{\sin (m + n)x}{2(m + n)} + C;\ m^2 \neq n^2$

50. $\displaystyle\int \cos mx \cos nx\, dx = \frac{\sin (m - n)x}{2(m - n)} + \frac{\sin (m + n)x}{2(m + n)} + C;\ m^2 \neq n^2$

Reduction Formulas for Trigonometric Functions

51. $\displaystyle\int \cos^n x\, dx = \frac{1}{n}\cos^{n-1} x \sin x + \frac{n - 1}{n}\int \cos^{n-2} x\, dx$

52. $\displaystyle\int \sin^n x\, dx = -\frac{1}{n}\sin^{n-1} x \cos x + \frac{n - 1}{n}\int \sin^{n-2} x\, dx$

53. $\displaystyle\int \tan^n x\, dx = \frac{\tan^{n-1} x}{n - 1} - \int \tan^{n-2} x\, dx;\ n \neq 1$

54. $\displaystyle\int \cot^n x\, dx = -\frac{\cot^{n-1} x}{n - 1} - \int \cot^{n-2} x\, dx;\ n \neq 1$

55. $\displaystyle\int \sec^n x\, dx = \frac{\sec^{n-2} x \tan x}{n - 1} + \frac{n - 2}{n - 1}\int \sec^{n-2} x\, dx;\ n \neq 1$

56. $\displaystyle\int \csc^n x\, dx = -\frac{\csc^{n-2} x \cot x}{n - 1} + \frac{n - 2}{n - 1}\int \csc^{n-2} x\, dx;\ n \neq 1$

57. $\displaystyle\int \sin^m x \cos^n x\, dx = -\frac{\sin^{m-1} x \cos^{n+1} x}{m + n} + \frac{m - 1}{m + n}\int \sin^{m-2} x \cos^n x\, dx;\ m \neq -n$

58. $\displaystyle\int \sin^m x \cos^n x\, dx = \frac{\sin^{m+1} x \cos^{n-1} x}{m + n} + \frac{n - 1}{m + n}\int \sin^m x \cos^{n-2} x\, dx;\ m \neq -n$

59. $\displaystyle\int x^n \sin ax\, dx = -\frac{x^n \cos ax}{a} + \frac{n}{a}\int x^{n-1} \cos ax\, dx;\ a \neq 0$

60. $\displaystyle\int x^n \cos ax\, dx = \frac{x^n \sin ax}{a} - \frac{n}{a}\int x^{n-1} \sin ax\, dx;\ a \neq 0$

Integrals Involving $a^2 - x^2$; $a > 0$

61. $\displaystyle\int \sqrt{a^2 - x^2}\, dx = \frac{x}{2}\sqrt{a^2 - x^2} + \frac{a^2}{2}\sin^{-1}\frac{x}{a} + C$

62. $\displaystyle\int \frac{dx}{x\sqrt{a^2 - x^2}} = -\frac{1}{a}\ln\left|\frac{a + \sqrt{a^2 - x^2}}{x}\right| + C$

63. $\displaystyle\int \frac{dx}{x^2\sqrt{a^2 - x^2}} = -\frac{\sqrt{a^2 - x^2}}{a^2 x} + C$

64. $\displaystyle\int x^2\sqrt{a^2 - x^2}\, dx = \frac{x}{8}(2x^2 - a^2)\sqrt{a^2 - x^2} + \frac{a^4}{8}\sin^{-1}\frac{x}{a} + C$

65. $\displaystyle\int \frac{\sqrt{a^2 - x^2}}{x^2}\, dx = -\frac{1}{x}\sqrt{a^2 - x^2} - \sin^{-1}\frac{x}{a} + C$

66. $\displaystyle\int \frac{x^2}{\sqrt{a^2 - x^2}}\, dx = -\frac{x}{2}\sqrt{a^2 - x^2} + \frac{a^2}{2}\sin^{-1}\frac{x}{a} + C$

67. $\displaystyle\int \frac{dx}{a^2 - x^2} = \frac{1}{2a}\ln\left|\frac{x + a}{x - a}\right| + C$

Integrals Involving $x^2 - a^2$; $a > 0$

68. $\displaystyle\int \sqrt{x^2 - a^2}\, dx = \frac{x}{2}\sqrt{x^2 - a^2} - \frac{a^2}{2}\ln\left|x + \sqrt{x^2 - a^2}\right| + C$

69. $\displaystyle\int \frac{dx}{\sqrt{x^2 - a^2}} = \ln\left|x + \sqrt{x^2 - a^2}\right| + C$

70. $\displaystyle\int \frac{dx}{x^2\sqrt{x^2 - a^2}} = \frac{\sqrt{x^2 - a^2}}{a^2 x} + C$

71. $\displaystyle\int x^2\sqrt{x^2 - a^2}\, dx = \frac{x}{8}(2x^2 - a^2)\sqrt{x^2 - a^2} - \frac{a^4}{8}\ln\left|x + \sqrt{x^2 - a^2}\right| + C$

72. $\displaystyle\int \frac{\sqrt{x^2 - a^2}}{x^2}\, dx = \ln\left|x + \sqrt{x^2 - a^2}\right| - \frac{\sqrt{x^2 - a^2}}{x} + C$

73. $\displaystyle\int \frac{x^2}{\sqrt{x^2 - a^2}}\, dx = \frac{a^2}{2}\ln\left|x + \sqrt{x^2 - a^2}\right| + \frac{x}{2}\sqrt{x^2 - a^2} + C$

74. $\displaystyle\int \frac{dx}{x^2 - a^2} = \frac{1}{2a}\ln\left|\frac{x - a}{x + a}\right| + C$

75. $\displaystyle\int \frac{dx}{x(x^2 - a^2)} = \frac{1}{2a^2}\ln\left|\frac{x^2 - a^2}{x^2}\right| + C$

Integrals Involving $a^2 + x^2$; $a > 0$

76. $\displaystyle\int \sqrt{a^2 + x^2}\, dx = \frac{x}{2}\sqrt{a^2 + x^2} + \frac{a^2}{2}\ln\left(x + \sqrt{a^2 + x^2}\right) + C$

77. $\displaystyle\int \frac{dx}{\sqrt{a^2 + x^2}} = \ln\left(x + \sqrt{a^2 + x^2}\right) + C$

78. $\displaystyle\int \frac{dx}{x\sqrt{a^2 + x^2}} = \frac{1}{a}\ln\left|\frac{a - \sqrt{a^2 + x^2}}{x}\right| + C$

79. $\displaystyle\int \frac{dx}{x^2\sqrt{a^2 + x^2}} = -\frac{\sqrt{a^2 + x^2}}{a^2 x} + C$

80. $\displaystyle\int x^2\sqrt{a^2 + x^2}\, dx = \frac{x}{8}(a^2 + 2x^2)\sqrt{a^2 + x^2} - \frac{a^4}{8}\ln\left(x + \sqrt{a^2 + x^2}\right) + C$

81. $\displaystyle\int \frac{\sqrt{a^2 + x^2}}{x^2}\, dx = \ln\left|x + \sqrt{a^2 + x^2}\right| - \frac{\sqrt{a^2 + x^2}}{x} + C$

82. $\displaystyle\int \frac{x^2}{\sqrt{a^2 + x^2}}\, dx = -\frac{a^2}{2}\ln\left(x + \sqrt{a^2 + x^2}\right) + \frac{x\sqrt{a^2 + x^2}}{2} + C$

83. $\displaystyle\int \frac{\sqrt{a^2 + x^2}}{x}\, dx = \sqrt{a^2 + x^2} - a\ln\left|\frac{a + \sqrt{a^2 + x^2}}{x}\right| + C$

84. $\displaystyle\int \frac{dx}{(a^2 + x^2)^{3/2}} = \frac{x}{a^2\sqrt{a^2 + x^2}} + C$

85. $\displaystyle\int \frac{dx}{x(a^2 + x^2)} = \frac{1}{2a^2}\ln\left(\frac{x^2}{a^2 + x^2}\right) + C$

Integrals Involving $ax \pm b$; $a \neq 0, b > 0$

86. $\displaystyle\int (ax + b)^n\, dx = \frac{(ax + b)^{n+1}}{a(n + 1)} + C;\ n \neq -1$

87. $\displaystyle\int (\sqrt{ax + b})^n\, dx = \frac{2}{a}\frac{(\sqrt{ax + b})^{n+2}}{n + 2} + C;\ n \neq -2$

88. $\displaystyle\int \frac{dx}{x\sqrt{ax - b}} = \frac{2}{\sqrt{b}}\tan^{-1}\sqrt{\frac{ax - b}{b}} + C$

89. $\displaystyle\int \frac{dx}{x\sqrt{ax + b}} = \frac{1}{\sqrt{b}}\ln\left|\frac{\sqrt{ax + b} - \sqrt{b}}{\sqrt{ax + b} + \sqrt{b}}\right| + C$

90. $\displaystyle\int \frac{x}{ax + b}\, dx = \frac{x}{a} - \frac{b}{a^2}\ln|ax + b| + C$

91. $\displaystyle\int \frac{x^2}{ax + b}\, dx = \frac{1}{2a^3}\left((ax + b)^2 - 4b(ax + b) + 2b^2\ln|ax + b|\right) + C$

92. $\displaystyle\int \frac{dx}{x^2(ax + b)} = -\frac{1}{bx} + \frac{a}{b^2}\ln\left|\frac{ax + b}{x}\right| + C$

93. $\displaystyle\int x\sqrt{ax + b}\, dx = \frac{2}{15a^2}(3ax - 2b)(ax + b)^{3/2} + C$

94. $\displaystyle\int \frac{x}{\sqrt{ax + b}}\, dx = \frac{2}{3a^2}(ax - 2b)\sqrt{ax + b} + C$

95. $\displaystyle\int x(ax + b)^n\, dx = \frac{(ax + b)^{n+1}}{a^2}\left(\frac{ax + b}{n + 2} - \frac{b}{n + 1}\right) + C;\ n \neq -1, -2$

96. $\displaystyle\int \frac{dx}{x(ax + b)} = \frac{1}{b}\ln\left|\frac{x}{ax + b}\right| + C$

Integrals with Exponential and Trigonometric Functions

97. $\displaystyle\int e^{ax}\sin bx\, dx = \frac{e^{ax}(a\sin bx - b\cos bx)}{a^2 + b^2} + C$

98. $\displaystyle\int e^{ax}\cos bx\, dx = \frac{e^{ax}(a\cos bx + b\sin bx)}{a^2 + b^2} + C$

Integrals with Exponential and Logarithmic Functions

99. $\displaystyle\int \frac{dx}{x\ln x} = \ln|\ln x| + C$

100. $\displaystyle\int x^n \ln x\, dx = \frac{x^{n+1}}{n + 1}\left(\ln x - \frac{1}{n + 1}\right) + C;\ n \neq -1$

101. $\displaystyle\int xe^x\, dx = xe^x - e^x + C$

102. $\displaystyle\int x^n e^{ax}\, dx = \frac{1}{a}x^n e^{ax} - \frac{n}{a}\int x^{n-1}e^{ax}\, dx;\ a \neq 0$

103. $\displaystyle\int \ln^n x\, dx = x\ln^n x - n\int \ln^{n-1} x\, dx$

Miscellaneous Formulas

104. $\displaystyle\int x^n \cos^{-1} x\, dx = \frac{1}{n + 1}\left(x^{n+1}\cos^{-1} x + \int \frac{x^{n+1}dx}{\sqrt{1 - x^2}}\right);\ n \neq -1$

105. $\displaystyle\int x^n \sin^{-1} x\, dx = \frac{1}{n + 1}\left(x^{n+1}\sin^{-1} x - \int \frac{x^{n+1}\, dx}{\sqrt{1 - x^2}}\right);\ n \neq -1$

106. $\displaystyle\int x^n \tan^{-1} x\, dx = \frac{1}{n + 1}\left(x^{n+1}\tan^{-1} x - \int \frac{x^{n+1}dx}{x^2 + 1}\right);\ n \neq -1$

107. $\displaystyle\int \sqrt{2ax - x^2}\, dx = \frac{x - a}{2}\sqrt{2ax - x^2} + \frac{a^2}{2}\sin^{-1}\left(\frac{x - a}{a}\right) + C;\ a > 0$

108. $\displaystyle\int \frac{dx}{\sqrt{2ax - x^2}} = \sin^{-1}\left(\frac{x - a}{a}\right) + C;\ a > 0$

GRAPHS OF ELEMENTARY FUNCTIONS

Linear functions

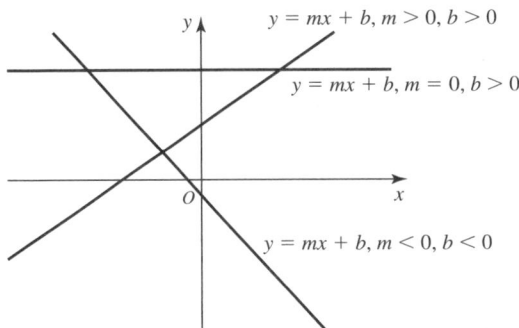

$y = mx + b, m > 0, b > 0$

$y = mx + b, m = 0, b > 0$

$y = mx + b, m < 0, b < 0$

Quadratic functions

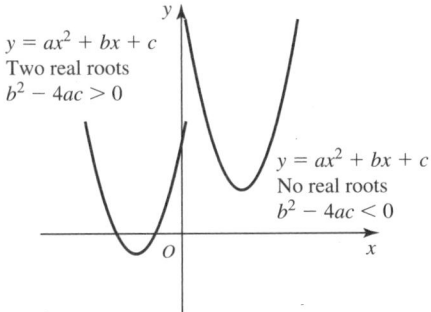

$y = ax^2 + bx + c$
Two real roots
$b^2 - 4ac > 0$

$y = ax^2 + bx + c$
No real roots
$b^2 - 4ac < 0$

Positive even powers

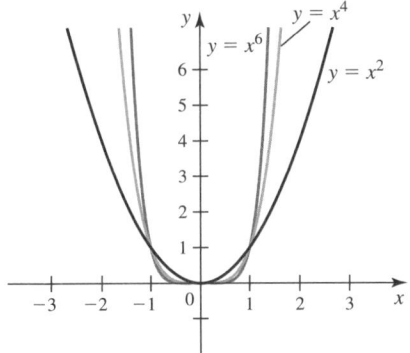

$y = x^6$ $y = x^4$ $y = x^2$

Positive odd powers

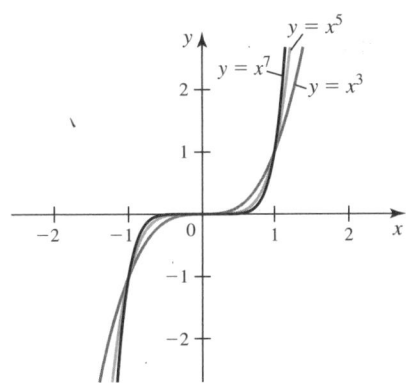

$y = x^5$ $y = x^7$ $y = x^3$

Negative even powers

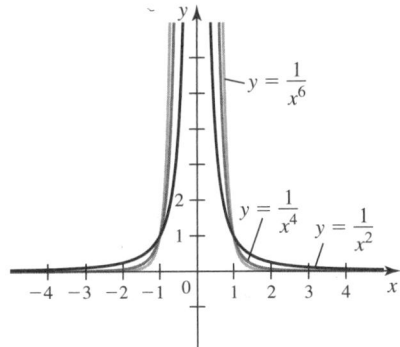

$y = \frac{1}{x^6}$ $y = \frac{1}{x^4}$ $y = \frac{1}{x^2}$

Negative odd powers

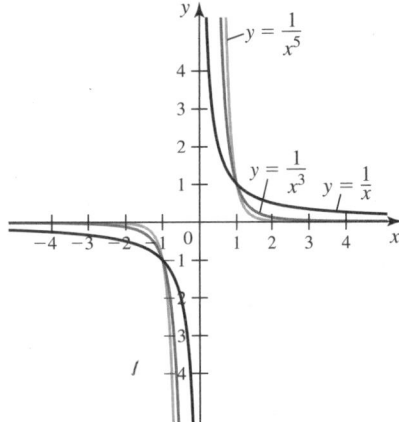

$y = \frac{1}{x^5}$ $y = \frac{1}{x^3}$ $y = \frac{1}{x}$

Exponential functions

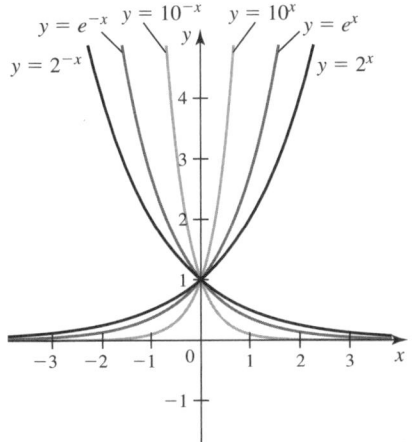

$y = e^{-x}$ $y = 10^{-x}$ $y = 10^x$ $y = e^x$
$y = 2^{-x}$ $y = 2^x$

Natural logarithmic and exponential functions

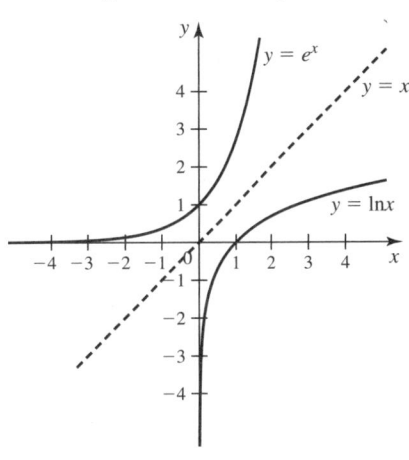

$y = e^x$ $y = x$ $y = \ln x$

DERIVATIVES

General Formulas

$$\frac{d}{dx}(c) = 0$$

$$\frac{d}{dx}(cf(x)) = cf'(x)$$

$$\frac{d}{dx}(f(x) + g(x)) = f'(x) + g'(x)$$

$$\frac{d}{dx}(f(x) - g(x)) = f'(x) - g'(x)$$

$$\frac{d}{dx}(f(x)g(x)) = f'(x)g(x) + f(x)g'(x)$$

$$\frac{d}{dx}\left(\frac{f(x)}{g(x)}\right) = \frac{g(x)f'(x) - f(x)g'(x)}{(g(x))^2}$$

$$\frac{d}{dx}(x^n) = nx^{n-1}, \text{ for real numbers } n$$

$$\frac{d}{dx}\big(f(g(x))\big) = f'(g(x)) \cdot g'(x)$$

Trigonometric Functions

$$\frac{d}{dx}(\sin x) = \cos x$$

$$\frac{d}{dx}(\cos x) = -\sin x$$

$$\frac{d}{dx}(\tan x) = \sec^2 x$$

$$\frac{d}{dx}(\cot x) = -\csc^2 x$$

$$\frac{d}{dx}(\sec x) = \sec x \tan x$$

$$\frac{d}{dx}(\csc x) = -\csc x \cot x$$

Inverse Trigonometric Functions

$$\frac{d}{dx}(\sin^{-1} x) = \frac{1}{\sqrt{1 - x^2}}$$

$$\frac{d}{dx}(\cos^{-1} x) = -\frac{1}{\sqrt{1 - x^2}}$$

$$\frac{d}{dx}(\tan^{-1} x) = \frac{1}{1 + x^2}$$

$$\frac{d}{dx}(\cot^{-1} x) = -\frac{1}{1 + x^2}$$

$$\frac{d}{dx}(\sec^{-1} x) = \frac{1}{|x|\sqrt{x^2 - 1}}$$

$$\frac{d}{dx}(\csc^{-1} x) = -\frac{1}{|x|\sqrt{x^2 - 1}}$$

Exponential and Logarithmic Functions

$$\frac{d}{dx}(e^x) = e^x$$

$$\frac{d}{dx}(b^x) = b^x \ln b$$

$$\frac{d}{dx}(\ln |x|) = \frac{1}{x}$$

$$\frac{d}{dx}(\log_b |x|) = \frac{1}{x \ln b}$$

Hyperbolic Functions

$$\frac{d}{dx}(\sinh x) = \cosh x$$

$$\frac{d}{dx}(\cosh x) = \sinh x$$

$$\frac{d}{dx}(\tanh x) = \text{sech}^2 x$$

$$\frac{d}{dx}(\coth x) = -\text{csch}^2 x$$

$$\frac{d}{dx}(\text{sech } x) = -\text{sech } x \tanh x$$

$$\frac{d}{dx}(\text{csch } x) = -\text{csch } x \coth x$$

Inverse Hyperbolic Functions

$$\frac{d}{dx}(\sinh^{-1} x) = \frac{1}{\sqrt{x^2 + 1}}$$

$$\frac{d}{dx}(\cosh^{-1} x) = \frac{1}{\sqrt{x^2 - 1}} \quad (x > 1)$$

$$\frac{d}{dx}(\tanh^{-1} x) = \frac{1}{1 - x^2} \quad (|x| < 1)$$

$$\frac{d}{dx}(\coth^{-1} x) = \frac{1}{1 - x^2} \quad (|x| > 1)$$

$$\frac{d}{dx}(\text{sech}^{-1} x) = -\frac{1}{x\sqrt{1 - x^2}} \quad (0 < x < 1)$$

$$\frac{d}{dx}(\text{csch}^{-1} x) = -\frac{1}{|x|\sqrt{1 + x^2}} \quad (x \neq 0)$$